TEACHER WRAPAROUND EDITION

Albert Kaskel Paul J. Hummer, Jr. Lucy Daniel

Glencoe
McGraw-Hill

New York, New York Columbus, Ohio Woodland Hills, California Peoria, Illinois

## A MERRILL BIOLOGY PROGRAM

**Biology: An Everyday Experience,** *Student Edition*
**Biology: An Everyday Experience,** *Teacher Edition*
**Biology: An Everyday Experience,** *Teacher Resource Package*
**Biology: An Everyday Experience,** *Study Guide*
**Biology: An Everyday Experience,** *Transparency Package*
**Biology: An Everyday Experience,** *Laboratory Manual, Student Edition*
**Biology: An Everyday Experience,** *Laboratory Manual, Teacher Edition*
**Biology: An Everyday Experience,** *Computer Test Bank*
**Biology: An Everyday Experience,** *Tech Prep Applications*

### PHOTO CREDITS

**4T,** Claude Steelman/Tom Stack & Assoc.; **5T,** Mark Thayer Studios; **6T,** Mark Newman, Tom Stack & Assoc.; **8T, 9T,** Mark Burnett/ Glencoe; **11T,** Bob Daemmrich; **12T, 13T,** Doug Martin; **18T,** Ted Rice; **20T, 21T,** Bob Daemmrich; **26T,** First Image; **27T,** Bob Daemmrich; **31T,** Gaillarde/Gamma Liaison Network; **31T,** Dr. E.R. Degginger/Color-Pic, Inc.; **31T,** Bill Ross/Allstock; **33T,** Doug Martin; **35T,** Chip Clark; **36T,** Doug Martin; **41T,** Ted Rice.

*Glencoe/McGraw-Hill*

*A Division of The* **McGraw-Hill** *Companies*



Send all inquiries to:
Glencoe/McGraw-Hill
936 Eastwind Drive
Westerville, OH 43081

ISBN 0-02-825686-7
Printed in the United States of America.

2 3 4 5 6 7 8 9 10 11 12 071/046 06 05 04 03 02 01 00 99

## TEACHER GUIDE TABLE OF CONTENTS

# Philosophy and Themes

The philosophy of *Biology: An Everyday Experience* is implied in its title. Biology is the study of living things, and living things are part of everyone's everyday experiences. An understanding of biology, therefore, results in a better understanding and appreciation of life. *Biology: An Everyday Experience* thus proceeds with the philosophy that to begin understanding life and life processes depends on a selective yet comprehensive introduction to applied biology.

The approach of this text is interesting, understandable, and practical, but most importantly it aids students who need an introduction to major biological concepts in an easy-to-read form. Difficult concepts are taught using analogies and examples with which students are familiar. Major concepts that may be unfamiliar can thus be learned through understanding rather than through memorization of countless facts. All concepts are presented with the most up-to-date information available with many opportunities for expansion as the needs of individuals or classes dictate. Topics that might be controversial are treated in a balanced and fair manner. In addition, practical applications of biological concepts throughout the text help make biology more "real" for students.

The approach emphasizes the fundamental unity in the diversity of life forms. The text focuses on major life processes. In so doing, each process is discussed using a variety of examples from all five kingdoms of organisms but with particular attention given to humans. In this way, students gain insight into the idea that all organisms, including themselves, carry out the same life functions.

With this philosophy and these ideas in mind, the text is built upon the following unifying themes:

1. There are many different kinds of living things, all of which are alike in some basic ways.
2. Living things depend on body systems that maintain and control body functions.
3. Living things reproduce and pass their traits to their offspring. These traits help a living thing survive in its environment.
4. Living things interact with each other and their environments.

*Biology: An Everyday Experience* is divided into eight units, organizing the text into major areas of biological study. Unit 1 introduces features of living things and classification. Unit 2 presents each of the five kingdoms. Units 3 and 4 present the structures and functions of animal systems. Unit 5 deals with plants, their parts, and their importance. Ways that living things reproduce, develop, change, and inherit traits are stressed in Units 6 and 7. Unit 8 presents the relationships of living and nonliving things in the environment.

The eight units are divided into 32 chapters that subdivide the units into more specific biological topics. Each chapter is further divided into numbered sections. The numbered sections allow the student to locate specific topics easily and aid the teacher in arranging class and homework assignments. The divisions and subdivisions of the text allow for adaptation to individual classroom needs and make the text useful in situations that require individualized instruction and flexible scheduling.

# Meeting National Science Standards

In the past decade, educators, public policy makers, corporate America, and parents have recognized the need for reform in science education. These groups have united in a call to action to solve this national problem. The *National Science Education Standards,* recently published by the National Research Council and representing the contribution of thousands of educators and scientists, offers a comprehensive vision of a scientifically literate society. The standards not only describe what students should know, but they also offer guidelines for science teaching and assessment. If you are using, or plan to use, the standards to guide changes in your science curriculum, you can be assured that ***Biology: An Everyday Experience*** aligns exceedingly well with *The National Science Education Standards.* More importantly, we believe it will help students succeed in science so that they will want to continue learning about science through high school and into adulthood.

## Science Content Standards

The National Science Content Standards for Grades 9-12 have been correlated to each major section of *Biology: An Everyday Experience.* You will find these correlations in the first column of the Objectives chart on the interleaf pages preceding each chapter. Correlations are designated according to the numbering system in the table of science content standards at the right.

## National Science Content Standards Unifying Concepts and Processes

**UCP.1** Systems, order, and organization
**UCP.2** Evidence, models, and explanation
**UCP.3** Change, constancy, and measurement
**UCP.4** Evolution and equilibrium
**UCP.5** Form and function

**Science as Inquiry**
**A.1** Abilities necessary to do scientific inquiry
**A.2** Understandings about scientific inquiry

**Physical Science**
**B.1** Structure of atoms
**B.2** Structure and properties of matter
**B.3** Chemical reactions
**B.4** Motions and forces
**B.5** Conservation of energy and increase in disorder
**B.6** Interactions of energy and matter

**Life Science**
**C.1** The cell
**C.2** Molecular basis of heredity
**C.3** Biological evolution
**C.4** Interdependence of organisms
**C.5** Matter, energy, and organization in living systems
**C.6** Behavior of organisms

**Earth and Space Science**
**D.1** Energy in the earth system
**D.2** Geochemical cycles
**D.3** Origin and evolution of the earth system
**D.4** Origin and evolution of the universe

**Science and Technology**
**E.1** Abilities of technological design
**E.2** Understandings about science and technology

**Science in Personal and Social Perspectives**
**F.1** Personal and community health
**F.2** Population growth
**F.3** Natural resources
**F.4** Environmental quality
**F.5** Natural and human-induced hazards
**F.6** Science and technology in local, national, and global challenges

**History and Nature of Science**
G.1 Science as a human endeavor
G.2 Nature of scientific knowledge
G.3 Historical perspectives

# Text Features

## Readability

Biological content with close attention to readability is a major feature of *Biology: An Everyday Experience.* The readability has been designed to promote comprehension, student interest, and student involvement in the subject area. In addition, the readability aids in teaching the material and in classroom management.

In controlling the reading level, several things have been given careful attention. The vocabulary in the text is consistent with the level of the students for whom the text is designed. Words with large numbers of syllables have been avoided wherever possible. Important terms have been printed in boldface type, and in many cases, are followed by phonetic spellings. Each term is clearly defined and reinforced throughout the discussion. Sentence construction and paragraph structure have also been carefully controlled to aid in readability. Readability is also aided by the use of many color photographs and illustrations that tie directly to the printed words. All of the aspects that aid in readability are enhanced by the typeface and type size in which the text is printed.

## Photographs and Illustrations

As an aid to comprehension and reinforcement, more than 850 color photographs and illustrations have been included in *Biology: An Everyday Experience.* Each chapter opening includes two photographs that are introduced by the opening paragraphs of the text. The introduction serves to stimulate student interest in the material to be presented in the chapter. In addition, the text photographs and illustrations reinforce and clarify concepts in adjacent paragraphs by directly relating to the printed words. The color artwork included in each Lab serves either as an illustration of the procedure students will follow in doing the lab or as artwork the student will use to complete the lab.

## Student Notes

Student notes are placed in the margin to aid in review as well as to serve as a reading aid. Notes are correlated with section objectives and provide thought-provoking questions for students.

## Technology

Each unit contains a Technology feature. These features are optional sections that highlight new breakthroughs and applications of the biology found in the chapters in which they occur. The Technology feature can be used to motivate students to seek additional information concerning current issues such as eye "fingerprints," seed banks, and angioplasty.

## Applying Technology

The Applying Technology features help to develop scientific and technological literacy in students by emphasizing problem solving, critical thinking skills, and teamwork. These performance-based activities consist of eight two-page presentations of a problem and a lab procedure that can be followed to solve the problem. Each feature includes a technology connection and points out several relevant career choices, which serve to motivate students to investigate the topics further. The margins of the Applying Technology pages of the *Teacher Edition* contain answers to the questions in the feature, as well as additional information about the problem.

## Science and Society

Each unit has a full-page Science and Society feature that introduces students to controversial issues in the sciences. Students are given some background information and then presented with three case studies related to the topic. At the end of each case study are thought-provoking questions that have no right or wrong answers, but invite students to develop and express their own opinions.

## Careers

Each unit contains a Career Close-Up feature. Each highlights a career possibility for students to investigate and gives information about the training and schooling required. Each Career Close-Up is contained within the chapter to which it most closely relates. Several of the careers are not typically thought of as biological, but do require some knowledge of biology. The margins of the Career Close-Up pages of the *Teacher Edition* contain additional career information.

## Unit Introductions and Closures

The unit-opening pages get students going with large, eye-catching photos and a stimulating introductory paragraph entitled "What would happen if ..." The introductory paragraph encourages students to stretch their imaginations, while integrating science, technology, and society themes.

The unit closures are titled "Connections, Biology in Your World." These pages integrate biology with consumer, leisure, art, history, and geography topics. A single theme unites the connections with a unit. For example, Unit 1 shows the connection between scientific method and Alex Haley's search for his family roots, continental drift, how to build a terrarium, and how to select a bicycle.

## Labs

Each chapter of *Biology: An Everyday Experience* contains two pages of labs that enhance student learning of biological concepts through seeing and doing. Each lab includes a materials list, a step-by-step procedure, a skills list, and several questions to test understanding of the lab performed. The end of each lab provides an extension for further exploration of concepts reinforced in the lab. The accompanying color illustrations either show the steps of the procedures or are art that is used by the students in the lab. Additional information about lab safety, care of living materials, and equipment and supplies can be found in this Teacher Guide.

## Skill Reinforcement

Skills are reinforced throughout *Biology: An Everyday Experience*. To pique student interest in learning new skills, two kinds of features have been included, Skill Checks and Mini Labs. Skill Checks are quick exercises that students can do in class or as homework assignments. Mini Labs are brief activities that require simple materials that can be found in the classroom or at home. To give students further support with skills, a Skill Handbook is provided at the end of the book. All Mini Labs and Skill Checks provide references to the Skill Handbook for students who need extra help.

## Idea Maps

Idea maps outline major concepts in diagram form. They provide a visual summary of the ideas students need to focus on learning. Idea maps can be used as study guides to major concepts and for review and reinforcement. Before students begin a section, have them study the idea map to see what the main ideas of the section are. Then, when they have finished reading the section, have them go over the idea map again. You may wish to have them use the major ideas presented in the idea map and write connecting sentences.

## Section- and Chapter-end Materials

Following each numbered section of the text are Check Your Understanding questions. These questions are designed to allow students to check their understanding of previous material. The first three questions correspond to the section objectives. The fourth question stimulates critical thinking, and the fifth question integrates biology with reading, writing, or math. Answers to these questions are provided in the margin of the *Teacher Edition*.

The chapter-end material begins with a Summary that concisely reviews the major concepts and principles of the chapter. Each Summary statement is numbered and corresponds to the section objectives.

The Testing Yourself section of the chapter-end material includes various types of questions. Using Words tests vocabulary understanding. Finding Main Ideas reinforces the reading and locating of important text material. Using Main Ideas requires students to recall important principles and concepts presented in the chapter by asking various types of questions. Skill Review questions review the skills to which students have been introduced in the Skill Checks, Mini Labs, and Labs.

Finding Out More presents Critical Thinking and Application questions to lead the student beyond the material covered in the text. This section is designed to provide a number of open-ended activities such as reports, projects, and experiments.

## Appendices

There are four appendices that may be used to expand students' learning or application of concepts. Appendix A shows SI units that students may commonly encounter. A system for easy conversion from one unit to another is described. You may use this section to teach students how to correctly and easily convert SI units. Appendix B includes the classification of living things. Organisms covered in Chapters 4-8 are included in the classification scheme. Appendix C covers root words, prefixes, and suffixes of commonly used biological terms. Appendix D lists rules for safety in the science classroom. Also provided is a chart of safety symbols that appear in the labs.

## Glossary and Index

The glossary provides students with a quick reference to key terms and their pronunciations within the text. Page references are provided so students can easily locate the page on which a word is defined. The index allows for easy finding of text material because it is complete and cross-referenced.

# Supplementary Materials

## Teacher Resource Package

The *Teacher Resource Package* for *Biology: An Everyday Experience* provides background information and comprehensive teaching material to aid in the effective teaching of biology. The following worksheet masters are provided: Application, Enrichment, Focus, Critical Thinking/Problem Solving, Skill, Reteaching, Lab, Study Guide, Transparency, Chapter Review, Chapter Test, and Unit Test.

Each chapter includes Application, Enrichment, Focus, Critical Thinking/Problem Solving, and Skill worksheets. The Application worksheets tie together biological concepts and safety, health, consumer, leisure, and environmental topics. Enrichment worksheets are designed to take students beyond material learned in the text. Focus masters provide five-minute activities to grab the attention of students and get them involved in the lesson. Focus activities require no prior knowledge of the subject matter on the part of the student. Critical Thinking/Problem Solving worksheets help to develop critical thinking skills while applying important concepts learned in the classroom to new situations. Skill worksheets are designed to reinforce process skills.

For each numbered section, a Reteaching master is provided. Reteaching masters are in the form of idea maps, transparencies, and worksheets for students to fill out. Idea maps in the *Teacher Resource Package* coordinate with those found in the *Student Edition* and serve as review and reinforcement aids. Transparencies can be made into overhead masters or passed out to students.

Each lab in the *Student Edition* has an accompanying lab worksheet. The worksheet reproduces the lab page and also includes rules and an expanded data table for students to write in their answers.

Over 160 Transparency masters and worksheets are provided to help you introduce and reinforce the lessons in *Biology: An Everyday Experience.* You can have students fill out their worksheets as you go over the master using the overhead, or the worksheets can be used for homework assignments.

Also included in the *Teacher Resource Package* is an evaluation program. There are two pages of review questions for each text chapter and three pages of test questions. Eight unit tests are included. Each test includes questions and problems of varying difficulty focused on different levels of learning. The reviews and tests can be used not only for assessment but also as learning tools to help students improve their knowledge of biology.

## Review and Reinforcement

For your convenience, the Study Guide worksheets in the *Teacher Resource Package* have been bound into a separate booklet.

The *Study Guide* emphasizes the text material by including charts and diagrams for the student to complete. The program concentrates on improving reading skills by making it necessary for the student to reread certain text sections in order to complete the exercises in the *Study Guide.* Besides emphasizing the text material and improved reading, the program offers the student additional reinforcement of important concepts and ideas.

The *Study Guide* can offer flexibility to the teacher. It can be used for in-class work as well as for homework. In addition, the *Study Guide* allows the teacher an additional means for student evaluation.

## Laboratory Program

*Biology: Laboratory Manual* and *Biology: Laboratory Manual (Teacher Annotated Edition)* is a learning through doing program of experiments and activities designed to reinforce important concepts. The experiments are written for ease of understanding and the illustrations show the student what to do. Because this program requires inexpensive equipment and materials along with minimum teacher preparation time, the program is easily implemented in a variety of classrooms that differ in sophistication of facilities.

## Transparency Package

A transparency package consisting of 48 four-color transparencies accompanied by a Teacher Guide are also available. The transparencies feature original art, text art, or unlabeled diagrams with labels on an overlay.

## Computer Test Bank

The *Computer Test Bank* is provided to help you construct chapter and unit tests. The package comes with a convenient teacher guide that includes a printout of all test questions.

## Cooperative Learning

The *Cooperative Learning Book* includes worksheets for students and valuable teacher information to help you organize your classes into cooperative groups. Each worksheet reinforces a different biological concept while exposing students to various cooperative learning strategies.

## Tech Prep Applications

The *Tech Prep Applications* book helps to fulfill the goals of a tech prep education. This supplement includes activities and projects that focus on real-life circumstances, problem solving, teamwork, and community resources.

# Teaching Cycle

What is the edge for which all teachers search when making a new textbook selection? Teachers anticipate a total program that imparts a wealth of experience they can translate into maximum efficiency, thereby avoiding any teacher's worst enemy—wasted time. The *Teacher Edition* of *Biology: An Everyday Experience* delivers the collective teaching experience of its authors, consultants, and reviewers to you so that you will have the treasure trove of experience teachers search for. By furnishing you with an effective teaching model, this book saves you preparation time and energy. You, in turn, are free to spend that time and energy on your most important responsibility—your students.

As a professional, you will be pleased to find that this program provides you with readily available activities that will engage your students for the entire class period. Each numbered section of every chapter may be considered an individual lesson and includes a comprehensive four-part teaching cycle that, when utilized consistently, will result in better cognitive transfer for your students.

Each chapter begins with two main parts that precede the teaching cycle, a CHAPTER OVERVIEW and a section called GETTING STARTED. The CHAPTER OVERVIEW highlights Key Concepts, Key Science Words, and Skill Development. In addition, it offers a tip for Bridging concepts to be learned with previous lessons. GETTING STARTED offers help on Using the Photos that open the chapter, teaching the MOTIVATION/Try This! activity, and a Chapter Preview. In addition, common student Misconceptions are pointed out to help you get your students off to a good start.

## PREPARATION

Before you begin any lesson, thorough PREPARATION is essential. In this step, you gather the Materials Needed to teach the lesson and preview the Key Science Words to be introduced and the Process Skills to be reinforced. With PREPARATION completed, you will have the foundation for keeping your lesson instructionally sound.

## 1 FOCUS

FOCUS, or the anticipatory set as it is sometimes called, is the first step of the *Biology: An Everyday Experience* teaching cycle that involves student contact. FOCUS helps prepare the students for learning the day's lesson. FOCUS hints may be in the form of Demonstrations, Audiovisuals, Brainstorming, or other innovative ideas that help focus the attention of the class so the lesson can begin. The function of a FOCUS activity is not to teach all of the concepts for the class period, but merely to pique student interest. A FOCUS item may also provide time for students to interact with you, thus enabling you to modify the lesson to fit what students already know. The FOCUS activity usually can be completed within the first five minutes of class and should be a direct lead-in to the concepts that will be presented. By focusing your students, you provide the structural framework for what is to come.

## 2 TEACH

TEACH, the second step of the teaching cycle, is the heart of any lesson. The primary aim of the *Biology: An Everyday Experience* teaching cycle is to give you the tools to accomplish the admittedly difficult task of getting biological concepts across to your students. Effective teaching strategies as well as alternate ways to teach the lesson are provided in this part of the teaching cycle. MOTIVATION and ACTIVITY ideas appear throughout this step. These ideas can help you develop student interest in the concepts being taught. To help you monitor and adjust your teaching to what the students appear to be learning, Check for Understanding hints are provided. For students who are having trouble, a Reteach suggestion immediately follows the Check for Understanding hint. Reteaching is a way to teach the same concepts or facts differently so they will be more amenable to students' individual learning styles. Also in the TEACH step you will find at least one Enrichment, Challenge, Cooperative Learning, Misconception, Software, and Audiovisual tip.

The authors of *Biology: An Everyday Experience* view teaching as a creative art. As you are aware, teachers work at making the transfer of knowledge as efficient and productive as it can be. One finding of educational researchers is that teachers who closely guide their students during a lesson are usually more effective. To this end, Guided Practice ideas are given. Guided practice takes place when you walk around the room and actively help students begin a task. Independent practice, on the other hand, occurs when students work alone.

Throughout the TEACH step and the rest of the teaching cycle are references to the *Teacher Resource Package* worksheets that go with the lesson. Also included are references to the *Lab Manual* and color *Transparency Package*. With the materials presented in the TEACH step, your efficiency and productivity as a teacher will increase. The TEACH step brings together the major elements that form a sound teaching approach.

Students can better understand complex biological processes and structures when they see them in action. Located throughout the TEACH step are numerous references to Glencoe and National Geographic Society videodiscs and videotapes. References to *Biology: The Dynamics of Life* and *The Secret of Life* videodiscs, *The Secret of Life* videotapes, and *National Geographic Society* videodisc series can be found under TEACH, and bar codes for the videodisc segments are displayed in the margin of the page at the point of use.

## 3 APPLY

APPLY is the third step of the teaching cycle. In the APPLY step, students have an opportunity for the concepts they have just learned to gel, to become firmer and clearer. In this step, hints may include suggestions for Convergent Questions, Divergent Questions, or Connections. Once students have been taken through the APPLY step, they are ready for the final step, closure.

## 4 CLOSE

CLOSE is the fourth and final step of the teaching cycle. An excellent CLOSE can be as short as the final minute or two of class, or it can last up to several minutes. There are several purposes to closure. The most important of these is to give students an opportunity to reflect upon what they have learned. One of several kinds of hints may be given in this step. They include Audiovisual suggestions and suggestions for how to Summarize the information.

Answers to Check Your Understanding questions are also presented under CLOSE. Each set of Check Your Understanding questions includes three questions pertaining to concept development, one Critical Thinking question, and one question that connects biology to reading, writing, or math.

As you review the *Teacher Edition* to *Biology: An Everyday Experience,* you will discover that you and your students are considered very important. With the enormous number of teaching strategies provided by the *Teacher Edition*, you should be able to accomplish the goals of your curriculum with a minimum of time and effort. This is true if you are teaching biology for the first time, or are a veteran teacher. The materials allow adaptability and flexibility so that student needs and curricular needs can be met.

# Reading Science

## Introduction

Scientific writing can be difficult for some students to read. Scientific material is crammed with details, concepts, and interrelationships. Understanding the intricacies of one concept may require an understanding of the subtleties of another. Often, the science textbook is the first experience a student has with trying to read such intensely detailed material. Earlier reading experience may have consisted entirely of descriptive and narrative works. So, the concentration of elements within scientific material may seem overwhelming to the reader who encounters it for the first time.

Science reading requires that students be able to sense relationships and think critically about what is already known and what is currently being read. Students need to possess a degree of competency with study skills in order to read scientific information. They must be able to use the text as a learning tool. They need to be able to interpret not only the printed page, but also the charts, pictures, tables, diagrams, formulas, and experiments. Students who experience difficulty in reading need to be walked through text material. They need to be taught how to read a science book.

## Chapter Preview

A chapter preview will serve to acquaint the student with new vocabulary and the organization of the chapter, and will give the student a purpose for reading. The section entitled *Biology and You* at the front of the text instructs students to preview parts of their text and to become familiar with the use of these parts. You may choose to go through the *Biology and You* section step-by-step with your class. Understanding how to use their texts will make students better able to study effectively throughout the school year.

After your class has read and completed *Biology and You,* you still may want to complete a chapter preview to reinforce proper use of the chapter's text material. Students should complete a teacher-directed step-by-step procedure before they thoroughly read a chapter.

Chapter 2, Features of Life and the Cell, is used to exemplify the procedure.

1. Direct students to turn to the table of contents and read the title of Chapter 2.
2. Have students list the numbered sections in order. Point out that 2:1 is the first numbered section of the chapter, 2:2 the second, and so on. Numbered sections are followed by subsections that contain information related to the numbered section. Numbered sections and subsections can be used as a skeletal outline of the chapter.
3. Starting from the beginning of Chapter 2, have students study the list of Key Science Words. Also have them read and become familiar with the section objectives. Have them star the known words and define the others as they read the chapter.
4. Since visual aids are distractors for some readers, the purpose for their inclusion should be discussed before the chapter is read. In this text the visual aids are directly tied to the text words. Have students read the labels that are included on pictures.
5. Read the titles of all charts, tables, or diagrams. Remind students to focus on these visual aids as the chapter is read.
6. Reading the Check Your Understanding sections first can provide a framework that helps students know what to look for in the section.
7. A Summary of Chapter 2 is found on page 44. Have students read this summary before reading the chapter. It gives an overview and identifies the important points.
8. Have students read the list of Key Science Words on page 44. Students may choose other words that are unfamiliar and add them to their personal word lists for study.
9. Guide students to read through the rest of the end matter on pages 44 and 45. The section Using Main Ideas has page references after the questions. If students are expected to complete Finding Out More, a teacher-made study guide could be provided to help students organize and report the information. The questions and activities at the end of the chapter will establish a framework that students can use to study the materials as they read.
10. After the preview, direct students to carefully read the chapter or work with students in reading through the chapter, paying special attention to key terms, visual aids, and information flow.

## Special Techniques

You may need to employ special teaching techniques for students who have reading problems. Some teachers prefer to have students read aloud in class. Each student should then restate orally what was just read. You may choose to read to your students for variety, or tape record the text readings and have students use the tapes individually.

Poor reading skills can contribute to other problems. Short attention span, inability to complete task, poor preparation for class, carelessness, easy frustration, and low academic self-concept can all result from reading difficulties. Methods used to help the student with these problems should be designed to help the student achieve some success (no matter how small). To a student accustomed to failure, even a small amount of success is a major first step in improving his or her motivation.

Many teachers find it advisable to work more slowly and cover less material than they would normally. Repetition of main ideas is important. Assignments should be short and practical. Sometimes, working through an assignment step-by-step is beneficial. A student may develop a sense of responsibility for an assignment if the teacher gives each student personal attention and specifically asks for the work.

Making students recopy sloppy work can also help the student to feel more respect for the work he or she has done. These are just a few techniques you may find helpful in trying to develop a feeling of success within the student.

## Visual Aids

Visual aids can be considered distracting from the printed text. In many cases, students ignore visual aids and don't understand their relationship to the printed words. The authors of this textbook use pictures, charts, tables, and diagrams to illustrate and condense information. Visual aids are an essential part of a chapter and should be read as carefully as the text. In this textbook, photographs and art are very closely tied to the text words.

At the appropriate time, several questions may be asked about a visual aid to focus attention on it. The authors have included several such questions in the text. Or the teacher may ask questions not provided by the authors.

## Paragraph Structure

In order to read science effectively, the student must be aware of the different paragraph forms commonly found in scientific writing. The introductory paragraphs at the beginning of a chapter establish the purpose of the chapter. Notice in Chapter 11 how the first paragraphs introduce the chapter. By using introductory paragraphs as a guide to reading the chapter, the reader can understand the author's purpose for writing.

Definition paragraphs are easy to recognize and can be most helpful for vocabulary clarification. Such paragraphs are particularly important to future understanding of terms when it is the assumption that their meaning is known to the reader.

In scientific writing, paragraphs that present a sequence of events or a chronology are often enumerated. Section 11:2 shows an example of this type of writing. Students many need to be helped to see the relationships and interrelationships of each point.

The transition paragraph is inserted in the discourse to call attention to a shift in time or to change the subject, or just to present an example.

There are many other paragraph forms common to scientific writing, but the authors of this text chose to use these forms most frequently.

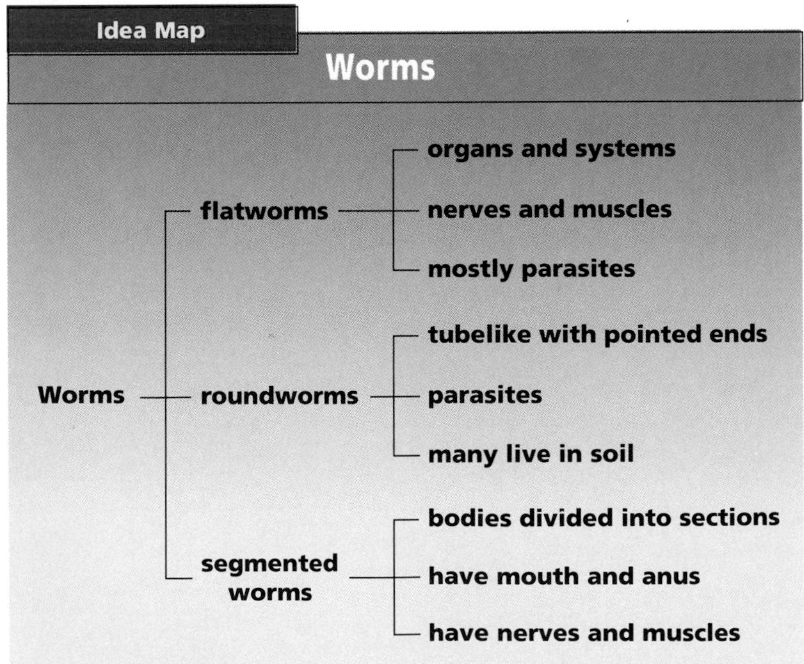


**Study Tip:** Use this idea map as a study guide to worms. The traits of each worm group are shown.

## Note Taking

Note taking can be a difficult and frustrating experience for some students. They often are not aware of how many notes to take. There are several ways note taking can be simplified. Adhering to the numbered sections of the chapter, the teacher can provide a skeletal outline for a chapter. This framework can be provided until students are confident with the outlining process. Completion of the outline should also be teacher-directed. If abbreviations are used, make a key (phy=phylum) to avoid confusion when referring to the outline. The skeletal framework also provides an organizational framework from which students can study.

Other forms of note taking also help students organize their notes, see relationships between ideas, and sort out essential information. A table outline condenses information and is easy to complete once the "features" are identified.

An outline in the form of an idea map provides a visual overview of a chapter or section. With this type of diagram, students can see the relationship of one topic to another. Numerous idea maps are found throughout the chapters. Have students use these as study guides, or have them construct their own.

## Vocabulary

The vocabulary of science consists of two types of words: technical and general. The authors of this text have used various methods to define technical vocabulary. Direct definition is characterized by the following format: vocabulary word, linking verb (is), followed by complete definition. This is the method that the authors use most frequently. Students should be directed to the glossary and cautioned that it presents words only as they are used in the text material.

In addition to technical vocabulary, students must be aware of general vocabulary. Certain words in our language have little or no meaning by themselves, but their use in sentences or paragraphs is of the utmost importance. These vocabulary words will be referred to as signal words. The first group is classified as "go" words. These direct the student to continue reading. The most common examples of these signal words are *and, next, in addition, first, second,* and *another.* "Caution" signals alert the reader to pay careful attention to the next point. The most common caution signals are *thus, therefore,* and *as a result.*

The last group of signal words are "turn" signals. These words indicate a different view, or an opposing idea. The most common words in this group are *nevertheless, however,* and *although.*[1]

The teacher's task is to teach the function of these words when they occur in context. It is advisable for the teacher to underline these words in the text so that he or she is alert to their occurrence. It is important to convey their function to the student.

[1] H. Alan Robinson, *Teaching Reading and Study Strategies: The Content Areas* (Boston Massachusetts: Allyn and Bacon, Inc., 1978), pp. 104-106.

# Teaching Special Students

| | DESCRIPTION | SOURCES OF HELP/INFORMATION |
|---|---|---|
| **Learning Disabled** | All learning disabled students have an academic problem in one or more areas, such as academic learning, language, perception, social-emotional adjustment, memory, or attention. | *Journal of Learning Disabilities*<br>*Learning Disability Quarterly* |
| **Behaviorally Disabled** | Students with behavior disorders deviate from standards or expectations of behavior and impair the functioning of others and themselves. These students may also be gifted or learning disabled. | *Exceptional Children*<br>*Journal of Special Education* |
| **Physically Disabled** | Students who are physically disabled fall into two categories—those with orthopedic impairments and those with other health impairments. Orthopedically impaired students have the use of one or more limbs severely restricted, so the use of wheelchairs, crutches, or braces may be necessary. Students with other health impairments may require the use of respirators or have other medical equipment. | Batshaw, M.L., and M.Y. Perset. *Children with Handicaps: A Medical Primer.* Baltimore: Paul H. Brooks, 1981.<br>Hale, G. (Ed.). *The Source Book for the Disabled.* NY: Holt, Rinehart & Winston, 1982.<br>*Teaching Exceptional Children* |
| **Visually Disabled** | Students who are visually disabled have partial or total loss of sight. Individuals with visual impairments are not significantly different from their sighted peers in ability range or personality. However, blindness can affect cognitive, motor, and social development, especially if early intervention is lacking. | *Journal of Visual Impairment and Blindness*<br>*Education of Visually Handicapped*<br>American Foundation for the Blind |
| **Hearing Impaired** | Students who are hearing impaired have partial or total loss of hearing. Individuals with hearing impairments are not significantly different from their hearing peers in ability range or personality. However, deafness can affect cognitive, motor, and social development if early intervention is lacking. Speech development also is often affected. | *American Annals of the Deaf*<br>*Journal of Speech and Hearing Research*<br>*Sign Language Studies* |
| **Multicultural and/or Bilingual** | Multicultural and/or bilingual students often speak English as a second language or not at all. The customs and behavior of people in the majority culture may be confusing for some of these students. Cultural values may inhibit some of these students from full participation. | *Teaching English as a Second Language Reporter*<br>Jones, R.L. (Ed.), *Mainstreaming and the Minority Child.* Reston, VA: Council for Exceptional Children, 1976. |
| **Gifted** | Although no formal definition exists, these students can be described as having above average ability, task commitment, and creativity. Gifted students rank in the top 5% of their class. They usually finish work more quickly than other students, and are capable of divergent thinking. | *Journal for the Education of the Gifted*<br>*Gifted Child Quarterly*<br>*Gifted/Creative/Talented* |

## TIPS FOR INSTRUCTION

1. Provide support and structure: clearly specify rules, assignments, and duties.
2. Establish situations that lead to success: use simple vocabulary.
3. Practice skills frequently—use games and drills to help maintain student interest.
4. Allow students to record answers on tape and allow extra time to complete tests and assignments.
5. Provide outlines or tape lecture material.
6. Pair students with peer helpers, and provide classtime for pair interaction.

---

1. Provide a carefully structured environment with regard to scheduling, rules, and room arrangement.
2. Clearly outline objectives and how you will help students obtain objectives. Seek input from them about their strengths, weaknesses, and goals.
3. Reinforce appropriate behavior and model it for students.
4. Do not expect immediate success. Instead, work for long-term improvement.
5. Balance individual needs with group requirements.

---

1. Assume that students understand more than they may be able to communicate.
2. Openly discuss with student any uncertainties you have about when to offer aid.
3. Ask parents or therapists what special devices or procedures are needed, and if any special safety precautions need to be taken.
4. Allow physically disabled students to do everything their peers do, including participating in field trips, special events, and projects.
5. Help nondisabled students and adults understand the characteristics of physically disabled students.

---

1. Help the student become independent—make the student accountable for assignments.
2. Teach classmates how to serve as guides.
3. Eliminate unnecessary noise in the classroom.
4. Encourage students to use their sense of touch. Provide tactile models whenever possible.
5. Describe people and events as they occur in the classroom.
6. Provide taped lectures and reading assignments.
7. Team the student with a sighted peer for laboratory work.

---

1. Seat the students where they can see your lip movements easily, and avoid visual distractions.
2. Avoid standing with your back to a window or light source.
3. Avoid moving around the room or writing on the board while speaking. Instead, use an overhead projector, which allows you to maintain eye contact while writing.
4. Encourage students to face the speaker, even if students must move around during class discussions.
5. Write all assignments on the board, or hand out handwritten instructions.
6. If the student has a manual interpreter, allow both student and interpreter to select the most favorable seating arrangements.

---

1. Do not allow the language a student brings to school to influence your expectations of the student's academic performance; however, beware of reverse discrimination.
2. Try to incorporate the student's language into your instruction. The help of a bilingual aide may be effective.
3. Include information about different cultures in your curriculum to aid students' self-image—avoid cultural stereotypes.
4. Encourage students to share their culture in the classroom.

---

1. Make arrangements for students to take selected subjects early.
2. Let students express themselves in art forms such as drawing, creative writing, or acting.
3. Make public services available through a catalog of resources, such as agencies providing free and inexpensive materials, community services and programs, and people in the community with specific expertise.
4. Ask "what if" questions to develop high-level thinking skills.
5. Emphasize concepts, theories, ideas, relationships, and generalizations.

# Assessment

## Assessment

What criteria do you use to assess your students as they progress through a course? Do you rely on formal tests and quizzes? To assess students' achievement in science, you need to measure not only their knowledge of the subject matter, but also their ability to handle apparatus; to organize, predict, record, and interpret data; to design experiments; and to communicate orally and in writing. ***Biology: An Everyday Experience*** has been designed to provide you with a variety of assessment tools, both formal and informal, to help you develop a clearer picture of your students' progress.

## Performance Assessment

Performance assessments are becoming more common in today's schools. Science curriculums are being revised to prepare students to cope with change and with futures that will depend on their ability to think, learn, and solve problems. Although learning fundamental concepts will always be important in the science curriculum, the concepts alone are no longer sufficient in a student's scientific education. Performance assessments differ in formality and complexity, but in most cases, the teacher observes a pupil or group of pupils involved in an activity and rates the performance and/or the products that result from the activity. Background information and specific examples of performance assessment are included in Glencoe's ***Alternate Assessment in the Science Classroom. Biology: An Everyday Experience*** provides numerous opportunities to observe student behavior both in informal and formal settings. The laboratory activities present many instances where you can informally observe students and evaluate their understanding of both concepts and process skills. Each **Lab** and **Mini Lab** contains suggestions that will enable you to assess students' understanding of how the lab relates to the concepts presented in the text.

Another approach for assessing student mastery of concepts and skills in the laboratory is provided in Glencoe's ***Performance Assessment in the Biology Classroom.*** It features 30 laboratory exercises that enable you to evaluate students' skills in handling laboratory equipment and students' knowledge of laboratory processes.

## Group Performance Assessment

Recent research has shown that cooperative learning structures produce improved student learning outcomes for students of all ability levels. ***Biology: An Everyday Experience*** provides many opportunities for cooperative learning and, as a result, many opportunities to observe group work processes and products. Glencoe's ***Cooperative Learning Resource Guide*** provides strategies and resources for implementing and evaluating group activities. In cooperative group assessment, all members of the group contribute to the work process and the products it produces. For example, if a mixed ability, four-member laboratory work group conducts an activity, you can use a rating scale or checklist to assess the quality of both group interaction and work skills. An example, along with information about evaluating cooperative work, is provided in the booklet ***Alternate Assessment in the Science Classroom.*** All four members of the group are expected to review and agree on the data sheet produced by the group. You can require each member to certify the group's results by signing the data sheet or lab report. In this approach, all members of the group receive the same grade on the work product. Research shows that cooperative group assessment is as valid as individual assessment. Additionally, it reduces the marking and grading work load of the teacher.

## Portfolios: Putting It All Together

The purpose of a student or cooperative group portfolio is to present examples of the individual's or group's work in a "non-testing" environment. A portfolio is simply a method for assembling and presenting selected examples of work products. The process of assembling the portfolio should be both integrative (of process and content) and reflective. The performance portfolio is *not* a complete collection of all worksheets and other assignments for a grading period. At its best, the portfolio should include integrated performance products that show growth in concept attainment and skill development. You can structure the portfolio development process by establishing categories and other limiting specifications. An essential component in portfolio development is the composition of a submission letter or reflective paper that lists the contents of the portfolio and discusses growth in knowledge, attitudes, and skills.

***Biology: An Everyday Experience*** presents a wealth of opportunities for performance portfolio development. Each chapter contains projects, enrichment activities, investigations, skill reviews, library research opportunities, and connections with life, society, and literature. Each of these student activities results in a product. A mixture of these products can be used to document student growth during the grading period.

Finally, ***Biology: An Everyday Experience*** strongly suggests the use of student journals. Students are encouraged to write observations, descriptions, and reflections in their journals. They are also encouraged to include diagrams and drawings. Excerpts from the student journal can be included in the individual or group portfolio. Additionally, as many writers have discovered, the journal will be an excellent resource for developing the reflective submission letter or paper.

## Content Assessment

While new and exciting performance skill assessments are emerging, paper-and-pencil tests are still a mainstay of student evaluation. Students must learn to conceptualize, process, and prepare for traditional content assessments. Presently and in the foreseeable future, students will be required to pass pencil-and-paper tests to exit high school, and to enter college, trade schools, and other training programs.

Traditional content assessment forms such as matching, multiple choice, and short essay items are effective in sampling content and can be quickly marked and scored. ***Biology: An Everyday Experience*** contains numerous strategies and formative checkpoints for evaluating student progress toward mastery of science concepts. Throughout the chapters in the ***Student Edition,*** **Check Your Understanding** questions and application tasks are presented. This spaced review process helps build learning bridges that allow all students to confidently progress from one lesson to the next.

After instruction for the chapter is completed, a summation of the major concepts is presented. Small groups of students can research the major concepts in the chapter and present restatements of their meaning in writing and as oral reports to the class.

After the main idea presentations, the formal review process for the written content assessment can begin. ***Biology: An Everyday Experience*** presents a two-page **Chapter Review** at the end of each chapter. Individual students or cooperative groups of three to five students can respond to these items to check their understanding of science terms, concepts, critical thinking skills, and problem solving techniques. By evaluating the student responses to this extensive review, you can determine if any substantial reteaching is needed.

For the formal content assessment, a two-page chapter review and a three-page test are provided in the ***Biology: An Everyday Experience Evaluation*** booklet for each chapter in the student text. If your individual assessment plan requires a test that differs from the **Chapter Test** in the resource package, customized tests can be easily produced using the ***Computer Test Bank.***

# Performance Objectives

## What Is a Good Performance Objective?

Statements of performance objectives can usually be divided into three distinct parts. First is a description of the conditions under which the desired behavior will be observed. Here are some examples that describe the condition:

"When presented with pictures of various protists...."

"When asked to list the parts of the plant...."

The second part of a well-stated performance objective is a clear description of exactly what behavior you are looking for. A good performance objective would avoid using such terms as "know about," "appreciate," or "sense the relationship between."

In other words, a good performance objective uses verbs that express some type of observable action. Here are some examples of action verbs that describe observable behaviors: states orally, identifies, watches, matches, lists, states, distinguishes, manipulates, hypothesizes, measures, constructs.

The third part of a well-stated performance objective is a description of some level of performance or criteria that may help you to know if the student or class has performed to the degree that you hoped for.

## Where Do Performance Objectives Come From?

Statements of performance objectives come in many forms, and no given number of them constitute a biology curriculum. You could never say that students, having achieved the objectives contained in this *Teacher Edition,* will then "know" biology. The performance objectives for each school, class, group of students, or individual student will vary according to the students, the teacher, the resources available in the learning environment, and the task at hand.

Ultimately, the task of selecting, composing, and evaluating performance objectives is up to you. You should draw on the performance objectives in this *Teacher Edition* and on other sources or other people who could help you. Students are good, if not the best, sources of performance objectives. When they can express to you their questions, their interests, and their goals, then you can cooperatively decide what objectives should be reached.

Any performance objective for teaching science should be based on one or more of the goals of science education. You should be able to say that the reason you're working to accomplish this particular objective is because it's one step or one part of a long-range goal of science education. For example, if one of the goals of science education is to develop the student's ability to utilize some of the methods and processes by which problems are solved scientifically, then one of the purposes of a lesson might be to develop an understanding of what an experiment is. If a student understands what an experiment is:

When asked to define an experiment, he/she will state that it is a test or a way of proving some theory or principle.

When asked what he/she would do to find out what effect exercise has on blood pressure or heartbeat rate, he/she will describe what to do and use the term "experiment" in the description.

When asked for examples of experiments that have been conducted so far this semester, he/she will cite at least three.

Specific performance objectives, such as the above, are only examples of one small part of that larger, broader goal for science education. However, each performance objective should be consistent with one or more of those goals so that you can justify what you are teaching, and so that you can see students progress toward that goal.

Performance objectives are listed for each chapter of *Biology: An Everyday Experience* in the margin and on the interleaf pages preceding each chapter. They are related to the four broad goals of science education.

1. **Attitudes** are behaviors that you might observe as students increase their enjoyment of science activities, and as they demonstrate an interest in the scientific objects and events in their environment. Few of these behaviors are prompted by the teacher. They are voluntary behaviors that indicate an affinity for, an inclination toward, and a preference for science.

2. **Process** goals are behaviors that you might observe as students use the thinking processes of analyzing, experimenting, applying, hypothesizing, theorizing, comparing and contrasting, classifying, and observing.

3. **Knowledge** goals are behaviors that you might observe as students demonstrate that they have gained an understanding of and can use certain terms, concepts, and generalizations of science. Generally, a knowledge objective is one that calls on students to recall or explain information, concepts, or principles that they learned in the past or that were presented in the text and other materials.

4. **Skill** goals are behaviors that you might observe as students gain skill in using scientific equipment correctly and organizing, recording, or reporting information accurately.

You will find that the criterion or level of performance for each objective is not included. The intent was to create guidelines and to state levels of expectancies for you. Only you can decide if 100% or only 50% of your students should be able to accomplish each task. You will need to develop and insert these criterion levels for your own students based on your knowledge of their abilities, the available materials, and your own situation.

## How Do You Know You've Achieved Your Objectives?

When you try to find out if students have learned anything, you probably give some form of written test—a true-false, completion, matching, or essay type of examination. You might also be exploring some of the assessment techniques given on pages 18T and 19T. All of these forms of evaluation are important in determining whether or not students have achieved some of the objectives.

# Cooperative Learning

Cooperative learning groups *learn* things together, not just do things together. Studies show that cooperative learning experiences promote more learning than competitive or individual learning experiences regardless of student age, subject matter, or learning activity. Harder learning tasks, such as problem solving, critical thinking, and conceptual learning, come out far ahead when cooperative strategies are used. Studies show that in classroom settings, adolescents learn more from each other about subject matter than they do from the teacher.

The basic elements of a cooperative learning group are as follows: (1) The students must perceive that they "sink or swim together." (2) Students are responsible for everyone else in the group, as well as for themselves, in learning the assigned material. (3) Students must see that they all have the same goal, that they need to divide up the tasks and share the responsibilities equally, and that they will be given one evaluation or reward that will apply to all the members of the group.

## Getting Started

1. *Arrange your room.* Students in groups should face each other as they work together. It is helpful to number the tables so you can refer to groups by number rather than by a student's name. Or, have each group choose a name.

2. *Decide on the size of the group.* Groups work best when teams are composed of two to five students. The materials available might dictate the size of the group.

3. *Assign students to groups.* Groups should reflect real-life situations. The group should be mixed socially, racially, ethnically, sexually, and by learning abilities. If students are allowed to choose their own groups there may be less task-oriented behavior, and the homogeneity that usually results will not allow them the opportunity to hear views that may differ from their own. If a group finds that things are not working out, members may complain that they want to switch groups. If students are told that they can change groups after they have proven their ability to work effectively together for a given time, maybe two weeks, they often decide they don't want to switch. Occasionally a student may insist on working alone. This student may change his or her mind upon seeing that everyone else's grades are better and that the groups are having more fun.

4. *Change groups.* Some teachers keep their groups together for a quarter, a semester, or for a teaching unit. The only criterion would be to have groups stay together long enough to experience success as a group. The advantage of changing groups is that students will have more opportunities to deal with a variety of classmates.

5. *Prepare students for cooperation.* This is a critical step. Tell students about the rationale, procedures, and expected outcomes of this method of instruction. Students need to know that you are not trying to force them to be friends with other people, but asking them to work together as they will later on in life with people who come together for a specific purpose.

Start small. You do not have to incorporate all of this information into your planning for your first cooperative groups. Some ideas you will find easy to adapt to your style while others may be more difficult.

6. *Explain group tasks.* For most activities, each group will need someone to take notes, someone to obtain supplies, someone to summarize as the group progresses, and someone who makes sure everyone is involved. Classes that have less experience using these methods may need to assign these roles to group members at the beginning of each activity. These jobs should be done by different students each time, so that one student does not feel that he or she is being burdened with one job all the time.

7. *Explain the day's lesson.* Develop a transparency that includes the following information or write this information on the chalkboard each day.

- ▶ **Form Groups:** list the number of students in each group.
- ▶ **Topic of the Day:** general academic topic.
- ▶ **Task:** title of lab or other activity. At this time, go over specific instructions, give background, or relate the work to previous learning.
- ▶ **Goal:** indicate whether students will all do individual work from which you will select one paper or product to grade, or whether they will produce only one product per group. Students' signatures on a product indicate that they will accept the grade.
- ▶ **Cooperative Skills:** list the specific group skills you will be checking. Start with one or two basic skills. The following skills start from basic and move to more advanced.

a. Stay with your group.
b. Encourage each other to participate.
c. Use each other's names. Use eye contact.
d. Ask your teacher for help only after you have decided as a group that you all need help.
e. Stress that all student contributions are valuable.

After students have some experience working in cooperative groups, you will expect higher-level cooperation skills.

a. Express support and acceptance, both verbal and nonverbal, through eye contact, enthusiasm, and praise.
b. Ask for clarification about what is happening.
c. Suggest new ideas.
d. Use appropriate humor that stays on task.

e. Describe feelings using "I messages."
f. Summarize and elaborate on what others have contributed.
g. Develop memory aids and analogies that are clever ways of remembering important points in the assignment.
h. Criticize ideas, not people.
i. Go beyond the first answer to a question.

## What the Teacher Does During Cooperative Group Work

1. *Monitor student behavior.* Just because students have been placed in groups and given instructions does not mean that they will automatically work cooperatively. When you first begin, you will want to observe very closely to see that things are off to a good start. Use a formal observation sheet to count the number of times you observe the behavior expected on that particular assignment for each group. Do not try to count too many different behaviors at the beginning. Reassure students that some of the things you are expecting them to say may sound contrived. Share your observations with each group.

2. *Provide assistance with the task.* Go ahead and clarify instructions, review concepts, or answer questions. But, avoid interfering in the group's process. You have now delegated authority to groups to carry out an assignment. It is of critical importance to let them make decisions *on their own.* Even though they are accountable to you for their work, you must let go and allow the groups to work things through for themselves. Students must be allowed to make mistakes, then evaluate themselves to see where they went wrong. You will find that because of their past experience, students who see you hovering nearby will automatically start asking you questions. Your first response should be, "Have you asked everyone in your group?" Explain to students that they will hear that question often from you.

Students will be doing many of the things you usually do. They will be asking each other questions, keeping each other on task, and encouraging everyone to participate. Your role will be supportive supervisor rather than direct supervisor. For example, as a supportive supervisor, you will help a group that has gotten stuck and is experiencing a high level of frustration. You might do this by asking a few open-ended questions. In a conflict situation, you might ask the group what they think the difficulty is and ask them to come up with some strategies for handling the conflict.

3. *Intervene to teach cooperative skills.* As you observe that some groups have more problems than others with learning cooperative skills, you may wish to intervene by asking the group to think about why they are not being effective and asking them to come up with a solution. If you notice that your groups seem to be having a difficult time coming up with ways of encouraging each other, you might ask them to come up with a list of things to say that would be encouraging and then post it in the room.

4. *Provide closure for the lesson.* Students should be asked to summarize what they have learned and should be able to relate it to what they have previously studied. You may want to review the main points and ask students to give examples and answer any final questions.

5. *Evaluate the group process.* In order for groups to be aware of their progress in learning to work together, they must be given time to evaluate how they are working together. Give a few minutes at the end of the lesson to decide if they achieved the criterion you set up for that particular lesson. Have them rate themselves as a group on a scale of one to ten and write down specified ways they could improve.

6. *Evaluate student learning.* Evaluation tools may include traditional tests and quizzes, group quizzes, and extra points for groups achieving overall high marks.

Keep in mind that cooperative learning doesn't just "happen." The first few days may seem like bedlam, with some students upset, others mistrustful, and others off task. Both you and your students will make mistakes. Just be patient and keep at it. You may wish to read more about cooperative learning groups in the references listed below.

Johnson, David W., and Roger T. Johnson. *Cooperation and Competition, Theory and Research.* Edina, MN: Interaction Book Co., 1989.

Kagan, Spencer. *Cooperative Learning, Resources for Teachers.* Laguna Niguel, CA: Resources for Teachers, 1989.

Slavin, Robert. *Cooperative Learning, Student Teams,* 2nd ed. Washington, DC: National Education Association, 1987.

# Planning Guide

*Biology: An Everyday Experience* is designed with built-in flexibility that both beginning and experienced teachers can appreciate. A planning guide has been provided to assist you with both long- and short-range planning. It is designed to help you provide students with the best possible biology program.

The planning guide presents each chapter and the number of class sessions suggested for that chapter. The entire course assumes 180 class sessions divided into six-week periods. This would be equivalent to 160 hours of combined class and laboratory time for the year (50 minutes per day). For alternative block scheduling planning, see the *Lesson Plans* book in the *Teacher Resource Package.*

The planning guide classifies the sections of each chapter into two categories: those of primary importance and those of secondary importance. Most biology curricula include those sections designated as being of primary importance. The sections listed in this category provide a minimum program that you can develop more fully as time permits by using the sections designated as being of secondary importance. In some cases, the material designated as being of secondary importance is information that helps relate the information in the chapter to students' experiences. For this reason, it is important to include this information whenever possible to enhance students' appreciation and understanding of biology.

## Planning Guide for *Biology: An Everyday Experience*

| UNIT | CHAPTER | CLASS SESSIONS | PRIMARY IMPORTANCE | SECONDARY IMPORTANCE |
|---|---|---|---|---|
| 1 | 1 | 6 | All sections | |
| 1 | 2 | 6 | All sections | |
| 1 | 3 | 6 | All sections | |
| 2 | 4 | 7 | All sections | |
| 2 | 5 | 5 | All sections | |
| 2 | 6 | 5 | All sections | |
| 2 | 7 | 7 | All sections | |
| 2 | 8 | 7 | All sections | |
| 3 | 9 | 7 | All sections | |
| 3 | 10 | 4 | All sections | |
| 3 | 11 | 5 | 11:1, 11:2, 11:3 | 11:4 |
| 3 | 12 | 5 | 12:1, 12:2, 12:4 | 12:3 |
| 3 | 13 | 5 | 13:1, 13:2, 13:4 | 13:3 |
| 3 | 14 | 5 | 14:1, 14:2 | 14:3 |
| 4 | 15 | 5 | 15:1, 15:2, 15:3 | 15:4 |
| 4 | 16 | 5 | 16:1, 16:2 | 16:3 |
| 4 | 17 | 4 | All sections | |
| 4 | 18 | 8 | All sections | |
| 5 | 19 | 7 | 19:1, 19:2 | 19:3 |
| 5 | 20 | 5 | All sections | |
| 5 | 21 | 6 | 21:1, 21:2 | 21:3 |
| 6 | 22 | 6 | 22:1, 22:2 | 22:3 |
| 6 | 23 | 5 | All sections | |
| 6 | 24 | 5 | All sections | |
| 6 | 25 | 5 | All sections | |
| 7 | 26 | 6 | All sections | |
| 7 | 27 | 5 | 27:1, 27:2 | 27:3 |
| 7 | 28 | 6 | All sections | |
| 7 | 29 | 7 | All sections | |
| 8 | 30 | 6 | 30:1, 30:2, 30:3 | 30:4 |
| 8 | 31 | 5 | 31:1, 31:3 | 31:2 |
| 8 | 32 | 6 | 32:1, 32:3 | 32:2 |

# Alternate Sequence Approach

You may choose to teach the subject matter in a sequence other than that which is presented in Chapters 1-32. When using an altered sequence, students may encounter minor vocabulary and concept difficulties. However, these gaps can easily be remedied by the teacher. The following alternate approach is provided merely as a guide and should not be construed as a rigid plan. Areas of emphasis are included along with the alternate sequence. Selection of topics should depend on the interests and ability levels of the classes you teach.

## Ecological Approach

| Chapters | Emphasis |
|---|---|
| 1-3 | Life characteristics<br>Cell anatomy<br>Classification of life forms into five kingdoms |
| 19, 30-32 | Plants and photosynthesis<br>How life forms are dependent upon one another<br>Role of energy in ecosystems<br>Recycling of needed chemicals<br>Population dynamics<br>Human role in disturbing the environment<br>Human problems resulting from a damaged environment<br>Future and current solutions to a damaged environment |
| 22, 26-29 | Role of mitosis and meiosis<br>Heredity principles<br>How life forms change as a consequence of natural selection<br>Role of changing environment as selection agent for evolution |
| 4-6<br>20-21, 23 | Survey of life forms (not including animals)<br>Protists, monera, fungi, plants |
| 7-16, 24-25 | Survey of animals<br>Major animal systems |
| 17-18 | Animal behavior<br>Drugs and behavior |

# Skill Development and Reinforcement

The development and reinforcement of thinking and manipulative skills is a major goal of *Biology: An Everyday Experience.* Throughout the text, students are involved in many activities that test and strengthen their ability to think and operate in a science setting. Specific skill-related exercises can be found throughout the book in Labs, Check Your Understanding questions, Mini Labs, Skill Checks, Skill Review questions, and the Skill Handbook.

## Labs

A lab is designed to reinforce and develop a variety of skills through hands-on activities that follow a scientific procedure. Each lab in *Biology: An Everyday Experience* includes a problem statement, a list of process skills, a list of materials, numbered procedure statements, tables, and two sets of questions: those that ask students about the data and observations collected during the lab, and those that ask students to analyze the data they collected and apply the information learned in the investigation to other situations. Many labs require the formation of a hypothesis. More complex process skills will be developed if students understand that hypotheses are neither right nor wrong and that science processes do not produce correct answers, but rather statements that are either supported or not supported. Finally, there is an extension that asks students to go beyond the lab by designing further experiments or participating in some other type of creative activity.

## Check Your Understanding

Within each chapter, five review questions appear at the close of each numbered section. In each instance, the fourth question is labeled Critical Thinking. A critical thinking question requires a student to utilize critical thinking skills to answer a question about information presented in the text. The fifth question makes a connection between biology and reading, writing, or math skills.

## Mini Labs and Skill Checks

Mini Labs and Skill Checks are sprinkled throughout each chapter. These activities give students an opportunity to reinforce skills such as sequencing, classifying, observing, inferring, or constructing a table. A reference to the Skill Handbook is provided with each Skill Check and Mini Lab for students who need further help. Mini Labs use simple materials that are readily available at home or in the classroom. Each chapter has a Skill Check on understanding science words. In this skill, science words are broken down into root words, prefixes, and suffixes. Extra help with biology words is found in Appendix C, which lists root words, prefixes, and suffixes that are common in biology.

## Skill Review

Within each Chapter Review are four Skill Review questions that review the skills first practiced in the Skill Checks, Mini Labs, and Labs. Students are given one more opportunity to practice valuable skills while learning important concepts in biology.

## Skill Handbook

For help with Mini Labs, Skill Checks, and Skill Review questions, students using *Biology: An Everyday Experience* may refer to the Skill Handbook on pages 704-719 of the textbook. The handbook contains instructions on skills and an example of each skill type. The skills in the Skill Handbook are:

- *Practicing Scientific Method
- *Reading Science
- *Using a Microscope
- *Organizing Information
- *Using Numbers

Each skill includes several subcategories. For example, Reading Science includes skills on sequencing, understanding science words, interpreting diagrams, and formulating models.

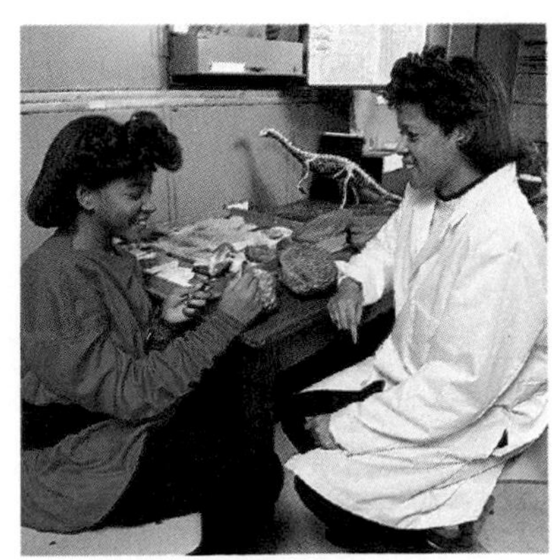

# Developing Process Skills

*Biology: An Everyday Experience* provides experiences that help students both develop and apply critical thinking process skills. These experiences are provided by the use of higher level divergent questions, and lab activities. Students are given the opportunity to use and apply both basic and complex process skills, thus facilitating their advancement toward higher levels of cognitive thinking.

Students able to think in terms of several solutions to a single problem are not limited to concrete activities. Instead, these students can formulate hypotheses as testable ideas in their minds and can demonstrate deductive patterns of thought. Through integration of the important critical thinking process skills listed below, students develop the ability to think logically and abstractly.

## Basic Process Skills

**Observing:** Students use one or more senses to increase their perceptions in order to learn more about objects and events.

**Classifying:** Students group objects or events based on common properties and/or categorize based on existing relationships among objects or events.

**Inferring:** Students propose interpretations, explanations, and causes from observed events and collected data.

**Communicating:** Students convey information verbally, in both oral and written forms, and visually through graphs, charts, pictures, and diagrams.

**Recognizing and Using Spatial Relationships:** Students estimate the relative positions of moving and non-moving objects.

**Measuring:** Students identify and order length, area, volume, mass, and temperature to describe and quantify objects or events.

**Predicting:** Students propose possible results or outcomes of future events based on observations and inferences drawn from previous events.

**Using Numbers:** Students transfer or apply ordering, counting, adding, subtracting, multiplying, and dividing to quantify data where appropriate in investigations or experiments.

## Complex Process Skills

**Interpreting Data:** Students explain the meaning of information gathered in scientific situations.

**Forming Hypotheses:** Students make an assumption in order to draw out and test its logical consequences.

**Separating and Controlling Variables:** Students recognize the many factors that affect the outcome of events and understand the relationship of the factors to one another so that one factor (variable) can be manipulated while the others are controlled.

**Experimenting:** Students test hypotheses or predictions under conditions where variables are both controlled and manipulated.

**Formulating Models:** Students construct mental, verbal, or physical representations of ideas, objects, or events. The models are then used to clarify explanations or to demonstrate relationships.

**Defining Operationally:** Students form a working definition that is based upon actual experience in which the student has participated.

## Process Skills Used in Labs

| Lab | Process Skills |
|---|---|
| 1-1 | observe, recognize and use spatial relationships |
| 1-2 | form hypotheses, interpret data, separate and control variables |
| 2-1 | make and use tables, interpret data, infer, form hypotheses |
| 2-2 | observe, compare, use a microscope |
| 3-1 | classify, observe |
| 3-2 | classify, communicate, make a table |
| 4-1 | observe, recognize and use spatial relationships, formulate a model |
| 4-2 | make and use tables, observe, measure in SI, use a microscope |
| 5-1 | make and use tables, interpret data, observe, classify, use a microscope |
| 5-2 | observe, infer, use a microscope, design an experiment |
| 6-1 | make and use tables, interpret data, use a microscope, measure in SI, infer |
| 6-2 | make and use tables, observe, use a microscope |
| 7-1 | observe, make and use tables, interpret data |
| 7-2 | observe, infer, predict, make and use table |
| 8-1 | make and use tables, observe, infer |
| 8-2 | interpret data, observe, infer |
| 9-1 | use numbers, experiment, measure in SI, make and use tables |
| 9-2 | form a hypothesis, experiment, make and use tables |
| 10-1 | separate and control variables, form a hypothesis, observe, design an experiment |
| 10-2 | classify, interpret diagrams |
| 11-1 | form a hypothesis, calculate, interpret data, make and use tables |
| 11-2 | formulate models, form a hypothesis, measure in SI |
| 12-1 | use a microscope, measure in SI, compare |
| 12-2 | use a microscope, compare, classify |
| 13-1 | form a hypothesis, measure in SI, interpret data |
| 13-2 | observe, interpret data, make and use tables, infer |
| 14-1 | form a hypothesis, experiment, measure in SI |
| 14-2 | formulate a model, experiment, infer |
| 15-1 | form a hypothesis, calculate, infer |
| 15-2 | experiment, interpret data |
| 16-1 | formulate a model, observe, measure in SI |
| 16-2 | predict, interpret data, make and use tables |

| Lab | Process Skills |
|---|---|
| 17-1 | predict, separate and control variables, observe, design an experiment |
| 17-2 | form hypotheses, observe, make and use tables |
| 18-1 | observe, interpret data, experiment |
| 18-2 | observe, interpret data, experiment |
| 19-1 | measure in SI, interpret diagrams, classify |
| 19-2 | observe, infer, experiment |
| 20-1 | observe, formulate a model, infer, design an experiment |
| 20-2 | observe, classify, form hypotheses, design an experiment |
| 21-1 | measure in SI, form hypotheses, interpret data, experiment |
| 21-2 | observe, form hypothesis, interpret data |
| 22-1 | observe, formulate a model, sequence |
| 22-2 | use numbers, make and use tables, form hypotheses, interpret data |
| 23-1 | form a hypothesis, experiment, observe |
| 23-2 | experiment, observe, measure in SI, infer |
| 24-1 | make and use tables, measure in SI, recognize and use spatial relationships, make scale drawings |
| 24-2 | make and use graphs, infer, relate cause and effect |
| 25-1 | measure in SI, use numbers, infer, make and use graphs |
| 25-2 | interpret data, measure in SI, compare |
| 26-1 | observe, predict, interpret data |
| 26-2 | observe, formulate a model, form hypotheses |
| 27-1 | make and use tables, formulate a model, calculate, design an experiment |
| 27-2 | observe, make and use tables, interpret data |
| 28-1 | formulate a model, interpret diagrams, observe |
| 28-2 | observe, infer, interpret data |
| 29-1 | form hypotheses, experiment, interpret data, infer, calculate |
| 29-2 | formulate a model, observe, infer |
| 30-1 | form hypotheses, interpret data, infer, design an experiment |
| 30-2 | observe, infer, design an experiment |
| 31-1 | observe, form hypotheses, interpret data, infer, design an experiment |
| 31-2 | observe, form hypotheses, interpret data, design an experiment |
| 32-1 | observe, experiment, interpret data, classify |
| 32-2 | observe, form hypotheses, interpret data, design an experiment |

# Problem Solving

A major challenge faced by teachers is teaching students how to approach problems. When asked to apply a concept or interpret an everyday situation in light of the concepts they have learned, many students become puzzled and confused. To aid you in helping your students develop problem solving skills, each unit of *Biology: An Everyday Experience* includes an Applying Technology feature. In addition to their value in the Tech Prep curriculum, these features help students (1) become aware of the process they use to solve problems, (2) identify the stages in that process, and (3) identify and choose strategies for problem solving.

Below are some misconceptions most students hold about problem solving.

*A way to solve the problem should be obvious when you read a problem.
*You can't write anything down unless you have the solution worked out in your head.
*No trial-and-error attempts or "playing around" with different approaches to a problem are allowed.
*There is only one "right" way to solve a problem.
*You can't change a problem to make it simpler.
*Problem solving is a step-by-step process that always proceeds in a linear, logical manner.

When faced with a problem, most students focus their attention on getting the "right" answer. Relationships among problems are rarely seen. Instead, each problem is looked on as a completely new, unique situation. Fear of beginning is also an obstacle to problem solving. You may want to present the following overall plan as one way to attack a problem.

1. *Verbalize the problem*—state the problem out loud and believe that you can solve it.
2. *Define the problem*—analyze the information given and determine what you know and what you need to know in order to solve the problem.
3. *Explore*—brainstorm strategies for organizing the given information and finding out what you need to know.
4. *Plan*—decide on a strategy or group of strategies and list steps and substeps of the strategies.
5. *Carry out the plan.*
6. *Assess*—decide what was done, whether or not you are closer to solving the problem, and why or why not.

Once students have an overall plan, they can begin to develop strategies for problem solving. A *strategy* is an organized approach to a problem that breaks down the task of obtaining and organizing information into stages. Having a group of strategies to choose from will help students over the initial panic felt when faced with a problem. Several strategies are listed here.

*construct a table, graph, or figure
*list all possible solutions
*look for a pattern
*talk to an expert
*act out the problem
*make a model
*guess and check
*work backwards
*make a drawing
*solve a simpler or similar problem

Remember that the attitude of students toward problem solving is a big factor in their success or failure. No matter how many models or strategies are given to students, if they think they are no good at solving problems, they will probably fail. Try to overcome this fear of the unknown by including problem solving in your daily teaching. Talk about problem solving as an important skill. Use the words "problem solving" often and consistently to describe the problem solving process. Finally, focus on helping students progress in using the process rather than in getting the "right" answer.

**Additional References**

Covington, M.V., et. al. *Productive Thinking Program.* Columbus, OH: Merrill Publishing Company, 1984.

deBono, E. *deBono's Thinking Course.* BBC Publication, 35 Marylebone High St., London, WIM 4AA, 1982.

Meiring, Steven P. *Problem Solving . . . A Basic Mathematics Goal: Part 1, Becoming a Better Problem Solver.* Palo Alto, CA: Dale Seymour Publications.

Meiring, Steven P. *Problem Solving . . . A Basic Mathematics Goal: Part 2, A Resource for Problem Solving.* Palo Alto, CA: Dale Seymour Publications.

# Cultural Diversity

"Multicultural education is an idea stating that all students, regardless of the groups to which they belong, such as those related to gender, ethnicity, race, culture, social class, religion, or exceptionality, should experience education equality in the schools."—James Banks

American classrooms reflect the rich and diverse cultural heritages of the American people. Students come from different ethnic backgrounds and different cultural experiences into a common classroom that must assist all of them to learn. The diversity itself is an important focus of learning experience.

Diversity can be repressed, creating a hostile environment; ignored, creating an indifferent environment; or appreciated, creating a receptive and productive environment. Responding to diversity and approaching it as a part of every curriculum is challenging to a teacher, experienced or not. The goal of science is understanding. The goal of multicultural education is to promote an understanding of how people from different cultures approach and solve the basic problems all humans have in living and learning. *Biology: An Everyday Experience* addresses this issue. In the Cultural Diversity features of the *Teacher Wraparound Edition,* information is provided about people and groups who have traditionally been misrepresented or omitted. Each feature includes a short student activity. The intent is to build awareness and appreciation for the global community in which we all live.

By providing these opportunities, *Biology: An Everyday Experience* is helping to meet the five major goals of multicultural education:

1. promoting the strength and value of cultural diversity
2. promoting the human rights of and respect for those who are different from oneself
3. promoting alternative life choices for people
4. promoting social justice and equal opportunity for all people
5. promoting equity in the distribution of power among groups

Three books that provide additional information on multicultural education are:

Banks, James A. (with Cherry A. McGee Banks). *Multicultural Education: Issues and Perspectives.* Boston: Allyn and Bacon, 1989.

Banks, James A. (with others). *Curriculum Guidelines for Multiethnic Education.* Washington, DC: National Council for the Social Studies, 1977.

Selin, Helaine. *Science Across Cultures: An Annotated Bibliography of Books on Non-Western Science, Technology, and Medicine.* New York: Garland, 1992.

# Using the Internet

## Using the Internet in the Classroom

The Internet is the world's largest computer network. It is a network of networks linked by high-speed data lines and wireless systems that freely exchange information. It provides access to individuals, corporations, educational institutions, government, and other groups.

To access the Internet, you need a computer with communication software, a modem, and a phone line. Many computers come equipped with an internal modem. To access the Internet, you also must have an account with an Internet service provider (server). Although these commercial providers can be expensive, it is possible in some cities to access the Internet without charge through university or local library systems.

## Internet for Teachers

After you have access to the Internet, you can use it to enhance your teaching in a variety of ways. You will be able to obtain numerous lessons, activities, project ideas, and labs on any topic you choose. Learn the locations of useful teacher contacts. When you find good ideas that you would like to use, simply download, print out, and hand out to your students.

The Internet can provide you with easy access to the most up-to-date information on any topic you may wish to research. There are thousands of free sources for graphics, videos, animation sequences, and shared software. By using the electronic mail–E-mail–capabilities of your server, you can exchange ideas with teachers worldwide, consult with research scientists, and order materials.

## Internet for Students

Students will also benefit from using the Internet. They can use the Internet to compare and contrast lab data with students in the next town or across the globe. As they use the Internet for information to enhance their lab reports and other classroom activities, they develop their search-and-retrieval strategies, which require careful analysis, evaluation, and application. They soon learn to express themselves unambiguously when they send messages to scientists, teachers, or friends over the Internet. Students who become familiar now with the technology of computers and the new method of communicating by asking questions and searching for answers on the Internet will be better prepared for the future in business, technology, education, and leisure.

## The World Wide Web

To access a wider source of multimedia information, you and your students can access the World Wide Web (WWW) by using a web browser. Many servers now provide the use of a web browser. The first page you see when you enter a WWW site is the homepage. From the homepage, you can obtain complete texts and graphics from books, clips from movies, or graphical replicas of art from museums. You can browse through books, magazines, and scientific periodicals. Go to the homepages of scientific-supply companies and order your lab supplies, equipment, books, or posters.

## The Glencoe Homepage

The Glencoe Homepage provides up-to-date links to useful web sites on the Internet. Simply go to the homepage at **http://www.glencoe.com**
and follow the links to access additional material for the chapter.

## Some Other Useful Web Sites

**AskERIC The Q & A Service**
http://ericir.syr.edu
This is an ongoing project that is building a digital library for educational information. You use it as you would a library.

**National Renewable Energy Laboratory**
http://www.nrel.gov/
The U.S. Department of Energy Lab offers information about renewable energy, including energy data, resource maps, publications, and job opportunities.

**U.S. Department of Education**
http://www.ed.gov/
This site provides information about various programs in the Department of Education.

**Webcrawler**
http://webcrawler.com
This is a search engine with useful links to high school biology homepages.

**Yahoo**
http://www.bham.wednet.edu
Another search engine with an extensive inventory of Web sites. Type in the key words *Biology Teacher* and it will list many useful resources for biology.

## Words of Caution

You might wish to caution your users that any scientific information they may read on the Internet has not been reviewed and authenticated by the usual peer-review system practiced in textbook and journal publishing. You will also have to carefully supervise and monitor student use of the Internet. Because there is no censorship on the Internet, students can easily find material that you or their parents find objectionable. However, with these cautions in mind, you will soon discover the use of the computer in the classroom to be an exciting and rewarding addition to your lesson plans.

# Tech Prep Education

Tech Prep is a rigorous and focused program of study that aims to create a workforce in the United States that is technically literate. It is designed to prepare students enrolled in a general curriculum for the demands of further education or for employment by providing them with essential academic and technical foundations, along with problem-solving, group-process, and lifelong-learning skills. These goals can be achieved by integrating vocational study with higher-level academic study.

## What Are the Characteristics of the Tech Prep Curriculum?

In 1990, Congress passed the Carl D. Perkins Vocational and Applied Technology Act to set aside funds for the development and administration of Tech Prep programs. The criteria outlined in Title III of the Perkins Act specify that Tech Prep programs take place during the last two years of high school, followed by two years of post-secondary occupational education, and that this education culminate in a certificate or associate degree. The Secretary's Commission on Achieving Necessary Skills (SCANS), an arm of the U.S. Department of Labor, published a report in June 1991 that outlined several competencies that characterize successful workers. The Tech Prep curriculum seeks to address these competencies, which include:

- the ability to use resources productively.
- the ability to use interpersonal skills effectively, including fostering teamwork, teaching others, serving customers, leading, negotiating, and working well with individuals from culturally diverse backgrounds.
- the ability to acquire, evaluate, interpret, and communicate data and information.
- the ability to understand social, organizational, and technological systems.
- the ability to apply technology to specific tasks.

## What Is the Role of the Community in Tech Prep Education?

The key to implementing a Tech Prep course of study is a partnership among high schools, post-secondary educational institutions, businesses, industry, and labor. Teachers and schools can enlist the support of their communities in a number of ways to enhance tech prep education in the science classroom. They include corporate partnerships, inviting scientists into the classroom, enlisting the aid of industrial tutors, and encouraging students to become involved in community service projects.

## How Does *Biology: An Everyday Experience* Address Tech Prep Issues?

*Biology: An Everyday Experience* helps you develop scientific and technological literacy in your students through a variety of performance-based activities that emphasize problem solving, critical thinking skills, and teamwork. In the *Student Edition,* many applications of biology are used throughout the text, in the Bio Tips, and in the features to illustrate the relevance of biology to everyday life and the world of work. Labs and Mini Labs, and Applying Technology features provide opportunities for practical applications of biological concepts. The Finding Out More section of the Chapter Review provides extra activities that will give the student practical experience with problem solving and that can be used as extensions for the interested student. Teaching strategies that are particularly suitable for the tech prep student are listed at the bottom of each chapter opener of the *Teacher Wraparound Edition. The Teacher Resource Package* includes a *Tech Prep Applications* booklet that gives additional activities and career ideas designed to fulfill the goals of tech prep education.

# Safety in the Laboratory

Safety is of prime importance in every classroom. However, the need for safety is even greater when science is taught. Students need to become aware of safety at all times, for themselves and their classmates.

Teachers and students using *Biology: An Everyday Experience* will find that safety awareness is emphasized throughout the book and the laboratory manual. In Labs where potential safety problems might arise, specific safety symbols are used to caution students. Safety symbols that are used in all Glencoe science programs are shown on page 33T. These symbols appear in Labs wherever appropriate. If you take the time to make students aware of them now at the beginning of the course, then they will know immediately what precautions are to be taken as soon as they see a particular symbol. Overall, safety problems can be avoided and accidents prevented with planning on the part of both teacher and student.

## In the Laboratory

1. Chemicals in the laboratory.
   Responsible storage, use, and disposal of chemicals in a science lab ensures the safety of you and your students. Follow the guidelines outlined on page 32T for safe and productive behaviors when working with chemicals.
2. Equipment in the laboratory.
   (a) Clean and dry all equipment before storing.
   (b) Label and organize all equipment.
   (c) Protect electronic equipment and microscopes from dust, humidity, and extremes in temperature.
   (d) Be sure all glassware is a heat-treated type that will not shatter.
3. Provide adequate work space and ventilation for activities.
4. Post safety and evacuation guidelines and safety symbols.
5. Check to ensure that safety equipment is accessible and working properly.
6. Provide containers for disposing of chemicals, broken glass, other waste products, and biological specimens. Disposal methods should meet your local guidelines.
7. Use a hot plate for any procedure requiring a heat source. If lab burners are used, a central shutoff valve for the gas supply should be accessible to you. Never use open flames when a flammable solvent is in the same room.

## Before the First Lab

1. Distribute and discuss safety rules, first aid guidelines, and safety contract found in the *Teacher Resource Package.*
2. Review the safe use of equipment, chemicals, and living organisms. Review the safety symbols.
3. Review use and location of fire extinguisher.
4. Discuss safe disposal of materials and lab cleanup policy.
5. Have students sign safety contract and return it. Test students' knowledge of safety symbols.

## Before Each Lab

1. Perform each activity yourself ahead of time to determine where students may need help.
2. Arrange the lab so that equipment and supplies are clearly labeled and easily accessible.
3. Have available only equipment and supplies needed to complete the assigned Lab.
4. Review the written procedure with students.
5. Be sure all students understand any precautions to be taken.

## During the Lab

1. Make sure the lab floors and counters are clean.
2. Insist that all students wear goggles and aprons.
3. Do not allow students to work alone.
4. Insist on caution when students use cutting devices.
5. Shield systems under pressure or vacuum.
6. Caution students not to point the open end of a heated test tube toward anyone.
7. Remove broken or chipped glassware and clean up any spills immediately. Dilute spills with water before removing.
8. Remind students that heated glassware *looks* cool.
9. Prohibit eating and drinking in the lab.

## Following the Lab

1. Be certain that students have disposed of broken glassware and chemicals properly.
2. Be sure all equipment is turned off.
3. Insist that each student wash his or her hands when lab work is completed.

# Chemical Storage and Disposal

## General Guidelines

Be sure to store all chemicals properly. The following are guidelines commonly used. Your school, city, county, or state may have additional requirements for handling chemicals. It is the responsibility of each teacher to become informed as to what rules or guidelines are in effect in his or her area.

1. Separate chemicals by reaction type. Strong acids should be stored together. Likewise, strong bases should be stored together and should be separated from acids. Oxidants should be stored away from easily oxidized materials and so on.
2. Be sure all chemicals are stored in labeled containers indicating contents, concentration, source, date purchased (or prepared), any precautions for handling and storage, and expiration date.
3. Dispose of any outdated or waste chemicals properly according to accepted disposal procedures.
4. Do not store chemicals above eye level.
5. Wood shelving is preferable to metal. All shelving should be firmly attached to walls and have antiroll edges.
6. Store only those chemicals that you plan to use.
7. Hazardous chemicals require special storage containers and conditions. Be sure to know what those chemicals are and the accepted practices for your area. Some substances must even be stored outside the building.
8. When working with chemicals or preparing solutions, observe the same general safety precautions that you would expect from students. These include wearing an apron and goggles. Wear gloves and use the fume hood when necessary. Students will want to do as you do whether they admit it or not.
9. If you are a new teacher in a particular laboratory, it is your responsibility to survey the chemicals stored there and to be sure they are stored properly or disposed of. Consult the rules and laws in your area concerning what chemicals can be kept in your classroom. For disposal, consult up-to-date disposal information from the state and federal governments.

## Disposal of Chemicals

Local, state, and federal laws regulate the proper disposal of chemicals. These laws should be consulted before chemical disposal is attempted. Although most substances encountered in high school biology can be flushed down the drain with plenty of water, it is not safe to assume that is always true. It is recommended that teachers who use chemicals consult the following books from the National Research Council.

*Prudent Practices for Handling Hazardous Chemicals in Laboratories,* Washington, DC: National Academy Press, 1981.

*Prudent Practices for Disposal of Chemicals from Laboratories,* Washington, DC: National Academy Press, 1983.

These books are useful and still in print, although they are several years old. Current law in your area would, of course, supersede the information in these books.

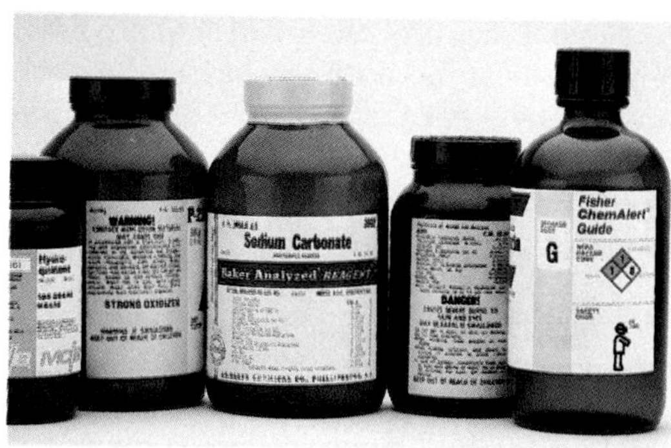

# Care of Living Organisms

Think about what made you decide to study biology. Was it curiosity about leaves and bugs? Maybe it was the teacher who hung all those birds' nests from the ceiling. Or was it an afternoon in a lab looking through a microscope at live amoebas? Many people become fascinated with biology as a result of a laboratory experience where they have to grow plants, make wet mount slides of living specimens, or track data on living organisms. From germinating bean seeds to keeping pet turtles, the biology classroom and curriculum comes to life with living things. The student who is bored by the textbook can be revived with a drop of pond water.

Teaching biology offers an opportunity for you to develop in your students a sense of excitement and old-fashioned wonder. Use of live organisms will enhance this wonder and help you reach at least one objective—students will be enthusiastic about coming to class. Many animal groups are described and illustrated in *Biology: An Everyday Experience.* You will not be able to introduce living specimens of all these groups to the classroom, but a few might stimulate student interest. Students will have to understand, of course, that they have an important role in the responsible care of these organisms. A few will probably even be willing to share that responsibility.

In addition, respect of the animals as classroom pets or as organisms to be observed in labs must be fostered. The authors of *Biology: An Everyday Experience* recommend the National Association of Biology Teachers "Guidelines for the Use of Live Animals." These guidelines are published in *The American Biology Teacher,* Vol. 48., No. 2, Feb. 1986. The guidelines are written in clear terms and can be used to stimulate a classroom discussion on the proper care and use of animals in society. The Labs throughout *Biology: An Everyday Experience* avoid the use of dissection and make use of observation of protists, fungi, and arthropods whenever possible. Students who wish to use animals for science fair projects or advanced study projects should use safe and respectful procedures that have been discussed thoroughly with you ahead of time. All work should be supervised for the well-being of both student and project animal.

Many times, students want to bring live animals to class. If facilities are adequate, this can be a great learning tool. However, students will have to understand the importance of providing food, space, fresh water, and adequate light and ventilation for the animals.

Weekends and vacations present problems. All living organisms will require regular feeding and a fresh supply of water. If a custodian cannot help, then students have to realize up front that animal day care is a necessary part of learning, and a regular group of volunteer caretakers will be needed. Students can take part in planning for this. Encourage students to set up teams of volunteers for regular care of organisms.

## Equipment and Supplies List

A complete list of equipment and expendables for all the labs in the text is given below. The list will help you in ordering equipment for the entire school year. The amounts listed are for 15 pairs of students. The labs are designed so that simple apparatus can be used. Whenever possible, use materials that can be obtained locally.

| Equipment | Quantity | Lab |
|---|---|---|
| aquarium (large) | 1 | 13-1 |
| aquarium (small) | 15 | 8-1 |
| balance | 2 | 1-2, 14-1 |
| beaker, 100 mL | 30 | 1-2, 14-1, 23-1, 23-2 |
| beaker, 200 mL | 30 | 9-1 |
| beaker, 250 mL | 30 | 13-1 |
| bolt (with 2 nuts), approximately 3.7 cm x 0.7 cm | 30 | 4-1 |
| bowl | 15 | 30-2 |
| cheese cloth, 20 cm x 20 cm | 15 | 9-1 |
| coins | 30 | 27-1 |
| coverslips | 1 box | 1-1, 2-2, 5-1, 6-1, 17-1, 30-2 |
| dish (small) | 15 | 7-1, 23-2 |
| dissecting pan | 15 | 7-2 |
| dropper | 320 | 1-1, 2-1, 5-1, 6-1, 6-2, 7-1, 9-1, 9-2, 13-2, 15-1, 17-1, 18-1, 18-2, 20-2, 32-1, 32-2 |
| envelopes (large)<br>Materials to be placed in each envelope:<br>bolt, brass fastener, button, glass slide, hairpin, match, nail, metal nut, paper clip, rubber band, safety pin, toothpick, washer, yarn (small piece) | 15 | 3-1 |
| feather, down | 15 | 8-2 |
| feather, wing or tail | 15 | 8-2 |
| flashlight | 15 | 7-1, 7-2 |
| forceps | 15 | 1-1, 6-1, 20-2, 30-2, 32-1 |
| funnel | 15 | 9-1 |
| glass tubing | 30 | 11-2 |
| graduated cylinder, 250 mL | 15 | 1-2, 9-1 |
| 100 mL | 15 | 29-2 |
| hand lens | 15 | 1-2, 5-2, 6-1, 7-1, 7-2, 8-2, 16-1, 25-2, 31-2 |
| hot plate | 15 | 9-1 |
| jars, baby food | 75 | 2-1 |
| lamp | 15 | 31-1 |
| map, state | 15 | 3-2 |
| metric ruler | 15 | 4-2, 6-1, 8-2, 15-1, 16-1, 19-1, 23-1, 24-1, 24-2, 25-2, 28-2, 29-2, 30-1, 12-1 |
| meter stick | 15 | 11-2 |
| microscopes, light | 15 | 1-1, 2-2, 4-2, 5-1, 5-2,10-1, 10-2, 17-1, 22-2, 24-1, 30-1 |
| microscope slides, glass | 2 boxes | 1-1, 2-2, 5-1, 6-1, 6-2, 9-2, 13-2, 15-2, 17-1, 18-1, 18-2, 30-2, 32-1 |
| mirror, plane surface | 15 | 17-2 |
| model of foot | 15 | 14-2 |
| net (aquarium fish) | 15 | 8-1, 13-1 |
| pan | 15 | 11-2 |
| paper clips | 1 box | 2-1 |
| paper punch | 15 | 14-2 |
| pencils, red, blue, green, yellow, purple | 15 | 10-2, 16-2, 24-2 |
| petri dish (with cover) | 15 | 4-2, 11-1, 12-2, 21-2, 24-2, 26-1 |
| plastic dish | 15 | 29-2 |

| | | |
|---|---|---|
| plastic stick | 15 | 4-2 |
| plastic squeeze bottle | 15 | 11-2 |
| polystyrene ball | 15 | 4-1 |
| razor blade, single edge | 15 | 21-1, 23-2, 9-2 |
| rubber stopper | | |
| two hole to fit plastic bottle | 15 | 11-2 |
| solid to fit test tubes | 90 | 17-1, 31-1 |
| scalpel | 15 | 5-2, 6-2 |
| scissors | 15 | 8-2, 14-2, 16-1, 23-1, 28-1, 28-2 |
| stereoscopic microscope | 15 | 6-2, 7-1 |
| test tube rack | 15 | 13-2, 15-2, 17-1, 31-1, 32-2 |
| test tubes, 12 cm | 90 | 13-2, 15-2, 17-1, 31-1, 32-2 |
| thermal glove | 15 | 9-1 |
| thermometer | 1 | 13-1 |
| watch (wall clock) | 1 | 11-1, 13-1, 17-2 |
| wire, #22 gauge, 14 cm long | 30 | 4-1 |

**Expendables**

| | | |
|---|---|---|
| Aluminum foil | 1 roll | 28-2 |
| cardboard piece | 15 | 30-2 |
| cheese cloth | 1 package | 9-1 |
| chicken bone, a piece | 15 | 14-1 |
| cotton swab | 15 | 7-2 |
| crayons: red, blue, green, yellow | 15 | 28-1 |
| cow bone, a piece | 15 | 14-1 |
| detergent | 50 mL | 32-2 |
| digestive system diagrams | 90 | 10-2 |
| file card | 1 package | 14-2 |
| filter paper strips, 13 cm x 4 cm | 15 | 19-2 |
| food samples: | 15 of each | 9-2 |
| apple juice, honey | | |
| sweet potato, molasses | | |
| maple syrup | | |
| gelatin dessert | 2 packages | 10-1 |
| glucose test paper | 2 rolls | 13-2, 15-2 |
| glue | 1 bottle | 30-2 |
| graph paper | 30 | 21-1, 24-2, 30-1 |
| ice | 2 cups | 13-1 |
| India ink, black | 1 bottle | 21-1 |
| labels | 200 | 13-1, 17-1, 23-2, 28-2 |
| magazine print letter "e" | 15 | 1-1 |
| marking pen | 15 | 21-2, 31-2 |
| marking pencil | 15 | 12-2, 22-2, 31-1 |
| metal fasteners | 1 box | 14-2 |
| milk, whole | 100 mL | 9-1 |
| milk cartons | 30 | 23-1 |
| owl pellet | 15 | 30-2 |
| paper bags | 30 | 26-2 |
| paper cup | 120 | 10-1, 27-1 |
| paper, black | 15 | 16-1, 29-1 |
| paper, heavy | 60 | 28-1 |
| paper, white | 105 | 16-1, 16-2, 17-2, 27-2, 28-1, 29-1 |
| paper dots | | |
| black | 450 | 29-1 |
| white | 450 | 29-1 |

| | | |
|---|---|---|
| paper towel | 5 rolls | 1-2, 6-1, 7-2, 16-2, 21-1, 21-2, 22-2, 26-1, 29-1 |
| pencil | 15 | 6-2, 17-2, 22-1, 23-1, 27-2 |
| permanent marker | 15 | 23-1 |
| pH chart | 30 | 32-1 |
| pH paper, wide range | 2 packages | 32-1 |
| pig bone, a piece | 15 | 14-1 |
| pipecleaners, 2-cm lengths | 360 pieces | 4-1 |
| plastic bag, self-sealing | 30 | 1-2, 21-1 |
| plastic cup, clear | 70 | 28-2, 31-2 |
| plastic sheet, clear | 1 | 12-2, 22-2 |
| plastic spoons | 60 | 10-1 |
| rubber bands | 15 | 23-1 |
| salt | 10 g | 1-2 |
| salt water, varying amounts | | 13-2, 31-2 |
| soil | 40 lb | 23-1, 28-2 |
| sticks, for stirring | 1 box | 10-1, 19-2 |
| straw, drinking | 15 | 2-1 |
| string | 1 roll | 14-2 |
| sugar, granulated | 10 g | solution preparation |
| tape, masking | 2 rolls | 1-2, 19-2, 27-1 |
| tape, transparent | 1 roll | 14-2, 16-1, 21-2, 23-1, 28-1 |
| tissues | 1 box | 22-2 |
| toothpicks, flat | 2 boxes | 2-2, 7-1, 21-1, 31-2 |
| tracing paper | 60 | 28-1 |
| turkey bone, a piece | 15 | 14-1 |
| water | | |
| distilled, varying amounts | | 13-2, 32-2 |
| rain, 300 mL | | 32-1 |
| tap, varying amounts | | 1-1, 1-2, 2-2, 7-2, 9-2, 10-1, 14-1, 18-1, 18-2, 26-1, 28-2, 29-2, 23-2, 30-2 |
| wax pencil | 15 | 2-1, 15-2, 18-1, 18-2, 26-1 |
| yeast, dead mixture | 150 mL | 32-2 |
| yeast, dry | 4 packets | 2-1, 32-2 |

**Chemicals**

| | | |
|---|---|---|
| acetone | 100 mL | solution preparation |
| alcohol | 100 mL | 18-2 |
| alcohol testing chemical | 10 mL | 18-2 |
| ammonia | 300 mL | 32-1 |
| aspirin | 1 small bottle | 18-1 |
| aspirin testing chemical | 10 mL | 18-1 |
| baking soda (sodium bicarbonate) | 50 g | 32-1 |
| bromthymol blue | 1 g | 32-2 |
| cola | 300 mL | 32-1 |
| copper chloride, CuCl | 100 g | solution preparation |
| enzyme | 1 | 9-1 |
| ethyl alcohol | 100 mL | 6-2 |
| ferric nitrate nonahydrate, $Fe(NO_3)_3\ 9H_2O$ | 200 mL | solution preparation |
| glucose | 100 mL | 13-2 |
| hydrogen peroxide, 3% solution | 1 bottle | 32-2 |
| lemon juice | 300 mL | 32-1 |
| methyl cellulose | 25 g | solution preparation |
| methylene blue stain | 25 g | 2-2 |

| | | |
|---|---|---|
| nitric acid, concentrated | 100 mL | solution preparation |
| oil, vegetable | 1/2 L | mini lab |
| painkiller test solutions | 50 mL | 18-1 |
| petroleum ether | 1 L | solution preparation |
| phenol red | 1 g | 2-1 |
| plastic powder (type 2 Jeltrate) | 1 L | 29-2 |
| potassium iodide | 10 g | solution preparation |
| silver nitrate | 10 g | 13-2 |
| sodium chloride | 100 g | solution preparation |
| sugar test paper | 100 pieces | 9-2 |
| test solutions of over-the-counter drugs and household items | 50 mL each | 18-2 |
| vinegar | 3/4 liter | 7-2, 32-1 |

**Biological Supplies**

| | | |
|---|---|---|
| *Anabaena,* prepared slides | 15 | 7-2 |
| bacteria, prepared slide | 15 | 4-2 |
| beans, red | 600 | 26-2 |
| beans, white | 600 | 26-2 |
| bean seedlings, 3 days old | 15 | 21-1 |
| blood, frog, prepared | 15 | 12-1 |
| blood, human, prepared | 15 | 12-1, 12-2 |
| brine shrimp or daphnia | 15 | 7-1 |
| brine shrimp eggs | 15 | 31-2 |
| carrot, root | 15 | 20-2, 23-2 |
| *Coleus* plant | 15 leaves | 28-2 |
| corn seeds | 900 | 21-2, 23-1, 26-1 |
| crayfish, live | 15 | 8-1 |
| cultures of: | | |
| amoeba | 1 | 5-1 |
| euglena | 1 | 5-1 |
| hydra | 1 | 7-1 |
| earthworms, live | 15 | 7-2, 11-1 |
| *Elodea,* several sprigs | 45 | 2-2, 31-1 |
| Enzyme cultures | 1 | 10-1 |
| fern fronds | 15 | 6-2 |
| garlic bulb | 15 | 23-2 |
| grasshopper, stages of metamorphosis | 15 | 25-2 |
| guppy | 15 | 2-1 |
| leaf | 15 | 29-2 |
| leaves, 15 types | 45 each | 19-1 |
| moss, preserved or live | 15 | 6-1 |
| moth, stages of metamorphosis | 15 | 25-2 |
| mushroom | 15 | 5-2 |
| mushroom basidium, prepared slide | 15 | 5-2 |
| radish root | 15 | 20-2 |
| starfish egg, prepared slide | 15 | 24-1 |
| starfish sperm, prepared slide | 15 | 24-1 |
| urine sample—see Preparation of Solutions | | |
| abnormal | | 13-2, 15-2 |
| normal | | 13-2, 15-2 |
| vinegar eels | 1 | 17-1 |
| yam root | 15 | 20-2 |

## Preparation of Solutions

Abnormal urine—Add 2 teaspoons honey to a normal urine mixture.

Alcohol-testing solution—Add 20 g potassium dichromate powder to a glass beaker. Pour 20 mL concentrated sulfuric acid into the beaker. Stir with a glass rod to dissolve most of the powered. *Slowly* add 60 mL distilled water and continue to stir. *Solution becomes very hot.* Allow to cool. Powder may precipitate out after cooling. Pour only the liquid portion of the solution into dropping bottles for student use. Solution has a shelf life of one year.

Aspirin testing solution—Add 2 g ferric nitrate nonahydrate, $Fe(NO_3)_3\ 9H_2O$ to 50 mL water. Then add 1 mL concentrated nitric acid to this solution.

Aspirin solution—Dissolve 1 aspirin in 50 mL water.

Blue liquid—Add 0.1 g bromthymol blue to 2000 mL water.

Dead yeast mixture—Boil 100 mL live yeast for at least 5 minutes on a hot plate.

Glucose solution—Add 1 teaspoon glucose to 250 mL tap water.

Green liquid—Add 0.1 g bromthymol blue to 2000 mL water. Then exhale into the liquid until a green color appears.

Hydrochloric acid—Pour 85 mL concentrated hydrochloric acid slowly into 1000 mL water. Do not pour water into concentrated acid.

Iodine solution—Dissolve 10 g potassium iodine and 3 g iodine in 1 L water. Store in the dark or in a brown bottle.

Liquid solvent—Combine 92 parts petroleum ether with 8 parts acetone. An alternative solvent is ethyl alcohol.

Live yeast mixture—Add 1 tablespoon each of dry yeast and sugar to 200 mL water.

Methyl cellulose—Add 10 g methyl cellulose powder to 90 mL warm distilled water. Stir gently to avoid bubbles.

Methylene blue stain—Dissolve 1.5 g methylene blue in 100 mL ethyl alcohol. Dilute by adding 10 mL of solution to 90 mL water.

Normal urine—Add 1 teaspoon salt and 4 drops yellow food coloring to 500 mL tap water.

Phenol red—Mix 0.4 g phenol red in 100 mL water to make stock solution. Transfer 1 mL stock solution to 100 mL water to yield a 0.00004% phenol red solution for use.

Salt water—Add 1 teaspoon sodium chloride to 500 mL tap water.

Silver nitrate—Add 4 g silver nitrate to 250 mL distilled water. Store in a brown bottle away from light.

Urine sample—Add 6 drops yellow food coloring to 1000 mL tap water.

Urine sample with glucose—Add 6 drops yellow food coloring to 1000 mL tap water, then add 2 teaspoons glucose or honey.

Yeast solution—Mix 1 teaspoon sugar and 1 teaspoon dry yeast to 200 mL water.

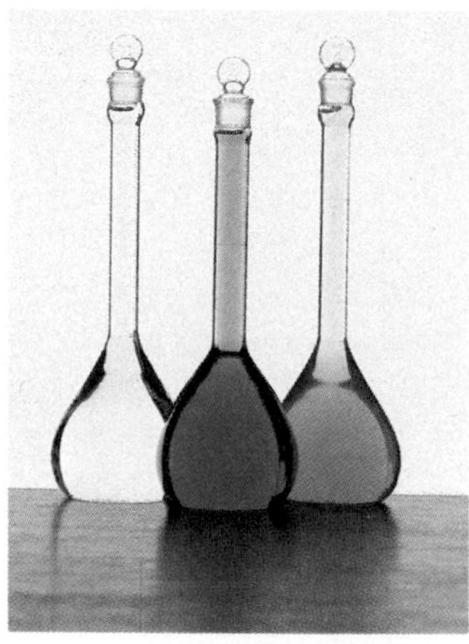

# Equipment Suppliers

Carolina Biological Supply Co.
2700 York Rd.
Burlington, NC 27215

Central Scientific Co.
11222 Melrose Ave.
Franklin Park, IL 60131

Fisher Scientific Co.
4901 W. LeMoyne
Chicago, IL 60615

LaPine Scientific Company
6001 S. Knox Ave.
Chicago, IL 60629

McKilligan Supply Corp.
435 Main St.
Johnson City, NY 13790

Nasco
901 Janesville Ave.
Fort Atkinson, WI 53538

Sargent-Welch Scientific Co.
7300 N. Linder
Skokie, IL 60077

Sargent-Welch Scientific of Canada Ltd.
285 Garyray Dr.
Weston, Ontario, Canada M9L 1P3

Science Kit and Boreal Labs
777 E. Park Dr.
Tonawanda, NY 14150

Ward's Natural Science Establishment, Inc.
P.O. Box 1712
Rochester, NY 14603

# Materials List by Lab

**Lab 1-1**
- microscope
- slides
- coverslips
- water
- dropper
- forceps
- small piece magazine print

**Lab 1-2**
- 2 100 mL beakers
- 40 pinto beans
- balance
- salt
- water
- 2 paper towels
- 50 mL graduated cylinder
- 2 self-sealing plastic bags
- masking tape

**Lab 2-1**
- 5 baby food jars
- red liquid
- wax pencil
- guppy
- yeast suspension
- dropper
- paper clip
- drinking straw

**Lab 2-2**
- microscope
- 2 slides
- 2 coverslips
- dropper
- methylene blue stain
- toothpicks, flat type
- *Elodea* leaf
- water

**Lab 3-1**
- envelopes with 12 items

**Lab 3-2**
- state maps

**Lab 4-1**
- bolt
- 2 nuts to fit bolt
- 2 pieces of wire
- polystyrene ball
- pipecleaners

**Lab 4-2**
- microscope
- prepared slide of *Anabaena*
- prepared slide of bacteria
- petri dish
- metric ruler

**Lab 5-1**
- microscope
- 3 slides
- 3 coverslips
- 3 droppers
- cultures of:
  - amoeba
  - euglena
  - paramecium

**Lab 5-2**
- microscope
- hand lens
- scalpel
- prepared mushroom slide
- mushroom gill

**Lab 6-1**
- stereomicroscope or hand lens
- 3 slides
- cover slip
- dropper
- forceps
- metric ruler
- paper towels
- moss

**Lab 6-2**
- stereomicroscope
- 2 slides
- scalpel
- dropper
- ethyl alcohol
- whole fern with spore cases
- pencil
- paper

**Lab 7-1**
- stereomicroscope
- hand lens
- dropper
- small dish
- flashlight
- hydra culture
- daphnia or brine shrimp

**Lab 7-2**
- hand lens
- shallow pan
- flashlight
- toothpick
- beaker of water
- cotton swab
- paper towels
- vinegar
- live earthworms

**Lab 8-1**
- small aquarium with crayfish

**Lab 8-2**
- hand lens
- metric ruler
- scissors
- wing or tail feather
- down feather

**Lab 9-1**
- 2 200-mL beakers
- dropper
- funnel
- hot plate
- stirring rod
- whole milk
- enzyme
- 250-mL graduated cylinder
- thermal glove
- cheese cloth

**Lab 9-2**
- slides
- droppers
- razor blade
- food samples for sugar test
- water
- sugar test paper

**Lab 10-1**
- 4 paper cups
- gelatin
- water
- 4 applicator sticks
- enzyme A
- enzyme B

**Lab 10-2**
- 5 diagrams of digestive system
- colored pencils

**Lab 11-1**
- petri dish
- water
- wall clock
- live earthworm

**Lab 11-2**
- plastic squeeze bottle
- pan or sink
- water
- meterstick
- rubber stopper and tube assembly

**Lab 12-1**
- microscope
- prepared slides of frog and human blood
- petri dish cover
- metric ruler

**Lab 12-2**
- sheet of clear plastic
- marking pencil
- microscope
- prepared slide of human blood

**Lab 13-1**
- 250-mL beaker
- goldfish in aquarium
- thermometer
- net
- watch or wall clock
- ice

**Lab 13-2**
- 4 test tubes
- test tube rack
- droppers
- labels
- distilled water
- salt water
- glucose
- glucose test tape
- silver nitrate
- normal artificial urine
- abnormal artificial urine
- 4 slides

**Lab 14-1**
- 100-mL graduated cylinder
- balance
- water
- bones of:
  - chicken
  - cow
  - pig
  - turkey

**Lab 14-2**
- scissors
- string
- tape
- paper punch
- metal fastener
- index card

**Lab 15-1**
- metric ruler

**Lab 15-2**
- test tube
- slides
- 8 droppers
- wax pencil
- normal urine sample
- urine sample from diabetic
- urine samples in test tubes marked C-E
- glucose test paper

**Lab 16-1**
- metric ruler
- scissors
- hand lens
- clear tape
- white paper
- black paper

**Lab 16-2**
- white paper
- red and green pencils

**Lab 17-1**
- microscope
- glass slide
- coverslip
- dropper
- test-tube rack
- 2 test tubes, small
- 2 stoppers to fit tubes
- 2 labels
- vinegar eels

**Lab 17-2**
- mirror
- watch with second hand
- pencil
- paper

**Lab 18-1**
- 4 glass slides
- wax marking pencil
- aspirin solution
- 6 test solutions of pain killers
- dropper tube
- aspirin-testing solution in dropper bottle

**Lab 18-2**
- 4 slides
- wax marking pencil
- 8 droppers
- water
- alcohol
- alcohol testing chemical in dropper bottle
- 6 test solutions of over-the-counter drugs and household chemicals

**Lab 19-1**
- 15 leaves of different plants
- metric ruler

**Lab 19-2**
- plant leaf
- coin
- small jar
- metric ruler
- filter paper strip
- liquid solvent
- applicator stick
- masking tape

**Lab 20-1**
- pencil
- paper towel cut into 10 x 15 cm and 1 x 10 cm strips
- stapler
- large plastic cup

**Lab 20-2**
- hand lens
- carrot root
- yam root
- radish root
- razor blade
- 3 petri dish halves
- iodine solution
- dropper
- forceps

**Lab 21-1**
- toothpick
- plastic bag
- paper towel
- graph paper
- razor blade
- India ink
- 2 bean seedlings

**Lab 21-2**
- petri dish
- marking pen
- 2 paper towels
- transparent tape
- 4 pre-soaked corn seeds

**Lab 22-1**
- pencil
- paper

**Lab 22-2**
- clear plastic sheet
- marking pencil
- facial tissue
- microscope

**Lab 23-1**
- metric ruler
- rubber band
- small beaker
- pencil
- tape
- permanent marker
- 2 milk cartons
- soil
- 20 corn seeds
- water
- stapler
- scissors

**Lab 23-2**
- garlic bulb
- carrot
- labels
- water
- toothpicks
- small beaker
- metric ruler
- razor blade
- shallow dish

**Lab 24-1**
- microscope
- prepared slide of starfish eggs
- prepared slide of starfish sperm
- metric ruler
- petri dish for drawing circles

**Lab 24-2**
- metric ruler
- graph paper
- colored pencils

**Lab 25-1**
- metric ruler

**Lab 25-2**
- metric ruler
- hand lens
- grasshopper stages of metamorphosis
- moth stages of metamorphosis

**Lab 26-1**
- 20 corn seeds
- paper towels
- wax pencil
- petri dish
- water

**Lab 26-2**
- 40 white beans
- 40 red beans
- 2 paper bags

**Lab 27-1**
- 2 coins
- paper cup
- masking tape

**Lab 27-2**
- pencil
- paper

**Lab 28-1**
- scissors
- tracing paper
- heavy paper
- tape
- blank paper
- crayons

**Lab 28-2**
- scissors
- clear plastic cup
- aluminum foil
- *Coleus* plant
- soil
- water
- label
- metric ruler

**Lab 29-1**
- white paper
- black paper
- 30 white dots
- 30 black dots
- damp paper towels

**Lab 29-2**
- plastic dish
- 100-mL graduated cylinder
- warm water
- metric ruler
- leaf
- plastic powder

**Lab 30-1**
- metric ruler
- graph paper

**Lab 30-2**
- light microscope
- slide
- coverslip
- bowl
- forceps
- cardboard
- glue
- owl pellet
- water

**Lab 31-1**
- 4 test tubes
- wax pencil
- 4 stoppers
- lamp
- blue test liquid
- green test liquid
- 2 *Elodea* sprigs
- 2 test-tube racks

**Lab 31-2**
- 3 small plastic cups
- marking pen
- hand lens
- flat toothpick
- distilled water
- brine shrimp eggs
- strong salt solution
- weak salt solution

**Lab 32-1**
- pH paper
- pH chart
- forceps
- marking pencil
- 7 slides
- 7 droppers
- ammonia
- baking soda
- cola
- vinegar
- 2 samples of rainwater
- distilled water

**Lab 32-2**
- 4 test tubes
- marking pencil
- test-tube rack
- 5 droppers
- blue test liquid
- dead yeast mixture
- live yeast mixture
- detergent
- hydrogen peroxide
- distilled water

# Audiovisual and Software Suppliers

AETNA Life and Casualty, Film Library, 151 Farmingham Ave., Hartford, CT 06115

Aims, 626 Justin Ave., Glendale, CA 91201

Almanac Films, Inc., 29 E. Tenth St., New York, NY 10003

American Cancer Society, 777 Third Ave., New York, NY 10017

American Film Registry, 832 S. Wabash, Chicago, IL 60605

American Heart Association, 44 E. 23 St., New York, NY 10010 (or contact local chapter)

Appalachian Hardwood Association, Room 408 NCNB Bldg., High Point, NC 27260

Association Films, Inc., 866 Third Avenue, New York, NY 10022

Audio Productions, Inc., 708 Third Avenue, New York, NY 10017

Audio Visual Center, Indiana University, Bloomington, IN 47405

Bailey Films, Inc., 6509 DeLongpre Ave., Hollywood, CA 90028

Barr Films, P.O. Box 5667, 3490 East Foothill Blvd., Pasadena, CA 91107

Bausch and Lomb, Inc., Film Service, 1400 N. Goodman St., Rochester, NY 14602

Beacon Films, P.O. Box 575, Norwood, MA 02062

Bergwall, Inc., 106 Charles Lindberg Blvd., Uniondale, NY 11553

BFA Educational Media, 2211 Michigan Ave., P.O. Box 1795, Santa Monica, CA 90406

Brancon Films, Inc., 200 W. 57th St., New York, NY 10019

Bray Studios, Inc., (BRAY), 630 Ninth Avenue, New York, NY 10036

Bureau of AV Services, University of Arizona, Tucson, AZ 85721

Carolina Biological Supply Company, Burlington, NC 27215

Chevron Chemical Co., Ortho Distribution (Film Lab), 200 Bush St., San Francisco, CA 94104

CCM Films, Inc., 5547 Ravenswood Ave., Chicago, IL 60640

Centron Educational Films, 108 Wilmont Rd., Deerfield, IL 60015

Churchill Films, 662 Robertson Blvd., Los Angeles, CA 90069

Clearvue, Inc., 6465 Avondale Ave., Chicago, IL 60631

Colonial Films, 71 Walton St., N.W., Atlanta, GA 30303

COMPress, A Div. of Wadsworth Inc., P.O. Box 102, Wentworth, NH 03282

CONDUIT, University of Iowa Oakdale Campus, Iowa City, IA 52242

Conservation Foundation, 1717 Massachusetts Ave. N.W., Washington, DC 20036

Contemporary Films, Inc., 267 W. 25 St., New York, NY 10001

Coronet Films, 108 Wilmont Rd., Deerfield, IL 60015

Cox Enterprises, 2900 S. Sawtelle Blvd., Los Angeles, CA 90064

Cross Educational Software, P.O. Box 1536, Ruston, LA 71270

DM, Inc., One DLM Park, Allen, TX 75002

Davidson & Associates, Inc., 3135 Kashiwa St., Torrence, CA 90505

Denoyer-Geppert Audiovisuals, 355 Lexington Ave., New York, NY 10017

Design Ware, Inc./EduWare Services, Inc., 185 Berry St., San Francisco, CA 94107

Walt Disney Productions, 500 S. Buena Vista Ave., Burbank, CA 91505

Doubleday Multimedia, Box C-19518, 1371 Reynolds, Irvine, CA 92713

EME, Old Mill Plain Road, Box 2805, Danbury, CT 06430

Earling Corp., 2225 Massachusetts Ave., Cambridge, MA 02140

Eastman Kodak Co., Audiovisual Service, 348 State St., Rochester, NY 14608

Education Dimensions Group, P.O. Box 126, Stamford, CT 06904

Educational Activities, 19377 Grand Ave., Baldwin, NY 11510

Educational Development Corp., Learning Resources Division, 4404 S. Florida Ave., Lakeland, FL 33803

Educational Film Library Association, 250 W. 57 St., New York, NY 10019

Educational Images Ltd., P.O. Box 3456 West Side, Elmira, NY 14905

Encyclopedia Brittanica Educational Corp., 425 N. Michigan Ave., Chicago, IL 60611

Eye Gate Media, P.O. Box 303, Jamaica, NY 11435

Farm Film Foundation, 1616 H Street, NW, Washington, DC 20006

Films, Inc, ATTN: Public Media, Inc., 5547 Ravenswood Ave., Chicago, IL 60640

Films for the Humanities and Sciences, P.O. Box 2053, Princeton, NJ 08543

Films of the Nation Distributors, Inc., 305 E. 86 St., New York, NY 10036

Filmstrip House, 6633 W. Howard St., Niles, IL 60648

Focus Media, Inc., 839 Stewart Ave., Box 865, Garden City, NY 11530

Frey Scientific Co., 905 Hickory Lane, Mansfield, OH 44905

Gamco Industries, Box 1911, Big Spring, TX 79720

Gateway Productions, Inc., 1859 Powell St., San Francisco, CA 94133

Golden Key Productions, Inc., Film Distributors, 1921 Hillhurst Ave., Hollywood, CA 90027

Guidance Associates, Inc., Communications Park, Box 3000, Mount Kisco, NY 10549

HRM Software, 175 Tompkins Ave., Pleasantville, NY 10570

Handel Films, 8730 Sunset Blvd., West Hollywood, CA 90069

Hartley Courseware, Inc., 133 Bridge, Dimondale, MI 48821

Hayden Software, 600 Suffolk St., Lowell, MA 01853

Heidenkamp Nature Pictures, 872 Old Hickory Rd., Pittsburgh, PA 15243

Human Relations Media, 175 Tompkins Ave., Pleasantville, NY 10570

IBM, P.O. Box 2150, Atlanta, GA 30055

Ideal Pictures Corp., 58 E. South Water St., Chicago, IL 60648

Indiana University Films, Audio-Visual Center, Bloomington, IN 47405

Insight Media, 121 West 85th Street, New York, NY 10024

Institutional Cinema Service, Inc., 915 Broadway, New York, NY 10075

Instructional Films, Inc., 1150 Wilmette Ave., Wilmette, IL 60091

International Film Bureau, 332 S. Michigan Ave., Chicago, IL 60604

International Film Bureau, Inc., 57 E. Jackson Blvd., Chicago, IL 60604

International Film Foundation, Inc., 200 W. 72nd St., Room 64, New York, NY 10023

JLM Visuals, 920 7th Avenue, Grafton, WI 53024

Jam Handy Organization, 28221 E. Grand Blvd., Detroit, MI 48211

Library Films, Inc., 79 Fifth Ave., New York, NY 10011

McGraw-Hill Text Films, 110 Fifteenth St., Del Mar, CA 92014

McGraw-Hill Publishing Co., Films Division, 330 W. 42 St., New York, NY 10036

Minnesota Educational Computing Corp./MECC, 3490 Lexington Ave., North, St. Paul, MN 55126

Modern Learning Aids, Division of Ward's Natural Science, P.O. Box 1712, Rochester, NY 14603

Modern Talking Picture Service, Inc., 45 Rockefeller Plaza, New York, NY 10020

Moonlight Productions, 2243 Old Middlefield Way, Mountain View, CA 94043

Nasco, 901 Janesville Ave., Fort Atkinson,WI 53538

Nasco West, Inc., P.O. Box 3837, Modesto, CA 95352

National Audubon Society, 1130 Fifth Ave., New York, NY 10028

National Film Board of Canada, 1251 Avenue of the Americas, New York, NY 10020

National Geographic Society, Dept. 86, Washington, DC 20036

National Safety Council, 425 N. Michigan Ave., Chicago, IL 60611

National Wildlife Federation, 1412 16th St. N.W., Washington, DC 20036

NET Film Service, Audio-Visual Center, Indiana University, Bloomington, IN 47401 (also 10 Columbus Circle, New York, NY 10019)

Opportunities for Learning, 20417 Nordhoff St., Chattsworth, CA 91311

Photo Lab, Inc., 3825 Georgia Ave., N.W., Washington, DC 20011

Prentice-Hall Media, 150 White Plains Rd., Tarrytown, NY 10591

Queue, Inc., 338 Commerce Dr., Fairfield, CT 06430

Edwin Shapiro Co., 43-55 Kissena Blvd., Flushing, NY 11355

Scholastic Software, 730 Broadway, New York, NY 10003

Science and Mankind, Inc., Communications Park, Box 200, White plains, NY 10602

Science Software Systems, Inc., 11899 W. Pico Blvd., West Los Angeles, CA 90064

Sensible Software, 210 South Woodward, Ste. 229, Birmingham, MI 48011

Shell Film Library, 1433 Sadlier Circle West Drive, Indianapolis, IN 46239

Society for Visual Education, Inc., 1345 W. Diversey Pkwy, Chicago, IL 60614

Sunburst Communications, Inc., 39 Washington Ave., Pleasantville, NY 10570

Time-Life Films, 100 Eisenhower Dr., Paramus, NJ 07652

United World Films, Inc., 221 Park Ave. South, New York, NY 10003

University of Illinois, Visual Aids Service, Champaign, IL 61820

Vernard Organization, 702 S. Adams St., Peoria, IL 61602

Ward's Natural Science Establishment, P.O. Box 1712, Rochester, NY 14603

Wilner Films and Slides, P.O. Box 231, Cathedral Station, New York, NY 10025

**U.S. Government Agencies**

Council for Agricultural Science and Technology, 137 Lynn Ave., Ames, IA 50010-7120

Fish and Wildlife Service, U.S. Dept. of the Interior, P.O. Box 128, College Park, MD 20740

National Aeronautics and Space Administration, Washington, DC 20546

National Oceanic and Atmospheric Administration, U.S. Dept. of Commerce, Motion Picture Service, 12231 Welkin Ave., Rockville, MD 20852

Soil Conservation Service, Washington, DC 20251

U.S. Atomic Energy Commission, Washington, DC 20545

U.S. Department of Agriculture, Motion Picture Service, Washington, DC 20251

U.S. Forest Service, U.S. Department of Agriculture, Washington, DC 20251

U.S. Navy Department, Office of Administration, Pentagon Bldg., Washington, DC 20350

U.S. Public Health Service, 200 Independence Avenue SW, Washington, DC 20201

U.S. Public Health Service, Communicable Disease Center, Atlanta, GA 30333

**Canada**

Audio-Visual Supply Co., Toronto General Trusts Bldg., Winnipeg, Manitoba

Canadian Film Institute, 142 Sparks St., Ottawa, Ontario

General Films Limited, 15334 13th Ave., Regina, Saskatchewan

Radio-Cinema, 5011 Verdun Ave., Montreal, Quebec

## Safety Symbols

These safety symbols are used to indicate possible hazards in the activities. Each activity has appropriate hazard indicators.

|  |  |  |  |
|---|---|---|---|
| | **DISPOSAL ALERT** This symbol appears when care must be taken to dispose of materials properly. |  | **ANIMAL SAFETY** This symbol appears whenever live animals are studied and the safety of the animals and the students must be ensured. |
|  | **BIOLOGICAL HAZARD** This symbol appears when there is danger involving bacteria, fungi, or protists. |  | **RADIOACTIVE SAFETY** This symbol appears when radioactive materials are used. |
|  | **OPEN FLAME ALERT** This symbol appears when use of an open flame could cause a fire or an explosion. |  | **CLOTHING PROTECTION SAFETY** This symbol appears when substances used could stain or burn clothing. |
|  | **THERMAL SAFETY** This symbol appears as a reminder to use caution when handling hot objects. |  | **FIRE SAFETY** This symbol appears when care should be taken around open flames. |
|  | **SHARP OBJECT SAFETY** This symbol appears when a danger of cuts or punctures caused by the use of sharp objects exists. |  | **EXPLOSION SAFETY** This symbol appears when the misuse of chemicals could cause an explosion. |
|  | **FUME SAFETY** This symbol appears when chemicals or chemical reactions could cause dangerous fumes. |  | **EYE SAFETY** This symbol appears when a danger to the eyes exists. Safety goggles should be worn when this symbol appears. |
|  | **ELECTRICAL SAFETY** This symbol appears when care should be taken when using electrical equipment. |  | **POISON SAFETY** This symbol appears when poisonous substances are used. |
|  | **SKIN PROTECTION SAFETY** This symbol appears when use of caustic chemicals might irritate the skin or when contact with microorganisms might transmit infection. |  | **CHEMICAL SAFETY** This symbol appears when chemicals used can cause burns or are poisonous if absorbed through the skin. |

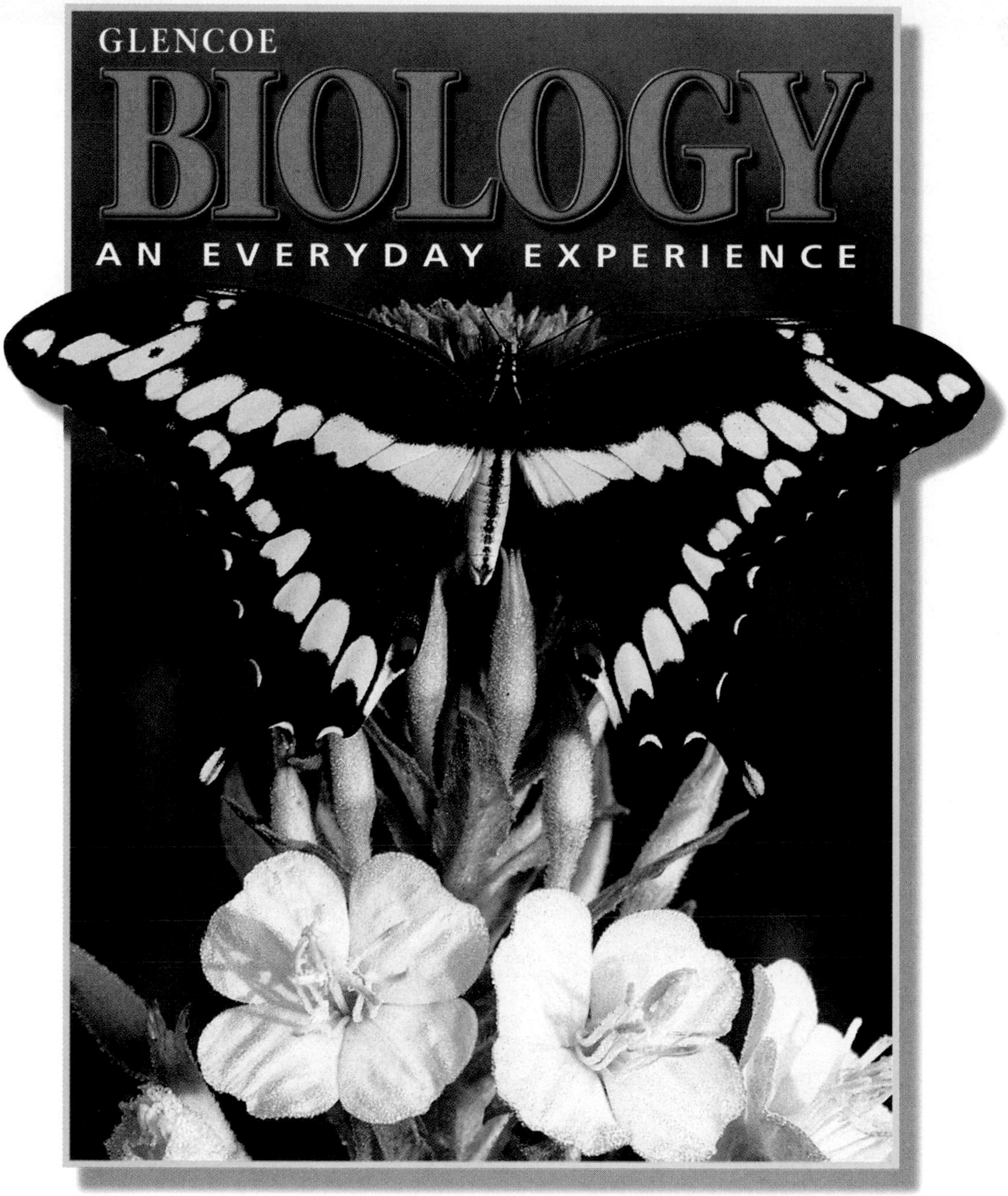

GLENCOE

# BIOLOGY

## AN EVERYDAY EXPERIENCE


Albert Kaskel Paul J. Hummer, Jr. Lucy Daniel

New York, New York Columbus, Ohio Woodland Hills, California Peoria, Illinois

## A MERRILL BIOLOGY PROGRAM

**Biology: An Everyday Experience,** *Student Edition*
**Biology: An Everyday Experience,** *Teacher Edition*
**Biology: An Everyday Experience,** *Teacher Resource Edition*
**Biology: An Everyday Experience,** *Study Guide*
**Biology: An Everyday Experience,** *Transparency Package*
**Biology: An Everyday Experience,** *Laboratory Manual, Student Edition*
**Biology: An Everyday Experience,** *Laboratory Manual, Teacher Edition*
**Biology: An Everyday Experience,** *Computer Test Bank*
**Biology: An Everyday Experience,** *Tech Prep Applications*

### CONTENT CONSULTANTS


**David M. Armstrong, Ph.D.**
*Director, University of Colorado Museum*
University of Colorado
Boulder, CO

**Mary D. Coyne, Ph. D.**
*Professor of Biological Sciences*
Department of Biological Sciences
Wellesley College
Wellesley, MA

**Joe W. Crim, Ph.D.**
*Associate Professor of Zoology*
Department of Zoology
University of Georgia
Athens, GA

**Marvin Druger**
*Professor of Biology and Science Education*
Department of Biology
Syracuse University
Syracuse, NY

**David G. Futch, Ph.D.**
*Associate Professor of Biology*
Department of Biology
San Diego State University
San Diego, CA

**Carl Gans, Ph.D.**
*Professor of Biology*
Department of Biology
University of Michigan
Ann Arbor, MI

**John Just, Ph.D.**
*Associate Professor of Biology*
School of Biological Sciences
University of Kentucky
Lexington, KY

**Richard Storey, Ph. D.**
*Associate Professor of Biology*
Department of Biology
Colorado College
Colorado Springs, CO

**James F. Waters, Ph.D.**
*Professor of Zoology*
Department of Biology
Humboldt State University
Arcata, Ca


### READING CONSULANT


**Barbara S. Pettegrew, Ph.D.**
*Director of Reading/Study Center*
*Assistant Professor of Education*
Otterbein College
Westerville, OH


### SPECIAL CONSULTANT


**Alton Biggs**
Allen High School
Allen, Texas


### REVIEWERS


**John A. Beach**
Fairless High School
Navarre, OH

**Tony Beasley**
*Science Supervisor*
Davidson County School Board
Nashville, TN

**Brenda Carrillo**
McCollum High School
San Antonio, TX

**Renee M. Carroll**
Taylor County High School
Perry, FL

**Dixie Duncan**
Williams Township School
Whiteville, NC

**Margorae Freimuth**
Argenta-Oreanna High School
Argenta, IL

**Raymond P. Gipson**
Blue Ridge High School
Morgan Hill, CA

**Karen S. Hewitt**
Coldspring High School
Coldspring, TX

**Marilyn B. Jacobs**
Huffman Eastgate High School
Huffman, TX

**Rex J. Kartchner**
St. David High School
St. David, AZ

**Barbara B. Kruse**
Alamosa High School
Alamosa, CO

**Lynn M. Smith**
Waterville Hill School
Waterville, ME

**Ouida E. Thomas**
B.F. Terry High School
Rosenberg, TX



*Glencoe/McGraw-Hill*

A Division of The McGraw·Hill Companies





Send all inquiries to"
Glencoe/McGraw-Hill
936 Eastwind Drive
Westerville, OH 43081

ISBN 0-02-825685-9
Printed in the United States of America.

1 2 3 4 5 6 7 8 9 10 11 12 071/043 06 05 04 03 02 01 00 99 98

# AUTHORS

**Albert Kaskel** has thirty-one years experience teaching science. He has extensive teaching experience in the city of Chicago and Evanston Township High School, Evanston, Illinois. His teaching experience includes all ability levels of biological science, physical science, and chemistry. He holds an M.Ed. degree from DePaul University. He received the Outstanding Biology Teacher Award for the State of Illinois in 1984.

**Paul J. Hummer, Jr.** taught science for twenty-eight years in the Frederick County, Maryland schools. He is currently a biology teacher at Hood College, Frederick, Maryland. He received his B.S.Ed. from Lock Haven State University, Lock Haven, PA, and his M.A.S.T. from Union College in Schenectady, NY. He has experience teaching various ability levels of biology as well as general science and physics. He received the Presidential Award for Excellence in Science and Mathematics Teaching in 1984.

**Lucy Daniel** taught biology at Rutherfordton-Spindale High School, Rutherfordton, North Carolina. She has thirty-five years of teaching experience in biology, life science, and general science. She holds a B.S. degree from the University of North Carolina at Greensboro and a M.A.S.E. from Western Carolina University at Cullowhee. She received the Presidential Award for Excellence in Science and Mathematics Teaching in 1984.



iii

## Biology *is* an Everyday Experience

**This is your life.** Biology is not just another science class. It's a subject you already know well, because it's about life.

**It's 10:00, do you know where your dinner is?** Every day, you eat and drink to stay alive. Where do your food and water come from, and how does your body use them?

**Appreciate the environment—it's the only one you have.** Trees don't just stand there, they help supply you with oxygen, prevent soil erosion, and make some of your food.

**The wonderful world of technology.** The field of medicine is closely tied to biology. Advances in medical technology may present you with difficult decisions.

**Biology happens!** Biology is about every living thing in your world and the relationships among them. The more you learn about biology, the more you will realize that biology *is* an everyday experience.

## Using *Biology: An Everyday Experience*—*a quick tour of your textbook*

*Biology: An Everyday Experience* not only presents information, it asks thought-provoking questions. Labs bring the text to life as you use scientific methods to solve problems. You will see how biology affects you as a consumer and learn about careers in biology. Take time now to see what your textbook offers.

*from beginning to end, you'll see how biology connects to the world around you*

**1** What would happen if. . . there were no mosquitoes? Have you ever thought about it? Each unit opener begins with a thought-provoking question like this. The unit introduction then shows you how even small differences in the relationships between living things can change your world in dramatic ways. So, what would happen if there were no mosquitoes? Read the opener to Unit 8 to find out.

1 Unit 8

CONTENTS

Chapter 30 Populations and Communities 630

Chapter 31 Ecosystems and Biomes 652

Chapter 32 Solving Ecological Problems 672

628

**Relationships in the Environment**

What would happen if...

there were no mosquitoes? You may have memories of hot, itchy summers when you thought the world would be better off without mosquitoes. But, would it? The fish in the photo depend on mosquito larvae for food. Many birds, in turn, depend on the fish. Without the mosquito, many fish would starve to death. Many birds would then starve. The balance of nature would be upset by the lack of mosquitoes.

What is the effect of spraying chemicals to kill mosquitoes? In many communities, trucks that spray to kill mosquitoes are a regular sight. The spray that kills mosquitoes kills honeybees, too. What would be the effect of killing honeybees? What is the cost to the environment of getting rid of mosquitoes? Is the cost too high?

629

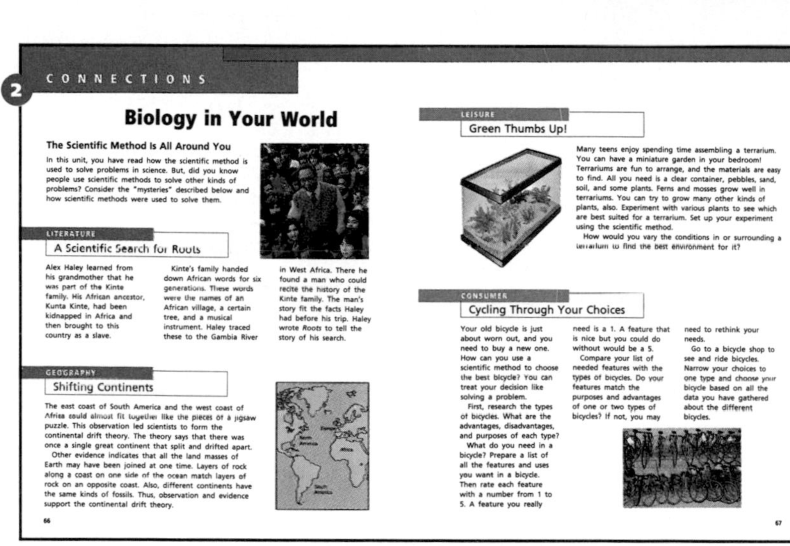

2 CONNECTIONS

**Biology in Your World**

**The Scientific Method Is All Around You**

In this unit, you have read how the scientific method is used to solve problems in science. But, did you know people use scientific methods to solve other kinds of problems? Consider the "mysteries" described below and how scientific methods were used to solve them.

LITERATURE

A Scientific Search for Roots

Alex Haley learned from his grandmother that he was part of the Kinte family. His African ancestor, Kunta Kinte, had been kidnapped in Africa and then brought to this country as a slave.

Kinte's family handed down African words for six generations. These words were the names of an African village, a certain tree, and a musical instrument. Haley traced these to the Gambia River in West Africa. There he found a man who could recite the history of the Kinte family. The man's story fit the facts Haley had before his trip. Haley wrote *Roots* to tell the story of his search.

GEOGRAPHY

Shifting Continents

The east coast of South America and the west coast of Africa could almost fit together like the pieces of a jigsaw puzzle. This observation led scientists to form the continental drift theory. The theory says that there was once a single great continent that split and drifted apart.

Other evidence indicates that all the land masses of Earth may have been joined at one time. Layers of rock along a coast on one side of the ocean match layers of rock on an opposite coast. Also, different continents have the same kinds of fossils. Thus, observation and evidence support the continental drift theory.

66

LEISURE

Green Thumbs Up!

Many teens enjoy spending time assembling a terrarium. You can have a miniature garden in your bedroom! Terrariums are fun to arrange, and the materials are easy to find. All you need is a clear container, pebbles, sand, soil, and some plants. Ferns and mosses grow well in terrariums. You can try to grow many other kinds of plants, also. Experiment with various plants to see which are best suited for a terrarium. Set up your experiment using the scientific method.

How would you vary the conditions in or surrounding a terrarium to find the best environment for it?

CONSUMER

Cycling Through Your Choices

Your old bicycle is just about worn out, and you need to buy a new one. How can you use a scientific method to choose the best bicycle? You can treat your decision like solving a problem.

First, research the types of bicycles. What are the advantages, disadvantages, and purposes of each type?

What do you need in a bicycle? Prepare a list of all the features and uses you want in a bicycle. Then rate each feature with a number from 1 to 5. A feature you really need is a 1. A feature that is nice but you could do without would be a 5.

Compare your list of needed features with the types of bicycles. Do your features match the purposes and advantages of one or two types of bicycles? If not, you may need to rethink your needs.

Go to a bicycle shop to see and ride bicycles. Narrow your choices to one type and choose your bicycle based on all the data you have gathered about the different bicycles.

67

**2** What do biology and Alex Haley have in common? A lot, as you'll discover when you read the close to Unit 1. Each unit is closed with mini essays that make connections between biology and consumer issues, leisure activities, art, literature, and history.

*clearly organized to get you started and keep you going*

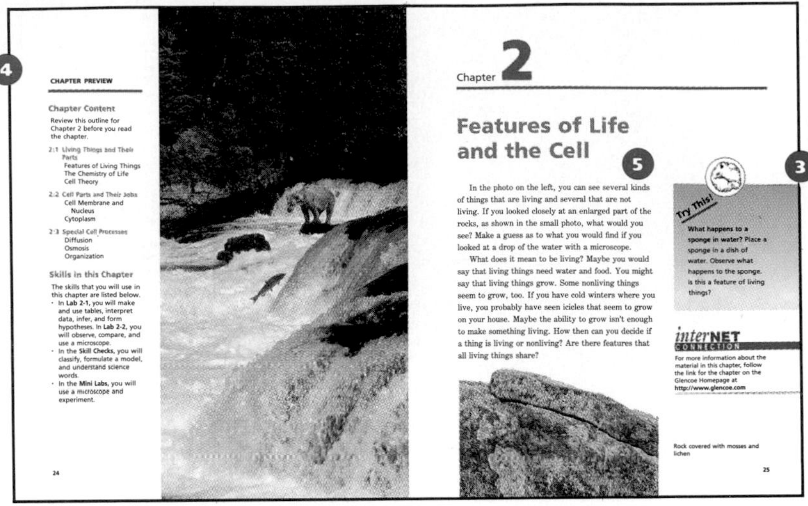

CHAPTER PREVIEW

Chapter Content

Review this outline for Chapter 2 before you read the chapter.

2:1 Living Things and Their Parts
Features of Living Things
The Chemistry of Life
Cell Theory

2:2 Cell Parts and Their Jobs
Cell Membrane and Nucleus
Cytoplasm

2:3 Special Cell Processes
Diffusion
Osmosis
Organization

Skills in this Chapter

The skills that you will use in this chapter are listed below.
- In **Lab 2-1**, you will make and use tables, interpret data, infer, and form hypotheses. In **Lab 2-2**, you will observe, compare, and use a microscope.
- In the **Skill Checks**, you will classify, formulate a model, and understand science words.
- In the **Mini Labs**, you will use a microscope and experiment.

24

Chapter 2

Features of Life and the Cell

In the photo on the left, you can see several kinds of things that are living and several that are not living. If you looked closely at an enlarged part of the rocks, as shown in the small photo, what would you see? Make a guess as to what you would find if you looked at a drop of the water with a microscope.

What does it mean to be living? Maybe you would say that living things need water and food. You might say that living things grow. Some nonliving things seem to grow, too. If you have cold winters where you live, you probably have seen icicles that seem to grow on your house. Maybe the ability to grow isn't enough to make something living. How then can you decide if a thing is living or nonliving? Are there features that all living things share?

Try This!

What happens to a sponge in water? Place a sponge in a dish of water. Observe what happens to the sponge. Is this a feature of living things?

interNET CONNECTION

For more information about the material in this chapter, follow the link for the chapter on the Glencoe Homepage at http://www.glencoe.com

Rock covered with mosses and lichen

25

**3** **Try This!**
Each chapter begins with an easy activity to do right at your desk, or at home. It gets you ready for learning.

**4** Listed for you in the **Chapter Preview** are the chapter contents. They tell what topics are covered and how they are organized. Study this before you dive into the chapter material. Also listed are skills that you will practice. A skill is something you get better at with practice. *Biology: An Everyday Experience* gives you all the practice you need to master skills that are important for success in biology, your other classes, and your everyday life.

**5** When was the last time you thought about what it means to be alive? Do small living things have the same life processes as large ones? These are the kinds of ideas you will ponder as you read the chapter openers. Each chapter opener has two photographs that are talked about in the introduction. As you read, think about what the photos mean and how they relate to the chapter.

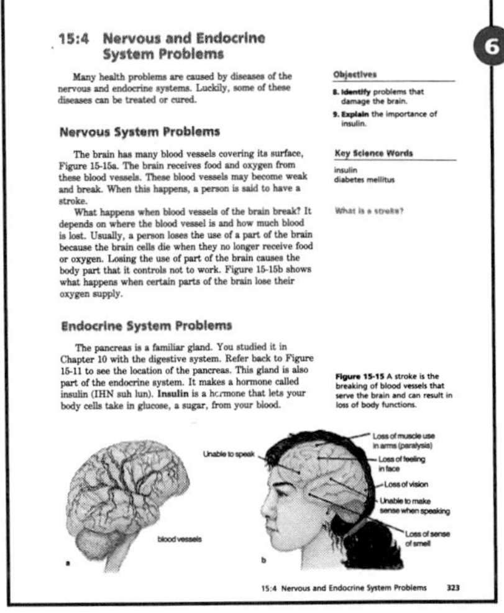

15:4 Nervous and Endocrine System Problems

Many health problems are caused by diseases of the nervous and endocrine systems. Luckily, some of these diseases can be treated or cured.

Objectives

**8. Identify** problems that damage the brain.
**9. Explain** the importance of insulin.

Key Science Words

insulin
diabetes mellitus

Nervous System Problems

The brain has many blood vessels covering its surface, Figure 15-15a. The brain receives food and oxygen from these blood vessels. These blood vessels may become weak and break. When this happens, a person is said to have a stroke.

What is a stroke?

What happens when blood vessels of the brain break? It depends on where the blood vessel is and how much blood is lost. Usually, a person loses the use of a part of the brain because the brain cells die when they no longer receive food or oxygen. Losing the use of part of the brain causes the body part that it controls not to work. Figure 15-15b shows what happens when certain parts of the brain lose their oxygen supply.

Endocrine System Problems

The pancreas is a familiar gland. You studied it in Chapter 10 with the digestive system. Refer back to Figure 15-11 to see the location of the pancreas. This gland is also part of the endocrine system. It makes a hormone called insulin (IHN suh lun). **Insulin** is a hormone that lets your body cells take in glucose, a sugar, from your blood.

**Figure 15-15** A stroke is the breaking of blood vessels that serve the brain and can result in loss of body functions.



15:4 Nervous and Endocrine System Problems 323

**6** Chapters are organized into two to four numbered sections. Each numbered section has several subsections that have red headings. The Objectives at the beginning of the numbered section tell you what major topics you'll be covering and what you should expect to learn about them. The Key Science Words are also listed in the order in which they appear in the section.

vi

*lots of ways to help you master important ideas and skills*

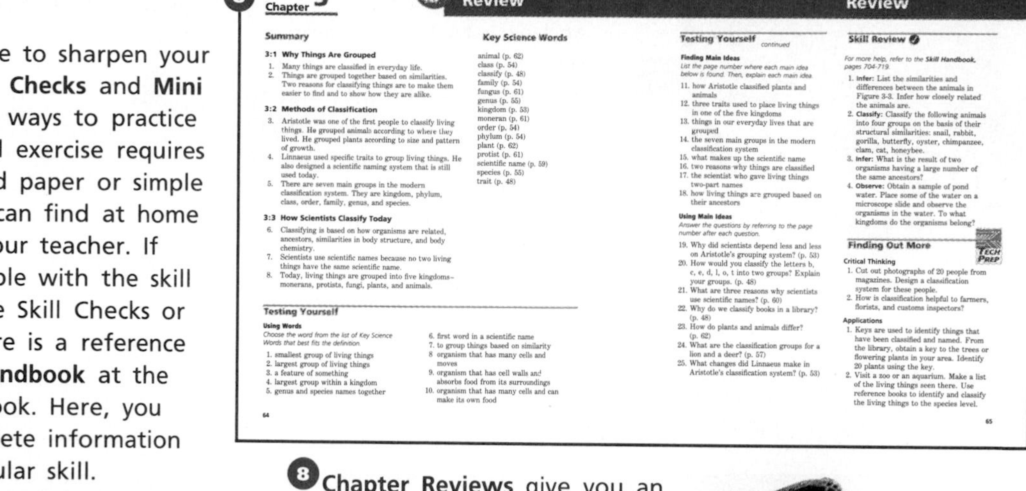

8

Chapter 3 Review

Summary

**3:1 Why Things Are Grouped**

1. Many things are classified in everyday life.
2. Things are grouped together based on similarities. Two reasons for classifying things are to make them easier to find and to show how they are alike.

**3:2 Methods of Classification**

3. Aristotle was one of the first people to classify living things. He grouped animals according to where they lived. He grouped plants according to size and pattern of growth.
4. Linnaeus used specific traits to group living things. He also designed a scientific naming system that is still used today.
5. There are seven main groups in the modern classification system. They are kingdom, phylum, class, order, family, genus, and species.

**3:3 How Scientists Classify Today**

6. Classifying is based on how organisms are related, ancestors, similarities in body structure, and body chemistry.
7. Scientists use scientific names because no two living things have the same scientific name.
8. Today, living things are grouped into five kingdoms–monerans, protists, fungi, plants, and animals.

Key Science Words

animal (p. 62)
class (p. 54)
classify (p. 48)
family (p. 54)
fungus (p. 61)
genus (p. 55)
kingdom (p. 53)
moneran (p. 61)
order (p. 54)
phylum (p. 54)
plant (p. 62)
protist (p. 61)
scientific name (p. 59)
species (p. 55)
trait (p. 48)

Testing Yourself

**Using Words**

*Choose the word from the list of Key Science Words that best fits the definition.*

1. smallest group of living things
2. largest group of living things
3. a feature of something
4. largest group within a kingdom
5. genus and species names together
6. first word in a scientific name
7. to group things based on similarity
8. organism that has many cells and moves
9. organism that has cell walls and absorbs food from its surroundings
10. organism that has many cells and can make its own food

64

Review

Testing Yourself *continued*

**Finding Main Ideas**

*List the page number where each main idea below is found. Then, explain each main idea.*

11. how Aristotle classified plants and animals
12. three traits used to place living things in one of the five kingdoms
13. things in our everyday lives that are grouped
14. the seven main groups in the modern classification system
15. what makes up the scientific name
16. two reasons why things are classified
17. the scientist who gave living things two-part names
18. how living things are grouped based on their ancestors

**Using Main Ideas**

*Answer the questions by referring to the page number after each question.*

19. Why did scientists depend less and less on Aristotle's grouping system? (p. 53)
20. How would you classify the letters b, c, e, d, l, o, t into two groups? Explain your groups. (p. 48)
21. What are three reasons why scientists use scientific names? (p. 60)
22. Why do we classify books in a library? (p. 48)
23. How do plants and animals differ? (p. 62)
24. What are the classification groups for a lion and a deer? (p. 57)
25. What changes did Linnaeus make in Aristotle's classification system? (p. 53)

Skill Review

*For more help, refer to the* ***Skill Handbook,*** *pages 704-719.*

1. **Infer:** List the similarities and differences between the animals in Figure 3-3. Infer how closely related the animals are.
2. **Classify:** Classify the following animals into four groups on the basis of their structural similarities: snail, rabbit, gorilla, butterfly, oyster, chimpanzee, clam, cat, honeybee.
3. **Infer:** What is the result of two organisms having a large number of the same ancestors?
4. **Observe:** Obtain a sample of pond water. Place some of the water on a microscope slide and observe the organisms in the water. To what kingdoms do the organisms belong?

Finding Out More

TECH PREP

**Critical Thinking**

1. Cut out photographs of 20 people from magazines. Design a classification system for these people.
2. How is classification helpful to farmers, florists, and customs inspectors?

**Applications**

1. Keys are used to identify things that have been classified and named. From the library, obtain a key to the trees or flowering plants in your area. Identify 20 plants using the key.
2. Visit a zoo or an aquarium. Make a list of the living things seen there. Use reference books to identify and classify the living things to the species level.

65

**7** Here's a chance to sharpen your skills. The **Skill Checks** and **Mini Labs** are good ways to practice skills. Each skill exercise requires only pencil and paper or simple materials you can find at home or get from your teacher. If you have trouble with the skill exercises in the Skill Checks or Mini Labs, there is a reference to the **Skill Handbook** at the back of the book. Here, you can find complete information about a particular skill.

**8** **Chapter Reviews** give you an opportunity to reinforce and apply your knowledge.

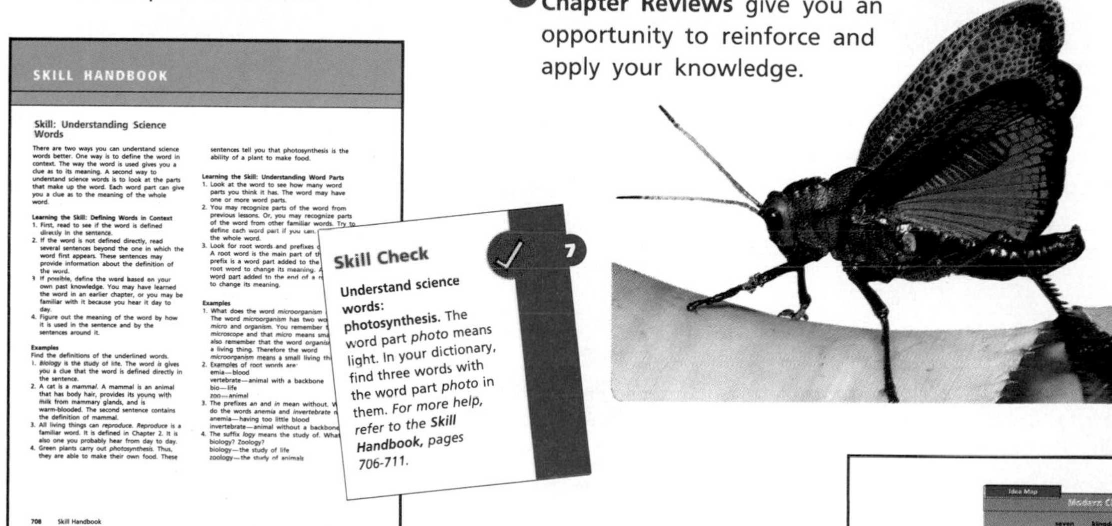

SKILL HANDBOOK

Skill: Understanding Science Words

There are two ways you can understand science words better. One way is to define the word in context. The way the word is used gives you a clue as to its meaning. A second way to understand science words is to look at the parts that make up the word. Each word part can give you a clue as to the meaning of the whole word.

**Learning the Skill: Defining Words in Context**

1. First, read to see if the word is defined directly in the sentence.
2. If the word is not defined directly, read several sentences beyond the one in which the word first appears. These sentences may provide information about the definition of the word.
3. If possible, define the word based on your own past knowledge. You may have learned the word in an earlier chapter, or you may be familiar with it because you hear it day to day.
4. Figure out the meaning of the word by how it is used in the sentence and by the sentences around it.

**Examples**

Find the definitions of the underlined words.

1. *Biology* is the study of life. The word *is* gives you a clue that the word is defined directly in the sentence.
2. A cat is a *mammal*. A mammal is an animal that has body hair, provides its young with milk from mammary glands, and is warm-blooded. The second sentence contains the definition of mammal.
3. All living things can *reproduce*. *Reproduce* is a familiar word. It is defined in Chapter 2. It is also one you probably hear from day to day.
4. Green plants carry out *photosynthesis*. Thus, they are able to make their own food. These sentences tell you that photosynthesis is the ability of a plant to make food.

**Learning the Skill: Understanding Word Parts**

1. Look at the word to see how many word parts you think it has. The word may have one or more word parts.
2. You may recognize parts of the word from previous lessons. Or, you may recognize parts of the word from other familiar words. Try to define each word part if you can. ... the whole word.
3. Look for root words and prefixes ... A root word is the main part of th... prefix is a word part added to the ... root word to change its meaning. ... word part added to the end of a r... to change its meaning.

**Examples**

1. What does the word *microorganism* ... The word *microorganism* has two wo... *micro* and *organism*. You remember ... *microscope* and that *micro* means sma... also remember that the word *organis*... a living thing. Therefore the word *microorganism* means a small living th...
2. Examples of root words are:
emia—blood
vertebrate—animal with a backbone
bio—life
zoo—animal
3. The prefixes *an* and *in* mean without. W... do the words *anemia* and *invertebrate* ...
anemia—having too little blood
invertebrate—animal without a backbone
4. The suffix *logy* means the study of. What... biology? Zoology?
biology—the study of life
zoology—the study of animals

708 Skill Handbook

Skill Check 7

**Understand science words: photosynthesis.** The word part *photo* means light. In your dictionary, find three words with the word part *photo* in them. *For more help, refer to the* ***Skill Handbook,*** *pages 706-711.*

**9** If you have trouble remembering and understanding a lesson, help is on the way. Each major concept has an Idea Map that you can use as a study guide. Ideas are summarized in an easy-to-read format. The Idea Maps will help make your study of biology a success.

**10** At the end of each major section are five **Check Your Understanding** questions. The first three questions reinforce what you have learned in the section. The fourth question challenges you to think critically about what you have read. The fifth question connects biology with reading, writing, or math.

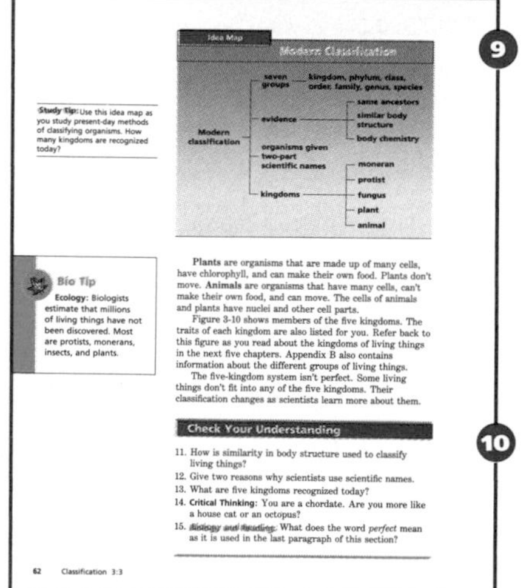

Idea Map

Modern Classification



**Study Tip:** Use this idea map as you study present-day methods of classifying organisms. How many kingdoms are recognized today?

**Bio Tip**

**Ecology:** Biologists estimate that millions of living things have not been discovered. Most are protists, monerans, insects, and plants.

**Plants** are organisms that are made up of many cells, have chlorophyll, and can make their own food. Plants don't move. **Animals** are organisms that have many cells, can't make their own food, and can move. The cells of animals and plants have nuclei and other cell parts.

Figure 3-10 shows members of the five kingdoms. The traits of each kingdom are also listed for you. Refer back to this figure as you read about the kingdoms of living things in the next five chapters. Appendix B also contains information about the different groups of living things.

The five-kingdom system isn't perfect. Some living things don't fit into any of the five kingdoms. Their classification changes as scientists learn more about them.

Check Your Understanding

11. How is similarity in body structure used to classify living things?
12. Give two reasons why scientists use scientific names.
13. What are five kingdoms recognized today?
14. **Critical Thinking:** You are a chordate. Are you more like a house cat or an octopus?
15. **Biology and Reading:** What does the word *perfect* mean as it is used in the last paragraph of this section?

62 Classification 3:3



vii

*really experience biology by observing, experimenting, asking questions*

11 Every chapter has two step-by-step labs. Procedures are clear and easy to follow. Sample data tables are given to help you organize the information you collect. At the end of each lab are questions that help to reinforce what you learned in the lab. Doing a lab has never been so easy!

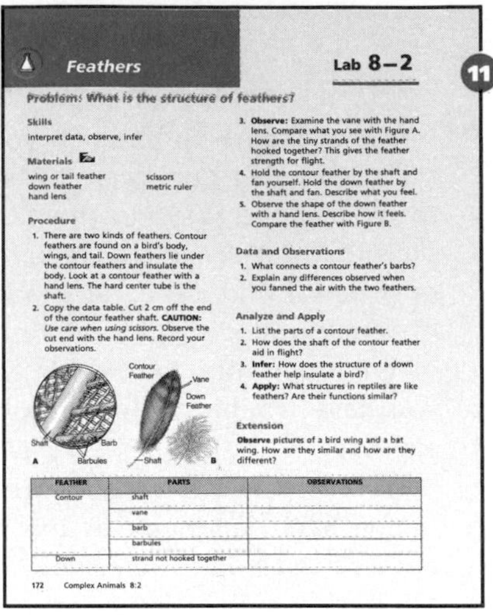

**Feathers** **Lab 8–2** 11

**Problem: What is the structure of feathers?**

**Skills**

interpret data, observe, infer

**Materials**

wing or tail feather
down feather
hand lens
scissors
metric ruler

**Procedure**

1. There are two kinds of feathers. Contour feathers are found on a bird's body, wings, and tail. Down feathers lie under the contour feathers and insulate the body. Look at a contour feather with a hand lens. The hard center tube is the shaft.
2. Copy the data table. Cut 2 cm off the end of the contour feather shaft. **CAUTION:** *Use care when using scissors.* Observe the cut end with the hand lens. Record your observations.
3. **Observe:** Examine the vane with the hand lens. Compare what you see with Figure A. How are the tiny strands of the feather hooked together? This gives the feather strength for flight.
4. Hold the contour feather by the shaft and fan yourself. Hold the down feather by the shaft and fan. Describe what you feel.
5. Observe the shape of the down feather with a hand lens. Describe how it feels. Compare the feather with Figure B.



**Data and Observations**

1. What connects a contour feather's barbs?
2. Explain any differences observed when you fanned the air with the two feathers.

**Analyze and Apply**

1. List the parts of a contour feather.
2. How does the shaft of the contour feather aid in flight?
3. **Infer:** How does the structure of a down feather help insulate a bird?
4. **Apply:** What structures in reptiles are like feathers? Are their functions similar?

**Extension**

**Observe** pictures of a bird wing and a bat wing. How are they similar and how are they different?

| FEATHER | PARTS | OBSERVATIONS |
|---|---|---|
| Contour | shaft | |
| | vane | |
| | barb | |
| | barbules | |
| Down | strand not hooked together | |

172 Complex Animals 8:2

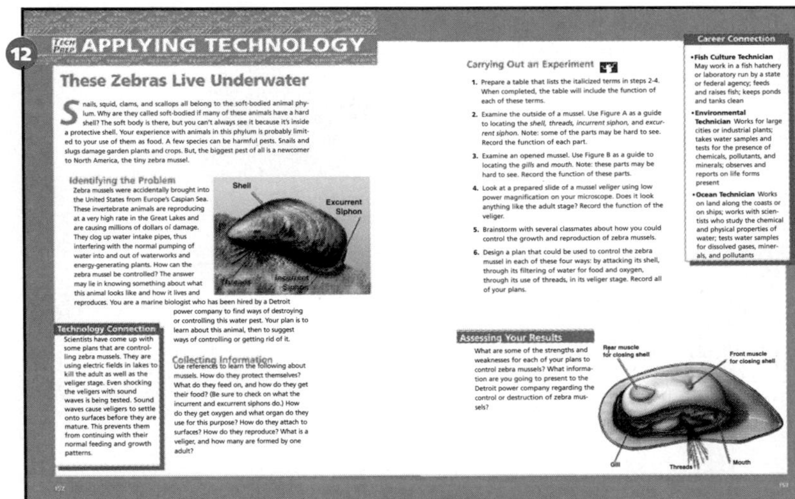

12 **APPLYING TECHNOLOGY**

**These Zebras Live Underwater**

Snails, squid, clams, and scallops all belong to the soft-bodied animal phylum. Why are they called soft-bodied if many of these animals have a hard shell? The soft body is there, but you can't always see it because it's inside a protective shell. Your experience with animals in this phylum is probably limited to your use of them as food. A few species can be harmful pests. Snails and slugs damage garden plants and crops. But, the biggest pest of all is a newcomer to North America, the tiny zebra mussel.

**Identifying the Problem**

Zebra mussels were accidentally brought into the United States from Europe's Caspian Sea. These invertebrate animals are reproducing at a very high rate in the Great Lakes and are causing millions of dollars of damage. They clog up water intake pipes, thus interfering with the normal pumping of water into and out of waterworks and energy-generating plants. How can the zebra mussel be controlled? The answer may lie in knowing something about what this animal looks like and how it lives and reproduces. You are a marine biologist who has been hired by a Detroit power company to find ways of destroying or controlling this water pest. Your plan is to learn about this animal, then to suggest ways of controlling or getting rid of it.



**Technology Connection**

Scientists have come up with some plans that are controlling zebra mussels. They are using electric fields in lakes to kill the adult as well as the veliger stage. Even shocking the veligers with sound waves is being tested. Sound waves cause veligers to settle onto surfaces before they are mature. This prevents them from continuing with their normal feeding and growth patterns.

**Collecting Information**

Use references to learn the following about mussels. How do they protect themselves? What do they feed on, and how do they get their food? (Be sure to check on what the incurrent and excurrent siphons do.) How do they get oxygen and what organ do they use for this purpose? How do they attach to surfaces? How do they reproduce? What is a veliger, and how many are formed by one adult?

**Carrying Out an Experiment**

1. Prepare a table that lists the italicized terms in steps 2-4. When completed, the table will include the function of each of these terms.
2. Examine the outside of a mussel. Use Figure A as a guide to locating the *shell, threads, incurrent siphon,* and *excurrent siphon.* Note: some of the parts may be hard to see. Record the function of each part.
3. Examine an opened mussel. Use Figure B as a guide to locating the *gills* and *mouth.* Note: these parts may be hard to see. Record the function of these parts.
4. Look at a prepared slide of a mussel *veliger* using low power magnification on your microscope. Does it look anything like the adult stage? Record the function of the veliger.
5. Brainstorm with several classmates about how you could control the growth and reproduction of zebra mussels.
6. Design a plan that could be used to control the zebra mussel in each of these four ways: by attacking its shell, through its filtering of water for food and oxygen, through its use of threads, in its veliger stage. Record all of your plans.

**Career Connection**

- **Fish Culture Technician** May work in a fish hatchery or laboratory run by a state or federal agency; feeds and raises fish; keeps ponds and tanks clean
- **Environmental Technician** Works for large cities or industrial plants; takes water samples and tests for the presence of chemicals, pollutants, and minerals; observes and reports on life forms present
- **Ocean Technician** Works on land along the coasts or on ships; works with scientists who study the chemical and physical properties of water; tests water samples for dissolved gases, minerals, and pollutants

**Assessing Your Results**

What are some of the strengths and weaknesses for each of your plans to control zebra mussels? What information are you going to present to the Detroit power company regarding the control or destruction of zebra mussels?



12 Each unit has a two-page Applying Technology feature. Read the short background paragraph, do the activity, and read how it relates to technology and careers.

viii

*explore how biology impacts technology and news-making issues and offers career choices*

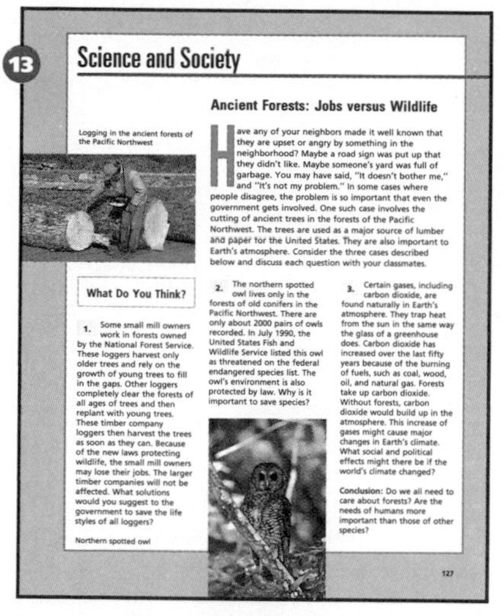

## Science and Society

### Ancient Forests: Jobs versus Wildlife

Logging in the ancient forests of the Pacific Northwest

Have any of your neighbors made it well known that they are upset or angry by something in the neighborhood? Maybe a road sign was put up that they didn't like. Maybe someone's yard was full of garbage. You may have said, "It doesn't bother me," and "It's not my problem." In some cases where people disagree, the problem is so important that even the government gets involved. One such case involves the cutting of ancient trees in the forests of the Pacific Northwest. The trees are used as a major source of lumber and paper for the United States. They are also important to Earth's atmosphere. Consider the three cases described below and discuss each question with your classmates.

**What Do You Think?**

1. Some small mill owners work in forests owned by the National Forest Service. These loggers harvest only older trees and rely on the growth of young trees to fill in the gaps. Other loggers completely clear the forests of all ages of trees and then replant with young trees. These timber company loggers then harvest the trees as soon as they can. Because of the new laws protecting wildlife, the small mill owners may lose their jobs. The larger timber companies will not be affected. What solutions would you suggest to the government to save the life styles of all loggers?

2. The northern spotted owl lives only in the forests of old conifers in the Pacific Northwest. There are only about 2000 pairs of owls recorded. In July 1990, the United States Fish and Wildlife Service listed this owl as threatened on the federal endangered species list. The owl's environment is also protected by law. Why is it important to save species?

3. Certain gases, including carbon dioxide, are found naturally in Earth's atmosphere. They trap heat from the sun in the same way the glass of a greenhouse does. Carbon dioxide has increased over the last fifty years because of the burning of fuels, such as coal, wood, oil, and natural gas. Forests take up carbon dioxide. Without forests, carbon dioxide would build up in the atmosphere. This increase of gases might cause major changes in Earth's climate. What social and political effects might there be if the world's climate changed?

**Conclusion:** Do we all need to care about forests? Are the needs of humans more important than those of other species?

Northern spotted owl

127

13 Are animal experiments necessary? Who decides which person in need of an organ transplant gets one? As you read the Science and Society features, you'll find that the answers to these and other questions are not so easy. The Science and Society features bring you closer to current issues and let you see the impact of technology on society. They prepare you for the day when you may need to participate in making decisions that affect your community and your environment.

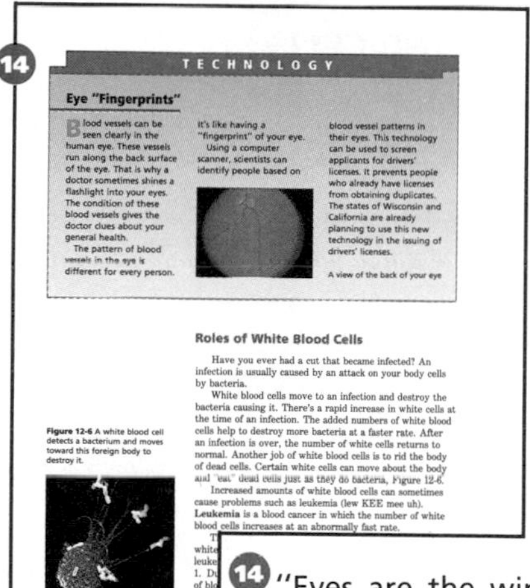

TECHNOLOGY

### Eye "Fingerprints"

Blood vessels can be seen clearly in the human eye. These vessels run along the back surface of the eye. That is why a doctor sometimes shines a flashlight into your eyes. The condition of these blood vessels gives the doctor clues about your general health.

The pattern of blood vessels in the eye is different for every person. It's like having a "fingerprint" of your eye.

Using a computer scanner, scientists can identify people based on blood vessel patterns in their eyes. This technology can be used to screen applicants for drivers' licenses. It prevents people who already have licenses from obtaining duplicates. The states of Wisconsin and California are already planning to use this new technology in the issuing of drivers' licenses.

A view of the back of your eye

### Roles of White Blood Cells

Have you ever had a cut that became infected? An infection is usually caused by an attack on your body cells by bacteria.

White blood cells move to an infection and destroy the bacteria causing it. There's a rapid increase in white cells at the time of an infection. The added numbers of white blood cells help to destroy more bacteria at a faster rate. After an infection is over, the number of white cells returns to normal. Another job of white blood cells is to rid the body of dead cells. Certain white cells can move about the body and "eat" dead cells just as they do bacteria, Figure 12-6.

Increased amounts of white blood cells can sometimes cause problems such as leukemia (lew KEE mee uh). **Leukemia** is a blood cancer in which the number of white blood cells increases at an abnormally fast rate.

**Figure 12-6** A white blood cell detects a bacterium and moves toward this foreign body to destroy it.

250 Blood 12:2

14 "Eyes are the windows to the soul." Did you know that your eyes are also like fingerprints? The patterns made by the vessels in the back of your eye are unique and can be used to identify you. No other person has eyes like you. This and other recent discoveries appear in the Technology features. The Technology features tie together biology and applications of the most recent research in biology. They make biology meaningful to *you.*

15

*Career Close-Up*

**Wildlife Photographer**

John's biology teacher invited a wildlife photographer to visit his class. The photographer brought slides and prints of many living things. Many photographs of flowers and insects were made with a close-up lens or a zoom lens. The photographer showed the students how to use a camera attachment on a light microscope. She told them that her work often took her to outdoor settings.

Students wanted to know about the training needed to be a photographer. She told them that she had taken photography and natural science courses in high school. Then she had taken several courses at a community college. She explained that in this field success depends on a person knowing the subject matter well.

Other students wanted to know where her work was used. She showed them magazines and books that contained her photographs.

Photographing wildlife is often a challenging occupation.

Biologists classify living things. Doing so puts organisms in order. It also shows how they are alike. There are over one and one-half million known kinds of living

15 You may never have thought about a career in biology, but think again. What kinds of interesting careers are there for you in biology? What about wildlife photography? How about being an athletic trainer? Each career feature focuses on one career, telling you the daily ins and outs of the job. Even if you want to work as soon as you graduate from high school, you'll find there are many careers that will let you do just that. Information about career training is given so that if you're interested, you can start planning now.



ix

# CONTENTS

x

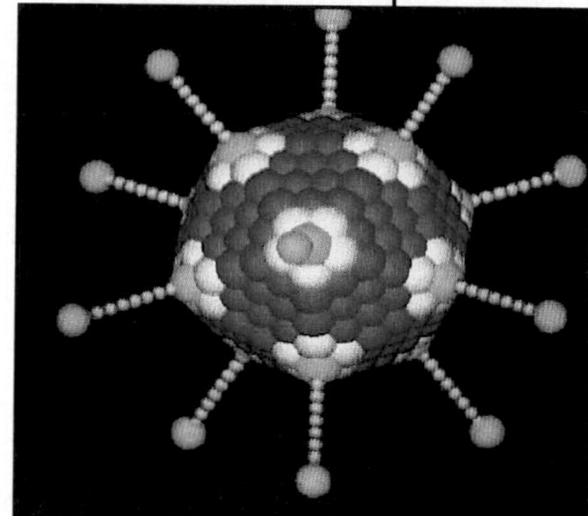

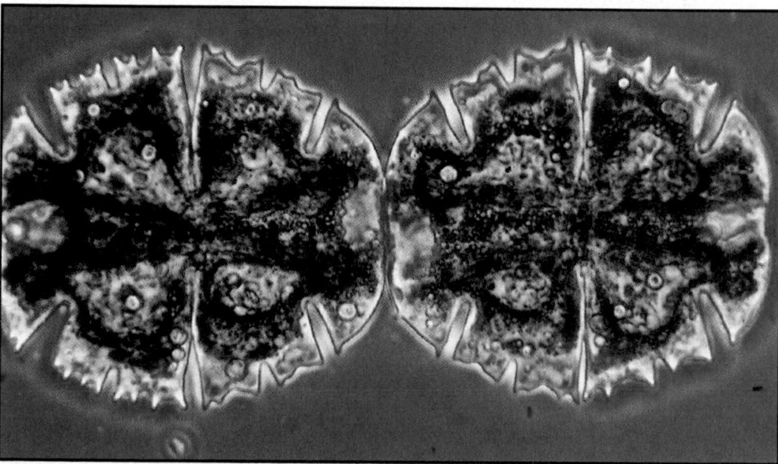



xi

xii

xiii

xiv

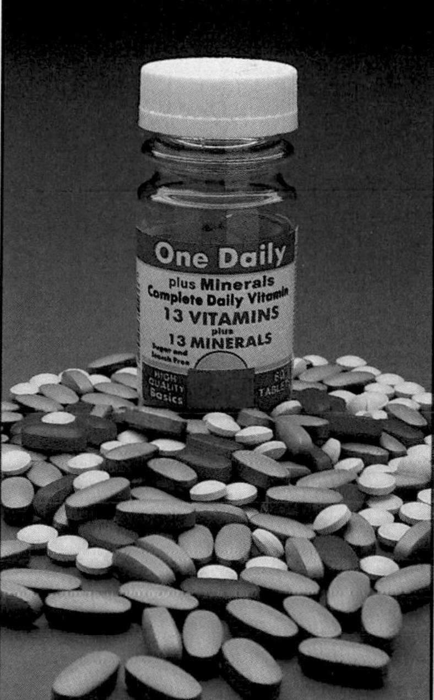

One Daily
plus Minerals
Complete Daily Vitamin
13 VITAMINS
plus
13 MINERALS

xv

xvii

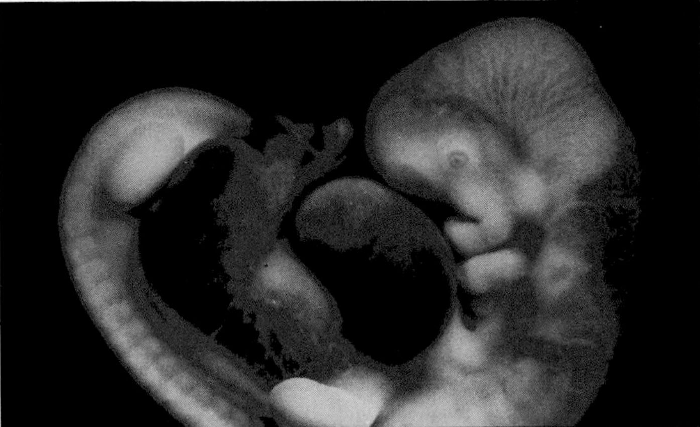



xviii

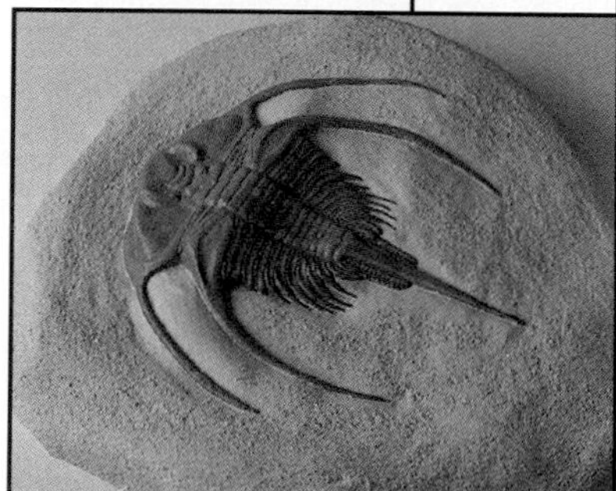

xix

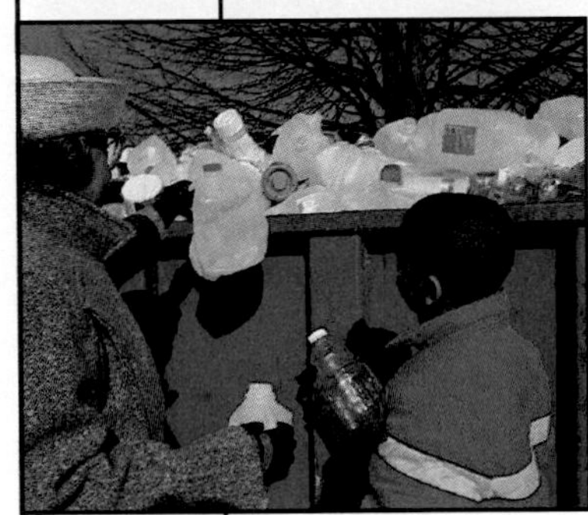

## CHAPTER 31 ECOSYSTEMS AND BIOMES 652



## CHAPTER 32 SOLVING ECOLOGICAL PROBLEMS 672



## APPENDICES



## SKILL HANDBOOK





xx

# LABS

Have you ever seen an animal capture and eat its prey using tentacles that surround its mouth? You may do this in one of the 64 **Labs** found in your textbook. In another **Lab,** you may watch as protein is digested. In other **Labs,** you test painkillers for the presence of aspirin, build models of viruses, and test the effects of pollution on yeast cells. **Labs** let you use what you have learned to *do* biology.

xxi

## APPLYING TECHNOLOGY

Can crops be watered with ocean water? Does light butter have less fat than regular butter? These are examples of **Applying Technology** features found in your textbook. You may investigate the problem of zebra mussels in the Great Lakes, or you may try to date rock layers from the fossils they contain. In the **Applying Technology** features, you can apply technology to solving biological problems and perhaps find a future career in the process.

## SCIENCE AND SOCIETY

Today it is possible for a couple to choose the sex of their child. Should this be done? How will a country's population be affected if all couples chose the sexes of their children? This is an example of a problem in which science and society interact as people line up on both sides of the debate. **Science and Society** features discuss real cases and present questions to help you form your *own* opinion about a current topic of debate in biology. Should animals be used to test the safety of new drugs? Should athletes use steroids to help them reach their goals? These and many other issues involving the environment, health, and conservation are presented in **Science and Society.**

**xxii**

## TECHNOLOGY

What do biologists do with new information about living things? The chemicals in the cells of each human are unique. No two people have the exact same chemical make-up. Learning to put that fact to practical use in solving crimes is one example of technology. Your textbook's **Technology** features show examples of biological knowledge at work. In **Technology,** you learn how bacteria can clean up oil spills and how eye-blinking behavior might help pilots.

## CAREER CLOSE-UP

What will you do with everything you learn in your biology course? Maybe you'll begin thinking about a career in a field involving biology. **Career Close-ups** are designed to give you some examples of career paths you might follow. You might become a wildlife photographer and travel around the world photographing living things. Or, you could be an animal breeder working to develop more nutritious beef. Perhaps you would like to become a mushroom farmer. The number of career choices in biology is almost endless.

xxiii

## MINI LABS

Would you like to learn how to make a bird feeder or try to design a child-proof drug package? The **Mini Lab** features in your textbook will tell you how.



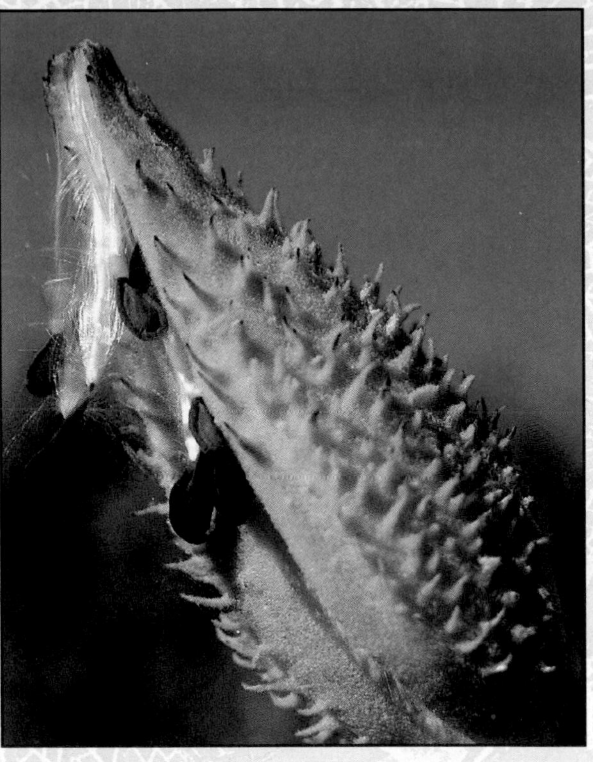

## SKILL CHECKS

What does the word *photosynthesis* mean? What other words are made of the same word parts? The **Skill Check** features will help you understand science words, as well as give you practice using other skills.

# Unit 1
# Kinds of Life

## UNIT OVERVIEW

Unit 1 provides background information for the study of biology. Chapter 1 deals with how biology is used in the present-day world, the tools used by biologists, scientific measurements, and scientific method. The features of living things, the cell as a basic unit of life, and the classification of living things are discussed in Chapters 2 and 3.

## Advance Planning

Audiovisual and computer software suggestions are located within each specific chapter under the heading TEACH.

Arrange a field trip to a nearby college for students to see an electron microscope.

Obtain materials for setting up a classroom aquarium to use throughout the year.

**To prepare for:**

| | |
|---|---|
| Lab 1-1: | Make sure that the light microscopes have been checked and are ready to use. Collect magazines from which to cut the letter *e* and pieces of color photographs. |
| Lab 1-2: | Make sure that you have pinto beans, salt, and self-sealing plastic bags. |
| Mini Lab: | Collect plant cuttings and aspirin. |
| Lab 2-1: | Ask students to bring baby food jars for this lab. Prepare phenol red solution and yeast solution. If you do not have a classroom aquarium, guppies will have to be purchased. |
| Lab 2-2: | Prepare methylene blue stain. Order *Elodea.* |
| Mini Labs: | Cut thin slices of cork. Obtain vanilla extract, a balloon, and a box. |
| Lab 3-1: | Collect the objects to be classified and place them in envelopes. |
| Lab 3-2: | Order state maps. |

# Unit 1

CONTENTS

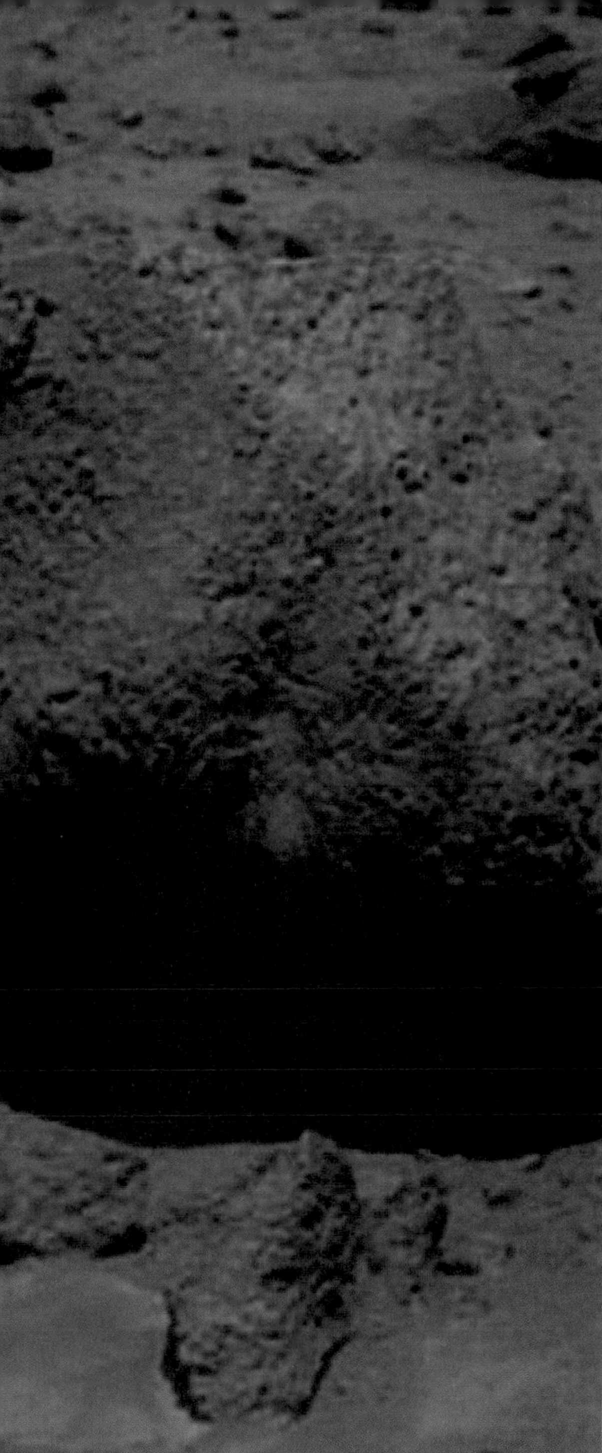

# Kinds of Life

**What would happen if...**

there were life on Mars? Living things as we know them need water, warmth, and protection from radiation from the sun. Does Mars provide these things? Life-forms on Mars would have to survive without any water because all of the water on Mars is frozen. Mars is farther away from the sun than Earth. As a result, the temperature on Mars is always below freezing. The atmosphere of Mars is too thin to block out harmful radiation from the sun. Thus any life on Mars would have to be able to withstand radiation.

The Pathfinder probe, carrying a surface rover named Sojourner, landed on the surface of Mars on July 4th, 1997. Sojourner, shown here, is studying Martian soil and rocks, looking for different chemicals and for living things.

If life exists on Mars, it probably is not life as we know it. But are there other types or definitions of life?

3

UNIT OVERVIEW

## Photo Teaching Tip

Have students read the passage that accompanies this pair of unit opening photographs. Open a discussion that will invite students to compare the conditions on Earth and Mars. Have them consider the possibility of the existence of other life-forms in the universe. Explain that scientists need to be open-minded in order to make hypotheses and explore the unknown. New and unusual ideas can often result in scientific breakthroughs. Have students write their own new definitions of life as they imagine it to exist on other planets.

## Theme Development

Each organism is made up of cells that help it maintain a balance within its environment. The themes of ecology, homeostasis, and diversity are developed in this unit.

**CULTURAL DIVERSITY**

See the Cultural Diversity feature located on the Connections page at the end of this unit.

For Tech Prep activities, see the Applying Technology feature in this unit, the Finding Out More applications questions in the Chapter Reviews, and the Tech Prep Applications booklet in the Teacher Resource Package.

# The Study of Life

## PLANNING GUIDE

| CONTENT | TEXT FEATURES | TEACHER RESOURCE PACKAGE | OTHER COMPONENTS |
|---|---|---|---|
| (1/2 day)<br>1:1 Biology in Use<br>Who Uses Biology<br>Tools of Biology | Skill Check: *Understand Science Words,* p. 7<br>Skill Check: *Make and Use Tables,* p. 8<br>Lab 1-1: *Microscopes,* p. 9<br>Check Your Understanding, p. 10 | Enrichment: *Fields of Science,* p. 1<br>Reteaching: *Parts of the Microscope,* p. 1<br>Focus: *How Does Biology Contribute to the Working of a Supermarket?* p. 1<br>Study Guide: *Biology in Use,* pp. 1-2<br>Lab 1-1: *Microscopes,* pp. 1-2<br>Transparency Master: *Tools Used in Biology,* p. 1 | **Laboratory Manual:**<br>*How Is the Light Microscope Used?* p. 1<br>Color Transparency 1: *The Microscope*<br>**STVS:** *Using Bacteria to Detect Carcinogens,* Plants and Simple Organisms (Disc 4, Side 1)<br>**STVS:** *Selective Breeding in Micro-Pigs,* Plants and Simple Organisms (Disc 4, Side 1) |
| (1 1/2 days)<br>1:2 Measurements Used in Biology<br>Length<br>Volume<br>Weight and Mass<br>Time and Temperature | Skill Check: *Calculate,* p. 12<br>Check Your Understanding, p. 14 | Application: *Safety in the Biology Laboratory,* p. 1<br>Reteaching: *Measurements Used in Biology,* p. 2<br>Skills: *Measure in SI,* p. 1<br>Study Guide: *Measurements Used in Biology,* pp. 3-4 | **Laboratory Manual:**<br>*How Are SI Length Measurements Made?* p. 5<br>**STVS:** *Research in the Pinelands,* Ecology (Disc 6, Side 1) |
| (2 days)<br>1:3 Scientific Method<br>Recognizing and Researching the Problem<br>Forming a Hypothesis<br>Testing a Hypothesis<br>Drawing Conclusions<br>Theory<br>Technology | Mini Lab: *Does Aspirin Help Plants Grow?* p. 16<br>Lab 1-2: *Scientific Method,* p. 17<br>Science and Society: *Technology: Helpful or Harmful?* p. 21<br>Idea Map, p. 15<br>Check Your Understanding, p. 20 | Reteaching: *How Can You Use a Scientific Method to Solve a Problem?* p. 3<br>Critical Thinking/Problem Solving: *Can You Spot the Scientific Method?* p. 1<br>Study Guide: *Scientific Method,* p. 5<br>Study Guide: *Vocabulary,* p. 6<br>Lab 1-2: *Scientific Method,* pp. 3-4 | **STVS:** *Animal Models of Human Physiology,* Animals (Disc 5, Side 2) |
| Chapter Review | Summary<br>Key Science Words<br>Testing Yourself<br>Finding Main Ideas<br>Using Main Ideas<br>Skill Review | **ASSESSMENT RESOURCES**<br>Chapter Review, pp. 1-2<br>Chapter Test, pp. 3-5<br>Performance Assessment in the Biology Classroom<br>Alternate Assessment in the Science Classroom<br>Computer Test Bank | |

## GLENCOE TECHNOLOGY

**Infinite Voyage,** *The Future of the Past* and *The Great Dinosaur Hunt*
**Science and Technology Videodisc Series,** *Using Bacteria to Detect Carcinogens,* Plants and Simple Organisms (Disc 4, Side1)
*Selective Breeding in Micro-Pigs,* Plants and Simple Organisms (Disc 4, Side 1)
*Research in the Pinelands,* Ecology (Disc 6, Side 1)
*Animal Models of Human Physiology,* Animals (Disc 5, Side 2)
**The Secret of Life,** *On the Brink: Portraits of Modern Science*

## MATERIALS NEEDED

| LAB 1-1, p. 9 | LAB 1-2, p. 17 | MARGIN FEATURES |
|---|---|---|
| microscope<br>microscope slide<br>coverslips<br>water<br>dropper<br>forceps<br>small piece of magazine print | 2 100-mL beakers<br>40 pinto beans<br>balance<br>salt<br>water<br>2 paper towels<br>50-mL graduated cylinder<br>2 self-sealing plastic bags<br>masking tape | Skill Check, p. 7<br>dictionary<br>Skill Check, p. 8<br>pencil<br>paper<br>Skill Check, p. 12<br>calculator<br>meter stick<br>Mini Lab, p. 16<br>plants<br>aspirin<br>water |

## OBJECTIVES

For more information about National Science Standards, see page 5T.

| SECTION | OBJECTIVE | CORRELATION of QUESTIONS to OBJECTIVES: CHECK YOUR UNDERSTANDING | CHAPTER REVIEW | TRP CHAPTER REVIEW | TRP CHAPTER TEST |
|---|---|---|---|---|---|
| 1:1 National Science Stds: UCP.3, A.2, E.2, G.1, G.2 | 1. **Describe** three jobs that depend on the use of biology. | 1, 4 | 16, 23 | 3, 26, 36 | 9, 35, 36, 37, 38, 39 |
| | 2. **List** five tools of biologists. | 2 | 19, 29 | 10, 11, 29 | 46 |
| | 3. **Compare** the features of the light microscope, stereomicroscope, and electron microscope. | 3, 5 | 6, 12, 24 | 16, 17, 18, 19, 20, 21, 22, 23, 24, 25 | 1, 3, 15, 16, 17, 18, 19, 20, 21, 22, 23 |
| 1:2 National Science Stds: UCP.3, B.2 | 4. **Explain** how SI units are grouped. | 6, 9 | 1, 21 | 2, 4, 32 | 8, 31, 32, 34 |
| | 5. **List** the SI units of measure for length, volume, mass, time, and temperature. | 7 | 2, 4, 13, 25 | 1, 9, 27, 31, 38 | 4, 7, 30, 31, 33, 34, 41 |
| | 6. **Measure** objects using SI units. | 8, 10 | 17, 18, 27 | 28, 35 | 6, 24, 25, 28, 43, 45 |
| 1:3 National Science Stds: UCP.2, UCP.3, A.1, A.2, E.1, E.2, F.4, F.5, F.6, G.1, G.2 | 7. **Explain** the steps of the scientific method. | 11, 12, 15 | 3, 7, 10, 11, 14, 15, 20, 22, 28 | 5, 6, 7, 8, 12, 13, 14, 15, 30, 33, 39 | 2, 10, 11, 13, 14, 37, 40, 42 |
| | 8. **Compare** the difference between a hypothesis and a theory. | 13 | 8 | 34 | 12 |
| | 9. **Explain** how technology is used to solve everyday problems. | 14 | 9, 26 | 37 | 5, 47 |

# The Study of Life

## CHAPTER OVERVIEW

### Key Concepts

In this chapter, students will study who uses biology, the tools that are used to study it, and how measurements are made using SI. Students also will learn the steps in scientific method and use those steps to solve problems.

### Key Science Words

| | |
|---|---|
| biology | light microscope |
| Celsius | meter |
| control | scientific method |
| data | stereo microscope |
| experiment | technology |
| hypothesis | theory |
| International System of Units | variable |
| kilogram | volume |

### Skill Development

In Lab 1-1, students will **observe** and **recognize and use spatial relationships** while learning to use the microscope. In Lab 1-2, they will **form hypotheses, interpret data**, and **separate and control variables** while learning to use scientific method. In the Skill Check on page 7, students will **understand** the **science word** *microscope.* In the Skill Check on page 8, they will **make** and **use tables** to compare microscope magnifications. In the Skill Check on page 12, they will **calculate** area. In the Mini Lab on page 16, they will **design an experiment** to see if plant cuttings will grow faster in aspirin water than plain water.

### Bridging

Students know the differences between things that are living and things that are not living. Discuss these differences and how an understanding of living things is used by many people. For instance, you may want to discuss the work that people in the community do. Students will see how biology is used in everyday life.

**CHAPTER PREVIEW**

### Chapter Content

Review this outline for Chapter 1 before you read the chapter.



### Skills in this Chapter

The skills that you will use in this chapter are listed below.

- In **Lab 1-1,** you will observe and recognize and use spatial relationships. In **Lab 1-2,** you will form hypotheses, interpret data, and separate and control variables.
- In the **Skill Checks,** you will make and use tables, calculate, and understand science words.
- In the **Mini Labs,** you will design an experiment.

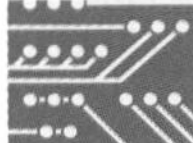

## TECH PREP

For Tech Prep activities in this chapter of the Teacher Wraparound Edition, see especially the Connection, Portfolio, and Student Journal on page 7, and the Motivation on page 15.
See also the Glencoe Homepage at **www.glencoe.com**

The following Glencoe resources provide additional opportunities for integrating science and technology.

**Communication Technology: Today and Tomorrow**
Section I: Introduction to Communication Technology
Basic Activity 1: Interrelationship of Technologies

**Production Systems Technology**
Section II: Resources for Production
Activity 6: The Metric System

Chapter **1**

# The Study of Life

Look at the photo of the classroom on the left. How many things can you name that are living or were once living? Think about the living or once-living things in your own classroom. The desks, pencils, and books in your classroom are plant products. Chalk comes from tiny living things found in the ocean. Plastics contain materials that come from the remains of plants and animals. Chairs, pens, rulers, and calculators are some of the products that can be made from plastics.

Consider the things that came from living or once-living things in the other parts of the school, such as the gym, cafeteria, or library. You can see many things that you need and use every day are living or were once living. **Biology** is the study of living and once-living things. *Bio-* means life and *-ology* means the study of. A person who studies living and once-living things is called a biologist.

**Try This!**

**How are an animal and a candle flame alike?** Observe a living animal and a candle flame. Record how they are similar and how they are different. How is the candle flame like the living animal? How is it different?

**interNET CONNECTION**

For more information about the material in this chapter, follow the link for the chapter on the Glencoe Homepage at **http://www.glencoe.com**



**GETTING STARTED**

## Using the Photos

Students may also consider their bedrooms or other rooms in their homes and try to come up with a list of living or once-living things found there. With the widespread use of plastics, there is almost nothing in the home or school environment that is not living or once living.

## MOTIVATION/Try This!

**How are an animal and a candle flame alike?** In the chapter-opening activity, students will **observe** a living and nonliving thing and **record** their observations. Students will see that the candle flame moves, uses energy from the candle, and responds to air currents. They will discover that nonliving things may appear to be living.

## Chapter Preview

Have students study the chapter outline before they begin to read the chapter. The chapter introduces the tools, measurements, and problem solving methods of biologists.

## Misconception

All people use biology every day. Students may think of people who use biology as those who wear white coats and work in laboratories surrounded by test tubes and beakers.

**GLENCOE TECHNOLOGY**

**Videodisc**

**Biology: The Dynamics of Life**
Movie: *Biologists at Work*
Disc 1, Side 1, Ch. 2

**ASSESSMENT PLANNER**

**Portfolio**
Strategies on the following pages represent student products that can be placed into a best-work portfolio: pp. 7, 17.

**PERFORMANCE ASSESSMENT**
Skill Check, pp. 8, 12
Mini Lab, p. 16
Lab, pp. 9, 17

**CONTENT ASSESSMENT**
Check for Understanding, pp. 10, 13, 14, 18, 20
Chapter Review, pp. 22-23

**GROUP ASSESSMENT**
Opportunities for group assessment occur with Cooperative Learning Strategies.

**Student Journal**
Strategies on the following pages represent opportunities for writing in a Student Journal: pp. 7, 11.

## 1:1 Biology in Use

## PREPARATION

### Materials Needed

Make copies of the Focus, Enrichment, Transparency, Reteaching, Study Guide, and Lab worksheets in the *Teacher Resource Package*.

▶ Gather a Petri dish, Duco cement, and a few pencil shavings for the Focus demonstration. Cut letter e's from a magazine for Lab 1-1.

### Key Science Words

biology
light microscope
stereomicroscope

### Process Skills

In Lab 1-1, students will observe and recognize and use spatial relationships. In the Skill Checks, students will understand science words and make and use tables.

## 1 FOCUS

▶ The objectives are listed on the student page. Remind students to preview these objectives as a guide to this numbered section.

### MOTIVATION/Demonstration

Fill a Petri dish with water. Place it on the overhead projector. Add 3 or 4 drops of Duco cement. Then add 2 or 3 pencil shavings. Ask students to observe and record their observations. Discuss what life processes can be seen (movement, eating).

### Guided Practice

Have students write down their answers to the margin question: to analyze evidence.

**Focus**/*Teacher Resource Package*, p. 1. Use the Focus worksheet shown at the bottom of this page as an introduction to the uses of biology.

### Objectives

1. **Describe** three jobs that depend on the use of biology.
2. **List** five tools of biologists.
3. **Compare** the features of the light microscope, stereomicroscope, and electron microscope.

### Key Science Words

biology
light microscope
stereomicroscope

**How do the police use biology?**

## 1:1 Biology in Use

The use of biology is not limited to biologists. Many people use biology daily in their hobbies and work. We will look at how you and other people use biology.

### Who Uses Biology?

Look at the photos. Are the people shown in the photos using biology in any way?

The beekeeper in Figure 1-1a is removing honey from the hive. A beekeeper must know how bees work together to produce honey. He must know where to place the hives to get the most honey.

Crime lab technicians (tek NISH uhnz) work with different kinds of evidence. Evidence may include cloth fibers, bits of paper, hair, skin, and blood. These things must be studied under a microscope.

If you wear eyeglasses or contact lenses, an optometrist (ahp TAHM uh trust) probably fitted them. The optometrist has studied the eyes and how they work.

Mushrooms are grown in buildings with carefully controlled temperature, moisture, and air flow. Mushroom growers must choose the best soil conditions for their crops. They must know how to identify and treat diseases of mushrooms.

Many people use biology in different ways. You use biology if you have ever asked yourself any of these questions: How can I tell when my plant needs water? What can I do about the fleas on my dog? How can I tell if this plant is poison ivy? The answers to these questions have to do with living things. You use biology all the time and may not even realize it.

**Figure 1-1** Beekeepers (a) and optometrists (b) use biology in their work.

a

b

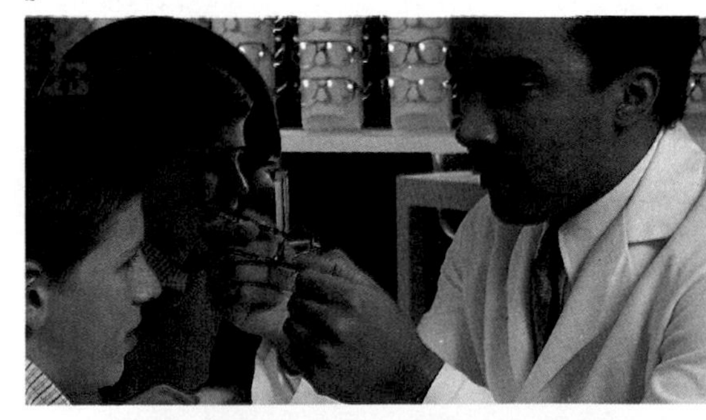



## OPTIONS

### Science Background

The first microscopes were built around 1600. Galileo observed insects with a compound microscope he had made. Two Dutch spectacles makers, Jan and Zacharias Janssen, are also credited with having developed compound microscopes.

### Science Background

Magnets in the electron microscope focus the electron beam onto a fluorescent screen, which produces an image of the specimen.

**FOCUS 1**

FOCUS CHAPTER 1

Name ______ Date ______ Class ______

HOW IS BIOLOGY USED IN A SUPERMARKET?

1. How do you think biology is useful to the meat and fish department of a supermarket? Explain.
2. How is biology useful to the produce department? Explain.
3. How might the salad and cheese department depend on a biologist who studies tiny living things and the diseases they can cause?

## Tools of Biology

Almost every day you use tools to measure and observe. You may use some of the same tools that a biologist uses. Have you ever used a ruler or a magnifying glass? Maybe you have used a pair of binoculars or a microscope.

A hand lens is a type of magnifying glass. It makes an object appear three to five times larger than it is.

A light microscope is one of the most important tools of biology, Figure 1-2. It magnifies small objects similar to the way binoculars magnify far-away objects. Both light microscopes and binoculars have lenses for magnifying objects.

In a **light microscope**, light passes through the object being looked at and then through two or more lenses. What you look at appears larger.

### Skill Check

**Understand science words: microscope.** The word part *micro* means small. In your dictionary, find three words with the word part *micro* in them. *For more help, refer to the **Skill Handbook**, pages 706-711.*

Eyepieces
Arm
Low-power objective
Revolving nosepiece
Stage clips
High-power objectives
Coarse adjustment
Stage
Fine adjustment
Diaphragm
Light source

**Figure 1-2** A light microscope magnifies small objects.

7

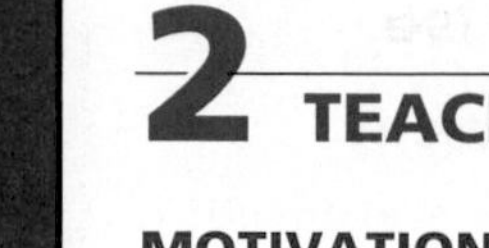

## 2 TEACH

### MOTIVATION/Brainstorming

Ask students why a paramedic, crime lab technician, or optician needs knowledge of biology.

### Concept Development

▶ You may want to introduce the section on tools of biology by showing students a hand lens, a pair of binoculars, and a microscope.

### Connection: Language Arts

Have students look through the classified section of the newspaper and circle jobs that use biology.

### Independent Practice

**Study Guide**/*Teacher Resource Package*, p. 1. Use the Study Guide worksheet shown at the bottom of this page for independent practice.

**Portfolio**

Optical instruments provide advances in areas of science other than biology. Have students research how microscopes are used in other areas of science, how cameras are used in other areas of science, and how telescopes are used in astronomy.

**Student Journal**

Have students make a list of tools that extend their ability to observe. Begin with the microscope. Explain how each instrument would extend a person's ability to observe.

### ENRICHMENT 1

ENRICHMENT CHAPTER 1

Name ______ Date ______ Class ______

Use after Section 1:1.

FIELDS OF SCIENCE

Scientists study and investigate nearly every aspect of life on Earth. They work in different fields of science. The names of many of these fields end in the suffix *-logy*. Find this suffix in the dictionary.

What is its meaning? **the science of or study of**

A suffix, such as *-logy*, must be combined with another word part to complete its meaning. When the suffix *-logy* is added to the prefix *bio-*, what does the resulting word mean? **the study of life**

*Use a dictionary to find the meanings of the prefixes in the name of each field of science listed below. In the second column of the chart, write a description of what is studied in each field listed in the first column.*

| Word | Meaning of word |
|---|---|
| Anthropology | the study of humans, their origins and cultures |
| Zoology | the study of animals |
| Ecology | the study of living organisms and their environment |
| Virology | the study of viruses |
| Bacteriology | the study of bacteria |
| Cytology | the study of cell structure |
| Entomology | the study of insects |
| Embryology | the study of the formation and development of embryos |
| Ethology | the study of animal behavior |
| Herpetology | the study of reptiles |
| Microbiology | the study of microscopic organisms |
| Paleontology | the study of plant and animal fossils |
| Oceanology | the study of marine resources and technology |
| Physiology | the study of how living things function |
| Ichthyology | the study of fish |

### STUDY GUIDE 1

STUDY GUIDE CHAPTER 1

Name ______ Date ______ Class ______

BIOLOGY IN USE

*In your textbook, read about people who use biology in Section 1:1. Then, answer the questions below.*

1. Many people use biology every day. What do the people shown below need to know about biology?

A florist needs to know how to care for plants.

A coach needs to know about bones and muscles.

A pharmacist needs to know how medicines affect the body.

A farmer needs to know how to raise plants and animals.

A veterinarian needs to know how to keep animals healthy.

A lifeguard needs to know about lungs and breathing.

2. Many people use biology in their hobbies. The column on the right lists ways biology can be used in hobbies. In front of the hobbies on the left, write the letter or letters of the way people use biology in each hobby.

a, c 1. growing houseplants — a. know about different kinds of plants
d 2. jogging — b. know about the habits of birds
b, e 3. bird watching — c. know how to make plants grow
d 4. swimming — d. know about bones and muscles
a, e 5. identifying wild flowers — e. know about classifying things

## OPTIONS

**How Is the Light Microscope Used?**/*Lab Manual*, pp. 1-4. Use this lab to give additional practice with the microscope.

**Challenge:** Have students expand the list of ways people use biology. Ask if biology might supply the answers to: How could I stay underwater longer? Why do I yawn?

**Enrichment**/*Teacher Resource Package*, p. 1. Use the Enrichment worksheet shown here to give more information on fields of biology.

# TEACH

## Concept Development

▶ Explain that the stereoscopic microscope has an ocular lens and an objective lens for each eye. This arrangement of lenses provides a three-dimensional view of the object's surface.

▶ Point out that the electron microscope produces a picture in a manner similar to a television.

▶ Explain to students that a microscope does two things: it magnifies and provides resolving power.

▶ Explain to students that spaces are used to group digits in long numbers because, in some countries, a comma indicates a decimal point.

## Guided Practice

Have students write down their answers to the margin questions: objects through which light will not pass.

**Cooperative Learning:** Divide the class into twos and have students in each group check each other on the parts and functions of the microscope.

## Skill Check

Use the Skill Check to make sure that students understand how to find the total magnification of a microscope.

## Independent Practice

**Study Guide**/*Teacher Resource Package*, p. 2. Use the Study Guide worksheet shown at the bottom of this page for independent practice.

a

b

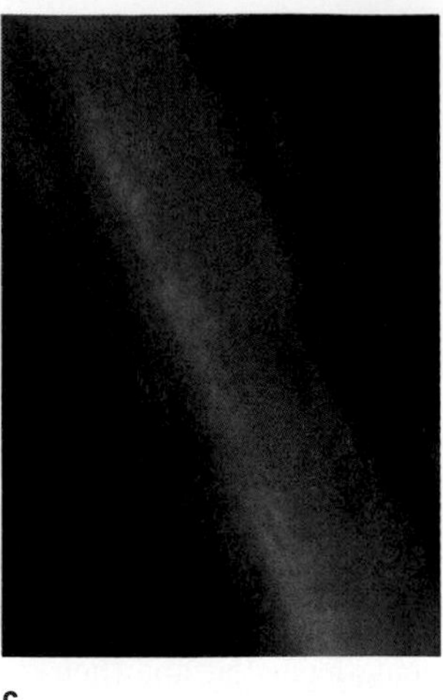

c

**Figure 1-3** A hair is shown magnified 10 times (a), 100 times (b), and 430 times (c).

### Skill Check

**Make and use tables:** Make a table that compares the magnifying powers of a hand lens, a stereomicroscope, low power and high power of a light microscope, and an electron microscope. *For more help, refer to the* ***Skill Handbook****, pages 715-717.*

**What kinds of objects would you view with a stereomicroscope?**

The microscope has a lens in the eyepiece. The eyepiece lens magnifies 10 times and is marked 10 ×. If you looked through just the eyepiece, a hair would look like the one in Figure 1-3a.

In addition to the eyepiece lens, a light microscope has one or more objective lenses that magnify the object further. One objective is a low-power lens. It magnifies 10 times and is marked 10 ×. To find the *total magnification* when looking through the eyepiece lens and the objective lens, you multiply the magnification of the two lenses together.

eyepiece lens × objective lens = total magnification
10 times × 10 times = 100 times larger

If you looked at the same hair with the low-power objective and the eyepiece together, you would see something that looks like Figure 1-3b. This hair is enlarged 100 times.

The second objective lens is a high-power lens. High-power objectives can magnify 40, 43, or 100 times. How would an objective that magnifies 43 times be marked? If you looked at the hair with a high-power objective and the eyepiece together, it might look like the hair in Figure 1-3c. This hair is magnified 430 times. The highest-powered light microscopes can magnify objects 2000 times.

Stereomicroscopes (STER ee oh MI kruh skohps) are often used in biology classes. A **stereomicroscope** is used for viewing large objects and things through which light cannot pass, such as insects. The stereomicroscope has an eyepiece lens for each eye, as shown in Figure 1-4.

8

## OPTIONS

**Enrichment:** Investigate preparation techniques for electron microscopy. Investigate the process of taking photographs through a microscope. If you have access to standard photographic equipment, you might try photomicrography.

**Transparency Master**/*Teacher Resource Package*, p. 1. Use the Transparency Master shown here to teach students about tools used in biology.

**STUDY GUIDE 2**

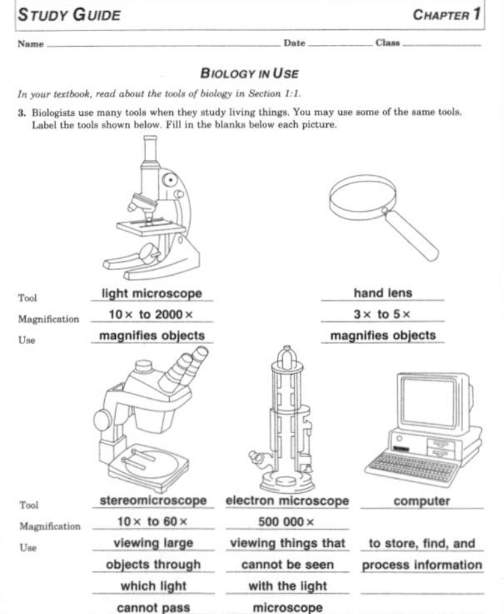

STUDY GUIDE CHAPTER 1

Name ______ Date ______ Class ______

BIOLOGY IN USE

*In your textbook, read about the tools of biology in Section 1:1.*

3. Biologists use many tools when they study living things. You may use some of the same tools. Label the tools shown below. Fill in the blanks below each picture.

| | | |
|---|---|---|
| Tool | light microscope | hand lens |
| Magnification | 10× to 2000× | 3× to 5× |
| Use | magnifies objects | magnifies objects |

| | | | |
|---|---|---|---|
| Tool | stereomicroscope | electron microscope | computer |
| Magnification | 10× to 60× | 500 000× | |
| Use | viewing large objects through which light cannot pass | viewing things that cannot be seen with the light microscope | to store, find, and process information |

**TRANSPARENCY 1**

TRANSPARENCY MASTER CHAPTER 1

TOOLS USED IN BIOLOGY

hand lens, wire gauze, baby food jar, balance, thermometer, culture dish, coverslip, microscope slide, rubber stopper, petri dish, dropper, metric ruler, graduated cylinder, test tube rack, forceps, dissecting needle, probe, scalpel, dissecting scissors, test tube, safety goggles, beaker, dissecting tray

# Lab 1–1 Microscopes

## Problem: How do you use the microscope?

### Skills

observe, recognize spatial relationships

### Materials

microscope
microscope slide
coverslip
water
dropper
forceps
small piece of magazine print

### Procedure

1. Place a drop of water in the center of a glass slide.
2. With forceps, place a small piece of magazine print in the drop of water. Make sure the piece of magazine has a letter *e* in it.
3. Hold a clean coverslip by the edges at one side of the drop of water. Carefully lower the coverslip onto the drop of water. You have just made a wet mount slide.
4. Place the slide on the stage of the microscope. Put the stage clips into place. Move the slide so that the magazine print is directly over the hole in the stage.
5. Turn the low-power objective into place.
6. Looking from the side, turn the coarse adjustment so that the low-power objective is near the coverslip.
7. Open the diaphragm so that the most light enters the microscope. If your microscope has a mirror, adjust it until you see a circle of light. This circle of light is called the field of view.
8. Look through the eyepiece. Try to keep both eyes open. Raise the body tube until you can see the letters. **CAUTION:** *Never lower the body tube while looking through the eyepiece.*
9. Use the fine adjustment to bring the letters into sharp focus. Find a letter e.
10. Adjust the slide so that the letter e is in the center of your field of view.
11. On a separate paper, make a drawing of the letter e. Label your drawing with your name and the magnification.
12. **Recognize spatial relationships:** Move the slide to the left, to the right, toward you, and away from you. Note the direction in which the *e* appears to move.
13. Turn the nosepiece to bring the high-power objective into position. Focus with the fine adjustment. **CAUTION:** *Use only the fine adjustment with the high-power objective.*
14. Draw the letter *e* as it appears under high power. Label your drawing with your name and the magnification.
15. Remove the wet mount. Rinse and dry the slide and coverslip. Set your microscope on low power and put it away.

### Data and Observations

1. What happens when you move the slide to the left? To the right? Toward you? Away from you?
2. Describe the appearance of the letter *e* under high power.

### Analyze and Apply

1. What is the purpose of the coverslip?
2. Why should the coverslip be held by the edges?
3. **Apply:** Why must a specimen be very thin to be viewed under the microscope?

### Extension

**Observe:** View other objects, such as a hair or thread under the microscope.



## *ANSWERS*

### Data and Observations

1. appeared to move to the right; appeared to move to the left; appeared to move away; appeared to move toward observer
2. Students may observe that the edges of the e are ragged looking; only part of it can be seen.

### Analyze and Apply

1. It prevents the objects being viewed from moving and protects the lenses.
2. to prevent fingerprints and smudges
3. Light must pass through it in order for it to be seen.

## Lab 1-1 Microscopes

### Overview

In this lab, students will use a microscope and observe its magnifying and resolving powers.

**Objectives:** Upon completing this lab students will be able to (1) **prepare** a wet mount slide, (2) **compare** observations made under high and low powers, (3) **observe** a microscope's magnifying power and resolving power.

**Time Allotment:** 40 minutes

### Preparation

▶ Assemble microscope slides, coverslips, droppers, and forceps. Cut letter e's from magazine print.

**Lab 1-1 worksheet**/*Teacher Resource Package*, pp. 1-2.

### Teaching the Lab

▶ Review with the students the correct procedures for carrying, using, and storing a microscope.

▶ **Troubleshooting:** Make sure the lowercase e is used, not capital E. E is not asymmetrical.

**Cooperative Learning:** Divide the class into groups of two students. For more information, see pp. 22T-23T in the Teacher Guide.

**The Microscope**/*Transparency Package,* number 1. Use color transparency number 1 as you teach this lab.

### ASSESSMENT

**Performance:** Have students look at a newspaper through the curved side of an empty glass, the flat bottom of an empty glass, a bowl filled with water, and a drop of water on waxed paper. Ask what they observed and how this relates to the microscope.

## TEACH

### Guided Practice

Have students write down their answers to the margin question: biologists use cameras to record an animal's daily habits; they use nets to capture animals and observe them; they use gardens to grow plants and observe them.

### Check for Understanding

Have students respond to the first three questions in Check Your Understanding.

### Reteach

**Reteaching**/*Teacher Resource Package*, p. 1. Use this worksheet to give students additional practice with the parts of the microscope.

**Extension:** Assign Critical Thinking, Biology and Math, or some of the **OPTIONS** available with this lesson.

## 3 APPLY

Show a videocassette (or filmstrip) on using the microscope.

## 4 CLOSE

Ask students to list the tools used by biologists studied in this lesson. Then have them give other examples of tools that can detect what the human senses cannot. Radio receivers, infrared and ultraviolet detectors, telescopes, and photographic film are some examples.

**Figure 1-4** A stereomicroscope is used for viewing objects through which light will not pass.

How does having two eyepiece lenses help? The advantage of having a lens for each eye is that it makes the object appear three-dimensional. A disadvantage of the stereomicroscope is that its magnifying power is not as great as that of a light microscope. The stereomicroscope magnifies only 10 to 60 times.

An electron microscope is a very high-powered microscope that uses a beam of electrons instead of light to magnify objects. The magnified object is seen on a monitor similar to a TV screen. An advantage of an electron microscope is that it can magnify more than 500 000 times. Objects can be viewed in great detail. A disadvantage is that objects can't be viewed in color. Also, objects must be stained with expensive dyes before they can be viewed under the electron microscope. The staining process requires a great deal of skill and training.

New information about biology is published every day. How could a biologist remember all of it? Many biologists use computers to store, find, and process information. The computer helps biologists search for information and organize it. It saves many hours of library research.

How do biologists use cameras, nets, and gardens?

Biologists use many other tools that you may not associate with biology. For example, biologists use cameras. A wildlife biologist may take photographs of deer to understand the deer's daily habits. Biologists who study fish need boats to take them to where the fish are. They also use nets to catch the fish so they can observe them. Did you ever imagine that a garden could be a biologist's tool? Biologists who study how plants grow use gardens all the time for their work. As you see, biologists use more than laboratory tools to study living things.

### Check Your Understanding

1. How does a beekeeper use biology?
2. What are three tools used by biologists?
3. A light microscope has an eyepiece that magnifies 10 × and an objective that is marked 43 ×. What is the total magnification for these two lenses?
4. **Critical Thinking:** How would a farmer use biology to grow corn?
5. **Biology and Math:** If a microscope eyepiece magnifies 4× and the total magnification is 400×, how much does the objective lens magnify?



### Answers to Check Your Understanding

1. A beekeeper must know how bees work together to produce honey; he must also know where to place the hives in order to get the most honey.
2. hand lens, ruler, microscope, stereomicroscope, binoculars, computers
3. 430 (10 x 43)
4. The farmer must know the best soil conditions for the corn, know how to identify and treat diseases of corn, and what kinds of fertilizers to use.
5. 100 (400 ÷ 4)

**RETEACHING 1**

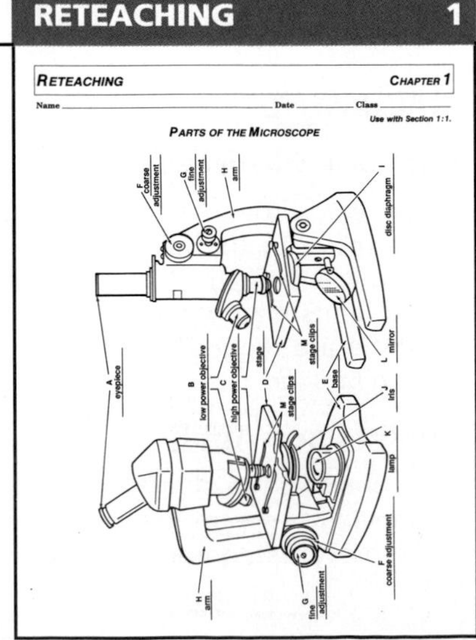
RETEACHING CHAPTER 1

Name ______ Date ______ Class ______

Use with Section 1:1.

PARTS OF THE MICROSCOPE

# 1:2 Measurements Used in Biology

It is 40 kilometers to the nearest city. The soft drink is in a 2-liter bottle. The mass of the bag of potato chips is 184 grams. A football field is 300 feet long. A photographer uses a 35 millimeter camera. The medicine comes in 25-milligram tablets.

In which of these sentences is the unit of measurement different? The football field measurement is different. It is in English units. The other measurements are in metric units. Scientists today use a modern form of the metric system. They use the International System of Units, or SI system. The **International System of Units** is a measuring system based on units of 10. Scientists all over the world use it. Measurements in the SI system are similar to measurements in our money system. Let's see how the two systems compare.

### Objectives

4. **Explain** how SI units are grouped.
5. **List** the SI units of measure for length, volume, mass, time, and temperature.
6. **Measure** objects using SI units.

### Key Science Words

International System of Units
meter
volume
kilogram
Celsius

## Length

Length is the distance from one point to another. What does our money system have in common with a system for measuring length? The basic unit in our money system is the dollar. As you know, the dollar can be divided into 10 dimes. Each dime can be divided into 10 cents as shown in Figure 1-5.

The SI unit of length is the **meter** (m). Figure 1-5 shows that like a dollar, a meter can be divided into 10 smaller units. The meter is divided into 10 units called decimeters (DES uh meet urz) (dm). Each decimeter can be divided into 10 centimeters (SENT uh meet urz) (cm).

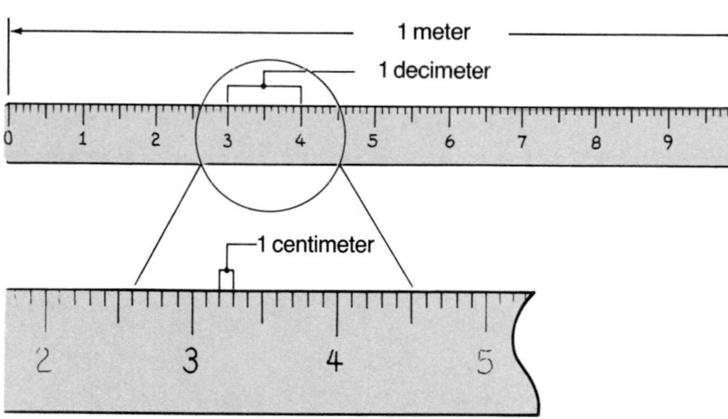


**Figure 1-5** The SI system is based on units of 10, just as our money system is.

## 1:2 Measurements Used in Biology

## PREPARATION

### Materials Needed

Make copies of the Application, Skill, Reteaching, and Study Guide worksheets in the *Teacher Resource Package*.

▶ Gather measuring instruments for the Focus demonstration.

▶ Gather metric rulers for the Skill Check.

### Key Science Words

International System of Units
meter
volume
kilogram
Celsius

### Process Skills

In the Skill Check, students will calculate.

## 1 FOCUS

▶ The objectives are listed on the student page. Remind students to preview these objectives as a guide to this numbered section.

### MOTIVATION/Demonstration

Show students measuring instruments such as a clock, rain gauge, barometer, galvanometer, and pedometer. Discuss what each measures in units.

### Student Journal

Have students estimate the sizes, in SI, of items in the classroom such as desks, windows, the chalkboard, etc. Then have students measure the objects and compare these measurements with their estimates.

### APPLICATION 1

APPLICATION: SAFETY — CHAPTER 1

Name ______ Date ______ Class ______

*Use after Section 1:1.*

SAFETY IN THE BIOLOGY LABORATORY

Working in a biology laboratory can be dangerous if safety rules are not followed.

*To prepare yourself for a safe year in the laboratory, complete the exercises below.*

1. The phrases below describe student behavior in the laboratory. In the space before each phrase, write *safe* if the behavior is safe or *unsafe* if the behavior is unsafe.

   unsafe holding a frog in your hand while you dissect it
   safe tying back long hair to keep it away from chemicals, burners, or other equipment
   unsafe eating your lunch while waiting for a solution to boil
   safe wearing goggles and a laboratory apron when working with chemicals
   unsafe turning the mirror of your microscope directly toward the sun to get more light
   unsafe identifying a chemical by tasting it

2. Draw a diagram of your laboratory classroom. Show the location of the *fire blanket, fire extinguishers, eyewash station, first aid kit, sand, safety shower*, and *exits*.

   exit, fire extinguisher, fire blanket, sand, exit, first aid kit, eyewash station, safety shower

   Diagrams will depend on the classroom floor plan.

3. Two laboratory accidents are described below. Explain what action should be taken and tell how the accident could have been avoided.

   a. A student's loose sleeve catches on fire as she reaches across a lighted burner.
   Wrap the arm in a fire blanket and get medical help. To avoid such an accident, tie back loose sleeves; do not reach across burners.

   b. A chemical splashes into a student's eye as he heats the chemical in a test tube.
   Wash the eye at the eyewash station and get medical help. To avoid the accident, wear goggles and point the test tube away from anyone.

## OPTIONS

### Science Background

SI is the form of the metric system used by the scientific community. SI units are metric units. SI is used by scientists all over the world and in everyday life in many countries. SI stands for the French Le Système International d'Unités.

**Application**/*Teacher Resource Package*, p. 1. Use the Application worksheet shown here to teach safety in the lab.

## TEACH

### MOTIVATION/Brainstorming

Ask students to list where measurements are made using the SI system. Examples may include sports, soft drink bottles, parts for foreign-made cars.

### Concept Development

▶ You will find added information on the SI system included in Appendix A.

▶ Avoid having students convert between the metric and English systems. Encourage students to "think metric."

### Connection: Consumer

Ask students to bring in food labels that show metric measurements. Have them convert the measurements on the labels to equivalents in larger or smaller units.

### Skill Check

You may have to remind students that area is determined by multiplying length by width.

### Guided Practice

Have students answer the margin question by measuring the width of their hands and writing down their answers.

### Independent Practice

**Study Guide**/*Teacher Resource Package*, p. 3. Use the Study Guide worksheet shown at the bottom of this page for independent practice.

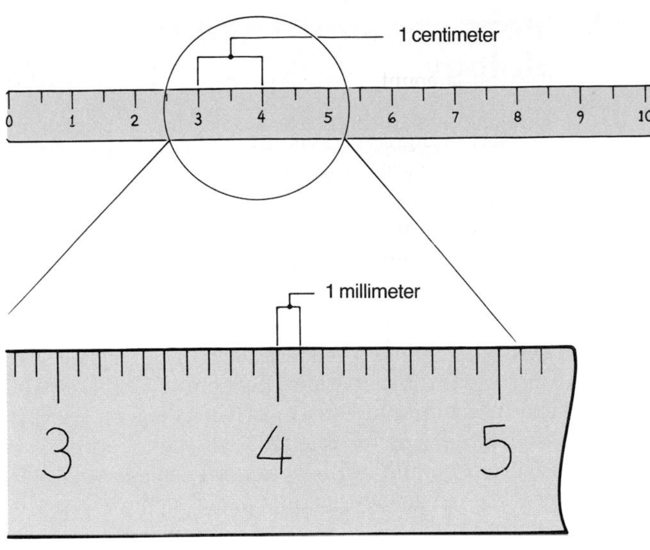


**Figure 1-6** There are 10 millimeters in one centimeter.

### Skill Check

**Calculate:** Use a calculator and a meter stick to find the area of your classroom and your desk. *For more help, refer to the **Skill Handbook**, pages 718-719.*

**What is the width of your hand in millimeters?**

In our money system, the smallest amount of money is one cent. However, the smallest length you can measure is not one centimeter. Figure 1-6 shows that each centimeter can be further divided into 10 units called millimeters (MIHL uh meet urz) (mm). Very small measurements are made in millimeters.

Do we have any units in our money system that are *larger* than the dollar? Yes, we group dollars together into tens, hundreds, and thousands. We also put meters into larger groups. Table 1-1 shows how metric units are grouped. The prefixes shown have special meanings. Notice that the first three prefixes change meters to larger units of measurement. For example, 1000 meters make one *kilo*meter (KIHL uh meet ur) (km). Ten *deci*meters make one meter. One hundred *centi*meters make one meter. The second three prefixes change meters to smaller units of measurement. How many *milli*meters make one meter?

**TABLE 1–1. HOW METRIC UNITS ARE GROUPED**

| Prefix (word part) | Symbol | Meaning |
|---|---|---|
| kilo- | k | 1000 |
| hecto- | h | 100 |
| deka- | da | 10 |
| deci- | d | 0.1 (1/10) |
| centi- | c | 0.01 (1/100) |
| milli- | m | 0.001 (1/1000) |



## OPTIONS

**Enrichment:** Explain that the cent is divided into 10 mills and relate that to the division of 1 cm into 10 mm. (The mill is used in tax levies.)

**How Are SI Length Measurements Made?**/*Lab Manual*, pp. 5-8. Use This lab to give further practice with SI measurements.

**Skill**/*Teacher Resource Package*, p. 1. Use the Skill worksheet shown here to give students additional practice measuring in SI.

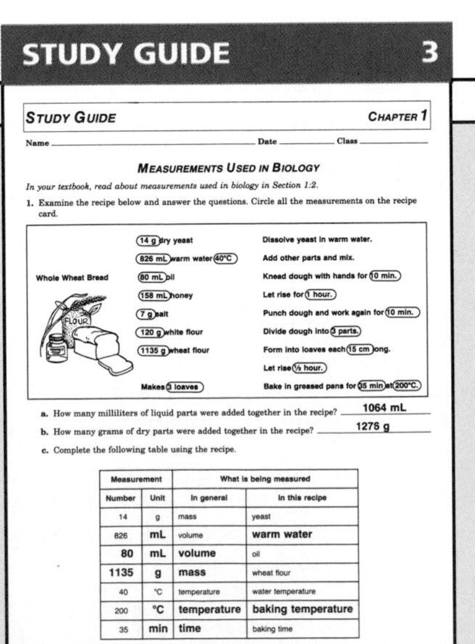

**STUDY GUIDE 3**

STUDY GUIDE — CHAPTER 1

Name ______ Date ______ Class ______

**MEASUREMENTS USED IN BIOLOGY**

*In your textbook, read about measurements used in biology in Section 1:2.*

1. Examine the recipe below and answer the questions. Circle all the measurements on the recipe card.



Whole Wheat Bread

14 g dry yeast
826 mL warm water 40°C
80 mL oil
158 mL honey
7 g salt
120 g white flour
1135 g wheat flour
Makes 3 loaves

Dissolve yeast in warm water.
Add other parts and mix.
Knead dough with hands for 10 min.
Let rise for 1 hour.
Punch dough and work again for 10 min.
Divide dough into 3 parts.
Form into loaves each 15 cm long.
Let rise ½ hour.
Bake in greased pans for 35 min at 200°C.

a. How many milliliters of liquid parts were added together in the recipe? 1064 mL

b. How many grams of dry parts were added together in the recipe? 1276 g

c. Complete the following table using the recipe.

| Measurement | | What is being measured | |
|---|---|---|---|
| Number | Unit | In general | In this recipe |
| 14 | g | mass | yeast |
| 826 | mL | volume | warm water |
| 80 | mL | volume | oil |
| 1135 | g | mass | wheat flour |
| 40 | °C | temperature | water temperature |
| 200 | °C | temperature | baking temperature |
| 35 | min | time | baking time |

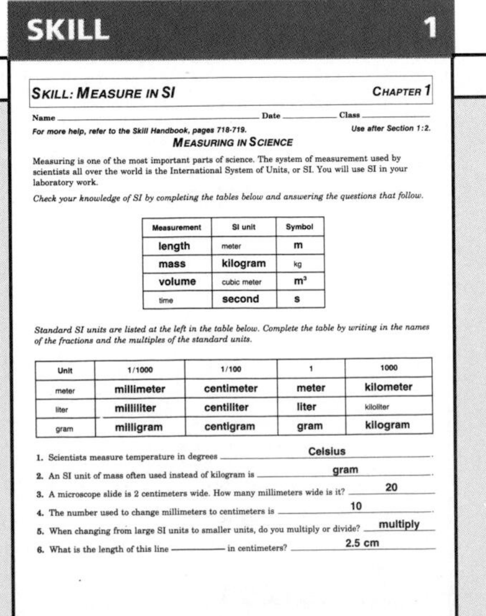

**SKILL 1**

SKILL: MEASURE IN SI — CHAPTER 1

Name ______ Date ______ Class ______

*For more help, refer to the Skill Handbook, pages 718-719.* — *Use after Section 1:2.*

**MEASURING IN SCIENCE**

Measuring is one of the most important parts of science. The system of measurement used by scientists all over the world is the International System of Units, or SI. You will use SI in your laboratory work.

*Check your knowledge of SI by completing the tables below and answering the questions that follow.*

| Measurement | SI unit | Symbol |
|---|---|---|
| length | meter | m |
| mass | kilogram | kg |
| volume | cubic meter | $m^3$ |
| time | second | s |

*Standard SI units are listed at the left in the table below. Complete the table by writing in the names of the fractions and the multiples of the standard units.*

| Unit | 1/1000 | 1/100 | 1 | 1000 |
|---|---|---|---|---|
| meter | millimeter | centimeter | meter | kilometer |
| liter | milliliter | centiliter | liter | kiloliter |
| gram | milligram | centigram | gram | kilogram |

1. Scientists measure temperature in degrees Celsius.
2. An SI unit of mass often used instead of kilogram is gram.
3. A microscope slide is 2 centimeters wide. How many millimeters wide is it? 20
4. The number used to change millimeters to centimeters is 10.
5. When changing from large SI units to smaller units, do you multiply or divide? multiply
6. What is the length of this line ——— in centimeters? 2.5 cm

## Volume

**Volume** is the amount of space a substance occupies. Units of volume are based on units of length. If we multiply length by width by height, we get volume. The SI unit of volume is the cubic meter ($m^3$). This volume is rather large for everyday use, so scientists often use the liter (L) instead. A liter is 1/1000 as large as a cubic meter. You use the liter when you do labs in science class, or when you buy soft drinks at the grocery store. One liter equals 1000 milliliters (mL).

Like meters, liters can be put into large groups. One kiloliter means 1000 liters. What does five kiloliters mean?

**Into how many smaller units can a liter be divided?**

## Weight and Mass

The photos in Figure 1-7 show an astronaut on Earth and an astronaut on the moon. The astronaut does not weigh the same in both places. Weight is a measure of the force of gravity on an object. The force of gravity on the moon is less than that on Earth. Therefore, an astronaut weighs less on the moon than on Earth. Even though the astronaut weighs less on the moon than on Earth, his or her mass does not change. Mass is how much matter is in something. Let's look at mass in another way. Consider the masses of two different objects, a box of bricks and a box of cotton. Which has more mass? Mass is measured by comparing an object of unknown mass to an object of known mass. The instrument used to measure mass is a balance.

a

**Figure 1-7** An astronaut's mass is the same on Earth (a) as it is on the moon (b).

b

## TEACH

### Concept Development

▶ Try to give students the idea of the volume of a liter without converting it to the English system (soft drink bottles are an example).

▶ Give students the metric masses of several common items. Examples: loaf of bread-450 g; pair of shoes-1000 g or 1 kg; pencil-10 g.

### Check for Understanding

Make flash cards with examples of metric units. When an example is shown, students should respond with the unit of measurement.

### Reteach

Show students different pieces of lab equipment. Have them identify the quantity measured and the unit of metric measurement.

**Cooperative Learning:** Divide the class into cooperative groups of four. Have each group invent a new system for measuring length. Remind each group that a system of measurement must have a standard.

### Guided Practice

Have students write down their answers to the margin question: 1000 milliliters.

### Independent Practice

**Study Guide**/*Teacher Resource Package*, p. 4. Use the Study Guide worksheet shown at the bottom of this page for independent practice.

## GLENCOE TECHNOLOGY

**Videodisc**

**Biology: The Dynamics of Life**
Movie: *The Light Microscope*
Disc 1, Side 1, Ch. 22

## STUDY GUIDE 4

STUDY GUIDE — CHAPTER 1

Name ______ Date ______ Class ______

**MEASUREMENTS USED IN BIOLOGY**

*In your textbook, read about measuring volume in Section 1:2.*

2. Look at the graduated cylinder on the right. Note the lines and numbers printed on its sides. The graduated cylinder is filled to the 45 mL mark. Read the level of the liquid in each of the graduated cylinders below. Write the correct volume on the line below each graduated cylinder.

45 mL

65 mL — 80 mL — 100 mL

850 mL — 750 mL — 1 L (1000 mL)

You may need to explain how to read a meniscus. Students may wonder why the art shows the liquid level curved.

*In your textbook, read about measuring temperature in Section 1:2.*

3. In each space, write one of the following phrases that best matches the temperature shown. Use these phrases—water for bath, water freezes, water boils, a cool day, and human with fever.

100° water boils
55° water for bath
40° human with fever
20° a cool day
0° water freezes

## OPTIONS

**Enrichment:** Ask students what their approximate weights would be on the moon. Moon weights would be about one-sixth of Earth weights. Ask how their weights would be affected on Jupiter. Jupiter is much larger, and the weights would be much greater. Then ask if their masses would change on the moon and on Jupiter. The answer is no.

The SI unit of mass is the **kilogram** (kg). There are 1000 grams (g) in a kilogram. Grams are used to measure small objects. An egg has a mass of about 60 grams. A gram may be divided into 1000 units called milligrams. Milligrams are often used to measure medicines or ingredients in foods such as breakfast cereals.

**For what kind of measurement is a kilogram used?**

## Time and Temperature

Time is the period between two events. The SI unit for measuring time is the second (s). A stopwatch or a clock with a second hand is used to measure time.

Temperature is a measure of the amount of heat in something. The SI scale for measuring temperature is the Kelvin scale (K). Scientists measure temperature with the **Celsius** (SEL see us) scale because it is easier to use. Temperature scales are divided into units called degrees. On the Celsius scale, the freezing point of water is 0 degrees and the boiling point is 100 degrees. Figure 1-8 shows the temperature in Celsius degrees of other familiar things.

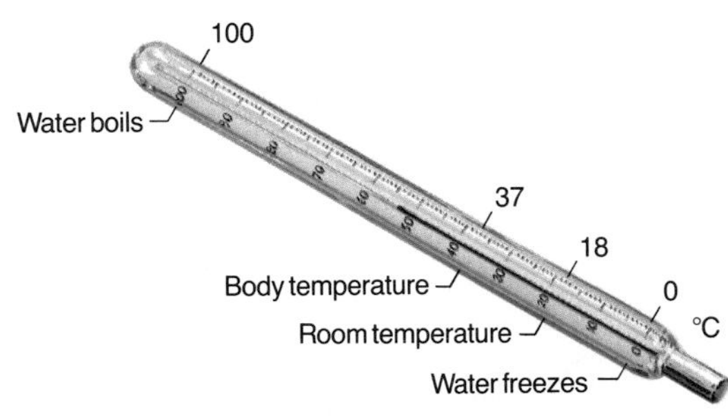


**Figure 1-8** The Celsius scale

## Check Your Understanding

6. What does the prefix *centi* mean?
7. What are some units used for measuring volume?
8. What is your body temperature in degrees Celsius?
9. **Critical Thinking:** How many centimeters make one kilometer?
10. **Biology and Math:** Suppose a cheetah can run 30 meters per second. How far can it run in 8 seconds?



# TEACH

### ACTIVITY/Hands-On

Have students use Celsius thermometers to measure room temperature, outdoor temperature, ice water, and boiling water. Caution students about the safety factors regarding boiling water.

### Guided Practice

Have students write down their answers to the margin question: mass.

### Check for Understanding

Have students respond to the first three questions in Check Your Understanding.

### Reteach

**Reteaching**/*Teacher Resource Package*, p. 2. Use this worksheet to give students additional practice with measuring in SI.

**Extension:** Assign Critical Thinking, Biology and Math, or some of the **OPTIONS** available with this lesson.

# 3 APPLY

### Software

*The Metric System*, Queue, Inc.

# 4 CLOSE

Have students identify the most useful units for measuring the things they can see in the classroom.

### Answers to Check Your Understanding

6. 1/100 of something
7. liters and milliliters
8. about 37°
9. 100 centimeters in a meter x 1000 meters in a kilometer = 100 000 cm
10. 240 m (30 x 8)

## GLENCOE TECHNOLOGY

**Videodisc**

**The Secret of Life**

*Cells: Microscopy*

## RETEACHING 2

**RETEACHING: IDEA MAP** — CHAPTER 1

Name ______ Date ______ Class ______

*Use with Section 1:2.*

**MEASUREMENTS USED IN BIOLOGY**

*Complete the idea map by using the following numbers to fill in the blanks: 1, 10, 200, 1000, 5000.*

metric equivalents
- 1 kilometer = **1000** meter(s)
- 100 milliliters = **1** liter(s)
- 5 kiloliters = **5000** liters(s)
- 1 meter = **1000** millimeters(s)
- 2 meters = **200** centimeters(s)
- 1 decimeter = **10** centimeters(s)

1. In what way is a dollar similar to a meter in the SI system? **A dollar can be divided into units of 10, just as a meter can.**

2. Which metric unit would you use to measure the following?
   a. the area of your classroom — **meter**
   b. the volume of liquid in a soft drink — **liter**
   c. the distance to a city in another state — **kilometer**
   d. your height — **meter or centimeter**
   e. the temperature of a bird — **degree Celsius**
   f. the mass of a vitamin pill — **milligram**
   g. the mass of two paper clips — **gram**
   h. time — **second**

# 1:3 Scientific Method

Much of the work of biology is to solve problems. Problems are not solved by flipping a coin or taking a guess as to the outcome. Scientists use a series of steps called a **scientific method** to solve problems. The following steps are often used: recognizing the problem, researching the problem, forming a hypothesis (hi PAHTH uh sus), testing the hypothesis, and drawing conclusions.

**Objectives**

7. **Explain** the steps of the scientific method.
8. **Compare** the difference between a hypothesis and a theory.
9. **Explain** how technology is used to solve everyday problems.

**Key Science Words**

scientific method
hypothesis
experiment
variable
control
data
theory
technology

## Recognizing and Researching the Problem

Have you ever tried to turn on a light and found that it didn't work? Maybe you have turned the key in a car's ignition and the car didn't start. If you have had problems like those, you probably have used a scientific method to solve them.

A biology class raised guppies for a class project. Paula thought the guppies would have more young if the light in the aquarium was turned off part of the time. Other students thought the light should be left on all the time. But, Paula had observed that her guppies at home had more young than the ones at school. She also knew that she kept her aquarium light turned off part of the time. The class had a problem. What should they do about the aquarium light?

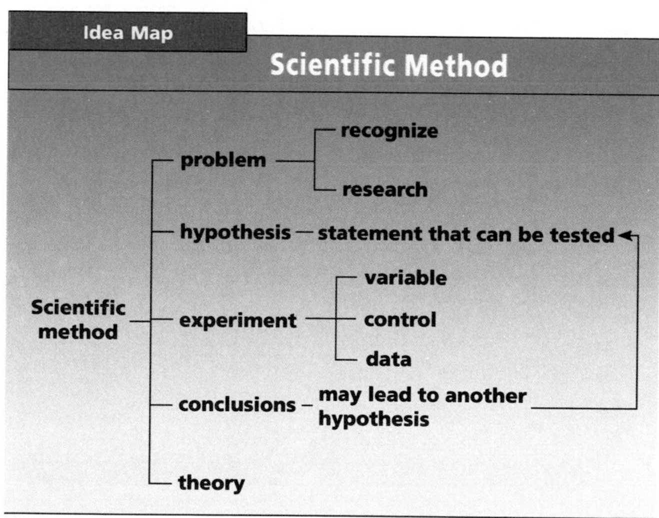


**Study Tip:** Use this idea map as a study guide to scientific method. Notice that a conclusion may lead to a second hypothesis. The second hypothesis then leads to more experiments and further conclusions.



## OPTIONS

### Science Background

Scientists obtain information by asking questions, making observations, and trying things out in a systematic way. They use a process called scientific method. A general model of scientific method is presented here.

# 1:3 Scientific Method

## PREPARATION

### Materials Needed

Make copies of the Critical Thinking/Problem Solving, Reteaching, Study Guide, and Lab worksheets in the *Teacher Resource Package*.

▶ For the Focus demonstration, get a slide projector and a screen.

▶ Gather cuttings of ivy, *Coleus*, and *Zebrina*, as well as aspirins for the Mini Lab.

▶ Gather pinto beans, salt, and self-sealing plastic bags for Lab 1-2.

### Key Science Words

| | |
|---|---|
| scientific method | control |
| hypothesis | data |
| experiment | theory |
| variable | technology |

### Process Skills

In Lab 1-2, students will form hypotheses, interpret data, and separate and control variables. In the Mini Lab, students will design an experiment.

## 1 FOCUS

▶ The objectives are listed on the student page. Remind students to preview these objectives as a guide to this numbered section.

### MOTIVATION/Demonstration

Use a slide projector to set up a problem for students to investigate. Turn on the projector. No light reaches the screen. Have students solve the problem of the malfunctioning projector by using scientific method.

### Idea Map

Have students use the idea map as a study guide to the major concepts of this section.

# 2 TEACH

## MOTIVATION/Brainstorm

Ask students to think of ways they use scientific method in their everyday activities. Answers may include analyzing plays in sports, learning to drive a car, learning to read. Ask students to tell how they use their investigative skills. See if any have tried to find out what was in a package before opening it. What procedures did they use?

### Concept Development

▶ Explain that using scientific method is similar to using common sense.

▶ Stress that there is no one scientific method.

▶ Point out that scientific method has no set order. The order will depend upon the investigation.

### Mini Lab

Have students work in groups of four. Provide cuttings of ivy, *Coleus,* and *Zebrina.* Also provide aspirins and 2-4 small beakers per group.

### ✓ ASSESSMENT

**Performance:** A shampoo commercial claims that Brand A cleans hair better and makes it more shiny than Brand B. Have students design an experiment to test this claim.

### Guided Practice

Have students write down their answers to the margin question: a statement that can be tested.

### Independent Practice

**Study Guide**/*Teacher Resource Package*, p. 5. Use the Study Guide worksheet shown at the bottom of this page for independent practice.

---

**Mini Lab**

**Does Aspirin Help Plants Grow?**

**Design an experiment:** Your aunt tells you that plant cuttings will grow faster if aspirin is added to the water. Design and conduct an experiment to find out if this is true. *For more help, refer to the* ***Skill Handbook,*** *pages 704-705.*

Paula was able to recognize the problem. Before Paula could work on the solution to the problem, she needed to do some research at the library. She found out what temperature to keep the water and the amount of water needed. However, there was little information about how much light a guppy needs.

## Forming a Hypothesis

A **hypothesis** is a statement that can be tested. Paula's hypothesis was: "If the light is turned off part of the time, then the guppies will have more young." Paula knew that her guppies received less light than the ones at school. She also observed two important things. Her tank at home had the same amount of water in it as the one at school. Also, the water temperature was the same in each tank. Paula was pretty sure that the conditions for the guppies were the same at home and at school, except for the amount of light. However, she needed to do a test to be absolutely sure.

## Testing a Hypothesis

What is a hypothesis?

Paula's next step was to test her hypothesis. Testing a hypothesis using a series of steps with controlled conditions is called an **experiment**. In her experiment, Paula divided 16 guppies into two equal groups. She put the groups into separate tanks and labeled them group A and group B. She kept the temperature in the tanks the same. She put the same amount of water in each tank. The light for group B was left on 24 hours a day. The light for group A was turned on for 12 hours and off for 12 hours each day.

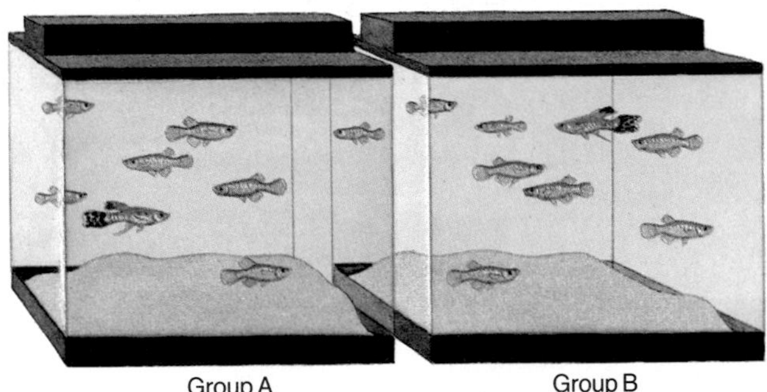


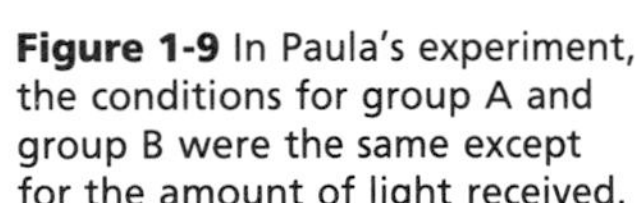

**Figure 1-9** In Paula's experiment, the conditions for group A and group B were the same except for the amount of light received.



## OPTIONS

**Challenge:** Have students find out about people such as Salk, Sabin, Fleming, Jenner, Pasteur, and Koch. Encourage them to investigate the ways in which scientists solved problems that have affected our lives.

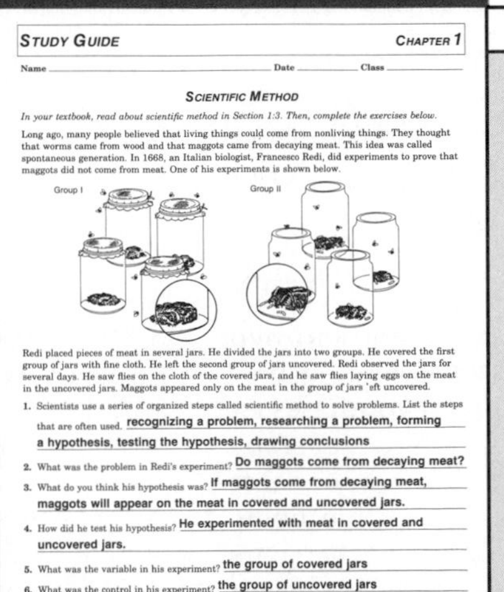

**STUDY GUIDE 5**

STUDY GUIDE — CHAPTER 1

Name ________ Date ________ Class ________

SCIENTIFIC METHOD

*In your textbook, read about scientific method in Section 1:3. Then, complete the exercises below.*

Long ago, many people believed that living things could come from nonliving things. They thought that worms came from wood and that maggots came from decaying meat. This idea was called spontaneous generation. In 1668, an Italian biologist, Francesco Redi, did experiments to prove that maggots did not come from meat. One of his experiments is shown below.

Group I — Group II

Redi placed pieces of meat in several jars. He divided the jars into two groups. He covered the first group of jars with fine cloth. He left the second group of jars uncovered. Redi observed the jars for several days. He saw flies on the cloth of the covered jars, and he saw flies laying eggs on the meat in the uncovered jars. Maggots appeared only on the meat in the group of jars 'eft uncovered.

1. Scientists use a series of organized steps called scientific method to solve problems. List the steps that are often used. recognizing a problem, researching a problem, forming a hypothesis, testing the hypothesis, drawing conclusions
2. What was the problem in Redi's experiment? Do maggots come from decaying meat?
3. What do you think his hypothesis was? If maggots come from decaying meat, maggots will appear on the meat in covered and uncovered jars.
4. How did he test his hypothesis? He experimented with meat in covered and uncovered jars.
5. What was the variable in his experiment? the group of covered jars
6. What was the control in his experiment? the group of uncovered jars
7. What do you think Redi's conclusion was? Maggots do not come from decaying meat.

# Lab 1–2 Scientific Method

## Problem: How is a scientific method used to solve problems?

### Skills

form hypotheses, interpret data, separate and control variables

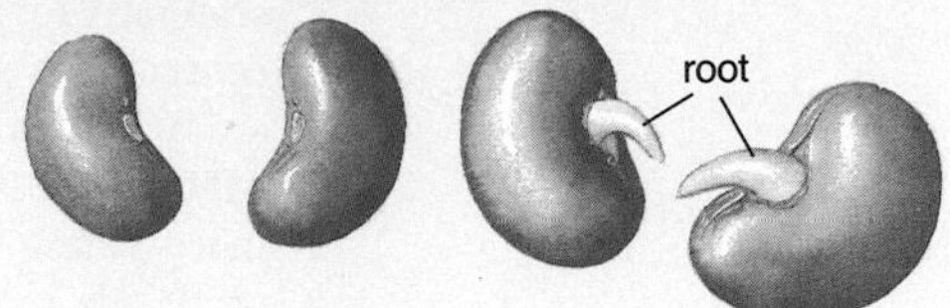


### Materials

| | |
|---|---|
| 2 100-mL beakers | 2 paper towels |
| 40 pinto beans | 50-mL graduated cylinder |
| balance | 2 self-sealing plastic bags |
| salt | masking tape |
| water | |

### Procedure

1. Solve this problem using a scientific method: Do seeds grow better when soaked in plain water or salt water?
2. Research the problem. Find out what seeds need in order to grow.
3. Develop a **hypothesis** based on the problem.
4. Conduct the experiment.
   a. Copy the data table.
   b. Label two beakers A and B. Fill each beaker with 50 mL of water.
   c. Add 20 beans to beaker A. Add 20 beans and 2 g salt to beaker B. Label each beaker with your name and set them aside overnight.
   d. On the following day, pour out the water from both beakers.
   e. Wrap the seeds from beaker A with a damp paper towel. Place the paper towel into a plastic bag and close the bag. Label the bag A.
   f. Repeat step e with the seeds in beaker B.

| | SEEDS USED | SEEDS THAT GREW |
|---|---|---|
| Water | 20 | 17 |
| Salt water | 20 | 0 |

5. Collect the data.
   a. On the next day check the seeds for growth. Look for small white roots.
   b. Count the number of seeds that are growing in each bag. Record these numbers in your data table.
6. Draw conclusions. Based on your data, decide if you need a new hypothesis.
7. If soil and containers are available, you may want to plant the seeds that are growing.

### Data and Observations

1. Under what condition, plain water or salt water, did more seeds grow?
2. Under what condition, plain water or salt water, did fewer seeds grow?

### Analyze and Apply

1. **Check your hypothesis:** Is your original hypothesis supported by your data? Why or why not?
2. **Infer:** How can you be sure that the conclusion you reached is correct?
3. What variable was being tested in this experiment? What was the control?
4. **Apply:** How do salt water and plain water affect seed growth?

### Extension

**Design an experiment** using distilled water in place of the salt water. Compare your results.



## ANSWERS

### Data and Observations

1. plain water
2. salt water

### Analyze and Apply

1. Students who hypothesized that more seeds would grow in plain water will say their hypotheses were supported.
2. Run another experiment to see if the results are the same.
3. The variable being tested was whether seeds grow better or worse with salt water; the seeds grown with plain water.
4. Salt water slows seed growth. Plain water does not slow seed growth.

## Lab 1-2 Scientific Method

### Overview

In this lab, students use scientific method to solve a problem.

**Objectives:** Upon completing this lab, students will be able to (1) **measure** volume and mass in SI units, (2) **form and test hypotheses**, (3) **record** observations.

**Time Allotment:** 20 minutes day 1, 15 minutes day 2

### Preparation

▶ **Alternate Materials:** Other types of seed may be used. Assemble beakers, pinto beans, salt, balances, paper towels, graduated cylinders, and self-sealing plastic bags.

**Lab 1-2 worksheet/***Teacher Resource Package*, pp. 3-4.

### Teaching the Lab

▶ **Troubleshooting:** The seeds may not germinate overnight. If not, allow 2 days before concluding the lab.

**Cooperative Learning:** Divide the class into groups of four students. For more information, see pp. 22T-23T in the Teacher Guide.

### ASSESSMENT

**Performance:** Have students hypothesize what would happen to plants that grow in darkness. Students can design an experiment to test the hypothesis. They should consider how darkness would affect the amount of food plants produce and plant growth, color, and overall health. Students will need the germinated seed from Lab 1-2, soil, and containers.

### *Portfolio*

Students can write an evaluation of Lab 1-2 and the performance assessment.

## TEACH

### Concept Development

▶ Point out that usually the purpose of an experiment is to test the effect of varying a single condition in an otherwise constant environment. This condition is the independent variable or experimental factor. A control sample is handled exactly the same way as the experimental sample, except that the variable to be tested is either held at a fixed level or omitted. Comparing the results of the control with the results of the experimental sample shows the effect of the variable.

### Guided Practice

Have students write down their answers to the margin question: recognizing and researching the problem, forming a hypothesis, testing the hypothesis, drawing conclusions.

### Check for Understanding

Have students discuss what would happen if scientific experiments did not include a control. Would the findings of such experiments be valid?

**Reteach**

Place a control tray of bean seedlings in the light and another identical tray in the dark. Ask students what the variable factor is and what must be kept constant. Have students follow scientific method in observing the effect of sunlight on growth.

**What are five steps of the scientific method described in this section?**

A **variable** is something that causes the changes observed in an experiment. Some variables in Paula's experiment were water temperature, amount of food given to each group, and number of hours the light was left on. For meaningful results, the effect of only one variable can be tested at a time. Paula tested the effect of having the light on for different lengths of time.

An experiment must also have a control. A **control** is a standard for comparing results. In Paula's experiment, the control was group B, in which the light was left on 24 hours a day. Paula planned to compare her results for group A with her results for group B to see if group A produced more young.

Paula kept a record of the information she gathered. Recording information is called collecting data. **Data** are the recorded facts or measurements from an experiment.

Paula recorded the number of adult guppies she put into each tank. Every day, she recorded the temperature of the water in each of the tanks. Once a week for four weeks, she recorded the number of young produced by each group, Figure 1-11. She removed the new young each week so she could count more easily.

## Drawing Conclusions

Paula's last step was to conclude whether or not the problem was solved. At the end of her experiment, the guppies in group A had more young than the guppies in group B, Figure 1-11. Could Paula conclude that 12 hours

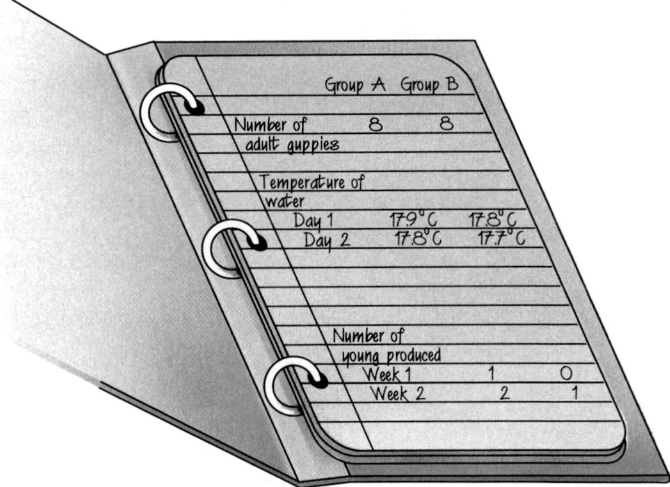


**Figure 1-10** Paula recorded the data that she collected.



## OPTIONS

**Critical Thinking/Problem Solving/** *Teacher Resource Package*, p. 1. Use the Critical Thinking/Problem Solving worksheet shown here as an extension to scientific method.

**CRITICAL THINKING 1**

CRITICAL THINKING/PROBLEM SOLVING — CHAPTER 1

Name ________ Date ________ Class ________

*Use after Section 1:3.*

**CAN YOU SPOT THE SCIENTIFIC METHOD?**

*Each sentence below describes a step of the scientific method. Match each sentence with a step of the scientific method listed below.*

A. recognize a problem
B. form a hypothesis
C. test the hypothesis with an experiment
D. draw conclusions

B 1. Stephen predicted that seeds would start to grow faster if an electric current traveled through the soil in which they were planted.

B 2. Susan said, "If I fertilize my geranium plants, they will blossom."

D 3. Jonathan's data showed that household cockroaches moved away from raw cucumber slices.

C 4. Rene grew bacteria from the mouth on special plates in the laboratory. She placed drops of different mouthwashes on bacteria on each plate.

C 5. Kathy used a survey to determine how many of her classmates were left-handed and how many were right-handed.

A 6. Dana wanted to know how synthetic fibers were different from natural fibers.

A 7. Jose saw bats catching insects after dark. He asked, "How do bats find the insects in the dark?"

A 8. Justin wondered if dyes could be taken out of plant leaves, flowers, and stems.

C 9. Arjulia soaked six different kinds of seeds in water for 24 hours. Then she planted the seeds in soil at a depth of 1 cm. She used the same amount of water, light, and heat for each kind of seed.

A 10. Bob read about growing plants in water. He wanted to know how plants could grow without soil.

B 11. Kevin said, "If I grow five seedlings in red light, I think the plants will grow faster than the five plants grown in white light."

D 12. Angela's experiment proved that earthworms move away from light.

B 13. Scott said, "If acid rain affects plants in a particular lake, it might affect small animals, such as crayfish, that live in the same water."

D 14. Michael fed different diets to three groups of guinea pigs. His experiment showed that guinea pigs need vitamin C and protein in their diets.

D 15. Kim's experiment showed that chicken egg shells were stronger when she gave the hen feed to which extra calcium had been added.

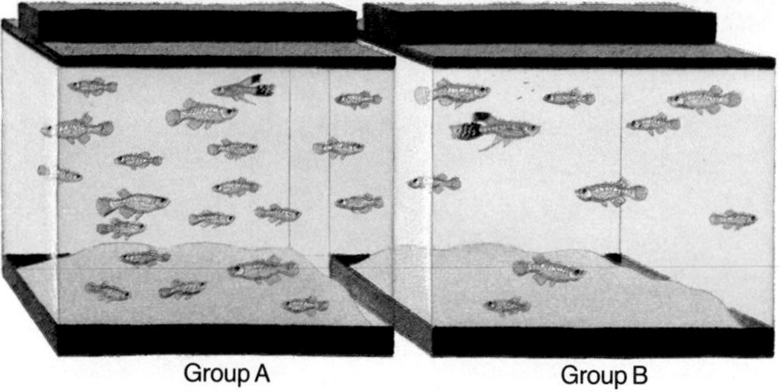


**Figure 1-11** At the end of the experiment, group A had more young.

of light and 12 hours of dark caused the difference? She might. One way to be sure of a conclusion is to repeat the experiment. Repeating experiments is how scientists confirm that their hypotheses are correct. To confirm her hypothesis, Paula repeated the experiment three times with different groups of guppies. Each time the same thing happened. The guppies had more young when they were exposed to 12 hours of light and 12 hours of dark each day.

Paula used a scientific method to solve a problem. Let's review the steps that she used.

1. Recognize the problem.
2. Research the problem.
3. Form a hypothesis.
4. Test the hypothesis with an experiment. The experiment should have one variable being tested and a control group.
5. Draw conclusions. If your first hypothesis is wrong, you may need to form another hypothesis and test that.

## Theory

If a hypothesis is tested with further experiments and the results are similar, it may become a theory. A **theory** is a hypothesis that has been tested again and again by many scientists, with similar results each time. You may have said at some time, "I think I know why the bus is late. My theory is that it got into a traffic jam on Fifth Street." But, this statement is not a theory. It hasn't been tested. A scientific theory is not a guess. It is the best explanation science has to offer about a problem. Theories can be used to predict future events. In the next chapter you will learn about the cell theory.

### Bio Tip

**Science:** Nonscientists usually use *theory* to mean an idea or thought as opposed to a fact. Scientists use *theory* to describe an idea that has so much supporting evidence that it is almost certainly true.



## TEACH

### Concept Development

▶ Emphasize the importance of repeating an experiment. A hypothesis cannot be supported or disproved with just one experiment.

▶ Explain to students that the word theory is used outside of science to mean speculation. Many people incorrectly assume that scientists mean the same thing when they speak of scientific theory. A scientific theory is an established, accepted idea that scientists do not usually question until new evidence is presented.

### Connection: Social Studies

Stress current events by making a bulletin board entitled Biology News Stories. Place high-interest articles and news stories, such as sports medicine topics, new breakthroughs in cancer research, the effects of steroids, the effects of drugs, and articles on sexually transmitted diseases, on the board. Ask students to bring in articles. Maintain the bulletin board throughout the year.

### Independent Practice

**Study Guide**/*Teacher Resource Package*, p. 6. Use the Study Guide worksheet shown at the bottom of this page for independent practice.

## GLENCOE TECHNOLOGY

**Videotape**

The Secret of Life

*On the Brink: Portraits of Modern Science*

## STUDY GUIDE 6

STUDY GUIDE — CHAPTER 1

Name ______ Date ______ Class ______

VOCABULARY

*Review the new words in Chapter 1 of your textbook. Then, complete this puzzle.*

ACROSS
9. a statement that can be tested
10. SI unit of mass—1000 grams
11. a microscope that has an eyepiece lens and an objective lens for each eye
12. use of scientific discoveries to solve everyday problems
13. something that causes changes in an experiment
14. a hypothesis that has been tested over and over again
15. temperature scale used by scientists
16. recorded facts or measurements from an experiment

DOWN
1. study of living and once-living things
2. International System of Units (abbreviation)
3. method used to solve a problem
4. SI unit of length
5. the standard for comparing results in an experiment
6. the amount of space a substance occupies
7. to test a hypothesis
8. microscope used in class

## OPTIONS

**Challenge:** An example of a well-tested theory is the germ theory of disease, developed from the work of Louis Pasteur. Have students report on the germ theory of disease.

## TEACH

### Guided Practice

Have students write down their answers to the margin question: they transplant organs and use artificial devices to replace limbs or organs.

### Check for Understanding

Have students respond to the first three questions in Check Your Understanding.

### Reteach

**Reteaching**/*Teacher Resource Package*, p. 3. Use this worksheet to give students additional practice using scientific method.

**Extension:** Assign Critical Thinking, Biology and Reading, or some of the **OPTIONS** available with this lesson.

## 3 APPLY

### Skill

Have students infer which fields of science are responsible for technologies that affect their lives.

## 4 CLOSE

Refer to a detective television program with which students are familiar. Ask what the detective must do to solve the problem, how he or she finds clues, what is done with the information in order to reach conclusions.

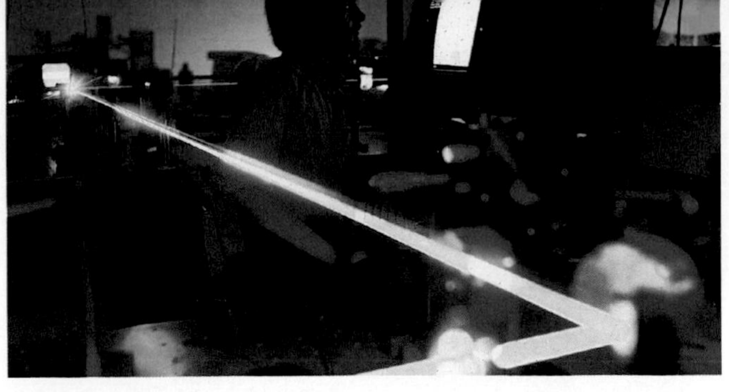

**Figure 1-12** Lasers are products of technology.

### Bio Tip

**Consumer:** A laser is a tool that produces a narrow, intense beam of light that can be used as a kind of scalpel. It can be used to remove unhealthy cells and in certain types of eye surgery.

How do doctors use technology to solve medical problems?

## Technology

Scientists have played a part in solving many problems in farming, industry, and medicine. The use of scientific discoveries to solve everyday problems is called **technology**. Doctors transplant human organs based on what they know about the human body. Over two million artificial devices are used each year to take the place of limbs or body organs lost to illnesses and accidents. These devices are products of technology.

Improved running shoes, laser surgery, and drugs that clear fats from the walls of blood vessels are also products of technology. These kinds of technological advances promote well-being. Other kinds of advances may be as helpful, but have harmful effects as well. Fertilizers help farmers grow better crops, but they may contribute to water pollution. Chemicals added to food keep foods fresh, but may have harmful effects on people if consumed for long periods of time.

### Check Your Understanding

11. What is the variable in an experiment? What is the control?
12. What are five steps in the scientific method described in this section?
13. Why is a theory more accepted than a hypothesis?
14. **Critical Thinking:** Give an example of how you have used technology to solve an everyday problem.
15. **Biology and Reading:** What three steps of the scientific method described in this section should be done before performing an experiment?



### Answers to Check Your Understanding

11. The variable is something that causes changes in an experiment. The control is the standard for comparing.
12. recognize the problem, research the problem, form a hypothesis, test the hypothesis with an experiment, draw conclusions
13. A theory has been tested again and again by many scientists with similar results each time.
14. Some students might include getting X rays and vaccinations or using commercial fertilizer to help house plants grow, and so on.
15. recognize the problem, research the problem, form a hypothesis

### RETEACHING 3

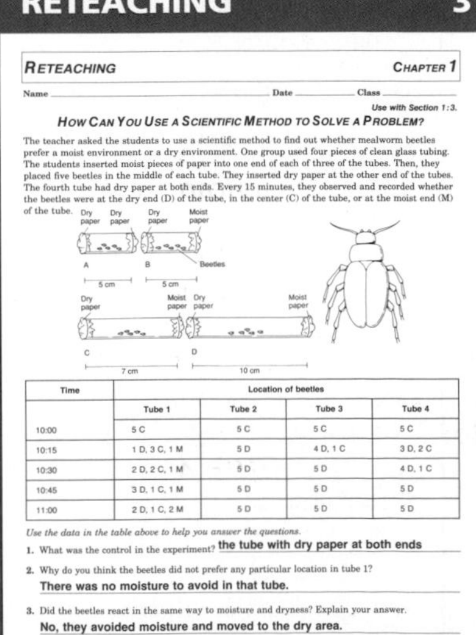

RETEACHING CHAPTER 1

Name ______ Date ______ Class ______

*Use with Section 1:3.*

HOW CAN YOU USE A SCIENTIFIC METHOD TO SOLVE A PROBLEM?

The teacher asked the students to use a scientific method to find out whether mealworm beetles prefer a moist environment or a dry environment. One group used four pieces of clean glass tubing. The students inserted moist pieces of paper into one end of each of three of the tubes. Then, they placed five beetles in the middle of each tube. They inserted dry paper at the other end of the tubes. The fourth tube had dry paper at both ends. Every 15 minutes, they observed and recorded whether the beetles were at the dry end (D) of the tube, in the center (C) of the tube, or at the moist end (M) of the tube.



| Time | Location of beetles | | | |
|---|---|---|---|---|
| | Tube 1 | Tube 2 | Tube 3 | Tube 4 |
| 10:00 | 5 C | 5 C | 5 C | 5 C |
| 10:15 | 1 D, 3 C, 1 M | 5 D | 4 D, 1 C | 3 D, 2 C |
| 10:30 | 2 D, 2 C, 1 M | 5 D | 5 D | 4 D, 1 C |
| 10:45 | 3 D, 1 C, 1 M | 5 D | 5 D | 5 D |
| 11:00 | 2 D, 1 C, 2 M | 5 D | 5 D | 5 D |

*Use the data in the table above to help you answer the questions.*

1. What was the control in the experiment? **the tube with dry paper at both ends**
2. Why do you think the beetles did not prefer any particular location in tube 1? **There was no moisture to avoid in that tube.**
3. Did the beetles react in the same way to moisture and dryness? Explain your answer. **No, they avoided moisture and moved to the dry area.**

# Science and Society

## Technology: Helpful or Harmful?

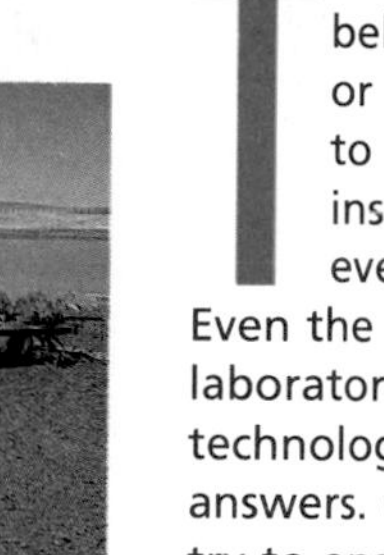

Technology has provided many products that we believe make our lives better. Imagine walking three or more miles to and from school each day or trying to keep clean without soap and toothpaste. Look inside your cabinets and closets at home. Almost everything in them is a product of recent technology. Even the clothes you wear may contain fibers invented in a laboratory. Besides providing useful products, our use of technology has raised questions that do not have easy answers. Consider the three situations described below and try to answer the questions.

### What Do You Think?

1. Nuclear energy is an important source of power. Many communities use nuclear energy to generate electricity for lighting homes and businesses. Nuclear power plants do not pollute the air as coal-burning power plants do. Use of nuclear energy, however, does present problems. There is no guaranteed safe way to dispose of used nuclear fuel. Also, many nuclear power plants are being shut down because they are not considered safe. Who is responsible for disposing of nuclear wastes safely? Should we continue to use and build nuclear power plants?

2. Farmers use fertilizers to grow more and better crops. More crops mean a better living for the farmer, a better food supply, and lower prices at the grocery store. Without fertilizers, we might not be able to buy many kinds of fruits and vegetables. Fertilizers, however, can drain from farmland into lakes and streams. Fish can't live in water with fertilizers in it. In some areas, people can't drink the water because of the fertilizers in it. Should laws be passed on when and how much fertilizers can be used?

3. A person is rushed to the hospital after an automobile accident. The doctors do not know if there is brain damage, or what the chances of survival are. A quick decision must be made about whether or not to use a life-support system called a respirator. A respirator is a device that helps a person breathe when that person can't breathe on his or her own. The respirator is used. Later, it is discovered that the person shows little brain activity. The person remains in a coma for months. There is little hope for recovery, and the family wants the respirator disconnected. Who should decide when life-support systems are needed?

**Conclusion:** How does technology help people? Is it possible to misuse technology?

Nuclear power plant



Science and Society

## Technology: Helpful or Harmful?

### Background

Currently, the Department of Energy (DOE) is planning the construction of a nuclear waste repository that will be used as a storage site for nuclear wastes. Discuss with students alternate energy sources: solar, wind, and geothermal.

A document called the Living Will states that in the event that a person becomes incapacitated, extraordinary life support measures are not used.

### Teaching Suggestions

**Cooperative Learning:** Have students work in cooperative groups of four to compose a letter about each issue and ideas on how to solve the problems.

### What Do You Think?

1. Students may argue that there needs to be a uniform set of regulations for disposing of nuclear wastes.
2. Any regulation necessitates monitoring, which creates both cost and bureaucracy.
3. Many students may feel that the patient should not have to pay. Every year, however, many people come out of comas, many of whom were never expected to do so.

**Conclusion:** Technology is only good as long as it is used to make our lives better and the environment no worse.

### Suggested Readings


Caplan, Arthur. "An Improved Future?" *Scientific American*, September 1995, pp. 142-143.

Caplan, Arthur. *Ethical Issues in Medicine and the Life Sciences.* New York: John Wiley & Sons, 1995.

Furth, Harold P. "Fusion." *Scientific American*, September 1995, pp. 174-177.

Horgan, John. "Seeking a Better Way to Die." *Scientific American*, May 1997, pp. 100-105.

Luck, Robert W. "What Technology Alone Cannot Do." *Scientific American*, September 1995, pp. 204-205.

Plucknett, Donald, and Donald Winkelmann. "Technology for Sustainable Agriculture." *Scientific American*, September 1995, pp. 182-186.

Sancton, T.A. "What on Earth Are We Doing?" *Time,* Jan. 2, 1989, pp. 24-30.

Whipple, Chris G. "Can Nuclear Waste Be Stored Safely at Yucca Mountain?" *Scientific American*, June 1996, pp. 72-79.

## CHAPTER 1 REVIEW

### Summary

Summary statements can be used by students to review the major concepts of this chapter.

### Key Science Words

All boldfaced terms from the chapter are listed.

### Testing Yourself

#### Using Words

1. kilogram
2. volume
3. scientific method
4. meter
5. biology
6. light microscope
7. data
8. hypothesis
9. technology
10. experiment
11. variable
12. stereomicroscope

#### Finding Main Ideas

13. p. 14; Time is measured in seconds.
14. p. 15; A scientific method is a series of steps used to solve a problem.
15. p. 16; An experiment is the testing of a hypothesis.
16. p. 6; People may use biology in hobbies and work, for example beekeeper, crime lab technician, optician, mushroom grower.
17. p. 13; An instrument used to measure mass is a balance.
18. p. 13; Multiply length by width by height to find volume.
19. p. 7; Tools a biologist uses include hand lens, light microscope, electron microscope, and computer.
20. p. 16; A hypothesis is a statement that can be tested.
21. p. 12; Prefixes are combined with a unit of measure to change that unit of measure.
22. p. 18; The part of an experiment that is not changed is the control group.



# Chapter 1 Review

## Summary

### 1:1 Biology in Use

1. Many jobs depend on the use of biology.
2. Tools that biologists use include microscopes, binoculars, cameras, nets, and computers.
3. The light microscope, stereomicroscope, and electron microscope are important tools in biology.

### 1:2 Measurements Used in Biology

4. The International System of Units (SI) is a measuring system based on units of 10 that all scientists use.
5. Meter, cubic meter, kilogram, second, and Celsius degrees are SI measurements.
6. SI units are used to measure length, volume, mass, time, and temperature.

### 1:3 Scientific Method

7. The steps of the scientific method are: recognizing the problem, researching the problem, forming a hypothesis, testing the hypothesis, and drawing conclusions.
8. A theory is a hypothesis that has been tested again and again by many scientists. Results are the same each time.
9. Technology is used to solve many everyday problems.

## Key Science Words

biology (p. 5)
celsius (p. 14)
control (p. 18)
data (p. 18)
experiment (p. 16)
hypothesis (p. 16)
International System of Units (p. 11)
kilogram (p. 14)
light microscope (p. 7)
meter (p. 11)
scientific method (p. 15)
stereomicroscope (p. 8)
technology (p. 20)
theory (p. 19)
variable (p. 18)
volume (p. 13)

## Testing Yourself

### Using Words

*Choose the word from the list of Key Science Words that best fits the definition.*

1. one thousand grams
2. measured in liters
3. series of steps in solving a problem
4. can be divided into 100 units called centimeters
5. study of living things
6. magnifies objects 430 ×
7. recorded measurements in an experiment
8. a statement that can be tested
9. using discoveries in science to solve everyday problems
10. steps used to test a hypothesis
11. something that causes changes in an experiment
12. magnifies objects through which light does not pass

# Review

## Testing Yourself *continued*

### Finding Main Ideas

*List the page number where each main idea below is found. Then, explain each main idea.*

13. how time is measured
14. what a scientific method is
15. what an experiment is
16. how people use biology
17. instrument used to measure mass
18. how to find volume
19. tools a biologist uses
20. what a hypothesis is
21. how prefixes are used
22. part of an experiment that is not changed

### Using Main Ideas

*Answer the questions by referring to the page number after each question.*

23. What must an optometrist and a mushroom grower know about biology? (p. 6)
24. How is a stereomicroscope different from a light microscope? (p. 8)
25. How are mass and weight different? (p. 13)
26. How are biology and technology related? (p. 20)
27. Which SI units could you use to measure
    (a) the distance to your friend's house? (p. 11)
    (b) the volume of one serving of orange juice? (p. 13)
    (c) the mass of a can of soup? (p. 14)
28. How are a variable and a control different? (p. 18)
29. Why do biologists use computers? (p. 10)

## Skill Review ✔

*For more help, refer to the **Skill Handbook,** pages 704-719.*

1. **Calculate:** An ostrich has a mass of 155 kg. A hummingbird has a mass of 10 g. How many times greater than the mass of the hummingbird is the mass of the ostrich?
2. **Design an experiment:** Design an experiment to show the effect of sunlight on the growth of bean seedlings.
3. **Observe:** Look around your classroom and name the tools that you observe.
4. **Make and use tables:** Make a table of the different kinds of measurements and the SI units for each.

## Finding Out More

### Critical Thinking

1. Design a controlled experiment to test the following hypothesis: Radish seeds will germinate faster in light than in darkness.
2. Why is it important to use both a number and a unit when measuring an object?

### Applications

1. Report to your class on new technologies in medicine, farming, or industry by reading the newspaper or watching television.
2. Find out which metric units are used by a hospital or pharmacy in preparing medicines. List several examples.



### Using Main Ideas

23. An optometrist must know how eyes work, the correct lens to prescribe. A mushroom grower must know soil conditions and mushroom diseases.
24. It does not magnify as much and it allows objects to be viewed that light cannot pass through.
25. The mass of an object remains the same. The weight of an object varies with gravitational pull.
26. Discoveries in biology are used to solve problems in the areas of farming, environment, and disease.
27. The measurements needed are (a) meter or kilometer, (b) milliliters, (c) gram.
28. The control remains the same; the variable is the factor that is being tested.
29. Biologists use computers to keep up with new information about biology.

## Skill Review

1. 10 g = .01 kg
   155 kg / .01 kg = 15 500 times as large
2. Have two containers containing the same number of bean seedlings. Keep the containers the same, except one will be kept in the light, one in the dark.
3. Tools can include microscopes (both light and stereo-), balance, graduated cylinders, beakers, and others.
4. Students could list SI units across the top and measurements along the side and then fill in the table.

## Finding Out More

### Critical Thinking

1. Have two containers with the same number of seeds in each. All the conditions must be kept the same except that one container will be placed in the dark and one will be placed in the light.
2. A number is used to show the amount and a unit is used to tell what is being measured.

### Applications

1. Amniocentesis, ultrasound, artificial hearts, organ transplants, insulin pumps, fuel from plants, biological control of insect pests can be used for reports.
2. Volume measurements and mass measurements (liters or milliliters and grams or milligrams or kilograms) are used. Some medications are available in powder form, which must have water added to it. Medications are measured in milligrams per dose.

**Chapter Review**/*Teacher Resource Package*, pp. 1-2.

**Chapter Test**/*Teacher Resource Package*, pp. 3-5.

**Chapter Test**/*Computer Test Bank*

# Features of Life and the Cell

## PLANNING GUIDE

| CONTENT | TEXT FEATURES | TEACHER RESOURCE PACKAGE | OTHER COMPONENTS |
|---|---|---|---|
| (1 1/2 days)<br>2:1 Living Things and Their Parts<br>Features of Living Things<br>The Chemistry of Life<br>Cell Theory | Skill Check: *Classify,* p. 27<br>Skill Check: *Formulate a Model,* p. 30<br>Mini Lab: *What Makes Up Cork?* p. 31<br>Lab 2-1: *Respiration,* p. 28<br>Check Your Understanding, p. 31 | Enrichment: *The Chemistry of Life,* p. 2<br>Reteaching: *Living Things and Their Parts,* p. 4<br>Focus: *Testing for Evidence of Life on Mars,* p. 3<br>Transparency Master: *Features of Life,* p. 5<br>Study Guide: *Living Things and Their Parts,* pp. 7-8<br>Lab 2-1: *Respiration,* pp. 5-6 | **STVS:** *Energy-Integrated Farm,* Ecology (Disc 6, Side 2) |
| (1 1/2 days)<br>2:2 Cell Parts and Their Jobs<br>Cell Membranes and Nucleus<br>Cytoplasm | Skill Check: *Understand Science Words,* p. 35<br>Lab 2:2 *Cells,* p. 36<br>Idea Map, p. 34<br>Check Your Understanding, p. 37 | Application: *A Cell Is Like a City,* p. 2<br>Reteaching: *Plant Cell and Animal Cell Structure,* p. 5<br>Skill: *Interpret Diagrams,* p. 2<br>Study Guide: *Cell Parts and Their Jobs,* p. 9<br>Lab 2:2 *Cells,* pp. 7-8 | **Laboratory Manual:**<br>*What Cell Parts Can You See with the Microscope?* p. 13<br>Color Transparency 2a: *Plant and Animal Cells*<br>**STVS:** *Genetic Engineering in Barley,* Plants and Simple Organisms (Disc 4, Side 1) |
| (1 day)<br>2:3 Special Cell Processes<br>Diffusion<br>Osmosis<br>Organization | Mini Lab: *What Is Diffusion?* p. 39<br>Check Your Understanding, p. 41<br>Applying Technology: *Salty Plants—How Will They Grow?* p. 42 | Reteaching: *Cell Processes,* p. 6<br>Critical Thinking/Problem Solving: *What Happened to Dinner?* p. 2<br>Study Guide: *Special Cell Processes,* pp. 10-11<br>Study Guide: *Vocabulary,* p. 12 | **Laboratory Manual:**<br>*What Are Diffusion and Osmosis?* p. 9<br>Color Transparency 2b: *Organization in Living Things*<br>**STVS:** *Natural Time-Release Capsules,* Human Biology (Disc 7, Side 1) |
| Chapter Review | Summary<br>Key Science Words<br>Testing Yourself<br>Finding Main Ideas<br>Using Main Ideas<br>Skill Review | **ASSESSMENT RESOURCES**<br>Chapter Review, pp. 6-7<br>Chapter Test, pp. 8-10<br>Performance Assessment in the Biology Classroom<br>Alternate Assessment in the Science Classroom<br>Computer Test Bank | |

## GLENCOE TECHNOLOGY

**Infinite Voyage,** *The Champion Within*
**Science and Technology Videodisc Series,** *Energy-Integrated Farm,* Ecology (Disc 6, Side 2)
*Genetic Engineering in Barley,* Plants and Simple Organisms (Disc 4, Side 1)
*Natural Time-Release Capsules,* Human Biology (Disc 7, Side 1)
**The Secret of Life,** *What's in Stetter's Pond?: The Basics of Life*

## MATERIALS NEEDED

| LAB 2-1, p. 28 | LAB 2-2, p. 36 | APPLYING TECHNOLOGY, p. 42 | MARGIN FEATURES |
|---|---|---|---|
| 5 small baby food jars<br>red liquid<br>wax pencil<br>guppy<br>yeast suspension<br>dropper<br>paper clip<br>drinking straw | light microscope<br>2 microscope slides<br>2 cloverslips<br>dropper<br>methylene blue stain<br>toothpicks, flat type<br>*Elodea* leaf<br>water | 2 trays of young plants<br>labels<br>salt water | Skill Check, p. 27<br>pencil paper<br>Skill Check, p. 30<br>polystyrene balls toothpicks<br>labels<br>Skill Check, p. 35<br>dictionary<br>Mini Lab, p. 31<br>microscope cork slice<br>Mini Lab, p. 39<br>box vanilla balloon |

## OBJECTIVES

For more information about National Science Standards, see page 5T.

| SECTION | OBJECTIVE | CORRELATION of QUESTIONS to OBJECTIVES: CHECK YOUR UNDERSTANDING | CHAPTER REVIEW | TRP CHAPTER REVIEW | TRP CHAPTER TEST |
|---|---|---|---|---|---|
| 2:1 National Science Stds: UCP.1, UCP.2, UCP.3, UCP.4, A.2, B.2, B.3, B.6, C.1, C.5, G.3 | 1. **Describe** eight features common to all living things. | 1, 4, 5 | 7, 10, 14, 20 | 4, 19, 11, 12, 13, 14, 28 | 14, 15, 16, 17, 18, 19, 20, 21, 23, 28 |
| | 2. **List** the elements that make up living things. | 2 | 16 | 15,16, 17, 18, 19 | 5, 6, 10, 12, 13 |
| | 3. **State** the major ideas of the cell theory. | 3 | 6, 17 | 1, 2 | 1, 2, 25, 27, 31 |
| 2:2 National Science Stds: UCP.1, UCP.2, UCP.3, UCP.5, B.6, C.1, C.5 | 4. **List** the parts of the cell. | 6, 7 | 1, 4, 23 | 5, 20, 21, 22, 23, 24, 25, 26, 27, 30, 32 | 31, 32, 33, 34, 35, 36, 37, 38, 39, 40, 48, 49 |
| | 5. **Describe** the functions of the cell parts. | 8, 9 | 2, 12, 18, 23, 26 | 6, 7, 33, 35, 37, 38, 39 | 7, 9, 11, 48, 49, 50, 51, 52, 53, 54, 55, 56 |
| 2:3 National Science Stds: UCP.1, UCP.2, UCP.3, UCP.4, UCP.5, B.2, B.4, B.5, C.1, C.5 | 6. **Explain** the process of diffusion in a cell. | 11 | 5, 13, 24, 25 | 8 | 26 |
| | 7. **Describe** the process of osmosis in a cell. | 12, 14 | 9, 19, 22, 24 | 36 | 3, 28, 30 |
| | 8. **Communicate** how cells, tissues, organs, and organ systems are organized. | 13, 15 | 3, 8, 11 | 3, 10 | 4, 8, 22 |

# Features of Life and the Cell

## CHAPTER OVERVIEW

### Key Concepts

In this chapter, students will study the eight characteristics that define living things. Students also will learn about cell structure and function and the processes by which cells adjust to their environments.

### Key Science Words

adaptation
cell
cell membrane
cellular respiration
cell wall
centriole
chloroplast
chromosome
consumer
cytoplasm
development
diffusion
mitochondria
nuclear membrane
nucleolus
nucleus
organ
organism
organ system
osmosis
producer
reproduce
ribosome
tissue
vacuole

### Skill Development

In Lab 2-1, students will **make and use tables, interpret data, infer**, and **form hypotheses** to discover different things that give off carbon dioxide. In Lab 2-2, students **use a microscope** as they **observe** and **compare** plant and animal cells. In the Skill Check on page 27, students will **classify** living and once-living things. In the Skill Check on page 30, they will **formulate a model** of a simple molecule. On page 35, they will **understand** the **science word** *chloroplast.* In the Mini Lab on page 31, students will **use a microscope** to examine cork. In the Mini Lab on page 39, they will **experiment** to learn about diffusion.

### Bridging

Students have already learned how scientific measurements are made in Chapter 1. The features of living things can be measured. For example, evidence that a living thing has grown is an increase in mass and length.

**CHAPTER PREVIEW**

### Chapter Content

Review this outline for Chapter 2 before you read the chapter.

**2:1 Living Things and Their Parts**
Features of Living Things
The Chemistry of Life
Cell Theory

**2:2 Cell Parts and Their Jobs**
Cell Membrane and Nucleus
Cytoplasm

**2:3 Special Cell Processes**
Diffusion
Osmosis
Organization

### Skills in this Chapter

The skills that you will use in this chapter are listed below.

- In **Lab 2-1,** you will make and use tables, interpret data, infer, and form hypotheses. In **Lab 2-2,** you will observe, compare, and use a microscope.
- In the **Skill Checks,** you will classify, formulate a model, and understand science words.
- In the **Mini Labs,** you will use a microscope and experiment.

24

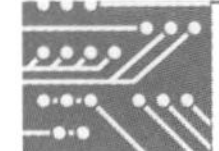

## TECH PREP

For Tech Prep activities in this chapter of the Teacher Wraparound Edition, see especially the Cooperative Learning activity on page 30, the Portfolio on page 31, and the Assessment on page 39.
See also the Glencoe Homepage at **www.glencoe.com**

The following Glencoe resources provide additional opportunities for integrating science and technology.

**Production Systems Technology**
Section II: Resources for Production
Activity 1: Working with Fluid Power

Chapter **2**

# Features of Life and the Cell

In the photo on the left, you can see several kinds of things that are living and several that are not living. If you looked closely at an enlarged part of the rocks, as shown in the small photo, what would you see? Make a guess as to what you would find if you looked at a drop of the water with a microscope.

What does it mean to be living? Maybe you would say that living things need water and food. You might say that living things grow. Some nonliving things seem to grow, too. If you have cold winters where you live, you probably have seen icicles that seem to grow on your house. Maybe the ability to grow isn't enough to make something living. How then can you decide if a thing is living or nonliving? Are there features that all living things share?

**Try This!**

**What happens to a sponge in water?** Place a sponge in a dish of water. Observe what happens to the sponge. Is this a feature of living things?

**interNET CONNECTION**

For more information about the material in this chapter, follow the link for the chapter on the Glencoe Homepage at **http://www.glencoe.com**

Rock covered with mosses and lichen



## GETTING STARTED

### Using the Photos

In looking at the photos, students should try to identify factors that characterize nonliving things. Have students attempt to define the word *life* without using the word *living* in the definition. It is easier to use traits of living things to define life.

### MOTIVATION/Try This!

**What happens to a sponge in water?** Upon completing the chapter-opening activity, students will have gained practice in developing **operational definitions.** As a followup, invite a doctor or nurse to speak about the difficulties of determining when life ends. Students will discover that the definition of a living thing is not always easy to formulate.

### Chapter Preview

Have students study the chapter outline before they begin to read the chapter. They should note the overall organization of the chapter, that living things are characterized first, that a feature of living things—the cell—is discussed second, and that cell functions are discussed third.

### Misconception

Respiration, the chemical reaction that yields energy within the cell, is one feature of living things. Explain to students that respiration is not the same as breathing.

## ASSESSMENT PLANNER

**Portfolio**

Strategies on the following pages represent student products that can be placed into a best-work portfolio: pp. 27, 31.

**PERFORMANCE ASSESSMENT**

Skill Check, pp. 27, 30
Mini Lab, pp. 31, 39
Lab, pp. 28, 36

**CONTENT ASSESSMENT**

Check for Understanding, pp. 29, 31, 35, 37, 40, 41
Chapter Review, pp. 44-45

**GROUP ASSESSMENT**

Opportunities for group assessment occur with Cooperative Learning Strategies.

**Student Journal**

Strategies on the following pages represent opportunities for writing in a Student Journal: pp. 33, 39, 40.

## 2:1 Living Things and Their Parts

## PREPARATION

### Materials Needed

Make copies of the Focus, Transparency Master, Study Guide, Enrichment, Reteaching, and Lab worksheets in the *Teacher Resource Package*.

▶ Prepare phenol red and yeast solutions and obtain small jars, straws, and guppies for Lab 2-1.

▶ Purchase polystyrene balls and toothpicks for the Skill Check.

▶ Prepare slices of cork for the Mini Lab.

### Key Science Words

| | |
|---|---|
| reproduce | cellular respiration |
| development | cell |
| consumer | adaptation |
| producer | |

### Process Skills

In Lab 2-1, students will make and use tables, interpret data, infer, and form hypotheses. In the Skill Checks, students will classify and formulate a model. In the Mini Lab, students will use a microscope.

## 1 FOCUS

▶ The objectives are listed on the student page. Remind students to preview these objectives as a guide to this numbered section.

### MOTIVATION/Bulletin Board

Display a periodic table of the elements and point out common elements and their symbols. Explain that elements can join together to form compounds.

Focus/*Teacher Resource Package*, p. 3. Use the Focus transparency shown at the bottom of this page as an introduction to living things.



**Objectives**

1. **Describe** eight features common to all living things.
2. **List** the elements that make up living things.
3. **State** the major ideas of the cell theory.

**Key Science Words**

reproduce
development
consumer
producer
cellular respiration
cell
adaptation

## 2:1 Living Things and Their Parts

The word *living* is not easy to define. However, biologists recognize that all living things share certain features. Let's look at some of those features.

### Features of Living Things

**Feature 1 Living things reproduce.**

**Reproduce** means to form offspring similar to the parents. If living things could not reproduce, there might be nothing alive on Earth today.

Living things reproduce in different ways. For example, a one-celled living thing may split to produce two living things, like the amoeba in Figure 2-1.

Sunflowers grow from seeds made by the parent plants. A sunflower produces many seeds, each of which can become a new sunflower. Bear cubs have a male and a female parent and are born alive.

**Feature 2 Living things grow.**

A living thing grows by using materials and energy from its environment to increase its size. Young oak saplings grow into large oak trees over a period of many years. A kitten grows into a cat.

**Feature 3 Living things develop.**

**Development** is all the changes that occur as a living thing grows. When a tadpole first begins to develop, it does not look like an adult frog, Figure 2-2a. It has no front or hind legs. As the animal grows and develops, legs begin to

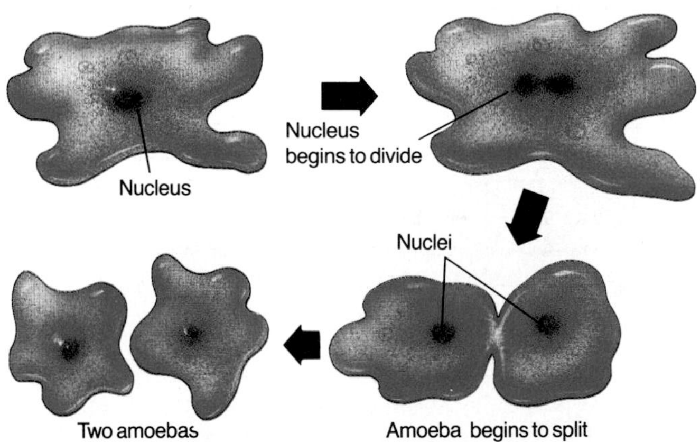


**Figure 2-1** An amoeba divides to reproduce.



## OPTIONS

### Science Background

All the features of living things are needed for the survival of the individual organism, except for reproduction. Reproduction is necessary for the survival of the species.

**FOCUS 3**

FOCUS CHAPTER 2

Name Date Class

TESTING FOR EVIDENCE OF LIFE ON MARS

Viking Lander Spacecraft on Mars

1. The Viking Lander took soil samples from Mars' surface. Radioactive nutrients were added to the soil.
2. If living things were present in the soil, they would use the radioactive nutrients to grow.

1. Explain how this experiment could prove to scientists that living things were present in the Martian soil.
2. The Lander did not find materials that make up living things. Do these results prove that life doesn't exist on Mars now? Do they prove that life never existed on Mars? Explain.

a

b

Figure 2-2 A tadpole (a) does not look like an adult frog. A developing bean seed (b) does not look like a young bean plant.

form. A young plant in a seed looks different from a fully developed plant, as shown by Figure 2-2b. The young plant does not have roots, stems, or flowers. Many of these parts develop as the plant grows.

**Feature 4 Living things need food.**

Many animals get food by hunting and eating other animals. Some animals eat only plants. Other animals eat both plants and animals. Animals are consumers. **Consumers** are living things that eat, or consume, other living things. Green plants make their own food. Green plants are producers. **Producers** are living things that make, or produce, their own food.

**Feature 5 Living things use energy.**

Energy is the ability to do work. Moving your arms or legs, blinking your eyes, breathing, and even sleeping require energy. Animals get energy from the food they eat. Plants get energy from the food they make during photosynthesis.

How do living things get energy from food? All living things carry out cellular respiration. **Cellular respiration** is the process by which food is broken down and energy is released. Many living things use oxygen in the process of cellular respiration. Water and a gas called carbon dioxide are given off. The energy given off during cellular respiration is used for other life processes. For example, energy is used to keep human body temperature at 37°C.

**Feature 6 Living things are made of cells.**

The **cell** is the basic unit of all living things. Some living things have many cells. Others have only one cell.

### Skill Check

**Classify:** Look around you and make a list of things that are living or were once living. How are living things different from things that were once living? *For more help, refer to the **Skill Handbook,** pages 715-717.*

**What is given off during cellular respiration?**



## 2 TEACH

### Concept Development

▶ Have students give examples of what consumers eat.

▶ Emphasize that both producers and consumers use food. Some animals eat many times their body weight each day.

▶ You may wish to introduce the concept that energy moves through the living world in one direction. Matter that makes up living things, however, is recycled. These concepts are important in the study of biology.

▶ Emphasize that energy is required for the other features of life.

### Skill Check

Living things reproduce, grow, develop, need food, use energy, are made of cells, respond, and adapt. Nonliving things do not have all of these features.

### Guided Practice

Have students write down their answers to the margin question: water and carbon dioxide gas.

**Portfolio**

Students can list and give examples of the eight features that distinguish living things from nonliving things. Ask students to rank these features in the order of their importance. Have students justify the order they chose. Find pictures from magazines that illustrate each feature.

**TRANSPARENCY 5**

TRANSPARENCY MASTER CHAPTER 2

FEATURES OF LIFE

| reproduce | are adapted |
|---|---|
| develop | grow |
| need food | use energy |
| made of cells | respond |

### OPTIONS

**Transparency Master**/*Teacher Resource Package*, p. 5. Use the Transparency Master shown here to teach the features of life.

### GLENCOE TECHNOLOGY

**Videodisc**

**The Secret of Life**

*Introduction: Question*

*Introduction: Response*

## Lab 2-1 Respiration

### Overview

In this lab, students use the scientific method to observe that living things give off $CO_2$. They also learn about humane treatment of animals.

**Objectives:** Upon completing the lab, students will be able to (1) **state** the purpose of an experimental control, (2) **recognize** that chemical indicators show that chemical reactions are taking place, (3) **observe** that living things give off $CO_2$.

**Time Allotment:** 30 minutes

### Preparation

▶ **Alternate Materials:** You may wish to substitute a goldfish or a germinating bean seed for the guppy.

▶ For phenol red indicator, mix 0.4 g phenol red in 100 mL water to make stock solution. Use 1 mL stock solution to 100 mL water to yield a 0.00004% solution of phenol red.

▶ For yeast solution, mix 1 teaspoon sugar and 1 teaspoon dry yeast in 200 mL water.

**Lab 2-1 worksheet**/*Teacher Resource Package*, pp. 5-6.

### Teaching the Lab

▶ Phenol red changes from red to yellow or orange in the presence of carbon dioxide.

▶ **Troubleshooting:** Activate the yeast by mixing it with lukewarm water. Do not use hot or cold water.

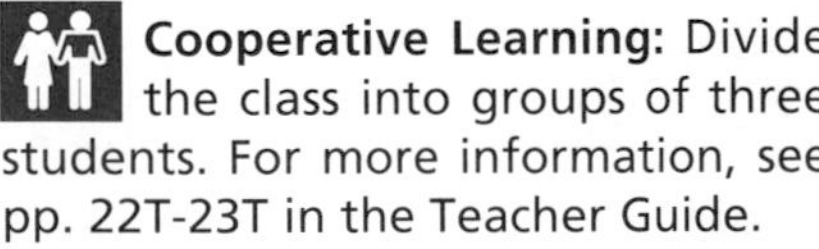
**Cooperative Learning:** Divide the class into groups of three students. For more information, see pp. 22T-23T in the Teacher Guide.

### ✓ ASSESSMENT

**Performance:** Have students repeat the lab using bromothymol blue. Ask if bromothymol blue indicates the presence of carbon dioxide. Ask students to explain their observations.

# Respiration

**Lab 2–1**

## Problem: Do all things give off carbon dioxide?

### Materials

5 small baby food jars
red liquid
wax pencil
guppy
yeast suspension
dropper
paper clip
straw

### Skills

make and use tables, interpret data, infer, form hypotheses

### Procedure

1. Copy the data table.
2. Number the five baby food jars 1 to 5. Fill each of the jars 2/3 full with red liquid.
3. Record in your notebook the color in each jar at the start of the activity.
4. Add the things listed here to the jars. Jar 1—Add nothing. Jar 2—Use the straw to blow gently into the red liquid for 2 minutes. Jar 3—Add 1 guppy. **CAUTION:** *Always use care when handling live animals.* Jar 4—Add 5 drops of yeast solution. Jar 5--Add 1 paper clip. Record what is added to each jar on your table.
5. In your notebook, write a **hypothesis** about which things you think will give off carbon dioxide. You will know when carbon dioxide is given off because the red liquid will turn orange or yellow.
6. Wait 15 minutes. Record in your table the color of the liquid in each jar using red, orange, or yellow as the choices.
7. Return the guppy to your teacher.

### Data and Observations

1. In which jars did the color change?
2. In which jars was carbon dioxide present?

### Analyze and Apply

1. **Check your hypothesis:** Is your hypothesis supported by your data? Why or why not?
2. What was the purpose of Jar 1?
3. **Interpret data:** Which things gave off carbon dioxide?
4. Which things didn't give off carbon dioxide?
5. What process in living things causes carbon dioxide to be given off?
6. **Apply:** Which of these things give off carbon dioxide: a tree? a bicycle? a fly? a shoe?

### Extension

**Infer:** Look around your classroom. Make a table listing the things that now give off carbon dioxide, things that gave off carbon dioxide when they were living, and things that never gave off carbon dioxide.

| JAR | 1 | 2 | 3 | 4 | 5 |
|---|---|---|---|---|---|
| What was added | nothing | breath | guppy | yeast | paper clip |
| Color at start | red | red | red | red | red |
| Color of liquid after 15 minutes | red | | yellow | orange | red |
| Color of liquid in Jar 2 after 2 minutes | | yellow | | | |



## ANSWERS

### Data and Observations

1. 2, 3, 4
2. 2, 3, 4

### Analyze and Apply

1. Students who hypothesized that jars 2, 3, and 4 would show the presence of carbon dioxide will say that their hypotheses were supported.
2. control
3. person, guppy, yeast
4. paper clip
5. cellular respiration
6. tree, fly

**Figure 2-3** Plants respond to light by growing toward it.

### Feature 7 Living things respond.

The plants in Figure 2-3 are responding to the light by growing toward it. When you call your dog, it responds to the sound of your voice. It comes to you. These are examples of living things responding to changes in the environment. The environment is made up of all the living and nonliving things that surround another living thing.

You respond to changes in the environment, too. When you get cold, you shiver. When you run, your heart beats faster. When your body gets warm, you sweat. These are examples of how your body responds to changes.

### Feature 8 Living things are adapted to their environments.

A trait that makes a living thing better able to survive is called an **adaptation.** Polar bears are adapted to living in the Arctic. They have thick fur and layers of fat to keep warm. Squirrels are adapted to living in trees. Their bushy tails help them keep their balance.

How can you tell if something is living? It must have all of the features just described. If it does not, it is not living. For example, a burning candle has some features that living things have, Figure 2-4. It gives off carbon dioxide gas and water. It uses oxygen and wax and changes shape. A candle responds. The candle flame flickers when a draft passes over it.

Does having these features mean that a burning candle is alive? The answer is no. A candle does not reproduce. It does not have cells. It does not grow. It is not adapted to its environment. The candle is not living, even though it has a few of the features that a living thing has.

**Figure 2-4** A burning candle has some features of living things. What are they? uses energy, responds

## STUDY GUIDE 7

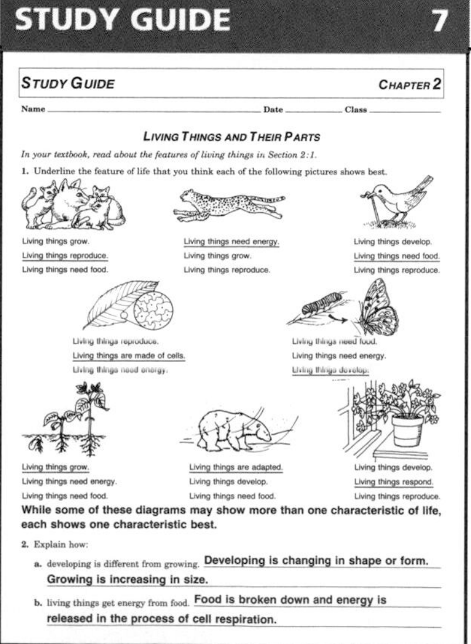

STUDY GUIDE CHAPTER 2

Name ______ Date ______ Class ______

LIVING THINGS AND THEIR PARTS

*In your textbook, read about the features of living things in Section 2:1.*

1. Underline the feature of life that you think each of the following pictures shows best.

Living things grow. / Living things reproduce. / Living things need food.

Living things need energy. / Living things grow. / Living things reproduce.

Living things develop. / Living things need food. / Living things reproduce.

Living things reproduce. / Living things are made of cells. / Living things need energy.

Living things need food. / Living things need energy. / Living things develop.

Living things grow. / Living things need energy. / Living things need food.

Living things are adapted. / Living things develop. / Living things need food.

Living things develop. / Living things respond. / Living things reproduce.

**While some of these diagrams may show more than one characteristic of life, each shows one characteristic best.**

2. Explain how:
   a. developing is different from growing. **Developing is changing in shape or form. Growing is increasing in size.**
   b. living things get energy from food. **Food is broken down and energy is released in the process of cell respiration.**

## STUDY GUIDE 8

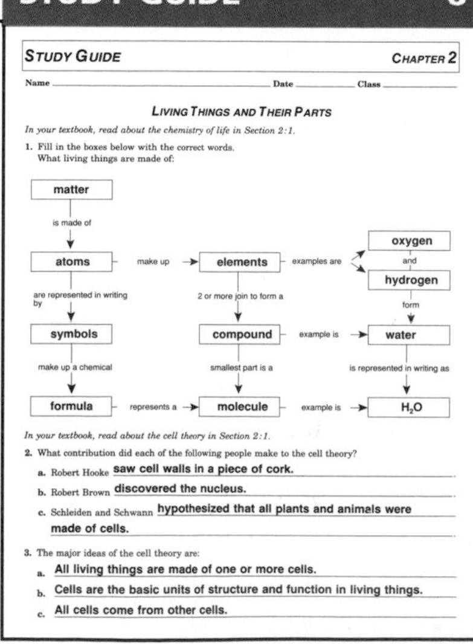

STUDY GUIDE CHAPTER 2

Name ______ Date ______ Class ______

LIVING THINGS AND THEIR PARTS

*In your textbook, read about the chemistry of life in Section 2:1.*

1. Fill in the boxes below with the correct words.
   What living things are made of:



*In your textbook, read about the cell theory in Section 2:1.*

2. What contribution did each of the following people make to the cell theory?
   a. Robert Hooke **saw cell walls in a piece of cork.**
   b. Robert Brown **discovered the nucleus.**
   c. Schleiden and Schwann **hypothesized that all plants and animals were made of cells.**

3. The major ideas of the cell theory are:
   a. **All living things are made of one or more cells.**
   b. **Cells are the basic units of structure and function in living things.**
   c. **All cells come from other cells.**

# TEACH

## Concept Development

▶ Responses in plants are usually less obvious than those in animals. A plant that shows rapid response is the mimosa. If you touch a mimosa leaf, all the small leaflets on the branch fold upward.

▶ Point out that organisms are adapted in terms of structure, function, and behavior.

▶ Discuss examples of human adaptations and how each example aids survival.

## Independent Practice

**Study Guide**/*Teacher Resource Package*, pp. 7-8. Use the Study Guide worksheets shown at the bottom of this page for independent practice.

## Check for Understanding

Make flash cards by writing one of the eight characteristics of living things on one side and an example of the characteristic on the other. Show an example and have students respond with a characteristic, and vice versa.

## Reteach

Have students name the features of living things that can be observed in a classroom aquarium.

## TEACH

### Concept Development

▶ Point out that even though Earth is filled with billions of things, these things are made of a limited number of elements.

▶ Point out that a compound consists of atoms that are joined together. A chemical formula represents the number and kinds of atoms in a compound.

### Skill Check

Have students find a biology or chemistry book that has diagrams of biological molecules. They may use gum drops instead of polystyrene balls.

### Guided Practice

Have students write down their answers to the margin question: $H_2O$.

**Cooperative Learning:** Divide the class into three groups and have each group make a display of objects made of only one particular element. For example, one group could collect objects made entirely of aluminum; another, objects made entirely of iron; and the third, objects made entirely of copper. Explain that pure gold and silver are too soft for jewelry so other elements are added to give them strength.

**GLENCOE TECHNOLOGY**

**Videodisc**

**The Secret of Life**

*Composition of Living Things*

### Skill Check

**Formulate a model:** Construct a model of a simple molecule, such as water or carbon dioxide. Use polystyrene balls and toothpicks. Label each atom in the molecule. *For more help, refer to the* ***Skill Handbook,*** *pages 706-711.*

## The Chemistry of Life

Living things are made of matter. Matter is anything that has mass and takes up space. Matter is made of tiny particles called atoms. Atoms make up elements. An element is a substance that is made up of only one kind of atom. Oxygen and hydrogen are examples of elements.

Matter is often made of two or more different elements joined together to form a compound. Joining different elements can change the way they act. For example, water is a compound made of the elements oxygen and hydrogen. Oxygen and hydrogen are gases. Yet, when they join they make water. As you know, water is a liquid. The smallest part of a compound is a molecule. A molecule of water is the smallest amount of water you can have and still have water.

**What is the chemical formula for water?**

Scientists use a certain symbol for each different element. When written together, the symbols make a chemical formula. A chemical formula is a way to write the name of a compound using symbols. The symbol for oxygen is O. The symbol for hydrogen is H. The chemical formula for a molecule of water is $H_2O$. The number 2 shows that there are two hydrogen atoms for every oxygen atom in a molecule of water.

Seven elements make up over 99 percent of the matter in living things. Those seven elements are carbon, hydrogen, oxygen, nitrogen, phosphorus (FAHS fuh rus), calcium, and sulfur, Figure 2-5. They are often called the building blocks of living matter.

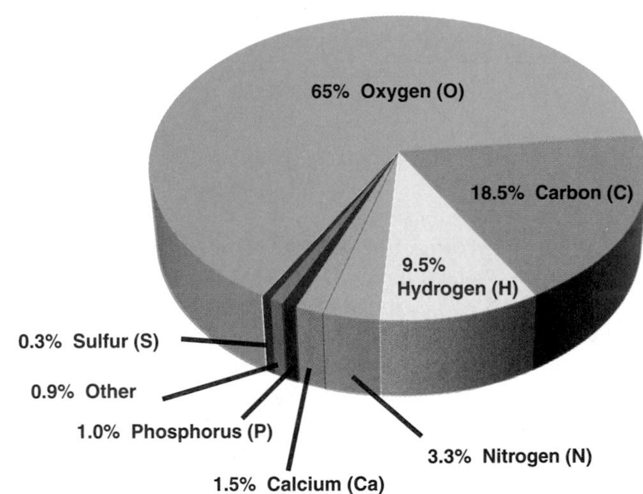


**Figure 2-5** Living things are made up of the elements shown here. Which element makes up most of a living thing?



## OPTIONS

**Enrichment:** Discuss the three common states of matter. A solid has a definite shape and occupies a definite volume. A liquid has a definite volume but changes shape to fill its container. A gas has no definite shape or volume but spreads out to fill whatever space is available.

**Enrichment**/*Teacher Resource Package,* p. 2. Use the Enrichment worksheet shown here to teach the chemistry of life.

**ENRICHMENT 2**

ENRICHMENT CHAPTER 2

Name ______ Date ______ Class ______

*Use after Section 2:1.*

THE CHEMISTRY OF LIFE

Living matter is composed of many different elements. Elements are made up of small units called atoms. Each atom has a nucleus that is surrounded by particles called electrons. Each electron has a negative charge. A nucleus is usually made up of two types of particles, protons and neutrons. Each proton has a positive charge. Neutrons have no charge. The number of protons in the nucleus of an atom is its atomic number. The number of protons and the number of neutrons added together is the atomic mass. The table below shows facts about the six elements that make up most of your body.

*Use the information about atoms and the facts in the table below to answer the questions that follow.*

| Element | Element symbol | Atomic number | Atomic mass | Percent of human body |
|---|---|---|---|---|
| hydrogen | H | 1 | 1 | 9.5 |
| carbon | C | 6 | 12 | 18.5 |
| nitrogen | N | 7 | 14 | 3.0 |
| oxygen | O | 8 | 16 | 65.0 |
| phosphorus | P | 15 | 31 | 1.0 |
| sulfur | S | 16 | 32 | 0.3 |

1. What is the most abundant element in the human body? oxygen
2. What is the symbol of the element in question 1? What is its atomic number? O, 8
3. What element makes up 18.5 percent of the body? What is its symbol? carbon, C
4. Write the atomic number of phosphorus. How many protons does it have? 15, 15
5. Fill in the blanks in this word equation:

   Atomic mass = the number of protons + the number of neutrons.
6. What is the atomic mass of nitrogen? 14
7. How many neutrons are in the nucleus of a nitrogen atom? 7
8. Write the symbol of the element found in 0.3 percent of the human body. S
9. Which element in the chart has no neutrons? hydrogen
10. How many positive charges are in the nucleus of a carbon atom? 6

## Cell Theory

In 1665, an English scientist named Robert Hooke looked at thin slices of cork under a microscope. Hooke saw that cork had a lot of empty spaces. Hooke used the word *cells* to describe the empty spaces in cork.

Today, biologists know that Hooke did not see living cells. He saw the walls of cells that were alive at one time.

By the nineteenth century, microscopes had been improved. Scientists could see that the cell had parts. First, Robert Brown discovered the central part of the cell, the nucleus (NEW klee us). Then, two German biologists, Matthias Schleiden and Theodor Schwann, did experiments to see what kinds of living things had cells. They formed hypotheses that all plants and animals were made of cells.

The experiments of Schleiden, Schwann, and other scientists led to the development of the cell theory. The major ideas of the cell theory are listed below.

1. All living things are made of one or more cells.
2. Cells are the basic units of structure and function in living things.
3. All cells come from other cells.

### Mini Lab

**What Makes Up Cork?**

**Use a microscope:** To see what Robert Hooke saw, slice a thin section from a piece of cork and look at it under the microscope. What do you see in the center of each cell? *For more help refer to the* ***Skill Handbook****, pages 712-714.*

**What events led to the cell theory?**

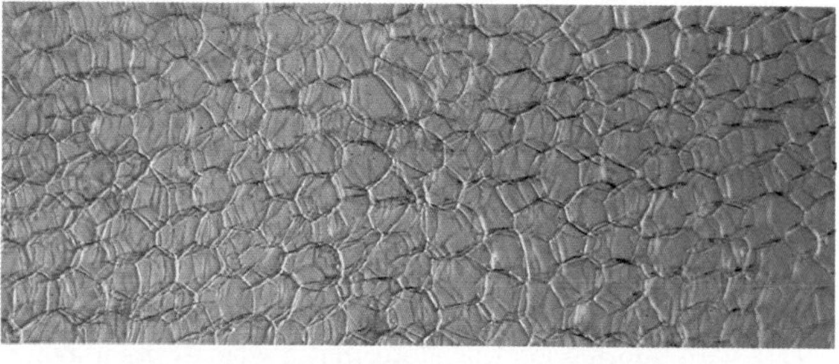

**Figure 2-6** These cork cells are similar to what Robert Hooke saw.

### Check Your Understanding

1. Name eight features of living things.
2. What six elements make up most of living matter?
3. What is the cell theory?
4. **Critical Thinking:** Why is cellular respiration necessary to living things?
5. **Biology and Writing:** Write a paragraph about a nonliving thing not mentioned in this section. Describe the features of living things that the nonliving thing has.



## TEACH

### Mini Lab

Students will see cell walls and empty space inside each cell.

### ASSESSMENT

**Performance:** Have students use a hand lens to look at objects around them, such as a chair, desk, hand, a hair, paper, etc. Ask students to explain how a hand lens changes what they observed.

### Guided Practice

Have students write down their answers to the margin question: experiments of scientists.

### Portfolio

Ask students to write a paragraph explaining the relationship between technology and the development of the cell theory. (Technology has produced better microscopes that have increased the ability to see smaller parts of cells.)

### Check for Understanding

Have students respond to the first three questions in Check Your Understanding.

### Reteach

**Reteaching**/*Teacher Resource Package*, p. 4.

**Extension:** Assign Critical Thinking, Biology and Writing, or some of the **OPTIONS** available with this lesson.

## 3 APPLY

### Connection: History

Have students write reports on Matthias Schleiden, Theodor Schwann, and the cell theory.

## 4 CLOSE

Show the video *The Living Cell*, Clearvue, Inc.

### RETEACHING 4

**RETEACHING: IDEA MAP** CHAPTER 2

Name ______ Date ______ Class ______

*Use with Section 2:1.*

**LIVING THINGS AND THEIR PARTS**

Complete the idea map by using the following phrases to fill in the blanks: changes in the environment, to carry out life processes, their environment, size, change form, offspring, cells, cellular respiration.

Living things
- reproduce — form **offspring**
- grow — increase in **size**
- develop — **change form**
- need food — **to carry out life processes**
- get energy — from **cellular respiration**
- are made — of **cells**
- respond — to **changes in the environment**
- are adapted — to **their environment**

*Answer the following questions or complete the sentences.*

1. Living things grow and develop. How are these two features of life different? **Growth is an increase in size. Development is all the changes that occur as a living thing grows.**
2. How do living things get energy from food? **by cellular respiration**
3. When you get cold, you shiver. This is an example of **homeostasis**.
4. Cacti can live in the desert. This is an example of **an adaptation**.
5. Something is living when it has **all the features of a living thing**.
6. What are the six elements that make up 97 percent of the matter in living things? **carbon, hydrogen, oxygen, nitrogen, phosphorus, and sulfur**

### Answers to Check Your Understanding

1. They reproduce, grow, develop, need food, use energy, are made of cells, respond, and adapt.
2. carbon, hydrogen, oxygen, nitrogen, sulfur, phosphorus
3. All living things are made of one or more cells. Cells are the basic units of structure and function in living things. All cells come from other cells.
4. Food is broken down and energy released.
5. Examples may be a car or a match.

# 2:2 Cell Parts and Their Jobs

## PREPARATION

### Materials Needed

Make copies of the Skill, Study Guide, Application, Reteaching, and Lab worksheets in the *Teacher Resource Package*.

▶ Have methylene blue stain and gather *Elodea* and toothpicks for Lab 2-2.

### Key Science Words

| | |
|---|---|
| cell membrane | ribosome |
| nucleus | mitochondria |
| nuclear membrane | vacuole |
| nucleolus | centriole |
| chromosome | chloroplast |
| cytoplasm | cell wall |

### Process Skills

In Lab 2-2, students will observe, compare, and use a microscope.

## 1 FOCUS

▶ The objectives are listed on the student page. Remind students to preview these objectives as a guide to this numbered section.

### MOTIVATION/Bulletin Board

Make a bulletin board showing the parts of an animal cell and a plant cell. As each cell part is studied, identify the part by placing its label on the bulletin board.

### Guided Practice

Have students write down their answers to the margin question: cytoplasm.

**Oral:** Ask students what would happen to a cell if the nucleus became damaged. (The cell would no longer function correctly because the nucleus controls all the activities of the cell.)

## 2:2 Cell Parts and Their Jobs

**Objectives**

4. **List** the parts of the cell.
5. **Describe** the functions of the cell parts.

**Key Science Words**

cell membrane
nucleus
nuclear membrane
nucleolus
chromosome
cytoplasm
ribosome
mitochondria
vacuole
centriole
chloroplast
cell wall

Cells are microscopic units that make up all living things. Cells are alive. They do everything needed to stay alive. They carry on cellular respiration. They grow and reproduce. A cell has many different parts to do all of these jobs. As you study the parts of the cell, refer to Figure 2-7. Figure 2-7 shows an animal cell and a plant cell and the cell parts of each.

### Cell Membrane and Nucleus

All cells are surrounded by a cell membrane. The **cell membrane** gives the cell shape and holds the cytoplasm. It also helps control what moves into and out of the cell.

The **nucleus** is the cell part that controls most of the cell's activities. It determines how and when proteins will be made. Proteins are complex substances with several different jobs. Some form cell parts. Others regulate activities of the cell. The nucleus also passes traits from parents to offspring.

The **nuclear membrane** is a structure that surrounds the nucleus and separates it from the rest of the cell. The nuclear membrane has openings that allow certain materials to move into and out of the nucleus.

Inside the nucleus is a smaller body called the nucleolus (new KLEE uh lus). The **nucleolus** is the cell part that helps make ribosomes (RI buh sohmz). You will read about ribosomes in the next section. Some cells have more than one nucleolus.

Also inside the nucleus are threadlike structures called chromosomes (KROH muh sohmz). **Chromosomes** are cell parts with information that determines what traits a living thing will have. Examples of traits are hair color, eye color, and sizes and shapes of leaves.

### Cytoplasm

**What is found between the cell membrane and the nucleus?**

The clear, jellylike material between the cell membrane and the nucleus that makes up most of the cell is called **cytoplasm** (SITE uh plaz um). Most of the cell's chemical reactions take place in the cytoplasm. Cytoplasm is mostly water, but it also has other chemicals. In addition, other cell parts that carry on special functions are found in the cytoplasm.



## OPTIONS

### Science Background

The nucleus determines which proteins will be made and when. Proteins, in turn, regulate most of the other chemical processes in the cell. In this way, the nucleus controls cell activity. Not all cells have nuclei. The red blood cells of mammals have nuclei when they are first formed, but they lose their nuclei before they enter the bloodstream.

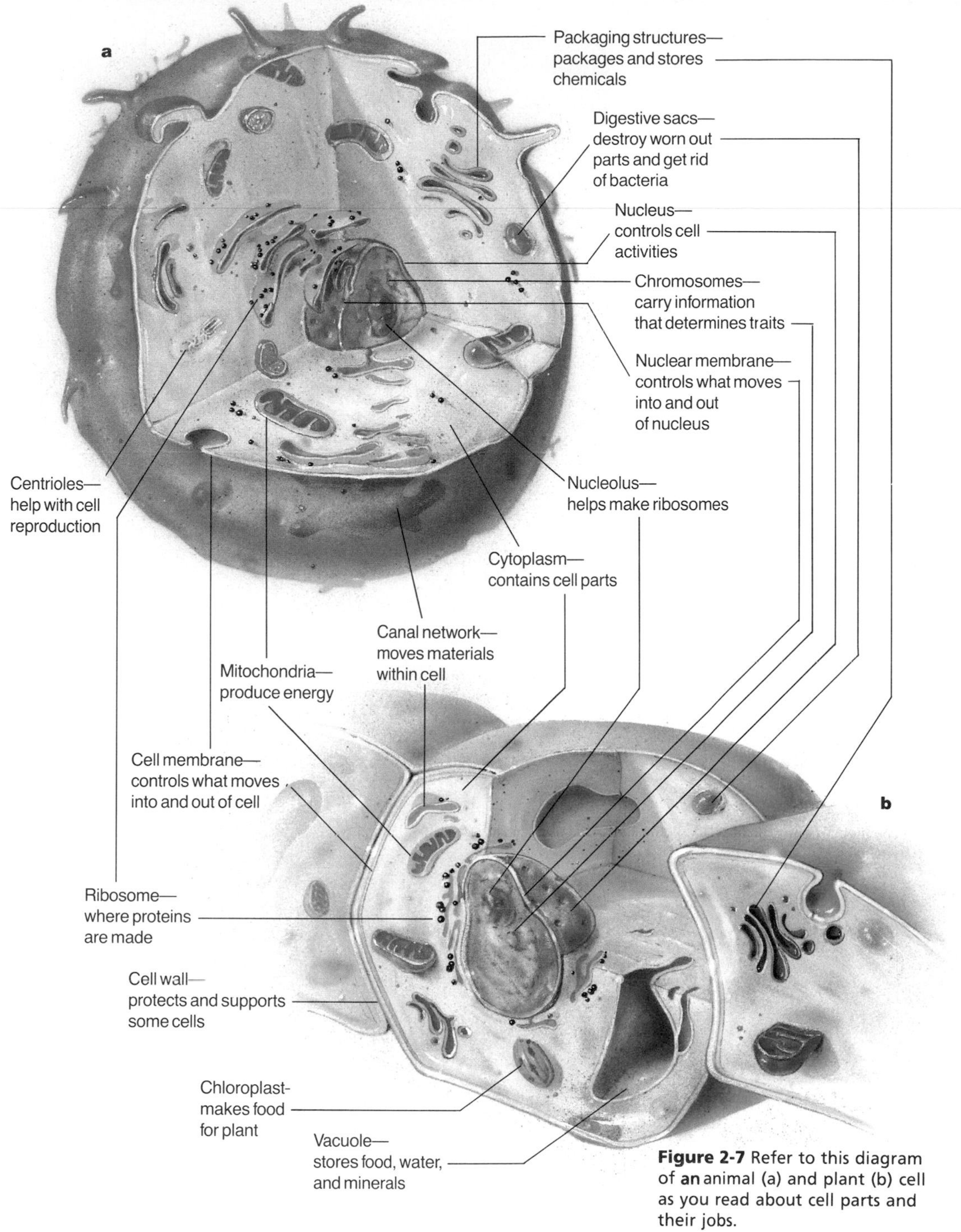


**Figure 2-7** Refer to this diagram of **an** animal (a) and plant (b) cell as you read about cell parts and their jobs.



# 2 TEACH

## MOTIVATION/Software

*The Cell: Examination, Structure, and Function,* Queue, Inc.

## MOTIVATION/Demonstration

Use a projector to show cell parts of both a plant and an animal cell. Prepared slides of frog blood and onion root tip are usually available.

## TEACH

### Concept Development

▶ Some students may believe a cell wall and a cell membrane to be the same structure. Distinguish between the two, and continue to reinforce the difference as cells are studied.

▶ Explain that a cell wall does not determine what may enter or leave a cell.

▶ Point out that cells of fungi and some one-celled organisms also have cell walls.

▶ Explain that chromosomes carry the hereditary information from one generation of cells to the next.

**Cooperative Learning:** Divide the class into cooperative groups of six. Have each group be responsible for learning the functions of two cell parts. Once each group has learned its cell parts, have the groups reshuffle and form new groups. Each person in the new group is responsible for explaining the functions of the cell parts he or she has become an "expert" on.

### Student Journal

Students should research the work of Ernest E. Just and write a report on his contributions to the study of cell parts.

## SKILL 2

**SKILL: INTERPRET DIAGRAMS** CHAPTER 2

Name ______ Date ______ Class ______

*For more help, refer to the Skill Handbook, pages 706-711.* *Use with Section 2:2.*

**PLANT AND ANIMAL CELLS**

Diagrams in a textbook give information just as words do. They help make the meanings of words clear. Studying Figure 2-7 will help you understand Section 2:2.

*Use Section 2:2 and Figure 2-7 to complete the table below.*

In the first column, list all the words in Section 2:2 that are printed in bold type. In the second column, tell where in the cell each cell part is located. In the third column, list the function(s) of each cell part.

| Cell part | Location | Function(s) |
|---|---|---|
| cell membrane | around the cell | gives the cell shape, controls what moves into and out of the cell |
| nucleus | near the center | controls most of the cell's activities |
| nuclear membrane | around the nucleus | allows materials to move into and out of the nucleus |
| nucleolus | in the nucleus | makes ribosomes |
| chromosomes | in the nucleus | determine what traits a living thing will have |
| cytoplasm | throughout the cell | makes up most of the cell |
| ribosomes | in the cytoplasm | where proteins are made |
| mitochondria | in the cytoplasm | produce energy |
| vacuole | in the cytoplasm | stores food, water, and minerals |
| centrioles | in the cytoplasm | help cell reproduction |
| chloroplasts | in the cytoplasm of plant cells | trap energy for food making |
| cell wall | surrounds the plant cell | protects and supports the plant cell |

## OPTIONS

**Plant and Animal Cells/***Transparency Package*, number 2a. Use color transparency number 2a as you teach the parts of cells.

**Skill/***Teacher Resource Package*, p. 2. Use the Skill worksheet shown here for students to interpret diagrams of plant and animal cells.

# TEACH

### Concept Development

▶ Explain that cytoplasm consists of many types of proteins and other large molecules.

▶ Point out that mitochondria are most numerous in cells that are active in some way, such as muscle cells and liver cells. In plants, mitochondria are numerous in cells that transport water against the force of gravity.

### Idea Map

Have students use the idea map as a study guide to the major concepts of this section.

### Guided Practice

Have students write down their answers to the margin question: They are the cell parts where proteins are made.

## GLENCOE TECHNOLOGY

**Videodisc**

**The Secret of Life**

*Animal Cell*

*Plant Cell*

*The Nucleus*

*Ribosomes*

**Study Tip:** Use this idea map as you study cell parts and their jobs. Remember that only plant cells have cell walls and chloroplasts, and only animal cells have centrioles. All cells have the other cell parts shown in the idea map.

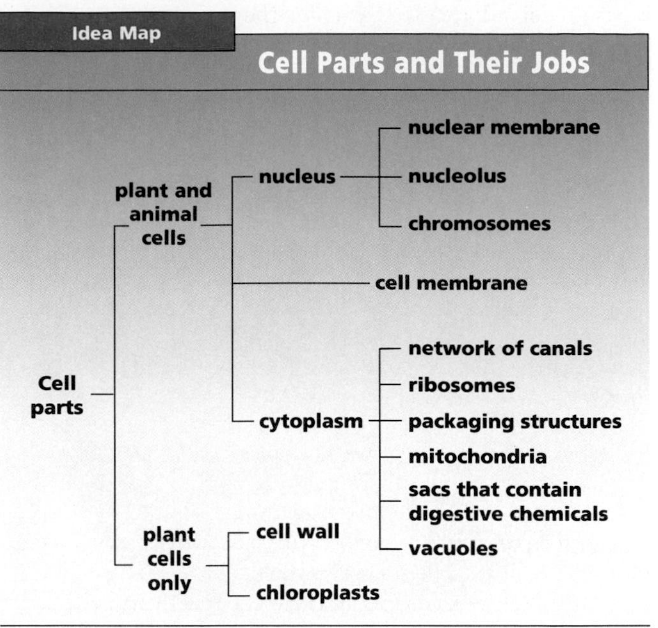


What is the job of the ribosomes?

First, the cytoplasm contains a network of canals that help move material around inside the cell. The canals connect the nuclear membrane and the cell membrane.

A second cell part found in the cytoplasm is the ribosome. **Ribosomes** are cell parts where proteins are made. Large numbers of ribosomes can be found along the canal network, where they are made. Ribosomes can also be found throughout the cytoplasm.

A third cell part found in the cytoplasm is a structure that packages and stores chemicals to be released from the cell. Large numbers of these packaging structures are found in cells that make saliva. Why do you suppose this is so? Large amounts of saliva are needed to break down the foods you eat.

Fourth, the cytoplasm contains rod-shaped bodies called mitochondria (mite uh KAHN dree uh). The **mitochondria** are cell parts that produce energy from food that has been digested. Mitochondria are often called "powerhouses" of the cell because they produce so much energy.

Small sacs that contain digestive chemicals are a fifth structure found in the cytoplasm. The chemicals made in these sacs break down large molecules. They get rid of

**Bio Tip**

**Health:** Hair is not made of cells. It is made of protein that is secreted by hair follicle cells. Hair follicle cells surround the root of the hair.



## OPTIONS

**What Cell Parts Can You See With the Microscope?**/*Lab Manual,* pp. 13-16. Use this lab as an extension to studying the parts of cells.

disease-causing bacteria that enter the cell. They also destroy worn-out cell parts and form products that can be used again.

Sixth, most cells have vacuoles (VAK yuh wolz) within the cytoplasm. A **vacuole** is a liquid-filled space that stores food, water, and minerals. Vacuoles also store wastes until the cell is ready to get rid of them. In most plant cells, the vacuole takes up a large amount of space within the cell. The fluid inside the vacuole helps to support the plant.

Centrioles (SEN tree olhz), a seventh structure within the cytoplasm, are located near the nucleus in animal cells but not in plant cells. **Centrioles** are cell parts that help with cell reproduction. They exist in pairs in the cell.

The cytoplasm of plant cells contains an eighth cell part, chloroplasts (KLOR uh plasts). **Chloroplasts** are cell parts that contain the green pigment, chlorophyll. Chlorophyll traps energy from the sun. Plants use this energy to make food. Chloroplasts give plants their green color.

The cells of plants, algae, fungi, and some bacteria have cell walls. Animal cells do not have cell walls. The **cell wall** is a thick outer covering outside the cell membrane. It protects and supports the cell.

The cell wall often remains after the rest of the cell has died. Wood is made of the walls of dead cells. What did Robert Hooke see when he looked at cork cells?

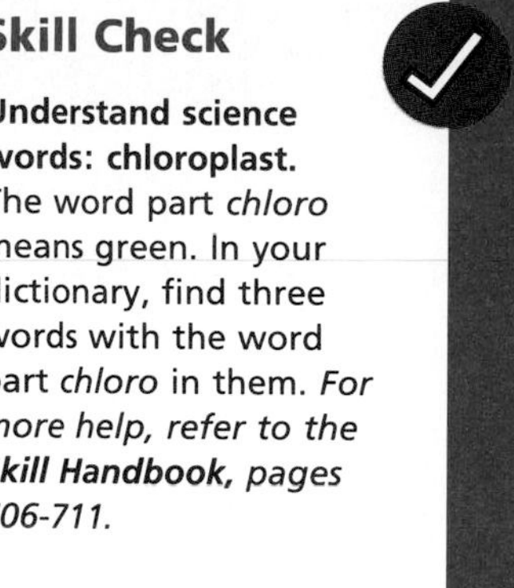

### Skill Check

**Understand science words: chloroplast.** The word part *chloro* means green. In your dictionary, find three words with the word part *chloro* in them. *For more help, refer to the* ***Skill Handbook****, pages 706-711.*

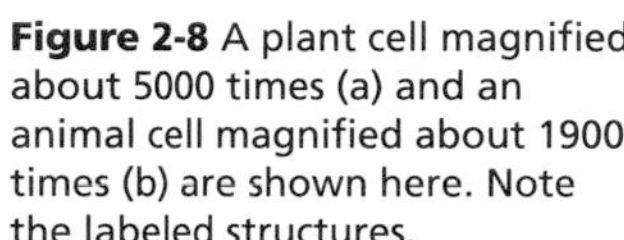

**Figure 2-8** A plant cell magnified about 5000 times (a) and an animal cell magnified about 1900 times (b) are shown here. Note the labeled structures.

a

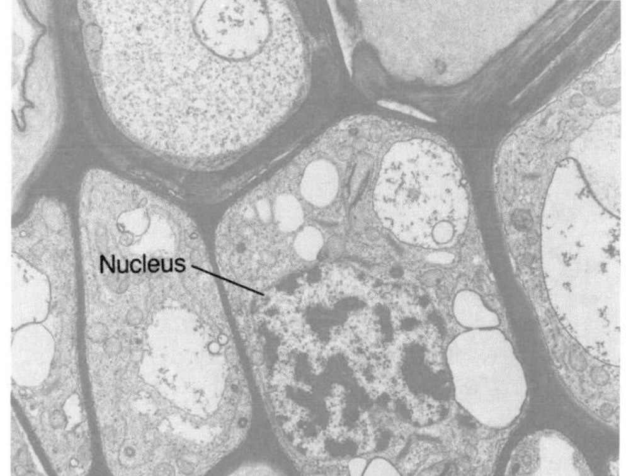


b

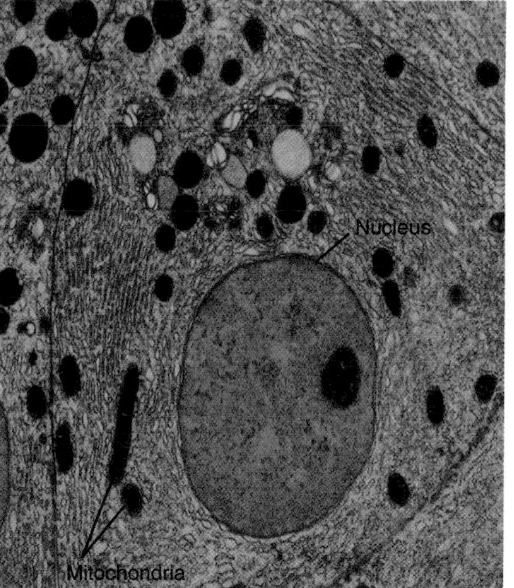




## TEACH

### Concept Development

▶ Explain that in unicellular organisms, food may be digested within certain vacuoles.

### Check for Understanding

Use the overhead projector and a transparency of a typical cell to point out each cell part and where it is located in the cell. Have students identify each part and give its function.

### Reteach

Write the function of each cell part on a 3 x 5 card. Have a student draw a card, read the function, name the cell part, and tell where the cell part is located in the cell.

### Independent Practice

**Study Guide**/*Teacher Resource Package*, p. 9. Use the Study Guide worksheet shown at the bottom of this page for independent practice.

**GLENCOE TECHNOLOGY**

**Videotape**

**The Secret of Life**

*What's in Stetter's Pond: The Basics of Life*

## APPLICATION 2

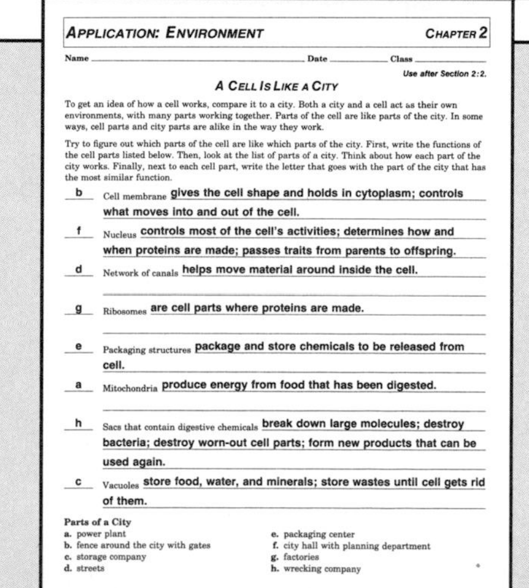

APPLICATION: ENVIRONMENT — CHAPTER 2

Name ______ Date ______ Class ______

*Use after Section 2:2.*

*A CELL IS LIKE A CITY*

To get an idea of how a cell works, compare it to a city. Both a city and a cell act as their own environments, with many parts working together. Parts of the cell are like parts of the city. In some ways, cell parts and city parts are alike in the way they work.

Try to figure out which parts of the cell are like which parts of the city. First, write the functions of the cell parts listed below. Then, look at the list of parts of a city. Think about how each part of the city works. Finally, next to each cell part, write the letter that goes with the part of the city that has the most similar function.

b Cell membrane gives the cell shape and holds in cytoplasm; controls what moves into and out of the cell.

f Nucleus controls most of the cell's activities; determines how and when proteins are made; passes traits from parents to offspring.

d Network of canals helps move material around inside the cell.

g Ribosomes are cell parts where proteins are made.

e Packaging structures package and store chemicals to be released from cell.

a Mitochondria produce energy from food that has been digested.

h Sacs that contain digestive chemicals break down large molecules; destroy bacteria; destroy worn-out cell parts; form new products that can be used again.

c Vacuoles store food, water, and minerals; store wastes until cell gets rid of them.

**Parts of a City**

a. power plant
b. fence around the city with gates
c. storage company
d. streets
e. packaging center
f. city hall with planning department
g. factories
h. wrecking company

## STUDY GUIDE 9

STUDY GUIDE — CHAPTER 2

Name ______ Date ______ Class ______

*CELL PARTS AND THEIR JOBS*

*In your textbook, read about cell parts and their jobs in Section 2:2.*

1. Label the parts of these two cells in the spaces provided.

mitochondrion
vacuole
cytoplasm
nucleolus
chromosome
nuclear membrane
canals
centriole
cell wall
packaging structures
cell membrane
ribosome
chloroplast

Cell A — Cell B

2. Read the descriptions of cell parts below and write in the name of the cell part. Use the color indicated to shade the pictures above.
   a. Use red for the part that gives the cell shape and holds the cytoplasm. cell membrane
   b. Use green for parts that make food. chloroplasts
   c. Use brown for the thick outer covering that protects and supports the cell. cell wall
   d. Use blue for the part that stores substances. vacuole
   e. Use black for parts that get energy from food. mitochondria
   f. Use purple for parts that carry hereditary information. chromosomes
   g. Use pink for the cell part that helps with cell reproduction. centriole
   h. Use orange for the parts that package and store chemicals. packaging structures
3. List two cell parts found only in a plant cell. chloroplasts, cell wall
4. Where in a cell do most chemical reactions take place? in the cytoplasm

## OPTIONS

**Application**/*Teacher Resource Package*, p. 2. Use the Application worksheet shown here to teach how a cell works.

# Lab 2-2 Cells

## Overview

In this lab, students use the light microscope to compare plant and animal cells. They will stain and observe cell parts.

**Objectives:** Upon completing this lab, students will be able to (1) **use a microscope** more efficiently, (2) **prepare** stained cells, (3) **compare** the parts of plant and animal cells that are visible with the light microscope.

**Time Allotment:** 40 minutes

## Preparation

▶ **Alternate Materials:** Lettuce can be substituted for *Elodea.*

**Lab 2-2 worksheet**/*Teacher Resource Package*, pp. 7-8.

## Teaching the Lab

▶ Caution students to rub their cheeks gently when obtaining cheek cells with the toothpicks. Also instruct them to properly dispose of the toothpick after use.

**Cooperative Learning:** Have students work in pairs. For more information, see pp. 22T-23T in the Teacher Guide.

▶ **Troubleshooting:** Too many cheek cells on the slide will result in not being able to see the cells.

**ASSESSMENT**

**Performance:** Provide students with prepared slides of other plant and animal cells. Ask students to observe the slides under low power and then high power. Students should determine which slides are of plants and which are of animals. Ask students to support their choices.

# Cells

# Lab 2–2

## Problem: How do animal and plant cells differ?

### Materials

light microscope
2 glass slides
2 coverslips
dropper
methylene blue stain
toothpick, flat type
*Elodea* leaf
water

| CELL PART | CHEEK CELL PARTS PRESENT | *ELODEA* CELL PARTS PRESENT |
|---|---|---|
| Cytoplasm | ✓ | ✓ |
| Nucleus | ✓ | ✓ |
| Chloroplast | | ✓ |
| Cell wall | | ✓ |
| Cell membrane | ✓ | ✓ |

### Skills

observe, compare, use a microscope

### Procedure

1. Copy the data table.
2. Put a drop of stain on a slide. Gently scrape the inside of your cheek with a toothpick. **CAUTION:** *Do not scrape hard enough to injure your cheek.*
3. Rub the toothpick in the stain. Break the toothpick in half and discard it as your teacher directs.
4. Cover the slide with a coverslip.
5. **Use a microscope:** Look at the cheek cells under low power, then high power.
6. Locate the nucleus, cytoplasm, and cell membrane. Fill in the table by putting a check mark in the box if the cell part can be seen.
7. Draw and label the nucleus, cytoplasm, and cell membrane of a cheek cell.
8. Prepare a slide of an *Elodea* leaf. Put an *Elodea* leaf in a drop of water on a slide. Add a coverslip.
9. Look at the *Elodea* cells under low power, then high power.
10. Locate the cell wall, chloroplasts, nucleus, and cytoplasm. Fill in the table.
11. Draw and label the cell wall, chloroplasts, nucleus, and cytoplasm of an *Elodea* cell.

### Data and Observations

1. Describe the shape of a cheek cell.
2. Describe the shape of an *Elodea* cell.
3. **Compare:** What parts did you see in both cells?
4. What parts are found in plant cells that are absent in animal cells?

### Analyze and Apply

1. What do the cell parts found only in plant cells do?
2. Is the nucleus always found in the center of the cell?
3. Which part of an animal cell gives shape to the cell?
4. Which parts of a plant cell give shape to the cell?
5. Why are stains such as methylene blue used when observing cells under the microscope?
6. **Apply:** Why don't animal cells have chloroplasts? (HINT: How do animals get energy?)

### Extension

**Observe** other plant and animal cells under the microscope. How are they different from cheek cells and *Elodea* cells? How are they alike?



## ANSWERS

### Data and Observations

1. round and uneven
2. rectangular
3. cytoplasm, nucleus, cell membrane
4. chloroplast, cell wall

### Analyze and Apply

1. Chloroplasts trap energy from the sun and make food; the cell wall protects the cell and gives it support.
2. no
3. cell membrane
4. cell wall
5. to make the cell parts visible
6. Animals get energy from the food they eat.

Your cells and the cells of every living thing around you are complex structures with many parts. Each part has a function that is important to the life of the cell. All of the living things in Figure 2-9 are made of cells. What other living things can you name that are made of cells?

**Figure 2-9** The maple tree (a), ladybug (b), horse (c), and snake (d) are all made of cells.

## Check Your Understanding

6. Which cell part is being described?
   (a) helps keep cytoplasm inside
   (b) controls most of the cell's activities
   (c) a liquid-filled space for storage
   (d) green parts of plants that trap energy from the sun
   (e) clear, jellylike material in which most of the cell's chemical reactions take place
7. Name two cell parts found in plant cells that are not found in animal cells.
8. Why are mitochondria called "powerhouses" of the cell?
9. **Critical Thinking:** How do mitochondria and chloroplasts differ?
10. **Biology and Writing:** Write three sentences describing the cell parts that make up a wooden table.



## TEACH

### Check for Understanding

Have students respond to the first three questions in Check Your Understanding.

### Reteach

*Reteaching/Teacher Resource Package*, p. 5. Use this worksheet to give students additional practice comparing plant cell and animal cell structures.

**Extension:** Assign Critical Thinking, Biology and Writing, or some of the **OPTIONS** available with this lesson.

## 3 APPLY

### Connection: Language Arts

Have students list the cell parts and use a dictionary to find the origin and meaning of each word.

## 4 CLOSE

### Make a Model

Have students work in pairs with clay and paints to make a model of either a plant or an animal cell.

### Answers to Check Your Understanding

6. (a) cell membrane, (b) nucleus, (c) vacuole, (d) chloroplast, (e) cytoplasm
7. cell wall and chloroplast
8. they produce so much energy
9. Chloroplasts trap energy from the sun and change it into food energy. Mitochondria produce energy from food that has been digested.
10. Answers will vary. Students should include cell walls in their answers.

## RETEACHING 5

RETEACHING CHAPTER 2

Name Date Class

Use with Section 2:2.

PLANT CELL AND ANIMAL CELL STRUCTURES

chloroplast, cell wall, cell membrane, cytoplasm, vacuole, mitochondrion, ribosome, canals, chromosome, nuclear membrane, nucleus, nucleolus, sacs with digestive chemicals, structures that package chemicals, centriole, Plant Cell, Animal Cell

## GLENCOE TECHNOLOGY

**Videodisc**

**The Secret of Life**

*Diffusion*

*Osmosis*

## 2:3 Special Cell Processes

### PREPARATION

#### Materials Needed

Make copies of the Study Guide, Critical Thinking/Problem Solving, and Reteaching worksheets in the *Teacher Resource Package*.

▶ Gather beakers and instant coffee for the Focus activity. Have balloons and vanilla extract on hand for the Mini Lab.

#### Key Science Words

| | |
|---|---|
| diffusion | organ |
| osmosis | organ system |
| tissue | organism |

#### Process Skills

In the Mini Lab, students will experiment.

### 1 FOCUS

▶ The objectives are listed on the student page. Remind students to preview these objectives as a guide to this numbered section.

#### ACTIVITY/Hands-On

To see diffusion, divide the class into groups of two and give each group a beaker of water and a small amount of instant coffee. Have students add the coffee to the water, leave it undisturbed, and observe what happens.

#### Guided Practice

Have students write down their answers to the margin question: movement of a substance from where there is a larger amount to where there is a smaller amount.

## 2:3 Special Cell Processes

**Objectives**

6. **Explain** the process of diffusion in a cell.
7. **Describe** the process of osmosis in a cell.
8. **Communicate** how cells, tissues, organs, and organ systems are organized.

**Key Science Words**

diffusion
osmosis
tissue
organ
organ system
organism

Cells need certain substances to stay alive. These substances include food and oxygen. How does a substance, such as oxygen, get through the cell membrane and into the cell?

### Diffusion

Imagine you have a box of marbles. The box is divided in half by a strip of cardboard with an opening. Figure 2-10 shows the box with marbles in it. Notice that all of the marbles are on one side of the cardboard strip.

The marbles are packed very tightly into the box. Some of the marbles roll to the other side of the box through the opening in the cardboard strip. The marbles keep rolling until each side of the box has about equal numbers of marbles. How did this happen?

If you walk near the chemistry lab in your school, you might smell a gas made during a chemistry experiment. How did the gas move from the lab into the hall where you are? The gas moved from where there was a large amount of it, in the lab, to where there was a small amount of it, in the hall. Gas molecules, like marbles, can move about. The movement of a substance from where there is a large amount of it to where there is a small amount of it is called **diffusion** (dif YEW zhun).

What is diffusion?

How does oxygen diffuse into cells? Doesn't the cell membrane get in the way? Think about the box divided by the cardboard strip. Marbles could roll to the other side of the box through the opening in the cardboard strip.

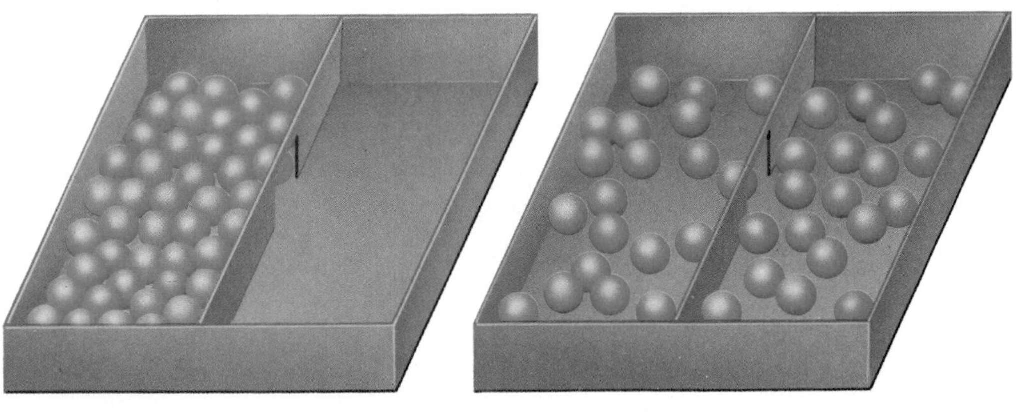

**Figure 2-10** Before the marbles roll through the opening in the cardboard strip, they are all on one side of the box (left). The box on the right shows what happens after the marbles roll from one side to the other.



### OPTIONS

#### Science Background

Cell membranes are selectively permeable. Water passes through the membrane but a substance dissolved in the water may or may not pass through. Carrier molecules in the cell membrane permit specific molecules on one side of the membrane to pass through to the other side. The process of diffusion is made easier, or facilitated, by the carrier molecules, so the process is called facilitated diffusion.

**What Are Diffusion and Osmosis?**/*Lab Manual,* pp. 9-12. Use this lab as an extension to teaching cell processes.

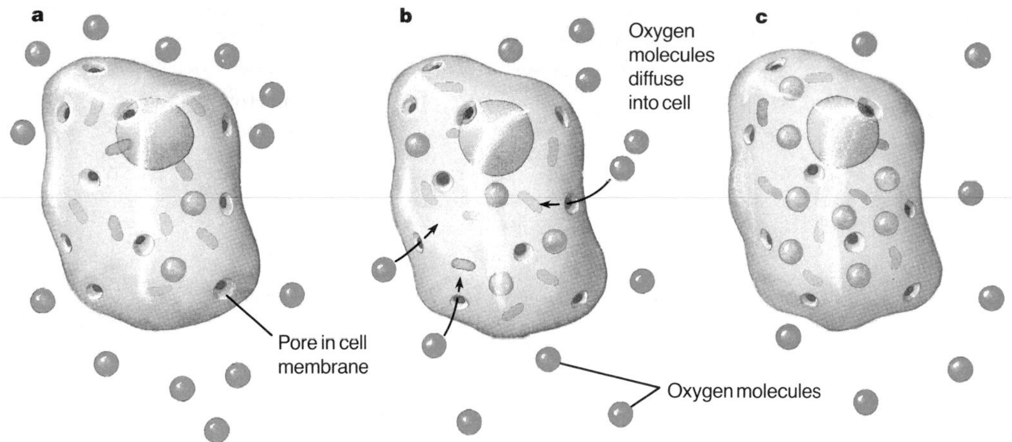


Membranes also have openings, called pores. Some molecules, such as oxygen, can pass through these pores, Figure 2-11a. The movement of molecules through the pores of a membrane is like the movement of marbles through the opening in the cardboard strip.

In Figure 2-11:

1. There is a large number of oxygen molecules outside the cell.
2. There is a small number of oxygen molecules inside the cell.
3. The oxygen moves from the outside to the inside by diffusion, as shown in Figure 2-11b. The oxygen moves from where there is a large amount of it to where there is a small amount of it. If we look at the cell later, we see about equal numbers of oxygen molecules inside and outside the cell, Figure 2-11c. Remember that at a later time, there were about equal numbers of marbles on both sides of the box.

**Figure 2-11** Oxygen diffuses into the cell through pores in the membrane.

## Osmosis

Water molecules also move across membranes. Like oxygen, they move from where there is a large number of molecules to where there is a small number of molecules. The movement goes on until the number of water molecules is about the same on both sides of the membrane. The movement of water across the cell membrane is called **osmosis** (ahs MOH sus).

### Mini Lab

**What Is Diffusion?**

**Experiment:** Put 10 drops of vanilla extract inside a balloon. Blow up the balloon, tie it, and place it inside a box. Close the box. After two hours, open the box. What do you smell? *For more help, refer to the* ***Skill Handbook,*** *pages 704-705.*



# 2 TEACH

### MOTIVATION/Demonstration

Place one slice of cucumber, potato, carrot, or celery in salt water and one slice in plain water. Have students observe how the slice in salt water wilts or softens.

### Independent Practice

**Study Guide**/*Teacher Resource Package*, p. 10. Use the Study Guide worksheet shown at the bottom of this page for independent practice.

### Mini Lab

Students will smell vanilla in the box. The vanilla diffuses through the balloon and into the box.

### ASSESSMENT

**Performance:** Have students examine the effects of high and low temperatures on diffusion. Provide hot water, cold water, and food coloring. Have students add the food coloring to the different temperatures of water. Students should time how long it takes for the molecules to become evenly distributed. This data can be recorded in a graph.

### Student Journal

Have students write a paragraph that explains diffusion and osmosis and make a diagram to illustrate the processes.

### CRITICAL THINKING 2

CRITICAL THINKING/PROBLEM SOLVING — CHAPTER 2

Name ______ Date ______ Class ______

**WHAT HAPPENED TO DINNER?** *Use after Section 2:3.*

Jeff's family planned to attend his brother's basketball game on a school night. His mother cooked spaghetti sauce. Jeff offered to make the salad and cook the spaghetti. First, Jeff washed the lettuce. Then, he sliced tomatoes, cucumbers, and radishes. Next, Jeff put some pepper, herbs, oil, and vinegar on the vegetables. Finally, he tossed the salad, covered it, and placed it in the refrigerator. Jeff read the directions for cooking a pound of spaghetti. He was surprised to see that he needed five quarts of water to boil the spaghetti. Jeff read the label on the box of spaghetti. He found that the food was made from ground-up plants. Jeff slid the dry spaghetti into boiling water and waited while it cooked. He drained the spaghetti. It took up much more space now than it had before it was cooked. Almost all the water in the pot was gone. What had happened to the water and the dry spaghetti? Jeff took the salad from the refrigerator. The vegetables were no longer crisp. There was more liquid in the salad bowl than he had added. What had happened to the vegetables?

**Analyzing the Problem**

1. How did the dry spaghetti increase in size? **The spaghetti absorbed water.**
2. Where did the liquid in the salad bowl come from? **from the vegetables**

**Solving the Problem**

1. When Jeff made the salad, were more water molecules inside or outside the plant cells? **inside the cells**
2. What happens when the number of water molecules is greater inside a cell than outside? **Water moves out of the cell.**
3. How does your answer to number 2 help to explain what happened to the vegetables in the salad? **Water moved out of the cells by osmosis because there was more water inside the cells of the plants making up the salad than outside.**
4. When Jeff put dry spaghetti into boiling water, was there more water inside or outside the food? **outside the food**
5. What happens when the number of water molecules is greater in one area than another area? **Water moves from the area in which there are more water molecules to the area in which there are fewer water molecules.**
6. How does your answer to number 5 help to explain what happened to the spaghetti? **The water molecules moved from where there were more of them (the water in the pot) to where there were fewer of them (spaghetti).**

### STUDY GUIDE 10

STUDY GUIDE — CHAPTER 2

Name ______ Date ______ Class ______

**SPECIAL CELL PROCESSES**

*In your textbook, read about diffusion and osmosis in Section 2:3.*

1. The first picture below, labeled *Before*, shows a cell surrounded by oxygen molecules before diffusion takes place. Each of the small black dots represents an oxygen molecule. Which of the three pictures labeled *After* shows where these oxygen molecules would be found after diffusion takes place? Circle your answer. **Students are being asked to show *net* diffusion. Diffusion continues after equilibrium is reached, but net diffusion does not.**

Before | After | After | After

2. What is diffusion? **the movement of a substance from where there is a large amount of it to where there is a small amount of it**
3. How do molecules get through the cell membrane? **The membrane has pores, or small openings, through which the molecules can pass.**
4. What is osmosis? **Osmosis is the movement of water across the cell membrane.**
5. Which way would the water molecules move in the following situations?
   a. cucumber slice is placed in salt water **out of the cucumber**
   b. salt is poured on a snail **out of the snail**
   c. vegetables are sprinkled with water **into the vegetables**
   d. potato slice is placed in pure water **into the potato**
6. Circle the letter in front of the sentence that best explains the process of osmosis.
   (a.) Osmosis is the movement of water into or out of a cell from where it is in large amounts to where it is in small amounts.
   b. Osmosis is the movement of water into or out of a cell from where it is in small amounts to where it is in large amounts.
   c. Osmosis is the movement of salt into or out of a cell from where it is in large amounts to where it is in small amounts.

## OPTIONS

**Critical Thinking/Problem Solving/** *Teacher Resource Package*, p. 2. Use the Critical Thinking/Problem Solving worksheet shown here for students to understand that foods are affected by osmosis and diffusion.

## TEACH

### Concept Development

▶ Point out that the concentration of water inside the cucumber would be greater than the concentration of water in the salt water.

▶ Remind students that the cell is the simplest level of organization that is alive.

### Guided Practice

Have students write down their answers to the margin question: Cells are surrounded by water molecules. Osmosis allows movement in and out so that the number of molecules inside and the number outside are about the same.

### Check for Understanding

Have students list examples of osmosis and diffusion. Suggest smells from a bakery and spraying fruits and vegetables with water.

### Reteach

Place a few raisins in a beaker of water and let the beaker stand overnight. Have students explain the process observed.

### Independent Practice

**Study Guide**/*Teacher Resource Package*, p. 11. Use the Study Guide worksheet shown at the bottom of this page for independent practice.

### Student Journal

Have students draw a diagram to show the relationship among cells, tissues, organs, and organ systems in a plant.

### GLENCOE TECHNOLOGY

**Videodisc**

Biology: The Dynamics of Life

Animation: *Passive Transport*

Disc 1, Side 1, Ch. 24

**Figure 2-12** Water molecules move into and out of cells by osmosis. If there are more water molecules inside than out, the water molecules will move out of the cell.

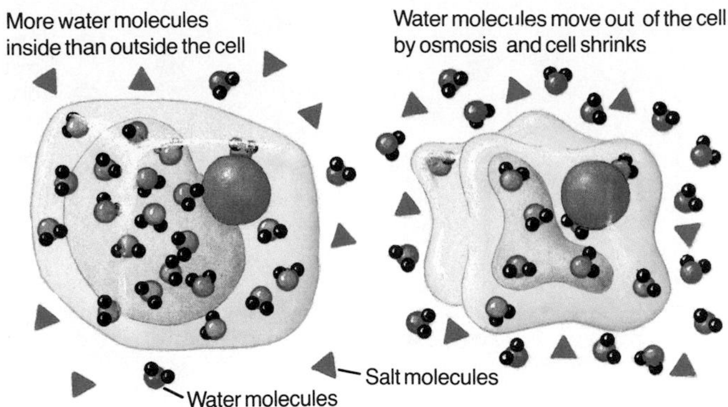


Why is osmosis important to cells?

Osmosis is important to cells because they are surrounded by water molecules. The number of water molecules inside and outside the cell must be about the same. When there are more water molecules outside the cell, they move into the cell. When there are more water molecules inside the cell, they move out.

Suppose you placed some cucumber slices into salt water. What do you think would happen? (HINT: Cucumber cells have nearly pure water inside. Salt water is not pure water because it has salt in it. So, there are more water molecules in a cucumber than in salt water.) The water molecules inside the cucumber would move out, as shown in Figure 2-12. When cells of a plant lose too much water, the plant wilts. What do you think would happen if your cells lost too much water?

**Bio Tip**

**Consumer:** Osmosis affects the taste of foods. When steak is salted before it is cooked, water moves out of its cells, making the steak dry and tasteless.

## Organization

Living things are organized in special ways. Some living things are made of only one cell. That one cell carries out all the activities necessary for life. It responds to the environment, reproduces, and uses energy.

In living things made of many cells, the cells are organized into groups, Figure 2-13. The cells of each group do special jobs. For example, cells that line the small intestine make chemicals for digestion. Groups of cells, such as those that line the intestine, are called tissues. A **tissue** is a group of similar cells that work together to carry out a special job. Bone, muscle, blood, and nerve are kinds of tissues found in animals. Bark and outer surfaces of leaves are kinds of tissues in plants.



## OPTIONS

### Science Background

Explain that pressure within a plant cell is called turgor pressure. Turgor pressure is the chief means of support in most nonwoody plants. When plant structures lose too much water, they collapse. The loss of turgor pressure, called plasmolysis, causes plants to wilt.

**Organization in Living Things**/*Transparency Package*, number 2b. Use color transparency 2b as you teach how living things are organized.

### STUDY GUIDE 11

STUDY GUIDE — CHAPTER 2

Name ________ Date ________ Class ________

SPECIAL CELL PROCESSES

*In your textbook, read about the organization of cells, tissues, and organs in Section 2:3.*

7. Place a check mark in the correct column for each phrase or diagram on the left.

| | Tissue | Organ | Organ system |
|---|---|---|---|
| Groups of cells all doing the same job | ✓ | | |
| Stomach, mouth, and intestines all working together | | | ✓ |
| Made up of a group of tissues | | ✓ | |
| [diagram] | ✓ | | |
| Small intestine | | ✓ | |
| [diagram] | | ✓ | |
| Leaf | | ✓ | |
| Muscle | ✓ | | |
| Bark | ✓ | | |
| Outer surface of a leaf | ✓ | | |

8. How is an organ different from a tissue? A tissue is made up of only one kind of cell. An organ is made up of many different kinds of tissues that work together to perform a job. Because the different tissues in an organ are each made up of a different kind of cell, an organ is made up of different kinds of cells.

9. What is an organism? An organism is a living thing.

10. Compare a one-celled organism with a many-celled organism. In a one-celled organism, a single cell carries out all the activities of life. In a many-celled organism, cells are organized into groups and the cells of each group do special jobs.

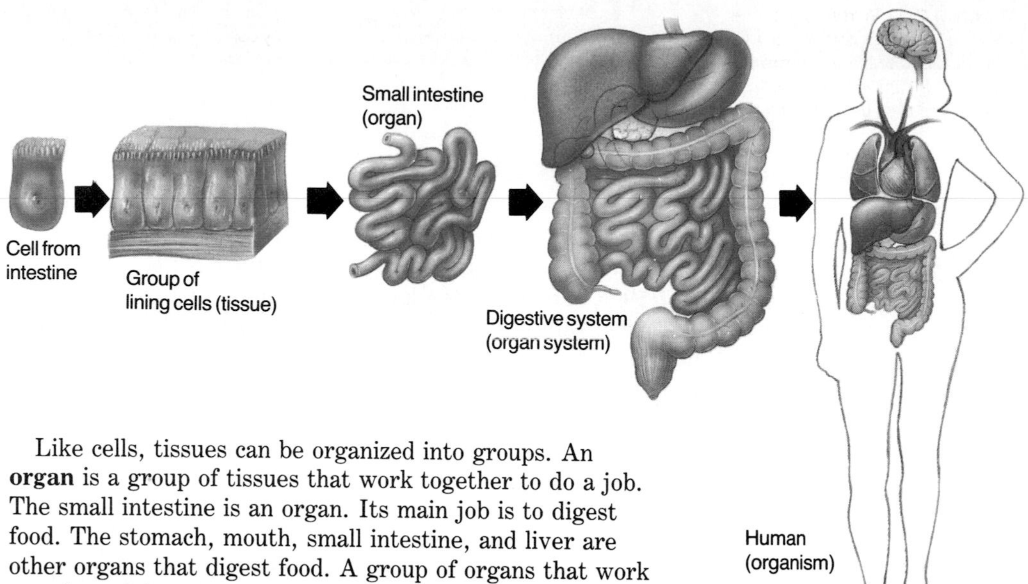


Figure 2-13 Living things (organisms) are organized into cells, tissues, organs, and organ systems.

Like cells, tissues can be organized into groups. An **organ** is a group of tissues that work together to do a job. The small intestine is an organ. Its main job is to digest food. The stomach, mouth, small intestine, and liver are other organs that digest food. A group of organs that work together to do a certain job is an **organ system.** The organs that work together to digest food are called the digestive system.

All the organ systems working together make up an organism (OR guh niz um). An **organism** is a living thing. Many organisms are made of only one cell. Yet, they have all the features of living things. Other organisms have organs and organ systems. Humans are organisms with several organ systems. Among other systems, humans have circulatory, reproductive, and nervous systems. These systems will be explained in Chapters 11, 15, and 24.

What is the relationship among cells, tissues, organs, and organ systems?

## Check Your Understanding

11. How does diffusion make it possible for you to smell different odors?
12. In osmosis, which substance diffuses?
13. How are tissues and organs different?
14. **Critical Thinking:** What would happen if you placed a slice of cucumber in distilled water? (HINT: Distilled water is purer than the water inside the cucumber cells.)
15. **Biology and Reading:** What are the living parts that make up an organ system?

## RETEACHING 6

**RETEACHING** — CHAPTER 2

Name ______ Date ______ Class ______

*Use with Section 2:3.*

### CELL PROCESSES

The cytoplasm of *Elodea* cells is composed of about 70 percent water molecules and 30 percent other kinds of molecules.

*Read the three examples given below.*

A. *Elodea* cells are put into a liquid that is 50 percent water.
B. *Elodea* cells are put into a liquid that is 70 percent water.
C. *Elodea* cells are put into a liquid that is 100 percent water.

50% — A; 70% — B; 100% — C; Elodea cells

*Study the predictions below and select the example above that matches the prediction. On the line to the left, write the letter of the example you select.*

A 1. The *Elodea* cells in the liquid will shrink.
C 2. The *Elodea* cells in the liquid will swell.
B 3. The *Elodea* cells will not change in size.

The diagram to the right shows a cellophane bag containing molasses. The bag is in a beaker of water. An open glass tube extends from the bag. The molasses comes up a short distance in the tube.

Glass tube; Cellophane bag; Water; Molasses

*Use the diagram to answer the questions below.*

a 1. Movement of water molecules across the membrane is the result of (a) osmosis. (b) breathing. (c) the glass tube. (d) air pressure.
b 2. After 24 hours, the size of the cellophane bag will probably be (a) smaller. (b) larger. (c) the same.
a 3. After 24 hours, the beaker will contain (a) water only. (c) both water and molasses. (b) molasses only. (d) neither water nor molasses.
c 4. After 24 hours, the cellophane bag will contain (a) water only. (c) both water and molasses. (b) molasses only. (d) neither water nor molasses.
b 5. After 24 hours, the level of liquid in the glass tube (a) will be lower. (b) will be higher. (c) will be the same.

## STUDY GUIDE 12

**STUDY GUIDE** — CHAPTER 2

Name ______ Date ______ Class ______

### VOCABULARY

*Review the new words in Chapter 2 of your textbook. Then complete the puzzle. The letters in the dark boxes will summarize what you are studying in Chapter 2.*

1. CYTOPLASM
2. CHLOROPLAST
3. REPRODUCE
4. MITOCHONDRIA
5. TISSUE
6. NUCLEAR MEMBRANE
7. CELLULAR RESPIRATION
8. PRODUCER
9. DIFFUSION
10. CELL MEMBRANE
11. ORGAN SYSTEM
12. CELL
13. OSMOSIS
14. VACUOLE
15. CENTRIOLE
16. DEVELOPMENT
17. ORGANISM
18. CELL WALL
19. ADAPTATION
20. CHROMOSOME
21. RIBOSOME
22. NUCLEUS
23. ORGAN
24. CONSUMER
25. NUCLEOLUS

1. clear jellylike material in cells
2. cell part that contains chlorophyll
3. form offspring similar to parents
4. cell parts that give off energy
5. group of similar cells carrying out a job
6. structure that surrounds the nucleus
7. process by which food is broken down and energy is released
8. living thing that makes its own food
9. movement of a substance from where there is a large amount to where there is a small amount of it
10. controls what moves into cells
11. group of organs working together
12. basic unit of all living things
13. movement of water across a membrane
14. stores water, food, and minerals
15. part near the nucleus that helps with reproduction
16. all the changes that occur as a living thing grows
17. living thing
18. thick outer covering of a plant cell
19. trait that helps a living thing survive
20. cell part with information that determines a living thing's traits
21. cell part where protein is made
22. controls the activities of the cell
23. group of tissues that work together
24. living thing that eats other living things
25. smaller body inside the nucleus that helps make ribosomes

### Guided Practice

Have students write down their answers to the margin question: cells make up tissues, tissues make up organs, and organs make up organ systems.

### Independent Practice

**Study Guide**/*Teacher Resource Package*, p. 12. Use the Study Guide worksheet shown at the bottom of this page for independent practice.

### Check for Understanding

Have students respond to the first three questions in Check Your Understanding.

### Reteach

**Reteaching**/*Teacher Resource Package*, p. 6. Use this worksheet to give students additional practice with cell processes.

**Extension:** Assign Critical Thinking, Biology and Reading, or some of the **OPTIONS** available with this lesson.

## 3 APPLY

### Brainstorming

Ask students how drying, smoking, and salting preserve foods.

## 4 CLOSE

### Convergent Question

Ask why people cannot use ocean water for drinking.

### Answers to Check Your Understanding

11. The molecules diffuse from where they are in large amounts to where they are in small amounts.
12. water
13. An organ is made up of tissues.
14. The slice of cucumber would swell up as water enters.
15. cells, tissues, and organs

APPLYING TECHNOLOGY

### Purpose

Students will determine whether bean plants can be grown when watered with salt water. They will use their understanding of osmosis to predict and explain their results.

### Process Skills

Interpret data, experiment, form hypotheses, separate and control variables, observe, infer

### Time Required

The initial marking of trays, examining of plants, and recording their appearance the first day will take 15 minutes. Daily observing, recording, and watering will take 10 minutes.

### Possible Problems

Be sure that students use the proper solution when watering each tray. Overwatering may result in mold growth destroying the young plants. Weekends pose a problem with plants drying out. To prevent this from happening, remove the trays from their light source during the weekend, and place trays in a cool area of the room. Add a mulch of peat moss over the top of the soil for the weekend or make sure that plants are well watered on Friday.

### Teaching Strategies

- You may want students to work in small groups of three or four to conserve materials.
- Start the young plants from seed at least ten days prior to the start of this activity. Plant in suitable small trays. Disposable peat moss pots are available from garden shops and are inexpensive.
- Popcorn, sunflower (in pet food section of a supermarket), mustard (in spice section of a supermarket), or grass are four easily obtained types of seed for planting. Seeds are also available from biological supply houses (use radish or pea).
- If time does not permit you to start young plants from seeds, use mature plants such as grass from a vacant lot or forest area or young *Coleus* plants from a garden shop.

TECH PREP APPLYING TECHNOLOGY

# Salty Plants— How Will They Grow?

All plants need water in order to live and grow. Farmers usually depend on rainfall to provide water to their growing crops. But, what if there is too little rain available? Then, farmers must supply their crops with water through irrigation. The problem with irrigation, however, is that sometimes there isn't a large enough water supply available. Solution: there is plenty of water in Earth's oceans. Why not use it to irrigate plants?

## Identifying the Problem

You are a botanist who has been hired by a large group of farmers along the west coast of the United States. Your job is to determine whether they can use water from the Pacific Ocean to water their crops. Your plan is to conduct an experiment to see whether bean plants can survive when watered with salt water. You will use only one variable, which will be the adding of salt water (seawater) to one group of plants. Also, remember that you will need a control. It will be the adding of freshwater to a second group of plants. Form a hypothesis as to how you believe salt water will affect the growth of plants.

### Technology Connection

The need for freshwater has resulted in technology that removes salt from ocean water. Desalination plants built near oceans are helping to supply freshwater to areas that are in need. Cities along the coast of California, as well as in Israel and Saudi Arabia, have already built these plants. They remove the salt from ocean waters by either distillation or reverse osmosis techniques.

## Collecting Information

Do some library research to determine what the difference is between "fresh" water and "sea or salt" water. Once you have this information, you are almost ready to begin your experiment. One last item to think about during your research is whether osmosis might be occurring between the cells of plants and the water within the soil in which they are growing.

42

### Technology Connection

Desalination is based on distillation of seawater. The process is expensive because it requires large amounts of energy to heat and cool the seawater in order to evaporate and then condense it. A newer technology called reverse osmosis is also being used to convert seawater to freshwater. In reverse osmosis, salt water is forced through a series of membranes. The membranes allow water but not the salt to pass through. Again, energy is required. However, less energy is needed than with distillation.

## Carrying Out an Experiment

1. Obtain two trays containing young plants.
2. Label one tray *experimental* and the other tray *control.*
3. Design a data table in which you record your daily observations. Your table should include the number of days that you conduct the experiment. Mark today as *Day 1.*
4. Add only salt water to the experimental tray as needed and freshwater (tap water) to the control tray as needed. Keep the soil in both trays moist.
5. Place the trays in a lighted area.
6. Observe the plants each day for a total of at least ten days. Record all observations in your table. Diagrams may be helpful to show any differences between the two groups.

## Assessing Your Results

Compare and contrast the size and general appearance of plants receiving freshwater with those receiving salt water. Based on your observations, what conclusions might you make about watering or irrigating bean plants with salt water? Based on your observations, what advice will you give to the farmers who hired you to solve their problem? Based on your understanding of osmosis, what conclusions might you make about the effect of salt water on plants? What was your control in this experiment and what was the variable? Why did you need a control? Was your hypothesis supported by your experimental findings? Explain.

### Career Connection

- **Agricultural Supply Salesperson** Sells farming equipment, agricultural chemicals, and other supplies to farmers; works in stores or in large dealerships
- **Crop Scientist** Works to increase the yield of crops and to improve farming methods. Their concerns may be about weed control, watering problems, and crop development.
- **Water Resources Engineer** Ensures that water is available for drinking, farming, and industrial use; works on problems associated with conserving water and preventing flooding

### Career Connection

**Career Path**

**Agricultural Supply Salesperson** A high school diploma is needed; experience in 4-H would be helpful.

**Crop Scientist** A four-year college degree is needed; the degree could be in agronomy, biology, or chemistry.

**Water Resources Engineer** A college degree beyond four years is required; courses in engineering and science are needed.

**Career Issue**

When freshwater is pumped from below ground to irrigate fields, salt water often flows in to replace the lost freshwater. This is a major problem along ocean coastal areas. This process is called saltwater intrusion. Have students find out and explain why it can be harmful to crops.

▶ Seawater is 3.5% salt. Prepare solutions of "seawater" for students by mixing 3.5 g of table salt with 96.5 mL of tap water. You may want to have students prepare their own salt water solutions. If balances are in short supply, determine the mass of 3.5 grams of salt, and convert this mass to teaspoons. Advise students to use this amount of salt per 100 mL of water.

▶ Soil type is not critical. Sand may be substituted for soil. Window light will also be adequate for the length of time needed to complete the experiment.

### Conclusions

Students will observe differences between those plants watered with freshwater and those receiving salt water. Plants receiving salt water will appear stunted and shriveled compared to those plants receiving freshwater. Many of the plants receiving salt water will die.

### Answers to Questions

Student answers may vary. In general, those plants receiving salt water may fail to grow or will look sick. Salt water is not recommended for watering of plants. It damages plants because of osmosis. Salt water has a lower concentration of water than water within a plant's cells. Thus, water moves from high concentration (within the plant) toward an area of lower concentration (outside the plant), resulting in damage to the plant. The control was the use of freshwater; the variable was the use of salt water. A control will allow for a basis of comparison and will show that no harmful effects are due to the water.

### ASSESSMENT

**Process** Have students write out the procedural steps that would be needed to determine experimentally the lowest possible salt solution concentration that will still allow for growth of plants.

# CHAPTER 2 REVIEW

## Summary

Summary statements can be used by students to review the major concepts of the chapter.

## Key Science Words

All boldfaced terms from the chapter are listed.

## Testing Yourself

### Using Words

1. chromosome
2. mitochondria
3. organ
4. chloroplast
5. diffusion
6. cell
7. cellular respiration
8. organ system
9. osmosis

### Finding Main Ideas

10. p. 29; Living things respond to changes in the environment.
11. p. 4l; An organ system is a group of organs working together.
12. p. 35; Centrioles help with cell reproduction.
13. p. 39; Molecules pass through pores in a membrane.
14. p. 27; Cellular respiration is the process by which food is broken down and energy is released.
15. p. 30; A molecule is the smallest part of a compound.
16. p. 30; The elements carbon, hydrogen, oxygen, nitrogen, sulfur, and phosphorus make up 97 percent of living matter.
17. p. 31; Robert Hooke is the scientist who first observed cells.
18. p. 32; Chromosomes carry information that determines what traits a living thing will have.
19. p. 40; The water molecules are concentrated inside the cell and move out into the salt water where there is less concentration of water molecules.

# Chapter 2

# Review

## Summary

### 2:1 Living Things and Their Parts

1. Living things have eight features in common. They reproduce, grow, develop, need food, use energy, have cells, respond, and are adapted to their environments.
2. Living things are made of matter. Six elements make up over 97 percent of the matter in living things.
3. The cell theory states: all living things are made of one or more cells; cells are the basic units of structure and function in living things; all cells come from other cells.

### 2:2 Cell Parts and Their Jobs

4. The main parts of the cell are the cell membrane, nucleus, chromosomes, cytoplasm, ribosomes, mitochondria, and vacuoles.
5. The jobs of the cell parts include protection, the making of energy, the moving of materials into and out of the cell, and reproduction.

### 2:3 Special Cell Processes

6. Materials move into or out of a cell by diffusion.
7. The movement of water molecules across the cell membrane is osmosis.
8. Living things are organized into cells, tissues, organs, and organ systems.

## Key Science Words

adaptation (p. 29)
cell (p. 27)
cell membrane (p. 32)
cellular respiration (p. 27)
cell wall (p. 35)
centriole (p. 35)
chloroplast (p. 35)
chromosome (p. 32)
consumer (p. 27)
cytoplasm (p. 32)
development (p. 26)
diffusion (p. 38)
mitochondria (p. 34)
nuclear membrane (p. 32)
nucleolus (p. 32)
nucleus (p. 32)
organ (p. 41)
organism (p. 41)
organ system (p. 41)
osmosis (p. 39)
producer (p. 27)
reproduce (p. 26)
ribosome (p. 34)
tissue (p. 40)
vacuole (p. 35)

## Testing Yourself

### Using Words

*Choose the word from the list of Key Science Words that best fits the definition.*

1. cell part with information that determines a living thing's trait
2. cell parts that produce energy when food is broken down
3. tissues that work together to do the same job
4. green part inside a plant cell
5. movement of substances, such as oxygen, into a cell
6. basic unit of living things
7. process in which food is broken down and energy is released
8. group of organs working together
9. diffusion of water into or out of a cell

# Review

## Testing Yourself *continued*

**Finding Main Ideas**

*List the page number where each main idea below is found. Then, explain each main idea.*

10. what living things respond to
11. what an organ system is
12. the function of centrioles
13. how molecules pass through a membrane
14. how living things get energy from food
15. what a molecule is
16. the six elements that make up over 97 percent of living matter
17. the scientist who first observed cells
18. the job of chromosomes
19. why cucumbers wilt in salt water

**Using Main Ideas**

*Answer these questions by referring to the page number after the question.*

20. How are growing and developing different? (p. 26)
21. Why are humans called organisms? (p. 41)
22. Why will a stalk of celery put in salt water begin to wilt? (p. 40)
23. How do your school and a cell compare? Match the job of each cell part with a part of your school that has a similar job. (pp. 32-35)

| | |
|---|---|
| (a) nucleus | 1. main office |
| (b) vacuole | 2. cafeteria |
| (c) chloroplasts | 3. furnace room |
| (d) mitochondria | 4. lockers |

24. How is osmosis a special kind of diffusion? (p. 39)
25. What is the function of pores in a cell membrane? (p. 39)
26. What characteristics of living things does a cell have? (pp. 26, 27, 29)

## Skill Review

*For more help, refer to the **Skill Handbook**, pages 704-719.*

1. **Make and use tables:** Make a table to show the features of living things shown by a cat, guppy, bird, and maple tree. In your table give examples for each feature of living things shown by these organisms.
2. **Classify:** Classify the following as elements or compounds: hydrogen, oxygen, water, carbon, carbon dioxide, nitrogen, and phosphorus.
3. **Formulate a model:** Construct and label a three-dimensional model of a plant or animal cell.
4. **Infer:** Compare the growth of a single-celled organism with that of a many-celled organism. What is the difference between the two types of growth?

## Finding Out More

**Critical Thinking**

1. Explain why some cells have more mitochondria than others.
2. Make a time line showing the events that led to the cell theory.

**Applications**

1. What kinds of tissues and organs are you eating when you eat a green salad?
2. Report on ways that desert organisms such as cacti and camels conserve and use water as adaptations for survival.



### Using Main Ideas

20. Growing is increasing in size. Developing is changing in shape and form.
21. The human body is a living thing made up of organ systems such as the circulatory, reproductive, and nervous systems.
22. Water will move across a membrane from where it is in large amounts to where it is in small amounts.
23. (a) Nucleus and (1) main office have similar jobs, as do (b) vacuole and (4) lockers, (c) chloroplasts and (2) cafeteria, and (d) mitochondria and (3) furnace room.
24. Diffusion is the movement of molecules, but osmosis is the movement of water molecules.
25. Pores in a membrane allow molecules to pass through.
26. A cell has all the characteristics of living things, including respiration, growth, and reproduction.

### Skill Review

1. Have students use the living things as headings across the top, and list the features of living things down the side.
2. Hydrogen, oxygen, carbon, nitrogen, and phosphorus are elements; water and carbon dioxide are compounds.
3. Students should be creative and may use clay, yarn, pipe cleaners, and even edible items. Models should have a key to the cell parts.
4. Many-celled organisms grow by increasing the number of cells; single-celled organisms do not.

### Finding Out More

#### Critical Thinking

1. Cells that move use more energy and have more mitochondria. A muscle cell would have more mitochondria than a skin cell.
2. The time line should begin with 1665. Students should research the cell theory to fill in the time line.

#### Applications

1. Leaves (lettuce, spinach) and roots (carrots) are organs made up of several tissues.
2. Cacti spines are modified leaves that give off almost no water. Camels are able to store quantities of water within their bodies.

 **Chapter Review**/*Teacher Resource Package*, pp. 6-7.

 **Chapter Test**/*Teacher Resource Package*, pp. 8-10.

**Chapter Test**/*Computer Test Bank*

# Classification

## PLANNING GUIDE

| CONTENT | TEXT FEATURES | TEACHER RESOURCE PACKAGE | OTHER COMPONENTS |
|---|---|---|---|
| (1/2 day)<br>3:1 Why Things Are Grouped<br>Classifying in Everyday Life<br>How Grouping Helps Us | Lab 3-1: *Classifying,* p. 49<br>Career Close-Up: *Wildlife Photographer,* p. 50<br>Check Your Understanding, p. 50 | Reteaching: *Grouping Living Things,* p. 7<br>Study Guide: *Why Things Are Grouped,* pp. 13-14<br>Lab 3-1: *Classifying,* pp. 9-10 | **Laboratory Manual:**<br>*How Can Paper Objects Be Grouped?* p. 17<br>**STVS:** *Insect Museum,* Animals (Disc 5, Side 1) |
| (1 day)<br>3:2 Methods of Classification<br>Early Classification<br>The Beginning of Modern Classification | Skill Check: *Understand Science Words,* p. 55<br>Mini Lab: *How Did Aristotle Classify Animals?* p. 53<br>Lab 3-2: *Common Names,* p. 52<br>Idea Map, p. 53<br>Check Your Understanding, p. 55 | Focus: *Classifying Dinosaurs,* p. 5<br>Reteaching: *Methods of Classification,* p. 8<br>Study Guide: *Methods of Classification,* p. 15<br>Lab 3-2: *Common Names,* pp. 11-12<br>Transparency Master: *Classification of Living Things,* p. 11 | **STVS:** *Naming Fish,* Animals (Disc 5, Side 2) |
| (2 1/2 days)<br>3:3 How Scientists Classify Today<br>Classifying Based on How Organisms Are Related<br>Other Evidence Used in Classifying<br>Scientific Names Come from Classification<br>Why Scientific Names Are Used<br>Classification of Kingdoms | Skill Check: *Classify,* p. 60<br>Mini Lab: *What Kingdom?* p. 58<br>Technology: *Six Kingdoms Instead of Five,* p. 61<br>Idea Map, p. 62<br>Check Your Understanding, p. 62 | Application: *Many Ways to Classify,* p. 3<br>Enrichment: *The Classification of Animals,* p. 3<br>Reteaching: *The Five Kingdoms,* p. 9<br>Critical Thinking/Problem Solving: *What Traits Can You Use to Classify Organisms?* p. 3<br>Skill: *The Five Kingdoms of Living Things,* p. 3<br>Study Guide: *How Scientists Classify,* pp. 16-17<br>Study Guide: *Vocabulary,* p. 18 | Color Transparency 3: *Five Kingdoms*<br>**Laboratory Manual:**<br>*How Can Living Things Be Grouped?* p. 21<br>**STVS:** *Simple Forms of Life in the Antarctic,* Plants and Simple Organisms (Disc 4, Side 1) |
| Chapter Review | Summary<br>Key Science Words<br>Testing Yourself<br>Finding Main Ideas<br>Using Main Ideas<br>Skill Review | **ASSESSMENT RESOURCES**<br>Chapter Review, pp. 11-12<br>Chapter Test, pp. 13-15<br>Performance Assessment in the Biology Classroom<br>Alternate Assessment in the Science Classroom<br>Computer Test Bank | |

### GLENCOE TECHNOLOGY

**Infinite Voyage,** *To the Edge of the Earth*
**Science and Technology Videodisc Series,** *Insect Museum,* Animals (Disc 5, Side 1)
*Naming Fish,* Animals (Disc 5, Side 2)
*Simple Forms of Life in the Antarctic,* Plants and Simple Organisms (Disc 4, Side 1)

**The Secret of Life,** *Gone Before You Know It: The Biodiversity Crisis*

## MATERIALS NEEDED

| LAB 3-1, p. 49 | LAB 3-2, p. 52 | MARGIN FEATURES |
|---|---|---|
| envelope with 12 different items | state map | Skill Check, p. 55<br>dictionary<br>Skill Check, p. 60<br>10 leaves<br>key or field guide<br>Mini Lab, p. 53<br>pencil<br>paper<br>Mini Lab, p. 58<br>pencil<br>paper |

## OBJECTIVES

For more information about National Science Standards, see page 5T.

| SECTION | OBJECTIVE | CORRELATION of QUESTIONS to OBJECTIVES | | | |
|---|---|---|---|---|---|
| | | CHECK YOUR UNDERSTANDING | CHAPTER REVIEW | TRP CHAPTER REVIEW | TRP CHAPTER TEST |
| 3:1 National Science Stds: UCP.1, G.1 | 1. **Give** examples of items in daily life that are grouped. | 2, 3 | 13, 22 | 1 | 50 |
| | 2. **Explain** how and why we classify things. | 1, 4 | 3, 7, 16, 20 | 3, 7, 9, 13, 14,15, 16, 19, 21, 26, 28 | 2, 11, 12, 13, 14, 15, 16, 17, 18, 19, 20 |
| 3:2 National Science Stds: UCP.1, G.1, G.2, G.3 | 3. **Explain** Aristotle's classification of living things. | 6, 9 | 11, 19 | 2 | 3, 7, 29 |
| | 4. **Compare** Linnaeus' system of classification with Aristotle's. | 7, 10 | 1, 2, 4, 17, 25 | 4 | 8 |
| | 5. **Name** the levels of classification that are used today. | 8 | 14 | 17, 24 | 5, 21, 39 |
| 3:3 National Science Stds: UCP.1, C.3, C.5, G.1, G.2, G.3 | 6. **List** traits used in classifying living things. | 11, 14 | 18, 24 | 10, 12 | 9, 33 |
| | 7. **State** the function of scientific names. | 12 | 5, 6, 15, 21 | 5, 8, 20, 22 | 1, 4, 6, 10, 35, 36, 37, 38, 40 |
| | 8. **List** the five kingdoms scientists recognize today. | 13 | 8, 9, 10, 12, 23 | 6, 18, 23 | 22, 23, 24, 25, 26, 27, 28, 30, 31, 32, 34 |

# Classification

## CHAPTER OVERVIEW

### Key Concepts

In this chapter, students will study why things are grouped in everyday life. They will learn about the history of classification, how organisms are classified, how scientific names come from classification, and why scientific names are used. They will be introduced to the five kingdom classification system.

### Key Science Words

animal
class
classify
family
fungus
genus
kingdom
moneran
order
phylum
plant
protist
scientific name
species
trait

### Skill Development

In Lab 3-1, students will **classify** and **observe** some common objects. In Lab 3-2, they will **classify, communicate,** and **make a table** of name places from a map. In the Skill Check on page 55, students will **understand** the **science word** *phylum.* In the Skill Check on page 60, they will use a key to **classify** leaves. In the Mini Lab on page 53, students will **infer** using Aristotle's classification scheme. In the Mini Lab on page 58, they will **classify** an imaginary organism.

### Bridging

Living organisms are classified according to similarities and differences. Bridge this chapter with the section on scientific method in Chapter 1, and relate the classifying of living organisms to data collection. Cell structure is also used in classification. Bridge this section with cells in Chapter 2.

**CHAPTER PREVIEW**

### Chapter Content

Review this outline for Chapter 3 before you read the chapter.



### Skills in this Chapter

The skills that you will use in this chapter are listed below.

- In **Lab 3-1,** you will classify and observe. In **Lab 3-2,** you will classify, communicate, and make a table.
- In the **Skill Checks,** you will classify and understand science words.
- In the **Mini Labs,** you will infer and classify.

Tubeworms

46

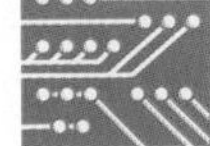

## TECH PREP

For Tech Prep activities in this chapter of the Teacher Wraparound Edition, see especially the Student Journal on page 48, and the Focus Activity on page 51. See also the Glencoe Homepage at **www.glencoe.com**

Chapter 3

# Classification

What features do biologists use to group living things? You know that most plants are green and do not move around. You also know that most animals are not green and do move around. The tubeworm on the left lives in the water attached to one spot. It is not green. Is a tubeworm a plant or an animal? The euglena in the smaller photo is green and moves around in the water. Is it a plant or an animal, or does it belong to some other group?

Biologists have a system for grouping living things. Each living thing has a specific name that biologists all over the world understand. In this chapter, you will learn why we group living things and about the system used to group them.

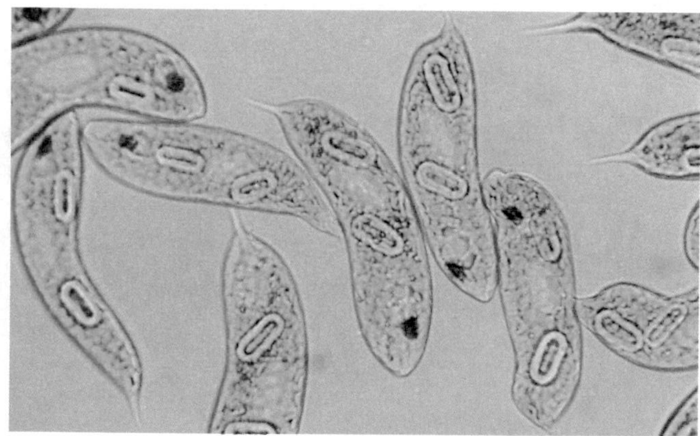

**Try This!**

**Why do we have two names?** Choose a first name that you hear often. Make a list of all the people you know who have the same first name. Can you see why a two-name system is needed?

**interNET CONNECTION**

For more information about the material in this chapter, follow the link for the chapter on the Glencoe Homepage at **http://www.glencoe.com**

*Euglena*, 450x



**GETTING STARTED**

## Using the Photos

By looking at the photos, students will see that classifying organisms is more complicated than they thought. Have them try to classify each organism as plant or animal.

## MOTIVATION/Try This!

**Why do we have two names?** When students have completed the chapter opening activity, they will be able to **infer** why two names are needed to identify a person. Then they will apply this to the two-name system used in classification.

## Chapter Preview

Have students study the chapter outline to see how the chapter is organized. This chapter discusses how grouping is used. This discussion is followed by the history of classification. The last section explains how scientists classify and name living organisms.

## Misconception

Students often think that all organisms fit perfectly into one of the five kingdoms. Point out examples of organisms that have traits of more than one kingdom, such as euglena and slime molds.

**ASSESSMENT PLANNER**

**Portfolio**

Strategies on the following page represent student products that can be placed into a best-work portfolio: p. 63.

**PERFORMANCE ASSESSMENT**

Skill Check, p. 60
Mini Lab, pp. 53, 58
Lab, pp. 49, 52

**CONTENT ASSESSMENT**

Check for Understanding, pp. 54, 55, 62
Chapter Review, pp. 64-65

**GROUP ASSESSMENT**

Opportunities for group assessment occur with Cooperative Learning Strategies.

**Student Journal**

Strategies on the following pages represent opportunities for writing in a Student Journal: pp. 48, 56.

# 3:1 Why Things Are Grouped

## PREPARATION

### Materials Needed

Make copies of the Study Guide, Reteaching, and Lab worksheets in the *Teacher Resource Package*.

▶ Prepare envelopes for Lab 3-1.

### Key Science Words

classify
trait

### Process Skills

In Lab 3-1, students will classify and observe.

## 1 FOCUS

▶ The objectives are listed on the student page. Remind students to preview these objectives as a guide to this numbered section.

### MOTIVATION/Demonstration

To introduce how items can be classified, ask each student to place one shoe in a pile in the middle of the floor. Begin sorting the shoes into two piles by using a feature such as right or left shoes or running and nonrunning shoes. Explain that you are using a certain trait to do the sorting. When someone knows what the trait is, have that person suggest another way to sort the shoes. Divide each of the piles further, one pile at a time, and make a simple key on the chalkboard.

### Guided Practice

Have students write down their answers to the margin questions: in the produce and bread sections; by a system of numbering.

**Student Journal**

Have students look through the classified section of a newspaper and determine how these data are organized.



**Objectives**

1. **Give examples** of items in daily life that are grouped.
2. **Explain** how and why we classify things.

**Key Science Words**

classify
trait

# 3:1 Why Things Are Grouped

How often do you classify things? You probably classify things more often than you think. To **classify** means to group things together based on similarities.

## Classifying in Everyday Life

If you were opening a sports store, how would you group the equipment in Figure 3-1? You might put the shoes in one place and the balls in another. You might want items for a certain sport grouped together.

Where would you find lettuce and bread in a grocery store?

Many things in our daily lives have been grouped for us. Think about how food is grouped in a grocery store. Frozen foods, meats, produce, bakery items, and canned foods are found in separate areas.

What subjects are you taking this year? Don't you group your courses? Spanish, French, and German are language courses. Typing and bookkeeping are business courses.

## How Grouping Helps Us

There are several reasons to classify things. One reason is to put things in order so that they become easier to find.

How are books arranged in a library?

Your school library must have thousands of books. They are arranged by a system of numbering that makes it easier to find a certain book. Think of trying to find a classroom with the room numbers out of order. Could you find a phone number in a phone book if the names were not in alphabetical order?

A second reason we classify things is to show that they share certain traits. A **trait** is a feature that a thing has. In a library, you see biographies grouped together. What trait do they share?

**Figure 3-1** Many everyday objects, such as this sports equipment, are classified.

## OPTIONS

### Science Background

The science of classifying living things is called taxonomy. More than one and one-half million organisms have been classified. Organisms are classified based on evidence that indicates how closely related they are.

**How Can Paper Objects Be Grouped?**/*Lab Manual,* pp. 17-20. Use this lab as an extension to classifying everyday objects.

# Lab 3–1 Classifying

## Problem: How can some common objects be classified?

### Skills

classify, observe

### Materials

envelope that contains 12 items

### Procedure

1. Copy the diagram.
2. Place the objects from your envelope on your desk.
3. Make a list of the objects. Your teacher will help you with their names.
4. These objects will be called Kingdom Objects. List the objects under Kingdom Objects in your diagram.
5. Sort the 12 objects into two groups on your desk. The objects in each group must have a common trait.
6. Call each of these two groups a phylum. Use the common trait as the name of the phylum. Place the phylum name for each group on your diagram.
7. On the diagram, write the name of each object that you placed in Phylum 1 and in Phylum 2.
8. Sort the objects of Phylum 1 into two groups. Each of the two groups must have a common trait.
9. Call each of these two groups a class. Use the common trait as the name of the class. Place the class names for Phylum 1 on your diagram under Classes A and B.
10. Sort the objects in Phylum 2 into two groups. Each of these two groups must have a common trait.
11. Each of these groups represents a class. Place the class names for Phylum 2 on your diagram under Classes C and D.
12. Put the items back into the envelope and return the envelope to your teacher.

Kingdom Objects ______
- Phylum 1 ______
  - Class A ______
  - Class B ______
- Phylum 2 ______
  - Class C ______
  - Class D ______

### Data and Observations

1. What traits did you use to classify the kingdom objects into two phyla?
2. What is the name of Phylum 1? Phylum 2?
3. What are the names of Class A and Class B?
4. What are the names of Class C and Class D?
5. What trait do all four classes have in common?
6. What traits do Classes A and B have in common?
7. What traits do Classes C and D have in common?

### Analyze and Apply

1. At what level of classification, phylum or class, did all objects share more traits?
2. At what level of classification, phylum or class, did all objects share fewer traits?
3. **Apply:** How could you use the classification method in this lab to group a collection of compact discs?

### Extension

**Classify** the items in your bedroom closet using the same method that you used in this lab.



## ANSWERS

### Data and Observations

1. metals or nonmetals, flat or not flat, pointed or not pointed
2. same as number 1
3. If the phylum was metals, the class names could be round, not round; steel, not steel.
4. If the phylum was nonmetals, the class names could be white, not white; glass, not glass.
5. All are objects.
6. If the phylum was metals, then classes A and B would both be metals and so on.
7. If the phylum was nonmetals, then classes C and D would both be nonmetals, and so on.

### Analyze and Apply

1. phylum
2. class
3. Students should classify their discs first by type of music, then by artist, and so on.

## Lab 3-1 Classifying

### Overview

In this lab, students will make their own classification system using traits to group common items.

**Objectives:** Upon completing this lab, students will be able to (1) **devise** a classification system (2) **infer** that any classification system can be correct if proper criteria are followed in formulating it.

**Time Allotment:** 45 minutes

### Preparation

▶ Have 1 envelope for each lab group containing:

| | |
|---|---|
| hairpin | bolt |
| paper clip | brass fastener |
| toothpick | rubber band |
| nail | yarn (2 cm) |
| button | glass slide |
| washer | seed |
| nut | key |

▶ **Alternate Materials:** Similar objects may be substituted.

**Lab 3-1 worksheet**/*Teacher Resource Package*, pp. 9-10.

### Teaching the Lab

▶ Be sure students understand that there is no one correct classification for their objects.

▶ **Troubleshooting:** Make sure all objects in Phylum 1 are separated before students work on Phylum 2.

**Cooperative Learning:** Divide the class into groups of two students. For more information, see pp. 22T-23T in the Teacher Guide.

### ASSESSMENT

**Performance:** Have students classify six or more family members or classmates, using such characteristics as sex, eye color, hair color, and weight. Each individual's last name can be the person's genus and first name the species.

## Career Close-Up

### Wildlife Photographer

#### Background

The training for a wildlife photographer usually includes training after high school. Success in this field depends on some knowledge of biology.

#### Related Careers

wildlife illustrator, wildlife biologist

#### Job Requirements

photography training, an artistic eye, patience

#### For More Information

Professional Photographers
of America
1090 Express Way
Des Plaines, IL 60018

### Independent Practice

**Study Guide**/*Teacher Resource Package*, pp. 13-14. Use the Study Guide worksheets shown at the bottom of this page for independent practice.

## 3 APPLY

Compare the Library of Congress classification system with the Dewey decimal system.

## 4 CLOSE

### ACTIVITY/Brainstorming

Have students give examples of things they classify.

### Answers to Check Your Understanding

1. feature of something; mammals/hair; birds/feathers
2. languages, sports, subjects
3. to make things easier to find; to show how things are alike
4. round: nucleus, nucleolus; long and thin: chromosomes
5. 1 500 000; $1.5 \times 10^6$

## Career Close-Up

### Wildlife Photographer

Photographing wildlife is often a challenging occupation.

John's biology teacher invited a wildlife photographer to visit his class. The photographer brought slides and prints of many living things. Many photographs of flowers and insects were made with a close-up lens or a zoom lens. The photographer showed the students how to use a camera attachment on a light microscope. He told them that his work often took him to outdoor settings.

Students wanted to know about the training needed to be a photographer. He told them that he had taken photography and natural science courses in high school. Then he had taken several courses at a community college. He explained that in this field success depends on a person knowing the subject matter well.

Other students wanted to know where his work was used. He showed them magazines and books that contained his photographs.

Biologists classify living things. Doing so puts organisms in order. It also shows how they are alike. There are over one and one-half million known kinds of living things. It would be impossible to find information about them if they were not grouped in some way.

### Check Your Understanding

1. What is meant by a trait? Give two examples of traits.
2. Give an example of something at school that is classified.
3. Give two reasons why things are classified.
4. **Critical Thinking:** Group the parts of an animal cell based on their shapes.
5. **Biology and Math:** What are two other ways to write the number one and one-half million?

## STUDY GUIDE 13

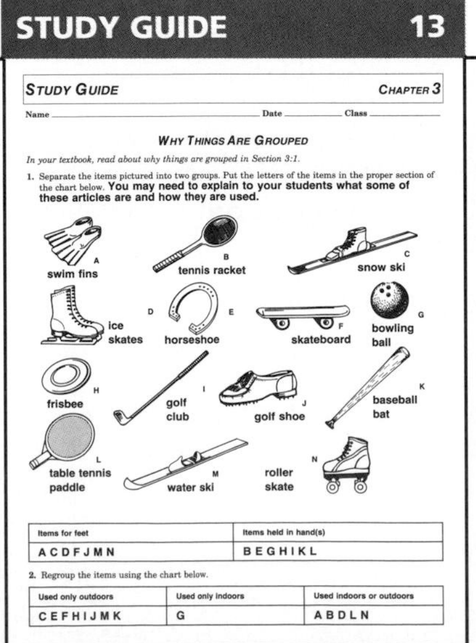

STUDY GUIDE — CHAPTER 3

Name ______ Date ______ Class ______

WHY THINGS ARE GROUPED

*In your textbook, read about why things are grouped in Section 3:1.*

1. Separate the items pictured into two groups. Put the letters of the items in the proper section of the chart below. **You may need to explain to your students what some of these articles are and how they are used.**



| Items for feet | Items held in hand(s) |
|---|---|
| A C D F J M N | B E G H I K L |

2. Regroup the items using the chart below.

| Used only outdoors | Used only indoors | Used indoors or outdoors |
|---|---|---|
| C E F H I J M K | G | A B D L N |

## STUDY GUIDE 14

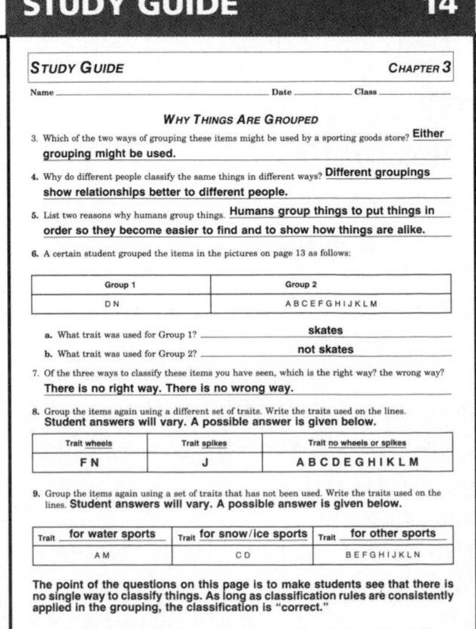

STUDY GUIDE — CHAPTER 3

Name ______ Date ______ Class ______

WHY THINGS ARE GROUPED

3. Which of the two ways of grouping these items might be used by a sporting goods store? **Either grouping might be used.**
4. Why do different people classify the same things in different ways? **Different groupings show relationships better to different people.**
5. List two reasons why humans group things. **Humans group things to put things in order so they become easier to find and to show how things are alike.**
6. A certain student grouped the items in the pictures on page 13 as follows:

| Group 1 | Group 2 |
|---|---|
| D N | A B C E F G H I J K L M |

a. What trait was used for Group 1? **skates**

b. What trait was used for Group 2? **not skates**

7. Of the three ways to classify these items you have seen, which is the right way? the wrong way? **There is no right way. There is no wrong way.**
8. Group the items again using a different set of traits. Write the traits used on the lines. **Student answers will vary. A possible answer is given below.**

| Trait wheels | Trait spikes | Trait no wheels or spikes |
|---|---|---|
| F N | J | A B C D E G H I K L M |

9. Group the items again using a set of traits that has not been used. Write the traits used on the lines. **Student answers will vary. A possible answer is given below.**

| Trait for water sports | Trait for snow/ice sports | Trait for other sports |
|---|---|---|
| A M | C D | B E F G H I J K L N |

**The point of the questions on this page is to make students see that there is no single way to classify things. As long as classification rules are consistently applied in the grouping, the classification is "correct."**

## 3:2 Methods of Classification

For hundreds of years, people have been grouping living things. The job has not been easy. What makes the job of classifying hard is that scientists do not always agree on how to group living things.

**Objectives**

3. **Explain** Aristotle's classification of living things.
4. **Compare** Linnaeus' system of classification with Aristotle's.
5. **Name** the levels of classification that are used today.

### Early Classification

Over 2000 years ago, a Greek scientist named Aristotle (AIR uh staht ul) was one of the first people to classify living things. He noticed that living things fit into two main groups—plant and animal. Most plants were green and didn't move. Most animals weren't green and did move.

Aristotle next divided all animals into three groups. He based his groups on where the animals lived as shown in Figure 3-2. Animals that lived in water went into one group. Animals that lived on land went into a second group. Animals that lived in the air and could fly made up a third group.

Aristotle then worked out a system for grouping plants. He based it on size of the plant and pattern of growth. You can see in Figure 3-2 that plants were placed into three smaller groups. Tall plants with one trunk were put into a tree group. Medium plants with many trunks were put into a shrub group. A privet hedge is an example of a shrub. Small plants with soft stems went into an herb group. Herbs included grasses and wildflowers.

**Key Science Words**

kingdom
phylum
class
order
family
genus
species

**How did Aristotle divide his animal group?**

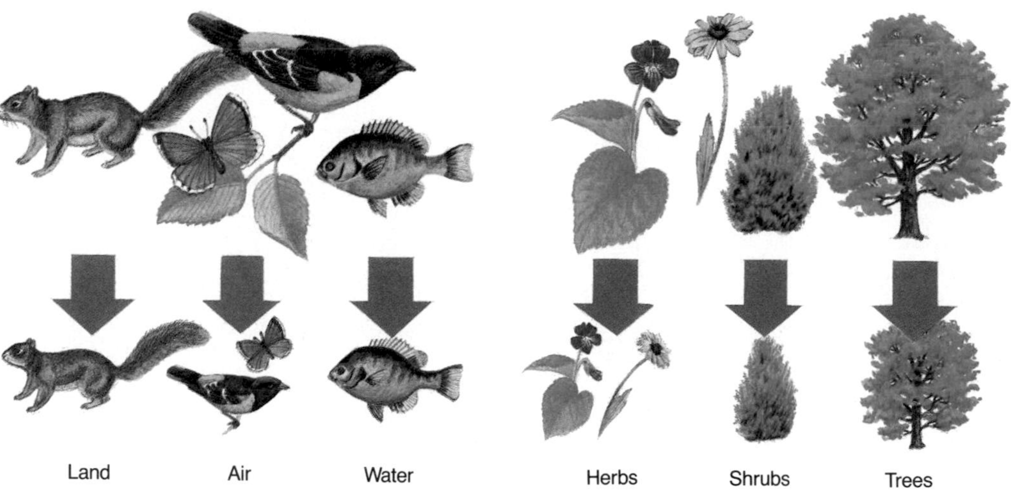


**Figure 3-2** Aristotle grouped animals into three groups and plants into three groups.



**FOCUS 5**

FOCUS CHAPTER 3

Name ____________ Date ________ Class ________

CLASSIFYING DINOSAURS

1. *Allosaurus* 2. *Brachiosaurus* 3. *Plateosaurus*
4. *Coelophysis* 5. *Psittacosaurus* 6. *Triceratops*
7. *Parasaurolophus* 8. *Diplodocus* 9. *Tyrannosaurus*

Find a trait you can see in the pictures that you could use to classify the dinosaurs shown into two groups. Put the number for each dinosaur into Group A or Group B. Then, find a trait you could use to divide Group A into Groups $A_1$ and $A_2$, and a trait that could divide Group B into Groups $B_1$ and $B_2$. Put the number for each dinosaur in the appropriate group.

## OPTIONS

### Science Background

Aristotle's classification system remained in use almost 2000 years. In the 16th and 17th centuries, there was renewed interest in classification when European explorers brought back many new unidentified plant and animal specimens from other lands. In the 17th century, John Ray developed an improved classification system. He classified plants according to the structure of the seed and gave each organism a Latin name.

## 3:2 Methods of Classification

## PREPARATION

### Materials Needed

Make copies of the Focus, Study Guide, Reteaching, and Lab worksheets in the *Teacher Resource Package*.

▶ Gather maps for Lab 3-2.

### Key Science Words

| | |
|---|---|
| kingdom | family |
| phylum | genus |
| class | species |
| order | |

### Process Skills

In Lab 3-2, students will classify, communicate, and make a table. In the Mini Lab, students will infer. In the Skill Check, students will understand science words.

## 1 FOCUS

▶ The objectives are listed on the student page. Remind students to preview these objectives as a guide to this numbered section.

### ACTIVITY/Brainstorming

In groups of five, give students two minutes to brainstorm ways that newspapers, magazines, motor vehicles, sports, and restaurants make use of classification.

Focus/*Teacher Resource Package*, p. 5. Use the Focus transparency shown at the bottom of this page to introduce classification of organisms.

### Guided Practice

Have students write down their answers to the margin question: Animals were divided into three groups — animals that lived on the ground, animals that lived in water, and animals that lived in the air and flew.

## Lab 3-2 Common Names

### Overview

In this lab, students will learn that some places are named for plants and animals. They will see that common names are used.

**Objectives:** Upon completing this lab, students will be able to: (1) **recognize** that names of living things are used to name places and (2) **relate** that common names are used more by people who are not scientists.

**Time Allotment:** 35 minutes

### Preparation

▶ **Alternate Materials:** Any map may be used.

**Lab 3-2 worksheet**/*Teacher Resource Package,* pp. 11-12.

### Teaching the Lab

▶ Spend some time with students to develop map-reading skills. Point out highways, railroads, rivers, elevations, and parks.

▶ You might want to make an overhead transparency of a map of your area.

**Cooperative Learning:** Divide the class into groups of two students. For more information, see pp. 22T-23T in the Teacher Guide.

▶ **Troubleshooting:** Check the map to make sure there are places named for plants and animals.

**Skill:** Have students use references to find the scientific names of three plants and three animals. Students should determine if the scientific name of each is similar to the common name.

# Common Names

**Lab 3–2**

## Problem: How many places are named for animals or plants?

### Skills

classify, communicate, make a table

### Materials

state map

### Procedure

1. **Make a table** in your notebook in which to record your observations. Use the headings Name and Type of Place in your table.
2. Use the map on this page or your state map to find each city or town that has the name of a plant or animal as part of its name. Record the names in your table.
3. Place a (C) after the name so you will know it is a city or town.
4. Use the map to find state or national parks, rivers, lakes, and historical places that have a plant or animal name as part of their names.
5. Record the names of the parks, rivers, lakes, and historical places in your table. Put a (P) after parks, (R) after rivers, (H) after historical places, and (L) after lakes.

### Data and Observations

1. Were animal names or plant names used more often on your map?
2. Which animal name appeared most often on your map?
3. Which plant name appeared most?

### Analyze and Apply

1. Why do you suppose humans use certain animal and plant names for cities, rivers, and parks?
2. Do cities, parks, or bodies of water seem to have more plant and animal names?
3. Why do humans use the common names of plants and animals rather than the scientific names to name places?
4. **Apply:** Would you expect to find a town called Giraffesville on your map? Explain.

### Extension

**Communicate:** Make a list of other common themes that are used in naming places, for example Indian names. Share your list with the class.

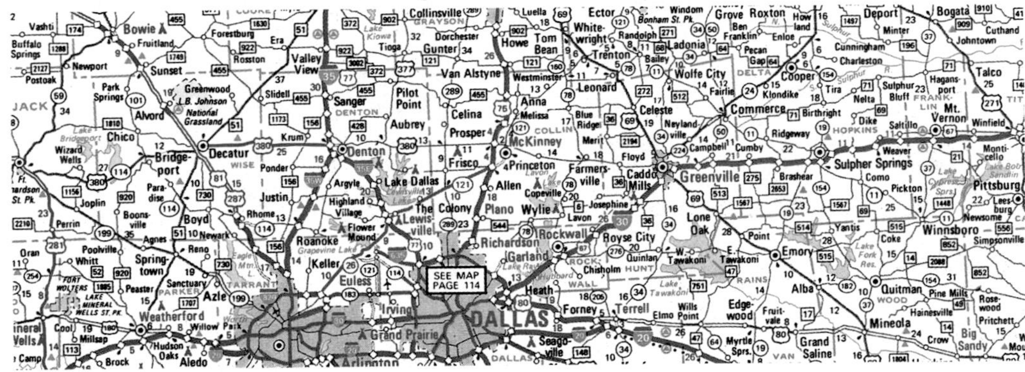




## ANSWERS

### Data and Observations

1. Answers for this activity will vary according to the map used.
2. Answers will vary according to map used.
3. Answers will vary according to map used.

### Analyze and Apply

1. Students may answer that people name places after things with which they are most familiar.
2. Answers may vary.
3. Common names are more familiar.
4. No. Giraffes are not found in the area shown on the map.

## The Beginning of Modern Classification

Scientists used Aristotle's system of classification for hundreds of years. However, as scientists found more and more living things, Aristotle's system became less useful. Many of the newly discovered living things did not fit into his system. Many scientists began to question the system because different traits were used to group plants and animals. A useful classification system should use the same traits for classifying all groups.

In 1735, Carolus Linnaeus (lin NAY us) developed a new classification system. He first placed living things into two main groups. He called these groups kingdoms. A **kingdom** is the largest group of living things. Plants made up one kingdom. Animals made up the other kingdom. Linnaeus placed living things with similar traits into the same group and called this group a species. He used very specific traits for his groups. For example, he used flower parts to group plant species. He placed similar species into a larger group called a genus (JEE nus).

Through his work, Linnaeus made a number of important changes in Aristotle's system.

1. He classified plants and animals into more groups.
2. He based his system on specific traits.
3. He gave organisms names that described their traits. These names had two parts. All living things still have two-part names.

**How do the classification systems of Aristotle and Linnaeus compare?**

### Mini Lab

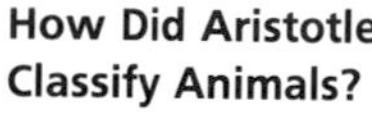

**How Did Aristotle Classify Animals?**

**Infer:** Make a list of animals that you have seen or that you know about from reading magazines and watching television. Classify the animals the way Aristotle would have. *For more help, refer to the* ***Skill Handbook,*** *pages 706-711.*

**Idea Map**

**Early Classification**

- Early classification
  - Aristotle
    - plant — trees, shrubs, herbs
    - animal — water, land, air
  - Linnaeus
    - plant kingdom — genus — species
    - animal kingdom — genus — species

**Study Tip:** Use this idea map as you study the development of classification systems. Aristotle and Linnaeus developed early classification systems.



# 2 TEACH

### MOTIVATION/Field Trip

Visit a local zoo. Ask students to classify the animals into a phylum. Compare answers in the classroom.

### Concept Development

▶ Point out that living things are first placed into large groups. Each large group is then divided into smaller groups.

▶ Emphasize that Linnaeus first based his classification system on flower structures only. He later included other plant structures.

### Guided Practice

Have students write down their answers to the margin question: Linnaeus' system is more precise than Aristotle's. Linnaeus used specific traits to group organisms.

### Mini Lab

Students should place the animals in groups based on whether they live in water, on land, or fly.

### ASSESSMENT

**Content:** Have students compare the classification system used by Aristotle with the one used by Linnaeus. Students should describe how these classification systems compare with the one scientists use today.

### Idea Map

Have students use the idea map as a study guide to the major concepts of this section.

### Independent Practice

**Study Guide**/*Teacher Resource Package,* p. 15. Use the Study Guide worksheet shown at the bottom of this page for independent practice.

**STUDY GUIDE 15**

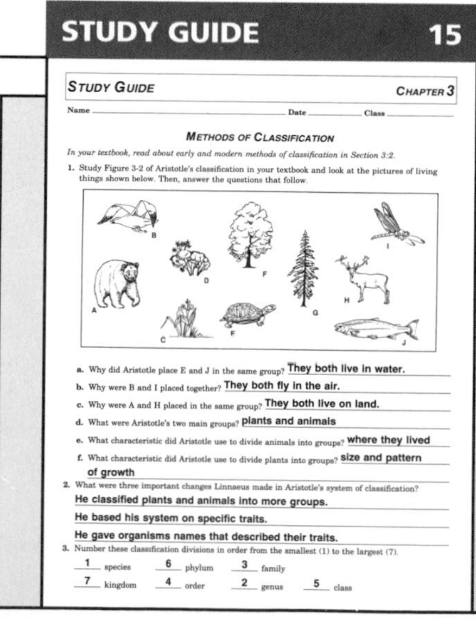

STUDY GUIDE CHAPTER 3

Name ______ Date ______ Class ______

METHODS OF CLASSIFICATION

*In your textbook, read about early and modern methods of classification in Section 3:2.*

1. Study Figure 3-2 of Aristotle's classification in your textbook and look at the pictures of living things shown below. Then, answer the questions that follow.



a. Why did Aristotle place E and J in the same group? **They both live in water.**
b. Why were B and I placed together? **They both fly in the air.**
c. Why were A and H placed in the same group? **They both live on land.**
d. What were Aristotle's two main groups? **plants and animals**
e. What characteristic did Aristotle use to divide animals into groups? **where they lived**
f. What characteristic did Aristotle use to divide plants into groups? **size and pattern of growth**

2. What were three important changes Linnaeus made in Aristotle's system of classification?
**He classified plants and animals into more groups.**
**He based his system on specific traits.**
**He gave organisms names that described their traits.**

3. Number these classification divisions in order from the smallest (1) to the largest (7).
**1** species **6** phylum **3** family
**7** kingdom **4** order **2** genus **5** class

## OPTIONS

### ACTIVITY/Demonstration

Present students with at least one representative of each kingdom. For the Kingdom Monera, this will need to be done with a picture. Ask students to write descriptions of what they can see from the specimens and pictures in order to describe the kingdom.

## TEACH

### Check for Understanding

Give students examples of plants and ask which Aristotle would have grouped together. Then ask which Linnaeus would have grouped together.

### Reteach

Have students outline how Aristotle divided living things into groups and then compare his system with that of Linnaeus. Outlines should show living things divided into two groups—plants and animals. The animal division should be subdivided into water, land, and air (where they live); the plant division should be subdivided into tall plants, medium plants, and small plants (size). Linnaeus classified plants and animals into more groups based on specific traits. He gave them names that described their traits.

### Guided Practice

Have students write down their answers to the margin question: Classification groups and addresses can be arranged from more general to more specific.

### GLENCOE TECHNOLOGY

**Videodisc**

Biology: The Dynamics of Life
Movie: *Museum Collections*
Disc 1, Side 2, Ch. 8

**TABLE 3–1. COMPARING CLASSIFICATION GROUPS AND ADDRESS INFORMATION**

| Address | Classification |
|---|---|
| Country | Kingdom |
| State | Phylum |
| County | Class |
| Town | Order |
| Neighborhood | Family |
| Street | Genus |
| House number | Species |

Today, there are seven groups for classifying organisms—kingdom, phylum, class, order, family, genus, and species. Why so many? Having more groups makes it easier to place an organism in the proper group. Think of how difficult it would be to find a long-lost cousin's house if all you knew was the country your cousin lived in. Table 3-1 shows the different kinds of address information that could be helpful in finding your cousin's house. How do the classification groups compare to the different kinds of address information?

How can classification groups be compared to addresses?

The kingdom is still the largest group of living things, just as it was in Linnaeus' time. A kingdom can be compared to the country your cousin lives in. Kingdoms are divided into groups called phyla (FI luh) (singular phylum). A **phylum** is the largest group within a kingdom. A phylum can be compared to a state. Phyla are divided into even smaller groups called classes. A **class** is the largest group within a phylum. Classes are divided into groups called orders. An **order** is the largest group within a class. Orders are divided into families. A **family** is the largest group within an order. Notice in Table 3-1 that classes can be compared to counties, orders to towns, and families to neighborhoods.

a

b

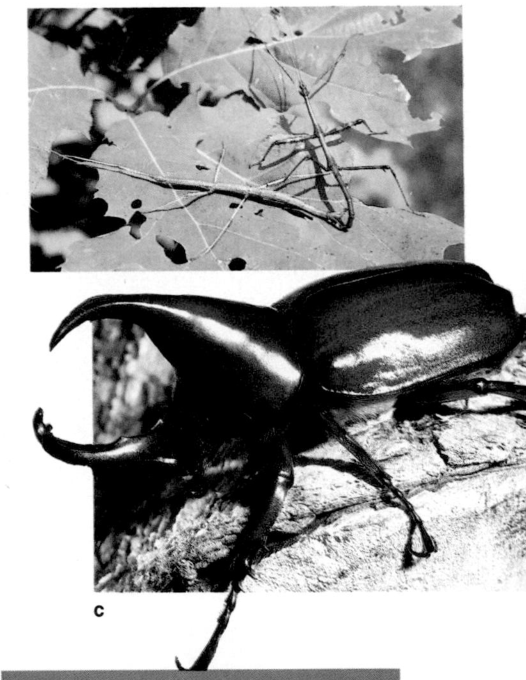

c

**Figure 3-3** The praying mantis (a), walking stick (b), and rhinoceros beetle (c) have names that describe their appearance.



**TRANSPARENCY 11**

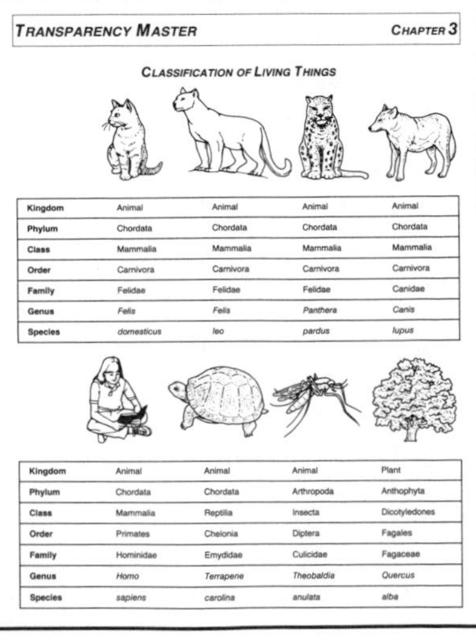

TRANSPARENCY MASTER — CHAPTER 3

CLASSIFICATION OF LIVING THINGS

| Kingdom | Animal | Animal | Animal | Animal |
|---|---|---|---|---|
| Phylum | Chordata | Chordata | Chordata | Chordata |
| Class | Mammalia | Mammalia | Mammalia | Mammalia |
| Order | Carnivora | Carnivora | Carnivora | Carnivora |
| Family | Felidae | Felidae | Felidae | Canidae |
| Genus | *Felis* | *Felis* | *Panthera* | *Canis* |
| Species | *domesticus* | *leo* | *pardus* | *lupus* |

| Kingdom | Animal | Animal | Animal | Plant |
|---|---|---|---|---|
| Phylum | Chordata | Chordata | Arthropoda | Anthophyta |
| Class | Mammalia | Reptilia | Insecta | Dicotyledones |
| Order | Primates | Chelonia | Diptera | Fagales |
| Family | Hominidae | Emydidae | Culicidae | Fagaceae |
| Genus | *Homo* | *Terrapene* | *Theobaldia* | *Quercus* |
| Species | *sapiens* | *carolina* | *anulata* | *alba* |

### OPTIONS

**Transparency Master**/*Teacher Resource Package*, p.11. Use the Transparency Master shown here to teach the classification groups of living things.

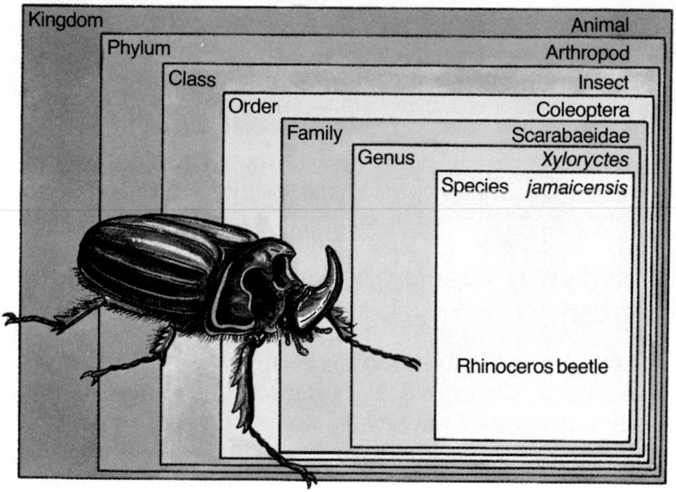


**Figure 3-4** Each classification group is a subset of the next-largest group. Of what group is the kingdom a subset? all living things

The names given to the different groups often describe the living things in them. Let's look at an example at the family level. Members of three different insect families are shown in Figure 3-3. Each insect belongs to a different family.

Another way to think of a kingdom is as a set. Think of the other groups, from largest to smallest, as subsets. All seven groups are shown in Figure 3-4. You can see that like classes and orders, families are divided into groups. The largest group within a family is a **genus.** The smallest group of living things is a **species.**

## Skill Check

**Understand science words: phylum.** The word part *phyl* means tribe or race. In your dictionary, find three words with the word part *phyl* in them. *For more help, refer to the* ***Skill Handbook,*** *pages 706-711.*

## Check Your Understanding

6. What traits were used by Aristotle to divide plants into groups?
7. Linnaeus made important changes in the way living things are classified. Name two.
8. There are seven main classification groups. Name these groups in order from largest to smallest.
9. **Critical Thinking:** A euglena is a one-celled organism that moves on its own and has chloroplasts. Why would Aristotle have trouble classifying this organism?
10. **Biology and Reading:** Who developed a two-kingdom system of classification for plants and animals?



## TEACH

### Check for Understanding

Have students respond to the first three questions in Check Your Understanding.

### Reteach

**Reteaching**/*Teacher Resource Package,* p. 8. Use this worksheet to give students additional practice with methods of classification.

**Extension:** Assign Critical Thinking; Biology and Reading, or some of the **OPTIONS** available with this lesson.

## 3 APPLY

### ACTIVITY/Game

Play "What's My Kingdom?". Read a description of an organism without naming it. Have students try to classify it in the proper kingdom.

## 4 CLOSE

### Audiovisual

Show the slide program *The Earth's Five Kingdoms,* JLM Visuals.

### Answers to Check Your Understanding

6. size and pattern of growth
7. He classified plants and animals into more groups, based his system on specific traits, and gave organisms names that described their traits.
8. kingdom, phylum, class, order, family, genus, species
9. It has both animal-like characteristics (movement) and plantlike characteristics (chloroplasts).
10. Linnaeus

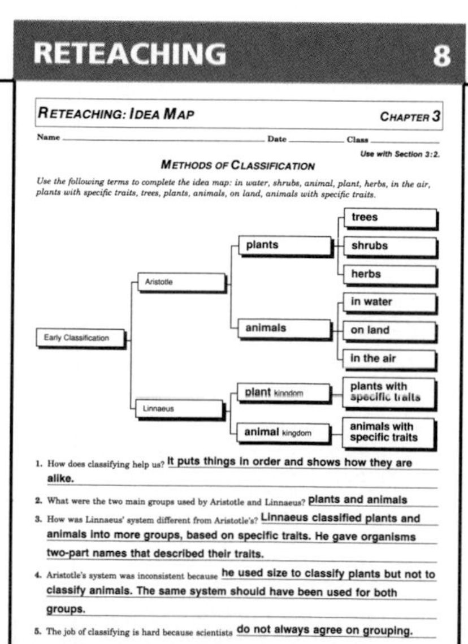

**RETEACHING 8**

**RETEACHING: IDEA MAP** — **CHAPTER 3**

Name ______ Date ______ Class ______

*Use with Section 3:2.*

**METHODS OF CLASSIFICATION**

*Use the following terms to complete the idea map: in water, shrubs, animal, plant, herbs, in the air, plants with specific traits, trees, plants, animals, on land, animals with specific traits.*

- Early Classification
  - Aristotle
    - plants
      - trees
      - shrubs
      - herbs
    - animals
      - in water
      - on land
      - in the air
  - Linnaeus
    - plant kingdom
      - plants with specific traits
    - animal kingdom
      - animals with specific traits

1. How does classifying help us? It puts things in order and shows how they are alike.
2. What were the two main groups used by Aristotle and Linnaeus? plants and animals
3. How was Linnaeus' system different from Aristotle's? Linnaeus classified plants and animals into more groups, based on specific traits. He gave organisms two-part names that described their traits.
4. Aristotle's system was inconsistent because he used size to classify plants but not to classify animals. The same system should have been used for both groups.
5. The job of classifying is hard because scientists do not always agree on grouping.

## 3:3 How Scientists Classify Today

## PREPARATION

### Materials Needed

Make copies of the Enrichment, Application, Critical Thinking, Skill, Study Guide, and Reteaching worksheets in the *Teacher Resource Package*.

▶ Use bird seed mix to make packets for the Focus activity.

▶ Gather leaves and keys or field guides for the Skill Check.

### Key Science Words

| | |
|---|---|
| scientific name | fungus |
| moneran | plant |
| protist | animal |

### Process Skills

In the Skill Check and Mini Lab, students will classify.

## 1 FOCUS

▶ The objectives are listed on the student page. Remind students to preview these objectives as a guide to this numbered section.

### ACTIVITY/Hands-on

Have students work in pairs for this activity. Give each pair a small packet of bird seed, a hand lens, and a metric ruler. Have them divide the seeds into two groups, then divide the two groups into smaller groups based on traits such as color, size, and shape.

### Student Journal

Have students research the contributions of Ernst Haeckel, Robert Whittaker, and Lynn Margulis to classification. Students can write a paragraph on the work of each one.

#### Objectives

6. **List** traits used in classifying living things.
7. **State** the function of scientific names.
8. **List** the five kingdoms scientists recognize today.

#### Key Science Words

scientific name
moneran
protist
fungi
plant
animal

## 3:3 How Scientists Classify Today

How do scientists know to which groups an organism belongs? They look at many traits. They compare the traits of one organism with those of another. Scientists also compare organisms living today with those that lived long ago. Let's look at how scientists classify living things.

### Classifying Based on How Organisms Are Related

Living things that are closely related are in many of the same classification groups. For example, if two plants are closely related, they will be in almost all of the same groups. If they are not closely related, they will not be in very many of the same groups.

Table 3-2 shows a list of groups to which the house cat shown in Figure 3-5 belongs. It also shows the traits for each group. You may not recognize some of the group names. Many are from the Greek or Latin language.

Compare the lion shown in Figure 3-6 to the house cat. In how many groups are they found together?

**TABLE 3–2. CLASSIFYING THE HOUSE CAT**

| Group | Group name | Group trait |
|---|---|---|
| Kingdom | Animal | Has many cells; eats food |
| Phylum | Chordate | Rodlike structure along the back for support |
| Class | Mammal | Nurses young; has hair |
| Order | Carnivore | Eats flesh; has large teeth |
| Family | Felidae | Sharp claws; large eyes |
| Genus | *Felis* | Small cats |
| Species | *catus* | Tame |

**Figure 3-5** To what classification groups does a house cat belong?



## OPTIONS

### Science Background

In 1866, Ernst Haeckel proposed a third kingdom called Protista. Robert Whittaker, a biologist at Cornell University, developed the five-kingdom system. It received little support until the 1970s, when extensive work in classifying was done by Lynn Margulis.

**Enrichment**/*Teacher Resource Package*, p. 3. Use the Enrichment worksheet shown here as an extension to how animals are classified today.

### ENRICHMENT 3

ENRICHMENT CHAPTER 3

Name ____________ Date ________ Class ________

Use with Section 3:3.

THE CLASSIFICATION OF ANIMALS

*Use this information and information from your textbook to complete the scientific names of the animals below and to fill in the chart.*

Shown below are pictures of animals with their common names and scientific names. Why do the scientific names of all the animals shown begin with the same word? **The animals all belong to the same genus.**

The genus for dogs and doglike animals is *Canis*. Dogs belong to the family Canidae and to the same order as cats.

| Common name: | coyote | wolf | dog |
|---|---|---|---|
| Scientific name: | *Canis latrans* | *Canis lupus* | *Canis familiaris* |

| | | | |
|---|---|---|---|
| Kingdom | Animal | Animal | Animal |
| Phylum | Chordate | Chordate | Chordate |
| Class | Mammal | Mammal | Mammal |
| Order | Carnivore | Carnivore | Carnivore |
| Family | Canidae | Canidae | Canidae |
| Genus | *Canis* | *Canis* | *Canis* |
| Species | *Canis latrans* | *Canis lupus* | *Canis familiaris* |

| TABLE 3–3. COMPARING THE HOUSE CAT AND LION | | | |
|---|---|---|---|
| | **House Cat** | **Lion** | **Comparison** |
| Kingdom | Animal | Animal | Same kingdom |
| Phylum | Chordate | Chordate | Same phylum |
| Class | Mammal | Mammal | Same class |
| Order | Carnivore | Carnivore | Same order |
| Family | Felidae | Felidae | Same family |
| Genus | *Felis* | *Panthera* | Different genus |
| Species | *catus* | *leo* | Different species |

Table 3-3 shows that five of the seven groups for these two animals are the same. Only the genus and species groups are different. The lion and the house cat are very similar. They have many of the same traits.

Look at the classifications in Table 3-4 for the house cat and deer. These two animals do not share many traits. Notice that only their first three classification groups are the same.

| TABLE 3–4. COMPARING THE HOUSE CAT AND DEER | | | |
|---|---|---|---|
| | **House Cat** | **Deer** | **Comparison** |
| Kingdom | Animal | Animal | Same kingdom |
| Phylum | Chordate | Chordate | Same phylum |
| Class | Mammal | Mammal | Same class |
| Order | Carnivore | Artiodactyla | Different order |
| Family | Felidae | Cervidae | Different family |
| Genus | *Felis* | *Odocoileus* | Different genus |
| Species | *catus* | *virginianus* | Different species |

Let's compare a house cat with an octopus. They have fewer groups in common than house cats and deer. House cats and deer have backbones and hair. What about an octopus? Table 3-5 shows that for the house cat and the octopus, only the kingdom is the same. The house cat and the octopus are both animals. After the kingdom level, these animals belong to different groups.

| TABLE 3–5. COMPARING THE HOUSE CAT AND OCTOPUS | | | |
|---|---|---|---|
| | **House Cat** | **Octopus** | **Comparison** |
| Kingdom | Animal | Animal | Same kingdom |
| Phylum | Chordate | Mollusk | Different phylum |
| Class | Mammal | Cephalopod | Different class |
| Order | Carnivore | Octopoda | Different order |
| Family | Felidae | Octopodidae | Different family |
| Genus | *Felis* | *Octopus* | Different genus |
| Species | *catus* | *vulgaris* | Different species |

**Figure 3-6** How are a lion (top), deer (middle), and octopus (bottom) like a house cat? all are animals; the lion and house cat are classified the same, except for genus and species; the deer and house cat belong to the same phylum and class



# 2 TEACH

## MOTIVATION/Display

Display skeletons of a frog, bird, bat, chicken, cat (whichever are available). Have students identify body parts that have the same structure and origin. Ask what traits a bird and a bat have that are similar and why they are placed in different classes.

## Concept Development

▶ Discuss why the classification of living organisms is more difficult than other kinds of classification. There are more different kinds of organisms with many different kinds of traits. Classification is based on the ways organisms are alike and the ways organisms are different.

**Cooperative Learning:** Point out to students that using a mnemonic device is an easy way to remember the classification system developed by Linnaeus. One example is King Phillip came over for graduation Sunday. Have students work in groups of three to come up with another mnemonic device that could be used.

## Independent Practice

**Study Guide**/*Teacher Resource Package,* p. 16. Use the Study guide worksheet shown at the bottom of this page for independent practice.

**STUDY GUIDE 16**

STUDY GUIDE — CHAPTER 3

Name ________ Date ________ Class ________

**HOW SCIENTISTS CLASSIFY**

*In your textbook, read about how scientists classify today in Section 3:3. Then, examine the classification of the three animals in the chart below.*

| | Animal A | Animal B | Animal C |
|---|---|---|---|
| Kingdom | Animal | Animal | Animal |
| Phylum | Chordate | Chordate | Chordate |
| Class | Mammal | Aves | Aves |
| Order | Primates | Passeriformes | Passeriformes |
| Family | Pongidae | Fringillidae | Fringillidae |
| Genus | *Pan* | *Spizella* | *Serinus* |
| Species | *troglodytes* | *passerina* | *canarius* |

1. Comparing Animals A and D, how many groups are the same? **2**
2. Comparing Animals A and C, how many groups are the same? **2**
3. Comparing Animals B and C, how many groups are the same? **5**
4. Which two animals (A and B, A and C, or B and C) are most alike in classification? **B and C**
5. Which two animals have more of the same traits? **B and C**
6. Which two animals have more of the same body parts? **B and C**
7. Which two animals are most closely related? **B and C**
8. What is the scientific name of Animal A? ***Pan troglodytes***
9. What does the first word of the scientific name represent? **genus**
10. What does the second word of the scientific name represent? **species**
11. What are three reasons for using scientific names? **No mistake can be made about what living thing is being described. Scientific names of living things seldom change. Scientists throughout the world use the same language to name living things.**

**APPLICATION 3**

APPLICATION: CONSUMER — CHAPTER 3

Name ________ Date ________ Class ________

*Use after Section 3:1.*

**MANY WAYS TO CLASSIFY**

People find many ways to classify and organize things around them. A person who runs a business might use a classification system to keep track of his or her goods.

1. Imagine that you own a pet store. How would you classify the animals in your store? List three different ways you could group the animals. **Answers will vary. Students might suggest grouping by species, by whether it is a land or water animal.**
2. How would having a classification system help you run your store? **It could help you and your customers find certain animals; it could help you more easily care for and feed the animals; it could help you separate animals that would fight with each other and so on.**
3. How do you think having a classification system is useful to scientists? **It helps scientists find information about organisms.**
4. Different scientists and other people who work with plants and animals have different ways to classify organisms. For each occupation listed at the left, state two ways to classify the organisms that people in that occupation work with.

| Occupation | Organisms they work with | Classification system |
|---|---|---|
| Doctor | disease-causing bacteria | by how they react to medicines<br>by what disease they cause |
| Pest exterminator | household insect pests | by where they are found in house<br>by how to get rid of them |
| Florist | flowering plants | by cost<br>by season they are available |
| Veterinarian | pets | by type of animal<br>by owner's last name |
| Dairy farmer | cows | by which cows are giving milk<br>by age |

## OPTIONS

**Application**/*Teacher Resource Package,* p.3. Use the Application worksheet shown here to teach students practical uses for classifying.

# TEACH

## Concept Development

▶ Make sure that students understand the meaning of the word *ancestor.*

▶ Other kinds of evidence are used in classifying. Chromosome structure and embryology are two examples.

▶ Explain that in the classification system, the number of groups increases from kingdom to species. However, the number of kinds of organisms in each group decreases. At the kingdom level, there are five groups that contain every kind of organism. At the species level, there are millions of groups but only one kind of organism in each group. The more groups that organisms share, the more related the organisms are, and the more ancestors they have in common.

▶ The evolutionary history of a species is called phylogeny.

**Misconception:** Students may think that the classification system for organisms never changes. The system changes as scientists gain new information.

### Mini Lab

Students need to include the major features of the group chosen.

### ✓ ASSESSMENT

**Performance:** Have students prepare a key that can be used to identify the five kingdoms. Students should verify that the key works by allowing other classmates to try it. (Review the proper organization and construction of a key before giving this assignment.)

## Guided Practice

Have students write down their answers to the margin question: Horseshoe crabs and spiders have similar body chemistry.

## Independent Practice

**Study Guide**/*Teacher Resource Package,* p. 17. Use the Study Guide worksheet shown at the bottom of this page for independent practice.

### Mini Lab

**What Kingdom?**

**Classify:** Write a description of an imaginary organism so that a person reading the description could classify the organism as a moneran, protist, fungus, plant, or animal. *For more help, refer to the **Skill Handbook,** pages 715-717.*

## Other Evidence Used in Classifying

Classification can be based on a living thing's ancestors. An ancestor is a related organism that lived some time in the past. For example, horses and donkeys have many of the same ancestors. They have more of the same ancestors than horses and goats do. Horses and goats have more of the same ancestors than horses and fish do. Of these pairs, horses and donkeys have the most ancestors in common and are the most related.

Similar body structures often show that living things have common ancestors. This is important in classification. You can see in Figure 3-7 that the front limbs of a human (a), a cat (b), a horse (c), a bird (d), and a bat (e) are similar in their bone structures. These similarities show a common ancestor. Compare the front limbs of animals (a) through (e) with the front limbs of an animal that lived long ago (f). All the limbs are similar even though they do different things. They have similar bones arranged in similar patterns.

**Why is the horseshoe crab classified with spiders?**

Another way to group living things is by body chemistry. A good example is the horseshoe crab. At first, the horseshoe crab was thought to be like other crabs. The blood of the horseshoe crab, however, is more like that of the spider. As a result, the horseshoe crab is now grouped with spiders.

For many years scientists have debated about how to classify the giant panda. Some classified it with raccoons, while others classified it with bears. Now its body chemistry shows it to be more closely related to bears.

**Figure 3-7** Similarity in body structures shows that living things may have had a common ancestor.

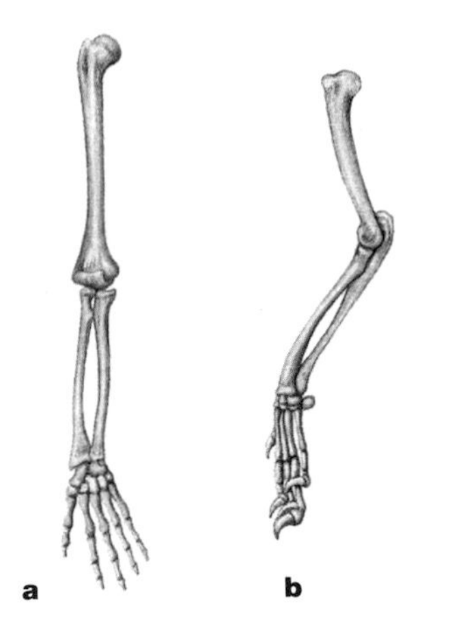


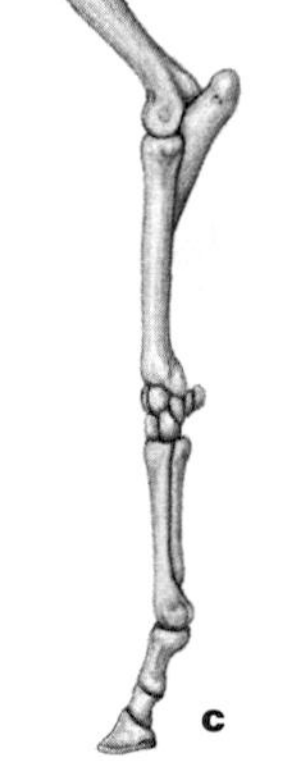


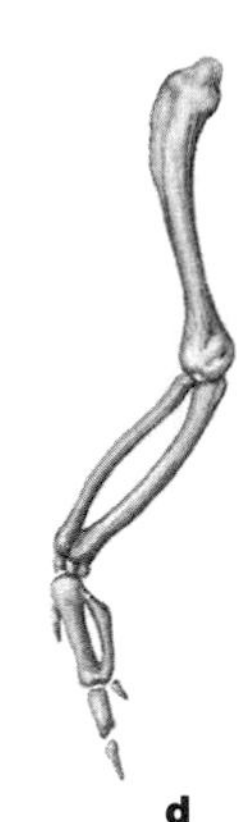


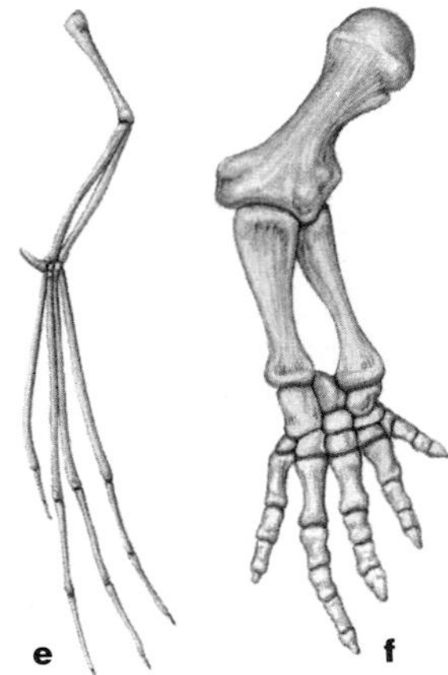




**STUDY GUIDE 17**

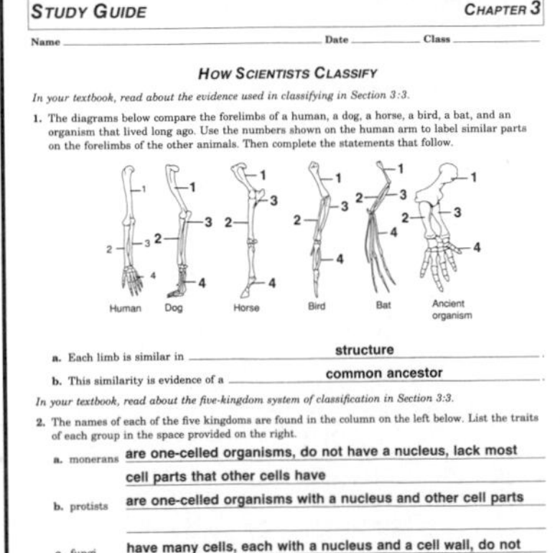

STUDY GUIDE — CHAPTER 3

Name ____ Date ____ Class ____

HOW SCIENTISTS CLASSIFY

*In your textbook, read about the evidence used in classifying in Section 3:3.*

1. The diagrams below compare the forelimbs of a human, a dog, a horse, a bird, a bat, and an organism that lived long ago. Use the numbers shown on the human arm to label similar parts on the forelimbs of the other animals. Then complete the statements that follow.



a. Each limb is similar in **structure**.

b. This similarity is evidence of a **common ancestor**.

*In your textbook, read about the five-kingdom system of classification in Section 3:3.*

2. The names of each of the five kingdoms are found in the column on the left below. List the traits of each group in the space provided on the right.

a. monerans **are one-celled organisms, do not have a nucleus, lack most cell parts that other cells have**

b. protists **are one-celled organisms with a nucleus and other cell parts**

c. fungi **have many cells, each with a nucleus and a cell wall, do not have chlorophyll, cannot make their own food**

d. plants **have many cells, have chlorophyll, can make their own food, do not move**

e. animals **have many cells, cannot make their own food, can move**

## Scientific Names Come from Classification

Tables 3-2 to 3-5 showed that living things are classified down to genus and species. The genus and species names together make up the **scientific name.** In Section 3:2, you learned that Linnaeus was the first to give organisms a scientific name with two parts.

The scientific name for a cat is *Felis catus* (FEE lus • CAT us). The name comes from the genus *Felis* and species *catus*. Notice that the genus is always capitalized and the species is not. Both the genus and the species are in italics.

A wolf's scientific name is *Canis lupus* (KAY nus • LEW pus). The name includes the genus *Canis* and species *lupus*. The scientific name for humans is *Homo sapiens*. What are the genus and the species to which humans belong?

Have you ever called a plant or an animal by its scientific name? You have, but you may not have realized it. Figure 3-8 shows some living things that you know about. Table 3-6 shows a list of their scientific names. You probably have used parts of these scientific names many times.

**What names make up a scientific name?**

**TABLE 3–6. SCIENTIFIC NAMES OF SOME PLANTS AND ANIMALS**

| Scientific name | Name you probably use |
|---|---|
| *Pinus sylvestris* | Scotch pine |
| *Rosa carolina* | rose |
| *Elephas maximus* | Asian elephant |
| *Gorilla gorilla* | gorilla |
| *Giraffa camelopardalis* | giraffe |

**Figure 3-8** Many living things have common names that sound like their scientific names.

## TEACH

### Concept Development

▶ Provide students with some scientific names that are easily converted to a common name. Ask the students to name the organisms based on the term which sounds similar.

▶ Point out that Linnaeus' naming system is still used today. His naming system is called binomial nomenclature. It is a system by which each species is given a two-word Latin name.

▶ The second word of a scientific name is an adjective. Often this term is used to describe the species.

▶ Greek and Latin are the languages used for scientific names. Scientists around the world may speak other languages but they all use Latin or Greek for scientific names.

▶ Students need not worry about proper pronunciation or meaning of Greek and Latin names. Some students will enjoy guessing the meanings of some Latin names.

**Analogy:** An analogy to two-word naming can be found in Chinese names. The first word in a Chinese name is that of the family, and the second and third words are those of the individual. American and European names represent the same idea in reverse order.

### Guided Practice

Have students write down their answers to the margin question: The genus and species names make up the scientific name.

**CRITICAL THINKING 3**

CRITICAL THINKING/PROBLEM SOLVING — CHAPTER 3

Name ______ Date ______ Class ______

*Use after Section 3:3.*

**WHAT TRAITS CAN YOU USE TO CLASSIFY ORGANISMS?**

*Tell how you would decide into what kingdom each organism described below fits.*

Organism A: You are looking at a drop of pond water through a microscope. You discover something you have never seen before. It does not swim around. It is made up of a string of green cells. Each cell is shaped like a shoebox, and has a large nucleus.

Organism B: You are walking along a beach looking at tidepools. In the water you find a soft organism that looks like a thick pink flower. It is about 3 centimeters high. The bottom part is a cylinder. As you watch, a small fish swims toward an opening in the top of the organism. The flowerlike parts catch the fish and slowly push it into the opening.

Organism C: You look at a large bush in the schoolyard. You know that it has been growing in the same place for more than a year. It has green leaves.

**Analyzing the Problem**

1. What traits help you classify organisms? moves or does not move or swim; green color; takes in food; made up of few or many cells; looks like something I already know
2. If you make a table called Characteristics of Kingdoms, what will you place in these columns:

| | |
|---|---|
| Protist | microscopic or small; plantlike or animallike |
| Animal | moves, takes in food |
| Plant | does not move, makes own food, has many cells |
| Fungus | do not move, are not green, can't make their own food |

**Solving the Problem**

1. In what kingdom will you classify Organism A? Explain your answer. protist because it is green, small, does not move, and has a nucleus
2. In what kingdom will you classify Organism B? Explain. animal because it takes in food and moves its body
3. In what kingdom will you classify Organism C? Explain. plant because it is green, does not move, has many cells

## OPTIONS

**Critical Thinking/Problem Solving/** *Teacher Resource Package*, p. 3. Use the Critical Thinking/Problem Solving worksheet shown here to give students additional practice classifying unknown organisms.

## TEACH

### Concept Development

▶ Point out that pill bug, sow bug, and wood louse are all names for the same animal. Sweet pepper, bell pepper, mango, and green pepper are all names for *Capsicum frutescens.*

▶ Write the names of organisms on the chalkboard that are given names that are misleading. Examples are ringworm (fungus), acorn worm (vertebrate), sea cucumber (echinoderm), poison oak (herb), ladybug (beetle).

▶ Scientists discovered microscopic organisms that did not seem to fit in the plant or animal kingdom. The microscope encouraged the development of the Protist kingdom.

### Skill Check

Provide students with field guides to classify leaves.

### Guided Practice

Have students list fish that aren't true fish. Answers could include: crayfish, starfish, jellyfish, and silverfish. Have them write the reasons scientific names are used.

### Independent Practice

**Study Guide**/*Teacher Resource Package,* p. 18. Use the Study Guide worksheet shown at the bottom of this page for independent practice.

## *GLENCOE* TECHNOLOGY

**Videotape**

The Secret of Life

*Gone Before You Know It: The Biodiversity Crisis*

---

### Skill Check

**Classify:** Collect 10 leaves from trees in your neighborhood. Bring the leaves to class and use a key or field guide to find the scientific names of the trees from which the leaves came. *For more help, refer to the* ***Skill Handbook,*** *pages 715-717.*

## Why Scientific Names Are Used

You call a robin, seal, apple tree, and house cat by their common names. Common names are names used in everyday language. Scientists prefer to use scientific names. There are several reasons for using scientific names instead of common names.

1. No mistake can be made about which living thing is being described. That's because two different living things don't have the same scientific name. Two different living things, however, may have the same common name. For example, hawk is the common name for several kinds of birds. Only one kind of hawk, however, is named *Buteo jamaicensis* (BOO tee oh • juh may uh KEN sis). This hawk has broad wings and a reddish tail. You may know it as the red-tailed hawk.
2. Scientific names seldom change.
3. Scientific names are written in the same language around the world. Using scientific names allows scientists to communicate no matter what their everyday language is. The language in which scientific names are written is Latin. Latin is used because it does not change.

**Figure 3-9** Only one of these hawks is *Buteo jamaicensis* (a).

a

b

c

## Classification of Kingdoms

Remember that early scientists grouped living things into two kingdoms—plant and animal. As scientists learned about more living things, they found that some living things did not fit into either kingdom. A new system of classification was needed to group all the living things being discovered.

60 Classification 3:3

---

## *OPTIONS*

### ACTIVITY/**Project**

Have interested students make an insect collection, classifying their organisms to order. The collection can be displayed in class.

**How Can Living Things Be Grouped?**/*Lab Manual,* pp. 21-24. Use this lab as an extension to the classification of living things.

### STUDY GUIDE 18

STUDY GUIDE CHAPTER 3

Name ______ Date ______ Class ______

VOCABULARY

*Review the new words used in Chapter 3 of your textbook. Then, complete the puzzle and definitions below.*

1. Use these words to fill in the blanks of the puzzle.

class, genus, order, scientific name, traits, kingdom, species, phylum, family

Animal K I N G D O M
Chordate P H Y L U M
Mammal C L A S S
Carnivore O R D E R
Felidae F A M I L Y
*Felis* G E N U S
*domesticus* S P E C I E S

*Felis domesticus* is the S C I E N T I F I C N A M E of the house cat. The name comes from the T R A I T S that the animal has.

2. Define each of the following words or phrases.

a. trait feature that a thing has
b. order largest group within a class
c. classify to group things together based on similarities
d. family largest group within an order
e. class largest group within a phylum
f. species smallest group of living things
g. phylum largest group within a kingdom
h. genus largest group within a family

3. A kingdom is the largest group of living things. List the five kingdoms into which scientists classify living things today. moneran, protist, fungus, plant, animal

## TECHNOLOGY

### Six Kingdoms Instead of Five

Scientists use classification as a tool and kingdom as a category for organizing life forms. Do scientists always agree as to the number of kingdoms to be used? No! They cannot agree as to whether there should be five or six kingdoms, thus creating a controversy.

The problem lies with the moneran kingdom, which includes different kinds of bacteria. Some scientists wish to put all bacteria into Kingdom Monera. Others want to divide bacteria into two kingdoms called Archaea and Eubacteria.

What is the basis for this two-kingdom grouping? Archaea live in very unusual places on Earth, such as in deep sea vents, salt lakes, and the hot sulfur springs shown here. These life forms have unique cell walls and membranes, different kinds of proteins and fats, and they don't need oxygen to live. Many give off smelly marsh gas. The life forms that would be placed into the eubacteria kingdom do not show most of these traits and would die if exposed to some of the same harsh conditions under which the Archaea now thrive.

Hot sulfur spring

Today most scientists use this system to classify living things into five kingdoms. The five kingdoms include: monerans (muh NIHR uns), protists (PROH tihsts), fungi, plants, and animals.

**Monerans** are one-celled organisms that don't have a nucleus. They lack most of the cell parts that other cells have. The kingdom has only two phyla, bacteria and blue-green bacteria.

**Protists** are mostly single-celled organisms that have a nucleus and other cell parts. Some have chlorophyll and can make their own food. Others must take in food from the surroundings. Some can move and some can't. The organisms in this kingdom are difficult for scientists to classify because they are so different from one another and from most other living things.

Mushrooms, molds, and yeasts are examples of fungi, (singular fungus). **Fungi** are organisms that have cell walls and absorb food from their surroundings. Fungi don't have chlorophyll, so they can't make their own food.

**What are two differences between monerans and protists?**



## SKILL 3

SKILL: CLASSIFY — CHAPTER 3

Name ______ Date ______ Class ______

*For more help, refer to the Skill Handbook, pages 715-717.* *Use after Section 3:3.*

THE FIVE KINGDOMS OF LIVING THINGS

The pictures at the bottom of the page show organisms that belong in each of the five kingdoms of living things. Classify the organisms in the pictures by filling out the table below. Indicate which organisms belong to which kingdoms by writing the numbers from the appropriate diagrams in the second column. In the third column, list two or more traits of the organisms in each kingdom.

| Kingdom | Organisms | Traits of kingdom |
|---|---|---|
| Moneran | 4, 5 | one-celled, do not have a nucleus |
| Protist | 2, 7, 9 | mostly one-celled, have a nucleus and other cell parts |
| Fungus | 6, 8 | have cell walls, cannot make their own food |
| Plant | 3, 11 | many-celled, do not move, make their own food |
| Animal | 1, 10, 12 | many-celled, can move, cannot make their own food |

1. Starfish 2. Paramecium 3. Moss 4. Blue-green bacteria
5. Bacteria 6. Bread mold 7. Amoeba 8. Mushroom
9. Euglena 10. Butterfly 11. Cactus 12. Earthworm

## OPTIONS

**Skill**/*Teacher Resource Package* p. 3. Use the Skill worksheet shown here to reinforce the skill of classifying.

## TECHNOLOGY

### Six Kingdoms Instead of Five?

#### Background

The controversy over the number of kingdoms can be used to illustrate how new information allows scientists to reorganize and review old ideas. Taxonomy has always been an area that has undergone change, with the discovery of new organisms and now with the ability to detect molecular similarities and differences among organisms.

#### Discussion

1. From what you have read in the chapter, has the number of kingdoms used by scientists ever changed? Give an example. (Yes, Aristotle's and Linnaeus's two-kingdom systems)
2. What type of information is used by scientists who want to add a new kingdom of bacteria? (unusual habitats, no need for oxygen, different structures)
3. Construct an idea map that illustrates the six-kingdom system being proposed. (Idea map should list the five kingdoms discussed plus Kingdom Archaea.)

#### References


Flannery, Maura C. "Back to Bacteria." *American Biology Teacher,* June 1997, pp. 370-373.

Madigan, Michael T. "Extremophiles." *Scientific American,* April 1997, pp. 82-87.


## TEACH

### Guided Practice

Have students write down their answers to the margin question: Monerans don't have a nucleus or most of the cell parts other cells have. Protists have a nucleus and other cell parts.

# TEACH

## Idea Map

Have students use the idea map as a study guide to the major concepts of this section.

## Check for Understanding

Have students respond to the first three questions in Check Your Understanding.

### Reteach

**Reteaching**/*Teacher Resource Package,* p. 9. Use this worksheet to give students additional practice with classification.

**Extension:** Assign Critical Thinking, Biology and Reading, or some of the **OPTIONS** available with this lesson.

## 3 APPLY

Have each student describe and sketch an imaginary organism that would require a new kingdom. Display the sketches.

## 4 CLOSE

### ACTIVITY/Challenge

Have students make a list of several common living things such as the horsefly, poison ivy, cockroach, etc. Have the students use references to find their scientific names.

### Answers to Check Your Understanding

11. Similar body structures often show that living things have common ancestors.
12. Mistakes can be made if common names are used. Scientific names do not usually change. All countries use the same scientific names.
13. Monera, Protist, Fungus, Animal, Plant
14. house cat
15. The word *perfect* means that the present five-kingdom system would never need to be changed.

**Study Tip:** Use this idea map as you study present-day methods of classifying organisms. How many kingdoms are recognized today?

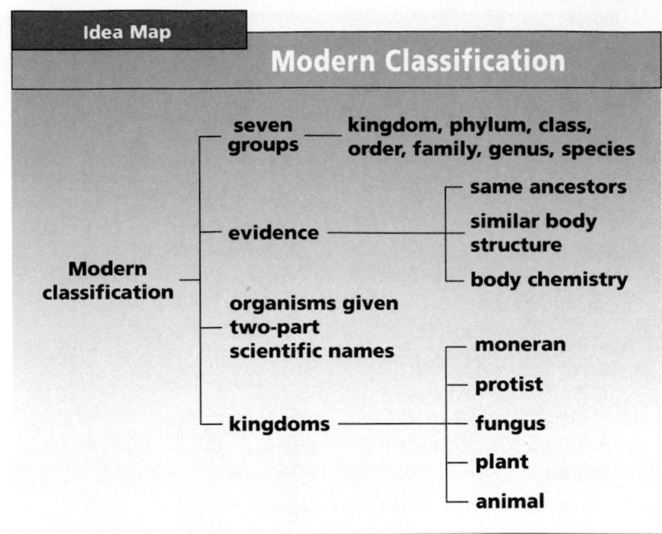


### Bio Tip

**Ecology:** Biologists estimate that millions of living things have not been discovered. Most are protists, monerans, insects, and plants.

**Plants** are organisms that are made up of many cells, have chlorophyll, and can make their own food. Plants don't move. **Animals** are organisms that have many cells, can't make their own food, and can move. The cells of animals and plants have nuclei and other cell parts.

Figure 3-10 shows members of the five kingdoms. The traits of each kingdom are also listed for you. Refer back to this figure as you read about the kingdoms of living things in the next five chapters. Appendix B also contains information about the different groups of living things.

The five-kingdom system isn't perfect. Some living things don't fit into any of the five kingdoms. Their classification changes as scientists learn more about them.

## Check Your Understanding

11. How is similarity in body structure used to classify living things?
12. Give two reasons why scientists use scientific names.
13. What are five kingdoms recognized today?
14. **Critical Thinking:** You are a chordate. Are you more like a house cat or an octopus?
15. **Biology and Reading:** What does the word *perfect* mean as it is used in the last paragraph of this section?



**RETEACHING 9**

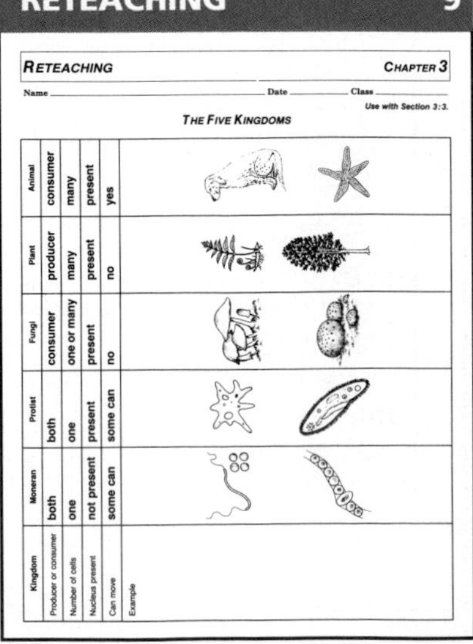

RETEACHING CHAPTER 3

Name ____ Date ____ Class ____

Use with Section 3:3.

THE FIVE KINGDOMS

| Kingdom | Producer or consumer | Number of cells | Nucleus present | Can move | Example |
|---|---|---|---|---|---|
| Moneran | both | one | not present | some can | |
| Protist | both | one | present | some can | |
| Fungi | consumer | one or many | present | no | |
| Plant | producer | many | present | no | |
| Animal | consumer | many | present | yes | |

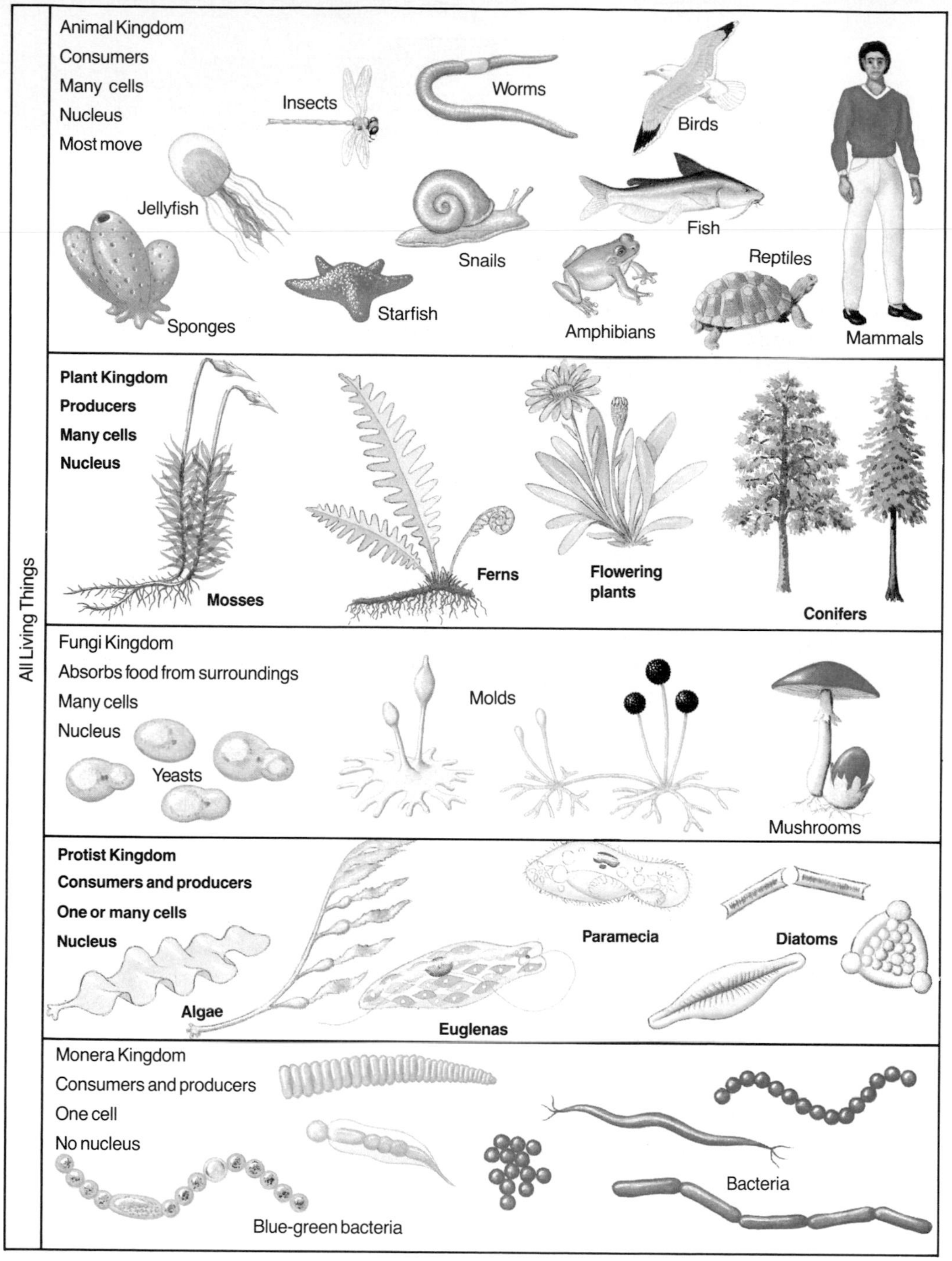

**Figure 3-10** Living things are classified into five kingdoms.



**Portfolio**

Have students prepare a chart that summarizes the similarities and differences between monerans, protists, fungi, plants, and animals. Each chart should include information regarding size, color, cell organization, habitat, and classification. Pictures from magazines can be used to illustrate each group.

## GLENCOE TECHNOLOGY

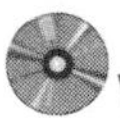

**Videodisc**

The Secret of Life

*Six Kingdoms*

*Order at Every Level*

*Classifying Organisms*

## OPTIONS

### ACTIVITY/Enrichment

Use the microprojector to show prepared slides of bacteria, blue-green bacteria, paramecia, and euglena. This will help students understand the traits used in placing organisms into kingdoms. Observe plants, fungi, and guppies to see the differences in how organisms get food.

**Five Kingdoms**/*Transparency Package,* number 3. Use color transparency number 3 as you introduce the five kingdoms.

## CHAPTER 3 REVIEW

### Summary

Summary statements can be used by students to review the major concepts of the chapter.

### Key Science Words

All boldfaced terms from the chapter are listed.

### Testing Yourself

#### Using Words

1. species
2. kingdom
3. trait
4. phylum
5. scientific name
6. genus
7. classify
8. animal
9. fungus
10. plant

#### Finding Main Ideas

11. p. 51; Aristotle classified plants based on how they grew and their size, and animals by where they lived.
12. pp. 61-62; Three traits used to place living things in one of the five kingdoms are number of cells, cell nucleus, and how the living things get food.
13. p. 48; Answers will vary, but some examples of everyday things that are grouped include sports items, groceries, classrooms, classes, and books.
14. p. 54; The seven main groups in the modern classification system are kingdom, phylum, class, order, family, genus, and species.
15. p. 59; The scientific name is made up of the genus and species.
16. pp. 48-50; We classify things to put them in order, to make them easier to find, and to show how things are alike.
17. p. 53; Linnaeus gave living things two-part names.
18. p. 58; The ancestors of two living things give a clue as to how related they are.

# Chapter 3 Review

## Summary

### 3:1 Why Things Are Grouped

1. Many things are classified in everyday life.
2. Things are grouped together based on similarities. Two reasons for classifying things are to make them easier to find and to show how they are alike.

### 3:2 Methods of Classification

3. Aristotle was one of the first people to classify living things. He grouped animals according to where they lived. He grouped plants according to size and pattern of growth.
4. Linnaeus used specific traits to group living things. He also designed a scientific naming system that is still used today.
5. There are seven main groups in the modern classification system. They are kingdom, phylum, class, order, family, genus, and species.

### 3:3 How Scientists Classify Today

6. Classifying is based on how organisms are related, ancestors, similarities in body structure, and body chemistry.
7. Scientists use scientific names because no two living things have the same scientific name.
8. Today, living things are grouped into five kingdoms–monerans, protists, fungi, plants, and animals.

## Key Science Words

animal (p. 62)
class (p. 54)
classify (p. 48)
family (p. 54)
fungus (p. 61)
genus (p. 55)
kingdom (p. 53)
moneran (p. 61)
order (p. 54)
phylum (p. 54)
plant (p. 62)
protist (p. 61)
scientific name (p. 59)
species (p. 55)
trait (p. 48)

## Testing Yourself

### Using Words

*Choose the word from the list of Key Science Words that best fits the definition.*

1. smallest group of living things
2. largest group of living things
3. a feature of something
4. largest group within a kingdom
5. genus and species names together
6. first word in a scientific name
7. to group things based on similarity
8. organism that has many cells and moves
9. organism that has cell walls and absorbs food from its surroundings
10. organism that has many cells and can make its own food

64

#### Using Main Ideas

19. Many of the newly discovered living things did not fit into Aristotle's grouping system.
20. They could all be classified into tall letters and short letters. T, b, l, and d are tall letters, and c, e, and o are short letters.
21. Scientists use scientific names so that no mistake can be made about what living thing is being described, because they seldom change, and because scientists all over the world use the same Latin names.
22. Books in a library are classified so they will be easier to find.
23. Plants are producers and animals are consumers.
24. Lion – animal kingdom, chordate phylum, mammal class, carnivore order, family Felidae, genus *Panthera,* species *leo;* deer — animal kingdom, chordate phylum, mammal class, order Artiodactyla, family Cervidae, genus *Odocoileus,* species *virginianus.*
25. Linnaeus used more groups, based his system on specific traits, and gave organisms names that described their traits.

# Review

## Testing Yourself *continued*

**Finding Main Ideas**

*List the page number where each main idea below is found. Then, explain each main idea.*

11. how Aristotle classified plants and animals
12. three traits used to place living things in one of the five kingdoms
13. things in our everyday lives that are grouped
14. the seven main groups in the modern classification system
15. what makes up the scientific name
16. two reasons why things are classified
17. the scientist who gave living things two-part names
18. how living things are grouped based on their ancestors

**Using Main Ideas**

*Answer the questions by referring to the page number after each question.*

19. Why did scientists depend less and less on Aristotle's grouping system? (p. 53)
20. How would you classify the letters b, c, e, d, l, o, t into two groups? Explain your groups. (p. 48)
21. What are three reasons why scientists use scientific names? (p. 60)
22. Why do we classify books in a library? (p. 48)
23. How do plants and animals differ? (p. 62)
24. What are the classification groups for a lion and a deer? (p. 57)
25. What changes did Linnaeus make in Aristotle's classification system? (p. 53)

## Skill Review 

*For more help, refer to the **Skill Handbook**, pages 704-719.*

1. **Infer:** List the similarities and differences between the animals in Figure 3-3. Infer how closely related the animals are.
2. **Classify:** Classify the following animals into four groups on the basis of their structural similarities: snail, rabbit, gorilla, butterfly, oyster, chimpanzee, clam, cat, honeybee.
3. **Infer:** What is the result of two organisms having a large number of the same ancestors?
4. **Observe:** Obtain a sample of pond water. Place some of the water on a microscope slide and observe the organisms in the water. To what kingdoms do the organisms belong?

## Finding Out More

TECH PREP

**Critical Thinking**

1. Cut out photographs of 20 people from magazines. Design a classification system for these people.
2. How is classification helpful to farmers, florists, and customs inspectors?

**Applications**

1. Keys are used to identify things that have been classified and named. From the library, obtain a key to the trees or flowering plants in your area. Identify 20 plants using the key.
2. Visit a zoo or an aquarium. Make a list of the living things seen there. Use reference books to identify and classify the living things to the species level.



## Skill Review

1. The three insects have different shapes. Two are long and thin, one is round and heavy bodied. All have six legs, all are insects. However, the two long and thin ones are probably more closely related.
2. Students could use the following classes:
   no legs, no wings - snail, oyster, clam
   2 legs, no wings – gorilla, chimpanzee
   4 legs, no wings – rabbit, cat
   6 legs, wings – butterfly, honeybee
3. They share more of the classification categories.
4. Students should see organisms that belong to protist, plant, and animal kingdoms.

## Finding Out More

### Critical Thinking

1. Race, size, age, hair color, and gender are some traits that can be used in classifying people.
2. Each must be able to classify and identify plants.

### Applications

1. Pocket keys with pictures or diagrams are helpful in identifying organisms in the field.
2. Students may ask the zoo or aquarium keeper how the information necessary to classify living things is obtained.

 **Chapter Review**/*Teacher Resource Book,* pp. 11-12.

 **Chapter Test**/*Teacher Resource Book,* pp. 13-15.

**Chapter Test**/*Computer Test Bank*

 **Unit Test**/*Teacher Resource Package,* pp. 16-17.

# Unit 1
# Kinds of Life

CONNECTIONS

## Biology in Your World

In order to study kinds of life or any other topic, scientists need a plan of action. The scientific method provides a structured way to answer scientific questions and solve problems. This method can be applied to all areas of life, from everyday mysteries to scientific phenomena.

### Literature Connection

Alex Haley's (1921- ) novel *Roots* documents his family history. Haley managed to trace his family to the village of Juffre in Gambia, West Africa. His great-great-great-great-great-grandfather, Kunta Kinte, was abducted from West Africa in 1767. Kunta was taken on a slave ship to Maryland, where he was sold to a Virginia plantation owner. *Roots* chronicles Haley's family from Kunta to Haley himself. The book was dramatized on television in 1977, and the last episode drew one of the largest audiences in television history.

### Geography Connection

According to Alfred Wegener's theory of continental drift, all the continents were once a single land mass called Pangaea. Around 200 million years ago, Pangaea split into two large land masses and these later broke into the present-day continents. The continents drifted about 2.5 cm per year to their present positions.

Wegener showed that plants that once grew in Greenland may be related to plants now growing in the tropics. This finding supports his theory of continental drift.

Today, the theory of plate tectonics is used to explain continental drift. According to this theory, Earth's outer crust is made of rigid plates. Convection currents in Earth's mantle cause the plates to move; thus, the continents drift over the years. Plate movement causes earthquakes, volcanoes, and the formation of mountains and ocean ridges.

CONNECTIONS

# Biology in Your World

### The Scientific Method Is All Around You

In this unit, you have read how the scientific method is used to solve problems in science. But, did you know people use scientific methods to solve other kinds of problems? Consider the "mysteries" described below and how scientific methods were used to solve them.

LITERATURE

### A Scientific Search for Roots

Alex Haley learned from his grandmother that he was part of the Kinte family. His African ancestor, Kunta Kinte, had been kidnapped in Africa and then brought to this country as a slave.

Kinte's family handed down African words for six generations. These words were the names of an African village, a certain tree, and a musical instrument. Haley traced these to the Gambia River in West Africa. There he found a man who could recite the history of the Kinte family. The man's story fit the facts Haley had before his trip. Haley wrote *Roots* to tell the story of his search.

GEOGRAPHY

### Shifting Continents

The east coast of South America and the west coast of Africa could almost fit together like the pieces of a jigsaw puzzle. This observation led scientists to form the continental drift theory. The theory says that there was once a single great continent that split and drifted apart.

Other evidence indicates that all the land masses of Earth may have been joined at one time. Layers of rock along a coast on one side of the ocean match layers of rock on an opposite coast. Also, different continents have the same kinds of fossils. Thus, observation and evidence support the continental drift theory.

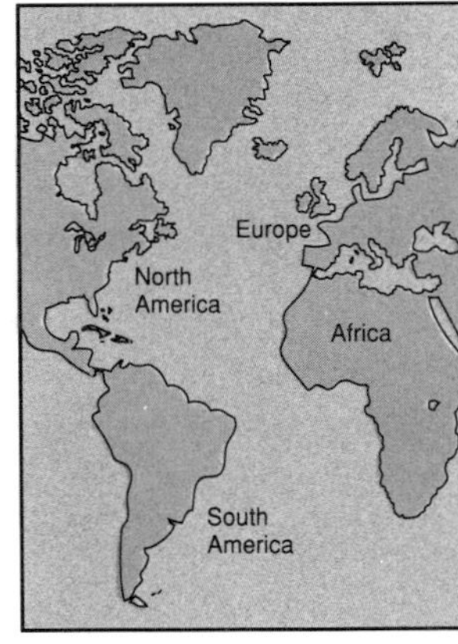


66

**GLENCOE TECHNOLOGY**

**Videodisc**

**Biology: The Dynamics of Life**
Animation: *Plate Tectonics*
Disc 1, Side 2, Ch. 3

LEISURE

## Green Thumbs Up!

Many teens enjoy spending time assembling a terrarium. You can have a miniature garden in your bedroom! Terrariums are fun to arrange, and the materials are easy to find. All you need is a clear container, pebbles, sand, soil, and some plants. Ferns and mosses grow well in terrariums. You can try to grow many other kinds of plants, also. Experiment with various plants to see which are best suited for a terrarium. Set up your experiment using the scientific method.

How would you vary the conditions in or surrounding a terrarium to find the best environment for it?

CONSUMER

## Cycling Through Your Choices

Your old bicycle is just about worn out, and you need to buy a new one. How can you use a scientific method to choose the best bicycle? You can treat your decision like solving a problem.

First, research the types of bicycles. What are the advantages, disadvantages, and purposes of each type?

What do you need in a bicycle? Prepare a list of all the features and uses you want in a bicycle. Then rate each feature with a number from 1 to 5. A feature you really need is a 1. A feature that is nice but you could do without would be a 5.

Compare your list of needed features with the types of bicycles. Do your features match the purposes and advantages of one or two types of bicycles? If not, you may need to rethink your needs.

Go to a bicycle shop to see and ride bicycles. Narrow your choices to one type and choose your bicycle based on all the data you have gathered about the different bicycles.

## CULTURAL DIVERSITY

### Cultural Adaptations to the Environment

Humans occupy almost all types of habitats. People adapt to Earth's varying environments in many ways. For example, humans are able to meet their nutritional needs by eating a varied diet and by growing many different kinds of crops. People have designed clothing suited to virtually all climate conditions—from heavy rainfall to subzero temperatures. Architectural designs that make use of available materials help people create shelters suitable for a particular environment. For example, the Inuit of North America created housing using their most available resources: snow and ice. Peoples of the southwestern United States often built their houses into mountainsides using a mud-clay mixture called adobe. Have students research and report on other examples of shelters or clothing designed for a particular environment.

CONNECTIONS

### Leisure Connection

Terrariums can provide plants and animals with ideal growing conditions. If kept moist and given indirect sunlight, a covered terrarium preserves moisture and heat to create a warm, humid environment. Plants that grow well together, such as ferns and ivy, are good choices for terrariums. Small animals such as lizards, toads, salamanders, and small snakes can also thrive in this atmosphere.

### Consumer Connection

In listing and rating features of a bicycle, students must consider how they will use a bike. Will it be used for racing, touring, off-road riding, or pleasure riding? Do they want to ride alone or with a partner? Will they carry a lot with them when they ride? Do they want hand or pedal brakes? Do they want a headlight for night-time riding?

Other considerations are the cost, comfort, and look of the bicycle.

### References


Haley, Alex. *Roots.* New York: Doubleday & Co., 1976.

Hershey, David R. "Doctor Ward's Accidental Terrarium." *The American Biology Teacher*, May 1996, pp. 276-281.

Taylor, S. Ross, and Scott M. McLennan. "The Evolution of Continental Crust." *Scientific American*, January 1996, pp. 76-81.

# Unit 2 Kingdoms

## UNIT OVERVIEW

Unit 2 is a survey of the organisms classified in the moneran, protist, fungus, plant, and animal kingdoms. Viruses are also discussed. The major characteristics used in classifying the organisms are identified and the features that distinguish them from other organisms are explained.

## Advance Planning

Arrange a field trip to a zoo or a botanical garden. Obtain materials for setting up a classroom terrarium.

Audiovisual and computer software suggestions are located within each chapter under the heading TEACH.

**To prepare for:**

Lab 4-1: Purchase 3.7 x 0.7 cm bolts and nuts, #22 gauge wire, 4.5 cm polystyrene balls and pipe cleaners.

Lab 4-2: Arrange to have prepared slides of *Anabaena* and bacteria, petri dishes, and metric rulers.

Mini Labs: Assemble beef broth, salt, sugar, vinegar, and water. Obtain yogurt and prepare methylene blue.

Lab 5-1: Order cultures of: amoeba, euglena, and paramecium.

Lab 5-2: Arrange to have prepared slides of mushroom basidia.

Mini Lab: Gather resealable plastic bags.

Lab 6-1: Collect or order moss plants with spore cases.

Lab 6-2: Collect or order bracken ferns with spore cases.

Mini Labs: Purchase red food coloring, carnations, and plastic bags. Collect a selection of open and closed pine cones.

Lab 7-1: Order hydra culture and food (*Daphnia* or brine shrimp).

Lab 7-2: Order or collect live earthworms.

Mini Labs: Collect rocks and gather suction cups. Gather wire mesh, funnels, and jars.

Lab 8-1: Order or collect crayfish.

Lab 8-2: Arrange to have tail or wing feathers, down feathers.

Mini Labs: Purchase flour and locate a dew spider web. Assemble wood, a hammer, bottle caps, nails, peanut butter, and bird seed.

# Unit 2

CONTENTS

68

# Kingdoms

## What would happen if...

there were no fungi? Fungi use and break down substances in rotting logs and other dead organisms. This action releases nutrients into the soil for plants to use. Without fungi, the nutrients in the soil would soon be used up. Then, no plants would grow. Without plants, animals such as the elk in the photo would have no food to eat.

Fungi are also needed to produce many things we've come to depend on. Yeasts, molds, and mushrooms are all fungi. Carbon dioxide gas produced by yeast causes bread to rise. Without yeast, our bread would be flat. Without a certain type of mold, there would be no penicillin. This important antibiotic keeps many deadly diseases under control. Other molds are used to flavor some types of cheeses.

Fungi may be simple and small, but they are a necessary part of our world.

69

UNIT OVERVIEW

### Photo Teaching Tip

Ask students to read the passage and discuss the possible outcomes of life without fungi. Lead students to the understanding that all organisms interact and have important roles in the balance of nature. The discussion may result in the conclusion that all life-forms are valuable and worth protecting.

### Theme Development

There are many different kinds of living things, all of which are alike in some basic ways. This unit describes the five kingdoms of living things and how all organisms interact with each other. The themes of diversity and ecology are developed in this unit.

**CULTURAL DIVERSITY**

See the Cultural Diversity feature located on the Connections page at the end of this unit.

For Tech Prep activities, see the Applying Technology feature in this unit, the Finding Out More applications questions in the Chapter Reviews, and the Tech Prep Applications booklet in the Teacher Resource Package.

# Viruses and Monerans

## PLANNING GUIDE

| CONTENT | TEXT FEATURES | TEACHER RESOURCE PACKAGE | OTHER COMPONENTS |
|---|---|---|---|
| (1 day)<br>4:1 Viruses<br>Traits of Viruses<br>Viruses and Disease<br>Controlling Viruses | Skill Check: *Interpret Diagrams,* p. 75<br>Lab 4-1: *Viruses,* p. 76<br>Idea Map, p. 77<br>Check Your Understanding, p. 78 | Reteaching: *Virus Life Cycles,* p. 10<br>Critical Thinking/Problem Solving: *Why Are Medicines Not Used to Treat Virus Diseases?* p. 4<br>Study Guide: *Viruses,* pp. 19-20<br>Lab 4-1: *Viruses,* pp. 13-14 | **STVS:** *Mapping a Virus,* Plants and Simple Organisms (Disc 4, Side 1) |
| (3 days)<br>4:2 Monera Kingdom<br>Traits of Bacteria<br>Bacteria and Disease<br>Helpful Bacteria<br>Controlling Bacteria<br>Blue-green Bacteria | Skill Check: *Understand Science Words,* p. 85<br>Mini Lab: *What Prevents the Growth of Bacteria?* p. 83<br>Mini Lab: *What Makes Yogurt?* p. 84<br>Technology: *Uses of Bacteria,* p. 86<br>Lab 4-2: *Monerans,* p. 89<br>Idea Map, p. 87<br>Check Your Understanding, p. 88 | Application: *Ways to Control Bacteria,* p. 4<br>Enrichment: *Viruses, Monerans, and Disease,* p. 4<br>Reteaching: *Monerans,* p. 11<br>Skill: *Make and Use Tables,* p. 4<br>Focus: *How Do Bacteria Affect Our Lives?* p. 7<br>Transparency Master: *Comparing Viruses and Monerans,* p. 15<br>Study Guide: *Monera Kingdom,* pp. 21-23<br>Study Guide: *Vocabulary,* p. 24<br>Lab 4-2: *Monerans,* pp. 15-16 | **Laboratory Manual:**<br>*How Can You Test for Bacteria in Foods?* p. 25<br>**Laboratory Manual:**<br>*Does Soil Contain Bacteria?* p. 27<br>**Laboratory Manual:**<br>*What Are the Traits of Blue-green Bacteria?* p. 31<br>Color Transparency 4: *Bacterial and Animal Cells*<br>**STVS:** *Anticavity Vaccine,* Plants and Simple Organisms (Disc 4, Side 1)<br>**STVS:** *Bacterial Control of Mosquitoes,* Plants and Simple Organisms (Disc 4, Side 1) |
| Chapter Review | Summary<br>Key Science Words<br>Testing Yourself<br>Finding Main Ideas<br>Using Main Ideas<br>Skill Review | **ASSESSMENT RESOURCES**<br>Chapter Review, pp. 18-19<br>Chapter Test, pp. 20-22<br>Performance Assessment in the Biology Classroom<br>Alternate Assessment in the Science Classroom<br>Computer Test Bank | |

### GLENCOE TECHNOLOGY

**Infinite Voyage,** *Secrets from a Frozen World*
**Science and Technology Videodisc Series,** *Mapping a Virus,* Plants and Simple Organisms (Disc 4, Side1)
*Anticavity Vaccine,* Plants and Simple Organisms (Disc 4, Side1)
*Bacterial Control of Mosquitoes,* Plants and Simple Organisms (Disc 4, Side1)

**The Secret of Life,** *Nothing to Sneeze At: Viruses*

## MATERIALS NEEDED

| LAB 4-1, p. 76 | LAB 4-2, p. 89 | MARGIN FEATURES |
|---|---|---|
| bolt<br>2 nuts to fit bolt<br>2 pieces wire<br>polystyrene ball<br>pipecleaners, cut in 2-cm lengths | microscope<br>prepared slide of *Anabaena*<br>prepared slide of bacteria<br>petri dish<br>metric ruler | Skill Check, p. 75<br>pencil<br>paper<br>Skill Check, p. 85<br>dictionary<br>Mini Lab, p. 83<br>4 clean beakers<br>4 mL beef broth<br>1 t. salt<br>1 t. sugar<br>1 t. vinegar<br>boiling water<br>Mini Lab, p. 84<br>microscope<br>microscope slide<br>coverslip<br>yogurt<br>water<br>methylene blue stain |

## OBJECTIVES

For more information about National Science Standards, see page 5T.

| SECTION | OBJECTIVE | CORRELATION of QUESTIONS to OBJECTIVES: CHECK YOUR UNDERSTANDING | CHAPTER REVIEW | TRP CHAPTER REVIEW | TRP CHAPTER TEST |
|---|---|---|---|---|---|
| 4:1 National Science Stds: UCP.1, UCP.2, UCP.5, C.1, C.4, F.1 | 1. **Identify** the traits of viruses. | 1 | 10, 11, 21 | 4, 8, 11, 12 | 2, 8, 9, 17, 21, 25, 30 |
| | 2. **Tell** how viruses reproduce. | 2 | 2, 22 | 5 | 7, 27 |
| | 3. **List** examples of how viruses affect living things. | 3, 4, 5 | 1, 12, 13, 23, 24 | 7, 18, 25, 28 | 5, 6, 25, 29, 32, 43, 49 |
| 4:2 National Science Stds: UCP.1, UCP.2, UCP.5, A.2, C.1, C.4, C.5, C.6, D.2, E.2, F.1, F.4, F.5, G.3 | 4. **Identify** the traits of bacteria. | 6 | 4, 8, 9, 14, 15, 18, 25, 26 | 2, 6, 9, 10, 13, 21, 29, 30, 31 | 1, 3, 4, 12, 17 |
| | 5. **Explain** how bacteria affect other living things. | 7, 9 | 3, 5, 16, 17, 19, 29, 30 | 3, 15, 16, 17, 19, 20, 22, 23, 24, 26, 27, 28, 32, 33, 34, 35 | 10, 11, 13, 21, 26, 31, 33, 34, 35, 36, 37, 38, 39, 40, 41, 42, 44, 45, 46, 47, 48, 50 |
| | 6. **Compare** the traits of bacteria and blue-green bacteria. | 8 | 7, 20, 31 | 1, 14, 15, 30 | 18, 19, 22, 23, 26 |

# Viruses and Monerans

## CHAPTER OVERVIEW

### Key Concepts

In this chapter, students will study the traits of viruses and bacteria, how they cause disease, and how they can be controlled. They will also study how bacteria are helpful. Students will study the traits of blue-green bacteria and how they are helpful and harmful.

### Key Science Words

| | |
|---|---|
| antibiotic | endospore |
| asexual reproduction | fission |
| bacteria | flagellum |
| biotechnology | host |
| blue-green bacteria | interferon |
| capsule | Koch's postulates |
| colony | parasite |
| communicable disease | pasteurization |
| decomposer | saprophyte |
| | vaccine |
| | virus |

### Skill Development

In Lab 4-1, students will **observe, recognize and use spatial relationships**, and **formulate a model** to understand the shape of a virus. In Lab 4-2, students will **make and use tables, observe, measure in SI,** and **use a microscope** to compare bacteria with blue-green bacteria. In the Skill Check on page 75, students will **interpret diagrams** to describe how a virus causes disease. On page 85, students will **understand** the **science word** *biotechnology*. In the Mini Lab on page 83, students will **infer** that salt and vinegar retard growth of bacteria. In the Mini Lab on page 84, students will **use a microscope** to observe bacteria in yogurt.

### Bridging

Review the features of living things and the parts of typical cells in Chapter 2 to help students understand that viruses do not have cell parts and do not have all the features of living things. Bridge this section to bacteria, showing how bacteria are different from other cells.

**CHAPTER PREVIEW**

### Chapter Content

Review this outline for Chapter 4 before you read the chapter.

**4:1 Viruses**
- Traits of Viruses
- Viruses and Disease
- Controlling Viruses

**4:2 Monera Kingdom**
- Traits of Bacteria
- Bacteria and Disease
- Helpful Bacteria
- Controlling Bacteria
- Blue-green Bacteria

### Skills in this Chapter

The skills that you will use in this chapter are listed below.
- In **Lab 4-1**, you will observe, recognize and use spatial relationships, and formulate a model. In **Lab 4-2**, you will make and use tables, observe, measure in SI, and use a microscope.
- In the **Skill Checks**, you will interpret diagrams and understand science words.
- In the **Mini Labs**, you will infer and use a microscope.

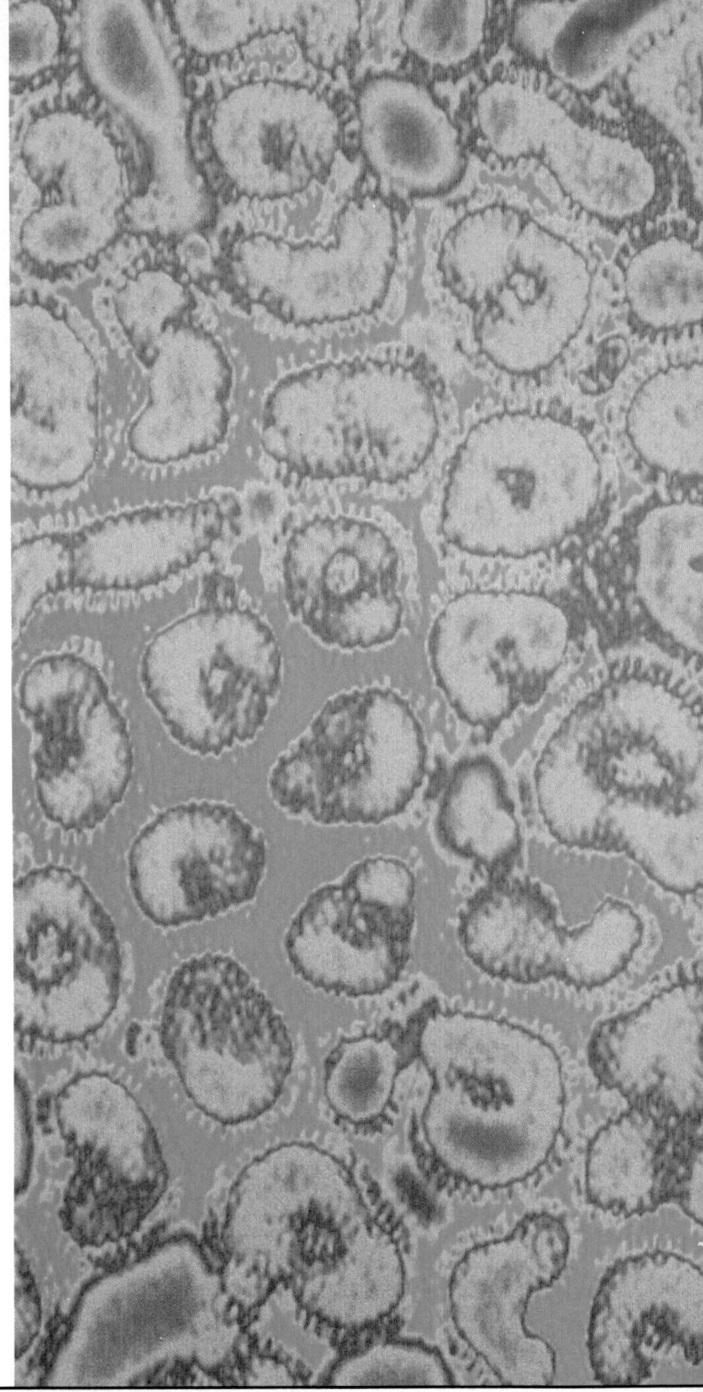

Flu virus

70

## TECH PREP

For Tech Prep activities in this chapter of the Teacher Wraparound Edition, see especially the Cooperative Learning activities on pages 77 and 85, the Portfolio on page 78, the Activity on page 83, and the Motivations on pages 79, 83, and 84. See also the Glencoe Homepage at **www.glencoe.com**

The following Glencoe resources provide additional opportunities for integrating science and technology.

**Technology: Science and Math in Action, Book Two**
Module 1: Recycling

Chapter 4

# Viruses and Monerans

Have you ever suffered from the flu? If so, you have had a disease caused by a virus. There are several kinds of viruses that cause the flu. You can get the flu more than once a year because you can be infected by different flu viruses. The photo on the left shows one of the viruses that can cause the flu. It is enlarged 120 000 times. What does this tell you about the size of viruses? In the smaller photo you can see bacteria that cause pneumonia. They are enlarged 75 000 times. Are bacteria or viruses larger?

In this chapter you will read about viruses and monerans. Viruses and monerans affect organisms that are billions of times larger than they are.

Pneumonia bacteria

### Try This!

**What methods do people use to fight viruses?** Interview several people to learn about folk remedies used to treat the common cold. Make a list of the folk remedies. After you have studied viruses, decide which folk remedies would be the most helpful in fighting a cold.

### interNET CONNECTION

For more information about the material in this chapter, follow the link for the chapter on the Glencoe Homepage at **http://www.glencoe.com**



## ASSESSMENT PLANNER

**Portfolio**

Strategies on the following pages represent student products that can be placed into a best-work portfolio: pp. 78, 80, 83.

**PERFORMANCE ASSESSMENT**

Skill Check, p. 75
Mini Lab, pp. 83, 84
Lab, pp. 76, 89

**CONTENT ASSESSMENT**

Check for Understanding, pp. 78, 87, 88
Chapter Review, pp. 90-91

**GROUP ASSESSMENT**

Opportunities for group assessment occur with Cooperative Learning Strategies.

**Student Journal**

Strategies on the following pages represent opportunities for writing in a Student Journal: pp. 73, 77, 81.

## GETTING STARTED

### Using the Photos

In looking at the photos, students will think at first that bacteria are smaller than viruses. Have students note the different degrees of magnification. By comparing the difference, they will be able to determine that viruses are smaller than bacteria.

### MOTIVATION/Try This!

**What methods do people use to fight viruses?** Upon completing the chapter-opening activity, students will have **communicated** with a number of persons and collected data that they can use in making a decision.

### Chapter Preview

Have students study the chapter outline before they begin to read the chapter. The structure of viruses, the role of viruses in causing disease, and how viruses are controlled are studied first. The unique features of the two phyla in the moneran kingdom and their helpful and harmful effects are then discussed.

### Misconception

Students may think that medicines cure colds. Medicines provide relief from symptoms, but not a cure. A cure would destroy the cold virus or disable it. We suffer from a cold until we develop antibodies. The medicines simply make us comfortable.

## 4:1 Viruses

### PREPARATION

#### Materials Needed

Make copies of the Critical Thinking/Problem-Solving, Study Guide, Reteaching, and Lab worksheets in the *Teacher Resource Package*.

▶ Obtain a basketball, ping-pong ball, and BB shot for the MOTIVATION/Demonstration.

▶ Purchase bolts, nuts, wire, polystyrene balls, and pipecleaners for Lab 4-1.

#### Key Science Words

virus
host
parasite
interferon
vaccine

#### Process Skills

In Lab 4-1, students will observe, formulate a model, and recognize and use spatial relationships. In the Skill Check, students will interpret diagrams.

### 1 FOCUS

▶ The objectives are listed on the student page. Remind students to preview these objectives as a guide to this numbered section.

#### ACTIVITY/Word Association

The word *virus* comes from the Latin word for poison. Ask students to describe their most recent viral attacks. This will help the class remember the word-root association.



**Objectives**

1. **Identify** the traits of viruses.
2. **Tell** how viruses reproduce.
3. **List** examples of how viruses affect living things.

**Key Science Words**

virus
host
parasite
interferon
vaccine

## 4:1 Viruses

Viruses are neither living nor nonliving things, and yet they have some traits of both. Viruses are difficult for scientists to classify. What are these unusual things and what are their unusual traits?

### Traits of Viruses

Viruses are so small that they can be seen only with an electron microscope. Figure 4-1 shows three kinds of viruses highly magnified. Viruses have many different shapes. Some are round and some are rod-shaped. Other viruses are many-sided.

A **virus** is made of a chromosome-like part surrounded by a protein coat. The chromosome-like part carries the hereditary material. The protein coat is responsible for the different shapes of viruses. The structures of some viruses are shown in Figure 4-1.

Viruses are not made of cells and have no cell parts. They do not grow or respond to changes in the environment as living things do. The trait that viruses share with living things is the ability to reproduce. However, viruses reproduce only inside living cells.

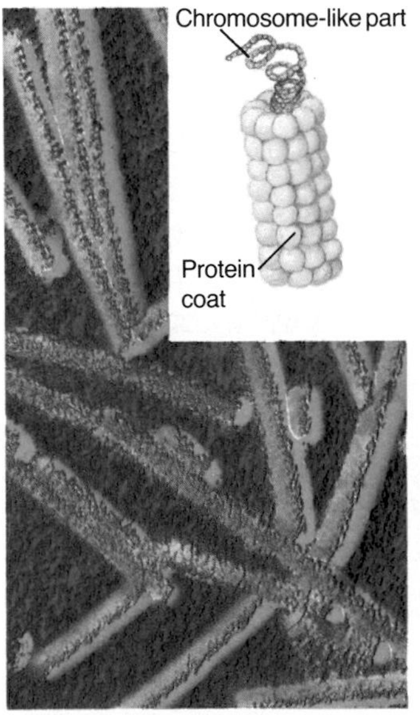


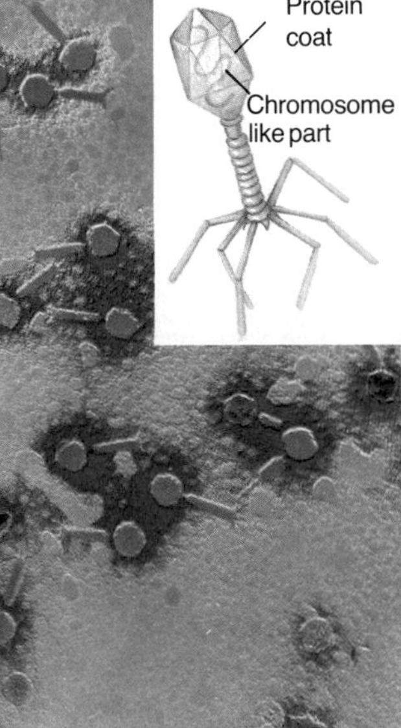


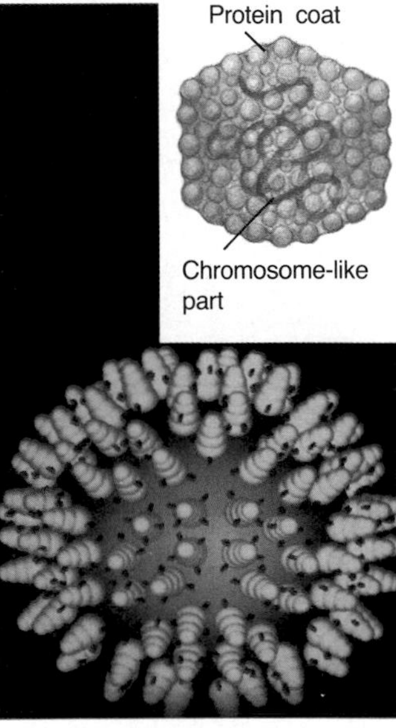


**Figure 4-1** All viruses are made of a chromosome-like part and a protein coat.



### OPTIONS

#### Science Background

In 1892, Dimitri Ivanowski, a Russian botanist, discovered that something smaller than bacteria could cause tobacco mosaic disease. In 1953, Dr. Wendell Stanley isolated the first virus, tobacco mosaic virus, and showed that viruses could be crystallized.

Parts of viruses that look like crystals were stored in labs for as long as 50 years. Scientists then placed the viruses inside living cells. The viruses grew and reproduced. You can see why some scientists consider viruses to be living and others consider them nonliving. For this reason, viruses are not grouped into any kingdom.

One trait used to group viruses is the kind of cell they infect. Plants, animals, fungi, monerans, and protists all serve as hosts to different kinds of viruses. A **host** is an organism that provides food for a parasite (PAR uh site). A **parasite** is an organism that lives in or on another living thing and gets food from it. In this example, viruses act like parasites.

Each kind of virus infects a certain host. For example, the tobacco mosaic virus in Figure 4-2 will infect tobacco plants, but not corn or wheat plants. Some viruses will infect only certain parts of their hosts. The rabies virus will infect only the nervous system of mammals. Common cold viruses will infect only the cells along the air pathway to the lungs. The shape of the protein coat and size of the virus are two other traits that scientists use to classify viruses.

Some scientists think that viruses came into being before cells existed. Others say that viruses are parts of early cells. Still another idea is that they may have come from disease-causing bacteria.

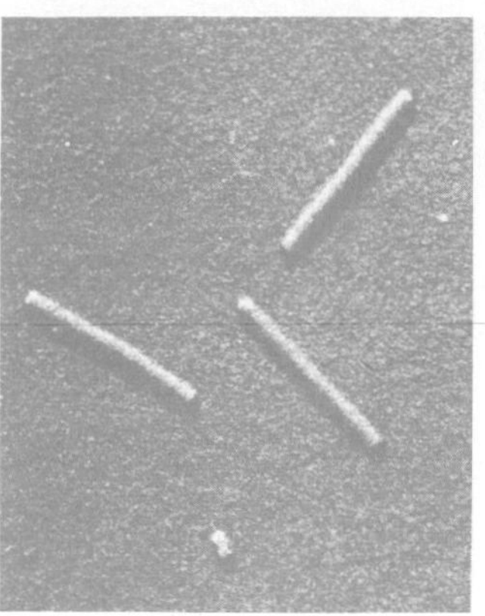

**Figure 4-2** Tobacco mosaic virus, magnified 100 000 times

What traits are used to classify viruses?

## Viruses and Disease

In spite of their small size, viruses cause many serious diseases in humans and other living things. Some of these diseases are listed in Table 4-1. Viruses may spread from one infected organism to another. In humans, viruses are spread by insects, air, water, food, and other people.

**TABLE 4–1. DISEASES CAUSED BY VIRUSES**

| Animal Diseases | Human Diseases | Plant Diseases |
|---|---|---|
| Rabies in dogs | Common cold, flu | Mosaic disease in tobacco |
| Foot and mouth disease in cattle | German measles, mumps | Bushy stunt in tomato |
| Newcastle disease in chickens | Measles, chickenpox | Maize dwarf mosaic |
| Distemper in dogs | Mononucleosis, cold sores | Mosaic disease in alfalfa |
| Cowpox in cows | Hepatitis, warts | Curly top in sugar beet |
| Feline leukemia in cats | Polio, smallpox | Dwarfism in rice |
| | Herpes, AIDS | |

**Bio Tip**

**Health:** An influenza virus can survive five minutes on human skin and 1.5 days on a kitchen counter top.



## OPTIONS

### Science Background

Until the discovery of viroids, viruses were thought to be the smallest things to cause disease. Viroids are short, naked strands of RNA. They are only a fraction of the size of viruses. They lack the protein coat of viruses. As with viruses, they reproduce only within a living cell. They are known to cause several plant diseases. They may also cause some diseases in animals and humans.

## 2 TEACH

### MOTIVATION/Demonstration

Show students a basketball, a ping-pong ball, and a BB shot. Tell them that one represents a human cheek cell, one represents a bacterium, and one represents a virus. Ask which each one represents.

### MOTIVATION/Software

*Microbiology Techniques*, EME.

### Concept Development

▶ Before starting this section, you may want to have students discuss whether viruses are living or nonliving. List their ideas on the chalkboard under the headings "Living" and "Nonliving."

▶ Mention that many different viruses are responsible for the large variety of colds and flus that infect humans.

▶ Encourage your students to bring in newspaper and magazine articles concerning viruses.

▶ Point out that scientists could not see viruses until after the electron microscope came into use in the 1940s.

▶ Point out the number of colds a person gets per year tends to decrease with age. It is possible that older people have more antibodies against most of the cold viruses.

### Guided Practice

Have students write down their answers to the margin question: the shape of the protein coat and size of the virus.

### Student Journal

Students should use reference books to learn more about and list the contributions made by Jenner, Pasteur, and Iwanoski to the understanding of viruses.

## TEACH

### Concept Development

▶ You may want to mention that some viruses may be considered to be beneficial. A variety of tulip has been cultivated because of the ornamental coloring produced by a virus that infects the tulip. Some viruses are used in research.

### Guided Practice

Have students write down their answers to the margin question: Insects break through the cell walls of the plant cells and let the virus enter.

### MOTIVATION/Demonstration

Use the overhead projector or the chalkboard to draw diagrams to explain viral replication. The reproduction of viruses is often called viral replication to distinguish it from the reproduction of cells.

**Videodisc**

**STV: Human Body**

*AIDS virus 1*

**48833**

*AIDS virus 2*

**48836**

*Cell bursting, releasing cold virus*

**48789**

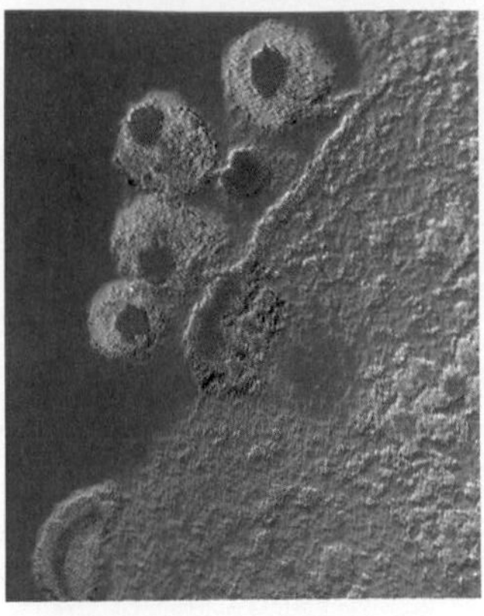

**Figure 4-3** AIDS virus being released from an infected cell. The viruses are color enhanced in red and green.

Acquired immune deficiency syndrome (AIDS) is a viral disease that destroys the body's immune system. When the immune system does not work properly, the body can't fight infections. People with AIDS die from diseases that most others can resist. Even a common cold can be life-threatening. The AIDS virus, Figure 4-3, is spread in four known ways. It is spread by sexual intercourse, blood products, the sharing of contaminated needles, and from a pregnant woman to her developing baby. Scientists studying AIDS have not yet discovered a cure.

Plant viruses may be spread by the wind or by insects. Chewing or sucking insects, such as aphids (AY fudz), whiteflies, and beetles, spread most plant viruses. Many plant viruses are unable to enter a plant without the help of an insect. While feeding, the insect breaks through the cell wall of the plant. This break in the cell wall lets the virus get into the host cell. Viruses may infect one or two leaves or an entire plant. Insects or wind then carry the viruses to other plants.

**How do viruses get into plant cells?**

How do viruses cause disease? Some reproduce rapidly and cause the cell to break open and release viruses, as shown in Figure 4-5. When this type of virus comes in

**Figure 4-4** Chewing and sucking insects break cell walls and allow viruses to enter plants.



## *OPTIONS*

### Science Background

Traditionally, plant and animal viruses have been named for the diseases they cause (e.g., smallpox, rabies, tobacco mosaic virus), while bacteriophages have been assigned number and letter codes (e.g., T2, T4, etc.).

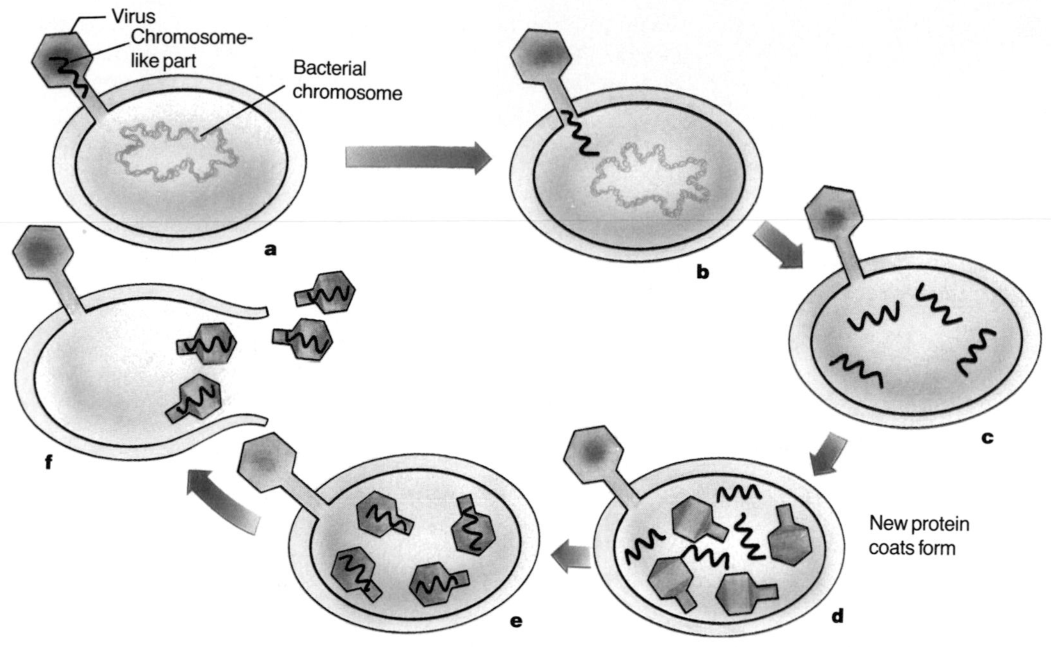


**Figure 4-5** Life cycle of a bacterial virus

contact with a host cell, it attaches itself to the cell (a). The chromosome-like part enters the cell (b). Then the chromosome-like part takes over the cell (c). The virus changes the hereditary material in the host cell so that the host cell produces more viruses instead of performing its usual work (d-e). The cell breaks open and releases the new viruses, which then invade other cells (f). Tissue damage and disease result. Viruses that cause polio, mumps, rabies, and flu in humans act this way.

Other viruses remain hidden in the cell a long time without reproducing. Cold sores are blisters around the lips. They are caused by a virus. No symptoms appear until something such as a fever or sunburn causes the virus to become active. The cold sores may disappear for long periods of time, but the virus remains in the body. As long as the virus does not reproduce, no cold sores appear. When the virus begins reproducing again, the cold sores reappear.

Other viruses cause the host cells to reproduce both themselves and the viruses. Groups of infected cells become lumps called tumors. Some tumors are harmless. A wart is an example of a harmless tumor. Examples of harmful tumors are some kinds of cancer.

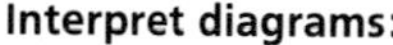

### Skill Check

**Interpret diagrams:** Look at Figure 4-5 and tell how a virus causes disease. Describe what is taking place in each lettered step. *For more help, refer to the* ***Skill Handbook,*** *pages 706-711.*



## TEACH

### Concept Development

▶ Point out that viruses that cause lysis of their host are called virulent phages. A phage may complete a life cycle in 25 to 45 minutes and give off several hundred new phages.

### Skill Check

Point out to students that the letters in the text correspond to the letters in the diagram.

### Independent Practice

**Study Guide**/*Teacher Resource Package*, p. 19. Use the Study Guide worksheet shown at the bottom of this page for independent practice.

## *GLENCOE* TECHNOLOGY

**Videodisc**

**Biology: The Dynamics of Life**
Animation: *The Lytic Cycle*
Disc 1, Side 2, Ch. 9

Animation: *The Lysogenic Cycle*
Disc 1, Slde 2, Ch. 10

**Videotape**

**The Secret of Life**
*Nothing to Sneeze at: Viruses*

## STUDY GUIDE 19

STUDY GUIDE — CHAPTER 4

Name ______ Date ______ Class ______

VIRUSES

*In your textbook, read about viruses in Section 4:1.*

1. Use the words *chromosome-like part* and *protein coat* to label the bacterial virus shown below.

protein coat
chromosome-like part

2. List six traits of viruses.
   a. Viruses are so small they can only be seen with an electron microscope.
   b. Viruses have different shapes.
   c. Viruses are made of a chromosome-like part surrounded by a protein coat.
   d. Viruses are not cells and have no cell parts.
   e. Viruses reproduce only in living cells.
   f. Viruses act like parasites.

3. Study the incorrect statements about viruses below. Change the underlined word or words to make each sentence correct. Write the correct new word or words on the line to the right.
   a. Viruses cause disease *only in humans*. — in plants, animals, and humans
   b. Each kind of virus infects *many* hosts. — only certain
   c. The rabies virus will infect only the *digestive* system of mammals. — nervous
   d. Cold sores are caused by a virus that remains *active*. — hidden
   e. Viruses are always *larger* than cells they infect. — smaller
   f. Viruses reproduce *outside* of living cells. — inside
   g. Viruses *do not change* the hereditary material in the host cell. — change

## OPTIONS

### Science Background

The viral disease rubella, also called German measles, is generally mild in children. But it can cause a serious problem if a woman has the disease during the fourth through the twelfth weeks of pregnancy. The developing fetus can suffer damage to the heart, the lens of the eye, the inner ear, and the brain.

# Lab 4-1 Viruses

## Overview

In this lab, students assemble three-dimensional models of two kinds of viruses to learn the structure of viruses. The lack of cell parts in the models will enable them to learn what viruses are not.

**Objectives:** Upon completing the lab, students will be able to (1) **describe** the shapes of viruses, (2) **relate** the basic parts of viruses, (3) **demonstrate** how a virus attacks a healthy cell.

**Time Allotment:** 45 minutes

## Preparation

▶ **Alternate Materials:** Toothpicks may be used for pipecleaners. Other sizes of bolts and nuts may be used.

▶ Cut #22 gauge wire in 14-cm lengths. Cut pipecleaners in 2-cm lengths.

**Lab 4-1 worksheet**/*Teacher Resource Package*, pp. 13-14.

## Teaching the Lab

▶ **Troubleshooting:** If the wire is not strong enough to hold the model up, fold the wire, twist it, and then place it on the bolt.

**Cooperative Learning:** Have students work in pairs to do the lab. For more information, see pp. 22T-23T in the Teacher Guide.

**Skill:** Have students compare a virus to a cell. (A virus is made of a chromosome-like part surrounded by a protein coat. A cell has cell parts, grows, and responds to changes in the environment.) Students should compare the structure of a virus to the model made in the activity.

# Viruses

# Lab 4–1

## Problem: What are some shapes of viruses?

### Skills

observe, formulate a model, recognize and use spatial relationships

### Materials

bolt, 3.7 cm × 0.7 cm
2 nuts to fit bolt
2 pieces of wire, 14 cm long (#22 gauge)
polystyrene ball, 4.5 cm in diameter
pipecleaners, cut in 2-cm lengths

### Procedure

1. **Observe:** Look at Figure A of the bacterial virus enlarged 260 000 times. Note the three parts labeled for you.
2. **Formulate a Model:** To build a model of the virus, attach two nuts onto the bolt as shown in Figure C.
3. Twist the wires around the bolt as in Figure D. Make them as tight as possible.
4. Fold down the wire ends and bend them as in Figure E.
5. Figure B shows a flu virus enlarged 300 000 times. Build a model of this virus using the polystyrene ball and pipecleaners.

### Data and Observations

1. Which virus part is represented by the top of the bolt and the nuts in your model?
2. Which virus part is represented by the threaded part of the bolt and the wires?

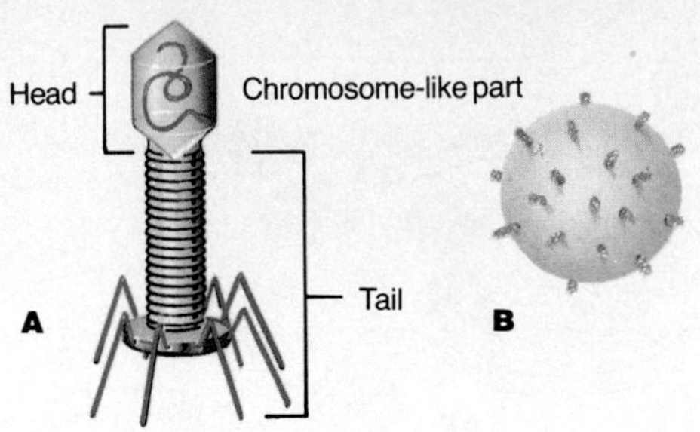


### Analyze and Apply

1. What chemical in a real virus would make up the threaded part of the bolt and the wires?
2. Which part of a virus would attach to the host's cell membrane?
3. Which parts would not enter a cell?
4. Which part does enter a cell?
5. If your flu virus model were a real virus, what would you expect to find inside the ball?
6. What chemical would you expect to make up the covering on the ball?
7. What are the basic parts of a virus?
8. **Apply:** How does the structure of a virus differ from that of a bacterial cell?

### Extension

**Formulate a model:** Using references, look up the shapes of some other viruses. Build models with common household items.

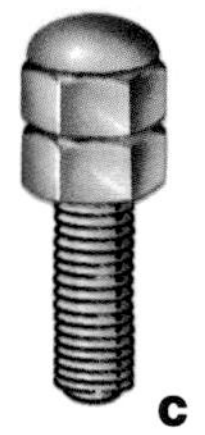


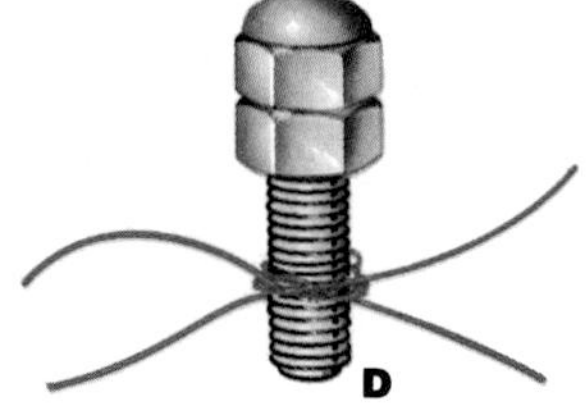


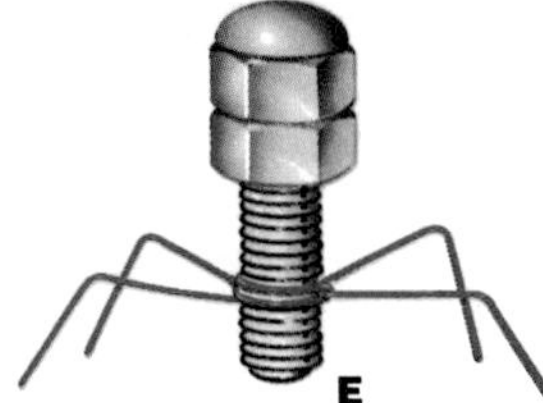




## ANSWERS

### Data and Observations

1. head
2. tail

### Analyze and Apply

1. protein
2. tail
3. the head and tail
4. the chromosome-like part
5. the chromosome-like part
6. protein
7. head, tail, and chromosome-like part
8. Viruses are smaller and do not have cell walls or other cell parts like bacteria.

## Controlling Viruses

Diseases caused by viruses are hard to treat or cure. There are no known drugs that destroy viruses. However, humans and other animals are protected against some viruses in several ways. Certain white blood cells can surround and destroy a virus. If the virus is not captured and destroyed by these white blood cells, other white blood cells make chemicals called antibodies (ANT ih bohd eez). Antibodies help destroy viruses or harmful bacteria by attaching to them. The virus or bacterium may be destroyed directly by the antibody or it may be held captive by the antibody until white blood cells can surround and destroy it. Each antibody acts only on one specific kind of virus or bacteria.

When human cells are first attacked by a virus, the cells produce interferon (ihnt ur FIHR ahn). **Interferon** is a chemical substance that interferes with the way viruses reproduce. When infected cells burst and release more viruses, they also release interferon. It is believed that the interferon warns nearby cells that a viral infection is taking place. The cells then produce their own chemicals to fight the viruses. Unlike antibodies, interferon will affect any type of virus that invades the body. The interferon produced in one species will work only in that species. For example, interferon produced by a horse will not work in humans.

What does a human cell do when it is attacked by a virus?

**Idea Map**

**Viruses**

- **Viruses**
  - **structure**
    - **chromosome-like part**
    - **protein coat**
  - **traits**
    - **reproduce only in living cells**
    - **each infects a certain host**
    - **causes disease**
  - **control**
    - **white blood cells - antibodies**
    - **interferon**
    - **vaccines**

**Study Tip:** Use this idea map as a study guide to viruses. The features of viruses and how viruses are controlled are shown.



## TEACH

### Concept Development

▶ Point out that scientists now believe that some people may be prone to cancer because their bodies do not produce enough interferon.

### Guided Practice

Have students write down their answers to the margin question: It produces interferon.

### Idea Map

Have students use the idea map as a study guide to the major concepts of this section.

**Cooperative Learning:** Divide the class into groups of four. Have each group research and report on the topics: culturing of viruses in tissues, such as the chick embryo; how interferon works; Burkitt's lymphoma; viroids; and viruses.

### Independent Practice

**Study Guide**/*Teacher Resource Package*, p. 20. Use the Study Guide worksheet shown at the bottom of this page for independent practice.

**Student Journal**

The body manufactures protein substances called interferon to fight viruses. Scientists are now experimenting with artificial interferon to fight viruses and other diseases. Have students research and write a report on antiviral drugs such as interferon, AZT, ara-A, and acyclovir.

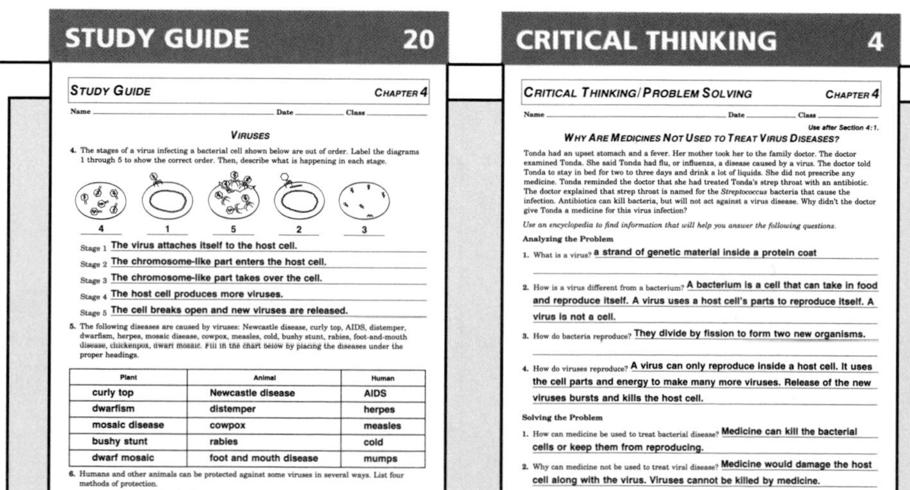

**STUDY GUIDE 20**

STUDY GUIDE CHAPTER 4

Name ______ Date ______ Class ______

*VIRUSES*

4. The stages of a virus infecting a bacterial cell shown below are out of order. Label the diagrams 1 through 5 to show the correct order. Then, describe what is happening in each stage.

4 1 5 2 3

Stage 1 The virus attaches itself to the host cell.
Stage 2 The chromosome-like part enters the host cell.
Stage 3 The chromosome-like part takes over the cell.
Stage 4 The host cell produces more viruses.
Stage 5 The cell breaks open and new viruses are released.

5. The following diseases are caused by viruses: Newcastle disease, curly top, AIDS, distemper, dwarfism, herpes, mosaic disease, cowpox, measles, cold, bushy stunt, rabies, foot-and-mouth disease, chickenpox, dwarf mosaic. Fill in the chart below by placing the diseases under the proper headings.

| Plant | Animal | Human |
|---|---|---|
| curly top | Newcastle disease | AIDS |
| dwarfism | distemper | herpes |
| mosaic disease | cowpox | measles |
| bushy stunt | rabies | cold |
| dwarf mosaic | foot and mouth disease | mumps |

6. Humans and other animals can be protected against some viruses in several ways. List four methods of protection.

a. white blood cells b. antibodies
c. interferon d. vaccines

7. How are viruses spread in humans? Viruses are spread by insects, air, water, food, and other people.

**CRITICAL THINKING 4**

CRITICAL THINKING/PROBLEM SOLVING CHAPTER 4

Name ______ Date ______ Class ______

*Use after Section 4:1.*

*WHY ARE MEDICINES NOT USED TO TREAT VIRUS DISEASES?*

Tonda had an upset stomach and a fever. Her mother took her to the family doctor. The doctor examined Tonda. She said Tonda had flu, or influenza, a disease caused by a virus. The doctor told Tonda to stay in bed for two to three days and drink a lot of liquids. She did not prescribe any medicine. Tonda reminded the doctor that she had treated Tonda's strep throat with an antibiotic. The doctor explained that strep throat is named for the *Streptococcus* bacteria that cause the infection. Antibiotics can kill bacteria, but will not act against a virus disease. Why didn't the doctor give Tonda a medicine for this virus infection?

*Use an encyclopedia to find information that will help you answer the following questions.*

**Analyzing the Problem**

1. What is a virus? a strand of genetic material inside a protein coat
2. How is a virus different from a bacterium? A bacterium is a cell that can take in food and reproduce itself. A virus uses a host cell's parts to reproduce itself. A virus is not a cell.
3. How do bacteria reproduce? They divide by fission to form two new organisms.
4. How do viruses reproduce? A virus can only reproduce inside a host cell. It uses the cell parts and energy to make many more viruses. Release of the new viruses bursts and kills the host cell.

**Solving the Problem**

1. How can medicine be used to treat bacterial disease? Medicine can kill the bacterial cells or keep them from reproducing.
2. Why can medicine not be used to treat viral disease? Medicine would damage the host cell along with the virus. Viruses cannot be killed by medicine.
3. What things in the body destroy viruses? White blood cells, fever, antibodies of the immune system destroy viruses.

## OPTIONS

**Critical Thinking/Problem Solving/** *Teacher Resource Package*, p. 4. Use the worksheet shown here to reinforce why medicines are not used to treat viral diseases.

## Portfolio

Have students research and write a report on the type of vaccine Jonas Salk developed. Students could interview older adults about their memory of polio and the polio vaccine.

### Check for Understanding

Have students respond to the first three questions in Check Your Understanding.

### Reteach

**Reteaching**/*Teacher Resource Package*, p. 10. Use this worksheet to give students additional practice with the life cycle of a virus.

**Extension:** Assign Critical Thinking, Biology and Writing, or some of the **OPTIONS** available with this lesson.

## 3 APPLY

### Videocassette

Show the videocassette *Viruses*, Coronet.

## 4 CLOSE

### Overhead

Make transparencies of the following viral groups: papovaviruses, adenoviruses, and herpes viruses.

### Answers to Check Your Understanding

1. A virus has no cell parts and does not grow like a living thing.
2. inside a living cell
3. It produces antibodies, interferon, and certain white cells that destroy viruses.
4. In response to both infection and a vaccine, the body produces antibodies.
5. The mosquito and the horse are the hosts. The bacteria are parasites.

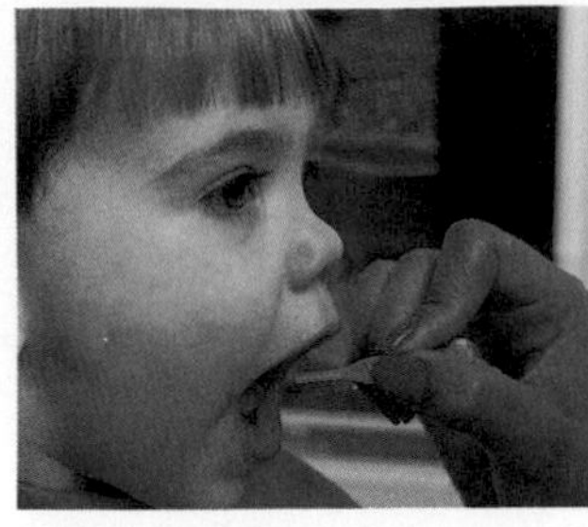

a

b

**Figure 4-6** A person takes polio vaccine for protection against the disease (a). Burning infected plants prevents the spread of plant diseases caused by viruses (b).

**Vaccines** are substances made from weakened or dead viruses. They can be used to treat some viral or bacterial diseases. There are vaccines for polio, rubella, measles, and influenza. Figure 4-6a shows a person getting the polio vaccine. The body reacts to a vaccine by producing antibodies to protect against the disease. In addition to people, pets and farm animals receive vaccines for protection against disease.

It is usually impossible to treat viral infections in plants. Therefore, farmers try to prevent the spread of viral infections by burning infected plants.

### Check Your Understanding

1. List two ways viruses differ from living things.
2. Where must viruses be found if they are to reproduce?
3. What are three ways the human body protects itself against viruses?
4. **Critical Thinking:** How is the body's reaction to a vaccine similar to its reaction when it is attacked by a virus?
5. **Biology and Writing:** A mosquito contains some viruses and bites a horse. The horse becomes sick with a disease caused by the viruses from the mosquito. Use a complete sentence to tell which two living things are the hosts. What is the parasite?



## OPTIONS

**Enrichment:** Explain the development of polio vaccines and the way that mass vaccinations were conducted in the 1960s.

**RETEACHING 10**

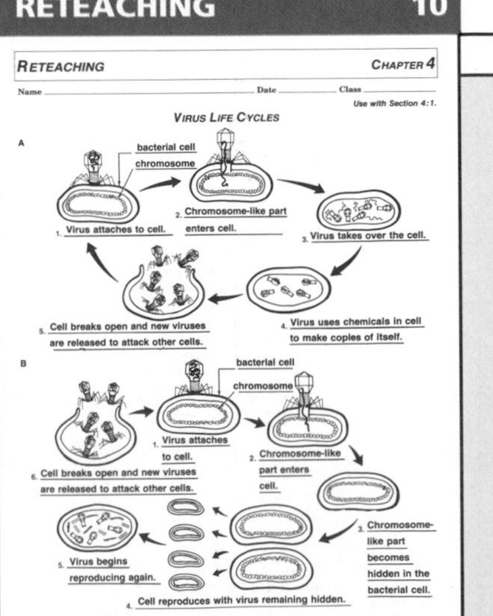

RETEACHING CHAPTER 4

Name Date Class

Use with Section 4:1.

VIRUS LIFE CYCLES

A

bacterial cell
chromosome

1. Virus attaches to cell.
2. Chromosome-like part enters cell.
3. Virus takes over the cell.
4. Virus uses chemicals in cell to make copies of itself.
5. Cell breaks open and new viruses are released to attack other cells.

B

bacterial cell
chromosome

1. Virus attaches to cell.
2. Chromosome-like part enters cell.
3. Chromosome-like part becomes hidden in the bacterial cell.
4. Cell reproduces with virus remaining hidden.
5. Virus begins reproducing again.
6. Cell breaks open and new viruses are released to attack other cells.

# 4:2 Monera Kingdom

Monerans are one-celled organisms that lack a nucleus, but do have nuclear material within the cell wall. They also lack many of the cell parts found in plant and animal cells. Two groups of monerans, bacteria and blue-green bacteria, are found everywhere around you. They live as single cells or in groups of cells.

**Objectives**

4. **Identify** the traits of bacteria.
5. **Explain** how bacteria affect other living things.
6. **Compare** the traits of blue-green bacteria and other bacteria.

**Key Science Words**

bacteria
colony
capsule
flagellum
fission
asexual reproduction
endospore
saprophyte
decomposer
Koch's postulates
communicable disease
antibiotic
biotechnology
pasteurization
blue-green bacteria

## Traits of Bacteria

What are bacteria? **Bacteria** are very small, one-celled monerans. They are larger than viruses, but are too small to be seen without a microscope. They are so small that 300 could fit side by side across the period at the end of this sentence.

Bacteria can be found almost everywhere. They live in water, air, soil, food, and on almost every object. They are on your skin and even inside your body. Bacteria have been found in arctic ice as well as in hot springs. Bacteria are so widespread because almost any material may be food for some kind of bacteria.

Bacteria are classified into three groups according to shape. Some bacteria are round. Some are shaped like rods. Others are spiral, Figure 4-7. Bacteria can be found as single cells, in pairs, or in clusters. Some singles and pairs may join together in a cluster or chain. The cells in a rapidly growing cluster or chain can make a colony. A **colony** is a group of similar cells growing next to each other that do not depend on each other.

a

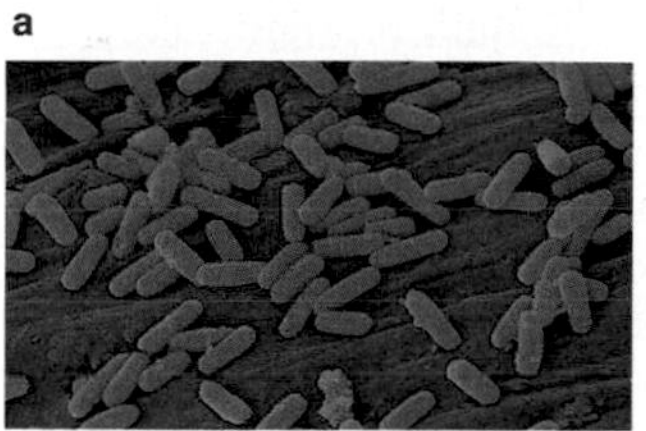

b

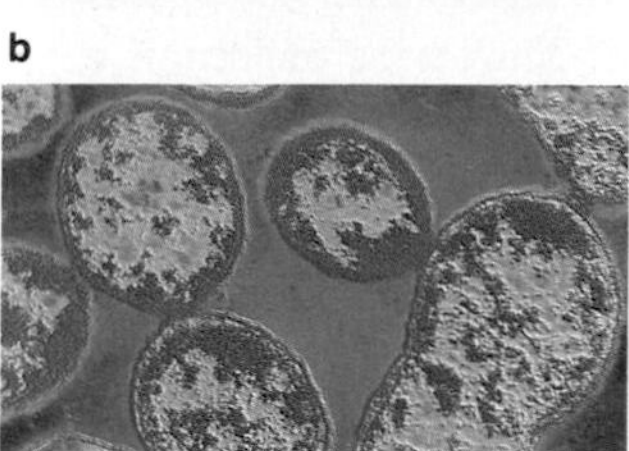

c

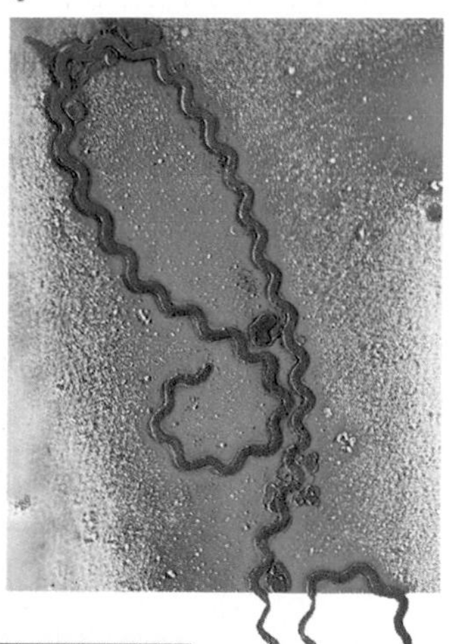

**Figure 4-7** Rod-shaped bacteria (a), round bacteria (b), and spiral bacteria (c) highly magnified



**FOCUS 7**

*FOCUS* CHAPTER 4

Name ______ Date ______ Class ______

*HOW BACTERIA AFFECT YOUR LIFE*

| Money spent because of harmful bacteria (in billions of dollars) | Money earned because of helpful bacteria |
|---|---|
| 78.6 Health-related costs: Hospitals, Drugs, Antibiotics | 311.2 Microbiological products: Drugs, Chemicals, Oil products |
| 65.9 Public-health costs: Epidemic control Purifying wate. , Pollution, Sewage | |
| 29.8 Treatment of food to prevent bacterial growth | 56.7 Foods and beverages |

1. Which is greater, the amount of money spent because of harmful bacteria or the amount earned because of helpful bacteria? Explain.
2. What surprised you most about this table?

## OPTIONS

### Science Background

Rod-shaped bacteria are called bacilli. Spherical bacteria are called cocci. Spiral-shaped bacteria are called spirilla.

**Bacterial and Animal Cells/** *Transparency Package,* number 4. Use color transparency number 4 as you teach the cell parts of bacteria.

# 4:2 Monera Kingdom

## PREPARATION

### Materials Needed

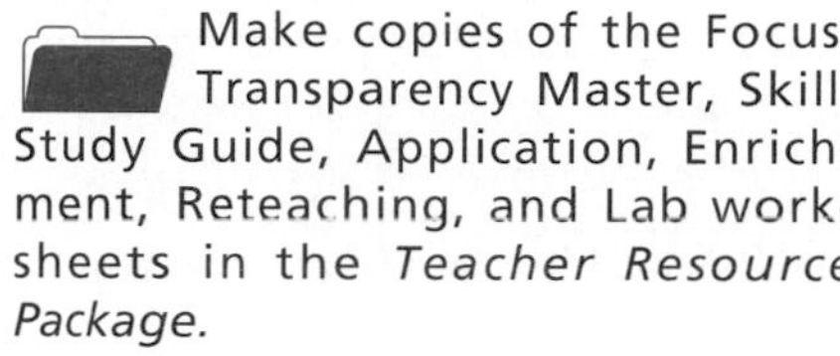

Make copies of the Focus, Transparency Master, Skill, Study Guide, Application, Enrichment, Reteaching, and Lab worksheets in the *Teacher Resource Package.*

▶ Obtain beef broth, salt, sugar, vinegar, and plain yogurt for the Mini Labs.

▶ Obtain prepared slides of *Anabaena* and bacteria for Lab 4-2.

### Key Science Words

bacteria
colony
capsule
flagellum
fission
asexual reproduction
endospore
saprophyte
decomposer
Koch's postulates
communicable disease
antibiotic
biotechnology
pasteurization
blue-green bacteria

### Process Skills

In Lab 4-2, students will make and use tables, observe, measure in SI, and use a microscope. In the Skill Check, students will understand science words. In the Mini Labs, students will infer and use a microscope.

## 1 FOCUS

▶ The objectives are listed on the student page. Remind students to preview these objectives as a guide to this numbered section.

### MOTIVATION/Bulletin Board

Put the title Monerans – Helpful and Harmful on a bulletin board. Display pictures of yogurt, cheese, tooth decay and crown gall. Have students decide in which column the pictures should be placed.

**Focus/***Teacher Resource Package,* p. 7. Use the Focus transparency shown at the bottom of this page as an introduction to how bacteria affect our lives.

## 2 TEACH

### MOTIVATION/Demonstration

Draw a bacterial cell and an animal cell on the chalkboard. Use colored chalk to draw in the cell parts. Have the students compare the two cells.

### MOTIVATION/Brainstorming

Have students name some skin diseases caused by bacteria (acne, pimples, boils).

### Concept Development

▶ Point out that the most unusual trait of bacteria is not size or shape, but chemical makeup. Because of their versatility, bacteria can be found almost everywhere.

▶ Explain that the interval from one fission to the next is highly variable. *E. coli,* a bacterium found in the human intestinal system, divides every 20 minutes under ideal conditions. *Treponema pallidum,* a bacterium that causes syphilis, can divide every 33 hours. Bacteria that cause leprosy reproduce only once every 12 days.

**Portfolio**

Have students make the following comparisons between viruses and bacteria: the structures present in each, the functions of each, and the relative sizes of each.

▶ Have students use references to compile a list of bacterial species. Students should categorize these species by shape (round, rod, or spiral) and regroup the bacteria into two categories, helpful or harmful. Students should determine if there is a correlation between a species' shape and whether it is harmful or helpful.

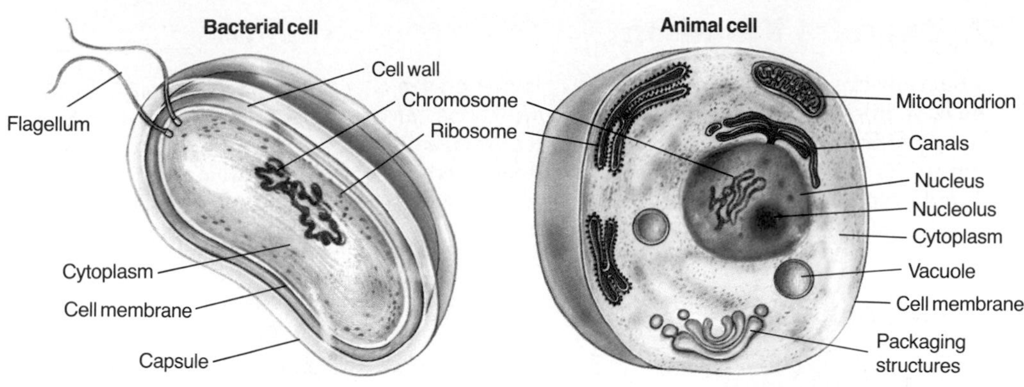


**Figure 4-8** Structures of bacterial and animal cells. What structures are found in a bacterial cell but not in an animal cell?

Under a microscope, bacteria look very different from other cells you have studied. Notice that the bacterial cell in Figure 4-8 does not have a nucleus. It has one main chromosome. The animal cell does have a nucleus. The bacterial cell also does not have most cell parts seen in the animal cell.

Even without certain cell parts, bacteria carry on the same jobs as other cells. Bacteria still reproduce, grow, and carry out cellular respiration.

Bacterial cells have a cell wall as well as a cell membrane. Some bacteria have a sticky outer layer called a **capsule,** Figure 4-8. The capsule keeps the cell from drying out and helps the cell stick to food and other cells. Some bacteria move with a long, whiplike thread called a **flagellum** (fluh JEL um).

Bacteria reproduce by fission (FIHSH un). **Fission** is the process of one organism dividing into two organisms. Fission is a kind of asexual (ay SEK shul) reproduction. **Asexual reproduction** is the reproducing of a living thing from only one parent. Bacteria reproduce by asexual reproduction. The circular chromosome of the bacterial cell makes a copy of itself, and the cell divides, Figure 4-9. The time that it takes a bacterial cell to grow and divide into two cells varies. If growing conditions are right, it can take only about 20 minutes.

What do bacteria need to live? They need moisture, a certain temperature, and food. A few bacteria can live at 0°C and others can live at temperatures of 75°C. Those that cause disease in humans live at 37°C, normal body temperature. Most bacteria grow best in darkness. Most need oxygen to live. Others cannot live in the presence of oxygen.

**Bio Tip**

**Consumer:** In Southeast Asia, rice is grown without using fertilizers because certain bacteria take nitrogen gas out of the air and use it to make nutrients for the rice plants.

**Figure 4-9** A bacterial cell divides to form two new cells. What is this form of reproduction called?

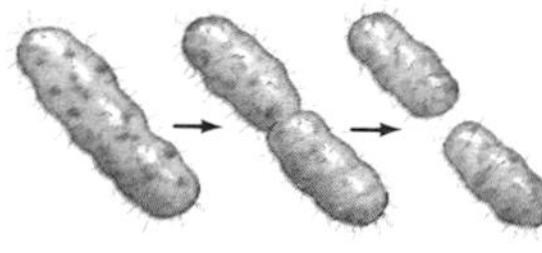



## OPTIONS

### Science Background

The classification of monerans is controversial. Kingdom Monera includes the archaebacteria or "ancient" bacteria and the Eubacteria, or "true" bacteria. Only two groups of Eubacteria are included here.

**Transparency Master/***Teacher Resource Package,* p. 15. Use the Transparency Master shown here to help students compare viruses and monerans.

**TRANSPARENCY 15**

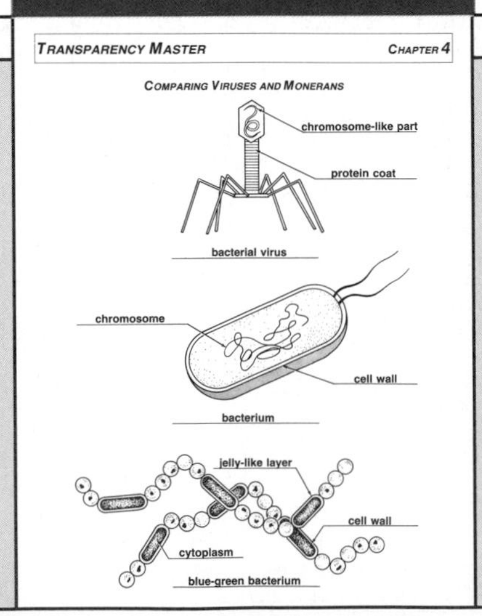

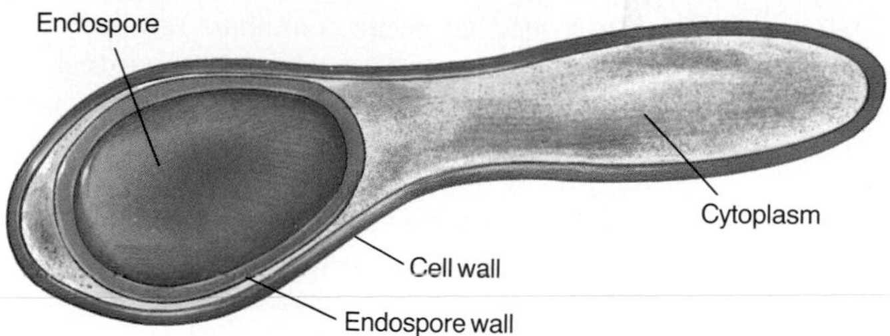


**Figure 4-10** An endospore has a thick, protective wall that allows the bacterium to withstand harsh conditions.

If living conditions are not right for bacteria to grow, they can survive by forming endospores (EN duh sporz), Figure 4-10. An **endospore** is a thick-walled structure that forms inside the cell, enclosing all the nuclear material and some cytoplasm. Some endospores can survive for many years. They can withstand boiling, freezing, and extremely dry conditions without damage. When conditions return to normal, endospores develop into bacteria. Endospores are not used for reproduction. They let bacteria survive when living conditions are not ideal.

What do bacteria use for food?

Most bacteria feed on other living things or on dead things. Bacteria that feed on living things are parasites. Bacteria that live on dead things are called saprophytes (SAP ruh fites). **Saprophytes** are organisms that use dead materials for food. They get energy by breaking down, or decomposing, dead materials. **Decomposers** are living things that get their food from breaking down dead matter into simpler chemicals. Decomposers are important because they return minerals and other materials to the soil, where other organisms can use them.

Some bacteria can make their own food. Some use the energy of the sun to make food. Others use the energy in substances containing iron and sulfur to make food.

## Bacteria and Disease

Some bacteria are parasites in humans. They can cause a variety of diseases. Human diseases caused by bacteria include strep throat, tuberculosis, certain kinds of pneumonia (noo MOH nyuh), gonorrhea (gahn uh REE uh), and meningitis.

Plants and animals other than humans also may have bacterial diseases. Fire blight and crown gall, Figure 4-11, are plant diseases caused by bacteria. The bacteria that cause fire blight are spread by rain.

**Figure 4-11** Crown gall is a plant disease caused by bacteria.

## OPTIONS

**Does Soil Contain Bacteria?/*Lab Manual*, pp. 27-30. Use this lab as an extension to where bacteria are found.**

## TEACH

### Concept Development

▶ Point out that most bacteria that form endospores are harmless. However, some such as bacteria that cause tetanus and botulism are extremely dangerous.

▶ You may wish to mention that some bacteria make their own food by chemosynthesis. Chemosynthesis is similar to photosynthesis except that sunlight is not necessary.

▶ Point out that many bacteria obtain food by secreting digestive enzymes through their cell walls. The digestive enzymes break down food molecules into smaller molecules that can be absorbed by the bacteria.

▶ You might mention that other bacterial diseases include toxic shock syndrome, impetigo, sinusitis, ear infections, scarlet fever, tetanus, food poisoning, whooping cough, and leprosy.

### Guided Practice

Have students write down their answers to the margin question: other living things or dead things.

### Student Journal

Students can make a list of all the diseases they know about. They should classify the diseases into two groups and explain why they classified the diseases as they did.

### GLENCOE TECHNOLOGY

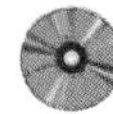

**Videodisc**

**Biology: The Dynamics of Life**
Movie: *Binary Fission*
Disc 1, Side 2, Ch. 11

# TEACH

## Concept Development

▶ Explain how Koch's postulates are being used today to identify disease-causing organisms. Have students bring in newspaper and magazine articles on disease research.

## Guided Practice

Have students write down their answers to the margin question: They show that bacteria taken from one host causes the same disease in a second host.

**GLENCOE TECHNOLOGY**

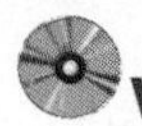

**Videodisc**

**The Secret of Life**

*Prokaryotic Cell*

*Prokaryotes/Phylogeny*

NATIONAL GEOGRAPHIC SOCIETY

**Videodisc**

**STV: Human Body Vol. 3**

*Immune System*

Unit 1, Side 1

*Viruses and Vaccination*

**21672-31347**

Some bacterial diseases that occur in animals can be passed on to humans. Anthrax, a disease found in livestock, is one example. It is usually passed on to people who work with animals, such as butchers and handlers of wool and leather.

How do we know that bacteria cause disease? In 1876, a German doctor named Robert Koch used a scientific method to show that anthrax was caused by a bacterium. He made a hypothesis that a bacterium caused the disease. He experimented to test the hypothesis. The experiments supported his hypothesis, so he formed a theory about the cause of disease.

How can scientists prove that bacteria cause disease?

**Koch's postulates** (KAHKS • PAHS chuh lutz) are steps for proving that a disease is caused by a certain microscopic organism. Here are Koch's postulates.

1. The organism must be present in a living thing when the disease occurs.
2. The organism must be taken from the host and grown in the laboratory, Figure 4-12a, b.
3. When the organisms from the laboratory are injected into healthy hosts, they must cause the same disease in the healthy hosts, Figure 4-12c.
4. The organism must be removed from the new hosts, grown in the laboratory, and shown to be the same as the organism from the first host, 4-12d.

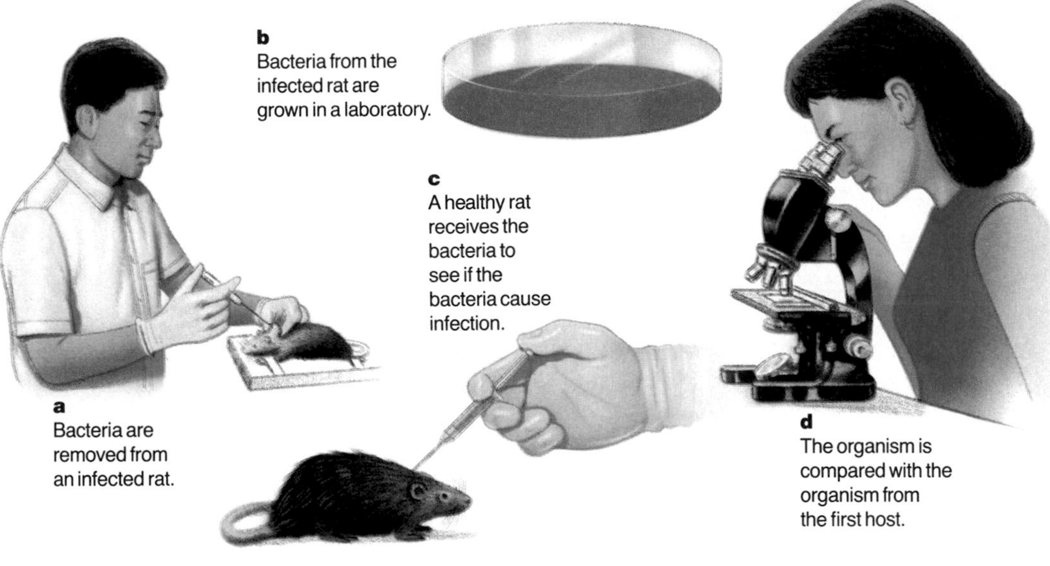


**Figure 4-12** Using Koch's postulates to prove the cause of a disease



## OPTIONS

### ACTIVITY/Challenge

Find out from the school nurse what diseases students are expected to be vaccinated against by the time they enter school. Have students research to determine which diseases are caused by bacteria and which are caused by viruses. Have them explain why vaccination for these diseases is important.

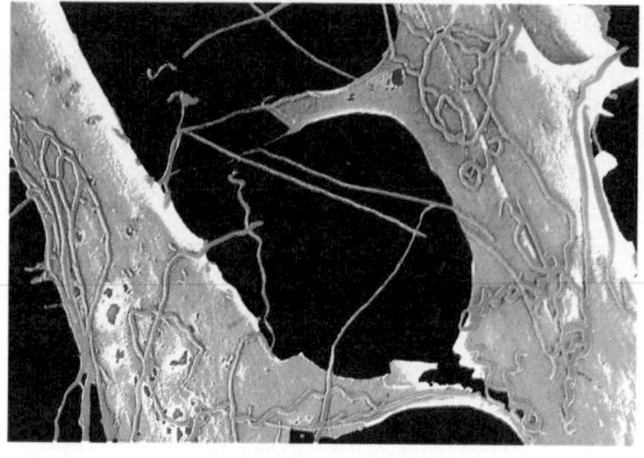

a

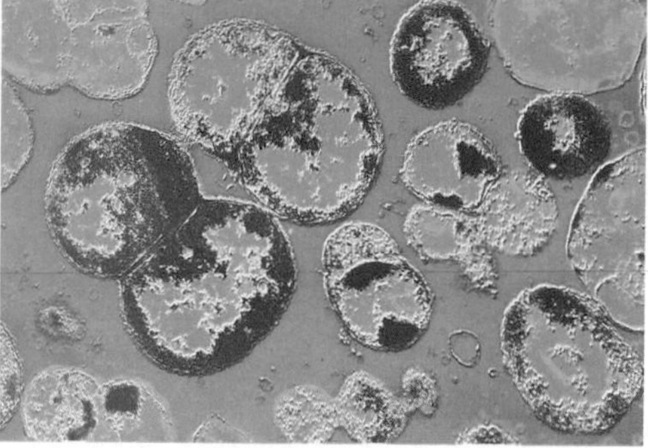

b

**Figure 4-13** The spiral bacteria that cause syphilis (a) and the round bacteria that cause gonorrhea (b)

Scientists still use Koch's postulates today. The procedure helps identify bacteria and other disease-causing organisms.

Some diseases caused by bacteria and viruses are called communicable (kuh MYEW nih kuh bul) diseases. **Communicable diseases** are ones that can be passed from one organism to another. They may be spread in several ways. Some diseases are spread by air when a person sneezes or coughs. Pneumonia, strep throat, and tuberculosis are spread through the air. Communicable diseases can also be spread by touching anything an infected organism has touched. Common items that might have bacteria are clothes, food, silverware, or toothbrushes. Another way of spreading diseases is by drinking water that contains bacteria. Some of the more serious diseases are those spread by sexual contact. Syphilis (SIHF uh lus) and gonorrhea are sexually transmitted diseases caused by bacteria, Figures 4-13a, b.

Insects also spread diseases. Flies, fleas, cockroaches, and mosquitoes carry disease organisms.

## Helpful Bacteria

You may be surprised to learn that there are more helpful bacteria than harmful ones. Most kinds of bacteria are helpful to humans. Bacteria are needed to decompose dead matter. Bacteria get energy from dead matter as they break it down into materials that other living things can use. In this way, the chemicals that are found in dead matter are recycled so that other organisms can use them to grow.

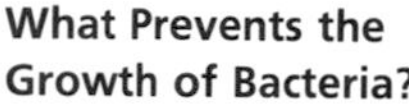

### Mini Lab

**What Prevents the Growth of Bacteria?**

**Infer:** Pour 100 mL beef broth into each of 4 beakers: A, B, C, D. Add 1 tsp. salt to B, 1 tsp. sugar to C, and 1 tsp. vinegar to D. Observe after 2 days. What prevents bacterial growth? *For more help, refer to the **Skill Handbook**, pages 706-711.*



## TEACH

### ACTIVITY/Guest Speaker

Invite someone from the local health department to discuss sexually transmitted diseases.

### MOTIVATION/Brainstorming

Brainstorm how bacterial diseases are spread. Discussions might center around why it is important to (1) wash hands after using the bathroom; (2) cover your mouth when sneezing or coughing; (3) see a doctor if bitten by an animal.

### Mini Lab

The beakers with salt and vinegar will be less cloudy than the control. The one with sugar will be more cloudy. Sugar provides food for bacterial growth.

### ASSESSMENT

**Performance:** Have students hypothesize what other additives may control the growth of bacteria. Students should design an experiment to test the hypothesis, then perform the experiment.

### Independent Practice

**Study Guide**/*Teacher Resource Package,* p. 21. Use the Study Guide worksheet shown at the bottom of this page for independent practice.

### Portfolio

Have students write a newspaper article on why the study of bacteria is important for everyone. Students should include the helpful effects of bacteria as well as the harmful ones.

### STUDY GUIDE 21

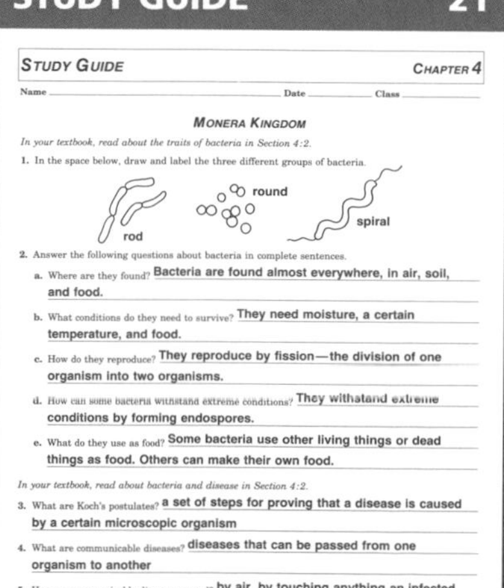

STUDY GUIDE — CHAPTER 4

Name ______ Date ______ Class ______

**MONERA KINGDOM**

*In your textbook, read about the traits of bacteria in Section 4:2.*

1. In the space below, draw and label the three different groups of bacteria.

rod — round — spiral

2. Answer the following questions about bacteria in complete sentences.
   a. Where are they found? Bacteria are found almost everywhere, in air, soil, and food.
   b. What conditions do they need to survive? They need moisture, a certain temperature, and food.
   c. How do they reproduce? They reproduce by fission—the division of one organism into two organisms.
   d. How can some bacteria withstand extreme conditions? They withstand extreme conditions by forming endospores.
   e. What do they use as food? Some bacteria use other living things or dead things as food. Others can make their own food.

*In your textbook, read about bacteria and disease in Section 4:2.*

3. What are Koch's postulates? a set of steps for proving that a disease is caused by a certain microscopic organism
4. What are communicable diseases? diseases that can be passed from one organism to another
5. How are communicable diseases spread? by air, by touching anything an infected organism has touched, by drinking water that contains bacteria, by sexual contact, by insects that carry disease organisms
6. What are two sexually transmitted diseases? syphilis and gonorrhea

## OPTIONS

### Science Background

Robert Koch discovered that the microorganisms causing African sleeping sickness are transmitted to humans by tsetse flies. He also discovered the tuberculosis bacterium and developed the first test to indicate if a person had ever been exposed to the bacterium.

## TEACH

### Concept Development

▶ Point out that millions of *E. coli* live in the human intestinal system where they carry out many beneficial functions such as aiding digestion and producing vitamins.

▶ Make sure students understand the difference between antibodies and antibiotics.

▶ Explain that antibiotics inhibit growth of bacteria by interfering with development of cell walls. Antibiotics are not effective against viruses. Remind students that viruses reproduce by using chemicals in living cells. Any substance that interferes with the virus would also kill the living cells.

▶ You may want to have a student contact a pharmacist or doctor to get a list of different antibiotics that are used today. Then discuss the use of the antibiotics.

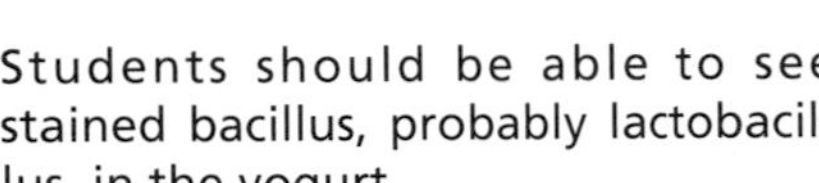

### Mini Lab

Students should be able to see stained bacillus, probably lactobacillus, in the yogurt.

### ASSESSMENT

**Performance:** Have students repeat the Mini Lab using a drop of cultured buttermilk. Students should determine if similar organisms are responsible for producing buttermilk.

### MOTIVATION/Display

Display food (containers may be used) made with bacteria – cottage cheese, yogurt, sour cream, buttermilk, sauerkraut, vinegar, pickles, coffee, and cocoa. Discuss the role of bacteria in making the foods.

### Mini Lab

**What Makes Yogurt?**

**Use a microscope:** Thin plain yogurt with water. Place a drop on a slide. Add a drop of methylene blue and a coverslip. Observe under high power. What do you see in the yogurt? *For more help, refer to the* ***Skill Handbook,*** *pages 712-714.*

Some bacteria grow in other living things and help them. Bacteria in the stomach of a cow break down grass and hay. In this way, the grass and hay can be used as food by both the bacteria and the cow. Without the action of bacteria, plants would have little food value for cows, sheep, and deer. Bacteria in your intestine make vitamins that you need.

Bacteria are used to make many products that are useful to us. Some bacteria break down the fibers of plants used to make linen and rope. Some are used in making leather from skins. Some of today's most useful antibiotics (an ti bi AHT iks) are produced by bacteria, Figure 4-14a. **Antibiotics** are chemical substances that kill or slow the growth of bacteria. Some antibiotics have a specific effect, working only on certain types of bacteria. Others destroy many types of bacteria.

Some foods owe their taste or texture to bacteria. Many dairy products are made by adding certain species of bacteria to milk. For example, cottage cheese, Swiss cheese, yogurt, sour cream, and buttermilk are all made by bacteria growing in milk. The flavors of coffee and cocoa are due to bacteria, Figure 4-14b. Sauerkraut is made by the action of bacteria on chopped cabbage. Bacteria also help change alcohol to vinegar.

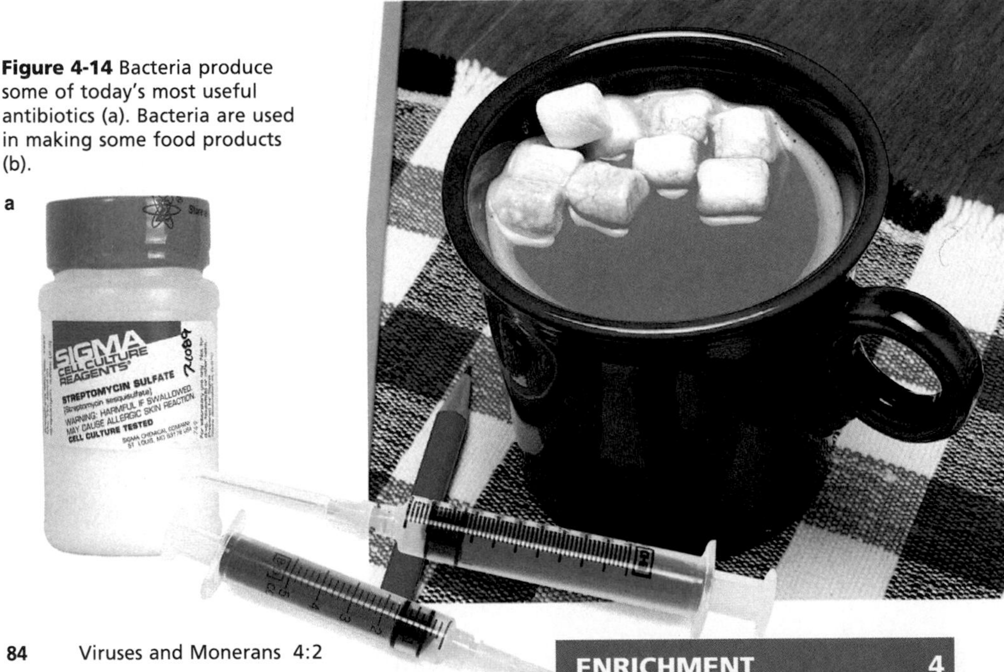


**Figure 4-14** Bacteria produce some of today's most useful antibiotics (a). Bacteria are used in making some food products (b).



### OPTIONS

**How Can You Test for Bacteria in Foods?**/*Lab Manual*, p. 25. Use this lab as an extension to studying bacteria in foods.

**Enrichment**/*Teacher Resource Package*, p. 4. Use the Enrichment worksheet shown here to help students understand whether a virus or moneran causes a certain disease and whether a vaccine is available for that disease.

### ENRICHMENT 4

ENRICHMENT CHAPTER 4

Name ______ Date ______ Class ______

Use after Section 4:2.

**VIRUSES, MONERANS, AND DISEASE**

Viruses and bacteria cause many human diseases. Some of these diseases are listed in the first column of the table below.

*Look up the diseases in your textbook or in a dictionary or an encyclopedia. Then fill in the table below. Tell whether the disease is caused by a virus or by a bacterium, tell what effect the disease has on the body, and tell whether there is a vaccine to prevent the disease.*

| Disease | Cause | Effect on the body | Vaccine? |
|---|---|---|---|
| rabies | virus | affects the nervous system, causing convulsions and death | yes |
| measles | virus | red, circular spots break out on the skin | yes |
| strep throat | bacterium | inflamed throat and high fever | no |
| influenza | virus | fever, aches, and pains | yes, for some strains |
| meningitis | bacterium | inflammation of the membranes around the brain | no |
| mononucleosis | virus | abnormal white blood cells | no |
| polio | virus | fever, paralysis, and wasting away of muscles | yes |
| chicken pox | virus | low-grade fever and rash of itching blisters on the skin | no |
| tuberculosis | bacterium | damages the lungs | no |
| AIDS | virus | destroys the immune system | no |

Bacteria are very important in the field of biotechnology (bi oh tek NAHL uh jee). **Biotechnology** is the use of living things to solve practical problems. Bacteria are being used to produce natural gas and detergents. Some are used to produce human insulin. Insulin is a chemical that the body normally makes to control the level of sugar in the blood. Some people cannot make insulin, so they must use the insulin produced by bacteria.

## Controlling Bacteria

If food is put into containers and heated to a high enough temperature, the bacteria inside can be destroyed. The containers can then be sealed while they are hot. Food can remain in them for long periods of time without spoiling, as long as the seals are not broken and the cans remain airtight. The process of sealing food in airtight cans or jars after killing the bacteria is called canning. Endospores can be killed by the canning process. They are a major cause of food poisoning.

The process of heating milk to kill harmful bacteria is called **pasteurization** (pas chuh ruh ZAY shun). The milk you buy in stores is pasteurized. Even pasteurized milk that has not been opened will finally spoil. Can you explain why?

Cooling food to a low temperature slows the growth of bacteria. You keep food in the refrigerator to slow bacterial growth. Freezing food is also used to slow bacterial growth. Frozen food can be stored safely for several months. Items kept in a freezer last longer than those kept in the refrigerator. Why?

### Skill Check

**Understand science words: biotechnology.** The word part *bio* means life. In your dictionary, find three words with the word part *bio* in them. *For more help, refer to the* ***Skill Handbook****, pages 706-711.*

**Figure 4-15** Canning is a way to prevent bacteria from spoiling food.



## TEACH

### Concept Development

▶ Point out that cans of food that bulge should be discarded. The bulge may be due to the presence of gases produced by botulism-causing bacteria growing inside.

▶ Discuss why refrigeration prevents food poisoning. Mention also the importance of cleanliness of food preparers and of the materials and equipment used in food preparation.

**Cooperative Learning:** Divide the class into groups of four and have each group obtain information on an example of biotechnology, such as using bacteria in genetic engineering, cleaning up toxic waste areas, or making insecticides, and share the information with the class.

### Independent Practice

**Study Guide**/*Teacher Resource Package*, p. 22. Use the Study Guide worksheet shown at the bottom of this page for independent practice.

**APPLICATION 4**

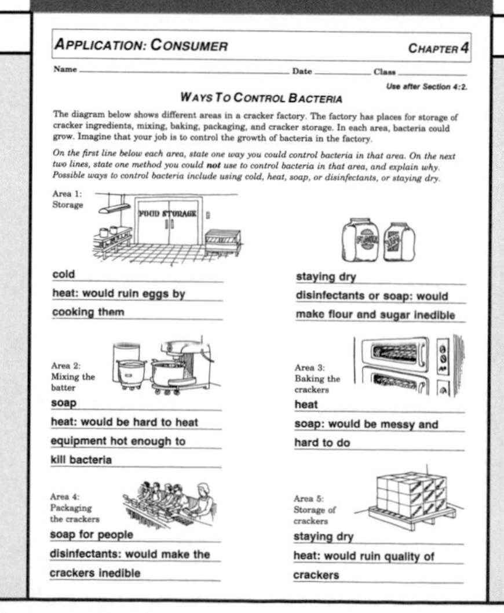

APPLICATION: CONSUMER — CHAPTER 4

Name ____________ Date ______ Class ______

Use after Section 4:2.

WAYS TO CONTROL BACTERIA

The diagram below shows different areas in a cracker factory. The factory has places for storage of cracker ingredients, mixing, baking, packaging, and cracker storage. In each area, bacteria could grow. Imagine that your job is to control the growth of bacteria in the factory.

*On the first line below each area, state one way you could control bacteria in that area. On the next two lines, state one method you could* ***not*** *use to control bacteria in that area, and explain why. Possible ways to control bacteria include using cold, heat, soap, or disinfectants, or staying dry.*

Area 1: Storage

cold
heat: would ruin eggs by
cooking them

staying dry
disinfectants or soap: would
make flour and sugar inedible

Area 2: Mixing the batter

soap
heat: would be hard to heat
equipment hot enough to
kill bacteria

Area 3: Baking the crackers

heat
soap: would be messy and
hard to do

Area 4: Packaging the crackers

soap for people
disinfectants: would make the
crackers inedible

Area 5: Storage of crackers

staying dry
heat: would ruin quality of
crackers

**STUDY GUIDE 22**

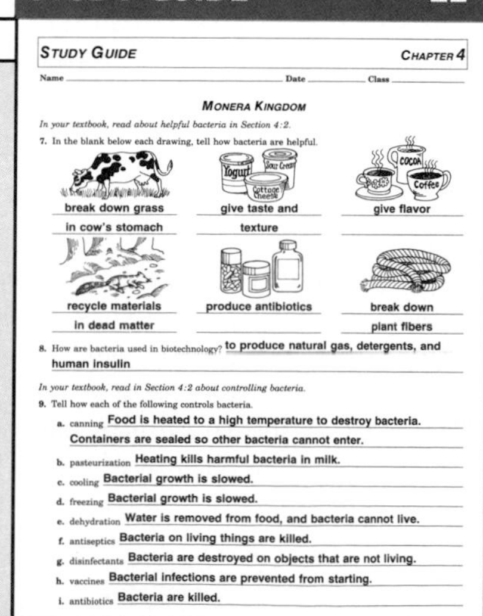

STUDY GUIDE — CHAPTER 4

Name ____________ Date ______ Class ______

MONERA KINGDOM

*In your textbook, read about helpful bacteria in Section 4:2.*

7. In the blank below each drawing, tell how bacteria are helpful.

break down grass in cow's stomach
give taste and texture
give flavor
recycle materials in dead matter
produce antibiotics
break down plant fibers

8. How are bacteria used in biotechnology? to produce natural gas, detergents, and human insulin

*In your textbook, read in Section 4:2 about controlling bacteria.*

9. Tell how each of the following controls bacteria.

a. canning Food is heated to a high temperature to destroy bacteria. Containers are sealed so other bacteria cannot enter.
b. pasteurization Heating kills harmful bacteria in milk.
c. cooling Bacterial growth is slowed.
d. freezing Bacterial growth is slowed.
e. dehydration Water is removed from food, and bacteria cannot live.
f. antiseptics Bacteria on living things are killed.
g. disinfectants Bacteria are destroyed on objects that are not living.
h. vaccines Bacterial infections are prevented from starting.
i. antibiotics Bacteria are killed.

## OPTIONS

### Science Background

Paul Ehrlich, a German scientist, was the first to find a chemical effective against a bacterial disease. In the early 1900s, he discovered an arsenic compound that destroyed the bacteria that caused syphilis.

**Application**/*Teacher Resource Package*, p. 4. Use the Application worksheet shown here to reinforce ways to control bacteria.

## TECHNOLOGY

### Uses of Bacteria

#### Background

Bacteria are being cultivated by many industrial companies to clean up hazardous pollutants. Because bacteria have rapid reproductive rates, they are able to adapt quickly when any new substance, even a toxic one, is added to their environment. Bacteria that become resistant survive and produce a new, tougher generation. Only the hardiest strains survive.

#### Discussion

1. How were the number of bacteria increased along Prince William Sound? Fertilizers were added to the water.
2. How does increasing the numbers of bacteria help in cleaning up oil spills and toxic wastes? It increases the chances of bacteria that are resistant to the poison surviving and reproducing.
3. Why are bacteria being cultured to clean up nitroglycerin? Not enough occur naturally, so they are allowed to reproduce in the laboratory and are returned to the soil.

#### References

Chollar, Susan. "The Poison Eaters." *Discover,* Apr. 1990, pp. 76-78.

Cultice, Curt. "Let's Do Lunch." *Business America*, Jan.-Feb. 1996, pp. 24-26.

Wickelgren, Ingrid. "Pay Dirt." *Popular Science*, March 1996, pp. 48-51.

## GLENCOE TECHNOLOGY

**Videodisc**

The Secret of Life

*Cyanobacteria*

## TECHNOLOGY

Bacteria are helpful in removing pollutants from soil and groundwater. After an oil spill in Prince William Sound in Alaska, the natural populations of oil-eating bacteria were not large enough to take care of the spill. Researchers added fertilizers to the oil-covered beaches to increase the number of bacteria that lived on the beaches.

At Los Alamos National Laboratory, bacteria that break down explosive wastes are being grown. A researcher has collected soil samples that contain TNT and nitroglycerin. The samples also contain bacteria that have adapted to the explosives and break them down. The bacteria are being cultured in large quantities and transferred back to the contaminated area.

Bacteria are used to mine gold, copper, and other valuable minerals from mines where the concentrations are so low they cannot be mined by traditional methods. Bacteria that release bound-up minerals are added to the ore. The minerals can then be mined.

Oil spill

**Figure 4-16** When water is removed from foods, bacteria cannot grow in them.

If water is taken from food, bacteria can't live in the food. Bacteria need the water for growth. For this reason, uncooked noodles and dry cereals can be left open without spoiling. These foods have very little water. Removing the water from food is called dehydration (dee hi DRAY shun). What other foods do you know of that are dehydrated?

Why do you put iodine, hydrogen peroxide, or alcohol on a cut, scrape, or other skin injury? These chemicals are called antiseptics (an ti SEP tiks). The word *antiseptic* means against infection. Antiseptics are chemicals that kill bacteria on living things. Disinfectants (dihs un FEK tunts) are stronger chemicals that are used to destroy bacteria on objects that are not living.

What do we do to get rid of bacteria on or in our bodies? We wash with soap, we brush our teeth, and we use mouthwash. Each time this is done, we wash, brush, kill, or rinse away millions of bacteria. The bacteria can cause body and breath odor and tooth decay.



## OPTIONS

### Science Background

Only a few species of bacteria produce spores, but their spores are nearly everywhere. Moist heat under pressure is a widely used method of sterilization used to kill spores.

**Skill**/*Teacher Resource Package*, p. 4. Use the Skill worksheet shown here to have students compare viruses, bacteria, and blue-green bacteria.

**SKILL 4**

**SKILL: MAKE AND USE TABLES** CHAPTER 4

Name ______ Date ______ Class ______

*For more help, refer to the Skill Handbook, pages 715-717.* *Use after Section 4:2.*

**VIRUSES, BACTERIA, AND BLUE-GREEN BACTERIA**

One way to study and understand information is to organize it into a table. By making a table, you can see at a glance how viruses, bacteria, and blue-green bacteria are the same and how they are different. The first column of the table below lists facts about viruses and monerans. To complete the table, decide whether the fact describes any of the organisms or viruses named in the other columns. Some facts describe more than one type of organism or virus.

*Place a check mark in a column if the fact describes that organism or virus. When the table is complete, use it to answer the questions below.*

| Fact | Viruses | Bacteria | Blue-green bacteria |
|---|---|---|---|
| Some move with a flagellum | | ✓ | |
| Have a jellylike layer | | | ✓ |
| Some have a capsule | | ✓ | |
| Reproduce by fission | | ✓ | ✓ |
| Live in colonies | | ✓ | ✓ |
| Reproduce only in host cells | ✓ | | |
| May form endospores | | ✓ | |
| All make their own food | | | ✓ |
| Have no nucleus | ✓ | ✓ | ✓ |
| Are single-celled | | ✓ | ✓ |
| Are not made of cells | ✓ | | |
| May cause disease | ✓ | ✓ | |
| Can spoil food | | ✓ | |
| Contain chlorophyll | | | ✓ |

1. How are viruses similar to bacteria? **They both can cause disease and both lack a nucleus.**
2. How are bacteria similar to blue-green bacteria? **They both reproduce by fission, live in colonies, lack a nucleus, and are single-celled.**

Some medicines help stop bacterial infections after you become sick. You may have been given an antibiotic or other medicine to fight a bacterial infection when you were sick. Vaccines prevent bacterial infections from happening. They protect the body from invading organisms.

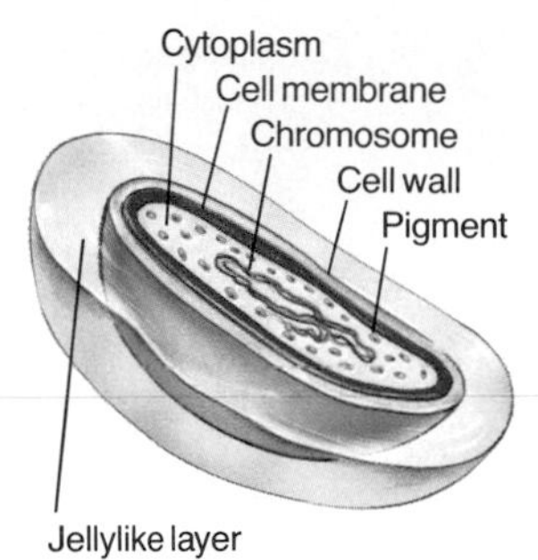


**Figure 4-17** Structure of a typical blue-green bacterium. Does it look more like a bacterial cell or an animal cell? bacterial cell

## Blue-green Bacteria

**Blue-green bacteria** are small, one-celled monerans that contain chlorophyll and can make their own food. For many years they were called blue-green algae and were classified as plants. Their cell structure is more like that of bacteria than of plants. They do not have nuclei or other typical plant cell parts, Figure 4-17. Each has a cell wall outside a cell membrane. The cell wall of a blue-green bacterium is made of materials different from the cell wall of a plant cell.

Blue-green bacteria contain colored pigments. They are usually a blue-green color, but some blue-green bacteria may appear red, black, brown, or purple. The blue-green bacteria in the Red Sea grow rapidly at certain times of the year, giving the sea its red color and its name.

Where do blue-green bacteria live? They are found in ponds, lakes, moist soils, swimming pools, and sometimes around leaky faucets. Some can even live in salt water and on snow. Others grow in the acid water of hot springs.

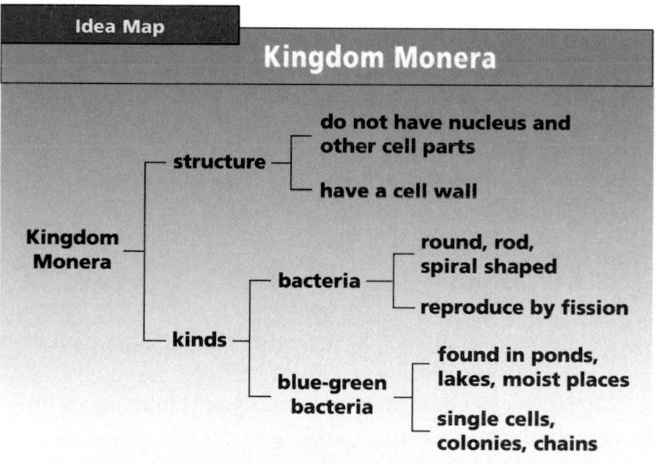


**Study Tip:** Use this idea map as a study guide to the monerans. Can you give an example of each kind of moneran?



## TEACH

### Concept Development

▶ Mention that blue-green bacteria are also called cyanobacteria. The color of blue-green bacteria is due to the presence of chlorophyll (green) and phycocyanin (blue).

▶ Point out that many blue-green bacteria take nitrogen from the air and change it to ammonia. Other kinds of bacteria change the ammonia to nitrates. The nitrates can be used by plants.

### Idea Map

Have students use the idea map as a study guide to the major concepts of this section.

### Check for Understanding

Show prepared slides of bacteria and blue-green bacteria. Have students make a list of characteristics of each. Discuss how they are similar and different.

### Reteach

Prepare a list of statements on 3" x 5" cards that describe bacteria and blue-green bacteria. Have each student select a card, read the statement, and tell which one it describes. Some statements may describe both.

### Independent Practice

**Study Guide**/*Teacher Resource Package*, pp. 23, 24. Use the Study Guide worksheet at the bottom of this page for independent practice.

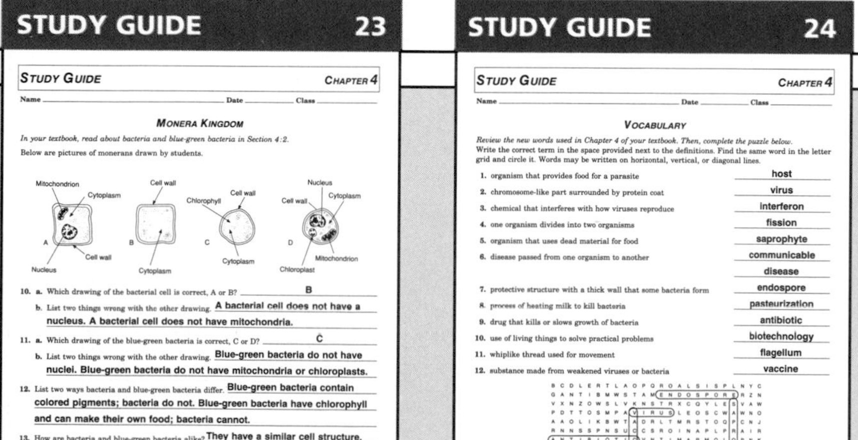

**STUDY GUIDE 23**

*STUDY GUIDE* — *CHAPTER 4*

Name ______ Date ______ Class ______

**MONERA KINGDOM**

*In your textbook, read about bacteria and blue-green bacteria in Section 4:2.*

Below are pictures of monerans drawn by students.



10. a. Which drawing of the bacterial cell is correct, A or B? **B**
    b. List two things wrong with the other drawing. **A bacterial cell does not have a nucleus. A bacterial cell does not have mitochondria.**
11. a. Which drawing of the blue-green bacteria is correct, C or D? **C**
    b. List two things wrong with the other drawing. **Blue-green bacteria do not have nuclei. Blue-green bacteria do not have mitochondria or chloroplasts.**
12. List two ways bacteria and blue-green bacteria differ. **Blue-green bacteria contain colored pigments; bacteria do not. Blue-green bacteria have chlorophyll and can make their own food; bacteria cannot.**
13. How are bacteria and blue-green bacteria alike? **They have a similar cell structure. They do not have nuclei or most other cell parts.**
14. How are blue-green bacteria helpful? **They are food for animals that live in the water. They produce oxygen. They recycle nitrogen, which is used by plants.**
15. How are blue-green bacteria harmful? **They grow rapidly in ponds and kill other living things. They make water unfit for drinking and swimming.**

**STUDY GUIDE 24**

*STUDY GUIDE* — *CHAPTER 4*

Name ______ Date ______ Class ______

**VOCABULARY**

*Review the new words used in Chapter 4 of your textbook. Then, complete the puzzle below.*
Write the correct term in the space provided next to the definitions. Find the same word in the letter grid and circle it. Words may be written on horizontal, vertical, or diagonal lines.

1. organism that provides food for a parasite **host**
2. chromosome-like part surrounded by protein coat **virus**
3. chemical that interferes with how viruses reproduce **interferon**
4. one organism divides into two organisms **fission**
5. organism that uses dead material for food **saprophyte**
6. disease passed from one organism to another **communicable disease**
7. protective structure with a thick wall that some bacteria form **endospore**
8. process of heating milk to kill bacteria **pasteurization**
9. drug that kills or slows growth of bacteria **antibiotic**
10. use of living things to solve practical problems **biotechnology**
11. whiplike thread used for movement **flagellum**
12. substance made from weakened viruses or bacteria **vaccine**

## OPTIONS

### ACTIVITY/Challenge

Ask students to find out why blue-green bacteria were once classified as plants and why they are now classified with bacteria (lack of cell parts and cell wall).

## TEACH

### Check for Understanding

Have students respond to the first three questions in Check Your Understanding.

### Reteach

**Reteaching**/*Teacher Resource Package,* p. 11. Use this worksheet to give students additional practice with monerans.

**Extension:** Assign Critical Thinking, Biology and Reading, or some of the **OPTIONS** available with this lesson.

## 3 APPLY

### Connection: Math

*E. coli* bacteria reproduce every 20 minutes. Have students calculate how many can be produced in an eight-hour period.

## 4 CLOSE

### Display

Display food items such as cereal, raisins, pickles, milk, canned food, and frozen food. Ask students to describe methods used to control bacterial growth.

### Answers to Check Your Understanding

6. Bacterial cells do not have a nucleus or other cell parts.
7. by air, drinking water, touching, sexual contact, and insects
8. They have chlorophyll and make their own food.
9. Bacteria need water to grow. The water has been removed from these foods.
10. *Cyano* is a prefix that means blue-green.

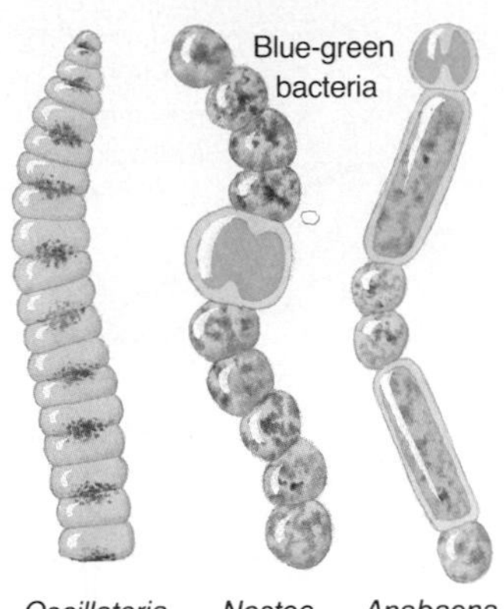

**Figure 4-18** Blue-green bacteria (a) can grow very rapidly and cover the surface of a pond (b).

Blue-green bacteria occur as single cells, colonies, and long, threadlike chains. Many have an outer jellylike layer that holds the cell to other cells. Some have gas bubbles that let the cells float at the surface, where they get sun.

Blue-green bacteria serve as food for animals that live in water. They produce oxygen as they make their own food. They are also important in recycling nitrogen that can then be used by plants.

Blue-green bacteria may grow too rapidly and cover an entire pond. When this happens, they can use up the oxygen in the pond and kill other living things. Many blue-green bacteria produce a bad odor. The odor is one clue that the water is unfit for drinking. Substances that pollute water are often used by blue-green bacteria for food. The amount of pollution in water is measured by counting the number of blue-green bacteria.

### Check Your Understanding

6. What are two ways that bacterial cells differ from animal cells?
7. What are three ways in which communicable diseases can be spread?
8. How do blue-green bacteria differ from other bacteria?
9. **Critical Thinking:** Why don't foods such as uncooked rice and raisins spoil?
10. **Biology and Reading:** The scientific name for blue-green bacteria is *cyanobacteria.* From the way these organisms are described, what do you think the prefix *cyano-* means?



## OPTIONS

**What Are the Traits of Blue-green Bacteria?**/*Lab Manual,* pp. 31-32. Use this lab as an extension to classifying blue-green bacteria.

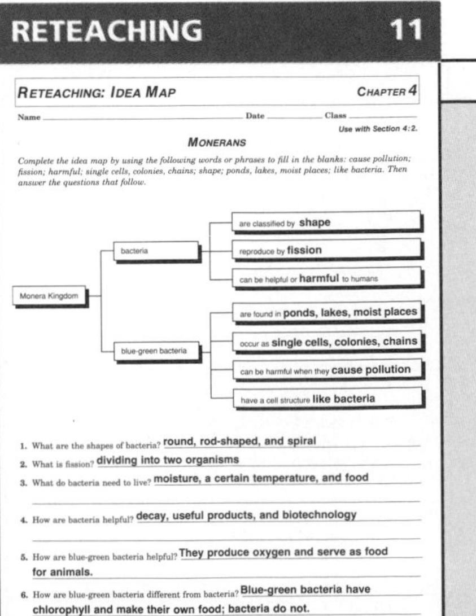
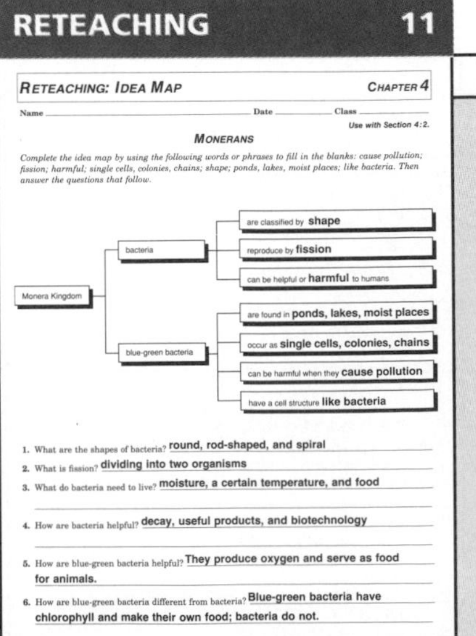

**RETEACHING 11**

*RETEACHING: IDEA MAP* — CHAPTER 4

Name ______ Date ______ Class ______

*Use with Section 4:2.*

**MONERANS**

*Complete the idea map by using the following words or phrases to fill in the blanks: cause pollution; fission; harmful; single cells, colonies, chains; shape; ponds, lakes, moist places; like bacteria. Then answer the questions that follow.*

- Monera Kingdom
  - bacteria
    - are classified by **shape**
    - reproduce by **fission**
    - can be helpful or **harmful** to humans
  - blue-green bacteria
    - are found in **ponds, lakes, moist places**
    - occur as **single cells, colonies, chains**
    - can be harmful when they **cause pollution**
    - have a cell structure **like bacteria**

1. What are the shapes of bacteria? **round, rod-shaped, and spiral**
2. What is fission? **dividing into two organisms**
3. What do bacteria need to live? **moisture, a certain temperature, and food**
4. How are bacteria helpful? **decay, useful products, and biotechnology**
5. How are blue-green bacteria helpful? **They produce oxygen and serve as food for animals.**
6. How are blue-green bacteria different from bacteria? **Blue-green bacteria have chlorophyll and make their own food; bacteria do not.**

# Lab 4–2 Monerans

## Problem: How do blue-green bacteria and bacteria compare?

### Skills

make and use tables, observe, measure in SI, use a microscope

### Materials

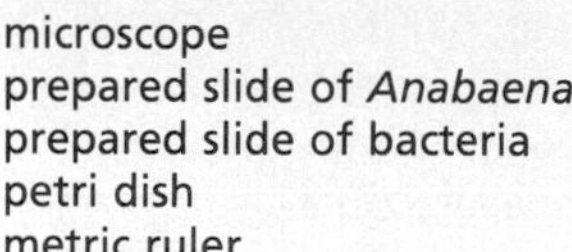

microscope
prepared slide of *Anabaena*
prepared slide of bacteria
petri dish
metric ruler

### Procedure

1. Copy the data table.
2. Use the low-power objective to locate a strand of *Anabaena*, Figure A.
3. **Observe:** Switch to the high-power objective of your microscope and observe the *Anabaena*. Look for cell parts inside the cells.
4. Complete the first three rows of the data table for *Anabaena*.
5. Using the petri dish as a guide, draw a circle on a sheet of paper.
6. Look through the eyepiece of the microscope. Draw one cell in the circle. (Pretend that the circle is the view through the eyepiece. Draw the cell the same size it appears through the eyepiece. This is called drawing to scale.)
7. **Measure in SI:** Measure the length of your *Anabaena* diagram in millimeters.
8. Record this number in your data table.
9. To find the actual length of the cell, multiply the length of your diagram by 0.0035. Use a calculator to help you. Record your answer.
10. Repeat steps 2 through 9 with the bacteria slide. See Figure B.

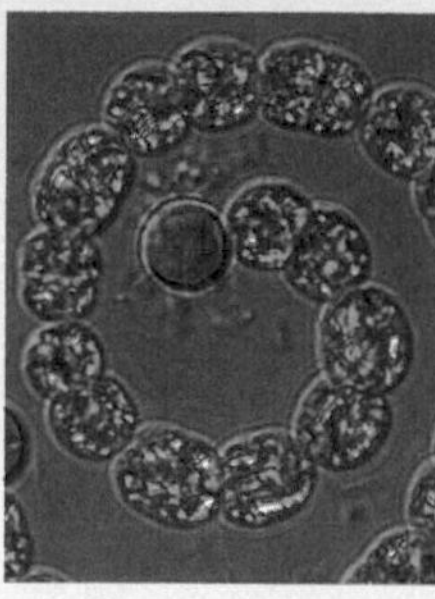

A

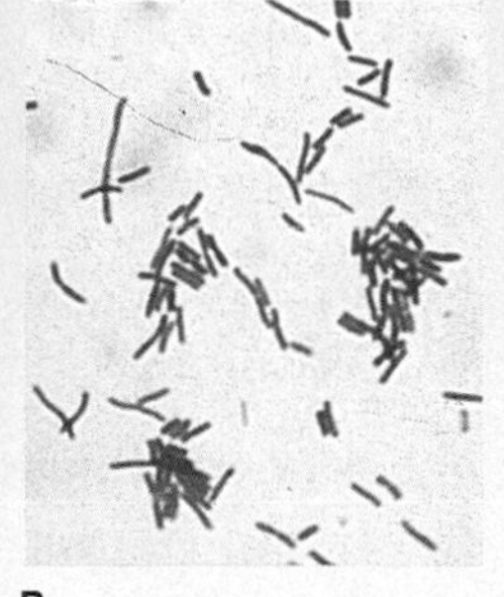

B

### Data and Observations

1. Which organism is smaller?
2. Which organism lives as a single cell?

| | *ANABAENA* | *BACTERIUM* |
|---|---|---|
| Shape of cells | round and oblong | rod |
| Single cell or filament | filament | single cell |
| Color | green | pink |
| Length of diagram | 10 mm | 4 mm |
| Actual length of cell | 0.035 | 0.014 |

### Analyze and Apply

1. Which organism can make its own food?
2. To which kingdom do *Anabaena* and bacteria belong? How do you know?
3. **Apply:** How do blue-green bacteria and other bacteria compare?

### Extension

**Observe** other bacteria and blue-green bacteria under the microscope. Compare these with *Anabaena* and the bacteria in this lab.



### COMPUTER PROGRAM VI

COMPUTER PROGRAM LAB 4-2 MONERANS

```
]POKE 33,33

]LIST

 10  GOSUB 390
 20  HOME
 30  PRINT "           ------------------------------";
 33  PRINT "           !   ANABAENA  ! BACTERIUM  !";
 35  PRINT "-----------------------------------------";
 40  PRINT "!LENGTH IN  !             !            !";
 50  PRINT "!MILLIMETERS!             !            !";
 60  PRINT "!OF DIAGRAM !             !            !";
 70  PRINT "-----------------------------------------";
 80  PRINT "!ACTUAL     !             !            !";
 90  PRINT "!LENGTH OF  !             !            !";
 100  PRINT "!CELL IN    !             !            !";
 110  PRINT "!MILLIMETERS!             !            !";
 120  PRINT "-----------------------------------------";
 130  PRINT "!SKIN   !       !           !          !";
 140  PRINT "-----------------------------------------"
 150  FOR X = 1 TO 2
 160  VTAB V(X): HTAB H(X): FLASH : PRINT " ";: NORMAL
 170  VTAB 15: HTAB 1: CALL  - 958
 180  PRINT "ENTER YOUR VALUE FOR THIS SPACE";
 190  INPUT ": ";A
 195 N(X) = A
 200  VTAB V(X): HTAB H(X): PRINT A
 210  NEXT
 220  VTAB 15: HTAB 1: CALL  - 958
 230  PRINT "ARE THESE NUMBERS                 ENTERED CORRECTLY? (
    Y/N): ";: GET A$
 240  IF A$ = "N" OR A$ = "n" THEN GOTO 20
 250  IF A$ < > "Y" AND A$ < > "y" THEN 220
 260  VTAB 15: HTAB 1: CALL  - 958: PRINT
 270  PRINT  "PRESS THE ";: INVERSE : PRINT "SPACE BAR";: NORMAL : PRINT "
     TO SEE              THE ACTUAL LENGTHS IN MILLIMETERS";: GET A$
 280  VTAB 16: HTAB 1: CALL  - 958
 290  FOR X = 3 TO 4
 300  VTAB V(X): HTAB H(X)
 310  PRINT N(X - 2) * .0035
 320  NEXT
 330  INVERSE : PRINT
 340  PRINT : PRINT "DO YOU WANT THIS PRINTED (Y/N)?;";: GET A$
 345  NORMAL
 348  VTAB 15: HTAB 1: CALL  - 958
 350  IF A$ = "N" OR A$ = "n" THEN  END
 360  IF A$ < > "Y" AND A$ < > "y" THEN GOTO 340
 370  GOSUB 2000
 380  END
 390  TEXT : HOME : PRINT "LAB 4-1                          MONERANS"
 400  PRINT : PRINT
 410  PRINT "YOU WILL BE ASKED TO TYPE IN YOUR      RESULTS FOR ONLY THE
     LAST TWO ROWS OF   THE DATA TABLE USED IN THE EXPERIMENT."
 430  PRINT
```

## ANSWERS

### Data and Observations

1. bacteria
2. bacteria

### Analyze and Apply

1. *Anabaena*
2. Monera; they do not have nuclei.
3. Bacteria are smaller. They are not green and do not live in colonies. The cells are single. *Anabaena* is green, has two kinds of cells, and lives in colonies.

## Lab 4-2 Monerans

### Overview

In this lab, students observe bacteria and blue-green bacteria to compare their traits. They will also make scale drawings and measure.

**Objectives:** Upon completing this lab, students will be able to (1) **determine** which is larger, bacteria or blue-green bacteria, (2) **determine** which makes its own food, (3) **recognize** traits of monerans.

**Time Allotment:** 25 minutes

### Preparation

▶ If your school does not have prepared slides of Anabaena and bacteria, have students use the photos on this page.

**Lab 4-2 worksheet**/*Teacher Resource Package*, pp. 15-16.

**Lab 4-2 Computer Program**/ *Teacher Resource Package*, p. vi. Use the computer program shown at the bottom of this page as you teach this lab.

### Teaching the Lab

▶ Explain to students that bacteria are normally colorless. The slides have been stained pink as a method of identifying the bacteria.

**Cooperative Learning:** Have students work in pairs for this lab. For more information, see pp. 22T-23T in the Teacher Guide.

**Performance:** Set up a lab practical of other one-celled organisms, including bacteria and protists. Have students draw and identify the cyanobacteria.

# CHAPTER 4 REVIEW

## Summary

Summary statements can be used by students to review the major concepts of the chapter.

## Key Science Words

All boldfaced terms from the chapter are listed.

## Testing Yourself

### Using Words

1. parasite
2. interferon
3. pasteurization
4. colony
5. biotechnology
6. antibiotic
7. blue-green bacteria
8. fission
9. flagellum
10. virus

### Finding Main Ideas

11. p. 73; They are classified by the kind of cell they infect.
12. p. 73; They are spread by insects, air, water, food, and other people.
13. pp. 77-78; Certain white blood cells destroy viruses, some white blood cells produce antibodies, vaccines are used to treat some diseases, and cells produce interferon.
14. p. 79; They are found almost everywhere–water, air, soil, food, on your skin, in your body.
15. p. 79; They are classified by their shape–round, rod, and spiral.
16. p. 83; They may be spread by air, touching anything an infected person has touched, drinking water that contains bacteria, and sexual contact.
17. p. 81; Fire blight and crown gall are plant diseases caused by bacteria.
18. pp. 85-87; High and low temperatures, canning, pasteurization, dehydration, antiseptics, washing, and medicines are all methods to control bacteria.
19. p. 83; They change dead matter into materials that other living things can use.
20. p. 88; They grow rapidly and cause water pollution in lakes and ponds.

# Chapter 4 Review

## Summary

### 4:1 Viruses

1. Viruses are not cells and have no cell parts. They are made of a chromosome-like part that carries the hereditary material. They are surrounded by a protein coat.
2. Viruses reproduce only inside living cells.
3. Viruses cause diseases in many living things.

### 4:2 Monera Kingdom

4. All bacteria are one-celled, have no nucleus, and are either round, rod-shaped, or spiral. Other bacteria cannot make their own food or make oxygen.
5. Some bacteria cause serious diseases. They may be spread by air, food, water, sexual contact, or insects.
6. Blue-green bacteria can make their own food and make oxygen. They occur as single cells, colonies, or long chains.

## Key Science Words

antibiotic (p. 84)
asexual reproduction (p. 80)
bacteria (p. 79)
biotechnology (p. 85)
blue-green bacteria (p. 87)
capsule (p. 80)
colony (p. 79)
communicable disease (p. 83)
decomposer (p. 81)
endospore (p. 81)
fission (p. 80)
flagellum (p. 80)
host (p. 73)
interferon (p. 77)
Koch's postulates (p. 82)
parasite (p. 73)
pasteurization (p. 85)
saprophyte (p. 81)
vaccine (p. 78)
virus (p. 72)

## Testing Yourself

### Using Words

*Choose the word from the list of Key Science Words that best fits the definition.*

1. organism that lives on or in another living thing and causes harm to it
2. substance that interferes with how a virus reproduces
3. process of heating milk to kill harmful bacteria
4. group of similar cells growing next to each other
5. the use of living things to solve practical problems
6. chemical that kills or slows the growth of bacteria
7. monerans that contain chlorophyll and can make their own food
8. dividing in half to form two new cells
9. whiplike thread used for movement
10. chromosome-like part surrounded by a protein coat

### Finding Main Ideas

*List the page number where each main idea below is found. Then, explain each main idea.*

11. how viruses are classified
12. how human viruses are spread
13. how humans and other animals protect themselves from viruses



### Using Main Ideas

21. Viruses are so small they can only be seen with an electron microscope. They have many different shapes. They are made of a chromosome-like part surrounded by a protein coat. They have no cell parts. They reproduce only inside living cells.
22. Chewing or sucking insects break through the plant cell wall and allow the virus to come in contact with the host cell.
23. (1) The chromosome-like part enters a host cell and takes over the cell. The virus makes copies of itself, breaks open, and sends the new viruses out to invade more cells. (2) Viruses remain hidden in cells without reproducing until something causes them to become active.
24. Interferon binds to the cell membrane of other body cells the first time a person has flu. It helps the cells resist other viruses.

# Review

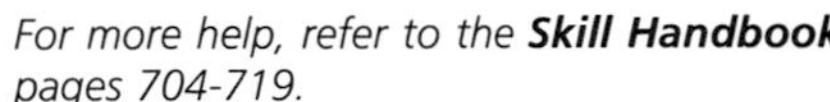

## Testing Yourself *continued*

14. where bacteria are found
15. how bacteria are classified
16. how communicable diseases are spread
17. two plant diseases caused by bacteria
18. methods by which bacteria can be controlled
19. how decay-causing bacteria are important
20. how blue-green bacteria are harmful

**Using Main Ideas**

*Answer the questions by referring to the page number after each question.*

21. What are some traits of viruses? (p. 72)
22. How do insects spread plant viruses? (p. 74)
23. In what two ways do viruses cause disease? (p. 74)
24. How does interferon act against viruses? (p. 77)
25. What are the differences between a bacterial cell and an animal cell? (p. 80)
26. How can bacteria survive when growing conditions are not favorable? (p. 81)
27. How do antiseptics and disinfectants differ from one another? (p. 86)
28. How are Koch's postulates used today? (p. 83)
29. In what ways are bacteria useful to humans? (p. 83)
30. How do blue-green bacteria differ from other bacteria? (p. 87)
31. Where can blue-green bacteria be found? (p. 87)

## Skill Review

*For more help, refer to the **Skill Handbook,** pages 704-719.*

1. **Formulate a model:** Place several green marbles in a bag filled with water. What organism does this represent?
2. **Interpret diagrams:** Look at Figure 4-8 and compare the parts of a bacterium with those of an animal cell. What cell parts does an animal cell have that a bacterium does not have?
3. **Infer:** Blue-green bacteria were once classified as algae. They are now classified as bacteria. The organisms have not changed. Why do you think their classification has changed?
4. **Make and use tables:** Make a table showing the cause of: measles, AIDS, pneumonia, cold, crown gall, flu, syphilis, warts, gonorrhea, tooth decay, fire blight, herpes.

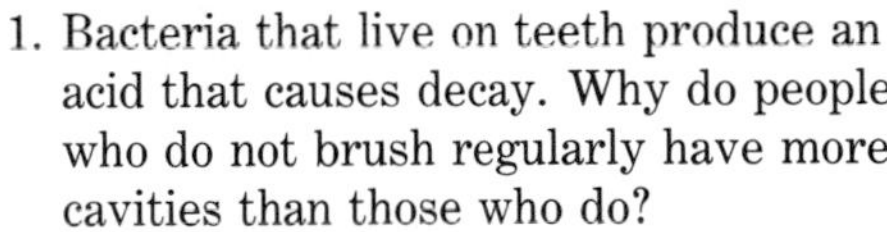


## Finding Out More

**Critical Thinking**

1. Bacteria that live on teeth produce an acid that causes decay. Why do people who do not brush regularly have more cavities than those who do?
2. Few people get the diseases measles and polio. Why are the vaccines still required in many states?

**Applications**

1. Report on the economic effects of viral diseases on crops.
2. Make a collage of ads for products such as deodorants, soaps, and toothpastes that control bacteria.



## Using Main Ideas

25. A bacterial cell does not have a nucleus or other cell parts. An animal cell has a nucleus and other cell parts.
26. They form endospores. A thick wall forms inside the cell, enclosing all the nuclear material and some cytoplasm.
27. Antiseptics are used to kill bacteria on living things. Disinfectants are used to kill bacteria on nonliving objects.
28. They are used to identify bacteria and other disease-causing organisms.
29. Bacteria break down dead matter; break down grass and hay so that it can be digested by animals; produce vitamins; make linen, rope, leather, antibiotics, and many foods.
30. They have chlorophyll and can make their own food.
31. They are found in ponds, lakes, moist soils, swimming pools, and sometimes around leaky faucets.

## Skill Review

1. a blue-green bacterium
2. nucleus, mitochondrion, vacuole, Golgi body, nucleolus, endoplasmic reticulum
3. Their cell structure is more like bacteria than algae.
4. Students may like to place headings of Virus and Bacteria across the top and list Human and Plant along the side, then fill in the diseases.

## Finding Out More

### Critical Thinking

1. Not brushing leaves particles on teeth that bacteria can use for food. This encourages bacterial growth.
2. If several people did not get the vaccine and contracted the diseases, the weakened viruses in the vaccines might not be effective against the active ones.

### Applications

1. Common plant viruses include tobacco mosaic disease, turnip yellow mosaic, and tomato bushy stunt.
2. Collect old magazines for students to use.

**Chapter Review**/*Teacher Resource Package,* pp. 18-19.

**Chapter Test**/*Teacher Resource Package,* pp. 20-22.

**Chapter Test**/*Computer Test Bank*

# Protists and Fungi

## PLANNING GUIDE

| CONTENT | TEXT FEATURES | TEACHER RESOURCE PACKAGE | OTHER COMPONENTS |
|---|---|---|---|
| (2 days)<br>5:1 Protist Kingdom<br>Animal-like Protists<br>Plantlike Protists<br>Funguslike Protists | Skill Check: *Understand Science Words,* p. 97<br>Skill Check: *Interpret Diagrams,* p. 100<br>Mini Lab: *Is There Life in a Mud Puddle?* p. 96<br>Lab 5-1: *Protists,* p. 101<br>Idea Map, p. 99<br>Check Your Understanding, p. 100 | Reteaching: *Protist Kingdom,* p. 12<br>Transparency Master: *Kinds of Protists,* p. 19<br>Study Guide: *Protist Kingdom,* pp. 25-26<br>Lab 5-1: *Protists,* pp. 17-18 | **Laboratory Manual:**<br>*Where Are Protists Found?* p. 33<br>**STVS:** *Soil Crusts in the Desert,* Plants and Simple Organisms (Disc 4, Side 1)<br>**STVS:** *Simple Forms of Life in the Antarctic,* Plants and Simple Organisms (Disc 4, Side 1) |
| (2 days)<br>5:2 Fungus Kingdom<br>Traits of Fungi<br>Kinds of Fungi<br>Lichens | Mini Lab: *What Do Fungi Need?* p. 104<br>Career Close-Up: *Mushroom Farmer,* p. 105<br>Lab 5-2: *Fungi,* p. 109<br>Check Your Understanding, p. 108 | Application: *Learning About Mushrooms,* p. 5<br>Enrichment: *Ten-pin Protists and Fungi,* p. 5<br>Reteaching: *Comparing Fungi,* p. 13<br>Critical Thinking/Problem Solving: *How Do Mushrooms Form Fairy Rings?* p. 5<br>Skill: *Sequence,* p. 5<br>Focus: *Mildew Makes Things Musty,* p. 9<br>Study Guide: *Fungus Kingdom,* pp. 27-29<br>Study Guide: *Vocabulary,* p. 30<br>Lab 5-2: *Fungi,* pp. 19-20 | **Laboratory Manual:**<br>*What Do Molds Need in Order to Grow?* p. 37<br>Color Transparency 5a: *Slime Mold Life Cycle*<br>Color Transparency 5b: *How Fungi Get Their Food*<br>**STVS:** *Fungal Collection for Research,* Plants and Simple Organisms (Disc 4, Side 1)<br>**STVS:** *Disease-Resistant Tomatoes,* Plants and Simple Organisms (Disc 4, Side 1) |
| Chapter Review | Summary<br>Key Science Words<br>Testing Yourself<br>Finding Main Ideas<br>Using Main Ideas<br>Skill Review | **ASSESSMENT RESOURCES**<br>Chapter Review, pp. 23-24<br>Chapter Test, pp. 25-27<br>Performance Assessment in the Biology Classroom<br>Alternate Assessment in the Science Classroom<br>Computer Test Bank | |

## GLENCOE TECHNOLOGY

**Infinite Voyage,** *Secrets from a Frozen World*
**Science and Technology Videodisc Series,** *Soil Crusts in the Desert,* Plants and Simple Organisms (Disc 4, Side1)
*Simple Forms of Life in the Antarctic,* Plants and Simple Organisms (Disc 4, Side 1)
*Fungal Collection for Research,* Plants and Simple Organisms (Disc 4, Side 1)
*Disease-Resistant Tomatoes,* Plants and Simple Organisms (Disc 4, Side 1)

**The Secret of Life,** *On the Brink: Portraits of Modern Science*

## MATERIALS NEEDED

| LAB 5-1, p. 101 | LAB 5-2, p. 104 | MARGIN FEATURES |
|---|---|---|
| microscope<br>3 microscope slides<br>3 coverslips<br>3 droppers<br>cultures of:<br>amoeba<br>euglena<br>paramecium | microscope<br>hand lens<br>scalpel<br>prepared mushroom slide<br>mushroom | Skill Check, p. 97<br>dictionary<br>Mini Lab, p. 96<br>microscope<br>microscope slide<br>coverslip<br>dropper<br>muddy water<br>Mini Lab, p. 104<br>bread<br>water<br>plastic bag |

## OBJECTIVES

For more information about National Science Standards, see page 5T.

| SECTION | OBJECTIVE | CORRELATION of QUESTIONS to OBJECTIVES: CHECK YOUR UNDERSTANDING | CHAPTER REVIEW | TRP CHAPTER REVIEW | TRP CHAPTER TEST |
|---|---|---|---|---|---|
| 5:1 National Science Stds: UCP.2, UCP.3, UCP.4, UCP.5, C.1, C.4, C.5, C.6, F.1, F.4, F.5 | 1. **Identify** the general characteristics of protists. | 1, 4 | 7, 14, 24, 32 | 18 | 5, 26 |
| | 2. **Compare** the traits of animal-like protists, plant-like protists, and fungus-like protists. | 3 | 5, 6, 9, 16, 26, 27, 29, 36 | 6, 7, 8, 30, 34 | 1, 3, 4, 6, 8, 10, 13, 23, 24, 25, 28, 29, 30, 31, 32, 41, 42, 55, 56, 57 |
| | 3. **Give an example** of each kind of protist. | 2 | 8, 11, 17, 21, 22, 23 | 9, 10, 11, 12, 13, 14, 15, 16, 17, 31, 32, 37 | 2, 36, 37, 38, 39, 40, 41, 42, 43 |
| 5:2 National Science Stds: UCP.1, UCP.2, UCP.3, UCP.4, UCP.5, C.1, C.4, C.5, C.6, D.2, F.1, F.3, F.4, F.5 | 4. **List** the traits of fungi and compare the traits of the major kinds of fungi. | 6, 8, 9 | 2, 3, 10, 12, 13, 15, 18, 19, 25, 28, 30, 33 | 2, 5, 19, 21, 22, 29, 33, 36, 38 | 9, 11, 12, 15, 17, 19, 43, 44, 45, 49, 50 |
| | 5. **Tell** how fungi can be harmful and helpful. | 7 | 4, 20, 31, 34, 35 | 1, 3, 4, 22, 25, 26, 27, 28, 29, 30, 31, 32, 33, 34, 35, 36, 37, 38, 39, 40, 41, 47 | 7, 14, 16, 18, 22, 46, 47, 48, 51, 54 |

# Protists and Fungi

## CHAPTER OVERVIEW

### Key Concepts

In this chapter, students will study the structure, traits, and economic importance of protists and fungi. The protists are divided into animal-like, plantlike, and fungilike groups. Students will learn about the kinds of fungi and their methods of reproduction

### Key Science Words

| | |
|---|---|
| algae | protozoan |
| budding | sac fungi |
| cilia | slime mold |
| club fungi | sporangia |
| hyphae | sporangium fungi |
| lichen | spore |
| multicellular | sporozoan |
| mutualism | |

### Skill Development

In Lab 5-1 students will **make and use tables, interpret data, observe, classify,** and **use a microscope** while observing the traits of protists. In Lab 5-2 they will **observe, infer, use a microscope** and **design an experiment** to study mushroom structure. In the Skill Check on page 97, they will **understand** the **science word** *multicellular.* In the Skill Check on page 100, they will **interpret diagrams** of slime molds. In the Mini Lab on page 96, students will **use a microscope** to observe life from a mud puddle. In the Mini Lab on page 104, they will **observe** the growth of fungi.

### Bridging

Students learned the basis for classifying organisms into five kingdoms in Chapter 3. In Chapter 4, they learned that monerans are single-celled. Bridge this to the fact that most protists are also single-celled, but they have cell parts. Bacteria and protists are important in decay.

**CHAPTER PREVIEW**

### Chapter Content

Review this outline for Chapter 5 before you read the chapter.

**5:1 Protist Kingdom**
- Animal-like Protists
- Plantlike Protists
- Funguslike Protists

**5:2 Fungus Kingdom**
- Traits of Fungi
- Kinds of Fungi
- Lichens

### Skills in this Chapter

The skills that you will use in this chapter are listed below.

- In **Lab 5-1,** you will make and use tables, interpret data, observe, classify, and use a microscope. In **Lab 5-2,** you will observe, infer, use a microscope, and design an experiment.
- In the **Skill Checks,** you will understand science words and interpret diagrams.
- In the **Mini Labs,** you will use a microscope and observe.

92

## TECH PREP

For Tech Prep activities in this chapter of the Teacher Wraparound Edition, see especially the Connection on page 96, the Student Journals on pages 98 and 106, and the Portfolio on page 106.
See also the Glencoe Homepage at **www.glencoe.com**

The following Glencoe resources provide additional opportunities for integrating science and technology.

**Technology: Science and Math in Action, Book Two**
Module 1: Recycling

Chapter 5

# Protists and Fungi

Many different protists live in streams, ponds, and oceans. Some have chlorophyll and can make their own food, while others must find food. Some have cell walls and others do not. Most protists are one-celled, but some seaweeds are several meters long. In the small photograph on this page, you can see some of the many kinds of protists that are found in a drop of water from a pond.

The mushroom growing near the pond on the left is a fungus. Mushrooms and other fungi were once grouped with plants. Then it was discovered that fungi have traits that are different from plants. Fungi do have cell walls, but they lack chlorophyll. They absorb food from their surroundings. This is why fungi are grouped in a separate kingdom.

**Try This!**

**Have you ever made mushroom prints?** Remove the stalks from several different mushrooms. Place each cap, open side down, on a separate sheet of paper. Cover each one with an inverted bowl or jar and leave overnight. Uncover and remove the caps. What do you see?

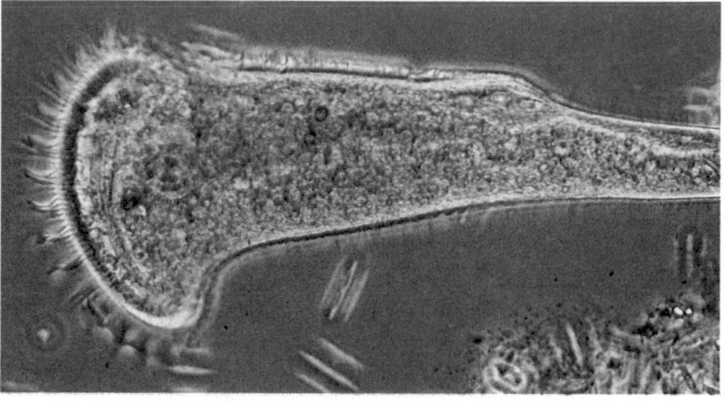

**interNET CONNECTION**

For more information about the material in this chapter, follow the link for the chapter on the Glencoe Homepage at **http://www.glencoe.com**

*Stentor* is a protist that lives in pond water.



**ASSESSMENT PLANNER**

**Portfolio**

Strategies on the following pages represent student products that can be placed into a best-work portfolio: pp. 97, 106.

**PERFORMANCE ASSESSMENT**

Skill Check, p. 100
Mini Lab, pp. 96, 104
Lab, pp. 101, 109

**CONTENT ASSESSMENT**

Check for Understanding, pp. 99, 100, 107, 108
Chapter Review, pp. 110-111

**GROUP ASSESSMENT**

Opportunities for group assessment occur with Cooperative Learning Strategies.

**Student Journal**

Strategies on the following pages represent opportunities for writing in a Student Journal: pp. 98, 106.

GETTING STARTED

## Using the Photos

In looking at the photos, students should compare the features of protists and fungi. Have them note features of fungi that are similar to those of plants.

## MOTIVATION/Try This!

**Have you ever made mushroom prints?** Upon completing the chapter-opening activity, students will **observe** that mushrooms have gills under the cap that release spores.

## Chapter Preview

Have students study the chapter outline before they begin to read this chapter. They should note that animal-like, plantlike, and funguslike protists are in the Protist Kingdom. The traits of fungi, the kinds of fungi, and lichens are studied in the Fungi Kingdom.

## Misconception

Students may think all wild mushrooms are poisonous. Very few species are poisonous, but because they look like the nonpoisonous species, wild mushrooms should never be eaten. Even experts have trouble telling the poisonous species from the nonpoisonous ones.

## 5:1 Protist Kingdom

### PREPARATION

#### Materials Needed

Make copies of the Transparency Master, Study Guide, Reteaching, and Lab worksheets in the *Teacher Resource Package.*

▶ Obtain different kinds of green algae for Focus activity.

▶ Collect water from a mud puddle for the Mini Lab.

▶ Obtain cultures of amoeba, paramecium, and euglena for Lab 5-1.

#### Key Science Words

protozoan
cilia
sporozoan
spore
algae
multicellular
slime mold

#### Process Skills

In Lab 5-1, students will make and use tables, interpret data, observe, classify, and use a microscope. In the Skill Checks, they will interpret diagrams and understand science words. In the Mini Lab, students will use a microscope.

### 1 FOCUS

▶ The objectives are listed on the student page. Remind students to preview these objectives as a guide to this numbered section.

#### MOTIVATION/Demonstration

Bring in green algae from flowerpots, ponds, and ditches. Make wet mounts of the samples for students to observe.

#### Guided Practice

Have students write down their answers to the margin question: one.



**Objectives**

1. **Identify** the general characteristics of protists.
2. **Compare** the traits of animal-like protists, plantlike protists, and funguslike protists.
3. **Give an example** of each kind of protist.

**Key Science Words**

protozoan
cilia
sporozoan
spore
algae
multicellular
slime mold

## 5:1 Protist Kingdom

Protists live almost everywhere. In Chapter 3, protists were described as one-celled organisms that have a nucleus and other cell parts with membranes. Protists are larger than bacteria, but most are so small they can be seen only with a microscope. Protists may be producers or consumers. Most can move about in search of food, light, or a place to live.

Although there are a number of different protist phyla, they can be put into three groups. You can see in Figure 5-1 an example of an animal-like protist, plantlike protist, and a funguslike protist.

### Animal-like Protists

**How many cells do protozoans have?**

Protozoans (proht uh ZOH uhnz) make up the largest group of protists. **Protozoans** are one-celled animal-like organisms with a nucleus. The name *protozoan* comes from the Greek words meaning "first animal." At one time, these organisms were classified in the animal kingdom because scientists thought they looked like tiny animals.

Protozoans, like animals, are consumers. The single cell takes in and digests food. It also reproduces. Most protozoans can move about. They are classified in different phyla depending on how they move. Table 5-1 lists the traits of four protozoan phyla.

**Figure 5-1** Protists can be grouped into animal-like protists (a), plantlike protists (b), and funguslike protists (c).

a

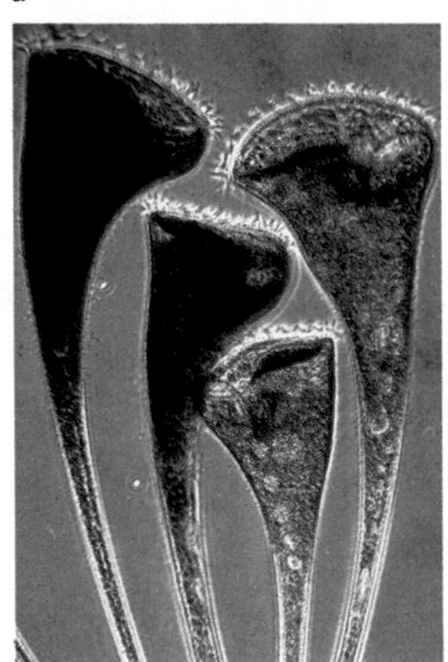

b

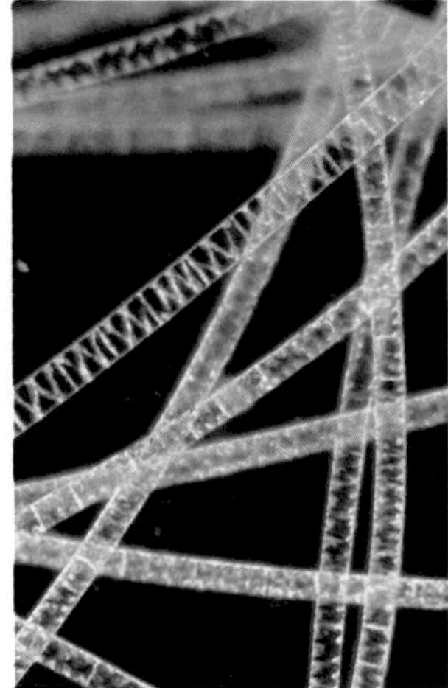

c

## OPTIONS

### Science Background

Green algae are classified on the basis of whether the plants are single cells, colonies, or multicellular. Most green algae live in fresh water. Some move by means of flagella. Brown algae are found in salt water. They grow attached to rocks, shells, or other algae.

| TABLE 5–1. ANIMAL-LIKE PROTISTS | |
|---|---|
| **Traits of Phylum** | **Example** |
| move with false feet | amoeba |
| move with cilia | *Paramecium* |
| do not move; all parasites | *Plasmodium* *(plaz MOH dee uhm)* |
| move with one or more flagella | trypanosome (trip AN uh some) |

Let's look at some of the protozoans. The protozoan in Figure 5-2 is an amoeba (uh MEE buh). The amoeba is a protist that moves by changing shape. It has parts made of cytoplasm that can reach out from the main body. These parts are called false feet. They help the amoeba move about and trap food. Figure 5-3 shows how. First, the amoeba's false feet surround a small protist or a bacterium. Soon the food is trapped inside the amoeba, where it will be digested.

**How does an amoeba move?**

Like bacteria, amoebas reproduce by fission. A parent cell divides into two new cells. Each of the cells has the same hereditary material as the parent cell.

Most kinds of amoebas are not harmful to humans, but some cause disease. One kind of amoeba causes a severe type of diarrhea called amoebic dysentery (uh MEE bihk • DIHS un ter ee). People get this disease by drinking water or eating food that contains the amoebas. This usually occurs in areas where conditions are not sanitary.

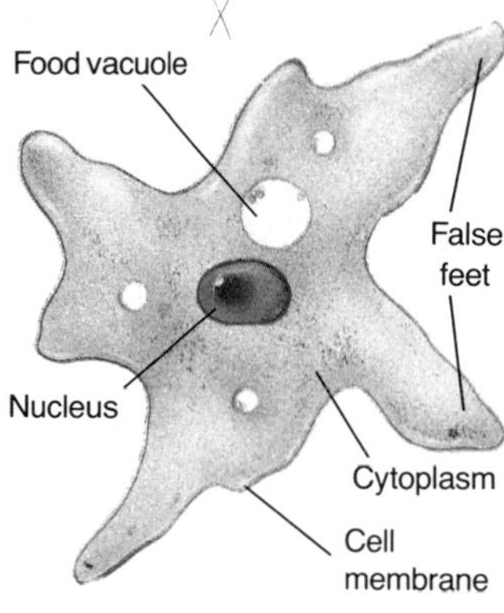


**Figure 5-2** Structures of an amoeba. Do amoebas have cell walls? no

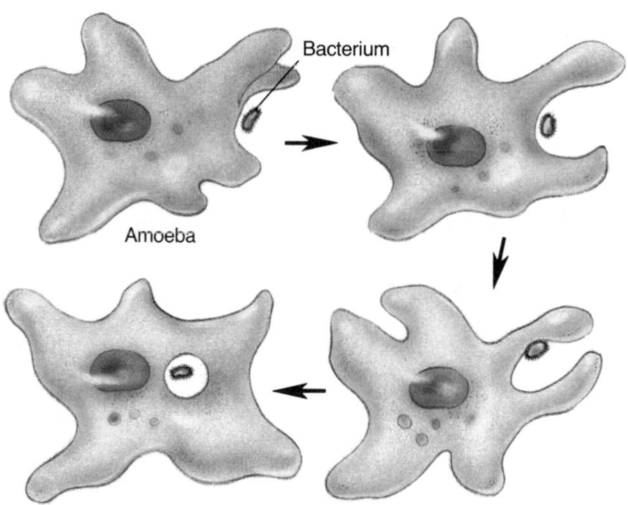


**Figure 5-3** Amoebas capture food with their false feet.



# 2 TEACH

## MOTIVATION/Videocassette

Show the videocassette *The Protozoa,* Opportunities for Learning, Inc.

## Concept Development

▶ Emphasize that protists occur wherever there is moisture. They may be found in the ocean, fresh water, soil, and bodies of other organisms (plants, animals, other protists).

▶ Point out that animal-like organisms are able to move, feed on other organisms, and have no chlorophyll.

▶ Explain that false feet are called pseudopods.

## Guided Practice

Have students write down their answers to the margin question: by changing its shape.

**GLENCOE TECHNOLOGY**

**Videodisc**

**Biology: The Dynamics of Life**
Movie: *Protists*
Disc 1, Side 2, Ch. 12

**Videodisc**

**STV: Human Body Vol. 3**
*Immune System*
Unit 1, Side 1
*Parasites, Cancer, Allergies*

**31350-38691**

**TRANSPARENCY 19**

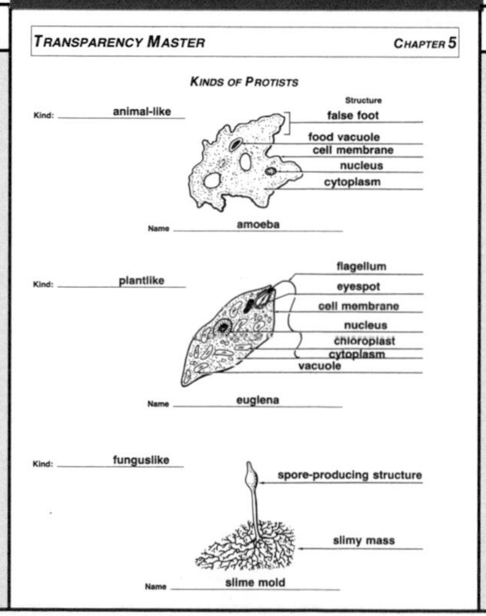


## OPTIONS

**Transparency Master**/*Teacher Resource Package*, p. 19. Use the Transparency Master shown here to teach the kinds of protists.

**Where Are Protists Found?**/*Lab Manual*, pp. 33-36. Use this lab as an extension to studying the Protist Kingdom.

# TEACH

## Concept Development

▶ Point out that cilia is plural for cilium, and flagella is the plural of flagellum.

▶ You may wish to review the term *parasite.*

▶ Explain that sporozoans are carried along by currents in the blood and other body fluids of their hosts.

## Mini Lab

Students should see different protists. Have them classify their drawings as animal-like or plantlike.

**Oral:** Ask students where the living organisms in the mud puddle came from. (Answers may include from endospores present in the soil and organisms present in the soil.)

## Independent Practice

**Study Guide**/*Teacher Resource Package,* p. 25. Use the Study Guide worksheet shown at the bottom of this page for independent practice.

## Guided Practice

Have students write down their answers to the margin question: sporozoans.

## Connection: Social Studies

Have students report to the class on protist-caused diseases such as toxoplasmosis, Chagas disease, and African sleeping sickness.

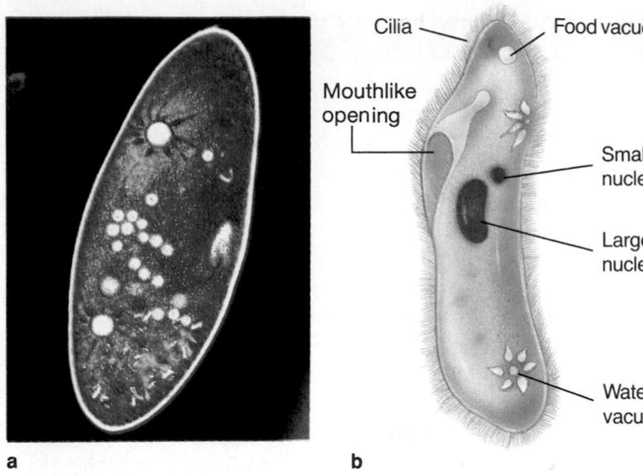


**Figure 5-4** A paramecium (a) and some of its structures (b). What structures does a paramecium use for movement? cilia

> ### Mini Lab
> **Is There Life in a Mud Puddle?**
>
> **Use a microscope:** Use a dropper to collect a sample of water from a mud puddle. Place a drop on a slide and add a coverslip. Focus on low power and then on high. Draw what you see. *For more help, refer to the **Skill Handbook,** pages 712-714.*

An example of another type of protozoan is the paramecium. A paramecium is a protozoan that has a definite shape and moves by means of cilia. **Cilia** are short, hairlike parts on the surface of the cell, as seen in Figure 5-4. The cilia move back and forth very fast, causing the organism to move through the water. Protists with cilia are found in most ponds and streams. Most members of this group do not cause diseases in humans.

When a paramecium finds food, such as a bacterium, it sweeps the food into its body with its cilia. The food enters through a mouthlike opening. A vacuole then forms around the food. The food is digested inside the food vacuole and used by the cell.

Some protozoans move by means of long, whiplike flagella. These protists are found in all kinds of water and in the soil. Some even live inside animals, such as the protists that live in the digestive systems of termites and digest wood for the termite.

Some protozoans are parasites and cause disease. Trypanosomes cause sleeping sickness in humans. They are carried from one person to another by a certain kind of fly that is found in Africa. Once inside the bloodstream, these protozoans produce poison that causes weakness or death in humans.

**Which protist reproduces by forming spores?**

**Sporozoans** (spor uh ZOH uhnz) are protozoans that reproduce by forming spores. **Spores** are special cells that develop into new organisms. Sporozoans do not have cilia, flagella, or false feet, so they can't move to get food. They live as parasites in humans and other animals. Remember that parasites cause harm to other living things.



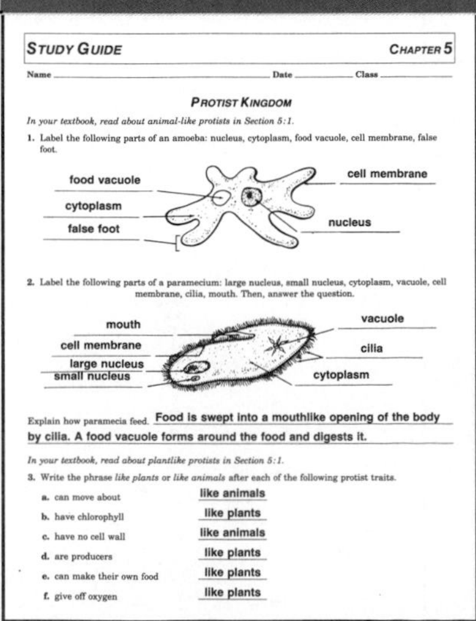

STUDY GUIDE 25

STUDY GUIDE CHAPTER 5

Name ______ Date ______ Class ______

PROTIST KINGDOM

*In your textbook, read about animal-like protists in Section 5:1.*

1. Label the following parts of an amoeba: nucleus, cytoplasm, food vacuole, cell membrane, false foot.

food vacuole
cytoplasm
false foot
cell membrane
nucleus

2. Label the following parts of a paramecium: large nucleus, small nucleus, cytoplasm, vacuole, cell membrane, cilia, mouth. Then, answer the question.

mouth
cell membrane
large nucleus
small nucleus
vacuole
cilia
cytoplasm

Explain how paramecia feed. Food is swept into a mouthlike opening of the body by cilia. A food vacuole forms around the food and digests it.

*In your textbook, read about plantlike protists in Section 5:1.*

3. Write the phrase *like plants* or *like animals* after each of the following protist traits.

a. can move about — like animals
b. have chlorophyll — like plants
c. have no cell wall — like animals
d. are producers — like plants
e. can make their own food — like plants
f. give off oxygen — like plants

Perhaps you have heard of malaria. The sporozoan named *Plasmodium* causes this disease. It is carried by female *Anopheles* (uh NAHF uh leez) mosquitoes. Humans get the disease when they are bitten by infected mosquitoes. The disease is common in tropical areas where these mosquitoes live.

## Plantlike Protists

**Algae** (AL jee) (singular, alga) are plantlike protists. They are like plants because they have chlorophyll and make their own food. Algae also have other pigments. The other pigments help capture light and pass its energy on to chlorophyll. Algae are producers. They produce large amounts of oxygen, which is used by other living things.

Algae live in fresh and salt water. Some also live in moist soils and on tree bark. Like animals, many algae can move about. They do not have a cell wall. Is this trait like that of a plant or an animal?

Algae may be one-celled or multicellular. **Multicellular** means that an organism has many different cells that do certain jobs for the organism. All algae cells have nuclei. Algae are grouped into different phyla depending on their color and their structures. Note the features of each phylum in Table 5-2.

### Skill Check

**Understand science words: multicellular.** The word part *multi* means many. In your dictionary, find three words with the word part *multi* in them. *For more help, refer to the* ***Skill Handbook****, pages 706-711.*

**TABLE 5–2. PLANTLIKE PROTISTS**

| Traits of Phylum | Example |
|---|---|
| move with flagellum, one-celled, no cell wall, found mostly in fresh water, green color | euglena |
| found mostly in salt water, one-celled, gold-brown color | diatom |
| found mostly in salt water, red or brown color, one-celled | dinoflagellate |
| green color, found in fresh and salt water and in soil, one-celled, multicellular, or in colonies | green algae |
| brown color, found in salt water, multicellular | kelp |

The euglena shown in Figure 5-5 lives in fresh water. A euglena is a one-celled alga that moves with a flagellum. The euglena swims by moving its flagellum back and forth in the water. This causes the cell to move in an unusual wiggling motion. Euglenas have a red eyespot near their front end. The eyespot is sensitive to light and helps the euglena find the light it needs to make food. Are euglenas producers or consumers?

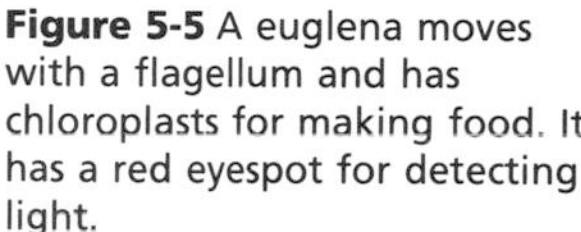

**Figure 5-5** A euglena moves with a flagellum and has chloroplasts for making food. It has a red eyespot for detecting light.

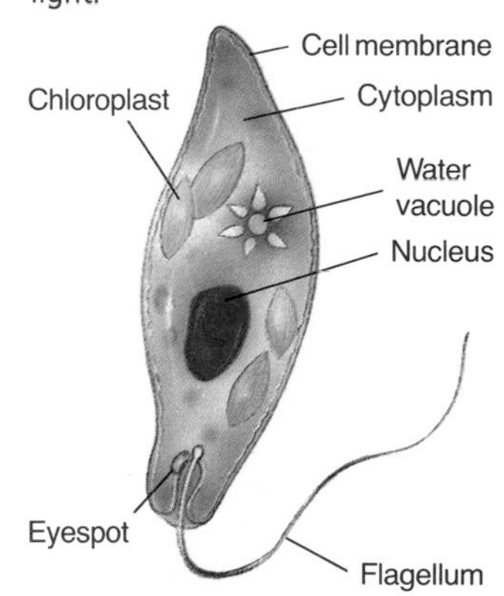




## TEACH

### Concept Development

▶ Point out that a vector is an animal such as the female *Anopheles* mosquito that carries a protist.

▶ Emphasize the importance of algae as producers. They are the beginning of most food chains in both salt and fresh water.

▶ All cells in one-celled algae are similar to one another and live more or less independently. In multicellular organisms, cells are specialized and dependent on one another.

▶ Stress that all algae contain chlorophyll. They capture light at various wavelengths and pass the energy on to chlorophyll.

▶ The euglena has a depression at the front end of the cell called the reservoir. Near the reservoir is a mass of red pigment called the eyespot. When the euglena senses light, it moves toward the light source. This is a helpful adaptation for a photosynthetic organism.

▶ The euglena reproduces by splitting lengthwise to form two daughter cells.

### ACTIVITY/Brainstorming

Have students brainstorm and decide which organism, amoeba, paramecium, or euglena, is the most complex. The paramecium has food and contractile vacuoles, micro- and macronuclei, and is capable of sexual and asexual reproduction.

**Portfolio**

Show students some cartoons about protists. Have students sketch their own cartoons about protists.

## OPTIONS

### Science Background

Red algae contain carrageenan which is used in puddings, preserves, and ice cream. Brown algae is used as animal feed and fertilizer in Northern Europe. Algin, a chemical obtained from brown algae, is often used in the manufacture of latex, ceramic glazes, cosmetics, and ice cream. There is an annual multimillion dollar market for algin in the United States alone.

**GLENCOE TECHNOLOGY**

**Videodisc**

**Biology: The Dynamics of Life**
Movie: *Kelp Forests*
Disc 1, Side 2, Ch. 13

## TEACH

### Concept Development

▶ Explain that diatoms store food in the form of oil.

▶ Point out that when boats, fish, or waves disturb some dinoflagellates at night, the water glows with a blue or green light. This is an example of bioluminescence.

▶ Species of dinoflagellates that cause red tides produce powerful nerve poisons that can kill fish. Shellfish such as oysters and clams are not affected by the poisons, but the poisons do collect in their bodies. Humans who eat the shellfish may be paralyzed or even killed.

### Independent Practice

**Study Guide**/*Teacher Resource Package,* p. 26. Use the Study Guide worksheet shown at the bottom of this page for independent practice.

### ACTIVITY/Demonstration

Demonstrate the structure of a diatom cell with a petri dish. The top and bottom halves represent the cell wall.

**Cooperative Learning:** Have students work in groups of four. Assign each group one kind of algae—green, red, or brown—to research. Have them find out what its role is as a food and oxygen producer and its economic importance.

### Student Journal

Have students find out where red tide occurs most often in the world. Research the most common environmental conditions under which red tide occurs. Include what methods, if any, are currently being used to fight red tide.

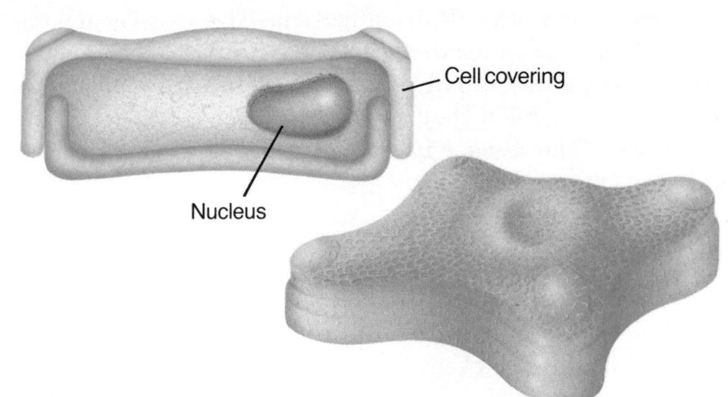


**Figure 5-6** Diatoms have many shapes. Each cell has a two-part cell covering.

Diatoms (DI uh tahmz) are beautiful, one-celled protists that have many shapes, Figure 5-6. They are shaped like boats, rods, disks, or triangles. Diatoms are producers. Would you expect them to have chlorophyll? Diatoms can move slowly through the water. They are one of the most important food sources for animals that live in water.

Diatoms have an interesting structure. They have a cell covering made of two parts. The covering overlaps like a box with a lid, as shown in Figure 5-6. It is made of the same material as glass. When diatoms die, they fall to the bottom of the lake or ocean. The coverings do not decay, but form thick layers. The coverings in these layers are used in toothpastes, scouring powders, and filters because they are gritty and are good cleansers.

Dinoflagellates (di noh FLAJ uh luhts) are algae that are usually found in oceans. Only a few types live in fresh water. They are usually red or brown. They have chlorophyll, but their red and brown pigments hide it. They move by beating their two flagella.

Red dinoflagellates are responsible for red tides. They reproduce in such large numbers that the ocean water turns red. Chemicals produced by these protists during red tides can kill thousands of fish. Humans can become ill if they eat shellfish that have absorbed these chemicals. Red tides are common off the coast of Florida and in other warm waters.

Green algae come in many different forms. They may be one-celled, multicellular, or they may form groups of cells that live together. *Volvox,* the organism shown in Figure 5-7, is a green alga that lives as a hollow ball containing hundreds of one-celled organisms with flagella. The cells are arranged in a single layer with their flagella facing outward. When the flagella beat, the ball of cells spins

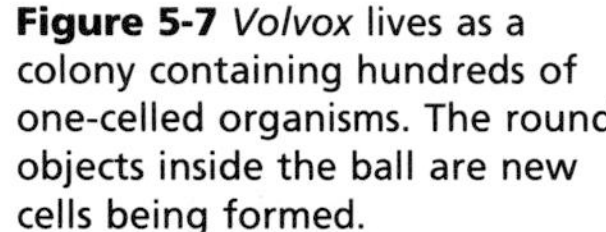

**Figure 5-7** *Volvox* lives as a colony containing hundreds of one-celled organisms. The round objects inside the ball are new cells being formed.

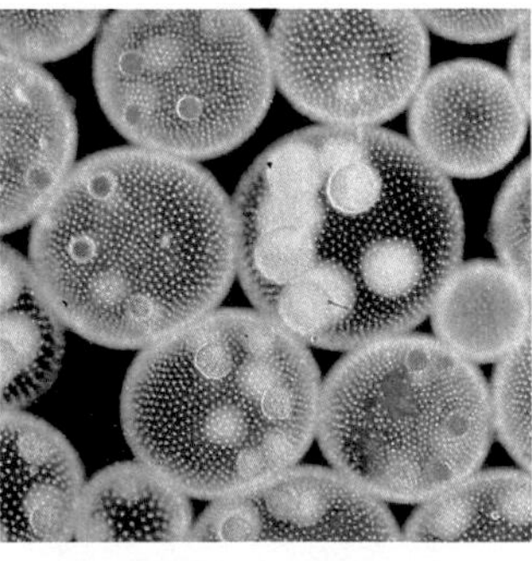



## OPTIONS

### Science Background

The cell walls of diatoms are composed of gelatinous material called pectin. In most diatoms there is also silica, a glassy material rich in silicon. When diatoms die, the outer shell of silica remains intact. The shells accumulate on the ocean floor, forming diatomaceous earth.

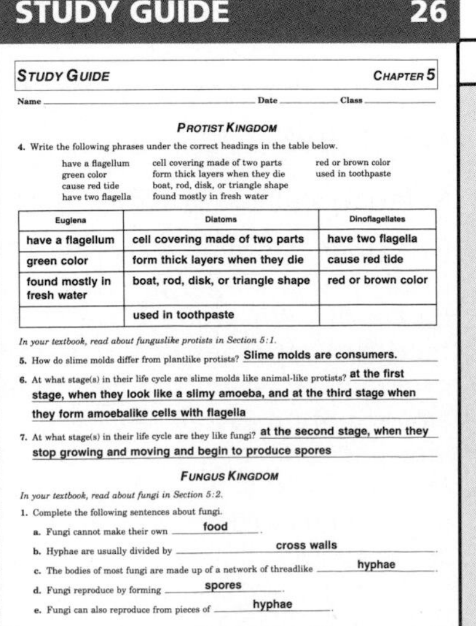

**STUDY GUIDE 26**

STUDY GUIDE CHAPTER 5

Name ______ Date ______ Class ______

PROTIST KINGDOM

4. Write the following phrases under the correct headings in the table below.

have a flagellum, green color, cause red tide, have two flagella, cell covering made of two parts, form thick layers when they die, boat, rod, disk, or triangle shape, found mostly in fresh water, red or brown color, used in toothpaste

| Euglena | Diatoms | Dinoflagellates |
|---|---|---|
| have a flagellum | cell covering made of two parts | have two flagella |
| green color | form thick layers when they die | cause red tide |
| found mostly in fresh water | boat, rod, disk, or triangle shape | red or brown color |
| | used in toothpaste | |

*In your textbook, read about funguslike protists in Section 5:1.*

5. How do slime molds differ from plantlike protists? Slime molds are consumers.
6. At what stage(s) in their life cycle are slime molds like animal-like protists? at the first stage, when they look like a slimy amoeba, and at the third stage when they form amoebalike cells with flagella
7. At what stage(s) in their life cycle are they like fungi? at the second stage, when they stop growing and moving and begin to produce spores

FUNGUS KINGDOM

*In your textbook, read about fungi in Section 5:2.*

1. Complete the following sentences about fungi.
   a. Fungi cannot make their own food.
   b. Hyphae are usually divided by cross walls.
   c. The bodies of most fungi are made up of a network of threadlike hyphae.
   d. Fungi reproduce by forming spores.
   e. Fungi can also reproduce from pieces of hyphae.

through the water. A *Volvox* ball reproduces when some of the cells in the ball come together to form small groups inside the ball. The ball bursts open and releases the small groups of cells.

Multicellular algae have many different shapes. Some form long strands that have many cells. Others have a leaflike shape that looks like lettuce.

Algae are important to us in many ways. They release oxygen into the water and air. The oxygen is used by other living things. Algae are important as food for animals such as fish, snails, and crayfish. Have you ever heard of a seaweed called kelp? Kelp is one kind of brown algae used by many people for food, Figure 5-8. Some scientists think that green algae will be an even more important source of food for humans in the near future.

Are algae products found in any foods that you eat? If you have eaten ice cream or jelly, you probably have eaten certain substances that come from algae. These substances are used to thicken foods.

**Figure 5-8** Kelp is a brown alga that many people eat.

## Funguslike Protists

**Slime molds** are funguslike protists that are consumers. They live in cool, damp places, such as the forest floor. They feed on bacteria growing on rotting logs and decaying leaves. A few slime molds are parasites.

**Bio Tip**

**Ecology:** Slime molds have been known to cover an entire golf course with their slimy masses. The masses creep over the ground, engulfing microorganisms and digesting them.

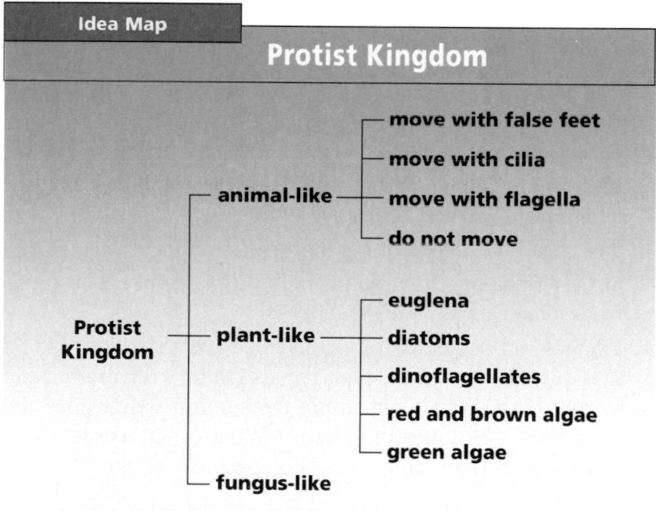


**Study Tip:** Use this idea map as a study guide to the protist kingdom. The features of the three protist groups are shown.



## TEACH

### Concept Development

▶ Slime molds are among the least familiar organisms to most students. Use actual specimens to acquaint the class with these organisms.

▶ Point out that slime molds are sometimes classified with fungi because of the way they reproduce (spores) and get their food (absorption).

### Check for Understanding

Quiz students on protists, methods of locomotion, how they obtain food, how they are classified and why they are important.

### Reteach

Have students make a table to show how protists in this chapter are classified and why they are important.

### Idea Map

Have students use the idea map as a study guide to the major concepts of this section.

**GLENCOE TECHNOLOGY**

**Videodisc**

**Biology: The Dynamics of Life**
Movie: *Slime molds*
Disc 1, Side 2, Ch. 14

## OPTIONS

**Slime Mold Life Cycle**/*Transparency Package*, number 5a. Use color transparency number 5a as you teach funguslike protists.

## TEACH

### Skill Check

The first stage looks like an amoeba and the second stage is like a fungus.

### Check for Understanding

Have students respond to the first three questions in Check Your Understanding.

### Reteach

**Reteaching**/*Teacher Resource Package*, p. 12. Use this worksheet to give students additional practice in understanding the Protist Kingdom.

**Extension:** Assign Critical Thinking, Biology and Reading, or some of the **OPTIONS** available with this lesson.

## 3 APPLY

### ACTIVITY/Software

*Protozoa,* Opportunities for Learning, Inc.

## 4 CLOSE

### Divergent Question

Ask students to make a list of how protists affect our daily lives.

### Answers to Check Your Understanding

1. flagella, cilia, flagellum
2. animal-like: *Paramecium,* amoeba, *Plasmodium,* trypanosome; plantlike: euglena, diatom, dinoflagellate, green algae
3. Most plantlike protists have chlorophyll and can make their own food. Animal-like protists cannot.
4. Protists have cell parts with membranes, and a nucleus. Monerans have chromosomes not contained in a nucleus and a cell wall. Viruses have a chromosome-like part surrounded by a protein coat. Viruses do not have cell parts.
5. It means animal. Other examples are zoo, zoology, zoological.

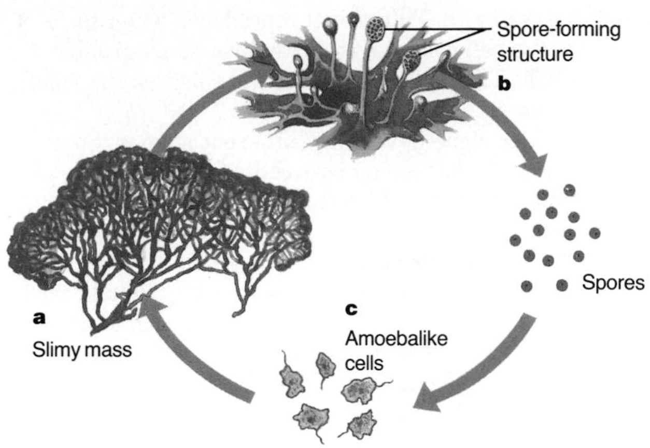


**Figure 5-9** The life cycle of a slime mold has three stages: the feeding and growth stage (a), the spore-forming stage (b), and the amoebalike stage (c).

### Skill Check

**Interpret diagrams:** Look at the slime mold in Figure 5-9. Identify the two stages that are like other organisms. What organisms in this chapter do the stages resemble? *For more help, refer to the **Skill Handbook**, pages 706-711.*

There are three stages in the life cycle of a slime mold, Figure 5-9. The first stage looks like a slimy mass. It moves much like an amoeba, by flowing across a surface. The mass may have beautiful colors such as red, yellow, or violet, or it may have no color.

In the second stage, the slime mold stops growing and moving. It produces spores inside a structure on a stalk. The slime mold in this second stage is like a fungus. This stage usually starts when the slime mold has no food.

During the third stage, the spores develop into little amoebalike cells with flagella. Then each of these cells loses its flagella and grows into a slimy mass again.

### Check Your Understanding

1. Explain how the following move: a *Volvox* colony, a paramecium, a euglena.
2. List two examples of animal-like protists and two examples of plantlike protists.
3. How are plantlike protists different from protozoans?
4. **Critical Thinking:** How do protists differ from viruses and from monerans?
5. **Biology and Reading:** As you learned from reading this section, the word *protozoan* comes from two Greek words. If the word part *proto-* means first, what does the word part *-zoan* mean in Greek? What other words can you think of that come from this same Greek word?



**RETEACHING 12**

**RETEACHING: IDEA MAP** **CHAPTER 5**

Name ________ Date ________ Class ________

*Use with Section 5:1.*

**PROTIST KINGDOM**

Use the following terms to complete the idea map: euglena, paramecium, dinoflagellate, amoeba, diatom, sporozoan, slime mold, trypanosome.

| Group | Trait | Protist |
|---|---|---|
| Animal-like protists | moves with false feet | amoeba |
| | moves with cilia | paramecium |
| | does not move | sporozoan |
| | moves with flagella | trypanosome |
| Plantlike protists | moves with flagella | euglena |
| | has two-part cell covering | diatom |
| | red or brown color | dinoflagellate |
| Funguslike protists | has three stages in life cycle | slime mold |

1. What traits do animal-like protists have? They are consumers; most move about; and all are one-celled.
2. How are the plantlike protists like plants? All have chlorophyll and make their own food. They produce large amounts of oxygen.
3. How are the funguslike protists like fungi? They live in damp places and feed on decaying material. They reproduce by spores.
4. How do amoebas differ from sporozoans in the way they reproduce? Amoebas reproduce by fission; sporozoans reproduce by forming spores.
5. How are plantlike protists classified? They are classified by their color and their structure.

# Lab 5–1 Protists

## Problem: What are the traits of protists?

### Skills

make and use tables, interpret data, observe, classify, use a microscope

### Materials

microscope
3 coverslips
3 droppers
3 glass slides

cultures of:
amoeba
paramecium
euglena

### Procedure

1. Copy the data table.
2. Put a drop of amoeba culture on a glass slide. Gently add a coverslip.

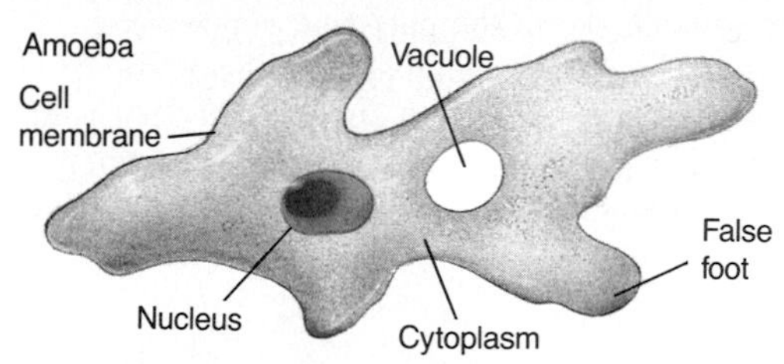


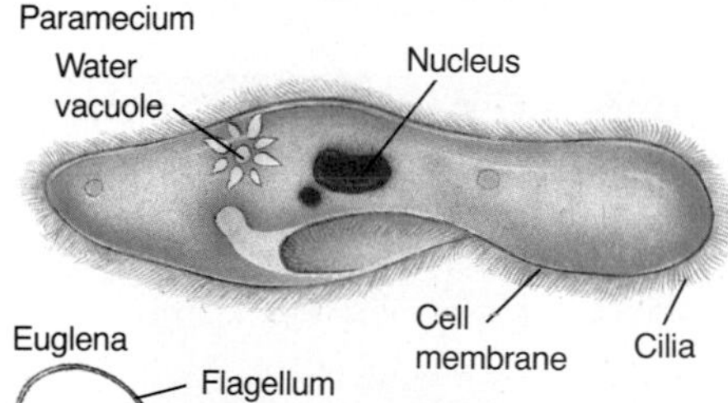


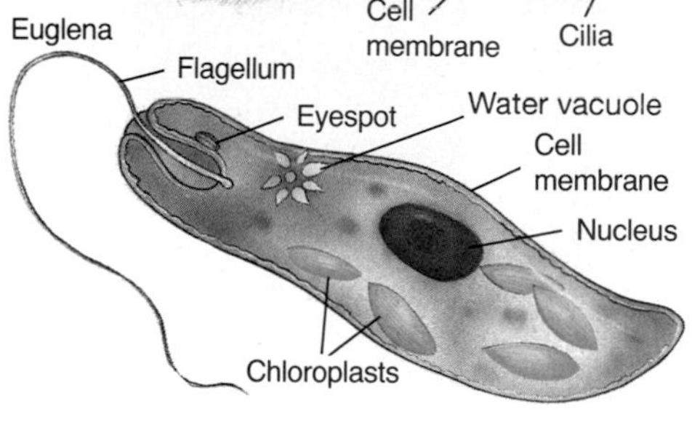


3. **Observe:** Examine the amoeba on low, then high power. Record in your table the parts you see.
4. Repeat steps 2 and 3 with the paramecium and euglena cultures.
5. Wash and dry the glass slides, coverslips, and droppers. Return them to their proper places.

### Data and Observations

1. How does the amoeba move?
2. Which protist is green?

| | Amoeba | Paramecium | Euglena |
|---|---|---|---|
| Cell membrane | yes | yes | yes |
| Parts used for movement | false feet | cilia | flagellum |
| Cell wall | no | no | no |
| Vacuoles | Answers | will vary. | |
| Nucleus | yes | yes | yes |
| Cytoplasm | yes | yes | yes |
| Green color (chlorophyll) | no | no | yes |

### Analyze and Apply

1. **Interpret data:** What traits do protists have in common?
2. Which protist moves the slowest? Which moves the fastest?
3. **Apply:** Which protists are like animals? Why? Which protists are like plants? Why?

### Extension

**Observe** a drop of pond water under the microscope. Identify the protists you see.



## *ANSWERS*

### Data and Observations

1. Cytoplasm flows in and out of false feet.
2. euglena

### Analyze and Apply

1. They have a cell membrane, nucleus, and cytoplasm. They all lack a cell wall.
2. Amoeba moves the slowest. Paramecium moves the fastest.
3. Amoeba and paramecium are like animals because they are consumers and don't make their own food. Euglena are like plants because they have chlorophyll.

## Lab 5-1 Protists

### Overview

In this lab, students use the microscope to observe the structure of representative protists. They also observe how protists move and any cell parts that are different in the three types.

**Objectives:** Upon completing this lab, students will be able to (1) **identify** common traits of protists and (2) **recognize** differences between protists.

**Time Allotment:** 45 minutes

### Preparation

▶ Methyl cellulose may be used to slow down paramecia. Add 10 g methyl cellulose powder to 90 mL warmed distilled water. Stir gently to avoid bubbles. Dispense in dropping bottles.

▶ **Alternate Materials:** Strands of cotton fibers may be used to slow protists instead of methyl cellulose.

**Lab 5-1 worksheet**/*Teacher Resource Package*, pp. 17-18.

### Teaching the Lab

**Cooperative Learning:** Have students work in pairs. For more information, see pp. 22T–23T in the Teacher Guide.

▶ **Troubleshooting:** Remind students never to focus down on a glass slide without watching from the side. The nucleus in euglena may be hard to find if there is an abundance of chloroplasts or if the organisms are very small.

### ASSESSMENT

**Performance:** Remove some scrapings from the side of a fish tank to make a slide of protists. Have students observe, identify, and draw what they see.

## 5:2 Fungus Kingdom

### PREPARATION

#### Materials Needed

Make copies of the Focus, Skill, Critical Thinking/Problem Solving, Application, Study Guide, Enrichment, Reteaching, and Lab worksheets in the *Teacher Resource Package.*

▶ Gather mildew, rust, puffballs, and mushrooms for the Focus activity.

▶ Obtain bread and resealable plastic bags for the Mini Lab.

▶ Obtain prepared slides of mushroom basidium and purchase mushrooms for Lab 5-2.

#### Key Science Words

| | |
|---|---|
| hyphae | sac fungi |
| sporangium fungi | budding |
| sporangia | mutualism |
| club fungi | lichen |

#### Process Skills

In the Mini Lab students will observe. In Lab 5-2, they will observe, infer, use a microscope, and design an experiment.

### 1 FOCUS

▶ The objectives are listed on the student page. Remind students to preview these objectives as a guide to this numbered section.

#### ACTIVITY/Display

Collect bracket fungi, mildew, rust, puffballs, and mushrooms. Place each in a covered container, label, and display. Use the display to show diversity in fungi. Point out that hyphae are not visible.

**Focus**/*Teacher Resource Package,* p. 9. Use the Focus transparency shown at the bottom of this page as an introduction to Fungus Kingdom.

## 5:2 Fungus Kingdom

**Objectives**

4. **List** the traits of fungi and compare the traits of the major kinds of fungi.
5. **Tell** how fungi can be harmful and helpful.

**Key Science Words**

hyphae
sporangium fungi
sporangia
club fungi
sac fungi
budding
mutualism
lichen

Mushrooms, molds, mildews, yeasts, rusts, and smuts are all fungi. Fungi are consumers and decomposers. They cannot make their own food. Some are parasites, but most of them are saprophytes. Remember from Chapter 4 that saprophytes live on dead matter. They break down waste and dead materials for food and return them to the soil.

### Traits of Fungi

You read in Chapter 3 that a fungus is an organism that has a cell wall and does not make its own food. Cells of fungi often have more than one nucleus. Fungi range in size from one-celled yeasts to large, multicellular mushrooms.

The bodies of most fungi are made of a network of threadlike structures called **hyphae** (HI fee). Hyphae contain cytoplasm and are usually divided by cross walls. The hyphae grow and branch until they cover and digest the food source on which the fungus is growing. Look at the bracket fungi growing on the tree in Figure 5-10b. The hyphae have spread throughout the tree bark. The part of the fungus that you can see is the part that reproduces.

Most fungi reproduce by forming spores. They are classified according to how they form spores. The spores of fungi are so small that you need a microscope to see them. Mushrooms and puffballs release large clouds of spores that are easy to see, as shown in Figure 5-10c. Each cloud contains millions of spores.

a

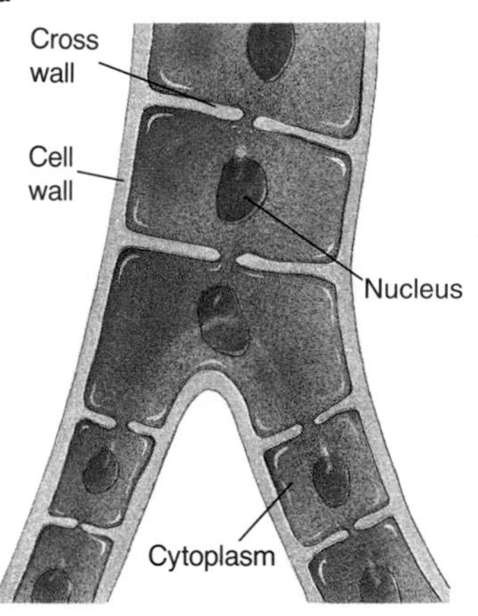


b

c

**Figure 5-10** Hyphae (a) of a fungus usually have cross walls. Bracket fungi (b) grow on trees. Mushrooms (c) reproduce by releasing spores.

### OPTIONS

#### Science Background

*Amanita muscaria* is one of the best known poisonous mushrooms. Its colorful yellow or red cap is covered by white scales. Toxins from *Amanita* become attached to and destroy cells of the liver. Symptoms appear 6 to 12 hours after eating. Death usually occurs in about 4 days. No antidote has been found.

**How Fungi Get Their Food**/*Transparency Package,* number 5b. Use color transparency number 5b as you teach the fungi.

**FOCUS 9**

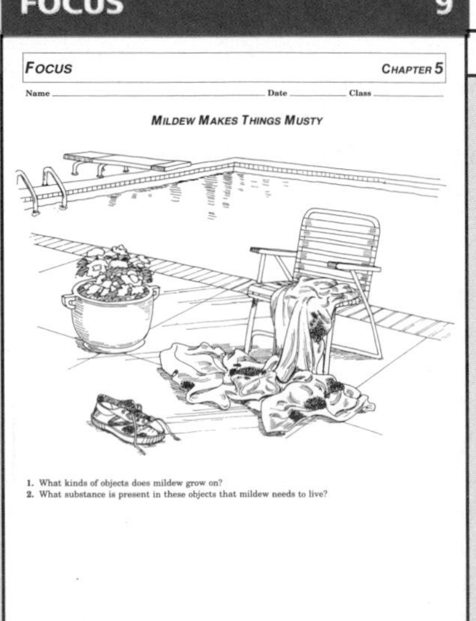

FOCUS CHAPTER 5

Name ______ Date ______ Class ______

MILDEW MAKES THINGS MUSTY

1. What kinds of objects does mildew grow on?
2. What substance is present in these objects that mildew needs to live?

**Figure 5-11** Fungi digest food outside their bodies.

Fungi also reproduce from pieces of hyphae. Sometimes, wind or water carry pieces of broken hyphae to new places. If enough moisture and food are present, the pieces grow into new fungi.

**What do fungi use for food?**

Most fungi can use once-living things for food. The bracket fungi in Figure 5-10b are feeding on a dead tree. Mildew grows on different items, using them as food. Have you ever seen mildew growing on old clothing? As fungi grow on a once-living thing, they decompose it.

A large amount of waste and dead material is deposited on Earth every day. If it were not for decomposers, the waste would build up and become a problem for all living things. Many of the decomposers that help break down dead materials and return them to the soil are fungi.

Fungi digest food outside their bodies. First, the hyphae release chemicals into the material surrounding them. The chemicals break down the food into small molecules. Then the hyphae absorb the digested food.

Not all fungi feed on dead material. Some feed on living things. They are parasites. Most fungi that are parasites grow on plants. Some attack crops such as corn, wheat, potatoes, and soybeans.

Some fungi are parasites of animals. They cause problems such as athlete's foot by living on human skin cells.

Many fungi grow in and on the roots of living plants, helping the plants get water and minerals. In return, the fungi receive food from the plant. Many trees and other plants can't live without these fungi.

### Bio Tip

**Health:** Ergotism is a serious fungal disease caused by eating fungus-infected rye. It causes severe abdominal pain, hallucinations, gangrene, and even death. The fungus is also a source of lysergic acid diethylamide, also known as LSD.



## 2 TEACH

### MOTIVATION/Audiovisual

Show the slides *Mushrooms and Other Fungi, Parts I and II,* JLM Visuals.

### Concept Development

▶ Discuss the nature of a spore as a single cell used for reproduction.

▶ Point out that fungal diseases of plants are still among the greatest threats to human welfare. Hundreds of millions of dollars are spent each year to protect crops with fungicides, to develop new fungicides, and to develop new strains of crops resistant to fungi.

▶ Discuss the amount of waste and dead material that would accumulate if there were no fungi to decompose the organic material.

▶ Explain that a fungus secretes enzymes into its food. These enzymes break large food molecules into molecules that are small enough to be absorbed by hyphae. After the food is absorbed, it travels to all parts of the fungus.

▶ Human fungal diseases cause infections of the skin and of internal organs. Fungus that causes a respiratory infection lives in the soil but becomes a parasite in the lungs if spores or hyphae are inhaled.

### Guided Practice

Have students write down their answers to the margin question: Most use once-living things.

### SKILL 5

**SKILL: SEQUENCE** — CHAPTER 5

Name ________ Date ________ Class ________

*For more help, refer to the Skill Handbook, pages 706-711.* *Use after Section 5:2.*

**LIFE CYCLES OF FUNGI**

The life cycles of two kinds of fungi are shown below. The diagrams show the fungi at different stages of development. They also indicate events or processes that take place at each stage of development.

*Study the life cycle of Rhizopus, the bread mold, shown below. Complete the sentences to explain the sequence, or the order, in which the events occur.*

In the picture the organisms labeled **A** are mature bread molds. They produce **spores** in the structures called **sporangia**. When the **spores** are released and land on a food source, they develop into **new bread molds**. To show the mold organisms in order from youngest to oldest, you would draw stage **B** first, then stage **C**, and then stage **A**.

*Study the life cycle of the mushroom below and then complete the sentences.*

A mushroom reproduces by means of **spores**, which are produced in club-shaped structures in the **gills** of the mushroom **cap**. These structures are shown in the picture labeled **A**.

*Some events in the life cycle of a mushroom are listed below. The events are not in order. Put the events in order by writing the letters of the sentences on the line.*

a. Mushroom becomes mature.
b. Spores are released.
c. Mushroom grows and develops.
d. Spore begins to grow.
e. Mushroom produces spores.
f. Spore lands on food source.

f, **d, c, a, e, b**

## OPTIONS

### ACTIVITY/Enrichment

Discuss the late blight fungus that caused the Irish potato famine, the chestnut tree fungus that has killed most of the chestnut trees in North America, and the fungus that causes Dutch elm disease.

**Skill**/*Teacher Resource Package,* p. 5. Use the Skill worksheet shown here for students to sequence the life cycles of fungi.

## GLENCOE TECHNOLOGY

**Videodisc**

**Biology: The Dynamics of Life**
Movie: *Fungal Decay*
Disc 1, Side 2, Ch. 15

# TEACH

## Concept Development

▶ Explain that sodium propionate is a chemical used by bakeries to inhibit growth of mold in bread. You may want to have students check bread packages for the term.

▶ Point out that an identifying feature of mushrooms is their spore print. A mushroom will drop spores in a pattern that shows the arrangement of gills.

▶ Remind students that the commercial mushrooms are an excellent source of vitamins and the minerals phosphorus and iron. They also contain some proteins.

▶ Hyphae growth rates can exceed 1 kilometer per day. This explains why mushrooms can spring up overnight.

### Mini Lab

The bread slice that was moistened will have mold growing on it. Molds need moisture for growth.

### ASSESSMENT

**Performance:** Have students repeat the experiment with dry cereal, instant potato flakes, fruit, or other foods and compare the results. Students should make a list of factors required for the growth of fungi.

### GLENCOE TECHNOLOGY

**Videodisc**

**Biology: The Dynamics of Life**
Movie: *Fungal Decay*
Disc 1, Side 2, Ch. 15

### Mini Lab

**What Do Fungi Need?**

**Observe:** Sprinkle a slice of bread with water. Place it in a plastic bag. Place a dry slice in a bag. Seal the bags and set in a warm place for a week. What do fungi need to grow? *For more help, refer to the* ***Skill Handbook,*** *pages 704-705.*

## Kinds of Fungi

Have you ever seen bread mold—a black, fuzzy substance growing on bread? If so, you have seen a sporangium (spuh RAN jee uhm) fungus. **Sporangium fungi** are fungi that produce spores in sporangia. **Sporangia** are structures, found on the tips of hyphae, that make spores. You can see that the phylum gets its name from these structures.

Let's return to our bread mold. Bread mold produces spores within sporangia that stick up above the bread, as shown in Figure 5-12. Each spore can form a new bread mold. Hyphae grow into the bread. The mold takes in food and water by means of its hyphae. Molds grow best in warm, moist, dark places, but they can grow also in cold places, such as the refrigerator.

Growth of molds on food can be prevented by cleanliness, drying, and the use of chemicals. Chemicals are added to breads to delay the growth of molds. Fruits such as raisins and apricots are preserved by drying.

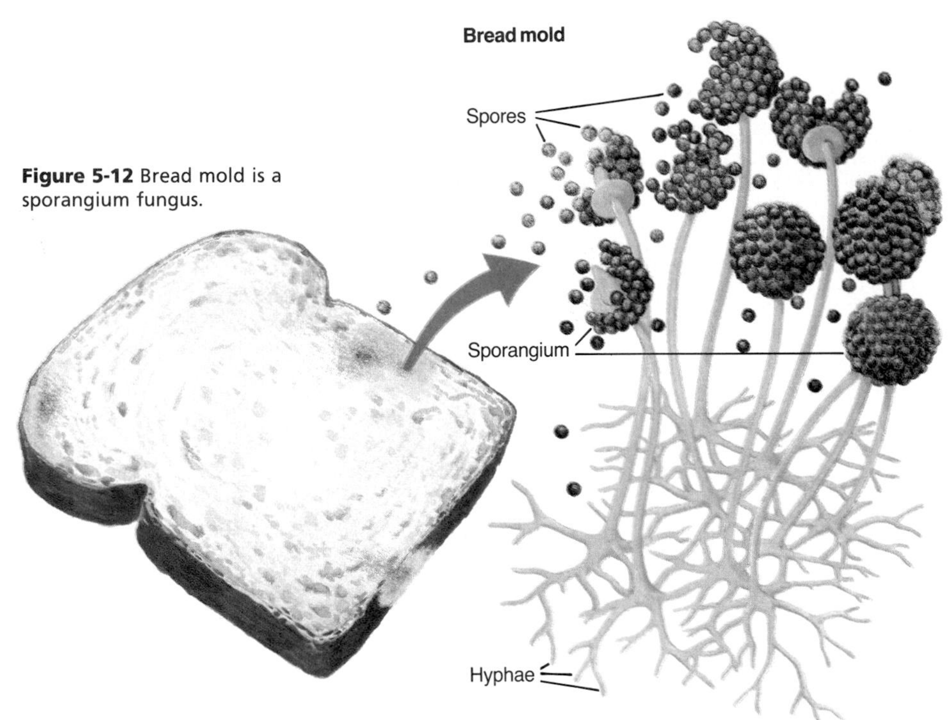


**Figure 5-12** Bread mold is a sporangium fungus.



## OPTIONS

### ACTIVITY/Challenge

Have some students set up an experiment with homemade bread and bakery bread to see which will grow mold faster.

**What Do Molds Need in Order to Grow?**/*Lab Manual,* pp. 37-40. Use this lab as an extension to the traits of fungi.

**Critical Thinking/Problem Solving/** *Teacher Resource Package,* p. 5. Use the Critical Thinking/Problem Solving worksheet shown here to teach how mushrooms grow.

### CRITICAL THINKING 5

CRITICAL THINKING/PROBLEM SOLVING — CHAPTER 5

Name ______ Date ______ Class ______

*Use after Section 5:2.*

**HOW DO MUSHROOMS FORM FAIRY RINGS?**

Jason looked out of his window one damp autumn morning and saw a circle of mushrooms. It had not been there the night before. Now, like the one in the picture below, the ring was growing in the grass in his yard. He remembered the fairy tale about elves and fairies leaving a ring where they have danced at night. However, he wanted to know how scientists explain fairy rings. He looked up the growing habits of mushrooms in his biology book.

Jason learned the life cycle of mushrooms. Each mushroom grows from a spore. The spore grows into threadlike structures called hyphae. The hyphae grow out in all directions forming a tangled network shaped like a circle. Around the edges of the circle, the cells use up the food in the soil. Then, some of the tips of the hyphae grow upright. They form a cap and a stalk, which is called a mushroom. When there is plenty of water, the cells expand greatly. The mushroom then produces spores and the cycle repeats itself.

**Analyzing the Problem**

1. Tell two ways mushrooms reproduce. **Hyphae grow and form a tangled mass. Spores form hyphae.**
2. Where is most of the mushroom located? **Most of the hyphae are under the ground.**
3. Where do the mushrooms develop? **The cap and stalk form when food is used up in the soil and the tips of the hyphae grow upright.**
4. When do mushrooms grow rapidly? **The cap and stalk grow up rapidly when the ground is very moist, on a damp morning or after rain.**

**Solving the Problem**

1. How do fairy rings form? **Mushroom hyphae grow from a spore out into a circle of hyphae. At the edges, mushrooms form. They expand when the ground is very moist, on a damp morning or after rain.**

## Career Close-Up

### Mushroom Farmer

Did you ever wonder where the mushrooms on your pizza came from? Most mushrooms are grown indoors, but some may be grown outdoors or in caves. Equipment is used to control temperature, humidity, and ventilation.

To grow mushrooms, a compost mixture is placed in shallow beds. Spores from mushroom caps are sown on the compost and covered with a thin layer of soil. A mass of hyphae will soon grow through the compost. In seven or eight weeks, mushrooms will appear.

After three months, the mushrooms can be harvested and put into boxes. They will be shipped to markets the same day.

No special schooling is needed for this job, but a knowledge of fungi is very helpful.

Mushroom farmers grow mushrooms in shallow beds of compost.

Fungi with club-shaped parts that produce spores are called **club fungi,** Figure 5-13. They form networks of branched hyphae underground. Mushrooms are club fungi. Mushrooms get food in much the same way as bread mold.

Mushrooms are an important food crop. Some, however, cause illness and a few types cause death. You should never eat wild mushrooms unless an expert says they are safe.

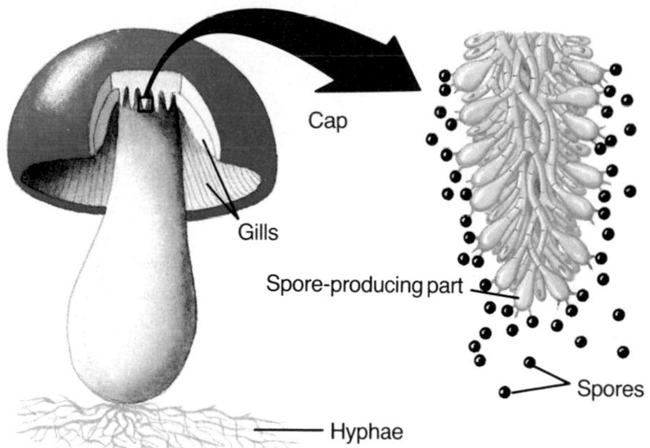


**Figure 5-13** Club fungi produce spores in club-shaped structures within gills. Where are the gills located? in the cap



## Career Close-Up

### Mushroom Farmer

#### Background

Mushrooms have been used in foods for thousands of years. They grow wild in fields. It was not until the late 1800s that mushrooms were grown commercially in the United States. Mushrooms are eaten fresh, canned, or dried.

#### Related Careers

plant farmers

#### Job Requirements

A high school education and community college courses in fungi, business management, and marketing are helpful. Books may be obtained from the local library on the conditions necessary for growing mushrooms and how to start a small business.

#### For More Information

Mushroom Growers Association
18 South Water Market
Chicago, IL 60608

## TEACH

### Independent Practice

**Study Guide**/*Teacher Resource Package,* p. 27. Use the Study Guide worksheet shown at the bottom of this page for independent practice.

### APPLICATION 5

**APPLICATION: SAFETY** — CHAPTER 5

Name ________ Date ________ Class ________

*Use after Section 5:2.*

**LEARNING ABOUT MUSHROOMS**

Mushrooms provide a food source for people, but some mushrooms are poisonous. Even one bite of a white mushroom called the "destroying angel" can be fatal. Many mushrooms look alike. Some very poisonous species look like common edible species. For this reason, you should never eat a mushroom from the wild unless an adult who knows mushrooms tells you it is safe.

Identifying mushrooms can be difficult. Many plants are identified by their leaves, stems, flowers, or fruits. Mushrooms do not have these parts, so they must be identified in a different way. The special parts of mushrooms can be used to identify them.

1. *Describe each part of the mushroom listed below.*

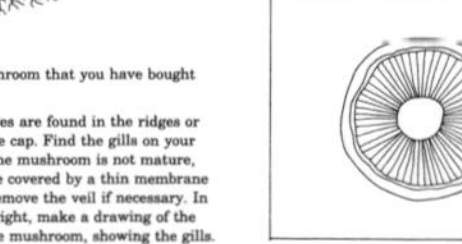

Cap the round top of a mushroom

Hyphae network of threadlike structures

Spores small reproductive cells

2. Observe a mushroom that you have bought at a grocery.

   Mushroom spores are found in the ridges or "gills" under the cap. Find the gills on your mushroom. If the mushroom is not mature, the gills may be covered by a thin membrane called a veil. Remove the veil if necessary. In the box at the right, make a drawing of the underside of the mushroom, showing the gills.

3. As the spores on the gills mature, the gills become darker. Once the spores are fully developed, they fall from the gills onto the ground. Most spores are hard to see without a magnifying glass. However, you can see a collection of spores by making a spore print. Use a mature mushroom with dark gills. Place the mushroom on a white sheet of paper, gill side down. Cover the mushroom and the paper with a glass jar or bowl to keep the spores from blowing away. Leave it there overnight. The next morning, remove the glass, and carefully lift the mushroom off the paper. You should be able to see a print of the spores as they have fallen from the gills. Share your results with your class.
4. Try to find more than one kind of mushroom in the supermarket. See if the spore patterns are the same or different.

### STUDY GUIDE 27

**STUDY GUIDE** — CHAPTER 5

Name ________ Date ________ Class ________

**FUNGUS KINGDOM**

2. Review the following words from Chapter 4. Then, define them in the spaces below.
   a. saprophyte organism that uses dead material for food
   b. parasite organism that lives in or on another living thing
3. Next to each picture, on the first line write the word *parasite* or *saprophyte*, to describe the kind of fungus shown. On the lower line, explain how you know that the fungus is a saprophyte or a parasite.

   saprophyte — feeds off dead material

   saprophyte — feeds off dead material

   parasite — feeds off living things

   parasite — feeds off living things

4. Put a check mark in the correct place on the table that indicates which traits are found in the different kinds of fungi.

| Trait | Sporangium fungi | Club fungi | Sac fungi |
|---|---|---|---|
| reproduce by spores | ✓ | ✓ | ✓ |
| spores made in a saclike part | | | ✓ |
| cause Dutch elm disease | | | ✓ |
| spores made in a sporangium | ✓ | | |
| mushrooms belong here | | ✓ | |
| yeast belongs here | | | ✓ |
| spores made in club-shaped part | | ✓ | |

## OPTIONS

**Application**/*Teacher Resource Package,* p. 5. Use the Application worksheet shown here for students to learn safety with mushrooms.

# TEACH

## Concept Development

▶ Point out that Roquefort, Brie, and Camembert are cheeses that get their special flavors from specific sac fungi used in the production process. The best soy sauces are made by fermenting boiled soybeans and wheat with a sac fungus. The use of fungi to increase the nutrient content of foods is very important in many countries.

▶ Point out that *Penicillium* is one of the six genera of molds from which researchers make antibiotics. *Penicillium* produces penicillins. Molds are used in making well over 1000 different antibiotics.

## Guided Practice

Have students write down their answers to the margin question: They are used for baking bread and making alcohol.

## Independent Practice

**Study Guide**/*Teacher Resource Package,* p. 28. Use the Study Guide worksheet shown at the bottom of this page for independent practice.

**Portfolio**

Have students research antibiotics produced from fungi. Students should look for sources of antibiotics, the methods of purifying them, and the development of new kinds of antibiotics.

**Student Journal**

The Irish potato blight of the mid-1800s was caused by the fungus *Phytophthora infestans.* Have students research the social and economic effects of this agricultural problem.

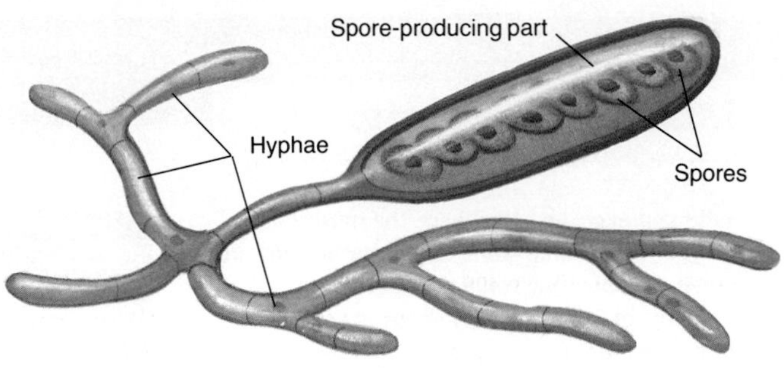


**Figure 5-14** Sac fungi produce spores in saclike structures.

Shelf fungi, rusts, smuts, and puffballs are club fungi, too. Many of these fungi are harmful to plants. Rusts and smuts can destroy entire crops of corn, wheat, and oats.

Yeasts, cup fungi, and powdery mildews are called sac fungi. **Sac fungi** produce spores in saclike structures, Figure 5-14. Yeasts are unusual sac fungi because they have only one cell. Yeasts reproduce by budding. **Budding** is reproduction in which a small part of the parent grows into a new organism. The bud grows out of the parent as shown in Figure 5-15b. Budding produces offspring that are identical to the parent. If conditions are not right for growth, the yeast cell can form a spore. When growing conditions improve, the spore forms a new yeast cell.

How are yeasts helpful to humans?

Sac fungi are useful to humans. Yeasts are used for making bread and alcohol. *Penicillium* (pen uh SIL ee uhm) is a fungus used to make the antibiotic penicillin. It is shown in Figure 5-15a. One species of *Penicillium* is used

a

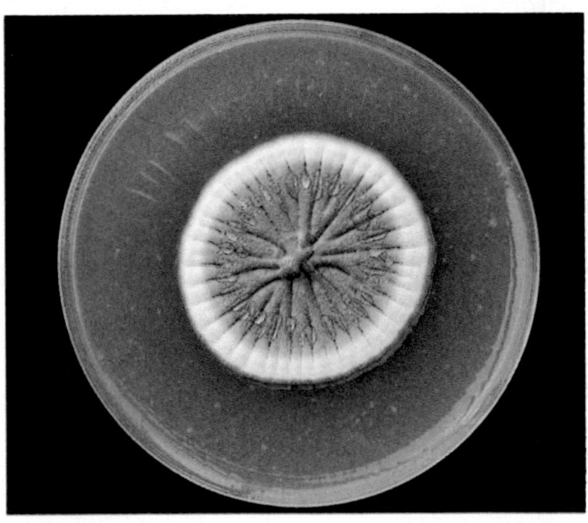

b

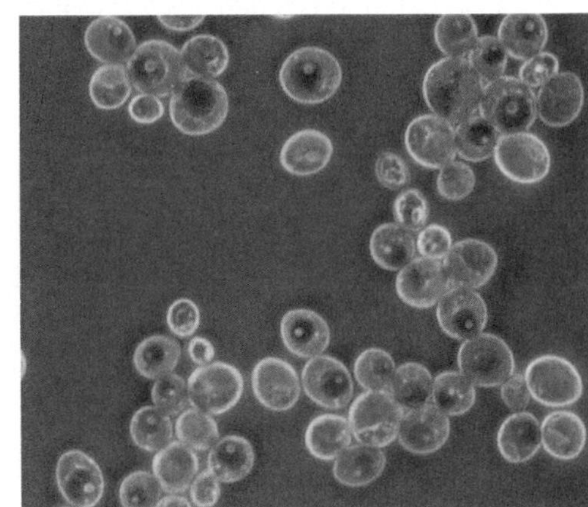

**Figure 5-15** *Penicillium* (a) is a sac fungus used to make penicillin. Yeasts reproduce by budding (b). Is budding sexual or asexual reproduction? asexual



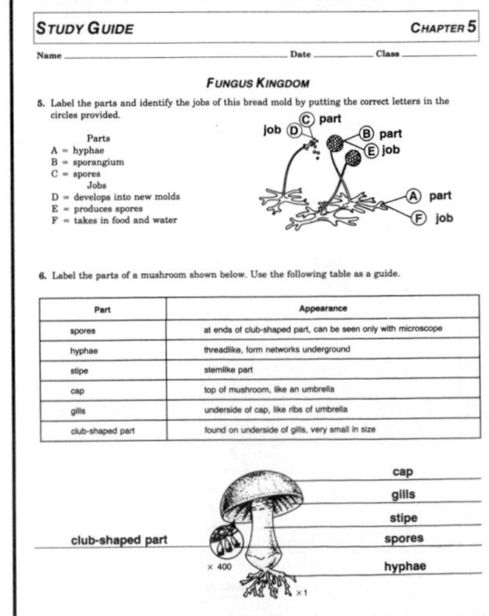

STUDY GUIDE 28

STUDY GUIDE CHAPTER 5

Name ________ Date ________ Class ________

FUNGUS KINGDOM

5. Label the parts and identify the jobs of this bread mold by putting the correct letters in the circles provided.

Parts
A = hyphae
B = sporangium
C = spores
Jobs
D = develops into new molds
E = produces spores
F = takes in food and water



6. Label the parts of a mushroom shown below. Use the following table as a guide.

| Part | Appearance |
|---|---|
| spores | at ends of club-shaped part, can be seen only with microscope |
| hyphae | threadlike, form networks underground |
| stipe | stemlike part |
| cap | top of mushroom, like an umbrella |
| gills | underside of cap, like ribs of umbrella |
| club-shaped part | found on underside of gills, very small in size |

| TABLE 5–3. HELPFUL AND HARMFUL FUNGI |
|---|
| **Helpful Fungi** |
| 1. Used as food—mushrooms |
| 2. Used to make blue and Roquefort cheese |
| 3. Used to make wine, beer, and whiskey—yeast |
| 4. Used to make bread rise—yeast |
| 5. Used to make soy sauce |
| 6. Used to make penicillin and other antibiotics |
| 7. Help to break down materials and get rid of wastes |
| 8. Enrich the soil |
| **Harmful Fungi** |
| 1. Cause foods, such as fruit and bread, to spoil |
| 2. Cause plant diseases, such as rusts, smuts, Dutch elm disease, and mildew |
| 3. Cause human diseases, such as athlete's foot, ringworm, thrush, and lung infections |
| 4. Destroy leather, fabrics, plastics |

to make blue cheeses. While many sac fungi are helpful, one causes Dutch elm disease. This disease has destroyed thousands of elm trees in the United States. Table 5-3 shows some examples of how fungi are helpful and harmful.

## Lichens

Some fungi live neither as parasites nor as saprophytes. They get food from other organisms without causing harm. In turn, they may give support or protection. A living arrangement in which both organisms benefit is called **mutualism.**

A **lichen** (LI kun) is a fungus and an organism with chlorophyll that live together. The organism with chlorophyll can be either a green alga or a blue-green bacterium. A lichen looks like a single organism but it is not. The two organisms in the lichen are so closely tangled together that they cannot be easily separated. You may have seen a British-soldier lichen, as shown in Figure 5-16. The green alga provides the food for the fungus. The fungus provides support, and holds water and minerals for the alga. The alga uses the water and minerals to make food for both itself and the fungus. Both organisms benefit.

Lichens are very sensitive to changes in the environment. If the air becomes too polluted, the green organism can't live. If the green organism dies, the fungus also dies.

### Bio Tip

**Consumer:** Most lemon flavoring (citric acid) today is made from a black mold called *Aspergillus niger.*

**Figure 5-16** British soldier lichen

# TEACH

## Concept Development

▶ Point out that lichens can grow in places where few other organisms can live because of their unusual living arrangement.

▶ Explain that most lichens reproduce by fragmentation, in which small pieces break off and blow away.

▶ Point out that the death of an increasing number of lichens in an area may indicate an increase in air pollution.

▶ Many of the chemicals produced by lichens have long been prized as beautiful dyes. Harris tweed is produced with lichen dyes. Litmus, a dye made from lichens, is used to determine whether a solution is an acid or a base.

## Check for Understanding

Provide students with diagrams of bread mold, a mushroom, and yeast cells. The diagram should have a place for students to label the fungi, hyphae, and spore-producing structure of each, to add examples of each kind of fungi, and to list how each is helpful or harmful.

### Reteach

Discuss why fungi are classified as fungi and the effects of fungi on the environment.

## Independent Practice

**Study Guide**/*Teacher Resource Package*, p. 29. Use the Study Guide worksheet shown at the bottom of this page for independent practice.

## ENRICHMENT 5

ENRICHMENT — CHAPTER 5

Name ______ Date ______ Class ______

*Use after Section 5:2.*

**TEN-PIN PROTISTS AND FUNGI**

Imagine you are bowling. Each of the ten "pins" shown below is a letter in a word related to protists or fungi. The object is to make a word with each group of ten pins. You may form a second word with any letters not used the first time. Score one point for each pin or letter used. If you make a ten-letter word, you have a strike. If you make two words and use all the letters, you have a spare.

*Write the word or words in the spaces beneath the pins. Write your score on the score sheet below.*

Z N O A / R S O / P O / S — sporozoans

M L M I / E D O / S L / S — slime molds

A R P M / U E C / M I / A — paramecium

O S Q M / E T O / I S / U — mosquitoes

M A I A / E K L / I N / L — animal-like

T N A O / Z O P / R O / S — protozoans

O O T E / S A P / T T / H — toothpaste

L C I I / N P E / I L / N — penicillin

N S I O / F C N / T E / I — infections

G M I U / N R A / O P / S — sporangium

1 2 3 4 5 6 7 8 9 10 Final score

Scores will depend on student answers.

## STUDY GUIDE 29

STUDY GUIDE — CHAPTER 5

Name ______ Date ______ Class ______

**FUNGUS KINGDOM**

7. Refer to Table 5-3 on page 107 of your textbook, and fill in the following blanks.
   a. List three ways in which fungi can be helpful. **Student answers may vary, but may indicate that fungi are used as food, in making food, to make medicine, or to enrich the soil.**
   b. List ways fungi can be harmful to things you own. **Fungi can destroy plants, leather, fabric, and plastic.**
   c. List ways fungi can be harmful to people. **They can cause food to spoil and can cause diseases in humans.**

*In your textbook, read about lichens in Section 5:2.*

8. What is mutualism? **Mutualism is a living arrangement in which the things living together benefit.**
9. What is a lichen? **A lichen is a fungus and an organism with chlorophyll that live together.**
   a. List two examples of lichens. **British soldier, reindeer moss**
   b. What is the role of the organism with chlorophyll? **It provides food for the fungus.**
   c. What is the role of the fungus? **It provides support and holds water and minerals for the other organism.**
10. Where do lichens live? **on bare rocks, on trees, and on Arctic ice**
   a. How do they help form soil? **Lichens release acids that break down rock. The broken down rock and the dead lichens form soil.**
   b. How do lichens help some animals? **Lichens provide food for some animals.**

## OPTIONS

**Enrichment**/*Teacher Resource Package*, p. 5. Use the Enrichment worksheet shown here to give student practice with words from this chapter.

## TEACH

### Idea Map

Have students use the idea map as a study guide to the major concepts of this section.

### Independent Practice

**Study Guide**/*Teacher Resource Package,* p. 30. Use the Study Guide worksheet shown at the bottom of this page for independent practice.

### Check for Understanding

Have students respond to the first three questions in Check Your Understanding.

### Reteach

**Reteaching**/*Teacher Resource Package,* p. 13. Use this worksheet to give students additional practice in comparing fungi.

**Extension:** Assign Critical Thinking, Biology and Writing, or some of the **OPTIONS** available with this lesson.

## 3 APPLY

### ACTIVITY/Filmstrip

Show the filmstrip *The Impact of Fungi on Man and His Environment,* Carolina Biological.

## 4 CLOSE

Ask how fungi are used in pizza making.

### Answers to Check Your Understanding

6. spores; from pieces of hyphae
7. sporangium fungi produce spores in sporangia; club fungi produce spores in club-shaped parts
8. to make bread, alcohol, penicillin
9. In fission, bacterium divides into two equal-sized bacteria. In budding, part of the parent grows into a new organism.
10. Fungi may be saprophytes, parasites, or mutualistic.

**Study Tip:** Use this idea map as a study guide to the fungus kingdom. The features and kinds of fungi are shown.

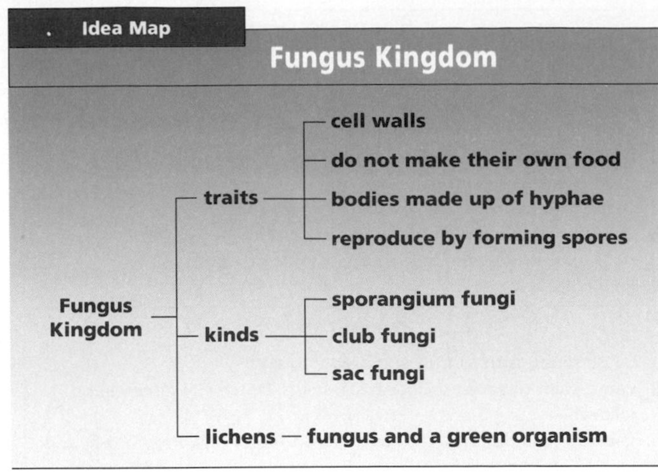


Lichens grow in places where neither fungi nor organisms with chlorophyll can be found alone. They grow on bare rocks, trees, and even on Arctic ice. They are often the first organisms to appear on rocks, such as lava, after a volcano erupts. Lichens release acids that break down the rock. Soil begins to form when the broken down rock mixes with lichens that have died. This process provides a place where other living things can grow.

Some types of lichens provide food for animals. In the Arctic, there is very little plant life. Reindeer moss is eaten by reindeer and caribou (KAIR uh boo), much as cattle graze on grass elsewhere. Caribou are deerlike animals that live in the Arctic.

### Check Your Understanding

6. List two ways in which fungi reproduce.
7. How do sporangium fungi differ from club fungi?
8. Name three uses of sac fungi.
9. **Critical Thinking:** How does budding in yeast differ from fission in bacteria?
10. **Biology and Writing:** Combine the following three simple sentences into one compound sentence. Some fungi are saprophytes. Some fungi are parasites. Some fungi are mutualistic.

**STUDY GUIDE 30**

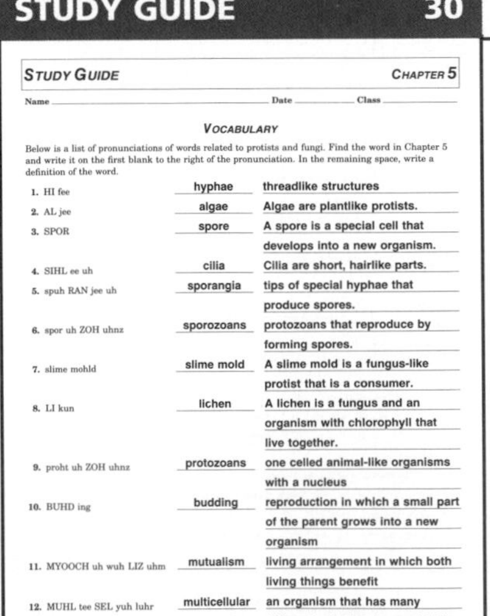

STUDY GUIDE — CHAPTER 5

Name ______ Date ______ Class ______

VOCABULARY

Below is a list of pronunciations of words related to protists and fungi. Find the word in Chapter 5 and write it on the first blank to the right of the pronunciation. In the remaining space, write a definition of the word.

| | Pronunciation | Word | Definition |
|---|---|---|---|
| 1. | HI fee | hyphae | threadlike structures |
| 2. | AL jee | algae | Algae are plantlike protists. |
| 3. | SPOR | spore | A spore is a special cell that develops into a new organism. |
| 4. | SIHL ee uh | cilia | Cilia are short, hairlike parts. |
| 5. | spuh RAN jee uh | sporangia | tips of special hyphae that produce spores. |
| 6. | spor uh ZOH uhnz | sporozoans | protozoans that reproduce by forming spores. |
| 7. | slime mohld | slime mold | A slime mold is a fungus-like protist that is a consumer. |
| 8. | LI kun | lichen | A lichen is a fungus and an organism with chlorophyll that live together. |
| 9. | proht uh ZOH uhnz | protozoans | one celled animal-like organisms with a nucleus |
| 10. | BUHD ing | budding | reproduction in which a small part of the parent grows into a new organism |
| 11. | MYOOCH uh wuh LIZ uhm | mutualism | living arrangement in which both living things benefit |
| 12. | MUHL tee SEL yuh luhr | multicellular | an organism that has many different cells that do special jobs. |

**RETEACHING 13**

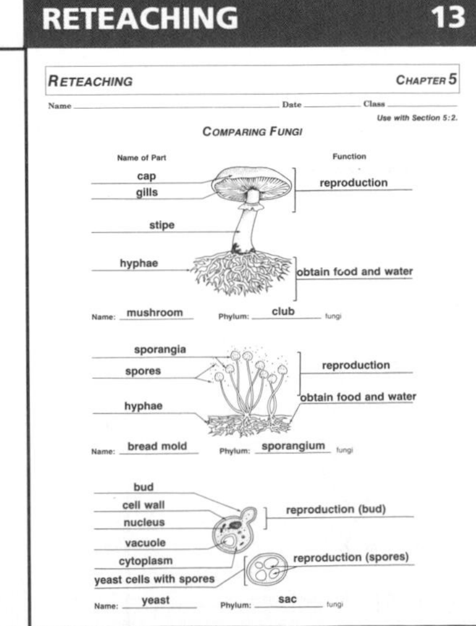

RETEACHING — CHAPTER 5

Name ______ Date ______ Class ______

Use with Section 5:2.

COMPARING FUNGI

Name of Part — Function

cap, gills — reproduction

stipe

hyphae — obtain food and water

Name: mushroom Phylum: club fungi

sporangia, spores — reproduction

hyphae — obtain food and water

Name: bread mold Phylum: sporangium fungi

bud, cell wall, nucleus, vacuole — reproduction (bud)

cytoplasm, yeast cells with spores — reproduction (spores)

Name: yeast Phylum: sac fungi

# Lab 5–2 Fungi

## Problem: Why is a mushroom a fungus and not a plant?

### Skills

observe, infer, use a microscope, design an experiment

### Materials

scalpel
microscope
hand lens
mushroom
prepared slide of mushroom gill

### Procedure

1. Obtain a mushroom and study its structure. Use Figure A to find the parts of the mushroom.
2. Draw and label the parts in your notebook.
3. Cut the mushroom lengthwise through the stipe and cap. **CAUTION:** *Always be careful when using a scalpel.*
4. **Observe:** Examine the cut areas with a hand lens. Locate the hyphae that form the reproductive part. Figure B will help.
5. Use the hand lens to observe the gills on the underside of the cap.
6. **Use a microscope:** Obtain a prepared slide of a mushroom gill. Examine it under low power on the microscope. Locate the structures that produce spores.
7. Locate the spores.
8. Draw and label the spores and the structures that produce them.

### Data and Observations

1. What mushroom part supports the cap?
2. Where are the gills located?
3. Where are the reproductive parts of the mushroom located?
4. Describe the color of the mushroom.

### Analyze and Apply

1. **Infer:** How would the location of the mushroom's reproductive parts help spread the spores?
2. What is a mushroom's food source?
3. Describe how a mushroom gets its food. Is it a parasite or a saprophyte?
4. **Apply:** How do mushrooms appear in your yard year after year when there is no sign of them between seasons? (HINT: Think about how mushrooms grow.)

### Extension

**Design an experiment** to show what foods a fungus can use for growth.

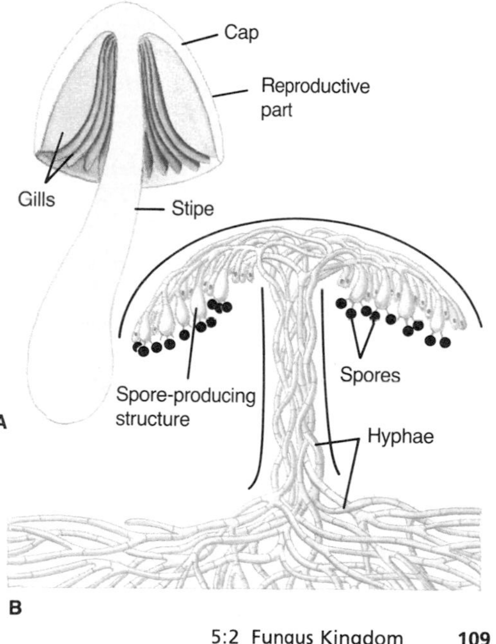




## Lab 5-2 Fungi

### Overview

In this lab, students will study the structure of a mushroom and determine why it is classified as a fungus and not a plant.

**Objectives:** Upon completing this lab, students will be able to (1) **identify** the parts of a mushroom and (2) **determine** why a mushroom is a fungus.

**Time Allotment:** 35 minutes

### Preparation

▶ The fresh mushrooms at the grocery store are usually the brown and white varieties of *Agaricus.*

**Lab 5-2 worksheet**/*Teacher Resource Package,* pp. 19-20.

### Teaching the Lab

**Cooperative Learning:** Have students work in pairs to complete the lab. For more information, see pp. 22T–23T in the Teacher Guide.

▶ **Troubleshooting:** Students may have difficulty in finding all the mushroom parts. Explain that the hyphae of mushrooms are underground. The mushroom is a fruiting body which bears spores. The reproductive structures are located on the gills.

### ASSESSMENT

**Performance:** Provide students with mushrooms from the grocery store that have aged until the undersides are brown. Have students remove the stems and arrange the mushrooms with the undersides of the caps facing down on a white paper and leave them overnight. Students should compare what they observe with how plants reproduce.

## ANSWERS

### Data and Observations

1. the stipe
2. underneath the cap
3. on the gills
4. brown or gray

### Analyze and Apply

1. The spores can easily be dispersed by the wind or fall to the ground.
2. decaying material
3. It digests its food outside its body. The hyphae release chemicals that break down the food into small molecules, which are then absorbed by the hyphae. Most mushrooms are saprophytes.
4. The spores may be left from the mushrooms present the year before, or hyphae may be left underground. These hyphae can produce new mushrooms.

# CHAPTER 5 REVIEW

## Summary

Summary statements can be used by students to review the major concepts of the chapter.

## Key Science Words

All boldfaced terms from the chapter are listed.

## Testing Yourself

### Using Words

1. mutualism
2. hyphae
3. multicellular
4. lichen
5. spore
6. budding
7. cilia
8. algae
9. protozoan
10. club fungi
11. slime mold
12. sac fungi
13. sporangia
14. sporozoan

### Finding Main Ideas

15. p. 106; Yeasts are one-celled while other fungi are multicellular.
16. p. 96; Sporozoans are protists that do not move.
17. p. 98; A diatom has a two-part cell covering of glasslike material that overlaps.
18. p. 102; Fungi are made of a network of threadlike hyphae.
19. pp. 104-106; The three kinds of fungi are sporangium fungi, club fungi, and sac fungi.
20. p. 106; Sac fungi are used to make bread, alcohol, and antibiotics.
21. p. 98; Red tides are caused by dinoflagellates.
22. p. 100; A slime mold is sometimes a slimy mass of material that flows like an amoeba. In a second stage, it stops flowing and produces spores. In a third stage, the spores become small amoebalike organisms with flagella.
23. p. 99; Most slime molds consume bacteria on rotting logs and decaying leaves.
24. pp. 95-97; Some human disease caused by protists are amoebic dysentery, sleeping sickness, and malaria.

### Using Main Ideas

25. The green alga provides food and the fungus provides support.
26. An amoeba traps its food with its false feet.
27. A person can get malaria from a sporozoan carried by female *Anopheles* mosquitoes.



# Chapter 5 Review

## Summary

### 5:1 Protist Kingdom

1. Protists are usually one-celled, have a nucleus and other cell parts with membranes, and may be either producers or consumers.
2. There are three kinds of protists—animal-like protists, plantlike protists, and funguslike protists. They have different methods of movement, feeding, and reproduction.
3. Amoebas, paramecia, and trypanosomes are animal-like protists. Euglena, diatoms, and *Volvox* are plantlike protists. Slime molds are funguslike protists.

### 5:2 Fungus Kingdom

4. Fungi have many cells with cell walls and nuclei. They are consumers and feed as saprophytes or parasites. They form mutualistic relationships with green organisms. There are three phyla of fungi. They are classified by how and where they produce spores.
5. Fungi can be used in many ways that are helpful to humans. They also cause diseases in plants and animals.

## Key Science Words

algae (p. 97)
budding (p. 106)
cilia (p. 96)
club fungi (p. 105)
hyphae (p. 102)
lichen (p. 107)
multicellular (p. 97)
mutualism (p. 107)
protozoan (p. 94)
sac fungi (p. 106)
slime mold (p. 99)
sporangia (p. 104)
sporangium fungi (p. 104)
spore (p. 96)
sporozoan (p. 96)

## Testing Yourself

### Using Words

*Choose the word from the list of Key Science Words that best fits the definition.*

1. a living arrangement in which both things benefit
2. threadlike structures of fungi that contain cytoplasm and have cross walls
3. having many cells that do different jobs
4. fungus and organism with chlorophyll that live together
5. special cell that forms a new organism
6. a way of producing offspring that are identical to the parent
7. tiny, hairlike parts on a cell surface
8. plantlike protist group
9. animal-like protist group
10. fungi with club-shaped parts
11. funguslike protist
12. fungi that have saclike structures
13. structures, on tips of hyphae, that make spores
14. a protozoan that forms spores

# Review

## Testing Yourself *continued*

**Finding Main Ideas**

*List the page number where each main idea below is found. Then, explain each main idea.*

15. how yeasts are different from other fungi
16. the name of protists that do not move
17. the structure of a diatom
18. what fungi are made of
19. the three kinds of fungi
20. why sac fungi are important
21. the cause of red tides
22. the life cycle of a slime mold
23. what most slime molds use for food
24. two human diseases caused by protists

**Using Main Ideas**

*Answer the questions by referring to the page number after each question.*

25. What is the role of each organism in a lichen? (p. 107)
26. How does an amoeba get food? (p. 95)
27. How does a person get malaria? (p. 97)
28. How do fungi digest food? (p. 103)
29. How are plantlike protists different from funguslike protists? (pp. 97, 99)
30. How do bracket fungi get food from trees? (p. 102)
31. Why should you not eat wild mushrooms? (p. 105)
32. How do protozoans get food? (p. 94)
33. Where are lichens found? (p. 108)
34. How can growth of molds on food be prevented? (p. 104)
35. How are lichens helpful? (p. 108)
36. How do certain dinoflagellates cause red tides? (p. 98)

## Skill Review

*For more help, refer to the* ***Skill Handbook,*** *pages 704-719.*

1. **Make and use tables:** Make a table to show the three kinds of fungi, where they produce spores, and examples of each.
2. **Classify:** Make a list of the traits of animal-like protists, plantlike protists, and funguslike protists. How are these traits used to classify protists?
3. **Infer:** Mushrooms often appear on a lawn soon after a rain. Using your knowledge of fungi, infer how this happens.
4. **Observe:** Observe a small amount of blue cheese under a stereomicroscope. What makes the cheese blue?

## Finding Out More

TECH PREP

**Critical Thinking**

1. What advantages does a lichen have over its fungus and green-organism parts?
2. How would life on Earth be affected if all the algae died?

**Applications**

1. Make a list of things you can do to keep fungi from growing in your home. Where are fungi most likely to be found in the home?
2. Use the library to learn about fungicides. Report on how they are used to keep fruits and vegetables safe from fungal infection.



### Using Main Ideas

28. Fungi digest food outside their bodies. They release chemicals that break down the food into small molecules. Then the fungi absorb the digested food.
29. Plantlike protists have chlorophyll and can make their own food, and funguslike protists are consumers.
30. The hyphae spread throughout the tree bark and absorb the food.
31. Some wild mushrooms cause illness and a few cause death.
32. The single cell takes in and digests food.
33. Lichens are found on bare rocks, trees, and Arctic ice.
34. Growth of molds can be prevented by cleanliness, drying, and the use of chemicals.
35. Lichens help to form soil and provide food for animals.
36. They reproduce in such large numbers that the water turns red.

### Skill Review

1. Have students list the three types of fungi across the top of their table and list the features under each.
2. Animal-like protists: animal-like, one-celled, cell nucleus, consumers, move about; plantlike protists: plantlike, one-celled or multicellular, cell nucleus, have chlorophyll, producers, some can move about; funguslike protists: funguslike, multicellular, cell nucleus, consumers or parasites, can move about during part of their life
3. The rain causes spores to be released and creates conditions right for new mushrooms to develop.
4. the sac fungi

## Finding Out More

### Critical Thinking

1. Lichens can make their own food and have a surface on which to grow.
2. The consumers that depend on algae for food would die. Less carbon dioxide would be removed from the atmosphere, and less oxygen would be placed into the atmosphere.

### Applications

1. Fungi can be kept from growing by keeping surfaces dry and clean. Fungi are most likely to be found in damp, dark, places such as wet basements, refrigerators, and bathrooms, and where there are water leaks such as under cabinets.
2. Fungicides are chemical substances used to kill fungi. They are sprayed or dusted on plants to kill rusts, mildew, smuts, and molds.

**Chapter Review**/*Teacher Resource Package*, pp. 23-24.

**Chapter Test**/*Teacher Resource Package*, pp. 25-27.

**Chapter Test**/*Computer Test Bank*

# Plants

## PLANNING GUIDE

| CONTENT | TEXT FEATURES | TEACHER RESOURCE PACKAGE | OTHER COMPONENTS |
|---|---|---|---|
| (2 days)<br>6:1 Plant Classification<br>Plant Features<br>How Are Plants Grouped? | Skill Check: *Understand Science Words,* p. 115<br>Mini Lab: *How Does Water Move Up in a Plant?* p. 116<br>Idea Map, p. 116<br>Check Your Understanding, p. 116 | Reteaching: *Plant Classification,* p. 14<br>Focus: *How We Use Plants,* p. 11<br>Study Guide: *Plant Classification,* pp. 31-32<br>Reteaching: *Life of a Moss,* p. 15<br>Study Guide: *Nonvascular Plants,* p. 33 | **Laboratory Manual:**<br>*What Are the Traits of Vascular and Nonvascular Plants?* p. 41<br>**STVS:** *City Trees,* Plants and Simple Organisms (Disc 4, Side 2) |
| (2 days)<br>6:2 Nonvascular Plants<br>Mosses and Liverworts<br>The Life Cycle of a Moss | Lab 6-1: *Nonvascular Plants,* p. 119<br>Idea Map, p. 118<br>Check Your Understanding, p. 120 | Reteaching: *Life of a Moss,* p. 15<br>Skill: *Outline,* p. 6<br>Study Guide: *Nonvascular Plants,* p. 33<br>Lab 6-1: *Nonvascular Plants,* pp. 21-22 | |
| (2 days)<br>6:3 Vascular Plants<br>Ferns<br>Conifers<br>Flowering Plants | Skill Check: *Interpret Diagrams,* p. 122<br>Mini Lab: *What Conditions Cause a Pine Cone to Open?* p. 126<br>Lab 6-2: *Ferns,* p. 123<br>Idea Map, p. 121<br>Science and Society: *Ancient Forests: Jobs Versus Wildlife,* p.127<br>Check Your Understanding, p. 129 | Application: *Poisonous Plants,* p. 6<br>Enrichment: *Leaf Adaptations,* p. 6<br>Reteaching: *Comparing Vascular Plants,* p. 16<br>Critical Thinking/Problem Solving: *What Happened to the Fern?* p. 6<br>Transparency Master: *Conifer Life Cycle,* p. 25<br>Study Guide: *Vascular Plants,* pp. 34-35<br>Study Guide: *Vocabulary,* p. 36<br>Lab 6-2: *Ferns,* pp. 23-24 | **Laboratory Manual:**<br>*How Can Conifers Be Identified?* p. 45<br>Color Transparency 6a: *Fern Life Cycle*<br>Color Transparency 6b: *Conifer Life Cycle*<br>**STVS:** *Water Hyacinth,* Plants and Simple Organisms (Disc 4, Side 2)<br>**STVS:** *Purifying with Plants,* Plants and Simple Organisms (Disc 4, Side 2) |
| Chapter Review | Summary<br>Key Science Words<br>Testing Yourself<br>Finding Main Ideas<br>Using Main Ideas<br>Skill Review | **ASSESSMENT RESOURCES**<br>Chapter Review, pp. 28-29<br>Chapter Test, pp. 30-32<br>Performance Assessment in the Biology Classroom<br>Alternate Assessment in the Science Classroom<br>Computer Test Bank | |

## GLENCOE TECHNOLOGY

**Infinite Voyage,** *Life in the Balance*
**Science and Technology Videodisc Series,** *City Trees,* Plants and Simple Organisms (Disc 4, Side 2)
*Water Hyacinth,* Plants and Simple Organisms (Disc 4, Side 2)
*Purifying with Plants,* Plants and Simple Organisms (Disc 4, Side 2)

**The Secret of Life,** *Gone Before You Know It: The Biodiversity Crisis*

## MATERIALS NEEDED

| LAB 6-1, p. 119 | LAB 6-2, p. 123 | MARGIN FEATURES |
|---|---|---|
| stereomicroscope or hand lens<br>light microscope<br>3 microscope slides<br>coverslip<br>dropper<br>forceps<br>metric ruler<br>paper towel<br>moss | stereomicroscope<br>2 microscope slides<br>scalpel<br>dropper<br>ethyl alcohol<br>whole fern for class<br>fern leaves with spore cases<br>pencil<br>paper | Skill Check, p. 115<br>dictionary<br>Skill Check, p. 122<br>pencil<br>paper<br>Mini Lab, p. 116<br>beaker<br>red food coloring<br>white carnation<br>Mini Lab, p. 126<br>2 plastic bags<br>open and closed pine cones |

## OBJECTIVES

For more information about National Science Standards, see page 5T.

| SECTION | OBJECTIVE | CORRELATION of QUESTIONS to OBJECTIVES: CHECK YOUR UNDERSTANDING | CHAPTER REVIEW | TRP CHAPTER REVIEW | TRP CHAPTER TEST |
|---|---|---|---|---|---|
| 6:1 National Science Stds: UCP.1, UCP.2, UCP.3, UCP.4, UCP.5, C.1, C.5 | 1. **Describe** two features common to most plants. | 1 | 1, 23 | 18, 25, 28, 33 | 1, 14, 21 |
| | 2. **Compare** vascular and nonvascular plants. | 2, 3, 15 | 3, 10, 13, 20 | 10, 12, 13 | 2, 9, 18, 19, 25, 26, 27, 29, 31 |
| 6:2 National Science Stds: UCP.1, UCP.2, UCP.5, C.1, C.5, F.3 | 3. **Compare** mosses and liverworts with other plants. 3A | 6, 7 | 14, 18 | 7, 17, 22 | 5, 16, 33, 35, 42, 49 |
| | 4. **Sequence** the steps in the life cycle of a moss. | 8 | 2, 17 | 1, 3, 4, 21 | 3, 4, 7, 23, 44 |
| | 5. **List** three ways mosses are important to other living things. | 9 | 22 | 35 | 6, 37, 39 |
| 6:3 National Science Stds: UCP.1, UCP.2, UCP.3, UCP.5, C.1, C.4, C.5, D.1, D.2, F.3 | 6. **Compare** the traits of ferns, conifers, and flowering plants. | 11 | 7, 19, 21 | 2, 6, 11, 14, 20, 23, 24, 26, 27, 29, 30 | 8, 10, 11, 12, 17, 20, 22 |
| | 7. **Explain** how ferns, conifers, and flowering plants reproduce. | 12 | 6, 8, 9, 12, 24 | 5, 9, 15, 16, 19, 20, 21, 24, 26, 36, 37, 39 | 7, 13, 15, 28, 41, 43, 44 |
| | 8. **Describe** how conifers and flowering plants are important to other living things. | 13, 14 | 15, 16 | 9 | 32, 34, 36, 38, 40 |

# PLANTS

## CHAPTER OVERVIEW

### Key Concepts

In this chapter, students will study the common traits of plants and the characteristics used to group them. Examples of nonvascular plants—mosses and liverworts—and examples of vascular plants—ferns, conifers, and flowering plants—are described. The economic importance of plants is discussed.

### Key Science Words

chlorophyll
conifer
egg
embryo
fern
fertilization
flower
flowering plant
moss
nonvascular plant
phloem
photosynthesis
pollen
seed
sexual reproduction
sperm
vascular plant
xylem

### Skill Development

In Lab 6-1, students will **make and use tables, interpret data, use a microscope, measure in SI,** and **infer** to find traits of nonvascular plants. In Lab 6-2, students will **make and use tables, observe,** and **use a microscope** to study the traits of ferns. In the Skill Check on page 115, students will **understand** the **science word** *photosynthesis.* In the Skill Check on page 122, students will **interpret diagrams** to understand the life cycle of a fern. In the Mini Lab on page 116, students will **experiment** with flowers to learn that vascular stems carry water. In the Mini Lab on page 126, students will **observe** the conditions necessary for pine cones to open.

### Bridging

Students have already learned how plants are classified in Chapter 3. Refer students to Figure 3-10 that shows the characteristics used to classify plants.

## CHAPTER PREVIEW

### Chapter Content

Review this outline for Chapter 6 before you read the chapter.



### Skills in this Chapter

The skills that you will use in this chapter are listed below.

- In **Lab 6-1,** you will make and use tables, interpret data, use a microscope, measure in SI, and infer. In **Lab 6-2,** you will make and use tables, observe, and use a microscope.
- In the **Skill Checks,** you will understand science words and interpret diagrams.
- In the **Mini Labs,** you will experiment and observe.

112

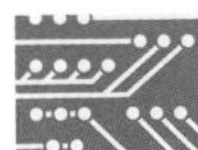

## TECH PREP

For Tech Prep activities in this chapter of the Teacher Wraparound Edition, see especially the Student Journals on pages 117, 124, and 128, the Connection on page 118, the Motivation on page 121, and the Portfolio on page 124.
See also the Glencoe Homepage at **www.glencoe.com**

The following Glencoe resources provide additional opportunities for integrating science and technology.

**Technology: Science and Math in Action, Book One**
Module 1: Hydroponics

Chapter 6

# Plants

If you were to walk through the forest pictured on the left, you would first see the giant redwood trees. Because of their small size, you might not notice the plants growing around the roots of the trees. Now, look at the tiny water plants in the photo below. These plants live on the surfaces of lakes or ponds.

There are many kinds of plants. They vary in size, shape, color, and life span. Some, like the small water plants, are only a few millimeters long. Others, such as the redwood tree, can grow to 100 meters tall. Some plants change colors and lose their leaves. Others stay green for many years. Some plants live for just a few months. Plants grow almost everywhere on Earth. They are adapted to many kinds of environments. In this chapter, you will read about plants and where they grow.

**Try This!**

**Where does the moisture come from?** Cover a small houseplant with a wide-mouth jar. Observe the plant for a day or two. What do you see in the jar? What is its source?

**interNET CONNECTION**

For more information about the material in this chapter, follow the link for the chapter on the Glencoe Homepage at **http://www.glencoe.com**

Duckweed

113

## ASSESSMENT PLANNER

**Portfolio**

Strategies on the following pages represent student products that can be placed into a best-work portfolio: pp. 124, 128.

**PERFORMANCE ASSESSMENT**

Skill Check, p. 122
Mini Lab, pp. 116, 126
Lab, pp. 119, 123

**CONTENT ASSESSMENT**

Check for Understanding, pp. 116, 120, 128, 129
Chapter Review, pp. 130-131

**GROUP ASSESSMENT**

Opportunities for group assessment occur with Cooperative Learning Strategies.

**Student Journal**

Strategies on the following pages represent opportunities for writing in a Student Journal: pp. 117, 124, 128.

## GETTING STARTED

### Using the Photos

In looking at the photos, students should compare the size of the redwood trees with the size of the small plants at their bases. They can mark off a few millimeters on a metric ruler to visualize how small water plants can be. Have them look at a meterstick and imagine 100 of them stacked end to end to visualize the height of a redwood tree.

### MOTIVATION/Try This!

**Where does the moisture come from?** When students have completed the chapter-opening activity, they will have **observed** moisture in the jar. The moisture comes from the leaves of the plant.

### Chapter Preview

Have students study the chapter outline before they begin to read the chapter. Students will note that plant features and the grouping of plants are studied before nonvascular and vascular plants. Examples of different kinds of plants within each group are discussed.

### Misconception

Students may think that plants get everything they need to make food from the soil. Carbon dioxide from the air, energy from sunlight, and chlorophyll are other things that are needed.

# 6:1 Plant Classification

## PREPARATION

### Materials Needed

Make copies of the Focus, Study Guide, and Reteaching worksheets in the *Teacher Resource Package.*

▶ Collect plants for the Focus motivation.

▶ Gather beakers, red food coloring, and white carnations for the Mini Lab.

### Key Science Words

chlorophyll
photosynthesis
vascular plant
nonvascular plant

### Process Skills

In the Mini Lab, students will experiment. In the Skill Check, students will understand science words.

## 1 FOCUS

▶ The objectives are listed on the student page. Remind students to preview these objectives as a guide to this numbered section.

### MOTIVATION/Display

Display a variety of interesting plants such as cacti, medicine plant *(Aloe vera)*, sensitive plant *(Mimosa pudica)*, living stones *(Lithops)*, and Madagascar sprouting leaf *(Kalanchoe)*.

Focus/*Teacher Resource Package* pp. 11-12. Use the Focus transparency shown at the bottom of this page as an introduction to how we use plants.

### Guided Practice

Have students write down their answers to the margin question: a stiff cell wall.

## 6:1 Plant Classification

**Objectives**

1. **Describe** two features common to most plants.
2. **Compare** vascular and nonvascular plants.

**Key Science Words**

chlorophyll
photosynthesis
vascular plant
nonvascular plant

There are over 350 000 different kinds of plants. How could you ever know them all? Scientists have grouped plants to make it easier to study and learn them. Just as with all other living things, plants are classified based on their traits.

### Plant Features

Although it may seem all plants are different, most have two traits in common. First, almost all plants have green cell parts called chloroplasts (KLOHR uh plasts). A chloroplast is shown in Figure 6-1. Chloroplasts are green because they contain chlorophyll (KLOHR uh fihl). **Chlorophyll** is a chemical that gives plants their green color and traps light energy. Plants trap light energy for photosynthesis (foht oh SIHN thuh sus). **Photosynthesis** is the process in which plants use water, carbon dioxide, and energy from the sun to make food. In the process, they release oxygen. This food-making process of plants occurs only in the chloroplasts.

What feature of cells is common to all plants?

Photosynthesis is the main process that separates plants from animals. Animals can't make their own food. They don't have chlorophyll. Unlike animals, plants can't move about to search for food. They have to take in the materials and energy needed for photosynthesis.

The second trait of plants is a stiff cell wall. The cell wall gives structure to each cell. The cell wall helps to support a plant on land. Fungi, monerans, and many protists also have cell walls, but animal cells don't.

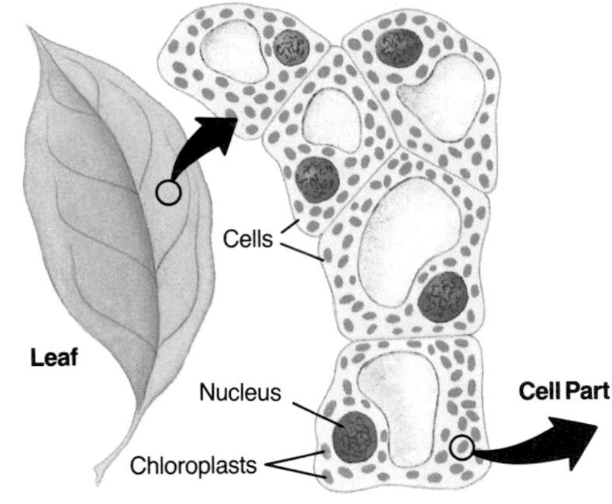


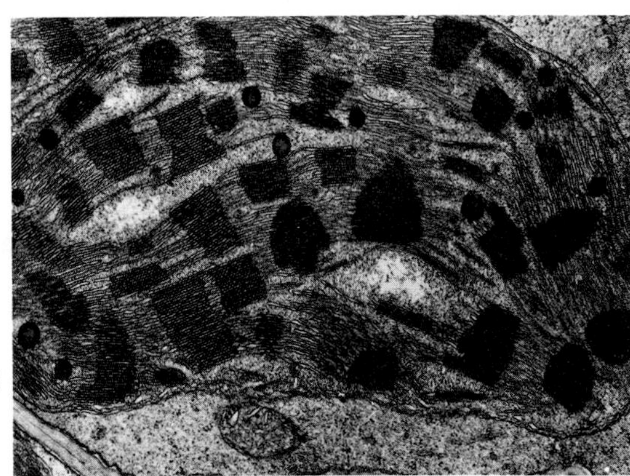

**Figure 6-1** A common cell part of almost all plants is the chloroplast, shown in the photo magnified 15 000 times.



## OPTIONS

### Science Background

The word *division* is often used for phylum in the plant kingdom. Algae make up the first three divisions. Mosses, liverworts, and hornworts are a fourth division. All vascular plants are in a fifth division.

**What Are the Traits of Vascular and Nonvascular Plants?**/*Lab Manual,* pp. 41-44. Use this lab as an extension to plant classification.

**FOCUS 11**

## How Are Plants Grouped?

Plants are grouped in a way similar to how you might group office buildings. If you were to group office buildings by whether they have elevators, you would be able to put them into two groups. Some office buildings have elevators, some don't.

Notice that the taller building in Figure 6-2 has an elevator. The building with only three floors does not have an elevator. Why do we build tall buildings with elevators rather than just stairways? People can get to offices in a tall building more easily if there are elevators in the building. Supplies can also be moved more easily from one floor to another.

Just as there are two groups of buildings, there are also two groups of plants. The two groups of plants are based on whether or not they have cells that form tubes through the length of the plant. Some plants have tubelike cells and some don't. The tubelike cells join end to end and look like water pipes or soda straws joined together. The function of the tubelike cells in plants is much like that of the elevators in a tall building. The elevators carry people up and down within the building. The people can easily reach all the offices. One set of tubelike cells allows water and minerals to move up and down within a plant. A separate set of tubelike cells allows food made by the leaves to move to other parts of the plant. The tubelike cells run through the roots, stems, and leaves of this kind of plant, Figure 6-2. They carry supplies up and down the plant. This is the way that water and food reach all the plant's cells.

### Skill Check

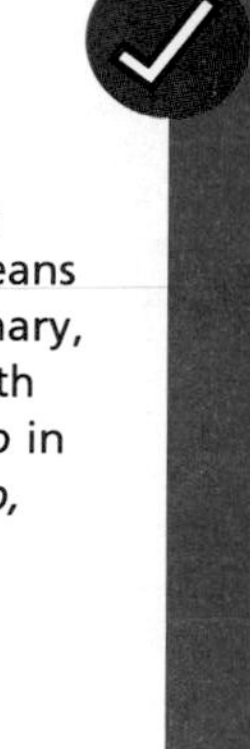

**Understand science words: photosynthesis.** The word part *photo* means light. In your dictionary, find three words with the word part *photo* in them. *For more help, refer to the **Skill Handbook**, pages 706-711.*

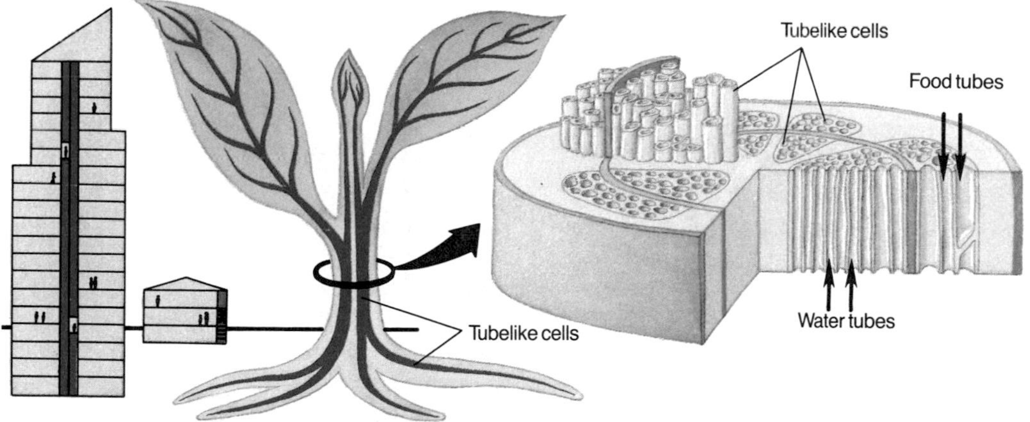


**Figure 6-2** The elevator in a tall building acts like the tubelike cells in vascular plants.

### STUDY GUIDE 31

STUDY GUIDE CHAPTER 6

Name ______ Date ______ Class ______

**PLANT CLASSIFICATION**

*In your textbook, read about plant features in Section 6:1.*

1. Shade green those parts of the living things shown below that are usually green.
2. Draw an X through each living thing that does not contain chlorophyll.
3. Circle each living thing that can make its own food.

Green; A height 30 cm; B width 12 cm; C length 2 cm; D height 30 m; Green; F 28 cm; Green

4. Which of the diagrams (A, B, C, D, or E) are plants? **A, D, E**
5. List four features of plants that are different from the features of animals.
   a. **Plants have green parts called chloroplasts that contain chlorophyll.**
   b. **Plants carry on photosynthesis.**
   c. **Plants can't move about.**
   d. **Plants have stiff cell walls.**
6. What is the main difference between plants and animals? **Plants can make their own food, animals can't.**
7. What is photosynthesis? **Photosynthesis is a process in which plants use energy from the sun, carbon dioxide, and water to make food.**
8. Where does photosynthesis occur in plants? **Photosynthesis occurs in the chloroplasts.**

### STUDY GUIDE 32

STUDY GUIDE CHAPTER 6

Name ______ Date ______ Class ______

**PLANT CLASSIFICATION**

9. Fill in the blanks below with the following words or phrases.

light energy; chloroplasts; photosynthesis; green color; chlorophyll; plants

**plants** — have cell parts called → **chloroplasts** — that contain a chemical called → **chlorophyll** — that gives plants their → **green color** — and traps → **light energy** — for use in the process of → **photosynthesis** — a process which separates animals from → **plants**

10. What are vascular plants? **Vascular plants are plants that have tubelike cells in their roots, stems, and leaves to carry food and water.**
11. What are nonvascular plants? **Nonvascular plants are plants that don't have tubelike cells in their roots, stems, or leaves.**
12. Why must nonvascular plants grow close to the ground? **They depend on osmosis to move water to all parts of the plant.**
13. Define the following words.
    a. root **plant structure that anchors plants in the ground and takes in water and minerals from the soil**
    b. stem **plant structure that carries water to all parts of the plant and holds leaves up to the sunlight**
    c. leaf **plant structure that makes food**

# 2 TEACH

### MOTIVATION/Software

*Project Classify: Plants,* National Geographic Society.

### ACTIVITY/Field Trip

Take a walk around the school or go on a field trip to look at plants. Have students identify the different plants they know.

### Concept Development

▶ Review chlorophyll and chloroplasts from Chapter 2.

▶ Point out that mosses and liverworts don't have tubelike cells to carry water and food. Discuss why these plants don't need tubelike cells. Mosses and liverworts are small plants that grow in moist environments.

**Using Science Words:** You may wish to introduce the terms *xylem* and *phloem* here. Tubelike cells that carry water are called xylem. Tubelike cells that carry food are called phloem.

### Independent Practice

**Study Guide**/*Teacher Resource Package,* pp. 31-32. Use the Study Guide worksheets shown at the bottom of this page for independent practice.

**GLENCOE TECHNOLOGY**

**Videodisc**

**The Secret of Life**
*Stem Structure*

## TEACH

### Idea Map

Have students use the idea map as a study guide to the major concepts of this section.

### Mini Lab

The flower will turn a shade of red. Tubelike structures take in water and carry it to the flower.

**Performance:** Have students design an experiment using celery stalks with leaves in water with food coloring to observe the tubelike structures through which water moves.

### Guided Practice

Have students write down their answers to the margin question: roots.

### Check for Understanding

Have students respond to the first three questions in Check Your Understanding.

### Reteach

**Reteaching**/*Teacher Resource Package*, p. 14. Use this worksheet to give students additional practice with plant classification.

**Extension:** Assign Critical Thinking, Biology and Reading, or some of the **OPTIONS** available with this lesson.

## 3 APPLY

### Demonstration

Set up a bog terrarium. Plant butterwort and cobra lily at one end. Add peat moss to the other end and plant pitcher plant, sundew, and Venus's flytrap.

## CLOSE

### Brainstorming

Have students name as many plant species as they can and group them.

**Study Tip:** Use this idea map as a study guide to the traits of plants. Notice how one trait is used to classify plants into two main groups.

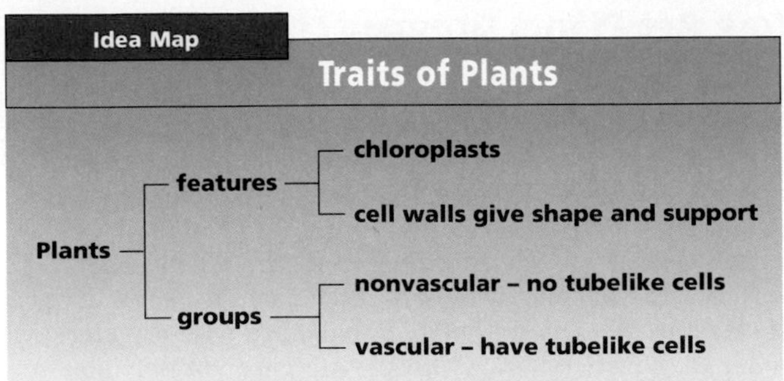


### Mini Lab

**How Does Water Move Up in a Plant?**

**Experiment:** Fill a beaker with water. Add red food coloring. Place a white carnation in the beaker. Explain what happens. *For more help, refer to the* ***Skill Handbook****, pages 704-705.*

What plant part common to vascular plants is absent in nonvascular plants?

Plants that have tubelike cells in their roots, stems, and leaves to carry food and water are **vascular** (VAS kyuh lur) **plants.** Roots, stems, and leaves are three kinds of organs. In vascular plants, roots anchor plants in the ground and take in water and minerals from the soil. Stems carry water to all parts of the plant and hold the leaves up to the sunlight. Leaves are the main organ for food making. Most of the chloroplasts are in the leaves of a plant.

Plants that don't have tubelike cells are shorter, just as office buildings without elevators are shorter than those with elevators. Plants that don't have tubelike cells in their stems and leaves are **nonvascular** (nahn VAS kyuh lur) **plants.** Nonvascular plants grow close to the ground in moist areas. They don't have roots. They take up water by osmosis through hairlike cells. You have read in Chapter 2 that osmosis is the movement of water across a cell membrane. The water also carries minerals from the soil to the plant.

### Check Your Understanding

1. Why is chlorophyll important to plants?
2. What is the main difference between vascular and nonvascular plants?
3. Why are vascular plants usually taller than nonvascular plants?
4. **Critical Thinking:** How does photosynthesis differ from cellular respiration?
5. **Biology and Reading:** Go back and read again the paragraph that introduces the word *photosynthesis*. Then read the paragraph that follows. What is the main idea of these two paragraphs?



### Answers to Check Your Understanding

1. It is used to make food.
2. Vascular plants have tubelike cells; nonvascular plants don't.
3. Vascular plants have tubelike cells that can move water over great distances; nonvascular plants grow close to the ground because they don't have roots.
4. Photosynthesis: carbon dioxide is used, oxygen is released, food is made; cellular respiration: oxygen is used, carbon dioxide is released, food is broken down.
5. Answers will vary.

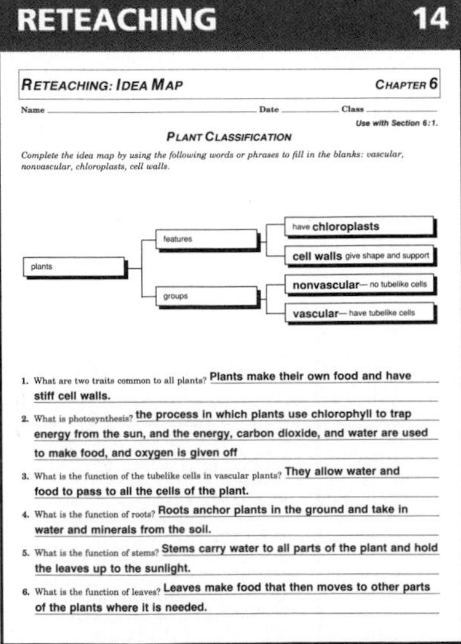

RETEACHING 14

*RETEACHING: IDEA MAP* CHAPTER 6

Name ______ Date ______ Class ______

*Use with Section 6:1.*

PLANT CLASSIFICATION

*Complete the idea map by using the following words or phrases to fill in the blanks: vascular, nonvascular, chloroplasts, cell walls.*

plants — features — have **chloroplasts**; **cell walls** give shape and support
plants — groups — **nonvascular**— no tubelike cells; **vascular**— have tubelike cells

1. What are two traits common to all plants? **Plants make their own food and have stiff cell walls.**
2. What is photosynthesis? **the process in which plants use chlorophyll to trap energy from the sun, and the energy, carbon dioxide, and water are used to make food, and oxygen is given off**
3. What is the function of the tubelike cells in vascular plants? **They allow water and food to pass to all the cells of the plant.**
4. What is the function of roots? **Roots anchor plants in the ground and take in water and minerals from the soil.**
5. What is the function of stems? **Stems carry water to all parts of the plant and hold the leaves up to the sunlight.**
6. What is the function of leaves? **Leaves make food that then moves to other parts of the plants where it is needed.**

# 6:2 Nonvascular Plants

You may have hiked on the mossy banks of a stream. Mosses are one kind of nonvascular plant. Most nonvascular plants live near water. They grow close to the ground where they can take up water even though they don't have tubelike cells.

**Objectives**

3. **Compare** mosses and liverworts with other plants.
4. **Sequence** the steps in the life cycle of a moss.
5. **List** three ways mosses are important to other living things.

**Key Science Words**

moss
sexual reproduction
egg
sperm
fertilization

## Mosses and Liverworts

You may have seen mosses as a lush, green, leafy mat on the floor of a forest or on rocks by a stream. A **moss** is a small, nonvascular plant that has both stems and leaves but no roots. If you looked closer to the ground you may have seen smaller patches of green plants that looked wet and slippery. These are liverworts, Figure 6-3.

Mosses and liverworts (LIV uhr wuhrts) are two very similar kinds of nonvascular plants. Unlike vascular plants, mosses and liverworts don't have roots. They are fixed to the surface of the ground or tree trunks by hairlike cells that take up water. Perhaps you have seen mosses and liverworts. If you have, you may have noticed that they are only a few centimeters tall and are common in wet or damp areas.

Mosses have fine, soft stems that often grow upright in mats. These mosses look like short-cropped hair. Some mosses form mats but have stems that creep along the ground. These mosses look like longer hair that is tangled and wavy. The leaves of mosses are only one or two cells thick and will soon dry out if taken from their moist environment. Moss leaves grow from all sides of the stems. Leaf shapes of mosses vary and moss species can be classified by their leaves.

**How do leaves and stems of mosses and liverworts differ from those in vascular plants?**

a

b

**Figure 6-3** Liverworts (a) and mosses (b) are nonvascular land plants.



## *OPTIONS*

### Science Background

Not all plants called mosses are true mosses. Irish moss is a red marine alga. Reindeer moss is a lichen. Spanish moss is a flowering plant. *Protococcus,* a moss that grows on trees, is a green algae. Club mosses, *Selaginella* and *Lycopodium,* are vascular plants.

## 6:2 Nonvascular Plants

## PREPARATION

### Materials Needed

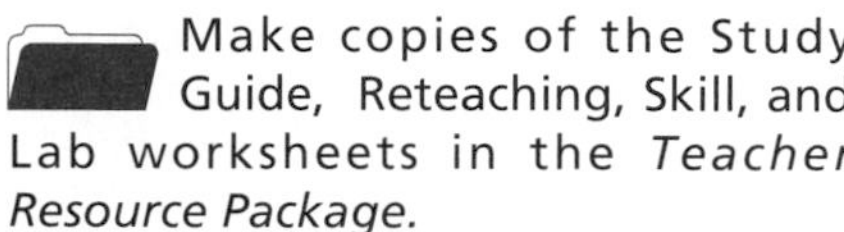

Make copies of the Study Guide, Reteaching, Skill, and Lab worksheets in the *Teacher Resource Package.*

▶ Collect mosses and liverworts for the Focus motivation.

▶ Obtain mosses for Lab 6-1.

### Key Science Words

moss
sexual reproduction
egg
sperm
fertilization

### Process Skills

In Lab 6-1, students will make and use tables, interpret data, use a microscope, measure in SI, and infer.

## 1 FOCUS

▶ The objectives are listed on the student page. Remind students to preview these objectives as a guide to this numbered section.

### MOTIVATION/Display

Bring in some mosses and liverworts for students to examine.

### Guided Practice

Have students write down their answers to the margin question: They are soft, only one or two cells thick, and don't have tubelike cells.

### *Student Journal*

Peat mosses were used to dress wounds during World War I. Have students research what physical properties make peat moss a useful dressing for wounds. (absorbs liquid and has antibiotic properties)

# 2 TEACH

### MOTIVATION/Display

Bring in some sphagnum moss and some peat moss. Place them in separate bowls and allow students to touch and look carefully at each.

### Concept Development

▶ Explain that peat is the remains of sphagnum moss that lived long ago.

### Idea Map

Have students use the idea map as a study guide to the major concepts of this section.

### Guided Practice

Have students write down their answers to the margin question: They are food.

### Connection: Math

Obtain some sphagnum moss from a florist or garden store. Have students find the mass of the dry moss. Then place it in water. Have students find the mass of the sphagnum moss after it has been in water for 30 or 40 minutes, and then calculate how much water it absorbed.

### Independent Practice

Study Guide/*Teacher Resource Package*, p. 33. Use the Study Guide worksheet shown at the bottom of this page for independent practice.

**Study Tip:** Use this idea map as a study guide to nonvascular plants. What do mosses and liverworts have in common?

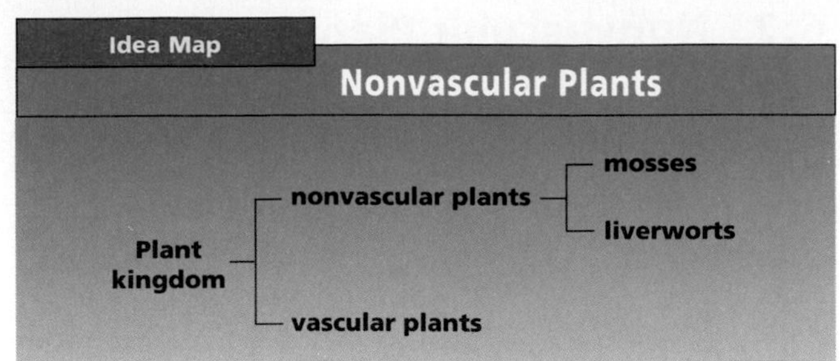


Many liverworts don't have roots, stems, or leaves. The body of these liverworts is often a flat, slippery layer of green cells that lies close to the ground. In other species, liverworts look more like mosses. They have creeping stems and small leaves. The main difference between mosses and these leafy liverworts is the arrangement of the leaves. In liverworts, the leaves grow in two or three flattened rows along the stem. Remember that moss leaves grow all around the stems.

**How are mosses important to small animals?**

Mosses and liverworts are useful to other living things, including you. They are food for some animals such as worms and snails. They also help hold the soil in place to keep it from washing away. Some mosses that live on rocks cause them to break down to form soil.

You may have used sphagnum (SFAHG num) moss in hanging baskets or flowers, Figure 6-4. Have you ever seen a gardener using sphagnum moss? Sometimes called peat moss, sphagnum moss is added to soil to increase the amount of water the soil can hold. Sphagnum moss is an upright moss that forms large, deep mats in bogs. In some areas, peat is cut, dried, and burned as fuel.

**Figure 6-4** Moss helps keep moisture around potted flowers.

## The Life Cycle of a Moss

Mosses and liverworts need a constant supply of water to survive. They also need water for sexual reproduction. **Sexual reproduction** is the forming of a new organism by the union of two reproductive cells. A female reproductive cell called an **egg** joins with a male reproductive cell called a **sperm.** The joining of the egg and sperm is called **fertilization.** Sexual reproduction occurs in most plants and animals. In many living things, the sperm swims to the egg. The sperm of mosses and liverworts must swim to the eggs to fertilize them. Sperm and eggs of mosses form at the tips of leafy stems.

## OPTIONS

Skill/*Teacher Resource Package*, p. 6. Use the Skill worksheet shown here to give students practice with outlining.

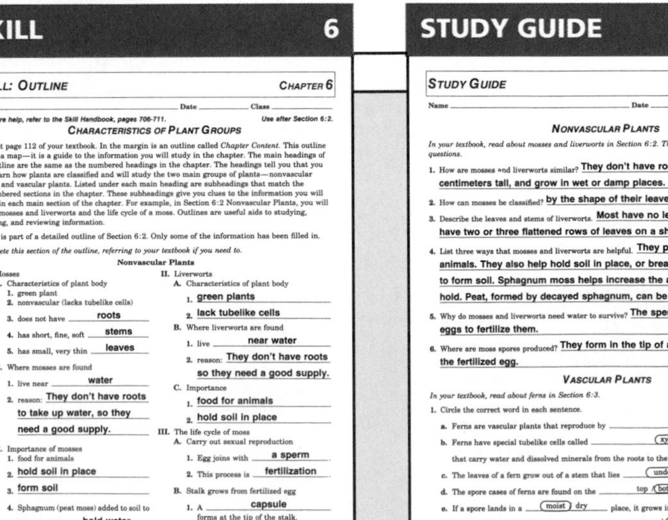

**SKILL 6**

SKILL: OUTLINE — CHAPTER 6

Name ______ Date ______ Class ______

*For more help, refer to the Skill Handbook, pages 706-711.* *Use after Section 6:2.*

CHARACTERISTICS OF PLANT GROUPS

Look at page 112 of your textbook. In the margin is an outline called *Chapter Content*. This outline is like a map—it is a guide to the information you will study in the chapter. The main headings of the outline are the same as the numbered headings in the chapter. The headings tell you that you will learn how plants are classified and will study the two main groups of plants—nonvascular plants and vascular plants. Listed under each main heading are subheadings that match the unnumbered sections in the chapter. These subheadings give you clues to the information you will study in each main section of the chapter. For example, in Section 6:2 Nonvascular Plants, you will study mosses and liverworts and the life cycle of a moss. Outlines are useful aids to studying, learning, and reviewing information.

Below is part of a detailed outline of Section 6:2. Only some of the information has been filled in.

*Complete this section of the outline, referring to your textbook if you need to.*

Nonvascular Plants

I. Mosses
  A. Characteristics of plant body
    1. green plant
    2. nonvascular (lacks tubelike cells)
    3. does not have roots
    4. has short, fine, soft stems
    5. has small, very thin leaves
  B. Where mosses are found
    1. live near water
    2. reason: They don't have roots to take up water, so they need a good supply.
  C. Importance of mosses
    1. food for animals
    2. hold soil in place
    3. form soil
    4. Sphagnum (peat moss) added to soil to help the soil hold water or it can be dried and used as fuel

II. Liverworts
  A. Characteristics of plant body
    1. green plants
    2. lack tubelike cells
  B. Where liverworts are found
    1. live near water
    2. reason: They don't have roots so they need a good supply.
  C. Importance
    1. food for animals
    2. hold soil in place

III. The life cycle of moss
  A. Carry out sexual reproduction
    1. Egg joins with a sperm.
    2. This process is fertilization.
  B. Stalk grows from fertilized egg
    1. A capsule forms at the tip of the stalk.
    2. The capsule produces spores that develop into new plants.

**STUDY GUIDE 33**

STUDY GUIDE — CHAPTER 6

Name ______ Date ______ Class ______

NONVASCULAR PLANTS

*In your textbook, read about mosses and liverworts in Section 6:2. Then answer the following questions.*

1. How are mosses and liverworts similar? They don't have roots, are only a few centimeters tall, and grow in wet or damp places.
2. How can mosses be classified? by the shape of their leaves
3. Describe the leaves and stems of liverworts. Most have no leaves or stems. Some have two or three flattened rows of leaves on a short, creeping stem.
4. List three ways that mosses and liverworts are helpful. They provide food for some animals. They also help hold soil in place, or break down rocks helping to form soil. Sphagnum moss helps increase the amount of water soil can hold. Peat, formed by decayed sphagnum, can be burned as fuel.
5. Why do mosses and liverworts need water to survive? The sperm must swim to the eggs to fertilize them.
6. Where are moss spores produced? They form in the tip of a stalk that grows from the fertilized egg.

VASCULAR PLANTS

*In your textbook, read about ferns in Section 6:3.*

1. Circle the correct word in each sentence.
  a. Ferns are vascular plants that reproduce by seeds / (spores).
  b. Ferns have special tubelike cells called (xylem) / phloem that carry water and dissolved minerals from the roots to the leaves.
  c. The leaves of a fern grow out of a stem that lies (underground) / above ground.
  d. The spore cases of ferns are found on the top / (bottom) side of the leaves.
  e. If a spore lands in a (moist) / dry place, it grows into a small, heart-shaped plant.
  f. The heart-shaped plant produces spores / (sperm and eggs).
  g. Ferns (do) / don't have a vascular system.

# Lab 6—1 Nonvascular Plants

## Problem: What are some traits of a nonvascular plant?

### Skills

make and use tables, interpret data, use a microscope, measure in SI, infer

### Materials

moss
light microscope
3 microscope slides
coverslip
hand lens or stereomicroscope
forceps
paper towel
metric ruler
dropper

### Procedure

1. Copy the data table.
2. Place the moss plant on the paper towel and examine it with the hand lens.
3. **Measure in SI:** Find the hairlike cells that act as roots for the moss. Measure their lengths in millimeters. Complete the table.
4. Observe the stem part with the hand lens. Complete the table for this part.
5. Observe the leaf parts with the hand lens. Complete the table for these parts.
6. Use the forceps to remove one of the leaves. Make a wet mount slide and observe the cells under low power.
7. Find the chloroplasts in the cells. Determine how thick the leaf is by moving the fine focus up and down.

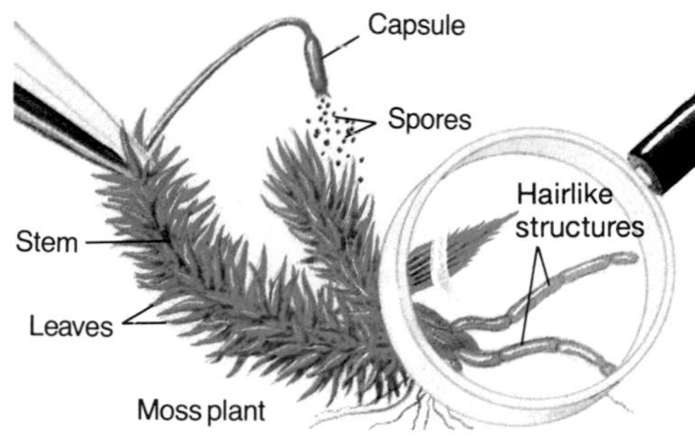


8. Find a spore capsule at the end of a brownish stalk. Observe the spore capsule with the hand lens. Complete the table.
9. Place the spore capsule on a slide. Add two drops of water. Place a second slide on top of the capsule. Press firmly down with your thumb on the top slide so that the capsule is crushed. Carefully remove the top slide. Place a coverslip over the crushed capsule.
10. Examine the released spores under low power. Draw and label the spores.

### Data and Observations

1. Which parts of the moss are green?
2. What shapes are moss leaves?

| PART | COLOR | SHAPE | LENGTH (mm) |
|---|---|---|---|
| Rootlike hairs | brown | hairlike | 1—10 |
| Stem | brown | threadlike | 20—80 |
| Leaf | green | round or oval | 1—5 |
| Spore case | brown | rounded | 2—6 |

### Analyze and Apply

1. **Infer:** How are the green parts used by a moss?
2. Describe the overall size of the plant.
3. How many cell layers are in a moss leaf?
4. **Apply:** Why do moss plants live in moist areas?

### Extension

**Design an experiment** to find out if mosses grow better in wet or dry places.



## ANSWERS

### Data and Observations

1. the leaves
2. round or oval

### Analyze and Apply

1. for photosynthesis
2. about 25 to 100 mm long
3. one or two
4. They don't have vascular tissues.

## Lab 6-1 Vascular Plants

### Overview

In this lab, students observe the parts of a moss to determine why mosses grow in moist environments and why they do not grow very tall.

**Objectives:** Upon completing the lab, students will be able to (1) **determine** that mosses do not have vascular tissue, (2) **recognize** that because of the absence of vascular tissue, mosses are incapable of growing tall.

**Time Allotment:** 35 minutes

### Preparation

▶ Mosses collected locally can be kept in cottage cheese containers covered with plastic wrap to prevent water loss. You may wish to place the moss specimens on trays from the supermarket to study.

**Lab 6-1 worksheet**/*Teacher Resource Package*, pp. 21-22.

### Teaching the Lab

▶ **Troubleshooting:** Students may need some help in focusing down through the leaf. Use the overhead projector and a petri dish. Write "bottom" on the bottom and "top" on the top of the dish. When one word is in focus on the overhead, the other will not be. Point out that the same is true with the layers of cells in the leaf.

**Cooperative Learning:** Have students work in pairs, taking turns using the microscope. For more information, see pp. 22T-23T in the Teacher Guide.

**Performance:** Provide students with other living specimens of mosses and liverworts. Have students identify the parts of each and describe the functions of each part.

## TEACH

### Guided Practice

Have students write down their answers to the margin question: spores.

### Check for Understanding

Have students respond to the first three questions in Check Your Understanding.

### Reteach

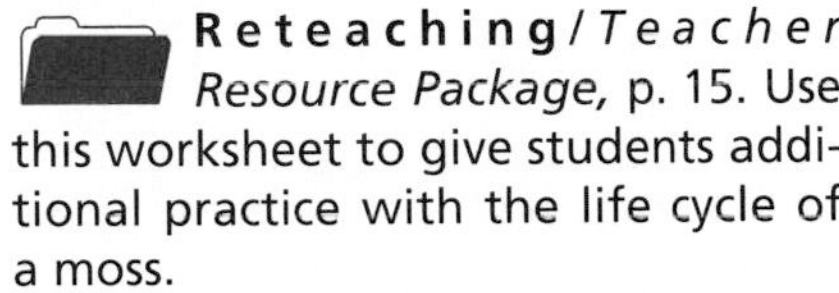

**Reteaching**/*Teacher Resource Package*, p. 15. Use this worksheet to give students additional practice with the life cycle of a moss.

**Extension:** Assign Critical Thinking, Biology and Reading, or some of the **OPTIONS** available with this lesson.

## 3 APPLY

### Brainstorming

Ask students whether the saying "mosses grow only on the north side of a tree" is a true statement.

## 4 CLOSE

### Audiovisual

Show the slide set *The Plant Kingdom*, Educational Images, Ltd.

### Answers to Check Your Understanding

6. by osmosis through hairlike cells that anchor the plant
7. Mosses are food for many animals, help to form soil, help to hold water in soil, and help hold soil in place.
8. sexually by sperm and eggs to form spores
9. Mosses provide a ground cover and help to prevent soil erosion.
10. A drier climate would reduce the number of mosses and liverworts because they require constant water to survive and for reproduction.

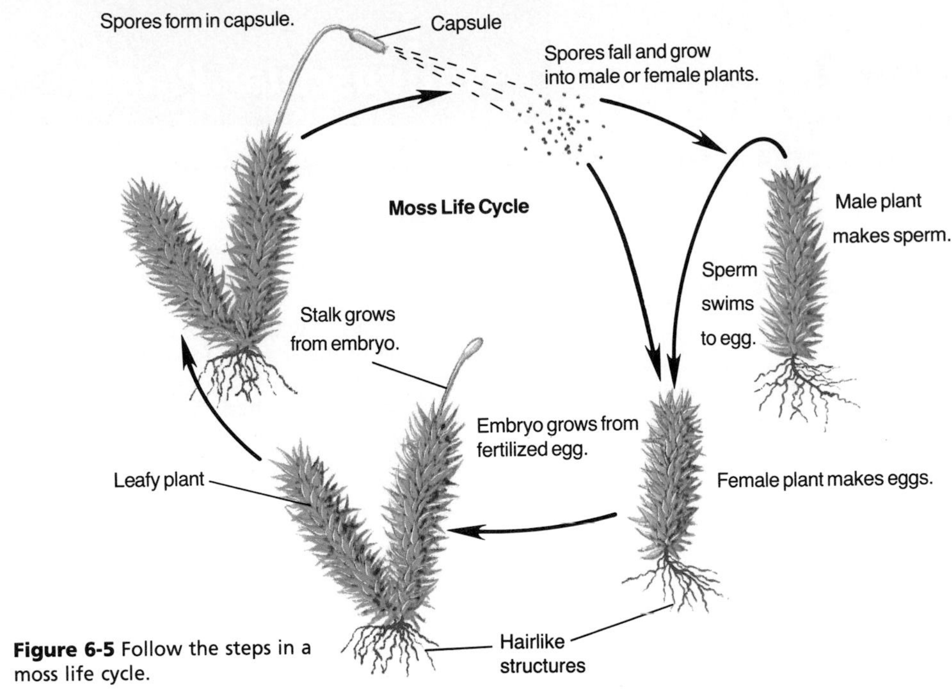


**Figure 6-5** Follow the steps in a moss life cycle.

Follow the arrows in Figure 6-5, which shows the steps in the life cycle of a moss. Notice in the figure that after fertilization, a stalk grows from the fertilized egg. Brown capsules form at the tip of each stalk. The capsules contain spores. Many plants make spores. Moss spores are as small and light as powdered sugar. Spores are blown away from the parent plant by wind, and where they land they grow into new leafy plants.

**What forms inside a moss capsule?**

### Check Your Understanding

6. How do mosses and liverworts take up water?
7. How are mosses important?
8. How do mosses reproduce?
9. **Critical Thinking:** How might a mat of mosses protect a forest floor?
10. **Biology and Reading:** From what you have learned about mosses and liverworts, what would be the effect on these plants if the climate became drier?



## GLENCOE TECHNOLOGY

**Videodisc**

**Biology: The Dynamics of Life**
Animation: *Life Cycle of a Moss*
Disc 1, Side 2, Ch. 17

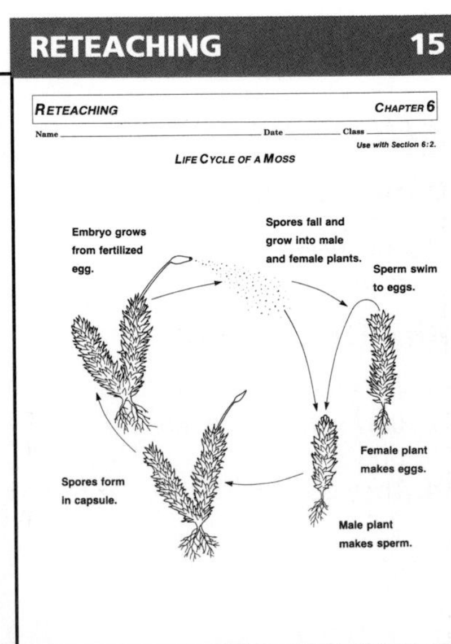

**RETEACHING 15**

RETEACHING CHAPTER 6

Name ______ Date ______ Class ______

Use with Section 6:2.

LIFE CYCLE OF A MOSS

## 6:3 Vascular Plants

Most of the plants that you see every day are vascular plants. You have read that vascular plants are plants with tubelike cells in roots, stems, and leaves. The tubelike cells carry food and water throughout the plant. There are two types of tubelike cells in vascular plants. **Xylem** (ZI lum) cells carry water and dissolved minerals from the roots to the leaves. **Phloem** (FLOH em) cells carry food that is made in the leaves to all parts of the plant.

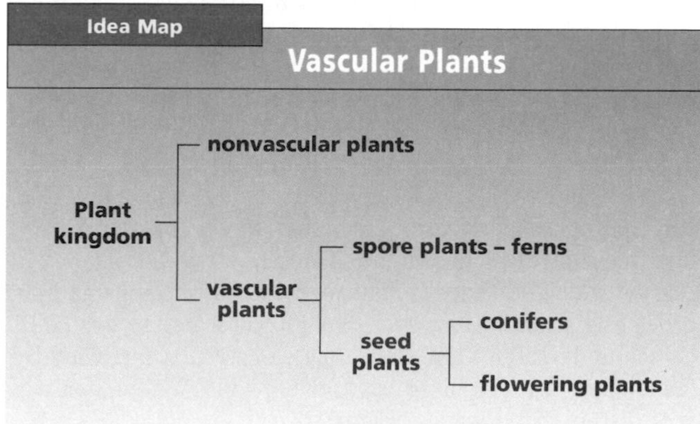


**Objectives**

6. **Compare** the traits of ferns, conifers, and flowering plants.
7. **Explain** how ferns, conifers, and flowering plants reproduce.
8. **Describe** how conifers and flowering plants are important to other living things.

**Key Science Words**

xylem
phloem
fern
seed
embryo
conifer
pollen
flowering plant
flower

**Study Tip:** Use this idea map as a study guide to vascular plants. How do ferns differ from other vascular plants?

### Ferns

A **fern** is a vascular plant that reproduces with spores. Because of their vascular system, ferns can grow taller than mosses and liverworts. However, one stage of the life cycle of a fern does not have a vascular system at all. For this reason ferns often grow in moist, shaded areas. Some ferns have adapted to much drier climates. Some live in water. Some species called tree ferns live in tropical forests and can grow to a height of 25 meters. There are over 12 000 known species of ferns. About 8000 of these species live in tropical regions.

**How are ferns different from conifers and flowering plants?**

You might think that a tree fern is unusual. However, over 300 million years ago, most of the plants that grew on Earth were tree ferns or other vascular spore plants. We know that these plants lived at that time because scientists have found ancient, preserved pieces of their stems. These ancient tree ferns lived in swamps. When they died and fell into the swamps, they formed the coal that we use for fuel today.



## *OPTIONS*

### Science Background

Seed plants are divided into two groups, based on where the seeds develop. Gymnosperms are plants whose seeds do not develop within ovaries. The largest and most familiar group of gymnosperms is the conifers. The second group of seed plants is angiosperms. They are the flowering plants whose seeds develop within ovaries.

## 6:3 Vascular Plants

## PREPARATION

### Materials Needed

Make copies of the Critical Thinking/Problem Solving, Study Guide, Application, Transparency, Enrichment, Reteaching, and Lab worksheets.

▶ Obtain bracken ferns for Lab 6-2.

▶ Collect both open and closed pine cones for the Mini Lab.

### Key Science Words

| | |
|---|---|
| xylem | conifer |
| phloem | pollen |
| fern | flowering plant |
| seed | flower |
| embryo | |

### Process Skills

In the Skill Check, students will interpret diagrams. In the Mini Lab, students will observe. In Lab 6-2, students will make and use tables, observe, and use a microscope.

## 1 FOCUS

▶ The objectives are listed on the student page. Remind students to preview these objectives as a guide to this numbered section.

### MOTIVATION/Bulletin Board

Arrange pictures of products obtained from plants under the headings: Wood Products, Food, Medicine, and Recreation. Use the bulletin board to discuss the value of plants.

### Idea Map

Have students use the idea map as a study guide to the major concepts of this section.

### Guided Practice

Have students write down their answers to the margin question: They reproduce with spores.

# 2 TEACH

### MOTIVATION/Display

Display a fern, a conifer with pine cones, and a flowering plant. Point out the fern's lacy leaves and where spores form. Emphasize that cones contain seeds, not spores, and that flowering plants have flowers and seeds that develop within a fruit.

### Concept Development

▶ Club mosses and horsetails are not discussed here. You may want to show pictures of both so that students will know they are vascular plants.

▶ Point out that a fern leaf is called a frond. Each frond grows from an underground stem called a rhizome.

▶ Make sure that students understand that spores are not seeds.

▶ Explain that the Christmas fern is believed to have gotten its name from its evergreen nature and the shape of its leaf parts that resemble Christmas stockings.

### Guided Practice

Have students write down their answers to the margin question: the small, heart-shaped plant.

**Misconception:** Many people mistake the spore cases on the underside of fern leaves for a disease.

### Skill Check

Use Figure 6-7 to make sure students understand the life cycle of a fern.

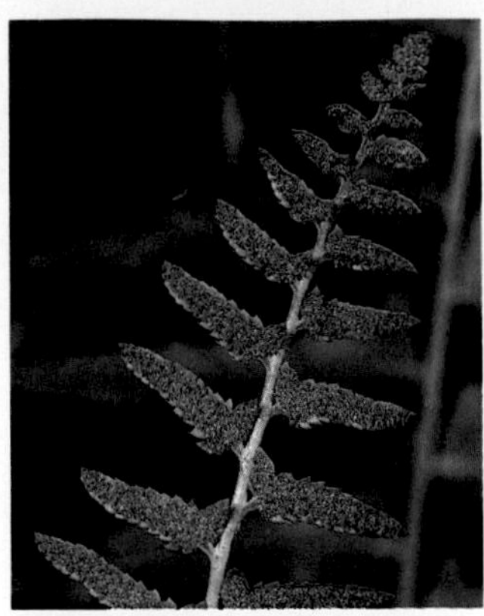

**Figure 6-6** A fern produces thousands of spores each year.

What part of a fern plant produces eggs and sperm?

The leaves of a fern grow from a stem that is horizontal and lies underground. The stem stores food and water. Small roots grow down from the underground stem. The roots anchor the plant and take up water and minerals from the soil into the xylem cells.

A fern leaf is often divided into many tiny leaflets that give it a feathery appearance. Many ferns lose their leaves at the end of a growing season. The underground stems remain and produce new leaves each spring. Ferns can be named by looking at the shapes of their leaves.

Ferns are similar to nonvascular plants in that they reproduce with spores. The spores of ferns are found on the underside of the leaves, Figure 6-6. If you have a fern in your school or home, look for brown or orange spots on the leaves. These roundish structures are called spore cases. The spore cases hold the spores.

Figure 6-7 shows the life cycle of a fern. Look for the spore cases on the fern leaf. When a spore case opens, the tiny spores are carried by wind and water. If a spore lands in a moist place, it grows into a small, flat, heart-shaped plant. The heart-shaped plant produces sperm cells and egg cells. The sperm cells swim through the water to egg cells on the underside of the plant. Each egg cell is fertilized by one sperm. The fertilized egg develops into a new fern with roots, stems, and leaves, Figure 6-7.

### Skill Check

**Interpret diagrams:** Look at the fern life cycle in Figure 6-7. What does the spore grow into? Where do the sperm and eggs develop? How do the sperm get to the eggs? *For more help, refer to the **Skill Handbook**, pages 706-711.*

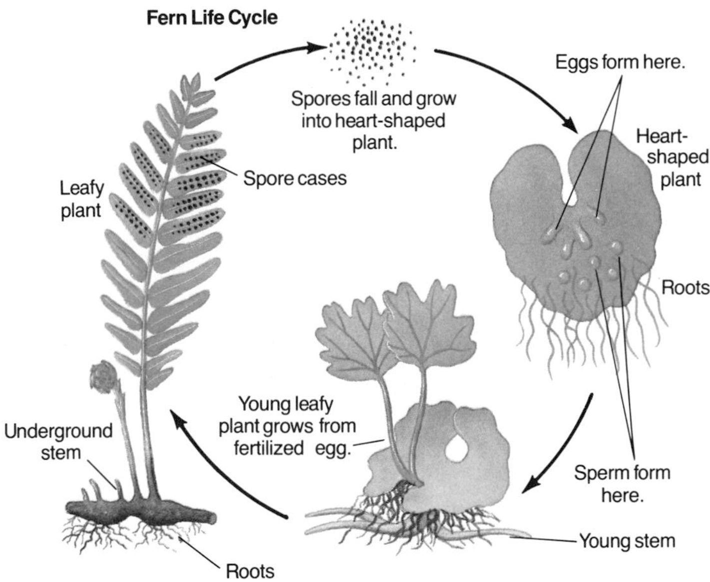


**Figure 6-7** Follow the steps in the fern life cycle.



## OPTIONS

**Fern Life Cycle**/*Transparency Package*, number 6a. Use color transparency number 6a as you teach the life cycle of a fern.

**Critical Thinking/Problem Solving**/*Teacher Resource Package*, p. 6. Use the Critical Thinking/Problem Solving worksheet shown here to reinforce traits of ferns.

**CRITICAL THINKING 6**

CRITICAL THINKING/PROBLEM SOLVING — CHAPTER 6

Name ______ Date ______ Class ______

*Use after Section 6:3.*

**WHAT HAPPENED TO THE FERN?**

A neighbor of Jennie's kept beautiful ferns on her shady porch all summer. Jennie had admired the plants and watched her neighbor care for them. When fall came, Jennie's neighbor carried the plants inside. She had some extra plants, so she gave Jennie two plants to grow in her room. Jennie wondered how much light a fern needs.

Jennie placed one plant in a brightly-lit window. She placed the other plant on the other side of the room. She only watered the plants when the soil was dry. After a few weeks, the leaves of the fern in the window were drooping and turning brown. The fern in the shade looked healthy, but there were powdery, brown spots on the undersides of the leaves. What was happening? Jennie looked at the section on vascular plants in her biology text.

**Analyzing the Problem**

1. How did Jennie test how much light is needed by ferns? She put one plant in bright light and one in shade.
2. Where do most ferns grow? Ferns usually grow in damp shady woods.
3. Do the powdery, brown spots on the underside of the fern leaf indicate the plant is not healthy? What are they? No, the plant is healthy. The powdery spots are spore cases.

**Solving the Problem**

1. What can Jennie conclude about how much light ferns need? Ferns need low light. They do not grow well in brighter light.
2. What can Jennie conclude about the brown spots on the underside of the fern leaves? They are part of the fern life cycle. The fern is growing normally.
3. What can Jennie conclude about light conditions for ferns in houses and in the forest? Ferns need low light for healthy growth.

## Lab 6–2 Ferns

### Problem: What are some traits of ferns?

**Skills**

make and use tables, observe, use a microscope

**Materials**

stereomicroscope
2 microscope slides
ethyl alcohol
whole fern plant (for the class)
fern leaf with spore cases
scalpel
dropper
paper and pencil

**Procedure**

1. **Observe:** Look at a whole fern plant. Locate the horizontal stem at the base.
2. Note how the leaves grow from the stem.
3. Look for veins that indicate xylem and phloem cells in the leaf.
4. Look at the underside of each leaf for brownish-yellow spots. Each spot is made up of a group of spore cases.
5. Draw a table like the one shown. Draw and label your fern plant. Label all the parts as shown in Figure A.
6. Remove a fern leaf. Notice how the cells of the veins connect the stem to the leaf.
7. Obtain a part of a leaf with spore cases.
8. Place a drop of water on a clean slide.
9. Use the scalpel to scrape one of the brownish-yellow spots into the drop of water on the slide. **CAUTION:** *Always cut away from yourself when using a scalpel.*
10. **Use a microscope:** Observe the slide under low power of the stereomicroscope. Look for club-shaped spore cases like the one shown in Figure B.
11. Draw and label a single spore case.
12. Now, make a slide of spores by removing one or two spore cases from your first slide to a clean, dry slide. Add a drop of ethyl alcohol.
13. Observe the slide under low power. Watch what happens to the spore cases.
14. Draw a few of the spores like those shown in Figure C.

**Data and Observations**

1. What kinds of cells made it difficult to tear fern leaves from the plant?
2. What do fern roots grow from?

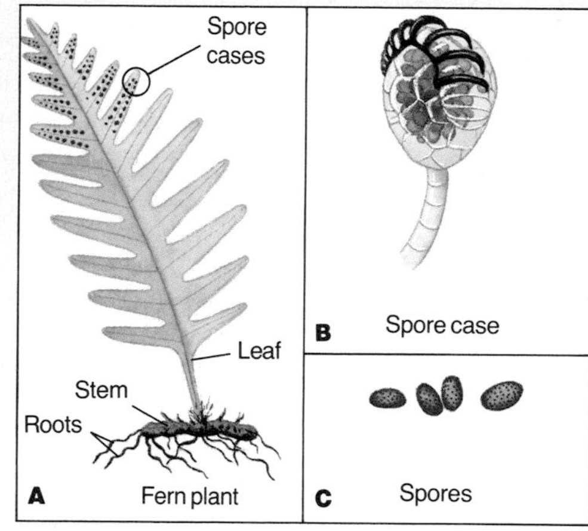


**Analyze and Apply**

1. What parts of the fern carry on photosynthesis?
2. When are spores produced in a fern life cycle?
3. Why are ferns usually larger than mosses and liverworts?
4. **Apply:** How are the leaves of ferns similar to the leaves of flowering plants?

**Extension**

**Compare** the roots of ferns with the rootlike hairs in mosses and liverworts.



## ANSWERS

### Data and Observations

1. xylem and phloem
2. an underground stem

### Analyze and Apply

1. the leaves
2. after the leaves have developed
3. They have a vascular system.
4. They have chlorophyll and veins.

## Lab 6-2 Ferns

### Overview

In this lab, students examine the structure of a fern and observe spore cases and spores.

**Objectives:** Upon completing the lab, students will be able to (1) **identify** the structures of a fern plant, (2) **explain** how ferns reproduce.

**Time Allotment:** 35 minutes

### Preparation

▶ **Alternate Materials:** Other kinds of ferns with spore cases may be used.

**Lab 6-2 worksheet**/*Teacher Resource Package*, pp. 23-24.

### Teaching the Lab

**Cooperative Learning:** Have the students work in pairs. One can prepare slides while the other uses the microscope. For more information, see pp. 22T-23T in the Teacher Guide.

▶ **Troubleshooting:** If too many spore cases are placed on the slide, students will not be able to see the spores. Spores eject from the spore cases when the alcohol is added.

**Skill:** Have students draw a diagram showing how a fern differs from a moss.

**GLENCOE TECHNOLOGY**

**Videodisc**

**Biology: The Dynamics of Life**
Movie: *Fern Development*
Disc 1, Side 2, Ch. 18

## TEACH

### Concept Development

▶ Point out that conifers make up about one-third of modern forests. The giant sequoias are conifers and are the largest trees.

### Independent Practice

**Study Guide**/*Teacher Resource Package,* p. 34. Use the Study Guide worksheet shown at the bottom of this page for independent practice.

**Portfolio**

Have students prepare an informational bulletin board about seeds used in cooking and the nutritional value of the seeds. Seeds that may be investigated include walnuts, peanuts, coconuts, sunflower seeds, pine nuts, cashews, and cocoa beans.

**Student Journal**

The chemicals in conifers make them valuable sources of tar, turpentine, and resin. Have students find out what parts of conifers are used to make these products.

**GLENCOE TECHNOLOGY**

**Videodisc**

**Biology: The Dynamics of Life**
Movie: *Giant Redwoods*
Disc 1, Side 2, Ch. 19

Animation: *Life Cycle of a Pine*
Disc 1, Side 2, Ch. 20

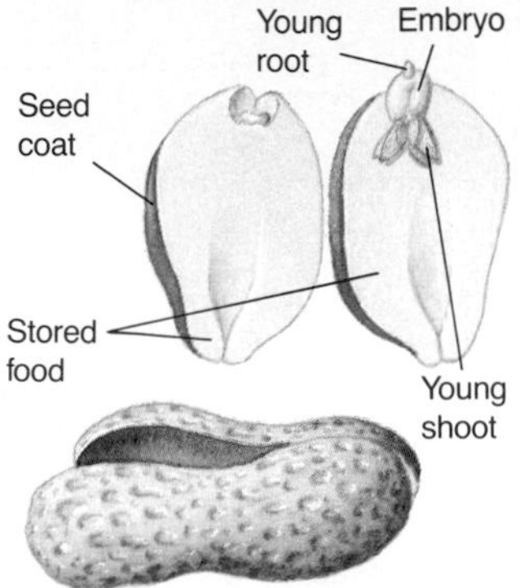


**Figure 6-8** A seed has a seed coat to protect the embryo with its food supply. What kind of seed is shown here? a peanut seed

## Conifers

The most common land plants are seed plants. Seed plants reproduce by forming seeds. A **seed** is the part of a plant that contains a new, young plant and stored food. The young plant in a seed is the embryo (EM bree oh). An **embryo** is an organism in its earliest stages of growth. A seed has a hard outer covering called the seed coat that protects the embryo, Figure 6-8. The food supply and the seed coat help the embryo survive for long periods of time when conditions are not right for growth.

There are two kinds of seed plants. One kind, called a **conifer** (KAHN uh fur), is a plant that produces seeds in cones. Conifers generally keep their leaves throughout the year. This is why they are sometimes called evergreens. A long time ago there were many different kinds of conifers. Conditions on Earth changed and many of the plants died out. Today, conifers are found mainly in northern areas of the world.

You have probably seen many examples of conifers. Most conifers are evergreen trees with small, needle-shaped leaves. An evergreen tree sheds its leaves in the same way that fur falls from the coat of a dog or cat—not all at once! If you are familiar with pine, spruce, and fir trees and think they are evergreen conifers, you are correct. Check your knowledge of these conifers with Figure 6-9. The larch, dawn redwood, and bald cypress are conifers that lose all their leaves each fall.

**Figure 6-9** The pine (a), the spruce (b), and the fir (c) are all conifers with needlelike leaves.

a

b

c

## OPTIONS

### Science Background

Conifers are examples of evolutionary success by several criteria. They make up about one-third of modern forests. They are the largest living plants. The giant redwoods grow as high as 100 meters. A sequoia contains sufficient wood to build a village of 50 six-room houses. A sequoia is not mature until it is 300 years old.

**Conifer Life Cycle**/*Transparency Package,* number 6b. Use color transparency number 6b as you teach the life cycle of a conifer.

**STUDY GUIDE 34**

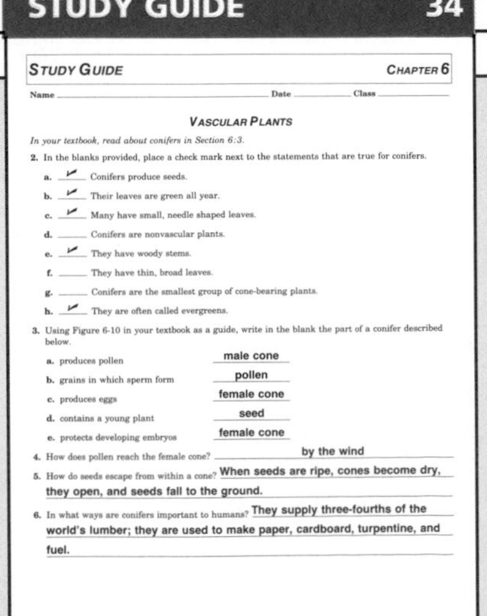

STUDY GUIDE — CHAPTER 6

Name ______ Date ______ Class ______

VASCULAR PLANTS

*In your textbook, read about conifers in Section 6:3.*

2. In the blanks provided, place a check mark next to the statements that are true for conifers.
   a. ✓ Conifers produce seeds.
   b. ✓ Their leaves are green all year.
   c. ✓ Many have small, needle shaped leaves.
   d. ____ Conifers are nonvascular plants.
   e. ✓ They have woody stems.
   f. ____ They have thin, broad leaves.
   g. ____ Conifers are the smallest group of cone-bearing plants.
   h. ✓ They are often called evergreens.
3. Using Figure 6-10 in your textbook as a guide, write in the blank the part of a conifer described below.
   a. produces pollen — male cone
   b. grains in which sperm form — pollen
   c. produces eggs — female cone
   d. contains a young plant — seed
   e. protects developing embryos — female cone
4. How does pollen reach the female cone? by the wind
5. How do seeds escape from within a cone? When seeds are ripe, cones become dry, they open, and seeds fall to the ground.
6. In what ways are conifers important to humans? They supply three-fourths of the world's lumber; they are used to make paper, cardboard, turpentine, and fuel.

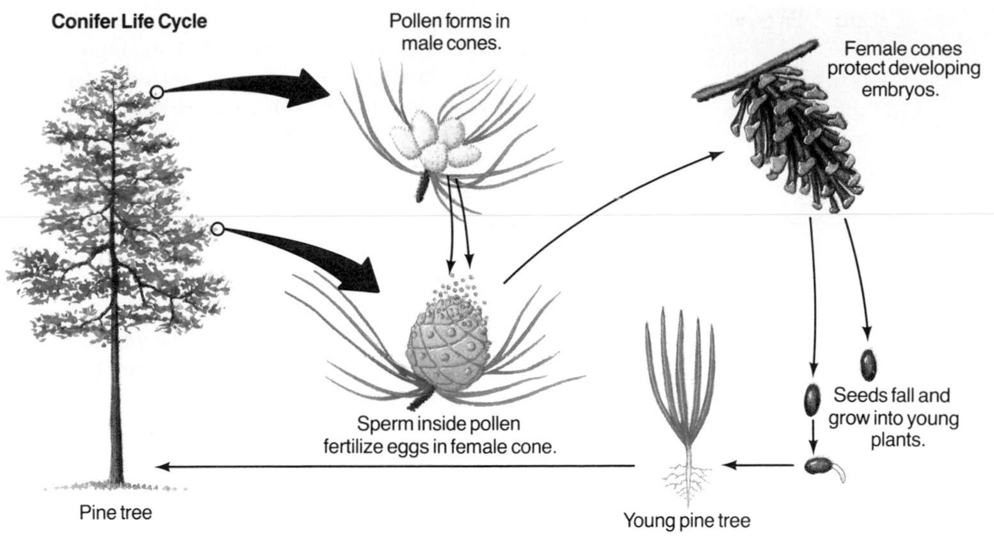


Figure 6-10 Follow the steps in the life cycle of a conifer.

Let's look at the life cycle of a pine tree. Have you ever noticed that two different kinds of cones grow on pine trees? Pine trees produce male and female cones. The small cones shown in Figure 6-10 are male cones, which produce pollen. **Pollen** are the tiny grains of seed plants in which sperm develop. You see male cones early in the spring. They open and shed pollen into the air. Wind carries the pollen to the larger female cones. The female cones, as shown in Figure 6-10, contain egg cells. The sperm cells fertilize the egg cells and the seeds containing the embryo form between the woody scales of the cone.

When the seeds are ripe, the cones become dry and the woody scales open. The seeds fall to the ground. When conditions are right, each seed grows into a young plant, Figure 6-10. Some conifers don't have woody cones. The conifer known as a yew produces its seeds inside red, fleshy cuplike parts. Birds eat these juicy cuplike parts and later drop the seeds away from the tree.

The roots, stems, and leaves of conifers are not like the soft stems of ferns. The roots and stems of conifers are hard and woody. The xylem cells have thicker cell walls. Also, the leaves of conifers are tough and needlelike. Examine the different kinds of conifer leaves shown in Figure 6-11. Can you see why some of them are called needles? Other leaves of conifers are scalelike in their shape, but feel quite soft.

**Figure 6-11** Conifer leaves are either needlelike as in pines, flattened needles as in hemlock, or scalelike as in arborvitae, a common garden shrub.

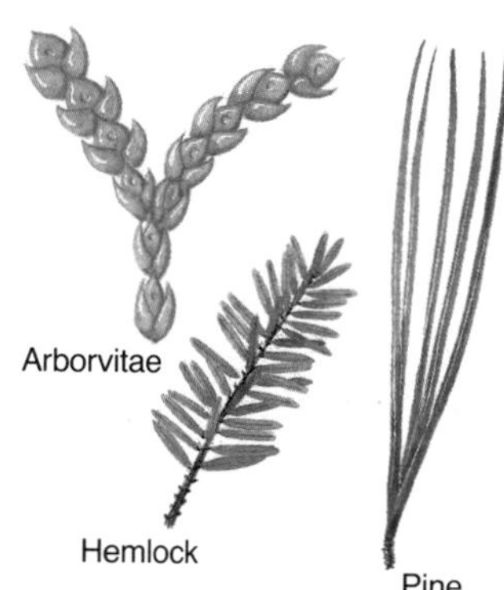




## TEACH

### Concept Development

▶ Mention that a few conifers, such as the yew and the juniper, do not have cones but have individual seeds that develop on branches.

### Guided Practice

Give students a photocopied diagram of the life cycle of a pine tree and have them label it.

### ACTIVITY/Demonstration

Bring cones of different conifers to class. Indicate that pollen (male) cones are found in clusters at the ends of branches in the spring. The seed (female) cones are larger and more woody than pollen cones and are used to identify various species of conifers. Show the location of seeds on the cone scales. Pull off a cone scale. Two seeds are attached.

**TRANSPARENCY 25**

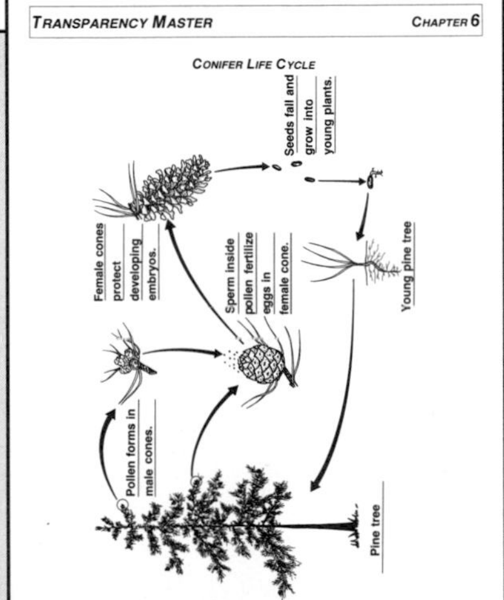


### OPTIONS

**Transparency Master**/*Teacher Resource Package*, p. 25. Use the Transparency Master shown here to explain the life cycle of a conifer.

**How Can Conifers Be Identified?**/*Lab Manual*, pp. 45-48. Use this lab as an extension to studying conifers.

# TEACH

## Concept Development

▶ Point out that many conifers grow well in poor or shallow soil where nutrients are scarce, and where there are often long cold or dry periods.

### Mini Lab

Cones that detect moisture will close, and those in a warm, dry place will open. Warm, dry conditions are needed for cones to open. If seeds were released in rainy weather, they would all fall at the bottom of the tree. They are released when the weather is dry so that they will be carried to a suitable spot for germination.

**Performance:** Provide students with a packet of different kinds of seeds. Have students observe each seed, record their observations (shape of seed, size, color, weight, texture), and hypothesize how each seed is dispersed.

### GLENCOE TECHNOLOGY

**Videodisc**

Biology: The Dynamics of Life

Movie: *Blooming Flowers*

Disc 1, Side 2, Ch. 21

**Figure 6-12** A bald cypress is a conifer that lives in swamps.

Conifers can live in many more places than most other plants. Some can live in very wet areas. For example, the bald cypress shown in Figure 6-12 lives in swamps. Other conifers live in very dry places. Junipers are common conifers of the desert. Conifers grow also on high mountain slopes or close to the sea.

Take a look around at everything that is made of wood. Wood is a plant material. Wood is made up mainly of xylem cells. Your pencil, desk, and many parts of your school are probably made of wood. What about your home? What parts of your home are made of wood? Conifers supply three-fourths of the lumber that is used in the world.

You are reading words that are printed on plant material. You pay for things with plant material. You even write on plant material. Why are all these statements true? Paper is made from wood. Cardboard is also made of paper. Think of all the things you buy that are in cardboard boxes. Almost all of the world's paper comes from conifers, Figure 6-13. Conifers also are a source of turpentines, disinfectants, and fuel.

Conifers are very important to other animals for food and shelter. The bark, buds, and seeds of conifers are eaten by insects, birds, squirrels, rabbits, and many other animals.

### Mini Lab

**What Conditions Cause Pine Cones to Open?**

**Observe:** Place open and closed pine cones into two plastic bags. Add a little water to one. Leave the other in a warm, dry place. Which conditions caused the cones to open? *For more help, refer to the* ***Skill Handbook,*** *pages 704-705.*

**Figure 6-13** Conifers supply most of the world's paper products. Which of these products have you used? Students will probably say all of them.



## OPTIONS

### ACTIVITY/Challenge

Have students research and illustrate the papermaking process from trees to paper.

# Science and Society

## Ancient Forests: Jobs versus Wildlife

Logging in the ancient forests of the Pacific Northwest

Have any of your neighbors made it well known that they are upset or angry by something in the neighborhood? Maybe a road sign was put up that they didn't like. Maybe someone's yard was full of garbage. You may have said, "It doesn't bother me," and "It's not my problem." In some cases where people disagree, the problem is so important that even the government gets involved. One such case involves the cutting of ancient trees in the forests of the Pacific Northwest. The trees are used as a major source of lumber and paper for the United States. They are also important to Earth's atmosphere. Consider the three cases described below and discuss each question with your classmates.

### What Do You Think?

1. Some small mill owners work in forests owned by the National Forest Service. These loggers harvest only older trees and rely on the growth of young trees to fill in the gaps. Other loggers completely clear the forests of all ages of trees and then replant with young trees. These timber company loggers then harvest the trees as soon as they can. Because of the new laws protecting wildlife, the small mill owners may lose their jobs. The larger timber companies will not be affected. What solutions would you suggest to the government to save the life styles of all loggers?

2. The northern spotted owl lives only in the forests of old conifers in the Pacific Northwest. There are only about 3000 pairs of owls recorded. In July 1990, the United States Fish and Wildlife Service listed this owl as threatened on the federal endangered species list. The owl's environment is also protected by law. Why is it important to save species?

Northern spotted owl

3. Certain gases, including carbon dioxide, are found naturally in Earth's atmosphere. They trap heat from the sun in the same way the glass of a greenhouse does. Carbon dioxide has increased over the last fifty years because of the burning of fuels, such as coal, wood, oil, and natural gas. Forests take up carbon dioxide. Without forests, carbon dioxide would build up in the atmosphere. This increase of gases might cause major changes in Earth's climate. What social and political effects might there be if the world's climate changed?

**Conclusion:** Do we all need to care about forests? Are the needs of humans more important than those of other species?



**ENRICHMENT 6**

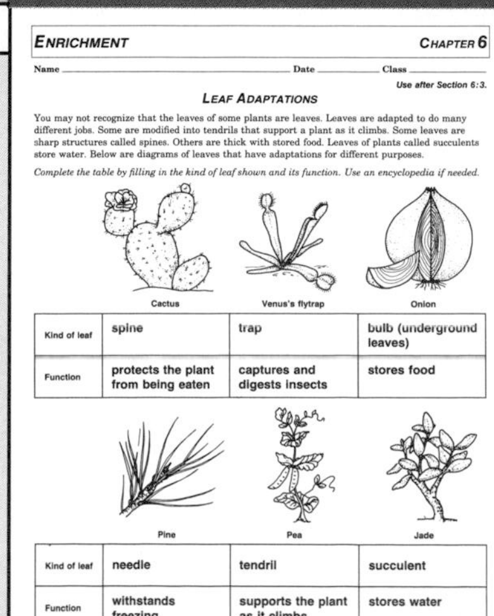

ENRICHMENT CHAPTER 6

Name ______ Date ______ Class ______

Use after Section 6:3.

LEAF ADAPTATIONS

You may not recognize that the leaves of some plants are leaves. Leaves are adapted to do many different jobs. Some are modified into tendrils that support a plant as it climbs. Some leaves are sharp structures called spines. Others are thick with stored food. Leaves of plants called succulents store water. Below are diagrams of leaves that have adaptations for different purposes.

*Complete the table by filling in the kind of leaf shown and its function. Use an encyclopedia if needed.*

| | Cactus | Venus's flytrap | Onion |
|---|---|---|---|
| Kind of leaf | spine | trap | bulb (underground leaves) |
| Function | protects the plant from being eaten | captures and digests insects | stores food |

| | Pine | Pea | Jade |
|---|---|---|---|
| Kind of leaf | needle | tendril | succulent |
| Function | withstands freezing | supports the plant as it climbs | stores water |

## OPTIONS

**Enrichment**/*Teacher Resource Package*, p. 6. Use the Enrichment worksheet shown here to help students understand leaf adaptations.

## Science and Society

### Ancient Forests: Jobs versus Wildlife

### Background

The U.S. Fish and Wildlife Service often becomes involved in controversies that concern the protection of threatened wildlife. If the U.S. Fish and Wildlife Service decides that a species is endangered, it can prohibit activity in a particular area. In the 1970s, construction of the Tellico Dam in Tennessee was halted while environmentalists studied how to protect the snail darter fish.

### Teaching Suggestions

The questions that appear in the student text should provide the basis for an excellent class debate.

### What Do You Think?

There are no incorrect answers to these questions. Students should attempt to answer them based on their own feelings, opinions, or background.

**Conclusion:** Students should have some understanding of how all living things are dependent on each other. Some may feel that humans are more important, but others will realize that humans can't survive without other living things and need to prevent endangering any species.

### Suggested Readings


Bicak, Charles J. "The Application of Ecological Principles in Establishing an Environmental Ethic." *The American Biology Teacher*, April 1997, pp. 200-206.

Karl, Thomas R., Neville Nicholls, and Jonathan Gregory. "The Coming Climate." *Scientific American*, May 1997, pp. 78-83.

Rice, Richard E., Raymond E. Gullison, and John W. Reid. "Can Sustainable Management Save Tropical Forests?" *Scientific American*, April 1997, pp. 44-49.

Satchell, Michael. "The Endangered Logger." *U.S. News and World Report*, June 25, 1990, pp. 27-29.

## TEACH

### Independent Practice

**Study Guide**/*Teacher Resource Package*, p. 35. Use the Study Guide worksheet shown at the bottom of this page for independent practice.

**Cooperative Learning:** Have students work in groups of four and prepare an illustrated report on plant adaptations.

### Check for Understanding

Take students on a short walk around the school or bring in pictures of a variety of plants. Ask them about each plant. Ask them if the plant is vascular or nonvascular.

### Reteach

Show slides or pictures of ferns, conifers, and flowering plants. Have students name the traits each plant has that make it a fern, a conifer, or a flowering plant.

### *Portfolio*

Have students collect several kinds of leaves from flowering plants and identify the plant from which each leaf was taken. Students should label the parts of each leaf and describe how the shape of the leaf helps the plant survive in its environment.

### *Student Journal*

Have students make a list of 20 items in their homes that are made from plant products. Students should determine whether the item came from a conifer or a flowering plant.

**Figure 6-14** A flower is the reproductive part of a flowering plant.

## Flowering Plants

The second kind of seed plant produces flowers and forms fruits. This seed plant is a flowering plant. A **flowering plant** is a vascular plant that produces seeds inside a flower. The **flower** is the reproductive part of the plant, Figure 6-14. Flowers have male parts that produce pollen. They also have female parts that produce eggs. A sperm cell produced in a pollen grain must join with an egg for a seed to form. Pollen is carried to the egg cells by wind, insects, or other animals. The female flower parts develop into a fruit that protects the seeds.

There are many more kinds of flowering plants in the world than nonflowering plants. Why might that be? The way the seeds are protected in flowering plants helps to make sure that the new plants survive. Flowering plants are adapted to live in many different environments.

Just because there are many more kinds of flowering plants than nonflowering plants, don't think that you will see mostly flowers when you take a walk outside. Not all flowering plants have big, sweet-smelling flowers. Many flowering plants, such as the maple in Figure 6-15a, have small, non-showy flowers. You may not have known they were flowering plants. Many broadleaved trees, vegetables, grasses, roadside weeds, and thorn bushes have flowers. So you see that not just roses and lilies are examples of flowering plants. Think about grass in lawns, fields, and parks. All grasses have flowers. The flowers on a grass plant are hard to see.

Another reason for not seeing flowers everywhere when you take a walk is that most flowering plants produce their

### Bio Tip

**Consumer:** Did you know that cauliflower and broccoli are heads of tightly packed flower buds? Next time you buy these vegetables, look more closely to find the parts of the flowers.

a

b

**Figure 6-15** The flowers of maples (a) are not as showy as those of magnolia trees (b).

128

## *OPTIONS*

### ACTIVITY/Enrichment

Provide a variety of seeds for students to examine and compare their shapes and adaptations for dispersal.

**Application**/*Teacher Resource Package*, p. 6. Use the Application worksheet shown here to explain poisonous plants.

**STUDY GUIDE 35**

STUDY GUIDE — CHAPTER 6

Name ______ Date ______ Class ______

VASCULAR PLANTS

7. Decide which of the following sentences are true and which are false. Write the word *true* in the blank next to the sentences that are true. For sentences that are false, replace the underlined word with a word that would make the sentence true. Write that word in the blank.

true a. Seeds of flowering plants are formed inside a flower.
vascular b. Flowering plants are nonvascular.
true c. The flower is involved in reproduction.
male d. Sperm cells are formed within female flower parts.
seed e. In a flowering plant, a fertilized egg forms a flower.
Female f. Male flower parts develop into an embryo.
largest g. Flowering plants make up the smallest group of all types of plants.
true h. Seeds contain a supply of food for the developing embryo.

8. Complete the following table. Put a check mark in the column of each kind of plant that has the trait listed.

| Trait | Moss and liverwort | Fern | Conifer | Flowering plant |
|---|---|---|---|---|
| Nonvascular | ✓ | | | |
| Vascular | | ✓ | ✓ | ✓ |
| Has no tubelike cells | ✓ | | | |
| Has tubelike cells | | ✓ | ✓ | ✓ |
| Has roots, stems, leaves | | ✓ | ✓ | ✓ |
| Doesn't have roots | ✓ | | | |
| Spores | ✓ | ✓ | | |
| Seeds | | | ✓ | ✓ |
| Cones | | | ✓ | |
| Flowers | | | | ✓ |
| Most common of plants | | | | ✓ |
| Gives off oxygen | ✓ | ✓ | ✓ | ✓ |

**APPLICATION 6**

APPLICATION: SAFETY — CHAPTER 6

Name ______ Date ______ Class ______

Use after Section 6:3.

POISONOUS PLANTS

Some plants have parts that are poisonous if they are touched or eaten. You can safely explore in woods and gardens if you know which plants to avoid.

*Use an encyclopedia or book about plants to find information on the plants listed below. Draw a picture of each plant. On the first line below your picture, tell if the plant is a tree, shrub, vine, or small flowering plant. On the next line, name a part of the plant that you could use to recognize the plant, such as star-shaped flowers or fuzzy leaves. On the last two lines, tell which parts of the plant are poisonous. Include whether the plant is poisonous to the touch or poisonous if swallowed.*

Poison ivy
vine or shrub
leaflets in threes; red leaves in fall
all parts; touching plant causes rash and itching

Rhubarb
small flowering plant
large leaves, reddish leafstalks
leaves; poisonous if swallowed (leafstalks are edible)

Foxglove
small flowering plant
flowers shaped like tubular bells
all parts, especially leaves; can be fatal if swallowed

Mistletoe
small plant, parasite on trees
white berries, evergreen leaves
berries; poisonous if swallowed

How does being poisonous help? Being poisonous could protect a plant from being eaten by animals.

flowers only at certain times of the year. For example, many forest trees bloom only in the spring. The garden lily and tomato plants flower in the summer. Some plants even flower at times other than spring and summer, Figure 6-16.

How are flowering plants important to us? You know that showy flowers are often used for decoration, but plants have many more uses. Oranges and potatoes are plant parts you can eat. Bread and cereals are foods that are made from plants. Without plants, there would be no fruits, vegetables, breads, or cereals to eat. What parts of the plants in Figure 6-17 have you eaten?

If there were no plants, could you eat meat instead? No, if there were no plants, there would be no animals. Animals depend on plants or other producers to live.

In what other ways do all animals need plants? Remember that plants make their own food by photosynthesis. Oxygen given off by the plant is needed by both plants and animals for cellular respiration. Much of the oxygen gas that we have on Earth is produced by plants. In turn, the carbon dioxide produced during cellular respiration in plants and animals is needed by plants to make food. You can now see why every living thing depends on every other living thing.

**Figure 6-16** Chrysanthemums bloom in late summer. Name some flowers that bloom in the spring. Answers may include daffodils and crocuses.

**In what ways do animals need plants?**

**Figure 6-17** Which part of a plant do you eat—stems, leaves, seeds, or fruits? All of these parts

## Check Your Understanding

11. What are three examples of conifers?
12. Where are sperm and egg cells formed in seed plants?
13. How are flowering plants important to us?
14. **Critical Thinking:** Which gas is more important to life on Earth, carbon dioxide or oxygen?
15. **Biology and Reading:** From reading the previous section, what are the three major groups of vascular plants?

## RETEACHING 16

*Reteaching* — Chapter 6

Name ______ Date ______ Class ______

*Use with Section 6:3.*

**Comparing Vascular Plants**

The chart on this page lists traits found in one or more groups of plants.

*If a trait is present in a plant group, place a check mark in the appropriate column.*

| Traits of land plants | Ferns | Conifers | Flowering plants |
|---|---|---|---|
| Tubelike cells are present. | ✓ | ✓ | ✓ |
| Is the largest group of plants. | | | ✓ |
| Are made up of cells with a nucleus. | ✓ | ✓ | ✓ |
| Are usually found in moist places. | ✓ | | |
| Spores form in a spore case. | ✓ | | |
| Seeds develop in a cone. | | ✓ | |
| Have stiff cell walls. | ✓ | ✓ | ✓ |
| Spores form in a capsule. | | | |
| Many species have waxy, needlelike leaves. | | ✓ | |
| Do not have tubelike cells. | | | |
| Cells contain chloroplasts. | ✓ | ✓ | ✓ |
| Require water for sperm to swim to egg. | ✓ | | |
| Seeds are enclosed in a fruit. | | | ✓ |
| Seeds contain an embryo surrounded by a protective coat. | | ✓ | ✓ |

## STUDY GUIDE 36

*Study Guide* — Chapter 6

Name ______ Date ______ Class ______

**Vocabulary**

*Review the new words used in Chapter 6 of your textbook. Then, complete the exercise. Fill in the blank in each of the sentences below with the correct word from the following list.*

| | | | |
|---|---|---|---|
| chlorophyll | flowers | sexual reproduction | nonvascular plant |
| xylem | photosynthesis | egg | fertilization |
| embryo | conifer | vascular plants | moss |
| sperm | phloem | fern | flowering plant |
| seed | pollen | | |

1. The chemical that gives plants their green color is **chlorophyll**.
2. **Nonvascular** plants don't have tubelike cells in their stems and leaves.
3. A small nonvascular plant that has both stems and leaves but no roots is a **moss**.
4. An **egg** is a female reproductive cell.
5. **Vascular plants** have tubelike cells in their roots, stems, and leaves.
6. A **conifer** produces seeds in cones.
7. A male reproductive cell is a **sperm**.
8. A **flowering plant** produces seeds inside a flower.
9. A **fern** is a vascular plant that reproduces with spores.
10. **Phloem** cells carry food from the leaves to all parts of the plant.
11. **Pollen** are the tiny grains of seed plants in which sperm develop.
12. **Flowers** are the reproductive parts of flowering plants.
13. A **seed** contains a new, young plant and stored food.
14. The young plant in a seed is the **embryo**.
15. **Xylem** cells carry water from the roots to the leaves.
16. In **sexual reproduction**, a new organism forms from the union of two reproductive cells.
17. The joining of the egg and the sperm is **fertilization**.
18. **Photosynthesis** is the process in which plants use water, carbon dioxide, and energy from the sun to make food.

# TEACH

## Guided Practice

Have students write down their answers to the margin question: for food and oxygen.

## Independent Practice

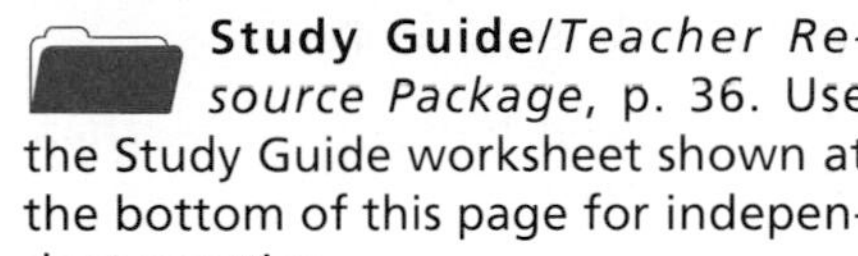

**Study Guide**/*Teacher Resource Package*, p. 36. Use the Study Guide worksheet shown at the bottom of this page for independent practice.

## Check for Understanding

Have students respond to the first three questions in Check Your Understanding.

## Reteach

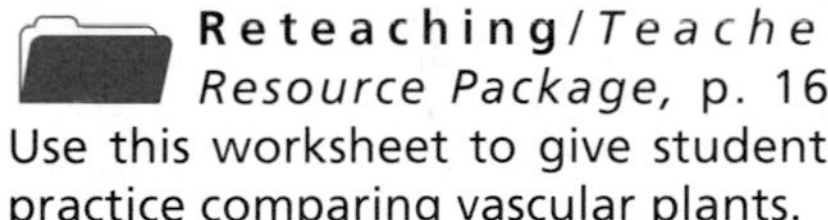

**Reteaching**/*Teacher Resource Package*, p. 16. Use this worksheet to give students practice comparing vascular plants.

**Extension:** Assign Critical Thinking, Biology and Reading, or some of the **OPTIONS** available with this lesson.

# 3 APPLY

## Connection: Language Arts

Make an illustrated booklet of tree leaves. Use both conifers and flowering plants. Have students look up simple and compound leaves and venation patterns.

# 4 CLOSE

**Cooperative Learning:** Have students work in groups of four to make a list of uses of plants.

## Answers to Check Your Understanding

11. pines, spruce, and fir trees
12. Sperm form in the male cone and male part of the flower; eggs form in the female cone and female part of the flower.
13. They are used as decoration and food, and they produce oxygen.
14. Answers will vary, but should show that carbon dioxide and oxygen are equally important.
15. ferns, conifers, and flowering plants

# CHAPTER 6 REVIEW

## Summary

Summary statements can be used by students to review the major concepts of the chapter.

## Key Science Words

All boldfaced terms from the chapter are listed.

## Testing Yourself

### Using Words

1. chlorophyll
2. fertilization
3. vascular plant
4. sexual reproduction
5. egg
6. seed
7. conifer
8. flowering plant
9. pollen

### Finding Main Ideas

10. p. 115; Water and food are carried in tubelike cells.
11. p. 114; Plants use sunlight, carbon dioxide, and water to make food.
12. p. 121; Ferns reproduce by spores.
13. pp. 115-116; Plants are grouped into vascular and nonvascular plants by whether or not they have tubelike cells to carry food and water.
14. p. 117; Mosses have no roots, but have stems and leaves that grow upright in mats; they look like short-cropped hair.
15. p. 129; They are used for decoration and food, and they give off oxygen.

### Using Main Ideas

16. Conifers produce three-fourths of the world's lumber, almost all of the world's paper, provide food and shelter for animals, and provide turpentine, resins, and fuel.
17. Mosses reproduce sexually by forming eggs and sperm. They also reproduce asexually by spores.
18. They need a constant supply of water to survive, and for sexual reproduction.



# Chapter 6

# Review

## Summary

### 6:1 Plant Classification

1. Most plants have chlorophyll in chloroplasts that traps light energy for photosynthesis. All plants have stiff cell walls.
2. Nonvascular plants do not have tubelike cells. Vascular plants have tubelike cells that carry food and water.

### 6:2 Nonvascular Plants

3. Mosses and liverworts are nonvascular plants. They have stems and leaves but no roots.
4. Mosses reproduce by sperm and eggs formed at the tips of leafy stems. The fertilized eggs develop into a stalk with a capsule. Spores are produced in a capsule. New moss plants grow from the spores.
5. Mosses are food for animals. They help form soil and keep it moist. Mosses are used in gardens.

### 6:3 Vascular Plants

6. Ferns, conifers, and flowering plants all have xylem and phloem cells. Conifers have seeds inside cones. Flowering plants have flowers and seeds.
7. Ferns reproduce by eggs and sperm and by spores on the leaves. Conifers produce sperm in pollen. Pollen and eggs are formed in cones. Flowering plants produce sperm in pollen and eggs in flowers.
8. Conifers are used for lumber and paper. All living things depend on plants for food and oxygen.

## Key Science Words

chlorophyll (p. 114)
conifer (p. 124)
egg (p. 118)
embryo (p. 124)
fern (p. 121)
fertilization (p. 118)
flower (p. 128)
flowering plant (p. 128)
moss (p. 117)
nonvascular plant (p. 116)
phloem (p. 121)
photosynthesis (p. 114)
pollen (p. 125)
seed (p. 124)
sexual reproduction (p. 118)
sperm (p. 118)
vascular plant (p. 116)
xylem (p. 121)

## Testing Yourself

### Using Words

*Choose the word from the list of Key Science Words that best fits the definition.*

1. chemical that gives plants their green color
2. when a sperm and an egg cell join
3. kind of plant with tubelike cells to carry water and food
4. type of reproduction when a sperm fertilizes an egg
5. female sex cell

# Review

## Testing Yourself *continued*

6. a structure that contains a small, new plant and a supply of stored food
7. plant that grows from the seeds of woody cones
8. vascular plant that produces seeds inside a flower
9. powderlike structures in seed plants in which sperm develop

### Finding Main Ideas

*List the page number where each main idea below is found. Then, explain each main idea.*

10. what's carried in tubelike cells
11. what plants use to make food
12. how ferns reproduce
13. how plants are grouped
14. what mosses look like
15. how flowering plants are important to other organisms

### Using Main Ideas

*Answer the questions by referring to the page number after each question.*

16. What are several ways that conifers are important? (p. 126)
17. How do mosses reproduce? (p. 118)
18. Why do mosses and liverworts grow in wet or damp places? (p. 117)
19. What are two reasons you may not find flowers on flowering plants? (p. 128)
20. How do nonvascular plants survive without xylem and phloem? (p. 116)
21. What is one way of grouping vascular plants? (pp. 121, 124)
22. What are the uses of mosses? (p. 118)
23. How are plants different from animals? (p. 114)
24. How are spores different from pollen? (pp. 120, 125)

## Skill Review 

*For more help, refer to the **Skill Handbook**, pages 704-719.*

1. **Infer:** Why are the spores of a fern more likely to be carried farther by wind than the spores of a moss?
2. **Observe:** Place a flower on a sheet of white paper. Carefully take the flower apart and compare it with the one in Figure 6-14.
3. **Interpret diagrams:** Look at the diagram in Figure 6-10. Why must pollen be formed before fertilization?
4. **Make and use tables:** Make a table to show the importance of mosses and liverworts, ferns, conifers, and flowering plants.

## Finding Out More

TECH PREP

### Critical Thinking

1. How do you think certain medicines from plants were discovered?
2. Compare the methods of reproduction between spore plants and flowering plants.

### Applications

1. In an old aquarium or glass jar, set up a terrarium with mosses, liverworts, and small ferns.
2. Report on common poisonous plants. Explain what part of the plant is poisonous, the effects of the poison, and whether there are any known antidotes.



### Using Main Ideas

19. Some flowers are small and not showy, some are hard to see, and some plants produce flowers only at certain times of the year.
20. They live close to the ground and take in water by osmosis through hairlike cells.
21. One way of grouping vascular plants is by whether they are flowering or nonflowering.
22. Mosses are food for some animals, they hold soil in place, and they help break up rock to form soil.
23. Plants can make their own food, but animals can't; plants can't move around, but animals can; plants have stiff cell walls, but animals don't.
24. Spores are formed in capsules and form into new plants. Pollen are grains of seed plants. Sperm develop in pollen and then fertilize egg cells, which can then develop into new plants.

## Skill Review

1. Mosses grow very close to the ground, but ferns grow tall.
2. Students should be able to see petals and, depending on the type of flower, male and female reproductive parts.
3. Sperm develop in pollen and the sperm fertilize egg cells.
4. Use Mosses and Liverworts, Ferns, Conifers, and Flowering Plants as heads and list the importance of each in columns underneath.

## Finding Out More

### Critical Thinking

1. Answers will vary, but most students will agree that trial-and-error methods determined that some plants had medicinal value before more scientific methods were developed.
2. Both use sexual reproduction. In spore plants, the fertilized eggs develop into a stalk with a capsule where spores develop. The fertilized eggs in flowering plants produce seeds protected in a fruit.

### Applications

1. A five-gallon aquarium is ideal. Plants can be collected from wooded areas, provided they are not protected.
2. Common poisonous plants include the nightshades, yew berries, poinsettia, philodendron, poison ivy, and poison oak.

 **Chapter Review**/*Teacher Resource Package*, pp. 28-29.

 **Chapter Test**/*Teacher Resource Package*, pp. 30-32.

**Chapter Test**/*Computer Test Bank*

# Simple Animals

## PLANNING GUIDE

| CONTENT | TEXT FEATURES | TEACHER RESOURCE PACKAGE | OTHER COMPONENTS |
|---|---|---|---|
| (1 1/2 days)<br>7:1 Animal Classification<br>Traits of Animals<br>How Animals Are Classified | Skill Check: *Understand Science Words,* p. 134<br>Skill Check: *Interpret Diagrams,* p. 135<br>Check Your Understanding, p. 136 | Reteaching: *Animal Classification,* p. 17<br>Skill: *Interpret Diagrams,* p. 7<br>Focus: *Finding Relationships Among Simple Animals,* p. 13<br>Study Guide: *Animal Classification,* p. 37 | **STVS:** *Zooplankton,* Animals (Disc 5, Side 1) |
| (1 1/2 days)<br>7:2 Sponges and Stinging-cell Animals<br>Sponges<br>Stinging-cell Animals | Mini Lab: *How Does a Sea Anemone Attach Itself?* p. 140<br>Lab 7-1: *Stinging-cell Animals,* p. 141<br>Check Your Understanding, p. 140 | Reteaching: *Sponges and Stinging-cell Animals,* p. 18<br>Study Guide: *Sponges and Stinging-cell Animals,* p. 38<br>Lab 7-1: *Stinging-cell Animals,* pp. 25-26 | **Laboratory Manual:**<br>*What Traits Does a Sponge Have for Living in Water?* p. 49<br>**Laboratory Manual:**<br>*What Are the Parts of a Stinging-cell Animal?* p. 57 |
| (2 days)<br>7:3 Worms<br>Flatworms<br>Roundworms<br>Segmented Worms | Skill Check: *Infer,* p. 143<br>Mini Lab: *Where Do You Find Roundworms?* p. 145<br>Idea Map, p. 147<br>Check Your Understanding, p. 148 | Application: *Tapeworm Life Cycle,* p. 7<br>Enrichment: *Leeches,* p. 7<br>Reteaching: *Anatomy of Worms,* p. 19<br>Study Guide: *Worms,* pp. 39-40<br>Study Guide: *Vocabulary,* p. 42 | Color Transparency 7a: *Life Cycle of a Pork Tapeworm*<br>**STVS:** *Nematodes,* Animals (Disc 5, Side 1)<br>**STVS:** *Leeches,* Animals (Disc 5, Side 1) |
| (1 day)<br>7:4 Soft-bodied Animals<br>Traits of Soft-bodied Animals<br>Classes of Soft-bodied Animals | Check Your Understanding, p. 151<br>Applying Technology: *These Zebras Live Underwater,* p. 152<br>Lab 7-2: *Earthworms,* p. 148 | Reteaching: *Classes of Soft-bodied Animals,* p. 20<br>Critical Thinking/Problem Solving: *What Is Destroying the Tomato Plants?* p. 7<br>Transparency Master: *Simple Animals,* p. 31<br>Study Guide: *Soft-bodied Animals,* p. 41<br>Lab 7-2: *Earthworms,* pp. 27-28 | **Laboratory Manual:**<br>*What Are the Parts of a Squid?* p. 53<br>Color Transparency 7b: *How a Squid Moves*<br>**STVS:** *Conch Farming,* Animals (Disc 5, Side 1) |
| Chapter Review | Summary<br>Key Science Words<br>Testing Yourself<br>Finding Main Ideas<br>Using Main Ideas<br>Skill Review | **ASSESSMENT RESOURCES**<br>Chapter Review, pp. 33-34<br>Chapter Test, pp. 35-37<br>Performance Assessment in the Biology Classroom<br>Alternate Assessment in the Science Classroom<br>Computer Test Bank | |

## GLENCOE TECHNOLOGY

**Infinite Voyage,** *To the Edge of the Earth*
**Science and Technology Videodisc Series,** *Zooplankton,* Animals (Disc 5, Side 1)
*Nematodes,* Animals (Disc 5, Side 1)
*Leeches,* Animals (Disc 5, Side 1)
*Conch Farming,* Animals (Disc 5, Side 1)

## MATERIALS NEEDED

| LAB 7-1, p. 141 | LAB 7-2, p. 148 | APPLYING TECHNOLOGY, p. 152 | MARGIN FEATURES |
|---|---|---|---|
| stereomicroscope<br>hand lens<br>dropper<br>small dish<br>flashlight<br>hydra culture<br>daphnia or brine shrimp | hand lens<br>shallow pan<br>flashlight<br>toothpick<br>cotton swab<br>paper towels<br>beaker of water<br>vinegar<br>live earthworms | mussels<br>prepared slide of mussel veliger<br>microscope | Skill Check, p. 134<br>dictionary<br>Skill Check, p. 135<br>pencil<br>paper<br>Mini Lab, p. 140<br>small rock<br>suction cup<br>Mini Lab, p. 145<br>funnel<br>wire mesh<br>jar<br>alcohol<br>moist soil |

## OBJECTIVES

For more information about National Science Standards, see page 5T.

| | | CORRELATION of QUESTIONS to OBJECTIVES | | | |
|---|---|---|---|---|---|
| SECTION | OBJECTIVE | CHECK YOUR UNDERSTANDING | CHAPTER REVIEW | TRP CHAPTER REVIEW | TRP CHAPTER TEST |
| 7:1 National Science Stds: UCP.1, UCP.2, UCP.5, C.4, C.5, C.6, G.1 | 1. **List** four traits of animals. | 1, 4 | 26, 30 | 6, 9, 27, 28, 35 | 1, 2, 3 |
| | 2. **Identify** nine major phyla of animals and give an example of each. | 2 | 9, 13, 15 | 3, 15, 32 | 5, 6, 7, 17, 28 |
| 7:2 National Science Stds: UCP.1, UCP.2, UCP.5, C.1, C.4, C.6 | 3. **List** the traits of sponges. | 6, 10 | 3, 19, 25 | 2 | 10, 14, 21, 23, 24, 26, 31 |
| | 4. **Describe** the traits of stinging-cell animals. | 7, 8 | 5, 17, 18, 27 | 4, 11, 30 | 8, 11, 12, 15, 19, 22 |
| 7:3 National Science Stds: UCP.1, UCP.2, UCP.5, C.4, C.6, F.1 | 5. **List** the three main phyla of worms and give an example of each. | 11, 12 | 1, 7, 12, 20, 33 | 8, 12, 13, 14, 29, 33 | 9, 34 |
| | 6. **Identify** the main features of flatworms, round worms, and segmented worms. | 13, 15 | 2, 8, 10, 11, 16, 22, 24, 28, 29 | 1, 5, 7, 16, 17, 18, 19, 20, 21, 22, 23, 24, 25, 26 | 13, 36, 37, 38, 39, 40, 41, 42, 43, 44, 45, 46, 47, 48, 49, 50, 51, 52, 53 |
| 7:4 National Science Stds: UCP.1, UCP.2, UCP.5, C.6 | 7. **Identify** the major features of soft-bodied animals. | 17, 18 | 6, 21, 23, 32 | 31, 34 | 16, 18, 20, 25, 27, 29, 30, 32, 33, 35 |
| | 8. **Explain** how soft-bodied animals are classified. | | 14 | 10 | 4 |

# Simple Animals

## CHAPTER OVERVIEW

### Key Concepts

This chapter serves as an introduction to the animal kingdom. The traits of the animals in the first six phyla (sponges, stinging-cell animals, flatworms, roundworms, segmented worms, and soft-bodied animals) are examined. Students should see that animals increase in complexity from the first phylum on.

### Key Science Words

anus
cyst
flatworm
hookworm
invertebrate
mantle
planarian
pore
roundworm
segmented worm
soft-bodied animal
sponge
stinging-cell animal
symmetry
tapeworm
tentacle
vertebrate

### Skill Development

In Lab 7-1, students will **observe, make and use tables,** and **interpret data** dealing with the traits of stinging-cell animals. In Lab 7-2, they will **observe, infer, predict,** and **make and use tables** in relation to the traits of earthworms. In the Skill Check on page 134, they will **understand** the **science word** *invertebrate.* In the Skill Check on page 135, students will **interpret diagrams** of symmetry. In the Skill Check on page 143, they will **infer** why tapeworms affect a person's general health. In the Mini Lab on page 140, students will **infer** how sea anemones attach themselves to rocks. In the Mini Lab on page 145, they will **observe** roundworms.

### Bridging

In Chapter 3, students learned how the classification scheme for all life forms is organized. Traits of animals were introduced in a very general way. This chapter expands on these traits and gives the student an opportunity to look at various phyla in detail.

**CHAPTER PREVIEW**

### Chapter Content

Review this outline for Chapter 7 before you read the chapter.



### Skills in this Chapter

The skills that you will use in this chapter are listed below.

- In **Lab 7-1,** you will observe, make and use tables, and interpret data. In **Lab 7-2,** you will observe, infer, predict, and make and use tables.
- In the **Skill Checks,** you will understand science words, interpret diagrams, and infer.
- In the **Mini Labs,** you will infer and observe.

132

## TECH PREP

For Tech Prep activities in this chapter of the Teacher Wraparound Edition, see especially the Motivation on page 137, the Portfolios on pages 138 and 146, and the Activity on page 147.

See also the Glencoe Homepage at **www.glencoe.com**

Chapter 7

# Simple Animals

In the photo on the opposite page, you can see some of the beautiful and unusual animals that are found in the sea. For many years people thought that sponges were plants. The sponges didn't move around by themselves and they didn't seem to capture food, both animal-like traits. Can you tell from the photo how a sponge captures its food? Early scientists couldn't either.

The small photo shows the mouth of a coral. It is surrounded by structures that contain the stinging cells the animal uses to capture food. With careful observation, scientists were able to watch such animal-like traits as food-getting. Today there is no doubt that these organisms are animals.

**Try This!**

**How does a squid move?** Inflate a balloon and release it. Observe the movement and direction of the balloon. Compare it with the movement of a squid.

**interNET CONNECTION**

For more information about the material in this chapter, follow the link for the chapter on the Glencoe Homepage at **http://www.glencoe.com**

The mouth of a coral is surrounded by tentacles.



**ASSESSMENT PLANNER**

***Portfolio***

Strategies on the following pages represent student products that can be placed into a best-work portfolio: pp. 134, 138, 146.

**PERFORMANCE ASSESSMENT**

Skill Check, pp. 135, 143

Mini Lab, pp. 140, 145

Lab, pp. 141, 148

**CONTENT ASSESSMENT**

Check for Understanding, pp. 136, 139, 146, 147, 150, 151

Chapter Review, pp.154-155

**GROUP ASSESSMENT**

Opportunities for group assessment occur with Cooperative Learning Strategies.

***Student Journal***

Strategies on the following pages represent opportunities for writing in a Student Journal: pp. 134, 146, 150.

**GETTING STARTED**

## Using the Photos

In looking at the photos, students should note the obvious traits of sponges. Have them discuss real sponges that they may have seen or used. They should try to speculate on how a sponge gets its food.

## MOTIVATION/Try This!

**How does a squid move?** Students will **observe** that air leaving the balloon forces it to move in a direction opposite escaping air. A squid moves in a direction opposite escaping water. It squirts water and uses this force to move either forward or backward.

## Chapter Preview

Have students study the chapter outline before they begin to read the chapter. They should note the overall organization of the chapter, that the general traits of animals and how they are classified is covered first. Next, the phyla making up sponges and stinging-cell animals are examined. Three different phyla commonly called worms are studied next. Finally, traits and the different classes of soft-bodied animals are studied.

## Misconception

Students usually believe that animals having the largest number of different species would be those most familiar to them. This would include vertebrates and, specifically, mammals. However, invertebrates far outnumber the vertebrates. They simply are less obvious to humans because of either their small size or the fact that many live in water habitats.

# 7:1 Animal Classification

## PREPARATION

### Materials Needed

Make copies of the Focus, Study Guide, Skill, and Reteaching worksheets in the *Teacher Resource Package.*

▶ Obtain slides for the Focus motivation.

### Key Science Words

symmetry
vertebrate
invertebrate

### Process Skills

In the Skill Checks, students will interpret diagrams and understand science words.

## 1 FOCUS

▶ The objectives are listed on the student page. Remind students to preview these objectives as a guide to this numbered section.

### MOTIVATION/Demonstration

Obtain prepared slides for all animal phyla. Ask students to group the samples using size, color, where found, and harmful or helpful .

**Focus**/*Teacher Resource Package,* p. 13. Use the Focus transparency shown at the bottom of this page as an introduction to relations among simple animals.

**Student Journal**

Have students make a list of five similarities and five differences between plants and animals.

**Portfolio**

Have students use the worksheet on page 7 of the *Skill* booklet to interpret diagrams of animals with symmetry.

**Objectives**

1. **List** four traits of animals.
2. **Identify** nine major phyla of animals and give an example of each.

**Key Science Words**

vertebrate
invertebrate
symmetry

# 7:1 Animal Classification

Scientists have discovered and classified about 1 1/2 million different kinds of living things. Of this number, about one million are animals. The different kinds of animals are grouped according to traits they have in common.

## Traits of Animals

What are some of the traits that make animals different from other living things? First, animals are consumers and can't make their own food. They must take in food from their surroundings. Some animals eat plants, some eat other animals, and others eat both plants and animals. Most animals digest and store food in their bodies.

Second, most animals can move from place to place. Moving around helps them find food. Animals that don't move have developed other ways to get food.

Third, animals are multicellular organisms. The cells of most animals are organized into tissues and organs that form systems. Many animals have nervous systems, digestive systems, and reproductive systems. Nearly all animals have muscles and systems for getting rid of wastes.

All animals belong to one of two groups—the vertebrates and the invertebrates (in VERT uh brayts). **Vertebrates** are animals with backbones. People and the animals most closely related to us, such as fish, amphibians, reptiles, birds, and other mammals, are vertebrates. **Invertebrates** are animals without backbones. Worms and insects are examples of invertebrates.

Finally, most animals have some kind of symmetry (SIH muh tree). **Symmetry** is the balanced arrangement of body parts around a center point or along a center line. Only a few very simple animals do not show symmetry. These animals grow in a variety of shapes. The sponge in Figure 7-1a does not show symmetry.

Some invertebrates show radial (RAYD ee ul) symmetry. In radial symmetry, the body parts are arranged in a circle around a center point. The sea anemone (uh NEM uh nee) in Figure 7-1b is an example of an animal that has radial symmetry.

All vertebrates and some invertebrates have bilateral (bi LAT uh rul) symmetry. In an animal with bilateral symmetry, the body can be divided lengthwise into two equal sides, a right side and a left side. The bee in Figure

**Skill Check**

**Understand science words: invertebrate.** The word part *in* means not. In your dictionary, find three words with the word part *in* in them. *For more help, refer to the **Skill Handbook**, pages 706-711.*



**FOCUS 13**

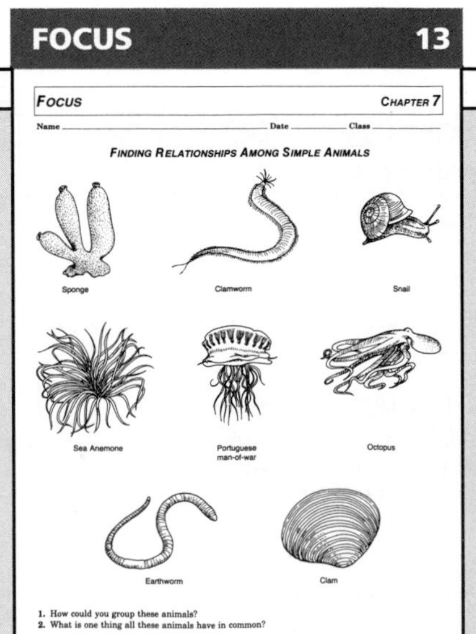


## OPTIONS

### Science Background

Phylum names used in this chapter are general terms. Technical terms used in taxonomy to describe the phyla are: Sponges - Porifera; Stinging-cell animals - Coelenterata or Cnidaria; Flatworms - Platyhelminthes; Roundworms - Nematoda; Segmented worms - Annelida; Soft-bodied animals - Mollusca; Jointed-leg animals - Arthropoda; Spiny-skin animals - Echinodermata; Chordates - Chordata.

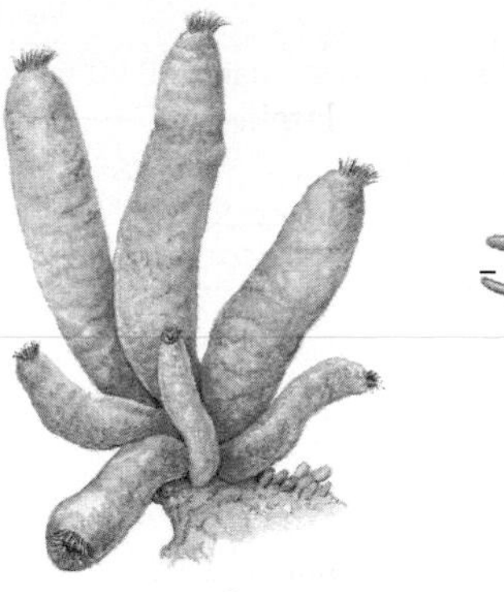

a

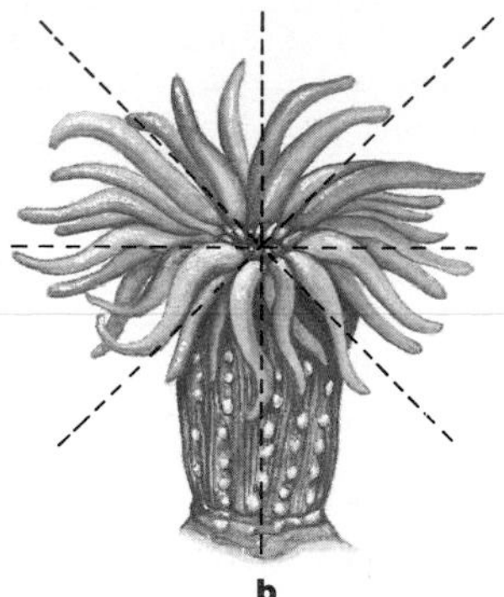

b

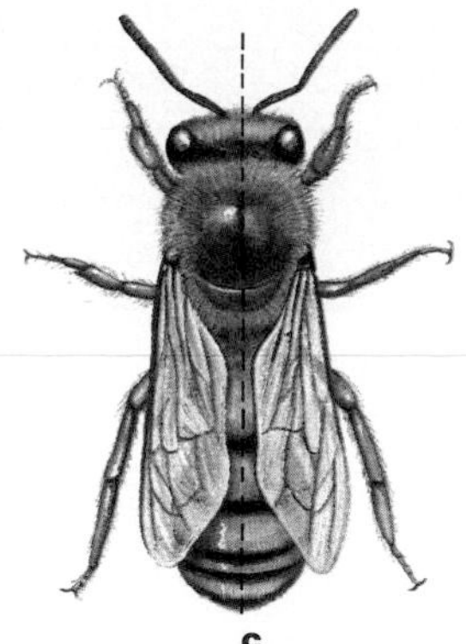

c

7-1c has bilateral symmetry. Notice that it has a head end and a back end. The head end studies where the animal is going and finds food. The bee also has an upper and a lower side. The lower side moves the animal along a surface.

Let's review the traits of animals.

1. Animals can't make food, but must catch and eat it.
2. Most animals can move from place to place.
3. Animals have many cells. The cells make up tissues and organs that form systems.
4. Most animals have symmetry.

**Figure 7-1** A sponge does not have symmetry (a). A sea anemone has radial symmetry (b). A bee has bilateral symmetry (c).

How are an animal's cells organized?

## How Animals Are Classified

Scientists group living things into one of five kingdoms. If an organism shows most of the traits just listed, then it is placed in the animal kingdom. If not, then it is placed in one of the other four kingdoms.

How do scientists know to which phylum an animal belongs? Consider the sponges. They have certain traits. If an animal shows these traits, it belongs in the phylum with sponges. If it does not, then it belongs in a different phylum.

Here is an example with which you are familiar. A new student arrives at your school. How does the office staff know whether the student should be placed in the ninth, tenth, eleventh, or twelfth grade? The student can be placed in a certain grade only if he or she has passed a certain number of courses. If the student has passed the number of courses required in the ninth grade, the student is put in the tenth grade. The number of courses passed is the "trait" used to place students into the proper grades.

### Skill Check

**Interpret diagrams:** Make a list of the invertebrates pictured in this chapter. Indicate the type of symmetry shown by each one. *For more help, refer to the* ***Skill Handbook****, pages 706-711.*



## 2 TEACH

### MOTIVATION/Brainstorming

Ask students to describe how animals differ from plants. Ask students how plants and animals are alike.

### Concept Development

▶ Ask students to compare the number of plant species (350 000) with the number of animal species.

**Cooperative Learning:** Divide the class into nine groups. Provide each group with one sample animal. You might use pictures, plastic-embedded, preserved, etc. Each animal should be from a different phylum. Have each group compile a list of the traits that they can see. Have a spokesperson for each group report to the class on their findings.

### Guided Practice

Have students write down their answers to the margin question: cells make up tissues and organs that form systems.

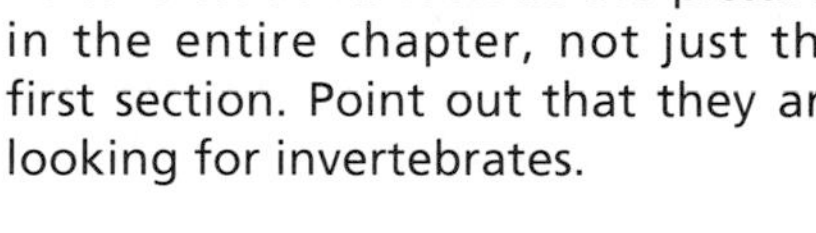

### Skill Check

Be sure students look at the pictures in the entire chapter, not just the first section. Point out that they are looking for invertebrates.

### Independent Practice

**Study Guide**/*Teacher Resource Package,* p. 37. Use the Study Guide worksheet shown at the bottom of this page for independent practice.

**SKILL 7**

*SKILL: INTERPRET DIAGRAMS: SYMMETRY* — *CHAPTER 7*

Name ________ Date ________ Class ________

*For more help, refer to the Skill Handbook, pages 706-711.* *Use after Section 7:1.*

*THE SYMMETRY OF ANIMALS*

*Review symmetry in Section 7:1 of your textbook. Then, decide whether each animal in the table has radial symmetry or bilateral symmetry. Write your answer in the second column. In the last column, explain how you decided the type of symmetry the animal has. Then, answer the questions.*

| Animal | Symmetry | Reason |
|---|---|---|
| | radial | has body parts around a central point |
| | bilateral | has two equal sides, a right side and a left side |
| | radial | has body parts around central point |
| | bilateral | has two equal sides, a right side and a left side |
| | bilateral | has two equal sides, a right side and a left side |
| | bilateral | has two equal sides, a right side and a left side |

1. What are the advantages of bilateral symmetry? **An animal can move in a specific direction.**
2. What are the advantages of radial symmetry? **An animal may be able to sense food or the approach of other animals from all sides.**

**STUDY GUIDE 37**

*STUDY GUIDE* — *CHAPTER 7*

Name ________ Date ________ Class ________

*ANIMAL CLASSIFICATION*

*In your textbook, read about how animals are classified in Section 7:1.*

1. What is the difference between vertebrates and invertebrates? **Vertebrates have a backbone, invertebrates do not.**
2. What is symmetry? **Symmetry is the balanced arrangement of body parts around a center point or along a center line.**
3. What is the difference between radial and bilateral symmetry? **In radial symmetry, the body parts are arranged in a circle around a center point. In bilateral symmetry, the body can be divided lengthwise into two equal sides, a right side and a left side.**
4. List four traits of animals.
   a. **Animals can't make their own food.**
   b. **Most animals can move from place to place.**
   c. **Animals have many cells.**
   d. **Most animals have symmetry.**
5. The animal kingdom is one of five kingdoms into which scientists group living things. Scientists group animals into nine major groups.
   a. What are these groups called? **phyla**
   b. Which group contains the largest number of different kinds of animals? **animals with jointed legs**
   c. Which group contains the smallest number of different kinds of animals? **sponges**
   d. Which group contains the simplest animals? **sponges**
   e. Which group contains the most complex animals? **chordates**
6. What main group of animals makes up the chordates phylum? **vertebrates**
7. What main group of animals makes up the rest of the animal kingdom? **invertebrates**

## OPTIONS

### ACTIVITY/Project

Have students prepare posters with pictures of animals. Have them divide the animals into vertebrates and invertebrates. There should be at least 20 examples in each category.

**Skill**/*Teacher Resource Package,* p. 7. Use the Skill worksheet shown here to have students interpret diagrams of animals with symmetry.

## TEACH

### Guided Practice

Have students write down their answers to the margin question: sponges.

### Check for Understanding

Have students respond to the first three questions in Check Your Understanding.

### Reteach

**Reteaching**/*Teacher Resource Package*, p. 17. Use this worksheet to give students additional practice in animal classification.

**Extension:** Assign Critical Thinking, Biology and Reading, or some of the **OPTIONS** available with this lesson.

## APPLY

### Summary

Have students discuss animal phyla as being either vertebrates or invertebrates.

## CLOSE

### Convergent Question

Ask students why scientists classify animals.

### Answers to Check Your Understanding

1. Animals are consumers, can move about, are multicellular, have symmetry, and are vertebrates or invertebrates.
2. jointed-leg animals; insects, crayfish
3. It can be divided into two equal parts on any plane.
4. It can't be classified as an animal because it makes its own food.
5. An amoeba has no symmetry line; a starfish will have lines from a center point on the body through each part; the person will have a line running lengthwise down the middle.

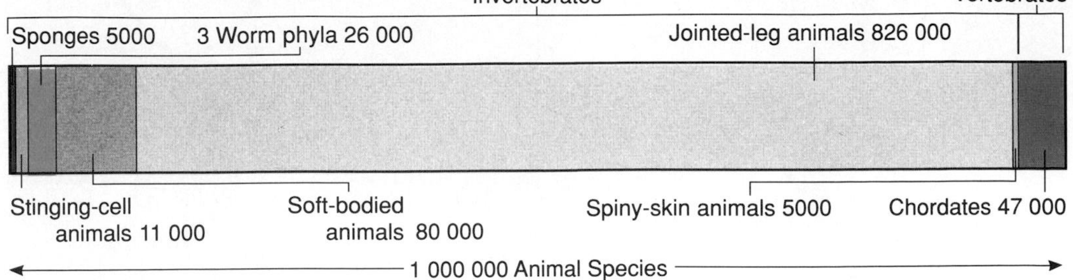


**Figure 7-2** Animals are placed in nine major phyla. Which phylum has the most complex animals? chordates

**Which phylum contains the simplest animals?**

The nine major groups into which animals are placed are shown in Figure 7-2. Each major group is a phylum. Also shown is the number of animal species in each phylum.

The figure shows that some phyla include many different kinds of animals. Other phyla are small. Animals with jointed legs make up the largest phylum. Insects, spiders, and crayfish belong to this phylum. Sponges make up the smallest of the phyla shown.

Figure 7-2 also shows the simplest phyla at the left and the most complex at the right. Sponges are the simplest of all animals. Chordates (KOR dayts) are the most complex. Most chordates are vertebrates. Eight of the nine phyla shown are grouped together as invertebrates. You can see from Figure 7-2 that there are many more invertebrates than vertebrates on Earth.

### Check Your Understanding

1. List four traits of animals.
2. Which animal phylum is the largest? Give two examples of members of this phylum.
3. How can an animal body be divided if it has radial symmetry?
4. **Critical Thinking:** An astronaut discovers an organism on a distant planet. It can move around, is multicellular, has bilateral symmetry, and makes its own food. Can this new organism be classified as an animal? Explain.
5. **Biology and Reading:** An amoeba has no symmetry. A starfish has radial symmetry. A human has bilateral symmetry. Draw an amoeba, a starfish, and a person. Draw the line or lines of symmetry for each.



## GLENCOE TECHNOLOGY

**Videodisc**

**The Secret of Life**

*Evolution of Symmetry*

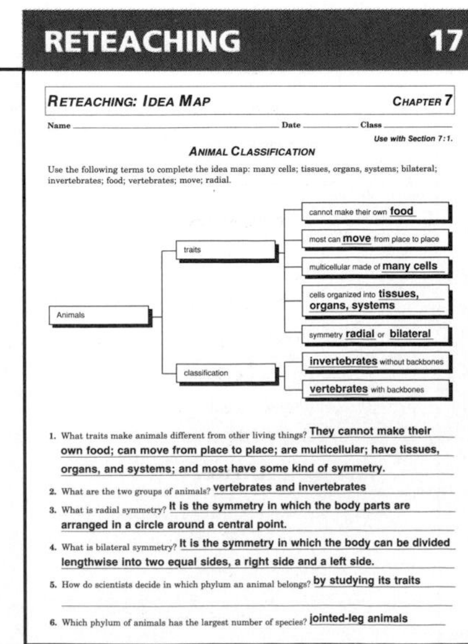

RETEACHING 17

RETEACHING: IDEA MAP — CHAPTER 7

Name ____ Date ____ Class ____

Use with Section 7:1.

ANIMAL CLASSIFICATION

Use the following terms to complete the idea map: many cells; tissues, organs, systems; bilateral; invertebrates; food; vertebrates; move; radial.

Animals
- traits
  - cannot make their own **food**
  - most can **move** from place to place
  - multicellular made of **many cells**
  - cells organized into **tissues, organs, systems**
  - symmetry **radial** or **bilateral**
- classification
  - **invertebrates** without backbones
  - **vertebrates** with backbones

1. What traits make animals different from other living things? **They cannot make their own food; can move from place to place; are multicellular; have tissues, organs, and systems; and most have some kind of symmetry.**
2. What are the two groups of animals? **vertebrates and invertebrates**
3. What is radial symmetry? **It is the symmetry in which the body parts are arranged in a circle around a central point.**
4. What is bilateral symmetry? **It is the symmetry in which the body can be divided lengthwise into two equal sides, a right side and a left side.**
5. How do scientists decide in which phylum an animal belongs? **by studying its traits**
6. Which phylum of animals has the largest number of species? **jointed-leg animals**

# 7:2 Sponges and Stinging-cell Animals

Sponges and stinging-cell animals do not have backbones. They are invertebrates. Sponges are the simplest invertebrates.

**Objectives**

3. **List** the traits of sponges.
4. **Describe** the traits of stinging-cell animals.

**Key Science Words**

sponge
pore
stinging-cell animal
tentacle

## Sponges

**Sponges** are simple invertebrates that have pores. A **pore** is a small opening. You may know a sponge as an object that you use when you take a bath or wash cars. Look at the sponges in Figure 7-3. Some look like small trees. Others look like vases. Most sponges have no definite shape. You will also notice that they come in many colors.

What are the traits of sponges? Unlike most other animals, sponges do not move about freely on their own. Most live attached to rocks in shallow oceans. Some sponges live in fresh water. The body of a sponge has many pores. Water enters through these pores and leaves through an opening in the top center of the sponge's body. Many canals carry the water throughout the sponge's body.

**How is water carried through the walls of a sponge's body?**

Since sponges can't move, how do they get food? All sponges must live in water. Water has small organisms in it that sponges can use for food. The pores in the sponge allow water and any food in the water to flow into the animal. Once inside, the food is trapped by food-getting cells.

**Figure 7-3** Sponges come in a variety of sizes, shapes, and colors.



## OPTIONS

### Science Background

There are three classes of sponges based on their skeleton composition. These are: Class Calcarea, skeleton of calcium carbonate spicules; Class Hexactinellida, skeleton of glass; Class Demospongiae, skeleton of protein called spongin.

There are three classes of stinging–cell animals. These are: Class Hydrozoa, hydroids such as Hydra and Portuguese man-of-war; Class Scyphozoa, jellyfish; Class Anthozoa, sea anemones and coral.

# 7:2 Sponges and Stinging-cell Animals

## PREPARATION

### Materials Needed

Make copies of the Study Guide, Reteaching, and Lab worksheets in the *Teacher Resource Package.*

▶ Have on hand a natural sponge and a plastic sponge for the Focus motivation.

▶ Obtain hydra culture, daphnia or brine shrimp, and flashlights for Lab 7-1.

▶ Obtain a suction cup and small rock for the Mini Lab.

### Key Science Words

sponge
pore
stinging-cell animal
tentacle

### Process Skills

In the Mini Lab, students will infer. In Lab 7-1, they will observe, make and use tables, and interpret data.

## 1 FOCUS

▶ The objectives are listed on the student page. Remind students to preview these objectives as a guide to this numbered section.

### MOTIVATION/Demonstration

Show students a piece of natural sponge and have them compare it to a commercial plastic type. Have them point out similarities and differences. You may want to prepare wet mounts of these two sponges for further comparison on a microscopic level.

### Guided Practice

Have students write down their answers to the margin question: through canals.

## 2 TEACH

### MOTIVATION/Bulletin Board

Arrange pictures of sponges, stinging–cell animals, worms, and mollusks on the bulletin board. Have students place the organisms under the appropriate phylum as the animals are studied. As the traits of the animals are discussed, look for the traits in the pictures.

### Concept Development

▶ Explain that sponges are filter feeders, organisms that get their food by filtering it out of the water around them.

▶ The mouth of the sponge is the osculum or excurrent pore.

▶ Explain that the special cells inside sponges for trapping food are collar cells. The collar cells have flagella like those of protists.

▶ The pores of the sponge are incurrent openings. Water is drawn into the sponge and circulated in the central cavity by the beating of the collar cells. Water passes out of the sponge through the osculum.

▶ Point out that most sponges that are now sold are synthetically made. Some places, such as Tarpon Springs, Florida, still catch live sponges for sale.

▶ The sponge is asymmetrical because it lacks symmetry.

**Portfolio**

Sponges in the ocean are becoming more and more scarce. Artificial sponges have been developed for human use. Have students find out how artificial sponges are made and what materials are used to make them. What are the ecological advantages of artificial sponges?

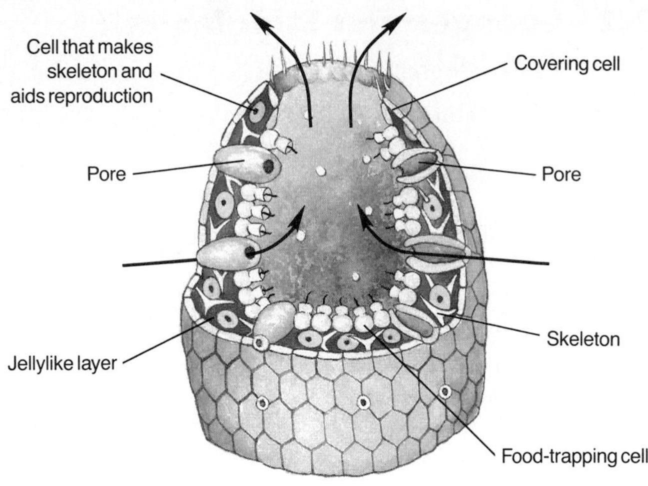


**Figure 7-4** Arrows show the direction in which water flows through a sponge. Sponges have three kinds of cells.

**Bio Tip**

**Consumer:** Most of the sponges sold today are artificial. They are made from cellulose or synthetic materials. Natural sponges hold much more water than artificial sponges, but they are not commonly used because they are expensive.

Sponges are only two cell layers thick, have no muscle or nerve cells, and have no tissues, organs, or organ systems. Sponges have three main kinds of cells. Each cell type has a certain job. One cell type traps food and moves water through the sponge by beating its flagellum. Other cells cover and protect the sponge. The third cell type produces chemicals that make the skeleton. This third cell type also aids in reproduction and carries food to other cells. The three kinds of cells are shown in Figure 7-4.

Sponges can reproduce both sexually and asexually. In sexual reproduction, the same sponge produces both sperm and eggs, but at different times. It takes two sponges for sexual reproduction to take place. Having both sperm and eggs ensures that sponges near each other can reproduce. Sperm cells produced by one sponge are carried to another sponge by the water. There, the sperm fertilize the eggs. New sponges usually develop on rocks on the ocean floor.

Sponges also reproduce asexually. Small pieces of a sponge's body may break off and form separate, new sponges. Sponges may also form buds. Each bud can develop into a new sponge.

You may have used a natural sponge for cleaning and washing because they hold a lot of water. The skeletons of sponges have many fibers. When the animal dies and decays, the fibers remain. The spaces between these fibers can be filled with water. Because the fibers are elastic, the water can be squeezed out.



### OPTIONS

**What Traits Does a Sponge Have for Living in Water?**/*Lab Manual,* pp. 49-52. Use this lab as an extension to studying sponges.

## Stinging-cell Animals

**Stinging-cell animals** are animals with stinging cells and hollow, saclike bodies that lack organs. Most of these animals live in the ocean. A few live in freshwater lakes and streams. Many stinging-cell animals are beautiful and delicate. They look very different from the animals you may see every day.

It is hard at first to compare hydras, corals, sea fans, jellyfish, and sea anemones with one another. Look at the animals in Figure 7-6. Animals in the stinging-cell animal phylum do not look much alike. Yet, all of them share certain traits.

Many stinging-cell animals have armlike parts called **tentacles** (TENT ih kulz) that surround the mouth. The jellyfish in Figure 7-6a has tentacles. If you were to look closely at corals and sea fans, you would see that these animals also have tentacles. A coral is really a group of many small animals. The same is true of sea fans. A hard structure on the outside of the animals supports and protects them and is made of chemicals produced by the animals. The chemicals harden into different shapes.

Figure 7-5 shows that stinging-cell animals have radial symmetry and saclike bodies made of two cell layers. A jellylike layer lies between the two cell layers. Inside the body of each animal is a body cavity. The cavity has one opening called the mouth. The only way into and out of the body is through the mouth.

Many stinging-cell animals fasten themselves to the ocean bottom or to rocks with a structure called a disc. They do not move from that place. How then do they get food? They catch it with their tentacles.

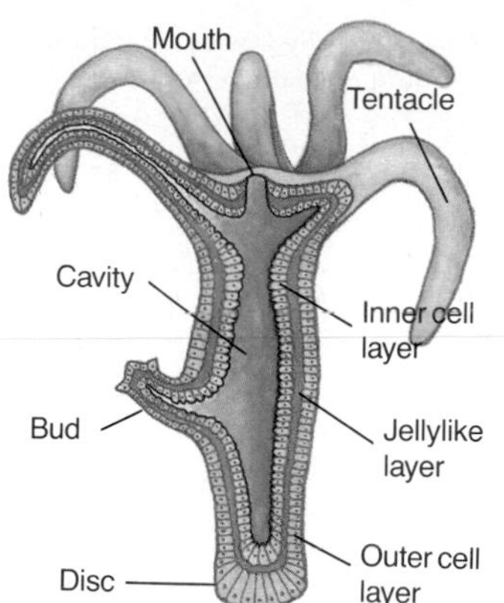


**Figure 7-5** Stinging-cell animals have sock-shaped bodies with two cell layers.

**How many cell layers does a stinging-cell animal's body contain?**

**Figure 7-6** Some of the many stinging-cell animals are a jellyfish (a), coral (b), and sea fan (c).

a

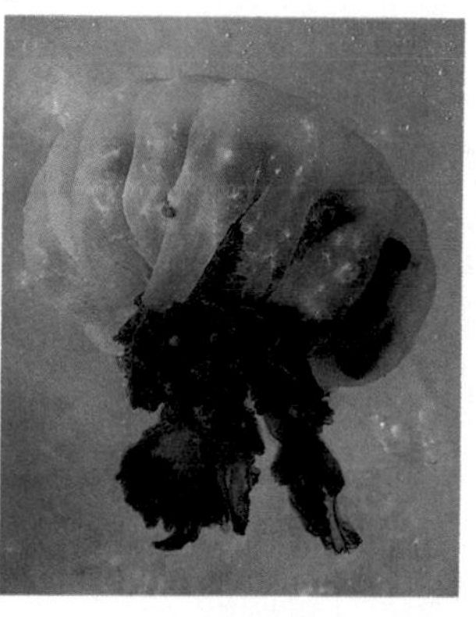

b

c

## TEACH

### Concept Development

▶ The stinging cells are called nematocysts and release a protein poison that has paralyzing action. Some sea anemones release only a sticky substance.

▶ Explain that common names for animals are often misleading. The jellyfish is not a fish.

▶ The mouth also serves as the anal opening through which wastes are extruded.

### Guided Practice

Have students write down their answers to the margin question: two.

### Check for Understanding

Have students list the headings Sponges and Stinging-cell animals. Have them list the traits of each under the proper headings.

### Reteach

Ask students to prepare a key that will enable them to determine if an animal is a sponge or a stinging-cell animal. Review the process of key construction and provide an example.

### Independent Practice

**Study Guide**/*Teacher Resource Package,* p. 38. Use the Study Guide worksheet shown at the bottom of this page for independent practice.

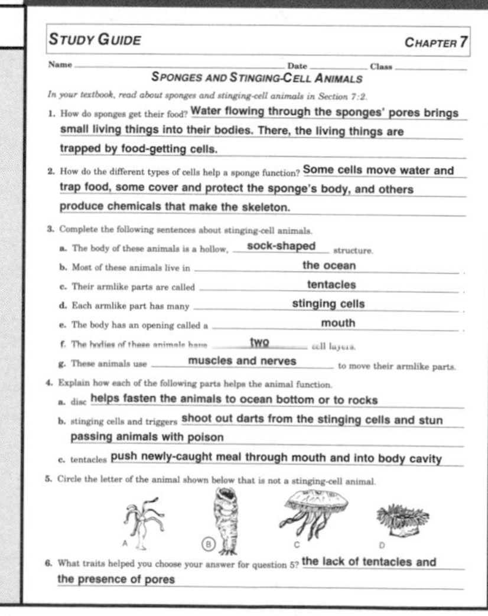

**STUDY GUIDE 38**

*STUDY GUIDE* — CHAPTER 7

Name ______ Date ______ Class ______

**SPONGES AND STINGING-CELL ANIMALS**

*In your textbook, read about sponges and stinging-cell animals in Section 7:2.*

1. How do sponges get their food? **Water flowing through the sponges' pores brings small living things into their bodies. There, the living things are trapped by food-getting cells.**
2. How do the different types of cells help a sponge function? **Some cells move water and trap food, some cover and protect the sponge's body, and others produce chemicals that make the skeleton.**
3. Complete the following sentences about stinging-cell animals.
   a. The body of these animals is a hollow, **sock-shaped** structure.
   b. Most of these animals live in **the ocean**.
   c. Their armlike parts are called **tentacles**.
   d. Each armlike part has many **stinging cells**.
   e. The body has an opening called a **mouth**.
   f. The bodies of these animals have **two** cell layers.
   g. These animals use **muscles and nerves** to move their armlike parts.
4. Explain how each of the following parts helps the animal function.
   a. disc **helps fasten the animals to ocean bottom or to rocks**
   b. stinging cells and triggers **shoot out darts from the stinging cells and stun passing animals with poison**
   c. tentacles **push newly-caught meal through mouth and into body cavity**
5. Circle the letter of the animal shown below that is not a stinging-cell animal.

   A (B) C D
6. What traits helped you choose your answer for question 5? **the lack of tentacles and the presence of pores**

## OPTIONS

### ACTIVITY/Challenge

Have some students set up a saltwater aquarium. Supplies may be purchased from a biological supply house or from aquarium stores.

**What Are the Parts of a Stinging-cell Animal?**/*Lab Manual,* pp. 57-59. Use this lab as an extension to studying stinging-cell animals.

**GLENCOE TECHNOLOGY**

**Videodisc**

**Biology: The Dynamics of Life**
Movie: *Ocean Cnidarians*
Disc 1, Side 2, Ch. 31

## TEACH

### Mini Lab

You may substitute a medicine dropper top (rubber part) and a small pebble for the suction cup and rock.

### ASSESSMENT

**Content:** Have students use the dictionary to find the meaning of the word *sessile.* Students should relate this term to the invertebrates in this section.

### Check for Understanding

Have students respond to the first three questions in Check Your Understanding.

### Reteach

**Reteaching**/*Teacher Resource Package,* p. 18. Use this worksheet to give students additional practice with sponges and stinging-cell animals.

**Extension:** Assign Critical Thinking, Biology and Reading, or some of the **OPTIONS** available with this lesson.

## 3 APPLY

### ACTIVITY/Interview

You are a sponge and are being interviewed. Explain what and how you eat, where you live, and what you do.

## 4 CLOSE

### Audiovisual

Show the filmstrip *Survey of the Animal Kingdom, Part I - Sponges, Anemones, Corals and Flatworms,* Educational Images Ltd.

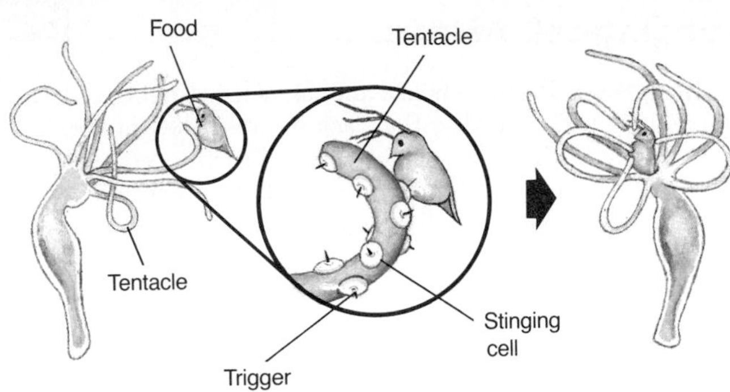


**Figure 7-7** Hydras catch food by stinging a small animal with a poison dart. The stunned animal is then stuffed into the hydra's mouth.

### Mini Lab

**How Does a Sea Anemone Attach Itself?**

**Infer:** Wet a suction cup and press it against a small rock. Pick the rock up by holding the suction cup. How are you able to lift the rock with the suction cup? *For more help, refer to the* ***Skill Handbook,*** *pages 706-711.*

The tentacles have thousands of stinging cells with hairlike triggers. When an animal brushes against the trigger, a stinging dart containing poison shoots into the water. The dart hits the animal and stuns it with the poison. The tentacles then push the newly caught meal through the mouth and into the body cavity, where it is digested. Undigested food leaves the animal through the mouth. Figure 7-7 shows a hydra with food it has caught.

Stinging cells of a jellyfish can cause a very painful sting in people. Anyone who lives near an ocean probably knows not to bother jellyfish that have washed onto shore.

Stinging-cell animals have muscle cells and nerve cells. Animals in this phylum use muscles and nerves to move their tentacles. Jellyfish use muscles to swim.

Stinging-cell animals reproduce sexually by forming eggs and sperm. These are released into the water, where the sperm fertilize the eggs. Some stinging-cell animals, like the hydra, can also reproduce asexually by forming buds. These buds break off and become separate animals.

### Check Your Understanding

6. List three traits of sponges.
7. How do stinging-cell animals get food?
8. List three traits of stinging-cell animals.
9. **Critical Thinking:** Why is radial symmetry useful to a stinging-cell animal?
10. **Biology and Reading:** Most of the sponges sold today are made from cellulose, a plant material. How do you think this kind of sponge got its name?



### Answers to Check Your Understanding

6. do not move on their own, full of pores, two cell layers thick
7. stun it with darts containing poison and capture it with tentacles
8. have tentacles, stinging-cells, and one opening
9. It allows them to catch food from any direction.
10. probably from the holes, or pores throughout the sponges

### RETEACHING 18

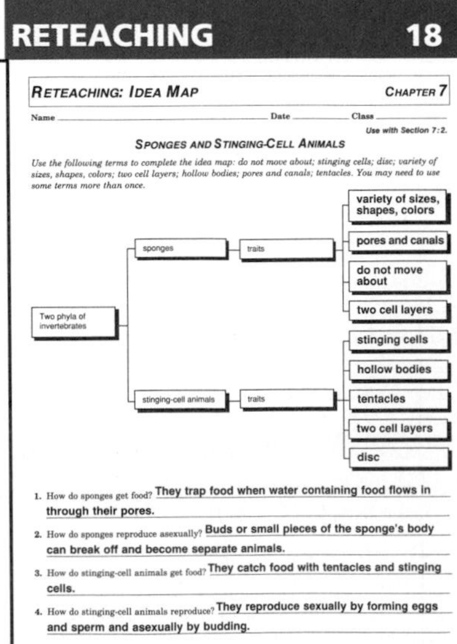

RETEACHING: IDEA MAP — CHAPTER 7

Name ______ Date ______ Class ______

Use with Section 7:2.

SPONGES AND STINGING-CELL ANIMALS

*Use the following terms to complete the idea map: do not move about; stinging cells; disc; variety of sizes, shapes, colors; two cell layers; hollow bodies; pores and canals; tentacles. You may need to use some terms more than once.*

Two phyla of invertebrates
- sponges — traits: variety of sizes, shapes, colors; pores and canals; do not move about; two cell layers
- stinging-cell animals — traits: stinging cells; hollow bodies; tentacles; two cell layers; disc

1. How do sponges get food? **They trap food when water containing food flows in through their pores.**
2. How do sponges reproduce asexually? **Buds or small pieces of the sponge's body can break off and become separate animals.**
3. How do stinging-cell animals get food? **They catch food with tentacles and stinging cells.**
4. How do stinging-cell animals reproduce? **They reproduce sexually by forming eggs and sperm and asexually by budding.**

# Lab 7–1 Stinging-cell Animals 

## Problem: What are the traits of a stinging-cell animal?

### Skills

observe, make and use tables, interpret data

### Materials

hydra culture
dropper
stereomicroscope
daphnia or brine shrimp
small dish
hand lens
flashlight

### Procedure

1. Copy the data tables.
2. Use a dropper to place a hydra and some water from the culture into a small dish.
3. **Observe:** Examine the hydra with a hand lens. Count the number of tentacles. Find the disc by which the animal attaches itself. Find the mouth. Record your observations in Table 1.
4. Tap the dish gently with your finger. Observe how the hydra moves and changes shape. Record your observations.
5. Place the dish on the stage of a stereomicroscope and focus on the hydra.
6. Draw the hydra and label the structures.
7. Drop a daphnia or a small amount of brine shrimp into the dish. Observe and record how the hydra takes in food.
8. Shine a flashlight on the dish. Observe and record the hydra's reaction.
9. Return the hydra to the culture.

### Data and Observations

1. How does the hydra react to light?
2. Where is the disc located?

### Analyze and Apply

1. How does the hydra capture food?
2. **Interpret data:** Why does the hydra react the way it does when you tap the dish?
3. What is the advantage to the hydra of having a mouth surrounded by tentacles?
4. **Apply:** How does a hydra differ from a sponge in food-getting?

### Extension

**Observe** a hydra when it is near other small animals and protists to determine what a hydra will eat.

| TABLE 1. | |
|---|---|
| **PART** | **OBSERVATION** |
| Location of mouth | on top, surrounded by tentacles |
| Number of tentacles | varies; may be 6 or more |
| Location of disc | at bottom |

| TABLE 2. | |
|---|---|
| **STIMULUS** | **REACTION** |
| Touch (tapping on dish) | contracts body and tentacles |
| Food | surrounds the food with tentacles |
| Light | contracts body and tentacles |



## ANSWERS

### Data and Observations

1. It contracts its body and tentacles.
2. at the bottom of the hydra

### Analyze and Apply

1. It surrounds the food with tentacles. Students may not see the discharge of stinging cells, but they should note that the food animal stops moving as it is stunned.
2. It is protecting itself.
3. The mouth is in a central location for receiving food from any of the tentacles.
4. Sponges get food in the water that enters their bodies through the pores. Special cells trap the food inside the body. Hydras have stinging cells and tentacles that trap food outside the body and then put it in the mouth. Sponges have no mouth.

## Lab 7-1 Stinging-cell Animals

### Overview

In this lab, students will observe a stinging-cell animal and will be able to study its feeding habits.

**Objectives:** Upon completion of this lab, students will be able to, (1) **describe** the general shape and characteristics of a hydra, (2) **relate** this animal's response to the presence of a nervous system, and (3) **describe** the method used by hydra in feeding.

**Time Allotment:** 35 minutes

### Preparation

▶ **Alternate Materials:** If prepared slides of hydra are available, then steps 3 and 6 may be done without use of living materials.

▶ Order hydra and daphnia, or brine shrimp in advance from a biological supply house. Purchase from any pet or aquarium shop and follow directions on container for hatching.

 **Lab 7-1 worksheet**/*Teacher Resource Package*, pp. 25-26.

### Teaching the Lab

▶ Place several coverslip pieces into the container of hydra overnight. The hydra will stick to the glass. You can remove the glass pieces with attached hydra using tweezers.

### ASSESSMENT

**Performance:** Provide students with a piece of coral to observe. Coral was once thought to be a plant. It is now classified in the animal kingdom. Have students give reasons why they think coral should be placed in the animal kingdom.

# 7:3 Worms

## PREPARATION

### Materials Needed

 Make copies of the Application, Study Guide, Enrichment, Lab worksheets, and Reteaching worksheets in the *Teacher Resource Package.*

▶ Gather wire mesh, funnels, and alcohol for the Mini Lab.

▶ Order earthworms for Lab 7-2.

### Key Science Words

flatworm, tapeworm, cyst, planarian, roundworm, hookworm, anus, segmented worm

### Process Skills

In the Skill Check, students will infer. In the Mini Lab, they will observe. In Lab 7-2, they will observe, predict, and make and use tables.

## 1 FOCUS

▶ The objectives are listed on the student page. Remind students to preview these objectives as a guide to this numbered section.

### ACTIVITY/Demonstration

Mark off a distance of 30 meters on the classroom floor. Ask students to imagine an animal of this length inside their body. Advise them that tapeworms have reached this length within human hosts.

**GLENCOE TECHNOLOGY**

**Videodisc**

The Secret of Life

*Flatworm Cross Section*

## 7:3 Worms

**Objectives**

5. **List** the three main phyla of worms and give an example of each.
6. **Identify** the main features of flatworms, roundworms, and segmented worms.

**Key Science Words**

flatworm
tapeworm
cyst
planarian
roundworm
hookworm
anus
segmented worm

When you hear the word *worm,* you probably think of an earthworm, but there are many different kinds of worms. Worms are invertebrates. They are classified into three main phyla—flatworms, roundworms, and segmented worms, based on their structures.

### Flatworms

Worms are more complex than sponges and stinging-cell animals. **Flatworms** are the simplest worms. They have a flattened body and three layers of cells. These layers include an outer layer, an inner layer, and a thick, middle layer. Organs and systems develop from cells of the middle layer.

Most flatworms are parasites. Remember that parasites live in or on other living things and get food from them. The organism that provides the food is the host. Some flatworms can cause serious problems.

A **tapeworm** is a kind of flatworm that has a flattened, ribbonlike body divided into sections. Tapeworms can live in the intestine of almost any kind of vertebrate. A tapeworm has suckers and hooks on one end that are used to hold onto the intestine of its host. Tapeworms have no mouth or organs to digest food. They absorb food that the host has already digested.

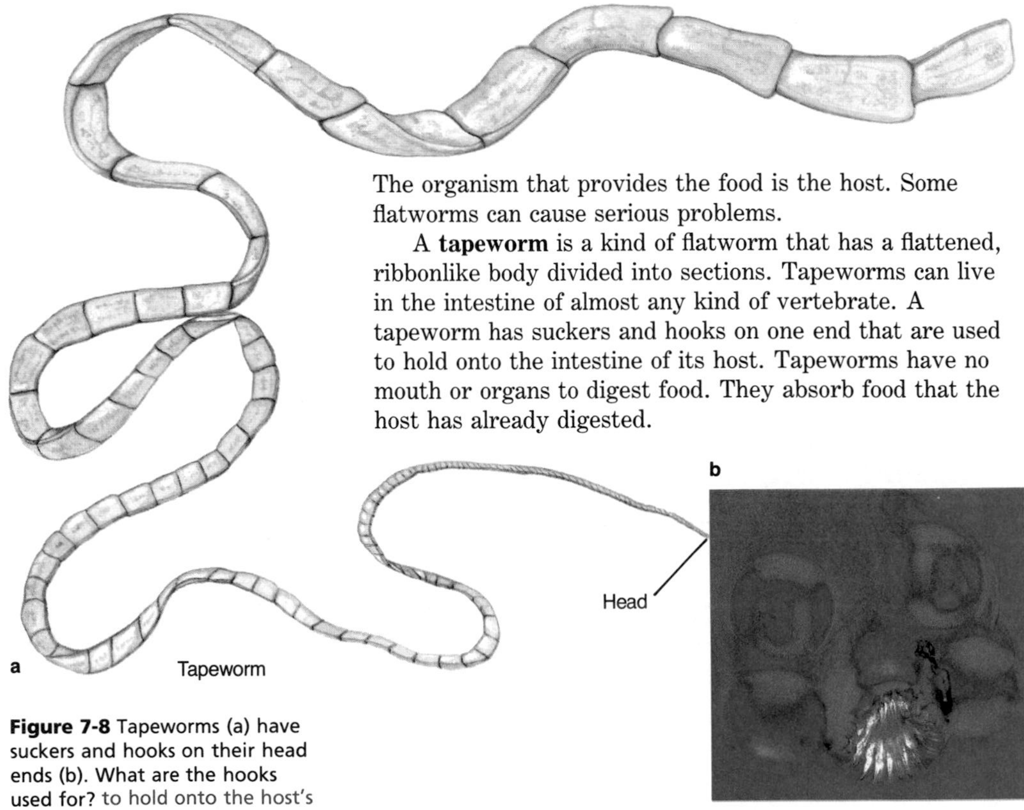


**Figure 7-8** Tapeworms (a) have suckers and hooks on their head ends (b). What are the hooks used for? to hold onto the host's intestine



## OPTIONS

### Science Background

The following classification for each worm phylum is: Phylum Platyhelminthes (flatworms); Class Turbellaria, (planaria); Class Cestoda, (beef tapeworm); Class Trematoda, (flukes); Phylum Nematoda (roundworms); Phylum Annelida (segmented worms); Class Polychaeta (clamworms); Class Oligochaeta (earthworms); Class Hirudinea (leech).

**Life Cycle of a Pork Tapeworm/** *Transparency Package,* number 7a. Use color transparency 7a as you teach flatworms.

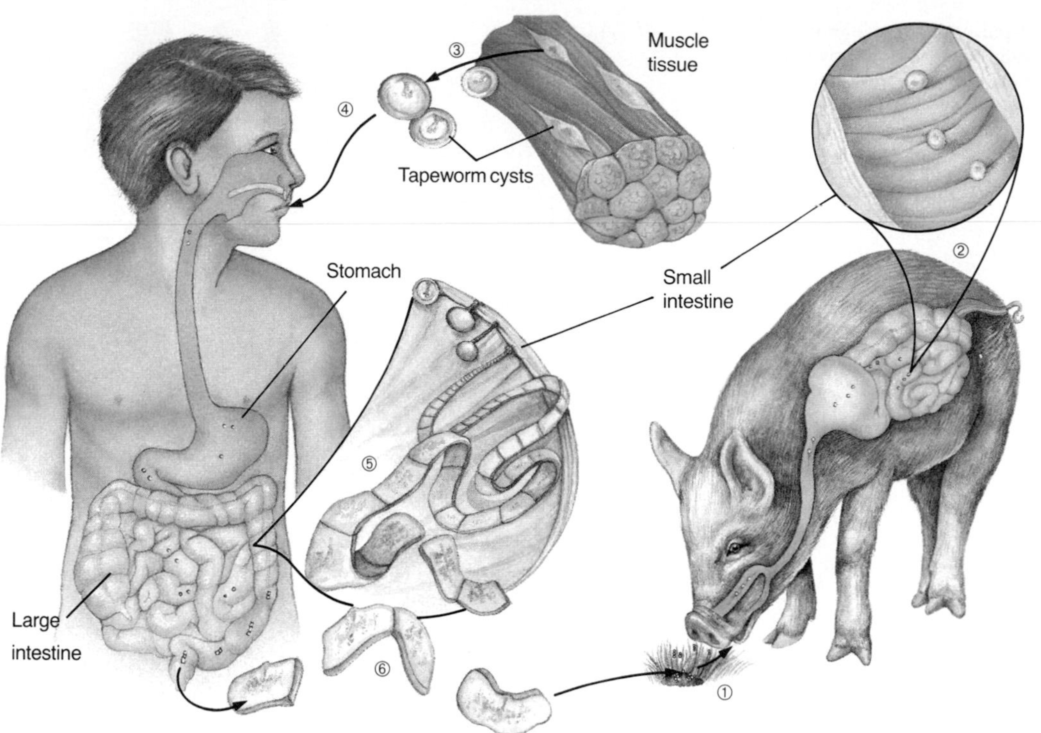


The tapeworm in Figure 7-8 is shown life size. Although some are much smaller, this one happens to be very large. Is it any wonder that a worm of this size uses a lot of the host's food?

Tapeworms have life cycles with many stages. Follow the steps of the pork tapeworm life cycle in Figure 7-9.

1. A pig eats tapeworm eggs that are on the ground. The eggs hatch in the pig's intestine.
2. The young worms enter the pig's bloodstream. Then they travel to the muscles and burrow into them.
3. The young worms form cysts (SIHSTZ) in the muscles. A **cyst** is a young worm with a protective covering.
4. If a person eats raw or undercooked meat that contains the cysts, the tapeworms get inside the person's intestine. The worms come out of their cysts, attach to the inside of the intestine, and begin to grow.
5. Tapeworms produce both eggs and sperm in each of their body sections. The sperm fertilize the eggs.
6. The body sections containing fertilized eggs break off and leave the host's body in solid waste through the intestine.

**Figure 7-9** Life cycle of a pork tapeworm

### Skill Check

**Infer:** Someone with tapeworms eats a lot of food but still feels hungry, tired, and loses weight. What is the cause? *For more help, refer to the* ***Skill Handbook****, pages 706-711.*

# 2 TEACH

### MOTIVATION/Demonstration

Show the class a preserved specimen of a tapeworm. Lay the worm on a table so students can get a sense of its size.

### Concept Development

▶ Point out that tapeworms cause malnutrition or mild vitamin deficiencies in some of their hosts. Some tapeworms cause pernicious anemia because they remove most of the Vitamin $B_{12}$ from the host's diet.

▶ Medicines have been given to humans who have tapeworms to cause expulsion of the worm.

▶ Several human tapeworms are transmitted by infected pork, beef, or wild game that has not been thoroughly cooked.

▶ One tapeworm section may contain as many as 100 000 eggs.

▶ The tapeworm body is made of sections called proglottids. New sections grow from the head region and push the older ones further back from the head.

### Skill Check

Tapeworms absorb nutrients from their host. This person is suffering from lack of necessary nutrients.

### Independent Practice

**Study Guide**/*Teacher Resource Package*, p. 42. Use the Study Guide worksheet shown at the bottom of this page for independent practice.

## APPLICATION 7

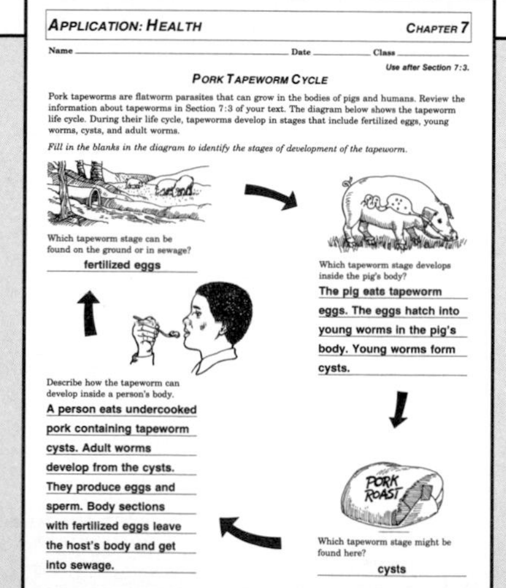

**APPLICATION: HEALTH** — **CHAPTER 7**

Name ______ Date ______ Class ______

*Use after Section 7:3.*

**PORK TAPEWORM CYCLE**

Pork tapeworms are flatworm parasites that can grow in the bodies of pigs and humans. Review the information about tapeworms in Section 7:3 of your text. The diagram below shows the tapeworm life cycle. During their life cycle, tapeworms develop in stages that include fertilized eggs, young worms, cysts, and adult worms.

*Fill in the blanks in the diagram to identify the stages of development of the tapeworm.*

Which tapeworm stage can be found on the ground or in sewage? **fertilized eggs**

Which tapeworm stage develops inside the pig's body? **The pig eats tapeworm eggs. The eggs hatch into young worms in the pig's body. Young worms form cysts.**

Which tapeworm stage might be found here? **cysts**

Describe how the tapeworm can develop inside a person's body. **A person eats undercooked pork containing tapeworm cysts. Adult worms develop from the cysts. They produce eggs and sperm. Body sections with fertilized eggs leave the host's body and get into sewage.**



## STUDY GUIDE 42

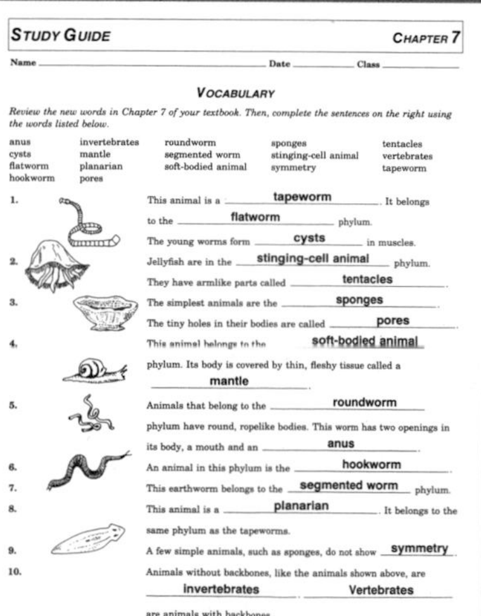

**STUDY GUIDE** — **CHAPTER 7**

Name ______ Date ______ Class ______

**VOCABULARY**

*Review the new words in Chapter 7 of your textbook. Then, complete the sentences on the right using the words listed below.*

| | | | | |
|---|---|---|---|---|
| anus | invertebrates | roundworm | sponges | tentacles |
| cysts | mantle | segmented worm | stinging-cell animal | vertebrates |
| flatworm | planarian | soft-bodied animal | symmetry | tapeworm |
| hookworm | pores | | | |

1. This animal is a **tapeworm**. It belongs to the **flatworm** phylum. The young worms form **cysts** in muscles.
2. Jellyfish are in the **stinging-cell animal** phylum. They have armlike parts called **tentacles**.
3. The simplest animals are the **sponges**. The tiny holes in their bodies are called **pores**.
4. This animal belongs to the **soft-bodied animal** phylum. Its body is covered by thin, fleshy tissue called a **mantle**.
5. Animals that belong to the **roundworm** phylum have round, ropelike bodies. This worm has two openings in its body, a mouth and an **anus**.
6. An animal in this phylum is the **hookworm**.
7. This earthworm belongs to the **segmented worm** phylum.
8. This animal is a **planarian**. It belongs to the same phylum as the tapeworms.
9. A few simple animals, such as sponges, do not show **symmetry**.
10. Animals without backbones, like the animals shown above, are **invertebrates**. **Vertebrates** are animals with backbones.

## OPTIONS

**Application**/*Teacher Resource Package*, p. 7. Use the Application worksheet shown here to teach the life cycle of a tapeworm.

## TEACH

### Concept Development

▶ Stress the importance of cooking meat thoroughly even though meat inspection prevents most infected meat from reaching the market.

▶ The ventral side of a planarian is covered with cilia. This allows the planarian to glide over a mucous path secreted by glands on its ventral surface.

▶ Emphasize that a planarian's digestive system has only one opening through which both food is taken in and wastes are expelled.

▶ Relate the digestive system of a planarian to that in a stinging-cell animal. Ask students if this feature should be considered simple or complex and why.

### Guided Practice

Have students write down their answers to the margin question: planarian.

### Independent Practice

**Study Guide**/*Teacher Resource Package*, p. 39. Use the Study Guide worksheet shown at the bottom of this page for independent practice.

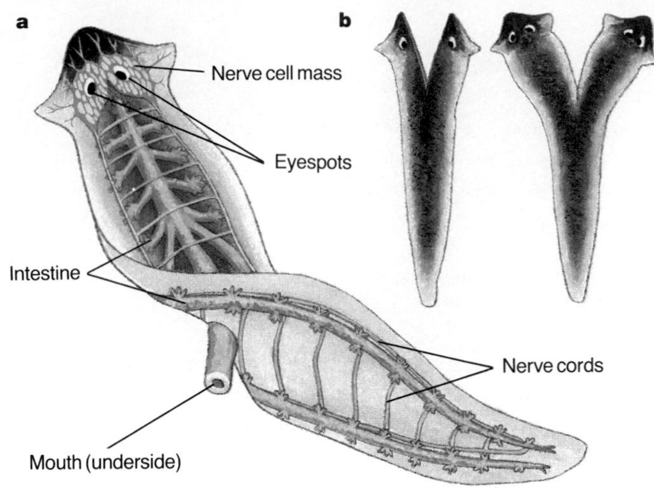


**Figure 7-10** A planarian (a) can reproduce asexually by regrowing new parts (b). How many parents are involved in this type of reproduction? one

Tapeworms in humans are not as common in the United States as in some other countries. In the United States, human wastes are treated with chemicals at sewage treatment plants. Meat is inspected for cysts by government inspectors. However, completely cooking meat is the best way to keep from getting tapeworms.

Although most flatworms are parasites, some aren't. Those that are not parasites are said to be free-living. Some of these flatworms live in water, but a few are found on land. A **planarian** (pluh NAIR ee un) is a common freshwater flatworm that is not a parasite. It is less than one centimeter long.

A planarian has the features shown in Figure 7-10a.

What flatworm has a mouth?

1. It has a flat body with muscles.
2. It has a triangular head with two spots that look like eyes. The eyespots have nerve cells that detect light.
3. It has two nerve cords. It also has two masses of nerve cells near the head. These structures allow the planarian to respond to the environment.
4. The mouth is near the middle of the body on the underside.
5. Planarians have an intestine that breaks down food. Undigested food leaves the intestine through the mouth.
6. Each animal has both male and female reproductive organs and produces both sperm and eggs. It reproduces sexually by exchanging sperm with another planarian. Planarians also reproduce asexually by pinching into two parts. Each part then forms a new animal, Figure 7-10b.



**STUDY GUIDE 39**

STUDY GUIDE — CHAPTER 7

Name ______ Date ______ Class ______

#### WORMS

*In your textbook, read about flatworms in Section 7:3.*

1. Study the steps of the pig tapeworm life cycle in Figure 7-9 of your textbook. Then, answer the following questions.
   a. What two host animals are part of the tapeworm's life cycle? **pig and human**
   b. Which animal will contain cysts of the worm? **pig**
   c. In which animal will the worm reproduce? **human**
   d. In which animal will the worm's body sections break off and leave the host's body in its solid waste? **human**
   e. In which animal will young worms travel to the muscles and burrow into them? **pig**
   f. In which animal will worms attach to the inside of the intestines? **human**
2. Why aren't tapeworms common in the United States? **These worms aren't common because in the United States human wastes are treated with chemicals at sewage plants, and meats are inspected for cysts.**
3. On the blanks below, label the parts of this planarian and briefly describe the job each part does.

   part **eyespots** — job **detect light**
   part **intestine** — job **breaks down food**
   part **nerve cords**
   part **nerve cell mass** — job **respond to environment**
   part **mouth** — job **takes in and gets rid of undigested food from the body**
4. How do planarians differ from most other flatworms? **Most flatworms are parasites. Planarians are free-living.**

## Roundworms

**Roundworms** are worms that have long bodies with pointed ends. They have three layers of cells. Many of these worms are small and can't be seen without a microscope. Many are parasites. Their hosts are people, dogs, cats, or even plants.

Some roundworms live in great numbers in the soil. A square meter of soil may contain four million worms! These worms are not animal parasites, but some of them may be plant parasites. Most roundworms, however, are free-living.

Hookworms are found in soil in the southeastern part of the United States. A **hookworm** is a roundworm that is a parasite of humans. Hookworms enter the body through the skin of the feet. Once inside, they move to the intestine. There they attach themselves and feed on the host's blood.

Let's take a closer look at a roundworm. Look at Figure 7-11 as you read about the features of this group.

1. These worms have long, rounded bodies. Each end is pointed. A set of muscles runs the length of the body. When roundworms move, they whip about.
2. One end of a roundworm has a mouth. The other end has an anus (AY nus). An **anus** is an opening through which undigested food leaves the body.
3. A roundworm has a tube within its body that connects the mouth and the anus. The tube is an intestine that digests food. Food and wastes do not mix as they do in the stinging-cell animals and flatworms, which have only one opening.
4. Males and females are separate animals.

### Mini Lab

**Where Do You Find Roundworms?**

**Observe:** Cover a funnel with wire mesh. Stand it in a jar of alcohol. Place moist soil on the mesh. What do you see in the alcohol when the soil begins to dry out? Why? *For more help, refer to the* ***Skill Handbook,*** *pages 704-705.*

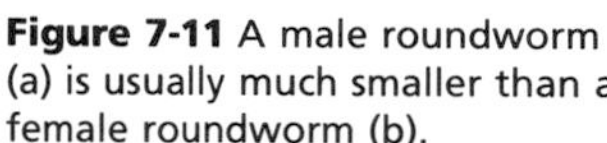

**Figure 7-11** A male roundworm (a) is usually much smaller than a female roundworm (b).

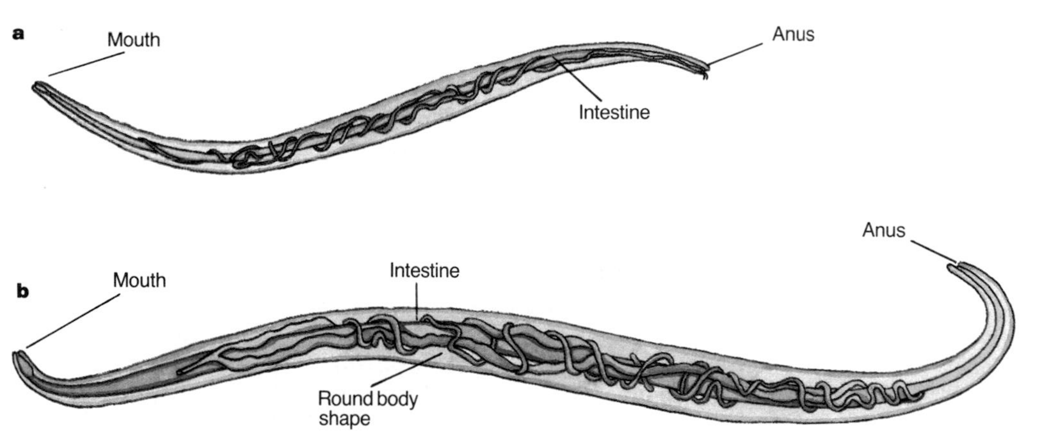




## TEACH

### Concept Development

▶ Mention that there are over 10 000 species of roundworms.

▶ Mention that hookworms and other roundworms can rupture the intestinal wall causing hemorrhages, infections, and allergic reactions. Sometimes larger worms can burrow into the host causing peritonitis and blocking ducts within the body.

### Mini Lab

Any type of alcohol may be used. A light over the soil will speed up the process of soil drying. Students may need a stereoscope or hand lens to examine the alcohol for roundworms.

### ✓ ASSESSMENT

**Skill:** Have students make a list of places where earthworms are found.

### Independent Practice

**Study Guide**/*Teacher Resource Package,* p. 40. Use the Study Guide worksheet shown at the bottom of this page for independent practice.

### GLENCOE TECHNOLOGY

**Videodisc**

**The Secret of Life**

*Earthworm*

*Earthworm Segment*

### STUDY GUIDE 40

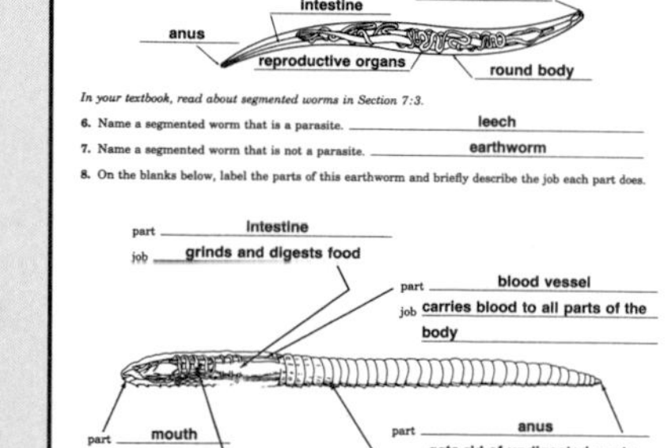

STUDY GUIDE — CHAPTER 7

Name ______ Date ______ Class ______

WORMS

*In your textbook read about roundworms in Section 7:3.*

5. On the blanks provided, label the parts of this roundworm.



*In your textbook, read about segmented worms in Section 7:3.*

6. Name a segmented worm that is a parasite. leech
7. Name a segmented worm that is not a parasite. earthworm
8. On the blanks below, label the parts of this earthworm and briefly describe the job each part does.



## OPTIONS

### ACTIVITY/Challenge

Have students prepare wet mounts of the roundworm *Turbatrix aceti* (vinegar eel) to observe roundworm movement. These may be purchased from a biological supply house.

## TEACH

### Concept Development

▶ Explain that segmentation is internal as well as external.

▶ Explain that a leech can ingest many times its own body weight of blood in one feeding.

▶ Explain that earthworms come near the surface to deposit wastes and to mate, usually at night.

▶ Point out that earthworms have bilateral symmetry.

▶ The bristles are called setae. Birds often have difficulty pulling earthworms out of the ground because the worms' setae dig into the soil.

### Check for Understanding

Have students prepare a chart with the headings Tapeworm, Planarian, Hookworm, and Earthworm across the top. Under each heading have them list the phylum name and traits of each.

### Reteach

Discuss students' charts from Check for Understanding. Have them fill in any omissions and correct any errors.

### *Student Journal*

Have students make a chart that compares flatworms, roundworms, and segmented worms. Students should include size, shape, structure, method of movement, method of respiration, and method of reproduction in their charts.

### *Portfolio*

Have students research medical uses for leeches and write a report. In the 1800s, leeches were applied to the temples of persons with headaches. Leeches also were used to treat mental illness, gout, skin diseases, and whooping cough.

a

b

**Figure 7-12** Earthworms (a) and leeches (b) are segmented worms.

### Bio Tip

**Health:** Leeches are used in medicine to drain excess blood from reattached body parts following surgery. Leeches produce a chemical that prevents blood from clotting. This chemical is used for treating cardiovascular disorders.

## Segmented Worms

The **segmented worms** are worms with bodies divided into sections called segments. These worms have three cell layers. They are the most complex of all the worms.

You may be familiar with some of the segmented worms. They live almost everywhere—in fresh water, salt water, and on land. Examples are shown in Figure 7-12. The leech is a segmented worm that is a parasite. Most leeches live in streams and lakes. They attach to the skin of host animals and suck their blood.

The common earthworm is a segmented worm. Figure 7-13 shows some of the earthworm's features. Look at this figure as you read about the earthworm's traits.

1. The body wall has layers of muscles. Each segment has bristles that help the earthworm move by gripping the soil.
2. Earthworms have a mouth and an anus.
3. The intestine is a tube. It has parts for holding food, grinding food, and digesting food.
4. Most segments have organs to get rid of wastes.
5. Two blood vessels run along the body. They meet to form five pairs of simple hearts. The hearts pump blood through the body, carrying oxygen and food to all cells.

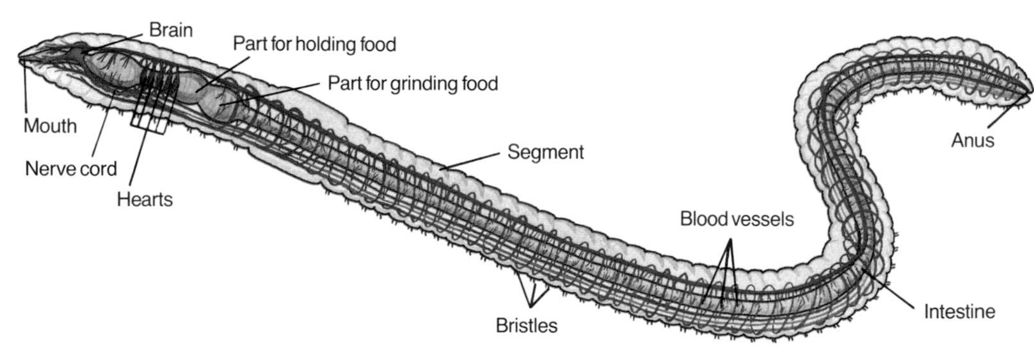


**Figure 7-13** Segmented worms, such as earthworms, have a complex structure.



## OPTIONS

### Science Background

Most of an earthworm's 100 to 150 segments are identical, except for the first and last. A swelling called the clitellum is found between segments 32 and 37. It is important in reproduction.

**Enrichment**/*Teacher Resource Package*, p. 7. Use the Enrichment worksheet shown here to teach students about leeches.

### ENRICHMENT 7

*ENRICHMENT* CHAPTER 7

Name ______ Date ______ Class ______

*Use after Section 7:3.*

#### LEECHES

*Read the information below about leeches. Then, answer the questions that follow.*

Imagine having four or five leeches attached to your arms, legs, or face. As the leeches suck your blood, they become fatter and fatter. Leeches are segmented worms related to earthworms. Not so long ago, people used leeches on their bodies in just this way. They thought leeches could cure diseases by removing "bad" blood.

People used leeches for removing blood—or bloodletting—for thousands of years. At first, leeches were believed to remove evil spirits from the body. In the 1800s, doctors also used leeches to treat a variety of illnesses, including gout, whooping cough, and mental illness. Millions of these leeches were used each year. They were in such high demand that people began raising leeches on farms. With time, the medical use of leeches became less popular. Some people had died from the overuse of leeches. By the late 1800s, doctors had learned that microorganisms cause diseases. New methods of treating diseases were also developed at that time. For most of the twentieth century, leeches were not used often for medical purposes.

Today, however, people are finding new uses for leeches. For example, doctors are using leeches in plastic surgery. In one case, a boy's ear was reattached after being bitten off by a dog. Leeches were applied around the ear to help drain excess blood. Without draining the blood, the reattached blood vessels would not heal. Researchers have discovered valuable chemicals in the saliva of the giant Amazon leech. One of these chemicals, called hementin, dissolves blood clots. Researchers think hementin might be useful for dissolving blood clots and preventing heart attacks and strokes in humans.

1. Why were leeches in high demand in the 1800s? **Leeches were used to treat many diseases, such as gout, whooping cough, and mental illness.**
2. Why did leeches become less popular in the late 1800s? **Many people had died from the overuse of leeches, doctors were learning more about the causes of diseases, and new methods of treating diseases were being developed.**
3. How are leeches being used in plastic surgery today? **Leeches are used to drain excess blood near the area of the surgery, thereby allowing reattached blood vessels to heal.**
4. What is hementin, and how might it be useful to humans? **Hementin is a chemical found in the saliva of the giant Amazon leech. Since it dissolves blood clots, it might be useful in treating blood clots and possibly preventing heart attacks and strokes in humans.**

Idea Map

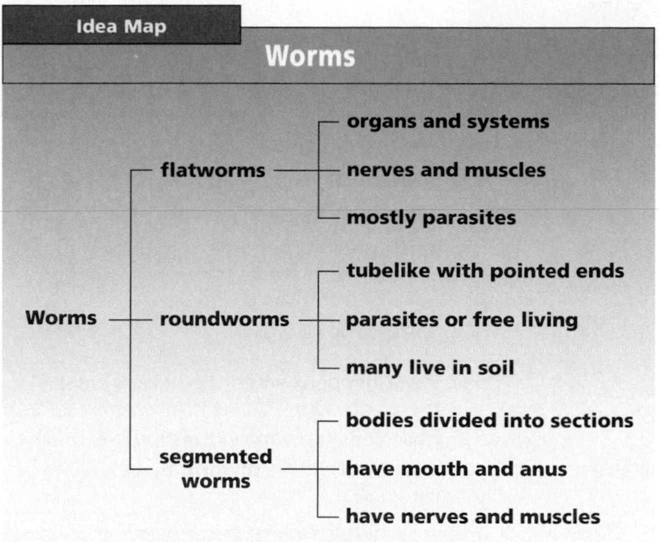


**Study Tip:** Use this idea map as a study guide to worms. The traits of each worm group are shown.

6. Nerves run the length of the body. There is a simple brain at the front end.
7. Each earthworm has both male and female sex organs. Earthworms reproduce when two worms exchange sperm. Other segmented worms may have separate sexes.

Earthworms take in soil through their mouths. Soil contains decayed matter, such as dead leaves, insects, and seeds. These things are food for the earthworm. The soil itself is not food and passes through the animal's intestine. Soil leaves the intestine through the anus. Earthworms move large amounts of soil from place to place by passing it through their bodies. They enrich the soil and loosen it, helping plants grow.

**Why do earthworms seem to eat soil?**

## Check Your Understanding

11. Give two examples of flatworms.
12. In what way are tapeworms harmful?
13. List three traits of roundworms.
14. **Critical Thinking:** How can you prevent hookworms from entering your body?
15. **Biology and Reading:** List two things that earthworms do that are helpful.



**RETEACHING 19**

RETEACHING CHAPTER 7

Name Date Class

Use with Section 7:3.

ANATOMY OF WORMS

*Label the parts of the planarian and the earthworm.*

planarian: nerve cord, masses of nerve cells, mouth, intestine

earthworm: blood vessel, male sex organs, anus, hearts, brain, mouth, sperm storage sacs, female sex organs, intestine, blood vessel, nerve cord, segment

# TEACH

## Idea Map

Have students use the idea map as a study guide to the major concepts of this section.

## Guided Practice

Have students write down their answers to the margin question: They take in soil through the mouth and absorb decayed matter from it.

## Check for Understanding

Have students respond to the first three questions in Check Your Understanding.

## Reteach

**Reteaching**/*Teacher Resource Package,* p. 19. Use this worksheet to give students additional practice in understanding the anatomy of worms.

**Extension:** Assign Critical Thinking, Biology and Reading, or some of the **OPTIONS** available with this lesson.

# 3 APPLY

## ACTIVITY/Challenge

Have students research types of worms that are parasites of humans.

# 4 CLOSE

## Divergent Question

Have students list ways of avoiding parasite infection.

## Answers to Check Your Understanding

11. tapeworms and planarians
12. They are parasites and use food of the host animal.
13. long, round bodies; mouth and anus; long digestive tube; separate sexes
14. by wearing shoes outside
15. enrich and loosen soil

## Lab 7-2 Earthworms

### Overview

In this lab, students examine the external structures of an earthworm and observe how the animal behaves when subjected to stimuli.

**Objectives:** Upon completion of this lab, students will be able to (1) **describe** structures that appear on the external surface of the animal, (2) **describe** how an earthworm responds to light, touch, and chemicals, and (3) **predict** how its features and behavior aid in its way of life.

**Time Allotment:** 1 class period

### Preparation

▶ Maintain earthworms in a polystyrene container with at least 6-8 cm of soil in the bottom. The soil should be collected at the same site as the earthworms. Keep the soil moist and the container covered. Temperature may range from 5°C to 15°C. Earthworms may be purchased from a biological supply house or purchased from a fish bait shop.

**Lab 7-2 worksheet**/*Teacher Resource Package,* pp. 27-28.

### Teaching the Lab

▶ Stress careful and gentle handling of the animals.

▶ Have students wash their hands when finished with this lab.

**ASSESSMENT**

**Performance:** Provide students with a stereoscopic microscope so they can examine a live vinegar eel. Have students compare its structure with that of an earthworm.

## Earthworms

## Lab 7–2

### Problem: What traits does an earthworm have that help it live in soil?

#### Skills

observe, infer, predict, make and use tables

#### Materials

live earthworms in a covered container
hand lens
shallow pan
toothpick
paper towels
beaker of water
vinegar
cotton swab
flashlight

#### Procedure

1. Copy the data table.
2. Open a container of earthworms and shine the flashlight on the worms. What do the worms do? Record what you see.
3. Wet your hands. Carefully place an earthworm on a moist paper towel in a shallow pan. **CAUTION:** *Do not let the worm dry out.* Keep your hands wet and moisten the worm by sprinkling it with water. Hold the worm gently between your thumb and forefinger. Observe its movements and record them in your table.
4. Rub your fingers gently along the body. Examine the bristles along the earthworm with a hand lens. See the figure for help.
5. **Observe:** With a toothpick, gently touch the earthworm on the front and back ends. Record your observations.
6. **Make and use tables:** Dip a cotton swab in vinegar. Place it in front of the earthworm on the paper towel. Record your observations.

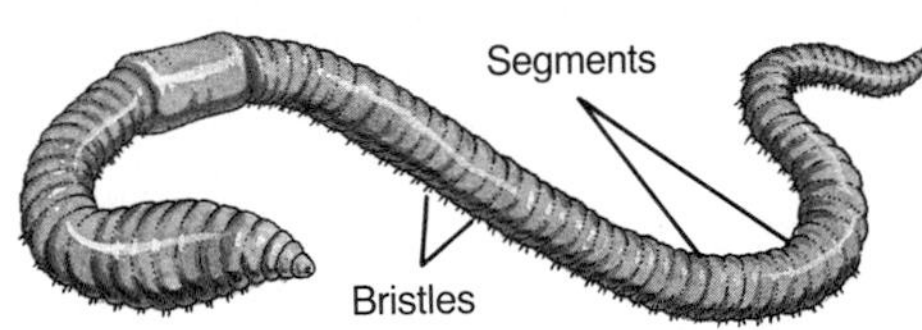


7. Return the earthworm to the covered container.

#### Data and Observations

1. What happens when the light is shined on the earthworms?
2. How does the earthworm react to touch?
3. How does the earthworm react to vinegar?

| CONDITION | REACTION |
|---|---|
| Light | moves away |
| Holding | tries to free itself |
| Touch | moves away |
| Vinegar | moves away |

#### Analyze and Apply

1. **Infer:** How does the earthworm's reaction to light help it to live in the soil?
2. How does the earthworm's reaction to being touched help it to live in the soil?
3. **Predict:** How would you expect an earthworm to react to chemicals like vinegar in the soil?
4. How is the front end of the earthworm helpful for living in soil?
5. How is the long, soft body helpful for living in the soil?
6. How are the bristles useful for living in the soil?
7. **Apply:** How is an earthworm similar to a hookworm?

#### Extension

**Observe** a planarian's response to light and touch.



## ANSWERS

### Data and Observations

1. They move away.
2. It moves away.
3. It moves away.

### Analyze and Apply

1. It keeps the worm under the moist soil and away from the light, which has a drying effect.
2. If the worm comes across something in its path, it can turn away.
3. If the worm senses a harmful chemical, it will turn away.
4. It is round and pointed to help the worm move through the soil.
5. It can move with the shape of the soil and move around things easily.
6. They help in movement. They also prevent the worm from being pulled out of the soil.
7. They both live in the soil. Both worms have long bodies.

## 7:4 Soft-bodied Animals

The soft-bodied animals are more complex than the other invertebrates that you have studied. Soft-bodied animals come in all sizes and live in many different environments. They are placed together in one phylum because of the traits they share.

**Objectives**

7. **Identify** the major features of soft-bodied animals.
8. **Explain** how soft-bodied animals are classified.

**Key Science Words**

soft-bodied animal
mantle

### Traits of Soft-bodied Animals

**Soft-bodied animals** are animals with a soft body that is usually protected by a hard shell. Animals in the soft-bodied animal phylum may be familiar to you, no matter where you live. Some live along the ocean on rocks, buried in the sand, or in the water. You may have gathered their shells while walking along a beach, Figure 7-14a. Some of these animals are found around streams and ponds. Others live on land.

Soft-bodied animals have certain traits. The body is covered by a thin, fleshy tissue called a **mantle.** The mantle makes the shell. The shell is the outermost covering. It protects the soft body. Soft-bodied animals have a muscular foot for moving from place to place, Figure 7-14b. Most have a head with a mouth. Many soft-bodied animals have a structure like a tongue covered with teeth inside the mouth. This structure scrapes food from surfaces, such as rocks. A snail has this structure.

**What are the traits of soft-bodied animals?**

The soft-bodied animals are grouped into classes based on three traits. The three traits are the kind of foot, whether or not a shell is present, and the number of shells.

**Figure 7-14** The shells of soft-bodied animals can be found on many beaches (a). Clams use a muscular foot for movement (b).



## 7:4 Soft-bodied Animals

### PREPARATION

#### Materials Needed

Make copies of the Transparency Master, Critical Thinking/Problem Solving, Study Guide, and Reteaching worksheets in the *Teacher Resource Package.*

#### Key Science Words

soft-bodied animal
mantle

### 1 FOCUS

▶ The objectives are listed on the student page. Remind students to preview these objectives as a guide to this numbered section.

#### ACTIVITY/Brainstorming

Have students name any seafood that they would enjoy eating. List their suggestions on a transparency for the overhead projector. When finished, place a check mark after those foods that are members of the soft-bodied phylum.

#### Guided Practice

Have students write down their answers to the margin question: soft body, mantle, shell, muscular foot, head with mouth.

**GLENCOE TECHNOLOGY**

**Videodisc**

**Biology: The Dynamics of Life**
Movie: *Squid Locomotion*
Disc 1, Side 2, Ch. 33

**TRANSPARENCY 31**

TRANSPARENCY MASTER — CHAPTER 7

SIMPLE ANIMALS

| Phylum | Traits | Example |
|---|---|---|
| Sponges | 1. no definite shape<br>2. body has canals and pores<br>3. do not move about<br>4. no tissues or organs<br>5. have three kinds of cells | |
| Stinging-cell animals | 1. stinging cells and hollow bodies<br>2. tentacles that surround a mouth<br>3. two cell layers<br>4. have muscles and nerve cells | |
| Flatworms | 1. three layers of cells<br>2. flattened bodies<br>3. one body opening<br>4. many parasites, some free-living | |
| Roundworms | 1. three layers of cells<br>2. round bodies with pointed ends<br>3. tubes within their bodies<br>4. two body openings<br>5. males and females are separate | |
| Segmented worms | 1. three layers of cells<br>2. bodies divided into segments<br>3. body wall has muscles<br>4. segments have organs for getting rid of wastes<br>5. blood vessels<br>6. nerves | |
| Soft-bodied animals | 1. soft body usually protected by shell<br>2. body covered by mantle<br>3. have a muscular foot | |

### OPTIONS

#### Science Background

The classification for the phylum Mollusca (soft-bodied animals) is as follows: Class Polyplacophora, chitons; Class Scaphopoda, tooth shells; Class Gastropoda, snails and slugs; Class Bivalvia, clams, mussels, oysters; Class Cephalopoda, squid, octopuses.

**Transparency Master**/*Teacher Resource Package,* p. 31. Use the Transparency Master shown here to reinforce the traits of simple animals.

# 2 TEACH

### MOTIVATION/Audiovisual

Show the video *Biovideo: Introduction to Invertebrates,* Carolina Biological Supply Co.

### Concept Development

▶ Soft-bodied animals are commonly called mollusks.

▶ Explain that clams, oysters, and scallops are filter feeders.

▶ Squids reach speeds of about 20 km per hour.

### Connection: Language Arts

Have students read the poem "The Chambered Nautilus" by Oliver Wendell Holmes.

### Guided Practice

Have students write down their answers to the margin question: snails and slugs; clams, oysters, scallops; octopus, squid, cuttlefish.

### Check for Understanding

Have students prepare an outline of Section 7:4. Have them use section titles as the major divisions.

### Reteach

Put the headings snails and slugs; clams, oysters, scallops; octopus, squid on the chalkboard. Have students give traits for each.

### Student Journal

Have students make a drawing of a squid and label its structures. They should use arrows to indicate the way a squid moves.

## Classes of Soft-bodied Animals

The first class of soft-bodied animals includes snails and slugs, shown in Figure 7-15a. They live on land and in the water. They glide slowly along by means of a wide, muscular foot. The foot puts down a trail of slime that helps the animal glide. Snails have a single shell and slugs have no shell. Sea and land slugs look like snails without shells. Notice the two pairs of tentacles on the snail's head in Figure 7-15a. They are sense organs. Each of the larger tentacles has an eye that detects light.

The second class of soft-bodied animals includes clams, oysters, and scallops. See Figure 7-14b. Unlike the snails, animals in the second class have two shells that fit together. Many clams spend their lives under water buried in sand or mud. Their foot is shaped like a shovel and is used to burrow in the sand.

How do buried clams get food? Figure 7-16 shows you. These animals usually have two tubes that stick out of the sand into the water. The animal takes in water through one tube. Any food in the water is filtered out by the animal. The water is then forced out through its other tube.

The octopus in Figure 7-15b and squid and cuttlefish belong to the third class of soft-bodied animals. Both squid and cuttlefish have shells inside their bodies. The octopus has no shell. In these animals, the muscular foot is divided into arms, or tentacles, surrounding the head. These animals have very well-developed eyes. They are rapid swimmers and move by shooting a jet of water.

a

b

**Figure 7-15** A snail (a) has a muscular foot. The foot of an octopus (b) is divided into eight arms that surround the head.

**What are the three classes of soft-bodied animals?**

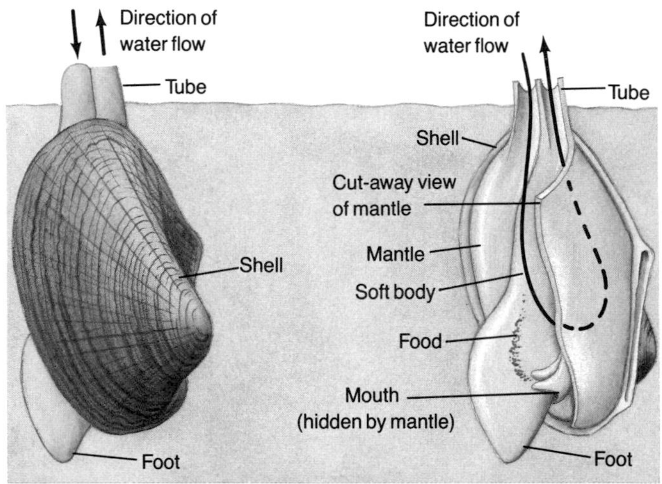


**Figure 7-16** Clams filter food out of water. The water is taken into the clam through one tube and forced out through another tube.



## OPTIONS

**What Are the Parts of the Squid?**/*Lab Manual,* pp. 53-56. Use this lab as an extension to studying soft-bodied animals.

**How a Squid Moves**/*Transparency Package,* 7b. Use color transparency 7b to teach the movement of a squid.

**Critical Thinking/Problem Solving**/*Teacher Resource Package,* p. 7. Use the Critical Thinking/Problem Solving worksheet shown here to explain how simple animals can destroy plants.

### CRITICAL THINKING 7

CRITICAL THINKING/PROBLEM SOLVING — CHAPTER 7

Name ______ Date ______ Class ______

*Use after Section 7:4.*

**WHAT IS DESTROYING THE TOMATO PLANTS?**

In the spring, Julio's family planted a garden. Family members worked together preparing the soil and planting seeds for corn, beans, squash, and lettuce. When the temperature became warm, they bought tomato plants and pepper plants at the garden center. Julio chose to set out the six tomato plants. He cut the bottoms from three cardboard milk cartons and put them around three of the plants to protect the plants. He had no cartons for the three remaining plants.

Julio watered and fertilized the plants and watched them grow taller. One morning before school, he found two of the unprotected plants lying on the ground. The stems were cut and the leaves were ragged. A shiny streak of slime led to the plants. What could be destroying the tomato plants?

**Analyzing the Problem**

1. During what part of the day were the plants destroyed? night
2. None of the plants surrounded by milk cartons were destroyed. How is this information helpful in solving the problem? The plants must be cut by something that creeps along the ground. It does not fly or jump over the height of a milk carton.
3. What organism leaves a slimy streak? a snail or slug
4. What living things eat plants? animals
5. To what phylum of animals do snails and slugs belong? They belong to the phylum of soft-bodied animals.

**Solving the Problem**

1. What could have destroyed the tomato plants? a creeping animal, such as a snail or slug
2. Julio has performed an experiment that tests safe growing conditions. What is the variable being tested? presence or absence of milk carton

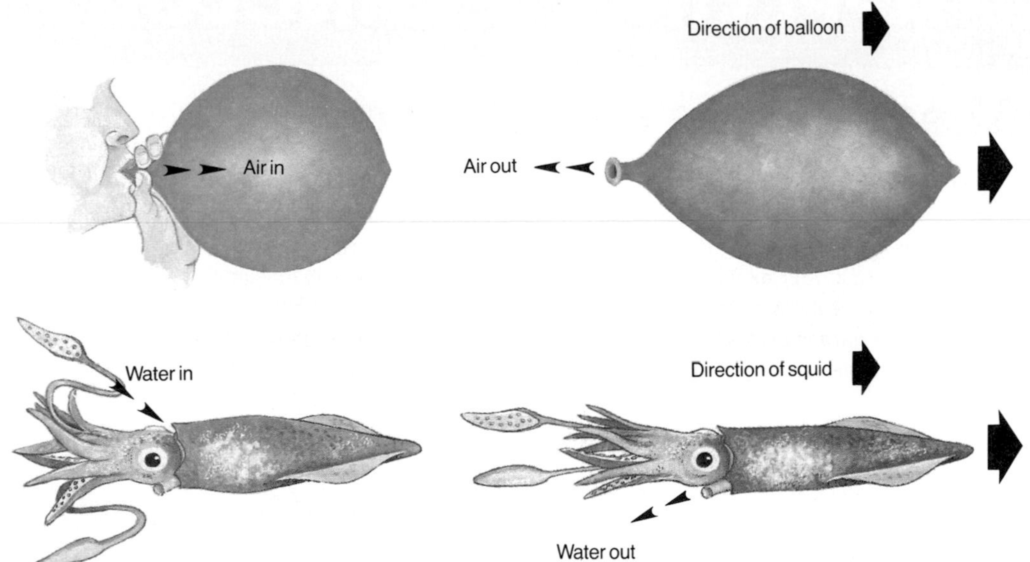


**Figure 7-17** Squid move rapidly by shooting a jet of water out of their bodies. Can you think of something else that moves this way? rockets and jet planes

The way a squid moves is shown in Figure 7-17. The movement is similar to that of a balloon. If you let go of a balloon filled with air, it flies in a direction opposite to that of the air coming out of it. The squid also swims in a direction opposite to that of the water coming out of it.

Let's review the features of soft-bodied animals.

1. All are invertebrates. No backbone is present.
2. They have a soft body covered by a mantle.
3. Most have one or two external shells or an internal shell. A few have no shell.
4. Most have a foot by which they move about.

## Check Your Understanding

16. What are four traits of soft-bodied animals?
17. How do clams get food?
18. What traits are used to classify soft-bodied animals?
19. **Critical Thinking:** How does the shell of a snail differ from that of a clam? How does the shell of an octopus differ from that of a clam?
20. **Biology and Math:** Giant squid may grow as long as 49 feet. Estimate how many meters long these squid grow.

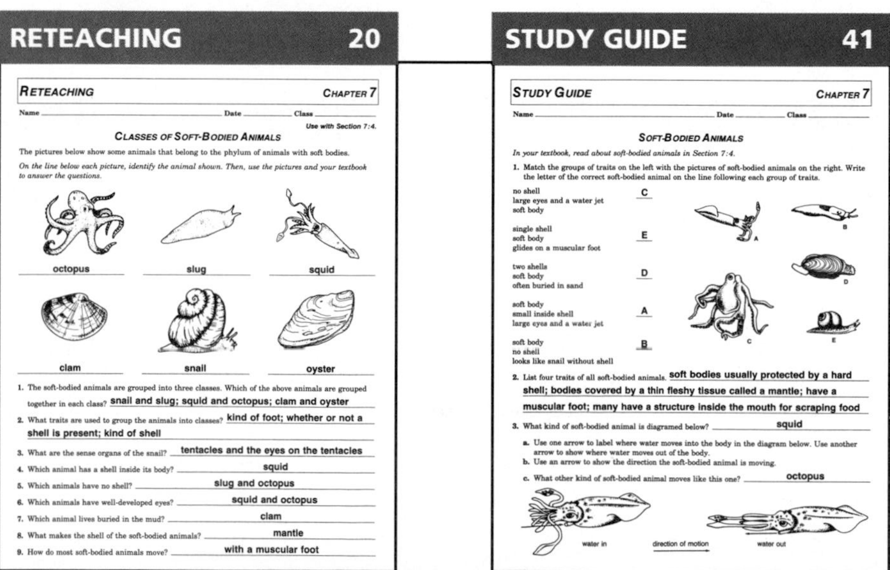

**RETEACHING 20**

RETEACHING — CHAPTER 7

Name ______ Date ______ Class ______

Use with Section 7:4.

CLASSES OF SOFT-BODIED ANIMALS

The pictures below show some animals that belong to the phylum of animals with soft bodies.

*On the line below each picture, identify the animal shown. Then, use the pictures and your textbook to answer the questions.*

octopus — slug — squid

clam — snail — oyster

1. The soft-bodied animals are grouped into three classes. Which of the above animals are grouped together in each class? snail and slug; squid and octopus; clam and oyster
2. What traits are used to group the animals into classes? kind of foot; whether or not a shell is present; kind of shell
3. What are the sense organs of the snail? tentacles and the eyes on the tentacles
4. Which animal has a shell inside its body? squid
5. Which animals have no shell? slug and octopus
6. Which animals have well-developed eyes? squid and octopus
7. Which animal lives buried in the mud? clam
8. What makes the shell of the soft-bodied animals? mantle
9. How do most soft-bodied animals move? with a muscular foot

**STUDY GUIDE 41**

STUDY GUIDE — CHAPTER 7

Name ______ Date ______ Class ______

SOFT-BODIED ANIMALS

*In your textbook, read about soft-bodied animals in Section 7:4.*

1. Match the groups of traits on the left with the pictures of soft-bodied animals on the right. Write the letter of the correct soft-bodied animal on the line following each group of traits.

no shell, large eyes and a water jet, soft body — C

single shell, soft body, glides on a muscular foot — E

two shells, soft body, often buried in sand — D

soft body, small inside shell, large eyes and a water jet — A

soft body, no shell, looks like snail without shell — B

2. List four traits of all soft-bodied animals. soft bodies usually protected by a hard shell; bodies covered by a thin fleshy tissue called a mantle; have a muscular foot; many have a structure inside the mouth for scraping food
3. What kind of soft-bodied animal is diagramed below? squid
   a. Use one arrow to label where water moves into the body in the diagram below. Use another arrow to show where water moves out of the body.
   b. Use an arrow to show the direction the soft-bodied animal is moving.
   c. What other kind of soft-bodied animal moves like this one? octopus

water in — direction of motion — water out

## TEACH

### Independent Practice

**Study Guide**/*Teacher Resource Package,* p. 41. Use the Study Guide worksheet shown at the bottom of this page for independent practice.

### Check for Understanding

Have students respond to the first three questions in Check Your Understanding.

### Reteach

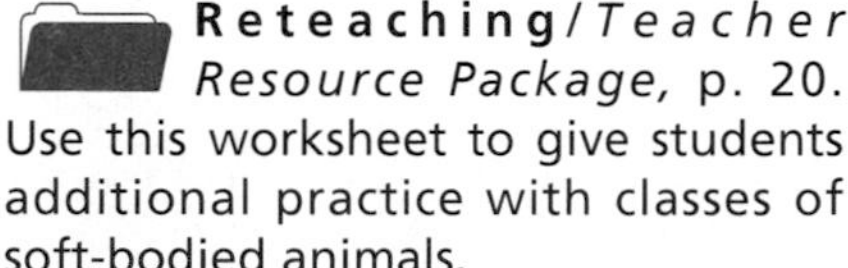

**Reteaching**/*Teacher Resource Package,* p. 20. Use this worksheet to give students additional practice with classes of soft-bodied animals.

**Extension:** Assign Critical Thinking, Biology and Math, or some of the **OPTIONS** available with this lesson.

## 3 APPLY

### Software

*The Animal Kingdom, Part I and II,* Queue, Inc.

## 4 CLOSE

### ACTIVITY/Brainstorming

Ask students to explain the importance of soft-bodied animals.

### Answers to Check Your Understanding

16. All are invertebrates; have a soft body covered by a mantle; most have one or two shells; most have a foot.
17. It filters food from water.
18. The kind of foot, whether a shell is present, number of shells.
19. (a) A snail's shell is all one part; a clam's shell has two parts. (b) The octopus has no outside shell; the clam has an outside shell with two parts.
20. Accept 14-16 meters.

## APPLYING TECHNOLOGY

### Purpose

Students will study the anatomy, physiology, and reproductive nature of the zebra mussel. They will then devise several methods that might be used to control or eradicate these pests.

### Process Skills

Observe, communicate, infer, formulate models

### Time Required

The entire activity can be completed in one class period.

### Possible Problems

The observation of an already-opened fresh or preserved mussel will be most unrewarding for students. It will be difficult for them to actually identify most of the organs listed. As an alternative, simply use the diagrams provided in the activity.

### Teaching Strategies

- Mussels (not zebra mussels, but a larger, similar species) are available from biological supply houses. A cheaper supply can be found in your local supermarket fresh fish department. Again, these will not be zebra mussels. Some stores may carry packages of frozen mussels.
- You may wish to discuss the mussels' anatomy, physiology, and life cycle with students in one large group so you do not need reference materials for the entire class. Films, videotapes, and filmstrips could be used to cover these same topics. Photocopy suitable reference materials for student use in the classroom.
- An excellent and understandable reference text for student use is *Animals Without Backbones* by Ralph Buchsbaum, The University of Chicago Press.
- Prepared slides of veligers are available from biological supply houses. To reduce the cost of veliger prepared slides, purchase only one. Project it with a microprojector onto

# TECH PREP APPLYING TECHNOLOGY

## These Zebras Live Underwater

Snails, squid, clams, and scallops all belong to the soft-bodied animal phylum. Why are they called soft-bodied if many of these animals have a hard shell? The soft body is there, but you can't always see it because it's inside a protective shell. Your experience with animals in this phylum is probably limited to your use of them as food. A few species can be harmful pests. Snails and slugs damage garden plants and crops. But, the biggest pest of all is a newcomer to North America, the tiny zebra mussel.

### Identifying the Problem

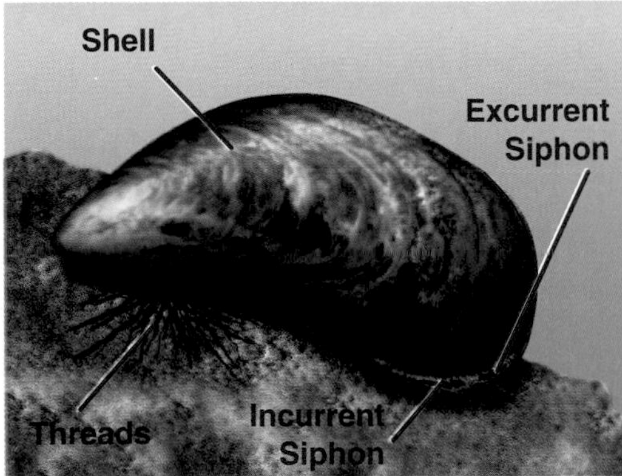


Zebra mussels were accidentally brought into the United States from Europe's Caspian Sea. These invertebrate animals are reproducing at a very high rate in the Great Lakes and are causing millions of dollars of damage. They clog up water intake pipes, thus interfering with the normal pumping of water into and out of waterworks and energy-generating plants. How can the zebra mussel be controlled? The answer may lie in knowing something about what this animal looks like and how it lives and reproduces. You are a marine biologist who has been hired by a Detroit power company to find ways of destroying or controlling this water pest. Your plan is to learn about this animal, then to suggest ways of controlling or getting rid of it.

#### Technology Connection

Scientists have come up with some plans that are controlling zebra mussels. They are using electric fields in lakes to kill the adult as well as the veliger stage. Even shocking the veligers with sound waves is being tested. Sound waves cause veligers to settle onto surfaces before they are mature. This prevents them from continuing with their normal feeding and growth patterns.

### Collecting Information

Use references to learn the following about mussels. How do they protect themselves? What do they feed on, and how do they get their food? (Be sure to check on what the incurrent and excurrent siphons do.) How do they get oxygen and what organ do they use for this purpose? How do they attach to surfaces? How do they reproduce? What is a veliger, and how many are formed by one adult?



#### Technology Connection

The use of electricity to kill the adult and veliger stage has been somewhat successful. It is very energy intensive, however, and freshwater is not a good conductor of electricity. Ultraviolet radiation is now being used on an experimental basis.

## Carrying Out an Experiment

1. Prepare a table that lists the italicized terms in steps 2-4. When completed, the table will include the function of each of these terms.
2. Examine the outside of a mussel. Use Figure A as a guide to locating the *shell, threads, incurrent siphon,* and *excurrent siphon.* Note: some of the parts may be hard to see. Record the function of each part.
3. Examine an opened mussel. Use Figure B as a guide to locating the *gills* and *mouth.* Note: these parts may be hard to see. Record the function of these parts.
4. Look at a prepared slide of a mussel *veliger* using low power magnification on your microscope. Does it look anything like the adult stage? Record the function of the veliger.
5. Brainstorm with several classmates about how you could control the growth and reproduction of zebra mussels.
6. Design a plan that could be used to control the zebra mussel in each of these four ways: by attacking its shell, through its filtering of water for food and oxygen, through its use of threads, in its veliger stage. Record all of your plans.

### Career Connection

- **Fish Culture Technician** May work in a fish hatchery or laboratory run by a state or federal agency; feeds and raises fish; keeps ponds and tanks clean
- **Environmental Technician** Works for large cities or industrial plants; takes water samples and tests for the presence of chemicals, pollutants, and minerals; observes and reports on life forms present
- **Ocean Technician** Works on land along the coasts or on ships; works with scientists who study the chemical and physical properties of water; tests water samples for dissolved gases, minerals, and pollutants

## Assessing Your Results

What are some of the strengths and weaknesses for each of your plans to control zebra mussels? What information are you going to present to the Detroit power company regarding the control or destruction of zebra mussels?

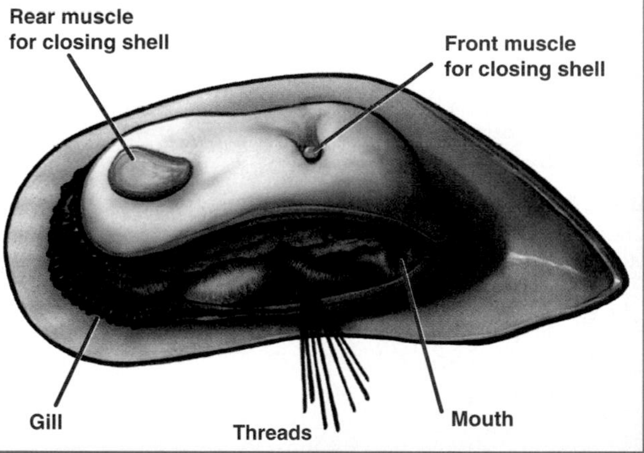


153

### Career Connection

**Career Path**
**Fish Culture Technician** Requires a high-school diploma or training on the job.
**Environmental Technicians** High-school diploma and several college courses in chemistry and biology
**Ocean Technician** Minimum of two years of college after graduation from high school; certain colleges offer a two-year Associates degree.
**Career Issue**
Ask students to explore the problem of how to get rid of or find a use for all the waste shells that might collect in lakes once the zebra mussels are killed.

the class screen or use a microvideo camera, if available.

▶ Have students work in groups of two or three to reduce the cost of specimens.

▶ Review the life cycle of bivalve mollusks (mussels, clams, oysters) with students.

▶ If you live near an ocean, you may wish to collect live mussels for classroom use.

▶ Advise students NOT to open the mussels provided. Opening of the shell is difficult and should be done carefully by the instructor.

### Conclusions

Students will be able to describe the role of various organs of a mussel and to understand its life cycle. They will be able to design several methods for controlling or eradicating the animal based on their knowledge of how it functions and reproduces.

### Answers to Questions

Mussels are protected by their shells, which are held together tightly by muscles to prevent predators from opening them up. The incurrent siphon pulls water, along with oxygen and food, into the body. Water passes over the gills, which remove oxygen for the mussel's use. The excurrent siphon expels water from the body. Mussel threads are used to attach the young animal to a surface such as a pipe or boat hull. Mussels shed their eggs or sperm into the water, where fertilization occurs. Fertilized eggs then develop into the veliger, which eventually forms the young adult.

Plans to destroy or control the animal will vary. Suggested answers may be: oils to block the incurrent siphon, poisons that kill the veliger stage, introduction of predators that feed on zebra mussels.

### ASSESSMENT

**Portfolio** Ask students to prepare a report that could be sent to a Detroit power company outlining the steps that might be taken to rid the Great Lakes of zebra mussels.

## CHAPTER 7 REVIEW

### Summary

Summary statements can be used by students to review the major concepts of the chapter.

### Key Science Words

All boldfaced terms from the chapter are listed.

### Testing Yourself

#### Using Words

1. cyst
2. flatworm
3. sponge
4. symmetry
5. tentacle
6. mantle
7. segmented worm
8. tapeworm
9. invertebrate
10. planarian

#### Finding Main Ideas

11. p. 136; The phylum with the largest number of animal types is the jointed-leg phylum.
12. p. 144; Tapeworms can be avoided by completely cooking meat.
13. p. 144 ; A planarian has eyespots made of nerve cells to detect light.
14. pp. 139-140; Hydra and sea anemones capture their food with tentacles that have stinging cells.
15. p. 138; Sponges have spaces between the many fibers that can fill with water.
16. p. 145; An anus is an opening through which undigested food can pass.
17. p. 151; Water pushed out of the squid moves the squid in a direction opposite to the direction of the water.
18. p. 145; A hookworm enters the body through the skin of the feet.
19. p. 150; Clams have two tubes. They take in water through one tube, filter the food from the water, and then force the water out through the other tube.
20. p. 147; Earthworms eat decayed matter in soil.

# Chapter 7

# Review

## Summary

### 7:1 Animal Classification

1. Animals can't make their own food. They can move about, are multicellular, and usually have cells organized into tissues, organs, and organ systems.
2. Animals are divided into nine major phyla based on traits they have in common.

### 7:2 Sponges and Stinging-cell Animals

3. Sponges are invertebrates that live attached to a surface in water.
4. Stinging-cell animals have stinging cells and hollow bodies with one opening.

### 7:3 Worms

5. Worms are classified into three major phyla—flatworms, roundworms, and segmented worms.
6. Flatworms have flattened bodies. Roundworms have rounded bodies. Segmented worms have segmented bodies.

### 7:4 Soft-bodied Animals

7. Soft-bodied animals are invertebrates that have a soft body covered by a mantle. They have one, two, or no shells and usually have a foot for movement.
8. Soft-bodied animals are grouped into classes by the kind of foot, whether or not a shell is present, and the number of shells.

## Key Science Words

anus (p. 145)
cyst (p. 143)
flatworm (p. 142)
hookworm (p. 145)
invertebrate (p. 134)
mantle (p. 149)
planarian (p. 144)
pore (p. 137)
roundworm (p. 145)
segmented worm (p. 146)
soft-bodied animal (p. 149)
sponge (p. 137)
stinging-cell animal (p. 139)
symmetry (p. 134)
tapeworm (p. 142)
tentacle (p. 139)
vertebrate (p. 134)

## Testing Yourself

### Using Words

*Choose the word from the list of Key Science Words that best fits the definition.*

1. young tapeworm with protective covering
2. the simplest worm
3. simplest animal
4. the arrangement of body parts
5. armlike part that surrounds a stinging-cell animal's mouth
6. thin, fleshy tissue that makes a shell
7. earthworm phylum
8. flatworm that lives on the digested food of another animal



#### Using Main Ideas

21. Sponges do not move about, live attached, and have a body with a series of pores or canals.
22. Animals are consumers, can move about, and are multicellular.
23. They cause painful stings.
24. In radial symmetry, body parts are arranged in a circle around a central point. In bilateral symmetry, the body can be divided lengthwise into two equal halves.
25. They enrich the soil and loosen it.
26. Leeches belong to the segmented worm phylum.
27. They attach to the skin of animals and suck blood.
28. Soft-bodied animals are grouped by the kind of foot, whether or not a shell is present, and the number of shells.

# Review

## Testing Yourself *continued*

9. animal without a backbone
10. flatworm that is not a parasite

**Finding Main Ideas**

*List the page number where each main idea below is found. Then, explain each main idea.*

11. the phylum that has the largest number of animal types
12. how to avoid getting a tapeworm
13. what a planarian uses to detect light
14. how animals such as the hydra and sea anemone capture their food
15. why sponges hold water easily
16. what an anus is
17. how a squid moves
18. how a hookworm enters the body
19. how clams buried in the sand get their food
20. what an earthworm eats

**Using Main Ideas**

*Answer the questions by referring to the page number after each question.*

21. How do you determine if a living thing belongs to the sponge phylum? (p. 137)
22. What features make animals different from other living things? (p. 134)
23. How are some stinging-cell animals harmful to humans? (p. 140)
24. How does radial symmetry differ from bilateral symmetry? (p. 134)
25. Why are earthworms important to farmers? (p. 147)
26. In which phylum are leeches found? (p. 146)
27. Why are leeches considered to be parasites? (p. 146)
28. What are the three traits used to group soft-bodied animals? (p. 149)

## Skill Review 

*For more help, refer to the Skill Handbook, pages 704-719.*

1. **Observe:** Use a hand lens to observe a natural sponge and a sponge made by humans. Predict which one will hold more water.
2. **Infer:** What can you infer about the fact that a tapeworm has no mouth?
3. **Interpret diagrams:** Look at Figure 7-9. Describe how a human might get tapeworms from a pig.
4. **Make and use tables:** Make a table that compares the traits of flatworms, roundworms, and segmented worms.

## Finding Out More

**Critical Thinking**

1. What advantage does a hydra have over a sponge in getting food?
2. Why is a tapeworm able to live without a digestive system?

**Applications**

1. Make a poster to illustrate and describe the life cycle of the dog heartworm. List the symptoms of a heartworm infection.
2. Write a report on how pearls are made by oysters. Explain the difference between natural pearls and cultured pearls.

**Chapter Review**/*Teacher Resource Package*, pp. 33-34.

**Chapter Test**/*Teacher Resource Package*, pp. 35-37.

**Chapter Test**/*Computer Test Bank*

## Skill Review

1. A natural sponge will hold more water because it has more holes.
2. It must absorb its food through the body wall.
3. If they eat the meat of the pig uncooked or undercooked.
4. Have students use the headings Flatworms, Roundworms, and Segmented Worms. Have them list the traits under each heading.

## Finding Out More

### Critical Thinking

1. The hydra can catch food with its stinging cells and tentacles and bring the food to it. A sponge must rely on the water current to bring it food. Hydra can digest larger food items than sponges.
2. It lives in the digestive system of its host and absorbs food that has already been digested.

### Applications

1. The larvae of dog heartworms are transmitted to dogs by mosquitoes. The worms develop in the dog's muscles and travel through the veins to the heart. The female reproduces in the heart and causes heart damage. Symptoms include listlessness and heart rhythm abnormalities.
2. Pearls are made when an irritant, such as a grain of sand, gets into the mantle area of an oyster. The mantle secretes the pearl substance around the irritant. For a cultured pearl, the irritant is placed in the oyster; for a natural pearl, the irritant gets into the oyster naturally.

# Complex Animals

## PLANNING GUIDE

| CONTENT | TEXT FEATURES | TEACHER RESOURCE PACKAGE | OTHER COMPONENTS |
|---|---|---|---|
| (1 1/2 days)<br>8:1 Complex Invertebrates<br>Jointed-leg Animals<br>Spiny-skin Animals | Mini Lab: *How Can You Catch a Spider Web?* p. 160<br>Lab 8-1: *Crayfish,* p. 162<br>Idea Map, p. 164<br>Check Your Understanding, p. 164 | Reteaching: *Complex Invertebrates,* p. 21<br>Skill: *Observe,* p. 8<br>Focus: *Body Coverings,* p. 15<br>Study Guide: *Invertebrates,* p. 43<br>Lab 8-1: *Crayfish,* pp. 29-30 | **Laboratory Manual:**<br>*What Are the Traits of Certain Jointed-leg Animals?* p. 59<br>**STVS:** *Sea Urchins and Power Plants,* Animals (Disc 5, Side 1)<br>**STVS:** *Horseshoe Crab,* Animals (Disc 5, Side 1) |
| (2 1/2 days)<br>8:2 Vertevbrates<br>Chordates<br>Characteristics of fish<br>Jawless Fish<br>Cartilage Fish<br>Bony Fish<br>Amphibians<br>Reptiles<br>Birds<br>Mammals | Skill Check: *Understand Science Words,* p. 169<br>Skill Check: *Infer,* p. 170<br>Mini Lab: *How Do You Make a Bird Feeder?* p. 171<br>Lab 8-2: *Feathers,* p. 172<br>Science and Society: *Are Animal Experiments Needed?* p. 175<br>Check Your Understanding, p. 174 | Application: *Beak Adaptations,* p. 8<br>Enrichment: *Orders of Mammals,* p. 8<br>Reteaching: *Vertebrates,* p. 22<br>Critical Thinking/Problem Solving: *What Can a Scientific Illustration Tell You?* p. 8<br>Transparency Master: *Chordate Classes,* p. 37<br>Study Guide: *Vertebrates,* pp. 44-47<br>Study Guide: *Vocabulary,* p. 48<br>Lab 8-2: *Feathers,* pp. 31-32 | **Laboratory Manual:**<br>*How Do Fish Classes Compare?* p. 63<br>**STVS:** *Studying Sharks,* Animals (Disc 5, Side 2)<br>**STVS:** *Sexing Birds,* Animals (Disc 5, Side 2) |
| Chapter Review | Summary<br>Key Science Words<br>Testing Yourself<br>Finding Main Ideas<br>Using Main Ideas<br>Skill Review | **ASSESSMENT RESOURCES**<br>Chapter Review, pp. 38-39<br>Chapter Test, pp. 40-42<br>Performance Assessment in the Biology Classroom<br>Alternate Assessment in the Science Classroom<br>Computer Test Bank | |

## GLENCOE TECHNOLOGY

**Infinite Voyage,** *Insects: The Ruling Class*
**Science and Technology Videodisc Series,** *Sea Urchins and Power Plants,* Animals (Disc 5, Side1)
*Horseshoe Crab,* Animals (Disc 5, Side 1)
*Studying Sharks,* Animals (Disc 5, Side 2)
*Sexing Birds,* Animals (Disc 5, Side 2)
**The Secret of Life,** *In the Land of Milk and Money: Biotechnology*

## MATERIALS NEEDED

| LAB 8-1, p. 162 | LAB 8-2, p. 172 | MARGIN FEATURES |
|---|---|---|
| small aquarium with crayfish | hand lens<br>metric ruler<br>scissors<br>wing or tail feather<br>down feather | Skill Check, p. 169<br>dictionary<br>Mini Lab, p. 160<br>black paper<br>spider web with dew<br>flour<br>Mini Lab, p. 171<br>wood<br>nails<br>bottle caps<br>hammer<br>peanut butter<br>bird seed<br>wire or string |

## OBJECTIVES

For more information about National Science Standards, see page 5T.

| | | CORRELATION of QUESTIONS to OBJECTIVES | | | |
|---|---|---|---|---|---|
| SECTION | OBJECTIVE | CHECK YOUR UNDERSTANDING | CHAPTER REVIEW | TRP CHAPTER REVIEW | TRP CHAPTER TEST |
| 8:1 National Science Stds: UCP.1, UCP.2, UCP.5, C.4, C.6 | 1. **Identify** the major traits of jointed-leg animals. | 1 | 11, 18, 30 | 1, 2, 11, 14, 15, 17, 33, 38 | 3, 5, 6, 15, 25 |
| | 2. **Compare** insects with other jointed-leg animals. | 2, 4 | 17, 25, 33 | 6, 9 | 1, 2, 4, 7, 31, 34, 35, 36, 37, 38, 39, 40, 41, 42, 43, 44 |
| | 3. **Describe** the traits of spiny-skin animals. | 3 | 12, 13, 27 | 7, 16, 18 | 19, 28 |
| 8:2 National Science Stds: UCP.1, UCP.2, UCP.3, UCP.5, C.4, C.6 | 4. **Compare** the traits of jawless fish, cartilage fish, and bony fish. | 6 | 3, 6, 15, 19, 22, 31 | 10, 23, 24, 26, 31, 36, 37 | 8, 16, 17, 24, 29, 33, 45, 46, 47, 48, 49, 50, 51, 52, 53, 54 |
| | 5. **Describe** the major traits of amphibians, reptiles, and birds. | 7 | 1, 4, 8, 10, 14, 16, 20, 21, 24, 26, 32 | 5, 12, 20, 21, 22, 27, 28, 29, 32, 34 | 9, 10, 11, 12, 14, 18, 20, 26, 30, 32 |
| | 6. **Identify** the characteristics of mammals. | 8 | 2, 5, 9, 23, 26, 29 | 3, 4, 8, 12, 13, 19, 25, 30, 35 | 13, 21, 23, 25, 27 |

# Complex Animals

## CHAPTER OVERVIEW

### Key Concepts

This chapter surveys the last three animal phyla. Students will learn that chordates are the most complex of animals. They also will learn which features are characteristic of chordates. Two of the phyla to be studied are invertebrates (jointed-leg animals and spiny-skin animals). Chordates are the only phylum studied that are classified as vertebrates.

### Key Science Words

| | |
|---|---|
| amphibian | hibernation |
| antennae | jawless fish |
| appendage | jointed-leg animal |
| bony fish | mammal |
| cartilage | mammary gland |
| cartilage fish | molting |
| chordate | reptile |
| cold-blooded | spiny-skin animal |
| compound eye | tube feet |
| endoskeleton | warm-blooded |
| exoskeleton | |
| gill | |

### Skill Development

In Lab 8-1, students will **make and use tables, observe,** and **infer** about the traits of jointed-leg animals. In Lab 8-2, they will **interpret data, observe,** and **infer** in relation to the structure of feathers. In the Skill Check on page 169, students will **understand** the **science word** *amphibian.* In the Skill Check on page 170, they will **infer** why reptiles lay fewer eggs than amphibians or fish. In the Mini Lab on page 160, students will **observe** a spider web. On page 171, they will **observe** birds at a bird feeder.

### Bridging

Chapter 7 introduced students to the invertebrate phyla. Chapter 8 continues with this theme. Two remaining invertebrate phyla are covered, and then vertebrates are introduced.

**CHAPTER PREVIEW**

### Chapter Content

Review this outline for Chapter 8 before you read the chapter.

**8:1 Complex Invertebrates**
- Jointed-leg Animals
- Spiny-skin Animals

**8:2 Vertebrates**
- Chordates
- Characteristics of Fish
- Jawless Fish
- Cartilage Fish
- Bony Fish
- Amphibians
- Reptiles
- Birds
- Mammals

### Skills in this Chapter

The skills that you will use in this chapter are listed below.

- In **Lab 8-1,** you will make and use tables, observe, and infer. In **Lab 8-2,** you will interpret data, observe, and infer.
- In the **Skill Checks,** you will understand science words and infer.
- In the **Mini Labs,** you will observe.

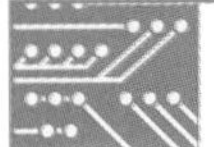

## TECH PREP

For Tech Prep activities in this chapter of the Teacher Wraparound Edition, see especially the Activity on page 158, the Motivation and the Student Journal on page 160, the Connection on page 167, and the Assessment on page 172.

See also the Glencoe Homepage at **www.glencoe.com**

Chapter 8

# Complex Animals

The photograph on the left shows one of the largest animals alive today. How can the elephant grow so big? Now look at the smaller photograph on this page. It is a daphnia, or water flea, magnified many times. Daphnia are found in lakes and ponds. Even though a daphnia is only 1 mm long, it is still a very complex animal. Looking into its transparent body, you can see that it is made up of many parts.

How are the daphnia and the elephant alike? In this chapter you will see what these two animals have in common, and how they differ. You will learn the traits of the more complex animals and how they are grouped.

**Try This!**

**What does a fish scale look like?** Place a dried fish scale on dark paper. Use a hand lens to count the wide, lighter bands. What do the bands represent?

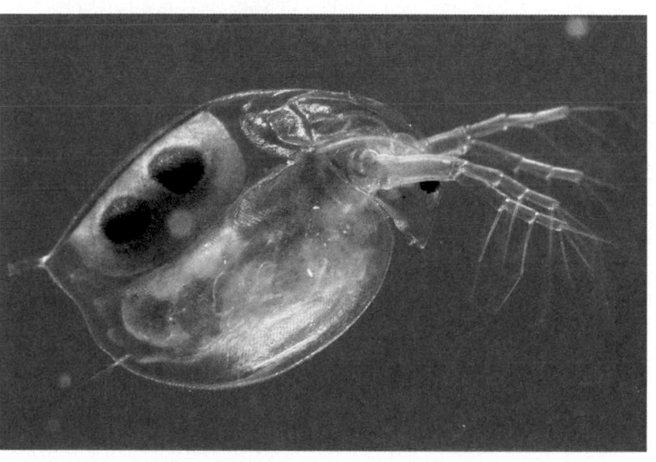

Daphnia

**interNET CONNECTION**

For more information about the material in this chapter, follow the link for the chapter on the Glencoe Homepage at **http://www.glencoe.com**



**ASSESSMENT PLANNER**

***Portfolio***
Strategies on the following page represent student products that can be placed into a best-work portfolio: p.174.

**PERFORMANCE ASSESSMENT**
Skill Check, p. 170
Mini Lab, pp. 160, 171
Lab, pp. 162, 172

**CONTENT ASSESSMENT**
Check for Understanding, pp. 164, 168, 173, 174
Chapter Review, pp. 176-177

**GROUP ASSESSMENT**
Opportunities for group assessment occur with Cooperative Learning Strategies.

***Student Journal***
Strategies on the following pages represent opportunities for writing in a Student Journal: pp. 160, 174.

**GETTING STARTED**

## Using the Photos

Elephants have internal skeletons, a trait that allows them to attain massive size. Daphnia have external skeletons, which limits their size. Daphnia have sense organs, digestive systems, and reproductive systems. So do elephants. However, the most significant trait that separates the two is the presence of a backbone in elephants and the absence of one in Daphnia.

## MOTIVATION/Try This!

**What does a fish scale look like?** Obtain fish scales from preserved fish, or by purchasing frozen fish from your local supermarket. Ask students to **observe** and try to determine what they represent. Bands represent rings of growth.

## Chapter Preview

Have students study the chapter outline before they begin to read the chapter. They should note the overall organization of the chapter, that the first section still deals with two invertebrate phyla. Next, vertebrates are discussed with only one phylum serving as an example of this group. They should note, however, that the Chordate phylum is divided into seven different classes.

## Misconception

There are many misconceptions about animals. Many people think that snakes are slimy when they are really covered with scales and feel cool and dry to the touch. Many believe that bats are blind. All bats can see and a few have better vision than people. Most bats, however, don't have an exceptionally keen sense of sight. They rely on their ability to echolocate to navigate. Most people believe that camels store water in their humps. A camel's hump contains fat as an energy reserve that helps it survive in the desert when food and water supplies are low.

## 8:1 Invertebrates

## PREPARATION

### Materials Needed

Make copies of the Focus, Skill, Study Guide, Reteaching, and Lab worksheets in the *Teacher Resource Package*.

▶ Make sure that living crayfish have been ordered for Lab 8-1 and that containers are available.

▶ Locate webs and obtain black paper and flour for the Mini Lab.

▶ Obtain 35-mm slides of representative arthropods.

### Key Science Words

jointed-leg animal
appendage
exoskeleton
molting
antennae
compound eye
spiny-skin animal
tube feet

### Process Skills

In the Mini Lab, students will observe. In Lab 8-1, students will make and use tables, observe, and infer.

## 1 FOCUS

▶ The objectives are listed on the student page. Remind students to preview these objectives as a guide to this numbered section.

### ACTIVITY/Brainstorming

Ask students to list the advantages and disadvantages of a suit of armor. Have them draw parallels to animals having a skeleton on the outside of their body (armor).

**Focus**/*Teacher Resource Package*, p. 15. Use the Focus transparency shown at the bottom of this page as an introduction to complex animals.

### Guided Practice

Have students write down their answers to the margin question: They allow quick movement.



## 8:1 Complex Invertebrates

**Objectives**

1. **Identify** the major traits of jointed-leg animals.
2. **Compare** insects with other jointed-leg animals.
3. **Describe** the traits of spiny-skin animals.

**Key Science Words**

jointed-leg animal
appendage
exoskeleton
molting
antennae
compound eye
spiny-skin animal
tube feet

The animals in the two invertebrate phyla in this chapter are more complex than the animals you studied in Chapter 7. One phylum, containing insects and spiders, is familiar to everyone. Insects are found almost everywhere. It is hard to think of life without insects.

### Jointed-leg Animals

Have you ever watched an army of ants moving food back to its nest? Hundreds of these animals work together to carry food that is much larger than they are. Ants crawl along in single file, following a trail that they marked on the way to the food. Back in the nest, hungry ants wait for the hunters to return. Ants are amazing animals!

Ants are jointed-leg animals. A **jointed-leg animal** is an invertebrate with an outside skeleton, bilateral symmetry, and jointed appendages (uh PEN dihj uz). An **appendage** is a structure that grows out of an animal's body. Jointed appendages bend to allow quick movement. Legs, antennae (an TEN ee), and wings are all appendages.

**What is an advantage to jointed appendages?**

The phylum of jointed-leg animals includes insects, spiders, and crayfish. It has more animal types than any other animal phylum. The graph on page 136 shows that over 80 percent of the known animal types on Earth are jointed-leg animals.

Jointed-leg animals have a segmented body. In many of these animals, the segments are fused into three parts—the head, the thorax (THOR aks), and the abdomen (AB duh mun). Look for these parts in the pictures in this section.

**Figure 8-1** Ants work together to carry food back to the nest.



## OPTIONS

### Science Background

The scientific name for the jointed-leg phylum is Arthropoda. There are six classes within this phylum. They include: horseshoe crabs; spiders, mites, and scorpions; crayfish, barnacles, and shrimp; insects such as ants, bees, and moths; centipedes; millipedes. The spiny-skin animals belong to the phylum Echinodermata. Classes in this phylum include: sea stars (starfish); sea urchins; sea cucumbers; feather stars.

FOCUS 15

Focus CHAPTER 8

Name Date Class

BODY COVERINGS

Notice that all these animals have different kinds of body coverings. How do you think each kind of body covering helps the animal?

Jointed-leg animals have an exoskeleton. An **exoskeleton** (EK soh skel uht uhn) is a skeleton on the outside of the body. It is made of a hard, waterproof, nonliving substance. It protects the body from drying and injury, and provides a place for muscles to attach.

How does an animal with an exoskeleton grow? The animal must shed its exoskeleton from time to time. Shedding the exoskeleton is called **molting.** After the old skeleton is shed, the body is quite soft. The animal swells by taking in extra water or air while the new skeleton hardens. This swelling gives the animal growing room inside the new skeleton. The animal is not very well protected from its enemies, so it stays hidden until its skeleton hardens.

There are five classes of jointed-leg animals. Crayfish, shrimp, lobsters, crabs, water fleas, and pill bugs are in one class, Figure 8-2. Crayfish and water fleas are found in fresh water. Pill bugs live in damp places on land. The rest of these animals live in the ocean.

Figure 8-3 shows the traits of animals in the first class. Animals in this class have mouthparts that hold, cut, and crush food. They have two pairs of antennae. **Antennae** are appendages of the head that are used for sensing smell and touch. These animals also have compound eyes for seeing. **Compound eyes** are eyes with many lenses. In contrast, you have simple eyes. They each have only one lens.

Most animals in this class have only two body sections. The head and the thorax are fused to make up one section. The abdomen is the second section. These animals usually have five pairs of legs for walking. A clawlike pair of legs at the head grabs and holds food.

The animals in this class are food for many fish and whales. Humans also eat lobsters, shrimp, and crabs.

a

b

**Figure 8-2** Crabs (a) and pill bugs (b) belong to the first class of jointed-leg animals.

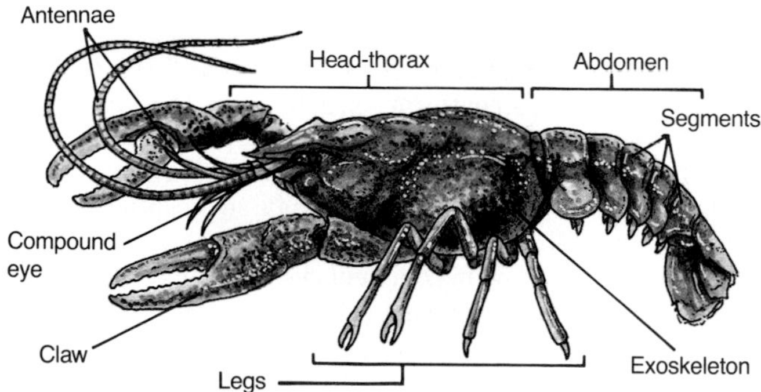


**Figure 8-3** Traits of a crayfish



# 2 TEACH

## MOTIVATION/Audiovisual

Show students 35 mm slides of some representative arthropods (any insect, crayfish or shrimp, spider, millipede, centipede). Ask them to list: (a) traits that are shared or are similar in these animals. (b) traits that are different in these animals.

## Concept Development

▶ The arthropods are the only invertebrate phylum with members adapted to life in air, land, soil, fresh water, or salt water.

▶ Explain that the jointed-leg animals are the first animals studied so far that have appendages. The mollusk foot is not a true appendage.

▶ The heavy exoskeleton is a greater problem for land animals than it is for those living in water. The buoyancy of the water helps support the exoskeleton of animals that live in water.

▶ Point out that jointed-leg animals also have an open circulatory system, dorsal heart, and ventral nerve cord, and most have a head, thorax, and abdomen.

▶ Point out that most jointed-leg animals undergo four to seven molts before reaching full adult size.

## Bridging

Ask students how the body segments of a jointed-leg animal differ from segments in a segmented worm and flatworm.

**SKILL 8**

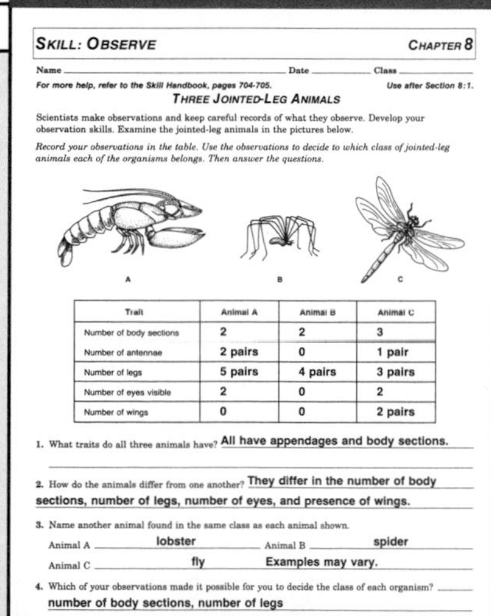

SKILL: OBSERVE — CHAPTER 8

Name ____ Date ____ Class ____

*For more help, refer to the Skill Handbook, pages 704-705.* *Use after Section 8:1.*

THREE JOINTED-LEG ANIMALS

Scientists make observations and keep careful records of what they observe. Develop your observation skills. Examine the jointed-leg animals in the pictures below.

*Record your observations in the table. Use the observations to decide to which class of jointed-leg animals each of the organisms belongs. Then answer the questions.*

A B C

| Trait | Animal A | Animal B | Animal C |
|---|---|---|---|
| Number of body sections | 2 | 2 | 3 |
| Number of antennae | 2 pairs | 0 | 1 pair |
| Number of legs | 5 pairs | 4 pairs | 3 pairs |
| Number of eyes visible | 2 | 0 | 2 |
| Number of wings | 0 | 0 | 2 pairs |

1. What traits do all three animals have? All have appendages and body sections.
2. How do the animals differ from one another? They differ in the number of body sections, number of legs, number of eyes, and presence of wings.
3. Name another animal found in the same class as each animal shown.
   Animal A lobster Animal B spider
   Animal C fly Examples may vary.
4. Which of your observations made it possible for you to decide the class of each organism? number of body sections, number of legs

## OPTIONS

**Traits of Jointed-leg Animals**/*Transparency Package,* number 8. Use color transparency number 8 as you teach about jointed-leg animals.

**Skill**/*Teacher Resource Package,* p. 8. Use the Skill worksheet shown here to give students practice with traits of jointed-leg animals.

## GLENCOE TECHNOLOGY

**Videodisc**

**Biology: The Dynamics of Life**

Movie: *Arthropods*
Disc 1, Side 2, Ch. 35

Movie: *Web-Spinning Spider*
Disc 1, Side 2, Ch. 36

## TEACH

### Concept Development

▶ Point out that the venom of a scorpion is a neurotoxin that causes paralysis of respiratory and cardiac muscles.

▶ Point out that most spiders are beneficial because they feed on insect pests.

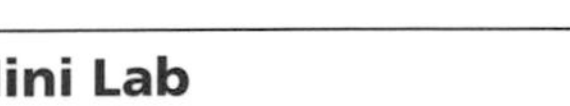

### Mini Lab

If students lightly dust the black paper with flour beforehand, they will be better able to catch and observe a spiderweb.

**Performance:** Have students observe the spiderweb and explain how an insect can be captured by it while the spider can walk on it to the insect. (The spoke threads on which the spider walks are not sticky.)

### MOTIVATION/Display

Display pictures of the black widow spider and the brown recluse spider. Explain that the bites of these two organisms can make a person extremely ill. While some deaths have been reported from spider bites, they usually result from complications rather than from the bite itself. Review first aid for spider bites and other animals' bites and stings. The First Aid Handbook of the American Red Cross is a good reference.

### Independent Practice

**Study Guide**/*Teacher Resource Package,* p. 43. Use the Study Guide worksheet shown below for independent practice.

**Student Journal**

Have students write a report on the tarantula. Students should include its structure, sense organs, method of respiration, how it obtains food, and whether it is dangerous to humans.

**Figure 8-4** Traits of a spider

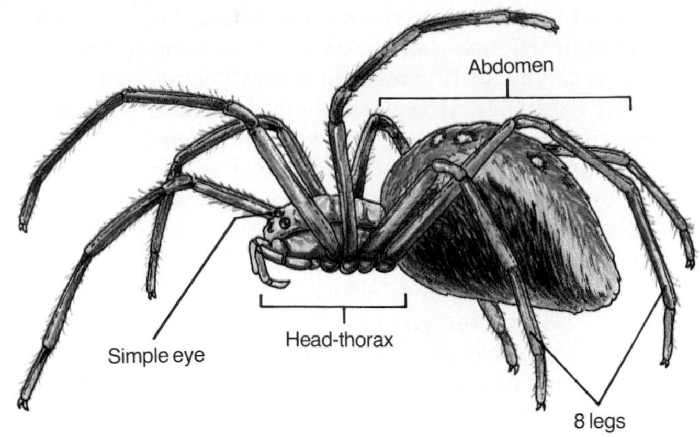


> ### Mini Lab
>
> **How Can You Catch a Spider Web?**
>
> **Observe:** Find a spider web covered with dew. Catch the web by placing a piece of black paper against it. Sprinkle some flour over the web on the paper and tap off the excess. What do you see? *For more help, refer to the **Skill Handbook,** pages 704-705.*

The second class of jointed-leg animals includes spiders, scorpions (SKOR pee uhnz), ticks, and harvestmen, also called daddy longlegs. Each has four pairs of walking legs, simple eyes, no antennae, and a body with two sections.

Spiders feed on insects. Most spiders trap insects in webs. The head of a spider has a pair of hollow fangs that connect to poison glands. When a spider traps an insect and bites into it, the poison stuns the insect. You may think that all spiders are dangerous to people. However, only a few spiders, like the black widow, have poison that causes serious sickness in people. Scorpions have a pointed stinger on the end of their abdomen. The stinger contains poison. Scorpions use their stinger for protection against enemies, but they can harm people, too.

The third and fourth classes of jointed-leg animals include the centipedes (SENT uh peedz) and millipedes (MIHL uh peedz). Centipedes and millipedes live on land under rocks or wood. They both have heads, long segmented bodies, and many legs.

**Figure 8-5** Centipedes have one pair of legs on each segment.

The name *centipede* suggests that these animals have one hundred legs. Remember that *centi-* means 100. *Pede* means foot. Actually centipedes usually have no more than 30 legs. You can see in Figure 8-5 that each of their segments has one pair of walking legs. The appendages on the first segment are poison claws. They help capture food. Centipedes usually eat insects.

Millipedes are slow-moving animals. They have two pairs of legs on most segments. They do not really have a thousand legs, as the name has you believe. Millipedes do not have poison claws. They usually eat decaying plants.



## OPTIONS

### ACTIVITY/Challenge

Have students prepare an illustrated report on the various hunting methods of spiders. Include the wolf spider, the trap-door spider, and the bird-eating spider.

**What Are the Traits of Certain Jointed-leg Animals?**/*Lab Manual,* pp. 59-62. Use this lab as an extension to the lesson on jointed-leg animals.

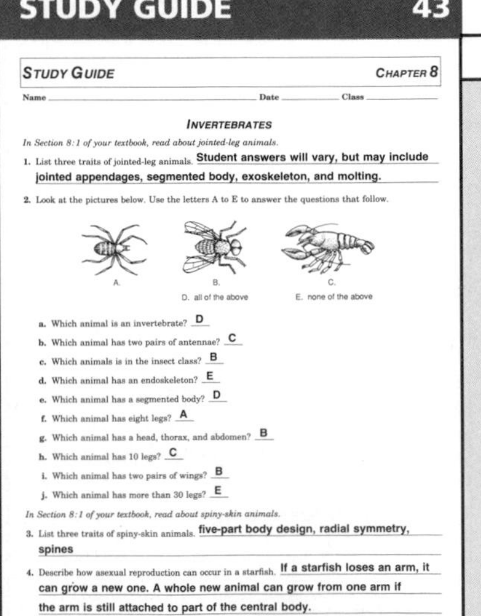

**STUDY GUIDE 43**

STUDY GUIDE CHAPTER 8

Name ______ Date ______ Class ______

**INVERTEBRATES**

*In Section 8:1 of your textbook, read about jointed-leg animals.*

1. List three traits of jointed-leg animals. **Student answers will vary, but may include jointed appendages, segmented body, exoskeleton, and molting.**

2. Look at the pictures below. Use the letters A to E to answer the questions that follow.

A. B. C.

D. all of the above E. none of the above

a. Which animal is an invertebrate? **D**
b. Which animal has two pairs of antennae? **C**
c. Which animals is in the insect class? **B**
d. Which animal has an endoskeleton? **E**
e. Which animal has a segmented body? **D**
f. Which animal has eight legs? **A**
g. Which animal has a head, thorax, and abdomen? **B**
h. Which animal has 10 legs? **C**
i. Which animal has two pairs of wings? **B**
j. Which animal has more than 30 legs? **E**

*In Section 8:1 of your textbook, read about spiny-skin animals.*

3. List three traits of spiny-skin animals. **five-part body design, radial symmetry, spines**

4. Describe how asexual reproduction can occur in a starfish. **If a starfish loses an arm, it can grow a new one. A whole new animal can grow from one arm if the arm is still attached to part of the central body.**

Insects are the fifth class of jointed-leg animals. There are more kinds of insects than all other animals combined. They live almost everywhere. They live deep in the ocean and high on mountain tops. They live in the air and on the ground, in the tropics and even at the North and South poles. They come in many shapes and colors. They can eat almost anything because they have special mouthparts for chewing, sucking, or lapping.

How can insects eat so many different kinds of food?

How can you tell if an organism is an insect? An insect's body has three main parts as shown in Figure 8-6. The head has two compound eyes and usually three simple eyes, one pair of antennae, and several mouthparts. The thorax has three pairs of walking legs and usually two pairs of wings, so the animal can fly. Insects are the only invertebrates that can fly.

Insects reproduce sexually, with separate sexes producing eggs and sperm. The fertilized eggs are laid in large numbers on leaves or branches of plants.

Insects can be helpful or harmful. Some insects destroy food crops. For example, Mediterranean fruit flies can destroy all of the fruit in an orchard. Moths can eat holes in your clothing when they feed on oils in the fabric or food particles you have left there. Termites destroy wood in buildings and fences. Why do people try to prevent houseflies from touching food? Houseflies carry bacteria that cause diseases.

Some insects help farmers by eating harmful insects. Ladybird beetles, or ladybugs, eat aphids that feed on many types of plants. Bees and other insects carry pollen from flower to flower. Crops must be pollinated in order to reproduce. There are insects that produce food and other useful materials. Honeybees, for example, produce honey.

### Bio Tip

**Health:** Deer ticks carry bacteria that cause Lyme disease. A tick picks up bacteria when it bites a mouse and sucks blood containing the bacteria. Then, when the tick bites a human, the bacteria are transferred. Lyme disease causes rashes and heart damage. It is treated with antibiotics.

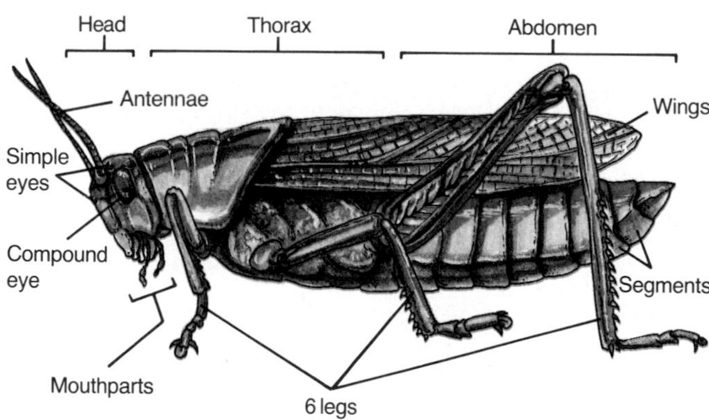


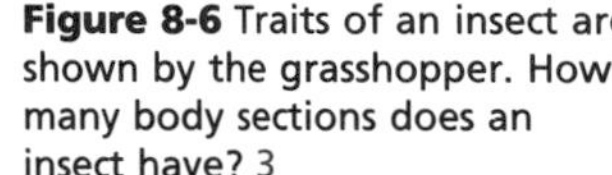

**Figure 8-6** Traits of an insect are shown by the grasshopper. How many body sections does an insect have? 3



## TEACH

### Concept Development

▶ Stress that insects are both helpful and harmful. Students may think that all insects are a nuisance, which is not true.

▶ Point out that mosquitoes do not pass the AIDS virus to people even though they do carry malaria from one person to another.

### Guided Practice

Compare the following three jointed-leg animal classes: insect, crayfish, spider. The comparison should consist of a chart with these headings: number of leg pairs, number of body sections, presence of exoskeleton, shows molting, antennae present, eyes present, wings present, lives on land, lives in water. When students are finished, ask them to place a checkmark after the traits shared by all animals. These will be phylum traits.

### Guided Practice

Have students write down their answers to the margin question: They have mouthparts for chewing, sucking, or lapping.

### ACTIVITY/Demonstration

Students may wish to actually follow the stages of insect development. This can be done by purchasing kits from biological supply houses of cricket, fly, butterfly, and moth life cycles.

## *OPTIONS*

### ACTIVITY/Challenge

Pheromones are being used to control many insect pests. Have students find out how they are being used to control fire ants.

### ACTIVITY/Project

Prepare a report on the many human diseases carried by insects. Include in the report if the disease is caused by a bacterium, virus, or protist.

## Lab 8-1 Crayfish

### Overview

Students will become familiar with the traits of animals in the jointed-leg phylum. They will study a crayfish as the representative animal for this phylum.

**Objectives:** Upon completion of the lab, students will be able to (1) **describe** the main traits for the jointed-leg phylum, (2) **relate** the structure of specific body parts to their function, (3) **state** the adaptive advantages of certain body structures.

**Time Allotment:** 30 minutes

### Preparation

▶ Living crayfish must be ordered in advance from a biological supply house. Make sure that you have sufficient small aquaria (battery jars) to house the animals while students observe them.

▶ **Alternate Materials:** It will be possible to complete this lab using preserved specimens rather than live animals. This may be a cheaper alternative because the same animals are used year after year.

**Lab 8-1 worksheet**/*Teacher Resource Package*, pp. 29-30.

### Teaching the Lab

▶ **Troubleshooting:** Wrap rubber bands around the front claws to prevent pinching. Do this before class.

**Cooperative Learning:** Divide the class into groups of two to three students. For more information, see pp. 22T-23T in the Teacher Guide.

**Performance:** Have students hypothesize how water temperature affects crayfish. Students should design an experiment to test the hypothesis, carry out the experiment, and record the data.

# Crayfish

**Lab 8–1**

## Problem: What are the traits of a jointed-leg animal?

### Skills

make and use tables, observe, infer

### Materials

small aquarium with live crayfish

### Procedure

1. Make a table similar to the one shown and record your observations as you make them. Use the figure to help you.
2. You will receive a crayfish in a small aquarium from your teacher. **CAUTION:** *Use care when working with live animals.* Leave the crayfish in the aquarium while you make your observations.
3. **Observe:** Touch the crayfish. Observe how it feels. Count the number of body sections.
4. Locate the two pairs of antennae. Are all the antennae the same length?
5. Look closely at the compound eyes. Notice that each eye is located at the top of a short stalk.
6. Locate the first pair of walking legs or claws. On what body section are they found?
7. Observe the other four pairs of walking legs. How are they different from the first pair?
8. Observe the five pairs of swimmerets. On which body section are they found?
9. Find the flipper at the end of the crayfish. How do you think it is used?
10. Return the crayfish to the place your teacher has assigned. Wash your hands.

| BODY SECTIONS | PARTS ATTACHED | FUNCTION |
|---|---|---|
| Fused head-thorax | 1 pr. small antennae | sensory |
| | 1 pr. large antennae | sensory |
| | 2 eyes | sensory |
| | 1 pr. claws | get food |
| | 4 pr. walking legs | movement |
| Abdomen | 5 pr. swimmerets | swim |
| | flipper | movement |

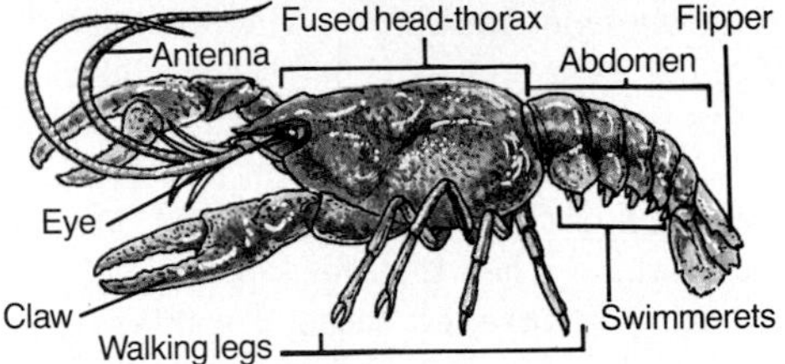


### Data and Observations

1. What appendages are attached to the first body section?
2. What appendages are attached to the abdomen?

### Analyze and Apply

1. Describe the exoskeleton of the crayfish.
2. What is the function of the exoskeleton?
3. How does the location of the eyes permit the crayfish to look in different directions?
4. **Infer:** How does the structure of the claws aid in getting food?
5. Which parts does the crayfish use to get food?
6. **Apply:** In what way are swimmerets in crayfish and fins in fish similar?

### Extension

**Observe** a spider. Identify the traits that spiders and crayfish have in common.



## ANSWERS

### Data and Observations

1. the antennae, eyes, claws, and walking legs
2. the swimmerets and flipper

### Analyze and Apply

1. The exoskeletion is a hard, waterproof, armorlike structure on the outside of a crayfish.
2. protection from drying and injury; and provides a place for muscles to attach
3. The eyes are on stalks and can move.
4. They are large claws for grabbing and holding food.
5. the claws
6. They are both used for steering the animal through the water.

a

b

c

Figure 8-7 A sea urchin (a), starfish (b), and sand dollar (c) are spiny-skin animals. They all show a five-part body design.

## Spiny-skin Animals

The animal in Figure 8-7a looks like a living pincushion. It is a sea urchin. The sea urchin, starfish, and sand dollar belong to the phylum of spiny-skin animals. A **spiny-skin animal** is an invertebrate with a five-part body design, radial symmetry, and spines. Count the arms of the starfish in Figure 8-7b. Sand dollars and sea urchins also show the five-part body design. You may have walked along a beach and seen these animals. They are found only in the oceans. They are very common on rocky shores.

What is the body design of a sand dollar?

What are some other traits of spiny-skin animals? If you look closely at the bottom of a starfish you will see dozens of tube feet, like those in Figure 8-8. **Tube feet** are parts like suction cups that help the starfish move, attach to rocks, and get food. If you ever try to pull a starfish from a rock, you will see how tightly it holds onto the rock with its tube feet. A tube foot works like a medicine dropper. If you squeeze the bulb of a dropper filled with water, you force out the water. If you squeeze the bulb and then place your finger against the open end, you will feel suction when you let go of the bulb. A starfish grips slippery rocks in much the same way. The starfish takes in ocean water and passes it through a series of canals to the tube feet. As this water moves in and out of the tube feet, suction is made and released.

Figure 8-8 Tube feet help a starfish move, attach to rocks, and get food.

## TEACH

### Concept Development

▶ *Echinoderm* means spiny skin. The term *spiny skin* refers to the spines of calcium found on the outside of these invertebrates. The spines are projections from an endoskeleton that is covered by a layer of skin.

▶ The starfish has radial symmetry, characteristic of spiny-skin animals. Other phylum traits include: no head region or body segmentation, presence of an endoskeleton, and are marine.

**Analogy:** You might compare the vascular system of the starfish to a hydraulic system.

### Guided Practice

Have students write down their answers to the margin question: five-part body design.

**Misconception:** Students usually believe that all starfish have five arms. Point out that some species have more or fewer than five. You may show pictures of sun stars that have eight or more arms, or the sunflower star of the California coast, which may have up to 20 arms. Emphasize that regardless of the number, the arms are not appendages but outgrowths or lobes of the body.

### GLENCOE TECHNOLOGY

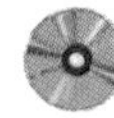

**Videodisc**

Biology: The Dynamics of Life
Movie: *Starfishes*
Disc 1, Side 2, Ch. 39

## OPTIONS

### Science Background

Because of the spiny external covering, the spiny-skin animals are not usually a source of food for humans. In the West Indies, sea urchin eggs are used for food, and in parts of Asia, dried sea cucumbers are used to make soup.

**What Are the Parts of a Starfish?**/*Lab Manual*, pp. 67-70. Use this lab as an extension to spiny-skin animals.

## TEACH

### Idea Map

Have students use the idea map as a study guide to the major concepts of this section.

### Check for Understanding

Have students respond to the first three questions in Check Your Understanding.

### Reteach

Reteaching/*Teacher Resource Package,* p. 21. Use this worksheet to give students additional practice in understanding complex invertebrates.

**Extension:** Assign Critical Thinking, Biology and Reading, or some of the **OPTIONS** available with this lesson.

## 3 APPLY

### ACTIVITY/Bridging

Prepare a list of traits that are similar in segmented worms and jointed-leg animals. Ask students if the list tends to support the idea that both phyla may be related. Are jointed-leg animals more closely related by similar traits to segmented worms or to spiny-skin animals?

## 4 CLOSE

### Audiovisual

Show the videocassette *Spiders; Aggression and Mating,* National Geographic Society Educational Service.

### Answers to Check Your Understanding

1. body divided into three parts, jointed appendages, exoskeleton
2. by molting
3. The tube feet act like suction cups.
4. They can live almost everywhere, eat anything, and can fly and walk.
5. Arthritis refers to a disease of the joints.

**Figure 8-9** Starfish can grow back missing arms. What type of reproduction is this? asexual

If a starfish loses an arm it can grow a new one, as seen in Figure 8-9. A whole new animal can grow from one arm if the arm is still attached to part of the central body. This is a form of asexual reproduction. Starfish also reproduce sexually with separate sexes producing eggs and sperm.

**Study Tip:** Use this idea map as a study guide to the invertebrates in this chapter. The features of jointed-leg animals and spiny-skin animals are shown.

**Idea Map**

**Complex Invertebrates**

- Complex invertebrates
  - jointed-leg animals
    - jointed appendages
    - segmented bodies
    - exoskeleton
  - spiny-skin animals
    - five-part body
    - spines
    - tube feet

### Check Your Understanding

1. List three traits of the jointed-leg animals.
2. How does an animal with an exoskeleton grow?
3. How do starfish use their tube feet for gripping surfaces?
4. **Critical Thinking:** Why are insects able to live in so many more different environments than other animals?
5. **Biology and Reading:** The scientific name for the jointed-leg animals is Arthropoda. The word *arthropoda* is made from two Greek words. The prefix *arthro-* means jointed. The suffix *-poda* means foot. What part of the body do you think the disease arthritis refers to?



## OPTIONS

### ACTIVITY/Challenge

You discover a "new" animal. Assume that it can talk, understands your questions, and is curled up so you cannot see any of its traits easily. Upon interviewing it, you want to determine to what phylum it belongs. What are some key questions that you might ask it in order to solve the phylum question?

### RETEACHING 21

**RETEACHING: IDEA MAP** — CHAPTER 8

Name ______ Date ______ Class ______

Use with Section 8:1.

**COMPLEX INVERTEBRATES**

*Use the following words or phrases to complete the idea map: spines, exoskeleton, five-part body, jointed appendages, tube feet, segmented bodies.*

- complex invertebrates
  - traits of jointed-leg animals
    - jointed appendages
    - segmented bodies
    - exoskeleton
  - traits of spiny-skin animals
    - five-part body
    - spines
    - tube feet

1. What is an appendage? a structure attached to the body
2. What is a segmented body? a body divided into sections
3. What is an exoskeleton? a skeleton on the outside of the body
4. How do tube feet help the spiny-skin animals move and get food? They are like suction cups allowing the animals to grip slippery rocks and food.
5. What main body parts do most jointed-leg animals have? head, thorax, abdomen
6. What happens to the exoskeleton when an invertebrate grows? The animal sheds the exoskeleton and grows another one.
7. How many legs does each of the following animals have? spider four pairs (8) insect three pairs (6) centipedes one pair per segment millipedes two pairs per segment

## 8:2 Vertebrates

You have been learning about some of the phyla that make up the invertebrate animals, those with no backbone. Animals that have a backbone are called vertebrates. All vertebrates belong to the chordate phylum. Vertebrates are the most complex organisms in the animal kingdom.

**Objectives**

4. **Compare** the traits of jawless fish, cartilage fish, and bony fish.
5. **Describe** the major traits of amphibians, reptiles, and birds.
6. **Identify** the characteristics of mammals.

**Key Science Words**

chordate
endoskeleton
cold-blooded
gill
jawless fish
cartilage
cartilage fish
bony fish
amphibian
hibernation
reptile
warm-blooded
mammal
mammary gland

### Chordates

The chordate phylum is probably the most familiar to you. The structures and ways of life of these animals are most like yours. They are your food, your pets, farm animals, the animals you see all around you. Chordates live in water as well as on land. Some chordates, such as bats and birds, are able to fly. You, too, belong to this phylum.

How do you identify a chordate? A **chordate** is an animal that, at some time in its life, has a tough, flexible rod along its back. The chordate phylum is named for this trait.

The chordate phylum contains the largest animals on Earth, one of which is shown in Figure 8-10. How are chordates able to grow so large? They have an endoskeleton (EN doh skel uht uhn). An **endoskeleton** is a skeleton on the inside of the body. An exoskeleton limits growth because it is on the outside of the body. Animals with endoskeletons don't have this problem.

**Figure 8-10** A hippopotamus can grow so large because it has an endoskeleton.



**TRANSPARENCY 37**

TRANSPARENCY MASTER — CHAPTER 8

CHORDATE CLASSES

| Class | Traits | Example |
|---|---|---|
| Jawless fish | 1. tubelike bodies covered with slime<br>2. do not have paired fins or jaws<br>3. round mouth lined with teeth<br>4. skeleton of cartilage<br>5. cold-blooded | |
| Cartilage fish | 1. skeleton of cartilage<br>2. jaws<br>3. toothlike scales on skin<br>4. paired fins<br>5. rows of sharp teeth<br>6. cold-blooded | |
| Bony fish | 1. skeleton of bone<br>2. skin with scales<br>3. gill covers<br>4. jaws<br>5. cold-blooded | |
| Amphibians | 1. lay eggs in water<br>2. young have gills, adults have lungs<br>3. broad mouths with a sticky tongue<br>4. cold-blooded | |
| Reptiles | 1. dry, scaly skin<br>2. well-developed lungs<br>3. two pairs of legs and clawed toes (except snakes)<br>4. egg has tough shell<br>5. cold-blooded | |
| Birds | 1. feathers and wings<br>2. hollow bones and powerful muscles<br>3. warm-blooded<br>4. have beaks, no teeth | |
| Mammals | 1. hair<br>2. feed milk to young<br>3. warm-blooded | |

## *OPTIONS*

### Science Background

The Chordate phylum consists of the following: Phylum Chordata; Subphylum Urochordata, or tunicates; Subphylum Cephalochordata, or Amphioxus; Subphylum Vertebrata.

**Transparency Master**/*Teacher Resource Package*, p. 37. Use the Transparency Master shown here to introduce the seven chordate classes.

## 8:2 Vertebrates

### PREPARATION

### Materials Needed

Make copies of the Transparency Master, Critical Thinking/Problem Solving, Application, Enrichment, Study Guide, Reteaching, and Lab worksheets in the *Teacher Resource Package*.

▶ Gather feathers for Lab 8-2.

▶ Obtain materials for bird feeders for the Mini Lab. Collect bird guides.

### Key Science Words

| | |
|---|---|
| chordate | bony fish |
| endoskeleton | amphibian |
| cold-blooded | hibernation |
| gill | reptile |
| jawless fish | warm-blooded |
| cartilage | mammal |
| cartilage fish | mammary gland |

### Process Skills

In Lab 8-2, they will interpret data, infer and observe. In the Skill Checks, they will understand science words and infer. In the Mini Lab, students will observe.

### 1 FOCUS

▶ The objectives are listed on the student page. Remind students to preview these objectives as a guide to this numbered section.

### ACTIVITY/Brainstorming

Ask students to compare their traits to those of a frog.

# 2 TEACH

### MOTIVATION/Demonstration

Show the class vertebral columns or skeletons of several vertebrates. Point out that vertebrates are distinguished from all other animals by an internal skeleton (endoskeleton) of cartilage and/or bone.

### Concept Development

▶ The tough flexible rod present in all chordates is called a notochord. In humans, it was present during embryonic development but was replaced by bony vertebrae. The word *chordate* refers to the chord in notochord.

▶ Many animals that are cold-blooded have some control over their body temperatures because they can move to warmer or cooler places. For example, lizards often sit in the sun. Their bodies warm while they are doing this.

### ACTIVITY/Display

Set up numbered specimens of the seven vertebrate classes around the room. Have students write the name of the organisms, its class, and the traits of the class for each specimen.

### ACTIVITY/Field Trip

To introduce fish, arrange a field trip to an aquarium. If you have a local aquarium, ask for a guided tour of the back rooms to see how the tanks are maintained and the fish are monitored.

**Using Science Word:** Explain that the plural form, when referring to more than one species, is fishes and when referring to more than one individual animal is fish.

### Independent Practice

**Study Guide**/*Teacher Resource Package*, p. 44. Use the Study Guide worksheet shown at the bottom of this page for independent practice.

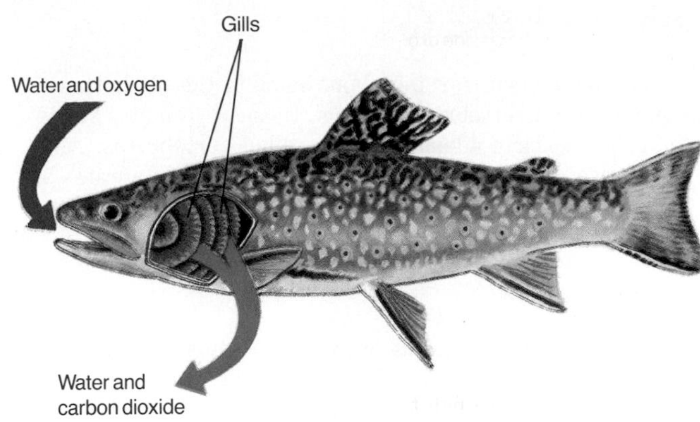


**Figure 8-11** Fish have gills that remove oxygen from water. What gas is removed from the fish by the gills? carbon dioxide

Vertebrates are chordates. In most vertebrates, the rod along the back is replaced by a backbone. Vertebrates have well-developed body systems. They have a circulatory system with a heart and blood vessels, a digestive system to change food into a useful form, a skeletal system for support, a respiratory system for gas exchange, and a nervous system for control. Vertebrates have large brains, well-developed senses, and are very intelligent animals. The seven vertebrate classes are jawless fish, cartilage fish, bony fish, amphibians, reptiles, birds, and mammals.

## Characteristics of Fish

Three of the seven vertebrate classes are fish. All fish have certain traits in common. Fish are cold-blooded vertebrates that live in water and breathe with gills. **Cold-blooded** means having a body temperature that changes with the temperature of the surroundings. Fish have gills on each side of the throat region. A **gill** is a structure used to breathe in water. The fish pumps water into its mouth and out through the gills, Figure 8-11. The gills pick up oxygen from the water as it passes through.

Most fish have scales that cover and protect their bodies. They also have fins that help them swim. The fins steer the fish as it moves its body from side to side through the water. A lateral line runs along each side of the body. The lateral line is an important sense organ for fish because it detects water movement and the presence of objects.

There are three different classes of fish in the chordate phylum. They are jawless fish, cartilage fish, and bony fish.



## OPTIONS

### Science Background

Jawless fish, cartilage fish, and bony fish are sometimes grouped into the superclass Pisces. This category denotes movement by body fins.

### STUDY GUIDE 44

STUDY GUIDE — CHAPTER 8

Name ______ Date ______ Class ______

**VERTEBRATES**

*In Section 8:2 of your textbook, read about chordates.*

1. What trait identifies a chordate? A chordate is an animal that, at some point in its life, has a tough, flexible rod along its back.
2. In most vertebrates, what replaces the rod along its back? a backbone

*In Section 8:2 of your textbook, read about characteristics of fish.*

3. Put a checkmark in the column of each kind of fish that has the trait listed.

| Trait | Jawless fish | Cartilage fish | Bony fish |
|---|---|---|---|
| have skeletons mostly of bone | | | ✓ |
| have toothlike scales | | ✓ | |
| are parasites | ✓ | | |
| group includes perch and trout | | | ✓ |
| have skeletons of cartilage | ✓ | ✓ | |
| group includes lamprey | ✓ | | |
| have smooth, bony scales | | | ✓ |
| have uncovered gills | ✓ | ✓ | |
| group includes sharks and rays | | ✓ | |
| have paired fins | | ✓ | ✓ |
| most have a swim bladder | | | ✓ |
| have tubelike bodies | ✓ | | |

4. Fill in the correct answers in the blanks provided below each picture.

a. Class bony fish Phylum chordate
b. Class jawless fish Phylum chordate
c. Class cartilage fish Phylum chordate

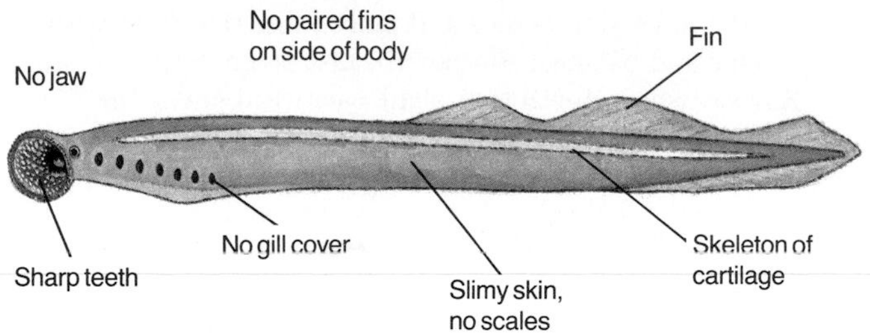


**Figure 8-12** Traits of jawless fish are shown by a lamprey.

## Jawless Fish

**Jawless fish** are fish that have no jaws and are not covered with scales. The skeletons of jawless fish are made of cartilage (KART ul ihj). **Cartilage** is a tough, flexible tissue that supports and shapes the body. Your ears and nose tip are made up of cartilage.

The lamprey (LAM pree) shown in Figure 8-12 is a jawless fish. Lampreys have tubelike bodies covered with slime that protects the skin. They don't have paired fins. Lampreys swim by waving their body from side to side.

The lamprey's mouth does not have jaws. It is a round opening lined with toothlike structures. How does the lamprey eat? Many lampreys attach themselves to the sides of other fish with their sharp, toothlike structures. The lamprey cuts a hole through the skin and sucks out the blood and body fluids. Are lampreys parasites?

## Cartilage Fish

**Cartilage fish** are fish in which the entire skeleton is made of cartilage. They have no bone. Unlike the jawless fish, cartilage fish have jaws, toothlike scales, and paired fins. Sharks and rays are cartilage fish. The shark in Figure 8-13 shows these traits.

**What kind of skeleton do cartilage fish have?**

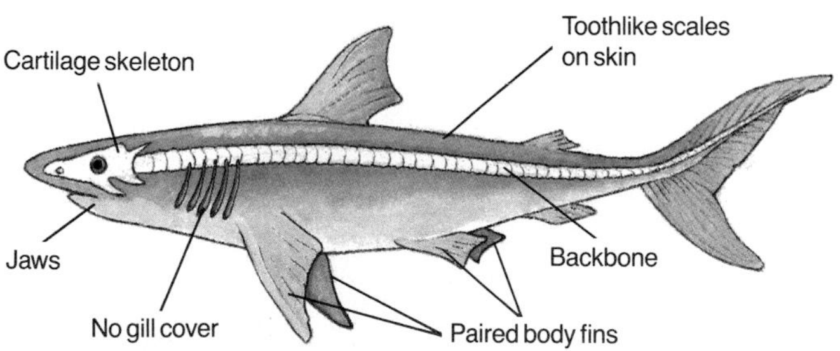


**Figure 8-13** Traits of cartilage fish are shown by a shark.



## TEACH

### Concept Development

▶ Lampreys have one heart, hagfishes have at least four, Hagfish have 5 to 15 pairs of gill openings and mucus-secreting glands over their bodies. The hagfishes have been omitted here.

▶ Point out that sharks and rays can be distinguished by the placement of their gill slits. The gill slits of a shark are always on the side of its head and the ray's gill slits are underneath its wide fins.

▶ Cartilage fish also have 2-chambered hearts.

### Connection: Language Arts

Make a list of everyday phrases that describe certain characteristics in terms of animal traits. Ask students to discuss the origin of the phrases and to decide whether the comparisons are valid. Use such phrases as "sly as a fox," "slippery as an eel," "dog tired," "blind as a bat," "eats like a bird," "a memory like an elephant," "free as a bird," and "bee line."

### Guided Practice

Have students write down their answers to the margin question: an internal skeleton of cartilage.

### GLENCOE TECHNOLOGY

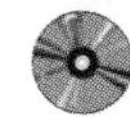

**Videodisc**

**Biology: The Dynamics of Life**
Movie: *Fish Schooling*
Disc 2, Side 1, Ch. 2

### CRITICAL THINKING 8

CRITICAL THINKING/PROBLEM SOLVING — CHAPTER 8

Name ______ Date ______ Class ______

*Use after Section 8:2.*

**WHAT CAN A SCIENTIFIC ILLUSTRATION TELL YOU?**

Science concepts can be shown with words and with pictures. Some things can be explained best in words, such as experimental results. Illustrations are more helpful for showing structures. For example, many fish have unusual adaptations to their environments.

*Use the descriptions below to match the fish to the illustrations. Write the name of each fish next to its picture. Then write how the adaptation might help the fish survive.*

Lionfish: Attacks other fish with its sharp fins, which look like feathers. Fins release a poison.

Sargassum Fish: Climbs seaweed with its pawlike fins. Its shaggy body looks like the seaweed.

Porcupine fish: Can fill itself with water and swell up in size. Its skin is covered with spines.

Flounder: Has a flat body that changes color to match its background. Both eyes are on one side of the body.

Flounder — Matches sea floor. Eyes see above.

Porcupine Fish — Puffs up too big to be swallowed.

Sargassum Fish — Is disguised like seaweed.

Lionfish — Attacks food fish with poison fins.

### OPTIONS

**Enrichment:** Mention that in the 1950s one species of sea lamprey threatened the entire Great Lakes fishing industry. The lamprey was partially controlled by chemicals that killed newly hatched larvae.

**How Do Fish Classes Compare?**/*Lab Manual,* pp. 63-66. Use this lab as an extension to the lesson on fish.

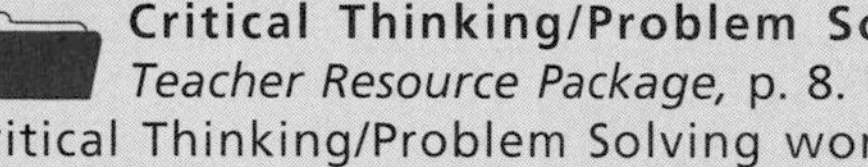

**Critical Thinking/Problem Solving/** *Teacher Resource Package,* p. 8. Use the Critical Thinking/Problem Solving worksheet shown here as an extension to fish adaptation.

## TEACH

### Concept Development

▶ Experiments have shown that sharks use their senses of smell and touch to detect and locate prey. The eyes are used only as they approach the prey.

▶ Students may be interested in knowing that the cartilage fish have no swim bladder. They begin to sink when they stop swimming.

▶ Mention that bony fish also have two-chambered hearts.

### ACTIVITY/Demonstration

You may demonstrate camouflage in bony fish by placing a few minnows in each of two beakers. Place one beaker on white cloth and place one on black cloth. Have students observe. Switch the beakers and observe again.

### ACTIVITY/Hands-On

You may want to have students dissect a dogfish shark to show the body systems of a cartilage fish. They are available at biological supply houses.

### Check for Understanding

Ask students to compare the three fish classes as to: (a) where they live; (b) what makes up the skeleton of each; (c) what they use to move about; (d) number of fins; (e) skin covering; (f) flap covering the gills.

### Reteach

Make flash cards of the traits of the three fish classes. On the back of each card, write the class(es) that have the trait. Show students the traits and have them name the corresponding classes.

### Independent Practice

**Study Guide**/*Teacher Resource Package,* p. 45. Use the Study Guide worksheet shown at the bottom of this page for independent practice.

**Figure 8-14** A ray has a flat body that is adapted for living on the ocean floor.

Sharks have slim bodies and paired fins that help with movement and balance. Sharks are fast swimmers.

Rows of sharp teeth that slant backward cover the shark's jaws. The teeth help sharks hold and cut up their food. Most sharks eat other animals that live in the ocean. However, the whale shark eats only protists.

Rays are flat and live on the ocean bottom, Figure 8-14. They eat fish and invertebrates in the ocean. Most rays are harmless to humans, but some stingrays have whiplike tails with stingers that can cause a painful wound.

## Bony Fish

Most fish known today belong to the class of bony fish. **Bony fish** have skeletons made mostly of bone, not of cartilage. Many fish that you eat, such as perch, bass, flounder, and trout, are in this group. You may have seen some of these fish in a supermarket.

Figure 8-16 shows the important traits of bony fish. Most bony fish have smooth, bony scales that are covered with a slimy coating, but some, like the catfish in Figure 8-15, have slime-covered leathery skin instead. The slime and scales protect the fish from infections and from enemies. The slime also makes it easier for the fish to glide through the water.

**Figure 8-15** Catfish have leathery skin instead of scales.

Bony fish have a flap that covers and protects the gills. Jawless fish and cartilage fish do not have this flap. Most bony fish have a swim bladder. A swim bladder is a baglike structure that fills with gases and helps the fish float and go up and down in the water. Fish can change the amount of gas in the swim bladder. As the bladder fills with gas, the fish rises. As the gas is let out, the fish goes deeper.

Most fish need water to reproduce. The female lays large numbers of eggs in the water. The male fish deposits sperm in the water, and some of the eggs are fertilized when the sperm swim to the eggs.

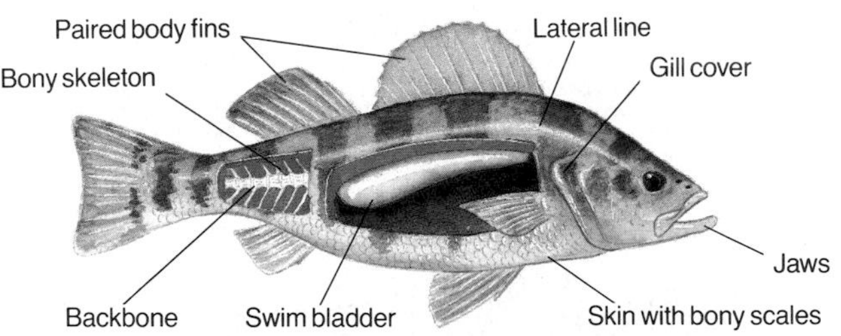
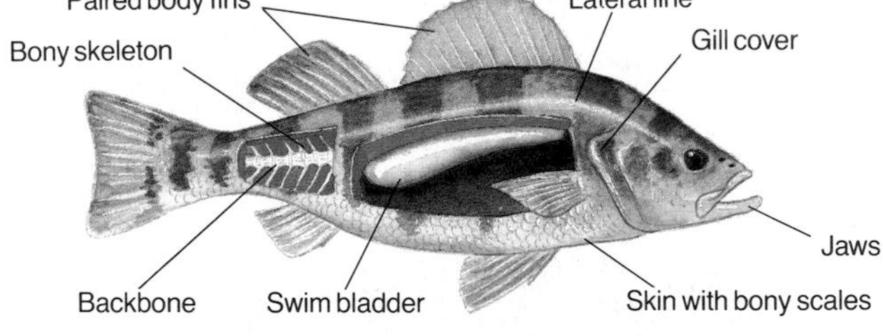


**Figure 8-16** Traits of a bony fish. What is the function of the swim bladder? It helps the fish go up and down in the water.



## OPTIONS

### ACTIVITY/Challenge

Have students report on the most dangerous shark waters in the world. Have them include those shark species present.

### STUDY GUIDE 45

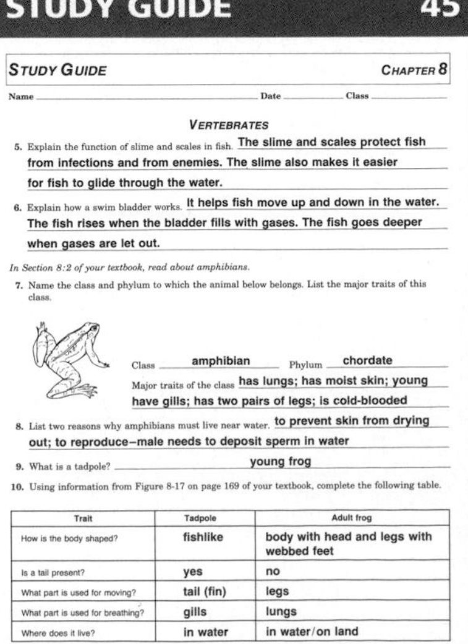

STUDY GUIDE CHAPTER 8

Name ______ Date ______ Class ______

VERTEBRATES

5. Explain the function of slime and scales in fish. The slime and scales protect fish from infections and from enemies. The slime also makes it easier for fish to glide through the water.

6. Explain how a swim bladder works. It helps fish move up and down in the water. The fish rises when the bladder fills with gases. The fish goes deeper when gases are let out.

*In Section 8:2 of your textbook, read about amphibians.*

7. Name the class and phylum to which the animal below belongs. List the major traits of this class.

Class amphibian Phylum chordate

Major traits of the class has lungs; has moist skin; young have gills; has two pairs of legs; is cold-blooded

8. List two reasons why amphibians must live near water. to prevent skin from drying out; to reproduce–male needs to deposit sperm in water

9. What is a tadpole? young frog

10. Using information from Figure 8-17 on page 169 of your textbook, complete the following table.

| Trait | Tadpole | Adult frog |
|---|---|---|
| How is the body shaped? | fishlike | body with head and legs with webbed feet |
| Is a tail present? | yes | no |
| What part is used for moving? | tail (fin) | legs |
| What part is used for breathing? | gills | lungs |
| Where does it live? | in water | in water/on land |

## Amphibians

Amphibians (am FIHB ee uns) include frogs, toads, and salamanders. An **amphibian** is an animal that lives part of its life in water and another part of its life on land. Usually, young amphibians live in water and adult amphibians live mostly on land. Amphibians that live mostly on land still need water. Without water, an amphibian's skin dries out. Amphibian eggs are laid in water so they won't dry out. The female lays large numbers of eggs. The male deposits sperm in the water. Only a few eggs are fertilized.

Amphibian young look very different from the adults. Notice that the tadpole in Figure 8-17b is different from an adult frog in many ways. Tadpoles must live in water and breathe with gills. The adults have lungs and can take in oxygen from the air. They also take in oxygen through their moist skins and the linings of their mouths.

Frogs and toads are probably familiar to you. Figure 8-17a shows some of their traits. Most frogs and toads do not have tails. They have broad mouths with long, sticky tongues for catching insects. They have two pairs of legs. The hind legs are much larger and more powerful than the front legs, so the animal can jump great distances. Frogs and toads have webbed feet for swimming. Their eyes stick out from their heads. This trait allows them to hide under the water with only their eyes showing above the surface. When an insect flies by, the frog flicks out its sticky tongue and catches the insect.

### Skill Check

**Understand science words: amphibian.** The word part *amphi* means both. In your dictionary, find three words with the word part *amphi* in them. *For more help, refer to the* ***Skill Handbook****, pages 706-711.*

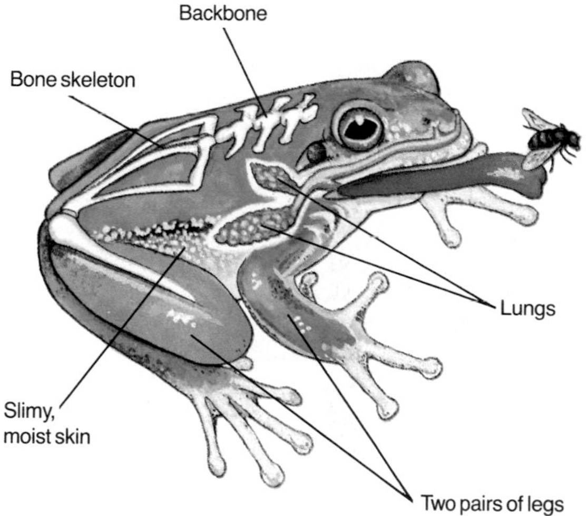

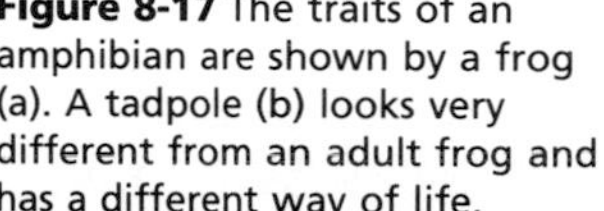

**Figure 8-17** The traits of an amphibian are shown by a frog (a). A tadpole (b) looks very different from an adult frog and has a different way of life.



## TEACH

### Concept Development

▶ Point out that many amphibians have poison glands beneath their outer skin. Their poison serves as a protection against natural enemies. Generally these poisons are harmless to humans. However, the South American tree frog Dendrobates secretes a powerful poison used by natives to make poison-tipped arrows.

▶ Remind students that all the amphibian traits are not necessarily present in every animal of each class. For example, *Xenopus laevis,* the South African aquatic frog, has no tongue and rarely leaves the water. There are exceptions for each class (and phylum) that is presented.

### ACTIVITY /Demonstration

Use a goldfish or guppy to demonstrate the flow of water across a fish's gills. Have students watch the opening and closing of the mouth and gill covers. Place a drop of food coloring in front of the fish's mouth and see how quickly the water comes out of the gills.

### GLENCOE TECHNOLOGY

**Videodisc**

**Biology: The Dynamics of Life**
Movie: *Frog Behavior*
Disc 2, Side 1, Ch. 3

## OPTIONS

### Science Background

Amphibians, reptiles, birds, and mammals are sometimes grouped into the superclass Tetrapoda. This category denotes movement by legs.

## TEACH

### Concept Development

▶ Point out that lizards (reptile) and salamanders (amphibian) have similar body shapes but are very different animals.

▶ Emphasize to students that prehistoric reptiles were the dominant animals of their day, just as mammals are dominant today.

### ACTIVITY/Demonstration

You might want to introduce the section by displaying a small turtle and having students note the traits.

### ACTIVITY/Display

Display pictures of reptiles such as the glass lizard, horned toad, gecko, different kinds of snakes, crocodiles, and alligators. Have students discuss how they are similar.

### Skill Check

Reptile eggs have leathery protective coverings. More reptile eggs will hatch and survive; therefore, fewer eggs need to be laid.

### Independent Practice

Study Guide/*Teacher Resource Package*, p. 46. Use the Study Guide worksheet at the bottom of this page for independent practice.

**GLENCOE TECHNOLOGY**

**Videodisc**

**Biology: The Dynamics of Life**
Movie: *Penguins*
Disc 2, Side 1, Ch. 7

a

b

**Figure 8-18** A cave salamander (a) and a snake (b) are adapted for living on land.

Salamanders are amphibians with a tail. Their two pairs of legs are about the same size, Figure 8-18a. They live only in moist places. Some salamanders keep their gills throughout their lives and live in water, even as adults.

Amphibians are cold-blooded vertebrates. Their body temperature drops as the temperature of their surroundings drops. Because they are cold-blooded, amphibians become inactive during cold weather. The state of being inactive during cold weather is called **hibernation.** Animals that hibernate eat no food and use only a little oxygen. Their energy needs are met with stored fat.

Amphibians help control insects. Amphibians also are used in life science studies and in medical research. They are eaten by snakes, turtles, birds, and mammals.

## Reptiles

You may have seen a snake at one time and thought it looked slimy. If you touched the snake, you probably were surprised to find that it was not slimy at all. Snakes are reptiles. A **reptile** is an animal that has dry, scaly skin and can live on land. Reptiles were the first vertebrates to live and reproduce entirely on land. Snakes, lizards, turtles, crocodiles, and alligators are all reptiles.

What are the traits of reptiles? Look at Figure 8-19. Reptiles are cold-blooded vertebrates with a backbone and an endoskeleton. Their scaly skin protects them, prevents water loss, and keeps them from drying out. Some reptiles are covered by hard plates instead of scales. They have well-developed lungs for breathing air. Most reptiles have two pairs of legs and clawed toes, but snakes and some lizards don't have legs. Most reptiles can move quickly.

### Skill Check

**Infer:** Compare reproduction in reptiles with reproduction in fish and amphibians. Explain why reptiles lay far fewer eggs than amphibians or fish. *For more help, refer to the **Skill Handbook**, pages 706-711.*

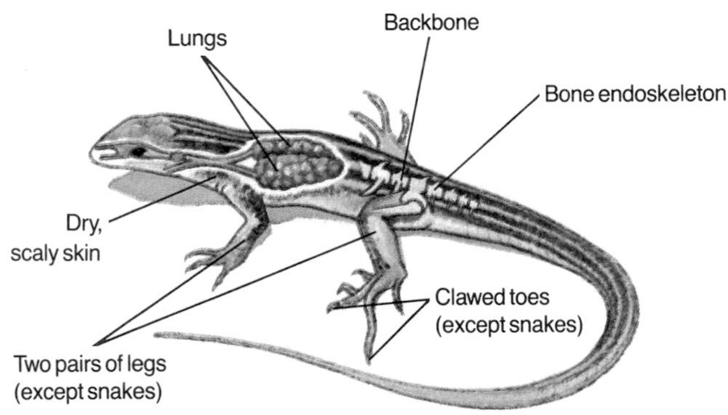


**Figure 8-19** A lizard shows the traits of a reptile.

170 Complex Animals 8:2

## OPTIONS

### Science Background

Reptiles have hearts with atria and a single (three-chambered heart) ventricle. Crocodiles have two ventricles (a four-chambered heart).

**STUDY GUIDE 46**

STUDY GUIDE CHAPTER 8

Name ______ Date ______ Class ______

VERTEBRATES

11. How are cold-blooded animals affected by the temperature in their surroundings? The temperature of cold-blooded animals drops as the temperature of their surroundings drop.

12. How does hibernation help amphibians survive? Animals that are inactive during cold weather eat no food and use only a little oxygen.

*In Section 8:2 of your textbook, read about reptiles.*

13. Name the class and phylum to which the animal below belongs. List the major traits of this class.

Class reptile Phylum chordate

Major traits of the class cold-blooded; has dry, scaly skin or hard plates; has lungs; has tough, leathery-shelled egg

14. List five kinds of reptiles. snakes, lizards, alligators, crocodiles, turtles

15. List three reasons why a reptile can live entirely on land. It has dry, scaly skin that prevents it from drying out, well-developed lungs for breathing air, and a tough, leathery-shelled egg that doesn't need to be laid in water.

*In Section 8:2 of your textbook, read about birds.*

16. Compare the following traits of birds with those of reptiles. Using the words *same* or *different*, fill in the blanks provided with the answer that applies.

a. warm-blooded different
b. scaly legs same
c. shelled eggs same
d. toes with claws same
e. feathers different
f. lungs same
g. beaks different
h. hollow bones different

They use their claws for running and climbing, and for digging nests in the soil. A reptile egg has a tough, leathery shell that protects it and keeps it from drying out. Because reptile eggs have shells, they can be laid on land instead of in the water. Even though reptiles do not need to live in water, many still do. Alligators, crocodiles, and turtles, for example, spend a good part of their lives in or near the water.

Reptiles eat insects and pests such as rats and mice. In some areas of the world, people eat reptiles and their eggs.

**How are reptile eggs protected?**

## Birds

Birds are vertebrates that have wings, a beak, two legs, and a covering of feathers over most of their bodies. Some of the traits of birds are shown in Figure 8-20. Like reptiles, birds have scales, but the scales are only on their legs. Birds also have claws on their toes. They have well-developed lungs. Female birds, like reptiles, lay eggs with shells from which the young develop.

Most birds are well adapted for flying. They have hollow bones, which makes them light in weight. Birds have powerful muscles to move their wings. Some birds, such as the ostrich, do not fly. Ostriches, however, can run as fast as 48 km an hour, about the speed of a racing bike.

Birds are warm-blooded. **Warm-blooded** means controlling the body temperature so that it stays about the same no matter what the temperature of the surroundings. Feathers help the body keep a constant body temperature.

### Mini Lab

**How Do You Make a Bird Feeder?**

**Observe:** Nail bottle caps onto a piece of wood. Fill the caps with peanut butter or birdseed. Hang the feeder from a tree with string. Observe the birds that come to the feeder. *For more help, refer to the **Skill Handbook,** pages 704-705.*

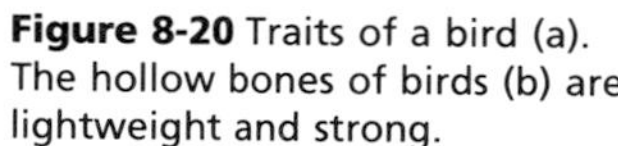

**Figure 8-20** Traits of a bird (a). The hollow bones of birds (b) are lightweight and strong.

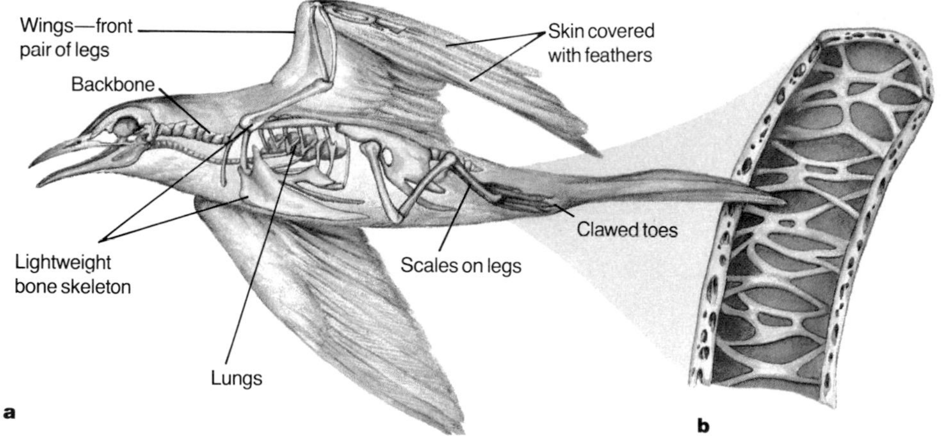


171

## TEACH

### Concept Development

▶ Bird skin secretes an oily substance that a bird uses to coat its feathers. This behavior is called preening. Bird feathers are replaced at least once a year by molting. Feathers are usually lost and replaced in pairs, preserving balance for flight.

▶ Point out that warm-blooded animals require more energy to heat their bodies than do cold-blooded animals. Birds must constantly seek food to provide energy to keep their bodies warm.

### Mini Lab

Depending on the area in which you live, students should be able to observe sparrows, finches, and woodpeckers. Woodpeckers (and squirrels) are especially attracted to peanut butter.

### ✓ ASSESSMENT

**Skill:** Provide students with a photocopied page of birds showing their beaks and feet. Have them observe the diagrams and describe how each bird obtains food.

### Guided Practice

Have students write down their answers to the margin question: by a tough, leathery shell.

### Independent Practice

**Study Guide**/*Teacher Resource Package,* p. 47. Use the Study Guide worksheet shown at the bottom of this page for independent practice.

**STUDY GUIDE 47**

STUDY GUIDE — CHAPTER 8

Name ______ Date ______ Class ______

VERTEBRATES

17. Explain the meaning of *warm-blooded.* Warm-blooded means controlling the body temperature so that it stays about the same no matter what the temperature of the surroundings.

18. What do birds have that helps them keep a constant body temperature? feathers

19. How are birds well adapted for flying? They have hollow bones, which makes them light in weight. They have powerful muscles that move their wings.

20. How are the beaks of birds adapted for getting food? The kind (shape and size) of beak they have is related to the kind of food they eat.

*In Section 8:2 of your textbook, read about mammals.*

21. Name the class and phylum to which the animal below belongs. List the major traits of this class.

Class mammal Phylum chordate

Major traits of the class has hair; is warm-blooded; has mammary glands, nurses young; young are born alive.

22. a. What is a mammal? A mammal is an animal that has hair and feeds milk to its young.

b. Are you a mammal? yes

23. a. What are mammary glands? body parts that produce milk

b. How are male mammary glands different from female mammary glands? Male mammary glands don't produce milk.

24. List two groups of mammals whose young do not develop inside the mother's body. pouched mammals, egg-laying mammals

**APPLICATION 8**

APPLICATION: LEISURE — CHAPTER 8

Name ______ Date ______ Class ______

*Use after Section 8:2.*

BEAK ADAPTATIONS

Birdwatching is a hobby that many people enjoy. Knowing what kind of food a bird eats can help a birdwatcher know where to find the bird.

To obtain food, a bird uses its beak like a tool. The shape of a bird's beak is related to the type of food the bird eats. Compare the function of each bird's beak to the function of the tools shown below. On the first line next to each bird picture, write the name of the tool whose function most closely resembles that of the bird's beak. Then, consider what kind of food each bird could obtain using its beak. On the second line, write the type of food you think each bird would eat. Refer to the foods and tools listed below. You may use a tool or food choice more than once.

| Bird | Tool | Food |
|---|---|---|
| woodpecker | awl | insects |
| parrot | nutcracker | seeds |
| flamingo | strainer | tiny organisms |
| vulture | meat knife | fleshy animals |
| hummingbird | soda straw | nectar |
| crossbill | nutcracker | seeds |
| hawk | meat knife | fleshy animals |
| cardinal | nutcracker | seeds |
| pelican | scoop | fish |

Tools: nutcracker, meat knife, soda straw, strainer, awl (for puncturing), scoop

Foods: seeds, nectar (in flowers), fleshy animals, tiny organisms (in water), insects (under tree bark), fish

## OPTIONS

**Application**/*Teacher Resource Package,* p. 8. Use the worksheet shown here to teach students about beak adaptations.

## Lab 8-2 Feathers

### Overview

In this lab, students become familiar with the two different types of feathers. They will be able to relate these differences in appearance to the function of each type of feather.

**Objectives:** Upon completion of the lab, students will be able to (1) **name** the different parts of a typical feather, (2) **list** differences between contour and down feathers, (3) **relate** the differences in structure of the two feather types to the particular function.

**Time Allotment:** 20 minutes

### Preparation

▶ Feathers are available from your local grocery store, local farmers, or some biological supply houses.

▶ Down feathers are available from old discarded pillows or comforters.

▶ Both feather types are available from biological supply houses as prepared slides.

▶ Caution students against collecting feathers found outdoors. It is illegal to collect bird eggs and feathers.

**Lab 8-2 worksheet**/*Teacher Resource Package*, pp. 31-32.

### Teaching the Lab

**Cooperative Learning:** Divide the class into groups of two to three students. For more information, see pp. 22T-23T in the Teacher Guide.

### ASSESSMENT

**Performance:** Provide students with synthetic materials that are used to make warm coats and down feathers. Have them do the following experiment. Wrap some of the synthetic material around a can. Then, put the down feathers in a cloth bag and wrap this bag around another can. Put a thermometer in each can. Apply light directly to each can. See which can heats up faster.

# Feathers

**Lab 8–2**

## Problem: What is the structure of feathers?

### Skills

interpret data, observe, infer

### Materials

| | |
|---|---|
| wing or tail feather | scissors |
| down feather | metric ruler |
| hand lens | |

### Procedure

1. There are two kinds of feathers. Contour feathers are found on a bird's body, wings, and tail. Down feathers lie under the contour feathers and insulate the body. Look at a contour feather with a hand lens. The hard center tube is the shaft.
2. Copy the data table. Cut 2 cm off the end of the contour feather shaft. **CAUTION:** *Use care when using scissors.* Observe the cut end with the hand lens. Record your observations.

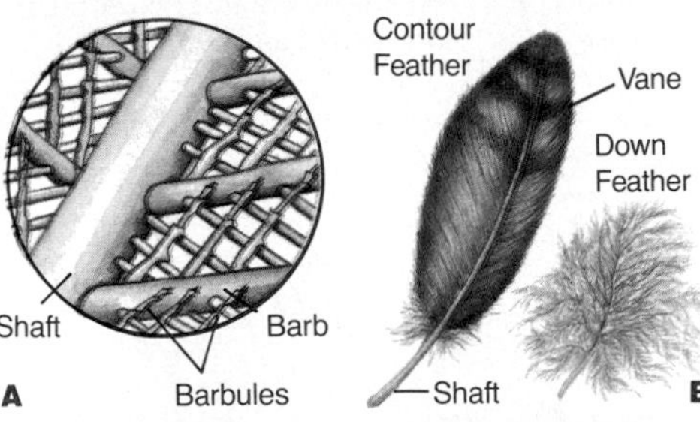


3. **Observe:** Examine the vane with the hand lens. Compare what you see with Figure A. How are the tiny strands of the feather hooked together? This gives the feather strength for flight.
4. Hold the contour feather by the shaft and fan yourself. Hold the down feather by the shaft and fan. Describe what you feel.
5. Observe the shape of the down feather with a hand lens. Describe how it feels. Compare the feather with Figure B.

### Data and Observations

1. What connects a contour feather's barbs?
2. Explain any differences observed when you fanned the air with the two feathers.

### Analyze and Apply

1. List the parts of a contour feather.
2. How does the shaft of the contour feather aid in flight?
3. **Infer:** How does the structure of a down feather help insulate a bird?
4. **Apply:** What structures in reptiles are like feathers? Are their functions similar?

### Extension

**Observe** pictures of a bird wing and a bat wing. How are they similar and how are they different?

| FEATHER | PARTS | OBSERVATIONS |
|---|---|---|
| Contour | shaft | hard, smooth, hollow |
| | vane | tiny strands on either side of the shaft |
| | barb | make up the vanes |
| | barbules | hook the barbs together |
| Down | strand | soft, small, strands not hooked together |



## ANSWERS

### Data and Observations

1. barbules
2. The down feather was not strong or stiff enough to fan air effectively.

### Analyze and Apply

1. shaft, barbs, barbules
2. It is strong and stiff and can push air during flight.
3. Because it is soft and fluffy, it can trap air close to the body.
4. Scales; Feathers have the same functions as scales. They have additional functions related to flight.

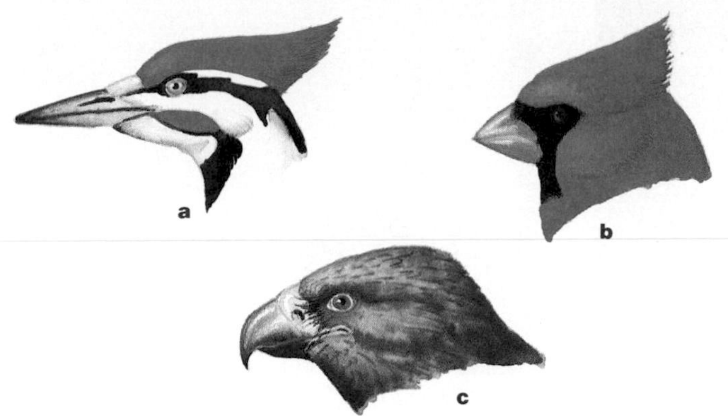


**Figure 8-21** The shape of a bird's beak is related to the kind of food it eats. Woodpecker (a), cardinal (b), and hawk (c). What food does each bird eat? (a) insects, (b) seeds, (c) meat

Birds have no teeth. Instead, they have beaks that they use to get food. The kind of beak a bird has is related to the kind of food it eats. Figure 8-21 shows some examples. Woodpeckers have beaks that cut through tree bark to expose insects. Sparrows and cardinals have thick beaks that crack seeds. Hawks have sharp beaks that tear meat.

Birds are important to us in many ways. They help farmers by eating insects and the seeds of weeds. Some birds eat rats and mice. Chickens and ducks provide eggs and meat. Some people have birds as pets. Many people enjoy watching and feeding wild birds. The bald eagle is the national bird in the United States. Do you know the state bird for your state?

### Bio Tip

**Leisure:** Millions of people in the United States enjoy bird watching. Each year, bird lovers spend over $500 million on birdseed, $100 million on bird feeders, boxes, and baths; and $18 million on field guides.

## Mammals

What traits are found in the animals in Figure 8-22? First, they have hair and are warm-blooded. The hair helps keep a constant body temperature. Second, the females have milk glands with which they nurse their young. These animals are mammals (MAM ulz). A **mammal** is an animal that has hair and feeds milk to its young. You are a mammal.

Most animals that you have studied hatch from eggs. Mammals usually develop inside the mother's body and are born alive. After birth, young mammals feed on milk produced by the mother's mammary glands. **Mammary** (MAM uh ree) **glands** are body parts that produce milk. Male mammals also have mammary glands, but they don't produce milk. Besides providing milk, all mammals care for their young.

**Figure 8-22** Mammals feed their young with milk from their mammary glands.

## TEACH

### Concept Development

▶Introduce this section with pictures and specimens of as many different species of mammals as possible.

**Cooperative Learning:** Divide the class into 7 groups. Each group is to be assigned one chordate class. The task for each group is to prepare five true-false or multiple choice questions that could be used in designing a test. One representative from each group will read off their questions to the class and ask students from other groups to supply the correct answers.

### Check for Understanding

Have students compare bird and mammal classes as to: (a) whether cold-or warm blooded; (b) type of body covering; (c) presence of a backbone; (d) appearance of front legs; (e) presence of teeth; (f) if young are provided milk; (g) if mammary glands are present.

### Reteach

Make flash cards of the traits of birds and mammals. On the back of each card, write the class(es) that have the trait. Show students the traits and have them name the corresponding classes.

### Independent Practice

**Study Guide**/*Teacher Resource Package*, p. 48. Use the Study Guide worksheet shown at the bottom of this page for independent practice.

**STUDY GUIDE 48**

STUDY GUIDE — CHAPTER 8

Name ______ Date ______ Class ______

**VOCABULARY**

*Review the new words used in Chapter 8 of your textbook. Then, fill in the blanks.*

Crayfish, insects, and spiders belong to the phylum of **jointed-leg animals**. These animals have **compound eyes** with many lenses. They also have legs, wings, and **antennae** which are used for sensing smell and touch. These body parts are called **appendages**. Animals in this phylum have a skeleton, called an **exoskeleton**, that is outside their bodies. As the animals grow, they shed their outer skeletons. This process is called **molting**.

The phylum of **spiny-skin animals** includes sea urchins, starfish, and sand dollars. These animals use their **tube feet** to move, attach to rocks, and get food.

The phylum of **chordates** includes animals that have a rod along their back at some time in their lives. Their skeleton, called an **endoskeleton**, is inside their body.

Fish live in water and breathe with **gills**. They are **cold-blooded** animals because their body temperature changes with their surroundings. **Jawless fish** have no jaws, scales, or bones. Their skeleton is made of **cartilage**. **Cartilage fish** have jaws, scales, and a cartilage skeleton. **Bony fish** have jaws, scales, and a bony skeleton.

**Amphibians** live part of their life in water and part on land.

**Hibernation** helps animals survive cold weather.

**Reptiles** have dry, scaly skin, can live on land, and breathe with lungs.

Birds have a beak, wings, and feathers. Their body temperature does not change with the temperature of their surroundings. Thus, they are **warm-blooded** animals.

**Mammals** also have a constant body temperature. They have hair on their bodies. The females produce milk from their **mammary glands**.

**ENRICHMENT 8**

ENRICHMENT — CHAPTER 8

Name ______ Date ______ Class ______

*Use after Section 8:2.*

**ORDERS OF MAMMALS**

Thousands of species of mammals are grouped into orders, each with specific traits. The traits of some of the mammal orders are given in the table below.

*Use a dictionary or an encyclopedia to find examples of animals in each of the orders listed. Then, complete the table.*

| Order name | Traits of the order | Examples |
|---|---|---|
| Monotremata | only mammal that lays egg | duckbilled platypus, echidna (spiny anteater) |
| Marsupiala | young are born as embryos; they develop in a pouch while nursing | opossum, kangaroo, koala bear |
| Insectivora | eat insects and burrow in the ground; have pointed teeth and long snouts | shrew, mole |
| Rodentia | have large incisors for gnawing; incisors grow continuously | mouse, rat, gopher, guinea pig |
| Edentata | have long tubular snout with sticky tongue; have reduced teeth or lack teeth altogether; some have bony plates in their skin | anteater, sloth, armadillo |
| Carnivora | have large canine teeth and sharp claws; prey on other animals | dog, cat, lion, fox |
| Chiroptera | forelimbs developed for flying; some fly at night using a sonarlike mechanism to find their way | bats |
| Lagomorpha | have a double row of incisors; incisors grow continuously | rabbit, hare, pika |
| Cetacea | live in fresh water or salt water; forelimbs are flippers; have no hind limbs | whale, dolphin, porpoise |
| Proboscidae | have long muscular nose (trunk) and large paired tusks | elephant |

## OPTIONS

**Enrichment**/*Teacher Resource Package*, p. 8. Use the Enrichment worksheet shown here to teach students the mammal orders.

## TEACH

### Guided Practice

Have students write down their answers to the margin question: They crawl into the pouch and attach to mammary glands there.

### Check for Understanding

Have students respond to the first three questions in Check Your Understanding

### Reteach

**Reteaching**/*Teacher Resource Package*, p. 22. Use this worksheet to give students additional practice understanding vertebrates.

### *Portfolio*

Have students make a list of mammals that have hooves. How are all of these animals alike? (similar teeth and eat plants) Draw pictures of the hooves of a cow and a horse and explain how the hooves are different.

**Extension:** Assign Critical Thinking, Biology and Reading, or some of the **OPTIONS** available with this lesson.

## 3 APPLY

### Audiovisual

Show the videocassette *Reptiles,* National Geographic Society Educational Services.

## 4 CLOSE

### ACTIVITY/Software

*Project Classify: Mammals,* National Geographic Society Educational Services.

### *Student Journal*

Have students write a paragraph comparing the similarities and differences of birds and mammals.

a

b

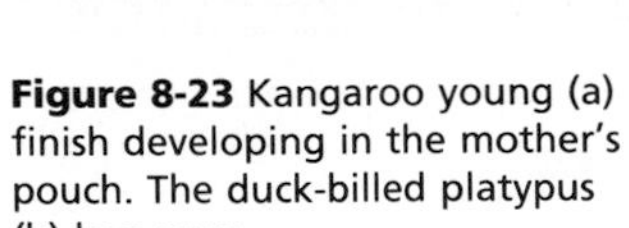

**Figure 8-23** Kangaroo young (a) finish developing in the mother's pouch. The duck-billed platypus (b) lays eggs.

**How do opossum young finish developing?**

| TABLE 8–1. CLASSIFICATION OF HUMANS |
|---|
| Kingdom: animal |
| Phylum: chordate |
| Subphylum: vertebrate |
| Class: mammal |
| Order: primate |
| Family: hominid |
| Genus: *Homo* |
| Species: *sapiens* |

Two groups of mammals whose young do not develop inside the mother's body are the pouched mammals and the egg-laying mammals. Kangaroos and opossums are pouched mammals. The young of kangaroos and opossums are not fully developed at birth. After birth, they crawl into a pouch outside the mother's body. There they attach themselves to the mother's mammary glands and feed on milk until they are more fully developed. Egg-laying mammals, such as the duck-billed platypus shown in Figure 8-23b, lay eggs like those of reptiles.

Now that you have studied the entire animal kingdom, let's see where you fit in. Remember that you are also a chordate and a mammal. You can see the smaller subgroups in Table 8-1. The name of your genus is *Homo.* Your species name is *sapiens.* These names make up the scientific name of humans, *Homo sapiens.* The Latin word *Homo* means man and *sapiens* means wise. Together *Homo sapiens* means wise man.

### Check Your Understanding

6. Compare the scales of jawless fish, cartilage fish, and bony fish.
7. How do birds get food?
8. What does being warm-blooded mean?
9. **Critical Thinking:** Why do mammals have a relatively small number of young?
10. **Biology and Reading:** Why can chordates grow larger than any other animals? What limits the size of animals that are not chordates?



### Answers to Check Your Understanding

6. jawless fish–no scales; cartilage fish–toothlike scales; bony fish–smooth, bony scales
7. with beaks and feet
8. having an internal temperature that remains about the same no matter what the temperature of the surroundings
9. The young must be nursed by the mother and only a small number of young can be nursed at one time.
10. Chordates have an endoskeleton that allows them to grow larger. Some invertebrates do not have the necessary support that would allow them to grow to very large sizes. Other invertebrates have an exoskeleton that limits their size.

**RETEACHING 22**

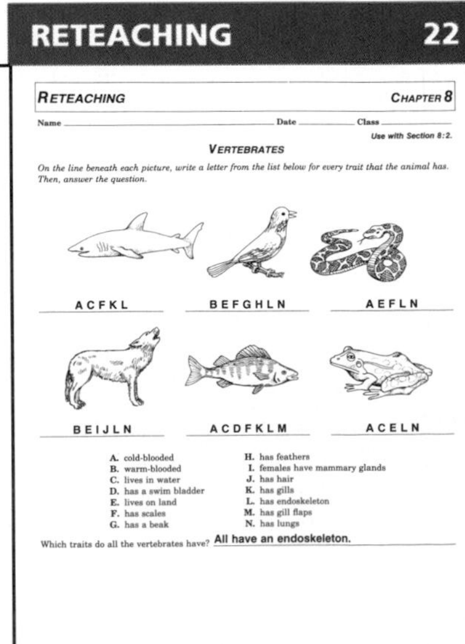

RETEACHING CHAPTER 8

Name ______ Date ______ Class ______

Use with Section 8:2.

VERTEBRATES

*On the line beneath each picture, write a letter from the list below for every trait that the animal has. Then, answer the question.*

ACFKL BEFGHLN AEFLN

BEIJLN ACDFKLM ACELN

A. cold-blooded
B. warm-blooded
C. lives in water
D. has a swim bladder
E. lives on land
F. has scales
G. has a beak
H. has feathers
I. females have mammary glands
J. has hair
K. has gills
L. has endoskeleton
M. has gill flaps
N. has lungs

Which traits do all the vertebrates have? All have an endoskeleton.

# Science and Society

## Are Animal Experiments Needed?

Millions of animals are used each year for research conducted in the United States and for dissections in high school classrooms. In most cases, the animals are treated with great care. In other cases, however, the animals are mistreated and many die. Sometimes the research itself is fatal and the animals die. Rats and mice are the animals used most often, but dogs, cats, pigs, rabbits, and frogs are also used. Monkeys and apes are used to study diseases such as cancer and AIDS because these animals are similar to humans. Many high school students dissect frogs, worms, fish, and even mammals. These animals are purchased from companies that usually raise them for the purpose of scientific study, although some animals are captured from the wild.

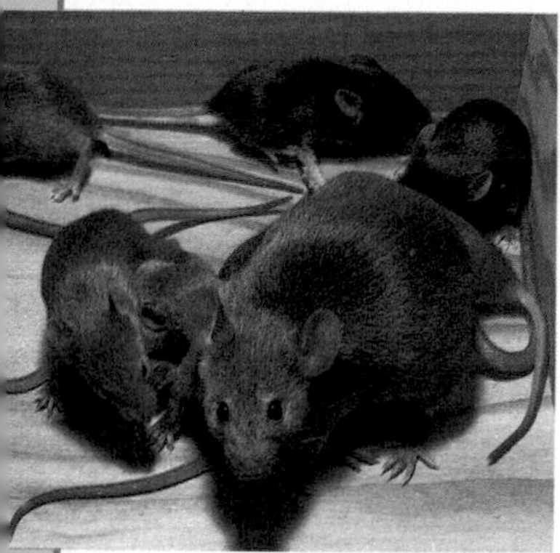

### What Do You Think?

1. Many people in the United States protest the use of animals in experiments and in dissections. These people argue that it is cruel and the animals are made to suffer unnecessarily. They also argue that animals are not enough like humans to give useful results. Should animals be used in laboratory experiments? Should experiments be done if they cause the animals to suffer? Should there be strict laws governing the use of animals in labs?

2. Many people defend the use of animals in lab experiments. Doctors studying living animals say the results of experiments allow many human lives to be saved. Vaccines that protect people from diseases were tested on animals before being given to humans. Animals are used to test the safety of drugs and are being used to develop artificial body parts. Do you think human welfare should be placed over animal welfare? If animals are not used, how should new drugs be tested?

3. Many teachers feel that biology students should dissect animals to learn the structure of body systems and how they work. These teachers say that students cannot learn these things by looking at pictures in textbooks. People who disagree say that this is not a good reason for killing animals. Do you think animal dissections are needed for teaching biology students? Does it make a difference whether the animals used are wild animals captured for scientific use or animals raised for sale to schools?

**Conclusion:** If animals are not used in laboratories, what are the alternatives?

Animal rights activists




Botting, Jack H., and Adrian R. Morrison. "Animal Research Is Vital to Medicine." *Scientific American*, February 1997, pp. 83-85.

Howard, Walter E. "An Ecologist's View of Animal Rights." *The American Biology Teacher*, April 1994, pp. 202-205.

Miller, Harlan B. and William H. Williams, Eds. *Ethics and Animals.* Clifton, N.J.: Humana Press Inc., 1983.

Mukerjee, Madhusree. "Trends in Animal Research." *Scientific American*, February 1997, pp. 86-93.


Science and Society

## Are Animal Experiments Needed?

### Background

Aristotle stated, "Plants exist for the sake of animals and the other animals for the good of man". How have attitudes toward animals changed since Aristotle's time?

### Teaching Suggestions

Discuss how nonhuman animals are similar to and different from humans. The discussion will hinge on whether or not students feel that nonhuman animals have rights, equal to those of humans.

### What Do You Think?

1. Most students will think that it is moral to use animals in experiments that don't cause suffering. In experiments that cause suffering, students will disagree. Most will agree that strict laws are needed.
2. Explain that many drugs and products can now be tested on cells rather than on whole animals. Discuss whether testing on animals other than humans can give reliable results that translate directly to humans.
3. Students will disagree as to whether or not animals should be dissected for biology class. Many schools are replacing animal dissections with computer simulations and models.

**Conclusion:** Students may suggest that new products be tested on a representative group of people who would be paid for the experiment. Others may suggest that new drugs be used on a few individuals, with their permission, who need a cure for a disease. Stress the importance of testing new products and drugs before their general widespread use.

### Suggested Readings


Barnard, Neal D., and Stephan R. Kaufman. "Animal Research Is Wasteful and Misleading." *Scientific American*, February 1997, pp. 80-82.

## CHAPTER 8 REVIEW

### Summary

Summary statements can be used by students to review the major concepts of the chapter.

### Key Science Words

All boldfaced terms from this chapter are listed.

### Testing Yourself

#### Using Words

1. amphibian
2. warm-blooded
3. cartilage
4. molting
5. mammary gland
6. gill
7. endoskeleton
8. cold-blooded
9. chordate
10. reptile
11. compound eye
12. spiny-skin animal

#### Finding Main Ideas

13. p. 163; Tube feet are used like suction cups to help starfish move.
14. p. 169; Frogs have long, sticky tongues to catch insects.
15. p. 167; A lamprey is a jawless fish that lives as a parasite.
16. p. 159; An animal molts to grow.
17. p. 165; The chordate phylum has the biggest animals.
18. p. 159; An exoskeleton is a skeleton on the outside of the body.
19. p. 168; Sharks eat other animals that live in the sea or protists.
20. p. 169; Amphibian eggs must be laid in water to keep them from drying out.
21. p. 171; Feathers help a bird keep a constant body temperature.
22. p. 168; Bony fish have a swim bladder that helps them to float.
23. p. 173; A mammary gland is a milk gland with which the young are nursed.
24. p. 171; A reptile has an egg with a tough shell that protects it and keeps it from drying out.

# Chapter 8 Review

## Summary

### 8:1 Complex Invertebrates

1. Jointed-leg animals have jointed appendages, segmented bodies, and exoskeletons.
2. Insects are jointed-leg animals with one pair of antennae, special mouthparts, and three pairs of walking legs. There are more kinds of insects than all other animals combined.
3. Spiny-skin animals have tube feet and a body design showing five parts.

### 8:2 Vertebrates

4. There are three classes of fish. Fish have gills. Most fish have scales. The jawless fish and cartilage fish have skeletons made of cartilage. Bony fish have skeletons made of bone. Most bony fish have a swim bladder that helps them go up and down in the water.
5. Amphibians live both in water and on land. Reptiles are animals with dry, scaly skin that can live on land. Birds are warm-blooded. They have feathers, wings, and light, hollow bones.
6. Mammals are warm-blooded and have hair. The females have milk glands with which they nurse the young.

## Key Science Words

amphibian (p. 169)
antennae (p. 159)
appendage (p. 158)
bony fish (p. 168)
cartilage (p. 167)
cartilage fish (p. 167)
chordate (p. 165)
cold-blooded (p. 166)
compound eye (p. 159)
endoskeleton (p. 165)
exoskeleton (p. 159)
gill (p. 166)
hibernation (p. 170)
jawless fish (p. 167)
jointed-leg animal (p. 158)
mammal (p. 173)
mammary gland (p. 173)
molting (p. 159)
reptile (p. 170)
spiny-skin animal (p. 163)
tube feet (p. 163)
warm-blooded (p. 171)

## Testing Yourself

### Using Words

*Choose the word from the list of Key Science Words that best fits the definition.*

1. animal that lives part of its life in water and another part on land
2. body temperature that stays the same no matter what the temperature of the surroundings
3. tough, flexible tissue that supports and shapes the body
4. shedding of the exoskeleton
5. structure that produces milk
6. used by fish to breathe
7. skeleton inside the body
8. body temperature that changes with the temperature of the environment
9. animal that, sometime in its life, has a stiff rod along the back
10. has dry, scaly skin and lives on land
11. has many lenses
12. has a body design with five parts

176

#### Using Main Ideas

25. Insects have two compound eyes, three pairs of legs, antennae, and a body in three sections. Spiders have eight legs, no compound eyes, no antennae, and a body in two sections.
26. They all have an endoskeleton and a backbone.
27. A starfish grips slippery rocks with its tube feet.
28. An endoskeleton is on the inside of the body. An exoskeleton is on the outside of the body.
29. They have the largest brains and well developed senses.
30. Centipedes have one pair of legs on each body segment, poison claws, and feed on insects. Millipedes have two pairs of legs on most segments, no poison claws, and eat decaying plants.
31. Sharks have rows of teeth that slant backward and help them hold their food.
32. Tadpoles must live in water because they breathe with gills.
33. Some insects eat harmful insects. Insects carry pollen from flower to flower, and some insects produce food and other useful materials.

# Review

## Testing Yourself *continued*

**Finding Main Ideas**

*List the page number where each main idea below is found. Then, explain each main idea.*

13. how tube feet are used
14. why frogs have long, sticky tongues
15. what a lamprey is
16. why an animal molts
17. which phylum has the biggest animals
18. what an exoskeleton is
19. what sharks eat
20. why amphibian eggs must be laid in water
21. how feathers help a bird
22. what helps a bony fish float
23. what a mammary gland is
24. what kind of egg a reptile has

**Using Main Ideas**

*Answer the questions by referring to the page number after the question.*

25. How do insects differ from spiders? (pp. 160, 161)
26. What are two ways that amphibians, reptiles, birds, and mammals are alike? (pp. 169-174)
27. How does a starfish grip slippery rocks? (p. 163)
28. How does an endoskeleton differ from an exoskeleton? (p. 165)
29. Why are vertebrates believed to be the most intelligent animals on Earth? (p. 166)
30. How do centipedes differ from millipedes? (p. 160)
31. What helps sharks hold their food? (p. 168)
32. Why must tadpoles live in water? (p. 169)
33. How are insects helpful? (p. 161)

## Skill Review ✓

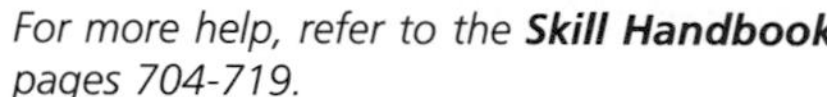

*For more help, refer to the **Skill Handbook,** pages 704-719.*

1. **Infer:** Frogs have light coloring on their bottom sides and dark coloring on their top sides. Infer how this coloring can be an advantage to the frog.
2. **Observe:** Examine Figures 8-3 and 8-6. How do crayfish and grasshoppers differ from each other?
3. **Infer:** Animals whose eggs are fertilized outside the body must reproduce in water. Why is this so?
4. **Make and use tables:** Make a table to show the traits of the seven vertebrate classes and give examples of each class.

## Finding Out More

TECH PREP

**Critical Thinking**

1. What problems might occur if all the insects in the world were to die?
2. Considering the advantages and disadvantages of endoskeletons and exoskeletons, why do you think there are more kinds of insects than mammals?

**Applications**

1. Set up a bird feeder at home. Learn what you can do to attract different kinds of birds to your feeder.
2. Write a report about how honeybees communicate with one another. Discuss how communication helps bees make honey.



## Skill Review

1. When a predator looks up from underwater, it is looking toward daylight and the light underside of the frog provides camouflage. When a predator looks down into water, it is looking away from the daylight, and the dark top of the frog provides camouflage.
2. Crayfish have two body sections, five pairs of walking legs, and no wings. Insects have three body sections, three pairs of walking legs, and wings.
3. The sperm needs a liquid in which to swim to the egg. Outside the body, water provides that liquid.
4. Students could place the headings Class, Trait, and Example across the top, and the names of the classes along the side. Then they could fill in the rest of the table.

## Finding Out More

### Critical Thinking

1. There would be widespread famine as crops failed because insects are needed for pollination of many food crops. Many fish, amphibians, reptiles, birds, and mammals would become extinct because they eat only insects.
2. There are more kinds of insects because the exoskeleton can come in many shapes and sizes. The endoskeleton is basically in one pattern and all vertebrates evolved from the same stock.

### Applications

1. Various kinds of wild bird seed will attract seed-eaters. Suet balls and pine cones covered with peanut butter will attract insect eaters, and woodpeckers.
2. Bees communicate through a dance that tells the other bees where and how far away a food source is. Encourage students to read *The Dance Language and Orientation of Bees,* by Karl von Frisch.

**Chapter Review**/*Teacher Resource Package,* pp. 38-39.

**Chapter Test**/*Teacher Resource Package,* pp. 40-42.

**Chapter Test**/*Computer Test Bank*

**Unit Test**/*Teacher Resource Package,* pp. 43-44.

# Unit 2 Kingdoms

CONNECTIONS

## Biology in Your World

In this unit, students learn how organisms are alike and how they are different. Studies of plant and animal groups help students make sense of their world.

### Literature Connection

While in Africa, Dian Fossey developed several techniques for observing mountain gorillas. Fossey found that if she stood or walked upright in the gorillas' line of sight, the animals would act frightened. She also found that looking directly at a male gorilla seemed threatening to the animal. The gorillas were most at ease when she imitated their behavior. She tried walking on her hands and knees and staying seated while observing them. When sitting, Fossey would pretend to feed and imitate the sounds gorillas make when content. She then discovered that if she partially hid herself, the animals would come out in the open to get a better view.

### Geography Connection

The Great Barrier Reef of Australia sustains 350 species of coral alone. About 1500 species of fish live in the water surrounding the reef. Sea turtles, birds, worms, algae, crabs, and starfish also make the reef their home. Scientists benefit by being able to study this wide variety of life in one place.

CONNECTIONS

# Biology in Your World

## The Animal World

Classification is used to make sense out of all the many kinds of living things in the world. In this unit, you have seen how organisms are similar. You have also looked at some of the relationships among organisms—from simple, one-celled monerans to complex animals.

LITERATURE

## Primate to Primate

Dian Fossey left her home in Kentucky in 1966 to go to Zaire (ZI uhr) and Rwanda (roo AHN duh) in Africa. There, she studied the endangered mountain gorilla in the Virunga (vuh RUN guh) Mountains.

Fossey learned to accept the animals on their own terms. She was able to learn how gorillas form family groups. She saw behaviors that scientists didn't know existed in gorillas.

During her studies, poachers were a constant threat. A poacher is a person who hunts or takes plants or animals illegally. Fossey fought a never-ending battle to protect the gorillas and their natural habitat.

Fossey studied gorillas for 13 years. Her experiences are discussed in her book, *Gorillas in the Mist.*

GEOGRAPHY

## Coral Reefs

Coral reefs are found in tropical oceans. Coral reefs are built by animals called coral polyps. These creatures are related to jellyfish. They live in colonies and make hard limestone skeletons around their soft bodies. As old colonies die, new ones grow on top of the old skeletons.

The largest coral reef is the Great Barrier Reef off the coast of Australia. It is over 1900 kilometers long.

Divers often see crabs, worms, fish, octopuses, and sponges living within a coral reef. Coastline development, oil spills, and pollution threaten these fragile and beautiful communities.

178

**GLENCOE TECHNOLOGY**

**Videodisc**

**Biology: The Dynamics of Life**
Movie: *Squid Locomotion*
Disc 1, Side 2, Ch. 33

## LEISURE

### Collecting Fossils

You can have fun searching for fossils of organisms that lived long ago. Fossils are often found in rocks such as shale, clay, sandstone, and limestone that were formed underwater. Since fossils are found in layers, try to find exposed areas such as cliffs, quarries, riverbanks, and road or railway cuts. You will need a shovel, hammer, chisel, notebook, and bags for specimens. You may find fossils of sea lilies, which are related to starfish. Fossil clams, snails, and insects are also common.

Take careful notes, and label and wrap all your specimens. You just might turn up something unusual!

## CONSUMER

### Tasty Morsels from the Sea

Do you enjoy eating lobster, crab, or shrimp? They are all jointed-leg animals. Scallops, clams, oysters, mussels, and snails are soft-bodied animals. To some people, sea urchins and sea cucumbers, both relatives of the starfish, are delicacies. Fish, such as cod, haddock, and tuna, which are vertebrates, are also enjoyed by many people.

You don't have to live near the ocean or go to a fancy restaurant to enjoy good seafood. Many grocery stores have a seafood department that receives fresh or frozen shipments regularly.

Seafood is best fresh and should be properly cooked. Ask at the store when shipments are received. To be considered fresh, seafood should have been received within the last 24 hours. Keep seafood cold and plan to serve it right away. You can freeze it if it hasn't been previously frozen.

Do you like to help in the kitchen? Try a simple recipe, and surprise your family with a seafood dinner.

179

## CONNECTIONS

### Leisure Connection

Fossils are common in almost every state. They can be found anywhere sedimentary rock is present. Often, fossils are unearthed by construction crews digging highway or building foundations. In North America, most fossils of mammals are found west of the Mississippi River.

Once a fossil is discovered, scientists classify it by comparing its features to those of other known species. Only the hard parts of an organism remain, so fossils are grouped by the shapes of the shells, teeth, and skeletons.

### Consumer Connection

When choosing seafood, price is often a consideration. Fresh or frozen fish is generally the least expensive seafood. Shrimp, crab, and lobster cost more than fish. Exotic or unusual foods, such as sea cucumbers or squid, also cost more. Seafood can be purchased ready to eat, and that preparation adds to the price, too.

### References


Arnold, Caroline. *A Walk on the Great Barrier Reef.* Minneapolis: Carolrhonda Books, Inc., 1988.

Fenton, Carroll Lane, and Mildred Adams Fenton. *The Fossil Book.* New York: Doubleday, 1989.

Fossey, Dian. *Gorillas in the Mist.* Boston: Houghton Mifflin, 1983.

Fredman, Catherine, "The Maine Event." *Travel-Holiday*, September 1996, pp. 78-84.

Luoma, John.R. "Reef Madness." *Audubon*, Nov.-Dec. 1996, pp. 24-27.

Musgrove, Mary. "Fossil Collecting." *Lapidary Journal*, August 1996, pp. 88-91.

Tourtellot, Jonathon B. "Keeping Our Coral Reefs Intact." *National Geographic Traveler*, Jan.-Feb. 1995, pp. 16-18.


## CULTURAL DIVERSITY

### Algae Harvesting in Japan

Discuss with students the nutritional benefits of algae and their use in cooking in many areas of the world, especially Asia. Point out that algae contain many nutrients, including protein and various vitamins and minerals. Algae are eaten fresh, boiled, or fried in many Asian recipes. They are used in soup and as part of sushi. One of the more common types of algae used in cooking is *Porphyra,* a red alga called nori. Since the 17th century, Japanese aquaculturalists have developed a huge industry around harvesting this alga from Tokyo Bay. Have students research Japanese algae harvesting techniques or bring in samples of Japanese foods that contain algae for the class to sample.

# Unit 3 Body Systems— Maintaining Life

## UNIT OVERVIEW

Unit 3 examines animals in detail, with emphasis on humans as the representative animal. A general discussion of nutrients (Chapter 9) is covered first. Then the digestive system (Chapter 10) is explored. Chapters 11 and 12 cover circulation and blood respectively. Chapters 13 and 14 explain the respiratory, urinary, skeletal, and muscular systems.

## Advance Planning

Audiovisual and computer software suggestions are located within each chapter under the heading TEACH.

**To prepare for:**

Lab 9-1: Gather honey, molasses, maple syrup, apple juice, and sweet potato. Obtain sugar test paper.

Lab 9-2: Order enzyme.

Mini Labs: Obtain tomatoes, tomato juice, potatoes, and potato chips.

Lab 10-1: Prepare enzymes A and B. Collect gelatin, plastic spoons, paper cups, and sticks.

Lab 10-2: Make copies of the transparency master of the digestive system. Provide colored pencils.

Lab 11-1: Order or collect live earthworms.

Lab 11-2: Collect plastic squeeze bottles.

Mini Lab: Arrange to borrow a stethoscope.

Lab 12-1: Order prepared slides of human and frog blood. Gather petri dish covers and metric rulers.

Lab 12-2: Order prepared slides of human blood. Provide sheets of clear plastic and marking pencils.

Mini Lab: Provide cooking oil, water, and red food coloring.

Lab 13-1: Order goldfish.

Lab 13-2: Prepare normal and abnormal artificial urine samples and salt water. Prepare silver nitrate and obtain glucose test tape.

Mini Labs: Prepare bromthymol blue and gather drinking straws. Order rubber tubing and filter paper.

Lab 14-1: Collect bones of pigs, cows, chickens, and turkeys.

Lab 14-2: Collect string, metal fasteners, index cards, and a paper punch.

Mini Labs: Purchase cooked chicken legs (thigh and drumstick attached). Prepare 6% hydrochloric acid and obtain chicken bones.

# Unit 3

CONTENTS

180

# Body Systems—Maintaining Life

**What would happen if...**

we could live forever? What would it mean? Would we still grow old, or would we stay young? How would everyone's living forever affect the environment? The population would increase at an even greater rate than it does now. We might soon run out of food, space, and resources.

Better medical care and advances in preventing disease have increased the human life span. Further advances may increase it more. However, we can't live forever. In the laboratory, human cells die after a set number of divisions.

Should we think more about improving the quality of life than increasing the length of life?

181

UNIT OVERVIEW

## Photo Teaching Tip

Ask students to read the unit opening passage and make a list of all the possible problems that might develop as a result of life without death. Initiate a discussion that first focuses the students' thoughts on your own community and later expand the discussion to include the impact of a growing population on the health of the planet. You may wish to suggest that students write an essay for extra credit on how they would like to use science to improve the quality of life.

## Theme Development

Animal life depends on the animal systems that maintain and control regulated functions. This unit uses the human as its major animal example. However, the various systems present in other animal phyla are also referenced. Thus, students will receive a cohesive picture of how animal systems perform needed functions throughout the entire animal kingdom. The theme of homeostasis is developed in this unit. Students will also gain an understanding as to how the various body systems evolved within the animal kingdom.

**CULTURAL DIVERSITY**

See the Cultural Diversity feature located on the Connections page at the end of this unit.

For Tech Prep activities, see the Applying Technology feature in this unit, the Finding Out More applications questions in the Chapter Reviews, and the Tech Prep Applications booklet in the Teacher Resource Package.

# Nutrition

## PLANNING GUIDE

| CONTENT | TEXT FEATURES | TEACHER RESOURCE PACKAGE | OTHER COMPONENTS |
|---|---|---|---|
| (3 1/2 days)<br>9:1 What Are the Nutrients in Food?<br>Food<br>Proteins, Fats, and Carbohydrates<br>Vitamins<br>Minerals<br>Water<br>Supplying Nutrients to Your Body | Skill Check: *Understand Science Words,* p. 185<br>Mini Lab: *How Does Salt Content Change in Processed Foods?* p. 189<br>Mini Lab: *What Are Processed Food Costs?* p. 191<br>Lab 9-1: *Milk Nutrients,* p. 187<br>Idea Map, p. 188<br>Check Your Understanding, p. 192 | Application: *Comparing Nutrients in Food,* p. 9<br>Enrichment: *Calculating RDA,* p. 9<br>Reteaching: *Nutrient Needs,* p. 23<br>Critical Thinking/Problem Solving: *Which Common Foods Provide the Most Nutrients?* p. 9<br>Skill: *Make a Bar Graph,* p. 9<br>Focus: *Sugar-Free Products,* p. 17<br>Transparency Master: *Reading Food Labels,* p. 41<br>Study Guide: *What Nutrients Are in Food?* pp. 49-51<br>Lab 9-1: *Nutrients,* pp. 33-34 | **Laboratory Manual:**<br>*What Are the Tests for Fats and Proteins?* p. 71<br>**Laboratory Manual:**<br>*How Do You Test for Vitamin C in Foods?* p. 75<br>Color Transparency 9: *Daily Nutrient Requirements*<br>**STVS:** *Measuring Calcium Deficiency,* Human Biology (Disc 7, Side 2) |
| (1 1/2 days)<br>9:2 Calories<br>Energy in Food<br>Calorie Content of Food<br>Using Calories | Skill Check: *Infer,* p. 195<br>Skill Check: *Interpret Data,* p. 197<br>Lab 9-2: *Energy Foods,* p. 194<br>Check Your Understanding, p. 198<br>Applying Technology: *Get the Fat Out,* p. 198 | Reteaching: *Calories Intake and Output,* p. 24<br>Study Guide: *Calories,* pp. 52-53<br>Study Guide: *Vocabulary,* p. 54<br>Lab 9-2: *Energy Foods,* pp. 35-36 | **STVS:** *Obesity and Heredity,* Human Biology (Disc 7, Side 2)<br>**STVS:** *Bulimia,* Human Biology (Disc 7, Side 2) |
| Chapter Review | Summary<br>Key Science Words<br>Testing Yourself<br>Finding Main Ideas<br>Using Main Ideas<br>Skill Review | **ASSESSMENT RESOURCES**<br>Chapter Review, pp. 45-46<br>Chapter Test, pp. 47-49<br>Performance Assessment in the Biology Classroom<br>Alternate Assessment in the Science Classroom<br>Computer Test Bank | |

### GLENCOE TECHNOLOGY

**Infinite Voyage,** *A Taste of Health*
**Science and Technology Videodisc Series,** *Measuring Calcium Deficiency,* Human Biology (Disc 7, Side 2)
*Obesity and Heredity,* Human Biology (Disc 7, Side 2)
*Bulimia,* Human Biology (Disc 7, Side 2)

## MATERIALS NEEDED

| LAB 9-1, p. 187 | LAB 9-2, p. 194 | APPLYING TECHNOLOGY, p. 198 | MARGIN FEATURES |
|---|---|---|---|
| 2 200-mL beakers<br>thermal glove<br>cheese cloth<br>dropper<br>funnel<br>hot plate<br>stirring rod<br>whole milk<br>enzyme<br>250-mL graduated cylinder | glass slides<br>droppers<br>razor blade<br>food samples for sugar test<br>water<br>sugar test paper | butter, regular<br>butter, light<br>test tubes<br>labels<br>beaker<br>hot plate<br>pan balance | Skill Check, p. 197<br>pencil<br>paper<br>Skill Check, p. 185<br>dictionary<br>Mini Lab, p. 189<br>canned tomato juice |

## OBJECTIVES

For more information about National Science Standards, see page 5T.

| SECTION | OBJECTIVE | CORRELATION of QUESTIONS to OBJECTIVES: CHECK YOUR UNDERSTANDING | CHAPTER REVIEW | TRP CHAPTER REVIEW | TRP CHAPTER TEST |
|---|---|---|---|---|---|
| 9:1 National Science Stds: UCP.3, B.6, C.5, F.1 | 1. **List** six important nutrients that the body needs. | 1, 4 | 5, 10, 19, 28 | 3, 9, 11, 15, 24, 25, 27 | 1, 2, 10 |
| | 2. **Explain** how each of the six important nutrients is used by the body. | 2 | 1, 3, 8, 13, 14, 16, 23 | 1, 4, 5, 7, 12, 13, 14, 20, 23, 26, 28 | 4, 6, 8, 9, 11, 12, 16, 17, 18, 19, 20, 21, 22, 28, 29, 30, 31, 32, 33, 34, 35, 36, 37, 38, 40, 41, 42, 43 |
| | 3. **Define** a balanced diet. | 3 | 4, 6, 15, 20, 24, 26 | 6 | 12, 13 |
| 9:2 National Science Stds: UCP.2, UCP.3, B.6, C.5, F.1 | 4. **Describe** how a Calorie is used by the body. | 6 | 2, 18, 25 | 2, 16, 17, 18, 21 | 5, 7, 14 |
| | 5. **Compare** the number of Calories found in different nutrients. | 7 | 7, 9, 12, 17, 27 | 22 | 3, 23, 24, 25, 26, 27 |
| | 6. **Compare** the number of Calories used in different activities. | 9, 10 | 11, 21, 22 | 8, 19 | 15 |

# Nutrition

## CHAPTER OVERVIEW

### Key Concepts

In this chapter, students are first introduced to the role and source of the six categories of nutrients. The concept of % Daily Value is described as it relates to the way one maintains a balanced diet. Caloric intake and output are related to gaining or losing weight.

### Key Science Words

| | |
|---|---|
| balanced diet | Percent Daily Value |
| Calorie | protein |
| carbohydrate | recommended daily allowance |
| fat | vitamin |
| mineral | |
| nutrient | |
| nutrition | |

### Skill Development

In Lab 9-1, students will **use numbers, experiment, measure in SI,** and **make and use tables** in determining nutrients in milk. In Lab 9-2, they will **form a hypothesis, experiment,** and **make and use tables** in determining which foods contain starch and sugar. In the Skill Check on page 185, students will **understand** the **science word** *carbohydrate.* In the Skill Check on page 195, they will infer the number of Calories in fat. In the Skill Check on page 197, they will **interpret data** on Calories used in activities. In the Mini Lab on page 189, students will **interpret data** to compare salt content in fresh and processed foods. In the Mini Lab on page 191, they will **interpret data** to compare the cost of raw and processed foods.

### Bridging

In Chapter 2, students learned about the cell and features of living things. In this chapter, students will be able to associate the fact that the foods that we eat supply nutrients to the cells.

**CHAPTER PREVIEW**

### Chapter Content

Review this outline for Chapter 9 before you read the chapter.



### Skills in this Chapter

The skills that you will use in this chapter are listed below.

- In **Lab 9-1,** you will use numbers, experiment, measure in SI, and make and use tables. In **Lab 9-2,** you will form a hypothesis, experiment, and make and use tables.
- In the **Skill Checks,** you will infer, interpret data, and understand science words.
- In the **Mini Labs,** you will interpret data.

182

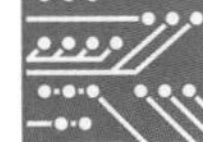

## TECH PREP

For Tech Prep activities in this chapter of the Teacher Wraparound Edition, see especially the Portfolios on pages 185 and 195, the Assessments on pages 189 and 191, the Cooperative Learning activity on page 189, and the Student Journals on pages 194 and 196.
See also the Glencoe Homepage at **www.glencoe.com**

The following Glencoe resources provide additional opportunities for integrating science and technology.

**Technology: Science and Math in Action, Book Two**
Module 5: Solar Power

**Production Systems Technology**
Section II: Resources for Production
Activity 1: Working with Fluid Power

Chapter 9

# Nutrition

What influences the types of foods people buy or eat? The photo on the left will give you some clues. It shows foods that can be prepared in a microwave oven. Usually these types of foods can be cooked quickly. Speed of preparation is one thing that influences the kinds of foods we buy. Is price important? Does advertising influence our buying habits? Even ethnic background plays a role in the types of food you eat. You may prefer Greek food as shown in the photo below. Don't stores today have food sections marked ethnic foods?

No matter what influences your choice of food, it's important to your body for several reasons. You will find out how and why food is important as you read this chapter.

**Try This!**

**Where do foods come from?** Do most foods that you eat each day come from plants or animals? Make a list of the foods that you ate yesterday. Categorize each food as either "plant" or "animal." What conclusion can be drawn from your list?

**interNET CONNECTION**

For more information about the material in this chapter, follow the link for the chapter on the Glencoe Homepage at **http://www.glencoe.com**

A Greek salad



GETTING STARTED

## Using the Photos

Discuss convenience, price, and ethnic preference as important considerations when it comes to choosing foods. Ask students to suggest other reasons people have for choosing food.

## MOTIVATION/Try This!

**Where do foods come from?** Students will **observe** that most foods come directly from plants (potatoes, rice, vegetables, fruits). Meats also come from plants because the animals used plants as their original food source.

## Chapter Preview

Have students study the chapter outline before they begin to read the chapter. They should note the overall organization of the chapter, that nutrients are studied first, and how the body uses these nutrients is studied next.

## Misconception

Students may believe that vitamins supply you with energy, and the more vitamins you take, the healthier you will look and feel. Both are false. No Calories are present in vitamins; therefore, no energy is obtained from them. Certain vitamins, such as the fat-soluble vitamins, in high doses will cause health problems.

**ASSESSMENT PLANNER**

***Portfolio***
Strategies on the following pages represent student products that can be placed into a best-work portfolio: pp. 185, 195.

**PERFORMANCE ASSESSMENT**
Skill Check, p. 197
Mini Lab, pp. 189, 191
Lab, pp. 187, 194

**CONTENT ASSESSMENT**
Check for Understanding, pp. 191, 192, 196, 197
Chapter Review, pp. 200-201

**GROUP ASSESSMENT**
Opportunities for group assessment occur with Cooperative Learning Strategies.

***Student Journal***
Strategies on the following pages represent opportunities for writing in a Student Journal: pp. 194, 196.

## 9:1 What Nutrients Are in Food?

## PREPARATIONS

### Materials Needed

Make copies of the Focus, Study Guide, Skill, Transparency Master, Enrichment, Application, Critical Thinking/Problem Solving, Reteaching, and Lab worksheets in the *Teacher Resource Package.*

▶ Obtain enzyme (rennilase), whole milk, and cheese cloth for Lab 9-1.

▶ Obtain starch and sucrose for the Focus activity.

### Key Science Words

nutrient
nutrition
protein
fat
carbohydrate
vitamin
recommended daily allowance
mineral
Percent Daily Value
balanced diet

### Process Skills

In Lab 9-1, they will use numbers, experiment, measure in SI, and make and use tables. In the Skill Check, they will understand science words. In the Mini Labs, students will interpret data.

## 1 FOCUS

▶ The objectives are listed on the student page. Remind students to preview these objectives as a guide to this numbered section.

### Guided Practice

Have students write down their answers to the margin question: carbohydrates, proteins, fats, vitamins, minerals, and water.

**Focus**/*Teacher Resource Book*, p. 17. Use the Focus transparency shown at the bottom of this page as an introduction to nutrients and Calories in food.

## 9:1 What Are the Nutrients in Food?

### Objectives

1. **List** six important nutrients that the body needs.
2. **Explain** how each of the six important nutrients is used by the body.
3. **Define** a balanced diet.

### Key Science Words

nutrient
nutrition
protein
fat
carbohydrate
vitamin
recommended daily allowance
Percent Daily Value
mineral
balanced diet

If a car runs out of gas, its motor stops. If you don't add fuel to your body, your motor also stops. Food is the fuel for living things.

### Food

The cells of your body must be supplied with food or they stop working. Cells use food for growth and repair. Food must be supplied to your body regularly in the right amount and balance. Your body needs fuel to keep working, just as a motor needs fuel to keep running. You and the motor need the right balance of fuel and oxygen to keep working. Without the proper balance of these substances, your body and the motor will slow down or stop.

**What are the six nutrients in food?**

To keep all your body cells running properly, you must supply them with the right kinds and amounts of nutrients (NEW tree unts). **Nutrients** are the chemicals in food that cells need. The study of nutrients and how your body uses them is called **nutrition** (new TRISH un). The six different nutrients in food are proteins (PROH teenz), fats, carbohydrates (kar boh HI drayts), vitamins, minerals, and water. Each of these six nutrients will be described in the following sections.

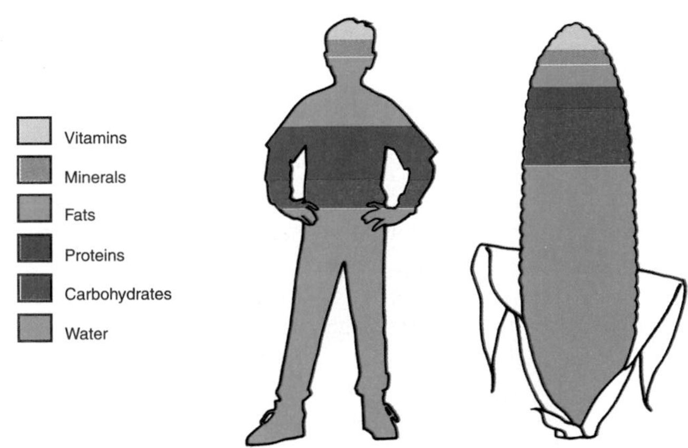


**Figure 9-1** The chemicals that make up the human body come from the exact same chemicals found in the foods you eat.



## OPTIONS

### Science Background

Both animal and plant foods supply protein. In general, protein from animals has more nutritional value because it has more of the essential amino acids. Meat, fish, poultry, milk, eggs, legumes, and cereal grains provide most of our protein needs.

FOCUS 17

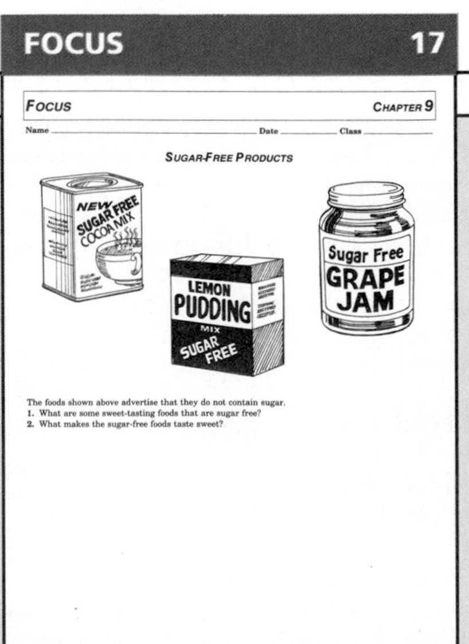

**Figure 9-2** These foods are excellent sources of proteins, fats, and carbohydrates.

## Proteins, Fats, and Carbohydrates

**Proteins** are nutrients that are used to build and repair body parts. Proteins make up large parts of tissues such as bone, muscle, and skin. Foods such as meat, eggs, fish, nuts, and chicken supply you with protein.

**Fats** are nutrients that are used as a source of energy by your body. Fats are compounds that store large amounts of energy. Salad dressing, butter, and cooking oils are foods high in fat.

**Carbohydrates** are nutrients that also supply you with energy. What then is the difference between fats and carbohydrates? The difference is that your body may store fats. It uses them as an energy source after it first uses up all of your carbohydrate supply. Foods containing starches and sugars, such as bread and fruit, supply you with carbohydrates.

How much of each nutrient is present in your body? Water is an important nutrient that makes up a large percentage of the human body. Figure 9-3 shows that a male's body is 60 percent water. How does this amount compare with what is present in a female? A male has 18 percent fat. Does a female have more or less of this nutrient? Look at Figure 9-3 and compare amounts of carbohydrate for males and females.

How much of each nutrient do you need each day? You might guess that you need more protein than carbohydrate because protein makes up more of your body. You might also guess that you need very little carbohydrate. After all, this nutrient makes up little of your body.

### Skill Check

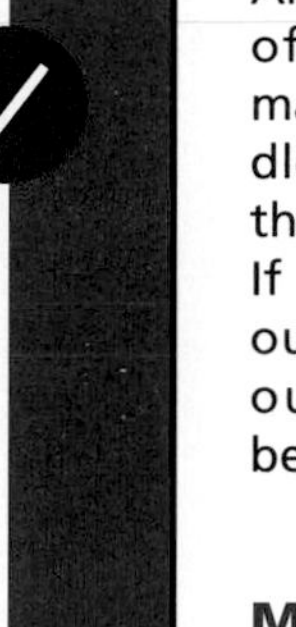

**Understand science words: carbohydrate.** The word part *carbo* means coal. In your dictionary, find three words that contain the word part *carbo. For more help, refer to the **Skill Handbook**, pages 706-711.*

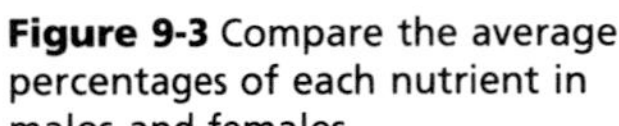

**Figure 9-3** Compare the average percentages of each nutrient in males and females.

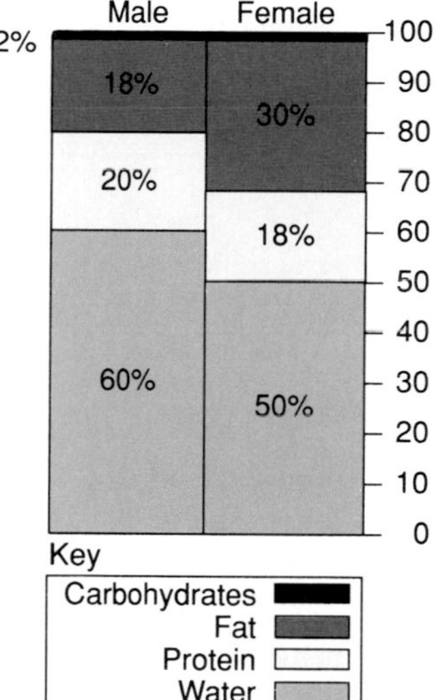




# 2 TEACH

### MOTIVATION/Demonstration

An analogy of the need for a supply of both food and oxygen may be made by burning a candle. The candle wax is the fuel. While burning, the candle uses oxygen from the air. If the candle uses all its fuel, it goes out. If air is excluded, it also goes out. Demonstrate by placing a beaker over the burning candle.

### MOTIVATION/Audiovisual

Show the video *Nutrition to Grow On*, Focus Media Inc.

### Concept Development

▶ Have students give other examples of protein foods that they are familiar with. Do the same with fats and carbohydrates. You may want to write the students' suggestions on the chalkboard under the proper headings.

▶ Take a survey of what some students had for breakfast or for supper the night before. A typical survey will show that more carbohydrates are eaten than any other food nutrient.

### Independent Practice

**Study Guide**/*Teacher Resource Package*, p. 49. Use the Study Guide worksheet shown at the bottom of this page for independent practice.

### Portfolio

Have students prepare a chart comparing the amount of fat in five food products that are advertised as being low-fat to a similar product that is not low-fat. Students should explain why a person would want to reduce fat intake when fats are an energy source.

**STUDY GUIDE 49**

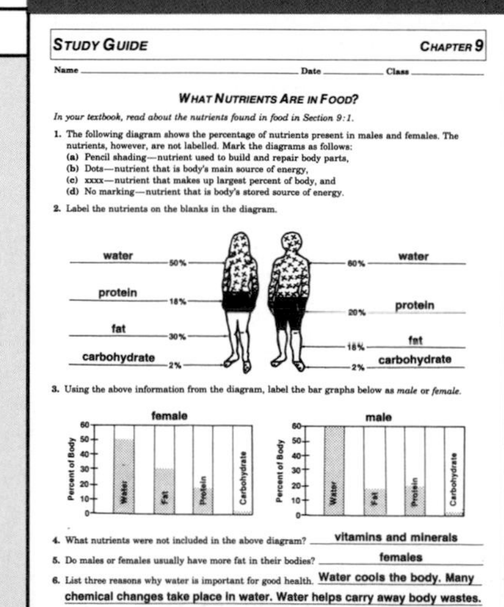

STUDY GUIDE — CHAPTER 9

Name ________ Date ________ Class ________

**WHAT NUTRIENTS ARE IN FOOD?**

*In your textbook, read about the nutrients found in food in Section 9:1.*

1. The following diagram shows the percentage of nutrients present in males and females. The nutrients, however, are not labelled. Mark the diagrams as follows:
   (a) Pencil shading—nutrient used to build and repair body parts,
   (b) Dots—nutrient that is body's main source of energy,
   (c) xxxx—nutrient that makes up largest percent of body, and
   (d) No marking—nutrient that is body's stored source of energy.
2. Label the nutrients on the blanks in the diagram.



3. Using the above information from the diagram, label the bar graphs below as *male* or *female*.



4. What nutrients were not included in the above diagram? **vitamins and minerals**
5. Do males or females usually have more fat in their bodies? **females**
6. List three reasons why water is important for good health. **Water cools the body. Many chemical changes take place in water. Water helps carry away body wastes.**

## OPTIONS

**Daily Nutrient Requirements**/*Transparency Package*, number 9. Use color transparency number 9 as you teach the nutrients your body needs.

**What Are the Tests for Fats and Proteins?**/*Lab Manual*, pp. 71-74. Use this lab as an extension to teaching the nutrients in food.

## TEACH

### Concept Development

▶ Have students use Figure 9-4 to determine how much of each nutrient they need. You may wish to have students construct a chart that summarizes the information.

▶ The graph in Figure 9-4 shows the number of grams of each major nutrient that a person should take in daily. Fiber is included in the graph because it now is recognized as an important component to the diet even though it does not contribute any nutrient value. Fiber consists of cellulose from the cell walls of fruits and vegetables, as well as hemicellulose present in grains. All fiber is indigestible in humans.

### Connection: Social Studies

Have students determine foods that are mainly fat, protein, and carbohydrate sources in different countries.

### GLENCOE TECHNOLOGY

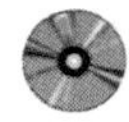

**Videodisc**

**The Secret of Life**

*Saturated Fatty Acids*

*Unsaturated Fatty Acid*

*Monosaccharide*

*Disaccharide*

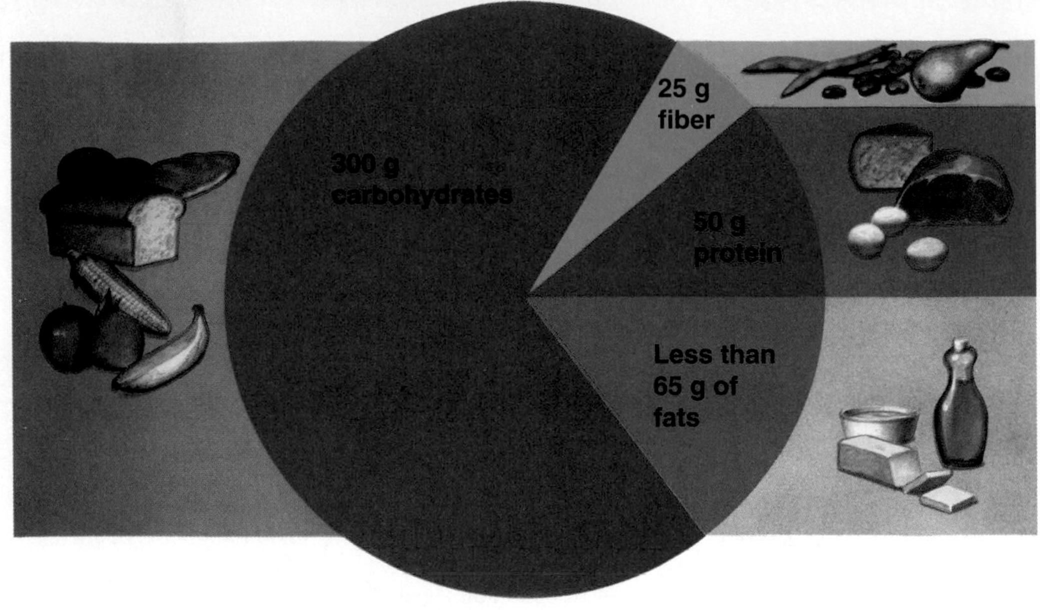


**Figure 9-4** This graph shows how much of each nutrient your body needs each day.

Something quite different is true. You need more carbohydrate than protein each day. Why? The body uses carbohydrates quickly. Carbohydrates are the body's main source of energy. They are not stored for a long time. Fats and proteins can be stored for a long time.

Look at Figure 9-4. Which nutrient is needed in the greatest amount each day? Except for fiber, which nutrient is needed in the smallest amount each day? Carbohydrates are needed in the greatest amount and protein is needed in the smallest amount. A person can remain healthy only if he or she takes in the correct amounts of each nutrient. Of course, the amount of each nutrient needed daily may differ slightly for each person. The amount of each nutrient you take in must be balanced with what your body uses. One way to stay healthy is to eat foods that will supply you with the correct amount of each nutrient. Your diet should be from 55 to 65 percent carbohydrates, less than 30 percent fats, and from 10 to 15 percent protein each day for you to stay healthy.

What parts of your body contain or store nutrients? Protein makes up the cytoplasm in your cells. All the body organs are mostly protein. Fats are stored under skin and around body organs. Carbohydrates are stored in the liver and blood.

**Mini Lab**

**Leisure:** Planning a meal for a group of friends can be an enjoyable way to learn about nutrients.



### OPTIONS

**Skill**/*Teacher Resource Package*, p. 9. Use the Skill worksheet shown here for students to make a bar graph of nutrients in cereals.

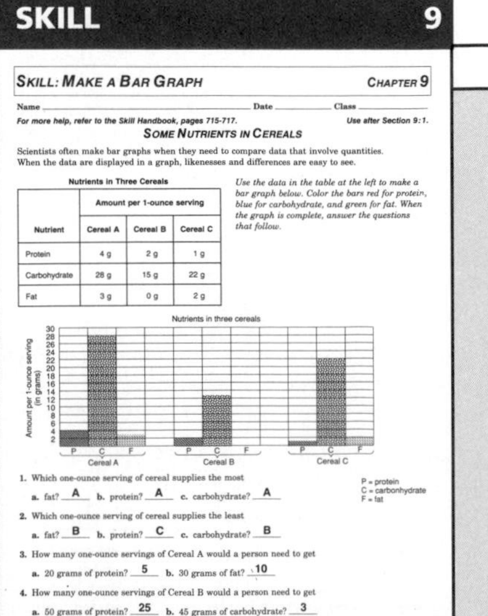

**SKILL 9**

**SKILL: MAKE A BAR GRAPH** CHAPTER 9

Name ________ Date ______ Class ______

*For more help, refer to the Skill Handbook, pages 715-717.* *Use after Section 9:1.*

**SOME NUTRIENTS IN CEREALS**

Scientists often make bar graphs when they need to compare data that involve quantities. When the data are displayed in a graph, likenesses and differences are easy to see.

**Nutrients in Three Cereals**

| Nutrient | Amount per 1-ounce serving | | |
|---|---|---|---|
| | Cereal A | Cereal B | Cereal C |
| Protein | 4 g | 2 g | 1 g |
| Carbohydrate | 28 g | 15 g | 22 g |
| Fat | 3 g | 0 g | 2 g |

*Use the data in the table at the left to make a bar graph below. Color the bars red for protein, blue for carbohydrate, and green for fat. When the graph is complete, answer the questions that follow.*



1. Which one-ounce serving of cereal supplies the most
   a. fat? A b. protein? A c. carbohydrate? A
2. Which one-ounce serving of cereal supplies the least
   a. fat? B b. protein? C c. carbohydrate? B
3. How many one-ounce servings of Cereal A would a person need to get
   a. 20 grams of protein? 5 b. 30 grams of fat? 10
4. How many one-ounce servings of Cereal B would a person need to get
   a. 50 grams of protein? 25 b. 45 grams of carbohydrate? 3

# Lab 9–1 Milk Nutrients

## Problem: What nutrients are present in milk?

### Skills

use numbers, experiment, make and use tables, measure in SI

### Materials

whole milk
hot plate
medicine dropper
enzyme
stirring rod
250-mL graduated cylinder
200-mL beakers (2)
thermal glove
cheese cloth
funnel

### Procedure

1. Copy the data table.
2. **Measure in SI:** Measure 100 mL of whole milk in the graduated cylinder.
3. Pour the milk into a beaker.
4. Place the beaker of milk onto a hot plate and warm the milk. **CAUTION:** *Use a thermal glove when handling hot objects.*
5. Add 20 drops of the enzyme to the milk.
6. Use a stirring rod to mix the milk and enzyme for several minutes.
7. Continue to stir until the milk separates into a solid white part and a liquid part.
8. Line a funnel with a single layer of wet cheese cloth. Position the funnel so that a beaker is below it, as shown in the figure.
9. Pour the milk into the funnel. Wait 2 to 3 minutes until all the liquid has drained through the cheese cloth.

10. The liquid that drains into the beaker is water. Use a graduated cylinder to measure the volume.
11. Record this volume in the data table.
12. The solid material left in the cheese cloth is protein. Record its color and texture in your data table.

### Data and Observations

1. What is the percent water in your milk sample?
2. What is the percent protein in your milk sample?

| STATE OF MILK | OBSERVATIONS |
|---|---|
| Original volume of milk used | |
| Volume of water in milk sample | |
| Color and texture of protein | |
| Color of liquid | |

### Analyze and Apply

1. Why did you add the enzyme to the milk?
2. What two essential nutrients did you study in this lab?
3. Milk can be separated into solid curds and liquid whey. In which step did you do this?
4. **Apply:** If you had tested skim milk instead of whole milk, would nutrients be present in different amounts in the two types of milk?

### Extension

**Experiment** with skim milk and 2-percent milk to measure the same nutrients as those for whole milk.



## *ANSWERS*

### Data and Observations

1. 80-85%
2. 15-20%

### Analyze and Apply

1. The enzyme helps to separate the protein out of the liquid.
2. protein and water
3. step 7 after enzyme was added in step 5
4. They would be present in the same amount. Skim milk has less fat, and this nutrient wasn't tested for in this lab.

## Lab 9-1 Milk Nutrients

### Overview

Students will experimentally measure the amount and/or presence of water and protein in milk.

**Objectives:** Upon completion of this lab, students will be able to (1) **state** that milk contains water, protein, and carbohydrates, (2) **describe** the lab technique used to separate these nutrients, (3) **identify** protein and water that have been separated from milk.

**Time Allotment:** 30 minutes

### Preparation

▶ **Alternate Materials:** Vinegar may be substituted in place of rennilase (about 1 teaspoon/100 ml of milk).

▶ Rennilase is available from a biological supply house.

**Lab 9-1 worksheet**/*Teacher Resource Package,* pp. 33-34.

### Teaching the Lab

▶ Review use and proper reading of a graduated cylinder before starting this lab.

▶ Protein may be verified as follows: concentrated nitric acid added to protein will result in a yellow color. *Do as a teacher demo.*

**Cooperative Learning:** Have students work in groups of two and list their data on the chalkboard. For more information, see pp. 22T-23T in the Teacher Guide.

### ASSESSMENT

**Performance:** Provide students with two samples of milk marked *A* and *B*. Have students determine which sample has the most protein and the most water. (Sample *A* is regular milk; Sample *B* is the same milk but diluted with 25 percent more water.)

## TEACH

### Concept Development

▶ Obtain food labels from a variety of foods. Make photocopies for your students and a transparency, for the overhead projector. Go over the transparency, pointing out the vitamins to familiarize students with the names.

▶ When describing the quantity of a nutrient needed each day for a balanced diet, RDA is used. When describing the amount of a specific nutrient present in one serving of a food, % DV is used. % DV is found on product labels for consumer use and is based on the dietary needs for a person taking in 2000 Calories each day. The value of 2000 is an average between 1600 Calories per day needed for women and 2400 Calories per day needed for men.

### Guided Practice

Have students write down their answers to the margin question: They are needed for growth and tissue repair.

### Idea Map

Have students use the idea map as a study guide to the major concepts of this section.

**GLENCOE TECHNOLOGY**

**Videodisc**

**The Secret of Life**

*Composition of Living Organisms*

### Bio Tip

**Consumer:** Which has more vitamin C, an orange or a red bell pepper? An orange contains 70 milligrams, while a red bell pepper contains 140 milligrams.

## Vitamins

Are you a label reader? Have you ever read a label on a cereal box or bread wrapper? Labels give you useful information about the foods you eat. For example, labels tell you which vitamins are in foods. **Vitamins** are chemical compounds needed in very small amounts for growth and tissue repair of the body. A well-balanced diet will supply you with the vitamins needed each day.

The vitamins have chemical names. For example, ascorbic acid is more commonly known as vitamin C. The chemical name for vitamin A is retinol. When you read a food label, you will see vitamins listed as A, $B_1$,$B_2$, $B_3$, C, and D, or sometimes by their chemical names. Look at Table 9-1 for the names of the vitamins you need each day.

Why are vitamins important?

Certain diseases may appear if too much or not enough of a vitamin is in your diet. Taking too much vitamin A, for example, can cause loss of hair or liver problems. Without enough vitamins, many chemical changes cannot take place within your cells. Table 9-1 lists some vitamins that appear on food labels. The table shows how the body uses the vitamins and describes what happens if you do not have enough of them. For example, if you were to go for many months without fresh fruit and vegetables, you may find your mouth sore, your skin rough, and that you bruise easily.

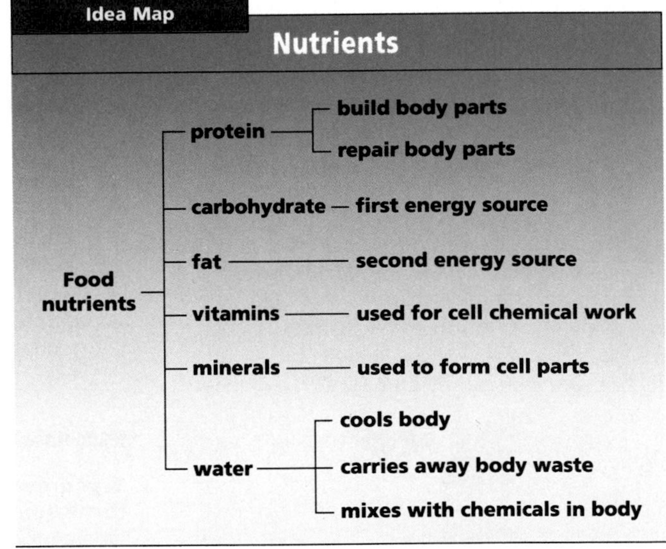


**Study Tip:** Use this idea map as a study guide to comparing how the body uses each of the six food nutrients.



## OPTIONS

**Transparency Master/*Teacher Resource Package*,** p. 41. Use the Transparency Master shown here to teach how to read food labels.

**TRANSPARENCY 41**

*TRANSPARENCY MASTER* — *CHAPTER 9*

*READING FOOD LABELS*

| Specific label information | What the information means |
|---|---|
| Cereal flakes<br>serving size = 1 ounce 28.4 g<br><br>servings per package: 16 | serving size = amount of food usually eaten in one serving<br>**28.4g**<br># servings per package = number of servings available in entire package; in this example, there are<br>**16 servings** |
| 1 oz serving<br>Calories 90<br><br>Protein 3 g<br>Carbohydrate 22 g<br>Fat 0 g<br>Sodium 220 mg<br>Potassium 160 mg | number of calories in 1 serving of 28.4 g<br><br>listed according to their mass; values listed are for 1 serving |
| Percentage of U.S. Recommended Daily Allowance (RDA) in a 1 oz serving<br><br>Protein 6<br><br>Vitamin A 15<br>Vitamin C *<br>Thiamine 25<br>Niacin 25<br>Vitamin D 25<br><br>Calcium *<br>Iron 100<br>Zinc 25<br>Copper 10<br><br>*Less than 2% of RDA | amount of nutrient needed each day to stay in good health, expressed as a percentage<br>These values are for one serving only.<br><br>protein = 6%; a person must get the other 94% from other foods to meet 100% RDA<br><br>these nutrients are all vitamins<br><br>these nutrients are all minerals<br>**Review the RDA still needed from other foods for vitamin C, thiamine, calcium, and zinc.** |
| Ingredients: wheat bran, sugar, corn syrup, salt, malt flavoring, preservative BHT | The ingredient listed first is present in the greatest amount; the last ingredient is the least amount. BHT is a preservative. The ingredient listed in greatest amount (wheat bran) should match nutrient present in greatest mass (carbohydrate). |

| TABLE 9–1. VITAMINS | | | | |
|---|---|---|---|---|
| **Vitamin** | **How Used in Body** | **Problems if Not Enough** | **Foods** | **RDA** |
| A (retinol) | vision, healthy skin | night blindness, rough skin | liver, broccoli, carrots | 1000 μg* |
| $B_1$ (thiamine) | allows cells to use carbohydrates | digestive problems, muscle paralysis | ham, eggs, raisins | 1.5 mg |
| $B_2$ (riboflavin) | allows cells to use carbohydrates and proteins | eye problems, cracking skin | milk, yeast, eggs | 1.7 mg |
| $B_3$ (niacin) | allows cells to carry out respiration | mental problems, skin rash, diarrhea | peanuts, tuna, chicken | 20.0 mg |
| C (ascorbic acid) | healthy membranes, wound healing | sore mouth and bleeding gums, bruises | green peppers oranges, lemons, tomatoes | 60.0 mg |
| D (calciferol) | bone growth | bowed legs, poor teeth | egg yolk, shrimp, milk, yeast | 10 μg* |

*μg = microgram (1 microgram is 1/1 000 000 of a gram)

Notice the column marked RDA in Table 9-1. RDA stands for recommended daily allowance. The **recommended daily allowance** is the amount of each vitamin and mineral a person needs each day to stay in good health. Some of the units in Table 9-1 are listed as mg, or milligrams, and μg, or micrograms. Remember from Chapter 1 that one milligram is one thousandth of a gram. A strand of hair has a mass of more than one milligram. You can see that vitamins are needed in very small amounts.

Let's look at a real food label. Figure 9-5 shows a label from orange juice. The vitamins are highlighted. Each vitamin is listed as a percent. The percent stands for % Daily Value (% DV). **Percent Daily Value** is the percent of nutrient found in one serving of a food compared to 100%. Remember, 100% of each nutrient is needed each day. For example, you eat a cereal that has a % DV of 15% for Vitamin A. This means that you still need 85% more of Vitamin A from other foods to complete the 100% total for that day. % DVs are based on the amount of food that an average person eats in one day.

## Mini Lab

**How Does Salt Content Change in Processed Foods?**

**Interpret data:** Find out how much salt is present in a tomato. Compare this with the salt in canned tomato juice. How do these amounts compare? *For more help, refer to the* ***Skill Handbook****, pages 704-705.*

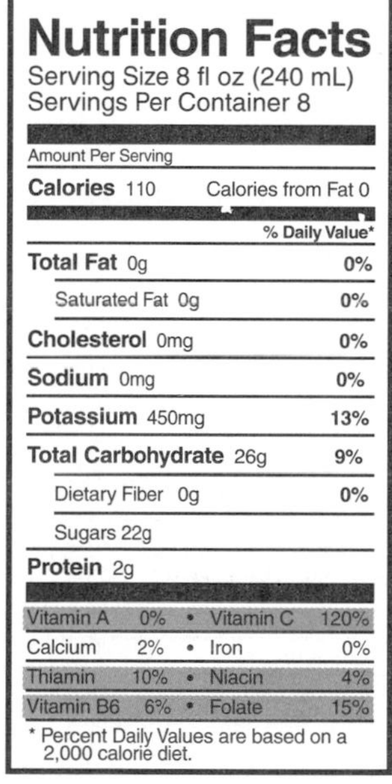

**Nutrition Facts**
Serving Size 8 fl oz (240 mL)
Servings Per Container 8

Amount Per Serving

| | |
|---|---|
| **Calories** 110 | Calories from Fat 0 |
| | % Daily Value* |
| **Total Fat** 0g | 0% |
| Saturated Fat 0g | 0% |
| **Cholesterol** 0mg | 0% |
| **Sodium** 0mg | 0% |
| **Potassium** 450mg | 13% |
| **Total Carbohydrate** 26g | 9% |
| Dietary Fiber 0g | 0% |
| Sugars 22g | |
| **Protein** 2g | |

| | | | |
|---|---|---|---|
| Vitamin A | 0% | Vitamin C | 120% |
| Calcium | 2% | Iron | 0% |
| Thiamin | 10% | Niacin | 4% |
| Vitamin B6 | 6% | Folate | 15% |

* Percent Daily Values are based on a 2,000 calorie diet.

**Figure 9-5** Food labels tell which vitamins are present. They are listed in %DV.



## TEACH

### Concept Development

▶ Remind students that 1000 milligrams equals 1 gram. Show them the mass of 1 gram of a clay ball. Tell them to imagine having to divide this ball into 1000 equal units to help them visualize the size of a milligram.

### Mini Lab

Processed foods generally have more salt. 1 tomato (120 g) contains approximately 4 mg of sodium. 1 cup of tomato juice (240 g) contains about 480 mg of sodium.

### ASSESSMENT

**Performance:** Provide students with photocopies of labels from a variety of foods, such as potato chips, pretzels, popcorn, canned vegetables, and milk. Have students list and rank the foods from highest to lowest in salt content *per average serving*. Then have students calculate and record the total amount of salt in the entire can, carton, or package of food.

**Cooperative Learning:** Divide the class into groups of 4 students. Have each group research the following information for a different vitamin: (a) its RDA, (b) food sources that are high and low in the vitamin, (c) how it is used by the body, (d) signs of vitamin deficiency, (e) side effects due to excess dosage.

## ENRICHMENT 9

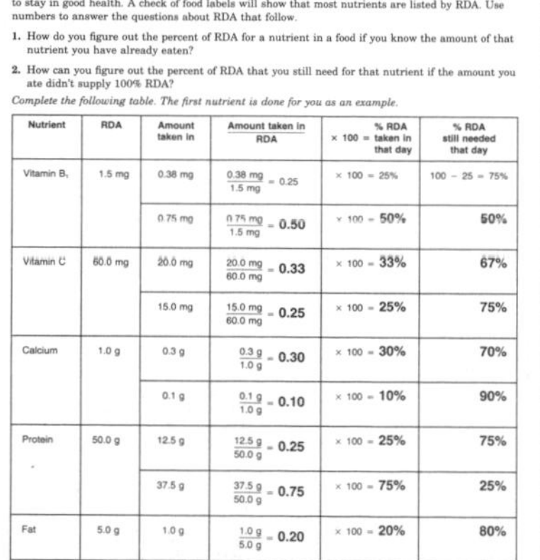

ENRICHMENT CHAPTER 9

Name ______ Date ______ Class ______

*Use after Section 9:1.*

**CALCULATING RDA**

The term *Recommended Daily Allowance* (RDA) is the amount of a nutrient a person needs each day to stay in good health. A check of food labels will show that most nutrients are listed by RDA. Use numbers to answer the questions about RDA that follow.

1. How do you figure out the percent of RDA for a nutrient in a food if you know the amount of that nutrient you have already eaten?
2. How can you figure out the percent of RDA that you still need for that nutrient if the amount you ate didn't supply 100% RDA?

*Complete the following table. The first nutrient is done for you as an example.*

| Nutrient | RDA | Amount taken in | Amount taken in / RDA | × 100 = % RDA taken in that day | % RDA still needed that day |
|---|---|---|---|---|---|
| Vitamin $B_1$ | 1.5 mg | 0.38 mg | $\frac{0.38\text{ mg}}{1.5\text{ mg}} = 0.25$ | × 100 = 25% | 100 − 25 = 75% |
| | | 0.75 mg | $\frac{0.75\text{ mg}}{1.5\text{ mg}} = 0.50$ | × 100 = 50% | 50% |
| Vitamin C | 60.0 mg | 20.0 mg | $\frac{20.0\text{ mg}}{60.0\text{ mg}} = 0.33$ | × 100 = 33% | 67% |
| | | 15.0 mg | $\frac{15.0\text{ mg}}{60.0\text{ mg}} = 0.25$ | × 100 = 25% | 75% |
| Calcium | 1.0 g | 0.3 g | $\frac{0.3\text{ g}}{1.0\text{ g}} = 0.30$ | × 100 = 30% | 70% |
| | | 0.1 g | $\frac{0.1\text{ g}}{1.0\text{ g}} = 0.10$ | × 100 = 10% | 90% |
| Protein | 50.0 g | 12.5 g | $\frac{12.5\text{ g}}{50.0\text{ g}} = 0.25$ | × 100 = 25% | 75% |
| | | 37.5 g | $\frac{37.5\text{ g}}{50.0\text{ g}} = 0.75$ | × 100 = 75% | 25% |
| Fat | 5.0 g | 1.0 g | $\frac{1.0\text{ g}}{5.0\text{ g}} = 0.20$ | × 100 = 20% | 80% |

## OPTIONS

**Enrichment**/*Teacher Resource Package*, p. 9. Use the Enrichment worksheet shown here to have students calculate RDA.

**How Do You Test for Vitamin C in Foods?**/*Lab Manual*, pp. 75-78. Use this lab as an extension to teaching about vitamins.

# TEACH

## Concept Development

▶ It may be difficult for students to distinguish between vitamins and minerals when using the chemical names. Use a periodic table of the elements for reference. If a periodic table in chart form is not available, photocopies of the table can be made from any chemistry textbook.

▶ Point out that all the minerals are also elements present on the periodic table.

▶ Ask students to list the minerals present in various foods by reading food labels. They should use the periodic table as a guide.

## Independent Practice

**Study Guide**/*Teacher Resource Package,* p. 50. Use the Study Guide worksheet shown at the bottom of this page for independent practice.

## GLENCOE TECHNOLOGY

**Videodisc**

The Secret of Life
*Water Molecule*

Biology: The Dynamics of Life
Movie: *Properties of Water*
Disc 1, Side 1, Ch. 19

### Bio Tip

**Health:** The American Heart Association recommends that you use only 500 mg of salt each day.

**TABLE 9-2. MINERALS**

| Mineral | How Used in Body | Problems if Not Enough | Foods | RDA |
|---|---|---|---|---|
| Iron | helps form blood cells, helps blood carry oxygen | anemia, feeling tired | liver, egg yolk, peas, enriched cereals, whole grains | 10-15 mg |
| Calcium | helps form bones and teeth | bones and teeth become weak or brittle | milk, cheese, sardines, nuts, whole-grain cereals | 800-1200 mg |
| Magnesium | helps form bones and teeth | muscles twitch | potatoes, fruits, whole-grain cereals | 325 mg |
| Iodine | helps make thyroid gland chemical | causes thyroid gland to enlarge | seafoods, eggs, milk, iodized table salt | 150 μg* |
| Sodium | muscle contractions, nerve messages | dizziness, tired feeling, cramps | bacon, butter, table salt | less than 2400 mg |

*μg = microgram (one microgram is 1/1 000 000 of a gram)

**Figure 9-6** Food labels tell which minerals are present in one serving. They are listed in %DV and in grams or milligrams.

## Minerals

**Minerals** are nutrients needed to help form different cell parts. Minerals, like vitamins, are chemicals. Your body needs minerals in very small amounts. A diet with a variety of foods provides needed minerals.

Table 9-2 lists five minerals and how they are used by your body.

The table also tells you what can happen when these minerals are missing from your diet. For example, calcium is a mineral needed to form strong bones and teeth. Lack of calcium may cause bones and teeth to become weak and even brittle.

Figure 9-6 shows an actual food label from crackers. The minerals are highlighted. Minerals are shown two ways on labels. First, the quantity of a mineral present in one serving of the food is listed in grams or milligrams, as labeled *A* in Figure 9-6. Second, the % DV is also given for minerals, as labeled *B* in Figure 9-6.

190

## OPTIONS

**Application**/*Teacher Resource Package,* p. 9. Use the Application worksheet shown here to have students compare nutrients in foods.

## STUDY GUIDE 50

**STUDY GUIDE** CHAPTER 9

Name ______ Date ______ Class ______

### WHAT NUTRIENTS ARE IN FOOD?

8. Complete the following table by putting check marks in the correct columns.

| | Fat | Protein | Carbohydrate |
|---|---|---|---|
| Present in butter, oils | ✓ | | |
| Body's main energy source | | | ✓ |
| Present in bread and fruit | | | ✓ |
| Makes up bone and muscle | | ✓ | |
| Need the most of daily | | | ✓ |
| Present in meat, fish | | ✓ | |
| Used most quickly in body | | | ✓ |
| 10% of your diet each day | | ✓ | |
| Found as cell membranes | ✓ | ✓ | |
| Stored under skin | ✓ | | |
| Stored in blood and liver | | | ✓ |

9. Examine these food labels. Then, answer the questions below.

Cereal (1 serving)

| Nutrient | RDA |
|---|---|
| Vitamin A | 20 |
| Vitamin C | 10 |
| Riboflavin | 30 |
| Thiamine | 10 |
| Niacin | 25 |
| Calcium | 40 |
| Iron | 60 |
| Potassium | 1 |

Breakfast drink (1 serving)

| Nutrient | RDA |
|---|---|
| Vitamin A | 30 |
| Vitamin C | 5 |
| Niacin | 2 |
| Riboflavin | 40 |
| Thiamine | 5 |
| Calcium | 50 |
| Iron | 4 |
| Zinc | 1 |
| Phosphorus | 1 |

a. Circle the vitamins on each label. b. Underline the minerals on each label.

c. Which of the two foods supplies more Vitamin C in one serving? cereal

## APPLICATION 9

**APPLICATION: CONSUMER** CHAPTER 9

Name ______ Date ______ Class ______

Use after Section 9:2.

### COMPARING NUTRIENTS IN FOODS

Visit a grocery store, and look at the labels of six different breakfast cereals. Look at cereals that are made to appeal more to children and cereals made to appeal more to adults. On each box, find the table that gives the nutrient content and recommended daily allowance (RDA) figures.

*Record the information listed below for each cereal. Be sure to use the figures that are given for the cereal alone, without milk added.*

| Name of cereal (Answers will vary) | % RDA Protein | % RDA Vitamin A | Grams: Carbohydrates |
|---|---|---|---|
| a. Cheerios | 6 | 25 | 20 |
| b. Special K | 10 | 15 | 20 |
| c. Puffed Rice | 0 | 0 | 13 |
| d. Quaker 100% Natural | 4 | 0 | 18 |
| e. Cap'n Crunch | 2 | 0 | 24 |
| f. Corn Flakes | 4 | 15 | 24 |

Student answers will vary depending on cereals used.

1. Use the information in your table to finish the bar graphs below.

% RDA Protein — Cereal a b c d e f; % RDA Vitamin A — Cereal a b c d e f; Grams Carbohydrates — Cereal a b c d e f

2. Which cereal do you think is the most healthful? least healthful? Explain. Answers will vary. Students might judge cereals based on their amounts of protein or vitamins.

**Figure 9-7** You need to replace water that is lost from the body as sweat.

## Water

Water is an important nutrient for good health. You learned earlier that the human body is made up of 50 to 60 percent water.

There are three reasons why water is important. First, water cools the body. You see or feel this happening when you sweat, Figure 9-7. Second, many chemicals within the body can combine only with water. Therefore, many chemical changes can take place only in water. Third, water helps carry away body wastes.

The average adult needs about two liters of water each day. Most people don't drink that much water but still get enough of it. Many foods contain water, Figure 9-8. In fact, some foods are over 90 percent water.

### Mini Lab

**What Are Processed-food Costs?**

**Interpret data:** Find the cost per gram of a raw potato. Compare this with the cost per gram of potato chips. *For more help, refer to the **Skill Handbook,** pages 704-705.*

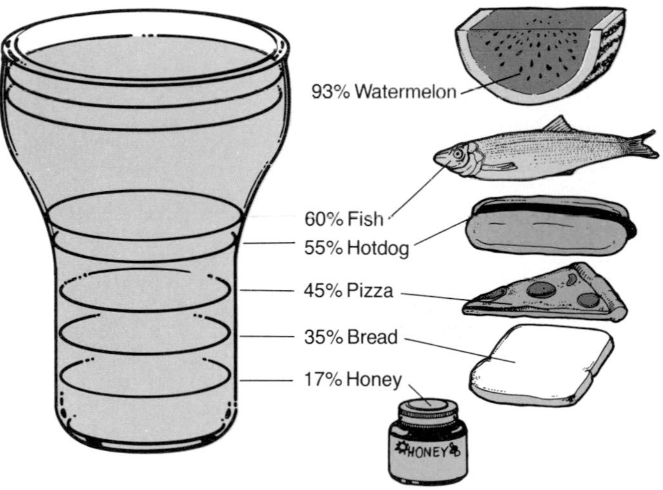


**Figure 9-8** Compare the percentage of water in some foods.



## TEACH

### Concept Development

▶ Ask students how much water they drink in one day. Show them the volume of 2 liters with a graduated cylinder.

### Mini Lab

You may actually want to supply students with a potato and have them mass the potato in class. Provide a balance, the cost of the potato, and the cost of a bag of potato chips and the mass of potatoes in the bag.

### ASSESSMENT

**Performance:** Provide students with a taffy apple, a balance, paper towels, the cost of the taffy apple, and cost and the mass of a bag of apples. Have students calculate the cost per gram of a plain apple versus the cost per gram of a taffy apple. (Students should deduct 3 grams from the mass of the taffy apple to account for the stick and coating mass.)

### Check for Understanding

Prepare a transparency of a food label for the overhead. Use the food label to discuss the needs for fat, protein, and carbohydrate; the food group to which the food belongs; the vitamins and minerals present; and the % DV for all vitamins and minerals shown.

### Reteach

Prepare a chart with the following headings: Nutrients, Amount in one serving, Role in body, Percent in body of male/female. Supply a copy of a food label. Have students complete the chart using the label and their text as a reference.

## CRITICAL THINKING 9

CRITICAL THINKING/PROBLEM SOLVING — CHAPTER 9

Name ______ Date ______ Class ______

*Use after Section 9:1.*

**WHICH COMMON FOODS PROVIDE THE MOST NUTRIENTS?**

A balanced diet contains the right percentages of carbohydrate, fat, protein, and water. Find out which foods provide the highest percentage of each. Look at the table below.

| Food | Percentage of water | Percentage of carbohydrate | Percentage of fat | Percentage of protein |
|---|---|---|---|---|
| Milk, whole | 87 | 5 | 4 | 4 |
| Bread, white | 36 | 52 | 4 | 8 |
| Lettuce | 96 | 3 | 0 | 1 |
| Sugar | 1 | 99 | 0 | 0 |
| Corn oil | 0 | 0 | 100 | 0 |
| Egg | 74 | 1 | 12 | 13 |
| Beef, ground | 54 | 0 | 20 | 26 |
| Potato | 75 | 21 | 0 | 4 |
| Cheese, cheddar | 38 | 4 | 33 | 25 |
| Grain cereal | 1 | 84 | 0 | 15 |
| Margarine | 16 | 1 | 82 | 1 |
| Potato chips | 4 | 50 | 42 | 4 |
| Spaghetti | 7 | 77 | 2 | 14 |
| Tomato | 94 | 5 | 0 | 1 |
| Orange juice | 88 | 10 | 1 | 1 |

**Analyzing the Problem**

1. Look at the table above. Fill in the foods that provide the highest percentage of each nutrient:
   water **lettuce, tomato, orange juice**
   carbohydrate **sugar, grain cereal**
   fat **corn oil, margarine**
   protein **ground beef, cheddar cheese, grain cereal, spaghetti**
2. People who need to lose weight for health reasons are advised to consume a lower percentage of fats. Which foods might they avoid? **corn oil, margarine, potato chips**
3. Protein is very important in the diet. Which three foods in the table provide high protein without providing high fat content? **grain cereal, spaghetti**

**Solving the Problem**

1. Can you conclude that a diet of potato chips and fruit juices is all you need to eat to be healthy? Explain. **No, a balanced diet must contain enough of all nutrients.**

## OPTIONS

**Critical Thinking/Problem Solving/** *Teacher Resource Package,* p. 9. Use the Critical Thinking/Problem Solving worksheet shown here to demonstrate which common foods provide the most nutrients.

## TEACH

### Guided Practice

Have students write down their answers to the margin question: a diet with the right amount of each nutrient.

### Independent Practice

**Study Guide**/*Teacher Resource Package,* p. 51. Use the Study Guide worksheet shown at the bottom of this page for independent practice.

### Check for Understanding

Have students respond to the first three questions in Check Your Understanding.

### Reteach

**Reteaching**/*Teacher Resource Package,* p. 23. Use this worksheet to give students additional practice with nutrient needs.

**Extension:** Assign Critical Thinking, Biology and Math, or some of the **OPTIONS** available with this lesson.

## 3 APPLY

### Convergent Question

Ask where nutrients come from that are needed by the body if one or two meals are skipped.

## 4 CLOSE

### Software

*Buying Nutritious Food,* Queue, Inc.

### Answers to Check Your Understanding

1. fats and carbohydrates
2. They help form certain cell parts.
3. by choosing foods from the six food groups in the correct proportions
4. Carbohydrates are not stored for a long time and are used quickly.
5. two paper clips

Pg. 192 patch correction

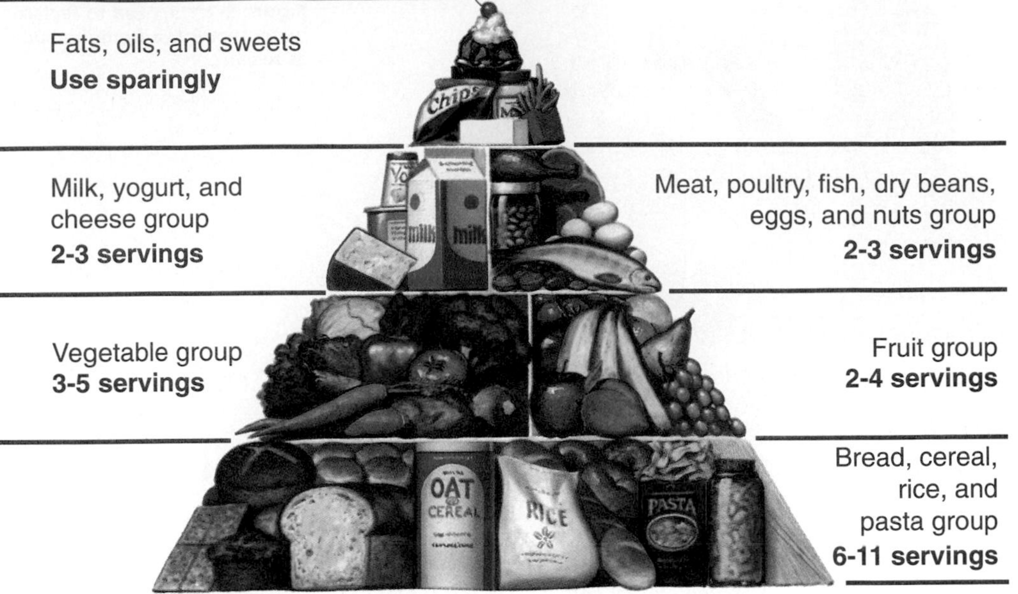


**Figure 9-9** A food pyramid shows that most foods each day should come from the base of the pyramid, with very few foods each day coming from the tip of the pyramid.

**What is a balanced diet?**

## Supplying Nutrients to Your Body

Most foods contain several nutrients. Whole milk is four percent protein, four percent fat, and five percent carbohydrate. It also has vitamins, minerals, and water.

How do you know which foods will give you a balanced diet? A **balanced diet** is a diet with the right amount of each nutrient. Figure 9-9 shows foods placed into six groups. Eating the number of servings for each food group every day will help you eat a balanced diet.

### Check Your Understanding

1. Name two nutrients that supply the body with energy.
2. How are minerals used by the body?
3. How can you be sure that your diet is balanced?
4. **Critical Thinking:** Why does the body need more carbohydrates than fats or proteins?
5. **Biology and Math:** The %DV of sodium is 2400 mg. Which of the following has about the same mass as 2400 mg: two paper clips, a quarter, a pencil, or a biology book?

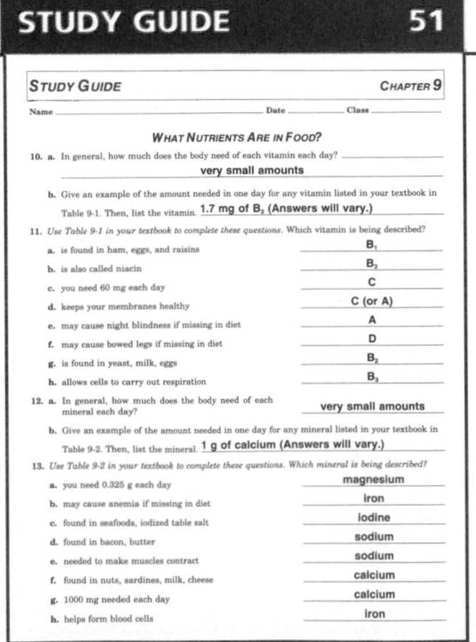

**STUDY GUIDE 51**

*STUDY GUIDE* — *CHAPTER 9*

Name ______ Date ______ Class ______

*WHAT NUTRIENTS ARE IN FOOD?*

10. a. In general, how much does the body need of each vitamin each day? **very small amounts**
    b. Give an example of the amount needed in one day for any vitamin listed in your textbook in Table 9-1. Then, list the vitamin. **1.7 mg of $B_2$ (Answers will vary.)**
11. *Use Table 9-1 in your textbook to complete these questions.* Which vitamin is being described?
    a. is found in ham, eggs, and raisins **$B_1$**
    b. is also called niacin **$B_3$**
    c. you need 60 mg each day **C**
    d. keeps your membranes healthy **C (or A)**
    e. may cause night blindness if missing in diet **A**
    f. may cause bowed legs if missing in diet **D**
    g. is found in yeast, milk, eggs **$B_2$**
    h. allows cells to carry out respiration **$B_3$**
12. a. In general, how much does the body need of each mineral each day? **very small amounts**
    b. Give an example of the amount needed in one day for any mineral listed in your textbook in Table 9-2. Then, list the mineral. **1 g of calcium (Answers will vary.)**
13. *Use Table 9-2 in your textbook to complete these questions. Which mineral is being described?*
    a. you need 0.325 g each day **magnesium**
    b. may cause anemia if missing in diet **iron**
    c. found in seafoods, iodized table salt **iodine**
    d. found in bacon, butter **sodium**
    e. needed to make muscles contract **sodium**
    f. found in nuts, sardines, milk, cheese **calcium**
    g. 1000 mg needed each day **calcium**
    h. helps form blood cells **iron**

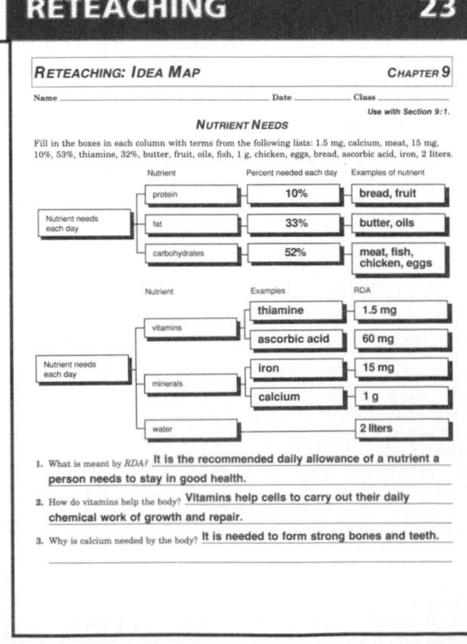

**RETEACHING 23**

*RETEACHING: IDEA MAP* — *CHAPTER 9*

Name ______ Date ______ Class ______

*Use with Section 9:1.*

*NUTRIENT NEEDS*

Fill in the boxes in each column with terms from the following lists: 1.5 mg, calcium, meat, 15 mg, 10%, 53%, thiamine, 32%, butter, fruit, oils, fish, 1 g, chicken, eggs, bread, ascorbic acid, iron, 2 liters.

Nutrient needs each day

| Nutrient | Percent needed each day | Examples of nutrient |
|---|---|---|
| protein | **10%** | **bread, fruit** |
| fat | **33%** | **butter, oils** |
| carbohydrates | **52%** | **meat, fish, chicken, eggs** |

Nutrient needs each day

| Nutrient | Examples | RDA |
|---|---|---|
| vitamins | **thiamine** | **1.5 mg** |
| | **ascorbic acid** | **60 mg** |
| minerals | **iron** | **15 mg** |
| | **calcium** | **1 g** |
| water | | **2 liters** |

1. What is meant by *RDA?* **It is the recommended daily allowance of a nutrient a person needs to stay in good health.**
2. How do vitamins help the body? **Vitamins help cells to carry out their daily chemical work of growth and repair.**
3. Why is calcium needed by the body? **It is needed to form strong bones and teeth.**

# 9:2 Calories

Did you ever have a soft drink or candy bar when you wanted to eat something that would give you a lift? Besides supplying your body with nutrients, food gives your body energy. You are able to talk, play sports, and read this book because of the energy food gives you. Let's look at how the energy in food is measured and how the body uses food energy.

**Objectives**

4. **Describe** how a Calorie is used by the body.
5. **Compare** the number of Calories found in different nutrients.
6. **Compare** the number of Calories used in different activities.

**Key Science Words**

Calorie

## Energy in Food

Have you ever noticed how often TV food commercials mention Calories? A **Calorie** is a measure of the energy in food. Foods high in Calories provide a lot of energy. Low-Calorie foods provide less energy. Diet soft drinks usually contain only one Calorie. A candy bar might contain over 200 Calories. Most of those Calories are in the form of fat and sugar.

Food, like wood or coal, can be burned. The heat given off by burning food can change the temperature of water. Scientists describe a Calorie as the amount of heat it takes to raise the temperature of 1000 g of water 1°C, Figure 9-10. If a slice of bread has 70 Calories, what does it mean? It means that if the slice of bread were burned, it would give off enough heat to raise the temperature of 1000 g of water 70°C.

**What is a Calorie?**

Of course, in your body, food energy is not used to heat water. Food energy is used to keep your body temperature close to 37°C. It is also used to move your muscles, pump your blood, and send messages along your nerves. Food energy is released when your cells carry on respiration.

**Figure 9-10** This procedure shows how one Calorie can be measured.

**a** Start with 1000 g of water at 19°C.

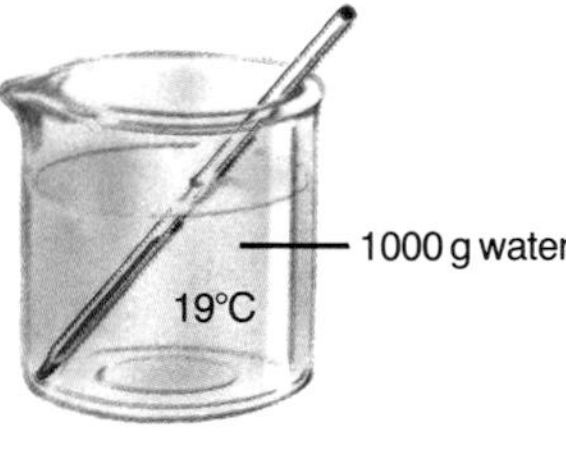


**b** Burn food to heat the water.

**c** When one Calorie of food has burned, the temperature of the water is one degree higher.

## OPTIONS

### Science Background

Calories are units of energy. They are arrived at by taking food samples and burning them. The amount of temperature change that occurs during burning is measured and converted into Calories. Food Calories are sometimes called kilocalories and are always capitalized.

## 9:2 Calories

## PREPARATION

### Materials Needed

Make copies of the Study Guide, Reteaching, and Lab worksheets in the *Teacher Resource Package.*

▶ Gather food samples needed for Lab 9-2. Purchase sugar (glucose) test paper.

▶ Bring in taco chips for the Focus demonstration.

### Key Science Words

Calorie

### Process Skills

In Lab 9-2, students will form a hypothesis, experiment, and make and use tables. In the Skill Checks, they will infer and interpret data.

## 1 FOCUS

▶ The objectives are listed on the student page. Remind students to preview these objectives as a guide to this numbered section.

### Guided Practice

Have students write down their answers to the margin question: a measure of the amount of energy in food.

## Lab 9-2 Energy Foods

### Overview

Students will test foods known to contain sugar. They will then test a number of unknown foods.

**Objectives:** Upon completing this lab, students will be able to (1) **test** for sugar in a variety of foods, (2) **determine** the purpose of testing for sugar in food.

**Time Allotment:** 30 minutes

### Preparation

▶ Sugar test paper is available in the diabetic needs section of your local drugstore or from a biological supply house.

▶ All sugars such as honey, molasses, and maple syrup must be diluted in water, about 5 parts water to 1 part food, in order for glucose test paper to register a green color. Do not test cooked potato or rice with sugar test paper. They will test positive for sugar.

**Lab 9-2 worksheet**/*Teacher Resource Package,* pp. 35-36.

### Teaching the Lab

▶ **Troubleshooting:** All sugars are carbohydrates but there are many different types of sugars. The test paper being used in this experiment will detect only glucose, a specific sugar type. Table sugar is not glucose. It is sucrose and is not detected with this test paper.

**Performance:** Provide students with samples of "unknowns" marked A, B, C, and D. Students should test and record whether or not sugar (glucose) is present in the samples provided. (Unknowns A and B = water, C and D = honey mixed with water.)

### Student Journal

Have students write copy for an ad for a nutritious snack food—a piece of fruit. Students should research the benefits to the body of the fruit they choose, and include this information in their copy.

## Energy Foods

## Lab 9–2

### Problem: Which foods contain sugar?

#### Skills

form a hypothesis, experiment, make and use tables

#### Materials

glass slides
razor blade
sugar test paper
food samples for sugar test
droppers
water

#### Procedure

1. Copy the data table.
2. **Form a hypothesis:** Look at the foods you are to test for sugar. In your notebook, write a hypothesis about which foods contain sugar and which don't.
3. Using a dropper, put a drop of honey on a glass slide. Honey is a food that contains sugar.
4. Touch a piece of sugar test paper to the honey. Wait one minute. When the test paper turns green, this shows that sugar is present.
5. Using a different dropper, put a drop of water on a glass slide. Touch a piece of sugar test paper to the water. Wait one minute. No green color shows that sugar is not present.
6. Using a razor blade, cut a small slice from a sweet potato and touch a piece of sugar test paper to the inside of the slice. **CAUTION:** *Always cut away from yourself when using a razor blade.*
7. Record in your table if a green color appears and if sugar is present.
8. Repeat steps 3 and 4 with maple syrup, molasses, and apple juice. Use a different dropper and piece of sugar test paper for each food tested.
9. Dispose of the food samples according to your teacher's directions.

#### Data and Observations

1. Which foods contained sugar?
2. Which foods did not contain sugar?

| FOOD | GREEN COLOR? | SUGAR PRESENT? |
|---|---|---|
| Honey | yes | yes |
| Water | no | no |
| Sweet potato | yes | yes |
| Maple syrup | yes | yes |
| Molasses | yes | yes |
| Apple juice | yes | yes |

#### Analyze and Apply

1. Explain how you can tell if a food
   (a) contains sugar.
   (b) does not contain sugar.
2. What type of nutrient is sugar?
3. Why was it important to test water for sugar?
4. **Check your hypothesis:** Is your hypothesis supported by your data? Why or why not?
5. To which of the six food groups do each of your food samples belong?
6. **Apply:** What is the role of sugar in the diet?

#### Extension

**Experiment** using the same foods and iodine solution to test for the presence of starch.

194 Nutrition 9:2

## ANSWERS

### Data and Observations

1. In the samples suggested, honey, sweet potato, maple syrup, molasses, and apple juice contain sugar.
2. Water was the only suggested sample with no sugar.

### Analyze and Apply

1. (a) Touch a food sample with sugar test paper. After one minute, the presence of sugar will turn the paper a green color.
   (b) If the paper doesn't turn green, there's no sugar present.
2. carbohydrate
3. Water acted as a control.
4. Students who hypothesized correctly which foods contain sugar will answer that their hypotheses were correct.
5. Answers will vary depending on sample chosen. The samples suggested here belong mainly to the fats, oils, and sweets group and to the fruit and vegetable groups.
6. Sugar is a carbohydrate that supplies your body with energy.

## Calorie Content of Food

Foods differ in the amount of energy, or Calories, they contain. How many Calories do you think are in a ham sandwich, a small salad, and a glass of milk?

To find the answer, add the Calories next to the list of foods in Table 9-3.

**TABLE 9–3. CALORIES IN CERTAIN FOODS**

| Foods Eaten | Calories |
|---|---|
| 2 slices of white bread | 140 (70 in each slice) |
| 1 spoonful of mayonnaise | 70 |
| 2 slices of ham | 148 (74 in each slice) |
| 1/8 head of lettuce | 10 |
| 1 slice tomato | 6 |
| 2 spoonfuls of salad dressing | 120 (60 in each spoonful) |
| 1 glass milk | 150 |

This lunch would supply 644 Calories.

You can see from the table that not all foods provide the same number of Calories. However, we have not been comparing equal masses of foods to one another. When we do compare equal masses, we find that fats supply the most Calories. For example, let's compare butter with bread. Look at Figure 9-11b as you read. Butter is mostly fat, and bread is mostly carbohydrate. What would happen if you ate an equal mass of butter and bread? The butter would give your body two and one-half times the number of Calories as you would get from the bread.

Equal masses of protein and carbohydrate are both lower in Calories than fats. Let's compare bread with ham, Figure 9-11a. Remember that bread is mostly carbohydrate. Ham is mostly protein. There are 70 Calories in a slice of bread and 74 Calories in a slice of ham of the same mass. They have almost equal numbers of Calories.

a — Bread, Ham — 70 Calories, 74 Calories

b — Bread, Butter — 70 Calories, 175 Calories

**Figure 9-11** Food of equal masses can have very different numbers of Calories.

### Skill Check

**Infer:** Cooking oil and lard are both fats. How will the numbers of Calories found in equal masses of these substances compare? *For more help, refer to the **Skill Handbook**, pages 706-711.*

**Which foods supply the most Calories?**



## 2 TEACH

### MOTIVATION/Demonstration

Illustrate the concept of equal masses of food by bringing to class the number of slices of bread needed to equal the mass of one stick of butter. Find the mass of bread by checking the nutrition label on the package. The mass of one slice will be listed. Calculate the number of slices that equal the mass of one stick of butter. Then the Calories in both foods can be compared.

### Concept Development

▶ Food is not burned in the true sense. It is, however, changed chemically and energy is released during cellular respiration.

### Guided Practice

Have students write down their answers to the margin question: foods that are high in fat.

### Independent Practice

**Study Guide**/*Teacher Resource Package*, p. 52. Use the Study Guide worksheet shown at the bottom of this page for independent practice.

### Portfolio

Have students prepare a chart comparing the number of Calories in one ounce each of ten high-fat foods and ten low-fat foods. Remind students that many labels will list the Calories in an average serving rather than in ounces. What do students conclude following this activity?

### STUDY GUIDE 52

STUDY GUIDE — CHAPTER 9

Name ______ Date ______ Class ______

**CALORIES**

*In your textbook, read about the energy in food in Section 9:2.*

1. Define *Calorie.* **A Calorie is a measure of the energy in food.**
2. What does it mean if someone says that a food contains 50 Calories? **It means that if this food were burned, it would give off enough heat to raise the temperature of 1000 g of water 50° C.**
3. What is the energy in food used for in your body? **cell work, moving muscles, pumping blood, sending messages along nerves**
4. Complete the table below by determining the number of Calories in each food sample and writing the number in the column at the right.

| Food sample | Amount of water in beaker | Starting temperature of water before burning food | Final temperature of water after burning food | Number of Calories in food |
|---|---|---|---|---|
| A | 1000 g | 18°C | 19°C | 1 |
| B | 1000 g | 10°C | 15°C | **5** |
| C | 1000 g | 0°C | 100°C | **100** |
| D | 1000 g | 55°C | 72°C | **17** |

5. A student ate the foods in the table below in one day. Complete the table by determining the total Calories taken in. Then, answer the questions on the following page.

| Food | Calories × | Amount eaten = | Total |
|---|---|---|---|
| Egg | 80 | 2 | 160 |
| Milk | 80 | 2 glasses | **160** |
| Bread | 70 | 4 slices | **280** |
| Bologna | 60 | 4 slices | **240** |
| Cola | 145 | 4 glasses | **580** |
| Hamburger | 250 | 2 | **500** |
| French fries | 100 | 1 serving | **100** |

## OPTIONS

### ACTIVITY/Enrichment

Illustrate how to measure the heat energy in food. Place a beaker filled with 1000 mL of water on a metal tripod. Position a regular-sized marshmallow below the beaker using a long needle whose opposite end is stuck into a cork. Record the temperature of the water in the beaker using a Celsius scale. Light the marshmallow with a match and allow it to burn completely. Remeasure the water temperature. The change in temperature indicates the number of Calories in the marshmallow.

## TEACH

### Concept Development

▶ Explain that runners deplete their stored carbohydrates after about 18-20 miles. They then begin to utilize stored fat for energy. Nonrunners will deplete their carbohydrate stores in about 48 hours if no food is eaten.

### Check for Understanding

Have students calculate their Caloric intake for a 24-hour period. Calorie charts are available from your home economics teacher or from any nutrition textbook. This activity works best if students are told in advance to keep a list of the foods they consume and the quantity of each food eaten. Calculators may be helpful.

**Reteach**

Have students discuss why their Calorie needs are different from those of other students. Have them discuss ways for reducing weight.

### Guided Practice

Have students write down their answers to the margin question: A person gains weight.

### Independent Practice

**Study Guide**/*Teacher Resource Package*, p. 53. Use the Study Guide worksheet shown at the bottom of this page for independent practice.

### *Student Journal*

Have students write a paragraph for the Health Page of the school newspaper telling which exercises use the most and which exercises use the fewest Calories in a one-hour time period.

### Bio Tip

A typical calculation for the Bio Tip on page 197: If a student weighs 185 pounds and is 70 inches tall, $185 \div 70 \div 70 \times 705 = 26.6$. This BMI is over 25. Therefore, the student is overweight.

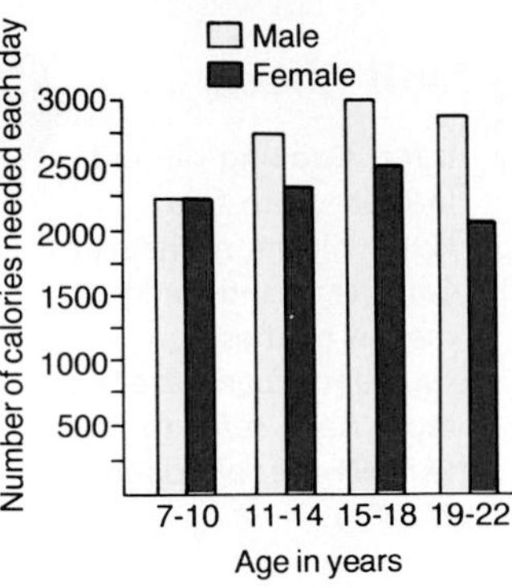


**Figure 9-12** Compare the Caloric needs for different ages and sexes. What age group needs the most Calories daily? 15-18

## Using Calories

Your body needs a certain number of Calories to keep a proper weight. How many Calories does your body need each day? The answer to that question is not simple. Figure 9-12 shows that age and sex of a person are important in figuring Calorie needs. The easiest way to receive the needed number of Calories is to eat different foods from each of the six groups shown on page 192.

How are Calories used by different people? How a person uses Calories depends on his or her size and on how active he or she is. For example, a larger person uses more Calories than a smaller person doing the same activity. Table 9-4 shows that a person who is 73 kg uses 240 Calories walking for one hour. Someone who is 54 kg uses 180 Calories walking for the same amount of time.

The more energy it takes to do an activity, the more Calories a person uses for that activity. For example, a 63-kg person uses 112 Calories standing for one hour. The same person uses 420 Calories playing tennis for one hour. Playing tennis uses more energy than standing.

**What happens if more Calories are taken into the body than are used?**

Taking in too many Calories or too little exercise may result in overweight. If the number of Calories taken in are equal to the number of Calories used, body weight stays about the same. If more Calories are taken in than used, weight is gained. What becomes of the extra Calories? They may be stored in the body as fat.

What happens if the body takes in less Calories than it uses? The body makes up the difference by using stored Calories, and the person loses weight. Another way of using stored Calories is by exercising.

**Figure 9-13** Taking in more or fewer Calories than are needed can result in overweight or underweight. Neither condition is healthy for the body.



**STUDY GUIDE 53**

*STUDY GUIDE* CHAPTER 9

Name ______ Date ______ Class ______

CALORIES

6. Which food gave the student the most Calories? **cola**
7. Were the servings and mass of each food equal? **no**
8. When comparing equal masses of food, which nutrient type will provide the most Calories? **fat**

*In your textbook, read about using Calories in Section 9:2.*

9. Look at Table 9-4 in your textbook. Use the information in the table and in your textbook to help you answer the following questions.
   a. How many Calories do you need each day? **Answers will vary.**
   b. What two things listed on the graph are important in determining how many Calories a person needs? **age and sex**
   c. How do Calorie needs differ as a person ages from 7 to 18? **Calorie needs increase with age.**
   d. How do Calorie needs differ between males and females above the age of 10? **Males need more Calories than females.**
   e. If a typical female age 18 takes in 1500 Calories each day for one month, will she gain or lose weight? Why? **She will lose weight because she needs 2500 Calories each day. Taking in less Calories than one needs over time will result in weight loss.**
   f. If a typical male age 15 takes in 3500 Calories each day for one month, will he gain or lose weight? Why? **He will gain weight because he needs 3000 Calories each day. Taking in more Calories than one needs over time will result in weight gain.**
   g. How do Calorie needs vary with different kinds of activities a person does? **The more energy it takes to do an activity, the more Calories a person uses to do that activity.**

| TABLE 9–4. CALORIES USED IN 1 HOUR | | | |
|---|---|---|---|
| **Type of Activity** | **Mass of Person** | | |
| | **54 kg** | **63 kg** | **73 kg** |
| Sleeping | 48 | 56 | 64 |
| Sitting | 72 | 84 | 96 |
| Eating | 84 | 98 | 112 |
| Standing | 96 | 112 | 123 |
| Walking | 180 | 210 | 240 |
| Playing tennis | 380 | 420 | 460 |
| Bicycling fast | 500 | 600 | 700 |
| Running | 700 | 850 | 1000 |

## Skill Check

**Interpret data:** Study Table 9-4. What happens to the number of Calories used as the type of activity becomes more difficult? *For more help, refer to the **Skill Handbook**, pages 704-705.*

**Figure 9-14** The number of Calories you will burn during a one-hour walk depends on your body mass.

## Check Your Understanding

6. How are Calories important to the body?
7. Which nutrient supplies the body with the most Calories?
8. Who needs more Calories, a 16-year-old boy or a 16-year-old girl?
9. **Critical Thinking:** A 73-kg person and a 54-kg person run at the same speed for one hour. Which person will use more Calories?
10. **Biology and Math:** Tom has a mass of 63 kg. In a typical day, Tom sleeps for 8 hours, sits for 6 hours, plays tennis for 2 hours, eats for 2 hours, walks for 3 hours, and stands for 3 hours. If Tom consumed 2500 Calories on a given day, would he have gained, maintained, or lost weight?

## Bio Tip

**Health:** Are You Overweight? Calculate your Body Mass Index (BMI). Divide your weight in pounds by your height in inches. Divide the answer again by your height in inches and multiply by 705. If your BMI is over 25, you are considered overweight.

## RETEACHING 24

RETEACHING — CHAPTER 9

Name ______ Date ______ Class ______

**CALORIE INTAKE AND OUTPUT** — Use after Section 9:2.

*A person will gain or lose weight depending on how many Calories he or she takes in and uses up. More Calories taken in over a period of time will result in weight gain.*

*In the tables below, calculate the number of Calories taken in and used up in one day. Then, answer the questions.*

| Calories taken in for one day | | | |
|---|---|---|---|
| Food | Calories | × Amount eaten | = Total |
| orange juice | 110 per glass | × 2 glasses | = 220 |
| muffin | 140 each | × 2 | = 280 |
| chicken | 180 per piece | × 3 pieces | = 540 |
| ham | 250 per slice | × 4 slices | = 1000 |
| potato | 100 each | × 1 | = 100 |
| butter | 60 per pat | × 3 pats | = 180 |
| bread | 70 per slice | × 4 slices | = 280 |
| total calories taken in for one day | | | = 2600 |

| Calories used up in one day | | | |
|---|---|---|---|
| Activity | Hours spent in activity | × Calories used each hour | = Total Calories used |
| standing | 3 | × 110 | = 330 |
| sitting | 8 | × 80 | = 640 |
| walking | 1 | × 200 | = 200 |
| playing tennis | 2 | × 400 | = 800 |
| eating | 1 | × 90 | = 90 |
| sleeping | 9 | × 50 | = 450 |
| total calories used up in one day | | | = 2510 |

1. Which was greater, total Calories taken in or Calories used up? Calories taken in
2. Did the person gain or lose weight that day? gained

## STUDY GUIDE 54

STUDY GUIDE — CHAPTER 9

Name ______ Date ______ Class ______

**VOCABULARY**

*Review the new words used in Chapter 9 of your textbook. Then, complete this crossword puzzle.*

Across: 1. CARBOHYDRATE 4. VITAMINS 6. NUTRITION 7. FATS 8. PROTEIN 10. BALANCED DIET
Down: 1. CALORIES 2. RDA 3. NUTRIENTS 5. MINERALS 9. WATER

**Across**
1. nutrient that supplies energy
4. nutrients that help cells carry on daily chemical work
6. study of how body uses food
7. nutrient that supplies energy if body's first energy source is used up
8. main nutrient in meat
10. the right amount of each nutrient (two words)

**Down**
1. measure of energy in food
2. abbreviation for Recommended Daily Allowance
3. chemicals in food needed by body
5. iron and calcium are examples
9. makes up 50 to 60% of your body

# TEACH

### Independent Practice

**Study Guide**/*Teacher Resource Package,* p. 54. Use the Study Guide worksheet shown at the bottom of this page for independent practice.

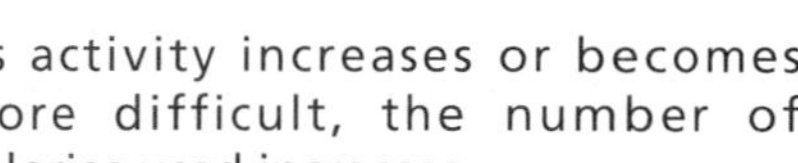

### Skill Check

As activity increases or becomes more difficult, the number of Calories used increases.

### Check for Understanding

Have students respond to the first three questions in Check Your Understanding.

### Reteach

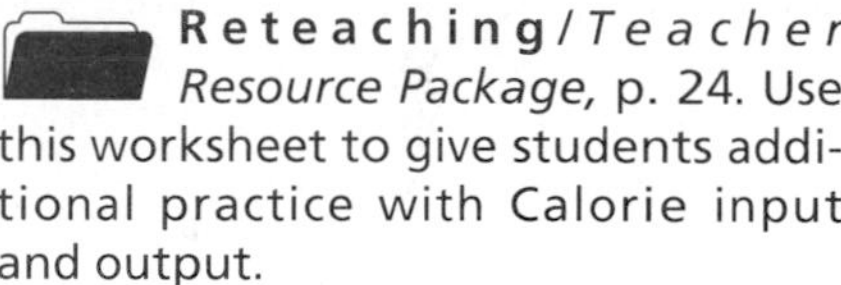

**Reteaching**/*Teacher Resource Package,* p. 24. Use this worksheet to give students additional practice with Calorie input and output.

**Extension:** Assign Critical Thinking, Biology and Math, or some of the **OPTIONS** available with this lesson.

# 3 APPLY

### Convergent Question

Have students give reasons why they need food Calories.

# 4 CLOSE

### ACTIVITY/Software

*Food, Fitness, and Appearance,* Queue, Inc.

### Answers to Check Your Understanding

6. They are a measure of the amount of energy in food. The body needs energy to do all its daily activities.
7. fat
8. a 16-year-old boy
9. the 73-Kg person
10. He would lose weight.

## APPLYING TECHNOLOGY

### Purpose

Students will determine how foods labeled "light" butter and "regular" butter differ in the amount of water and fat present.

### Process Skills

Classify, experiment, infer, predict, measure, use numbers, interpret data, observe, define operationally

### Time Required

Setting up the water bath and weighing and melting the samples could take a full class period, although students will not have to spend all their time watching the melting process. Cooling of the samples will take overnight. Students can observe the test tubes the following day in class.

### Possible Problems/Safety Concerns

Students must avoid mixing up the samples when weighing and placing the samples into test tubes. Stress proper labeling of all glassware and samples when being weighed. It is suggested that waxed paper or aluminum foil be placed on the pan of the balance to provide ease in cleaning up. The paper or foil can be marked as to sample type. Goggles are to be worn in the event that there is splattering during melting. Popsicle sticks or plastic spoons may help students in putting the butter samples into the test tubes.

### Teaching Strategies

▶ Students may wish to use "regular" and "light" margarine rather than butter.
▶ Allow students to work in groups of three or four to reduce the amount of materials needed.
▶ If balances are in short supply, the mass of butter needed (30 grams) can be estimated by having students use about 1 tablespoon of

# TECH PREP APPLYING TECHNOLOGY

## Get the Fat Out

Fat is a nutrient needed by humans. So why all the fuss about reducing fat intake in the diet? It is known that a diet high in fat contributes to obesity. It also is a major cause of heart disease. How much fat should you include in your diet if you are trying to eat healthy? Check the labels on any food package, and you will see that fat should be limited to less than 65 grams each day. Labels also tell you that you should limit your saturated-fat intake to less than 20 grams of the 65-gram total. Saturated fat is a type of fat that differs chemically from another type, called unsaturated fat. An ideal diet, therefore, should consist of less than 45 grams of unsaturated fat and less than 20 grams of saturated fat each day.

### Identifying the Problem

You are a nutritionist. People come to you for advice about the kinds of foods they should and should not be eating. A high school student asks you the following questions. Are "light" fats, such as those found in "light" butter, really light? Do they contain less fat than "regular" fats? Do they really help to cut down on the amount of fat a person eats each day? You want to experiment with "regular" and "light" butter to see if there is any difference between the two product types. This way, as a nutritionist, you will be better able to answer the questions raised earlier by the student.

**Nutrition Facts**
Serving Size 1 Tbsp. (14g)
Servings Per Container about 32

| Amount Per Serving | |
|---|---|
| Calories 100 Calories from Fat 100 | |
| | % Daily Value* |
| Total Fat 11g | 17% |
| Saturated Fat 8g | 38% |
| Cholesterol 30mg | 10% |
| Sodium 0mg | 0% |
| Total Carbohydrate 0g | 0% |
| Protein 0g | |
| Vitamin A 8% | |

Not a significant source of dietary fiber, sugars, vitamin C, calcium, and iron.

*Percent Daily Values are based on a 2,000 calorie diet.

#### Technology Connection

Nutritionists and scientists are trying to design new artificial foods that look and taste like fats but aren't fats. Their goal is to find a chemical that provides few Calories, but tastes like fat. The newest chemical is a protein product called Simpless. It is being used in "fat-free butter" spreads. Another fat-free chemical currently being used is a product called Olestra. If you were to design a fat substitute, what are some of the qualities that this artificial food would have to show? Explain why these qualities might be needed.

### Collecting Information

Visit your local grocery store. Look in the dairy section for "regular" and "light" butter. Butter is a fat. See if the food labels on these products give you a clue as to how the "light" butter differs from the "regular" butter. Also, see if the two butter types



#### Technology Connection

Simpless is formed through a process called microparticulation. Milk and egg protein are heated and blended into a mistlike substance. Simpless is digested as it passes through the human digestive system. Olestra has a sucrose molecule taking the place of glycerol and six fatty acid molecules instead of the usual three in fats. It is not digested by humans. It can be used for cooking and frying, whereas Simpless cannot. Student answers regarding the qualities of a fat substitute might include answers such as: it must feel like butter, it must have the texture of butter, it must taste like a fat, it must give fried foods the same crunch and taste.

differ in the amounts of saturated and unsaturated fats. Make a hypothesis as to how the makers of these products can make butter "light."

## Carrying Out an Experiment 

1. Weigh out equal masses of both butter types, about 30 grams of each.
2. Place each sample into a separate labeled test tube.
3. *Put on protective goggles.* Place the tubes into a beaker half-filled with water. Place the beaker and tubes onto a hot plate and heat the water until the butter melts.
4. When each sample is totally melted, turn off the hot plate. Leave the tubes in the water bath and allow the tubes to cool overnight. Examine the tubes the next day in class.
5. Make a prediction as to what the different layers of solids and liquids are that can be seen in the test tubes. Record any measurements, observations, or other data that you feel might be important in explaining how "light" and "regular" butters differ. Diagrams may also be made as part of your data reporting.

## Assessing Your Results

Compare and contrast the two butter samples. Explain how "light" and "regular" butter may differ. Based on your experimental results, how would you answer the student's original questions? Do "light"-fat foods really contain less fat than "regular"-fat foods? Could "light"-fat foods help to cut down on the amount of fat a person uses each day? Explain how.

### Career Connection

- **Food Processing Technician** May work as an inspector for a state or federal agency; helps with quality control by testing for the presence of bacteria, impurities, or toxic substances in foods
- **Food Scientist** Works to improve processes of canning, freezing, storing, packaging, and distributing of foods; may also work in the research and development areas of food processing companies
- **Dietetic Technician** Assists dietitians; work is usually in hospitals, day care centers, nursing homes, or schools; may supervise the ordering, storing, preparing, and serving of food

regular butter and about 1½ tablespoons of light butter. Light butter is whipped and, therefore, requires more volume.

▶ Fats can be melted in a microwave oven if you have access to one.

▶ Cleanup of test tubes with butter can be accomplished by melting the fats and pouring into an empty plastic soda pop bottle. Seal and discard in the garbage. DO NOT pour melted butter down a sink drain.

▶ Explain to students that fat molecules consist of one glycerol and three fatty acid molecules. Fatty acids are long chains of carbon and hydrogen atoms. If there is one or more double bonds along the carbon chain, the fat is unsaturated. If no double bonds exist, the fat is saturated. Saturated fats are typically solid, are found in animals, and are implicated in the production of cholesterol. Unsaturated fats are typically liquid and are found in plants.

### Conclusions

Students will see that "light" butter contains more than twice the amount of water found in "regular" butter. The fat content in both butter types, however, remains rather high.

### Answers to Questions

The companies making "light" fats add water to the product to reduce the amount of fat per serving. This reduces the Calories for each serving of fat. However, the amount of fat remains high even in the "light" varieties.

**Process** Have students write out the procedural steps they might follow to determine experimentally how and if "light" peanut butter differs from "regular" peanut butter.

### Career Connection

**Career Path**

**Food Processing Technician** High-school diploma and several years of college; two-year college degree may be needed; some on-the-job training.

**Food Scientist and Dietetic Technician** Both careers require a high-school diploma and a college degree.

**Career Issue**

Have students find out what a new fat substitute such as Olestra must go through in order to be approved by the Food and Drug Administration.

## CHAPTER 9 REVIEW

### Summary

Summary statements can be used by students to review the major concepts of the chapter.

### Key Science Words

All boldfaced terms from the chapter are listed.

### Testing Yourself

#### Using Words

1. nutrition
2. Calorie
3. protein
4. balanced diet
5. nutrient
6. recommended daily allowance
7. fat
8. vitamin
9. carbohydrate
10. mineral

#### Finding Main Ideas

11. p. 196; Calories used depends on a person's size as well as how active the person is.
12. p. 195; Multiply the mass of each food by the number of Calories in a known mass, and then add those numbers together.
13. p. 189; Vitamin C is used for healthy membranes and wound healing. Vitamin D is used for bone growth.
14. p. 191; Water is important for cooling the body, for dissolving chemicals, and for carrying away body wastes.
15. p. 192; Dairy foods contain carbohydrates, fat, protein, water, minerals, and vitamins.
16. p. 196; They are stored as fat.
17. pp. 188-190; Vitamins are needed for growth and tissue repair, while minerals are needed to form cell parts.

#### Using Main Ideas

18. Vitamin C is supplied by the fruit-vegetable group; carbohydrates are supplied by all groups except meat; water is supplied by all the food groups; magnesium is supplied by the fruit-vegetable and grain groups.



# Chapter 9 Review

## Summary

### 9:1 What Are the Nutrients in Food?

1. The body needs six different nutrients to stay healthy—proteins, fats, carbohydrates, vitamins, minerals, and water.
2. Fats and carbohydrates are needed for energy. Proteins are needed to build and repair body parts. Vitamins are needed for chemical work in the cell. Minerals are needed to form certain cell parts. Water cools the body. Chemical changes take place in water. Water carries away body wastes.
3. Choosing a variety of foods from the six nutrients will provide you with a balanced diet.

### 9:2 Calories

4. A Calorie is a measure of the energy in food.
5. When comparing equal masses of food, fats provide more Calories than any other nutrient.
6. How a person uses Calories depends on his or her age, sex, size, and the kind and amount of activity a person does. Using less Calories than the amount taken in may result in weight gain. Using more Calories than the amount taken in may result in weight loss.

## Key Science Words

balanced diet (p. 192)
Calorie (p. 193)
carbohydrate (p. 185)
fat (p. 185)
mineral (p. 190)
nutrient (p. 184)
nutrition (p. 184)
Percent Daily Value (p. 189)
protein (p. 185)
recommended daily allowance (p. 189)
vitamin (p. 188)

## Testing Yourself

### Using Words

*Choose the word from the list of Key Science Words that best fits the definition.*

1. study of how your body uses food
2. measure of food energy
3. nutrient used to build and repair body parts, such as skin and bone
4. eating the proper amount of all nutrients each day
5. type of chemical in food needed by all living cells
6. amount of a vitamin needed each day
7. energy food, such as oil or butter
8. nutrient, such as niacin, that aids a cell with its chemical work
9. body's main source of energy
10. nutrients, such as calcium or iron

### Finding Main Ideas

*List the page number where each main idea below is found. Then, explain each main idea.*

11. why the number of Calories used during an activity is different for each person

# Review

## Testing Yourself *continued*

12. how to find the total number of Calories in a meal
13. the roles of vitamins C and D
14. why water is important
15. the nutrients present in dairy foods
16. what happens to unused Calories in the body
17. ways in which vitamins differ from minerals

### Using Main Ideas

*Answer the questions by referring to the page number after each question.*

18. Which food group supplies the body with vitamin C, carbohydrates, water, and magnesium? (pp. 190, 192)
19. How do the numbers of Calories used change as activity increases? (p. 197)
20. Which nutrients are used for growth and repair of skin and bones? (p. 185)
21. What does it mean if a certain food gives you 20 percent of your %DV for niacin? (p. 189)
22. What happens to body weight when
    (a) more Calories are used than taken in? (p. 196)
    (b) more Calories are taken in than used? (p. 196)
23. Which foods would you suggest a person with anemia and muscle twitches eat in an attempt to correct the condition? (p. 190)
24. How many Calories are in equal masses of
    (a) fat and carbohydrate? (p. 195)
    (b) fat and protein? (p. 195)
    (c) carbohydrate and protein? (p. 195)
25. What are the six important nutrients found in food? (p. 184)

## Skill Review ✓

*For more help, refer to the **Skill Handbook**, pages 704-719.*

1. **Infer:** Males and females have different amounts of fat in their bodies. What does this tell you about how different sexes store this nutrient?
2. **Use numbers:** Calculate the cost of one gram of canned, cooked kidney beans compared with the cost of an equal amount of plain, uncooked kidney beans. (A can costs $0.85 for 200 gm, while a bag of beans costs $1.25 for 1000 gm.)
3. **Measure in SI:** Use a pan balance to determine the masses of two foods listed in Table 9-3 that have similar Caloric values.
4. **Interpret data:** List the number of Calories that a 63-kg person will use up in one hour of walking and running.

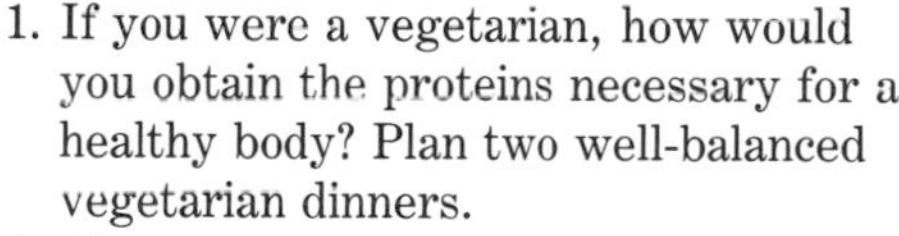


## Finding Out More

### Critical Thinking

1. If you were a vegetarian, how would you obtain the proteins necessary for a healthy body? Plan two well-balanced vegetarian dinners.
2. If a person takes vitamin and mineral supplements daily, do they still need a variety of foods? Explain your answer.

### Applications

1. Build a tin can calorimeter. Measure the Calories in different foods.
2. Use food charts to plan two different dinners—one low in sodium, and one low in carbohydrates.



### Using Main Ideas

19. As activity increases, the number of Calories used increases.
20. Proteins are nutrients that build and repair body parts.
21. The food supplies 20 percent of the amount of niacin needed each day to stay in good health.
22. (a) If more Calories are used than taken in, a person will lose weight. (b) If more Calories are taken in than used, a person will gain weight.
23. A person with anemia should eat more liver, eggs, peas, enriched cereals, and whole grains. A person with muscle twitching should eat more potatoes, fruits, and whole grain cereals.
24. (a) Fat has more Calories than an equal mass of carbohydrate. (b) Fat has more Calories than an equal mass of protein. (c) Equal masses of protein and carbohydrate have almost equal numbers of Calories.
25. The six important nutrients found in food are proteins, carbohydrates, fats, water, minerals, and vitamins.

## Skill Review

1. Females store more fat than males do.
2. A bag of beans costs less per gram.
3. Answers will vary depending on the foods chosen.
4. walking = 210 Calories; running = 850 Calories

## Finding Out More

### Critical Thinking

1. Students should use Figure 9–9, Table 9-1, and Table 9-2 for help.
2. Yes. Nutrients other than vitamins and minerals are needed for good health.

### Applications

1. Remove the top of a tin can. Make wedge-shaped cuts along the bottom edge and a round opening in the center of the bottom for a water-filled test tube. Place food on a wire stand, turn the can over the stand, and place a tube through the hole. Record the water temperature, ignite the food, and record the final water temperature.
2. Fresh foods will be lower in sodium than processed foods. Dairy and meat products will be lower in carbohydrates than grains and produce.

**Chapter Review**/*Teacher Resource Package,* pp. 45-46.

**Chapter Test**/*Teacher Resource Package,* pp. 47-49.

**Chapter Test**/*Computer Test Bank*

# Digestion

## PLANNING GUIDE

| CONTENT | TEXT FEATURES | TEACHER RESOURCE PACKAGE | OTHER COMPONENTS |
|---|---|---|---|
| (1 1/2 days)<br>10:1 The Process of Digestion<br>Breakdown of Food<br>Physical and Chemical Changes | Skill Check: *Experiments,* p. 205<br>Lab 10-1: *Digestion,* p. 207<br>Idea Map, p. 205<br>Check Your Understanding, p. 206 | Reteaching: *Digestion,* p. 25<br>Critical Thinking/Problem Solving: *What Can Teeth Reveal About Diet?* p. 10<br>Focus: *Breaking Down Food,* p. 19<br>Study Guide: *The Process of Digestion,* p. 55<br>Lab 10-1: *Digestion,* pp. 37-38 | **STVS:** *Measuring Body Fat,* Human Biology (Disc 7, Side 2) |
| (4 days)<br>10:2 The Human Digestive System<br>Nutrients Are Digested<br>A Trip Through the Digestive System<br>Moving Digested Food Into Body Cells<br>Digestion in Other Animals<br>Problems of the Digestive System | Skill Check: *Understand Science Words,* p. 212<br>Mini Lab: *How Long Is the Digestive System?* p. 210<br>Mini Lab: *How Do Villi Aid Absorption?* p. 214<br>Career Close-Up: *Dietician,* p. 211<br>Lab 10-2: *Digestive System,* p. 213<br>Idea Map, p. 212<br>Check Your Understanding, p. 217 | Application: *Gallstones, a Problem of the Digestive System,* p. 10<br>Enrichment: *Digesting Different Nutrients,* p. 10<br>Reteaching: *Digestive Systems of Animals,* p. 26<br>Reteaching: *Problems of the Digestive System,* p. 27<br>Skill: *Sequence,* pp. 10-11<br>Transparency Master: *Human Digestive System,* p. 45<br>Study Guide: *The Human Digestive System,* pp. 56-58<br>Study Guide: *Problems of the Digestive System,* p. 59<br>Study Guide: *Vocabulary,* p. 60<br>Lab 10-2: *Digestive System,* pp. 39-40 | **Laboratory Manual:**<br>*How Do Digestive System Lengths Compare?* p. 79<br>**Laboratory Manual:**<br>*How Does the Fish Digestive System Work?* p. 83<br>Color Transparency 10: *The Human Digestive System*<br>**STVS:** *Bulimia,* Human Biology (Disc 7, Side 2) |
| Chapter Review | Summary<br>Key Science Words<br>Testing Yourself<br>Finding Main Ideas<br>Using Main Ideas<br>Skill Review | **ASSESSMENT RESOURCES**<br>Chapter Review, pp. 50-51<br>Chapter Test, pp. 52-54<br>Performance Assessment in the Biology Classroom<br>Alternate Assessment in the Science Classroom<br>Computer Test Bank | |

## GLENCOE TECHNOLOGY

**Infinite Voyage,** *A Taste of Health*
**Science and Technology Videodisc Series,**
*Measuring Body Fat,* Human Biology (Disc 7, Side 2)
*Bulimia,* Human Biology (Disc 7, Side 2)

## MATERIALS NEEDED

| LAB 10-1, p. 207 | LAB 10-2, p. 213 | MARGIN FEATURES |
|---|---|---|
| 4 paper cups<br>gelatin<br>water<br>4 applicator sticks<br>4 plastic spoons<br>enzyme A<br>enzyme B<br>refrigerator (optional) | five diagrams of the digestive system<br>colored pencils: red, blue, green, yellow, and purple | Skill Check, p. 205<br>2 sugar cubes<br>glass<br>warm water<br>Skill Check, p. 212<br>dictionary<br>Mini Lab, p. 210<br>pencil<br>paper<br>Mini Lab, p. 214<br>water<br>paper towels<br>beaker |

## OBJECTIVES

For more information about National Science Standards, see page 5T.

| SECTION | OBJECTIVE | CORRELATION of QUESTIONS to OBJECTIVES: CHECK YOUR UNDERSTANDING | CHAPTER REVIEW | TRP CHAPTER REVIEW | TRP CHAPTER TEST |
|---|---|---|---|---|---|
| 10:1 National Science Stds: UCP.1, UCP.2, UCP.3, UCP.4, UCP.5, B.2, B.3, B.6, C.1, C.5 | 1. **Relate** the importance of the digestive system. | 1 | 3 | 28 | 7, 47 |
| | 2. **Compare** a physical change and a chemical change in the digestive system. | 2, 4 | 1, 5, 6, 13, 21 | 8, 33 | 10, 31, 32, 33, 34, 46, 48 |
| | 3. **Explain** the role of enzymes in a chemical change. | 3 | 4, 10, 12 | 11, 12, 13, 14, 15, 16, 17, 30 | 44 |
| 10:2 National Science Stds: UCP.1, UCP.2, UCP.3, UCP.4, UCP.5, B.3, B.4, B.6, C.1, C.5, F.1 | 4. **Trace** the path of food from the mouth to the body cells. | 6, 8 | 2, 8, 9, 11, 14, 17, 20 | 1, 3, 6, 9, 10, 11, 12, 13, 14, 15, 16, 17, 18, 19, 20, 21, 22, 23, 24, 25, 26, 27, 28, 29, 31, 32, 34, 35 | 1, 2 , 3, 6, 8, 11, 12, 13, 14, 15, 16, 17, 18, 19, 20, 21, 22, 23, 24, 25, 26, 35, 36, 37, 38, 43, 45 |
| | 5. **Compare** the human digestive system with those of other animals. | 8, 10 | 15, 18, 19 | 4, 5, 7 | 4, 9, 27, 28, 29, 30 |
| | 6. **Identify** the problems of the digestive system. | 7, 9 | 16, 22 | 2 | 5, 41, 42 |

# Digestion

## CHAPTER OVERVIEW

### Key Concepts

This chapter examines the role of the digestive system and some of its common problems. A meal is followed through the human digestive tract.

### Key Science Words

appendix
bile
chemical change
digestion
digestive system
enzyme
esophagus
gallbladder
hydrochloric acid
large intestine
liver
mucus
pancreas
physical change
saliva
salivary gland
small intestine
stomach
villi

### Skill Development

In Lab 10-1, students will **separate and control variables, form a hypothesis, observe,** and **design an experiment.** In Lab 10-2, students will **classify** and **interpret diagrams.** In the Skill Check on page 205, students will **experiment** with dissolving sugar. On page 212, they will **understand** the **science word** *appendix.* In the Mini Lab on page 210, students will **use numbers** to diagram and label digestive organs. In the Mini Lab on page 214, they will experiment with towels absorbing water.

### Bridging

Review digestion by fungi in Chapter 5, by simple invertebrates in Chapter 7, and by complex invertebrates in Chapter 8. Bridge this chapter with these earlier chapters to show how the digestive process becomes more complex as the complexity of the animals increases.

**CHAPTER PREVIEW**

### Chapter Content

Review this outline for Chapter 10 before you read the chapter.

**10:1 The Process of Digestion**
Breakdown of Food
Physical and Chemical Changes

**10:2 The Human Digestive System**
Nutrients Are Digested
A Trip Through the Digestive System
Moving Digested Food into Body Cells
Digestion in Other Animals
Problems of the Digestive System

### Skills in this Chapter

The Skills that you will use in this chapter are listed below.

- In **Lab 10-1,** you will separate and control variables, form a hypothesis, observe, and design an experiment. In **Lab 10-2,** you will classify and interpret diagrams.
- In the **Skill Checks,** you will experiment and understand science words.
- In the **Mini Labs,** you will use numbers and experiment.

202

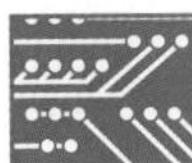

## TECH PREP

For Tech Prep activities in this chapter of the Teacher Wraparound Edition, see especially the Student Journals on pages 211 and 212, the Activity on page 214, the Connection on page 215, and the Cooperative Learning and Motivation activities on page 216.

See also the Glencoe Homepage at **www.glencoe.com**

Chapter 10

# Digestion

Have you ever watered a very dry houseplant? If so, you probably sprinkled water around the base of the plant rather than flooding one spot with water. Look at the photo on the left. By wetting a large area of soil, the amount of water immediately absorbed by the plant is increased.

Your body needs to absorb nutrients from food just as a houseplant needs to absorb water. This is the job of various organs of your digestive system. One such digestive organ is the small intestine. The picture below shows an enlarged view of the lining of the small intestine. How does the irregular surface of this organ aid in the digestion of food? As you read this chapter, you will discover the answer to this question.

**Try This!**

**Does particle size affect the rate at which a substance dissolves?** Predict which would dissolve faster, a copper sulfate crystal or an equal mass of powdered copper sulfate. Check your prediction by placing these substances in two beakers, each with 100 mL of warm water.

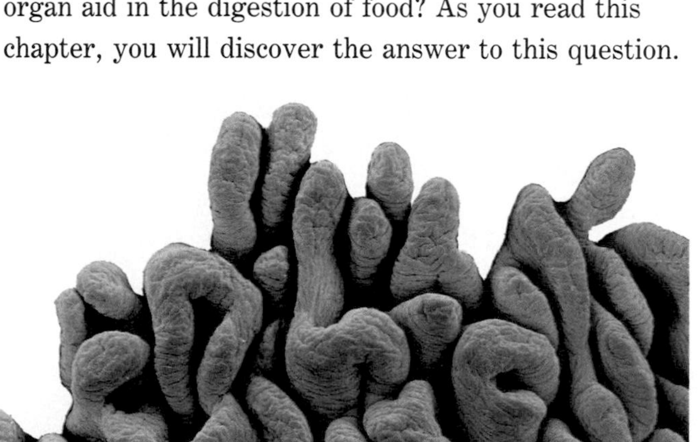

**interNET CONNECTION**

For more information about the material in this chapter, follow the link for the chapter on the Glencoe Homepage at **http://www.glencoe.com**

High power view inside the wall of the small intestine



**ASSESSMENT PLANNER**

**Portfolio**
Strategies on the following pages represent student products that can be placed into a best-work portfolio: pp. 210, 215.

**PERFORMANCE ASSESSMENT**
Skill Check, p. 205
Mini Lab, pp. 210, 214
Lab, pp. 207, 213

**CONTENT ASSESSMENT**
Check for Understanding, pp. 206, 215, 217
Chapter Review, pp. 218-219

**GROUP ASSESSMENT**
Opportunities for group assessment occur with Cooperative Learning Strategies.

**Student Journal**
Strategies on the following pages represent opportunities for writing in a Student Journal: pp. 211, 212.

**GETTING STARTED**

## Using the Photos

The digestive system changes food both chemically and physically into a form suitable for cell use. The lining of the small intestine has a very large surface area. This structure is designed for rapid absorption of digested food.

## MOTIVATION/Try This!

**Does particle size affect the rate at which a substance dissolves?** Upon completing the chapter opening activity, students will **infer** that smaller particles are broken down faster than larger pieces. Chewing breaks food down physically into smaller pieces. This provides a greater surface area on which enzymes can work. Saliva is able to do a more efficient job with small chunks than with large chunks.

## Chapter Preview

Have students study the chapter outline before they begin to read the chapter. They should note the overall organization of the chapter, that the process of digestion is studied first, that nutrients are then covered along with the pathway through the system, and that problems associated with this system are covered last.

## Misconception

Ask any student what the most important organ of the digestive system is and they will answer "the stomach." Actually, a person can live without a stomach. It is not a critical organ for digestion.

## 10:1 The Process of Digestion

### PREPARATION

#### Materials Needed

Make copies of the Focus, Critical Thinking/Problem Solving, Study Guide, Reteaching, and Lab worksheets in the *Teacher Resource Package.*

▶ Gather sugar cubes and warm water for the Skill Check.

▶ Purchase paper cups, gelatin, applicator sticks, plastic spoons, and enzymes for Lab 10-1.

#### Key Science Words

digestive system
digestion
physical change
chemical change
enzyme

#### Process Skills

In the Skill Check, students will experiment. In Lab 10-1, students will separate and control variables, form a hypothesis, observe, and design an experiment.

### 1 FOCUS

▶ The objectives are listed on the student page. Remind students to preview these objectives as a guide to this numbered section.

#### ACTIVITY/Analogy

On an assembly line, certain tasks are completed in a certain area. Certain steps of digestion are completed in certain areas of the digestive tract.

#### Guided Practice

Have students write down their answers to the margin question: so that it is in a form that the body cells can use.

**Focus**/*Teacher Resource Package*, p. 19. Use the Focus transparency shown at the bottom of this page as an introduction to physical digestion.



## 10:1 The Process of Digestion

**Objectives**

1. **Relate** the importance of the digestive system.
2. **Compare** a physical change and a chemical change in the digestive system.
3. **Explain** the role of enzymes in a chemical change.

**Key Science Words**

digestive system
digestion
physical change
chemical change
enzyme

When food is first eaten, it is not in a form that can be used by cells in the body. Food must be broken down into a form that cells can use. All cells need food for energy, growth, and repair. How does the body break down food into a form that is usable to cells? The body changes food into a usable form by means of the digestive system. The **digestive system** is a group of organs that take in food and change it into a form the body can use.

### Breakdown of Food

**Why does food need to be digested?**

Much of your digestive system is hollow. Food moves through it just as materials pass through a tube. To understand how the digestive system works, we can compare it to a factory.

Raw materials are brought into a factory. The materials are transported to different locations in the factory. At each location, the materials are changed. The final products that leave the factory are quite different from the original raw materials. For example, iron ore enters a factory as a raw material. As it moves through the factory, the iron ore is changed to steel. The steel is later changed into new products. Figure 10-1 shows these events.

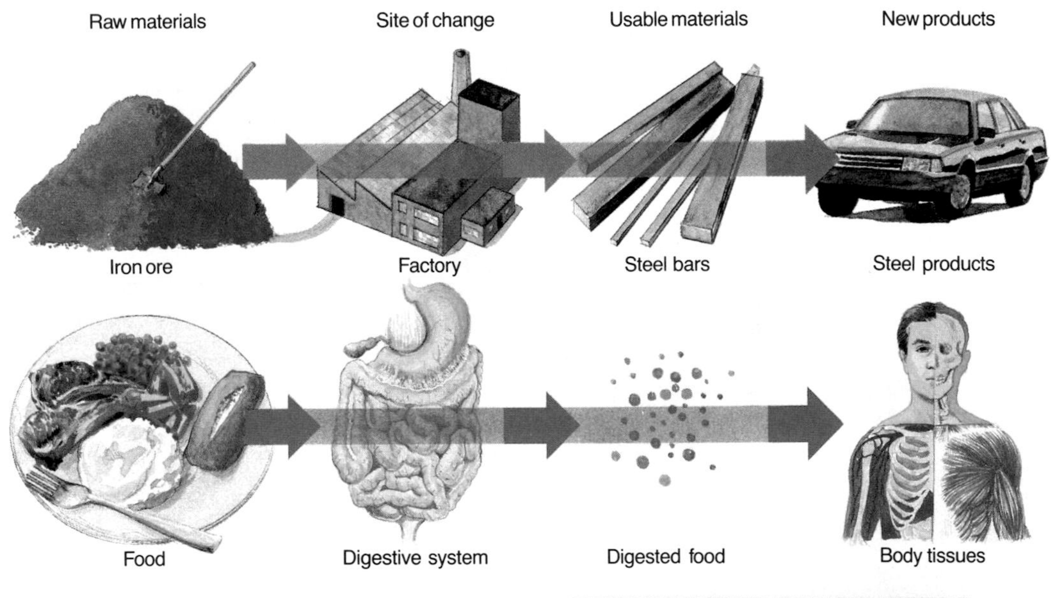


**Figure 10-1** The human digestive system is like a factory. Raw materials are used to make products.



### OPTIONS

#### Science Background

Enzymes do not actually change a chemical into new molecules, they simply speed up the reaction. They act as catalysts. It would take years to convert starch to glucose without a catalyst. Within our digestive system, it takes only minutes.

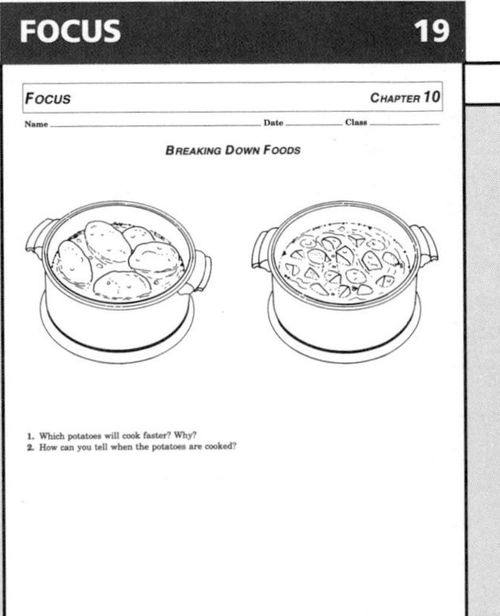
FOCUS 19

FOCUS CHAPTER 10

Name ______ Date ______ Class ______

BREAKING DOWN FOODS

1. Which potatoes will cook faster? Why?
2. How can you tell when the potatoes are cooked?

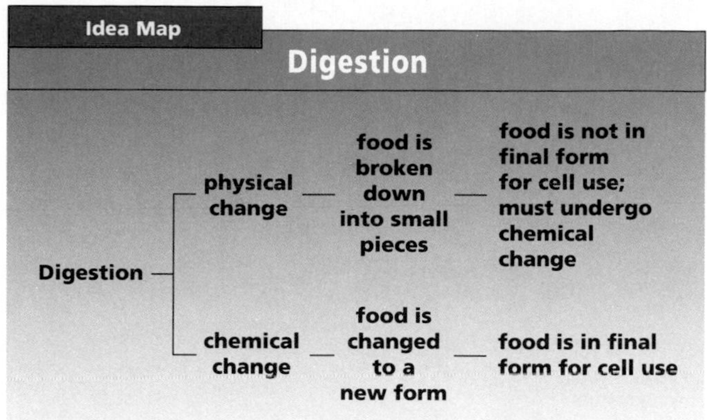


**Study Tip:** Use this idea map as a study guide to identifying the differences between physical and chemical changes. Which type of change turns food into a form the body can use?

Your digestive system works in a way similar to the factory. Food is the raw material of your digestive "factory." It enters your digestive system through your mouth. As soon as food enters your body, it begins to change form. The changing of food into a usable form is called **digestion.** As food leaves your mouth, it enters the long tube that makes up your digestive system. The food continuously changes form as it passes through the tube. Finally, the food is in a form that can be used to supply the body with energy or to help make bone, skin, or muscle cells. These are the products of your digestive factory.

### Skill Check

**Experiment:** Compare how long it takes for a sugar cube and a crushed sugar cube to dissolve in a glass of warm water, with or without stirring the water. *For more help, refer to the **Skill Handbook,** pages 704-705.*

## Physical and Chemical Changes

The digestive system is like a long tube. It is narrow in some places and wide in others. Food is broken down as it moves through this sometimes wide, sometimes narrow tube.

How is food broken down as it passes through the digestive tube? There are two ways. They are physical changes and chemical changes. Physical and chemical changes to food occur at different times and places along the digestive tube.

A **physical change** occurs when large food pieces are broken down into smaller pieces. The food is still in the same form. Only the size and shape of food particles are different. Chewing by the teeth causes a physical change in food. Grinding and mixing also cause a physical change. These physical changes occur farther on down the long tube of the digestive system.



# 2 TEACH

### MOTIVATION/Videocassette

Show the videocassette *Eating to Live,* Films for the Humanities.

### Idea Map

Have students use the idea map as a study guide to the major concepts of this section.

### ACTIVITY/Demonstration

A demonstration of the differences between physical and chemical change may be helpful. Physical change—Break a wood splint into two or four small pieces. Have students examine the wood splint before and after breaking. The starting and final products are exactly the same. Chemical change—Burn the splint. Have students examine the starting and final products. The starting and final products are no longer similar.

### Skill Check

The crushed sugar cube should dissolve much more rapidly than the whole sugar cube.

### Guided Practice

Have students write down their answers to the margin question: chewing and grinding food with the teeth.

**CRITICAL THINKING 10**

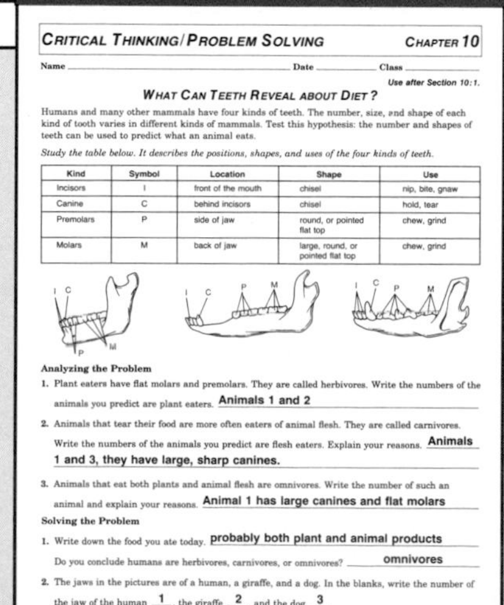

CRITICAL THINKING/PROBLEM SOLVING — CHAPTER 10

Name ______ Date ______ Class ______

*Use after Section 10:1.*

**WHAT CAN TEETH REVEAL ABOUT DIET?**

Humans and many other mammals have four kinds of teeth. The number, size, and shape of each kind of tooth varies in different kinds of mammals. Test this hypothesis: the number and shapes of teeth can be used to predict what an animal eats.

*Study the table below. It describes the positions, shapes, and uses of the four kinds of teeth.*

| Kind | Symbol | Location | Shape | Use |
|---|---|---|---|---|
| Incisors | I | front of the mouth | chisel | nip, bite, gnaw |
| Canine | C | behind incisors | chisel | hold, tear |
| Premolars | P | side of jaw | round, or pointed flat top | chew, grind |
| Molars | M | back of jaw | large, round, or pointed flat top | chew, grind |



**Analyzing the Problem**

1. Plant eaters have flat molars and premolars. They are called herbivores. Write the numbers of the animals you predict are plant eaters. **Animals 1 and 2**
2. Animals that tear their food are more often eaters of animal flesh. They are called carnivores. Write the numbers of the animals you predict are flesh eaters. Explain your reasons. **Animals 1 and 3, they have large, sharp canines.**
3. Animals that eat both plants and animal flesh are omnivores. Write the number of such an animal and explain your reasons. **Animal 1 has large canines and flat molars**

**Solving the Problem**

1. Write down the food you ate today. **probably both plant and animal products**
   Do you conclude humans are herbivores, carnivores, or omnivores? **omnivores**
2. The jaws in the pictures are of a human, a giraffe, and a dog. In the blanks, write the number of the jaw of the human **1**, the giraffe **2**, and the dog **3**.

## OPTIONS

**Critical Thinking/Problem Solving/** *Teacher Resource Package,* p. 10. Use the worksheet shown here to reinforce how teeth are involved in physical digestion.

**Videodisc**

**STV: Human Body Vol. 1**
*Digestive System*
Unit 2, Side 2
*Digestive System (in its entirety)*

**655-30380**

## TEACH

### Independent Practice

**Study Guide**/*Teacher Resource Package*, p. 55. Use the Study Guide worksheet shown at the bottom of this page for independent practice.

### Guided Practice

Have students write down their answers to the margin question: They speed up the rate of change.

### Check for Understanding

Have students respond to the first three questions in Check Your Understanding.

### Reteach

**Reteaching**/*Teacher Resource Package*, p. 25. Use this worksheet to give students practice in understanding digestion.

**Extension:** Assign Critical Thinking, Biology and Writing, or some of the **OPTIONS** available with this lesson.

## 3 APPLY

### Audiovisual

Show the videocassette *Breakdown*, Films for the Humanities.

## 4 CLOSE

### ACTIVITY/Brainstorming

Ask: How and where does the digestive system change a slice of bread?

### Answers to Check Your Understanding

1. to take in food and change it into a form the body can use
2. physical and chemical changes
3. chemicals that change the rate of a chemical reaction
4. Cells cannot use the starch in bread. It must be broken down into glucose molecules.
5. Answers may include that a factory is not-living and digestion occurs in living things.

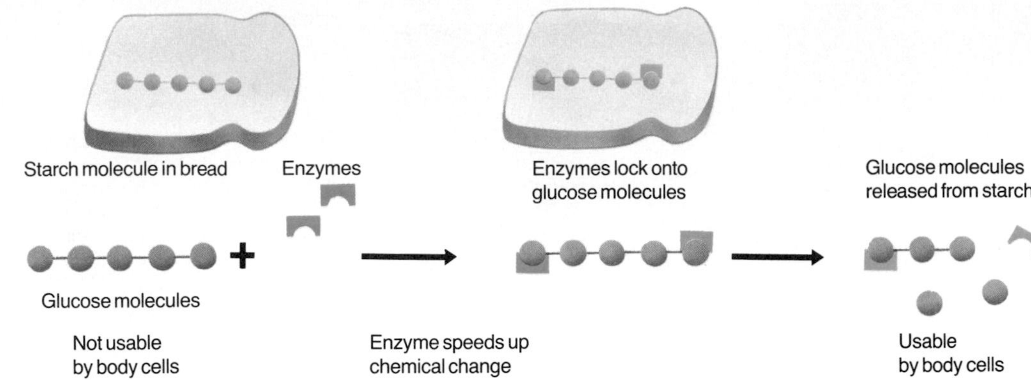


**Figure 10-2** Starch is a large molecule made of many glucose molecules strung together. An enzyme breaks apart the starch molecule, releasing glucose molecules that are usable by the body.

**How do enzymes help chemical changes?**

A chemical change occurs when food changes form. How does a chemical change differ from a physical change? A **chemical change** turns food into a form that cells can use. Your digestive system makes chemicals that help with the chemical changes. These chemicals are added to food as it moves through the organs of your digestive system. The chemicals are called enzymes (EN zimes). **Enzymes** are chemicals that speed up the rate of chemical change. Figure 10-2 shows how enzymes work.

Bread is made of a carbohydrate called starch. Bread itself is not usable to cells. It must be digested first. The starch in bread is made of many molecules of a chemical called glucose (GLEW kohs). Glucose is a type of sugar molecule. An enzyme speeds up the change of starch by removing glucose molecules from the starch. Once the glucose molecules are separated, they are in a form your body cells can use.

### Check Your Understanding

1. What is the main job of the digestive system?
2. Identify two ways that food is broken down.
3. What are enzymes?
4. **Critical Thinking:** Explain why bread must be digested.
5. **Biology and Writing:** You have learned that the process of digestion is similar to the work of a factory. Write a paragraph that describes the differences between a real factory and digestion.

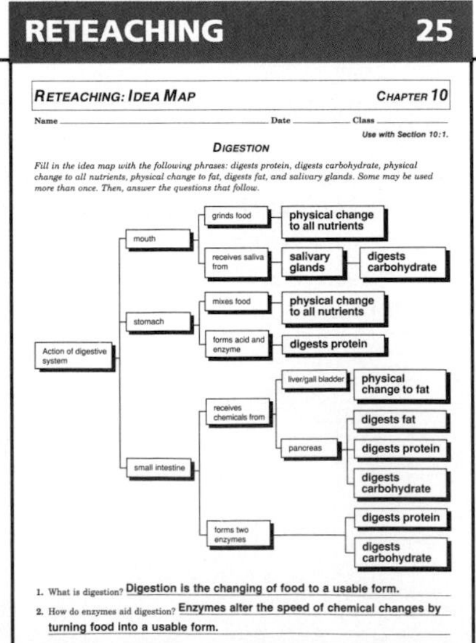

**RETEACHING 25**

RETEACHING: IDEA MAP — CHAPTER 10

Name ______ Date ______ Class ______

*Use with Section 10:1.*

*DIGESTION*

*Fill in the idea map with the following phrases: digests protein, digests carbohydrate, physical change to all nutrients, physical change to fat, digests fat, and salivary glands. Some may be used more than once. Then, answer the questions that follow.*

- Action of digestive system
  - mouth
    - grinds food — **physical change to all nutrients**
    - receives saliva from — **salivary glands** — **digests carbohydrate**
  - stomach
    - mixes food — **physical change to all nutrients**
    - forms acid and enzyme — **digests protein**
  - small intestine
    - receives chemicals from
      - liver/gall bladder — **physical change to fat**
      - pancreas — **digests fat**, **digests protein**, **digests carbohydrate**
    - forms two enzymes — **digests protein**, **digests carbohydrate**

1. What is digestion? **Digestion is the changing of food to a usable form.**
2. How do enzymes aid digestion? **Enzymes alter the speed of chemical changes by turning food into a usable form.**

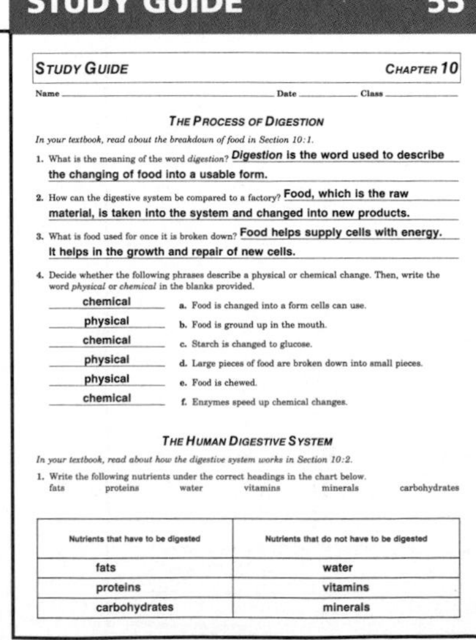

**STUDY GUIDE 55**

STUDY GUIDE — CHAPTER 10

Name ______ Date ______ Class ______

*THE PROCESS OF DIGESTION*

*In your textbook, read about the breakdown of food in Section 10:1.*

1. What is the meaning of the word *digestion*? **Digestion is the word used to describe the changing of food into a usable form.**
2. How can the digestive system be compared to a factory? **Food, which is the raw material, is taken into the system and changed into new products.**
3. What is food used for once it is broken down? **Food helps supply cells with energy. It helps in the growth and repair of new cells.**
4. Decide whether the following phrases describe a physical or chemical change. Then, write the word *physical* or *chemical* in the blanks provided.

**chemical** a. Food is changed into a form cells can use.
**physical** b. Food is ground up in the mouth.
**chemical** c. Starch is changed to glucose.
**physical** d. Large pieces of food are broken down into small pieces.
**physical** e. Food is chewed.
**chemical** f. Enzymes speed up chemical changes.

*THE HUMAN DIGESTIVE SYSTEM*

*In your textbook, read about how the digestive system works in Section 10:2.*

1. Write the following nutrients under the correct headings in the chart below.
fats proteins water vitamins minerals carbohydrates

| Nutrients that have to be digested | Nutrients that do not have to be digested |
|---|---|
| fats | water |
| proteins | vitamins |
| carbohydrates | minerals |

# Lab 10–1 Digestion

## Problem: How are proteins digested?

### Skills

separate and control variables, form a hypothesis, observe, design an experiment

### Materials

4 paper cups
gelatin
water
4 applicator sticks
refrigerator (optional)
4 plastic spoons
enzyme A
enzyme B

### Procedure

1. Copy the data table.
2. Number four paper cups 1 through 4.
3. **Separate and control variables:** Fill each cup as follows:
   Cup 1—2 spoonfuls of gelatin
   This is the control in your experiment.
   Cup 2—2 spoonfuls of gelatin and 1 spoonful of water
   This setup is a variable to show the effect of water on the gelatin.
   Cup 3—2 spoonfuls of gelatin and 1 spoonful of enzyme A
   This setup is a variable to show the effect of enzyme A.
   Cup 4—2 spoonfuls of gelatin and 1 spoonful of enzyme B
   This setup will show the effect of enzyme B.
4. Mix the contents of each cup with a different applicator stick.
5. **Form a hypothesis:** Which cup's contents will show digestion of the gelatin? Write your **hypothesis** in your notebook.
6. Wait 20 minutes. (Or place cups in a refrigerator for 20 minutes.)

| CUP | CONTENTS OF CUP | SOLID OR LOOSE? | DIGESTION OCCURRED? |
|---|---|---|---|
| 1 | Gelatin | solid | no |
| 2 | Gelatin + water | solid | no |
| 3 | Gelatin + enzyme A | loose | yes |
| 4 | Gelatin + enzyme B | solid | no |

7. **Observe:** After waiting, record in your table whether or not the gelatin is solid or loose and whether digestion occurred. (NOTE: If gelatin is *solid,* enzyme *did not* digest gelatin. If gelatin is loose, enzyme *did* digest gelatin.)
8. Dispose of the gelatin as indicated by your teacher.

### Data and Observations

1. Which cup showed gelatin digestion?
2. What is your evidence?
3. What was added to this cup?

### Analyze and Apply

1. What is an enzyme?
2. Where are digestive enzymes usually made?
3. What is your evidence that water is not an enzyme?
4. **Check your hypothesis:** Is your hypothesis supported by your data? Why or why not?
5. **Apply:** Gelatin is a protein. What body organs form enzymes that help digest this nutrient?

### Extension

**Design an experiment** to determine whether saliva contains an enzyme that can digest gelatin. Try your experiment and report the results.



## ANSWERS

### Data and Observations

1. 3
2. the gelatin became loose
3. enzyme A

### Analyze and Apply

1. a chemical that helps to speed up the rate of digestion
2. in the digestive system
3. Cup 2 with water and gelatin alone showed no change in the gelatin.
4. Students who hypothesized that cup 3 would show digestion will answer yes.
5. stomach, pancreas, small intestine

## Lab 10-1 Digestion

### Overview

In this lab, students will compare visually the action of enzymes on a certain food type.

**Objectives:** When students have completed this lab, they will have learned that (1) gelatin may be chemically changed when enzymes are added to it, (2) not all enzymes work on the same food.

**Time Allotment:** 45 minutes

### Preparation

▶ Enzyme A is pineapple juice (must be fresh or frozen, *cannot be canned*).

▶ Enzyme B is diastase –mix 10 g diastase with 90 mL water.

▶ Gelatin preparation–add only half the amount of water indicated on the gelatin package.

**Lab 10-1 worksheet**/*Teacher Resource Package,* pp. 37-38.

### Teaching the Lab

▶ **Troubleshooting:** It may take overnight for gelatin to solidify.

**Performance:** Provide students with paper and colored pencils. Have them design and label models that depict the shape of an enzyme and protein both before being mixed together and after being mixed together. Suggest to students that they check Figure 10-2 first and advise them that protein does not exactly resemble a starch or glucose molecule.

## 10:2 The Human Digestive System

## PREPARATION

### Materials Needed

 Make copies of the Transparency Master, Application, Enrichment, Reteaching, Skill, Study Guide, and Lab worksheets in the *Teacher Resource Package.*

▶ Photocopy diagrams of the digestive system and gather colored pencils for Lab 10-2.

### Key Science Words

| | |
|---|---|
| saliva | liver |
| salivary gland | bile |
| esophagus | gallbladder |
| stomach | large intestine |
| hydrochloric acid | appendix |
| small intestine | villi |
| pancreas | mucus |

### Process Skills

In the Mini Labs, students will use numbers and experiment. In Lab 10-2, they will classify and interpret diagrams. In the Skill Check, students will understand science words.

## 1 FOCUS

▶ The objectives are listed on the student page. Remind students to preview these objectives as a guide to this numbered section.

### ACTIVITY/Brainstorming

An operation performed on very obese people involves removing a section of the small intestine. Have your students speculate as to why this kind of operation helps with weight loss. Less intestine means less absorption of food.

### Objectives

4. **Trace** the path of food from the mouth to body cells.
5. **Compare** the human digestive system with those of other animals.
6. **Identify** the problems of the digestive system.

### Key Science Words

saliva
salivary gland
esophagus
stomach
hydrochloric acid
small intestine
pancreas
liver
bile
gallbladder
large intestine
appendix
villi
mucus

## 10:2 The Human Digestive System

Humans provide a good example of how digestive systems work. If you understand digestion in humans, you should be able to understand it in other animals.

### Nutrients Are Digested

Many nutrients must be digested before they can be used, while others are already in a usable form. Table 10-1 shows which nutrients are already usable and which are not.

Water, vitamins, and minerals can move directly from your digestive system into body cells without being changed. Fats, proteins, and carbohydrates must be acted upon by enzymes in your digestive system. There are different kinds of enzymes for each nutrient that must be digested. For example, enzymes that speed up changes in fats cannot change proteins and carbohydrates. Enzymes that speed up changes in proteins cannot help with the breakdown of fats or carbohydrates. The digestive system, therefore, must make different kinds of enzymes for each different nutrient.

**TABLE 10–1. NUTRIENTS**

| Nutrient | Already in Usable Form |
|---|---|
| Water | yes |
| Vitamins | yes |
| Minerals | yes |
| Fat | no |
| Protein | no |
| Carbohydrate | no |

### A Trip Through the Digestive System

Let's take a trip through the human digestive system to see how it works. To make it a little more interesting, we will look at what happens to a hamburger. Remember that ground meat is mostly protein, mayonnaise is mostly fat, and the bun is mostly carbohydrate. The entire trip of the hamburger through the digestive system takes about 21 hours. Figure 10-3 shows how the entire digestive system looks from one end to the other.

The hamburger enters the system through the mouth. In the mouth, teeth break and grind the food. The breaking and grinding cause the hamburger to change physically. **Saliva** is a liquid that is formed in the mouth and that contains an enzyme. Saliva speeds up chemical changes in the carbohydrates of the bun. It has no chemical effect on proteins or fats. Figure 10-3 shows that saliva is made in the salivary (SAL uh ver ee) glands. **Salivary glands** are three pairs of small glands located under the tongue and behind the jaw. Saliva passes from the glands through small tubes leading to the mouth.



## OPTIONS

### Science Background

Think of the digestive system as a long tube through which food passes. Each organ of the system makes up the tube or is attached to the tube and helps with the digestion of different food types.

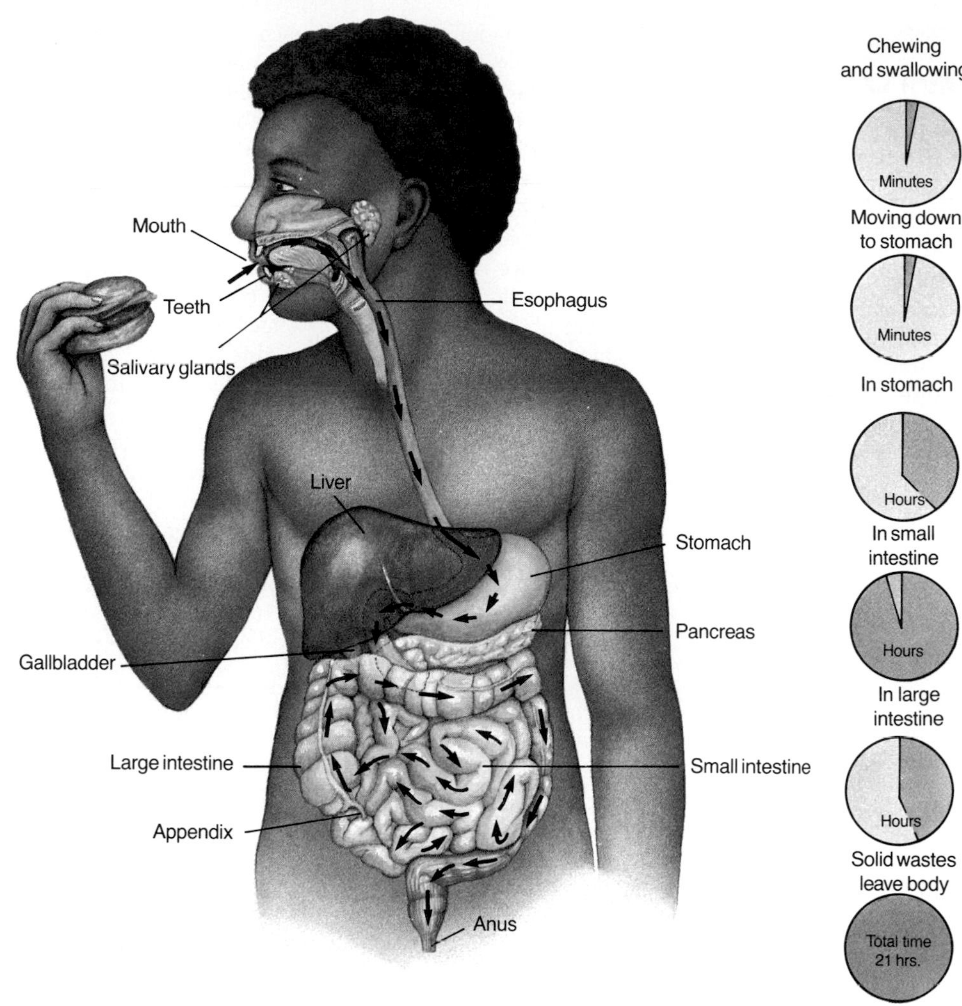


**Figure 10-3** Follow the path of a meal through the digestive system.

The broken down hamburger remains in the mouth for about one minute. Swallowing moves it from the mouth into the esophagus (ih SAHF uh gus). The **esophagus** is a tube that connects the mouth to the stomach. Muscles in the esophagus push the food toward the stomach. These events are shown in Figure 10-3. This part of the trip takes less than one minute.

As it leaves the esophagus, the food enters the stomach. The **stomach** is a baglike, muscular organ that mixes and chemically changes protein. It can hold about one liter of liquid and food.

**How does food reach your stomach?**

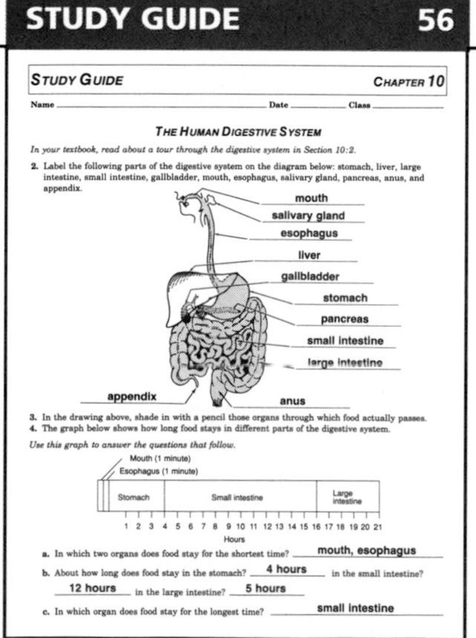

STUDY GUIDE 56

STUDY GUIDE CHAPTER 10

Name ________ Date ________ Class ________

*THE HUMAN DIGESTIVE SYSTEM*

*In your textbook, read about a tour through the digestive system in Section 10:2.*

2. Label the following parts of the digestive system on the diagram below: stomach, liver, large intestine, small intestine, gallbladder, mouth, esophagus, salivary gland, pancreas, anus, and appendix.



3. In the drawing above, shade in with a pencil those organs through which food actually passes.
4. The graph below shows how long food stays in different parts of the digestive system.

*Use this graph to answer the questions that follow.*



a. In which two organs does food stay for the shortest time? **mouth, esophagus**

b. About how long does food stay in the stomach? **4 hours** in the small intestine? **12 hours** in the large intestine? **5 hours**

c. In which organ does food stay for the longest time? **small intestine**

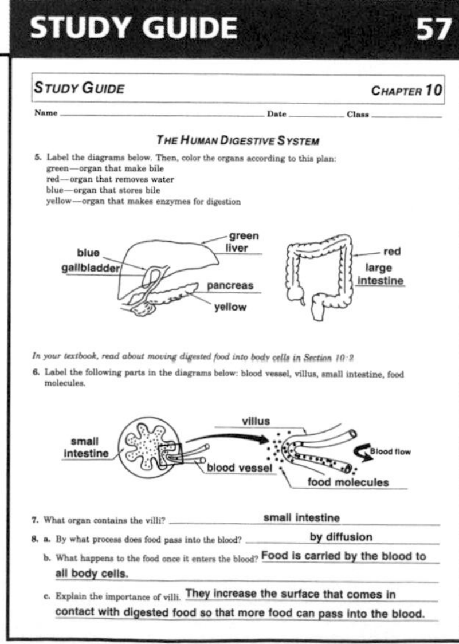

STUDY GUIDE 57

STUDY GUIDE CHAPTER 10

Name ________ Date ________ Class ________

*THE HUMAN DIGESTIVE SYSTEM*

5. Label the diagrams below. Then, color the organs according to this plan:
green—organ that make bile
red—organ that removes water
blue—organ that stores bile
yellow—organ that makes enzymes for digestion



*In your textbook, read about moving digested food into body cells in Section 10:2*

6. Label the following parts in the diagrams below: blood vessel, villus, small intestine, food molecules.



7. What organ contains the villi? **small intestine**

8. a. By what process does food pass into the blood? **by diffusion**

b. What happens to the food once it enters the blood? **Food is carried by the blood to all body cells.**

c. Explain the importance of villi. **They increase the surface that comes in contact with digested food so that more food can pass into the blood.**

## 2 TEACH

### MOTIVATION/Demonstration

Have students observe prepared slides of the intestine (cross section) under the microscope. This will allow students to see the villi. A 35-mm slide projected on the classroom screen would also be an effective way for them to see the extensive surface area of this organ.

### Concept Development

▶ The human body may form 250 to 1500 mL of saliva per day.

▶ Point out that the digestive system produces eleven different enzymes. Each enzyme changes a specific nutrient into a usable form.

### Guided Practice

Have students write down their answers to the margin question: Muscles in the esophagus move food toward the stomach.

### MOTIVATION/Enrichment

Point out that food is prevented from entering the windpipe by the closure of the epiglottis during swallowing. This closure is an example of a reflex action. The muscle contraction called peristalsis is a second example of a reflex. The muscles that line the esophagus contract to push food down into the stomach.

### Independent Practice

**Study Guide**/*Teacher Resource Package,* pp. 56-57. Use the Study Guide worksheets shown at the bottom of this page for independent practice.

## TEACH

### Concept Development

▶ Explain that food passes directly from the stomach to the small intestine, but the liver, gallbladder, and pancreas pour their chemicals into the small intestine. It is important to emphasize that food does not enter these organs. Figure 10-4, which shows this relationship, should be pointed out to students.

▶ Bile consists of bile salts, cholesterol, and bilirubin (a hemoglobin derivative). The liver produces 500 mL of bile each day.

▶ Mention that the liver has many roles in addition to that of forming bile. You may want to mention that the pancreas also has a dual role and will be discussed in a later chapter.

### Mini Lab

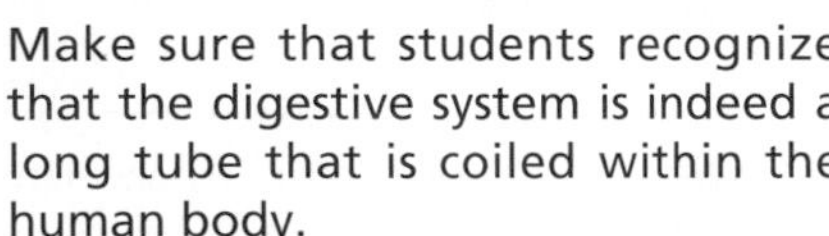

Make sure that students recognize that the digestive system is indeed a long tube that is coiled within the human body.

**ASSESSMENT**

**Skill:** Provide students with a diagram that is a long, twisted tube marked off and labeled into sections of the esophagus, stomach, small intestine, and large intestine. Students are to measure the lengths of each organ in cm and prepare a bar graph of their data. Lengths of organs should be in proportion to the values used in the Mini Lab.

**Portfolio**

Copy the diagram on page 210. Through a series of short arrows, trace the pathway that bile will take from the liver to the gallbladder and to the intestines. Use a different color series of arrows to show the pathway of pancreas enzymes. Use a third color series of arrows to show the pathway of food. Add labels to all arrows explaining what they represent.



### Mini Lab

**How Long Is the Digestive System?**

**Use numbers:** Diagram and label the lengths and names of the following digestive organs: esophagus (25 cm), stomach (20 cm), small intestine (700 cm), and large intestine (150 cm). *For more help, refer to the* ***Skill Handbook,*** *pages 718-719.*

Cells on the inside of the stomach make two chemicals that help in digestion. One is an enzyme. The enzyme speeds up the chemical change of protein or meat in the hamburger. The other chemical made by the stomach is hydrochloric (hi druh KLOR ik) acid. **Hydrochloric acid** is a chemical often called stomach acid.

The muscular walls of the stomach mix and churn the food. Mixing is a physical part of digestion. Look at Figure 10-3 and notice the small clock next to the stomach. It shows how long foods remains in the stomach.

After about four hours, the stomach pushes food into the small intestine. The **small intestine** is a long, hollow, tubelike organ. Most of the chemical digestion of food takes place in the small intestine.

Before describing what the small intestine does, we must take a short detour. Figure 10-4 shows three different organs that are part of the digestive system. These organs are the pancreas (PAN kree us), liver, and gallbladder. Food doesn't pass through these organs.

The **pancreas** makes three different enzymes. One kind of enzyme helps to break down fats, one kind helps to break down proteins, and the third kind helps to break down carbohydrates. The enzymes pass through a small tube from the pancreas to the small intestine.

The **liver,** the largest organ in the body, makes a chemical called bile. **Bile** is a green liquid that breaks large fat droplets into small fat droplets. This change is physical and takes place in the small intestine. Bile is delivered from the liver to the gallbladder. The **gallbladder** is a small, baglike part located under the liver. It stores bile until it is needed by the small intestine.

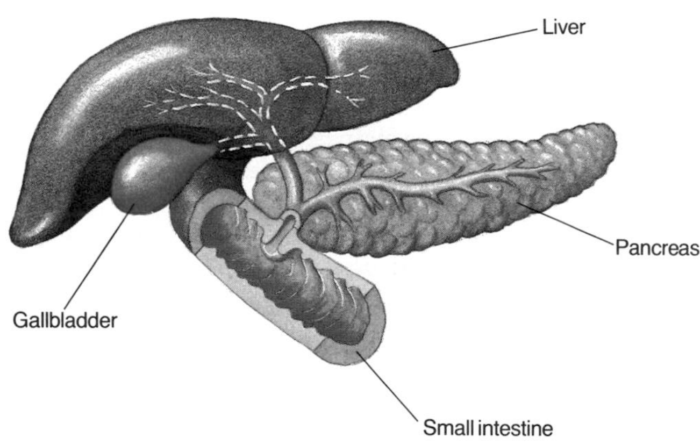


**Figure 10-4** Bile from the gallbladder breaks up fat. The pancreas makes three enzymes that help digest fat, protein, and carbohydrates.



## OPTIONS

**Application**/*Teacher Resource Package,* p. 10. Use the Application worksheet shown here to introduce the concept of digestive system disorders.

**APPLICATION 10**

APPLICATION: HEALTH — CHAPTER 10

Name ______ Date ______ Class ______

*Use after Section 10:3.*

GALLSTONES, A PROBLEM OF THE DIGESTIVE SYSTEM

Bile is formed by the liver and stored in the gallbladder until it is needed to help digest fat. While in the gallbladder, bile sometimes forms hard masses called gallstones. If gallstones move down the tube leading to the small intestine, they cause problems. They block the path through which bile flows into the small intestine. They may even block the flow of enzymes from the pancreas into the small intestine. In Figure 1, arrows show the normal route of bile from the liver and of enzymes from the pancreas. Figure 2 shows the positions of three different gallstones.

*Study the diagrams to answer the following questions.*

1. Will gallstone A cause any problem with fat digestion? Explain. **No, bile can still enter and leave the gallbladder.**
2. What problems, if any, will gallstone B cause? **Bile will not be able to pass from the gallbladder into the small intestine.**
3. What problems, if any, will gallstone C cause? **Neither bile from the gallbladder nor enzymes from the pancreas will be able to enter the small intestine.**

Sometimes the gallbladder must be removed to prevent other gallstones from forming. Study Figure 3 to see the new pathway that bile takes after this operation.

4. Does bile still reach the small intestine if the gallbladder is removed? Explain. **Yes, it goes from the liver to the small intestine.**
5. Can enzymes from the pancreas still enter the small intestine? Explain. **Yes, the pathway is not changed.**

Figure1 — Tube from liver to small intestine; Gallbladder; Enzyme passes out of pancreas; Pancreas; Section of small intestine

Figure 2 — A; B; C

Figure 3 — Gall bladder removed

## Career Close-Up

### Dietitian

**S**teve was interested in a career as a dietitian. As part of his career day project at school, he decided to interview a dietitian to find out more about the job.

**Steve:** What is a dietitian?

**Dietitian:** A dietitian is one who understands nutrition. A dietitian plans diets for groups or individuals. These diets may be regular, balanced meals or diets for patients with special needs.

**Steve:** What type of training is needed?

**Dietitian:** Usually a four-year college degree is needed. This background will lead to a Registered Dietitian Degree. A six- to eighteen-month internship in a hospital is required.

**Steve:** Where are you most likely to find a job?

**Dietitian:** Hospitals are one place. You could also work for a company, school, university, or nursing home that has its own employee or patient dining hall. You might even work for an eating disorder clinic. Large food companies and drug companies hire dietitians.

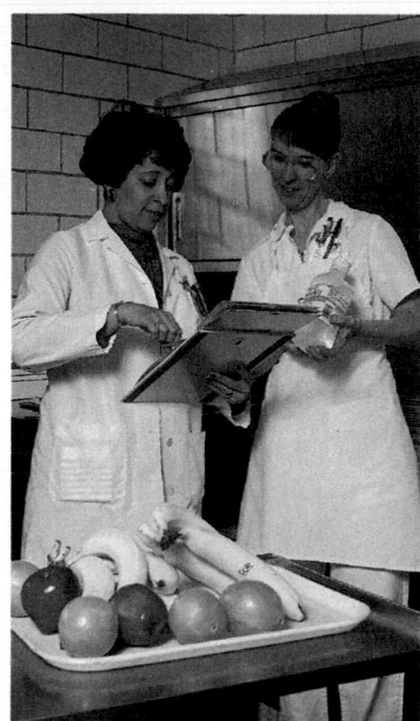

A dietitian approves the meals to be served to hospital patients.

Let's continue our tour and follow the food into the small intestine. Remember, the pancreas and gallbladder have already emptied their chemicals into the small intestine. In addition, the small intestine itself makes several enzymes. Cells that line the small intestine make enzymes that help with digestion of proteins in the hamburger meat and carbohydrates in the bun. All foods that enter the small intestine are finally changed chemically into a form that is usable by the body. The main goal of the digestive system has now been carried out. Figure 10-3 shows these changes. It also shows you that food spends about 12 hours in the small intestine. Why should food spend so much time in this part of the digestive system?

This is where chemical digestion and absorption are completed.



## Career Close-Up

### Dietician

#### Background

This field is expanding as the population ages. More and more senior citizen housing centers are employing licensed dieticians to help with the meal planning for these individuals.

#### Related Careers

Food services in nursing homes, hospitals, and rehabilitation centers are in need of people to prepare, cook, and serve food. A degree as a dietician is not required.

#### Job Requirements

Becoming a licensed dietician usually requires a two- or four-year college degree and licensing by the state or local government.

#### For More Information

Check with your school counselor, home economics teacher, or school librarian for more information.

## TEACH

### Concept Development

▶ The small intestine releases about 1500 mL of liquid per day in the form of digestive enzymes and mucus. It absorbs about 8500 mL of liquid per day.

### Independent Practice

**Study Guide**/*Teacher Resource Package,* p. 58. Use the Study Guide worksheet shown at the bottom of this page for independent practice.

### Student Journal

Have students pretend they are a sandwich that was eaten at noon. Have students keep a diary noting where they would most likely be in the body at 12:01, 12:03, and every hour on the hour after 12:00. Students should report what their surroundings might be like for each entry in the diary.

### STUDY GUIDE 58

STUDY GUIDE — CHAPTER 10

Name ______ Date ______ Class ______

THE HUMAN DIGESTIVE SYSTEM

9. Complete the table below.

| Organ | Enzyme formed here? (yes or no) | Food passes through this organ? (yes or no) | Name of nutrient acted upon or digested (fat, protein, carbohydrate, none) | Food absorbed here? (yes or no) | Water absorbed here? (yes or no) | Name of chemical made here |
|---|---|---|---|---|---|---|
| Mouth and salivary glands | yes | yes (mouth) | carbohydrate | no | no | saliva |
| Esophagus | no | yes | none | no | no | none |
| Stomach | yes | yes | protein | no | no | hydrochloric acid, enzyme |
| Liver and gallbladder | no | no | fat | no | no | bile |
| Pancreas | yes | no | fat, protein, carbohydrate | no | no | enzymes |
| Small intestine | yes | yes | fat, protein, carbohydrate | yes | no | enzymes |
| Large intestine | no | yes | none | no | yes | none |

10. Decide whether the following statements are true or false. If true, write *true* in the left-hand column. If false, change the underlined word to one that will make the statement true and write the correct word in the blank.

long a. Plant-eating animals usually have short digestive systems.

two b. An earthworm has one opening in its digestive system.

true c. A tapeworm has no digestive system.

short d. Meat-eating animals usually have long digestive systems.

true e. Humans are among the animals that have a complex digestive system.

## OPTIONS

**The Human Digestive System**/*Transparency Package,* number 10. Use color transparency number 10 as you teach the digestive system.

## TEACH

### Concept Development

▶ Explain that the large intestine aids in the reabsorption of sodium into the bloodstream. The large intestine is composed of three main sections: the ascending colon, the transverse colon, and the descending colon.

### ACTIVITY/Display

Obtain a model of the human body (torso). Most models have detachable organs. Ask students to identify all detached organs that compose the digestive system. Have them attempt to reassemble the model correctly.

### Idea Map

Have students use the idea map as a study guide to the major concepts of this section.

### *Student Journal*

Tell students to pretend they have been hired to check to see if the human digestive system is properly changing foods both physically and chemically. Students should write a preliminary report on what they expect to find in the mouth, esophagus, stomach, and small and large intestines. Students can decide if they want to include the liver and appendix in the report and explain why.

**Idea Map**

**Digestive Organs**

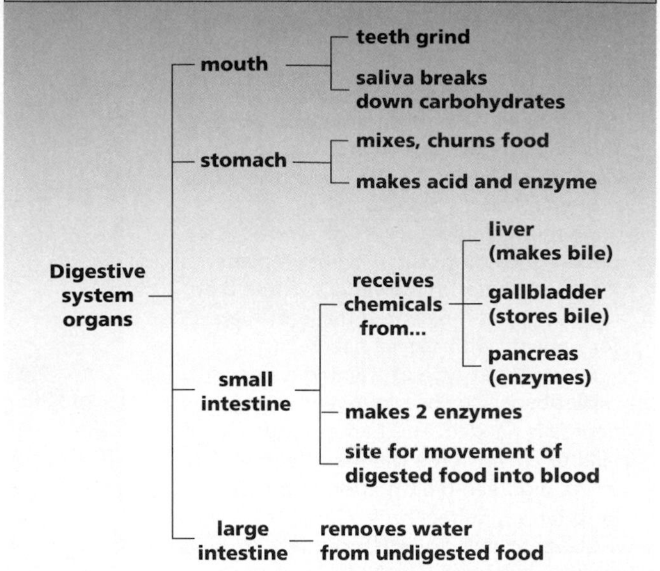

- **Digestive system organs**
  - **mouth**
    - **teeth grind**
    - **saliva breaks down carbohydrates**
  - **stomach**
    - **mixes, churns food**
    - **makes acid and enzyme**
  - **small intestine**
    - **receives chemicals from...**
      - **liver (makes bile)**
      - **gallbladder (stores bile)**
      - **pancreas (enzymes)**
    - **makes 2 enzymes**
    - **site for movement of digested food into blood**
  - **large intestine**
    - **removes water from undigested food**

**Study Tip:** Use this idea map as a study guide to understanding the job of each digestive organ.

After passing through the small intestine, food enters the large intestine. The **large intestine** is a tubelike organ at the end of the digestive tract. It is called the large intestine because of its width. The large intestine is about 5 cm wide, and the small intestine is only 2.5 cm wide. Not much digestion occurs in the large intestine. Its main job is to remove water from undigested food. Water is then returned to the bloodstream. The clock in Figure 10-3 tells you that food spends about five hours in the large intestine. Not all of the food is digested when it reaches the end of the large intestine. It then leaves the body as solid waste through the anus.

One last part of the digestive system should be mentioned. This part is the appendix (uh PEN dihks). The **appendix** is a small fingerlike part found where the small and large intestines meet. The appendix in our body does not take part in the digestion of food.

Altogether your digestive system forms a tube about 900 cm long. The average height of an adult is about 170 cm. Your digestive system is about five times longer than your body! How can such a long tube fit inside of you? Why must it be so long?

### Skill Check

**Understand science words: appendix.** The word part *pend* means to hang. In your dictionary, find three words with the word part *pend* in them. *For more help, refer to the* ***Skill Handbook****, pages 706-711.*

212

## *OPTIONS*

**Enrichment**/*Teacher Resource Package*, p. 10. Use the Enrichment worksheet shown here to help students.

**Transparency Master**/*Teacher Resource Package*, p. 45. Use the Transparency Master shown here to help students learn the parts of the digestive system.

**ENRICHMENT 10**

ENRICHMENT CHAPTER 10

Name ______ Date ______ Class ______

Use after Section 10:2.

**DIGESTING DIFFERENT NUTRIENTS**

*Use the information below to complete the tables for carbohydrate digestion and protein digestion. Key words for the tables are underlined.*

Salivary glands produce the enzyme salivary amylase. The pancreas produces pancreatic amylase. Both enzymes bring about chemical changes, changing carbohydrates to double sugars. Salivary glands release enzymes into the mouth. The pancreas releases enzymes into the small intestine. Glands in the small intestine make carbohydrases, enzymes that bring about the chemical breakdown of double sugars to single sugars. This action occurs in the small intestine.

Proteins are changed into polypeptides in the stomach. An enzyme called pepsin and a chemical called hydrochloric acid, both made by glands in the stomach, aid this chemical change. Trypsin, an enzyme made in the pancreas, is released into the small intestine. Trypsin helps in the chemical breakdown of polypeptides into peptides. Glands in the small intestine release the enzyme erepsin (ee REP sihn). This enzyme chemically changes peptides into amino acids.

**Carbohydrate Digestion**

| Steps in nutrient digestion | Enzyme name | Organ making enzyme | Organ where change occurs | Chemical or physical |
|---|---|---|---|---|
| Carbohydrate<br>Double sugar | salivary amylase and pancreatic amylase | salivary glands and pancreas | mouth and small intestine | chemical |
| Single sugar | carbohydrases | small intestine | small intestine | chemical |

**Protein Digestion**

| Steps in nutrient digestion | Enzyme or chemical name | Organ making enzyme or chemical | Organ where change occurs | Chemical or physical |
|---|---|---|---|---|
| Protein<br>Polypeptide | pepsin and hydrochloric acid | stomach | stomach | chemical |
| Peptide | trypsin | pancreas | small intestine | chemical |
| Amino acids | erepsin | small intestine | small intestine | chemical |

**TRANSPARENCY 45**

TRANSPARENCY MASTER CHAPTER 10

**HUMAN DIGESTIVE SYSTEM**

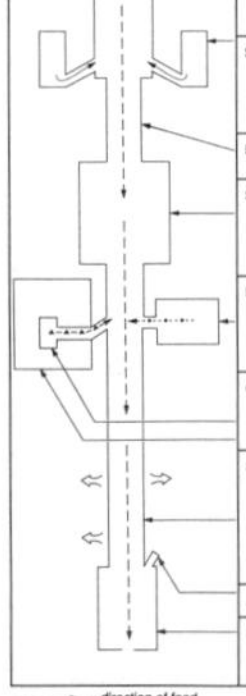

| Organ name | Main functions, chemical made, and foods acted upon |
|---|---|
| Mouth | 1. Chewing food begins digestive process.<br>2. Food enters digestive system. |
| Salivary gland | 1. An enzyme in saliva starts digestion of carbohydrates.<br>2. Food is moistened for ease in swallowing. Saliva moves into mouth by way of small tube. |
| Esophagus | 1. Food is moved by muscle contraction to the stomach. |
| Stomach | 1. Enzymes and hydrochloric acid are formed here.<br>2. Protein digestion begins.<br>3. Muscle movement of stomach mixes food. |
| Pancreas | 1. Three enzymes are formed here for digestion of fats, proteins, and carbohydrates.<br>2. All enzymes from pancreas are added to the small intestine. |
| Gallbladder and liver | 1. Liver forms bile that is stored in the gallbladder.<br>2. Bile causes fats to break up.<br>3. Bile is added to the small intestine. |
| Small intestine | 1. Two enzymes are formed here. One digests protein, and other carbohydrates.<br>2. Final digestion of fats, protein, and carbohydrates occurs here.<br>3. All digested foods are absorbed into bloodstream. |
| Appendix | 1. Performs no known digestive job. |
| Large intestine | 1. Absorbs large amounts of water.<br>2. Undigested food given off as solid waste. |

(Column "Human digestive system": diagram)

# Lab 10–2 Digestive System 

## Problem: What are the jobs of the digestive system organs?

### Skills

classify, interpret diagrams

### Materials

five diagrams of the digestive system
colored pencils: red, blue, green, yellow, purple

### Procedure

1. Label your five diagrams A, B, C, D, and E.
2. **Classify:** On diagram A, label the liver, esophagus, large intestine, mouth, small intestine, gallbladder, pancreas, salivary gland, stomach, anus, and appendix.
3. With a regular pencil, shade only the parts through which food actually passes.
4. Label the diagram "Human Digestive System and Food Pathway."
5. On diagram B, label the organs that aid the chemical change of carbohydrates. Color these organs red.
6. Label the diagram "Organs that Help to Digest Carbohydrates."
7. On diagram C, label the organs that aid the chemical change of protein. Color these organs blue.
8. Label the diagram "Organs that Help to Digest Protein."

9. On diagram D, label the organs that help with chemical and physical changes of fat. Color these organs green.
10. Label the diagram "Organs that Help to Digest Fat."
11. On diagram E, label the organs of the digestive system where diffusion of digested food into blood occurs. Color these organs yellow.
12. Using the same diagram, E, label the organs where diffusion of water occurs. Color these organs purple.
13. Label the diagram "Organs that Help Remove Digested Food and Water."

### Data and Observations

1. Which organs did you color red?
2. Which organs did you color blue?
3. Which organs did you color green?
4. Which organs did you color yellow?
5. Which organs did you color purple?

### Analyze and Apply

1. **Interpret diagrams:** Which two organs are the most important in digestion? Explain your answer.
2. Which organs help digest carbohydrates, protein, and fat?
3. **Apply:** Diagram the digestive path of a slice of cheese and pepperoni pizza. Indicate where the crust, cheese, and pepperoni are digested.

### Extension

**Draw a diagram** of the digestive system. Create a color code to indicate where chemical, physical, both physical and chemical, and no changes occur. Then, color your diagram accordingly.



## ANSWERS

### Data and Observations

1. mouth, pancreas, small intestine
2. stomach, pancreas, small intestine
3. pancreas, liver, gallbladder
4. small intestine
5. large intestine

### Analyze and Apply

1. the pancreas and small intestine, because most of the digestive enzymes are produced there
2. carbohydrates – mouth, pancreas, and small intestine; protein – stomach, pancreas, and small intestine; fat – pancreas, liver, and gallbladder
3. On a diagram, students will show the crust digested in the mouth and small intestine, the cheese digested in the stomach and small intestine, and the pepperoni digested in the stomach and small intestine.

## Lab 10-2 Digestive System

### Overview

Using diagrams, students will identify the organs responsible for digestion and absorption in the human digestive system.

**Objectives:** After completing this lab, students will be able (1) to **name** and **describe** the organs of the human digestive system, (2) to **indicate** the organs where absorption of food and water occur, (3) to **name** the specific organs in which carbohydrate, fat, and protein digestion occur.

**Time Allotment:** 1-2 class periods (See the Cooperative Learning note.)

### Preparation

▶ Each student will need five copies of the digestive system. This diagram is available in the *Transparency Masters* book, p. 47. (See the Cooperative Learning note.)

 **Lab 10-2 worksheet**/*Teacher Resource Package,* pp. 39-40.

### Teaching the Lab

**Cooperative Learning:** To reduce the amount of time and materials needed, allow students to work in groups of five. Each student would then be responsible for only one diagram and the entire group would turn in one completed set. For more information, see pp. 22T-23T in the Teacher Guide.

### ASSESSMENT

**Skill:** Provide students with an outline map of the digestive system and two colored pencils. Have students shade the organs that make enzymes that would aid in digestion of a hamburger (having both fat and protein). Students should indicate which color represents which nutrient. If enzymes from the same organ are responsible for both fat and protein, only half the organ should be shaded with each corresponding color.

## TEACH

**Using Science Words:** Relate the terms *absorption* and *absorb.* Ask students to describe the action of a sponge or paper towel as it soaks up water. What term is often used to describe this action? The towel or sponge absorbs. Relate the term *absorption* to osmosis and diffusion.

### ACTIVITY/Brainstorming

Ask students why dieters often increase the amount of vegetables they eat. They may see the connection between eating a lot of foods high in bulk, and not being able to digest as much of one's food. Few nutrients are acquired from these types of foods.

### Mini Lab

The correlation must be made that villi help to increase the surface area of the intestines and aid in the absorption of digested food.

### ASSESSMENT

**Performance:** Provide students with a square of cheesecloth, sand and coarse stones or pebbles, and a shallow dish or pan. Have students design a model that illustrates how the intestines will absorb digested foods but not undigested foods. Have them explain their model to you.

### GLENCOE TECHNOLOGY

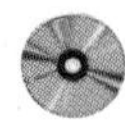

**Videodisc**

**Biology: The Dynamics of Life**
Movie: *X Ray of Swallowing*
Disc 2, Side 1, Ch. 31

## Moving Digested Food into Body Cells

Food digested in the small intestine is ready to be used by the body cells. Food cannot stay in the small intestine and do the body any good. It must be carried to all body cells by the blood. How does most food get out of the small intestine and into the blood? Food gets out of the small intestine and into the blood mainly by diffusion. Diffusion is a process that you studied in Chapter 2.

The inside surface of the small intestine helps absorb food molecules. The small intestine is a long, hollow tube much like a garden hose. The main difference between your small intestine and a hose is the inside lining. The lining of your small intestine is not smooth. Figure 10-5 shows that the small intestine has many tiny, fingerlike parts covering its entire inside surface. The fingerlike parts on the lining of the small intestine are called **villi** (VIHL i). Each villus contains blood vessels that carry digested food.

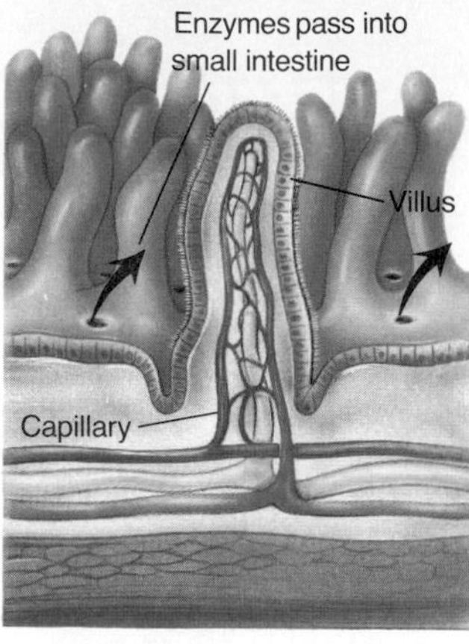


**Figure 10-5** The surface area of the small intestine is increased by many hundreds of villi.

Once inside your blood vessels, digested food is carried by the blood to all body cells. How does digested food get from your blood vessels into your body cells? This happens through a special kind of diffusion. Figure 10-6 shows how food gets into the blood vessels.

The villi allow your small intestine to absorb more digested food than if it were smooth. Why? Villi increase the intestinal surface that comes in contact with digested food. With more intestinal surface, more digested food can pass into the blood.

### Mini Lab

**How Do Villi Aid Absorption?**

**Experiment:** Compare how one, two, three, and four folded paper towels absorb water. Dip each paper towel into a beaker of water. Record the volume of water left in the beaker. *For more help, refer to the* ***Skill Handbook,*** *pages 704-705.*

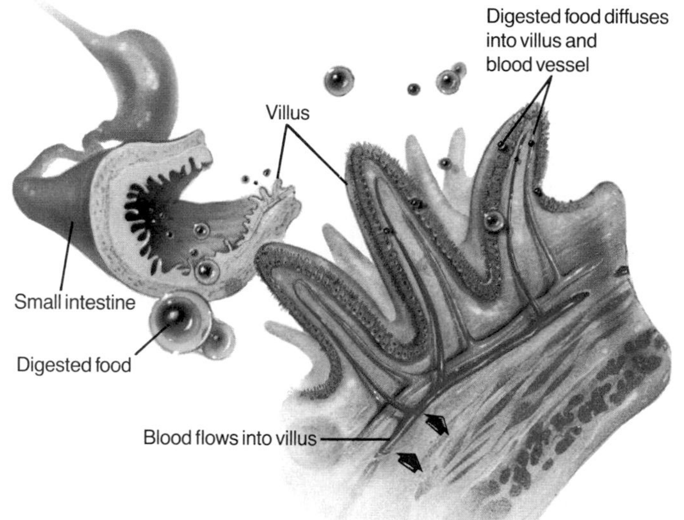


**Figure 10-6** Digested food particles pass into the blood through the walls of the villi.



### OPTIONS

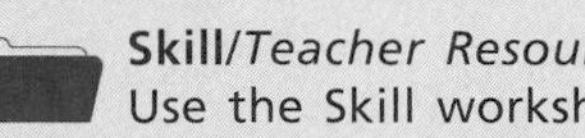

**Skill**/*Teacher Resource Package,* p. 10. Use the Skill worksheet shown here to reinforce digestion of nutrients.

**How Do Digestive System Lengths Compare?**/ *Lab Manual,* pp. 79-82. Use this lab as an extension to the digestive system

**SKILL 10**

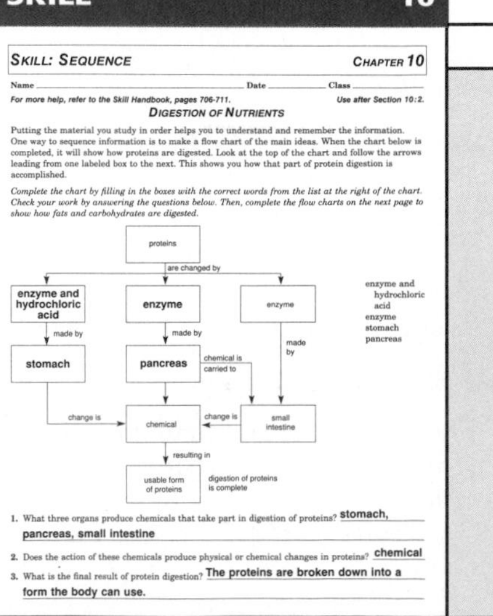

**SKILL: SEQUENCE** — **CHAPTER 10**

Name ______ Date ______ Class ______

*For more help, refer to the Skill Handbook, pages 706-711.* *Use after Section 10:2.*

**DIGESTION OF NUTRIENTS**

Putting the material you study in order helps you to understand and remember the information. One way to sequence information is to make a flow chart of the main ideas. When the chart below is completed, it will show how proteins are digested. Look at the top of the chart and follow the arrows leading from one labeled box to the next. This shows you how that part of protein digestion is accomplished.

*Complete the chart by filling in the boxes with the correct words from the list at the right of the chart. Check your work by answering the questions below. Then, complete the flow charts on the next page to show how fats and carbohydrates are digested.*



1. What three organs produce chemicals that take part in digestion of proteins? **stomach, pancreas, small intestine**
2. Does the action of these chemicals produce physical or chemical changes in proteins? **chemical**
3. What is the final result of protein digestion? **The proteins are broken down into a form the body can use.**

Consider this example. Trapeze artists may use a small net during their act. If they should fall, the chances of landing in this net are small. If they use a larger net, however, they have a better chance of landing in the net. A similar thing happens with the small intestine. The larger the surface of the intestine, the better the chances are that food molecules will come into contact with it. The more food that contacts the surface of the intestine, the more food that passes through the intestine and is picked up by the blood.

## Digestion in Other Animals

How does the human digestive system compare with those of other animals? Different kinds of animals eat different kinds of food. The kind of food eaten is related to the animal's digestive system. Animals that eat plants, such as cows and rabbits, usually have long digestive systems. Animals that eat meat, such as cats and wolves, usually have shorter digestive systems. Plant eaters have longer digestive systems because plants are harder to digest than meat. The longer digestive system gives the food more time to change into a usable form.

**Why do plant eaters have long digestive systems?**

Like humans, many other animals have digestive systems with two openings. For example, the earthworm also has a mouth and an anus, Figure 10-7. Also like humans, the earthworm's digestive system has different organs that have different jobs. Some of the organs are the sites of chemical changes. Other organs produce physical changes in foods. Dogs, birds, snakes, fish, insects, and squid also have digestive systems with two openings.

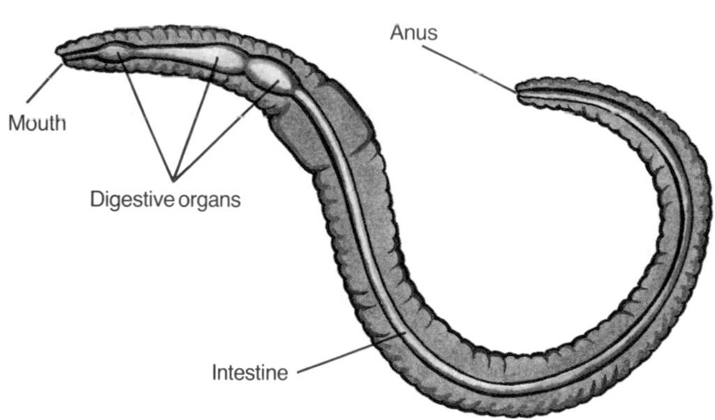


**Figure 10-7** There are different organs in the earthworm's digestive system that cause physical or chemical changes in food.



## TEACH

### Guided Practice

Have students write down their answers to the margin question: Plants are difficult to digest.

### Check for Understanding

Ask: Did all the animal phyla studied in Chapters 7 and 8 have a digestive system? Which phyla had one or two openings to the system?

### Reteach

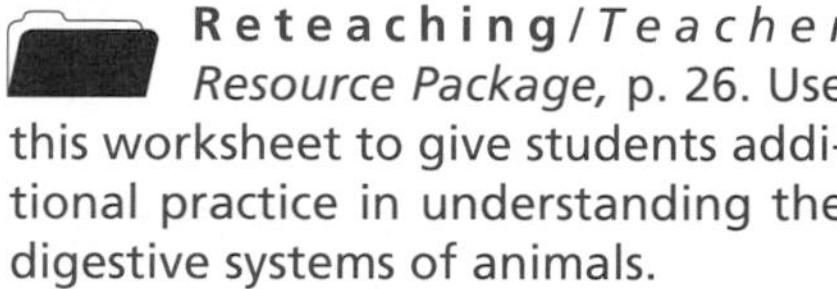

**Reteaching**/*Teacher Resource Package,* p. 26. Use this worksheet to give students additional practice in understanding the digestive systems of animals.

### Connection: Math

Have students make a pie graph showing the amount of time food spends in each digestive organ. Times in minutes are: mouth–2; esophagus–1; stomach–240; small intestine–720; large intestine–240.

Have students divide each time by the total time (1200 minutes) and multiply the result by 360 (total number of degrees in the pie). Students can use a protractor to mark the degrees around the circumference, then they can draw a line from each mark to the center.

### Portfolio

Prepare two diagrams. Show an animal on each – one with one opening in the digestive system and another with two. Have students use Figures 10-7 and 10-8 as a guide and use a series of short arrows to trace the pathway of food in and out of these animals. They should use colored pencils to draw arrows that show the pathway of digested foods into those cells that absorb digested food. Title the diagrams and label the arrows.

## RETEACHING 26

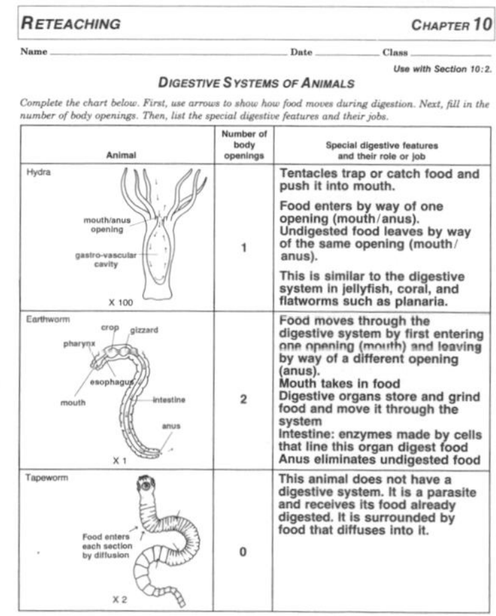

RETEACHING CHAPTER 10

Name ______ Date ______ Class ______

Use with Section 10:2.

DIGESTIVE SYSTEMS OF ANIMALS

*Complete the chart below. First, use arrows to show how food moves during digestion. Next, fill in the number of body openings. Then, list the special digestive features and their jobs.*

| Animal | Number of body openings | Special digestive features and their role or job |
|---|---|---|
| Hydra (mouth/anus opening; gastro-vascular cavity; X 100) | 1 | Tentacles trap or catch food and push it into mouth. Food enters by way of one opening (mouth/anus). Undigested food leaves by way of the same opening (mouth/anus). This is similar to the digestive system in jellyfish, coral, and flatworms such as planaria. |
| Earthworm (crop; gizzard; pharynx; esophagus; mouth; intestine; anus; X 1) | 2 | Food moves through the digestive system by first entering one opening (mouth) and leaving by way of a different opening (anus). Mouth takes in food Digestive organs store and grind food and move it through the system Intestine: enzymes made by cells that line this organ digest food Anus eliminates undigested food |
| Tapeworm (Food enters each section by diffusion; X 2) | 0 | This animal does not have a digestive system. It is a parasite and receives its food already digested. It is surrounded by food that diffuses into it. |

## OPTIONS

### Science Background

Bacteria within the digestive systems of animals such as cows, sheep, camels, and deer enable these animals to digest plant matter or the cellulose of plant cell walls. Thus, these animals are able to receive considerable nutrition from these foods. Humans cannot benefit from plant matter because we lack these bacteria.

**How Does the Fish Digestive System Work?**/*Lab Manual,* pp. 83-86. Use this lab as a comparison to the human digestive system.

**Videodisc**

**STV: Human Body**

*Stomach Lining 1*

**48630**

## TEACH

### Independent Practice

Study Guide/*Teacher Resource Package,* p. 59. Use the Study Guide worksheet shown at the bottom of this page for independent practice.

**Cooperative Learning:** Divide the class into two groups: patients and doctors. Have each patient describe his/her digestive complaint to a doctor. The doctor should then diagnose the problem and suggest treatment.

### MOTIVATION/Brainstorming

How many commercials on TV advertise products to cure digestive problems? Have students name the products and then discuss them.

### ACTIVITY/Demonstration

Add a few drops of dilute hydrochloric acid to several thicknesses of paper. Have students note how rapidly the paper is destroyed by the acid. This simulates the nature of an ulcer.

Repeat the demonstration only this time cover the paper with petroleum jelly (vaseline) before adding the acid. Note the slower action time for the acid. Petroleum jelly simulates the mucus lining of the stomach.

### Guided Practice

Have students write down their answers to the margin question on page 217: Excess food from the stomach can back up into the esophagus.

### Concept Development

▶ Explain to students that new research suggests that stomach ulcers are caused by a certain type of bacteria. If this is true, ulcers may be treatable with antibiotics.

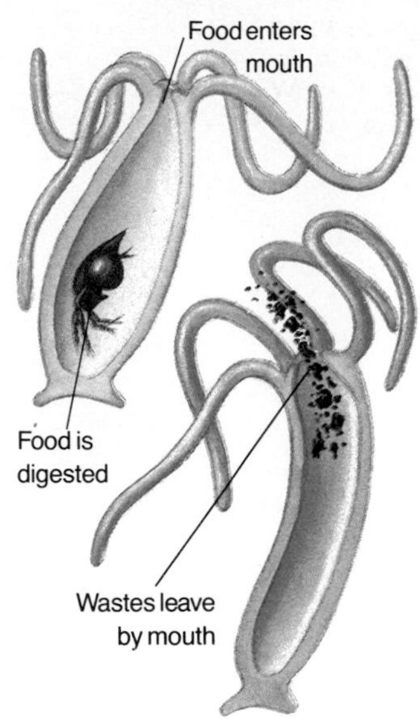


**Figure 10-8** The single opening in a hydra acts as both mouth and anus. How are nutrients absorbed by a hydra? into the body cells by diffusion

Some animals have digestive systems with only one opening. The hydra is a small, simple animal with only one opening. Its digestive process is simple. The opening serves as both a mouth and an anus as shown in Figure 10-8. Its stomach is just a hollow sac within the animal's body. What other animals you studied in Chapter 7 have one opening to the digestive system? jellyfish and planarian

A few animals have no digestive system. Animals such as tapeworms are examples. Tapeworms live inside the digestive systems of other animals. Digested food passes into their bodies by diffusion. It may not be a very good arrangement for the animal in whose body the tapeworm lives, but it certainly works for the tapeworm.

## Problems of the Digestive System

The digestive system, like any body system, may have problems. You may have heard of ulcers or heartburn. An ulcer (UL sur) is a sore or hole on the inside of either the stomach or the small intestine. Figure 10-9 shows a cut-away view of the inside of a normal stomach and the same view of a stomach with an ulcer. The ulcer is caused by the stomach lining being digested, or "eaten" away. Remember that enzymes and stomach acids are found in the stomach. These enzymes and acids cause the ulcer.

Normally, however, the stomach or intestine does not digest itself. There is usually a chemical covering on the inside. This covering is called mucus (MYEW kus). **Mucus**

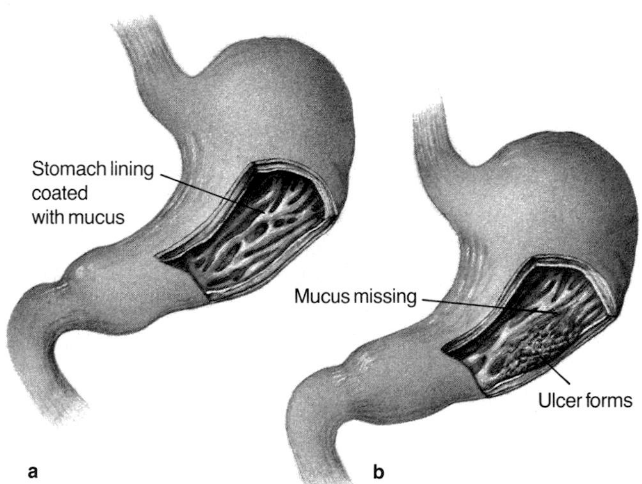


**Figure 10-9** A healthy stomach (a) is coated with mucus. A stomach with an ulcer (b) can be very painful.



## OPTIONS

### ACTIVITY/Challenge

Have students research other problems of the digestive system (appendicitis, liver cancer, gallstones, diarrhea).

### Science Background

Ulcers and heartburn are only two of the more common problems associated with digestion. Neither is very serious unless the ulcer results in internal bleeding or perforates the stomach wall resulting in infection of the peritoneal cavity.

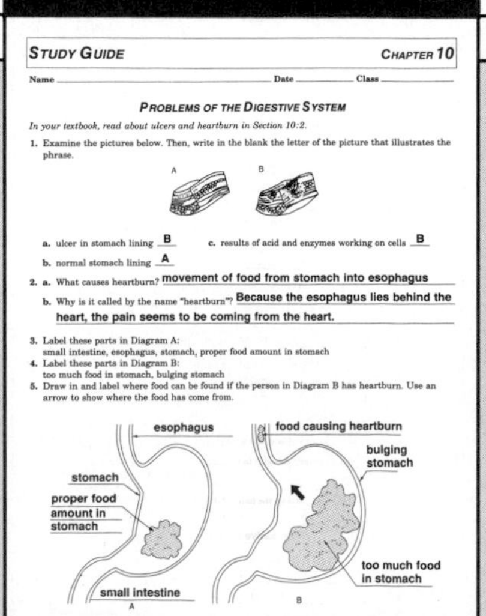

**STUDY GUIDE 59**

STUDY GUIDE — CHAPTER 10

Name ______ Date ______ Class ______

PROBLEMS OF THE DIGESTIVE SYSTEM

*In your textbook, read about ulcers and heartburn in Section 10:2.*

1. Examine the pictures below. Then, write in the blank the letter of the picture that illustrates the phrase.

A B

a. ulcer in stomach lining B
b. normal stomach lining A
c. results of acid and enzymes working on cells B

2. a. What causes heartburn? movement of food from stomach into esophagus
   b. Why is it called by the name "heartburn"? Because the esophagus lies behind the heart, the pain seems to be coming from the heart.
3. Label these parts in Diagram A: small intestine, esophagus, stomach, proper food amount in stomach
4. Label these parts in Diagram B: too much food in stomach, bulging stomach
5. Draw in and label where food can be found if the person in Diagram B has heartburn. Use an arrow to show where the food has come from.

esophagus
stomach
proper food amount in stomach
small intestine
A
food causing heartburn
bulging stomach
too much food in stomach
B

**Figure 10-10** Eating too many rich or acidic foods can cause heartburn. Which food is more acidic, a pickle, or a potato? a pickle

is a thick, sticky material that protects the stomach and intestinal linings from enzymes and stomach acid. It's the same sticky, thick material that lines your nose.

Heartburn is a problem caused by stomach acids moving into the esophagus. Most of the time, stomach acids stay in the stomach. They then pass into the small intestine. When stomach acids move into the esophagus, they cause a burning feeling. However, nothing is being "burned." The problem also has nothing to do with the heart. The esophagus lies behind the heart. The pain feels like it is coming from the heart, even though it isn't.

Eating too much at one time may result in heartburn, Figure 10-10. The stomach can hold only so much food. If too much is eaten, some may back up into the esophagus.

### Bio Tip

**Health:** Aspirin and ethyl alcohol are absorbed directly into the bloodstream through the stomach.

**Why does eating too much cause heartburn?**

## Check Your Understanding

6. Beginning with the mouth, describe the path food takes as it moves through the digestive system.
7. What causes an ulcer?
8. Identify the differences between the digestive process of a human and a hydra.
9. **Critical Thinking:** How do antacids reduce heartburn?
10. **Biology and Reading:** Name three organs that are not part of the digestive tube. How do these organs aid the digestive process?

## STUDY GUIDE 60

**STUDY GUIDE** CHAPTER 10

Name ___ Date ___ Class ___

**VOCABULARY**

*Review the new words used in Chapter 10. Use the terms below to fill in the blanks in the sentences that follow.*

| | | | | |
|---|---|---|---|---|
| mucus | digestive system | digestion | stomach | saliva |
| villi | physical change | esophagus | small intestine | enzymes |
| bile | chemical change | pancreas | gall bladder | hydrochloric |
| liver | salivary glands | appendix | large intestine | acid |

1. Changing of food into a usable form is called **digestion**.
2. **Enzymes** speed up the rate of chemical change.
3. **Saliva** helps digest carbohydrates in the mouth.
4. **Bile** helps break down fats.
5. The tube that connects the mouth to the stomach is the **esophagus**.
6. **Salivary glands** are located under the tongue and behind the jaw.
7. The **pancreas** is an organ that makes three different enzymes.
8. The **gallbladder** stores bile.
9. Most digestion and absorption of food takes place in the **small intestine**.
10. The **large intestine** removes water from undigested food.
11. The **appendix** is a small fingerlike part located where the small and large intestines meet.
12. The breaking of food into small pieces is **physical change**.
13. Protein digestion begins in the **stomach**.
14. Bile is made in the **liver**.
15. Fingerlike parts on the lining of the small intestine are called **villi**.
16. **Chemical change** turns food into a form that cells can use.
17. The **digestive system** is a group of organs that take in food and change it into a form the body can use.
18. The chemical **hydrochloric acid** is made by the stomach.
19. **Mucus** protects the stomach and intestinal linings.

## RETEACHING 27

**RETEACHING** CHAPTER 10

Name ___ Date ___ Class ___

*Use after Section 10:3.*

**PROBLEMS OF THE DIGESTIVE SYSTEM**

*Label the following items on the drawings below: mucus layer, stomach ulcer, esophagus, stomach, and small intestine. Then answer the questions.*

esophagus
small intestine
stomach
mucus layer
stomach ulcer

1. What food does the stomach digest? **proteins**
2. What kinds of chemicals does the stomach produce? **enzymes and hydrochloric acid**
3. What is the function of the mucus lining in the stomach? **The mucus protects the lining of the stomach from enzymes and stomach acid.**
4. What is an ulcer? **An ulcer is a sore, or hole, on the inside of the stomach or intestine.**
5. What causes a person with an ulcer to feel pain? **The acids and enzymes begin to digest the stomach lining where the ulcer is because the lining is not protected by the mucus there.**
6. What is heartburn? **Heartburn occurs when stomach acids move up into the esophagus.**

## TEACH

### Independent Practice

**Study Guide**/*Teacher Resource Package,* p. 60. Use the Study Guide shown at the bottom of this page for independent practice.

### Check for Understanding

Have students respond to the first three questions in Check Your Understanding.

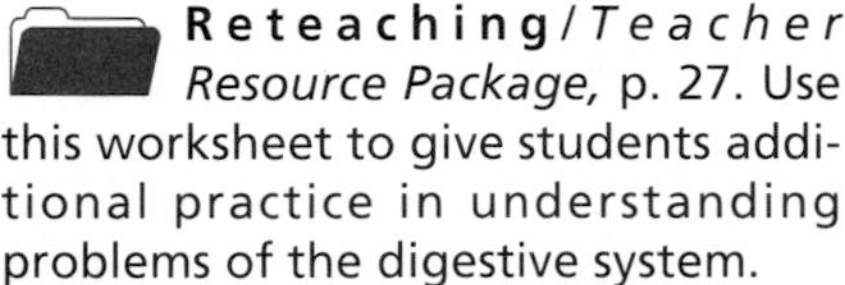

### Reteach

**Reteaching**/*Teacher Resource Package,* p. 27. Use this worksheet to give students additional practice in understanding problems of the digestive system.

**Extension:** Assign Critical Thinking, Biology and Reading, or some of the **OPTIONS** available with this lesson.

## 3 APPLY

### Software

*The Digestion Simulator,* Cambridge Development Laboratories.

## 4 CLOSE

### Audiovisual

Show the filmstrip *The Digestive System,* Clearvue, Inc.

### Answers to Check Your Understanding

6. mouth, esophagus, stomach, small intestine, large intestine, and anus
7. enzymes and stomach acid digest the stomach lining
8. human – two openings, specialized organs; hydra – one opening, no digestive organs
9. They neutralize the stomach acid.
10. salivary glands – produce digestive enzymes and mucus; pancreas – produces digestive enzymes; liver – produces digestive enzymes and bile; gallbladder – stores bile made by the liver

# CHAPTER 10 REVIEW

## Summary

Summary statements can be used by students to review the major concepts of the chapter.

## Key Science Words

All boldfaced terms from the chapter are listed.

## Testing Yourself

### Using Words

1. physical change
2. esophagus
3. digestion
4. pancreas
5. saliva
6. bile
7. gallbladder
8. villi
9. salivary gland
10. enzyme

### Finding Main Ideas

11. pp. 210-212; Food spends four hours in the stomach, twelve hours in the small intestine, and five hours in the large intestine.
12. p. 208; Only one type of enzyme can digest each food type.
13. pp. 204-205; In a factory, raw materials are brought in and changed as they pass from one place to the next. New products are formed. The digestive system does the same thing. The digestive system is like a long tube. Food moves through it in one direction.
14. p. 208; Teeth grind the food, causing a physical change. Salivary glands add saliva to the food, causing chemical changes.
15. p. 215; Different organs can carry out different jobs. Certain organs can carry out physical changes, while others can carry out chemical changes.
16. p. 216; An ulcer is formed when the stomach lining is digested by enzymes and acid.

### Using Main Ideas

17. Food passes through the mouth, esophagus, stomach, small intestine, large intestine, and anus.
18. The stomach makes enzymes and hydrochloric acid.



## Chapter 10

## Review

### Summary

**10:1 The Process of Digestion**

1. The digestive system—a long, hollow tube—takes in food and changes it into a form usable to body cells.
2. Physical changes grind and break down food. Chemicals called enzymes speed up the changing of food into a usable form.
3. Enzymes help to change fats, proteins, and carbohydrates into usable forms by speeding up chemical changes.

**10:2 The Human Digestive System**

4. Food passes from mouth to esophagus to stomach to small intestine to large intestine. Villi in the small intestine increase surface area and help with the movement of digested food into body cells.
5. Digestive systems in animals differ in length, number of openings, and specialization.
6. An ulcer is formed by a part of the stomach lining being digested. Mucus protects the stomach and intestine linings from enzymes and acid. Heartburn is caused by stomach acids that enter the esophagus. Eating too much at one time may result in heartburn.

### Key Science Words

appendix (p. 212)
bile (p. 210)
chemical change (p. 206)
digestion (p. 205)
digestive system (p. 204)
enzyme (p. 206)
esophagus (p. 209)
gallbladder (p. 210)
hydrochloric acid (p. 210)
large intestine (p. 212)
liver (p. 210)
mucus (p. 217)
pancreas (p. 210)
physical change (p. 205)
saliva (p. 208)
salivary gland (p. 208)
small intestine (p. 210)
stomach (p. 209)
villi (p. 214)

### Testing Yourself

**Using Words**

*Choose the word from the list of Key Science Words that best fits the definition.*

1. breaking of large food pieces into small pieces
2. connects mouth to stomach
3. process of changing food into a usable form
4. organ that makes three different enzymes
5. changes carbohydrates in the mouth
6. liquid that causes a physical change in fats
7. organ that stores bile
8. fingerlike parts on the lining of the small intestine
9. makes saliva
10. chemical that speeds up the changing of food into a usable form



19. A tapeworm has no openings for digestion. A hydra has one opening for digestion, and a human has two openings for digestion.
20. Mucus protects the stomach and intestinal linings from enzymes and stomach acid.
21. A physical change is the breaking of large food pieces into smaller food pieces by the grinding and chewing of the teeth in the mouth. A chemical change is the changing of starch in bread into glucose by enzymes in saliva.
22. Heartburn is caused by eating more food than the stomach can hold. The extra food and stomach acid back up into the esophagus.

# Review

## Testing Yourself *continued*

**Finding Main Ideas**

*List the page number where each main idea below is found. Then, explain each main idea.*

11. how much time food spends in each of the following: the stomach, small intestine, and large intestine
12. why so many different enzymes are needed for digestion
13. how the digestive system is like a factory
14. the body parts that help with physical and chemical changes in the mouth
15. how a digestive system with two openings is more specialized than a digestive system with one opening
16. what causes a stomach ulcer

**Using Main Ideas**

*Answer the questions by referring to the page number after each question.*

17. What are the organs through which food passes on its trip through the digestive system? (pp. 209, 210)
18. What two chemicals are made in the stomach? (p. 210)
19. What is an example of an animal with no openings to its digestive system? One opening? Two openings? (pp. 215, 216)
20. What is the function of mucus in the digestive system? (p. 217)
21. What is an example of a physical change and a chemical change that takes place in the digestive system? (pp. 205, 206)
22. What causes heartburn? (p. 217)

## Skill Review

*For more help, refer to the **Skill Handbook,** pages 704-719.*

1. **Form a hypothesis:** Which food would enzymes work on more rapidly: a slice of bread or an equal mass of bread crumbs? Explain by using the term *surface area.*
2. **Use numbers:** Determine from Figure 10-3 how long it takes for a meal to pass into the stomach and out into the large intestine.
3. **Interpret diagrams:** Use the information shown in Figure 10-3 to make a bar graph illustrating the amount of time food generally remains in the major digestive organs.
4. **Classify:** Make a list of five foods you have eaten during the past 18 hours. Classify each food according to the main nutrient it contains.

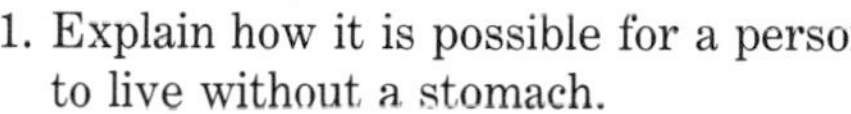


## Finding Out More

**Critical Thinking**

1. Explain how it is possible for a person to live without a stomach.
2. Explain why it is impossible for a person to live without a pancreas.

**Applications**

1. How do the number and shape of teeth compare among adult animals, such as dogs, cats, cows, and horses? Relate tooth shape with diet.
2. Plants contain the substance cellulose. Report on the relationship between this substance and the length of a plant eater's digestive system.

**Chapter Review**/*Teacher Resource Package,* pp. 50-51.

**Chapter Test**/*Teacher Resource Package,* pp. 52-54.

**Chapter Test**/*Computer Test Bank*

## Skill Review

1. bread crumbs; The bread crumbs are already in smaller pieces than a slice of bread and have more surface area on which the enzymes can react.
2. into the stomach – less than 1 minute; stomach – 4 hours; small intestine – 12 hours; digestion in the small intestine takes the longest time.
3. Student graphs should show: stomach – 4 hours; small intestine – 12 hours; large intestine – 5 hours.
4. Students should classify foods as protein, fat, carbohydrate, vitamin, mineral, or water.

## Finding Out More

### Critical Thinking

1. Most of the absorption of food takes place in the small intestine. If the person eats foods that are already partially broken down, he or she can live without a stomach.
2. Enzymes from the pancreas are necessary to digest carbohydrates, proteins, and fats in the small intestine.

### Applications

1. Animals that eat plants have a few large, flat teeth with broad grinding surfaces. Meat eaters have a larger number of smaller teeth with pointed surfaces for ripping and tearing flesh. Animals that eat both plants and animals have both pointed teeth and grinding molars.
2. Cellulose is very difficult to break down. The more cellulose a plant eater eats, the longer its digestive system.

# Circulation

## PLANNING GUIDE

| CONTENT | TEXT FEATURES | TEACHER RESOURCE PACKAGE | OTHER COMPONENTS |
|---|---|---|---|
| (1 day)<br>11:1 The Process of Circulation<br>Pickup and Delivery<br>Circulation in Animals | Idea Map, p. 222<br>Check Your Understanding, p. 223 | Reteaching: *Circulation in Animals,* p. 28<br>Study Guide: *The Process of Circulation,* p. 61 | **Laboratory Manual:**<br>*How Do Hearts of Different Animals Compare?* p. 91<br>**STVS:** *Modeling Blood Flow,* Human Biology (Disc 7, Side 1) |
| (3 days)<br>11:2 The Human Heart<br>Heart Structure<br>The Pumping of the Heart<br>Heart Valves<br>The Jobs of the Heart | Lab 11-1: *Pulse Rate,* p. 224<br>Skill Check: *Understand Science Words,* p. 227<br>Mini Lab: *How Do the Pulses of Males and Females Compare?* p. 228<br>Check Your Understanding, p. 231 | Enrichment: *Tracing a Heartbeat,* p. 11<br>Reteaching: *Heartbeat Sequence and Events,* p. 29<br>Critical Thinking/Problem Solving: *How Does Circulation Vary with Different Activities?* p. 11<br>Focus: *Electrocardiograph,* p. 21<br>Study Guide: *The Human Heart,* pp. 62-63<br>Lab 11-1: *Pulse Rate,* pp. 41-42 | **Laboratory Manual:**<br>*How Does the Heart Work?* p. 87<br>Color Transparency 11: *Blood Flow Between Heart and Lungs*<br>**STVS:** *Testing Heart Valves,* Human Biology (Disc 7, Side 1) |
| (2 days)<br>11:3 Blood Vessels<br>Arteries<br>Veins<br>Capillaries | Lab 11-2: *Blood Pressure,* p. 235<br>Check Your Understanding, p. 234 | Reteaching: *Blood Vessels and Blood Flow,* p. 30<br>Skill: *The Circulatory System,* p. 12<br>Study Guide: *Blood Vessels,* p. 64<br>Lab 11-2: *Blood Pressure,* pp. 43-44 | **STVS:** *Measuring Blood Pressure,* Human Biology (Disc 7, Side 1) |
| (2 days)<br>11:4 Problems of the Circulatory System<br>High Blood Pressure<br>Heart Attack<br>Preventing Heart Problems | Technology: *Angioplasty,* p. 238<br>Check Your Understanding, p. 239 | Application: *Blood Pressure,* p. 11<br>Reteaching: *Circulatory System Problems,* p. 31<br>Study Guide: *Problems of the Circulatory System,* p. 65<br>Study Guide: *Vocabulary,* p. 66<br>Transparency Master: *Circulation Pathway in Humans,* p. 49 | **STVS:** *Through-the-Skin Heart Drug,* Human Biology (Disc 7, Side 1) |
| Chapter Review | Summary<br>Key Science Words<br>Testing Yourself<br>Finding Main Ideas<br>Using Main Ideas<br>Skill Review | **ASSESSMENT RESOURCES**<br>Chapter Review, pp. 55-56<br>Chapter Test, pp. 57-59<br>Performance Assessment in the Biology Classroom<br>Alternate Assessment in the Science Classroom<br>Computer Test Bank | |

### GLENCOE TECHNOLOGY

**Science and Technology Videodisc Series,** *Modeling Blood Flow,* Human Biology (Disc 7, Side1)
*Testing Heart Valves,* Human Biology (Disc 7, Side1)
*Measuring Blood Pressure,* Human Biology (Disc 7, Side1)
*Through-the-Skin Heart Drug,* Human Biology (Disc 7, Side1)

## MATERIALS NEEDED

| LAB 11-1, p. 224 | LAB 11-2, p. 235 | MARGIN FEATURES |
|---|---|---|
| petri dish<br>water<br>wall clock<br>live earthworm | plastic squeeze bottle<br>pan or sink<br>water<br>meterstick<br>rubber stopper and tube assembly | Skill Check, p. 227<br>dictionary<br>Mini Lab, p. 228<br>pencil<br>paper |

## OBJECTIVES

For more information about National Science Standards, see page 5T.

| SECTION | OBJECTIVE | CORRELATION of QUESTIONS to OBJECTIVES: CHECK YOUR UNDERSTANDING | CHAPTER REVIEW | TRP CHAPTER REVIEW | TRP CHAPTER TEST |
|---|---|---|---|---|---|
| 11:1 National Science Stds: UCP.1, UCP.5, B.4, C.1, C.5 | 1. **Identify** the circulatory system. | 1, 2, 4 | 7, 13, 21 | 1, 28 | 5, 12 |
| | 2. **Compare** the circulatory systems of earthworms, insects, and humans. | 3, 5 | 13 | 4 | 10, 13, 14 |
| 11:2 National Science Stds: UCP.1, UCP.2, UCP.3, UCP.4, UCP.5, B.4, C.1 | 3. **Describe** how blood is pumped through the heart. | 6, 9 | 4, 8, 17, 25 | 29, 37 | 45, 48, 49, 50 |
| | 4. **Explain** what causes the sounds the heart makes. | 7 | 14, 18, 19 | 2, 21, 24, 25, 30 | 7, 24, 46, 47 |
| | 5. **Trace** the pathway of blood through the left and right sides of the heart. | 8, 10 | 5, 16, 23, 26 | 3, 15, 16, 17, 18, 19, 20, 22, 23, 26, 27, 32, 38 | 1, 16, 17, 18, 19, 21, 22, 23, 25, 26, 27, 28, 29, 30, 31, 32, 39, 40 |
| 11:3 National Science Stds: UCP.1, UCP.2, UCP.3, UCP.5, B.4, C.1, C.5 | 6. **Discuss** the importance of arteries. | 11 | 3, 11 | 9, 10, 13, 31 | 6, 33, 34, 35, 38, 39, 41, 43 |
| | 7. **List** four traits of veins. | 12 | 1, 19 | 8, 12 | 2, 36, 44 |
| | 8. **Give the function** of capillaries. | 13 | 2 | 11, 14 | 4, 37, 38, 42 |
| 11:4 National Science Stds: UCP.2, UCP.3, UCP.4, UCP.5, B.4, F.1 | 9. **Explain** how blood pressure is measured. | 16, 19 | 24 | 7, 35, 36 | 8, 9, 11 |
| | 10. **Trace** the events before and after a heart attack. | 17, 20 | 6, 9, 15 | 6, 33, 34 | 3 |
| | 11. **Relate** ways to help prevent and correct heart problems. | 18 | 20, 22 | 5 | 15 |

# Circulation

## CHAPTER OVERVIEW

### Key Concepts

The role of the circulatory system is described using examples from different animal phyla. The anatomy of the human circulatory system is presented. Comparisons are made in the role of the heart, arteries, veins, and capillaries, as well as the blood pressure within them. Problems in the human circulatory system and the role of modern technology in correcting them are also considered.

### Key Science Words

| | |
|---|---|
| aorta | hypertension |
| artery | pulmonary artery |
| atrium | pulmonary vein |
| bicuspid valve | semilunar valve |
| blood pressure | tricuspid valve |
| capillary | valve |
| cholesterol | vein |
| circulatory system | vena cava |
| coronary vessel | ventricle |
| heart attack | |

### Skill Development

In Lab 11-1, students will **form a hypothesis, calculate, interpret data,** and **make and use tables** in comparing worm and human pulse rates. In Lab 11-2, they will **formulate models, form a hypothesis,** and **measure in SI** as they study blood pressure in veins and arteries. In the Skill Check on page 227, students will **understand** the **science word** *semilunar.* In the Mini Lab on page 228, students will **design an experiment** to compare pulse rates of males and females.

### Bridging

The fate of digested food as studied in the previous chapter can now be followed. Once food passes out of the small intestine, it moves through the body via the bloodstream.

**CHAPTER PREVIEW**

### Chapter Content

Review this outline for Chapter 11 before you read the chapter.



### Skills in this Chapter

The skills that you will use in this chapter are listed below.

- In **Lab 11-1,** you will form a hypothesis, calculate, interpret data, and make and use tables. In **Lab 11-2,** you will formulate models, form a hypothesis, and measure in SI.
- In the **Skill Checks,** you will understand science words.
- In the **Mini Labs,** you will design an experiment.

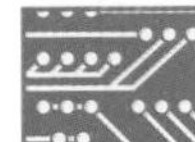

## TECH PREP

For Tech Prep activities in this chapter of the Teacher Wraparound Edition, see especially the Portfolios on pages 222 and 231, the Motivation on page 226, the Connection on page 229, the Activity on page 237, and the Student Journal on page 238.

See also the Glencoe Homepage at **www.glencoe.com**

Chapter 11

# Circulation

Look at the photo of the crowded highway shown on the left. All the cars and trucks may have special pickup or delivery jobs. People in cars may be on their way to work, to school, or even going on vacation. Trucks may be picking up or delivering goods to homes, warehouses, or stores.

The photograph below shows a close-up view of human blood moving through a blood vessel. Blood cells are round. Notice that the walls of a blood vessel are very thin.

How are the jobs of cars and trucks on a highway similar to the jobs of blood cells? How is the job of the highway similar to the job of a blood vessel? Can you think of ways in which they are different?

**Try This!**

**What kinds of sounds does the heart make?** Use a stethoscope to listen to the sounds your heart makes. How many different sounds do you hear? Is there a pattern to the sounds? How many sounds do you hear during a one-minute period?

*inter*NET CONNECTION

For more information about the material in this chapter, follow the link for the chapter on the Glencoe Homepage at **http://www.glencoe.com**

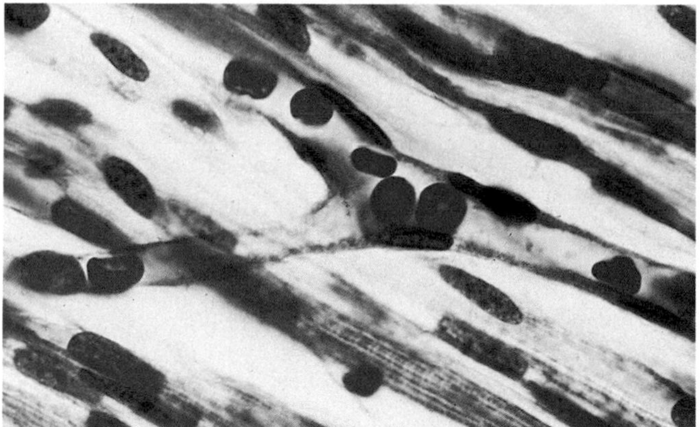

Blood vessel



## ASSESSMENT PLANNER

**Portfolio**

Strategies on the following pages represent student products that can be placed into a best-work portfolio: pp. 222, 231.

**PERFORMANCE ASSESSMENT**

Skill Check, p. 227

Mini Lab, p. 228

Lab, pp. 224, 235

**CONTENT ASSESSMENT**

Check for Understanding, pp. 223, 230, 231, 234, 238, 239

Chapter Review, pp. 240-241

**GROUP ASSESSMENT**

Opportunities for group assessment occur with Cooperative Learning Strategies.

**Student Journal**

Strategies on the following pages represent opportunities for writing in a Student Journal: pp. 230, 238.

## GETTING STARTED

### Using the Photos

Use the photo analogy to explain the functions of delivery and pickup performed by the circulatory system. Direct students' attention to the fact that the highway system and the extensive network of blood vessels within the body are very similar in terms of function.

### MOTIVATION/Try This!

**What kinds of sounds does the heart make?** Allow students to listen to their heart sounds for several minutes with stethoscopes. If stethoscopes are not available, students can feel their heartbeat using their hand placed over their chest. Student observations will differ. However, the pattern they **observe** will consist of two distinct sounds occurring close together in a continuous rhythm.

### Chapter Preview

Have students study the chapter outline before they begin to read the chapter. They should note the overall organization of the chapter, that the process of circulation is studied first in a very general way, then the human heart is covered in detail. Blood vessel structure and function is next, and finally, problems of the circulatory system and how they may be solved or reduced are studied.

### Misconception

Most students do not realize that the human heart is: (a) a muscle (b) hollow inside (c) receives no nourishment or oxygen from the blood being pumped through it (d) divided into two separate sides.

## 11:1 The Process of Circulation

## PREPARATION

### Materials Needed

Copy the Study Guide and Reteaching worksheets in the *Teacher Resource Package.*

▶ Order earthworms for Lab 11-1.

### Key Science Words

circulatory system

### Process Skills

In Lab 11-1, students will form a hypothesis, calculate, interpret data, and make and use tables.

## 1 FOCUS

▶ The objectives are listed on the student page. Remind students to preview these objectives as a guide to this numbered section.

### Idea Map

Have students use the idea map as a study guide to major concepts.

### Guided Practice

Have students write down their answers to the margin question: blood, blood vessels, and heart.

### Independent Practice

**Study Guide**/*Teacher Resource Package,* p. 61. Use the Study Guide worksheet shown at the bottom of this page for independent practice.

**Portfolio**

Have students draw diagrams to show how they would assemble two models: one illustrating a closed circulatory system and the other an open circulatory system. Students should title their diagrams and label what parts in each model correspond to a real animal.

## 11:1 The Process of Circulation

**Objectives**

1. **Identify** the circulatory system.
2. **Compare** the circulatory systems of earthworms, insects, and humans.

**Key Science Words**

circulatory system

Your body is made of billions of cells. Each cell is like a tiny factory that must be supplied with certain chemicals. You have a pickup and delivery system in your body to supply these chemicals.

### Pickup and Delivery

**What makes up the circulatory system?**

Your body's pickup and delivery system is the circulatory system. Your **circulatory system** is made of your blood, blood vessels, and heart. Blood delivers needed materials, such as oxygen, water, and food, to your cells. Blood picks up the cells' waste products, such as carbon dioxide gas.

Blood travels through a series of tubes called blood vessels. Your heart serves as a pump to help move, or circulate, blood through these vessels.

### Circulation in Animals

Not all animals have circulatory systems. Simple animals, such as a sponge, sea anemone, or hydra, do not have circulatory systems. The bodies of these animals are just a few cells thick. Water moves freely in and out of their bodies. Nearly every cell in the animal's body comes in contact with the water. Oxygen and nutrients in the water diffuse into the cells. Wastes, such as carbon dioxide, diffuse out of the cells into the water. The water in which these simple animals live acts as a pickup and delivery service for them.

**Study Tip:** Use this idea map as a study guide to comparing circulation patterns in different animals.

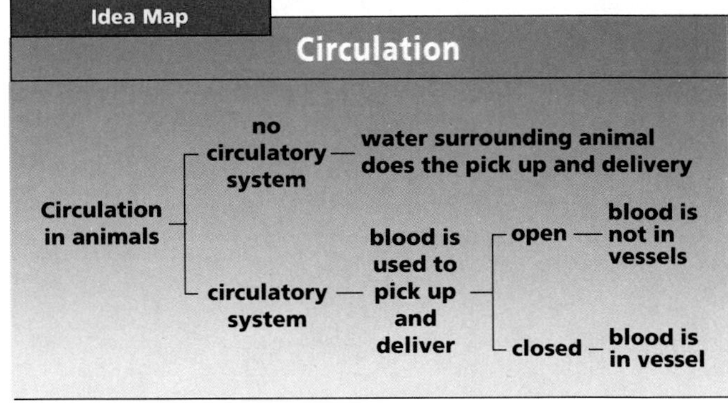




## OPTIONS

### Science Background

Point out that circulatory systems evolved as body size increased. With a circulatory system, all internal cells can come in contact with blood. The animal is less dependent on an aqueous environment for diffusion of materials into and out of the animal.

**STUDY GUIDE 61**

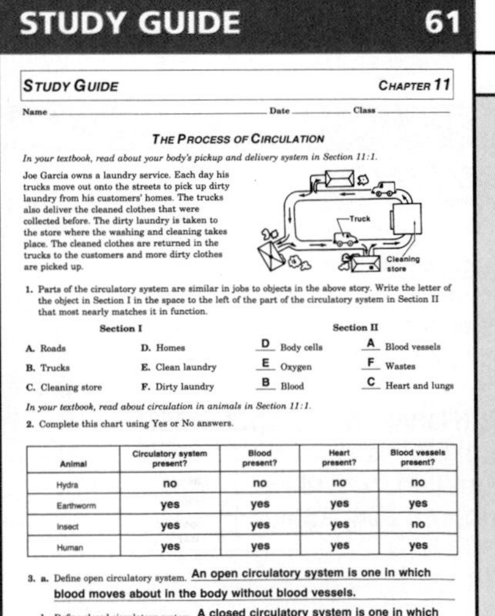

**STUDY GUIDE** CHAPTER 11

Name ______ Date ______ Class ______

**THE PROCESS OF CIRCULATION**

*In your textbook, read about your body's pickup and delivery system in Section 11:1.*

Joe Garcia owns a laundry service. Each day his trucks move out onto the streets to pick up dirty laundry from his customers' homes. The trucks also deliver the cleaned clothes that were collected before. The dirty laundry is taken to the store where the washing and cleaning takes place. The cleaned clothes are returned in the trucks to the customers and more dirty clothes are picked up.



1. Parts of the circulatory system are similar in jobs to objects in the above story. Write the letter of the object in Section I in the space to the left of the part of the circulatory system in Section II that most nearly matches it in function.

| Section I | | Section II | |
|---|---|---|---|
| A. Roads | D. Homes | D Body cells | A Blood vessels |
| B. Trucks | E. Clean laundry | E Oxygen | F Wastes |
| C. Cleaning store | F. Dirty laundry | B Blood | C Heart and lungs |

*In your textbook, read about circulation in animals in Section 11:1.*

2. Complete this chart using Yes or No answers.

| Animal | Circulatory system present? | Blood present? | Heart present? | Blood vessels present? |
|---|---|---|---|---|
| Hydra | no | no | no | no |
| Earthworm | yes | yes | yes | yes |
| Insect | yes | yes | yes | no |
| Human | yes | yes | yes | yes |

3. a. Define open circulatory system. An open circulatory system is one in which blood moves about in the body without blood vessels.

b. Define closed circulatory system. A closed circulatory system is one in which blood moves about in the body within blood vessels.

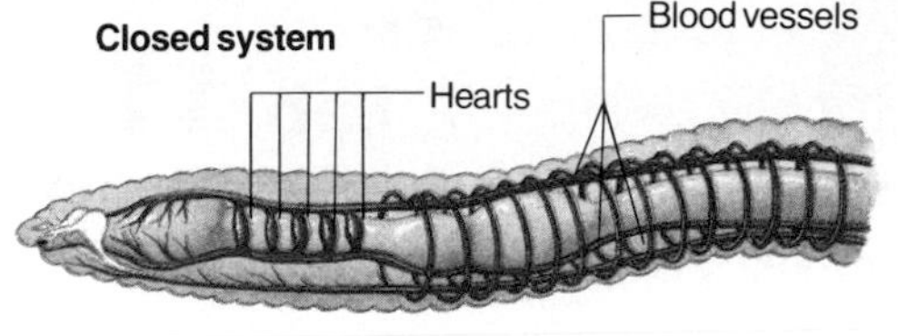


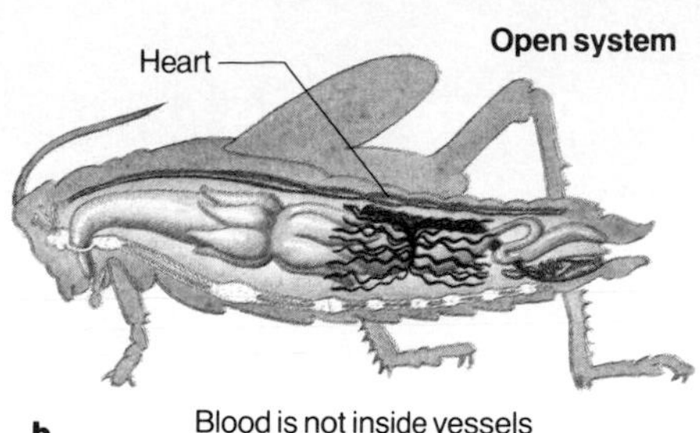


Complex animals, such as earthworms and insects, do have circulatory systems. These animals have many layers of cells and tissues. Oxygen and nutrients would not reach each cell of the animal's body if there were no circulatory system. Look carefully at Figure 11-1 and you will notice an important difference between the circulatory systems of earthworms and insects. Notice that the earthworm has blood vessels and hearts, Figure 11-1b. The circulatory system in the earthworm is said to be closed. Blood is inside vessels. The insect has a heart but no blood vessels contained within. Blood in the insect's body moves about without traveling in vessels. The insect's circulatory system is said to be open. Think about your own body. Your body is made up of many complex systems of organs. You have a closed circulatory system to supply all your cells with the nutrients they need.

**Figure 11-1** Earthworms have a closed circulatory system (a). Insects have an open circulatory system (b).

**What kind of circulatory system does an insect have?**

## Check Your Understanding

1. What are the two main jobs of the circulatory system?
2. How do animals that lack a circulatory system get their needed materials?
3. (a) How are the circulatory systems of insects and earthworms alike?
   (b) How are the systems different?
4. **Critical Thinking:** What advantage does a closed circulatory system have over an open circulatory system? (HINT: What would happen if an animal with an open circulatory system got a cut?)
5. **Biology and Reading:** Describe the relationship between the complexity of an animal's body and the type of circulatory system it has.



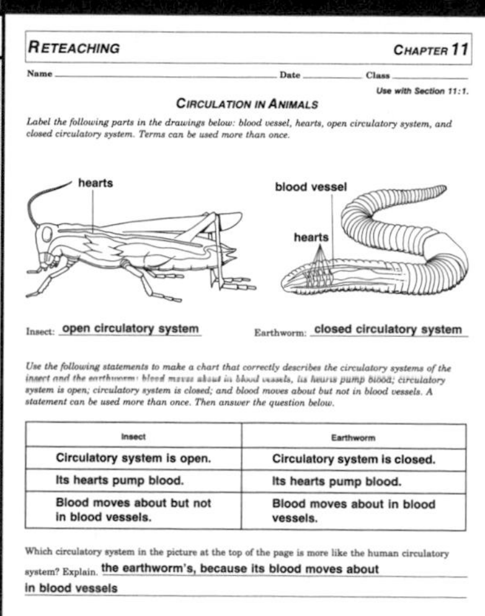

**RETEACHING 28**

**RETEACHING** **CHAPTER 11**

Name ____ Date ____ Class ____

Use with Section 11:1.

**CIRCULATION IN ANIMALS**

*Label the following parts in the drawings below: blood vessel, hearts, open circulatory system, and closed circulatory system. Terms can be used more than once.*

hearts

blood vessel

hearts

Insect: open circulatory system

Earthworm: closed circulatory system

*Use the following statements to make a chart that correctly describes the circulatory systems of the insect and the earthworm: blood moves about in blood vessels; its hearts pump blood; circulatory system is open; circulatory system is closed; and blood moves about but not in blood vessels. A statement can be used more than once. Then answer the question below.*

| Insect | Earthworm |
|---|---|
| Circulatory system is open. | Circulatory system is closed. |
| Its hearts pump blood. | Its hearts pump blood. |
| Blood moves about but not in blood vessels. | Blood moves about in blood vessels. |

Which circulatory system in the picture at the top of the page is more like the human circulatory system? Explain. the earthworm's, because its blood moves about in blood vessels

## OPTIONS

**How Do Hearts of Different Animals Compare?**/*Lab Manual*, pp. 91-94. Use this lab to give students additional practice with circulation in other animals.

# 2 TEACH

### Guided Practice

Have students write down their answers to the margin question: open; a heart but no blood vessels.

### Check for Understanding

Have students respond to the first three questions in Check Your Understanding.

### Reteach

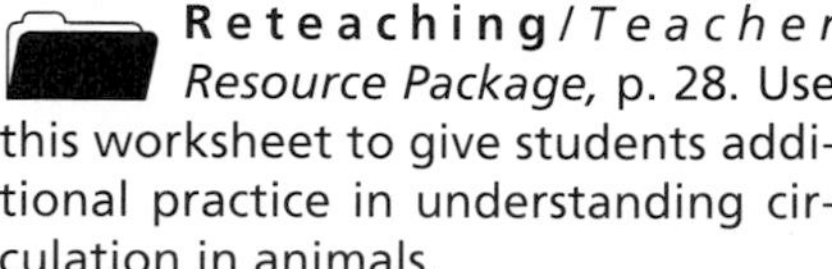

**Reteaching**/*Teacher Resource Package*, p. 28. Use this worksheet to give students additional practice in understanding circulation in animals.

**Extension:** Assign Critical Thinking, Biology and Reading, or some of the **OPTIONS** available with this lesson.

# 3 APPLY

### Divergent Question

You discover a very small 10-celled animal living in the ocean. Explain why you are certain that this animal has no circulatory system.

# 4 CLOSE

### Summary

Prepare a list of what exactly is delivered to and picked up from your cells by the circulatory system.

### Answers to Check Your Understanding

1. It carries oxygen and nutrients to and removes wastes from cells.
2. Nutrients are obtained from the water by diffusion.
3. (a) They both have the same job and have hearts. (b) Insects have an open system with no blood vessels, while earthworms have a closed system with blood vessels.
4. A closed circulatory system protects the flow of blood through an organism's body.
5. Organisms with closed circulatory systems are more complex.

## Lab 11-1 Pulse Rate

### Overview

Students will be able to observe the pulsing of blood in an earthworm. This will allow them to count and compare the pulse of an earthworm to their own.

**Objectives:** Upon completion of the lab, students will be able to (1) **state** that the circulatory system for worms and humans is closed, (2) **calculate** an average pulse for earthworms and humans, (3) **compare** the pulse of an earthworm to their own.

**Time Allotment:** 1 class period

### Preparation

▶ Earthworms may be ordered from a supply house, dug from the soil, or purchased locally from a bait shop.

▶ **Alternate Materials:** Any shallow dish may be used for petri dishes.

**Lab 11-1 worksheet**/*Teacher Resource Package,* pp. 41-42.

### Teaching the Lab

▶ The blood vessel in the earthworm can be observed near the tail.

▶ **Troubleshooting:** You may want to review averaging with students before starting the lab.

**Cooperative Learning:** Divide the class into groups of two or three. For more information, see pp. 22T-23T in the Teacher Guide.

**Lab 11-1 Computer Program**/*Teacher Resource Package,* p. viii. Use the computer program shown at the bottom of this page as you teach this lab.

### ASSESSMENT

**Performance:** Have students use their average pulse rate and the following average rates per minute to construct a bar graph: hummingbird = 400, elephant = 24. Students should use the graph to describe a pattern between pulse and body size.

# Pulse Rate

# Lab 11-1

## Problem: How do a worm's pulse and your pulse compare?

### Skills

form a hypothesis, calculate, interpret data, make and use tables

### Materials

live earthworm
petri dish with cover
water, room temperature
wall clock

### Procedure

1. Copy the data table.
2. Place the live earthworm in a petri dish. **CAUTION:** *Keep the earthworm moist by stroking now and again with wet fingers.*
3. Look for the blood vessel along the top surface of the earthworm, Figure A.
4. Count and record the pulse of the earthworm for exactly 15 seconds.
5. Multiply your answer by 4 to get the worm's pulse for one minute. Record your result in the data table.
6. Repeat steps 4 and 5 three more times.
7. **Calculate** an average pulse for the earthworm for one minute. (Add all 4 pulse rates and then divide the total by 4.)
8. Return the earthworm to your teacher.
9. **Form a hypothesis** about how your average pulse rate will compare with that of the earthworm.
10. Use Figure B as a guide or ask your teacher to help you locate your pulse. Take your own pulse for one minute.
11. Record your results in your data table.
12. Repeat step 10 three more times and calculate an average pulse for yourself.

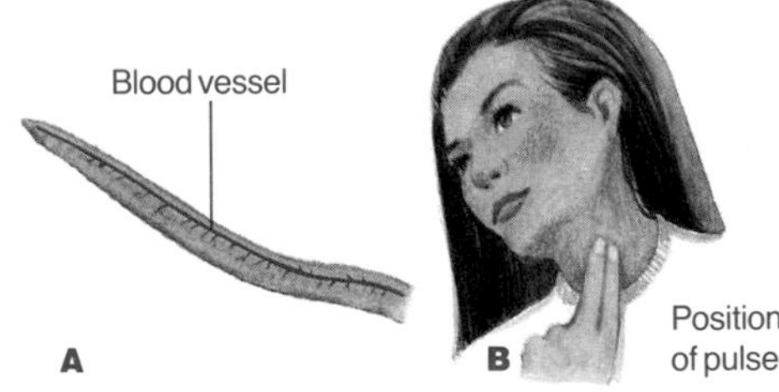


| TRIAL | WORM PULSE FOR 15 SECONDS | WORM PULSE FOR 1 MINUTE | YOUR PULSE FOR 1 MINUTE |
|---|---|---|---|
| 1 | 8 | 32 | 82 |
| 2 | 6 | 24 | 80 |
| 3 | 6 | 24 | 78 |
| 4 | 7 | 28 | 78 |
| Total | — | 108 | 318 |
| Average | — | 27 | 79.5 |

### Data and Observations

1. What was the earthworm's average pulse over one minute?
2. What was your average pulse?

### Analyze and Apply

1. **Check your hypothesis:** Is your hypothesis supported by your data? Why or why not?
2. What is a pulse?
3. Why is it important to repeat measurements in an experiment?
4. **Apply:** Your heartbeat rate slows down when you relax. Explain how this change will affect your pulse rate.

### Extension

**Design an experiment** to determine the effect different kinds of exercise have on your pulse rate.



## ANSWERS

### Data and Observations

1. about 27 pulses a minute
2. about 79.5 pulses a minute

### Analyze and Apply

1. Students who hypothesized that their pulse rates would be higher will answer yes.
2. the rhythmic movement of blood through an artery
3. Measurements vary with each sample taken. The more measurements taken, the less chance of error.
4. As your heartbeat rate slows down, so will your pulse rate. Blood is pumped from your heart into your arteries where you check your pulse rate.

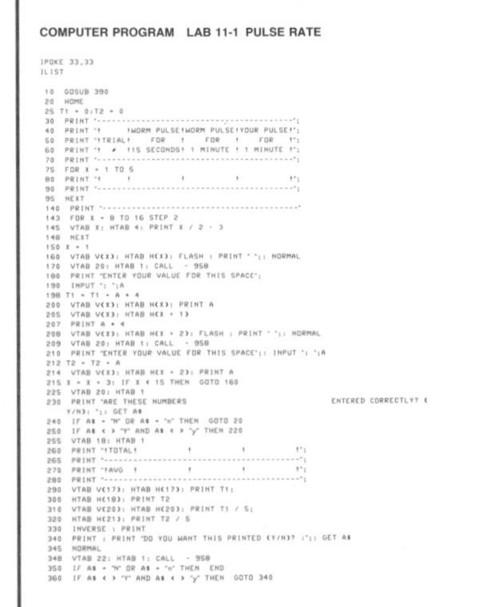

COMPUTER PROGRAM VIII

COMPUTER PROGRAM LAB 11-1 PULSE RATE

```
]POKE 33,33
]LIST

10  GOSUB 390
20  HOME
25 T1 = 0:T2 = 0
30  PRINT "--------------------------------------";
40  PRINT "!      !WORM PULSE!WORM PULSE!YOUR PULSE!";
50  PRINT "!TRIAL!    FOR    !    FOR    !   FOR    !";
60  PRINT "!  #   !15 SECONDS! 1 MINUTE ! 1 MINUTE !";
70  PRINT "--------------------------------------";
75  FOR X = 1 TO 5
80  PRINT "!      !          !          !          !";
90  PRINT "--------------------------------------";
95  NEXT
140  PRINT "--------------------------------------"
143  FOR X = 8 TO 16 STEP 2
145  VTAB X: HTAB 4: PRINT X / 2 - 3
148  NEXT
150 X = 1
160  VTAB V(X): HTAB H(X): FLASH : PRINT " ";: NORMAL
170  VTAB 20: HTAB 1: CALL  - 958
180  PRINT "ENTER YOUR VALUE FOR THIS SPACE";
190  INPUT ": ";A
198 T1 = T1 + A * 4
200  VTAB V(X): HTAB H(X): PRINT A
205  VTAB V(X): HTAB H(X + 1)
207  PRINT A * 4
208  VTAB V(X): HTAB H(X + 2): FLASH : PRINT " ";: NORMAL
209  VTAB 20: HTAB 1: CALL  - 958
210  PRINT "ENTER YOUR VALUE FOR THIS SPACE";: INPUT ": ";A
212 T2 = T2 + A
214  VTAB V(X): HTAB H(X + 2): PRINT A
215 X = X + 3: IF X < 15 THEN  GOTO 160
225  VTAB 20: HTAB 1
230  PRINT "ARE THESE NUMBERS                    ENTERED CORRECTLY? (
    Y/N): ";: GET A$
240  IF A$ = "N" OR A$ = "n" THEN  GOTO 20
250  IF A$ < > "Y" AND A$ < > "y" THEN 220
255  VTAB 18: HTAB 1
260  PRINT "!TOTAL!          !          !          !";
265  PRINT "--------------------------------------";
270  PRINT "!AVG  !          !          !          !";
280  PRINT "--------------------------------------";
290  VTAB V(17): HTAB H(17): PRINT T1;
300  HTAB H(18): PRINT T2
310  VTAB V(20): HTAB H(20): PRINT T1 / 5;
320  HTAB H(21): PRINT T2 / 5
330  INVERSE : PRINT
340  PRINT : PRINT "DO YOU WANT THIS PRINTED (Y/N)? :";: GET A$
345  NORMAL
348  VTAB 22: HTAB 1: CALL  - 958
350  IF A$ = "N" OR A$ = "n" THEN  END
360  IF A$ < > "Y" AND A$ < > "y" THEN  GOTO 340
```

# 11:2 The Human Heart

Can you believe there is a pump within your chest that started working before you were born? It sounds impossible, but that is what your heart is.

**Objectives**

3. **Describe** how blood is pumped through the heart.
4. **Explain** what causes the sounds the heart makes.
5. **Trace** the pathway of blood through the left and right sides of the heart.

**Key Science Words**

atrium
ventricle
artery
vein
valve
tricuspid valve
bicuspid valve
semilunar valve
vena cava
pulmonary artery
pulmonary vein
aorta

## Heart Structure

The heart is a muscle that pumps blood through the body. Figure 11-2b shows a photograph of the human heart.

Use Figure 11-2a to learn the parts of the human heart. There are three things to note about this diagram of the heart.

1. It is a cut-away view of the inside of the heart.
2. The colors bright red and blue represent blood within the heart.
3. The uncolored parts are heart muscle.

Because it has two sides, you can think of the heart as two separate pumps, one on the left and one on the right. Figure 11-2a is divided into left and right sides.

You might think that the sides of the heart are backward. The sides are labeled as if the heart were inside someone facing you. So, the heart diagram's right and left sides are opposite to your right and left sides.

Each side of the heart has a small chamber on the top and a large chamber on the bottom. The small, top chambers of the heart are **atria** (AY tree uh). The atria are divided into the right atrium and the left atrium. The large bottom chambers of the heart are called the **ventricles** (VEN trih kulz). The ventricles also are divided into right and left sections.

a

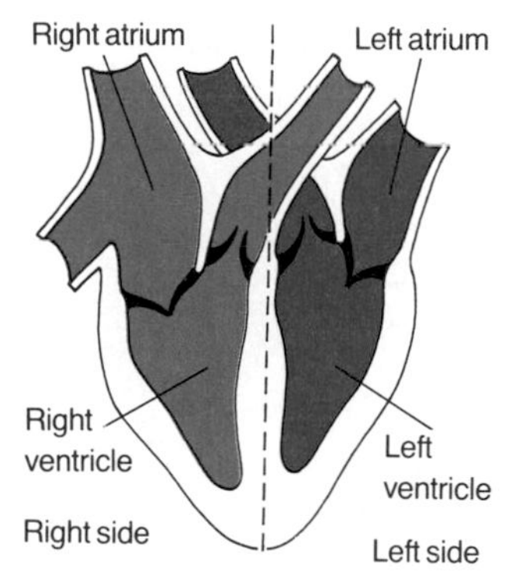


b

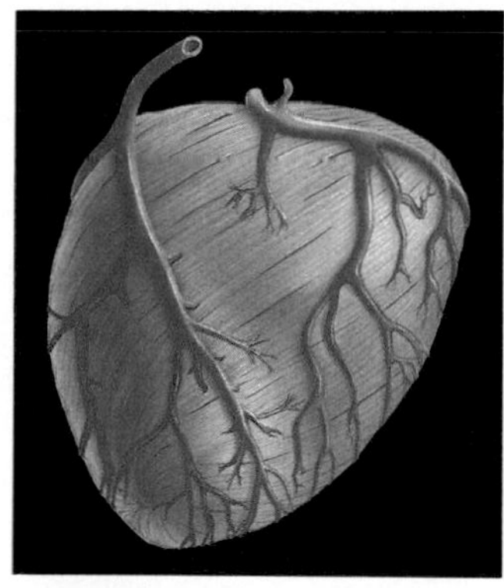

**Figure 11-2** The human heart has four compartments. Which compartments of the heart does blood enter from veins? the atria



## 11:2 The Human Heart

## PREPARATION

### Material Needed

Make copies of the Focus, Enrichment, Critical Thinking/Problem Solving, Study Guide, Reteaching, and Lab worksheets in the *Teacher Resource Package.*

▶ Order a preserved heart from a biological supply house for the Focus activity.

### Key Science Words

atrium
ventricle
artery
vein
valve
tricuspid valve
bicuspid valve
semilunar valve
vena cava
pulmonary artery
pulmonary vein
aorta

### Process Skills

In the Skill Check, they will understand science words.

## 1 FOCUS

▶ The objectives are listed on the student page. Remind students to preview these objectives as a guide to this numbered section.

### ACTIVITY/Display

Purchase a preserved heart (sheep or pig) from a supply house or use a model to introduce students to the size and general anatomy of this organ.

**Focus**/*Teacher Resource Package,* p. 21. Use the Focus transparency shown at the bottom of this page as an introduction to the action of the human heart.

**FOCUS 21**

FOCUS — CHAPTER 11

Name ______ Date ______ Class ______

ELECTROCARDIOGRAPH

A heartbeat begins with a rhythmical electrical discharge in one part of the heart. The electric current then spreads throughout the heart during the rest of the heartbeat. An instrument called an electrocardiograph can record this current. The patient in the picture has wires attached to his chest and arms and to an electrocardiograph.

What might a doctor find out about a patient from his or her electrocardiograph?

## OPTIONS

### Science Background

The heart is composed of cardiac muscle. The middle and main part of the heart is the myocardium. The inner part is the endocardium. The sac that surrounds the heart is the pericardium.

# 2 TEACH

### MOTIVATION/Demonstration

Have students take their pulse for one minute while resting. Put their pulse rates on the chalkboard and note the wide variation seen within the class. Point out to students that pulse rate is a reflection of heartbeat. One heartbeat corresponds to one surge of blood through the arteries. We call this surge or wave of blood a pulse.

### Concept Development

▶ Remind students that when one looks at the front view of a person, the left and right sides are reversed.

▶ Remind students that all diagrams in this chapter are front views.

▶ The first letter of *artery* and the first letter of *away* both begin with an a. This may help students to remember that arteries carry blood away from the heart while veins carry blood in the opposite direction.

**Analogy:** Ask students if they have ever trapped water between their fists and shot it out at someone. This illustrates the idea graphically because squeezing together the fists requires the action of muscles.

### Guided Practice

Have students write down their answers to the margin question: from the atria.

**Videodisc**

STV: Human Body Vol. 1
*Circulatory and Respiratory Systems*
Unit 1, Side 1
*Circulatory and Respiratory Systems (in its entirety)*

**1064-30932**

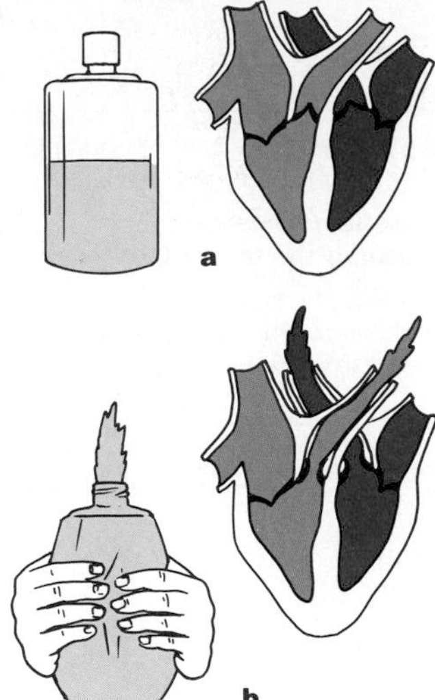


**Figure 11-3** When a bottle or heart is not squeezed, no liquid is pushed out. When a bottle or heart is squeezed, liquid is squirted out.

**How do the ventricles receive blood?**

## The Pumping of the Heart

How does the heart work as a pump? Figure 11-3 shows the heart's squeezing action by comparing it to the squeezing of a plastic bottle. The top picture shows the plastic bottle when it is not squeezed. No liquid is pushed out. The same is true for the heart. If the heart muscle does not squeeze, no blood is pumped. The bottom picture shows the plastic bottle being squeezed and liquid squirting out. The bottom diagram of the heart shows that when muscles of the ventricles squeeze, blood is pushed out.

While a person is resting, the heart pumps 60 to 80 times each minute. After running, the heart may pump 150 times in one minute. Each pump of the heart is called a beat. The structure of the heart allows blood to move through the circulatory system in only one direction. Your heart muscle will pump for more than two billion times during your lifetime!

Figure 11-4 shows several large blood vessels called arteries (ART uh reez) and veins. An **artery** is a blood vessel that carries blood away from the heart. A **vein** is a blood vessel that carries blood back to the heart. The arrows in the diagram show the direction of blood flow.

1. Blood from the veins enters the heart's right and left atria, Figure 11-4a. Neither chamber is pumping.
2. Figure 11-4b shows the atria beginning to pump or squeeze. Right and left ventricles are relaxed and are not pumping. These chambers receive blood from the atria.
3. Figure 11-4c shows the right and left ventricles squeezing blood into two large arteries leading to the body and the lungs. When the ventricles are pumping, the atria are relaxed. Blood returning from the body enters the atria and the entire pumping cycle begins again.

**Figure 11-4** Follow the steps that show the pumping action of the heart.

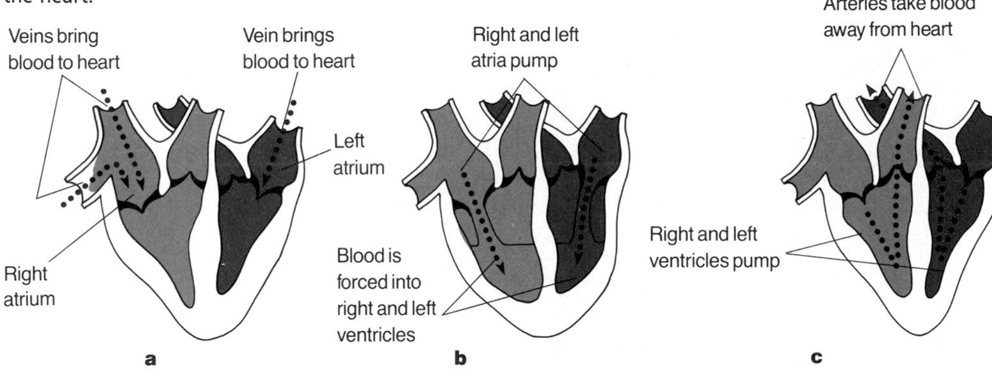




## OPTIONS

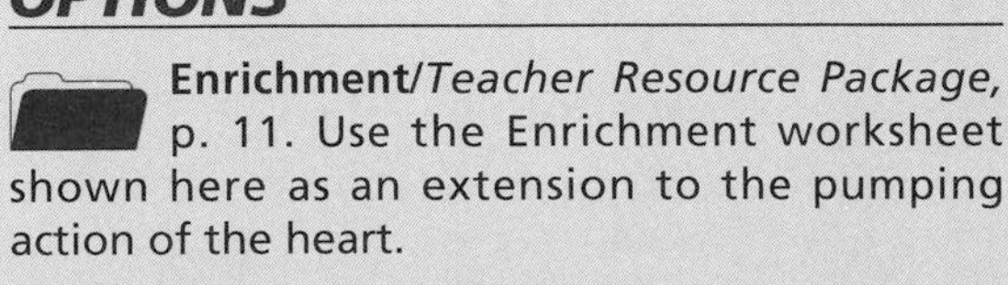

**Enrichment**/*Teacher Resource Package*, p. 11. Use the Enrichment worksheet shown here as an extension to the pumping action of the heart.

**ENRICHMENT 11**

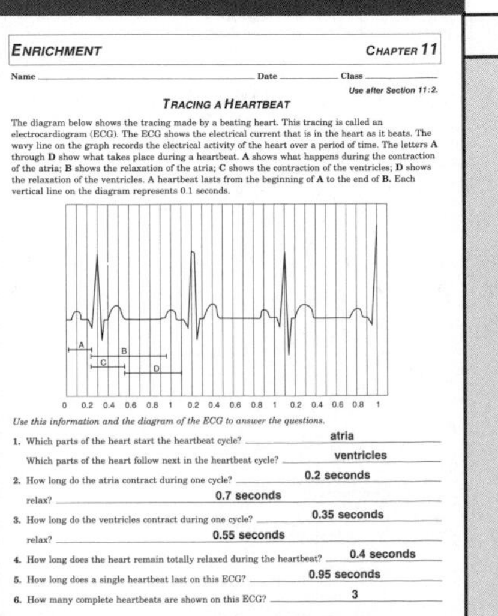

ENRICHMENT CHAPTER 11

Name ____ Date ____ Class ____

*Use after Section 11:2.*

TRACING A HEARTBEAT

The diagram below shows the tracing made by a beating heart. This tracing is called an electrocardiogram (ECG). The ECG shows the electrical current that is in the heart as it beats. The wavy line on the graph records the electrical activity of the heart over a period of time. The letters **A** through **D** show what takes place during a heartbeat. **A** shows what happens during the contraction of the atria; **B** shows the relaxation of the atria; **C** shows the contraction of the ventricles; **D** shows the relaxation of the ventricles. A heartbeat lasts from the beginning of **A** to the end of **B.** Each vertical line on the diagram represents 0.1 seconds.



*Use this information and the diagram of the ECG to answer the questions.*

1. Which parts of the heart start the heartbeat cycle? atria
   Which parts of the heart follow next in the heartbeat cycle? ventricles
2. How long do the atria contract during one cycle? 0.2 seconds
   relax? 0.7 seconds
3. How long do the ventricles contract during one cycle? 0.35 seconds
   relax? 0.55 seconds
4. How long does the heart remain totally relaxed during the heartbeat? 0.4 seconds
5. How long does a single heartbeat last on this ECG? 0.95 seconds
6. How many complete heartbeats are shown on this ECG? 3

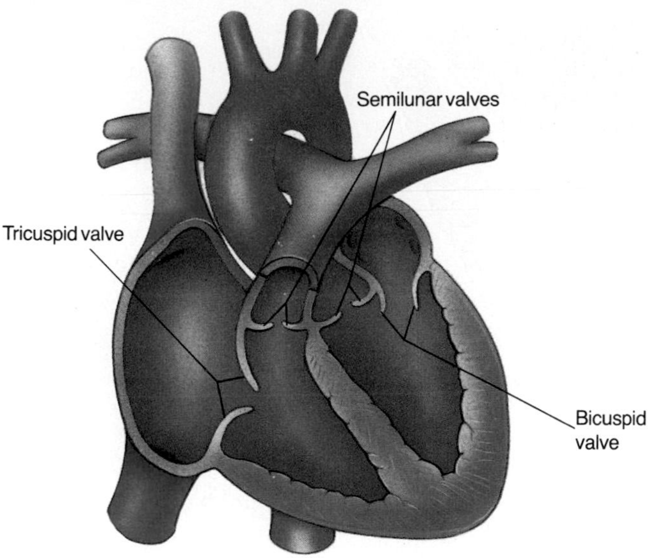


**Figure 11-5** Locate the valves of the heart.

## Heart Valves

If a pump is to work well, it must keep liquid moving in only one direction. A pump has valves that help with this task. So does the heart. **Valves** are flaps in the heart that keep blood flowing in one direction.

Figure 11-5 shows the location of the valves. Notice that there are two sets of valves. One set is between the atria and ventricles. The valve between the right atrium and right ventricle is called the **tricuspid** (tri KUS pud) **valve.** The valve between the left atrium and the left ventricle is called the **bicuspid** (bi KUS pud) **valve.** Note that these two valves are like one-way doors. They only open downward into the ventricles to keep blood flowing in one direction.

The other valves are the semilunar (sem ih LEW nur) valves. **Semilunar valves** are located between the ventricles and their arteries. Note that these two valves are also like one-way doors. They open only in an upward direction away from the ventricles.

Have you ever heard your heart beat? Do you know what causes the sounds you hear? Some people think they are hearing the heart muscle as it pumps. Actually, the sounds you hear are your heart valves closing. These valves, like doors, make sounds when they close. Doesn't a door make a loud sound when it slams shut?

### Skill Check

**Understand science words: semilunar.** The word part *semi* means half. In your dictionary, find three words with the word part *semi* in them. *For more help, refer to the* ***Skill Handbook,*** *pages 706-711.*



## TEACH

### Concept Development

▶ Point out to students that blood on the right and left sides of the heart never mixes directly. Also, it can be noted that the pathway of blood, once within the heart, follows a regular pattern. Blood enters the top chambers, is pumped into the lower chambers, and is then pumped out of the lower chambers to body organs.

▶ Point out to students that a cusp is a flap. The prefix *tri-* refers to three, while the prefix *bi-* refers to two. There are three cusps or sections to the tricuspid valve. The bicuspid valve has two sections or cusps. The term *semilunar* is related to the fact that these valves have a shape similar to a half-moon.

### GLENCOE TECHNOLOGY

**Videodisc**

**Biology: The Dynamics of Life**
Animation: *One-Way Valves*
Disc 2, Side 1, Ch. 34

Animation: Recording Heart Rhythms
Disc 2, Side 1, Ch. 35

**The Secret of Life**
*Circulatory System*

### CRITICAL THINKING 11

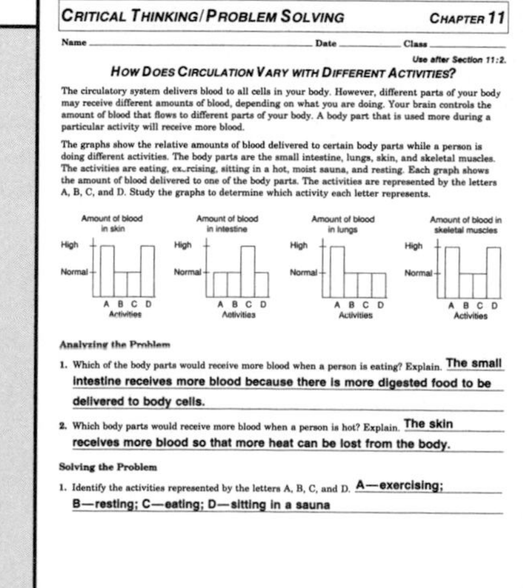

CRITICAL THINKING/PROBLEM SOLVING CHAPTER 11

Name ______ Date ______ Class ______

Use after Section 11:2.

**HOW DOES CIRCULATION VARY WITH DIFFERENT ACTIVITIES?**

The circulatory system delivers blood to all cells in your body. However, different parts of your body may receive different amounts of blood, depending on what you are doing. Your brain controls the amount of blood that flows to different parts of your body. A body part that is used more during a particular activity will receive more blood.

The graphs show the relative amounts of blood delivered to certain body parts while a person is doing different activities. The body parts are the small intestine, lungs, skin, and skeletal muscles. The activities are eating, ex_rcising, sitting in a hot, moist sauna, and resting. Each graph shows the amount of blood delivered to one of the body parts. The activities are represented by the letters A, B, C, and D. Study the graphs to determine which activity each letter represents.



**Analyzing the Problem**

1. Which of the body parts would receive more blood when a person is eating? Explain. The small intestine receives more blood because there is more digested food to be delivered to body cells.
2. Which body parts would receive more blood when a person is hot? Explain. The skin receives more blood so that more heat can be lost from the body.

**Solving the Problem**

1. Identify the activities represented by the letters A, B, C, and D. A—exercising; B—resting; C—eating; D—sitting in a sauna

## OPTIONS

**Critical Thinking/Problem Solving/** *Teacher Resource Package*, p. 11. Use the Critical Thinking/Problem Solving worksheet shown here as an extension to structure and function of heart valves.

## TEACH

### Concept Development

▶ A murmur is an indication that the heart is not working efficiently. If the deficiency is great, heart enlargement may result.

▶ Explain to students that valve closure is not due directly to muscle action. Valves close due to blood pressure changes occurring within the chambers. As the ventricles contract, they force blood up against the underside of the bicuspid and tricuspid valves, thus forcing them to close. As the ventricles relax, blood in the aorta and pulmonary artery falls back toward the ventricles. The semilunar valves fill with blood, which results in their closure.

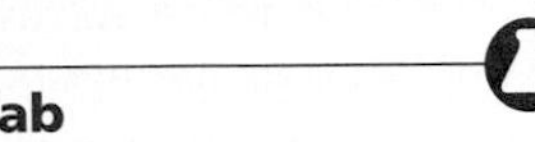

### Mini Lab

Students should include a large enough sample size and should make sure the subjects are all engaging in the same activity (sitting).

### ✓ ASSESSMENT

**Performance:** Have students make a graph comparing the resting heart rates of physically active and nonactive people in the class. Students should list benefits of regular exercise.

### Guided Practice

Have students write down their answers to the margin question: the heart valves closing.

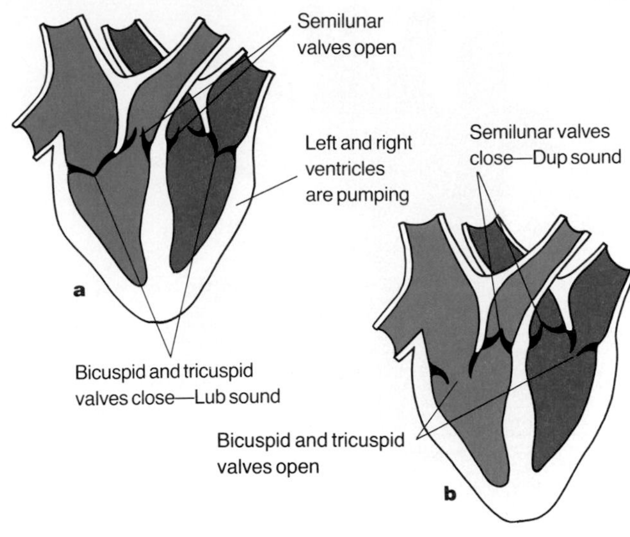


**Figure 11-6** The lub and dup sounds of the heart are caused by the closing of the heart's valves.

### Mini Lab

**How Do the Pulses of Males and Females Compare?**

**Design an experiment:** Design an experiment to compare the pulse rates of males and females. Carry out the experiment. Write a report. *For more help, refer to the* ***Skill Handbook,*** *pages 704-705.*

Because you have two sets of valves in your heart that close at different times, your heart makes two sounds. The first sound is caused by the closing of your bicuspid and tricuspid valves. Figure 11-6a shows these valves closed. The blood can no longer pass from the atria to the ventricles. When your ventricles squeeze, these valves close and cause a "lub" sound. The second sound occurs when your semilunar valves close. These valves close when your ventricles stop squeezing. A "dup" sound is now heard. Figure 11-6b shows the semilunar valves closed. The blood can no longer pass from the ventricles out of the main arteries.

The two heart sounds heard together make a "lub-dup" sound. The tricuspid and bicuspid valves close harder than the semilunar valves. That is why the sounds differ. A doctor listening to your heart can tell if your heart is working normally. A "lub-dup" sound means your valves are working normally. If the doctor hears a "lub-swish-dup" sound, the bicuspid or tricuspid valves are not closing properly. Blood leaking past the valves causes the "swish" sound. The blood moves in the opposite direction from which it should be going. Blood flowing backward through the heart when valves are not closing tightly is called a heart murmur. Which valves are not closing properly if a doctor hears a "lub-dup-swish" sound? Since the "swish" sound follows the "dup" sound, it must be caused by the semilunar valves not closing properly.

**What makes the "lub-dup" sound in your heart?**



## OPTIONS

### ACTIVITY/Challenge

Purchase a preserved heart and allow students to dissect it under your guidance. Make sure that they observe external as well as internal anatomy.

**How Does the Heart Work?**/*Lab Manual,* pp. 87-90. Use this lab to give students additional practice with heart function.

## The Jobs of the Heart

So far we've only seen how your heart pumps. Now let's look at your blood vessels and see how the rest of your circulatory system works. Remember, your heart has a right and left side. Each side has its own job of pumping blood around the body.

**Right Side of the Heart**

The right side of your heart pumps blood only to your lungs. The path the blood follows can be seen by matching numbers 1 through 5 in Figure 11-7 with statements 1 through 5.

1. Blood enters the heart's right side at the right atrium from a large vein called the vena cava. The **vena cava** (VEE nuh • KAY vuh) is the largest vein in the body and carries blood from the body back to the heart. The blue color is used to show that at this point blood contains a lot of carbon dioxide gas and little oxygen gas.
2. Blood is pumped into the right ventricle.
3. From here, blood is pumped into a large artery. This artery is called the pulmonary (PUL muh ner ee) artery. The **pulmonary artery** carries blood away from the heart to the lungs. Because you have two lungs, the pulmonary artery divides in two. Blood will be pumped to both lungs.
4. Blood enters both lungs. In the lungs, the blood picks up oxygen and loses carbon dioxide. Notice that the color of the blood is now bright red. Blood containing a lot of oxygen is shown in bright red.
5. Blood returns to the heart through the pulmonary veins. **Pulmonary veins** carry blood from the lungs to the left side of the heart. The total time needed for blood to travel from your heart to your lungs and back again is only about 10 seconds.

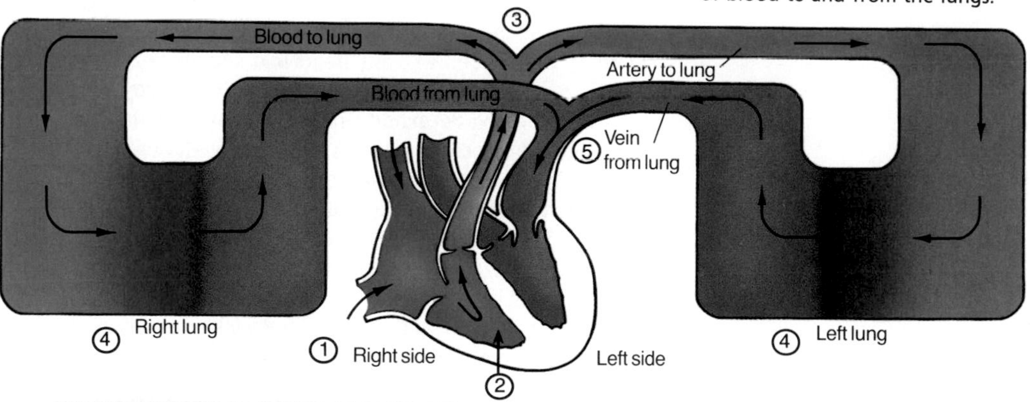


**Figure 11-7** Trace the pathway of blood to and from the lungs.



## TEACH

### Concept Development

▶ Stress the following:
1. Blood enters the heart through a vein.
2. The blood in the right side of the heart contains much carbon dioxide. Pictures of the heart always show blood in this condition as being colored blue.
3. The right side of the heart pumps blood only to the lungs.
4. Blood leaves the heart through an artery.
5. Blood loses its supply of carbon dioxide and gains a supply of oxygen in the lungs.

### Connection: Math

Have students calculate the volume of blood pumped by the average heart in one minute, one hour, and one day. Use the following information: The left ventricle pumps about 70 mL of blood every time it contracts; the heart beats 70 times in one minute.

**Misconception:** Emphasize that deoxygenated blood is not blue; it is dark red. Blue is used to illustrate deoxygenated blood because it contrasts well with red.

**Cooperative Learning:** Divide the class into groups of two students. Give them the following lists of questions: (a) For the right side of the heart, name the chambers, valves, blood vessels that enter the chamber, blood vessels that leave the chamber, condition of blood entering the chamber, condition of blood leaving the chamber, where blood is coming from as it enters the chamber, where blood is going as it leaves the chamber. (b) Give them a second list of questions for the left side of the heart. One team member will ask questions about the left side, the second about the right side. Have the teams continue to ask the questions until all are answered correctly.

### Independent Practice

**Study Guide**/*Teacher Resource Package*, p. 62. Use the Study Guide worksheet shown at the bottom of this page for independent practice.

## OPTIONS

**Blood Flow Between Heart and Lungs**/*Transparency Package*, number 11. Use color transparency number 11 as you teach pulmonary circulation.

**STUDY GUIDE 62**

STUDY GUIDE — CHAPTER 11

Name ____ Date ____ Class ____

THE HUMAN HEART

*In your textbook, read about the structure of the human heart in Section 11:2.*

1. Label the diagram below using these words: right atrium, right ventricle, left atrium, left ventricle, tricuspid valve, semilunar valve (used twice), bicuspid valve, right side of heart, left side of heart

right atrium; left atrium; tricuspid valve; semilunar valve; bicuspid valve; semilunar valve; right ventricle; left ventricle; right side of heart; left side of heart

*In your textbook, read about how the heart works as a pump in Section 11:2.*

2. Complete the following chart by putting checkmarks in the correct columns.

| | A | B |
|---|---|---|
| Atria are relaxed | | ✓ |
| Ventricles are relaxed | ✓ | |
| Atria are pumping | ✓ | |
| Ventricles are pumping | | ✓ |
| Ventricles are receiving blood from atria | ✓ | |
| Atria are receiving blood returning to heart | | ✓ |
| Blood is pumped to body | | ✓ |

## TEACH

### Concept Development

▶ Stress the following:
1. Blood enters the heart through a vein.
2. The blood in the left side of the heart contains much oxygen.
3. The left side of the heart pumps blood to all body organs except the lungs.
4. Blood leaves the heart through an artery.
5. Blood loses its supply of oxygen and gains a supply of carbon dioxide from the body cells.

### Check for Understanding

Provide students with a diagram similar to Figure 11-8. Have them label the following: heart chambers, blood vessels, heart valves. Ask them to draw in arrows that show the direction of blood flow.

### Reteach

Have students draw a circle and divide it into four quarters. Each quarter will represent the chambers of the human heart. Ask students to label all four chambers and draw in the location of bicuspid and tricuspid valves.

### Guided Practice

Have students write down their answers to the margin question: It comes from the lungs where it picked up oxygen.

### Independent Practice

**Study Guide**/*Teacher Resource Package*, p. 63. Use the Study Guide worksheet shown at the bottom of this page for independent practice.

### *Student Journal*

Have students trace a drop of blood within a human body starting and ending at the right atrium. Students should list what pathways the blood will follow and what happens to the blood throughout the journey.

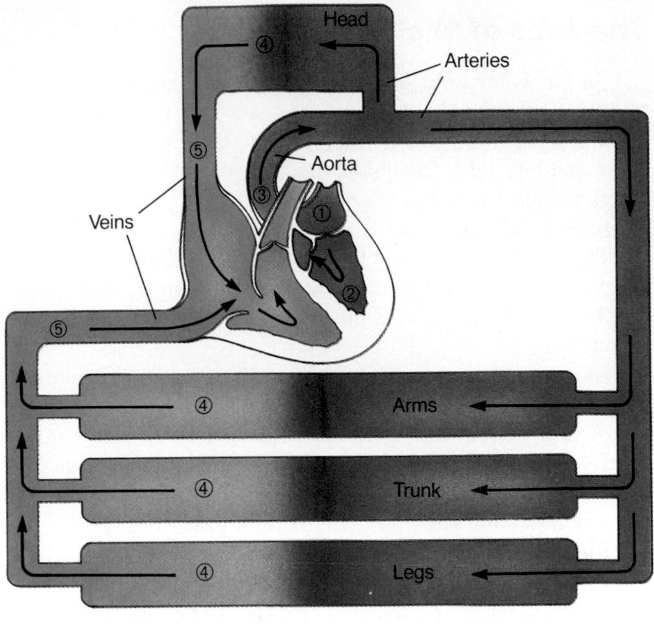


**Figure 11-8** Trace the pathway of blood through the body.

### Left Side of the Heart

The left side of the heart pumps blood to all parts of the body. The path the blood follows can be seen by matching numbers 1 through 5 in Figure 11-8 with statements 1 through 5.

**How is oxygen-rich blood supplied to the left side of the heart?**

1. Blood rich in oxygen has just arrived at the left atrium from the lungs.
2. The blood is pumped from the left atrium into the left ventricle.
3. From the left ventricle, blood is pumped out of the heart into the aorta (ay ORT uh). The **aorta** is the largest artery in the body. It carries blood away from the heart's left side. Then it branches and goes to the body or head.
4. Body parts such as your head and organs receive blood that is rich in oxygen as shown by the bright red color.
5. Finally, blood returns to the heart's right side. Notice that blood has lost its oxygen and is now shown as a blue color. Why does this change occur?

Remember, one main job of blood is to carry two gases, oxygen and carbon dioxide. As blood passes through the lungs, it picks up oxygen. This oxygen is then delivered to all body cells. Carbon dioxide is produced as a waste product by all body cells. The blood picks up this carbon dioxide and drops it off at the lungs.

**Bio Tip**

**Health:** A resting heart will pump 5 liters of blood through the body each minute.



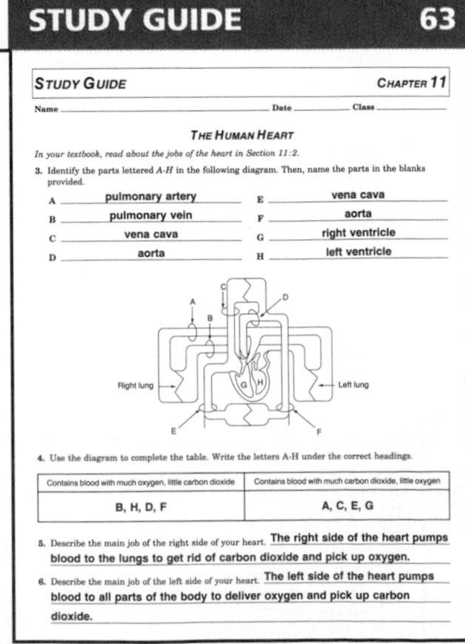

**STUDY GUIDE 63**

STUDY GUIDE — CHAPTER 11

Name ______ Date ______ Class ______

THE HUMAN HEART

*In your textbook, read about the jobs of the heart in Section 11:2.*

3. Identify the parts lettered *A-H* in the following diagram. Then, name the parts in the blanks provided.

A pulmonary artery
B pulmonary vein
C vena cava
D aorta
E vena cava
F aorta
G right ventricle
H left ventricle



4. Use the diagram to complete the table. Write the letters A-H under the correct headings.

| Contains blood with much oxygen, little carbon dioxide | Contains blood with much carbon dioxide, little oxygen |
|---|---|
| B, H, D, F | A, C, E, G |

5. Describe the main job of the right side of your heart. The right side of the heart pumps blood to the lungs to get rid of carbon dioxide and pick up oxygen.

6. Describe the main job of the left side of your heart. The left side of the heart pumps blood to all parts of the body to deliver oxygen and pick up carbon dioxide.

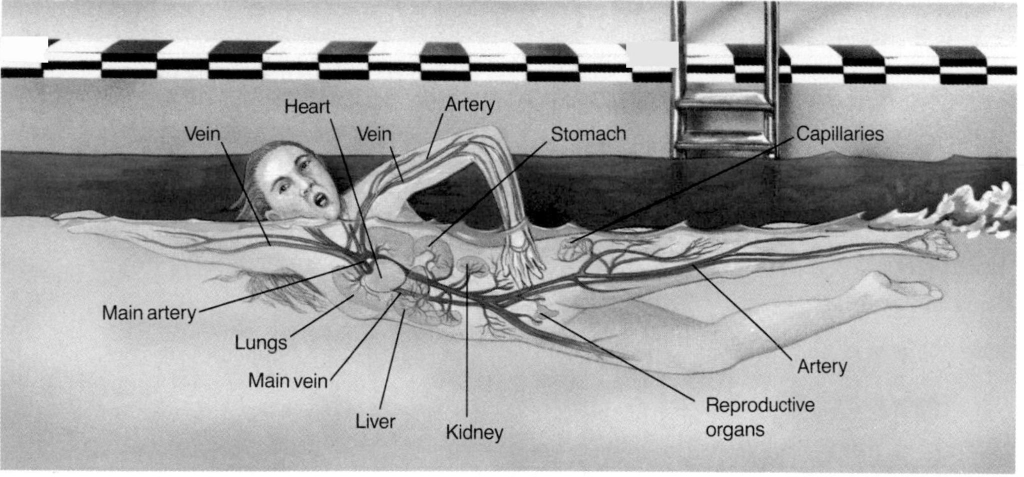


There is a very important difference between the left and right sides of your heart. Blood in the right side of your heart contains a lot of carbon dioxide. Blood in the left side of your heart contains a lot of oxygen. Which side of the heart pumps blood only to your lungs? The side with more carbon dioxide. Which side of the heart pumps blood only to your body? The left side. Figure 11-9 helps to show how all the blood vessels supply all the organs of your body.

**Figure 11-9** Oxygen is delivered to all the organs of your body by arteries. Veins pick up blood that has carbon dioxide in it and return it to the lungs.

## Check Your Understanding

6. (a) Are ventricles relaxed or pumping when they fill? Describe the condition of the atria as the ventricles fill. (b) Are ventricles relaxed or pumping when they empty? Describe the condition of the atria as the ventricles empty.
7. What causes the sounds of the heart?
8. Trace the pathway of blood once it leaves the (a) left atrium, (b) right atrium, (c) left ventricle, and (d) right ventricle.
9. **Critical Thinking:** An amphibian heart has one ventricle and two atria. How would the blood circulation through an amphibian heart differ from that in a human heart?
10. **Biology and Math:** A person's normal heartbeat rate is 70 beats each minute. About how many times does a person's heart beat in a day?



## TEACH

Have students respond to the first three questions in Check Your Understanding.

### Check for Understanding

Have students respond to the first three questions in Check Your Understanding.

### Reteach

**Reteaching/*Teacher Resource Package*, p. 29.** Use this worksheet to give students additional practice in understanding heartbeat sequence and events.

**Extension:** Assign Critical Thinking, Biology and Math, or some of the **OPTIONS** available with this lesson.

### Portfolio

Have students cut and paste the two diagrams together. Copy for students Figures 11-7 and 11-8. Explain to students that due to a congenital birth defect, a baby may be born with no wall between the left and right ventricles. With colored pencils, have students shade in the expected color of blood through such a baby's body. Explain how this problem might affect the baby's development.

## 3 APPLY

### ACTIVITY/Software

*The Heart Simulator,* Cambridge Development Laboratory, Inc.

## 4 CLOSE

### ACTIVITY/Software

*The Heart,* Queue, Inc.

## RETEACHING 29

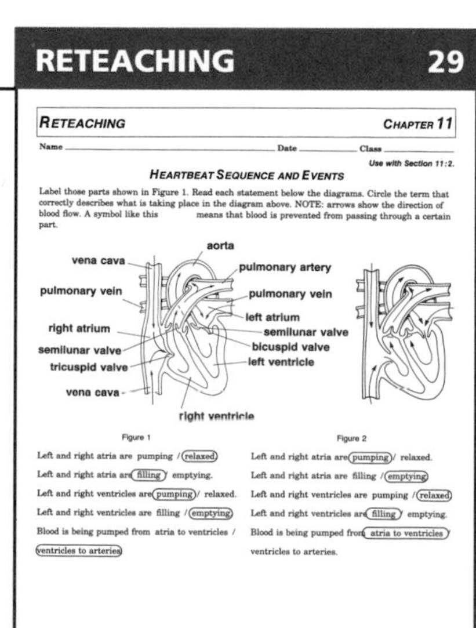

RETEACHING — CHAPTER 11

Name ______ Date ______ Class ______

Use with Section 11:2.

HEARTBEAT SEQUENCE AND EVENTS

Label those parts shown in Figure 1. Read each statement below the diagrams. Circle the term that correctly describes what is taking place in the diagram above. NOTE: arrows show the direction of blood flow. A symbol like this means that blood is prevented from passing through a certain part.



Figure 1

Left and right atria are pumping / (relaxed)

Left and right atria are (filling) / emptying.

Left and right ventricles are (pumping) / relaxed.

Left and right ventricles are filling / (emptying)

Blood is being pumped from atria to ventricles / (ventricles to arteries)

Figure 2

Left and right atria are (pumping) / relaxed.

Left and right atria are filling / (emptying)

Left and right ventricles are pumping / (relaxed)

Left and right ventricles are (filling) / emptying.

Blood is being pumped from (atria to ventricles) / ventricles to arteries.

## Answers to Check Your Understanding

6. (a) When the ventricles fill, they are relaxed and the atria are pumping. (b) When the ventricles empty, they are pumping and the atria are relaxed.
7. the closing of the heart valves
8. (a) It enters the left ventricle. (b) It enters the right ventricle. (c) It is pumped into the aorta. (d) It is pumped into the pulmonary artery.
9. Blood from each atrium flows into the same ventricle. There is a mixing in the ventricle of oxygenated and deoxygenated blood.
10. 70 beats per minute × 60 minutes per hour × 24 hours = 100 800 beats

## 11:3 Blood Vessels

## PREPARATION

### Materials needed

Make copies of the Skill, Study Guide, Reteaching and Lab worksheets in the *Teacher Resource Package.*

▶Build the rubber stopper and tube assembly needed for Lab 11-2.

### Key Science Words

blood pressure capillary

### Process Skills

In Lab 11-2, students will formulate models, form a hypothesis, and measure in SI.

## 1 FOCUS

▶ The objectives are listed on the student page. Remind students to preview these objectives as a guide to this numbered section.

### ACTIVITY/Demonstration

Fill a graduated cylinder with 35 mL of red-colored water. Tell students that this is the volume of blood that fills each ventricle when it is relaxed. Therefore, the heart ejects a total of 70 mL of blood per heartbeat. Have students trace the ejected blood. What type of blood vessel does the blood enter when ejected from the heart?

### Guided Practice

Have students write down their answers to the margin question: by the pumping action of the heart.

**Objectives**

6. **Discuss** the importance of arteries.
7. **List** four traits of veins.
8. **Give the function** of capillaries.

**Key Science Words**

blood pressure
capillary

## 11:3 Blood Vessels

Blood vessels are tubelike structures through which blood moves. You have about 96 000 kilometers of blood vessels. If they were all placed end-to-end, they would go around Earth about two and one-half times. The three types of blood vessels in your body are arteries, veins, and capillaries (KAP uh ler eez).

### Arteries

Remember that arteries carry blood away from the heart. The aorta and pulmonary artery are two arteries you have studied. In what direction is blood being carried within these vessels?

As you can see in Figure 11-11, arteries are round and have thick walls made of many muscle cells. Arteries, such as the aorta, are quite wide in diameter. The aorta can be as wide as a garden hose. Arteries branch into smaller and narrower vessels as they carry blood away from the heart. Look again at Figure 11-9 on page 231 to see what the entire network of arteries might look like.

**How is pressure created within arteries?**

One important feature of arteries is blood pressure. What do you think would happen if you squeezed the water-filled plastic bottle shown in Figure 11-10? Water would squirt out of the hollow tube. Why does this happen? When you squeeze the bottle, you are reducing its volume. You give the liquid less room inside the bottle. The liquid has to go somewhere, so it squirts out of the tube. The water in the bottle is put under high pressure because of the squeezing. The water moves into the tube where the pressure is lower.

**Figure 11-10** If a bottle is squeezed, high pressure forces water to squirt out.

The heart's pumping action on blood is similar to the plastic bottle example. Blood is squeezed by the heart muscle just as water was squeezed out of the bottle by your hand. Just as water moved into the tube, blood is forced into the arteries. The force created when blood pushes against the walls of vessels is called **blood pressure.** This pressure drives the blood through the blood vessels. When you feel your pulse, you are feeling your blood moving through your arteries. Your pulse rate agrees with your heartbeat rate.

The three important traits of arteries are (1) they carry blood away from the heart (2) they carry blood under high pressure (3) they are round in shape and have thick, muscular walls.



## OPTIONS

### Science Background

Arteries carry blood away from heart, regardless of condition of blood within them. Veins carry blood to heart, regardless of condition of blood within them.

**Skill**/*Teacher Resource Package,* p. 12. Use the Skill worksheet shown here to give additional practice with the circulatory system.

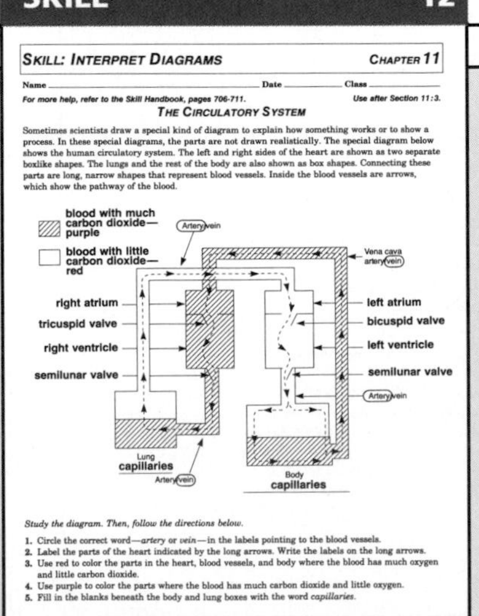

**SKILL 12**

SKILL: INTERPRET DIAGRAMS CHAPTER 11

Name ______ Date ______ Class ______

*For more help, refer to the Skill Handbook, pages 706-711.* *Use after Section 11:3.*

THE CIRCULATORY SYSTEM

Sometimes scientists draw a special kind of diagram to explain how something works or to show a process. In these special diagrams, the parts are not drawn realistically. The special diagram below shows the human circulatory system. The left and right sides of the heart are shown as two separate boxlike shapes. The lungs and the rest of the body are also shown as box shapes. Connecting these parts are long, narrow shapes that represent blood vessels. Inside the blood vessels are arrows, which show the pathway of the blood.



*Study the diagram. Then, follow the directions below.*

1. Circle the correct word—*artery* or *vein*—in the labels pointing to the blood vessels.
2. Label the parts of the heart indicated by the long arrows. Write the labels on the long arrows.
3. Use red to color the parts in the heart, blood vessels, and body where the blood has much oxygen and little carbon dioxide.
4. Use purple to color the parts where the blood has much carbon dioxide and little oxygen.
5. Fill in the blanks beneath the body and lung boxes with the word *capillaries.*

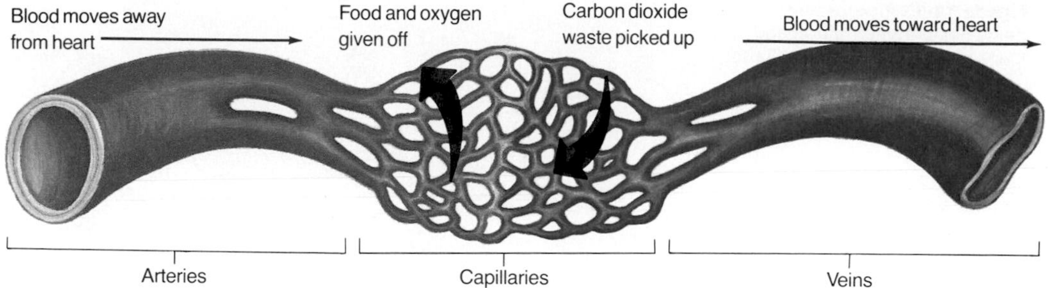


**Figure 11-11** The three types of blood vessels are arteries, veins, and capillaries. Where do pickups and deliveries occur? in the capillaries

## Veins

Blood must be brought back to the heart in veins. The vena cava and pulmonary vein are two veins you have studied. In what direction is blood being carried within these vessels?

Look again at Figure 11-11. You can see how veins differ from arteries. Veins have less muscle than arteries. Their shape is rather flat, but they may be much wider than arteries. Veins have thinner walls than arteries. Veins in your arms and legs have many one-way valves inside them. These valves keep blood flowing in one direction, toward the heart. Figure 11-12 shows what happens to the valves if blood begins to flow backward away from the heart. The valves are helpful because blood in veins is under low pressure. Without the valves, how would blood in your leg or arm veins return to your heart without flowing back into your feet or hands?

**In what direction does blood in the veins travel?**

The four important traits of veins are (1) they carry blood to the heart (2) they carry blood under low pressure (3) they are flat in shape and have little muscle (4) they have many one-way valves to keep blood flowing in one direction.

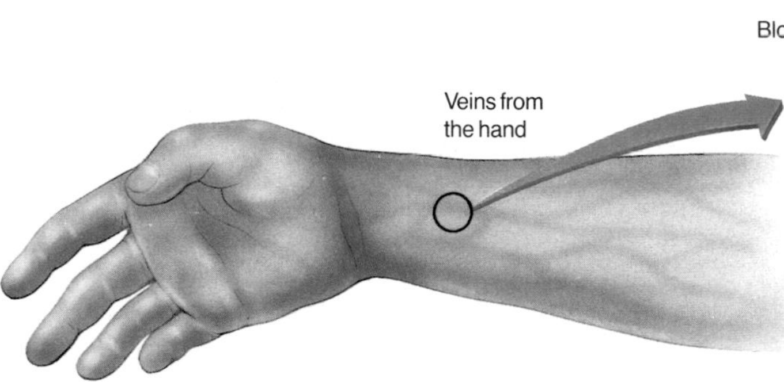


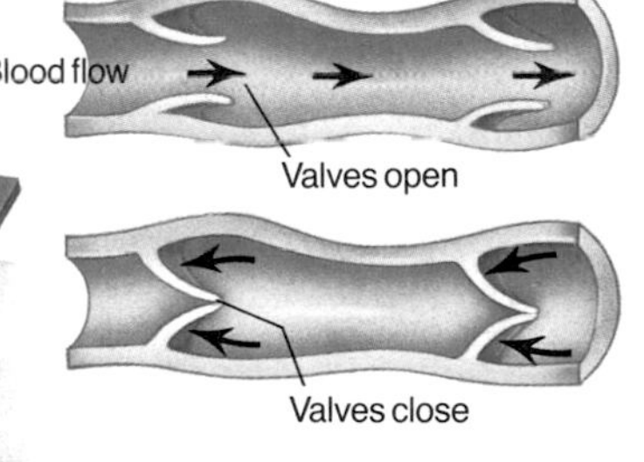


**Figure 11-12** One-way valves in veins keep blood flowing in one direction toward the heart.



## STUDY GUIDE 64

STUDY GUIDE CHAPTER 11

Name ______ Date ______ Class ______

**BLOOD VESSELS**

*In your textbook, read about arteries, veins, and capillaries in Section 11:3.*

1. Circle the word or words that make the statement true.
   a. Arteries carry blood toward / (away from) the heart.
   b. Arteries carry blood under (high) / low pressure.
   c. Arteries have thin / (thick) muscular walls.
   d. An example of an artery is the vena cava / (aorta).
   e. When your feel your pulse, you are feeling blood moving through your veins / (arteries).
2. What is the main job of veins? **to carry blood back to the heart**
3. Compare the amount of muscle in veins and arteries. **Veins have thin walls with little muscle. Arteries have thick, muscular walls.**
4. Compare the pressure within veins and arteries. **Veins carry blood under low pressure. Arteries carry blood under high pressure.**
5. Explain how the one-way valves in veins help maintain circulation. **Since blood in these vessels is under low pressure, valves help keep blood flowing in one direction toward the heart.**
6. Place a checkmark next to those statements below that are true.
   ___ a. Blood pressure is higher in capillaries than in arteries.
   ✓ b. Capillary walls are only one cell thick.
   ___ c. The human body has more veins than capillaries.
   ✓ d. Capillaries bring blood close to all body cells.
   ✓ e. All the pickups of carbon dioxide and other cell wastes, and all the deliveries of oxygen, food, and other materials, occur in the capillaries.

## OPTIONS

**Science Background:** Varicose veins are related to nonfunctioning, poorly functioning, or nonexisting valves along the veins in the legs. As a consequence, blood pools in these vessels, causing them to bulge.

# 2 TEACH

### MOTIVATION/Videocassette

Show the videocassette, *Life Under Pressure,* Films for the Humanities and Sciences, Inc.

### Concept Development

▶ Remind students that all the "work" of delivering and picking up is taking place in the capillaries. Blood cells can move single file through the capillaries.

**Misconception:** Students may believe that all arteries carry blood with much oxygen, while all veins carry blood with much carbon dioxide. Point out that the pulmonary artery and pulmonary vein are exceptions to this rule.

### ACTIVITY/Brainstorming

Make sure students understand which side of the heart must push the blood out under greater pressure. Ask them which side must push blood farther in order to get it where it must go. You may want to point out that the left side of the heart is a much thicker muscle than the right side. Ask students to explain why.

### Guided Practice

Have students write down their answers to the margin question: in one direction back to the heart.

### Independent Practice

**Study Guide**/*Teacher Resource Package,* p. 64. Use the Study Guide worksheet shown at the bottom of this page for independent practice.

## TEACH

### Guided Practice

Have students write down their answers to the margin question: oxygen and nutrients diffuse into body cells and waste diffuses out of body cells.

### Check for Understanding

Have students respond to the first three questions in Check Your Understanding.

### Reteach

**Reteaching**/*Teacher Resource Package,* p. 30. Use this worksheet to give students additional practice in understanding blood vessels and blood flow.

**Extension:** Assign Critical Thinking, Biology and Math, or some of the **OPTIONS** available with this lesson.

## 3 APPLY

### ACTIVITY/Demonstration

Use a variety of tubelike materials to represent arteries and veins.

## 4 CLOSE

### ACTIVITY/Brainstorming

Ask students: Why do you always use an artery to feel your pulse?

### Answers to Check Your Understanding

11. They carry blood away from the heart, blood is carried under high pressure, and they have thick, muscular walls.
12. They carry blood to the heart, blood is carried under low pressure, and they have one-way valves to keep blood flowing in one direction.
13. They bring blood cells carrying oxygen and nutrients close to body cells.
14. an artery, because the blood is under high pressure
15. 96 000 km/2.5 = 38 400 km

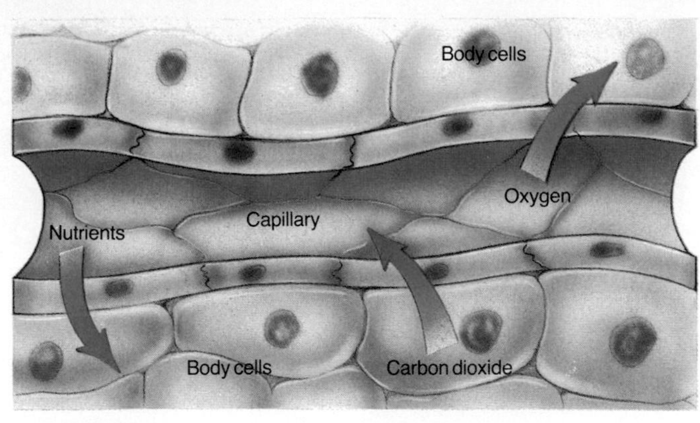


**Figure 11-13** Pickups and deliveries between blood and body cells occur in the capillaries.

## Capillaries

What occurs in the capillaries?

There are blood vessels in your body that become very narrow until they can no longer be called arteries or veins. A **capillary** is the smallest kind of blood vessel. The wall of a capillary is only one cell thick and may be thinner than a single page in this book. Some blood cells can easily pass between the cells of the capillary wall. You have more capillaries in your body than any other blood vessel type.

Capillaries have a very important job. They bring blood close to every body cell. All the pickups and deliveries occur in the capillaries. Blood in the capillaries can deliver oxygen, food, or other needed materials to all body cells. At the same time, waste chemicals made in body cells, such as carbon dioxide, are picked up by the blood in capillaries. Figure 11-13 shows these events.

### Check Your Understanding

11. List three traits of arteries.
12. List four traits of veins.
13. What is the main job of capillaries?
14. **Critical Thinking:** What would be more life-threatening, a cut in an artery or a cut in a vein? Why?
15. **Biology and Math:** You have read that the total length of all your blood vessels is about 96 000 kilometers, enough to go around Earth about two and one-half times. Use this information to calculate the circumference of, or distance around, Earth.

234 Circulation 11:3

## GLENCOE TECHNOLOGY

**Videodisc**

**Biology: The Dynamics of Life**
Movie: *Capillaries*
Disc 2, Side 1, Ch. 33

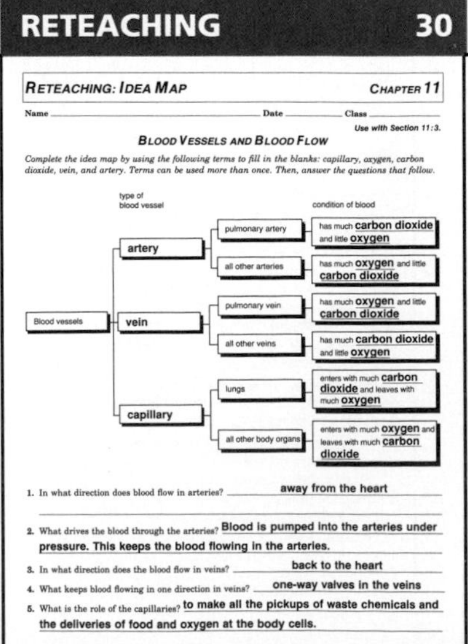

## RETEACHING 30

**RETEACHING: IDEA MAP** CHAPTER 11

Name ______ Date ______ Class ______

*Use with Section 11:3.*

**BLOOD VESSELS AND BLOOD FLOW**

*Complete the idea map by using the following terms to fill in the blanks: capillary, oxygen, carbon dioxide, vein, and artery. Terms can be used more than once. Then, answer the questions that follow.*

| | type of blood vessel | | condition of blood |
|---|---|---|---|
| Blood vessels | artery | pulmonary artery | has much carbon dioxide and little oxygen |
| | | all other arteries | has much oxygen and little carbon dioxide |
| | vein | pulmonary vein | has much oxygen and little carbon dioxide |
| | | all other veins | has much carbon dioxide and little oxygen |
| | capillary | lungs | enters with much carbon dioxide and leaves with much oxygen |
| | | all other body organs | enters with much oxygen and leaves with much carbon dioxide |

1. In what direction does blood flow in arteries? away from the heart
2. What drives the blood through the arteries? Blood is pumped into the arteries under pressure. This keeps the blood flowing in the arteries.
3. In what direction does the blood flow in veins? back to the heart
4. What keeps blood flowing in one direction in veins? one-way valves in the veins
5. What is the role of the capillaries? to make all the pickups of waste chemicals and the deliveries of food and oxygen at the body cells.

# Lab 11-2 Blood Pressure

## Problem: How does blood pressure in arteries and veins compare?

### Skills

formulate models, form a hypothesis, measure in SI

### Materials

plastic squeeze bottle
pan or sink
rubber stopper and tube assembly
water
meterstick

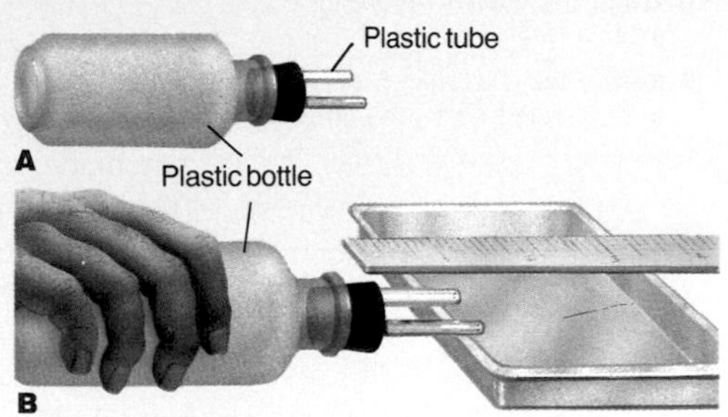


### Procedure

1. Copy the data table.
2. Fill the plastic bottle with water.
3. Insert the rubber stopper and tube assembly into the plastic bottle, Figure A.
4. Position the bottle and meterstick over a pan or sink as shown in Figure B.
5. **Form a hypothesis** about which tube the water will squirt from farther. One tube is glass and the other is plastic.
6. Give the bottle one firm squeeze.
7. Note the distance that water squirts from each tube.
8. **Measure in SI:** Measure in centimeters the distance that water squirts from each tube. Record the distance in your table.
9. Repeat steps 2 through 4 and 6 through 7 two more times.
10. Calculate the average distance water squirted for each tube. (Add the 3 distances and divide the total by 3.)

| TRIAL | DISTANCE WATER SQUIRTS (cm) | |
|---|---|---|
| | GLASS TUBE | PLASTIC TUBE |
| 1 | 42 | 36 |
| 2 | 38 | 34 |
| 3 | 37 | 36 |
| Total | 117 | 106 |
| Average | 39 | 35.3 |

### Data and Observations

1. Which tube squirted water farther?
2. Which tube squirted water the shorter distance?

### Analyze and Apply

1. (a) What body organ does the plastic bottle represent?
   (b) What body liquid does the water represent?
2. Arteries are firmer than veins.
   (a) Which tube represents an artery?
   (b) Which tube represents a vein?
3. Which tube was under higher pressure?
4. Which tube was under lower pressure?
5. How does blood pressure in arteries and veins compare?
6. **Check your hypothesis:** Is your hypothesis supported by your data? Why or why not?
7. **Apply:** Why are arteries firmer than veins?

### Extension

**Design an experiment** to show the effect of less salt intake on blood pressure.



## Lab 11-2 Blood Pressure

### Overview

In this lab, students measure the distance that water squirts when forced through two different tubes. This model will be compared with arteries and veins.

**Objectives:** Upon completion of the lab, students will be able to (1) **relate** the distance water squirts from two tubes to the pressure within them. (2) **compare** the pressure within arteries and veins.

**Time Allotment:** 1 class period

### Preparation

▶ For the rubber stopper and tube assembly use the following: glass tube 22 cm long, rubber or plastic tube 18 cm long (inner diameter 4 mm), 2-hole rubber stopper (size 6 to 7), lubricant, and cloth towel.

**Lab 11-2 worksheet**/*Teacher Resource Package,* pp. 43-44.

### Teaching the Lab

▶ For safety reasons, the teacher should build the rubber stopper assembly.

**Cooperative Learning:** Divide the class into groups of two to three students. For more information, see pp. 22T-23T in the Teacher Guide.

**Lab 11-2 Computer Program**/*Teacher Resource Package,* p. x. Use the computer program shown at the bottom of this page as you teach this lab.

### ✓ ASSESSMENT

**Performance:** Have students use a squeeze bottle to demonstrate how blood pressure would differ in wide versus narrow arteries. Students should predict the results, then try the experiment with the model.

### COMPUTER PROGRAM X

COMPUTER PROGRAM LAB 11-2 BLOOD PRESSURE

```
]POKE 33,33
]LIST

 40  DIM GLASS(5),PLASTIC(5)
 70  GOSUB 470
 90  HOME
 100  GOSUB 670
 110  GOSUB 430
 120  PRINT "!  TRIAL  !DISTANCE WATER SQUIRTS (CM) !";
 130  PRINT "!         !----------------------------";
 140  PRINT "!         !GLASS TUBE   !PLASTIC TUBE  !";
 150  GOSUB 430
 160 S1 = 0:S2 = 0
 170  FOR X = 1 TO 5
 180  VTAB (X * 2 + 5): PRINT "!    ";X;"    !";
 190  HTAB 18: INPUT "";GLASS(X)
 200  VTAB (X * 2 + 5): HTAB 25: PRINT "!    ";
 210  INPUT "";PLASTIC(X)
 220  VTAB (X * 2 + 5): HTAB 40: PRINT "!";
 230  GOSUB 430
 240 S1 = S1 + GLASS(X):S2 = S2 + PLASTIC(X)
 250  NEXT X
 270  PRINT "ARE THESE NUMBERS CORRECT? (Y/N)";
 280  GET SPACE$
 290  IF SPACE$ = "N" THEN 80
 300  IF SPACE$ < > "Y" THEN 280
 310  VTAB  PEEK (37) + 1: HTAB 1: CALL  - 868
 320  PRINT "! TOTAL   !";
 330  VTAB 17: HTAB 18: PRINT S1;: HTAB 25: PRINT "!    ";S2;
 [illegible]  HTAB 18: PRINT [illegible]
 350  GOSUB 430
 360  PRINT "! AVERAGE !          "
 370  VTAB 19: HTAB 18: PRINT S1 / 5;: HTAB 25: PRINT "!    ";S2 / 5;
 380  HTAB 40: PRINT "!";
 390  GOSUB 430
 400  PRINT "DO YOU WANT TO PRINT THIS (Y/N)?";: GET AN$: IF AN$ = "Y" THEN VTA
B  PEEK (37) + 1: HTAB 1: CALL  - 868: GOSUB 1000: END
 410  IF AN$ < > "N" THEN VTAB  PEEK (37) + 1: HTAB 1: CALL  -868: GOTO 400
 420  END
 430  HTAB 1
 440  PRINT "---------------------------------------";
 450  RETURN
 470  TEXT : HOME
 480  GOSUB 670
 490  PRINT : PRINT
 500  PRINT "INSTRUCTIONS:"
 510  PRINT
 580  PRINT : PRINT "YOU WILL BE ASKED TO TYPE IN YOUR       RESULTS FOR THE 5 T
RIALS ONTO A DATA    TABLE ";
 590  PRINT "THAT IS EXACTLY LIKE THE ONE USED IN THE EXPERIMENT."
 600  PRINT
 610  PRINT "AFTER TYPING EACH NUMBER, PRESS THE     RETURN KEY.  AFTER TYPING Y
```

## ANSWERS

### Data and Observations

1. glass tube
2. plastic tube

### Analyze and Apply

1. (a) the heart (b) blood
2. (a) glass tube (b) plastic tube
3. glass tube
4. plastic tube
5. Blood pressure is higher in arteries than in veins.
6. Students who hypothesized that water will squirt farther out of the glass tube will answer yes.
7. They need to be firmer in order to withstand the higher blood pressure.

## 11:4 Problems of the Circulatory System

## PREPARATION

### Materials Needed

Make copies of the Application, Transparency Master, Study Guide, and Reteaching worksheets in the *Teacher Resource Package.*

### Key Science Words

hypertension
cholesterol
coronary vessel
heart attack

## FOCUS

▶ The objectives are listed on the student page. Remind students to preview these objectives as a guide to this numbered section.

### ACTIVITY/Brainstorming

Ask why the heart needs its own blood vessels when so much blood is passing through it? All pick-ups and deliveries are accomplished in the capillaries. Thus, the heart is no different than any other organ in the body.

## GLENCOE TECHNOLOGY

**Videodisc**

**The Secret of Life**
*Plaque-Clogged Artery*

# 11:4 Problems of the Circulatory System

**Objectives**

9. **Explain** how blood pressure is measured.
10. **Trace** the events before and after a heart attack.
11. **Relate** ways to help prevent and correct heart problems.

**Key Science Words**

hypertension
cholesterol
coronary vessel
heart attack

Like any body part, the circulatory system can develop problems. Many of these problems can be corrected.

## High Blood Pressure

When your blood pressure is taken, two measurements are made. One is taken when the ventricles are contracting. This upper pressure is the greater measurement. When the ventricles are not pumping, a lower pressure is taken. The normal range of blood pressure for young adults is 110 to 140 units for the upper pressure and 65 to 90 units for the lower pressure. Blood pressure is recorded as a fraction such as $\frac{120}{80}$. In general, blood pressure increases gradually as a person gets older.

Arteries are partly muscle, which gives them the ability to contract. When the artery contracts, the inside of the artery gets smaller. Blood must now pass through a narrower opening than before. The result is blood pushing through at higher pressure.

Let's use an example with a garden hose representing an artery. Water inside the hose can represent blood.

What happens if you squeeze the hose while the water is on? Water squirts out farther and with more force, Figure 11-14. You have raised the pressure of the water by making it pass through a narrower opening. The same thing happens with arteries.

**Hypertension** (HI pur ten chun) occurs when blood pressure is extremely high. Hypertension may also refer to a disease caused by high blood pressure. It occurs when arteries are too narrow for easy movement of blood.

Hypertension affects millions of people. Unlike many diseases, hypertension does not always make a person feel ill. Many people with high blood pressure do not know that they have it.

Hypertension can cause damage to body organs. The high pressure makes the heart work harder, which can cause heart failure. It can also cause a blood vessel to burst, which can cause a stroke. For this reason, many people have their blood pressure checked regularly, Figure 11-15. If the blood pressure is high, a doctor may suggest a change in diet or activities. In some cases, medicines are also given to help reduce high blood pressure.

**Figure 11-14** The water pressure in a hose is higher when it is squeezed.

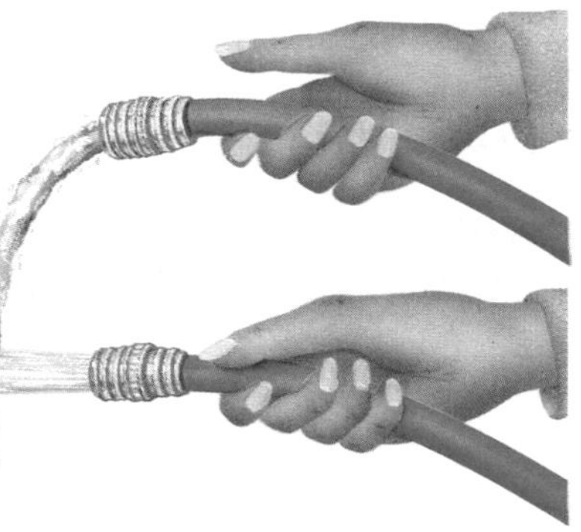



## OPTIONS

### Science Background

Heart disease does not have a single cause, but has many contributing factors. Heart disease is difficult to control because of these many contributing factors, which are covered in this section.

**Application**/*Teacher Resource Package,* p. 11. Use the Application worksheet shown here to teach students about blood pressure.

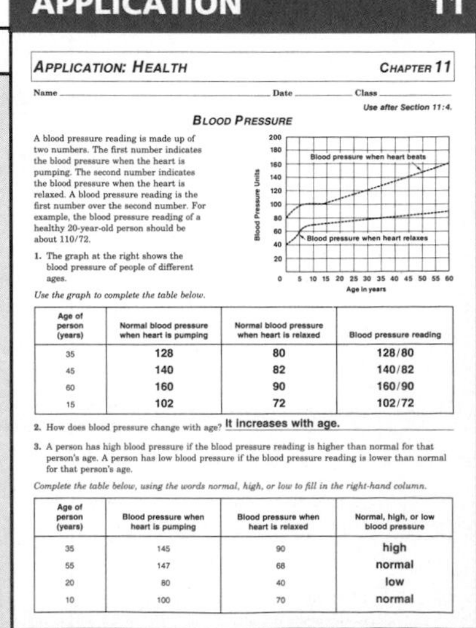

**APPLICATION 11**

APPLICATION: HEALTH CHAPTER 11

Name ______ Date ______ Class ______

*Use after Section 11:4.*

BLOOD PRESSURE

A blood pressure reading is made up of two numbers. The first number indicates the blood pressure when the heart is pumping. The second number indicates the blood pressure when the heart is relaxed. A blood pressure reading is the first number over the second number. For example, the blood pressure reading of a healthy 20-year-old person should be about 110/72.

1. The graph at the right shows the blood pressure of people of different ages.



*Use the graph to complete the table below.*

| Age of person (years) | Normal blood pressure when heart is pumping | Normal blood pressure when heart is relaxed | Blood pressure reading |
|---|---|---|---|
| 35 | 128 | 80 | 128/80 |
| 45 | 140 | 82 | 140/82 |
| 60 | 160 | 90 | 160/90 |
| 15 | 102 | 72 | 102/72 |

2. How does blood pressure change with age? It increases with age.

3. A person has high blood pressure if the blood pressure reading is higher than normal for that person's age. A person has low blood pressure if the blood pressure reading is lower than normal for that person's age.

*Complete the table below, using the words normal, high, or low to fill in the right-hand column.*

| Age of person (years) | Blood pressure when heart is pumping | Blood pressure when heart is relaxed | Normal, high, or low blood pressure |
|---|---|---|---|
| 35 | 145 | 90 | high |
| 55 | 147 | 68 | normal |
| 20 | 80 | 40 | low |
| 10 | 100 | 70 | normal |

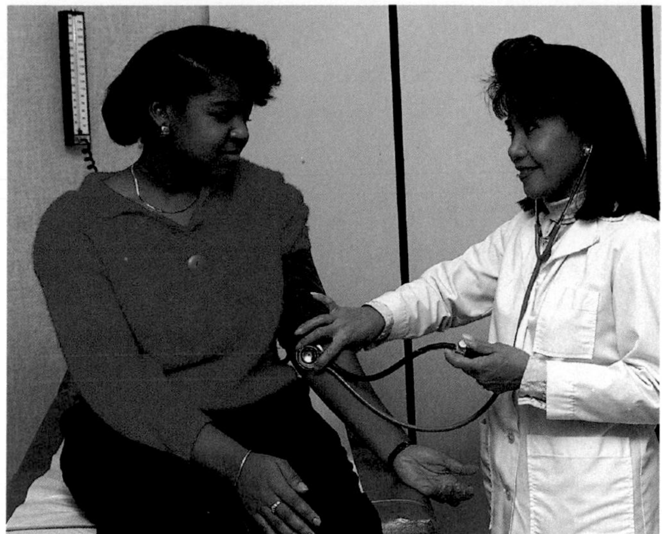

**Figure 11-15** Your blood pressure is checked regularly by your doctor.

High blood pressure may be worsened by a person's diet. For example, too much salt in the diet may make hypertension worse. **Cholesterol** (kuh LES tuh ral) is a fatlike chemical found in certain foods. Cholesterol can coat the inside of arteries and cause them to become narrower. This narrowing raises the blood pressure, just as squeezing the hose did.

### Bio Tip

**Consumer:** Foods that come from plants, such as fruit and vegetables, do not contain cholesterol. Natural foods contain much less salt (sodium) than processed foods. The amount of salt in a box of uncooked rice is much less than the amount in a frozen rice dish.

## Heart Attack

The heart is made of muscle cells. These cells, just like other cells in the body, must receive oxygen and nutrients. How does the heart receive oxygen and nutrients? The heart has its own supply of blood vessels. The blood vessels that carry blood to and from the heart itself are called **coronary** (KOR uh ner ee) **vessels.** Oxygen and nutrients are carried to the heart by the blood in the coronary vessels.

Coronary vessels, like other blood vessels, can become clogged by cholesterol or a blood clot. When the vessels are clogged, blood cannot reach a part of the heart. This part of the heart then begins to die because of a lack of nutrients and oxygen. The death of a section of heart muscle is called a **heart attack.** If too much heart muscle dies, the heart is unable to pump properly. When blood doesn't deliver enough oxygen to body tissues, the person could die.

What is a heart attack?

# 2 TEACH

### MOTIVATION/Demonstration

This is an excellent place for the use of a heart model or a preserved heart. The coronary blood vessels are usually very prominent on both. Trace the intricate pathway of these blood vessels around the heart.

### Concept Development

▶ Students may ask why there is blood pressure even when the heart ventricles are not pumping. The blood is confined to a narrow vessel (the aorta and other major arteries) and still retains pressure from this confinement. Also, the arteries themselves are capable of muscle contraction and contribute to the maintaining of blood pressure.

▶ Coronary blood vessels are normally small in diameter and are highly susceptible to cholesterol deposition. Minor narrowing of these blood vessels can impede blood flow to the heart muscle.

### ACTIVITY/Guest Speaker

Ask the school nurse to visit your classroom and demonstrate the use of a sphygmomanometer. If possible, allow several students to take each other's pressure.

**Analogy:** Compare the buildup of cholesterol in arteries with the accumulation of rust or minerals inside a pipe. The buildup of cholesterol deposits restricts blood flow. The buildup of rust in a pipe reduces water flow.

### Guided Practice

Have students write down their answers to the margin question: the death of a section of heart muscle.

### Independent Practice

**Study Guide/***Teacher Resource Package,* pp. 65-66. Use the Study Guide worksheets shown at the bottom of this page for independent practice.

**STUDY GUIDE 65**

STUDY GUIDE — CHAPTER 11

Name ________ Date ________ Class ________

PROBLEMS OF THE CIRCULATORY SYSTEM

*In your textbook, read about high blood pressure and heart attacks in Section 11:4.*

1. The following diagrams show cross sections of three arteries. Choices A, B, and C may be used more than once and in combinations when answering these questions.

A (normal) B (contracted) C (filled with cholesterol)

a. Which diagram or diagrams *might* have an upper blood pressure value of 160? B, C
Explain why. A narrow opening raises blood pressure.

b. Which diagram or diagrams *might* have an upper blood pressure value of 110? A
Explain why. Diagram A shows a normal opening through which blood should be able to push at a normal pressure value of 110.

2. What are coronary blood vessels? blood vessels that carry blood to and from the heart

3. What can happen if these vessels become clogged? Since blood can't reach a part of the heart, that part begins to die due to lack of nutrients and oxygen. This can result in a heart attack.

*In your textbook, read about preventing heart problems in Section 11:4.*

4. How can diet help prevent heart problems? Eating a balanced diet can prevent becoming overweight and can reduce cholesterol intake.

5. Why is a smoker more likely to have heart problems than a nonsmoker? Nicotine in tobacco smoke causes blood vessels to narrow in size. This causes the heart to work harder to pump blood.

**STUDY GUIDE 66**

STUDY GUIDE — CHAPTER 11

Name ________ Date ________ Class ________

VOCABULARY

*Review the new words used in Chapter 11 of your textbook. Then, read the statements below.*

For each statement, write TRUE or FALSE on the first blank. Make a false statement true by changing the underlined term. Write the correct term on the second blank.

| | | | |
|---|---|---|---|
| 1. | TRUE | | Valves are flaps in the heart and vein. |
| 2. | TRUE | | Arteries carry blood away from the heart. |
| 3. | TRUE | | The circulatory system is made up of blood, blood vessels, and the heart. |
| 4. | FALSE | vena cava | The pulmonary vein is the largest vein. |
| 5. | FALSE | blood vessels | Blood pressure is the force of blood pushing against the walls of the heart. |
| 6. | TRUE | | Atria are the top chambers of the heart. |
| 7. | FALSE | semilunar | The tricuspid valves are located between the ventricles and their arteries. |
| 8. | FALSE | high | Hypertension is caused by low blood pressure. |
| 9. | FALSE | capillary | A vena cava is the smallest kind of blood vessel. |
| 10. | TRUE | | Coronary vessels carry blood to the heart. |
| 11. | FALSE | Pulmonary veins | Capillaries carry blood from the lungs to the left side of the heart. |
| 12. | FALSE | tricuspid | The semilunar valve is located between the right atrium and right ventricle. |
| 13. | TRUE | | A heart attack is the death of a heart muscle. |
| 14. | FALSE | pulmonary artery | The aorta carries blood from the heart to the lungs. |
| 15. | TRUE | | Veins carry blood back to the heart. |
| 16. | TRUE | | The bicuspid valve is located between the left atrium and the left ventricle. |
| 17. | TRUE | | Cholesterol can narrow arteries. |
| 18. | FALSE | aorta | The pulmonary artery is the largest artery. |
| 19. | TRUE | | Ventricles are the bottom chambers of the heart. |

## TECHNOLOGY

### Angioplasty

#### Background

This technique does not require the recuperation time associated with open heart surgery, so it is less costly and traumatic for the patient.

#### Discussion

Cholesterol is not removed from the blood vessels. It is simply pressed up against the inside of the arteries. Therefore, the procedure must be repeated as cholesterol buildup and subsequent narrowing returns.

#### References


Bittl, John A. "Bypass or Angioplasty: Which One When?" *Patient Care*, March 30, 1996, pp. 18-28.

"Heart Procedures and Survival." *The New York Times*, April 16, 1997 p. C8.


## TEACH

#### Check for Understanding

Ask students to name the part of the circulatory system being affected by: too much cholesterol, too little exercise, too much body fat, smoking.

#### Reteach

Ask students to prepare an idea map for problems of the circulatory system and their solutions.

#### Guided Practice

Have students write down their answers to the margin question: nicotine causes blood vessels to narrow in size.

#### *Student Journal*

Have students write a specific exercise and nutrition program that will help an adult prevent heart problems. Students also should provide details about the general health condition of the person for whom the program is written.

## TECHNOLOGY

### Angioplasty

Angioplasty is a way of opening blocked arteries using a balloon.

1. A deflated balloon is inserted into an artery.
2. The balloon is inflated and pushes cholesterol against the artery wall.
3. The balloon is deflated and removed. The artery now has an open pathway and blood can flow freely in the vessel.

This type of operation is easier than removing a blocked blood vessel. Patients who have angioplasty may return to normal activities in a few days.

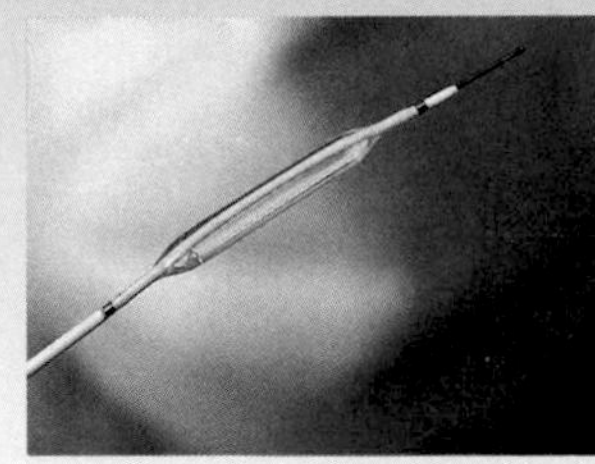

Not all patients can be treated this way. For many patients, replacing clogged vessels is the best form of treatment. A problem with angioplasty is that the cholesterol that caused the problem is still present.

A deflated balloon is prepared for angioplasty.

### Preventing Heart Problems

What can a person do to try to prevent heart problems? Scientific research has shown three things that may help. They are exercise, proper diet, and not smoking.

**Why is smoking harmful to the heart?**

Why is exercise important? Just as leg or arm muscles need exercise, the heart muscle also needs exercise. Physical exercise makes your heart muscle stronger because your heart beats more often when you exercise. Increasing the heartbeat rate regularly is good exercise for your heart.

Diet is also important. Eating a balanced diet can help a person avoid being overweight. Being overweight puts a strain on the heart because fat forms around the heart. The heart has to work harder to pump blood throughout the body. Eating a balanced diet also may help to reduce your cholesterol intake. Table 11-1 shows the cholesterol content of popular foods. How many foods that are high in cholesterol do you eat?

Studies show that a person who smokes is more likely to have heart problems than someone who doesn't smoke. Nicotine, a chemical in tobacco smoke, causes blood vessels to narrow in size. The heart must work harder to pump blood through these narrow vessels.

**TABLE 11–1. CHOLESTEROL VALUES FOR CERTAIN FOODS**

| Food | Cholesterol in One Normal Serving (in milligrams) |
|---|---|
| Chicken liver | 630 |
| Egg yolk | 252 |
| Pork spareribs | 120 |
| Ham | 100 |
| Chicken | 90 |
| Flounder | 70 |
| Pizza | 60 |
| Milk (whole) | 34 |
| Butter (1 tbsp.) | 20 |
| Milk (skim) | 5 |
| Egg white | 0 |



## *OPTIONS*

#### Science Background

Cholesterol is a naturally-occurring compound in the body, so it is impossible to reduce the presence of cholesterol to zero.

**Transparency Master**/*Teacher Resource Package*, p.49. Use the Transparency Master shown here as you teach about the effects of cholesterol on the circulatory system.

**TRANSPARENCY 49**

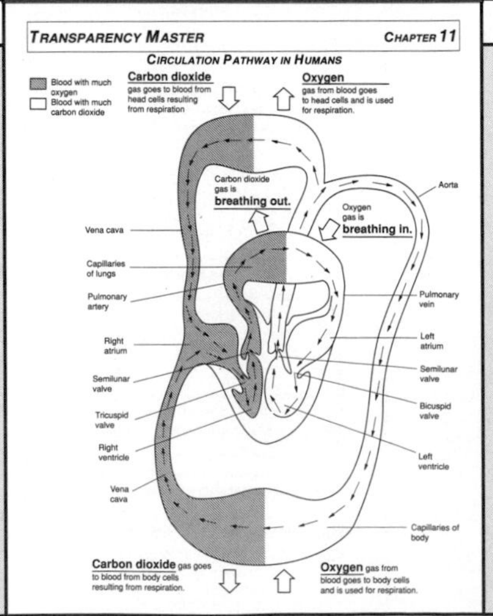

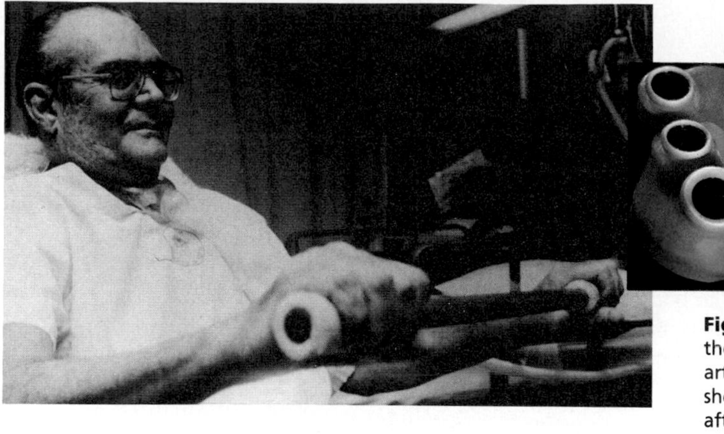

**Figure 11-16** Barney Clark was the first person to receive an artificial heart like the one shown (inset). He died in 1983 after living 112 days with the plastic and metal heart.

What can be done for the person who can't be helped with exercise or diet? If blocked coronary vessels are the problem, unclogged vessels from another part of the patient's body can replace those in the heart. There are drugs that dissolve blood clots inside blood vessels. Heart transplant operations are also possible. Hearts are donated by families of persons who die.

An artificial heart made of plastic and metal has been used for heart patients in an emergency until a human heart is donated. There were 3698 people on the waiting list for a donor heart at the end of 1996. Figure 11-16 shows what this heart looks like. This type of heart has kept a person alive for over 600 days.

## Check Your Understanding

16. What two measurements are made when a person's blood pressure is taken?
17. What causes a section of heart muscle to die?
18. What are three ways of preventing heart problems?
19. **Critical Thinking:** An overweight person's blood pressure is usually higher than the blood pressure of a person of average weight. Why does blood pressure increase with body weight?
20. **Biology and Reading:** A heart attack occurs when blood vessels inside the heart muscle are damaged. The scientific name for a heart attack comes from the name for these blood vessels. What is this name?



## TEACH

### Check for Understanding

Have students respond to the first three questions in Check Your Understanding.

### Reteach

**Reteaching**/*Teacher Resource Package,* p. 31. Use this worksheet to give students additional practice in understanding circulatory problems.

**Extension:** Assign Critical Thinking, Biology and Reading, or some of the **OPTIONS** available with this lesson.

## 3 APPLY

### ACTIVITY/Brainstorming

Why it is possible to remove large veins from the body without harm to the person?

## 4 CLOSE

### Convergent Question

Discuss the problem of eating foods that are high in cholesterol. Is it a problem that teenagers should be concerned with now? (Yes, many diseases that appear later in life take many years to develop.)

### Answers to Check Your Understanding

16. upper, when the ventricles are contracting, and lower, when the ventricles are at rest
17. if it does not receive enough nutrients and oxygen
18. exercise, proper diet, and not smoking
19. The heart must work harder to pump the blood through the entire body.
20. coronary or coronary attack

## RETEACHING 31

**RETEACHING: IDEA MAP** — **CHAPTER 11**

Name ________ Date ________ Class ________

*Use with Section 11:4.*

**CIRCULATORY SYSTEM PROBLEMS**

*Use the following terms to complete the idea map below: proper diet, artificial heart, not smoking, drugs to dissolve blood clots, replace blocked vessels, exercise, heart transplants. Then, answer the questions that follow.*

- Heart problems
  - solution for many
    - exercise
    - proper diet
    - not smoking
  - solution for some
    - drugs to dissolve blood clots
    - replace blocked vessels
    - heart transplants
    - artificial heart

1. What is the normal upper blood pressure range for young adults? **110-140**
   the normal lower blood pressure range? **65-90**
2. What are some causes of hypertension? **too much salt and too much cholesterol in the diet**
3. Which of the following would be a healthier food in terms of the amount of cholesterol it contains—scrambled eggs or pizza? **pizza**
4. Why is a person who smokes more likely to have heart problems than someone who doesn't? **Nicotine in tobacco smoke causes blood vessels to narrow. The heart has to work harder to pump blood through the narrow vessels.**

## *OPTIONS*

### ACTIVITY/Challenge

Have students design a "public information" poster that advises people on how to reduce the risk of heart attack.

# CHAPTER 11 REVIEW

## Summary

Summary statements can be used by students to review the major concepts of the chapter.

## Key Science Words

All boldfaced terms from this chapter are listed.

## Testing Yourself

### Using Words

1. valve
2. capillary
3. aorta
4. bicuspid valve
5. pulmonary artery
6. heart attack
7. circulatory system
8. ventricle
9. hypertension

### Finding Main Ideas

10. p. 228; The heart sounds are caused by the closing of the heart valves.
11. p. 232; Their round shape and thick, muscular walls aid the movement of blood under high pressure.
12. p. 229; Blood leaving the lungs is high in oxygen and low carbon dioxide.
13. p. 222; Simple animals with no circulatory system live in a liquid environment where each body cell is in contact with the liquid.
14. p. 228; A heart murmur is caused by blood leaking past the heart valves when they don't close properly.
15. p. 236; High blood pressure occurs when the arteries are too narrow for free movement of blood.
16. pp. 229-230; The right side of the heart pumps blood only to the lungs. The left side of the heart pumps blood to all parts of the body except the lungs.
17. pp. 225, 227; The human heart consists of two atria, two ventricles, a tricuspid valve, a bicuspid valve, and semilunar valves.

# Chapter 11 Review

## Summary

### 11:1 The Process of Circulation

1. The circulatory system works as a pickup and delivery system for the body.
2. Insects have an open circulatory system. Earthworms and humans have closed systems.

### 11:2 The Human Heart

3. Your heart is a muscle with four chambers. Blood is pumped from top to bottom chambers.
4. The closing of heart valves produces heart sounds.
5. The heart's right side pumps blood only to the lungs. The heart's left side supplies blood to the rest of the body.

### 11:3 Blood Vessels

6. Arteries have thick walls and carry blood away from the heart. Blood in arteries is under high pressure.
7. Veins have valves and carry blood to the heart.
8. Capillaries are very small blood vessels.

### 11:4 Problems of the Circulatory System

9. Blood pressure is caused by blood pushing against the walls of the blood vessels.
10. A heart attack is caused by blockage of the coronary blood vessels.
11. Many heart problems can be avoided by exercise, proper diet, and not smoking.

## Key Science Words

aorta (p. 230)
artery (p. 226)
atrium (p. 225)
bicuspid valve (p. 227)
blood pressure (p. 232)
capillary (p. 234)
cholesterol (p. 237)
circulatory system (p. 222)
coronary vessel (p. 237)
heart attack (p. 237)
hypertension (p. 236)
pulmonary artery (p. 229)
pulmonary vein (p. 229)
semilunar valve (p. 227)
tricuspid valve (p. 227)
valve (p. 227)
vein (p. 226)
vena cava (p. 229)
ventricle (p. 225)

## Testing Yourself

### Using Words

*Choose the word from the list of Key Science Words that best fits the definition.*

1. flap in a vein that keeps blood flow going in one direction
2. smallest blood vessel
3. largest artery in body
4. valve between left atrium and ventricle
5. carries blood away from heart to lungs
6. death of a section of heart muscle
7. your heart, blood vessels, and blood
8. bottom heart chamber
9. disease caused by high blood pressure

240

### Using Main Ideas

18. The heartbeat is the closing of the tricuspid and bicuspid valves of the heart.
19. Veins carry blood toward the heart, blood in veins is under low pressure, and veins are flat with many one-way valves that keep blood from flowing backwards.
20. Most heart attacks are caused by the blocking of the coronary arteries that supply oxygen and nutrients to the heart.
21. The circulatory system picks up oxygen from the lungs and nutrients from the digestive system and delivers them to the body cells. It also picks up carbon dioxide and wastes from the body cells and delivers them to the lungs and excretory system.
22. Heart attacks can be prevented by exercising, a proper diet, and by not smoking.
23. The tricuspid valve is between the right atrium and the right ventricle. The bicuspid valve is between the left atrium and the left ventricle. The semilunar valves are between the ventricles and their arteries.
24. The upper blood pressure is measured when the ventricles contract. The lower blood pressure is measured when the ventricles are relaxed.

# Review

## Testing Yourself *continued*

**Finding Main Ideas**

*List the page number where each main idea below is found. Then, explain each main idea.*

10. what the "lub-dup" sounds are
11. how the traits of arteries aid blood movement
12. the condition of blood leaving the lungs
13. how simple animals can live without a circulatory system
14. what causes a heart murmur
15. what causes high blood pressure
16. what special jobs the heart's right and left sides have
17. what the main parts of the human heart are

**Using Main Ideas**

*Answer the questions by referring to the page number after each question.*

18. What is a heartbeat? (p. 227)
19. What are four traits of veins? (p. 233)
20. What is the cause of most heart attacks? (p. 237)
21. What does your circulatory system pick up and deliver? (p. 222)
22. How can people prevent heart problems? (p. 238)
23. What are the locations and names of all heart valves? (p. 227)
24. What occurs in the heart when one measures the upper and lower blood pressure? (p. 236)
25. How is the pumping action of the heart similar to the squeezing of a plastic bottle? (p. 232)
26. What is the pathway that blood takes as it leaves your heart's right and left ventricles? (p. 226)

## Skill Review 

*For more help, refer to the* ***Skill Handbook,*** *pages 704-719.*

1. **Design an experiment:** Design an experiment to find out whether the average pulse rate of a young child is different from that of a teenager.
2. **Understand science words:** What do the word parts *bi* and *tri* mean?
3. **Formulate a model:** Construct and label a three-dimensional model of the heart and the blood vessels that lead to and from it.
4. **Calculate:** The total sodium (salt) content of a bag of popcorn is 680 mg. The total sodium content of a bag of potato chips is 1140 mg. The popcorn bag contains 4 servings while the bag of potato chips contains 6 servings. Determine the sodium content of a single serving of each snack. Which snack has the lower sodium content?

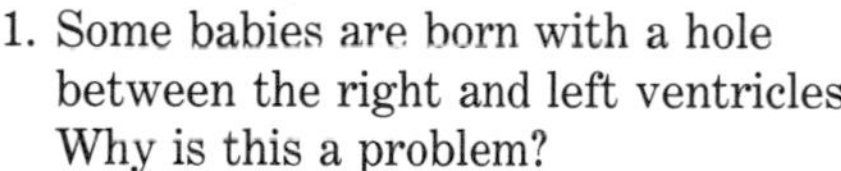


## Finding Out More

**Critical Thinking**

1. Some babies are born with a hole between the right and left ventricles. Why is this a problem?
2. Suppose you are asked to design an artificial heart. What technical problems might you need to solve?

**Applications**

1. Design meals for three days that would be suitable for a person who wishes to reduce his cholesterol.
2. Explain why your heartbeat rate speeds up during exercise.



### Using Main Ideas

25. When a plastic bottle is squeezed, pressure inside the bottle causes the liquid to be pushed out. When the heart chambers contract, pressure inside the chambers causes blood to be pushed out.
26. When blood leaves the right ventricle, it goes to the lungs through the pulmonary artery. When blood leaves the left ventricle, it goes to the body cells through the aorta.

### Skill Review

1. The blood pressure of the teenager should be slightly higher than that of a young child.
2. *Bi* means two, and *tri* means three.
3. The model should contain two atria, two ventricles, a vena cava, an aorta, coronary arteries, a pulmonary artery, and pulmonary veins. There should be valves between the atria and ventricles and between the ventricles and arteries.
4. popcorn–680 mg/4 = 170 mg per serving; potato chips–1140 mg/6 = 190 mg per serving; Popcorn has a lower sodium content per serving than potato chips.

## Finding Out More

### Critical Thinking

1. The hole between the atria allows high-oxygen blood to mix with low-oxygen blood, making the transport of oxygen to the body cells less efficient.
2. The major problem with an artificial heart is finding a material that won't be rejected by the body. Other concerns include leaking, keeping the pressure, force, and timing of the contractions correct; and preventing the formation of blood clots within the heart.

### Applications

1. Answers will vary. The meals should be low in animal fat and high in grains and fruits and vegetables.
2. The heart must pump blood faster to the muscle cells as oxygen is used up during exercise.

**Chapter Review**/*Teacher Resource Package*, pp. 55-56.

**Chapter Test**/*Teacher Resource Package*, pp. 57-59.

**Chapter Test**/*Computer Test Bank*

# Blood

## PLANNING GUIDE

| CONTENT | TEXT FEATURES | TEACHER RESOURCE PACKAGE | OTHER COMPONENTS |
|---|---|---|---|
| (1 day)<br>12:1 The Role of Blood<br>Blood Functions | Check Your Understanding, p. 245 | Reteaching: *Functions of Blood,* p. 32<br>Study Guide: *The Role of Blood,* p. 67 | **STVS:** *Blood Clot Treatment,* Human Biology (Disc 7, Side 1) |
| (2 days)<br>12:2 Parts of Human Blood<br>Plasma<br>Red Blood Cells<br>Roles of Red Blood Cells<br>White Blood Cells<br>Roles of White Blood Cells<br>Platelets | Lab 12-1: *Red Blood Cells,* p. 246<br>Skill Check: *Understand Science Words,* p. 252<br>Mini Lab: *How Is Human Blood Made Up?* p. 248<br>Lab 12-2: *Human Blood Cells,* p. 253<br>Technology: *Eye "Fingerprints,"* p. 250<br>Check Your Understanding, p. 252 | Enrichment: *Red and White Blood Cell Numbers,* p. 12<br>Reteaching: *Parts of the Blood,* p. 33<br>Critical Thinking/Problem Solving: *How Are Active and Passive Immunity Different?* p. 12<br>Study Guide: *Parts of Human Blood,* p. 68<br>Lab 12-1: *Red Blood Cells,* pp. 45-46<br>Lab 12-2: *Human Blood Cells,* pp. 47-48 | **Laboratory Manual:**<br>*How Can Blood Diseases Be Identified?* p. 95<br>**STVS:** *White Cells in Action,* Human Biology (Disc 7, Side 1) |
| (1 day)<br>12:3 Blood Types<br>Blood Types Differ<br>Mixing Blood Types | Check Your Understanding, p. 255 | Reteaching: *Blood Types,* p. 34<br>Study Guide: *Blood Types,* p. 69 | **Laboratory Manual:**<br>*What Blood Types Can Be Mixed?* p. 99<br>Color Transparency 12a: *Blood Types*<br>**STVS:** *Natural Time-Release Capsules,* Human Biology (Disc 7, Side 1) |
| (1 day)<br>12:4 Immunity<br>Your Immune System<br>How the Immune System Works<br>How Shots Keep You Healthy<br>AIDS and the Immune System | Skill Check: *Infer,* p. 258<br>Check Your Understanding, p. 259 | Application: *How Does AIDS Cause Death?* p. 12<br>Reteaching: *Immunity,* p. 35<br>Skill: *Outline,* p. 13<br>Focus: *How Can Someone Be Infected with the AIDS Virus?* p. 23<br>Study Guide: *The Immune System,* pp. 70-71<br>Study Guide: *Vocabulary,* p. 72<br>Transparency Master: *AIDS and the Immune System,* p. 53 | Color Transparency 12b: *The Immune System*<br>**STVS:** *Combined Immune Deficiency,* Human Biology (Disc 7, Side 1) |
| Chapter Review | Summary<br>Key Science Words<br>Testing Yourself<br>Finding Main Ideas<br>Using Main Ideas<br>Skill Review | **ASSESSMENT RESOURCES**<br>Chapter Review, pp. 60-61<br>Chapter Test, pp. 62-64<br>Performance Assessment in the Biology Classroom<br>Alternate Assessment in the Science Classroom<br>Computer Test Bank | |

## GLENCOE TECHNOLOGY

**Infinite Voyage,** *The Champion Within*
**The Secret of Life,** *Nothing to Sneeze At: Viruses*

## MATERIALS NEEDED

| LAB 12-1, p. 246 | LAB 12-2, p. 253 | MARGIN FEATURES |
|---|---|---|
| microscope<br>prepared slides of human and frog blood<br>petri dish cover<br>metric ruler | sheet of clear plastic<br>marking pencil<br>microscope<br>prepared slide of human blood | Skill Check, p. 252<br>dictionary<br>Mini Lab, p. 248<br>oil<br>water<br>red food coloring |

## OBJECTIVES

For more information about National Science Standards, see page 5T.

| SECTION | OBJECTIVE | CORRELATION of QUESTIONS to OBJECTIVES: CHECK YOUR UNDERSTANDING | CHAPTER REVIEW | TRP CHAPTER REVIEW | TRP CHAPTER TEST |
|---|---|---|---|---|---|
| 12:1 National Science Stds: UCP.5, B.4, C.5 | 1. **Identify** the pickup and delivery jobs of the blood. | 1, 2 | 11, 12 | 3 | 1, 7, 11, 12 |
| | 2. **Explain** the pickup and delivery system in animals without blood. | 3 | 14, 22 | 5 | 13 |
| 12:2 National Science Stds: UCP.1, UCP.2, UCP.3, UCP.5, C.1, C.5, F.1 | 3. **Describe** the living and nonliving parts of blood. | 6 | 3, 9, 18 | 1, 11, 12, 13, 14, 25 | 18, 19, 20, 21, 22, 23, 24, 30, 35 |
| | 4. **Give the functions** of the blood cells and cell-like parts. | 7, 10 | 1, 2, 19 | 4, 6, 9, 15, 16, 17, 18, 19, 20, 21, 22, 23, 26, 27 | 2, 5, 6, 10, 22, 23, 24, 25, 28, 29, 30, 31, 33, 46 |
| | 5. **Explain** the problems that may occur with blood cells and platelets. | 8, 9 | 4, 5, 8 | 29 | 9, 26, 34 |
| 12:3 National Science Stds: UCP.1, UCP.2, UCP.5, C.1 | 6. **Compare** the red cells and plasma proteins in the four main blood types. | 11, 12 | 10, 21 | 7, 8, 24 | 8, 41, 42, 43 |
| | 7. **Explain** why blood types can't be mixed. | 13, 14 | 15 | 9 | 4, 44, 45 |
| 12:4 National Science Stds: UCP.1, UCP.2, UCP.3, UCP.4, UCP.5, C.1, E.2, F.1, F.6 | 8. **Describe** how the immune system works. | 16, 20 | 7, 13, 17 | 10 | 3, 35, 36, 37, 38, 39, 40 |
| | 9. **Explain** how shots prevent disease. | 17 | 20, 23 | 2 | 47 |
| | 10. **Discuss** the AIDS virus. | 18, 19 | 6, 16 | 28 | 14, 48 |

# Blood

## CHAPTER OVERVIEW

### Key Concepts

This chapter discusses the specific functions of blood and its component parts. Blood types of the AB series are covered. The immune system and how it protects us against disease is discussed.

### Key Science Words

| | |
|---|---|
| AIDS | immune system |
| anemia | immunity |
| antibody | leukemia |
| antigen | plasma |
| bone marrow | platelet |
| hemoglobin | red blood cell |
| hemophilia | white blood cell |

### Skill Development

In Lab 12-1, students will **use a microscope, measure in SI,** and **compare** red blood cells. In Lab 12-2, they will **use a microscope, compare,** and **classify** human blood cells. In the Skill Check on p. 252, students will **understand** the **science word** *hemophilia.* On page 258, they will **infer** about immunity. In the Mini Lab on p. 248, students will **formulate a model** of blood.

### Bridging

The circulatory system was described as a pickup and delivery system. Blood does the actual work of carrying all the chemicals to be picked up and delivered, while the circulatory system pumps the blood.

**CHAPTER PREVIEW**

### Chapter Content

Review this outline for Chapter 12 before you read the chapter.



### Skills in this Chapter

The skills that you will use in this chapter are listed below.

- In **Lab 12-1,** you will use a microscope, measure in SI, and compare. In **Lab 12-2,** you will use a microscope, compare, and classify.
- In the **Skill Checks,** you will infer and understand science words.
- In the **Mini Lab,** you will formulate a model.

242

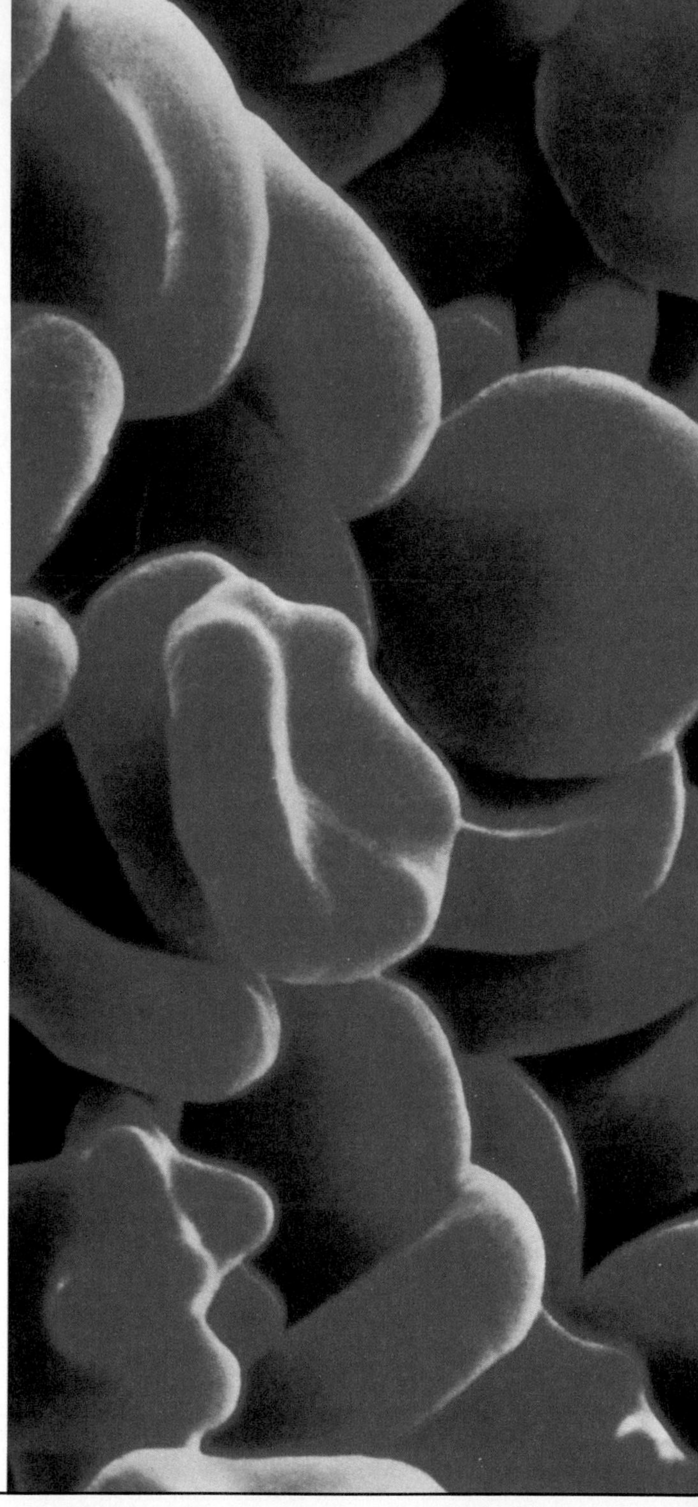

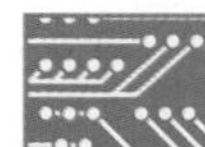

## TECH PREP

For Tech Prep activities in this chapter of the Teacher Wraparound Edition, see especially the Brainstorming activities on pages 235 and 247, the Options Activity on page 248, the Cooperative Learning activity on page 251, the Connection on page 255, and the Portfolio on page 258.

See also the Glencoe Homepage at **www.glencoe.com**

Chapter 12

# Blood

Suppose you were asked to classify blood as either a solid or a liquid. If you have ever had a bloody nose or deep cut, you would probably classify blood as a liquid. But if you viewed blood through a microscope, you might classify it as a solid. The large photograph on the left shows what your blood looks like magnified by an electron microscope. Notice the solid particles. These round, red parts are cells. The small photograph below shows bottles of fruit juices with about the same volume of liquid as you have blood in your body, or about five liters. You may be asking yourself: if blood is made of cells, how can it be liquid in form? After reading this chapter, you will know the answer to this puzzling question.

**Try This!**

**How much blood does the body contain?** The body of an adult human contains about 5 liters of blood. List three ways of showing this quantity. For example, your body contains as much blood as 5 quarts of milk. (1 liter is a little more than 1 quart.)

For more information about the material in this chapter, follow the link for the chapter on the Glencoe Homepage at **http://www.glencoe.com**

Your body holds five liters of blood.



GETTING STARTED

## Using the Photos

Students can discover that blood is composed of cells (pink in color, round). This is apparent only when blood is magnified. It is considered a tissue because it is composed of different cell types.

## MOTIVATION/Try This!

**How much blood does the body contain?** Ways to **measure** could include 5 1-liter soda bottles, a gallon plus a quart, a pail with 5 liters of water in it.

## Chapter Preview

Have students study the chapter outline before they begin to read the chapter. They should note the overall organization of the chapter. First the general role of blood is covered, then the various parts of blood are studied, followed by a description of blood types, and finally a study of the immune system.

## Misconception

Blood in arteries is thought to be red in color, while blood in veins is thought to be blue. Blood in most arteries is bright red due to the presence of oxygen, while blood in most veins is deep red due to the lack of oxygen.

**ASSESSMENT PLANNER**

***Portfolio***

Strategies on the following pages represent student products that can be placed into a best-work portfolio: pp. 254, 256, 258.

**PERFORMANCE ASSESSMENT**

Skill Check, pp. 252, 258
Mini Lab, p. 248
Lab, pp. 246, 253

**CONTENT ASSESSMENT**

Check for Understanding, pp. 245, 252, 255, 259
Chapter Review, pp. 260-261

**GROUP ASSESSMENT**

Opportunities for group assessment occur with Cooperative Learning Strategies.

***Student Journal***

Strategies on the following pages represent opportunities for writing in a Student Journal: pp. 244, 248.

## 12:1 The Role of Blood

## PREPARATION

### Materials Needed

Make copies of the Reteaching, Study Guide, and Lab worksheets in the *Teacher Resource Package.*

▶ Obtain petri dish covers, rulers, and prepared slides for Lab 12-1.

### Key Science Words

No new science words will be introduced in this section.

### Process Skills

In Lab 12-1, students will use a microscope and measure in SI.

## 1 FOCUS

▶ The objectives are listed on the student page. Remind students to preview these objectives as a guide to this numbered section.

### MOTIVATION/Brainstorming

What do you think is the most important role of blood?

### Guided Practice

Have students write down their answers to the margin questions: Carbon dioxide waste goes to the lungs and chemical waste goes to the kidneys; blood and seawater carry nutrients to the cell of the organism.

**Student Journal**

Have students pretend they are a human body cell. They should list their daily needs and explain how these needs are fulfilled. Have them describe how waste is removed and write a similar paragraph for a body cell in a jellyfish.

## 12:1 The Role of Blood

**Objectives**

1. **Identify** the pickup and delivery jobs of the blood.
2. **Explain** the pickup and delivery system in animals without blood.

In Chapter 11 you learned that blood is part of your circulatory system. Blood has many important jobs. You will learn about the roles of blood as you read this chapter.

### Blood Functions

Blood is a body tissue that is part liquid and part cells. What are the functions of blood? Because it's part of the circulatory system, the blood's functions are to pick up and deliver nutrients and wastes.

First, let's list some of the delivery jobs of blood.

1. Blood delivers digested nutrients from fats, proteins, and carbohydrates to all body cells.
2. Blood delivers oxygen to all body cells.
3. Blood delivers chemical messengers to some body cells.
4. Blood delivers water, minerals, and vitamins to cells.

Next, let's list the pickup jobs of blood.

1. Blood picks up carbon dioxide waste from cells and carries it to the lungs.
2. Blood picks up chemical waste from cells and carries it to the kidneys.
3. Blood moves excess body heat into the skin.

**Where does blood carry the wastes that it picks up from body cells?**

You may have noticed the red flush that appears on the skin of a person after heavy exercise. Exercise results in the buildup of excess body heat. The red color is due to the widening of blood capillaries in the skin, Figure 12-1.

a

b

**Figure 12-1** As you begin to exercise, the body heats up (a). The blood vessels in the skin get wider and excess heat is given off through the skin (b).



## OPTIONS

### Science Background

Spiders and their relatives have colorless blood. Crustaceans have amoebocytes in their blood to aid in transport of gases. Insects have colorless blood containing amoebocytes.

Blood in most animals consists mainly of plasma with respiratory pigments dissolved in it. These pigments may consist of hemoglobin or chlorocruorin. Hemoglobin makes the blood red; chlorocruorin causes blood to appear green.

**Figure 12-2** Many sea animals, such as these sea anemones, have no blood or blood vessels. Water brings oxygen and nutrients to their cells.

Other functions of blood include helping to fight diseases and helping to stop bleeding.

Not all animals have blood. Animals such as sponges, jellyfish, and flatworms have no blood or blood vessels. How do the cells of these animals get oxygen or food? How are they able to get rid of waste chemicals? Most animals without blood live in water. Water serves as their "blood." It brings needed oxygen and food to all cells. It also carries away any waste chemicals from the body.

**How is blood to a human like seawater to a jellyfish?**

## Check Your Understanding

1. What are six substances blood delivers to body cells?
2. What three substances are picked up by blood?
3. How do animals without blood get oxygen or food?
4. **Critical Thinking:** How does blood help a person maintain normal body temperature?
5. **Biology and Reading:** One of the following sentences is a fact. The other sentence can be made into a fact by changing the boldfaced word. Change the word and then write both facts on your paper.

The average adult body contains about five **liters** of blood.
Blood is a body **system** made up of liquid and cells.

**RETEACHING 32**

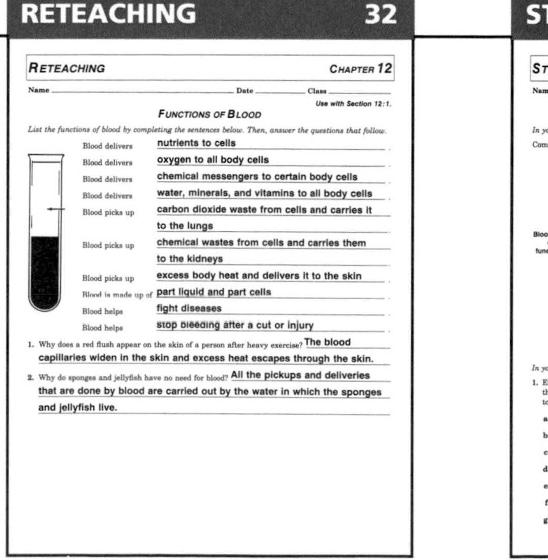

RETEACHING CHAPTER 12

Name ________ Date ______ Class ______

Use with Section 12:1.

FUNCTIONS OF BLOOD

*List the functions of blood by completing the sentences below. Then, answer the questions that follow.*

Blood delivers nutrients to cells.
Blood delivers oxygen to all body cells.
Blood delivers chemical messengers to certain body cells.
Blood delivers water, minerals, and vitamins to all body cells.
Blood picks up carbon dioxide waste from cells and carries it to the lungs.
Blood picks up chemical wastes from cells and carries them to the kidneys.
Blood picks up excess body heat and delivers it to the skin.
Blood is made up of part liquid and part cells.
Blood helps fight diseases.
Blood helps stop bleeding after a cut or injury.

1. Why does a red flush appear on the skin of a person after heavy exercise? The blood capillaries widen in the skin and excess heat escapes through the skin.
2. Why do sponges and jellyfish have no need for blood? All the pickups and deliveries that are done by blood are carried out by the water in which the sponges and jellyfish live.

**STUDY GUIDE 67**

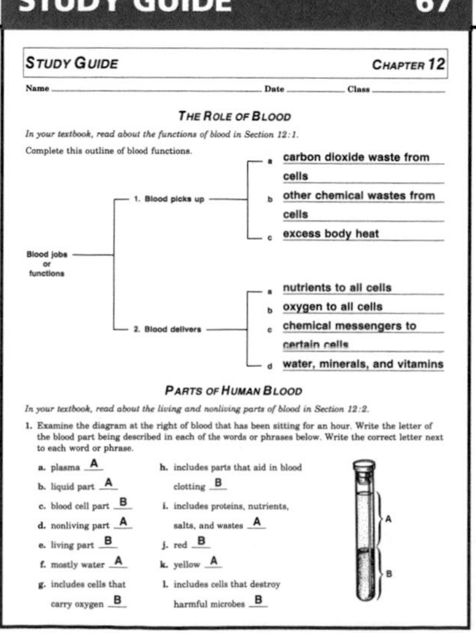

STUDY GUIDE CHAPTER 12

Name ________ Date ______ Class ______

THE ROLE OF BLOOD

*In your textbook, read about the functions of blood in Section 12:1.*

Complete this outline of blood functions.

Blood jobs or functions
- 1. Blood picks up
  - a carbon dioxide waste from cells
  - b other chemical wastes from cells
  - c excess body heat
- 2. Blood delivers
  - a nutrients to all cells
  - b oxygen to all cells
  - c chemical messengers to certain cells
  - d water, minerals, and vitamins

PARTS OF HUMAN BLOOD

*In your textbook, read about the living and nonliving parts of blood in Section 12:2.*

1. Examine the diagram at the right of blood that has been sitting for an hour. Write the letter of the blood part being described in each of the words or phrases below. Write the correct letter next to each word or phrase.

a. plasma A
b. liquid part A
c. blood cell part B
d. nonliving part A
e. living part B
f. mostly water A
g. includes cells that carry oxygen B
h. includes parts that aid in blood clotting B
i. includes proteins, nutrients, salts, and wastes A
j. red B
k. yellow A
l. includes cells that destroy harmful microbes B



# 2 TEACH

### Independent Practice

**Study Guide**/*Teacher Resource Package,* p. 67. Use the Study Guide worksheet shown at the bottom of this page for independent practice.

### Check for Understanding

Have students respond to the first three questions in Check Your Understanding.

### Reteach

**Reteaching**/*Teacher Resource Package,* p. 32. Use this worksheet to give students additional practice with the functions of blood.

**Extension:** Assign Critical Thinking, Biology and Reading, or some of the **OPTIONS** available with this lesson.

# 3 APPLY

### Brainstorming

Invent a "fluid" that can be used as a substitute for blood. What would be some of its characteristics?

# 4 CLOSE

### Audiovisual

Show the filmstrip *Blood: River of Life, Mirror of Health,* Human Relations Media.

### Answers to Check Your Understanding

1. digested nutrients, oxygen, chemical messengers, water, minerals, and vitamins
2. carbon dioxide waste, chemical waste, and excess body heat
3. from surrounding water
4. Blood picks up excess body heat and moves that heat to the skin where the heat leaves the body.
5. The average adult body contains about five liters of blood. Blood is a body tissue made up of liquid and cells.

## Lab 12-1 Red Blood Cells

### Overview

This lab will test the observational skills of students and will allow them to compare and contrast blood cells from different animals.

**Objective:** Upon completing this lab, students will be able to (1) **differentiate** between red blood cells (2) **make** a drawing to scale, and (3) **calculate** cell size.

**Time Allotment:** 30 minutes

### Preparation

▶ Obtain prepared slides of human and frog blood smears.

**Lab 12-1 worksheet**/*Teacher Resource Package,* pp. 45-46.

### Teaching the Lab

▶ If a video camera is available, it can be attached to a microscope and the slides viewed on a television screen.

▶ Review the procedure for finding objects first under low power and then under high power.

▶ Assume that the photo of bird blood cells is shown under high power magnification. Thus, students can measure the cells directly from the page to obtain cell size.

▶ **Troubleshooting:** Students may have difficulty with the math and with measuring/drawing to scale. An example of the steps involved may first be shown on the overhead projector as a demonstration and review.

### ASSESSMENT

**Performance:** Have students look at slides of both human and frog red blood cells. Students should draw and label a diagram of what they see. Students also should prepare a chart that lists three similarities and three differences between the two types of cells.

**Cooperative Learning:** Divide the class into groups of three students. For more information, see pp. 22T-23T in the Teacher Guide.

# Red Blood Cells

## Lab 12–1

### Problem: How do red blood cells of different animals compare?

#### Skills

use a microscope, measure in SI, compare

#### Materials

microscope
petri dish cover
metric ruler
prepared slides of human and frog blood

#### Procedure

1. Copy the data table.
2. Use the cover of a petri dish to draw a circle on a piece of paper.
3. **Use a microscope:** Look at a prepared slide of human red blood cells under low power on the microscope. Change to high power. Most of the cells you see are red blood cells. Cells that appear blue are not red blood cells.
4. Draw a red blood cell in your circle. Assume your circle is the same as your field of view. Draw the cell to scale.
5. **Measure in SI:** Use a ruler to measure the size in millimeters of the blood cell in your drawing. Record your answer in your data table.
6. Repeat steps 3 through 5 with a prepared slide of frog blood.
7. On your diagrams, check for the presence of a nucleus and a cell membrane. Check these parts in your table if present.
8. Describe and record the shapes of the blood cells in your table. Use the following choices—round or oval.
9. Complete the rest of the data table using the photo of bird blood.

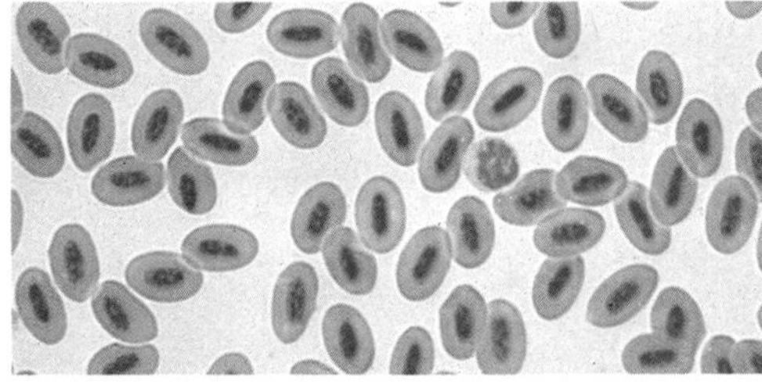

Answers will vary.

| BLOOD FROM | SIZE | SHAPE | PARTS PRESENT? NUCLEUS | CELL MEMBRANE |
|---|---|---|---|---|
| Human | 0.01 mm | round | — | √ |
| Frog | 0.02 mm | oval | √ | √ |
| Bird | — | oval | √ | √ |

#### Data and Observations

1. Which animals have red blood cells:
   (a) with a nucleus?
   (b) that are round?
   (c) that are oval?
2. Which animal has the larger red blood cells—frog or human?

#### Analyze and Apply

1. What is the main job of red blood cells?
2. In what blood vessels do red blood cells transport oxygen?
3. **Apply:** Based upon the appearance of red blood cells, which two animals are most closely related?

#### Extension

**Calculate:** The photo of bird red blood cells is magnified 1000 times. What is the actual measurement of the diameter of one blood cell?



## ANSWERS

### Data and Observations

1. (a) frog, bird
   (b) human
   (c) frog, bird
2. frog

### Analyze and Apply

1. They carry oxygen to body cells.
2. arteries
3. frog and bird

### Extension

0.007 mm

## 12:2 Parts of Human Blood

Blood is different from most body tissues. Not only is it able to flow, but it's also part living and part nonliving.

**Objectives**

3. **Describe** the living and nonliving parts of blood.
4. **Give the functions** of the blood cells and cell-like parts.
5. **Explain** the problems that may occur with blood cells and platelets.

**Key Science Words**

plasma
red blood cell
bone marrow
hemoglobin
anemia
white blood cell
leukemia
platelet
hemophilia

### Plasma

Let's look at two tubes of blood. The tube on the left in Figure 12-3a shows blood just removed from a person's vein. The blood is a deep red color. No special parts can be seen. Now look at the tube on the right. This tube has been sitting for about an hour. Notice how the blood has separated into two parts. Blood cells have settled to the bottom. A yellow liquid appears on top.

**Plasma** (PLAZ muh) is the nonliving, yellow liquid part of blood. Blood plasma is 92 percent water. The remaining eight percent includes blood proteins, nutrients, salts, and waste chemicals.

Most of the pickup and delivery jobs of blood are carried out by plasma.

### Red Blood Cells

**Red blood cells** are cells in the blood that carry oxygen to the body tissues. Red blood cells make up most of the blood. There are about five million red blood cells in a small drop of blood. Red blood cells are part of the living portion of blood. They give blood its red color.

Which substance makes up the nonliving part of blood?

Figure 12-3b shows red blood cells as seen under a light microscope. Look at this photograph as you read about red blood cells.

a

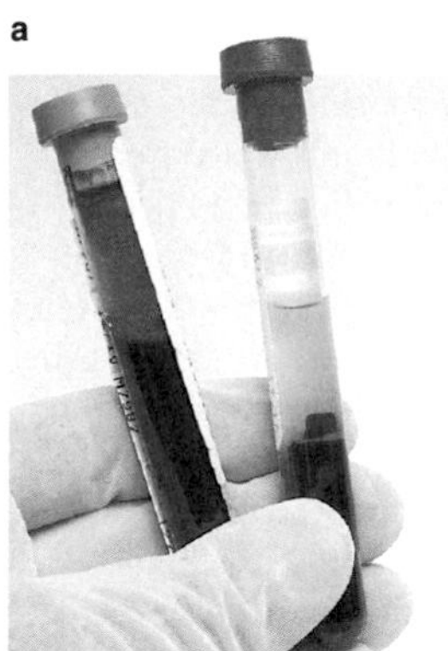

b

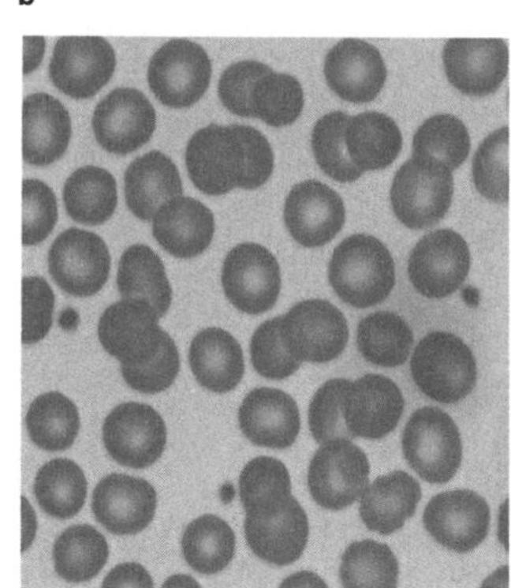

**Figure 12-3** When blood is left to stand (left tube) the red cells and plasma separate out (right tube) (a). Blood cells and plasma can be examined under a light microscope (b).



## OPTIONS

### Science Background

Blood is an example of a tissue that is rather highly specialized and diverse in function. Each component part, plasma and blood cells, has very specific functions relating to transport of materials as well as to immunity.

## 12:2 Parts of Human Blood

## PREPARATION

### Materials Needed

Make copies of the Study Guide, Enrichment, Critical Thinking/Problem Solving, Reteaching, and Lab worksheets in the *Teacher Resource Package.*

▶ Obtain plastic sheets, marking pencils, and prepared blood slides for Lab 12-2.

▶ Gather cooking oil and red food coloring for the Mini Lab.

### Key Science Words

plasma
red blood cell
bone marrow
hemoglobin
anemia
white blood cell
leukemia
platelet
hemophilia

### Process Skills

In Lab 12-2, students will use a microscope, compare, and classify. In the Skill Check, they will understand science words. In the Mini Lab, students will formulate a model.

## 1 FOCUS

▶ The objectives are listed on the student page. Remind students to preview these objectives as a guide to this numbered section.

### MOTIVATION/Brainstorming

Describe:
(a) as many different diseases or disorders connected with blood as you can.
(b) as many different single words that describe the appearance or general function of blood as you can.

### Guided Practice

Have students write down their answers to the margin question: Plasma is the non-living part of the blood.

# 2 TEACH

### MOTIVATION/Software

*The Circulatory System,* Queue, Inc.

### Concept Development

▶ Ask students to compare the appearance of red blood cells as seen in bone marrow to those seen in blood. Red cells in marrow are immature red blood cells that have not yet lost their nucleus.

### Idea Map

Have students use the idea map as a study guide to the major concepts of this section.

### Independent Practice

**Study Guide/***Teacher Resource Package,* p. 68. Use the Study Guide worksheet shown at the bottom of this page for independent practice.

### Guided Practice

Have students write down their answers to the margin question: Anemia is a condition in which there are too few red blood cells. Extra iron in the diet may help.

### Mini Lab

Blood that is mixed appears uniformly red. If allowed to stand, however, cells will settle and plasma will appear. This model allows for observing both conditions.

### ASSESSMENT

**Performance:** Have students redo the model to represent only plasma and its parts. Note: Water without coloring can still be seen when mixed with oil.

### Student Journal

Have students write a job description for a red blood cell. This description should include the important roles of a red blood cell in the body.



**Study Tip:** Use this idea map as a study guide for comparing the living and nonliving parts of blood.

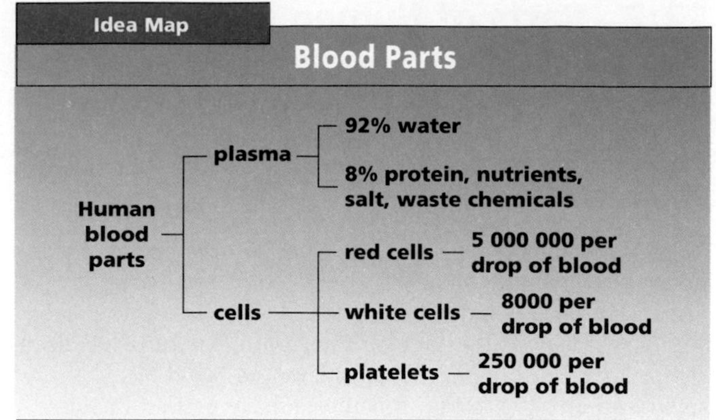


### Mini Lab

**How Is Blood Made Up?**

**Formulate a model:** Add 20 mL of cooking oil to 20 mL of water and a drop of red food coloring. Shake the tube. Allow the liquids to separate. What do the layers represent? *For more help, refer to the* ***Skill Handbook,*** *pages 706-711.*

Red blood cells are round. They have thin centers and thick edges. They look like doughnuts without holes. Red blood cells have a nucleus only when they are first formed. The nucleus is lost as the cell matures and begins to move through the blood vessels.

Red blood cells have a life span of about 120 days. The body must constantly make new red cells. Red blood cells are made primarily in your bone marrow. **Bone marrow** is the soft center part of the bone.

## Roles of Red Blood Cells

Red blood cells are the main part of blood that helps with the blood's job of pickup and delivery. They deliver oxygen to all body cells. Oxygen coming into the lungs is picked up by red blood cells as the cells pass through capillaries in the lungs. Red blood cells are then carried by the plasma to all body parts. Once these cells reach the cells of the body, oxygen is given up by the red blood cells.

How can red blood cells perform this job of oxygen pickup and delivery so well? The answer lies in a special chemical found in red blood cells. This chemical is called hemoglobin (HEE muh gloh bun). **Hemoglobin** is a protein in red blood cells that joins with oxygen and gives the red cells their color. Hemoglobin in blood cells carries oxygen to other body cells.

What is anemia? How is it treated?

Hemoglobin contains the mineral iron. If a person's diet is low in iron, the total number of red blood cells and hemoglobin in the blood decreases. This decrease in red blood cells and hemoglobin can lead to anemia. **Anemia** is a



## OPTIONS

### ACTIVITY/Challenge

Have students prepare a list of foods that are rich in iron. References can be obtained from a nutrition text. Good sources include: liver, kidney beans, prunes, oatmeal, blackstrap molasses, sweet potato, endive, spinach, figs, and tomato juice

### STUDY GUIDE 68

STUDY GUIDE CHAPTER 12

Name ______ Date ______ Class ______

**PARTS OF HUMAN BLOOD**

*In your textbook, read about blood cells and platelets in Section 12:2.*

2. Complete the table below by writing the phrases that follow in the correct column in the table. The first one is done for you.

| Red blood cells | White blood cells | Platelets |
|---|---|---|
| 5 million in a small drop of blood | 8000 in small drop of blood | 250 000 in a small drop of blood |
| contain hemoglobin | destroy microbes | not whole cells |
| transport oxygen | can move between capillaries and among body cells | aid in blood clotting |
| if number is low, person feels tired | increase during infections | life span of 5 days |
| life span of 120 days | remove dead cells | results in hemophilia if not working |
| cell with no nucleus | made in spleen | |
| look like doughnuts without holes | move like amoebas | |
| | life span of 10 days | |
| | increase abnormally during leukemia | |
| | cell with a nucleus | |

8000 in a small drop of blood
250 000 in a small drop of blood
5 million in a small drop of blood
not whole cells
destroy microbes
aid in blood clotting
can move between capillaries and among body cells
increase during infections
contain hemoglobin
remove dead cells
transport oxygen
made in spleen
if number is low, person feels tired
life span of 5 days
life span of 10 days
life span of 120 days
increase abnormally during leukemia
cell with a nucleus
cell with no nucleus
results in hemophilia if not working
look like doughnuts without holes

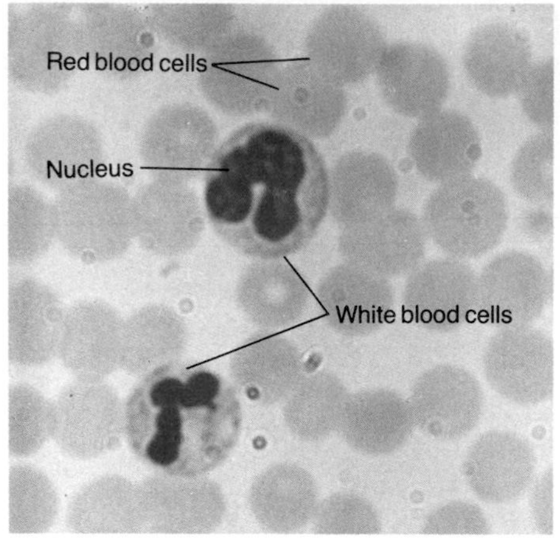


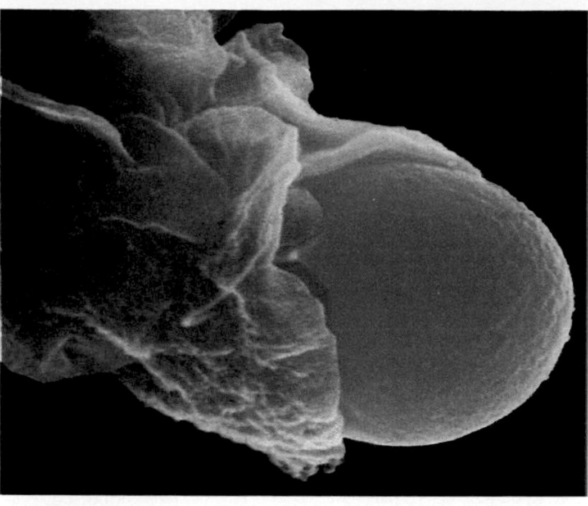

condition in which there are too few red blood cells in the blood. A person with anemia will usually feel weak, tired, and short of breath. These symptoms are caused by too little oxygen reaching body cells. Extra iron in the diet may help solve the problem. How much iron is needed each day? Look back at Table 9-2 on page 190 for the answer.

**Figure 12-4** White blood cells as seen under a light microscope (left) are larger than red blood cells. In fact, when a red blood cell dies, white cells surround and destroy it, as shown under an electron microscope (right).

## White Blood Cells

The second type of blood cell is the white blood cell. **White blood cells** are the cells in the blood that destroy harmful microbes, remove dead cells, and make proteins that help prevent disease.

White blood cells are part of the living portion of blood. Figure 12-4 shows white blood cells magnified under a light microscope and an electron microscope.

A white blood cell has a nucleus. The nucleus appears dark blue in both cells in Figure 12-4 (left). Also notice that white blood cells are larger than red blood cells. Certain white blood cells may live for months or years, but most have a life span of about 10 days. A healthy person has about 8000 white cells in a very small drop of blood.

Like red cells, white cells are made in bone marrow. Unlike red cells, however, white cells are also made in organs such as the spleen, thymus gland, and tonsils.

Some kinds of white blood cells can move out of capillaries and travel among the body cells. Figure 12-5 shows how white blood cells move like amoebas and at times even look like them.

**Figure 12-5** An important trait of white blood cells is their ability to move into body tissues from the capillaries.

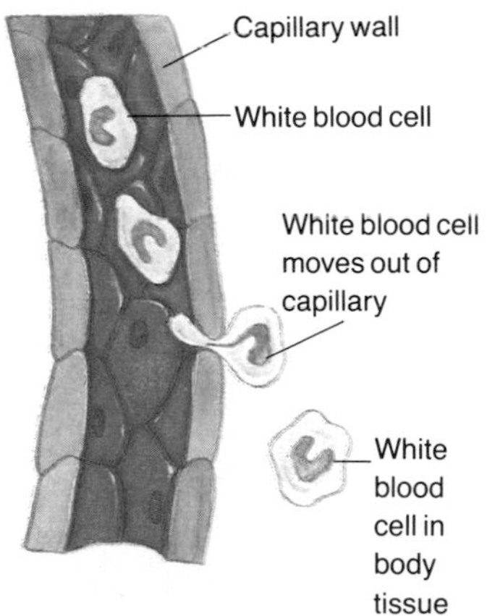




## TEACH

### Concept Development

▶ The daily iron requirement for females between the ages of 11 and 50 is 18 mg. The requirement for males to age 18 is the same, after which it drops to 10 mg. Ask students to speculate as to why this is so. A menstruating female will lose 1 mg of iron per day during menstruation.

▶ See if students can relate the symptoms of sickle-cell anemia, such as being tired, weak, and short of breath, with the corresponding lack of oxygen in body cells.

▶ Ask students to compare a photo of white cells to one of red cells. They will be able to pick out the major similarities and differences between these two cell types even before reading about them.

▶ The number of white blood cells is used as a diagnostic tool for determining if an infection exists. Ask your students what a doctor might be looking for when a blood test is done on a person who may have appendicitis. Appendicitis is an infection that occurs in the appendix. An elevated white cell count alerts the doctor to a possible infection.

**Videodisc**

**STV: Human Body Vol. 1**
*Circulatory and Respiratory Systems*
Unit 1, Side 1
*Blood and Circulation*

**9508-12959**

## ENRICHMENT 12

*ENRICHMENT* CHAPTER 12

Name ______ Date ______ Class ______

*Use after Section 12:2.*

*RED AND WHITE BLOOD CELL NUMBERS*

You can count the number of cells in a small drop of blood by using a microscope slide with a grid, called a counting chamber. By counting just a sample of blood, you can find out the number of cells in a drop of blood without counting all the cells.

1. Count all the red blood cells in the counting chambers below, marked for persons, A, B, and C. Record the totals in the chart below and calculate the total number of cells in each sample.

A B C

2. Use the same method to find out how many white blood cells are in the samples for persons D, E, and F. Record your results in the chart.

D E F

| Person | Red cell count | Add 5 zeros | Total | Person | White cell count | Add 2 zeros | Total |
|---|---|---|---|---|---|---|---|
| A | 73 | 00000 | 7 300 000 | D | 80 | 00 | 8 000 |
| B | 50 | 00000 | 5 000 000 | E | 173 | 00 | 17 300 |
| C | 32 | 00000 | 3 200 000 | F | 43 | 00 | 4 300 |

*Answer the following questions using information from the chart and from a reference book.*

3. Use your text and the results above to find out which persons have the normal number of red cells. B white cells? D

## OPTIONS

**Enrichment**/*Teacher Resource Package*, p. 12. Use the Enrichment worksheet shown here to help students understand blood counts.

## TECHNOLOGY

### Eye "Fingerprints"

#### Background

The system works similar to scanners in a supermarket that read bar codes. The eye scanner reads blood vessel patterns present at the back of the eye along the retina.

#### Discussion

Congress has already passed a law that will require truck drivers and bus drivers to submit to some unique identifying processes before obtaining a license. This will prevent a person from receiving a new license if his or her original has been revoked.

#### Reference

"Look Me in the Eye." *Discover,* Feb. 1990, pp. 8-11.

## TEACH

### Concept Development

▶ Discuss the possible treatment of leukemia with radiation. White cell production is decreased in the body as a result of exposure to radiation. Thus, the disease can be slowed through X rays and other radiation treatments.

▶ One way of determining if a person has been accidentally exposed to radiation is to do a white cell count. A very low number of white cells could indicate exposure to radiation.

**Misconception:** Students believe that white cells eat red cells in leukemia. They do not. The rise in white cell number in leukemia is at the expense of the red and platelet cells. This is because all blood cells, regardless of whether they are to become red, white, or platelet cells,have a common origin, stem cells that differentiate into the different blood cell types.

Blood will clot when it comes in contact with a foreign surface, not when it comes in contact with the air, as is commonly believed.

## TECHNOLOGY

### Eye "Fingerprints"

Blood vessels can be seen clearly in the human eye. These vessels run along the back surface of the eye. That is why a doctor sometimes shines a flashlight into your eyes. The condition of these blood vessels gives the doctor clues about your general health.

The pattern of blood vessels in the eye is different for every person. It's like having a "fingerprint" of your eye.

Using a computer scanner, scientists can identify people based on blood vessel patterns in their eyes. This technology can be used to screen applicants for drivers' licenses. It prevents people who already have licenses from obtaining duplicates. The states of Wisconsin and California are already planning to use this new technology in the issuing of drivers' licenses.

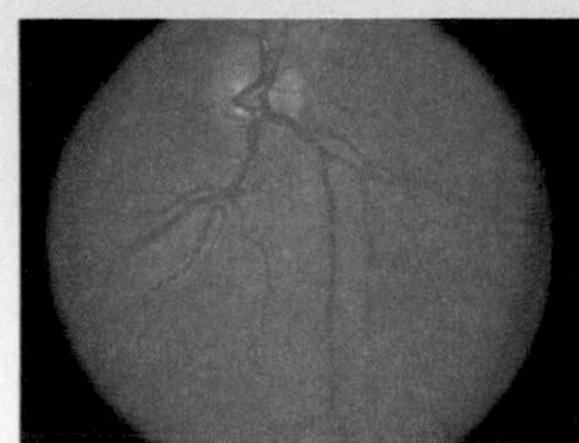

A view of the back of your eye

### Roles of White Blood Cells

Have you ever had a cut that became infected? An infection is usually caused by an attack on your body cells by bacteria.

White blood cells move to an infection and destroy the bacteria causing it. There's a rapid increase in white cells at the time of an infection. The added numbers of white blood cells help to destroy more bacteria at a faster rate. After an infection is over, the number of white cells returns to normal. Another job of white blood cells is to rid the body of dead cells. Certain white cells can move about the body and "eat" dead cells just as they do bacteria, Figure 12-6.

Increased amounts of white blood cells can sometimes cause problems such as leukemia (lew KEE mee uh). **Leukemia** is a blood cancer in which the number of white blood cells increases at an abnormally fast rate.

There are three main differences between the rise in white blood cell numbers during an infection and during leukemia.

1. During leukemia, the number of white cells in each drop of blood may reach 100 000 or more. During an infection, the number rarely goes above 30 000.

**Figure 12-6** A white blood cell detects a bacterium and moves toward this foreign body to destroy it.

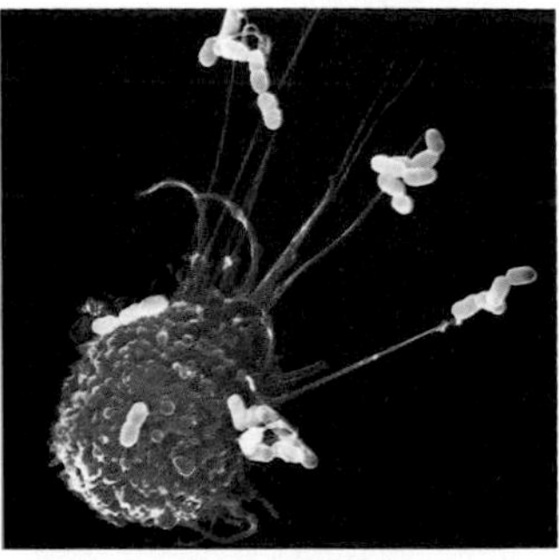



## OPTIONS

**How Can Blood Diseases Be Identified?/** *Lab Manual,* pp. 96-98. Use this lab as an extension to blood diseases.

2. With leukemia, the number of white cells does not return to normal.
3. The cells formed in such large numbers are not normal cells. They can't do the job performed by normal white blood cells.

## Platelets

Cuts and scrapes are common events in our lives. We usually don't worry about these injuries because we know that the flow of blood will stop quickly. A clot forms, and in a few days a scab appears. A blood clot is blood that has formed a plug to stop the bleeding. Figure 12-8 (left) shows a clot starting to form.

The forming of blood clots leads us to the last living part of blood, platelets (PLAYT lutz). **Platelets** are cell parts that aid in forming blood clots. Platelets are not complete cells. However, platelets do come from cells that break apart, so they are still a living part of blood. Notice in Figure 12-8 (right) that the platelets are much smaller than red cells.

When an injury occurs in the body, platelets break apart and release a chemical. This chemical starts the formation of a clot. Several chemicals are involved in clot formation.

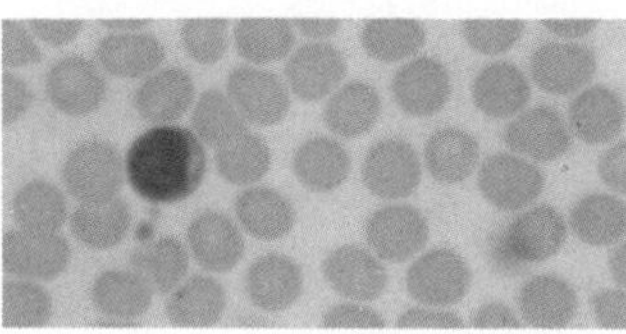

**Figure 12-7** Which photo (top or bottom) is from a person with leukemia? bottom

**What is the main job of platelets?**

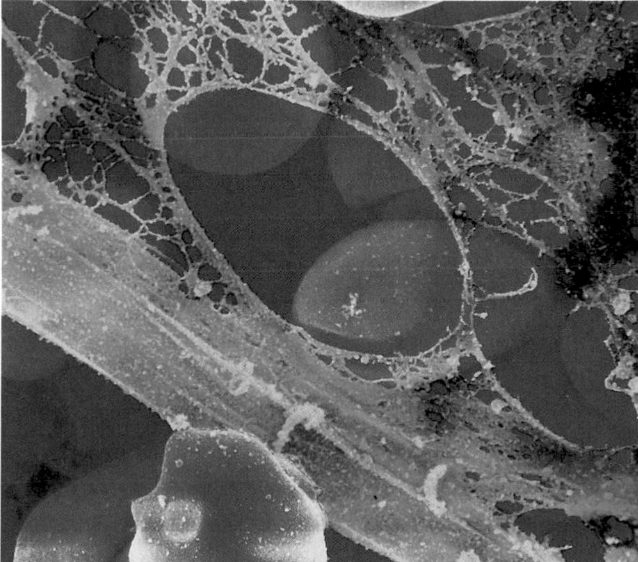

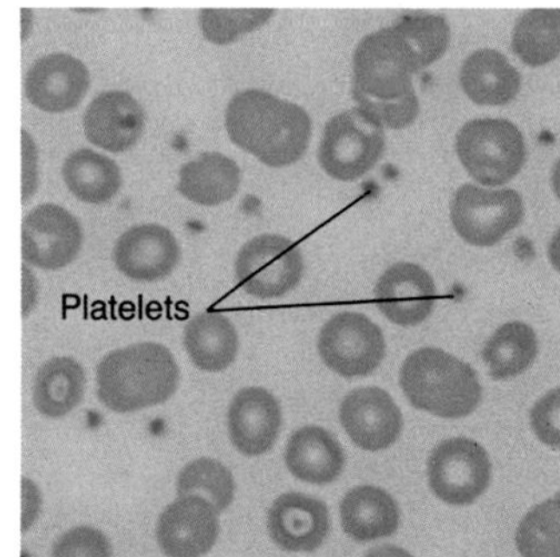


**Figure 12-8** Red blood cells caught in fibers form the beginning of a blood clot, as seen under an electron microscope (left). Platelets, as seen under a light microscope (right), produce a chemical that binds the fibers together.



## TEACH

### Concept Development

▶ Platelets are also called thrombocytes. Students may be familiar with the related terms *thrombus* or *thrombosis.* Both refer to a clot.

**Cooperative Learning:** Divide the class into 4 groups. Provide them with references that will enable them to gather information on the following diseases or disorders: pernicious anemia, sickle-cell anemia, leukemia, hemophilia. Group 1 determines the cause of each problem. Group 2 determines the symptoms for each problem. Group 3 determines the diagnosis for each problem. Group 4 determines the treatment or cure for each problem. Allow each group to exchange information with the others to gain a complete picture for each disease.

### Guided Practice

Have students write down their answers to the margin question: Platelets aid the blood in clotting.

**GLENCOE TECHNOLOGY**

**Videotape**

The Secret of Life

*Nothing to Sneeze at: Viruses*

**CRITICAL THINKING 12**

CRITICAL THINKING/PROBLEM SOLVING — CHAPTER 12

Name ______ Date ______ Class ______

*Use after Section 12:2.*

***HOW ARE ACTIVE AND PASSIVE IMMUNITY DIFFERENT?***

The students in a biology class studied immunity. Then they worked in small groups to find examples of active and passive immunity in their own lives. They also asked how the body reacts to foreign proteins.

The students learned about active and passive immunity. Active immunity occurs when a person has a disease and is later immune to it, or when vaccines that contain dead pathogens are introduced into the body. Active immunity is long lasting. Passive immunity results when antibodies are introduced into the body. This kind of immunity lasts only a short time.

*Read the report of each group and answer the questions below.*

The first group of students studied the inoculations given to young babies. They found diptheria, pertussis, and tetanus injections (DPT) are given to babies at 2, 4, 15, and 18 months old. Each injection contains dead bacteria. The baby develops immunity to the diseases. Most states require another booster before children start school.

The second group studied injections against measles. They found that measles is caused by a virus. The injection contains viruses that are inactive, so they cause a mild form of the disease. After children get a slight rash and fever once, they are not likely to catch that kind of measles again.

The third group studied the injection given after a person is cut by a dirty nail. The injection contains antibodies from the blood of horses and helps to prevent tetanus.

**Analyzing the Problem**

1. Define immunity. the ability of a person to resist the effects of a disease after the person has once contacted the disease or its cause
2. Define active immunity. long-lasting protection after a person has been infected with a disease-causing bacteria or virus
3. What is a vaccine? a preparation of disease-causing bacteria or viruses that have been treated so they are not able to infect a person with the disease
4. Define passive immunity. short-lasting protection after a person has received antibodies against a specific disease
5. What is an antigen? a protein that stimulates the body to produce antibodies

**Solving the Problem**

1. Is the DPT injection an example of active or passive immunity? active
2. Is the measles vaccine an example of active or passive immunity? active
3. Is the tetanus injection an example of active or passive immunity? passive

## OPTIONS

### Science Background

Clotting of blood is a complex series of enzyme reactions that takes place in the body. Not only are platelets needed for the process, but clotting proteins within the plasma are also essential.

**Critical Thinking/Problem Solving/** *Teacher Resource Package,* p. 12. Use the Critical Thinking/Problem Solving worksheet shown here to reinforce immunity.

## TEACH

### Independent Practice

**Study Guide**/*Teacher Resource Package,* p. 72. Use the Study Guide worksheet shown at the bottom of this page for independent practice.

### Check for Understanding

Have students respond to the first three questions in Check Your Understanding.

### Reteach

**Reteaching**/*Teacher Resource Package,* p. 33. Use this worksheet to give students additional practice with parts of the blood.

**Extension:** Assign Critical Thinking, Biology and Reading, or some of the **OPTIONS** available with this lesson.

## 3 APPLY

### Brainstorming

You have just cut yourself. What kinds of cells are trapped in the clot?

## 4 CLOSE

### Audiovisual

Show the videocassette *Internal Defenses,* Films for the Humanities and Sciences.

### Answers to Check Your Understanding

6. red blood cells, white blood cells, and platelets
7. destroy harmful microbes, remove dead cells, and make proteins that help prevent disease
8. During leukemia, the number of white blood cells in a drop of blood may reach 100 000 or more, the number of white blood cells does not return to normal, and the cells formed are abnormal.
9. Iron is one of the nutrients used to make red blood cells.
10. months or years; about 10 days

**Figure 12-9** A person with hemophilia needs a transfusion of blood clotting factor, even after a minor accident.

### Skill Check

**Understand science words: hemophilia.** The word part *hemo* means blood. In your dictionary, find three words that contain the word part *hemo. For more help, refer to the* ***Skill Handbook,*** *pages 706-711.*

A small drop of blood contains about 250 000 platelets. The life span of platelets is very short, only about five days. Like red cells and some white cells, platelets are made in the bone marrow.

Two problems can happen with platelets. First if the number of platelets is too low, it's hard for blood clots to form. Second, if platelets lack the clotting chemical, this causes hemophilia (hee muh FIHL ee uh). **Hemophilia** is a disease in which a person's blood won't clot. A minor cut or bruise can be very dangerous to a person with hemophilia because of the large amount of blood that might be lost. Hemophilia is a genetic disease. This means that a person inherits the disease from his or her parents.

### Check Your Understanding

6. What substances make up the living part of blood?
7. What are three jobs of white blood cells?
8. What are three ways the rise in white blood cell numbers during leukemia differs from the rise in white blood cell numbers during infection?
9. **Critical Thinking:** You have learned that increasing the amount of iron in one's diet is a way of treating anemia. What does this indicate about the relationship between iron and red blood cells?
10. **Biology and Reading:** What is the maximum life span of a white blood cell? What is the average life span of most white blood cells?

### RETEACHING 33

RETEACHING CHAPTER 12

Name ______ Date ______ Class ______

Use with Section 12:2.

PARTS OF THE BLOOD

Fill in the table below. Tell what role each type of cell plays and describe what the cells are like and how they live and do their jobs.

| Cell type | Role | Characteristics |
|---|---|---|
| Red cell | carries oxygen to body cells | 5 000 000 per drop of blood<br>round, doughnut-like in shape<br>has no nucleus in mature cell<br>life span of 120 days<br>made in bone marrow<br>contains hemoglobin |
| White cell | destroys harmful microbes and removes dead cells<br>makes antibodies | 8 000 per drop of blood<br>larger than red blood cell<br>has a nucleus<br>most live for 10 days<br>made in bone marrow, lymph glands, spleen, and thymus<br>can move out of blood capillaries |
| Platelet | aids in forming blood clots that stop bleeding | 250 000 per drop of blood<br>cell part, no nucleus, irregular shape<br>life span of 5 days<br>made in bone marrow |

### STUDY GUIDE 72

STUDY GUIDE CHAPTER 12

VOCABULARY

*Review the new words used in Chapter 12 of your textbook. Then, use the terms below to fill in the blanks in the sentences that follow.*

immunity, antigens, hemophilia, AIDS, bone marrow, immune system, red blood cells, white blood cells, plasma, anemia, leukemia, hemoglobin, platelets, antibodies

1. Plasma is the yellow, nonliving part of blood.
2. Red blood cells are the cells in the blood that carry oxygen to the tissues.
3. Hemoglobin is a protein in red blood cells that joins with oxygen.
4. When a person has anemia there are too few red blood cells or too little hemoglobin or both.
5. The cells in the blood that remove microbes and dead cells are white blood cells.
6. When a person has leukemia, the number of white blood cells increases abnormally.
7. Platelets are bloodlike parts important in blood clotting.
8. The blood of a person with hemophilia will not clot.
9. The immune system helps keep a person free of disease.
10. Antibodies are chemicals that destroy bacteria or viruses.
11. Antigens are foreign substances that invade the body and cause disease.
12. Immunity is the ability of a person who once had a disease to be protected from getting it again.
13. AIDS is a disease of the immune system.
14. Red blood cells are made in bone marrow, the soft center part of the bone.

# Lab 12–2 Human Blood Cells 

## Problem: How can human blood cells be counted, identified, and compared?

### Skills

use a microscope, compare, classify

### Materials

sheet of clear plastic
marking pencil
prepared slide of human blood
microscope

### Procedure

1. Copy the data table.
2. Figures A and B show what blood cells would look like under the microscope. Each figure represents blood from a healthy person. Place a piece of clear plastic over Figure A.
3. Count and record the number of red blood cells. Use the marking pencil to check the cells counted so that you do not count them twice.
4. Count and record the number of white blood cells.
5. Count and record the number of platelets.
6. Repeat steps 2 through 5 for Figure B.
7. **Use a microscope:** Look at a prepared slide of human blood under low power on the microscope. Change to high power.
8. Draw and label a red blood cell, white blood cell, and platelet.

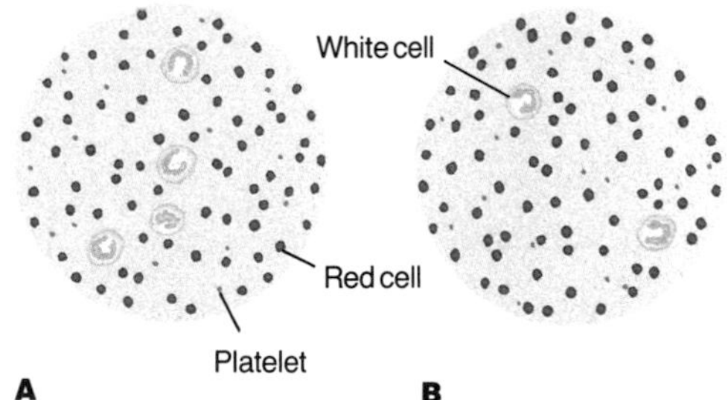


Answers will vary.

| NUMBER OF CELLS | FIGURE A | FIGURE B | PREPARED SLIDE |
|---|---|---|---|
| Red cells | 75 | 77 | 108 |
| White cells | 4 | 2 | 3 |
| Platelets | 14 | 16 | 22 |

9. Record the numbers of red blood cells, white blood cells, and platelets in an area similar to that in the figures.

### Data and Observations

1. According to your results with Figures A and B:
   (a) which blood cell type is most common?
   (b) which type is second-most common?
   (c) which type is least common?
2. According to your results with the prepared slide:
   (a) which blood cell type is most common?
   (b) which type is second-most common?
   (c) which type is least common?

### Analyze and Apply

1. **Classify:** List three ways that:
   (a) red cells differ from white cells.
   (b) red cells differ from platelets.
2. Which count of cells, from figures or from a prepared slide, is more accurate?
3. **Apply:** How is a cell count important when diagnosing an illness?

### Extension

**Compare:** Examine a prepared slide of frog blood under high power. Count the numbers of red blood cells and white blood cells. Compare frog and human blood.



## COMPUTER PROGRAM XII

COMPUTER PROGRAM LAB 12-2 HUMAN BLOOD CELLS

```
]POKE 33,33

]LIST

 10  GOSUB 390
 20  HOME
 30  PRINT "------------------------------------";
 40  PRINT "!  NUMBER    !   FIGURE   !   FIGURE   !";
 50  PRINT "! OF CELLS   !     A      !     B      !";
 70  PRINT "------------------------------------";
 80  PRINT "!RED CELLS   !            !            !";
 90  PRINT "------------------------------------";
 100  PRINT "!WHITE CELLS !            !            !";
 120  PRINT "------------------------------------";
 130  PRINT "!PLATELETS   !            !            !";
 140  PRINT "------------------------------------";
 150  FOR X = 1 TO 6
 160  VTAB V(X): HTAB H(X): FLASH : PRINT " ";: NORMAL
 170  VTAB 15: HTAB 1: CALL  - 958
 180  PRINT "ENTER YOUR VALUE FOR THIS SPACE";
 190  INPUT ": ";A
 200  VTAB V(X): HTAB H(X): PRINT A
 210  NEXT
 220  VTAB 15: HTAB 1: CALL  - 958
 230  PRINT "ARE THESE NUMBERS                 ENTERED CORRECTLY? (
     Y/N): ";: GET A$
 240  IF A$ = "N" OR A$ = "n" THEN  GOTO 20
 250  IF A$ < > "Y" AND A$ < > "y" THEN 220
 260  VTAB 15: HTAB 1: CALL  - 958: PRINT
 270  FLASH : PRINT "PRESS THE SPACE BAR TO SEE CORRECT DATA": GET A$
 278  NORMAL
 280  VTAB 16: HTAB 1: CALL  - 868
 290  FOR X = 1 TO 6
 300  VTAB V(X): HTAB H(X) + 6
 310  READ C: PRINT "(";C;")"
 320  NEXT
 330  INVERSE : PRINT
 340  PRINT : PRINT "DO YOU WANT THIS PRINTED (Y/N)? :";: GET A$
 345  NORMAL
 348  VTAB 14: HTAB 1: CALL  - 958
 350  IF A$ = "N" OR A$ = "n" THEN  END
 360  IF A$ < > "Y" AND A$ < > "y" THEN  GOTO 340
 370  GOSUB 2000
 380  END
 390  TEXT : HOME : PRINT "LAB 12-2              HUMAN BLOOD CELLS"
 400  PRINT : PRINT
 410  PRINT "YOU WILL BE ASKED TO TYPE IN YOUR       RESULTS FOR THE 2 HU
     MAN BLOOD SAMPLES  ONTO A DATA TABLE THAT IS EXACTLY LIKE  THE ONE
     USED IN THE EXPERIMENT."
 430  PRINT
 440  PRINT "AFTER TYPING EACH NUMBER, PRESS THE     RETURN KEY. AFTER TY
     PING YOUR DATA, YOU WILL BE ASKED IF THE NUMBERS ARE ENTEREDCORRECTL
     Y.IF CORRECT TYPE THE LETTER Y.IF WRONG, TYPE THE LETTER N."
 450  INVERSE
```

## ANSWERS

### Data and Observations

1. (a) red cells
   (b) platelets
   (c) white cells
2. (a) red cells
   (b) platelets
   (c) white cells

### Analyze and Apply

1. (a) red cells are red, have no nucleus, and are more numerous than white cells; (b) red cells are red, are larger, and are more numerous than platelets
2. from figures, because the number doesn't vary
3. Too few or too many red or white cells indicates particular illnesses such as anemia and leukemia.

## Lab 12-2 Human Blood Cells

### Overview

Students will identify and determine anatomical differences among the three blood cell types. They will also compare the relative numbers of each.

**Objectives:** Upon completing this lab, students should be able to (1) **distinguish** red, white, and platelet cells, (2) **compare** the numbers of each blood cell type present in normal human blood, and (3) **describe** the changes in blood cell types that take place during anemia and leukemia.

**Time Allotment:** 30 minutes

### Preparation

 **Lab 12-2 worksheet**/*Teacher Resource Package,* p. 47-48.

 **Lab 12-2 Computer Program/** *Teacher Resource Package,* p. xii. Use the computer program shown at the bottom of this page as you teach this lab.

### Teaching the Lab

▶ **Troubleshooting:** It may be necessary for students to move the slide slowly along the stage while viewing under high power so that several fields of view are noted. Not all fields will contain white blood cells.

### ASSESSMENT

**Content:** Have students redraw Figure B as if the blood cells were taken from a person with the following condition or disease: anemia, leukemia, and hemophilia. Have students explain their drawings and how they differ from the original Figure B.

## PREPARATION

### Material Needed

Make copies of the Reteaching and Study Guide worksheets in the *Teacher Resource Package.*

### Key Science Words

No new science words will be introduced in this section.

## 1 FOCUS

▶ The objectives are listed on the student page. Remind students to preview these objectives as a guide to this numbered section.

### MOTIVATION/Brainstorming

Ask students if they know their blood type. Frequently, parents will know this information.

### Guided Practice

Have students write down their answers to the margin questions: Proteins in the plasma of type A blood fit like a puzzle piece with the proteins of red cells in type B blood; mixing the red cell and plasma proteins could cause clumping.

**Portfolio**

Have students use Figures 12-10 and 12-11 as a guide to diagram the appearance of blood cells and plasma proteins when types A and AB are mixed. Students should do another diagram for types B and AB. Have students label all plasma proteins, blood cells, and red cell proteins.

# 12:3 Blood Types

**Objectives**

6. **Compare** the red cell and plasma proteins in the four main blood types.
7. **Explain** why blood types can't be mixed.

There are different kinds or types of blood. If you were to donate blood, your blood usually would be given to someone with the same type. If the wrong blood type were given, the person receiving the blood could have serious problems or even die.

## Blood Types Differ

**How is the plasma protein of type A blood like the red cell protein of type B blood?**

There are four main blood types: A, B, AB, and O. What makes each blood type different from the others? The difference is due to proteins found on red blood cells and in blood plasma. Look at Figure 12-10 and notice the different model shapes used to show red cell proteins and plasma proteins. If the protein is on a red blood cell, we shall call it a red cell protein. If it is in the plasma, we shall call it a plasma protein.

Note that the red cell protein shapes on types A and B blood cells are different. Also note that the red cell protein shapes on type AB are the same as both types A and B. Type O red blood cells have no red cell proteins.

Now notice the shapes of the plasma proteins. Type A plasma proteins are different from those in type B. Type AB has no plasma proteins, and type O has plasma proteins like types A and B.

**Figure 12-10** Blood types differ in their proteins. Each blood type has red cell and plasma proteins that don't fit each other.

| Blood type | Red cell protein present | Plasma protein present | Plasma protein and cell protein |
|---|---|---|---|
| A | Red cell, Protein | | No fit |
| B | | | No fit |
| AB | | None | No fit |
| O | None | | No fit |



## OPTIONS

**Blood Types**/*Transparency Package,* number 12a. Use color transparency number 12a as you teach blood types.

**What Blood Types Can Be Mixed?**/*Lab Manual,* pp. 99-102. Use this lab as an extension to blood types.

## Mixing Blood Types

Why is it safe to mix certain blood types and not others? The answer lies in the shapes of the red cell proteins and plasma proteins. Figure 12-10 is a model that shows how the shape of the red cell protein in type A blood can't fit into the shape of its own plasma protein. The red cell and plasma proteins have different shapes. The same is true for B, AB, and O types as well.

What happens when different blood types are mixed? Our models can help explain the result. Let's see what happens when a person with type A blood receives type B blood. Proteins in the plasma of type A blood fit like puzzle pieces into the red cell protein shape of type B blood. Proteins in the plasma of type B blood also fit like puzzle pieces into the red cell protein shape of type A blood. Note this happening in Figure 12-11.

Many type A and B blood cells now join together like all the completed pieces in a puzzle and large clumps of blood cells form. These clumps are too large to pass through capillaries. They plug these blood vessels and prevent the normal flow of blood. If these clumps block vessels in the heart, brain, or lungs, death can occur.

Why wouldn't it be safe for a person with type A blood to receive type AB blood? Why wouldn't it be safe for a person with type O blood to receive blood types A or B? Now you know why most people receive blood that is the same type as their own.

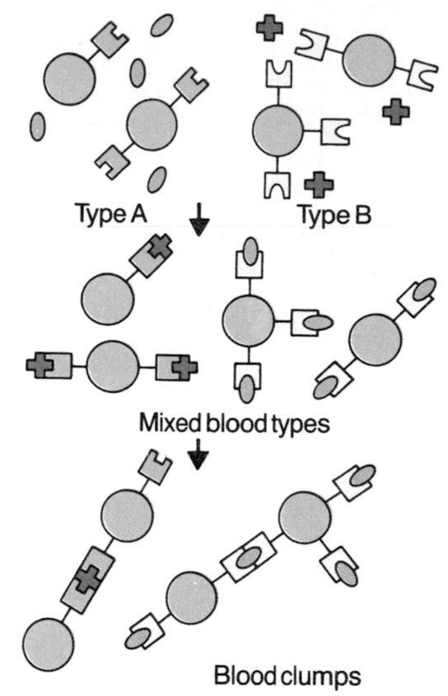


**Figure 12-11** Blood clots form when blood types A and B are mixed.

**Why can't blood types be mixed?** The plasma proteins in the type A blood would cause the AB blood to clump. The plasma proteins in the type O blood would cause A or B blood to clump.

### Check Your Understanding

11. What are the four main blood types?
12. Describe the model shapes of the red cell proteins and plasma proteins found in type O blood.
13. What happens to blood proteins and plasma when blood types A and B are mixed?
14. **Critical Thinking:** Sometimes AB plasma is transfused from one person to another. What trait of this plasma makes it safer to use in a transfusion than whole blood?
15. **Biology and Reading:** Look at Figure 12-10. Notice the shapes of plasma proteins for type O blood. Now, look at the shapes of the red cell proteins for type AB blood. What would happen if you mixed type O blood with type AB blood?

## 2 TEACH

### Independent Practice

**Study Guide**/*Teacher Resource Package*, p. 69. Use the Study Guide worksheet shown at the bottom of this page for independent practice.

### Check for Understanding

Have students respond to the first three questions in Check Your Understanding.

### Reteach

**Reteaching**/*Teacher Resource Package*, p. 34. Use this worksheet to give students additional practice with blood types.

**Extension:** Assign Critical Thinking, Biology and Reading, or some of the **OPTIONS** available with this lesson.

## 3 APPLY

### Connection: Math

Eighty-five percent of the population in the United States is Rh+ and 15% is Rh-. Ask students to calculate how many people out of 100 are O+ and O- if 50% of the population has blood type O. (Note: O- = .50 × .15 = .0765 or 7 people.)

## 4 CLOSE

Explain why you must wait at least a month before donating blood again.

### Answers to Check Your Understanding

11. A, B, AB, and O
12. Type O red blood cells have no red cell proteins while type O plasma has plasma proteins like those found in types A and B blood.
13. The blood cells join with the plasma proteins and form clumps.
14. Type AB plasma does not contain red blood cell proteins.
15. The blood would clump.

### STUDY GUIDE 69

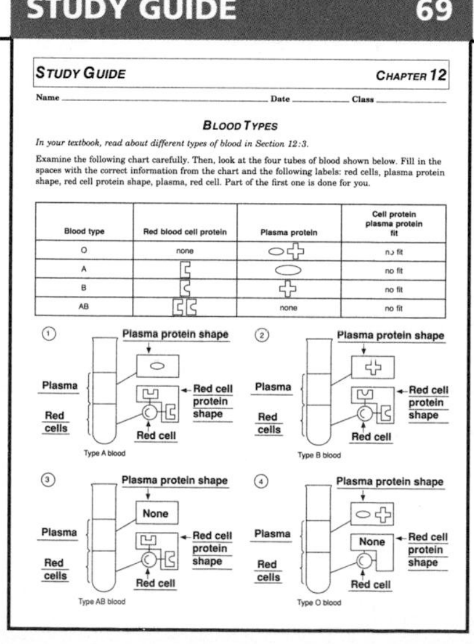

STUDY GUIDE — CHAPTER 12

Name ______ Date ______ Class ______

**BLOOD TYPES**

*In your textbook, read about different types of blood in Section 12:3.*

Examine the following chart carefully. Then, look at the four tubes of blood shown below. Fill in the spaces with the correct information from the chart and the following labels: red cells, plasma protein shape, red cell protein shape, plasma, red cell. Part of the first one is done for you.

| Blood type | Red blood cell protein | Plasma protein | Cell protein plasma protein fit |
|---|---|---|---|
| O | none | | no fit |
| A | | | no fit |
| B | | | no fit |
| AB | | none | no fit |



### RETEACHING 34

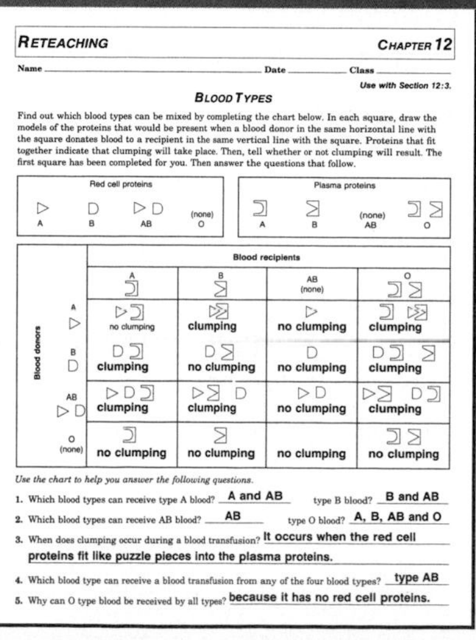

RETEACHING — CHAPTER 12

Name ______ Date ______ Class ______

*Use with Section 12:3.*

**BLOOD TYPES**

Find out which blood types can be mixed by completing the chart below. In each square, draw the models of the proteins that would be present when a blood donor in the same horizontal line with the square donates blood to a recipient in the same vertical line with the square. Proteins that fit together indicate that clumping will take place. Then, tell whether or not clumping will result. The first square has been completed for you. Then answer the questions that follow.



| Blood donors | Blood recipients: A | B | AB (none) | O |
|---|---|---|---|---|
| A | no clumping | clumping | no clumping | clumping |
| B | clumping | no clumping | no clumping | clumping |
| AB | clumping | clumping | no clumping | clumping |
| O (none) | no clumping | no clumping | no clumping | no clumping |

*Use the chart to help you answer the following questions.*

1. Which blood types can receive type A blood? A and AB type B blood? B and AB
2. Which blood types can receive AB blood? AB type O blood? A, B, AB and O
3. When does clumping occur during a blood transfusion? It occurs when the red cell proteins fit like puzzle pieces into the plasma proteins.
4. Which blood type can receive a blood transfusion from any of the four blood types? type AB
5. Why can O type blood be received by all types? because it has no red cell proteins.

## 12:4 Immunity

## PREPARATION

### Materials Needed

Make copies of the Focus, Application, Transparency Master, Study Guide, Skill, and Reteaching worksheets in the *Teacher Resource Package.*

### Key Science Words

immune system
antibody
antigen
immunity
Acquired Immune Deficiency Syndrome

### Process Skills

In the Skill Check, students will infer.

## 1 FOCUS

▶ The objectives are listed on the student page. Remind students to preview these objectives as a guide to this numbered section.

### MOTIVATION/Brainstorming

Ask students why they put a bandage over a cut or scrape? cover their mouth when coughing? receive shots for certain diseases?

Focus/*Teacher Resource Package,* p. 23. Use the Focus transparency shown at the bottom of this page as an introduction to AIDS.

**Portfolio**

Have students prepare a chart that lists the organs or cell types that perform these tasks: makes red cells, stores red cells, rids body of red cells, transports red cells, makes white cells, stores white cells, transports white cells, and makes antibodies.

**Objectives**

8. **Describe** how the immune system works.
9. **Explain** how shots prevent disease.
10. **Discuss** the AIDS virus.

**Key Science Words**

immune system
antibody
antigen
immunity
AIDS

## 12:4 Immunity

Another job of your blood helps you stay healthy. Certain blood cells play an important role in helping your body get rid of disease-causing viruses and bacteria.

### Your Immune System

Your immune system cures you of the flu, measles, mumps, or even boils. The **immune system** is made of proteins, cells, and tissues that identify and defend the body against foreign chemicals and organisms. It helps to keep you free of disease. Actually, the word *immune* means "to free."

Teardrops, mucus, and skin are part of your immune system. They prevent disease-causing organisms from entering your body. Other parts of the immune system are shown in Figure 12-12. Locate the parts of the immune system and examine their roles.

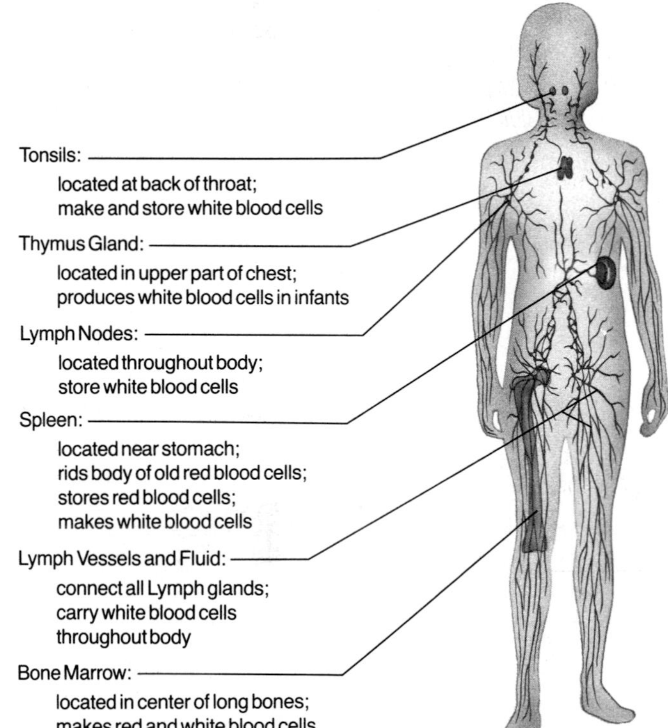


**Figure 12-12** In which organs of the immune system are white cells made? tonsils, thymus gland, spleen, and bones



## OPTIONS

### Science Background

Each specific antigen shape causes the formation of an antibody that has an almost complementary but opposite shape. The resulting combination of matching antigen and antibody shapes can be compared to a lock and key fit.

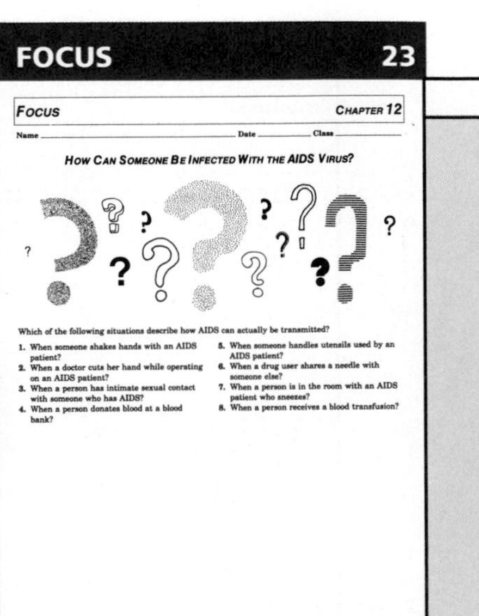
FOCUS 23

FOCUS CHAPTER 12

Name Date Class

HOW CAN SOMEONE BE INFECTED WITH THE AIDS VIRUS?

Which of the following situations describe how AIDS can actually be transmitted?

1. When someone shakes hands with an AIDS patient?
2. When a doctor cuts her hand while operating on an AIDS patient?
3. When a person has intimate sexual contact with someone who has AIDS?
4. When a person donates blood at a blood bank?
5. When someone handles utensils used by an AIDS patient?
6. When a drug user shares a needle with someone else?
7. When a person is in the room with an AIDS patient who sneezes?
8. When a person receives a blood transfusion?

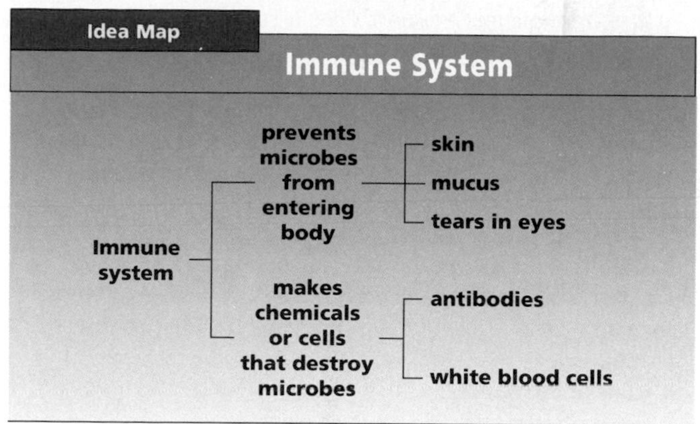


**Study Tip:** Use this idea map as a study guide for identifying the functions of the immune system.

## How the Immune System Works

There are many kinds of white blood cells. Each kind has its own job in the protection of the body. Some white blood cells have the job of making special chemicals called antibodies (ANT ih bohd eez). **Antibodies** are chemicals that help destroy bacteria or viruses. Bacteria, viruses, and other foreign substances are examples of antigens (ANT ih junz) that enter our bodies. **Antigens** are foreign substances, usually proteins, that invade the body and cause diseases.

Models can show how antibodies made by white cells get rid of antigens such as bacteria. Figure 12-13 shows a white blood cell with a model of an antibody on its cell membrane. The antibody looks like the letter *M*. Certain antibodies stay attached to the cell membrane. Other antibodies are released into the blood plasma. These released antibodies match a bacterium that has an exactly opposite "shape" on its surface as shown in Figure 12-13. Note that the antibody and antigen shapes fit together like a lock and key.

When the antigen and antibody on the surface of a white cell fit together correctly, the bacterium may be easily destroyed. First, the cell membrane of the bacterium may break open. When the cell membrane is broken, the bacterium will die. Second, with antibodies stuck to it, the bacterium can now be destroyed by other white blood cells. Each different type of virus or bacterium that enters the body has a different antigen shape on its surface. The white blood cells must have a similar number of different antibody shapes.

**Bio Tip**

**Health:** Have you had your shots? Check Table 12-2 on page 258 to see which shots you might need.

What are antigens?

**Figure 12-13** An antigen is destroyed when a white cell produces a matching antibody.

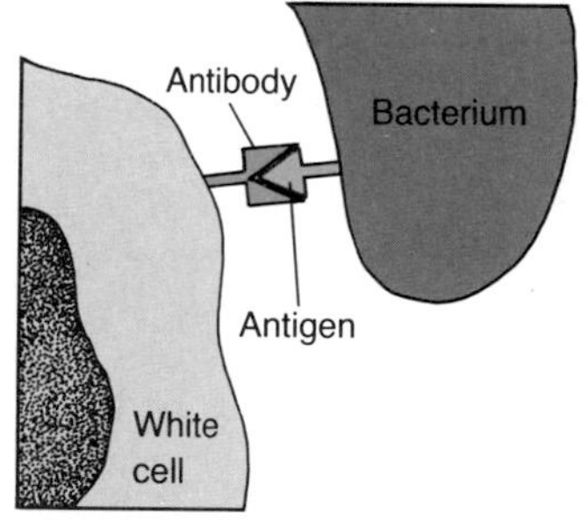




# 2 TEACH

### Concept Development

▶ Discuss the fact that the body may receive its immunity in several ways. Most common to students is naturally acquired immunity. The catching of a disease can lead to the formation of antibodies against a future exposure to the disease. Once infected, a person cannot be reinfected.

▶ The second way of obtaining immunity is through vaccines. This is artificially acquired immunity. A person is given a mild form of a disease to allow the immune system to make antibodies against the disease.

### Idea Map

Have students use the idea map as a study guide to the major concepts of this section.

### Guided Practice

Have students write down their answers to the margin question: Antigens are foreign substances that invade the body and cause diseases.

### Independent Practice

**Study Guide**/*Teacher Resource Package*, p. 70. Use the Study Guide worksheet shown at the bottom of this page for independent practice.

**APPLICATION 12**

**APPLICATION: HEALTH** CHAPTER 12

Name ______ Date ______ Class ______

*Use after Section 12:4.*

**HOW DOES AIDS CAUSE DEATH?**

A person with the disease AIDS does not die directly from the virus itself. Illness and death take place because the person's immune system is destroyed. The virus destroys one kind of white blood cell that helps fight disease-causing organisms. These special white blood cells are called T4 cells.

*Study the diagrams to compare what happens in the blood of a healthy person and of a person with AIDS. Answer the questions that follow.*

a. A disease-causing antigen (bacterium) enters the body.
b. T4 white blood cells detect bacterium.
c. T4 cells direct another kind of white blood cell, called B white blood cells, to make antibodies against the bacterium.
d. Antibodies join with bacterium.
e. Bacterium is destroyed and disease is prevented.

Healthy person

a. A disease-causing antigen (AIDS virus) enters the body.
b. AIDS virus enters T4 white blood cell.
c. AIDS virus reproduces in T4 cell and destroys it.
d. The new AIDS viruses from first T4 cell destroy other T4 cells.
e. B white blood cells receive no direction to make antibodies because T4 cells have been destroyed.
f. New antigen (a bacterium) enters body and is not destroyed.
g. Bacteria reproduce unchecked and may kill person.

Person with AIDS

1. Give examples of two types of antigens. **bacterium and AIDS virus**
2. What is the job of a T4 white blood cell? **to detect antigens and to direct B white blood cells to make antibodies against the antigens**
3. What is the job of a B white blood cell? **to make antibodies that destroy antigens**
4. Which white cells are destroyed by the AIDS virus? **T4 cells**
5. Why don't B cells do their job when the AIDS virus is present? **because T4 cells are destroyed and are not there to give B cells directions to make antibodies**

**STUDY GUIDE 70**

**STUDY GUIDE** CHAPTER 12

Name ______ Date ______ Class ______

**IMMUNITY**

*In your textbook, read about immunity in Section 12:4.*

1. Define immune system. **proteins, cells, and tissues that identify and defend the body against foreign chemicals and organisms**
2. a. Define antigen. **a foreign substance that causes disease and also antibodies to form**
   b. Are antigens helpful? Explain. **No, they invade the body and cause diseases.**
3. a. Define antibody. **a chemical that helps destroy bacteria or viruses**
   b. Are antibodies helpful? Explain. **Yes, they keep us healthy by getting rid of antigens that enter our bodies.**
4. Figures A, B, C, and D are labeled for you. You are to:
   a. draw the antigen onto the correct cell that normally has antigens on it.
   b. draw the antibody onto the correct cell that normally has antibodies on it.

A White cell — B Bacterium — C Antigen — D Antibody

5. How is the meeting of the white blood cell and the bacterium shown in question 4 like the meeting of a lock and a key? **The shape of the white blood cell's antibody will fit together with the shape of the bacterium's antigen.**
6. List two ways that this meeting can result in the death of the bacterium. **By combining with the white blood cell, the membrane of the bacterium may break open. With antibodies stuck to it, the bacterium can then be destroyed by other white blood cells.**

## OPTIONS

**The Immune System**/*Transparency Package*, number 12b. Use color transparency number 12b as you teach the immune system.

**Application**/*Teacher Resource Package*, p. 12. Use the Application worksheet shown here to reinforce the study of AIDS.

## TEACH

### Concept Development

▶ The hepatitis virus infects only liver cells and the cold virus infects only nasal cells. The AIDS virus infects only certain white blood cells.

### Guided Practice

Have students write down their answers to the margin questions on these two pages: The antigens in the vaccine trigger the immune system to make antibodies to the DPT vaccine. Later, if the person is exposed to the disease, the body is ready with antibodies to destroy the virus; AIDS destroys white cells that fight infections.

### Skill Check

The body has a memory for the measles immunity; however, since there are many flu viruses, a person can be invaded by a flu virus of which the body has no memory.

### *Portfolio*

Using Table 12-1 and Figure 12-13 as a guide, have students design and draw models to illustrate three different antigen and corresponding antibody shapes on white blood cells. Students should label the white cells *bacterium, antigen,* and *antibody.* Students should use their models to explain why the body needs to make different antibodies for each different antigen.

**TABLE 12–1. CAUSES OF SOME DISEASES**

| Disease | Cause |
|---|---|
| Diphtheria | Bacteria |
| Tetanus (Lock Jaw) | Bacteria |
| Pertussis (Whooping cough) | Bacteria |
| Polio (Infantile paralysis) | Virus |
| Measles | Virus |
| Rubella (German measles) | Virus |
| Mumps | Virus |

**What occurs in a person's body after receiving a DPT shot?**

### Skill Check

**Infer:** It's common for a person to get the measles only once during his or her lifetime. Yet, a person can get the flu many times during his or her lifetime. What does this indicate about the body's "memory" of these immunities? *For more help, refer to the **Skill Handbook,** pages 706-711.*

The immune system now does a remarkable thing. It begins to make many new white blood cells, each with the shape of the antibody just used to get rid of the antigen. It's as if the immune system has a memory. White blood cells will reproduce antibodies in the bloodstream for many months or years. Their job is to prevent future disease caused by the same type of bacteria. If the same type of bacteria gets into your body five years later, the white blood cells with the antibodies will quickly destroy the bacteria. This time, the bacteria are so quickly destroyed that you may not even know you were invaded by the bacteria.

## How Shots Keep You Healthy

When you receive shots to prevent a disease, your immune system's memory often is being refreshed. DPT shots against diphtheria (dif THIHR ee uh), pertussis (pur TUH suhs), and tetanus (TET nus) are good examples. The DPT shot is made up of proteins from these three bacterial diseases.

When you receive a DPT shot, it's as if you are getting antigens of these three diseases. Your immune system begins to make antibodies against these antigens. Or, if you have already come into contact with the disease, your body may already have a memory of the antigens. Later on, if the actual bacteria causing these diseases were to enter your body, your immune system would be ready and waiting. You will have gained immunity. Table 12-1 lists other diseases for which you may have received shots. **Immunity** is the ability of a person who once had a disease to be protected from getting the same disease again. Table 12-2 shows the schedule for receiving shots.

**TABLE 12–2. IMMUNIZATION SCHEDULE FOR PERSONS AGE 7 AND OLDER***

| Spacing | Vaccines | Explanation |
|---|---|---|
| 1st visit | Td (adult)<br>TOPV<br>M/M/R | Td—tetanus and diphtheria adult vaccine<br>TOPV—trivalent oral polio vaccine<br>M/M/R—one dose of measles/mumps/rubella combined vaccine |
| 2 months after first | Td (Adult) TOPV | |
| 6—12 months after 2nd | Td (Adult) TOPV | |

*who did not begin their immunization series before age 15 months



## OPTIONS

**Skill**/*Teacher Resource Package,* p. 13. Use the Skill worksheet shown here to have students outline the section on the immune system.

**Transparency Master**/*Teacher Resource Package,* p. 53. Use the Transparency Master worksheet shown here to help students with the role of white blood cells.

**TRANSPARENCY 53**

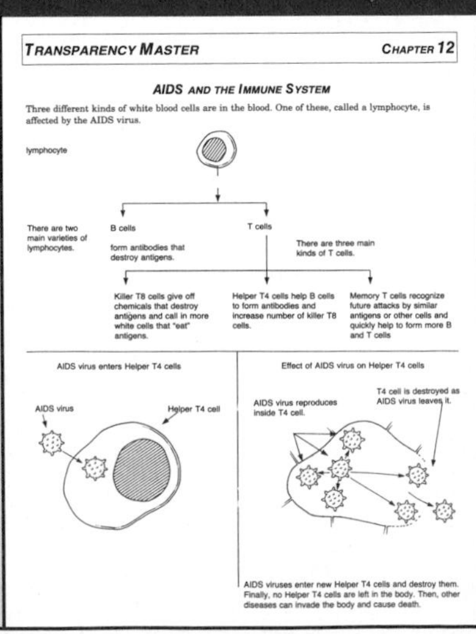
TRANSPARENCY MASTER — CHAPTER 12

AIDS AND THE IMMUNE SYSTEM

Three different kinds of white blood cells are in the blood. One of these, called a lymphocyte, is affected by the AIDS virus.



**SKILL 13**

SKILL: OUTLINE — CHAPTER 12

Name ______ Date ______ Class ______

*For more help, refer to the Skill Handbook, pages 706-711.* *Use after Section 12:4.*

THE IMMUNE SYSTEM

*Complete the outline below by filling in the blanks under the main ideas.*

A. What the immune system is
1. Define the immune system. **the proteins, cells, and tissues of the body that identify and defend the body against foreign chemicals and organisms**
2. Name the body parts of the immune system and tell the function of each part. **Tears, skin, and mucus keep organisms out of the body; tonsils, thymus gland, lymph nodes, and spleen store white blood cells; and bone marrow makes white blood cells.**
3. What do white blood cells do in the immune system? **Some white blood cells make antibodies.**
4. What is an antibody? **a chemical that helps destroy bacteria or viruses**

B. How the immune system works
1. What are antigens? **foreign substances that invade the body**
2. What do antigens do? **cause antibodies to form**
3. What happens when an antigen meets an antibody that matches it? **They join like a lock and key.**
4. Name two ways that an antigen such as a bacterium can be destroyed when it has joined an antibody. **The cell membrane of a bacterium may break open. White blood cells destroy the bacterium when antibodies are stuck to it.**

C. How immunity is obtained
1. Explain how the body has lasting immunity after a specific antigen has been destroyed. **White blood cells that produce antibodies against the antigen are formed for years to come.**
2. How does a shot help give lasting immunity? **A shot contains antigens of a specific disease. The antigens cause the immune system to form antibodies that will protect against the disease organisms.**

## AIDS and the Immune System

**Acquired Immune Deficiency Syndrome,** or **AIDS,** is a disease of the immune system. AIDS is caused by a virus that reproduces only inside one kind of white blood cell.

What happens when this type of white blood cell is totally destroyed? Other diseases can then easily invade the body. Diseases that the body's immune system would normally overcome can now cause death.

How can one be infected by the AIDS virus, Figure 12-14? The virus is present in body fluids, such as blood and semen. There are four known ways the virus can be passed from one person to another. One is during sexual intercourse, when body fluids containing the virus may be passed along through broken body tissues. Second, another way is by sharing needles during drug use. The needle may have a small amount of blood on it. If the blood has the AIDS virus in it, the virus can be injected into the next person's blood. Third, the AIDS virus may be passed from a pregnant woman to her unborn child. Fourth, it's possible to get AIDS from a blood transfusion if the blood is infected. Modern technology has made it possible to test and reject blood that has the AIDS virus.

AIDS is spreading rapidly and there is no known cure for it. Through mid-1996, there have been more than 510 000 recorded cases of AIDS in the United States. Of this number, more than 315 000, or over 60 percent, have died of AIDS.

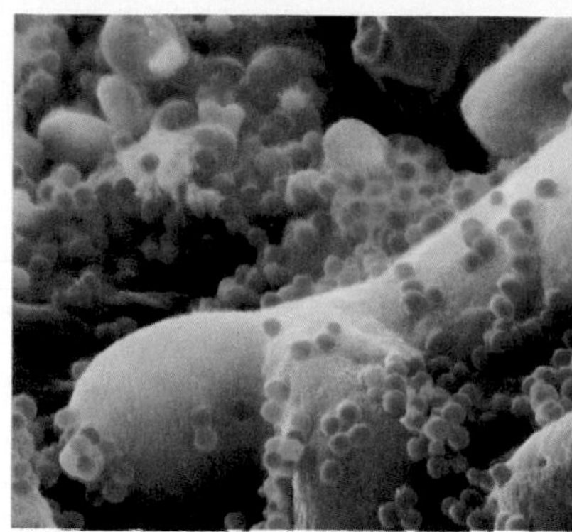

**Figure 12-14** AIDS virus particles attack white blood cells and destroy the immune system.

What effect does the AIDS virus have on a person's immune system?

### Check Your Understanding

16. Compare antigens and antibodies.
17. What is a DPT shot made up of?
18. What are four ways the AIDS virus can be passed from one person to another?
19. **Critical Thinking:** Why do you think there's not a shot against AIDS?
20. **Biology and Writing:** When a person has an organ transplant, the transplanted organ has antigens different from those of the person receiving the organ. Usually, the person receiving the organ is given drugs that lower the body's immune response. Explain how this treatment might help the transplant procedure.

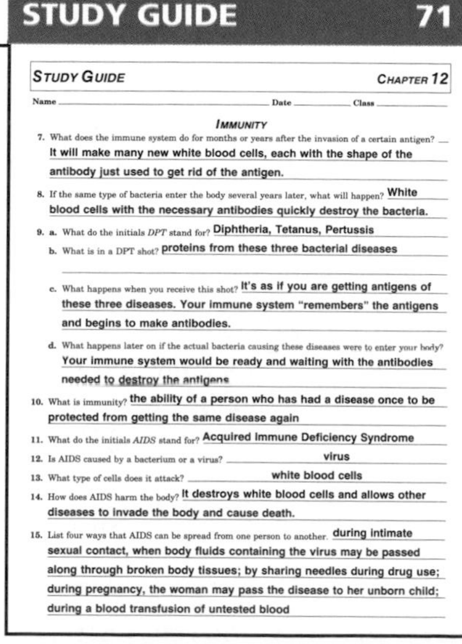

**STUDY GUIDE 71**

STUDY GUIDE — CHAPTER 12

Name ______ Date ______ Class ______

*IMMUNITY*

7. What does the immune system do for months or years after the invasion of a certain antigen? It will make many new white blood cells, each with the shape of the antibody just used to get rid of the antigen.
8. If the same type of bacteria enter the body several years later, what will happen? White blood cells with the necessary antibodies quickly destroy the bacteria.
9. a. What do the initials *DPT* stand for? Diphtheria, Tetanus, Pertussis
   b. What is in a DPT shot? proteins from these three bacterial diseases
   c. What happens when you receive this shot? It's as if you are getting antigens of these three diseases. Your immune system "remembers" the antigens and begins to make antibodies.
   d. What happens later on if the actual bacteria causing these diseases were to enter your body? Your immune system would be ready and waiting with the antibodies needed to destroy the antigens
10. What is immunity? the ability of a person who has had a disease once to be protected from getting the same disease again
11. What do the initials *AIDS* stand for? Acquired Immune Deficiency Syndrome
12. Is AIDS caused by a bacterium or a virus? virus
13. What type of cells does it attack? white blood cells
14. How does AIDS harm the body? It destroys white blood cells and allows other diseases to invade the body and cause death.
15. List four ways that AIDS can be spread from one person to another. during intimate sexual contact, when body fluids containing the virus may be passed along through broken body tissues; by sharing needles during drug use; during pregnancy, the woman may pass the disease to her unborn child; during a blood transfusion of untested blood

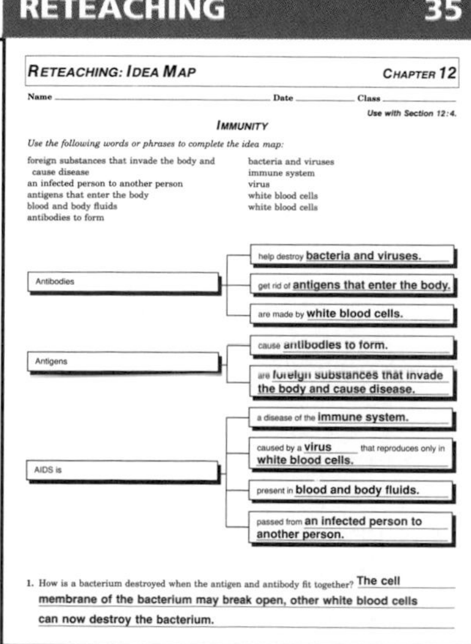

**RETEACHING 35**

RETEACHING: IDEA MAP — CHAPTER 12

Name ______ Date ______ Class ______

Use with Section 12:4.

*IMMUNITY*

*Use the following words or phrases to complete the idea map:*

foreign substances that invade the body and cause disease
an infected person to another person
antigens that enter the body
blood and body fluids
antibodies to form
bacteria and viruses
immune system
virus
white blood cells
white blood cells

Antibodies
- help destroy bacteria and viruses.
- get rid of antigens that enter the body.
- are made by white blood cells.

Antigens
- cause antibodies to form.
- are foreign substances that invade the body and cause disease.

AIDS is
- a disease of the immune system.
- caused by a virus that reproduces only in white blood cells.
- present in blood and body fluids.
- passed from an infected person to another person.

1. How is a bacterium destroyed when the antigen and antibody fit together? The cell membrane of the bacterium may break open, other white blood cells can now destroy the bacterium.

## TEACH

### Independent Practice

**Study Guide**/*Teacher Resource Package,* p. 71. Use the Study Guide worksheet shown at the bottom of this page for independent practice.

### Check for Understanding

Have students respond to the first three questions in Check Your Understanding.

### Reteach

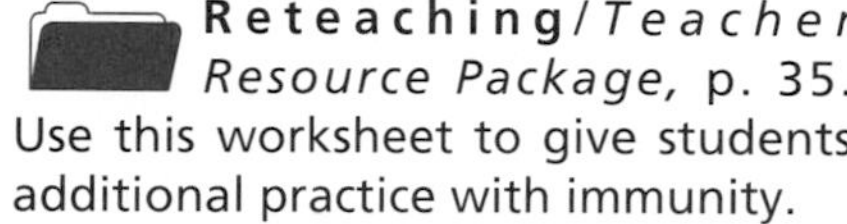

**Reteaching**/*Teacher Resource Package,* p. 35. Use this worksheet to give students additional practice with immunity.

**Extension:** Assign Critical Thinking, Biology and Writing, or some of the **OPTIONS** available with this lesson.

## 3 APPLY

Have students use references to determine and record which disease provides lasting immunity.

## 4 CLOSE

### Audiovisual

Show the videocassette *The Nature and Transmission of AIDS,* Films for the Humanities and Sciences.

### Answers to Check Your Understanding

16. Antigens are foreign substances that invade the body and cause disease. Antibodies are chemicals that destroy antigens.
17. proteins from the three different bacterial diseases—diphtheria, pertussis and tetanus.
18. during sexual intercourse, by sharing needles during drug use, during pregnancy, and by transfusion with infected blood
19. because the AIDS virus has many different forms
20. By receiving drugs that lower the body's immune system, the antibodies that would destroy the organ are not made and the organ is accepted.

# CHAPTER 12 REVIEW

## Summary

Summary statements can be used by students to review the major concepts of the chapter.

## Key Science Words

All boldfaced terms from this chapter are listed.

## Testing Yourself

### Using Words

1. platelets
2. hemoglobin
3. plasma
4. leukemia
5. anemia
6. AIDS
7. antibody
8. hemophilia
9. red blood cell

### Finding Main Ideas

10. p. 254; Blood types differ due to different types of proteins found in the plasma or on the red cells.
11. p. 244; Blood picks up waste chemicals and excess body heat.
12. p. 244; Blood delivers nutrients, oxygen, water, minerals, and chemical messengers.
13. p. 256; Parts of the immune system are tonsils, thymus gland, lymph nodes, vessels, and fluid; spleen, bone marrow. Tears, mucus, and skin can also be included.
14. p. 245; Animals that do not have blood include sponges, jellyfish, and flatworms.
15. p. 255; Blood cell proteins of type A will match with plasma proteins from type B. Blood cell proteins of type B will match with plasma proteins of type A. The result is clumping.
16. p. 259; A person can get AIDS from the exchange of body fluids during sexual intercourse, by sharing a needle used by someone with AIDS, and from a pregnant mother to her unborn child.
17. p. 257; Antigens are foreign substances that enter the body and cause disease. Antibodies are chemicals made by the white blood cells to destroy antigens.

# Chapter 12

# Review

## Summary

### 12:1 The Role of Blood

1. Blood is a body tissue that helps with delivery and pickup of chemicals, nutrients, gases, and wastes.
2. Animals without blood usually live in water. Water delivers oxygen and food to the cells and picks up wastes.

### 12:2 Parts of Human Blood

3. Blood plasma is a nonliving, yellow liquid that makes up 55 percent of blood. Red blood cells, white blood cells, and platelets are the living parts of blood.
4. Red blood cells carry oxygen in the blood. White blood cells destroy harmful microbes and remove dead cells. Platelets aid in blood clotting.
5. Health problems can occur in people who lack the proper amount of red blood cells, white blood cells, or platelets.

### 12:3 Blood Types

6. The four main blood types are A, B, AB, O.
7. Different blood types cannot be mixed.

### 12:4 Immunity

8. The immune system keeps the body free of most diseases. Antibodies destroy antigens that enter the body.
9. Shots help prevent disease.
10. AIDS is a serious disease caused by a virus.

## Key Science Words

AIDS (p. 259)
anemia (p. 248)
antibody (p. 257)
antigen (p. 257)
bone marrow (p. 248)
hemoglobin (p. 248)
hemophilia (p. 252)
immune system (p. 256)
immunity (p. 258)
leukemia (p. 250)
plasma (p. 247)
platelet (p. 251)
red blood cell (p. 247)
white blood cell (p. 249)

## Testing Yourself

### Using Words

*Choose the word from the list of Key Science Words that best fits the definition.*

1. helps form blood clots
2. gives red blood cells their color
3. part of blood that is nonliving, yellow liquid
4. disease in which white cell numbers increase



### Using Main Ideas

18. Plasma is 92% water and 8% other nutrients and salts.
19. Antigens and antibodies have shapes that fit together like lock and key. When they interlock, the bacterium can be easily destroyed by white cells eating it or making it break apart.
20. They act as if they have a memory when they produce antibodies that are capable of recognizing and destroying the same antigen many times, even years later.
21. Red cell protein shapes on types A and B blood cells are different. Type AB red cells have both A and B protein shapes. Type O red cells have no red cell proteins. Type A plasma proteins fit type B cell proteins. Type B plasma proteins fit type A cell proteins. Type AB has no plasma proteins, and type O has both A and B.
22. These animals live in water, which supplies these needs.
23. Shots tell the immune system to make antibodies against certain diseases.

# Review

## Testing Yourself *continued*

5. a problem resulting from too little hemoglobin in red cells
6. disease caused by a virus that destroys the immune system
7. chemicals that defend the body against foreign chemicals and organisms
8. a disease in which blood does not clot
9. cell with the job of carrying oxygen

### Finding Main Ideas

*List the page number where each main idea below is found. Then, explain each main idea.*

10. why blood types differ
11. what the pickup jobs of blood are
12. several delivery jobs of blood
13. the main parts of your immune system
14. which animal groups do not have blood
15. what happens when blood type A is mixed with type B
16. three ways of getting AIDS
17. how antibodies and antigens differ

### Using Main Ideas

*Answer the questions by referring to the page number after each question.*

18. What are the nutrients that make up blood plasma? (p. 247)
19. How do antibodies help get rid of bacteria? (p. 257)
20. When do white blood cells act as if they have a memory? (p. 258)
21. How do red cell proteins and plasma proteins differ in the four blood types? (p. 254)
22. How do animals with no blood get nutrients and oxygen? (p. 245)
23. How do shots help to prevent diseases? (p. 258)

## Skill Review 

*For more help, refer to the **Skill Handbook,** pages 704-719.*

1. **Infer:** The common cold is a viral disease that most people survive easily. Explain why a cold can be life-threatening to a person infected with the AIDS virus.
2. **Measure in SI:** The red blood cells in the large photo on page 242 are magnified about 6000 times. Measure the diameter of one cell and determine its actual size in micrometers. (1 mm = 1000 micrometers)
3. **Use a microscope:** Examine a prepared slide of human white blood cells. Compare the sizes of the different kinds of white cells and the different shapes of their nuclei.
4. **Understand science words:** The word part *anti* means against. Define *antibody* and *antigen* using the term *against* in your definitions.

## Finding Out More

### Critical Thinking

1. Trace the events that occur in your immune system after a mumps shot.
2. A person seriously needs a blood transfusion but none is available. What could be given temporarily to help?

### Applications

1. Research the progress of scientists' search for a cure for AIDS.
2. Interview a local dentist to determine how AIDS has altered even the most routine dental procedures.



## Skill Review

1. A person with an AIDS infection has an immune system that does not function properly. When this person gets a cold, the body cannot fight infection because the immune system does not work. The infection can therefore continue or get worse, possibly even cause death.
2. If the cell measures 65 mm, then the actual size is:
   $\frac{65 \times 1000}{6000}$ = 10.8 micrometers
3. Students may find five different kinds of white blood cells. Some white blood cells have large, round nuclei that fill the cell; others have lobed nuclei, and some are surrounded by small granules.
4. An antigen is something that fights against good health, such as a bacterium or virus. Antibodies fight against the antigen and help restore good health.

## Finding Out More

### Critical Thinking

1. When the mumps shot is given, the immune system detects that there is a foreign substance in the body. It identifies the substance, the mumps vaccine, and starts to make antibodies to destroy the antigen. The immune system destroys the antigen and then "remembers" the antibody necessary to do so. If sometime in the future the same antigen enters the body, the immune system kicks in and destroys the antigen again, using the "remembered" antibodies.
2. Type O plasma could be used until a blood match can be obtained, because type O plasma would not contain the red cell proteins.

### Applications

1. AIDS is a relatively new disease. Progress is being made, however, and drugs to slow the progression of the disease are being studied. A vaccine is being developed. There is no cure at this time.
2. The AIDS virus is present in blood and body fluids and only a small amount, as little as a drop, of blood is necessary to transmit the disease. Dentists do procedures that can cause bleeding. As a result, dentists take extra care to be sure there is no disease transmission by wearing surgical masks, gloves, and sterilizing all instruments after they are used.

 **Chapter Review**/*Teacher Resource Package,* pp. 60-61.

 **Chapter Test**/*Teacher Resource Package,* pp. 62-64.

**Chapter Test**/*Computer Test Bank*

# Respiration and Excretion

## PLANNING GUIDE

| CONTENT | TEXT FEATURES | TEACHER RESOURCE PACKAGE | OTHER COMPONENTS |
|---|---|---|---|
| (1 1/2 days)<br>13:1 The Role of Respiration<br>Gases of Respiration<br>Respiration in Animals | Idea Map, p. 264<br>Check Your Understanding, p. 265 | Reteaching: *Animal Respiratory Systems,* p. 36<br>Focus: *A Mammal That Can Breathe Like a Fish,* p. 25<br>Study Guide: *The Role of Respiration,* p. 73 | **Laboratory Manual:**<br>*How Does Breathing Occur?* p. 103<br>**STVS:** *Nicotine and the Lungs,* Human Biology (Disc 7, Side 2) |
| (2 days)<br>13:2 Human Respiratory System<br>Respiratory Organs<br>Gas Exchange<br>Breathing | Skill Check: *Understand Science Words,* p. 267<br>Mini Lab: *Is Carbon Dioxide Present?* p. 269<br>Mini Lab: *How Much Air Do You Breathe Out?* p. 271<br>Lab 13-1: *Breathing Rate,* p. 266<br>Idea Map, p. 271<br>Check Your Understanding, p. 271 | Enrichment: *The Epiglottis,* p. 13<br>Reteaching: *Human Respiratory System,* p. 37<br>Critical Thinking/Problem Solving: *Does Air Breathed In or Out Contain Carbon Dioxide?* p. 13<br>Skill: *Interpreting Diagrams,* p. 14<br>Study Guide: *Human Respiratory System,* p. 74<br>Lab 13-1: *Breathing Rate,* pp. 49-50 | Color Transparency 13a: *Breathing In and Out*<br>**STVS:** *Children and Smog,* Human Biology (Disc 7, Side 2) |
| (1/2 day)<br>13:3 Problems of the Respiratory System<br>A Breathing Problem<br>Respiratory Diseases | Skill Check: *Define Words in Context,* p. 273<br>Check Your Understanding, p. 273 | Reteaching: *Respiratory Problems,* p. 38<br>Study Guide: *Problems of the Respiratory System,* p. 75 | **STVS:** *Detecting Cystic Fibrosis,* Human Biology (Disc 7, Side 2) |
| (2 days)<br>13:4 The Role of Excretion<br>Waste Removal<br>Human Kidneys<br>Excretion and the Skin<br>Problems of the Excretory System | Skill Check: *Interpret Diagrams,* p. 277<br>Mini Lab: *What Does a Kidney Do?* p. 275<br>Lab 13-2: *Urine,* p. 278<br>Science and Society: *Organ Transplants,* p. 281<br>Check Your Understanding, p. 280 | Application: *The Artificial Kidney Machine,* p. 13<br>Reteaching: *Journey Through the Kidneys,* p. 39<br>Transparency Master: *Nephron Unit of the Kidney,* p. 59<br>Study Guide: *The Role of Excretion,* pp. 76-77<br>Study Guide: *Vocabulary,* p. 78<br>Lab 13-2: *Urine,* pp. 51-52 | **Laboratory Manual:**<br>*What Chemical Wastes Are Removed by the Lungs, Skin, and Kidneys?* p. 107<br>Color Transparency 13b: *Human Excretory System*<br>**STVS:** *Laser Treatment of Bladder Cancer,* Human Biology (Disc 7, Side 2) |
| Chapter Review | Summary<br>Key Science Words<br>Testing Yourself<br>Finding Main Ideas<br>Using Main Ideas<br>Skill Review | **ASSESSMENT RESOURCES**<br>Chapter Review, pp. 65-66<br>Chapter Test, pp. 67-69<br>Performance Assessment in the Biology Classroom<br>Alternate Assessment in the Science Classroom<br>Computer Test Bank | |

### GLENCOE TECHNOLOGY

**Science and Technology Videodisc Series,** *Nicotine and the Lungs,* Human Biology (Disc 7, Side 2)
*Children and Smog,* Human Biology (Disc 7, Side 2)
*Detecting Cystic Fibrosis,* Human Biology (Disc 7, Side 2)
*Laser Treatment of Bladder Cancer,* Human Biology (Disc 7, Side 2)

## MATERIALS NEEDED

| LAB 13-1, p. 266 | LAB 13-2, p. 278 | MARGIN FEATURES |
|---|---|---|
| 250-mL beaker<br>goldfish in aquarium<br>thermometer<br>net<br>watch or wall clock<br>ice | 4 test tubes<br>test tube rack<br>droppers<br>labels<br>distilled water<br>salt water<br>glucose<br>glucose test tape<br>silver nitrate in dropper bottle<br>normal artificial urine<br>abnormal artificial urine<br>4 slides | Skill Check, p. 267<br>dictionary<br>Mini Lab, p. 269<br>bromthymol blue<br>beaker<br>straw<br>Mini Lab, p. 271<br>1-L graduated cylinder<br>pan<br>rubber tube<br>water<br>Mini Lab, p. 275<br>funnel<br>filter paper<br>muddy water |

## OBJECTIVES

For more information about National Science Standards, see page 5T.

| SECTION | OBJECTIVE | CORRELATION of QUESTIONS to OBJECTIVES: CHECK YOUR UNDERSTANDING | CHAPTER REVIEW | TRP CHAPTER REVIEW | TRP CHAPTER TEST |
|---|---|---|---|---|---|
| 13:1 National Science Stds: UCP.1, UCP.3, UCP.4, B.3, B.6, C.1, C.5 | 1. **Define** the role of the respiratory system. | 1 | 12 | 9 | 11 |
| | 2. **Explain** why cells need oxygen and give off carbon dioxide. | 2 | 10 | 3 | 6, 7 |
| | 3. **Compare** the respiratory systems of different animals. | 3, 4, 5 | 18 | 1, 20 | 2, 12 |
| 13:2 National Science Stds: UCP.1, UCP.2, UCP.3, UCP.4, UCP.5, B.4, C.5 | 4. **Give the function** of different parts of the respiratory system. | 6, 10 | 3, 5, 7 | 2, 18, 27, 28, 29, 31, 32, 33, 34, 37 | 14, 15, 16, 17, 18, 19, 20, 21, 22 |
| | 5. **Compare** the pathways of gases as they enter and leave the body. | 7, 9 | 16, 19 | 12, 16, 30 | 8, 45, 46, 47, 48, 49 |
| | 6. **Sequence** the movement of the diaphragm and the rib cage during the breathing process. | 8, 14 | 13, 17 | 4, 11, 13, 14, 15, 38 | 39, 40, 41, 42, 43 |
| 13:3 National Science Stds: UCP.5, F.1, F.4, F.5 | 7. **Relate** the effects of carbon monoxide poisoning and how to avoid the gas. | 11 | 9, 20 | 5 | 5 |
| | 8. **Explain** the cause of pneumonia. | 12 | 2 | 36 | 9 |
| | 9. **Describe** emphysema. | 13, 15 | 15, 21 | 39 | 10 |
| 13:4 National Science Stds: UCP.1, UCP.2, UCP.3, UCP.4, UCP.5, E.1, F.1 | 10. **Discuss** the importance of the excretory system. | | 4, 8 | 10, 40 | 3 |
| | 11. **Sequence** the steps for filtering of blood in a kidney. | 16, 19, 20 | 1, 11, 20 | 6, 7, 17, 21, 22, 23, 24, 25, 26, 35, 39 | 1, 23, 24, 25, 26, 27, 28, 29, 30, 31, 32, 33 |
| | 12. **Describe** the skin as part of the excretory system. | 18 | 14 | 8, 19 | 4, 13 |

# Respiration and Excretion

**CHAPTER OVERVIEW**

## Key Concepts

In this chapter, students will study the anatomy and physiology of the human respiratory and excretory systems. They will learn the functions of the kidneys, skin, and lungs and the problems associated with these systems. Organ transplants are discussed.

## Key Science Words

| | |
|---|---|
| alveoli | respiratory system |
| bronchi | trachea |
| carbon monoxide | urea |
| diaphragm | ureter |
| emphysema | urethra |
| epiglottis | urinary bladder |
| excretory system | urine |
| nephron | |
| pneumonia | |

## Skill Development

In Lab 13-1, students will **form a hypothesis, measure in SI,** and **interpret data** while determining the effect of temperature on the breathing rate of fish. In Lab 13-2, they will **observe, interpret data, make and use tables,** and **infer** as they detect chemicals in urine. In the Skill Check on page 267, students will **understand** the **science word** *epiglottis.* On page 277, they will **interpret diagrams** while following the pathways of blood, wastes, and water. On page 273, they will **define words in context** while describing a communicable disease. In the Mini Lab on page 269, students will **experiment** to detect carbon dioxide. In the Mini Lab on page 271, they will **design an experiment** to measure air exhaled. In the Mini Lab on page 275, they will **formulate a model** to show how the kidneys work.

## Bridging

Chapter 2 introduced the term *cellular respiration.* Students will now have an opportunity to see how oxygen gas is supplied to cells and how carbon dioxide gas is removed from the body.

**CHAPTER PREVIEW**

### Chapter Content

Review this outline for Chapter 13 before you read the chapter.

**13:1 The Role of Respiration**
- Gases of Respiration
- Respiration in Animals

**13:2 Human Respiratory System**
- Respiratory Organs
- Gas Exchange
- Breathing

**13:3 Problems of the Respiratory System**
- A Breathing Problem
- Respiratory Diseases

**13:4 The Role of Excretion**
- Waste Removal
- Human Kidneys
- Excretion and the Skin
- Problems of the Excretory System

### Skills in this Chapter

The skills that you will use in this chapter are listed below.

- In **Lab 13-1,** you will form a hypothesis, measure in SI, and interpret data. In **Lab 13-2,** you will observe, interpret data, make and use tables, and infer.
- In the **Skill Checks,** you will understand science words, interpret diagrams, and define words in context.
- In the **Mini Labs,** you will experiment, formulate a model, and design an experiment.

262

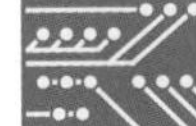

## TECH PREP

For Tech Prep activities in this chapter of the Teacher Wraparound Edition, see especially the Portfolios on pages 267 and 277, the Connection and the Activity on page 270, the Student Journals on pages 271 and 272, and the Activity on page 273.

See also the Glencoe Homepage at **www.glencoe.com**

Chapter 13

# Respiration and Excretion

If you were asked, "How much air can your lungs hold?" could you come up with the correct answer? The person in the photo on the left is snorkeling. Different people can stay under water for different lengths of time. This time can vary from a few seconds to several minutes. The small photo below shows one way to measure your own lung volume. What gases enter the lungs when you breathe in? What gases leave the lungs when you breathe out? Why is breathing important? What are some of the effects of breathing in polluted air? This chapter will help you answer these questions.

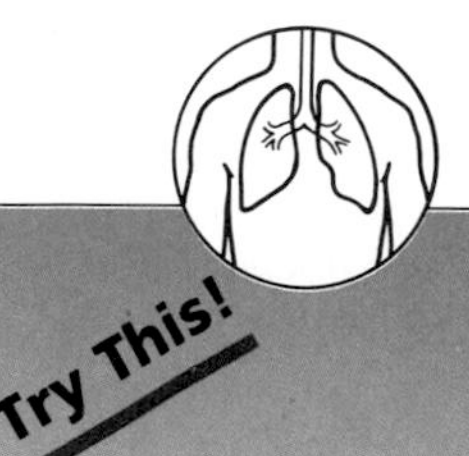

**Try This!**

**How many breaths do you take?** Count the number of times you breathe in and out for exactly one minute. What is your breathing rate?

**interNET CONNECTION**

For more information about the material in this chapter, follow the link for the chapter on the Glencoe Homepage at **http://www.glencoe.com**

Measuring the volume of air you breathe out



GETTING STARTED

## Using the Photos

Students will think that oxygen and carbon dioxide are the only gases they breathe in and out. Explain that other gases are present and that they may or may not affect breathing. The connection between breathing and cellular respiration needs to be emphasized. Discuss how lung volume varies with age, sex, and physical condition.

## MOTIVATION/Try This!

**How many breaths do you take?** Students will **observe** that their breathing rate is between 15 to 30 times per minute. A stopwatch will be needed. Students can **calculate** a class average.

## Chapter Preview

Have students study the chapter outline before they begin to read the chapter. They should note the overall organization of the chapter, that the role of respiration is presented first, the human respiratory system is discussed next, respiratory problems are then examined, and last, the excretory system is explained.

## Misconception

Students often confuse breathing with respiration. Breathing is a mechanical process that forces air into and out of the lungs. Respiration is a chemical process that occurs within all cells.

## ASSESSMENT PLANNER

**Portfolio**

Strategies on the following pages represent student products that can be placed into a best-work portfolio: pp. 267, 277.

**PERFORMANCE ASSESSMENT**

Skill Check, pp. 267, 273, 277
Mini Lab, pp. 269, 271, 275
Lab, pp. 266, 278

**CONTENT ASSESSMENT**

Check for Understanding, pp. 265, 269, 271, 273, 277, 280
Chapter Review, pp. 282-283

**GROUP ASSESSMENT**

Opportunities for group assessment occur with Cooperative Learning Strategies.

**Student Journal**

Strategies on the following pages represent opportunities for writing in a Student Journal: pp. 271, 272.

# 13:1 The Role of Respiration

## PREPARATION

### Materials Needed

Make copies of the Focus, Study Guide, and Reteaching worksheets in the *Teacher Resource Package.*

### Key Science Words

respiratory system

## 1 FOCUS

▶ The objectives are listed on the student page. Remind students to preview these objectives as a guide to this numbered section.

### MOTIVATION/Brainstorming

Ask students why it is harder to breathe at high altitudes.

### Guided Practice

Have students write down their answers to the margin questions: to exchange oxygen and carbon dioxide; carbon dioxide.

### Idea Map

Have students use the idea map as a study guide to the major concepts of this section.

**Focus**/*Teacher Resource Package,* pp. 25-26. Use the Focus transparency shown at the bottom of this page as an introduction to respiration.



## 13:1 The Role of Respiration

**Objectives**

1. **Define** the role of the respiratory system.
2. **Explain** why cells need oxygen and give off carbon dioxide.
3. **Compare** the respiratory systems of different animals.

**Key Science Words**

respiratory system

When you talk of respiration, many people think of breathing. Is there more to respiration than this? Yes, respiration involves more body parts besides the nose and mouth.

### Gases of Respiration

**What is the role of the respiratory system?**

**What waste gas is given off by all cells?**

Chapter 12 described how your blood delivers oxygen to all your body cells. It also explained how blood removes carbon dioxide from your cells.

Oxygen is brought into your body by the respiratory (RES pruh tor ee) system. Carbon dioxide is carried out of your body by the respiratory system. Your **respiratory system** is made of body parts that help with the exchange of gases. It brings in oxygen and removes carbon dioxide.

What happens to oxygen in cells? Oxygen is a gas used by cells in cellular respiration. During this chemical process, sugar reacts with oxygen, and energy is released for all of the cells' activities. At the same time, carbon dioxide and water are formed. Carbon dioxide is a waste gas given off by all cells. Water is an important nutrient for all cell processes. Remember that cellular respiration occurs in cells of all organisms, including plants. Cellular respiration can be shown as an equation:

$$\underset{\text{Oxygen}}{O_2} + \underset{\text{sugar}}{C_6H_{12}O_6} \xrightarrow{\text{Energy}} \underset{\text{carbon dioxide}}{CO_2} + \underset{\text{water}}{H_2O}$$

**Study Tip:** Use this idea map as a study guide to the role of the respiratory system. What is the process that releases energy from food?

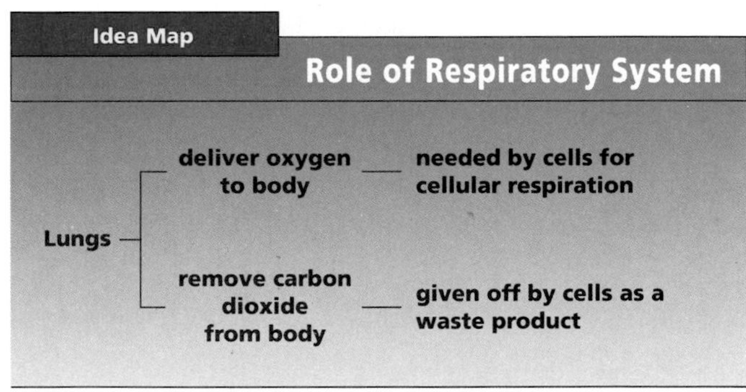




## OPTIONS

### Science Background

Certain life forms, such as bacteria and yeast, and body cells, such as muscle cells, can carry on respiration without oxygen. This is called anaerobic respiration. This process yields little available energy for the cell and results in food such as glucose being degraded into either ethyl alcohol or lactic acid. Most life forms, however, carry on aerobic respiration. This process requires oxygen and is more energy efficient than anaerobic respiration.

**FOCUS 25**

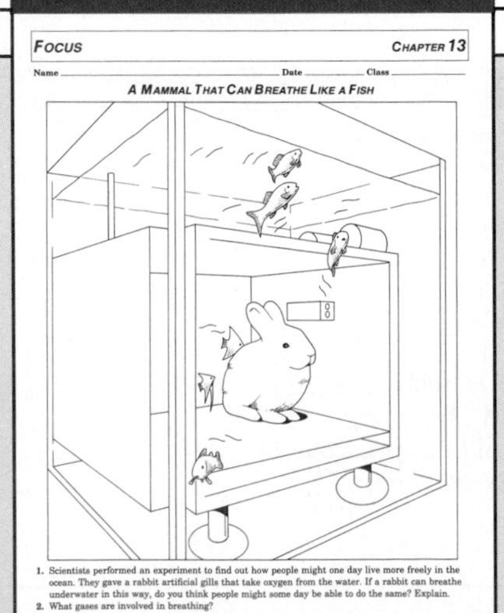
FOCUS CHAPTER 13

Name ________ Date ________ Class ________

A MAMMAL THAT CAN BREATHE LIKE A FISH

1. Scientists performed an experiment to find out how people might one day live more freely in the ocean. They gave a rabbit artificial gills that take oxygen from the water. If a rabbit can breathe underwater in this way, do you think people might some day be able to do the same? Explain.
2. What gases are involved in breathing?

## Respiration in Animals

Most animals have respiratory systems to help with the exchange of gases. In Figure 13-1, the blue shading shows the respiratory systems of some animals. Earthworms exchange gases through their skins. Insects have a large network of small tubes that carry air to all body parts. Fish carry on gas exchange through organs called gills. Frogs exchange gases through two types of organs—lungs and skin. Humans, as you know, also have lungs.

How do fish exchange gases?

The systems may look different in different animals, but all these systems are adapted for gas exchange. One important trait of all respiratory systems is that there is a large surface area over which air passes. This large area provides a surface for gases to diffuse into or out of the animals' bodies. In the earthworm, for example, the entire body surface acts as an area for gas exchange.

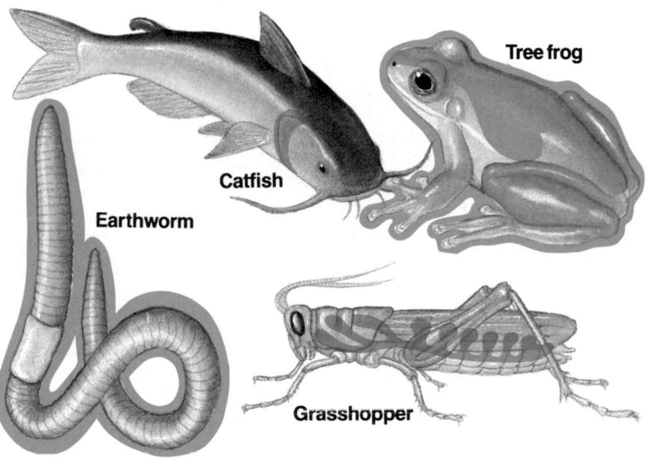


**Figure 13-1** Compare the respiratory systems of some animals. The shaded areas in blue show where animals exchange gases with air.

### Check Your Understanding

1. Which gases are exchanged by the respiratory system?
2. Why is oxygen needed by our cells?
3. Why do respiratory systems have a large surface area?
4. **Critical Thinking:** Why do birds need a more efficient respiratory system than humans?
5. **Biology and Writing:** Write a four to ten line paragraph that describes one kind of respiratory system.

## RETEACHING 36

**RETEACHING: IDEA MAP** — CHAPTER 13

Name ____________ Date ______ Class ______

*Use with Section 13:1.*

**ANIMAL RESPIRATORY SYSTEMS**

Use the following terms to complete the idea map below: skin, gills, insect, frog, fish, earthworm, skin and lungs, and network of small tubes.

Respiratory System in Animals

| Animal | Organ(s) Involved |
|---|---|
| earthworm | skin |
| insect | network of small tubes |
| fish | gills |
| frog | skin and lungs |

1. What respiratory trait is shared by all animals? large surface area for gas exchange
2. What two gases are exchanged during respiration? Oxygen is taken in and carbon dioxide is given off.
3. What products are formed during cellular respiration? carbon dioxide and water
4. Where does cellular respiration take place? Cellular respiration takes place in the cells of all organisms.
5. In what organ does gas exchange take place in humans? Gas exchange takes place in the lungs.
6. Why is a large surface area for gas exchange important? It provides a place where gas can diffuse into or out of the animals' blood or body cells.
7. What chemical process takes place in cellular respiration? Sugar reacts with oxygen, and energy is released.

## STUDY GUIDE 73

**STUDY GUIDE** — CHAPTER 13

Name ____________ Date ______ Class ______

**THE ROLE OF RESPIRATION**

*In your textbook, read about respiration in Section 13:1.*

1. What is the function of the respiratory system? to help with the exchange of gases; to bring in oxygen and to remove carbon dioxide
2. Explain why a large surface area is needed in a respiratory system. This large area provides a surface for gases to diffuse into or out of the blood or body cells.
3. Match the main methods of getting oxygen with the animals below. Write the letter of the method below each animal.
A = skin B = skin and lungs C = lungs D = gills E = small tubes
B A D E C

**HUMAN RESPIRATORY SYSTEM**

*In your textbook, read about respiratory system organs in Section 13:2.*

1. Label the parts on the diagram below using these words: bronchus, nose, nasal chamber, lung, trachea, epiglottis, alveoli.



2. Place the letter of the description below in the correct circle next to the labeled part of the diagram.
A = place where air is warmed or cooled
B = clusters of air sacs where gas exchange occurs
C = place where air first enters respiratory system
D = the windpipe
E = one of two major organs of respiration
F = carries air from trachea into lung
G = flap that prevents food from entering trachea

# 2 TEACH

### Guided Practice

Have students write down their answers to the margin question: through gills.

### Independent Practice

**Study Guide**/*Teacher Resource Package,* p. 73. Use the Study Guide worksheet shown at the bottom of this page for independent practice.

### Check for Understanding

Have students respond to the first three questions in Check Your Understanding.

### Reteach

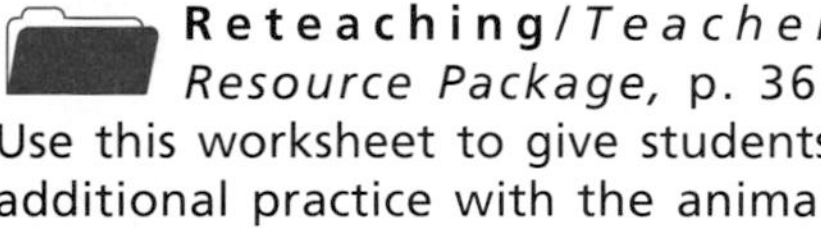

**Reteaching**/*Teacher Resource Package,* p. 36. Use this worksheet to give students additional practice with the animal respiratory system.

**Extension:** Assign Critical Thinking, Biology and Writing, or some of the **OPTIONS** available with this lesson.

# 3 APPLY

### Demonstration

Lift up the gill covering on a fresh fish to expose the extensive surface area of the respiratory system.

# 4 CLOSE

**Using Science Words:** Have students explain how respiration and breathing differ.

### Answers to Check Your Understanding

1. oxygen and carbon dioxide
2. to release energy
3. The larger the surface area, the more gases can be exchanged.
4. They have a higher rate of metabolism than humans.
5. An example is earthworms exchanging gases through skin.

## Lab 13-1 Breathing Rate

### Overview

In this lab, students will determine if the respiration rate of a fish changes as the amount of oxygen dissolved in the water varies.

**Objectives:** Upon completing this lab, students will have (1) **measured** the rate at which a fish uses oxygen, (2) **observed** that fish breathe faster in warm water than in cold water.

**Time Allotment:** 1 class period

### Preparation

▶ Pet shops stock feeder goldfish that are very reasonably priced.

**Lab 13-1 worksheet**/*Teacher Resource Package*, pp. 49-50.

**Lab 13-1 Computer Program**/*Teacher Resource Package*, p. xiv. Use the computer program shown at the bottom of this page as you teach this lab.

### Teaching the Lab

▶ Fish may live in water that is totally covered by ice because the very cold water contains much dissolved oxygen.

**ASSESSMENT**

**Content:** Have students prepare a graph of their average data. Students should place gill opening values along the bottom and temperature values along the left side of the graph. Students can extend the plotted line on their graphs to predict the expected number of gill openings at 10° and 40° Celsius.

**Cooperative Learning:** Students can work in groups. Each student in the group should make separate counts and average his or her counts. This will increase observer reliability.

# Breathing Rate

# Lab 13–1

## Problem: Does temperature affect the breathing rate of fish?

### Skills

form a hypothesis, measure in SI, interpret data

### Materials

250-mL beaker
net
ice
goldfish in aquarium tank
thermometer
watch or wall clock

### Procedure

1. Copy the data table.
2. Fill the beaker with water from the aquarium.
3. Use a net to transfer one goldfish from the tank to the beaker. **CAUTION:** *Handle animals with care.*
4. Locate the fish's gill cover. Note that the gill cover opens and closes as the fish breathes in and out.
5. **Measure in SI:** Use a thermometer to measure the water temperature in the beaker. Record this temperature in your table.
6. Count the number of times the fish opens its gill cover during one minute. Record this number in your data table.
7. Repeat step 6 two more times, recording your results in the table as trials 2 and 3.
8. Add a small piece of ice to the beaker to slowly drop the water temperature. Continue to add ice until the temperature of the water is 10 degrees lower than at the start. Wait at least 5 minutes before going on to step 9. As you wait, write a **hypothesis** as to whether you think the fish's breathing rate will speed up, slow down, or remain the same with the cooler water.
9. Repeat steps 5 through 7.
10. Complete the data table.
11. Allow the water temperature to rise to room temperature. Then, use the net to return your fish to the aquarium.

### Data and Observations

1. What were you measuring when you counted gill cover openings?
2. **Interpret data:** Were there more or fewer gill openings with cold water when compared with warm water?

| WATER TEMPERATURE: 24°C | | WATER TEMPERATURE: 14°C | |
|---|---|---|---|
| TRIAL | NUMBER OF GILL OPENINGS | TRIAL | NUMBER OF GILL OPENINGS |
| 1 | 63 | 1 | 44 |
| 2 | 58 | 2 | 46 |
| 3 | 65 | 3 | 45 |
| Total | 186 | Total | 135 |
| Average | 62 | Average | 45 |

### Analyze and Apply

1. Cold water contains more oxygen than warm water. Explain why a fish in cold water breathes more slowly than a fish in warm water.
2. **Check your hypothesis:** Is your hypothesis supported by your data? Explain.
3. **Apply:** Why might a fish die if the water temperature gets too warm?

### Extension

**Predict** the number of gill openings at 5 degrees lower than the water was at the start. Experiment to test your prediction.



## ANSWERS

### Data and Observations

1. the rate of breathing
2. fewer

### Analyze and Apply

1. Oxygen is supplied to the blood faster in colder water.
2. Students who hypothesized that the breathing rate will slow down will answer yes. The oxygen supply is greater in colder water.
3. Warm water has less oxygen and a fish must breathe faster. Eventually a fish may use energy faster than it can take in oxygen, and will die.

## COMPUTER PROGRAM XIV

COMPUTER PROGRAM LAB 13-1 BREATHING RATE

```
]LIST

 5  DIM YES(5),NO(5)
 8  HOME
 10  GOSUB 480
 40  HOME : VTAB 12: INPUT "ENTER FIRST WATER TEMPERATURE: ";T1
 50  INPUT "ENTER SECOND WATER TEMPERATURE: ";T2
 60  HOME : GOSUB 450
 70 T1$ =  STR$ (T1)
 80  IF  LEN (T1$) < 3 THEN T1$ = "0" + T1$: GOTO 80
 90 T2$ =  STR$ (T2)
 100  IF  LEN (T2$) < 3 THEN T2$ = "0" + T2$: GOTO 100
 110  PRINT "! WATER TEMP    "T1$" ! WATER TEMP   "T2$" !";
 120  GOSUB 450
 130  PRINT "!       ! NUMBER OF !       ! NUMBER OF!";
 140  PRINT "! TRIAL !   GILL    ! TRIAL !   GILL   !";
 150  PRINT "!       ! OPENINGS  !       ! OPENINGS !";
 160  GOSUB 450
 170 S1 = 0:S2 = 0
 180  FOR X = 1 TO 3
 190  VTAB (X * 2 + 8): PRINT "!   ";X;"   !";
 200  HTAB 14: INPUT "";NO(X)
 210  VTAB (X * 2 + 8): HTAB 21: PRINT "!   ";X;"   !   ";
 220  INPUT "";YES(X)
 230  VTAB (X * 2 + 8): HTAB 40: PRINT "!";
 240  GOSUB 450
 250 S1 = S1 + NO(X):S2 = S2 + YES(X)
 260  NEXT X
 270  GOSUB 450: INPUT "ARE THESE NUMBERS CORRECT? ";S$
 280  IF S$ = "N" OR S$ = "n" THEN 80
 290  VTAB  PEEK (37) - 1: HTAB 1
 300  PRINT "! TOTAL !           ! TOTAL !          !";
 310  VTAB 16: HTAB 14: PRINT S1;: HTAB 34: PRINT S2
 320  GOSUB 450
 330  PRINT "!AVERAGE!           !AVERAGE!          !";
 340 A1 =  VAL ( LEFT$ ( STR$ (S1 / 3),6))
 350 A2 =  VAL ( LEFT$ ( STR$ (S2 / 3),6))
 360  VTAB 18: HTAB 14: PRINT A1;: HTAB 34: PRINT A2;
 370  VTAB 18: HTAB 40: PRINT "!";
 380  GOSUB 450
 390  IF SPACE$ = "N" THEN 80
 400  IF SPACE$ = "Y" THEN  END
 410  PRINT "DO YOU WANT THIS PRINTED (Y/N)? ";: GET A$
 420  IF A$ <  > "Y" AND A$ <  > "N" THEN VTAB 20: HTAB 1: GOTO 410
 430  IF A$ = "Y" THEN VTAB 20: HTAB 1: CALL  - 868: GOSUB 670
 440  END
 450  HTAB 1:
 460  PRINT "+---------------------------------------";
 470  RETURN
 480  TEXT : HOME
 490  GOSUB 600
 500  PRINT : PRINT "YOU WILL BE ASKED TO TYPE IN YOUR WATER TEMPERATURE
     AND RESULTS FOR THE 3 TRIALSONTO A DATA TABLE ";
 510  PRINT "THAT IS EXACTLY LIKE  THE ONE USED IN THE EXPERIMENT."
 520  PRINT
```

## 13:2 Human Respiratory System

What are the parts of the human respiratory system and how do they work? When you breathe, all you may feel is the slight movement of your chest. What else is involved in breathing?

**Objectives**

4. **Give the function** of different parts of the respiratory system.
5. **Compare** the pathways of gases as they enter and leave the body.
6. **Sequence** the movement of the diaphragm and rib cage during the breathing process.

**Key Science Words**

epiglottis
trachea
bronchi
alveoli
diaphragm

### Respiratory Organs

Use Figure 13-2 to tour the respiratory system and study its parts.

1. Our tour starts with the nose. As you breathe in, air enters the nose. Hairs inside the nose trap dust particles and prevent them from entering the lungs.
2. The nasal chamber is a large space above the roof of your mouth. It warms or cools the entering air. Mucus is formed inside the nose. Mucus keeps the air moist as it passes through the nose.
3. Air enters the nose and passes to the windpipe. The windpipe is covered by the epiglottis (ep uh GLAHT uhs). The **epiglottis** is a small flap that closes over the windpipe when you swallow. This closing prevents food or liquid from entering the windpipe when you are eating. If food passed into your windpipe, you would choke.
4. The air passes into your windpipe, which is called the trachea (TRAY kee uh). The **trachea** is a tube about 15 centimeters long that carries air to two shorter tubes that lead into your lungs.
5. **Bronchi** (BRAUN ki) are two short tubes that carry air from the trachea to the left and right lung.

**Skill Check**

**Understand science words: epiglottis.** The word part *glott* means language. In your dictionary, find three words with the word part *glott* in them. *For more help, refer to the* ***Skill Handbook,*** *pages 706-711.*

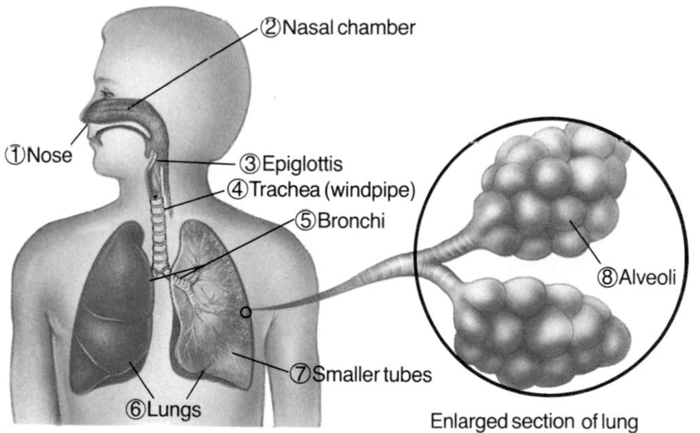


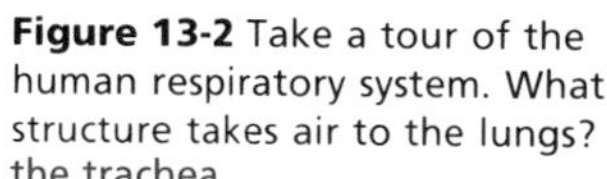

**Figure 13-2** Take a tour of the human respiratory system. What structure takes air to the lungs? the trachea



## 13:2 Human Respiratory System

### PREPARATION

#### Materials Needed

Make copies of the Reteaching, Skill, Critical Thinking/Problem Solving, Enrichment, Study Guide, and Lab worksheets in the *Teacher Resource Package*.

▶ Have goldfish on hand for Lab 13-1.

▶ Gather straws and rubber tubes for the Mini Labs.

#### Key Science Words

epiglottis
trachea
bronchi
alveoli
diaphragm

#### Process Skills

In Lab 13-1, students will form a hypothesis, measure in SI, and interpret data. In the Mini Labs, students will experiment and design an experiment. In the Skill Check, they will understand science words.

### 1 FOCUS

▶ The objectives are listed on the student page. Remind students to preview these objectives as a guide to this numbered section.

#### ACTIVITY/Brainstorming

Ask students if lungs are made of muscle. Establish that they are not, and ask how lungs pump air into and out of the body.

**Portfolio**

Provide students with two straws, scissors, tape, two twist ties, and two balloons. Have students construct a model of the human respiratory system showing the trachea, epiglottis, bronchi, and lungs.

### OPTIONS

#### Science Background

The human respiratory system consists of a series of hollow tubes and air sacs that terminate in and actually form the lungs. Capillaries surround these air sacs and serve as the site for gas exchange between air and blood. Blood carries oxygen to all body cells while carrying carbon dioxide away from the body cells to lung air sacs.

# 2 TEACH

## MOTIVATION/Demonstration

Use a 1- or 2-liter empty plastic soda bottle to show that pressure influences the movement of air and that air does occupy space. Ask students why air is pushed out of the bottle when the sides are squeezed together and why air rushes back in when the sides return to normal.

## Concept Development

▶ Ask students what happens when the epiglottis fails to close quickly. They have all experienced coughing that follows an attempt to clear the windpipe.

▶ Mucus lining the nasal passage and chamber traps dust, foreign matter, and bacteria. The mucus is brought up from the trachea and is either swallowed or spit out.

▶ Nitrogen gas is present in the blood in the same concentration as in the air. It is not utilized in the body but is absorbed into the bloodstream.

**Analogy:** The branches of the respiratory system can be compared to a tree with many smaller and smaller branches radiating out from the central trunk.

## Guided Practice

Have students write down their answers to the margin question: Gases are exchanged in the alveoli.

**Cooperative Learning:** Divide the class into three groups and have each design an idea map on the respiratory system.

**Critical Thinking/Problem Solving**/*Teacher Resource Package*, p. 13. Use the Critical Thinking/Problem Solving worksheet shown here to give students practice with components of breathed air.

**Skill**/*Teacher Resource Package*, p. 14. Use the Skill worksheet shown here to give students practice in interpreting diagrams of the respiratory system.

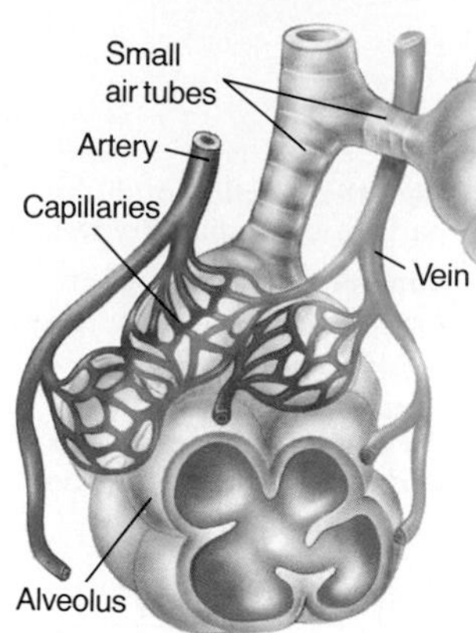


**Figure 13-3** In the alveoli, gas exchange takes place in less than one-tenth of a second.

6. The two lungs are major organs in your respiratory system. Lungs are large, soft organs in which oxygen and carbon dioxide are exchanged.
7. As the bronchi enter each lung, they branch into thousands of smaller and smaller tubes. Each tiny tube finally leads into tiny air sacs.
8. There are about 300 million alveoli (al VEE uh li) found in each lung. **Alveoli** are the tiny air sacs of the lungs. The alveoli look like clusters of grapes.

Each air sac, or alveolus, is surrounded by tiny blood vessels called capillaries, Figure 13-3. Oxygen in the air sacs moves through the cell membranes into the blood. Carbon dioxide moves out of the blood into the air sacs. If all the air sacs in a person's lungs were opened and spread out, they would cover an area equal to about 1/5 of the area of a basketball court, Figure 13-4. The air sacs of your lungs provide a large surface area for gases to pass into and out of your blood.

Why are alveoli important?

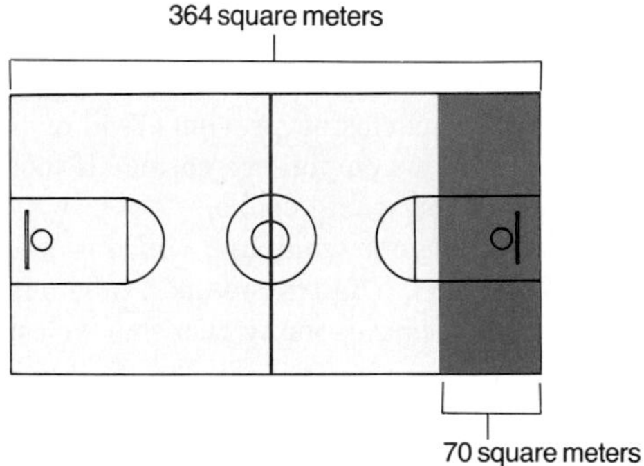


**Figure 13-4** The area covered by a person's air sacs equals about one-fifth the area of a basketball court.

## Gas Exchange

The function of your lungs is to exchange gases. Table 13-1 shows how much oxygen ($O_2$), carbon dioxide ($CO_2$), and nitrogen ($N_2$) gas we breathe in and out.

**TABLE 13–1. MAKEUP OF AIR WE BREATHE**

| Gas | Chemical Symbol | Air Entering Lungs | Air Leaving Lungs |
|---|---|---|---|
| Oxygen | $O_2$ | 19.97% | 16.00% |
| Carbon Dioxide | $CO_2$ | .03% | 4.00% |
| Nitrogen | $N_2$ | 80.00% | 80.00% |

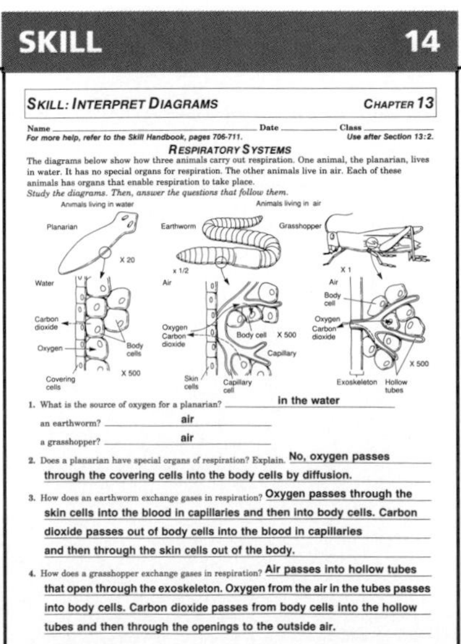

**SKILL 14**

SKILL: INTERPRET DIAGRAMS — CHAPTER 13

Name ____________ Date ______ Class ______

*For more help, refer to the Skill Handbook, pages 706-711.* *Use after Section 13:2.*

**RESPIRATORY SYSTEMS**

The diagrams below show how three animals carry out respiration. One animal, the planarian, lives in water. It has no special organs for respiration. The other animals live in air. Each of these animals has organs that enable respiration to take place.

*Study the diagrams. Then, answer the questions that follow them.*



1. What is the source of oxygen for a planarian? **in the water**
   an earthworm? **air**
   a grasshopper? **air**
2. Does a planarian have special organs of respiration? Explain. **No, oxygen passes through the covering cells into the body cells by diffusion.**
3. How does an earthworm exchange gases in respiration? **Oxygen passes through the skin cells into the blood in capillaries and then into body cells. Carbon dioxide passes out of body cells into the blood in capillaries and then through the skin cells out of the body.**
4. How does a grasshopper exchange gases in respiration? **Air passes into hollow tubes that open through the exoskeleton. Oxygen from the air in the tubes passes into body cells. Carbon dioxide passes from body cells into the hollow tubes and then through the openings to the outside air.**

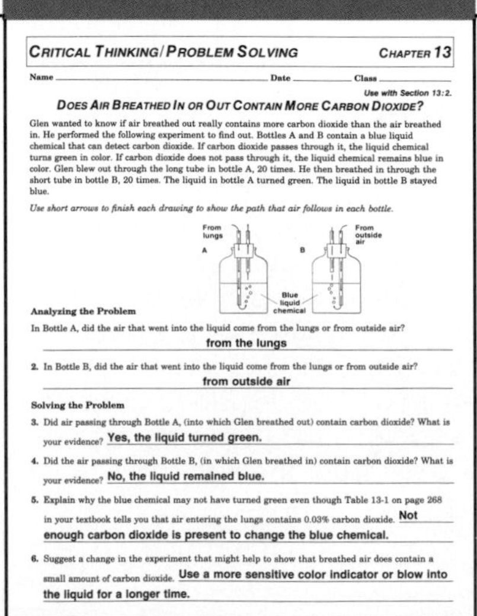

**CRITICAL THINKING 13**

CRITICAL THINKING/PROBLEM SOLVING — CHAPTER 13

Name ____________ Date ______ Class ______

*Use with Section 13:2.*

**DOES AIR BREATHED IN OR OUT CONTAIN MORE CARBON DIOXIDE?**

Glen wanted to know if air breathed out really contains more carbon dioxide than the air breathed in. He performed the following experiment to find out. Bottles A and B contain a blue liquid chemical that can detect carbon dioxide. If carbon dioxide passes through it, the liquid chemical turns green in color. If carbon dioxide does not pass through it, the liquid chemical remains blue in color. Glen blew out through the long tube in bottle A, 20 times. He then breathed in through the short tube in bottle B, 20 times. The liquid in bottle A turned green. The liquid in bottle B stayed blue.

*Use short arrows to finish each drawing to show the path that air follows in each bottle.*



**Analyzing the Problem**

In Bottle A, did the air that went into the liquid come from the lungs or from outside air? **from the lungs**

2. In Bottle B, did the air that went into the liquid come from the lungs or from outside air? **from outside air**

**Solving the Problem**

3. Did air passing through Bottle A, (into which Glen breathed out) contain carbon dioxide? What is your evidence? **Yes, the liquid turned green.**
4. Did the air passing through Bottle B, (in which Glen breathed in) contain carbon dioxide? What is your evidence? **No, the liquid remained blue.**
5. Explain why the blue chemical may not have turned green even though Table 13-1 on page 268 in your textbook tells you that air entering the lungs contains 0.03% carbon dioxide. **Not enough carbon dioxide is present to change the blue chemical.**
6. Suggest a change in the experiment that might help to show that breathed air does contain a small amount of carbon dioxide. **Use a more sensitive color indicator or blow into the liquid for a longer time.**

Blood in the capillaries around the alveoli is low in oxygen. As you breathe in, air and the oxygen in it fill each alveolus. Oxygen passes out of the air sacs into the blood of the capillaries by diffusion, Figure 13-5.

Red blood cells carry oxygen to all body cells. The oxygen diffuses into the body cells, Figure 13-6. Why, then, does less oxygen move out of the lungs than goes in as shown in Table 13-1? Less moves out because some oxygen is transported to all your body cells.

Now, why is more carbon dioxide given off than is taken in by the lungs? Carbon dioxide is formed in all body cells as a waste gas during cellular respiration. The gas moves by diffusion from body cells into nearby capillaries, Figure 13-6. In time, blood loaded with carbon dioxide will travel back to the lungs. When blood arrives at the lungs, carbon dioxide diffuses out of the blood in the capillaries and into the alveoli, Figure 13-5. All the carbon dioxide gas given off by your body cells during cellular respiration is carried by your blood to your lungs. Once there, it is given off in the air you breathe out. Do the data in Table 13-1 show this? Why do you think the amount of nitrogen that enters and leaves the lungs stays the same? It's because nitrogen gas is not used by body cells.

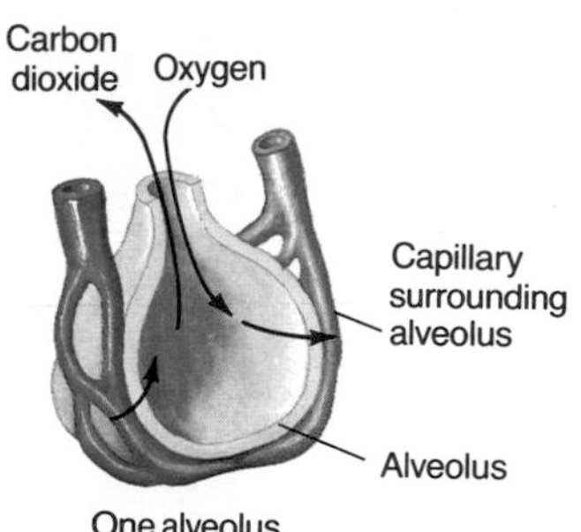


**Figure 13-5** The pathways of oxygen and carbon dioxide between the alveolus and its blood capillaries

**Why is more carbon dioxide given off than is taken in?**

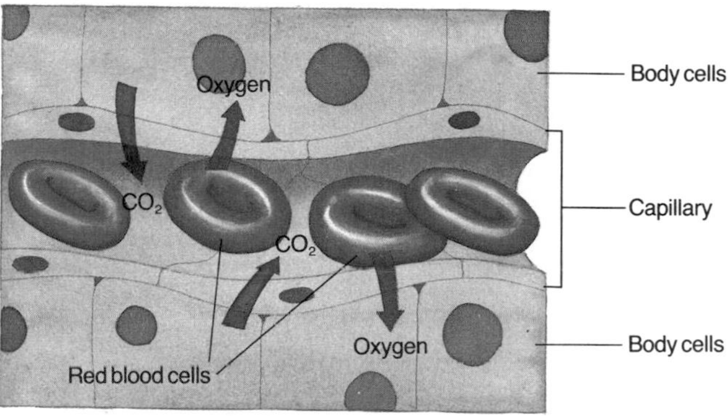


**Figure 13-6** Oxygen moves by diffusion from red blood cells into body cells. Carbon dioxide ($CO_2$) diffuses into capillaries from the cells.

## Mini Lab

**Is Carbon Dioxide Present?**

**Experiment:** You can detect carbon dioxide with bromthymol blue. Exhale 40 times through a straw into a beaker filled with the chemical. What color does the chemical turn? *For more help, refer to the* ***Skill Handbook****, pages 704-705.*

## Breathing

Usually, you don't think about your breathing. If you did think about it, however, you would probably notice that you breathe in and out about 12 to 16 times in one minute. How does your breathing work to bring air in and push air out of your lungs? Your rib cage and a muscle at the bottom of your chest help you to breathe.



## TEACH

### Guided Practice

Have students write down their answers to the margin question: Carbon dioxide is formed as a waste gas in all body cells.

### Check for Understanding

Have students compare and contrast the (a) sites where oxygen and carbon dioxide first enter the blood (alveoli/body capillary); (b) sites where oxygen and carbon dioxide leave blood (body capillary/alveoli); (c) sources of oxygen and carbon dioxide (air/body cells).

### Reteach

Have students summarize the events studied by making a table that lists the following across the top: Breathing out, Breathing in. List the following along the side: Direction of diaphragm movement, Condition of diaphragm, Direction of rib movement.

### Mini Lab

The Mini Lab will require a solution of 0.1 g bromthymol blue to 2 L of water. The solution will turn green with the presence of $CO_2$.

### ASSESSMENT

**Performance:** Provide students with a beaker of bromthymol blue and a football pump. Have students demonstrate how they can determine that inhaled air does not contain carbon dioxide gas.

### Independent Practice

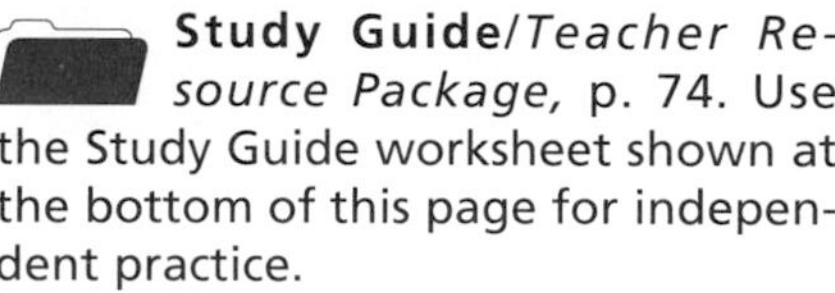

**Study Guide**/*Teacher Resource Package,* p. 74. Use the Study Guide worksheet shown at the bottom of this page for independent practice.

STUDY GUIDE 74

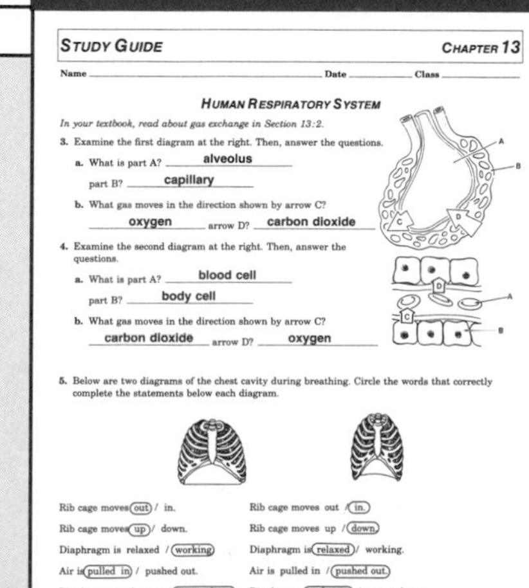

STUDY GUIDE — CHAPTER 13

Name ______ Date ______ Class ______

HUMAN RESPIRATORY SYSTEM

*In your textbook, read about gas exchange in Section 13:2.*

3. Examine the first diagram at the right. Then, answer the questions.
   a. What is part A? alveolus
   part B? capillary
   b. What gas moves in the direction shown by arrow C? oxygen arrow D? carbon dioxide
4. Examine the second diagram at the right. Then, answer the questions.
   a. What is part A? blood cell
   part B? body cell
   b. What gas moves in the direction shown by arrow C? carbon dioxide arrow D? oxygen
5. Below are two diagrams of the chest cavity during breathing. Circle the words that correctly complete the statements below each diagram.

| | |
|---|---|
| Rib cage moves (out) / in. | Rib cage moves out / (in.) |
| Rib cage moves (up) / down. | Rib cage moves up / (down.) |
| Diaphragm is relaxed / (working.) | Diaphragm is (relaxed) / working. |
| Air is (pulled in) / pushed out. | Air is pulled in / (pushed out.) |
| Diaphragm pushes up / (moves down.) | Diaphragm (pushes up) / moves down. |
| Lungs get squeezed / (expand.) | Lungs (get squeezed) / expand. |
| Person is breathing (in) / out. | Person is breathing in / (out.) |

## OPTIONS

**How Does Breathing Occur?**/*Lab Manual,* pp. 103-106. Use this lab as an extension to the human respiratory system.

# TEACH

## Concept Development

▶ Explain that the diaphragm is striated muscle that is normally involuntary, but that may be controlled while singing, holding one's breath, or consciously taking a deep breath.

## Divergent Question

Ask students if breathing is usually automatic. Does it speed up during exercise? When may it not be automatic? (breath holding)

## Connection: Math

Students can calculate the amount of air they breathe in during one minute and one hour. Assume that humans breathe in about 500 mL of air each time they breathe in. Have students (a) count the number of times they breathe in during one minute, (b) calculate the volume of air breathed in by multiplying their answer from step (a) by 500 mL, (c) calculate the volume of air breathed in during one hour by multiplying their answer from step (b) by 60.

## Guided Practice

Have students write down their answers to the margin question: The lungs expand and you breathe in.

## ACTIVITY/Enrichment

Have students build a lung model using a plastic soda pop bottle with the bottom cut off, balloons, a Y tube, rubber sheeting, and a rubber stopper.

**Videodisc**

**STV: Human Body Vol. 1**
*Circulatory and Respiratory Systems*
Unit 1, Side 1
*Circulatory and Respiratory Systems (in its entirety)*

**1064-30932**

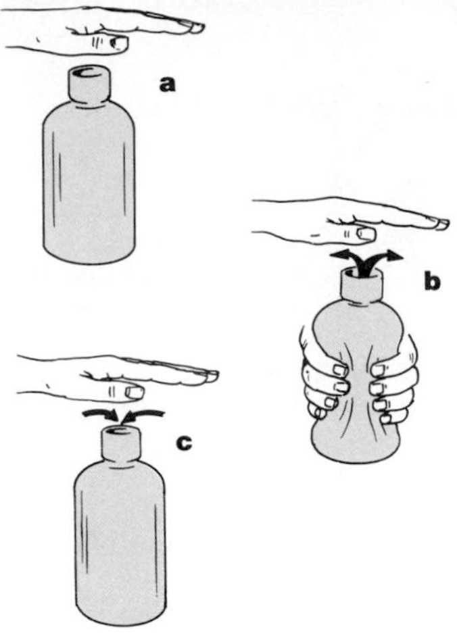


**Figure 13-7** No movement of the bottle's sides results in no air moving in or out (a). Pushing the bottle's sides in causes air to move out (b). The bottle sides snap back and air is pulled in (c).

Hold your hand over a plastic bottle such as the one shown in Figure 13-7. You will feel no air moving in or out of the bottle opening. Squeeze the bottle as shown in Figure 13-7. You will feel air rushing out. Why? As the sides of the bottle are pushed in, air is pushed out. Now allow the sides of the bottle to snap back to their original shape, as in Figure 13-7. You can feel air rushing back into the bottle.

Your chest is like the plastic bottle. Its shape and size change, allowing you to breathe in and out. The size of your chest changes due to the action of a muscle located along the bottom of your chest. This muscle is your diaphragm (DI uh fram). The **diaphragm** is a sheetlike muscle that separates the inside of your chest from the intestines and other organs of your abdomen. Read the following statements and match them with parts of the diagram in Figure 13-8.

**What happens when the diaphragm flattens and moves downward?**

1. The diaphragm is relaxed. When it relaxes, it is domelike and pushes upward.
2. There is space between the lungs and the inside of the chest wall. The diaphragm pushes up against the lungs, and the space gets smaller.
3. Your lungs are soft. They are squeezed as the space grows smaller.
4. Air is pushed out of the alveoli as the lungs are squeezed. You breathe out.
5. The diaphragm begins to contract. As it tightens up, it flattens and moves downward.
6. The space between the lungs and the inside of the chest wall becomes larger.

**Figure 13-8** Notice the changes in the diaphragm, lungs, rib cage, and chest cavity during breathing out (a) and breathing in (b).

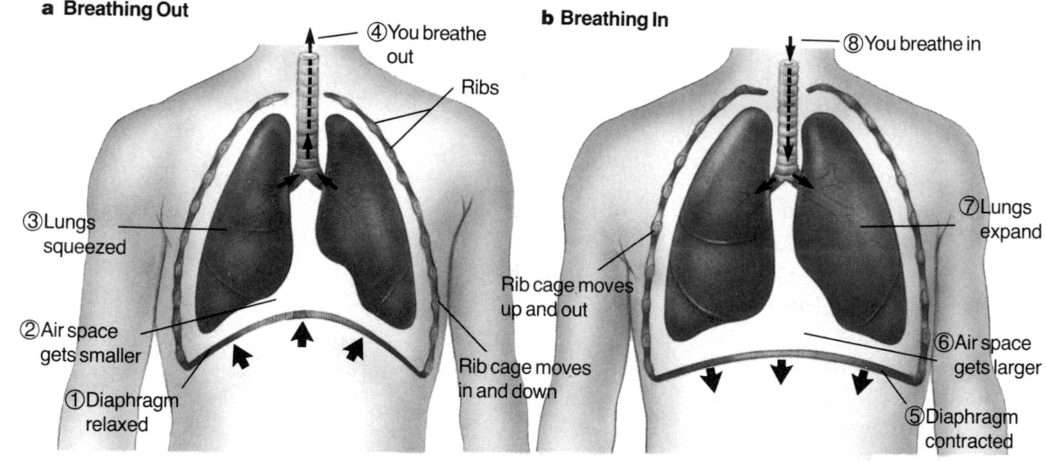




## OPTIONS

**Breathing In and Out**/*Transparency Package*, number 13a. Use color transparency 13a as you teach breathing.

**Enrichment**/*Teacher Resource Package*, p. 13. Use the Enrichment worksheet shown here to teach the epiglottis.

**ENRICHMENT 13**

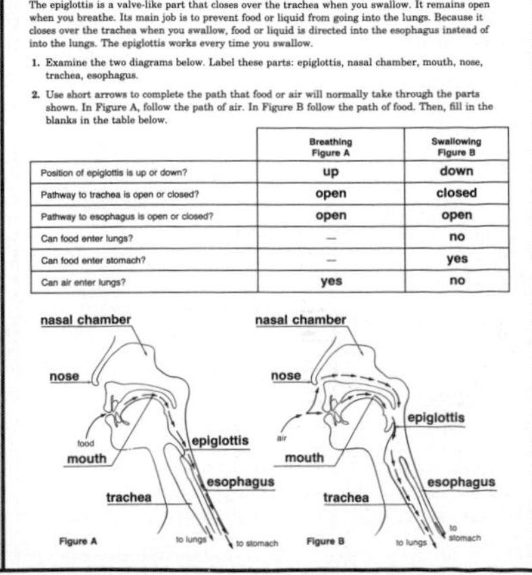

ENRICHMENT — CHAPTER 13

Name ______ Date ______ Class ______

*Use after Section 13:3.*

THE EPIGLOTTIS

The epiglottis is a valve-like part that closes over the trachea when you swallow. It remains open when you breathe. Its main job is to prevent food or liquid from going into the lungs. Because it closes over the trachea when you swallow, food or liquid is directed into the esophagus instead of into the lungs. The epiglottis works every time you swallow.

1. Examine the two diagrams below. Label these parts: epiglottis, nasal chamber, mouth, nose, trachea, esophagus.
2. Use short arrows to complete the path that food or air will normally take through the parts shown. In Figure A, follow the path of air. In Figure B follow the path of food. Then, fill in the blanks in the table below.

| | Breathing Figure A | Swallowing Figure B |
|---|---|---|
| Position of epiglottis is up or down? | up | down |
| Pathway to trachea is open or closed? | open | closed |
| Pathway to esophagus is open or closed? | open | open |
| Can food enter lungs? | — | no |
| Can food enter stomach? | — | yes |
| Can air enter lungs? | yes | no |

## Idea Map

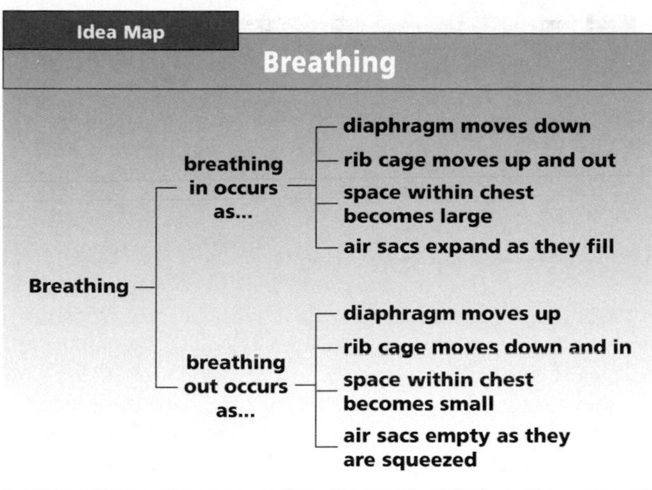


**Study Tip:** Use this idea map as a study guide to breathing. Does breathing occur in your chest or abdomen?

7. Your lungs expand because the space surrounding the lungs has increased.
8. Air is pulled into the alveoli as the lungs expand. This causes you to breathe in.

Your rib cage also helps with breathing. The rib cage moves in and out as well as up and down, Figure 13-8. The muscles between each rib help with this movement. The ribs can be pulled closer together or moved apart. When your rib cage moves in and slightly down, the ribs close up slightly and the space within your chest gets smaller. Air is squeezed out and you breathe out. As your rib cage moves out and slightly up, the ribs move apart and the space within your chest expands. Air rushes in from the outside and you breathe in.

### Mini Lab

**How Much Air Do You Breathe Out?**

**Design an experiment:** Use a 1-L graduated cylinder, rubber tube, pan, and water to measure how much air you breathe out. Check with your teacher before doing it. *For more help, refer to the **Skill Handbook**, pages 704-705.*

### Check Your Understanding

6. What are the functions of the trachea, bronchi, lungs, and alveoli?
7. How is carbon dioxide formed in your cells, and how does it pass out of your body?
8. In which directions do your rib cage and diaphragm move as you breathe in?
9. **Critical Thinking:** What causes a person to choke?
10. **Biology and Reading:** Where does air go after it leaves the smallest tubes of the bronchi?



## TEACH

### Mini Lab

Water displacement measurements will be between 350-500 mL.

### ASSESSMENT

**Performance:** Have students design and do an experiment to test the hypothesis that students with a larger body frame exhale more air during a normal exhalation than do students with a smaller body frame.

### Check for Understanding

Have students respond to the first three questions in Check Your Understanding.

### Reteach

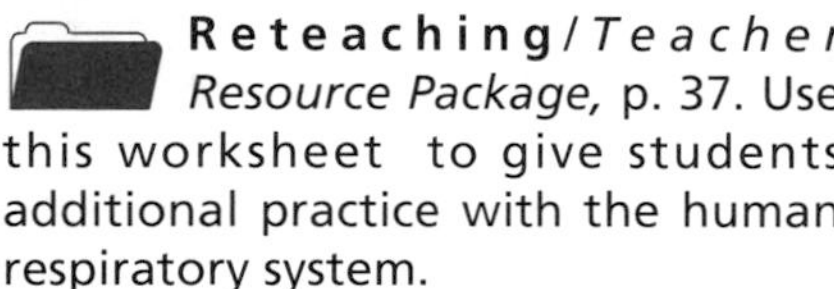

**Reteaching**/*Teacher Resource Package,* p. 37. Use this worksheet to give students additional practice with the human respiratory system.

**Extension:** Assign Critical Thinking, Biology and Reading, or some of the **OPTIONS** available with this lesson.

### Student Journal

A newborn baby has no air space between lungs and the inside of the chest wall. Therefore, breathing is more difficult. Have students alter the seven steps shown in Figure 13-8 to describe what occurs in a newborn's chest.

## 3 APPLY

### /Interview

Interview a molecule of oxygen that just returned from a trip through the human body. Ask it to describe its journey.

## 4 CLOSE

### Audiovisual

Show the videocassette *Breath of Life,* Films for the Humanities and Sciences.

### RETEACHING 37

RETEACHING — CHAPTER 13

Name ______ Date ______ Class ______

Use with Section 13:2.

→ Oxygen pathway
⇢ Carbon dioxide pathway

HUMAN RESPIRATORY SYSTEM

mouth
esophagus
enlarged view of alveoli
X 2000

| Respiratory system parts and functions |
|---|
| **nose**<br>• air enters<br>• hairs block or trap dust and unwanted objects |
| **nasal chamber**<br>• warms air and adds moisture as it enters |
| **epiglottis**<br>• flap that closes over trachea during swallowing<br>• keeps food and liquid out of lungs |
| **trachea (windpipe)**<br>• tube that carries air from nasal chamber to bronchi |
| **bronchus**<br>• short tube that carries air from trachea to right and left lungs |
| **lung**<br>• main organ of respiratory system<br>• tiny air sacs and blood capillaries where oxygen moves into blood and carbon dioxide moves out |
| **diaphragm**<br>• muscle at bottom of chest cavity<br>• moves up when relaxed and helps with breathing out<br>• moves down when contracted and helps with breathing in |
| **capillary**<br>• blood vessel that surrounds alveoli<br>• site of gas exchange between blood and air |
| **alveoli**<br>• air sacs<br>• site of gas exchange between blood and air |

### Answers to Check Your Understanding

6. trachea — brings air from nasal chamber to bronchi; bronchi — carry air from trachea to lungs; lungs — help exchange oxygen and carbon dioxide; alveoli — sites of gas exchange between lungs and blood
7. forms during cellular respiration and leaves the body when you breathe out.
8. Rib cage moves up and out as diaphragm moves down.
9. If a person eats when talking, food may fall into the open windpipe.
10. into the alveoli

## 13:3 Problems of the Respiratory System

## PREPARATION

### Materials Needed

Make copies of the Reteaching and Study Guide worksheets in the *Teacher Resource Package*.

### Key Science Words

carbon monoxide
pneumonia
emphysema

### Process Skills

In the Skill Check, students will define words in context.

## 1 FOCUS

▶ The objectives are listed on the student page. Remind students to preview these objectives as a guide to this numbered section.

### MOTIVATION/Audiovisual

Show the filmstrip *Partners for Life: The Human Heart and Lungs, Part 2 Disease,* Human Relations Media.

### Guided Practice

Have students write down their answers to the margin question: They contain carbon monoxide, a poisonous gas.

### Independent Practice

**Study Guide**/*Teacher Resource Package,* p. 75. Use the Study Guide worksheet shown at the bottom of this page for independent practice.

### Student Journal

Have students trace the pathway that molecules of carbon monoxide and nitrogen would follow through the body after being inhaled into the lungs. Students should explain why one gas can cause serious problems and why the other gas has no harmful effects.



### Objectives

7. **Relate** the effects of carbon monoxide poisoning and how to avoid the gas.
8. **Explain** the cause of pneumonia.
9. **Describe** emphysema.

### Key Science Words

carbon monoxide
pneumonia
emphysema

# 13:3 Problems of the Respiratory System

Problems of the respiratory system can be caused by diseases of the respiratory organs. Some problems, however, are caused by the quality of the air we breathe.

## A Breathing Problem

A person can live for many days without food. Without oxygen, however, a person can live for only a few minutes. Cells can't live without oxygen.

Some gases are poisonous. If you breathe these gases, they act by taking the place of oxygen in your blood. One poisonous gas is carbon monoxide (CO). **Carbon monoxide** is an odorless, colorless gas sometimes found in the air. If carbon monoxide is inhaled, it is more easily picked up by your red blood cells than oxygen.

**Why are exhaust fumes dangerous?**

Where does carbon monoxide come from? You know that when fuels are burned, oxygen is used up and carbon dioxide is released—just as in cellular respiration. Carbon monoxide is also given off in small amounts when fuels are burned. For example, carbon monoxide is given off in exhaust fumes when gasoline is burned in a car. The small amount of gas mixes with the surrounding air and is not a health problem. It becomes a serious problem, however, if the exhaust is given off in a closed area, such as a garage. A common cause of carbon monoxide pollution is from the smoke of cigarettes, Figure 13-9.

### Bio Tip

**Leisure:** Swimming is an excellent sport for people with asthma. It provides exercise for the body at the same time as helping a person relax.

**Figure 13-9** A person who breathes too much carbon monoxide can't deliver enough oxygen to his or her cells.



## OPTIONS

### Science Background

There are several different organisms capable of causing pneumonia. These include bacterial, viral, and protist organisms. The most common form, however, is bacterial. This form of pneumonia is caused by the bacterium *Streptococcus pneumoniae*. The form of pneumonia associated with AIDS is caused by a protozoan, *Pneumocystis carinii.*

### STUDY GUIDE 75

STUDY GUIDE — CHAPTER 13

Name ______ Date ______ Class ______

**PROBLEMS OF THE RESPIRATORY SYSTEM**

*In your textbook, read about respiratory problems in Section 13:3.*

A B C D

1. Drawing A shows the normal pathway of 2 oxygen molecules (black dots) going from the alveoli into a lung capillary. Finish the drawing by showing the path that the other 4 oxygen molecules will take in a normal person.
2. Drawing B shows alveoli filled with oxygen and carbon monoxide (squares). Finish the drawing by showing the path that the gas molecules will take when in the lungs of a person with carbon monoxide poisoning.
3. Explain why a person with the problem in diagram B could die. **Carbon monoxide gas is picked up more easily than oxygen by red blood cells. If too much is inhaled, too little oxygen will be delivered to the person's cells.**
4. Drawing C shows alveoli filled with pus, liquid, and mucus.
   a. What disease may cause this to occur? **pneumonia**
   b. Can oxygen (black dots) reach the capillaries surrounding these alveoli? **no** Explain. **The alveoli are blocked with pus, liquid, and mucus. This prevents oxygen from getting into the blood.**
   c. Is this disease communicable or noncommunicable? **communicable** Explain. **Pneumonia is caused by bacteria or viruses and can be passed from one person to another.**
5. Drawing D shows the alveoli of a person with emphysema.
   a. How do these alveoli compare with those of a normal person? **The alveoli of a normal person have more surface area. The alveoli of a person with emphysema are broken down.**
   b. How is emphysema harmful? **A person with emphysema is out of breath because there are not enough alveoli to get oxygen into the blood.**

## Respiratory Diseases

**Pneumonia** (noo MOH nyuh) is a lung disease caused by bacteria, a virus, or both. These microbes invade the alveoli and multiply. Their presence causes the lung tissue to form extra mucus. Mucus and pus from microbes fill the alveoli and cause difficulty in breathing. The blocked lungs prevent oxygen from getting into the blood. Because pneumonia is caused by bacteria or viruses, it is communicable. You read in Chapter 4 that a communicable disease is passed from one person to another.

Why is pneumonia communicable?

Some diseases are not passed on from person to person. Emphysema (em fuh SEE muh) is a respiratory disease that is not communicable. **Emphysema** is a lung disease that results in the breakdown of alveoli. A person with emphysema always seems to be out of breath because there are not enough alveoli to get oxygen into the blood. A common cause of emphysema is the breathing in of chemicals while smoking or from air pollution, Figure 13-10.

What happens to a person's lungs with emphysema?

### Skill Check

**Define words in context:** If you say that you are "catching a cold," would this describe a communicable disease? *For more help, refer to the* ***Skill Handbook,*** *pages 706-711.*

**Figure 13-10** More people suffer from respiratory diseases in polluted cities than in cleaner, rural areas.

### Check Your Understanding

11. Why is carbon monoxide poisonous?
12. What causes pneumonia?
13. What is emphysema?
14. **Critical Thinking:** A spirometer measures lung capacity and volume. Why are these measurements important?
15. **Biology and Reading:** What happens to the surface area of the lung when alveoli are lost as a result of emphysema?



## 2 TEACH

### Guided Practice

Have students write down their answers to the margin questions: it is caused by bacteria and viruses; they have difficulty with breathing.

### Check for Understanding

Have students respond to the first three questions in Check Your Understanding.

### Reteach

**Reteaching**/*Teacher Resource Package,* p. 38. Use this worksheet to give students additional practice with respiratory problems.

**Extension:** Assign Critical Thinking, Biology and Reading, or some of the **OPTIONS** available with this lesson.

## 3 APPLY

### ACTIVITY/Challenge

Have students report on the cause of and problems associated with asthma.

## 4 CLOSE

### Software

*Respiratory Diseases and Disorders,* Queue, Inc.

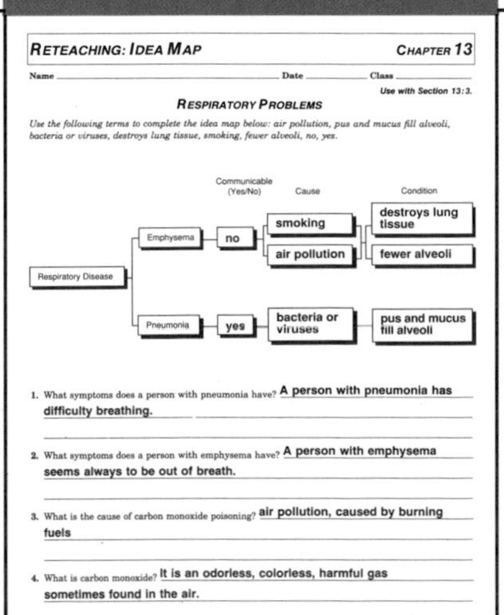

RETEACHING 38

RETEACHING: IDEA MAP CHAPTER 13

Name ______ Date ______ Class ______

Use with Section 13:3.

RESPIRATORY PROBLEMS

*Use the following terms to complete the idea map below: air pollution, pus and mucus fill alveoli, bacteria or viruses, destroys lung tissue, smoking, fewer alveoli, no, yes.*



1. What symptoms does a person with pneumonia have? A person with pneumonia has difficulty breathing.
2. What symptoms does a person with emphysema have? A person with emphysema seems always to be out of breath.
3. What is the cause of carbon monoxide poisoning? air pollution, caused by burning fuels
4. What is carbon monoxide? It is an odorless, colorless, harmful gas sometimes found in the air.

### Answers to Check Your Understanding

11. Not enough oxygen is delivered to cells when CO is breathed.
12. bacteria or a virus
13. breakdown of alveoli
14. can show if you have a lung disease
15. Surface area decreases.

### GLENCOE TECHNOLOGY

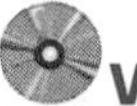

**Videodisc**

**Biology: The Dynamics of Life**
Animation: *Mechanics of Breathing*
Disc 2, Side 1, Ch. 32

## 13:4 The Role of Excretion

### PREPARATION

#### Materials Needed

Make copies of the Application, Reteaching, Transparency Master, Study Guide, and Lab worksheets in the *Teacher Resource Package*.

▶ Have muddy water for the Mini Lab.

▶ Prepare artificial normal and abnormal urine, silver nitrate, glucose, glucose test tape, and salt water for Lab 13-2.

#### Key Science Words

excretory system
urea
ureter
urinary bladder
urethra
nephron
urine

#### Process Skills

In the Mini Lab, students will formulate a model. In Lab 13-2, they will observe, interpret data, make and use tables, and infer. In the Skill Check, they will interpret diagrams.

### 1 FOCUS

▶ The objectives are listed on the student page. Remind students to preview these objectives as a guide to this numbered section.

#### ACTIVITY/Demonstration

Use a human torso model to examine and illustrate the number, size, and location of kidneys and the connection of kidneys to blood supply and urinary bladder.

#### Guided Practice

Have students write down their answers to the margin question: Urea is a poisonous waste product from the breakdown of proteins.



## 13:4 The Role of Excretion

**Objectives**

10. **Discuss** the importance of the excretory system.
11. **Sequence** the steps for filtering of blood in a kidney.
12. **Describe** the skin as part of the excretory system.

**Key Science Words**

excretory system
urea
ureter
urinary bladder
urethra
nephron
urine

Chemical wastes are formed by living cells. Carbon dioxide is removed from your body by your lungs. Other organs, such as your skin and kidneys, help in getting rid of other waste materials.

### Waste Removal

Your blood contains many different chemical wastes. These wastes are chemicals that are not needed by the body and may be harmful. If wastes weren't removed from the body, the tissues would fill with poisonous waste products. The wastes would destroy cells and tissues. Fever, poisoning, or even death can result from a buildup of wastes in the tissues.

Wastes are either made by your body cells or are the remains of undigested food in your diet. Getting rid of wastes is called excretion. In Chapter 10, you read that solid wastes are removed by the digestive system. Liquid wastes are removed by the excretory (EK skruh tor ee) system. The **excretory system** is made up of those organs that rid the body of liquid wastes.

**Why does urea need to be removed from the body?**

**Urea** (yoo REE uh) is a waste that results from the breakdown of body protein. It's poisonous and must be removed from the body. Urea is picked up from the cells by the blood and carried to the kidneys for removal. The kidneys are the most important organs of the excretory system.

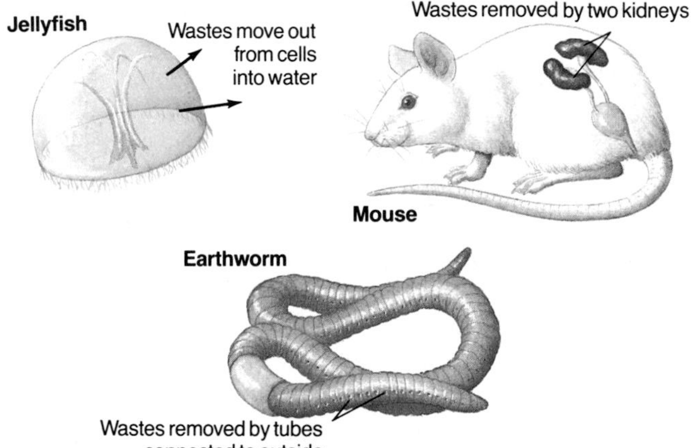


**Figure 13-11** Compare the excretory organs in a jellyfish, earthworm, and mouse.



### OPTIONS

#### Science Background

Kidneys, lungs, and skin are all involved with excretion of waste chemicals. However, only the kidneys can eliminate urea. Only the lungs can eliminate carbon dioxide. Water and salts are eliminated from several different organs.

**Human Excretory System**/*Transparency Package,* number 13b. Use color transparency 13b as you teach the role of excretion.

Do all other animals have kidneys as we do? Many animals, such as mice, fish, and frogs, have kidneys. Simpler animals without kidneys, however, must still get rid of wastes. Simple animals such as sponges and jellyfish don't need kidneys. Wastes made by their cells pass out of their cells directly into the water in which they live. Earthworms excrete their wastes through a pair of tubes in each body segment. These tubes connect the inside of the animal to the outside. Compare the organs of excretion in the animals in Figure 13-11.

## Human Kidneys

The human body has two kidneys, each about as big as a fist. Kidneys could be described as blood filters. To understand the job of your kidneys, let's look at how a filter works, Figure 13-12. Let's assume you have some muddy water and you want to separate the mud from the water. The muddy water can be poured through filter paper. The mud will be caught by the paper, while the water will pass through. Your kidneys work in a way similar to the filter paper. They filter wastes from your blood. During one day, your kidneys filter up to 200 liters of blood.

If wastes were not removed from your body, they would build up in your blood and act as a poison. In order to carry out their job of removing wastes, your kidneys are hooked up to your blood vessels. Match the following numbered steps to parts of the human excretory system shown in Figure 13-13.

### Mini Lab

**What Does a Kidney Do?**

**Formulate a model:** Filter some muddy water to model how the kidneys work. *For more help, refer to the* ***Skill Handbook****, pages 706-711.*

a

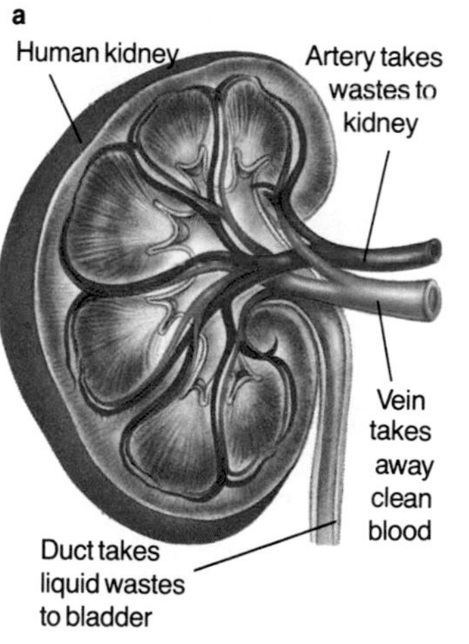


b

**Figure 13-12** The human kidney (a) is a filter for blood. The filter paper catches mud. Clear water passes through the filter (b).



## STUDY GUIDE 76

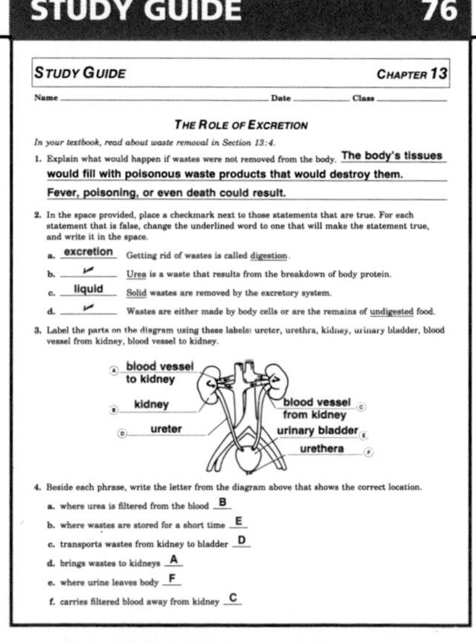

STUDY GUIDE CHAPTER 13

Name ____ Date ____ Class ____

THE ROLE OF EXCRETION

*In your textbook, read about waste removal in Section 13:4.*

1. Explain what would happen if wastes were not removed from the body. The body's tissues would fill with poisonous waste products that would destroy them. Fever, poisoning, or even death could result.

2. In the space provided, place a checkmark next to those statements that are true. For each statement that is false, change the underlined word to one that will make the statement true, and write it in the space.
   a. excretion Getting rid of wastes is called digestion.
   b. ✓ Urea is a waste that results from the breakdown of body protein.
   c. liquid Solid wastes are removed by the excretory system.
   d. ✓ Wastes are either made by body cells or are the remains of undigested food.

3. Label the parts on the diagram using these labels: ureter, urethra, kidney, urinary bladder, blood vessel from kidney, blood vessel to kidney.



4. Beside each phrase, write the letter from the diagram above that shows the correct location.
   a. where urea is filtered from the blood B
   b. where wastes are stored for a short time E
   c. transports wastes from kidney to bladder D
   d. brings wastes to kidneys A
   e. where urine leaves body F
   f. carries filtered blood away from kidney C

# 2 TEACH

### MOTIVATION/Filmstrip

Show the filmstrip *Your Kidneys – Living Filters,* Clearvue Inc.

### Concept Development

▶ Explain that the kidneys selectively filter from the blood excess water, urea, certain ions or salts, and bile pigments resulting from hemoglobin breakdown. The bile pigment gives urine its yellow color.

### Mini Lab

Coffee filters may be used if filter paper is not available. Obtaining clear water may require two filterings or double filters. Do the demonstration using water and sand rather than muddy water to hasten the process of filtration. Let students examine the water to be filtered. Make sure that they see the filter paper and its contents, and the pure water that has separated.

**Performance:** Have students redesign the kidney model to illustrate how an abnormal, diseased, or injured kidney that cannot properly filter would work.

### Independent Practice

**Study Guide/***Teacher Resource Package,* p. 76. Use the Study Guide worksheet shown at the bottom of this page for independent practice.

## GLENCOE TECHNOLOGY

**Videodisc**

**Biology: The Dynamics of Life**
Animation: *Nephron Filtration*
Disc 2, Side 1, Ch. 36

**Videotape**

**The Secret of Life**
*Human Excretory System*

## TEACH

### Convergent Question

Explain that the kidney is considered to be a body or blood filter. Ask: What does it filter from the blood? What does it not filter from the blood?

### MOTIVATION/Audiovisual

Show the filmstrip *Urinary System,* Carolina Biological Supply Co.

### Independent Practice

**Study Guide**/*Teacher Resource Package*, p. 77. Use the Study Guide worksheet shown at the bottom of this page for independent practice.

### Guided Practice

Have students write down their answers to the margin question: Wastes diffuse out of the blood into nephrons inside the kidneys. Urea, excess salts, and excess water pass out of the kidneys via the ureters to the urinary bladder and then out of the body via the urethra.

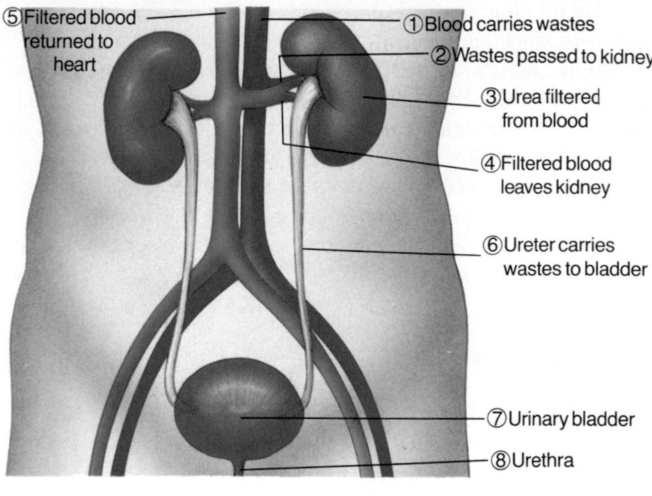


**Figure 13-13** Follow the path of blood as it is cleaned by the human excretory system.

1. Blood carrying wastes moves through the body's arteries.
2. Small arteries bring blood to be filtered into each kidney.
3. The kidneys filter urea from the blood.
4. The filtered blood leaves the kidney through a vein. The blood is now free of wastes.
5. The blood in the veins is taken to the heart and other parts of the body.
6. Wastes from the blood leave each kidney through a ureter (YOOR ut ur). A **ureter** is a tube that carries wastes from a kidney to the urinary bladder.
7. The **urinary bladder** is a sac that stores liquid wastes removed from the kidneys.
8. The **urethra** (yoo REE thruh) is a tube that carries liquid wastes from the urinary bladder to outside the body.

**What is the path of wastes from the blood through the body?**

How do the kidneys do their job of filtering blood? Look at Figure 13-14. It shows an enlarged view of a nephron (NEF rahn), which is a small part of a kidney. A **nephron** is a tiny filter unit of the kidney. Each kidney has about one million nephrons that allow the kidneys to filter your blood every day.

Use the figure and the numbered steps to follow the pathway of blood through a nephron.

1. Blood with blood cells, salts, sugar, urea, and water enters each nephron.
2. The blood first passes into a tightly coiled capillary inside a cuplike part of the nephron.



## OPTIONS

**Application**/*Teacher Resource Package*, p. 13. Use the Application worksheet shown here to give students practice in understanding the operation of an artificial kidney.

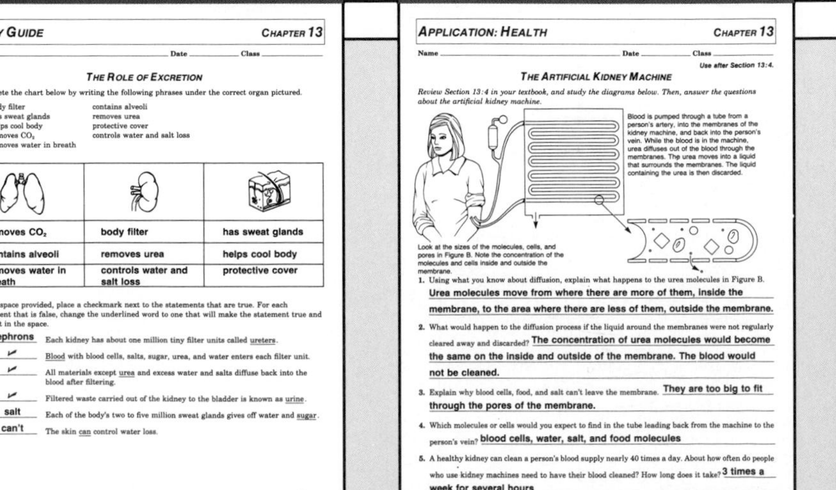

**STUDY GUIDE 77**

STUDY GUIDE — CHAPTER 13

Name ______ Date ______ Class ______

*THE ROLE OF EXCRETION*

5. Complete the chart below by writing the following phrases under the correct organ pictured.

body filter; has sweat glands; helps cool body; removes $CO_2$; removes water in breath; contains alveoli; removes urea; protective cover; controls water and salt loss

| (lungs) | (kidney) | (skin) |
|---|---|---|
| removes $CO_2$ | body filter | has sweat glands |
| contains alveoli | removes urea | helps cool body |
| removes water in breath | controls water and salt loss | protective cover |

6. In the space provided, place a checkmark next to the statements that are true. For each statement that is false, change the underlined word to one that will make the statement true and write it in the space.

a. nephrons — Each kidney has about one million tiny filter units called ureters.
b. ✓ — Blood with blood cells, salts, sugar, urea, and water enters each filter unit.
c. ✓ — All materials except urea and excess water and salts diffuse back into the blood after filtering.
d. ✓ — Filtered waste carried out of the kidney to the bladder is known as urine.
e. salt — Each of the body's two to five million sweat glands gives off water and sugar.
f. can't — The skin can control water loss.

**APPLICATION 13**

APPLICATION: HEALTH — CHAPTER 13

Name ______ Date ______ Class ______

*Use after Section 13:4.*

*THE ARTIFICIAL KIDNEY MACHINE*

*Review Section 13:4 in your textbook, and study the diagrams below. Then, answer the questions about the artificial kidney machine.*

Blood is pumped through a tube from a person's artery, into the membranes of the kidney machine, and back into the person's vein. While the blood is in the machine, urea diffuses out of the blood through the membranes. The urea moves into a liquid that surrounds the membranes. The liquid containing the urea is then discarded.

Look at the sizes of the molecules, cells, and pores in Figure B. Note the concentration of the molecules and cells inside and outside the membrane.

1. Using what you know about diffusion, explain what happens to the urea molecules in Figure B. Urea molecules move from where there are more of them, inside the membrane, to the area where there are less of them, outside the membrane.
2. What would happen to the diffusion process if the liquid around the membranes were not regularly cleared away and discarded? The concentration of urea molecules would become the same on the inside and outside of the membrane. The blood would not be cleaned.
3. Explain why blood cells, food, and salt can't leave the membrane. They are too big to fit through the pores of the membrane.
4. Which molecules or cells would you expect to find in the tube leading back from the machine to the person's vein? blood cells, water, salt, and food molecules
5. A healthy kidney can clean a person's blood supply nearly 40 times a day. About how often do people who use kidney machines need to have their blood cleaned? How long does it take? 3 times a week for several hours

3. Water, salts, sugar, and urea pass into the cuplike part of the nephron. Everything but blood cells is forced out of the blood.
4. Follow the flow shown by the dashed black arrows that indicate the movement of these materials into a long tube.
5. Notice that the capillary with the original blood cells from step 2 now twists itself around the long tube. The movement of blood cells is shown by the dashed red arrows.
6. All materials except urea, water, and salts diffuses back into the blood. This movement is shown by the solid black arrows. Water passes back into the blood by osmosis.
7. The cleaned blood now leaves the kidney and returns to the heart.
8. The filtered wastes stay in the long tube and are carried out of the kidney through the ureter to the bladder. Waste liquid that reaches the ureter is called **urine.** Urine passes out of the body.

Taking in more water or salt means more of these nutrients will be excreted. Less water or salt taken in means less will be excreted. Kidneys regulate the amount of water and salts kept within the body. A healthy person can excrete about one liter of urine each day.

### Skill Check

**Interpret diagrams:** What feature of Figure 13-14 makes it easier for you to follow the pathways of blood, wastes, and water? *For more help, refer to the* ***Skill Handbook,*** *pages 706-711.*

**Figure 13-14** A section of a kidney (a) shows the position of a nephron (b). Follow the pathway of materials through a nephron (c).

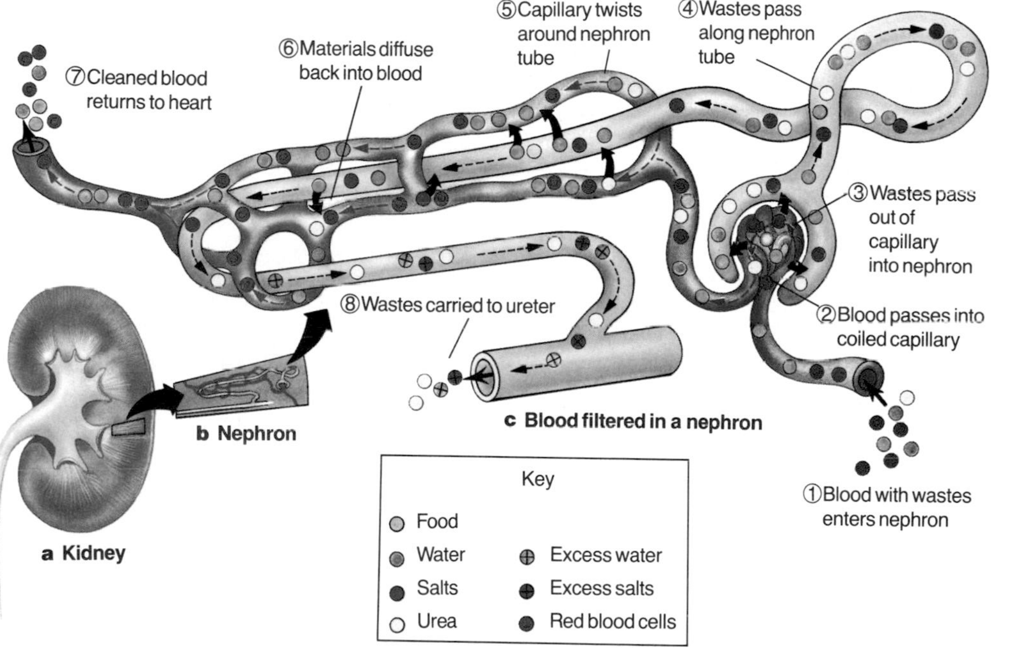




## TEACH

### Concept Development

▶ Go over the sequence of events in Figure 13-14. Ask students to follow the pathway of one chemical at a time through the nephron. Be sure to specify whether water and salt are to be considered in excess or as chemicals needed and retained by the body. The pathway that needed water and salts follow will be different from the pathway for excess salt and water.

### Guided Practice

Provide students with a list of the following parts: artery entering kidney, vein leaving kidney, kidney, ureter, urethra, nephron, urinary bladder. Ask them to indicate the number of these structures present in the human body.

### Check for Understanding

Have students prepare a chart with the following headings: Blood components, Enters kidney by way of artery, Leaves kidney by way of vein, Leaves kidney by way of ureter, Leaves kidney as urine. Under Blood components, list the following items: red cell, needed water, excess water, urea, glucose (food), needed salt, excess salt. Have students complete the chart by placing check marks in the proper columns.

### Reteach

Ask students which blood parts or chemicals are needed by the body and which are not needed. Have them use information from Check for Understanding.

### TRANSPARENCY 59

TRANSPARENCY MASTER CHAPTER 13

NEPHRON UNIT OF KIDNEY

all chemicals except blood cells are squeezed out from coiled capillary and caught here
artery entering nephron
cup-like part
tightly coiled capillary
artery leaving coiled capillary
food is taken out of tube and put back into blood
long tube
blood returning to body from nephron
tube leads from nephron
capillary twists itself around long tube
water is taken out of tube and put back into blood
excess
urine
excess
salt is taken out of tube and put back into blood
chemicals in water leave nephron as urine
Key: Red blood cells = ○ Salts = ▲ ▲ extra Food = ● Urea = ■ Water = ● ● extra

## OPTIONS

**Transparency Master**/*Teacher Resource Package*, p. 59. Use the Transparency Master shown here to teach the nephron unit of the kidney.

### Portfolio

Medicines such as aspirin follow a pathway through the kidneys much like the pathway shown for urea. Copy Figure 13-14 for students leaving out all the small circles that represent cells and molecules. Have students trace the pathway that aspirin would follow through the diagram. Students can use circles to represent aspirin, and describe the events that take place using numbered steps 1 through 8 as a guide.

## Lab 13-2 Urine

### Overview

In this lab, students are given an opportunity to check the chemical composition of normal and abnormal urine.

**Objectives:** Upon completing this lab, students will have learned (1) **to test** for and recognize if salt is present in a liquid, (2) **to test** for and recognize if glucose is present in a liquid, (3) **to test** for the presence or absence of salt and glucose in urine.

**Time Allotment:** 1 class period

### Preparation

▶ For normal urine – add 1 teaspoon sodium chloride (table salt) and 4 drops yellow food coloring to 500 mL tap water.

▶ For abnormal urine – same as above with 1 teaspoon glucose added.

▶ For silver nitrate – add 4 g silver nitrate to 250 mL distilled water.

▶ For glucose – add 1 teaspoon glucose to 250 mL tap water.

▶ For salt water – add 1 teaspoon sodium chloride to 500 mL tap water.

**Lab 13-2 worksheet**/*Teacher Resource Package,* pp. 51-52.

### Teaching the Lab

▶ Remind students to use silver nitrate cautiously.

▶ Use distilled water (available in grocery store) for tube 1 when testing for salt. Some tap water will contain high enough concentrations of salt to yield a false positive test.

▶ Discuss abnormal urine as being like that from a diabetic.

**ASSESSMENT**

**Performance:** Provide students with four tubes of artificial urine marked A, B, C, and D. Have students determine which tubes contain urine with and without salt and which tubes contain urine with and without glucose. Supply students with labels, test tubes, silver nitrate solution, glucose test tape, and glass slides.

# Urine

**Lab 13–2**

## Problem: What chemicals can be detected in urine?

### Skills

observe, interpret data, make and use tables, infer

### Materials

4 test tubes
test-tube rack
distilled water
salt water
normal artificial urine
abnormal artificial urine
silver nitrate in bottle with dropper
labels
glucose
glucose test tape
droppers
4 glass slides

### Procedure

1. Copy the data table.
2. Use labels to number four test tubes 1, 2, 3, and 4 and place them in a rack.
3. Half fill each tube with the following– 1: water, 2: salt water, 3: normal artificial urine, 4: abnormal artificial urine.
4. Add five drops of silver nitrate to each tube. If a white haze appears, this means that salt is present. **CAUTION:** *Do not spill silver nitrate on skin or clothing. Rinse with water if spillage occurs.*
5. **Make and use tables:** Record your before and after results in the table.
6. Use labels to number four slides 1, 2, 3, and 4.
7. Add one drop of the following to each slide– 1: water, 2: glucose, 3: normal artificial urine, 4: abnormal artificial urine.
8. Touch a small piece of glucose test tape to each drop. If a green color appears on the tape, it means that the sugar glucose is present.
9. Record your before and after results in the table.
10. Dispose of your samples as indicated by your teacher.

### Data and Observations

1. How can you tell if a liquid contains salt?
2. How can you tell if a liquid contains glucose?

| TUBE | TUBE CONTENTS | HAZE PRESENT BEFORE? | HAZE PRESENT AFTER? | SALT PRESENT? |
|---|---|---|---|---|
| 1 | water | no | no | no |
| 2 | salt water | no | yes | yes |
| 3 | normal urine | no | yes | yes |
| 4 | abnormal urine | no | yes | yes |
| **SLIDE** | **SLIDE CONTENTS** | **COLOR OF PAPER BEFORE** | **COLOR OF PAPER AFTER** | **GLUCOSE PRESENT?** |
| 1 | water | yellow | yellow | no |
| 2 | glucose | yellow | green | yes |
| 3 | normal urine | yellow | yellow | no |
| 4 | abnormal urine | yellow | green | yes |

### Analyze and Apply

1. Using your data, does the normal artificial urine contain glucose or salt?
2. Using your knowledge, would urine from your body contain glucose or salt? Why?
3. Using your data, does the abnormal urine contain glucose?
4. **Apply:** Why is it useful to know if urine contains salt or glucose?

### Extension

**Experiment:** Add table sugar to urine and test for glucose.



## ANSWERS

### Data and Observations

1. Silver nitrate makes a liquid hazy in the presence of salt.
2. Glucose test tape turns green in the presence of glucose.

### Analyze and Apply

1. salt
2. It would contain salt because excess salt in the blood is excreted. It wouldn't contain glucose because glucose is absorbed back into the blood.
3. yes
4. If urine doesn't contain salt or if it does contain glucose, then the nephrons in the kidneys may not be functioning correctly.

## Excretion and the Skin

Skin has three main jobs. First, it protects the body from infection. Second, it is a sense organ. Third, it helps in the excretion of water and salts.

Figure 13-15 shows what a slice through skin looks like. The top layer of this organ is made of a layer of living cells covered with dead cells that are constantly flaking off. A lower layer of skin has capillaries and sweat glands. The sweat glands open onto the surface of the skin. Hair grows from this bottom layer. This layer also has many different types of nerve cells. You will learn in Chapter 16 how nerves in the skin make it an important sense organ.

Your body has two to five million sweat glands like those shown in Figure 13-15. Each gland gives off water and salts much as the kidneys do. Unlike kidneys, however, your skin can't control water or salt loss. It's possible for you to lose half a liter of water each day through the skin without noticing it, unless you sweat heavily.

Does loss of water through the skin help the body? This water loss is one way the body cools itself. You usually sweat more when the weather is warm or when you exercise. Water moving onto the skin from the sweat glands evaporates. The evaporation cools the body.

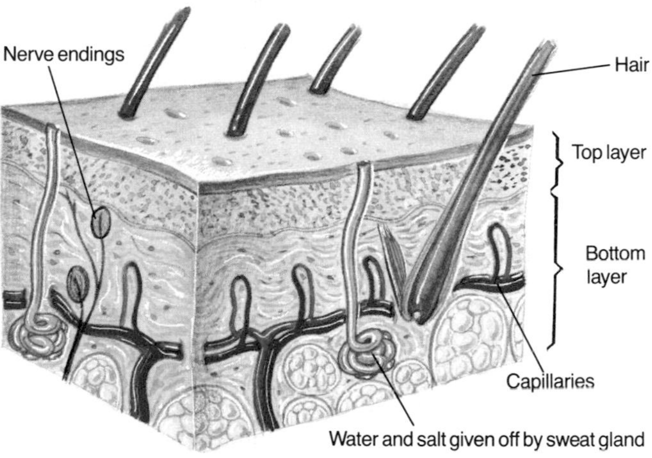


**Figure 13-15** The sweat glands are a part of the excretory system. Find these glands in this cross section of skin.

## Problems of the Excretory System

Like other organs, the kidneys may become diseased. The kidneys can be damaged over time if a person has high blood pressure. Another cause of kidney damage is infection caused by bacteria.

### Bio Tip

**Consumer:** Sweating helps cool the body. Antiperspirants block the sweat glands and stop sweating. Deodorants, however, just mask the odor of sweat.

**How does sweat help the body?**



## TEACH

### ACTIVITY/Brainstorming

Ask students what the skin protects against. Answers should include preventing disease-causing organisms, such as bacteria, from entering the body and loss of water.

### ACTIVITY/Demonstration

Have a student exhale onto a piece of metal or a mirror. Explain that moisture in exhaled air condenses on a cooler surface. Explain that students can see their breath on a cold morning because the water vapor in warm breath condenses in cold air as water droplets.

### Guided Practice

Have students write down their answers to the margin question: Sweat cools the surface of the skin, which cools the body.

### Independent Practice

**Study Guide**/*Teacher Resource Package,* p. 78. Use the Study Guide worksheet shown at the bottom of this page for independent practice.

**Videodisc**

**STV: Human Body Vol. 3**

*Immune System*

Unit 1, Side 1

*First Defense—The Skin*

**07510-10160**

### STUDY GUIDE 78

STUDY GUIDE — CHAPTER 13

Name ________ Date ________ Class ________

**VOCABULARY**

*Review the new words used in Chapter 13 of your textbook. Then, complete this puzzle.*

1. Put the following words into two groups by adding them to this table under the proper heading: urethra, diaphragm, bronchi, ureter, trachea, nephron, alveoli, epiglottis, urinary bladder, respiratory system, excretory system.

| Parts of the respiratory system | Parts of the excretory system |
|---|---|
| lungs | kidney |
| rib cage | sweat glands |
| diaphragm | urethra |
| bronchi | ureter |
| trachea | nephron |
| alveoli | urinary bladder |
| epiglottis | |

2. Put the following words into two groups by writing them in this table under the proper heading: urea, diaphragm, urine, emphysema, excess salt, pneumonia.

| Respiration | Excretion |
|---|---|
| emphysema | urea |
| pneumonia | urine |
| diaphragm | excess salt |

## OPTIONS

**What Chemical Wastes Are Removed by the Lungs, Skin, and Kidneys?**/*Lab Manual,* pp. 107-110. Use this lab as an extension to the role of excretion.

## TEACH

### Guided Practice

Have students write down their answers to the margin question: Both remove wastes from a person's blood.

### Check for Understanding

Have students respond to the first three questions in Check Your Understanding.

### Reteach

**Reteaching**/*Teacher Resource Package*, p. 39. Use this worksheet to give students additional practice on the route wastes take through the kidneys.

**Extension:** Assign Critical Thinking, Biology and Reading, or some of the **OPTIONS** available with this lesson.

## 3 APPLY

### Summary

Have students name wastes that are eliminated by the skin and kidneys.

## 4 CLOSE

### Discussion

Discuss why one would test urine for drug use.

### Answers to Check Your Understanding

16. ureters — carry liquid waste (urine) from kidneys to urinary bladder; urethra — carries liquid waste from bladder to outside the body
17. excess water, urea, and excess salts
18. sense organ, excretion of salts and water, protects body from infection
19. When the concentration of urea is higher in the tube of the nephron than in the blood, it will pass out of the tube into the blood.
20. the kidney

**Figure 13-16** Each session on an artificial kidney machine can take up to five hours.

**How does a kidney machine compare with a kidney?**

Humans are lucky in that they can live with only one kidney. Thus, if one kidney is damaged, we still have the other one to help with excretion. But, what happens when both kidneys are damaged? In such cases, either a kidney machine must be used, or a kidney transplant operation should be performed.

An artificial kidney machine, Figure 13-16, removes urea from the blood when the kidneys can't function. The artificial kidney machine works by passing a person's blood through a very long tube made from a thin membrane. This membrane is surrounded by liquid. Urea in the blood diffuses through the tube membrane and into the liquid. Fresh liquid is constantly added and the used liquid is constantly removed. Blood without urea is then returned to the person. The kidney machine is designed to act just like your nephrons. A person may need up to three sessions a week on the kidney machine.

### Check Your Understanding

16. What are the jobs of your ureters and urethra?
17. Name three things that kidneys filter out of the blood.
18. What are the three main jobs of the skin?
19. **Critical Thinking:** How does the process of diffusion allow materials in a nephron to be returned to the blood?
20. **Biology and Reading:** There's a pair of tubes in each body segment of an earthworm that removes wastes made by the cells. What organ in humans has the same function as these tubes in earthworms?



**RETEACHING 39**

RETEACHING — CHAPTER 13

Name ______ Date ______ Class ______

*Use with Section 13:4.*

*JOURNEY THROUGH THE KIDNEYS*

*Describe the path of the blood as it travels through the kidneys. Use these terms to fill in the blanks: all body parts, urinary bladder, small arteries, ureters, urea, vein, and urethra.*

1. Wastes in blood enter through **small arteries**.
2. Kidneys filter **urea** the blood.
3. Blood leaves a kidney through a **vein**.
4. Veins take clean blood to **all body parts**.
5. Wastes leave kidneys through **ureters**, which carry wastes from the kidneys to the **urinary bladder**.
6. The **urethra** is a tube that carries liquid wastes to outside the body.

*Describe the path of the blood as it travels through a single nephron. Fill in the blanks below.*

1. The tiny filter units of the kidney are the **nephrons**.
2. Blood enters the capillary of each nephron bringing **blood cells, salts, food, urea, and water**.
3. The **blood cells** remain in the capillary.
4. The **water, salts, food, and urea** pass into the cuplike part of the nephron and flow through a long tube.
5. The **capillary** twists around the long tube.
6. The **food, needed water, and needed salts** pass back into the blood.
7. The **blood cells, needed water, needed salts, and food** leave the nephron and return to the body.
8. The **urea, excess water, and salts** stay in the long tube and are carried out of the body as **urine**.

*Answer the following questions.*

1. How does loss of water through sweat glands help the body? **It cools the body by lowering the temperature by evaporation.**
2. How does the skin work as an organ of excretion? **It excretes water and salt.**

# Science and Society

## Organ Transplants

After a kidney transplant operation, a person has a new chance to live a long and healthy life.

The first successful kidney transplant operation was performed in the 1960s. A kidney was donated from one identical twin to the other. Up to that time, the major problem with organ transplants was that tissues from another person's body were usually rejected. Today, we use drugs that reduce rejection. Kidney transplants are now rather common. Why is the kidney such a good organ for transplants? Remember, each person has two of these organs and could survive with only one. It is not unusual for a healthy person to donate a kidney to another member of his or her family or to an unrelated person. Healthy kidneys are also available from people who have agreed to donate their organs when they die.

### What Do You Think?

1. There are a greater number of people needing kidney transplants than there are available kidneys. Thousands of patients are now waiting for a healthy kidney donor. The demand is much greater than the supply. Where should the supply of organs come from? Could they be taken from animals such as baboons, terminally ill patients, or from prisoners on death row? At the present time, the most common source is from the bodies of people in fatal auto accidents. Should everyone be expected to carry around a permission card to donate their kidneys?

2. Kidney and other transplant operations are very expensive. Should only those who can afford to pay for the operations be able to have them? If the government or insurance companies pay for the transplant operations, who will decide who gets an organ? Is the life of a four-year-old child worth more than that of a sixty-year-old adult? How should these choices be controlled?

3. An illegal market exists that encourages healthy people to sell one of their kidneys to the highest bidder. Prices as high as $13 000 for a kidney have been reported. Many of these kidneys come to the United States from foreign countries. These organs may not be in a state suitable for transplanting. A person may spend a lot of money and then discover the organ to be useless. Should the United States pass laws that prevent the importing of human organs?

**Conclusion:** What can be done to assure that anybody who needs a transplant can have one?

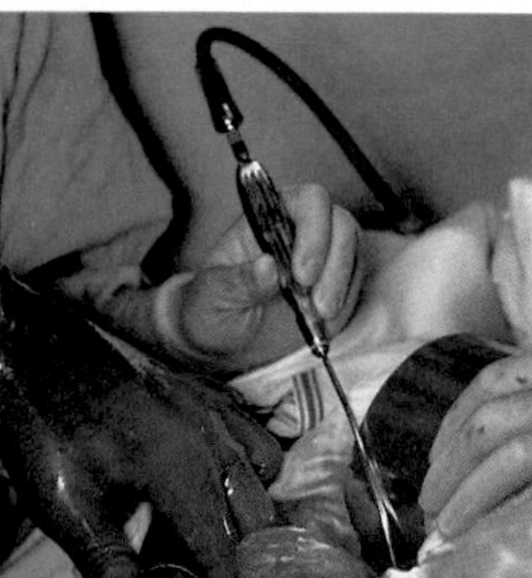

A kidney operation



## GLENCOE TECHNOLOGY

**Videotape**

**The Secret of Life**

*In the Land of Milk and Money: Biotechnology*

Science and Society

Organ Transplants

### Background

The types of transplant operations performed today in decreasing order are: cornea, kidney, bone marrow, skin, liver, lung, heart, and endocrine. The major problem with transplants is the rejection reaction that results from action taken by the recipient's immune system. A drug called cyclosporin A is able to suppress the immune system, so the success rate of transplants has increased in recent years.

### Teaching Suggestions

The questions that appear in the student text should provide the basis for an excellent debate. You may wish to assign, for extra credit, the task of defending or rejecting the question being raised.

### What Do You Think?

There are no incorrect answers to these questions. Students should attempt to answer them based on their own feelings or opinions.

**Conclusion:** Students may suggest that the government would have to become more involved in order to keep transplants fair for all people.

### Suggested Reading

Baker, Stephen. "Who Gets a Liver-And Who Doesn't." *Business Week*, December 9, 1996, p. 153.

Johnson, Lynn, and Glenn Dowling. "Rondie's Gift: A Life-and-Death Choice." *Life*, March 1997, pp. 54-63.

Novitt-Moreno, Anne. "Medical Miracles: Are We Reaching the Bionic Age?" *Current Health 2*, December 1996, pp. 6-13.

Wedemeyer, Lisa. "The Right Thing to Do: By Donating Bone Marrow." *Newsweek*, April 28, 1997, p. 14.

# Chapter 13 Review

## Summary

### 13:1 The Role of Respiration

1. Oxygen is taken in and carbon dioxide is given off in the respiratory system.
2. Oxygen is used by cells to convert food to energy. Carbon dioxide is given off as waste.
3. Some animals have no respiratory system. Animals with a respiratory system use skin, gills, lungs, or pairs of tubes for gas exchange.

### 13:2 Human Respiratory System

4. The respiratory system is made up of tubes that supply air to the alveoli in the lungs.
5. Oxygen moves from lungs to blood to body cells. Carbon dioxide moves from body cells to the lungs.
6. Your diaphragm relaxes and the rib cage moves in when you breathe out. Your diaphragm contracts and the rib cage moves out when you breathe in.

### 13:3 Problems of the Respiratory System

7. Carbon monoxide is a poisonous gas.
8. Pneumonia is caused by bacteria and viruses.
9. Emphysema causes breakdown of lung tissues.

### 13:4 The Role of Excretion

10. Kidneys filter wastes such as urea, excess water, and salts from the blood.
11. Blood, carrying wastes from body cells, is filtered in nephrons in the kidney.
12. Your skin excretes water and cools the body.

## Key Science Words

alveoli (p. 268)
bronchi (p. 267)
carbon monoxide (p. 272)
diaphragm (p. 270)
emphysema (p. 273)
epiglottis (p. 267)
excretory system (p. 274)
nephron (p. 276)
pneumonia (p. 273)
respiratory system (p. 264)
trachea (p. 267)
urea (p. 274)
ureter (p. 276)
urethra (p. 276)
urinary bladder (p. 276)
urine (p. 277)

## Testing Yourself

### Using Words

*Choose the word from the list of Key Science Words that best fits the definition.*

1. tiny filter unit of kidney
2. communicable disease of lungs
3. flap that keeps food out of trachea
4. tube that carries urine from kidney to bladder
5. tiny air sacs of lungs
6. sheetlike muscle that separates chest from rest of body



# CHAPTER 13 REVIEW

## Summary

Summary statements can be used by students to review the major concepts of the chapter.

## Key Science Words

All boldfaced terms from the chapter are listed.

## Testing Yourself

### Using Words

1. nephron
2. pneumonia
3. epiglottis
4. ureter
5. alveoli
6. diaphragm
7. bronchi
8. urethra
9. carbon monoxide

### Finding Main Ideas

10. p. 264; Carbon dioxide is formed in all body cells as a waste gas during cellular respiration. Carbon dioxide diffuses into the blood and is given off in the air as you breathe out.
11. p. 277; Urea, sugar, water, and salts diffuse out of nephron capillaries .
12. p. 269; Oxygen diffuses into cells, and carbon dioxide diffuses out of cells.
13. p. 270; The domelike diaphragm moves upward and squeezes trapped air in the chest wall. The pressure of the air increases, the lungs are squeezed, and the space gets smaller.
14. p. 279; Water loss through the skin is one way the body cools itself.
15. p. 273; A person with emphysema does not have enough alveoli to get oxygen into the blood.
16. p. 269; Oxygen diffuses into the alveoli from the bronchi and then into blood capillaries surrounding alveoli. Carbon dioxide diffuses out of capillaries into alveoli and then out of alveoli into the bronchi.

### Using Main Ideas

17. The diaphragm moves down as it contracts, allowing the chest wall to enlarge. Air pressure in the chest wall decreases, and air rushes in to inflate the lungs.
18. Earthworms exchange gases by diffusion across the skin. Fish carry on gas exchange across organs called gills.
19. Air enters the nose and passes into the trachea. From the trachea, the air moves into two bronchi that lead to the right and left lung.
20. Urea is the waste that results from the breakdown of body protein, while urine is the liquid waste, containing urea, that passes out of the body. The ureters carry liquid waste from the kidneys to the urinary bladder, while the urethra carries liquid waste from the urinary bladder to outside the body. Carbon dioxide is given off as a waste product of cellular respiration. Carbon monoxide is given off in the burning of fossil fuels.
21. Emphysema can be prevented by not smoking and by avoiding air pollution.

# Review

## Testing Yourself *continued*

7. tubes that carry air from trachea to both lungs
8. tube that carries urine to the outside of your body
9. poisonous gas in exhaust fumes

**Finding Main Ideas**

*List the page number where each main idea below is found. Then, explain each main idea.*

10. why more carbon dioxide is given off than taken in
11. which chemicals are forced out of the capillaries of the nephrons
12. which two gases are exchanged during cellular respiration
13. what makes the space around your lungs get smaller as you breathe out
14. why water loss through the skin can be helpful
15. why a person with emphysema is always short of breath
16. the events that take place in the alveoli

**Using Main Ideas**

*Answer the questions by referring to the page number after each question.*

17. Why do your lungs expand when your diaphragm contracts? (p. 271)
18. How do the respiratory systems of earthworms and fish compare? (p. 265)
19. What is the sequence of organs through which air passes, starting with the nose? (pp. 267, 268)
20. How do the following pairs of words differ–urea and urine, ureter and urethra, and carbon dioxide and carbon monoxide? (pp. 268, 272, 274, 276, 277)
21. What can be done to lower the chances of getting emphysema? (p. 273)

## Skill Review ✓

*For more help, refer to the **Skill Handbook**, pages 704-719.*

1. **Experiment:** Blow up a round balloon for 15 seconds. Tie it closed. Run on the spot for one minute, then blow up a second round balloon for 15 seconds. Tie it closed. Compare the volume of each balloon by measuring their circumferences. How did exercise affect your breathing efficiency?
2. **Formulate a model:** Make a model of your lungs using everyday household materials.
3. **Understand science words:** Use the dictionary to find two words closely related to trachea that are used in biology. Give their meanings.
4. **Infer:** Why can you see your breath on a cold morning? What is being excreted?

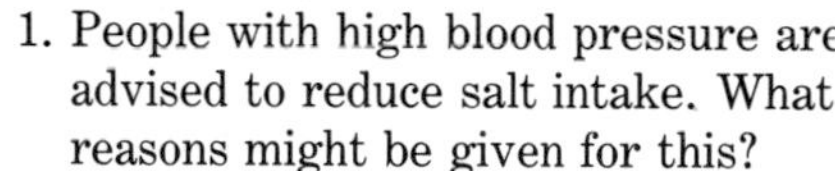

## Finding Out More

TECH PREP

**Critical Thinking**

1. People with high blood pressure are advised to reduce salt intake. What reasons might be given for this?
2. Compare the chemicals in blood entering and leaving the kidney.

**Applications**

1. Place your hand just below your ribs and feel the movement of your diaphragm and rib cage as you breathe in and out.
2. Which organs of the respiratory system are used when you cough, cry, hiccup, laugh, yawn, and sing?



## Skill Review

1. Exercise should increase your breathing efficiency.
2. Plastic squeeze bottles can be used for lungs, and straws for bronchi and trachea.
3. Examples: tracheid – water-conducting tubes in plants; tracheitis – inflammation of the trachea
4. Warm water vapor from the nose meets the cold air and fogs. Water vapor is being excreted.

## Finding Out More

### Critical Thinking

1. In high blood pressure, the kidneys retain salt and water instead of passing them out. An elevated salt level in the blood causes constriction of blood vessels and less blood gets to kidneys and cells.
2. Blood entering the kidney contains salts, water, blood cells, proteins, urea, and sugar. The blood leaving the kidneys contains blood cells, proteins, and sugar. It also contains salts and water, but in a lower concentration than the blood entering.

### Applications

1. The diaphragm will move down as you breathe in and up when you breathe out.
2. The organs include the epiglottis, trachea, lungs, and diaphragm.

**Chapter Review**/*Teacher Resource Package*, pp. 65-66.

**Chapter Test**/*Teacher Resource Package*, pp. 67-69.

**Chapter Test**/*Computer Test Bank*

# Support and Movement

## PLANNING GUIDE

| CONTENT | TEXT FEATURES | TEACHER RESOURCE PACKAGE | OTHER COMPONENTS |
|---|---|---|---|
| (2 days)<br>14:1 The Role of the Skeleton<br>Functions of the Skeleton<br>Bones in the Skeleton<br>Bone Growth<br>Bone Structure<br>Joints | Mini Lab: *What Happens to Bone When Calcium is Removed?* p. 287<br>Mini Lab: *How Strong Can a Hollow Bone Be?* p. 289<br>Lab 14-1: *Bone Density,* p. 290<br>Idea Map, p. 286<br>Check Your Understanding, p. 291 | Enrichment: *Human Skeletons,* p. 14<br>Reteaching: *Bone and Skeleton,* p. 40<br>Study Guide: *The Role of the Skeleton,* pp. 79-80<br>Lab 14-1: *Bone Density,* pp. 53-54 | **Laboratory Manual:**<br>*How Do Male and Female Skeletons Differ?* p. 115<br>Color Transparency 14: *The Human Skeleton*<br>**STVS:** *Orthopedic Implants,* Human Biology (Disc 7, Side 2) |
| (2 days)<br>14:2 The Role of Muscles<br>Movement<br>Human Muscle Types<br>How Muscles Work<br>Muscles Work in Pairs | Skill Check: *Classify,* p. 294<br>Skill Check: *Infer,* p. 296<br>Career Close-up: *Athletic Trainer,* p. 293<br>Check Your Understanding, p. 297 | Application: *Exercise and Skeletal Muscles,* p. 14<br>Reteaching: *Muscles,* p. 41<br>Critical Thinking/Problem Solving: *What Are the Jobs of Different Muscles?* p. 14<br>Skill: *Understand Science Words,* p. 15<br>Transparency Master: *Muscle Types,* p. 65<br>Study Guide: *The Roles of Muscles,* pp. 81-82 | **STVS:** *Tricycle for the Handicapped,* Human Biology (Disc 7, Side 2) |
| (2 days)<br>14:3 Bone and Muscle Problems<br>Problems of the Skeletal System<br>Problems of the Muscular System<br>Making Products Fit People | Lab 14-2: *Muscles,* p. 298<br>Skill Check: *Understand Science Words,* p. 299<br>Check Your Understanding, p. 301 | Focus: *When Muscles No Longer Work: Stephen Hawking,* p. 29<br>Reteaching: *Bone and Muscle Problems,* p. 42<br>Study Guide: *Bone and Muscle Problems,* p. 83<br>Study Guide: *Vocabulary,* p. 84<br>Lab 14-2: *Muscles,* pp. 55-56 | **Laboratory Manual:**<br>*What Causes Sports Injuries?* p. 111<br>**STVS:** *Arthritis Research,* Human Biology (Disc 7, Side 2) |
| Chapter Review | Summary<br>Key Science Words<br>Testing Yourself<br>Finding Main Ideas<br>Using Main Ideas<br>Skill Review | **ASSESSMENT RESOURCES**<br>Chapter Review, pp. 70-71<br>Chapter Test, pp. 72-74<br>Performance Assessment in the Biology Classroom<br>Alternate Assessment in the Science Classroom<br>Computer Test Bank | |

### GLENCOE TECHNOLOGY

**Infinite Voyage,** *The Champion Within*
**Science and Technology Videodisc Series,** *Orthopedic Implants,* Human Biology (Disc 7, Side 2)
*Tricycle for the Handicapped,* Human Biology (Disc 7, Side 2)
*Arthritis Research,* Human Biology (Disc 7, Side 2)

## MATERIALS NEEDED

| LAB 14-1, p. 290 | LAB 14-2, p. 298 | MARGIN FEATURES |
|---|---|---|
| 100-mL graduated cylinder<br>balance<br>water<br>bones of chicken, cow, pig, turkey | scissors<br>string<br>tape<br>paper punch<br>metal fastener<br>index card | Skill Check, p. 299<br>dictionary<br>Mini Lab, p. 287<br>beaker<br>tweezers<br>chicken bone<br>6% hydrochloric acid<br>Mini Lab, p. 289<br>tape<br>paper<br>Mini Lab, p. 295<br>cooked chicken leg (drumstick attached) |

## OBJECTIVES

For more information about National Science Standards, see page 5T.

| SECTION | OBJECTIVE | CORRELATION of QUESTIONS to OBJECTIVES | | | |
|---|---|---|---|---|---|
| | | CHECK YOUR UNDERSTANDING | CHAPTER REVIEW | TRP CHAPTER REVIEW | TRP CHAPTER TEST |
| 14:1 National Science Stds: UCP.1, UCP.2, UCP.5, B.4 | 1. **Explain** the functions and growth of the skeleton. | 1, 5 | 11, 16, 19 | 3, 5, 7, 8, 11, 15, 16, 19, 33 | 2, 6, 7, 15, 18, 37, 38, 39 |
| | 2. **Define** the six tissues of bones. | 2 | 5, 7, 14 | 13, 23, 34 | 4, 9, 11, 13 |
| | 3. **Identify** three types of joints. | 3, 4 | 10 | 9, 20, 22, 24, 25, 31 | 1, 20, 22, 23, 43, 44, 45, 46, 47, 48 |
| 14:2 National Science Stds: UCP.1, UCP.2, UCP.3, UCP.5, B.4, C.1 | 4. **Compare** three kinds of muscles. | 6 | 2, 3, 21 | 2, 18, 32, 35, 36 | 12, 26, 27, 28, 29, 30, 31, 32, 40, 41 |
| | 5. **Explain** how muscles work to move body parts. | 7, 8, 9 | 6, 23 | 1, 12, 14, 17, 21 | 10, 17, 21, 24, 25, 26, 33 |
| | 6. **Relate** how muscles work in pairs. | 10 | 8, 15, 22 | 26, 27, 28, 29, 30 | 3, 16, 49, 50, 51, 52, 53, 54 |
| 14:3 National Science Stds: UCP.5, B.4, E.1, E.2, F.1 | 7. **Discuss** problems of the skeletal system. | 11 | 4, 12, 20 | 4, 6 | 8, 35, 42 |
| | 8. **Give examples** of problems of the muscular system. | 12, 14, 15 | 24 | 36 | 5, 19, 36 |
| | 9. **Explain** the purpose of new designs for products. | 13 | 17 | 10 | 14 |

# Support and Movement

## CHAPTER OVERVIEW

### Key Concepts

In this chapter, students will study the structure and function of the skeletal and muscular systems. The anatomy of bones and joints is examined. Muscles, muscle pairing, and their relation to movement are discussed. Several problems of the skeletal and muscular systems are examined.

### Key Science Words

arthritis
ball-and-socket joint
cardiac muscle
fixed joint
hinge joint
involuntary muscle
ligament
muscular dystrophy
muscular system
skeletal muscle
skeletal system
smooth muscle
solid bone
spongy bone
sprain
tendon
voluntary muscle

### Skill Development

In Lab 14-1, students will **interpret data, form a hypothesis, experiment,** and **measure in SI** while comparing mammal and bird bone density. In Lab 14-2, they will **formulate a model, experiment,** and **infer** while examining muscle function. In the Skill Check on page 294, they will classify muscle types. In the Skill Check on page 296, they will **infer** where muscle tissue goes when it contracts. In the Skill Check on page 299, they will **understand** the **science word** *arthritis.* In the Mini Lab on page 287, they will **experiment** with bone calcium. In the Mini Lab on page 289, they will **formulate a model** of hollow bone. In the Mini Lab on page 295, they will **observe** a chicken thigh.

### Bridging

The human heart is muscle. The lungs are able to move air in and out as muscles contract. Food moves through the digestive system due to the action of muscles in the stomach and intestine. This chapter will show how these muscles actually work to move the body, blood, air, and food.



**CHAPTER PREVIEW**

### Chapter Content

Review this outline for Chapter 14 before you read the chapter.



### Skills in this Chapter

The skills that you will use in this chapter are listed below.

- In **Lab 14-1,** you will interpret data, form a hypothesis, experiment, and measure in SI. In **Lab 14-2,** you will formulate a model, experiment, and infer.
- In the **Skill Checks,** you will infer, classify, and understand science words.
- In the **Mini Labs,** you will formulate a model, observe, and experiment.

## TECH PREP

For Tech Prep activities in this chapter of the Teacher Wraparound Edition, see especially the Activities on pages 286, 288, and 296, the Connection on page 287, and the Options Activity on page 297.

See also the Glencoe Homepage at **www.glencoe.com**

The following Glencoe resources provide additional opportunities for integrating science and technology.

**Experience Technology**
Section III: Using Resources to Solve Technical Problems
Activity Brief: Using Technology to Overcome Physical Handicaps

Chapter **14**

# Support and Movement

All your body systems tackle mighty jobs each day. They help you walk, sit, write, tie your shoelaces, and even watch a TV show. But what about the extra work that you give your body? Do you jog, dance, do judo, lift heavy weights, swim, cycle, or play baseball? These are not activities you need to do to survive. So, how can your body put up with these extra demands?

Your bones and muscles give your body support and strength, just as the steel frame supports the bridge shown in the photo on the left. The strength of your skeleton and muscles can be relied on in most exercises, such as the one shown in the photo below. In this chapter, you will see how the pieces and parts of the skeleton and muscles work together.

**Try This!**

**How big is your muscle?** Use a tape measure to determine the size (diameter) of your upper arm when relaxed. Remeasure as you bend your arm and make a fist. Muscles can shorten. When did your muscle shorten?

*inter***NET** CONNECTION

For more information about the material in this chapter, follow the link for the chapter on the Glencoe Homepage at **http://www.glencoe.com**



**ASSESSMENT PLANNER**

***Portfolio***

Strategies on the following pages represent student products that can be placed into a best-work portfolio: pp. 294, 296.

**PERFORMANCE ASSESSMENT**

Skill Check, pp. 294, 296, 299
Mini Lab, pp. 287, 289, 295
Lab, pp. 290, 298

**CONTENT ASSESSMENT**

Check for Understanding, pp. 289, 291, 297, 300, 301
Chapter Review, pp. 302-303

**GROUP ASSESSMENT**

Opportunities for group assessment occur with Cooperative Learning Strategies.

***Student Journal***

Strategies on the following pages represent opportunities for writing in a Student Journal: p. 300.

**GETTING STARTED**

## Using the Photos

Have students give examples of other types of support systems, living and nonliving. Living support systems include exoskeletons, the hydrostatic skeleton of an earthworm, and cell walls of plants. Nonliving supports include the chassis of a car and the frame of a house. Ask students what all these systems have in common.

## MOTIVATION/Try This!

**How big is your muscle?** A strip of clay that is pushed in from either side will end up thicker in the center. Muscles tend to show the same effect as they shorten. Students will observe that muscle tissue bunches up and as a result, the muscle appears to thicken.

## Chapter Preview

Have students study the chapter outline before they begin to read the chapter. They should note the overall organization of the chapter, that the role of the skeleton is covered first, that the study of muscles is next, and that a discussion of bone and muscle problems ends the chapter.

## Misconception

Students are not aware of the fact that bone is living tissue. Yet, they realize that a fracture of this body part will heal and repair itself in time. They are also not sure of the fact that the bulk of internal organs is muscle tissue.

# 14:1 The Role of the Skeleton

## PREPARATION

### Materials Needed

Make copies of the Enrichment, Reteaching, Study Guide, and Lab worksheets in the *Teacher Resource Package*.

▶ Obtain paper, tape, chicken bones, and hydrochloric acid for the Mini Labs.

▶ Gather bones from the butcher shop for Lab 14-1.

▶ Obtain old X rays from the local hospital for the demonstrations.

### Key Science Words

skeletal system
solid bone
spongy bone
ligament
hinge joint
ball-and-socket joint
fixed joint

### Process Skills

In Lab 14-1, students will interpret data, form a hypothesis, experiment, and measure in SI. In the Mini Labs, they will experiment and formulate a model.

## 1 FOCUS

▶ The objectives are listed on the student page. Remind students to preview these objectives as a guide to this numbered section.

### ACTIVITY/Demonstration

Obtain outdated X rays from your local hospital. Project them onto the classroom screen using an overhead projector. Ask students if they can identify certain body areas or bones from the X rays.

### Idea Map

Have students use the idea map as a study guide to the major concepts of this section.

# 14:1 The Role of the Skeleton

**Objectives**

1. **Explain** the functions and growth of the skeleton.
2. **Define** the six tissues of bones.
3. **Identify** three types of joints.

**Key Science Words**

skeletal system
solid bone
spongy bone
ligament
hinge joint
ball-and-socket joint
fixed joint

All vertebrates, from fish to mammals, have a skeleton. The skeleton provides support to their bodies. It also has other jobs such as protection, storage of materials, production of blood cells, and helping with movement.

## Functions of the Skeleton

You know that all body systems are made of organs. Bones are the organs in the skeletal system. The **skeletal system** is the framework of bones in your body. As a framework, the skeleton helps to support the entire body. A skeleton also has other jobs. The skeleton helps to protect certain body organs. Your brain, heart, and lungs are organs with soft tissues. They could easily be damaged. Your skull protects your brain against injury. Ribs form a cage that protects your heart and lungs.

Many bones in your body make blood cells. You learned in Chapter 12 that bone marrow is the soft center part of bone. Blood cells are made within the bone marrow. You may have eaten a ham steak. A ham steak is the muscle of a pig's thigh, and the bone in the steak is a section of the pig's thigh bone. At the center of the bone is the marrow.

**What part of bone produces blood cells?**

Another job of bone is to store calcium (Ca). Remember that calcium is a mineral used by the body and is part of all bones. Calcium gives bones their strength. Without calcium, your bones would become weak or brittle.

The last job of your skeleton is to provide a place for muscles to attach. If muscles were not attached to your bones, you would not be able to move.

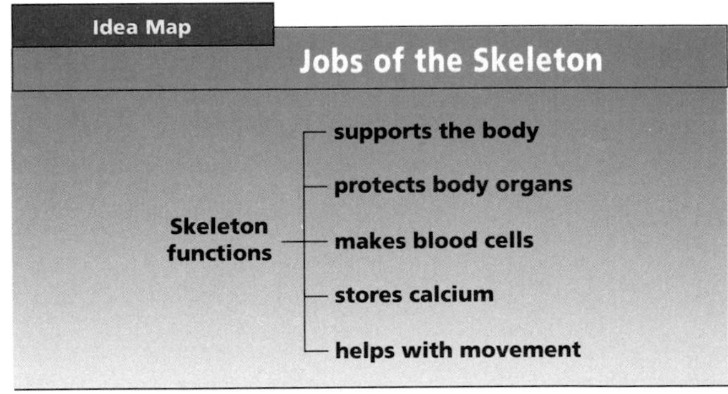


**Study Tip:** Use this idea map as a study guide to the functions of the skeleton. What bones allow you to write the answer to this question?



## *OPTIONS*

### Science Background

The interdependence of the muscular and skeletal systems is developed in this chapter. Muscles contract to move body parts that are usually attached to some part of the skeletal system. This is, of course, particularly true for skeletal muscle.

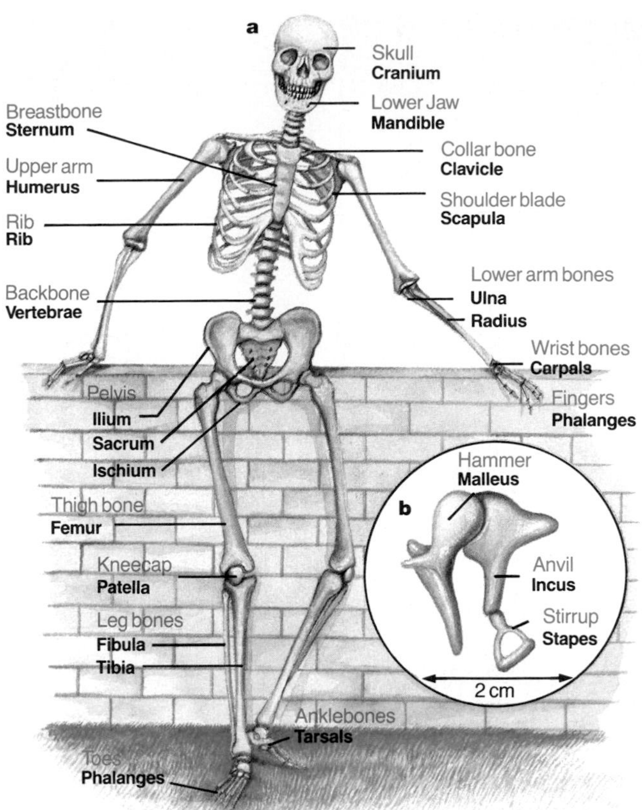


**Figure 14-1** The skeleton is made of flat bones in the skull and ribs, irregular bones in the backbone, long bones in the arms and legs, and short bones in the hands and feet (a). The smallest bones are in the ear (b).

## Bones in the Skeleton

If you counted all the bones in an adult human skeleton, you would find 206 of them. Examine the human skeleton in Figure 14-1. All the bones can't be counted in this diagram because not all of them are shown. Also, some bones are joined with other bones, making them look like one. For example, there are 22 different bones in the human skull. You can see only a few of these in the diagram.

Notice that the common name of a bone is usually different from the medical name. The common names are printed in blue. The medical names are printed in black. The three small ear bones shown to the bottom right are the body's smallest bones. Together, they measure only two centimeters long. They look so much like a hammer, an anvil, and a stirrup that these are their common names.

### Mini Lab

**What Happens to Bone When Calcium Is Removed?**

**Experiment:** Place a chicken bone into a beaker with dilute HCl. Leave the bone overnight. Rinse carefully to remove any acid. Now, describe the bone. *For more help, refer to the **Skill Handbook**, pages 704-705.*



# 2 TEACH

### MOTIVATION/Demonstration

A skeleton (either real or plastic) would illustrate the major functions being described here.

### Concept Development

▶ The top ten rib pairs are attached to the sternum. The lowest two pairs are not attached and are called "floating" ribs.

### Guided Practice

Have students write down their answers to the margin question on page 286: marrow.

### Connection: Social Studies

Have students investigate some of the benefits and limitations of armor.

### Mini Lab

Caution students to use care with hydrochloric acid.

### ASSESSMENT

**Performance:** The shell of an egg is almost pure calcium. Have students design an experiment to illustrate that an acid, such as vinegar, rather than water only reacts with calcium. (Note: Pieces of eggshell are suitable for this assessment rather than the entire egg itself. Students also should design some method for judging shell size before and after treatments.)

### Independent Practice

**Study Guide**/*Teacher Resource Package*, p. 79. Use the Study Guide worksheet shown at the bottom of this page for independent practice.

**The Human Skeleton**/*Transparency Package*, number 14. Use color transparency number 14 as you teach the bones of the skeleton.

**STUDY GUIDE 79**

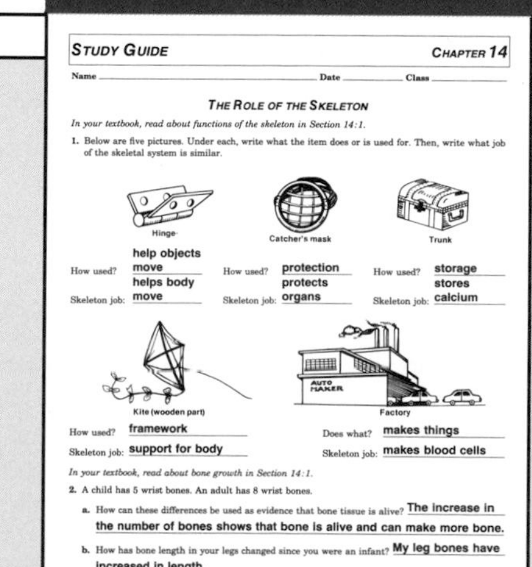

STUDY GUIDE — CHAPTER 14

Name ________ Date ________ Class ________

THE ROLE OF THE SKELETON

*In your textbook, read about functions of the skeleton in Section 14:1.*

1. Below are five pictures. Under each, write what the item does or is used for. Then, write what job of the skeletal system is similar.

Hinge — How used? help objects move — Skeleton job: helps body move

Catcher's mask — How used? protection — Skeleton job: protects organs

Trunk — How used? storage — Skeleton job: stores calcium

Kite (wooden part) — How used? framework — Skeleton job: support for body

Factory — Does what? makes things — Skeleton job: makes blood cells



*In your textbook, read about bone growth in Section 14:1.*

2. A child has 5 wrist bones. An adult has 8 wrist bones.
   a. How can these differences be used as evidence that bone tissue is alive? The increase in the number of bones shows that bone is alive and can make more bone.
   b. How has bone length in your legs changed since you were an infant? My leg bones have increased in length.
   c. How can this change be used as evidence that bone tissue is alive? Something which is living can grow in size.

## OPTIONS

### ACTIVITY/Challenge

Ask students to use references (anatomy books) in order to determine the regions and number of bones associated with the spinal cord. An "idea map" could be designed from their findings. The five regions of the spine include:

| | |
|---|---|
| neck or cervical region | 7 |
| thoracic | 12 |
| lumbar | 5 |
| sacrum (5 bones fused) | 5 |
| tail or coccyx | 4 |

# TEACH

## Concept Development

▶ Bone is a combination of organic matter (35%) and mineral salts (65%). The organic part is composed of cells and a protein called collagen. The organic matter makes bone flexible. The mineral salts, such as calcium carbonate or calcium phosphate, make bone hard.

## ACTIVITY/Demonstration

Find examples of X rays showing wrists and ankles of a young person and an older person. Place the X rays on an overhead projector and reduce the light in the room. An opaque projector will also work. Compare specific infant bones to the same bones of an older person. Have students compare both the size and number of bones for young and old people.

## Guided Practice

Have students write down their answers to the margin question: cartilage, outer membrane, solid bone, spongy bone, marrow, ligaments.

## ACTIVITY/Project

Have students test for the presence of calcium in a variety of skeleton-like parts such as snail shells, chicken cartilage, beef bone. A 6 percent solution of hydrochloric acid added to these various tissues will form bubbles if calcium is present.

**Enrichment**/*Teacher Resource Package*, p. 14. Use the Enrichment worksheet shown here to teach students more about the parts of the skeleton.

## ACTIVITY/Demonstration

Bring a beef bone into class for students to examine. If the bone is fresh, students will be able to see the periosteum and to pull off portions of it. Chicken bones show cartilage very well. A beef knee joint with bone above and below the joint will show the many different ligaments holding the bones in place. A long beef bone cut open lengthwise from your butcher will show solid, spongy, and marrow regions.



**Figure 14-2** The hand and wrist bones of an infant (a) and an adult (b) indicate an increase in numbers of bones with age.

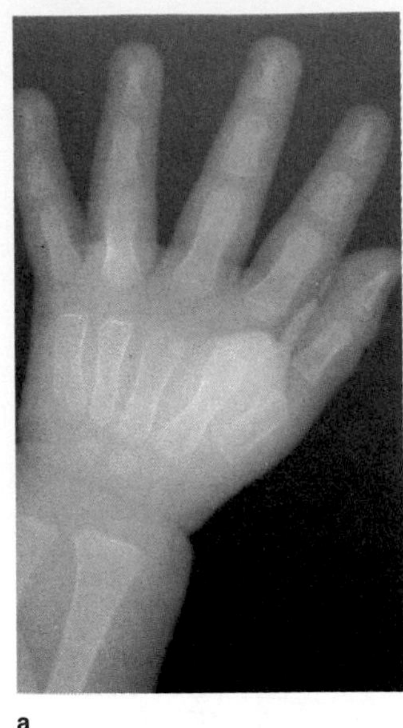


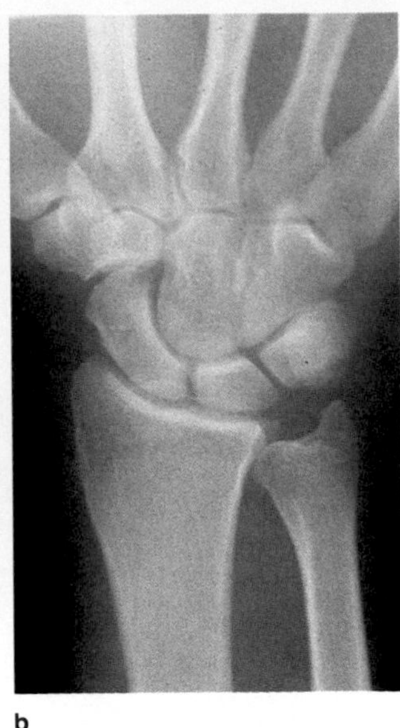


## Bone Growth

Bone tissue is alive and is made of cells, just as in other organs. Because bone cells are living, they can reproduce and make more bone. This results in bone growth. You know that bones grow because you are taller now than when you were younger.

Not only does bone size change as you grow, the number of your bones increases. Figure 14-2a shows an X ray of the hand and wrist bones of a three-year-old child. The X-ray photo on the right shows the hand and wrist bones of an adult. It is clear that the size of the bones has increased during growth. Has the number of bones increased? Count the number of bones shown in the wrist of each X ray. An infant's wrist has five bones. An adult's wrist has eight bones.

**Bio Tip**

**Consumer:** Many foods provide you with a natural supply of calcium. They include milk, milk products, green leafy vegetables, oranges, dried peas and beans, and shellfish.

## Bone Structure

What are six tissues of bone?

Most bones are made of six kinds of tissues. Figure 14-3 shows the positions of these six tissues in a bone.

You read in Chapter 8 that cartilage is a tough, flexible tissue that supports and shapes the body. This tissue acts as a cushion where bones come together. Cartilage doesn't



## OPTIONS

### Science Background

Cartilage is a special tissue consisting of cells called chondrocytes and a protein matrix made up of collagen. Cartilage has no mineral salts such as calcium. No blood vessels run through cartilage. This is why cartilage is so slow to heal if torn or injured.

**How Do Male and Female Skeletons Differ?**/*Lab Manual*, pp. 115-118. Use this lab as an extension to skeleton structure.

**ENRICHMENT 14**

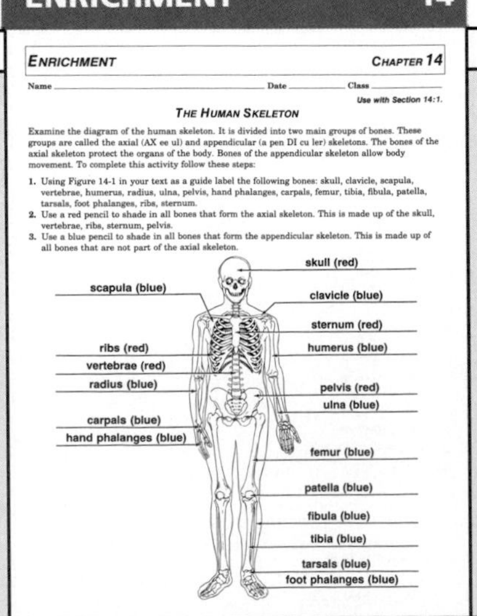

ENRICHMENT CHAPTER 14

Name ______ Date ______ Class ______

Use with Section 14:1.

THE HUMAN SKELETON

Examine the diagram of the human skeleton. It is divided into two main groups of bones. These groups are called the axial (AX ee ul) and appendicular (a pen DI cu ler) skeletons. The bones of the axial skeleton protect the organs of the body. Bones of the appendicular skeleton allow body movement. To complete this activity follow these steps:

1. Using Figure 14-1 in your text as a guide label the following bones: skull, clavicle, scapula, vertebrae, humerus, radius, ulna, pelvis, hand phalanges, carpals, femur, tibia, fibula, patella, tarsals, foot phalanges, ribs, sternum.
2. Use a red pencil to shade in all bones that form the axial skeleton. This is made up of the skull, vertebrae, ribs, sternum, pelvis.
3. Use a blue pencil to shade in all bones that form the appendicular skeleton. This is made up of all bones that are not part of the axial skeleton.

store calcium, so it is much softer than bone. Your ears and the tip of your nose can be bent from side to side. They are shaped by cartilage. When eating chicken, you may have noticed cartilage on the ends of the bones. Cartilage in foods such as chicken and beef is often called gristle.

Bone is covered by a thin sheet of tissue that forms an outer membrane. This membrane has many nerves and blood vessels that supply the bone with blood. This is the part that hurts when a bone gets bumped or bruised. It is also the growth area for new bone.

The inside of a bone shows other tissues. **Solid bone** is the very compact or hard part of a bone. It is usually found along the outer edges of bones. This part of the bone is very strong. Calcium is stored in solid bone and makes it strong.

**Spongy bone** is the part of a bone that has many empty spaces, much like those in a sponge. Spongy bone is usually found toward the ends of bones. It is strong like solid bone and gives the bone strength. Spongy bone also stores calcium but is lightweight because of the many empty spaces. Bird skeletons are lightweight because much of their bone tissue is spongy bone. How is this feature helpful to birds?

Remember that marrow is the soft center in bone. Blood cells and platelets are made in the marrow.

**Ligaments** (LIGH uh munts) are tough fibers that hold one bone to another. They are usually found at bone joints where two bones come together.

## Mini Lab

**How Strong Can a Hollow Bone Be?**

**Formulate a model:** Curl a sheet of paper into a tube about 3 cm in diameter. Tape it so it does not unwind. How many books can be supported on top of the cylinder? *For more help, refer to the* ***Skill Handbook,*** *pages 706-711.*

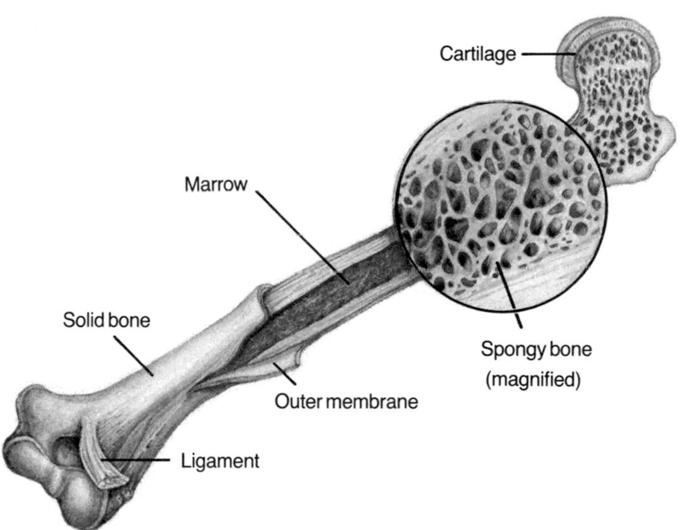


**Figure 14-3** The six tissues of bone are shown in this cut-away diagram.



## STUDY GUIDE 80

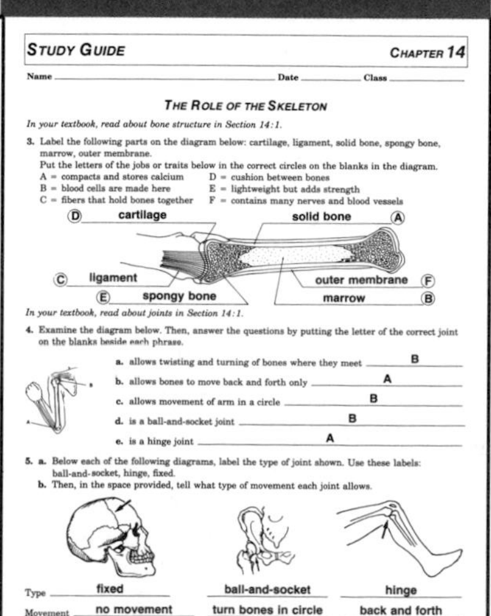

STUDY GUIDE CHAPTER 14

Name ______ Date ______ Class ______

*THE ROLE OF THE SKELETON*

*In your textbook, read about bone structure in Section 14:1.*

3. Label the following parts on the diagram below: cartilage, ligament, solid bone, spongy bone, marrow, outer membrane.
Put the letters of the jobs or traits below in the correct circles on the blanks in the diagram.
A = compacts and stores calcium
B = blood cells are made here
C = fibers that hold bones together
D = cushion between bones
E = lightweight but adds strength
F = contains many nerves and blood vessels

(D) cartilage
(A) solid bone
(C) ligament
(F) outer membrane
(E) spongy bone
(B) marrow

*In your textbook, read about joints in Section 14:1.*

4. Examine the diagram below. Then, answer the questions by putting the letter of the correct joint on the blanks beside each phrase.
a. allows twisting and turning of bones where they meet B
b. allows bones to move back and forth only A
c. allows movement of arm in a circle B
d. is a ball-and-socket joint B
e. is a hinge joint A

5. a. Below each of the following diagrams, label the type of joint shown. Use these labels: ball-and-socket, hinge, fixed.
b. Then, in the space provided, tell what type of movement each joint allows.

| | | | |
|---|---|---|---|
| Type | fixed | ball-and-socket | hinge |
| Movement | no movement | turn bones in circle | back and forth |

**Videodisc**

STV: Human Body Vol. 2
*Muscular and Skeletal Systems*
Unit 1, Side 1
*Muscular and Skeletal Systems (in its entirety)*

**1107-33550**

## TEACH

### Mini Lab

The strip of folded paper may require some additional support initially. The weight of the book will usually crumple it before students realize what has occurred. Books will have to be centered onto the cylinders, otherwise they will fall off. Hollow bones, as illustrated by the cylinder of paper, have more strength than solid bones.

### ASSESSMENT

**Performance:** Have students determine if there is a correlation between tube strength and tube diameter. Allow them to form a hypothesis, design and conduct their experiment, record their findings, and indicate if their hypothesis was or was not supported by their data.

### Guided Practice

Provide students with a diagram of a cross section of bone. Ask them to label the parts that have been studied.

### Check for Understanding

Ask students to name parts of bone that illustrate the following skeletal functions: support, protection, form blood, store calcium, place for muscle attachment.

### Reteach

Certain bone parts have specialized functions. Have students classify the following as to whether they add strength, are a covering, or provide cushioning: cartilage, spongy bone, solid bone, periosteum.

### Independent Practice

**Study Guide**/*Teacher Resource Package*, p. 80. Use the Study Guide worksheet shown at the bottom of this page for independent practice.

## Lab 14-1 Bone Density

### Overview

Students will learn how to determine the density of bone and use the technique to compare the densities of mammal and bird bones.

**Objectives:** Upon completion of the lab, students will be able to (1) **calculate** density, (2) **compare** the density of mammal and bird bone, (3) **relate** the lower density of bird bones to a bird's lifestyle.

**Time Allotment:** 1 class period

### Preparation

▶ Use similar (homologous) bones for comparison of densities, for example ribs (beef ribs, pork chops, chicken and turkey breast).

**Lab 14-1 worksheet**/*Teacher Resource Package*, pp. 53-54.

### Teaching the Lab

▶ Review the math involved with density calculations.

**Performance:** Provide students with a piece of cartilage (obtainable from the local butcher shop or from dinner scraps). Have students determine the density of this tissue and compare it to that of mammal bone.

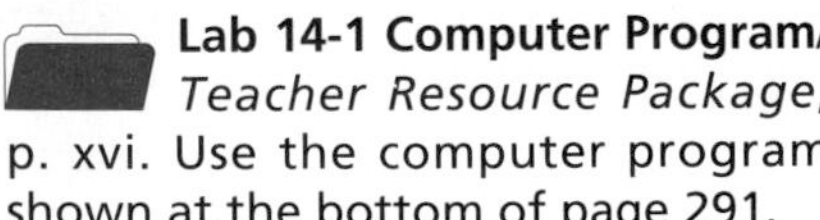

**Cooperative Learning:** Divide the class into groups of three students. For more information, see pp. 22T-23T in the Teacher Guide.

**Lab 14-1 Computer Program/** *Teacher Resource Package*, p. xvi. Use the computer program shown at the bottom of page 291.

# Bone Density

## Lab 14–1

### Problem: How do bird and mammal bone densities compare?

#### Skills

interpret data, form a hypothesis, experiment, measure in SI

#### Materials

100-mL graduated cylinder
water
balance
bones of cow, pig, chicken, turkey

#### Procedure

1. Copy the data table.
2. **Measure in SI:** Use a balance to determine the mass in grams of a cow bone.
3. Record the mass of the bone.
4. Measure 50 mL of water in the graduated cylinder.
5. **Experiment:** Place the bone into the cylinder. Read the new volume of water. Remember to read the volume at eye level.
6. Subtract 50 from the new volume reading. This will give you the volume of the bone in millimeters.
7. Record this volume of the bone in your data table.
8. To calculate the density of the bone, divide the mass of the bone by its volume. (NOTE: Density is a measure of how compact a material is. The more compact it is, the higher its density. The less compact it is, the lower its density.)
9. Write a **hypothesis** in your notebook about which type of bone you think will have higher density, bird or mammal. Justify your hypothesis. (HINT: Read Section 14:1 again.)
10. Repeat steps 2 to 8 using a pig bone, a chicken bone, and a turkey bone.

| | MAMMAL BONE | | BIRD BONE | |
|---|---|---|---|---|
| | COW | PIG | CHICKEN | TURKEY |
| Mass | 38 g | 13.5 g | 34 g | 28 g |
| Volume | 23 mL | 10 mL | 30 mL | 25 mL |
| Density | 1.6 g/mL | 1.3 g/mL | 1.1 g/mL | 1.1 g/mL |

#### Data and Observations

1. Which kind of bone, mammal or bird, has the higher density?
2. Which kind of bone, mammal or bird, has the lower density?

#### Analyze and Apply

1. **Check your hypothesis:** Is your hypothesis supported by your data? Explain.
2. How would you describe a mammal bone in terms of its density? How would you describe a bird bone's density?
3. (a) Which material would be more dense, a piece of steel or a piece of wood?
   (b) What kind of experiment could you do to prove your answer correct?
4. **Interpret your data** to explain if mammal bone is mainly solid or spongy.
5. Using your data, is bird bone mainly solid or spongy? Explain.
6. **Apply:** Using your data, explain how the type of bone might help birds and mammals survive.

#### Extension

**Design an experiment** that would help you measure the density of each of the three tissues of bone: spongy bone, solid bone, and marrow.

290 Support and Movement 14:1

## ANSWERS

### Data and Observations

1. mammal
2. bird

### Analyze and Apply

1. Students who hypothesized that mammal bones are more dense will say yes. A mammal bone has more solid bone than a bird bone.
2. A mammal bone is more compact. A bird bone is less compact.
3. (a) a piece of steel
   (b) Compare the difference in volume of a piece of steel and a piece of wood by placing in water, recording the new volumes, and subtracting the original water volumes.
4. Mammal bone is mainly solid; the density is higher than bird bone.
5. Bird bone is spongy; the density is lower than mammal bone.
6. Spongy bone is light and would help birds fly. Solid bone is heavy and strong and would support the weight of mammals and absorb the shock of running on the ground.

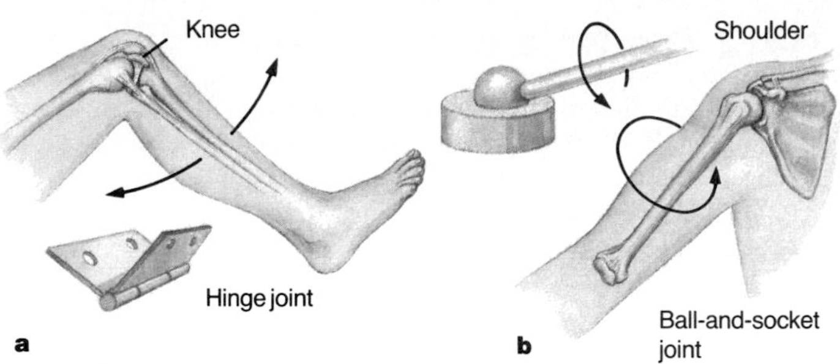


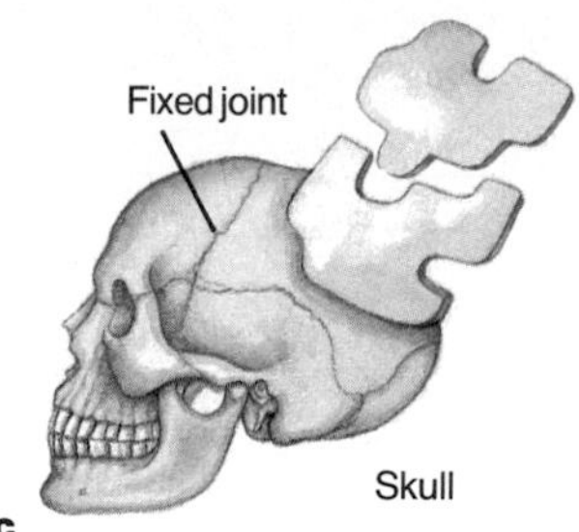


**Figure 14-4** Hinge joints allow only back and forth movements, as in the hinge of a door (a). Ball-and-socket joints allow twisting and turning movements, as in a radio antenna (b). Fixed joints in the skull allow no movement, as in a jigsaw puzzle (c).

## Joints

The place where bones come together is called a joint. There are several types of joints in the body. Most joints allow for different kinds of bone movement. Another type of joint allows no movement at all.

**Hinge joints** allow bones to move only back and forth. Your knee and elbow joints are examples of hinge joints, Figure 14-4a. Compare a knee and an elbow to a real hinge. Why do you think you can open your mouth? Your lower jaw is attached to the skull by a hinge joint.

**What joint allows only back and forth movement?**

Your upper arm bone meets the shoulder, and your upper leg bone meets the pelvis in ball-and-socket joints. A **ball-and-socket joint** allows you to twist and turn the bones in a circle where they meet. Figure 14-4b compares the movement of a ball-and-socket joint to an antenna.

The skull has joints where the bones come together. These joints don't move. Joints that don't move are called **fixed joints.** In Figure 14-4c, the zigzag lines on the skull are the fixed joints.

### Check Your Understanding

1. What mineral supplies strength to bone?
2. Describe the jobs of cartilage, ligaments, and marrow.
3. Describe the movements of a hinge joint and a ball-and-socket joint. Give examples of each.
4. **Critical Thinking:** What are some sports you couldn't do if your shoulder and upper arm were joined by a hinge joint? Explain your answer.
5. **Biology and Math:** Is the smallest bone in the body less than two centimeters long? How do you know?

## COMPUTER PROGRAM XVI

COMPUTER PROGRAM LAB 14-1 BONE DENSITY

```
5  HOME : GOSUB 1000
10  FOR X = 1 TO 4
20  READ ANIMAL$(X)
30  NEXT
40  FOR QUESTIN = 1 TO 4
50  PRINT "ENTER MASS OF "ANIMAL$(QUESTIN)" BONE : ";
60 A$ = "": HTAB 33
61  GET S$: IF S$ =  CHR$ (13) THEN 65
63 A$ = A$ + S$
64  PRINT S$;: GOTO 61
65 MASS(QUESTIN) = VAL (A$)
68  PRINT "g"
69 A$ = ""
70  PRINT "ENTER VOLUME OF "ANIMAL$(QUESTIN);" BONE : ";
75  HTAB 33
80  GET S$: IF S$ = CHR$ (13) THEN 84
81 A$ = A$ + S$
83  PRINT S$;: GOTO 80
84 VOLUME(QUESTIN) = VAL (A$)
85  PRINT "mL": PRINT
90  NEXT
100  PRINT : PRINT " ARE THESE NUMBERS CORRECT? (Y/N) :";: GET A$
110  IF A$ <  > "Y" AND A$ <  > "N" THEN  VTAB  PEEK (37): HTAB 1: GOTO
   100
120  IF A$ = "N" THEN  HOME : GOTO 40
130  HOME
140  FOR ANIMAL = 1 TO 4
150  PRINT "DENSITY OF "ANIMAL$(ANIMAL)" BONE IS:";
155 D = MASS(ANIMAL) / VOLUME(ANIMAL)
157 d =  INT (D * 100 + .5) / 100
170  HTAB 30: PRINT D"g/mL"
180  PRINT
190  NEXT
200  PRINT : PRINT
210  PRINT "DO YOU WANT THIS PRINTED? (Y/N) :";: GET A$
215  PRINT
220 IF A$ <  > "N" AND A$ <  > "Y" THEN  VTAB  PEEK (37): HTAB 1: GOTO
   210
230  IF A$ = "N" THEN  END
240  PR# 1
250  FOR ANIMAL = 1 TO 4
260  PRINT "MASS OF "ANIMAL$(ANIMAL)" BONE: "MASS(ANIMAL)"g"
270  PRINT "VOLUME OF "ANIMAL$(ANIMAL)" BONE: "VOLUME(ANIMAL)"mL"
280  PRINT "DENSITY OF "ANIMAL$(ANIMAL)" BONE: "; INT ((MASS(ANIMAL) / V
   OLUME(ANIMAL)) * 100 + .5) / 100"g/mL"
290  PRINT
300  NEXT
310  PR# 0: END
1000  PRINT "LAB 14-1                    BONE DENSITY"
1005  PRINT : INVERSE : PRINT "INSTRUCTIONS:": NORMAL : PRINT : PRINT
1010  PRINT "YOU WILL BE ASKED TO ENTER YOUR DATA    FOR THIS EXPERIMENT
   . A DATA CHART       DIFFERENT FROM THE ONE USED IN THE TEXT WILL AP
   PEAR ON THE SCREEN."
1030  VTAB 24: HTAB 5
```

## RETEACHING 40

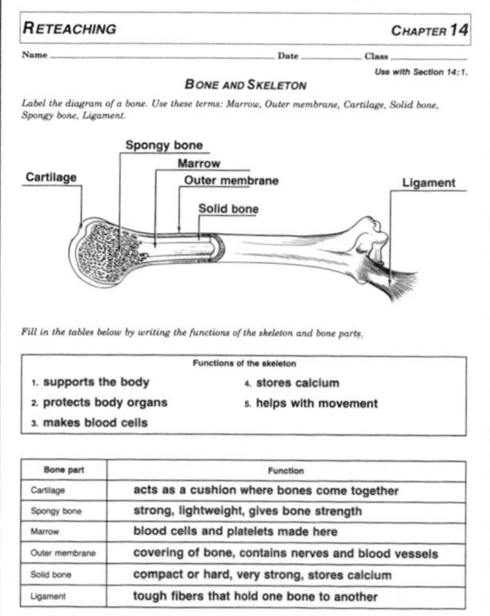
RETEACHING CHAPTER 14

Name ______ Date ______ Class ______

Use with Section 14:1.

BONE AND SKELETON

*Label the diagram of a bone. Use these terms: Marrow, Outer membrane, Cartilage, Solid bone, Spongy bone, Ligament.*



*Fill in the tables below by writing the functions of the skeleton and bone parts.*

Functions of the skeleton

1. supports the body
2. protects body organs
3. makes blood cells
4. stores calcium
5. helps with movement

| Bone part | Function |
|---|---|
| Cartilage | acts as a cushion where bones come together |
| Spongy bone | strong, lightweight, gives bone strength |
| Marrow | blood cells and platelets made here |
| Outer membrane | covering of bone, contains nerves and blood vessels |
| Solid bone | compact or hard, very strong, stores calcium |
| Ligament | tough fibers that hold one bone to another |

# TEACH

### Guided Practice

Have students write down their answers to the margin question: hinge joint.

### Check for Understanding

Have students respond to the first three questions in Check Your Understanding.

### Reteach

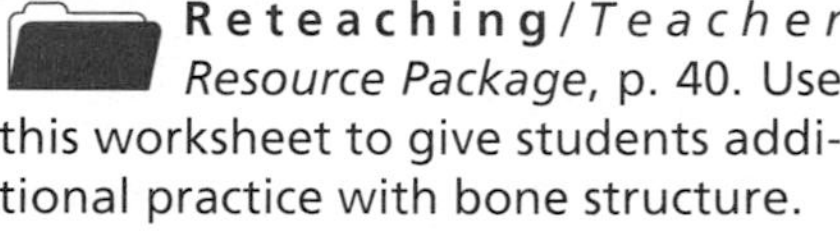
Reteaching/*Teacher Resource Package*, p. 40. Use this worksheet to give students additional practice with bone structure.

**Extension:** Assign Critical Thinking, Biology and Math, or some of the **OPTIONS** available with this lesson.

# 3 APPLY

### ACTIVITY/Enrichment

Obtain an owl pellet. Carefully remove all hair and feathers to expose the bones within.

# 4 CLOSE

### Software

*Human Systems, Series II, Skeletal,* Opportunities for Learning, Inc.

### Answers to Check Your Understanding

1. calcium
2. Cartilage acts as a cushion between bones; ligaments hold one bone to another; marrow forms blood cells.
3. hinge joint–moves back and forth, elbow or knee; ball-and-socket–moves in a circle, hips or shoulder
4. baseball, volleyball, tennis, football; you couldn't throw overhand with a hinge joint.
5. The smallest bone in the body must be less than two centimeters long because the three bones of the ear measure two centimeters.

# 14:2 The Role of Muscles

## PREPARATION

### Materials Needed

Make copies of the Transparency Master, Critical Thinking/Problem Solving, Skill, Application, Reteaching, Study Guide, and Lab worksheets in the *Teacher Resource Package*.

▶ Purchase chicken legs for the Mini Lab. Boil them at home prior to bringing them to school.

▶ Obtain file cards, string, tape, and metal fasteners for Lab 14-2.

### Key Science Words

muscular system
skeletal muscle
voluntary muscle
cardiac muscle
involuntary muscle
smooth muscle
tendon

### Process Skills

In Lab 14-2, students will formulate a model, infer, and experiment. In the Mini Lab, they will observe. In the Skill Checks, they will infer and classify.

## 1 FOCUS

▶ The objectives are listed on the student page. Remind students to preview these objectives as a guide to this numbered section.

### Bridging

Discuss the relationship between muscles and bones.

# 14:2 The Role of Muscles

**Objectives**

4. **Compare** three kinds of muscles.
5. **Explain** how muscles work to move body parts.
6. **Relate** how muscles work in pairs.

**Key Science Words**

muscular system
skeletal muscle
voluntary muscle
cardiac muscle
involuntary muscle
smooth muscle
tendon

The ball flies in your direction and you leap into the air to meet it. Your sister's pass from the other side of the volleyball net was a clever one, but you manage to return the ball. At the same time, you notice your shoelace is undone and you kneel to tie it. In these few seconds, you have used hundreds of different muscles in your body.

## Movement

Muscle is a body tissue that can change its shape and length and thus cause movement. The kinds of movements muscles carry out depend on where the muscles are attached. For example, arm and leg muscles allow you to move the bones of your arms and legs, such as when you leap to punch a volley ball, Figure 14-5. Your heart is an organ with very strong muscle tissue. Its main job is to help pump or move blood. Your stomach and intestines also have muscle tissues. The movement of muscles in these organs helps to push food along.

The **muscular system** is all the muscles in your body. Almost all animal groups have a muscular system, which allows them to move about.

**Figure 14-5** Muscles allow your body to perform a variety of different movements.



## OPTIONS

### Science Background

Muscle tissue is capable of contracting (shortening) when stimulated by a nerve message. These messages are either under voluntary or involuntary control. The actual physiology behind muscle contraction is rather complex but does involve the sliding of muscle protein fibers across one another, resulting in a shortening effect.

## Career Close-Up

### Athletic Trainer

**Are you interested in sports? You might want a job working with athletes. You might want to work with students in high school, but you could also train athletes in college. Maybe one day you will be training a professional team. To become an athletic trainer you will need to understand how muscles and bones make up the human body. Take courses in biology, health, human physiology, chemistry, and physics in order to start your career. Other useful courses include nutrition, drugs, first aid, health, and human anatomy.**

**In a typical day you may be asked to work directly with athletes. As the trainer, you will supervise exercises and you might be required to treat sprains, muscle cramps, or other minor body injuries. Other activities for an athletic trainer include meal planning, repairing of sports equipment, and ordering of training room supplies.**

**As an athletic trainer, your future is rewarding. With a college degree, the starting salary can be up to $26 000 per year.**

An athletic trainer understands the importance of warm-up exercises to avoid muscle damage.

## Human Muscle Types

The human body has three different kinds of muscle. They are skeletal, smooth, and cardiac (KAR dee ac) muscle. Each muscle type has a different structure and pattern that make it different from the other. Each type of muscle has a different job, and each has a different location in the body.

**Skeletal muscles** are muscles that move the bones of the skeleton. Your arms and legs are moved by this muscle type. Skeletal muscles make up the bulk of your body. Important skeletal muscles for athletic activity include those in the arms, the legs, the abdomen, the chest, and the shoulders.

What are the three different kinds of muscles?



## Career Close-Up

### Athletic Trainer

#### Background

Athletic trainers not only work with athletes, they may frequently consult with and discuss problems with the team coach and doctor.

#### Related Careers

A physical therapist, physical education teacher, or paramedic are careers related to athletic trainers.

#### Job Requirements

Athletic trainers must be certified by the National Athletic Trainers Association.

#### For More Information

The Athletic Institute
200 Castlewood Drive
North Palm Beach, FL 33408

## 2 TEACH

### MOTIVATION/Audiovisual

Project 35 mm color slides of all three muscle types onto the screen. Do not tell students what they are looking at. Ask them to describe similarities and differences.

### Concept Development

▶ Point out to students that their internal organs are mainly muscle.

### Bridging

Do a quick review of the animal kingdom with students. Point out that all phyla except sponges exhibit motion.

### Guided Practice

Have students write down their answers to the margin question: skeletal, cardiac, and smooth.

### Videocassette

*Muscle Power,* Films for the Humanities and Sciences.

## TRANSPARENCY 65

TRANSPARENCY MASTER — CHAPTER 14

MUSCLE TYPES

Muscle type—skeletal
Characteristics:
1. bandlike appearance from dark and light stripes
2. long fibers
3. separate cells hard to see
4. cell nucleus present
5. voluntary muscle
6. moves bones

Found in:
1. arms, leg muscles
2. any muscle you have control over

Muscle type—smooth
Characteristics:
1. no bandlike appearance
2. not connected to bones
3. individual cells are easier to see
4. nucleus present
5. involuntary muscle

Found in:
1. digestive organs (stomach, intestines)
2. blood vessels, ureters

Muscle type—cardiac
Characteristics:
1. bandlike appearance from dark and light stripes
2. long fibers that form a "weave" by joining together
3. individual cells hard to see
4. cell nucleus present
5. involuntary
6. not connected to bone

Found in:
heart only



## OPTIONS

### ACTIVITY/Challenge

Use prepared slides to show the differences among the three muscle types. An alternate suggestion is to obtain 35 mm colored slides. Or, if a video camera is available, project slides through the microscope onto a TV for class viewing.

**Transparency Master/***Teacher Resource Package*, p. 65. Use the Transparency Master shown here to teach students about muscle types.

## TEACH

**Misconception:** Ask students what they are eating when they eat hot dogs, hamburgers, roast, or steak. Students do not realize that meat purchased in a grocery or as a hamburger or hot dog is animal muscle.

**Cooperative Learning:** Divide the class into five groups. Have each group be responsible for researching and listing as many body organs as possible formed from skeletal, cardiac, and smooth muscle. Each group is to report verbally to the other groups on the findings. Offer some type of reward to the group with the longest list.

### Independent Practice

**Study Guide**/*Teacher Resource Package*, p. 81. Use the Study Guide worksheet shown at the bottom of this page for independent practice.

**Portfolio**

Prepare three microscopes with prepared slides of smooth, cardiac, and skeletal muscles. Cover the slide labels. Mark the scopes A, B, and C. Have students identify each muscle type by name and describe those features or traits that helped in their identification.

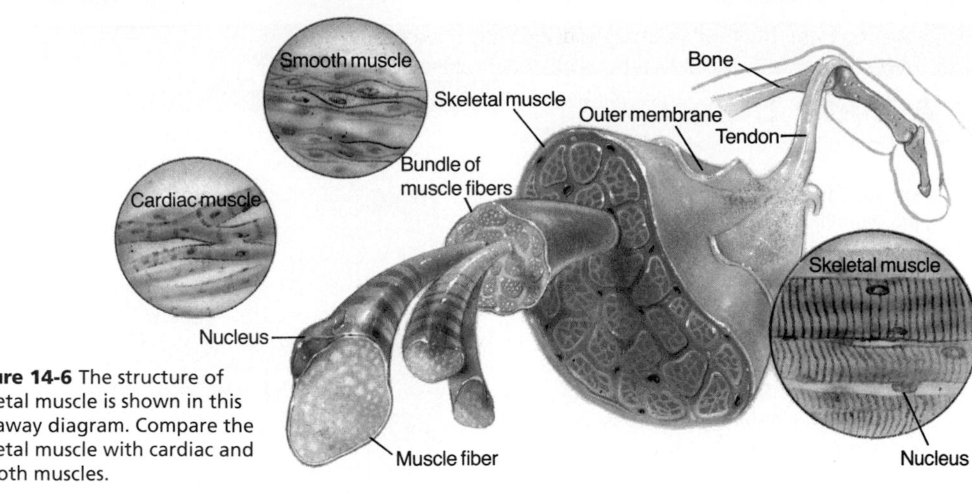


**Figure 14-6** The structure of skeletal muscle is shown in this cut-away diagram. Compare the skeletal muscle with cardiac and smooth muscles.

Figure 14-6 shows how skeletal muscle is made of bundles of fibers. Each fiber is a muscle cell. Notice that skeletal muscle looks striped. The stripes are caused by two kinds of proteins that make up the muscle fibers. The whole muscle is surrounded by a layer of fibrous tissue.

Skeletal muscles are voluntary. **Voluntary muscles** are muscles you can control. If you decide to lift something, you control the muscles needed for lifting. If you pull a rope, jump, look out the window, or point, you decide what you want your muscles to do. You control your arm, leg, head, and finger movements.

Most of the animal tissue that people eat is skeletal muscle. If you eat meat, you are usually eating muscle. For example, the parts of a fish that you eat are its muscles. You eat mostly muscle when you eat a chicken drumstick. Hamburger is the ground-up muscle of beef cows.

**Cardiac muscle** is the muscle that makes up the heart. The heart is the only body organ made of this muscle type. Cardiac muscle is not connected to any bones, Figure 14-7.

Cardiac muscle is made of bundles of fibers, just as in skeletal muscle. Again, this muscle type looks striped. The main difference between cardiac and skeletal muscles is that the bundles of cardiac muscle fibers form a tight weave. The tight weave makes cardiac muscle very strong.

Cardiac muscle is involuntary. **Involuntary muscles** are muscles you can't control. You don't think whether your heart needs to pump slow or fast. Your brain controls this movement as needed.

**Skill Check**

**Classify:** In what ways are skeletal, cardiac, and smooth muscle alike? How do they differ? *For more help, refer to the* ***Skill Handbook*** *pages 715-717.*

**Figure 14-7** The heart is the only organ with cardiac muscle.

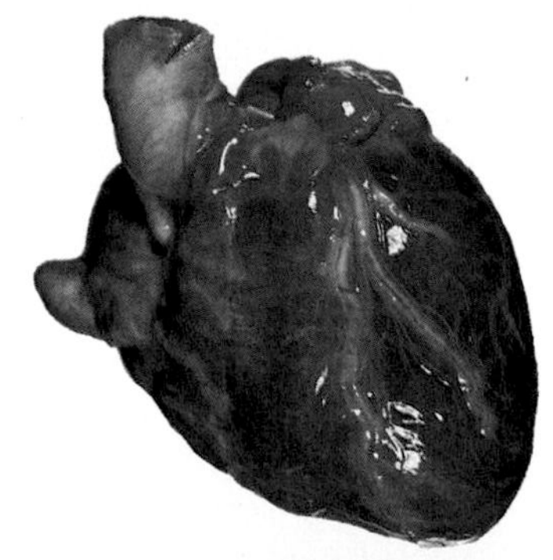

294

## OPTIONS

### ACTIVITY/Challenge

Ask students to prepare a wet mount of a very small amount of fresh hamburger meat. If a small enough sample is used, students will see the individual fibers and striations on the fibers. Proper lighting is critical.

**Critical Thinking/Problem Solving/** *Teacher Resource Package*, p. 14. Use the Critical Thinking/Problem Solving worksheet shown here as an extension to muscle types.

**CRITICAL THINKING 14**

CRITICAL THINKING/PROBLEM SOLVING — CHAPTER 14

Name ______ Date ______ Class ______

Use after Section 14:2.

**WHAT ARE THE JOBS OF DIFFERENT MUSCLES?**

Read the story below. As you read, think about the different types of muscles and what jobs they do. What kinds of activities do you do that depend on certain types of muscles?

It was 11:15 in the morning and time for gym. Alicia enjoyed gym class because she was a good athlete and probably the fastest *runner* (1) in her class. Today they were going outdoors to *run sprints* (2). Alicia *knew* (3) she could win if she did some *leg stretching* (4) exercises first.

It was very *warm outside* (5). After running the sprints and winning, Alicia was aware of how hard she was *breathing* (6). Her *rib cage* (7) was *moving in and out* (8) rapidly. She could even feel her *heart beating* (9). When she *placed her hand* (10) over her neck, she could feel her pulse. A pulse is caused by the pumping of blood through the *arteries* (11).

Usually she did not sweat much, but today she was *sweating a lot* (12). After *showering* (13) and *changing back into her school clothes* (14), Alicia was ready for lunch. She knew lunch was near because she was getting *hunger pangs* (15) in her *stomach* (16). Alicia asked her *gym teacher* (17) before leaving what caused the hunger pangs. The teacher told her that hunger pangs were the result of *movements or contractions* (18) of an empty stomach.

**Analyzing the Problem**

1. Write the numbers of the phrases that name organs made of muscle. **4, 7, 9, 11, 16**
2. Write the numbers of the phrases that name things that your body can do only because you have muscles. **1, 2, 4, 6, 8, 9, 10, 12, 13, 14, 15, 18**
3. Write the numbers of the phrases of body parts that best match the kind of muscle listed. Smooth: **11, 16** Skeletal: **4, 7** Cardiac: **9**
4. Write the numbers of the phrases that name things that your body can do under voluntary control (you can control it). **1, 2, 3, 4, 10, 13, 14**
5. Write the numbers of the phrases that name things that your body can do under involuntary control (you can't control it). **6, 8, 9, 12, 15, 18**

**Solving the Problem**

6. Think about the activities you do every day. Write a paragraph like the one above that describes your activities. Underline words that tell what you can do because of muscles. Circle words that list body parts that are made of muscle.

**Student answers will vary.**

**STUDY GUIDE 81**

STUDY GUIDE — CHAPTER 14

Name ______ Date ______ Class ______

**THE ROLE OF MUSCLES**

*In your textbook, read about human muscle types in Section 14:2.*

1. Identify these three drawings as smooth, skeletal, or cardiac muscle. Label them correctly. They are about 1500 times natural size.

**skeletal muscle** **cardiac muscle** **smooth muscle**

2. Complete the table by checking the correct column for each trait listed.

| Trait or location | Skeletal muscle | Cardiac muscle | Smooth muscle |
|---|---|---|---|
| Makes up your small and large intestines | | | ✓ |
| Makes up your heart | | ✓ | |
| Makes up your body muscles | ✓ | | |
| Has stripelike appearance | ✓ | ✓ | |
| Is voluntary | ✓ | | |
| Can be controlled | ✓ | | |
| Moves your bones | ✓ | | |
| Has no stripes | | | ✓ |
| Muscles form a tight weave | | ✓ | |
| Involuntary | | ✓ | ✓ |
| Is not connected to bone | | ✓ | ✓ |
| Can't be controlled | | ✓ | ✓ |
| Can contract | ✓ | ✓ | ✓ |
| Can shorten | ✓ | ✓ | ✓ |
| Most often eaten as meat | ✓ | | |

**Smooth muscle** is involuntary muscle that makes up the intestines, arteries, and many other body organs. You have no control over the working of your intestines or blood vessels. The cells of smooth muscle are long and spindly, but much shorter than the fibers of skeletal muscles. Smooth muscle does not have stripes. Like cardiac muscle, smooth muscle is not connected to any bones.

## How Muscles Work

A muscle works by changing its length. For a muscle to do its job, its fibers must shorten. The entire muscle shortens when the muscle fibers contract. The contraction of muscles allows you to move your body parts. Muscles don't lengthen to move body parts, they only shorten.

Let's look at the working of a skeletal muscle. Skeletal muscles are connected to bones by tendons (TEN duhns). A **tendon** is a tough, fibrous tissue that connects muscle to bone. The tendons cause the bone to be pulled when the muscle contracts.

How would your muscles work if you were to pull a boat in to dock with a rope? Look at Figure 14-8a to see the shape of your upper arm muscles before you begin to pull. They are relaxed. The muscle fibers have not contracted or shortened. Figure 14-8b shows how your arm muscles would look while you were pulling on the rope. Your muscle fibers are working. When they work, they contract or shorten. The bulging you see when a muscle is working is due to the thickening of muscle fibers as they shorten.

### Mini Lab

**What Is the Structure of a Chicken Leg?**

**Observe:** Examine a cooked chicken drumstick. Locate the muscles, tendons, cartilage, and bone. Why are so many tendons needed? Where do tendons lead to? *For more help, refer to the **Skill Handbook,** pages 704-705.*

How do skeletal muscles move body parts?

Muscle contracted

Muscle relaxed

b

a

**Figure 14-8** To pull a boat into dock you might use a rope. Your arm muscles would first be relaxed (a) and then contracted (b) as you pull on the rope.



## TEACH

### Mini Lab

Demonstrate muscle movement in a chicken leg by using a chicken thigh with the foot still attached. Most of the leg of a chicken is muscle. Tendons are very tough (almost bone-like) fibers connected to the muscle. Pulling on the various tendons will allow students to see the corresponding movements of the toes, as well as up-and-down movement of the foot.

### ASSESSMENT

**Performance:** On their palm-down diagram, students should label the location of the tendons extending to the fingers and thumbs. On their palm-up diagram, students should label the location of the tendons that extend from the palm to the upper wrist.

### Guided Practice

Have students write down their answers to the margin question: by contracting.

**Analogy:** Use a thick rubber band to illustrate muscle contraction.

**Using Science Words:** Make sure that students know the meaning of the words *contract* and *contraction.* To illustrate, describe someone rolling up into a ball as being contracted.

### Independent Practice

Study Guide/*Teacher Resource Package*, p. 82. Use the Study Guide worksheet at the bottom of this page for independent practice.

**SKILL 15**

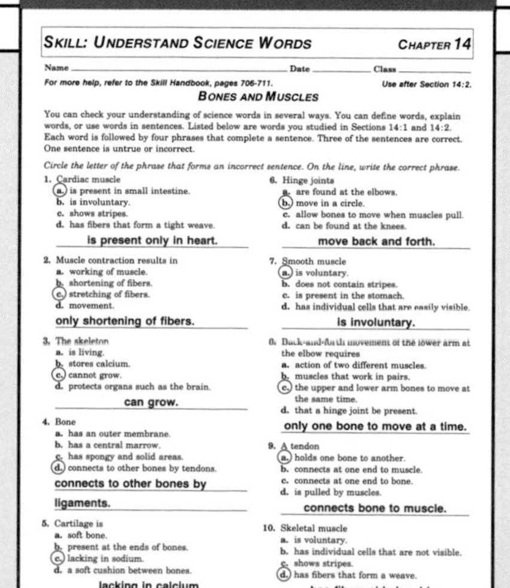

SKILL: UNDERSTAND SCIENCE WORDS — CHAPTER 14

Name ______ Date ______ Class ______

*For more help, refer to the Skill Handbook, pages 706-711.* *Use after Section 14:2.*

BONES AND MUSCLES

You can check your understanding of science words in several ways. You can define words, explain words, or use words in sentences. Listed below are words you studied in Sections 14:1 and 14:2. Each word is followed by four phrases that complete a sentence. Three of the sentences are correct. One sentence is untrue or incorrect.

*Circle the letter of the phrase that forms an incorrect sentence. On the line, write the correct phrase.*

1. Cardiac muscle
   a. is present in small intestine.
   b. is involuntary.
   c. shows stripes.
   d. has fibers that form a tight weave.
   **is present only in heart.**
2. Muscle contraction results in
   a. working of muscle.
   b. shortening of fibers.
   c. stretching of fibers.
   d. movement.
   **only shortening of fibers.**
3. The skeleton
   a. is living.
   b. stores calcium.
   c. cannot grow.
   d. protects organs such as the brain.
   **can grow.**
4. Bone
   a. has an outer membrane.
   b. has a central marrow.
   c. has spongy and solid areas.
   d. connects to other bones by tendons.
   **connects to other bones by ligaments.**
5. Cartilage is
   a. soft bone.
   b. present at the ends of bones.
   c. lacking in sodium.
   d. a soft cushion between bones.
   **lacking in calcium.**
6. Hinge joints
   a. are found at the elbows.
   b. move in a circle.
   c. allow bones to move when muscles pull.
   d. can be found at the knees.
   **move back and forth.**
7. Smooth muscle
   a. is voluntary.
   b. does not contain stripes.
   c. is present in the stomach.
   d. has individual cells that are easily visible.
   **is involuntary.**
8. Back-and-forth movement of the lower arm at the elbow requires
   a. action of two different muscles.
   b. muscles that work in pairs.
   c. the upper and lower arm bones to move at the same time.
   d. that a hinge joint be present.
   **only one bone to move at a time.**
9. A tendon
   a. holds one bone to another.
   b. connects at one end to muscle.
   c. connects at one end to bone.
   d. is pulled by muscles.
   **connects bone to muscle.**
10. Skeletal muscle
    a. is voluntary.
    b. has individual cells that are not visible.
    c. shows stripes.
    d. has fibers that form a weave.
    **has fibers side by side.**

**STUDY GUIDE 82**

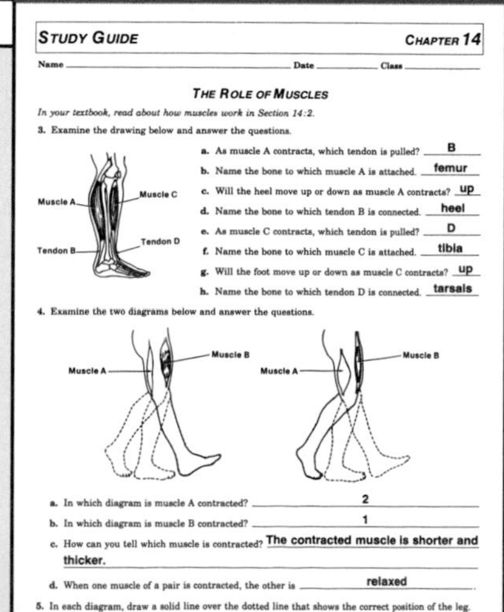

STUDY GUIDE — CHAPTER 14

Name ______ Date ______ Class ______

THE ROLE OF MUSCLES

*In your textbook, read about how muscles work in Section 14:2.*

3. Examine the drawing below and answer the questions.



a. As muscle A contracts, which tendon is pulled? **B**
b. Name the bone to which muscle A is attached. **femur**
c. Will the heel move up or down as muscle A contracts? **up**
d. Name the bone to which tendon B is connected. **heel**
e. As muscle C contracts, which tendon is pulled? **D**
f. Name the bone to which muscle C is attached. **tibia**
g. Will the foot move up or down as muscle C contracts? **up**
h. Name the bone to which tendon D is connected. **tarsals**

4. Examine the two diagrams below and answer the questions.



a. In which diagram is muscle A contracted? **2**
b. In which diagram is muscle B contracted? **1**
c. How can you tell which muscle is contracted? **The contracted muscle is shorter and thicker.**
d. When one muscle of a pair is contracted, the other is **relaxed**.

5. In each diagram, draw a solid line over the dotted line that shows the correct position of the leg.

## OPTIONS

Skill/*Teacher Resource Package*, p. 15. Use the Skill worksheet shown here to help students understand the terminology in this section.

### Science Background

The axons of motor nerves terminate in or on muscle tissue. When a stimulus moves down the nerve to the axon, a chemical is released from the axon. The chemical is called a transmitter substance and will stimulate the muscle tissue to contract. This chemical is usually acetylcholine.

## TEACH

**Misconception:** Exercise creates larger muscle cells, not more muscle cells. The number of muscle cells doesn't change.

### Skill Check

The muscle tissue forms a bulge where the contracting fibers overlap.

### ACTIVITY/Demonstration

Have students feel their calf muscles while standing on tiptoe. Remind them that standing on tiptoe requires that the foot be turned downward. Have them note the contraction of their calf muscles. Then, have them return to a normal standing position as they continue to concentrate on their calf muscles. Holding onto the front of the leg while the foot is turned up will allow them to see that different muscles are needed to raise the foot.

### Portfolio

Using Figure 14-9 as a guide, students should make a series of labeled diagrams that depict the changes that occur to the paired muscles of the arm as it moves up and down. Figures 14-1, 14-8, and 14-10 can be used as references for the muscles and bones of the arm.

**Videodisc**

**STV: Human Body Vol. 2**
*Muscular and Skeletal Systems*
Unit 1, Side 1
*Skeletal Muscles*

**21752-24658**

### Skill Check

**Infer:** If muscles shorten when they contract, where does the excess muscle tissue go? *For more help, refer to the **Skill Handbook,** pages 706-711.*

Bones with skeletal muscles also allow you to walk. Walking is an exercise that strengthens muscles because of the continual contracting and stretching of pairs of muscles. Let's look at this movement.

Figure 14-9a shows the muscles that are used to raise or lower your foot. All the muscles in this diagram are relaxed. This means that the foot is not moving up or down. Notice that one end of muscle A is attached to the top of the tibia (TIHB ee uh), your leg bone. This end won't move during contraction. The lower end of the tibia is attached by a tendon to the tops of the tarsals, the bones of your foot. This end moves during contraction. One end of muscle B is attached to the bottom of the femur, your thighbone. This end won't move during contraction. The other end is attached by a tendon to the back of your heel bone. This end moves.

Figure 14-9b shows that muscle A is relaxed. You can tell this because the muscle is the same length as muscle A shown in Figure 14-9a. Muscle B, however, is contracted. You can tell this because it is much shorter than it was in Figure 14-9a. As muscle B contracts, it pulls up the bottom of your heel. A hinge joint in the ankle allows for this movement.

How is your foot pulled up? A different muscle must be used this time. Muscle B is not used. Muscle B in Figure 14-9c is now relaxed. Muscle A is contracted. When it shortens, it pulls up the foot. You can now move forward.

**Figure 14-9** When you walk, you lift your feet. At first, your muscles are relaxed (a). Then, the heel is pulled up (b), and the foot is lifted off the ground (c). Why is cycling a good exercise for leg muscles? The muscles are continually stretched.

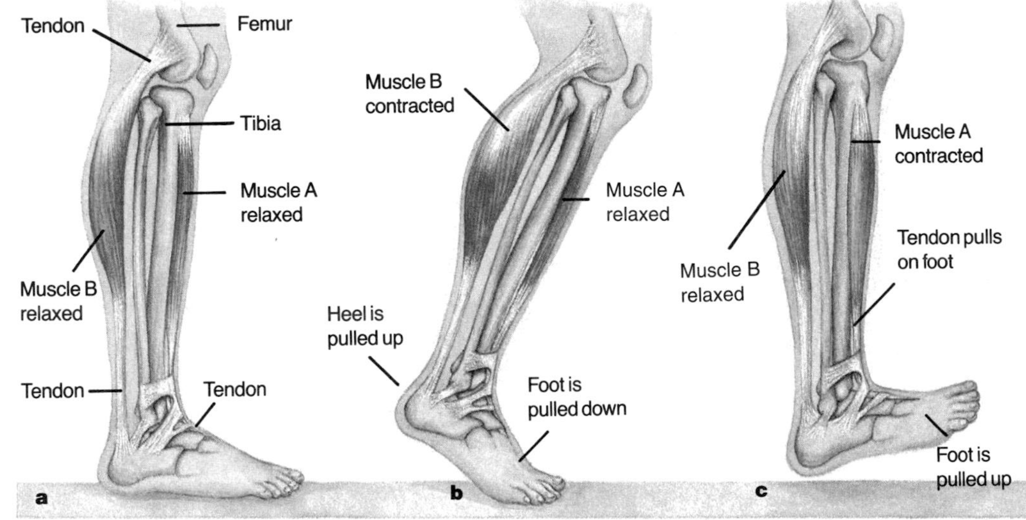




## OPTIONS

**Enrichment:** The pairing of muscles throughout the body is called antagonistic pairing. In the example being used, muscle A is a flexor. It raises the foot. Muscle B is called an extensor. It lowers the foot. Most major muscle pairings include a flexor and an extensor.

**Application**/*Teacher Resource Package*, p. 14. Use the Application worksheet shown here to show the connection between exercise and muscle contraction.

**APPLICATION 14**

APPLICATION: LEISURE — CHAPTER 14

Name ____ Date ____ Class ____

Use after Section 14:2.

EXERCISE AND SKELETAL MUSCLES

You have learned that skeletal muscles work in pairs. When you move a part of your body, one muscle contracts while another relaxes.

*In each of the exercises shown below, state which muscles are contracting. Next, state which muscles are relaxing (or stretching). Try out each exercise to see if your answer is correct.*

Side stretch
Contracting: muscles along ribs on the side with the arm reaching down
Relaxing or stretching: muscles along ribs on the side with the arm reaching up

Leg lift
Contracting: muscles on top of leg and stomach muscles
Relaxing or stretching: muscles on the back of leg

Touch toes
Contracting: stomach and chest muscles
Relaxing or stretching: neck and back muscles

Back bend
Contracting: back muscles and muscles on the back of legs
Relaxing or stretching: stomach muscles

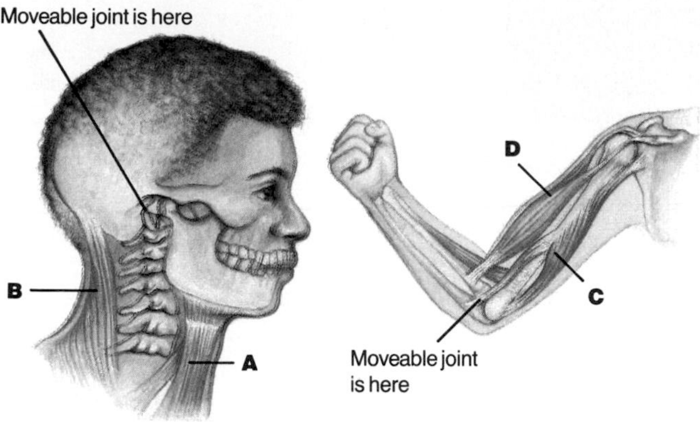


**Figure 14-10** Which way will the head and arm move as different muscles contract and relax?

## Muscles Work in Pairs

How do skeletal muscles work in pairs?

Because muscles can only shorten, it takes a pair of skeletal muscles to move bones back and forth. When one muscle contracts, the opposite muscle must relax. In the examples just used, one muscle was needed to pull your foot up. A different muscle was needed to lower your foot.

Look at these other examples in Figure 14-10. See if you can figure out which muscle has which job. Remember, muscles move body parts only when they shorten. The body part moves because muscles pull on a bone.

How does the head move when muscle A relaxes and muscle B contracts? The head turns up.

How does the arm move when muscle C contracts and muscle D relaxes? The arm straightens.

### Check Your Understanding

6. Give examples of the three types of muscles and where they are found in the body.
7. What is the job of tendons?
8. How does the length of a relaxed muscle compare with that of a contracted muscle?
9. **Critical Thinking:** Muscle cells have more mitochondria than other cells. Why?
10. **Biology and Reading:** You have some muscles in your eye that you can control. Other muscles in your eye work automatically. What two kinds of muscles must control movements in your eye?



## RETEACHING 41

RETEACHING CHAPTER 14

Name ______ Date ______ Class ______

*Use with Section 14:2.*

**MUSCLES**

Figures 1 and 2 show two simple diagrams of bones in the human body. The bones meet at a hinge joint and are connected by a pair of muscles. The figures show that when a muscle is relaxed and not working, it will be thin in width. When it is contracted and working, it becomes thicker in width.

Figure 1: Muscle A (relaxed), Bone A, Bone B, Hinge joint, Muscle B (relaxed)

Figure 2: Muscle A (contracted), Bone A, Bone B

*Use Figures 1 and 2 to answer the following questions. Circle the correct answer.*

1. Figure 1 shows bone A moving / (not moving) up.
2. Figure 1 shows bone B moving / (not moving) up.
3. Figure 1 shows muscle A thicker / (thinner) than in Figure 2.
4. Figure 1 shows muscle B thicker than / thinner than / (the same as) in Figure 2.
5. Figure 1 shows muscle A (relaxed) / contracted when compared with Figure 2.
6. Figure 1 shows muscle B (relaxed) / contracted.
7. Figure 1 shows muscle A working / (not working).
8. Figure 1 shows muscle B working / (not working).
9. Figure 2 shows bone A moving / (not moving) up.
10. Figure 2 shows bone B (moving) / not moving up.
11. Figure 2 shows muscle A (thicker) / thinner than in Figure 1.
12. Figure 2 shows muscle B thicker than / thinner than / (the same as) in Figure 1.
13. Figure 2 shows muscle A relaxed / (contracted) when compared with Figure 1.
14. Figure 2 shows muscle B (relaxed) / contracted.
15. Figure 2 shows muscle A (working) / not working.
16. Figure 2 shows muscle B working / (not working).

## OPTIONS

### ACTIVITY/Challenge

Ask students to diagram a muscle, bone, joint, and tendon and show the resulting bone movement when the muscle contracts. Ask students to modify the diagram by including another muscle that can move the bone in the opposite direction. (If Lab 14-2 has been done, ask students to use a set of muscles and bones other than the lower leg and foot).

## TEACH

### Guided Practice

Have students write down their answers to the margin question: because muscles can only shorten.

### Check for Understanding

Have students respond to the first three questions in Check Your Understanding.

### Reteach

*Reteaching/Teacher Resource Package*, p. 41. Use this worksheet to give students additional practice with muscle action.

**Extension:** Assign Critical Thinking, Biology and Reading, or some of the **OPTIONS** available with this lesson.

## 3 APPLY

### Software

*Human Systems, Series II Muscular,* Opportunities for Learning, Inc.

## 4 CLOSE

### ACTIVITY/Enrichment

Have students do a little research to find the answer to the following: Which type of muscle, skeletal, cardiac, or smooth, will fatigue the most rapidly? Least rapidly?

### Answers to Check Your Understanding

6. skeletal–arms, legs; cardiac–heart; smooth–stomach, intestines, blood vessels
7. Tendons connect muscle to bone.
8. When contracted, muscle is much shorter and thicker than when relaxed.
9. Muscle cells need more energy for movement than other cells. Mitochondria are where energy is produced from food.
10. You have control over the skeletal muscles that move the eye, but not the smooth muscles.

## Lab 14-2 Muscles

### Overview

Students will build a paper model of a foot and leg. Antagonistic muscle pairs are demonstrated.

**Objectives:** Upon completion of the lab, students will be able to (1) build a **model** of the human foot and leg and (2) **demonstrate** the action of muscle pairs in bringing about movement of the foot.

**Time Allotment:** 30 minutes

### Preparation

▶ Prepare patterns of the model foot and leg before class.

**Lab 14-2 worksheet**/*Teacher Resource Package*, pp. 55-56.

### Teaching the Lab

▶ Pulling up on the string means pulling toward the top of the model.

▶ Suggest that the model be taped to the desk along the upper edge of the leg piece. Make sure the tape does not block movement of the strings or foot.

▶ **Troubleshooting:** Remind students that tape must be positioned as shown. It cannot overlap onto other model parts.

### ASSESSMENT

**Performance:** Provide students with a file card, punch, tape, scissors, string, and a metal fastener. Ask them to design a model that illustrates the placement and action of those muscles needed to nod the head forward and backward. Refer them to Figure 14-10 if help is needed.

**Cooperative Learning:** Divide the class into groups of two students. For more help, refer to pp. 22T-23T in the Teacher Guide.

# Muscles

**Lab 14–2**

## Problem: Do muscles work in pairs?

### Skills

formulate models, experiment, infer

### Materials

index card
scissors
string
metal fastener
tape
paper punch

### Procedure

1. Copy the data table.
2. Cut one index card in half lengthwise. Attach one half of the card to the other half using the metal fastener as shown in Figure A. One half card is the foot, and the other half card is the leg.
3. Punch two holes in the top of the leg bone card, and thread a piece of string, 15 cm long, through each hole. Attach the strings to the base of the foot card as shown in Figure B.
4. **Experiment:** Pull up on string A. Record in your table how the foot part of the model moves. Also note if string B gets shorter or longer between the hole on the leg card and the tape on the foot card.
5. Record your findings in the table.
6. Pull up on string B. Record in your table how the foot part of the model moves. Also note if string A gets shorter or longer between the hole on the leg card and the tape on the foot card.

### Data and Observations

1. In your model:
   (a) what does the metal fastener represent?
   (b) what do strings A and B represent?
   (c) what do the pieces of tape represent?
2. When the foot moved down, which string got shorter? Which got longer?
3. When the foot moved up, which string got shorter? Which got longer?

| | FOOT MOVES UP OR DOWN? | STRING A LENGTHENS OR SHORTENS | STRING B LENGTHENS OR SHORTENS |
|---|---|---|---|
| String A pulled up | down | shortens | lengthens |
| String B pulled up | up | lengthens | shortens |

### Analyze and Apply

1. (a) Was a short string supposed to show a contracted or a relaxed muscle?
   (b) Was a long string supposed to show a contracted or a relaxed muscle?
2. Can the same string (muscle) move the foot up and down? Explain.
3. **Apply:** Which other sets of muscles in your body act the same way as those in your model?

### Extension

**Formulate a model** of a weight-lifting machine that would act to strengthen the thigh muscles while moving the leg forward and backward.



## ANSWERS

### Data and Observations

1. (a) a joint
   (b) a pair of muscles
   (c) tendons
2. A got shorter, B got longer
3. B got shorter, A got longer

### Analyze and Apply

1. (a) contracted
   (b) relaxed
2. No. Only a contracted muscle causes movement.
3. Answers may include lower arm, upper arm, hands, upper legs, and neck muscles.

# 14:3 Bone and Muscle Problems

From studying this chapter, you know that the skeletal and muscular systems work together. So, a problem in one system may cause a problem with the other. Modern science and technology is used to solve some of the problems that affect the skeletal and muscular systems.

**Objectives**

7. **Discuss** the problems of the skeletal system.
8. **Give examples** of problems of the muscular system.
9. **Explain** the purpose of new designs for products.

## Problems of the Skeletal System

Some problems with the skeletal system are caused by diseases of the joints. **Arthritis** (ar THRIT us) is a disease of bone joints. One type of arthritis results in breakdown of the cartilage at the joints. The result is pain and swelling at the diseased joint. In time, a person may not be able to bend or move the affected part of the body, Figure 14-11.

**Key Science Words**

arthritis
sprain
muscular dystrophy

Today, there is some help for the problems of arthritis. Artificial joints made of plastic or metal are sometimes used to replace diseased joints. Hip, ankle, and knee joints have been replaced by this technology. Figure 14-11 shows an artificial hip joint. It is made of metal and replaces the ball part on the femur, which is attached to the ball-and-socket joint in the hip.

**What are the effects of arthritis?**

Bones are strong because they store calcium. If the level of calcium is low, bones will break easily. Bones tend to become brittle as a person ages. Exercise and eating dairy foods helps to keep bones strong.

**Skill Check**

**Understand science words: arthritis.** The word part *arthr* means joint. In your dictionary, find three words with the word part *arthr* in them. *For more help, refer to the* ***Skill Handbook,*** *pages 706-711.*

**Figure 14-11** Arthritis can cause severe disability. The ball part of a ball-and-socket joint can be replaced with metal (inset).



## 14:3 Bone and Muscle Problems

### PREPARATION

**Materials Needed**

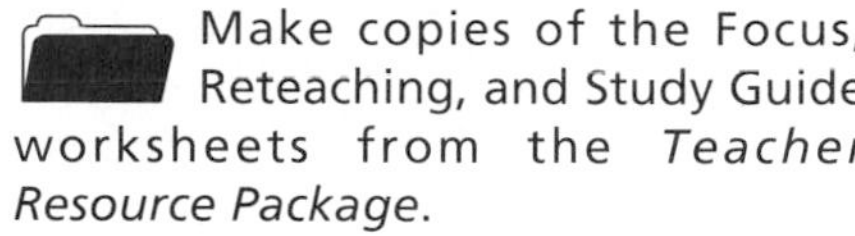

Make copies of the Focus, Reteaching, and Study Guide worksheets from the *Teacher Resource Package.*

▶ Obtain a foot or knee model from a local nursing school, hospital, or orthopedic surgeon for use in the demonstration of ligaments.

**Key Science Words**

arthritis
sprain
muscular dystrophy

**Process Skills**

In the Skill Check, students will understand science words.

### 1 FOCUS

▶ The objectives are listed on the student page. Remind students to preview these objectives as a guide to this numbered section.

Focus/*Teacher Resource Package*, p. 27. Use the Focus transparency shown at the bottom of this page as an introduction to bone and muscle problems.

**MOTIVATION/Demonstration**

Muscle has a very rich supply of capillaries. This is easily illustrated with a piece of fresh meat. Hold the meat up and blood will drain from the tissue.

**Guided Practice**

Have students write down their answers to the margin question: a breakdown of cartilage that causes pain and swelling at the joint.

**FOCUS 27**

FOCUS CHAPTER 14

Name ______ Date ______ Class ______

WHEN MUSCLES NO LONGER WORK: STEPHEN HAWKING

Normal muscle
Motor Neurons
Impulse from motor neuron causes muscle to contract
Stephen Hawking: Scientist with motor-neuron disease
Shriveled muscle
Damaged motor neuron cannot cause muscle to react

1. What do you think happens when the nerves that make muscles work degenerate or become useless? What kinds of activities are interfered with?
2. Why would a person with damaged nerves find it difficult to talk or eat?

### OPTIONS

**Science Background**

A number of joint replacements are now possible including hip, knee, and ankle. These replacements are often needed as a consequence of arthritis.

Arthritis is considered an autoimmune disease. That is, the body's immune system is responsible for causing the disease. Two general types are recognized: rheumatoid and osteoarthritis. Both result in deterioration of cartilage at the ends of bone.

# 2 TEACH

### Bridging

Remind students that a muscle cannot contract without some command. This command comes from the nerve cells associated with the muscle. The vast majority of apparent muscle problems are the result of nerve problems rather than muscle dysfunction.

### Check for Understanding

Discuss answers to these questions with students: (a) How does a fracture differ from a sprain? How does a muscle cramp differ from a muscle strain?

### Reteach

Prepare an idea map of problems associated with the skeleton and muscle.

### Guided Practice

Have students write down their answers to the margin question: a tense muscle due to low oxygen supply.

### Independent Practice

**Study Guide**/*Teacher Resource Package*, pp. 83-84. Use the Study Guide worksheets shown at the bottom of this page for independent practice.

### *Student Journal*

Picking up a pencil involves the use of both finger bones and muscles. Have students describe how the skeletal and muscular systems are dependent on one another when doing this action. Students should include the names and contributions of any other tissue types, such as tendons, periosteum, and cartilage, that also are used during this task.

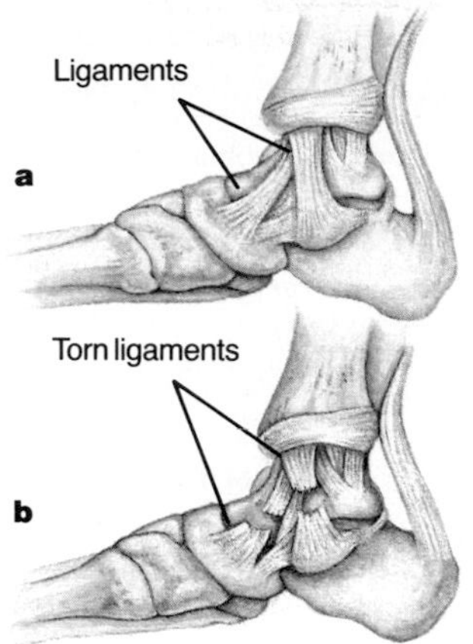


**Figure 14-12** Normal (a) and torn (b) ligaments of an ankle with a sprain

Most bones in the body are connected by ligaments. Have you ever twisted your ankle? If you have, the pain you felt was caused by injury to your ankle ligaments, Figure 14-12. **Sprains** are injuries that occur to your ligaments at a joint. A sprain results when the ligaments are torn and blood vessels are damaged.

## Problems of the Muscular System

Have you ever strained a muscle while lifting or after a sudden move? Have you ever had muscle cramps while swimming or during a long-distance run? Strains and cramps are two quite different problems.

**What is a cramp?**

You might strain a muscle if you haven't exercised for a while. A sudden use of a poorly exercised muscle may tear the fibers. Regular exercise will help you avoid straining your muscles. A muscle cramp is when the muscle contracts strongly and then can't relax. Muscle cramps result from a poor supply of oxygen. If you exercise one set of muscles too long, the oxygen supply may run low. When oxygen returns, the muscles will recover.

Skeletal muscles are controlled by nerve cells. A disease called muscular dystrophy (MUS kyuh lur • DIHS truh fee) blocks the nerve messages to muscles. **Muscular dystrophy** is a disease that causes the slow wasting away of skeletal muscle tissue. Muscular dystrophy can be inherited and is most common in males, Figure 14-13.

**Figure 14-13** By transplanting healthy muscle cells into a person with muscular dystrophy, scientists have given new hope to many people with this disease.



## *OPTIONS*

**What Causes Sports Injuries?**/*Lab Manual*, pp. 111-114. Use this lab as an extension to bone and muscle problems.

**STUDY GUIDE 83**

STUDY GUIDE — CHAPTER 14

Name ______ Date ______ Class ______

**BONE AND MUSCLE PROBLEMS**

*In your textbook, read about bone and muscle problems in Section 14:3.*

1. What is arthritis? **Arthritis is a disease of bone joints.**
2. One type of arthritis results in breakdown of the cartilage at the joints. How would this affect a person's life? **This breakdown causes pain and swelling at the diseased joint. In time, a person may not be able to bend or move the affected part of the body.**
3. What can be done to help people whose joints are severely affected by arthritis? **Diseased joints can be replaced with plastic or metal ones.**
4. What can happen to bones if a person doesn't get enough calcium in their diet? **The bones can become brittle and break easily.**
5. What are ligaments? **Ligaments are tough fibers that connect bones to each other.**
6. What happens to ligaments that are sprained? **They are torn and blood vessels are damaged.**
7. a. What is a muscle cramp? **A muscle cramp occurs when a muscle contracts strongly and can't relax.**
   b. When does it occur? **If you exercise some muscles too long, oxygen supply to those muscles may run low.**
8. a. What is muscular dystrophy? **It's a disease that blocks nerve messages to muscle tissue.**
   b. What effect does it have on muscles? **It causes muscles to waste away.**

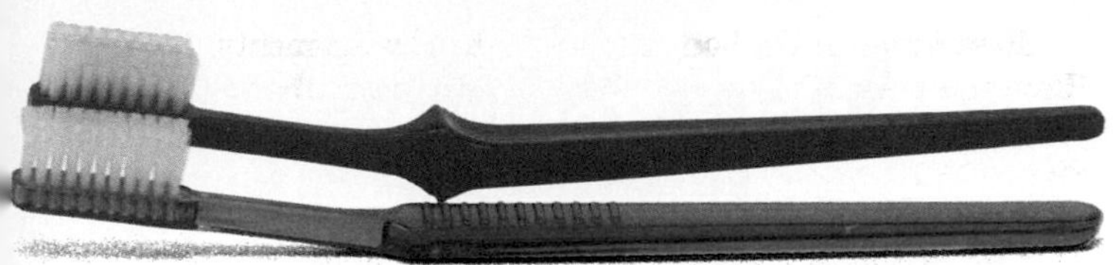

**Figure 14-14** How are these toothbrushes designed to fit your body?

## Making Products Fit People

How often have you sat in class and complained that the desk or chair was very uncomfortable? Backaches, muscle fatigue, and general discomfort are often the result of poor product design. Why? In the past, the most important thing for a new product was that it should look good. It wasn't important that a product fit the shape of the human body.

All of this is changing, however. New product design that fits the user's body shape is now becoming important. Automakers have put lumbar supports in their car seats. The lumbar region is the "small of the back" that gets tired after long periods of sitting.

Look at the toothbrushes in Figure 14-14. Will their handles reduce the chances of dropping them? Will they feel more comfortable in your hands? Will the bristles reach all the surfaces of your teeth and gums? Their design helps solve these kinds of problems.

### Bio Tip

**Consumer:** Exercise cycles are popular for home fitness programs. Many makes and models are advertised. You should select the cycle that best fits your needs and body shape.

**What kind of product design is becoming popular?**

### Check Your Understanding

11. What is arthritis?
12. What happens to ligaments when the ankle is sprained?
13. What new design in autos makes drivers more comfortable?
14. **Critical Thinking:** Low back pain and neck pains are often due to muscle spasms related to stress. What might be some ways to avoid these problems?
15. **Biology and Math:** There are several kinds of muscular dystrophy. In the kind of muscular dystrophy that occurs most often in the U.S., 3 out of every 10 000 boys will be expected to show symptoms of the disease. How many boys would be expected to show symptoms of muscular dystrophy in a city of 60 000?

## TEACH

### Guided Practice

Have students write down their answers to the margin question: one that fits the user's body shape.

### Check for Understanding

Have students respond to the first three questions in Check Your Understanding.

### Reteach

*Reteaching/Teacher Resource Package*, p. 42. Use this worksheet to give students additional practice with bone and muscle problems.

**Extension:** Assign Critical Thinking, Biology and Math, or some of the **OPTIONS** available with this lesson.

## 3 APPLY

### Divergent Question

The human body may use several systems in order to function normally. Ask students for an example of this type of cooperation using the movement of an arm or leg.

## 4 CLOSE

A person has a cast on his broken leg for three months. When the cast is removed, he notices that the muscle of the leg has shrunk. Explain why.

### Answers to Check Your Understanding

11. a disease of bone joints
12. ligaments are torn
13. a lumbar support that supports the lower back
14. exercising and relaxing
15. In a city of 60 000, 18 boys would be expected to have the disease.
    3/10 000 = x/60 000;
    180 000/10 000 = x;
    x = 18

**STUDY GUIDE 84**

*STUDY GUIDE* — *CHAPTER 14*

Name ______ Date ______ Class ______

*VOCABULARY*

*Review the new words used in Chapter 14 of your textbook. Then, complete the puzzle. Use the words or phrases to fill in the blanks in the sentences. Do not use any term more than once.*

| | | | |
|---|---|---|---|
| skeletal system | solid bone | fixed joints | ligaments |
| spongy bone | hinge joints | muscular system | ball-and-socket joint |
| skeletal muscles | sprain | voluntary muscles | muscular dystrophy |
| cardiac muscle | tendons | involuntary muscle | smooth muscle |
| arthritis | | | |

1. Cardiac muscle is found only in the heart.
2. Muscle you have no control over is involuntary muscle.
3. Bones are held together by ligaments. A sprain results when they are torn.
4. Spongy bone is usually found toward the ends of bones. The outer part of bones is usually solid bone.
5. The skeletal system is a framework for the body.
6. All the muscles in your body make up the muscular system.
7. Muscles are connected to bones by tendons.
8. Muscles you can control are voluntary muscles.
9. Many body organs, such as arteries, are made up of smooth muscle.
10. Arthritis is a disease of the joints.
11. Muscular dystrophy is a disease that causes the wasting away of muscle tissue.
12. Muscles that move the bones of the skeleton are skeletal muscles.
13. Your knees and elbows are hinge joints.
14. Bones can turn in a circle at a ball-and-socket joint.
15. Joints that don't move are called fixed joints.

**RETEACHING 42**

*RETEACHING: IDEA MAP* — *CHAPTER 14*

Name ______ Date ______ Class ______

*Use with Section 14:3.*

*BONE AND MUSCLE PROBLEMS*

*Use the following terms to complete the idea map below: sudden, strong contraction of muscle; stretching or tearing of ligaments; disease of the bone joints; slow wasting away of muscle tissue; small tears occur in muscles.*

| Problem | Descriptions |
|---|---|
| muscle cramp | sudden, strong contraction of muscles |
| sprain | stretching or tearing of ligaments |
| muscular dystrophy | slow wasting away of muscle tissue |
| arthritis | disease of the bone joints |
| strain | small tears occur in muscles |

1. What disease results in wasting away of muscle tissue? muscular dystrophy
2. How are arthritis patients sometimes helped? Artificial joints of plastic or metal are sometimes used to replace diseased joints.
3. What causes the pain when you twist your ankle? You stretch your ankle ligaments.
4. How is a chair that fits the user's body shape important? The person sitting in the chair is more comfortable and will be less fatigued.

# CHAPTER 14 REVIEW

## Summary

Summary statements can be used by students to review the major concepts of the chapter.

## Key Science Words

All boldfaced words from the chapter are listed.

## Testing Yourself

### Using Words

1. ball-and-socket joint
2. smooth muscle
3. cardiac muscle
4. arthritis
5. solid bone
6. tendon
7. ligament
8. skeletal muscle
9. skeletal system

### Finding Main Ideas

10. p. 291; A hinge joint is the elbow or knee. Fixed joints are the skull bones. A ball-and-socket joint is the upper arm and shoulder.
11. p. 299; Artificial joints of plastic or metal are used to replace diseased joints.
12. p. 286; It supports the body, protects organs, stores materials, makes blood cells, and helps with movement.
13. p. 300; A sprain is an injury to a ligament.
14. p. 300; Muscular dystrophy interrups nerve messages to muscles and causes the muscle tissue to waste away.
15. p. 289; The six tissues of bone are cartilage, outer membrane, solid bone, spongy bone, marrow, and ligaments.
16. p. 297; Muscle can only shorten. One muscle moves a bone in only one direction, and a second muscle is needed to move it in the opposite direction.
17. p. 287; The leg bones are the femur, tibia, and fibula. The arm bones are the humerus, radius, and ulna. The collarbone is the clavicle, and the shoulder blade is the scapula.
18. p. 301; Products can be designed that support the lower back and can be adjusted for height and angle.

# Chapter 14 Review

## Summary

### 14:1 The Role of the Skeleton

1. The skeleton serves as a framework, protects body organs, makes blood cells, stores calcium, and provides a place for muscle attachment. Bone is alive and grows.
2. The six tissues of bones include a fibrous membrane around solid bone, spongy bone, and marrow, as well as cartilage and ligaments.
3. Three types of joints are hinge, ball-and-socket, and fixed.

### 14:2 The Role of Muscles

4. Muscles bring about body movement. The three types of muscle are skeletal, cardiac, and smooth.
5. Muscles contract in order to work. They are connected to bone by tendons.
6. Muscles must work in pairs in order to bring about movement of body parts.

### 14:3 Bone and Muscle Problems

7. Arthritis, lack of calcium, and sprains are three common problems of the skeletal system.
8. Strains and cramps are two common muscle problems. Muscular dystrophy is a more serious and often inherited muscle problem.
9. New products are being designed to fit the shape of the human body.

## Key Science Words

arthritis (p. 299)
ball-and-socket joint (p. 291)
cardiac muscle (p. 294)
fixed joint (p. 291)
hinge joint (p. 291)
involuntary muscle (p. 294)
ligament (p. 289)
muscular dystrophy (p. 300)
muscular system (p. 292)
skeletal muscle (p. 293)
skeletal system (p. 286)
smooth muscle (p. 295)
solid bone (p. 289)
spongy bone (p. 289)
sprain (p. 300)
tendon (p. 295)
voluntary muscle (p. 294)

## Testing Yourself

### Using Words

*Choose the word from the list of Key Science Words that best fits the definition.*

1. allows arm to move in a full circle
2. involuntary muscle that makes up intestines and blood vessels
3. muscle that makes up the heart
4. a disease of bone joints
5. compact or hard part of bone
6. attaches muscle to bone
7. holds two bones together
8. type of muscle that works in pairs
9. framework of all body bones



### Using Main Ideas

19. The joint in the lower jaw is a hinge joint. A ball-and-socket joint is formed between the upper leg and pelvis, and in the shoulder.
20. The skull protects the brain against injury. The ribs form a cage around the heart and lungs.
21. The ligaments are torn.
22. Smooth muscle is found in the stomach and blood vessels. Cardiac muscle is found in the heart. Skeletal muscle is found in the arms.
23. The muscle in the front of the leg is attached at the top of the tibia and at the bottom to the foot bones. When this muscle contracts, the foot is pulled up. Muscle at the back of the leg is attached at the top to the femur and at the bottom to the underside of the heel. When the muscle contracts, the foot is pulled down.
24. You can control voluntary muscles. You can't control involuntary muscles.
25. Muscle cramps result.

# Review

## Testing Yourself *continued*

**Finding Main Ideas**

*List the page number where each main idea below is found. Then, explain each main idea.*

10. examples of the three types of joints in the human body
11. how the problem of arthritis is treated
12. five main jobs of the skeletal system
13. what a sprain is
14. a disease that causes the blocking of nerve messages to muscle
15. the six tissues of bone
16. why two different muscles are needed to move bones back and forth
17. the medical names of the bones of the leg, arm, collar, and shoulder blade
18. how products can be designed to fit the shape of the human body

**Using Main Ideas**

*Answer the questions by referring to the page number after each question.*

19. What type of joint is found in each of the following: lower jaw, upper leg and pelvis, and shoulder? (p. 291)
20. What are two examples of bones protecting body organs? (p. 286)
21. Why do you feel pain when you sprain your ankle? (p. 300)
22. What type of muscle is found in or as part of each of the following: stomach, blood vessels, heart, and arms? (pp. 294, 295)
23. How do you move your foot up and down? (p. 296)
24. What is the difference between voluntary and involuntary muscles? (p. 294)
25. What happens to muscle when the oxygen supply runs low? (p. 300)

## Skill Review ✓

*For more help, refer to the **Skill Handbook,** pages 704-719.*

1. **Infer:** When weight lifters flex their muscles, are the muscles shortening, relaxing, or stretching?
2. **Understand science words:** The word *involuntary* means without choice. The word part *in* means without. In your dictionary, find another meaning of the word part *in*.
3. **Classify:** Which words or statements are not correct for all muscles: voluntary, living, contract, made from cells, and form a tight weave?
4. **Infer:** A muscle that moves your leg up or down depends on what direction of action at your knee joint?

## Finding Out More

TECH PREP

**Critical Thinking**

1. Draw an imaginary animal. Build your new animal's skeletal system from balsa wood or paper. Hinge bones together with wire or thread. Explain to the class how your animal is adapted to survive.
2. How does the skeleton of a human compare with the skeleton of an alligator? Explain how each skeleton helps the organism survive.

**Applications**

1. Find out the causes and problems of rheumatoid arthritis, myasthenia gravis, and gout.
2. Make a list of the different types of bone fractures that can occur.

**Chapter Review**/*Teacher Resource Package*, pp. 70-71.

**Chapter Test**/*Teacher Resource Package*, pp. 72-74.

**Chapter Test**/*Computer Test Bank*

**Unit Test**/*Teacher Resource Package*, pp. 75-76.

## Skill Review

1. The muscles are shortening. A muscle bunches up when it contracts.
2. The word part in also means within or toward.
3. Not all muscles are voluntary or form a tight weave.
4. Movement is the result of a hinge joint.

## Finding Out More

### Critical Thinking

1. Students should describe how their new animal's skeleton would be beneficial for feeding and protection.
2. The human skeleton is upright and helps with walking. The hands have a thumb that helps with holding objects. The alligator skeleton is on four legs and low to the ground, which allows it to hide and move fast over the ground. It has a long tail that helps it swim.

### Applications

1. The cause of rheumatoid arthritis is unknown, but it is thought to be related to deficiencies in the autoimmune system. If severe, it can result in loss of use of the affected joints. Myasthenia gravis is a disease of the muscular system. A person becomes progressively weaker until movement is no longer possible. Gout is a metabolic disorder. The joints become inflamed and very painful.
2. Fractures include closed (simple) and open (compound).

# Unit 3 Body Systems—Maintaining Life

CONNECTIONS

## Biology in Your World

The human body, an incredible living system, has smaller systems within it that work to maintain life. Students learn in this unit that their mind, body, and behavior all contribute to their health and well-being.

### Literature Connection

Norman Cousins teaches future doctors about the biochemistry of the emotions. Cousins has overcome two serious illnesses and believes his positive outlook helped his recoveries. He also feels patients and doctors should work as a team when making decisions about treatment. If patients work together with their doctors, they are less apt to feel helpless. Cousins thinks helplessness and panic contribute to the deaths of some ill people.

### Art Connection

Rembrandt's painting *The Anatomy Lecture of Dr. Nicolaes Tulp* was the first of his many important group portraits. Usually, the artist chosen for a guild portrait had political connections. Because a guild member had ties to Rembrandt's art dealer, Rembrandt was selected to paint the autopsy. The surgeons liked the painting, and he was commissioned by many others for group portraits.

CONNECTIONS

# Biology in Your World

## An Amazing Machine

Even as you sleep, your digestive system is breaking down food. Your circulatory system is delivering oxygen and nutrients to cells. Eating well and getting plenty of exercise will keep your body systems running smoothly as they do their tasks.

LITERATURE

## Think Positive

Your heart really pumps life into your body. It is a remarkable, complex machine.

Norman Cousins was an editor and author. For many years he was editor of the magazine *Saturday Review.* He had a massive heart attack in 1980. Instead of giving up, he looked at his illness as a challenge. He began to take part in medical decisions concerning his health. Cousins knew that panic and depression could make his heart condition worse. He used positive thought, humor, and relaxation to avoid panic and depression so his heart could recover. In *The Healing Heart,* Cousins tells about his illness and recovery. He stresses that a person's attitude has a lot to do with recovering from an illness. Cousins died in 1990 at the age of 78.

ART

## The Inside Story

One of the ways medical students learn about the inside parts of the human body is through dissection. In the 1600s, the bodies of executed criminals were used. In 1632, the Dutch painter Rembrandt van Rijn was asked to paint a group portrait of the Amsterdam Guild of Surgeons. His painting, *The Anatomy Lecture of Dr. Nicolaes Tulp,* records a public dissection. In the painting, Dr. Tulp is showing the tendons of the arm. This painting was one of the first group portraits Rembrandt did in Amsterdam. It is considered one of the greatest dramatic group portraits.

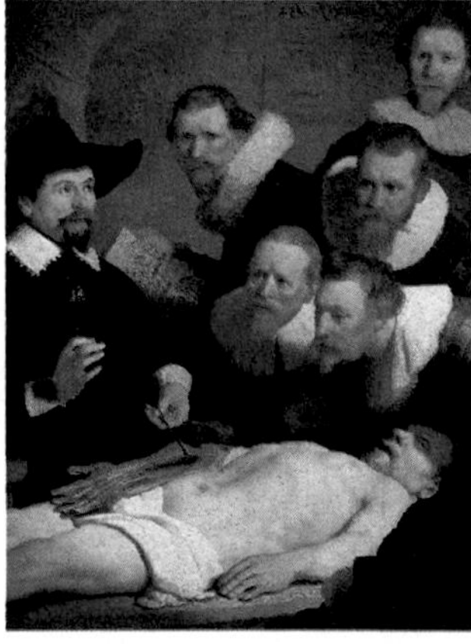

## LEISURE

## How Fit Are You?

Exercise can be an enjoyable way to spend leisure time. The benefits of exercise are long-lasting. Endurance exercises done on a regular basis will help your heart, lungs, and circulatory system work better. These exercises include walking, jogging, bicycling, and swimming. Lifting weights will build muscle strength, and stretching will increase flexibility.

To improve your overall fitness, plan an exercise program. Before you begin one, get advice from a doctor. Your program need not take large amounts of time or money. You may want to exercise with a friend.

Besides an exercise program, there may be other ways to get exercise. Are there daily activities you could change? Maybe you could walk to the store rather than ride in a car.

## CONSUMER

## Healthy Dieting

Many people diet to lose weight. What is a safe diet? A safe diet provides the nutrients the body needs every day, but with fewer calories. An effective diet includes an exercise program. Many doctors think a person should lose no more than two pounds per week. Before beginning a diet and exercise program, a person should see a doctor.

How safe are fad diets? Some fad diets promise weight loss without exercise. Others stress eating foods with a lot of a certain nutrient. Some diets promise a loss of several pounds a week. How safe and effective do you think these diets are? How well do they meet safe diet requirements?

## CULTURAL DIVERSITY

### Human Skeletal Variation

Variation exists in human body form, especially within the skeletal system. Differences in skeletal morphology are related to the geographical origins of populations. For example, a leaner body form is often observed in people who live in arid regions where greater skin-surface area in proportion to body weight facilitates heat loss. A stockier build is more adaptive for inhabitants of cold regions. Differences in body form and skin color have been used as a justification for racism. Have students write short essays about how their attitudes of racism have been affected by knowledge of the influences of human variation.

## CONNECTIONS

### Leisure Connection

Leisure exercise activities benefit the body in several ways. Physically fit people can do everyday chores without tiring and are less likely to injure themselves. A fit body resists disease and infection. Exercise reduces the risk of heart disease and can help control adult diabetes and high blood pressure.

### Consumer Connection

Many fad diets are nutritionally unbalanced. Some people lose weight quickly on these diets but at the expense of their health. Unbalanced diets can cause fatigue, gall stones, or even death.

The best way to lose or maintain weight is to eat a balanced diet and exercise moderately. Weight taken off slowly is easier to keep off and causes less strain on your body. Groups such as Weight Watchers offer support for dieters.

### References


Blonz, Edward R. "The World's Best-Kept Diet Secrets." *Better Homes and Gardens*, March 1994, pp. 52-53.

Krucoff, Carol. "Can Exercise Make You Smarter?" *Saturday Evening Post*, March-April, 1996, p. 16.

Neimark, Jill. "On the Frontline of Alternative Medicine." *Psychology Today*, Jan.-Feb. 1997, pp. 52-60

Wardlaw, Gordon, and Paul Insel. *Perspectives in Nutrition.* St. Louis: Times Mirror/Mosby College Publishers, 1990.

White, Christopher. *Rembrandt.* London: Thames and Hudson Ltd., 1984.

# Unit 4 Body Systems—Controlling Life

## UNIT OVERVIEW

Unit 4 ties together the concepts of coordination and communication among living things. Chapter 15 focuses on the nervous and endocrine systems. Sense organs are discussed in Chapter 16. Chapter 17 concentrates on animal behavior and how it relates to the nervous system. Chapter 18 deals with the effect of drugs upon behavior.

## Advance Planning

Audiovisual and computer software suggestions are located within each specific chapter under the heading TEACH. Photocopies of specific items for teaching the chapter or an activity should be prepared in advance.

Prepare a time schedule of chapters in this unit so that any chemicals with a short shelf life may be prepared when needed.

**To prepare for:**

| | |
|---|---|
| Lab 15-1: | Gather metric rulers. |
| Lab 15-2: | Prepare normal and diabetic "urine" samples. Gather droppers, test tubes, wax pencils, and glucose test paper. |
| Mini Lab: | Arrange to have flashlights available. |
| Lab 16-1: | Gather black and white paper, tape, scissors, and a hand lens. |
| Lab 16-2: | Assemble red and green colored pencils. |
| Lab 17-1: | Order vinegar eels. |
| Lab 17-2: | Collect mirrors. |
| Mini Labs: | Order or collect live earthworms. Ask students to bring in shoe boxes. Obtain black paper. Purchase onions and collect knives. |
| Lab 18-1: | Prepare 6 test solutions of painkillers and aspirin-testing solution in dropper tubes. |
| Lab 18-2: | Prepare 6 test solutions of over-the-counter drugs, household items, and alcohol testing chemical. |
| Mini Lab: | Order bacteria culture and agar. Treat paper disks with onion and garlic. |

Unit **4**

**CONTENTS**



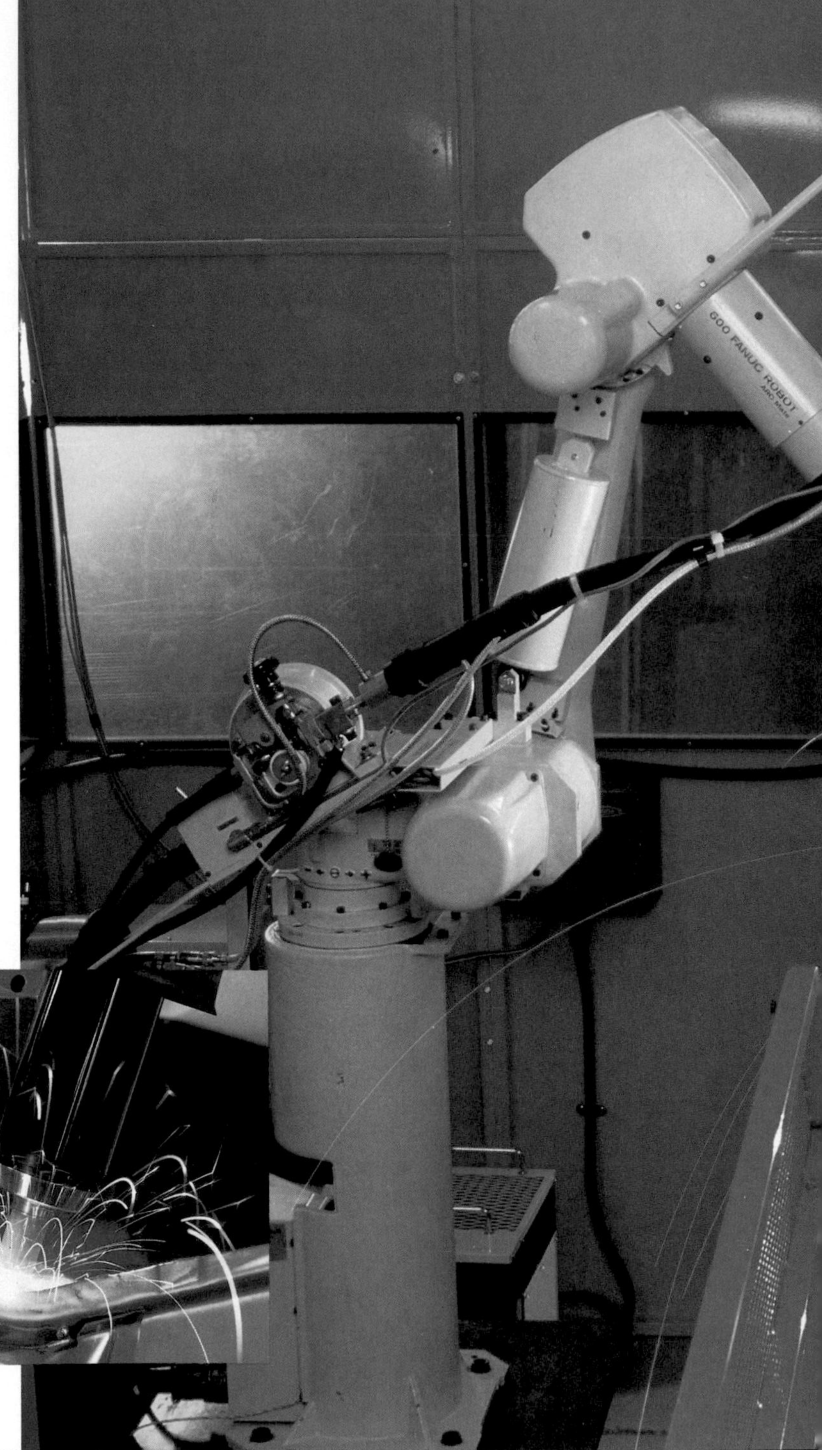

# Body Systems—Controlling Life

**What would happen if...**

robots could think and make decisions? So far, robots can't think, guess, imagine, create, or make decisions. But researchers hope to produce computer programs that mimic human thought.

With such computer programs, the possible uses of robots would increase greatly. More and more jobs could be taken over by robots.

Being able to run a factory or business without people has many advantages. Robots are not paid, and they could work 24 hours a day. Dirty, boring, and dangerous jobs could be done by robots. But, would people who lose their jobs to robots find other jobs or would they become unemployed? What would be the impact on society of increased unemployment?



**UNIT OVERVIEW**

## Photo Teaching Tip

After students have read the passage that accompanies these unit opening photographs of a robot, ask them to make a list of all the things they do each day that they would like to have a personal robot to carry out for them. Have students consider the choice between personal control and being controlled. This could lead the discussion to the misuse and abuse of drugs, which can result in a loss of control over one's own choices. Point out that technology can be used to help us gain more control over our environment and that as scientists we have the choice to use or abuse this control.

## Theme Development

Animal life depends on the animal systems that maintain and regulate body functions. This unit, as did the previous one, continues to use the human as its major animal example. The theme of homeostasis is developed in this unit. The nervous and endocrine systems plus the sense organs of various animal phyla other than Chordates are referenced in this unit. Control of life functions is finalized, with special emphasis given to animal behavior and the effect of drugs on behavior.

**CULTURAL DIVERSITY**

See the Cultural Diversity feature located on the Connections page at the end of this unit.

For Tech Prep activities, see the Applying Technology feature in this unit, the Finding Out More applications questions in the Chapter Reviews, and the Tech Prep Applications booklet in the Teacher Resource Package.

# Nervous and Chemical Control

## PLANNING GUIDE

| CONTENT | TEXT FEATURES | TEACHER RESOURCE PACKAGE | OTHER COMPONENTS |
|---|---|---|---|
| (1 day)<br>15:1 The Role of the Nervous System<br>Response<br>Nervous Systems of Animals | Check Your Understanding, p. 311 | Reteaching: *Nervous Systems of Invertebrates,* p. 43<br>Study Guide: *The Role of the Nervous System,* p. 85 | **STVS:** *Detecting the Body's Magnetic Fields,* Human Biology (Disc 7, Side 1) |
| (3 days)<br>15:2 Human Nervous System<br>Nerve Cells<br>Pathways of Messages<br>Nervous System Parts<br>Jobs of the Brain<br>Reflexes | Skill Check: *Interpret Diagrams,* p. 314<br>Skill Check: *Understand Science Words,* p. 316<br>Mini Lab: *What Is an Eye Reflex?* p. 317<br>Lab 15-1: *Reaction Time,* p. 319<br>Idea Map, p. 317<br>Check Your Understanding, p. 318 | Application: *Reflexes Protect You,* p. 15<br>Reteaching: *Parts of the Brain,* p. 44<br>Focus: *What You Can Do Without Thinking,* p. 29<br>Transparency Master: *Neuron Parts and Pathways,* p. 69<br>Study Guide: *The Human Nervous System,* pp. 86-87<br>Lab 15-1: *Reaction Time,* pp. 57-58 | **Laboratory Manual:**<br>*Which Brain Side is Dominant?* p. 119<br>**Laboratory Manual:**<br>*What Are the Functions of the Brain?* p. 123<br>Color Transparency 15a: *Movement of a Message Across a Synapse*<br>**STVS:** *Brain Development,* Human Biology (Disc 7, Side 1) |
| (1 1/2 days)<br>15:3 The Role of the Endocrine System<br>Chemical Control<br>The Pituitary Gland<br>The Thyroid Gland | Idea Map, p. 322<br>Check Your Understanding, p. 322 | Enrichment: *How Pancreas Hormones Work Together,* pp. 15-16<br>Reteaching: *Endocrine Glands,* p. 45<br>Critical Thinking/Problem Solving: *How Does the Brain Affect the Amount of Urine Made by the Kidneys?* p. 15<br>Study Guide: *The Role of the Endocrine System,* p. 88<br>Study Guide: *Vocabulary,* p. 90 | Color Transparency 15b: *The Endocrine System*<br>**STVS:** *Insulin Pills,* Human Biology (Disc 7, Side 2) |
| (1 1/2 days)<br>15:4 Nervous and Endocrine System Problems<br>Nervous System Problems<br>Endocrine System Problems | Lab 15-2: *Diabetes,* p. 325<br>Check Your Understanding, p. 324<br>Applying Technology: *Get a Kick Out of This,* p. 326 | Reteaching: *Endocrine System Problems,* p. 46<br>Skill: *Summarize,* p. 16<br>Study Guide: *The Role of the Endocrine System,* p.89<br>Lab 15-2: *Diabetes,* pp. 59-60 | **STVS:** *Insulin Pump,* Human Biology (Disc 7, Side 2)<br>**STVS:** *Brain By-Pass Surgery,* Human Biology (Disc 7, Side 1) |
| Chapter Review | Summary<br>Key Science Words<br>Testing Yourself<br>Finding Main Ideas<br>Using Main Ideas<br>Skill Review | **ASSESSMENT RESOURCES**<br>Chapter Review, pp. 77-78<br>Chapter Test, pp. 79-81<br>Performance Assessment in the Biology Classroom<br>Alternate Assessment in the Science Classroom<br>Computer Test Bank | |

## MATERIALS NEEDED

| LAB 15-1, p. 319 | LAB 15-2, p. 325 | APPLYING TECHNOLOGY, p. 326 | MARGIN FEATURES |
|---|---|---|---|
| metric ruler | test tube rack<br>microscope slides<br>8 droppers<br>wax pencils<br>normal urine sample<br>urine sample from diabetic<br>urine samples in test tubes marked C-E<br>glucose test paper | hammer, soft<br>lab table or desk | Skill Check, p. 316 dictionary<br>Mini Lab, p. 317 flashlight |

## OBJECTIVES

For more information about National Science Standards, see page 5T.

| SECTION | OBJECTIVE | CORRELATION of QUESTIONS to OBJECTIVES: CHECK YOUR UNDERSTANDING | CHAPTER REVIEW | TRP CHAPTER REVIEW | TRP CHAPTER TEST |
|---|---|---|---|---|---|
| 15:1 National Science Stds: UCP.1, UCP.2, UCP.5, C.6 | 1. **Explain** how animals keep in touch with their body parts and their surroundings. | 1, 2, 4 | 1, 20 | 9 | 12, 13, 47 |
| | 2. **Compare** the nervous systems of common animals. | 3 | 16 | 10 | 14, 48 |
| 15:2 National Science Stds: UCP.1, UCP.2, UCP.3, UCP.4, UCP.5, C.1, C.6 | 3. **Explain** how nerve cells carry messages through the body. | 6, 10 | 3, 6, 18, 25 | 4, 15, 28, 31 | 3, 5, 9, 10, 11, 39, 40, 41, 42, 43, 44, 45 |
| | 4. **State** the functions of the three major parts of the nervous system. | 7, 9 | 4, 12, 17, 22 | 1, 3, 6, 8, 11, 17, 20, 23, 24, 32 | 2, 6, 15, 17, 18, 19, 20, 21, 22, 23, 24, 25, 46, 53 |
| | 5. **Describe** a reflex action. | 8 | 10, 15 | 14, 27 | 7, 16, 49, 56 |
| 15:3 National Science Stds: UCP.1, UCP.2, UCP.5, F.1 | 6. **Explain** the function of the endocrine system. | 11 | 9 | 5, 12, 13, 16, 18, 19, 21, 22, 25, 26 | 1, 8, 26, 27, 28, 29, 30, 31, 32, 35, 37, 52 |
| | 7. **Relate** the importance of the pituitary and thyroid glands. | 12, 14, 15 | 5, 8, 11, 14, 24 | 29, 33 | 33, 34, 38, 51, 57 |
| 15:4 National Science Stds: UCP.2, UCP.5, F.1 | 8. **Identify** problems that damage the brain. | 16, 20 | 21 | 2 | 54 |
| | 9. **Explain** the importance of insulin. | 17, 18 | 1, 17, 13, 23 | 7, 30, 34 | 4, 36, 50, 55 |

# Nervous and Chemical Control

## CHAPTER OVERVIEW

### Key Concepts

This chapter looks at the body's two major systems of communication, the nervous and endocrine systems. The neuron is described and its role in carrying messages through the nervous system is analyzed. The most common endocrine glands are described, and the way hormones keep themselves in balance is discussed. Certain problems of the nervous and endocrine systems are also examined.

### Key Science Words

| | |
|---|---|
| axon | nerve |
| brain | nervous system |
| cerebellum | neuron |
| cerebrum | pituitary gland |
| dendrite | reflex |
| diabetes mellitus | spinal cord |
| endocrine system | synapse |
| hormone | thyroid gland |
| insulin | thyroxine |
| medulla | |

### Skill Development

In Lab 15-1, students will **form a hypothesis, calculate**, and **infer** to determine reaction time. In Lab 15-2, they will **experiment** and **interpret data** to determine if glucose is present in urine of a diabetic. In the Skill Check on page 314, students will **interpret diagrams** of synapses. In the Skill Check on page 316, they will **understand** the **science word** *cerebrum.* In the Mini Lab on page 317, they will **observe** eye reflexes.

### Bridging

Skeletal muscles contract on a voluntary basis and heart muscle contracts on an involuntary basis. From where within the body do these messages that bring about contraction originate? The nervous system's role in muscle contraction can now be related to an earlier study of muscle action.

**CHAPTER PREVIEW**

### Chapter Content

Review this outline for Chapter 15 before you read the chapter.

**15:1 The Role of the Nervous System**
- Response
- Nervous Systems of Animals

**15:2 Human Nervous System**
- Nerve Cells
- Pathways of Messages
- Nervous System Parts
- Jobs of the Brain
- Reflexes

**15:3 The Role of the Endocrine System**
- Chemical Control
- The Pituitary Gland
- The Thyroid Gland

**15:4 Nervous and Endocrine System Problems**
- Nervous System Problems
- Endocrine System Problems

### Skills in this Chapter

The skills that you will use in this chapter are listed below.

- In **Lab 15-1,** you will form a hypothesis, calculate, and infer. In **Lab 15-2,** you will experiment and interpret data.
- In the **Skill Checks,** you will interpret diagrams and understand science words.
- In the **Mini Lab,** you will observe.

308

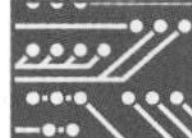

## TECH PREP

For Tech Prep activities in this chapter of the Teacher Wraparound Edition, see especially the Motivation on page 313, the Portfolio on page 314, the Connection on page 318, and the Interview on page 324.

See also the Glencoe Homepage at **www.glencoe.com**

Chapter 15

# Nervous and Chemical Control

The larger photo on the left is a photograph of a circuit board from a computer. The smaller photo below shows the human brain. It is often said that a computer and the human brain are very much alike. The brain is made of living cells. There are billions of cells in the human brain. Each cell sends and receives messages from nearby cells. These messages are interpreted as sights, sounds, smells, or thoughts. A computer has thousands of computer chips. Each chip sends and receives messages from the other chips in the computer. Both brains and computers process information. How is the processing of information by the brain important? In what other ways are messages carried through the body?

**Try This!**

**Can you control your reflexes?** Put on a pair of safety goggles. Have someone toss a paper wad toward your eyes. Do you blink? Can you keep yourself from blinking? What is this involuntary action called?

*inter*NET CONNECTION

For more information about the material in this chapter, follow the link for the chapter on the Glencoe Homepage at **http://www.glencoe.com**

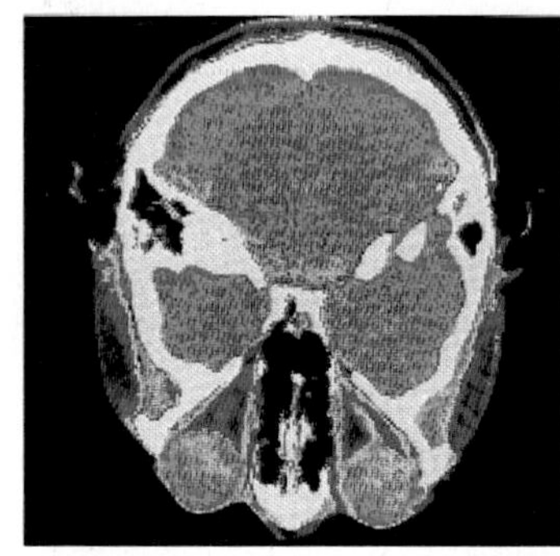

CAT scan of the human brain



ASSESSMENT PLANNER

***Portfolio***

Strategies on the following pages represent student products that can be placed into a best-work portfolio: pp. 314, 323.

**PERFORMANCE ASSESSMENT**

Skill Check, pp. 314, 316
Mini Lab, p. 317
Lab, pp. 319, 325

**CONTENT ASSESSMENT**

Check for Understanding, pp. 311, 318, 322, 324
Chapter Review, pp. 328-329

**GROUP ASSESSMENT**

Opportunities for group assessment occur with Cooperative Learning Strategies.

***Student Journal***

Strategies on the following pages represent opportunities for writing in a Student Journal: p. 318.

GETTING STARTED

## Using the Photos

In looking at the photos, students should try to identify similarities and differences between the brain and a circuit board. Both process information. The brain is more complex and can do more than a computer.

## MOTIVATION/Try This!

**Can you control your reflexes?** Students are **experimenting** with a reflex. This type of action is not under voluntary control and will occur each time the same stimulus (wad of paper moving toward the eyes) is detected.

## Chapter Preview

Have students study the chapter outline before they begin to read the chapter. They should note the overall organization of the chapter, that the introduction describes the general role of the nervous system and introduces students to the system found in simpler animal phyla. Next, it describes the nature of the human nervous system. The role of the endocrine system is examined next with a detailed study of certain endocrine glands. Last, problems that arise with the endocrine and nervous system are described.

## Misconception

Ask students the cause of diabetes and they will tell you that it is related to the kidneys (presence of sugar in urine). In reality, the problem is associated with an endocrine gland and the lack of the hormone insulin.

## 15:1 The Role of the Nervous System

## PREPARATION

### Materials Needed

 Make copies of the Reteaching and Study Guide worksheets in the *Teacher Resource Package.*

### Key Science Words

nervous system

## 1 FOCUS

▶ The objectives are listed on the student page. Remind students to preview these objectives as a guide to this numbered section.

### Convergent Question

Ask students what animals would be unable to do if they had no nervous system. Students will realize that almost all body functions are related directly or indirectly to the nervous system.

### Guided Practice

Have students write down their answers to the margin questions: The hydra has a nerve net; animals are able to respond to changes by means of their nervous systems.

**Videodisc**

STV: Human Body Vol. 2
*Nervous System*
Unit 2, Side 2
*Nervous System (in its entirety)*

1066-30655



## 15:1 The Role of the Nervous System

**Objectives**

1. **Explain** how animals keep in touch with their body parts and their surroundings.
2. **Compare** the nervous systems of common animals.

**Key Science Words**

nervous system

Most animals have special cells and organs that help them keep in touch with parts of their own bodies. They are able to send messages to all body parts and receive messages from those body parts. These special cells also allow an animal to receive messages from its surroundings.

### Response

Most animals can quickly detect changes that take place around them. Usually, they respond quickly to these changes. A response is the action of an organism because of a change in its environment.

Let's use an example to show how quickly animals respond to changes around them. Have you ever seen a dog chase a cat? It seems impossible for the dog to catch the cat. Why? First, the cat usually sees the dog. The cat's brain gets the message that something nearby is moving toward it. The brain responds by sending messages to the cat's leg muscles. The cat runs up a tree to protect itself.

The cat sees and responds to the dog because it has a nervous system. A **nervous system** is made of cells and organs that let an animal detect changes and respond to them. This system is made of nerve cells, sense organs, and usually a brain.

### Nervous Systems of Animals

The simplest animals to have a nervous system are the stinging-cell animals. Figure 15-1a shows what this system looks like in the hydra. Hydras have a very simple nervous

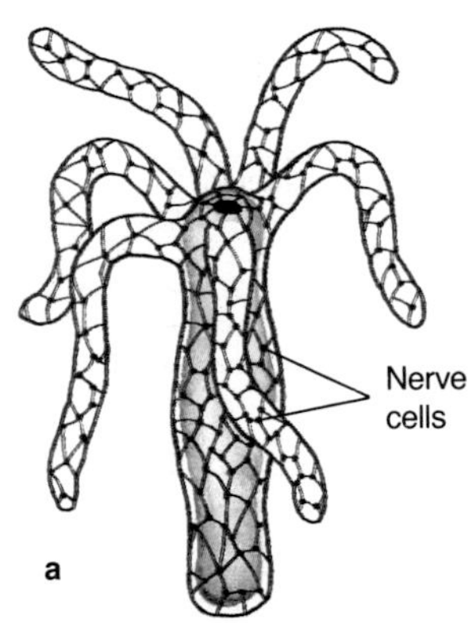


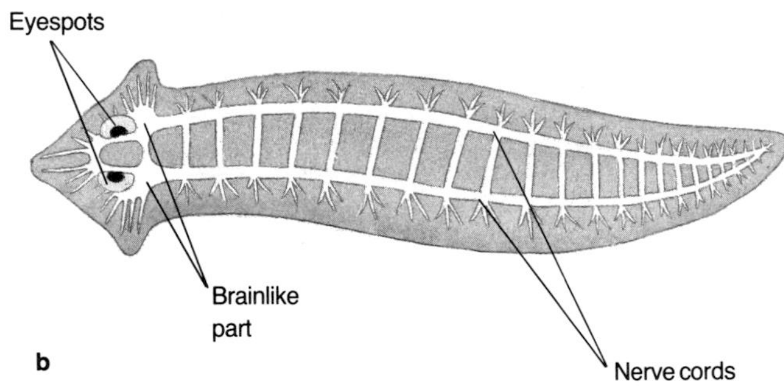


**Figure 15-1** Hydras (a) and planarians (b) have simple nervous systems.



## OPTIONS

### Science Background

A nerve net transmits impulses in all directions. Impulses are weaker, and response time is slower. A nervous system transmits impulses along discrete nerve pathways. Impulses are stronger, and response time is faster.

system made up of nerve cells that form a net throughout the body. This simple grouping of nerve cells lets these animals detect and respond to changes around them.

A flatworm has a nervous system that is more complex than that of the hydra. The nervous system of a flatworm has several parts. You can see in Figure 15-1b that a planarian has three parts that a hydra doesn't have. The planarian has a brainlike part, eyespots, and two nerve cords. Planarians can respond to changes in light with their eyespots. Because they have a brainlike part and nerve cords, they also can respond more quickly than a hydra.

Jointed-leg animals have a more complex nervous system than flatworms. Figure 15-2 shows the inside of a spider. A spider has four pairs of eyes. It has a brain, nerve cord, and many nerves going to its body parts. Spiders can respond even more quickly than planarians.

**What kind of nervous system does a hydra have?**

**How are animals able to respond to changes around them?**

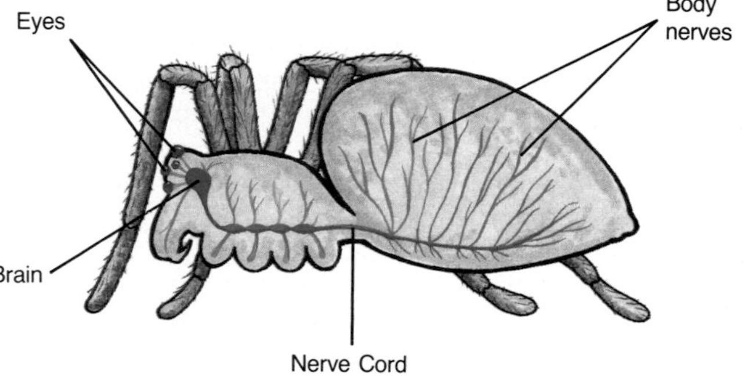


**Figure 15-2** Nervous system of a spider

## Check Your Understanding

1. What is meant by *response?* Give an example not found in this chapter.
2. What is a nervous system? What parts form a nervous system?
3. Compare the nervous systems of a hydra, planarian, and spider.
4. **Critical Thinking:** What cell parts do protists use to respond to changes in the environment?
5. **Biology and Writing:** A planarian has a nervous system. Do you think a planarian can learn? Write a paragraph of at least three sentences to explain your answer.

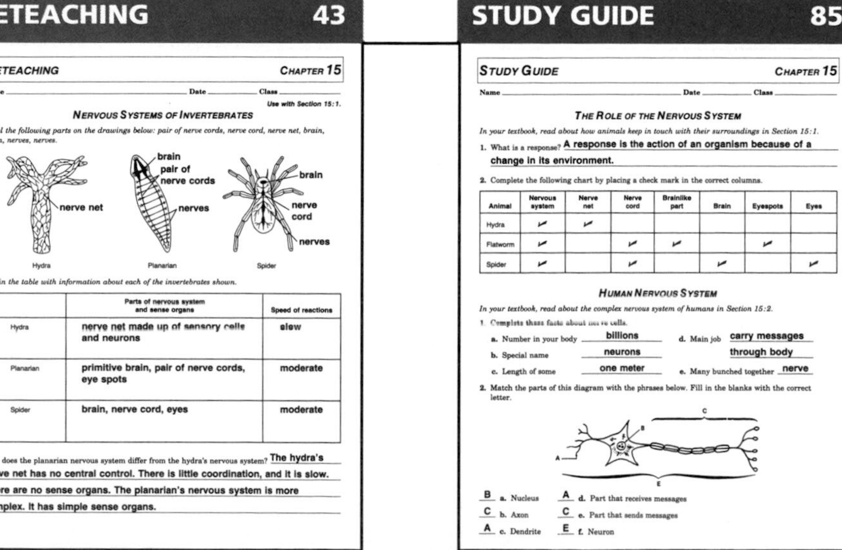

**RETEACHING 43**

RETEACHING — CHAPTER 15

Name ______ Date ______ Class ______

Use with Section 15:1.

NERVOUS SYSTEMS OF INVERTEBRATES

*Label the following parts on the drawings below: pair of nerve cords, nerve cord, nerve net, brain, brain, nerves, nerves.*



*Fill in the table with information about each of the invertebrates shown.*

| | Parts of nervous system and sense organs | Speed of reactions |
|---|---|---|
| Hydra | nerve net made up of sensory cells and neurons | slow |
| Planarian | primitive brain, pair of nerve cords, eye spots | moderate |
| Spider | brain, nerve cord, eyes | moderate |

How does the planarian nervous system differ from the hydra's nervous system? The hydra's nerve net has no central control. There is little coordination, and it is slow. There are no sense organs. The planarian's nervous system is more complex. It has simple sense organs.

**STUDY GUIDE 85**

STUDY GUIDE — CHAPTER 15

Name ______ Date ______ Class ______

THE ROLE OF THE NERVOUS SYSTEM

*In your textbook, read about how animals keep in touch with their surroundings in Section 15:1.*

1. What is a response? A response is the action of an organism because of a change in its environment.
2. Complete the following chart by placing a check mark in the correct columns.

| Animal | Nervous system | Nerve net | Nerve cord | Brainlike part | Brain | Eyespots | Eyes |
|---|---|---|---|---|---|---|---|
| Hydra | ✓ | ✓ | | | | | |
| Flatworm | ✓ | | ✓ | ✓ | | ✓ | |
| Spider | ✓ | | ✓ | | ✓ | | ✓ |

HUMAN NERVOUS SYSTEM

*In your textbook, read about the complex nervous system of humans in Section 15:2.*

1. Complete these facts about nerve cells.
   a. Number in your body billions
   b. Special name neurons
   c. Length of some one meter
   d. Main job carry messages through body
   e. Many bunched together nerve
2. Match the parts of this diagram with the phrases below. Fill in the blanks with the correct letter.



B a. Nucleus
C b. Axon
A c. Dendrite
A d. Part that receives messages
C e. Part that sends messages
E f. Neuron

## 2 TEACH

### Independent Practice

**Study Guide**/*Teacher Resource Package,* p. 85. Use the Study Guide worksheet shown at the bottom of this page for independent practice.

### Check for Understanding

Have students respond to the first three questions in Check Your Understanding.

### Reteach

**Reteaching**/*Teacher Resource Package,* p. 43. Use this worksheet to give students additional practice with the nervous systems of invertebrates.

**Extension:** Assign Critical Thinking, Biology and Writing, or some of the **OPTIONS** available with this section.

## 3 APPLY

### ACTIVITY/Challenge

Have students research the nervous system in other animal phyla.

## 4 CLOSE

Show the videocassette *Nervous System,* National Geographic.

### Answers to Check Your Understanding

1. It is how an animal reacts to a change in its environment. A robin responds to a worm by grabbing it in its beak.
2. cells and organs that let an animal detect changes and respond to them; nerve cells, sense organs, and possibly a brain
3. The hydra has a nerve net. Flatworms have a brain-like part, eyespots, and two nerve cords. Spiders have a brain, eyes, body nerves, and a nerve cord.
4. nucleus, vacuole, cell membrane
5. Answers will vary but should include that a planarian has a brainlike part.

## 15:2 Human Nervous System

### PREPARATION

#### Materials Needed

Make copies of the Focus, Transparency Master, Application, Study Guide, Reteaching, and Lab worksheets in the *Teacher Resource Package.*

▶ Obtain penlights for the Mini Lab on an eye reflex.

▶ Gather metric rulers for Lab 15-1.

#### Key Science Words

neuron
nerve
dendrite
axon
synapse
brain
spinal cord
cerebrum
cerebellum
medulla
reflex

#### Process Skills

In Lab 15-1, students will form a hypothesis, calculate, and infer. In the Skill Checks, students will understand science words. In the Mini Lab, students will observe and interpret diagrams.

### 1 FOCUS

▶ The objectives are listed on the student page. Remind students to preview these objectives as a guide to this numbered section.

#### MOTIVATION/Brainstorming

Ask students to name activities that would be impossible if we had no nervous system and to offer a possible reason for their answer. Ask if it would be possible to remain alive without a nervous system.

**Focus**/*Teacher Resource Package,* p. 29. Use the Focus Transparency shown at the bottom of this page as an introduction to reflexes.

## 15:2 Human Nervous System

**Objectives**

3. **Explain** how nerve cells carry messages through the body.
4. **State** the functions of the three major organs of the nervous system.
5. **Describe** a reflex action.

**Key Science Words**

neuron
nerve
dendrite
axon
synapse
brain
spinal cord
cerebrum
cerebellum
medulla
reflex

Humans have one of the most complex nervous systems. If you understand how the human nervous system works, then you will have an idea of how nervous systems work in many other animals. Let's look at parts of the human nervous system in more detail.

### Nerve Cells

Nerve cells are the main part of any nervous system. They carry messages through the nervous system. Your body contains billions of nerve cells. Nerve cells are called **neurons** (NOO rahnz).

Neurons are often compared to electrical wires. Wires are very thin and long. Wires carry messages and have a covering of insulation around them. Wires usually connect two things, such as an electric outlet and a lamp. Neurons are very thin and can be very long. They are as long as one meter in some parts of your body. Neurons also have a covering. Neurons carry messages through your body. They make it possible for body parts to keep in touch with each other. Figure 15-3 shows how a wire and a neuron compare.

We can also compare a cable with a nerve. A cable is made of many wires bunched together. Cables are thick. A **nerve** is many neurons bunched together. Nerves can be thick, too. Figure 15-3 compares a nerve with a cable. Notice how thick the nerve is compared to the single neuron.

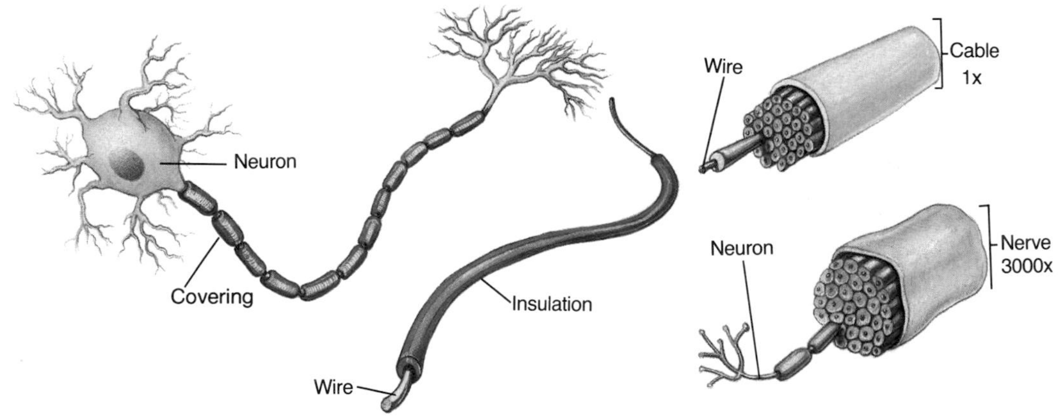


**Figure 15-3** Neurons and nerves can be compared to wires and cables.



### OPTIONS

#### Science Background

Many neurons have a fat-containing sheath (myelin sheath) that surrounds the axon. The myelin sheath helps to increase the speed of nerve impulses and insulates the cell. Myelinated neurons make up the white matter in the brain and spinal cord. Unmyelinated neurons make up gray matter.

FOCUS 29

FOCUS CHAPTER 15

Name Date Class

WHAT YOU CAN DO WITHOUT THINKING

1. The people shown are doing things without thinking about them. What are they doing?
2. What other things can you do without thinking about them?

A neuron, shown in Figure 15-4, has many of the same parts that other cells have. There are a few differences, however, between neurons and other cells. Two differences are the length and the shape of the neuron. The long, thin shape helps the neuron do its job well. The third difference between neurons and other cells is in the shape of the ends. Notice the branching shape of both ends of the neuron. One end is called the dendrite. **Dendrites** are parts of the neuron that receive messages from nearby neurons. The other end, usually longer, is called the axon (AK sahn). The **axon** is the part of the neuron that sends messages to surrounding neurons or body organs.

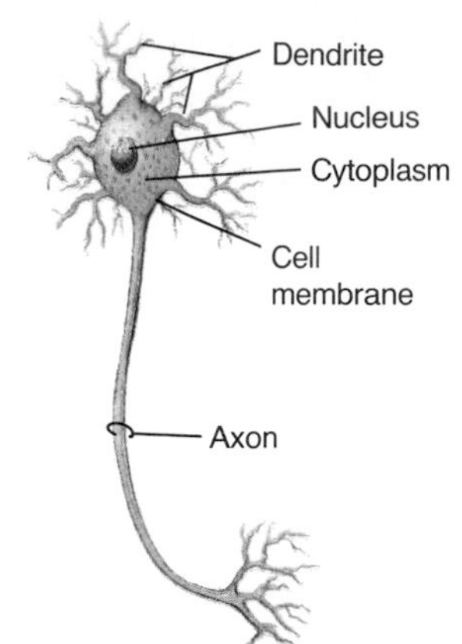

**Figure 15-4** The long, thin shape of a neuron helps it do its job of carrying messages.

## Pathways of Messages

How do nerves carry messages from one part of your body to another? Figure 15-5 shows what the path of nerves would look like. There are three important things that you should see in the diagram.

1. The pathway that carries messages from brain to hand is not the same pathway that carries messages from hand to brain. The different colored arrows in this diagram show the directions in which messages move along the pathways.
2. The neurons that make up a long pathway do not touch each other. There is a very small space called a synapse (SIN aps) between one neuron and the next. A **synapse** is a small space between the axon of one neuron and the dendrite of a nearby neuron.
3. The dendrites of one neuron are always next to the axon of another neuron.

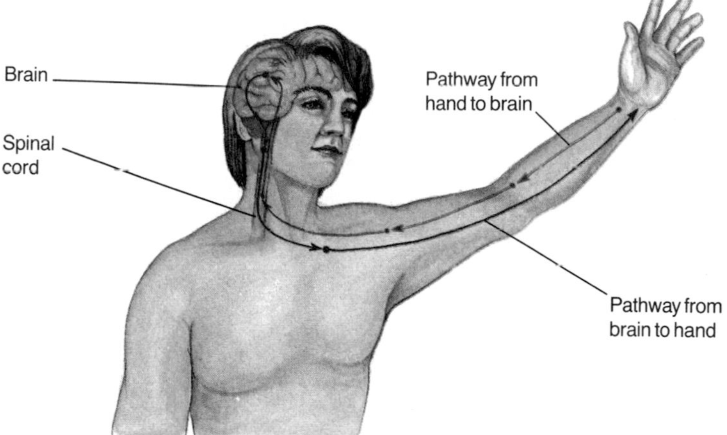

**Figure 15-5** The nerve pathways that carry messages away from the brain are different from the nerve pathways that carry messages to the brain.



# 2 TEACH

### MOTIVATION/Demonstration

See if your physics department has a wave motion apparatus. It consists of a series of thin metal tubes fastened to a thin metal strip. The motion of the tubes will simulate a wave pattern when the strip is twisted. Such a demonstration will help students visualize how a nerve message is propagated along the length of the nerve.

### Concept Development

▶ Understanding the terms *axon, dendrite,* and *synapse* is critical to understanding how the nervous system works. Make sure that students understand these terms before moving on.

▶ The three points listed in the diagram reference are important. Ask students to redraw the diagram of neuron pathways on the chalkboard, making sure that they observe the three rules in their drawings.

▶ Ask students to label the axon and dendrite ends of the neurons that were drawn on the chalkboard. It is important that axon and dendrite labels do appear next to one another, but are labeled on different neurons. Add labels to each synapse.

**Analogy:** Have students read the comparison between nerve cells and a wire. Ask them to describe differences between the two. Possible answers are that nerves are composed of cells, that they are composed of living matter, and that some nerves can repair themselves, while wires cannot. A cable cut in cross section may be obtained from your shop teacher. It would serve as a good visual aid in making the comparison between nerves and cables more tangible.

**TRANSPARENCY 69**

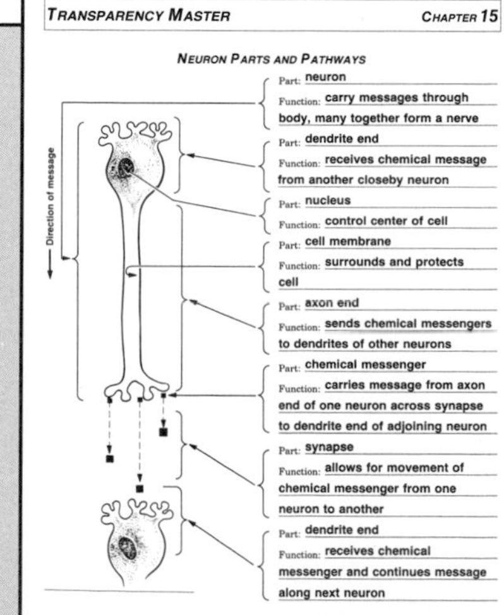


## *OPTIONS*

**Enrichment:** The speed of messages along a nerve is about 100 meters per second. If a person could run at that speed, he or she could cover the length of a football field in about one second.

**Transparency Master**/*Teacher Resource Package,* p. 69. Use the Transparency Master shown here as you teach about neuron pathways.

## TEACH

**Analogy:** An analogy can be made between a radio station transmitting tower, a home radio receiver, and a neuron's axon/dendrite ends. The tower sends out the signal (axon end), while the radio receives the message (dendrite end). Radio signals represent the chemical given off by the axon. Continue the analogy of radio waves by asking if a typical home radio receiver can send messages back to the radio station. It can't. Thus, the rule established for one-way messages along neuron pathways is further illustrated.

### Guided Practice

Have students write down their answers to the margin question: Message moves along neuron from dendrite to axon. Message reaches tip of axon. Chemical is given off by axon when message arrives. Chemical passes across synapse and reaches dendrite of next neuron. Chemical restarts the message and message continues along the new neuron.

### Skill Check

Messages travel from one neuron to the next by means of chemical messengers that cross the synapse.

**Portfolio**

Provide students with two 15-cm lengths of wire, scissors, tape, and heavy paper. Have them prepare a model of two adjoining neurons. Have them tape their models to the paper and use arrows to show message direction. Have them label the following: synapse, dendrite, axon, neuron, insulation, and site of chemical release.

Messages move along a neuron from one end to the other. The message is an electrical charge that moves along the axon just as electricity moves along a wire. It flows along the neuron from dendrite end to axon end. How can this message travel across the synapse between one neuron and the next?

**How do messages move across a synapse?**

Figure 15-6 shows how messages move across a synapse. First, the message moves along a neuron from the dendrite to the axon, Figure 15-6a. Next, the message reaches the tip of the axon, Figure 15-6b. Notice that a chemical is given off by the axon when the message arrives there. This chemical then passes across the synapse and reaches the dendrite of the next neuron as shown in Figure 15-6c. The chemical restarts the message and the message continues along the new neuron.

Now you know why different pathways are needed to carry messages from the brain to the hand and from the hand to the brain. Messages do not travel in both directions along the same neuron. Only the axon of the neuron gives off the chemical that crosses the synapse.

### Skill Check

**Interpret diagrams:** Look at Figure 15-6. How does the message travel from one neuron to the next? *For more help, refer to the **Skill Handbook**, pages 706-711.*

## Nervous System Parts

Nerve cells make up three important parts or areas of your nervous system. These parts are the brain, spinal cord, and body nerves.

The first major organ of the human nervous system is the brain. The **brain** is the organ that sends and receives messages to and from all body parts. It also records and interprets these messages. It is made of billions of neurons.

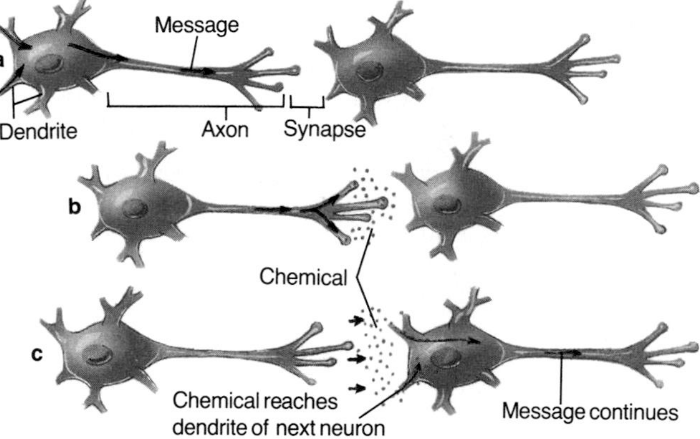


**Figure 15-6** Messages move from one neuron to the next with the help of a chemical messenger that crosses the synapse.



## OPTIONS

### ACTIVITY/Enrichment

You can use a skull to show how well it serves as a protective structure. Most skulls are hinged on top so that students can actually measure the bone thickness of different regions of the skull.

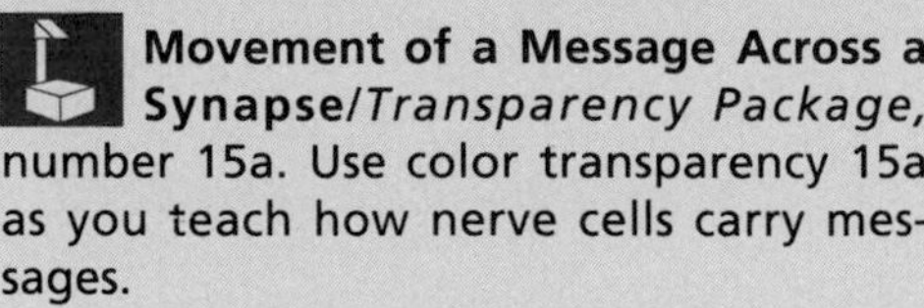

**Movement of a Message Across a Synapse**/*Transparency Package*, number 15a. Use color transparency 15a as you teach how nerve cells carry messages.

The second major part of the human nervous system is the spinal cord. Your **spinal cord** is the part that carries messages from the brain to body nerves or from body nerves to the brain. Messages travel up the spinal cord on the way to your brain and down the spinal cord to body nerves. The spinal cord is made of millions of neurons. Figure 15-7 shows the location of your spinal cord. It runs down the center of your back. The spinal cord can be compared to the power lines that enter or leave a power plant. The power plant could be compared to the brain.

Figure 15-8 shows how well the brain and spinal cord are protected against injury. Both are covered by bone and membranes. Both are also surrounded by fluid that cushions them. The brain is covered by your skull. The spinal cord is covered by bones called vertebrae. The word *vertebrate*, used in Chapter 7, sounds very similar. What do *vertebrae* and *vertebrate* mean?

The third part of your nervous system is the group of nerves that enter and leave the spinal cord. Body nerves connect organs, muscles, and skin to the spinal cord. Messages move along body nerves from organs or muscles to the spinal cord and then to the brain. Messages also move from the brain to the spinal cord and then along body nerves to organs or muscles. Body nerves are like the wires that lead from the power lines to your house.

**What are the three major parts of the nervous system?**

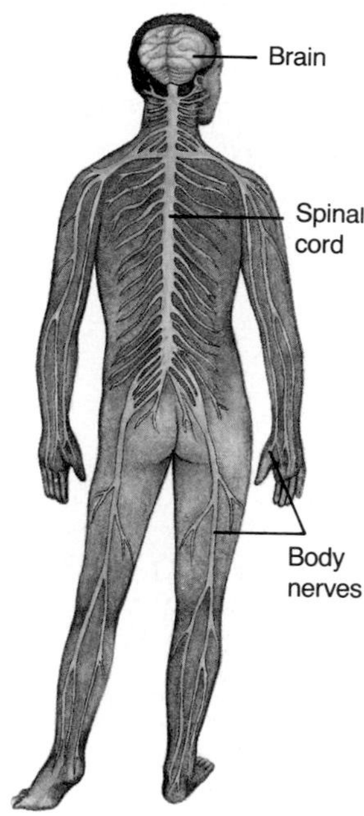


**Figure 15-7** The parts of the nervous system include the brain, spinal cord, and body nerves.

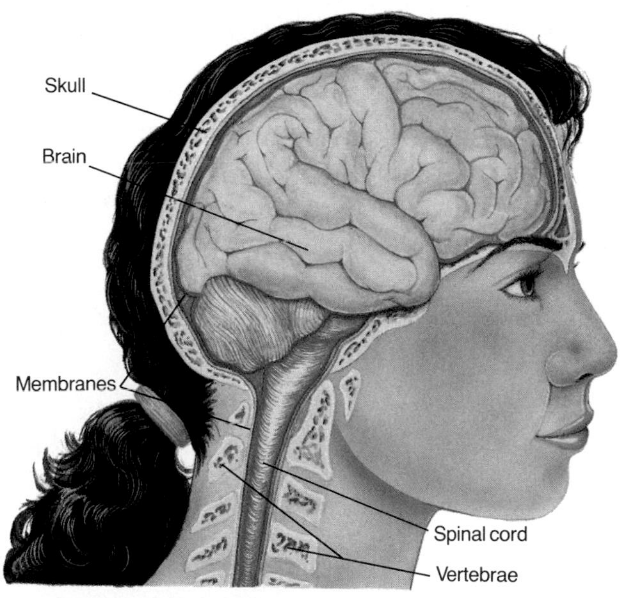


**Figure 15-8** The brain and spinal cord are well protected.



## STUDY GUIDE 86

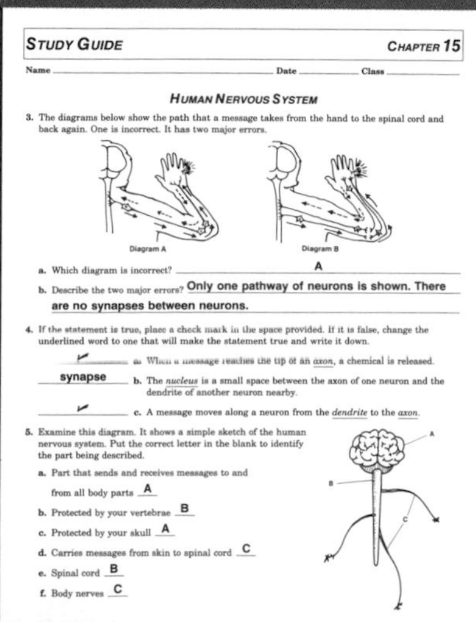

*STUDY GUIDE* *CHAPTER 15*

Name ________ Date ________ Class ________

*HUMAN NERVOUS SYSTEM*

3. The diagrams below show the path that a message takes from the hand to the spinal cord and back again. One is incorrect. It has two major errors.

Diagram A Diagram B

a. Which diagram is incorrect? **A**

b. Describe the two major errors? **Only one pathway of neurons is shown. There are no synapses between neurons.**

4. If the statement is true, place a check mark in the space provided. If it is false, change the underlined word to one that will make the statement true and write it down.

✓ a. When a message reaches the tip of an <u>axon</u>, a chemical is released.

**synapse** b. The <u>*nucleus*</u> is a small space between the axon of one neuron and the dendrite of another neuron nearby.

✓ c. A message moves along a neuron from the <u>*dendrite*</u> to the <u>*axon*</u>.

5. Examine this diagram. It shows a simple sketch of the human nervous system. Put the correct letter in the blank to identify the part being described.

a. Part that sends and receives messages to and from all body parts **A**

b. Protected by your vertebrae **B**

c. Protected by your skull **A**

d. Carries messages from skin to spinal cord **C**

e. Spinal cord **B**

f. Body nerves **C**

# TEACH

## Concept Development

▶ Explain to students that when the spinal cord is injured, nerve function is affected below the point of injury. Incoming messages from sensory neurons cannot reach the brain, and messages from the brain cannot reach the motor neurons. The degree to which nerve function is diminished is related to the severity of the injury. A severed spinal cord results in complete paralysis below the point of injury. Neurons in the central nervous system do not repair themselves.

▶ You may want to introduce the terms *central nervous system* and *peripheral nervous system* at this point. The brain and spinal cord comprise the central nervous system. Body nerves comprise the peripheral nervous system. Some students may have heard of peripheral vision. Peripheral vision refers to vision toward the side. Peripheral nerves are those that extend from the spinal cord toward both sides of the body.

## Guided Practice

Have students write down their answers to the margin question: Three major parts of the nervous system are the brain, spinal cord, and body nerves.

## Independent Practice

**Study Guide**/*Teacher Resource Package*, p. 86. Use the Study Guide worksheet shown at the bottom of this page for independent practice.

## GLENCOE TECHNOLOGY

**Videodisc**

**Biology: The Dynamics of Life**

Animation: *Impulse Transmission: Neuron*

Disc 2, Side 1, Ch. 37

Animation: *Impulse Transmission: Synapse*

Disc 2, Side 1, Ch. 38

# TEACH

## Concept Development

▶ Have students speculate as to the advantage of the brain having a folded surface rather than a flat surface. Surface area is greatly expanded by the folds.

▶ Review the terms *voluntary* and *involuntary.* Ask students why we say "control your tongue" or "control your emotions." These actions are voluntary, and therefore we can control them.

▶ A summary of cerebrum anatomy and function at this point should emphasize that the cerebrum has left and right sides. Each side controls the opposite side of the body. The cerebrum is also divided into front and back halves.

## Demonstration

Using a brain model or a preserved brain, locate for students the areas of the brain: cerebrum, cerebellum, medulla. As you point out these areas, remind students of their functions.

## Independent Practice

**Study Guide**/*Teacher Resource Package,* p. 87. Use the Study Guide worksheet shown at the bottom of this page for independent practice.

## ACTIVITY/Software

*The Nervous System and the Sense Organs,* Queue, Inc.

## Connection: Social Studies

Have students determine the nature of the following White House occupants' nervous or endocrine system problems: F. D. Roosevelt (polio), John Kennedy (Addisons disease), Barbara Bush (thyroid problems).

### Skill Check

**Understand science words: cerebrum.** The word part *cerebr* means brain. In your dictionary, find three words with the word part *cerebr* in them. *For more help, refer to the **Skill Handbook**, pages 706-711.*

## Jobs of the Brain

The brain, like other body organs, has several different parts. Each part has a different job. The human brain has three main parts. Let's look at each part. They are the cerebrum (suh REE brum), cerebellum (ser uh BEL uhm), and medulla (muh DUL uh). The **cerebrum** is the brain part that controls thought, reason, and the senses. Look at Figure 15-9. Most of what you see is the cerebrum. It is the largest part of the brain. It looks like a huge walnut because of its folds.

The cerebrum has many jobs. One of its jobs is to store messages. We call stored messages memory. The cerebrum receives messages from all the sense organs. For example, sounds may be interpreted as music, laughter, or a whistle. Sights may be interpreted as brightly colored flowers or dark thunder clouds. The cerebrum is also the center for muscle control. Messages for moving the arms and legs start in the cerebrum. Messages about pain or touch end up in the cerebrum. The cerebrum also controls personality.

Many jobs of the cerebrum are voluntary. Remember from Chapter 14 that voluntary means you have control. For example, you decide if you want to move a toe or foot.

Figure 15-9 also shows a top view of the cerebrum. It is divided into left and right sides. The left side of the cerebrum controls the right side of the body. The left side also receives messages coming from the right side of the body. The right side of the cerebrum controls and receives messages from the left side of the body.

**Figure 15-9** The three main parts of the brain are the cerebrum, cerebellum, and medulla. Which part controls involuntary jobs only? the medulla

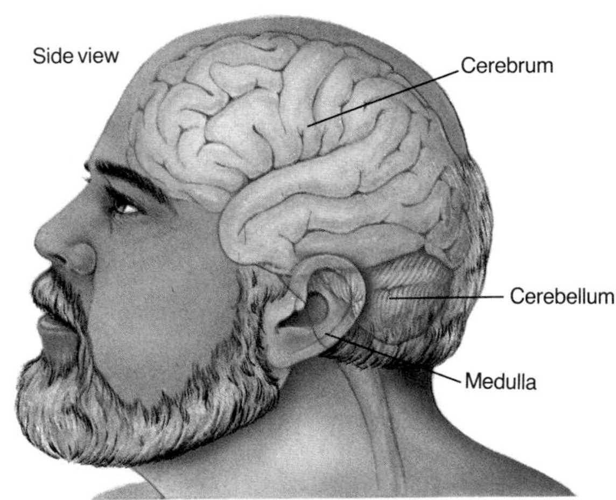


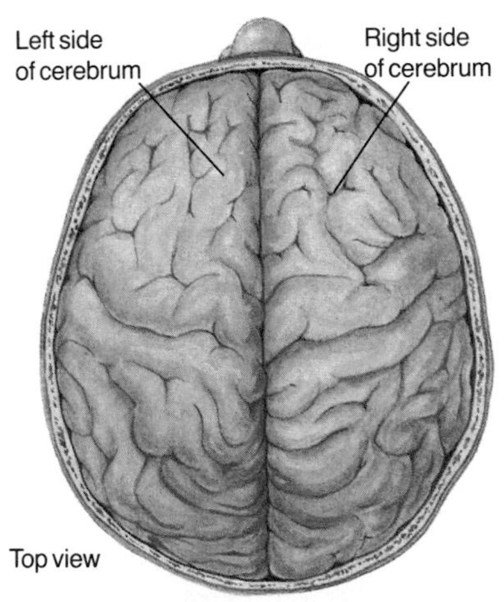




## OPTIONS

**Enrichment:** The human cerebrum has four main regions. They are the frontal lobe, which controls speech and sorting of senses; parietal lobe, which controls body awareness; occipital lobe, which controls vision; temporal lobe, which controls hearing, reading, and smell.

**Which Side of the Brain Is Dominant?**/*Lab Manual*, pp. 119-122. Use this lab as an extension to brain anatomy and function.

**STUDY GUIDE 87**

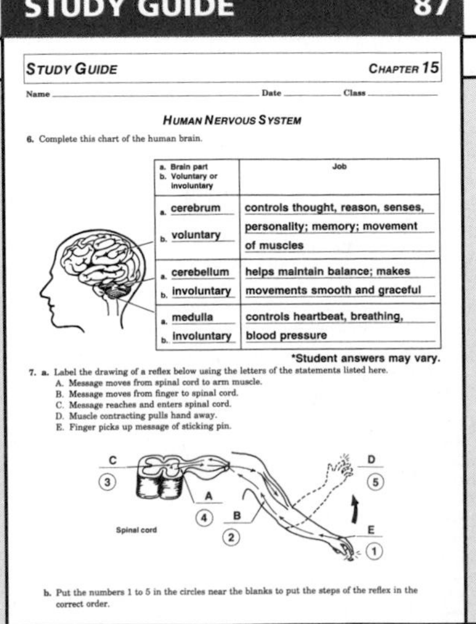

STUDY GUIDE — CHAPTER 15

Name ______ Date ______ Class ______

HUMAN NERVOUS SYSTEM

6. Complete this chart of the human brain.

| a. Brain part<br>b. Voluntary or involuntary | Job |
|---|---|
| a. cerebrum<br>b. voluntary | controls thought, reason, senses, personality; memory; movement of muscles |
| a. cerebellum<br>b. involuntary | helps maintain balance; makes movements smooth and graceful |
| a. medulla<br>b. involuntary | controls heartbeat, breathing, blood pressure |

*Student answers may vary.

7. a. Label the drawing of a reflex below using the letters of the statements listed here.
   A. Message moves from spinal cord to arm muscle.
   B. Message moves from finger to spinal cord.
   C. Message reaches and enters spinal cord.
   D. Muscle contracting pulls hand away.
   E. Finger picks up message of sticking pin.



b. Put the numbers 1 to 5 in the circles near the blanks to put the steps of the reflex in the correct order.

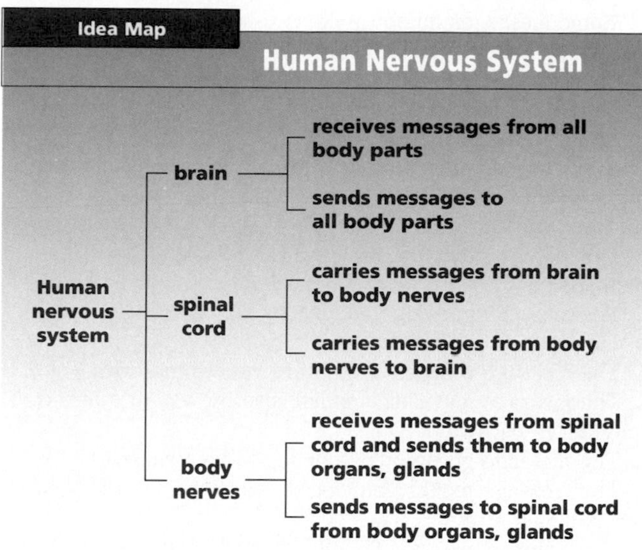


**Study Tip:** Use this idea map as you study the three main parts of the human nervous system. What are the main jobs of the brain, spinal cord, and body nerves?

Locate the cerebellum in Figure 15-9. The **cerebellum** is the brain part that helps make your movements smooth and graceful, rather than robotlike. How does the size of the cerebellum compare to that of the cerebrum? Where is the cerebellum located in relation to the front and back of your head? All nerves that enter or leave the brain on their way to and from the muscles deliver messages to the cerebellum. The cerebellum helps you keep your balance. The cerebellum's actions are involuntary. This means that you can't control them.

The third part of the brain is the medulla. The medulla may look as if it were part of the spinal cord, but its job is very different. The **medulla** is the brain part that controls heartbeat, breathing, and blood pressure. All jobs handled by the medulla are involuntary. The medulla works without your thinking about it. When was the last time you thought about keeping your heart beating?

## Mini Lab

**What Is an Eye Reflex?**

**Observe:** Have a partner cover one eye. Shine a small flashlight into the other eye. Uncover the first eye. Are the pupils the same size? *For more help, refer to the **Skill Handbook,** pages 704-705.*

## Reflexes

In most cases, any message received by your body must get to the brain before you can react to it. Think of what happens when you see a ball flying toward you. You raise your arms or duck after your brain gets the message that the ball is coming toward you.

What is a reflex?



**APPLICATION 15**

*APPLICATION: SAFETY* *CHAPTER 15*

Name ______ Date ______ Class ______

*REFLEXES PROTECT YOU* *Use after Section 15:2.*

To ride a bike safely, take part in sports, work with tools, or drive a car, you need to react quickly to changes around you. Your nervous system sends messages to your brain to tell you about the changes. Then, your brain sends messages to your body to tell it what to do. Sometimes, you need to react even more quickly. You know that some messages can cause you to react without going to the brain. These reflexes go to the spinal cord and back to the muscles.

1. You blink or duck when a ball or other object flies toward your face. How does this reflex protect you? **Blinking or ducking prevents the object from striking the eyes and injuring them.**
2. Pulling away from a sharp object or from something hot are also reflexes. How does this protect you? **prevents injury or burns**
3. What are some other reflexes? How do they protect you or help keep you safe? **Answers will vary.**
4. You can experience some reflex responses. Sit on a chair with one leg crossed over the other like the girl in the picture. With your fingers, find the soft place below your kneecap. Hit the spot gently but firmly with the edge of your hand as shown.
   a. Describe what happens to your leg. **the leg kicks upward**
   b. How could this reflex help keep you safe? **Answers will vary but students might suggest that the action helps avoid injury from a blow to the knee or allows a quick response.**
5. Tickle your nostrils gently with a fluffy ball of clean cotton.
   a. Describe your response. **Response will probably be a sneeze.**
   b. How could this response protect you? **The sneeze prevents the object that caused it from entering the nose.**

## OPTIONS

**What Are the Functions of the Brain?**/*Lab Manual,* pp. 123-126. Use this lab as an extension to the brain and its functions.

**Application**/*Teacher Resource Package,* p. 15. Use the Application worksheet shown here to teach students how reflexes are protective.

# TEACH

## Idea Map

Have students use the idea map as a study guide to the major concepts of this section.

## Mini Lab

Both eyes will respond to the stimulus even though only one eye is receiving the stimulus. As a variation, the classroom lights can be turned off and then turned back on instead of using the flashlight.

**ASSESSMENT**

**Skill:** Have students cover one eye and look to the far left with the uncovered eye. Have them remove the cover and observe the covered eye. In which direction is it looking? Why is this *not* a reflex?

## Guided Practice

Have students write down their answers to the margin question: Reflexes are involuntary, they happen quickly, they may not involve the brain. Most reflexes are helpful.

**Cooperative Learning:** Ask students to complete the following, working in groups: one student should be designated the "artist or recorder" and one, the presenter. The task is as follows: (a) draw a diagram of the brain onto a transparency, (b) label the three brain areas, (c) indicate which areas are under voluntary or involuntary control, (d) list functions for each brain area. Have the presenter of each group show the group's transparency to the rest of the class via the overhead projector.

## Guided Practice

Ask students to list as many functions as possible under the following four nervous system headings: Spinal Cord, Cerebrum, Cerebellum, Medulla. Allow them to use their textbook as a reference.

## TEACH

### Check for Understanding

Have students respond to the first three questions in Check Your Understanding.

### Reteach

**Reteaching**/*Teacher Resource Package,* p. 44. Use this worksheet to give students additional practice with parts of the brain.

**Extension:** Assign Critical Thinking, Biology and Reading, or some of the **OPTIONS** available with this section.

### *Student Journal*

Have students sequence the events that occur when a finger is stuck with a pin. They may use steps 1-6 as a guide. Have them also outline the path that the pain message would take to the brain. Note: The pain center is located in the cerebrum.

## 3 APPLY

### Connection: Health

Have a health professional such as nurse or doctor speak to the class about head injuries.

## 4 CLOSE

### Videocassette

Show the videocassette *Our Talented Brain,* Films for the Humanities and Sciences.

### *GLENCOE* TECHNOLOGY

**Videodisc**

The Secret of Life

*Reflex Arc—Anatomy*

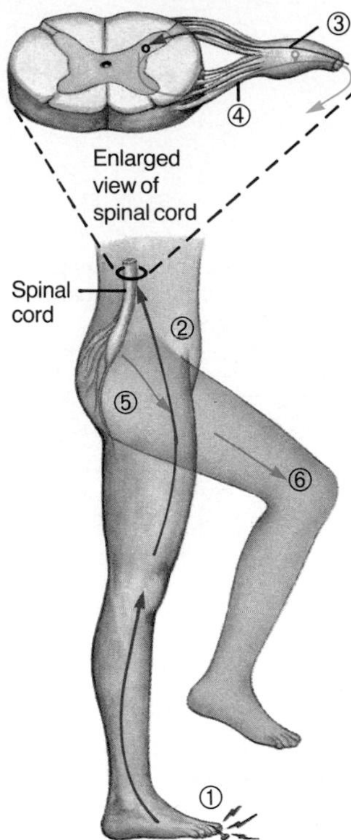


**Figure 15-10** In a reflex, the message moves from the body part to the spinal cord and back to the body part.

Some messages do not make it to the brain. They go into the spinal cord and quickly back out to the muscles. The body is able to react in a very short time. Quick, protective reactions that occur within the nervous system are called **reflexes.**

How does a reflex work? Think of what happens when you accidentally step on a tack. Follow the numbered steps in Figure 15-10.

1. The skin on your foot receives the message that it has been stuck.
2. The message goes up your leg by way of a body nerve pathway.
3. The message reaches and enters your spinal cord.
4. The message leaves the spinal cord by way of a different body nerve pathway.
5. The message goes down your leg to your muscles.
6. The message makes your leg muscle contract and your foot is pulled away.

Four things are true for all reflexes. First, they are involuntary. Second, they happen very quickly. The events in Figure 15-10 took a fraction of a second. Third, reflexes may or may not involve the brain. In the tack example, your brain may have received a pain message. But, the message reached the brain *after* you pulled your foot away. If your brain had to receive the pain message before you could react, it would take more time. Injury could result or be more severe. This brings us to the fourth point about reflexes. Most reflexes are helpful. They help protect you from further harm. Coughing, blinking, and swallowing are reflexes. How does coughing or blinking protect you?

### Check Your Understanding

6. Compare the jobs of dendrites and axons.
7. Describe the locations and jobs of the brain's three parts.
8. How do reflexes help? Give two examples.
9. **Critical Thinking:** In humans, the cerebrum is very large and folded. How does having a large cerebrum help humans?
10. **Biology and Reading:** You have learned that neurons are like electrical wires. You also learned that neurons have coverings, just as electrical wires do. From this analogy, what are the coverings on neurons for?



### Answers to Check Your Understanding

6. Dendrites receive messages from neurons. Axons send messages to neurons.
7. cerebrum—top of brain, controls thought, reason, memory, muscles, and senses; cerebellum—back of brain, makes muscle movements smooth and graceful; medulla—above spinal cord, controls breathing, heartbeat, and blood pressure
8. Reflexes are quick, protective, and involuntary. Coughing and blinking are examples.
9. It allows humans to think and reason.
10. The coverings insulate neurons so they don't touch one another.

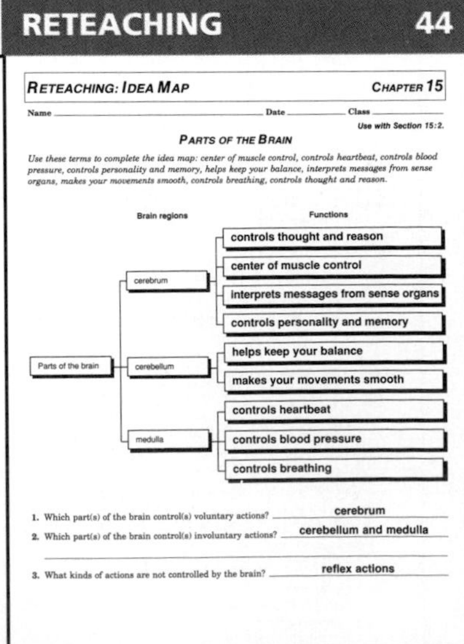

RETEACHING 44

RETEACHING: IDEA MAP CHAPTER 15

Name ____________ Date ________ Class ________

*Use with Section 15:2.*

PARTS OF THE BRAIN

*Use these terms to complete the idea map: center of muscle control, controls heartbeat, controls blood pressure, controls personality and memory, helps keep your balance, interprets messages from sense organs, makes your movements smooth, controls breathing, controls thought and reason.*

| Brain regions | | Functions |
| --- | --- | --- |
| Parts of the brain | cerebrum | controls thought and reason |
| | | center of muscle control |
| | | interprets messages from sense organs |
| | | controls personality and memory |
| | cerebellum | helps keep your balance |
| | | makes your movements smooth |
| | medulla | controls heartbeat |
| | | controls blood pressure |
| | | controls breathing |

1. Which part(s) of the brain control(s) voluntary actions? cerebrum
2. Which part(s) of the brain control(s) involuntary actions? cerebellum and medulla
3. What kinds of actions are not controlled by the brain? reflex actions

# Lab 15–1 Reaction Time

## Problem: What can change reaction time?

### Skills

form a hypothesis, calculate, infer

### Materials

metric ruler

### Procedure

Reaction time is how long it takes for a message to travel along your nerve pathways.

1. Copy the data table.
2. Have your partner hold a metric ruler at the end with the highest number.
3. Place the thumb and first finger of your left hand close to, but not touching, the end with the lowest number.
4. When your partner drops the ruler, try to catch it between your thumb and finger.
5. Record where the top of your thumb is when you catch the ruler.
6. Repeat steps 2 to 5 three more times.
7. State if you think the ruler will fall farther if you catch it with your right hand. Write your **hypothesis** in your notebook.
8. Repeat steps 2 to 5 four more times using your right hand to catch the ruler.
9. Switch roles and drop the ruler for your partner.
10. **Calculate:** To complete your data table, calculate the time in seconds needed for the ruler to fall. Multiply the distance by 0.01. Use a calculator if you have one.
11. Find the average for each column.

### Data and Observations

1. Did you catch the ruler faster with your left hand or your right hand?
2. Which hand is your writing hand?

### Analyze and Apply

1. **Check your hypothesis:** Is your hypothesis supported by your data? Why or why not?
2. **Infer:** Compare your results with several classmates. Which hand was faster at catching the ruler? Why? (HINT: Which hand do most people use to write?)
3. Why was it a good idea to run several trials?
4. **Apply:** Explain why a message moving along nerve pathways takes time.

### Extension

**Design an experiment** to show how age affects reaction time.

| TRIAL | LEFT HAND | | RIGHT HAND | |
|---|---|---|---|---|
| | DISTANCE RULER FALLS (cm) | TIME IN SECONDS | DISTANCE RULER FALLS | TIME IN SECONDS |
| 1 | 27 | 0.23 | 25 | 0.22 |
| 2 | 26 | 0.23 | 25 | 0.22 |
| 3 | 28 | 0.24 | 23 | 0.21 |
| 4 | 30 | 0.24 | 26 | 0.23 |
| Total | 111 | 0.94 | 99 | 0.88 |
| Average | 27.75 | 0.24 | 24.75 | 0.22 |



## ANSWERS

### Data and Observations

1. Answers will vary.
2. Most students will say their right hand.

### Analyze and Apply

1. Students who are right-handed and hypothesized the ruler would not fall as far when they tried to catch it with their right hand will probably say their hypotheses were supported.
2. In general, best reaction times will correlate with handedness of student.
3. Times may vary somewhat, and it is best to take an average.
4. Time is needed for the message to move from one neuron to the next.

## Lab 15-1 Reaction Time

### Overview

This lab will provide students with a simple means of measuring reaction time.

**Objectives:** Upon completion of this lab, students will be able to (1) **state** the meaning and causes of reaction time, (2) **measure** reaction time under different conditions, (3) **recognize** that reaction times are not always the same.

**Time Allotment:** 30 minutes

### Preparation

**Lab 15-1 worksheet**/*Teacher Resource Package*, pp. 57-58.

**Lab 15-1 Computer Program**/*Teacher Resource Package*, p. xviii.

### Teaching the Lab

▶ During the prelab have students practice holding the ruler correctly.

▶ If a student is unable to catch the ruler in time, record this as 35 cm.

▶ A more accurate but more complex method of calculating elapsed time may be used with students who have better math skills. It is as follows: (1) Multiply the distance in cm by 2. (2) Divide the result in step 1 by 1000. (3) Calculate the square root of the result in step 2.

**Cooperative Learning:** Divide the class into groups of two students. For more information, see pp. 22T-23T in the Teacher Guide.

### ASSESSMENT

**Performance:** Have students prepare an enlarged diagram of Figure 15-7 with these three changes: (1) make the drawing from the waist up, (2) leave the head off, (3) use Figure 15-8 for the head portion. Use arrows to trace the pathway of nerves that were used in this lab.

## 15:3 The Role of the Endocrine System

## PREPARATION

### Materials Needed

Make copies of the Enrichment, Critical Thinking, Study Guide, and Reteaching worksheets in the *Teacher Resource Package.*

### Key Science Words

endocrine system
hormone
pituitary gland
thyroid gland
thyroxine

## 1 FOCUS

▶ The objectives are listed on the student page. Remind students to preview these objectives as a guide to this numbered section.

### Brainstorming

Pose the following questions to the class: (a) If messages from the endocrine system do not use a nerve pathway, what other pathway for distribution may be available in the body? (b) How might the speed of messages not carried by nerves compare to those carried by nerves?

### Guided Practice

Have students write down their answers to the margin question: Hormones are chemicals made in one part of an organism that affect other parts of the organism.

**GLENCOE TECHNOLOGY**

**Videodisc**
The Secret of Life
*Endocrine System—Glands*

## 15:3 The Role of the Endocrine System

**Objectives**

6. **Explain** the function of the endocrine system.
7. **Relate** the importance of the pituitary and thyroid glands.

**Key Science Words**

endocrine system
hormone
pituitary gland
thyroid gland
thyroxine

What are hormones?

Many animals have an additional system for sending messages through their bodies. This system does not use nerve cells. It uses chemicals formed in special glands.

### Chemical Control

The second system that allows different parts of your body to keep in touch is called the endocrine (EN duh krin) system. The **endocrine system** is made of small glands that make chemicals for carrying messages through the body. Endocrine glands are found throughout the body. The chemicals made by endocrine glands are called hormones (HOR mohnz). **Hormones** are chemicals made in one part of an organism that affect other parts of the organism. Hormones are released into the blood. Once in the blood, hormones travel to different organs of the body. Changes take place in the organs when they receive the chemical messages that hormones carry.

Figure 15-11 shows the main endocrine glands in the human body. Note that their jobs, or functions, are also given. Some of these glands will be studied in greater detail in the next few sections.

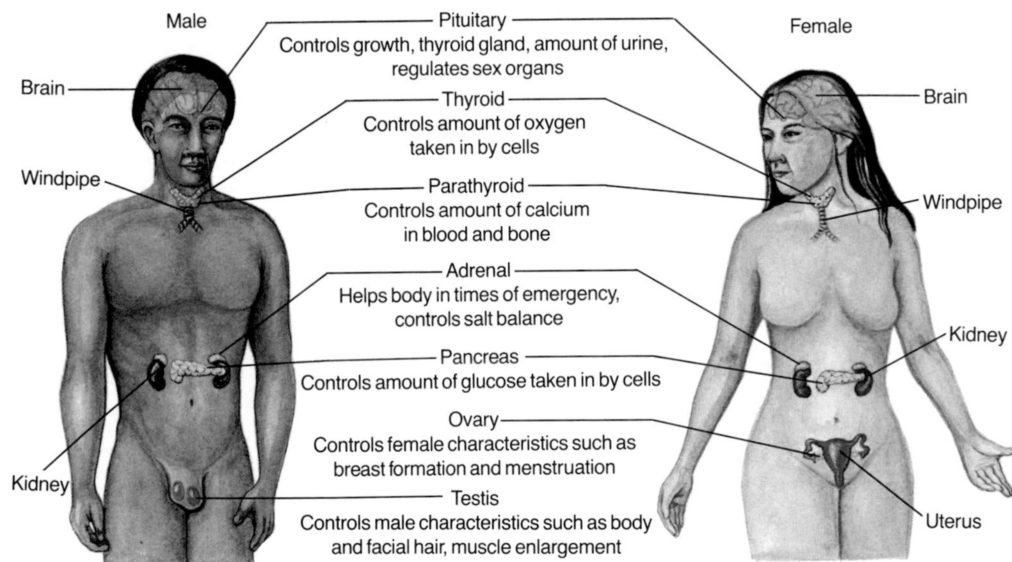


**Figure 15-11** The endocrine system is made up of several glands and organs that control different body functions.



## OPTIONS

### Science Background

The endocrine system consists of a series of ductless glands. Hormones move directly into fluid that bathes body cells. Chemicals made by ducted glands, such as salivary glands, are sent directly to the target organ. Ducted glands are called exocrine glands.

**Enrichment**/*Teacher Resource Package,* p. 15. Use the Enrichment worksheet shown here as an extension to endocrine system functions.

**ENRICHMENT 15**

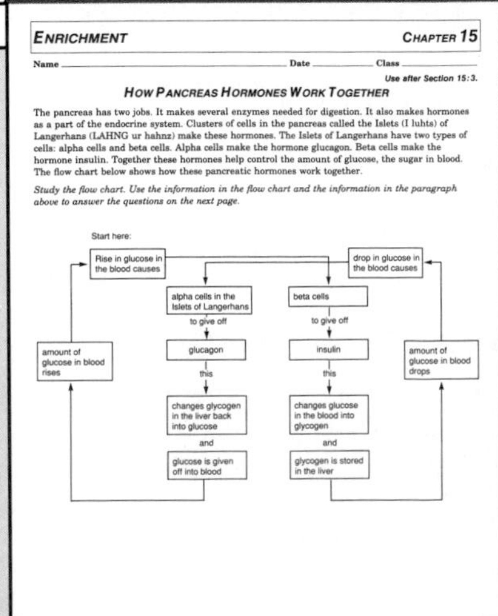

ENRICHMENT CHAPTER 15

Name ______ Date ______ Class ______

*Use after Section 15:3.*

**HOW PANCREAS HORMONES WORK TOGETHER**

The pancreas has two jobs. It makes several enzymes needed for digestion. It also makes hormones as a part of the endocrine system. Clusters of cells in the pancreas called the Islets (I luhts) of Langerhans (LAHNG ur hahnz) make these hormones. The Islets of Langerhans have two types of cells: alpha cells and beta cells. Alpha cells make the hormone glucagon. Beta cells make the hormone insulin. Together these hormones help control the amount of glucose, the sugar in blood. The flow chart below shows how these pancreatic hormones work together.

*Study the flow chart. Use the information in the flow chart and the information in the paragraph above to answer the questions on the next page.*

Start here:
Rise in glucose in the blood causes
drop in glucose in the blood causes
alpha cells in the Islets of Langerhans
beta cells
to give off
to give off
glucagon
insulin
this
this
changes glycogen in the liver back into glucose
changes glucose in the blood into glycogen
and
and
glucose is given off into blood
glycogen is stored in the liver
amount of glucose in blood rises
amount of glucose in blood drops

## The Pituitary Gland

The **pituitary** (puh TEW uh ter ee) **gland** is an endocrine gland that forms many different hormones. It is often called the master gland because it makes hormones that regulate other endocrine glands. The pituitary forms more hormones than any other endocrine gland in the body. These hormones control many organs, other endocrine glands, and different body functions. Figure 15-12 shows where the pituitary gland is located in the human body. It is located just below the brain.

As a master gland, the pituitary has many jobs. For example, it makes hormones that control body growth. It causes a person to reach sexual maturity, also called puberty. This gland helps to control the amount of urine that you form.

Let's take a closer look at one hormone made by the pituitary gland. This hormone controls body growth and is called growth hormone. How did biologists find out that the pituitary controls body growth? They used the scientific method. Their experiments are shown in Figure 15-13. Look over their work.

Rats with their pituitary glands removed did not grow. Those with a pituitary gland did grow. The rats that had their pituitary glands removed and received injections of pituitary gland hormone grew like normal rats.

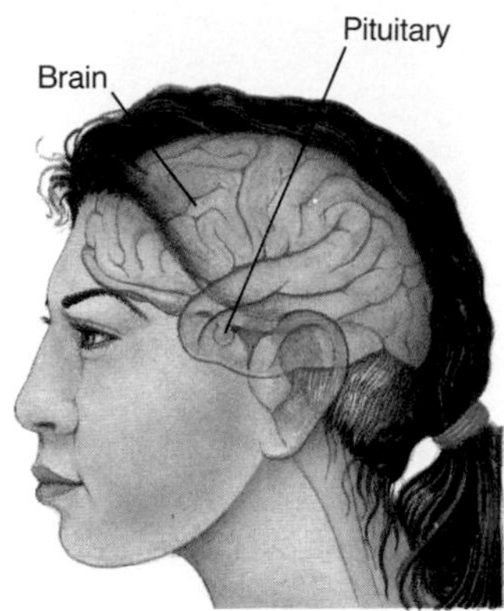


**Figure 15-12** The pituitary gland is located below the brain.

**How is body growth controlled?**

**Figure 15-13** The experiment pictured here showed that growth hormone from the pituitary gland controlled growth.

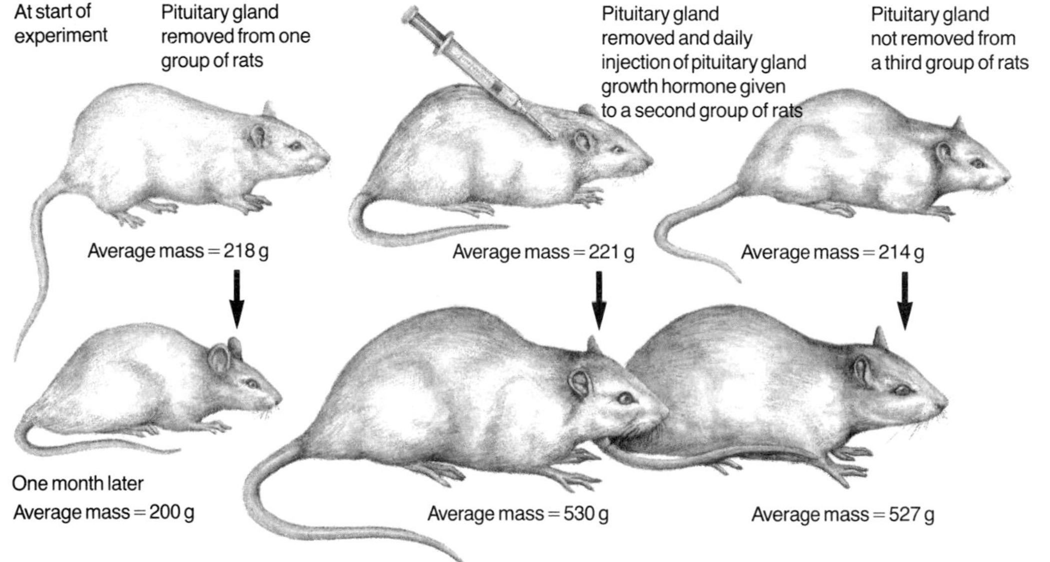




# 2 TEACH

### MOTIVATION/Brainstorming

Ask students to list similarities and differences between the endocrine and nervous systems. Suggest similarities in function, composition, and location. Differences could be related to size, speed of action, and how messages are carried.

### Concept Development

▶ Explain that the term *endocrine* comes from Greek, meaning inside (*endo-*) and to separate (*crine*). The endocrine glands are within the body and are separate tissue.

▶ Remind students that even though the pituitary forms eight different hormones, and each one is distributed by the blood, each hormone has a different body part or organ that it influences.

### Guided Practice

Have students write down their answers to the margin question: Body growth is controlled by hormones made in the pituitary gland.

### MOTIVATION/Demonstration

A mannequin would help show the locations of the endocrine glands.

### Independent Practice

**Study Guide**/*Teacher Resource Package*, p. 88. Use the Study Guide worksheet shown at the bottom of this page for independent practice.

**CRITICAL THINKING 15**

CRITICAL THINKING/PROBLEM SOLVING — CHAPTER 15

Name ______ Date ______ Class ______

*Use after Section 15:3.*

***HOW DOES THE BRAIN AFFECT THE AMOUNT OF URINE MADE BY THE KIDNEYS?***

Your pituitary gland produces a hormone called antidiuretic hormone, or ADH. The word *antidiuretic* (ANT i DI yew RET ihk) means "against water loss." The following diagrams explain how this hormone works. The diagrams show how ADH keeps enough water in the body.

*Study the diagrams and answer the questions that follow.*

Person drinks very little water for 24 hours. Pituitary gland. Brain detects blood is low in water. ADH. Result: Pituitary releases ADH into blood. ADH causes kidney to return more water to blood. Low urine output results.

Person drinks very much water for 24 hours. Pituitary gland. Brain detects blood contains much water. Result: Pituitary does not release ADH into blood. Kidney returns little water to blood. High urine output results.

1. What gland produces ADH? pituitary
2. What organ does ADH affect? kidney
3. When will ADH be released? Circle the correct answer.
   a. when the amount of water drunk is (very low) / very high
   b. when the blood contains more / (less) water than usual
4. How does the body respond to ADH?
   a. Kidneys respond to ADH by returning more water to the blood.
   b. This results in a urine output that is low.
   c. Is body water being conserved? yes
5. How does the body respond to a lack of ADH?
   a. Kidneys respond to a lack of ADH by returning less water to the blood.
   b. This results in a urine output that is high.
   c. Is body water being conserved? no

**STUDY GUIDE 88**

STUDY GUIDE — CHAPTER 15

Name ______ Date ______ Class ______

***THE ROLE OF THE ENDOCRINE SYSTEM***

*In your textbook, read about the endocrine system in Section 15:3.*

1. On the blanks below, write the name and job of the glands being shown.

brain, windpipe, kidney, uterus, Female, Male, scrotum

a. pituitary — controls growth; makes hormones that effect kidney and sex organs; makes eight different hormones
b. thyroid — makes thyroxine; controls how fast cells use food
c. parathyroid — controls balance of calcium
d. adrenal — helps body in time of emergency
e. pancreas — controls the amount of sugar in the blood; makes insulin
f. ovary — controls female characteristics
g. testis — controls male characteristics

## OPTIONS

**Critical Thinking/Problem Solving**/*Teacher Resource Package*, p. 15. Use the Critical Thinking/Problem Solving worksheet shown here to give students practice solving problems regarding the pituitary gland.

**The Endocrine System**/*Transparency Package*, number 15b. Use color transparency number 15b as you teach the functions of the endocrine system.

## TEACH

### Idea Map

Have students use the idea map as a study guide to the major concepts of this section.

### Independent Practice

**Study Guide**/*Teacher Resource Package,* p. 90. Use the Study Guide worksheet shown at the bottom of this page for independent practice.

### Check for Understanding

Have students respond to the first three questions in Check Your Understanding.

### Reteach

**Reteaching**/*Teacher Resource Package,* p. 45. Use this worksheet to give students additional practice with the endocrine glands.

**Extension:** Assign Critical Thinking, Biology and Reading, or some of the **OPTIONS** available with this section.

## 3 APPLY

Have students research what mineral is present in thyroxine and why this mineral is important in the diet.

## 4 CLOSE

### Videocassette

Show the video *Messengers,* Films for the Humanities and Sciences.

### Answers to Check Your Understanding

11. It sends hormone messages through the body.
12. It releases human growth hormone, which triggers growth.
13. The thyroid gland is an endocrine gland that forms the hormone thyroxine.
14. the thyroid; If thyroxine production is low, the person may gain weight.
15. In this case, master means a gland that controls other glands.

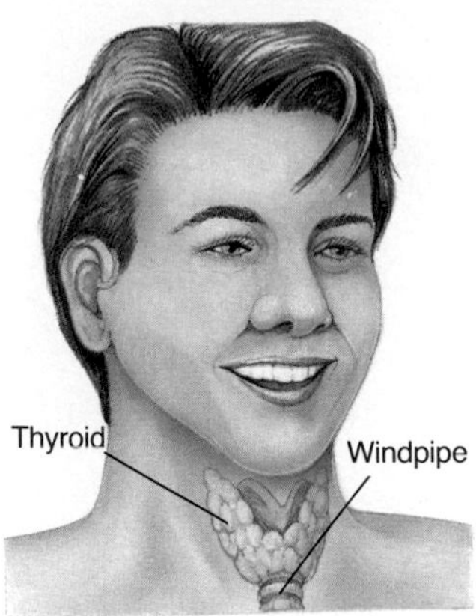


**Figure 15-14** The thyroid gland makes a hormone that controls how fast your cells release energy from food.

## The Thyroid Gland

The **thyroid** (THI royd) **gland** is an important endocrine gland found near the lower part of your neck. It lies in front of the windpipe and is about the size of your ear. Figure 15-14 shows its location. The thyroid's job is to make thyroxine (thi RAHK sun). **Thyroxine** is the hormone that controls how fast your cells release energy from food.

Sometimes the body may form too little or too much of a hormone. If the thyroid makes too little thyroxine, a person may gain weight and feel tired. How is this problem related to what thyroxine does? Could it be that food does not release enough energy? If the thyroid makes too much thyroxine, a person may lose weight and feel nervous. How could this problem be related to thyroxine's job?

**Study Tip:** Use this idea map as you study the human endocrine system. What are the jobs of the endocrine system glands?

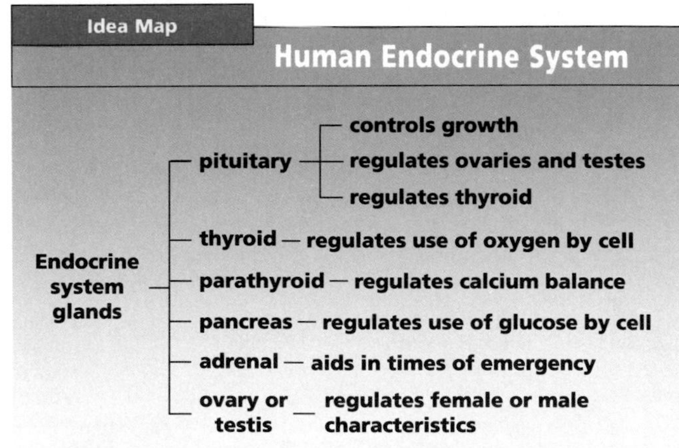


### Check Your Understanding

11. What is the job of your endocrine system?
12. How is the pituitary gland important for growth?
13. How are the thyroid gland and thyroxine related?
14. **Critical Thinking:** If a person is overweight and can't lose weight through diet and exercise, what endocrine gland should he or she have checked? Why?
15. **Biology and Reading:** You learned that the pituitary is sometimes called the master gland. What do you think the word *master* means when used in this way?

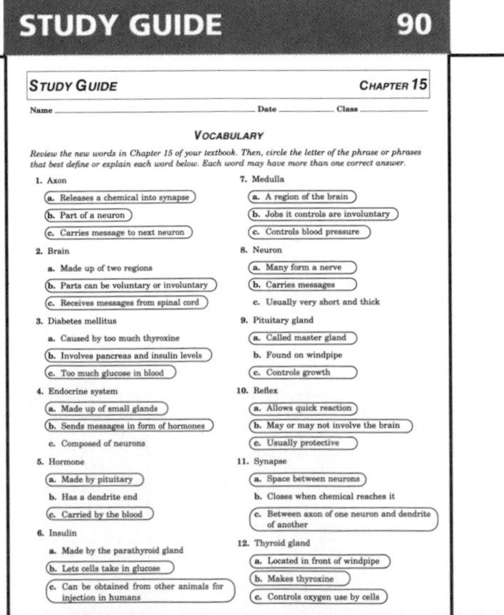

**STUDY GUIDE 90**

**STUDY GUIDE** CHAPTER 15

Name ____ Date ____ Class ____

**VOCABULARY**

*Review the new words in Chapter 15 of your textbook. Then, circle the letter of the phrase or phrases that best define or explain each word below. Each word may have more than one correct answer.*

1. Axon
   a. Releases a chemical into synapse
   b. Part of a neuron
   c. Carries message to next neuron
2. Brain
   a. Made up of two regions
   b. Parts can be voluntary or involuntary
   c. Receives messages from spinal cord
3. Diabetes mellitus
   a. Caused by too much thyroxine
   b. Involves pancreas and insulin levels
   c. Too much glucose in blood
4. Endocrine system
   a. Made up of small glands
   b. Sends messages in form of hormones
   c. Composed of neurons
5. Hormone
   a. Made by pituitary
   b. Has a dendrite end
   c. Carried by the blood
6. Insulin
   a. Made by the parathyroid gland
   b. Lets cells take in glucose
   c. Can be obtained from other animals for injection in humans
7. Medulla
   a. A region of the brain
   b. Jobs it controls are involuntary
   c. Controls blood pressure
8. Neuron
   a. Many form a nerve
   b. Carries messages
   c. Usually very short and thick
9. Pituitary gland
   a. Called master gland
   b. Found on windpipe
   c. Controls growth
10. Reflex
    a. Allows quick reaction
    b. May or may not involve the brain
    c. Usually protective
11. Synapse
    a. Space between neurons
    b. Closes when chemical reaches it
    c. Between axon of one neuron and dendrite of another
12. Thyroid gland
    a. Located in front of windpipe
    b. Makes thyroxine
    c. Controls oxygen use by cells

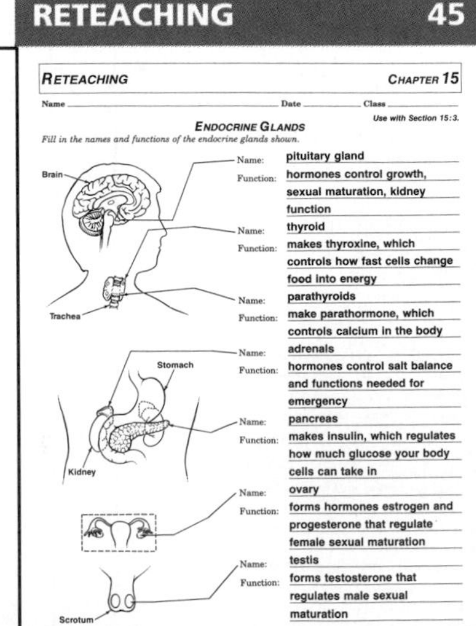

**RETEACHING 45**

**RETEACHING** CHAPTER 15

Name ____ Date ____ Class ____

*Use with Section 15:3.*

**ENDOCRINE GLANDS**

*Fill in the names and functions of the endocrine glands shown.*



Name: pituitary gland
Function: hormones control growth, sexual maturation, kidney function

Name: thyroid
Function: makes thyroxine, which controls how fast cells change food into energy

Name: parathyroids
Function: make parathormone, which controls calcium in the body

Name: adrenals
Function: hormones control salt balance and functions needed for emergency

Name: pancreas
Function: makes insulin, which regulates how much glucose your body cells can take in

Name: ovary
Function: forms hormones estrogen and progesterone that regulate female sexual maturation

Name: testis
Function: forms testosterone that regulates male sexual maturation

# 15:4 Nervous and Endocrine System Problems

Many health problems are caused by diseases of the nervous and endocrine systems. Luckily, some of these diseases can be treated or cured.

### Objectives

**8. Identify** problems that damage the brain.

**9. Explain** the importance of insulin.

### Key Science Words

insulin
diabetes mellitus

## Nervous System Problems

The brain has many blood vessels covering its surface, Figure 15-15a. The brain receives food and oxygen from these blood vessels. These blood vessels may become weak and break. When this happens, a person is said to have a stroke.

What is a stroke?

What happens when blood vessels of the brain break? It depends on where the blood vessel is and how much blood is lost. Usually, a person loses the use of a part of the brain because the brain cells die when they no longer receive food or oxygen. Losing the use of part of the brain causes the body part that it controls not to work. Figure 15-15b shows what happens when certain parts of the brain lose their oxygen supply.

## Endocrine System Problems

The pancreas is a familiar gland. You studied it in Chapter 10 with the digestive system. Refer back to Figure 15-11 to see the location of the pancreas. This gland is also part of the endocrine system. It makes a hormone called insulin (IHN suh lun). **Insulin** is a hormone that lets your body cells take in glucose, a sugar, from your blood.

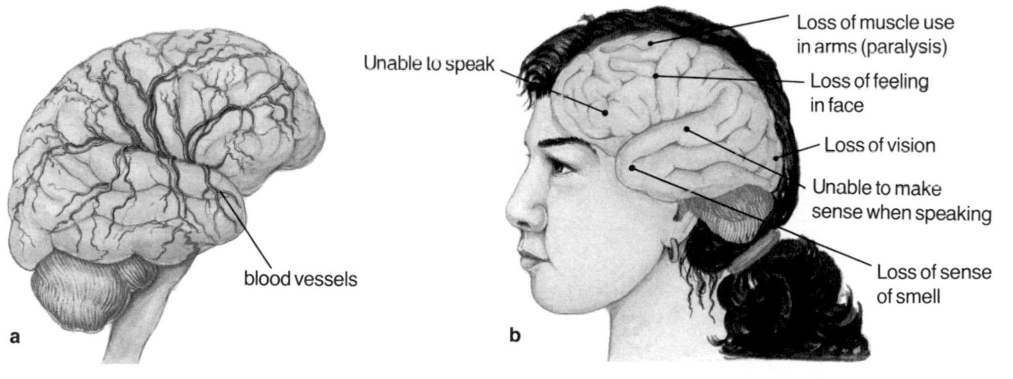


**Figure 15-15** A stroke is the breaking of blood vessels that serve the brain and can result in loss of body functions.



**SKILL 16**

*SKILL: SUMMARIZE* — CHAPTER 15

Name ______ Date ______ Class ______

*For more help, refer to the Skill Handbook, pages 706-711.* *Use after Section 15:4.*

THE ENDOCRINE SYSTEM

Summarizing is a good way to remember information, check your understanding of a subject, and review what you have studied. After you answer the following questions, you will have a three-part summary—(1) a comparison of the body's two systems of communication, (2) the functions of some organs of the endocrine system, and (3) some effects on the body when certain endocrine glands do not function properly.

*Complete the following statements by filling in the blanks with words that make the sentences correct.*

1. The body has two communication systems. One is the __nervous__ system. The other is the __endocrine__ system.
2. The nervous system uses __nerves (neurons)__ to detect changes and respond to them. The endocrine system uses __hormones__, which are chemical messengers.
3. The nervous system is made up of organs such as the __brain (spinal cord)__. The endocrine system is made up of __glands__.
4. The nervous system sends messages along __neuron__ pathways. The endocrine system sends messages through the __blood__.

*Use the diagram at the right to answer the questions below. Write the letter of the correct choice in the blank.*

5. Gland __B__ makes thyroxine.
6. Gland __C__ controls calcium balance.
7. Gland __A__ controls sexual maturity.
8. Gland __B__ controls the speed at which cells use food.
9. Gland __A__ controls body growth.

*Complete the following statements by writing **too much** or **too little** in the blanks.*

10. A person gains weight if __too little__ thyroxine is made.
11. A person may feel nervous if __too much__ thyroxine is made.
12. A person may grow __too little__ if the pituitary gland does not work.
13. Glucose may not pass into cells if __too little__ insulin is made.

## OPTIONS

### Science Background

Strokes may be caused by or brought on by hypertension, blood clots, and/or atherosclerosis of vessels associated with the cerebrum.

**Skill**/*Teacher Resource Package,* p. 16. Use the Skill worksheet shown here to give students practice summarizing concepts.

# 15:4 Nervous and Endocrine System Problems

## PREPARATION

### Materials Needed

Make copies of the Skill Study Guide, Reteaching, and Lab worksheets in the *Teacher Resource Package.*

▶ Purchase glucose test paper and prepare urine samples for Lab 15-2.

### Key Science Words

insulin
diabetes mellitus

### Process Skills

In Lab 15-2, students will experiment and interpret data.

## 1 FOCUS

▶ The objectives are listed on the student page. Remind students to preview these objectives as a guide to this numbered section.

### MOTIVATION/Demonstration

Use a brain model to review the brain's control of the body.

### Guided Practice

Have students write down their answers to the margin question: the breaking of a blood vessel that serves the brain.

### *Portfolio*

Have students trace Figure 15-8 and label the three brain areas. Then, have them add and label the pituitary gland. Using information from Figure 15-15 (b) as a guide, students should label and color code the areas responsible for: speech, muscle control, production of growth hormone, smell, vision, control of urine production, speech coordination, and detecting sensation.

## 2 TEACH

### Guided Practice

Have students write down their answers to the margin question: Insulin lets the body cells take in glucose from the blood.

### Independent Practice

**Study Guide**/*Teacher Resource Package*, p. 89. Use the Study Guide worksheet shown at the bottom of this page for independent practice.

### Check for Understanding

Have students respond to the first three questions in Check Your Understanding.

### Reteach

**Reteaching**/*Teacher Resource Package*, p. 46. Use this worksheet to give students additional practice with endocrine system problems.

**Extension:** Assign Critical Thinking, Biology and Reading, or some of the **OPTIONS** available with this section.

## 3 APPLY

### Interview

Have students research treatment and symptoms of diabetes mellitus.

## 4 CLOSE

### Software

*The Human System Series 2*, Focus Media, Inc.

### Answers to Check Your Understanding

16. Brain cells die.
17. Insulin opens cell membranes and allows sugar to enter cells.
18. failure of the pancreas to make enough insulin
19. test the urine for glucose
20. because vessels that carry food and oxygen to it have broken

**Figure 15-16** Insulin helps cells take in glucose from the blood.

What is insulin?

### Bio Tip

**Health:** Scientists can grow microbes that make human insulin. The scientists insert the chromosome material that controls the making of human insulin into the microbes. The microbes then multiply and produce large amounts of human insulin.

Figure 15-16a shows how insulin works. The blue circles stand for glucose. The red figures stand for insulin. Insulin opens the cell membrane, allowing glucose to enter the cells. Once inside your cells, glucose can be used as food.

What if the pancreas stops making insulin? If insulin is missing, glucose can't get into the cells. It remains in the blood. As a result, a person may lose weight. Not treating this problem may lead to blindness, heart disease, or death.

The problem just described has a name. **Diabetes mellitus** (di uh BEET us • MEL uht us) is a disease that results when the pancreas doesn't make enough insulin. People with this disease have too much glucose in their blood and not enough glucose in their cells. The excess glucose that is in the blood leaves the body in the urine.

### Check Your Understanding

16. What happens when the brain loses its oxygen supply?
17. How does insulin do its job?
18. What is the cause of diabetes mellitus?
19. **Critical Thinking:** If you were a doctor, how would you test someone for diabetes?
20. **Biology and Reading:** Why does the part of the brain damaged by a stroke no longer receive food or oxygen?

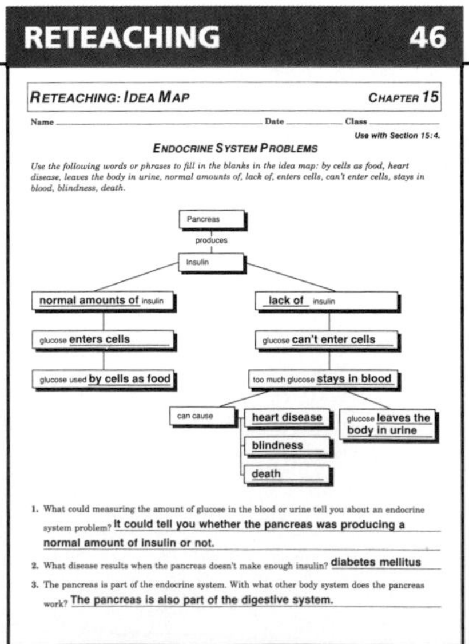

**RETEACHING 46**

*RETEACHING: IDEA MAP* — *CHAPTER 15*

Name ______ Date ______ Class ______

*Use with Section 15:4.*

*ENDOCRINE SYSTEM PROBLEMS*

*Use the following words or phrases to fill in the blanks in the idea map: by cells as food, heart disease, leaves the body in urine, normal amounts of, lack of, enters cells, can't enter cells, stays in blood, blindness, death.*



1. What could measuring the amount of glucose in the blood or urine tell you about an endocrine system problem? It could tell you whether the pancreas was producing a normal amount of insulin or not.
2. What disease results when the pancreas doesn't make enough insulin? diabetes mellitus
3. The pancreas is part of the endocrine system. With what other body system does the pancreas work? The pancreas is also part of the digestive system.

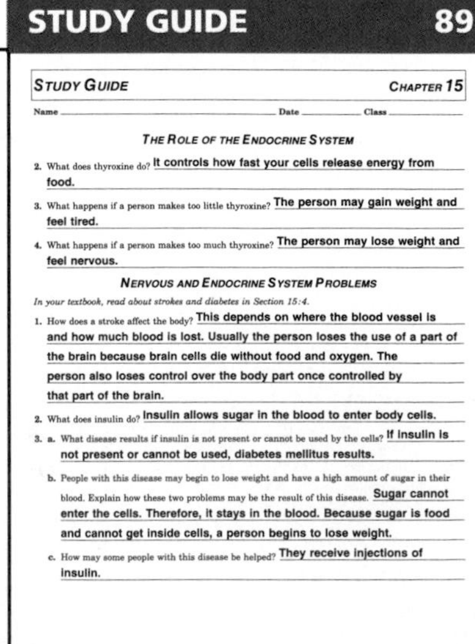

**STUDY GUIDE 89**

*STUDY GUIDE* — *CHAPTER 15*

Name ______ Date ______ Class ______

*THE ROLE OF THE ENDOCRINE SYSTEM*

2. What does thyroxine do? It controls how fast your cells release energy from food.
3. What happens if a person makes too little thyroxine? The person may gain weight and feel tired.
4. What happens if a person makes too much thyroxine? The person may lose weight and feel nervous.

*NERVOUS AND ENDOCRINE SYSTEM PROBLEMS*

*In your textbook, read about strokes and diabetes in Section 15:4.*

1. How does a stroke affect the body? This depends on where the blood vessel is and how much blood is lost. Usually the person loses the use of a part of the brain because brain cells die without food and oxygen. The person also loses control over the body part once controlled by that part of the brain.
2. What does insulin do? Insulin allows sugar in the blood to enter body cells.
3. a. What disease results if insulin is not present or cannot be used by the cells? If insulin is not present or cannot be used, diabetes mellitus results.
   b. People with this disease may begin to lose weight and have a high amount of sugar in their blood. Explain how these two problems may be the result of this disease. Sugar cannot enter the cells. Therefore, it stays in the blood. Because sugar is food and cannot get inside cells, a person begins to lose weight.
   c. How may some people with this disease be helped? They receive injections of insulin.

# Lab 15–2 Diabetes

## Problem: Is glucose found in the urine of a person with diabetes?

### Skills

experiment, interpret data

### Materials

glucose test paper
glass slides
normal urine sample (A)
urine sample from diabetic (B)
urine samples in test tubes marked C-E
test-tube rack
wax pencil
8 droppers

### Procedure

1. Copy the data table.
2. Use a wax pencil to draw two circles on a glass slide. Mark the circles A and B.
3. Add two drops of normal urine to the circle marked A. Add two drops of urine from a diabetic to the circle marked B.
4. Touch a small piece of glucose test paper to each urine sample. A green color means glucose is present. A yellow color means that no glucose is present. Record the color of the paper in your data table.
5. Draw one circle on each of three slides. Label the slides C to E.
6. Using a dropper, put two drops of urine from test tube C into the circle on slide C. Do the same with test tubes D and E.
7. Test each urine sample with glucose test paper. Record the color of the paper in your table. Complete your data table.
8. Dispose of your samples as directed by your teacher.

### Data and Observations

1. In which samples was glucose present?
2. In which samples was glucose not present?

### Analyze and Apply

1. **Interpret data:** Which samples tested could be from a normal person? A person with diabetes? Explain.
2. Explain why glucose is present in the urine of a person with diabetes.
3. Name the endocrine gland not working if a person has diabetes.
4. **Apply:** What kinds of foods should a person with diabetes avoid? Why?

### Extension

**Experiment** to test if glucose test paper will detect different amounts of glucose.

| SAMPLE | COLOR OF TEST PAPER | GLUCOSE PRESENT? | DIABETES PRESENT? |
|---|---|---|---|
| Normal urine (A) | yellow | no | no |
| Urine from diabetic (B) | green | yes | yes |
| C* | green | yes | yes |
| D | green | yes | yes |
| E | yellow | no | no |

*Answers will vary depending on samples used.



## ANSWERS

### Data and Observations

1. C, D
2. E

### Analyze and Apply

1. E; C, D; E has no glucose in it but C, D have glucose in them.
2. Glucose can't get into the body cells because there isn't enough insulin produced.
3. pancreas
4. foods high in sugars; The body cells won't be able to take in the sugars because insulin is lacking.

## Lab 15-2 Diabetes

### Overview

Students will test for the presence of glucose in urine.

**Objectives:** Upon completion of the lab, students will be able to (1) **test** for the presence of glucose in urine, (2) **recognize** glucose in urine as a symptom of diabetes mellitus.

**Time Allotment:** 30 minutes

### Preparation

▶ Precut glucose test paper into 1 cm strips.

▶ Urine samples without glucose—for every 1000 mL tap water, use six drops yellow food coloring.

▶ Urine samples with glucose—prepare as above and add 2 teaspoons of glucose or honey.

**Lab 15-2 worksheet**/*Teacher Resource Package*, pp. 59-60.

### Teaching the Lab

▶ Explain that a person with diabetes has excess glucose in his or her blood.

▶ **Troubleshooting:** Caution students not to mix droppers and to rinse and dry slides before reusing them.

**Cooperative Learning:** Divide the class into groups of two or three. For more information, see pp. 22T-23T in the Teacher Guide.

### ASSESSMENT

**Performance:** Prepare a series of unknowns marked A-D. Provide students with glucose test paper, slides, and a wax pencil. Have them determine which unknowns are from a person with diabetes.

APPLYING TECHNOLOGY

### Purpose

Students will learn about the nature of reflexes. They will have an opportunity to experiment with the knee-jerk reflex and learn that it cannot be consciously controlled.

### Process Skills

Observe, experiment, form hypotheses, communicate, interpret data, define operationally

### Time Required

One class period for using references and having the entire class do the activity when sharing one hammer. Less time will be needed to complete the activity if there are several hammers available.

### Possible Problems

Advise students to strike the area below the knee cap GENTLY. A lot of force is not required to elicit the response.

### Teaching Strategies

▶ Reflex hammers (percussion hammers) can be purchased from biological supply houses or borrowed from the school nurse or a local doctor. If only one or two hammers are available, either have one or two groups demonstrate the activity for the class or have groups rotate using the hammer while others do the reference work.

▶ You may want to find references and bring them or copies to class for student use.

▶ Reviewing the pathway of a reflex on the overhead projector or chalkboard may be helpful for some students.

▶ Ask your school nurse to visit class and discuss briefly the significance of patients who show no knee-jerk response.

TECH PREP APPLYING TECHNOLOGY

# Get a Kick Out of This

Reflexes are usually protective. This means that the body is somehow protected as a result of an occurring reflex. Think about coughing. Coughing is a reflex that you don't have to learn or think about. It occurs when food or liquid gets into your lungs. You start to cough in an attempt to bring the food or liquid up and out of the lungs. Did you have to think about starting the process of coughing? Can you stop yourself from coughing once it starts?

## Identifying the Problem

You may have heard of a reflex called the knee-jerk reflex. If lightly struck with a soft object just below the knee, your leg kicks upward. The question you are trying to answer is as follows: Can you prevent this reflex from happening if you think about it and concentrate on it not happening? Form a hypothesis that will guide you in your answering of this question.

## Collecting Information

Review the meaning of a reflex. Check on the pathway that a reflex follows through the nervous system. Be sure that you understand the role of the spinal cord in a reflex. Use a reference to check on the meaning of the terms *motor neuron* and *sensory neuron.* Study the figure on this page, and locate the motor and sensory neurons that are

**Technology Connection**

The spinal cord carries messages from the brain to all body parts and from all body parts back to the brain. When cut or damaged, neurons in the spinal cord do not repair themselves, and a person is left paralyzed. There is some new hope for people with this type of injury. Researchers have found that if spinal cord neuron cells from embryo rats are placed into the spinal cord of adult rats with spinal cord damage, the adult rats show improvement in movement. This supports the idea that some repair of spinal cord neurons has occurred.

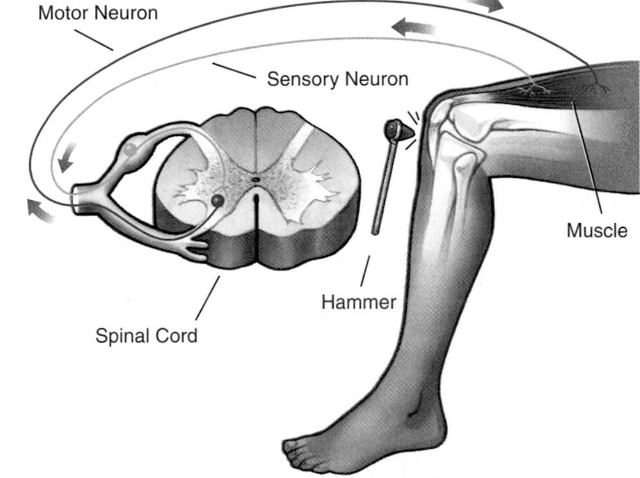




**Technology Connection**

The spinal cord of the central nervous system contains cells that produce proteins which inhibit the new growth of axon ends of neurons. Embryo cells that are transplanted into adult rats tend to ignore these proteins and thus are able to form new neuron connections.

used in a knee-jerk reflex. Describe the job of the motor and sensory neurons in this figure.

## Carrying Out an Experiment

You will need to work in a group of three to carry out this activity. Student A will record results. Student B will be the subject, and Student C will start the reflex by using the soft hammer.

1. Student B is to sit on a desk or lab table with his or her feet dangling, that is, not touching the floor.
2. Student C *lightly* taps the area just below the kneecap of Student B, using a soft hammer. See the diagram below.
3. Student A notes the reaction of Student B's leg and records the observation. It may be necessary to tap Student B several times before the reflex is seen. Record each tap as a separate trial.
4. To test your hypothesis, Student B should concentrate on not allowing the leg to jerk up when struck with the hammer. Again, Student A should record what happens. Note: use several trials rather than just one.
5. Change roles so that all students in your group have a chance to try steps 1-4.

## Assessing Your Results

Describe the knee-jerk reflex. Explain the role of the following in the knee-jerk reflex: soft hammer, sensory neuron, motor neuron, spinal cord. Did a knee jerk occur with each and every trial? Offer an explanation for why this may have happened. Were you or your teammates able to prevent the knee-jerk reflex from occurring? Offer a reason why this might be so. Was your hypothesis supported or not? Trace the pathway of the message that passes through your body during a knee-jerk reflex.

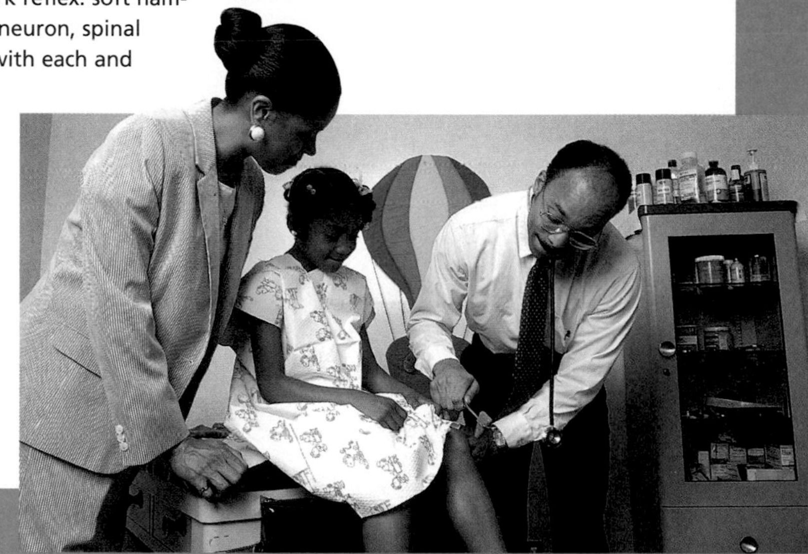

### Career Connection

- **Home Health Aide** Helps people recovering at home after a hospital stay or aids the handicapped; gives baths, massages, changes bandages and bedding; may give medicines and help with household chores
- **Nurse's Aide** Cares for patients in a hospital or nursing home; takes blood pressure, temperature, and pulse of patients
- **Radiologic Technician** Works in a hospital, doctor's office, or private laboratory; takes X rays, MRIs, and CAT scans; develops X-ray film

### Conclusions

Students will be able to explain the role of motor and sensory neurons in a reflex. They will be able to trace the path of a simple reflex through the body. They will observe a reflex and determine experimentally that it cannot be controlled consciously.

### Answers to Questions

Sensory neurons detect a stimulus (the hammer striking the knee area) and carry the message to the spinal cord. Spinal cord neurons connect the sensory neuron directly to a motor neuron. Motor neurons carry the message from the spinal cord to a muscle (the thigh muscle) causing it to contract, resulting in the leg kicking up. A knee jerk may not have occurred each time because the stimulus was too light. Students are not able to consciously control the knee-jerk reflex. Reflexes are not controlled by the brain; they occur automatically. The pathway is as follows: sensory neuron carries message to spinal cord, spinal cord sends message to leg muscle by way of motor neuron.

### ASSESSMENT

**Performance** Have students draw the pathway that a reflex would take through the body when a person touches a hot stove and the hand is immediately pulled away. Labels should be added to the diagram, and arrows should show the direction in which the message flows.

### Career Connection

**Career Path**
**Home Health Aide and Nurse's Aide** Both require a high-school diploma and on-the-job training.
**Radiologic Technician** High-school diploma plus either two or four more years of college course work and continued refresher courses for updating the newest technology.

**Career Issue**
Have students investigate whether neurons not located in the brain or spinal cord can regrow after being cut or injured. Ask how this differs from injuries to the spinal cord or brain.

## CHAPTER 15 REVIEW

### Summary

Summary statements can be used by students to review the major concepts of the chapter.

### Key Science Words

All boldfaced terms from the chapter are listed.

### Testing Yourself

#### Using Words

1. reflex
2. diabetes mellitus
3. neuron
4. medulla
5. thyroxine
6. dendrite
7. insulin
8. thyroid
9. endocrine

#### Finding Main Ideas

10. p. 318; They are involuntary, happen very quickly, may or may not involve the brain, and most are helpful.
11. p. 320; The adrenal gland controls salt balance and helps the body in times of emergency. The pituitary controls growth and is a master gland that controls other glands. The parathyroid controls the amount of calcium in the blood.
12. p. 316; The cerebrum is voluntary; the cerebellum and medulla are involuntary.
13. p. 324; It acts like a gatekeeper. Insulin "opens" the cell membrane allowing glucose to pass into the cell.
14. p. 321; Three rat groups were used. One group (control) had no treatment. The second group had the pituitary removed but received daily injections of the pituitary hormone. The third group had the pituitary removed. The first two groups grew at a normal rate. The third group did not grow at all.
15. p. 318; Skin receives message that tack sticks toe—message travels by way of neurons up leg to spinal cord—message leaves spinal cord and travels down leg to muscle—toe is pulled away.

# Chapter 15 Review

## Summary

### 15:1 The Role of the Nervous System

1. The nervous system allows animals to receive messages from their surroundings.
2. Nervous systems become more complex as animal groups become more complex.

### 15:2 Human Nervous System

3. Neurons are nerve cells. They have ends called dendrites and axons. Neurons make up nerve pathways.
4. The brain, spinal cord, and body nerves form the human nervous system.
5. Reflexes are involuntary, quick reactions of the nervous system that protect the body.

### 15:3 The Role of the Endocrine System

6. The endocrine system consists of glands that send hormones throughout the body.
7. The pituitary gland makes many hormones that control different body functions. Thyroxine is the hormone made by the thyroid gland.

### 15:4 Nervous and Endocrine System Problems

8. Blood vessels that supply the brain with food and oxygen may break, resulting in a stroke.
9. Insulin is a hormone made by the pancreas. Insulin controls glucose movement into your cells.

## Key Science Words

axon (p. 313)
brain (p. 314)
cerebellum (p. 317)
cerebrum (p. 316)
dendrite (p. 313)
diabetes mellitus (p. 324)
endocrine system (p. 320)
hormone (p. 320)
insulin (p. 323)
medulla (p. 317)
nerve (p. 312)
nervous system (p. 310)
neuron (p. 312)
pituitary gland (p. 321)
reflex (p. 318)
spinal cord (p. 315)
synapse (p. 313)
thyroid gland (p. 322)
thyroxine (p. 322)

## Testing Yourself

**Using Words**

*Choose the word from the list of Key Science Words that best fits the definition.*

1. quick reaction that is protective
2. disease caused by lack of insulin
3. special name for a nerve cell
4. brain area that controls heartbeat
5. chemical made by the thyroid gland
6. neuron end that receives messages
7. hormone made by the pancreas
8. gland found in the neck
9. system that sends chemical messages through the body



#### Using Main Ideas

16. Stinging-cell animals have nerve cells spread over their body. Complex animals such as spiders have a brain, nerve cord, and sense organs.
17. Right side of cerebrum receives and controls messages from the left side of body.
18. A message reaches the axon end of a neuron. A chemical is given off. The chemical moves across the space and restarts the message on the dendrite end of the neuron.
19. The pituitary gland is located at the base of the cerebrum; the thyroid is over the windpipe; the parathyroid is located on the thyroid; the pancreas is located below the stomach and even with the kidneys; the adrenal is on top of the kidney; the ovary is in the lower part of the body near the uterus.
20. The nervous system is made of cells that allow an animal to detect changes around it. This same system will allow an animal to respond to these changes.
21. The person loses the use of a part of the brain and the body part controlled by that part of the brain.
22. The brain receives and sends messages to all body parts. These messages enter and

# Review

## Testing Yourself *continued*

**Finding Main Ideas**

*List the page number where each main idea below is found. Then, explain each main idea.*

10. the four main features of a reflex
11. the functions of the adrenal, pituitary, and parathyroid glands
12. the brain regions that are voluntary
13. how insulin works
14. how to prove with experiments that the pituitary gland controls growth
15. the pathway that a reflex follows

**Using Main Ideas**

*Answer these questions by referring to the page number after each question.*

16. How do the nervous systems of simple and more complex animal groups differ? (p. 311)
17. Which side of the cerebrum receives and controls messages from the left side of your body? (p. 316)
18. How does a message move from one neuron to the next? (p. 313)
19. Name the locations of the pituitary, thyroid, parathyroid, pancreas, adrenal, and ovary. (p. 320)
20. How do animals keep in touch with their surroundings? (p. 310)
21. What can happen when someone has a stroke? (p. 323)
22. How do the brain and spinal cord work together? (p. 315)
23. What happens if the pancreas does not produce enough insulin? (p. 324)
24. What happens if the thyroid produces too much thyroxine? (p. 322)
25. Why are different pathways needed to carry messages from hand to brain and brain to hand? (p. 313)

## Skill Review

*For more help, refer to the **Skill Handbook,** pages 704-719.*

1. **Interpret data:** Which rat in the experiment in Figure 15-13 was the control? Why?
2. **Interpret diagrams:** Study the diagrams of neurons in this chapter. How does the shape of a neuron help it do its job?
3. **Infer:** What would happen to messages traveling along the spinal cord if the spinal cord were cut? How could this type of injury affect body parts?
4. **Observe:** Examine slides of thyroid tissue. Diagram what you see.

## Finding Out More

TECH PREP

**Critical Thinking**

1. Find out what the following diseases are and what their causes are: multiple sclerosis, cerebral palsy, epilepsy, polio.
2. What are the symptoms of diabetes mellitus?

**Applications**

1. Design and build a model of a neuron, a neuron pathway of at least three neurons, and a nerve. (HINT: Spaghetti might work well.)
2. Prepare a large diagram that shows the locations and jobs of the different areas of the cerebrum.



### Using Main Ideas

leave the brain by way of the spinal cord.

23. Diabetes mellitus may result.
24. Too much thyroxine will result in weight loss and feeling nervous.
25. Only the axon end of neurons can give off a chemical that keeps the messages moving from one neuron to the next. Thus, messages cannot travel in both directions along the same pathway.

### Skill Review

1. Rats in which the pituitary gland was not removed and no growth hormone was added were the control.
2. Its long thin shape enables electrical impulses to travel quickly along its length. The ends (axon and dendrite) are close together to allow the impulses to go from one neuron end to the next neuron easily.
3. They would be disrupted because nerves do not repair themselves, unlike other cells. When a body part is not used, it shrinks (atrophies).
4. The diagram will have open spaces surrounded by small cells that will stain dark.

### Finding Out More

**Critical Thinking**

1. Multiple sclerosis results when the covering of nerve cells is somehow destroyed. Cerebral palsy is due to brain damage at birth and may result when oxygen is cut off to the brain during delivery. Epilepsy occurs when the electrical patterns of the central nervous system are disrupted. Polio is caused by a virus that damages the protective covering of nerve cells.
2. Symptoms of diabetes mellitus include: excess sugar in blood and urine, thirst, hunger, and weight loss.

**Applications**

1. Students could lay spaghetti end to end and glue it to construction paper.
2. Students should include the cerebrum, cerebellum, and medulla.

**Chapter Review**/*Teacher Resource Package,* pp. 77-78.

**Chapter Test**/*Teacher Resource Package,* pp. 79-81.

**Chapter Test**/*Computer Test Bank*

# Senses

## PLANNING GUIDE

| CONTENT | TEXT FEATURES | TEACHER RESOURCE PACKAGE | OTHER COMPONENTS |
|---|---|---|---|
| (1 day)<br>16:1 Observing the Environment<br>Senses Aid Survival<br>Sense Organs of Animals | Check Your Understanding, p. 333 | Reteaching: *Animal Sense Organs,* p. 47<br>Study Guide: *Observing the Environment,* pp. 91-92 | **STVS:** *How Bats Hear,* Human Biology (Disc 5, Side 2) |
| (3 1/2 days)<br>16:2 Human Sense Organs<br>The Eye<br>The Tongue and Nose<br>The Ear<br>The Skin | Skill Check: *Interpret Diagrams,* p. 341<br>Skill Check: *Understand Science Words,* p. 343<br>Mini Lab: *What Is an Afterimage?* p. 335<br>Lab 16-1: *The Eye,* p. 337<br>Lab 16-2: *The Senses,* p. 338<br>Career Close-Up: *Licensed Practical Nurse,* p. 342<br>Idea Map, p. 340<br>Check Your Understanding, p. 343 | Enrichment: *Animal Sense Organs,* p. 17<br>Reteaching: *The Human Eye,* p. 48<br>Critical Thinking/Problem Solving: *How Are a Camera and a Human Eye Similar?* p. 16<br>Focus: *An Electrical Sense Organ,* pp. 31-32<br>Study Guide: *Human Sense Organs,* pp. 93-94<br>Lab 16-1: *The Eye,* pp. 61-62<br>Lab 16-2: *The Senses,* pp. 63-64<br>Transparency Master: *Human Sense Organs I and II,* pp. 75-76 | **Laboratory Manual:**<br>*How Are Your Senses Sometimes Fooled?* p. 131<br>**Laboratory Manual:**<br>*How Can You Test Your Senses?* p. 127<br>**Laboratory Manual:**<br>*What Are the Parts of an Eye?* p. 135<br>Color Transparency 16: *Pathway of Light Through the Eye*<br>**STVS:** *Sounds Made by the Ear,* Human Biology (Disc 7, Side 2) |
| ( 1 1/2 days)<br>16:3 Problems With Sense Organs<br>Correcting Vision Problems<br>Protecting Against Hearing Loss<br>Correcting Hearing Problems | Check Your Understanding, p. 347 | Application: *Skin Disorders,* p. 16<br>Reteaching: *Problems with Sense Organs,* p. 49<br>Skill: *Infer,* p. 17<br>Study Guide: *Problems With Sense Organs,* p. 95<br>Study Guide: *Vocabulary,* p. 96 | **STVS:** *Hearing by Touch,* Human Biology (Disc 7, Side 2)<br>**STVS:** *Low Vision Clinic,* Human Biology (Disc 7, Side 2) |
| Chapter Review | Summary<br>Key Science Words<br>Testing Yourself<br>Finding Main Ideas<br>Using Main Ideas<br>Skill Review | **ASSESSMENT RESOURCES**<br>Chapter Review, pp. 82-83<br>Chapter Test, pp. 84-86<br>Performance Assessment in the Biology Classroom<br>Alternate Assessment in the Science Classroom<br>Computer Test Bank | |

## GLENCOE TECHNOLOGY

**Infinite Voyage,** *Miracles by Design*
**Science and Technology Videodisc Series,** *How Bats Hear,* Human Biology (Disc 5, Side 2)
*Sounds Made by the Ear,* Human Biology (Disc 7, Side 2)
*Hearing by Touch,* Human Biology (Disc 7, Side 2)
*Low Vision Clinic,* Human Biology (Disc 7, Side 2)

## MATERIALS NEEDED

| LAB 16-1, p. 337 | LAB 16-2, p. 338 | MARGIN FEATURES |
|---|---|---|
| ruler<br>scissors<br>clear tape<br>white paper<br>black paper<br>top part of hand lens | white paper<br>red and green pencils | Skill Check, p. 343<br>dictionary<br>Mini Lab, p. 335<br>pencil<br>paper |

## OBJECTIVES

For more information about National Science Standards, see page 5T.

| SECTION | OBJECTIVE | CORRELATION of QUESTIONS to OBJECTIVES: CHECK YOUR UNDERSTANDING | CHAPTER REVIEW | TRP CHAPTER REVIEW | TRP CHAPTER TEST |
|---|---|---|---|---|---|
| 16:1 National Science Stds: UCP.1, UCP.5, C.6 | 1. **Relate** ways a planarian and an earthworm sense light. | 2, 4 | 12, 22 | 2 | 3 |
| | 2. **Compare** the sense organs of a cricket and a snake. | 3 | 1, 8, 9, 11, 12, 13, 20, 23, 24, 25 | 5 | 10, 16, 23 |
| 16:2 National Science Stds: UCP.1, UCP.2, UCP.3, UCP.4, UCP.5, B.2 | 3. **State** the functions of the parts of the eye. | 6, 9, 10 | 4, 5, 6, 15, 16, 23 | 3, 9, 25, 26, 27, 28, 29, 30, 31, 32, 36, 38 | 5, 6, 11, 14, 17, 20, 21, 25, 28, 29, 43, 44, 45, 46, 47, 48, 49, 50, 51, 52 |
| | 4. **Explain** the jobs of the nose, tongue, and skin as sense organs. | 8 | 2, 3, 10, 17, 19, 23 | 7, 8, 34, 35, 39 | 4, 8, 9, 13, 18, 24, 30 |
| | 5. **Describe** the parts of the outer, middle, and inner ear. | 7 | 7, 18, 21 | 4, 11, 12, 13, 14, 15, 16, 17, 18, 19, 20, 21, 22, 23, 33, 37 | 1, 7, 26, 31, 32, 33, 34, 35, 36, 37, 38, 39, 40, 41, 42 |
| 16:3 National Science Stds: UCP.3, UCP.4, UCP.5, E.1, E.2, F.1 | 6. **Describe** how common vision problems are corrected. | 11, 14, 15 | 14 | 10, 40 | 19, 28 |
| | 7. **Discuss** reasons for protecting your ears against loud noises. | 12 | 14 | 1 | 2 |
| | 8. **Explain** ways of correcting hearing problems. | 13 | | 6 | 12, 22, 27 |

# Senses

## CHAPTER OVERVIEW

### Key Concepts

This chapter begins by pointing out the importance of sense organs. An overview of certain sense organs in the animal kingdom is presented, followed by a detailed look at human sense organs, including the eye, ear, skin, tongue, and nose. Finally, the chapter looks at some of the more familiar problems associated with the senses.

### Key Science Words

auditory nerve
cochlea
cone
cornea
dermis
eardrum
epidermis
farsighted
iris
lens
lens muscle
nearsighted
olfactory nerve
optic nerve
pupil
retina
rod
sclera
semicircular canals
sense organ
taste bud
vitreous humor

### Skill Development

In Lab 16-1, students will **formulate a model, observe,** and **measure in SI** as they learn how the eye works. In Lab 16-2, they will **predict, interpret data,** and **make and use tables** as they test their sense of vision. In the Skill Check on page 341, students will **interpret a diagram** of the ear. In the Skill Check on page 343, they will **understand** the **science word** *epidermis.* In the Mini Lab on page 335, they will **observe** an afterimage.

### Bridging

The nervous system allows animals to monitor their surroundings. This chapter explains to students how the nervous system gets its input or information about the surroundings. Sense organs are an extension of the nervous system, which students have just read about in Chapter 15.

**CHAPTER PREVIEW**

### Chapter Content

Review this outline for Chapter 16 before you read the chapter.



### Skills in this Chapter

The skills that you will use in this chapter are listed below.

- In **Lab 16-1,** you will formulate a model, observe, and measure in SI. In **Lab 16-2,** you will interpret data and make and use tables.
- In the **Skill Checks,** you will interpret diagrams and understand science words.
- In the **Mini Lab,** you will observe.

330

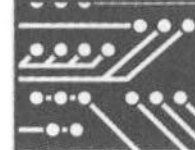

## TECH PREP

For Tech Prep activities in this chapter of the Teacher Wraparound Edition, see especially the Motivations on pages 335, 339, 341, 344, and 345, the Apply Demonstration and the Options Challenge on page 343, the Student Journal on page 344, and the Connection on page 345.

See also the Glencoe Homepage at **www.glencoe.com**

The following Glencoe resources provide additional opportunities for integrating science and technology.

**Experience Technology**
Section III: Using Resources to Solve Technical Problems
Activity Brief: Using Technology to Overcome Physical Handicaps

Chapter **16**

# Senses

Think about all the things you can enjoy because of your senses. You can enjoy the aroma of baking pizza because of your sense of smell. You can savor the rich sweetness of a chocolate chip cookie because of your sense of taste. Your senses let you experience the richness of the world around you.

But, did you realize there are things your senses can't detect? What do you think an insect can see? Insects can see ultraviolet light from the sun that you can't see. The flower on the left is what you see when you look at a flower. The flower on the right is what an insect sees. This flower was photographed in ultraviolet light. The insect's ability to see the middle of the flower as a darkened area helps guide the insect to the flower. Like you, the insect's senses allow it to keep in touch with the world around it.

**Try This!**

**Did you hear that?** Listen to the sounds you make as you speak, breathe, and chew. Plug your ears with your fingers and listen to these sounds again. How do the sounds differ? Can you explain why?

**interNET CONNECTION**

For more information about the material in this chapter, follow the link for the chapter on the Glencoe Homepage at **http://www.glencoe.com**

Flowers photographed under natural light (left) and ultraviolet light (right)



## ASSESSMENT PLANNER

**Portfolio**

Strategies on the following pages represent student products that can be placed into a best-work portfolio: pp. 341, 344.

**PERFORMANCE ASSESSMENT**

Skill Check, pp. 341, 343
Mini Lab, p. 335
Lab, pp. 337, 338

**CONTENT ASSESSMENT**

Check for Understanding, pp. 333, 343, 347
Chapter Review, pp. 348-349

**GROUP ASSESSMENT**

Opportunities for group assessment occur with Cooperative Learning Strategies.

**Student Journal**

Strategies on the following pages represent opportunities for writing in a Student Journal: pp. 332, 344.

**GETTING STARTED**

### Using the Photos

The ability to see colors is found in many animals including fish, reptiles, most birds, and some mammals. An insect's eye is made up of thousands of lenses and produces a picture that is somewhat like an image produced by a color TV. Both are composed of small dots. Have students compare the colors of the flowers as they appear to humans and insects. An insect can see UV light but we can't. However, an insect can't distinguish between red and black but we can.

### MOTIVATION/Try This!

**Did you hear that?** Hearing is based on detection of molecules that vibrate or move. These vibrations are usually air molecules in motion; however, they may be vibrations of a solid. Sound heard when the ears are plugged occurs as a result of vibrations through bone. Upon completion of this activity, students will be able to compare sound as a result of air molecule motion with sound as a result of bone vibration.

### Chapter Preview

Have students study the chapter outline before they begin to read the chapter. They should note the overall organization of the chapter, that the introduction describes the sense organs of various animals. Next it explains the parts and functions of the human sense organs. Last, it deals with problems associated with sense organs and how certain problems are solved or prevented.

### Misconception

Mention an eye transplant operation and most students will assume that the whole eye is removed and a new one from a donor is inserted. What actually is transplanted is only the cornea, the outer covering of the eye. At this time, one cannot cut a major nerve such as the optic nerve and have it regrow. (See the reference to this research in Chapter 15.)

## 16:1 Observing the Environment

### PREPARATION

#### Materials Needed

Make copies of the Focus, Reteaching, and Study Guide worksheets in the *Teacher Resource Package*.

▶ Obtain plastic mounts or slides for the demonstration.

#### Key Science Words

sense organ

### 1 FOCUS

▶ The objectives are listed on the student page. Remind students to preview these objectives as a guide to this numbered section.

#### MOTIVATION/Demonstration

Show students plastic mounts or 35 mm slides of a planarian, earthworm, cricket, and snake.

#### Guided Practice

Have students write down their answers to the margin questions: A planarian senses light with its two eyespots; a snake picks up odor molecules with its tongue.

**Focus**/*Teacher Resource Package*, p. 31. Use the Focus transparency shown at the bottom of this page as an introduction to sense organs.

**Student Journal**

Have students imagine they are snakes teaching biology to their snake students. The topic is "sense organs." Tell students to describe how they would compare their vision and olfactory organs to those used by humans.

## 16:1 Observing the Environment

**Objectives**

1. **Relate** ways a planarian and an earthworm sense light.
2. **Compare** the sense organs of a cricket and a snake.

**Key Science Words**

sense organ

Living things must know what is going on around them. They may use eyes. What if they do not have eyes? They will have other ways to keep in contact with their surroundings.

### Senses Aid Survival

The human body has many sense organs. **Sense organs** are parts of the nervous system that tell an animal what is going on around it. Eyes, ears, nose, tongue, and even our skin are sense organs.

Could you stay alive for very long without sense organs? It would be very hard. What about other animals? How well would they survive without any sense organs?

### Sense Organs of Animals

The sense organs of four different animals will be studied in this section. These animals are a flatworm, a segmented worm, an insect, and a snake.

**How does a planarian sense light?**

A planarian is a simple, small animal that lives in ponds or streams. Figure 16-1a shows two of its sense organs. Eyespots of a planarian are simple. They can tell only light from dark. A planarian has knobs on its front end that detect the direction of water currents and food.

Earthworms are a little more complex than planarians. If you look at an earthworm's body closely, you cannot see anything that looks like an eye. Yet, the animal can detect light. How? Earthworms have nerve cells just below the skin that detect light.

Figure 16-1b shows three different sense organs of a cricket. The cricket has parts for detecting sound on the

**Figure 16-1** A planarian (a) and a cricket (b) have sense organs that help them keep in touch with their surroundings.

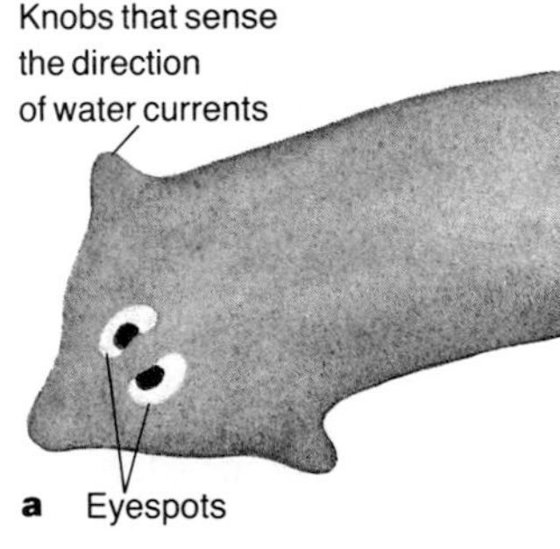


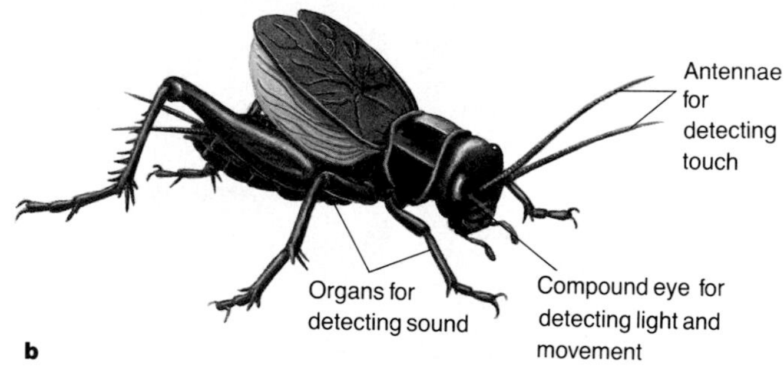




### OPTIONS

#### Science Background

The vast array of different sense organs parallels animal complexity. Simple animals tend to have poorly developed nervous systems and correspondingly poorly developed sense organs. The more complex the nervous system, the greater the complexity and variety of sense organs seen in the animal kingdom.

**FOCUS 31**

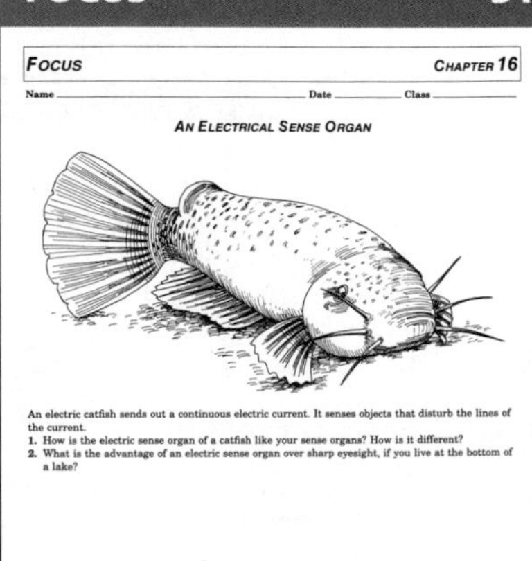

FOCUS CHAPTER 16

Name ______ Date ______ Class ______

AN ELECTRICAL SENSE ORGAN

An electric catfish sends out a continuous electric current. It senses objects that disturb the lines of the current.

1. How is the electric sense organ of a catfish like your sense organs? How is it different?
2. What is the advantage of an electric sense organ over sharp eyesight, if you live at the bottom of a lake?

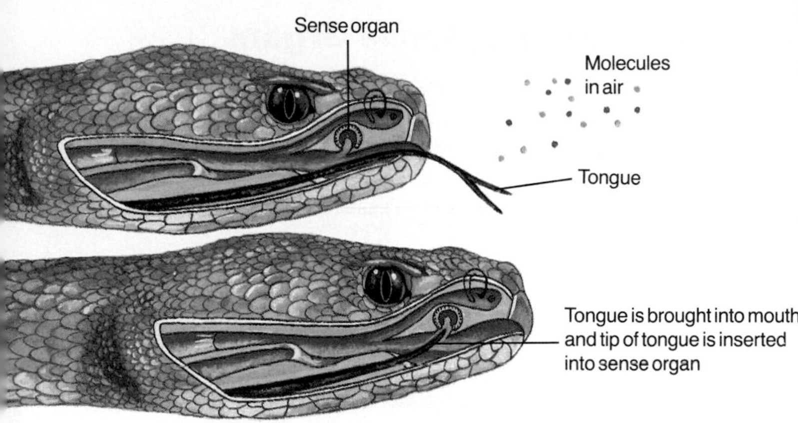

Figure 16-2 A snake has a sense organ in the roof of its mouth for sensing odors.

front legs. The antennae detect touch and chemicals. The eyes detect movement and light and dark. Insects have compound eyes. In this case, the word *compound* means many. A compound eye is made of hundreds or even thousands of very small eyelike parts. These eyelike parts are grouped together to form the compound eye.

The sense organs of snakes and other reptiles are more complex than those of insects. Snakes have eyes similar to human eyes. Their eyes have many of the same parts that human eyes have.

How does a snake smell?

Snakes have a very good sense of smell. Have you ever noticed how a snake darts its tongue in and out? A snake picks up odor molecules in the air with its tongue, then brings it back into its mouth. The snake then sticks the tip of its tongue into special sense organs found on the roof of its mouth. These sense organs look like two curled grooves as shown in Figure 16-2. Nerve cells line these grooves. The nerve cells pick up the molecule messages brought in by the tongue. Messages are then sent to the snake's brain.

## Check Your Understanding

1. Name five different sense organs in the human body.
2. Explain how an insect can detect light.
3. How does a snake use its tongue as a sense organ?
4. **Critical Thinking:** Since earthworms live underground, why do they need to detect light?
5. **Biology and Reading:** How do senses aid survival?

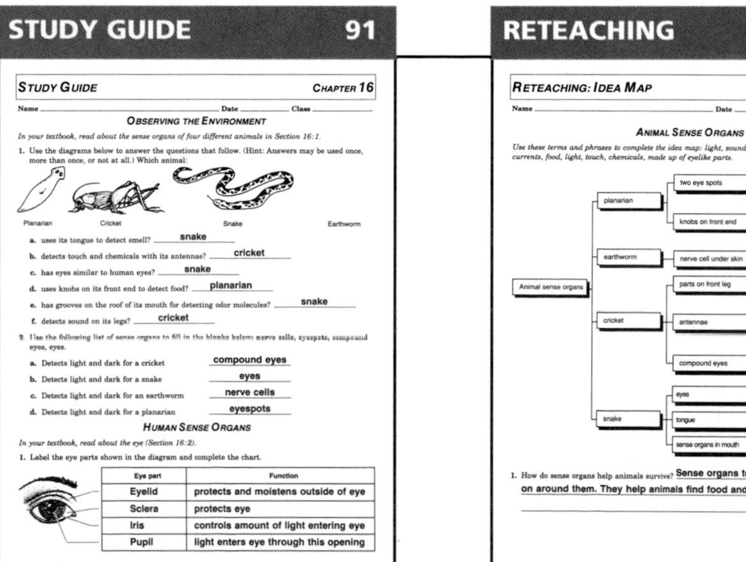

**STUDY GUIDE 91**

STUDY GUIDE — CHAPTER 16

Name ______ Date ______ Class ______

OBSERVING THE ENVIRONMENT

*In your textbook, read about the sense organs of four different animals in Section 16:1.*

1. Use the diagrams below to answer the questions that follow. (Hint: Answers may be used once, more than once, or not at all.) Which animal:

Planarian Cricket Snake Earthworm

a. uses its tongue to detect smell? snake
b. detects touch and chemicals with its antennae? cricket
c. has eyes similar to human eyes? snake
d. uses knobs on its front end to detect food? planarian
e. has grooves on the roof of its mouth for detecting odor molecules? snake
f. detects sound on its legs? cricket

2. Use the following list of sense organs to fill in the blanks below: nerve cells, eyespots, compound eyes, eyes.

a. Detects light and dark for a cricket compound eyes
b. Detects light and dark for a snake eyes
c. Detects light and dark for an earthworm nerve cells
d. Detects light and dark for a planarian eyespots

HUMAN SENSE ORGANS

*In your textbook, read about the eye (Section 16:2).*

1. Label the eye parts shown in the diagram and complete the chart.

| Eye part | Function |
|---|---|
| Eyelid | protects and moistens outside of eye |
| Sclera | protects eye |
| Iris | controls amount of light entering eye |
| Pupil | light enters eye through this opening |

**RETEACHING 47**

RETEACHING: IDEA MAP — CHAPTER 16

Name ______ Date ______ Class ______

Use with Section 16:1.

ANIMAL SENSE ORGANS

*Use these terms and phrases to complete the idea map: light, sound, objects, odor, odor, water currents, food, light, touch, chemicals, made up of eyelike parts.*

Animal sense organs
- planarian
  - two eye spots — light
  - knobs on front end — water currents; food
- earthworm
  - nerve cell under skin — light
- cricket
  - parts on front leg — sound
  - antennae — touch; chemicals
  - compound eyes — made up of eyelike parts
- snake
  - eyes — objects
  - tongue — odor
  - sense organs in mouth — odor

1. How do sense organs help animals survive? Sense organs tell animals what is going on around them. They help animals find food and protect themselves.

## 2 TEACH

### Independent Practice

**Study Guide**/*Teacher Resource Package,* p. 91. Use the Study Guide worksheet shown at the bottom of this page for independent practice.

### Check for Understanding

Have students respond to the first three questions in Check Your Understanding.

### Reteach

**Reteaching**/*Teacher Resource Package,* p. 47. Use this worksheet to give students additional practice with animal sense organs.

**Extension:** Assign Critical Thinking, Biology and Reading, or some of the **OPTIONS** available with this lesson.

## 3 APPLY

### ACTIVITY/Demonstration

Show students the lateral line along the side of a preserved fish and explain that it detects vibrations in the water.

## 4 CLOSE

Discuss how difficult it is to catch a fly. Relate this to the efficiency of the insect eye in detecting motion.

### Answers to Check Your Understanding

1. eyes, ears, nose, tongue, skin
2. with its compound eyes.
3. Nerve cells pick up messages brought in by the tongue and transfer them to the brain.
4. They need to detect light because when they do go to the surface, they have to stay out of strong light to avoid drying out.
5. They detect changes in the organism's environment. The organism can then react to those changes, some of which could threaten survival.

## 16:2 Human Sense Organs

## PREPARATION

### Materials Needed

Make copies of the Transparency Masters, Reteaching, Critical Thinking/ Problem Solving, Enrichment, Study Guide, and Lab worksheets in the *Teacher Resource Package.*

▶ Obtain black and white paper for Lab 16-1. Obtain white paper and green and red pencils for Lab 16-2.

▶ Gather ear plugs, eye and ear models, and PTC paper for the demonstrations.

### Key Science Words

| | |
|---|---|
| sclera | optic nerve |
| iris | taste bud |
| pupil | olfactory nerve |
| cornea | eardrum |
| lens | cochlea |
| lens muscle | auditory nerve |
| retina | semicircular canal |
| vitreous humor | epidermis |
| rod | dermis |
| cone | |

### Process Skills

In Lab 16-1, students will observe, measure in SI, and formulate a model. In Lab 16-2, students will predict, interpret data, and make and use tables. In the Skill Checks, students will interpret diagrams and understand science words. In the Mini Lab, students will observe.

## 1 FOCUS

▶ The objectives are listed on the student page. Remind students to preview these objectives as a guide to this numbered section.

### MOTIVATION/Brainstorming

Ask students to name as many sense organs in the human body as they can.



### Objectives

3. **State** the functions of the parts of the eye.
4. **Explain** the jobs of the nose, tongue, and skin as sense organs.
5. **Describe** the parts of the outer, middle, and inner ear.

### Key Science Words

| | |
|---|---|
| sclera | cone |
| iris | optic nerve |
| pupil | taste bud |
| cornea | olfactory nerve |
| lens | eardrum |
| lens muscle | cochlea |
| retina | auditory nerve |
| vitreous humor | semicircular canal |
| rod | epidermis |
| | dermis |

## 16:2 Human Sense Organs

The human body has a variety of sense organs. Each sense organ is specialized as to what it can detect. Eyes detect light. The nose and tongue detect chemical molecules. Ears detect moving air molecules. Skin detects heat, cold, pain, touch, and pressure.

### The Eye

**What are the parts of the eye?**

How do our eyes give us information about our surroundings? What are the parts of the human eye? Figure 16-3 shows a front view and a side view of the eye. Let's look at what the parts of the eye do.

The eyelid protects and moistens the outside of the eye. Each time you blink, liquid spreads over the front of your eye.

The **sclera** (SKLER uh) is the tough, white outer covering of the eye. The sclera protects the eye.

The **iris** is a muscle. The iris controls the amount of light entering the eye. It is also the part that gives the eye its color. When you say that someone has blue or brown eyes, you are describing the color of the iris.

The **pupil** is an opening in the center of the iris. Light enters the eye through the pupil. As the amount of light changes, the size of both the iris and the pupil changes. The pupil becomes larger in dim light and smaller in bright light. How does a larger pupil help you see in the dark?

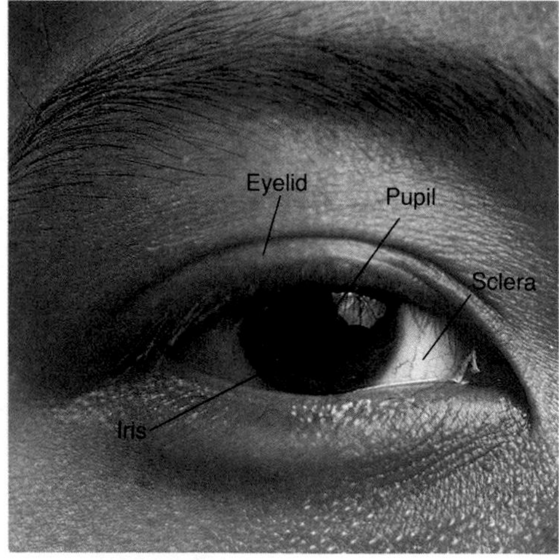


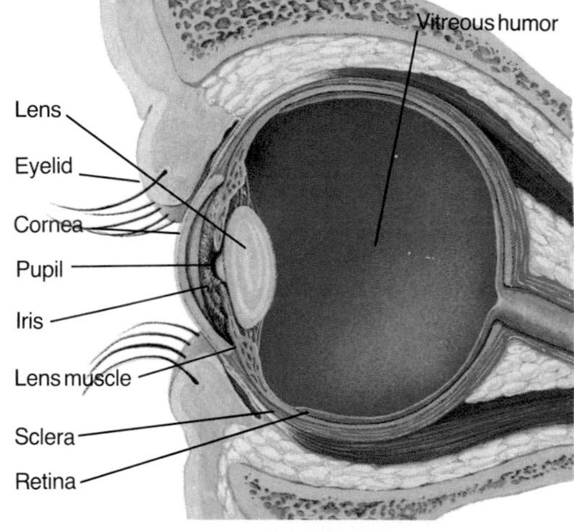


**Figure 16-3** The parts of the human eye are shown here.



## OPTIONS

### Science Background

Human sense organs are capable of detecting a variety of stimuli. These stimuli can be categorized into 3 groups. The eyes detect light energy. The ears detect vibrational energy. The nose detects molecules in the air, and the tongue detects molecules dissolved in water.

**Enrichment**/*Teacher Resource Package,* p. 17. Use the Enrichment worksheet shown here to help students understand animal sense organs.

**ENRICHMENT 17**

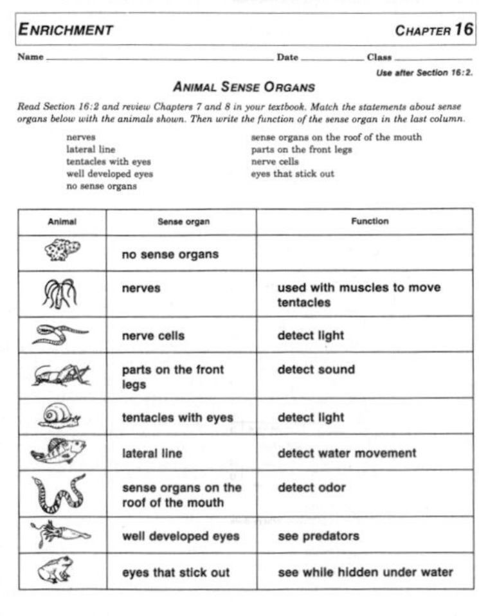

ENRICHMENT CHAPTER 16

Name ______ Date ______ Class ______

*Use after Section 16:2.*

**ANIMAL SENSE ORGANS**

*Read Section 16:2 and review Chapters 7 and 8 in your textbook. Match the statements about sense organs below with the animals shown. Then write the function of the sense organ in the last column.*

nerves
lateral line
tentacles with eyes
well developed eyes
no sense organs
sense organs on the roof of the mouth
parts on the front legs
nerve cells
eyes that stick out

| Animal | Sense organ | Function |
|---|---|---|
| | no sense organs | |
| | nerves | used with muscles to move tentacles |
| | nerve cells | detect light |
| | parts on the front legs | detect sound |
| | tentacles with eyes | detect light |
| | lateral line | detect water movement |
| | sense organs on the roof of the mouth | detect odor |
| | well developed eyes | see predators |
| | eyes that stick out | see while hidden under water |

The clear, outer covering at the front of the eye is the **cornea** (KOR nee uh). It protects the eye and allows light to enter the eye through the pupil. The cornea also helps to focus the light by bending the light rays as they enter the eye.

The **lens** is a clear part of the eye that changes shape as you view things at different distances. The lens muscle attaches to the lens. The **lens muscle** pulls on the lens and changes its shape. This helps to focus objects that are close up or far away.

The **retina** (RET nuh) is a structure at the back of the eye made of light-detecting nerve cells. The retina is often compared to film in a camera.

The **vitreous** (VI tree us) **humor** is a jelly-like material inside the eye. It pushes out on the eye parts. Its main job is to keep the eye round in shape.

Now that we have looked at the parts of the eye, we can follow the path that light takes through the eye. Follow the numbered steps in Figure 16-4.

1. Light rays enter the eye through the cornea. Notice that the cornea slightly bends the light rays. The light rays are bent toward each other.
2. Light passes through the pupil and then through the lens. Light is bent again as it passes through the lens. Again, the light rays are bent toward each other.
3. Because the vitreous humor is clear, light continues to pass through the eye.

### Mini Lab

**What Is an Afterimage?**

**Observe:** Stare out of the classroom window for at least 30 seconds. Then, close your eyes for 15 seconds. What you see is called an afterimage. What does the afterimage look like? *For more help, refer to the **Skill Handbook,** pages 704-705.*

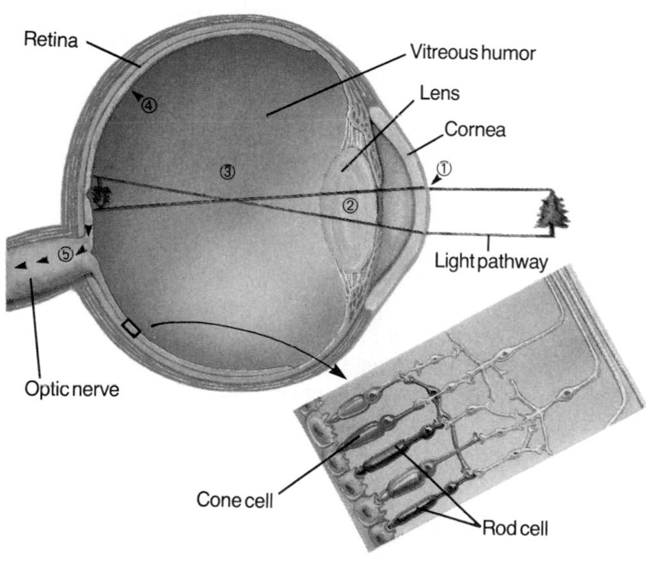


**Figure 16-4** Light passes through the cornea, lens, and vitreous humor before it strikes the retina.



# 2 TEACH

### MOTIVATION/Demonstration

A simple demonstration will show that the pupils of the eyes dilate in dim light. Have pairs of students sit so they can see each other's eyes. Darken the classroom as much as possible and leave lights off for at least 30 seconds. Turn them back on and have students note the change in the size of their partner's pupils. It may be necessary to repeat this demonstration several times.

### Guided Practice

Have students write down their answers to the margin question: The parts of the eye are sclera, iris, pupil, cornea, lens, retina, and vitreous humor.

### Mini Lab

Afterimages are related to the over stimulation of and chemical changes occurring in retinal cells.

### ASSESSMENT

**Skill:** Have students sketch the view through the window of the classroom. Then, have them diagram the same view that appears as the afterimage. They should label both diagrams and describe how they are alike and how they differ.

### Concept Development

▶ Explain that parts of the eye through which light passes are almost devoid of blood vessels, as blood vessels would impair vision. All other parts, however, do contain blood vessels. "Bloodshot" eyes are due to the dilation of capillaries located in the conjunctiva, a thin layer covering the sclera.

▶ Color vision is dependent on three different cone types. Each cone contains either a green, blue, or red receptor. The vast array of colors is due to the many possible combinations of the three basic receptors.

### CRITICAL THINKING 16

CRITICAL THINKING/PROBLEM SOLVING CHAPTER 16

Name ______ Date ______ Class ______

Use after Section 16:2.

HOW ARE A CAMERA AND THE HUMAN EYE SIMILAR?

Glass plate cover (fits over front of camera when not being used) — Camera body — Glass plate — Iris opening — Iris — Lens — Camera interior — Film

The diagram to the left shows the side view of a camera. Certain parts have been labeled for you. The function or job of each part is listed in the table below. Complete the last two columns in the table.

*Match the part of the eye with the camera part that has the same function. Write the function of the eye part in the space provided. Then answer the questions that follow.*

| Camera part | Function | Eye part | Function |
|---|---|---|---|
| Iris | adjusts for bright and dim light | iris | adjusts for bright and dim light |
| Iris opening | allows light to enter camera | pupil | allows light to enter eye in varying amounts |
| Glass plate | protects iris and lens | cornea | protects iris and lens |
| Lens | adjusts for close-up and distant objects | lens | adjusts for close-up and distant viewing |
| Camera body | holds parts together | sclera | forms outer protective layer and gives shape |
| Glass plate cover | protects glass plate, iris, and lens from dust | eyelid | protects cornea from dust and objects |
| Camera interior | allows light to pass through to film | vitreous humor | passes light to retina and gives shape to eye |
| Film | is sensitive to light; picture appears upside down | retina | is sensitive to light; image is upside down on retina |

**Analyzing the Problem**

1. Camera film is developed in a darkroom. Where does developing take place for the eye? in the brain
2. What cells in the retina are like black and white film? like color film? rods; cones

**Solving the Problem**

1. How are a camera and the human eye alike? Both receive light and form "pictures" of what is focussed upon.

## OPTIONS

**Critical Thinking/Problem Solving/** *Teacher Resource Package,* p. 16. Use the worksheet shown here to help students compare a camera and the human eye.

**Pathway of Light Through the Eye/** *Transparency Package,* number 16. Use color transparency number 16 as you teach how the eye functions.

**What Are the Parts of an Eye?/***Lab Manual,* pp. 135-140. Use this lab to reinforce what students learn about the parts of an eye.

# TEACH

## Concept Development

▶ Students will often see "floaters" within their eyes when looking into bright light. It is believed that these are blood cells that are floating in the vitreous humor.

▶ An easy way of remembering which cell type does what is to relate the letter c in "cone" to the letter c in "color."

▶ The prefix *opt-* is present in such words as *optician, optometrist, optics.* Ask students if they can name some words sharing the prefix *opt-*. This exercise becomes a useful tool for students when trying to remember the word *optic.*

## Independent Practice

**Study Guide**/*Teacher Resource Package,* p. 92. Use the Study Guide worksheet shown at the bottom of this page for independent practice.

## Idea Map

Have students use the idea map as a study guide to the major concepts of this section.

**Videodisc**

STV: Human Body Vol. 2
*Nervous System*
Unit 2, Side 2
*Sensory Receptors*

**9346-12793**

STV: Human Body
*Eye's Rods and Cones Magnified 10,000X*

**48694**

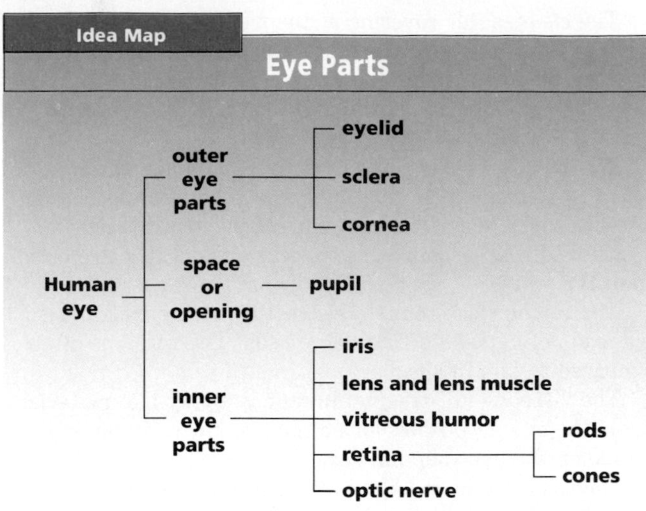


**Study Tip:** Use this idea map as you study the parts of the human eye and their functions. Write down the function of each eye part.

4. Finally, the light rays strike the retina. Notice that the retina is made of two types of nerve cells. These cells are called rods and cones because of their shapes. **Rods** are nerve cells that detect motion and help us to tell if an object is light or dark. They also help us to detect the shapes of objects. We use our rods when looking at something in dim light.

**Cones** are nerve cells that can detect color. There are three types of cones. Each type detects only one color: red, green, or blue. All other colors that we see are the result of two or three of the different types of cones acting together. Figure 16-5 shows a section of the retina enlarged 1000 times. Notice the many rods and cones.

Look closely again at the retina in Figure 16-4. When the image of the tree forms on the retina, it is upside down. It appears upside down because the light rays crossed over one another when they were bent. Which two eye parts bend light?

5. The message sent by the light leaves the retina and enters the optic (AHP tihk) nerve. The **optic nerve** is a nerve that carries messages from the retina to the brain. Once a message arrives at the vision center of the cerebrum, two things happen. One, the message is interpreted. In our example, the brain decodes the message and we see a tree. Two, the brain interprets the message as an object right side up. We see the tree in its normal position.

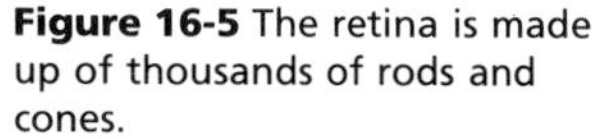

**Figure 16-5** The retina is made up of thousands of rods and cones.

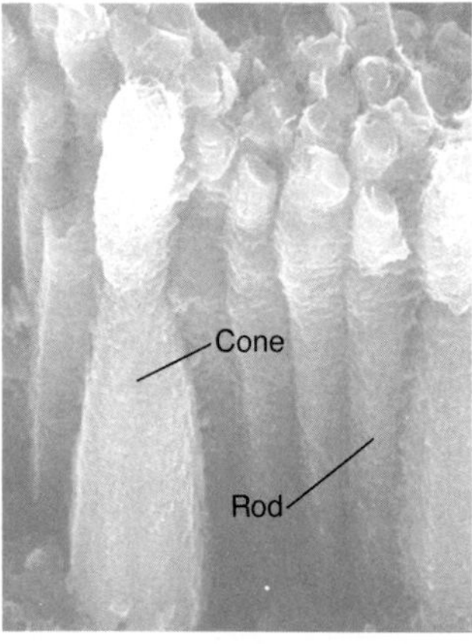




## OPTIONS

### Science Background

Tears contain an enzyme (lysozyme) that destroys bacteria. Lysozyme helps to fight off bacterial infections and lubricates the eye.

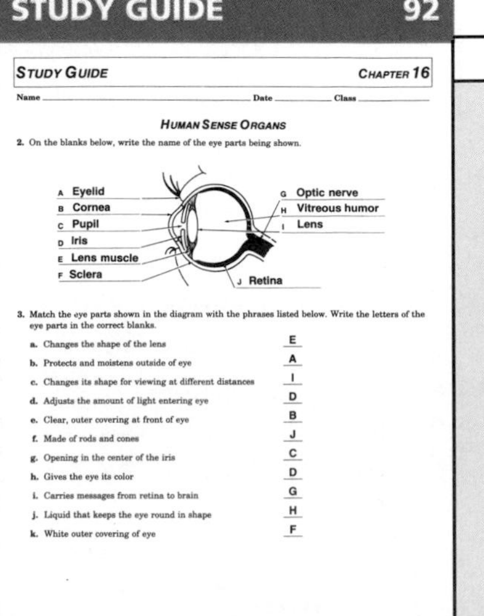

STUDY GUIDE 92

STUDY GUIDE CHAPTER 16

Name ________ Date ________ Class ________

HUMAN SENSE ORGANS

2. On the blanks below, write the name of the eye parts being shown.

A Eyelid
B Cornea
C Pupil
D Iris
E Lens muscle
F Sclera
G Optic nerve
H Vitreous humor
I Lens
J Retina

3. Match the eye parts shown in the diagram with the phrases listed below. Write the letters of the eye parts in the correct blanks.

a. Changes the shape of the lens E
b. Protects and moistens outside of eye A
c. Changes its shape for viewing at different distances I
d. Adjusts the amount of light entering eye D
e. Clear, outer covering at front of eye B
f. Made of rods and cones J
g. Opening in the center of the iris C
h. Gives the eye its color D
i. Carries messages from retina to brain G
j. Liquid that keeps the eye round in shape H
k. White outer covering of eye F

# Lab 16-1 The Eye

## Problem: How does the eye work?

### Skills

observe, measure in SI, formulate a model

### Materials

black paper
ruler
clear tape
white paper
scissors
top part of a hand lens

### Procedure

1. Make a data table in which to diagram your results.
2. Use a ruler to trace a letter *L* onto black paper. Make the letter 15 cm high and 8 cm long. Make each leg of the *L* 3 cm wide.
3. Cut out the letter and tape it onto the classroom window right side up.
4. Hold a piece of white paper as shown in the figure. The paper should be 40 to 60 cm away from the window.
5. With your back toward the window, look at the white paper.
6. Hold the top of the hand lens about 4 cm from the paper. Move the lens slowly toward and away from the paper until you see the letter *L* clearly on the paper.

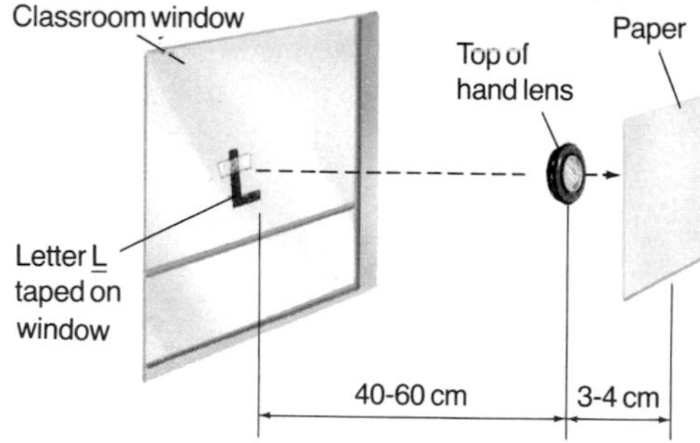


7. Complete the data table by making a diagram of the letter *L* as it appears on the window and on the paper.
8. **Measure in SI:** Measure the distance between the lens and paper in cm. Record this distance in your data table.
9. Determine the shape of the lens by rubbing your fingers over both sides of it. Diagram the shape of the lens in your data table.

### Data and Observations

1. What was the appearance of the letter *L* on the window?
2. What was the appearance of the letter *L* on the paper?
3. Which of the following better describes the shape of the lens: (a) thicker toward the edges and thinner toward the center or (b) thinner toward the edges and thicker toward the center?

### Analyze and Apply

1. Which part of the human eye is represented by the:
   (a) lens?
   (b) white paper?
2. (a) In a real eye, what structures would make up the paper?
   (b) In a real eye, what shape would the lens have?
3. **Apply:** In this lab, you moved the lens back and forth to obtain a clear view of the letter *L* on the paper. What does the action of moving the lens back and forth represent in the real eye?

### Extension

**Formulate a model** to show what causes nearsightedness. How does the distance between the lens and white paper change?



## ANSWERS

### Data and Observations

1. right side up and frontwards
2. upside down and backwards
3. b

### Analyze and Apply

1. a) lens
   b) retina
2. a) rods and cones
   b) thinner toward the edges and thicker toward the middle
3. focusing

## Lab 16-1 The Eye

### Overview

Students will construct a simple eye model. This model illustrates the role and shape of the lens. Students will discover that the image that falls upon the retina is inverted.

**Objectives:** Upon completing the lab, students will be able to (1) **compare** the shape of a hand lens to that of the eye lens, (2) **recognize** the role of the lens and retina, (3) **observe** that the image that falls on the retina is inverted.

**Time Allotment:** 30 minutes

### Preparation

▶ **Alternate Materials:** You may use a hand-held magnifying lens in place of the hand lens.

**Lab 16-1 worksheet**/*Teacher Resource Package,* pp. 61-62.

### Teaching the Lab

▶ **Troubleshooting:** The distance between the lens and the paper is going to be rather short (3-5 cm), as indicated in the diagram.

▶ Alert students to the fact that the image of the letter is indeed inverted.

**Cooperative Learning:** It will be necessary for students to work in groups of two so as to hold the lens and paper in proper position and measure the distance between them at the same time. For more information, see pp. 22T-23T in the Teacher Guide.

### ASSESSMENT

**Performance:** Have students modify the model so that an object, such as the letter *L*, would appear on the paper right-side up.

## Lab 16-2 The Senses

### Overview

Students will determine if their sense organs are 100 percent reliable. They will test their blind spots and color afterimaging. They should conclude that the eye is not totally reliable.

**Objectives:** Upon completing the lab, students will be able to (1) **test** for the blind spot and (2) **describe** an afterimage.

**Time Allotment:** 30 minutes

### Preparation

**Lab 16-2 worksheet**/*Teacher Resource Package*, pp. 63-64.

### Teaching the Lab

▶ To find the blind spot, have students start with the page at arm's length and move the paper slowly toward them. The blind spot will become apparent when the page is about 20 cm from the eye.

▶ **Troubleshooting:** Students may need to repeat the steps before they see the afterimage. Do not tell them what they are supposed to see.

**Cooperative Learning:** Divide the class into groups of two or three students. For more information, see pp. 22T-23T in the Teacher Guide.

**ASSESSMENT**

**Skill:** Have students diagram a model of the human eye like Figure 16-4. Have them illustrate how the cross that was viewed during the blind-spot test would appear as light rays from it pass through the various eye parts and fall on the blind spot of the retina. Note: The blind spot is located on the retina at the point where it forms the optic nerve.

## The Senses

**Lab 16-2**

### Problem: How reliable is your sense of vision?

### Skills

interpret data, make and use tables

### Materials

white paper
green and red pencils

### Procedure

1. Copy the data table.
2. Close your left eye. With your right eye, stare at the cross in the figure for about 10 seconds.
3. Very slowly move the page toward you. At a certain distance, the dot in the figure will disappear. When it disappears, light from the dot is falling on your blind spot. The blind spot is an area on the retina with no rods or cones.
4. Close your right eye. Repeat steps 2 and 3, only this time stare at the circle.
5. Record in your table whether each eye has a blind spot.
6. Stare at the triangles for 30 seconds. Then, look at a blank piece of paper. You will see a similar figure on the blank paper. Record in your table what colors you see in the outer triangle and the inner triangle in the second figure.

| TEST | | OBSERVATIONS |
|---|---|---|
| Blind spot | Right eye | blind spot present |
| | Left eye | blind spot present |
| Colors of triangles | Outer triangle | red |
| | Inner triangle | green |

### Data and Observations

1. Is your blind spot present in both eyes?
2. What happened to the colors of the triangles when you looked at the white paper?

### Analyze and Apply

1. Why aren't you usually aware of your blind spot?
2. Which parts of the eye detect the colors of the triangles?
3. **Apply:** Was the change in color you observed in the triangles taking place in your eyes or in your brain? Explain your answer.

### Extension

The size of an object determines how far away the object must be before the light from it falls on your blind spot. **Design an experiment** to test this idea.

## ANSWERS

### Data and Observations

1. yes
2. They were reversed.

### Analyze and Apply

1. An object must be in a certain position in front of the eyes for the blind spot to be noticed.
2. cones
3. The change was taking place inside the brain. The rods and cones don't interpret messages, the brain does.

## The Tongue and the Nose

The tongue is the major sense organ for taste. The tongue detects molecules dissolved in water. The tongue senses different molecules and you taste the sweetness of a candy bar or the sourness of a lemon.

Notice in Figure 16-6a that the surface of the tongue has many small bumps. On each of these bumps are tiny pits that contain taste buds. **Taste buds** are nerve cells in the tongue that detect chemical molecules. You have about 10 000 taste buds. Figure 16-6a shows a greatly magnified view of several taste buds.

There are four different kinds of taste buds on the tongue. Each kind is located in a different area of the tongue. Each area of the tongue detects a certain kind of molecule, or taste. The four different tastes are sour, salty, sweet, and bitter. You know that there are more than just four tastes. What you taste are different combinations of the four basic tastes.

Have you ever had trouble tasting things when you had a cold? That's because the senses of taste and smell work together. They both detect molecules. Your nose detects gas molecules in the air. These gas molecules are then interpreted by the brain as odors.

Figure 16-6b shows how your nose works. Molecules in the form of a gas are detected by neurons that line the top of the nasal chamber. There are seven different types of nerve cells in the nose. Each type detects one kind of odor. The **olfactory** (ohl FAK tree) **nerve** then carries the message to your brain. The cerebrum interprets the message as smoke or perfume or some other odor.

### Bio Tip

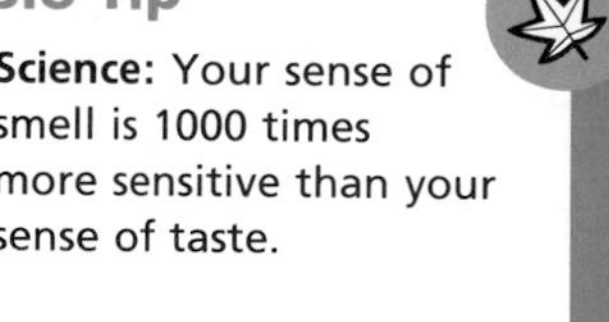

**Science:** Your sense of smell is 1000 times more sensitive than your sense of taste.

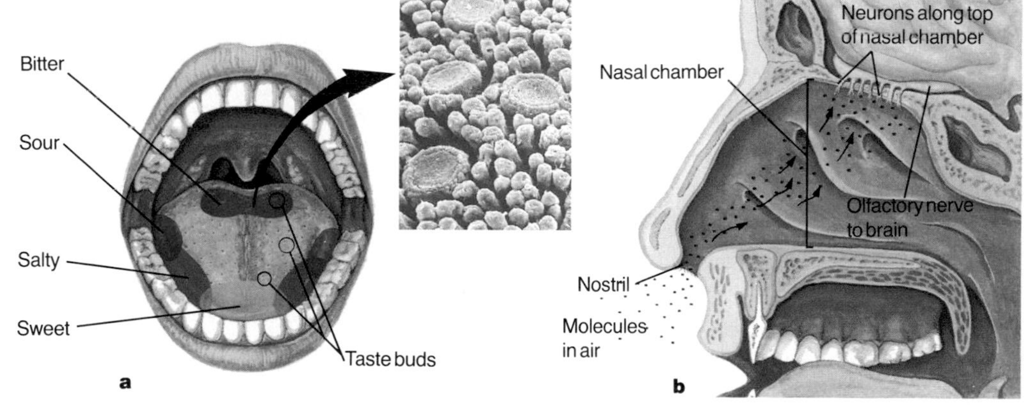


**Figure 16-6** Taste buds (a) detect four kinds of tastes. The photograph shows the surface of the tongue magnified 1000 times. The large, round structures are tastebuds. The nose (b) detects odors in the form of gases.



## TEACH

### Concept Development

▶ Review the term *molecule* with your students. Then ask if it is molecules that one detects on the taste buds. The answer is yes.

### MOTIVATION/Demonstration

Ask students if each one perceives the same exact taste for all types of food. Why do people have likes and dislikes when it comes to food? The perception of taste may have some genetic basis. Tell students that the ability to taste certain chemicals is a trait that they may inherit much like eye, skin, or hair color. To illustrate this idea, provide each student with a small piece of PTC paper. (This paper is available from most biological supply houses.) Ask them to taste the paper. About 80 percent of the class will describe its taste as bitter. The remaining 20 percent will describe it as having no taste. Ask the students who tasted nothing to raise their hand.

### GLENCOE TECHNOLOGY

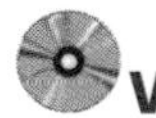

**Videodisc**

**Biology: The Dynamics of Life**

Animation: *Sense of Sight*
Disc 2, Side 1, Ch. 39

Animation: *Sense of Hearing*
Disc 2, Side 1, Ch. 40

### TRANSPARENCY 75

TRANSPARENCY MASTER — CHAPTER 16

HUMAN SENSE ORGANS I

lens, vitreous humor, eyelid, optic nerve, cornea, pupil, iris, retina, lens muscle, sclera, The Eye

bitter, sour, salty, sweet, The Tongue

olfactory nerve, nerve cells, nasal chamber, nostril, The Nose

### OPTIONS

**Transparency Master**/*Teacher Resource Package*, p. 75. Use the Transparency Master shown here to help students study the human sense organs.

**How Can You Test Your Senses?**/*Lab Manual*, pp. 127-130. Use this lab as an extension to learning about senses.

## TEACH

### Concept Development

▶ The small middle-ear bones are called the malleus (hammer), incus (anvil), and stapes (stirrup).

▶ A model of the human ear would help students to visualize the locations and the size relationships of parts being described.

▶ Compare the outer ear (pinna) to a megaphone or an old-fashioned ear trumpet. Ask why people who are hard of hearing often cup their hands in back of their ears when trying to hear. They are trying to direct more sound waves into the outer ear.

### Idea Map

Have students use the idea map as a study guide to the major concepts of this section.

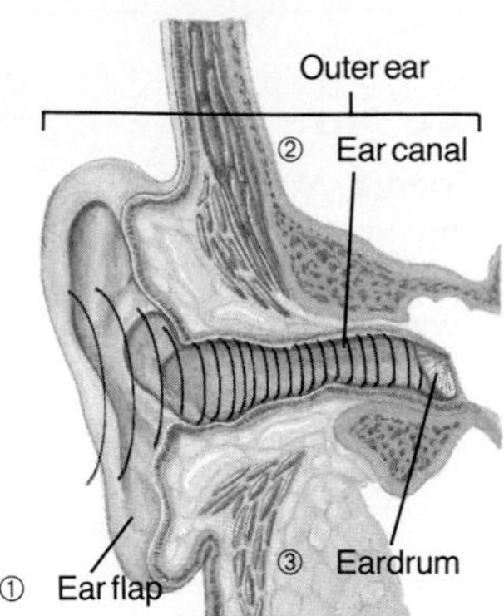


**Figure 16-7** Sound waves are directed into the ear by the ear flap and ear canal.

## The Ear

Ears are sense organs for detecting sound waves. Sound waves are air molecules in motion. How does the ear detect moving air molecules and allow you to hear?

The human ear is divided into three areas. These areas are called the outer, middle, and inner ear. The following description tells how these parts work together and enable you to hear.

Figure 16-7 shows the three parts of the outer ear. Look at this figure as you read steps 1 through 3.

1. The only part of your ear that you can see is the ear flap. The ear flap helps direct sound waves into a narrow tube leading into the ear. The ear flap is made of cartilage and connective tissue. It is soft and easily bent.
2. The narrow tube leading into the ear is called the ear canal. The ear canal carries sound waves to the middle ear.
3. Sound waves bump against a thin membrane called the eardrum. The **eardrum** is the membrane that vibrates at the end of the ear canal. The vibrations, or back-and-forth movements, are caused by sound waves striking the eardrum. A fatlike chemical, earwax, is made by cells in the ear canal. Earwax helps to keep insects and other foreign matter out of the ear and keeps the eardrum soft.

Figure 16-8 shows the middle ear. It has two main parts. Look at this figure as you read steps 4 and 5.

4. Three small bones make up one of the main parts of the

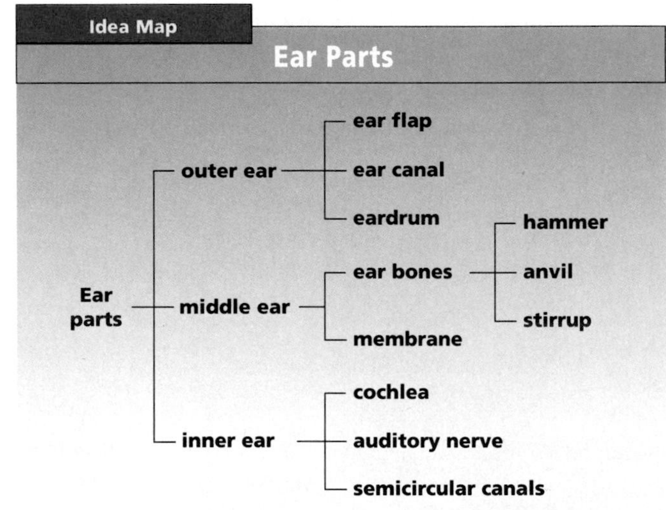


**Study Tip:** Use this idea map as you study the parts of the human ear and their functions. Write down the function of each ear part.



### OPTIONS

**Transparency Master**/*Teacher Resource Package*, p. 77. Use the Transparency Master shown here to help students study the human sense organs.

**TRANSPARENCY 77**

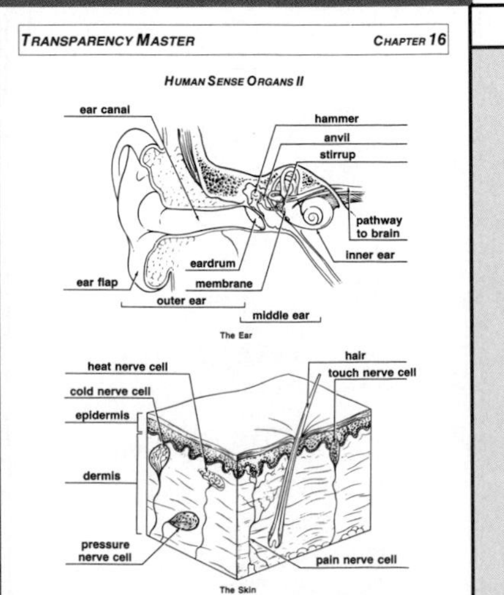

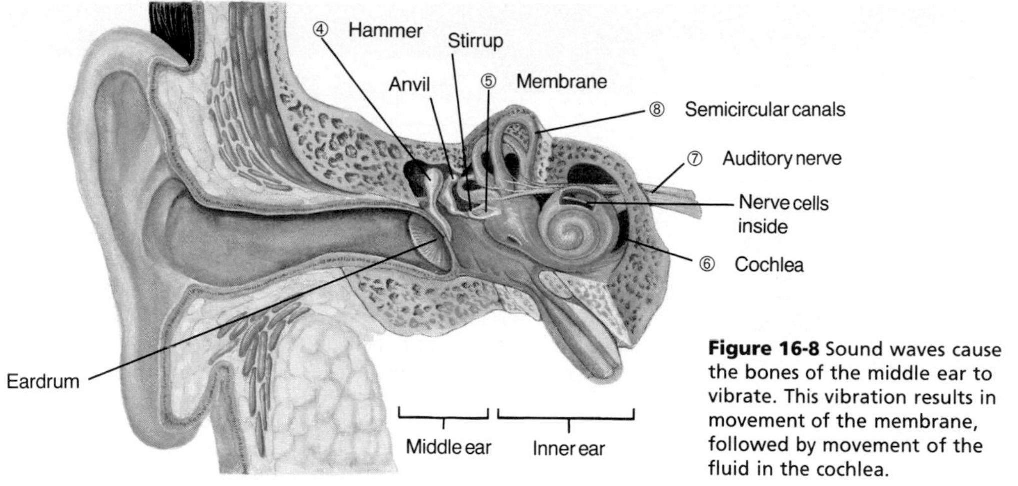


**Figure 16-8** Sound waves cause the bones of the middle ear to vibrate. This vibration results in movement of the membrane, followed by movement of the fluid in the cochlea.

middle ear. Because of their shapes, the bones are called the hammer, anvil, and stirrup. These bones are quite small and are connected to one another. The hammer is also connected to the eardrum. When the eardrum vibrates, the hammer moves. The movement of the hammer causes the other two bones to move.

5. The stirrup is connected to a membrane in the middle ear that vibrates with the motion of the ear bones.

The inner ear also has three main parts. Look at Figure 16-8 and read steps 6, 7, and 8 to see what they do.

6. This part is called the cochlea (KAHK lee uh). *Cochlea* is Latin for "snail shell." The **cochlea** is a liquid-filled, coiled chamber in the ear that contains nerve cells. When the middle ear membrane vibrates, it makes the liquid in the cochlea move. The nerve cells in the cochlea detect this movement.

How does the cochlea work?

7. Each nerve cell in the cochlea is connected to a large nerve, the auditory (AHD uh tor ee) nerve. The **auditory nerve** carries messages of sound to the brain.

Once messages reach the cerebrum of your brain, they are interpreted as particular sounds. You hear a whistle, a barking dog, or some other sound.

8. The semicircular canal has a very special job not related to hearing. The **semicircular canals** are inner ear parts that help us keep our balance. Nerve messages pass from these ear parts to your brain's cerebellum. When these messages arrive in the brain, they send messages to body muscles to contract or relax as needed to maintain balance.

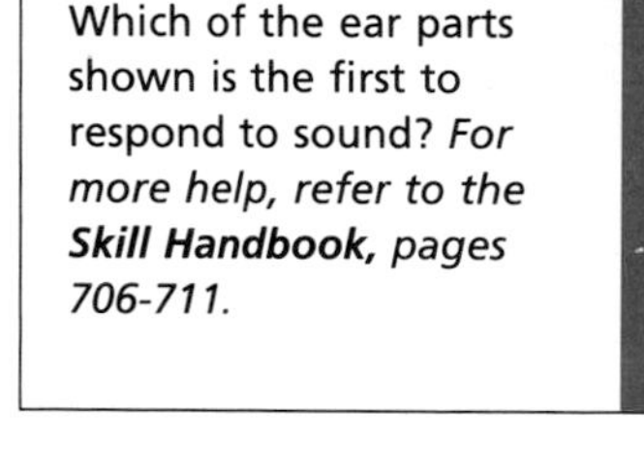

**Skill Check**

**Interpret diagrams:** Study Figure 16-7. Which of the ear parts shown is the first to respond to sound? *For more help, refer to the* ***Skill Handbook,*** *pages 706-711.*



# TEACH

## MOTIVATION/Demonstration

You can demonstrate the role of the inner ear bones with a hypodermic syringe (piston removed). As the piston is inserted into the cylinder and pushed inward, the pressure within the cylinder increases. In the analogy, the cylinder represents the inner ear and the piston represents the ear bones. These bones increase the pressure of the sound vibration 22 times. The oval window vibrates more intensely from the increased pressure than did the eardrum when the sound struck it.

## Guided Practice

Have students write down their answers to the margin question: The cochlea contains liquid and nerve cells. When the middle ear membrane vibrates, the liquid in the cochlea moves. The nerve cells in the cochlea detect this movement. The nerve cells are connected to the auditory nerve, which carries messages of sound to the brain.

## Skill Check

The eardrum is the first ear part that responds to sound.

## MOTIVATION/Software

*The Ear,* Queue, Inc.

## Independent Practice

**Study Guide**/*Teacher Resource Package,* p. 93. Use the Study Guide worksheet shown at the bottom of this page for independent practice.

## Portfolio

Prepare an enlarged copy of Figure 16-8 but omit the semicircular canals. Have students use colored pencils to color code the different ear parts according to the following scheme: shade red the parts that transmit sound through the air (a gas), shade blue the parts that transmit sound through solid structures, and shade green the parts that transmit sound through liquid.

STUDY GUIDE 93

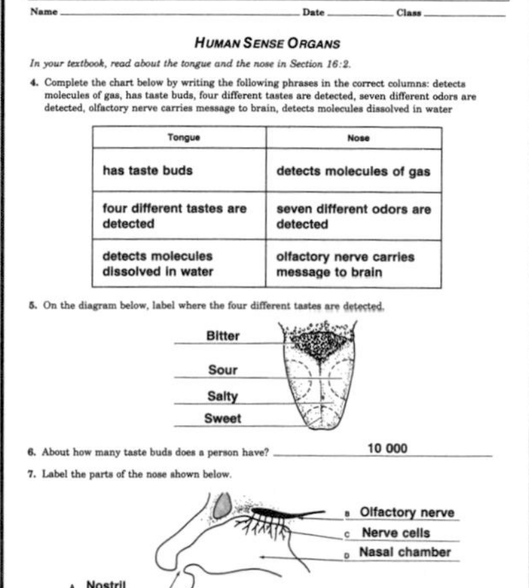

STUDY GUIDE CHAPTER 16

Name ______ Date ______ Class ______

HUMAN SENSE ORGANS

*In your textbook, read about the tongue and the nose in Section 16:2.*

4. Complete the chart below by writing the following phrases in the correct columns: detects molecules of gas, has taste buds, four different tastes are detected, seven different odors are detected, olfactory nerve carries message to brain, detects molecules dissolved in water

| Tongue | Nose |
|---|---|
| has taste buds | detects molecules of gas |
| four different tastes are detected | seven different odors are detected |
| detects molecules dissolved in water | olfactory nerve carries message to brain |

5. On the diagram below, label where the four different tastes are detected.

Bitter
Sour
Salty
Sweet

6. About how many taste buds does a person have? 10 000

7. Label the parts of the nose shown below.

A. Nostril
B. Olfactory nerve
C. Nerve cells
D. Nasal chamber

## OPTIONS

**How Are Your Senses Sometimes Fooled?**/*Lab Manual,* pp. 131-134. Use this lab as an extension to understanding the senses.

## Career Close-Up

### Licensed Practical Nurse

#### Background

A practical nurse works with patients in hospitals and nursing homes, and provides in-home care. A practical nurse is involved with many aspects of patient care and works closely with registered nurses and doctors.

#### Related Careers

Related careers are registered nurse and nurse's aide.

#### Job Requirements

In order to be a licensed practical nurse, a person must be a high school graduate and take special training at a junior college or hospital.

#### For More Information

National Federation of Licensed Practical Nurses, Inc., Box 11038, Durham, NC 27703.

## TEACH

### Guided Practice

Have students write down their answers to the margin question: The skin has nerve cells for detecting pain, pressure, touch, heat, and cold.

### Independent Practice

**Study Guide**/*Teacher Resource Package,* p. 94. Use the Study Guide worksheet shown at the bottom of this page for independent practice.

## Career Close-Up

### Licensed Practical Nurse

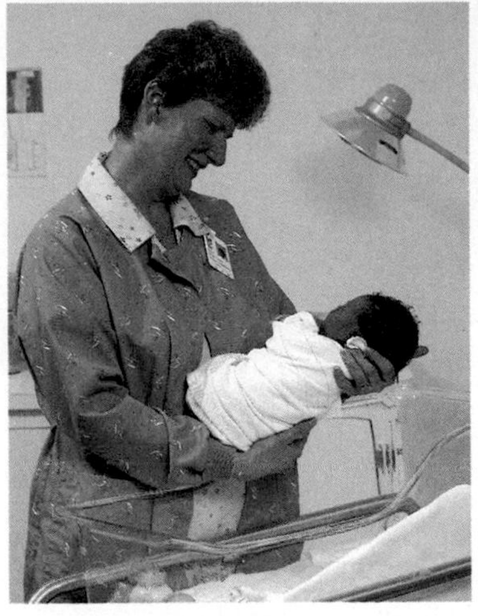
An LPN cares for patients.

**Did you know that you can be a nurse without going to college for four years? You can if you become a licensed practical nurse, or LPN. An LPN must graduate from high school and take a year of special training at a junior college or hospital. When the LPN finishes training, he or she can go right to work helping doctors and registered nurses do their jobs.**

**A practical nurse works directly with patients. The LPN takes temperatures, changes bandages, and feeds and bathes patients. Practical nurses are also licensed to give drugs to patients. They learn about how drugs affect patients and the proper use of drugs.**

**Practical nurses may work in hospitals, nursing homes, and homes for the elderly. Practical nurses are licensed by the state in which they work.**

Have you ever experienced motion sickness? It's a terrible feeling! Motion sickness results when the eyes and semicircular canals receive conflicting messages. For example, when you are in a car, your inner ear will be telling you that you are moving. If for some reason you can't look out of the car to see where you are going, your eyes will be telling you that you aren't moving. The messages are mixed up and you feel ill. One way to prevent motion sickness is to look out the front window of the car. Most drivers don't experience motion sickness because they are always looking out the front window.

## The Skin

How does the skin function as a sense organ?

You may not think of your skin as a sense organ. It does, however, have many different kinds of nerve cells that detect changes around the body. Before we describe these nerve cells, let's look at the skin itself. Figure 16-9



## OPTIONS

### Science Background

One part of the body that has a large number of pain neurons close to the surface is the cornea of the eye. When even a very small object gets into the eye, it causes severe pain.

**STUDY GUIDE 94**

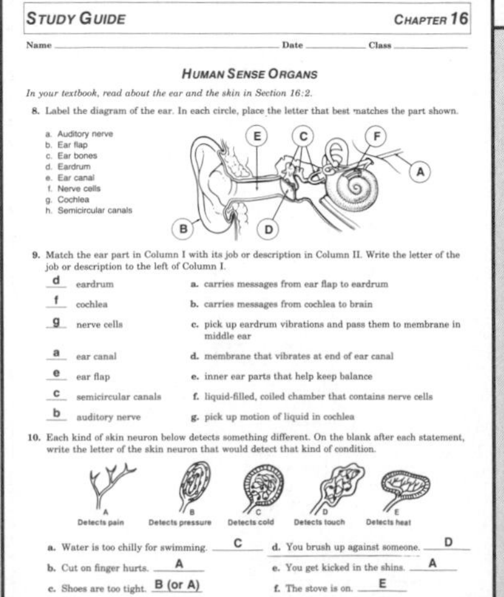
STUDY GUIDE CHAPTER 16

Name ______ Date ______ Class ______

HUMAN SENSE ORGANS

*In your textbook, read about the ear and the skin in Section 16:2.*

8. Label the diagram of the ear. In each circle, place the letter that best matches the part shown.

a. Auditory nerve
b. Ear flap
c. Ear bones
d. Eardrum
e. Ear canal
f. Nerve cells
g. Cochlea
h. Semicircular canals



9. Match the ear part in Column I with its job or description in Column II. Write the letter of the job or description to the left of Column I.

d eardrum — a. carries messages from ear flap to eardrum
f cochlea — b. carries messages from cochlea to brain
g nerve cells — c. pick up eardrum vibrations and pass them to membrane in middle ear
a ear canal — d. membrane that vibrates at end of ear canal
e ear flap — e. inner ear parts that help keep balance
c semicircular canals — f. liquid-filled, coiled chamber that contains nerve cells
b auditory nerve — g. pick up motion of liquid in cochlea

10. Each kind of skin neuron below detects something different. On the blank after each statement, write the letter of the skin neuron that would detect that kind of condition.



a. Water is too chilly for swimming. C
b. Cut on finger hurts. A
c. Shoes are too tight. B (or A)
d. You brush up against someone. D
e. You get kicked in the shins. A
f. The stove is on. E

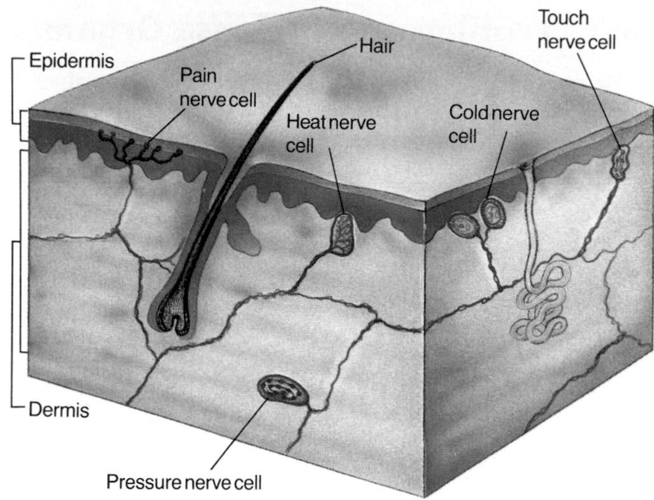


**Figure 16-9** The skin contains five kinds of nerve cells.

shows a closeup of a small section of skin. Two different layers form the skin. The **epidermis** (ep uh DUR mus) is the outside layer of cells of an organism. The **dermis** is a thick layer of cells that form in the inner part of the skin.

Notice in Figure 16-9 that five nerve cell types are shown. Each nerve cell detects a different condition. The nerve cells detect pain, pressure, touch, heat, and cold. Note that most nerve cells are found in the dermis. Only nerve cells that detect pain are found in both the epidermis and the dermis.

Like messages from other sense organs, messages from the nerve cells in the skin also travel to the cerebrum. There they are interpreted as hot, cold, pain, pressure, or touch messages.

### Skill Check

**Understand science words: epidermis.** The word part *epi* means outer. In your dictionary, find three words with the word part *epi* in them. *For more help, refer to the* ***Skill Handbook****, pages 706-711.*

## Check Your Understanding

6. What is the relationship between the iris and the pupil?
7. How do the bones of the middle ear enable you to hear?
8. How are taste and smell related? How do they differ?
9. **Critical Thinking:** Many animals are active only at night. Would these night creatures have larger or smaller pupils than humans? Why?
10. **Biology and Reading:** Where are the nerve cells of the eye found? What are they called?



## TEACH

### Check for Understanding

Have students respond to the first three questions in Check Your Understanding.

### Reteach

**Reteaching**/*Teacher Resource Package,* p. 48. Use this worksheet to give students additional practice with the human eye.

**Extension:** Assign Critical Thinking, Biology and Reading, or some of the **OPTIONS** available with this lesson.

## 3 APPLY

**Demonstration:** Provide two or three students with sets of earplugs, such as those used by swimmers. Have them wear the earplugs for the first 20 minutes of class. Then, have them report on what took place during the 20 minutes. Discuss the importance of the senses.

## 4 CLOSE

Discuss why: (a) foods taste so bland when you have a cold, (b) your ears plug up when riding up or down in an elevator, (c) the entire eye cannot be replaced.

### Answers to Check Your Understanding

6. The iris is a muscle in the eye that controls the size of the pupil.
7. When the eardrum vibrates, the hammer moves and causes the other two bones to move. The vibrations are transferred through the inner ear to the auditory nerve, which sends the impulse to the brain, and sound is perceived.
8. Sense organs for taste and smell detect molecules. To detect taste, molecules must be dissolved in water. To detect odor, molecules must be in gas form.
9. larger; to allow in more light
10. Nerve cells of the eye are found in the retina and are called rods and cones.

### RETEACHING 48

RETEACHING CHAPTER 16

Name ______ Date ______ Class ______

Use with Section 16:2.

THE HUMAN EYE

*Label the parts of the human eye shown below. Use the following terms and phrases to list the name and functions of each part: lens, iris, cornea, retina, vitreous humor, sclera, bends light so objects will be in focus, adjusts the amount of light entering the eye, changes shape to help you see close and far, has light detecting surface like the film of a camera, keeps the eye round in shape, protects eye and gives it shape.*

Image
Object
lens — changes shape to help you see close and far
cornea — bends light so objects will be in focus
iris — adjusts the amount of light entering the eye
vitreous humor — keeps the eye round in shape
sclera — protects eye and gives it shape
retina — has light detecting surface llike the film of a camera

1. What is the function of the rods in the retina? The rods detect motion and help us to see in dim light.
2. What is the function of the cones? Each cone can detect one color: red, green, or blue.
3. What is the function of the optic nerve? It carries messages from the retina to the vision center of the cerebrum.

## OPTIONS

**Challenge:** Use reference books to research some eye problems associated with improper muscle action.

**Challenge:** Interview a teacher of the hearing impaired. Find out the types of problems experienced by hearing impaired students. Ask to learn several words used in signing.

## 16:3 Problems with Sense Organs

## PREPARATION

### Materials Needed

Make copies of the Skill, Application, Reteaching, and Study Guide worksheets in the *Teacher Resource Package*.

▶ Obtain an eye chart for testing students' vision. Also obtain a tape recorder for the demonstration.

### Key Science Words

nearsighted
farsighted

## 1 FOCUS

▶ The objectives are listed on the student page. Remind students to preview these objectives as a guide to this numbered section.

### MOTIVATION/Demonstration

Use an eye chart to demonstrate that a student volunteer who wears glasses has no trouble reading the chart with glasses but has difficulty reading the same chart at the same distance without glasses. NOTE: If an eye chart is not available, print up a series of block letters that measure exactly 1 cm in height. Have the student stand 20 feet from the chart.

### Student Journal

Have students imagine they are optometrists and have just given a patient an eye exam. They tell the patient her vision is 20/80 rather than 20/20. The patient asks for an explanation. Have students explain, using diagrams if necessary, what is meant by 20/20 and 20/80 vision.

### Portfolio

Have students use Figures 16-4 and 16-10 as guides and illustrate the pathway that light would follow through the eye. They should describe the possible problems that would result if: (a) the cornea did not bend light and (b) the lens did not bend light.



**Objectives**

6. **Describe** how common vision problems are corrected.
7. **Discuss** reasons for protecting your ears against loud noises.
8. **Explain** ways of correcting hearing problems.

**Key Science Words**

nearsighted
farsighted

# 16:3 Problems with Sense Organs

Many people have lost their sense of hearing or vision completely. There are many more people with hearing or vision problems. What causes some of these problems? What can be done to correct them?

## Correcting Vision Problems

Have you ever wondered what is meant by 20/20 vision? A person with 20/20 vision is said to have normal vision. Figure 16-10a tells you what the numbers mean. Notice that the units used by eye doctors are not in metric or SI. A person with 20/20 vision sees certain things clearly at 20 feet. A person with 20/60 vision sees clearly at 20 feet what a person with 20/20, or normal, vision would see clearly at 60 feet.

Not everyone has 20/20 vision. Prove it to yourself by looking around the classroom. How many of your classmates are wearing glasses or contact lenses?

Why do some people have normal vision and others have problems seeing clearly? Figures 16-10b-d show you. The lines from the letter *T* to the retina show the path of light into the eye. Notice that the lines in Figure 16-10b meet on the retina. This eye has 20/20 vision. The eye in Figure 16-10c is nearsighted. **Nearsighted** means being able to see clearly close up but not far away. The path of light meets in front of the retina, instead of on it. This is because the nearsighted eye is longer from front to back than the normal eye.

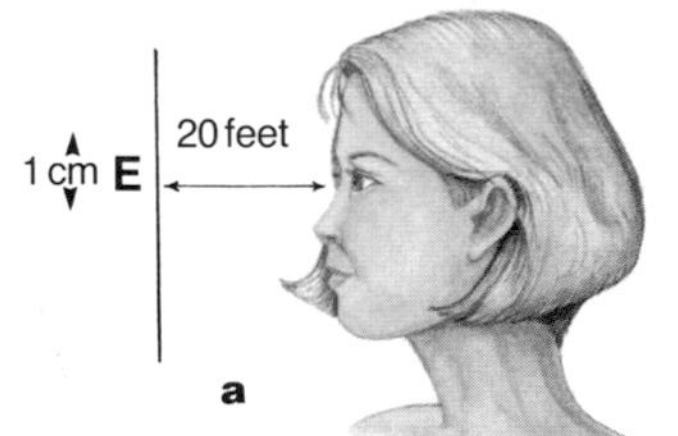


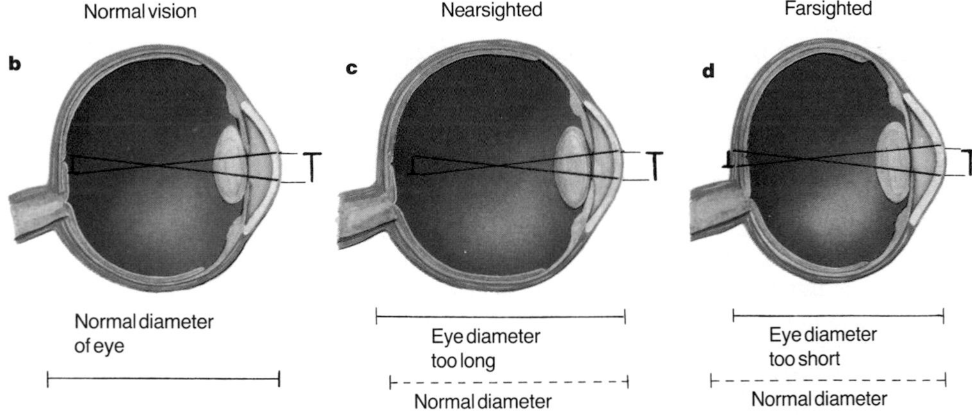


**Figure 16-10** A person with 20/20 vision can see a 1 cm high letter clearly at 20 feet (a). In a normal eye (b), light rays meet on the retina. In a nearsighted eye (c), light rays meet in front of the retina. In a farsighted eye (d), light rays meet behind the retina.



## OPTIONS

### Science Background

Nearsightedness is also called myopia. It results from improper bending of light by the cornea, by the lens, or by an abnormally long eye. Farsightedness is also called hypermetropia. It results from improper bending of light by the cornea, by the lens, or by an abnormally short eye.

**Skill**/*Teacher Resource Package*, p. 17. Use the Skill worksheet shown here to help students relate cause and effect.

**SKILL 17**

**SKILL: INFER** CHAPTER 16

Name ______ Date ______ Class ______

*For more help, refer to the Skill Handbook, pages 706-711.* *Use after Section 16.3.*

**SENSE ORGANS**

Scientists look for patterns among their observations. One type of pattern they search for is cause and effect. The cause is the reason something happens. The effect is what happens.

*In the table below, consider each statement at the left to be a cause. Think about the probable effect of each cause. Write your response in the second column.*

| Cause | Effect |
|---|---|
| A virus affects the function of your semicircular canals. | You do not have normal balance. |
| You have difficulty hearing a speech and cup your hand behind your ear. | Sound waves are directed into the ear and you hear better. |
| You get a dirt particle in your eye. | You blink and liquid spreads over the front surface of the eye. |
| You go from bright sunshine into a darkened theater. | The pupils of your eyes become larger. |
| Particles of perfume reach neurons in your nasal chamber. | The olfactory nerve carries the message to the brain. The cerebrum interprets the message as the odor of perfume. |
| You have a stuffy head cold when you eat a slice of pizza. | You cannot taste the pizza as well as you normally would. |
| You touch your tongue to a slice of lemon. | Your taste buds detect a sour taste. |
| The middle ear bones cannot move back and forth. | Deafness occurs. |
| Rays of light meet in front of the retina instead of on the retina. | Vision is blurred. The condition is nearsightedness. |

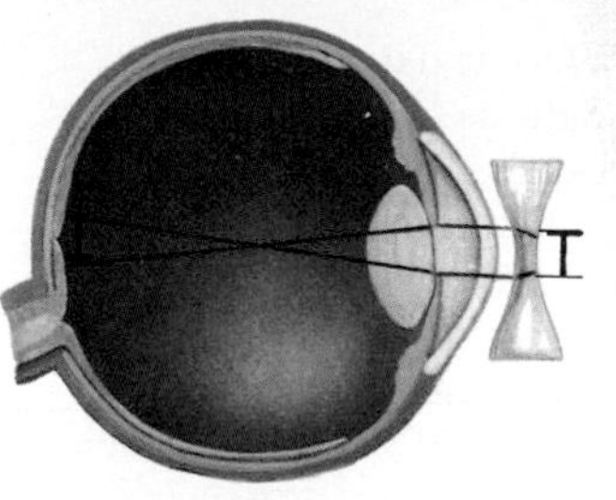

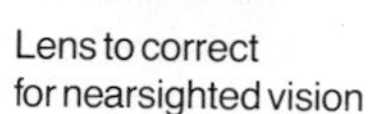

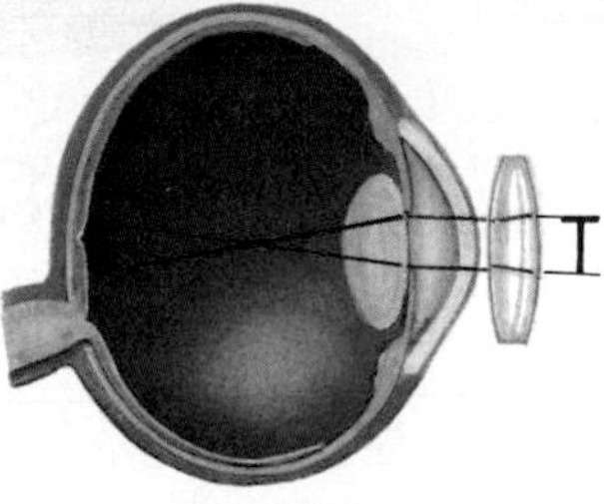

Lens to correct for nearsighted vision

Lens to correct for farsighted vision

**Figure 16-11** Lenses bend the light entering the eye and cause it to fall on the retina.

Figure 16-10d shows what happens within the eye when a person is farsighted. **Farsighted** means being able to see objects clearly far away but not close up. When the path of light enters the eye, it meets in back of the retina, instead of on it. Notice in the figure that the farsighted eye is shaped differently than the normal eye. The farsighted eye is shorter from front to back.

How are these kinds of vision problems corrected? They are corrected with lenses people can wear. Figure 16-11 shows how the lenses of glasses or contacts help correct vision problems. The lenses bend the light before it enters the eye. This extra bending allows the light rays to meet on the retina as they do in a normal eye. As a result, objects appear clear and sharp instead of fuzzy or blurry.

**Bio Tip**

**Consumer:** Tanning parlors increase the risk of eye cataracts. A cataract is a clouding of the lens. It results in loss of vision and blindness if not treated.

How does wearing glasses correct nearsightedness?

## Protecting Against Hearing Loss

Can listening to loud music damage your hearing? The answer to this question is yes. Listening to any kind of loud noise over a long period of time can cause permanent damage to your hearing.

How does listening to loud noises damage hearing? Loud noises damage the inside of the cochlea. The cochlea contains fluid and thousands of hairlike cells. The hairlike cells vibrate as the fluid within the cochlea is set into motion by sound waves. The movement of the hairlike cells starts a message in the auditory nerve leading to the brain. Whenever the ear is exposed to loud noises, some of the hairlike cells become flattened. Once the hairlike cells are flattened, they can no longer send messages to the auditory nerve. If enough hairlike cells are damaged, hearing loss results. Also, the hairlike cells can't repair themselves. Hearing that is damaged this way can't be restored.

## 2 TEACH

### MOTIVATION/Demonstration

Borrow an earmuff type hearing protection device from a student's parent. Allow the class to try it on with loud music playing. This will help them better understand the degree of protection given by these devices.

### Concept Development

▶ Figure 16-10 may require some added explanation. If a person can see a 1 cm high letter clearly at 20 feet, their vision is said to be 20/20. If the smallest letter a person can see clearly at 20 feet is 1.3 cm high, their vision is said to be 20/30.

▶ Obtain a tape recorder and play some music for the students at a normal sound level. Slowly increase the volume. Ask students to raise their hand when the volume becomes irritating to them.

### Connection: Consumer

Have students compare the magnifications of binoculars offered for sale in mail order catalogues or stores. What are the ranges of powers recommended for bird watching and opera viewing?

### Independent Practice

Study Guide/*Teacher Resource Package,* pp. 95-96. Use the Study Guide worksheets shown at the bottom of this page for independent practice.

### Guided Practice

Have students write down their answers to the margin question: Wearing glasses bends the light entering the eye and allows the light rays to meet on the retina.

**STUDY GUIDE 95**

STUDY GUIDE — CHAPTER 16

Name ______ Date ______ Class ______

PROBLEMS WITH SENSE ORGANS

*In your textbook, read about problems with sense organs in Section 16:3.*

1. Examine the diagrams below. Determine what vision problem is shown. In the space provided, name and describe the problem.

Farsighted means being able to see clearly far away but not close up. The path of light meets in back of the retina instead of on it.

Nearsighted means being able to see clearly close up but not far away. The path of light meets in front of the retina, instead of on it.

2. How are these problems corrected? They are corrected with special lenses that cause the path of light to meet on the retina.

3. What are two causes of deafness or hearing loss and how is each one treated?
   a. The nerve cells in the cochlea do not work. An electronic ear can help.
   b. The ear bones cannot move. The stirrup is removed and replaced with a small piece of plastic that vibrates properly.

**STUDY GUIDE 96**

STUDY GUIDE — CHAPTER 16

Name ______ Date ______ Class ______

VOCABULARY

*Review the new words used in Chapter 16 of your textbook. Then, use the terms below to fill in the blanks in the sentences that follow.*

| | | | |
|---|---|---|---|
| auditory nerve | cochlea | eardrum | rods |
| olfactory nerve | cones | epidermis | sclera |
| lens muscles | cornea | nearsighted | pupils |
| vitreous humor | irises | farsighted | retina |
| optic nerve | lenses | taste buds | dermis |
| semicircular canals | | | |

1. The sclera is a tough, white covering that protects the eye.
2. Your taste buds allow you to taste sweet, salty, bitter, and sour.
3. The olfactory nerve carries messages from neurons in the nose to the brain.
4. When you focus on distant objects, your lens muscles change the shape of the lenses in your eyes.
5. When sound waves enter your ear, they cause the eardrum to vibrate.
6. The vitreous humor keeps your eyes round in shape.
7. The skin is made of two layers: the dermis and the epidermis.
8. The retina is made of two kinds of neurons: rods that detect motion and cones which detect color.
9. The cornea is a clear covering over the pupil.
10. The semicircular canals in your inner ear help you keep your balance.
11. Your irises determine the color of your eyes.
12. A farsighted person might need to wear glasses to read a book. A nearsighted person might need to wear glasses to drive a car.
13. The nerve cells in the cochlea send messages to the auditory nerve, which carries messages of sound to the brain.
14. When you enter a dark room, your pupils become larger.
15. The optic nerve carries messages from the retina to the brain.

## TEACH

### Concept Development

▶ Not all hearing takes place because of sound waves entering the outer ear. Sound conduction by the skull also plays an important role. Sound bypasses the middle ear and stimulates the cochlea. The ability to hear sound conducted by bone is used to determine if a person is suffering from nerve deafness or middle ear deafness.

### Guided Practice

Have students write down their answers to the margin questions: A sound of 85 decibels or more can damage hearing. An electronic ear has a wire that is placed inside the cochlea and carries messages from outside the ear to the cochlea.

**How loud a sound can damage your hearing?**

How loud must a sound be before it damages your hearing? Table 16-1 shows the loudness of different sounds that you hear frequently. The loudness of sound is measured in units called decibels (DEH suh bulz). The softer a sound is, the lower its decibel value will be. The louder a sound is, the higher its decibel value will be. Listening to sounds that have decibel values over 85 can damage your hearing. Notice in the table that the decibel values for a motorcycle, jet, and rock concert are over 85.

What can be done to prevent hearing loss? Avoiding the sources of loud noises is one way. If loud noises can't be avoided, then hearing protection can be used. People who work at airports wear earmuffs to protect their hearing. People who work in factories often use earplugs made of plastic or foam rubber. It is important to prevent hearing loss. Once the hearing has been damaged by loud noises, it can't be restored.

**TABLE 16–1. DECIBEL VALUES OF COMMON SOUNDS**

| Decibels | Sound | |
|---|---|---|
| 20 | rustling of leaves | |
| 30 | whisper | |
| 50 | light traffic | |
| 60 | air conditioner | |
| 70 | loud talk | |
| | television | |
| | vacuum cleaner | |
| 85 | food blender | |
| 90 | motorcycle at 10 meters | hearing damage may result |
| | shouting | hearing damage may result |
| | power mower | hearing damage may result |
| 100 | snowmobile | hearing damage may result |
| | jet overhead | hearing damage may result |
| 110 | portable radio with headphones | hearing damage may result |
| 120 | thunderclap | hearing damage may result; painful to the ear |
| | rock concert | hearing damage may result; painful to the ear |
| 140 | jet nearby | hearing damage may result; painful to the ear |
| 160 | shotgun | hearing damage may result; painful to the ear |

## Correcting Hearing Problems

There are many causes of deafness or hearing loss. Sometimes the problem lies with the auditory nerve or cochlea. Hearing problems can happen if the middle ear bones don't move smoothly. Let's look at some new ways of treating these problems.

A person may become deaf if the nerve cells in the cochlea don't work. An electronic ear can help, Figure 16-12. The electronic ear has a long wire that is placed inside the cochlea. The wire carries messages from outside



## OPTIONS

**Challenge:** Have students research how a hearing aid works. A hearing aid will help with conduction deafness or hearing loss due to the inability of the bones to vibrate and conduct sounds to the cochlea. In contrast, an implant helps with nerve deafness.

**Application**/*Teacher Resource Package*, p. 16. Use the Application worksheet shown here to help students learn about skin disorders.

**APPLICATION 16**

APPLICATION: HEALTH — CHAPTER 16

Name ______ Date ______ Class ______

SKIN DISORDERS — *Use after Section 16:3.*

The largest organ of the human body is the skin. Many disorders affect the skin.

*Use a dictionary or encyclopedia to find the answers to the following questions.*

1. The disorders listed below are caused by infections. For each disorder, tell whether it is caused by a bacteria, virus, fungus, insect, or insect-like creature. Then, describe the symptoms.

   Boils **bacteria; painful red lumps with pus**

   Warts **virus; small, hard lump (Virus gets into skin cells and causes them to multiply rapidly.)**

   Ringworm **fungus; patches of ring-shaped, red, itchy, scaly skin**

   Impetigo **bacteria; thin blisters that burst and form crusts (Occurs mainly in children.)**

   Athlete's foot **fungus; cracks form in the skin between the toes, peeling skin**

2. Burns may be caused by heat from fire, chemicals, electric shock, or overexposure to sunlight or other radiation. Describe the symptoms of the three types of burns below.

   First-degree burns **skin turns red; damages only the epidermis**

   Second-degree burns **skin blisters; damages epidermis and part of the dermis; may leave scars**

   Third-degree burns **skin blisters and/or turns black; damage to all skin layers; scarring; may require skin grafts; may be life-threatening**

3. Tumors are abnormal growths of cells. Some types of tumors form on the skin. They may be malignant or benign. Define these terms.

   Malignant **cancerous; can spread through the body, causing illness or death**

   Benign **not cancerous; will not spread throughout the body**

4. Describe the symptoms of skin cancer caused by overexposure to the sun. **One or more tumors may form on the skin, often on the face. The tumor may look like a pink lump or an ulcer with a crust. In one kind of skin cancer, a new mole may form or an old mole may become sore, itchy, or crusty. It may enlarge or bleed. The cancer can spread and cause death.**

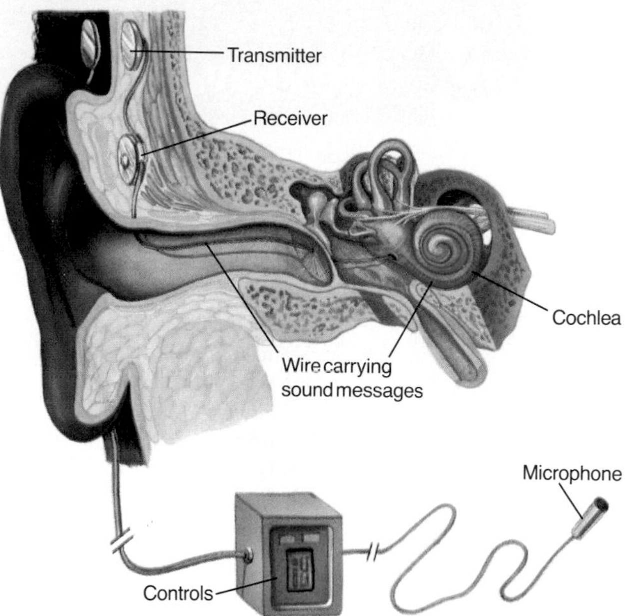


**Figure 16-12** An electronic ear helps to restore certain kinds of hearing loss.

the ear to the cochlea. The electronic ear lets people hear sounds that they couldn't hear without it.

Remember that the job of middle ear bones is to move back and forth. Deafness may result if the ear bones can't move. The ear bone that most often gets stuck is the stirrup. When it does, the middle ear membrane can't move, and messages no longer reach the cochlea. To treat this problem, the stirrup is removed during an operation. It is replaced with a small piece of plastic. The plastic stirrup vibrates properly and moves the membrane.

**How does an electronic ear work?**

## Check Your Understanding

11. Explain what causes a person to be nearsighted or farsighted.
12. How does loud noise cause hearing loss?
13. Describe two ways of correcting hearing problems.
14. **Critical Thinking:** How would damage to the optic nerve affect vision?
15. **Biology and Reading:** What system of measurement do eye doctors use?



## TEACH

### Check for Understanding

Have students respond to the first three questions in Check Your Understanding.

### Reteach

**Reteaching**/*Teacher Resource Package,* p. 49. Use this worksheet to give students additional practice with vision problems.

**Extension:** Assign Critical Thinking, Biology and Reading, or some of the **OPTIONS** available with this lesson.

## 3 APPLY

### Brainstorming

Have students consider the following question: Why might damage to the retina or optic nerve be the most difficult problem to correct?

## 4 CLOSE

### ACTIVITY/Audiovisual

Show the videocassette *Eyes and Ears,* Films for the Humanities and Sciences.

### Answers to Check Your Understanding

11. Nearsightedness is caused by an eyeball that is too long. Light focuses in front of the retina. Farsightedness is caused by an eyeball that is too short. Light focuses at the back of the retina.
12. Loud noises damage the inside of the cochlea by flattening the hairlike cells in the cochlea. Once the hairlike cells are flattened, they can't send messages to the auditory nerve.
13. An electronic ear may be used. A stirrup may be replaced with one made of plastic.
14. The person would probably be blind.
15. Doctors who study the eye use standard English measurements.

## RETEACHING 49

RETEACHING CHAPTER 16

Name ____ Date ____ Class ____

*Use with Section 16:3.*

PROBLEMS WITH SENSE ORGANS

| Vision Problems | Description | Cause |
|---|---|---|
| Farsighted | able to see clearly far away but not close up | path of light meets in back of the retina instead of on it |
| Nearsighted | able to see clearly close up but not far away | path of light meets in front of the retina instead of on it. |

*Label the drawings as: normal, farsighted, nearsighted.*

A nearsighted B normal C farsighted

1. How can vision problems be corrected? They can be corrected by wearing lenses that bend the light entering the eye. This bending allows the light rays to meet on the retina.
2. What can help a person when the nerve cells in the cochlea do not work? An electronic ear, consisting of a long wire is placed in the cochlea. The wire carries messages from outside the ear to the cochlea. The person can hear sounds he or she could not hear without it.
3. What can help a person who is deaf because the stirrup can't move? The stirrup is replaced with a small piece of plastic that vibrates properly and moves the oval window.

## OPTIONS

**Enrichment:** Ask students to research one of the following questions. Are decibels and hearing range the same thing? How do they differ? What is the hearing range for humans, dogs, dolphins, bats?

**Enrichment:** Ask students to design a poster that alerts teens to the danger of continuous loud music.

## CHAPTER 16 REVIEW

### Summary

Summary statements can be used by students to review the major concepts of this chapter.

### Key Science Words

All boldfaced terms from the chapter are listed.

### Testing Yourself

#### Using Words

1. vitreous humor
2. eardrum
3. cochlea
4. olfactory nerve
5. dermis
6. taste bud
7. nearsighted
8. iris
9. cone
10. semicircular canal
11. lens

#### Finding Main Ideas

12. p. 332; Planaria have two eye spots that detect light and dark. Insects have compound eyes made up of many eyelike parts. Humans have a complex eye that can change for viewing close and distant objects. It can detect colors and motion. Only one lens is present.
13. p. 334; The parts that protect the eye are the eyelid, cornea, and sclera. The iris and pupil adjust the amount of light allowed to enter the eye.
14. p. 347; Causes of deafness can be middle-ear problems, such as immobile bones, or nerve problems involving the inner ear or auditory nerve.
15. p. 343; Nerve cells for pressure and touch are found in the dermis; the nerve cells for pain are found in both the dermis and epidermis.
16. p. 339; Various tastes are a result of combinations of the basic tastes.
17. p. 341; Nerve cells in the cochlea connect to the auditory nerve.
18. pp. 344-345; A nearsighted person sees clearly close up but not far away as a result of an eye that is longer from front to back than a normal eye. A farsighted person sees clearly far away but not close up as a result of an eye that is shorter from front to back than a normal eye.

#### Using Main Ideas

19. Sound waves cause these parts to vibrate.
20. (a) iris and lens muscle, (b) eyelid, sclera, and cornea, (c) vitreous humor, (d) retina and optic nerve, (e) cornea and lens
21. A person standing 20 feet from an eye chart can see what a person with normal vision can see at 60 feet.
22. Sound is detected by parts on its front legs (ears). Touch is detected by antennae on its head and movement is detected by eyes on its head.
23. The olfactory nerve carries messages from nose to brain, the auditory nerve from ear to brain, and the optic nerve from eye to brain.
24. A compound eye has many eyelike parts. A human eye has a single lens.
25. The message is interpreted and messages are flipped right side up.

# Chapter 16

# Review

## Summary

### 16:1 Observing the Environment

1. Sense organs differ among animals. Planaria have eyespots. Earthworms have light-sensitive skin.
2. Crickets detect sound with their front legs, touch and chemicals with antennae, and light with compound eyes. Snakes smell with their tongues.

### 16:2 Human Sense Organs

3. The human eye has parts for reducing or increasing the amount of light entering the eye, focusing, and for seeing color or black and white.
4. The tongue detects molecules dissolved in water. It senses sour, salty, sweet, and bitter. The human nose detects molecules or odors in the form of gases. Skin contains nerves that detect pain, pressure, touch, heat, and cold.
5. The human ear detects air molecules in motion. It is divided into an outer, middle, and inner area.

### 16:3 Problems with Sense Organs

6. Nearsighted and farsighted vision are corrected with glasses or contact lenses to provide 20/20 vision.
7. Loud noise causes hearing loss by damaging the nerve cells in the cochlea.
8. Certain types of deafness can be helped with a plastic stirrup or electronic ear.

## Key Science Words

auditory nerve (p. 341)
cochlea (p. 341)
cone (p. 336)
cornea (p. 335)
dermis (p. 343)
eardrum (p. 340)
epidermis (p. 343)
farsighted (p. 345)
iris (p. 334)
lens (p. 335)
lens muscle (p. 335)
nearsighted (p. 344)
olfactory nerve (p. 339)
optic nerve (p. 336)
pupil (p. 334)
retina (p. 335)
rod (p. 336)
sclera (p. 334)
semicircular canals (p. 341)
sense organ (p. 332)
taste bud (p. 339)
vitreous humor (p. 335)

## Testing Yourself

### Using Words

*Choose the word from the list of Key Science Words that best fits the definition.*

1. liquid inside the eye
2. membrane at the end of the ear canal
3. liquid-filled chamber of the ear
4. carries messages from the nose to the brain
5. inner layer of the skin
6. detects molecules on the tongue
7. can see clearly close up but not far away
8. adjusts amount of light entering eye
9. nerve cell that detects color
10. helps us keep our balance
11. changes shape for viewing distant and nearby objects

348

# Review

## Testing Yourself *continued*

**Finding Main Ideas**

*List the page number where each main idea below is found. Then, explain each main idea.*

12. how the eyes of planaria, insects, and humans differ
13. the eye parts that protect the eye and the eye parts that adjust light
14. two causes of deafness
15. layer of skin in which you find many pressure, touch, and pain nerves
16. how you are able to taste more than just four types of taste
17. how the cochlea aids hearing
18. how nearsighted and farsighted vision differ

**Using Main Ideas**

*Answer the questions by referring to the page number after the question.*

19. What effect do sound waves have on the eardrum, ear bones, and middle ear membrane? (p. 341)
20. Which eye parts (a) are muscles, (b) protect the eye, (c) give the eye shape, (d) are made of nerves, and (e) bend light? (pp. 334-335)
21. What is meant by 20/60 vision? (p. 344)
22. How does a cricket detect sound, touch, and movement? (pp. 332-333)
23. What is the job and the sense organ associated with the following nerves?
    (a) olfactory (p. 339)
    (b) auditory (p. 341)
    (c) optic (p. 336)
24. How does a compound eye differ from a human eye? (p. 333, pp. 334-336)
25. Describe two jobs of the cerebrum when messages from the eye reach it. (p. 336)

## Skill Review 

*For more help, refer to the* ***Skill Handbook,*** *pages 704-719.*

1. **Interpret diagrams:** Study Figure 16-6. Where on the tongue are the four tastes detected?
2. **Predict:** How would puncturing the eardrum affect the ability to hear?
3. **Observe:** Look at your pupils in dim light, then in bright light. How does the size of the pupil change?
4. **Formulate a model:** Strike a tuning fork with a rubber mallet. Note the sound it makes. Place the tuning fork into a glass of water. What happens to the water? How does this model show what happens within the cochlea?

## Finding Out More

**Critical Thinking**

1. Why is it impossible to transplant the retina or optic nerve? Why is it possible to transplant the cornea?
2. Why is the pupil black?

**Applications**

1. Design an experiment to test which sense is better developed in most people, taste or smell.
2. Using a large ball and strips of cloth, construct a model of the human eye. Use the cloth to represent the muscles located on the outside of the eye. Determine the job of each muscle pair. Indicate these jobs on your model.



## Skill Review

1. Sweet taste is detected on the tip of the tongue, sour on the sides toward the back, bitter on the far back center, and salt on the sides toward the front.
2. Puncturing the eardrum would cause sounds to be distorted, since the eardrum would not be functioning properly. It would not vibrate correctly and that would cause sounds to be perceived differently.
3. The size of the pupil gets larger in dim light, smaller in brighter light. The iris closes partially in order to make the pupil smaller, opens partially to make the pupil larger.
4. The water in the glass will form waves similar to those in a pond when a stone is dropped into it. In the same manner, when a sound wave (vibration) hits the fluid in the cochlea, the liquid sends out waves that are sensed by the nerve cells in the cochlea, which send the message to the auditory nerve.

## Finding Out More

### Critical Thinking

1. It is impossible to transplant the retina or optic nerve because both are nerve tissue, and nerve tissue, once it is severed, does not regenerate. Corneas, however, can be transplanted because that part of the eye will repair itself.
2. The pupil appears black because light striking the retina is absorbed, not reflected.

### Applications

1. Various solutions of different concentrations can be used. Starting with the weakest solution, students could tell at which point they taste or smell the solution.
2. The ball represents the eyeball. Strips of cloth on the top and bottom of the ball would represent the muscles that move the eye upward and downward. Those on the sides would move the eye from side to side.

 **Chapter Review**/*Teacher Resource Package,* pp. 82-83.

 **Chapter Test**/*Teacher Resource Package,* pp. 84-86.

**Chapter Test**/*Computer Test Bank*

# Animal Behavior

## PLANNING GUIDE

| CONTENT | TEXT FEATURES | TEACHER RESOURCE PACKAGE | OTHER COMPONENTS |
|---|---|---|---|
| (3 days)<br>17:1 Behavior<br>Steps of Behavior<br>Innate Behavior<br>Learned Behavior | Skill Check: *Understand Science Words,* p. 353<br>Lab 17-1: *Innate Behavior,* p. 355<br>Technology: *Blinking Reveals Behavior Secrets,* p. 356<br>Idea Map, p. 353<br>Check Your Understanding, p. 357<br>Mini Lab: *Do Onions Make You Cry?* p. 356 | Application: *Driving a Car,* p. 17<br>Reteaching: *Innate and Learned Behavior,* p. 50<br>Critical Thinking/Problem Solving: *Where Can You Observe a Circadian Rhythm?* p. 17<br>Focus: *Animal Behavior,* p. 33<br>Study Guide: *Behavior,* pp. 97-98<br>Lab 17-1: *Innate Behavior,* pp. 65-66 | **Laboratory Manual:**<br>*What Happens When You Learn Through Trial and Error?* p. 145<br>Color Transparency 17: *Stimulus and Response*<br>**STVS:** *Rattlesnakes,* Animals (Disc 5, Side 2) |
| (3 days)<br>17:2 Special Behaviors<br>Behaviors for Reproduction<br>Behaviors for Finding Food<br>Protective Behaviors<br>Migration<br>Parental Behavior | Lab 17-2: *Behavior,* p. 358<br>Skill Check: *Form Hypotheses,* p. 360<br>Mini Lab: *How Do Earthworms Behave?* p. 362<br>Idea Map, p. 364<br>Check Your Understanding, p. 365 | Enrichment: *Sleep Behavior,* p. 18<br>Reteaching: *Special Types of Behavior,* p. 51<br>Skill: *Interpret Data,* p. 18<br>Transparency Master: *Steps of Behavior,* p. 81<br>Study Guide: *Special Behaviors,* pp. 99-101<br>Study Guide: *Vocabulary,* p. 102<br>Lab 17-2: *Behavior,* pp. 67-68 | **Laboratory Manual:**<br>*What Self-protecting Behaviors Do Isopods Show?* p. 141<br>**STVS:** *March of the Spiny Lobster,* Animals (Disc 5, Side 1)<br>**STVS:** *Songbird Study,* Animals (Disc 5, Side 2) |
| Chapter Review | Summary<br>Key Science Words<br>Testing Yourself<br>Finding Main Ideas<br>Using Main Ideas<br>Skill Review | **ASSESSMENT RESOURCES**<br>Chapter Review, pp. 87-88<br>Chapter Test, pp. 89-91<br>Performance Assessment in the Biology Classroom<br>Alternate Assessment in the Science Classroom<br>Computer Test Bank | |

## GLENCOE TECHNOLOGY

**Infinite Voyage,** *The Living Clock*
**Science and Technology Videodisc Series,** *Rattlesnakes,* Animals (Disc 5, Side 2)
*March of the Spiny Lobster,* Animals (Disc 5, Side 1)
*Songbird Study,* Animals (Disc 5, Side 2)
**The Secret of Life,** *Gone Before You Know It: The Biodiversity Crisis*

## MATERIALS NEEDED

| LAB 17-1, p. 355 | LAB 17-2, p. 358 | MARGIN FEATURES |
|---|---|---|
| microscope<br>glass slide<br>coverslip<br>dropper<br>test tube rack<br>2 test tubes, small<br>2 stoppers to fit tubes<br>2 labels<br>vinegar eels | mirror<br>watch with second hand<br>pencil<br>paper | Skill Check, p. 353<br>dictionary<br>Mini Lab, p. 362<br>shoe box<br>black paper<br>live earthworm<br>Mini Lab, p. 356<br>onion<br>knife |

## OBJECTIVES

For more information about National Science Standards, see page 5T.

| SECTION | OBJECTIVE | CORRELATION of QUESTIONS to OBJECTIVES | | | |
|---|---|---|---|---|---|
| | | CHECK YOUR UNDERSTANDING | CHAPTER REVIEW | TRP CHAPTER REVIEW | TRP CHAPTER TEST |
| 17:1 National Science Stds: UCP.2, C.6 | 1. **Explain** the difference between a stimulus and a response. | 1 | 4, 9, 16, 20 | 1, 3, 9, 17, 29, 33 | 2, 5, 7, 30, 31, 32, 33, 34, 35 |
| | 2. **Explain** innate behavior and give examples. | 2, 4 | 1, 2, 3, 12, 15 | 18, 20, 23, 24, 32, 37 | 15, 17, 19, 21, 23, 25, 27, 28, 29 |
| | 3. **Describe** how behavior is learned and give examples of learned behavior. | 3 | 1, 3, 18 | 2, 7, 17, 21, 22, 28 | 3, 6, 16, 18, 20, 22, 24, 25, 27 |
| 17:2 National Science Stds: UCP.1, C.6, G.1 | 4. **Explain** how courting behaviors help an animal species reproduce. | 8 | 5, 10, 19, 23, 24 | 5, 11, 15, 30, 34 | 1, 4, 10, 39, 40, 41 |
| | 5. **Describe** how behavior helps an animal find food, shelter, and protection from enemies. | 6, 9, 10 | 7, 8, 14, 22 | 4, 6, 10, 12, 14, 16, 19, 26, 31 | 9, 11, 12, 13, 26, 37, 38, 42, 43, 44 |
| | 6. **Compare** reproduction in animals that give parental care with reproduction in animals that give no care. | 7 | 6, 11, 21 | 8, 13, 25 | 8, 14, 45, 46, 47, 48, 49 |

# Animal Behavior

## CHAPTER OVERVIEW

### Key Concepts

The chapter develops the concepts of stimulus and response. A distinction is made between innate and learned behavior. Special behaviors that promote survival within the animal kingdom are detailed, with a number of specific examples provided.

### Key Science Words

| | |
|---|---|
| behavior | migration |
| courting behavior | parental care |
| innate behavior | pheromone |
| instinct | social insect |
| learned behavior | stimulus |

### Skill Development

In Lab 17-1, students will **predict, separate and control variables, observe,** and **design an experiment** to show vinegar eels' response to gravity. In Lab 17-2, they will **form hypotheses, observe,** and **make and use tables** in a study of innate and learned behavior. In the Skill Check on page 353, students will **understand** the **science word** *innate.* On page 360, they will **form hypotheses** about moth behavior. In the Mini Lab on page 356, students will **infer** if crying from cut onions is learned or innate behavior. On page 362, they will **observe** earthworm protective behavior.

### Bridging

Animal behavior is the application of the nervous system in animals. Students will see a correlation between how one acts and the control that the nervous system has over these actions.

**CHAPTER PREVIEW**

### Chapter Content

Review this outline for Chapter 17 before you read the chapter.

**17:1 Behavior**
- Steps of Behavior
- Innate Behavior
- Learned Behavior

**17:2 Special Behaviors**
- Behaviors for Reproduction
- Behaviors for Finding Food
- Protective Behaviors
- Migration
- Parental Behavior

### Skills in this Chapter

The skills that you will use in this chapter are listed below.

- In **Lab 17-1,** you will predict, separate and control variables, observe, and design an experiment. In **Lab 17-2,** you will form hypotheses, observe, and make and use tables.
- In the **Skill Checks,** you will understand science words and form hypotheses.
- In the **Mini Labs,** you will infer and observe.



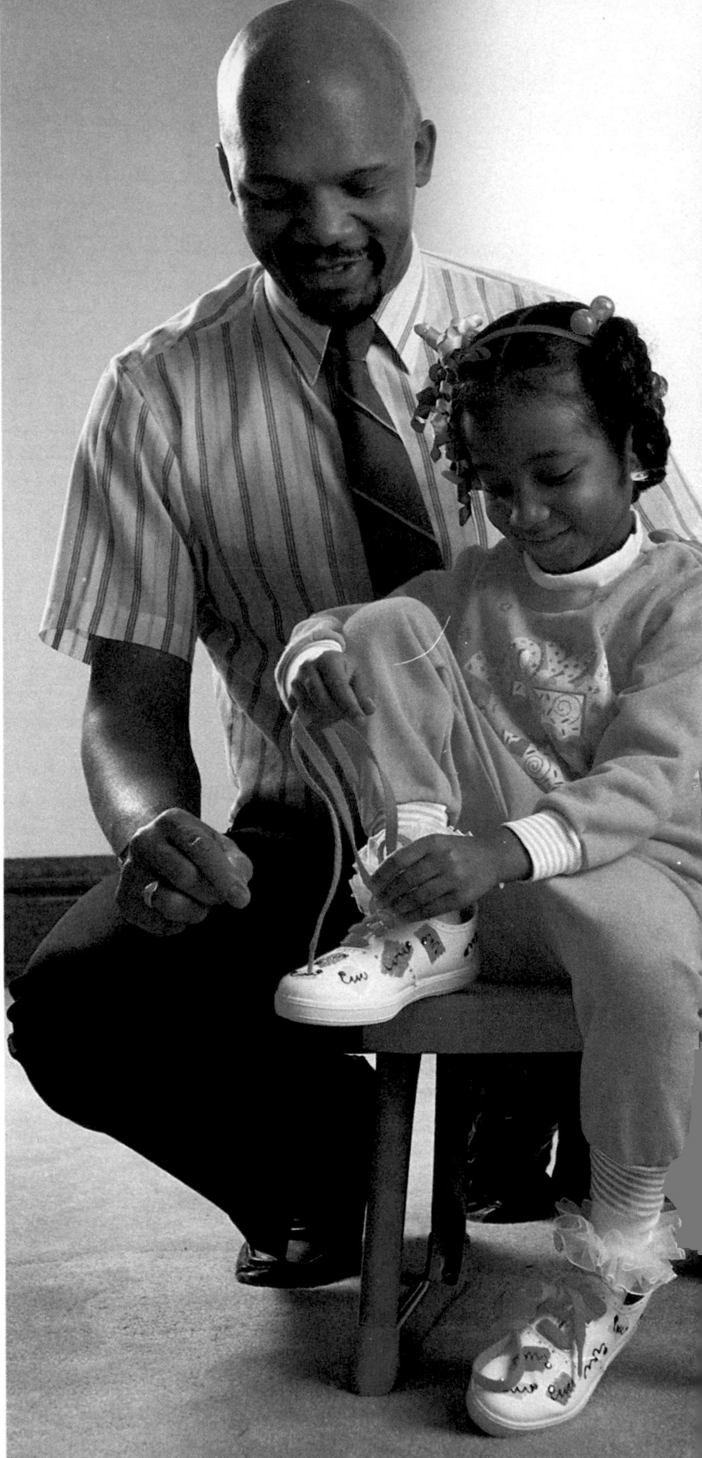

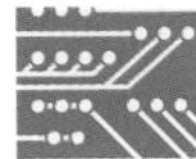

## TECH PREP

For Tech Prep activities in this chapter of the Teacher Wraparound Edition, see especially the Motivations on page 354, the Cooperative Learning activity on page 357, the Activities on pages 359 and 361, the Portfolio on page 361, and the Student Journal on page 362.

See also the Glencoe Homepage at **www.glencoe.com**

Chapter 17

# Animal Behavior

Children must learn how to tie their shoelaces. This skill is learned at home or in school. The child in the photo on the opposite page is being taught to tie her shoelaces. After a lot of practice, she will have learned this skill.

A bird building a nest has never gone to school to learn nest-building. How does the bird shown in the photograph below know how to build a nest without learning about it first? How does learning how to tie shoelaces differ from nest-building? These are some of the questions you will explore in this chapter.

**Try This!**

**How do you learn?** Find a book on origami, or paper folding. Follow directions on folding an animal, while you time how long it takes you to do this. Repeat the folding of the same animal several more times. Note how long each new trial takes. Does the length of time change as you practice?

**interNET CONNECTION**

For more information about the material in this chapter, follow the link for the chapter on the Glencoe Homepage at **http://www.glencoe.com**

Birds don't have to learn how to build nests.



**ASSESSMENT PLANNER**

***Portfolio***

Strategies on the following pages represent student products that can be placed into a best-work portfolio: pp. 361, 364.

**PERFORMANCE ASSESSMENT**

Skill Check, pp. 353, 360
Mini Lab, pp. 356, 362
Lab, pp. 355, 358

**CONTENT ASSESSMENT**

Check for Understanding, pp. 357, 365
Chapter Review, pp. 366-367

**GROUP ASSESSMENT**

Opportunities for group assessment occur with Cooperative Learning Strategies.

***Student Journal***

Strategies on the following pages represent opportunities for writing in a Student Journal: pp. 353, 362.

**GETTING STARTED**

## Using the Photos

The two photos illustrate two kinds of behavior—learned and innate. Have students brainstorm other behaviors that they think belong to one or the other category. Another example of learned behavior you might wish to demonstrate is the tying of a bowtie. Swallowing and salivating are two more examples of innate behavior you might discuss.

## MOTIVATION/Try This!

**How do you learn?** Students will **observe** that the time it takes to repeat the folding process should decrease as the task is repeated and learned. Locate library books that deal with origami for the paper folding.

## Chapter Preview

Have students study the chapter outline before they begin to read the chapter. They should note the overall organization of the chapter, that the general nature of behavior is covered first. Next, specific types of behaviors seen in the animal kingdom are covered.

## Misconception

Ask students to describe any animal behavior and they will typically describe it in human terms using such words as *pain, emotions,* and *desire.* We have a tendency to relate all animal behavior in terms of human reactions. We do this for all animals, simple or complex. Caution students not to think of animal behavior in terms of how we see it.

## 17:1 Behavior

### PREPARATION

#### Materials Needed

Make copies of the Focus, Critical Thinking/Problem Solving, Study Guide, Application, Reteaching, and Lab worksheets in the *Teacher Resource Package.*

▶ Order vinegar eels (*Turbatrix aceti*) from a biological supply house for Lab 17-1.

▶ Obtain mirrors from your physics teacher or alert students to bring them to class for Lab 17–2.

▶ Obtain onions for the Mini Lab.

#### Key Science Words

behavior
stimulus
innate behavior
instinct
learned behavior

#### Process Skills

In the Mini Lab, students will infer. In Lab 17–1, they will predict, separate and control variables, observe and design an experiment. In Lab 17–2, they will form hypotheses, observe, and make and use tables. In the Skill Check, they will understand science words.

### 1 FOCUS

▶ The objectives are listed on the student page. Remind students to preview these objectives as a guide to this numbered section.

#### ACTIVITY/Demonstration

Bang a textbook on the desk without warning the students. Discuss their reaction to the noise. Is this something we learned to do? Ask if a newborn child would be startled by the same noise.

**Focus**/*Teacher Resource Package,* p. 33. Use the Focus transparency shown at the bottom of this page as an introduction to animal behavior.

## 17:1 Behavior

#### Objectives

1. **Explain** the difference between a stimulus and a response.
2. **Explain** innate behavior and give examples.
3. **Describe** how behavior is learned and give examples of learned behavior.

#### Key Science Words

behavior
stimulus
innate behavior
instinct
learned behavior

How do animals know what is happening around them? How do they react to events occurring in their environment? Animals react to changes around them in many different ways. However, they all use their sense organs, nervous systems, and muscles when they react.

### Steps of Behavior

Everything that an animal does is part of its behavior. **Behavior** is the way an animal acts. The way an animal acts depends on what is going on around it.

Let's use an example to see how behavior works. The feeding behavior of an owl can be broken down into steps. Follow the steps in Figure 17-1 as you read.

1. The owl sees a mouse on the ground.
2. Messages pass along nerves from the bird's eyes to its brain.
3. The brain sends out new messages along nerve pathways to muscles.
4. Messages arriving at muscles allow the owl to grab the mouse.

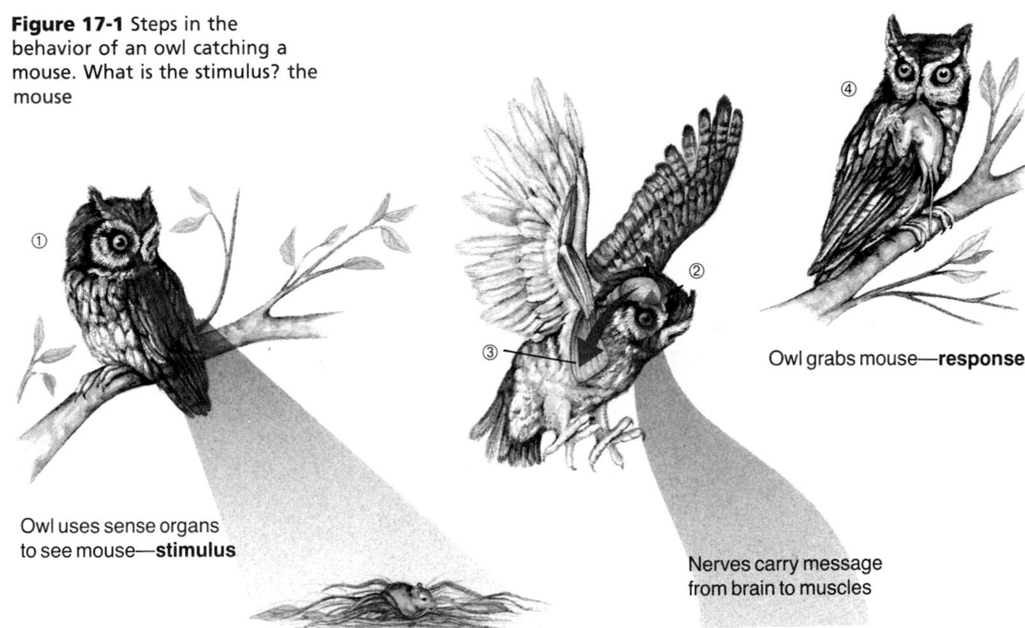


**Figure 17-1** Steps in the behavior of an owl catching a mouse. What is the stimulus? the mouse



### OPTIONS

#### Science Background

The study of behavior is called ethology. Ethology comes from the Greek word *ethos* meaning character or custom.

**Stimulus and Response**/*Transparency Package,* number 17. Use color transparency number 17 as you teach the steps of behavior.

FOCUS 33

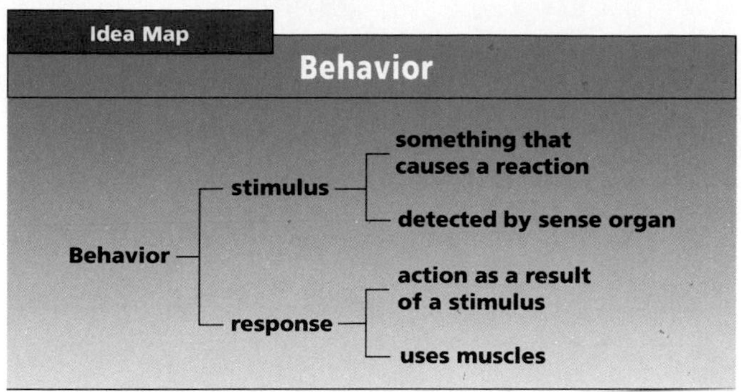


**Study Tip:** Use this idea map as a study guide to stimulus and response. Can you give an example of a stimulus and its response?

There are two important parts to all behavior. They are stimulus and response. A **stimulus** is something that causes a reaction in an organism. In the owl example, the mouse was the stimulus. Seeing it led to the owl's behavior.

**What are the two parts of behavior?**

Moving to get the mouse was the owl's response. You read in Chapter 15 that a response is the action of an organism as a result of a change, in this case a stimulus. Notice that the bird's response used muscles. How were they used?

Here is a different example. You are standing in a dark room and you think you are alone. You don't see anyone, but someone else is in the room with you and taps you on the arm. You jump and scream. What was the stimulus in this example? Which sense organs detected the stimulus? What were the responses? Which muscles were used to respond?

## Innate Behavior

There are two main types of behavior in animals—innate and learned. When a robin builds a nest, it is using innate behavior. **Innate behavior** is a way of responding that does not require learning. You did not learn to grab for objects, yawn, smile at your mother, or cry as a baby. All of these behaviors are innate. One type of innate behavior is an instinct. An **instinct** is a complex pattern of behavior that an animal is born with. Nest-building and mouse-catching are instincts. They are more complex than simple innate behavior, such as a yawn. In nest-building, for example, the bird must search for the proper material, gather it, and build the nest. The behavior may take several days to complete.

### Skill Check

**Understand science words: innate.** The word part *in* means inside. In your dictionary, find three words with the word part *in* in them. *For more help, refer to the **Skill Handbook**, pages 706-711.*



## 2 TEACH

### MOTIVATION/Brainstorming

Provide students with the following written paragraph: You approach what looks like a friendly dog tied to a leash. The dog snarls and bares its teeth. It begins barking at you and strains at its leash as it tries to reach you. You back away and leave. The dog stops barking.

Ask students:

1. What was the stimulus that started the dog's behavior?
2. What was the dog's response?
3. Did the dog use sense organs to detect the stimulus?

### Concept Development

▶ Instinctive behaviors are practiced by an entire species or a large fraction of the population. For example, male fruit flies all vibrate their wings when courting a female.

**Using Science Words:** Familiarity with the terms *stimulus* and *response* is essential to understanding this chapter. Review the definitions and then give students several examples in which they identify the stimulus and the response.

### Idea Map

Have students use the idea map as a study guide to the major concepts of this section.

### Guided Practice

Have students write down their answers to the margin question: stimulus and response.

### *Student Journal*

Have students use the steps of behavior on p. 352 as a guide, and list the steps of behavior for a cat catching a mouse. The steps must include the correct use of the terms *stimulus* and *response.*

### CRITICAL THINKING 17

CRITICAL THINKING/PROBLEM SOLVING — CHAPTER 17

Name ______ Date ______ Class ______

*Use after Section 17:1.*

**WHERE CAN YOU OBSERVE A CIRCADIAN RHYTHM?**

Each day, members of a species of crabs called *Sesarma* run along the sea bottom during high tide. They are hunting for food. At low tide, the crabs stay in their burrows. The crabs are showing a circadian (sir KAY dee un) rhythm. An animal that does the same thing at the same time every day is showing a circadian rhythm. The graph in Figure 1 below shows the number of crabs that are out of their burrows at different times during the day. Figure 2 shows the times of high and low tides in the ocean where the crabs live.

A scientist observed the behavior of the crabs in the laboratory. Figure 3 shows what he observed. The graph shows the number of crabs that are out of their burrows at different times of day. The arrows show the times of high tides in the ocean from which the crabs were taken. Next, the scientist changed the conditions in the laboratory. He kept the light on day and night. The crabs still followed the pattern of behavior shown in Figure 3. Does this indicate that the crabs were showing a circadian rhythm?

**Analyzing the Problem**

1. How does the crabs' cycle of behavior in the ocean relate to the cycle of high and low tides? The cycles match exactly.
2. How might a cycle of light and dark influence the crabs' cycle? The crabs might be attracted to light or to darkness.
3. Did the crabs' behavior in the laboratory relate to high and low tides in the same way as when the crabs were in the ocean? yes
4. Did the crabs' behavior change when the lights were on all the time? no

**Solving the Problem**

1. Do the crabs seem to have a "built in" clock that tells them when high or low tide is taking place? Explain. Yes, their cycle was the same in the laboratory as in the ocean.
2. Explain how the crabs' behavior cycle fits the definition of circadian rhythm. The crabs are out of their burrows at the same times every day.

Figure 1 — Crabs out of burrow; 1000; 0 (zero); 6 p.m.; 12 midnight; 6 a.m.

Figure 2 — Cycle of tides; Low tide; High tide; 6 p.m.; 12 midnight; 6 a.m.

Figure 3 — Crabs out of burrows (in laboratory); 50; 0 (zero); High tide; 6 p.m.; 12 midnight; 6 a.m.

## OPTIONS

**Critical Thinking/Problem Solving/** *Teacher Resource Package,* p. 17. Use the worksheet shown here to help students understand circadian rhythms.

## TEACH

### MOTIVATION/Analogy

Use a tape recorder in class to demonstrate the analogy of innate behavior. Play a short section of a current song. Stop the tape and rewind it. Play the section again. Then ask these questions:

1. (a) Was the song already on the tape when I bought it? (b) Will the same song play each time I start the tape at the same place? (c) Will the tape change if I wait a month? A year?
2. (a) Is a message already in the nervous system when an animal is born? (b) Will the same message be played back with the same response? (c) Does innate behavior ever change?

### Independent Practice

**Study Guide**/*Teacher Resource Package,* pp. 97–98. Use the Study Guide worksheets shown at the bottom of this page for independent practice.

### Guided Practice

Have students write down their answers to the margin question: no, innate behavior can't be changed.

### MOTIVATION/Brainstorming

Ask students to give examples of other innate behaviors. Some possibilities are: yawning, coughing, sneezing, blinking, crying; growling or snapping in a dog, grooming in a cat.

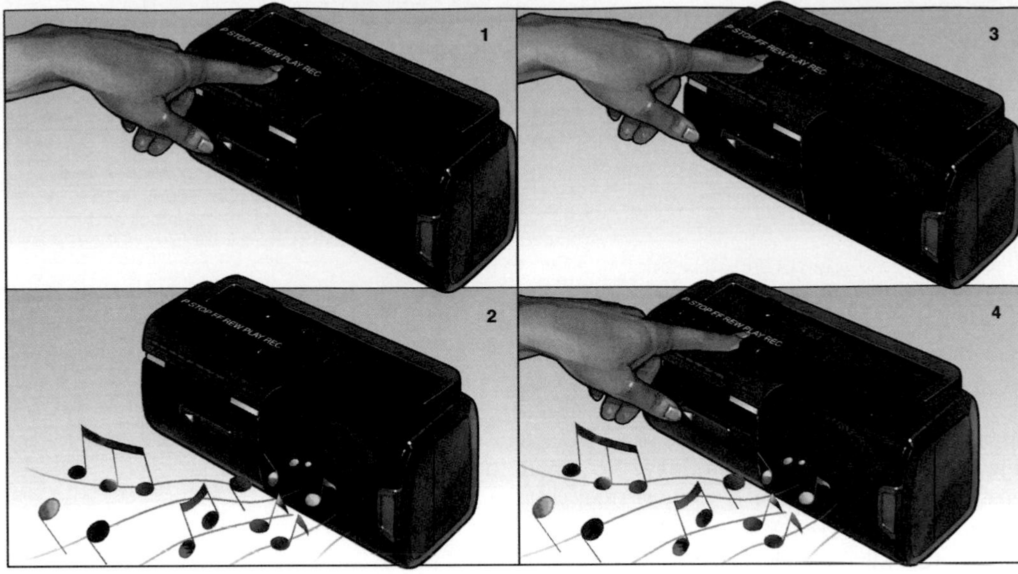


**Figure 17-2** Innate behavior can be compared to the playing of a prerecorded tape.

**Can innate behavior be changed?**

Let's compare innate behavior to running a tape recorder. Follow the steps in Figure 17-2 as you read.
1. You put a tape of your favorite song into a tape recorder. Then you press the play button.
2. The recorder plays the song.
3. You rewind the tape and press play again.
4. The recorder plays the same song as long as you continue to play the same tape.

Innate behavior is similar to playing this pre-recorded tape. Putting the tape into the recorder is like putting the stimulus into the nervous system. The playing song is like the response to the stimulus.

Each time a certain stimulus for an innate behavior occurs, the animal has the same response. Innate behaviors cannot be changed. Wasn't this also the case with our tape recorder? Let's look at a behavior that you have experienced but never had to learn. You swallow, and food goes down your trachea, or windpipe, instead of your esophagus. Your response is to cough. Coughing brings the food up and out of your windpipe. Why is coughing innate behavior? The same thing happens each time. You don't have to think about it and you can't change it.

Instincts, too, are automatic. A robin does not have to think about what kind of nest to build. All robins build the same shape nests, made out of the same materials. They can't change the kind of nest they build.

> **Bio Tip**
>
> **Health:** An innate behavior can be seen in infants when they grasp a rattle. The middle finger always closes on the handle first.

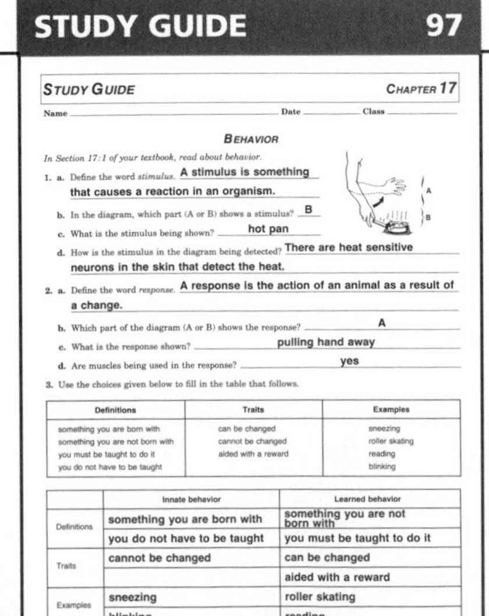

**STUDY GUIDE 97**

STUDY GUIDE — CHAPTER 17

Name ______ Date ______ Class ______

**BEHAVIOR**

*In Section 17:1 of your textbook, read about behavior.*

1. a. Define the word *stimulus.* A stimulus is something that causes a reaction in an organism.
   b. In the diagram, which part (A or B) shows a stimulus? B
   c. What is the stimulus being shown? hot pan
   d. How is the stimulus in the diagram being detected? There are heat sensitive neurons in the skin that detect the heat.
2. a. Define the word *response.* A response is the action of an animal as a result of a change.
   b. Which part of the diagram (A or B) shows the response? A
   c. What is the response shown? pulling hand away
   d. Are muscles being used in the response? yes
3. Use the choices given below to fill in the table that follows.

| Definitions | Traits | Examples |
|---|---|---|
| something you are born with<br>something you are not born with<br>you must be taught to do it<br>you do not have to be taught | can be changed<br>cannot be changed<br>aided with a reward | sneezing<br>roller skating<br>reading<br>blinking |

| | Innate behavior | Learned behavior |
|---|---|---|
| Definitions | something you are born with | something you are not born with |
| | you do not have to be taught | you must be taught to do it |
| Traits | cannot be changed | can be changed |
| | | aided with a reward |
| Examples | sneezing | roller skating |
| | blinking | reading |

**STUDY GUIDE 98**

STUDY GUIDE — CHAPTER 17

Name ______ Date ______ Class ______

**BEHAVIOR**

4. Determine if each of the following human behaviors is innate or learned. Write the word *innate* or *learned* on the blank beside each.
   a. speaking a foreign language learned
   b. coughing innate
   c. feeling pain innate
   d. sweating innate
   e. tying shoes learned
   f. biting fingernails learned
   g. having a fever innate
   h. shivering innate
   i. riding a bicycle learned
5. Determine if each of the following behaviors is innate or learned behavior for a dog. Write the word *innate* or *learned* on the blank beside each.
   a. fetching a newspaper learned
   b. nursing as a puppy innate
   c. scratching innate
   d. panting innate
   e. wagging tail innate
   f. "speaking" for supper learned
6. Place a checkmark next to the human behaviors that would get better with practice.
   a. speaking a foreign language ✓
   b. feeling pain ______
   c. tying shoelaces ✓
   d. sweating ______
   e. eating with a fork ✓
   f. riding a bicycle ✓
   g. dreaming ______
   h. having a fever ______
7. Place a check next to the dog behaviors below that would get better with practice.
   a. sleeping ______
   b. rolling over for a biscuit ✓
   c. fetching a newspaper ✓
   d. panting ______
   e. sitting up ✓
8. Using your answers from questions 6 and 7, explain if it is possible to:
   a. improve innate behaviors No, these behaviors are automatic and cannot be changed.
   b. improve learned behaviors Yes, learned behaviors can be improved by repetition or practice.

# Lab 17–1 Innate Behavior

## Problem: How do vinegar eels respond to gravity?

### Skills

predict, separate and control variables, observe, design an experiment

### Materials

| | |
|---|---|
| microscope | 2 test tubes, small |
| dropper | 2 stoppers to fit tubes |
| glass slide | 2 labels |
| coverslip | test-tube rack |
| vinegar eels | |

### Procedure

1. Copy the data table.
2. Use a dropper to place a drop of liquid containing vinegar eels onto a glass slide. Gently add a coverslip.
3. **Observe:** Observe the eels under low power of your microscope. Note their movement.
4. Draw several eels in the circle in the table.
5. **Predict:** Certain animals move toward gravity, while others move against gravity. Predict whether the vinegar eels will move toward gravity, against gravity, or will show no response to gravity.
6. Fill a small test tube with the liquid containing vinegar eels. Stopper the tube. If there is any air in the tube, remove the stopper and add a few more drops of liquid. Replace the stopper.
7. Fill a second tube half full with the liquid containing vinegar eels. Stopper the tube.
8. Place both tubes in the test-tube rack and label them with your name. Let the tubes stand overnight.
9. The following day, note carefully the location of the vinegar eels in each tube. **NOTE:** *Do not move or touch the tubes.*
10. Record on the tube outlines in the table where most of the eels are found.
11. Return the eels to their original container.

| OBSERVATIONS OF VINEGAR EELS | |
|---|---|
| Vinegar eels, 100X | |

### Data and Observations

1. Describe the appearance and movement of vinegar eels under the microscope.
2. If vinegar eels move toward gravity, they will be found near the bottom of the test tube. If they move away from gravity, they will be found near the top of the test tube. If they do not show any behavior in response to gravity, they will be found evenly throughout the tube. What type of behavior do vinegar eels show toward gravity?

### Analyze and Apply

1. What evidence do you have that the animals are responding to gravity and not to the air at the top of the tube?
2. Is the behavior of vinegar eels in response to gravity innate or learned? Explain.
3. **Apply:** How might vinegar eels benefit from the type of behavior that they show in response to gravity?

### Extension

**Design an experiment** to show the type of behavior vinegar eels might show in response to light and dark.



## ANSWERS

### Data and Observations

1. Vinegar eels are very small, roundworms that show a jerky movement.
2. They move away from gravity. Most eels will be found near the top of the test tube.

### Analyze and Apply

1. Vinegar eels in both tubes showed the same response.
2. The response is innate behavior. It does not change.
3. Food or oxygen might be more plentiful at the top.

## Lab 17-1 Innate Behavior

### Overview

Students will use vinegar eels (members of the phylum Nematoda, or roundworms) to study the behavior of these animals to gravity.

**Objectives:** Upon completion of the lab, students will be able to (1) **describe** the appearance of these animals (2) **describe** how one can experimentally study behavior of an animal.

**Time Allotment:** Setup time 20 minutes, following day 10 minutes

### Preparation

**Lab 17-1 worksheet**/*Teacher Resource Package,* pp. 65-66.

### Teaching the Lab

▶ Remind students on the following day to look carefully at the tube to determine where the animals are concentrated. A hand lens or magnifying lens might help in observing the vinegar eels.

▶ **Troubleshooting:** The adding of a viscous liquid (methyl cellulose) to the slide will slow movement of the animals.

**Cooperative Learning:** Students could work in teams of two to conserve on equipment. For more information, see pp. 22T-23T in the Teacher Guide.

### ✓ ASSESSMENT

**Skill:** Have students diagram the locations of vinegar eels in two test tubes–one completely full and the other half full–that are placed on their sides overnight.

TECHNOLOGY

### Blinking Reveals Behavior Secrets

**Background**

According to scientist John Stern of Washington University, blinking is related to behavior states such as excitement, fatigue, and anxiety.

**Discussion**

Have students perform the following experiment: Determine the average number of blinks per minute for several students in the class who claim they are not tired. Determine the average number of blinks for several students who claim they are very tired. Is there any difference in the averages of the two groups?

**References**

Chollar, Susan. "In the Blink of an Eye." *Psychology Today*, Mar. 1988, p. 8.

## TEACH

**Mini Lab**

This is a good example of behavior that cannot be changed, controlled, or altered with experience.

**ASSESSMENT**

**Skills:** Allow students to smell freshly ground pepper and note their reactions. Have them evaluate the taste of a lemon drop. Ask them if the responses to smelling pepper and tasting lemon are learned or innate behaviors.

***GLENCOE* TECHNOLOGY**

**Videodisc**

**Biology: The Dynamics of Life**
Movie: *Elephant Behavior*
Disc 1, Side 1, Ch. 3

Movie: *Territorial Behavior*
Disc 2, Side 1, Ch. 25

TECHNOLOGY

## Blinking Reveals Behavior Secrets

Blinking your eyes is an innate behavior. When you blink your eyes, you are protecting them from dust or smoke, and you are providing moisture for the cornea to prevent it from drying out. You don't have to think about blinking. It happens by itself.

Normally, people blink about 10 to 20 times each minute. Scientists have found that if a person is lying, he or she blinks more often.

Scientists also found that a person who is tired will blink about 30 or 40 times per minute. This information would be helpful in alerting pilots to the fact that they are too tired to be flying airplanes. Scientists could design an instrument that would tell the pilot if his or her blinking has increased beyond the normal rate. Thanks to people who study blinking, flying may be safer in the future.

Airplane pilot

## Learned Behavior

The second main type of behavior is learned behavior. **Learned behaviors** are behaviors that must be taught. These behaviors must be practiced until a certain stimulus results in a certain response. Behaviors such as tying your shoelaces or writing your name are learned behaviors.

Let's look at how learned behavior compares to recording a tape. Follow the steps in Figure 17-3 as you read.

1. You put a blank tape into a tape recorder. Then you press the play button. You hear nothing.
2. You rewind the tape and record a song.
3. You play back the tape. This time the recorder plays the song you just recorded.

You can continue to record the song until it sounds good to you. You can correct mistakes and record again.

Learned behavior is like recording a tape. There may be no response the first time a stimulus is received. The same kind of thing happens with a blank tape. It plays nothing.

Responding over and over to a stimulus can lead to learned behavior. Recording over and over can lead to a song that sounds good to you. Just as you recorded on the blank tape, the nervous system can "record" a behavior.

### Mini Lab

**Do Onions Make You Cry?**

**Infer:** Cut an onion. Hold the cut edge close to your face. Describe what happens. Could you have prevented this behavior? Is this behavior innate or learned? Why? *For more help, refer to the **Skill Handbook**, pages 706-711.*



### *OPTIONS*

**Challenge:** Have students relate some behaviors their pets have learned.

**Application**/*Teacher Resource Package*, p. 17. Use the Application worksheet shown here to show students the importance of learning safety skills for driving a car.

**What Happens When You Learn Through Trial and Error?**/*Lab Manual*, pp. 145-148. Use this lab as an extension to learning behavior.

**APPLICATION 17**

APPLICATION: SAFETY CHAPTER 17

Name ______ Date ______ Class ______

*Use after Section 17:1.*

LEARNED BEHAVIOR: DRIVING A CAR

You have learned that behaviors that must be taught are called learned behaviors. Many learned behaviors are made up of a series of small steps. To learn to drive a car safely, many behaviors must be learned.

*Under each heading below, list the steps that are part of learning to drive. Observe a driver in your family if necessary.*

1. To start a car, you would need to learn:
   to have the car in the parking gear
   where to put the key
   how to turn the key in the ignition
2. To make the car move in a straight line at a slow speed, you would also need to learn:
   how to shift gears
   how to work the gas pedal and control your speed
   how to work the brake
   how to hold the steering wheel to make the car go straight
3. To drive on streets and go around corners, you would also need to learn:
   how to stay in your lane
   how to make turns and use your turn signals
   how to read street signs
4. To drive on highways, you would also need to learn:
   how to drive at a high speed
   how to enter a highway on the entrance ramps
   how to shift lanes
5. Many behaviors that involve several steps must be taught in a certain order. Is this true for learning to drive a car? Explain. Yes; a person would have to learn the basic things, like starting a car, before trying something harder, like driving in traffic.

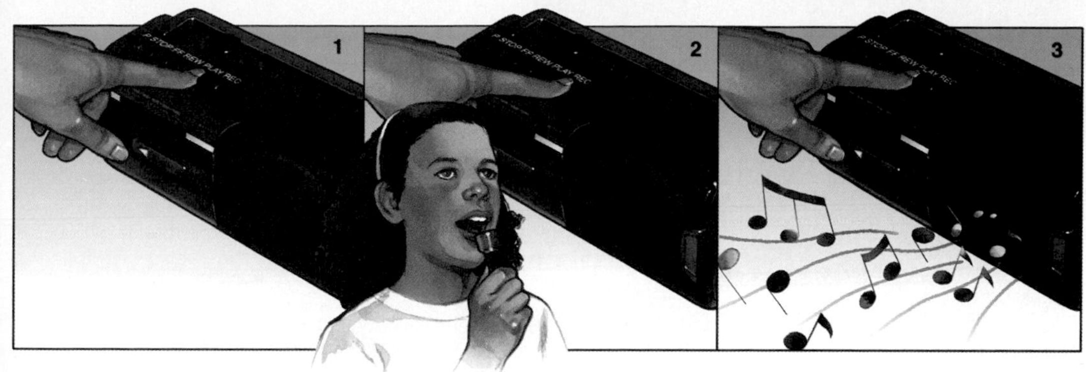

**Figure 17-3** Learned behavior can be compared to the recording of a tape.

Suppose you want to teach a dog to raise its paw. When you first tell the dog to raise its paw, nothing happens. Repeat the message over and over and pick up the dog's paw each time. The dog will learn what to do. Then, each time the dog hears the stimulus word, "paw," it will respond the same way. It lifts its paw.

Can learned behavior change? Yes, you can teach a dog to lift its paw and bark at the same time. You can even teach the dog to lift its paw, bark, and then roll over.

Most animals learn new behaviors faster if there is a reward. Animal trainers often use food as a reward when an animal shows the correct behavior. The next time you see dolphins jump through a hoop at an aquarium, look for the reward of fish they receive from their trainer. What do dog trainers give dogs as a reward for correct behaviors?

**What helps animals learn faster?**

## Check Your Understanding

1. What is the difference between stimulus and response? Give an example of each.
2. How does an instinct differ from simple innate behavior?
3. What type of behavior must be practiced?
4. **Critical Thinking:** Why would you expect the behavior of a honeybee to be more complex than that of a sponge?
5. **Biology and Reading:** Read the following sentences. Write the one that summarizes the information in this section.
   a. Behavior is how an animal knows what is happening.
   b. Behavior is everything that an animal does and is made up of unlearned and learned responses.
   c. Behavior is all of the complex learned responses to stimuli by animals that have brains.



## TEACH

### Guided Practice

Have students write down their answers to the margin question: animals learn faster if there is a reward.

### Check for Understanding

Have students respond to the first three questions in Check Your Understanding.

### Reteach

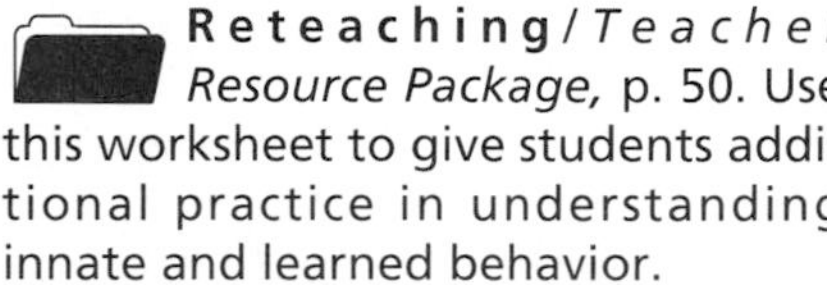

Reteaching/*Teacher Resource Package,* p. 50. Use this worksheet to give students additional practice in understanding innate and learned behavior.

**Extension:** Assign Critical Thinking, Biology and Reading, or some of the **OPTIONS** available with this lesson.

## 3 APPLY

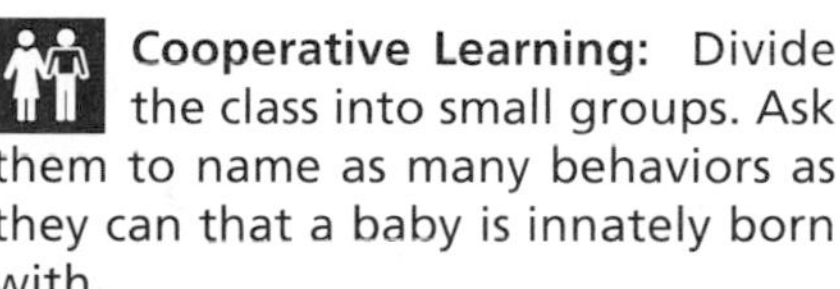

**Cooperative Learning:** Divide the class into small groups. Ask them to name as many behaviors as they can that a baby is innately born with.

## 4 CLOSE

### ACTIVITY/Videocassette

Show the videocassette *Weave and Spin,* Carolina Biological Supply Co.

### Answers to Check Your Understanding

1. A stimulus causes a reaction to start or a behavior to begin. A response is the behavior done as a result of the stimulus. Example: stimulus is a bird seeing a worm; response is bird catching the worm.
2. An instinct is a complex pattern of innate behavior.
3. learned behavior
4. A honeybee has a more complex brain and muscular system than a sponge and is capable of both innate and learned behavior.
5. b

## RETEACHING 50

RETEACHING — CHAPTER 17

Name ______ Date ______ Class ______

*Use with Section 17:1.*

*INNATE AND LEARNED BEHAVIOR*

I. Innate behavior

Stimulus: A spider sees a fly land in a web or feels the movement of web strands.

Response: The spider's muscles contract as it moves toward the fly.

What part of the nervous system detects the stimulus? sense organs
What part of the body responds to the stimulus? muscles
Was the animal taught this behavior? no
Was the animal born with this response? yes
Can the behavior be changed? no
Did the animal practice this behavior? no
Does a reward help the animal learn this behavior? no

II. Learned behavior

Stimulus: Dolphin sees hoop above water.

Response: Muscles contract as dolphin jumps through hoop.

What part of the nervous system detects the stimulus? sense organs
What part of the body responds to the stimulus? muscles
Was the animal taught this behavior? yes
Was the animal born with this response? no
Can the behavior be changed? yes
Did the animal practice this behavior? yes
Does a reward help the animal learn this behavior? yes

## Lab 17-2 Behavior

### Overview

Students will use change in pupil size as an example of an innate behavior and copying mirror images as an example of learned behavior.

**Objectives:** Upon completion of the lab, students will be able to (1) **give** pupil size as **an example** for innate human behavior, (2) **recognize** that innate behavior cannot be changed or modified with increasing numbers of trials, (3) **identify** copying a mirror image as a learned behavior that can change.

**Time Allotment:** 40 minutes

### Preparation

Lab 17-2 worksheet/*Teacher Resource Package*, pp. 67-68.

### Teaching the Lab

**Cooperative Learning:** Students can be paired up for the entire experiment. For more information, see pp. 22T-23T in the Teacher Guide.

▶ **Troubleshooting:** Advise students that the change in pupil size will not take very long.

**ASSESSMENT**

**Performance:** Provide students with the following data from two experiments:

| Length of time needed to complete task | | |
|---|---|---|
| Trial | Experiment 1 | Experiment 2 |
| 1 | 22 minutes | 5 seconds |
| 2 | 18 minutes | 4 seconds |
| 3 | 12 minutes | 5 seconds |

Have students decide which experimental data illustrates innate behavior and which illustrates learned behavior and have them explain why.

# Behavior

## Lab 17-2

### Problem: How do innate and learned behavior differ?

#### Skills

form hypotheses, observe, make and use tables

#### Materials

watch with second hand
pencil and paper
mirror

#### Procedure

Part A

1. Copy the data table.
2. **Observe:** Look at your eyes in the mirror and note the size of your pupils with the lights on and the lights off.
3. The teacher will turn off the lights. When the lights are turned on, look at the watch. Time how long it takes your pupils to change size. Record your data.
4. Write a **hypothesis** to tell whether the length of time it takes your pupils to change size will increase, decrease, or stay the same with repeated trials.
5. Repeat step 3 four more times.

Part B

1. Write your name on a piece of paper.
2. Position a mirror as shown in Figure B.
3. While looking *only* into the mirror, try to trace your name. Have a partner time you. Record the data in the table.

| TIME IN SECONDS | | |
|---|---|---|
| Trial | Pupil Change | Trace Name |
| 1 | Time will | Time will |
| 2 | not change. | decrease. |
| 3 | | |
| 4 | | |
| 5 | | |



4. Write a **hypothesis** to tell whether the length of time it takes you to trace your name will change with repeated trials.
5. Repeat steps 2 and 3 four more times.

#### Data and Observations

1. Did the pupils take the same amount of time to change with repeated trials?
2. Did the time needed to trace your name change with repeated trials?

#### Analyze and Apply

1. **Check your hypotheses:** Are your **hypotheses** supported by your data? Why or why not?
2. Was the behavior in Part A innate or learned? Was the behavior in Part B innate or learned? How do you know?
3. Compare innate and learned behaviors.
4. **Apply:** How does the behavior in Part A help humans with their vision?

#### Extension

**Design an experiment** to show if learned behavior is forgotten if it is not practiced.



## ANSWERS

### Data and Observations

1. The pupils took the same amount of time to change with repeated trials.
2. The time needed to trace a name became shorter with each trial.

### Analyze and Apply

1. Answers will vary. Students who hypothesized that the time for Part A would remain the same and the time for Part B would decrease will say that their hypotheses were supported.
2. The behavior in Part A is innate because it did not change. The behavior in Part B is learned because it is learned with practice.
3. Innate behavior cannot be changed and is automatic. Learned behavior improves with practice, is not automatic, and can change.
4. A wider pupil allows more light to enter the eye in dim surroundings. A narrow pupil reduces the amount of light that enters the eye, thus preventing possible injury to the retina.

# 17:2 Special Behaviors

How do certain behaviors help animals? Some types of behavior help animals find food. Other types may help an animal find a mate. Certain behaviors protect animals from their enemies. Finding a place to live while having young is another type of special behavior.

### Objectives

**4. Explain** how courting behaviors help an animal species reproduce.

**5. Describe** how behavior helps an animal find food, shelter, and protection from enemies.

**6. Compare** reproduction in animals that give parental care with reproduction in animals that give no care.

### Key Science Words

courting behavior
pheromone
social insects
migration
parental care

## Behaviors for Reproduction

Male frogs and crickets make sounds. You may have heard these sounds on a summer night. The males use these sounds to attract females. This behavior brings males and females together for mating. The sounds are the stimulus, and the female responds by finding the male. Does this type of behavior use sense organs or a nervous system? The answer to this question is that both are used. The female must hear the sounds and her brain must send the proper signals to her muscles so she can respond. Is a frog's croaking or a cricket's chirping innate or learned behavior? Why do you think so?

Some animals, such as birds and fish, have complex courting behaviors when they reproduce. **Courting behaviors** are behaviors used by males and females to attract one another for mating. Figure 17-4 shows a peacock with its tail feathers spread. The feather display is a courting behavior. The male uses it to attract the female. The colorful neck flap of certain lizards helps attract females and threaten other males. Figure 17-4 shows the neck flap on an anole lizard.

### Bio Tip

**Leisure:** People enjoy watching fireflies. Males of different species make special flash patterns when they look for mates. A female of the same species flashes back the same signal. He flies down to mate with her. If all species had the same flash, males would mate with the wrong females.

**Figure 17-4** Peacocks and anole lizards have colorful courting behavior.



## STUDY GUIDE 99

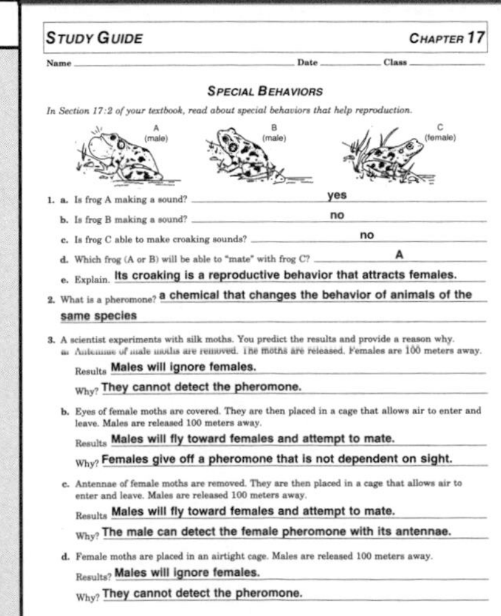

STUDY GUIDE — CHAPTER 17

Name ______ Date ______ Class ______

SPECIAL BEHAVIORS

*In Section 17:2 of your textbook, read about special behaviors that help reproduction.*



1. a. Is frog A making a sound? **yes**
   b. Is frog B making a sound? **no**
   c. Is frog C able to make croaking sounds? **no**
   d. Which frog (A or B) will be able to "mate" with frog C? **A**
   e. Explain. **Its croaking is a reproductive behavior that attracts females.**
2. What is a pheromone? **a chemical that changes the behavior of animals of the same species**
3. A scientist experiments with silk moths. You predict the results and provide a reason why.
   a. Antennae of male moths are removed. The moths are released. Females are 100 meters away.
   Results **Males will ignore females.**
   Why? **They cannot detect the pheromone.**
   b. Eyes of female moths are covered. They are then placed in a cage that allows air to enter and leave. Males are released 100 meters away.
   Results **Males will fly toward females and attempt to mate.**
   Why? **Females give off a pheromone that is not dependent on sight.**
   c. Antennae of female moths are removed. They are then placed in a cage that allows air to enter and leave. Males are released 100 meters away.
   Results **Males will fly toward females and attempt to mate.**
   Why? **The male can detect the female pheromone with its antennae.**
   d. Female moths are placed in an airtight cage. Males are released 100 meters away.
   Results? **Males will ignore females.**
   Why? **They cannot detect the pheromone.**

## OPTIONS

### Science Background

Behaviorists will categorize learned behavior into five main types—habituation, imprinting, classical conditioning, operant conditioning, and insight. This section emphasizes the types of behavior that aid an animal's survival. These examples involve the five types listed above.

# 17:2 Special Behaviors

## PREPARATION

### Materials Needed

Make copies of the Study Guide, Transparency Master, Enrichment, Skill, and Reteaching worksheets in the *Teacher Resource Package.*

▶ Earthworms will be needed for the Mini Lab. Order or obtain locally in bait shops or dig them yourself from soil.

### Key Science Words

courting behavior
pheromone
social insect
migration
parental care

### Process Skills

In the Mini Lab, students will observe. In the Skill Check, they will form hypotheses.

## 1 FOCUS

▶ The objectives are listed on the student page. Remind students to preview these objectives as a guide to this numbered section.

### ACTIVITY/Brainstorming

What would happen if a newborn child were to fall into deep water? Would it hold its breath and attempt to swim? What would happen if an animal such as a very young cat were placed onto a high ledge? Would it fall off or would it back away from the ledge? Are the behaviors of both the young cat and baby examples of protective behaviors?

### Independent Practice

**Study Guide**/*Teacher Resource Package,* p. 99. Use the Study Guide worksheet shown at the bottom of this page for independent practice.

# 2 TEACH

### MOTIVATION/Brainstorming

Have students look at the diagrams in Figure 17-5. Assuming this is an example of an innate behavior influenced by a pheromone, have students suggest ways other than removing the antennae that would prove that the male is responding to the specific female pheromone. They may suggest using pheromones from other insects, or using other female species. Ask students what type of control might have been used in this experiment. They should suggest that the male moths should have been tested to see if they would fly to the female when she was not releasing pheromones.

### Concept Development

▶ Croaking or chirping is innate. Frogs and crickets are born with this ability. Suggest to students that an animal may inherit certain nerve pathways at birth just as it inherits all other traits.

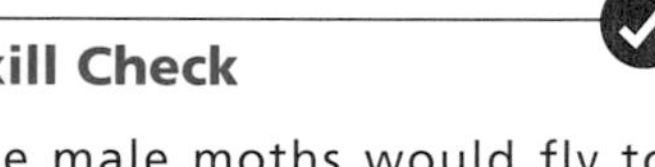

### Skill Check

The male moths would fly to the paper with pheromone. They would also fly to where the female moth had been before she was put into the cage.

### Guided Practice

Have students write down their answers to the margin question: a chemical that attracts animals of the same species.

## *GLENCOE* TECHNOLOGY

**Videodisc**

**Biology: The Dynamics of Life**
Movie: *Bees*
Disc 2, Side 1, Ch. 27

Movie: *Bird Courtship*
Disc 2, Side 1, Ch. 24

What is a pheromone?

Not all male animals make sounds or have displays to attract their mates. Some animals give off chemicals with distinct odors that do the same job. Chemicals that affect the behavior of members of the same species are called **pheromones** (FER uh mohnz).

Many female insects use pheromones to attract mates. One example is the silk moth. The female silk moth attracts the male by giving off a pheromone that is carried by wind currents. Males can detect the odor of the chemical with their antennae from many kilometers away. They respond by flying toward the odor. When the males reach the female, she chooses one male and allows him to mate with her.

How did scientists prove that a chemical attracts the male moth? Find the answer in Figure 17-5. Notice that it doesn't matter whether the males can see the females. They aren't attracted to the females by sight. Males are attracted to the odor of the pheromone, which is given off by the females.

How could you prove that the male senses the chemical with its antennae? You could remove the antennae from the males and note their responses. Scientists did this, and the male moths flew in all different directions. They could not smell the stimulus and find the female.

### Skill Check

**Form hypotheses:** What would happen to the behavior of the male moths in Figure 17-5 if the pheromone were added to the paper before the female moth was placed in the airtight cage? *For more help, refer to the **Skill Handbook**, pages 704-705.*

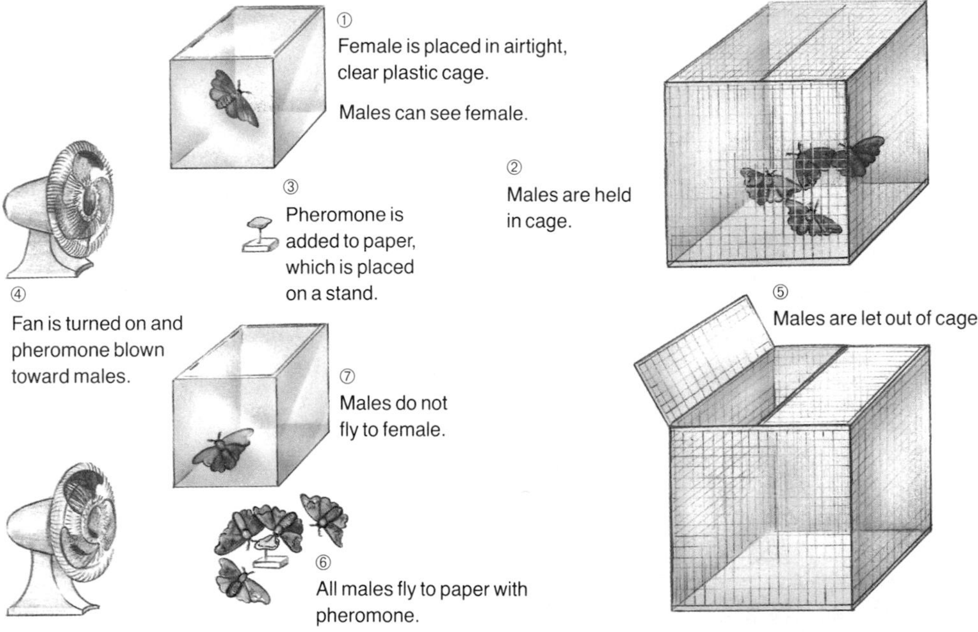


**Figure 17-5** Male moths are attracted to the female's pheromone, as shown in this experiment.



## *OPTIONS*

### Science Background

The female silk moth, *Bombyx* releases a pheromone called bombykol. It is detected by the antennae of the male from distances as far away as one mile. Seventy percent of the male's olfactory receptors detect only this molecule.

**Transparency Master**/*Teacher Resource Package,* p. 81. Use the Transparency Master shown here to help students understand the steps of behavior.

**TRANSPARENCY 81**

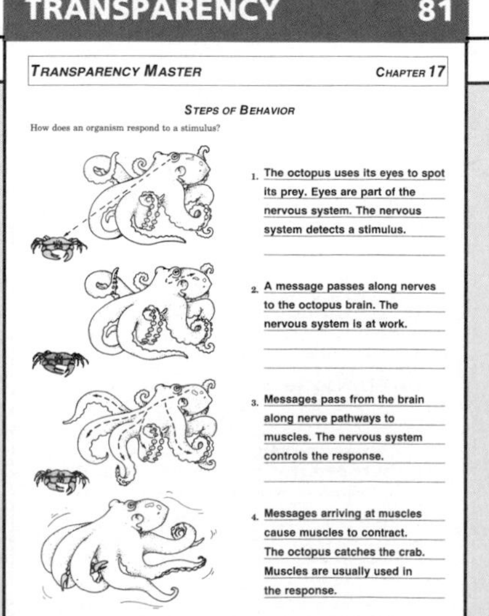

TRANSPARENCY MASTER — CHAPTER 17

STEPS OF BEHAVIOR

How does an organism respond to a stimulus?

1. The octopus uses its eyes to spot its prey. Eyes are part of the nervous system. The nervous system detects a stimulus.
2. A message passes along nerves to the octopus brain. The nervous system is at work.
3. Messages pass from the brain along nerve pathways to muscles. The nervous system controls the response.
4. Messages arriving at muscles cause muscles to contract. The octopus catches the crab. Muscles are usually used in the response.

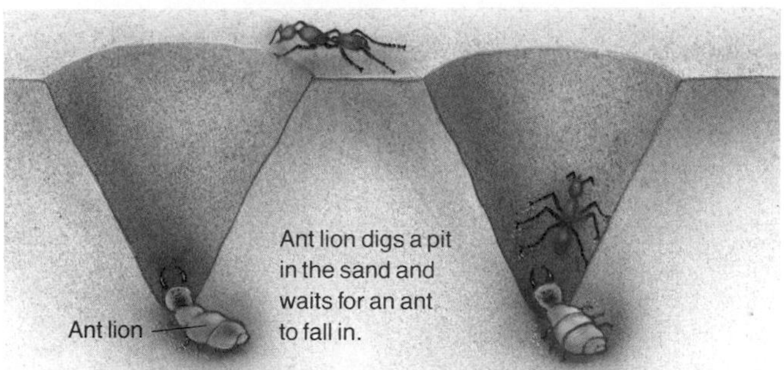


**Figure 17-6** An ant lion digs a pit to help in getting food.

## Behaviors for Finding Food

How would you like to walk along a beach and fall into a pit? Falling into the pit might not be too bad in itself. What if a hungry animal were waiting for you at the bottom? Things could go downhill very quickly.

Ant lions catch food by building a pit. Figure 17-6 shows how this behavior works. They wait for animals to fall into their pits. When an animal such as an ant falls in, it becomes a meal for the ant lion. Do you think that pit-building is innate or learned? Why do you think so?

Honeybees have complex behavior that helps them find food. They "talk" to one another and tell each other where food is located. The way bees "talk" is explained in Figure 17-7. When bees "talk," they move in a figure eight pattern. The "talking" bee faces the direction of the food when it is in the middle part of the figure eight. How do bees tell one another how far the food is? The "talking" bee wags its abdomen. The number of wags tells the other bees how far the food is from the hive.

Some spiders build webs to trap food. Each species of spider makes a web with a special design, as shown in Figure 17-8. Do you think this behavior is innate or learned?

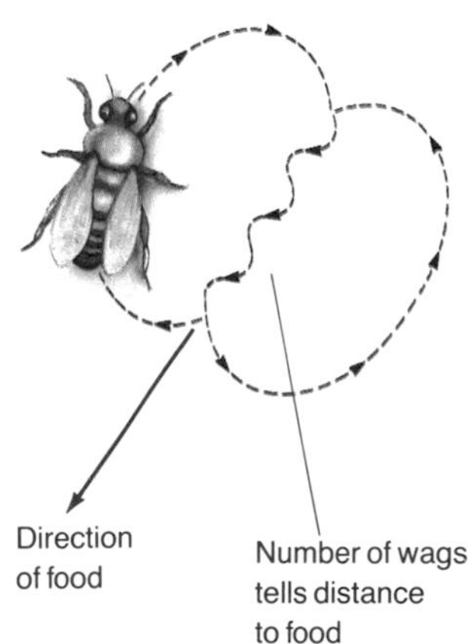


**Figure 17-7** Honeybees use special behavior in "talking" to other bees.

**Figure 17-8** Spiders use their webs to trap food. Each species of spider builds a web of a unique design.



## TEACH

### Concept Development

▶ The archer fish, *Toxotes jaculatrix,* spits droplets of water onto insects sitting on twigs that overhang streams and ponds. The droplets of water knock the insects into the water where they become easy prey for the fish.

### ACTIVITY/Brainstorming

Ask students to name other animal behaviors that illustrate methods for catching food. They may mention spiders that spin webs, trap door spiders, scorpions, ants that follow trails left by other ants (trails of formic acid).

### Independent Practice

**Study Guide/***Teacher Resource Package,* p. 100. Use the Study Guide worksheet at the bottom of this page for independent practice.

### ACTIVITY/Demonstration

Bring a compass into class. Ask how it works and whether it would be useful in determining the direction to follow to a certain location. Point out that a bee uses a "built-in compass" (its nervous system) to find food location, and is able to communicate the direction to other bees. Also point out that a compass is limited to telling direction. A bee can communicate direction and distance.

### ACTIVITY/Project

Ask students to build a maze suitable for a mouse. Have students determine how many trials and the length of time for each trial that it takes for the animal to learn the maze in search of food. Have them report their findingsto the class.

### *Portfolio*

Ask students to use Figure 17-7 as a guide and prepare diagrams to show the dance that bees would perform under the following circumstances: (a) food is 1000 m from the hive and is directly to the right of the hive entrance, (b) food is 200 m from the hive and directly to the left of the hive entrance. Note: The fewer the number of wags, the closer the food.

## STUDY GUIDE 100

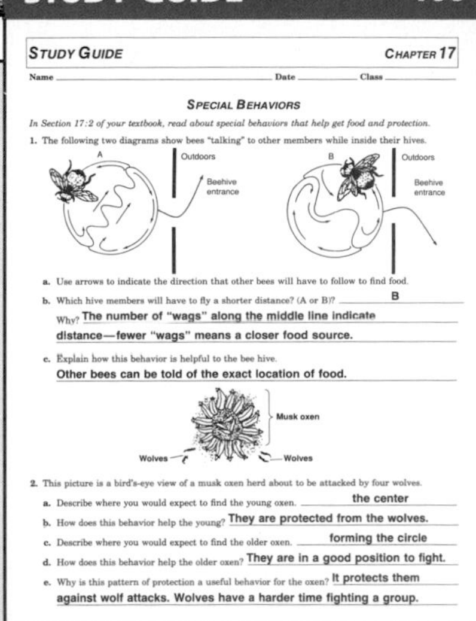

STUDY GUIDE — CHAPTER 17

Name ______ Date ______ Class ______

SPECIAL BEHAVIORS

*In Section 17:2 of your textbook, read about special behaviors that help get food and protection.*

1. The following two diagrams show bees "talking" to other members while inside their hives.



a. Use arrows to indicate the direction that other bees will have to follow to find food.

b. Which hive members will have to fly a shorter distance? (A or B)? **B**

Why? **The number of "wags" along the middle line indicate distance—fewer "wags" means a closer food source.**

c. Explain how this behavior is helpful to the bee hive.

**Other bees can be told of the exact location of food.**



2. This picture is a bird's-eye view of a musk oxen herd about to be attacked by four wolves.

a. Describe where you would expect to find the young oxen. **the center**

b. How does this behavior help the young? **They are protected from the wolves.**

c. Describe where you would expect to find the older oxen. **forming the circle**

d. How does this behavior help the older oxen? **They are in a good position to fight.**

e. Why is this pattern of protection a useful behavior for the oxen? **It protects them against wolf attacks. Wolves have a harder time fighting a group.**

## *OPTIONS*

### ACTIVITY/Project

You are to design a short filmstrip that includes both a tape and pictures. Write the script and suggest pictures that could be used to describe any of the special behaviors just studied. (You may want students to work on this project in small groups) Note: You can purchase blank 35 mm filmstrip from Carolina Biological Supply (catalog # 60-4420). Students can draw directly on the film with pencil creating their own filmstrip.

## TEACH

### Concept Development

▶ Defense behavior in the bombardier beetle consists of spraying its adversary with a chemical from special glands in the abdomen. Reactions that take place at the time of spraying raise the temperature of the chemical to 100°C.

▶ Antipredator, or protection, behaviors are exhibited by many species. Marmosets issue shrill calls to warn other members of the group about possible predators. When pursued by predators, many birds flock together. Fish form schools and antelopes form herds. Herding makes it difficult for the predator to single out one victim.

### ACTIVITY/Brainstorming

Most students are familiar with the fact that fish often swim in groups called schools. Ask why this is considered a protective behavior. When a fish swims alone, it is more vulnerable to being caught.

### Mini Lab

This Mini Lab will illustrate a protective behavior carried out by earthworms. Ask students to try and analyze why the earthworm avoids bright light. Can they name other animals that tend to do the same thing?

### ✓ ASSESSMENT

**Performance:** Ask students to prepare an experiment to determine whether earthworms prefer a wet or dry soil for burrowing. Suggest that the experiment be conducted while earthworms are in the dark.

### Guided Practice

Have students write down their answers to the margin question: for protection.

### Student Journal

Tell students they have been asked to prepare a job description for young worker bees, explaining to the bees what life as a worker bee is like. Have students include the daily chores that will need attention. Also have them include whether or not worker bees will be required to attend class to learn how to carry out their chores.

### Mini Lab

**How Do Earthworms Behave?**

**Observe:** Place an earthworm in a box. Cover half the box with black paper. Note if the worm moves toward or away from light. What behavior is this? How does it help a worm? *For more help, refer to the* ***Skill Handbook,*** *pages 704-705.*

## Protective Behaviors

Musk oxen, such as in Figure 17-9a, are mammals that live in the far north of Alaska and Canada. They are never found living alone. They are always found in groups. Living in a group gives them protection.

Imagine a wolf pack moving toward a herd of musk oxen. These wolves hunt the oxen for food. The oxen see, smell, or hear the wolves. They set up a defense against the wolves by forming a ring, as shown in Figure 17-9b. The adults stand on the outside. The young stay inside the ring. If a battle takes place, the older and larger oxen are in a good position to fight. The wolves can't get to the young inside the ring. The wolves also have a harder time fighting a group than they would a single animal. This type of innate group behavior helps to protect all the animals in the herd.

Many social insects have innate protective behaviors. **Social insects** are insects that live in groups, with each individual doing a certain job. Bees, wasps, ants, and termites are social insects. Bees have workers, drones, and a queen. The queen is the only female in the hive to reproduce. Drones are males. They mate with the queen. Workers are females that can't reproduce. Workers gather food, keep the hive clean, and protect it and the queen.

Ants and termites have an additional job. They have soldiers that help to protect the entire group. If an enemy enters an ant or a termite nest, the soldiers respond. They destroy the enemy by cutting it apart with their sharp jaws. A social insect lives in a group for protection.

**Why do some animals live in groups?**

**Figure 17-9** Musk oxen (a) form a ring (b) to defend their young against wolves.

a

b



## OPTIONS

**Enrichment**/*Teacher Resource Package,* p. 18. Use the Enrichment worksheet shown here to help students understand sleep behavior.

**What Self-protecting Behaviors Do Isopods Show?**/*Lab Manual,* pp. 141-144. Use this lab as an extension to protective behaviors.

### ENRICHMENT 18

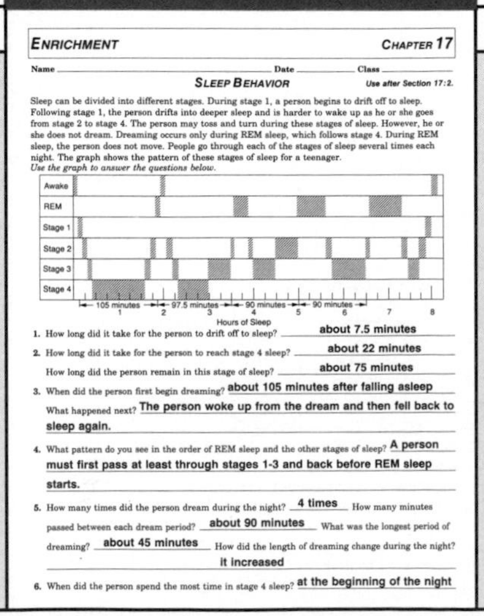

ENRICHMENT CHAPTER 17

Name ______ Date ______ Class ______

SLEEP BEHAVIOR *Use after Section 17:2.*

Sleep can be divided into different stages. During stage 1, a person begins to drift off to sleep. Following stage 1, the person drifts into deeper sleep and is harder to wake up as he or she goes from stage 2 to stage 4. The person may toss and turn during these stages of sleep. However, he or she does not dream. Dreaming occurs only during REM sleep, which follows stage 4. During REM sleep, the person does not move. People go through each of the stages of sleep several times each night. The graph shows the pattern of these stages of sleep for a teenager.

*Use the graph to answer the questions below.*



1. How long did it take for the person to drift off to sleep? **about 7.5 minutes**
2. How long did it take for the person to reach stage 4 sleep? **about 22 minutes**
   How long did the person remain in this stage of sleep? **about 75 minutes**
3. When did the person first begin dreaming? **about 105 minutes after falling asleep**
   What happened next? **The person woke up from the dream and then fell back to sleep again.**
4. What pattern do you see in the order of REM sleep and the other stages of sleep? **A person must first pass at least through stages 1-3 and back before REM sleep starts.**
5. How many times did the person dream during the night? **4 times** How many minutes passed between each dream period? **about 90 minutes** What was the longest period of dreaming? **about 45 minutes** How did the length of dreaming change during the night? **It increased**
6. When did the person spend the most time in stage 4 sleep? **at the beginning of the night**

a

## Migration

Many animals move from one place to another. **Migration** is a kind of behavior in which animals move from place to place in response to the season of the year. Many birds, such as the Atlantic golden plover in Figure 17-10b, migrate. The plovers' path is shown in Figure 17-11. Plovers spend spring, summer, and fall in northern Canada. As winter comes, the plovers migrate to South America. They fly a total distance of 10 000 kilometers, more than twice the distance across the United States.

When plovers migrate, they find more food. Food is plentiful during the spring through fall in Canada. As winter arrives, food supplies disappear. The plovers migrate to where there is more food.

b

**Figure 17-10** The fur seal (a) and the Atlantic golden plover (b) migrate to find a better place to live in winter.

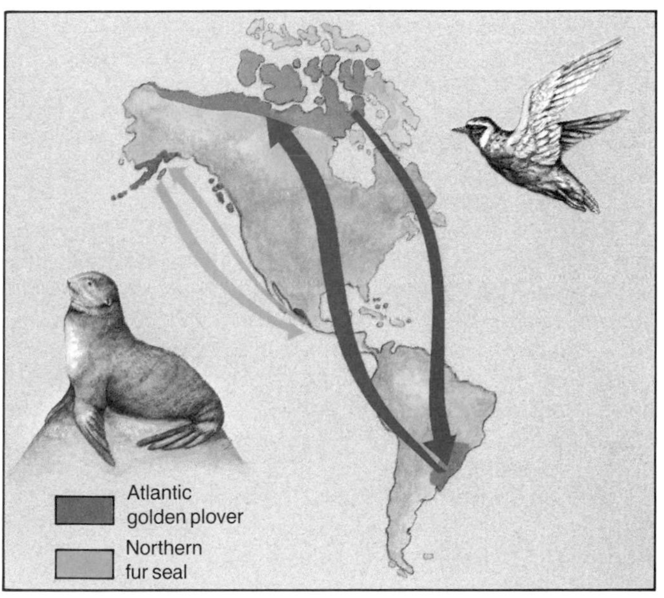


**Figure 17-11** Migration routes of the Atlantic golden plover and the Northern fur seal. Why do these animals migrate? to find food and a warmer place



## TEACH

### Concept Development

▶ Ask where robins go in the winter. Students should realize that these birds migrate to another area (South America). If you live along the Pacific coast, the migration of whales is an annual event. The mechanism these animals use to guide themselves will vary depending on species.

▶ Many birds migrate at night, using the stars to aid in navigation much as the sun is used during the day. Experiments using night migratory birds placed in a planetarium showed this to be true.

▶ Distinguish between migration and homing. Migration occurs seasonally, while homing occurs whenever an animal is moved from its original home.

### Connection: Math

The distance between New York and San Francisco is 4800 kilometers. Below is a list of animals and the distance they migrate each year. Calculate how many times these animals could have gone from one coast to the other.

| | |
|---|---|
| Ruby-throated hummingbird | 800 km |
| Arctic terns | 17 600 km |
| Leatherback turtle | 4800 km |
| Monarch butterfly | 2900 km |

### Independent Practice

**Study Guide**/*Teacher Resource Package*, p. 102. Use the Study Guide worksheet at the bottom of this page for independent practice.

**SKILL 18**

**SKILL: INTERPRET DATA** — CHAPTER 17

Name ______ Date ______ Class ______

*For more help, refer to the Skill Handbook, pages 704-705.* *Use after Section 17:2.*

**MIGRATION OF SALMON**

When salmon are ready to mate and reproduce, they return from a lake or ocean to the river in which they hatched two years earlier. How do they know when they reach the river from which they hatched? One scientist thought that the fish use their sense of smell to recognize the river. He decided to test this hypothesis.

Six thousand salmon were hatched in large tanks. After three months, the fish were divided equally among three tanks. Chemical A had been added to the water in one tank and chemical B to another tank. The water in the third tank was not treated. The fish in the untreated tank formed the control group. All the fish were marked to show which group they belonged in. When the fish were ready to reproduce, they were released into Lake Michigan. Six small rivers flowed into the lake. Earlier, the scientist had added chemical A to Kewanee River and chemical B to Manitowac River. He had not added chemicals to the other rivers. The scientist and his helpers captured marked salmon from all of the rivers. They counted the number of fish from each river that had the mark of each of the groups in the experiment. The table shows the data they collected.

Number of fish recovered

| River | Group A | Group B | Control group |
|---|---|---|---|
| Stony Creek | 1 | 0 | 47 |
| Annapee River | 6 | 1 | 78 |
| Kewanee River | 1559 | 20 | 173 |
| Molash Creek | 0 | 0 | 24 |
| Manitowac River | 4 | 1311 | 121 |
| Pigeon River | 5 | 1 | 204 |

*Interpret the data in the table and answer the questions below.*

1. To which river did most of the salmon raised with chemical A return? Kewanee
2. To which river did most of the salmon raised with chemical B return? Manitowac
3. Did the salmon in the control group show a strong preference for a particular river? no
4. Do the data support the hypothesis? Explain why or why not. Yes; almost all the fish raised with chemical A or with chemical B went to the river containing the chemical they were raised with. The controls had no special preference.
5. What type of behavior—learned or innate—did the salmon show? learned

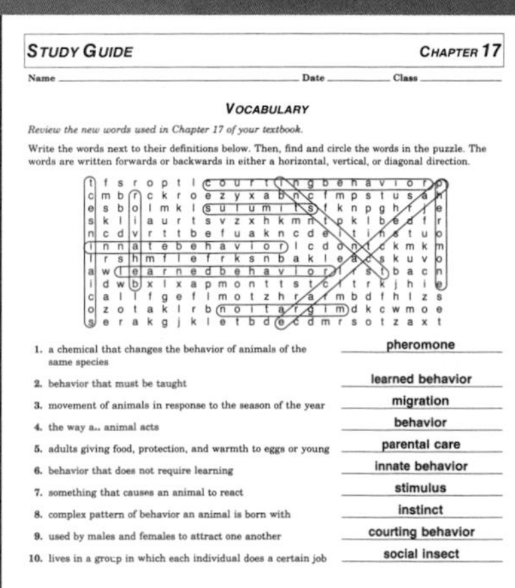

**STUDY GUIDE 102**

**STUDY GUIDE** — CHAPTER 17

Name ______ Date ______ Class ______

**VOCABULARY**

*Review the new words used in Chapter 17 of your textbook.*

Write the words next to their definitions below. Then, find and circle the words in the puzzle. The words are written forwards or backwards in either a horizontal, vertical, or diagonal direction.

1. a chemical that changes the behavior of animals of the same species — pheromone
2. behavior that must be taught — learned behavior
3. movement of animals in response to the season of the year — migration
4. the way a.. animal acts — behavior
5. adults giving food, protection, and warmth to eggs or young — parental care
6. behavior that does not require learning — innate behavior
7. something that causes an animal to react — stimulus
8. complex pattern of behavior an animal is born with — instinct
9. used by males and females to attract one another — courting behavior
10. lives in a group in which each individual does a certain job — social insect

## OPTIONS

### Science Background

Ruby-throated hummingbirds migrate 800 km twice a year across the Gulf of Mexico. Arctic terns migrate 35 000 km each year from the Arctic Circle to South America. Leatherback turtles migrate 5000 km from French Guiana to Africa. Monarch butterflies migrate 3000 km from southern Canada to Mexico.

**Skill**/*Teacher Resource Package*, p. 18. Use the Skill worksheet shown here to help students practice interpreting data.

## TEACH

### Concept Development

▶ Circadian rhythm refers to changes in behavior that are based on a 24-hour clock. Circannual rhythm occurs on a yearly basis. Both types of behavior may be the result of an internal biological clock mechanism.

▶ Whales spend considerable time in the food-rich cold waters of northern oceans. In the fall, they migrate from the cold waters to more temperate regions where calving takes place.

▶ Animals use a variety of navigational devices to find their way. Polarized light is a directional cue that honeybees use. The positions of the stars are also used by many species, including migratory birds.

### ACTIVITY/Brainstorming

Find out what students know about homing pigeons and their homing abilities. Discuss possible experiments to find out how pigeons find their way around. What are the various senses that pigeons use? (Pigeons may be sensitive to different orientations of the sun and stars, or they may be influenced by magnetic fields surrounding Earth.)

### Idea Map

Have students use the idea map as a study guide to the major concepts of this section.

### Independent Practice

**Study Guide**/*Teacher Resource Package,* p. 101. Use the Study Guide worksheet at the bottom of this page for independent practice.

**Portfolio**

Have students prepare a vertical bar graph that illustrates the relationship between number of offspring produced and number that usually survive. Compare fish to birds to humans.

**Idea Map**

**Special Behaviors**

Special behaviors
- aid in reproduction
  - croaking of frogs
  - pheromones
- finding food — bee language
- protection — social insects protect hive
- migration — migration of seals and birds
- parental care — feeding and protection of young

**Study Tip:** Use this idea map as a study guide to special behaviors. How many other examples can you give?

Fur seals, shown in Figure 17-10a, migrate every year. They swim from the coast of Alaska to the coast of Mexico and back again. That's a total distance of about 7000 kilometers each year, the distance between Chicago and England! Their path of migration is also shown in Figure 17-11.

The fur seals' migration helps the animals to reproduce. By moving south, seals find better places to reproduce and care for their young. Seals mate and raise their young on land. It is easier to raise young where it is warm.

How do animals that migrate find their way? This question is of great interest to scientists. Some of the answers can be found in studies made with penguins. Adelie (uh DAY lee) penguins, shown in Figure 17-12, live in the Antarctic. They have nests along the coast, where they live and reproduce. Could these animals find their way back to their nests from miles away?

Scientists moved the penguins far away from their nests and followed them. They found that the penguins used the sun for direction. On sunny days, the penguins moved in a straight line toward their original nests. On cloudy days, the penguins did not move in the correct direction.

Is this type of behavior innate or learned? Here is a clue. Scientists moved baby penguins to a new location. The babies also used the sun to find their way back. They did not have the adults to follow.

How do birds that migrate at night find their way? Scientists found that some birds use the stars for migration. Other birds seem to have built-in "compasses" that tell them the direction of the magnetic north pole.

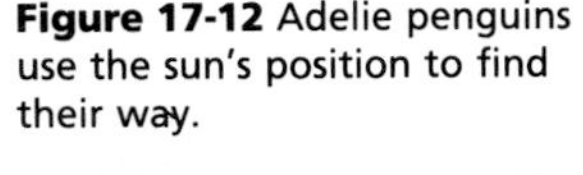

**Figure 17-12** Adelie penguins use the sun's position to find their way.

**GLENCOE TECHNOLOGY**

**Videodisc**

**Biology: The Dynamics of Life**
Movie: *Salmon Migration*
Disc 2, Side 1, Ch. 26

**STUDY GUIDE 101**

STUDY GUIDE — CHAPTER 17

Name ________ Date ________ Class ________

**SPECIAL BEHAVIORS**

*In Section 17:2 of your textbook, read about migration and parental care.*

1. Define migration. **A behavior in which animals move from place to place in response to the season of the year.**
2. Give two examples of animals that migrate. **Atlantic golden plover, fur seals**
3. Give two reasons why animals migrate. **Animals move to new areas as their food supply drops and to find safe places to reproduce and raise young.**
4. Using your answer to question 3, explain why the:
   a. plover migrates. **It migrates to find food.**
   b. fur seal migrates. **They find warmer and safer places to reproduce and care for their young.**
5. Using Figure 17-11 in your text, describe the path of migration for:
   a. plover **from northern Canada to South America**
   b. fur seal **from Alaska to Mexico**
6. What seems to be the main way in which adult Adelie penguins can find their way back to their nests? **using the sun as a guide**
7. a. Does migration seem to be innate or learned? **innate**
   b. What experimental proof can be given using Adelie penguins as an example? **Baby penguins could find their way home, never having been taught by parents.**
8. What is meant by parental care? **behavior in which adults give food, protection, and warmth to eggs or young**
9. Is parental care as important to most fish as it is to most mammals? Explain. **No, parental care is more important to animals that produce fewer young. Parental care gives the small number of offspring of mammals a better chance to survive. Most fish produce hundreds or thousands of eggs at one time.**

## Parental Behavior

After you were born, many things had to be done for you. You could not get your own food. You could not find a safe place to live. You could not protect yourself in any way. You would have died if your parents or other adults hadn't taken care of you. Are humans the only animals that care for their young? Most other mammals and birds care for their young. Some fish, amphibians, and reptiles provide protection for their eggs until they hatch.

**Parental care** is a behavior in which adults give food, protection, and warmth to eggs or young, Figure 17-13. The female parent is often the one to give parental care. The male parent also may protect the eggs or young. Male penguins often care for eggs while the female is feeding.

Parental care becomes more important in animals that produce fewer young. Most fish produce hundreds or thousands of eggs at a time, but do not provide any care at all. Some eggs are eaten, some never hatch, and some survive and grow to adulthood. Large numbers of eggs must be produced to be sure that some will survive.

Animals that provide better parental care can produce fewer young and have them survive. Wolves, squirrels, and birds give lots of parental care to their young. These animals usually have no more than six young at a time. Most of the young live because of the care they receive from their parents.

**Figure 17-13** Many animals, such as birds, feed and protect their young. What is this kind of behavior called? parental care

**Why is parental care important in animals that produce few young?**

### Check Your Understanding

6. How do crickets and silk moths differ in courting behavior?
7. Describe the path of migration for Atlantic golden plovers, and tell why this behavior helps the animals.
8. Why must an animal that has few young give parental care?
9. **Critical Thinking:** What advantages might a blackbird have by living in a flock of blackbirds rather than by living alone?
10. **Biology and Reading:** In this section you learned that social insects live in groups and have protective behaviors. What word means the opposite of *social* as used in the sentence above? Name an insect that can be described using this word.



## TEACH

### Guided Practice

Have students write down their answers to the margin question: it is important that all or most of the young survive.

### Check for Understanding

Have students respond to the first three questions in Check Your Understanding.

### Reteach

**Reteaching**/*Teacher Resource Package,* p. 51. Use this worksheet to give students additional practice in understanding special types of behavior.

**Extension:** Assign Critical Thinking, Biology and Reading, or some of the **OPTIONS** available with this lesson.

## 3 APPLY

**Misconception:** Many students may not think of human behavior as animal behavior. Show the filmstrip *Experiments in Human Behavior,* Human Relations Media.

## 4 CLOSE

### ACTIVITY/Videocassette

Show the videocassette *Never Cry Wolf,* Walt Disney Studios. Discuss the behaviors seen in the video.

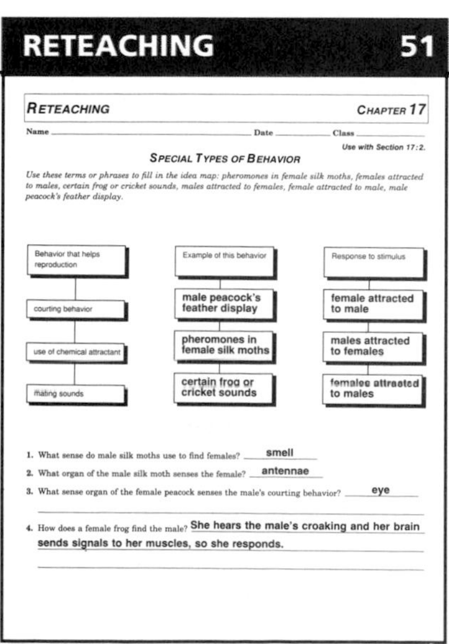

**RETEACHING 51**

*RETEACHING* CHAPTER 17

Name ______ Date ______ Class ______

*Use with Section 17:2.*

SPECIAL TYPES OF BEHAVIOR

*Use these terms or phrases to fill in the idea map: pheromones in female silk moths, females attracted to males, certain frog or cricket sounds, males attracted to females, female attracted to male, male peacock's feather display.*

| Behavior that helps reproduction | Example of this behavior | Response to stimulus |
| --- | --- | --- |
| courting behavior | male peacock's feather display | female attracted to male |
| use of chemical attractant | pheromones in female silk moths | males attracted to females |
| mating sounds | certain frog or cricket sounds | females attracted to males |

1. What sense do male silk moths use to find females? smell
2. What organ of the male silk moth senses the female? antennae
3. What sense organ of the female peacock senses the male's courting behavior? eye
4. How does a female frog find the male? She hears the male's croaking and her brain sends signals to her muscles, so she responds.

### Answers to Check Your Understanding

6. Crickets use sound to attract mates while silk moths use pheromones.
7. Atlantic golden plovers fly from Canada to South America and back again each year. Migration allows them to move to where there is more food in the winter.
8. It must be sure that those offspring survive.
9. It would get protection from the flock and be able to find food more easily.
10. The opposite of social is solitary. *Single, alone, by itself,* or similar terms may also be used. Solitary insects include a fly, butterfly, grasshopper, beetle.

### GLENCOE TECHNOLOGY

**Videodisc**

**The Secret of Life**

*Starlings/migration*

## CHAPTER 17 REVIEW

### Summary

Summary statements can be used by students to review the major concepts of the chapter.

### Key Science Words

All boldfaced terms from the chapter are listed.

### Testing Yourself

#### Using Words

1. learned behavior
2. innate behavior
3. behavior
4. instinct
5. courting behavior
6. parental care
7. migration
8. social insect
9. stimulus
10. pheromone

#### Finding Main Ideas

11. p. 365; Animals that give parental care have fewer young, but more of them survive. Animals that do not give parental care have more young, but fewer of them survive.
12. p. 361; A bee dances for the other bees to tell them where and how far away the food is.
13. p. 356; The first time a stimulus occurs, no response occurs, like a blank tape. Responding over and over to a stimulus can lead to learned behavior that records the behavior on the nervous system, like information is recorded on a tape.
14. pp. 363-364; Migration allows animals to move to areas where there is more food or better breeding sites.
15. p. 354; A prerecorded tape plays the same song every time it is put into a recorder. An innate response produces the same behavior each time the stimulus occurs.
16. p. 353; Sense organs detect the stimulus and muscles perform the behavior response.
17. p. 364; Penguins moved far away from their nests used the sun for direction and found their way back. Baby penguins that did not have the chance to learn from the adults could also find their way back.
18. p. 357; New behaviors are learned faster if there is a reward.

#### Using Main Ideas

19. If the antennae are removed from the male moths, they cannot find the source of the pheromone.
20. The bird sees the mouse and messages pass along nerves from the bird's eyes to the brain. The brain sends out messages along nerves to the muscles causing the bird to move and catch the mouse.
21. birds caring for eggs; humans feeding their young
22. Some birds use the stars or Earth's magnetic field to migrate at night.
23. A male peacock spreading its tail feathers, the spreading of the colorful neck flap in certain lizards, and the pheromones produced by female silk moths are all courting behaviors.
24. Yes, courting behavior is a complex pattern of behavior that does not have to be learned.

# Chapter 17

# Review

## Summary

### 17:1 Behavior

1. Behavior is the way an animal acts. All behavior is a response to a stimulus.
2. Innate behavior does not require learning. An animal is born with it. Instincts are complex patterns of innate behavior.
3. Learned behavior must be taught. Animals are not born with it. Learned behavior must be practiced, can be changed, and can be learned more quickly if a reward is given.

### 17:2 Special Behaviors

4. Courting behavior and response to pheromones are innate behaviors that help reproduction.
5. Many animals use special innate behaviors for catching or finding food. Some animals protect themselves from attack by staying in large groups. Social insects live in groups and have specialized behavior. Animals migrate in response to seasonal changes.
6. Certain animals care for their young. Young animals that receive parental care have a better chance of surviving than those that do not.

## Key Science Words

behavior (p. 352)
courting behavior (p. 359)
innate behavior (p. 353)
instinct (p. 353)
learned behavior (p. 356)
migration (p. 363)
parental care (p. 365)
pheromone (p. 360)
social insects (p. 362)
stimulus (p. 353)

## Testing Yourself

### Using Words

*Choose the word from the list of Key Science Words that best fits the definition.*

1. behavior that must be taught
2. behavior that does not have to be learned
3. the way an animal acts
4. a complex pattern of responding to a stimulus that is not learned
5. behavior used to attract males or females for mating
6. adults giving food or protection to young
7. moving from place to place in response to the season
8. insects that may have workers, drones, soldiers, and a queen
9. something that causes a reaction
10. chemical that affects the behavior of other members of the same species

# Review

## Testing Yourself *continued*

**Finding Main Ideas**

*List the page number where each main idea below is found. Then, explain each main idea.*

11. how the number of young produced and the number that live compare between animals that give parental care and those that do not
12. how a bee tells others where food is
13. why learned behavior may be compared to recording a tape
14. why animals migrate great distances each year
15. why innate behavior may be compared to playing a prerecorded tape
16. how sense organs and muscles are related to stimulus and response
17. the experimental evidence used to show that penguins use innate behavior to find their nests
18. why a reward might be given when teaching a new behavior

**Using Main Ideas**

*Answer the questions by referring to the page number after the question.*

19. What experimental evidence shows that male silk moths use their antennae to detect pheromones? (p. 360)
20. What is the sequence of stimulus and response when a bird sees and catches a mouse? (p. 352)
21. What are two examples of parental care? (p. 365)
22. How do birds find their way while migrating at night? (p. 364)
23. What are three examples of courting behavior in animals? (p. 359)
24. Is courting behavior an instinct? Explain your answer. (p. 353)

## Skill Review 

*For more help, refer to the* ***Skill Handbook,*** *pages 704-719.*

1. **Infer:** Why was pheromone used to attract the male moths in Figure 17-5?
2. **Infer:** How do you know that Adelie penguins use the sun to find the way?
3. **Form hypotheses:** You train goldfish to come to the top of a fish tank to eat when you turn on the light. The next time you feed them, you tap the tank instead of turning on the light. Hypothesize what the fish will do.
4. **Make and use tables:** Make a table with five examples each of human innate and learned behaviors.

## Finding Out More

TECH PREP

**Critical Thinking**

1. Decide if the following human baby behaviors are innate or learned: crying, suckling, grasping, crawling, toilet training, yawning, putting on clothes. Explain your answers.
2. Male killer bees are attracted by a pheromone given off by a female. How might the number of bees be reduced?

**Applications**

1. Design a plan to teach a dog to fetch the newspaper. Include the following terms: stimulus, nervous system, brain, response, sense organs, reward, practice.
2. Make a list of animals that give parental care and those that don't. Find out how many offspring each produces. Relate amount of care to the number and survival rate of offspring.



## Skill Review

1. to show that it was the pheromone and not the sight of the female that was the stimulus
2. because they can't find their nests on cloudy days
3. The fish will not come to the top of the tank because the light is their stimulus.
4. Examples may include: innate–coughing, sneezing, yawning, hiccups, jerking your hand back after touching something hot; learned–tying your shoes, reading, writing, solving problems, driving a car

## Finding Out More

### Critical Thinking

1. crying–innate; suckling–innate; grasping–innate; crawling–learned; toilet training–learned; yawning–innate; putting on clothes–learned. Crying, suckling, grasping and yawning are performed perfectly the first time. Crawling, toilet training, and putting on clothes are all learned because although the initial ability may be innate, the behaviors improve with practice.
2. Traps baited with female pheromones can be placed out to attract male killer bees.

### Applications

1. Answers will vary. Accept all suitable answers.
2. Animals that provide more infant care will be more complex than those that provide little care. Those animals that provide little care usually produce more offspring.

 **Chapter Review**/*Teacher Resource Package,* pp. 87-88.

 **Chapter Test**/*Teacher Resource Package,* pp. 89-91.

**Chapter Test**/*Computer Test Bank*

# Drugs and Behavior

## PLANNING GUIDE

| CONTENT | TEXT FEATURES | TEACHER RESOURCE PACKAGE | OTHER COMPONENTS |
|---|---|---|---|
| (2 days)<br>18:1 An Introduction to Drugs<br>What Is a Drug?<br>Obtaining Drugs Legally<br>Reading Drug Labels<br>Speed of Drug Action | Mini Lab: *Are Garlic and Onions Antibodies?* p. 371<br>Mini Lab: *Can a Drug Package Be Made Childproof?* p. 372<br>Lab 18-1: *Aspirin,* p. 373<br>Check Your Understanding, p. 375 | Reteaching: *Drugs in the Body,* p. 52<br>Study Guide: *An Introduction to Drugs,* pp. 103-104<br>Lab 18-1: *Aspirin,* pp. 69-70 | **Laboratory Manual:**<br>*How Do Drugs Affect the Ability of Seeds to Grow Into Plants?* p. 153<br>Color Transparency 18: *Path of a Swallowed Drug*<br>**STVS:** *Natural Time-Release Capsules,* Human Biology (Disc 7, Side 1) |
| (2 1/2 days)<br>18:2 How Drugs Affect Behavior<br>Stimulants<br>Depressants<br>Psychedelic Drugs | Skill Check: *Sequence,* p. 377<br>Idea Map, p. 379<br>Check Your Understanding, p. 379 | Reteaching: *How Drugs Work on Neurons,* p. 53<br>Study Guide: *How Drugs Affect Behavior,* p. 105 | **STVS:** *Panic Disorder and Brain Activity,* Human Biology (Disc 7, Side 1) |
| (1 1/2 days)<br>18:3 Uses of Over-the-Counter Drugs<br>Antihistimines<br>Cough Suppressants<br>Antacids | Skill Check: *Understand Science Words,* p. 380<br>Mini Lab: *What Determines the Price of a Drug?* p. 381<br>Check Your Understanding, p. 382 | Reteaching: *Over-the-Counter Drugs,* p. 54<br>Study Guide: *Uses of Over-the-Counter Drugs,* p. 106 | **Laboratory Manual:**<br>*Which Antacid Works Best?* p. 149<br>**STVS:** *Treating Motion Sickness,* Human Biology (Disc 7, Side 2) |
| (2 days)<br>18:4 Careless Drug Use<br>Drug Misuse and Abuse<br>Cocaine<br>Caffeine and Nicotine<br>Alcohol | Skill Check: *Make and Use Tables,* p. 388<br>Lab 18-2: *Alcohol,* p. 387<br>Science and Society: *Steroids,* p. 389<br>Check Your Understanding, p. 388 | Enrichment: *Sleep Behavior,* p. 18<br>Application: *Smoking and Lung Cancer,* p. 18<br>Reteaching: *Careless Use of Drugs,* p. 55<br>Critical Thinking/Problem Solving: *How is Blood Alcohol Concentration Useful?* p. 18<br>Skill: *Classify,* p. 19<br>Focus: *Attempts to Give Up Smoking,* pp. 35-36<br>Transparency Master: *Review of Drugs,* p. 85<br>Study Guide: *Careless Drug Use,* p. 107<br>Study Guide: *Vocabulary,* p. 108<br>Lab 18-2: *Alcohol,* pp. 71-72 | **STVS:** *Nicotine and the Lungs,* Human Biology (Disc 7, Side 2) |
| Chapter Review | Summary<br>Key Science Words<br>Testing Yourself<br>Finding Main Ideas<br>Using Main Ideas<br>Skill Review | **ASSESSMENT RESOURCES**<br>Chapter Review, pp. 92-93<br>Chapter Test, pp. 94-96<br>Performance Assessment in the Biology Classroom<br>Alternate Assessment in the Science Classroom<br>Computer Test Bank | |

## MATERIALS NEEDED

| LAB 18-1, p. 373 | LAB 18-2, p. 387 | MARGIN FEATURES |
|---|---|---|
| 4 glass slides<br>8 dropers<br>wax marking pencil<br>water<br>aspirin solution<br>6 test solutions of pain killers<br>aspirin-testing solution in dropper bottle | 4 glass slides<br>wax marking pencil<br>8 droppers<br>water<br>alcohol<br>alcohol-testing chemical in dropper bottle<br>6 test solutions of over-the-counter drugs and household chemicals | Skill Check, p. 377<br>pencil<br>paper<br>Skill Check, p. 380; Mini Lab, p. 381<br>dictionary<br>Mini Lab, p. 371<br>petri dish<br>agar<br>bacteria culture<br>paper disks treated with onion and garlic<br>Mini Lab, p. 372<br>paper<br>pencil |

## OBJECTIVES

For more information about National Science Standards, see page 5T.

| SECTION | OBJECTIVE | CORRELATION of QUESTIONS to OBJECTIVES | | | |
|---|---|---|---|---|---|
| | | CHECK YOUR UNDERSTANDING | CHAPTER REVIEW | TRP CHAPTER REVIEW | TRP CHAPTER TEST |
| 18:1 National Science Stds: UCP.2, UCP.3, UCP.4, F.1 | 1. **Compare** prescription and over-the-counter drugs. | 2, 4 | 2, 6, 10, 14 | 2, 7, 10, 11, 26 | 10, 12, 26, 27 |
| | 2. **Relate** the importance of information listed on drug labels. | 3, 5 | 3, 4, 5 | 4, 8, 13, 14, 17, 20 | 1, 4, 7, 19, 43, 44, 45, 46, 47 |
| | 3. **Describe** how drugs stop pain. | 1 | 16 | 3, 12, 14 | 11, 15 |
| 18:2 National Science Stds: UCP.2, UCP.3, UCP.4, UCP.5, C.1, C.6, F.1 | 4. **Describe** the effects of stimulants and depressants on the body. | 6, 9, 10 | 7, 9, 19, 22 | 16, 18 | 3, 17, 25, 28, 29, 31, 33 |
| | 5. **Discuss** how psychedelic drugs and inhalants affect the body. | 7 | 15, 17, 26 | 1, 9, 16 | 14, 30 |
| 18:3 National Science Stds: UCP.5, B.3, B.4, F.1 | 6. **Give the function** of antihistimines and cough suppressants. | 11 | 12, 24, 25 | 25 | 2, 14, 20, 21 |
| | 7. **Explain** the role of antacids in the body. | 13 | 11, 25 | 22 | 5, 9, 24 |
| 18:4 National Science Stds: UCP.3, UCP.4, C.6, F.1, F.5, F.6 | 8. **List** reasons for not using other people's drugs. | 16 | 20, 21, 23 | 5, 15, 19, 23 | 8, 16, 18, 23, 31, 32, 33, 34 |
| | 9. **Discuss** the problems caused by using caffeine, nicotine, cocaine, and alcohol. | 17, 18, 19, 20 | 1, 8, 13, 18 22 | 6, 16, 21, 24 | 6, 13, 15, 22, 35, 36, 37, 38, 39, 40, 41, 42 |

# Drugs and Behavior

**CHAPTER OVERVIEW**

## Key Concepts

This chapter looks at the physiological action of drugs on the human body. Useful consumer information is provided about drug dose, reading drug labels, and common drugs such as antacids and antihistamines. Stimulants, depressants, and commonly abused drugs are also examined.

## Key Science Words

| | |
|---|---|
| antacid | inhalant |
| antihistamine | nicotine |
| caffeine | overdose |
| cocaine | over-the-counter drug |
| controlled drug | prescription drug |
| dependence | psychedelic drug |
| depressant | side effect |
| dosage | stimulant |
| drug | symptom |
| drug abuse | |
| ethyl alcohol | |

## Skill Development

In Lab 18-1, they will **observe, interpret data,** and **experiment** in determining which medicines contain aspirin. In Lab 18-2, they will **observe, interpret data,** and **experiment,** in determining the presence of alcohol in over-the-counter drugs and household items. In the Skill Check on page 377, students will **sequence** the message path from neuron to neuron. On page 380, they will **understand** the **science word** *antihistamine.* In the Skill Check on page 388, they will **make and use tables** to show alcohol slows down body activities. In the Mini Lab on page 371, students will experiment with an antibiotic. On page 372, they will **observe** drug packaging. One page 381, they will infer the cost of drugs.

**CHAPTER PREVIEW**

### Chapter Content

Review this outline for Chapter 18 before you read the chapter.

**18:1 An Introduction to Drugs**
What Is a Drug?
Obtaining Drugs Legally
Reading Drug Labels
Speed of Drug Action

**18:2 How Drugs Affect Behavior**
Stimulants
Depressants
Psychedelic Drugs

**18:3 Uses of Over-the-Counter Drugs**
Antihistamines
Cough Suppressants
Antacids

**18:4 Careless Drug Use**
Drug Misuse and Abuse
Cocaine
Caffeine and Nicotine
Alcohol

### Skills in this Chapter

The skills that you will use in this chapter are listed below.

- In **Labs 18-1** and **18-2,** you will observe, interpret data, and experiment.
- In the **Skill Checks,** you will sequence, understand science words, and make and use tables.
- In the **Mini Labs,** you will experiment, observe, and infer.

368

## Bridging

Chapter 15 described the synapse as a space between adjoining neurons. A chemical messenger released from one neuron would move across the synapse and stimulate the adjoining neuron, thus completing a pathway for the nervous system. Students will now be taught the fact that stimulants and depressants operate by influencing the chemicals that move across the synapse, thus allowing a drug to either speed up or slow down messages.

For Tech Prep activities in this chapter of the Teacher Wraparound Edition, see especially the many Motivations, Activities, Portfolios, Connections, and Student Journals throughout the chapter.

See also the Glencoe Homepage at **www.glencoe.com**

Chapter **18**

# Drugs and Behavior

The plant in the photo on the opposite page is foxglove, a member of the genus *Digitalis*. The small photo on this page shows several pills. What do the plant and the pills have in common? Foxglove leaves are used to produce a drug called digoxin. Digoxin is used to treat heart failure, a disease that is caused by damaged heart muscle. The digoxin pills in the small photo help to increase the force of the heart contractions so the weakened heart can pump blood to all parts of the body.

**Try This!**

**What drugs have you bought?** Prepare a list of all the different types of drugs that you have purchased in a drugstore this past year. Describe why you bought each drug.

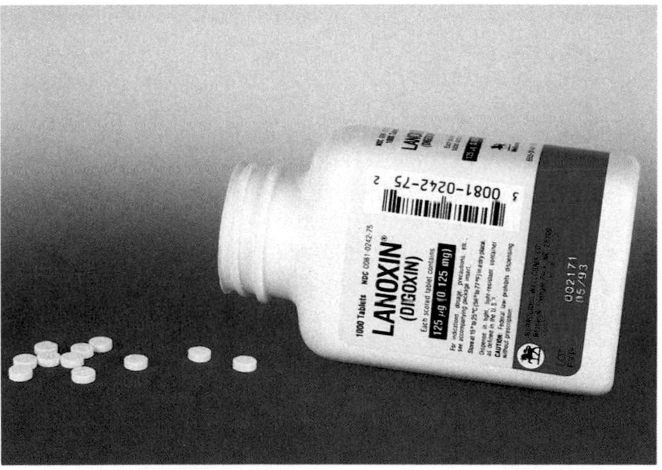


***inter*NET CONNECTION**

For more information about the material in this chapter, follow the link for the chapter on the Glencoe Homepage at **http://www.glencoe.com**

Digoxin tablets



**ASSESSMENT PLANNER**

***Portfolio***

Strategies on the following pages represent student products that can be placed into a best-work portfolio: pp. 372, 374, 385.

**PERFORMANCE ASSESSMENT**
Skill Check, pp. 377, 380, 388
Mini Lab, pp. 371, 372, 381
Lab, pp. 373, 387

**CONTENT ASSESSMENT**
Check for Understanding, pp. 375, 378, 379, 382, 388
Chapter Review, pp. 390-391

**GROUP ASSESSMENT**
Opportunities for group assessment occur with Cooperative Learning Strategies.

***Student Journal***

Strategies on the following pages represent opportunities for writing in a Student Journal: pp. 378, 386.

**GETTING STARTED**

## Using the Photos

Use the photos to point out to students that many of our modern medicines originally were derived from parts of plants. Bark from the Peruvian Cinchong tree was probably the most valuable of medical discoveries. The drug quinine extracted from the bark of the tree, is effective against malaria which was widespread disease in the 17th century. You may wish to discuss how the search for new drugs has been one of the major forces in history behind world exploration. Discuss how a country's economy may be controlled by exports of drugs such as tobacco and caffeine.

## MOTIVATION/Try This

**What drugs have you bought?** As a society, we are drug oriented and students will **observe** that there seems to be a legal drug available (over-the-counter) for almost every different ailment as well as for every different behavioral change being sought.

## Chapter Preview

Have students study the chapter outline before they begin to read the chapter. They should note the overall organization of the chapter, that the first topic described is a general introduction to drugs. This is followed by a section dealing with how drugs affect behavior. Next, a section deals with the use of over-the-counter drugs. Last, the chapter discusses careless drug use.

## Misconception

Students hear and read mainly about illegal drug use. They are not aware of the drug industries' major contribution to our well being and health. Pharmaceutical companies spend a large portion of their profits on researching and developing drugs to fight as well as to counteract diseases and their symptoms.

## 18:1 An Introduction to Drugs

### PREPARATION

#### Materials Needed

Make copies of the Study Guide, Lab, and Reteaching worksheets in the *Teacher Resource Package.*

▶ Gather materials needed for Lab 18-1. Prepare aspirin testing solution.

▶ Prepare petri dishes and broth cultures for the Mini Lab.

▶ Obtain garlic and onion bulbs for the Mini Lab

#### Key Science Words

drug
symptom
controlled drug
prescription drug
over-the-counter drug
side effect
dosage
overdose

#### Process Skills

In the Mini Lab, students will experiment. In Lab 18-1, they will observe, interpret data, and experiment.

### 1 FOCUS

▶ The objectives are listed on the student page. Remind students to preview these objectives as a guide to this numbered section.

#### MOTIVATION/Videocassette

Show the videocassette *Altering Our Living Chemistry,* Human Relations Media.

**Objectives**

1. **Compare** prescription and over-the-counter drugs.
2. **Relate** the importance of information listed on drug labels.
3. **Describe** how drugs stop pain.

**Key Science Words**

drug
symptom
controlled drug
prescription drug
over-the-counter drug
side effect
dosage
overdose

## 18:1 An Introduction to Drugs

Where do most drugs come from? The answer may surprise you. Most drugs come from plants. Aspirin comes from the willow tree. Castor oil comes from the seeds of the castor bean. Marijuana comes from the hemp plant. Some drugs come from fungi. In Chapter 5 you learned that the drug penicillin comes from a fungus called *Penicillium.* Other drugs come from bacteria, and a few come from animals.

### What Is a Drug?

What is a drug? A **drug** is a chemical that changes the way a living thing functions when it is taken into the body. Drugs are used for the treatment, cure, and prevention of disease. Drugs may be familiar chemicals such as those found in tea, coffee, or cigarettes. Did you brush your teeth today? If you did, you probably used a drug called fluoride in your toothpaste.

Why do people use drugs? Most drugs are used for medical reasons. These drugs are often called medicines. Doctors use certain drugs to kill organisms that cause diseases. Have you ever had strep throat? This disease is caused by bacteria. Doctors prescribe drugs like penicillin to cure strep throat.

People use other drugs to treat disease symptoms. **Symptoms** are changes that occur in your body as a result of a disease. You may know you have strep throat when your throat is so sore that it hurts to swallow. A sore throat is a symptom. Coughing and rashes are other examples of symptoms. These symptoms can be treated with cough medicines and ointments.

Most drugs are used to help people. If drugs are used incorrectly, however, they can harm people. For this reason, the manufacture and sale of many drugs that might be dangerous if used incorrectly are controlled by law. Drugs that are controlled by law are called **controlled drugs.** Marijuana, cocaine, and tranquilizers are all controlled drugs. Many people use controlled drugs illegally to change their behavior.

If you need a controlled drug for medical reasons, a doctor will direct you to take it, and you will be able to buy it or get it legally from the doctor. If you use a controlled drug without getting it from a doctor or drugstore, you are using the drug illegally.

**Figure 18-1** Medicines, and the fluoride in toothpaste, are drugs.

### OPTIONS

#### Science Background

Examples of drugs derived from plants include: pectin from lemon peel and used in antidiarrheal medicines; digitalis from foxglove and used in treating heart disorders; aloe from Aloe vera and used as an ointment for sunburn; castor oil from seeds of castor bean and used as a cathartic.

**How Do Drugs Affect the Ability of Seeds to Grow into Young Plants?**/*Lab Manual,* pp. 153-156. Use this lab as an extension to the effects of drugs.

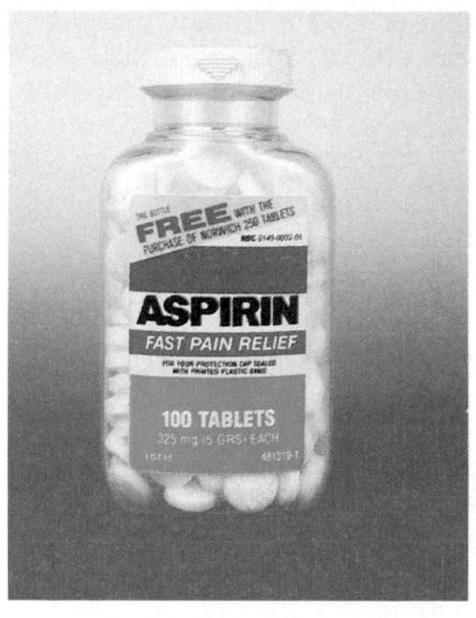


**Figure 18-2** Of all the over-the-counter drugs, aspirin is bought most often.

## Obtaining Drugs Legally

A legal drug is a drug used legally to treat a disease or its symptoms. There are two ways to obtain drugs legally. One way is with a prescription, and the other way is by buying them over the counter without a prescription.

What is a prescription drug?

A **prescription drug** is one that a doctor must tell you to take. The word *prescription* means direction. The doctor directs you to take a certain drug. Drugs obtained this way are legal.

Why do you need a doctor to tell you which drug to take? Do you know which drug will help your specific illness? Do you know how much of the drug to take? How would you know of any problems that could result from using the drug? You can see why a doctor must prescribe certain drugs for you. Some drugs are harmful if used incorrectly.

**Over-the-counter drugs** are those that you can buy legally without a prescription. They are also called nonprescription drugs. Figure 18-2 shows the huge selection of over-the-counter drugs in a supermarket. It includes everything from cough medicine to sunburn ointment to aspirin.

### Mini Lab

**Are Garlic and Onions Antibiotics?**

**Experiment:** Apply bacteria to a petri dish of agar. Add paper disks treated with garlic or onion juice. Use a plain disk as a control. Observe growth after two days. What do you see? *For more help, refer to the **Skill Handbook,** pages 704-705.*

## Reading Drug Labels

What can you learn from a drug label? Most drug labels tell you what the drug can be used for. The label also may have special instructions on what to do before using the drug.



## 2 TEACH

### MOTIVATION/Story

Relate the discovery of penicillin by Dr. Alexander Fleming in 1928. He noted that a colony of mold appearing by accident upon one of his petri dishes prevented the growth of bacteria near the mold. His observation of this unexpected result led to the discovery of penicillin and all future antibiotics.

### Guided Practice

Have students write down their answers to the margin question: a drug that a doctor must tell you to take.

### Mini Lab

Purchase sterile agar plates and broth from a supply house or prepare your own. Inoculate broth with *E. Coli* available from supply houses. Apply *E. Coli* from the broth cultures to the agar plates. Onions and garlic juice may be added to paperdiscs. These discs are positioned on inoculated petri dishes. A circle of inhibition will be seen surrounding the disks after 24 hours of incubation.

### ASSESSMENT

**Skill:** Provide students with diagrams that depict disks placed onto agar dishes. Show bacterial growth as shaded regions around the disk. Make the diameter of clear zones (free of bacterial growth) in each diagram of different diameters. Ask students to rank the diagrams in order of most to least effective in preventing bacterial growth.

### OPTIONS

**Science Background**

Over-the-counter (OTC) drugs are also called nonprescription drugs. There are about 350 000 OTC products with annual sales of $5 billion. OTC drugs that are subject to abuse include cough syrups, sleeping aids, and appetite suppressants.

**ACTIVITY/Challenge**

Have students prepare a brief table/chart on one over-the-counter drug of their choice. Their table/chart should include the following information: (a)drug use, (b)any special directions prior to drug use, (c)how often and how much of a drug to take (dosage), (d)any listed warning and (e)any listed cautions.

### GLENCOE TECHNOLOGY

**Videotape**

The Secret of Life

*In the Land of Milk and Money: Biotechnology*

## TEACH

### Concept Development

▶ Remind students that the stomach contains an acid and enzymes that could alter the nature of a drug. This is why certain drugs are injected.

### ACTIVITY/Demonstration

Make an overhead transparency of some common drug labels to help students locate the information on the label as it is being presented.

### Mini Lab

Students should be encouraged to use whatever design they feel is suitable for their product. Ask them to not duplicate those types of tamperproof packaging already available and used.

### ASSESSMENT

**Skill:** Provide students with a variety of childproof drug containers. Ask students to: (a) suggest ways in which a child may still be able to open or break open the container, (b) suggest ways in which the container may be improved.

### Guided Practice

Have students write down their answers to the margin question: what the drug can be used for, special instructions for using the drug, special side effects, cautions, dosage.

### ACTIVITY/Display

Ask your druggist for a copy of the warnings placed on drugs.

### Independent Practice

**Study Guide**/*Teacher Resource Package,* p. 103. Use the Study Guide worksheet shown at the bottom of this page for independent practice.

**Portfolio**

Have students copy all the information provided on an over-the-counter drug label of their choice. They should circle and label the following: warnings, use of drug, dose, special instructions, cautions.

**Figure 18-3** Warnings, special instructions, and side effects are present on all legal drug labels.

Shake well before using
For the relief of diarrhea
Directions:
Adult— 2 tablespoons every 4 hours, and no more than 8 doses in 24 hours
Children— 1 tablespoon every 4 hours, no more than 6 doses in 24 hours
WARNING— Do not take this drug with aspirin. May cause ringing in ears.
CAUTION— Do not take this drug if you have gout or diabetes.

Special instructions
Use of drug
Dose
Warning gives side effect and what other drug not to use
Caution

### Mini Lab

**Can a Drug Package Be Made Childproof?**

**Observe:** Survey the methods used to make over-the-counter drugs tamperproof and childproof. Design a childproof package for a product that you now use. *For more help, refer to the **Skill Handbook**, pages 704-705.*

Look at the drug label in Figure 18-3. One of the most important parts of a label is the warning. Warnings tell about possible side effects. A **side effect** is a change other than the expected change caused by a drug. This change often is not wanted or desirable. What is the side effect that may take place with use of the drug in Figure 18-3?

Warnings also tell you which foods or other drugs must not be taken with the drug. Mixing drugs with certain other drugs or foods can be dangerous. Mixing drugs can cause an unexpected and unwanted change in the body. For example, some cold medicines help get rid of cold symptoms. If used with certain diet pills, a steep rise in blood pressure, confusion, and nervousness can result.

Drug labels also include cautions. A caution may tell you to keep the drug away from children.

What do drug labels tell you?

Another very important thing a drug label tells you is the dosage. The **dosage** is how much and how often to take a drug. "Take one pill every four hours" is an example of a drug dosage.

Age influences drug dosage. Very young children must receive smaller amounts of a drug than adults. It is important to make sure that a child is not given too much of a drug. Drug dosage also is related to body size. Most drugs taken into the body leave it quickly or are changed into a different form by the body. The kidney, lungs, and liver are three organs that help the body get rid of drugs. For a drug to do its job, a certain amount of it must be in the body for a certain length of time. There must be enough drug taken in to make up for the amount leaving. This amount varies with body size.



## OPTIONS

### ACTIVITY/Challenge

Ask students to discuss the problem of drug interactions with a druggist and share the information with the class. They should prepare a report, citing specific examples.

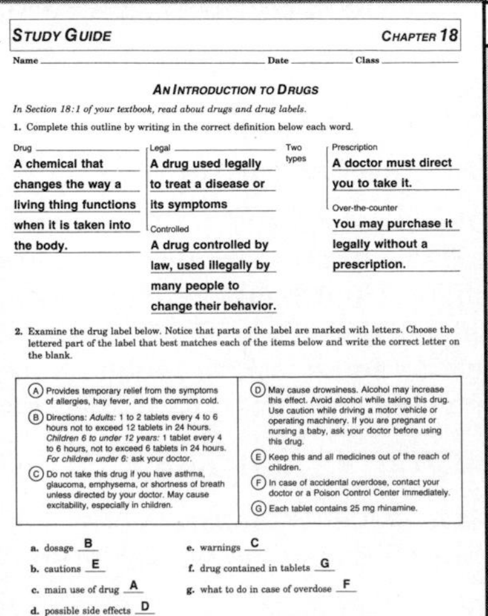
STUDY GUIDE 103

STUDY GUIDE CHAPTER 18

Name ______ Date ______ Class ______

AN INTRODUCTION TO DRUGS

*In Section 18:1 of your textbook, read about drugs and drug labels.*

1. Complete this outline by writing in the correct definition below each word.

Drug: **A chemical that changes the way a living thing functions when it is taken into the body.**

Two types:
- Legal: **A drug used legally to treat a disease or its symptoms**
  - Prescription: **A doctor must direct you to take it.**
  - Over-the-counter: **You may purchase it legally without a prescription.**
- Controlled: **A drug controlled by law, used illegally by many people to change their behavior.**

2. Examine the drug label below. Notice that parts of the label are marked with letters. Choose the lettered part of the label that best matches each of the items below and write the correct letter on the blank.

(A) Provides temporary relief from the symptoms of allergies, hay fever, and the common cold.
(B) Directions: *Adults:* 1 to 2 tablets every 4 to 6 hours not to exceed 12 tablets in 24 hours. *Children 6 to under 12 years:* 1 tablet every 4 to 6 hours, not to exceed 6 tablets in 24 hours. *For children under 6:* ask your doctor.
(C) Do not take this drug if you have asthma, glaucoma, emphysema, or shortness of breath unless directed by your doctor. May cause excitability, especially in children.
(D) May cause drowsiness. Alcohol may increase this effect. Avoid alcohol while taking this drug. Use caution while driving a motor vehicle or operating machinery. If you are pregnant or nursing a baby, ask your doctor before using this drug.
(E) Keep this and all medicines out of the reach of children.
(F) In case of accidental overdose, contact your doctor or a Poison Control Center immediately.
(G) Each tablet contains 25 mg rhinamine.

a. dosage **B**
b. cautions **E**
c. main use of drug **A**
d. possible side effects **D**
e. warnings **C**
f. drug contained in tablets **G**
g. what to do in case of overdose **F**

# Lab 18–1 Aspirin

## Problem: Which medicines contain aspirin?

### Skills

observe, interpret data, experiment

### Materials

4 glass slides
wax marking pencil
aspirin-testing chemical, in dropper bottle
6 test solutions of painkillers
8 droppers
water
aspirin solution

### Procedure

1. Copy the data table.
2. Use a marking pencil to draw two circles on a glass slide. Label the circles 1 and 2.
3. Add one drop of water to circle 1. Use a clean dropper to add one drop of aspirin solution to circle 2.
4. Add one drop of aspirin-testing chemical to each circle. **CAUTION:** *Aspirin-testing chemical is poisonous and can burn the skin.*
5. **Observe:** Wait one minute. Note the color in each circle. A violet color means that aspirin is present. Record your results.
6. Draw two circles on each of three more glass slides. Label the circles A through F.
7. Record in your data table the name of each medicine being tested.
8. Add a drop of a different test solution to each circle. Use a clean dropper for each solution.
9. Add one drop of aspirin-testing chemical to each circle.
10. Wait one minute. Record the color that appears in each circle. Complete the last column of your data table.
11. Dispose of your solutions and wash your glassware.

### Data and Observations

1. Which circles showed a violet color?
2. Which circles did not show a violet color?

| CIRCLE | CONTENTS | COLOR WITH TEST CHEMICAL | ASPIRIN PRESENT? |
|---|---|---|---|
| 1 | water | clear | no |
| 2 | aspirin | violet | yes |
| A | Advil | clear | no |
| B | Excedrin | clear | no |
| C | Nuprin | clear | no |
| D | Bufferin | violet | yes |
| E | Midol | clear | no |
| F | Tylenol | clear | no |

### Analyze and Apply

1. What was the purpose of the slide labeled circle 1 and circle 2?
2. Where should aspirin be listed if it is present in a medicine?
3. Ask your teacher to show you the labels from the medicines tested. Which medicines contain aspirin? Does this agree with your data?
4. **Apply:** Why should aspirin be listed as an ingredient if it is present in a medicine?

### Extension

Caffeine is often added to aspirin products because the aspirin works more rapidly if taken with caffeine. Check the labels of aspirin products available at a drugstore to see which products contain caffeine. **Make a table** of the products that contain both drugs and those that contain only aspirin.



## ANSWERS

### Data and Observations

1. Answers will vary. Bufferin and other aspirin products will show violet color.
2. Answers will vary. Advil, Excedrin, Nuprin, Midol, and Tylenol will not show a violet color.

### Analyze and Apply

1. to show that aspirin causes a violet color and that the test chemical does not turn violet without aspirin
2. on the drug label under "ingredients"
3. Answers will vary. Products that gave a positive test will have aspirin in them; products that did not turn violet with the test chemical will not contain aspirin.
4. Some people might want to avoid taking aspirin because they are allergic to it.

## Lab 18-1 Aspirin

### Overview

Students will test a known aspirin and a known nonaspirin solution with an aspirin-detecting chemical. They will then determine if aspirin is or is not present in a number of painkilling drugs. NOTE: Aspirin-testing chemical is an acid solution. Advise students to use caution and avoid spillage. Rinse with water if spillage occurs on skin or clothing.

**Objectives:** Upon completion of this lab, students will be able to: (1) **determine** how to test for the presence or absence of aspirin in a drug, (2) **test** a variety of painkilling drugs for the presence of aspirin.

**Time Allotment:** 30 minutes

### Preparation

**Lab 18-1 worksheet**/*Teacher Resource Package*, pp. 69-70.

▶ To make the aspirin solution, dissolve one aspirin tablet in 50 mL of warm water.

▶ To make the aspirin-detecting solution, add 2 g ferric nitrate nonahydrate, $Fe(NO_3)_3 \cdot 9H_2O$, to 50 mL water. Then add 1 mL concentrated nitric acid. **CAUTION:** *Nitric acid is very corrosive.*

### Teaching the Lab

▶ **Troubleshooting:** (a) Placing aspirin-detecting solution and aspirin solutions in marked dropping bottles will reduce spillage accidents.

▶ Use fresh solutions of drugs. Advil may give a false positive.

### ✓ ASSESSMENT

**Performance:** Provide students with four unknowns marked 1-4. Have them prepare a data table and determine which unknowns contain aspirin. Have students record their data.

# TEACH

## ACTIVITY/Demonstration

A demonstration can be done with a funnel to illustrate proper dose and overdose. Pour water into a funnel at such a rate that as much water enters as leaves the funnel. This corresponds to the proper dose. To demonstrate an overdose, pour water more rapidly than it flows out, so that it eventually spills over the edge of the funnel.

## Guided Practice

Have students prepare an idea map that compares the two ways of taking drugs, swallowing and injection.

## Independent Practice

**Study Guide**/*Teacher Resource Package,* p. 104. Use the Study Guide worksheet shown at the bottom of this page for independent practice.

## ACTIVITY/Brainstorming

Ask your students how a drug is able to move from the blood to the nervous system. Drugs can diffuse from cell to cell, eventually diffusing into nerve cells.

**Portfolio**

Figure 18-4 uses an analogy to describe the way in which the body reacts to a normal drug dose and a drug overdose. Have students redraw Figures **a** and **b** with a kidney in place of the sink. They should label and use arrows to show the flow of drugs into and out of the kidney. Have them refer to Figure 13-12 (a) for help.

**Videodisc**

**STV: Human Body**
*Prescription Drugs*

**48837**

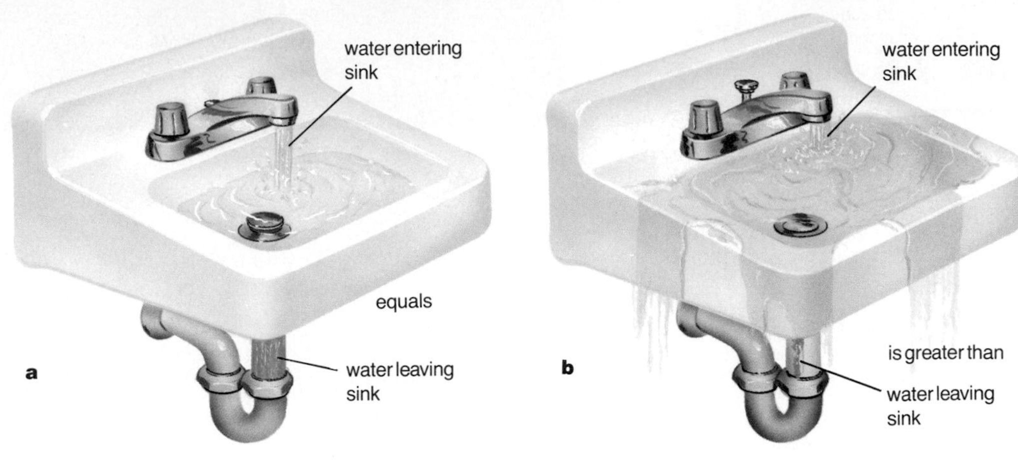


**Figure 18-4** Normal drug dose (a) – the amount of drug entering the body equals the amount leaving the body. Overdose (b) – the amount of drug entering the body is greater than the amount leaving the body.

Figure 18-4a uses a leaking sink to help explain how the body balances the amount of drug entering and leaving it. Think of the water going into the sink as a drug entering the body. The leaking stopper stands for the organ removing the drug. If the water entering the sink is equal to the water leaving it, the water will not overflow. Nor will the sink become empty. The same type of thing happens with the proper drug dosage. The amount of drug entering the body equals the amount leaving it.

What happens if a person does not take the correct drug dose? Not taking enough of a drug may result in the drug not doing its job. Taking too much of a drug can result in a drug overdose. An **overdose** is the result of too much of a drug in the body. Let's look at the sink example again. Figure 18-4b shows what might lead to a drug overdose. Too much of a drug is added to the body. The body can't get rid of the drug fast enough, and a drug overdose results. Drug overdoses are dangerous because they prevent the body from working properly.

**Bio Tip**

**Health:** Do not store any medicine in the glove compartment of a car. High temperatures may alter the chemicals in the drug.

## Speed of Drug Action

There are two ways that people commonly take drugs. These ways are swallowing or getting a shot. Suppose a pain comes from your toe. How can swallowing a pill or getting a shot stop that pain? Look at Figure 18-5 as you read the following steps.



## *OPTIONS*

### Science Background

A drug that is injected is generally fast-acting, and dosage is more accurately controlled. A drawback to drug injection is that sterile conditions must be maintained.

**STUDY GUIDE 104**

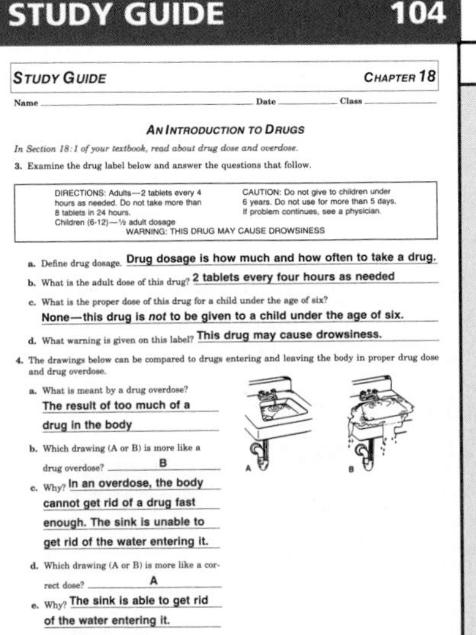

STUDY GUIDE — CHAPTER 18

Name ______ Date ______ Class ______

*AN INTRODUCTION TO DRUGS*

*In Section 18:1 of your textbook, read about drug dose and overdose.*

3. Examine the drug label below and answer the questions that follow.

DIRECTIONS: Adults—2 tablets every 4 hours as needed. Do not take more than 8 tablets in 24 hours. Children (6-12)—½ adult dosage

CAUTION: Do not give to children under 6 years. Do not use for more than 5 days. If problem continues, see a physician.

WARNING: THIS DRUG MAY CAUSE DROWSINESS

a. Define drug dosage. **Drug dosage is how much and how often to take a drug.**

b. What is the adult dose of this drug? **2 tablets every four hours as needed**

c. What is the proper dose of this drug for a child under the age of six? **None—this drug is *not* to be given to a child under the age of six.**

d. What warning is given on this label? **This drug may cause drowsiness.**

4. The drawings below can be compared to drugs entering and leaving the body in proper drug dose and drug overdose.

a. What is meant by a drug overdose? **The result of too much of a drug in the body**

b. Which drawing (A or B) is more like a drug overdose? **B**

c. Why? **In an overdose, the body cannot get rid of a drug fast enough. The sink is unable to get rid of the water entering it.**

d. Which drawing (A or B) is more like a correct dose? **A**

e. Why? **The sink is able to get rid of the water entering it.**

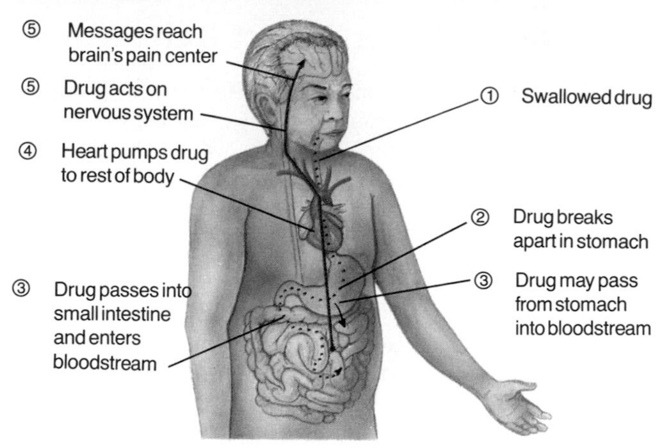


**Figure 18-5** Path followed by a swallowed drug. What body system is responsible for pain relief? nervous system

1. A person swallows a drug.
2. The drug reaches the stomach and begins to break up.
3. Some of the drug may pass from the stomach directly into the bloodstream. The rest will continue into the small intestine and then into the bloodstream.
4. Once the drug is in the blood, the heart pumps it to all parts of the body.
5. The drug acts on the nervous system. The nervous system sends messages to the brain and shuts off the brain's pain center.

**How does a painkiller stop pain after it is swallowed?**

Drugs taken by mouth don't work as quickly as injected drugs. Injected drugs go directly into blood capillaries. They don't have to spend any time passing through the digestive system first.

## Check Your Understanding

1. How is buying a prescription drug different from buying an over-the-counter drug?
2. Name four things that a drug label tells you.
3. Describe how a drug stops pain in the body.
4. **Critical Thinking:** If you are very sick, the doctor may give you a shot of antibiotic in the office and then give you pills to take later. Why do you think he or she gave you both a shot and pills? Why not just pills?
5. **Biology and Reading:** Look at the drug label in Figure 18-3. Why should this drug be taken? How often should the person take this drug?



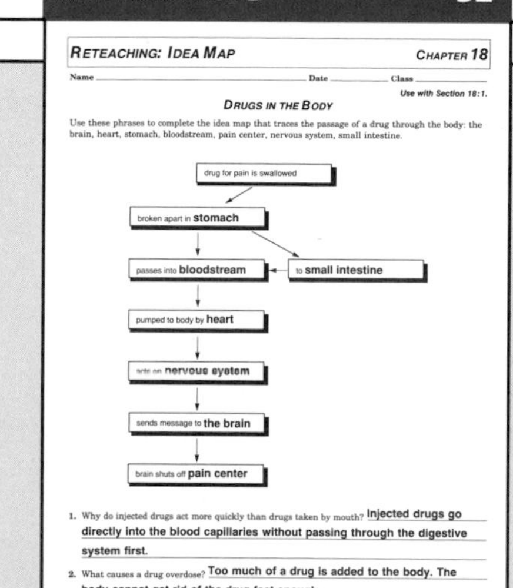
RETEACHING 52

*RETEACHING: IDEA MAP* CHAPTER 18

Name ______ Date ______ Class ______

*Use with Section 18:1.*

*DRUGS IN THE BODY*

Use these phrases to complete the idea map that traces the passage of a drug through the body: the brain, heart, stomach, bloodstream, pain center, nervous system, small intestine.

drug for pain is swallowed → broken apart in **stomach** → passes into **bloodstream** (← to **small intestine**) → pumped to body by **heart** → acts on **nervous system** → sends message to **the brain** → brain shuts off **pain center**

1. Why do injected drugs act more quickly than drugs taken by mouth? **Injected drugs go directly into the blood capillaries without passing through the digestive system first.**
2. What causes a drug overdose? **Too much of a drug is added to the body. The body cannot get rid of the drug fast enough.**

## OPTIONS

**Path of a Swallowed Drug**/*Transparency Package,* number 18. Use color transparency number 18 as you teach the speed of drug action.

## TEACH

### Guided Practice

Have students write down their answers to the margin question: It blocks the movement of messengers at the synapse in the brain.

### Check for Understanding

Have students respond to the first three questions in Check Your Understanding.

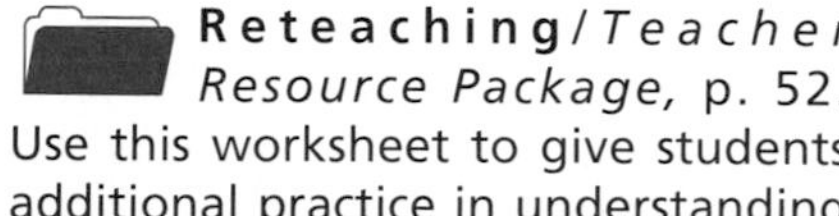

### Reteach

**Reteaching**/*Teacher Resource Package,* p. 52. Use this worksheet to give students additional practice in understanding special types of behavior.

**Extension:** Assign Critical Thinking, Biology and Reading, or some of the **OPTIONS** available with this lesson.

## 3 APPLY

### ACTIVITY/Videocassette

Show the videocassette *Chemicals in a Chemical Body,* Human Relations Media.

## 4 CLOSE

### Software

*Drugs: Who's in Control?* Opportunities for Learning.

### Answers to Check Your Understanding

1. A doctor must prescribe a prescription drug.
2. drug use, special instructions on what to do before taking the drug, warnings on side effects or mixing with certain foods or other drugs, drug dosage
3. The drug stops the movement of the chemical messenger at the synapse in the brain.
4. The shot works faster than the pills. Pills give longer-lasting relief.
5. for the relief of diarrhea; every four hours

## 18:2 How Drugs Affect Behavior

### PREPARATION

#### Materials Needed

Make copies of the Study Guide and Reteaching worksheets in the *Teacher Resource Package*.

#### Key Science Words

stimulant
depressant
psychedelic drug
inhalant

#### Process Skills

In the Skill Check, students will sequence.

### 1 FOCUS

▶ The objectives are listed on the student page. Remind students to preview these objectives as a guide to this numbered section.

#### ACTIVITY/Word Association

Ask students to take the four words listed under Key Science Words and
a) define the terms in their own words.
b) list dictionary words that are derived from or are similar to them.

#### Bridging

Review nerve physiology, particularly the role and location of axon, synapse, and dendrite ends.

**Videodisc**
**STV: Human Body Vol. 2**
*Nervous system*
Unit 2, Side 2
*Neurotransmission*

**15941-21903**

## 18:2 How Drugs Affect Behavior

#### Objectives

4. **Describe** the effects of stimulants and depressants on the body.
5. **Discuss** how psychedelic drugs and inhalants affect the body.

#### Key Science Words

stimulant
depressant
psychedelic drug
inhalant

Three of the ways in which drugs affect the body are to speed up the body's activities, slow down the body's activities, or affect a person's senses or way of thinking. These changes in the body cause changes in behavior.

### Stimulants

A **stimulant** (STIHM yuh lunt) is a drug that speeds up body activities that are controlled by the nervous system. Many stimulants are controlled drugs. Examples of stimulants that are controlled drugs are cocaine and amphetamines (am FEHT uh meenz). Caffeine and nicotine are also stimulants, but they are not controlled drugs.

How does a stimulant speed up the body's activities? To answer this question, we must look at nerve cells in the brain.

In Chapter 15 you learned how neurons carry messages. Chemical messengers given off by the axon end of one neuron move across the synapse and are picked up by the dendrite end of the next neuron. Usually these chemical messengers are destroyed after crossing the synapse. This prevents the original message from going on continuously. Figure 18-6 shows how this works.

How do stimulants change this pattern? They may cause the axon of a neuron to give off more of the chemical

> **Bio Tip**
> **Consumer:** An average of four prescription drugs per year, at a yearly cost of $22, are used by people age 16 or younger. Eighteen prescription drugs per year, at a yearly cost of $138, are used by those 65 or older.

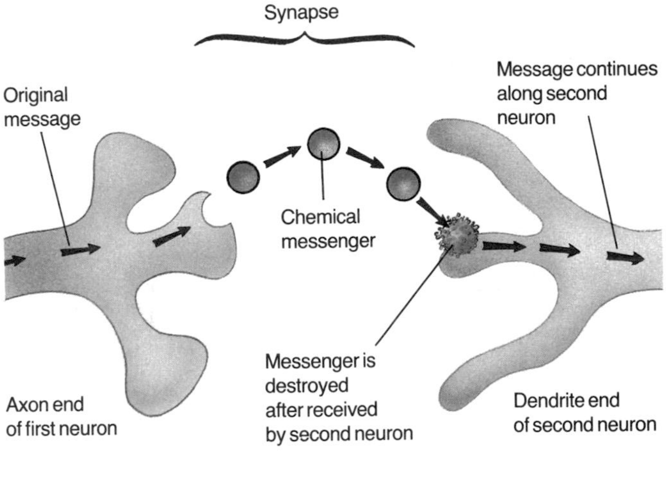


**Figure 18-6** Messages move from one neuron to the next across a synapse using a chemical messenger.



### OPTIONS

#### Science Background

LSD stands for lysergic acid diethylamide. Users of LSD report hallucinations, visions, great euphoria, panic, fear, depression. PCP stands for phencyclidine. PCP has been labeled as the most devastating of the hallucinogens because of its severe, long-lasting effects. With chronic use, memory loss, slurred speech, nervousness, anxiety, severe depression, overt aggressive behavior, delirium, convulsions, muscle rigidity, and coma may be experienced.

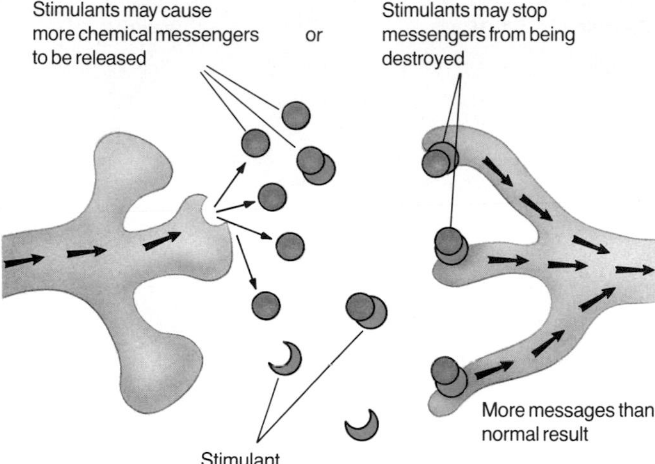


**Figure 18-7** There are two ways that stimulants can change the chemical messengers in a synapse.

messenger than normal. Or, stimulants may prevent the chemical messenger from being destroyed once it reaches the dendrite of the next neuron. In both cases, the second neuron keeps receiving chemical messengers. Figure 18-7 shows changes caused by stimulants. With stimulants, messages move from one neuron to the next for a longer time.

How does a stimulant affect the nervous system?

What are some effects of stimulants on the body? Stimulants speed up the heart rate. They increase blood pressure. There is a decrease in appetite. A person taking stimulants has a speeded-up feeling. He or she may feel very alert. Over time, these changes may harm the body because they use up the body's supply of nutrients and energy too quickly.

## Depressants

The word *depress* means to push down. A **depressant** (dih PRES unt) is a drug that slows down messages in the nervous system. Depressants slow down the body the way brakes slow down a car. Depressants are controlled drugs.

The main roles of depressants are to calm behavior, reduce pain, and help people sleep. Many mild sleep aids are sold over the counter, even though they are controlled drugs. Codeine (KOH deen), morphine, and barbiturates (bahr BICH uh ruhts) are other depressants. These are legal drugs that may be prescribed by doctors.

### Skill Check

**Sequence:** Draw the path that a message takes to go from one neuron to another. Then draw the same path to show what happens if a depressant is present. *For more help, refer to the* ***Skill Handbook,*** *pages 706-711.*



## OPTIONS

### Science Background

Examples of neurotransmitters are acetylcholine, dopa, dopamine, epinephrine, norepinephrine also known as adrenaline and noradrenaline, glutamic acid, serotonin, and gamma-aminobutyric acid.

## 2 TEACH

### MOTIVATION/Brainstorming

Ask students to name the general region of the nervous system where a stimulant is working if it is used to prevent sleep.

**Analogy:** Use the analogy of a phone message to reinforce normal nerve physiology. Think of what happens with a telephone. You call someone and talk for a while. When you hang up, the message stops moving across the telephone line. You must make another call if you wish to continue the conversation.

### Concept Development

▶ Make sure students understand Figure 18-7. A stimulant will continue to stimulate the dendrite ends of the adjoining neurons for a long period of time, in spite of the fact that no new messages are passing along the nerve. Use the telephone analogy again. Think of what happens if you hang up the phone and the line remains open. Your phone is still connected. You just pick up the phone if you wish to talk to the same person.

### ACTIVITY/Brainstorming

Have students identify the axon and dendrite ends of the nerves in Figure 18-6. Ask which nerve end would be blocked from giving off the chemical messenger.

### Guided Practice

Have students write down their answers to the margin question: It causes the neuron to give off more chemical messenger than normal or prevents the chemical messenger frombeing destroyed once it reaches the next neuron.

### Skill Check

Student drawings should be similar to the sequences shown in Figures 18-6 and 18-8.

# TEACH

## Concept Development

▶ Inhalants result in euphoria, hallucinations, irritability, anxiety, panic, heart arrhymias, fatigue, depression, and liver failure.

## Check for Understanding

Have students prepare a table that lists the following: (a) Side effects on the body when using stimulants, depressants, psychedelics. (b) Change brought about at the synapse when using stimulants, depressants.

## Reteach

Have students define the words *stimulate, depress,* and *psychedelic.* Ask them to write three different sentences using the three words in a nondrug-related context. Ask students to give examples of the three drug types. Ask them to speculate on how the synapse is being influenced by psychedelics.

## Independent Practice

**Study Guide**/*Teacher Resource Package,* p. 105. Use the Study Guide worksheet shown at the bottom of this page for independent practice.

## Student Journal

Have students imagine the following: You are a chemical messenger that is planning a trip across a neuron synapse. Describe where your trip must begin, where your travels take you, and where your trip must end. Also explain how your trip would differ if stimulants and depressants were present, but not at the same time.

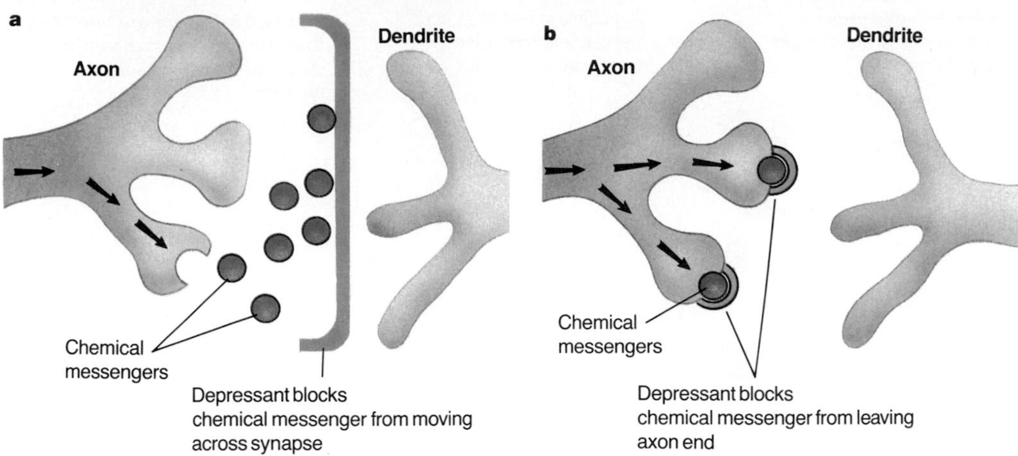


**Figure 18-8** Depressants change chemical messengers at a synapse in two ways.

Let's see how a depressant stops pain. Suppose you had a bad toothache. Neurons carry the pain message to your brain. At the brain, the message enters the brain's pain center. Then you feel the discomfort called pain.

Painkillers work in your brain at the synapse, just as stimulants do. Certain depressants may block the movement of chemical messengers across the synapse. This is shown in Figure 18-8a. Some depressants block the axon of a neuron from giving off the chemical messenger. This is shown in Figure 18-8b. Painkillers don't change what causes the pain. They simply keep you from feeling pain by stopping the chemical messenger from doing its job.

**Figure 18-9** The leaves, stems, flowers, and pollen of marijuana contain several drugs.

## Psychedelic Drugs

A third group of drugs that change behavior is the psychedelic (si kuh DEL ihk) drugs. A **psychedelic drug** is one that alters the way the mind works and changes the signals we receive from our sense organs. Hearing, seeing, and thinking are changed. For example, someone may say that an object appears brighter than it really is. Sounds become louder. Senses blend together. Someone may report "tasting colors" or "seeing music."

There are two general groups of psychedelic drugs. One group is found in nature. Natural psychedelic drugs are those found in certain kinds of mushrooms, cactus plants, marijuana, and the leaves of other desert and jungle plants. The second group of psychedelic drugs is synthetic, or made by humans. Drugs in both groups are controlled.



## OPTIONS

### Science Background

Inhalants include the following chemical compounds: toluene, benzene, kerosene, gasoline, and Freon (a fluorocarbon found in refrigerators and air conditioners as a coolant).

Inhalants typically act on the nervous system as does alcohol. They "short circuit" the nervous system's normal pathways causing neurons to send messages spontaneously without their receiving of normal messages from sense organs.

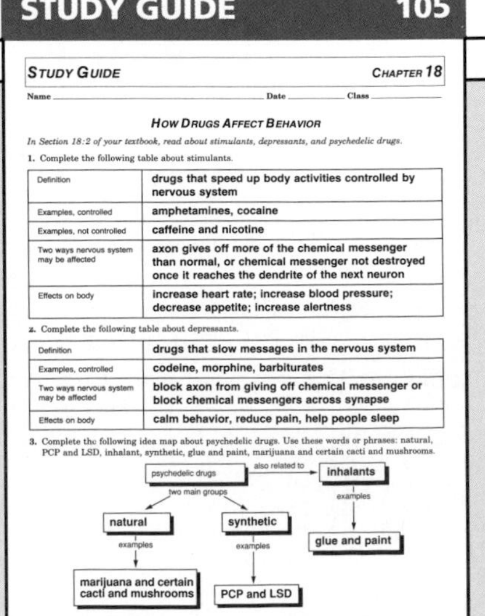

STUDY GUIDE 105

STUDY GUIDE CHAPTER 18

Name ______ Date ______ Class ______

HOW DRUGS AFFECT BEHAVIOR

*In Section 18:2 of your textbook, read about stimulants, depressants, and psychedelic drugs.*

1. Complete the following table about stimulants.

| | |
|---|---|
| Definition | drugs that speed up body activities controlled by nervous system |
| Examples, controlled | amphetamines, cocaine |
| Examples, not controlled | caffeine and nicotine |
| Two ways nervous system may be affected | axon gives off more of the chemical messenger than normal, or chemical messenger not destroyed once it reaches the dendrite of the next neuron |
| Effects on body | increase heart rate; increase blood pressure; decrease appetite; increase alertness |

2. Complete the following table about depressants.

| | |
|---|---|
| Definition | drugs that slow messages in the nervous system |
| Examples, controlled | codeine, morphine, barbiturates |
| Two ways nervous system may be affected | block axon from giving off chemical messenger or block chemical messengers across synapse |
| Effects on body | calm behavior, reduce pain, help people sleep |

3. Complete the following idea map about psychedelic drugs. Use these words or phrases: natural, PCP and LSD, inhalant, synthetic, glue and paint, marijuana and certain cacti and mushrooms.

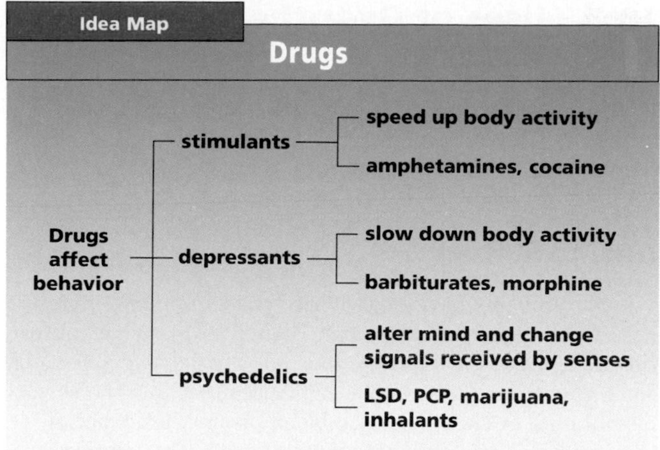


**Study Tip:** Use this idea map as a study guide to drugs that affect behavior. How many other examples can you give?

Synthetic drugs that alter the mind include PCP and LSD. PCP is also known as "angel dust." Use of PCP causes a variety of physical changes. These include high blood pressure, difficulties with walking or standing, and numbness. PCP also causes changes in behavior. Users of PCP may become very violent. They may also have a loss in memory or try to harm themselves.

What problems can psychedelic drugs cause?

Related to the psychedelics are inhalants. An **inhalant** is a drug breathed in through the lungs in order to cause a behavior change. The chemicals in glues, paints, and typewriter correction fluid are inhalants. Their use can cause irregular heartbeat and liver damage. The heart may stop beating, and death may occur.

## Check Your Understanding

6. Describe the action of a stimulant on the nervous system.
7. What behavior changes are caused by psychedelic drugs?
8. What are the main roles of depressants in the body?
9. **Critical Thinking:** How do diet pills work at the nerve level? Are they stimulants or depressants?
10. **Biology and Reading:** According to what you learned in this section, what word means the opposite of depressant?



## TEACH

### Idea Map

Have students use the idea map as a study guide to the major concepts of this section.

### Guided Practice

Have students write down their answers to the margin question: they change the way the mind works and change the signals received from the sense organs.

### Check for Understanding

Have students respond to the first three questions in Check Your Understanding.

### Reteach

**Reteaching**/*Teacher Resource Package,* p. 53. Use this worksheet to give students additional practice in understanding drugs in the body.

**Extension:** Assign Critical Thinking, Biology and Reading, or some of the **OPTIONS** available with this lesson.

## 3 APPLY

### ACTIVITY/Filmstrip

Show the filmstrip *A Capsule Look at What Drugs Do to You,* Carolina Biological Supply Co.

## 4 CLOSE

### Audiovisual

Show the filmstrip *Alcohol: America's Drug of Choice,* Carolina Biological Supply Co.

**RETEACHING 53**

RETEACHING — CHAPTER 18

Name — Date — Class

Use with Section 18:2.

HOW DRUGS WORK ON NEURONS

### Answers to Check Your Understanding

6. It causes the neuron to give off more chemical messenger than normal or prevents the chemical messenger from being destroyed once it reaches the next neuron.
7. Hearing, seeing, and thinking are changed. Objects may look brighter than normal. Senses blend together.
8. to stop pain, calm behavior, or help people sleep
9. They are stimulants. Messages move from one neuron to the next for a longer time. They decrease the appetite.
10. stimulant

# 18:3 Uses of Over-the-counter Drugs

## PREPARATION

### Materials Needed

Make copies of the Study Guide and Reteaching worksheets in the *Teacher Resource Package.*

▶ Obtain sodium bicarbonate (baking soda), vinegar, and litmus paper for the Focus motivation.

### Key Science Words

antihistamine
antacid

### Process Skills

In the Mini Lab, students will infer. In the Skill Check, they will understand science words.

## 1 FOCUS

▶ The objectives are listed on the student page. Remind students to preview these objectives as a guide to this numbered section.

### MOTIVATION/Demonstration

The action of sodium bicarbonate or any other antacid can be demonstrated. Fill a small beaker with vinegar (or any other dilute acid). Use a chemical indicator that will show whether a solution is acidic or basic. Add the sodium bicarbonate to the acid. A foam will appear, which indicates production of carbon dioxide gas. The indicator will change color showing that acid is no longer present in the beaker. If using litmus paper, the neutral solution that results will neither change blue nor red litmus. Repeat the demonstration with other antacids purchased from a drugstore.

## 18:3 Uses of Over-the-Counter Drugs

**Objectives**

6. **Give the function** of antihistamines and cough suppressants.
7. **Explain** the role of antacids in the body.

**Key Science Words**

antihistamine
antacid

There are thousands of over-the-counter drugs available. They relieve minor symptoms of illnesses, but they do not cure diseases. How do they work?

### Antihistamines

You probably have experienced the symptom of a stuffy nose when you have had a cold, Figure 18-10. What causes this? When you have a cold, your body is fighting a virus of some kind. Blood goes to the capillaries that line the nasal membranes. A clear fluid, the blood plasma, leaks out of the capillaries into the surrounding tissue. This plasma causes the tissues in your nose to swell. The swelling is what you feel when you have a stuffy nose.

Do you use a nasal spray or nose drops when your nose is stuffy? Nasal sprays and nose drops contain chemicals called antihistamines (ANT i HIHS tuh meenz). An **antihistamine** is a drug that reduces swelling of the tissues by stopping the leaking of blood plasma from capillaries. Most nasal sprays cause the leaking and swelling to stop. Then the stuffy feeling goes away. Antihistamines are also found in pills. They are used to reduce symptoms of hay fever, bee stings, and colds.

**Skill Check**

**Understand science words: antihistamine.** The word part *anti* means against. In your dictionary, find three words with the word part *anti* in them. *For more help, refer to the* ***Skill Handbook,*** *pages 706-711.*

**Figure 18-10** Antihistamines relieve the stuffy feeling of a cold. How do antihistamines work? they stop plasma from leaking out of capillaries



## OPTIONS

### Science Background

An invasion of microbes into the body stimulates certain white blood cells to release a chemical called histamine. This chemical causes blood capillary dilation and permeability. The increased permeability results in fluid loss and swelling. In a cold, histamine causes added mucus production and the typical runny nose associated with a cold. Antihistamines counteract the action of histamines.

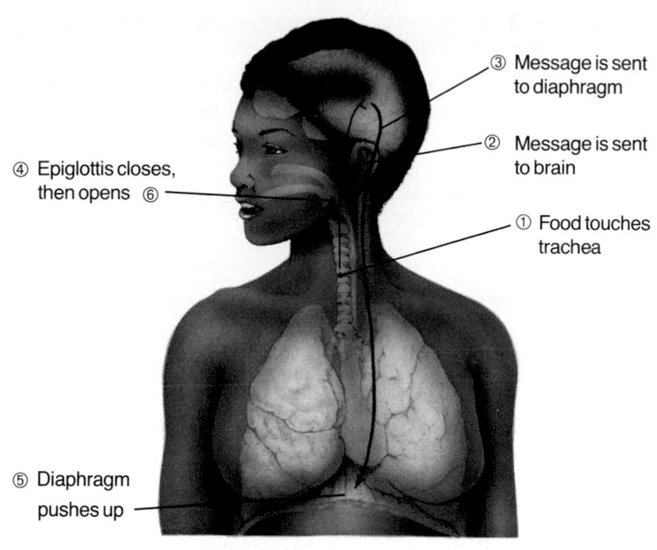


**Figure 18-11** How and why you cough

## Cough Suppressants

Coughing is your body's way of getting rid of whatever blocks your windpipe or lungs. Figure 18-11 shows why you cough. Something, such as food, may be touching the sides of your trachea. Nerves in the trachea sense the food and send a message to the part of the brain that controls coughing. The brain then sends the message to your diaphragm and your epiglottis, the flap that covers the opening of your windpipe. The epiglottis closes. Your diaphragm pushes up into your chest. Pressure builds up within your chest and the epiglottis opens. A burst of air under high pressure is pushed from your lungs, and you cough.

What is the role of cough suppressants in coughing? Suppressants are medicines that slow down or suppress the part of the brain that controls coughing. Cough medicines usually contain depressants. These drugs work on the chemical messenger of the synapse in the same way as other depressants.

### Mini Lab

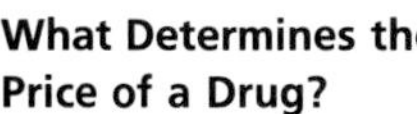

**What Determines the Price of a Drug?**

**Infer:** Compare the price of a "name brand" drug with a less known brand. Determine the cost per dose for each. Name some factors that might affect the price. *For more help, refer to the **Skill Handbook**, pages 706-711.*

**What kind of drug do cough suppressants contain?**

**Figure 18-12** Cough suppressants are available in many different forms.



**STUDY GUIDE 106**

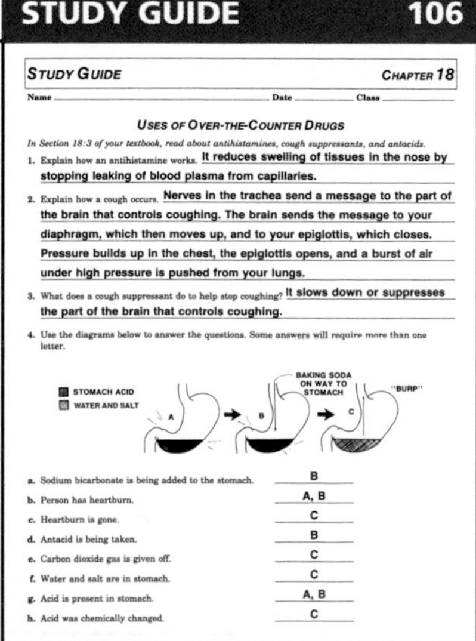

STUDY GUIDE CHAPTER 18

Name ______ Date ______ Class ______

USES OF OVER-THE-COUNTER DRUGS

*In Section 18:3 of your textbook, read about antihistamines, cough suppressants, and antacids.*

1. Explain how an antihistamine works. It reduces swelling of tissues in the nose by stopping leaking of blood plasma from capillaries.
2. Explain how a cough occurs. Nerves in the trachea send a message to the part of the brain that controls coughing. The brain sends the message to your diaphragm, which then moves up, and to your epiglottis, which closes. Pressure builds up in the chest, the epiglottis opens, and a burst of air under high pressure is pushed from your lungs.
3. What does a cough suppressant do to help stop coughing? It slows down or suppresses the part of the brain that controls coughing.
4. Use the diagrams below to answer the questions. Some answers will require more than one letter.

STOMACH ACID
WATER AND SALT
BAKING SODA ON WAY TO STOMACH
"BURP"

a. Sodium bicarbonate is being added to the stomach. B
b. Person has heartburn. A, B
c. Heartburn is gone. C
d. Antacid is being taken. B
e. Carbon dioxide gas is given off. C
f. Water and salt are in stomach. C
g. Acid is present in stomach. A, B
h. Acid was chemically changed. C

# 2 TEACH

### ACTIVITY/Brainstorming

Ask your students if coughing is voluntary or involuntary. It can be both. The term *reflex* best describes involuntary coughing.

### Concept Development

▶ Most students have experienced swelling of the nose membranes. The swelling is due to the formation of histamine.

▶ Histamine is a chemical that is released from cells when they are injured. In the case of a cold, the cells of the nasal chamber membranes are injured by the cold virus.

▶ Histamine not only causes tissue swelling, but it also stimulates the secretion of mucus.

### Mini Lab

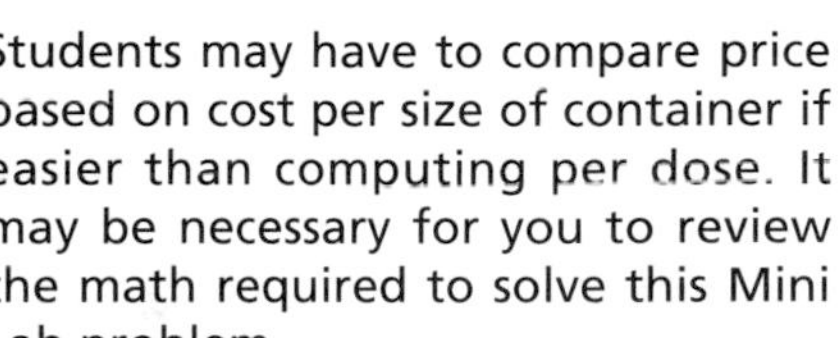

Students may have to compare price based on cost per size of container if easier than computing per dose. It may be necessary for you to review the math required to solve this Mini Lab problem.

### ASSESSMENT

**Skill:** Provide students with the following data: cost of three name-brand drugs, cost of these same three drugs but from lesser known brands, and normal dose for each drug. Have them calculate the cost per dose for each drug.

### Independent Practice

**Study Guide**/*Teacher Resource Package,* p. 106. Use the Study Guide worksheet shown at the bottom of this page for independent practice.

### Guided Practice

Have students write down their answers to the margin question: a depressant.

# TEACH

### Guided Practice

Have students write down their answers to the margin question: It changes the acid to water and a salt.

### Check for Understanding

Have students respond to the first three questions in Check Your Understanding.

### Reteach

Reteaching/*Teacher Resource Package*, p. 54. Use this worksheet to give students additional practice in understanding how drugs work on neurons.

**Extension:** Assign Critical Thinking, Biology and Reading, or some of the **OPTIONS** available with this lesson.

## 3 APPLY

### ACTIVITY/Videocassette

Show the videocassette *Over-the-counter Drugs and Valium,* Human Relations Media.

## 4 CLOSE

### ACTIVITY/Brainstorming

Ask students to force a cough. Have them note the changes that take place in the throat, epiglottis, diaphragm, and chest wall. Ask what normally triggers a cough. Is it voluntary or involuntary? How did the forcing of a cough differ from normal coughing in terms of being voluntary or involuntary?

### Answers to Check Your Understanding

11. Antihistamines stop blood plasma from leaking out of capillaries. Tissue swelling decreases.
12. depressants
13. It combines with stomach acid to form water and salt.
14. usually involuntary; reflex pathway
15. depressants

**Figure 18-13** Baking soda can change an acid to water and a salt.

## Antacids

Sodium bicarbonate (bi KAR buh nate) is the chemical name for baking soda. Many people have it in their homes. It is used to make cakes rise during baking. It is also useful as a drug. Let's see how. The stomach normally makes acid for digestion. When the stomach makes too much acid, you might get heartburn. Baking soda gets rid of the extra acid by causing a chemical change. You see this change in Figure 18-13. In the figure, baking soda has changed vinegar, an acid, to water and a salt. It does the same thing to stomach acid. The chemical change releases carbon dioxide gas. You burp the gas. The pain goes away because water and salt do not irritate the stomach as the acid did.

**What does an antacid do to an acid?**

Drugs like baking soda are called antacids (ant AS udz). An **antacid** is a drug that changes acid into water and a salt. The word *antacid* means against acid.

### Check Your Understanding

11. Describe the changes that take place in your nose when you use an antihistamine during a cold.
12. What type of drug is present in cough medicines?
13. How does an antacid help relieve heartburn?
14. **Critical Thinking:** Is coughing voluntary or involuntary? What nerve pathway is involved in coughing?
15. **Biology and Reading:** What class of drug that you read about in Section 18:2 is similar to the suppressants that you read about in this section?



## *OPTIONS*

### ACTIVITY/Challenge

Ask students to use a chemistry text to determine the chemical equation for the antacid reaction. This is a neutralization (acid-base) reaction. The acid and the sodium bicarbonate form carbon dioxide gas and a salt.

**Which Antacid Works Best?**/*Lab Manual,* pp. 149-152. Use this lab as an extension to understanding antacids.

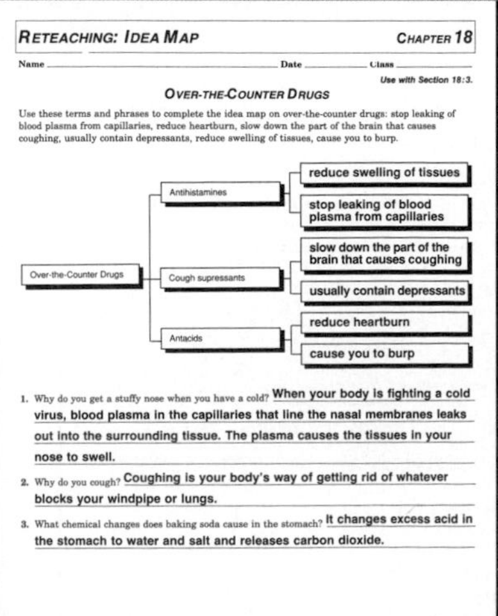

RETEACHING 54

RETEACHING: IDEA MAP CHAPTER 18

Name ______ Date ______ Class ______

*Use with Section 18:3.*

OVER-THE-COUNTER DRUGS

Use these terms and phrases to complete the idea map on over-the-counter drugs: stop leaking of blood plasma from capillaries, reduce heartburn, slow down the part of the brain that causes coughing, usually contain depressants, reduce swelling of tissues, cause you to burp.

Over-the-Counter Drugs
- Antihistamines
  - reduce swelling of tissues
  - stop leaking of blood plasma from capillaries
- Cough supressants
  - slow down the part of the brain that causes coughing
  - usually contain depressants
- Antacids
  - reduce heartburn
  - cause you to burp

1. Why do you get a stuffy nose when you have a cold? When your body is fighting a cold virus, blood plasma in the capillaries that line the nasal membranes leaks out into the surrounding tissue. The plasma causes the tissues in your nose to swell.
2. Why do you cough? Coughing is your body's way of getting rid of whatever blocks your windpipe or lungs.
3. What chemical changes does baking soda cause in the stomach? It changes excess acid in the stomach to water and salt and releases carbon dioxide.

## 18:4 Careless Drug Use

Most people use drugs in a wise and careful way. When drug use becomes careless, serious problems can result.

### Objectives

8. **List** reasons for not using other people's drugs.
9. **Discuss** the problems caused by using caffeine, nicotine, cocaine, and alcohol.

### Key Science Words

drug abuse
dependence
cocaine
caffeine
nicotine
ethyl alcohol

### Drug Misuse and Abuse

"Not feeling well? Why don't you try some of my pills?" Have you ever heard someone say that? Should you use drugs that a doctor has told someone else to use? The answer is *no*. Using someone else's drugs is dangerous.

Why is using someone else's drugs dangerous? First, a doctor prescribed the drug for a health problem someone else had. You may not have the same problem.

Second, the dose prescribed for someone else may not be correct for you. Dosage varies with a person's weight and age.

Third, you may be allergic (uh LUR jik) to the drug. If you have a drug allergy (AL ur gee), you are sensitive to a particular drug. You could end up with a rash, itchy eyes, or a runny nose. Sometimes the results are much more serious. An allergic reaction to a drug can make you unable to breathe, or it might cause a drop in blood pressure. These kinds of allergic reactions may cause death.

What are three reasons for not taking another person's drugs?

Using a drug for a health purpose, but using it in the wrong way or amount, is called drug misuse. Using drugs when they are not needed at all is another problem. For example, cocaine and morphine are painkillers. They often end up being abused. **Drug abuse** is the incorrect or improper use of a drug. Taking a controlled drug illegally is drug abuse.

Drug abuse can lead to dependence. **Dependence** means needing a certain drug in order to carry out normal daily activities. Many drugs cause dependence. Morphine, heroin, codeine, alcohol, and even the drugs in coffee, tobacco, and some soft drinks are drugs that people may become dependent upon.

If a person becomes dependent on a drug and stops using it, that person will suffer from withdrawal. Withdrawal sickness causes loss of appetite, vomiting, stomach pains, and other symptoms. These symptoms continue until the person's nervous system has recovered from the effects of the drug. Withdrawal symptoms can be so severe that sudden withdrawal from certain drugs can cause death.

**Figure 18-14** The flower of the opium poppy is the source of morphine and codeine.

**FOCUS 35**

*Focus* CHAPTER 18

Name ______ Date ______ Class ______

*Attempts at Giving Up Smoking: A Study of 5 389 Smokers*

One study had the following results:
Only 4 percent of smokers succeeded in quitting smoking for six months to a year.

Almost 35 percent of those who gave up smoking for six months took up smoking again in the next six months.

1. What do the results of this study reveal about smoking?
2. Cigarettes contain the drug nicotine. Based on the study, do you think smokers are addicted to the drug in cigarettes?

### OPTIONS

#### Science Background

Dependence may be physical or psychological. A person can be either or both. Dependence can develop as a result of periodic or prolonged use of a drug.

## 18:4 Careless Drug Use

### PREPARATION

#### Materials Needed

Make copies of the Focus, Enrichment, Application, Skill, Transparency Master, Critical Thinking/Problem Solving, Study Guide, Lab, and Reteaching worksheets in the *Teacher Resource Package*.

▶ Prepare test solutions and alcohol-testing chemical for Lab 18-2.

#### Key Science Words

| | |
|---|---|
| drug abuse | caffeine |
| dependence | nicotine |
| cocaine | ethyl alcohol |

#### Process Skills

In Lab 18-2, they will observe, interpret data, and experiment. In the Skill Check they will make and use tables.

### 1 FOCUS

▶ The objectives are listed on the student page. Remind students to preview these objectives as a guide to this numbered section.

#### ACTIVITY/Filmstrip

Show the filmstrip *Teen Substance Abusers*, Carolina Biological Supply Co.

#### Guided Practice

Have students write down their answers to the margin question: a drug is prescribed for a certain health problem that you may not have, the dosage may not be correct for you, you may be allergic to the drug.

**Focus**/*Teacher Resource Package*, p. 35. Use the Focus transparency shown at the bottom of this page as an introduction to careless drug use in relation to smoking.

# 2 TEACH

### MOTIVATION/Videocassette

Show the videocassette *Cocaine and Crack: Formula for Failure,* Human Relations Media.

### Concept Development

▶ The most harmful effect of cocaine use may be that this drug is a potential local anesthetic. It has been the cause of heart stoppage due to the blocking of messages to heart muscle, preventing the heart from contracting.

### Independent Practice

**Study Guide**/*Teacher Resource Package,* p. 108. Use the Study Guide worksheet shown at the bottom of this page for independent practice.

### Guided Practice

Have students write down their answers to the margin question: a strong form of cocaine.

## Cocaine

The leaves of a coca bush are used to make a drug called cocaine. **Cocaine** is a controlled drug used for its stimulant effects. Cocaine is a major cause of drug-related deaths in the United States. It may be injected with a needle, inhaled through the nose, or smoked. When inhaled, the drug enters capillaries in the nose and then goes through the circulatory system to the brain.

Upon first taking cocaine, a person may become more active. Blood pressure rises, and breathing and heart rate speed up. If too much of the drug is taken, the reverse may occur. The activity of the medulla slows down greatly. Remember from Section 15:2 that the medulla controls involuntary actions, such as breathing and heart rate. The slowing of the activity of the medulla can cause breathing or heart rate to slow or stop suddenly. It is not known how large a dose is needed for this to happen, and the drug acts differently in different people.

What is crack?

Crack is a very strong form of cocaine. Crack is made from cocaine that has been formed into a paste and then allowed to dry and harden. When smoked, the drug enters the blood by way of the lungs. In this way, it acts on the body in a matter of seconds. Use of crack can cause heart attacks and lung damage. Crack is so dangerous that it can cause death, even after the first use.

Crank, also called "ice," is another drug that is smoked. Crank is even more dangerous than crack because it causes quicker dependence than crack. Crank often causes death.

**Figure 18-15** Police often find large quantities of illegal drugs (a) when they make a raid (b).

a

b

### OPTIONS

**Enrichment**/*Teacher Resource Package,* p. 19. Use the Enrichment worksheet shown here to help students understand how the body rids itself of drugs.

**STUDY GUIDE 108**

STUDY GUIDE — CHAPTER 18

Name ______ Date ______ Class ______

VOCABULARY

*Review the new words in Chapter 18 of your textbook. Then, write the term that each of the phrases below describes on the blank that follows the phrase.*

1. a drug that slows messages in the nervous system **depressant**
2. too much of a drug **overdose**
3. will form carbon dioxide in the stomach **antacid**
4. a drug in cigarettes **nicotine**
5. a chemical that changes the way a living thing functions when it is taken into the body **drug**
6. a drug that speeds up body activities controlled by the nervous system **stimulant**
7. the incorrect or improper use of a drug **drug abuse**
8. a change that is not expected caused by a drug **side effect**
9. drug that relieves a stuffy nose **antihistamine**
10. a drug you can buy legally without a prescription **over-the-counter drug**
11. a change that takes place in the body due to disease **symptom**
12. needing a certain drug in order to carry out normal daily activities **dependence**
13. a drug in cola, tea, and coffee **caffeine**
14. a drug that a doctor must tell you to take **prescription drug**
15. drug made from the leaves of the coca bush **cocaine**
16. drugs that are controlled by law **controlled drug**
17. how much and how often to take a drug **dosage**
18. drug that changes the signals from the sense organs **psychedelic drug**
19. drug breathed in through the lungs **inhalant**
20. drug found in alcoholic drinks **ethyl alcohol**

**ENRICHMENT 19**

ENRICHMENT — CHAPTER 18

Name ______ Date ______ Class ______

*Use after Section 18:4.*

HOW THE BODY RIDS ITSELF OF DRUGS

The body has a number of ways of ridding itself of different kinds of drugs. The table shows the different organs or body fluids involved in the process.

*Read the table. Use your textbook to review how the body systems or organs work. Then answer the questions below.*

| | |
|---|---|
| Kidneys | Some drugs are water soluble. This type of drug will end up in the blood when taken into the body. Water-soluble drugs are filtered from blood as the blood moves through the kidneys. The drugs pass out of the body in the urine. |
| Liver | Some drugs are fat soluble. The liver can change fat-soluble drugs into water-soluble drugs. Once changed into water-soluble drugs, the kidneys filter them from the blood. They pass out of the body in the urine. The liver also breaks down alcohol into simpler substances. |
| Bile | Some drugs dissolve in bile. Bile is a fluid made in the liver. Drugs that are changed in the liver may leave the liver in bile. Bile passes into the intestine and is eliminated with undigested foods through the anus. |
| Lungs | Some gases taken in through the lungs, such as those used to put a patient to sleep during surgery, pass into the blood. They are given off from the blood back into the lungs, where they are exhaled. Ethyl alcohol also passes from the blood to the lungs where it is exhaled. |

1. How does the body rid itself of water-soluble drugs? **Water-soluble drugs are filtered from the blood as it moves through the kidneys. The drugs pass from the body in the urine.**
2. How does the body rid itself of fat-soluble drugs? **The liver changes fat-soluble drugs into water-soluble drugs. The drugs are filtered out of the blood by the kidneys and pass from the body in the urine.**
3. How does bile help rid the body of drugs? **Some drugs dissolve in bile. The bile passes out of the liver into the intestine. Eventually the bile is eliminated from the body through the anus with other digestive wastes.**
4. How do the lungs help rid the body of drugs? **Some drugs move from the blood to the lungs, where they pass out of the body when air is exhaled.**

**Figure 18-16** Coffee beans are the seeds of the coffee tree and the source of the stimulant caffeine. People in the United States drink about 400 million cups of coffee each day.

## Caffeine and Nicotine

**Caffeine** is a drug found in coffee and tea. It also is found in cocoa, chocolate, and some soft drinks. Some over-the-counter drugs, such as cold remedies and diet pills, have caffeine.

How does caffeine affect a person? You can almost guess by what people say about coffee. "It helps me wake up." "It gives me a lift." Caffeine is a stimulant.

How much caffeine is found in certain foods? Table 18-1 shows you.

**TABLE 18–1. CAFFEINE CONTENT OF SOME FOODS**

| Food or drug | Milligrams of caffeine |
|---|---|
| cup of coffee | 80 |
| cup of tea | 40 |
| regular cola soft drink, 1 can | 40 |
| chocolate bar | 10 |
| chocolate chip cookie | 4 |

How much caffeine causes a change in behavior? As little as 100 milligrams may cause a change. A person may feel more awake and alert. He or she will notice that the heart rate speeds up. Caffeine may make it hard to sleep. Caffeine also causes dependence.

**Nicotine** is a stimulant found in tobacco. Nicotine speeds up the heart and increases blood pressure. Cigarettes, cigars, snuff, and chewing tobacco have nicotine.

### Bio Tip

**Health:** In the United States, there are 359 000 tobacco-related deaths a year. This is equal to 1000 deaths each day.



## TEACH

### Concept Development

▶ Caffeine is one of the most widely used stimulants in the world. It affects the central nervous system and the heart.

▶ Nicotine is a central nervous system stimulant. It also stimulates the adrenal glands to release epinephrine. The results are symptoms associated with the fight or flight mechanism. These symptoms include increased blood pressure, increased heart rate, and increased blood flow to muscles.

▶ Nicotine is classified chemically as an alkaloid, as are caffeine, opium, and cocaine. This means that it contains nitrogen, usually in the form of an amine.

▶ Review *milligram* with students, and point out that the amounts listed in Table 18-1 are small. Ask students if small amounts of a drug are capable of causing obvious changes in behavior.

### Connection: History

Have students research and report on early history and use of cocaine.

### Portfolio

Have students prepare an idea map using the following terms: *stimulant, depressant, cocaine, ethyl alcohol, nicotine, caffeine, coca bush, coffee, yeast, tobacco, drug abuse.* Other words may be used to further explain any of the terms provided.

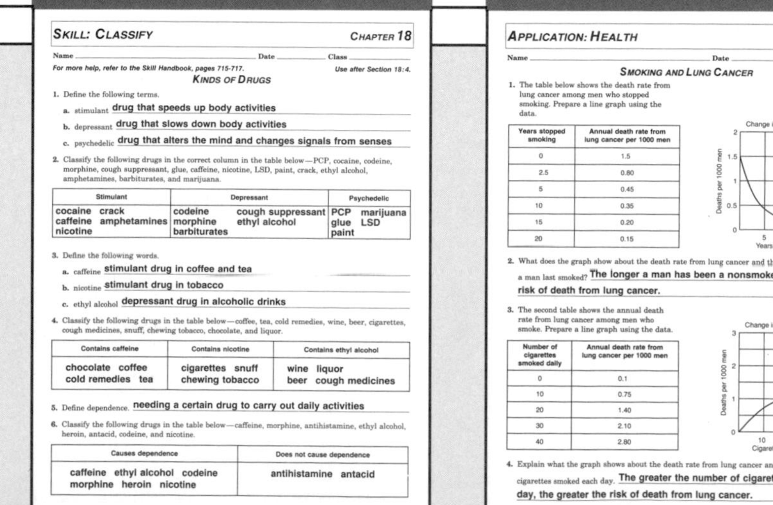

**SKILL 19**

**SKILL: CLASSIFY** — CHAPTER 18

Name ______ Date ______ Class ______

*For more help, refer to the Skill Handbook, pages 715-717.* — *Use after Section 18:4.*

**KINDS OF DRUGS**

1. Define the following terms.
   a. stimulant **drug that speeds up body activities**
   b. depressant **drug that slows down body activities**
   c. psychedelic **drug that alters the mind and changes signals from senses**
2. Classify the following drugs in the correct column in the table below—PCP, cocaine, codeine, morphine, cough suppressant, glue, caffeine, nicotine, LSD, paint, crack, ethyl alcohol, amphetamines, barbiturates, and marijuana.

| Stimulant | Depressant | Psychedelic |
|---|---|---|
| cocaine crack caffeine amphetamines nicotine | codeine cough suppressant morphine ethyl alcohol barbiturates | PCP marijuana glue LSD paint |

3. Define the following words.
   a. caffeine **stimulant drug in coffee and tea**
   b. nicotine **stimulant drug in tobacco**
   c. ethyl alcohol **depressant drug in alcoholic drinks**
4. Classify the following drugs in the table below—coffee, tea, cold remedies, wine, beer, cigarettes, cough medicines, snuff, chewing tobacco, chocolate, and liquor.

| Contains caffeine | Contains nicotine | Contains ethyl alcohol |
|---|---|---|
| chocolate coffee cold remedies tea | cigarettes snuff chewing tobacco | wine liquor beer cough medicines |

5. Define dependence. **needing a certain drug to carry out daily activities**
6. Classify the following drugs in the table below—caffeine, morphine, antihistamine, ethyl alcohol, heroin, antacid, codeine, and nicotine.

| Causes dependence | Does not cause dependence |
|---|---|
| caffeine ethyl alcohol codeine morphine heroin nicotine | antihistamine antacid |

**APPLICATION 18**

**APPLICATION: HEALTH** — CHAPTER 18

Name ______ Date ______ Class ______

**SMOKING AND LUNG CANCER** — *Use after Section 18:4.*

1. The table below shows the death rate from lung cancer among men who stopped smoking. Prepare a line graph using the data.

| Years stopped smoking | Annual death rate from lung cancer per 1000 men |
|---|---|
| 0 | 1.5 |
| 2.5 | 0.80 |
| 5 | 0.45 |
| 10 | 0.35 |
| 15 | 0.20 |
| 20 | 0.15 |



2. What does the graph show about the death rate from lung cancer and the number of years since a man last smoked? **The longer a man has been a nonsmoker, the lower the risk of death from lung cancer.**
3. The second table shows the annual death rate from lung cancer among men who smoke. Prepare a line graph using the data.

| Number of cigarettes smoked daily | Annual death rate from lung cancer per 1000 men |
|---|---|
| 0 | 0.1 |
| 10 | 0.75 |
| 20 | 1.40 |
| 30 | 2.10 |
| 40 | 2.80 |



4. Explain what the graph shows about the death rate from lung cancer and the number of cigarettes smoked each day. **The greater the number of cigarettes smoked each day, the greater the risk of death from lung cancer.**

## OPTIONS

**Application**/*Teacher Resource Package,* p. 18. Use the Application worksheet shown here to help students understand the effects of smoking on health.

**Skill**/*Teacher Resource Package,* p. 19. Use the Skill worksheet shown here to help students with classifying drugs.

## TEACH

### ACTIVITY/Brainstorming

Ask students how snuff, chewing tobacco, and cigarettes differ in the way nicotine enters the body. Students may not realize that many drugs can be absorbed into the bloodstream by way of the small capillaries that line the mouth and cheeks.

### ACTIVITY/Demonstration

Use the alcohol-testing chemical from Lab 18-2 to determine if alcohol is formed by yeast. Prepare a small flask of water (50 mL), table sugar (1 teaspoon), and packet of dry yeast. Mix and allow to stand overnight. Add one drop of liquid to a glass slide and test for presence of alcohol. See Lab 18-2 for details in interpretation of any color changes. Use water as a control to show that no color change occurs.

### Student Journal

Have students describe the journey that a drug such as a painkiller taken for a toothache would follow if: (a) taken by mouth, (b) taken by injection. They should explain the similarities and differences between this drug's journey and that of ethyl alcohol. They may use Figures 18-5 and 18-18 for reference.

**Figure 18-17** Tobacco products all contain nicotine.

What problems are caused by using tobacco? First, the nicotine in tobacco causes dependence. Second, the other chemicals in tobacco cause lung, throat, and mouth cancer. Third, smoking is strongly linked with heart disease. Warning labels appear on all cigarette packages. Fourth, second-hand smoke can harm people who are not smoking but who breathe other people's smoke. Tobacco smoke contains poisonous gases.

## Alcohol

Beer, wine, and whiskey contain a drug called ethyl (ETH ul) alcohol. **Ethyl alcohol** is a drug formed from sugars by yeast and is found in all alcoholic drinks. Some over-the-counter drugs, such as cough medicines, contain ethyl alcohol. There are other types of alcohol, such as rubbing alcohol, but they are not used in drugs because they are poisonous.

Figure 18-18 shows the path of swallowed alcohol in the body. Alcohol enters the blood from the stomach. Once in the blood, alcohol acts on the nervous system.

Many people think that alcohol is a stimulant because it speeds up heart and breathing rates. Alcohol is actually a depressant because it slows down the nervous system, leading to dangerous changes.

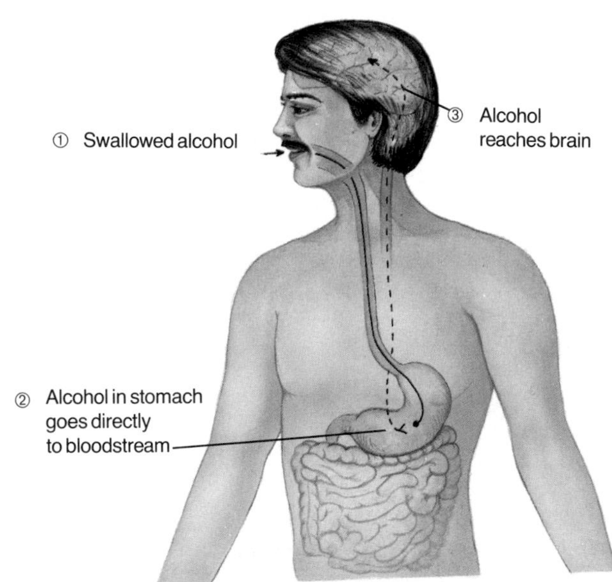


**Figure 18-18** Swallowed alcohol goes quickly into the bloodstream and reaches the brain. What kind of drug is alcohol? depressant



## OPTIONS

**Transparency Master**/*Teacher Resource Package*, p. 85. Use the Transparency Master shown here to help students review drugs.

**Critical Thinking/Problem Solving/** *Teaching Resource Package*, p. 18. Use the worksheet shown here to help students understand unsafe blood alcohol levels.

### TRANSPARENCY 85

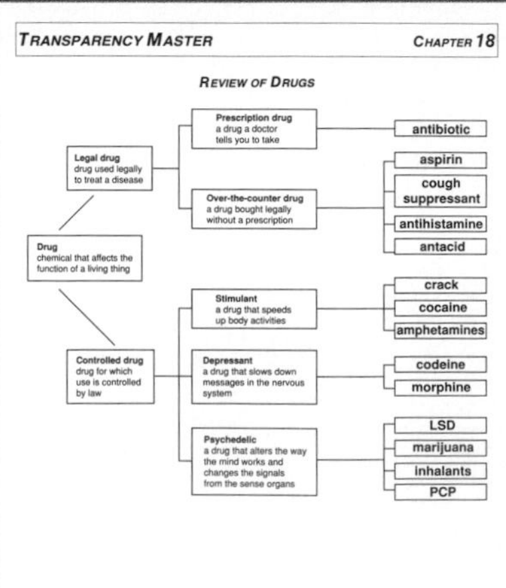


### CRITICAL THINKING 18

**CRITICAL THINKING/PROBLEM SOLVING** — **CHAPTER 18**

Name ______ Date ______ Class ______

*Use after Section 18:4.*

**HOW IS BLOOD ALCOHOL CONCENTRATION USEFUL?**

Alcohol has many effects on a person's body. Some of these effects impair the ability to drive a car safely. The same amount of alcohol consumed can affect different people differently. Blood alcohol concentration (BAC) describes the amount of alcohol in 100 mL of blood. Blood alcohol concentration can show how a person's behavior will be affected by alcohol. Factors that affect the blood alcohol concentration include how much and how quickly alcohol is consumed, the person's body mass, and whether there is food in the stomach. Blood alcohol concentration is used to determine when a person is legally drunk.

Table 1

| Blood alcohol concentration (BAC) in % | | | |
|---|---|---|---|
| Number of drinks in 1 hour | Body weight 45 kg | 64 kg | 82 kg |
| 1 | 0.03 | 0.02 | 0.02 |
| 2 | 0.06 | 0.04 | 0.03 |
| 3 | 0.10 | 0.07 | 0.05 |
| 4 | 0.13 | 0.09 | 0.07 |

Table 2

| Blood alcohol concentration (BAC) in % | | | |
|---|---|---|---|
| Number of drinks in 3 hours | Body weight 45 kg | 64 kg | 82 kg |
| 1 | 0.01 | 0.005 | 0.005 |
| 2 | 0.02 | 0.01 | 0.01 |
| 3 | 0.06 | 0.03 | 0.01 |
| 4 | 0.10 | 0.06 | 0.03 |

**Analyzing the Problem**

1. What is the BAC of a 45 kg person having two drinks in one hour and a 45 kg person having two drinks in three hours. one hour: 0.06%; three hours: 0.02%
2. How does time affect BAC? BAC is lower when drinks are consumed over a longer period of time.
3. Compare the BAC if a 45 kg person has two drinks in one hour with the BAC if the same person has four drinks in one hour. Two drinks results in a BAC of 0.06%; four drinks results in a BAC of 0.13%. Blood alcohol concentration is twice as high when twice as much alcohol is consumed.

**Solving the Problem**

1. In most states, a person is considered legally drunk when the blood alcohol concentration is 0.1%. However, a BAC as low as 0.01% can interfere with a person's ability to drive safely. When does the table suggest it might be safe for a person to drive after consuming one drink? It might be safe for a person weighing 64 kg or more to drive three hours after consuming one drink.

Lab **18–2**

## Alcohol

### Problem: Is alcohol present in over-the-counter drugs and household items?

#### Skills

observe, interpret data, experiment

#### Materials

4 glass slides
wax marking pencil
water
alcohol
8 droppers
alcohol-testing chemical, in dropper bottle
6 test solutions of over-the-counter drugs and household items

#### Procedure

1. Copy the data table.
2. Use a marking pencil to draw two circles on a slide. Label the circles 1 and 2.
3. Add one drop of water to circle 1. Use a clean dropper to add one drop of alcohol to circle 2.
4. Add one drop of alcohol-testing chemical to each circle. **CAUTION:** *Rinse immediately with water if alcohol-testing chemical is spilled on skin or clothing.*
5. **Observe:** Wait five minutes. Look for a color change. A yellow to orange color means alcohol *is not present.* A green, deep green, or blue color means that alcohol *is present.* Record your results.
6. Draw two circles on each of three glass slides. Label the circles A through F.
7. Record in your data table the name of each drug or household item being tested.
8. Add a drop of a different test solution to each circle. Use a clean dropper for each.
9. Add one drop of alcohol-testing chemical to each circle.
10. Wait five minutes. Record any color changes. Complete the last column of your data table.
11. Dispose of your solutions and wash your glassware.

| CIRCLE | CONTENTS | COLOR WITH TEST CHEMICAL | ALCOHOL PRESENT? |
|---|---|---|---|
| 1 | water | orange | no |
| 2 | alcohol | blue | yes |
| A | cough syrup | blue | yes |
| B | rubbing alcohol | blue | yes |
| C | mouthwash 1 | orange | no |
| D | mouthwash 2 | blue | yes |
| E | soft drink | orange | no |
| F | Geritol | blue | yes |

#### Data and Observations

1. Which products showed a color change to blue or green?
2. Which products did not show a color change to blue or green?

#### Analyze and Apply

1. What was the purpose of circles 1 and 2?
2. Ask your teacher to show you the labels from the products tested. Do your data agree with the labels?
3. Did all the products that contained alcohol also show alcohol listed on their labels?
4. What does your answer to question 3 tell you about the accuracy of these labels?
5. **Apply:** Why is it important that alcohol be listed if it is present in a product?

#### Extension

There are many different kinds of alcohol (rubbing, grain, wood, etc.). Will the alcohol-testing chemical detect all types of alcohol? **Design an experiment** to test this.



## ANSWERS

### Data and Observations

1. Answers will vary. Products that contain alcohol will show a color change to blue or green.
2. Products that do not contain alcohol will show a yellow or orange color.

### Analyze and Apply

1. to show what a positive alcohol test looks like and to show that the test chemical does not turn blue or green unless alcohol is present
2. Answers will vary with the products tested.
3. Products that contain alcohol should list alcohol on the label.
4. Drug labels must be accurate for safety and health reasons.
5. A person may not be able to tolerate alcohol or may have to avoid alcohol for health reasons.

## Lab 18-2 Alcohol

### Overview

In this lab, students will test a known alcohol and a known nonalcohol solution with a chemical. They will then determine if alcohol is present in other solutions.

**Objectives:** Upon completion of this lab, students will be able to (1) **determine** how to test for the presence of alcohol (2) **test** a variety of items for alcohol.

**Time Allotment:** 30 minutes

### Preparation

▶ Alcohol: use ethyl or denatured alcohol from a supply house or use rubbing alcohol from a drugstore.

▶ Alcohol-testing chemical: add 20 g potassium dichromate powder to a glass beaker. Pour 20 mL concentrated sulfuric acid to the beaker. Stir with glass rod to dissolve most of powder. *Slowly* add 60 mL distilled water and continue to stir. *Solution becomes very hot.* Allow to cool. Powder may precipitate out after cooling. Pour only liquid portion of solution into dropping bottles for students use. Solution has a shelf life of one year.

**Lab 18-2 worksheet**/*Teacher Resource Package,* pp. 71-72.

### Teaching the Lab

▶ **Troubleshooting:** Placing all liquids in marked dropping bottles will reduce spillage and contamination.

**Performance:** Provide students with samples of different types of alcohol (methyl, rubbing alcohol, ethyl alcohol). Have them determine if the reagent used in the experiment is specific for ethyl alcohol, or if it detects other alcohol types as well.

## TEACH

### Skill Check

Table 18-2 clearly shows that with the increase in blood alcohol concentration there is an increase in the lack of body control, even up to the point of death.

### Independent Practice

**Study Guide**/*Teacher Resource Package,* p. 107. Use the Study Guide shown at the bottom of page 389 for independent practice.

### Check for Understanding

Have students respond to the first three questions in Check Your Understanding.

### Reteach

**Reteaching**/*Teacher Resource Package,* p. 55. Use this worksheet to give students additional practice in understanding the careless use of drugs.

**Extension:** Assign Critical Thinking, Biology and Reading, or some of the **OPTIONS** available with this lesson.

## 3 APPLY

### Software

*Alcohol,* Opportunities for Learning.

## 4 CLOSE

### Audiovisual

Show the video *Maybe I Am: The Story of a Teenage Alcoholic,* Human Relations Media.

### Answers to Check Your Understanding

16. You may not have the same health problem, the dosage may not be correct for you, and you may be allergic to the drug.
17. addiction, breathing and heart rate problems, death
18. coffee, tea, cocoa, soft drink
19. Answers will vary. A person can get help from a peer group.
20. Answers will vary—to warn people not to use drugs.

**Figure 18-19** About one half of all car accident deaths involve the use of alcohol.

Table 18-2 shows what happens to the body as the amount of alcohol in the blood increases. The letters *BAC* mean *b*lood *a*lcohol *c*oncentration. The BAC is a way of describing the amount of alcohol in 100 mL of blood.

**TABLE 18–2. EFFECTS OF ALCOHOL ON THE BODY**

| % BAC | Effect |
|---|---|
| 0.01 – 0.05 | heart and breathing rates increase, judgment decreases |
| 0.06 – 0.10 | alertness, coordination, and judgment decrease |
| 0.11 – 0.15 | reaction time is slower, speech is slurred, balance is upset |
| 0.16 – 0.29 | frequent staggering or falling, loss of awareness |
| 0.30 – 0.39 | stupor, as if under an anesthetic |
| 0.40 and up | unconsciousness, breathing stops, death |

### Skill Check

**Make and use tables:** Alcohol is a depressant. How does Table 18-2 show that drinking alcohol slows down body activities? *For more help, refer to the* ***Skill Handbook,*** *pages 715-717.*

Another problem is the effect of alcohol on unborn babies. Alcohol passes from mother to baby during pregnancy. The baby can be born with very serious health problems such as fetal alcohol syndrome. These babies are smaller than normal, may be mentally retarded, and have heart defects. Many are born dependent on alcohol.

## Check Your Understanding

16. Give three reasons a person should not use someone else's drugs.
17. Describe problems caused by using cocaine and crack.
18. Name four foods that contain caffeine.
19. **Critical Thinking:** Think of a way that might help people stop abusing alcohol or tobacco.
20. **Biology and Reading:** From what you learned in this section, what do you believe was the author's purpose for writing about drugs?

### RETEACHING 55

RETEACHING CHAPTER 18

Name ________ Date ________ Class ________

*Use with Section 18:4.*

**CARELESS USE OF DRUGS**

1. List the reasons why using drugs prescribed for someone else is dangerous:
   a. You may not have the same problem.
   b. The dose prescribed for someone else may not be correct for you.
   c. You may be allergic to the drug, which could give you a rash, itchy eyes, or a runny nose. It might make you unable to breathe or cause your blood pressure to drop.
2. Place a checkmark in the blank next to the drugs(s) that may cause dependence.
   morphine ✓ alcohol ✓
   heroin ✓ caffeine in coffee ✓
   codeine ✓ nicotine in tobacco ✓
3. What are two opposite effects of cocaine? At first a person's blood pressure rises and heart rate speeds up; later the action of the medulla slows down greatly, so breathing and heart rate may slow or stop suddenly.
4. What are two dangerous effects of crack? It may cause a heart attack and lung damage.
5. How can caffeine affect the body? It can make a person feel alert and may make it hard to sleep.
6. What are the effects of nicotine on the body? It causes lung, throat, and mouth cancer. It is also strongly linked with heart disease.
7. What is meant by fetal alcohol syndrome? When alcohol passes from mother to baby during pregnancy, the baby may be smaller than normal, mentally retarded, and have heart defects.

# Science and Society

## Steroids

Testes in the male have two jobs. Not only do they make sperm cells, but they also make a hormone called testosterone (teh STAHS tuh rohn). Testosterone causes the male voice to deepen, hair to grow on the face, and muscles to increase in size. All of this is normal when it occurs at puberty. It is now possible to increase the amount of this hormone in the body by taking testosterone-like drugs. These drugs are called anabolic steroids (an uh BAHL ik • STIHR oydz). They cause a person's muscles to become larger and stronger, and they increase endurance. These changes occur faster than through normal exercise. Some athletes see these drugs as helpful.

### What Do You Think?

**1.** Athletes who use steroids are finding that the drugs can make them the strongest, fastest, or biggest person around. Many of these athletes train for years, and sports are their lives. Taking steroids is as much a part of their training routine as lifting weights or running. Taking the drugs is helping them to win medals. Many athletes, however, have been kicked off teams or stripped of their medals when it was discovered they were taking steroids. Should an athlete be allowed to use drugs and then compete? Would it be fair to you if you were in a contest with a person who was using anabolic steroids?

**2.** Many sports organizations, high schools, and colleges test all participating athletes for the presence of drugs in their urine. If drugs are found, the athlete is usually suspended for a length of time and prevented from competing. These drug tests are required of all professional athletes. The United States Constitution guarantees all citizens the right to privacy. Do you think these drug tests violate the right to privacy? What would happen if an athlete refused to be tested?

**3.** Steroids cause major health problems. Their use causes liver damage, heart disease, high blood pressure, and sterility. They can cause baldness, growth of breasts, and personality changes in males. Females may develop facial hair, a deep voice, stopping of the menstrual cycle, or a decrease in breast size. Many of these changes are not reversed when the athlete stops taking the drug. Who will pay the athletes' medical bills?

**Conclusion:** Should laws be passed that prevent or limit the use of steroids by athletes? Is it a government's business what a person does with his or her body?

Bodybuilders may be tested for drugs before they can compete.



**STUDY GUIDE 107**

**STUDY GUIDE** **CHAPTER 18**

Name ______ Date ______ Class ______

**CARELESS DRUG USE**

*In Section 18:4 of your textbook, read about careless drug use.*

Place a checkmark in front of those sentences that are true.

____ 1. It is safe to use the drugs that a physician has told someone else to use.

____ 2. The dose of a drug prescribed by a physician for someone else is always the same as the dose that you would need.

✓ 3. A drug allergy means that you are sensitive to a particular drug.

____ 4. Allergic drug reactions can never cause death.

✓ 5. Drug abuse is the incorrect or improper use of a drug.

____ 6. Drug abuse can never lead to drug dependence.

✓ 7. Codeine, alcohol, and heroin are examples of drugs that may cause dependence.

8. Complete the following table by placing checkmarks in the correct columns.

| | Cocaine | Caffeine | Nicotine | Alcohol |
|---|---|---|---|---|
| May cause breathing or heart to stop suddenly | ✓ | | | |
| Causes feeling of alertness and difficulty in falling asleep | | ✓ | | |
| Stimulant | ✓* | ✓ | ✓ | |
| Depressant | | | | ✓ |
| Leads to dependence | ✓ | ✓ | ✓ | ✓ |
| When BAC is 0.4-0.5, unconsciousness or coma can occur | | | | ✓ |
| Causes cancer and linked to heart disease | | | ✓ | |
| One form, crack, can cause death with first use | ✓ | | | |
| Can lead to withdrawal symptoms if stopped after dependent use | ✓ | ✓ | ✓ | ✓ |

*Large doses can result in severe depressant effects.

### Suggested Readings


Culotte, Elizabeth. "She'll Pump You Up." *Science World*, January 13, 1995. pp. 11-15.

Hoberman, John M. and Charles E. Yesalis. "The History of Testosterone." *Scientific American*, February, 1995, pp. 76-81.

King, Peter. "We Can Clean It Up." *Sports Illustrated,* July 9, 1990, pp. 34-38.

Nightengale, Bob. "Steroids in Baseball? Say It Isn't So, Bud." *The Sporting News*, July 24, 1995, p. 16.

Rozin, Skip. "Steroids and Sports: What Price Glory?" *Business Week*, October 17, 1994, pp. 176-177.

Silverman, Rick. "I Was a Teenage Science Project." *Muscle and Fitness*, April 1997, p. 200.


## Science and Society

### Steroids

#### Background

Steroids are hormones composed of complex rings of carbon and hydrogen. The steroid hormones include sex hormones and those secreted by the adrenal cortex. Steroids are used legally to treat disorders such as allergies and skin disorders, but can't be used for long periods of time because of side effects. The majority of abuse is with anabolic (tissue-building) steroids.

#### Teaching Suggestions

Emphasize that taken under prescription, steroids are a legitimate medicine.

#### What Do You Think?

1. Students will not think it is fair that some use steroids and others don't. Most will feel that steroids should be banned to make the sport fair to everyone.
2. Most students will probably say that they agree with drug testing just to make sports competitions fair.However, they may then discuss how drug testing of innocent people is not fair.
3. Most students will feel that the medical costs are the responsibility of the person who takes the steroid because he or she has a choice.

**Conclusion:** The subject of government intervention may cause some controversy. Some students may feel that the government shouldn't have any right to say what a person does with his or her body.

## CHAPTER 18 REVIEW

### Summary

Summary statements can be used to review the major concepts of the chapter.

### Key Science Words

All boldfaced terms from the chapter are listed.

### Testing Yourself

#### Using Words

1. drug abuse
2. over-the-counter drug
3. side effect
4. overdose
5. dosage
6. controlled drug
7. stimulant
8. caffeine
9. depressant
10. prescription drug

#### Finding Main Ideas

11. p. 382; Baking soda changes acid into water and salt.
12. p. 380; An antihistamine stops the leaking of plasma from blood capillaries in the nasal tissue.
13. p. 388; Heart and breathing rate increase, alertness and coordination decrease, reaction time decreases, the person staggers and falls. Increased use can cause stupor, depression of breathing, and death.
14. p. 371; A doctor must tell you to take a prescription drug, while no prescription is needed for over-the-counter drug.
15. p. 379; Blood pressure rises, the person may be unable to walk or stand, experience numbness, and behavior may become violent.
16. p. 375; The drug reaches the stomach, moves from the stomach to the small intestine, and then into the bloodstream. The drug reaches all the neurons and changesthe messengers being sent to the brain. The pain center in the brain is shut off.
17. p. 383; The nervous system is still being affected by the drug until the effects wear off.
18. p. 372; The kidney and the liver get rid of drugs.
19. p. 376; A stimulant causes a neuron to produce more chemical messenger or prevents the messenger from being destroyed after it has done its job.

#### Using Main Ideas

20. They may have a rash, itchy eyes, a runny nose, not be able to breathe and a drop in blood pressure.
21. The age and body size of a person.
22. Stimulants include caffeine, nicotine, amphetamines, and cocaine. Depressants include ethyl alcohol, morphine, and barbiturates.
23. Drug dependence means that a person must have a certain drug in order to carry out normal daily activities. Morphine, heroine, codeine, nicotine, and alcohol cause dependence.
24. Something may be touching your trachea. Nerves in the trachea sense this and send messages to the brain's cough center. Messages are then sent to the diaphragm and epiglottis and you cough.
25. Antacids change stomach acid to water and a salt. Antihistamines reduce swelling and plasma loss from blood. Cough medicine acts as a depressant on the cough center of the brain.

# Chapter 18 Review

## Summary

### 18:1 An Introduction to Drugs

1. Prescription drugs are obtained through a doctor. Over-the-counter drugs are obtained without a prescription.
2. Drug labels give the use, dosage, and warnings against side effects.
3. Drugs that are injected work faster in the body than those that are swallowed.

### 18:2 How Drugs Affect Behavior

4. Stimulants increase the amount of chemical messenger given off into a synapse. Depressants keep messages from moving across a synapse.
5. Psychedelic drugs and inhalants alter the mind by changing messages received by sense organs.

### 18:3 Uses of Over-the-Counter Drugs

6. Antihistamines stop plasma loss from capillaries. Cough suppressants slow the brain's cough center.
7. Antacids change acid into water and a salt.

### 18:4 Careless Drug Use

8. Using someone else's drugs can be dangerous.
9. Cocaine is a controlled stimulant. Caffeine and nicotine are both stimulants. Alcohol is a depressant. All of these drugs cause dependence.

## Key Science Words

antacid (p. 382)
antihistamine (p. 380)
caffeine (p. 385)
cocaine (p. 384)
controlled drug (p. 370)
dependence (p. 383)
depressant (p. 377)
dosage (p. 372)
drug (p. 370)
drug abuse (p. 383)
ethyl alcohol (p. 386)
inhalant (p. 379)
nicotine (p. 385)
overdose (p. 374)
over-the-counter drug (p. 371)
prescription drug (p. 371)
psychedelic drug (p. 378)
side effect (p. 372)
stimulant (p. 376)
symptom (p. 370)

## Testing Yourself

### Using Words

*Choose the word from the list of Key Science Words that best fits the definition.*

1. using drugs incorrectly or improperly
2. drug bought legally without a prescription
3. unexpected change while using a drug
4. having too much of a drug in the body
5. how much and how often a drug should be used
6. drug whose manufacture and sale are controlled by law
7. drug that speeds up the nervous system
8. drug found in coffee or tea

390

# Review

## Testing Yourself *continued*

9. drug that slows the nervous system
10. drug that a doctor must prescribe

**Finding Main Ideas**

*List the page number where each main idea below is found. Then, explain each main idea.*

11. how baking soda works in the stomach
12. how an antihistamine works
13. problems that result as blood alcohol concentration increases
14. the difference between prescription and over-the-counter drugs
15. changes that a psychedelic drug, such as PCP, may cause
16. the path that a swallowed drug takes to reach the brain's pain center
17. what causes withdrawal sickness
18. how the body gets rid of drugs
19. how a stimulant speeds up the body

**Using Main Ideas**

*Answer these questions by referring to the page number after each question.*

20. What happens when a person has an allergy to a drug? (p. 383)
21. What two factors are used to determine a drug dose? (p. 372)
22. Is each of the following drugs a stimulant or a depressant? caffeine, nicotine, morphine, cocaine, amphetamine (pp. 376-377)
23. What is dependence and what drugs can you become dependent on? (p. 383)
24. Why and how do you cough? (p. 381)
25. How do each of the following drugs work? antacids (p. 382); antihistamines (p. 380); cough suppressants (p. 381)
26. What is a controlled drug and why must it be controlled? (p. 370)

## Skill Review 

*For more help, refer to the* ***Skill Handbook,*** *pages 704-719.*

1. **Make and use tables:** Using Table 18-1, estimate the average amount of caffeine you consume in a day. How can you decrease this amount?
2. **Infer:** Use the sink and running water idea to infer what happens when a child takes an adult dose of a drug rather than the correct child dose.
3. **Interpret data:** If you do the Mini Lab on page 371, how do you know that the paper disk itself does not prevent bacteria from growing?
4. **Sequence:** List the steps that lead to a cough. What step is blocked by a cough suppressant?

## Finding Out More

TECH PREP

**Critical Thinking**

1. Drivers suspected of drinking while intoxicated blow up a balloon, and the police test the air in the balloon for alcohol. How does alcohol get into the suspect's breath?
2. Explain why urine can be tested to determine if a person has used drugs.

**Applications**

1. Examine the labels of two over-the-counter drugs. List and compare the following for each drug: use, dosage, how often it should be used, when to use, warnings.
2. Make a report on fetal alcohol syndrome or the effect of crack on an unborn baby.



### Using Main Ideas

26. A controlled drug is one in which the manufacture and distribution is controlled by law. Drugs that may be dangerous if used incorrectly are controlled drugs.

### Skill Review

1. Answers will vary depending on the amounts of caffeine thatindividual students consume. Students might suggest that they could consume more decaffeinated beverages and eat foods with less chocolate.
2. If the child takes a children's dosage of the drug, the amount of the drug entering the body equals the amount leaving, like a leaking sink. If the child takes an adult dosage, the amount entering is greater than the amount leaving and the sink overflows.
3. You must have a control. One disk should not be treated with the onion juice.
4. (a) irritation to the trachea, (b) nerves in the trachea send messages to the brain's cough center, (c) messages are received by the cough center, (d) messages are sent from the brain to the diaphragm and epiglottis; (e) you cough. Step (c) is suppressed.

### Finding Out More

#### Critical Thinking

1. The lungs help get rid of drugs.
2. The kidneys help get rid of drugs.

#### Applications

1. Answer will vary according to the drug labels used.
2. Fetal alcohol syndrome can cause low birth weight babies, premature birth, brain damage, and alcohol dependence in the fetus. Crack can cause the same symptoms in a fetus as alcohol, but the effects are more acute.

 **Chapter Review**/*Teacher Resource Package,* pp. 92-93.

 **Chapter Test**/*Teacher Resource Package,* pp. 94-96.

**Chapter Test**/*Computer Test Bank*

 **Unit Test**/*Teacher Resource Package,* pp. 97-98.

# Unit 4 Body Systems—Controlling Life

CONNECTIONS

## Biology in Your World

The body systems and body parts that control our bodies are also involved in how we receive and respond to stimuli. The sense organs receive information from the environment. Students need to understand the importance of their sense organs and how to protect them.

### Literature Connection

With the help of Anne Sullivan, Helen Keller (1880-1968) learned the manual alphabet and how to read and write in Braille in just three years. Keller then began speech lessons and eventually spoke well enough to attend high school and college.

After college, she worked to improve conditions for the blind. She traveled, lectured, raised money, and wrote books and articles for her cause. Keller wrote seven books, including *The Story of My Life.*

### Art Connection

George Seurat (1859-1891) was influenced by the impressionist painters' subjects and use of bright colors. Impressionists tended to paint what the eye sees at a glance. However, Seurat did not favor this loose style. Seurat used a more scientific approach to painting. He studied theories on color and light to develop pointillism. His paintings look smooth and unbroken from a distance, but up close they are a series of unconnected dots of color. Seurat's style influenced many painters of the early 1900s.

CONNECTIONS

# Biology in Your World

## Appreciating Our Senses

In this unit, you learned how body systems are controlled. Our bodies can receive information from outside our bodies. Our ears, eyes, nose, mouth, and skin collect this outside information. Factors such as disease and injury can cause damage to any of our sense organs.

LITERATURE

## A Silent and Dark World?

In 1881, illness left 18-month-old Helen Keller unable to see, hear, or talk. Most people thought she would not be able to learn. For the next six years, Keller lived in a world she described as a "no-world."

When Keller was 7 years old, a young teacher named Anne Sullivan came to live with her. Sullivan opened up the world for Keller. Using the manual alphabet and braille, Keller was able to communicate. In the manual alphabet, finger positions stand for letters. Braille is a method of reading and writing using raised dots.

At age 22, Keller wrote *The Story of My Life.* Her writings have helped educators design training programs for blind and deaf people. They have also inspired many blind and deaf people to overcome their handicaps.

ART

## Optical Painting

Must an artist mix paints to get certain shades? No, mixing of colors can take place in the brain. George Seurat (suh ROH) was a French painter who wanted his paintings to show the colors of nature in a scientific way. He covered his canvases with tiny dots of pure color. Because the dots are so small and regular, the eye does the mixing. This technique, called pointillism, is seen in Seurat's 1885 painting, *Le Bec du Hoc, Grandcamp.* Look closely at the colors to see for yourself.

392

LEISURE

## World Without Sound

All during your waking hours, and sometimes in your sleep, sounds surround you. How much do you depend on sound? How would a hearing loss affect your activities? To find out, wear earplugs while at home some evening or on a Saturday. Be sure to insert the earplugs carefully. Do not participate in an activity that would be unsafe without your hearing. As you participate in an indoor hobby, sport, or relaxation activity, make a list of the ways your "hearing loss" affects the activity. You may find that you can't do some of your usual activities. You may find that you don't enjoy the activity as much with reduced hearing. Did you also find that you could do some things better?

You may notice that your "hearing loss" affects other people. Did you have to turn the television volume up so high that it became too loud for others? Did others have to speak more loudly to get your attention?

CONSUMER

## Protect Your Sight

With many types of sunglasses available, you can be sure to make a fashion statement. But you also need to make choices when it comes to the level of protection sunglasses offer. It is important to protect your eyes from the effects of harmful ultraviolet (UV) radiation. Scientists think that too much UV radiation can lead to a clouding of the lens of the eye.

Sunglasses do not have to be expensive to be effective. Some inexpensive sunglasses provide good UV protection. Several grades of UV blockage are available. Look for a stamp on the lens that tells you the grade.

The next time you consider buying sunglasses, remember that your eyes are worth protecting—you only have one pair.

393

## CULTURAL DIVERSITY

### Acupuncture

The Chinese have used acupuncture as a system of medicine for thousands of years. Specific locations on the body are punctured by thin needles as long as 23 cm. The needles are twisted to elicit results that range from cancer cures to pain relief. A modern use of acupuncture in China is in providing anesthesia for surgery, although only three percent of all Chinese surgery uses acupuncture. Acupuncture has been slow to gain acceptance in Western medical practices, although most scientists agree that acupuncture has merit. Have students make a poster showing the human body. They should locate some of the many acupuncture points and point out what part of the body each affects.

CONNECTIONS

### Leisure Connection

Many hearing-impaired people live safely and independently. Hearing aids can help most people with hearing loss. For people with more severe hearing loss, telephones can be attached to a video screen and keyboard. A decoder connected to a television set makes close-captioning visible. Doorbells and alarm clocks can be connected to lights that flash.

Other hearing-impaired persons own hearing dogs. The dogs are trained in much the same way as seeing eye dogs. They alert their owners to doorbells, alarm clocks, a crying baby, or a smoke alarm.

### Consumer Connection

Besides the possibility of causing clouding of the lenses, overexposure to U.V. radiation can cause swelling and inflammation of the eyes. Also, ulcers and tumors may develop in the eyes.

Another consideration in eye protection is glare. Polarized lenses block much of the light reflected from such surfaces as snow and water. With less glare, the eye can see objects more clearly.

### References


Brink, Susan. "Sun Struck." *U.S. News and World Report,* June 24, 1996, pp. 62-68.

Burke, James. "Impressions." *Scientific American*, September 1996, pp. 182-183.

Gard, Carolyn J. "Eye See." *Current Health 2*, January 1997, pp. 30-32.

Keller, Helen. *The Story of My Life.* New York: Doubleday & Co., 1905.

Rewald, John. *Seurat: A Biography.* New York: Abrams, 1990.

Silverstein, Alvin, and Virginia B. Silverstein. *Glasses and Contact Lenses: Your Guide to Eyes, Eyewear, and Eye Care.* New York: J.B. Lippincott, 1989.

# Unit 5 Plant Systems And Functions

## UNIT OVERVIEW

Unit 5 focuses on plant organs and tissues and how these function. Chapter 19 deals with leaf structure and photosynthesis. Chapter 20 discusses roots and stems and the ways in which they support the plant and transport food and water. Chapter 21 deals with tropisms caused by hormones, growth of plant parts, and some diseases and insects that affect plants.

## Advance Planning

Audiovisual and software suggestions are located within each specific chapter under the heading TEACH.

**To prepare for:**

| | |
|---|---|
| Lab 19-1: | Collect leaves for student teams. These can be found and dried weeks in advance. |
| Lab 19-2: | Collect leaves and cut filter paper into strips. |
| Mini Lab: | Purchase waxed paper and collect leaves. |
| Lab 20-1: | Cut paper towels into strips prior to the lab. |
| Lab 20-2: | Purchase yams, radishes, and carrots. Prepare iodine solution. |
| Mini Lab: | Find a tree stump or log. |
| Lab 21-1: | Germinate the bean seeds for at least 4 days prior to the lab. |
| Lab 21-2: | Germinate the corn seeds for at least 4 days prior to the lab. |
| Mini Lab: | Purchase radish seeds. |

# Unit 5

CONTENTS

394

# Plant Systems and Functions

**What would happen if...**

there were no forests? Without forests there would be more land available for farming. Then, perhaps enough food could be produced to feed the world's population. Also, more land would be usable for housing, factories, and other types of development.

Without forests, though, there would be no shelter or food for many animals. There would not be enough lumber and other wood products for our needs. Plants use carbon dioxide and give off oxygen during photosynthesis. Without forests, the amount of carbon dioxide might increase and there might not be enough oxygen.

Can we keep our forests and still meet our food and space needs? Scientists are looking for ways to better manage our resources so we can meet our needs without upsetting the environment.

395

**UNIT OVERVIEW**

## Photo Teaching Tip

The passage that introduces this unit may not have as much impact on your students if you live in a prairie or desert region of the United States as it will on students from a more forested region. First, point out that forests cover a major part of the tropics as well as the northern, coastal, and more mountainous regions of the United States. Then, have students list the disadvantages of a treeless landscape. Make students aware that a forest is not just a collection of trees but that it provides a home, food, and shelter for countless other organisms.

## Theme Development

Living things depend on the many body parts that maintain and control functions of the entire organism. The theme of homeostasis is developed in this unit. Students will study the different parts of a plant and how they maintain and control the function of the organism. Students may not be aware that plants, like other living things, also have parts that carry out numerous processes necessary for life. The main organs of a plant are leaf, root, and stem.

**CULTURAL DIVERSITY**

See the Cultural Diversity feature located on the Connections page at the end of this unit.

For Tech Prep activities, see the Applying Technology feature in this unit, the Finding Out More applications questions in the Chapter Reviews, and the Tech Prep Applications booklet in the Teacher Resource Package.

# The Importance of Leaves

## PLANNING GUIDE

| CONTENT | TEXT FEATURES | TEACHER RESOURCE PACKAGE | OTHER COMPONENTS |
|---|---|---|---|
| (1 1/2 days)<br>19:1 The Structure of Leaves<br>Leaf Traits<br>Cells of the Leaf<br>Water Loss in Plants | Skill Check: *Understand Science Words,* p. 403<br>Mini Lab: *How Does a Leaf Epidermis Do It's Work?* p. 402<br>Mini Lab: *Why Do Leaves Wilt?* p. 404<br>Lab 19-1: *Leaves,* p. 400<br>Check Your Understanding, p. 404 | Enrichment: *The Effects of Sunlight and Water on Leaves,* p. 20<br>Reteaching: *Cells in a Leaf,* p. 56<br>Critical Thinking/Problem Solving: *How Do Drops of Water Form on Leaves?* p. 19<br>Transparency Master: *Water Loss in Plant,* p. 89<br>Study Guide: *The Structure of Leaves,* pp. 109-111<br>Lab 19-1: *Leaves,* pp. 73-74 | **Laboratory Manual:**<br>*What Do the Inside Parts of Leaves Look Like?* p. 157<br>Color Transparency 19: *Cross Section of a Leaf*<br>**STVS:** *Managing a Forest,* Plants and Simple Organisms (Disc 4, Side 2) |
| (1 1/2 days)<br>19:2 Leaves Make Food<br>Building From Raw Materials<br>Changing Raw Materials into Sugar<br>How Is Sugar Used? | Skill Check: *Infer,* p. 408<br>Idea Map, p. 407<br>Check Your Understanding, p. 409 | Application: *The Greenhouse Effect,* p. 19<br>Reteaching: *Photosynthesis,* p. 57<br>Skill: *Make and Use a Line Graph,* p. 20<br>Focus: *Effects of Sunlight on Growth,* p. 37<br>Study Guide: *Leaves Make Food,* p. 112 | **STVS:** *Water Hyacinth,* Plants and Simple Organisms (Disc 4, Side 2) |
| (1 1/2 days)<br>19:3 Leaves for Food<br>Animals Depend on Leaves<br>Practical Uses of Leaves<br>Changes in Leaves | Lab 19-2: *Pigments,* p. 415<br>Science and Society: *Farms of the Future,* p. 413<br>Idea Map, p. 410<br>Check Your Understanding, p. 414 | Reteaching: *Features of Leaves,* p. 58<br>Study Guide: *Leaves for Food,* p. 113<br>Study Guide: *Vocabulary,* p. 114<br>Lab 19-2: *Pigments,* pp. 75-76 | **Laboratory Manual:**<br>*Where in a Leaf Does Photosynthesis Take Place?* p. 161<br>**STVS:** *Farming Indoors,* Plants and Simple Organisms (Disc 4, Side 2) |
| Chapter Review | Summary<br>Key Science Words<br>Testing Yourself<br>Finding Main Ideas<br>Using Main Ideas<br>Skill Review | **ASSESSMENT RESOURCES**<br>Chapter Review, pp. 99-100<br>Chapter Test, pp. 101-103<br>Performance Assessment in the Biology Classroom<br>Alternate Assessment in the Science Classroom<br>Computer Test Bank | |

## GLENCOE TECHNOLOGY

**Infinite Voyage,** *The Keepers of Eden*
**Science and Technology Videodisc Series,** *Managing a Forest,* Plants and Simple Organisms (Disc 4, Side 2)
*Water Hyacinth,* Plants and Simple Organisms (Disc 4, Side 2)
*Farming Indoors,* Plants and Simple Organisms (Disc 4, Side 2)

## MATERIALS NEEDED

| LAB 19-1, p. 400 | LAB 19-2, p. 415 | MARGIN FEATURES |
|---|---|---|
| 15 leaves of different plants<br>metric ruler | plant leaf<br>coin<br>small jar<br>metric ruler<br>filter paper strip<br>liquid solvent<br>applicator stick<br>masking tape | Skill Check, p. 403<br>dictionary<br>Mini Lab, p. 402<br>water<br>waxed paper<br>paper towel<br>leaf<br>Mini Lab, p. 404<br>leaf<br>paper towel |

## OBJECTIVES

For more information about National Science Standards, see page 5T.

| SECTION | OBJECTIVE | CORRELATION of QUESTIONS to OBJECTIVES: CHECK YOUR UNDERSTANDING | CHAPTER REVIEW | TRP CHAPTER REVIEW | TRP CHAPTER TEST |
|---|---|---|---|---|---|
| 19:1 National Science Stds: UCP.1, UCP.2, UCP.3, UCP.4, UCP.5, C.1, F.5 | 1. **Examine** and **compare** the parts of a leaf in different plants. | 1 | 3, 6, 19 | 1, 2, 17, 18, 19, 28, 35 | 1, 5, 7, 10, 26, 30, 47, 52 |
| | 2. **Give the functions** of the cells in a leaf. | 4, 5 | 1, 7, 16 | 15, 20, 21, 22, 23, 24, 25, 26, 29, 30, 31, 32, 34 | 4, 8, 11, 12, 14, 15, 16, 17, 18, 27, 29, 32, 34, 35, 36, 37, 38, 39, 40, 41, 42, 43, 44, 45, 46, 48, 49, 50, 51 |
| | 3. **Explain** the process of transpiration. | 2, 3 | 2, 4, 5, 10, 18 | 3, 7, 27, 33 | 3, 53, 55 |
| 19:2 National Science Stds: UCP.2, UCP.3, UCP.4, B.2, B.3, B.5, B.6, C.5, D.1 | 4. **Describe** the process of photosynthesis. | 6 | 8, 9, 24 | 13, 16 | 24, 25, 33, 54 |
| | 5. **Explain** the chemical equation for photosynthesis. | 10 | 11, 20, 22, 23, 27 | 11, 12 | 21, 22, 23, 28 |
| | 6. **Compare** the importance of sugar in photosynthesis and respiration. | 7 | 12, 15 | 4, 5, 14 | 19 |
| 19:3 National Science Stds: UCP.3, UCP.4, B.5, B.6, C.4, C.5, D.1, F.3, F.4 | 7. **Determine** why plants are important to other things. | 8, 9, 11, 14 | 14, 21 | 6, 9 | 2 |
| | 8. **List** different uses of leaves. | 12, 15 | 13, 18 | 8 | 9, 13 |
| | 9. **Explain** why leaves change color. | 13 | 24 | 10 | 6, 20, 31 |

# The Importance of Leaves

## CHAPTER OVERVIEW

### Key Concepts

In this chapter, students will study photosynthesis as the chief function of leaves. They will compare the structure and internal parts of the leaves. Students will also study the importance of leaves as food for animals and as food and medicines for humans.

### Key Science Words

blade
epidermis
guard cell
midrib
palisade layer
spongy layer
stoma
transpiration
wilting

### Skill Development

In Lab 19-1, students will **classify, interpret diagrams,** and **measure in SI** to determine the traits of different leaves. In Lab 19-2, students will **experiment, observe,** and **infer** to discover pigments found in leaves. In the Skill Check on page 403, students will **understand** the **science word** *transpiration.* In the Skill Check on page 408, students will **infer** what happens to plants that are covered. In the Mini Lab on page 402, students will **experiment** with leaf epidermis. In the Mini Lab on page 404, students will **observe** a leaf wilting.

### Bridging

Students learned about the parts of plant cells in Chapter 2 and plant classification in Chapter 6. In this chapter, they will learn the names of specific cells in a leaf and how to classify plants by the features of their leaves.

**CHAPTER PREVIEW**

### Chapter Content

Review this outline for Chapter 19 before you read the chapter.

**19:1 The Structure of Leaves**
- Leaf Traits
- Cells of the Leaf
- Water Loss in Plants

**19:2 Leaves Make Food**
- Building from Raw Materials
- Changing Raw Materials into Sugar
- How Is Sugar Used?

**19:3 Leaves for Food**
- Animals Depend on Plants
- Practical Uses of Leaves
- Changes in Leaves

### Skills in this Chapter

The skills that you will use in this chapter are listed below.

- In **Lab 19-1,** you will measure in SI, interpret diagrams, and classify. In **Lab 19-2,** you will observe, infer, and experiment.
- In the **Skill Checks,** you will understand science words and infer.
- In the **Mini Labs,** you will observe and experiment.

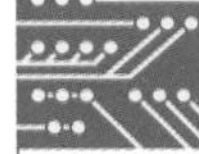

## TECH PREP

For Tech Prep activities in this chapter of the Teacher Wraparound Edition, see especially the Cooperative Learning activity on page 401, the Analogy and Student Journal on page 406, the Motivations on pages 410 and 411, the Activities on pages 412 and 414, and the Portfolio on page 412.

See also the Glencoe Homepage at **www.glencoe.com**

The following Glencoe resources provide additional opportunities for integrating science and technology.

**Technology: Science and Math in Action, Book One**
Module 1: Hydroponics

Chapter **19**

# The Importance of Leaves

Have you ever wondered where your energy to run, cycle, or play sports comes from? How were you able to grow from the size of a book bag as a baby to the size you are today? The answer to both of these questions is food. The baker in the photo to the left and the leaves in the photo below are both making food. Energy is needed to bake the bread and for the leaves to make food. Where does the energy come from to bake the bread? Where do the leaves get the energy to make food?

The baker is making food for himself and other humans. The leaves are making food for the leaf cells and other plant parts. Why do you need food? Why do plants need food? They don't move about as you do, but they do grow.

**Try This!**

**How do leaves vary?** Search around your area for the largest and the smallest leaf. Compare the colors, shapes, and sizes of the two leaves.

For more information about the material in this chapter, follow the link for the chapter on the Glencoe Homepage at **http://www.glencoe.com**



GETTING STARTED

## Using the Photos

In looking at the photos, students should relate how baking bread and photosynthesis in leaves are both food-making processes. Point out that carbon dioxide gas, which they cannot see, and water on a leaf are two raw materials needed for photosynthesis. Both processes use energy; bread making uses heat, and photosynthesis uses light.

## MOTIVATION/Try This!

**How do leaves vary?** Upon completing the chapter-opening activity, students will have gained practice in **observing** similarities and differences among leaves.

## Chapter Preview

Have students study the chapter outline before they begin to read the chapter. They should note that the structure of leaves is discussed first and that the food-making process of leaves is discussed second. The uses of and changes in leaves is discussed in the last section.

## Misconception

Students are aware that plants take in water through the roots but may fail to realize that plants wilt because they lose water through the leaves.

### ASSESSMENT PLANNER

***Portfolio***

Strategies on the following pages represent student products that can be placed into a best-work portfolio: pp. 403, 412.

**PERFORMANCE ASSESSMENT**
Skill Check, pp. 403, 408
Mini Lab, pp. 402, 404
Lab, pp. 400, 415

**CONTENT ASSESSMENT**
Check for Understanding, pp. 402, 404, 408, 409, 412, 414
Chapter Review, pp. 416-417

**GROUP ASSESSMENT**
Opportunities for group assessment occur with Cooperative Learning Strategies.

***Student Journal***

Strategies on the following pages represent opportunities for writing in a Student Journal: pp. 406, 409.

# 19:1 The Structure of Leaves

## PREPARATION

### Materials Needed

Make copies of the Study Guide, Critical Thinking/Problem Solving, Enrichment, Transparency Master, Reteaching, and Lab worksheets in the *Teacher Resource Package*.

▶ Gather leaves for Lab 19-1.

▶ Have a potted plant on hand for the Mini Labs.

### Key Science Words

blade
midrib
epidermis
palisade layer
spongy layer
stoma
guard cell
transpiration
wilting

### Process Skills

In Lab 19-1, students will classify, interpret diagrams, and measure in SI. In the Skill Check, students will understand science words. In the Mini Labs, students will experiment and observe.

## 1 FOCUS

▶ The objectives are listed on the student page. Remind students to preview these objectives as a guide to this numbered section.

### MOTIVATION/Brainstorming

Have students make a list of leaves they eat. Compile a list on the chalkboard.

## 19:1 The Structure of Leaves

**Objectives**

1. **Examine and compare** the parts of a leaf in different plants.
2. **Give the function** of the cells in a leaf.
3. **Explain** the process of transpiration.

**Key Science Words**

blade
midrib
epidermis
palisade layer
spongy layer
stoma
guard cell
transpiration
wilting

Look around your neighborhood for plants. Apart from flowers, the plant parts that most catch your eye are probably leaves. Leaves are usually green but may show other colors. A plant often can be named just from looking at one of its leaves. Each species has a distinct set of leaf traits.

### Leaf Traits

Most leaves are flat and green. The thin, flat part of a leaf is the **blade.** The blades of leaves can be round, heart-shaped, long and narrow, or short and broad. There are leaf blades that don't even look like leaves. The leaf blades of pine trees look like bundles of needles. The leaf blades of cedar trees look like the scales on a fish. Some leaves are adapted to one type of environment. The fleshy leaves of a jade plant are adapted for life in a dry, hot desert. The leaf blade is the main part of a plant that makes food.

A thin stalk usually joins the leaf to the stem. Stalks can be many different lengths and thicknesses. The part of the celery plant that you eat is the leaf stalk. It may be four to eight centimeters wide at the bottom. Many leaf stalks are no larger around than your pencil lead. Some plants, such as corn and grass, have leaves that are not attached to the stem by stalks. The leaf blades are joined directly to the stem. Figure 19-1 shows the stalk and the blade of a leaf.

Remember from Chapter 6 that there are two kinds of cells in the vascular tissue of plants. The stalk of a leaf contains veins with phloem and xylem vessels. These

**Bio Tip**

**Health:** Humans get poison ivy rashes when their skin comes in contact with the leaves of the poison ivy plant. You can identify poison ivy by its leaves that have three leaflets.

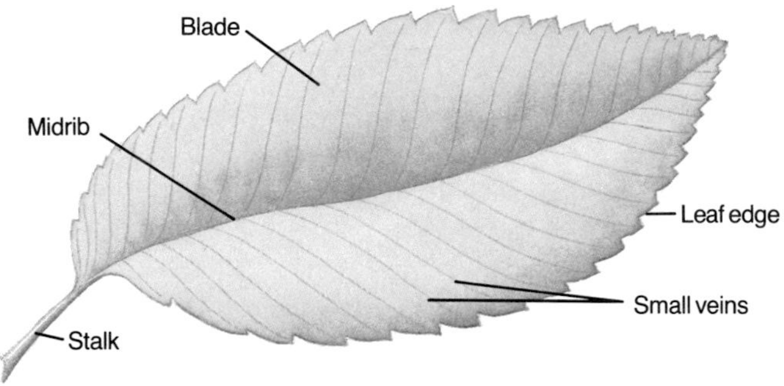


**Figure 19-1** A leaf has two main parts—a stalk and a blade.



## OPTIONS

### Science Background

Scientists showed that plants produce oxygen during photosynthesis by doing the following experiment. A plant was placed in one jar, a mouse in a second jar, and a mouse and plant in a third jar. The mouse in the second jar died from lack of oxygen, but the mouse in the third jar survived.

a

b

**Figure 19-2** A royal water lily leaf (a) can be large enough to support a baby, and a water fern (b) has leaves smaller than your fingernail.

vessels, just like the blood vessels in your body, transport nutrients around the tissues of the plant. In the leaf stalk, vessels carry food and water between stem and leaf.

The stalk of a leaf appears to run into the blade and continues to the tip of the leaf. This part of the leaf blade is the midrib. The **midrib** is the main vein of the leaf. Notice in Figure 19-1 the smaller veins branching from the midrib. The midrib and smaller veins are made of phloem and xylem vessels that carry food and water in the leaf. The midribs and veins form different patterns in different leaves.

**What leaf traits do scientists use to identify plants?**

Scientists use leaf traits to identify and group plants. They use leaf shape, kind of edge, vein pattern, and the arrangement on the stem. It would be difficult to show all the different sizes and shapes of leaves. Look at Figure 19-2. Compare the size of the royal water lily leaf from South America with that of a water fern leaf from North America. Some tropical palm trees have leaves that are as long as a three-story building is high!

How would you describe the shape of grass leaves? Bladelike? Think of some other plants that have leaves shaped like grass leaves. Look at the different leaf shapes in Figure 19-3. Which are shaped like grass leaves?

The leaves in Figure 19-3 also show different kinds of edges. Leaves may have smooth edges or a toothed edge. There may be hundreds of teeth along the edge of a leaf. Notice also in Figure 19-3 that the edges of some leaves curve in and out again to form "fingers" or lobes. The numbers and shapes of lobes are used to identify some plants.

**Figure 19-3** Leaves have many shapes and vein patterns. Which of these leaves have parallel veins? grass and corn

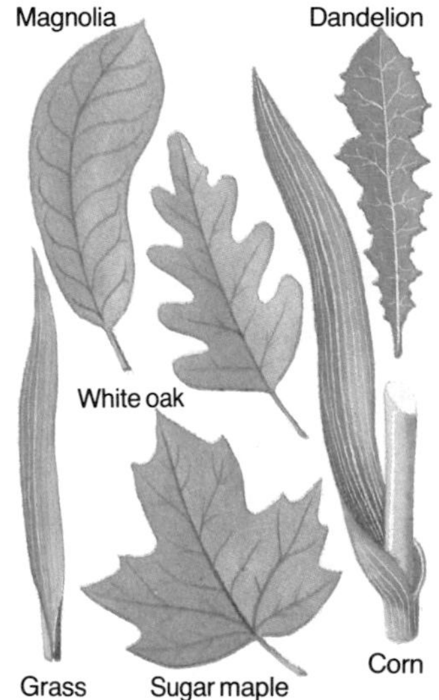




# 2 TEACH

### MOTIVATION/Demonstration

Place dried, pressed leaves on an overhead projector. Point out the differences in shapes and edges of the leaves.

### Concept Development

▶ Explain to students that the spines on a cactus plant and the red petal-like flowers of a poinsettia are modified leaves.

▶ Emphasize that veins are found in roots, stems, and leaves because they are the pipeline for water and food for the plant.

▶ Emphasize that the edges, veins, and shape are the main features of a leaf blade used to identify a particular species of plants. Students will use the same features when they do Lab 19-1.

### Guided Practice

Have the student make a drawing of a leaf showing the features of the edge, size, and vein pattern. Have them label the appropriate parts.

### Independent Practice

**Study Guide**/*Teacher Resource Package*, p. 109. Use the Study Guide worksheet shown at the bottom of this page for independent practice.

### Guided Practice

Have students write down their answers to the margin question: shapes, edges, patterns of veins, and arrangement on stem.

### STUDY GUIDE 109

STUDY GUIDE — CHAPTER 19

Name ________ Date ________ Class ________

THE STRUCTURE OF LEAVES

*In your textbook, read about leaf traits in Section 19:1.*

1. Label the following parts of this diagram: stalk, midrib, blade, smaller vein.

stalk
midrib
smaller vein
blade

*In your textbook, read about the cells in a leaf in Section 19:1.*

2. Label the following parts of this diagram of a leaf section: air spaces, upper epidermis, lower epidermis, wax layer, guard cell, wax layer, vein, stoma, spongy layer, palisade layer.

wax layer (yellow)
upper epidermis (red)
palisade layer (green)
spongy layer (green)
air spaces
vein (blue)
lower epidermis (red)
guard cell (green)
stoma
wax layer (yellow)

3. Using these colors, shade in the following parts on the diagram above:
Green—leaf cells that make food
Blue—leaf cells that carry water and food
Red—leaf cells that protect
Yellow—waterproof layer

## OPTIONS

### ACTIVITY/Enrichment

Bring in some fresh spinach, kale, or other greens that have a stalk attached to the leaf. Point out how the veins of the stalk run into the blade. Show students how to trace the vein from the stalk to the blade.

## Lab 19-1 Leaves

### Overview

In this lab, students will observe 15 different leaves and note their shapes, edges, and vein patterns. They will try to find out which of the patterns are the most common.

**Objectives:** Upon completing the lab, students will be able to: (1) **measure** various leaf parts, (2) **compare** the leaf parts with pictures.

**Time Allotment:** 45 minutes

### Preparation

▶ Use previously collected leaves.

**Lab 19-1 worksheet**/*Teacher Resource Package*, pp. 73-74.

### Teaching the Lab

**Cooperative Learning:** Divide the class into teams of two students. For more information, see pp. 22T-23T in the Teacher Guide.

▶ **Troubleshooting:** Make sure the students handle the specimens carefully as leaves can break easily when dry.

### ASSESSMENT

**Performance:** Have students classify the 15 leaves into four to five distinct groups based on the common properties they share. Tell students to use the data chart and leaf samples to classify the groups of leaves. They are to give names to the groups by using the physical properties they share. For example, one group may be called "small, round-toothed, spear-shaped leaves."

# Leaves

## Lab 19—1

### Problem: What are some traits of leaves?

#### Skills

classify, interpret diagrams, measure in SI

#### Materials

15 leaves of different plants
metric ruler

#### Procedure

1. Copy the data table and make a row for each one of your 15 leaves.
2. **Measure in SI:** Choose a leaf. Measure the length of the leaf blade in centimeters from the tip of the leaf blade to its base. For help, see Figure A.
3. Compare the leaf with the shapes given in Figure B. Record the leaf shape.
4. Compare the leaf edge with those in Figure C. Record the leaf edge.
5. Compare and record the vein pattern on your leaf with the vein patterns in Figure D.
6. Repeat steps 2 through 5 with the other 14 leaves.

| TRAITS OF LEAVES | | | | |
|---|---|---|---|---|
| Leaf | Blade Length (cm) | Shape | Edge | Veins |
| 1 | 17.5 | oval | no teeth | feather |
| 2 | 13 | star | small teeth | palm |

#### Data and Observations

1. Which leaves were the longest?
2. How many of your leaves had parallel veins? How many had palm veins?

#### Analyze and Apply

1. Which type of leaf edge is most common in your leaf samples?
2. These are the traits of two leaves: (a) oval, seven pointed lobes, small toothed edge, feather veins; (b) star shape, small teeth, palm veins. Are these two leaves the same species? Explain your answer.
3. **Apply:** How are leaf traits important?

#### Extension

**Classify** the 15 leaves by using a field guide.

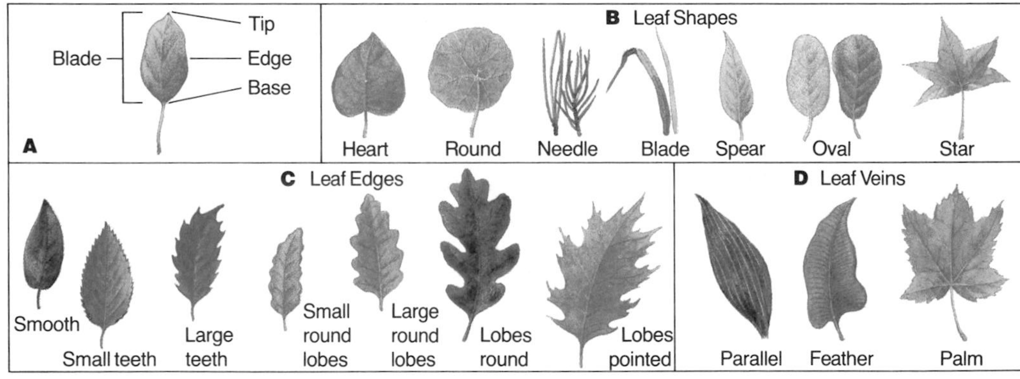


400 The Importance of Leaves 19:1

## ANSWERS

### Data and Observations

1. Answers will vary depending on leaf selection.
2. Answers will vary. Examine each student's data table.

### Analyze and Apply

1. Answers will vary. Examine student data.
2. The two leaves are very different in leaf shape, leaf edge, and leaf veins. They are probably two separate species.
3. Leaf traits help in the classifying of plants.

Leaves are arranged in different ways on the stems. Figure 19-4 shows three main leaf arrangements along a stem. Some plants have pairs of leaves opposite each other along the stem. Some plants have one leaf at each point along the stem. They alternate from one side to the other up the stem. In some plants, there are three or more leaves attached around one point on the stem to form a circle or whorled pattern. No matter how they are arranged, the leaves seldom overlap. Each leaf is arranged on the stem to catch the most sunlight it can.

**Figure 19-4** Leaves are arranged in an opposite, alternate, or whorled pattern on the stem.

## Cells of the Leaf

If you cut a leaf blade in half, you would see its cells. The cells inside a leaf blade are arranged in layers as shown in Figure 19-5. Most leaves are covered with a waxy layer. The waxy layer protects the leaf from water loss and from feeding insects. This layer is not made of cells. Did you ever notice how water forms beads on a newly waxed car? This layer of wax forms a protective coat for the paint on a car. Water cannot pass through the waxy layer on a car, nor can water leak through the waxy layer of a plant. The layer of wax on leaves makes them appear shiny, just as the wax makes a car shiny.

Beneath the waxy layer of a leaf is a layer of cells called the epidermis (ep uh DUR mus). The **epidermis** is the outer layer of cells of a plant. The epidermis is like a skin and is usually only one cell thick.

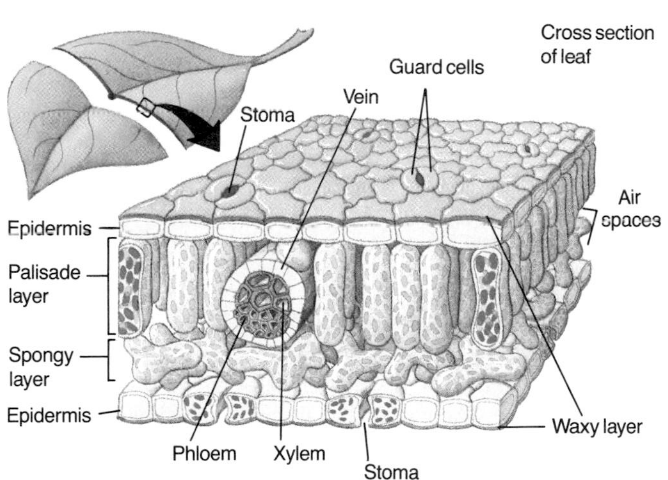


**Figure 19-5** The cells of a leaf are arranged in layers. In which layer are there air spaces? the spongy layer



## TEACH

### Concept Development

▶ Explain that only a few trees have the whorled pattern on the stem.

▶ Point out that the leaf has two layers of epidermis and both of these layers protect the leaf.

▶ Inform students that some leaf epidermis has small hairs projecting through the waxy layer and may not appear smooth if they were to look at it through a hand lens.

**Cooperative Learning:** Have students in groups of two or three collect and press leaves in a herbarium press or large catalog. After a week of drying and pressing, have students mount their leaves on construction paper and identify the leaves using common names.

**Misconception:** Explain to students that the waxy layer is the only non-living layer on a leaf. Students may believe that it is a living layer because it is attached to the leaf.

### GLENCOE TECHNOLOGY

**Videodisc**

**The Secret of Life**

*Cross-section of Leaf*

### CRITICAL THINKING 19

CRITICAL THINKING/PROBLEM SOLVING CHAPTER 19

Name ______ Date ______ Class ______

*Use after Section 19:2.*

*HOW DO DROPS OF WATER FORM ON LEAVES?*

As Anita walked into school she noticed the small grass plants growing in the cracks of the pavement and in a nearby vacant lot. Small drops of water shone on the surfaces of each blade of grass. She wondered how the drops formed. She could think of several hypotheses. The water might come from inside the leaf. Or it might be dew condensing from the air. Or, the water might be evaporating from the soil. In her biology class, the teacher suggested she and a team might study transpiration.

Anita and her team decided to try several experiments using plants in pots to answer their questions. First, they would observe the time of day when the droplets formed. Second, they would cover the soil and pot, to prevent evaporation. Also, they would cover plants with glass jars and find out if drops of water still formed. Finally, they would look at the leaves with a microscope. Would they see tiny holes in the leaves?

**Analyzing the Problem**

1. The students observed droplets at several different times of day. Did this information help them decide if the water came from dew or from the plant cells? **Yes, dew is found in the morning, so the drops are probably not dew.**
2. The students covered the pots and the soil of several plants with plastic, to prevent evaporation from the soil. Then they covered the whole pot and plant with a larger jar. In the morning, the plants inside the jars had drops of water on the leaves. What could they hypothesize about the drops on the plants in jars? **The drops of water came from inside the plants, not from evaporation from the soil.**
3. The students looked at the leaf under the microscope. They could see a large number of pores on the leaves. There were more on the underside than on the tops of the leaves. They compared them to a picture in their textbook. What are the pores called? **stomata**
   In what form does water enter and leave through the stomata? **water vapor**
   If the air outside is cooler, what can happen to the water vapor? **It can condense into water droplets on the leaves.**

**Solving the Problem**

1. Now the students could decide where the water drops came from, and how water passes out of a plant. Write your conclusion here. **The water forms inside the plant cells and passes out as vapor through the stomata. Then, it condenses on the leaves.**

### OPTIONS

**Critical Thinking/Problem Solving/** *Teacher Resource Package*, p. 19. Use the Critical Thinking/Problem Solving worksheet shown here for students to learn how drops of water form on leaves.

**Cross Section of a Leaf/***Transparency Package*, number 19. Use color transparency 19 as you teach the cells found in a leaf.

# TEACH

### Concept Development

▶ Point out that some leaves may have as many as 200 stomata per square millimeters.

### Guided Practice

Have students write down their answers to the margin question: layer of long, green cells in a leaf just below the upper epidermis.

### Mini Lab

The water was absorbed by the towel. The waxed paper and leaf have a layer of wax that keeps water from being absorbed.

### ✓ ASSESSMENT

**Oral:** Ask students to make a hypothesis as to why plants living in dry regions of the world have thicker or thinner waxy layers on the epidermis.

### Independent Practice

**Study Guide**/*Teacher Resource Package*, p. 110. Use the Study Guide worksheet shown at the bottom of this page for independent practice.

### Check for Understanding

Have the students make a list of the cells in the leaf from top to bottom and write a function for the parts.

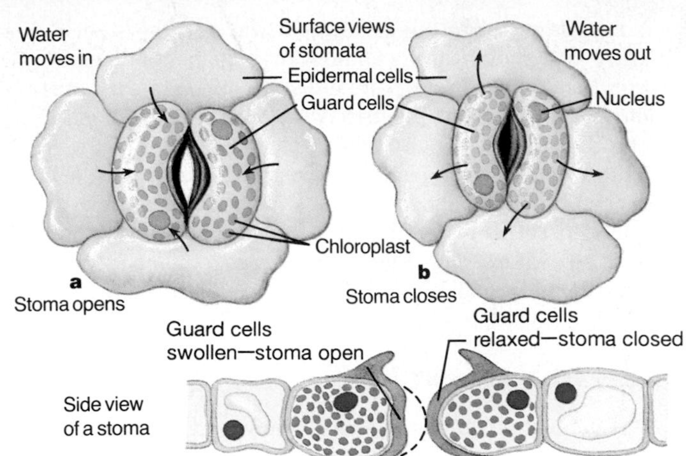


**Figure 19-6** Water enters the guard cells and causes them to swell. The stoma opens (a) because the guard cells push apart. The stoma closes (b) when water leaves the guard cells. Below is a cross section through a stoma, as seen from the edge of the leaf.

**What is the palisade layer?**

Examine Figure 19-5 again. Look at the layer of cells below the epidermis. The layer of long, green cells below the upper epidermis of a leaf is the **palisade layer.** Cells of the palisade layer make most of the food for the plant. These cells contain a lot of chloroplasts. Remember that chlorophyll is the green pigment in chloroplasts that helps the plant make food.

Directly below the palisade layer of a leaf is a layer of round, green cells called the **spongy layer.** These cells are loosely arranged and have spaces between filled with water vapor and air. The cells of the spongy layer also make food for the leaf. Leaf veins are often found among the palisade and spongy layers of cells.

Below the spongy layer is another layer of epidermis with its waxy layer. Find a stoma (STOH muh) in the lower epidermis of Figure 19-5. A **stoma** is a small pore or opening in the epidermis of a leaf. The plural of stoma is stomata. Leaves have stomata on both the upper and lower epidermis. The stoma acts as a doorway for gases, including water vapor, to enter and leave the leaf. Stomata are usually open during the day to take in carbon dioxide for photosynthesis. They are closed at night when carbon dioxide isn't needed.

The size of a stoma changes. How does this happen? Around the stoma are two bean-shaped cells called guard cells. **Guard cells** are green cells that change the size of the stoma in a leaf. The size of the stoma changes as the guard cells swell or shrink when they take in or let out water by osmosis, Figure 19-6.

### Mini Lab

**How Does a Leaf Epidermis Do Its Work?**

**Experiment:** Put two drops of water on waxed paper, on a paper towel, and on a leaf from a potted plant. Compare what happened to the drops of water.

*For more help, refer to the* ***Skill Handbook,*** *pages 704-705.*



## OPTIONS

**Enrichment**/*Teacher Resource Package*, p. 20. Use the Enrichment worksheet shown here to teach the effect of sunlight and water on leaves.

**What Do the Inside Parts of Leaves Look Like?**/*Lab Manual*, pp. 157-160. Use this lab as an extension to the cellular structure of leaves.

### STUDY GUIDE 110

STUDY GUIDE — CHAPTER 19

Name ______ Date ______ Class ______

**THE STRUCTURE OF LEAVES**

4. *The chart below describes several different kinds of leaves and gives an example of each. Read each description and example. Examine the diagrams carefully. Then, write the letter of the leaf type on the blank next to the diagram that it best matches.*

| Shape | Other traits | Example | |
|---|---|---|---|
| Fan | | Ginkgo | (a) |
| Heart | Edges have teeth. | Cottonwood | (b) |
| Heart | Edges are smooth. Tip is very pointed. | Catalpa | (c) |
| Heart | Edges are smooth. Tip is not very pointed. | Redbud | (d) |
| Oval | Edges are smooth. | Magnolia | (e) |
| Oval | Edges have a few large teeth. | Holly | (f) |
| Oval | Edges have many small teeth. | Elm | (g) |
| Needle | Needles are in twos. | Virginia pine | (h) |
| Needle | Needles are in threes. | Pitch pine | (i) |
| Needle | Needles are in fives. | White pine | (j) |
| 5-part | All leaflets attach at same point. | Buckeye | (k) |
| 5-part | Three leaflets attach at top, two near bottom. | Shagbark hickory | (l) |
| More than 5-parts | Edges have teeth. Leaflets are oppositely attached. | Sumac | (m) |
| More than 5-parts | Edges are smooth. Leaflets are oppositely attached. | White ash | (n) |
| Oval | Edges are toothed. Tip is very pointed. | Hackberry | (o) |
| Wavy lobed | Lobes are pointed. | Pin oak | (p) |
| Wavy lobed | Lobes are rounded. | White oak | (q) |

### ENRICHMENT 20

ENRICHMENT — CHAPTER 19

Name ______ Date ______ Class ______

*Use after Section 19:2.*

**THE EFFECT OF SUNLIGHT AND WATER ON LEAVES**

The amount of sunlight and water a plant gets affects the cells of its leaves. They may develop more layers of cells or fewer layers of cells depending on the amount of sunlight and water they receive.

Study the drawings of leaf sections below. The levels of sunlight and water each leaf gets is given below each cross section. Compare the drawings of the leaf sections, then answer the questions that follow.

Normal sun Normal water — Most sun Normal water — Normal sun Most water — Most sun Little water

1. What type of cells cause Leaf B to be thicker than Leaf A? **The extra palisade cells cause Leaf B to be thicker.**
2. Does Leaf B get more or less sun than the other leaves? **Leaf B gets more sun than Leaf A and Leaf C.**
3. What cell layer is thickest in Leaf C? Is Leaf C getting more water than Leaf A? **The spongy layer is thickest. Leaf C is getting more water than Leaf A.**
4. What nutrient do the extra cells in Leaf C store? **These cells store water.**
5. What is this stored nutrient used for in Leaf C? **Water is used in making food by photosynthesis.**
6. Leaf D gets little water but lives in very bright sunlight. What differences do you notice in Leaf D? **Leaf D has a thicker epidermis and fewer stomata than the other leaves.**
7. How is Leaf D helped by having these differences? **The thicker epidermis and fewer stomata prevent too much water loss.**
8. Where might you find a plant with leaves like Leaf D? **A plant with leaves similar to Leaf D would most likely be found in hot, dry places.**

**Figure 19-7** Transpiration is most obvious over the treetops of a tropical forest.

## Water Loss in Plants

What is transpiration?

Plants lose water daily through their stomata. The process of water passing out through the stomata of leaves is **transpiration** (trans puh RAY shun). Each day, a plant may lose up to 90 percent of the water it takes up through its roots.

Plant cells are mostly water. Water in a plant keeps the cells firm. Sometimes, a plant loses too much water and wilts. **Wilting** is when a plant loses water faster than it can be replaced. A punctured tire losing air is similar to a plant losing water. The tire goes flat if the escaping air is not replaced at the same rate. The plant shown in Figure 19-8 wilted when it didn't get watered for several days.

During hot, dry days, water loss is greater than usual and the stomata close. On cooler, damp days, water loss slows down and the stomata remain open.

### Skill Check

**Understand science words: transpiration.** The word part *trans* means through. In your dictionary, find three words with the word part *trans* in them. *For more help, refer to the* ***Skill Handbook,*** *pages 706-711.*

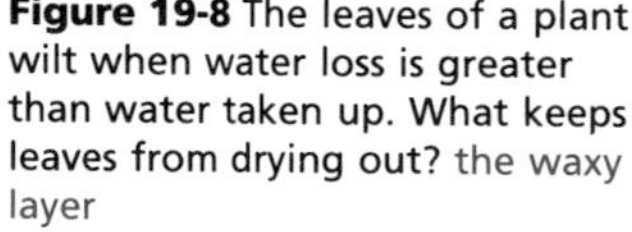

**Figure 19-8** The leaves of a plant wilt when water loss is greater than water taken up. What keeps leaves from drying out? the waxy layer



## TEACH

### Concept Development

▶ Point out that curling of a leaf is an adaptation that helps prevent a leaf from losing water. A curled leaf loses less water than a flat or rigid leaf.

▶ Explain that water can be lost through the stems, but most of the water loss in plants is through the leaves.

▶ Emphasize that temperature, wind, humidity, and light affect transpiration.

▶ Tell students that a single corn plant can transpire about two liters of water a day.

▶ Point out that the number of stomata on a leaf affects the amount of water transpired through a leaf.

### Guided Practice

Have students write down their answers to the margin question: process of water passing out through the stomata of leaves.

### Reteach

Have students relate shape and size of cells to their function. (epidermal–brick-shaped, protection; palisade–long, balloonlike, make food; spongy cells–like wet sponge with air holes, make food; guard cell–bean-shaped, lets water in or out).

### Portfolio

Have students solve the following problem and place it in their portfolios: Each day of the growing season, a corn plant transpires 2 L of water. A bathtub holds 150 L of water. How long would it take the corn plant to transpire enough water to fill the bathtub? Show the steps you used to solve the problem. Make a drawing if you think it is appropriate.

### TRANSPARENCY 89

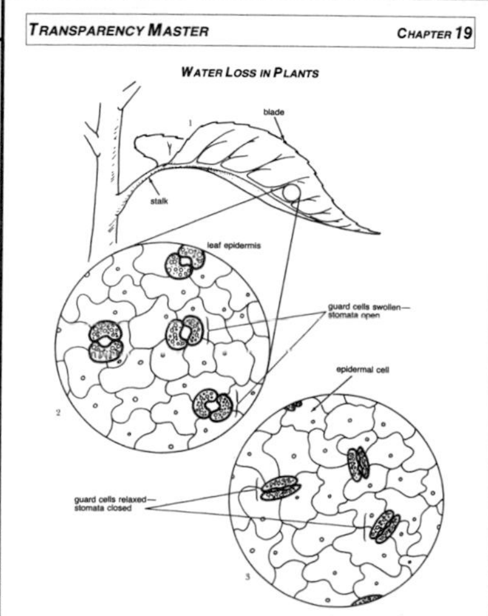


### OPTIONS

**Transparency Master**/*Teacher Resource Package*, p. 89. Use the Transparency Master shown here to teach water loss in plants.

## TEACH

### Mini Lab

The leaf is firm at the start, and limp as it loses water.

### ASSESSMENT

**Content:** The produce section of a grocery store sprinkles water frequently on some kinds of produce but not others. Give an explanation as to why this is done.

### Check for Understanding

Have students respond to the first three questions in Check Your Understanding.

### Reteach

Reteaching/*Teacher Resource Package,* p. 56. Use this worksheet to give students additional practice with cells in a leaf.

## 3 APPLY

### Demonstration

Show water loss in a plant by placing a clear plastic bag over a well-watered plant. Place in a lighted area and later point out condensation on the side of the bag.

## 4 CLOSE

### Connection: Writing

Have students write a paragraph on the differences between leaves in cool and warm climates.

### Answers to Check Your Understanding

1. stalk, blade, veins
2. guard cells
3. A plant wilts when water is lost faster than it can be replaced.
4. To avoid losing too much water, a desert plant has fewer stomata than a forest plant.
5. 1 percent

**Figure 19-9** A greenhouse provides a warm, wet environment for plants.

### Mini Lab

**Why Do Leaves Wilt?**

**Observe:** Remove a leaf from a potted plant and place it on a paper towel at the start of class. What is the texture of the leaf at the start and end of the period? Explain what happened to the leaf. *For more help, refer to the **Skill Handbook,** pages 704-705.*

Do you have a houseplant at home? If you placed this plant outdoors, would it survive? It's unlikely the houseplant would be adapted to your local climate. For example, many ferns couldn't grow where it is hot and dry. Houseplants are often grown first in greenhouses before they are sold to you, Figure 19-9. If you have ever been inside a greenhouse, you may have noticed how humid and warm it is. The temperature is kept the same at all times. The air is kept moist so plants lose less water through their stomata.

Enough water passes through the plants in a small field of corn in 100 days to fill a bathtub at least 10 000 times. Each corn plant may lose up to 500 liters of water in one season. Now you may understand why farmers need to be concerned about the weather. Without rain, their corn crops would suffer.

### Check Your Understanding

1. What three parts of most leaves can you see without a microscope?
2. What cells of the epidermis control sizes of stomata?
3. Explain what causes a plant to wilt.
4. **Critical Thinking:** Would a desert plant have as many stomata in its leaves as a forest plant? Explain.
5. **Biology and Math:** Plants normally lose about 99 percent of the water they absorb. What part of the water they absorb is used for growth?

**STUDY GUIDE 111**

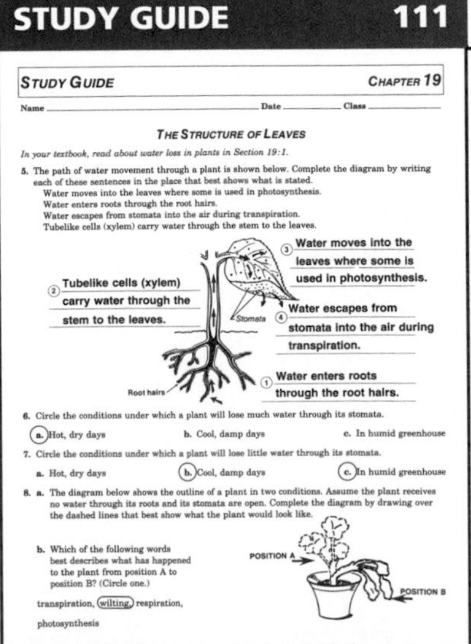

STUDY GUIDE CHAPTER 19

Name ______ Date ______ Class ______

THE STRUCTURE OF LEAVES

*In your textbook, read about water loss in plants in Section 19:1.*

5. The path of water movement through a plant is shown below. Complete the diagram by writing each of these sentences in the place that best shows what is stated.
   Water moves into the leaves where some is used in photosynthesis.
   Water enters roots through the root hairs.
   Water escapes from stomata into the air during transpiration.
   Tubelike cells (xylem) carry water through the stem to the leaves.



6. Circle the conditions under which a plant will lose much water through its stomata.
   a. Hot, dry days   b. Cool, damp days   c. In humid greenhouse
7. Circle the conditions under which a plant will lose little water through its stomata.
   a. Hot, dry days   b. Cool, damp days   c. In humid greenhouse
8. a. The diagram below shows the outline of a plant in two conditions. Assume the plant receives no water through its roots and its stomata are open. Complete the diagram by drawing over the dashed lines that best show what the plant would look like.
   b. Which of the following words best describes what has happened to the plant from position A to position B? (Circle one.)
   transpiration, wilting, respiration, photosynthesis



**RETEACHING 56**

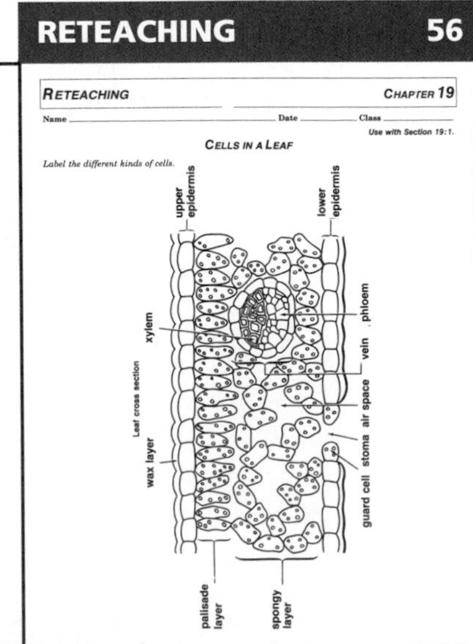

RETEACHING CHAPTER 19

Name ______ Date ______ Class ______

Use with Section 19:1.

CELLS IN A LEAF

*Label the different kinds of cells.*

## 19:2 Leaves Make Food

Plants use materials from their environment to make their own food. The food made by plants is also used by other living things. In meadows, deserts, forests, oceans, and rivers, animals depend on green plants for food.

### Objectives

4. **Describe** the process of photosynthesis.
5. **Explain** the chemical equation for photosynthesis.
6. **Compare** the importance of sugar in photosynthesis and respiration.

### Building from Raw Materials

At one time, people thought that plants got their food from the soil. Experiments have since shown that as a plant grows taller and heavier, the amount of soil around its roots hardly changes. Today, we know that plants make their own food. The leaf is the main part of a plant that makes food.

Figure 19-10 shows the frame of a house. The way a green leaf makes food by photosynthesis can be compared to the way a carpenter builds a house. Both processes build a product from raw materials.

**How is building a house like photosynthesis?**

A carpenter starts out by using raw materials such as lumber, nails, and shingles. As the carpenter builds, these raw materials begin to take on the shape of a building. Energy is needed to build a house. Human energy is used to carry, arrange, and join materials, and to drive nails into wood. Electrical energy is used to run power saws or drills.

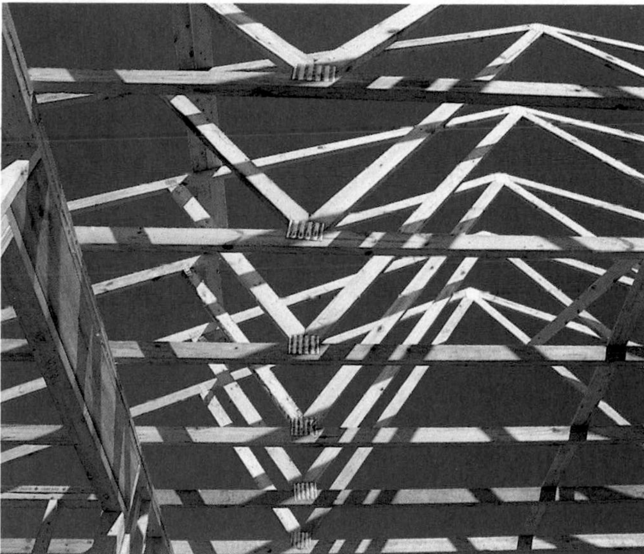

**Figure 19-10** The raw materials for building a house include wood and nails.



## 19:2 Leaves Make Food

## PREPARATION

### Materials Needed

Make copies of the Focus, Study Guide, Application, Skill, and Reteaching worksheets in the *Teacher Resource Package*.

### Key Science Words

No new science words will be introduced in this section.

### Process Skills

In the Skill Check, students will infer.

## 1 FOCUS

▶ The objectives are listed on the student page. Remind students to preview these objectives as a guide to this numbered section.

### Convergent Question

Ask students to make a list of materials a plant needs to get its food.

### Guided Practice

Have students write down their answers to the margin question: They both use energy to combine raw materials to make a product and produce wastes.

Focus/*Teacher Resource Package*, p. 37. Use the Focus Transparency shown at the bottom of this page as an introduction to the effect of sunlight on growth.

**FOCUS 37**

FOCUS CHAPTER 19

Name ______ Date ______ Class ______

EFFECT OF SUNLIGHT ON GROWTH

| Nov | Dec | Jan | Feb | Mar | Apr | May | Jun | Jul | Aug | Sep | Oct |
|---|---|---|---|---|---|---|---|---|---|---|---|
| Daily hours of sunlight | | | | | | | | | | | |
| 9 | 8 | 8 | 9-10 | 10-12 | 13-14 | 15 | 16 | 15 | 14-13 | 12-10 | 10-9 |

Hairy willow herb

Chrysanthemum

Short days → Long days → Short days

Plants need sunlight to make food. They also need light to grow and develop flowers. Study the chart above. List how much sunlight each of the plants in the illustration need to flower.

## OPTIONS

### Science Background

Many different processes occur in the leaf but photosynthesis is the most important one because food is made for use by the live cells of other plant parts. All chemical energy used by animals and humans started in the plants.

# 2 TEACH

## MOTIVATION/Video

Show the video *Introduction to Photosynthesis,* Insight Media.

### Concept Development

▶ Explain that soil actually supplies plants with minerals that are needed for many life processes.

▶ Emphasize that a specific amount of light, carbon dioxide, and water, as well as a suitable temperature are necessary factors for photosynthesis.

▶ Point out that oxygen, the waste product of photosynthesis, is needed by both plants and animals for respiration.

▶ Explain that plants are dependent on animals for carbon dioxide for photosynthesis.

▶ Explain that plants also get carbon dioxide from the weathering of rocks and volcanoes.

**Misconception:** Students may think plants get food by taking it in through roots. They are confusing uptake of minerals and water with the food making process of leaves.

**Analogy:** Make sure that students understand that building a house is a physical process of putting raw materials together; photosynthesis is a chemical process of putting raw materials together. Stress that both require raw materials and end with a usable product.

**Student Journal**

Tell students to reread section 19:2. Have them develop a chart in parallel outline form showing the similarities between the steps used to build a house and how a plant makes food. Tell them they should decide how many steps to add to the outline.

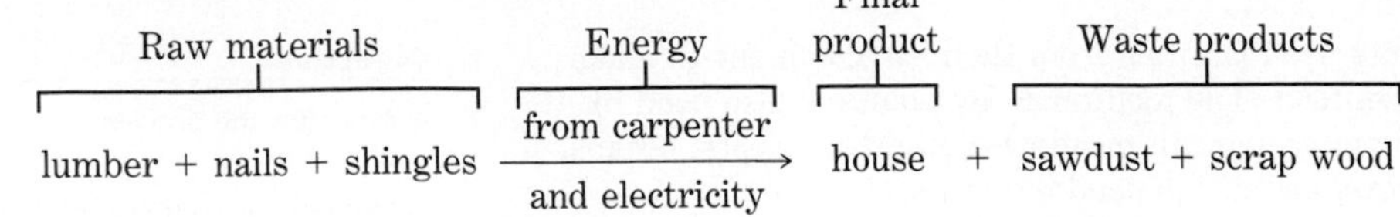


**Bio Tip**

**Consumer:** Fertilizer for houseplants is often called plant food. Look at a label on a packet of fertilizer to find out what it really contains.

When a house is built, some waste products are formed. Sawdust and scrap wood are waste products. If collected, the sawdust or wood scraps may be used for some other purpose. The equation above shows a formula for building a house. Notice how the raw materials are combined using energy to give a product and waste materials.

Keeping the example of house building in mind, let's think about the raw materials needed for photosynthesis. This process of food-making uses water and carbon dioxide gas as raw materials. Use Figure 19-11 to trace the path of carbon dioxide and water as they enter a plant. Plant roots take water from the soil. Water passes from the roots, up the stem, and into the leaves. Carbon dioxide in the air enters the leaves through the stomata.

## Changing Raw Materials into Sugar

The product of photosynthesis is food in the form of sugar. Just as in house building, plants also have waste products from building food.

How does a plant make sugar? Remember that the chemical formula for carbon dioxide is $CO_2$. The letter C stands for the element carbon. The letter O stands for the element oxygen. The number 2 means there are two atoms of oxygen in each molecule of carbon dioxide.

**Figure 19-11** The raw materials for photosynthesis are the waste products of cellular respiration.

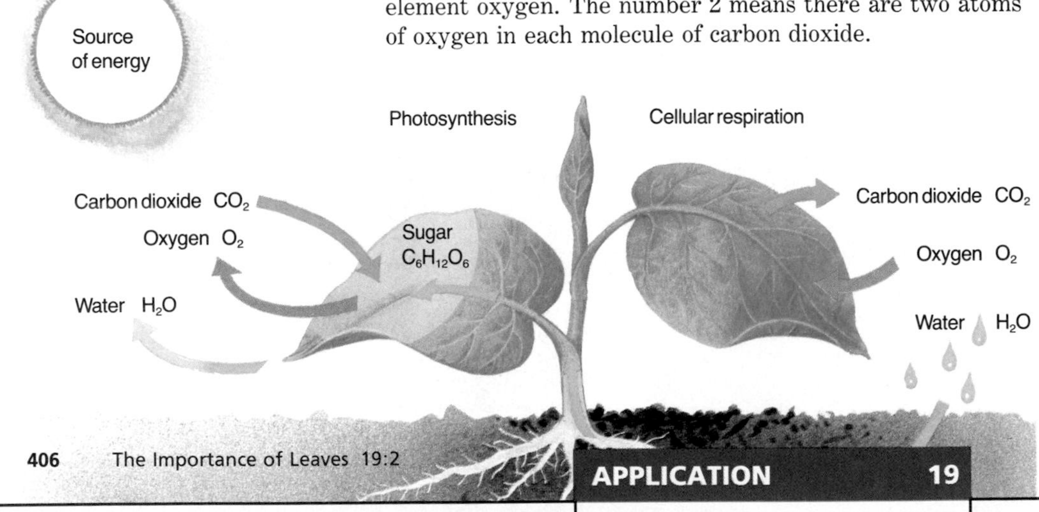




## OPTIONS

**Application**/*Teacher Resource Package*, p. 19. Use the Application worksheet shown here to teach the greenhouse effect.

**APPLICATION 19**

*APPLICATION: ENVIRONMENT* — *CHAPTER 19*

Name ______ Date ______ Class ______

*Use after Section 19:2.*

### THE GREENHOUSE EFFECT

Earth's atmosphere acts like the glass walls and roof of a greenhouse. It lets in the light of the sun. The light energy then changes into heat energy. The heat energy does not move easily out through the atmosphere, just as heat does not escape easily from a greenhouse. As a result, Earth, like the greenhouse, stays warm. The way the atmosphere helps warm Earth is called the greenhouse effect.

Perhaps you have heard that the greenhouse effect can cause a problem if too much heat is trapped. Extra heat is trapped because Earth's atmosphere is changing. The amount of carbon dioxide in the atmosphere is increasing. Some of the extra carbon dioxide is produced by burning fuels. Some of these fuels include wood, coal, oil, and natural gas. Also, when trees and other plants are destroyed, the problem gets worse. Living plants use up some carbon dioxide through photosynthesis. When the plants are destroyed, the carbon dioxide level continues to rise.

1. What are two ways that people could help lessen the greenhouse effect? **People can burn less fuel and destroy fewer trees and other plants.**
2. Why is the burning of forests a double problem? **The burning creates carbon dioxide, and trees, which could have used up carbon dioxide, are destroyed.**
3. How do you think the recycling of paper could help lessen the greenhouse effect? **Fewer trees would be cut down to make paper.**
4. Does Earth benefit from the greenhouse effect? Explain. **Yes, Earth is kept warm enough for life to exist. Without the atmosphere to hold in heat, Earth would be too cold.**
5. Would planting trees help lessen the greenhouse effect? Explain. **Yes, more trees would use up more carbon dioxide through photosynthesis.**
6. Look for a news report on the greenhouse effect in a newspaper or magazine. Report your findings. **Answers will vary.**

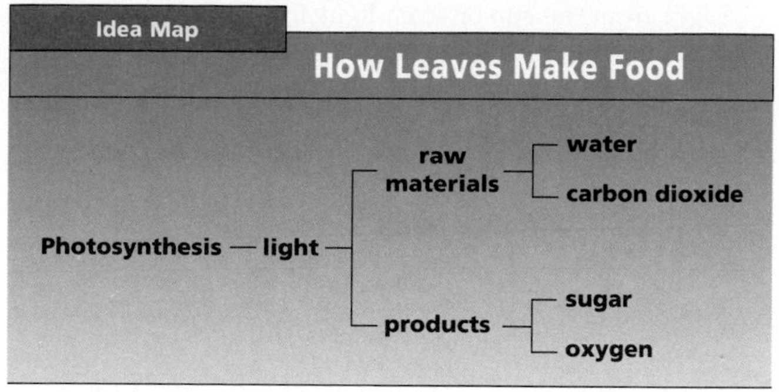


**Study Tip:** Use this idea map as a study guide to how leaves make food. From where does a plant get its raw materials?

The chemical formula for water is $H_2O$. The letter H stands for hydrogen. The formula tells us that a water molecule is made of two atoms of hydrogen and one atom of oxygen. The three elements found in carbon dioxide and water—carbon, oxygen, and hydrogen—are the same raw materials that plants need to make sugar, Figure 19-11.

**What three elements make up sugar?**

The general formula for sugar is $C_6H_{12}O_6$. This is a simple sugar called glucose. The three elements taken in by the plant are arranged in different ways in the leaf to make sugars. All sugars are made of carbon, hydrogen, and oxygen. During photosynthesis, plants first make glucose. This simple sugar is used by the plant to make other more complex carbohydrates such as starch. When food is moved around the plant, another sugar called sucrose is made. Sucrose is the same as your table sugar.

Leaves use six molecules of water and six molecules of carbon dioxide to make one molecule of sugar. How can a leaf cause molecules to change?

Remember how the carpenter used human and electric energy to build the house. Plants also use energy to make sugar. This energy comes from light, Figure 19-12.

The chemical equation for photosynthesis is:

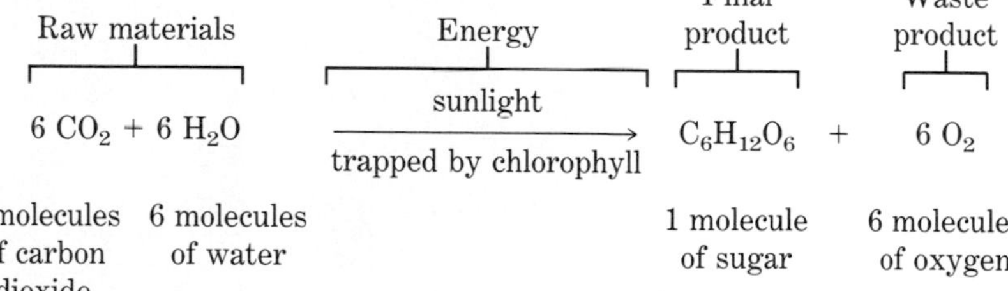




## STUDY GUIDE 112

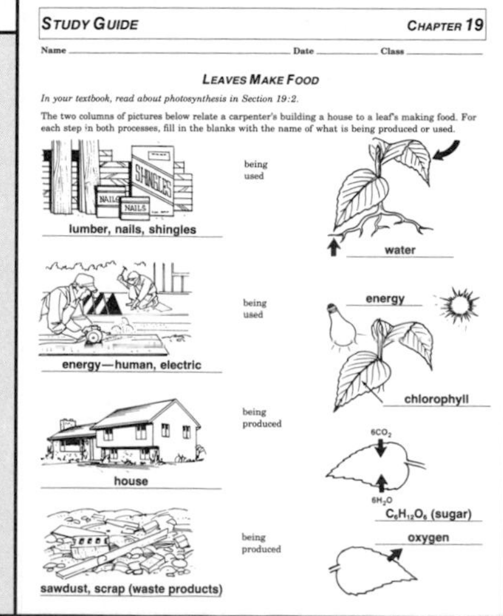
STUDY GUIDE CHAPTER 19

Name ______ Date ______ Class ______

LEAVES MAKE FOOD

*In your textbook, read about photosynthesis in Section 19:2.*

The two columns of pictures below relate a carpenter's building a house to a leaf's making food. For each step in both processes, fill in the blanks with the name of what is being produced or used.



## OPTIONS

### ACTIVITY/Challenge

Have students demonstrate that oxygen is made during photosynthesis. Have them place *Elodea* in a test tube filled with water and place the tube near a bright light source. After several minutes, students will see bubbles coming from the plant. Allow them to explain their observations.

## TEACH

### Concept Development

▶ Explain that photosynthesis occurs only in cells that have chlorophyll, and those cells are like factories.

### Connection: History

Have students investigate and report on Van Helmont's experiments with plant growth. Students can also make a time line to show how many discoveries led up to the current understanding we have about photosynthesis.

### Idea Map

Have students use the idea map as a study guide to the major concepts of this section.

### Guided Practice

Have students write down their answers to the margin question: carbon, hydrogen, and oxygen.

### Independent Practice

Study Guide/*Teacher Resource Package*, p. 112. Use the Study Guide worksheet shown at the bottom of this page for independent practice.

## GLENCOE TECHNOLOGY

**Videodisc**

**The Secret of Life**

*Photosynthesis Demonstration Segment*

*Photosynthesis*

## TEACH

### Concept Development

▶ Point out that plants use only about 1 percent of the light that strikes them during photosynthesis.

▶ Explain that leaves are too small to store all the food made during photosynthesis, but plants also store food in roots, stems, and seeds.

### MOTIVATION/Brainstorming

Point out that 90 percent of the photosynthesis in the world occurs in algae. Ask how oxygen gets from water to land.

### Check for Understanding

Have students make a parallel chart showing the processes of building a house and photosynthesis. Have them label the raw materials and products.

### Reteach

Have students make posters of the two processes, building a house and photosynthesis, and display them. Have some explain their posters.

### Guided Practice

Have students draw a plant and trace the paths of water and carbon dioxide through a plant.

### Skill Check

**Infer:** Plants growing beneath other plants sometimes turn yellow. Infer what will happen to these plants if they continue to be covered. *For more help, refer to the **Skill Handbook**, pages 706-711.*

Light from the sun or even light from a light bulb is the source of energy used by plants. How does a plant use light energy to make sugar? You have read in Chapter 6 that plants have chlorophyll. The chlorophyll in the leaves is able to trap light energy. Once trapped, the energy is used by the plant to make sugar.

The word formula for photosynthesis is:

$$\text{Carbon dioxide} + \text{Water} \xrightarrow[\text{by chlorophyll}]{\text{Light trapped}} \text{Sugar} + \text{Oxygen}$$

Oxygen is a waste product of photosynthesis. Just as the carpenter can recycle the waste products sawdust and wood scraps after building a house, a plant can recycle the oxygen from photosynthesis. Plants and other living things use the oxygen for cellular respiration.

## How Is Sugar Used?

Suppose you wanted to tear down an old house. Some of the materials in the old house could be reused to build other things such as a shed or garage. The sugar made during photosynthesis is also broken down and used to build other molecules for growth in the plant. When sugar is broken down, energy is released. Remember from Chapter 2 that this process is called cellular respiration. The energy released when sugar is broken down is used for life processes by all living things.

You have read that plants use carbon dioxide and water to make sugar, with oxygen as a waste product. In

a

b

**Figure 19-12** A carpenter uses electrical energy to build a house (a). Plants use light energy to make sugars (b).



## OPTIONS

Skill/*Teacher Resource Package*, p. 20. Use the Skill worksheet shown here and have students make a line graph on light and photosynthesis.

**SKILL 20**

**SKILL: MAKE AND USE A LINE GRAPH** CHAPTER 19

Name ____________ Date ________ Class ________

*For more help, refer to the Skill Handbook, pages 715-717.* *Use after Section 19:2.*

**LIGHT AND PHOTOSYNTHESIS**

Some students used the setup shown in Figure 1 to study photosynthesis. They put an Elodea plant in a jar of water and placed the jar 3 cm from a bright light. First, they observed bubbles begin to rise from the plant. The bubbles were oxygen that formed as a result of photosynthesis. The students counted the number of bubbles that formed in one minute and recorded what they saw. Next, the students moved the jar 6 cm from the light. After a few minutes, they counted the bubbles that formed in one minute. They continued to move the jar farther from the light and to record their observations. The table below shows the data they collected.

| Distance from light in cm | 3 | 6 | 9 | 12 | 15 | 18 | 21 | 24 |
|---|---|---|---|---|---|---|---|---|
| Number of bubbles each minute | 23 | 16 | 11 | 8 | 6 | 3 | 0 | 0 |

Number of bubbles each minute
0 2 4 6 8 10 12 14 16 18 20 22 24 26
0 2 2 6 8 10 12 14 16 18 20 22 24
Distance from light in cm

*Plot the data from the data table onto the grid above. Connect the data points with a ruler to make a line graph. Then, answer the questions below.*

1. What happened to the number of bubbles as the jar was moved farther from the light? Fewer oxygen bubbles were released.
2. Look at the line graph. How many oxygen bubbles are released if the distance between the jar and the light is 5 cm? 18 10 cm? 10 13 cm? 7
3. Explain how the experiment shows that light is needed for photosynthesis. The farther the jar is from the light, the less light the plant receives for photosynthesis. A smaller number of bubbles are released, indicating that less photosynthesis is occurring.

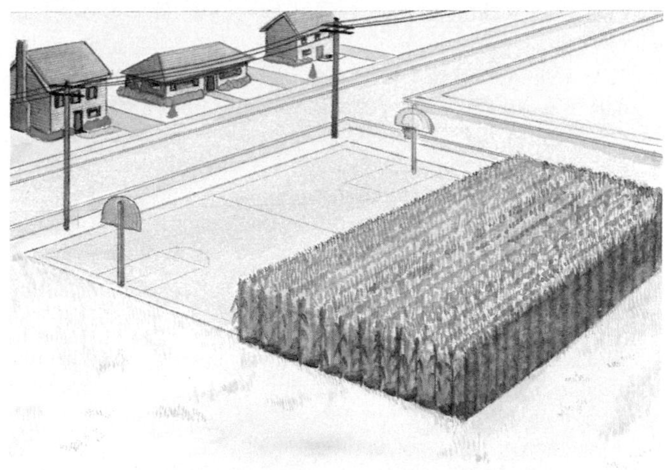

**Figure 19-13** If the energy stored in this field of corn could be converted to electrical energy, you and two of your neighbors would have enough electricity for your homes for one year.

respiration, plants and animals use oxygen to break down sugar. Carbon dioxide and water are the waste products. Notice in Figure 19-11 that the raw materials of photosynthesis are the waste products in cellular respiration. The raw materials for respiration are the products in photosynthesis.

Plants store large amounts of energy in the sugar they make during photosynthesis. Suppose plants could change the chemical energy in the food they make into electrical energy. In 100 days, a cornfield the size of a basketball court could make enough electricity in three homes to run their lights, appliances, and heating systems for one year! Plants are a major source of energy for all living things.

**How is energy stored in plants?**

## Check Your Understanding

6. What is the energy source for photosynthesis?
7. What chemical elements are found in sugar made by plants? What chemical elements are used by a plant for photosynthesis?
8. How is the sugar made during photosynthesis important to the plant and to animals?
9. **Critical Thinking:** Why can't animals use the energy from sunlight to make sugar?
10. **Biology and Math:** The formula for sugar is $C_6H_{12}O_6$. How many atoms of each kind are in the formula?



## TEACH

### Guided Practice

Have students write down their answers to the margin question: in the sugar they make.

### Check for Understanding

Have students respond to the first three questions in Check Your Understanding.

### Reteach

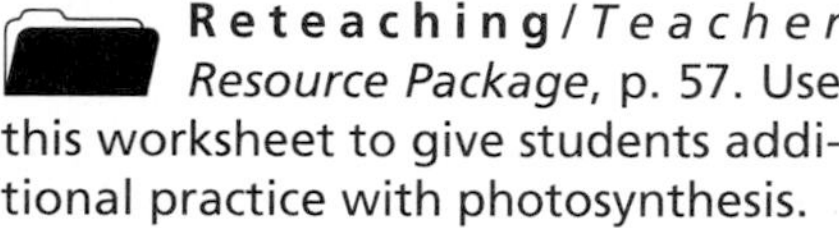

Reteaching/*Teacher Resource Package*, p. 57. Use this worksheet to give students additional practice with photosynthesis.

**Extension:** Assign Critical Thinking, Biology and Math, or some of the **OPTIONS** available with this lesson.

## 3 APPLY

### Divergent Question

Ask students to predict what would happen if all the plants on Earth were to die tomorrow.

## 4 CLOSE

### Software

*Photosynthesis and Transport,* Queue, Inc.

### Student Journal

Suppose you trace the path of a leaf miner in a leaf for several weeks. The tunnel always appears narrower where the miner started its tunnel and wider several weeks later. Have the students write how they would find out why the tunnel gets wider and place the written material in their journals.

### RETEACHING 57

**RETEACHING: IDEA MAP** CHAPTER 19

Name ________ Date ________ Class ________

*Use with Section 19:2.*

**PHOTOSYNTHESIS**

Use the following terms to help you complete the idea map: trapped by chlorophyll, carbon dioxide, sunlight, oxygen, water, sugar, waste, useful.

Photosynthesis
- energy
  - sunlight
  - trapped by chlorophyll
- raw materials
  - water
  - carbon dioxide
- products
  - useful — sugar
  - waste — oxygen

1. What is the useful product of photosynthesis? **sugar**
2. What are the raw materials of photosynthesis? **water and carbon dioxide**
3. What elements make up water? **hydrogen and oxygen**
4. What elements make up carbon dioxide? **carbon and oxygen**
5. How does chlorophyll aid in photosynthesis? **It helps trap sunlight.**

### Answers to Check Your Understanding

6. light from the sun
7. carbon, oxygen, and hydrogen; carbon, oxygen, and hydrogen
8. Plants break down sugar to release energy. When animals eat plants, they break down the sugar stored in the plants, releasing energy.
9. They don't have chloroplasts with chlorophyll.
10. There are 6 atoms of carbon, 12 atoms of hydrogen, and 6 atoms of oxygen.

## 19:3 Leaves for Food

## PREPARATION

### Materials Needed

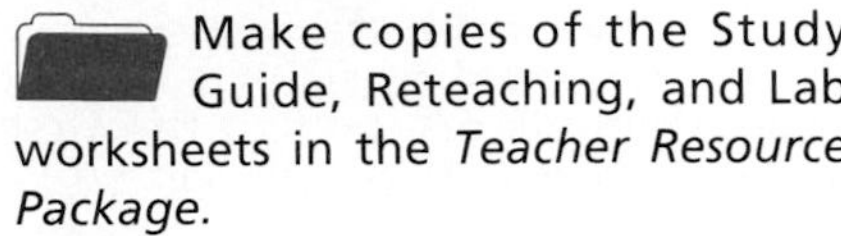

Make copies of the Study Guide, Reteaching, and Lab worksheets in the *Teacher Resource Package*.

▶ Gather plant leaves, coins, small jars, filter paper, liquid solvent, applicator sticks, and masking tape for Lab 19-2.

▶ Purchase produce for the Focus display.

### Key Science Words

No new science words will be introduced in this section.

### Process Skills

In Lab 19-2, students will experiment, observe, and infer.

## 1 FOCUS

▶ The objectives are listed on the student page. Remind students to preview these objectives as a guide to this numbered section.

### MOTIVATION/Display

Display some produce from the grocery store that has edible leaves. Have students tell you how they know the materials are leaves.

### Guided Practice

Have students write down their answers to the margin question: food and energy.

### Idea Map

Have students use the idea map as a study guide to the major concepts of this section.

## 19:3 Leaves for Food

**Objectives**

7. **Determine** why plants are important to other living things.
8. **List** different uses of leaves.
9. **Explain** why leaves change color.

You live on a planet that is sometimes called the green planet. Why? Flying a few hundred miles above Earth, you would see it as green. Much of the green color comes from plant leaves. The gases exchanged by all these leaves make Earth the perfect environment for humans and all other living things that need oxygen for respiration.

### Animals Depend on Plants

What do leaves provide to most living things?

Without photosynthesis, most living things would disappear from Earth. Mice, rabbits, owls, and fish as well as humans would die. Why? Leaves are the major food-producing organs of plants. Food produced by leaves and stored throughout a plant is eaten by animals. Plant eaters digest the plant food and use the released energy for all their life processes.

Many insects are consumers of plant tissues. Figure 19-14a shows insects called leaf miners eating their way through the inside of a leaf. Note the odd-shaped tunnels they make. The tunnels show where they have eaten through the inside of the leaf. Which cells of the leaf have they eaten? All the layers of cells with green chlorophyll have been eaten. These are the cells that produce sugar.

In Figure 19-14b notice the caterpillar eating a leaf. Does a caterpillar eat a leaf in the same way as a leaf miner? What cells are being eaten by the caterpillar?

Animals that eat plants are a source of energy for animals that feed only on other animals. Figure 19-14c shows this kind of consumer eating another animal. The

**Study Tip:** Use this idea map as a study guide to the importance of leaves. What are two ways that both humans and other animals depend on plants?

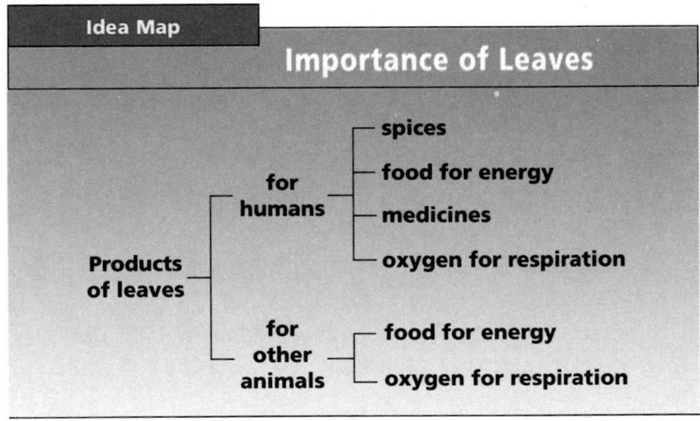




## OPTIONS

### Science Background

Leaves are important to all other living things on Earth because they are the food source for the plant eaters. The plant eaters are food for the animal eaters. Leaves also supply humans with food, medicines, and numerous products that can be extracted from leaves.

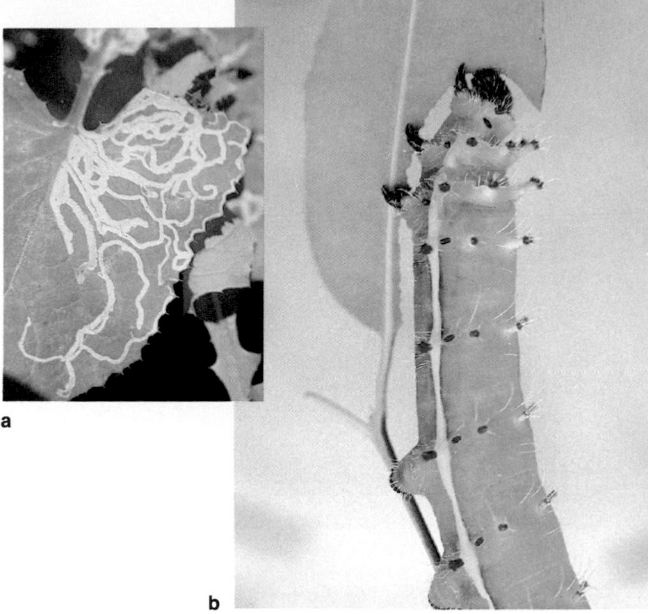

Figure 19-14 Leaf miners tunnel through leaves (a), caterpillars eat whole leaves (b), and birds eat insects that feed on leaves (c).

bird uses the energy from the caterpillar for its own growth and other life processes. The energy from the caterpillar came originally from the leaf of the plant. In turn, the leaf trapped the energy from sunlight. Where do you think your energy for running or cycling comes from? If you follow the chain of feeding back from your hamburger or hot dog you will come to a lot of plants. These plants trap the energy that you use each day.

What if all plants died? Can you predict what would happen to all the plant-eating consumers? They would die because the plants that they depend on for food would be dead. If there were no insects or other plant eaters, such as mice and rabbits, other consumers that eat animals would die. If there were no leaves, there would be less oxygen. Most living things depend on oxygen for respiration.

## Practical Uses of Leaves

Many animals, including humans, use leaves for food. You know that cattle, sheep, and other grazing animals eat the leaves of grass. What leaves have you eaten? You may have tried the leaves of cabbage, lettuce, spinach, or onions. In the produce section of a grocery store, you could make a list of many of the leaves people eat.

**Bio Tip**

**Leisure:** Develop a vegetable garden. You can grow tomatoes, lettuce, carrots, and radishes in an area less than ten square feet and with very little effort.



## 2 TEACH

**MOTIVATION/Brainstorming**

**Concept Development**

▶ Remind students that the palisade, guard, and spongy cells are the cells in a leaf that contain chlorophyll.

▶ Stress that plants are the beginning link in any food chain.

▶ Reinforce the fact that plants as well as animals need oxygen for respiration.

**GLENCOE TECHNOLOGY**

**Videodisc**

**Biology: The Dynamics of Life**
Movie: *How Organisms Interact*
Disc 1, Side 1, Ch. 4

Movie: *Deep Sea Rifts*
Disc 1, Side 1, Ch. 27

***OPTIONS***

**Where in a Leaf Does Photosynthesis Take Place?**/*Lab Manual,* pp. 161-164. Use this lab as an extension to ways in which animals depend on plants for food.

## TEACH

### ACTIVITY/Hands-On

Have students place sprouted bean plants in each of two small cups of soil. Add water and place one in a lighted area and the other in a dark area. Have students observe at daily intervals and record changes in the color of the leaves. The plants in the dark should develop yellow leaves.

### Check for Understanding

Have students explain the conditions that cause a leaf to turn from green to another color.

### Reteach

Divide students into teams and give each team a *Coleus* plant. Have them write a paragraph explaining why the plant has three different colors on the leaf.

### Independent Practice

Study Guide/*Teacher Resource Package,* p. 113. Use the Study Guide worksheet shown at the bottom of the page for independent practice.

### Guided Practice

Have students write down their answers to the margin question: to produce the chemical digitalis that is used to treat heart disease.

### *Portfolio*

Give students a list of common spices found in the grocery store. Tell them to put a check mark next to those spices that are made from leaves and tell how they are used in the food we eat. If you wish, you could have students visit ethnic food stores and generate their own lists.

**Figure 19-15** Leaves have many uses.

Some leaves are used for flavorings or for spices. If you drink tea, you are drinking boiled water in which tea leaves have been soaked. Peppermint and spearmint are two other flavorings from leaves. Sage, bay, and parsley are leaves used as spices in cooking. Looking at the spice shelves of a grocery store would help you think of other leaves used as spices. Which of the products from plant leaves in Figure 19-15 have you used?

**What are foxglove leaves used for?**

Many plant leaves contain drugs that are useful in treating diseases. The foxglove is a flowering plant that produces the chemical digitalis (dihj uh TAL us) used as a drug to treat heart disease. As you can see, leaves are very useful to humans.

## Changes in Leaves

Have you ever seen grass that has been growing under a board for several days? The grass may have changed color from green to yellow. What happened to the grass under the board? Grass leaves and many other plant leaves contain pigments other than chlorophyll. Two other common pigments are yellow and red in color. Usually leaves are green because the green pigments in chlorophyll cover up any other pigments present. Chlorophyll forms only when there is a supply of light. When the grass leaves were covered by the board, no chlorophyll was made and the green color was lost.



**STUDY GUIDE 113**

STUDY GUIDE CHAPTER 19

Name ______ Date ______ Class ______

LEAVES FOR FOOD

*In your textbook, read about leaf color changes in Section 19:3.*

1. These two diagrams show sections of a leaf. The left leaf has been in the light for 10 days. The right leaf has been in the dark for 10 days. Color the cells of the leaves either green or yellow depending on what they would look like to us. Color only those cells within the brackets. **Students should not color cells of vascular bundles.**

Light — all green
Dark — all yellow

2. Label the two diagrams with the leaf parts you can see. Use colored pencils to shade in what color a red oak leaf will be during each season shown.

Summer: stem, stalk, blade, midrib, smaller vein, green
Autumn: blade, stalk, midrib, stem, smaller vein, red

# Science and Society

## Farms of the Future

Growing plants without soil

Farmland that once was used to grow crops for food is now being sold for housing or industry. Less land for farming and growing crops could mean less food for humans. Scientists are now growing plants in the laboratory without soil. They found that as long as light was available, plants could grow while supported only in water that contained minerals. Can this technique of growing plants without soil be used to grow food crops for humans? One plant factory in Illinois grows plants such as lettuce and spinach in four weeks. Containers of sprouted seeds are placed on a slow-moving conveyer belt and given a certain amount of light, water, and minerals. Minerals are added as they are used by the plants. A sprout of lettuce uses 2.5 liters of water and the water is recycled. The same lettuce planted in the field would take six to nine weeks to grow and would use 25 liters of water.

### What Do You Think?

1. In an artificial environment, there are no pests, such as weeds or insects, to harm the crops. If the plants were outside, the farmer would have to use pesticides. Inside the factory, the farmer needs to supply the plants with heating, cooling, and lighting. All of these things need electricity. In the fields outside, the farmer would use the energy from the sun. Is it more economical to grow plants in factories or in fields?

Which apple would you eat?

2. If we can't afford to build more plant factories, will we be able to grow enough food? If you look around where you live, you might notice that there are more houses, hotels, and offices than there were ten years ago. Did this land once have farm crops? As the human population expands each year, we will build more houses and use up more crop land. What will we all eat? Will we eat more meat?

3. When you buy a head of lettuce, do you put it back if you see a bruise or leaves with holes? If you bought factory grown plants, they wouldn't have holes from insects. If you bought plants grown in the fields and not sprayed with pesticides, you would probably see some insect damage. How important is it for your vegetables to be perfect? Would you put up with insect-damaged fruits if you knew you were helping to reduce use of chemicals in the environment?

**Conclusion:** Could plant factories completely change farming all over the world?



## Suggested Readings


Adler, Tina. "Black-Eyed Peas Go to Mars." *Science News*, Dec. 2, 1995, pp. 376-377.

Collett, R.K. "Eyes on the Prize." *Flower and Garden Magazine*, August-September 1994, p. 74.

Hershey, David R. "Solution Culture Hydroponics." *The American Biology Teacher*, February 1994, p. 111.

Klinkenborg, Verlyn. "A Farming Revolution: Sustainable Agriculture." *National Geographic*, December 1995, pp. 60-88.


Science and Society

## Farms of the Future

### Background

Growing plants in greenhouses and without soil has been done for many decades and is called hydroponics. Growing them in a windowless building is a new approach. Light, minerals, and water are carefully controlled. Basil, bibb lettuce, and watercress are the three main crops produced.

### Teaching Suggestions

Have the students make a flow chart to show (1) how the raw materials are supplied for photosynthesis, and (2) how any of the raw materials are recycled. Have students find out how raw materials are supplied and recycled in a regular greenhouse. Have them compare a greenhouse with a hydroponics building.

### What Do You Think?

1. Answers will vary, but many will say that field-grown produce is more economical.
2. Answers will vary, but some will mention improving methods of food production to make better use of land. Point out that raising beef cattle and other meat sources also requires large tracts of land.
3. Answers will vary. Students who are more ecology-minded will want to reduce the use of chemicals.

**Conclusion:** Students may see these factories as an answer to areas with drought, little land suitable for farming, and severe insect problems. Have students consider the cost of building large factories and the number of factories necessary to supply world populations. Have them consider the amount and cost of electricity and minerals needed to grow the crops. Remind students that countries suffering from lack of food are already poor and probably could not afford these factories.

## TEACH

### Guided Practice

Have students write down their answers to the margin question: It breaks down.

### Independent Practice

**Study Guide**/*Teacher Resource Package,* p. 114. Use the Study Guide worksheet shown at the bottom of this page for independent practice.

### Check for Understanding

Have students respond to the first three questions in Check Your Understanding.

### Reteach

**Reteaching**/*Teacher Resource Package,* p. 58. Use this worksheet to give students additional practice with the features of leaves.

**Extension:** Assign Critical Thinking, Biology and Reading, or some of the **OPTIONS** available with this lesson.

## 3 APPLY

### ACTIVITY/Project

Have students list useful leaf products. Have them put product names on tag board strips.

## 4 CLOSE

### Summary

Use the tag board strips for review.

### Answers to Check Your Understanding

11. Leaves are the major organs of food producing. Animals eat plants and use the released energy for life processes.
12. for food, spices, flavorings, and medicines
13. Chlorophyll breaks down and other pigments show through.
14. Plants wouldn't photosynthesize, and all living things would die.
15. sage, bay, and parsley

a

b

**Figure 19-16** Some plants have pigments that make them show colors other than green (a). Maple trees in the northeastern parts of the United States give a spectacular show of colors in the autumn (b).

**What happens to chlorophyll in the autumn?**

Some leaves are always a color other than green. Examine the *Coleus* plant in Figure 19-16. How do you explain the colors of the *Coleus* leaves? Have you noticed other plants with colored leaves?

In the area where you live, the leaves of trees may change colors in autumn. When chlorophyll breaks down in autumn, the other pigments in the leaves can be seen. If a leaf contains red and yellow pigments, it will turn red or orange when the chlorophyll is gone.

Why do leaves change their colors in autumn? In the fall, the flow of sap in the stem slows down. The temperature drops and there are fewer hours of sunlight in the autumn. With fewer nutrients and less light, chlorophyll breaks down and the other colored pigments in the leaves show through.

### Check Your Understanding

11. Why would most living things disappear from Earth if plants didn't undergo photosynthesis?
12. In what ways are leaves useful to people?
13. Why do some leaves change color in autumn?
14. **Critical Thinking:** If Earth's climate changed so that there was always a thick cover of clouds, what might happen to plants and other living things?
15. **Biology and Reading:** List three leaves that you read about that are used as spices.

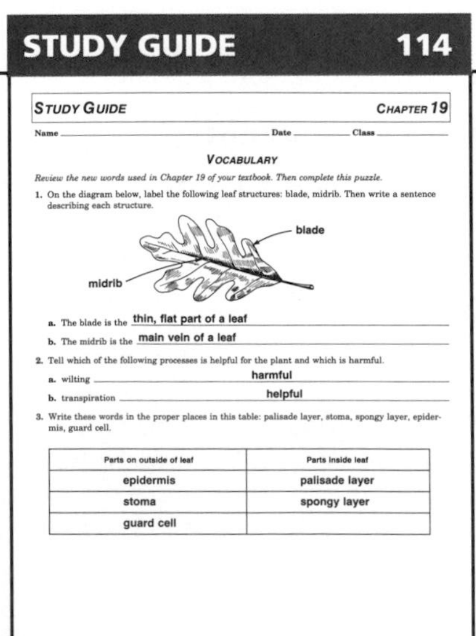

STUDY GUIDE 114

STUDY GUIDE — CHAPTER 19

Name ______ Date ______ Class ______

VOCABULARY

*Review the new words used in Chapter 19 of your textbook. Then complete this puzzle.*

1. On the diagram below, label the following leaf structures: blade, midrib. Then write a sentence describing each structure.



a. The blade is the thin, flat part of a leaf

b. The midrib is the main vein of a leaf

2. Tell which of the following processes is helpful for the plant and which is harmful.

a. wilting harmful

b. transpiration helpful

3. Write these words in the proper places in this table: palisade layer, stoma, spongy layer, epidermis, guard cell.

| Parts on outside of leaf | Parts inside leaf |
|---|---|
| epidermis | palisade layer |
| stoma | spongy layer |
| guard cell | |

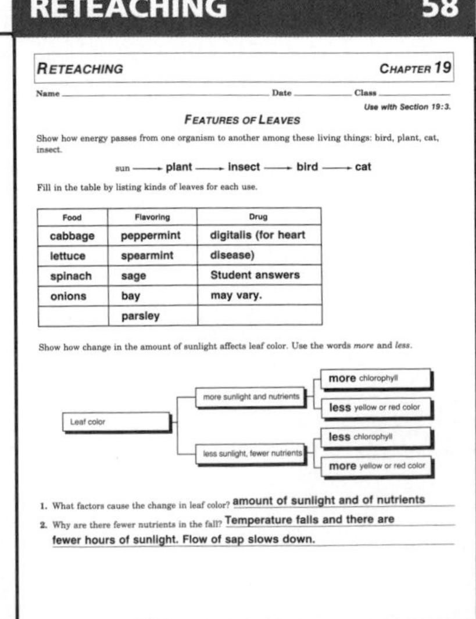

RETEACHING 58

RETEACHING — CHAPTER 19

Name ______ Date ______ Class ______

*Use with Section 19:3.*

FEATURES OF LEAVES

Show how energy passes from one organism to another among these living things: bird, plant, cat, insect.

sun → plant → insect → bird → cat

Fill in the table by listing kinds of leaves for each use.

| Food | Flavoring | Drug |
|---|---|---|
| cabbage | peppermint | digitalis (for heart |
| lettuce | spearmint | disease) |
| spinach | sage | Student answers |
| onions | bay | may vary. |
| | parsley | |

Show how change in the amount of sunlight affects leaf color. Use the words *more* and *less.*



1. What factors cause the change in leaf color? amount of sunlight and of nutrients
2. Why are there fewer nutrients in the fall? Temperature falls and there are fewer hours of sunlight. Flow of sap slows down.

# Lab 19–2 Pigments

## Problem: What pigments are found in a leaf?

### Skills

experiment, observe, infer

### Materials

plant leaf
coin
small jar
metric ruler
filter paper strip (13 x 2 cm)
liquid solvent
applicator stick
masking tape



### Procedure

1. Lay the leaf across the filter paper, 2 cm from one end. Rub the coin back and forth over the leaf as shown in Figure A. This action will make a dark green line at the 2-cm mark.
2. Pour the liquid solvent into the jar up to a depth of 1 cm. **CAUTION:** *Fumes from liquid solvent are dangerous. Do not breathe in the fumes. Do not use near a flame. Use a fume hood if you have one.*
3. Attach the strip of filter paper to the stick with masking tape, as shown in Figure B. The end with the leaf rubbing should hang down.
4. Hang the end of the paper strip with the leaf-rubbing line on it down into the solvent as shown in Figure B. Make sure the solvent level does not come above the green line. The paper strip may touch the bottom of the jar as long as the solvent level does not cover the leaf-rubbing line.
5. **Experiment:** Allow solvent to move up the paper until it is 2 cm from the top. Watch it carefully so that it doesn't run off the top of the strip.
6. Remove the paper and allow it to dry. Use a fume hood if you have one.
7. Dispose of the solvent as indicated by your teacher.

### Data and Observation

1. What color was the paper strip at the leaf-rubbing line before you put the strip in the solvent?
2. What happened to the color of the paper strip at the leaf-rubbing line when the solvent reached it?
3. What colors showed up on the paper strip between the starting and ending lines?

### Analyze and Apply

1. Did you see all of the colors the leaf contained before the paper strip was put in the solvent? Explain your answer.
2. **Infer:** Is there more than one shade of any color? If so, what might this mean?
3. Do you think chlorophyll was present? How do you know?
4. **Apply:** How could you find out whether other leaves have the same colors as the one you used in this activity?

### Extension

**Experiment** with a multi-colored leaf and compare the results with those here.



## *ANSWERS*

### Data and Observations

1. dark green
2. It began to break up; feather, and move up with the solvent.
3. Answers will depend on leaf used. Many will show green and yellow.

### Analyze and Apply

1. No, the different colors separated out after the solvent passed through them.
2. Yes, there are probably more than one kind of chlorophyll and of other pigments.
3. Yes, one of the pigment colors was green. Chlorophyll is a green pigment.
4. Carry out the same experiment using different leaves.

## Lab 19-2 Pigments

### Overview

In this lab, students will experiment to find out what color pigments are found in a leaf.

**Objectives:** Completing the lab, students will be able to: (1) **determine** the number of pigments in a leaf, (2) **analyze** the pigments in a leaf.

**Time Allotment:** 45 minutes

### Preparation

▶ Use rolls of filter paper, 2 cm wide.

▶ Most labs will not have fume hoods, so you may want to pour the solvent in the chemistry lab hood prior to the class period.

▶ To avoid pressure buildup on the stopper, don't stopper or cap too tightly.

▶ **Alternate Materials:** You may wish to substitute fingernail polish remover for the solvent.

**Lab 19-2 worksheet**/*Teacher Resource Package,* pp. 75-76.

### Teaching the Lab

▶ As the alcohol goes up the paper strip the students will see different color bands of pigments.

▶ **Troubleshooting:** Make sure that the leaf rubbing is above the solvent level.

**Cooperative Learning:** Divide the class into teams of two students. For more information, see pp. 22T–23T in the Teacher Guide.

### ASSESSMENT

**Skill:** Have students repeat the activity using ethyl alcohol in place of the ether-acetone solvent. Have them compare the color patterns on both filter-paper strips and write an explanation for the similarities and differences in the patterns.

# CHAPTER 19 REVIEW

## Summary

Summary statements can be used by students to review the major concepts of the chapter.

## Key Science Words

All boldfaced terms from the chapter are listed.

## Testing Yourself

### Using Words

1. palisade layer
2. stoma
3. blade
4. wilting
5. transpiration
6. midrib
7. epidermis

### Finding Main Ideas

8. p. 407; Plants need carbon, oxygen, hydrogen, and energy to make sugar.
9. p. 407; Carbon dioxide and water are raw materials that are changed into sugar.
10. p. 403; Wilting is when a plant loses water too fast.
11. p. 408; The products of photosynthesis are sugar and oxygen.
12. pp. 407-408; Sugar is broken down and energy is released during cellular respiration.
13. pp. 411-412; Leaves are used by people for food, spices, and medicines.
14. p. 410; Plants are the major source of energy for all living things.

### Using Main Ideas

15. It is broken down and energy is released.
16. The layers of leaf cells that carry on photosynthesis are the palisade and spongy layers.
17. Both are processes in which a plant loses water.
18. The chemical digitalis is made from foxglove.
19. Leaves can be identified by their edges, size, shape, and the way they are arranged on a stem.
20. The elements carbon, oxygen, and hydrogen make up sugar.



# Chapter 19

# Review

## Summary

### 19:1 The Structure of Leaves

1. Leaves are the main plant parts that make food. They are usually green and are made up of a stalk, a blade, a midrib, and veins. Leaves differ in arrangement, shape, size, and leaf edges.
2. Leaf cells form an epidermis, palisade layer, spongy layer, xylem, and phloem.
3. Plants lose water by transpiration. Wilting occurs when water is lost faster than it is replaced.

### 19:2 Leaves Make Food

4. In photosynthesis, leaves with chlorophyll use sunlight energy, carbon dioxide, and water to make sugar. This sugar is used as food by living things.
5. The raw materials for photosynthesis are carbon dioxide and water. The products are sugar and oxygen, a waste product. The energy source is sunlight, which is trapped by chlorophyll.
6. Sugar is broken down by cellular respiration into carbon dioxide and water. A large amount of energy is released during respiration.

### 19:3 Leaves for Food

7. Plants are necessary to supply animals, including humans, with food and oxygen.
8. Leaves are used by people for flavorings, spices, foods, and medicines.
9. Less light and cooler temperatures in autumn cause less chlorophyll to be made in leaves.

## Key Science Words

blade (p. 398)
epidermis (p. 401)
guard cell (p. 402)
midrib (p. 399)
palisade layer (p. 402)
spongy layer (p. 402)
stoma (p. 402)
transpiration (p. 403)
wilting (p. 403)

## Testing Yourself

### Using Words

*Choose the word from the list of Key Science Words that best fits the definition.*

1. layer of long, thin cells in a leaf
2. small pore in a leaf's epidermis
3. thin, flat part of a leaf
4. process in which plants lose water faster than it can be replaced
5. process in which plants lose water

# Review

## Testing Yourself *continued*

6. main vein of a leaf
7. cells on the surface of a leaf blade

**Finding Main Ideas**

*List the page number where each main idea below is found. Then, explain each main idea.*

8. what plants need to make sugar
9. what raw materials are changed into sugar in a leaf
10. the result of plants losing water too fast
11. the products of photosynthesis
12. what happens to the sugar made during photosynthesis
13. the uses people have for leaves
14. why animals depend on plants

**Using Main Ideas**

*Answer the questions by referring to the page number after each question.*

15. What happens to the sugar that a plant makes? (p. 408)
16. What two layers of leaf cells can carry on photosynthesis? (p. 402)
17. What is the relationship between wilting and transpiration? (p. 403)
18. What is the drug from foxglove leaves? (p. 412)
19. What features can you use to identify different kinds of leaves? (p. 398)
20. What elements make up sugar? (p. 407)
21. How are plants, insects, and birds connected by the process of photosynthesis? (p. 410)
22. What are the raw materials needed for photosynthesis? (p. 407)
23. Other than raw materials, what else do leaves need for photosynthesis? (p. 408)
24. Why do leaves change color? (p. 414)

## Skill Review 

*For more help, refer to the* ***Skill Handbook,*** *pages 704-719.*

1. **Classify:** Classify the leaves of plants near your home by shape, edge, and vein pattern.
2. **Measure in SI:** Grow a bean plant from a seed. Measure the volume of water that you use to water the plant each week. After four weeks, find the total volume of water the plant was given. Be careful not to overwater the plant.
3. **Understand science words:** Use a dictionary to find the meaning of the word part *derm*. Find two other words that have this word part.
4. **Interpret diagrams:** Look at Figure 19-6. Which of the drawings shows the stoma in cross section?

## Finding Out More

**Critical Thinking**

1. If plants require water for photosynthesis, how is it possible for desert plants to make food during the dry season?
2. Explain why the number of humans that can live on Earth is related to the rate of photosynthesis in plants.

**Applications**

1. Make a collection of leaves. Press the leaves in a catalog for two weeks, then use a field guide to identify them.
2. Visit the produce and spice areas of a grocery store. List the leaves or leaf parts that are for sale as food.



### Using Main Ideas

21. Insects gain energy from eating plants, and birds gain energy from eating insects.
22. Carbon, oxygen, and hydrogen are raw materials needed for photosynthesis.
23. Leaves use energy from light in photosynthesis.
24. In the fall, there is less light and chlorophyll breaks down, allowing other pigments to show through.

### Skill Review

1. Students may want to group leaves within each classification.
2. Students will need to make a chart to keep track of the amount of water used.
3. *Dermis* means skin. Dermatosis is a disease of the skin, *dermal* means derived from skin.
4. The lower drawing shows a cross section through a stoma.

### Finding Out More

#### Critical Thinking

1. Desert plants store water for use during dry seasons.
2. Plants produce oxygen that is necessary for human respiration.

#### Applications

1. Students can collect grass and other small plant leaves as well as tree leaves.
2. Students will find that the green spices come from leaves, but other colors of spices come from different plant parts.

**Chapter Review/** *Teacher Resource Package*, pp. 99-100.

**Chapter Test/***Teacher Resource Package*, pp. 101-103.

**Chapter Test/***Computer Test Bank*

# Plant Support and Transport

## PLANNING GUIDE

| CONTENT | TEXT FEATURES | TEACHER RESOURCE PACKAGE | OTHER COMPONENTS |
|---|---|---|---|
| (1 1/2 days)<br>20:1 Stem Structure<br>Herbaceous Stems<br>Woody Stems<br>Stem Growth | Mini Lab: *How Can You Find the Age of a Woody Stem?* p. 423<br>Career Close-Up: *Nursery Worker,* p. 422<br>Check Your Understanding, p. 423 | Enrichment: *How a Twig Grows,* p. 21<br>Reteaching: *Important Words for Stems,* p. 59<br>Skill: *Interpret Diagrams,* p. 21<br>Focus: *The History of a Tree's Growth,* p. 39<br>Study Guide: *Stem Structure,* pp. 115-116 | **Laboratory Manual:**<br>*What Does a Woody Stem Look Like Inside?* p. 165<br>Color Transparency 20: *Cells of a Woody Stem*<br>**STVS:** *Oil from Wood,* Plants and Simple Organisms (Disc 4, Side 2) |
| (1 1/2 days)<br>20:2 The Jobs of Stems<br>Transport<br>Storage<br>How Are Stems Used? | Skill Check: *Make and Use Tables,* p. 427<br>Technology: *Newer Medicines From Plants,* p. 426<br>Check Your Understanding, p. 427 | Reteaching: *The Jobs of Stems,* p. 60<br>Transparency Master: *Cells in Herbaceous Stems,* p. 93<br>Study Guide: *The Job of Stems,* p. 117 | **STVS:** *Sound of Thirsty Plants,* Plants and Simple Organisms (Disc 4, Side 2) |
| (1 day)<br>20:3 Root Structure<br>Taproots and Fibrous Roots<br>Cells of a Root<br>Root Growth | Skill Check: *Understand Science Words,* p. 429<br>Check Your Understanding, p. 430 | Reteaching: *Cells in a Root,* p. 60<br>Study Guide: *Root Structure,* p. 118 | **Laboratory Manual:**<br>*What Do the Inside Parts of a Root Look Like?* p. 169 |
| (2 days)<br>20:4 The Jobs of Roots<br>Absorption<br>Anchorage and Storage<br>How Are Roots Used? | Skill Check: *Sequence,* p. 431<br>Lab 20-1: *Absorption,* p. 432<br>Lab 20-2: *Storage,* p. 434<br>Idea Map, p. 433<br>Check Your Understanding, p. 435 | Application: *Importance of Stems and Roots,* p. 20<br>Reteaching: *The Jobs of Roots,* p. 62<br>Critical Thinking/Problem Solving: *What Caused the Trees to Die?* p. 20<br>Study Guide: *The Job of Roots,* p. 119<br>Study Guide: *Vocabulary,* p. 120<br>Lab 20-1: *Absorption,* pp. 77-78<br>Lab 20-2: *Storage,* pp. 79-80 | **STVS:** *Growing Plants in Space,* Plants and Simple Organisms (Disc 4, Side 2) |
| Chapter Review | Summary<br>Key Science Words<br>Testing Yourself<br>Finding Main Ideas<br>Using Main Ideas<br>Skill Review | **ASSESSMENT RESOURCES**<br>Chapter Review, pp. 104-105<br>Chapter Test, pp. 106-108<br>Performance Assessment in the Biology Classroom<br>Alternate Assessment in the Science Classroom<br>Computer Test Bank | |

## GLENCOE TECHNOLOGY

**Science and Technology Videodisc Series,** *Oil from Wood,* Plants and Simple Organisms (Disc 4, Side 2)
*Sound of Thirsty Plants,* Plants and Simple Organisms (Disc 4, Side 2)
*Growing Plants in Space,* Plants and Simple Organisms (Disc 4, Side 2)

## MATERIALS NEEDED

| LAB 20-1, p. 432 | LAB 20-2, p. 434 | MARGIN FEATURES |
|---|---|---|
| pencil<br>paper towel strips:<br>1–10 x 15 cm<br>4–1 x 10 cm<br>stapler<br>large plastic cup | hand lens<br>carrot root<br>yam root<br>radish root<br>razor blade<br>3 petri dish halves<br>iodine solution<br>dropper<br>forceps | Skill Check, p. 427<br>pencil<br>paper<br>Skill Check, p. 429<br>dictionary<br>Skill Check, p. 431<br>pencil<br>paper<br>Mini Lab, p. 423<br>tree stump or log |

## OBJECTIVES

For more information about National Science Standards, see page 5T.

| SECTION | OBJECTIVE | CORRELATION of QUESTIONS to OBJECTIVES: CHECK YOUR UNDERSTANDING | CHAPTER REVIEW | TRP CHAPTER REVIEW | TRP CHAPTER TEST |
|---|---|---|---|---|---|
| 20:1 National Science Stds: UCP.1, UCP.2, UCP.3, UCP.4, UCP.5, C.1 | 1. **Discuss** the traits of herbaceous stems. | 3 | 8, 22 | 27 | 5, 8, 53 |
| | 2. **Describe** the five layers of a woody stem. | 1 | 1, 19 | 5, 7, 8, 11, 15, 16, 31 | 13, 14, 23, 24, 25, 27, 28, 29, 30, 31, 32, 46, 47, 48, 49, 50, 51 |
| | 3. **Explain** how stems grow. | 2 | 5, 6, 7, 18 | 9, 13, 14, 32, 34 | 10, 11, 12 |
| 20:2 National Science Stds: UCP.1, UCP.2, UCP.3, UCP.4, UCP.5, B.4, C.4 | 4. **Explain** how a stem serves as a transport system. | 7, 9 | 13, 19 | 12 | 26 |
| | 5. **Discuss** the storage function of stems. | 6 | 11, 12, 24 | 10 | 7 |
| | 6. **Identify** how stems are useful to humans. | 8 | 14 | 6 | 9 |
| 20:3 National Science Stds: UCP.1, UCP.2, UCP.3, UCP.5, G.1 | 7. **Compare** taproots and fibrous roots. | 11, 15 | 2 | 28, 33 | 2, 15, 16, 17, 18, 19, 20, 21, 22 |
| | 8. **Sequence** the layers of root cells. | 12, 14 | 4, 10, 20 | 17, 18, 19, 20, 21, 22, 26, 30 | 33, 34, 35, 36, 37, 38, 39, 40, 41, 42, 43, 44, 45, 52 |
| | 9. **Describe** how roots grow. | 13 | 16, 17 | 1 | 6 |
| 20:4 National Science Stds: UCP.1, UCP.5, B.4, C.4, F.3 | 10. **Describe** root hairs and their function. | 17 | 3, 15 | 4, 23, 29 | 3 |
| | 11. **Relate** how roots anchor the plant and serve as storage. | 16 | 23 | 24, 25 | 1 |
| | 12. **Discuss** the benefits and problems of roots. | 18, 20 | 21 | 2, 3 | 4 |

# Plant Support and Transport

## CHAPTER OVERVIEW

### Key Concepts

This chapter emphasizes the structures and functions of roots and stems. The chapter points out that some cells found in the roots are continuous throughout the stem and leaves. The economic importance of roots and stems to humans is also presented.

### Key Science Words

| | |
|---|---|
| annual ring | herbaceous stem |
| cambium | lateral bud |
| cork | root hair |
| cortex | taproot |
| endodermis | terminal bud |
| fibrous root | woody stem |

### Skill Development

In Lab 20-1, students will **observe, formulate a model, infer,** and **design an experiment** in relation to root hair function. In Lab 20-2, they will **observe, classify, form hypotheses,** and **design an experiment** while studying starch storage in roots. In the Skill Check on page 427, students will **make and use tables** of plants used by humans. In the Skill Check on page 429, they will **understand** the **science word** *fibrous root.* In the Skill Check on page 431, they will **sequence** plant parts through which water passes. In the Mini Lab on page 423, students will **use numbers** to find the age of a woody stem.

### Bridging

In Chapter 19, students learned about the features used to classify leaves, and the parts and functions of the external and internal parts of leaves. In this chapter, they will learn about the external and internal parts of stems and roots.

**CHAPTER PREVIEW**

### Chapter Content

Review this outline for Chapter 20 before you read the chapter.



### Skills in this Chapter

The skills that you will use in this chapter are listed below.

- In **Lab 20-1,** you will observe, formulate a model, infer, and design an experiment. In **Lab 20-2,** you will observe, classify, form hypotheses, and design an experiment.
- In the **Skill Checks,** you will sequence, make and use tables, and understand science words.
- In the **Mini Labs,** you will use numbers.

418

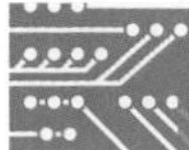

## TECH PREP

For Tech Prep activities in this chapter of the Teacher Wraparound Edition, see especially the Activities on pages 423, 427, and 433, the Motivation on page 425, the Portfolio and Options Activity on page 426, the Student Journal on page 431, and the Connection on page 433.

See also the Glencoe Homepage at **www.glencoe.com**

The following Glencoe resources provide additional opportunities for integrating science and technology.

**Technology: Science and Math in Action, Book One**
Module 1: Hydroponics

Chapter 20

# Plant Support and Transport

Plants, like animals, support themselves. Unlike animals, plants have no skeleton. How do you suppose the plants in the photo on the left stay upright without a skeleton? Which plant part supports them?

You may have noticed how pipes supply your home with water and gas. Look at the water pipes in the photo below. Like these pipes, the cells of a plant carry needed materials to all parts of the plant. Which parts of a plant supply the plant with these materials? Look again at the picture on the facing page. How do you suppose the plants get some of their needed materials? Can you tell where the root stops and the stem begins? How is a root different from a stem?

**Try This!**

**Do roots of plants grow only in soil?** Find a plant that grows up the sides of buildings. Some of these plants are called vines. Find out which part of the vine attaches to the building.

**interNET CONNECTION**

For more information about the material in this chapter, follow the link for the chapter on the Glencoe Homepage at **http://www.glencoe.com**

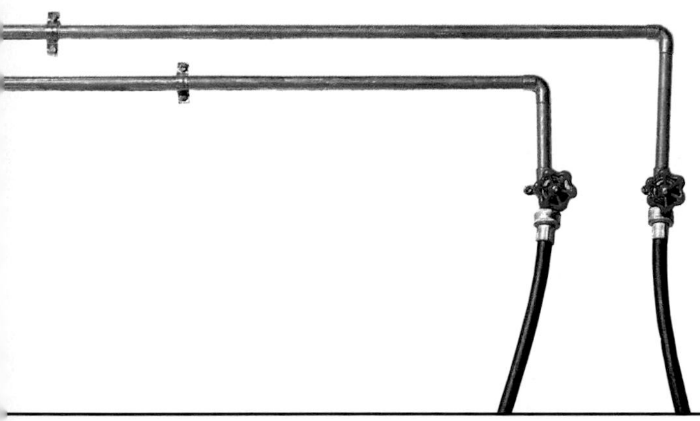

Water pipes supply a home.



**GETTING STARTED**

### Using the Photos

Students should be able to relate how the roots anchor the plants in soil and how the stems support the leaves. Point out that larger plants need a larger root system to carry out the plant's functions. Tell them that the roots spread out in soil as far as the leaves spread out above the ground.

### MOTIVATION/Try This!

**Do roots of plants grow only in soil?** Students will **observe** that some vines that grow up walls send out roots from the sides of the stems. These are called adventitious roots. Examples of plants with adventitious roots are ivy and grape vines.

### Chapter Preview

Have students study the chapter outline before they begin to read the chapter. They should note the overall organization of the chapter, that the first topic is stem structure followed by the jobs of stems and how stems are used by humans. These topics are followed by root structures, the jobs of roots, and lastly, by how roots are important to humans.

### Misconception

Students may believe falsely that sap rises in a tree in the spring and goes down to the roots in the fall. There is water in a stem all year.

**ASSESSMENT PLANNER**

***Portfolio***

Strategies on the following pages represent student products that can be placed into a best-work portfolio: pp. 426, 429.

**PERFORMANCE ASSESSMENT**

Skill Check, pp. 427, 429, 431
Mini Lab, p. 423
Lab, pp. 432, 434

**CONTENT ASSESSMENT**

Check for Understanding, pp. 423, 425, 427, 430, 435
Chapter Review, pp. 436-437

**GROUP ASSESSMENT**

Opportunities for group assessment occur with Cooperative Learning Strategies.

***Student Journal***

Strategies on the following pages represent opportunities for writing in a Student Journal: pp. 421, 428, 431.

## 20:1 Stem Structure

## PREPARATION

### Materials Needed

Make copies of the Focus, Enrichment, Skill, Reteaching, and Study Guide worksheets in the *Teacher Resource Package.*

▶ Obtain cornstalks for the demonstration activity.

▶ Obtain a tree stump or fireplace log for the Mini Lab.

### Key Science Words

| | |
|---|---|
| herbaceous stem | cambium |
| cortex | terminal bud |
| woody stem | lateral bud |
| cork | annual ring |

### Process Skills

In the Mini Lab, students will use numbers.

## 1 FOCUS

▶ The objectives are listed on the student page. Remind students to preview these objectives as a guide to this numbered section.

### ACTIVITY/Brainstorming

Ask students to give you the names of several plants they have seen on their way to school. Tell them to group their plants into two groups: those with green stems and those with bark. Ask them which plants are taller, those with green stems or those with bark. Ask them why they think one type of plant may grow taller than the other.

**Focus**/*Teacher Resource Package*, pp. 39-40. Use the Focus transparency shown at the bottom of this page as an introduction to woody stems.

### Guided Practice

Have students write down their answers to the margin question: they don't have enough xylem cells.

## 20:1 Stem Structure

**Objectives**

1. **Discuss** the traits of herbaceous stems.
2. **Describe** the five layers of woody stems.
3. **Explain** how stems grow.

**Key Science Words**

herbaceous stem
cortex
woody stem
cork
cambium
terminal bud
lateral bud
annual ring

A stem is a plant part that supports the plant and transports materials. Roots are on one end of a stem and leaves are attached along the stem's length. The external parts of stems have features that help identify plants. Think how the stem of a daisy differs from the stem of an oak tree.

### Herbaceous Stems

**Why aren't herbaceous stems tall?**

There are two stem types found in plants. They are either woody or herbaceous (hur BAY shus). **Herbaceous stems** are soft, green stems. Plants with herbaceous stems usually grow no taller than two meters. If they grew taller, they would probably fall over. Why? Herbaceous stems don't have many xylem cells. Xylem cells help support a plant stem. Recall from Chapter 6 that xylem tissue has tubelike cells that carry water and minerals throughout a plant.

Bean plants have herbaceous stems. If you slice the stem of a bean plant and look at a section under a microscope, it looks like Figure 20-1a. Many herbaceous vegetables have this pattern. Notice how the bundles of xylem and phloem cells are arranged in a circle. Other herbaceous stems, such as corn, lily, and wheat, have scattered bundles of xylem and phloem, as in Figure 20-1b.

Notice the cortex in the two herbaceous stems. The **cortex** is a food storage tissue in plants. Food is usually stored in plants in the form of starch. The epidermis of a herbaceous stem makes up its outer covering.

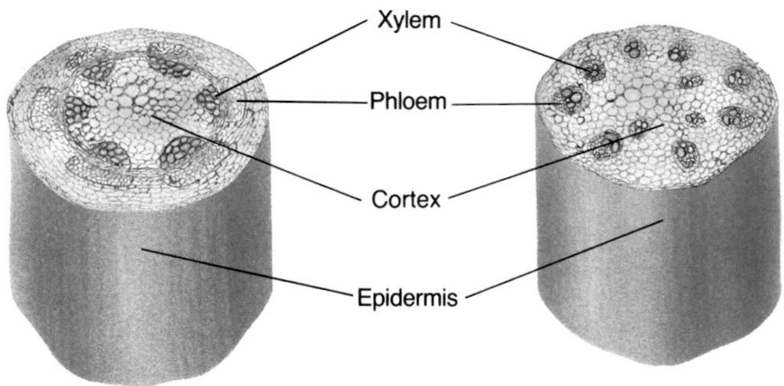


**Figure 20-1** There are two patterns of xylem and phloem bundles in herbaceous stems.



## OPTIONS

### Science Background

There are two kinds of stems, woody and herbaceous. The stems of herbaceous plants are soft and green, while woody stems are hard and not green. Some woody stems can grow to a height of 100 meters, while most herbaceous stems grow to a height of 0.5-2 meters. As they grow, stems increase in length and thickness.

**Cells of a Woody Stem**/*Transparency Package*, number 20. Use color transparency number 20 as you teach stem structure.

**FOCUS 39**

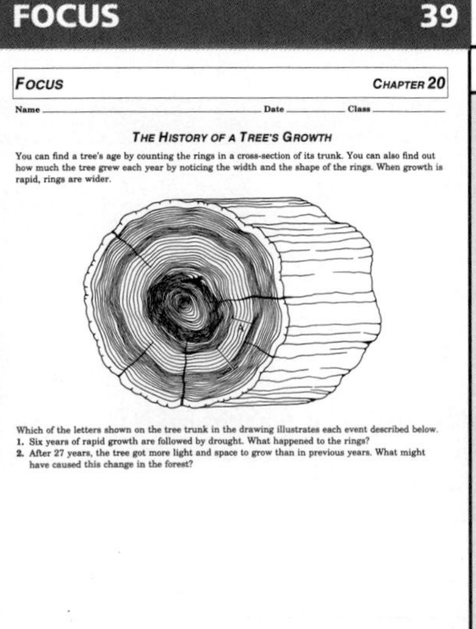

*Focus* — *Chapter 20*

Name ______ Date ______ Class ______

*The History of a Tree's Growth*

You can find a tree's age by counting the rings in a cross-section of its trunk. You can also find out how much the tree grew each year by noticing the width and the shape of the rings. When growth is rapid, rings are wider.

Which of the letters shown on the tree trunk in the drawing illustrates each event described below.

1. Six years of rapid growth are followed by drought. What happened to the rings?
2. After 27 years, the tree got more light and space to grow than in previous years. What might have caused this change in the forest?

## Woody Stems

Trees and shrubs have woody stems. A **woody stem** is a nongreen stem that grows to be thick and hard. The rough outer covering of a woody stem is called the bark. Bark protects a woody stem better than epidermis protects a herbaceous stem. If you ever look at a tree stump, look for the dark outer covering. This is the bark. Many trees can be identified by the traits of their bark. The bark of some trees is very smooth, Figure 20-2. Other trees have rough bark. The bark of some trees is scaly.

A woody stem is made up of five different cell layers, or tissues. Each layer has a certain job. By following the numbers in Figure 20-3, you will be able to see where these layers are found.

1. The outer layer of a woody stem is the **cork.** Cork is made of dead cells and protects the stem from insects, disease, and water loss.
2. Inside the cork layer is the layer of cells called the cortex. Remember that cortex cells store food.
3. Inside the cortex is a ring of phloem cells. Recall that phloem carries food from leaves to all parts of the plant.
4. The next layer is the cambium. **Cambium** is a thin layer of cells that divides to form new phloem on the outside and new xylem on the inside. Each year, the cambium produces new xylem and phloem cells that make the stem thicker. Thus, woody stems grow wider with age. The four layers—cork, cortex, phloem, and cambium—together make up the bark of a woody stem.

**Figure 20-2** Beech trees can be identified by their smooth bark.

What is the outer layer of a woody stem?

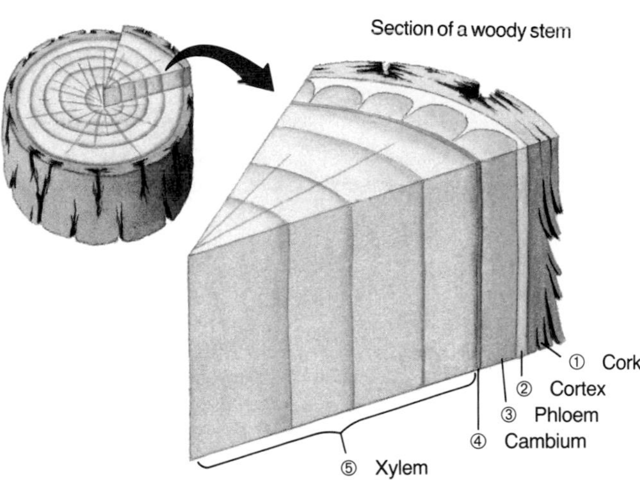


**Figure 20-3** A woody stem is made of five cell layers. How many layers make up bark? four



# 2 TEACH

### MOTIVATION/Demonstration

Obtain some dry cornstalks in the fall. You can use these to demonstrate strings or veins in plants.

### Guided Practice

Have students write down their answers to the margin question: bark.

### ACTIVITY/Display

Make photocopies of the cross sections of pieces of small tree trunks or limbs. These can be put on display with the question "Can you tell the age of these woody stems?"

### Independent Practice

**Study Guide**/*Teacher Resource Package*, p. 115. Use the Study Guide worksheet shown at the bottom of this page for independent practice.

### Student Journal

Have students reread Section 20:1 and make an idea map that compares cells and their functions in woody and herbaceous stems.

**SKILL 21**

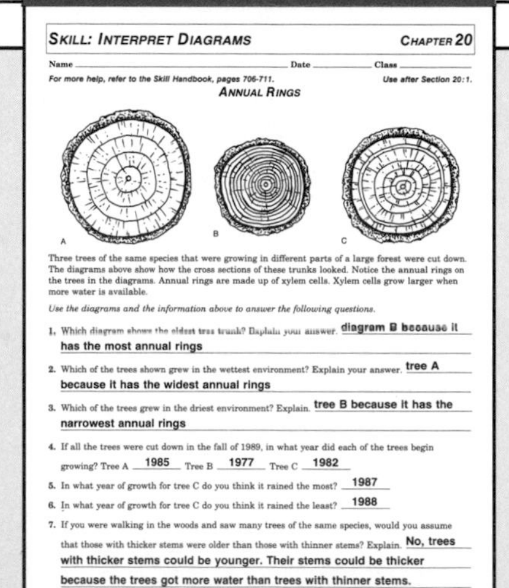

SKILL: INTERPRET DIAGRAMS — CHAPTER 20

Name ____ Date ____ Class ____

*For more help, refer to the Skill Handbook, pages 706-711.* *Use after Section 20:1.*

ANNUAL RINGS



Three trees of the same species that were growing in different parts of a large forest were cut down. The diagrams above show how the cross sections of these trunks looked. Notice the annual rings on the trees in the diagrams. Annual rings are made up of xylem cells. Xylem cells grow larger when more water is available.

*Use the diagrams and the information above to answer the following questions.*

1. Which diagram shows the oldest tree trunk? Explain your answer. **diagram B because it has the most annual rings**
2. Which of the trees shown grew in the wettest environment? Explain your answer. **tree A because it has the widest annual rings**
3. Which of the trees grew in the driest environment? Explain. **tree B because it has the narrowest annual rings**
4. If all the trees were cut down in the fall of 1989, in what year did each of the trees begin growing? Tree A **1985** Tree B **1977** Tree C **1982**
5. In what year of growth for tree C do you think it rained the most? **1987**
6. In what year of growth for tree C do you think it rained the least? **1988**
7. If you were walking in the woods and saw many trees of the same species, would you assume that those with thicker stems were older than those with thinner stems? Explain. **No, trees with thicker stems could be younger. Their stems could be thicker because the trees got more water than trees with thinner stems.**

**STUDY GUIDE 115**

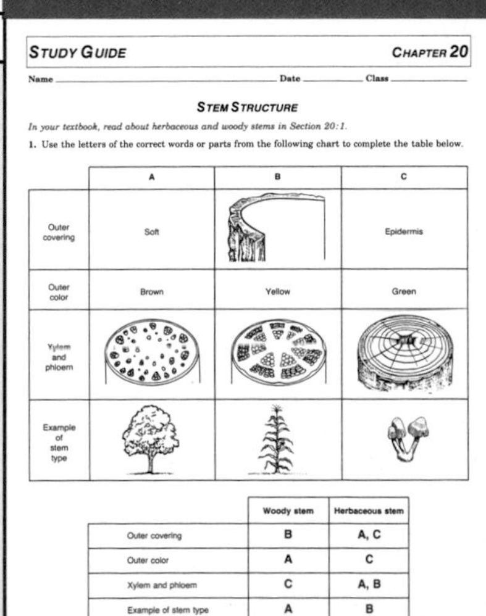

STUDY GUIDE — CHAPTER 20

Name ____ Date ____ Class ____

STEM STRUCTURE

*In your textbook, read about herbaceous and woody stems in Section 20:1.*

1. Use the letters of the correct words or parts from the following chart to complete the table below.

| | A | B | C |
|---|---|---|---|
| Outer covering | Soft | | Epidermis |
| Outer color | Brown | Yellow | Green |
| Xylem and phloem | | | |
| Example of stem type | | | |

| | Woody stem | Herbaceous stem |
|---|---|---|
| Outer covering | B | A, C |
| Outer color | A | C |
| Xylem and phloem | C | A, B |
| Example of stem type | A | B |

## OPTIONS

### ACTIVITY/Challenge

Have students write a statement to explain what would happen if sap did not rise in a woody stem each spring. Let several students read their statement to the class.

**Skill**/*Teacher Resource Package*, p. 21. Use the Skill worksheet shown here to give additional practice with aging woody stems.

## Career Close-Up

### Nursery Worker

**Background**

More people want to spend their leisure time away from the chores of yard work and depend on nurseries to solve their yard management problems.

**Related Careers**

lawn manager, horticulture technician, and landscape management

**Job Requirements**

high school diploma with on-the-job training

**For More Information**

Check the library for government publications listing careers.

## TEACH

### Independent Practice

**Study Guide**/*Teacher Resource Package,* p. 116. Use the Study Guide worksheet shown at the bottom of this page for independent practice.

### Guided Practice

Have students write down their answers to the margin question: the bud at the tip of stems.

## Career Close-Up

### Nursery Worker

A nursery worker must know how to select the best quality plants.

**Many people visit a nursery for gardening advice and to buy plants for their garden. In a large nursery, there are often several nursery workers who are there to help you.**

**The job of nursery worker includes planting a variety of plants, either at the nursery or a customer's home. Being able to understand the measurements in a landscape plan is an important skill for this job. The worker is responsible for preparing the soil and then setting out the plants.**

**When a nursery worker is not tending plants, he or she may spend some time repairing equipment. During colder months, he or she may help care for houseplants.**

**A high school education is important for a career as a nursery worker. You should have a general interest in caring for plants. As a nursery worker, you will need to learn about pruning, plant diseases, and insect pests. A class in horticulture (HOR tuh kul chur) will provide you with this knowledge. Most plant nurseries offer excellent on-the-job training.**

5. The innermost layer of a woody stem is xylem. The xylem cells have very thick cell walls that help support the plant. Xylem cells take water up through the stem. Dead xylem cells make up the wood of woody stems. This means that wood used for building homes or furniture came from the xylem of a tree.

### Stem Growth

Plants with herbaceous stems usually grow for only one year or one season. Plants with woody stems grow for more than one season. How much do plant stems grow in one year? Some, such as an oak tree, grow less than 15 cm while others, such as bamboo, grow nearly six meters.



## OPTIONS

### Science Background

Point out that tree rings tell something about environmental conditions. Thick annual rings denote good years of growth because the environmental factors were optimum. Thinner annual rings denote less than optimum factors.

**Enrichment**/*Teacher Resource Package,* p. 21. Use the Enrichment worksheet shown here as an extension to stem growth.

## STUDY GUIDE 116

STUDY GUIDE — CHAPTER 20

Name ______ Date ______ Class ______

**STEM STRUCTURE**

*In your textbook, read about stem growth in Section 20:1.*

2. Examine the diagram of a core sample taken from a tree. Then, answer the following questions.

Bark — One year's growth — Center of tree — A B C

a. What is counted to determine the age of a tree? **rings of xylem**
b. In what part of the year do the dark bands form? **summer**
c. Why are they dark? **The xylem cells are smaller (because growth is slower).**
d. In what part of the year do the light bands form? **spring**
e. Why are they light? **The xylem cells are larger (because growth is faster).**
f. How old was this tree when the core sample was taken? **8 years old**
g. Which band (A, B, or C) shows the poorest year of growth? **B**
h. Which band (A, B, or C) shows the year with the most rainfall? **A**

3. Examine the diagram below of a core sample taken in 1991.

**Students may need help identifying growth rings.**

Bark — 1991 — 1990 — 1989 — 1988 — Center of tree

a. How old was this tree in 1991? **6 years old**
b. What year was rainfall the least where this tree was growing? **1988**
c. What year did the tree grow the least? **1988**
d. What year was rainfall the most where this tree was growing? **1989**
e. What year did the tree grow the most? **1989**

## ENRICHMENT 21

ENRICHMENT — CHAPTER 20

Name ______ Date ______ Class ______

*Use after Section 20:1.*

**HOW A TWIG GROWS**

The picture shows a twig that is several years old. At the tip of the twig is the terminal bud. The twig grows from the tip. The cells just behind the terminal bud elongate. Bud scales surround and protect the terminal bud. At the end of the year the twig stops growing and the bud scales fall off. The bud scales leave a scar called a bud-scale scar. The bud-scale scar looks like a set of rings going all the way around the twig. Each year's growth is marked by the bud-scale scars. Leaves also make scars on the twig when they fall off in the autumn. These scars are called leaf scars. Lateral buds form on the sides of a twig. These buds show where new twigs will form.

*Label each of the structures mentioned above on the drawing of the twig. Then answer the questions that follow.*

Lateral buds — Leaf scars — Terminal bud — Bud scale — Bud-scale scar

1. How many centimeters did the twig grow in its first year? **5 cm** In its second year? **4 cm** How many centimeters did it grow last year? **3.5 cm**
2. How old is the twig? How can you tell? **4 years old. The bud-scale scars show 3-years growth and the terminal bud shows a fourth year.**
3. How many leaves grew on the twig last year? **1** How many leaves grew on the twig during its first year? **3** During which year did the twig have the most leaves? Explain how you know. **The second year. There are four leaf scars.**
4. How old is the smaller twig that is growing from the base of the larger twig? **3 years** Will the larger twig sprout any new twigs this year? Explain. **Yes. There are lateral buds just above the last bud-scale scar.**

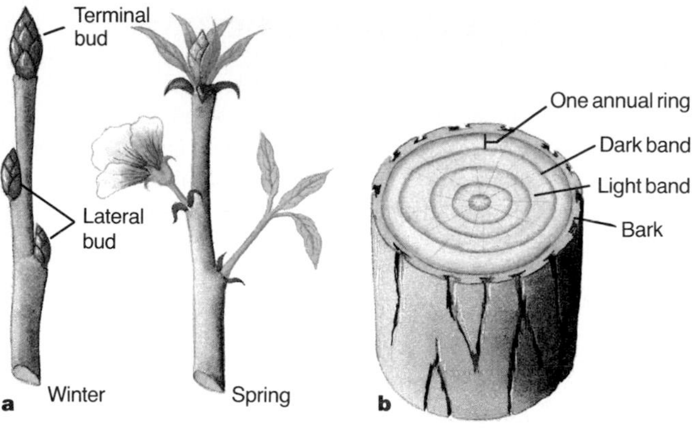


**Figure 20-4** Plants grow in length from terminal buds. New branches, leaves, or flowers grow from lateral buds (a). A tree trunk is made of annual rings of wood (b).

What is a terminal bud?

How do stems grow in length? Look at Figure 20-4a. Notice that there are two kinds of buds on the stem. The **terminal bud** is the bud at the tip of a stem. This bud is responsible for the plant's growth in length. Along the sides of a stem are **lateral buds** that give rise to new branches, leaves, or flowers.

A cross section of a tree trunk shows the trunk is made up of a series of rings, Figure 20-4b. Remember that the cambium forms a ring of new xylem cells each year. Each ring of xylem is called an **annual ring.** Each annual ring is often made up of a dark band and a light band of cells. The lighter band forms in the spring when there is plenty of rain and xylem growth is more rapid. The band looks lighter because the xylem cells are larger. Growth is slower in the summer when there is less rain. The band looks darker because the xylem cells are smaller.

### Mini Lab

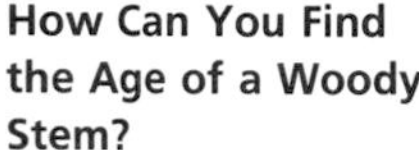

**How Can You Find the Age of a Woody Stem?**

**Use numbers:** Find a tree stump or fireplace log and look at the rings. Calculate the age of the woody stem. *For more help, refer to the **Skill Handbook,** pages 718-719.*

### Check Your Understanding

1. List two common plants with herbaceous stems.
2. How can woody stems be used to help identify trees?
3. How does the cambium cause a woody stem to grow?
4. **Critical Thinking:** If you drive a nail into a young tree trunk two feet from the ground, how high up the trunk do you think the nail will be in twenty years?
5. **Biology and Writing:** Plants have two ____ of stems. They are either woody or soft and green. Which word best fits in the blank? kinds, lengths, colors



## TEACH

### Mini Lab

Make sure that students count the light and dark rings as one year of growth.

### ASSESSMENT

**Skill:** Cut several 5- to 8-cm diameter sections of woody stems from wood found in a brush pile, a freshly cut tree, or fireplace wood. Tell students to count the number of light and dark rings to determine the age of the tree. Have them explain why some of the larger tree sections could be younger in age than those that are smaller in diameter.

### Reteach

Reteaching/*Teacher Resource Package,* p. 59. Use this worksheet to give students additional practice with parts of a stem.

## 3 APPLY

### ACTIVITY/Field Trip

Take students on a field trip around the school yard or a city park to compare herbaceous and woody stems.

## 4 CLOSE

### ACTIVITY/Challenge

Have students find out how they can tell the age of a tree without cutting it down.

### Answers to Check Your Understanding

1. Answers will vary. Examples are corn, lilies, and most vegetables.
2. the texture and kind of bark found on the tree
3. The cambium cells produce new layers of xylem cells. The added cells make the stems thicker.
4. The nail will still be 2 feet from the ground.
5. kinds

## RETEACHING 59

RETEACHING CHAPTER 20

Name ____ Date ____ Class ____

*Use with Section 20:1.*

IMPORTANT WORDS FOR STEMS

**Using Vocabulary**
Review the science words as you read Section 20:1. Study the definitions of the boldfaced words. Then complete the exercise below. Each science word has two choices that are correct.

*Write the letters of both correct choices on the line to the left.*

a, c 1. Cambium
(a) forms new xylem and phloem.
(b) is found only in leaves of herbaceous plants.
(c) is found between xylem and phloem.

a, c 2. Bark
(a) is found in plants with woody stems.
(b) is part of xylem in stems.
(c) is made of cork, phloem, cambium, and cortex cells.

b, c 3. Herbaceous stems
(a) are hard with rough bark.
(b) are usually soft and green.
(c) usually have less xylem than woody stems.

a, c 4. Woody stems
(a) are the thick hard stems of certain plants.
(b) are soft and green.
(c) are the nongreen stems of trees.

a, c 5. Cork
(a) forms the outer layer of a woody stem.
(b) is made of living cells.
(c) is made of dead cells that protect the stem.

a, c 6. A lateral bud
(a) forms a new leaf or flower.
(b) is found at the tip of all stems.
(c) is found along the sides of branches.

a, b 7. Cortex
(a) of an herbaceous stem is soft like a sponge.
(b) is the food storage area of the stem.
(c) is where food is made.

## OPTIONS

**What Does a Woody Stem Look Like Inside?**/*Lab Manual,* pp. 165-168. Use this lab as an extension to stem structure.

# 20:2 The Jobs of Stems

## PREPARATION

### Materials Needed

Make copies of the Transparency, Study Guide, and Reteaching worksheets in the *Teacher Resource Package*.

▶ Obtain a potted plant for the Focus activity.

▶ Obtain maple syrup and other plant products for the demonstration.

### Key Science Words

No new science words will be introduced in this section.

### Process Skills

In the Skill Check, students will make and use tables.

## 1 FOCUS

▶ The objectives are listed on the student page. Remind students to preview these objectives as a guide to this numbered section.

### ACTIVITY/Brainstorming

Show students a potted plant and ask them to work in groups of two or three and try to answer the questions: Where do roots end and stems start? How do you know? How could you find out?

### Guided Practice

Have students write down their answers to the margin question: xylem.

## 20:2 The Jobs of Stems

**Objectives**

4. **Explain** how a stem serves as a transport system.
5. **Discuss** the storage function of stems.
6. **Identify** how stems are useful to humans.

The stem is an important link between the roots and the leaves of a plant. Roots take up and store materials that leaves need. Leaves produce materials that roots and stems need. The stem is the part of the plant through which materials travel between the roots and leaves.

### Transport

Water and minerals are needed by plant parts for growth. Leaves, as you read in Chapter 19, make food for the plant by photosynthesis. Water enters the leaves from the stems. How does the stem get this water? The water is taken up by the roots of the plant.

**What cells in a stem carry water and minerals?**

Water taken in by the roots is transported up through the stem. Minerals from the soil are dissolved in this water and are also carried up the stem. All water and minerals are carried up in the long, tubelike cells called xylem.

What causes water to move through a plant from its roots to its leaves? Biologists have developed a theory that explains how water moves upward in a plant. One part of the theory is that water moves through a plant much as water moves into a paper towel, Figure 20-5. If you put a dry paper towel into water, the water is absorbed by the paper towel.

The second part of the theory explains how water passes up and out of a plant. Water moves up the plant through xylem cells and is lost from the leaves through the stomata. You read in Chapter 19 that transpiration is the loss of water through the stomata of leaves. This movement of water through a plant is like a thread being pulled through a straw, Figure 20-6a. The molecules of water stick together in a threadlike stream through a plant.

Figure 20-6b shows how important the stem is as a link between the leaves and roots. As water moves out of the leaves, more water is pulled into the leaves from the stem. The water in the stem is pulled up from the roots. New water enters the roots by osmosis.

Food made in leaves is also transported through stems. Food in the form of sugars moves down the stem through the tubelike cells called phloem. Why does food move downward? Root cells are alive and also need energy in order to stay alive. They would die without food. The sugars made in the leaf are used by the roots and all the other cells of the plant.

**Figure 20-5** Water enters a plant in a way that can be compared to how a paper towel absorbs water. diffusion

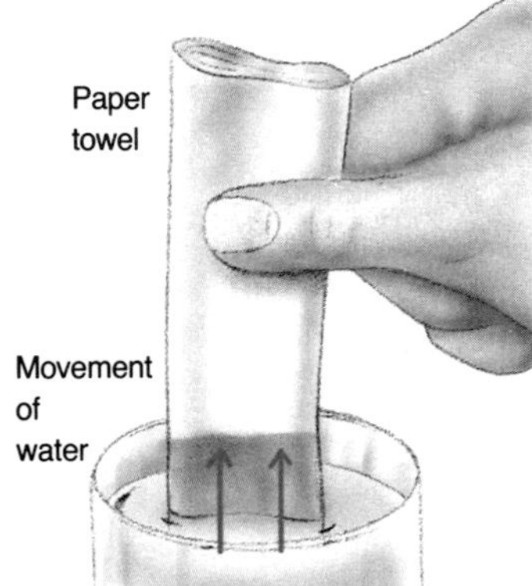




**TRANSPARENCY 93**

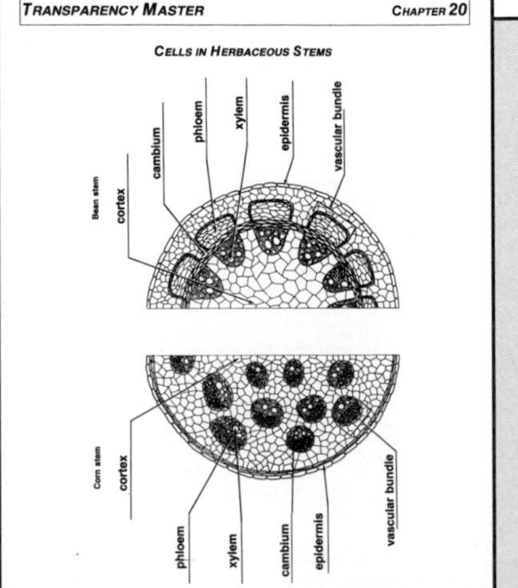


## OPTIONS

### Science Background

The stem connects the roots to the leaves. Water absorbed by the roots moves through the stem to the leaf for photosynthesis. The food made in the leaf is transported to the stem and roots for storage. This food can be used by all plant cells for respiration.

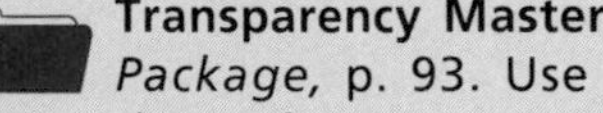
**Transparency Master**/*Teacher Resource Package*, p. 93. Use the Transparency Master shown here as you teach structure and function of stems.

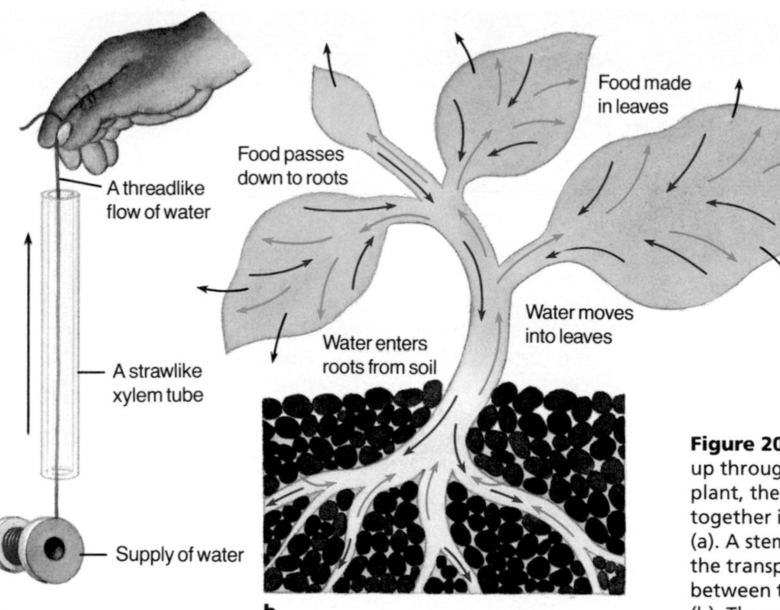


**Figure 20-6** When water moves up through xylem tubes in a plant, the molecules stick together in a continuous stream (a). A stem is the plant's organ for the transport of water and food between the roots and the leaves (b). The red arrows show the pathway of food. The blue arrows show the path of water, and the black arrows show water loss from leaves.

## Storage

Look at the food products stored in containers on the pantry shelf. A container is opened when food is needed. Stems also store materials that are used when needed. What types of materials are stored by the stem? Stems store water, and food in the form of starch. Water comes from the roots and the starch is made from sugars that are carried from the leaves.

**How does water storage help a plant?**

One benefit of water storage in herbaceous stems is that it prevents wilting. Water helps keep the cells of the stems stiff. You might say that these stiff cells support a plant like the skeleton supports an animal's body. Plants that live in dry areas have stems that store large amounts of water. A cactus stores a lot of water in the stem. This water is used for photosynthesis during the long, dry season.

Some stems store large amounts of starch. Starch is stored in stems until needed by a plant for its new growth. This food is a reserve supply for other plant parts. When needed, the starch is changed back into sugar. The sugar moves through the phloem cells from the stem to the root or the leaves. Many plants use much of their stored food for new growth each spring.

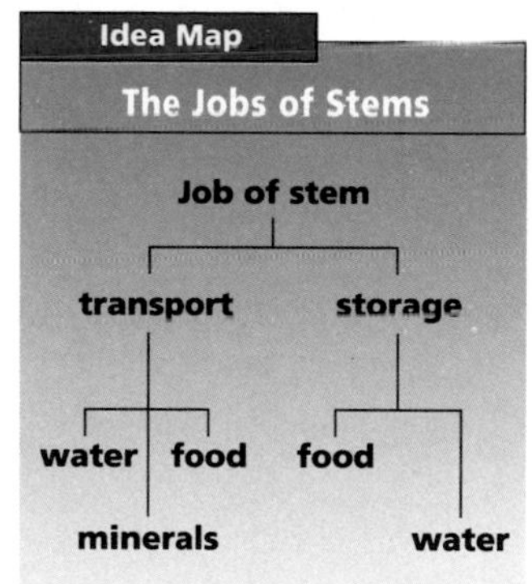


**Study Tip:** Use this idea map as a study guide to the jobs of stems. What cells transport water in stems?



## STUDY GUIDE 117

STUDY GUIDE CHAPTER 20

Name ______ Date ______ Class ______

THE JOBS OF STEMS

*In your textbook, read about transport and storage in Section 20:2.*

1. How do leaves get the water and minerals they need? **Water and minerals are taken in by the roots and transported up through the stem through long, tubelike cells called xylem.**
2. How does new water enter the roots? **by osmosis**
3. How is the movement of water through a plant like a thread being pulled through a straw? **The molecules of water stick together in a threadlike stream through the plant.**
4. In what form is sugar stored in plants? **as starch**
5. a. What kind of cells allow the transport of water upward in a plant? **xylem cells**
   b. What kind of cells allow the transport of food downward in a plant? **phloem cells**
6. Examine the diagrams below. Which one (A or B) best shows the path water takes through a stem? **A**

Leaf Stem Root A — Leaf Stem Root B

7. Explain what is wrong with the diagram that you did not choose as the answer to the last question. **The xylem and phloem do not connect from roots through the stem to the leaves in diagram B.**

# 2 TEACH

### MOTIVATION/Demonstration

Show students various products of plant stems. Suggestions include maple syrup, sugar, turpentine, asparagus. Have students give you the names of other materials they think come from stems.

### Concept Development

▶ Make sure students understand that most of a plant's water is lost through the leaves, while very little is lost through the stem.

### Idea Map

Have students use the idea map as a study guide to the major concepts of this section.

### Check for Understanding

Have students write a short paragraph explaining what materials are moved through the stem.

### Reteach

Give students a diagram of a plant with roots, stem, and leaves. Have them draw arrows to show direction of water and food movement.

### Guided Practice

Have students write down their answers to the margin question: It prevents wilting.

### Independent Practice

**Study Guide**/*Teacher Resource Package,* p. 117. Use the Study Guide worksheet shown at the bottom of this page for independent practice.

## GLENCOE TECHNOLOGY

**Videodisc**

The Secret of Life

*Stem Structure*

## TECHNOLOGY

### Newer Medicines from Plants

#### Background

Explorers to the New World returned with unknown plants that had useful medicinal properties. The explorers had bartered with the natives for information and the plants themselves. A closer study of plants was started as a result of this.

#### Discussion

Let students know that one of the most promising places that new medicinal plants are being found today is in the tropical rain forests of the world. Since the tropical forests are disappearing at a rapid rate, we may be losing some valuable sources of medicines.

#### References


Cox, Paul A. and Michael J. Balik. "The Ethnobotanical Approach to Drug Discovery." *Scientific American*, June 1994, pp. 82-87.

Nicolaou, K.C., Rodney K. Guy and Pierre Potier. "Taxoids: New Weapons Against Cancer." *Scientific American*, June 1996, pp. 94-99.

Norton, Rob. "Owls, Trees, and Ovarian Cancer." *Fortune*, February 5, 1996, p. 49.

Plotkin, Mark. *Tales of a Shaman's Apprentice: An Ethnobotanist Searches for New Medicines in the Amazon Rain Forest*. New York: Viking Penguin, 1994.


#### *Portfolio*

Direct students to do outside reading and make a list of plants and plant parts being used for medicines that have been discovered in the past 20 years. Have them state what diseases or medical problems these plant medicines are used for and also list the part of the world in which these plants are found. Other students can research what plants Native Americans used for medicines, what they were used for, and how effective they were.

## TEACH

#### Guided Practice

Have students write down their answers to the margin question: woody.



## TECHNOLOGY

### Newer Medicines from Plants

Several centuries ago, explorers brought plants from the tropical forests of the New World back to Europe. The explorers had discovered that these plants were being used for medicines by native peoples.

For example, quinine was found to be used as a cure for malaria. The natives of South America had made an extract of quinine by soaking the bark of the *Cinchona* (sihn CHOH nuh) tree in water. Drinking the water that contained the bark extract helped cure malaria.

Plants use the sugar from photosynthesis and minerals from the soil to make many different compounds that have medicinal properties. Many of the original plant medicines have since been made from synthetic chemicals. Now, researchers are going back to the tropical rain forests in search of new plant medicines. AIDS and cancer are two diseases for which we need new medicines.

Why is the tropical rain forest a good place to hunt for medicinal plants? There are more species of plants in the tropical rain forests than anywhere else in the world. Scientists believe that with all these undiscovered species, there must be some with new medicinal value.

Many plant scientists are involved in the search for new plants in the world's rain forests. The work is very time consuming. Will all of this effort help society? Imagine the benefits of discovering a plant that produces a chemical that can cure AIDS.

A chemical from the bark of the Peruvian Cinchona tree helps cure malaria all over the world.

### How Are Stems Used?

Does paper come from woody or herbaceous stems?

The most important product we get from stems is wood. When you write with a wooden pencil, you are using a product from a stem. Look around your home or classroom and you will find many places where wood is used. You read in Chapter 6 how conifers are important for paper products. Paper products you use come from the wood of these tree stems. This book, dollar bills, food packets, and many other paper products are part of your daily life, Figure 20-7a. Your life would be very different without stems.



### *OPTIONS*

#### ACTIVITY/Challenge

Have students find pictures in magazines of products that come from stems of plants. Have them make a collage of pictures on a large piece of poster board. Have them make a list of the products and how humans use them.

a

b

**Figure 20-7** Plants provide us with paper products (a) and sap or liquid products, such as maple syrup and chewing gum (b).

Stems are useful also as food. When you eat asparagus, cauliflower, and broccoli, you are eating the flower stems of these plants. People in some countries grow bamboo plants and eat the young, soft stems. Food seasonings such as cinnamon and dill are from stems. Some foods are made by taking sap from trees. Maple sugar is made from the sap of sugar maple trees. Sap is the liquid that flows up the stem in the xylem cells. In spring, the sap contains up to four percent sugar, which was stored in the roots over the winter.

Rubber, turpentine, and materials for making chewing gum are made from other liquids in the stems of different trees. After the liquids are collected from the trees, they are processed for our use, Figure 20-7b.

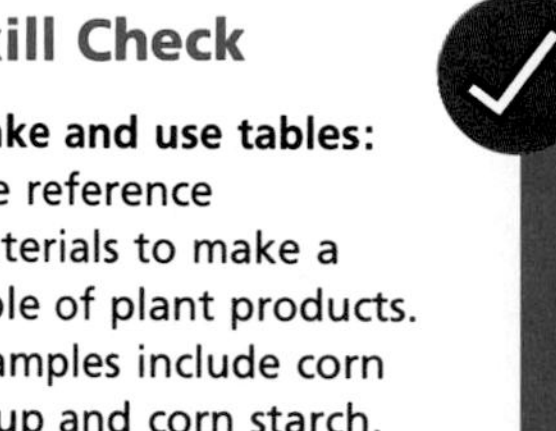

### Skill Check

**Make and use tables:** Use reference materials to make a table of plant products. Examples include corn syrup and corn starch. *For more help, refer to the **Skill Handbook**, pages 715-717.*

## Check Your Understanding

6. What happens to minerals in the soil before they can get into roots?
7. How does water storage help herbaceous stems?
8. What are five products we get from stems?
9. **Critical Thinking:** How would the rate of water movement through a plant on a rainy day compare with the rate on a hot, sunny day?
10. **Biology and Writing:** Make a complete sentence from each of the three sentence fragments.
    1. A stem supports a plant and transports
    2. are dissolved in water and carried through roots.
    3. One benefit of water storage in herbaceous stems



## TEACH

### Check for Understanding

Have students respond to the first three questions in Check Your Understanding.

### Reteach

**Reteaching**/*Teacher Resource Package*, p. 60. Use this worksheet to give students additional practice with jobs of stems.

**Extension:** Assign Critical Thinking, Biology and Writing, or some of the **OPTIONS** available with the lesson.

## 3 APPLY

### Software

*How Plants Grow: The Inside Story,* Opportunities for Learning.

## 4 CLOSE

### ACTIVITY/Interview

Have students visit a lumberyard and find out the names of some hardwoods and softwoods that are sold there. Have them find out why these woods are called hardwoods and softwoods.

### Answers to Check Your Understanding

6. They have to be dissolved in water.
7. It prevents the plant from wilting.
8. Answers will vary. Examples are rubber, turpentine, medicines, maple sugar, food, and paper.
9. The rate of water flow through a plant would be higher on a hot, sunny day because water would evaporate faster into the atmosphere.
10. 1. A stem supports a plant and transports materials.
    2. Materials from the soil are dissolved in water and are carried through the roots.
    3. One benefit of water storage in herbaceous stems is that it prevents wilting.

**RETEACHING 60**

**RETEACHING: IDEA MAP** CHAPTER 20

Name ______ Date ______ Class ______

*Use with Section 20:2.*

**THE JOBS OF STEMS**

*Use the following terms to complete the idea map: water, root, minerals, food, leaves, water.*

Stem — is a link between: root, leaves; transports: water, minerals; stores: food, water

1. How does water move through the plant during transpiration? **Water escapes through the stomata of leaves. As water moves out of the leaves, water moves up the stem to take its place. As water moves upward in the stem, water moves into the roots.**
2. Why is the stem called a link between parts of a plant? **The roots take in important materials that the leaves need. The leaves produce materials that roots need. Materials travel between roots and leaves through the stem.**
3. What keeps the stem stiff and prevents the plant from wilting? **water stored in the stem**
4. Through what parts of the stem do water and minerals travel? **through the xylem**

## GLENCOE TECHNOLOGY

### Videodisc

**Biology: The Dynamics of Life**
Animation: *Water Uptake in Roots*
Disc 1, Side 2, Ch. 22

## 20:3 Root Structure

## PREPARATION

### Materials Needed

Make copies of the Reteaching and Study Guide worksheets in the *Teacher Resource Package*.

### Key Science Words

taproot
fibrous root
root hair
endodermis

### Process Skills

In the Skill Check, students will understand science words.

## 1 FOCUS

▶ The objectives are listed on the student page. Remind students to preview these objectives as a guide to this numbered section.

### ACTIVITY/Discussion

Have students discuss what is meant by the word *root* in each phrase below. Have them discuss how the phrase ties in with plant roots.
family roots
root of a problem
rooting around

### Guided Practice

Have students write down their answers to the margin question: a large, simple root with smaller side roots.

### Student Journal

Have students reread Section 20:3 and make an idea map that compares the cells and their functions in woody and herbaceous roots. Tell them to compare this idea map to the one they did for stems.



## 20:3 Root Structure

**Objectives**

7. **Compare** taproots and fibrous roots.
8. **Sequence** the layers of root cells.
9. **Describe** how roots grow.

**Key Science Words**

taproot
fibrous root
root hair
endodermis

A root is the plant part that takes in water and minerals for the plant. The following sections will describe the structures of roots and how they grow.

### Taproots and Fibrous Roots

What is a taproot?

There are two basic types of root systems—taproots and fibrous roots. The part of a carrot that we eat is the root. In plants like the carrot, there is one main root called a taproot. A **taproot** is a large, single root with smaller side roots. The thick taproots of some plants store water and food during dry or cold periods. In the vegetable section of a grocery store, you can see many different kinds of taproots. Beets, carrots, and turnips are common vegetables that are taproots.

If you weed a garden or lawn, you know that some weeds such as dandelions are very hard to pull out of the soil, Figure 20-8a. Their long, thick taproots will usually break off before you can pull the plant from the soil.

Not all plants have taproots. Some plants have fibrous (FI brus) roots, Figure 20-8b. **Fibrous roots** are many-branched roots that grow in clusters. Unlike the taproot, there is no main, large root. As the name suggests, the roots are fiberlike. Fibrous roots spread out over a large area. Thus, fibrous roots collect water from a larger area than reached by taproots.

Oak trees and hickory trees have long, thick taproots. Some maple trees and beech trees have fibrous roots. Which kind of tree is more likely to blow over in a harsh windstorm?

a

b

**Figure 20-8** A taproot has stored food (a), and a fibrous root has many branches (b).



## OPTIONS

### Science Background

Roots are classified by structure and by the size and number of primary roots present. Taproots have a large primary root that grows straight down into the soil. Numerous secondary roots branch off this root. Fibrous roots have many stringlike primary roots with many more secondary roots branching from the primary roots. Some of the cells found in the stem are also found in the roots. Roots can grow in length and thickness like a stem.

**What Do the Inside Parts of a Root Look Like?**/*Lab Manual,* pp. 169-172. Use this lab as an extension to root structure.

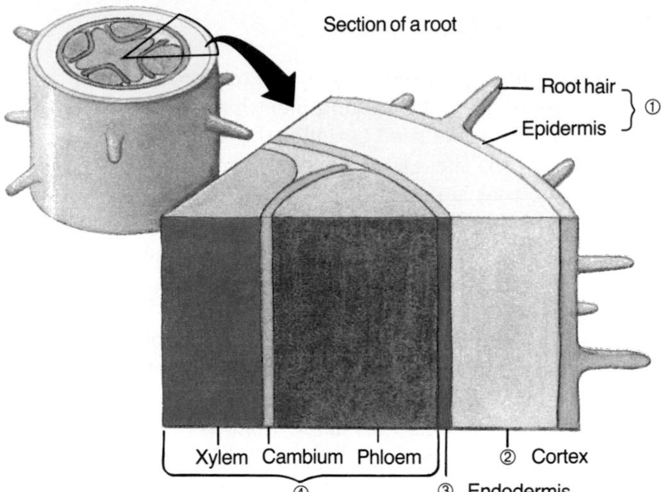


**Figure 20-9** A root is made of four main layers of cells. From which layer do root hairs grow? the epidermis

## Cells of a Root

The root is made of several layers of cells. The different layers have different jobs. Using Figure 20-9, let's examine the cell layers of a root.

1. The outside layer of a root is the epidermis, which protects the root. Growing from the epidermis are the root hairs. **Root hairs** are threadlike cells of the epidermis that absorb water and minerals for a plant.
2. Inside the epidermis is the layer of cells called the cortex. Just as in a stem, the cells of the cortex store food.
3. The next layer is the endodermis. The **endodermis** is a ring of waxy cells that surrounds the xylem in roots. This layer of cells helps the plant retain the water taken in by the xylem.
4. Inside the endodermis, notice the bundles of xylem. Between these xylem cells are phloem cells. You know that xylem and phloem carry materials in the plant. Xylem cells in the roots carry water and minerals up to the stem. Phloem cells bring food from the leaves.

If you eat a raw or cooked carrot, you might notice a difference in the feel and taste between the outer and inner layers of the carrot root. The outer ring is softer and sweeter. This is the cortex with its stored food. The inner part is tougher and not as sweet. What makes this layer taste different? This layer is made of the xylem and phloem. There are also more fibers in this ring.

**What is the outer layer of a root?**

### Skill Check

**Understand science words: fibrous root.** The word part *fibr* means threadlike. In your dictionary, find three words with the word part *fibr* in them. *For more help, refer to the* ***Skill Handbook,*** *pages 706-711.*



# 2 TEACH

### MOTIVATION/Demonstration

Cut a carrot into enough thin sections so that each student has one. Tell them to hold the carrot toward the light and locate the inner and outer rings. Point out that the xylem and phloem are thick-walled, smaller diameter cells in the inner ring, and the cortex is thinner-walled, large diameter cells in the outer ring.

### Concept Development

▶ Point out that cortex cells occupy most of the space in a root because they store food for the plant.

### Check Your Understanding

Have students list the cells found in a root from the outside to the center and write a function for each.

### Reteach

Set up several microscopes showing a cross section of a root. Place a specific cell at the end of the pointer. Have students identify the cell. Go over all the cell types, and review the size and shape of each according to its function.

### Guided Practice

Have students write down their answers to the margin question: epidermis.

### Independent Practice

**Study Guide**/*Teacher Resource Package,* p. 118. Use the Study Guide worksheet shown at the bottom of this page for independent practice.

### Portfolio

Tell students to write a paragraph in their portfolios explaining why smaller carrots may have a sweeter taste and not be as hard to chew as larger carrots. Have them focus on the internal parts of the carrot to help explain their answers.

**STUDY GUIDE 118**

STUDY GUIDE CHAPTER 20

Name ______ Date ______ Class ______

ROOT STRUCTURE

*In your textbook, read about root cells and growth in Section 20:3.*

1. Label the drawing below using these labels: root hair, cortex, xylem, phloem, epidermis, endodermis.

xylem; phloem; endodermis; cortex; epidermis; root hair

2. Label the diagram below using these labels: primary root, secondary root, root hairs.

Ⓐ secondary root
Ⓑ root hairs
primary root Ⓒ

3. From the diagram above, put the correct letters on the blanks below.

a. largest root of plant C
b. absorb water and minerals B
c. first root to form in plants C
d. would measure greatest distance if put in line B
e. forms from primary root A

**GLENCOE TECHNOLOGY**

**Videodisc**

**The Secret of Life**
*Dicot Root*

## TEACH

### Guided Practice

Have students write down their answers to the margin question: straight down from a seed.

### Check for Understanding

Have students respond to the first three questions in Check Your Understanding.

### Reteach

**Reteaching**/*Teacher Resource Package*, p. 61. Use this worksheet to give students additional practice with root structure.

**Extension:** Assign Critical Thinking, Biology and Reading, or some of the **OPTIONS** available with this section.

## APPLY

### ACTIVITY/Software

*Plant Biology, From the Ground Up,* Opportunities for Learning.

## CLOSE

### Filmstrip

Show the filmstrip *Plants: Parts and Processes—Roots, Stems, and Leaves,* National Geographic Society.

### Answers to Check Your Understanding

11. A taproot is a single, large root with many small secondary roots attached. A fibrous root has several large roots growing in a cluster.
12. epidermis–protection; cortex–food storage; xylem–carry water; phloem–carry food
13. They form from the simple primary root as branches of the primary root.
14. epidermis, cortex, xylem, phloem
15. tomato plant because it is less firmly rooted

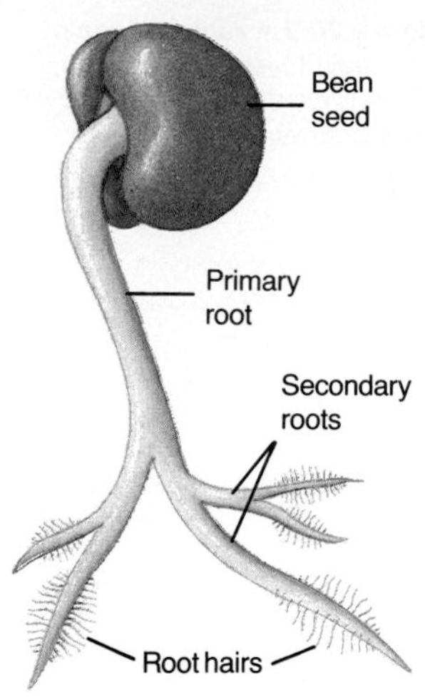


**Figure 20-10** The first root of a new plant is its primary root. Other roots are secondary roots.

**How does a primary root grow?**

## Root Growth

If you split open a bean seed, you will see the plant embryo inside. This new, young plant has leaves, a stem, and a root. When the seed sprouts, the first part to grow out of the seed is the root.

Only one root is present in the seed. When it grows out of the seed, it is the primary root. The word *primary* means first. The primary root is the first and largest root to form in a plant. It usually grows straight down, toward the pull of gravity.

Later, secondary roots form. Secondary roots are smaller roots that branch from the primary root. The thinner, shorter, threadlike parts near the tips of the secondary roots are the root hairs. Compare the primary roots and secondary roots in Figure 20-10.

Most roots keep growing as long as the plant lives. When roots grow, new cells are added to the tips of the roots. This is how the root grows down into the soil.

How large can a root system grow? In 1926, an American scientist asked the same question. To find the answer, he collected a lot of data. First he carefully removed all the fibrous roots of a rye plant, a grasslike cereal plant. He then measured the lengths of each of the roots and added all the lengths together. The results showed that if the roots were placed end to end, they could form a line 612 km long, as long as the state of Tennessee!

Next, the scientist measured the root hairs. The lengths of all the root hairs were added together. If every root hair were put end to end, the line would go from the east to the west coast of the United States and back again! You can see that the rye plant has a very large root system.

### Check Your Understanding

11. How does a taproot differ from a fibrous root?
12. List the layers of cells found in a root and describe the job of each.
13. How do secondary roots form?
14. **Critical Thinking:** Which layers of cells do roots and stems have in common?
15. **Biology and Reading:** Which would be easier to pull up—a dandelion that has a taproot, or a tomato plant that has a fibrous root system?



## *OPTIONS*

### ACTIVITY/Challenge

Have teams germinate and grow corn and bean seeds in different kinds of soil to see in which type of soil the roots appear to grow best. Use potting soil, clay soil, and sandy soil. Add the same amount of water to each type of soil.

### Science Background

Roots of woody plants can grow from 0.5 to 10 meters into the soil. Most herbaceous plants have roots that grow 0.3 to 0.5 meter into the soil.

**RETEACHING 61**

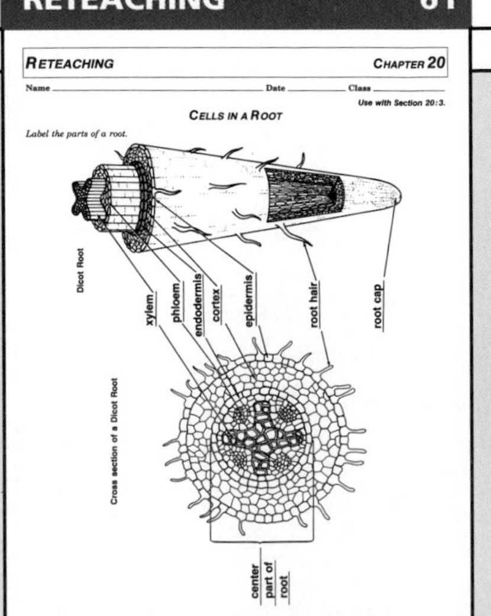
RETEACHING CHAPTER 20

Name Date Class

Use with Section 20:3.

CELLS IN A ROOT

*Label the parts of a root.*

Dicot Root

xylem, phloem, endodermis, cortex, epidermis, root hair, root cap

Cross section of a Dicot Root

center part of root

## 20:4 The Jobs of Roots

Is the root vital to a plant? The plant would die if its roots were removed. Roots anchor the plant in one place and supply a plant with nutrients.

### Objectives

**10. Describe** root hairs and their function.

**11. Relate** how roots anchor the plant and store nutrients.

**12. Discuss** the benefits and problems of roots.

### Absorption

Plants get needed materials for growth through their roots. Roots absorb water from their surroundings by osmosis. In Chapter 2, you read that osmosis is the movement of water molecules across a cell membrane. Roots also take in minerals from their surroundings with the water they absorb.

How does water enter a root?

Water enters the root through the root hairs. Root hairs increase the surface area of the root and so increase the amount of water that can be absorbed or taken in by the plant. When a gardener moves a plant from one part of a garden to another, the plant's root hairs may be torn off. The amount of water these plants can absorb is reduced. To protect the plant, the gardener digs up a ball of soil around the plant's roots and moves this soil along with the plant, Figure 20-11a. This protects the root hairs.

The root supplies the plant with raw materials. Water is a raw material for photosynthesis. Minerals are raw materials for the plant to make proteins and other chemicals such as chlorophyll.

Like all living cells, root cells must be supplied with oxygen. Root cells get oxygen from air in the spaces between soil particles. Gardeners and landscapers break up the soil around plants to ensure that air gets into the soil, Figure 20-11b. Sandy soils or soils rich in organic material have more air spaces than clay soils.

### Skill Check

**Sequence:** List the parts of a plant in the order in which water passes in and out of the plant. *For more help, refer to the **Skill Handbook**, page 706.*

**Figure 20-11** Roots are fragile and easily broken when a plant is moved (a). When the soil is broken up, air can circulate around plant roots (b).

a

b

## 20:4 The Jobs of Roots

## PREPARATION

### Materials Needed

Make copies of the Critical Thinking/Problem Solving, Application, Reteaching, Study Guide, and Lab worksheets in the *Teacher Resource Package*.

▶ Cut paper towel strips for Lab 20-1. Collect carrot, yam, and radish roots for Lab 20-2.

▶ Germinate the radish seeds before you do the demonstration.

### Key Science Words

No new science words will be introduced in this section.

### Process Skills

In Lab 20-1, students will observe, formulate a model, infer, and design an experiment. In Lab 20-2, they will observe, classify, form hypotheses, and design an experiment. In the Skill Check, they will sequence.

## 1 FOCUS

▶ The objectives are listed on the student page. Remind students to preview these objectives as a guide to this numbered section.

### Guided Practice

Have students write down their answers to the margin question: by osmosis through the root hairs.

### Student Journal

Have students find out how and why the soil is conditioned when shrubs and trees are planted in a park or new development. Plant nurseries usually provide information sheets explaining how to condition the soil. Have students relate what they found to the content in the text.

### CRITICAL THINKING 20

CRITICAL THINKING/PROBLEM SOLVING — CHAPTER 20

Name ________ Date ________ Class ________

*Use after Section 20:4.*

**WHAT CAUSED THE TREES TO DIE?**

Two students were walking up a street in a new housing development. Sewer lines had just been laid, and the street had just been paved. They noticed that some of the trees by the side of the road looked unhealthy. The leaves were wilted and falling, although it was still summer. The students wondered why this was happening. They also wondered if the trees could recover.

The students listed what could have damaged the trees. Had earthmoving equipment hit the trees? The trunks did not look as if they had been bumped. Perhaps the paving material was bad for the trees. The trees were a few meters from the road. The students had read that the root system grows as far out from the trunk as the branches spread out. They also knew that nutrients travel inside the trunk to the leaves. The students made a hypothesis that the roots and trunks were somehow affected by the new street development.

**Analyzing the Problem**

1. What are three functions of roots? Roots anchor the plant in the soil, absorb water and minerals, and store food made in the leaves.
2. What two materials do roots take in from the soil? Roots take in water and minerals.
3. By what processes do the materials get into the roots from the soil? absorption and diffusion
4. How are materials carried from the roots to the leaves? Materials travel up the trunk in tubes called xylem.

**Solving the Problem**

1. If soil and pavement were packed heavily on the roots how could the tree be affected? Water and dissolved minerals would diffuse into the roots more slowly.
2. What will happen to the support tissues in the trunk if they do not have enough water? They become dry and cannot carry water to the leaves.
3. Could the root system be disturbed by paving if the trees are a few meters from the road? Yes, because the roots spread out widely.
4. Can the root system repair itself with time? It might regrow under the road, because cell division happens actively at the tips of roots.

### OPTIONS

**Critical Thinking/Problem Solving/** *Teacher Resource Package,* p. 20. Use the Critical Thinking/Problem Solving worksheet shown here as an extension to root functions.

## Lab 20-1 Absorption

### Overview

Students will build a model of a root hair. They can observe how water enters a root hair and is further transported into the epidermal cell.

**Objectives:** After completing this lab, students will be able to: (1) **formulate a model** of a root hair, (2) **observe** how water moves in the root hair model, (3) **define** root hair.

**Time Allotment:** 45 minutes

### Preparation

▶ Prepare the paper towel strips prior to class. Use a thick paper towel like that found in most school lavatories as these towels will not fall apart during use.

 **Lab 20-1 worksheet**/*Teacher Resource Package*, pp. 77-78.

### Teaching the Lab

▶ Make sure students bend the paper towel as suggested. The students should be able to see the water being absorbed rapidly by the paper towel strips.

**Cooperative Learning:** Divide the class into groups of two students. For more information, see pp. 22T-23T in the Teacher Guide.

**ASSESSMENT**

**Performance:** Have students make an ink spot with a black overhead pen on each of the four strips of the root model. Tell them to put the ink on the part of the four strips that will be above the water level and that these spots will represent minerals on the outside of the root hairs. As the water is absorbed upward from the root hairs into the main root, students should notice the dissolving and movement of different color bands that make up the black ink. Students are to explain how the spreading ink pattern is similar to how the minerals are being dissolved and carried inside the roots.

# Absorption

## Lab 20–1

### Problem: How does a root hair work?

#### Skills

observe, formulate a model, infer, design an experiment

#### Materials

paper towel strips:
- 10 X 15 cm (1 strip)
- 1 X 10 cm (4 strips)

stapler
large plastic cup
pencil

#### Procedure

1. Staple the four small paper strips to the larger strip, Figure A. Bend the smaller strips so they are at right angles from the larger strip.
2. Fill the cup to within 2.5 cm of the top with water.
3. Lay a pencil across the rim of the cup. Lay the root model over the pencil. See Figure B. The smaller towel strips will hang down into the water.
4. **Observe** the root model for 20 minutes.
5. Remove the root model from the cup.

#### Data and Observations

1. What happened to the smaller strips when they were put in water?
2. Did the large strip absorb any water? If so, where did this water come from?

#### Analyze and Apply

1. Which part of your model represented the root hairs?
2. Which part of your model represented the main root from which root hairs grow?
3. In what way did the small paper strips act like root hairs?
4. Why are there hundreds of root hairs on a single root?
5. **Infer:** What would happen to a plant if its root hairs were damaged?
6. **Apply:** Would you expect desert plants or tropical rain forest plants to have more root hairs? Why?

#### Extension

**Design an experiment** that shows the rate of water absorption in a root. Use a carrot, water, and food coloring.

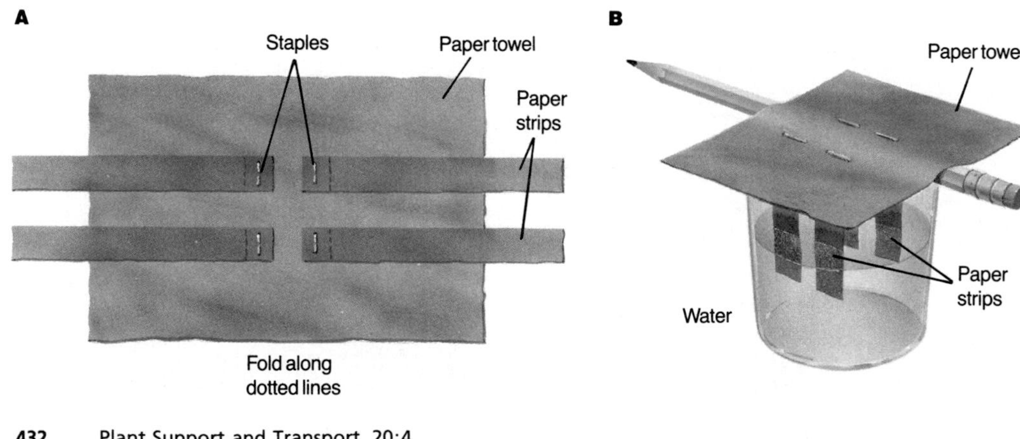




## ANSWERS

### Data and Observations

1. Water began to travel up the strips.
2. Yes, the water traveled up the smaller strips into the larger strips.

### Analyze and Apply

1. the smaller paper strips
2. the larger paper strips
3. They absorbed water from their surroundings.
4. to provide a larger surface area for water absorption into the plant
5. It would not be able to absorb water and so would wilt and then die.
6. desert plants, because the soil doesn't contain much water, and so the more root hairs, the greater the chance of contacting water

## Anchorage and Storage

A major job of a root is to anchor the plant. A root is similar to a boat anchor. An anchor catches on the rocks or soil under the water and holds the boat in one place. The roots of most plants spread through the particles of soil and hold the plant in one place, Figure 20-12a.

Many roots have large cells that store food. The usual food stored is starch. Compare the size of the storage cells with the other cells in Figure 20-12b. You have read that these storage cells make up the cortex. The food in these cells is used later by other cells in the plant. Large amounts of food are stored in taproots. Carrot and turnip roots are good examples of plants that have many food storage cells. The taproot of a dandelion stores food for the plant. You may have noticed that a dandelion will soon grow back if you don't pull up its whole root. It uses its stored food to grow new leaves, stems, and flowers.

Roots store food made in one growing season to use at the start of the next growing season. Many herbaceous plants depend on the stored food in the roots to survive the winter. If food were not stored in the roots for the winter, the plant could not grow new parts the next spring.

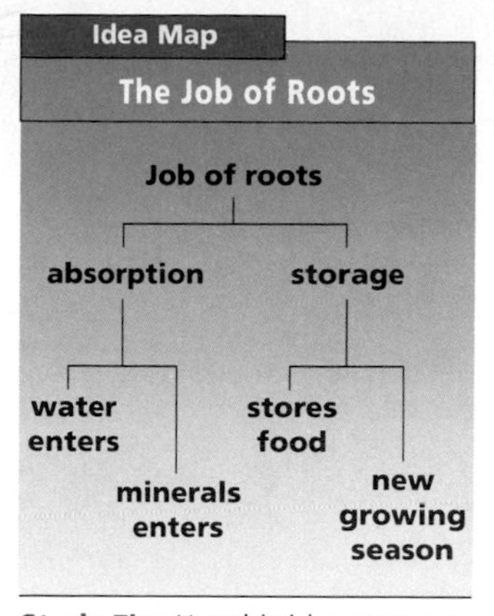


**Study Tip:** Use this idea map as a study guide to the jobs of roots. What two things do roots absorb?

**Why do roots store food?**

## How Are Roots Used?

You have already learned that people eat many kinds of roots such as beets, carrots, and sweet potatoes. Products made from the roots are also eaten. Sassafras (SAS uh fras) roots are used to make tea. Horseradish roots are ground and used as a sauce to season other foods. Roots are also used to make products such as perfumes and medicines.

**Figure 20-12** Roots anchor a plant in the soil (a). Many roots have stored food in the cells of the cortex (b).

a

b

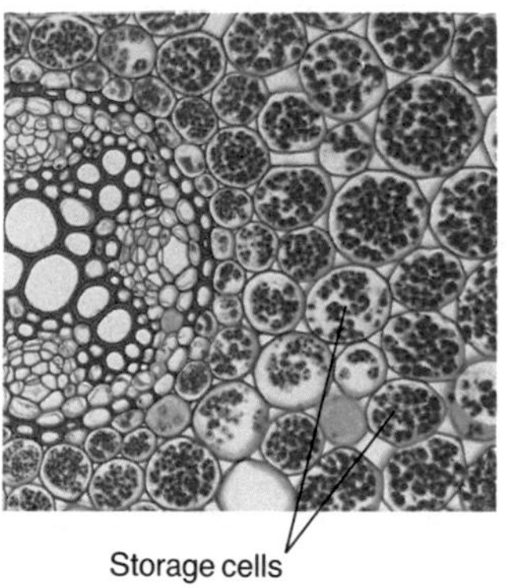




# 2 TEACH

### MOTIVATION/Videocassette

Show the videocassette *The Plant: An Operator's Manual*, Human Relations Media.

### ACTIVITY/Demonstration

Root hairs can be shown to students by sprouting radish seeds. Place 50 to 60 radish seeds on a paper towel. Fold the paper towel and wet it. Put the paper towel in a plastic bag. The seeds will sprout within 48 hours. Students can look at root hairs with a hand lens.

### Connection: Health

Have students make posters showing the importance of roots and stems to nutrition by cutting pictures from magazines.

### Idea Map

Have students use the idea map as a study guide to the major concepts of this section.

### Guided Practice

Have students write down their answers to the margin question: to use when needed for growth.

### Independent Practice

**Study Guide**/*Teacher Resource Package*, p. 119. Use the Study Guide worksheet shown at the bottom of this page for independent practice.

**APPLICATION 20**

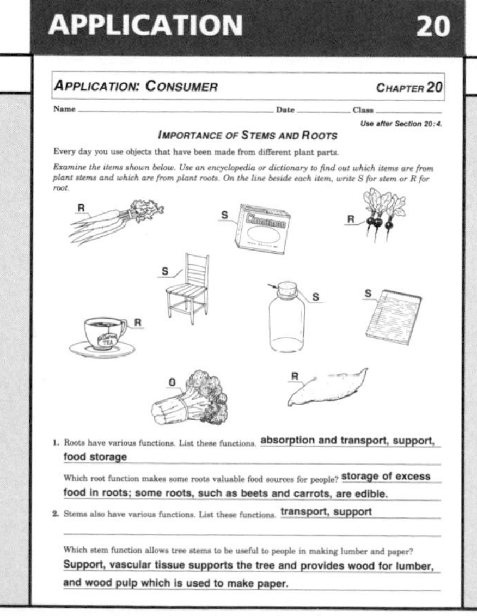

*APPLICATION: CONSUMER* — CHAPTER 20

Name ____ Date ____ Class ____

*Use after Section 20:4.*

*IMPORTANCE OF STEMS AND ROOTS*

Every day you use objects that have been made from different plant parts.

*Examine the items shown below. Use an encyclopedia or dictionary to find out which items are from plant stems and which are from plant roots. On the line beside each item, write S for stem or R for root.*

R, S, R, S, S, S, R, O, R

1. Roots have various functions. List these functions. **absorption and transport, support, food storage**

   Which root function makes some roots valuable food sources for people? **storage of excess food in roots; some roots, such as beets and carrots, are edible.**

2. Stems also have various functions. List these functions. **transport, support**

   Which stem function allows tree stems to be useful to people in making lumber and paper? **Support, vascular tissue supports the tree and provides wood for lumber, and wood pulp which is used to make paper.**

**STUDY GUIDE 119**

*STUDY GUIDE* — CHAPTER 20

Name ____ Date ____ Class ____

*ROOT STRUCTURE*

*In your textbook, read about taproots and fibrous roots in Section 20:4.*

4. Examine the drawings below. Then, put A or B on the blank after each phrase or word.
   a. taproot **B**
   b. fibrous roots **A**
   c. often spreads out **A**
   d. often grows very deep **B**

A B

*THE JOBS OF ROOTS*

*In your textbook, read about absorption, anchorage, and storage in Section 20:4.*

*Examine the pictures below. Beside each picture write a sentence describing what the object pictured does. Then, write a sentence describing a similar job of roots. The first one is done for you.*

Sponge, Suitcase, Hair net, Paper clip, Anchor, Paper towels, Stapler, Silo, Tent stake

1. A sponge absorbs water. Roots absorb water and minerals from the soil.
2. **A suitcase helps to store and move items. Roots store food and move water and minerals to the stem.**
3. **A hair net holds hair together. Roots hold soil particles together.**
4. **A paper clip holds things together. Roots hold soil particles together.**
5. **An anchor holds a boat in place. Roots hold a plant in place.**
6. **Paper towels absorb water. Roots absorb water and minerals from the soil.**
7. **A stapler holds papers together. Roots hold soil particles together.**
8. **A silo stores food. A root stores food.**
9. **A tent stake anchors a tent. Roots anchor a plant.**

## OPTIONS

### Science Background

**Enrichment:** Farmers sometimes plant grasses in their fields as a cover crop in the fall. The fibrous roots bind the soil particles together and prevent soil erosion. In the spring, the grass is plowed under and provides minerals for the new crops.

**Application**/*Teacher Resource Package*, p. 20. Use the Application worksheet shown here to emphasize the importance of stems and roots.

## Lab 20-2 Storage

### Overview

Students will discover that roots have a large area in which starch is stored. The iodine test verifies where the cortex, the starch-storage area of a root, is.

**Objectives:** After completing this lab, students will be able to: (1) **determine** the presence of starch using lab techniques, (2) **identify** where starch is stored in a root, (3) **compare** areas of starch storage in different roots.

**Time Allotment:** 45 minutes

### Preparation

▶ Purchase the roots from the produce section of a grocery store the day before you need them. You may wish to cut the roots prior to the class period.

▶ For *iodine solution;* dissolve 10 g potassium iodide and 3 g iodine in 1000 mL water. Store in a brown bottle.

Lab 20-2 worksheet/*Teacher Resource Package*, pp. 79-80.

### Teaching the Lab

▶ Caution the students about getting iodine on their skin. Have them flush the area of the skin with water.

**Cooperative Learning:** Divide the class into groups of three students. For more information, see pp. 22T-23T in the Teacher Guide.

**ASSESSMENT**

**Skill:** Have students repeat the iodine test with other root and stem parts, such as white potatoes, rutabaga, and celery. Direct students to allow 5 to 10 minutes for the starch-iodine reaction to occur in some of the plant parts. Have them compare the amount of starch present by the amount of the area that turns blue-black after adding iodine.

# Storage

## Lab 20–2

### Problem: What parts of roots store starch?

**Skills**

observe, classify, form hypotheses, design an experiment

**Materials**

| | |
|---|---|
| carrot root | 3 petri dish halves |
| yam root | iodine solution |
| radish root | dropper |
| razor blade | hand lens |
| forceps | |

**Procedure**

1. Copy the data table.
2. Using the razor blade, cut a very thin slice of each root as shown in the figure. **CAUTION:** *Use extreme care with the razor blade.*
3. Place each slice in a separate petri dish.
4. Write a **hypothesis** about which part of the carrot root will show the presence of starch. Will the other roots show the presence of starch? (HINT: Read pages 429 and 433 again.)
5. Add one dropperful of iodine to each root slice as shown in the figure. Make sure each slice has iodine on it at all times. **CAUTION:** *Iodine is poisonous and can stain and burn the skin.*

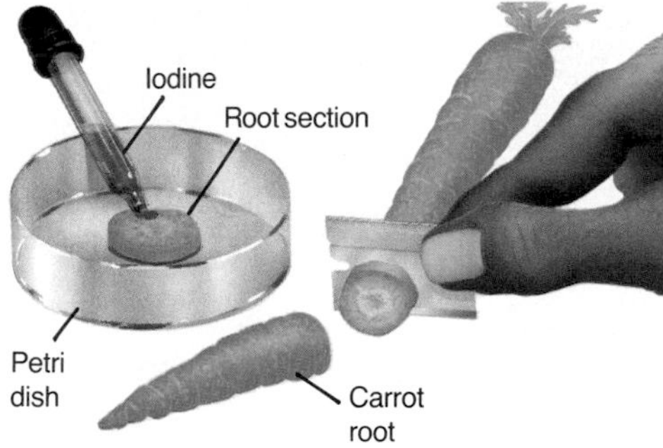


6. **Observe:** During the next 15 minutes, examine the root slices with the hand lens. Look for blue-black areas. A blue-black color means starch is present. Note in your data table in which areas starch is present.
7. Draw each root slice in your notebook. Label the cortex and xylem in the carrot drawing.

**Data and Observations**

1. Which roots showed starch was present? How could you tell?
2. Which part of each root slice showed more starch present?

| TYPE OF ROOT SLICE | BLUE-BLACK AREAS AFTER ADDING IODINE |
|---|---|
| Carrot | |
| Yam | |
| Radish | |

**Analyze and Apply**

1. What cells make up the center of the root? What do they do?
2. What cells make up the outer part of the root? What do they do?
3. How are the starch storage areas of the three roots alike?
4. **Classify:** Are the roots used taproots or fibrous roots? Explain your answer.
5. **Check your hypothesis:** Is your hypothesis supported by your data? Why or why not?
6. **Apply:** How is starch made by a plant?

**Extension**

**Design an experiment** to test different stems for stored food.

434 Plant Support and Transport 20:4

## ANSWERS

### Data and Observations

1. all roots, because each stained blue-black with iodine
2. See sample data in the data table.

### Analyze and Apply

1. xylem and phloem; carry water and food through the root
2. cells of the cortex; store starch
3. make up the largest part of the root
4. all taproots; they are one main root
5. Those students who hypothesized that the outer part of the carrot root contains starch will answer that their hypotheses were supported. The outer part of a carrot root is the cortex, the food-storage cells.
6. from sugar made in the leaves

a

b

**Figure 20-13** These root nodules contain bacteria that fix nitrogen for the plant (a). Growth of tree roots can sometimes damage sidewalks or sewage pipes (b).

Roots are beneficial to humans in other ways. If you were to pull up a small clump of grass growing in a yard, you would see soil sticking to the roots. Roots of plants help hold soil together. The soil is not easily carried away by rain, melting snow, or wind.

**How do roots help soil?**

The roots of many trees and plants, such as beans and clover, help make the soil rich in nitrogen. Some bacteria can change nitrogen in air to a form that plants can use. These bacteria live on plant roots, Figure 20-13a. Soil rich in nitrogen is good for growing corn and other food crops.

Sometimes roots are a nuisance, Figure 20-13b. The roots of some trees, like the willow and maple, grow into underground pipes. Roots can block the pipes that carry water and wastes away from houses. Root growth is so strong it can break pipes. It's a good idea to find out where pipes are laid before planting trees. Sometimes sidewalks or roads are also broken by root growth.

### Bio Tip

**Biology:** Many cities don't plant silver maples and willow trees because their roots cause damage to sewers.

## Check Your Understanding

16. Why are root hairs important?
17. What are three jobs of a root?
18. What are three benefits of roots?
19. **Critical Thinking:** You want to plant a shade tree in your yard. What things do you need to find out about the tree and your yard before you plant?
20. **Biology and Reading:** You have read that roots are sometimes a nuisance. What does it mean for a plant's roots to be a nuisance?

## RETEACHING 62

**RETEACHING: IDEA MAP** CHAPTER 20

Name ______ Date ______ Class ______

*Use with Section 20:4.*

**THE JOBS OF ROOTS**

*Use the following terms to complete the idea map: turnips, water and minerals, starch, carrots, holds plant in one place, in large cells.*

- Roots
  - absorption — water and minerals
  - storage
    - where? — in large cells
    - what? — starch
    - examples — carrots, turnips
  - anchorage — holds plant in one place

1. How do water and minerals enter the root? through the root hairs by osmosis
2. In what type of cell is a root hair found? epidermis
3. What is osmosis? Osmosis is the moving of water through a cell membrane from an area of greater concentration to an area of lesser concentration.
4. Why does photosynthesis stop when a plant has no water? Water is one of the raw materials of photosynthesis.

## STUDY GUIDE 120

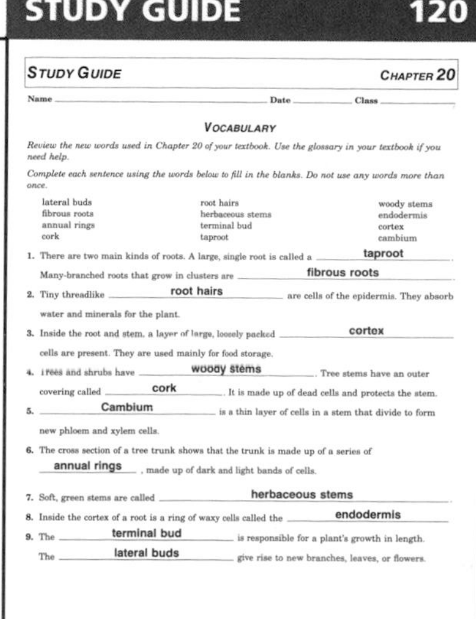

**STUDY GUIDE** CHAPTER 20

Name ______ Date ______ Class ______

**VOCABULARY**

*Review the new words used in Chapter 20 of your textbook. Use the glossary in your textbook if you need help.*

*Complete each sentence using the words below to fill in the blanks. Do not use any words more than once.*

| | | |
|---|---|---|
| lateral buds | root hairs | woody stems |
| fibrous roots | herbaceous stems | endodermis |
| annual rings | terminal bud | cortex |
| cork | taproot | cambium |

1. There are two main kinds of roots. A large, single root is called a taproot. Many-branched roots that grow in clusters are fibrous roots.
2. Tiny threadlike root hairs are cells of the epidermis. They absorb water and minerals for the plant.
3. Inside the root and stem, a layer of large, loosely packed cortex cells are present. They are used mainly for food storage.
4. Trees and shrubs have woody stems. Tree stems have an outer covering called cork. It is made up of dead cells and protects the stem.
5. Cambium is a thin layer of cells in a stem that divide to form new phloem and xylem cells.
6. The cross section of a tree trunk shows that the trunk is made up of a series of annual rings, made up of dark and light bands of cells.
7. Soft, green stems are called herbaceous stems.
8. Inside the cortex of a root is a ring of waxy cells called the endodermis.
9. The terminal bud is responsible for a plant's growth in length. The lateral buds give rise to new branches, leaves, or flowers.

## TEACH

### Guided Practice

Have students write down their answers to the margin question: they hold it together.

### Independent Practice

**Study Guide**/*Teacher Resource Package,* p. 120. Use the Study Guide worksheet shown at the bottom of this page for independent practice.

### Check for Understanding

Have students respond to the first three questions in Check Your Understanding.

### Reteach

**Reteaching**/*Teacher Resource Package,* p. 62. Use this worksheet to give students additional practice with roots.

**Extension:** Assign Critical Thinking, Biology and Reading, or some of the **OPTIONS** available with this lesson.

## 3 APPLY

### ACTIVITY/Field Trip

Have students collect weeds with taproots and fibrous roots.

## 4 CLOSE

### ACTIVITY/Filmstrip

Show the filmstrip *The Primary Plant Body,* Carolina Biological Supply Co.

### Answers to Check Your Understanding

16. Root hairs absorb water for the plant from the soil by osmosis.
17. The jobs are absorption, storage, transport, and anchorage.
18. They are used as food and medicines, and hold soil together.
19. You need to find out the location of any underground pipes and cables before you plant.
20. Nuisance means here that roots make problems for people.

## CHAPTER 20 REVIEW

### Summary

Summary statements can be used by students to review the major concepts of the chapter.

### Key Science Words

All boldfaced terms from the chapter are listed.

### Testing Yourself

#### Using Words

1. annual ring
2. taproot
3. root hair
4. cork
5. terminal bud
6. lateral bud
7. cambium
8. herbaceous stem
9. woody stem
10. endodermis
11. cortex

#### Finding Main Ideas

12. p. 425; Plants store food for the next growing season in the stems.
13. p. 424; Water moves into the roots and up through the stem through xylem cells.
14. p. 426; Stems are used to make paper products, as food, and to make products such as rubber and turpentine.
15. p. 431; Root hairs in the root take in water by osmosis.
16. p. 430; The primary root is the first root to form in a plant. It grows straight down into the soil.

#### Using Main Ideas

17. The primary root is already present in a seed. The secondary root grows from a primary root.
18. Cambium cells form new xylem cells each year; new xylem cells are formed as rings of new cells.
19. Leaves use water for photosynthesis.
20. The order of cell types in a root is phloem, endodermis, cortex, and epidermis.
21. They can grow into and block underground pipes that carry water and waste from houses. They can break sidewalks and roads as they grow.
22. Xylem and phloem are grouped together in clusters in a corn stem and in a circular pattern in the bean stem.
23. Roots are a good food source for animals because they store large amounts of starch.
24. Water is used for photosynthesis during the dry season.

# Chapter 20 Review

## Summary

### 20:1 Stem Structure

1. Herbaceous stems are soft and green.
2. Woody stems are thick and hard. From outside to inside, they are made up of cork, cortex, phloem, cambium, and xylem.
3. Stem growth in length occurs in terminal and lateral buds.

### 20:2 The Jobs of Stems

4. A stem transports materials from the roots to the leaves.
5. Stems store food and water in the cortex.
6. Stems provide humans with wood, food, seasonings, rubber, and turpentine.

### 20:3 Root Structure

7. A taproot is a large single root with smaller side roots. Fibrous roots are many-branched roots.
8. Roots are made up of an outer epidermis, cortex, xylem, and phloem.
9. The first root to grow out of a seed is the primary root. Secondary roots grow out from the primary root.

### 20:4 The Jobs of Roots

10. Root hairs absorb water and minerals by osmosis.
11. Roots anchor the plant and store starch.
12. Roots are used for food and to make medicines. Roots also help hold the soil together.

## Key Science Words

annual ring (p. 423)
cambium (p. 421)
cork (p. 421)
cortex (p. 420)
endodermis (p. 429)
fibrous root (p. 428)
herbaceous stem (p. 420)
lateral bud (p. 423)
root hair (p. 429)
taproot (p. 428)
terminal bud (p. 423)
woody stem (p. 421)

## Testing Yourself

### Using Words

*Choose the word from the list of Key Science Words that best fits the definition.*

1. a ring of xylem in a woody stem
2. root with a single, large root
3. threadlike cell of the root epidermis
4. dead cells on the outer part of a tree that partly make up bark
5. stem part that causes growth in length
6. gives rise to new branches, leaves, or flowers

# Review

## Testing Yourself *continued*

7. cells that form new xylem and phloem
8. soft, green stem
9. a nongreen, hard stem
10. ring of waxy cells around the xylem and phloem in the root
11. plant tissue that stores food

### Finding Main Ideas

*List the page number where each main idea below is found. Then, explain each main idea.*

12. stems storing food for the next growing season
13. how water lost in leaves is replaced
14. how stems are useful for paper products
15. how a root takes in water
16. how a primary root develops

### Using Main Ideas

*Answer the questions by referring to the page number after each question.*

17. How are primary and secondary roots different? (p. 430)
18. How does an annual ring form? (p. 423)
19. Why is transporting water to leaves an important function of stems? (p. 424)
20. Starting with the center, what is the order of these cell types as they would occur in a root: cortex, phloem, endodermis, and epidermis? (p. 429)
21. How can roots be a nuisance to humans? (p. 435)
22. How is the arrangement of xylem and phloem different in bean and corn stems? (p. 420)
23. Why are roots a good source of food for animals? (p. 433)
24. Why do plants in dry areas store much water in stems? (p. 425)

## Skill Review 

*For more help, refer to the **Skill Handbook,** pages 704-719.*

1. **Sequence:** What are the root cells and stem cells in order from the outside to the inside of a root and stem?
2. **Observe:** Observe the plants in the produce section of a grocery store. Make a list of those that are roots.
3. **Classify:** Make a collection of ten weeds from your yard or street. Use a field guide to name them. Group them as either woody or herbaceous.
4. **Use numbers:** Calculate the age of a tree that is 20 cm in diameter if each annual ring is 5 mm thick.

## Finding Out More

TECH PREP

### Critical Thinking

1. What causes knots in wooden boards?
2. What makes a tree ooze sap where limbs are removed? Explain why this happens more often during spring and summer.

### Applications

1. Find out which plant parts of onions and white potatoes are the stems.
2. Visit a furniture store and find out what kinds of wood are used in making furniture and why.



## Skill Review

1. root cells: epidermis, cortex, endodermis, xylem, and phloem; woody stem cells: cork, cortex, phloem, cambium, and xylem
2. Answers will vary. Some root foods are beets, carrots, sweet potatoes, turnips, sassafras, and horseradish.
3. Remind students that herbaceous stems are soft and green. Woody stems are harder.
4. 10 cm (1/2 the diameter) x 10 mm/cm x 1 yr/5mm = 20 years

## Finding Out More

### Critical Thinking

1. Knots are caused by the stems of the plant.
2. The fluids within the xylem cells are leaking out. The plant is transporting larger amounts of food and water during the spring and summer than during the winter.

### Applications

1. The parts that are eaten are the stems.
2. Oak, maple, cherry, and walnut are often used because they are hard and durable.

**Chapter Review**/*Teacher Resource Package*, pp. 104-105.

**Chapter Test**/*Teacher Resource Package*, pp. 106-108.

**Chapter Test**/*Computer Test Bank*

# Plant Response, Growth, and Disease

## PLANNING GUIDE

| CONTENT | TEXT FEATURES | TEACHER RESOURCE PACKAGE | OTHER COMPONENTS |
|---|---|---|---|
| (2 days)<br>21:1 Plant Response<br>Growing and Flowering<br>Tropisms<br>Other Plant Responses | Skill Check: *Observe,* p. 444<br>Skill Check: *Understand Science Words,* p. 446<br>Lab 21-1: *Root Growth,* p. 442<br>Lab 21-2: *Plant Responses,* p. 445<br>Check Your Understanding, p. 446 | Application: *Short-day and Long-day Plants,* p. 21<br>Reteaching: *Plant Responses,* p. 63<br>Critical Thinking/Problem Solving: *How Do Climbing Plants Respond to Sun?* p. 21<br>Transparency Master: *Soil Particles and Roots,* p. 97<br>Study Guide: *Plant Responses,* pp. 121-122<br>Lab 21-1: *Root Growth,* pp. 81-82<br>Lab 21-2: *Plant Responses,* pp. 83-84 | **Laboratory Manual:**<br>*What Tropisms Can Be Seen in Growing Plants?* p. 173<br>Color Transparency 21: *Plant Growth Responses*<br>**STVS:** *Growing Plants in Space,* Plants and Simple Organisms (Disc 4, Side 2) |
| (1 1/2 days)<br>21:2 Growth Requirements<br>Seasonal Growth<br>Light<br>Water<br>Minerals<br>Soil<br>Temperature | Mini Lab: *Do Plants Need Light?* p. 448<br>Idea Map, p. 448<br>Check Your Understanding, p. 451 | Enrichment: *Tolerance in Plants,* pp. 22-23<br>Skill: *Interpret Data,* p. 22<br>Reteaching: *Growth Requirements,* p. 64<br>Focus: *Plant Needs,* p. 41<br>Study Guide: *Growth Requirements,* pp. 123-124 | **Laboratory Manual:**<br>*Is Light an Important Growth Requirement for Plants?* p. 177<br>**STVS:** *Salt-Resistant Crops,* Plants and Simple Organisms (Disc 4, Side 2) |
| (1 day)<br>21:3 Plant Diseases and Pests<br>Bacteria and Viruses<br>Fungi<br>Insect Pests | Check Your Understanding, p. 453<br>Applying Technology: *How Much Nitrogen?* p. 454 | Reteaching: *Plant Diseases and Insects,* p. 65<br>Study Guide: *Plant Diseases and Insects,* p. 125<br>Study Guide: *Vocabulary,* p. 126 | **STVS:** *Insecticides from Desert Plants,* Plants and Simple Organisms (Disc 4, Side 2) |
| Chapter Review | Summary<br>Key Science Words<br>Testing Yourself<br>Finding Main Ideas<br>Using Main Ideas<br>Skill Review | **ASSESSMENT RESOURCES**<br>Chapter Review, pp. 109-110<br>Chapter Test, pp. 111-113<br>Performance Assessment in the Biology Classroom<br>Alternate Assessment in the Science Classroom<br>Computer Test Bank | |

### GLENCOE TECHNOLOGY

**Infinite Voyage,** *The Living Clock*
**Science and Technology Videodisc Series,** *Growing Plants in Space,* Plants and Simple Organisms (Disc 4, Side 2)
*Salt-Resistant Crops,* Plants and Simple Organisms (Disc 4, Side 2)
*Insecticides from Desert Plants,* Plants and Simple Organisms (Disc 4, Side 2)

## MATERIALS NEEDED

| LAB 21-1, p. 442 | LAB 21-2, p. 445 | APPLYING TECHNOLOGY, p. 454 | MARGIN FEATURES |
|---|---|---|---|
| toothpick<br>plastic bag<br>paper towel<br>graph paper<br>razor blade<br>India ink<br>2 bean seedlings | petri dish<br>marking pen<br>2 paper towels<br>transparent tape<br>4 pre-soaked corn seeds | plastic trays<br>bucket of soil<br>newspaper<br>grass seed<br>fertilizer<br>soil test kit | Skill Check, p. 444<br>pencil<br>pea seed<br>potted plant<br>Skill Check, p. 446<br>dictionary<br>Mini Lab, p. 448<br>2 paper cups<br>12 radish seeds<br>water<br>soil |

## OBJECTIVES

For more information about National Science Standards, see page 5T.

| SECTION | OBJECTIVE | CORRELATION of QUESTIONS to OBJECTIVES | | | |
|---|---|---|---|---|---|
| | | CHECK YOUR UNDERSTANDING | CHAPTER REVIEW | TRP CHAPTER REVIEW | TRP CHAPTER TEST |
| 21:1 National Science Stds: UCP.3, UCP.4, C.6 | 1. **Explain** how hormones affect plant growth. | 1 | 2, 14, 20, 23 | 2, 3, 8, 22 | 16, 20, 22, 23, 25 |
| | 2. **Compare** short-day, long-day, and day-neutral plants. | 2, 4 | 1, 6 | 10, 11, 27 | 5, 11, 24, 31, 32, 33, 35, 36, 37, 38, 40 |
| | 3. **Distinguish** among tropisms and other plant movements. | 3, 5 | 4, 7, 9, 13, 16, 22 | 1, 6, 7, 9, 13, 17, 18, 24, 28, 30 | 1, 8, 10, 12, 14, 15, 34, 39 |
| 21:2 National Science Stds: UCP.3, UCP.4, F.3, F.4 | 4. **Compare** annual, biennial, and perennial plants. | 6, 10 | 8, 10, 18 | 15, 25 | 2, 17 |
| | 5. **List** the growth requirements of plants. | 7, 8, 9 | 3, 12, 19, 25 | 4, 19, 20, 21, 26, 29 | 3, 4, 6, 7, 9, 13, 18, 19, 26, 27, 28, 29, 30, 48, 49, 50, 51, 52, 53 |
| 21:3 National Science Stds: UCP.3, UCP.4, C.4, F.4 | 6. **Explain** how bacteria and viruses affect plants. | 11, 15 | 15, 21 | 5, 12, 16 | 21, 41, 44, 47 |
| | 7. **Describe** how a fungus can kill a plant. | 12 | 17 | 12 | 21, 42, 43, 46, |
| | 8. **Relate** two ways that insects damage plants. | 13, 14 | 1, 2 | 12, 14, 23 | 13, 21, 45 |

# Plant Growth and Disease

## CHAPTER OVERVIEW

### Key Concepts

This chapter points out the role of hormones in various responses of plant growth and flower production. Four growth requirements for plants—light, water, minerals, and temperature—are discussed in terms of their effects on the development of a plant. Plant diseases, insects, and their effects on plant growth are discussed.

### Key Science Words

| | |
|---|---|
| annual | perennial |
| biennial | phototropism |
| day-neutral plant | short-day plant |
| fertilizer | thigmotropism |
| gravitropism | tropism |
| long-day plant | |

### Skill Development

In Lab 21-1 , students will **measure in SI, form hypotheses, interpret data,** and **experiment** with the growth region in a root. In Lab 21-2 , they will **observe, form hypotheses,** and **interpret data** in relation to gravitropism and root growth. In the Skill Check on page 444, students will **observe** thigmotropism. In the Skill Check on page 446, they will **understand** the **science word** *biennial.* In the Mini Lab on page 448, students will **experiment** with plant growth in response to light.

### Bridging

In Chapters 19 and 20 of this unit, students learned about the structures and functions of leaves, roots, and stems. In Chapter 15, they found out about the role of hormones in animals. In this chapter, students will learn how hormones bring about changes in plants. They will find out how hormones, growth requirements, insects, and diseases alter the growth of plants.

**CHAPTER PREVIEW**

### Chapter Content

Review this outline for Chapter 21 before you read the chapter.



### Skills in this Chapter

The skills that you will use in this chapter are listed below.

- In **Lab 21-1,** you will measure in SI, form hypotheses, interpret data, and experiment. In **Lab 21-2,** you will observe, form hypotheses, and interpret data.
- In the **Skill Checks,** you will observe and understand science words.
- In the **Mini Labs,** you will experiment.

438

## TECH PREP

For Tech Prep activities in this chapter of the Teacher Wraparound Edition, see especially the Student Journal on page 441, the Activities on pages 449 and 453, the Portfolios on pages 450 and 452, and the Connection on page 453.

See also the Glencoe Homepage at **www.glencoe.com**

The following Glencoe resources provide additional opportunities for integrating science and technology.

**Technology: Science and Math in Action, Book One**
Module 1: Hydroponics

Chapter 21

# Plant Growth and Disease

Plants, like animals, respond to changes in their environments. Plants don't have sense organs, nor do they move about. However, plants can still move in response to changes in their environments.

One well known response in plants can be seen in the photo to the left. If you live in the middle or northern states of the United States, you may have seen the new growth of plants in the spring. What causes a plant to grow in the spring, make flowers, or drop its leaves in the fall?

Another change in plants is caused by disease. Plants, like animals, can become sick as shown in the photo below. This chapter will explain how a plant responds to its environment, what it needs to grow, and some of the causes of plant diseases.

**Try This!**

**How can you see a plant move?** Place a houseplant in a window. Notice that the stems are straight. Look at the plant after two days. Has the plant moved? In what direction have the stems or leaves moved?

**interNET CONNECTION**

For more information about the material in this chapter, follow the link for the chapter on the Glencoe Homepage at **http://www.glencoe.com**

Diseased fruit is not edible.



**ASSESSMENT PLANNER**

**Portfolio**

Strategies on the following pages represent student products that can be placed into a best-work portfolio: pp. 449, 450, 452.

**PERFORMANCE ASSESSMENT**
Skill Check pp. 444, 446
Mini Lab, p. 448
Lab, pp. 442, 445

**CONTENT ASSESSMENT**
Check for Understanding, pp. 444, 446, 451, 453
Chapter Review, pp. 456-457

**GROUP ASSESSMENT**
Opportunities for group assessment occur with Cooperative Learning Strategies.

**Student Journal**

Strategies on the following pages represent opportunities for writing in a Student Journal: p. 441.

**GETTING STARTED**

## Using the Photos

Students may not notice that plants have responses because it takes several days to see some of the responses. Point out that the *Mimosa* plant shows a response to touch within a few seconds after the stimulus. The rapid response is caused by loss of water on the underneath side of the leaf. This causes the leaf to curl upward.

## MOTIVATION/Try This!

**How can you see a plant move?** Upon completing the chapter-opening activity, students will have **observed** that plants respond to light. You may want to bring in photos of plants, such as sunflowers, that show response to sunlight. Remind students that they studied stimulus and response in Chapter 17.

## Chapter Preview

Have students study the chapter outline before they begin to read the chapter. The first topic deals with how hormones affect plants and how plants respond to a stimulus. This is followed by a discussion of certain growth requirements. The final topic deals with plant diseases and insects that can affect plant growth.

## Misconception

Students may believe that the responses they see in a plant are caused by a nervous system or a "desire" on the part of the plant to respond to an environmental factor. Remind students that plants do not have nervous systems.

## 21:1 Plant Responses

### PREPARATION

### Materials Needed

Make copies of the Application, Study Guide, Transparency Master, Critical Thinking/Problem Solving, Reteaching, and Lab worksheets in the *Teacher Resource Package.*

▶ Sprout the seeds at least 3-4 days before you start Lab 21-1.

▶ Soak corn seeds the night before you do Lab 21-2.

▶ Obtain a pea seed and a pot of soil for the Skill Check.

### Key Science Words

short-day plant
long-day plant
day-neutral plant
tropism
phototropism
gravitropism
thigmotropism

### Process Skills

In Lab 21-1, students will measure in SI, form hypotheses, interpret data, and experiment. In Lab 21-2, students will observe, form hypotheses, and interpret data. In the Skill Checks, they will observe and understand science words.

### 1 FOCUS

▶ The objectives are listed on the student page. Remind students to preview these objectives as a guide to this numbered section.

### MOTIVATION/Overhead

Draw a graph on an overhead transparency to show the duration of sunlight in the various seasons. Use the graph to explain that increasing daily sunlight causes flowering in long-day plants and decreasing daily sunlight causes flowering in short-day plants.

### Guided Practice

Have students write down their answers to the margin question: A plant will grow bushy.



**Objectives**

1. **Explain** how hormones affect plant growth.
2. **Compare** short-day, long-day, and day-neutral plants.
3. **Distinguish** among tropisms and other plant movements.

**Key Science Words**

short-day plant
long-day plant
day-neutral plant
tropism
phototropism
gravitropism
thigmotropism

## 21:1 Plant Responses

If you took a picture of a bean plant early in the morning and then again later in the day, you would see that some parts of the plant had moved. Early in the morning, bean leaves droop down close to the stem. The photograph taken later in the day would show the leaves held out and away from the stem. Bean leaves respond to a change between night and day. Is the movement of bean leaves a growth response, or do they move by some other means? The following sections describe and explain various plant responses.

### Growth and Flowering

The photos in Figure 21-1 are of the same plant. The photos were taken in spring, just a week apart. In just a week, the plant has grown. Growth, flowering, and branching are all plant processes that are controlled by plant growth hormones. In Chapter 15, you read that hormones are chemicals made in one part of an organism that then affect another part of the organism.

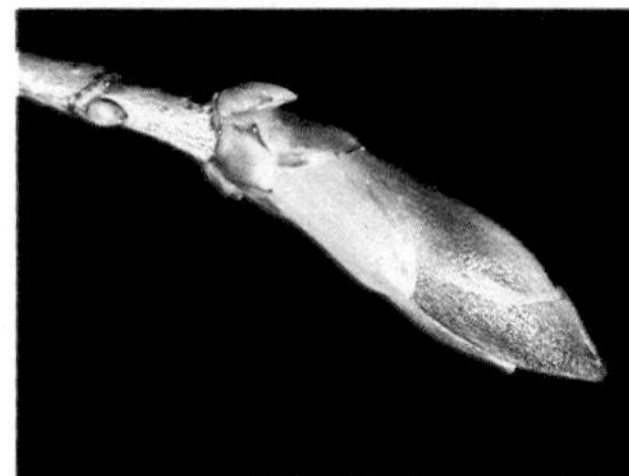

**Figure 21-1** A maple tree before (top) and after (bottom) its buds burst open in the spring

Hormones that affect growth are made in the cells of the growing regions of plants. Locate the growth regions of a stem tip and root tip in Figure 21-2a. Growth in the stem tip causes a plant to grow taller and branches to grow larger. Growth in the root tip allows the roots to grow longer. Notice that the cells producing the growth hormones make up only a small area in the growth regions at the very tips of both shoots and roots.

Just what happens when a plant grows? Look at Figure 21-2a. Growth hormones near the root and stem tips cause cells to increase in number. These cell divisions make the root and shoot grow. Farther back from the growing tip, the hormone causes each cell to get longer. The increase in length of the cells pushes the root downward into the soil or the stem higher into the air.

The hormones in the terminal buds allow the terminal bud to grow while preventing growth of the lateral bud. The side branches that usually develop from lateral buds grow more slowly, Figure 21-2b. If you want a tree or shrub to grow more dense and bushy, you could trim off the tips of a few of branches. The lateral buds on these branches will then take over the role of the terminal bud and grow faster. The plant will get bushy because of the new growth of lateral branches.

**What happens when you remove terminal buds?**



### OPTIONS

### Science Background

Plant hormones are produced in the same tissues they affect. Plant hormones are sometimes referred to as growth regulators because they can speed up or slow down the growth of different plant parts. The hormone that promotes growth causes cells to elongate in the root and stem tips and stimulates cells to divide. The hormones that inhibit growth cause dormancy in plants.

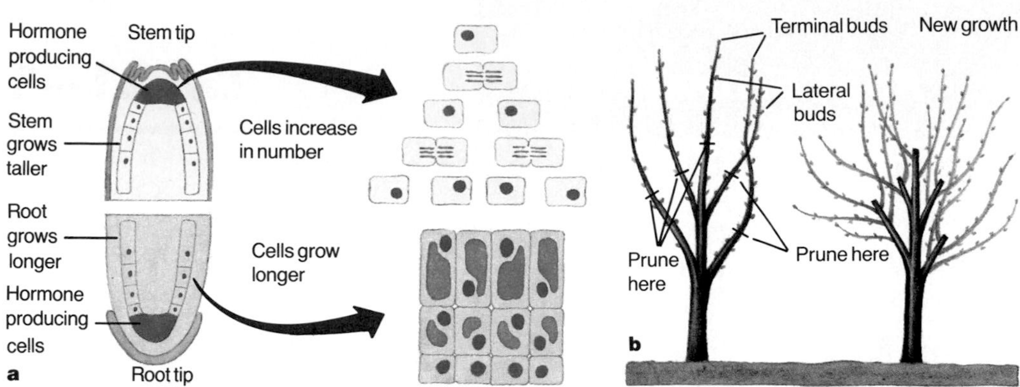


You may have noticed that plants don't flower all at the same time. Flowering is controlled by hormones and by the length of time a plant receives light and dark, Figure 21-3. Some plants bloom in the autumn when the days are getting shorter and nights are getting longer. As the days get shorter, there comes a time when a flowering hormone is produced. Plants that flower when the day length falls below 12 to 14 hours are called **short-day plants.** Ragweed, chrysanthemum, and poinsettia are well know short-day plants. Many spring wildflowers are short-day plants. The flower buds form in the fall as the days get shorter. They don't open up until the next spring.

Other plants bloom in the summer as the days get longer. **Long-day plants** are plants that flower when the day length rises above 12 to 14 hours. Long-day plants include lettuce, clover, and gladiolus. Some plants, such as roses and dandelions, can produce flowers at most times of the year. They don't depend on any particular number of hours of light to produce flowering hormones. Once they start flowering, they bloom until frost stops further growth. These plants are day-neutral. **Day-neutral plants** are plants in which flowering doesn't depend on day length.

**Figure 21-2** Cells in the growth regions of a stem and a root first increase in number at the tips and then get longer farther back (a). If terminal buds are removed, plants grow bushy (b).

**What plants bloom only in summer?**

Short-day plants
24 hours
Long-day plants
Under 12 hours dark
Long day
Over 12 hours dark
Short day

**Figure 21-3** Flowering is often controlled by the number of hours of light and dark that a plant receives.

# 2 TEACH

## MOTIVATION/Software

*How Plants Grow: The Inside Story,* Opportunities for Learning.

## Concept Development

▶ Point out that many weed killers are actually hormones.

▶ Mention that growth hormones are added to some plants to speed up the time it takes for growth. The grower can get the produce to market earlier in the season.

## Guided Practice

Have students write down their answers to the margin question: long-day plants.

## Independent Practice

**Study Guide/***Teacher Resource Package,* p. 121. Use the Study Guide worksheet shown at the bottom of this page for independent practice.

## Student Journal

Have students find out how florists growing plants in a greenhouse have white lilies bloom in time for the Easter season, and have chrysanthemums that normally bloom in the fall bloom during the entire year. Tell students to find out how florists use hormones and light to control the flowering of plants.

## APPLICATION 21

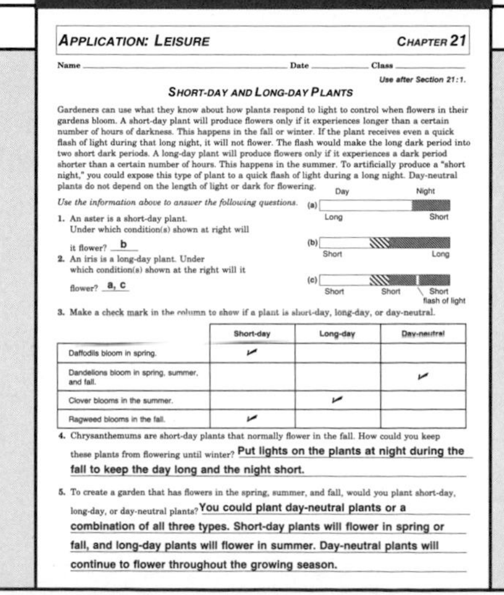

APPLICATION: LEISURE CHAPTER 21

Name ____ Date ____ Class ____

*Use after Section 21:1.*

SHORT-DAY AND LONG-DAY PLANTS

Gardeners can use what they know about how plants respond to light to control when flowers in their gardens bloom. A short-day plant will produce flowers only if it experiences longer than a certain number of hours of darkness. This happens in the fall or winter. If the plant receives even a quick flash of light during that long night, it will not flower. The flash would make the long dark period into two short dark periods. A long-day plant will produce flowers only if it experiences a dark period shorter than a certain number of hours. This happens in the summer. To artificially produce a "short night," you could expose this type of plant to a quick flash of light during a long night. Day-neutral plants do not depend on the length of light or dark for flowering.

*Use the information above to answer the following questions.*

Day Night
(a) Long Short
(b) Short Long
(c) Short Short Short flash of light

1. An aster is a short-day plant. Under which condition(s) shown at right will it flower? b
2. An iris is a long-day plant. Under which condition(s) shown at the right will it flower? a, c
3. Make a check mark in the column to show if a plant is short-day, long-day, or day-neutral.

| | Short-day | Long-day | Day-neutral |
|---|---|---|---|
| Daffodils bloom in spring. | ✓ | | |
| Dandelions bloom in spring, summer, and fall. | | | ✓ |
| Clover blooms in the summer. | | ✓ | |
| Ragweed blooms in the fall. | ✓ | | |

4. Chrysanthemums are short-day plants that normally flower in the fall. How could you keep these plants from flowering until winter? Put lights on the plants at night during the fall to keep the day long and the night short.
5. To create a garden that has flowers in the spring, summer, and fall, would you plant short-day, long-day, or day-neutral plants? You could plant day-neutral plants or a combination of all three types. Short-day plants will flower in spring or fall, and long-day plants will flower in summer. Day-neutral plants will continue to flower throughout the growing season.

## STUDY GUIDE 121

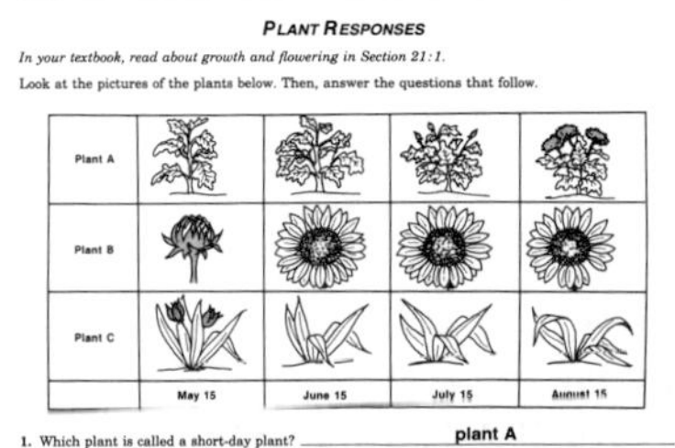

STUDY GUIDE CHAPTER 21

Name ____ Date ____ Class ____

PLANT RESPONSES

*In your textbook, read about growth and flowering in Section 21:1.*

Look at the pictures of the plants below. Then, answer the questions that follow.



1. Which plant is called a short-day plant? plant A
   Explain your answer. The plant forms flowers in the fall of the year when the days are getting shorter, thereby providing less sunlight each new day.
2. Which plant is called a long-day plant? plant C
   Explain your answer. The plant forms flowers in the spring of the year when the days are getting longer, thereby providing more sunlight each new day.
3. Which plant is called a day-neutral plant? plant B
   Explain your answer. The plant forms flowers at most times of the year; it does not depend on day length to produce flowers.

## OPTIONS

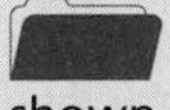

**Application/***Teacher Resource Package,* p. 21. Use the Application worksheet shown here to reinforce short-day and long-day plants.

## Lab 21-1 Root Growth

### Overview

In this lab, students will learn that the growth region of a root is located in the first 2 mm of the root tip. When they cut off this region in a seedling, the root does not continue to grow. In the seedling where the root tip was not removed, growth is continuous.

**Objectives:** After completing this lab, students will be able to (1) **measure** the distance between lines on a root, (2) **explain** why the distance between some lines changed, and (3) **infer** where growth takes place in a root.

**Time Allotment:** 45 minutes setup time and 5-10 minutes per day for 3 days

### Preparation

▶ Germinate the seeds 3-4 days prior to the lab.

**Lab 21-1 worksheet**/*Teacher Resource Package*, pp. 81-82.

### Teaching the Lab

▶ **Troubleshooting:** Make sure students give the ink a chance to dry before they put the seedlings in the plastic bag.

**Performance:** Have students make a hypothesis about the growth regions of the roots of other plants such as corn, peas, and squash. Direct students to test their hypothesis with one or more of the plants by germinating and marking the roots as done in this lab. Have students restate their hypothesis and tell why they accept or reject it.

# Root Growth

# Lab 21–1

## Problem: Where is the growth region in a root?

### Skills

measure in SI, form hypotheses, interpret data, experiment

### Materials

2 bean seedlings
black India ink
paper towel
1 mm graph paper
plastic bag
razor blade
toothpick

### Procedure

1. Copy the data table.
2. Place a bean seedling on the graph paper. The graph paper will serve as a ruler.
3. **Measure in SI:** Dip a toothpick in the ink. Starting from the tip of the root, draw three lines 2 mm apart, Figure A.
4. Repeat steps 2 and 3 with one more bean.
5. Choose one bean seedling and use a razor blade to remove 2 mm from the root tip. **CAUTION:** *Be careful when using a razor blade.*
6. Label two areas of a paper towel Bean 1 and Bean 2. Bean 2 will be the seedling with no root tip.
7. **Experiment:** Wet the paper towel. Place the two seedlings in place. Put the towel in a plastic bag.
8. Write a **hypothesis** in your notebook about which part of the root will grow the most in three days.
9. **Measure in SI:** Examine the seedlings for three days. Each day, use the graph paper to measure the distances between the ink marks. Record these data in your table.

| DAY | DAY 1 | | DAY 2 | | DAY 3 | |
|---|---|---|---|---|---|---|
| BEAN NO. | 1 | 2 | 1 | 2 | 1 | 2 |
| Tip to first mark | 3 mm | — | 11 mm | — | 16 mm | — |
| First to second mark | 3 mm | 2 mm | 3 mm | 2 mm | 3 mm | 2 mm |
| Second to third mark | 2 mm | 2 mm | 2 mm | 2 mm | 2 mm | 2 mm |

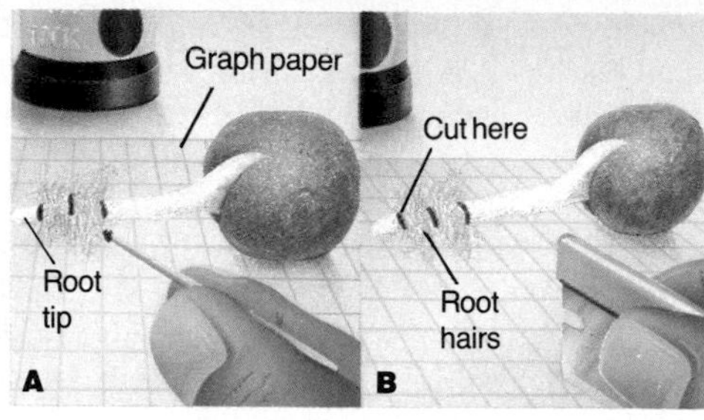


### Data and Observations

1. Which bean had the most root growth?
2. In which part of the root tip of Bean 1 was there the most root growth?

### Analyze and Apply

1. **Interpret data:** Why did the distance between some lines increase or stay the same?
2. Did growth take place in the bean with no root tip? Explain your answer.
3. **Check your hypothesis:** Is your hypothesis supported by your data? Why or why not?
4. **Apply:** What substance was present in the areas that caused the roots to grow?

### Extension

**Compare** how roots of other seedlings grow.



## *ANSWERS*

### Data and Observations

1. bean 1
2. between the root tip and the first line

### Analyze and Apply

1. When the distance between lines increased, it meant the cells were dividing more rapidly in these regions.
2. No, the tip is the place where growth hormone is produced.
3. Students who hypothesized that the roots will grow most at the tip will answer yes.
4. growth hormone

## Tropisms

Plants respond to stimuli such as light and gravity. Animals respond to stimuli because they have sense organs. How do plants respond to stimuli if they don't have a nervous system? Some plant responses, such as flowering and movement towards light, are caused by changes in growth patterns of cells.

A **tropism** (TROH pihz um) is a movement of a plant caused by a change in growth as a response to a stimulus. You may not have seen a tropism because most of the movements are so slow. When a root grows down or when a stem bends over, the plant is showing a tropism.

One tropism that is easy to recognize is phototropism (foh toh TROH pihz um). **Phototropism** is the growth of a plant in response to light. A plant showing phototropism bends toward the light. When leaf stalks bend toward the light, more light falls on the leaves. How is this useful to the plant? It's clear in Figure 21-4 that light is very important to the growth of sunflowers.

Hormones play a role in tropisms. The growth hormone in a stem tip is affected by light. Light causes the growth hormone to move to the dark side of the stem where it causes the cells to lengthen, Figure 21-5. Because the cells on the dark side of the stem become longer, the stem bends toward the light.

**Figure 21-4** Sunflowers were probably named for their obvious response to sunlight.

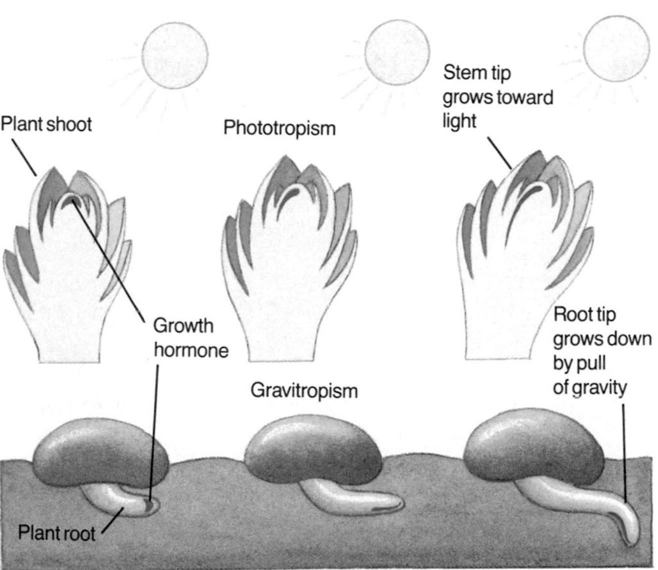


**Figure 21-5** Phototropism causes a stem to grow towards light, and gravitropism causes roots to grow down.



## TEACH

### Concept Development

▶ Explain to students that plant tropisms are actually movements that are caused by unequal growth of some of the plant parts.

▶ Point out that leaves of vines growing on walls or other plants turn outward in response to light.

▶ Explain that hormones also produce effects other than tropisms. A hormone in the leaf stalk helps hold the leaf on the tree during the growing season.

▶ Point out that a tropism is a plant's response to a stimulus. The prefixes *photo-*, *gravi-*, and *thigmo-* define the stimuli.

### GLENCOE TECHNOLOGY

**Videodisc**

**Biology: The Dynamics of Life**
Movie: *Plant Tropisms*
Disc 1, Side 2, Ch. 28

Movie: *Germination*
Disc 1, Side 2, Ch. 27

**The Secret of Life**
*Growing Root Tip*

### CRITICAL THINKING 21

CRITICAL THINKING/PROBLEM SOLVING — CHAPTER 21

Name ______ Date ______ Class ______

*Use after Section 21:1.*

**HOW DO CLIMBING PLANTS RESPOND TO THE SUN?**

Juan was walking past a chain-link fence in his neighborhood. He noticed that some plants called vines were growing up the fence, curled in and out of its wires. Vines such as ivy were attached to the fence by small rootlike structures. Some, including morning glory plants, had small stems that curled around and around the wire. He knew these were called tendrils.

Juan recalled that his parents put up thin stakes in rows in their garden. They planted ivy, bean, and morning glory plants near the stakes. He examined these plants, too, and found both rootlike structures and tendrils.

Juan also noticed that the morning glory flowers sometimes faced toward, and sometimes away from, the street. He made careful observations, and realized that the stem below the flower twisted around during the day. So did the tendrils. Juan used his textbook to understand how vines respond to their environment.

**Analyzing the Problem**

1. What is a tropism? the growth response of a plant to a stimulus in the environment
2. What do most plant tropisms involve? movement
3. What is the substance that travels in the stem of a plant and is involved in plant tropisms? a hormone

**Solving the Problem**

1. To what stimulus are the stems below the flowers responding when they turn during the day? sunlight
2. What is this tropism called? phototropism
3. To what stimulus are the vines responding when they turn to touch the wire or stake? contact
4. What is this tropism called? thigmotropism
5. How do tendrils become curled? As they grow, they twist around and around the wire or stake they touch.
6. How do stems become curled? As the flowers turn to follow the sun, the stems twist around and around.
7. What is the role of hormones in tropisms? Hormones stimulate some of the cells to grow and divide more quickly. This growth makes the stem or tendril bend.

### OPTIONS

**What Tropisms Can Be Seen in Growing Plants?**/*Lab Manual*, pp. 173-176. Use this lab as an extension to tropisms.

**Critical Thinking/Problem Solving/** *Teacher Resource Package*, p. 21. Use the Critical Thinking/Problem Solving worksheet shown here to reinforce how vines respond to the sun.

## TEACH

### ACTIVITY/Demonstration

Coil a piece of string around a wooden dowel to show how a stem of a climbing plant "holds on" to other objects.

### Guided Practice

Have students write down their answers to the margin question: a growth response to contact.

### Skill Check

Pea seeds can be purchased in garden or grocery stores. A foam cup can be used for planting the pea seed.

### Check for Understanding

Hand out a drawing of a bean plant that shows roots, stems, and leaves. Write the following on the chalkboard: gravitropism, phototropism, and thigmotropism. Tell the students to make a drawing to show how the plant responds to each tropism and write a one-sentence definition for each.

### Reteach

Use a sensitive plant to demonstrate its response to touch. Have students compare its response to the three tropisms studied in this chapter.

### Independent Practice

**Study Guide**/*Teacher Resource Package,* p. 122. Use the Study Guide worksheet shown at the bottom of this page for independent practice.

### Skill Check

**Observe:** Observe a climbing plant. Plant a pea seed. Place a pencil in the pot. Observe thigmotropism as the plant grows. *For more help, refer to the* ***Skill Handbook,*** *pages 704-705.*

**What is thigmotropism?**

Roots grow downward into the soil as a result of gravitropism (grav uh TROH pihz um). *Gravi* means having weight. **Gravitropism** is the response of a plant to gravity. As in the stem, the growth region of the root produces a growth hormone.

If you put a sprouting bean on its side on top of some potting soil, the root will grow downward into the soil within a few days. Large amounts of growth hormone in the root move through the cells to the lower side of the root, as if pulled down by their weight, Figure 21-5. This diffusion is in response to gravity. The hormone slows down the growth in length of the lower cells. The cells above continue to grow longer and the root grows down into the soil.

Grape vines, honeysuckle, and greenbriar are very common climbing plants in the forests of the United States. These climbing plants hold onto trees or fences by thigmotropism. *Thigma* means touch. **Thigmotropism** is a plant growth response to contact. Many climbing plants have twining parts called tendrils that respond to contact with a likely support by coiling around this object. The start of coiling can happen within ten minutes. Peas, beans, and squash are examples of food plants with tendrils that show thigmotropism, Figure 21-6.

**Figure 21-6** Thigmotropism helps a plant climb a fence.

444

## OPTIONS

**Transparency Master**/*Teacher Resource Package,* p. 97. Use the Transparency Master worksheet shown here to help students with root growth.

**Plant Growth Response**/*Transparency Package,* number 21. Use color transparency number 21 as you teach plant growth responses.

### STUDY GUIDE 122

STUDY GUIDE — CHAPTER 21

Name ______ Date ______ Class ______

PLANT RESPONSES

*In your textbook, read about tropisms in plants in Section 21:1. Then, read the description below and answer the questions that follow.*

Some plants in a greenhouse were tipped over and not set upright for several weeks. When the greenhouse manager found them, and stood them up, they looked like those in the diagram.

A B C

4. Which plant(s) were not tipped over? C Explain your answer. The plant roots are growing down into the soil and the stem is growing upward toward the light. This is the way plants normally grow.
5. Which plant(s) were tipped over? A and B Explain your answer. When a plant is turned on its side, the plant roots begin to grow downward and the stem begins to grow upward.
6. What is the name of the tropism that describes how the roots are responding? gravitropism
7. What is the name of the tropism that describes how the stem is responding? phototropism
8. Define each tropism. Gravitropism is the response of plants to gravity. Phototropism is the response of plants to light.
9. Which part of the plant responds to gravity? roots How can you tell? Roots grow toward Earth, where the pull of gravity is strongest.
10. Which part of the plant responds to light? stem How can you tell? Stems grow toward light.

### TRANSPARENCY 97

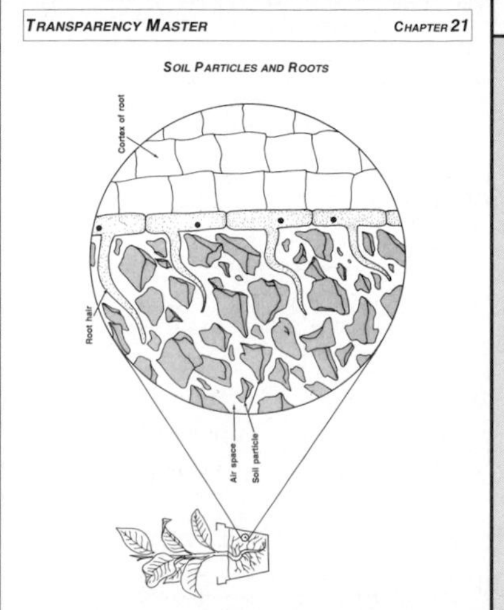

# Lab 21–2 Plant Responses 

## Problem: How does gravitropism affect the growth of roots?

### Skills

observe, form hypotheses, interpret data

### Materials

| | |
|---|---|
| petri dish | transparent tape |
| 2 paper towels | 4 presoaked corn seeds |
| marking pen | |

### Procedure

1. Copy the data table.
2. Soak two paper towels in water.
3. Wrinkle the towels and place them in the bottom half of the petri dish.
4. Place four presoaked seeds on top of the wet paper towels as shown in Figure A. Place the seeds so the narrow end is pointing toward the center of the dish.
5. Put the top on the petri dish. The lid will press the seeds into the wet towels.
6. Seal the lid with transparent tape.
7. Draw an arrow on the top of the petri dish with a marking pen as shown in Figure A. This will show the direction of the force of gravity.
8. Stand the petri dish on edge so that the arrow is pointing straight down.
9. Write a **hypothesis** in your notebook about which direction you think the roots will grow.
10. **Observe** the seeds for the next four days. Each day, fill in the data table with arrows to show the direction of root growth for each of the four seeds.

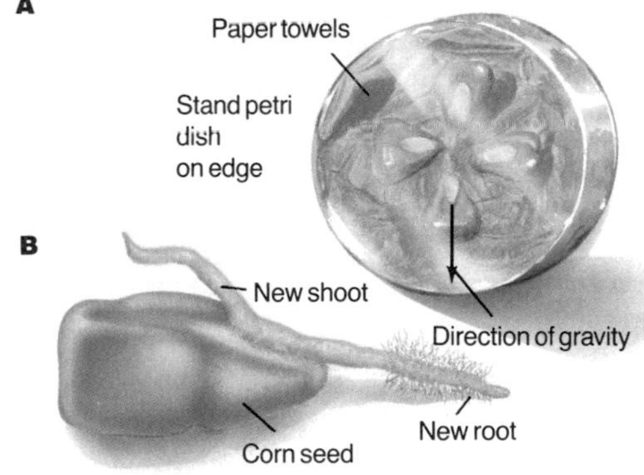


### Data and Observations

1. In what direction were the roots growing on Day 1?
2. What is the direction of root growth at the end of four days?

| SEED | DIRECTION OF GROWTH OF ROOT | | | |
|---|---|---|---|---|
| | Day 1 | Day 2 | Day 3 | Day 4 |
| | → | ↘ | ↓ | ↓ |
| | ↓ | ↓ | ↓ | ↓ |
| | ↗ | ↗ | → | ↓ |
| | ← | ↙ | ↙ | ↓ |

### Analyze and Apply

1. **Interpret data:** To what factor of the environment are the roots responding?
2. What is this response called?
3. **Check your hypothesis:** Is your hypothesis supported by your data? Why or why not?
4. **Apply:** If this same experiment were carried out in space, would the results be the same? Explain your answer.

### Extension

**Design an experiment** that shows the effect of phototropism.



## ANSWERS

### Data and Observations

1. straight towards the center of the dish
2. straight down in the direction of the arrow

### Analyze and Apply

1. the force of gravity
2. gravitropism
3. Students who hypothesized that the roots would all grow downward will answer yes.
4. No, the force of gravity is very weak in space. The roots would all grow straight.

## Lab 21-2 Plant Responses

### Overview

In this lab, students will conduct an experiment to find out how roots respond to gravity. Students will see that the roots grow downward shortly after germination.

**Objectives:** After completing this lab, students will be able to (1) **use** controls and variables in an experiment, (2) **interpret** data from an experiment, and (3) **explain** the effect of gravity on corn roots.

**Time Allotment:** 45 minutes, then 5-10 minutes per day for 4 days

### Preparation

▶ The night before you start this lab, soak the seeds in water. In the morning, place them between wet paper towels in a tray.

**Lab 21-2 worksheet**/*Teacher Resource Package,* pp. 83-84.

### Teaching the Lab

▶ **Troubleshooting:** Make sure students put enough wet paper towels in the petri dish to hold the seeds in place.

**Cooperative Learning:** Divide the students into teams of two. For more information, see pp. 22T-23T in the Teacher Guide.

### ✓ ASSESSMENT

**Oral:** Tell students to use their results from Lab 21-1 to predict what will happen to the roots in this lab if they were to cut 2 mm from the corn seedling root tips. Have students cut the root tips and place the corn seedlings in the dish as shown. They should report their findings to the class.

## TEACH

### Check for Understanding

Have students respond to the first three questions in Check Your Understanding.

### Reteach

**Reteaching**/*Teacher Resource Package*, p. 63. Use this worksheet to give students additional practice with plant responses.

**Extension:** Assign Critical Thinking, Biology and Reading, or some of the **OPTIONS** available with this lesson.

## 3 APPLY

### ACTIVITY/Demonstration

To demonstrate the bending of a stem, inflate a long, thin balloon with air. Draw a grid around the balloon. Let the balloon represent a stem, and the grid the cells of the stem. Bend the balloon (stem) to the right and ask students what happened to the cells on the left. They should respond that the cells became longer. Explain that the increase in length of the cells on one side of a stem causes the stem to bend toward light.

## 4 CLOSE

### Software

*The Plant Growth Simulator,* Queue, Inc.

### Answers to Check Your Understanding

1. The bud grows very slowly.
2. when the day length falls below 12 to 14 hours
3. In plants, phototropism is a growth response to light and thigmotropism is a growth response to contact.
4. Take them indoors and increase or decrease the amount of light they receive.
5. Most plant tropisms occur too slowly to be observed.

**Figure 21-7** Leaf and flower movements in these bloodroots (left) and movements of hairs on a sundew leaf (right) are controlled by changes in cell pressures. Where are the cells that control these movements? at the bases of leaves, petals, and hairs

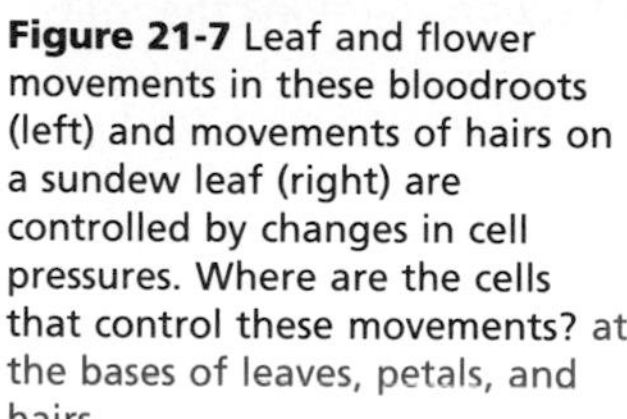

### Skill Check

**Understand science words: biennial.** The word part *bi* means two. In your dictionary, find three words with the word part *bi* in them. *For more help, refer to the **Skill Handbook,** pages 706-711.*

## Other Plant Responses

Not all responses of plants are caused by hormones. Some plant parts move because of changes in pressure inside the cells. You read in Chapter 19 that stomata in the leaves open and close for gas exchange. Stomata open and close when the water pressure inside the guard cells changes. A stoma opens when the guard cells are full with water. When water is low in the cell, the stoma closes.

There are many other plant movements that are caused by changes in cell pressure. For example, the movement of flowers and leaves shown in Figure 21-7 and described earlier for bean plant leaves are caused by changes in cell pressure. At night, the cells at the bases of petals and leaf stalks in some plants lose water and these parts fold up. The cells regain water in the daytime, and the leaves and flowers open out.

Some plants have leaves that trap insects, Figure 21-7. When the hairs on the leaves are touched, such as when an insect lands on them, the pressure in the cells at the base of hairs changes, causing the hairs to fold over rapidly and trap the insect.

### Check Your Understanding

1. What is the effect of a high level of hormone in a bud?
2. What causes a short-day plant to flower?
3. What is the difference between phototropism and thigmotropism?
4. **Critical Thinking:** How could you get plants to flower out of their usual flowering season?
5. **Biology and Reading:** Why is it difficult to observe a plant tropism?



## OPTIONS

### ACTIVITY/Enrichment

Have a few kinds of insect-eating plants to show students. You can feed the plants wingless fruit flies to see how they respond to the insects.

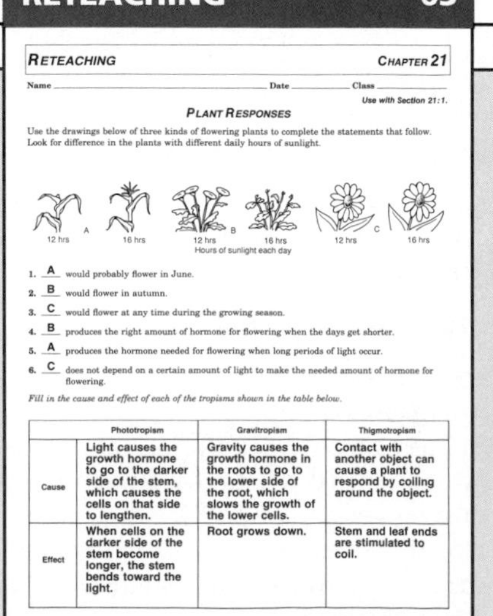

RETEACHING 63

RETEACHING CHAPTER 21

Name ________ Date ________ Class ________

*Use with Section 21:1.*

PLANT RESPONSES

Use the drawings below of three kinds of flowering plants to complete the statements that follow. Look for difference in the plants with different daily hours of sunlight.



1. A would probably flower in June.
2. B would flower in autumn.
3. C would flower at any time during the growing season.
4. B produces the right amount of hormone for flowering when the days get shorter.
5. A produces the hormone needed for flowering when long periods of light occur.
6. C does not depend on a certain amount of light to make the needed amount of hormone for flowering.

*Fill in the cause and effect of each of the tropisms shown in the table below.*

| | Phototropism | Gravitropism | Thigmotropism |
|---|---|---|---|
| Cause | Light causes the growth hormone to go to the darker side of the stem, which causes the cells on that side to lengthen. | Gravity causes the growth hormone in the roots to go to the lower side of the root, which slows the growth of the lower cells. | Contact with another object can cause a plant to respond by coiling around the object. |
| Effect | When cells on the darker side of the stem become longer, the stem bends toward the light. | Root grows down. | Stem and leaf ends are stimulated to coil. |

# 21:2 Growth Requirements

Plants can live from a few months to thousands of years. As a plant grows, the numbers of cells increase. Plants use the energy from the sun and water, minerals, and carbon dioxide from the environment to build these new cells. Factors needed by a plant for proper growth are called growth requirements. Growth requirements for plants include light, air, water, minerals, and the right temperature. Water and minerals are usually taken from the soil, so the condition of the soil is often another growth requirement. Growth may be slowed if any of these factors is not present in the right amount.

**Objectives**

4. **Compare** annual, biennial, and perennial plants.
5. **List** the growth requirements of plants.

**Key Science Words**

annual
biennial
perennial
fertilizer

**What are five growth requirements of plants?**

## Seasonal Growth

Most of the plants you know, such as oak trees, pine trees, poison ivy, daisies, lettuce, and onions reproduce by seeds. Seed-producing plants can grow from one to many years. Some herbaceous plants complete their growth, produce seeds, and then die within one year. Plants that complete their life cycle within one year are called **annual** (AN yul) plants. Garden plants like corn, peas, and lettuce are annual plants, Figure 21-8a.

Some herbaceous plants need two years to complete growth and produce seeds. These are called biennials, Figure 21-8b. **Biennial** (bi EN ee ul) plants produce seeds at the end of the second year of growth and then die. Cabbages and turnips are biennial plants. This is why you don't see flowers or seeds on a head of cabbage.

**When do biennial plants produce seeds?**

a

b

**Figure 21-8** Pea plants (a) are annuals, hollyhocks (b) are biennials. What are dandelions? perennials



## 21:2 Growth Requirements

## PREPARATION

### Materials Needed

Make copies of the Focus, Enrichment, Skill, Study Guide, and Reteaching worksheets in the *Teacher Resource Package.*

▶ Obtain radish seeds, cups, and soil for the Mini Lab.

### Key Science Words

| annual | perennial |
|---|---|
| biennial | fertilizer |

### Process Skills

In the Mini Lab, students will experiment.

## 1 FOCUS

▶ The objectives are listed on the student page. Remind students to preview these objectives as a guide to this numbered section.

### ACTIVITY/Brainstorming

Ask students to list factors that help a plant grow better.

### Guided Practice

Have students write down their answers to the margin questions: light, air, water, minerals, the right temperature; in the second year.

**Focus**/*Teacher Resource Package,* p. 41. Use the Focus transparency shown at the bottom of this page as an introduction to growth requirements.

**FOCUS 41**

FOCUS CHAPTER 21

Name ______ Date ______ Class ______

PLANT NEEDS

Plants need certain things for growth. The plant on top is healthy. The plant on the bottom is not. List the things you think the plant on the bottom needs.

## OPTIONS

### Science Background

Plants have many requirements for growth. The five listed in this section are the most important. Growth in a plant depends on the interaction of these five factors. Each of these requirements is needed in a specific amount by the plant. For instance, without light the plant would not be able to make chlorophyll even if the other factors were present in optimum amounts.

# 2 TEACH

### MOTIVATION/Filmstrip

Show the filmstrip *Germination and Plant Growth,* Clearvue, Inc.

### Concept Development

▶ Point out that light is the energy source for plants during photosynthesis.

▶ Point out that the amount of light a plant gets each day varies with the amount of clouds in the sky.

▶ Explain that some plants have a specific tolerance for shade or direct sunlight.

▶ Explain that biennials store in their roots food that was made during the first year of growth, and they use this food to produce flowers during the second year of growth.

▶ Point out that the annual plants seen in a field or meadow grow from new seeds each year.

### Idea Map

Have students use the idea map as a study guide to the major concepts of this section.

### Mini Lab

Get potting soil, radish seeds, and cups ready. The plants grown in the dark will be tall, yellow, and thin stemmed, and will not be as straight as those grown in the light.

### ASSESSMENT

**Skill:** Have student teams repeat the activity by covering the top of the cups with colored cellophane to find out what effect colored light has on the seedlings. The soil level should be 4 to 5 cm below the top of the cup to give the seedlings space to grow upward. Cellophane can be attached to the cup with rubber bands.

A plant that lives longer than two years is a perennial (puh REN ee ul). A **perennial** plant is one that doesn't die at the end of one or two years of growth. Perennials usually produce seeds year after year. Many woody plants, such as shrubs and trees, are perennials. Their woody aboveground parts don't die at the end of each growing season. Some herbaceous plants, such as daffodils, onions, and many spring wildflowers, are also perennials. The aboveground parts of herbaceous perennials do die at the end of each growing season. Herbaceous perennials have underground parts with stored food that is used to produce new growth in the spring. Daffodils, onions, and many spring wildflowers are herbaceous perennials.

**Study Tip:** Use this idea map as a study guide to plant growth requirements. Which two requirements are found in soil?

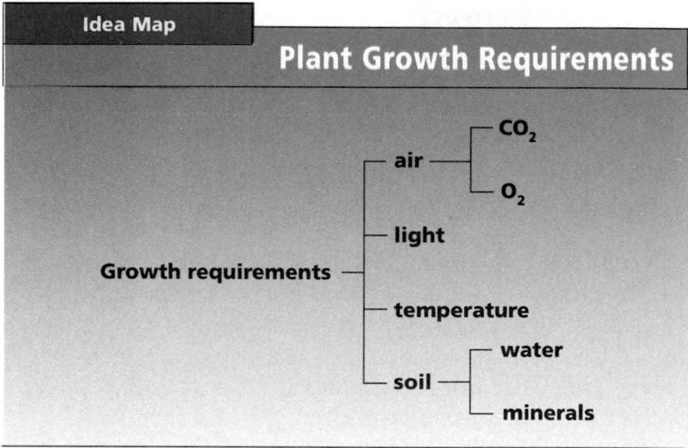


#### Mini Lab

**Do Plants Need Light?**

**Experiment:** Plant 12 radish seeds in moist soil in each of two paper cups. Place one cup in a well-lit area and the other in a cupboard. Compare the growth of the plants for a week. *For more help, refer to the **Skill Handbook,** pages 704-705.*

## Light

Plants need light for proper growth. The amount of light a plant gets controls much of its development. Production of chlorophyll, growth of buds, time of flowering, and ability to make food are all plant processes that need light.

Some plants need direct sunlight in order to grow well, others don't. Impatiens, popular bedding plants with brightly-colored flowers, grow best if they aren't planted in direct sunlight. Tomatoes and corn grow best in direct sunlight. Corn grown in the shaded part of a garden will produce short stalks and small ears. Tomatoes will take longer to ripen if grown in the shade. Plants grown in areas that are too shady will grow tall, be lighter green in color, and have fewer, but larger, leaves.



### OPTIONS

**Is Light an Important Growth Requirement for Plants?**/*Lab Manual,* pp. 177-180. Use this lab as an extension to growth requirements of plants.

**Enrichment**/*Teacher Resource Package,* p. 22. Use the Enrichment worksheet shown here to help students understand tolerance in plants.

### ENRICHMENT 22

ENRICHMENT CHAPTER 21

Name ______ Date ______ Class ______

TOLERANCE IN PLANTS

If you hiked up a mountain that was several thousand kilometers high, you might notice that plants growing at the bottom of the mountain are not all the same as those growing on top. The air and soil on top of the mountain are cooler than at the bottom. Some plants can grow only in warm weather, some can grow only in cold weather, and others can grow in warm and cold weather. The ability of a plant to live and grow in an unfavorable environment is called tolerance.

*Look at the drawing on the next page. Then, fill in the table below. Use check marks to show in which zones the plants in the drawing grow. Use the table and a dictionary or encyclopedia to help you answer the questions that follow.*

1. Which plant(s) grow in all four areas of the mountain? Explain your answer using the word *tolerance.* **Mosses have a tolerance for wide temperature differences because they can live in all zones.**
2. Which plant(s) grow in only one of the four areas? Explain your answer using the word *tolerance.* **Palm trees and live oak trees have a tolerance for only the warmest temperatures because they grow only in the warmest zones.**
3. Do spruce trees have the same tolerance as white oaks? Explain your answer. **No, spruce trees have tolerance for colder temperatures than white oaks. Spruce trees grow farther up the mountain than white oaks do.**

| Plants | Zone A | Zone B | Zone C | Zone D |
|---|---|---|---|---|
| mosses | ✓ | ✓ | ✓ | ✓ |
| white oak | ✓ | ✓ | | |
| birch | | ✓ | ✓ | |
| hickory | ✓ | ✓ | | |
| pine | ✓ | ✓ | ✓ | |
| live oak | ✓ | | | |
| palm | ✓ | | | |
| spruce | ✓ | ✓ | ✓ | |
| maple | ✓ | ✓ | | |

a

b

**Figure 21-9** The environments of a cactus (a) and a fern (b) are different in rainfall levels.

## Water

The two plants shown in Figure 21-9 have different water needs. The cactus can grow in a dry, sandy soil. The fern needs a moist, rich soil to grow well. A desert cactus takes up water from rain that comes only a few times each year. The cactus stores the water for use between rainfalls. The fern isn't adapted to store water. It needs to live in an area where the soil is damp most of the time.

How can some plants live in a desert with little rainfall?

Water has many important uses in plants. Water makes up 30 to 90 percent of a plant's mass. Plants use some of this water for photosynthesis. Minerals taken up by the roots are dissolved in water. Once inside a plant, this water carries the minerals throughout the plant. A farmer or gardener needs to water crops if there isn't enough rain to supply the needs of the plants.

## Minerals

Most plants grow in soil. Soil contains many minerals important to plant growth. Plants also use minerals for making chlorophyll and cell walls. Three important minerals used by plants are nitrogen, phosphorus, and potassium.

Plants may absorb more of one type of mineral than another. Why? Since plants need some minerals such as nitrogen, in larger amounts than others, the soil can become drained of that mineral. If the soil of a garden or field is low in one kind of mineral, fertilizer can be added. A **fertilizer** is a substance made of minerals that improves soil for plant growth. When you want to help your flowering plants bloom or your shrubs grow faster, you might apply different types of fertilizer. The labels on fertilizer packets give the percentages of nitrogen, phosphorus, and potassium that the fertilizers contain, Figure 21-10.

**Figure 21-10** Fertilizer can supply missing minerals to soil.

### STUDY GUIDE 123

STUDY GUIDE — CHAPTER 21

Name ______ Date ______ Class ______

GROWTH REQUIREMENTS

*In your textbook, read about growth requirements of plants in Section 21:2. Then, read the following description of a set of experiments on plant growth.*

Experiments were done to find what effects the amount of water, light, minerals, and soil had on some plants. In each experiment, the amounts of light, water, minerals, soil, and temperature were kept the same except where shown in the figure on the next page. The plants were grown for nearly two months. Answer the questions after you have studied the diagrams.

1. In experiment 1, which plant grew best? plant C How often was this plant watered? every three days
2. What would happen to the plant in answer 1 if it were planted in an area where it rained four days each week? The plant would not grow as well as it did in the experiment.
3. Does the kind of plant used in experiment 1 grow well with a great amount of water? no Explain your answer. This kind of plant does not grow well if watered too often, such as three times a day or even once a day.
4. What growth requirement is being tested in experiment 2? light
5. In experiment 2, which plant grew best? plant A How much sun did the plant receive? full sun
6. Suppose you bought several of these plants to put in your flower beds. Where would you *not* plant them? in shady areas Why? This kind of plant does not grow well in the shade.
7. What growth requirement is being tested in experiment 3? type of soil needed
8. In experiment 3, which plant grew best? plant B What type of soil was this plant grown in? potting soil
9. Could the type of plant in experiment 3 be grown in the desert? no Why or why not? The desert is mostly sandy soil. This plant does not grow well in sandy soil.

### STUDY GUIDE 124

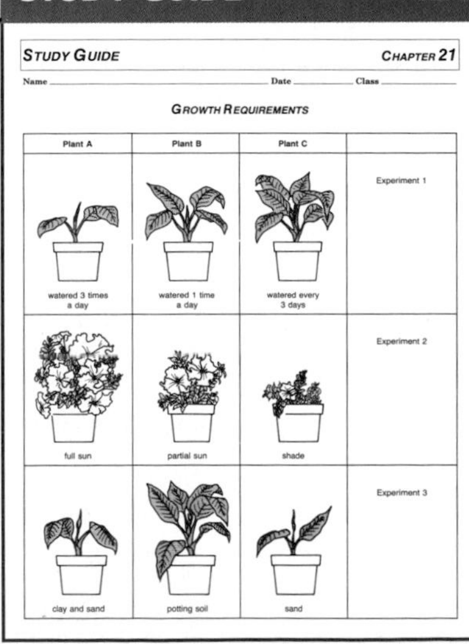

STUDY GUIDE — CHAPTER 21

Name ______ Date ______ Class ______

GROWTH REQUIREMENTS

| Plant A | Plant B | Plant C | |
|---|---|---|---|
| watered 3 times a day | watered 1 time a day | watered every 3 days | Experiment 1 |
| full sun | partial sun | shade | Experiment 2 |
| clay and sand | potting soil | sand | Experiment 3 |

# TEACH

### Concept Development

▶ Explain that some perennials do not bloom during their first year of growth.

### Bridging

Tie in Chapters 19, 20, and 21 by reviewing the importance of water loss in plants. Explain that certain plants, like the cactus, can store lots of water in their waxy coated stems so that they lose very little water. Plants living in drier environments have fewer stomata and smaller leaves to prevent excessive water loss.

### Portfolio

A person living in Washington, DC, decided to raise cactus plants and knew they had to be planted in sandy soil, but was not sure how often to add water to the growing plants. A friend told her that she should watch the weather forecast for the different cities in the United States, and when it rains in Phoenix, Arizona, she should add water to her cactus plants. Have students write a response to whether or not they think she received good advice.

### ACTIVITY/Display

Cut out sections of several fertilizer bags that list the mineral content. Display these for the students. List below the sections what each type of fertilizer is best used for.

### Guided Practice

Have students write down their answers to the margin question: The plants store water.

### Independent Practice

**Study Guide**/*Teacher Resource Package*, pp. 123-124. Use the Study Guide worksheets shown at the bottom of this page for independent practice.

## TEACH

### Concept Development

▶ Explain that the leaves and stems of plants that die in the fall are broken down into nutrients and recycled by other plants.

▶ Point out that rich soil contains a large amount of dead matter.

▶ Explain that dead matter in soil is food for earthworms and other invertebrates and increases the ability of the soil to hold water.

▶ Emphasize that soil provides oxygen and minerals for plants, and a substrate to which they can attach.

▶ Point out that temperature is the growth requirement that humans are most familiar with.

▶ Explain that temperature varies with the intensity of heat energy.

▶ Reinforce that many plants grow slower when the temperature is lower because chemical reactions are slowed down.

### Portfolio

Have students use a sheet of typing paper to design a flower garden that is 6 m × 2 m in size. The flower garden is to have different flowers blooming from spring through fall. Taller plants are to be placed in the back and smaller plants are in the front. Perennial plant catalogs will give data on the height of plants and the time of blooming. Students are to use symbols for each plant species (#, *, ∧, and so on) and make a key for which symbol represents each plant.

### Bio Tip

**Leisure:** Because flowers bloom at different times of the year, you can design a garden so that you will have flowers all year round.

## Soil

You have learned that soil supplies water and minerals to plants. Soil also gives the plant a place to set down its roots. Soil is a mixture that contains minerals, living organisms, dead or decaying matter, air, and water. The minerals come from rocks that have been worn down by wind and rain.

There are three main types of soil, Figure 21-11. Clay soil is hard because the soil particles are small and packed closely together. There are fewer air spaces. Other soil is crumbly because the soil particles are larger. There are more air spaces between the soil particles. Crumbly soil is better for plant growth. The air spaces in this type of soil give roots room to grow. The air spaces also keep roots dry. If roots are soaked too long, they may rot.

If the soil in your garden were like clay, you might first loosen the soil with a hoe. Also, you could add some sand and peat moss. Adding sand and dead organic matter to loosened clay soil increases the number of air spaces and adds nutrients.

Some soils are poor for plant growth. Pure sandy soil is an example of poor soil because it allows water to drain through too fast. The air spaces can't trap and hold water. Plants in desert environments often have enormous root systems to collect as much water as possible.

## Temperature

Which plant in Figure 21-12 will be most affected by colder temperatures? Some plants can live in cold climates, while others die when the temperature falls below freezing.

a

b

c

**Figure 21-11** Clay soil (a) is very fine; crumbly soil (b) is rich in organic matter; and sandy soil (c) is coarse and can't hold water.



## OPTIONS

### ACTIVITY/Challenge

Have students put a small spoonful of soil into a large test tube filled with water. Tell them to shake the tube until the clump of soil has broken apart. After the soil has settled overnight, they should notice that the layers from top to bottom are silt, clay, sand, and gravel.

**Skill**/*Teacher Resource Package,* p. 22. Use the Skill worksheet shown here to have students interpret data.

**SKILL 22**

**SKILL: INTERPRET DATA** CHAPTER 21

Name ______ Date ______ Class ______

*For more help, refer to the Skill Handbook, pages 704-705.* *Use after Section 21:2.*

**GROWTH OF FLOWERING PLANTS**

*Read the following paragraph. Then, study the data in the table and answer the questions.*

For their first flower garden, the Smiths bought several kinds of plants. Some of the plants would grow well in the sun and others would grow better in the shade. They also knew they had bought some annuals, some biennials, and some perennials. They planted some of each kind of plant in the shade and the rest in sunny areas. The table shows their record of how the plants grew.

| Name of plant | Where planted | | When plant bloomed | New growth each spring | Notes |
|---|---|---|---|---|---|
| | Sunny | Shady | | | |
| Impatiens | X | | June-October | No | Stem long, thin; plant droopy |
| | | X | June-October | No | Stems thick, green; many flowers |
| Mint | X | | | No | Stems bent over in a month; plants died. |
| | | X | May-June | Yes | Stems green, thick; many flowers |
| Chrysanthemum | X | | September-October | Yes | Stems green; many flowers |
| | | X | | No | Stems tall, yellow; died in two months |
| Foxglove | X | | June-July of second year | Only during second spring | Stems tall, green |
| | | X | | No | Stems and leaves yellow; died |

1. Which of the plants were annuals? Explain. **Impatiens, died after the first season.**
2. Which of the plants were biennials? Explain. **Foxglove, it produced flowers at the end of the second season and then died.**
3. Which of the plants were perennials? Explain. **Mint and chrysanthemum, they did not die at the end of one or two years.**
4. Which of the plants should grow best under trees and shrubs? Explain. **Impatiens and mint, they grow best in shade.**

a

b

Figure 21-12 Some plants live only in cold climates (a), whereas others, such as these palm trees, couldn't survive low temperatures (b).

Temperature affects the rate at which a plant will take up water, and the rates of photosynthesis and growth.

Farmers and gardeners know that they can plant most seeds or plants only after all danger of frost has passed. If they do their planting before this time, their plants will freeze or their seeds won't grow. Some seeds, however, need cold temperatures. For example, one kind of wheat called winter wheat is planted in northern states in the fall. It starts to grow until the temperature gets near freezing. The wheat is then protected by snow cover throughout the winter and begins to grow again in the spring. Some seeds, such as those of apples, will not grow unless they have first been frozen.

## Check Your Understanding

6. What are two ways in which annual and perennial plants differ?
7. What are three minerals important to the growth of plants?
8. Compare the light requirements of two different plants.
9. **Critical Thinking**: If you collected plants that were native to Texas and brought them home with you to Maine, what would you need to do to keep your plants alive?
10. **Biology and Reading**: Why don't you see flowers or seeds on a head of cabbage?

## STUDY GUIDE 126

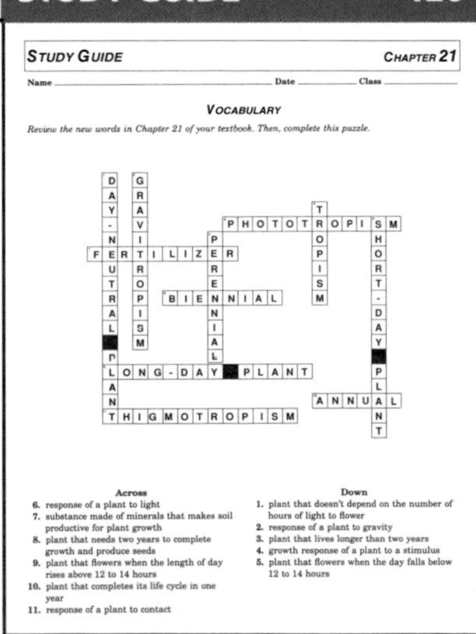

STUDY GUIDE — CHAPTER 21

Name ________ Date ________ Class ________

VOCABULARY

*Review the new words in Chapter 21 of your textbook. Then, complete this puzzle.*



Across
6. response of a plant to light
7. substance made of minerals that makes soil productive for plant growth
8. plant that needs two years to complete growth and produce seeds
9. plant that flowers when the length of day rises above 12 to 14 hours
10. plant that completes its life cycle in one year
11. response of a plant to contact

Down
1. plant that doesn't depend on the number of hours of light to flower
2. response of a plant to gravity
3. plant that lives longer than two years
4. growth response of a plant to a stimulus
5. plant that flowers when the day falls below 12 to 14 hours

## RETEACHING 64

RETEACHING — CHAPTER 21

Name ________ Date ________ Class ________

*Use with Section 21:2.*

GROWTH REQUIREMENTS

1. The growth requirements of plants are **light, water, nutrients, soil, and temperature**.
2. If any of the growth requirements are not just right, **growth may be slowed**.
3. Suppose a group of plants of the same type and age are grown—some of them in a lighted area and some in a dark cabinet.
   a. After seven days the plant growing in the lighted area would have **normal leaves and a normal stem**.
   b. The plant in the dark area would have a **yellowish stem and leaves**.
4. Corn grown in the shaded part of the garden will produce **short stalks and ears**.
5. Cactus plants **store water** to help them through the dry season.
6. Ferns need to live in an area where the **soil is damp most of the time**.
7. **Water** makes up 30 to 90 percent of a plant's mass.
8. Plants can absorb minerals from the soil because they are **dissolved in water**.
9. Plants use minerals for growth and for **making chlorophyll and cell walls**.
10. Two kinds of minerals that plants need are **nitrates and phosphates**.
11. When necessary, **fertilizers** can supply minerals that are missing from soil.
12. Loosening the soil around plants provides **larger air spaces around the roots**.
13. **Biennial** plants produce seeds at the end of the second year of growth.
14. **Perennial** plants usually do not die at the end of one or two years of growth, and they produce seeds from year to year.

# TEACH

## Independent Practice

**Study Guide**/*Teacher Resource Package*, p. 126. Use the Study Guide worksheet shown at the bottom of this page for independent practice.

## Check for Understanding

Have students respond to the first three questions in Check Your Understanding.

### Reteach

**Reteaching**/*Teacher Resource Package*, p. 64. Use this worksheet to give students additional practice with plant growth requirements.

**Extension:** Assign Critical Thinking, Biology and Reading, or some of the **OPTIONS** available with this lesson.

# 3 APPLY

## ACTIVITY/Audiovisual

Show the video *Plant Dynamics*, Carolina Biological Supply.

# 4 CLOSE

## Software

*Plants,* Opportunities for Learning.

## Answers to Check Your Understanding

6. Annual plants complete their life cycle and die within one year. Perennial plants live for more than two years and produce seeds year after year.
7. nitrogen, phosphorus, potassium
8. Some plants, such as tomatoes, need direct sunlight; others, such as impatiens, need shade.
9. You would need artificial light and controlled temperatures similar to those in Texas.
10. Cabbages are biennials and they are eaten in the first year. Flowers and seeds are not formed until the second year.

## 21:3 Plant Diseases and Pests

### PREPARATION

#### Materials Needed

Make copies of the Reteaching and Study Guide worksheets in the *Teacher Resource Package.*

#### Key Science Words

No new science words will be introduced in this section.

### 1 FOCUS

▶ The objectives are listed on the student page. Remind students to preview these objectives as a guide to this numbered section.

#### ACTIVITY/Display

You can preserve many fungi that are plant parasites by putting the plant part with the fungus attached into a preservative such as alcohol. These can be displayed and observed in the lab. Shelf fungi can be air dried. Excellent photos of plant diseases are available from your state Department of Agriculture.

#### Guided Practice

Have students write down their answers to the margin questions: a virus; it grows into and blocks xylem tissue.

#### *Portfolio*

Collect information sheets from plant nurseries or garden centers about plant diseases and pests. Have students use them to make lists of plant diseases and pests other than those found in Section 21:3. They are to have the same three headings that are found in the book. For diseases, they should list the name of the plant disease, the type of plant part it infects, and what the damage looks like on the plant. For insects, they should list the plant on which it is a pest and how it damages the plant. Have students report their findings to the class.

## 21:3 Plant Diseases and Pests

**Objectives**

6. **Explain** how bacteria and viruses affect plants.
7. **Describe** how a fungus can kill a plant.
8. **Relate** two ways that insects damage plants.

Diseases and insect pests can slow down or stop plant growth. Diseases affect all plants.

### Bacteria and Viruses

Bacteria cause many plant diseases. They can enter a plant through the stomata or small cuts. Once inside, the bacteria destroy plant cells when they invade the cytoplasm. If the bacteria spread throughout the plant, the plant might die. Bacteria are a common cause of blister spots on fruit and leaves, Figure 21-13a.

**What causes yellow spotting on leaves?**

Viruses also cause plant diseases. Garden plants and houseplants can be affected by a plant virus that causes small yellow spots on leaves. Eventually the yellow spots darken as the tissue dies. This kind of disease is called a mosaic disease, Figure 21-13b.

Viruses often cause the growth of tumors in leaves. The leaves become deformed. Plant viruses don't infect animals so it doesn't hurt you to touch these leaves.

### Fungi

**How does a fungus destroy an elm tree?**

Many plant diseases are caused by fungi, Figure 21-13c. A plant disease caused by a fungus can spread rapidly through entire fields of crops.

Dutch elm disease is caused by a fungus that has killed hundreds of thousands of American elms. A small bark beetle carries the spores of the fungus from infected to healthy trees. The fungus enters the stem, begins to grow, and eventually plugs up the xylem tissue.

**Figure 21-13** Plants can be infected by bacteria (a), viruses (b), or fungi (c).

a

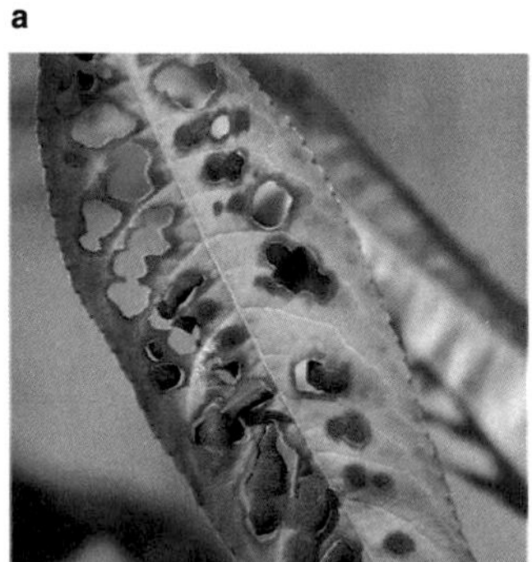

b

c

### *OPTIONS*

#### Science Background

The growth requirements studied in the last section are environmental factors. Living factors, like insects and disease-causing organisms, can also slow down or stop the growth of plants. There are about 200 bacteria, 400 viruses, and 1000 fungi recognized as plant pathogens. Many of these pathogens attack a single species or genus of plants. Insects are responsible for carrying plant diseases from one plant to another. Insects destroy plants by eating plant parts or entire plants. Since they can move about, they sometimes destroy large areas of plants at one time.

a

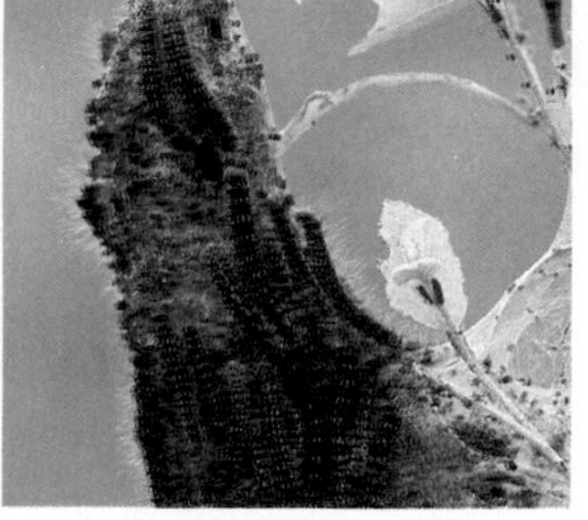

b

**Figure 21-14** An insect can transfer viruses from its mouthparts to uninfected plants (a). Tent caterpillars can eat all the leaves on a tree in a few days (b).

Other crop diseases caused by fungi include wheat rusts and corn smuts. There are also a wide variety of fungi that cause fruit and vegetable rot. Recall that spores are the reproductive structures of fungi. Many fungi are spread by the wind and rain that carry the spores.

## Insect Pests

**How do insects transfer viruses to plants?**

In general, insects are a great help to plants. They carry pollen from flower to flower. Sometimes, however, insects cause plant diseases. Suppose an insect visits a plant that has a virus, bacterium, or fungus in its cells. The insect takes in these microbes when it eats tissues of the plant, Figure 21-14a. When the insect visits the next plant, it transfers the disease microbe from its mouthparts to the uninfected plant. The disease soon develops in that plant.

Insects also can damage or even kill plants by eating too many leaves. A plant with damaged leaves can't make as much food. For example, the tent caterpillar, Figure 21-14b, builds a web in a tree and can eat all of the leaves on that tree in a few days.

### Bio Tip

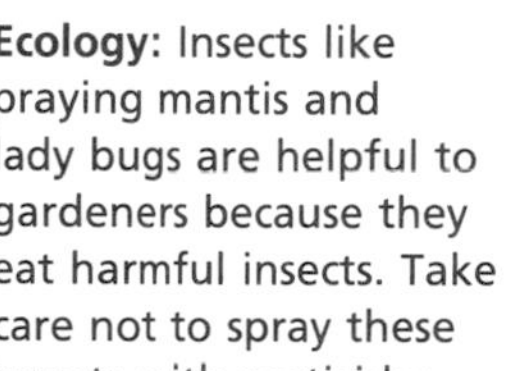

**Ecology:** Insects like praying mantis and lady bugs are helpful to gardeners because they eat harmful insects. Take care not to spray these insects with pesticides.

### Check Your Understanding

11. What is mosaic disease?
12. How do spores of Dutch elm disease enter a tree?
13. How can an insect reduce photosynthesis in a plant?
14. **Critical Thinking:** What is a way to control insect pests in your garden without spraying pesticides?
15. **Biology and Reading:** What two kingdoms contain organisms that cause plant diseases?

**STUDY GUIDE 125**

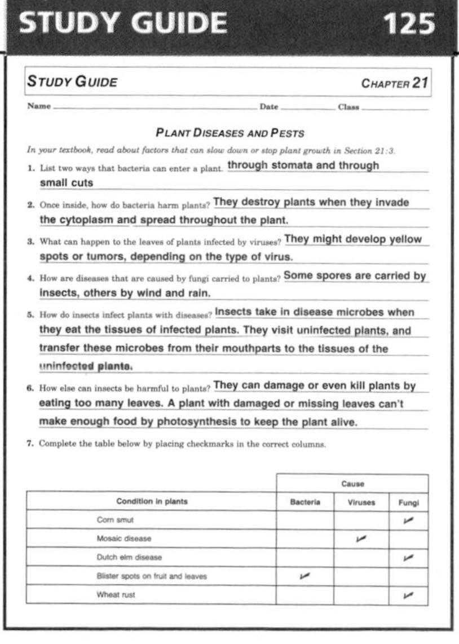

STUDY GUIDE — CHAPTER 21

Name ______ Date ______ Class ______

PLANT DISEASES AND PESTS

*In your textbook, read about factors that can slow down or stop plant growth in Section 21:3.*

1. List two ways that bacteria can enter a plant. **through stomata and through small cuts**
2. Once inside, how do bacteria harm plants? **They destroy plants when they invade the cytoplasm and spread throughout the plant.**
3. What can happen to the leaves of plants infected by viruses? **They might develop yellow spots or tumors, depending on the type of virus.**
4. How are diseases that are caused by fungi carried to plants? **Some spores are carried by insects, others by wind and rain.**
5. How do insects infect plants with diseases? **Insects take in disease microbes when they eat the tissues of infected plants. They visit uninfected plants, and transfer these microbes from their mouthparts to the tissues of the uninfected plants.**
6. How else can insects be harmful to plants? **They can damage or even kill plants by eating too many leaves. A plant with damaged or missing leaves can't make enough food by photosynthesis to keep the plant alive.**
7. Complete the table below by placing checkmarks in the correct columns.

| | Cause | | |
|---|---|---|---|
| Condition in plants | Bacteria | Viruses | Fungi |
| Corn smut | | | ✓ |
| Mosaic disease | | ✓ | |
| Dutch elm disease | | | ✓ |
| Blister spots on fruit and leaves | ✓ | | |
| Wheat rust | | | ✓ |

**RETEACHING 65**

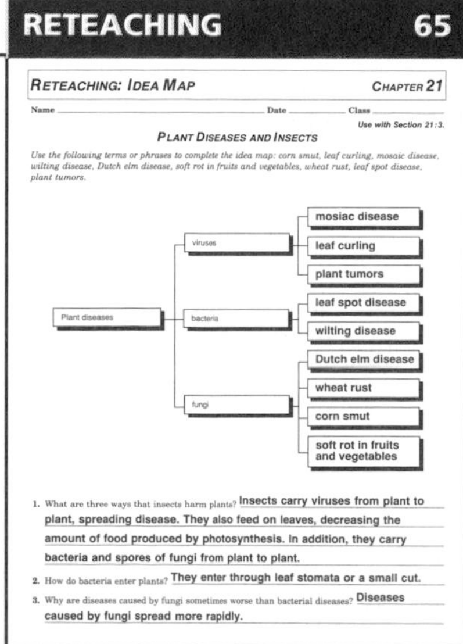

RETEACHING: IDEA MAP — CHAPTER 21

Name ______ Date ______ Class ______

Use with Section 21:3.

PLANT DISEASES AND INSECTS

*Use the following terms or phrases to complete the idea map: corn smut, leaf curling, mosaic disease, wilting disease, Dutch elm disease, soft rot in fruits and vegetables, wheat rust, leaf spot disease, plant tumors.*



1. What are three ways that insects harm plants? **Insects carry viruses from plant to plant, spreading disease. They also feed on leaves, decreasing the amount of food produced by photosynthesis. In addition, they carry bacteria and spores of fungi from plant to plant.**
2. How do bacteria enter plants? **They enter through leaf stomata or a small cut.**
3. Why are diseases caused by fungi sometimes worse than bacterial diseases? **Diseases caused by fungi spread more rapidly.**

## 2 TEACH

### Guided Practice

Have students write down their answers to the margin question: Insects pass disease microbes on their mouthparts to the next plant.

### Independent Practice

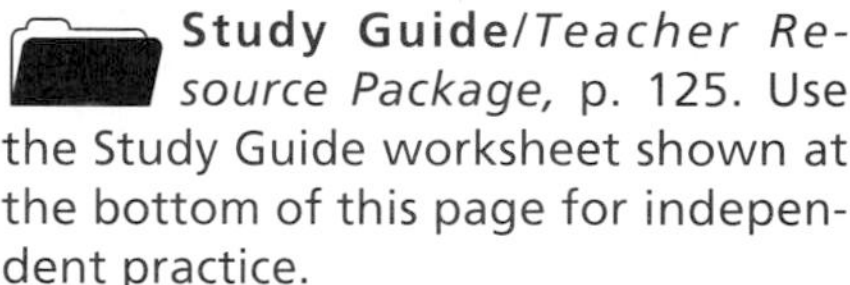

**Study Guide**/*Teacher Resource Package,* p. 125. Use the Study Guide worksheet shown at the bottom of this page for independent practice.

### Check for Understanding

Have students respond to the first three questions in Check Your Understanding.

### Reteach

**Reteaching**/*Teacher Resource Package,* p. 65. Use this worksheet to give students additional practice with plant diseases.

**Extension:** Assign Critical Thinking, Biology and Reading, or some of the **OPTIONS** available with this lesson.

## 3 APPLY

### Connection: Social Studies

Have students investigate plant diseases that affect humans.

## 4 CLOSE

### ACTIVITY/Guest Speaker

Have a county agricultural agent talk to your class about plant diseases and pests.

### Answers to Check Your Understanding

11. a disease caused by a virus that makes yellow spots on leaves
12. The fungus enters the stem.
13. It eats a large number of the plant's leaves.
14. Answers may include collecting pests by hand or using a detergent to wash the leaves.
15. Fungi and Monera

APPLYING TECHNOLOGY

### Purpose

Students will determine the amount of fertilizer needed to promote optimum growth of grass.

### Process Skills

Classify, measure, observe, infer, separate and control variables, interpret data, hypothesize, experiment

### Time Required

Soil preparation and testing will take 1 1/2 class periods; growing time will take approximately two weeks

### Possible Problems

If space is a limiting factor in your classroom, the same procedure can be attempted with smaller containers, available at the garden center. Guide students in choosing a type of grass that grows easily in your geographic region.

### Teaching Strategies

▶ You will need to have three or four shovels or trowels available for students to use to dig in the soil.
▶ Contact your local garden center, Agricultural Extension Office, or lab supply company for soil test kits.
▶ When mixing the soil and fertilizer, allow the soil to come to room temperature by placing the soil on a piece of newspaper in your classroom. The soil and fertilizer can be combined by hand in any kind of bucket or on the newspaper.
▶ Have students interview local farmers of various crops to find out what additives they use in their soil. Students should ask about the farmers' source for additives and how soil testing is done for large farming operations.

TECH PREP **APPLYING TECHNOLOGY**

# How Much Nitrogen?

One of the most important nutrients needed by plants is nitrogen, a substance that plants use to make proteins such as chlorophyll. Air is mostly nitrogen, but most plants can't make proteins from the nitrogen in the air. To solve this problem, nitrogen can be added to soil as fertilizer. Most fertilizers have varying amounts of nitrates, chemicals that contain nitrogen. Farmers, gardeners, and landscapers test for nutrients in the soil and add fertilizers to help their plants grow the best.

## Identifying the Problem

Your school is planning to plant new grass on its athletic fields. You are a landscaper who has been hired to do the work. As the landscaper, part of your job is to decide what changes should be made to the soil to make it best for growing grass. You will need to choose a type of grass that grows well in your climate and find out what soil conditions are best for that type of grass. After testing the soil on the playing field and experimenting with a couple of new grass seed mixtures, you'll be ready to suggest what type of fertilizer would help make the new grass grow the healthiest.

## Collecting Information

Go to a local lawn and garden center or farm supply store, and find out about different types of grass. Collect information about the growing conditions and nutritional needs of each type. Decide which type of grass you want to use in your experiment, one that would grow best in the conditions around your school. Think about ways you can find the best soil conditions for the type of grass you have chosen.

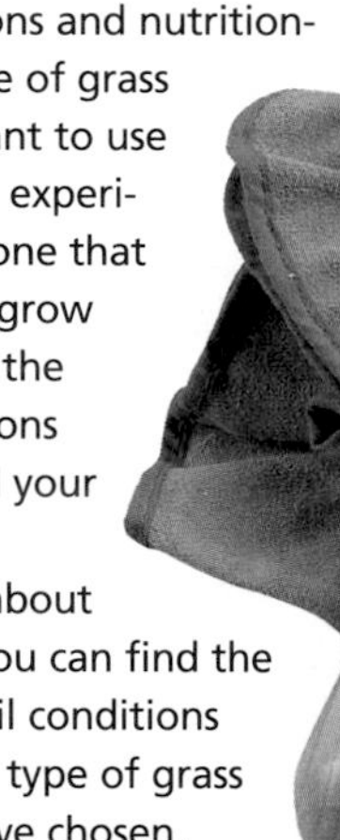

### Technology Connection

As plant experts learn more about the nutritional needs of plants, fertilizer companies must design new fertilizers and other products to meet those needs. If you were to design a new fertilizer, what are some of the qualities it would have to show? Explain why these qualities might be needed.



### Technology Connection

Fertilizers must have several physical qualities in order to be effective. The chemicals must be water soluble and must be in a form that can be taken up easily by plants. The fertilizer should also be in a form that can be applied easily to plants.

Agronomists have discovered several trace elements that are required by plants. Sodium ions are needed by all plants that utilize $C_4$ photosynthesis. Molybdenum and Cobalt are needed for nitrogen fixation.

## Carrying Out an Experiment

1. Collect several plastic trays, a bucket of soil from the area around your school, several sheets of newspaper, grass seed, fertilizer, and a soil test kit. Soil test kits are found in your local lawn and garden center. They can be used to test soil pH, nutrient content, and drainage.
2. Examine the soil from your school. What materials can you see in the soil? Is it sandy, or does it have clumps of clay? Is it wet or dry? Is it packed tightly or loosely?
3. Test the soil with the soil test kit. Record the results.
4. Predict how much fertilizer you should add to the soil to make the grass grow the best. See if the fertilizer label has information that is helpful to you.
5. Mix different amounts of fertilizer into samples of the school soil, and place them in the plastic trays. As a control, use a sample of the school soil with no added fertilizer. Make a recipe for each of the soils you mix. Record what and how much of each ingredient you add.
6. Test each of the soils with the soil test kit.
7. Plant the same amount of grass seed in each sample.
8. Place the trays near a window and keep the soil samples moist. Allow the grass to grow for two to three weeks. Record your results.

### Career Connection

- **Nursery Worker** Sells and takes care of nursery plants; gives advice to customers about plant care and diseases; delivers and plants trees and bushes
- **Lawn Care Technician** Applies fertilizers, herbicides, and insecticides to lawns; analyzes lawn problems and decides on treatments
- **Soil Specialist** Tests soil chemistry and recommends treatments to improve soil

## Assessing Your Results

How well did your plan work? What would you change if you were to do the experiment again? Compare and contrast the soils you mixed. Which of the soil samples had the best plant growth? Based on your results, what soil recipe would you suggest using for the athletic fields?

### Career Connection

**Career Path**
**Nursery Worker** High-school diploma and on-the-job training; volunteer work or summer job in nursery
**Lawn Care Technician** High-school diploma and on-the-job training; summer job or internship with lawn-care service company
**Soil Specialist** Technical school diploma; internship in soil chemistry lab
**Career Issue**
Fertilizers commonly wash into rivers and streams, where they cause algae blooms. Have students find out if these blooms pose a health risk to people.

### Conclusions

Students should see a difference in the grasses after ten days to two weeks. The grass that is overfertilized may not grow or will be yellow in color when it germinates. The soil that has been properly prepared should produce a full, thick tray of green grass in this time.

### Answers to Questions

Student answers will vary depending on the amount of fertilizer they used. Students may decide to use more or less fertilizer if the experiment is repeated.

### ASSESSMENT

**Performance** Have students collect results of other members of the class after two to three weeks and graph the results. Students should decide from the graph what amount of fertilizer is optimum for growth of this type of grass.

# CHAPTER 21 REVIEW

## Summary

Summary statements can be used by students to review the major concepts of the chapter.

## Key Science Words

All boldfaced terms from the chapter are listed.

## Testing Yourself

### Using Words

1. short-day plant
2. gravitropism
3. fertilizer
4. tropism
5. phototropism
6. long-day plant
7. thigmotropism
8. annual
9. day-neutral plant
10. perennial

### Finding Main Ideas

11. p. 453; Insects spread plant diseases with their mouthparts.
12. p. 449; Minerals such as nitrogen, phosphorus, and potassium are found in soil and used by plants to make chlorophyll and cell walls.
13. p. 443; Growth hormone causes stems to bend toward light.
14. p. 440; If the terminal buds are removed from a plant, bushy growth will result.
15. p. 452; Bacteria destroy the plant cells when they invade the cytoplasm.
16. p. 444; Climbing vines hold onto trees and fences by thigmotropism.
17. p. 452; Dutch elm disease is caused by a fungus.
18. p. 447; Biennial plants produce seeds at the end of the second year of growth.

### Using Main Ideas

19. The growth requirements of plants are light, water, soil, nutrients, and temperature.
20. Growth hormones are found in the growing regions of the plant: the terminal bud, lateral bud, and root tip.
21. Bacteria invade a plant through the stomata or a small leaf cut, and cause a disease by destroying the cell's cytoplasm.
22. Three kinds of tropisms in plants are phototropism, gravitropism, and thigmotropism.
23. The terminal bud produces a hormone that keeps the lateral bud from growing as fast as the cells near the terminal bud.
24. Caterpillars eat the leaves and slow down photosynthesis.
25. The three main minerals used by plants are nitrogen, phosphorus, and potassium.

# Chapter 21

# Review

## Summary

### 21:1 Plant Responses

1. Hormones in plants control growth and flowering.
2. Short-day plants flower when day length shortens. Long-day plants flower when there are longer periods of daylight. The flowering of day-neutral plants does not depend on the length of daylight.
3. Tropisms are growth responses to stimuli such as light, gravity, and touch. Other plant movements are caused by changes in cell pressure.

### 21:2 Growth Requirements

4. Annuals complete their life cycle in one year. Biennials produce seeds at the end of the second year. Perennial plants can live for many years.
5. Growth requirements for plants include varied amounts of light, water, minerals, air, and temperature. The type of soil is often important.
6. Different plants in different environments need different levels of light, temperature, water, and minerals.

### 21:3 Plant Diseases and Insects

7. Bacteria and viruses cause many plant diseases. They enter the plant through cuts, stomata, or roots.
8. Fungal diseases spread from plant to plant by spores.
9. Insects eat plants and damage them by carrying bacteria, viruses, and fungi from plant to plant.

## Key Science Words

annual (p. 447)
biennial (p. 447)
day-neutral plant (p. 441)
fertilizer (p. 449)
gravitropism (p. 444)
long-day plant (p. 441)
perennial (p. 448)
phototropism (p. 443)
short-day plant (p. 441)
thigmotropism (p. 444)
tropism (p. 443)

## Testing Yourself

### Using Words

*Choose the word from the list of Key Science Words that best fits the definition.*

1. plant that blooms in the autumn
2. plant response to gravity
3. substance made of nutrients that improves soil for plant growth
4. plant response to a stimulus
5. the response of a plant to light
6. plant that blooms in the summer
7. plant growth in response to touch
8. plant that dies at end of one growing season

# Review

## Testing Yourself *continued*

9. plant that blooms from early summer until frost
10. plant that produces seed year after year

### Finding Main Ideas

*List the page number where each main idea below is found. Then, explain each main idea.*

11. how insects spread plant diseases
12. minerals used by a plant to make cell walls and chlorophyll
13. what causes stems to bend toward light
14. how a person could cause bushy growth in plants
15. how plant diseases are caused by bacteria
16. how climbing vines hold onto trees and fences
17. what causes Dutch elm disease
18. when biennial plants produce seeds

### Using Main Ideas

*Answer the questions by referring to the page number after each question.*

19. What are the growth requirements for plants? (p. 447)
20. Where are growth hormones found in plants? (p. 440)
21. How do bacteria invade a plant and cause a disease? (p. 452)
22. What are three kinds of tropisms in plants? (pp. 443, 444)
23. What keeps lateral buds from growing faster than the terminal buds? (p. 440)
24. What effect does a tent caterpillar have on a tree? (p. 453)
25. What are three main minerals used by plants? (p. 449)

## Skill Review ✓

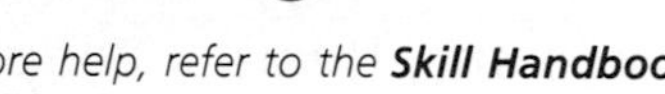

*For more help, refer to the* ***Skill Handbook,*** *pages 704-719.*

1. **Observe:** Observe plants that are growing in the shade of other plants. Explain how they differ from similar plants growing in the sun.
2. **Experiment:** Germinate two bean seeds with a moist paper towel. Experiment to find the effects of different stimuli on root and shoot growth.
3. **Measure in SI:** For five days, measure the height of a bean plant that you planted.
4. **Interpret data:** You have a garden of roses. One day, you notice some of the leaves have black spots. A week after a heavy rain, you notice that most of the leaves have black spots. The leaves then begin to drop off the plants. What's wrong with your roses?

## Finding Out More

TECH PREP

### Critical Thinking

1. Explain why cutting grass with a mower doesn't stop grass growing.
2. How would a scientist decide whether or not soil is missing some minerals needed for plant growth?

### Applications

1. Find out how the amounts of light and water provided for plants are controlled in a greenhouse.
2. Find out the common names of some hormones made by plants. Visit the garden section of a department store or hardware store for help.



## Skill Review

1. Plants growing in the shade will be smaller than similar plants growing in the sun.
2. Answers will vary depending on the stimuli chosen. As an example, roots will grow away from light and shoots will grow toward light.
3. The plant will grow higher and develop more leaves as it grows.
4. The roses are probably infected with a fungus whose spores were spread by the rain.

## Finding Out More

### Critical Thinking

1. Grass grows from the base; the blades of grass are merely leaves. When you mow the grass, you don't cut off the region of growth.
2. The scientist might test the soil to determine its composition, or try growing plants in it to see how well they grow.

### Applications

1. Students will find that light and water are controlled by the glass structure of the greenhouse.
2. Rootone, Hormodin, Quick-root, and Rootgro all contain indolebutyric acid, an auxin.

**Chapter Review**/*Teacher Resource Package,* pp. 109-110.

**Chapter Test**/*Teacher Resource Package,* pp. 111-113.

**Chapter Test**/*Computer Test Bank*

**Unit Test**/*Teacher Resource Package,* pp. 114-115.

# Unit 5 Plant Systems and Functions

**CONNECTIONS**

## Biology in Your World

In this unit, students learn that plants are essential to human existence. Farming is society's structured way of using plants to feed many people. Farmers constantly look for new ways to produce the most crops for the least amount of money. But, they also consider the effect on the environment and the safety of the food supply.

### Literature Connection

John Burroughs (1837-1921) was a famous author who wrote about outdoor life. He grew up on a New York farm, where he often read the works of Walt Whitman, Ralph Waldo Emerson, and John James Audubon. He worked for the U.S. Department of the Treasury and as a bank examiner until he moved to his own farm on the Hudson River. Burroughs wrote seven books and many essays about nature's beauty. His friends included Theodore Roosevelt, Henry Ford, and Thomas Edison.

### History Connection

From 1845 to 1847, much of Ireland's booming population lived in poverty, and potatoes were a staple of the diet. During the potato famine, 750 000 Irish died of starvation or disease. Hundreds of thousands moved to other countries. Most moved to the United States. Today, late blight can be controlled with fungicides.

**C O N N E C T I O N S**

# Biology in Your World

## Healthy Plants, Healthy People, Healthy Environment

In this unit, you learned about plants and their importance. Everyone depends on plants as a food source. Safe farming practices and pest control are needed for healthy food crops. A healthy environment and successful farming are goals of scientists and farmers.

**LITERATURE**

## Deep Woods

John Burroughs was a writer who loved nature and loved to write about it. Many of his essays and books are about his own experiences exploring the Catskill Mountains of New York, where he spent much of his life.

Burroughs brings the out-of-doors to life for his readers. His vivid, interesting writing style makes you feel as if you are with Burroughs in the woods or on a mountaintop.

*Deep Woods* is a collection of essays from several of Burroughs' books. Included are essays about the Catskills, Maine, Alaska, and Yosemite. Sit back and enjoy Burroughs' descriptions of the beauties and wonders of nature!

**HISTORY**

## The Potato Famine

For more than 100 years, potatoes had been the main part of the diet in Ireland. But in the fall of 1845, the potato crop rotted in about half of Ireland. Starvation and disease followed as crop after crop rotted over the next three years. We know now that a parasitic mold caused crops to rot. The mold had spread rapidly from North America to England, and then to Ireland.

Today, chemicals can be used to control the growth of molds. Use of these chemicals can reduce the loss of crops. The chemicals may also improve the quality of crops. Maybe famines like the one in Ireland will not occur again.

LEISURE

## Do You Have a Green Thumb?

You are never too old to enjoy getting dirty! Ask permission to cultivate a small, sunny area of your yard. You will need a spade, a rake, a hoe, and some seed catalogs. Choose several types of fruits and vegetables. All plants have slightly different requirements. You may have better luck with some than with others.

When you prepare your garden, make sure the soil is broken up well for proper drainage. If the soil is mostly clay, add peat moss, sand, or decayed leaves.

At harvest time, notice which fruits and vegetables did well. How can you improve your harvest next year? For each crop you grew, which part do you eat—leaf, stem, fruit, or root?

CONSUMER

## Using Chemicals Wisely

Are the fruits and vegetables you buy in grocery stores safe to eat? Aflatoxin is a chemical produced by a fungus. The fungus can grow on grains and other crops. Aflatoxin is a natural cancer-causing agent. It can affect peanuts, corn, and the milk of corn-fed cows.

In 1989, scientists tested corn for the presence of aflatoxin. They found unacceptable levels of the chemical in 6 percent of the tested samples.

A certain chemical works against aflatoxin. The chemical reduces the amount of aflatoxin present in the milk of cows that eat the treated grain.

Should farmers spray their crops with chemicals to kill fungi? Not all fungi are harmful to plants. One soil-living fungus helps plants remove needed nutrients from the soil. Thus, farmers should use caution when applying chemicals so that the chemicals don't enter the soil and kill the fungi there.

459

CULTURAL DIVERSITY

### Origin and Cultural Significance of Corn

Corn was one of the earliest crops to be domesticated and has been a major food source for Native Americans of both North and South America for about 7000 years. In addition to its importance as a food source, corn also occupies a symbolic place in the culture of many Native American groups. Ask students working in groups to research uses for corn other than as a food source. They might research the use of blue corn among the various Pueblo Indian groups in the American southwest. Ask each student group to prepare an illustrated essay of its findings.

CONNECTIONS

### Leisure Connection

Some students will not have access to a garden plot. Fortunately, some vegetables grow well in containers. Students should prepare containers as they would for a houseplant.

Dill, parsley, and other herbs do well in small pots on a sunny windowsill. Tomatoes, squash, and pumpkins grow nicely in large tubs. Cherry tomatoes and strawberries thrive in hanging baskets.

### Consumer Connection

Many foods contain very low levels of naturally occurring carcinogens. Unfortunately, the levels of aflatoxin found in milk were unsafe. A compound added to dairy cows' grain can control this problem.

However, some people think chemical residues in food are dangerous. Some farmers are reacting to this by producing organically grown food. Organic fruits and vegetables are now available in most supermarkets.

### References


Bruemmer, Fred. "My Life Among Wild Pinnipeds." *International Wildlife*, July-August 1996, pp. 10-22.

Burroughs, John. *Deep Woods.* Salt Lake City: Gibbs Smith, 1990.

Carlson, Shawn. "The Pleasures of Exploring Ponds." *Scientific American*, September 1996, pp. 169-171.

Daly, Douglas C. "The Leaf That Launched a Thousand Ships." *Natural History*, January 1996, pp. 24-33.

Fraser, Laura. "Homegrown Harvest." *Health*, May-June 1997, pp. 70-77.

Fry, Peter, and Fiona Somerset Fry. *A History of Ireland.* New York: Routledge London, 1988.

Long, Cheryl. "Compost Beats Disease Big Time." *Organic Gardening*, February 1996, pp. 12-13.

Oxford, Edward. "The Great Famine." *American History*, March-April 1996. pp. 52-62.

# Unit 6
# Reproduction And Development

## UNIT OVERVIEW

Unit 6 is based upon the concept that reproduction is a basic life process and examines both plant and animal reproduction. Meiosis and mitosis are discussed in Chapter 22. Chapter 23 deals with reproduction in plants. Sexual reproduction in animals is discussed in Chapter 24. Chapter 25 studies development in humans.

## Advance Planning

Audiovisual and computer software suggestions are located within each specific chapter under the heading TEACH.

Invite someone from the research community to speak about cancer and relate the lack of information on cell reproduction to the difficulty of finding a cure for the disease.

**To prepare for:**

| | |
|---|---|
| Lab 22-2: | Purchase a sheet of clear plastic. |
| Mini Lab: | Purchase radish seedlings. |
| Lab 23-1: | Purchase carrots and garlic bulbs. |
| Lab 23-2: | Purchase corn seeds and soak half of them for two days. |
| Mini Lab: | Soak the bean seeds for two days. |
| Lab 24-1: | Order prepared slides of starfish eggs and sperm. |
| Lab 24-2: | Gather colored pencils and graph paper. |
| Mini Labs: | Order sea urchin eggs and sperm. Order prepared slides of hydra. |
| Lab 25-2: | Order grasshoppers and moths in various stages of metamorphosis. |
| Mini Lab: | Prepare glucose solution and gather eggshells. |

# Unit 6

**CONTENTS**

# Reproduction and Development

**What would happen if...**

the present rate of population growth continued? The population of North America would increase to about 487 000 000 by the year 2000.

If a population of 487 000 000 people were spread out evenly over all of North America, there would be about 20 people per square kilometer. But the population is not spread out evenly. Few people live in desert and mountain areas. Most people are in cities along rivers, lakes, or oceans. The population density in cities can be hundreds of people per square kilometer.

The world population was about 5 880 000 000 in 1997. At a growth rate of 1.4%, the world population would be about 6 147 000 000 in the year 2000. That's about 41 people per square kilometer. How might this many people affect life on Earth?



UNIT OVERVIEW

## Photo Teaching Tip

Ask students to read the unit opening passage and look at the photographs. Have students discuss all the places they have visited where they wouldn't like to live. Maybe students have visited a volcano in Hawaii, the swamps of Georgia, or the peaks of the Rocky Mountains. Although these places may be nice to live near, not many people could live in or on them. Now, have students think what it would be like if all the people in the world came to visit their town because it was advertised as the nicest place on Earth. What would be some of the problems of this sudden overcrowding? Explain that this unit will deal with reproduction and that one of the consequences of reproduction in humans is overpopulation. Have students suggest some solutions to the problem of overpopulation.

## Theme Development

Living things reproduce and pass traits to offspring. Students will learn that living things reproduce and develop at different rates and go through various stages. They will study mitosis and meiosis, reproduction in various plants and animals, early development, and metamorphosis. Students will also learn about the reproductive system in humans and the development of the human fetus.

**CULTURAL DIVERSITY**

See the Cultural Diversity feature located on the Connections page at the end of this unit.

For Tech Prep activities, see the Applying Technology feature in this unit, the Finding Out More applications questions in the Chapter Reviews, and the Tech Prep Applications booklet in the Teacher Resource Package.

# Cell Reproduction

## PLANNING GUIDE

| CONTENT | TEXT FEATURES | TEACHER RESOURCE PACKAGE | OTHER COMPONENTS |
|---|---|---|---|
| (2 days)<br>22:1 Mitosis<br>Body Growth and Repair<br>An Introduction to Mitosis<br>Steps of Mitosis | Mini Lab: *How Rapid Is Mitosis?* p. 470<br>Lab 22-1: *Steps of Mitosis,* p. 467<br>Technology: *Hope for Spinal Cord Injury Patients,* p. 465<br>Idea Map, p. 466<br>Check Your Understanding, p. 470 | Application: *The Stages of Mitosis,* p. 22<br>Reteaching: *Mitosis in a Root Tip,* p. 66<br>Critical Thinking/Problem Solving: *How Can Students Make Slides of Dividing Cells?* p. 22<br>Focus: *Growing,* p. 43<br>Transparency Master: *The Steps of Mitosis,* p. 101<br>Study Guide: *Mitosis,* p. 127<br>Lab 22-1: *Steps of Mitosis,* pp. 85-86 | **Laboratory Manual:**<br>*What Happens When Cells Divide?* p. 181<br>Color Transparency 22a: *Steps of Mitosis*<br>**STVS:** *Nerve Regeneration in Garfish,* Animals (Disc 5, Side 2) |
| (2 1/2 days)<br>22:2 Meiosis<br>An Introduction to Meiosis<br>Steps of Meiosis<br>Sperm, Eggs, and Fertilization | Skill Check: *Understand Science Words,* p. 474<br>Idea Map, p. 474<br>Check Your Understanding, p. 475 | Enrichment: *Mitosis, Meiosis, and Number of Chromosomes,* p. 24<br>Reteaching: *Meiosis and Fertilization,* p. 67<br>Transparency Master: *The Steps of Meiosis,* p. 103<br>Study Guide: *Meiosis,* pp. 128-130 | Color Transparency 22b: *Meiosis: Halving the Chromosome Number*<br>Color Transparency 22c: *Steps of Meiosis*<br>Color Transparency 22d: *Meiosis in Humans*<br>**STVS:** *Raising Super Fish,* Animals (Disc 5, Side 2) |
| (1 1/2 days)<br>22:3 Changes in the Rate of Mitosis<br>Aging<br>Cancer | Skill Check: *Classify,* p. 479<br>Lab 22-2: *Rate of Mitosis,* p. 477<br>Check Your Understanding, p. 479 | Reteaching: *Changes in the Rate of Mitosis,* p. 68<br>Skill: *Form a Hypothesis,* p. 23<br>Study Guide: *Changes in the Rate of Mitosis,* p. 131<br>Study Guide: *Vocabulary,* p. 132<br>Lab 22-2: *Rate of Mitosis,* pp. 87-88 | **Laboratory Manual:**<br>*Are There More Dividing Cells or Resting Cells in a Root Tip?* p. 185<br>**STVS:** *Bitter Melon Cancer Treatment,* Plants and Simple Organisms (Disc 4, Side 2)<br>**STVS:** *Laser Treatment of Bladder Cancer,* Human Biology (Disc 7, Side 2) |
| Chapter Review | Summary<br>Key Science Words<br>Testing Yourself<br>Finding Main Ideas<br>Using Main Ideas<br>Skill Review | **ASSESSMENT RESOURCES**<br>Chapter Review, pp. 116-117<br>Chapter Test, pp. 118-120<br>Performance Assessment in the Biology Classroom<br>Alternate Assessment in the Science Classroom<br>Computer Test Bank | |

## GLENCOE TECHNOLOGY

**Infinite Voyage,** *The Living Clock*
**Science and Technology Videodisc Series,** *Nerve Regeneration in Garfish,* Animals (Disc 5, Side 2)
*Raising Super Fish,* Animals (Disc 5, Side 2)
*Bitter Melon Cancer Treatment,* Plants and Simple Organisms (Disc 4, Side 2)
*Laser Treatment of Bladder Cancer,* Human Biology (Disc 7, Side 2)

**The Secret of Life,** *On the Brink: Portraits of Modern Science*

## MATERIALS NEEDED

| LAB 22-1, p. 467 | LAB 22-2, p. 477 | MARGIN FEATURES |
|---|---|---|
| pencil<br>paper | clear plastic sheet<br>marking pencil<br>facial tissue<br>microscope | Skill Check, pp. 474, 479<br>dictionary<br>Mini Lab, p. 470<br>radish seedling<br>ruler |

## OBJECTIVES

For more information about National Science Standards, see page 5T.

| SECTION | OBJECTIVE | CORRELATION of QUESTIONS to OBJECTIVES | | | |
|---|---|---|---|---|---|
| | | CHECK YOUR UNDERSTANDING | CHAPTER REVIEW | TRP CHAPTER REVIEW | TRP CHAPTER TEST |
| 22:1 National Science Stds: UCP.1, UCP.2, UCP.3, UCP.4, UCP.5, B.4, C.1 | 1. **Relate** some benefits of mitosis. | 1, 4 | 3, 11, 17 | 3, 11, 18 | 3, 23, 25, 46, 50 |
| | 2. **Identify** cell parts involved in mitosis. | 3 | — | 17, 28 | 2, 12, 39 |
| | 3. **Trace** the steps of mitosis. | 2, 5 | 9, 13, 14 | 2, 6, 8, 20, 27, 29, 30, 31 | 1, 5, 10, 15, 18, 20, 28, 32, 36, 38, 40, 41, 42, 44, 48 |
| 22:2 National Science Stds: UCP.1, UCP.2, UCP.3, UCP.4, UCP.5, B.4, C.1 | 4. **Explain** the results of meiosis. | 7, 10 | 10, 16, 17 | 1, 4, 5, 7 | 4, 7, 9, 19, 22, 24, 26, 29, 30 |
| | 5. **Trace** the steps of meiosis. | 6 | 1 | 32, 33, 34, 35, 36, 37, 38, 39 | 14, 37, 43, 45, 47, 49 |
| | 6. **Compare** the sex cells of males and females. | 8, 9 | 4, 5, 6, 7 | 12, 14, 15, 16, 19, 20, 21, 22, 23, 24, 25, 26 | 11, 13, 17, 21, 27, 34 |
| 22:3 National Science Stds: UCP.1, UCP.3, UCP.4, C.1, F.1 | 7. **Explain** some effects of aging. | 11 | 12, 15 | 9 | 33 |
| | 8. **Describe** the causes and effects of cancer. | 12, 13, 15 | 2, 8, 18 | 10 | 6, 8, 16 |

# Cell Reproduction

## CHAPTER OVERVIEW

### Key Concepts

The two cell processes of mitosis and meiosis are described in detail. However, the chapter does not use the technical phase names. The role of mitosis in the aging process and its role in cancer are also discussed.

This chapter serves as the basis for the concepts of plant and animal reproduction and genetics.

### Key Science Words

body cells
cancer
meiosis
mitosis
ovary
polar body
puberty
sex cell
sister chromatids
testes

### Skill Development

In Lab 22-1, students will **observe, formulate models,** and **sequence** to learn the steps in mitosis. In Lab 22-2, they will **use numbers, make and use tables, form hypotheses,** and **interpret data.** In the Skill Check on page 474, students will **understand** the **science word** *ovary*. In the Skill Check on page 479, they will **classify** causes of cancer. In the Mini Lab on page 470, they will **measure in SI** to determine the amount of growth in a radish root.

### Bridging

Students learned previously that blood cells have a short life span. How does the human body actually replace these blood cells? This chapter looks at this specific process in detail.

Also, the concept of fertilization was introduced in Chapter 6. Why special cells (those with a reduced chromosome number) are needed for this process will become clear.

**CHAPTER PREVIEW**

### Chapter Content

Review this outline for Chapter 22 before you read the chapter.

**22:1 Mitosis**
- Body Growth and Repair
- An Introduction to Mitosis
- Steps of Mitosis

**22:2 Meiosis**
- An Introduction to Meiosis
- Steps of Meiosis
- Sperm, Eggs, and Fertilization

**22:3 Changes in the Rate of Mitosis**
- Aging
- Cancer

### Skills in this Chapter

The skills that you will use in this chapter are listed below.

- In **Lab 22-1,** you will observe, formulate a model, and sequence. In **Lab 22-2,** you will use numbers, make and use tables, form hypotheses, and interpret data.
- In the **Skill Checks,** you will classify and understand science words.
- In the **Mini Labs,** you will measure in SI.

462

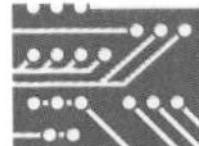

## TECH PREP

For Tech Prep activities in this chapter of the Teacher Wraparound Edition, see especially the Connection and Portfolio on page 469, the Apply Challenge on page 470, the Student Journal on page 473, and the Motivation on page 476.

See also the Glencoe Homepage at **www.glencoe.com**

The following Glencoe resources provide additional opportunities for integrating science and technology.

**Production Systems Technology**
Section V: Building the Future
Activity 1: Design for an Aging Population

Chapter **22**

# Cell Reproduction

Every part of your body is made up of cells. You and all other living things started out as single cells. In some plants and animals, that cell came from a part of one parent. In most plants and animals and in humans, the cell came from the joining of a cell from a male parent and a cell from a female parent. You have learned that these two origins of living things are the difference between asexual and sexual reproduction. Look at the difference in size between the parent and its young in the photo on the left. The larger animal has millions more cells than its offspring. Look at the tree seedling below. You know that the parent tree is much larger. How does one cell become millions of cells? Where do these new cells come from?

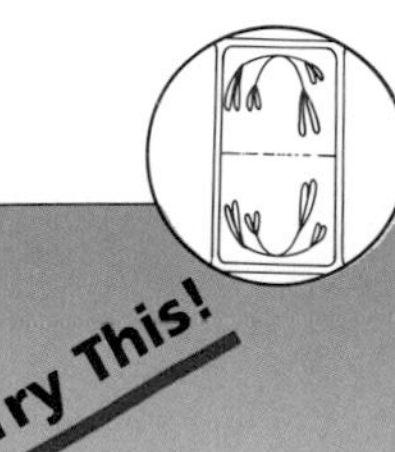

**Try This!**

**How many chromosomes do onion cells have?** Use a microscope to examine a prepared slide of onion root tip cells under high power. Do all the cells look like they are dividing?

A young tree

***inter*NET CONNECTION**

For more information about the material in this chapter, follow the link for the chapter on the Glencoe Homepage at **http://www.glencoe.com**



## ASSESSMENT PLANNER

**Portfolio**

Strategies on the following pages represent student products that can be placed into a best-work portfolio: pp. 469, 471.

**PERFORMANCE ASSESSMENT**
Skill Check, pp. 474, 479
Mini Lab, p. 470
Lab, pp. 467, 477

**CONTENT ASSESSMENT**
Check for Understanding, pp. 470, 475, 479
Chapter Review, pp. 480-481

**GROUP ASSESSMENT**
Opportunities for group assessment occur with Cooperative Learning Strategies.

**Student Journal**

Strategies on the following pages represent opportunities for writing in a Student Journal: pp. 473, 478.

**GETTING STARTED**

### Using the Photos

In looking at the photos, students will know that a larger animal or tree has grown more than a smaller one. They should be able to say that growth results from adding cells to the organism. Adding cells is a result of cell division.

### MOTIVATION/Try This!

**How many chromosomes do onion cells have?** Slides of longitudinal onion root tip are to be used. Students must **observe** the region of the onion root tip where cell reproduction is occurring. This region is next to the root cap. Only those cells that are reproducing by mitosis will show chromosomes. All others will show a typical nucleus, but no chromosomes will be visible.

A 35-mm slide of an onion root tip projected onto a screen would allow all students to see the same cellular material and would also allow you to point out those cells that do or do not show chromosomes within the nucleus.

### Chapter Preview

Have students study the chapter outline before they begin to read the chapter. They should note the overall organization of the chapter and that the first topic described is the process of mitosis. This is followed by a section dealing with meiosis. Lastly, the chapter deals with the changes associated with aging and cancer.

### Misconception

Every cell in a normal human body contains the same number of chromosomes (except for egg and sperm). This number is 46. Each parent has contributed exactly half of this number. That is, your father contributed 23, and your mother contributed 23. You have exactly 1/2 of your father's chromosomes and 1/2 of your mother's chromosomes.

## 22:1 Mitosis

### PREPARATION

#### Materials Needed

Make copies of the Focus, Application, Critical Thinking/Problem Solving, Transparency Master, Lab, Study Guide, and Reteaching worksheets in the *Teacher Resource Package.*

▶ Obtain radish seeds for the Mini Lab.

#### Key Science Words

mitosis
body cell
sister chromatids

#### Process Skills

They will measure in SI in the Mini Lab.

### 1 FOCUS

▶ The objectives are listed on the student page. Remind students to preview these objectives as a guide to this numbered section.

#### MOTIVATION/Brainstorming

Ask students how it is possible for their skin to replace itself when scraped or cut off during a fall or injury.

**Focus**/*Teacher Resource Package,* p. 43. Use the Focus transparency shown at the bottom of this page as an introduction to cell reproduction.

#### Guided Practice

Have students write down their answers to the margin question: for the body to grow and repair itself.

## 22:1 Mitosis

**Objectives**

1. **Relate** some benefits of mitosis.
2. **Identify** cell parts involved in mitosis.
3. **Trace** the steps of mitosis.

**Key Science Words**

mitosis
body cell
sister chromatids

You started life as a single cell. You now have millions of cells in your body. Somehow, you have grown. The new cells in your body came from cell reproduction.

### Body Growth and Repair

Living things grow. You were growing even before you were born. Living things also repair themselves when they are injured. As a child, you probably cut or scraped your skin often. You must have seen new skin grow back on your hands and knees many times. Your body must make new cells to grow and to repair itself. New cells are made by the process of cell reproduction.

**Why are new cells important to your body?**

One kind of cell reproduction in organisms is called mitosis (mi TOH sus). **Mitosis** is cell reproduction in which two identical cells are made from one cell. Each new cell grows in size until it too is ready to reproduce by mitosis. All body cells in humans are formed by mitosis. **Body cells** are cells that make up most of the body, such as the skin, blood, bones, and stomach. Figure 22-1 shows that all body cells don't live for the same length of time. Cells carry on mitosis at different rates in different organs to replace cells that are worn out. In which body organs do cells carry on mitosis most often? In which organs do cells carry on mitosis least often? In some body cells, such as muscle cells, mitosis never occurs after birth. You are born with all the muscle cells you will ever have.

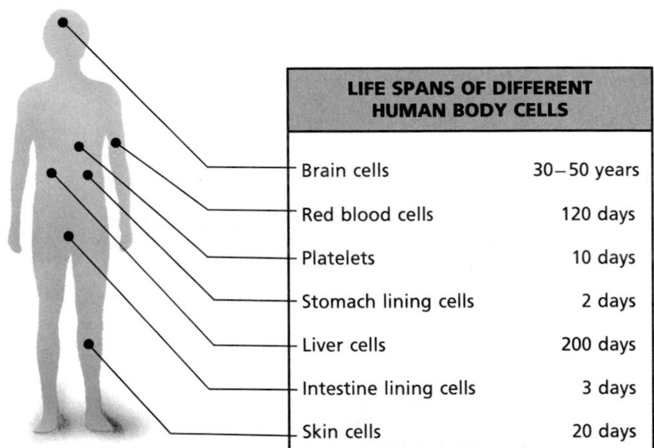

| LIFE SPANS OF DIFFERENT HUMAN BODY CELLS | |
|---|---|
| Brain cells | 30–50 years |
| Red blood cells | 120 days |
| Platelets | 10 days |
| Stomach lining cells | 2 days |
| Liver cells | 200 days |
| Intestine lining cells | 3 days |
| Skin cells | 20 days |

**Figure 22-1** The length of time that human cells live in the body depends on the type of cell.



### OPTIONS

#### Science Background

Prophase corresponds to step one. Step two is metaphase. Step three is anaphase. Step four is telophase.

**What Happens When Cells Divide?**/*Lab Manual,* p. 181. Use this lab to give additional hands-on practice with mitosis.

## TECHNOLOGY

### Hope for Spinal Cord Injury Patients

It has long been thought that nerve cells don't divide by mitosis after you are born. This used to explain why a person with a spinal injury was often permanently disabled. The nerve pathways from brain to body muscles were cut and would never heal.

Recent research shows that nerve cells can regrow and repair themselves. Scientists carried out the following experiment. The optic nerves leading from the eyes to the brains of several hamsters were cut and sections were removed. Normally, this would have meant the loss of sight for the animals. The removed section was then replaced with a piece of nerve taken from another part of the body of each animal. After seven weeks, nerve endings had regrown. The piece of body nerve that was added had somehow acted as a path for the original nerve to follow as it grew. Were the animals able to see again? The scientists have already found that the hamsters could detect changes in light intensity. This means that the optic nerve is working to some extent.

Will this type of scientific breakthrough lead to the healing of spinal cord injuries? As with most scientific research, a long time is needed to solve medical problems.

Hamsters

Mitosis goes on during a person's entire life. You are constantly forming new cells to replace those worn out or lost. Mitosis begins before you are born and continues up to the time you die.

## An Introduction to Mitosis

How does mitosis work? We could compare it to how a photocopy machine works. Look at Figure 22-2. Suppose you have a sketch of your pet. Your friend wants a copy. You copy the sketch on a photocopy machine. You now have two identical sketches of your pet.

The process of mitosis is similar. You start with one cell. The end result is two identical cells. The photocopy machine makes a copy in about five seconds. Mitosis may take several hours.

**Figure 22-2** How are photocopying and mitosis similar? They each result in an exact copy of the original.

## TECHNOLOGY

### Hope for Spinal Cord Injury Patients

#### Background

Mitosis occurs a limited number of times in certain tissues. Nerve tissue is limited as to the number of mitotic divisions it undergoes. Thus, after birth, brain tissue and spinal cord tissue cannot repair itself or regrow.

#### Discussion

Point out to students that a breakthrough in being able to regrow nerve tissue in the human body must be accomplished in many small steps. Research typically builds on previous findings and many "small steps" are needed before a final satisfactory solution is obtained.

#### References


Ditunno, John F., and Christopher Formal. "Chronic Spinal Cord Injury." *The New England Journal of Medicine*, February 24, 1994. pp. 550-557.

Dreher, Nancy. "The Devastating Effects of Spinal Cord Injuries." *Current Health* 2, May 1996, pp. 13-16.

Montgomery, Geoffrey. "A Brain Reborn." *Discover,* June 1990, pp. 48-53.

Sternberg, S., "Steps Toward Healing Damaged Spines." *Science News*, July 27, 1996, p. 52.


## 2 TEACH

### MOTIVATION/Demonstration

To demonstrate chromosome thickening, use a telephone cord detached from a phone. When stretched out, it is comparable to the chromosomes in a resting phase. When folded up upon itself, shortened and thickened, it is comparable to what happens during prophase.

## APPLICATION 22

**APPLICATION: HEALTH** CHAPTER 22

Name ______ Date ______ Class ______

*Use after Section 22:1.*

**THE STEPS OF MITOSIS**

To stay healthy, your body cells must undergo mitosis regularly. Dead or damaged cells are replaced with new cells.

*Label the diagrams below to show the stages in the four steps of mitosis.*

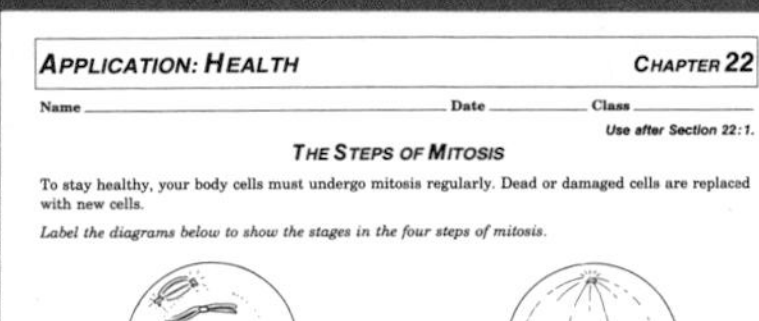

1. Sister chromatids shorten and thicken. Nuclear membrane breaks up. Centrioles move away from each other.
2. Centrioles move to opposite ends of the cell. Fibers attach to sister chromatids. Sister chromatids line up along the center of the cell.

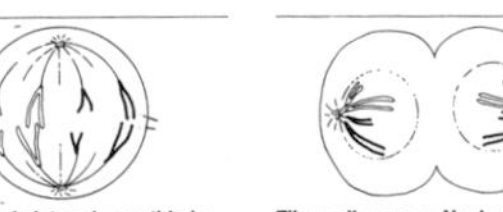

3. Each pair of sister chromatids is pulled apart by the fibers. Fibers pull each chromatid strand toward centrioles at opposite ends of the cell.
4. Fibers disappear. Nuclear membrane reforms. Two sides of the cell, each with a complete set of chromosomes, separate as the cell membrane pinches the cell in half.

## OPTIONS

**Enrichment:** Most cells are not actively engaged in mitosis all the time. They spend much of their time in interphase. Interphase is sometimes called a "resting" phase. In reality the cell is not resting. Chromosome replication is occurring during this phase. For example, skin cells will spend 15 to 20 days in interphase, while nerve cells spend 50 to 60 years.

**Application**/*Teacher Resource Package,* p. 22. Use the Application worksheet shown here to review the stages of mitosis.

## GLENCOE TECHNOLOGY

**Videodisc**

**The Secret of Life**

*Mitosis Segment*

## TEACH

### Concept Development

▶ *Chromosome* defines a nuclear structure before it makes a copy of itself (this copying process is covered in Chapter 28). *Sister chromatids* define the structures resulting from replication of the original chromosome.

▶ A key word in this section is *identical.* Identical refers to chromosome number as well as chromosomal makeup or gene traits. Each new cell formed during mitosis is a clone of every other cell in the body, having the same genetic material as every other cell.

▶ The sequence of events described here takes place during interphase. It is important for students to understand that during this phase, the chromosomes are making copies of themselves. The original cell described had only four chromosomes. Now, each chromosome has made a copy of itself, resulting in 8 sister chromatids.

### Idea Map

Have students use the idea map as a study guide to the major concepts of this section.

### Guided Practice

Have students write down their answers to the margin question: Each chromosome becomes doubled.

**GLENCOE TECHNOLOGY**

**Videodisc**

**Biology: The Dynamics of Life**
Animation: *The Cell Cycle*
Disc 1, Side 1, Ch. 28

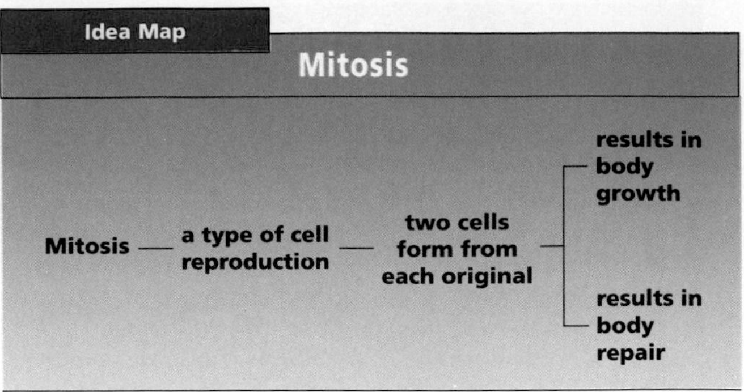


**Study Tip:** Use this idea map as a study guide to mitosis. If you cut your finger, how does mitosis help you?

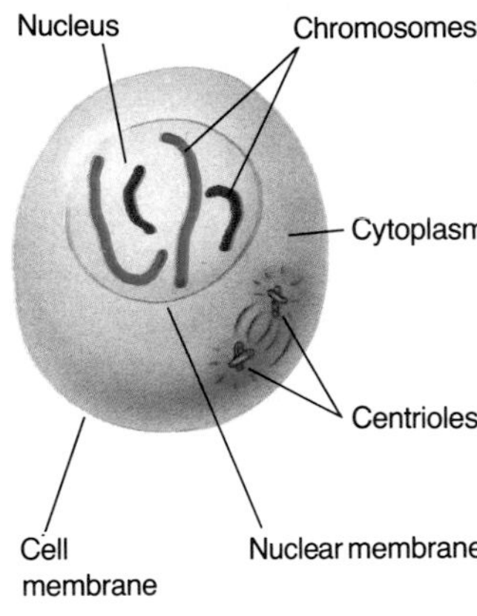


**Figure 22-3** Study the parts of an animal cell.

What cell parts are used in mitosis? Figure 22-3 shows what a typical animal cell might look like. Remember from Chapter 2 that the entire cell is surrounded by a cell membrane. Most of the material inside the cell is called cytoplasm. A cell nucleus is present in the cytoplasm. A nuclear membrane surrounds the nucleus. You have also read that inside the nucleus there are chromosomes. They may not be easy to see at this time, but they are there. Notice that there are also centrioles in the cytoplasm.

Most cells contain many chromosomes. To explain how a cell goes through mitosis, let's imagine a cell with just four chromosomes. An important step takes place in the nucleus of every cell before mitosis begins. Each chromosome becomes doubled. Figure 22-4 shows how this works. Notice how each chromosome looks after it has doubled. The two strands of each doubled chromosome are held together at one point. The two strands of a chromosome after it becomes doubled are called **sister chromatids** (KROH muh tidz). Each chromatid is an exact copy of the original chromosome. The centrioles also double just before mitosis begins.

**What does each chromosome do before mitosis begins?**

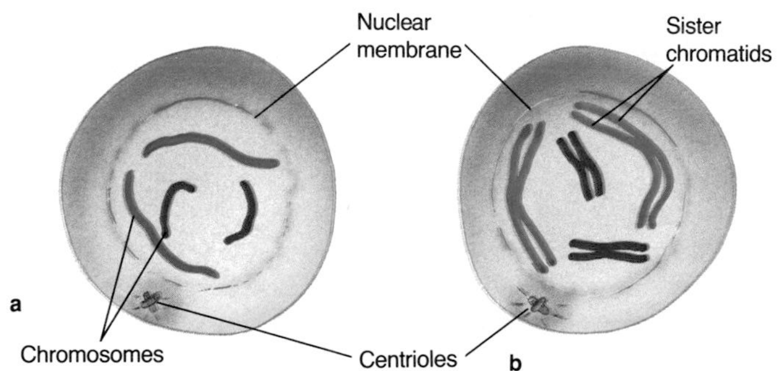


**Figure 22-4** Chromosomes become doubled before mitosis begins.



## OPTIONS

**Critical Thinking/Problem Solving/** *Teacher Resource Package,* p. 22. Use the Critical Thinking worksheet shown here to help students with problem-solving skills.

**CRITICAL THINKING 22**

CRITICAL THINKING/PROBLEM SOLVING — CHAPTER 22

Name ______ Date ______ Class ______

*Use after Section 22:1.*

***HOW CAN STUDENTS MAKE SLIDES OF DIVIDING CELLS?***

A class of biology students looked at prepared slides of mitosis and learned the names of the steps. Some of the slides were labeled "onion root tip mitosis." The students wondered how these slides were prepared. The teacher told them how to do a special project.

First, the students obtained six onions that were just beginning to sprout roots. They also collected six jars with mouths that were almost the same diameter as the onions. They filled the jars with water and balanced the onions on the jars. They put the jars on the windowsill. The onions sprouted roots in a few days.

When the roots were about a centimeter long, the students carefully cut off the end 2 to 3 millimeters. They laid each root tip on a glass slide. Then, they put a drop of blue stain on the root. They covered the drop with a coverslip and a cork, and pressed down heavily and carefully on the cork with a thumb. After several minutes, they put the slide under a microscope and looked for cells.

They saw many box-like cells with darkly stained dots in the middle. Some cells looked as if they were full of thread. Some cells looked as if they had a few colored threads in them. By comparing their observations to the figures in their textbook, they were able to identify the steps in mitosis.

**Analyzing the Problem**

1. What caused the onion roots to sprout? water and growth factors in the onion
2. Where did the onion roots get the energy to grow, divide, and sprout? from stored food
3. Why did the students add the blue stain to the roots? The stain dyes the nucleus and chromosomes a darker color than the rest of the cell.
4. Why did the students squash the root tip under the coverslip and cork? The root is thick, so that it is hard to see individual cells. Squashing spreads out the cells.

**Solving the Problem**

1. What part of the cell looked like tangled threads? chromosomes
2. What do the chromosomes look like when the cell is not undergoing mitosis? They are not easily seen.

# Lab 22–1 Steps of Mitosis

## Problem: What do the steps of mitosis look like?

### Skills

observe, formulate models, sequence

### Materials

paper
pencil

### Procedure

1. Look at Figure 22-4 on page 466 showing the stages of mitosis. Notice in Figure 22-4a that the chromosomes have not yet formed sister chromatids. In other words, they're not yet doubled.
2. **Formulate a model:** Trace the cell diagram in Figure 22-4a onto a piece of paper.
3. Draw a new cell that shows the next change before mitosis begins. Trace the diagram in Figure 22-4b.
4. **Sequence:** Use Figure 22-5 on pages 468 and 469 to help you draw the next four steps that would take place in this cell during mitosis.
5. In your diagrams, use the following words to label the parts:
   **Original cell**—cell membrane, nucleus, nuclear membrane, cytoplasm
   **Change before mitosis**—sister chromatids
   **Step 1**—sister chromatids, spindle fibers
   **Step 2**—fibers, sister chromatids at cell center
   **Step 3**—separation of sister chromatids, fiber, chromosomes
   **Step 4**—fibers disappearing, building a new cell wall, nuclear membrane reforms, chromosomes
   **Body cells**—chromosomes, nucleus, cell membrane, cytoplasm, body cells

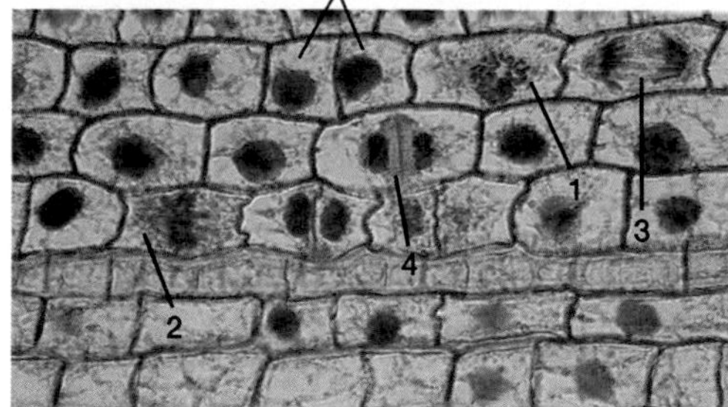


### Data and Observations

1. At what step in mitosis do the sister chromatids separate?
2. Would your diagrams differ if two chromosomes had been used? Explain.

### Analyze and Apply

1. What is the difference between chromosomes and sister chromatids?
2. How would your diagrams differ if this were meiosis in terms of:
   (a) the final number of cells?
   (b) the final number of chromosomes in each cell if you had started with 46 chromosomes?
   (c) the number of steps needed?
   (d) the labels on the last diagram?
3. **Apply:** Give two examples of human body organs where you would find:
   (a) mitosis taking place.
   (b) meiosis taking place.
4. How are fibers important to cell division?

### Extension

**Formulate a model:** Yarn and paper clips may be used to model the series of changes that occur to two chromosomes before and during mitosis.



## ANSWERS

### Data and Observations

1. step 3
2. No, only in the number of chromosomes. There will be two, not four, chromosomes in the two new cells. The chromosomes all go through the same steps in the same order, no matter how many chromosomes are involved.

### Analyze and Apply

1. Sister chromatids are the two strands in a chromosome after it has become doubled.
2. (a) There would be four, not two, cells. (b) There would be 23 chromosomes in each cell. (c) Six steps, not four, are needed for meiosis. (d) One label on the last diagram would be sex cells, not body cells.
3. (a) skin and liver
   (b) testes and ovaries
4. Fibers help move the chromosomes.

## Lab 22-1 Steps of Mitosis

### Overview

Students will review the sequential changes occurring in cells by diagraming and labeling stages of mitosis.

**Objectives:** Upon completion of this lab, students will be able to: (1) **identify** changes within cells as mitosis occurs, (2) **label** those cell organelles that appear during the mitotic process, (3) **compare** the changes occurring during mitosis and meiosis.

**Time Allotment:** 1 class period

### Preparation

**Lab 22-2 worksheet**/*Teacher Resource Package,* pp. 87-88.

### Teaching the Lab

▶ You may wish to provide photocopies of outline diagrams from the *Teacher Resource Package* to your students.

▶ Encourage students to use the diagrams in the text as a guide for completion of labels.

▶ **Troubleshooting:** You may want to delay having students answer any questions relating to meiosis until this concept has been covered in class. Students can answer all questions pertinent to mitosis at the conclusion of this lab.

**Cooperative Learning:** Divide the class into groups of three students. For more information, see pp. 22T-23T in the Teacher Guide.

### ASSESSMENT

**Skill:** The photograph in the lower-left corner shows cells before and during mitosis. Tell students to describe the events that are taking place in photos 1-5.

## TEACH

### Concept Development

▶ Point out to students the major changes and accomplishments of mitosis. Remind them of these facts:

1. Two new cells are formed.
2. Each new cell is identical in chromosome number to the other and to the original cell.
3. Each new cell will, in time, undergo mitosis after first entering a resting phase.
4. Body cells are formed by mitosis to replace worn out cells or to form new body cells.
5. Growth takes place as new cells are formed. Repair of an injury might also be occurring.

**Cooperative Learning:** Divide the class into 5 groups. Each group is to prepare a short written description and a labeled diagram on a transparency of major changes that take place within a cell during a specific stage of mitosis.

Group 1 = describe changes in the cell prior to mitosis.

Group 2 = describe changes during step one.

Group 3 = describe changes during step two.

Group 4 = describe changes during step three.

Group 5 = describe changes during step four.

### Guided Practice

Have students write down their answers to the margin question: step two.

## *GLENCOE* TECHNOLOGY

**Videotape**

**The Secret of Life**

*On the Brink: Portraits of Modern Science*

## Steps of Mitosis

Follow the numbers in Figure 22-5 as you read here about the four steps of mitosis.

**Step One**

1. Sister chromatids begin to shorten and thicken. If you were to look at a cell in this step of mitosis under a light microscope, you would be able to see the chromosomes. However, you might not be able to see the strands of each sister chromatid.
2. The nuclear membrane begins to break down. The pairs of sister chromatids now look like they are floating in the cytoplasm.
3. The centrioles move away from each other.
4. Fibers form between the centrioles. These fibers are strands of protein that form between the two ends of a cell.

**Step Two**

1. The centrioles move apart to opposite ends of the cell. The fibers look like they are stretched between the two ends of the cell.
2. Sister chromatids become attached to the fibers at the point where the two strands are joined to each other. The

**In which step of mitosis are the chromosomes pulled to the center of the cell?**

**Figure 22-5** Use these diagrams to help you follow the steps of mitosis as you read the text. How many times does the cell divide?

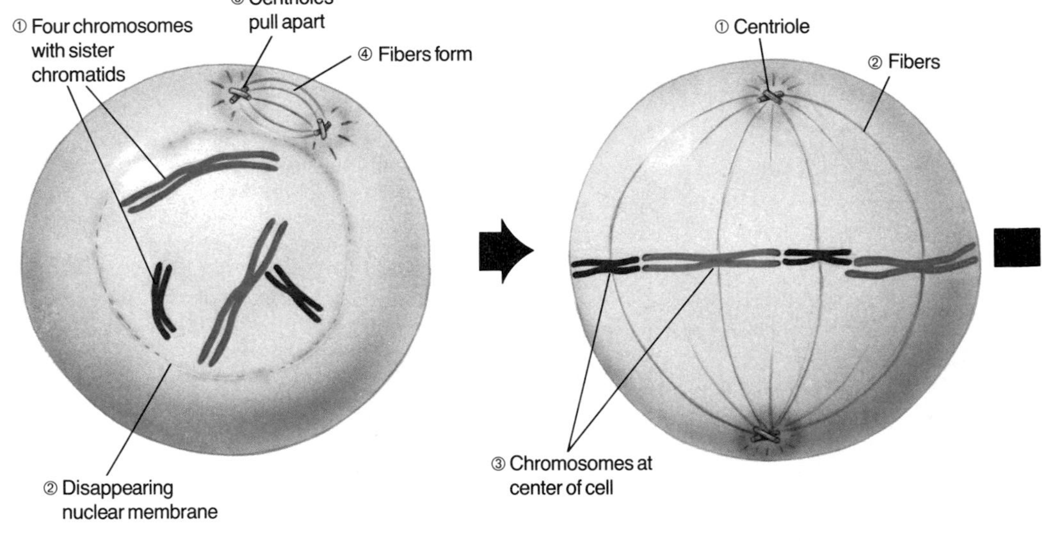




## *OPTIONS*

**Transparency Master**/*Teacher Resource Package,* p. 101. Use the transparency shown here as you teach mitosis.

**Challenge:** Have students use a reference to determine the exact name used to describe each phase of mitosis. Having determined these, have students find the meaning of each prefix being used. Each prefix will relate to the events taking place in a very general way.

**TRANSPARENCY 101**

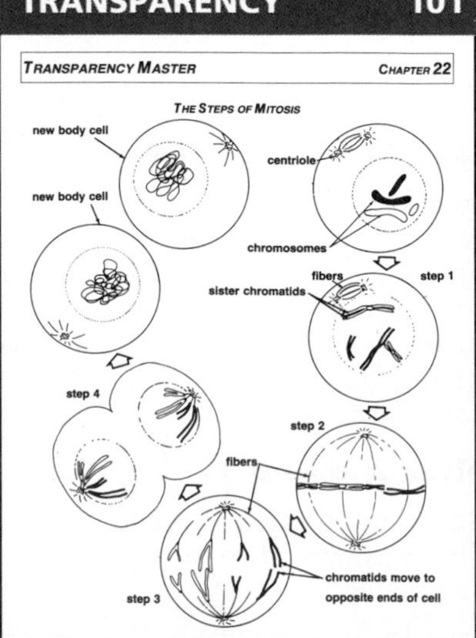

pairs of sister chromatids are pulled toward the center of the cell.
3. The four pairs of sister chromatids are now lined up at the center of the cell.

**Step Three**

1. The sister chromatids are pulled apart by the fibers. Each chromatid is now separate from its identical partner.
2. The fibers pull each chromatid strand toward the centrioles at opposite ends of the cell. Remember that each chromatid is an exact copy of one original chromosome.

**Step Four**

1. Each end of the cell now has a complete set of chromosomes. In this example, a complete set is four chromosomes. Notice that this is the same as the number we started with in Step One.
2. The fibers begin to disappear.
3. The nuclear membrane begins to reform.
4. The cell membrane begins to pinch in until the cytoplasm is divided in half. Two new cells have formed. They each have identical chromosomes. They also have the same number of chromosomes as the cell they originally came from.

### Bio Tip

**Consumer:** You may have seen advertisements that claim their creams will make your skin soft and as new as a baby's. Many of these creams work by adding moisture to your skin, not by causing new cells to grow.

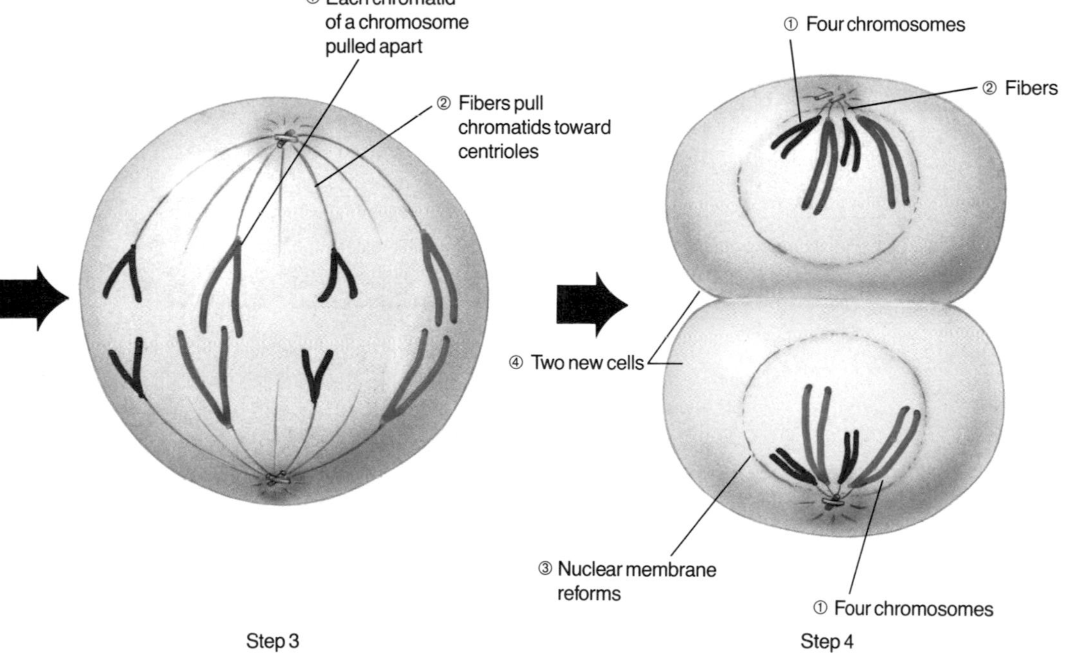




## TEACH

### Concept Development

▶ Students can be reminded that step four is almost a step one in reverse. Again, the telephone cord analogy can be used. This time, however, the cord will start out highly twisted and will be stretched out as the chromosomes become less visible.

### Guided Practice

Provide students with outline drawings of the cell. Review each step or phase on the overhead projector, writing in the details and drawing in the chromosomes and sister chromatids. Have students record the information and diagrams from the overhead screen.

### Connection: Math

Have students start with one cell of a type that undergoes mitosis once every hour and calculate the number of cells there would be at the end of 24 hours. The answer should be approximately 2 followed by 16 zeros ($1.7 \times 10^{16}$).

### Independent Practice

**Study Guide**/*Teacher Resource Package*, p. 127. Use the Study Guide worksheet shown at the bottom of this page for independent practice.

### Guided Practice

Quiz students concerning the number of chromosomes before and after mitosis and the number of sister chromatids during mitosis. Give them one number and have them fill in the other two. Example: If there are 10 chromosomes before mitosis, students should respond that there are 20 sister chromatids and 10 chromosomes in each of the resulting two cells.

### Portfolio

Provide students with pipe cleaners, scissors, half a petri dish for drawing circles, and tape. Have them construct a series of three models to show the changes that occur in a cell during steps 1, 2, and 3 of mitosis. Have them label each model.

## STUDY GUIDE 127

STUDY GUIDE — CHAPTER 22

Name ______ Date ______ Class ______

**MITOSIS**

*In your textbook, read about the steps of mitosis in Section 22:1.*

1. The following steps of mitosis are out of order. Place the numbers *1-5* in the blanks to show the correct order.

A 3 B 2 C 1 D 5 E 4

2. In the blanks below, write the letter of the diagram above that is being described.
   a. Two new identical cells are formed. D
   b. Cytoplasm begins to separate. E
   c. Sister chromatids are first pulled apart. A
   d. Chromosomes are completely separated and at opposite ends of the cell. E
   e. Sister chromatids can be seen for the first time. B
   f. This is what cells look like before going through mitosis. C
   g. Nuclear membrane begins to break down. C
   h. Sister chromatids move to the cell's center and line up on fibers. B
   i. A nuclear membrane begins to form around chromosomes. D

## OPTIONS

**Steps of Mitosis**/*Transparency Package*, number 22a. Use color transparency number 22a as you teach mitosis.

# TEACH

## Mini Lab

Germinate radish seeds by soaking the seeds in water overnight. Remove from water, wrap in damp paper toweling, and place in plastic bag. Radish seeds germinate within 24-36 hours.

## ASSESSMENT

**Skill:** Provide diagrams of roots emerging from seeds and labeled 1, 2, and 3. Each diagram should include an original and a 24-hour-old root. Vary the root length on the 24-hour root diagrams, but make all original roots of equal length. Have students determine which root is undergoing the least mitosis and which is undergoing the most mitosis, and have them explain their answers.

### Check for Understanding

Have students respond to the first three questions in Check Your Understanding.

### Reteach

**Reteaching**/*Teacher Resource Package,* p. 66 Use this worksheet to give students additional practice in this steps of mitosis.

**Extension:** Assign Critical Thinking, Biology and Math, or some of the **OPTIONS** available with this lesson.

## 3 APPLY

**Challenge:** Design an experiment that measures the rate of growth of fingernails or hair. Carry out the experiment. Relate this growth to mitosis.

## 4 CLOSE

Explain the following: hair and nails grow, skin flakes off, your body grows, lost blood is replaced, cuts and burns heal.

a

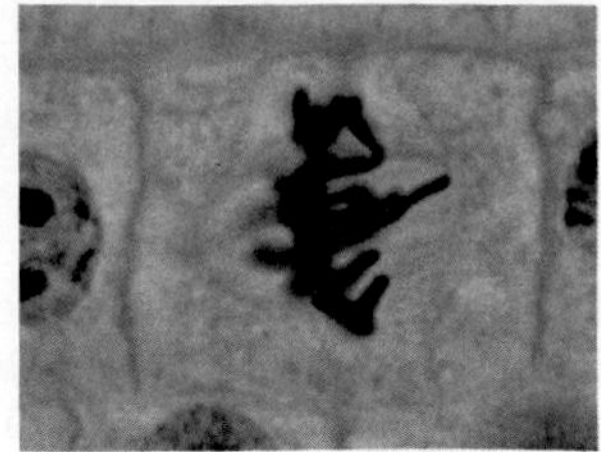

b

c

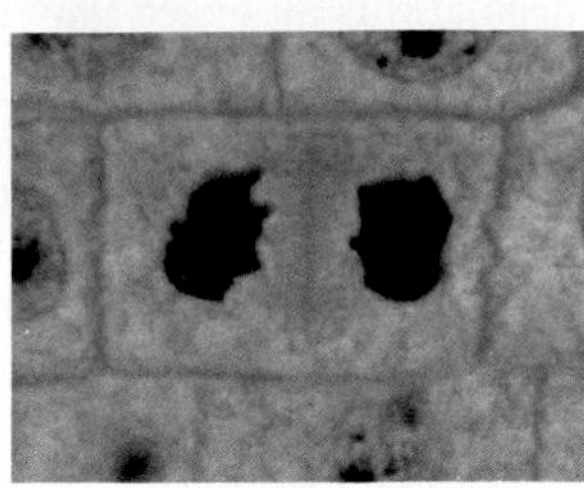

d

**Figure 22-6** Mitosis in plant cells: Step One (a), Step Two (b), Step Three (c), Step Four (d)

### Mini Lab

**How Rapid Is Mitosis?**

**Measure in SI:** Measure the amount of growth in the root and shoot of a radish seedling over a 24 hour period. *For more help, refer to the* ***Skill Handbook,*** *pages 718-719.*

The two new cells are smaller in size than the original. However, each new cell now begins to grow.

What are the benefits of mitosis? First, mitosis helps us grow by producing new cells. Second, mitosis replaces cells lost through cell death and injury, such as when you cut your finger.

Figure 22-6 shows photographs of plant cells in the four steps of mitosis. As you can see, mitosis in plant and animal cells is very similar. Mitosis in plant cells shows two differences from mitosis in animal cells. Plant cells lack centrioles, and at the end of cell division a cell wall is laid down.

### Check Your Understanding

1. Why is mitosis important to your body?
2. What are the jobs of protein fibers during mitosis?
3. What happens to chromosomes before the nuclear membrane breaks down in mitosis?
4. **Critical Thinking:** How does mitosis help a plant stem grow in width?
5. **Biology and Math:** Some cells divide every two hours. How many cells would result from mitosis in twenty-four hours if you started with one cell?



### Answers to Check Your Understanding

1. New cells are needed for growth and repair.
2. Fibers move sister chromatids to the center of the cell and pull them apart.
3. Chromosomes make copies of themselves, the centriole makes a copy of itself.
4. Plant growth in width is from mitosis in cells of the cambium.
5. 4096 cells

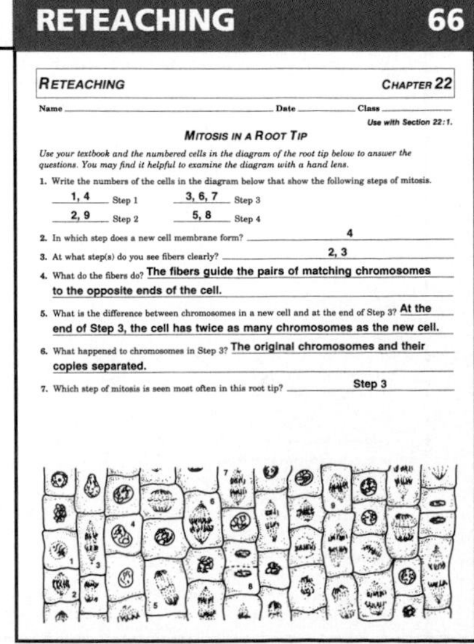

**RETEACHING 66**

RETEACHING CHAPTER 22

Name ______ Date ______ Class ______

Use with Section 22:1.

MITOSIS IN A ROOT TIP

*Use your textbook and the numbered cells in the diagram of the root tip below to answer the questions. You may find it helpful to examine the diagram with a hand lens.*

1. Write the numbers of the cells in the diagram below that show the following steps of mitosis.
   1, 4 Step 1 3, 6, 7 Step 3
   2, 9 Step 2 5, 8 Step 4
2. In which step does a new cell membrane form? 4
3. At what step(s) do you see fibers clearly? 2, 3
4. What do the fibers do? The fibers guide the pairs of matching chromosomes to the opposite ends of the cell.
5. What is the difference between chromosomes in a new cell and at the end of Step 3? At the end of Step 3, the cell has twice as many chromosomes as the new cell.
6. What happened to chromosomes in Step 3? The original chromosomes and their copies separated.
7. Which step of mitosis is seen most often in this root tip? Step 3

## 22:2 Meiosis

Not all cells reproduce by mitosis. A different type of cell reproduction is important to the passing on of traits to offspring.

**Objectives**

**4. Explain** the results of meiosis.

**5. Trace** the steps of meiosis.

**6. Compare** the sex cells of males and females.

### An Introduction to Meiosis

Besides body cells, most living things also have sex cells. **Sex cells** are reproductive cells produced in sex organs. Sperm are sex cells made by the male. An egg is a sex cell made by the female. Sex cells are made during meiosis (mi OH sus). **Meiosis** is a kind of cell reproduction that forms eggs and sperm.

**Key Science Words**

sex cell
meiosis
puberty
testes
ovary
polar body

To explain how a cell goes through meiosis, let's look again at our cell with four chromosomes, Figure 22-7. The four chromosomes make up two matching pairs.

In meiosis, a cell divides twice. When the cell divides for the first time, each chromosome in a pair moves away from its partner. Each chromosome of a pair goes to a different cell. The sister chromatids stay joined together. The two cells then divide again.

Look at Figure 22-7. How many chromosomes are in each of the four final cells? The number of chromosomes in each cell is one-half the original number. The original cell started with four chromosomes. Each new cell now has only two, or exactly one-half of the original number. Let's examine the process of meiosis in more detail to see how the chromosome number was halved.

**After meiosis, how does the number of chromosomes in a new cell compare with those in the original cell?**

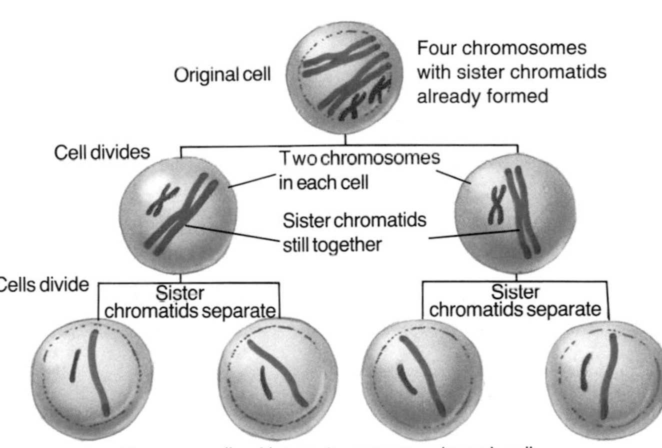


**Figure 22-7** Meiosis begins with one cell and results in four cells. Each new cell has one-half the original chromosome number.



**TRANSPARENCY 103**

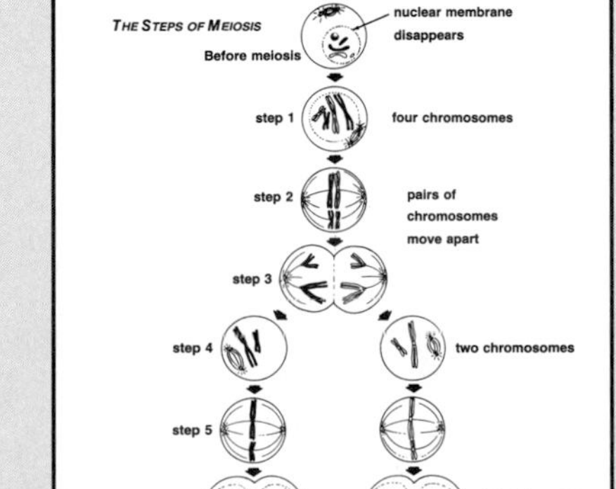


## OPTIONS

### Science Background

*Meiosis* means "to reduce" and refers to the fact that meiosis results in a reduction of the number of chromosomes from the diploid to the haploid number. In humans, the diploid number is 46 and the haploid number is 23.

**Transparency Master**/*Teacher Resource Package*, p. 103. Use the Transparency Master shown here to help you teach the steps of meiosis.

## 22:2 Meiosis

## PREPARATION

### Materials Needed

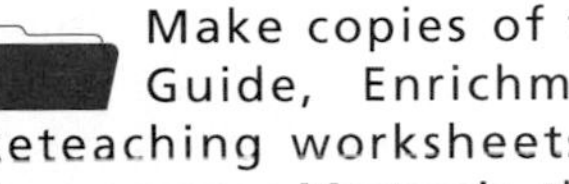

Make copies of the Study Guide, Enrichment, and Reteaching worksheets and the Transparency Master in the *Teacher Resource Package.*

### Key Science Words

| | |
|---|---|
| sex cell | testes |
| meiosis | ovary |
| puberty | polar body |

### Process Skills

In the Skill Check, students will understand science words.

## 1 FOCUS

▶ The objectives are listed on the student page. Remind students to preview these objectives as a guide to this numbered section.

### MOTIVATION/Brainstorming

Ask students what would be the result if we got 46 chromosomes from each parent. Chromosome number would double each generation.

### Guided Practice

Have students write down their answers to the margin question: there are half the original number of chromosomes in each of the four sex cells.

**Portfolio**

Copy Figure 22-6. Have students make these changes in the four diagrams· (a) enlarge the diagrams to show more detail, (b) assume these cells are from an animal rather than from a plant, (c) add labels to the important cell parts, (d) add a description of what is happening in each of the four figures.

# 2 TEACH

### MOTIVATION/Brainstorming

Ask students: How do the four final copies (new cells) differ from the original in terms of size and number of chromosomes present in original versus the number in each copy? They are smaller and have one-half the original number of chromosomes.

### Concept Development

- Provide students with the following problem: A cell is about to undergo meiosis. The cell has 20 chromosomes in its nucleus. Ask students to predict how many new cells will form from the original, as well as the chromosome number of each new cell.
- Have students contrast their answers to the effect of mitosis.
- Step one shows an important difference between meiosis and mitosis. Point out that no grouping of sister chromatids into fours took place during mitosis. This grouping into fours is called tetrad formation.
- Steps two and three differ from mitosis only in that there are groups of four rather than groups of two sister chromatids.
- Meiosis is divided into two major phases, meiosis I and meiosis II. In meiosis I, sets of four sister chromatids move together rather than only two, as in mitosis. Steps one through three are part of meiosis I. Steps four through six are part of meiosis II.

### Independent Practice

**Study Guide**/*Teacher Resource Package,* p. 128. Use the Study Guide worksheet shown at the bottom of this page for independent practice.

## Steps of Meiosis

Follow the numbers in Figure 22-8 as you read here about the six steps of meiosis. Just as in mitosis, each chromosome forms sister chromatids before meiosis begins.

**Step One**

1. The sister chromatids shorten and thicken.
2. The nuclear membrane begins to break down.
3. The centrioles begin to move away from one another and fibers form.
4. The matching chromosomes now come together to form pairs. Remember that each chromosome has two strands of sister chromatids. The pair of chromosomes now looks like a set of four strands and can be seen with a light microscope. Remember that in mitosis this pairing step doesn't take place.

**Step Two**

1. The centrioles have moved to opposite ends of the cell.
2. The sister chromatids become attached to the fibers.
3. Fibers move the two pairs of matching chromosomes to the center of the cell. Note that each chromosome is made up of sister chromatids that are still attached to one another. However, the paired chromosomes are not joined.

**Figure 22-8** Follow the steps of meiosis as you read the text. How many times do the cells divide?

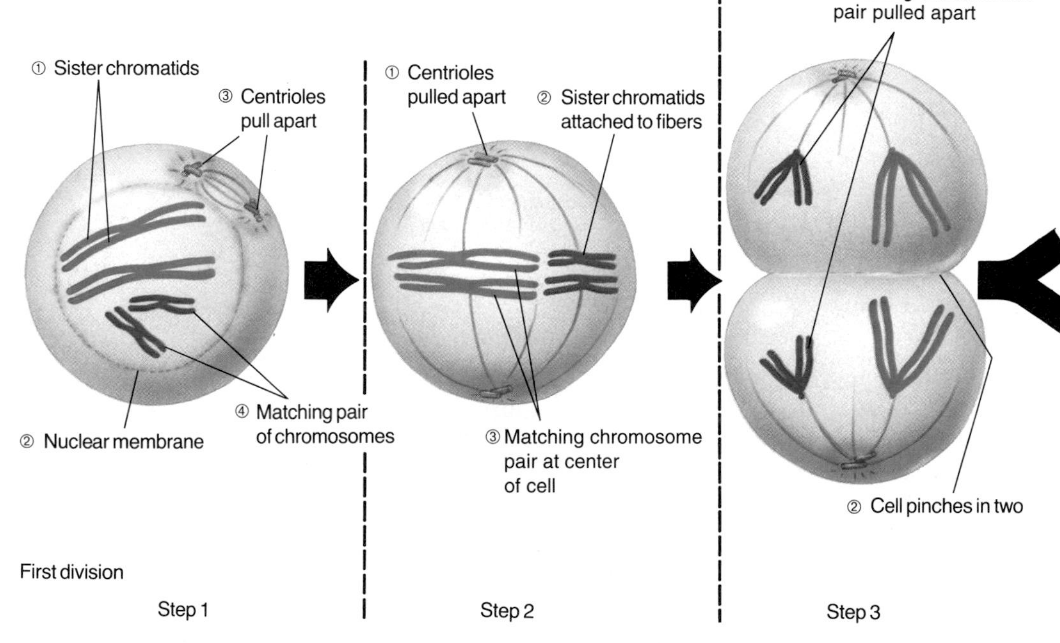




## OPTIONS

**Meiosis: Halving Chromosome Number/** *Transparency Package,* number 22b. Use color transparency number 22b as you teach meiosis.

**STUDY GUIDE 128**

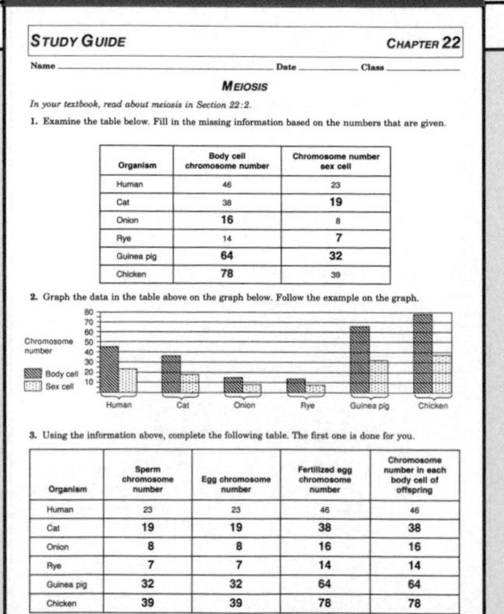

STUDY GUIDE CHAPTER 22

Name ______ Date ______ Class ______

MEIOSIS

*In your textbook, read about meiosis in Section 22:2.*

1. Examine the table below. Fill in the missing information based on the numbers that are given.

| Organism | Body cell chromosome number | Chromosome number sex cell |
|---|---|---|
| Human | 46 | 23 |
| Cat | 38 | 19 |
| Onion | 16 | 8 |
| Rye | 14 | 7 |
| Guinea pig | 64 | 32 |
| Chicken | 78 | 39 |

2. Graph the data in the table above on the graph below. Follow the example on the graph.



3. Using the information above, complete the following table. The first one is done for you.

| Organism | Sperm chromosome number | Egg chromosome number | Fertilized egg chromosome number | Chromosome number in each body cell of offspring |
|---|---|---|---|---|
| Human | 23 | 23 | 46 | 46 |
| Cat | 19 | 19 | 38 | 38 |
| Onion | 8 | 8 | 16 | 16 |
| Rye | 7 | 7 | 14 | 14 |
| Guinea pig | 32 | 32 | 64 | 64 |
| Chicken | 39 | 39 | 78 | 78 |

**Step Three**

1. Fibers move the matching chromosomes apart. Remember that in mitosis the sister chromatids came apart. In meiosis, the sister chromatids remain joined, but each matching pair of chromosomes separates.
2. The cell membrane begins to pinch the cell into two and divides the cytoplasm in half.

**Step Four**

1. Two new cells have now formed. Each cell has two chromosomes, each with sister chromatids.
2. The centrioles double and fibers form again.
3. A new nuclear membrane doesn't form at this time.

**Step Five**

The second division of meiosis now begins. Steps five and six now occur in both of the new cells. The movement of chromatids is similar to that in steps one to four of mitosis.

1. The centrioles move apart, and the fibers are formed between them.
2. The fibers connect to the sister chromatids at the point where the chromatids are joined together.
3. The sister chromatids are pulled to the center of each cell.

**In which step of meiosis do the pairs of sister chromatids separate?**

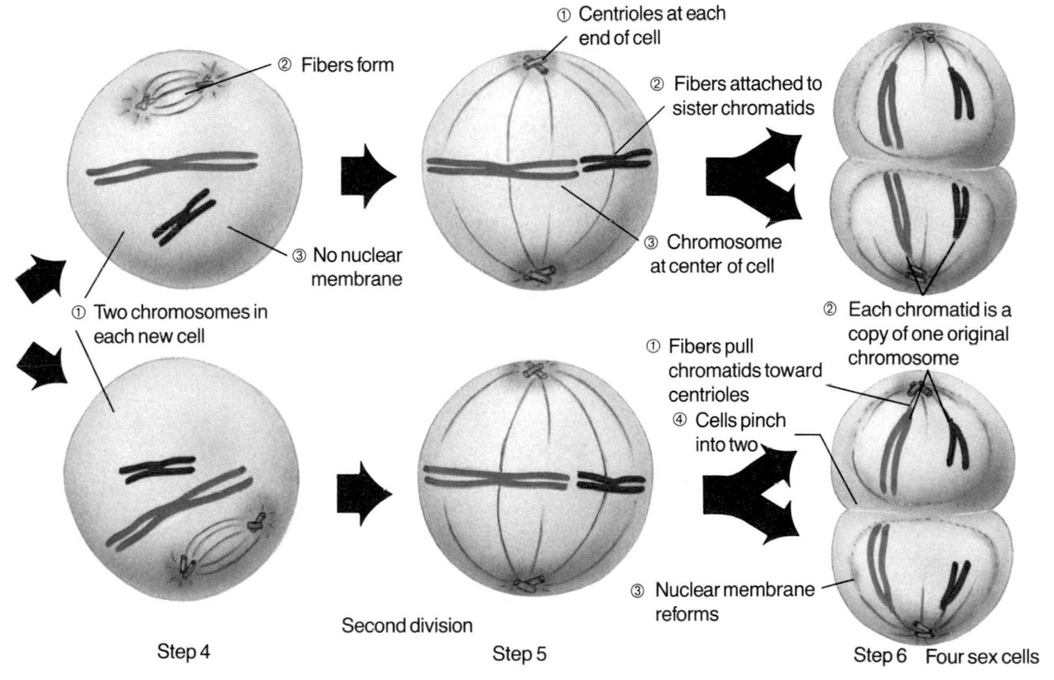




## TEACH

### Concept Development

▶ Point out that reduction in chromosome number occurs during steps one to three. At the end of meiosis I, each cell has only half the chromosome complement of the original cell.

▶ Step four is like mitosis beginning again. The important thing to stress here is that no new chromosome replication occurs.

### Guided Practice

Have students write down their answers to the margin question: step three.

### Guided Practice

Quiz students concerning mitosis and meiosis. Ask them to compare and contrast the following features: where and when the process happens, chromosome number before and after the process.

### Independent Practice

**Study Guide**/*Teacher Resource Package*, p. 129. Use the Study Guide worksheet shown at the bottom of this page for independent practice.

### Student Journal

Tell students the following: You are asked to describe to a student who has only general knowledge of mitosis and meiosis how they are alike and how they differ. Write down what you will tell them as you prepare your presentation of the two processes.

**STUDY GUIDE 129**

STUDY GUIDE CHAPTER 22

Name Date Class

MEIOSIS

*In your textbook, read about mitosis and meiosis in Section 22:2.*

4. The chart below shows the steps of mitosis and meiosis in the correct order. Complete the chart by answering the questions beside the diagrams.

a. Where are the chromosomes? in nucleus
b. How many are present in each cell? four
c. What happened to the chromosomes in each cell? They doubled forming sister chromatids.
d. Where are the sister chromatids? cell center
e. What is happening to the sister chromatids? They move apart. Pairs move apart in meiosis.
f. How many cells are formed in mitosis? two
g. Are the chromosomes single or double? single
h. Are the chromosomes single or double now in meiosis? double
i. What happens in meiosis now? Sister chromatids move to opposite ends of the cell.
j. How many cells are formed in meiosis? four

## OPTIONS

**Steps of Meiosis**/*Transparency Package*, number 22c. Use color transparency number 22c as you teach the steps of meiosis.

## GLENCOE TECHNOLOGY

**Videodisc**

**Biology: The Dynamics of Life**

Animation: *Meiosis*

Disc 1, Side 1, Ch. 30

# TEACH

## Concept Development

▶ You may wish to point out that no new cell formed by meiosis will have paired chromosomes.

▶ The fact that only one egg cell survives should not contradict all that has been learned regarding meiosis. Point out to students that the process of meiosis still formed four egg cells.

## Guided Practice

Have students write down their answers to the margin question: testes.

## Idea Map

Have students use the idea map as a study guide to the major concepts of this section.

## Independent Practice

**Study Guide**/*Teacher Resource Package,* p. 130. Use the Study Guide worksheet shown at the bottom of this page for independent practice.

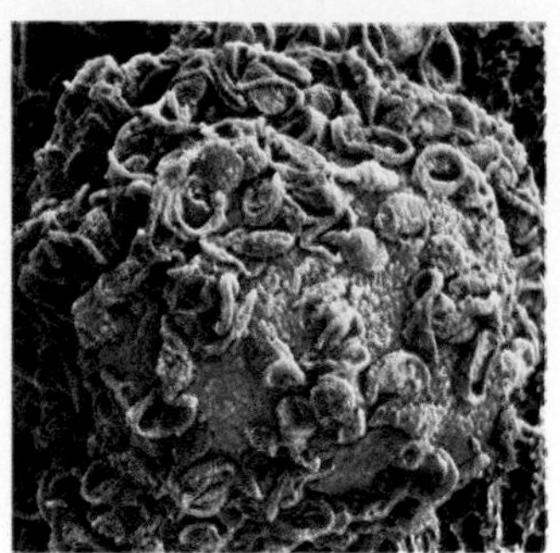

**Figure 22-9** Human egg and sperm cells

### Skill Check

**Understand science words: ovary.** The word part *ovi* means egg. In your dictionary, find three words with the word part *ovi* in them. *For more help, refer to the **Skill Handbook,** pages 706-711.*

**In which organs are sperm formed?**

**Step Six**

1. The fibers pull each strand of the sister chromatids apart and to opposite ends of the cell.
2. Each sister chromatid is an exact copy of just one of the original chromosomes.
3. The nuclear membrane begins to reform around each new set of chromosomes.
4. Cell membranes begin to pinch each cell in two along the center. The cytoplasm is divided between the new cells. The four new cells are sex cells.

Remember, that in our cell we started with four chromosomes. We now have two chromosomes in each of four cells. In a normal cell, there may be many more chromosomes. But, at the end of meiosis there will always be half the original number of chromosomes in each of four sex cells.

## Sperm, Eggs, and Fertilization

Figure 22-9 shows a photograph of human egg and sperm cells. How are the cells alike? How are they different? The cells are alike in four ways. One, both are sex cells. Two, both formed during meiosis. Three, each has half the number of chromosomes found in body cells. Four, in humans, both cells begin to develop by meiosis at puberty (PEW bur tee). **Puberty** is the stage in life when a person begins to develop sex cells. It takes place between the ages of 10 and 15.

How are sperm and eggs different? Eggs are much larger than sperm. Each sperm has a tail. The tail helps it move. Sperm form in the testes (TES teez). In animals, **testes** are the male sex organs that produce sperm. Eggs form in the ovaries (OHV uh reez). In animals and some plants, **ovaries** are the female sex organs.

**Study Tip:** Use this idea map as a study guide for comparing eggs and sperm. Which sex cells are smaller?

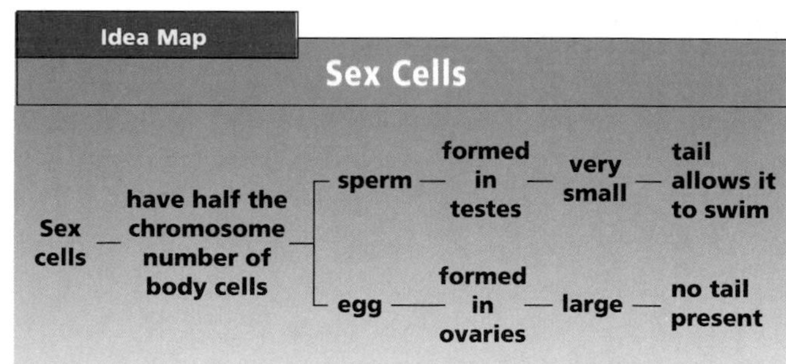




## OPTIONS

**Enrichment**/*Teacher Resource Package,* p. 24. Use the Enrichment worksheet shown here to give students practice with determining chromosome number after meiosis.

**Meiosis in Humans**/*Transparency Package,* number 22d. Use color transparency number 22d as you teach meiosis.

**STUDY GUIDE 130**

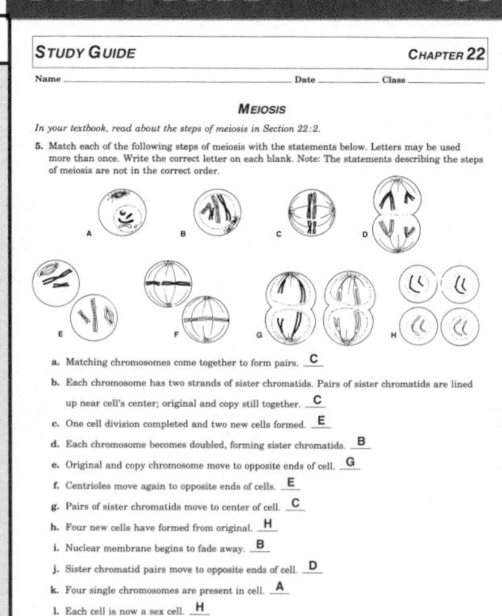

STUDY GUIDE CHAPTER 22

Name ______ Date ______ Class ______

MEIOSIS

*In your textbook, read about the steps of meiosis in Section 22:2.*

5. Match each of the following steps of meiosis with the statements below. Letters may be used more than once. Write the correct letter on each blank. Note: The statements describing the steps of meiosis are not in the correct order.



a. Matching chromosomes come together to form pairs. C
b. Each chromosome has two strands of sister chromatids. Pairs of sister chromatids are lined up near cell's center; original and copy still together. C
c. One cell division completed and two new cells formed. E
d. Each chromosome becomes doubled, forming sister chromatids. B
e. Original and copy chromosome move to opposite ends of cell. G
f. Centrioles move again to opposite ends of cells. E
g. Pairs of sister chromatids move to center of cell. C
h. Four new cells have formed from original. H
i. Nuclear membrane begins to fade away. B
j. Sister chromatid pairs move to opposite ends of cell. D
k. Four single chromosomes are present in cell. A
l. Each cell is now a sex cell. H
m. Nuclear membrane is reappearing. G

**ENRICHMENT 24**

ENRICHMENT CHAPTER 22

Name ______ Date ______ Class ______

*Use after Section 22:2.*

MITOSIS, MEIOSIS, AND NUMBER OF CHROMOSOMES

Organisms differ in the number of chromosomes in their cells. The table below shows the number of chromosomes in the body cells of different organisms.

*Use the table to help you answer the questions below. Refer to your textbook for the different steps of mitosis and meiosis.*

| Organism | Chromosome number | Organism | Chromosome number |
|---|---|---|---|
| alligator | 32 | amoeba | 50 |
| cat | 32 | chimpanzee | 48 |
| dog | 78 | eel | 36 |
| fly | 12 | human | 46 |
| lettuce | 18 | pigeon | 80 |

1. How many chromosomes in the body cells of an alligator? 32 an amoeba? 50 a dog? 78 a human? 46 a pigeon? 80 lettuce? 18 a fly? 12
2. How many chromosomes in the sex cells of a dog? 39 a cat? 16 a chimpanzee? 24
3. During Step One of mitosis how many pairs of sister chromatids are in the body cell of a human? 46 a cat? 32 lettuce? 18 an alligator? 32
4. During Step Three of mitosis how many chromosomes are in the dividing body cell of a human? 92 a dog? 156 a pigeon? 160 an alligator? 64 a cat? 64 a fly? 24
5. After mitosis is complete, how many chromosomes are in the resulting body cells of a dog? 78 a cat? 32 a human? 46 a pigeon? 80 an amoeba? 50
6. During Step One of meiosis how many pairs of sister chromatids are present in the forming sex cell of an amoeba? 50 an eel? 36 lettuce? 18
7. At the end of Step Three of meiosis how many pairs of sister chromatids are present in each of the forming cells of a human? 23 a cat? 16 a dog? 39 a fly? 6
8. When meiosis is complete how many chromosomes are in each resulting sex cell of an alligator? 16 a human? 23 an eel? 18
9. After fertilization takes place, how many chromosomes are in the fertilized cell of a dog? 78 a cat? 32 a human? 46 an amoeba? 50 an alligator? 32

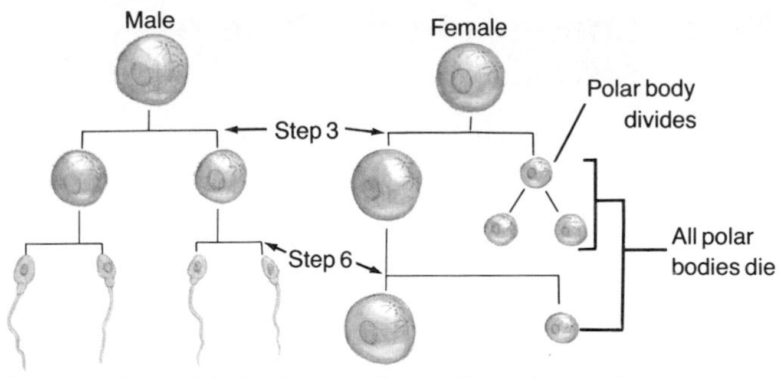


**Figure 22-10** Compare the products of meiosis between human males and females. Step 3 and Step 6 refer to the steps of meiosis shown on pages 472 and 473.

During meiosis, each original male cell becomes four sperm, Figure 22-10. Meiosis in males occurs all the time from the beginning of puberty.

Now, look at how eggs are formed, Figure 22-10. When the cell first pinches in half at the end of the first division, one cell is smaller than the other. The smaller cell is called a polar body. A **polar body** is a small cell formed during meiosis in a female. The polar body divides and then dies. The large cell that remains forms another polar body when the cell pinches in half again. This third polar body also dies. The large cell that remains becomes the egg. An egg is formed once a month from the onset of puberty.

When sperm and egg join, the chromosomes from each cell also come together. The new organism has a complete set of chromosomes in each one of its body cells. Half the chromosomes in the organism come from the father, and half come from the mother.

## Check Your Understanding

6. How do the number of chromosomes at the beginning and end of meiosis compare?
7. Describe changes in chromosomes during meiosis.
8. Compare the results of meiosis between human male and female sex cells.
9. **Critical Thinking:** What is the advantage of producing more sperm than eggs?
10. **Biology and Math:** If the body cells of an animal have 50 chromosomes, how many chromosomes would its sex cells have after meiosis?



## RETEACHING 67

RETEACHING CHAPTER 22

Name ______ Date ______ Class ______

Use with Section 22:2.

MEIOSIS AND FERTILIZATION

*Review the steps of meiosis in your textbook. Then draw the proper number of chromosomes in each diagram to show meiosis taking place. Use the same number and the same size of chromosomes as shown in Step 1 to complete the diagrams. You may want to use four different colored pencils to show the original and duplicated chromosomes.*

Meiosis
Fertilization
Step 1
Step 2
Step 3
Step 4
Step 5
Step 6
sperm cell
fertilized egg
mitosis

# TEACH

## Check for Understanding

Have students respond to the first three questions in Check Your Understanding.

### Reteach

**Reteaching**/*Teacher Resource Package*, p. 67. Use this worksheet to give students additional practice with meiosis and fertilization.

**Extension:** Assign Critical Thinking, Biology and Math, or some of the **OPTIONS** available with this lesson.

# 3 APPLY

## Divergent Question

Describe as many similarities and differences as you can between:
(a) sperm and egg cells
(b) mitosis and meiosis

# 4 CLOSE

## Audiovisual

Show the videocassette *Meiosis*, Carolina Biological Supply, Co.

## Answers to Check Your Understanding

6. The number is halved.
7. Chromosomes make copies of themselves, forming sister chromatids. Sister chromatids pair up with others of the same size. Sets of sister chromatids are moved to the center of the cell. Fibers separate sets of sister chromatids by moving them to opposite cell ends. Two cells then show fibers moving sister chromatids to the center of the cell. Sister chromatids are separated and move to opposite cell ends.
8. Female—one sex cell remains after meiosis. Male—four sex cells are formed.
9. More sperm are necessary because they leave the protection of the body and therefore have a lower chance of surviving.
10. 50/2 = 25

## 22:3 Changes in the Rate of Mitosis

## PREPARATION

### Materials Needed

Make copies of the Skill, Study Guide, Reteaching, and Lab worksheets in the *Teacher Resource Package.*

▶ Locate or purchase clear plastic sheets and marking pencils for Lab 22-2.

### Key Science Words

cancer

### Process Skills

In the Skill Check, students will classify. In the Lab 22-2, they will use numbers, make and use tables, form hypotheses, and interpret data.

## 1 FOCUS

▶ The objectives are listed on the student page. Remind students to preview these objectives as a guide to this numbered section.

### MOTIVATION/Brainstorming

Students are familiar with the term *cancer.* Ask them to list for you facts they know about cancer. Write their ideas on the chalkboard.

### Guided Practice

Have students write down their answers to the margin question: It slows down.

**Videodisc**

**STV: Human Body Vol. 3**
*Immune System*
Unit 1, Side 1
*Parasites, Cancer, Allergies*

**31350-38691**

## 22:3 Changes in the Rate of Mitosis

**Objectives**

7. **Explain** some effects of aging.
8. **Describe** the causes and effects of cancer.

**Key Science Words**

cancer

Changes often occur in the growth of cells. The rate of mitosis can speed up or slow down. Let's look at two effects caused by a change in the rate of mitosis.

### Aging

Aging is the process of becoming older. All living things age. Loss of hair, wrinkled skin, and loss of calcium in bones are some of the common signs of aging in humans.

**How does aging affect fingernail growth?**

Many changes that result from aging have something to do with mitosis. Let's take fingernail growth for example. Your fingernails grow by the process of mitosis. The new cells push out the old cells. Old cells harden, die, and form the fingernail. As a person ages, mitosis in cells of the fingernails slows down. Fingernail growth then slows down.

Changes in heart and body muscle also occur as you age. Muscle cells do not undergo mitosis. You are born with a certain number of muscle cells. As you grow, the muscle cells get bigger, but no new cells are made. As you become older, the muscle cells wear out, and no new ones replace them. This is why your heart weakens and cannot pump blood as well as you get older. Remember that your heart is made of muscle.

Table 22-1 shows a comparison between body functions at age 20 and age 70. Each difference may be the result of a slowing in the rate of cell mitosis. Mitosis is not a very well understood process. Scientists don't know why mitosis slows down with age.

**TABLE 22–1. CHANGES THAT OCCUR IN THE HUMAN BODY WITH AGE**

| Body System | Trait | 20-Year-Old | 70-Year-Old |
|---|---|---|---|
| Skin/Nails | Rate of fingernail growth | 1 mm/week | .6 mm/week |
| Nervous | Reaction time | .8 seconds | .95 seconds |
| Circulatory | Pumping action of heart | 3.7 liters/minute | 2.9 liters/minute |
| Nervous | Memory | 14 of 24 words recalled | 7 of 24 words recalled |
| Respiratory | Lung volume with deep breath | 5.5 liters/ inhalation | 3 liters/ inhalation |
| Muscular | % of body fat (male) | 15% | 30% |



## OPTIONS

### Science Background

Tissues subjected to "wear and tear" have the highest mitotic rates. Cells lining the digestive tract are constantly scraped by passing food and are surrounded by acidic liquids. Skin cells are constantly abraded or drying out.

**Skill**/*Teacher Resource Package,* p. 23. Use the Skill worksheet shown here to help students learn the process of developing a hypothesis.

**SKILL 23**

**SKILL: FORM A HYPOTHESIS** **CHAPTER 22**

Name ______ Date ______ Class ______

*For more help, refer to the Skill Handbook, pages 704-705.* *Use after Section 22:3.*

**GROWTH PATTERNS IN TWO FISH**

Joe and Rosa bought two goldfish that were the same color, size, and shape. They kept the fish in the same tank and fed them every day. They thought the fish were of the same species because they looked alike. Joe and Rosa had read that all cells in the body of a living thing go through mitosis at one rate. If the goldfish were of the same species, they thought the cells of both fish would all grow at the same rate. They made a hypothesis that the fish would grow at the same rate and would be about the same size a year later. To their surprise, one fish grew faster than the other. In six months, one fish was 2 cm longer than the other. What could be the explanation?

*Answer the following questions to find out why the hypothesis about the growth of the fish was incorrect. Develop a hypothesis of your own about the growth shown by the goldfish.*

1. What is mitosis? **cell reproduction in which two new cells are formed**
2. How does the size of the two new cells formed by mitosis compare with the size of the cell from which they were produced? **the new cells are smaller**
3. What happens to the size of each new cell soon after it has been formed? **Each cell grows in size.**
4. How does a living thing grow? **As cells reproduce by mitosis, the number of cells in a living thing increases. The increase results in growth.**
5. On what did the two students base their prediction about the growth of the two goldfish? **on the fact that the cells of the same kind of living thing go through mitosis at one certain rate**
6. State a hypothesis to explain why one fish grew faster than the other. **Answers will vary. Some possible answers are: the two goldfish may not be the same kind. One fish might have replaced more worn-out cells than the other fish. One fish might have eaten more than the other.**
7. Explain how you could test your hypothesis. **Answers will vary. A possible answer: Place two fish of the same kind and size in separate aquariums. Feed each fish the same amount of food each day. Measure both fish at least once a week. Record the measurements each time to keep track of any differences in length. Use a key to identify the species.**

# Lab 22–2 Rate of Mitosis

## Problem: Does mitosis occur in all cells all the time?

### Skills

use numbers, make and use tables, form hypotheses, interpret data

### Materials

clear plastic sheet
marking pencil
facial tissue
microscope

### Procedure

1. Copy the data table.
2. Lay a plastic sheet over the diagrams of tissue cells in the figure.
3. **Use numbers:** Count the total number of cells in each diagram. Use a marking pencil to check off each cell as you count. Record the number of cells.
4. Wipe off the plastic with the facial tissue. Place it over the diagrams again.
5. **Form a hypothesis:** Read the table in Figure 22-1 on page 464 again. State which of the three tissues you think will show the most cells in mitosis. Write your **hypothesis** in your notebook.
6. Repeat steps 1 and 2, only this time count the numbers of cells reproducing and the numbers of cells not reproducing by mitosis. Record these numbers.

### Data and Observations

1. **Interpret data:** Which tissue in the diagrams showed the most cells in mitosis?
2. Which tissue type in the diagrams showed the least number of cells in mitosis?

| TISSUE | TOTAL NUMBER OF CELLS | NUMBER OF CELLS NOT IN MITOSIS | NUMBER OF CELLS IN MITOSIS |
|---|---|---|---|
| Liver | 50 | 47 | 3 |
| Stomach Lining | 50 | 20 | 30 |
| Skin | 50 | 32 | 18 |

### Analyze and Apply

1. Which tissue type would:
   (a) repair itself the fastest?
   (b) repair itself the slowest?
2. **Check your hypothesis:** Is your hypothesis supported by your data? Why or why not?
3. **Apply:** Root tip cells reproduce faster than skin cells, but slower than liver cells. Draw 50 root tip cells. Show how many you think would be reproducing.

### Extension

**Use numbers:** Calculate the average number of dividing cells in one field of view of a prepared slide of onion root tip cells.

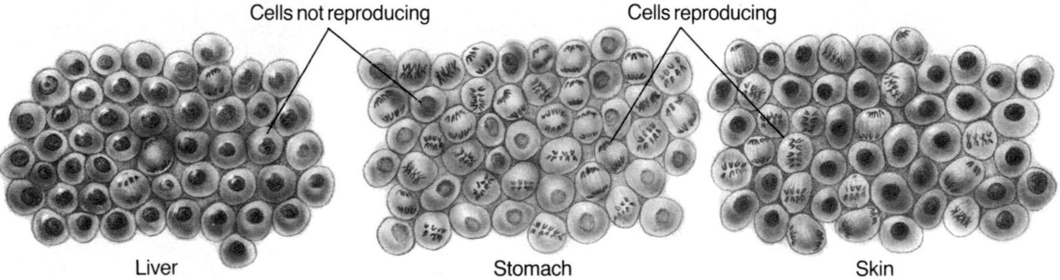




## COMPUTER PROGRAM XX

COMPUTER PROGRAM LAB 22-2 RATE OF MITOSIS

```
]POKE 33,33

]LIST

 10  GOSUB 390
 20  HOME
 30  PRINT "--------------------------------------";
 40  PRINT "!       ! TOTAL  ! NUMBER OF !NUMBER OF!";
 50  PRINT "!TISSUE ! NUMBER ! CELLS NOT !CELLS IN !";
 60  PRINT "!       !OF CELLS! IN MITOSIS!MITOSIS  !";
 70  PRINT "--------------------------------------";
 80  PRINT "!LIVER  !        !           !         !";
 90  PRINT "--------------------------------------";
 100  PRINT "!STOMACH!        !           !         !";
 110  PRINT "!LINING !        !           !         !";
 120  PRINT "--------------------------------------";
 130  PRINT "!SKIN   !        !           !         !";
 140  PRINT "--------------------------------------"
 150  FOR X = 1 TO 9
 160  VTAB V(X): HTAB H(X): FLASH : PRINT " ";: NORMAL
 170  VTAB 15: HTAB 1: CALL  - 958
 180  PRINT "ENTER YOUR VALUE FOR THIS SPACE";
 190  INPUT ": ";A
 200  VTAB V(X): HTAB H(X): PRINT A
 210  NEXT
 220  VTAB 15: HTAB 1: CALL  - 958
 230  PRINT "ARE THESE NUMBERS                    ENTERED CORRECTLY? (
     Y/N): ";: GET A$
 240  IF A$ = "N" OR A$ = "n" THEN  GOTO 20
 250  IF A$ < > "Y" AND A$  < > "y" THEN 220
 [illegible]  VTAB 15: HTAB 1: CALL  - 958: PRINT
 270  FLASH : PRINT "PRESS THE SPACE BAR TO SEE CORRECT DATA": GET A$
 278  NORMAL
 280  VTAB 16: HTAB 1: CALL  - 868
 290  FOR X = 1 TO 9
 300  VTAB V(X): HTAB H(X) + 3
 310  READ C: PRINT "(";C;")"
 320  NEXT
 330  INVERSE : PRINT
 340  PRINT : PRINT "DO YOU WANT THIS PRINTED (Y/N)? :";: GET A$
 345  NORMAL
 348  VTAB 15: HTAB 1: CALL  - 958
 350  IF A$ = "N" OR A$ = "n" THEN  END
 360  IF A$ < > "Y" AND A$  < > "y" THEN  GOTO 340
 370  GOSUB 2000
 380  END
 390  TEXT : HOME : PRINT "LAB 22-1                    MITOSIS RATE"
 400  PRINT : PRINT
 410  PRINT "YOU WILL BE ASKED TO TYPE IN YOUR      RESULTS FOR THE 3 TI
     SSUES ONTO A DATA   TABLE THAT IS EXACTLY LIKE THE ONE USED"
 420  PRINT "IN THE EXPERIMENT."
 430  PRINT
 440  PRINT "AFTER TYPING EACH NUMBER, PRESS THE     RETURN KEY. AFTER TY
     PING YOUR DATA, YOU WILL BE ASKED IF THE NUMBERS ARE ENTEREDCORRECTL
     Y.IF CORRECT TYPE THE LETTER Y.IF WRONG, TYPE THE LETTER N."
```

## ANSWERS

### Data and Observations

1. stomach
2. liver

### Analyze and Apply

1. (a) stomach (b) liver
2. Students who hypothesized that the stomach would show the most cells in mitosis will answer yes.
3. The diagram will show about five to ten cells.

## Lab 22-2 Rate of Mitosis

### Overview

In this activity, students will recognize that different tissues undergo mitosis at different rates.

**Objectives:** Upon completion of this lab, students will be able to: (1) **distinguish** between cells not undergoing mitosis, (2) **correlate** mitosis rate with growth and repair rate.

**Time Allotment:** 30 minutes

### Preparation

▶ **Alternate Materials:** Providing a photocopy of the diagrams to each student will eliminate the need for clear plastic sheets.

**Lab 22-2 worksheet**/*Teacher Resource Package,* pp. 85-86.

**Lab 22-2 Computer Program**/*Teacher Resource Package,* p. xx. Use the computer program shown at the bottom of this page as you teach this lab.

### Teaching the Lab

▶ **Troubleshooting:** Review with students which cells in the diagrams represent cells undergoing mitosis.

**Cooperative Learning:** Divide the class into groups of three students. For more information, see pp. 22T-23T in the Teacher Guide.

**Performance:** Provide the following data in table form:

Life Span of Different Plant Cells:

| | |
|---|---|
| Leaf cells | 30 days |
| Root cells | 23 days |
| Root tip cells | 2 days |
| Stem cells | 48 days |

Tell students to rank the plant cell types from most to least frequent occurrence of mitosis. Have them explain their criteria for ranking.

# 2 TEACH

### MOTIVATION/Brainstorming

Ask students to explain the significance of the data in Table 22-1 and if each change with age might be the result of cell loss. Might these changes be associated with lack of strength or sensory abilities such as vision and hearing? With aging, the process of mitosis slows down.

### Concept Development

▶ Make sure that students are correlating the events described on this page with Figure 22-5. The events are much more easily understood if this is done.

▶ The key concept in cancer is rapid cell mitosis in some tissue, which crowds out the tissue undergoing mitosis at a normal rate.

### Guided Practice

Have students write down their answers to the margin question: Lung tissue becomes crowded by abnormal cells.

### Independent Practice

**Study Guide**/*Teacher Resource Package*, pp. 131-132. Use the Study Guide worksheets shown at the bottom of this page for independent practice.

### *Student Journal*

Using Figure 22-11 as a guide, have students measure the width in millimeters of the normal (Figure a) and cancer tissues (Figures b and c). Tell them to describe the events that are occurring in Figures a-c. They should include the term *mitosis* as well as the measurements taken in millimeters as part of their description.

**Bio Tip**

**Health:** Exposure to radiation can affect the cell divisions in a person's sex cells. When you receive X rays at a hospital or dentist you should wear an apron made of lead for protection. X rays can't pass through lead.

**How does cancer in the lungs affect lung tissue?**

## Cancer

Healthy cells have regular rates of reproducing. For example, a skin cell undergoes mitosis once every 20 days. During the other 19 days, it doesn't reproduce. A liver cell may undergo mitosis once every 200 days.

**Cancer** is a disease in which body cells reproduce at an abnormally fast rate. Follow the steps in Figure 22-11 that show cancer forming in the lungs. The cells are shown about 250 times their real size. Figure 22-11a shows normal lung cells. A thin layer of cells covers the lungs. Notice the changes in Figure 22-11b. The outer layer of cells has changed from one cell layer to many cell layers.

In Figure 22-11c, the outer layer of cells is no longer thin and even. Rapid mitosis has increased the number of cells. The shapes of the cells and their nuclei have also changed. The abnormal cells begin to crowd the lung tissues inside. The outer cells have become cancer cells. They will continue to increase in number until they crowd out all normal lung tissue.

Why do some cells start reproducing abnormally and become cancer cells? Three well-known causes include chemicals, radiation, and viruses. Cells may become cancer cells if they are in contact with poisonous chemicals for a long time. For example, it has been shown that chewing tobacco may cause cancer of the mouth. Skin cells may become cancer cells if a person spends too much time in the sun. High levels of X rays may cause cancer in bones. Even viruses cause certain cancers.

**Figure 22-11** Follow the changes in lung tissue as the covering cells become cancer cells.

a

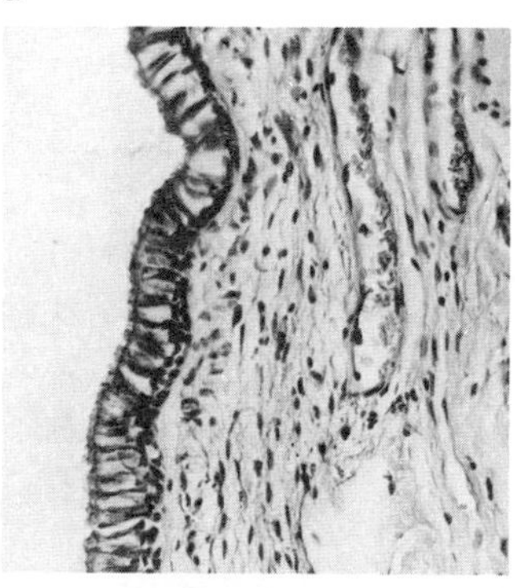

b

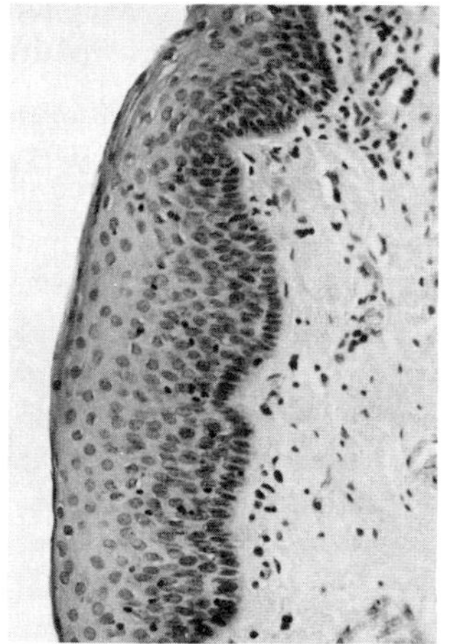

c

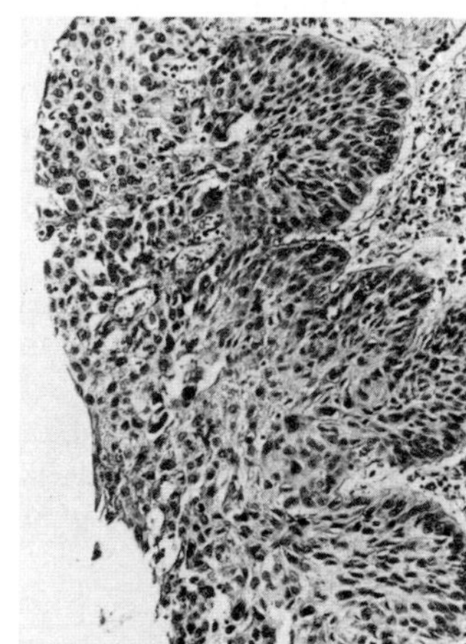

**STUDY GUIDE 131**

STUDY GUIDE — CHAPTER 22

Name ______ Date ______ Class ______

**MEIOSIS**

*In your textbook, read about sperm, eggs, and fertilization in Section 22:2.*

6. Look at the drawing below. Label the parts as egg, sperm, or chromosomes.

sperm / chromosomes / egg — A B C

a. How many chromosomes are in each human sex cell? **23**
b. How many pairs of chromosomes are in each sex cell? **none**
c. What process has occurred between diagram B and C? **fertilization**
d. How many chromosomes are in the egg cell in diagram C? **46**

**CHANGES IN THE RATE OF MITOSIS**

*In your textbook, read about aging and cancer in Section 22:3.*

1. What happens to fingernail growth as a person ages? Explain your answer. **Fingernail growth slows down because mitosis in cells of the fingernails slows down.**
2. Why is the heart weaker and less able to pump blood as a person ages? **As a person grows, muscle cells get bigger, but no new cells are made. As a person ages, the muscle cells wear out, and no new ones replace them.**
3. How are cancer cells different from normal cells? **They undergo rapid mitosis, and they no longer have their usual shape and nucleus.**
4. List three causes of cancer. **chemicals, radiation, and viruses**

**STUDY GUIDE 132**

STUDY GUIDE — CHAPTER 22

Name ______ Date ______ Class ______

**VOCABULARY**

*Review the new words used in Chapter 22 of your textbook. Then complete this exercise.*

1. Match one of these choices to each of the following statements. Write the proper choice on each blank: mitosis, meiosis.
   a. Four new cells are formed from each original. **meiosis**
   b. This process makes sperm cells. **meiosis**
   c. Two new cells are formed from each original. **mitosis**
   d. New skin cells are made this way. **mitosis**
   e. This type of cell reproduction helps you grow. **mitosis**
   f. This type of cell reproduction makes egg cells. **meiosis**
   g. This makes cells with half the original cell chromosome number. **meiosis**
   h. This makes cells with the same chromosome number as the original. **mitosis**
2. Match one of the choices to each of the following phrases. Write the proper choice on each blank: testes, ovaries, cancer, polar body, puberty, sex cells, body cells, sister chromatids
   a. stage when person produces sex cells **puberty**
   b. organs that produce eggs **ovaries**
   c. small cell formed and then dies during meiosis in female **polar body**
   d. organs that produce sperm **testes**
   e. egg and sperm **sex cells**
   f. cells that make up the skin, blood, bones, and stomach **body cells**
   g. the two strands of a doubled chromosome **sister chromatids**
   h. a disease in which body cells reproduce faster than normal **cancer**

**Figure 22-12** Chewing tobacco is a habit that may cause mouth cancer.

Cancer is caused by many things. There are probably many causes that scientists have not yet discovered. For these reasons, the problem of how to treat cancer is a very difficult one to solve. Over the past 30 years, regular medical treatment that includes doses of radiation has resulted in a doubling of the survival rates of cancer patients.

### Skill Check

**Classify:** The following are possible causes of cancer. Classify them as chemical, radiation, or other: smoking, radon, asbestos, smog, UV rays, and paint thinner. *For more help, refer to the* ***Skill Handbook,*** *pages 715-717.*

## Check Your Understanding

11. What are some of the changes that take place as a person ages?
12. What is the effect of cancer on mitosis?
13. List four possible causes of cancer.
14. **Critical Thinking:** Your skin produces a dark protective pigment when exposed to the sun's burning rays. There are a variety of suntan lotions. Some help you tan faster, some block the rays of the sun. Which suntan lotion should you choose to protect you from developing skin cancer?
15. **Biology and Reading:** Some cancers are inherited. Retinoblastoma (REH tihn uh blah STOH muh) is a kind of eye cancer that is inherited by about 1 in 20 000 children. What does the word *inherited* mean as used in this description?



## TEACH

### Skill Check

Chemical: smoking, smog, paint thinner
Radiation: radon, UV rays
Other: asbestos, AIDS virus

### Check for Understanding

Have students respond to the first three questions in Check Your Understanding.

### Reteach

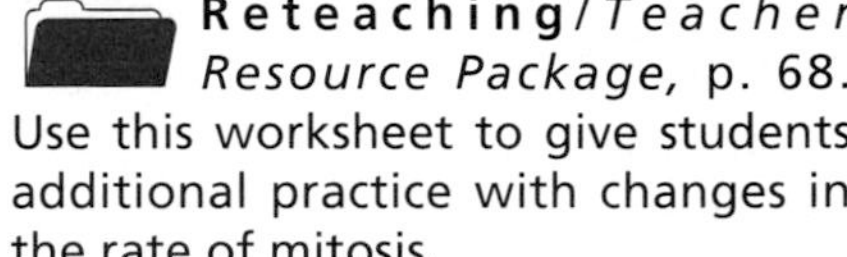

**Reteaching**/*Teacher Resource Package,* p. 68. Use this worksheet to give students additional practice with changes in the rate of mitosis.

**Extension:** Assign Critical Thinking, Biology and Reading, or some of the **OPTIONS** available with this lesson.

## 3 APPLY

Offer an explanation for the following: Warts are a type of tumor but are not cancerous. Explain why they are not harmful.

## 4 CLOSE

### Audiovisual

Show the videocassette *Living with Cancer,* Films for the Humanities.

### Answers to Check Your Understanding

11. loss of hair, wrinkled skin, changes in bones
12. Cancer causes abnormal mitosis in body cells.
13. contact with certain chemicals, X rays, viruses, too much sun
14. Suntan lotions that block the rays of the sun can help reduce the risk of skin cancer.
15. Inherited means that the disease is caused by chromosomes from the mother or father.

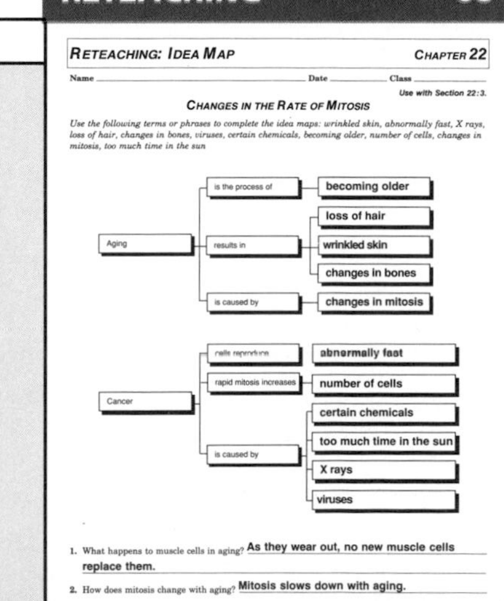

RETEACHING 68

RETEACHING: IDEA MAP — CHAPTER 22

Name ______ Date ______ Class ______

*Use with Section 22:3.*

CHANGES IN THE RATE OF MITOSIS

*Use the following terms or phrases to complete the idea maps: wrinkled skin, abnormally fast, X rays, loss of hair, changes in bones, viruses, certain chemicals, becoming older, number of cells, changes in mitosis, too much time in the sun*



1. What happens to muscle cells in aging? As they wear out, no new muscle cells replace them.
2. How does mitosis change with aging? Mitosis slows down with aging.

## OPTIONS

**Are There More Dividing Cells or Resting Cells in a Root Tip?**/*Lab Manual,* p. 185. Use this lab as an extension to the rate of mitosis.

# CHAPTER 22 REVIEW

## Summary

Summary statements can be used by students to review the major concepts of the chapter.

## Key Science Words

All boldfaced terms from the chapter are listed.

## Testing Yourself

### Using Words

1. meiosis
2. cancer
3. mitosis
4. ovary
5. puberty
6. polar body
7. testes

### Finding Main Ideas

8. p. 478; Covering cells are usually two cells thick. They form a thin even layer over connective tissue. As cancer begins, the covering layer begins to thicken and crowds out connective tissue. Covering cells are no longer normal.
9. p. 469; In step two, the sister chromatids become attached to the spindle fibers and are pulled toward the center of the cell. In step three, the sister chromatids are pulled apart by the spindle fibers and become separated.
10. p. 474; Egg and sperm cells are made, four cells are formed from each original cell, the number of chromosomes is half the number in the original cell, and each new cell has only one of each type of chromosome.
11. p. 464; Mitosis allows us to grow and to replace damaged or worn-out cells.
12. p. 476; Mitosis slows down as we age. Muscle cells wear out and no new ones replace them. As a result, for example, your heart is not as strong as when you were younger.
13. pp. 468, 472; In mitosis, the chromosomes are not paired; in meiosis, the matching chromosomes are paired up.

# Chapter 22

# Review

## Summary

### 22:1 Mitosis

1. Living things grow and repair themselves in a process of cell reproduction called mitosis.
2. Before mitosis begins, chromosomes become doubled and form sister chromatids. Each chromatid is an exact copy of the original chromosome.
3. During mitosis, two identical cells form from one original cell. Mitosis takes place in four steps. It results in two new body cells, each with the same number of chromosomes as in the original cell.

### 22:2 Meiosis

4. Sex cells form through a process called meiosis.
5. Meiosis involves two cell divisions. It results in four sex cells, each with half the number of chromosomes as the original cell.
6. In human males, four sperm result from meiosis. In females, only one egg results from meiosis. Egg and sperm join together at fertilization. Each supplies half the new cell's chromosomes. The new organism develops by mitosis.

### 22:3 Changes in the Rate of Mitosis

7. Aging is due to a slowing down of mitosis in some cells of the body.
8. Cancer is the result of abnormal mitosis. Cells reproduce too rapidly.
9. Cancer can be caused by chemicals, radiation or viruses.

## Key Science Words

body cell (p. 464)
cancer (p. 478)
meiosis (p. 471)
mitosis (p. 464)
ovary (p. 474)
polar body (p. 475)
puberty (p. 474)
sex cell (p. 471)
sister chromatids (p. 466)
testes (p. 474)

## Testing Yourself

### Using Words

*Choose the word from the list of Key Science Words that best fits the definition.*

1. cell reproduction that forms sex cells
2. abnormal cell mitosis
3. cell reproduction in which two identical cells are made
4. female sex organ
5. stage in life when sex cells start to undergo meiosis

480

### Using Main Ideas

14. step 6
15. The heart of an older person can't pump as much blood because some of the heart muscle cells have worn out and muscle cells are not replaced when worn out. It is therefore less efficient.
16. A cat would have 19 chromosomes in its sex cells.
17. (a) mitosis – two cells, meiosis – four cells; (b) mitosis – same as the original cell, meiosis – one-half the original cell; (c) mitosis – body cell, meiosis – sex cell.
18. Causes of cancer include chemicals, radiation, and viruses.

# Review

## Testing Yourself *continued*

6. small cell in the female that dies after being formed by meiosis
7. organs that produce sperm

### Finding Main Ideas

*List the page number where each main idea below is found. Then, explain each main idea.*

8. changes that take place in lung tissue with cancer
9. the role of the fibers during Steps Two and Three of mitosis
10. what is accomplished during meiosis
11. what the benefits of mitosis are
12. specific changes that take place between age 20 and age 70
13. how the chromosome grouping in Step Two of mitosis differs from that in Step Two of meiosis

### Using Main Ideas

*Answer the questions by referring to the page number after each question.*

14. In what step of meiosis do sister chromatids separate? (p. 474)
15. Why can't the heart of an older person pump as much blood as that of a younger person? (p. 476)
16. Cats have 38 chromosomes in their body cells. How many chromosomes would their sex cells have? (p. 471)
17. How do each of the following compare for mitosis and meiosis? (pp. 469, 474)
    (a) the number of cells formed
    (b) numbers of chromosomes in the new cells compared with number of chromosomes in the original cell
    (c) type of cell formed
18. What are three main causes of cancer? (p. 478)

## Skill Review 

*For more help, refer to the **Skill Handbook,** pages 704-719.*

1. **Classify:** Compare the following traits of eggs and sperm: size, number of chromosomes, process that makes them, polar bodies formed, tail present.
2. **Sequence:** What changes take place in the cell membrane and nuclear membrane before or during mitosis?
3. **Measure in SI:** A student makes 500 new bone cells and each is 0.015 mm in size. Assuming they all form in the same direction, how much did this student grow?
4. **Observe:** Examine a prepared slide of a lily anther and count the number of chromosomes in the young pollen cells.

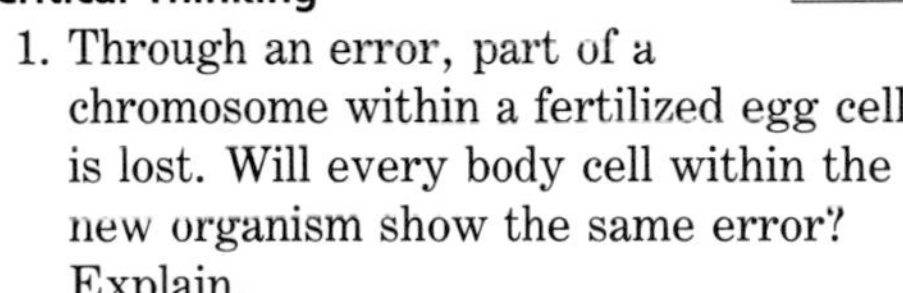


## Finding Out More

### Critical Thinking

1. Through an error, part of a chromosome within a fertilized egg cell is lost. Will every body cell within the new organism show the same error? Explain.
2. How might tanning parlors and cigarettes be related to cancer?

### Applications

1. Demonstrate how rapidly a living thing increases in size by calculating the number of cells that result from 10 generations of mitosis.
2. Build models of a cell's chromosomes showing the changes that take place during mitosis and meiosis.



## Skill Review

1. Traits of egg and sperm that are similar: number of chromosomes, process that makes them. Traits that are different: size, polar bodies formed, tail present.
2. During mitosis, the nuclear membrane breaks down, allowing the chromosomes to go to opposite ends of the cell. It then reforms around the nuclear material. The cell membrane pinches in two and separates the cell into two new identical cells.
3. 500 x 0.015 mm = 7.50 mm
4. Students should be able to see 12 chromosomes.

## Finding Out More

### Critical Thinking

1. Yes. All body cells within an organism have chromosomes that are exact duplicates of those in the original fertilized egg. If part of a chromosome is lost, this part cannot later be duplicated.
2. The tanning rays used in a tanning parlor and cigarettes are known to cause cancer.

### Applications

1. There will be 1024 cells after 10 generations of mitosis.
2. Thin and thick yarn of different colors or colored plastic-coated wire can be used. The wire can be coiled to show thickening.

 **Chapter Review**/*Teacher Resource Package,* pp. 116-117.

 **Chapter Test**/*Teacher Resource Package,* pp. 118-120.

**Chapter Test**/*Computer Test Bank*

# Plant Reproduction and Development

## PLANNING GUIDE

| CONTENT | TEXT FEATURES | TEACHER RESOURCE PACKAGE | OTHER COMPONENTS |
|---|---|---|---|
| (1 1/2 days)<br>23:1 Asexual Reproduction in Plants<br>Asexual Reproduction by Roots<br>Asexual Reproduction by Leaves<br>Asexual Reproduction by Stems | Idea Map, p. 485<br>Check Your Understanding, p. 487 | Reteaching: *Asexual Reproduction in Plants,* p. 69<br>Focus: *Seeds That Sprouted After Hundreds of Years,* p. 45<br>Study Guide: *Asexual Reproduction in Plants,* p. 133 | **STVS:** *Super Trees,* Plants and Simple Organisms (Disc 4, Side 2) |
| (1 1/2 days)<br>23:2 Sexual Reproduction in Plants<br>Flowers and Sexual Reproduction<br>Pollination and Fertilization | Idea Map, p. 488<br>Idea Map, p. 490<br>Check Your Understanding, p. 491 | Reteaching: *Fertilization in Flowers,* p. 70<br>Critical Thinking/Problem Solving: *Will Fertilization Take Place?* p. 23<br>Skill: *Recognize Spatial Relationships,* p. 24<br>Transparency Master: *Flower Anatomy and Pollination,* p. 113<br>Study Guide: *Sexual Reproduction in Plants,* pp. 134-135 | **Laboratory Manual:**<br>*What Are the Parts of a Flower?* p. 189<br>Color Transparency 23a: *Parts of a Flower*<br>**STVS:** *Bats as Pollinators,* Plants and Simple Organisms (Disc 4, Side 2) |
| (2 days)<br>23:3 Plant Development<br>Seeds and Fruits<br>Plant Development from Seeds<br>Comparing Kinds of Reproduction | Skill Check: *Understand Science Words,* p. 494<br>Mini Lab: *Will Seeds Grow Without Stored Food?* p. 492<br>Lab 23-1: *Germination,* p. 495<br>Lab 23-2: *Rooting,* p. 497<br>Technology: *Seed Banks,* p. 493<br>Check Your Understanding, p. 496<br>Applying Technology: *Are These Seeds Alive?* p. 498 | Application: *Flowers and Fruits,* p. 23<br>Enrichment: *Identifying Different Types of Fruits,* p. 25<br>Reteaching: *What Happens to Certain Flower Parts After Fertilization?* p. 71<br>Study Guide: *Plant Development,* pp. 136-137<br>Study Guide: *Vocabulary,* p. 138<br>Lab 23-1: *Germination,* pp. 89-90<br>Lab 23-2: *Rooting,* pp. 91-92 | **Laboratory Manual:**<br>*What Plant Part Are You Eating?* p. 193<br>Color Transparency 23b: *From Flower to Fruit*<br>Color Transparency 23c: *Germination*<br>**STVS:** *Seed Banks,* Plants and Simple Organisms (Disc 4, Side 2) |
| Chapter Review | Summary<br>Key Science Words<br>Testing Yourself<br>Finding Main Ideas<br>Using Main Ideas<br>Skill Review | **ASSESSMENT RESOURCES**<br>Chapter Review, pp. 121-122<br>Chapter Test, pp. 123-125<br>Performance Assessment in the Biology Classroom<br>Alternate Assessment in the Science Classroom<br>Computer Test Bank | |

## GLENCOE TECHNOLOGY

**Science and Technology Videodisc Series,** *Super Trees,* Plants and Simple Organisms (Disc 4, Side 2)
*Bats as Pollinators,* Plants and Simple Organisms (Disc 4, Side 2)
*Seed Banks,* Plants and Simple Organisms (Disc 4, Side 2)

## MATERIALS NEEDED

| LAB 23-1, p. 495 | LAB 23-2, p. 497 | APPLYING TECHNOLOGY, p. 498 | MARGIN FEATURES |
|---|---|---|---|
| metric ruler<br>rubber band<br>small beaker<br>pencil<br>tape<br>permanent marker<br>2 milk cartons<br>soil<br>20 corn seeds<br>water<br>stapler<br>scissors | garlic bulb<br>carrot<br>labels<br>water<br>toothpicks<br>small beaker<br>metric ruler<br>razor blade<br>shallow dish | 150 bean seeds<br>beaker<br>paper towels<br>plastic bag, self-seal<br>label<br>shallow dish<br>tetrazolium<br>forceps | Skill Check, p. 494<br>dictionary<br>Mini Lab, p. 492<br>soaked bean seeds<br>paper towel<br>petri dish |

## OBJECTIVES

For more information about National Science Standards, see page 5T.

| SECTION | OBJECTIVE | CORRELATION of QUESTIONS to OBJECTIVES | | | |
|---|---|---|---|---|---|
| | | CHECK YOUR UNDERSTANDING | CHAPTER REVIEW | TRP CHAPTER REVIEW | TRP CHAPTER TEST |
| 23:1 National Science Stds: UCP.1, E.2, G.1, G.2 | 1. **Explain** how plants reproduce asexually from roots. | 1 | 15, 23, 24 | 1, 33 | 26, 29, 30, 32, 34 |
| | 2. **Describe** how plants can reproduce asexually from leaves. | — | 9, 26 | 6 | 29, 30, 31, 32, 33, 34 |
| | 3. **Compare** asexual reproduction from runners, tubers, bulbs, cuttings, and grafting. | 2 | 2, 4, 10 | 3, 13,14, 18, 37 | 1, 4, 14, 25, 27, 28, 29, 30, 32, 33, 34 |
| 23:2 National Science Stds: UCP.1, UCP.2, UCP.5, C.4 | 4. **List** the parts of a flower and their functions. | 5, 6 | 5, 7, 18, 25 | 11, 22, 23, 24, 25, 26, 27, 28, 30, 31, 36 | 9, 10, 15, 16, 17, 18, 19, 21, 22, 23, 24, 38, 40, 41 |
| | 5. **Describe** two methods of pollination. | 7 | 8, 13, 21 | 32, 35 | 8 |
| | 6. **Sequence** the steps that lead to fertilization in flowering plants. | — | 16, 17, 22 | 8, 15, 19, 20, 29 | 5, 7, 20, 35 |
| 23:3 National Science Stds: UCP.1, UCP.2, UCP.3, UCP.4, UCP.5, C.1, C.4 | 7. **Describe** how fruits and seeds develop. | 10, 14 | 6, 11, 20 | 4, 7, 9 | 3, 11, 12, 36, 37, 39, 42 |
| | 8. **Explain** how seeds are scattered and grow into new plants. | 11, 13 | 3, 14 | 2, 5, 34 | 6, 13 |
| | 9. **Compare** asexual and sexual reproduction in plants. | 12 | 1, 12, 19 | 1, 10, 12, 16, 17, 38 | 2, 43, 44, 45, 46, 47, 48, 49, 50, 51, 52 |

# Plant Reproduction and Development

## CHAPTER OVERVIEW

### Key Concepts

In this chapter students will study sexual and asexual reproduction by plants. These two processes are correlated with mitosis and meiosis. The advantages and disadvantages of sexual and asexual reproduction are also examined.

### Key Science Words

| | |
|---|---|
| bulb | petal |
| clone | pistil |
| cross pollination | pollination |
| cutting | runner |
| fruit | self pollination |
| germination | sepal |
| grafting | stamen |
| ovule | tuber |

### Skill Development

In Lab 23-1, students will **form a hypothesis, experiment,** and **observe** to see if soaked seeds germinate faster than unsoaked seeds. In Lab 23-2, they will **experiment, observe, measure in SI,** and **infer** whether or not plants can grow from roots and bulbs. In the Skill Check on page 494, students will **understand** the **science word** *germination.* In the Mini Lab on page 492, students will experiment, to see if seeds will grow without food.

### Bridging

Plants exhibit the same life functions as animals. These functions were established in Chapter 2. One life function listed at that time was the ability of living things to reproduce. This chapter enables students to look at the process of plant reproduction in great detail.

Chapter 22 dealt with the topics of mitosis and meiosis. How plants actually utilize these processes during asexual and sexual reproduction will be emphasized.

**CHAPTER PREVIEW**

### Chapter Content

Review this outline for Chapter 23 before you read the chapter.



### Skills in this Chapter

The skills that you will use in this chapter are listed below.

- In **Lab 23-1,** you will form a hypothesis, experiment, and observe. In **Lab 23-2,** you will experiment, observe, measure in SI, and infer.
- In the **Skill Check,** you will understand science words.
- In the **Mini Lab,** you will experiment.

482

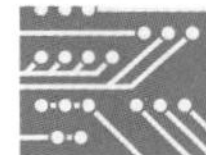

## TECH PREP

For Tech Prep activities in this chapter of the Teacher Wraparound Edition, see especially the Motivations on pages 485 and 489, the Activity and Options Activity on page 486, the Student Journal on page 493, and the Connection on page 496.

See also the Glencoe Homepage at **www.glencoe.com**

Chapter 23

# Plant Reproduction and Development

You have read that two of the features of living things are that they grow and reproduce. How do flowering plants grow and reproduce? Look at the flower in the photo on the left. You may have noticed that insects are attracted to flowers. Insects have important relationships with many different flowering plants. Many insects help flowers reproduce.

Look at the photo below. This is a fruit that developed from the flower on the left. How did an insect help the flower turn into a fruit? The flower, the fruit, and the insect all play a role in the sexual reproduction of flowering plants. Flowering plants may also reproduce asexually from roots, stems, and leaves.

**Try This!**

**Are fruits different from vegetables?** Examine several different fruits and vegetables. Decide how these two plant parts differ. Form a definition of these two terms based on your observations.

**interNET CONNECTION**

For more information about the material in this chapter, follow the link for the chapter on the Glencoe Homepage at **http://www.glencoe.com**

A fruit develops from a flower.

483

## ASSESSMENT PLANNER

**Portfolio**

Strategies on the following pages represent student products that can be placed into a best-work portfolio: pp. 489, 490.

**PERFORMANCE ASSESSMENT**
Skill Check, p. 494
Mini Lab, pp. 492, 493
Lab, pp. 495, 497

**CONTENT ASSESSMENT**
Check for Understanding, pp. 486, 487, 491, 496
Chapter Review, pp. 500-501

**GROUP ASSESSMENT**
Opportunities for group assessment occur with Cooperative Learning Strategies.

**Student Journal**

Strategies on the following pages represent opportunities for writing in a Student Journal: pp. 490, 493.

GETTING STARTED

### Using the Photos

Have students look at the photos and try to determine the role of an insect in plant reproduction. Students should know that an egg must be fertilized by a sperm. They may know that the egg is present in the pistil and pollen is present in the anthers. Lead the discussion so students realize that the insect transfers the pollen, which contains sperm cells, to the pistil. A fruit is the mature ovary.

### MOTIVATION/Try This!

**Are fruits different from vegetables?** Include several root examples (carrots, radish), stems (asparagus), and leaf parts (lettuce, brussel sprouts) as well as fruits (apples, cherry, squash). Students will **observe** on their own that all fruits contain seeds while other plant parts do not. They will resist calling a tomato or cucumber a fruit. However, if seeds are present within the plant part, it is a fruit.

### Chapter Preview

Have students study the chapter outline before they begin to read the chapter. They should note the overall organization of the chapter, that the first topic described deals with asexual reproduction, and that the second topic is sexual reproduction. The chapter ends by covering the topic of plant development.

### Misconception

Ask students the major purpose of a plant's flower and they will reply that it supplies us with beauty. Few students realize that it is the flower that is involved with plant reproduction. The flower is actually the reproductive organ for flowering plants.

## 23:1 Asexual Reproduction in Plants

## PREPARATION

### Materials Needed

Make copies of the Focus, Study Guide, and Reteaching worksheets in the *Teacher Resource Package.*

▶ Obtain the various plants or bulbs needed for the demonstrations and activities.

### Key Science Words

tuber
runner
bulb
cutting
grafting

## 1 FOCUS

▶ The objectives are listed on the student page. Remind students to preview these objectives as a guide to this numbered section.

Focus/*Teacher Resource Package,* p. 45. Use the Focus transparency shown at the bottom of this page as an introduction to plant reproduction.

### ACTIVITY/Demonstration

Purchase a *Kalanchoe* plant from a garden or floral shop. Pass the plant around, asking students to note the small notches located in the leaves of the plant. Have them look carefully at what is seen in the notches. (Small plants are usually present.) Ask students to speculate as to the function of these "plantlets."

### Objectives

1. **Explain** how plants reproduce asexually from roots.
2. **Describe** how plants can reproduce asexually from leaves.
3. **Compare** asexual reproduction from runners, tubers, bulbs, cuttings, and grafting.

### Key Science Words

tuber
runner
bulb
cutting
grafting

## 23:1 Asexual Reproduction in Plants

As you have read before, roots, stems, and leaves are plant parts with a variety of different jobs. One job of these plant parts is reproduction. Plant reproduction from roots, stems, and leaves is asexual reproduction. Sometimes they are also helpful in plant reproduction. Recall from Chapter 4 that reproduction from one parent is called asexual reproduction.

### Asexual Reproduction by Roots

Many plants have roots that can grow into whole new plants. The sweet potatoes that you buy in the grocery store are roots. If you place a sweet potato in water, it will soon grow many small, new plants. Figure 23-1 shows how these changes take place. Notice that each new plant has its own roots, stems, and leaves. Each small plant can be pulled off the sweet potato and planted.

The sweet potato plants formed in this way are produced by asexual reproduction. The new plants have grown by mitosis. The original plant part didn't produce eggs and sperm, and there was no fertilization.

Many flowering plants reproduce asexually from roots. Dandelions and morning glory roots can be cut into many small pieces. Each piece of root can be planted and in time will form a new plant. Many trees, such as poplars and magnolias, are not killed when they are cut down. If the roots are left in the ground, they will send up new shoots.

**Figure 23-1** A sweet potato is placed in water (a). Small plants begin to grow (b). Each new plant has its own roots, stems, and leaves (c).

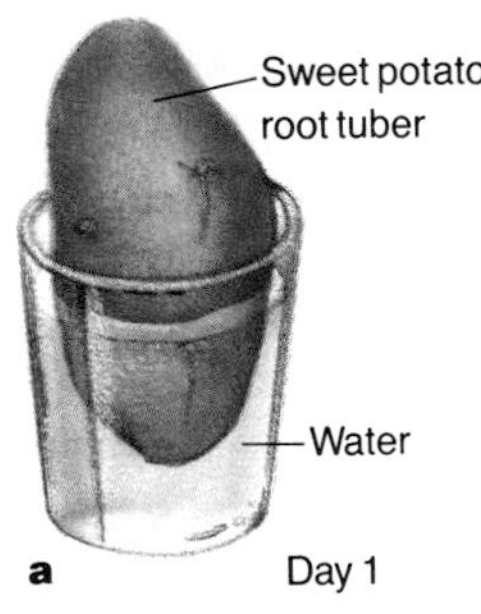


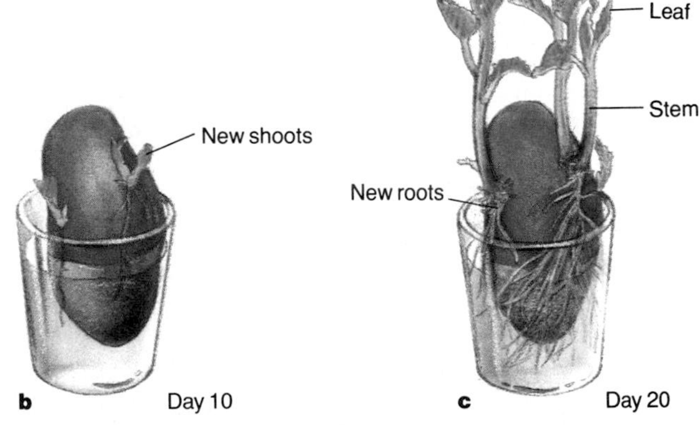


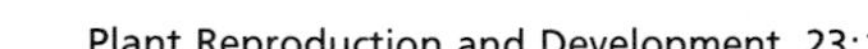


## OPTIONS

### Science Background

The emphasis in this section lies in the correlation between asexual reproduction and mitosis. During mitosis, body cells are formed. These cells are identical to all other cells within the plant in terms of chromosome number. These cells are not specialized gamete or sex cells.

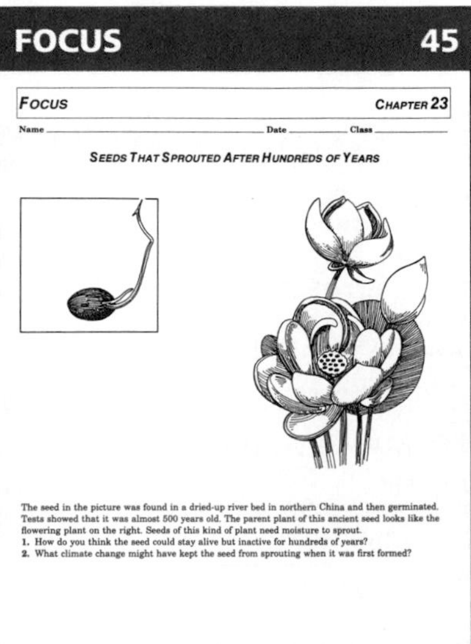
FOCUS 45

FOCUS CHAPTER 23

Name ______ Date ______ Class ______

SEEDS THAT SPROUTED AFTER HUNDREDS OF YEARS

The seed in the picture was found in a dried-up river bed in northern China and then germinated. Tests showed that it was almost 500 years old. The parent plant of this ancient seed looks like the flowering plant on the right. Seeds of this kind of plant need moisture to sprout.

1. How do you think the seed could stay alive but inactive for hundreds of years?
2. What climate change might have kept the seed from sprouting when it was first formed?

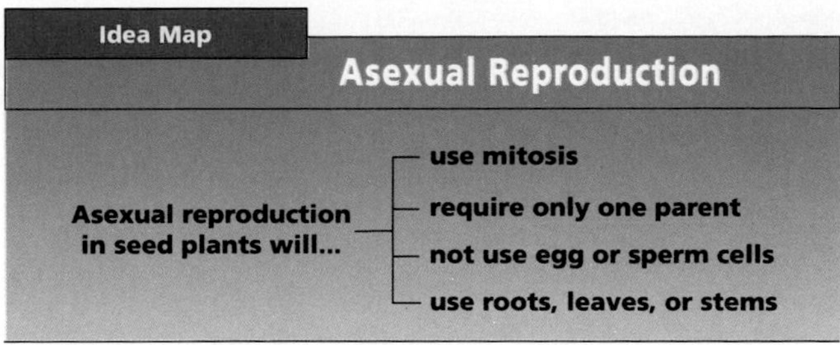


Study Tip: Use this idea map as a study guide to asexual reproduction in plants. Which plant parts can reproduce asexually?

Other plants that reproduce asexually from roots include wild carrots, lilac bushes, rose bushes, and apple trees. They send up new shoots from their root systems underground. Many plants like the sweet potato have swollen roots called tubers. A **tuber** is an underground root or stem swollen with stored food. Begonias and dahlias, common garden flowers, also have root tubers. They grow new plants every year from the food stored in the tubers.

**What kind of root is a sweet potato?**

## Asexual Reproduction by Leaves

Some plants can reproduce asexually from leaves. Some leaves, when removed from the main plant, will grow into an entire new plant. For example, a leaf of an African violet plant can be placed with its stalk in the soil. An entire new plant will grow from this single leaf. Figure 23-2 shows the steps in the growth of a new plant from the leaf of a snake plant.

**How can an African violet reproduce asexually?**

Remember that the process of making new plants from leaves, like roots, is asexual. There is only one parent. The new plants look just like the parent plants.

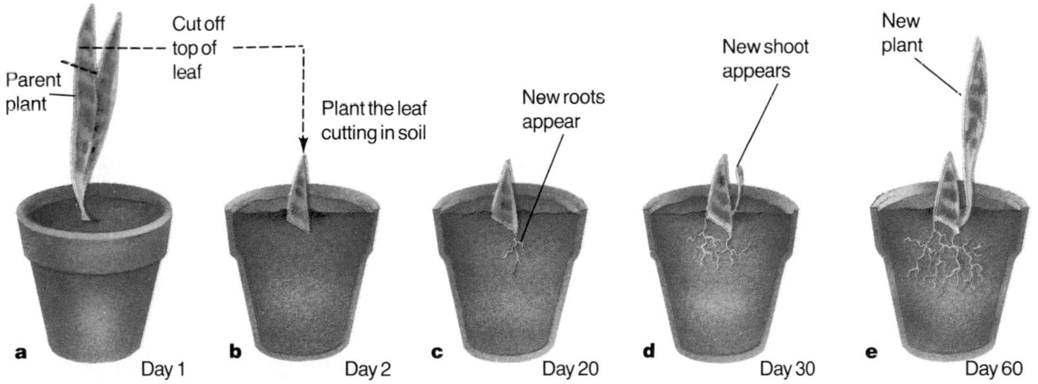


**Figure 23-2** These are the steps for producing a new snake plant asexually from a leaf of a parent plant. Which plant part grows first? the roots



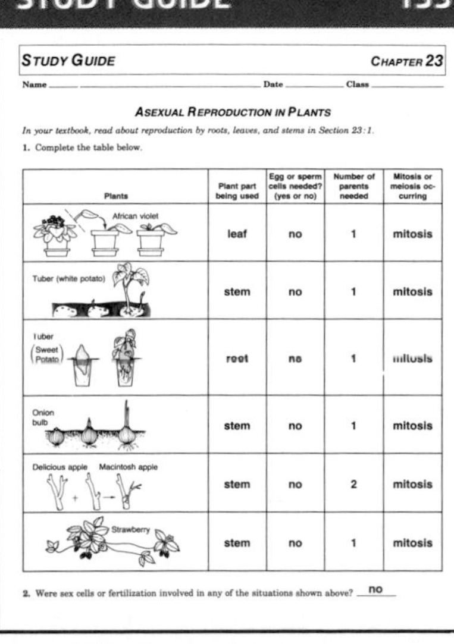

STUDY GUIDE 133

STUDY GUIDE CHAPTER 23

Name ______ Date ______ Class ______

ASEXUAL REPRODUCTION IN PLANTS

*In your textbook, read about reproduction by roots, leaves, and stems in Section 23:1.*

1. Complete the table below.

| Plants | Plant part being used | Egg or sperm cells needed? (yes or no) | Number of parents needed | Mitosis or meiosis occurring |
|---|---|---|---|---|
| African violet | leaf | no | 1 | mitosis |
| Tuber (white potato) | stem | no | 1 | mitosis |
| Tuber (Sweet Potato) | root | no | 1 | mitosis |
| Onion bulb | stem | no | 1 | mitosis |
| Delicious apple + Macintosh apple | stem | no | 2 | mitosis |
| Strawberry | stem | no | 1 | mitosis |

2. Were sex cells or fertilization involved in any of the situations shown above? no

# 2 TEACH

## MOTIVATION/Demonstration

Place some sweet potatoes in beakers of water as shown in Figure 23-1. Have students check them weekly and note how long it takes for new plants to appear. Allow students to remove the new plants and grow them in soil.

## Concept Development

▶ Some common examples of plants that can be propagated from leaves are chrysanthemum, begonia, geranium, African violet, and piggy-back plant.

▶ In leaf cuttings, the leaf blade, or the leaf blade and petiole (leaf stalk) are used in starting the new plant. Adventitious roots and shoots form from the leaf cutting.

▶ A sweet potato is classified as a root because of its internal structures. Its similarity in appearance to the stem of the white potato is coincidental.

## Idea Map

Have students use the idea map as a study guide to the major concepts of this section.

## Independent Practice

**Study Guide**/*Teacher Resource Package,* p. 133. Use the Study Guide worksheet shown at the bottom of this page for independent practice.

## Guided Practice

Have students write down their answers to the margin questions: root tuber; from its leaves.

## TEACH

### ACTIVITY/Demonstration

Bring a variety of plant stems such as white potato, onion, garlic, and shallot to class. Remind students that these are considered stem parts.

### Guided Practice

Have students write down their answers to the margin question: by cuttings and grafting.

### Check for Understanding

Have students explain the similarities and differences between:
- bulb and tuber
- cutting and runner
- asexual and sexual reproduction

### Reteach

Ask certain students to read their answers to the Check for Understanding questions. Discuss any problem areas in their responses.

### Guided Practice

Have students list examples of plants that reproduce asexually using the following parts: roots, leaves, stems.

**Figure 23-3** New spider plants form at the end of runners (a). Potatoes are stem tubers (b). Both plants are reproducing asexually from stems.

## Asexual Reproduction by Stems

**What are two ways that plants can be reproduced from stems?**

There are several ways that stems of flowering plants reproduce asexually. Two natural ways are by runners and underground stems.

Clover, strawberry, and spider plants grow small, new plants at the end of runners. A **runner** is a stem that grows along the ground or in the air and forms a new plant at its tip. Figure 23-3a shows an example of runners forming on a spider plant. Each new plant is a copy of the parent. In the forests of South Africa, where the spider plant grows naturally, these new plants would eventually break away from the parent plant. As houseplants, the small plants can be removed from the parent and placed in their own pots. They will then grow into new adult spider plants.

Many new plants grow asexually from underground stems. A potato is a stem tuber and an onion is a bulb. Both are underground stems. Have you ever found small plants growing from the sides of a potato? What you might have seen is shown in Figure 23-3b. Other plants with stem tubers are water lilies and caladium, a common houseplant with two-colored leaves shaped like arrowheads.

Many spring wildflowers have bulbs. A **bulb** is a short, underground stem surrounded by fleshy leaves that contain stored food. Look at the onion bulb that is cut in half, Figure 23-4. Notice the young shoot already formed at the center of the bulb. Garden flowers such as tulips, daffodils, and hyacinths (HI uh sihnths) are grown from bulbs.

**Figure 23-4** Notice the small stem at the base of an onion bulb.

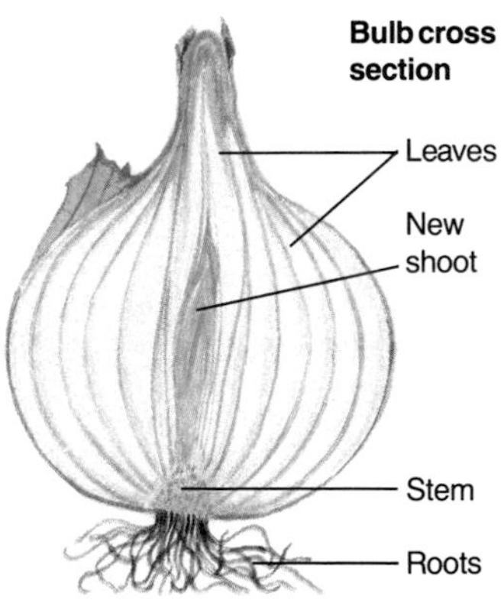


Gardeners use their knowledge of this ability of stems to grow new plants to grow plants from cuttings or graftings. Have you ever tried to grow new geranium



## OPTIONS

### ACTIVITY/Challenge

Snake plants and African violets are common species available in most plant stores. As a class project, have students asexually propagate these two plants. Plastic foam or paper cups make excellent inexpensive plant pots.

① Cut off a small branch close to the stem.

② Place the stem in water.

③ New roots appear in a few days.

Parent plant

Cut here

a

Cutting

b

Figure 23-5 These are the steps for producing a new plant from a stem cutting.

plants using the method shown in Figure 23-5? If so, you have grown a plant from a cutting. A **cutting** is a small section of plant stem that has been removed from a parent plant and planted to form a new plant.

**Grafting** is a process of joining a stem of one plant to the stem of another plant. For example, the stem of a Delicious apple tree may be joined or grafted to the stem of a MacIntosh apple tree. The original Delicious apple stem will continue to form Delicious apples. The original MacIntosh stem will continue to form MacIntosh apples. The result is a tree that will form both kinds of apples.

Why is grafting important to gardeners, farmers, and plant scientists? A gardener can grow a variety of fruits from one tree. The most common reason for grafting is to protect a plant from disease. For example, some kinds of rose plants or apple trees are more resistant to disease than other kinds. The kinds that have better flowers or fruits are grafted onto the roots of similar kinds that are resistant to diseases, Figure 23-6.

Figure 23-6 An old graft can be seen as a line where the two stems meet.

## Check Your Understanding

1. How does a root tuber help a plant reproduce?
2. How can an African violet reproduce asexually?
3. Name five different ways that a plant stem can help in asexual reproduction and give an example of each.
4. **Critical Thinking:** How might asexual reproduction help a plant survive a forest fire?
5. **Biology and Reading:** What does the prefix *a-* mean when it is attached to the word *sexual*?

## RETEACHING 69

RETEACHING: IDEA MAP — CHAPTER 23

Name ________ Date ______ Class ______

Use with Section 23:1.

ASEXUAL REPRODUCTION IN PLANTS

Use these terms or phrases to complete the idea map: tulip, spider plant, coleus, onion, strawberry, potato, Delicious apple, ivy.

stems
- cutting: coleus, ivy
- runner: strawberry, spider plant
- underground stem
  - tuber: potato
  - bulb: onion, tulip
- grafting: Delicious apple

1. How many parents are required for asexual reproduction? Explain. Only one parent is required because cells are produced by mitosis.
2. What is a clone? A clone is a group of living things that come from one parent and are identical to the parent.
3. What is grafting? Grafting is the joining of the stem of one plant to the stem of another plant.
4. Why is grafting important to farmers? The fruit formed on a grafted stem may have better size and quality than the fruit on the parent stem. The parent stem may be more resistant to diseases.



# TEACH

## Check for Understanding

Have students respond to the first three questions in Check Your Understanding.

### Reteach

**Reteaching/***Teacher Resource Package*, p. 69. Use this worksheet to give students additional practice in asexual reproduction in plants.

**Extension:** Assign Critical Thinking, Biology and Reading, or some of the **OPTIONS** available with this lesson.

# 3 APPLY

## ACTIVITY/Brainstorming

Do farmers use asexual reproduction methods in growing crops of corn, wheat, grapes, onions?

# 4 CLOSE

## Summary

Obtain a plant catalog. Copy one or two pages and have students list the plants that are available that reproduce asexually.

## Answers to Check Your Understanding

1. A tuber stores up food, which is used by the new plants that grow from it.
2. New plants can grow from a leaf.
3. cuttings–geranium; runners–strawberry; tubers–potato; bulbs– onion; grafting–apple trees
4. When all the plants in a forest have been burned to the ground, there would not be any flowers left to form seeds. Any underground plant parts that had a supply of stored food could grow and produce new plants.
5. The prefix *a-* means without or away from. *Asexual* means without sex.

## 23:2 Sexual Reproduction in Plants

## PREPARATION

### Materials Needed

Make copies of the Study Guide, Transparency Master, Skill, Critical Thinking/Problem Solving, and Reteaching worksheets in the *Teacher Resource Package.*

▶ Flowers may be purchased from a floral shop for the demonstration.

### Key Science Words

| | |
|---|---|
| sepal | ovule |
| petal | pollination |
| stamen | cross pollination |
| pistil | self pollination |

## 1 FOCUS

▶ The objectives are listed on the student page. Remind students to preview these objectives as a guide to this numbered section.

### ACTIVITY/Demonstration

Obtain a flower from a floral shop (preserved flowers may also be used). Prepare a slide of pollen grains from the anther for microscopic viewing. Allow students to view the slide and ask them to describe what they see, estimate the size of the individual cells, and offer a possible explanation of what these cells are.

### Idea Map

Have students use the idea map as a study guide to the major concepts of this section.

### Guided Practice

Have students write down their answers to the margin question: stamen.

# 23:2 Sexual Reproduction in Plants

**Objectives**

4. **List** the parts of a flower and their functions.
5. **Describe** two methods of pollination.
6. **Sequence** the steps that lead to fertilization in flowering plants.

**Key Science Words**

sepal
petal
stamen
pistil
ovule
pollination
cross pollination
self pollination

Flowering plants make our world beautiful to live in. Flowers come in an enormous variety of colors, shapes, and sizes. But, plants don't produce flowers for our pleasure. Flowers are important to the survival of the plant. Without flowers, many plants wouldn't be able to reproduce sexually.

## Flowers and Sexual Reproduction

Figure 23-7 is a diagram of a section through one kind of flower. Find the numbers on the diagram as you read the following statements that describe the different flower parts. In most flowers some flower parts are showy. These parts are neither male nor female. In most flowers some parts are female and other parts are male. Some flowers lack one or more of these parts. The flower in Figure 23-7 shows typical flower parts for a complete flower.

Parts 1 and 2 are sepals and petals. **Sepals** are often green, leaflike parts of a flower that protect the young flower while it is still a bud. **Petals** are often brightly colored and scented parts of a flower that attract insects to the flower.

Part 3 is the male flower part. The **stamen** is the male reproductive organ of a flower. Usually there are one to many stamens in each flower. Each stamen has a saclike part at its top in which pollen grains are formed. Figure 23-8a shows pollen grains under an electron microscope. You read in Chapter 6 that pollen grains are male structures that contain sperm cells.

**What flower part forms pollen grains?**

**Study Tip:** Use this idea map as a study guide to parts of a flower. Which structure in a flower contains ovules?

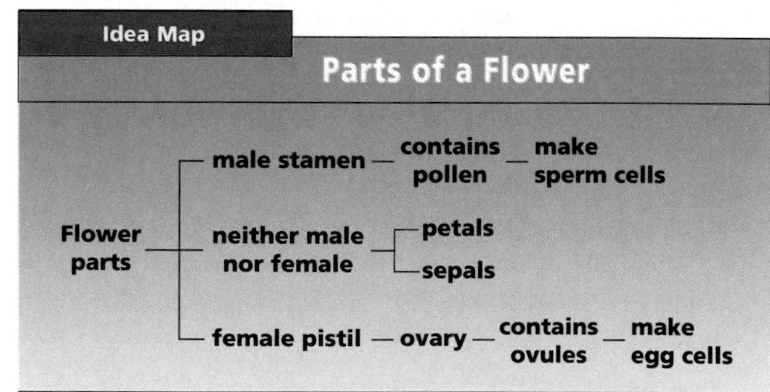




## OPTIONS

### Science Background

The emphasis within this section of the text lies in the correlation between sexual reproduction and meiosis. During meiosis, sex cells are formed. These cells differ from all other cells within the plant in terms of chromosome number. These cells are specialized gamete or sex cells.

**Transparency Master**/*Teacher Resource Package*, p. 113. Use the Transparency Master shown here to help students with flower anatomy.

**TRANSPARENCY 113**

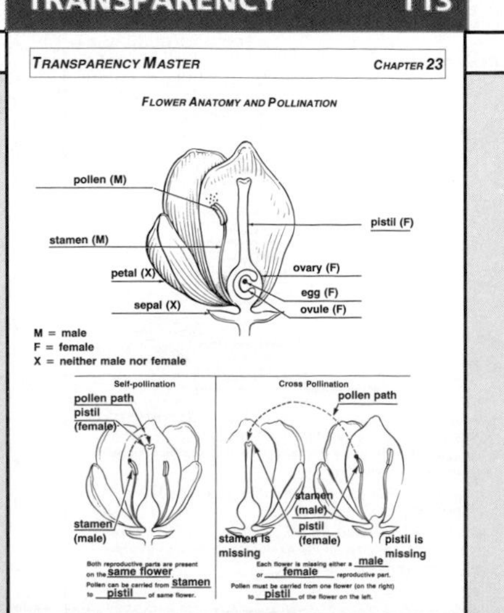

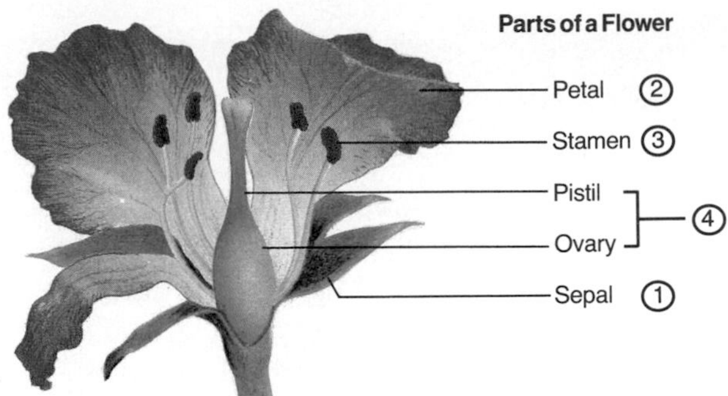


**Figure 23-7** Follow the text to read the functions of each part of a flower.

Part 4 in Figure 23-7 is the female flower part. The **pistil** (PIH stul) is the female reproductive organ of a flower. Often there is only one pistil in the center of a flower. Each pistil has a large, round ovary at its base. In Figure 23-8b this is cut open so you can see the ovules inside. Above the ovary is a stalk with a sticky tip. The sticky tip helps to trap pollen grains. Inside the ovary are tiny, round parts called ovules. **Ovules** contain the egg cells of a seed plant.

If you ever looked closely at a flower, such as a dandelion, a rose, a carnation, or a daffodil, you may have noticed that none of these has the structure of our typical flower in Figure 23-7. Some flowers don't have sepals, some have petals that are fused to form a tube, some have many tiny flowers all clustered together in a head, and some don't have any stamens. Flowers, just like all other organisms, show an enormous range of variation.

**Figure 23-8** Pollen grains under an electron microscope show wonderful surface patterns (a). A section through a pistil as seen under a light microscope shows the position of its ovules (b).

a

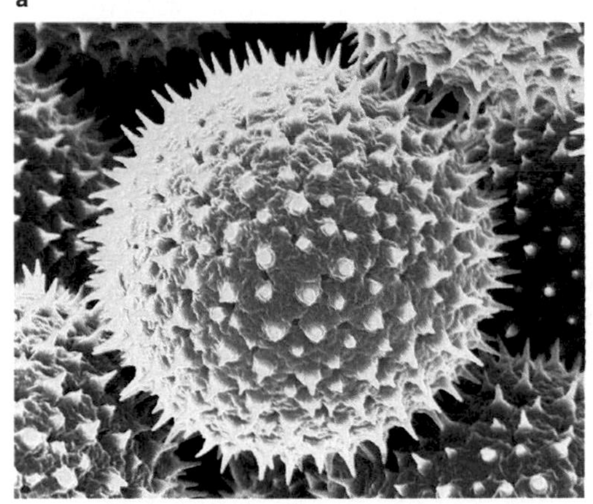

b

# 2 TEACH

## MOTIVATION/Demonstration

Obtain a large flower for student observation and dissection. Avoid composite flowers such as daisies, sunflowers, or mums. Day lily and iris are suitable. If the anthers are mature, a fine yellow powder can easily be seen. Ask students what this powder is.

## Independent Practice

**Study Guide**/*Teacher Resource Package*, p. 134. Use the Study Guide worksheet shown at the bottom of this page for independent practice.

### *Portfolio*

Have students copy Figure 23-7 in pencil and make the following changes or additions: (a) color code parts that are male as blue, female as red, neither male nor female as green; (b) add labels that describe the function of the important flower parts on the diagram.

## *GLENCOE* TECHNOLOGY

**Videodisc**

**Biology: The Dynamics of Life**
Animation: *Double Fertilization*
Disc 1, Side 2, Ch. 24

## SKILL 24

**SKILL: RECOGNIZE SPATIAL RELATIONSHIPS** CHAPTER 23

Name ______ Date ______ Class ______

*For more help, refer to the Skill Handbook, pages 706-711.* *Use after Section 23:2.*

**THE FLOWER**

When you use a microscope, you often observe a cross section of an organism or a part of an organism. To understand what you observe, you need to know how the cross section relates to the whole specimen. Look at the drawings of three objects below. The lines through the objects show where the cross section was made. Next to each object is a diagram of the cross section. Object C has two cross sections, made at different levels of the object.

A Cross section B Cross section C Cross section 1 Cross section 2

1. Draw cross sections of the objects below at levels 1, 2, and 3.

Cross section 1 Cross section 2 Cross section 3

2. On the left is a diagram of the ovary of a flower. A line through the ovary shows where a cross section was made. Fill in the diagram on the right to show what the cross section of the ovary looks like.

## STUDY GUIDE 134

**STUDY GUIDE** CHAPTER 23

Name ______ Date ______ Class ______

**SEXUAL REPRODUCTION IN PLANTS**

*In your textbook, read about flowers and sexual reproduction in Section 23:2.*

1. Label these four parts: pistil, petal, stamen, sepal.

(red) pistil
stamen (blue)
petal (yellow)
sepal (green)

2. Use the following colors to shade in the above drawing.
   a. yellow, for parts that attract insects
   b. green, for parts that protect flower in bud stage
   c. red, for parts that are female
   d. blue, for parts that are male
3. Label the parts shown here. Use these choices: sticky tip, pollen grains, pistil, ovary, stamen, ovule, saclike part, stalk.

saclike part, pollen grains, stamen, pistil, sticky tip, stalk, ovary, ovule

4. Name the part being described. Use these choices: saclike part, ovary, ovules, pollen grains, sticky tip, pistil, stamen.
   a. contains ovules ovary
   b. male flower part stamen
   c. contains pollen saclike part
   d. will form sperm pollen grains
   e. female flower part pistil
   f. traps pollen grains sticky tip
   g. hold female reproductive cells ovules

## OPTIONS

**Parts of a Flower**/*Transparency Package*, number 23a. Use color transparency number 23a as you teach the parts of a flower.

**What Are the Parts of a Flower?**/*Lab Manual*, pp. 189-192. Use this lab as an extension to flower anatomy.

**Skill**/*Teacher Resource Package*, p. 24. Use the Skill worksheet shown here to have students recognize sections of flower parts.

## TEACH

### Concept Development

▶ Emphasize that self pollination involves one flower, and cross pollination involves two flowers.

### Guided Practice

Have students write down their answers to the margin questions: pollination by pollen from the same flower or same plant; three.

**Cooperative Learning:** Divide the class into several groups. Assign the following task to all groups.

(a) draw two flowers that must undergo cross pollination. Show the transfer of pollen.

(b) draw a flower that depends on self pollination. Show the transfer of pollen.

One student should be assigned as recorder within each group.

### Idea Map

Have students use the idea map as a study guide to the major concepts of this section.

### *Portfolio*

Provide students with the following items: flower, bulb, leaf, tuber, prepared slide of pollen, and microscope. Have students list the items observed and describe whether each item is associated with asexual or sexual plant reproduction. For those items that are associated with asexual reproduction, students are to describe the method most likely to be used, such as cutting, bulb, runner, and so on.

### *Student Journal*

Tell students to imagine the following: You are a pollen grain that keeps a diary. You have just landed on a flower stigma. Your diary entries are made for each hour of the day rather than just one entry for each day. Record the hourly diary entries that explain what happens to you over the next 20 hours. Note: It takes a pollen tube about 12 hours to grow down through a style.

**Study Tip:** Use this idea map as a study guide to sexual reproduction in plants. What kind of cell division is necessary for sexual reproduction?

**Idea Map**

**Sexual Reproduction**

**Sexual reproduction in seed plants will...**
- **use meiosis**
- **require two parents**
- **use pollen and ovaries**
- **use egg and sperm cells**
- **form new plants not identical to parents**

## Pollination and Fertilization

You know that in sexual reproduction an egg and a sperm must join for fertilization. How does pollen get from the stamens of a plant to the pistil? Then, how do sperm cells inside the pollen reach the eggs within the ovules?

**What is self pollination?**

Pollen is transported from stamens to pistils by a process called pollination (pahl ih NAY shun). **Pollination** is the transfer of pollen from the male part of a seed plant to the female part.

In flowering plants, the design of the flower is very important in pollination. Many flowers with colorful petals have a strong scent that attracts insects. Insects climb over the flowers and pick up the pollen on their bodies. Many insects, such as bees, collect pollen for food. Some flowers also make a sugary chemical called nectar that bees and birds also use as food. Pollen catches on the hairs and feathers of their bodies while they feed on the nectar.

Some plants don't have large, colorful petals or attractive scents. These plants usually rely on wind to move the pollen from one plant to another. Grasses and many trees have this type of flower. Many trees, such as maples and willows, have long, hanging flowers that are yellow or green in color. These flowers are not attractive to insects. The pollen formed in these flowers is blown around by the wind.

Pollen is often carried by insects, birds, or wind between different plants. This is called cross pollination, Figure 23-9. **Cross pollination** is when pollen from the stamen of one flower is carried to the pistil of another flower on a different plant. Figure 23-9 also shows self pollination. **Self pollination** is when pollen moves from the stamen of one flower to the pistil of the same flower, or of another flower on the same plant.

**Figure 23-9** Compare two types of pollination. How many plants are needed for self pollination? one

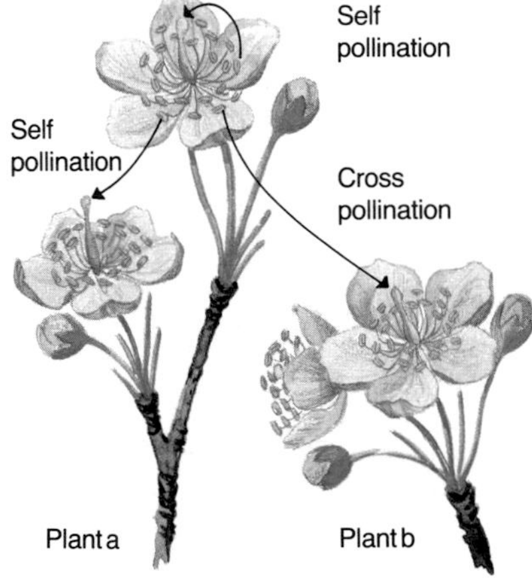




## *OPTIONS*

### ACTIVITY/Challenge

List the chromosome number of the cells that form these flower parts: petal, sepal, pollen, egg, sperm, ovule, stigma, style, anther. Assume that this plant has a body cell chromosome number of 12.

**Critical Thinking/Problem Solving/** *Teacher Resource Package,* p. 23. Use the worksheet shown here to reinforce fertilization in plants.

**CRITICAL THINKING 23**

CRITICAL THINKING/PROBLEM SOLVING — CHAPTER 23

Name ________ Date ________ Class ________

*Use after Section 23:2.*

**WILL FERTILIZATION TAKE PLACE?**

Below are diagrams of four imaginary flowers. Can the flowers shown in the column to the left be fertilized? In the second column, write your answer, yes or no. If your answer is no, write or draw in the third column what must be changed so that the flower can be fertilized.

| Flower | Can it be fertilized? | What must be changed so the flower can be fertilized? |
|---|---|---|
| 1. sepals. This flower must be cross-pollinated. | no | Sepals must be open. |
| 2. | no | Flower needs ovary and ovule with eggs, or a separate female flower. |
| 3. Flower is bee-pollinated. Nectar found in special glands. | no | A way is needed for a bee to contact the pistil after it brushes against the pollen on the anthers near its food. |
| 4. | yes, if wind or insect carries pollen from stamens to pistil | Flower needs wind or insect to carry pollen to pistil |

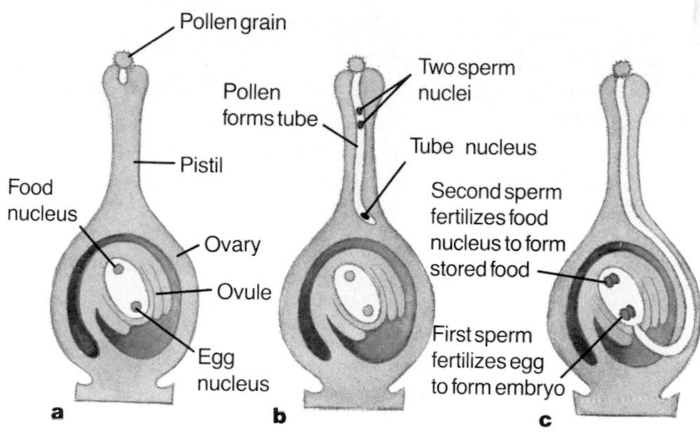


**Figure 23-10** Follow the sequence of steps in plant fertilization as shown in these sections of a pistil. How many sperm are needed to fertilize one ovule? two

How does sperm inside a pollen grain reach the egg inside an ovary? Follow the steps in Figure 23-10.

1. A pollen grain lands on the sticky tip of the pistil. It must somehow reach the egg within the ovule.
2. The pollen grain begins to grow a tube into the stalk of the pistil. This pollen tube grows all the way down to the ovary. Inside each pollen grain are three nuclei. One nucleus leads the way down the tube. The other two nuclei are sperm nuclei and these follow the first nucleus.
3. The tube reaches the ovule. A sperm nucleus inside the tube can now pass into the ovule and join with the egg. Fertilization takes place. The other sperm from the pollen tube joins with another nucleus in the ovule to form a food supply for the new seed.
4. A new plant or embryo now grows inside the ovule. As the ovule grows, it becomes a seed.

### Bio Tip

**Health:** A person who is a vegetarian gets most of his or her needed protein from plant seeds.

How many nuclei pass down a pollen tube?

## Check Your Understanding

6. List flower parts that are male, female, or neither.
7. What are two types of pollination?
8. How many sperm are needed to fertilize one ovule?
9. **Critical Thinking:** Hummingbirds and moths both feed on nectar of flowers. A hummingbird has a very long, thin beak and a moth has a very long mouthpart for feeding. What kinds of flowers would they most likely visit?
10. **Biology and Writing:** Are petals and sepals sexual parts in plants? Why or why not?

## TEACH

### Independent Practice

**Study Guide**/*Teacher Resource Package,* p. 135. Use the Study Guide worksheet shown at the bottom of this page for independent practice.

### Check for Understanding

Have students respond to the first three questions in Check Your Understanding.

### Reteach

**Reteaching**/*Teacher Resource Package,* p. 70. Use the Reteaching worksheet shown at the bottom of this page to give students additional practice with fertilization in flowers.

**Extension:** Assign Critical Thinking. Biology and Writing, or some of the **OPTIONS** available with this lesson.

## 3 APPLY

### ACTIVITY/Software

(1) *Plant Growth Simulator, Elementary Version;* (2) *Plant Growth Simulator, Secondary Version,* Focus Media, Inc.

## 4 CLOSE

### ACTIVITY/Videocassette

Show the videocassette *Sexual Encounters of the Floral Kind,* Carolina Biological Supply Co.

### Answers to Check Your Understanding

6. male–stamen; female–pistil; neither–petals and sepals
7. self pollination and cross pollination
8. two
9. flowers with very long tubular petals that contain a lot of nectar
10. Petals and sepals are not sexual parts in plants. Only those parts that produce sperm and egg are sexual.

**STUDY GUIDE 135**

STUDY GUIDE — CHAPTER 23

Name ______ Date ______ Class ______

SEXUAL REPRODUCTION IN PLANTS

*In your textbook, read about pollination and fertilization in plants in Section 23:2.*

pollen grains; sticky tip; saclike part; cross-pollination; pollen grains; saclike part; sticky tip; self-pollination

5. Examine the pictures shown above.
   a. Label the saclike part, pollen grains, and sticky tip in both diagrams.
   b. Use arrows to draw the pathway that pollen grains take in order for pollination to occur.
   c. On the line below each diagram, label the process self-pollination or cross-pollination.
6. The diagrams below show pollination and fertilization of a flower. However, the diagrams are out of order. Put the letter of each step below the diagram that it best matches

b d a c

   a. A pollen grain lands on the sticky tip of the pistil.
   b. A tube from the pollen grain grows down into the stalk.
   c. Two sperm nuclei and a tube nucleus travel down the tube.
   d. One sperm fertilizes the egg in the ovule. One sperm fertilizes a nucleus to form a food supply.
7. Examine the diagram above that has labels A and B on it.
   a. Which part (A or B) becomes the seed? A
   b. Which part (A or B) becomes the fruit? B
   c. Which part (A or B) contains the embryo? A

**RETEACHING 70**

RETEACHING — CHAPTER 23

Name ______ Date ______ Class ______

*Use with Section 23:2.*

FERTILIZATION IN FLOWERS

*Label the diagrams and fill in the blanks that describe what happens in fertilization.*

| | |
|---|---|
| A nucleus; pollen grain; pollen tube; pistil | Pollen lands on pistil during pollination. Top of pistil is sticky and traps the pollen. A small tube grows into the pistil from the pollen grain. |
| B pollen grain; pistil; pollen tube; sperm nuclei; tube nucleus | The pollen tube continues to grow down the pistil. The sperm nuclei and the tube nucleus move into pollen tube. |
| C pollen tube; sperm nuclei; egg; ovule; food nucleus; ovary | The pollen tube reaches the ovule. Two sperm nuclei enter the ovule. The ovule contains an egg and a food nucleus. |
| D egg; food nucleus; sperm; sperm | One sperm fertilizes the egg and one sperm fertilizes the food nucleus, forming future stored food. The ovule grows to form a seed. |

## 23:3 Plant Development

## PREPARATION

### Materials Needed

Make copies of the Application, Study Guide, Enrichment, Lab, and Reteaching worksheets in the *Teacher Resource Package.*

▶ Purchase corn seeds from a supply house (or popcorn from a grocery store) and collect milk cartons for Lab 23-1. Purchase carrots and garlic for Lab 23-2.

▶ Purchase beans for the Mini Lab.

▶ Obtain fruits and seeds for the demonstrations.

### Key Science Words

fruit
germination
clone

### Process Skills

In Lab 23-1, students will form hypotheses, experiment, and observe. In Lab 23-2, they will experiment, observe, measure in SI, and infer. In the Mini Lab, they will experiment. In the Skill Check, they will understand science words.

## 1 FOCUS

▶ The objectives are listed on the student page. Remind students to preview these objectives as a guide to this numbered section.

### ACTIVITY/Brainstorming

Ask students the value to a plant of having a:
(a) very large seed containing much stored food.
(b) very large fruit containing much stored food.

### Guided Practice

Have students write down their answers to the margin question: a seed.

## 23:3 Plant Development

### Objectives

7. **Describe** how fruits and seeds develop.
8. **Explain** how seeds are scattered and grow into new plants.
9. **Compare** asexual and sexual reproduction in plants.

### Key Science Words

fruit
germination
clone

In addition to reproduction, two other features of living things are that they grow and develop. Once a seed is formed, it's ready to grow and develop into a new plant. What kinds of changes occur when the seed begins to grow?

### Seeds and Fruits

After fertilization, changes take place in the flower. Use Figure 23-11 to follow these changes.
1. In this flower, many ovules have been fertilized by sperm cells.
2. Shortly after fertilization, the petals, stamens, and stalk of the pistil wither and die.
3. At the same time, the ovules and the ovary begin to grow.
4. Finally, the ovules mature into seeds. A seed is a plant part that contains a plant embryo. Remember from Chapter 6 that the embryo of a plant has a new young plant and stored food. The young plant and the stored food came from the joining of egg and sperm nuclei in the ovule.

As the ovary matures it becomes a fruit. A **fruit** is an enlarged ovary that contains seeds. You may have noticed that almost all the fruits you have ever eaten have contained seeds. If you use this definition of a fruit, can a tomato be a fruit? Is a squash, a green pepper, a cucumber, or a pumpkin a fruit? You must answer yes, because they all grew from ovaries and all have seeds inside.

### Mini Lab

**Will Seeds Grow Without Stored Food?**

**Experiment:** Remove only the embryo from several soaked bean seeds. Place the embryos on moist paper in a closed petri dish and wait to see if they grow. *For more help, refer to the* ***Skill Handbook,*** *pages 704-705.*

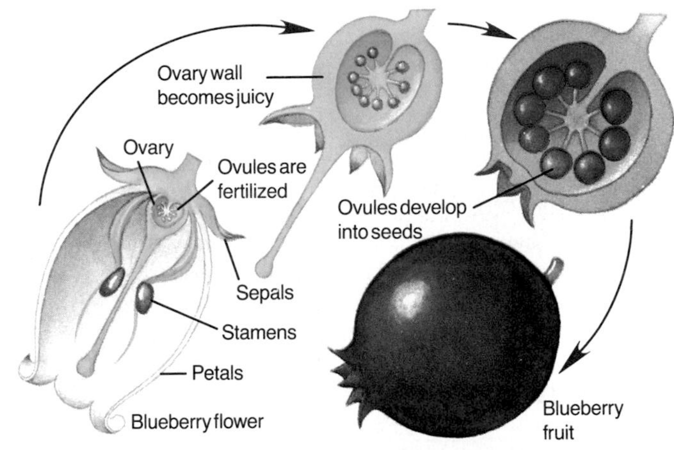


**Figure 23-11** Follow the steps that occur after fertilization in the forming of seeds in a fruit.



## OPTIONS

### Science Background

Plant hormones released by the fertilized ovule result in enlargement of the ovary and ovule to form a fruit and seed.

**From Flower to Fruit**/*Transparency Package*, number 23b. Use color transparency number 23b as you teach about fruit formation.

**Application**/*Teacher Resource Package*, p. 23. Use the Application worksheet shown here to reinforce the concept of fruits

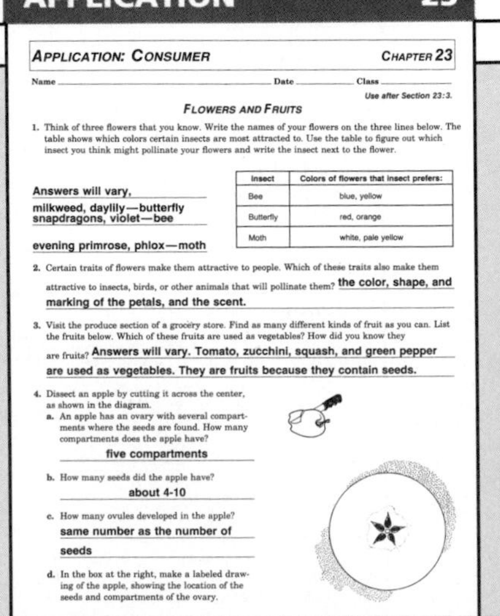

**APPLICATION 23**

*APPLICATION: CONSUMER* CHAPTER 23

Name ______ Date ______ Class ______

*Use after Section 23:3.*

*FLOWERS AND FRUITS*

1. Think of three flowers that you know. Write the names of your flowers on the three lines below. The table shows which colors certain insects are most attracted to. Use the table to figure out which insect you think might pollinate your flowers and write the insect next to the flower.

**Answers will vary,**
**milkweed, daylily—butterfly**
**snapdragons, violet—bee**
**evening primrose, phlox—moth**

| Insect | Colors of flowers that insect prefers: |
|---|---|
| Bee | blue, yellow |
| Butterfly | red, orange |
| Moth | white, pale yellow |

2. Certain traits of flowers make them attractive to people. Which of these traits also make them attractive to insects, birds, or other animals that will pollinate them? **the color, shape, and marking of the petals, and the scent.**

3. Visit the produce section of a grocery store. Find as many different kinds of fruit as you can. List the fruits below. Which of these fruits are used as vegetables? How did you know they are fruits? **Answers will vary. Tomato, zucchini, squash, and green pepper are used as vegetables. They are fruits because they contain seeds.**

4. Dissect an apple by cutting it across the center, as shown in the diagram.
   a. An apple has an ovary with several compartments where the seeds are found. How many compartments does the apple have? **five compartments**
   b. How many seeds did the apple have? **about 4-10**
   c. How many ovules developed in the apple? **same number as the number of seeds**
   d. In the box at the right, make a labeled drawing of the apple, showing the location of the seeds and compartments of the ovary.

## TECHNOLOGY

### Seed Banks

There are all kinds of banks. There are banks to hold your money, sperm banks, and even banks for body organs. Scientists also have formed a seed bank at a large laboratory in Fort Collins, Colorado. It is called the United States National Seed Storage Laboratory. The deposits made in this bank are exactly what you would expect from its name—seeds. In Fort Collins, seeds from all over the world are frozen in vats of liquid nitrogen. Someday, these living seeds will be removed from the vats and germinated.

Why do seeds need to be stored? Seeds are the link from one generation of plants to the next. One in 10 plant species is endangered. If a plant species becomes extinct, there is no way to get it back. Extinction is forever.

A seed bank is a means of saving plants that are endangered. Scientists of the future could cross-breed plants that have desirable traits with plants from the past.

Without plants for food, most living things can't survive. The "interest" the seed bank pays on seed deposits is that all living things benefit. The seed bank will help make the best seeds available for growing food plants for the future. Scientists will also be able to replace extinct plant species back into the wild when a new and suitable environment is found.

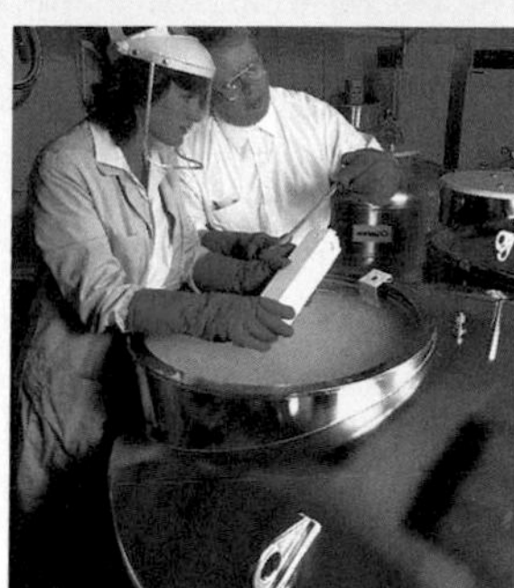

Seeds can be stored for a long time under controlled conditions.

If you could observe the development of a peanut or a walnut, you would conclude that they too were fruits. From these observations, it's clear that fruits are not just the juicy, fleshy fruits like oranges and grapes. Any plant part that develops from an ovary can be called a fruit. Fruits can also be nutty, and some aren't even edible by humans.

One seed develops from each fertilized ovule in an ovary. How many ovules were inside a cherry flower before fertilization? The answer is one. How do we know? Because a cherry fruit has only one seed. How many ovules are inside an orange flower? At least ten because there are about ten seeds in an orange, Figure 23-12. Have you ever counted the seeds in a watermelon? If you have, then you will know how many ovules were fertilized.

**What is the ovule in a fruit?**

**Figure 23-12** The seeds of an orange develop from ovules. The juicy fruit and rind of the orange develop from the ovary.

## STUDY GUIDE 136

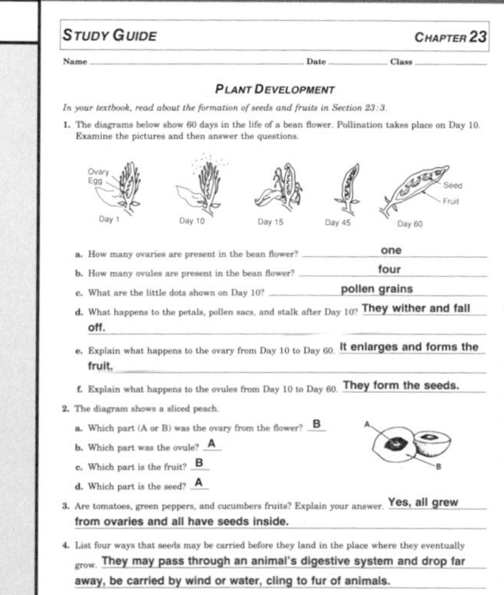

STUDY GUIDE — CHAPTER 23

Name ______ Date ______ Class ______

PLANT DEVELOPMENT

*In your textbook, read about the formation of seeds and fruits in Section 23:3.*

1. The diagrams below show 60 days in the life of a bean flower. Pollination takes place on Day 10. Examine the pictures and then answer the questions.



a. How many ovaries are present in the bean flower? one
b. How many ovules are present in the bean flower? four
c. What are the little dots shown on Day 10? pollen grains
d. What happens to the petals, pollen sacs, and stalk after Day 10? They wither and fall off.
e. Explain what happens to the ovary from Day 10 to Day 60. It enlarges and forms the fruit.
f. Explain what happens to the ovules from Day 10 to Day 60. They form the seeds.

2. The diagram shows a sliced peach.

a. Which part (A or B) was the ovary from the flower? B
b. Which part was the ovule? A
c. Which part is the fruit? B
d. Which part is the seed? A

3. Are tomatoes, green peppers, and cucumbers fruits? Explain your answer. Yes, all grew from ovaries and all have seeds inside.

4. List four ways that seeds may be carried before they land in the place where they eventually grow. They may pass through an animal's digestive system and drop far away, be carried by wind or water, cling to fur of animals.

## OPTIONS

**What Plant Part Are You Eating?**/*Lab Manual*, pp. 193-196. Use this lab as an extension to fruits.

## TECHNOLOGY

### Seed Banks

#### Background

Two thirds of harvested rice comes from only four species. Previously, over 100 different species of rice were grown. Those not being grown today are lost forever unless the seeds are saved in seed banks.

#### Discussion

Worldwide, plant species are disappearing at the rate of two a day. Discuss with students that this is a similar situation to that of endangered animal species.

#### Reference


Adler, Tina. "Sowing Hope." *Science News*, January 7, 1995, pp. 12-13.

Shell, E.R. "Seeds in the Bank Could Stave Off Disaster on the Farm." *Smithsonian,* Jan. 1990, pp. 94-100.


## TEACH

### Mini Lab

The process of removing the entire embryo from the seed will result in some embryos being broken or mutilated. Discard these damaged embryos.

### ASSESSMENT

**Skill:** Have students prepare a graph that predicts the amount of embryo growth that would result in three days if they performed the experiment with these three seeds: (a) an entire seed attached to the embryo, (b) half a seed attached to the embryo, (c) no seed attached to the embryo. Explain your graphed data.

### *Student Journal*

Tell students to imagine the following: You are a photographer who specializes in "before and after" photos. The following flower parts were taken by you as "before fertilization" photos: sepals, petals, style, stigma, ovary, ovule, stamens. Describe what your photos might look like one week after fertilization and one month later.

## TEACH

### ACTIVITY/Hands-On

Provide lima or pinto beans for the class to examine after soaking overnight. The next day students should be able to open the seeds with ease to observe the internal seed parts.

### Guided Practice

Have students write down their answers to the margin question: it scatters them away from the parent plant.

### Filmstrip

Show the filmstrip *Fruit and Seed Structure and Function,* Educational Images.

### ACTIVITY/Demonstration

That seeds remain viable for long periods of time can be demonstrated by germinating seeds bought from grocery store shelves. What appear to be dead seeds are, in reality, seeds that will germinate if supplied with water.

### Independent Practice

**Study Guide**/*Teacher Resource Package,* p. 137. Use the Study Guide worksheet shown at the bottom of this page for independent practice. Use Study Guide p. 138 for vocabulary review.

### Skill Check

**Understand science words: germination.** The word *germ* means a small part of living matter that can develop into a living organism. In your dictionary, find three words with the word part *germ* in them. *For more help, refer to the* ***Skill Handbook,*** *pages 706-711.*

## Plant Development from Seeds

What happens to the fruit and seeds after they are fully ripe? If the fruit is brightly colored and juicy it may be eaten by an animal. The seeds pass through the animal's digestive system and when they are dropped, they are often far from the parent plant. In most cases, however, the fruit remains on the parent plant and dries up as the seeds mature. Sometimes, the fruits eventually burst open and the seeds are thrown away from the parent plant. Some seeds are carried away by wind. Larger seeds may be carried away by water. Sometimes, the fruits are carried away with the seeds held inside. Dandelion seeds are scattered by the wind while still held in their tiny dried fruits. These one-seeded fruits have hairs that act like a parachute. Have you ever seen a sticky seed that clings to the fur of an animal? The wall of the fruit is covered with tiny hooks that catch in fur or clothing. These seeds are carried away in an animal's coat. The new plants can then grow far from the parent plant where there is more space, light, and water that they need to survive.

**How is the wind important to some seeds?**

What happens next? If a seed lands on soil that has moisture and proper temperature, it may germinate. **Germination** (jur muh NAY shun) is the first growth of a young plant from a seed. In many flowering plants, the stored food is the main part of the seed. This food is available for the growth of the new plant.

Let's look at a bean seed as it germinates. First the root and then the stem grow from the seed. In beans, the two seed halves are made mostly of stored food for the new plant. The stem that comes out of the seed has a small pair of leaves on it. Once out of the seed, the stem and leaves soon turn green and begin to make food by photosynthesis. When the food supply of the two seed halves is used up, they drop off the plant.

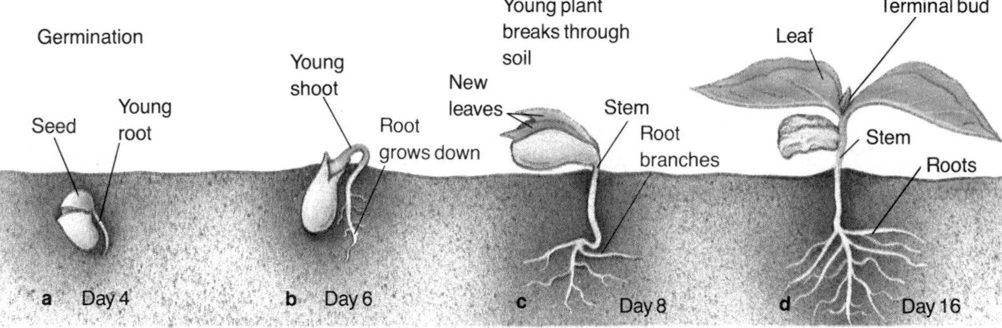


**Figure 23-13** Follow the sequence of steps in the development of a bean plant from a seed.



## OPTIONS

**Germination**/*Transparency Package,* number 23c. Use color transparency number 23c as you teach germination.

**Enrichment**/*Teacher Resource Package,* p. 25. Use the Enrichment worksheet shown here to help students understand fruits.

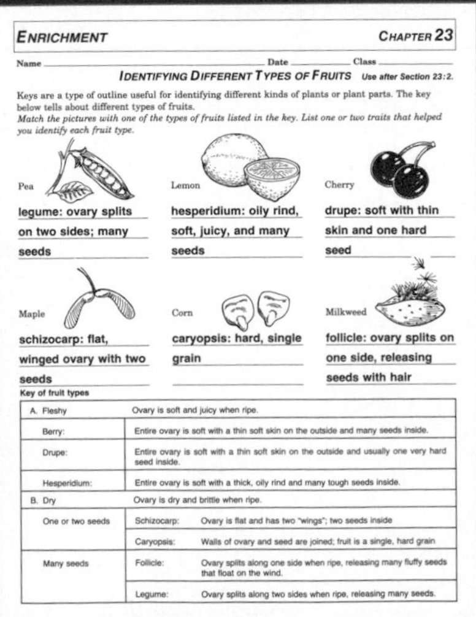

**ENRICHMENT 25**

ENRICHMENT — CHAPTER 23

Name ______ Date ______ Class ______

IDENTIFYING DIFFERENT TYPES OF FRUITS — Use after Section 23:2.

Keys are a type of outline useful for identifying different kinds of plants or plant parts. The key below tells about different types of fruits.

*Match the pictures with one of the types of fruits listed in the key. List one or two traits that helped you identify each fruit type.*

Pea: legume: ovary splits on two sides; many seeds

Lemon: hesperidium: oily rind, soft, juicy, and many seeds

Cherry: drupe: soft with thin skin and one hard seed

Maple: schizocarp: flat, winged ovary with two seeds

Corn: caryopsis: hard, single grain

Milkweed: follicle: ovary splits on one side, releasing seeds with hair

Key of fruit types

| | | |
|---|---|---|
| A. Fleshy | Ovary is soft and juicy when ripe. | |
| Berry: | Entire ovary is soft with a thin soft skin on the outside and many seeds inside. | |
| Drupe: | Entire ovary is soft with a thin soft skin on the outside and usually one very hard seed inside. | |
| Hesperidium: | Entire ovary is soft with a thick, oily rind and many tough seeds inside. | |
| B. Dry | Ovary is dry and brittle when ripe. | |
| One or two seeds | Schizocarp: | Ovary is flat and has two "wings"; two seeds inside |
| | Caryopsis: | Walls of ovary and seed are joined; fruit is a single, hard grain |
| Many seeds | Follicle: | Ovary splits along one side when ripe, releasing many fluffy seeds that float on the wind. |
| | Legume: | Ovary splits along two sides when ripe, releasing many seeds. |

**STUDY GUIDE 137**

STUDY GUIDE — CHAPTER 23

Name ______ Date ______ Class ______

PLANT DEVELOPMENT

*In your textbook, read about plant development from seeds in Section 23:3.*

5. What is germination? first growth of a young plant from a seed

6. Examine these diagrams of a developing plant. Then, fill in the blanks with the following labels: main root, leaves spread out and trap sunlight, secondary roots, first leaves wither and fall off, first leaves appear above ground, seed halves drop off plant.

Seed halves drop off plant.
Main root
First leaves appear above ground.
Secondary roots
Leaves spread out and trap sunlight.
First leaves wither and fall off

7. How does asexual reproduction help a plant? Only one parent is needed in this form of reproduction. Plants grow in the same place in a short time, and are exact clones of the parent so that helpful traits will be passed along.

8. How does sexual reproduction help a plant? Seeds can germinate after hundreds of years of being buried in the ground. This form of reproduction involves two parents so that a wide variety of traits can appear in the offspring, a fact that can improve survival chances if environmental conditions change.

# Lab 23–1 Germination

## Problem: Will soaked seeds germinate faster than unsoaked seeds?

### Skills

form a hypothesis, experiment, observe, interpret data

### Materials

| | |
|---|---|
| 2 milk cartons | metric ruler |
| soil | rubber band |
| 10 soaked corn seeds | small beaker |
| 10 unsoaked corn seeds | pencil |
| water | tape |
| stapler | permanent marker |
| scissors | |

### Procedure

1. Copy the data table.
2. Prepare milk carton planters as shown in the figure.
3. Fill each carton with moist soil.
4. Wrap a rubber band around a pencil exactly 2 cm from the end of the eraser. Using your pencil as a measuring stick, poke a row of ten holes exactly 2 cm deep in both cartons as shown.
5. **Experiment:** To the first carton, add one soaked corn seed to each hole. Use the marker to label the carton Soaked Corn Seeds. Add your name and the date.

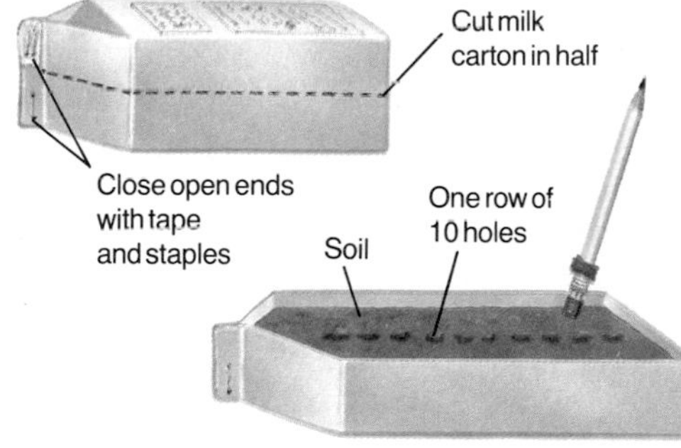


6. To the second carton, add one unsoaked corn seed to each hole. Label the carton Unsoaked Corn Seeds.
7. Gently cover the seeds with moist soil.
8. Read the problem question and write a **hypothesis** about it in your notebook.
9. **Observe:** Check each carton every day. Record the total number of plants that germinate each day.

### Data and Observations

1. How long did it take for the first unsoaked corn seed to germinate?
2. How long did it take for the first soaked seed to germinate?

| | NUMBER OF SEEDS GERMINATED | | | | | |
|---|---|---|---|---|---|---|
| Days from planting | 0 | 1 | 2 | 5 | 6 | 7 |
| Soaked | 0 | 0 | 1 | 6 | 8 | 9 |
| Not soaked | 0 | 0 | 0 | 2 | 3 | 5 |

### Analyze and Apply

1. What is the purpose of this experiment?
2. (a) What is meant by a control?
   (b) Which group of seeds were the control?
3. (a) What is meant by a variable?
   (b) What was the variable?
4. **Check your hypothesis:** Is your hypothesis supported by your data? Why or why not?
5. **Apply:** How could you find out if your results would be the same using different kinds of seeds?

### Extension

**Design an experiment** to test the effect of soaking the seeds for less than 24 hours.



## ANSWERS

### Data and Observations

1. about six days
2. about four days

### Analyze and Apply

1. to determine if soaking a seed affects the rate of germination
2. (a) a procedure used during an experiment in which one group being tested is left unchanged (b) those that were not soaked
3. (a) some type of change or treatment made to one group (b) the soaking of the seeds
4. Students who hypothesized that soaked seeds would germinate faster than unsoaked seeds will answer, yes.
5. by repeating the experiment using different kinds of seeds

## Lab 23-1 Germination

### Overview

Seeds that are planted are usually watered to promote germination. This lab allows students to determine whether the presoaking of seeds in water will speed up the process of germination.

**Objectives:** Upon completion of this lab, students will be able to: (1) **compare** the speed of seed germination of a control group to that of an experimental group, (2) **explain** the meaning of the term variable, (3) **interpret** data regarding the influence of watering on speed of germination.

**Time Allotment:** Day 1 – 30 minutes; Day 2 - 8 – 5 minutes

### Preparation

▶ **Alternate Materials:** Substitute polystyrene cups for milk carton and beakers, sand for soil. Substitute bean seeds for corn. Popcorn can be used in place of corn seeds.

**Lab 23-1 worksheet**/*Teacher Resource Package*, pp. 89-90.

### Teaching the Lab

▶ **Troubleshooting:** Do not allow soil to dry out, but reduce watering to avoid molding of seeds and young plants. If mold appears, water should be restricted even further.

**Skill:** Present the following data table to students:

| | Number of Seeds Germinated | | | | | | |
|---|---|---|---|---|---|---|---|
| Days from planting | 0 | 1 | 2 | 3 | 4 | 5 | 6 |
| Seeds X | 0 | 0 | 0 | 0 | 2 | 14 | 23 |
| Seeds Y | 0 | 0 | 0 | 0 | 0 | 2 | 4 |

Have students answer these questions:

(1) Which seeds, X or Y, were soaked before planting? Explain.
(2) Which seeds, X or Y, were not soaked before planting? Explain.

## TEACH

### Guided Practice

Have students write down their answers to the margin question: a group of living things that came from one parent.

### Check for Understanding

Have Students respond to the first three questions in Check Your Understanding.

### Reteach

**Reteaching**/*Teacher Resource Package*, p. 71. Use the Reteaching worksheet shown at the bottom of this page to give students additional practice in the events after fertilization.

**Extension:** Assign Critical Thinking. Biology and Reading, or some of the **OPTIONS** available with this lesson.

## 3 APPLY

### Connection: Social Studies

Have students use maps to determine the major crops for a specific region or country.

## 4 CLOSE

### ACTIVITY/Videocassette

Show the videocassette *Life Cycle of the Flowering Balsam,* Carolina Biological Supply Co.

### Answers to Check Your Understanding

11. Petals die; ovules form seeds; ovary becomes a fruit; stamens die.
12. the root
13. asexual—new plants will have the same traits as the parent; sexual—new plants will have traits from each parent
14. wind—always present; water—has buoyancy to keep seeds afloat; animals—can probably carry seeds the farthest
15. a fruit, ripened ovary

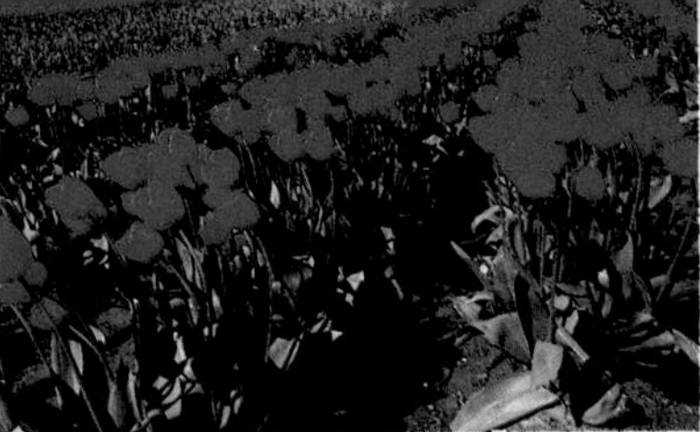

**Figure 23-14** Tulips of one variety all came from the same stock of bulbs and form a clone.

What is a clone?

### Bio Tip

**Leisure:** You can make an indoor garden from kitchen wastes. Grow a pineapple plant from the leafy top of a pineapple, an orange tree from an orange seed, a carrot from a carrot top, and an avocado plant from an avocado seed.

## Comparing Kinds of Reproduction

You have read about two ways a plant can reproduce. Let's now compare the advantages of both asexual and sexual reproduction in plants.

**Asexual reproduction**

1. In asexual reproduction only one parent is needed. A new plant can form from a root, stem, or leaf by mitosis. Reproduction by this method produces adult plants in a short time.
2. Some plants, like bananas, do not form seeds. Thus, the only way for them to reproduce is asexually.
3. All the new plants from asexual reproduction are clones of the parent plant. A **clone** (KLOHN) is a group of living things that come from one parent and are identical to the parent. This means that any trait or feature that is useful for survival will be kept in all offspring. Useful traits include large flowers, bright colors, or juicy fruits.

**Sexual reproduction**

1. Sexual reproduction requires two parents. Plants produced this way are made up of traits from both parents. No two plants will have the same combination of traits and so there will be a wide variety of traits in the offspring.
2. If proper growing conditions are not present, many seeds can survive for long periods of time. Some seeds have been known to germinate after hundreds of years of lying buried in the ground.
3. With a variety of traits, different plants will have a better chance to survive if conditions change.

### Check Your Understanding

11. Describe what becomes of each of the following flower parts after fertilization: petals, ovules, ovary, stamens.
12. Which part of a plant grows first after a seed begins to germinate?
13. How will plants produced by asexual reproduction and sexual reproduction compare with their parents?
14. **Critical Thinking:** Compare methods of seed scattering from a parent plant. Give advantages for each method and give reasons why one method might be best for a plant's survival.
15. **Biology and Reading:** If you eat an orange or a tomato, which part of the plant are you eating?



## GLENCOE TECHNOLOGY

**Videodisc**

**Biology: The Dynamics of Life**

Animation: *Fruit Formation*

Disc 1, Side 2, Ch. 25

Movie: *Seed Dispersal*

Disc 1, Side 2, Ch. 26

## RETEACHING 71

RETEACHING CHAPTER 23

Name ______ Date ______ Class ______

*Use with Section 23:3.*

**WHAT HAPPENS TO CERTAIN FLOWER PARTS AFTER FERTILIZATION?**

1. Use these terms to label the flower parts on this diagram: sepal, ovary, stamen, petal. On the line to the right of each flower part, explain what happens to each part after pollination and fertilization have taken place.

pistil — withers up and dies
stamen — withers up and dies
petal — withers up and dies
ovary — enlarges and forms fruit
sepal — withers up and dies

2. Use the following terms to label the parts of the diagram: egg, ovule, ovary. Explain what happens to each part after fertilization. Use the line to the right of each part you have labeled.

ovary — enlarges to become fruit
egg — becomes the embryo or future plant
ovule — enlarges and becomes future seed

3. Compare the kinds of reproduction by placing a checkmark in the correct column in the table below.

| | Asexual reproduction | Sexual reproduction |
|---|---|---|
| Two parents are needed. | | ✓ |
| Only one parent is needed. | ✓ | |
| Development is from seeds. | | ✓ |
| Seeds are not needed. | ✓ | |
| New plants are clones. | ✓ | |
| New plants have a variety of traits. | | ✓ |
| Adult plants are produced in a short time. | ✓ | |

# Lab 23–2 Rooting

## Problem: Can new plants be grown from roots and bulbs?

### Skills

experiment, observe, measure in SI, infer

### Materials

garlic bulb
carrot
labels
water
toothpicks
small beaker
metric ruler
razor blade
shallow dish

| DATE | | | | |
|---|---|---|---|---|
| garlic bulb | For | sample | data | |
| carrot | see | TRP | pages | 91-92 |

### Procedure

1. Copy the data table.
2. **Experiment:** Stick three toothpicks into a garlic bulb as shown in the figure.
3. Fill a small beaker with water. Label the beaker with your name and the date.
4. Balance the garlic in the water as shown. Make sure that the pointed end of the garlic is sticking up out of the water.
5. **Observe:** Use the data table to draw what the garlic looks like today. Mark the date on the table.
6. **Measure in SI:** Use a metric ruler to measure about 2 cm from the top of a carrot. Use a razor blade to cut off a slice of a carrot where you measured it.

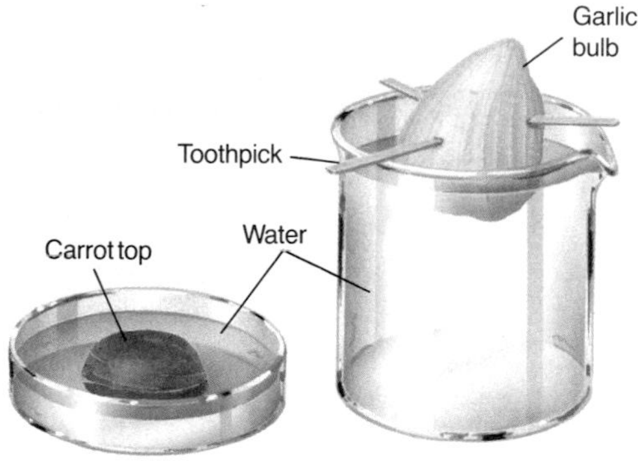


**CAUTION:** *Always cut away from yourself when using a razor blade.*

7. Place the carrot section into a shallow dish. The cut end should be facing down.
8. Label your dish. Add water to the dish.
9. Use the data table to draw what the root looks like today. Write in the data.
10. Check the bulb and root every day. Make a labelled diagram every four days on your data table.

### Data and Observations

1. What plant part of the garlic was used? Of the carrot?
2. Describe what formed on the garlic and carrot slice after several days.

### Analyze and Apply

1. Define asexual reproduction.
2. **Infer:** What did the parent plants of the garlic and carrot look like?
3. Using your data, how fast do garlic and carrot plants reproduce asexually?
4. **Apply:** How does asexual reproduction help a garlic plant survive?

### Extension

**Design an experiment** to test if garlic bulbs will grow better in the light than in the dark.



## ANSWERS

### Data and Observations

1. garlic—the bulb, which includes a stem and leaves; carrot—the root
2. garlic—roots from the base of the stem; carrot—roots from the top of the root

### Analyze and Apply

1. production of offspring from one parent
2. The parent of the garlic bulb will be the same as the bulb in this experiment; it was probably formed by asexual reproduction. The parent of the carrot may have been grown from seed and therefore may have been a little different.
3. Answers will vary. The garlic and carrot parts will probably show new growth after four to ten days.
4. A bulb is an underground storage organ that can survive over winter. Without the bulb, the garlic plant would not survive the cold temperatures above ground.

## Lab 23-2 Rooting

### Overview

Students will grow plants by asexual reproduction. This lab illustrates that new plants can be grown from plant parts such as bulbs or roots.

**Objectives:** Upon completion of this lab, students will be able to: (1) **observe** and record the events associated with asexual reproduction using a plant bulb, and root, (2) **compare** the rate at which asexual reproduction occurs when using different plant parts and different plants, (3) **determine** that new plants grown asexually are clones of the parent plant.

**Time Allotment:** Day 1 – 20 minutes
Days 2 to 3 – 5 minutes
Day 4 – 15 minutes
Days 5 to 7 – 5 minutes
Day 8 – 15 minutes

### Preparation

▶ **Alternate Materials:** Substitute onions for garlic, radish or beets for carrots, and paper cups for beaker.

**Lab 23-2 worksheet**/*Teacher Resource Package*, pp. 91-92.

### Teaching the Lab

▶ **Troubleshooting:** Garlic bulb and carrot root section must always be kept in water. On weekends, add sufficient water to dish and beaker, and cover with a plastic bag.

▶ Students must look carefully for new root growth occurring in garlic. Carrot tops must be cleared of original growth when first prepared so any new growth can be observed.

### ASSESSMENT

**Performance:** Provide a flower, onion bulb, seed, and potato tuber with "eyes" to students. Have them construct a chart that lists these plant parts along the top of their chart. The following categories should be listed down the left side: relies on asexual reproduction, relies on sexual reproduction, results in clone offspring, uses two parents, uses one parent. Have students complete the chart by placing check marks in the correct rows and columns.

## APPLYING TECHNOLOGY

### Purpose

Students will determine how to evaluate whether seeds purchased in a grocery store are alive.

### Process Skills

Experiment, observe, classify, infer, predict, use numbers, interpret data, formulate models

### Time Required

Soaking seeds will take several hours. Students can soak seeds overnight if necessary and use them the following day. Wrapping in wet towels will take 15 minutes. Observing and counting the seeds three days later will take 20 minutes.

### Possible Problems/Safety Concerns

Seeds used directly from grocery shelves may contain mold spores. Presoaking the seeds for two minutes in a 1 percent bleach solution (Add 10 mL of bleach to 990 mL of water) will kill any mold spores. Students may mistake mold growing on seeds as root growth. Students may have difficulty deciding whether or not a root is evident on the seeds. A root of any size should be taken as evidence that the seed is alive. Tetrazolium will stain hands. You may wish to have students wear plastic gloves.

### Teaching Strategies

▶ Allow students to work in groups of two or three.

▶ You may wish to purchase the seeds rather than relying on students to provide them.

▶ Make sure that the plastic bags are sealed to prevent the toweling and seeds from drying out and dying.

▶ A variety of seeds could be used. Peas, pinto bean, kidney bean, and black-eyed peas work well. Lima beans tend to fall apart during soaking.

TECH PREP **APPLYING TECHNOLOGY**

# Are These Seeds Alive?

Seeds serve as a food source for humans and other animals. They are also the most common method by which most plants reproduce. Farmers start the growing season by planting seeds for their spring and summer crops. They need to know what percentage of the seeds they plant are alive.

## Identifying the Problem

You are a farmer who is about to plant soybeans for the coming season. You want to know what percentage of the beans planted will grow into new plants. Another way of looking at the problem is to ask whether all the seeds that you are about to plant are alive. Certainly, dead seeds will not grow into new plants. Therefore, you don't want to waste your time and energy planting a batch of seeds that contains many dead seeds. The problem is that you can't tell if a seed is alive just by looking at it. You want to experiment with your bean seeds to see how many in the package are alive. Thus, as a farmer, you can better predict how many future plants might grow once planted.

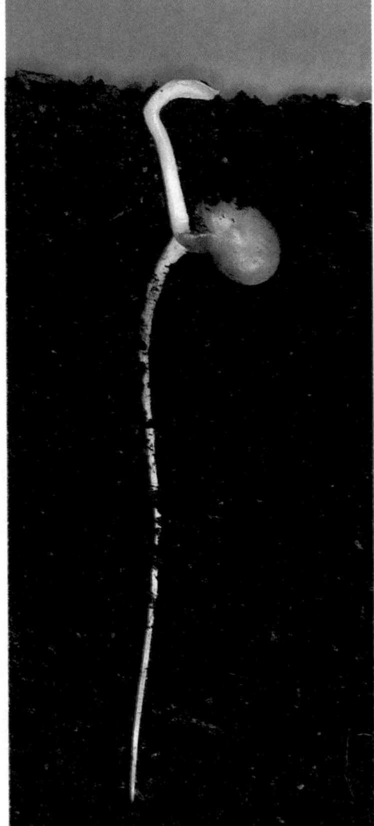

### Technology Connection

As a farmer, you may not have the time to test samples of all the different kinds of seeds that you are about to plant. Scientists have a quick test that can tell if a seed is alive. A chemical called tetrazolium causes living seeds to turn pink or red. Dead seeds do not change color. Soak another 50 bean seeds as before. Open each into two equal halves, and throw away one half of each seed. Place the remaining half of each seed into a shallow dish with the flat, inner side facing down. Add a small amount of tetrazolium to the dish. Wait 30 minutes. Using forceps, turn over each seed half and examine the underside for a red or pink color. Record the number of living and nonliving seeds. How do the results of the tetrazolium test compare with your results in the original bean seeds?

## Collecting Information

Check the packages of beans that your teacher has supplied. Choose a variety that you want to test. Record the bean type you have selected. Examine the seeds and try to

498

### Technology Connection

Tetrazolium is available from most biological supply houses as 2,3,5-triphenyl tetrazolium chloride. Prepare the solution by adding 1 g to 20 mL of distilled water. Use the solution within 5 hours of preparing. Tetrazolium is an indicator of the process of respiration. The color change occurs when the chemical reacts with hydrogen released during this process.

determine which seeds are alive. Record your observations. Try to estimate the percentage of seeds that are alive, and record this number. As a farmer or student, you should know that what you are doing is called sampling. You are checking a small sample of seeds rather than the entire batch to see if they are alive. Results from your small sample will tell you accurately what is true for the entire bag of seeds.

## Carrying Out an Experiment  

1. Soak 100 bean seeds in a beaker of water. Soak the seeds for several hours.
2. Remove the seeds from the water, and wrap them completely in several layers of wet paper towels.
3. Slip the towels into a self-seal plastic bag. Label your bag with your name and the date.
4. After three days, open the bag, and examine the seeds. If a seed is alive, a root will be seen growing from it. If no root can be seen, the seed is not alive.
5. Count and record the number of living and nonliving bean seeds that you observe.

## Career Connection

- **Farm Laborer** Carries out routine tasks such as caring for animals and crops; may be asked to clean pens, fertilize crops, or harvest crops when mature
- **Farm Office Worker** Gives clerical and technical support to farm owners; does accounting tasks such as payroll, handling billings, and mailings
- **Agronomist** Studies plants and soil in an effort to improve the process of growing farm crops; may work for large seed companies; aids in solving soil or crop problems

## Assessing Your Results

Calculate and record the percentage of living bean seeds in your original 100-bean sample. Based on your experimental results, how would you answer the original question about whether all the seeds that you are about to plant are alive?

## Conclusions

Students will be able to determine the number of seeds that are alive by observing the presence of roots. They may infer that water is needed for seed germination and that one cannot tell whether a seed is alive by looking at it. Seeds purchased from a supermarket may show close to 100 percent living seeds, but the percentage will vary with how long the seeds have been on the store shelf and the type of seed used.

## Answers to Questions

Students will be unable to tell whether a seed is alive by looking at it, so their estimates and percentages of living seeds will vary. The experiment should show that more than 90 percent of the seeds are living. If students do the tetrazolium test, the percentage of living seeds should be very close to that determined by germinating the seeds.

## ASSESSMENT

**Performance** Allow students to work with tetrazolium to determine the percentage of living seeds in their seed sample. Ask them to compare and contrast their results with those from the germination test. Have them explain any differences in percentage of living seeds when comparing the two methods.

## Career Connection

**Career Path**

**Farm Laborer** High-school diploma; on-the-job training

**Farm Office Worker** High-school diploma and some college helpful; several business and accounting courses in high school and college will be helpful.

**Agronomist** A college degree is needed; college courses in botany, chemistry, geology, soil science, and animal husbandry

**Career Issue**

Seeds planted by farmers usually have been pre-treated with a compound that gives them a red coloring. Have students find out the purpose of this treatment. What substance was used in the past? (Thiram is a fungicide that prevents mold growth once the seeds are planted. A red dye is added to identify these treated seeds. Mercury compounds were formerly used but are now banned because of their toxicity.)

## CHAPTER 23 REVIEW

### Summary

Summary statements can be used by students to review the major concepts of the chapter.

### Key Science Words

All boldfaced terms from the chapter are listed.

### Testing Yourself

#### Using Words

1. clone
2. tuber
3. germination
4. grafting
5. stamen
6. fruit
7. pistil
8. self pollination

#### Finding Main Ideas

9. p. 485; Some leaves, when removed from the main plant, will grow into an entire new plant when the leaf stalk is placed in soil.
10. p. 487; Grafting is important in producing fruit crops and in protecting plants from disease.
11. p. 492; Squash, walnuts, and dandelion seeds are enlarged ovaries that contain seeds.
12. p. 496; Asexual reproduction is reproduction by mitosis alone, where only one parent is needed, and where the new plants are clones of the parent plant. Sexual reproduction is reproduction by meiosis, where two parents are required, and where the new plants will have a variety of traits different from the parent plants.
13. p. 490; Cross pollination is when pollen from the stamen of one flower is carried to the pistil of another flower on a different plant. Self pollination is when the pollen moves from the stamen of one flower to the pistil of the same flower or another flower on the same plant.
14. p. 494; After germination, the root and then the stem grow from the seed. The stem and first leaves on the stem turn green and begin to make food by photosynthesis.
15. p. 485; Root tubers grow new plants each year from the food stored in the tubers.
16. p. 489; The pistil is the female reproductive organ of a flower and produces the egg cells of a seed plant.
17. p. 491; The pollen grain lands on the pistil and grows a tube down into the stalk of the pistil. Three nuclei make their way down the tube to the ovule. One of the nuclei enters the ovule and joins with the egg.
18. p. 488; Sepals protect the young flower while it is still a bud. Petals attract insects.

#### Using Main Ideas

19. Sexual reproduction produces a variety of traits in the offspring and makes it possible for germination to take place when growing conditions are favorable. Asexual reproduction preserves favorable traits in offspring, and since only one parent is needed, new plants can be produced in a short period of time.
20. A seed containing a young root and a young shoot, along with stored food, are found in an ovary.
21. The color, structure, and scent of flowers help to attract pollinators.

# Chapter 23 Review

## Summary

### 23:1 Asexual Reproduction in Plants

1. Asexual reproduction uses only one parent and produces identical offspring. Some plants reproduce from roots and root tubers.
2. Some plants reproduce asexually from leaves.
3. Some plants reproduce by underground stems, such as tubers and bulbs, or by runners. Some plants can be reproduced from cuttings or by grafting.

### 23:2 Sexual Reproduction in Plants

4. Sexual reproduction uses two parents and produces different offspring. The female part of the flower is the pistil. The male part is the stamen.
5. Pollination is the transfer of pollen from male to female flower parts. The two methods of pollination are self and cross pollination.
6. Pollen tubes grow down to ovules in the pistil. Sperm nuclei join with egg nuclei in the ovules.

### 23:3 Plant Development

7. After fertilization, a fruit forms from the ovary, and seeds form from ovules.
8. Seeds germinate and form new plants.
9. Asexual reproduction requires only one parent. Useful traits are kept in offspring. Sexual reproduction requires two parents and results in a wide variety of traits.

## Key Science Words

bulb (p. 486)
clone (p. 496)
cross pollination (p. 490)
cutting (p. 487)
fruit (p. 492)
germination (p. 494)
grafting (p. 487)
ovule (p. 489)
petal (p. 488)
pistil (p. 489)
pollination (p. 490)
runner (p. 486)
self pollination (p. 490)
sepal (p. 488)
stamen (p. 488)
tuber (p. 485)

## Testing Yourself

### Using Words

*Choose the word from the list of Key Science Words that best fits the definition.*

1. exact copies of a living thing
2. underground stem of potato
3. the growth of a young plant from a seed
4. joining a stem of one plant to the stem of another
5. male reproductive organ of flower
6. enlarged ovary with seeds
7. female reproductive organ of flower
8. carrying of pollen from stamen of one flower to pistil of same flower

# Review

## Testing Yourself *continued*

**Finding Main Ideas**

*List the page number where each main idea below is found. Then, explain each main idea.*

9. how to grow a new plant from a leaf
10. why grafting is important
11. why squash, walnuts, and dandelion "seeds" are fruits
12. the differences between asexual and sexual reproduction in plants
13. how self pollination and cross pollination differ
14. the changes or development that take place in a seed after germination
15. how plants reproduce from root tubers
16. what the function of a pistil is
17. how sperm in pollen reach the egg
18. the roles of petals and sepals in a flower

**Using Main Ideas**

*Answer the questions by referring to the page number after each question.*

19. What are two advantages of sexual and two advantages of asexual reproduction to a plant? (p. 496)
20. What are the parts that form inside an ovary? (p. 492)
21. How do the color, structure, and scent of flowers aid pollination? (p. 490)
22. What happens to sperm and egg nuclei at fertilization? (p. 491)
23. How may some trees reproduce asexually after they have been cut down? (p. 484)
24. How does a sweet potato reproduce asexually? (p. 484)
25. What is the job of a stamen? (p. 488)
26. What are two ways leaves produce new plants? (p. 485)

## Skill Review

*For more help, refer to the* ***Skill Handbook,*** *pages 704-719.*

1. **Understand science words:** Which terms don't fit the definition of asexual reproduction? Explain why. clone, egg germination, mitosis, one parent
2. **Experiment:** What part of an experiment could be the control if you compared the rates of germination of wheat seeds and bean seeds?
3. **Infer:** How do plants get onto islands?
4. **Form a hypothesis:** If you noticed lots of young oak seedlings beneath an old oak tree, what would you hypothesize about how they got there?

## Finding Out More

TECH PREP

**Critical Thinking**

1. *Aerodynamic* means designed for flight. What are some seeds that are carried away from the parent plant by wind with the help of aerodynamic fruits?
2. What is the advantage of a seed surviving for many years without germinating?

**Applications**

1. Examine pollen from different flowers under a microscope. Diagram what you see. How is pollen related to hay fever?
2. Prepare a list of 10 plants used by humans as food. Determine if these plants are grown by farmers using methods of asexual or sexual reproduction.



### Using Main Ideas

22. The sperm and egg nuclei form an embryo that grows inside the ovule.
23. After they are cut down, some trees will send up new shoots from the roots that will grow into a cluster of new trees.
24. A sweet potato reproduces asexually by growing new plants from the root tuber.
25. The stamen is the male reproductive organ of a flower and produces pollen grains that contain sperm cells.
26. A leaf can produce a new plant when the stalk is placed in soil. The leaves of some plants can form new leaves on the tips of each leaf. These new plants will drop off and become new plants.

### Skill Review

1. egg—is the result of meiosis; germination—seeds are formed by sexual reproduction
2. The rate of germination of one of the seeds must be known and used as a control to compare with the rate of germination of the other seeds.
3. Seeds are carried onto the island by the wind or by birds.
4. The oak tree dropped its seeds (acorns) below the tree.

### Finding Out More

**Critical Thinking**

1. Examples include milkweed, maple trees, cottonwood trees, dandelion, grasses, wheat, willow trees, ragweed, thistle, cattails
2. It can wait for the most favorable growing conditions and have a better chance of surviving as an adult plant.

### Applications

1. Pollen carried by wind lands in the nasal cavities and produces an allergic reaction.
2. Answers will vary. Examples are sexual—corn, apples, peaches, cherries, plums, cucumbers, tomatoes, pears, beans, peas; asexual—potatoes, sweet potatoes, apple trees, figs, strawberries, onions, artichoke, garlic, turnips, olive trees, plum trees.

**Chapter Review**/*Teacher Resource Package*, pp. 121-122.

**Chapter Test**/*Teacher Resource Package*, pp. 123-125.

**Chapter Test**/*Computer Test Bank*

# Animal Reproduction

## PLANNING GUIDE

| CONTENT | TEXT FEATURES | TEACHER RESOURCE PACKAGE | OTHER COMPONENTS |
|---|---|---|---|
| (1/2 day)<br>24:1 Asexual Reproduction<br>Review of Asexual Reproduction<br>Methods of Asexual Reproduction | Mini Lab: *What Do Hydra Buds Look Like?* p. 505<br>Idea Map, p. 504<br>Check Your Understanding, p. 506 | Reteaching: *Asexual Reproduction,* p. 72<br>Study Guide: *Asexual Reproduction,* p. 139 | **Laboratory Manual:**<br>*How Do Some Animals Reproduce Asexually?* p. 197<br>**STVS:** *Bacterial Waste Treatment,* Plants and Simple Organisms (Disc 4, Side 1) |
| (2 1/2 days)<br>24:2 Sexual Reproduction<br>Review of Sexual Reproduction<br>External and Internal Fertilization<br>Breeding Season | Mini Lab: *Are Sperm Attracted to Eggs?* p. 508<br>Lab 24-1: *Sex Cells,* p. 511<br>Idea Map, p. 507<br>Check Your Understanding, p. 510 | Application: *Reproduction in Animals,* p. 24<br>Enrichment: *Types of Animal Reproduction,* p. 26<br>Reteaching: *Sexual Reproduction,* p. 73<br>Critical Thinking/Problem Solving: *How Are Some Mammals Able to Birth More Than One Young at a Time?* p. 24<br>Skill: *Sequence,* p. 25<br>Study Guide: *Sexual Reproduction,* p. 140<br>Lab 24-1: *Sex Cells,* pp. 93-94 | **Laboratory Manual:**<br>*How Do Internal and External Reproduction Compare?* p. 201<br>**STVS:** *Alligator Courtship,* Animals (Disc 5, Side 2) |
| (2 days)<br>24:3 Reproduction in Humans<br>Human Reproductive System<br>Stages of Reproduction<br>The Menstrual Cycle<br>Diseases of the Reproductive System | Skill Check: *Sequence,* p. 518<br>Skill Check: *Understand Science Words,* p. 514<br>Lab 24-2: *Menstrual Cycle,* p. 512<br>Science and Society: *Sperm Banks,* p. 517<br>Check Your Understanding, p. 519 | Reteaching: *Egg Release and Attachment,* p. 74<br>Focus: *Test-Tube Tigers,* p. 47<br>Transparency Master: *Human Reproductive System,* p. 121<br>Study Guide: *Reproduction in Humans,* pp. 141-143<br>Study Guide: *Vocabulary,* p. 144<br>Lab 24-2: *Menstrual Cycle,* pp. 95-96 | **Laboratory Manual:**<br>*What Are the Stages of the Menstrual Cycle?* p. 205<br>Color Transparency 24a: *Stages of Reproduction*<br>Color Transparency 24b: *The Menstrual Cycle* |
| Chapter Review | Summary<br>Key Science Words<br>Testing Yourself<br>Finding Main Ideas<br>Using Main Ideas<br>Skill Review | **ASSESSMENT RESOURCES**<br>Chapter Review, pp. 126-127<br>Chapter Test, pp. 128-130<br>Performance Assessment in the Biology Classroom<br>Alternate Assessment in the Science Classroom<br>Computer Test Bank | |

### GLENCOE TECHNOLOGY

**Infinite Voyage,** *The Keepers of Eden*
**Science and Technology Videodisc Series,** *Bacterial Waste Treatment,* Plants and Simple Organisms (Disc 4, Side 1)
*Alligator Courtship,* Animals (Disc 5, Side 2)

## MATERIALS NEEDED

| LAB 24-1, p. 511 | LAB 24-2, p. 512 | MARGIN FEATURES |
|---|---|---|
| microscope<br>prepared slide of starfish eggs<br>prepared slide of starfish sperm<br>metric ruler<br>petri dish for drawing circles | metric ruler<br>graph paper<br>colored pencils | Skill Check, p. 514<br>dictionary<br>Skill Check, p. 518<br>pencil<br>paper<br>Mini Lab, p. 505<br>microscope<br>prepared hydra slide<br>Mini Lab, p. 508<br>microscope<br>dropper<br>sea urchin eggs<br>sea urchin sperm |

## OBJECTIVES

For more information about National Science Standards, see page 5T.

| SECTION | OBJECTIVE | CORRELATION of QUESTIONS to OBJECTIVES | | | |
|---|---|---|---|---|---|
| | | CHECK YOUR UNDERSTANDING | CHAPTER REVIEW | TRP CHAPTER REVIEW | TRP CHAPTER TEST |
| 24:1 National Science Stds: UCP.3, UCP.4, C.6 | 1. **Identify** the features of asexual reproduction. | 1, 4 | 10, 20 | 10 | 1, 21, 28, 31, 32, 33 |
| | 2. **Describe** types of asexual reproduction. | 2, 3, 5 | 10, 22 | 33 | 12, 15, 35, 37 |
| 24:2 National Science Stds: UCP.3, UCP.4, C.6 | 3. **Describe** the features of sexual reproduction. | 6, 10 | 21 | 2 | 5, 6, 19, 27, 28, 34 |
| | 4. **Compare** internal and external fertilization. | 7, 9 | 15, 18 | 1, 5, 11, 12, 13, 14, 15, 16, 37 | 3, 4, 8, 11, 12, 14, 18, 29, 36 |
| | 5. **Discuss** ways animals improve chances of fertilization. | 8 | 2, 5, 14, 20 | 6, 8 | 2, 17, 23, 30 |
| 24:3 National Science Stds: UCP.1, UCP.2, UCP.3, UCP.4, UCP.5, F.1 | 6. **Identify** the reproductive parts of humans and the stages of human reproduction. | 11 | 4, 7, 8, 12, 16 | 3, 4, 21, 22, 23, 24, 25, 26, 27, 28, 29, 30, 31, 34, 35, 36, 38 | 7, 16, 24, 39, 40, 41, 42, 43, 44, 45, 46, 47, 48, 49, 50, 51, 52, 53, 54, 55, 56, 57, 58 |
| | 7. **Compare** changes in the menstrual cycle with and without fertilization. | 12, 15 | 3, 9, 13, 19 | 9, 7, 18, 19, 20, 32 | 9, 10, 20, 22, 26 |
| | 8. **Discuss** symptoms and problems of sexually transmitted diseases. | 13, 14 | 17 | 7 | 25 |

# Animal Reproduction

## CHAPTER OVERVIEW

### Key Concepts

In this chapter, students will study methods of both sexual and asexual reproduction. The major emphasis is on the human reproductive system. Events of the menstrual cycle are described, as well as sexually transmitted diseases of the reproductive system.

### Key Science Words

| | |
|---|---|
| estrogen | regeneration |
| estrous cycle | reproductive system |
| external fertilization | sexually transmitted disease |
| internal fertilization | scrotum |
| menstrual cycle | uterus |
| menstruation | vagina |
| oviduct | vas deferens |
| penis | |

### Skill Development

In Lab 24-1, students will **make and use tables, measure in SI, recognize and use spatial relationships,** and **make scale drawings** in comparing sperm and egg cells. In Lab 24-2, they will **make and use graphs, infer,** and **relate cause and effect** in observing changes during the menstrual cycle. In the Skill Check on page 518, students will **sequence** events that take place during the menstrual cycle. On page 514, students will **understand** the **science word** *oviduct.* In the Mini Lab on page 505, students will **use a microscope** to examine hydra buds. In the Mini Lab on page 508, students will **observe** whether sperm are attracted to eggs.

### Bridging

In Chapter 22, students learned about the processes of mitosis and meiosis in detail. The practical aspects of that chapter can now be applied to asexual and sexual reproduction occurring within the animal kingdom and especially within humans.

**CHAPTER PREVIEW**

### Chapter Content

Review this outline for Chapter 24 before you read the chapter.



### Skills in this Chapter

The skills that you will use in this chapter are listed below.

- In **Lab 24-1,** you will make and use tables, measure in SI, recognize and use spatial relationships, and make scale drawings. In **Lab 24-2,** you will make and use graphs, infer, and relate cause and effect.
- In the **Skill Checks,** you will sequence and understand science words.
- In the **Mini Labs,** you will use a microscope and observe.

## TECH PREP

For Tech Prep activities in this chapter of the Teacher Wraparound Edition, see especially the Motivation on page 504, the Cooperative Learning activity and Student Journal on page 516, the Connection and Student Journal on page 518, and the Activity on page 519.

See also the Glencoe Homepage at **www.glencoe.com**

Chapter **24**

# Animal Reproduction

Look at the cloud of butterflies in the photo on the left. Where did they come from and where are they all going? The butterflies are migrating from Canada and the United States to Mexico. As the butterflies move southward, more and more butterflies join the flock. A flock of butterflies can number in the thousands. When the butterflies arrive in Mexico, they reproduce. In the spring, the butterflies will turn northward and head back to Canada and the United States.

How do butterflies reproduce? Like many other organisms, butterflies produce egg and sperm and have sexual reproduction. The joining together of the egg and sperm results in a new butterfly. In this chapter, you will study how different animals reproduce to form more of their own kind.

**Try This!**

**How do fish eggs and chicken eggs compare?** Examine some fish eggs and a chicken egg. What do fish eggs and chicken eggs have in common? How do they differ?

***inter*NET CONNECTION**

For more information about the material in this chapter, follow the link for the chapter on the Glencoe Homepage at **http://www.glencoe.com**

Adult monarch butterfly



**ASSESSMENT PLANNER**

***Portfolio***
Strategies on the following pages represent student products that can be placed into a best-work portfolio: p. 509.

**PERFORMANCE ASSESSMENT**
Skill Check, pp. 514, 518
Mini Lab, pp. 505, 508
Lab, pp. 511, 512

**CONTENT ASSESSMENT**
Check for Understanding, pp. 505, 506, 509, 510, 518, 519
Chapter Review, pp. 520-521

**GROUP ASSESSMENT**
Opportunities for group assessment occur with Cooperative Learning Strategies.

***Student Journal***
Strategies on the following pages represent opportunities for writing in a Student Journal: pp. 516, 518.

**GETTING STARTED**

## Using the Photos

In looking at the photos, students should note that butterflies are insects and, as do many organisms, they reproduce sexually in warm temperatures. Have students discuss why warm temperatures may be important for reproduction of some organisms.

## MOTIVATION/Try This!

**How do fish eggs and chicken eggs compare?** In completing the chapter opening activity, students will **observe** that both eggs are quite large. They are the reproductive cells of a female animal, have formed by meiosis, and are chordate eggs. They differ in size and appearance. Explain that the actual egg cell is difficult, if not impossible, to see. What is seen is mainly stored food.

## Chapter Preview

Have students study the chapter outline before they begin to read the chapter. They should note the overall organization of the chapter. The first topic deals with asexual reproduction in animals. This is followed with a discussion of sexual reproduction. The last part of the chapter covers human reproduction and sexually transmitted diseases.

## Misconception

Students may not be aware of the fact that most higher animals will release eggs regardless of whether they are or are not fertilized. Example: most students will assume that a chicken will lay eggs only if it has mated with a rooster. Egg production and release by a chicken is the same as egg production and release of an egg by a human. Most of the time the eggs are not fertilized.

# 24:1 Asexual Reproduction

## PREPARATION

### Materials Needed

Make copies of the Study Guide and Reteaching worksheets in the *Teacher Resource Package.*

▶ Obtain prepared slides of hydra showing budding for the Mini Lab.

### Key Science Words

regeneration

### Process Skills

In the Mini Lab, students will use a microscope.

## 1 FOCUS

▶ The objectives are listed on the student page. Remind students to preview these objectives as a guide to this major section.

### MOTIVATION/Story

Fishermen used to cut up starfish that they found in their fishing nets and throw them back into the sea. Each piece of starfish thrown back into the sea grew into another adult.

### Guided Practice

Have students write down their answers to the margin question: mitosis.

### Idea Map

Have students use the idea map as a study guide to the major concepts of the section.



## 24:1 Asexual Reproduction

**Objectives**

1. **Identify** the features of asexual reproduction.
2. **Describe** types of asexual reproduction.

**Key Science Words**

regeneration

All living things reproduce. Reproduction allows for the survival of different species of living things. For example, if butterflies didn't reproduce, there would be no more butterflies. The method or way in which living things reproduce differs from species to species. There are two main ways that animals reproduce. One of them is asexual reproduction.

### Review of Asexual Reproduction

**What type of cell reproduction occurs in asexual reproduction?**

Sometimes only one parent is needed in order to have young. When reproduction requires only one parent, it is called asexual reproduction. You first studied asexual reproduction in Chapter 4 when you studied reproduction in bacteria. Let's review some features of asexual reproduction.

1. Egg and sperm cells are not used. Therefore, there is no meiosis.
2. Organs such as ovaries and testes are not used because no egg and sperm are produced in asexual reproduction.
3. Mitosis is the type of cell reproduction involved.
4. Since there are no egg or sperm certain body cells must undergo mitosis to form the offspring.
5. Offspring are identical to one another and the parent. They have the same kind of chromosome material and the same traits. The offspring are clones.

**Study Tip:** Use this idea map as you study asexual reproduction. Note the different forms of asexual reproduction—budding and regeneration.

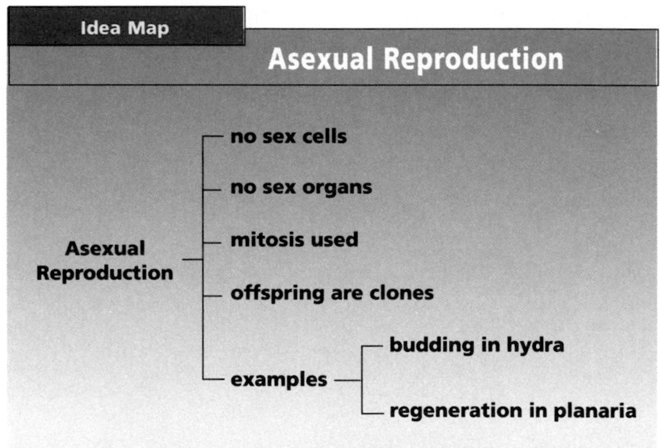




## OPTIONS

### Science Background

Asexual reproduction yields clones. All offspring produced by asexual reproduction are identical to the parent due to the fact that no genetic recombination takes place.

Asexual reproduction is sometimes said to be a simple form of reproduction. Since asexual reproduction is simple, do you suppose most animals use it? The answer is no. Simple animals, such as sponges, hydras, and flatworms, can reproduce this way. Complex animals, such as humans, birds, and fish, do not reproduce this way.

## Methods of Asexual Reproduction

Animals are able to reproduce asexually by budding and regeneration (rih jen uh RAY shun). You learned about budding in Chapter 5. It is reproduction in which a small part of the body grows into a new organism. A hydra is a good example. Follow the steps in Figure 24-1 to see how budding works.

The new hydra, or bud, is at first smaller than the parent. Its other features, such as body shape and presence of tentacles, are the same as the parent's. This is an important feature of asexual reproduction. All offspring are exact copies, or clones, of the parent.

Figure 24-1 shows a photograph of a hydra forming a bud. The photo is enlarged about 50 times. Did the bud form by mitosis or meiosis? Why?

A few animals can reproduce asexually by regeneration. **Regeneration** is reproduction in which the parent separates into two or more pieces and each piece forms a new organism. A planarian, which is a flatworm, is an animal that undergoes regeneration.

### Mini Lab

**What Do Hydra Buds Look Like?**

**Use a microscope:** Examine prepared slides of a hydra under low power. Diagram and label a hydra with buds. Does a hydra bud look like the parent? *For more help, refer to the **Skill Handbook,** pages 712-714.*

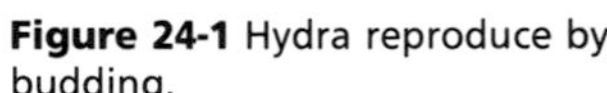

How is a bud of a hydra like the parent?

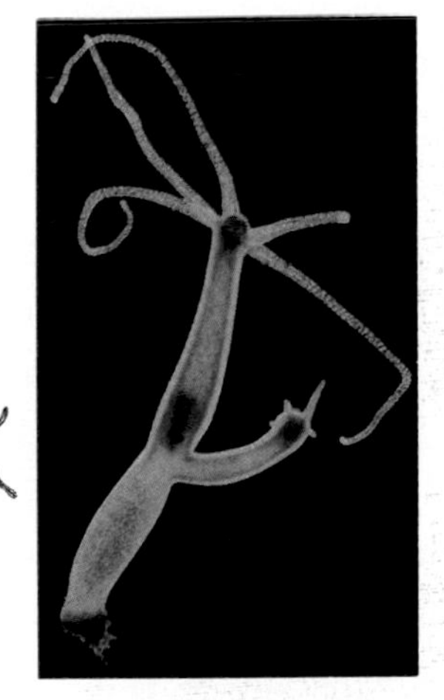


**Figure 24-1** Hydra reproduce by budding.

## OPTIONS

### ACTIVITY/Enrichment

Planaria are available from biological supply houses. You may want to purchase some and have students watch them regenerate. Cut the animals in half and observe the two halves each day. About 1 week is needed for visible changes.

**How Do Some Animals Reproduce Asexually?**/*Lab Manual,* pp. 197-200. Use this lab as an extension to asexual reproduction.

## 2 TEACH

### MOTIVATION/Audiovisual

Show the filmstrip *Asexual Reproduction,* Carolina Biological Supply Co.

### Concept Development

▶ Make sure that students thoroughly understand the five key features of asexual reproduction.

▶ Point out that if mitosis is responsible for budding, then all cells must be identical to the original parent cells. This is the characteristic of a clone.

### Mini Lab

Prepared slides of hydra are available from biological supply houses. Caution students to use only low power magnification with these slides.

### ✓ ASSESSMENT

**Performance:** Have students observe a prepared slide of a budding hydra. Ask them to diagram the bud and parent and to make a second diagram that shows what the bud and parent are expected to look like one month later. Both sets of diagrams should have labels that describe the difference in time as well as differences that appear in both parent and bud.

### Guided Practice

Have students write down their answers to the margin question: It is identical to the parent.

### Check for Understanding

Ask students the following: (1) What is asexual reproduction? (2) What process of cell reproduction is used? (3) How do the offspring look when compared to the parent, and what term describes this condition?

### Reteach

Use answers to Check for Understanding for class discussion.

## TEACH

### Independent Practice

**Study Guide**/*Teacher Resource Package*, p. 139. Use the Study Guide worksheet shown at the bottom of this page for independent practice.

### Check for Understanding

Have students respond to the first three questions in Check Your Understanding.

### Reteach

**Reteaching**/*Teacher Resource Package*, p. 72. Use this worksheet to give students additional practice with asexual reproduction.

**Extension:** Assign Critical Thinking, Biology and Reading, or some of the **OPTIONS** available with this lesson.

## 3 APPLY

### Brainstorming

What would the world population look like if humans could only reproduce by asexual reproduction?

## 4 CLOSE

Ask students which of the five features of asexual reproduction apply to the healing of a cut finger and which do not.

### Answers to Check Your Understanding

1. reproduction by one parent
2. reproduction in which the parent separates into two or more pieces and each piece forms a new organism
3. reproduction in which a small part of the body grows into a new organism
4. only one parent is needed
5. Starfish reproduce by regeneration, a kind of asexual reproduction.

**Figure 24-2** Planarians can reproduce by regeneration.

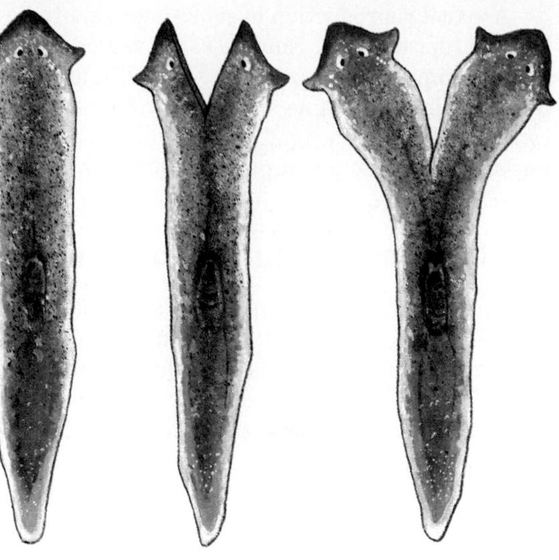

Figure 24-2 shows how a planarian regenerates. The planarian starts regenerating at its head or its tail end. Two heads or two tails form and then the animal begins to split along its body. Regeneration continues until what started out as a single animal ends up as two planarians.

The figure also shows a photograph of a planarian enlarged about 50 times. Note that the planarian appears to have two tail ends. In time, the two tails will separate to form two whole planarians.

### Check Your Understanding

1. What is asexual reproduction?
2. What is regeneration?
3. What is budding?
4. **Critical Thinking:** What would be an advantage of asexual reproduction?
5. **Biology and Reading:** People who fish for a living consider starfish pests. When starfish are caught in the fishing nets, they are sometimes cut up and tossed back into the water. Each piece grows into a new starfish. What kind of reproduction is this?

### STUDY GUIDE 139

STUDY GUIDE — CHAPTER 24

Name ______ Date ______ Class ______

**ASEXUAL REPRODUCTION**

*In your textbook, read about asexual reproduction in Section 24:1.*

1. Identify the type of asexual reproduction shown in each picture.

a. regeneration b. budding

2. On each blank write T if the statement is true or F if the statement is false.

F a. During asexual reproduction animals reproduce by meiosis.
F b. A hydra forms buds in its ovaries.
F c. An offspring produced by sexual reproduction is identical to its parent.
T d. Complex animals reproduce by sexual reproduction.
T e. Regeneration occurs when an organism separates into pieces and each piece forms a new organism.
F f. Organisms produce eggs and sperm when they reproduce asexually.
T g. A planarian regenerates by mitosis.
T h. A hydra is a simple animal.
F i. Asexual reproduction requires two parents.
T j. When a planarian regenerates, it can form two heads.
F k. All the offspring of a hydra are different from each other.

3. Answer the question below.

What is a clone? A clone is an offspring that is an exact copy of its parent.

### RETEACHING 72

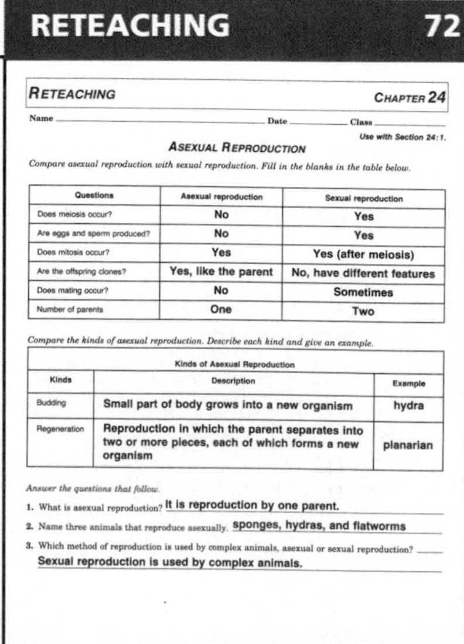

RETEACHING — CHAPTER 24

Name ______ Date ______ Class ______

Use with Section 24:1.

**ASEXUAL REPRODUCTION**

*Compare asexual reproduction with sexual reproduction. Fill in the blanks in the table below.*

| Questions | Asexual reproduction | Sexual reproduction |
|---|---|---|
| Does meiosis occur? | No | Yes |
| Are eggs and sperm produced? | No | Yes |
| Does mitosis occur? | Yes | Yes (after meiosis) |
| Are the offspring clones? | Yes, like the parent | No, have different features |
| Does mating occur? | No | Sometimes |
| Number of parents | One | Two |

*Compare the kinds of asexual reproduction. Describe each kind and give an example.*

| Kinds of Asexual Reproduction | | |
|---|---|---|
| Kinds | Description | Example |
| Budding | Small part of body grows into a new organism | hydra |
| Regeneration | Reproduction in which the parent separates into two or more pieces, each of which forms a new organism | planarian |

*Answer the questions that follow.*

1. What is asexual reproduction? It is reproduction by one parent.
2. Name three animals that reproduce asexually. sponges, hydras, and flatworms
3. Which method of reproduction is used by complex animals, asexual or sexual reproduction? Sexual reproduction is used by complex animals.

# 24:2 Sexual Reproduction

Can you name five animals that use sexual reproduction? You might name dogs, cats, horses, humans, and cattle. What about birds, insects, frogs, or fish? These animals also use sexual reproduction.

Most animals use sexual reproduction in forming offspring. Some simple animals like hydra and planarians can use sexual reproduction in addition to asexual reproduction.

**Objectives**

3. **Describe** the features of sexual reproduction.
4. **Compare** internal and external fertilization.
5. **Discuss** ways animals improve chances of fertilization.

**Key Science Words**

external fertilization
internal fertilization
estrous cycle

## Review of Sexual Reproduction

Let's review some of the features of sexual reproduction.

1. Sex cells, called egg and sperm, are used.
2. Organs are used to form the sex cells. Testes of the male form sperm cells. Ovaries of the female form egg cells.
3. Meiosis is the type of cell reproduction that forms the sex cells.
4. Mating may take place. Mating is the process of a male and female joining together to make sure that egg and sperm meet. Fertilization of egg by sperm is more likely to take place after mating.
5. Offspring formed by sexual reproduction will not look alike. They usually do not look exactly like either parent, although they may show some features of each parent.

Will offspring formed by sexual reproduction be clones? No. Sexual reproduction forms offspring with different features.

**Sexual reproduction uses what kind of cell reproduction?**

**Idea Map**

**Sexual Reproduction**

- Sexual Reproduction
  - sex cells
  - sex organs
  - meiosis used
  - mating
  - offspring different

**Study Tip:** Use this idea map as you study sexual reproduction. Note that in sexual reproduction, eggs can be fertilized outside the body or inside the body.



**SKILL 25**

SKILL: SEQUENCE — CHAPTER 24

Name ______ Date ______ Class ______

*For more help, refer to the Skill Handbook, pages 706-711.* *Use after Section 24:2.*

SEXUAL MATURITY OF ANIMALS

When animals are old enough to reproduce, they are sexually mature. Table 1 shows that the age of sexual maturity varies in different animals. Does the time needed to reach sexual maturity depend on the group of the animal?

*Look at the information in Table 1. Then, complete Table 2 by listing the animals from Table 1 in order of age of sexual maturity. Begin with the animal that takes the least amount of time. End with the one that takes the greatest amount of time. Then, answer the questions that follow the tables.*

Table 1

| Group | Animal | Age of sexual maturity |
|---|---|---|
| Soft-bodied | scallop | 4 years |
| Jointed-leg | water flea | 4 days |
| | cicada | 17 years |
| Spiny-skin | starfish | 1 year |
| Bony fish | eel | 8 years |
| | bass | 3 years |
| Amphibian | toad | 2 years |
| Reptile | garter snake | 2 years |
| | box turtle | 4 years |
| Bird | sparrow | 1 year |
| Mammal | whale | 2 years |
| | deer | 3 years |
| | elephant | 12 years |

Table 2

| Animal | Age of sexual maturity |
|---|---|
| water flea | 4 days |
| starfish | 1 year |
| sparrow | 1 year |
| toad | 2 years |
| garter snake | 2 years |
| whale | 2 years |
| bass | 3 years |
| deer | 3 years |
| scallop | 4 years |
| box turtle | 4 years |
| eel | 8 years |
| elephant | 12 years |
| cicada | 17 years |

1. Which animal takes the shortest time to reach sexual maturity? water flea
   Which takes the longest time? cicada
2. Do mammals take more or less time than the other animals to reach sexual maturity? Explain. Except for the elephant, most of the mammals reach sexual maturity in about the same time as the other animals.
3. Does the time needed to reach sexual maturity depend on the animal's group? no

## OPTIONS

### Science Background

Some animal species have separate sexes, as in the case of humans. This is known as dioecious. Some animals, such as flatworms and segmented worms, have combined sexes. They are monoecious animals capable of self-fertilization, but they generally exhibit cross-fertilization.

Skill/*Teacher Resource Package,* p. 25. Use the Skill worksheet shown here to sequence the sexual maturity of animals.

# 24:2 Sexual Reproduction

## PREPARATION

### Materials Needed

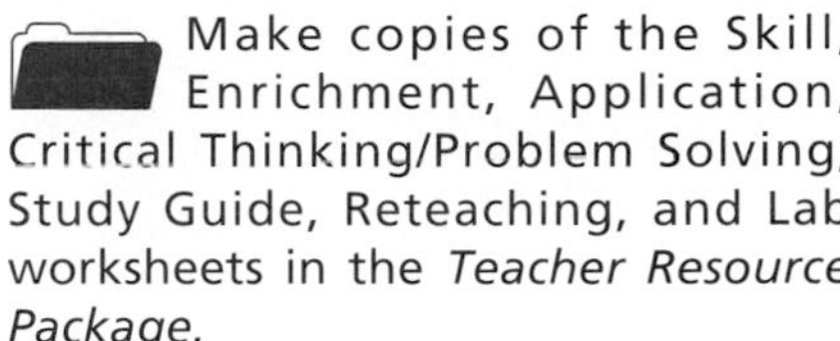

Make copies of the Skill, Enrichment, Application, Critical Thinking/Problem Solving, Study Guide, Reteaching, and Lab worksheets in the *Teacher Resource Package.*

▶ Purchase or collect sea urchin sperm and eggs for the Mini Lab.

▶ Obtain prepared slides of starfish egg and sperm for Lab 24-1.

### Key Science Words

external fertilization
internal fertilization
estrous cycle

### Process Skills

In the Mini Lab, students will observe. In Lab 24-1, students will make and use tables, measure in SI, recognize and use spatial relationships, and make scale drawings.

## 1 FOCUS

▶ The objectives are listed on the student page. Remind students to preview these objectives as a guide to this numbered section.

### MOTIVATION/Audiovisual

Show 35-mm color slides of sperm from a variety of animals. Have students note similarities and differences.

### Guided Practice

Have students write down their answers to the margin question: meiosis.

### Idea Map

Have students use the idea map as a study guide to the major concepts of this section.

# 2 TEACH

### Concept Development

▶ Point out how the five features of sexual reproduction differ from those of asexual reproduction.

▶ Suggest that fertilization is random and that breeding seasons increase chances for fertilization.

### Independent Practice

**Study Guide**/*Teacher Resource Package*, p. 140. Use the Study Guide worksheet shown at the bottom of this page for independent practice.

### Guided Practice

Have students write down their answers to the margin question: outside the body.

### Mini Lab

Purchase sea urchin eggs and sperm from a biological supply house or obtain locally if possible.

**✓ ASSESSMENT**

**Content:** Based on their observations of sperm and egg, have students answer the following: (a) Was this a demonstration of internal or external fertilization? Explain. (b) Would the same observations be seen if eggs from fish were used with sperm from fish? Explain. (c) Would the same observations be seen if eggs from a fish and sperm from a sea urchin were used? Explain.

**Figure 24-3** After the female fish deposits her eggs, the male will fertilize them.

## External and Internal Fertilization

In some animals the sex cells join outside the body. How is this possible? Animals like the female fish in Figure 24-3 release their eggs in water. Meanwhile, male fish also release their sperm in water. Then sperm and egg meet, and fertilization takes place.

Where does external fertilization take place?

The events just described are called external fertilization. **External fertilization** is the joining of egg and sperm outside the body. It's easy to remember because *external* means outside. The fertilized egg then forms a new animal. The young usually grow without care from their parents. This lack of care by the parents seems to be true for many animals that have external fertilization.

Most animals with external fertilization live in water or deposit their sex cells in water. Sperm must be able to swim to the egg. They could not do so unless they were in water.

How can animals with external fertilization make sure that egg and sperm meet? They can't always. They can, however, improve the chances in two ways. First, these animals give off thousands of sex cells at one time. The large number of sex cells increases the odds that one of the eggs will be fertilized. Second, these animals gather in large groups. You may have heard large groups of frogs croaking on a warm, rainy spring night. They gather in large groups to reproduce. When the sex cells are given off, the chance that they will meet is greater.

### Mini Lab

**Are Sperm Attracted to Eggs?**

**Observe:** Add a dropperful of sea urchin sperm to some sea urchin eggs. Observe what happens under the microscope. *For more help, refer to the* ***Skill Handbook****, pages 704-705.*



### OPTIONS

**Enrichment**/*Teacher Resource Package*, p. 26. Use the Enrichment worksheet shown here to help students learn types of animal reproduction.

**STUDY GUIDE 140**

STUDY GUIDE — CHAPTER 24

Name ________ Date ________ Class ________

**SEXUAL REPRODUCTION**

*In your textbook, read about external and internal fertilization in Section 24:2.*

1. Define external and internal fertilization.
   a. external fertilization: **External fertilization is the joining of egg and sperm outside the body.**
   b. internal fertilization: **Internal fertilization is the joining of egg and sperm inside the body.**
2. Where do most animals that have external fertilization live? **Most animals with external fertilization live in water.**
3. What would happen to sperm and eggs if land animals had external fertilization? **They would dry out.**
4. List two advantages of internal fertilization. **Internal fertilization increases the chance that an egg will become fertilized. It allows reproduction to take place out of water.**

*In your textbook, read about breeding seasons in Section 24:2.*

5. What is a breeding season? **A breeding season is a certain time of the year when animals reproduce.**
6. How does the breeding season of frogs help them? **The breeding season takes place when a lot of water is available and the temperature is warm enough for the young to survive.**
7. How does the breeding season of large mammals such as deer help them? **Since their young take longer to develop, the breeding season helps make sure that the young are born in the warm part of the year when they have a better chance to survive.**
8. What is an estrous cycle? **An estrous cycle is a time when a female is ready to mate.**

**ENRICHMENT 26**

ENRICHMENT — CHAPTER 24

Name ________ Date ________ Class ________

*Use after Section 24:2.*

**TYPES OF ANIMAL REPRODUCTION**

*Read the descriptions below of how different types of animals reproduce. Then, decide which method of reproduction is being described: budding, regeneration, external fertilization, or internal fertilization. Label each method as a form of sexual or asexual reproduction.*

1. A male dragonfly deposits its sperm in an opening near the female dragonfly's neck.
   **internal fertilization; sexual reproduction**
2. A sponge can reproduce by growing a miniature hollow cylinder along its side. This miniature cylinder can break off and form a new organism.
   **budding; asexual reproduction**
3. Male clams shed sperm into the water and female clams shed eggs into the water. Water currents bring the sperm and the eggs together.
   **external fertilization; sexual reproduction**
4. Earthworms mate, joining head to tail. Each exchanges sperm with the other. The sperm move into a pouch called the seminal receptacle (SEM un ul · ree SEP tih kul). A few days later the ovaries release eggs. The sperm join with the eggs inside the earthworm's body.
   **internal fertilization; sexual reproduction**
5. The female yellow perch lays eggs on the bottom of a lake. The male releases milt over the eggs. This is a fluid that contains sperm. Some of the eggs are fertilized by the sperm, and tiny fish hatch from the eggs.
   **external fertilization; sexual reproduction**
6. If a piece of a sponge breaks off from the main body of the sponge, the piece can grow into a new organism.
   **regeneration; asexual reproduction**
7. To attract a female, one kind of bird, the male vermillion flycatcher, dives in the air while singing a special song. Mating can take place in midair. The male presses his body near the female's so that the openings to their reproductive systems touch. Then, he releases sperm. The sperm travel into the female's body.
   **internal fertilization; sexual reproduction**
8. One of a starfish's five arms breaks off at the base. The arm can grow an entire new body.
   **regeneration; asexual reproduction**
9. Free-swimming, fully-formed jellyfish break away from the body of a stationary jellyfish.
   **budding; asexual reproduction**

Fish, starfish, sponges, and frogs are animals with external fertilization. These animals are found in or near water. Frogs can hop around on land, but they usually reproduce only in water.

If *external* means outside, then what does *internal* mean? It means inside. Some animals have internal fertilization. **Internal fertilization** is the joining of egg and sperm inside the female's body.

When mating takes place during internal fertilization, sperm are released inside the female's body. Releasing the sperm this way increases the chance that an egg will become fertilized. It also allows reproduction to take place out of water. Animals with internal fertilization can reproduce on land. For example, insects, reptiles, birds, and mammals have internal fertilization. In addition to reproducing on land, animals with internal fertilization usually protect and care for their young.

**What are two ways that animals improve chances for external fertilization?**

## Breeding Season

Most animals reproduce only during a certain time each year. For example, spring is the breeding season for frogs, Figure 24-4. A breeding season is a certain time of the year when particular kinds of animals reproduce.

The breeding season for frogs takes place when a lot of water is available and the temperature is warm enough. Why is water important for frog reproduction?

**Figure 24-4** Frogs breed only at certain times of the year.



## TEACH

### Concept Development

▶ Stress that breeding seasons occur in animals showing both types of fertilization.

### Guided Practice

Have students write down their answers to the margin question: by giving off many sex cells at one time and by gathering in large groups to mate.

### Check for Understanding

Ask students the following:
(1) What is sexual reproduction? (2) What process of cell reproduction is used? (3) How do the offspring look when compared to the parent? (4) What special cell types are needed and where are they made?

### Reteach

Use answers to Check for Understanding for class discussion.

### Portfolio

Have students prepare an idea map that illustrates the similarities and differences between internal and external fertilization. Their idea maps should begin with the phrase "Types of Fertilization."

## CRITICAL THINKING 24

**CRITICAL THINKING/PROBLEM SOLVING** — **CHAPTER 24**

Name ______ Date ______ Class ______

*Use after Section 24:2.*

**HOW CAN SOME MAMMALS BEAR MORE THAN ONE YOUNG AT A TIME?**

The students in a biology class watched a videotape about mammals. It showed a cow giving birth to a calf and a dog giving birth to four puppies. Some students were puzzled. They knew one sperm fertilized one egg. They thought most female mammals released only one egg during each estrous cycle. In the library, they found a table in a reference book about animal reproduction. The table showed the usual number of young born at one time, or the number in a litter. The table showed a range of numbers for several types of mammals.

| Name of mammal | Number of young in a litter |
|---|---|
| gorilla | 1 |
| pig | 4–6 |
| brown rat | 6–9 |
| whale | 1 |
| dog | 7 |

*Use the table and information from your textbook to answer the questions below.*

**Analyzing the Problem**

1. What is an estrous cycle? The stages in the reproductive cycle of a female mammal, including a time when the female is ready to mate.
2. In the table, which animals can give birth to the most young at one time? rat, dog, pig
3. Which animals have the least number of young in a litter? gorilla, whale
4. In order for a pig to give birth to 4–6 young in one litter, how many eggs are produced and fertilized in an estrous cycle? 4–6

**Solving the Problem**

1. Suppose you were given one female and one male of the same species. How could you test how many eggs the female releases at one time? How could you know how many eggs were fertilized? Let the animals mate and count the offspring in one litter. The number of eggs and sperm that join to form fertilized eggs equals the number of offspring.
2. Using the same animals, how could you determine the range of number of young in a litter? Let the animals have several litters and calculate an average.

## APPLICATION 24

**APPLICATION: ENVIRONMENT** — **CHAPTER 24**

Name ______ Date ______ Class ______

*Use after Section 24:2.*

**REPRODUCTION IN ANIMALS**

Different animals give birth to different numbers of offspring. Also, the time it takes for offspring to grow inside the mother varies for different animals. This period is called the gestation (je STAY shuhn) period. The number of offspring and gestation period for several animals is listed in the table below.

| Animal | Average number of offspring at each birth | Length of gestation period (months) |
|---|---|---|
| Gray squirrel | 9 | 1½ |
| Dog | 7 | 2 |
| Red fox | 6 | 2 |
| Lion | 3 | 3½ |
| Black bear | 2 | 7 |
| Zebra | 1 | 11½ |

Length of gestation period (months)

22 20 18 16 14 12 10 8 6 4 2

gray squirrel dog red fox lion black bear zebra

Number of offspring

9 8 7 6 5 4 3 2 1

gray squirrel dog red fox lion black bear zebra

1. Use the information in the table to fill in the bar graphs.

*Study the information in Section 24:2 of your text and the bar graphs above to answer the following questions.*

2. How does the number of offspring relate to the length of the gestation period of the animal? The animals with fewer offspring have longer gestation periods, and the animals with many offspring have shorter gestation periods.
3. Which season would you expect to be the breeding season for a gray squirrel? Explain. Spring, because a squirrel's gestation period is only 1½ months long. If it breeds in the spring, the babies would be born in the spring or summer, when the weather is good for young animals to survive.
4. For a zebra to give birth in the spring, when would its breeding season have to be? Explain. It would have to breed in the spring, because its gestation period is 11½ months long. It would give birth in the next spring.

## OPTIONS

**Application/***Teacher Resource Package,* p. 24. Use the Application worksheet shown here to teach reproduction in animals.

**Critical Thinking/Problem Solving/** *Teacher Resource Package,* p. 24. Use the Critical Thinking/Problem Solving worksheet shown here to help students understand how some mammals give birth to more than one young at a time.

## TEACH

### Check for Understanding

Have students respond to the first three questions in Check Your Understanding.

### Reteach

*Reteaching/Teacher Resource Package,* p. 73. Use this worksheet to give students additional practice with understanding sexual reproduction.

**Extension:** Assign Critical Thinking, Biology and Reading, or some of the **OPTIONS** available with this lesson.

## 3 APPLY

### ACTIVITY/Challenge

Have students use references to determine (a) the frequency of estrous cycles in different animals and (b) what controls whether or not animals mate only in the spring.

## 4 CLOSE

### Divergent Question

Ask students the following: What evidence can you provide to show that (a) humans do not have a breeding season and (b) dogs do have an estrous cycle?

### Answers to Check Your Understanding

6. egg and sperm cells
7. joining of egg and sperm outside the body of an animal
8. A breeding season is a certain time of year when particular kinds of animals reproduce; an estrous cycle is a cycle in which a female will mate only at certain times.
9. external fertilization; since offspring get no protection from parents, many die; more must be produced
10. The starfish undergo meiosis to produce sperm and eggs and reproduce by sexual reproduction.

**Figure 24-5** Large mammals, such as the deer, breed in the fall and have their young the following spring.

Most fish, birds, and many small mammals breed in the spring of the year. The young hatch from eggs, or are born during the late spring or summer of that same year. Many large mammals, such as the deer shown in Figure 24-5, breed in the fall of the year. The unborn young take longer to develop. The young are born during early spring of the next year. This breeding season helps make sure that the young are born in the warm part of the year when they have a better chance to survive.

Another way that animals improve the chance of fertilization is by having an estrous (ES trus) cycle. An **estrous cycle** is a cycle in which a female will mate only at certain times. It is only at these times that the female's eggs can be fertilized.

Mammals have estrous cycles. For example, rats are ready to mate every five days. Most dogs have two estrous cycles per year. Cows have a 20-day cycle.

### Check Your Understanding

6. Which cells are used in sexual reproduction?
7. What is meant by external fertilization?
8. What is the difference between a breeding season and an estrous cycle?
9. **Critical Thinking:** Which animals do you think produce more offspring, those with external fertilization or those with internal fertilization? Why?
10. **Biology and Reading:** What kind of reproduction is it when starfish release sperm and eggs into the water?



## OPTIONS

**How Do Internal and External Reproduction Compare?**/*Lab Manual,* pp. 201-204. Use this lab as an extension to studying sexual reproduction.

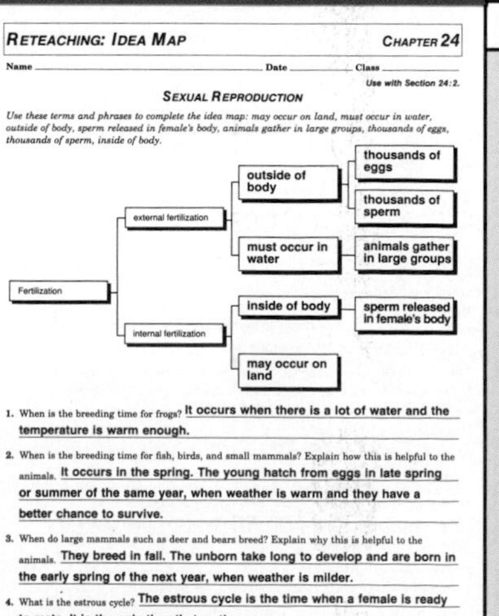

RETEACHING 73

RETEACHING: IDEA MAP — CHAPTER 24

Name ____ Date ____ Class ____

*Use with Section 24:2.*

SEXUAL REPRODUCTION

*Use these terms and phrases to complete the idea map: may occur on land, must occur in water, outside of body, sperm released in female's body, animals gather in large groups, thousands of eggs, thousands of sperm, inside of body.*



1. When is the breeding time for frogs? It occurs when there is a lot of water and the temperature is warm enough.
2. When is the breeding time for fish, birds, and small mammals? Explain how this is helpful to the animals. It occurs in the spring. The young hatch from eggs in late spring or summer of the same year, when weather is warm and they have a better chance to survive.
3. When do large mammals such as deer and bears breed? Explain why this is helpful to the animals. They breed in fall. The unborn take long to develop and are born in the early spring of the next year, when weather is milder.
4. What is the estrous cycle? The estrous cycle is the time when a female is ready to mate. It is the only time that mating occurs.

# Lab 24–1 Sex Cells

## Problem: How alike are egg and sperm cells?

### Skills

make and use tables, measure in SI, recognize and use spatial relationships, make scale drawings

### Materials

microscope
prepared slide of starfish eggs
prepared slide of starfish sperm
metric ruler
petri dish for drawing circles

### Procedure

1. Copy the data table.
2. Use your microscope to look at a prepared slide of starfish eggs under low power and high power.
3. Fill out the top row in your table.
4. Repeat step 2 with a prepared slide of starfish sperm.
5. Fill out the bottom row in your data table.
6. To measure the actual size of an egg cell, draw a circle on a sheet of paper. Use the petri dish as a guide.
7. **Recognize and use spatial relationships:** Using high power, draw one egg in the circle. Draw it to scale as seen through the microscope. Label it "starfish egg."
8. **Measure in SI:** Measure the diameter of your diagram in millimeters. Record this number in your table.
9. Multiply the number in step 8 by 0.004 to get the actual size of the cell. Record this number in your table.
10. Repeat steps 6 to 9 with the slide of sperm. Label your diagram "starfish sperm." Note that there are two parts to a sperm. Label the small, oval end "head." Label the long, threadlike part "tail."

### Data and Observations

1. Compare the actual sizes of egg and sperm.
2. Compare the shapes of egg and sperm.

### Analyze and Apply

1. Compare the following for starfish and humans:
   (a) type of reproduction—sexual or asexual
   (b) type of fertilization—internal or external
2. **Apply:** Explain how the shape of a sperm cell helps it swim.

### Extension

**Make scale drawings:** Prepare a wet mount of a small piece of magazine print. Locate a period at the end of a sentence. Using the same procedures as in the lab, measure the diameter of the period under high power. Compare the sizes of the starfish egg and sperm to that of the period.

*Answers will vary.

| | SHAPE OF CELL | BODY OR SEX CELL | HAS A TAIL | SWIMS | NUMBER SEEN UNDER HIGH POWER | LENGTH OF DIAGRAM (mm) | MULTIPLY BY | SIZE OF CELL (mm) |
|---|---|---|---|---|---|---|---|---|
| Egg | round | sex | no | no | 1 | 44 mm* | 0.004 | 0.176 mm* |
| Sperm | round with tail | sex | yes | yes | 30* | 2 mm* | 0.004 | 0.008 mm* |



## ANSWERS

### Data and Observations

1. An egg is about 0.16 mm in diameter; a sperm head is 0.003 mm.
2. Eggs are round; sperm have an oval head and a tail.

### Analyze and Apply

1. (a) both are sexual
   (b) starfish - external; human - internal
2. A sperm has a tail that moves back and forth to help the sperm swim.

## GLENCOE TECHNOLOGY

**Videodisc**

**Biology: The Dynamics of Life**
Animation: *Human Fertilization*
Disc 2, Side 1, Ch. 41

## Lab 24-1 Sex Cells

### Overview

In this lab, students are given an opportunity to look at prepared slides showing typical animal sex cells, thus reinforcing their skills with a microscope. After students have actually measured the sizes of both egg and sperm cells, the vast difference in size and the difference in numbers of sex cells produced by a male and female will become evident.

**Objectives:** Upon completing this lab, students will be able to (1) **distinguish** an egg cell from a sperm cell, (2) **compare** the size difference between these two cells, (3) **measure** and record the actual cell size of starfish sex cells.

**Time Allotment:** 20 minutes

### Preparation

▶ **Alternate Materials:** Sea urchin egg and sperm slides may be substituted. If petri dishes are not available, a circle with a diameter of 100 mm is all that is needed.

**Lab 24-1 worksheet**/*Teacher Resource Package,* pp. 93-94.

### Teaching the Lab

▶ If slides are not available, the activity may still be performed by students. Use the two photos provided with the activity. They are close to the size seen when viewed under high power magnification. Have students use them for their drawings to be made to scale on the 100-mm circle.

### ASSESSMENT

**Skill:** Provide students with a fish egg (caviar from the grocery store) and a metric ruler. Do not tell them that it is an egg cell. Ask them to decide if the cell is an egg or sperm. Then, have them complete the information called for in the chart at the bottom of this lab. They may omit the following categories: Number Seen, Length of Diagram, Multiply By. Have them state how they were able to decide initially if the cell was an egg or sperm.

# Lab 24-2 Menstrual Cycle

## Overview

In this lab, students are introduced to two of the female hormones responsible for controlling the events of the menstrual cycle. They will graph data presented in table form and will be able to correlate changes in uterus thickness with increasing or decreasing concentrations of these hormones.

**Objectives:** Upon completion of this lab, students will be able to (a) **construct a graph** on estrogen and progesterone hormone concentrations, (b) **construct a graph** on changes in uterus thickness, (c) **relate** the lowering and rising of hormone concentrations to a loss (menstruation) and thickening of uterus lining.

**Time Allotment:** 30 minutes

## Preparation

▶ You may wish to review basic graphing techniques with your students prior to starting this activity.

Lab 24-2 worksheet/*Teacher Resource Package,* pp. 95-96.

## Teaching the Lab

▶ The text describes estrogen but does not include any discussion of progesterone. You may wish to point out that both hormones influence uterine buildup.

▶ **Troubleshooting:** Check graphs to make sure that students have indeed used proper units and have prepared the graphs correctly.

**Skill:** Using completed graphs, have students fill in the expected data on a revised data table similar to the one in the lower-left corner. Their revised data table calls for estrogen units, progesterone units, and uterine thickness on days: 3, 7, 18, 22, and 26.

# Menstrual Cycle

**Lab 24–2**

## Problem: What causes uterus thickness to change during the menstrual cycle?

### Skills

make and use graphs, infer, relate cause and effect

### Materials

graph paper
colored pencil
ruler

### Procedure

1. Obtain a sheet of graph paper.
2. Draw and label two graphs as shown in the figure. *For more help, refer to the* ***Skill Handbook,*** *pages 715-717.*
3. Use the data in the table to complete the top graph.
   (a) Plot the estrogen units on the graph.
   (b) Connect the points for estrogen with a line and label the line "estrogen units."
   (c) Plot the progesterone units.
   (d) Use a colored pencil to connect the points for progesterone and label the line progesterone units.
4. Use the data in the table to complete the bottom graph.
   (a) Plot uterus thickness on the graph.
   (b) Connect the points on the graph with a line.

| DAY | AMOUNT OF HORMONE | | THICKNESS OF UTERUS LINING (mm) |
|---|---|---|---|
| | ESTROGEN | PROGESTERONE | |
| 1 | 50 | 5 | .5 |
| 5 | 65 | 5 | 1.5 |
| 10 | 200 | 5 | 2.25 |
| 15 | 75 | 40 | 3.0 |
| 20 | 100 | 150 | 4.0 |
| 25 | 50 | 100 | 5.0 |
| 27 | 50 | 30 | 4.75 |
| 1 | 50 | 5 | .5 |

### Data and Observations

1. How do the amounts of estrogen and progesterone change throughout the menstrual cycle?
2. How does the thickness of the uterus change throughout the menstrual cycle?

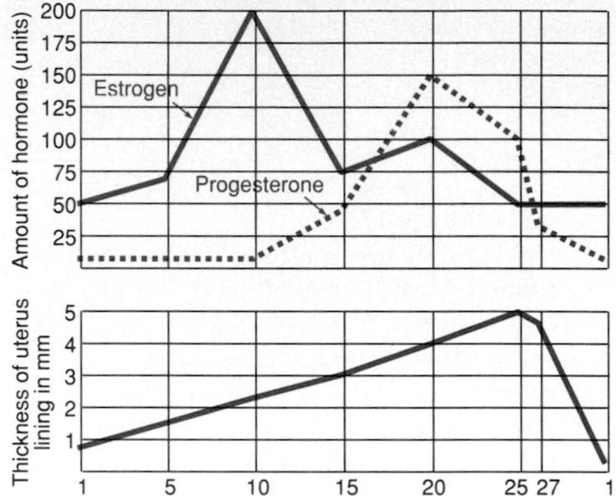


### Analyze and Apply

1. **Infer:** How is the change in the thickness of the uterine lining related to the change in estrogen amount for days 1-10?
2. Compare the change in thickness of the uterine lining with the change in progesterone amount for days 10-27.
3. **Apply:** (a) What happens to the uterine lining between day 27 and day 1?
   (b) What is this process called?

### Extension

**Relate cause and effect:** What happens to the egg on days 1-13, 14, and 15-27? How does this compare to the amounts of hormones and uterus thickness?



## ANSWERS

### Data and Observations

1. Estrogen increases until day 10, then decreases. Progesterone stays the same through day 10, then increases until day 20, then decreases.
2. The thickness of the uterus increases until day 25, then begins to decrease.

### Analyze and Apply

1. As the amount of estrogen goes up and down, the thickness of the uterus continues to increase.
2. As the amount of progesterone goes up, the thickness of the uterus goes up, until day 20. The amount of progesterone then begins to drop, but the thickness continues to increase for about 5 more days.
3. (a) The uterus decreases in thickness because it is being shed. (b) menstruation

# 24:3 Reproduction in Humans

Humans have reproductive systems with many parts. The reproductive organs of males and females each do special jobs. Let's look at the human reproductive system to see how the different parts work.

**Objectives**

6. **Identify** the reproductive parts of humans and the stages of human reproduction.
7. **Compare** changes in the menstrual cycle with and without fertilization.
8. **Discuss** symptoms and problems of sexually transmitted diseases.

**Key Science Words**

reproductive system
scrotum
penis
vas deferens
oviduct
uterus
vagina
menstrual cycle
estrogen
menstruation
sexually transmitted disease

## Human Reproductive System

The system used to produce offspring is called the **reproductive system.** What does this system look like in males and females?

Figure 24-6 shows the main parts of the male reproductive system. The job of each part is also listed. Study these jobs. Notice that this figure is a cut-away view from the side. Also notice that certain parts of the male reproductive system are also part of the excretory system studied in Chapter 13.

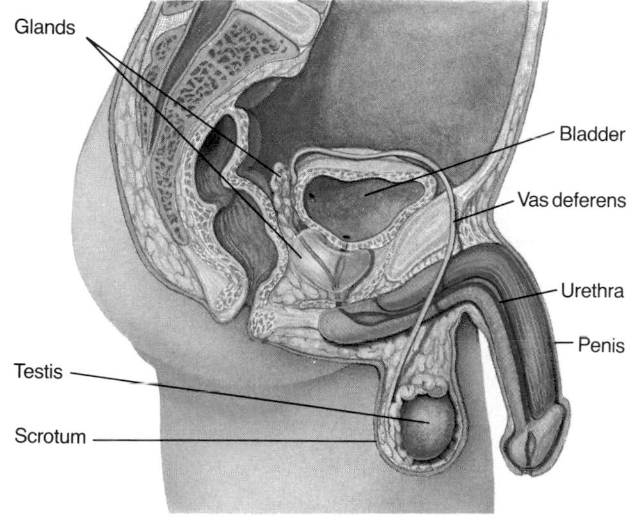


**Figure 24-6** The male reproductive system is shown here in a cut-away view from the side.

| PART | JOB |
|---|---|
| Testis | Produces sperm cells by meiosis |
| **Scrotum** (SKROH tum) | Sac that holds testes |
| **Penis** (PEE nus) | Places sperm in vagina during mating |
| **Vas deferens** (VAS·DEF uh runs) | Carries sperm from testes to urethra |
| Urethra (yoo REE thruh) | Carries sperm and urine out of body |
| Glands | Provides liquid in which sperm can swim |



# 24:3 Reproduction in Humans

## PREPARATION

### Materials Needed

Make copies of the Focus, Transparency Master, Study Guide, Reteaching, and Lab worksheets in the *Teacher Resource Package*.

▶ Obtain graph paper and colored pencils for Lab 24-2.

### Key Science Words

reproductive system
scrotum
penis
vas deferens
oviduct
uterus
vagina
menstrual cycle
estrogen
menstruation
sexually transmitted disease

### Process Skills

In Lab 24-2, students will make and use graphs, infer, and relate cause and effect. In the Skill Checks, students will sequence and understand science words.

## 1 FOCUS

▶ The objectives are listed on the student page. Remind students to preview these objectives as a guide to this numbered section.

### ACTIVITY/Audiovisual

Show the filmstrip *Human Body Series: Reproductive System*, Carolina Biological Supply Co.

**Focus**/*Teacher Resource Package*, p. 47. Use the Focus transparency shown at the bottom of this page as an introduction to the reproductive system.

FOCUS 47

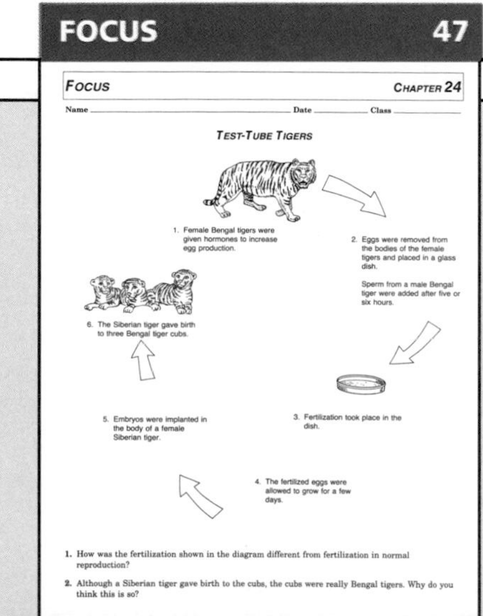


## OPTIONS

### Science Background

The menstrual cycle is under hormonal control. Hormones involved are gonadotropic releasing factor (stimulates anterior pituitary), follicle stimulating hormone (stimulates egg development), luteinizing hormone (stimulates egg release), and estrogen and progesterone (inhibit the release of FSH and LH, thus preventing more than one egg from maturing).

# 2 TEACH

### MOTIVATION/Brainstorming

Ask students to speculate as to the possible outcome if sperm are present and two different eggs are released at the same time. (Fraternal twins are possible.)

### Concept Development

▶ Ask students to pay particular attention to the information contained in Figure 24-6. It not only introduces several new terms but also identifies the functions of the parts of the male reproductive system.

▶ Be sure students study the functions of each part of the female reproductive system as shown in Figure 24-7.

▶ Review with students the nature of egg and sperm cells with regard to their chromosome number. Remind students that each human sex cell contains 23 pairs of chromosomes.

### Independent Practice

**Study Guide**/*Teacher Resource Package*, p. 141. Use the Study Guide worksheet shown at the bottom of this page for independent practice.

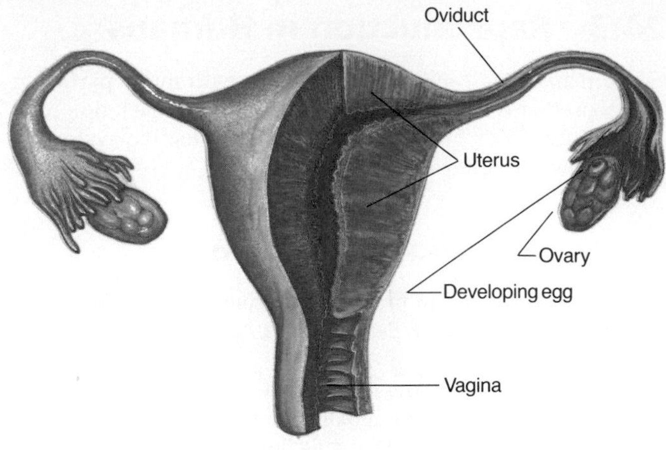


**Figure 24-7** The female reproductive system is shown here in a cut-away view from the front.

| PART | JOB |
|---|---|
| Ovary | Produces egg cells |
| Oviduct | Carries egg from ovary to uterus |
| Uterus | Place where fertilized egg develops |
| Vagina | Receives penis of male during mating |

The female reproductive system is shown in Figure 24-7. This figure shows a cut-away view from the front. Study the function of each part.

> **Skill Check**
>
> **Understand science words: oviduct.** The word part *ovi* means egg. In your dictionary, find three words with the word part *ovi* in them. *For more help, refer to the **Skill Handbook**, pages 706-711.*

## Stages of Reproduction

How does an egg become fertilized? What happens to the egg after it is fertilized? Look at Figure 24-8 and follow the numbers in the figure as you read.

1. Egg cells are formed in each ovary.
2. Each month, one ovary releases an egg. Usually, only one egg is released about every 28 days. The ovaries usually take turns releasing eggs. One ovary releases an egg one month, and the other ovary releases an egg the next month.

Figure 24-8 shows what happens next.

3. Once released from the ovary, the egg moves into a tube called an oviduct (OH vuh dukt). **Oviducts** are tubelike organs that connect the ovaries to the uterus. The **uterus** (YEWT uh rus) is a muscular organ in which the fertilized egg develops. The uterus is made of smooth muscle.

Figure 24-8 shows how fertilization takes place.

4. Sperm are released into the vagina during mating. The



## OPTIONS

**Stages of Reproduction**/*Transparency Package*, number 24a. Use color transparency number 24a as you teach the stages of reproduction.

**Transparency Master**/*Teacher Resource Package*, p. 121. Use the Transparency Master shown here to teach the human reproductive system.

**STUDY GUIDE 141**

STUDY GUIDE — CHAPTER 24

Name ______ Date ______ Class ______

REPRODUCTION IN HUMANS

*In your textbook, read about the human reproductive systems in Section 24:3.*

1. Label the diagram below using these words: oviduct, ovary, uterus, egg, vagina.

(E) oviduct; (C) ovary; (B) egg; (A) uterus; (D) vagina

2. In each circle on the diagram, write the letter of the job listed below that the body part does.
   A. Muscular organ where fertilized egg develops.
   B. Cell formed in the ovary.
   C. A place where the egg is formed.
   D. Muscular tube that leads to uterus.
   E. Tubelike parts that connect ovary to uterus.
3. Fill in the blanks below with the correct numbers.
   a. Ovaries in a human female 2
   b. Testes in a human male 2
   c. Eggs usually produced by human at one time 1
4. Label the diagram below using these words: testes, penis, vas deferens, urethra, scrotum, glands.
5. In each circle on the diagram, write the letter of the job listed below that the body part does.
   A. Sperm pass out of the body here.
   B. Sperm are made here.
   C. Sperm move from the testes to the penis here.
   D. Provides liquid for sperm.
   E. Carries sperm out of body.
   F. Sac that holds testes.

(C) vas deferens; glands (D); (E) urethra; scrotum (F); (A) penis; (B) testes

**TRANSPARENCY 121**

TRANSPARENCY MASTER — CHAPTER 24

HUMAN REPRODUCTIVE SYSTEM

Male — Side view: urinary bladder, glands, vas deferens, glands, urethra, penis, testes, scrotum. Front view: glands, vas deferens, glands, urethra, penis, testes, scrotum.

Female — Side view: oviduct, ovary, uterus, urinary bladder, vagina, urethra. Front View: oviduct, ovary, uterus, vagina.

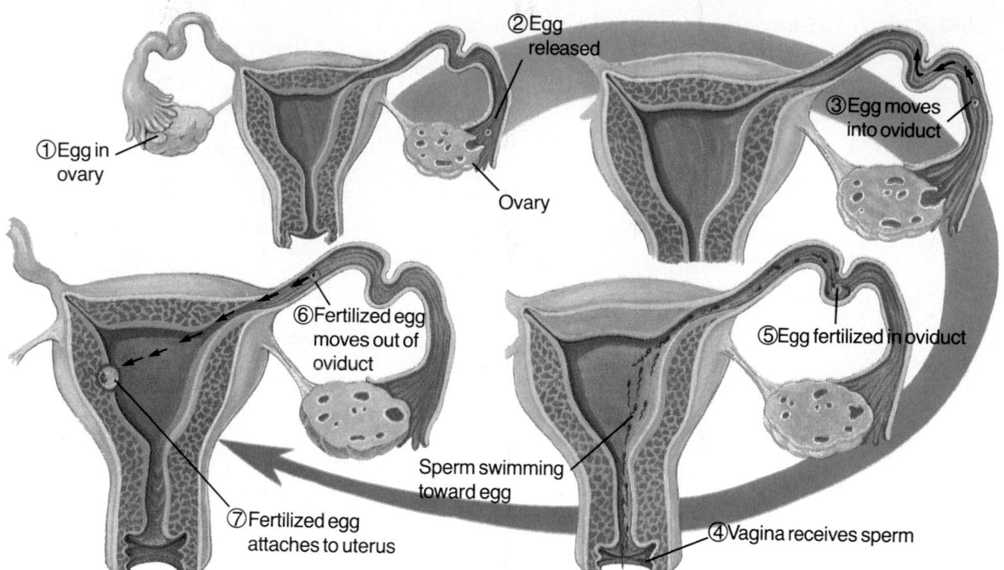


**Figure 24-8** The stages of reproduction in the female are shown here. Where is the egg fertilized?in the oviduct

**vagina** (vuh JI nuh) is a muscular tube that leads from outside the female's body to the uterus. Sperm swim from the vagina into the uterus and then into the oviducts.
5. If an egg is present, fertilization takes place.

Figure 24-8 shows what happens after fertilization.
6. As the fertilized egg moves down the oviduct, it divides several times and becomes an embryo.
7. The embryo then attaches itself to the wall of the uterus. Once attached, it will remain there for nine months as it develops into a baby.

**What happens to an egg after it is fertilized in the oviduct?**

## The Menstrual Cycle

You know that one egg is released by an ovary about every 28 days. What controls this timing? Many of the changes in the reproductive system are controlled by hormones. You learned in Chapter 15 that hormones are chemicals that affect certain body organs.

The monthly changes that take place in the female reproductive organs are called the **menstrual** (MEN struhl) **cycle.** The word *menstrual* comes from Latin and means monthly.

The menstrual cycle occurs about every month starting when a female is 10 to 13 years old. The monthly cycles continue for about 40 years.

### Bio Tip

**Science:** Three billion human eggs would fill a small bucket. The same number of human sperm would fit into a thimble.



## TEACH

### Concept Development

▶ Point out that the events described illustrate the major ideas connected with internal fertilization.

▶ For ease in organization, the menstrual cycle can be divided into the three phases: days 1-14, days 14-24, and days 25-28.

### Bridging

A review of Chapter 15 will refresh students' memories regarding hormones, their properties, where they are formed, and how they are carried through the body.

**Misconception:** Students usually are not aware of the fact that menstruation prepares the reproductive system to once again begin to form an egg and an environment conducive to fertilization and development of the fertilized egg.

### Guided Practice

Have students write down their answers to the margin question: It divides into a mass of cells and travels to the uterus where it becomes embedded.

### Independent Practice

**Study Guide**/*Teacher Resource Package,* p. 142. Use the Study Guide worksheet shown at the bottom of this page for independent practice.

**STUDY GUIDE 142**

STUDY GUIDE CHAPTER 24

Name ______ Date ______ Class ______

REPRODUCTION IN HUMANS

*In your textbook, read about stages of reproduction in Section 24:3.*

6. Examine the diagrams below and label the parts. Then, answer the questions.

| Diagram labels | Questions |
| --- | --- |
| egg, uterus, ovary | a. Why will fertilization not take place in this diagram? No sperm are present. Fertilization is the joining of egg and sperm. |
| sperm, vagina | b. Why will fertilization not take place in this diagram? No egg is present. |
| embryo | c. Why will fertilization not take place in this diagram? (HINT: Note that an embryo is already present.) Usually no eggs are released if fertilization and attachment within the uterus have occurred. |
| oviduct | d. Why will fertilization not take place in this diagram? There are no ovaries to produce eggs, and no eggs or sperm. |

## OPTIONS

**Challenge:** Have students note changes occurring in the ovary and uterus during days 1 to 14 and compare them to changes during days 15 to 28.

**The Menstrual Cycle**/*Transparency Package*, number 24b. Use color transparency number 24b as you teach the menstrual cycle.

**Videodisc**

**STV: Human Body Vol. 3**
*Reproductive Systems*
Unit 2, Side 2
*Reproductive Systems (in its entirety)*

**00667-36800**

## TEACH

### Concept Development

▶ Explain that the ovary is an egg-forming structure and an endocrine gland.

▶ Point out changes that occur within the uterus during the menstrual cycle.

▶ The major job of the uterus during the last 14 days of the menstrual cycle is to prepare for a fertilized egg.

▶ Animals that have an estrous cycle do not have menstruation. The thickened uterus is absorbed internally and, as a consequence, no obvious blood loss is detected.

### Independent Practice

**Study Guide**/*Teacher Resource Package*, p. 143. Use the Study Guide worksheet shown at the bottom of this page for independent practice.

### Guided Practice

Have students write down their answers to the margin question: The lining of the uterus is shed.

**Cooperative Learning:** Divide the class into 8 or 9 groups. Have each group prepare a table similar to Table 24-1 to represent changes occurring in the female menstrual cycle if fertilization *does* occur. Ask the groups to record their tables onto a transparency and present their tables to the class.

### Student Journal

Tell students the following: You are an endocrinologist, one who studies hormones and their effect on the human body. You are preparing a list that includes action and source for those hormones involved in the menstrual cycle. Prepare such a list and note that the hormone formed by the pituitary gland is called FSH or Follicle Stimulating Hormone.

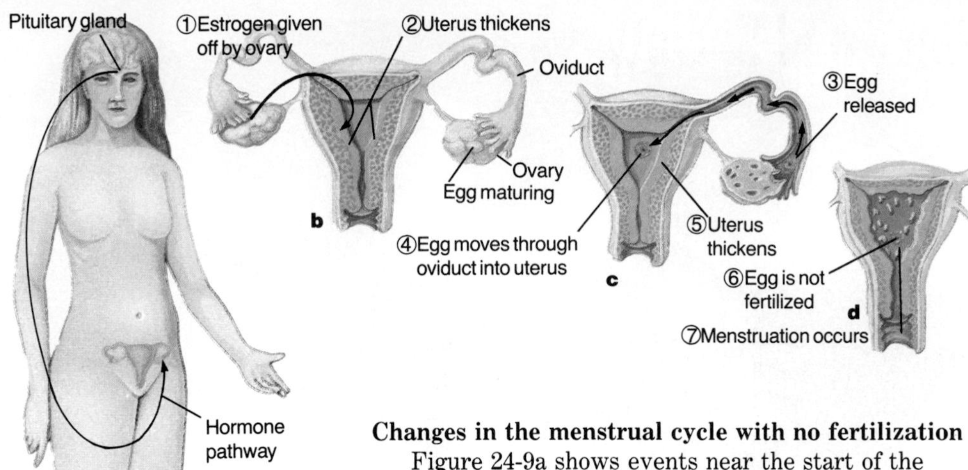


**Figure 24-9** The changes shown here take place during the menstrual cycle if no egg is fertilized.

**What happens in the uterus when an egg is not fertilized?**

#### Changes in the menstrual cycle with no fertilization

Figure 24-9a shows events near the start of the menstrual cycle. The pituitary gland gives off a hormone. This hormone travels by way of the blood to the ovaries. It causes an egg in the ovaries to mature.

Figure 24-9b shows the changes that follow. The numbered steps in the diagram match the numbered statements.

1. The ovary itself forms estrogen (ES truh jun). **Estrogen** is a female hormone. It is responsible for the changes seen in a female as she enters puberty.
2. Estrogen also causes the lining of the uterus to increase in thickness. This increase in thickness makes it possible for a fertilized egg to attach to the uterine lining.

Figure 24-9c shows the next set of events. No sperm are present, so fertilization of the egg will not occur.

3. The ovary releases an egg. After the egg is released, the ovary begins to produce progesterone (pruh JES tuh rohn), a second female hormone.
4. The egg moves through the oviduct and enters the uterus.
5. Meanwhile, the uterine lining continues to thicken from the progesterone being made by the ovary.

Figure 24-9d shows what happens next.

6. The egg has not been fertilized. Therefore, it will not attach to the uterus.
7. The thick uterine lining is no longer needed. It begins to break apart. The cells of the thickened uterine lining break off and leave the body through the vagina along with the unfertilized egg and a small amount of blood. This loss of cells from the uterine lining, blood, and egg is called **menstruation** (men STRAY shun).



## OPTIONS

### Science Background

Onset of menstruation and sexual maturity is called menarche. Cessation of egg release and menstruation is called menopause.

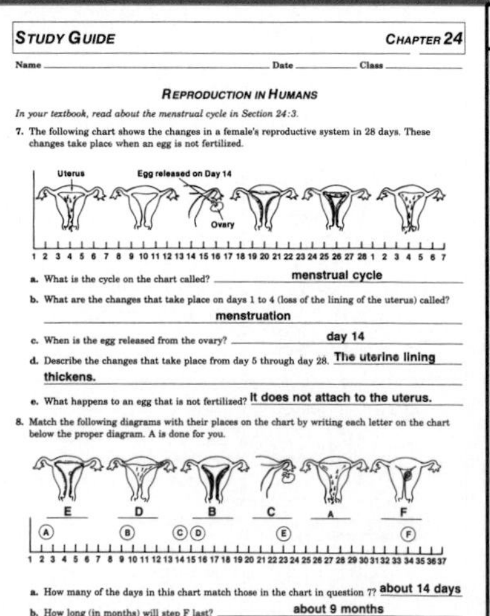

STUDY GUIDE 143

STUDY GUIDE CHAPTER 24

Name ______ Date ______ Class ______

REPRODUCTION IN HUMANS

*In your textbook, read about the menstrual cycle in Section 24:3.*

7. The following chart shows the changes in a female's reproductive system in 28 days. These changes take place when an egg is not fertilized.



a. What is the cycle on the chart called? menstrual cycle

b. What are the changes that take place on days 1 to 4 (loss of the lining of the uterus) called? menstruation

c. When is the egg released from the ovary? day 14

d. Describe the changes that take place from day 5 through day 28. The uterine lining thickens.

e. What happens to an egg that is not fertilized? It does not attach to the uterus.

8. Match the following diagrams with their places on the chart by writing each letter on the chart below the proper diagram. A is done for you.



a. How many of the days in this chart match those in the chart in question 7? about 14 days

b. How long (in months) will step F last? about 9 months

# Science and Society

## Sperm Banks

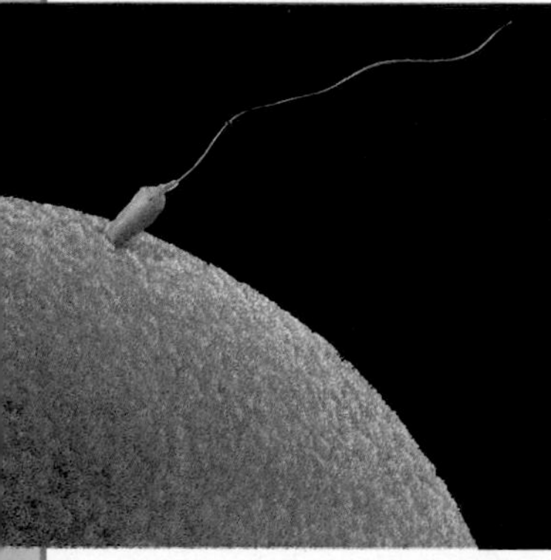

Technology has made it possible to store human sperm for years in sperm banks. Sperm are collected from donors and then frozen in liquid nitrogen. Chemicals are added to the sperm to keep them alive and prevent them from being damaged by the freezing temperatures. The sperm are kept frozen until they are needed.

How are the sperm banks used? For some couples, the husband is unable to produce sperm. These couples can receive sperm donated to the sperm bank by other males. Males who wish to donate sperm contact the sperm bank. The sperm is collected and stored. The stored sperm are used to fertilize one of the woman's eggs. The sperm are injected into the uterus with a syringe. This method of fertilizing eggs is called artificial insemination.

### What Do You Think?

1. Couples who use sperm banks often don't know who donated the sperm. The name of the donor isn't revealed to them. Thus, the couple doesn't know who the biological father is. The donor also doesn't know who receives his sperm. How is keeping the identities of donors and couples who receive sperm a secret helpful? Would it be better to reveal the identities?

2. Couples who use sperm banks can look at catalogs that list the traits of different sperm donors. In this way, they can choose some of the genetic qualities of their future child. What if the couple wanted their child to be like Albert Einstein or Bruce Springsteen? Who would be responsible if the child were born with a genetic disorder instead of being like Einstein?

3. Imagine you could create a person with beautiful hair and eyes, a great personality, and lots of physical ability. This "super person" would have the traits you find desirable. Over time, a large group of "super people" could be produced. But, your friend likes blond hair and brown eyes. You prefer brown hair and blue eyes. Who decides which traits are the most desirable ones to have? What kinds of problems does society face with the ability to create "super people"?

**Conclusion:** Do the benefits of sperm banks outweigh the possible problems?

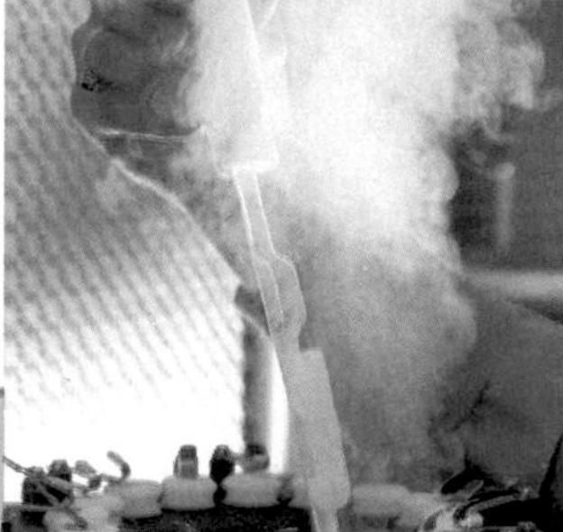

Sperm are stored in liquid nitrogen.



Peart, Karen N. "Can We Outdo Mother Nature?" *Scholastic Update*, September 2, 1994, pp. 20-24.

Seligson, Susan V. "Seeds of Doubt." *The Atlantic Monthly*, March 1995, pp. 28-32.



## Sperm Banks

### Background

The technique of freezing and thawing viable sperm has been around since 1938. The first successful artificial insemination of a female using sperm bank technology was reported in 1953.

A sample of donor sperm (typically the donors are screened and classified) is treated with glycerol to prevent ice crystals from forming. It is then cooled to -80°C for forty-five minutes. Then it is placed into a tank of liquid nitrogen for permanent storage (-196°C).

### Teaching Suggestions

Many different scenarios can be created by students as to advantages and disadvantages of using artificial insemination. Have students discuss some of their own ideas about this technique. Scenarios may include: birth defects resulting, not knowing the donor, storing sperm prior to a vasectomy.

### What Do You Think?

1. Genetic disease may be a major concern of parents.
2. and 3. Have students look up the meaning of the term *eugenics* to answer these questions.

**Conclusion:** Students may want to discuss the possibilities of a controlled population similar to something out of science fiction. Guide them to consider whether some couples would be happier with a child produced through artifical insemination or an adopted child. Have them consider that any child can be born with defects.

### Suggested Readings

Brownlee, Shannon. "The Baby Chase." *U.S. News & World Report*, December 5, 1994, pp. 84-91.

Fishman, Steve. "Inconceivable Conception." *Vogue*, December 1994, pp. 306-313.

## TEACH

### Concept Development

▶ Students should realize that a sperm cell that can swim on its own is needed for human fertilization. The trip through the female reproductive tract is rather long. Fertilization occurs in the oviducts, which are a distance from the vagina itself.

### Connection: History

Ask students to research:

1. the derivation of (a) the term *venereal disease* (The term itself is now replaced by STD; however, students are still familiar with the term VD. The term stems from Venus, the Goddess of Love.) and (b) hermaphrodite (a person who is both male and female, derived from Hermes and Aphrodite)
2. figures in history who died of or who were considered to have contracted some form of STD

### Check for Understanding

Have students compare the male and female reproductive systems by making these general comparisons: (1) names of parts and functions that are similar and different, (2) how sex cells are similar and different, (3) pathways that sperm and egg follow when leaving the reproductive system, and (4) main steps of the menstrual cycle and the reasons such a cycle is not needed in the male.

### Reteach

Prepare diagrams of the human male and female reproductive systems and have students label the major organs.

### Student Journal

Table 24-2 describes certain sexually transmitted diseases. Using only gonorrhea and the information provided for it, have students describe the organs of both the reproductive and excretory system that are affected by the organism that causes this problem. Also tell them to describe the type of organism responsible and whether or not this disease is curable.

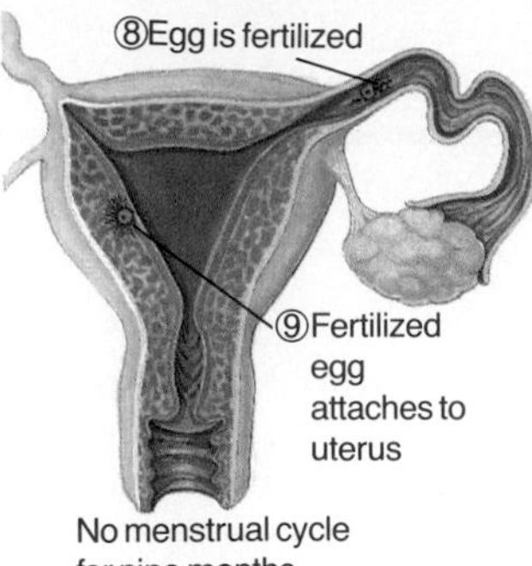


**Figure 24-10** If fertilization occurs, the fertilized egg attaches to the uterus.

**TABLE 24–1. MENSTRUAL CYCLE**

| Day in Cycle | Event Occurring if Egg Is Not Fertilized | Diagram |
|---|---|---|
| 1-4 | Menstruation—loss of egg, blood, uterus lining | 24-9d |
| 5-13 | Thickening of uterus lining by estrogen | 24-9a,b |
| 14 | Release of egg from ovary | 24-9c |
| 15-28 | Continued thickening of uterus lining | 24-9c |
| 1-4 | Menstruation if egg is not fertilized | 24-9d |

With menstruation, the cycle starts again. We could return to Figure 24-9a and follow the events for a second time. What event marks the beginning of a menstrual cycle? Table 24-1 shows you. It lists the days within the cycle when major events take place.

**Changes in the menstrual cycle if fertilization occurs**

What if fertilization does occur? Until step 5, the events are the same. After that, something different happens. Study Figure 24-10 as you read.

8. Sperm are present, and an egg is fertilized. Meanwhile, the uterine lining has thickened in preparation for receiving the embryo that develops from the fertilized egg.
9. The embryo attaches to the uterus. It remains attached for nine months while it develops into a baby. No new menstrual cycle will begin until the baby is born.

#### Skill Check

**Sequence:** Study Table 24-1 and Figure 24-9. Sequence the events that take place on days 5–28 of the menstrual cycle. *For more help, refer to the **Skill Handbook**, pages 706-711.*

## Diseases of the Reproductive System

Like all other body systems, the human reproductive system can be invaded by microbes. This invasion can result in diseases being passed from person to person during sexual contact. Diseases transmitted through sexual contact are called **sexually transmitted diseases,** or STDs.

Some common STDs are listed in Table 24-2. These diseases can damage the reproductive system as well as other body systems. What kinds of microbes cause STDs? The diseases gonorrhea (gawn uh REE uh), chlamydia (kluh MIH dee uh), and syphilis (SIF uh lus) are caused by bacteria. AIDS and genital herpes are caused by viruses.

Gonorrhea, chlamydia, and syphilis can be treated and cured with antibiotics. Genital herpes and AIDS aren't curable.



## OPTIONS

### Science Background

The specific organism responsible for each STD is as follows: gonorrhea, *Neisseria gonorrhoea*; chlamydia, *Chlamydia trachomatis*; syphilis, *Treponema pallidum;* genital herpes, *Herpes genitalis* or Herpes simplex virus II (HSV II); AIDS, HIV or Human Immunodeficiency Virus.

**TABLE 24-2. SEXUALLY TRANSMITTED DISEASES**

| Disease | Symptoms | Problems |
|---|---|---|
| Gonorrhea | females–puslike discharge from vagina, painful urination, no symptoms in many women<br>males–pus discharge from penis, painful urination | females–sterility, passed to newborn from infected mother during birth<br>males–infection of the testes |
| Chlamydia | (see gonorrhea), no symptoms in 70% of females, 10% of males | females–sterility, problem pregnancies<br>males–sterility |
| Syphilis | sore on vagina or penis followed by rash, sore throat, swollen glands | damage to heart and brain, passed from mother to fetus |
| Genital Herpes | blisterlike cut or sore on vagina or penis, may look like a rash | passed to newborn from infected mother during birth |
| AIDS | fever, night sweating, dry cough, weight loss, swollen glands, constant tiredness | destruction of the immune system, pneumonia, cancer, fatal after long illness |

How can a person reduce the chance of getting an STD? The best way is to avoid sexual contact. A second way is to use a condom during sexual contact. A third way is to have sexual contact with only one partner.

Which STDs can be cured?

In addition to sexually transmitted diseases, the reproductive system can be affected by other types of diseases. In males, the gland that supplies the liquid in which sperm swim can become swollen or develop tumors. In females, the tissue that surrounds the opening to the uterus can develop cancer. Both of these problems can be cured if they are detected early. Regular visits to a doctor are the best prevention.

## Check Your Understanding

11. Name the functions of the ovary and testes.
12. At about what age does the menstrual cycle first begin?
13. What kinds of organisms cause STDs?
14. **Critical Thinking:** How is it possible for a person to have a sexually transmitted disease for a long time without knowing it?
15. **Biology and Reading:** If nonidentical twins are produced, how many eggs must have been produced at the same time in the mother's ovaries? Is this normal?

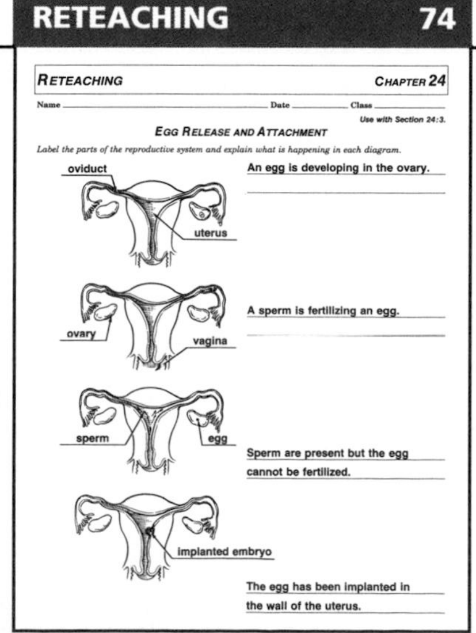

**RETEACHING 74**

*RETEACHING* CHAPTER 24

Name ______ Date ______ Class ______

Use with Section 24:3.

EGG RELEASE AND ATTACHMENT

*Label the parts of the reproductive system and explain what is happening in each diagram.*



An egg is developing in the ovary.

A sperm is fertilizing an egg.

Sperm are present but the egg cannot be fertilized.

The egg has been implanted in the wall of the uterus.

**STUDY GUIDE 144**

*STUDY GUIDE* CHAPTER 24

Name ______ Date ______ Class ______

VOCABULARY

*Review the new words used in Chapter 24 of your textbook. Then, fill in the blank next to each definition with the word or phrase it defines.*

penis 1. male reproductive organ that deposits sperm inside a female
scrotum 2. sac that holds testes
uterus 3. organ in which a fertilized egg will develop
vagina 4. tube leading from outside the female's body to the uterus
oviducts 5. tubelike organs that connect the ovaries to the uterus
regeneration 6. reproduction in which parent separates into two or more pieces, each of which forms a new organism
menstruation 7. loss of the uterine lining
estrous cycle 8. cycle in which the female will mate only at certain times
external fertilization 9. joining of egg and sperm outside the body
menstrual cycle 10. monthly cycle that takes place in female reproductive organs
internal fertilization 11. joining of egg and sperm inside the female's body
estrogen 12. a female hormone
sexually transmitted diseases 13. diseases transmitted through sexual contact
reproductive system 14. system used to produce offspring

# TEACH

### Guided Practice

Have students write down their answers to the margin question: gonorrhea, chlamydia, and syphilis.

### Independent Practice

**Study Guide**/*Teacher Resource Package,* p. 144. Use the Study Guide worksheet shown at the bottom of this page for independent practice.

### Check for Understanding

Have students respond to the first three questions in Check Your Understanding.

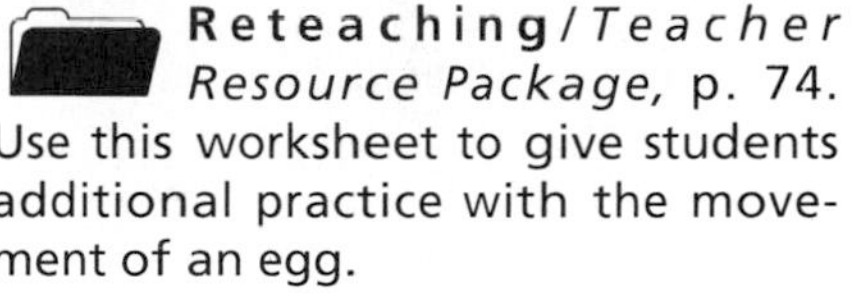

### Reteach

**Reteaching**/*Teacher Resource Package,* p. 74. Use this worksheet to give students additional practice with the movement of an egg.

**Extension:** Assign Critical Thinking, Biology and Reading, or some of the **OPTIONS** available with this lesson.

# 3 APPLY

### ACTIVITY/Challenge

Ask students to research and report on various types of birth control.

# 4 CLOSE

### Software

*Reproduction Organs,* Queue, Inc.

### Answers to Check Your Understanding

11. Ovaries produce egg cells, and testes produce sperm cells.
12. 10-13 years old
13. bacteria and viruses
14. Many STDs have no symptoms or very mild symptoms.
15. two eggs; twins are not very common

## CHAPTER 24 REVIEW

### Summary

Summary statements can be used by students to review the major concepts of the chapter.

### Key Science Words

All boldfaced terms from the chapter are listed.

### Testing Yourself

#### Using Words

1. estrogen
2. breeding season
3. menstruation
4. uterus
5. estrous cycle
6. external fertilization
7. oviduct
8. vagina
9. menstrual cycle
10. regeneration

#### Finding Main Ideas

11. p. 504; Egg and sperm cells are not used, special organs are not used, mitosis takes place, certain body cells undergo mitosis to form offspring, and the offspring are clones of the parent.
12. p. 515; Sperm swim up through the vagina into the uterus.
13. pp. 515-516; The egg, under hormone control, begins to mature within the ovary. The uterus lining, also under hormone control begins to thicken. The production of hormones decreases and the uterus lining is shed if no fertilization occurs.
14. p. 508; Thousands of sex cells are released. Also, the animals gather in large groups before releasing their sex cells.



# Chapter 24

# Review

## Summary

### 24:1 Asexual Reproduction

1. Asexual reproduction does not require sex cells, mating, or sex organs. It does use mitosis.
2. Budding and regeneration are two types of asexual reproduction.

### 24:2 Sexual Reproduction

3. Sexual reproduction uses mating and egg and sperm that are formed during meiosis.
4. In external fertilization, egg and sperm join outside the bodies of the parents. In internal fertilization, egg and sperm join inside the body of the female.
5. Having breeding seasons or an estrous cycle increases the chance of fertilization.

### 24:3 Reproduction in Humans

6. Eggs are formed in the ovaries and released into the oviducts. Fertilization of an egg occurs in the oviduct. The fertilized egg attaches itself to the uterine lining after it has divided several times.
7. The menstrual cycle prepares the uterus for a fertilized egg. If fertilization does not take place, menstruation occurs.
8. Sexually transmitted diseases are caused by bacteria and viruses and are spread by sexual contact.

## Key Science Words

estrogen (p. 516)
estrous cycle (p. 510)
external fertilization (p. 508)
internal fertilization (p. 509)
menstrual cycle (p. 515)
menstruation (p. 516)
oviduct (p. 514)
penis (p. 513)
regeneration (p. 505)
reproductive system (p. 513)
sexually transmitted disease (p. 518)
scrotum (p. 513)
uterus (p. 514)
vagina (p. 515)
vas deferens (p. 513)

## Testing Yourself

### Using Words

*Choose the word from the list of Key Science Words that best fits the definition.*

1. hormone found in females
2. certain time of the year when animals reproduce
3. loss of egg, uterine lining, and blood
4. organ in which a fertilized egg develops
5. cycle in which a female will mate only at certain times
6. egg and sperm joining outside body
7. tubelike organ connecting ovary to uterus
8. where sperm are released in the female
9. monthly changes in the female reproductive system

# Review

## Testing Yourself *continued*

10. what it is called when the parent breaks into two or more pieces to reproduce

**Finding Main Ideas**

*List the page number where each main idea below is found. Then, explain each main idea.*

11. what the main features of asexual reproduction are
12. the pathway followed by sperm within the vagina
13. changes that take place during the different stages of the menstrual cycle
14. how animals with external fertilization improve the chances of fertilization
15. why animals with internal fertilization can reproduce on land
16. the roles of testes and ovaries

**Using Main Ideas**

*Answer the questions by referring to the page number after each question.*

17. What are three ways you can reduce the chances of getting an STD? (p. 519)
18. Why do animals with external fertilization almost always live or reproduce in water? (p. 508)
19. What happens on or about day 14 of the menstrual cycle? (p. 518)
20. How does mating help animals reproduce? (p. 509)
21. How does asexual reproduction compare to sexual reproduction in terms of (p. 504, p. 507):
    (a) the process of cell division used?
    (b) whether sex cells are needed?
22. How is budding different from regeneration? (p. 505)

## Skill Review

*For more help, refer to the* ***Skill Handbook,*** *pages 704-719.*

1. **Make and use tables:** Make a table that compares the features of asexual and sexual reproduction.
2. **Sequence:** Sequence the steps involved in regeneration of a planarian.
3. **Understand science words:** What does the prefix *ex* mean in the word *external?*
4. **Make scale drawings:** An egg viewed under high power (430x) measures 30 mm when drawn to scale in a circle that has a diameter of 100 mm. What is the actual size of the egg?

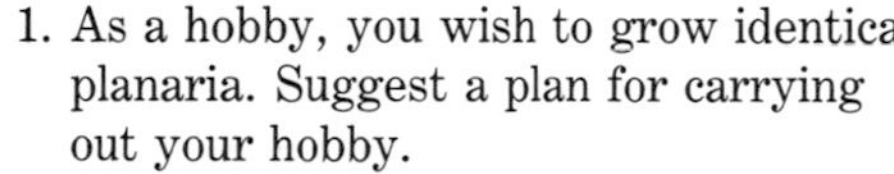

## Finding Out More



**Critical Thinking**

1. As a hobby, you wish to grow identical planaria. Suggest a plan for carrying out your hobby.
2. Each ovary takes a turn releasing one egg every 28 days. Yet, people do have fraternal (unalike) twins. Explain how this is possible.

**Applications**

1. Explain how a condom, diaphragm, fertility awareness, birth control pill, vasectomy, and tubal ligation actually work as birth control methods.
2. Find out which mammals have menstrual or estrous cycles and how long these cycles are. Prepare a chart of your findings for room display.



### Finding Main Ideas

15. p. 509; The sperm are released inside the female's body.
16. pp. 513-514; Testes produce sperm cells, and ovaries produce egg cells.

### Using Main Ideas

17. Avoid sexual contact, use a condom, and have sexual contact with only one partner.
18. Sperm cells must be able to swim to the egg, and this can only be done in water.
19. A mature egg moves into the uterus.
20. It increases the chance that an egg will be fertilized, and it allows fertilization to take place out of the water.
21. a. In asexual, cells undergo mitosis; in sexual, cells undergo meiosis.
    b. In asexual, no sex cells are needed; in sexual, the sex cells, egg and sperm, are needed.
22. In budding, a new organism develops from a small part of the body. In regeneration, the parent can split into two or more pieces. Each piece forms a new organism.

## Skill Review

1. List asexual and sexual as headings. List features in columns under each heading.
2. Sequence of steps is: regeneration begins at the head, two heads form; planarian continues to split along body, two separate planaria are formed.
3. *Ex* means out of or from
4. 0.0697 mm

## Finding Out More

### Critical Thinking

1. You could make a small vertical cut in the middle of the head of a planarian. It will then divide and form two identical planaria.
2. Two eggs are released.

### Applications

1. Condom, diaphragm, and vasectomy are methods that prevent sperm from reaching an egg. Tubal ligation prevents an egg from reaching the uterus. Birth control pills prevent an egg from maturing; fertility awareness is knowing the time an egg can be fertilized.
2. Answers will vary depending on the mammals that are researched. Examples of animals having estrous cycles are dogs and cats.

**Chapter Review**/*Teacher Resource Package*, pp. 126-127.

**Chapter Test**/*Teacher Resource Package*, pp. 128-130.

**Chapter Test**/*Computer Test Bank*

# Animal Development

## PLANNING GUIDE

| CONTENT | TEXT FEATURES | TEACHER RESOURCE PACKAGE | OTHER COMPONENTS |
|---|---|---|---|
| (2 days)<br>25:1 Development Inside the Female<br>Early Stages of Development<br>Needs of the Embryo<br>Human Development<br>Human Birth | Skill Check: *Interpret Digrams,* p. 526<br>Lab 25-1: *Development,* p. 531<br>Career Close-up: *Ultrasound Technician,* p. 527<br>Idea Map, p. 525<br>Check Your Understanding, p. 530 | Enrichment: *Comparing Weight in Different Stages of Mammal Development,* p. 27<br>Reteaching: *Vocabulary Puzzle,* p. 75<br>Transparency Master: *Cleavage of a Fertilized Egg,* p. 127<br>Study Guide: *Development Inside the Female,* pp. 145-147<br>Lab 25-1: *Development,* pp. 97-98 | **Laboratory Manual:**<br>*How Does a Human Fetus Change During Development?* p. 211<br>**Laboratory Manual:**<br>*What Changes Occur During Birth?* p. 215<br>Color Transparency 25a: *Cleavage*<br>**STVS:** *Brain Development,* Human Biology (Disc 7, Side 1) |
| (1/2 day)<br>25:2 Development Outside the Female<br>Eggs That Are Laid<br>Needs of the Embryo | Mini Lab: *What Can Pass Through an Egg Shell?* p. 533<br>Check Your Understanding, p. 533 | Reteaching: *Development Outside the Female,* p. 76<br>Skill: *Form a Hypothesis,* p. 26<br>Study Guide: *Development Outside the Female,* p. 148 | **STVS:** *Reptilian Sex Change,* Animals (Disc 5, Side 2) |
| (1 1/2 days)<br>25:3 Metamorphosis<br>Frog Metamorphosis<br>Insect Metamorphosis | Skill Check: *Understand Science Words,* p. 535<br>Lab 25-2: *Metamorphosis,* p. 534<br>Idea Map, p. 536<br>Check Your Understanding, p. 537 | Application: *Metamorphosis,* p. 25<br>Reteaching: *Insect Metamorphosis,* p. 77<br>Critical Thinking/Problem Solving: *What Happens to Insects in the Winter?* p. 25<br>Focus: *Metamorphosis* p. 49<br>Study Guide: *Metamorphosis,* p. 149<br>Study Guide: *Vocabulary,* p. 150<br>Lab 25-2: *Metamorphosis,* pp. 99-100 | Color Transparency 25b: *Frog Metamorphosis*<br>**STVS:** *Wasp Biological Control,* Ecology (Disc 6, Side 1) |
| Chapter Review | Summary<br>Key Science Words<br>Testing Yourself<br>Finding Main Ideas<br>Using Main Ideas<br>Skill Review | **ASSESSMENT RESOURCES**<br>Chapter Review, pp. 131-132<br>Chapter Test, pp. 133-135<br>Performance Assessment in the Biology Classroom<br>Alternate Assessment in the Science Classroom<br>Computer Test Bank | |

### GLENCOE TECHNOLOGY

**Infinite Voyage,** *The Living Clock*
**Science and Technology Videodisc Series,** *Brain Development,* Human Biology (Disc 7, Side 1)
*Reptilian Sex Change,* Animals (Disc 5, Side 2)
*Wasp Biological Control,* Ecology (Disc 6, Side 1)

**The Secret of Life,** *Sex and the Single Gene: Cell Development*

## MATERIALS NEEDED

| LAB 25-1, p. 531 | LAB 25-2, p. 534 | MARGIN FEATURES |
|---|---|---|
| metric ruler | metric ruler<br>hand lens<br>grasshopper stages of metamorphosis<br>moth stages of metamorphosis | Skill Check, p. 526<br>paper<br>pencil<br>Skill Check, p. 535<br>dictionary<br>Mini Lab, p. 533<br>beaker<br>eggshell<br>glucose |

## OBJECTIVES

For more information about National Science Standards, see page 5T.

| SECTION | OBJECTIVE | CORRELATION of QUESTIONS to OBJECTIVES | | | |
|---|---|---|---|---|---|
| | | CHECK YOUR UNDERSTANDING | CHAPTER REVIEW | TRP CHAPTER REVIEW | TRP CHAPTER TEST |
| 25:1 National Science Stds: UCP.1, UCP.2, UCP.3, UCP.4, UCP.5, C.5 | 1. **Sequence** the changes that take place in a fertilized egg before it attaches to the uterus. | 3, 5 | 6, 13 | 28 | 1, 29, 35 |
| | 2. **Explain** how the needs of an embryo are met as it develops. | 2, 4 | 2, 7, 9, 15, 19 | 15, 16, 18, 19, 20, 21 | 6, 7, 9, 11, 17, 18, 22, 25, 34, 48, 49, 50, 53, 54 |
| | 3. **Describe** the stages of human development from the first month until birth. | 1 | 1, 8, 14, 18, 20 | 3, 5, 17, 22, 23, 25 | 8, 12, 15, 23, 24, 27, 31, 32, 33, 36, 37, 38, 39, 40, 41, 42, 43, 44, 45, 46, 47, 51, 52 |
| 25:2 National Science Stds: UCP.1, UCP.2, UCP.3, UCP.4, UCP.5, C.1 | 4. **Compare** development in frogs, birds, and reptiles. | 6, 9, 10 | 13, 20 | 4 | 19, 20, 26, 30 |
| | 5. **Explain** how the needs of a chick embryo are met. | 7, 8 | 12, 16, 23 | 1, 2, 6 | 4, 13, 16, 21 |
| 25:3 National Science Stds: UCP.1, UCP.3, UCP.4, C.6 | 6. **Sequence** the stages of frog metamorphosis. | 11, 13 | 3, 10, 17, 21 | 24, 28 | 3, 28 |
| | 7. **Explain** incomplete metamorphosis. | 15 | 4, 10, 11, 22 | 12, 13, 14, 28 | 5, 14 |
| | 8. **Describe** complete metamorphosis. | 12 | 5, 10, 22 | 7, 8, 9, 10, 11, 26, 28 | 2, 3, 10 |

# Animal Development

## CHAPTER OVERVIEW

### Key Concepts

In this chapter, students will study the growth, development, and birth process of an embryo inside the body of a female animal. A human embryo is used as the representative example. Students will study growth and development outside the body using the chicken as an example. In the last sections, students will study both complete and incomplete animal metamorphosis.

### Key Science Words

amniotic sac
cleavage
complete metamorphosis
fetus
incomplete metamorphosis
labor
larva
metamorphosis
navel
nymph
placenta
pupa
umbilical cord

### Skill Development

In Lab 25-1, students will **measure in SI, use numbers, infer,** and **make and use graphs** to judge the age of a human fetus. In Lab 25-2, students will **interpret data, measure in SI,** and **compare** complete and incomplete metamorphosis. In the Skill Check on page 526, students will **interpret diagrams** as they study the path of oxygen from mother to embryo. On page 535, students will **understand** the **science word** *metamorphosis.* In the Mini Lab on page 533, students will **design an experiment** to test whether substances can pass through an egg shell.

### Bridging

Chapter 24 covered the events leading to fertilization of an egg. This chapter continues the story by explaining the changes that occur to a fertilized egg as it undergoes development into a fetus.

**CHAPTER PREVIEW**

### Chapter Content

Review this outline for Chapter 25 before you read the chapter.



### Skills in this Chapter

The skills that you will use in this chapter are listed below.

- In **Lab 25-1,** you will measure in SI, use numbers, infer, and make and use graphs. In **Lab 25-2,** you will interpret data, measure in SI, and compare.
- In the **Skill Checks,** you will interpret diagrams and understand science words.
- In the **Mini Labs,** you will design an experiment.

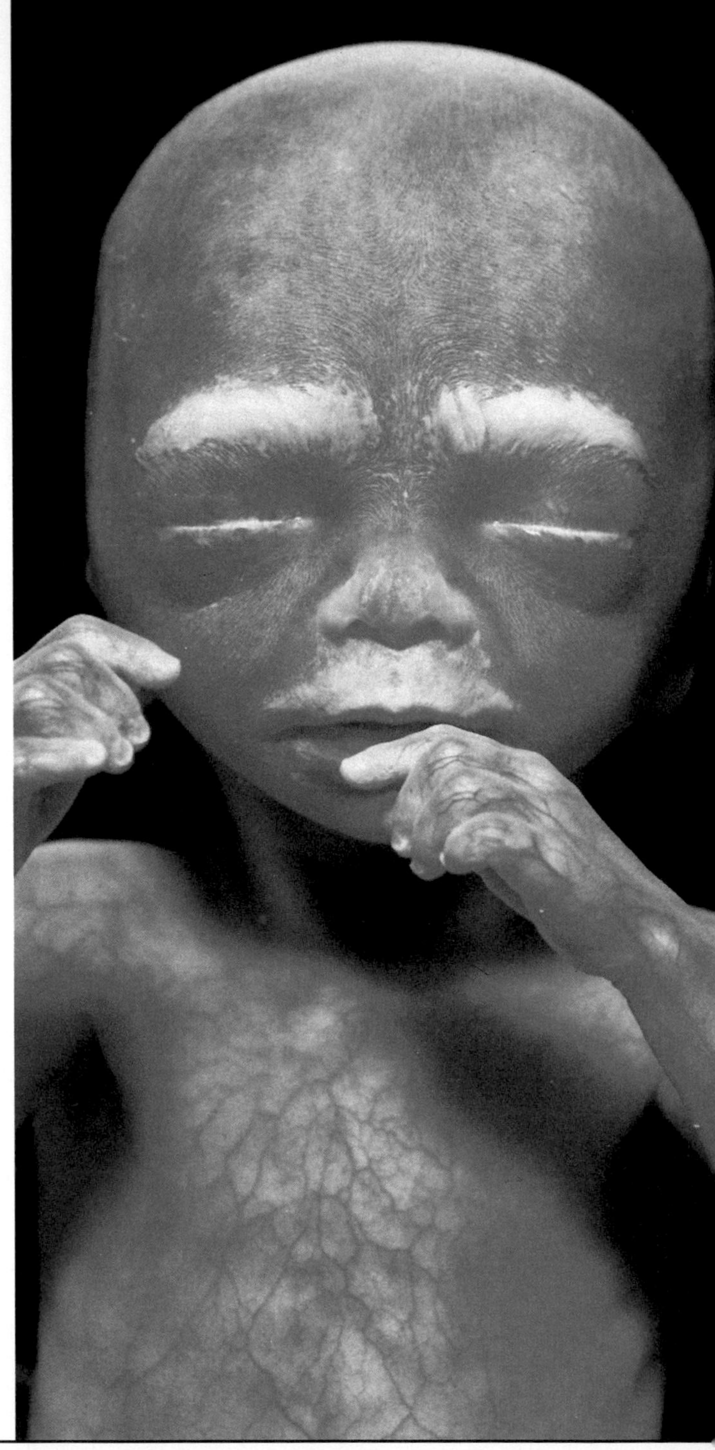

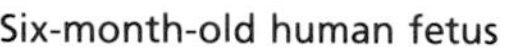

Six-month-old human fetus

## TECH PREP

For Tech Prep activities in this chapter of the Teacher Wraparound Edition, see especially the Activity on page 524, the Connection and Portfolio on page 528, the Motivation on page 536, and the Convergent Question on page 537.

See also the Glencoe Homepage at **www.glencoe.com**

Chapter 25

# Animal Development

A human female carries an unborn child for about 266 days, approximately 9 months. Other animals carry their young different lengths of time. Rats carry their young for only 22 days. In cats and dogs, the length of time for carrying young is about 60 days. In horses, it is about 336 days. Some animals don't carry their young at all. The young develop outside the female's body. Birds, frogs, and insects are some animals that develop outside the female.

The photo on the left shows a six-month-old unborn human as it appears within the mother's body. Has it formed all the parts it will need at birth? How does it get food and oxygen? You will discover the answers to these questions as you read this chapter.

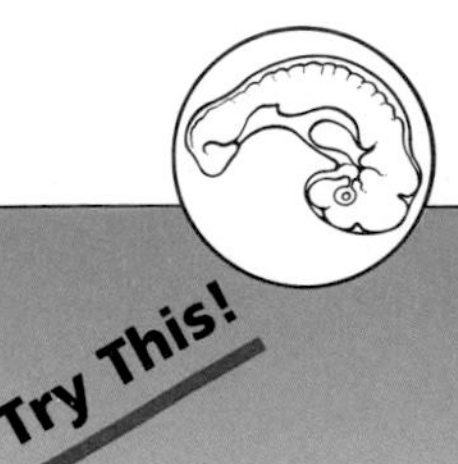

**Try This!**

**What changes take place during birth?** Use a compass to draw a circle 2 mm in diameter. This circle is the usual size of the opening to the uterus. Draw a circle 98 mm in diameter. This circle is the size of the opening to the uterus during birth. Why is this change in size needed?

**interNET CONNECTION**

For more information about the material in this chapter, follow the link for the chapter on the Glencoe Homepage at **http://www.glencoe.com**

523

**ASSESSMENT PLANNER**

**Portfolio**

Strategies on the following pages represent student products that can be placed into a best-work portfolio: pp. 528, 534.

**PERFORMANCE ASSESSMENT**

Skill Check, pp. 526, 535
Mini Lab, p. 532
Lab, pp. 531, 534

**CONTENT ASSESSMENT**

Check for Understanding, pp. 529, 530, 533, 537
Chapter Review, pp. 538-539

**GROUP ASSESSMENT**

Opportunities for group assessment occur with Cooperative Learning Strategies.

**Student Journal**

Strategies on the following pages represent opportunities for writing in a Student Journal: pp. 525, 536.

GETTING STARTED

## Using the Photos

In looking at the photos, students should try to identify the body parts that are visible in the embryo. Students may be surprised to find so many body parts already formed. Emphasize that body parts continue to grow and develop until birth.

## MOTIVATION/Try This!

**What changes take place during birth?** Upon completing the chapter-opening activity, students will have **formulated models.** The opening of the uterus must enlarge to accommodate the fetus during birth. The uterus, being smooth muscle, slowly changes in size during labor.

## Chapter Preview

Have students study the chapter outline before they begin to read the chapter. They should note the overall organization of the chapter, that the first topic described deals with development inside the body of a female animal. This is followed by development occurring outside the body of a female animal. Finally, the various changes taking place among frogs and insects during metamorphosis is covered.

## Misconception

Blood between mother and fetus does not mix directly during development. The placenta does connect fetus to mother, and the placenta contains a rich supply of blood vessels as does the uterus. However, the capillary networks of both organs (uterus and placenta) do not connect directly. Transfer of materials between placenta and uterus or vice versa occurs via diffusion. Molecules of food, oxygen, and blood proteins such as antibodies can diffuse across this barrier but blood cells do not.

# 25:1 Development Inside the Female

## PREPARATION

### Materials Needed

Make copies of the Transparency Master, Reteaching, Study Guide, Enrichment, and Lab worksheets in the *Teacher Resource Package*.

▶ Obtain clay for the Focus activity.

### Key Science Words

cleavage
amniotic sac
placenta
umbilical cord
fetus
labor
navel

### Process Skills

In the Skill Check, students will understand diagrams. In Lab 25-1, students will measure in SI, use numbers, infer, and make and use graphs.

## 1 FOCUS

▶ The objectives are listed on the student page. Remind students to preview these objectives as a guide to this numbered section.

### ACTIVITY/Hands-On

Provide each student with a small wad of clay about the size of a pea. Ask them to divide the clay into two equal halves, then each half into two equal halves, then again, and again so that they have sixteen pieces. The original wad of clay represents the mass of a fertilized egg. Each division into two equal halves represents the separation of cell contents during mitosis. Students will note that the number of cells increases but the mass of cellular material does not.

### Guided Practice

Have students write down their answers to the margin question: when it undergoes mitosis to make two cells.



## 25:1 Development Inside the Female

**Objectives**

1. **Sequence** the changes that take place in a fertilized egg until it attaches to the uterus.
2. **Explain** how the needs of an embryo are met as it develops.
3. **Describe** the stages of human development from the first month until birth.

**Key Science Words**

cleavage
amniotic sac
placenta
umbilical cord
fetus
labor
navel

You began as a fertilized egg and are now a young adult. You have gone through many changes. Many of these changes took place while you were still in your mother's body.

### Early Stages of Development

All living things go through development. You learned in Chapter 2 that development is all the changes that occur in a living thing as it grows. In humans, development begins when a single fertilized egg changes into many different body parts or tissues. Development results in the forming of bone, muscle, hair, nerve, and skin tissues.

What are the changes that can be seen in humans as they develop? In order to answer this question we must start with a single fertilized egg. The fertilized egg undergoes changes that take it from a one-celled stage to a many-celled stage. These changes are largely the result of mitosis.

Most of the early changes take place while the egg is still within the female's oviduct. Follow the numbered steps of Figure 25-1 as you continue to read. Also pay attention to how long it takes for each change to take place.

1. An egg is fertilized.
2. The fertilized egg undergoes mitosis to make two cells. It is now an embryo. How long did this step take?

**When does a fertilized egg become an embryo?**

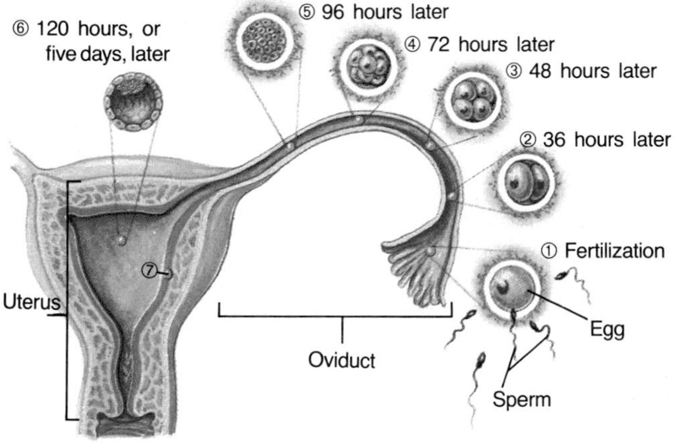


**Figure 25-1** The fertilized egg goes through cleavage in the oviduct. How long does it take for the embryo to become a solid ball of cells? 96 hours



## OPTIONS

### Science Background

In humans, during the weeks of early development, the formation of the three embryonic germ layers has begun to form with ectoderm, endoderm, and mesoderm now being present. Ectoderm forms the skin, nervous system, retina, cornea, lens, and inner ear. Endoderm forms the liver, pancreas, thyroid, and linings of the digestive and respiratory systems. Mesoderm forms muscles, bone, connective tissue, blood and blood vessels, kidneys, and reproductive organs.

3. Mitosis continues and two cells form four cells. Four cells then form eight. Eight cells will form sixteen cells.
4. Three days later, the embryo is sixteen cells in size.
5. Next, the embryo forms a solid ball of cells. Has it left the oviduct yet?
6. Finally, after five days (120 hours), the embryo moves out of the oviduct into the uterus. By this time, it is a hollow ball of cells.
7. Next, the embryo attaches itself to the lining of the uterus. This is where it develops for the next 37 weeks.

These early changes are called cleavage (KLEE vihj). **Cleavage** is the series of changes that take place to turn one fertilized egg into a hollow ball of many cells. The number of cells increases from one to about one hundred in just five days. The word *cleave* means to chop or divide. What divides during cleavage?

All the changes that take place during cleavage occur within the body of the female. Many animals, such as humans, cats, and whales, show this type of development.

## Needs of the Embryo

Embryos, just like adults or newborns, have certain needs. These needs include protection, food, oxygen, and getting rid of wastes. How are these needs met?

**Protection**

The embryo must be protected against injury. An amniotic (am nee AHT ik) sac forms around it as shown in Figure 25-2. An **amniotic sac** is a tissue filled with liquid that protects the embryo. The liquid around the embryo cushions it in the same way as packing material cushions the contents of a package.

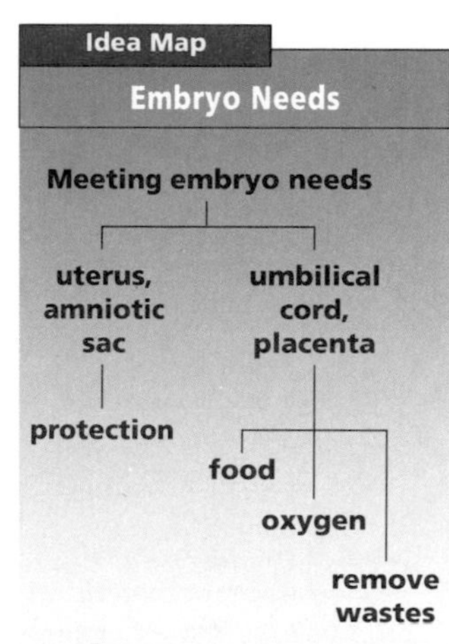


**Study Tip:** Use this idea map as a study guide to development inside the female. How are the needs of the embryo met?

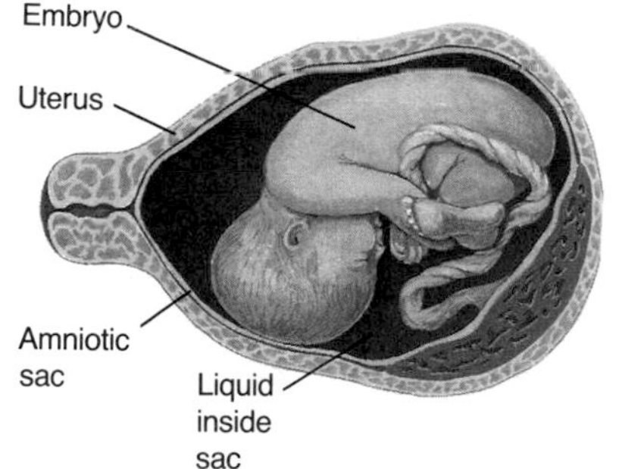


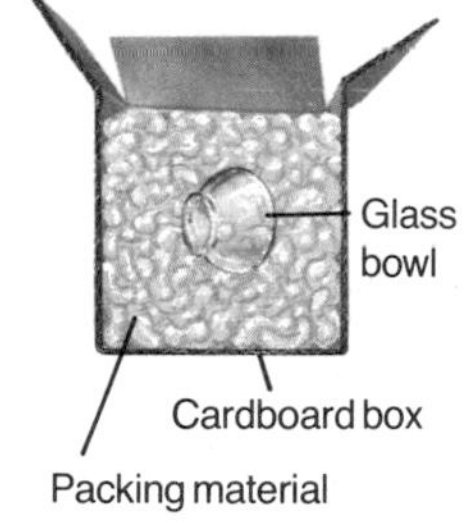


**Figure 25-2** The amniotic sac protects the embryo, much like the packing material protects the bowl.



# 2 TEACH

### MOTIVATION/Audiovisual

Show the video *The Course of Development*, Human Relations Media.

### Concept Development

▶ Make sure that students read the numbered statements and match them with the numbers appearing in Figure 25-1.

▶ If Chapter 24 was not studied, it may be helpful to review the anatomy of the female human reproductive system. Emphasis should be placed on the ovary, oviducts, and uterus.

▶ Point out that growth results from an increase in cells. Development refers to the actual changes responsible for forming new tissues.

▶ You may want to preview the four needs by simply listing them on the chalkboard before studying each one in detail. Or, try to elicit them from your students by asking what the needs might be.

**Using Science Words:** The word *embryo* in Greek means to swell. When a female becomes pregnant, a new human begins to grow from a fertilized egg. As it grows, there is a swelling of the female.

### Idea Map

Have students use the idea map as a study guide to the major concepts of this section.

### Student Journal

Set up the following scene for students: You are standing on top of an oviduct when you see an egg released from the ovary and move along the oviduct toward the uterus. No sperm are present. Report in writing what you observe. One month later, you see an egg released and this time sperm are present. Report in writing what you observe. Assume that you are standing on the oviduct for at least seven days.

**TRANSPARENCY 127**

TRANSPARENCY MASTER CHAPTER 25

CLEAVAGE OF A FERTILIZED EGG

## OPTIONS

**Cleavage**/*Transparency Package,* number 25a. Use color transparency number 25a as you teach early changes in a human embryo.

**Transparency Master**/*Teacher Resource Package,* p. 127. Use the Transparency Master shown here to teach cleavage of a fertilized egg.

# TEACH

## Concept Development

▶ Emphasize that no mixing of fetal and maternal blood occurs. Gases, food, and wastes pass through the placental barrier by means of diffusion.

▶ As a review, you may want to ask students why oxygen is needed by all cells. Terms such as *respiration, cellular respiration,* or *energy release* would be most acceptable.

▶ You may wish to relate ideas learned in Chapter 9 on nutrition to this section. Ask students the form of food as it passes through the placenta. (It is digested in the form of glucose, fatty acids, amino acids.)

## Bridging

Remind students that certain drugs and disease-causing organisms can also pass from the blood of mother to their embryos. This is why children can be born with AIDS, drug addiction, or syphilis.

## Skill Check

Students should follow the blue arrows in Figure 25-3 from the mother's blood in the uterus to the embryo.

## MOTIVATION/Brainstorming

Ask students to speculate on the kinds of problems that a fetus or embryo would face if for some reason the placenta did not make proper contact with the mother's uterus.

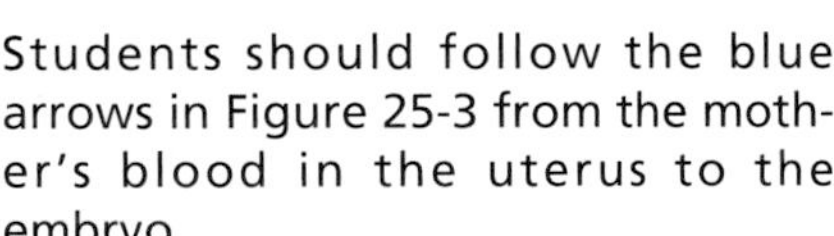

NATIONAL GEOGRAPHIC SOCIETY

**Videodisc**

**STV: Human Body Vol. 3**
*Reproductive Systems*
Unit 2, Side 2
*Embryonic Development*

**23244-31609**

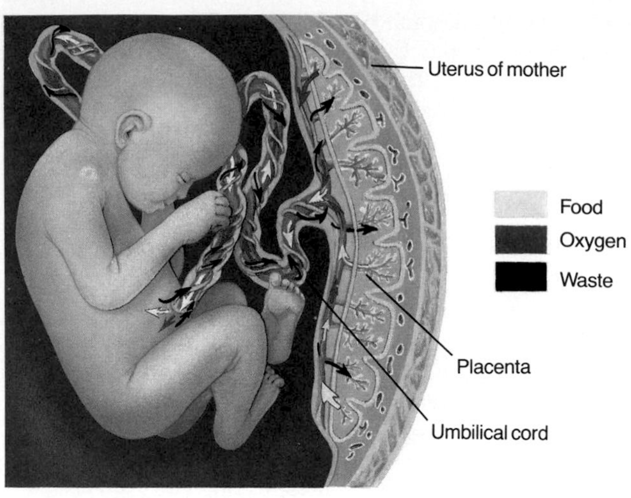


**Figure 25-3** The embryo receives food and oxygen and gets rid of wastes through the placenta.

### Skill Check

**Interpret diagrams:** Study Figure 25-3. What is the pathway of oxygen from mother to embryo? *For more help, refer to the **Skill Handbook,** pages 706-711.*

### Food

How does an embryo get food? Certainly it doesn't eat while inside the amniotic sac. Yet, it must be supplied with food in order to grow.

Before a human is born, food is supplied by the mother. The embryo's food is in the mother's blood. This food comes from food that the mother eats.

How does the food get from the mother's blood to the embryo? Food passes from an organ called the placenta (pluh SENT uh) to the embryo. The **placenta** is an organ that connects the embryo to the mother's uterus. The placenta has many blood vessels in it. Food passes from the mother's blood through the placenta into a cord, called the umbilical (uhm BIL ih kuhl) cord. The **umbilical cord** is made of blood vessels that connect the embryo to the placenta.

Figure 25-3 shows the placenta and umbilical cord in a human embryo. Notice the yellow arrows. They show the direction of food from mother to embryo.

### Oxygen

The embryo receives oxygen the same way as it receives food. Oxygen passes from the mother's blood in the uterus to the blood in the placenta. Once in the capillaries of the placenta, oxygen passes into the blood vessels of the umbilical cord. These blood vessels connect directly to the blood vessels of the embryo. Oxygen is then passed to the embryo. Figure 25-3 shows this pathway with blue arrows.



## OPTIONS

### Science Background

The umbilical cord consists of three blood vessels connecting fetus to placenta. These vessels include two veins and one artery. The vein brings oxygenated blood to the fetus, and the arteries carry away deoxygenated blood. The placenta provides a large surface area for exchange of materials between fetal and maternal circulatory systems.

## Career Close-Up

### Ultrasound Technician

**A woman is pregnant and her doctor suggests an ultrasound of the fetus. What actually is ultrasound? Who will take the ultrasound?**

**An ultrasound technician operates a machine that can take pictures of any organ inside the body. The machine uses sound waves that reflect off body organs. These sound waves are changed into pictures that allow a doctor to see inside the body. Unlike X rays, ultrasound pictures show movement. Ultrasound is also safer than X rays.**

**Ultrasound pictures will show if a woman is carrying twins. They help the doctor judge the age of the fetus. In this way, the doctor can give a better prediction as to when the child will be born. Abnormal development or birth defects may also be detected. Finally, the sex of the fetus can be determined.**

**To be an ultrasound technician a person needs a high school education. Two or more years of special training may be needed. After the proper education and training, most ultrasound technicians work in hospitals.**

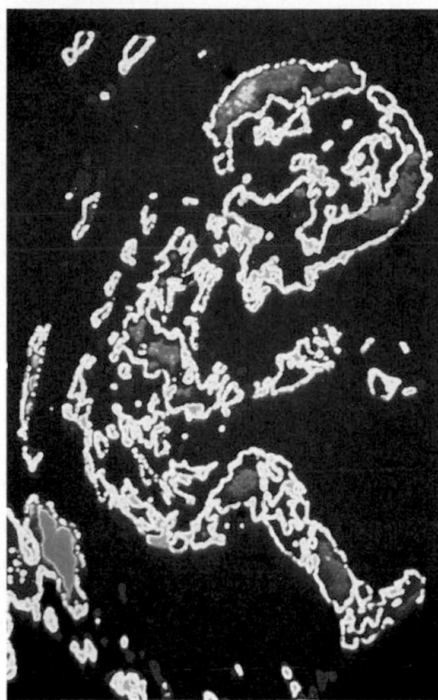

Ultrasound shows what a developing fetus looks like.

### Wastes

The cells of the embryo produce wastes. Carbon dioxide and urea are two such waste chemicals. The cord and placenta help get rid of these wastes. The embryo's blood carries these wastes from the embryo to the placenta, where the wastes enter the mother's bloodstream. Figure 25-3 shows this pathway. In which direction are the black arrows that carry wastes going? Why?

**How does an embryo get rid of wastes?**

A pregnant female eats for herself and the developing embryo. She also breathes for herself and the embryo. She must get rid of her wastes and those of the embryo.

Now you see why proper nutrition is important during pregnancy. Good health is also very important at this time.



## STUDY GUIDE 145

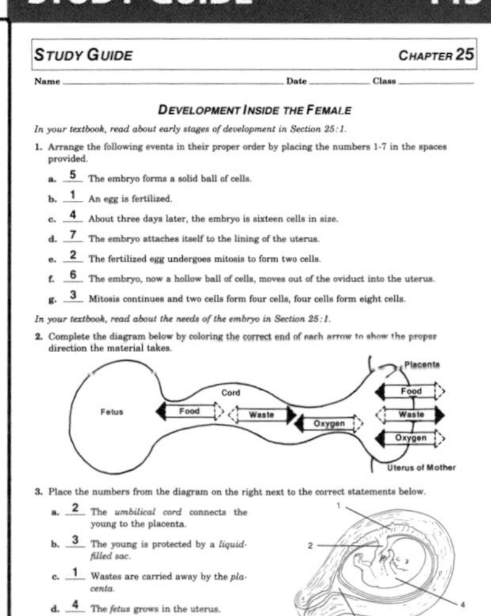

STUDY GUIDE CHAPTER 25

Name ______ Date ______ Class ______

DEVELOPMENT INSIDE THE FEMALE

*In your textbook, read about early stages of development in Section 25:1.*

1. Arrange the following events in their proper order by placing the numbers 1-7 in the spaces provided.
   a. 5 The embryo forms a solid ball of cells.
   b. 1 An egg is fertilized.
   c. 4 About three days later, the embryo is sixteen cells in size.
   d. 7 The embryo attaches itself to the lining of the uterus.
   e. 2 The fertilized egg undergoes mitosis to form two cells.
   f. 6 The embryo, now a hollow ball of cells, moves out of the oviduct into the uterus.
   g. 3 Mitosis continues and two cells form four cells, four cells form eight cells.

*In your textbook, read about the needs of the embryo in Section 25:1.*

2. Complete the diagram below by coloring the correct end of each arrow to show the proper direction the material takes.



3. Place the numbers from the diagram on the right next to the correct statements below.
   a. 2 The *umbilical cord* connects the young to the placenta.
   b. 3 The young is protected by a *liquid-filled sac*.
   c. 1 Wastes are carried away by the *placenta*.
   d. 4 The *fetus* grows in the uterus.



## GLENCOE TECHNOLOGY

**Videodisc**

The Secret of Life

*Development Segment*

## Career Close-Up

### Ultrasound Technician

#### Background

The technician directs high frequency sound waves by way of a transducer to a patient's body. The sound waves create an echo when reflected from body tissue, providing an image on a video monitor.

#### Related Careers

Many ultrasound technicians may also be certified as X-ray technicians, nuclear medicine technologists, or radiation therapy technologists.

#### Job Requirements

A two- or three-year course of study is required after high school graduation. Certification is then given by the American Registry of Radiologic Technologists.

#### For More Information

Society of Diagnostic Medical Sonographers, 12225 Greenville Avenue, Suite 434, Dallas, TX 75234, (214)-235-7367; or American Medical Association, Department of Allied Health Education and Accreditation, 515 North State Street, Chicago, IL 60610, (312)-464-5000

## TEACH

### Guided Practice

Have students write down their answers to the margin question: The embryo's blood carries wastes from the embryo and the placenta by way of the umbilical cord.

### Independent Practice

**Study Guide**/*Teacher Resource Package*, p. 145. Use the Study Guide worksheet shown at the bottom of this page for independent practice.

## TEACH

**Analogy:** Show objects having masses comparable to a fetus at the end of the second and third months. A small paper clip has a mass of about one gram. The length of the fetus from crown to rump after the third month is about the width of one's palm.

### Connection: Math

Provide students with calendars and give them an opportunity to determine some expected birth dates by listing the dates of a mother's last menstrual period. Have them add 280 days (40 weeks) to the first day of the last menstrual period. A second method of calculating a birth date is to count back three calendar months from the first day of the last menstrual period and then add one week. You may want students to try both methods to see if they end with the same delivery date.

### Guided Practice

Have students write down their answers to the margin question: two months.

**Portfolio**

Tell students to prepare a time line that shows the major events occurring during human development for weeks 1, 4, 8, 12, 16, and 38 (birth). The time line must also: (a) indicate when the change from embryo to fetus occurs, (b) include the approximate size and mass during each week shown.

**Videodisc**

**STV: Human Body Vol. 3**
*Reproductive Systems*
Unit 2, Side 2
*Fetal Development*

**31611-36800**

**Bio Tip**

**Health:** A fetus born at 20 weeks of development has about a 5 percent chance of survival.

## Human Development

What changes occur as the embryo continues to develop? When do these changes take place? Let's look at the human embryo to help answer these questions.

### First Month

Many changes take place during the first month. Figure 25-4a shows what the embryo looks like at one month. The stomach, brain, and heart start to form. Liver and ears begin to appear. The thyroid gland forms. The heart is already beating. It can be seen as a bulge just in front of the arms.

You can even see the start of eye development. The dark circle in the head will be the embryo's eye. Fingers can be seen. Can you see legs at this point in development?

**At what age does an embryo begin to move?**

How large is the embryo at this stage? To see how large the embryo is, measure the small figure in the box in millimeters. This figure is an embryo drawn to actual size. The embryo has a mass of 0.02 grams.

### Second Month

During the second month, arms and legs form. Fingers and toes appear. The eyes have eyelids. The mouth has lips and the nose has nostrils. Figure 25-4b shows these parts. Muscles and bone form. The embryo can even move at this age.

**Figure 25-4** A one-month-old embryo (a) and a two-month-old embryo (b) are shown enlarged and actual size.

a

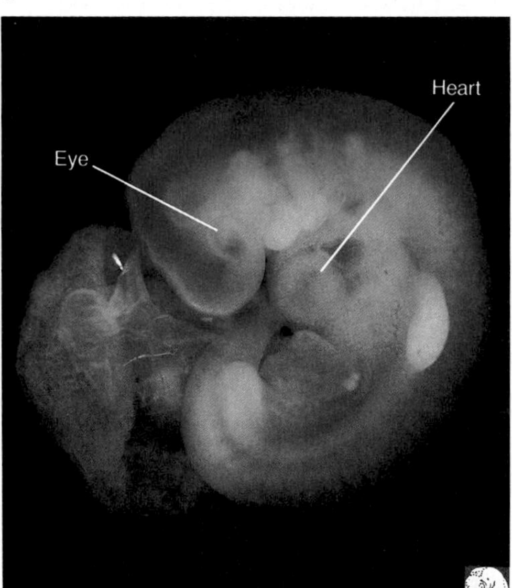


b

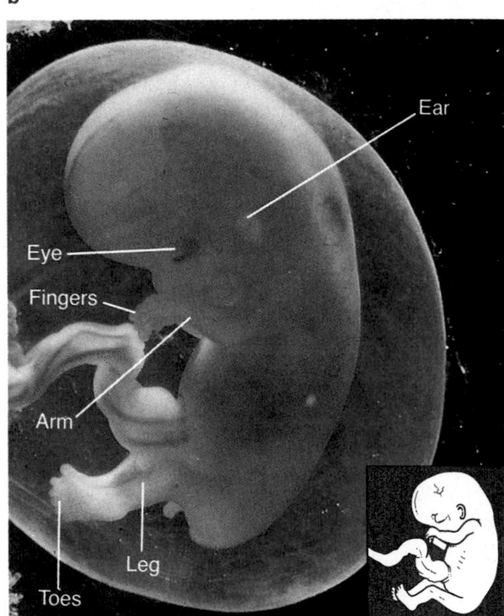

### OPTIONS

**Challenge:** Have your students investigate at what age a fetus may live if born prematurely. An age of 22 weeks is usually given as the age after which there is some chance for survival. However, fetuses born during the 26th to 28th weeks of development still have a difficult time surviving.

**Enrichment**/*Teacher Resource Package*, p. 27. Use the Enrichment worksheet shown here for students to compare weight in different stages of mammal development.

**ENRICHMENT 27**

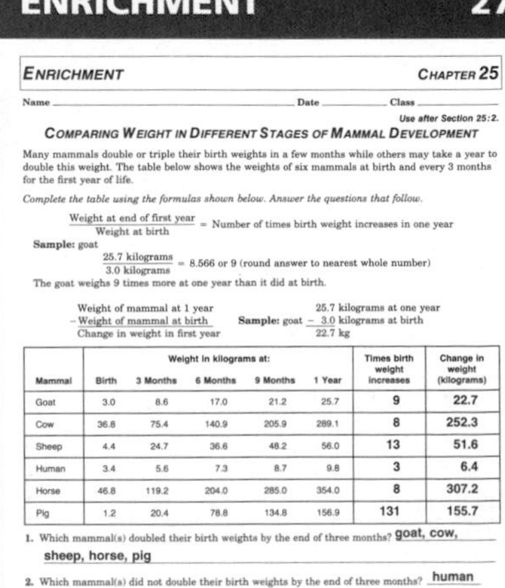

ENRICHMENT CHAPTER 25

Name ______ Date ______ Class ______

*Use after Section 25:2.*

COMPARING WEIGHT IN DIFFERENT STAGES OF MAMMAL DEVELOPMENT

Many mammals double or triple their birth weights in a few months while others may take a year to double this weight. The table below shows the weights of six mammals at birth and every 3 months for the first year of life.

*Complete the table using the formulas shown below. Answer the questions that follow.*

$$\frac{\text{Weight at end of first year}}{\text{Weight at birth}} = \text{Number of times birth weight increases in one year}$$

**Sample:** goat

$$\frac{25.7 \text{ kilograms}}{3.0 \text{ kilograms}} = 8.566 \text{ or } 9 \text{ (round answer to nearest whole number)}$$

The goat weighs 9 times more at one year than it did at birth.

Weight of mammal at 1 year
− Weight of mammal at birth
Change in weight in first year

**Sample:** goat 25.7 kilograms at one year − 3.0 kilograms at birth = 22.7 kg

| Mammal | Weight in kilograms at: Birth | 3 Months | 6 Months | 9 Months | 1 Year | Times birth weight increases | Change in weight (kilograms) |
|---|---|---|---|---|---|---|---|
| Goat | 3.0 | 8.6 | 17.0 | 21.2 | 25.7 | 9 | 22.7 |
| Cow | 36.8 | 75.4 | 140.9 | 205.9 | 289.1 | 8 | 252.3 |
| Sheep | 4.4 | 24.7 | 36.6 | 48.2 | 56.0 | 13 | 51.6 |
| Human | 3.4 | 5.6 | 7.3 | 8.7 | 9.8 | 3 | 6.4 |
| Horse | 46.8 | 119.2 | 204.0 | 285.0 | 354.0 | 8 | 307.2 |
| Pig | 1.2 | 20.4 | 78.8 | 134.8 | 156.9 | 131 | 155.7 |

1. Which mammal(s) doubled their birth weights by the end of three months? goat, cow, sheep, horse, pig
2. Which mammal(s) did not double their birth weights by the end of three months? human
3. Which mammal weighed the most at the end of the first year? horse
4. Which mammal weighed the least at the end of the first year? human
5. Which mammal increased its birth weight the greatest number of times in one year? pig
6. Which mammal increased its birth weight the least number of times in one year? human

You cannot tell yet whether the embryo is a boy or a girl. Its mass is 1.0 gram. Measure the small figure in the box in millimeters. Has the embryo doubled in size from the first month?

**Third Month**

At the start of the third month, the embryo is called a fetus (FEET us). The **fetus** is an embryo that has all of its body systems. The organs are not complete, but they are all present. You can tell whether the fetus is a boy or a girl. Its mass is 14 grams. Figure 25-5 shows a life-size fetus at three months. Measure the length of the fetus in millimeters.

**Fourth Month to Birth**

By the fourth month, the fetus is rather active. It kicks, bends, and turns. The mother begins to feel it moving. By the end of the fourth month, the fetus is 160 millimeters long and has a mass of 100 grams.

Development goes on for the next five months. After a total of 38 weeks, or about 266 days, birth will take place. The fetus will be about 500 millimeters and 3000 grams. How much larger is it than it was at one month?

Not all animals develop in 38 weeks. Table 25-1 lists how long certain animals take to develop before they are born. Dogs take nine weeks to develop. Elephants take 84 weeks. Does length of time needed for development appear to be related to animal size? In what way?

**TABLE 25–1. TIME FOR DEVELOPMENT**

| Animal | Weeks for Development |
|---|---|
| Rat | 3 |
| Rabbit | 4 |
| Cat | 9 |
| Dog | 9 |
| Sheep | 21 |
| Bear | 28 |
| Gorilla | 36 |
| Human | 38 |
| Horse | 48 |
| Elephant | 84 |

Amniotic sac
Placenta
Umbilical cord

**Figure 25-5** This three-month-old fetus is shown actual size.



## TEACH

### Check for Understanding

Have students copy the following terms: cleavage, amniotic sac, placenta, umbilical cord, fetus, labor, and navel. Ask them to properly use each term in a sentence.

### Reteach

Have students explain how a fertilized egg, embryo, and fetus differ in terms of stage of development, size, and the degree of change that has taken place.

**Cooperative Learning:** Divide the class into 4 groups. Have each group prepare a short oral report on one of the following topics: Early changes in development, Early needs of the developing embryo, Human development, or Human birth.

Each group should prepare any needed charts or diagrams to help with their presentation. These charts should be placed on a transparency for viewing by the entire class. There should be a spokesperson for each group who will present the report to the rest of the class.

### Independent Practice

**Study Guide**/*Teacher Resource Package*, pp. 146-147. Use the Study Guide worksheet shown at the bottom of this page for independent practice.

**STUDY GUIDE 146**

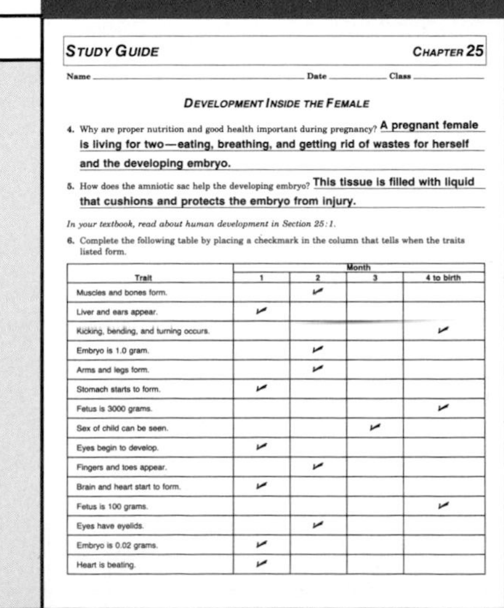

STUDY GUIDE — CHAPTER 25

Name ______ Date ______ Class ______

DEVELOPMENT INSIDE THE FEMALE

4. Why are proper nutrition and good health important during pregnancy? A pregnant female is living for two—eating, breathing, and getting rid of wastes for herself and the developing embryo.

5. How does the amniotic sac help the developing embryo? This tissue is filled with liquid that cushions and protects the embryo from injury.

*In your textbook, read about human development in Section 25:1.*

6. Complete the following table by placing a checkmark in the column that tells when the traits listed form.

| Trait | Month 1 | 2 | 3 | 4 to birth |
|---|---|---|---|---|
| Muscles and bones form. | | ✓ | | |
| Liver and ears appear. | ✓ | | | |
| Kicking, bending, and turning occurs. | | | | ✓ |
| Embryo is 1.0 gram. | | ✓ | | |
| Arms and legs form. | | ✓ | | |
| Stomach starts to form. | ✓ | | | |
| Fetus is 3000 grams. | | | | ✓ |
| Sex of child can be seen. | | | ✓ | |
| Eyes begin to develop. | ✓ | | | |
| Fingers and toes appear. | | ✓ | | |
| Brain and heart start to form. | ✓ | | | |
| Fetus is 100 grams. | | | | ✓ |
| Eyes have eyelids. | | ✓ | | |
| Embryo is 0.02 grams. | ✓ | | | |
| Heart is beating. | ✓ | | | |

**STUDY GUIDE 147**

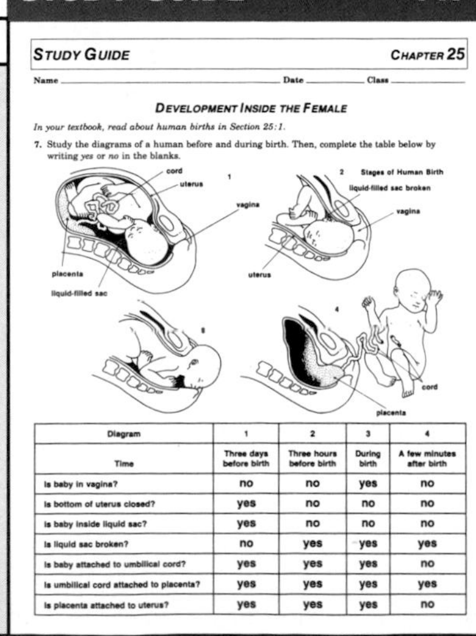

STUDY GUIDE — CHAPTER 25

Name ______ Date ______ Class ______

DEVELOPMENT INSIDE THE FEMALE

*In your textbook, read about human births in Section 25:1.*

7. Study the diagrams of a human before and during birth. Then, complete the table below by writing *yes* or *no* in the blanks.



| Diagram | 1 | 2 | 3 | 4 |
|---|---|---|---|---|
| Time | Three days before birth | Three hours before birth | During birth | A few minutes after birth |
| Is baby in vagina? | no | no | yes | no |
| Is bottom of uterus closed? | yes | no | no | no |
| Is baby inside liquid sac? | yes | no | no | no |
| Is liquid sac broken? | no | yes | yes | yes |
| Is baby attached to umbilical cord? | yes | yes | yes | no |
| Is umbilical cord attached to placenta? | yes | yes | yes | yes |
| Is placenta attached to uterus? | yes | yes | yes | no |

## OPTIONS

**How Does a Human Fetus Change During Development?**/*Lab Manual*, pp. 211-214. Use this lab as an extension to studying early human development.

## TEACH

### Check for Understanding

Have students respond to the first three questions in Check Your Understanding.

### Reteach

**Reteaching**/*Teacher Resource Package,* p. 75. Use this worksheet to give students additional practice with the vocabulary in this section.

**Extension:** Assign Critical Thinking, Biology and Reading, or some of the **OPTIONS** available with this lesson.

## 3 APPLY

### Demonstration

Use an X ray of a female pelvis to point out the opening in the pelvis (pelvic inlet).

## 4 CLOSE

### Software

*The Reproductive System: A Baby Is Born,* Queue, Inc.

### Answers to Check Your Understanding

1. (a) a change in form or appearance
   (b) inside the oviduct
2. amniotic sac
3. (a) the start of the third month (b) by end of first month (c) by third month (d) by third month
4. CO binds with blood cells so that they can't carry oxygen. The embryo can suffer brain damage because of lack of oxygen to the brain.
5. The embryo is a hollow ball of cells that attaches itself to the lining of the uterus when it leaves the oviduct.

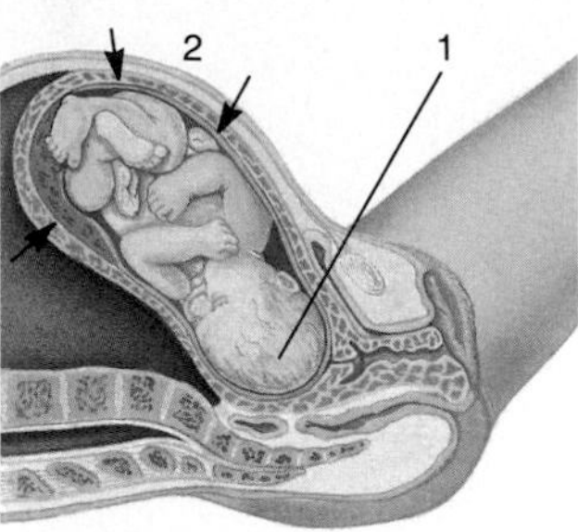


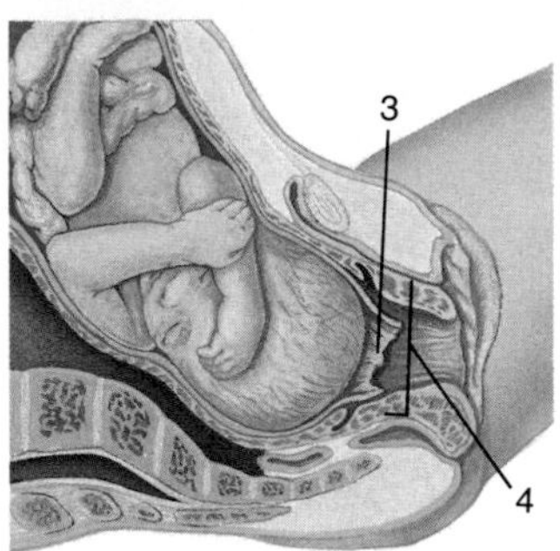


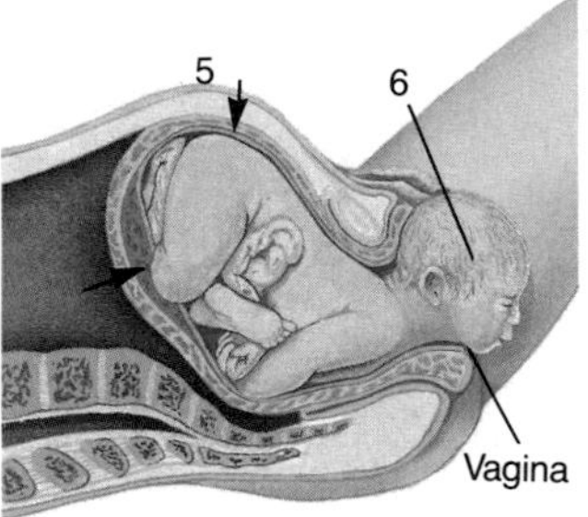


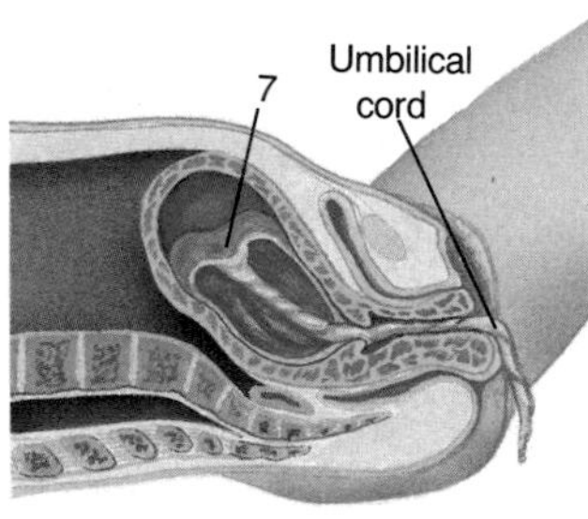


**Figure 25-6** Study the steps of human birth shown here as you read the text.

## Human Birth

Birth has one main purpose. It must push the fetus out of the uterus through the vagina. Follow the steps in Figure 25-6 as you read.

1. Before birth, the fetus usually turns head down. It will be born head first.
2. The uterus begins to contract. The uterus is smooth muscle. Is smooth muscle voluntary or involuntary? involuntary Contractions of the uterus are called **labor.**
3. The amniotic sac breaks. Liquid within the sac passes out through the mother's vagina.
4. The opening at the bottom of the uterus begins to widen. The fetus's head begins to move into this opening. Steps 2, 3, and 4 may take 8 to 12 hours.
5. More contractions take place. They help to push the fetus through the mother's vagina. Steps 4 and 5 might take several hours.
6. The fetus, or baby, is born.
7. The last step is getting rid of the placenta. The uterus contracts and pushes out the placenta and the amniotic sac. At this stage, the placenta is often called the afterbirth.

One evidence of your birth that still exists is your navel. Your **navel** is where the umbilical cord was attached to your body.

### Check Your Understanding

1. (a) What is development?
   (b) Where does early development take place in humans?
2. What part protects the embryo?
3. At what age:
   (a) is an embryo called a fetus?
   (b) does the heart begin to beat?
   (c) can you tell if the embryo is male or female?
   (d) are all organs present?
4. **Critical Thinking:** If a pregnant woman smokes, carbon monoxide can pass through the placenta to the fetus. What effect would this have on the fetus? (HINT: In Chapter 13, you studied the effects of carbon monoxide on the body.)
5. **Biology and Reading:** What is the shape of the embryo at the time it moves out of the oviduct and into the uterus? What happens to the embryo after it leaves the oviduct?



## OPTIONS

**What Changes Occur During Birth?**/*Lab Manual,* pp. 215-218. Use this lab as an extension to studying human development and birth.

### RETEACHING 75

**RETEACHING** CHAPTER 25

Name ______ Date ______ Class ______

Use with Section 25:1.

**VOCABULARY PUZZLE**

*The terms listed below are in Section 25:1 of your textbook. Unscramble each term and write a definition that relates to the way the term is used in Section 25:1.*

1. tidocuv — oviduct — where the embryo first develops
2. milabucil odcr — umbilical cord — contains blood vessels that connect the embryo to the placenta
3. topdelenemv — development — change in form or appearance
4. tonicima csa — amniotic sac — a tissue filled with liquid that protects the embryo
5. sefut — fetus — an embryo that has all its body systems
6. suture — uterus — where the fetus develops
7. vnale — navel — where the umbilical cord was attached to the body
8. eagcavel — cleavage — series of changes that take place to turn a fertilized egg into a ball of cells
9. barlo — labor — contractions of the uterus
10. letapanc — placenta — organ that connects the embryo to the mother's uterus

*Write sentences that show you understand the meaning of the following words.*

11. embryo — The embryo begins to develop in the oviduct and continues developing in the lining of the uterus.
12. placenta — Food and oxygen pass from the mother's blood through the placenta to the fetus.

## Lab 25–1 Development

### Problem: How do you judge the age of a human fetus?

#### Skills

measure in SI, use numbers, infer, make and use graphs

#### Materials

ruler

#### Procedure

1. Copy the data table below.
2. **Measure in SI:** The figure shows three fetuses 1/10 actual size. Measure the body length and foot length for each fetus in millimeters. Write the measurements in your data table.
3. Fill in all but the last column of your data table.
4. The data table on this page lists ages of a fetus. It also shows the average size of a fetus at each of these ages. Use the table and your measurements to determine the ages of fetuses A, B, and C.

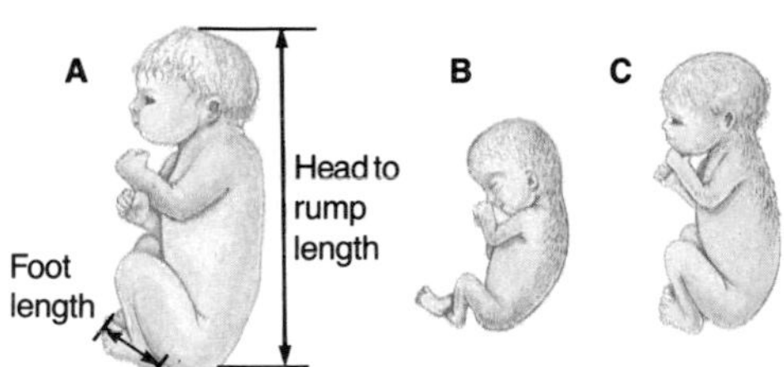


#### Data and Observations

1. At what age does the fetus open its eyes?
2. How much does the fetus grow between weeks 20 and 28?

| AGE IN WEEKS | AVERAGE BODY LENGTH (mm) | AVERAGE FOOT LENGTH (mm) | EYES | HAIR |
|---|---|---|---|---|
| 16 | 140 | 27 | closed | none |
| 20 | 210 | 40 | closed | body, head |
| 28 | 270 | 60 | open | body, head |
| 36 | 340 | 80 | open | head |

#### Analyze and Apply

1. How does a fetus obtain food and oxygen?
2. **Infer:** Could ultrasound tell you the sex of a fetus at six weeks of age? Why?
3. **Apply:** Assume you are an ultrasound technician and you report the age of fetus C to the doctor. The doctor tells you that the fetus should be 34 weeks old. The mother drinks a lot of alcohol. How would this affect the size of her fetus?

#### Extension

**Make a line graph** of the data for average body length.

| FETUS | BODY LENGTH | | FOOT LENGTH | | EYES | HAIR | AGE OF FETUS (WEEKS) |
|---|---|---|---|---|---|---|---|
| | MEASURED | ACTUAL (×10) | MEASURED | ACTUAL (×10) | | | |
| A | 33 | 330 | 7.2 | 72 | open | head | 36 |
| B | 21 | 210 | 3.6 | 36 | closed | body, head | 20 |
| C | 27 | 270 | 5.4 | 54 | open | body, head | 28 |



## ANSWERS

### Data and Observations

1. 28 weeks
2. 60 mm

### Analyze and Apply

1. through the placenta
2. No, sex can't be determined until the third month.
3. Alcohol would prevent the fetus from growing as fast. Therefore, it would be smaller than normal.

## Lab 25-1 Development

### Overview

Students will determine the age of three fetuses by comparing their measurements to a table of standard sizes. They will also use traits such as presence of body hair and open or closed eyes.

**Objectives:** Upon completion of this lab, students will be able to (1) **determine** how data obtained from ultrasound can be used to judge the approximate age of a fetus, (2) **define** the role of the placenta, (3) **state** at what age in development one can establish the sex of the fetus.

**Time Allotment:** 20 minutes

### Preparation

▶ Make copies of the data table for students.

▶ Show a transparent ruler on the overhead projector to remind students which units are millimeters.

▶ Explain to students that the data chart asks them to multiply their measurements by ten because the figures are 1/10th actual size.

**Lab 25-1 worksheet**/*Teacher Resource Package,* pp. 97-98.

### Teaching the Lab

▶ **Troubleshooting:** Students will have to use their data and observations with information provided in order to determine age of diagrams. Measurements of head to rump and foot length will not match exactly with the data in the table. Students must match their values with the nearest values in the table.

### ✓ ASSESSMENT

**Skill:** Have students prepare a graph of the average foot length of a human fetus using the data provided in the table at the top of the page. Tell them to predict the approximate age of a fetus if foot length is: (a) 32 mm, (b) 45 mm, (c) 70 mm.

## 25:2 Development Outside the Female

### PREPARATION

#### Materials Needed

Make copies of the Skill, Study Guide, and Reteaching worksheets in the *Teacher Resource Package*.

▶ Purchase chicken eggs for the Focus activity.

#### Key Science Words

No new science words will be introduced in this section.

#### Process Skills

In the Mini Lab, students will design an experiment.

### 1 FOCUS

▶ The objectives are listed on the student page. Remind students to preview these objectives as a guide to this numbered section.

#### ACTIVITY/Hands-On

Provide students with a chicken egg to examine its outer shape, inner shell surface, and inner contents. Break the eggs open into shallow dishes.

#### Guided Practice

Have students write down their answers to the margin question: in water.

#### ✓ ASSESSMENT

**Content:** The contents of an egg can be blown out without destroying the shell. Punch a small hole in each end of the egg and blow into one opening. Seal the holes with wax or clay. Fill a small beaker with water and hold the egg under the water. Have students state what the bubbles seen rising from the shell are and what this indicates about the nature of the shell itself.

## 25:2 Development Outside the Female

**Objectives**

4. **Compare** development in frogs, birds, and reptiles.
5. **Explain** how the needs of a chick embryo are met.

All animals undergo development. The female, however, may not have a uterus. In this case, development must take place elsewhere.

### Eggs That Are Laid

A female frog lays eggs outside her body. Birds and reptiles do the same thing. These eggs undergo development in a way similar to that in humans. One difference is that for frogs, birds, and reptiles, development takes place outside the female's body.

**Where do frog eggs develop?**

Fish, amphibians, and many invertebrates lay their eggs in water. These eggs undergo cleavage after they are fertilized, just as the fertilized human egg did. Figure 25-7 shows photos of cleavage taking place in a frog egg. Photo (a) is of one cell. Photo (b) is of two cells. How many cells can you see in photos (c) and (d)? 8 and 16

Bird and reptile eggs go through cleavage stages just as frog and human eggs do. Development in birds and reptiles takes place on land, not in water. Eggs of birds and reptiles can develop on land because they have shells. The shell keeps the embryo from drying out.

### Needs of the Embryo

What are the needs of an embryo that undergoes development in an egg that is laid? Let's use a chicken egg as an example. A chick embryo must have food and oxygen. It must get rid of waste chemicals. It must also be protected against drying out and injury. How are these needs met? Look at Figure 25-8 as you read.

a

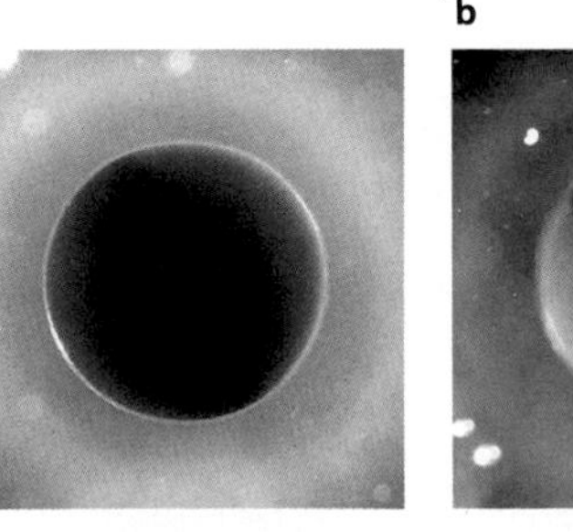

b

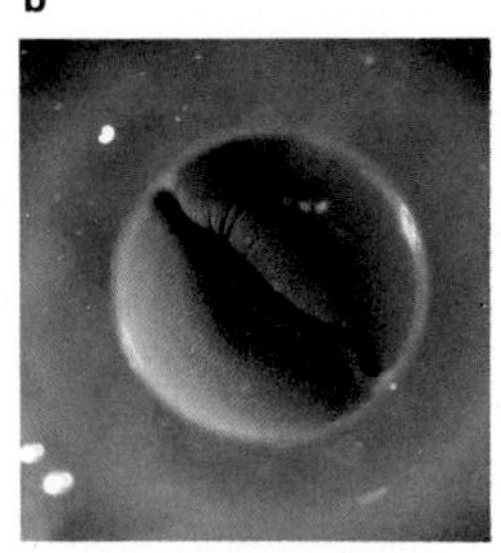

c

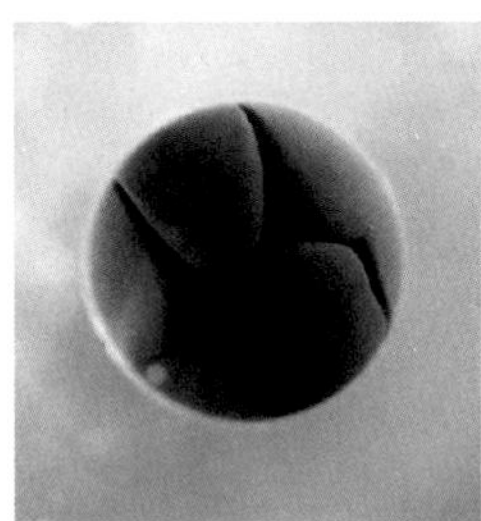

d

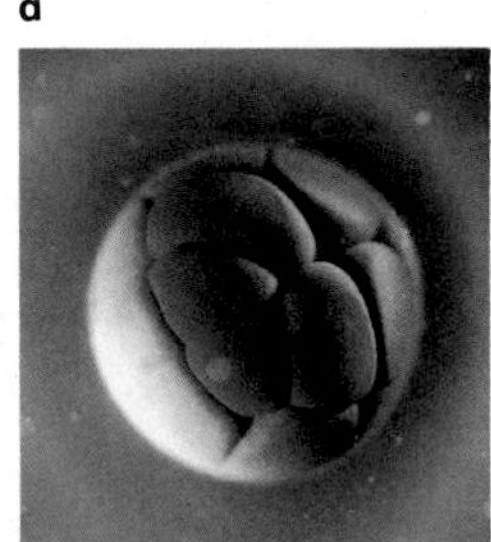

**Figure 25-7** A frog egg undergoes cleavage.



### OPTIONS

#### Science Background

Egg size in animals varies from the very large ostrich egg (200 mm) to the chicken egg (50 mm) to the mammalian egg (0.05 mm). The relatively large sizes of bird eggs and reptile eggs are due to the large amounts of yolk and albumen.

**Skill**/*Teacher Resource Package*, p. 26. Use the Skill worksheet shown here for students to predict when organisms will hatch or be born.

**SKILL 26**

**SKILL: FORM A HYPOTHESIS** CHAPTER 25

Name ______ Date ______ Class ______

*For more help, refer to the Skill Handbook, pages 704-705.* *Use after Section 25:2.*

**INCUBATION AND GESTATION**

After an egg is fertilized, the new organism has to go through a period of development either in the mother's uterus or inside the egg. The period of time that the young develops in the uterus before birth is called the gestation period. The period of time that the young develops in an egg before hatching is called the incubation period. The day that the young will be born or will hatch can be predicted if you know the time needed for gestation or incubation and if you know when mating of the parents took place.

*Table 1 below shows the incubation or gestation periods for some animals. Use the data in the table to predict when the young of each animal will hatch or will be born. Write the date you predict in the last column of the table. You can refer to Table 2 for the number of days in each month. When you have completed the table, answer the questions that follow.*

**Table 1 Incubation or gestation time for some animals**

| Animal | Day of mating | Type of development | Time needed | Day of birth |
|---|---|---|---|---|
| Red fox | Jan. 20 | Gestation | 52 days | Mar. 13 |
| Alligator | Jan. 25 | Incubation | 60 days | Mar. 26 |
| Rabbit | Mar. 15 | Gestation | 28 days | Apr. 12 |
| Opossum | Apr. 1 | Gestation | 13 days | Apr. 14 |
| Sparrow | Apr. 10 | Incubation | 13 days | Apr. 23 |
| Box turtle | Apr. 20 | Incubation | 88 days | July 17 |
| Bullfrog | July 3 | Incubation | 5 days | July 8 |
| Fin whale | Dec. 6 | Gestation | 360 days | Dec. 1 |

**Table 2 Number of days in each month**

| Jan. 31 | Mar. 31 | May 31 | July 31 | Sept. 30 | Nov. 30 |
|---|---|---|---|---|---|
| Feb. 28 | Apr. 30 | June 30 | Aug. 31 | Oct. 31 | Dec. 31 |

1. Which animal has the shortest period of development? bullfrog
2. Do the young of this animal develop inside or outside the female? outside
3. Which animal has the longest period of development? fin whale
4. Do the young develop inside or outside the female? inside

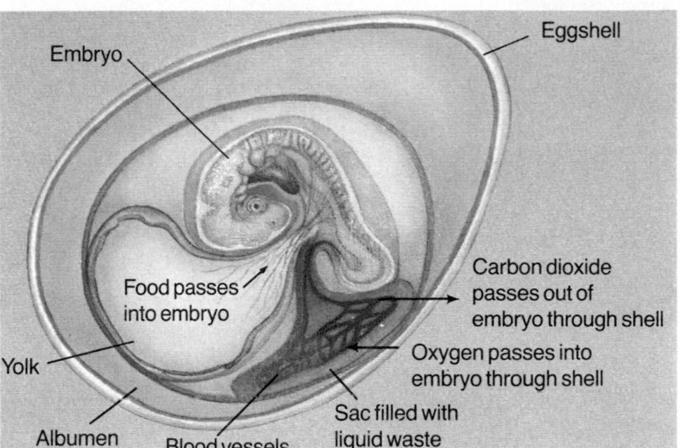


**Figure 25-8** Chick embryo

### Food Supply

You may not realize it, but you are familiar with the two parts that supply food to the embryo. They are the yolk and albumen (al BYEW mun). Yolk is the yellow part and is made up of protein and fat. Albumen is the white of the egg. Albumen also has protein in it. By hatching time, the chick has used up the yolk and albumen.

### Oxygen Supply

Oxygen diffuses through the eggshell from the air outside. Blood vessels that lie just below the shell pick up the oxygen. The oxygen then passes to the embryo.

### Waste Removal

Gas wastes leave the egg through the shell. Liquid wastes are stored in a sac within the egg until hatching.

### Protection

The shell protects the chick embryo. It protects the embryo from water loss and injury.

## Mini Lab

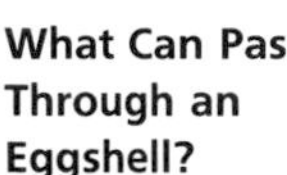

**What Can Pass Through an Eggshell?**

**Design an experiment:** Design an experiment to test if glucose can pass through an eggshell. What is the control in your experiment? *For more help, refer to the* ***Skill Handbook,*** *pages 704-705.*

**How is a chick embryo supplied with oxygen?**

## Check Your Understanding

6. Where do frog and bird development take place?
7. What does egg yolk supply to a developing chick?
8. What happens to wastes as a chick develops?
9. **Critical Thinking:** What is an advantage of developing outside the female's body? What is a disadvantage?
10. **Biology and Reading:** Which vertebrates lay eggs in water? Which vertebrates lay eggs on land?

## RETEACHING 76

*Reteaching: Idea Map* — *Chapter 25*

Name ______ Date ______ Class ______

*Use with Section 25:2.*

**Development Outside the Female**

*Use these terms and phrases to complete the idea map: from air through eggshell, yolk, eggshell prevents, water loss, albumen, injury, carbon dioxide, through eggshell, stored in sac in egg, liquid.*

Needs of chick embryo
- food supplied by — yolk; albumen
- oxygen source — from air through eggshell
- waste removal — carbon dioxide — through eggshell; liquid — stored in sac in egg
- protection — eggshell prevents — water loss; injury

1. Where does development usually take place in frogs, birds, and reptiles? **outside the female's body**
2. Name two kinds of vertebrates that lay their eggs in water. **fish and amphibians**
3. What feature allows birds and reptiles to develop on land? Explain why this feature is helpful. **Birds and reptiles have shells around their eggs that keep the embryo from drying out during development.**

## STUDY GUIDE 148

*Study Guide* — *Chapter 25*

Name ______ Date ______ Class ______

**Development Outside the Female**

*In your textbook, read about eggs and the needs of the young before hatching in Section 25:2.*

1. Write the names of the parts of the developing chicken in the correct blank.
   - A. **embryo**
   - B. **yolk**
   - C. **shell**
   - D. **sac with waste**
   - E. **albumen**

   A, B, C, D, E

2. Match the terms with their definitions by writing the correct letter from Column II on the blanks after the words in Column I.

| I | II |
|---|---|
| yolk **h** | a. place where liquid wastes are stored |
| blood vessels just below shell **e** | b. the protein in egg white |
| eggshell **g** | c. the young chicken |
| sac within egg **a** | d. leave the egg through the shell |
| albumen **b** | e. pick up oxygen that passes to embryo |
| embryo **c** | f. diffuses into shell from air outside |
| gas wastes **d** | g. protects embryo from water loss and injury |
| oxygen **f** | h. the yellow part of egg containing protein and stored fat |

3. How is development in frogs, birds, and reptiles similar to that in humans? **The fertilized eggs of all of these animals undergo cleavage.**
4. How is development in frogs, birds, and reptiles different from that in humans? **Only human fertilized eggs develop inside the female's body. The other animals listed lay eggs outside the female.**
5. How are the eggs of birds and reptiles suited to the environment in which they develop? **Since these eggs develop on land, the eggshells of these animals protect the young from drying out.**

## 2 TEACH

### Independent Practice

**Study Guide**/*Teacher Resource Package,* p. 148. Use the Study Guide worksheet shown at the bottom of this page for independent practice.

### Check for Understanding

Have students respond to the first three questions in Check Your Understanding.

### Reteach

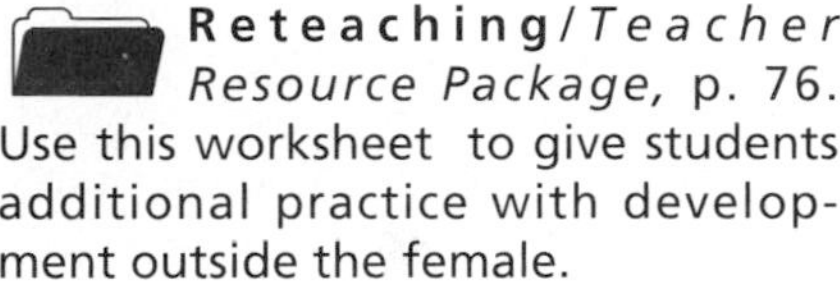

**Reteaching**/*Teacher Resource Package,* p. 76. Use this worksheet to give students additional practice with development outside the female.

**Extension:** Assign Critical Thinking, Biology and Reading, or some of the **OPTIONS** available with this lesson.

## 3 APPLY

### Divergent Question

Ask why a chicken egg is so much larger than a human egg.

## 4 CLOSE

### Audiovisual

Show the filmstrip *Frog Development,* Carolina Biological Supply Co.

### Answers to Check Your Understanding

6. outside the body; frog, in water; bird and reptile, on land
7. yolk–protein and fat
8. Gas wastes leave through the shell; liquids are stored in a sac.
9. Advantage: female produces more offspring; disadvantage: more offspring die without the protection of the female's body.
10. fish and amphibians, in water; reptiles and birds, on land.

## Lab 25-2 Metamorphosis

### Overview

Students will observe and compare stages of complete and incomplete insect metamorphosis.

**Objectives:** Upon completion of the lab, students will be able to (1) **identify** stages of complete metamorphosis, (2) **identify** stages of incomplete metamorphosis, (3) **compare** stages of incomplete and complete insect metamorphosis.

**Time Allotment:** 30 minutes

### Preparation

▶ Purchase insect stages as preserved specimens or plastic embedded mounts from a biological supply house.

▶ Collection of some insect stages may be undertaken by students.

 **Lab 25-2 worksheet**/*Teacher Resource Package*, pp. 99-100.

### Teaching the Lab

▶ **Troubleshooting:** Tell students which stage is capable of reproduction.

**Performance:** Provide students with an outline diagram of nymph and adult stages of an insect. Include legs but not wings on both diagrams. Label them as *nymph* and *adult.* Ask students to complete the diagrams as they believe they should actually appear (wings on adult but missing from nymph). Ask students to explain how they know that this insect carries out incomplete metamorphosis.

**Portfolio**

Tell students to copy the data chart from Lab 25-2 with the following changes. Include four rows; add the words *egg, young larva, mature larva,* and *adult* under Stage; delete the columns labeled Body Length and Can It Fly. Have them complete the chart for a frog using Figure 25-9 as a guide.

# Metamorphosis

**Lab 25–2**

### Problem: How do complete and incomplete metamorphosis compare?

#### Skills

interpret data, measure in SI, compare

#### Materials

hand lens
grasshopper stages of metamorphosis
moth stages of metamorphosis
metric ruler

#### Procedure

**Part A**

1. Copy the data table.
2. Using a hand lens, examine the eggs of a grasshopper.
3. Complete the data table for this stage.
4. Repeat steps 2 and 3 for the nymph and adult stages.

**Part B**

1. Using a hand lens, examine the eggs of a moth.
2. Complete the data table for this stage.
3. Repeat step 2 for the larva, pupa, and adult stages. Measure only body length for the adult. Do not include the wings.

#### Data and Observations

1. For the grasshopper, list two ways in which
   (a) the egg differs from the nymph or adult
   (b) the nymph and adult are alike
2. For the moth, list two ways in which
   (a) the pupa differs from the adult
   (b) the larva and adult differ

#### Analyze and Apply

1. **Interpret data:** Which type of metamorphosis do grasshoppers show—complete or incomplete?
2. Which type of metamorphosis do moths show–incomplete or complete?
3. **Apply:** Metamorphosis has made it possible for insects to live in almost every type of environment and use many different materials for food. Why do you think this is so?

#### Extension

**Compare** the stages of metamorphosis for a crayfish to those of a grasshopper.

| | STAGE | BODY LENGTH (mm) | LEGS PRESENT | NUMBER OF LEGS | WINGS PRESENT | CAN IT FLY? | MOUTH‡ PARTS PRESENT? | CAN IT FEED? | CAN IT REPRODUCE? |
|---|---|---|---|---|---|---|---|---|---|
| Part A | Egg | 3* | no | | no | no | no | no | no |
| | Nymph | 12* | yes | 6 | no | no | yes | yes | no |
| | Adult | 25* | yes | 6 | yes | yes | yes | yes | yes |
| Part B | Egg | 1* | no | | no | no | no | no | no |
| | Larva | 40* | yes | 6(16)** | no | no | yes | yes | no |
| | Pupa | 22* | no | | no | no | no | no | no |
| | Adult | 30* | yes | 6 | yes | yes | yes | yes | yes |

*Answers will vary. **Only the first three pairs of legs are true legs. ‡Parts may be difficult to see.



## ANSWERS

### Data and Observations

1. (a) The egg does not have mouthparts, legs, or wings.
   (b) Both have legs and mouthparts.
2. (a) The pupa does not have legs, wings, or mouthparts.
   (b) The larva is larger than adult and does not have wings.

### Analyze and Apply

1. incomplete
2. complete
3. The different stages of metamorphosis are adapted to withstanding different environmental conditions. Feeding on different materials prevents the adults and immature stages from using the same food sources and possibly not having enough food.

## 25:3 Metamorphosis

If you saw a puppy, you would be able to tell that it was a dog. If you saw a baby bird, you would be able to tell that it was a bird. Many young animals look like the adults, but some do not. These animals must go through another type of change before they look like an adult.

### Objectives

**6. Sequence** the stages of frog metamorphosis.
**7. Explain** incomplete metamorphosis.
**8. Describe** complete metamorphosis.

### Key Science Words

metamorphosis
larva
nymph
incomplete metamorphosis
complete metamorphosis
pupa

### Frog Metamorphosis

Many animals hatch and go through a series of changes before they look like their parents. Changes in appearance and lifestyle that occur between the young and the adult stages are called **metamorphosis** (met uh MOR fuh sus).

You may have seen tadpoles swimming in a pond or stream. These tadpoles will undergo metamorphosis to become adult frogs. Follow the changes in Figure 25-9 to see how a young frog changes into an adult.

1. The adult female frog lays eggs in the water. Eggs are fertilized outside the body. Notice that frog eggs do not have shells. They have a jellylike covering that protects them.
2. Early development takes place for about 12 days.

### Skill Check

**Understand science words: metamorphosis.** The word part *meta* means change. In your dictionary, find three words with the word part *meta* in them. *For more help, refer to the **Skill Handbook,** pages 706-711.*

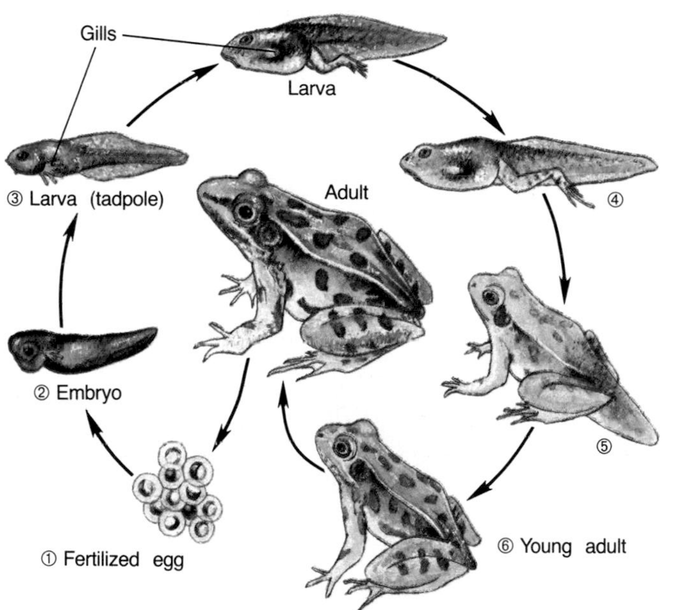


**Figure 25-9** The stages of metamorphosis in a leopard frog are shown here.

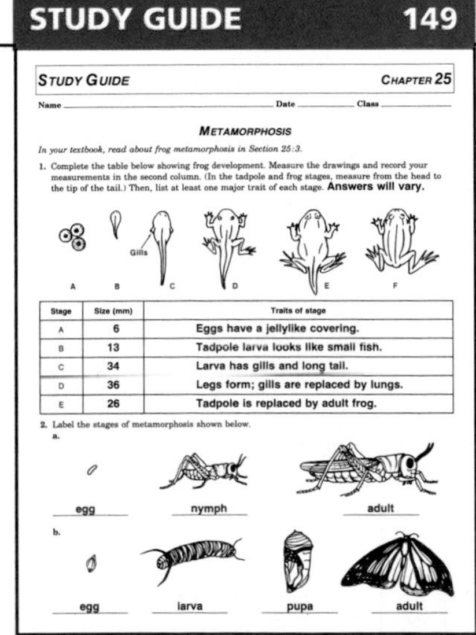

**STUDY GUIDE 149**

*STUDY GUIDE* CHAPTER 25

Name ______ Date ______ Class ______

*METAMORPHOSIS*

*In your textbook, read about frog metamorphosis in Section 25:3.*

1. Complete the table below showing frog development. Measure the drawings and record your measurements in the second column. (In the tadpole and frog stages, measure from the head to the tip of the tail.) Then, list at least one major trait of each stage. **Answers will vary.**

A B Gills C D E F

| Stage | Size (mm) | Traits of stage |
|---|---|---|
| A | 6 | Eggs have a jellylike covering. |
| B | 13 | Tadpole larva looks like small fish. |
| C | 34 | Larva has gills and long tail. |
| D | 36 | Legs form; gills are replaced by lungs. |
| E | 26 | Tadpole is replaced by adult frog. |

2. Label the stages of metamorphosis shown below.

a. egg nymph adult

b. egg larva pupa adult

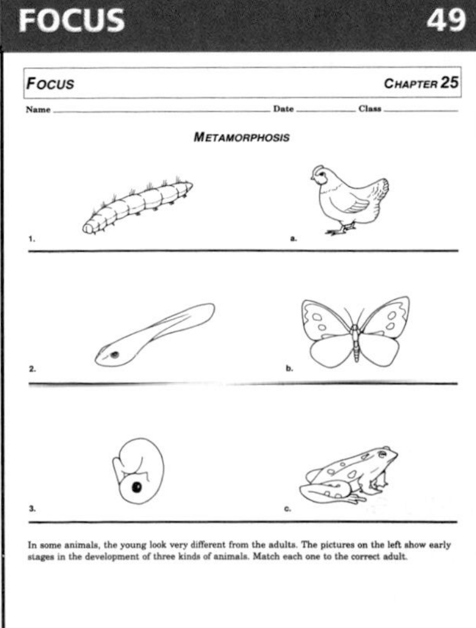

**FOCUS 49**

*FOCUS* CHAPTER 25

Name ______ Date ______ Class ______

*METAMORPHOSIS*

1. a.

2. b.

3. c.

In some animals, the young look very different from the adults. The pictures on the left show early stages in the development of three kinds of animals. Match each one to the correct adult.

## 25:3 Metamorphosis

## PREPARATION

### Materials Needed

Make copies of the Focus, Critical Thinking/Problem Solving, Study Guide, Application, Reteaching, and Lab worksheets in the *Teacher Resource Package.*

▶ Order preserved grasshopper and moth stages of metamorphosis for Lab 25-2.

### Key Science Words

metamorphosis
larva
nymph
incomplete metamorphosis
complete metamorphosis
pupa

### Process Skills

In Lab 25-2, students will interpret data, measure in SI, and compare. In the Skill Check, students will understand science words.

## 1 FOCUS

▶ The objectives are listed on the student page. Remind students to preview these objectives as a guide to this numbered section.

### Convergent Question

Ask students to describe maggots, caterpillars, and tadpoles.

**Focus**/*Teacher Resource Package,* p. 49. Use the Focus transparency at the bottom of this page as an introduction to metamorphosis.

### Independent Practice

**Study Guide**/*Teacher Resource Package,* p. 149. Use the Study Guide worksheet shown at the bottom of this page for independent practice.

# 2 TEACH

### MOTIVATION/Display

Purchase a few tadpoles so that students can observe and watch these animals undergoing metamorphosis.

### Concept Development

▶ Emphasize that the word *incomplete* suggests that something is missing. Incomplete metamorphosis lacks the larva and pupa stages.

### Guided Practice

Have students write down their answers to the margin questions: A tadpole has gills for obtaining oxygen and a tail for swimming. Its mouth and body are shaped differently from adults.

### Idea Map

Have students use the idea map as a study guide to the major concepts of this section.

### Student Journal

Tell students to imagine the following: You are an adult cricket who is looking through the family photo album that shows you as an infant cricket and all the stages of development that you have passed through. Describe these photos and explain how your photo album compares to one for your butterfly friend.

---

**How does a tadpole differ from an adult frog?**

3. A tadpole appears from the embryo. This tadpole is a larva (LAR vuh). A **larva** is a young animal that looks completely different from the adult. If you didn't know that a tadpole was a frog, you might mistake it for a small fish.
4. Note how the larva differs from the adult. Look at its mouth, body, and tail shape. The larva has gills for obtaining oxygen. Does the adult have the exact same shape?
5. The larva continues to change. Front and hind legs form. Gills are replaced by lungs.
6. Finally the larva, or tadpole, no longer exists. It has changed into a young adult frog.

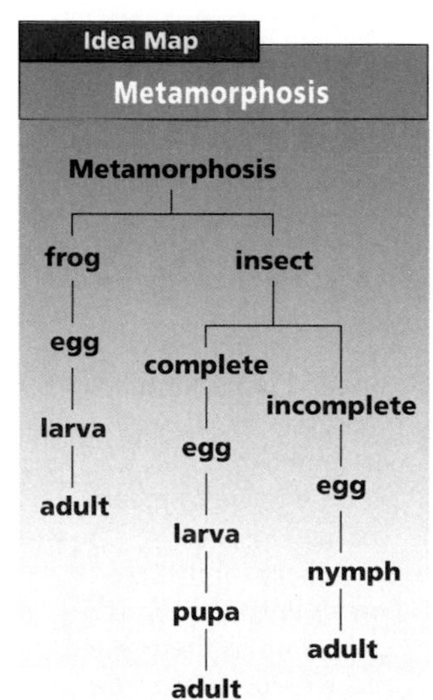


**Study Tip:** Use this idea map as a study guide to metamorphosis.

## Insect Metamorphosis

Insects also undergo metamorphosis. Unlike frogs, insects have two types of metamorphosis. They are called incomplete and complete metamorphosis.

**Incomplete metamorphosis**

When some insects hatch from eggs, they almost look like miniature adults. There are many differences, however. These young are called nymphs (NIHMFS). A **nymph** is a young insect that looks similar to the adult. Note in Figure 25-10 that when a nymph first hatches, it does not have wings like the adult. It is also sexually immature.

The nymph grows and changes until it reaches adult size and appearance. This type of change is called incomplete metamorphosis. **Incomplete metamorphosis** is a series of changes in an insect from nymph to adult. Grasshoppers and crickets have incomplete metamorphosis.

**Complete metamorphosis**

Many insects, such as butterflies and moths, have complete metamorphosis. **Complete metamorphosis** is a series of changes in an insect in which the young do not look like the adult. Figure 25-11 shows these changes.

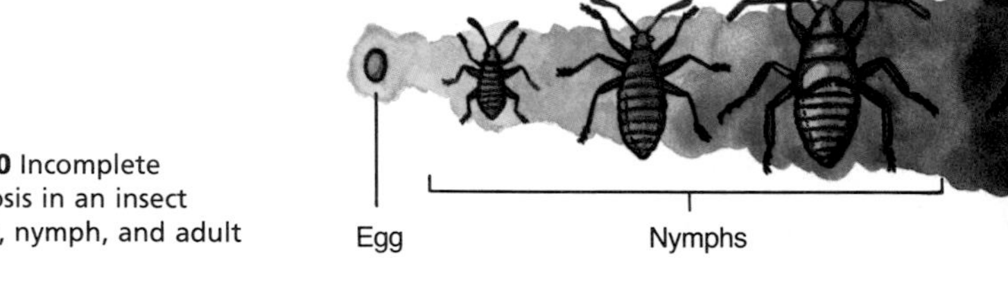


**Figure 25-10** Incomplete metamorphosis in an insect includes egg, nymph, and adult stages.



---

## OPTIONS

**Frog Metamorphosis**/*Transparency Package,* number 25b. Use color transparency number 25b as you teach frog metamorphosis.

**Critical Thinking/Problem Solving**/*Teacher Resource Package,* p. 25. Use the worksheet shown here to explain what happens to insects in winter.

**Application**/*Teacher Resource Package,* p. 25. Use the Application worksheet to teach metamorphosis of cicadas.

### CRITICAL THINKING 25

CRITICAL THINKING/PROBLEM SOLVING — CHAPTER 25

Name ________ Date ________ Class ________

*Use after Section 25:3.*

**WHAT HAPPENS TO INSECTS IN THE WINTER?**

Sam lives in a climate where the winter air temperature stays below zero degrees Celsius for several months. He knows that he sees few flying insects outdoors in winter. Sam's biology text says that insects are one of the kinds of animals that are cold-blooded. This means that their body temperature is like that of their environment. He wondered if this meant insects freeze during the winter. If so, how do insects reappear in the spring?

Sam spent several weeks in January making an insect collection. He found several life stages of insects. None of the insects were moving when he found them. He kept each insect in a jar with several holes in the lid. He observed the insects every day for several weeks. Then, he made the following table. Sam used it to answer his question: what happens to insects in the winter?

*Use the information in the table to answer the questions.*

| Stage of insect found | Changes observed over four weeks inside |
|---|---|
| cocoon on a twig | Cocoon did not move but changed to adult moth in two weeks. |
| larva, adult in grain bin | Mealworm stages moved when taken inside. |
| eggs in soil | After several weeks, crickets hatched from the eggs. |

**Analyzing the Problem**

1. What is metamorphosis? **the change from a larval to an adult form**
2. What is incomplete metamorphosis? **the changes in the life cycle of an insect in which the larval stages look like small adults**
3. What is complete metamorphosis? **the changes in the life cycle of an insect in which the larval stages look different from the adults**

**Solving the Problem**

1. What evidence did Sam have that the insects did not freeze and die in winter? **When brought into a warm room, they began to move around or develop and hatch.**
2. At which stages can insects survive the winter? **egg, larva, pupa, and adult**
3. Make a hypothesis about how insects survive the deep cold of winter. **They seem to stop or slow down their activity until they are in a warmer place.**

### APPLICATION 25

APPLICATION: ENVIRONMENT — CHAPTER 25

Name ________ Date ________ Class ________

*Use after Section 25:3.*

**METAMORPHOSIS**

Some insects take a long time to complete their metamorphosis. Certain types of cicadas take 17 years! The cycle begins when the adults mate, and the females lay eggs in the twigs of trees. The eggs develop into tiny nymphs. Then, the nymphs drop to the ground and burrow about one foot into the soil. They bite into the roots of grasses or trees and feed on the sap. The nymphs stay underground, slowly growing, for 17 years. Then, the nymphs emerge from the ground, molt, and become adults with wings. They mate, and the cycle begins again.

All of the cicadas within several square miles have the same cycle. Millions of cicadas emerge over a period of a few days. Several weeks later, the nymphs burrow into the ground. Then, no cicadas will be seen in that area for another 17 years.



*Use this information along with the diagram to answer the following questions.*

1. Does the 17-year cicada undergo complete or incomplete metamorphosis? **Incomplete** Explain. **It is incomplete because the nymph grows and changes until it reaches adult size and appearance. A cicada does not go through stages that do not look like the adult.**
2. Birds, cats, and other animals feed on cicadas when the cicadas emerge from the ground. Do you think cicadas make a reliable food source for these animals? Explain. **No. The cicadas would provide a lot of food, but only once every 17 years.**
3. What is the advantage to the cicada of a nymph stage that lives underground for so long? **While it is underground, it is safer from predators.**
4. Imagine that cicada nymphs burrowed into the ground under a tree in a vacant lot. What changes might happen in 17 years that would be harmful to the nymphs? **The tree might be cut down; the lot could be paved over; a house could be built there.**

**Figure 25-11** Complete metamorphosis includes egg (upper left), larva (upper right), pupa (lower left), and adult (lower right) stages.

1. The adult female butterfly lays fertilized eggs that develop into embryos.
2. The embryo hatches into a small larva 10 to 15 days later. This larva is also called a caterpillar.
3. The larva moves about and feeds on plant matter.
4. The larva stops feeding and enters a pupa (PYEW puh) stage. A **pupa** is a quiet, non-feeding stage that occurs between larva and adult. The pupa is surrounded by a protective covering. Many changes take place in the pupa.
5. After about one week, the pupa opens and an adult butterfly emerges.

**How do pupa and nymph stages differ?**

## Check Your Understanding

11. Name the stages of metamorphosis in a frog.
12. Arrange the following stages of development in their proper order: nymph, fertilized egg, adult.
13. How are frog eggs protected?
14. **Critical Thinking:** What form of a tapeworm is similar to the pupa stage of a butterfly? Why?
15. **Biology and Writing:** Is metamorphosis in a frog more like complete or incomplete metamorphosis? Write a paragraph that explains your answer.

**RETEACHING 77**

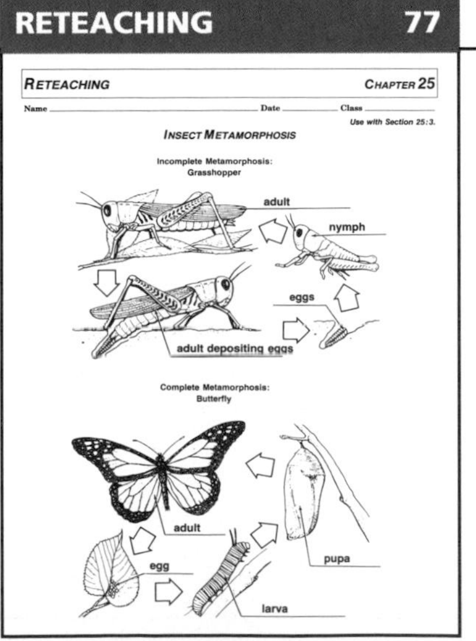
RETEACHING CHAPTER 25

Name ____ Date ____ Class ____

Use with Section 25:3.

INSECT METAMORPHOSIS



**STUDY GUIDE 150**

STUDY GUIDE CHAPTER 25

Name ____ Date ____ Class ____

VOCABULARY

*Review the new words in Chapter 25 of your textbook. Then, complete this puzzle.*

Below are five tables. List the following words in every table in which they are correct. Some words are used more than once.

amniotic sac breaks, complete metamorphosis, fetus, cleavage, incomplete metamorphosis, labor, larva, metamorphosis, navel, nymph, placenta, pupa, umbilical cord

| A change that takes place in all living things |
|---|
| cleavage |
| |
| |
| |

| Changes or stages that take place in insects |
|---|
| nymph |
| larva |
| incomplete metamorphosis |
| pupa |
| complete metamorphosis |

| Changes or stages that take place in frogs |
|---|
| larva |
| metamorphosis |
| |
| |

| Parts or stages in humans |
|---|
| placenta |
| umbilical cord |
| fetus |
| navel |

| Changes that occur during a human birth |
|---|
| amniotic sac breaks |
| labor |
| |
| |

## TEACH

### Guided Practice

Have students write down their answers to the margin questions: pupa—quiet, non-feeding stage; nymph—like a mini adult.

### Independent Practice

**Study Guide**/*Teacher Resource Package*, p. 150. Use the Study Guide worksheet shown at the bottom of this page for independent practice.

### Check for Understanding

Have students respond to the first three questions in Check Your Understanding.

### Reteach

**Reteaching**/*Teacher Resource Package*, p. 77. Use this worksheet to give students additional practice with insect metamorphosis.

**Extension:** Assign Critical Thinking, Biology and Writing, or some of the **OPTIONS** available with the lesson.

## 3 APPLY

### Convergent Question

Ask students what stage of the moth life cycle is most likely to damage wool clothing. (larva, because it is a feeding stage)

## 4 CLOSE

### Bridging

Have students compare human development and metamorphosis.

### Answers to Check Your Understanding

11. egg, embryo, larva, adult
12. fertilized egg, nymph, adult
13. They have a jellylike covering.
14. the cyst stage; both are quiet non-feeding stages
15. Answers will vary.

# CHAPTER 25 REVIEW

## Summary

Summary statements can be used by students to review the major concepts of the chapter.

## Key Science Words

All boldfaced terms from the chapter are listed.

## Testing Yourself

### Using Words

1. fetus
2. umbilical cord
3. larva
4. nymph
5. pupa
6. cleavage
7. placenta
8. labor
9. amniotic sac
10. metamorphosis

### Finding Main Ideas

11. pp. 536-537; Both start from eggs and end with adult stages. In incomplete metamorphosis, a nymph stage appears. In complete metamorphosis, larva and pupa stages appear.
12. p. 533; The shell protects the embryo from water loss and injury.
13. pp. 524-525; Mitosis occurs in a fertilized egg to make 2 cells, then 4, then 8, then 16 cells; it forms a solid ball, then a hollow ball, and attaches to the uterus.
14. p. 532; The development of both takes place outside the female's body and both undergo cleavage. Frogs develop in water, and birds develop on land.
15. p. 530; They push the fetus out through the mother's vagina. They also push the placenta out after the fetus is born.
16. pp. 525-527; The embryo receives protection from amniotic sac, receives oxygen and food from placenta through umbilical cord, and gets rid of wastes through umbilical cord to the placenta.
17. p. 533; An eggshell allows oxygen and carbon dioxide to pass through the shell and protects the embryo.
18. pp. 535-536; It differs in body, mouth, and tail shape, and has gills for obtaining oxygen.

### Using Main Ideas

19. At four months, the fetus is active. It continues to grow and develop until birth occurs at the end of the ninth month.
20. The uterus provides a place for development; the placenta connects the embryo to the uterus; the umbilical cord connects the embryo to the placenta.
21. Human development takes place inside the body, but birds, amphibians, and reptiles develop outside the body.

# Chapter 25 Review

## Summary

### 25:1 Development Inside the Female

1. Cleavage is a series of changes from a fertilized egg to a hollow ball of cells.
2. A human embryo receives protection, food, and oxygen, and gets rid of wastes while in the uterus.
3. Most major changes to the human embryo take place within the first three months. Development continues through the ninth month.

### 25:2 Development Outside the Female

4. Development occurs outside the female's body in frogs, birds, and reptiles.
5. The chicken egg supplies the chick embryo with protection, food, and a way of obtaining oxygen and getting rid of gas wastes.

### 25:3 Metamorphosis

6. During metamorphosis, frogs change from an egg to a larva to an adult.
7. Incomplete insect metamorphosis has nymph stages that look like the adult.
8. Complete metamorphosis has larva and pupa stages that don't look like the adult.

## Key Science Words

amniotic sac (p. 525)
cleavage (p. 525)
complete metamorphosis (p. 536)
fetus (p. 529)
incomplete metamorphosis (p. 536)
labor (p. 530)
larva (p. 536)
metamorphosis (p. 535)
navel (p. 530)
nymph (p. 536)
placenta (p. 526)
pupa (p. 537)
umbilical cord (p. 526)

## Testing Yourself

### Using Words

*Choose the word from the list of Key Science Words that best fits the definition.*

1. name given to the embryo at the start of the third month
2. blood vessels connecting the embryo to the placenta
3. young frog that looks completely different from the adult
4. young insect that looks similar to the adult
5 quiet, non-feeding stage during metamorphosis
6. series of changes that turn an egg into a hollow ball with many cells
7. organ connecting the embryo to the uterus
8. contractions of the uterus
9. liquid-filled tissue that protects the embryo
10. change in appearance that occurs between young and adult stages

# Review

## Testing Yourself *continued*

**Finding Main Ideas**

*List the page number where each main idea below is found. Then, explain each main idea.*

11. how incomplete and complete metamorphosis are alike and how they differ
12. how the shell of an egg allows development out of water
13. the changes that take place during cleavage
14. how early development of a frog and bird are alike, how they differ
15. how contractions of the uterus help during birth
16. the four main needs of an embryo that develops inside the female and how each need is met
17. the main jobs of an eggshell
18. how a larva differs from an adult frog

**Using Main Ideas**

*Answer the questions by referring to the page number after each question.*

19. What changes happen to a human fetus from four months to birth? (p. 529)
20. What are the jobs of the uterus, placenta, and umbilical cord? (pp. 525, 526, 530)
21. How does human development compare with bird, amphibian, and reptile development? (pp. 524, 532)
22. How does a frog egg obtain protection? (p. 535)
23. What are the stages that take place during incomplete and complete metamorphosis? (pp. 536, 537)
24. What are the jobs of the shell, sac, and yolk in a bird egg? (p. 533)

## Skill Review

*For more help, refer to the **Skill Handbook,** pages 704-719.*

1. **Design an experiment:** Design an experiment to determine if flies will undergo metamorphosis faster if kept in the dark. What would be your control in this experiment?
2. **Infer:** An embryo measures 15 mm in length. Is it older or younger than 1 month in age?
3. **Make and use graphs:** Make a line graph that shows the number of cells in a human embryo after 36, 48, 72, 96, and 120 hours.
4. **Understand diagrams:** Study Figure 25-1. What changes does the fertilized egg undergo in the oviduct?

## Finding Out More

TECH PREP

**Critical Thinking**

1. Babies are often born with diseases that the mother had during pregnancy. AIDS is an example. How might such a disease be passed to the fetus?
2. During pregnancy, the placenta sometimes pulls loose from the uterus. How might this affect the embryo?

**Applications**

1. How does the organ used for respiration differ between tadpoles and adult frogs? Relate this to where the tadpole and adult live.
2. Match the parts of a bird embryo with the following human parts and their jobs: kidney, skin, mouth, lungs. Explain your answer.



### Using Main Ideas

22. A jellylike covering protects the egg.
23. The stages in incomplete metamorphosis are egg, nymph, and adult. In complete metamorphosis, they are egg, larva, pupa, and adult.
24. The shell protects the embryo and keeps it from drying out, the sac holds liquid wastes, and the yolk provides protein and fat.

### Skill Review

1. A control would be to keep the same number of flies in natural light as those kept in complete darkness and those kept in constant light.
2. older
3. after 36 hours, 2 cells; 48, 4 cells; 72, 16 cells; 96, solid ball of cells; 120, hollow ball of cells
4. It divides into 2 cells, then 4, then 8, then 16 to form a solid ball of cells. After 5 days, it leaves the oviduct as a hollow ball of cells.

### Finding Out More

#### Critical Thinking

1. Diseases can pass from the mother's blood through the placenta to the umbilical cord and on to the embryo just as food does.
2. The placenta is the pathway between mother and embryo for food, oxygen, and waste removal. Without these, the embryo can't survive.

### Applications

1. The tadpole lives in water and has gills to obtain oxygen from the water. The adult frog lives on land, has lungs, and obtains oxygen from the air.
2. The sac is analogous to the kidney; the shell is analogous to skin and lungs; the yolk supplies food, as a mouth does.

**Chapter Review**/*Teacher Resource Package,* pp. 131-132.

**Chapter Test**/*Teacher Resource Package,* pp. 133-135.

**Chapter Test**/*Computer Test Bank*

**Unit Test**/*Teacher Resource Package,* pp. 136-137.

## Unit 6 Reproduction and Development

**CONNECTIONS**

### Biology in Your World

Some organisms are close to extinction, and other organisms have outgrown their natural boundaries and threaten other forms of life. The more we know abut how organisms reproduce and develop, the better we can control populations and restore nature's balance.

#### Literature Connection

Aldous Huxley (1894-1963) came from a family of scientists. His grandfather and brothers made significant contributions to zoology, biology, and physiology. Huxley's satirical novel, *Brave New World,* is about a society that worships science and machines at the expense of individual dignity. The novel expresses his concern about scientific discoveries and their effect on society's values. Today, advances in the area of human reproduction pose difficult legal and ethical questions for the scientific community.

#### Art Connection

English sculptor Henry Moore (1898-1986) was influenced early in his career by Mexican and African art. His works in stone, wood, and bronze have simple lines and many look weathered. Several of his sculptures depict families or mothers with their children. Another favorite theme was reclining figures. One of his best-known sculptures, *Reclining Figure,* is on display at Lincoln Center in New York City.

**CONNECTIONS**

# Biology in Your World

### Growth, Development, Change

Development is not the same for all animals. Some animals undergo distinct stages, changing form entirely. Other animals just grow larger. The following examples compare human development with that of other organisms.

**LITERATURE**

### A Crystal Ball?

In 1931, test-tube babies occurred only in science fiction. In *Brave New World,* author Aldous Huxley wrote about human eggs being fertilized outside the uterus. He also wrote about the eggs being made to divide into 96 identical embryos, or clones.

The process Huxley wrote about is a reality today. It is called *in vitro* fertilization. With *in vitro* fertilization, some childless couples are now able to have children. Huxley's prediction about clones has also come true. Animals such as frogs and mice have been cloned in the laboratory.

**ART**

### Family Group, by Henry Moore

For all of history, our survival has depended on our ability to reproduce. However, just producing babies is not enough. Without care and attention, even well-fed babies will not gain weight.

Early artwork of females stressed fertility because of the importance of reproduction. Some artists have gone beyond just showing fertility. Henry Moore, a 20th century English sculptor, is one such artist. In 1949 he did a bronze sculpture called "Family Group," which shows a loving family environment. The parents are protecting and caring for the children. How does Moore's sculpture relate to what is known about growth and development?

LEISURE

## See for Yourself!

In warm weather you can find frogs' eggs and tadpoles in ponds. The eggs look like small black beads surrounded by clear jelly. They can be collected using a long-handled dip net. The eggs will hatch in about one week at room temperature (don't leave them in the sun or near a heat source). Keep only eight tadpoles for each gallon of water. Use either pond water, bottled spring water, or tap water that has been sitting out for a few days. About a week after hatching, feed the tadpoles pond plants or boiled lettuce. In a few months, the tadpoles will develop hind legs. Front legs form later as the tails shorten. As the lungs develop, the young frogs will need a rock to rest on. If you plan to raise a frog, you must feed it live insects or mealworms. Good luck!

Return the frogs to the pond when you are finished observing them.

CONSUMER

## The Luxury of Silk

As you learned in this unit, a moth does not hatch from an egg completely formed. Instead, a caterpillar hatches from the egg. The caterpillar eventually forms a cocoon around itself. A moth then emerges from the cocoon.

Over 4000 years ago, the Chinese discovered that the cocoons of silk moths could be unwound into silk threads. Today, the caterpillars and cocoons of these moths are grown in factories. The silk fibers are woven into shiny cloth that is lightweight, yet strong.

Silk is used to make upholstery and fashionable clothing. Silk garments are often more expensive than those made from synthetic fibers. Why do you think silk is more expensive?



CULTURAL DIVERSITY

### Rites of Passage

In many cultures and religions, traditional celebrations mark the transition from childhood to adulthood. For example, in Mexican tradition, a girl celebrates her rite of passage on her fifteenth birthday in a celebration known as *quinceanera.* In the United States, Apache girls kneel on a sacred deerskin during a four-day celebration of their first menstrual period. Discuss with students examples of initiation rites in various cultures. You may wish to have students research topics by asking them to identify a culture and then determine whether that culture has a traditional rite of passage.

CONNECTIONS

### Leisure Connection

Frog eggs can be bought throughout the year from biological supply houses. Peeper eggs mature quickly.

An aquarium can provide an excellent environment for the eggs to develop. Aquatic plants provide oxygen and can be anchored in gravel at the bottom of the aquarium. If any frog eggs die, they will float to the surface. Be sure to remove dead eggs or animals. The best water temperature for development is 20°C. If the water is colder, the tadpoles take longer to develop.

### Consumer Connection

When the silkworm moth breaks out of its cocoon, it breaks the long silk thread into many pieces. These pieces can be spun into silk yarn and used as a fabric filler, but they are not as valuable as unbroken fibers. As a result, most silkworms are killed while in their cocoons and silk farmers have fewer moths to lay eggs.

Silk fabric woven with fillers such as spun silk is less expensive than pure silk. Silk blends retain many of the desirable qualities of silk, but cost less.

### References

Cowley, Geoffrey. "The Future of Birth." *Newsweek*, September 4, 1995, pp. 42-43.

Lewis, Ricki. "Unraveling the Weave of Spider Silk." *Bioscience*, October 1996, pp. 636-639.

Luoma, Jon R. "Vanishing Frogs." *Audubon*, May-June 1997, pp. 60-70.

Moore, Henry, and John Hedgecoe. *Henry Moore.* San Francisco: Chronicle Books, 1986.

Rennie, John. "Grading the Gene Tests." *Scientific American*, June 1994, pp. 88-97.

Shoemaker, Vaughn H. "Frogs and Toads in Deserts." *Scientific American*, March 1994, pp. 82-89.

# Unit 7
# Traits of Living Things

## UNIT OVERVIEW

Unit 7 focuses on how genetic continuity occurs in all living things. Chapter 26 deals with genes, chromosomes, and their relationship to traits of living things. Chapter 27 focuses on human genetics and covers sex-related traits, chromosome errors, and genetic diseases. Chapter 28 discusses the structure of DNA and how it works. Chapter 29 covers the topics of adaptation, natural selection, and isolation as they relate to the theory of evolution.

## Advance Planning

Audiovisual and computer software suggestions are located within each specific chapter under the heading TEACH.

**To prepare for:**

| | |
|---|---|
| Lab 26-1: | Order corn seeds and gather petri dishes. |
| Lab 26-2: | Provide 40 red and 40 white beans and 2 paper bags for each student team. |
| Mini Lab: | Purchase various types of dried beans. |
| Lab 27-1: | Put tape on 35 sets of coins in advance of this lab and use the coins for other classes. |
| Lab 28-1: | Purchase tracing paper, heavy paper, and red, blue, green, and yellow crayons. |
| Lab 28-2: | Purchase or order *Coleus* plants, clear plastic cups, and aluminum foil. |
| Mini Lab: | Purchase crayons and markers. |
| Lab 29-1: | Using a paper punch, punch out 30 black and 30 white dots per student team. Collect white and black paper. |
| Lab 29-2: | Order plastic powder and gather leaves. |
| Mini Labs: | Order petrified wood and collect pieces of unpetrified wood. Purchase 10 pea pods per team. |

# Unit 7

**CONTENTS**



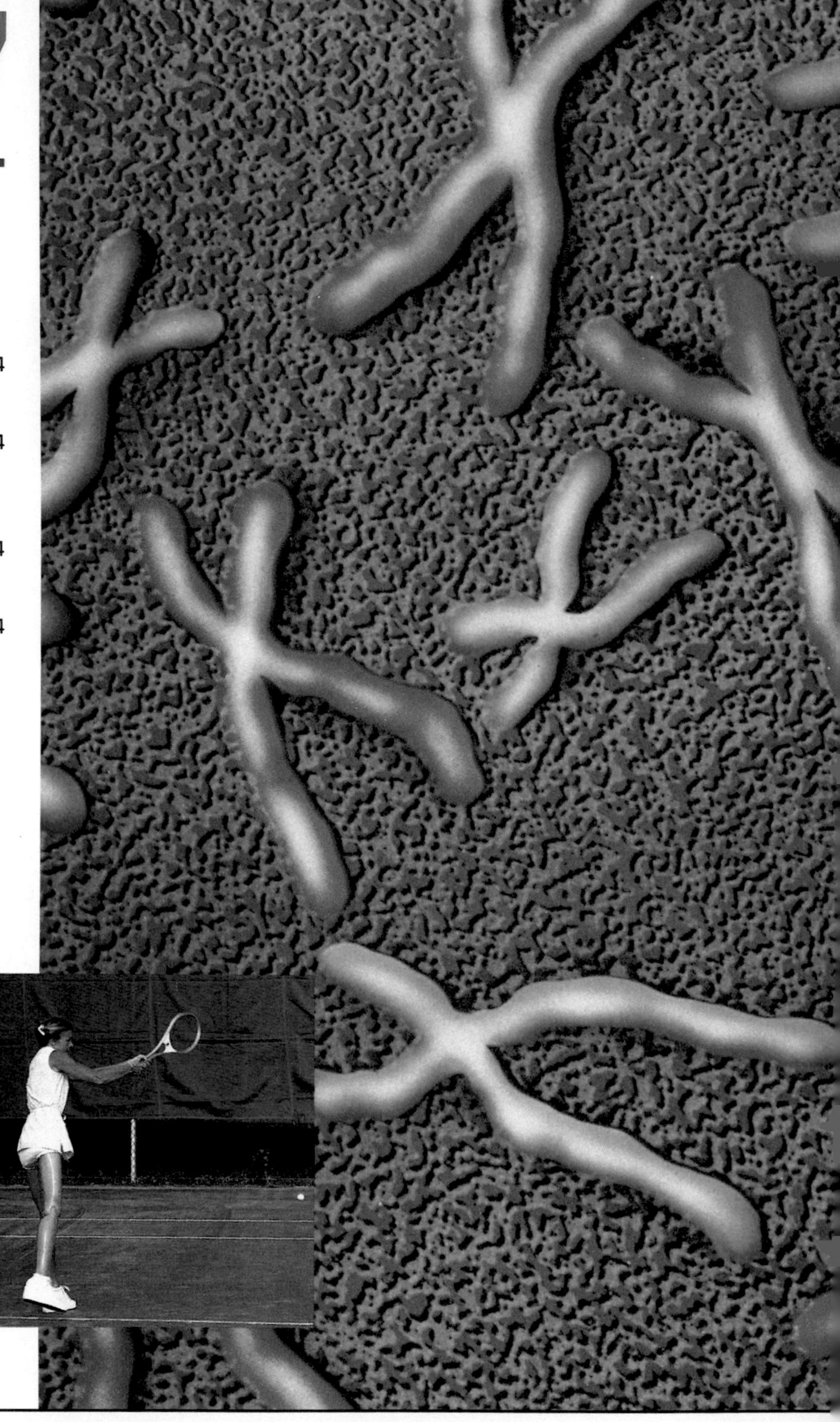

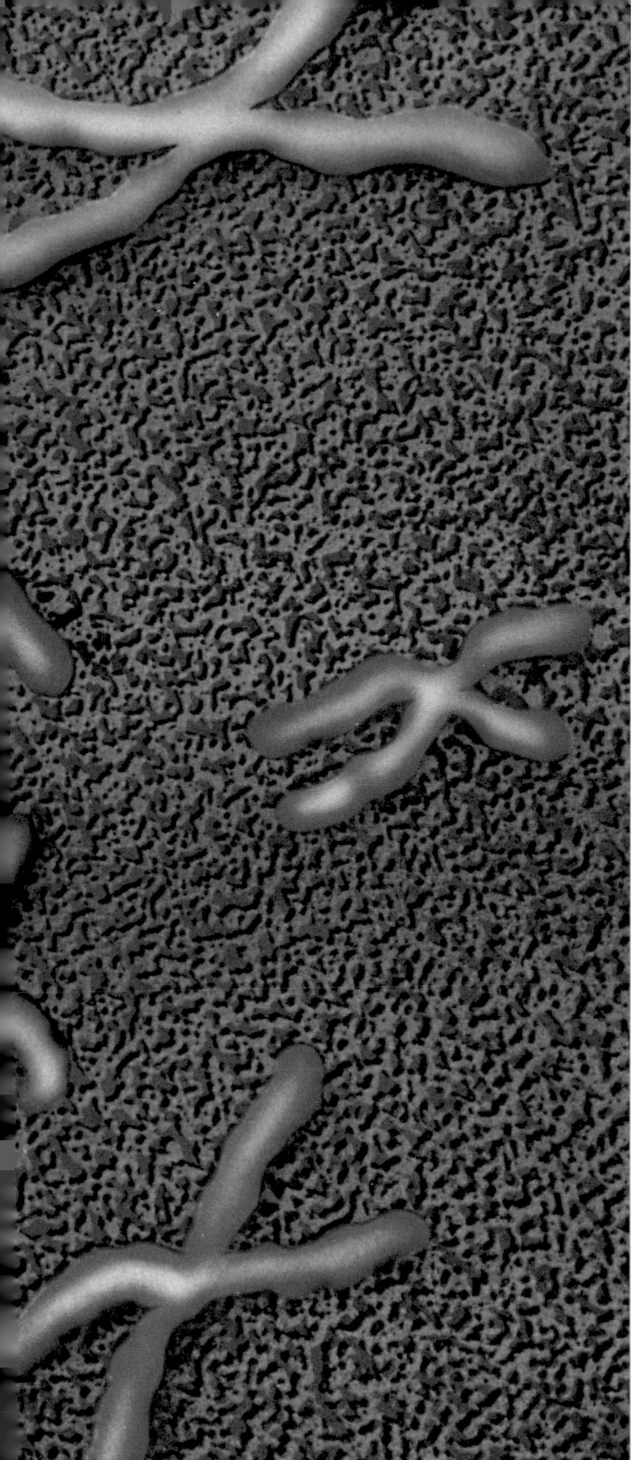

# Traits of Living Things

**What would happen if...**

humans could regrow body parts that were damaged or cut off? People who lose a body part in an accident or during surgery would grow a replacement. An infected arm or leg could be removed to stop the spread of the infection. A new arm or leg would grow in its place.

The growing back of body parts is controlled by genes. In a young child, a finger tip could regrow if one were cut off. At your age, only the skin and a few other body tissues regrow. For some reason, most body tissues lose the ability to regrow as humans age.

Some organisms can regrow body parts. Scientists are studying them to learn how body parts regrow. Maybe they will learn how to make the genes that control the ability of body parts to regrow.

543

**UNIT OVERVIEW**

## Photo Teaching Tip

Ask students to read the passage that accompanies the unit opening photographs and then discuss what it might be like to be without a hand, a foot, or another body part. Explain that the field of genetics is probably the fastest growing area of biology today. It is now possible to move parts of a chromosome from one cell to another. Explain that this modern technology is known as genetic engineering. Ask students what might be some of the consequences of this type of technology. What might be some benefits and what could be considered dangerous about changing the basic blueprint of an organism? Is genetic engineering the answer to all our disease or handicap problems?

## Theme Development

Living things reproduce and pass traits to offspring. Students have studied in prior chapters how cells reproduce and how other living things reproduce and develop through their life cycles. Now, students will find out how the various traits of the parents will be passed to and expressed in their offspring. They will learn how DNA controls traits and makes proteins. They will find out how the changes in the traits of an organism help to explain how evolution occurs.

**CULTURAL DIVERSITY**

See the Cultural Diversity feature located on the Connections page at the end of this unit.

For Tech Prep activities, see the Applying Technology feature in this unit, the Finding Out More applications questions in the Chapter Reviews, and the Tech Prep Applications booklet in the Teacher Resource Package.

# Inheritance of Traits

## PLANNING GUIDE

| CONTENT | TEXT FEATURES | TEACHER RESOURCE PACKAGE | OTHER COMPONENTS |
|---|---|---|---|
| (3 days)<br>26:1 Genetics, How and Why<br>Chromosomes<br>Genes on Chromosomes<br>Passing Traits to Offspring<br>Dominant and Recessive Genes<br>When Both Parents Are Heterozygous | Skill Check: *Understand Science Words,* p. 549<br>Mini Lab: *What Traits Do Some Seeds Share?* p. 547<br>Mini Lab: *What Type of Earlobes Do Most Students Have?* p. 548<br>Check Your Understanding, p. 551<br>Idea Map, p. 547 | Reteaching: *Traits of Plants and Animals,* p. 78<br>Study Guide: *Genetics: How and Why,* p. 151 | **Laboratory Manual:**<br>*How Can the Genes of Offspring Be Predicted?* p. 219<br>**STVS:** *Microinjecting Polygenes,* Plants and Simple Organisms (Disc 4, Side 1) |
| (2 days)<br>26:2 Expected and Observed Results<br>The Punnett Square<br>Expected Results<br>Observed Results<br>Mendel's Work | Skill Check: *Calculate,* p. 558<br>Lab 26-1: *Corn Genes,* p. 554<br>Lab 26-2: *Punnett Squares,* p. 556<br>Career Close-up: *Animal Breeder,* p. 555<br>Check Your Understanding, p. 561<br>Idea Map, p. 558 | Application: *Genetics Problems in Agriculture,* p. 26<br>Enrichment: *Incomplete Dominance,* p. 28<br>Reteaching: *Using the Punnett Square to Solve Problems,* p. 79<br>Critical Thinking/Problem Solving: *How Can Plants Show Unexpected Traits?* p. 26<br>Skill: *Use Numbers,* p. 27<br>Focus: *Crossbreeding Dogs,* p. 51<br>Transparency Master: *Punnett Square,* p. 133<br>Study Guide: *Expected and Observed Results,* pp. 152-155<br>Study Guide: *Vocabulary,* p. 156<br>Lab 26-1: *Corn Genes,* pp. 101-102<br>Lab 26-2: *Punnett Squares,* pp. 103-104 | **Laboratory Manual:**<br>*What Is a Test Cross?* p. 223<br>Color Transparency 26: *Offspring From Two Heterozygous Parents*<br>**STVS:** *Genetic Engineering in Barley,* Plants and Simple Organisms (Disc 4, Side 1) |
| Chapter Review | Summary<br>Key Science Words<br>Testing Yourself<br>Finding Main Ideas<br>Using Main Ideas<br>Skill Review | **ASSESSMENT RESOURCES**<br>Chapter Review, pp. 138-139<br>Chapter Test, pp. 140-142<br>Performance Assessment in the Biology Classroom<br>Alternate Assessment in the Science Classroom<br>Computer Test Bank | |

## GLENCOE TECHNOLOGY

**Infinite Voyage,** *The Geometry of Life*
**Science and Technology Videodisc Series,**
*Microinjecting Polygenes,* Plants and Simple Organisms (Disc 4, Side 1)
*Genetic Engineering in Barley,* Plants and Simple Organisms (Disc 4, Side 1)

**The Secret of Life,** *Sex and the Single Gene: Cell Development*

## MATERIALS NEEDED

| LAB 26-1, p. 554 | LAB 26-2, p. 556 | MARGIN FEATURES |
|---|---|---|
| 20 green: albino corn seeds<br>paper towels<br>wax pencil<br>petri dish<br>water | 40 white beans<br>40 red beans<br>2 paper bags | Skill Check, p. 549<br>dictionary<br>Skill Check, p. 558<br>pencil<br>paper<br>Mini Lab, p. 547<br>various kinds of dried beans<br>Mini Lab, p. 548<br>pencil<br>paper |

## OBJECTIVES

For more information about National Science Standards, see page 5T.

| SECTION | OBJECTIVE | CORRELATION of QUESTIONS to OBJECTIVES | | | |
|---|---|---|---|---|---|
| | | CHECK YOUR UNDERSTANDING | CHAPTER REVIEW | TRP CHAPTER REVIEW | TRP CHAPTER TEST |
| 26:1 National Science Stds: UCP.1, UCP.2, UCP.3, UCP.4, UCP.5, C.1, C.2 | 1. **Compare** the number of chromosomes in sex cells and body cells. | 1, 5 | 5, 9, 14, 18, 22, 25, 26 | 1, 2, 4 | 1, 2, 35, 37 |
| | 2. **Distinguish** between dominant and recessive genes. | 2 | 2, 7, 21 | 7, 9, 10, 13, 18, 19 | 4, 9, 18, 19, 20, 24, 25, 26, 28, 33 |
| | 3. **Describe** how different gene combinations result from fertilization and how traits are passed to offspring. | 3, 4, 10 | 1, 3, 4, 8, 11, 12, 17, 23 | 3, 5, 6, 14, 16, 17 | 3, 5, 6, 8, 29, 32, 34, 36 |
| 26:2 National Science Stds: UCP.1, UCP.2, UCP.3, UCP.4, A.2, G.1, G.2, G.3 | 4. **Discuss** the purpose of a Punnett square. | 7 | 6, 10, 15, 19 | 11, 12, 20 | 10, 11, 14, 15, 17 |
| | 5. **Compare** expected results and observed results. | 6, 9 | 16, 27 | 15 | 12, 13, 16, 17, 21, 22, 23, 27, 31, 38 |
| | 6. **Explain** the importance of Gregor Mendel's work. | 8 | 13, 20 | 8 | 7, 30 |

# Inheritance of Traits

## CHAPTER OVERVIEW

### Key Concepts

This chapter introduces genetics with a discussion and description of where chromosomes and genes are located and what their jobs are. This chapter also explains how traits are passed from parents to offspring and what dominant and recessive traits are. Students learn how to determine expected and observed results by using the Punnett square and will read about the accomplishments of Gregor Mendel.

### Key Science Words

| | |
|---|---|
| dominant gene | Punnett square |
| gene | pure dominant |
| genetics | pure recessive |
| heterozygous | recessive gene |

### Skill Development

In Lab 26-1, students will **observe, predict**, and **interpret data** with corn seeds that germinate into green plants. In Lab 26-2, students will **observe** and **formulate a model** of traits in heterozygous parents. They will **form hypotheses** to show possible combinations of traits in offspring. In the Skill Check on page 549, students will **understand** the **science word** *heterozygous.* In the Skill Check on page 558, students will **calculate** expected results of several coin tosses. In the Mini Lab on page 547, students will **classify** dried beans according to traits. In the Mini Lab on page 548, students will **make and use tables** to show numbers of students with particular traits.

### Bridging

In Unit 6, students learned how reproduction occurs in cells and organisms. That unit also focused on plant and animal development. In this unit, students will learn how all organisms pass on their genetic traits and how living things change through time by evolution.

**CHAPTER PREVIEW**

### Chapter Content

Review this outline for Chapter 26 before you read the chapter.



### Skills in this Chapter

The skills that you will use in this chapter are listed below.

- In **Lab 26-1,** you will observe, predict, and interpret data. In **Lab 26-2,** you will observe, formulate a model, and form hypotheses.
- In the **Skill Checks,** you will calculate and understand science words.
- In the **Mini Labs,** you will classify and make and use tables.

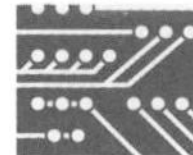

## TECH PREP

For Tech Prep activities in this chapter of the Teacher Wraparound Edition, see especially the Portfolio on page 546, the Activity on page 549, the Options Activities on pages 550 and 558, and the Student Journals on pages 553 and 555.

See also the Glencoe Homepage at **www.glencoe.com**

Chapter **26**

# Inheritance of Traits

The small photo on this page shows two parent rabbits. The large photo on the opposite page shows some baby rabbits. Which of the baby rabbits do you think belong to the two parents? What traits did you use to help you answer the question? What traits did the parents pass to the offspring?

How about the all-white baby rabbit? Did someone who was caring for several litters of rabbits put the white rabbit in the wrong container? Probably not. Perhaps the parents passed the trait for white fur to the baby white rabbit. Can offspring show a trait that is not seen in the parents? You should be able to answer this question after you read this chapter.

**Try This!**

**Are two peas in a pod alike?** Open a pea pod and look carefully at the peas. Make a list of ways the peas are the same and ways they are different. Each pea in a pod is like each child in a family. Children in a family show different traits.

*inter***NET** CONNECTION

For more information about the material in this chapter, follow the link for the chapter on the Glencoe Homepage at **http://www.glencoe.com**

Parent rabbits



## ASSESSMENT PLANNER

***Portfolio***

Strategies on the following pages represent student products that can be placed into a best-work portfolio: pp. 546, 560.

**PERFORMANCE ASSESSMENT**

Skill Check, pp. 549, 558
Mini Lab, pp. 547, 548
Lab, pp. 554, 556

**CONTENT ASSESSMENT**

Check for Understanding, pp. 550, 551, 558, 561
Chapter Review, pp. 562-563

**GROUP ASSESSMENT**

Opportunities for group assessment occur with Cooperative Learning Strategies.

***Student Journal***

Strategies on the following pages represent opportunities for writing in a Student Journal: pp. 553, 555, 559.

**GETTING STARTED**

### Using the Photos

Students will notice that the offspring of rabbits can be different in color, size, and other features when compared to the parents. Some students may assume that the white rabbit is from a different litter. Tell students they will learn why organisms can look different from either parent.

### MOTIVATION/Try This!

**Are two peas in a pod alike?** Upon completing this chapter-opening activity, students will learn to **compare** traits. Students can relate differences and similarities among peas to those among children in the same family.

### Chapter Preview

Have students study the chapter outline before they begin to read the chapter. Students should note that chromosomes are discussed first and then both dominant and recessive genes. The second section discusses the differences between expected and observed results of gene combinations.

### Misconception

Explain to students that offspring of animals do not need to look like both parents. Some traits of parents do not show up in offspring and are called recessive.

## 26:1 Genetics, How and Why

## PREPARATION

### Materials Needed

Make copies of the Study Guide and Reteaching worksheets in the *Teacher Resource Package*.

▶ Obtain different kinds of dried bean seeds for the Mini Lab.

### Key Science Words

genetics
gene
dominant gene
recessive gene
pure dominant
pure recessive
heterozygous

### Process Skills

In the Skill Check, students will understand science words. In the Mini Labs, students will classify and make and use tables.

## 1 FOCUS

▶ The objectives are listed on the student page. Remind students to preview these objectives as a guide to this numbered section.

### MOTIVATION/Bulletin Board

Set up a bulletin board that shows pictures of wild and domestic animals and their young. Ask students why most wild young look like their parents while domestic young have traits very different from their parents. Discuss how and why humans selectively breed animals.

### Portfolio

Have students make a construction paper model of a sex cell chromosome and a pair of body cell chromosomes. Direct students to make the chromosomes at least 20 cm long and place at least 20 genes on their model by using small pieces of knitting yarn glued to the chromosome. Have them write why there are some similarities and differences in the two kinds of chromosomes.

## 26:1 Genetics, How and Why

### Objectives

1. **Compare** the number of chromosomes in sex cells and in body cells.
2. **Distinguish** between dominant and recessive genes.
3. **Describe** how different gene combinations result from fertilization and how traits are passed to offspring.

### Key Science Words

genetics
gene
dominant gene
recessive gene
pure dominant
pure recessive
heterozygous

**Genetics** (juh NET ihks) is the study of how traits are passed from parents to offspring. Offspring usually show some traits of each parent. For a long time, scientists did not understand how this could happen. Later, they found that the traits of the parents were passed to offspring by sex cells.

### Chromosomes

Before you learn how traits are passed from parents to offspring, let's review something about cells. Look at a cell in Figure 26-1. The large, round part in the center is the nucleus. It has two main jobs. One is to direct the actions of other cell parts. The other is to allow the cell to reproduce.

Inside the nucleus are long, threadlike parts called chromosomes. Chromosomes can be seen best when a cell is ready to reproduce. During cell reproduction, the chromosomes become short and thick.

Suppose we look closer at two kinds of cells. One kind is called a body cell. Remember that body cells are cells that make up most of the tissues and organs in your body. Look at Figure 26-1a. Notice that the body cell has two of each kind of chromosome. The chromosomes in body cells are paired.

Another type of cell is a sex cell. The sex cell can be a sperm cell or an egg cell. Notice in Figure 26-1b that there is only one of each kind of chromosome present in the sex cell. Sex cells, then, have half as many chromosomes as body cells.

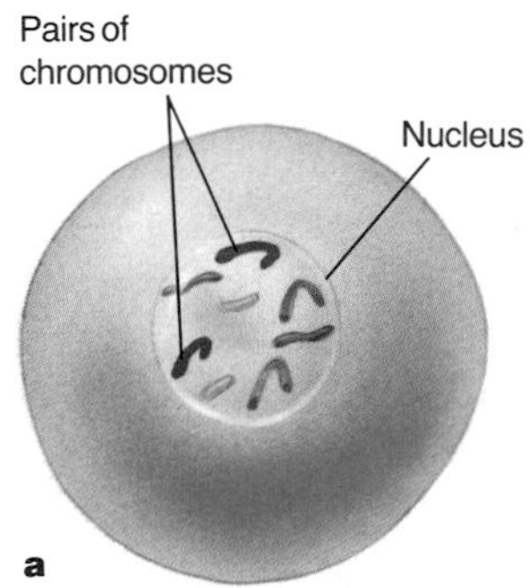


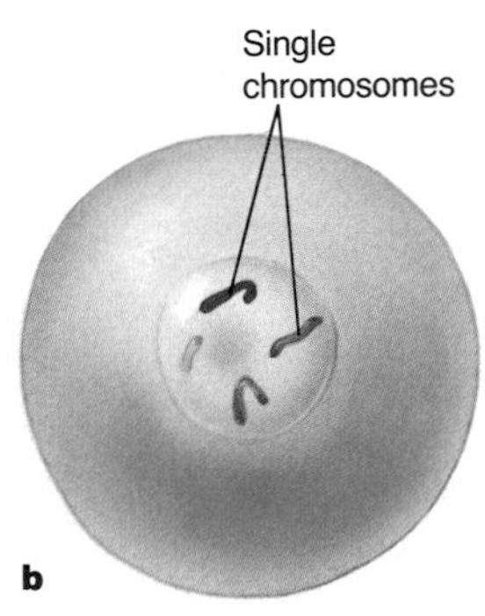


**Figure 26-1** Chromosomes are found in the nucleus. Body cells have two copies of each chromosome (a). Sex cells have one copy of each chromosome (b).



## OPTIONS

### Science Background

Inheritance of traits in all living things is controlled by genes. Genes are located on chromosomes. The amount of space that each gene takes up is different. Chromosomes are found in all body and sex cells of living things. The chromosome number in a sex cell is half that of a body cell. Genes are of two types: dominant and recessive. Combinations of these two genes in an organism give rise to three genotypes and two phenotypes.

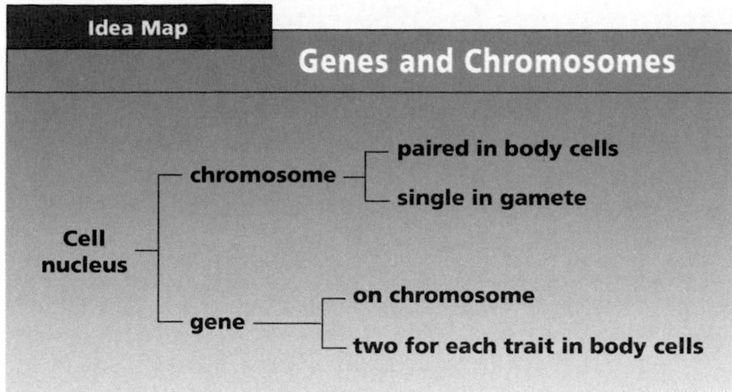


**Study Tip:** Use this idea map as a study guide to genes and chromosomes. Some features of each are shown.

## Genes on Chromosomes

All chromosomes contain genes (JEENZ). A **gene** is a small section of chromosome that determines a specific trait of an organism. Examples of traits are eye color, hair color, and shape of body parts such as ears. Chemical processes inside the body, which cannot be seen, are also traits. Organisms have thousands of different traits. Genetics is really a study of the genes that control all of these traits. Note that the word *genetics* contains the word *gene*.

Genes are arranged on a chromosome, one next to another, much like beads on a necklace. Each chromosome has different kinds of genes that control different traits. Figure 26-2 shows drawings of human chromosomes. The locations of several genes are shown.

Remember that chromosomes in body cells are paired. The genes on chromosomes in body cells are paired, too. There is one gene of each gene pair on each chromosome of a chromosome pair. You can see this pairing in Figure 26-2. Each trait we study will have one gene pair, or two genes that represent it. The two genes representing the trait are located on chromosomes that make a pair.

### Mini Lab

**What Traits Do Some Seeds Share?**

**Classify:** Look at the different dried beans found in the grocery store. Classify the beans into groups based on color, shape, and size. *For more help, refer to the* ***Skill Handbook****, pages 715-717.*

How many genes in a body cell represent each trait?

Sex Cell Chromosome

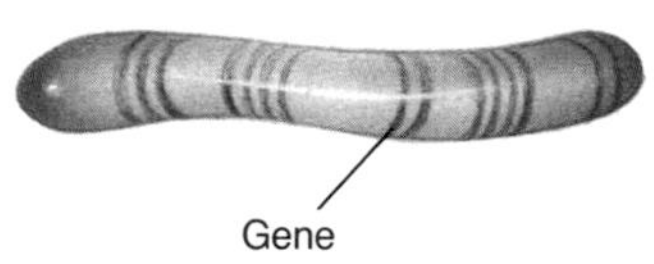


Body Cell Chromosome Pair

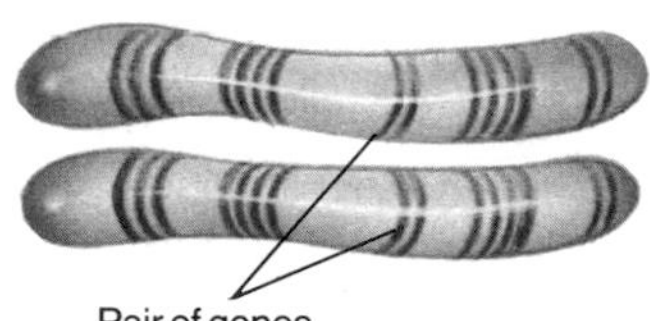


**Figure 26-2** Genes are arranged next to each other on a chromosome. How many copies of a gene are in a sex cell? one



## OPTIONS

### Science Background

Point out that *chromosome* means color body. Before the turn of the 20th century, chromosomes were not thought to have any role in heredity. Explain that chromosomes were seen as dark bodies inside a cell when stain was added. Chromosomes absorbed more stain than other cell parts—hence the name chromosomes.

## 2 TEACH

### MOTIVATION/Brainstorming

Show photos of some animals with their offspring–cats with kittens, etc.–and have students discuss traits the offspring have inherited from the parents. List the traits on the chalkboard. Ask how parents pass traits to their offspring.

### Concept Development

▶ Explain that some traits in living things may not show in either parent but can appear in offspring.

### Idea Map

Have students use the idea map as a study guide to the major concepts of this section.

### Mini Lab

Get bags of several different kinds of bean seeds at the grocery store. Mix the beans together in a large container and give each student a small cup of beans. Have them compare the size, color, and shapes of the beans.

### ASSESSMENT

**Oral:** Give student teams a small cup full of dried beans from a package of "Twelve Bean Soup" beans. Have students sort the beans and make a list of the traits they think each of the different beans inherited from its parents. Tell them that what they describe as the properties of the beans are most probably the traits that they inherited from their parents. They should report their findings to the class.

**Analogy:** Use the following analogy to compare chromosomes in body cells and sex cells: Have students imagine one closet with 23 pairs of shoes and a second closet with only 23 shoes, no pairs. A body cell is similar to the first closet; a sex cell is similar to the second closet.

### Guided Practice

Have students write down their answers to the margin question: two genes or one pair.

## TEACH

### Concept Development

▶ Point out that all genes on a chromosome are not the same size. Some cover a smaller or larger section of a chromosome than others.

▶ Make sure students understand that each chromosome can contain several thousand genes.

▶ Point out that one of the chromosomes in each body cell is a maternal chromosome because it came from the person's mother; the other is a paternal chromosome because it came from the father.

▶ Explain that humans inherit thousands of traits from their parents.

▶ Emphasize that the genes in the egg and sperm of parents are passed to offspring at the time of fertilization.

▶ Point out that during fertilization any sperm cell can fertilize any egg cell.

### Mini Lab

Make sure that students understand that a student with free earlobes can be pure dominant or heterozygous. Usually a class will have more students with the dominant trait because the recessive gene is hidden and not expressed in heterozygotes.

### ASSESSMENT

**Skill:** Have the students make another table with two columns: Free Earlobes; Attached Earlobes. Tell them to conduct a survey of their neighborhood to find out how many persons have free or attached earlobes. Tell them to calculate the percentage of persons who have free or attached earlobes in each of the two groups and compare the neighborhood data to the class data.

### MOTIVATION/Overhead

Make a transparency of Figure 26-3 showing the kinds of sex cells parents can make. Go over the diagram with care so that students understand this concept.

### Mini Lab

**What Type of Earlobes Do Most Students Have?**

**Make and use tables:** Make a table showing how many classmates have free or attached earlobes. Do more have the dominant or the recessive trait? *For more help, refer to the* ***Skill Handbook,*** *pages 715-717.*

## Passing Traits to Offspring

How are traits passed from parents to their offspring? To answer this question, let's use the trait of earlobe shape in humans as an example. A person can have attached earlobes or free earlobes. Figure 26-3 shows what these traits look like. What genes would you expect to find in the body cells of each of these people?

Suppose the mother shows the attached earlobe trait. Suppose the father shows the free earlobe trait. Which trait would appear in their children?

Figure 26-3 shows what kinds of sex cells the parents can make. The mother has the gene pair for attached earlobes. She can make eggs that have this gene. The father has the gene pair for free earlobes. He can make sperm that have this gene. What genes for earlobe shape will their child have? The child will have one gene for each trait.

In fertilization, one sperm will join with one egg. Which sperm and egg will join? We don't know. We can see that it would make no difference in this example. Any child produced will have a gene for attached earlobes and a gene for free earlobes. From which parent did the child receive the gene for attached earlobes? From which parent did the child receive the gene for free earlobes?

**Figure 26-3** Attached earlobe trait (a) and free earlobe trait (b). The trait for earlobe shape is passed from parents to their child.

a

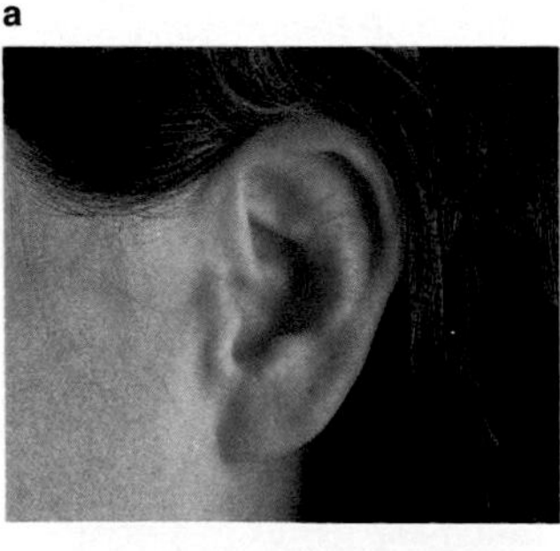

b

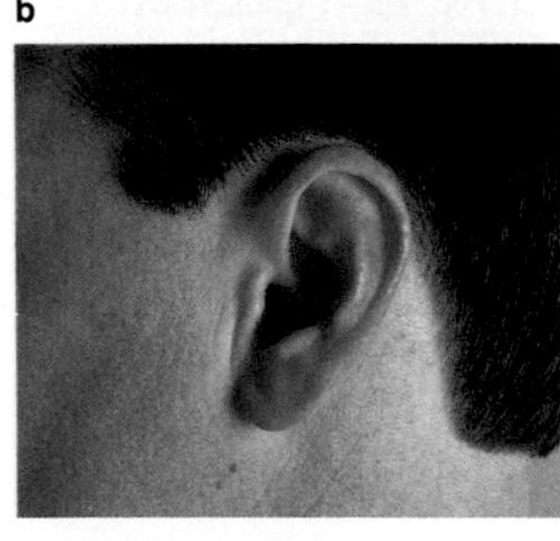

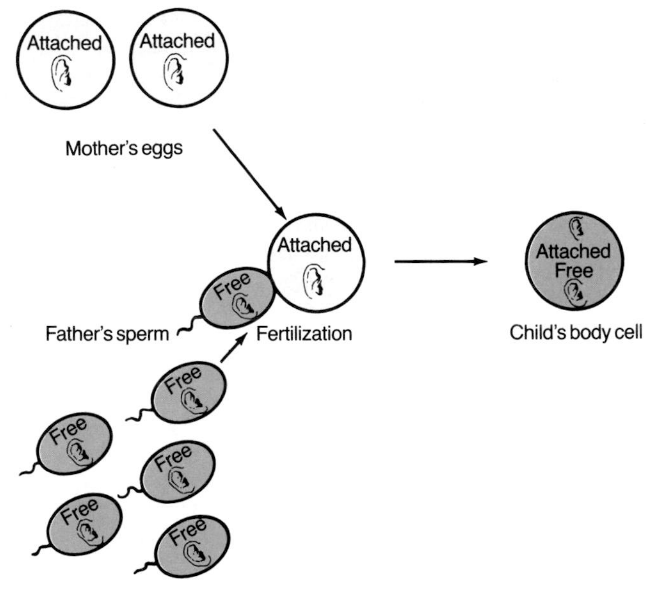

548 Inheritance of Traits 26:1

## OPTIONS

### ACTIVITY/Challenge

Have students cut out sperm and egg cells from construction paper and label one-half of them attached and one-half free. Have students show the different combinations that can occur during fertilization.

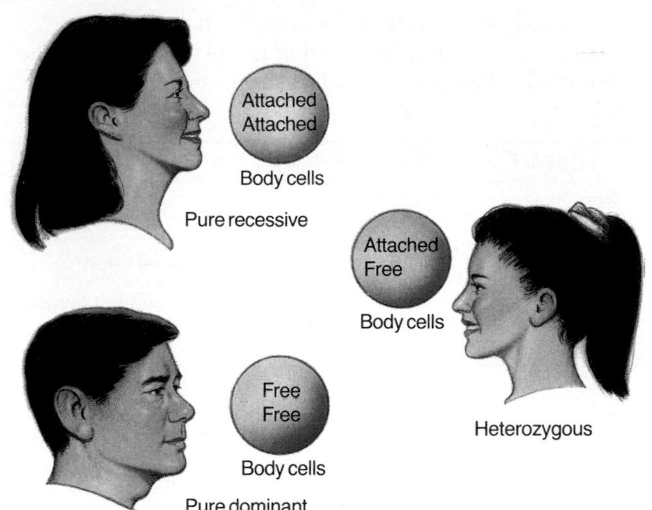


**Figure 26-4** A child with one gene for attached earlobes and one gene for free earlobes will have free earlobes.

## Dominant and Recessive Genes

What does a child with one gene for attached earlobes and one for free earlobes look like? Figure 26-4 shows you. Children born with both genes will have free earlobes. Why? Some genes can keep others from showing their traits. Genes that keep other genes from showing their traits are called **dominant** (DAHM uh nunt) **genes.** The genes that do not show their traits when dominant genes are present are called **recessive** (rih SES ihv) **genes.** In this example, the gene for free earlobes is dominant. The gene for attached earlobes is recessive.

An organism with two dominant genes for a trait is said to be **pure dominant.** Using the word *pure* means that both genes are the same. In our example, the father is pure dominant for free earlobes.

An organism with two recessive genes for a trait is said to be **pure recessive.** In our example, the mother is pure recessive for attached earlobes.

The child with a gene for attached earlobes and a gene for free earlobes has two different genes. The child is heterozygous (HET uh roh ZI gus). A **heterozygous** individual is one with a dominant and a recessive gene for a trait. Even though the heterozygous individual has the recessive gene, the recessive trait does not show. The trait of the dominant gene shows.

### Skill Check

**Understand science words: heterozygous.** The word part *hetero* means different. In your dictionary, find three words with the word part *hetero* in them. *For more help, refer to the* ***Skill Handbook,*** *pages 706-711.*

**What is a recessive gene?**



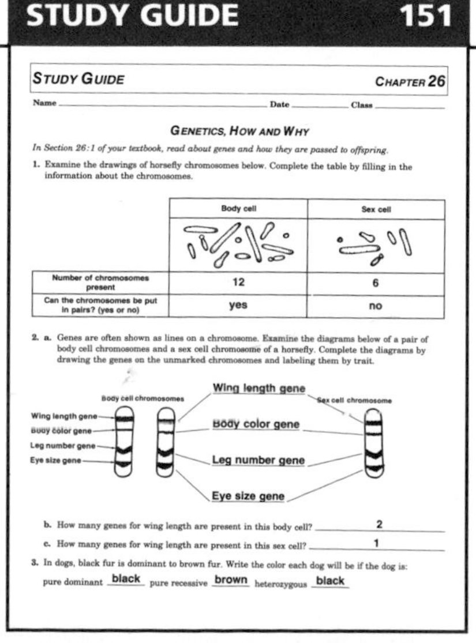

**STUDY GUIDE 151**

STUDY GUIDE — CHAPTER 26

Name ________ Date ________ Class ________

GENETICS, HOW AND WHY

*In Section 26:1 of your textbook, read about genes and how they are passed to offspring.*

1. Examine the drawings of horsefly chromosomes below. Complete the table by filling in the information about the chromosomes.

| | Body cell | Sex cell |
|---|---|---|
| Number of chromosomes present | 12 | 6 |
| Can the chromosomes be put in pairs? (yes or no) | yes | no |

2. a. Genes are often shown as lines on a chromosome. Examine the diagrams below of a pair of body cell chromosomes and a sex cell chromosome of a horsefly. Complete the diagrams by drawing the genes on the unmarked chromosomes and labeling them by trait.



b. How many genes for wing length are present in this body cell? 2

c. How many genes for wing length are present in this sex cell? 1

3. In dogs, black fur is dominant to brown fur. Write the color each dog will be if the dog is:
pure dominant black pure recessive brown heterozygous black

# TEACH

### Concept Development

▶ Point out that alternate forms of a given gene are called alleles.

▶ Make sure students understand that a person having one dominant and one recessive gene for a trait does not show a trait that is intermediate between the two.

▶ Reinforce that a dominant trait completely blocks the appearance of a recessive trait.

▶ Point out to students that not all traits are inherited as either a dominant or recessive trait, as the earlobe trait is. Other patterns of inherited traits will be discussed in the next chapter.

### ACTIVITY/Guest Speaker

Have a veterinarian come to your classroom to explain purebred animals.

### Independent Practice

**Study Guide**/*Teacher Resource Package,* p. 151. Use the Study Guide worksheet shown at the bottom of this page for independent practice.

### Guided Practice

Have students write down their answers to the margin question: genes that don't show their traits when dominant genes are present.

## GLENCOE TECHNOLOGY

**Videodisc**

**The Secret of Life**
*Question Segment*

*Answer Segment*

**The Secret of Life**
*Dominant Versus Recessive*

## TEACH

### Concept Development

▶ Stress that when fertilization does take place, it is usually random. The law of random fertilization states that any sperm cell can fertilize any egg cell.

▶ Emphasize that the dominant trait in a heterozygote is expressed over a recessive trait.

▶ Reinforce how gene combinations produce pure dominant, pure recessive, and heterozygous offspring.

### Check for Understanding

Have students form teams of three for a discussion group. After students discuss the following questions in their group, pull all of the teams together for a final wrap-up. (1) Where would you find a gene in the sperm and egg? (in the nucleus) (2) Where would you find a chromosome in the sperm and egg? (in the nucleus) (3) How many pairs of chromosomes would you find in a human sperm? (none) (4) How many pairs of chromosomes would you find in a human egg? (none) (5) How are sperm and eggs different from body cells? (Sex cells have half the number of chromosomes that a body cell has.) (6) How many chromosomes do we get from each of our parents? (23 or one of each pair)

### Reteach

Have students make models of a sperm and an egg using the following materials: cell–construction paper; chromosomes–string; and gene–piece of colored yarn glued to string. Tell them to put only three chromosomes in their model.

### Guided Practice

Have students write down their answers to the margin question: two.

Table 26-1 gives some traits in plants and animals. These traits are passed to offspring in the same way as the earlobe trait. They are dominant or recessive.

**TABLE 26–1. TRAITS OF PLANTS AND ANIMALS**

| Trait Found in | Dominant Trait | Recessive Trait |
|---|---|---|
| flies | long wings | short wings |
| pea plants | red flowers | white flowers |
| humans | can roll tongue | can't roll tongue |
| corn plants | normal height | dwarf |
| dogs | short hair | long hair |

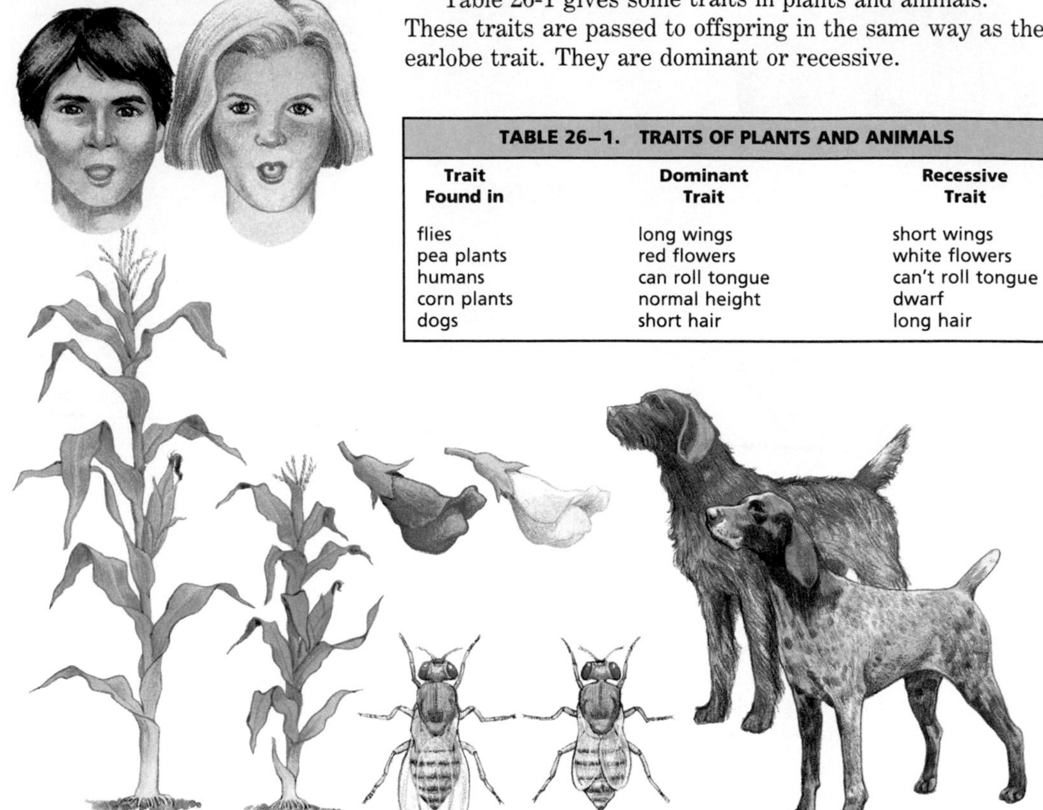

## When Both Parents Are Heterozygous

When one parent had both genes for attached earlobes and the other parent had both genes for free earlobes, all of their children had a combination of genes. The children had free earlobes. Now, let's suppose both parents have a combination of genes. What could their children look like?

**How many different kinds of sex cells can a heterozygous person make?**

If the mother's body cells have both genes, she can make two kinds of eggs. Each egg will have a gene for attached earlobes or a gene for free earlobes, but not both. If the father's body cells have both genes, he can make two kinds of sperm. Each sperm will have the gene for attached earlobes or the gene for free earlobes, but not both. Figure 26-5 shows the possible combinations when egg and sperm join. There are four possible combinations.



## OPTIONS

**How Can the Genes of Offspring Be Predicted?**/*Lab Manual,* pp. 219-222. Use this lab as an extension to predicting traits in offspring.

### ACTIVITY/Enrichment

Explain that there are many different varieties of dogs, cats, horses, and cows. Have students try to find pictures of the different varieties of these animals and give a class report on their findings. Pet food companies have excellent color drawings of dogs and cats and their ancestral lineage.

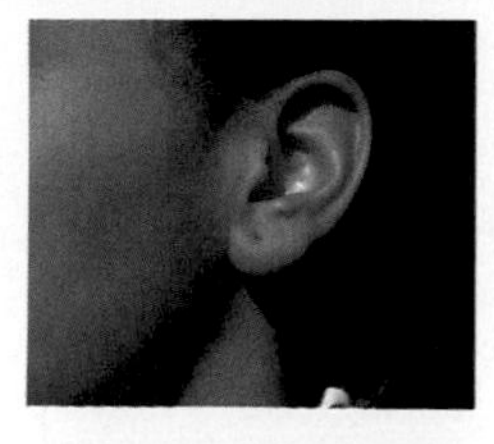

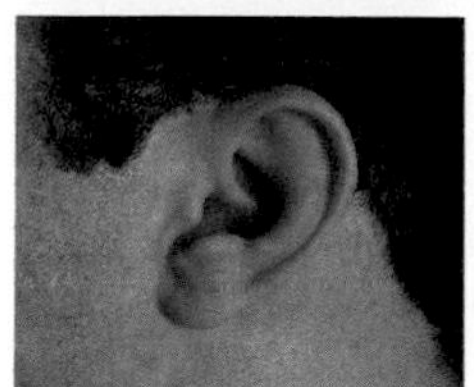

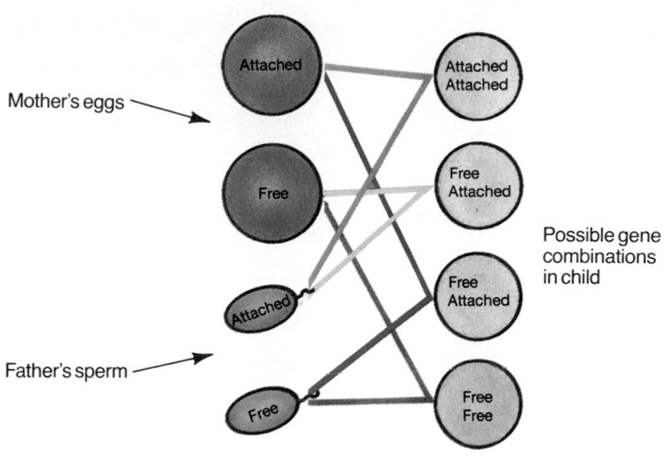


**Figure 26-5** What genes can a child have if both parents are heterozygous? ***FF, Ff, Ff, ff***

Table 26-2 shows what a child with each combination would look like. There are three chances in four that a child would have free earlobes and one chance in four that a child would have attached earlobes. We call this a 3 to 1 ratio.

| TABLE 26–2. GENE COMBINATIONS FOR EARLOBE TRAITS | | | |
|---|---|---|---|
| **With these genes from the egg:** | **sperm:** | **The child is:** | **The child has:** |
| free | free | pure dominant | free earlobes |
| free | attached | heterozygous | free earlobes |
| free | attached | heterozygous | free earlobes |
| attached | attached | pure recessive | attached earlobes |

## Check Your Understanding

1. How are chromosomes in sex cells different from those in body cells?
2. One person is pure dominant for a trait while another is heterozygous. How are their genes different?
3. How many different kinds of offspring could result if one parent is pure dominant and the other parent is pure recessive? How many different kinds of offspring could result if both parents are heterozygous?
4. **Critical Thinking:** Why does an individual usually carry only two genes for a certain trait?
5. **Biology and Math:** Why do living organisms have an even number of chromosomes in their body cells?



## RETEACHING 78

**RETEACHING: IDEA MAP** CHAPTER 26

Name ______ Date ______ Class ______

*Use with Section 26:1.*

**TRAITS OF PLANTS AND ANIMALS**

*Use the following terms to complete the idea map: recessive, dominant, heterozygous, pure recessive, pure dominant. You may need to use some terms more than once. Then answer the questions that follow.*

Genes
- dominant trait
  - pure dominant: dominant, dominant
  - heterozygous: dominant, recessive
- recessive trait
  - pure recessive: recessive, recessive

1. When is a living thing a pure dominant individual? **when it has two dominant genes for a trait**
2. What are dominant genes? **genes that keep other genes from showing their traits**
3. What are recessive genes? **genes that do not show their traits when dominant genes are present**
4. What is meant by a heterozygous individual? **A heterozygous individual has one dominant gene and one recessive gene for a trait.**
5. Could a child be born with a recessive trait if one parent is pure dominant for that trait? Explain. **No, the child would inherit a dominant gene from the parent who is pure dominant, so the recessive trait could not show.**
6. Could a child be born with a recessive trait if both parents are heterozygous for that trait? Explain. **Yes, there is one chance in four that the child will be born with the recessive trait.**

## TEACH

### Check for Understanding

Have students respond to the first three questions in Check Your Understanding.

### Reteach

**Reteaching**/*Teacher Resource Package*, p. 78. Use this worksheet to give students additional practice with traits of plants and animals.

**Extension:** Assign Critical Thinking, Biology and Math, or some of the **OPTIONS** available with this lesson.

## 3 APPLY

### Audiovisual

Show the video *Genetics: Observing Patterns of Inheritance,* Insight Media.

## 4 CLOSE

### Convergent Question

Pose the question: What would happen if all of the traits an organism inherited were recessive? (All of the organisms of the same type would have the same traits.)

### Answers to Check Your Understanding

1. Chromosomes are in pairs in body cells but there is only one chromosome of each pair in a sex cell.
2. A pure dominant person has two dominant genes; a heterozygous person has one dominant and one recessive.
3. one; three
4. An individual receives only one gene from each parent for a trait.
5. Chromosomes are paired, and two of any number is an even number.

# 26:2 Expected and Observed Results

## PREPARATION

### Materials Needed

Make copies of the Focus, Transparency Master, Study Guide, Skill, Critical Thinking/Problem Solving, Enrichment, Application, Reteaching, and Lab worksheets in the *Teacher Resource Package.*

▶ Obtain green: albino corn seeds for Lab 26-1.

▶ Purchase white and red kidney beans and paper bags for Lab 26-2.

### Key Science Words

Punnett square

### Process Skills

In Lab 26-1, students will observe, predict, and interpret data. In Lab 26-2 students will observe, formulate a model, and form hypotheses. In the Skill Check, they will calculate.

## 1 FOCUS

▶ The objectives are listed on the student page. Remind students to preview these objectives as a guide to this numbered section.

### MOTIVATION/Demonstration

To help students understand expected results, use a deck of 52 cards and have students predict how many times you could draw the following from the top of the deck: (a) an ace (4/52 or 1/13), (b) an ace of hearts (1/52), (c) a red ace (2/52 or 1/26).

### Guided Practice

Have students write down their answers to the margin question: which genes can combine when egg and sperm join.

Focus/*Teacher Resource Package*, p. 51. Use the transparency shown at the bottom of this page as an introduction to crossing traits in organisms.

# 26:2 Expected and Observed Results

**Objectives**

4. **Discuss** the purpose of a Punnett square.
5. **Compare** expected results and observed results.
6. **Explain** the importance of Gregor Mendel's work.

**Key Science Words**

Punnett square

How can knowing the types of genes that each parent has be helpful? You can predict what traits their children could have. Sometimes, however, the combinations of genes that you expect do not appear in the offspring. Let's find out why this is true.

## The Punnett Square

What does a Punnett square show?

We have seen how an egg and a sperm may combine to form an offspring. Each time, we figured out all the possible combinations of egg and sperm cells. There is an easier way to do this. This easy way is called the Punnett (PUH nuht) square. The **Punnett square** is a way to show which genes can combine when egg and sperm join. To make things easier, letters are used in place of genes. A large letter, such as ***F,*** is used for a dominant gene. The large ***F*** stands for free earlobe. A small letter, such as ***f,*** is used for a recessive gene. The small ***f*** stands for attached earlobe. A person with ***FF*** genes is pure dominant and has free earlobes. A person with ***Ff*** genes is heterozygous and has free earlobes. Notice that the large letter goes first in heterozygous organisms. A person with ***ff*** genes is pure recessive. What kind of earlobes does that person have?

Follow these six steps to determine the possible combinations of genes a child could have. In this example, both parents will be heterozygous. They will have a gene for free earlobes and a gene for attached earlobes. Use the letters ***Ff*** to stand for the gene pair in their body cells.

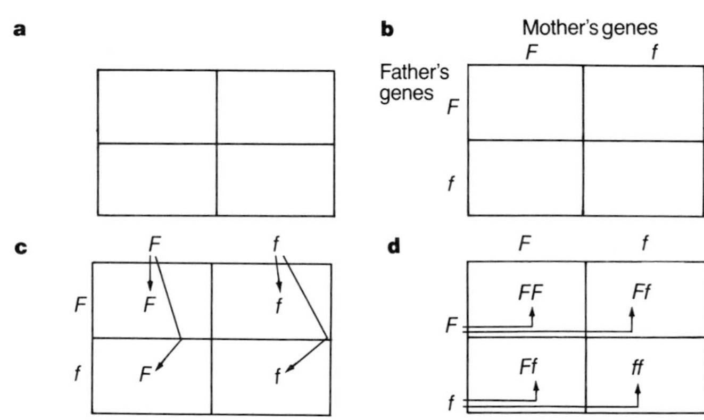


**Figure 26-6** How to make and use a Punnett square



## OPTIONS

### Science Background

Geneticists use a Punnett square to calculate the results of a genetic cross. The Punnett square was developed by Reginald Punnett in the early 1900s. In the Punnett square, the known male gametes are placed along either the top or left edge of the square and the eggs along the remaining side. The expected offspring appear in the boxes inside the square when all of the possible male and female gametes have been combined.

**FOCUS 51**

FOCUS CHAPTER 26

Name Date Class

CROSSBREEDING DOGS

If a short-haired pointer is crossed with an Irish setter, what kind of hair do you think the puppies will have? On what are you basing your answer? How could you find out if you are not sure?

1. Draw a Punnett square as shown in Figure 26-6a. Each small box stands for one possible combination of genes that could show up in the offspring. Each combination of genes comes from a sperm cell fertilizing an egg cell.
2. Decide what kinds of genes will be in the sex cells of each parent. Write the letters that stand for the genes in the mother's egg cells across the top of the square. Figure 26-6b shows you how. In this case, put the two genes for eggs, ***F*** and ***f***, across the top.
3. Now, write the letters that stand for genes in the father's sperm along the side of the square. There are two possible genes for the sperm. They are ***F*** (free earlobes) and ***f*** (attached earlobes).
4. Copy the letters that appear at the top of the square into the boxes below each letter. Figure 26-6c shows how.
5. Copy the letters that appear at the side of the square into the boxes next to each letter. Figure 26-6d shows how.
6. Look at the small boxes in the Punnett square. They show the possible combinations of eggs and sperm. When egg and sperm combine, a new organism develops. The boxes also show what combinations of genes the organism could have. In our example, a child could have one of the following combinations: ***FF***, ***Ff***, or ***ff***.

The Punnett square in Figure 26-7 shows you how the child could look. Remember, ***F*** is dominant over ***f***. There are two different combinations of genes that would result in a child with free earlobes—***FF*** and ***Ff***. Notice that three out of the four boxes contain either ***FF*** or ***Ff***. Three out of four gene combinations of egg and sperm would result in a child with free earlobes.

There is only one combination of genes that would result in a child with attached earlobes—***ff***. Only one out of four boxes in the Punnett square has ***ff***. On average, one out of four gene combinations would result in a child with attached earlobes.

## Expected Results

The Punnett square that you drew shows what kinds of traits offspring can have. It shows what to expect when the sperm and egg of two parents join. Expected results are what can be predicted in offspring based on the genetic traits of parents. We predicted that one out of four children would have attached earlobes. We could do this because we knew what genes the parents had.

### Bio Tip

**Consumer:** Sometimes an ear of white corn contains a yellow kernel of corn. A pollen grain carrying the dominant yellow gene was carried by the wind from a nearby garden. It fertilized the egg in one of the kernels of the white ear of corn.

**Figure 26-7** Three out of four gene combinations are for free earlobes.

| | *F* | *f* |
|---|---|---|
| *F* | *FF* Free earlobes | *Ff* Free earlobes |
| *f* | *Ff* Free earlobes | *ff* Attached earlobes |

553

# 2 TEACH

### MOTIVATION/Demonstration

Make a transparency of a Punnett square to show the earlobe problem step-by-step. Make sure that students realize that for the pair of genes appearing in each box of the Punnett square, one gene is from each parent.

### Independent Practice

**Study Guide**/*Teacher Resource Package*, p. 152. Use the Study Guide worksheet shown at the bottom of this page for independent practice.

### Student Journal

The Smiths and Taylors each had baby girls born in the same hospital on Tuesday. The Smiths think the babies were mixed up at birth because both of the Smiths have free earlobes but their baby has attached earlobes. They noticed that the Taylor baby and Mr. Taylor have free earlobes but Mrs. Taylor has attached earlobes. Since the Taylor baby has earlobes like both of the Smiths, the Smiths are sure the Taylor baby belongs to them rather than the Taylors. Since the Smith's baby has attached earlobes like Mrs. Taylor, the Smiths are sure this baby belongs to the Taylors. Develop a series of Punnett squares to explain why the Smiths could be wrong. Show how each of the two families could have a child with free earlobes.

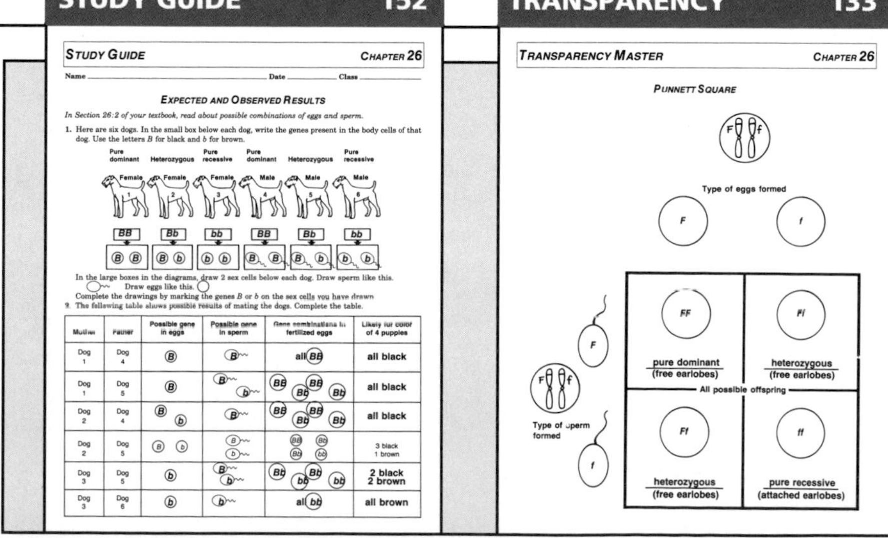

STUDY GUIDE 152

STUDY GUIDE CHAPTER 26

Name ______ Date ______ Class ______

EXPECTED AND OBSERVED RESULTS

*In Section 26:2 of your textbook, read about possible combinations of eggs and sperm.*

1. Here are six dogs. In the small box below each dog, write the genes present in the body cells of that dog. Use the letters *B* for black and *b* for brown.



In the large boxes in the diagrams, draw 2 sex cells below each dog. Draw sperm like this. Draw eggs like this.
Complete the drawings by marking the genes *B* or *b* on the sex cells you have drawn

2. The following table shows possible results of mating the dogs. Complete the table.

| Mother | Father | Possible gene in eggs | Possible gene in sperm | Gene combinations in fertilized eggs | Likely fur color of 4 puppies |
|---|---|---|---|---|---|
| Dog 1 | Dog 4 | B | B | all BB | all black |
| Dog 1 | Dog 5 | B | B b | BB BB Bb Bb | all black |
| Dog 2 | Dog 4 | B b | B | BB BB Bb Bb | all black |
| Dog 2 | Dog 5 | B b | B b | BB Bb Bb bb | 3 black 1 brown |
| Dog 3 | Dog 5 | b | B b | Bb Bb bb bb | 2 black 2 brown |
| Dog 3 | Dog 6 | b | b | all bb | all brown |

TRANSPARENCY 133

TRANSPARENCY MASTER CHAPTER 26



## OPTIONS

**Transparency Master**/*Teacher Resource Package*, p. 133. Use the Transparency Master shown here to teach the Punnett square.

## Lab 26-1 Corn Genes

### Overview

Students will find that germinating corn seeds will either have green or white leaves. From their data, they will see that not all corn seeds have genes for green leaves. Until germination, they cannot tell which genes the seeds have because all the seeds look alike.

**Objectives:** Students will be able to (1) **classify** offspring by genetic traits, (2) **infer** which trait, green or albino, is dominant, and (3) **predict** what would happen to future generations of offspring with the albino trait.

**Time Allotment:** 30 minutes the first day; 10 minutes each day for 4 days after sprouting

### Preparation

▶ Order green: albino corn seeds in advance of the activity. Get petri dishes, towels, and light sources ready. Corn seeds can be ordered from Carolina Biological Supply, Catalog Number 17-7130.

Lab 26-1 worksheet/*Teacher Resource Package*, pp. 101-102.

### Teaching the Lab

▶ If you have small trays you could have students grow the seeds in soil.

### ASSESSMENT

**Content:** Tell students that a seed company accidentally mixed two kinds of corn seeds together—one that grows green plants and the other that grows albino plants. These seeds were placed in one of 20 large storage bins and the seed company is not sure which of the bins has the mixed seeds. These seeds were to be sold to farmers who would want to grow green corn plants only. Have students use their data from this lab to outline the steps they would take to find out which bin has the mixed seeds. Have them explain why the farmers would not want to buy the mixed seeds at a reduced price.

# Corn Genes

## Lab 26–1

### Problem: Do all corn seeds have genes for becoming green plants?

#### Skills

observe, predict, interpret data

#### Materials

20 corn seeds
paper towels
petri dish
water
wax marking pencil

#### Procedure

1. Copy the data table.
2. Look at the seeds carefully. Can you tell which seeds will grow into green plants and which will grow into white plants?
3. Moisten a paper towel with water.
4. Place the towel in the bottom half of the petri dish, Figure A. Fold it to fit.
5. Place the seeds on the towel. Cover the dish. Write your name on the cover.
6. Place the dish under a light and check it every day. Add more water if needed.
7. **Observe:** When the first seeds begin to sprout, record the leaf color. This is day 1.
8. Check your corn seeds for four more days. Record the numbers of new plants with green and with white leaves each day.

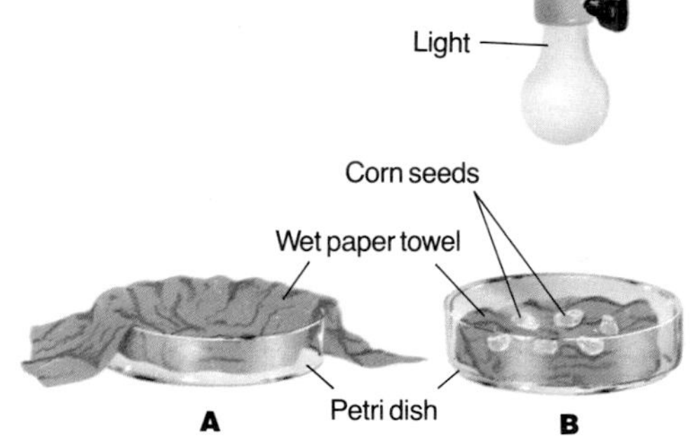




#### Data and Observations

1. What colors were the plants when they first sprouted?
2. Were there more green plants or white plants five days after sprouting?

| | NUMBER OF CORN SEEDS SPROUTED | NUMBER WITH GREEN LEAVES | NUMBER WITH WHITE LEAVES |
|---|---|---|---|
| Day 1 | 1 | 1 | 0 |
| Day 2 | 2 | 2 | 0 |
| Day 3 | 6 | 4 | 2 |
| Day 4 | 7 | 6 | 1 |
| Day 5 | 3 | 2 | 1 |
| Total | 19 | 15 | 4 |

#### Analyze and Apply

Corn plants that remain white several days after sprouting have the albino trait. These plants soon die because they cannot make food. These plants are missing chlorophyll.

1. Could you tell before the seeds sprouted which would form green leaves? Explain.
2. **Predict:** What results would you see if the seeds came from heterozygous parents?
3. Why do albino plants not live as long as green plants?
4. **Apply:** Do you think most plants in fields and forests would be albino or green? Explain your answer.

#### Extension

**Design an experiment** to find out if the green corn plants in this lab carry the albino gene as a recessive trait.

## ANSWERS

### Data and Observations

1. white
2. green

### Analyze and Apply

1. No, the trait was not visible in the seed; the trait is in the genes.
2. green plants and albino plants in a 3:1 ratio
3. They don't have chlorophyll.
4. Green; albino plants would die because they couldn't make food.

## Career Close-Up

### Animal Breeder

An animal breeder works with a veterinarian or animal breeding specialist to keep records on animals and record what offspring of animals look like. Egg and milk production, size, and coat color are but a few of the genetic traits that animals show. Records of the traits of parent animals can help the animal breeder predict what traits can show in the offspring.

Many animal breeders work for state universities. Others may work on large farms that specialize in breeding. For example, farms that raise hundreds of cows may use an animal breeder to follow weight gain of cattle or record milk production. Animal breeders may work on horse ranches that specialize in race horses. They can also work at kennels where dogs are bred to be sold.

Wherever an animal breeder works, much of the training is learned on the job. A knowledge of biology and genetics is useful in this career.

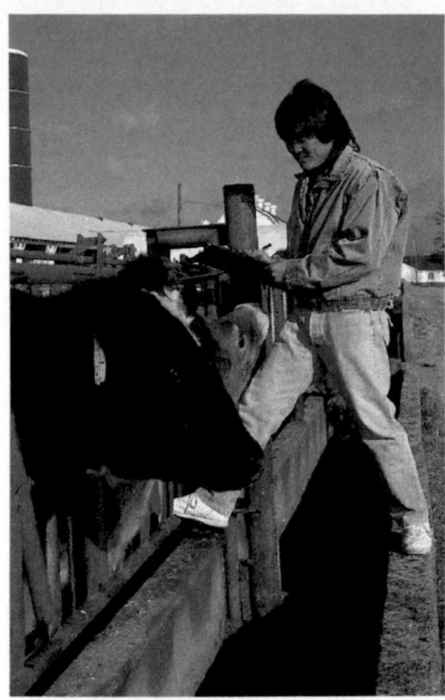

Some animal breeders record the milk production of cows.

The type of chin that a person has is also a genetic trait. A person with a cleft chin has a small indentation in the middle of the chin. A cleft chin is a dominant trait, Figure 26-8. A smooth chin is a recessive trait.

a

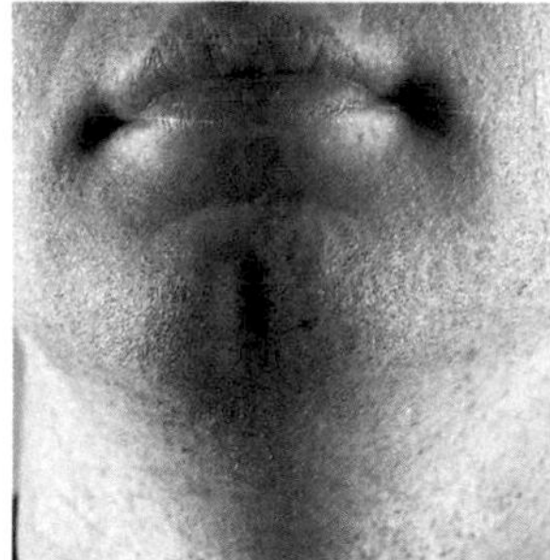

b

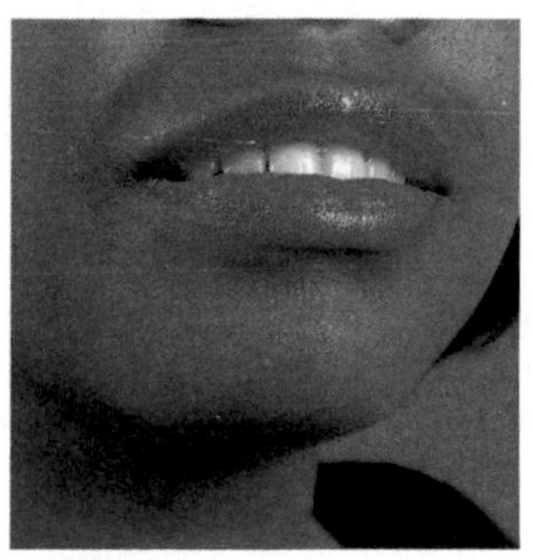

**Figure 26-8** Person with cleft chin (a) and smooth chin (b)



**SKILL 27**

SKILL: USE NUMBERS CHAPTER 26

Name ________ Date ________ Class ________

*For more help, refer to the Skill Handbook, pages 718-719.* *Use after Section 26:2.*

**FUR COLOR IN RABBITS**

Mary and Jake are breeding rabbits to sell to a pet shop. They will be paid $20.00 for each white rabbit and $8.00 for each rabbit of any other color. They did not know how the trait for fur color is inherited in rabbits. To find out, they put many mating pairs of rabbits together and kept careful records of the offspring. Each pair of rabbits produced four offspring.

*Use the table and your knowledge of heredity to answer the questions below.*

| Mating | Number of mating pairs | Fur color of parents | Fur color of offspring | |
|---|---|---|---|---|
| | | | White | Gray |
| A | 10 | white × white | 40 | 0 |
| B | 10 | gray × white | 0 | 40 |
| C | 10 | gray × gray | 10 | 30 |
| D | 10 | gray × gray | 0 | 40 |
| E | 10 | gray × white | 20 | 20 |
| Total | 50 | | 70 | 130 |

1. Complete the table above by filling in the numbers of mating pairs and the total number of offspring of each color.
2. How much money could Mary and Jake make by selling the white offspring? $1400
3. How much money could they make by selling the gray offspring? $1040
4. Suppose the pet shop changes its policy and will pay $20.00 for gray rabbits and $8.00 for white rabbits. How much money would Mary and Jake make by selling the offspring of the mating pairs? $2600 for the gray rabbits and $560 for the white rabbits, for a total of $3160
5. Which gene for fur color is dominant? The gene for gray fur.
6. Which mating will produce the most white rabbits? white × white

### OPTIONS

Skill/*Teacher Resource Package*, p. 27. Use the Skill worksheet shown here for students to use numbers in understanding breeding.

## Career Close-Up

### Animal Breeder

#### Background

The animal breeder has to keep accurate records about expected and observed traits of offspring and their parents. Today, many of the records are placed in a computer file for ease of storage and access. The animal breeder is expected to be able to cross parents of known genotypes to get a specific number of offspring of desired genotypes and phenotypes.

#### Related Careers

animal caretaker and pet store clerk

#### Job Requirements

A high school diploma and an interest in genetics and animals are necessary requirements for becoming an animal breeder. On-the-job training by the employer and some courses at a community college are helpful.

#### For More Information

Check the library for government publications listing career opportunities and also with the state department of agriculture office in your state.

### Student Journal

Have students make a chart listing the traits that animal breeders find beneficial in certain animals such as hunting dogs, pigs, beef cattle, and chickens. Students could interview hunters, farmers, veterinarians, or livestock ranchers. Direct the students to list why these persons think the traits are beneficial.

## Lab 26-2 Punnett Squares

### Overview

Students will use beans of two colors to simulate two genes for a trait found in sperm and eggs. By selecting a bean at random from one bag (potential egg cell) and a bean from a second bag (potential sperm cell), they will simulate fertilization of the egg.

**Objectives:** Students will be able to: (1) **demonstrate** how random fertilization works, (2) **simulate** a genetic cross using beans as sperm and egg cells, and (3) **compare** expected and observed data.

**Time Allotment:** 45 minutes

### Preparation

▶ Get the beans and paper bags for the activity.

**Lab 26-2 worksheet**/*Teacher Resource Package*, pp. 103-104.

### Teaching the Lab

▶ Beans can be stored in zip-lock type plastic bags for future years.

▶

| | *R* | *r* |
|---|---|---|
| *R* | *RR* | *Rr* |
| *r* | *Rr* | *rr* |

Expected ratios for offspring are 3 red: 1 white; 1*RR*: 2*Rr*: 1*rr*

▶ Optional: Paper bags can be marked male and female.

**Performance:** Have students remove five red beans from each of the two bags and add five white beans to each of the two bags. Direct the students to make a hypothesis about the number of pure dominant, heterozygous, and pure recessive offspring they think they will get if they do 40 trials of combining a sperm and an egg. Tell students to repeat the lab procedures to get 40 offspring and compare this data to the original lab activity. Have students restate their hypothesis and tell why they accept or reject it.

# Punnett Squares

**Lab 26–2**

## Problem: What determines how offspring will look?

### Skills

observe, formulate a model, form hypotheses

### Materials

40 white beans
40 red beans
2 paper bags

### Procedure

1. Copy the data table.
2. Place 20 white and 20 red beans into a paper bag. Label the bag female parent. These beans represent eggs.
3. Place 20 white and 20 red beans into a second paper bag. Label this bag male parent. These beans represent sperm.
4. Set up a Punnett square to show the parent gene types. Use ***R*** to stand for the dominant gene, red. Use ***r*** to stand for the recessive gene, white. Remember that each parent can produce sex cells carrying the red trait or the white trait. They are both heterozygous.
5. Complete your Punnett square to show the expected offspring of these parents.
6. Shake the bags.
7. Reach into each bag without looking and remove one bean. The two beans stand for the gene combination that results when sperm and egg join.
8. **Observe:** Look at the beans. Record the colors of the beans in your table next to trial 1. Use a check mark to record your results in the proper column.
9. Put the two beans back into the bags from which they came.
10. Write a **hypothesis** to show how many red/red ***(RR)***, red/white ***(Rr)***, and white/white ***(rr)*** pairs you will get in 40 trials. Use your Punnett square for help.
11. Repeat steps 6–9 for 39 more trials.
12. Using your table, find the totals of red/red (***RR***), red/white (***Rr***), and white/white (***rr***) combinations.

### Data and Observations

1. What gene pairs result in red offspring?
2. What gene pairs result in white offspring?

| | FEMALE PARENT | | MALE PARENT | | GENE PAIRS |
|---|---|---|---|---|---|
| TRIAL | RED BEANS | WHITE BEANS | RED BEANS | WHITE BEANS | |
| 1 | ✓ | | ✓ | | *RR* |
| 2 | | ✓ | | ✓ | *rr* |
| 3 | | ✓ | | ✓ | *rr* |
| 4 | ✓ | | | ✓ | *Rr* |
| 5 | | ✓ | ✓ | | *Rr* |
| | | | | | |
| | | | | | |
| 40 | | ✓ | | ✓ | *rr* |

### Analyze and Apply

1. Out of 40 offspring, how many did you expect to be pure dominant ***(RR)***, heterozygous ***(Rr)***, and pure recessive ***(rr)***?
2. Which of the three gene combinations did you expect in greatest number?
3. **Check your hypothesis:** Was your hypothesis supported by your data? Why or why not?
4. **Apply:** What determines how offspring will look?

### Extension

**Design an experiment** to show that the more trials you do, the closer your observed results come to your expected results.



## ANSWERS

### Data and Observations

1. ***RR, Rr***
2. ***rr***

### Analyze and Apply

1. 10 ***RR***; 20 ***Rr***; 10 ***rr***
2. ***Rr***
3. students who hypothesized that there will be 10 ***RR***, 20 ***Rr***, 10 ***rr*** in 40 trials will say that their hypotheses were supported.
4. the combination of genes that come together in the offspring

Let ***I*** stand for cleft chin and ***i*** stand for smooth chin. It might be easier to remember the symbols if you think of ***I*** standing for indentation. How might a child appear if one parent were ***Ii*** and the other were ***ii?*** Use the Punnett square in Figure 26-9 to get the answer. Two out of four gene combinations could be for children with cleft chins. Note that these children with cleft chins have ***Ii*** genes. The other two combinations could be for children with smooth chins, or children with ***ii*** genes. These results are what is expected. Would you expect these parents to have a pure dominant child? Why or why not?

| | *I* | *i* |
|---|---|---|
| *i* | *Ii* Cleft chin | *ii* Smooth chin |
| *i* | *Ii* Cleft chin | *ii* Smooth chin |

**Figure 26-9** Children can have two possible chin types if one parent is heterozygous and the other is pure recessive.

## Observed Results

We know that the results expected from the Punnett square do not always occur in every family. Look at Table 26-3. These data are from ten different families that show the cleft chin trait. In each family, one parent has the genes ***Ii*** and has a cleft chin. The other parent has the genes ***ii*** and has a smooth chin.

The table shows exactly what you would see if you looked at the children of these families. The traits actually seen in offspring when parents with certain genetic traits mate are the observed results. Using a Punnett square allows you to predict that half the children in these families could have cleft chins. Half the children could have smooth chins. The observed results, however, do not exactly match the expected results because you don't know which sperm and egg will join.

**TABLE 26–3. OFFSPRING AND CHIN TYPES**

| Family | Number of Children in Family | Number with Cleft Chin | Number with Smooth Chin |
|---|---|---|---|
| A | 2 | 0 | 2 |
| B | 1 | 1 | 0 |
| C | 5 | 3 | 2 |
| D | 4 | 2 | 2 |
| E | 2 | 1 | 1 |
| F | 3 | 1 | 2 |
| G | 1 | 1 | 0 |
| H | 6 | 4 | 2 |
| I | 2 | 1 | 1 |
| J | 3 | 1 | 2 |
| Totals | 29 | 15 | 14 |



## TEACH

### Concept Development

▶ Explain that each child born to two heterozygous parents has one out of four chances of having a smooth chin even though the first child has a smooth chin.

▶ Explain that observed results are not calculated.

▶ Point out that the expected offspring from each family shown in Table 26-3 is one child with cleft chin for each child having a smooth chin.

▶ Explain that the more data you have to look at, the closer the observed will be to the expected. (The totals on Table 26-3 show this.)

▶ Emphasize that families D and E are closer to expected results than other families.

### Independent Practice

**Study Guide**/*Teacher Resource Package,* p. 153. Use the Study Guide worksheet shown at the bottom of this page for independent practice.

## GLENCOE TECHNOLOGY

**Videodisc**

**Biology: The Dynamics of Life**

Animation: *Punnett Squares*

Disc 1, Side 1, Ch. 29

## STUDY GUIDE 153

STUDY GUIDE CHAPTER 26

Name ______ Date ______ Class ______

**EXPECTED AND OBSERVED RESULTS**

*In Section 26:2 of your textbook, read about solving genetics problems using the Punnett square.*

3. Examine the diagrams below. Each is a step in the Punnett square method. Put the steps in order by writing the numbers 1 to 4 below them on the correct blanks.

2 3 1 4

4. What do the letters outside the Punnett square stand for? genes in eggs and sperm

What do the letters inside each box stand for? genes in offspring

5. Examine the following Punnett squares and circle those that are correct.

6. Complete the following to determine the expected offspring.

## OPTIONS

### ACTIVITY/Challenge

Have students use the Punnett squares to determine the offspring from the following crosses of people having cleft or smooth chin:

a. ***II*** x ***II*** (all cleft chin)

b. ***ii*** x ***ii*** (all smooth chin)

c. ***Ii*** x ***II*** (1/2 ***Ii***: 1/2 ***II***; all cleft chin)

## TEACH

### Idea Map

Have students use the idea map as a study guide to the major concepts of this section.

### Skill Check

They should get about 1/2 heads and 1/2 tails. They are able to predict expected results.

### Guided Practice

Have students write down their answers to the margin question: The more results you observe, the closer observed results will be to expected results.

### Check for Understanding

Review expected and observed results. Make sure students understand that expected results can be calculated, but observed results are actually seen. Set up some simple problems that students can do by using the Punnett square. Do one problem as a class so they can learn from any mistakes.

**Reteach**

Have students use their Punnett squares from Check for Understanding and write out the expected results.

### Independent Practice

**Study Guide**/*Teacher Resource Package*, p. 154. Use the Study Guide worksheet shown at the bottom of this page for independent practice.

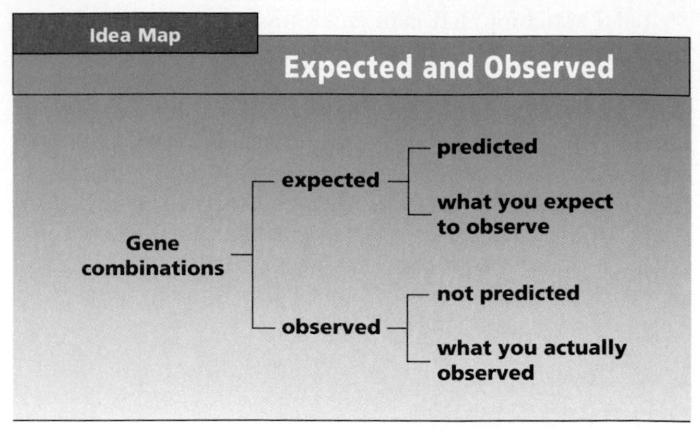


**Study Tip:** Use this idea map as a study guide to expected and observed results. The idea map shows how they are different.

**Skill Check**

**Calculate:** Calculate how many heads and tails you would get if you were to flip a coin 10 times and then 50 times. Explain which you are able to calculate, expected or observed results. *For more help, refer to the* ***Skill Handbook,*** *pages 718-719.*

**When are observed results similar to expected results?**

A little more than half the children have cleft chins. Nearly half the children have smooth chins. Each family by itself may not show the results you expect to see. If you add the results together, they are close to what you expect to see in a Punnett square. In fact, the more results you observe, the closer the observed results will be to the expected results.

Think about what happens when you flip a coin. You expect the coin to come up heads or tails. If you flip a coin two times, does it come up heads once and tails once? It may. It may come up heads twice or tails twice. Two heads or two tails are not what you expect. In fact, if you flip the coin four times you may even see four heads or four tails. Why?

Each time you flip a coin, you are starting over. If you flip the coin once and see heads, it does not mean that the next flip must come up tails. The coin can land either way each time you flip it. With just a few flips, you are less likely to see the exact same number of heads as tails. If you flip the coin many, many times, the number of heads and tails should be about equal. The observed results will be closer to the expected results.

## Mendel's Work

Let's go back in time to 1865. An Austrian monk named Gregor Mendel saw certain traits in the garden pea plants he grew in his garden. Mendel counted and recorded the



### OPTIONS

**ACTIVITY/Challenge**

Ask students to find out how many students in their neighborhood have a cleft chin and how many have a smooth chin. Ask which trait they expect to see more often and why.

**Critical Thinking/Problem Solving/** *Teacher Resource Package*, p. 26. Use the worksheet shown here to teach inherited traits.

**CRITICAL THINKING 26**

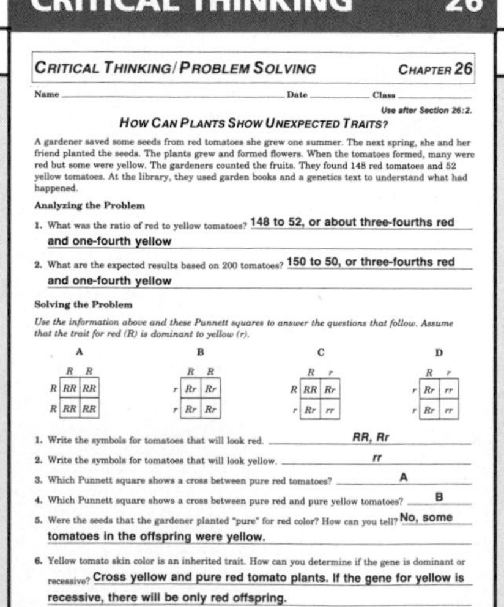

CRITICAL THINKING/PROBLEM SOLVING — CHAPTER 26

Name ______ Date ______ Class ______

Use after Section 26:2.

HOW CAN PLANTS SHOW UNEXPECTED TRAITS?

A gardener saved some seeds from red tomatoes she grew one summer. The next spring, she and her friend planted the seeds. The plants grew and formed flowers. When the tomatoes formed, many were red but some were yellow. The gardeners counted the fruits. They found 148 red tomatoes and 52 yellow tomatoes. At the library, they used garden books and a genetics text to understand what had happened.

**Analyzing the Problem**

1. What was the ratio of red to yellow tomatoes? **148 to 52, or about three-fourths red and one-fourth yellow**
2. What are the expected results based on 200 tomatoes? **150 to 50, or three-fourths red and one-fourth yellow**

**Solving the Problem**

*Use the information above and these Punnett squares to answer the questions that follow. Assume that the trait for red (R) is dominant to yellow (r).*

A

| | R | R |
|---|---|---|
| R | RR | RR |
| R | RR | RR |

B

| | R | R |
|---|---|---|
| r | Rr | Rr |
| r | Rr | Rr |

C

| | R | r |
|---|---|---|
| R | RR | Rr |
| r | Rr | rr |

D

| | R | r |
|---|---|---|
| r | Rr | rr |
| r | Rr | rr |

1. Write the symbols for tomatoes that will look red. **RR, Rr**
2. Write the symbols for tomatoes that will look yellow. **rr**
3. Which Punnett square shows a cross between pure red tomatoes? **A**
4. Which Punnett square shows a cross between pure red and pure yellow tomatoes? **B**
5. Were the seeds that the gardener planted "pure" for red color? How can you tell? **No, some tomatoes in the offspring were yellow.**
6. Yellow tomato skin color is an inherited trait. How can you determine if the gene is dominant or recessive? **Cross yellow and pure red tomato plants. If the gene for yellow is recessive, there will be only red offspring.**

**STUDY GUIDE 154**

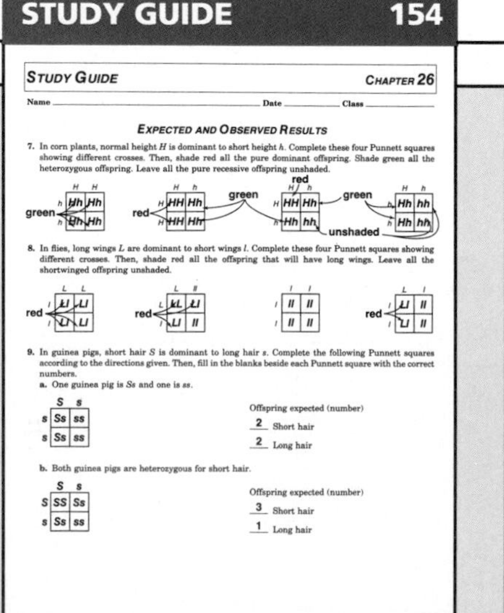

STUDY GUIDE — CHAPTER 26

Name ______ Date ______ Class ______

EXPECTED AND OBSERVED RESULTS

7. In corn plants, normal height *H* is dominant to short height *h*. Complete these four Punnett squares showing different crosses. Then, shade red all the pure dominant offspring. Shade green all the heterozygous offspring. Leave all the pure recessive offspring unshaded.



| | H | H |
|---|---|---|
| h | Hh | Hh |
| h | Hh | Hh |

| | H | h |
|---|---|---|
| H | HH | Hh |
| H | HH | Hh |

| | H | h |
|---|---|---|
| H | HH | Hh |
| h | Hh | hh |

| | H | h |
|---|---|---|
| h | Hh | hh |
| h | Hh | hh |

8. In flies, long wings *L* are dominant to short wings *l*. Complete these four Punnett squares showing different crosses. Then, shade red all the offspring that will have long wings. Leave all the shortwinged offspring unshaded.



| | L | L |
|---|---|---|
| l | Ll | Ll |
| l | Ll | Ll |

| | L | ll |
|---|---|---|
| L | LL | Ll |
| l | Ll | ll |

| | l | l |
|---|---|---|
| l | ll | ll |
| l | ll | ll |

| | L | l |
|---|---|---|
| l | Ll | ll |
| l | Ll | ll |

9. In guinea pigs, short hair *S* is dominant to long hair *s*. Complete the following Punnett squares according to the directions given. Then, fill in the blanks beside each Punnett square with the correct numbers.

a. One guinea pig is *Ss* and one is *ss*.

| | S | s |
|---|---|---|
| s | Ss | ss |
| s | Ss | ss |

Offspring expected (number)
**2** Short hair
**2** Long hair

b. Both guinea pigs are heterozygous for short hair.

| | S | s |
|---|---|---|
| S | SS | Ss |
| s | Ss | ss |

Offspring expected (number)
**3** Short hair
**1** Long hair

a

Round Wrinkled

b Green Yellow

**Figure 26-10** Gregor Mendel used a scientific method to study pea plants (a). Two of the traits he studied are shape and color of the seed (b).

traits he saw in the pea plants. He used a scientific method to do hundreds of experiments. Using the data he gathered and his knowledge of mathematics, he was able to explain some basic laws of genetics. Mendel explained what dominant and recessive traits were. He also showed how these traits passed from parent to offspring. How did he do it?

**Why is Mendel's work important?**

Mendel looked at several traits in the garden pea plant, Figure 26-10. One of those traits was height of the plant. He noticed that tall parent plants mated with short parent plants always produced tall plants, Figure 26-11. Remember that if ***T*** stood for the tall gene and ***t*** stood for the short gene, the parents would be ***TT*** and ***tt.*** What genes would the offspring have? Would the offspring be heterozygous or pure?

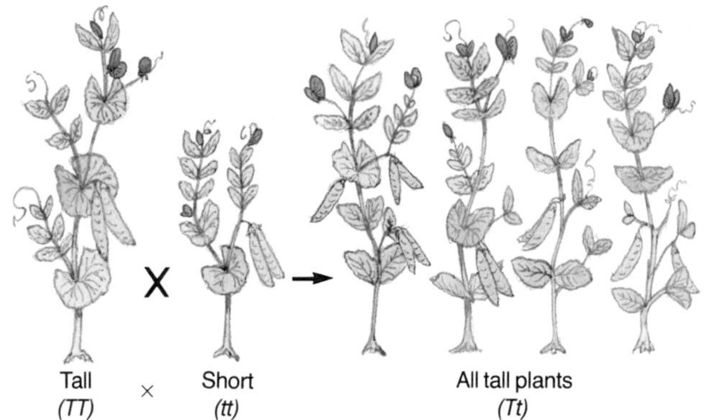


**Figure 26-11** Mendel's results when he mated a tall parent plant with a short parent plant



## TEACH

### Concept Development

▶ Point out that Mendel did not know about meiosis but assumed that traits, which he called unit characters, separated from each other when gametes were formed.

### Independent Practice

**Study Guide**/*Teacher Resource Package,* p. 155. Use the Study Guide worksheet shown at the bottom of this page for independent practice.

### Guided Practice

Have students write down their answers to the margin question: He explained basic laws of genetics and dominant and recessive traits.

### Connection: Social Studies

Have students give a report on the ideas of some scientists that extended Mendelian genetics. Suggest Carl Correns, Wilhelm Johannsen, W. E. Castle, or Hugo DeVries.

### *Student Journal*

Have students find out what scientists who lived at the same time as Mendel thought about how traits were inherited by living things. Tell them to use references in the library and focus their reading on the period of time from 1850-1875. Direct students to write a page or two in their journals about their findings.

## STUDY GUIDE 155

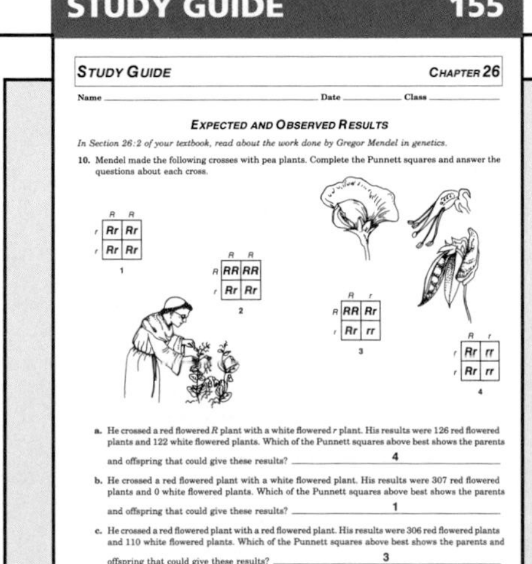

STUDY GUIDE CHAPTER 26

Name ______ Date ______ Class ______

**EXPECTED AND OBSERVED RESULTS**

*In Section 26:2 of your textbook, read about the work done by Gregor Mendel in genetics.*

10. Mendel made the following crosses with pea plants. Complete the Punnett squares and answer the questions about each cross.

1

| | R | R |
|---|---|---|
| r | Rr | Rr |
| r | Rr | Rr |

2

| | R | R |
|---|---|---|
| R | RR | RR |
| r | Rr | Rr |

3

| | R | r |
|---|---|---|
| R | RR | Rr |
| r | Rr | rr |

4

| | R | r |
|---|---|---|
| r | Rr | rr |
| r | Rr | rr |

a. He crossed a red flowered *R* plant with a white flowered *r* plant. His results were 126 red flowered plants and 122 white flowered plants. Which of the Punnett squares above best shows the parents and offspring that could give these results? **4**

b. He crossed a red flowered plant with a white flowered plant. His results were 307 red flowered plants and 0 white flowered plants. Which of the Punnett squares above best shows the parents and offspring that could give these results? **1**

c. He crossed a red flowered plant with a red flowered plant. His results were 306 red flowered plants and 110 white flowered plants. Which of the Punnett squares above best shows the parents and offspring that could give these results? **3**

d. He crossed a red flowered plant with a red flowered plant. His results were 300 red flowered plants and 0 white flowered plants. Which of the Punnett squares above best shows the parents and offspring that could give these results? **2**

## ENRICHMENT 28

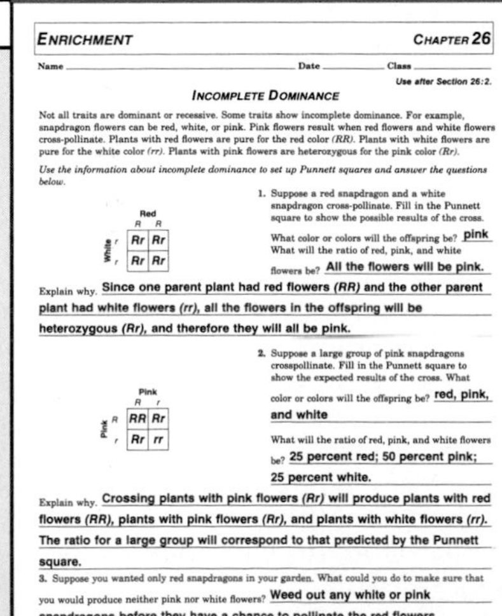

ENRICHMENT CHAPTER 26

Name ______ Date ______ Class ______

*Use after Section 26:2.*

**INCOMPLETE DOMINANCE**

Not all traits are dominant or recessive. Some traits show incomplete dominance. For example, snapdragon flowers can be red, white, or pink. Pink flowers result when red flowers and white flowers cross-pollinate. Plants with red flowers are pure for the red color *(RR)*. Plants with white flowers are pure for the white color *(rr)*. Plants with pink flowers are heterozygous for the pink color *(Rr)*.

*Use the information about incomplete dominance to set up Punnett squares and answer the questions below.*

Red (across), White (down)

| | R | R |
|---|---|---|
| r | Rr | Rr |
| r | Rr | Rr |

1. Suppose a red snapdragon and a white snapdragon cross-pollinate. Fill in the Punnett square to show the possible results of the cross.

What color or colors will the offspring be? **pink**
What will the ratio of red, pink, and white flowers be? **All the flowers will be pink.**

Explain why. **Since one parent plant had red flowers *(RR)* and the other parent plant had white flowers *(rr)*, all the flowers in the offspring will be heterozygous *(Rr)*, and therefore they will all be pink.**

Pink (across), Pink (down)

| | R | r |
|---|---|---|
| R | RR | Rr |
| r | Rr | rr |

2. Suppose a large group of pink snapdragons crosspollinate. Fill in the Punnett square to show the expected results of the cross. What color or colors will the offspring be? **red, pink, and white**

What will the ratio of red, pink, and white flowers be? **25 percent red; 50 percent pink; 25 percent white.**

Explain why. **Crossing plants with pink flowers *(Rr)* will produce plants with red flowers *(RR)*, plants with pink flowers *(Rr)*, and plants with white flowers *(rr)*. The ratio for a large group will correspond to that predicted by the Punnett square.**

3. Suppose you wanted only red snapdragons in your garden. What could you do to make sure that you would produce neither pink nor white flowers? **Weed out any white or pink snapdragons before they have a chance to pollinate the red flowers.**

## *OPTIONS*

**Enrichment**/*Teacher Resource Package,* p. 28. Use the Enrichment worksheet shown here to explain incomplete dominance.

## TEACH

**Cooperative Learning:** Put students into teams of four and have them use reference materials to answer questions below. Have a member of each team report their findings to the class. Have them answer the following:

1. How might Mendel's father's occupation have influenced Mendel? (His father was an orchardist and worked in plant husbandry.)
2. Why did Mendel's laws sit on a shelf for over 30 years before anyone used them? (Most people did not understand what Mendel had done and did not understand what math had to do with the inheritance of traits.)
3. How did scientists of that time think traits were inherited? (Look up blood theory of inheritance.)
4. Other than the laws of genetics, what else did Mendel accomplish in science? (He developed several new varieties of fruits and vegetables.)
5. How did Mendel know that he had true breeding strains of peas for his experiments? (He bought them from a "seedsman" who would guarantee the "pure breeding" quality of the seeds.)

**Portfolio**

Direct students to calculate how close Mendel's observed results were to the expected 3:1 ratio for each of the four traits shown in Table 26-4.

**GLENCOE TECHNOLOGY**

**Videodisc**

The Secret of Life

*Heredity in Mendel's Peas*

Mendel then took the tall offspring and mated them with each other. He noticed that about three-fourths of the plants they produced were tall and about one-fourth were short. Look at Figure 26-12. Can you see that Mendel's results came very close to what we now expect to see when heterozygous plants are mated? Mendel concluded that his plants were heterozygous after he saw the results of his experiments.

Remember that a heterozygous individual has a dominant gene and a recessive gene. Mendel's heterozygous plants would be ***Tt.*** Study the Punnett square in Figure 26-12 to see what happens when two heterozygous plants mate.

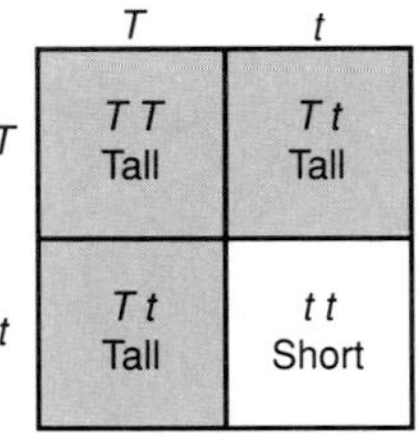


**Figure 26-12** Mendel's results when he mated two heterozygous plants. How many gene combinations are possible? three

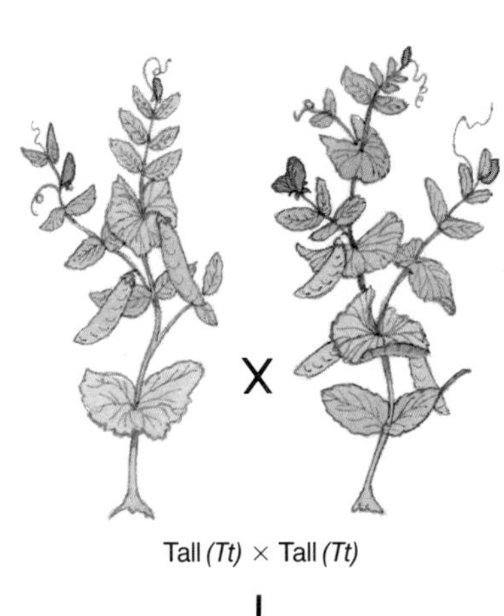


Tall *(Tt)* × Tall *(Tt)*

↓

3 tall plants *(1TT, 2Tt)*    1 short plant *(tt)*



## OPTIONS

**What Is a Test Cross?**/*Lab Manual*, pp. 223-226. Use this lab as an extension to crossing traits.

**Offspring From Two Heterozygous Parents**/*Transparency Package*, number 26. Use color transparency number 26 as you teach traits in the offspring of heterozygous parents.

**Application**/*Teacher Resource Package*, p. 26. Use the Application worksheet shown here to have students solve genetic problems in agriculture.

**APPLICATION 26**

APPLICATION: CONSUMER — CHAPTER 26

Name ______ Date ______ Class ______

*Use after Section 26:2.*

GENETICS PROBLEMS IN AGRICULTURE

*Fill in the Punnett squares to solve each of the following problems and answer the questions.*

1. Imagine that you raise guinea pigs to sell and that brown guinea pigs are your best sellers. In guinea pigs, black coat color, ***B***, is dominant to brown coat color, ***b***. Show the expected offspring from the mating of the following parents. From which pair could you raise the greatest number of brown guinea pigs? Black male *Bb* × brown female *bb*

(a) Black male ***Bb*** × brown female ***bb***

| | b | b |
|---|---|---|
| B | Bb | Bb |
| b | bb | bb |

Offspring will be:
Genes: 1 *Bb*:1 *bb*
Traits: 1 black and 1 brown

(b) Black male ***Bb*** × black female ***BB***

| | B | B |
|---|---|---|
| B | BB | BB |
| b | Bb | Bb |

Offspring will be:
Genes: 1 *BB*:1 *Bb*
Traits: All black

(c) Brown male ***bb*** × black female ***BB***

| | B | B |
|---|---|---|
| b | Bb | Bb |
| b | Bb | Bb |

Offspring will be:
Genes: All *Bb*
Traits: All black

2. In tomatoes, red fruit color, ***R***, is dominant to yellow fruit color, ***r***. A farmer has tomatoes that produce either red or yellow tomatoes. He has signed a contract with a large seed company to provide pure red, ***RR***, and pure yellow, ***rr***, seeds. The seed company does not want any heterozygous, ***Rr***, seeds. How could the farmer tell if his red tomatoes are pure or heterozygous? He could tell by crossing red tomatoes with yellow tomatoes. If the seeds produce both yellow and red tomatoes, the red parent plant was heterozygous (*Rr*). If the seeds produce only red tomatoes, the red parent plant was pure dominant (*RR*).

(a)

| | r | r |
|---|---|---|
| R | Rr | Rr |
| r | rr | rr |

Offspring will be:
Genes: 1 *Rr*:1 *rr*
Traits: 1 red:1 yellow

(b)

| | r | r |
|---|---|---|
| R | Rr | Rr |
| R | Rr | Rr |

Offspring will be:
Genes: All *Rr*
Traits: All red

| TABLE 26-4. MENDEL'S RESULTS | | | |
|---|---|---|---|
| | Number of Offspring Observed | | |
| Trait | Dominant Trait | Recessive Trait | Total |
| Color of pea pod | green–428 | yellow–152 | 580 |
| Shape of pea | round–5474 | wrinkled–1850 | 7324 |
| Color of pea | yellow–6022 | green–2001 | 8023 |
| Flower color | red–705 | white–224 | 929 |

Mendel studied seven traits in pea plants. Some traits that he looked at were the color of the pea pod, shape of the peas, color of the peas, and color of the flower. Table 26-4 shows that he got the same results whenever he mated heterozygous plants. About three-fourths of the offspring had the dominant trait and one-fourth had the recessive trait.

## Check Your Understanding

6. Why are observed results sometimes different from expected results?
7. Draw a Punnett square to show what genes a child would be expected to have if each parent were heterozygous for long eyelashes. Use ***L*** to stand for the dominant trait, long eyelashes. Use ***l*** to stand for the recessive trait, short eyelashes.
8. What laws of genetics did Mendel explain?
9. **Critical Thinking:** You mate a red-flowered plant with a white-flowered plant. You expect all the offspring to be red, but you find that half of them are white. Explain why.
10. **Biology and Writing:** Being an albino is a recessive trait in which no color is produced in the skin or eyes. Could an albino child be produced by two normally-pigmented parents? Explain.

## RETEACHING 79

**RETEACHING** — **CHAPTER 26**

Name ____________ Date ________ Class ________

*Use with Section 26:2.*

### USING THE PUNNETT SQUARE TO SOLVE PROBLEMS

*Use the Punnett squares provided to find the expected offspring from each of the matings (genetic crosses). In using the Punnett square, show the genes in the sex cells of each parent. Also show the genes that are possible in the offspring and what their traits will be.*

1. Garden pea plants can be either tall or short. Tall plants have at least one dominant gene present, *T*, while short plants are always pure recessive, *tt*. Show the results of the following crosses in the Punnett squares provided.

(a) *Tt* × *tt*

| | t | t |
|---|---|---|
| T | Tt | Tt |
| t | tt | tt |

Offspring will be:
Genes: 1 *Tt* :1 *tt*
Traits: 1 tall:1 short

(b) *Tt* × *Tt*

| | T | t |
|---|---|---|
| T | TT | Tt |
| t | Tt | tt |

Offspring will be:
Genes: 1 *TT* :2 *Tt* :1 *tt*
Traits: 3 tall:1 short

(c) *TT* × *tt*

| | t | t |
|---|---|---|
| T | Tt | Tt |
| T | Tt | Tt |

Offspring will be:
Genes: All *Tt*
Traits: All tall

2. Below is a table showing the traits of cucumber plants that were pollinated and the traits of the cucumbers produced. Use the information on the table to fill in the Punnett squares below. Use *G* for the dominant trait, green, and *g* for the recessive trait, striped.

| Traits of cucumbers crossed | Traits of cucumbers produced |
|---|---|
| (a) green × striped | all green |
| (b) striped × striped | all striped |
| (c) green × green | some green; some striped |

(a) green × striped

| | g | g |
|---|---|---|
| G | Gg | Gg |
| G | Gg | Gg |

Offspring will be:
Genes: All Gg
Traits: All green

(b) striped × striped

| | g | g |
|---|---|---|
| g | gg | gg |
| g | gg | gg |

Offspring will be:
Genes: All gg
Traits: All striped

(c) green × green

| | G | g |
|---|---|---|
| G | GG | Gg |
| g | Gg | gg |

Offspring will be:
Genes: 1 GG:2 Gg :1 gg
Traits: 3 green: 1 striped

## STUDY GUIDE 156

**STUDY GUIDE** — **CHAPTER 26**

Name ____________ Date ________ Class ________

### VOCABULARY

*Review the new words used in Chapter 26 of your textbook. Then, answer these questions.*

1. Below each of the following words are choices. Circle the choices that are examples of each of those words.
   a. Dominant gene
   (D) e k (L) (N) o (R) (S)
   b. Recessive gene
   M (n) (d) F G (i) (k) P
   c. Pure dominant
   (AA) Gg (KK) ll pp Rr (TT)
   d. Pure recessive
   (ee) Ff HH Oo (qq) Uu (ww)
   e. Offspring combinations in which dominant gene *must* show
   (AA) (Dd) (EE) ff (Jj) (RR) (Ss)
   f. Offspring combinations in which recessive gene *must* show
   (aa) Gg Ff KK Oo PP (ss) (tt)
2. Fill in the blanks below using these choices: dominant, genes, genetics, heterozygous, pure, recessive, chromosomes, Punnett square.
   a. Chromosomes have parts that determine traits. These parts are genes.
   b. A person having two genes that are alike is said to be pure.
   c. A gene that prevents others from showing is said to be dominant.
   d. A gene that may not show up even though it is there is said to be recessive.
   e. Long rod-shaped bodies inside a cell's nucleus are called chromosomes.
   f. One who studies how traits are passed on is studying genetics.
   g. A person with one dominant and one recessive gene for a trait is heterozygous.
   h. A way to show which genes can combine when an egg and sperm join is a Punnett square.

# TEACH

### Independent Practice

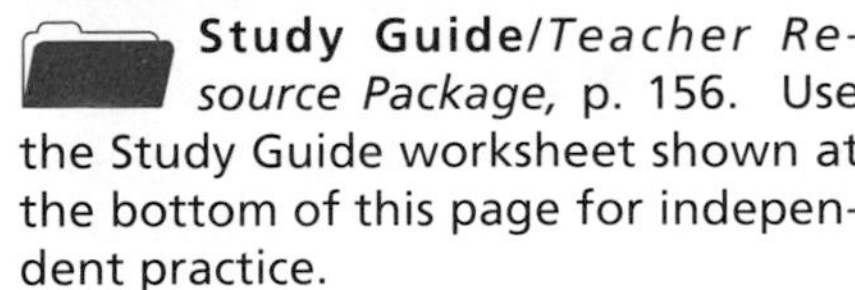

**Study Guide**/*Teacher Resource Package,* p. 156. Use the Study Guide worksheet shown at the bottom of this page for independent practice.

### Check for Understanding

Have students respond to the first three questions in Check Your Understanding.

### Reteach

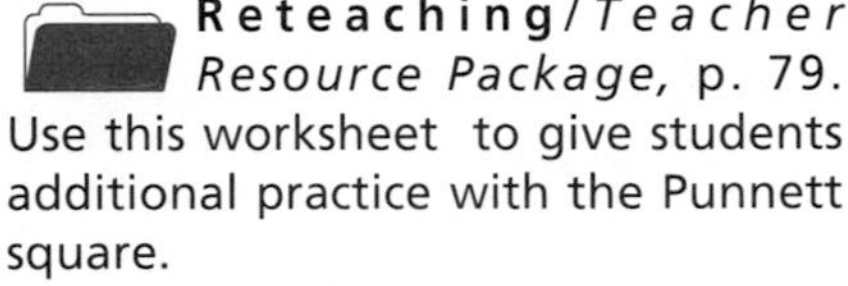

**Reteaching**/*Teacher Resource Package,* p. 79. Use this worksheet to give students additional practice with the Punnett square.

**Extension:** Assign Critical Thinking, Biology and Writing, or some of the **OPTIONS** available with this lesson.

# APPLY

### Convergent Question

Have students explain why some families in Table 26-3 do not have a 1:1 ratio of children with cleft and smooth chins.

# CLOSE

### Software

*Genetics*, Queue, Inc.

### Answers to Check Your Understanding

6. because you don't know which sperm and egg will join
7. 1 ***LL***: 2 ***Ll***: 1 ***ll***
8. what dominant and recessive traits were and how these traits passed from parent to offspring
9. The red parent plant is heterozygous.
10. yes, if both parents are heterozygous and carry the recessive gene for albinism

## CHAPTER 26 REVIEW

### Summary

Summary statements can be used by students to review the major concepts of the chapter.

### Key Science Words

All boldfaced terms from the chapter are listed.

### Testing Yourself

#### Using Words

1. heterozygous
2. dominant gene
3. pure dominant
4. pure recessive
5. gene
6. Punnett square
7. recessive gene
8. genetics

#### Finding Main Ideas

9. p. 547; They are arranged one next to another, like beads on a necklace.
10. pp. 552-553; Letters stand for genes. Letters for one parent go across the top and letters for the other parent go along the side. The letters are copied into the boxes beside each to show how they can combine.
11. p. 549; Dominant genes prevent the recessive traits from showing.
12. p. 547; Genes determine the traits.
13. p. 559; Mendel used the scientific method to do hundreds of experiments.
14. p. 546; They are best seen when a cell is ready to reproduce.
15. p. 553; The boxes represent the possible combinations of genes.
16. p. 557; The traits that offspring have are observed results.
17. pp. 550-551; Each parent can make two kinds of egg or sperm, which can join in four possible combinations.
18. pp. 546-547; There is only one of each kind of chromosome present in a sex cell, and each chromosome has different kinds of genes.

# Chapter 26 Review

## Summary

### 26:1 Genetics, How and Why

1. Chromosomes are threadlike parts in the nuclei of body cells and sex cells. Body cells have twice as many chromosomes as sex cells.
2. Dominant genes keep recessive genes from showing their traits. An organism may have two dominant genes for a trait, two recessive genes, or one of each.
3. An individual that is pure for a trait can make one kind of sex cells for a certain trait. A heterozygous individual can make two kinds of sex cells for a certain trait. The kind of offspring produced depends on which sex cells combine.

### 26:2 Expected and Observed Results

4. The Punnett square helps to predict the combinations of genes offspring can receive.
5. Expected results are the predicted results. Observed results are what you actually see.
6. Mendel reported how traits were inherited in garden pea plants. He explained basic principles of genetics.

## Key Science Words

dominant gene (p. 549)
gene (p. 547)
genetics (p. 546)
heterozygous (p. 549)
Punnett square (p. 552)
pure dominant (p. 549)
pure recessive (p. 549)
recessive gene (p. 549)

## Testing Yourself

### Using Words

*Choose the word from the list of Key Science Words that best fits the definition.*

1. has a dominant and a recessive gene for a trait
2. a gene that prevents a recessive gene from showing
3. has two dominant genes for a trait
4. having two recessive genes for the same trait
5. a small section of a chromosome that determines a trait
6. a way to show which genes combine when egg and sperm join
7. a gene that does not show when the dominant gene for the trait is present
8. study of how traits are passed from parents to offspring

### Finding Main Ideas

*List the page number where each main idea below is found. Then, explain each main idea.*

9. how genes are arranged on a chromosome
10. how a Punnett square can be used to predict combinations of genes when sperm and eggs combine



#### Using Main Ideas

19. It predicts the possible combinations of egg and sperm cells.
20. Mendel performed experiments with pea plants.

# Review

## Testing Yourself *continued*

11. why recessive traits do not show in heterozygous individuals
12. what determines an organism's traits
13. how traits in the garden pea plant were studied
14. when chromosomes in a cell are seen best
15. what the four small boxes of a Punnett square represent
16. the traits that offspring actually have
17. that heterozygous parents can have children with a dominant or recessive trait
18. that each sex cell has one gene for a given trait

### Using Main Ideas

*Answer the questions by referring to the page number after each question.*

19. What does a Punnett square predict? (p. 552)
20. Who discovered how garden pea plants inherit their traits? (p. 559)
21. What are genes that do not show even though they are present? (p. 549)
22. How are chromosomes and the nucleus related? (p. 546)
23. How is fertilization of an egg by a sperm like flipping a coin? (p. 558)
24. How are genes related to chromosomes? (p. 547)
25. What are two types of sex cells? (p. 546)
26. How do body cells and sex cells differ in chromosome number? (p. 546)
27. If Mendel crossed a heterozygous tall plant with a pure recessive short plant, what kinds of plants could he expect them to produce? (p. 553)

## Skill Review

*For more help, refer to the* ***Skill Handbook,*** *pages 704-719.*

1. **Calculate:** Use a calculator to find out how close Mendel's data (Table 26-4) are to showing a 3:1 ratio of dominant to recessive traits. Divide each trait by the total.
2. **Form hypotheses:** Form a hypothesis to explain why two offspring in a family have different traits.
3. **Interpret data:** Dogs with spots show a dominant trait and dogs without spots show a recessive trait. A dog breeder mates two spotted dogs. There are three dogs without spots in the first litter. Explain why.
4. **Classify:** Classify the following as cells or cell parts: sperm, gene, chromosome, egg, nucleus.

## Finding Out More

TECH PREP

### Critical Thinking

1. Suppose you planted flower seeds. The package stated that the plants grow two feet tall. You noticed that 30 percent of the plants grew less than one foot tall. What does this tell you about the tallness trait?
2. Explain why some children can resemble one parent more than the other.

### Applications

1. Find out which members of your family can roll their tongues and which can't.
2. Find out how seed companies produce flower seeds that will grow into plants with certain traits.



### Using Main Ideas

21. Genes that don't show up are recessive.
22. Chromosomes are threadlike parts found inside the nucleus.
23. Each time you flip a coin you are starting over, and each time an egg and sperm join you do not know exactly what to expect.
24. Genes are small sections along a section of chromosome.
25. Sperm and egg are two types of sex cells.
26. A sex cell has one of each kind of chromosome; a body cell has two of each kind of chromosome.
27. He could expect a ratio of 1 tall plant to 1 short plant.

## Skill Review

1. Students should see that all of Mendel's data shows a 3 dominant:1 recessive.
2. No two sperm or egg cells from the same person have the same gene combinations.
3. The dogs with spots are heterozygous and three dogs in the first litter received recessive traits from each parent.
4. Sperm and egg are cells; gene, chromosome, and nucleus are cell parts.

## Finding Out More

### Critical Thinking

1. The tallness trait is dominant, but the parent plants were heterozygous.
2. Children resemble the parent whose genes are dominant over the other's recessive genes.

### Applications

1. Students could record their findings on a chart and then compare their charts.
2. Students could write to seed companies to ask for information, or they can inquire at local greenhouses.

**Chapter Review**/*Teacher Resource Package,* pp. 138-139.

**Chapter Test**/*Teacher Resource Package,* pp. 140-142.

**Chapter Test**/*Computer Test Bank*

# Human Genetics

## PLANNING GUIDE

| CONTENT | TEXT FEATURES | TEACHER RESOURCE PACKAGE | OTHER COMPONENTS |
|---|---|---|---|
| (2 1/2 days)<br>27:1 The Role of Chromosomes<br>Chromosome Number<br>A Way to Tell Chromosome Number<br>Sex—A Genetic Trait | Skill Check: *Understand Science Words,* p. 568<br>Lab 27-1: *Sex Determination,* p. 571<br>Science and Society: *Can Couples Choose the Sexes of Their Children?* p. 569<br>Idea Map, p. 566<br>Check Your Understanding, p. 570 | Reteaching: *Chromosome Numbers,* p. 80<br>Focus: *What Do You Know About Human Genetics?* p. 53<br>Study Guide: *The Role of Chromosomes,* p. 157<br>Lab 27-1: *Sex Determination,* pp. 105-106 | Color Transparency 27: *Sex Determination*<br>**STVS:** *Breeding Fruit Flies,* Plants and Simple Organisms (Disc 4, Side 1) |
| (2 1/2 days)<br>27:2 Human Traits<br>Survey of Human Traits<br>Incomplete Dominance<br>Blood Types in Humans<br>Genes on the X Chromosome | Mini Lab: *What Traits Do Humans Share?* p. 574<br>Lab 27-2: *Human Traits,* p. 577<br>Idea Map, p. 573<br>Check Your Understanding, p. 576 | Application: *Knowing Your Blood Type,* p. 27<br>Reteaching: *Colorblindness,* p. 81<br>Critical Thinking/Problem Solving: *Could the Taylors' Son Be Colorblind?* p. 27<br>Study Guide: *Human Traits,* pp. 158-159<br>Lab 27-2: *Human Traits,* pp. 107-108<br>Transparency Master: *A Trait With Incomplete Dominance,* p. 137 | **Laboratory Manual:**<br>*What Do Normal and Sickled Cells Look Like?* p. 227<br>**Laboratory Manual:**<br>*How Are Traits of the Sex Chromosomes Inherited?* p. 231<br>**STVS:** *Obesity and Heredity,* Human Biology (Disc 7, Side 2) |
| (2 days)<br>27:3 Genetic Disorders<br>Errors in Chromosome Number<br>Genetic Disorders and Sex Chromosomes<br>Genetic Disorders and Autosomes<br>Genetic Counseling | Skill Check: *Formulate Models,* p. 579<br>Idea Map, p. 579<br>Check Your Understanding, p. 581 | Enrichment: *Tracing a Genetic Disorder in a Family,* p. 29<br>Reteaching: *Genetic Disorders,* p. 82<br>Skill: *Make a Line Graph,* p. 28<br>Study Guide: *Genetic Disorders,* pp. 160-161<br>Study Guide: *Vocabulary,* p. 162 | **STVS:** *Detecting Cystic Fibrosis,* Human Biology (Disc 7, Side 2) |
| Chapter Review | Summary<br>Key Science Words<br>Testing Yourself<br>Finding Main Ideas<br>Using Main Ideas<br>Skill Review | **ASSESSMENT RESOURCES**<br>Chapter Review, pp. 143-144<br>Chapter Test, pp. 145-147<br>Performance Assessment in the Biology Classroom<br>Alternate Assessment in the Science Classroom<br>Computer Test Bank | |

## GLENCOE TECHNOLOGY

**Infinite Voyage,** *The Geometry of Life*
**Science and Technology Videodisc Series,** *Breeding Fruit Flies,* Plants and Simple Organisms (Disc 4, Side 1)
*Obesity and Heredity,* Human Biology (Disc 7, Side 2)
*Detecting Cystic Fibrosis,* Human Biology (Disc 7, Side 2)
**The Secret of Life,** *Tinkering With Our Genes: Genetic Medicine*

## MATERIALS NEEDED

| LAB 27-1, p. 571 | LAB 27-2, p. 577 | MARGIN FEATURES |
|---|---|---|
| 2 coins<br>paper cup<br>masking tape | pencil<br>paper | Skill Check, p. 568<br>dictionary<br>Skill Check, p. 579<br>poster board<br>markers<br>paper<br>Mini Lab, p. 574<br>pencil<br>paper |

## OBJECTIVES

For more information about National Science Standards, see page 5T.

| SECTION | OBJECTIVE | CORRELATION of QUESTIONS to OBJECTIVES: CHECK YOUR UNDERSTANDING | CHAPTER REVIEW | TRP CHAPTER REVIEW | TRP CHAPTER TEST |
|---|---|---|---|---|---|
| 27:1 National Science Stds: UCP.2, UCP.3, UCP.4, UCP.5, C.1, C.2, E.2 | 1. **Compare** the chromosome numbers in body cells and sex cells. | 1, 5 | 1, 16 | 1, 18 | 19, 20 |
| | 2. **Describe** methods that doctors use to study chromosomes of a fetus. | 2 | 2, 11 | 20 | 21 |
| | 3. **Compare** the chromosomes of males and females. | 3, 4 | 5, 6, 9, 15 | 3, 7, 12, 19 | 2, 11, 12, 17 |
| 27:2 National Science Stds: UCP.2, UCP.3, UCP.4, UCP.5, C.2, F.1 | 4. **Compare** recessive and dominant traits with incomplete dominance. | 6 | 7, 12 | 4, 6, 10, 15, 22 | 3, 4, 5, 7, 8, 10 |
| | 5. **Describe** different ways human traits can be inherited. | 7, 8, 9 | 14, 17, 19, 20, 21 | 5, 11, 13, 14, 16, 17 | 1, 9, 13, 22, 23, 24, 25, 37, 38, 39, 40, 41 |
| 27:3 National Science Stds: UCP.2, UCP.3, UCP.4, A.2, F.1, G.1, G.2 | 6. **Describe** some genetic disorders in humans. | 12, 14 | 3, 8, 18 | 2, 8, 21 | 6, 16, 18 |
| | 7. **Give examples** of how genetic counseling can help families. | — | 4, 10 | 9 | 14, 15, 26, 27, 28, 29, 30, 31, 32, 33 |

# Human Genetics

## CHAPTER OVERVIEW

### Key Concepts

This chapter introduces students to the role of chromosomes in human genetics. Students will learn about several inherited traits and their patterns of inheritance. Students will also learn about health problems caused by chromosome errors.

### Key Science Words

amniocentesis
autosome
color blindness
dyslexia
genetic counseling
incomplete dominance
pedigree
sex chromosome
sickle-cell anemia
X chromosome
Y chromosome

### Skill Development

In Lab 27-1, students will **make and use tables, formulate a model, calculate,** and **design an experiment** to determine how many males and females could be expected in a family of four children. In Lab 27-2, students will **observe, make and use tables,** and **interpret data** to determine common traits. In the Skill Check on page 568, students will **understand** the **science word** *autosome.* On page 579, students will formulate a model to show how a dyslexic person sees letters. In the Mini Lab on page 574, students will **make and use tables** to show traits that they have inherited from their parents.

### Bridging

Students will focus on the traits that humans inherit. The major concepts that students learned in Chapter 26 will be used to reinforce and learn patterns of genetics as applied to humans. They will learn that the way humans inherit most of their traits is similar to that of other living things.

**CHAPTER PREVIEW**

### Chapter Content

Review this outline for Chapter 27 before you read the chapter.

**27:1 The Role of Chromosomes**
- Chromosome Number
- A Way to Tell Chromosome Number
- Sex—A Genetic Trait

**27:2 Human Traits**
- Survey of Human Traits
- Incomplete Dominance
- Blood Types in Humans
- Genes on the X Chromosome

**27:3 Genetic Disorders**
- Errors in Chromosome Number
- Genetic Disorders and Sex Chromosomes
- Genetic Disorders and Autosomes
- Genetic Counseling

### Skills in this Chapter

The skills that you will use in this chapter are listed below.

- In **Lab 27-1,** you will make and use tables, formulate a model, calculate, and design an experiment. In **Lab 27-2,** you will observe, make and use tables, and interpret data.
- In the **Skill Checks,** you will formulate a model, and understand science words.
- In the **Mini Labs,** you will make and use tables.

564

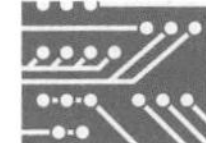

## TECH PREP

For Tech Prep activities in this chapter of the Teacher Wraparound Edition, see especially the Student Journals on pages 568 and 573, the Portfolio on page 572, the Activity on page 576, and the Connection and Activity on page 580.

See also the Glencoe Homepage at **www.glencoe.com**

Chapter 27

# Human Genetics

Did you ever notice how often the children in a family look alike? This is because they share many of the same traits they inherited from their parents. Some traits may not be the same in all the children, however. How can you explain these differences?

The girls in the large photo are sisters. What traits do they have in common? What traits can you see that are different in the two sisters? Can you tell anything about the traits of the parents without actually seeing them?

The girl in the smaller photo is a cousin of the two sisters. The father of the two sisters and the father of the cousin are brothers. What traits do the sisters share with the cousin? Why do sisters share more traits than two cousins?

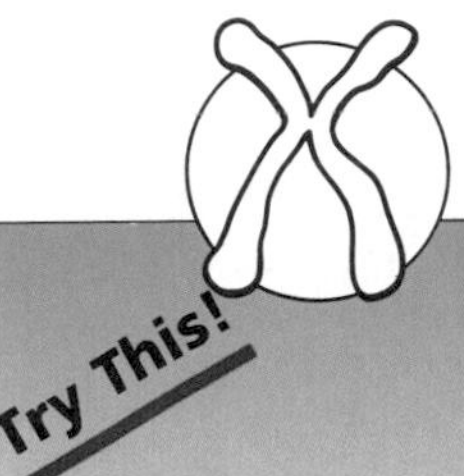

**Try This!**

**Are human traits different from traits of other animals?** Look at pictures of several animals. Make lists to show which traits or body parts are shared by humans and other animals and which are different.

**interNET CONNECTION**

For more information about the material in this chapter, follow the link for the chapter on the Glencoe Homepage at **http://www.glencoe.com**



**ASSESSMENT PLANNER**

**Portfolio**
Strategies on the following pages represent student products that can be placed into a best-work portfolio: pp. 572, 574.

**PERFORMANCE ASSESSMENT**
Skill Check, pp. 568, 579
Mini Lab, p. 574
Lab, pp. 571, 577

**CONTENT ASSESSMENT**
Check for Understanding, pp. 568, 570, 575, 576, 581
Chapter Review, pp. 582-583

**GROUP ASSESSMENT**
Opportunities for group assessment occur with Cooperative Learning Strategies.

**Student Journal**
Strategies on the following pages represent opportunities for writing in a Student Journal: pp. 568, 573.

**GETTING STARTED**

## Using the Photos

Students should notice that all three girls share some common family features. Point out that two sisters will share more similar traits than two cousins. Explain that sometimes cousins resemble each other more than two sisters or a sister and brother because of the way chromosomes separate during meiosis.

## MOTIVATION/Try This!

**Are human traits different from traits of other animals?** Upon completing the chapter-opening activity, students will have gained practice in learning to **classify.** Discuss the students' lists. Point out that geneticists think that humans inherit between 60 000 and 100 000 traits. Many of these traits control metabolic processes.

## Chapter Preview

Have students study the chapter outline before they begin to read the chapter. They should note that they will study the role of chromosomes first. Next, they will learn about genetic traits in humans. In the last section, they will study chromosome errors and genetic disorders.

## Misconception

Students are usually unaware of patterns of inheritance caused by traits other than Mendelian dominance or recessiveness. They also assume that they must show traits that are not different from either their mother or father.

## 27:1 The Role of Chromosomes

## PREPARATION

### Materials Needed

Make copies of the Focus, Study Guide, Reteaching, and Lab worksheets in the *Teacher Resource Package*.

▶ Have coins on hand for Lab 27-1.

### Key Science Words

amniocentesis
X chromosome
sex chromosome
Y chromosome
autosome

### Process Skills

In Lab 27-1, students will make and use tables, formulate models, calculate, and design an experiment. In the Skill Check, students will understand science words.

## 1 FOCUS

▶ The objectives are listed on the student page. Remind students to preview these objectives as a guide to this numbered section.

### MOTIVATION/Bulletin Board

Construct a bulletin board of human races. Use the title Humans—One Species, Many Differences. Use photos to show differences in skin color, hair color, height, etc.

Focus/*Teacher Resource Package*, p. 53. Use the Focus transparency shown at the bottom of this page as an introduction to human genetics.

### Guided Practice

Have students write down their answers to the margin question: in pairs.

### Idea Map

Have students use the idea map as a study guide to the major concepts of this section.

# 27:1 The Role of Chromosomes

**Objectives**

1. **Compare** the chromosome numbers in body cells and sex cells.
2. **Describe** methods that doctors use to study chromosomes of a fetus.
3. **Compare** the chromosomes of males and females.

**Key Science Words**

amniocentesis
X chromosome
sex chromosome
Y chromosome
autosome

**How are chromosomes arranged in body cells?**

All living things can pass their traits to their offspring. You learned in Chapter 26 that these traits can be dominant or recessive. The genes that control the traits are found on sections of the chromosomes.

## Chromosome Number

There are three things you should know about chromosome numbers in living things.
1. Each human sperm or egg has 23 chromosomes.
2. Each human body cell has 23 pairs of chromosomes. There are 46 chromosomes in each human body cell.
3. Different organisms have different numbers of chromosomes. The chromosomes in most living things are paired. A carrot plant has 18 chromosomes in each of its body cells. That's equal to nine pairs per cell. Table 27-1 gives the numbers of chromosomes in some other living things.

**TABLE 27–1. CHROMOSOME NUMBERS**

| Animal or Plant | Number of Chromosomes in Body Cells | Number of Chromosome Pairs in Body Cells |
|---|---|---|
| Red clover | 12 | 6 |
| Pea | 14 | 7 |
| Onion | 16 | 8 |
| Corn | 20 | 10 |
| White pine | 24 | 12 |
| Cat | 38 | 19 |
| Rabbit | 44 | 22 |
| Chicken | 78 | 39 |

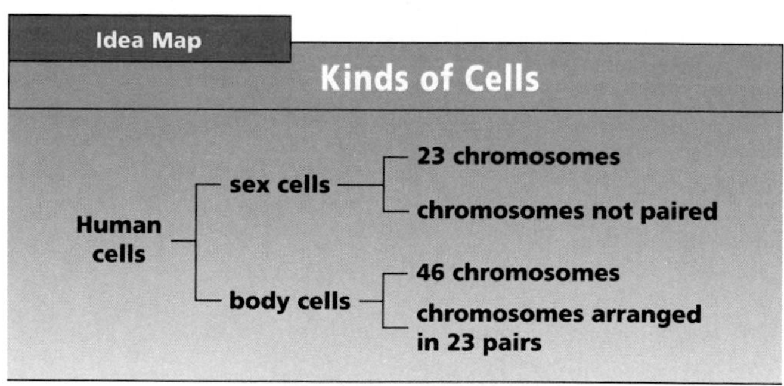


**Study Tip:** Use this idea map as a study guide to the chromosomes in human cells. The numbers of chromosomes in body and sex cells are shown.



## OPTIONS

### Science Background

It was not known until the start of the 20th century that genes and chromosomes were important in the study of heredity. Geneticists found sex chromosomes in many animals to be XX in females and XY in males. In some animals, the male has XX sex chromosomes while the female has only one X. The arrangement of chromosomes in pairs from largest to smallest is called a karyotype, and this system of classification was developed in the late 1950s.

**FOCUS 53**

FOCUS CHAPTER 27

Name ____________ Date ________ Class ________

**WHAT DO YOU KNOW ABOUT HUMAN GENETICS?**

The purpose of this survey is to find out what people really understand about human genetics. Read each statement and record whether you believe the statement to be true (T) or false (F). Record your own answers in column A. Find two other persons, preferably adults, to survey. Record their answers in the columns marked B and C.

| | A | B | C | |
|---|---|---|---|---|
| 1. | | | | 1. If two parents have brown eyes, then all of their children will have brown eyes. |
| 2. | | | | 2. Some people are born with inherited disorders. |
| 3. | | | | 3. A man's sperm cells determine whether a child will be a boy or a girl. |
| 4. | | | | 4. No one has ever seen a chromosome. |
| 5. | | | | 5. Persons of the same sex can donate blood to others of the same sex even if the blood types are different. |
| 6. | | | | 6. Each parent gives an equal number of chromosomes to his or her children. |
| 7. | | | | 7. Only the father's traits show in a male child. |
| 8. | | | | 8. Identical twins are always of the same sex. |
| 9. | | | | 9. Fraternal twins are more closely related to each other than to other children in the family. |
| 10. | | | | 10. Color blindness is more common in males than females. |
| 11. | | | | 11. A parent can pass traits to a child even though the parent does not show the trait. |
| 12. | | | | 12. All harmful traits are caused by recessive genes. |
| 13. | | | | 13. Identical twins have more of the same genes than other children in the family. |
| 14. | | | | 14. Some traits do not show in a person because the traits are recessive. |
| 15. | | | | 15. None of the grandparents' traits are passed on to the grandchildren. |

1. Did you and the people you surveyed have the same answers to the questions? Why do you think this is so?
2. What other traits of humans do you think are inherited?

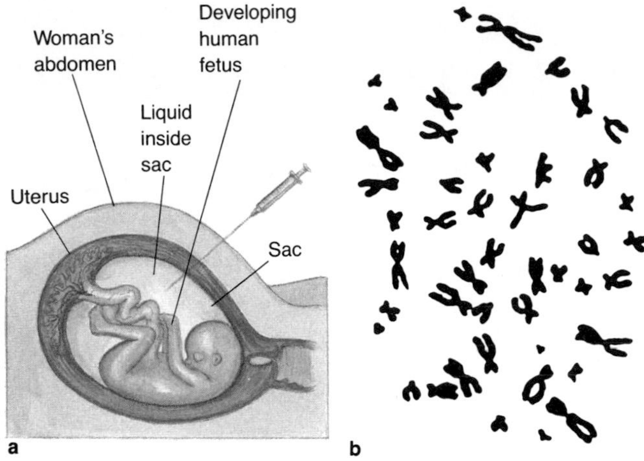


**Figure 27-1** Amniocentesis (a). Chromosomes taken from the sac in which the fetus develops (b).

## A Way to Tell Chromosome Number

Doctors today can tell if a child has the correct number of chromosomes even before it is born. They use a process called amniocentesis (am nee oh sen TEE sus). **Amniocentesis** is a way of looking at the chromosomes of a fetus.

Look at Figure 27-1a. A needle is inserted into the pregnant woman's abdomen. The needle enters the amniotic sac in which the fetus develops, and some liquid is removed through the needle. In this liquid are cells from the fetus's skin. The cells have rubbed off the fetus just as your skin cells do.

**How does a doctor perform amniocentesis?**

These cells are grown for about 10 days in a nutrient liquid that makes them divide. Remember that chromosomes can be seen best when the cell is dividing. The cells are then studied with a microscope and an enlarged photo like the one in Figure 27-1b is made. The chromosomes are cut out and the ones that match are arranged in pairs. A large chart is made showing the chromosome pairs. The lab technician looks at the chromosomes to make sure there are 46 of them and there are no missing or added parts.

Using a newer way to look at the chromosomes of a fetus, a doctor can take a small piece of the placenta surrounding the fetus. This area has cells with the same number and kind of chromosomes as the fetus. The chromosomes are counted and studied to see whether parts are missing or added.



# 2 TEACH

### MOTIVATION/Display

Make a transparency of Table 27-1 and have students compare the number of chromosomes in plants and animals. This table shows animals having more chromosomes than plants. Explain that some plants have more chromosomes than animals. For example, fruit flies have only four pairs of chromosomes. Point out that chromosome number has nothing to do with how complex an organism is. Some protozoa have 60 to 100 chromosomes.

### Concept Development

- Explain that sexual reproduction is important in that both parents contribute genetic material to their offspring. In this way the offspring shows some traits of each parent.
- Make sure students understand that amniocentesis is not a routine procedure, and is not given just to determine the sex of a child.
- Explain that amniocentesis is used to find out if an unborn baby has some abnormal chromosomes or some other genetic defect. A doctor can use amniotic fluid to test for certain genetic defects, such as Tay-Sachs disease.

### Guided Practice

Have students write down their answers to the margin question: by removing liquid from the amniotic sac and examining the cells in it.

## OPTIONS

### Science Background

Taking a small piece from the placenta is called chorionic villi sampling. This technique takes very little time, and the patient is able to resume normal activity within an hour. Chorionic sampling, as well as amniocentesis, does not involve major surgery.

**GLENCOE TECHNOLOGY**

**Videodisc**

**The Secret of Life**
*Amniocentesis*

# TEACH

## Concept Development

▶ Point out that other mammals have sex chromosomes similar to humans—females have XX and males have XY.

▶ To help students remember that the male contains the Y chromosome, draw attention to the male as determining the sex of the offspring.

## Check for Understanding

Have students set up a Punnett square with parents' sex chromosomes on the outside. Show each of the mother's sex cells by writing her X chromosomes across the top of the Punnett square. Show the father's sex cells by writing X and Y along the side of the Punnett square. Write the symbols for each parent's sex chromosomes in the small square. Each small square in the Punnett square stands for a way that the chromosomes could combine when the egg and sperm join.

### Reteach

Have students write a short paragraph to explain how humans inherit sex as a genetic trait.

## Guided Practice

Have students write down their answers to the margin question: female.

### Student Journal

In the last few years, geneticists have determined on what part of specific human chromosomes the genes for numerous traits are located. Their search was for those genes that produce traits harmful to humans. Tell students to do some library research and find the names of at least four genes that have been located over the past few years and tell what problems the genes cause. Have them tell why money is spent finding these genes rather than the genes for hair color, eye color, and height. Their findings can be written in their journals.

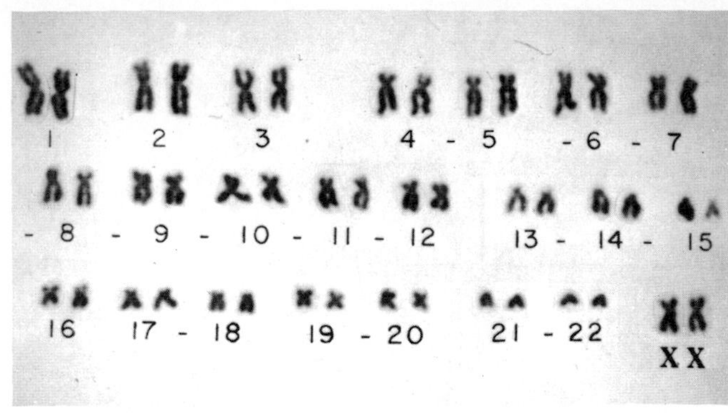


**Figure 27-2** These photos of chromosomes have been cut out and arranged in pairs. How many chromosomes does a human body cell have? 46

## Sex—A Genetic Trait

Whether you were born male or female depends on your chromosomes. In humans, a special pair of chromosomes determines sex.

**What sex is a person with two X chromosomes?**

Figure 27-2 shows the 23 pairs of chromosomes from a female. Notice that the chromosomes are numbered, except for those in the lower right corner. Those are marked XX. Each of the **X chromosomes** is a sex chromosome of the female. **Sex chromosomes** are chromosomes that determine sex. A human female has two X chromosomes in each body cell.

As you can see in Figure 27-3, the male's sex chromosomes are different from those of a female. The larger sex chromosome is an X chromosome. The smaller one is called a Y chromosome. The **Y chromosome** is a sex chromosome found only in males. A human male has one X and one Y chromosome in each body cell.

The numbered chromosomes in Figure 27-2 are called autosomes (AH toh sohmz). **Autosomes** are chromosomes that don't determine the sex of a person. They are also called nonsex or body chromosomes. The autosomes of a male are like those of a female.

### Skill Check

**Understand science words: autosome.** The word part *auto* means self. In your dictionary, find three words with the word part *auto* in them. *For more help, refer to the* ***Skill Handbook****, pages 706-711.*

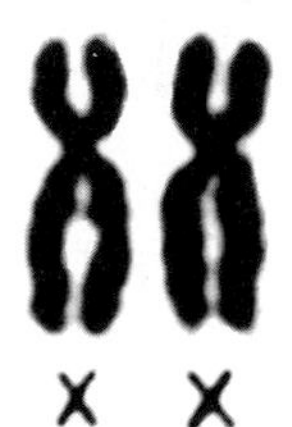

**Figure 27-3** A female has two X chromosomes. A male has one X and one Y chromosome.



### OPTIONS

**Sex Determination**/*Transparency Package*, number 27. Use color transparency number 27 as you teach how sex is determined.

# Science and Society

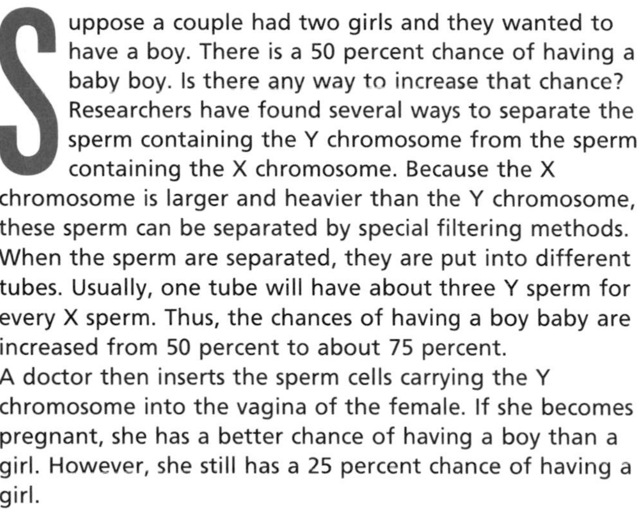

## Can Couples Choose the Sexes of Their Children?

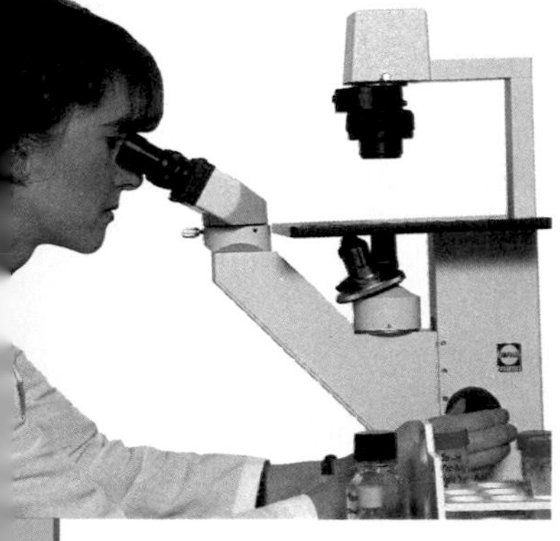

Suppose a couple had two girls and they wanted to have a boy. There is a 50 percent chance of having a baby boy. Is there any way to increase that chance? Researchers have found several ways to separate the sperm containing the Y chromosome from the sperm containing the X chromosome. Because the X chromosome is larger and heavier than the Y chromosome, these sperm can be separated by special filtering methods. When the sperm are separated, they are put into different tubes. Usually, one tube will have about three Y sperm for every X sperm. Thus, the chances of having a boy baby are increased from 50 percent to about 75 percent.
A doctor then inserts the sperm cells carrying the Y chromosome into the vagina of the female. If she becomes pregnant, she has a better chance of having a boy than a girl. However, she still has a 25 percent chance of having a girl.

### What Do You Think?

1. About 200 genetic disorders are controlled by the X chromosome. Many of these are fatal or result in lifelong health problems. These disorders can be prevented if a couple has only girls. Is it acceptable for couples to choose the sex of their children to prevent health problems?

2. Certain cultures prefer children of one sex over children of the other sex. Suppose that over the next 10 years, all couples in Japan decided to have only boys. Predict some problems that could result in the country.

3. Choosing the sex of a child can be very expensive. Not all people can afford to choose the sex of their children. If choosing the sex of children becomes popular, should the government or health insurance companies pay the medical costs of couples who cannot afford to pay?

**Conclusion:** Under what conditions, if any, should people be allowed to choose the sex of their children? If you knew that these methods guaranteed the sex of a child instead of just increasing the chances, how would you feel about them?

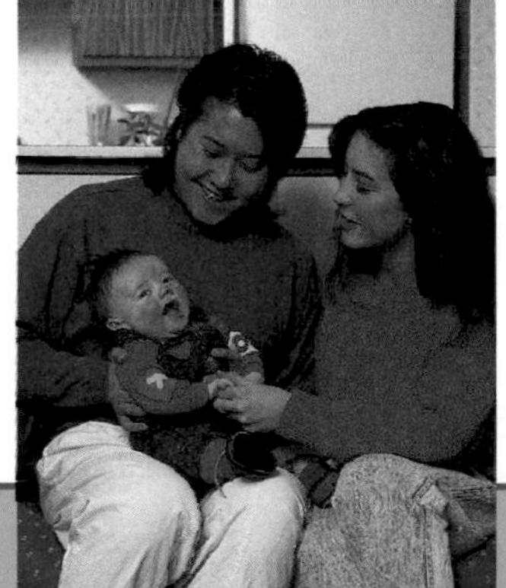

Some cultures prefer male children to female children.



## Science and Society

### Can Couples Choose The Sexes of Their Children?

#### Background

Scientists have placed X- and Y-bearing sperm cells in two different materials—serum albumin and a gel. They found that a larger number of Y-bearing sperm cells reached the bottom faster in the albumin. In the gel, a larger number of X-bearing sperm cells reached the bottom first. When artificially inseminated with cells from serum albumin, women conceived a larger number of male children. Those inseminated with the cells from the gel conceived a larger number of female children.

#### Teaching Suggestions

Point out that parents who have two girls or two boys and want another child of the opposite sex could benefit from this procedure.

Explain that parents who know of specific sex-linked traits in the family could select to have females who would be free of the harmful trait.

#### What Do You Think?

1. Some students will feel that this is a valid reason for choosing sex of offspring, but others may still feel it is not right.
2. Students may mention a decline in marriages and birth rates, or failure to fill some occupations or sports programs.
3. Those who are in favor of choosing the sex of a child will probably feel that there should be some help with the expense.

**Conclusion:** Some students may feel that these methods would be wrong for changing chance due to natural events. Others may even feel a bit repulsed by the idea of scientific control over reproduction. Some may feel that couples should have the right to make their own decisions. Students may feel, however, that the methods are acceptable for medical reasons.

#### Suggested Readings


Kelley, Richard L. and Mitzi I. Kuroda. "Equality for X Chromosomes." *Science*, December 8, 1995, pp. 1607-1611.

Leutwyler, Kristin, "X Marks the Spots." *Scientific American*, May 1996, pp. 16-17.

Martin, Julia and Natalie Jordet. "Having a Third." *Parents Magazine*, June 1996, pp. 107-110.

Steinbacher, Roberta and Ronald Ericsson. "Should Parents be Prohibited from Choosing the Sex of Their Child?" *Health*, March-April 1994, p. 24.


#### GLENCOE TECHNOLOGY

**Videotape**

**The Secret of Life**

*Sex and the Single Gene: Cell Development*

## TEACH

### Independent Practice

**Study Guide**/*Teacher Resource Package*, p. 157. Use the Study Guide worksheet shown at the bottom of this page for independent practice.

### Check for Understanding

Have students respond to the first three questions in Check Your Understanding.

### Reteach

**Reteaching**/*Teacher Resource Package*, p. 80. Use this worksheet to give students additional practice with chromosome numbers.

**Extension:** Assign Critical Thinking, Biology and Math, or some of the **OPTIONS** available with this lesson.

## 3 APPLY

### Convergent Question

Some members of royalty were known to divorce their wives because they gave birth to all girls. What could a study of genetics have told them?

## 4 CLOSE

### Summary

Have the students summarize the pros and cons to using amniocentesis.

### Answers to Check Your Understanding

1. A sex cell has 23 chromosomes; a body cell has 46.
2. a way to look at chromosomes in an unborn child
3. Each has one X chromosome. The female has a second X chromosome, while the male has a Y chromosome.
4. The number of chromosomes in the organism would double with each generation.
5. There is a 50 percent chance that her next child will be a boy.

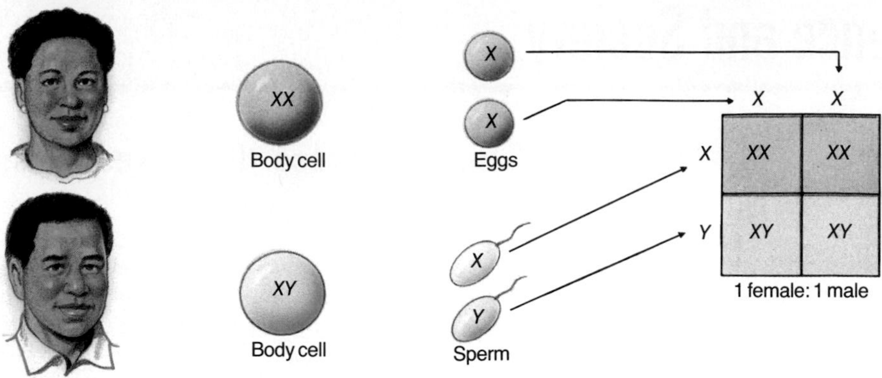


**Figure 27-4** A child receives an X chromosome from the mother and an X or a Y chromosome from the father.

What kind of sex chromosomes, X or Y, are found in eggs and sperm? Because females have only one kind of sex chromosome, they can make only one kind of egg. Each egg of a female has one X chromosome. Males have two different sex chromosomes, so they can make two different kinds of sperm. Each sperm of a male has either an X chromosome or a Y chromosome. As you can see from Figure 27-4, a child can receive only an X chromosome from its mother. The child can receive an X or Y chromosome from its father.

Notice that the father determines the sex of the children. If the child receives the father's X chromosome, the child will be a girl. If the child receives the father's Y chromosome, what will be the sex of the child? This Punnett square shows that we can expect half the children to be females and the other half to be males.

### Check Your Understanding

1. How do sex cells and body cells of humans differ in chromosome number?
2. What is amniocentesis?
3. How are the sex chromosomes of human males and females alike? How are they different?
4. **Critical Thinking:** What would happen if each sex cell had the same number of chromosomes as a body cell?
5. **Biology and Math:** Suppose a woman has three girls. What is the chance her next child will be a boy?

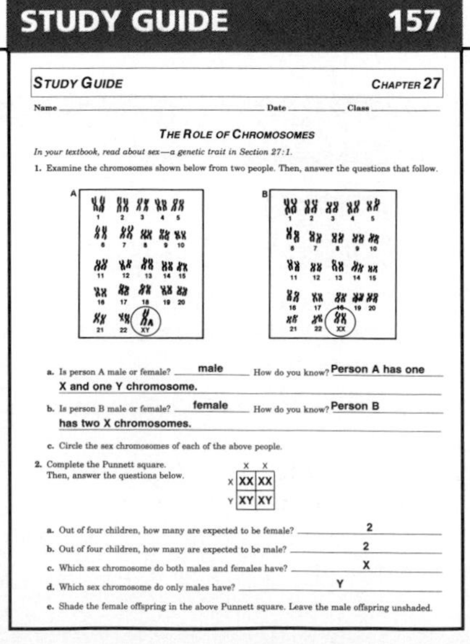

**STUDY GUIDE 157**

STUDY GUIDE — CHAPTER 27

Name ________ Date ________ Class ________

*THE ROLE OF CHROMOSOMES*

*In your textbook, read about sex—a genetic trait in Section 27:1.*

1. Examine the chromosomes shown below from two people. Then, answer the questions that follow.



a. Is person A male or female? **male** How do you know? **Person A has one X and one Y chromosome.**

b. Is person B male or female? **female** How do you know? **Person B has two X chromosomes.**

c. Circle the sex chromosomes of each of the above people.

2. Complete the Punnett square. Then, answer the questions below.

| | X | X |
|---|---|---|
| X | XX | XX |
| Y | XY | XY |

a. Out of four children, how many are expected to be female? **2**

b. Out of four children, how many are expected to be male? **2**

c. Which sex chromosome do both males and females have? **X**

d. Which sex chromosome do only males have? **Y**

e. Shade the female offspring in the above Punnett square. Leave the male offspring unshaded.

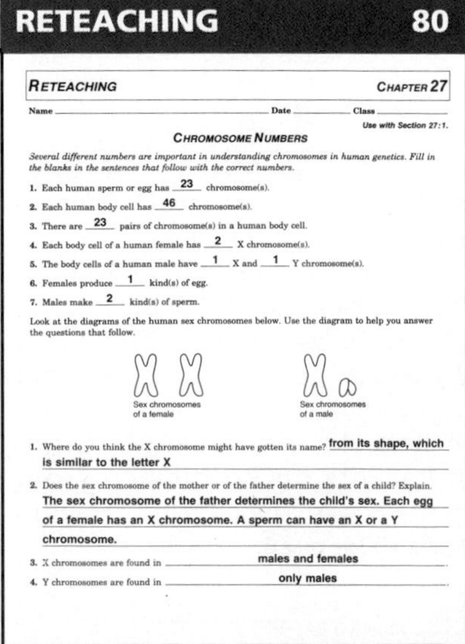

**RETEACHING 80**

RETEACHING — CHAPTER 27

Name ________ Date ________ Class ________

*Use with Section 27:1.*

*CHROMOSOME NUMBERS*

*Several different numbers are important in understanding chromosomes in human genetics. Fill in the blanks in the sentences that follow with the correct numbers.*

1. Each human sperm or egg has **23** chromosome(s).
2. Each human body cell has **46** chromosome(s).
3. There are **23** pairs of chromosome(s) in a human body cell.
4. Each body cell of a human female has **2** X chromosome(s).
5. The body cells of a human male have **1** X and **1** Y chromosome(s).
6. Females produce **1** kind(s) of egg.
7. Males make **2** kind(s) of sperm.

Look at the diagrams of the human sex chromosomes below. Use the diagram to help you answer the questions that follow.



1. Where do you think the X chromosome might have gotten its name? **from its shape, which is similar to the letter X**
2. Does the sex chromosome of the mother or of the father determine the sex of a child? Explain. **The sex chromosome of the father determines the child's sex. Each egg of a female has an X chromosome. A sperm can have an X or a Y chromosome.**
3. X chromosomes are found in **males and females**.
4. Y chromosomes are found in **only males**.

# Lab 27–1 Sex Determination 

## Problem: In a family of four children, how many will be girls and how many will be boys?

### Skills

make and use tables, formulate a model, calculate, design an experiment

### Materials

2 coins
paper cup
masking tape

### Procedure

1. Copy the data table 10 times.
2. Cover two coins with tape. Mark one coin with an X on each side. Mark the other coin with an X on one side and a Y on the other side.
3. Draw a Punnett square to show how many males and females you expect to see in one family with four children. Multiply the number of males by 10 and the number of females by 10. The totals are the numbers of males and females you would expect to see in ten families.
4. Put both coins in the paper cup. Shake the cup and drop the coins on your desk. Read the combination.
5. On table 1, make a check mark for your result on the first shake.
6. Shake, read, and mark your results three more times.
7. **Calculate:** Total the columns marked XX and XY.
8. Repeat steps 4 to 7 nine more times using a new table for each group of four shakes.
9. Add the number of females on all 10 tables. This is the total number of females you observed. Do the same for males.
10. Compare the observed totals in step 9 with the expected totals in step 3.

### Data and Observations

1. How many male and female children do you expect in each family? Explain your answer using the words *chromosome, X, Y, sperm,* and *egg.*
2. How many males and how many females did you expect in 10 families?

| | TABLE 1 RESULTS | |
|---|---|---|
| SHAKE | XX | XY |
| 1 | | |
| 2 | Answers will | |
| 3 | vary. | |
| 4 | | |
| Total | | |

### Analyze and Apply

1. Was the total number of males and females observed different from what you expected? If so, how can you explain the difference?
2. If you had made tables for 20 families, how many male children and how many female children would you expect?
3. If you had made tables for 100 families, would the observed results have been closer to the expected results? Explain your answer.
4. **Apply:** Suppose a doctor told a couple with three girls that their next child would be a boy. Is the doctor correct? Explain.

### Extension

**Design an experiment** to show how often families with four children will have two boys and two girls.



## COMPUTER PROGRAM XXII

COMPUTER PROGRAM LAB 27-1 SEX DETERMINATION

```
]POKE 33,33

]LIST

10  TEXT : HOME
20  PRINT "READ LAB 27-1 IN YOUR TEXT BOOK."
25  PRINT : PRINT
30  PRINT "THE  COMPUTER WILL TOSS THE COINS FOR YOU AND WILL ALSO RECORD YOUR RES
ULTS ON A  TABLE"
40  DIM Z(40),A(10,4)
50 X =  INT ( RND (1) * 60) + 1
60 X =  INT (X / 10)
70  IF X < 1 OR X > 5 THEN 50
80  IF X = 1 THEN XX = 20:XY = 20
90  IF X = 2 THEN XX = 21:XY = 19
100  IF X = 3 THEN XX = 22:XY = 18
110  IF X = 4 THEN XX = 19:XY = 21
120  IF X = 5 THEN XX = 18:XY = 22
130  FOR X = 1 TO XX
140 P =  INT ( RND (1) * 40) + 1
150  IF Z(P) <  > 0 THEN 140
160 Z(P) = 1
170  NEXT X
180  FOR X = 1 TO XY
190 P =  INT ( RND (1) * 40) + 1
200  IF Z(P) <  > 0 THEN 190
210 Z(P) = 2
220  NEXT X
230 SS = 0
240  FOR X = 1 TO 10
250  FOR Y = 1 TO 4
260 SS = SS + 1
270 A(X,Y) = Z(SS)
280  NEXT Y,X
290  PRINT : PRINT : PRINT "PRESS ANY KEY TO START";: GET ER$
300  TEXT : HOME
310  VTAB (5): PRINT "TABLE             RESULTS"
320  PRINT "-----------------------"
330  PRINT "SHAKE            XX    XY"
340  POKE 34,8
350  PRINT "-----------------------"
360  FOR X = 1 TO 10
370 R1 = 0:R2 = 0
380  FOR Y = 1 TO 4
390  VTAB (13): PRINT "-----------------------"
400  VTAB (5): HTAB (7): PRINT X
410  VTAB (21): PRINT "PRESS ANY KEY FOR EACH                    COIN TOSS OF TABL
E #";X;: GET ZZ$
420  VTAB (Y + 8): HTAB (3): PRINT Y;
430  IF A(X,Y) = 1 THEN  HTAB (17): PRINT "*":R1 = R1 + 1:RX = RX + 1
440  IF A(X,Y) = 2 THEN  HTAB (23): PRINT "*":R2 = R2 + 1:RY = RY + 1
450  IF R1 = 2 AND R2 = 2 THEN TB = TB + 1
460  NEXT Y
470  VTAB (14): PRINT "TOTAL";
480  HTAB (17): PRINT R1;
```

## ANSWERS

### Data and Observations

1. 2, There are 2 chances out of 4 that a sperm with a Y chromosome and an egg with an X chromosome will join.
2. 20 males; 20 females

### Analyze and Apply

1. Answers will vary; differences are due to chance.
2. 40 males; 40 females
3. Yes, the more results you observe, the closer the observed results will be to the expected.
4. No, there is 1 chance out of 2 that the baby will be a boy, and 1 chance out of 2 that it will be a girl.

## Lab 27-1 Sex Determination

### Overview

In this lab, students will determine by simulation how often certain combinations of males and females can occur in a family. They will compare answers predicted from a Punnett square to answers obtained by flipping coins.

**Objectives:** Upon completing the lab, students will be able to (1) **predict** expected data, (2) **experiment** to get observed data, and (3) **compare** the expected to observed data in order to infer what happened in the experiment.

**Time Allotment:** 45 minutes

### Preparation

▶ You may wish to put tape on the coins in advance of class.

**Lab 27-1 worksheet**/*Teacher Resource Package*, pp. 105-106.

### Teaching the Lab

▶ Explain to students that the results they get each time they drop the coins are strictly chance happenings. In the same manner, the sex of a child is usually a chance happening.

**Lab 27-1 Computer Program/** *Teacher Resource Package*, p. xxii. Use the computer program shown at the bottom of this page as you teach this lab.

### ✓ ASSESSMENT

**Skill:** Have students use their knowledge of the X and Y chromosomes to interpret the chart below. Have them explain why each of the six families has the number of boys and girls shown by making Punnett squares to help explain their results.

| Family | Number of Girls | Number of Boys | Total |
|---|---|---|---|
| A | 3 | 1 | 4 |
| B | 2 | 3 | 5 |
| C | 1 | 1 | 2 |
| D | 5 | 0 | 5 |
| E | 1 | 2 | 3 |
| F | 0 | 1 | 1 |

## 27:2 Human Traits

### PREPARATION

#### Materials Needed

Make copies of the Study Guide, Transparency Master, Application, Critical Thinking/Problem Solving, Reteaching, and Lab worksheets in the *Teacher Resource Package*.

#### Key Science Words

incomplete dominance
sickle-cell anemia
color blindness

#### Process Skills

In the Mini Lab, students will make and use tables. In Lab 27-2, students will observe, make and use tables, and interpret data.

### 1 FOCUS

▶ The objectives are listed on the student page. Remind students to preview these objectives as a guide to this numbered section.

#### Convergent Question

Ask students why humans are able to inherit traits as other living things do. Ask them what all living things share that enables them to pass on their traits.

#### Portfolio

Have students solve the problems on blood types in humans that appear below and place their answers in their portfolios.

A. One parent has blood type A and the second parent has blood type B; all of their three children have blood type O. Set up a Punnett square to show how the children inherited their blood type from their parents.

B. Two parents have blood type AB. A medical technician told them it would be impossible for them to have a child with blood type O or A. Write a paragraph in which you tell whether the parents were given correct information. Use Punnett squares to help explain your response.



#### Objectives

4. **Compare** recessive and dominant traits with incomplete dominance.
5. **Describe** different ways human traits can be inherited.

#### Key Science Words

incomplete dominance
sickle-cell anemia
color blindness

## 27:2 Human Traits

Have you ever gone "people watching"? The next time you are at a shopping center, movie theater, or football game, look at the people around you. They have many traits that you can see. These traits come from their parents. People also have many traits that you can't see. These traits direct things that go on inside the body.

### Survey of Human Traits

The girl in Figure 27-5a has freckles. She also has dimples in her cheeks. Both traits are caused by dominant genes. In both of these traits, one or both of the parents shows the dominant trait. Remember, only one dominant gene is needed to make a dominant trait show up. Two recessive genes are needed to make a recessive trait show up.

What kinds of traits are recessive? Attached earlobes is a recessive trait. Neither of the parents needs to have attached earlobes for their child to show this trait. Set up a Punnett square to show why that is true.

Other recessive traits are straight hair and not being able to roll the sides of your tongue. Can you guess what the dominant traits are? The ability to roll your tongue is dominant. Curly hair is dominant to straight hair. Short eyelashes are recessive. What is the dominant trait?

a

b

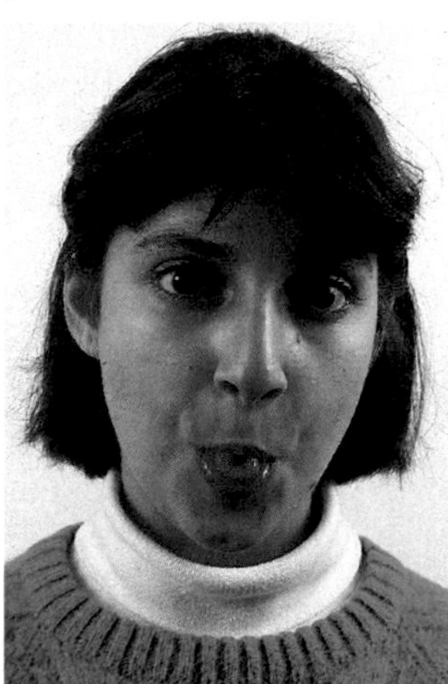

**Figure 27-5** Freckles and dimples are dominant traits (a). The ability to roll the tongue is a dominant trait (b).



### OPTIONS

#### Science Background

Between 8 and 13 percent of African Americans have the sickle-cell trait and are heterozygous with incomplete dominance. One out of every 500 African-American babies is born with sickle-cell anemia.

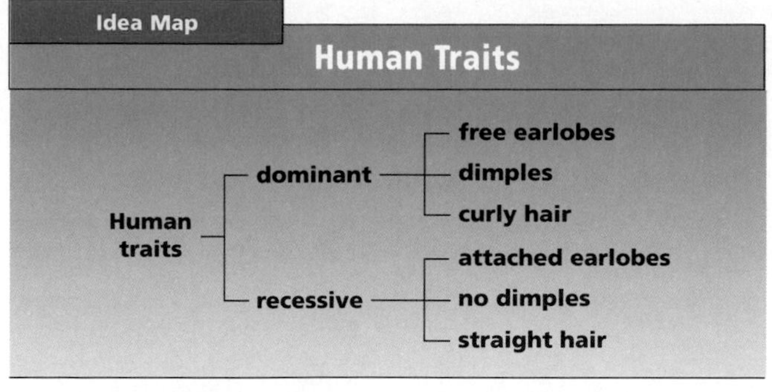


**Study Tip:** Use this idea map as a study guide to some human traits. How many other dominant and recessive traits can you think of?

## Incomplete Dominance

Some traits are neither totally dominant nor totally recessive. Scientists know of several genes that don't show total dominance. A case in which neither gene is totally dominant to the other is called **incomplete dominance.** If you placed a piece of blue glass over a piece of yellow glass and held it up to the light, what would you see? Figure 27-6 shows that the glass would look green. If you took apart the pieces of glass, you would see that they were still blue and yellow.

**What trait is seen in an individual that is heterozygous for an incompletely dominant trait?**

The same kind of thing happens with incomplete dominance. Remember that when a dominant and a recessive gene are together in a heterozygous organism, only the dominant trait is usually seen. In incomplete dominance, we see a new trait that is a blend of the dominant and recessive traits. When pure dominant red snapdragons are mated with pure recessive white snapdragons, the heterozygous offspring are all pink. It is important to know that the genes themselves do not combine. When they are separated, we see the original traits again. If two pink snapdragons are mated, they produce red, white, and pink offspring.

a

b

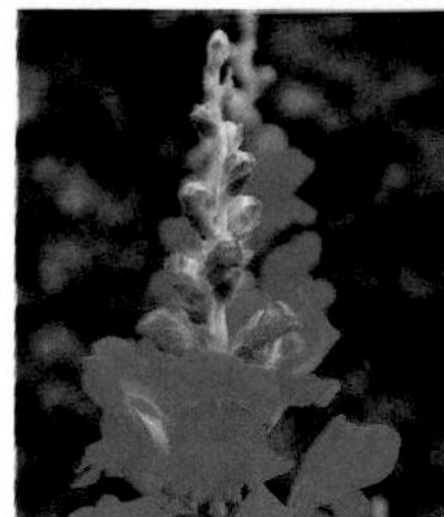

**Figure 27-6** In incomplete dominance, the combining of two genes gives a new trait, much as combining blue and yellow glass appears to make green glass (a). The combining of genes from red and white snapdragons gives offspring that are pink (b).



## 2 TEACH

### MOTIVATION/Filmstrip

Show the video *The Gene Machine,* Films for the Humanities and Science.

### Concept Development

▶ Make sure that students understand that some of their own traits may not be seen in either of their parents. These traits may be recessive, and each parent may have contributed the same recessive genes.

▶ Point out that recessive genes can skip several generations before they show up in offspring.

▶ Explain that geneticists know which of the human traits are dominant and recessive.

▶ Point out that the traits produced by incomplete dominance are sometimes called "intermediate" traits because their phenotypic effect is intermediate between the two traits.

### Idea Map

Have students use the idea map as a study guide to the major concepts of this section.

### Guided Practice

Have students write down their answers to the margin question: a new trait that is a combination of dominant and recessive traits.

### *Student Journal*

The manager of a new plant nursery decided to raise perennials from seeds. She wanted to continue producing plants that bred true each year. She was told that one kind of plant she wanted to raise showed incomplete dominance and could produce any one of three different colors. Have students write in their journals how the nursery manager could find out whether the trait for flower color was caused by incomplete dominance. Tell them to use Punnett squares to help explain their statements.

### *OPTIONS*

**How Are Traits on Sex Chromosomes Inherited?**/*Lab Manual,* pp. 231-234. Use this lab as an extension to studying human traits.

## TEACH

### Concept Development

▶ Tell students that persons with sickle-cell anemia can be given medication to lessen the effect of the disease; however, it cannot be cured.

### Independent Practice

**Study Guide**/*Teacher Resource Package,* p. 158. Use the Study Guide worksheet shown at the bottom of this page for independent practice.

### Guided Practice

Have students write down their answers to the margin question: the shape of red blood cells.

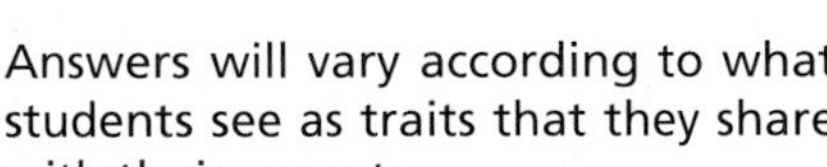

### Mini Lab

Answers will vary according to what students see as traits that they share with their parents.

**ASSESSMENT**

**Oral:** Direct students to report how they would determine if the traits they listed in their tables were dominant or recessive. Tell them to use Punnett squares and expected and observed results to explain their answers.

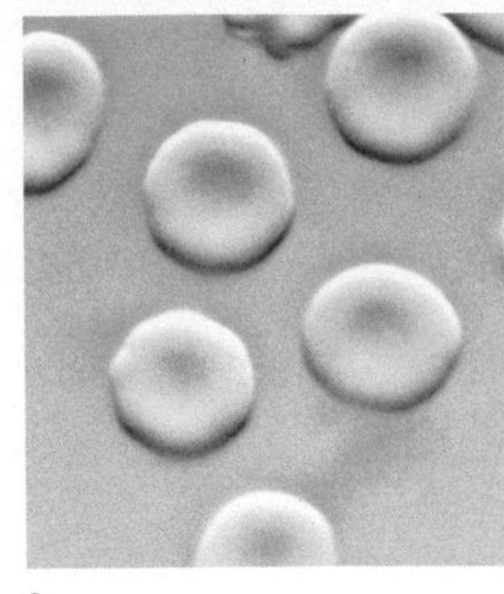

a

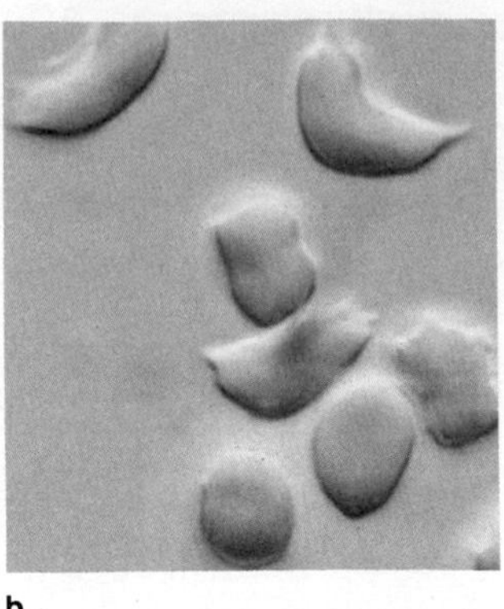

b

**Figure 27-7** Normal round (a) and sickled (b) red blood cells. Which kind of cells does a heterozygous person have? both

**What human trait shows incomplete dominance?**

Some traits in humans show incomplete dominance. One of these is the shape of red blood cells. In most people, red blood cells are round. Some people have red blood cells shaped like those in Figure 27-7b. These cells are called sickle cells. They are shaped like a sickle that is used to cut grass. Some people have both shapes of blood cells.

Let the letter $R$ stand for the gene for round blood cells. Let $R'$ stand for the gene for sickle cells. In incomplete dominance, two different letters or symbols are used for the genes controlling the trait. The $R$ gene is not totally dominant over the $R'$ gene. The $R'$ gene is not totally dominant over the $R$ gene. Capital letters can stand for both. A person with round cells has $RR$ genes. A person with all sickle cells has $R'R'$ genes. A person with both kinds of cells has $RR'$ genes.

Someone with $RR'$ genes usually doesn't have serious health problems. People who have some sickle cells, however, may not be as active as those who have all round cells. As you know, red blood cells carry oxygen to body cells. Sickle cells do not carry oxygen as well as round blood cells.

People with $R'R'$ genes have sickle-cell anemia. **Sickle-cell anemia** is a genetic disorder in which all the red blood cells are shaped like sickles. This disorder is much more common among African-Americans than among people of other races. People with sickle-cell anemia have serious health problems. Their lives may be shortened because of the disorder. The sickle cells cannot carry enough oxygen. Body tissues can be damaged because they don't receive enough oxygen. Sickle cells do not move easily through the blood capillaries because their shape causes the capillaries to become clogged.

> ### Mini Lab
> **What Traits Do Humans Share?**
>
> **Make and use tables:** Make a table of traits inherited from your parents. Compare your list with that of a classmate. How many traits do you have in common? *For more help, refer to the **Skill Handbook**, pages 715-717.*



## OPTIONS

**What Do Normal and Sickled Cells Look Like?**/*Lab Manual,* pp. 227-230. Use this lab as an extension to studying incomplete dominance.

**Transparency Master**/*Teacher Resource Package,* p. 137. Use the Transparency Master shown here to teach traits with incomplete dominance.

**STUDY GUIDE 158**

STUDY GUIDE — CHAPTER 27

Name ______ Date ______ Class ______

HUMAN TRAITS

*In your textbook, read about incomplete dominance in Section 27:2.*

1. Red blood cell shape shows incomplete dominance in humans. $R$ is the gene for round cell shape and $R'$ is the gene for sickle cell shape.
   a. Put checkmarks in the following table to show the shape of cells for persons with the genes listed.

| | (round cells) | (round and sickle cells) | (sickle cells) |
|---|---|---|---|
| R'R' | | | ✔ |
| RR' | | ✔ | |
| RR | ✔ | | |

   b. Which gene, $R$ or $R'$, is dominant? neither Which is recessive? neither
2. a. Describe the condition that a person with $R'R'$ genes has. The person has all sickled blood cells. Sickled red blood cells do not carry oxygen as well as normal red blood cells. Thus, people with all sickled red blood cells have serious health problems and their lives may be shortened
   b. What is the name of this disease? sickle-cell anemia
3. Human blood types show incomplete dominance as well as dominance. Fill in the table at the right showing possible genes a person with each blood type might have.

| Blood type | Possible genes |
|---|---|
| A | AA or AO |
| B | BB or BO |
| O | OO |
| AB | AB |

4. Which blood type genes are dominant to other blood type genes? A and B are dominant to O.
5. Which blood type genes show incomplete dominance to each other? A and B

**TRANSPARENCY 137**

TRANSPARENCY MASTER — CHAPTER 27

A TRAIT WITH INCOMPLETE DOMINANCE

Red blood cell shapes in humans.

Round — Sickle-shaped — Both round and sickle shaped

| | R | R' |
|---|---|---|
| R | RR<br>All red blood cells are round. | RR'<br>Some red blood cells are sickle-shaped. |
| R' | RR'<br>Some red blood cells are sickle-shaped. | R' R'<br>All red blood cells are sickle-shaped. Has sickle-cell anemia. |

If two parents have ***RR′*** genes, what would we expect the children to have? Look at the Punnett square in Figure 27-8. One out of four children is expected to have all round cells. Two out of four children are expected to have both round and sickle cells. The fourth child is expected to have all sickle cells. That child would have sickle-cell anemia.

| | *R* | *R′* |
|---|---|---|
| *R* | *RR* | *RR′* |
| *R′* | *RR′* | *R′R′* |

**Figure 27-8** How sickle-cell anemia can be inherited when both parents are heterozygous. How many children can be expected to have sickle-cell anemia? one-fourth of all the children

## Blood Types in Humans

Remember from Chapter 12 that there are four blood types—A, B, AB, and O. Human blood type is controlled in part by dominance of genes. Although three genes control blood types, each person has only two of them. The two genes that you have control your blood type.

The three genes that control blood type are ***A, B,*** and ***O.*** Both ***A*** and ***B*** are dominant to ***O. A*** and ***B,*** however, are not dominant to each other. Look at Table 27-2 as you read the next three paragraphs.

If you have type O blood, you have ***OO*** genes. If you have type AB blood, you have one ***A*** gene and one ***B*** gene.

If you have type A blood, you have ***AA*** or ***AO*** genes. If you have ***AO*** genes, the ***A*** gene is dominant to the ***O*** gene.

If you have type B blood, you have ***BB*** or ***BO*** genes. If you have ***BO*** genes, the ***B*** gene is dominant to the ***O*** gene.

**TABLE 27–2. BLOOD TYPES AND GENES**

| Blood Type | Genes |
|---|---|
| O | ***OO*** |
| AB | ***AB*** |
| A | ***AA*** or ***AO*** |
| B | ***BB*** or ***BO*** |

## Genes on the X Chromosome

Like other chromosomes, the sex chromosomes carry genes. Figure 27-9 shows some traits that are controlled by genes on the sex chromosomes. Notice that females have two genes for each of these traits. That is because females have two X chromosomes.

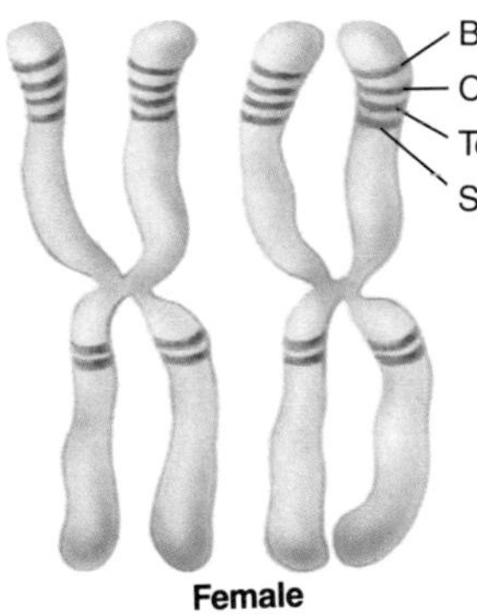


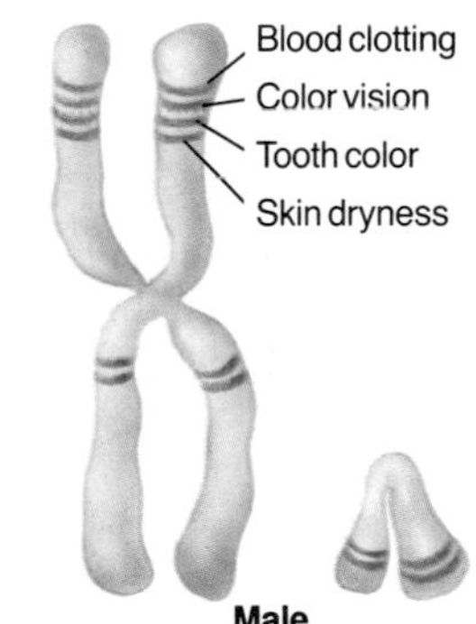


**Figure 27-9** Certain genes on the X chromosome are missing from the Y chromosome.



## TEACH

### Concept Development

▶ Point out that even though there are three genes for blood type, a human can inherit any combination of two of the three genes.

▶ Emphasize that the blood type of a child can be different from that of the parents.

▶ Stress that not all genes on the sex chromosomes carry undesirable traits, such as color blindness.

▶ Point out that most states give a test for color blindness when people take the test for a driver's license.

### Check for Understanding

Give students a short quiz in which they have to construct Punnett squares to solve at least two problems. Have transparencies with Punnett squares already drawn on them and ask some students to demonstrate how they got their answer.

### Reteach

As students work on the problems and check their answers, work with groups of students who are still having difficulty with a certain problem. Setting up a difficult problem on an overhead projector can help students understand how genes combine.

**Misconception:** Students may think Rh + or Rh- factor is the same as the A, B, and O blood groups. Tell students that the Rh factors are controlled by other genes.

**CRITICAL THINKING 27**

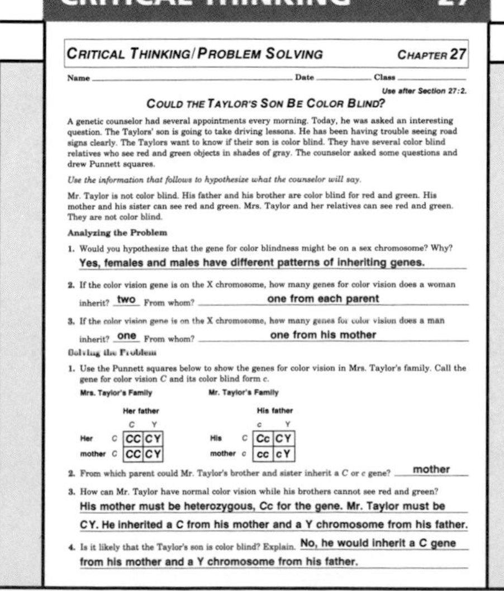

CRITICAL THINKING/PROBLEM SOLVING CHAPTER 27

Name ______ Date ______ Class ______

*Use after Section 27:2.*

**COULD THE TAYLOR'S SON BE COLOR BLIND?**

A genetic counselor had several appointments every morning. Today, he was asked an interesting question. The Taylors' son is going to take driving lessons. He has been having trouble seeing road signs clearly. The Taylors want to know if their son is color blind. They have several color blind relatives who see red and green objects in shades of gray. The counselor asked some questions and drew Punnett squares.

*Use the information that follows to hypothesize what the counselor will say.*

Mr. Taylor is not color blind. His father and his brother are color blind for red and green. His mother and his sister can see red and green. Mrs. Taylor and her relatives can see red and green. They are not color blind.

**Analyzing the Problem**

1. Would you hypothesize that the gene for color blindness might be on a sex chromosome? Why? Yes, females and males have different patterns of inheriting genes.
2. If the color vision gene is on the X chromosome, how many genes for color vision does a woman inherit? two From whom? one from each parent
3. If the color vision gene is on the X chromosome, how many genes for color vision does a man inherit? one From whom? one from his mother

**Solving the Problem**

1. Use the Punnett squares below to show the genes for color vision in Mrs. Taylor's family. Call the gene for color vision *C* and its color blind form *c*.

**Mrs. Taylor's Family** (Her father across the top, Her mother down the side)

| | *C* | Y |
|---|---|---|
| *C* | CC | CY |
| *C* | CC | CY |

**Mr. Taylor's Family** (His father across the top, His mother down the side)

| | *c* | Y |
|---|---|---|
| *C* | Cc | CY |
| *c* | cc | cY |

2. From which parent could Mr. Taylor's brother and sister inherit a *C* or *c* gene? mother
3. How can Mr. Taylor have normal color vision while his brothers cannot see red and green? His mother must be heterozygous, Cc for the gene. Mr. Taylor must be CY. He inherited a C from his mother and a Y chromosome from his father.
4. Is it likely that the Taylor's son is color blind? Explain. No, he would inherit a C gene from his mother and a Y chromosome from his father.

**APPLICATION 27**

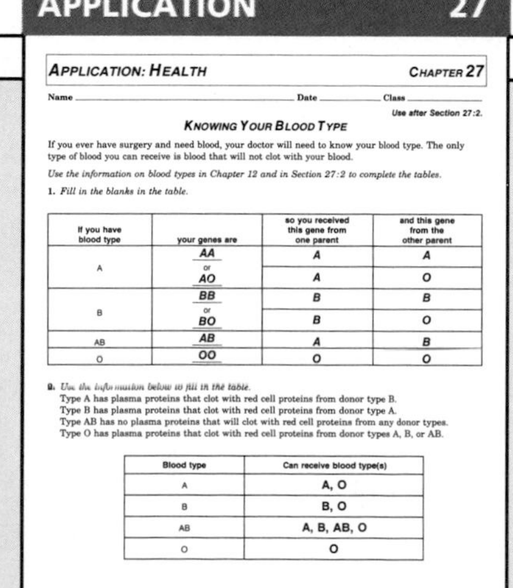

APPLICATION: HEALTH CHAPTER 27

Name ______ Date ______ Class ______

*Use after Section 27:2.*

**KNOWING YOUR BLOOD TYPE**

If you ever have surgery and need blood, your doctor will need to know your blood type. The only type of blood you can receive is blood that will not clot with your blood.

*Use the information on blood types in Chapter 12 and in Section 27:2 to complete the tables.*

1. *Fill in the blanks in the table.*

| If you have blood type | your genes are | so you received this gene from one parent | and this gene from the other parent |
|---|---|---|---|
| A | *AA* | *A* | *A* |
| | or *AO* | *A* | *O* |
| B | *BB* | *B* | *B* |
| | or *BO* | *B* | *O* |
| AB | *AB* | *A* | *B* |
| O | *OO* | *O* | *O* |

2. *Use the information below to fill in the table.*
Type A has plasma proteins that clot with red cell proteins from donor type B.
Type B has plasma proteins that clot with red cell proteins from donor type A.
Type AB has no plasma proteins that will clot with red cell proteins from any donor types.
Type O has plasma proteins that clot with red cell proteins from donor types A, B, or AB.

| Blood type | Can receive blood type(s) |
|---|---|
| A | A, O |
| B | B, O |
| AB | A, B, AB, O |
| O | O |

## OPTIONS

**Application**/*Teacher Resource Package*, p. 27. Use the Application worksheet shown here to help students understand blood types.

**Critical Thinking/Problem Solving/** *Teacher Resource Package*, p. 27. Use the Critical Thinking/Problem Solving worksheet shown here for students to learn who could be colorblind.

## TEACH

### Independent Practice

**Study Guide**/*Teacher Resource Package*, p. 159. Use the Study Guide worksheet shown at the bottom of this page for independent practice.

### Check for Understanding

Have students respond to the first three questions in Check Your Understanding.

### Reteach

**Reteaching**/*Teacher Resource Package*, p. 81. Use this worksheet to give students additional practice with color blindness.

**Extension:** Assign Critical Thinking, Biology and Reading, or some of the **OPTIONS** available with this lesson.

## 3 APPLY

### ACTIVITY/Challenge

Give students some simple genetics problems that deal with incomplete dominance, blood types, and sex-linked traits.

## 4 CLOSE

### ACTIVITY/Guest Speaker

Ask a local airplane pilot to explain why pilots need excellent color vision.

### Answers to Check Your Understanding

6. when neither gene is totally dominant over the other
7. AA or AO
8. The male doesn't have a second X chromosome to prevent recessive genes from being expressed.
9. No. He would get the nearsighted trait from his mother.
10. A male is hemizygous for traits on the X chromosome because he has only one X chromosome. The other sex chromosome of a male is the Y chromosome.

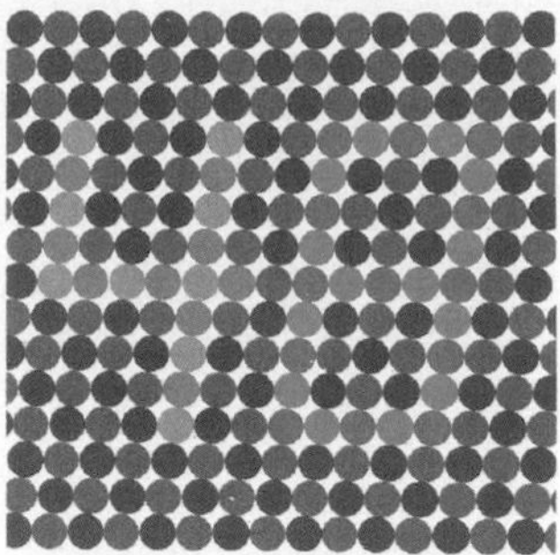

**Figure 27-10** To be tested for red-green color vision, a person may be shown a figure like this one.

Males have only one gene for each of these traits. The Y chromosome does not have the genes that are on the X chromosome. When the X and Y chromosomes are together, genes on the X chromosome control the traits.

Let's look at color blindness. **Color blindness** is a problem in which red and green look like shades of gray or other colors. Look at Figure 27-10. Someone who is color blind for red and green can't see the numbers in it.

Being able to see red and green as two separate colors is a dominant trait. Let's use ***C*** for this gene. Not being able to see red and green is a recessive trait. A ***c*** will be used for this gene.

A female who has ***CC*** genes will be able to see red and green. If she has ***Cc*** genes, she still will be able to see red and green. If she has ***cc*** genes, she will be color blind. Two of the three gene combinations for the female allow her to see red and green.

A male is more likely to be color blind. Why? Think about what genes he could have. Remember that the Y chromosome will not have a gene for the trait.

A male who has the ***C*** gene on his X chromosome will be able to see red and green. If he has the ***c*** gene on his X chromosome, he will not be able to see red and green. There are only two possible gene combinations for the male. One of them will allow him to see red and green. What would happen if a woman with ***Cc*** genes and a man with a ***c*** gene on his X chromosome had children? See if you can answer this question. 1 normal male: 1 color blind male: 1 heterozygous normal female: 1 color blind female

### Check Your Understanding

6. What is incomplete dominance?
7. What two gene combinations can a person with type A blood have?
8. Why are there more color blind males than color blind females?
9. **Critical Thinking:** The gene for nearsightedness in humans is found on the X chromosome. A boy has a nearsighted father. Does this mean that the boy will become nearsighted? Why or why not?
10. **Biology and Reading:** The word part *hemi* is found in the word *hemisphere*. *Hemi* means half. The scientific term for genes on the X chromosome in a male is hemizygous. Why?

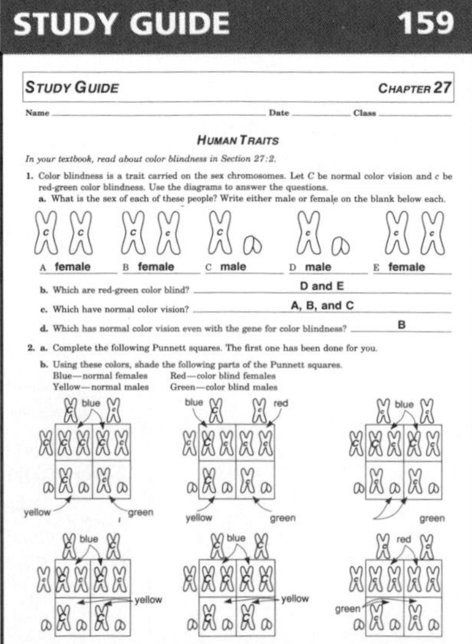

**STUDY GUIDE 159**

STUDY GUIDE — CHAPTER 27

Name ______ Date ______ Class ______

HUMAN TRAITS

*In your textbook, read about color blindness in Section 27:2.*

1. Color blindness is a trait carried on the sex chromosomes. Let *C* be normal color vision and *c* be red-green color blindness. Use the diagrams to answer the questions.
   a. What is the sex of each of these people? Write either male or female on the blank below each.
   A female B female C male D male E female
   b. Which are red-green color blind? D and E
   c. Which have normal color vision? A, B, and C
   d. Which has normal color vision even with the gene for color blindness? B
2. a. Complete the following Punnett squares. The first one has been done for you.
   b. Using these colors, shade the following parts of the Punnett squares.
   Blue—normal females Red—color blind females
   Yellow—normal males Green—color blind males



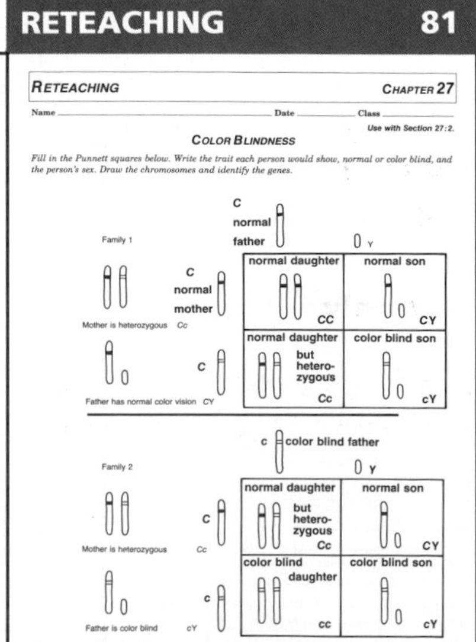

**RETEACHING 81**

RETEACHING — CHAPTER 27

Name ______ Date ______ Class ______

Use with Section 27:2.

COLOR BLINDNESS

*Fill in the Punnett squares below. Write the trait each person would show, normal or color blind, and the person's sex. Draw the chromosomes and identify the genes.*

# Lab 27–2 Human Traits

## Problem: Which traits are most common?

### Skills

observe, make and use tables, interpret data

### Materials

pencil
paper

### Procedure

1. Copy the data table.
2. **Make and use tables:** Have another student check you for each of the traits listed in the table. Record in your table with a check mark whether you show the dominant or recessive trait.
3. Reverse roles and check your partner for each trait.

| TRAIT | DOMINANT | RECESSIVE |
|---|---|---|
| Skin color | Answers will | vary. |
| Extra fingers or toes | | |
| Freckles | | |
| Earlobe | | |
| Tongue rolling | | |
| Shape of hairline | | |
| Hair on middle sections of fingers | | |
| Chin shape | | |

### Data and Observations

1. How many dominant traits do you have? How many recessive traits do you have?
2. How many other students in your class have the same traits as you?

### Analyze and Apply

1. Which statement best explains this lab?
   (a) All students in the class show the same dominant and recessive traits.
   (b) A person may show some dominant and some recessive traits.
   (c) It is easier to recognize a dominant trait in someone than a recessive trait.
2. **Apply:** Were there more dominant or more recessive traits in your class? How can you explain this?

### Extension

**Design an experiment** to determine if any of the traits are on the X chromosome.

| TRAIT | DOMINANT | RECESSIVE |
|---|---|---|
| Skin color | Dark colors | Light colors |
| Extra fingers or toes | Six or seven fingers or toes | Five fingers and toes |
| Freckles | Freckles | No Freckles |
| Earlobe | Free | Attached |
| Tongue rolling | Can roll edges | Cannot roll edges |
| Shape of hairline | Pointed in middle | Not pointed in middle |
| Hair on middle sections of fingers | Hair | No hair |
| Chin shape | Indentation in middle | No indentation |



## ANSWERS

### Data and Observations

1. Answers will vary.
2. Answers will vary.

### Analyze and Apply

1. b
2. Answers will vary. Some recessive genes will be found more frequently than dominant genes.

## GLENCOE TECHNOLOGY

**Videodisc**
The Secret of Life
*Blood Types*

## Lab 27-2 Human Traits

### Overview

Students will find out which inherited traits are most common among classmates by checking the phenotypes of another class member. Using their data, they will be able to determine whether there are more dominant or recessive traits visible in the members of the class. They will also find out that many members of the class have the same traits.

**Objectives:** Upon completing the lab, students will be able to (1) **determine** the dominant and recessive traits found in a population of students, (2) **state** which dominant and recessive traits they have, and (3) **use numbers** to show which traits were found more and less often.

**Time Allotment:** 45 minutes

### Preparation

**Lab 27-2 worksheet**/*Teacher Resource Package*, pp. 107-108.

### ASSESSMENT

**Performance:** Have the students make a table with six rows for the following traits: freckles, no freckles, can roll tongue, cannot roll tongue, hairline pointed, hairline not pointed. Include columns for male vs. female. Tell students to conduct a survey of 30-40 people in their neighborhood to find out which of the traits they show. Tell them to put a mark in the appropriate box for each trait. Have them calculate the percentage of people having each of the traits and compare the neighborhood data to the class data.

### Portfolio

Direct students to put their written reports from the assessment into their portfolios. Tell them to include their data charts and their analysis. Have them write a final paragraph telling what they learned about human genetics.

## 27:3 Genetic Disorders

## PREPARATION

### Materials Needed

Make copies of the Study Guide, Skill, Enrichment, and Reteaching worksheets in the *Teacher Resource Package*.

▶ Obtain poster board for the Skill Check.

### Key Science Words

dyslexia
genetic counseling
pedigree

### Process Skills

In the Skill Check, students will formulate models.

## 1 FOCUS

▶ The objectives are listed on the student page. Remind students to preview these objectives as a guide to this numbered section.

### MOTIVATION/Brainstorming

Ask students what the difference is between a genetic disorder and genetic error.

### Independent Practice

**Study Guide**/*Teacher Resource Package*, p. 160. Use the Study Guide worksheet shown at the bottom of this page for independent practice.

**GLENCOE TECHNOLOGY**

**Videodisc**

**Biology: The Dynamics of Life**
Animation: *Sex-Linked Traits*
Disc 1, Side 1, Ch. 35

**Objectives**

6. **Describe** some genetic disorders in humans.
7. **Give examples** of how genetic counseling can help families.

**Key Science Words**

dyslexia
genetic counseling
pedigree

## 27:3 Genetic Disorders

Each day, nearly 600 babies are born in the United States with some type of disorder. Some of the disorders are inherited. Let's look at a few of them.

### Errors in Chromosome Number

Some people are born with more or fewer than 46 chromosomes. This happens when the sperm or egg cell does not have 23 chromosomes. Figure 27-12 shows how this can come about.

During meiosis, the sister chromatids are supposed to pull apart from each other. Notice in Figure 27-12 that the sister chromatids of the short chromosome pulled apart and were separated into different cells. The sister chromatids of the longer chromosome stuck together. Both of them went into the same cell. That cell now has an extra chromosome.

If these cells were eggs, the one on the bottom would have one fewer chromosome, or 22. If this egg joined a sperm, a child with only 45 chromosomes would result. The egg on the top has an extra chromosome. If this egg joined a sperm, a child with 47 chromosomes would result.

Not pulling apart can happen to almost any pair of chromosomes. When it happens to the sex chromosomes, certain traits show up. Figure 27-11 shows sex chromosome patterns that sometimes show up in people. The O stands for a missing sex chromosome. Notice the boy and the girl who have only one sex chromosome. The sex chromosome from one parent is missing. What are some of the problems that result from having one of these chromosome patterns?

| Female | Male |
|---|---|
| XO | YO |
| | XXY |

**Figure 27-11** Sex chromosome patterns

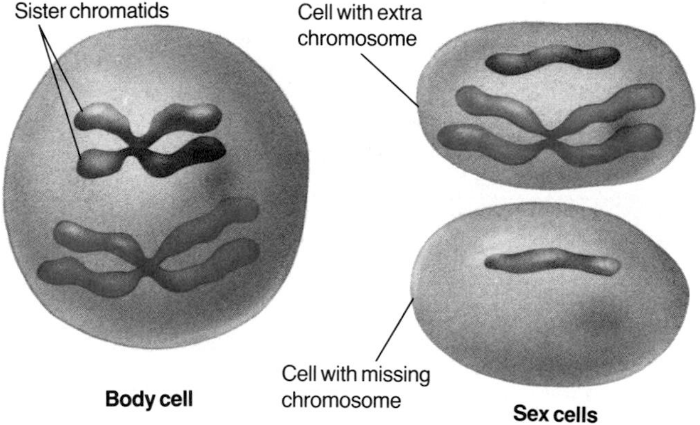


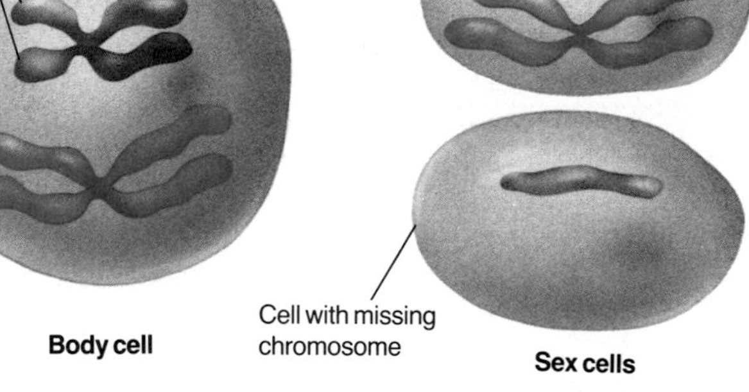

**Figure 27-12** When sister chromatids do not pull apart during meiosis, there can be an error in chromosome number in the child.



## OPTIONS

### Science Background

Geneticists refer to the sticking together of homologous chromosomes during meiosis as nondisjunction. Nondisjunction means failure to separate from one another.

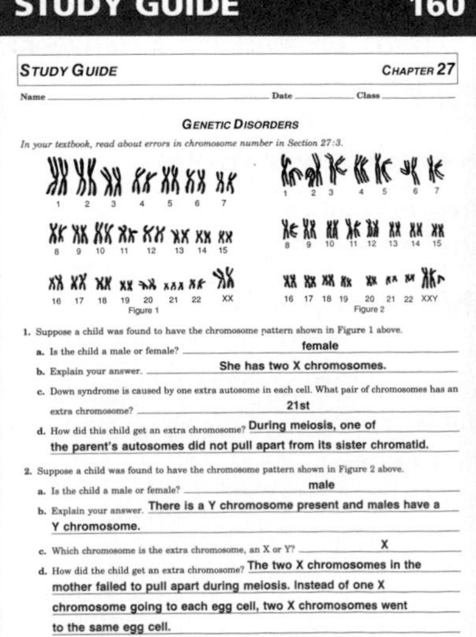

**STUDY GUIDE 160**

STUDY GUIDE — CHAPTER 27

Name ________ Date ______ Class ______

GENETIC DISORDERS

*In your textbook, read about errors in chromosome number in Section 27:3.*

Figure 1 (1–22, XX) — Figure 2 (1–22, XXY)

1. Suppose a child was found to have the chromosome pattern shown in Figure 1 above.
   a. Is the child a male or female? **female**
   b. Explain your answer. **She has two X chromosomes.**
   c. Down syndrome is caused by one extra autosome in each cell. What pair of chromosomes has an extra chromosome? **21st**
   d. How did this child get an extra chromosome? **During meiosis, one of the parent's autosomes did not pull apart from its sister chromatid.**
2. Suppose a child was found to have the chromosome pattern shown in Figure 2 above.
   a. Is the child a male or female? **male**
   b. Explain your answer. **There is a Y chromosome present and males have a Y chromosome.**
   c. Which chromosome is the extra chromosome, an X or Y? **X**
   d. How did the child get an extra chromosome? **The two X chromosomes in the mother failed to pull apart during meiosis. Instead of one X chromosome going to each egg cell, two X chromosomes went to the same egg cell.**

A YO male will die before it is born. An XO female and an XXY male cannot make sex cells, but they can live.

The child in Figure 27-13 was born with an extra autosome. During meiosis, one of his autosomes did not pull apart from its copy. A doctor would say he has Down syndrome. The child learns more slowly than most children his age. Children with Down syndrome often have heart problems. Researchers are learning more about problems caused by having an extra chromosome.

**Figure 27-13** Child with Down syndrome

## Genetic Disorders and Sex Chromosomes

You have studied traits controlled by genes on the sex chromosomes. The ability to see red and green was one of them. There are also serious genetic disorders controlled by genes on the X chromosome.

A certain recessive gene on the X chromosome can cause a rare disorder called hemophilia. Remember that hemophilia is a disorder in which a person's blood does not clot. Bleeding from a cut or bruise may take hours to stop. Hemophilia almost always shows up in male children. How would it be possible for a female child to inherit hemophilia? by inheriting one recessive gene from each parent

## Genetic Disorders and Autosomes

Genetic disorders can also be caused by genes found on the autosomes.

**Dyslexia** (dis LEHK see uh) is a genetic disorder that is also called word blindness. It is caused by a dominant gene. People with dyslexia see and write some letters of the alphabet or parts of words backward. They have trouble learning to read.

### Skill Check

**Formulate a model:** Make a model of the alphabet on a poster board showing how a person with dyslexia might see these letters. *For more help, refer to the **Skill Handbook**, pages 706-711.*

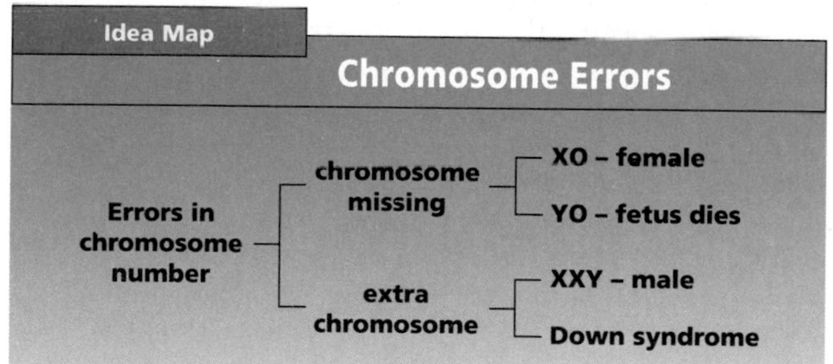


**Study Tip:** Use this idea map as a study guide to errors in chromosome number. Some examples of these errors are shown.

579

# 2 TEACH

### Software

*Genetics Counselor,* Queue, Inc.

### Concept Development

- Explain that scientists do not understand why the chromosomes fail to separate from each other.
- Explain that parts can be added or removed from a chromosome. This can cause a change in the traits a person inherits.
- Point out that viruses, chemicals, and radiation can cause a change in the structure of a chromosome.

### Independent Practice

**Study Guide**/*Teacher Resource Package*, p. 161. Use the Study Guide worksheet shown at the bottom of this page for independent practice.

### Skill Check

Tell students that some dyslexic people reverse the letters *b* for *d* and *p* for *q*. They also change the order of letters and numerals, such as was for saw, and 13 for 31.

### Idea Map

Have students use the idea map as a study guide to the major concepts of this section.

**Cooperative Learning:** Have teams of 3-4 write a news report on Down syndrome. Have each team give an oral report.

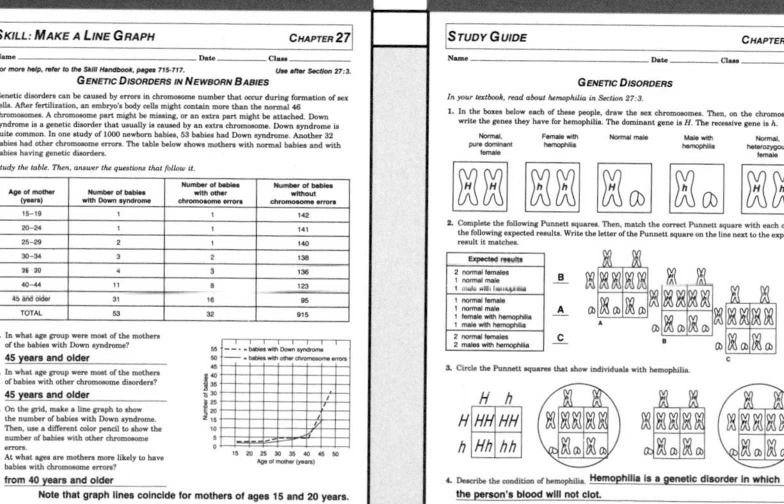

**SKILL 28**

**SKILL: MAKE A LINE GRAPH** — CHAPTER 27

Name ______ Date ______ Class ______

*For more help, refer to the Skill Handbook, pages 715-717.* *Use after Section 27:3.*

**GENETIC DISORDERS IN NEWBORN BABIES**

Genetic disorders can be caused by errors in chromosome number that occur during formation of sex cells. After fertilization, an embryo's body cells might contain more than the normal 46 chromosomes. A chromosome part might be missing, or an extra part might be attached. Down syndrome is a genetic disorder that usually is caused by an extra chromosome. Down syndrome is quite common. In one study of 1000 newborn babies, 53 babies had Down syndrome. Another 32 babies had other chromosome errors. The table below shows mothers with normal babies and with babies having genetic disorders.

*Study the table. Then, answer the questions that follow it.*

| Age of mother (years) | Number of babies with Down syndrome | Number of babies with other chromosome errors | Number of babies without chromosome errors |
|---|---|---|---|
| 15–19 | 1 | 1 | 142 |
| 20–24 | 1 | 1 | 141 |
| 25–29 | 2 | 1 | 140 |
| 30–34 | 3 | 2 | 138 |
| 35–39 | 4 | 3 | 136 |
| 40–44 | 11 | 8 | 123 |
| 45 and older | 31 | 16 | 95 |
| TOTAL | 53 | 32 | 915 |

1. In what age group were most of the mothers of the babies with Down syndrome? **45 years and older**
2. In what age group were most of the mothers of babies with other chromosome disorders? **45 years and older**
3. On the grid, make a line graph to show the number of babies with Down syndrome. Then, use a different color pencil to show the number of babies with other chromosome errors.
4. At what ages are mothers more likely to have babies with chromosome errors? **from 40 years and older**



**Note that graph lines coincide for mothers of ages 15 and 20 years.**

**STUDY GUIDE 161**

**STUDY GUIDE** — CHAPTER 27

Name ______ Date ______ Class ______

**GENETIC DISORDERS**

*In your textbook, read about hemophilia in Section 27:3.*

1. In the boxes below each of these people, draw the sex chromosomes. Then, on the chromosomes write the genes they have for hemophilia. The dominant gene is *H*. The recessive gene is *h*.



2. Complete the following Punnett squares. Then, match the correct Punnett square with each one of the following expected results. Write the letter of the Punnett square on the line next to the expected result it matches.

| Expected results | |
|---|---|
| 2 normal females, 1 normal male, 1 male with hemophilia | **B** |
| 1 normal female, 1 normal male, 1 female with hemophilia, 1 male with hemophilia | **A** |
| 2 normal females, 2 males with hemophilia | **C** |

3. Circle the Punnett squares that show individuals with hemophilia.

| | H | h |
|---|---|---|
| H | HH | HH |
| h | Hh | hh |

4. Describe the condition of hemophilia. **Hemophilia is a genetic disorder in which the person's blood will not clot.**

## OPTIONS

### Science Background

In 1910, Dr. Thomas Hunt Morgan discovered that traits governed by genes carried on the sex chromosomes showed a different pattern of inheritance than traits governed by autosomal genes.

**Skill**/*Teacher Resource Package*, p. 28. Use the Skill worksheet shown here to have students make a line graph on genetic disorders in newborn babies.

## TEACH

### Concept Development

▶ Genetic counselors are usually found in hospitals that do research in genetic diseases.

▶ Explain that geneticists collect data to predict how often these genes will show in a human population.

▶ Point out that one in twenty persons in the United States carries a recessive gene for cystic fibrosis.

### Guided Practice

Have students write down their answers to the margin questions: a genetic disorder in which some chemicals don't break down as they should; a diagram that shows how a certain trait is passed along in a family.

### Connection: Social Studies

Have students investigate which local and state agencies deal with health problems of genetic origin. Have them find out what these agencies do to help people and how much money is spent dealing with these kinds of health problems.

### ACTIVITY/Speaker

The local chapter of the Association for Retarded Citizens usually has a speakers bureau that can provide a free speaker to speak to your class about genetic counseling.

**GLENCOE TECHNOLOGY**

**Videotape**

**The Secret of Life**

*Tinkering with Our Genes: Genetic Medicine*

**What is PKU?**

Twenty years ago, a child born with PKU would spend its life in a hospital. PKU is a genetic disorder in which some chemicals in the body do not break down as they should. These chemicals can harm the brain cells. Today babies are tested for PKU soon after birth. A child born with PKU can be given a special diet, starting within the first few weeks of life. This diet has only small amounts of the chemicals that will not break down properly. The child can grow normally with this diet.

**Bio Tip**

**Health:** Cystic fibrosis patients lose large amounts of salt from their bodies, resulting in very salty sweat. For many years, having very salty sweat was used as a test for this disorder.

## Genetic Counseling

"It runs in the family." Have you heard this comment before? It means that many family members have a certain trait. The trait could be found in parents, grandparents, brothers, sisters, or other relatives. Family members can have the same hair color, eye color, or shape of nose.

Genetic disorders run in families, too. How can a couple find out whether a disorder in their family will show up in their children? They can seek genetic counseling. **Genetic counseling** is the use of genetics to predict and explain traits in children. A genetic counselor can tell whether a problem is caused by genes.

A genetic counselor asks a lot of questions about many members of a family. Then the counselor makes some conclusions. A genetic counselor can help answer the following questions:

1. How did their baby get the disorder?
2. If the baby is healthy, does it have a problem gene?
3. Is the trait dominant or recessive?
4. What will happen to the baby's health as it gets older?
5. What are the chances that future children will have the trait?

**Figure 27-14** A couple receiving genetic counseling

The baby of a young couple died when it was two years old. The doctor told the couple that their baby had cystic fibrosis (sis tik • fi BROH suhs). This is a genetic disorder in which the lungs and pancreas don't work the way they should. There are very serious breathing problems. Many children with this disorder die at an early age.

The baby's parents went to a genetic counselor, who told them that the disorder was caused by having two recessive genes, ***ff***. Both parents appeared normal, but they were each heterozygous for the dominant and recessive genes, ***Ff***.



### OPTIONS

**Enrichment:** Call the local chapter of the March of Dimes and request the booklet "A Guide to Human Chromosome Defects." This booklet gives good background information on karyotypes and chromosome errors, and the drawings and charts can be made into overhead transparencies.

**Enrichment**/*Teacher Resource Package*, p. 29. Use the Enrichment worksheet shown here for students to trace a genetic disorder in a family.

**ENRICHMENT 29**

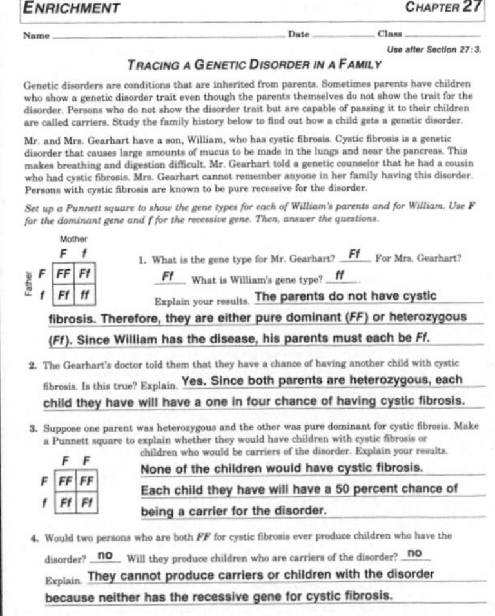

ENRICHMENT CHAPTER 27

Name ______ Date ______ Class ______

*Use after Section 27:3.*

*TRACING A GENETIC DISORDER IN A FAMILY*

Genetic disorders are conditions that are inherited from parents. Sometimes parents have children who show a genetic disorder trait even though the parents themselves do not show the trait for the disorder. Persons who do not show the disorder trait but are capable of passing it to their children are called carriers. Study the family history below to find out how a child gets a genetic disorder.

Mr. and Mrs. Gearhart have a son, William, who has cystic fibrosis. Cystic fibrosis is a genetic disorder that causes large amounts of mucus to be made in the lungs and near the pancreas. This makes breathing and digestion difficult. Mr. Gearhart told a genetic counselor that he had a cousin who had cystic fibrosis. Mrs. Gearhart cannot remember anyone in her family having this disorder. Persons with cystic fibrosis are known to be pure recessive for the disorder.

*Set up a Punnett square to show the gene types for each of William's parents and for William. Use **F** for the dominant gene and **f** for the recessive gene. Then, answer the questions.*

| Father \ Mother | F | f |
|---|---|---|
| F | FF | Ff |
| f | Ff | ff |

1. What is the gene type for Mr. Gearhart? Ff For Mrs. Gearhart? Ff What is William's gene type? ff. Explain your results. The parents do not have cystic fibrosis. Therefore, they are either pure dominant (FF) or heterozygous (Ff). Since William has the disease, his parents must each be Ff.

2. The Gearhart's doctor told them that they have a chance of having another child with cystic fibrosis. Is this true? Explain. Yes. Since both parents are heterozygous, each child they have will have a one in four chance of having cystic fibrosis.

3. Suppose one parent was heterozygous and the other was pure dominant for cystic fibrosis. Make a Punnett square to explain whether they would have children with cystic fibrosis or children who would be carriers of the disorder. Explain your results.

| | F | F |
|---|---|---|
| F | FF | FF |
| f | Ff | Ff |

None of the children would have cystic fibrosis. Each child they have will have a 50 percent chance of being a carrier for the disorder.

4. Would two persons who are both **FF** for cystic fibrosis ever produce children who have the disorder? no Will they produce children who are carriers of the disorder? no Explain. They cannot produce carriers or children with the disorder because neither has the recessive gene for cystic fibrosis.

Ff

Ff

Both parents are heterozygous

FF | Ff | Ff | ff

Normal children | Child with cystic fibrosis

**Figure 27-15** A family pedigree for cystic fibrosis

The counselor showed the parents a drawing of a family pedigree like the one in Figure 27-15. A **pedigree** is a diagram that can show how a certain trait is passed along in a family. A pedigree may be used to trace a family trait and to predict whether future generations will have the trait.

**What is a pedigree?**

The couple learned that there was a one in four chance of any of their children being born with the disorder and that tests such as amniocentesis can tell whether a fetus has the disorder. Talking to the counselor gave them an idea of what to expect if they decided to have more children.

## Check Your Understanding

11. How can an error in meiosis cause a sex cell to have the wrong number of chromosomes?
12. Name three genetic disorders and tell how each one of them is controlled by genes.
13. How can a genetic counselor use a pedigree to help families with genetic disorders?
14. **Critical Thinking:** Why do you think an XO female can survive but a YO male will die before birth?
15. **Biology and Math:** A genetic counselor sees a couple who is worried that their unborn child may have hemophilia. Tests show that the mother is heterozygous for the recessive gene, which is on the X chromosome. Will the baby have hemophilia? Why or why not?

## RETEACHING 82

**RETEACHING: IDEA MAP** CHAPTER 27

Name ______ Date ______ Class ______

Use with Section 27:3.

**GENETIC DISORDERS**

*Use the following terms to complete the idea map below: cystic fibrosis, dyslexia, hemophilia, color blindness, PKU.*

genetic disorder → sex chromosome → hemophilia, color blindness

genetic disorder → autosome → cystic fibrosis, dyslexia, PKU

1. What is meant by an error in chromosome number? **It means that a person is born with more than or fewer than 46 chromosomes.**
2. What is meant by a genetic disorder? **It means that a person inherits a gene that causes a disorder in body functions.**
3. How does color blindness affect a person? **The person cannot see red and green colors correctly.**
4. How does hemophilia affect a person? **It keeps the blood from clotting.**
5. What do the genetic disorders in 3 and 4 have in common? **They are controlled by genes found on the sex chromosome.**
6. How does dyslexia affect a person? **A person with dyslexia sees and writes the letters of the alphabet backward.**
7. How does cystic fibrosis affect a person? **The person's lungs and pancreas do not work properly.**

## STUDY GUIDE 162

**STUDY GUIDE** CHAPTER 27

Name ______ Date ______ Class ______

**VOCABULARY**

*Review the new words in Chapter 27 of your textbook. Then, complete this puzzle.*

**Across**

5. Use of genetics to predict and explain traits in children
6. Condition in which neither gene is dominant
7. Chromosome that does not determine sex
8. Disorder in which red blood cells are not round
9. Sex chromosomes of the female
10. Chromosomes that determine if child is male or female
11. Diagram that can show how a trait is passed along in a family

**Down**

1. Genetic disorder called word blindness
2. Disorder in which some colors are not seen as they should be
3. Way of looking at chromosomes of a fetus
4. Sex chromosomes of a male

# TEACH

### Independent Practice

**Study Guide**/*Teacher Resource Package*, p. 162. Use the Study Guide worksheet shown at the bottom of this page for independent practice.

### Check for Understanding

Have students respond to the first three questions in Check Your Understanding.

### Reteach

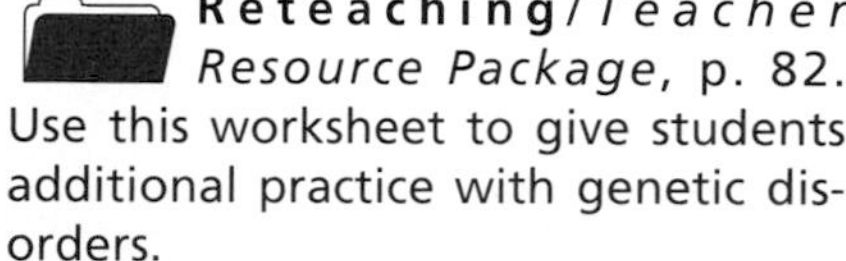

**Reteaching**/*Teacher Resource Package*, p. 82. Use this worksheet to give students additional practice with genetic disorders.

**Extension:** Assign Critical Thinking, Biology and Math, or some of the **OPTIONS** available with this lesson.

# APPLY

### ACTIVITY/Skill

Give students several genetics problems to evaluate their understanding of genetic disorders.

# CLOSE

### Filmstrip

Show the video *Heredity, Health, and Genetic Disorders,* Human Relations Media.

### Answers to Check Your Understanding

11. When sister chromatids both go into the same cell, the other cell will be missing a chromosome.
12. hemophilia—recessive gene on the X chromosome; dyslexia—dominant gene on autosome; cystic fibrosis—recessive gene on autosome
13. A pedigree traces how a trait is passed on in a family.
14. Some or all genes on the X chromosome are necessary for life.
15. No, only male children of this couple could have the disorder.

# CHAPTER 27 REVIEW

## Summary

Summary statements can be used by students to review the major concepts of the chapter.

## Key Science Words

All boldfaced terms from the chapter are listed.

## Testing Yourself

### Using Words

1. autosome
2. amniocentesis
3. dyslexia
4. genetic counseling
5. sex chromosome
6. X chromosome
7. incomplete dominance
8. color blindness
9. Y chromosome
10. pedigree

### Finding Main Ideas

11. p. 567; Amniocentesis is a way to look at cells of a fetus.
12. p. 573; A new trait in incomplete dominance is a combination of the dominant and recessive trait.
13. p. 578; A pair of chromosomes does not pull apart, and two chromosomes go into one cell.
14. p. 574; Body cells are damaged because sickle cells do not carry enough oxygen to them. Because of their shape, sickle cells clog articles.
15. p. 568; The X and Y chromosomes are chromosomes that determine sex.
16. p. 566; Human body cells contain 46 chromosomes, but sex cells have 23 chromosomes.

### Using Main Ideas

17. The male determines the sex of a child.
18. The parents can be heterozygous and carry the recessive gene for cystic fibrosis.
19. The parents can be heterozygous and both carry the gene for type O blood. The child inherited a pure recessive trait.



# Chapter 27 Review

## Summary

### 27:1 The Role of Chromosomes

1. Each human sex cell has 23 chromosomes. Each human body cell has 46 chromosomes arranged in pairs.
2. Amniocentesis is a way to look at the chromosomes of a fetus.
3. The sex of a person is determined by a pair of sex chromosomes. Human females have two X chromosomes. Human males have one X and one Y chromosome.

### 27:2 Human Traits

4. If a person has a dominant trait, one or both of the parents will also have the trait. If a person has a recessive trait, both of the parents will also carry the gene for the trait. Traits in which neither gene is totally dominant over the other show incomplete dominance.
5. Three genes control blood types in humans but each person has only two genes for this trait. Males show recessive traits located on the X chromosome more often than females do.

### 27:3 Genetic Disorders

6. If chromosome pairs do not pull apart during meiosis, errors in chromosome number result. Hemophilia, cystic fibrosis, and dyslexia are genetic disorders.
7. A genetic counselor uses knowledge of genetics to predict and explain disorders in children.

## Key Science Words

amniocentesis (p. 567)
autosome (p. 568)
color blindness (p. 576)
dyslexia (p. 579)
genetic counseling (p. 580)
incomplete dominance (p. 573)
pedigree (p. 581)
sex chromosome (p. 568)
sickle-cell anemia (p. 574)
X chromosome (p. 568)
Y chromosome (p. 568)

## Testing Yourself

*Choose the word from the list of Key Science Words that best fits the definition.*

1. chromosome not related to sex
2. way to observe cells from a fetus
3. word blindness
4. the use of genetics to predict problem traits in children
5. chromosome that determines sex of a child
6. female sex chromosome

# Review

## Testing Yourself *continued*

7. trait in which neither gene is totally dominant over the other
8. vision problem related to sex chromosome
9. male sex chromosome
10. diagram used to trace a family trait

#### Finding Main Ideas

*List the page number where each main idea below is found. Then, explain each main idea.*

11. how to see cells of a fetus
12. how a new trait appears when there is incomplete dominance
13. how errors in chromosome number occur
14. why people with sickle-cell anemia have serious health problems
15. what the X and Y chromosomes are
16. that sex cells contain half the number of chromosomes as body cells

#### Using Main Ideas

*Answer the questions by referring to the page number after each question.*

17. Which parent determines the sex of a child? (p. 570)
18. How can a child have cystic fibrosis when the parents do not? (p. 580)
19. How can two parents with type A blood have a child with type O blood? (p. 575)
20. A color-blind female marries a normal male. What could the color vision of their children be? (p. 576)
21. One parent is heterozygous for blood type A. The other is heterozygous for blood type B. What might the children's blood types be? (p. 575)

## Skill Review

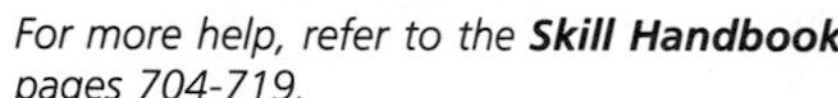

*For more help, refer to the **Skill Handbook,** pages 704-719.*

1. **Make and use tables:** Make a table to show the numbers of baby boys and girls born in a local hospital in one month. How does the ratio of boys to girls compare with what you expect?
2. **Observe:** Find out how many students can roll their tongue and how many cannot. Explain why there are more students showing one trait.
3. **Interpret data:** Two parents have four children, each of which has a different blood type. One child is type A, one is type B, one is type O, and one is type AB. How can you explain this?
4. **Calculate:** If the body cells of an animal contain 78 chromosomes, how many chromosomes does a sperm cell from that animal contain?

## Finding Out More

TECH PREP

#### Critical Thinking

1. Explain why geneticists would want to spend more time finding out about genetic disorders in humans rather than how height, hair color, and eye color are inherited.
2. Why do most states require that newborn babies be tested for PKU soon after they are born?

#### Applications

1. Find out how people with Down syndrome lead useful lives.
2. Find out the treatment for cystic fibrosis.



### Using Main Ideas

20. They could have color-blind sons and normal-vision daughters.
21. They could have one with AB, one with BO, one with AO, and one with OO.

## Skill Review

1. Depending on the source of the data, students may notice that one month more girls will be born and the next month more boys will be born. Point out that they should expect a 1:1 ratio.
2. Rolling the tongue is a dominant trait.
3. The parents are both heterozygous for blood type. One is AO and the other is BO.
4. 39

## Finding Out More

### Critical Thinking

1. Genetic disorders can cause serious problems and death.
2. If babies born with PKU are given a special diet within the first weeks of life, they can develop normally.

### Applications

1. Students will find that many people with Down syndrome lead productive lives by attending school and holding jobs.
2. Those with cystic fibrosis require frequent breathing and lung treatments.

**Chapter Review**/*Teacher Resource Package*, pp. 153-154.

**Chapter Test**/*Teacher Resource Package*, pp. 155-157.

**Chapter Test**/*Computer Test Bank*

# DNA—Life's Code

## PLANNING GUIDE

| CONTENT | TEXT FEATURES | TEACHER RESOURCE PACKAGE | OTHER COMPONENTS |
|---|---|---|---|
| (3 1/2 days)<br>28:1 The DNA Molecule<br>DNA Structure<br>DNA Chromosomes<br>Proof That DNA Controls Traits<br>How DNA Works<br>Making Proteins<br>How DNA Copies Itself | Mini Lab: *How Are Nitrogen Bases Paired?* p. 591<br>Lab 28-1: *DNA,* p. 588<br>Technology: *DNA Fingerprinting,* p. 593<br>Idea Map, p. 587<br>Idea Map, p. 589<br>Check Your Understanding, p. 594 | Enrichment: *The DNA Code,* p. 30<br>Reteaching: *DNA Structure and How DNA Copies Itself,* p. 83<br>Skill: *Form a Hypothesis,* p. 29<br>Transparency Master: *DNA Controls Traits,* p. 141<br>Study Guide: *The DNA Molecule,* pp. 163-165<br>Lab 28-1: *DNA,* pp. 109-110 | **Laboratory Manual:**<br>*How Does DNA Make Protein?* p. 235<br>Color Transparency 28a: *The Relationship Between DNA and the Cell*<br>Color Transparency 28b: *How DNA Copies Itself*<br>**STVS:** *Bacteriophage,* Plants and Simple Organisms (Disc 4, Side 1) |
| (2 1/2 days)<br>28:2 How The Genetic Message Changes<br>Mutations<br>Cloning<br>Plant and Animal Breeding<br>Splicing Genes and Gene Therapy | Skill Check: *Understand Science Words,* p. 601<br>Mini Lab: *How Does a Mutation Occur?* p. 596<br>Lab 28-2: *Cloning, p.* 598<br>Idea Map, p. 597<br>Check Your Understanding, p. 601 | Application: *Plant and Animal Breeding,* p. 28<br>Reteaching: *DNA Changes,* p. 84<br>Critical Thinking/Problem Solving: *How Can You Find Changes in the Genetic Code?* p. 28<br>Focus: *Twins,* p. 55<br>Study Guide: *How the Genetic Message Changes,* pp. 166-167<br>Study Guide: *Vocabulary,* p. 168<br>Lab 28-2: *Cloning,* pp. 111-112 | **Laboratory Manual:**<br>*How Can a Mutation in DNA Affect an Organism?* p. 239<br>**STVS:** *Microinjecting Polygenes,* Plants and Simple Organisms (Disc 4, Side 1) |
| Chapter Review | Summary<br>Key Science Words<br>Testing Yourself<br>Finding Main Ideas<br>Using Main Ideas<br>Skill Review | **ASSESSMENT RESOURCES**<br>Chapter Review, pp. 148-149<br>Chapter Test, pp. 150-152<br>Performance Assessment in the Biology Classroom<br>Alternate Assessment in the Science Classroom<br>Computer Test Bank | |

## GLENCOE TECHNOLOGY

**Infinite Voyage,** *The Geometry of Life*
**Science and Technology Videodisc Series,**
*Bacteriophage,* Plants and Simple Organisms (Disc 4, Side 1)
*Microinjecting Polygenes,* Plants and Simple Organisms (Disc 4, Side 1)

**The Secret of Life,** *In the Land of Milk and Money: Biotechnology*

## MATERIALS NEEDED

| LAB 28-1, p. 588 | LAB 28-2, p. 598 | MARGIN FEATURES |
|---|---|---|
| scissors<br>tracing paper<br>heavy paper<br>tape<br>blank sheet of paper<br>red, blue, green, and yellow crayons | scissors<br>clear plastic cup<br>aluminum foil<br>*Coleus* plant<br>soil<br>water<br>label<br>metric ruler | Skill Check, p. 601<br>dictionary<br>Mini Lab, p. 591<br>markers<br>paper<br>Mini Lab, p. 596<br>crayons<br>scissors<br>paper |

## OBJECTIVES

For more information about National Science Standards, see page 5T.

| SECTION | OBJECTIVE | CORRELATION of QUESTIONS to OBJECTIVES: CHECK YOUR UNDERSTANDING | CHAPTER REVIEW | TRP CHAPTER REVIEW | TRP CHAPTER TEST |
|---|---|---|---|---|---|
| 28:1 National Science Stds: UCP.1, UCP.2, UCP.3, UCP.4, UCP.5, A.1, A.2, B.2, C.1, C.2, G.1, G.2, G.3 | 1. **Describe** the structure of DNA. | 1, 5 | 3, 8, 18 | 1, 7, 8, 9, 16, 17, 19 | 1, 2, 4, 34, 35, 36, 37, 38, 39 |
| | 2. **Explain** how DNA controls genetic traits. | 2 | 1, 16, 27 | 2, 12, 25, 27 | 3, 8, 40 |
| | 3. **Describe** how DNA copies itself and works to make proteins. | 4 | 2, 9, 10, 14, 15, 21, 24 | 6, 10, 24 | 5, 6, 9, 12, 28, 41 |
| 28:2 National Science Stds: UCP.2, UCP.3, UCP.4, C.2, E.1, E.2, F.1, F.6 | 4. **Describe** how mutations occur. | 6 | 4, 12, 22 | 3, 28 | 7, 29, 42, 44 |
| | 5. **Describe** how cloning and breeding produce off-spring with desired traits. | 7, 9 | 2, 7, 11, 17, 25 | 5, 18, 19, 20, 21, 22, 23, 26 | 10, 11, 13, 15, 16, 17, 19, 20, 21, 23, 24, 26, 30, 31, 32, 43 |
| | 6. **Explain** how recombinant DNA and gene therapy can help humans. | 8, 10 | 5, 13, 20, 23 | 4, 11, 13, 14, 29 | 14, 18, 22, 25, 27, 33 |

# DNA – Life's Code

## CHAPTER OVERVIEW

### Key Concepts

In this chapter, students will examine the appearance and function of DNA. They will study the experimental evidence that shows DNA is the genetic material of the cell. Students also will study errors that occur in the DNA code and their consequences. They will learn about the modern applications of DNA technology.

### Key Science Words

| | |
|---|---|
| breeding | mutation |
| DNA | nitrogen base |
| fraternal twins | radiation |
| gene therapy | recombinant DNA |
| genetic code | RNA |
| identical twins | |

### Skill Development

In Lab 28-1, students will **formulate a model, interpret diagrams,** and **observe** to determine what a model of DNA looks like. In Lab 28-2, students will **observe, infer,** and **interpret data** to understand how a *Coleus* plant can be cloned. In the Skill Check on page 601, students will **understand** the **science word** *recombinant DNA*. In the Mini Lab on page 591, students will **formulate a model** of paired nitrogen bases. In the Mini Lab on page 596, students will **formulate a model** of normal DNA and a mutation.

### Bridging

Students learned in detail the steps of mitosis and meiosis in Chapter 22. This chapter will explain exactly how it is possible for a chromosome to copy itself during the forming of sister chromatids.

**CHAPTER PREVIEW**

### Chapter Content

Review this outline for Chapter 28 before you read the chapter.



### Skills in this Chapter

The skills that you will use in this chapter are listed below.

- In **Lab 28-1,** you will formulate a model, interpret diagrams, and observe. In **Lab 28-2,** you will observe, infer, and interpret data.
- In the **Skill Checks,** you will understand science words.
- In the **Mini Labs,** you will formulate models.

584

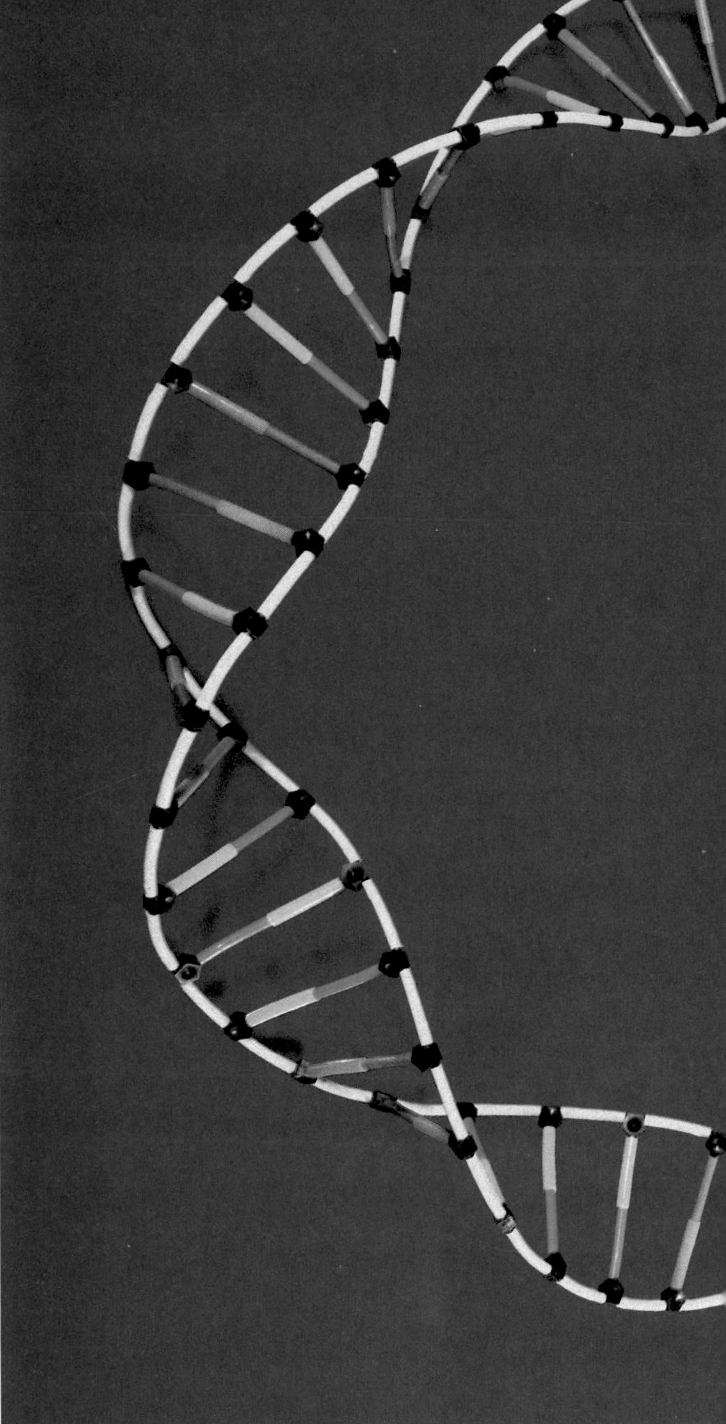

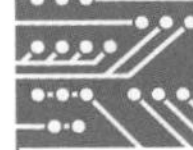

## TECH PREP

For Tech Prep activities in this chapter of the Teacher Wraparound Edition, see especially the Options Activities on pages 591 and 600, the Motivation on page 596, the Student Journal on page 597, and the Activity on page 601.

See also the Glencoe Homepage at **www.glencoe.com**

Chapter 28

# DNA—Life's Code

Many different kinds of codes are used for many different reasons. Braille is an example of a code. It is used by a visually impaired person to help him or her read. The photo on this page shows a book written in Braille. Another code is one that is found in each of your cells. It is a code that determines all of your traits. It directs your body to be you. What you see in the photo on the left is a model of a chemical molecule. How this molecule acts as a code to determine your body's traits is what this chapter is all about.

**Try This!**

**How many words can you make with only four letters?** Design an alphabet that has only four letters. Write as many words as possible using the four letters. You may use the same letter more than once.

***inter*NET CONNECTION**

For more information about the material in this chapter, follow the link for the chapter on the Glencoe Homepage at **http://www.glencoe.com**

Braille is a code.



**GETTING STARTED**

## Using the Photos

Have students think of other codes they are familiar with or perhaps have used. Have them relate written codes to chemical codes.

## MOTIVATION/Try This!

**How many words can you make with only four letters?** Upon completing the chapter-opening activity, students will demonstrate how only four major molecules in a cell code for all the variations in living things.

## Chapter Preview

Have students study the chapter outline before they begin to read the chapter. They should note the overall organization of the chapter, that the first topic is a description of the DNA molecule. This is followed by information that describes how the DNA code may change and how these changes may now be put to beneficial use.

## Misconception

Students have learned that the nucleus of a cell is the control center. In reality, the control of the cell lies not with the nucleus but with what is present within the nucleus. Chromosomes and their corresponding genes control the cell. Genes, in turn, are actually under the control of the DNA code.

**ASSESSMENT PLANNER**

***Portfolio***

Strategies on the following pages represent student products that can be placed into a best-work portfolio: pp. 591, 593.

**PERFORMANCE ASSESSMENT**

Skill Check, p. 601
Mini Lab, pp. 591, 596
Lab, pp. 588, 598

**CONTENT ASSESSMENT**

Check for Understanding, pp. 593, 594, 600, 601
Chapter Review, pp. 602, 603

**GROUP ASSESSMENT**

Opportunities for group assessment occur with Cooperative Learning Strategies.

***Student Journal***

Strategies on the following pages represent opportunities for writing in a Student Journal: pp. 590, 597.

## 28:1 The DNA Molecule

## PREPARATION

### Materials Needed

 Make copies of the Study Guide, Skill, Transparency Master, Enrichment, Reteaching, and Lab worksheets in the *Teacher Resource Package*.

▶ Have crayons, scissors, and tracing paper on hand for Lab 28-1.

▶ Purchase suggested candies for Check for Understanding.

### Key Science Words

DNA
nitrogen base
RNA
genetic code

### Process Skills

In the Mini Lab, students will formulate a model. In Lab 28-1, students will formulate a model, interpret diagrams, and observe.

## 1 FOCUS

▶ The objectives are listed on the student page. Remind students to preview these objectives as a guide to this numbered section.

### MOTIVATION/Convergent Question

Tell students that a certain chromosome has genes that control the following traits located on it: skin color, number of teeth, number of fingers, and blood type. Ask them what traits will be present on the sister chromatid when it is formed.

**GLENCOE TECHNOLOGY**

**Videodisc**
**The Secret of Life**
*DNA: Structure and Replication*

**Objectives**

1. **Describe** the structure of DNA.
2. **Explain** how DNA controls genetic traits.
3. **Describe** how DNA copies itself and works to make proteins.

**Key Science Words**

DNA
nitrogen base
RNA
genetic code

## 28:1 The DNA Molecule

Within each of your cells is a chemical that controls life. It is what we have been calling *genes* in the past few chapters. Now you will find out how this chemical makes up the genes, how it determines all your traits, and how it passes an organism's instructions from generation to generation. You will see why this chemical has been called the code of life.

### DNA Structure

The photograph on page 584 is a model of a DNA molecule. The letters DNA stand for *d*eoxyribo*n*ucleic (dee AHK sih ri boh noo klay ik) *a*cid. We shall use only the letters DNA each time we talk about this molecule. What is DNA? **DNA** is a molecule that makes up genes and determines the traits of all living things. Humans, birds, mushrooms, plants, protozoans, and bacteria have DNA. All living things contain DNA in their cells.

Scientists often use models to explain things that are very complex. We shall use a model of DNA to explain what DNA looks like and how it does its job.

Figure 28-1a shows a ladder. It is made up of two upright side pieces and many rungs. Figure 28-1b shows a small part of the DNA model. Note that it looks something like the ladder. An actual DNA molecule is much longer

**Figure 28-1** DNA is often compared to a ladder (a). A model of DNA is shown (b).

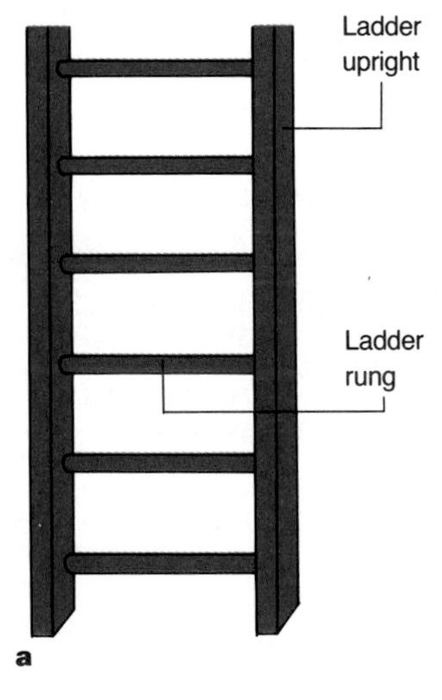


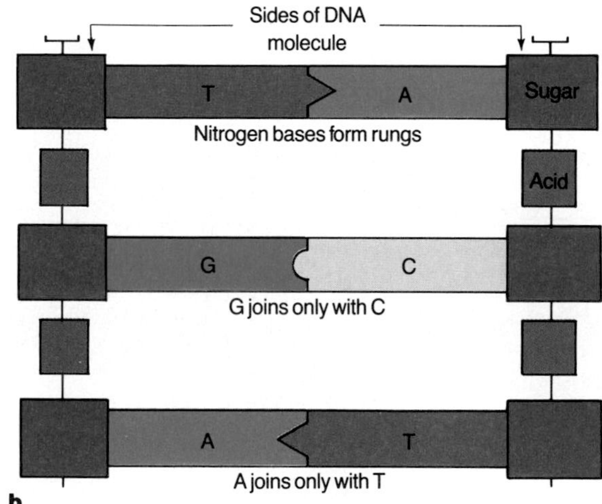




### OPTIONS

**Science Background**

DNA is a molecule present in all life forms. Its function and structure (except for a few organisms such as viruses) always is the same. Thus, one can use this similarity as evidence for common evolutionary ancestry of all life forms.

than the model. Scientists have estimated that a single DNA molecule in a human cell may contain about 100 million rungs.

There are six features of the DNA model.

1. DNA has two main sides. These sides are like the upright parts of a ladder.
2. The sides are made of two different chemicals. One is a sugar. The other is an acid. These two chemicals alternate along each side.
3. There are parts that connect the two sides together. These parts look like the rungs of a ladder.
4. **Nitrogen bases** form the rungs of a DNA molecule.
5. There are four different nitrogen bases in DNA. The letters A, T, C, and G stand for the four bases.
6. The four bases join each other in certain ways to form the rungs. Like pieces of a puzzle, shape A will fit only shape T. Shape C will fit only shape G. As a result, base A joins only with base T. Base C joins only with base G.

You should know one more thing about the shape of DNA. Imagine holding each end of a ladder and twisting it. You would end up with a structure like the one in Figure 28-2. This figure shows that DNA is twisted into a spiral.

**What are the parts of a DNA molecule?**

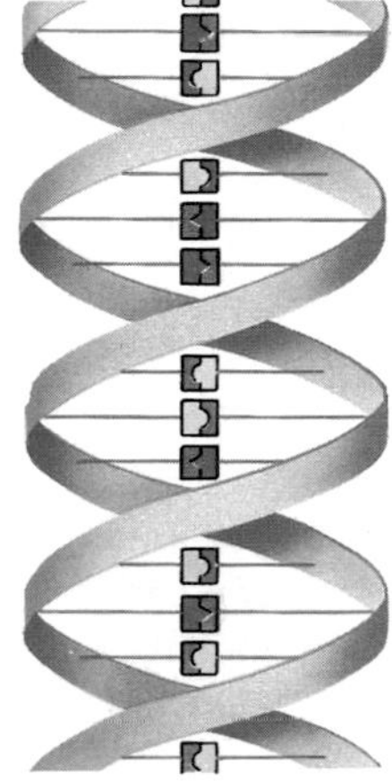

**Figure 28-2** DNA is twisted into a spiral.

## DNA and Chromosomes

DNA is in every cell in your body. It is also in every cell of all other living things. Where in these cells is it found? You probably know the answer already. DNA makes up parts of the chromosomes found in the nucleus.

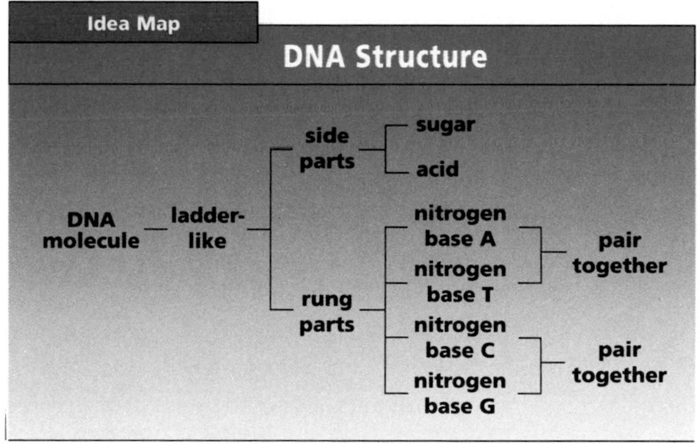


**Study Tip:** Use this idea map as a study guide to the structure of the DNA molecule. What is the shape of the DNA ladder?



# 2 TEACH

### ACTIVITY /Software

*DNA – The Basics*, Queue, Inc.

### Concept Development

▶ Sketch a ladder on the chalkboard and identify the upright sides of the ladder as well as the rungs. This will help students understand the terms *upright* and *rung.*

### Guided Practice

Have students write down their answers to the margin question: two sides made of alternating sugar and acid molecules and cross rungs made of nitrogen bases.

### Using Science Words

Show students where the initials *DNA* come from within the word *DeoxyriboNucleic Acid.* The *-nucleic* portion of the word is related to the part of the cell in which this chemical is located. *Deoxyribo-* is the stem for *deoxyribose,* the sugar molecule of DNA. *Acid* relates to phosphoric acid, another molecule of DNA.

### Brainstorming

Ask students the significance of the fact that all life forms have the same chemical responsible for regulating gene traits. (From an evolutionary point of view, this strongly suggests common origin because of common chemical makeup.)

### Idea Map

Have students use the idea map as a study guide to the major concepts of this section.

## OPTIONS

### ACTIVITY/Enrichment

Use puzzle pieces to emphasize complementarity of base pairs A-T and C-G. Place four puzzle pieces on the overhead projector. Two of the pieces should fit each other and the remaining two should also fit each other.

**The Relationship Between DNA and the Cell**/*Transparency Package,* number 28a. Use color transparency 28a as you teach where DNA is located.

## Lab 28-1 DNA

### Overview

Students will construct a model of the DNA molecule that demonstrates its major components and shows how nitrogen bases are joined together in a specific pattern.

**Objectives:** Upon completion of this lab, students will be able to (1) **state** which nitrogen bases form pairs, (2) **recognize** that the DNA molecule is ladder-like in appearance, (3) **explain** the value of using models.

**Time Allotment:** 1 class period

### Preparation

▶ Use manila folders or file cards for the heavy paper.

▶ Verify that students have four of each letter available.

▶ Suggest to the students that the nitrogen bases fit together like puzzle pieces.

▶ The completed DNA molecule from top to bottom should show this order of nitrogen bases: T-A, A-T, C-G, C-G, G-C, A-T, T-A, C-G.

**Lab 28-1 worksheet**/*Teacher Resource Package,* pp. 109-110.

### Teaching the Lab

▶ **Troubleshooting:** Suggest to students that it is okay to turn nitrogen bases (models) upside down.

**ASSESSMENT**

**Performance:** Have students prepare a model of a nucleotide that includes a nitrogen base (A, T, C, or G), a sugar, and an acid. Have them label the three parts.

**GLENCOE TECHNOLOGY**

**Videodisc**

**Biology: The Dynamics of Life**
Animation: *DNA Replication*
Disc 1, Side 1, Ch. 31

# DNA

## Lab 28–1

### Problem: What does a model of DNA look like?

#### Skills

formulate a model, interpret diagrams, observe

#### Materials

tracing paper
scissors
blank sheet of paper
heavy paper
red, blue, green, and yellow crayons
tape

#### Procedure

1. Trace the four parts shown in Figure A with tracing paper.
2. Cut out the four tracings and copy each part onto heavy paper four times. You should have 16 model parts.
   **CAUTION:** *Always be careful when using scissors.*
3. Label each part with its correct letter (A, T, C, G).
4. Cut out the 16 model parts. Label the other side of each part to match the first side.
5. Color only the parts on your figures that are nitrogen bases. Use this coloring code: A = red, T = blue, C = green, G = yellow. Color both sides of each part.
6. Arrange eight of the figures in any order on a blank sheet of paper. Use Figure B to help you.
7. Tape the eight figures in place.
8. Arrange the remaining eight figures according to how they fit the first eight figures. You can turn the parts in any direction. Tape these figures in place.
9. Compare your model with those of your classmates.

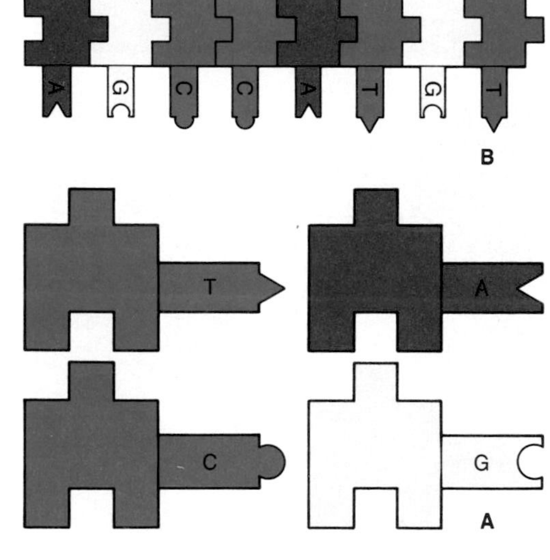


#### Data and Observations

1. What part, or letter, does the T fit? G? A? C? Explain these fits using your model.
2. In what ways is your model different from those of your classmates? In what ways is it alike?

#### Analyze and Apply

1. Where would you find this model if it were in a real cell?
2. Name the two chemicals that form the outer (uncolored) parts of your model.
3. **Interpret diagrams:** What chemical name describes the rungs that go across the middle?
4. **Apply:** Why do scientists use models?

#### Extension

**Make a model** to show the steps used by DNA as it copies itself. Use the parts from this lab.



## ANSWERS

### Data and Observations

1. A; C; T; G; In the model the pointed end of all Ts can only fit the matching triangle cut out of all As. The rounded end of all Cs can only fit the matching semicircular cutout of all Gs. They fit like puzzle pieces.
2. The sequence of bases will be different. The basic structure of the model, with the bases forming rungs, will be similar.

### Analyze and Apply

1. in the nucleus, as part of chromosomes
2. a sugar and an acid
3. nitrogen bases
4. A model may help to better explain complex things.

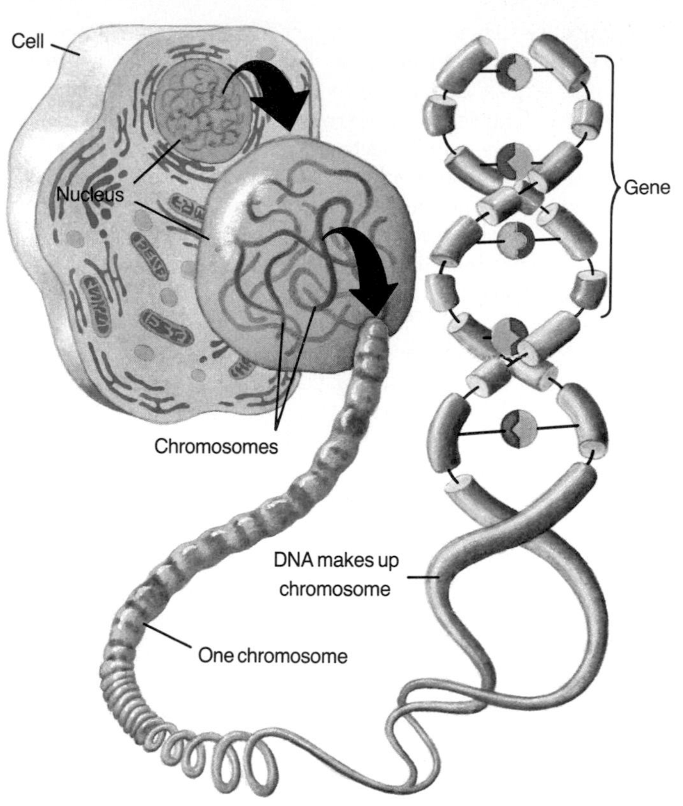


**Figure 28-3** How the cell, nucleus, chromosomes, genes, and DNA are related

Look carefully at Figure 28-3. It shows how DNA is related to chromosomes. Keep in mind that this drawing is many times larger than the real thing.

Now think about this question. Where in the cell can we find genes? Genes are parts of chromosomes. We can describe a gene in two ways. First, it is a short piece of DNA. Second, it is a certain number of bases (rungs on the ladder) on the DNA molecule.

In Chapter 26, a gene was defined as a small section of chromosome that determines traits. All three definitions of a gene are correct.

## Proof That DNA Controls Traits

How do scientists know that DNA controls our traits? To answer this question, let's look at some experiments done in 1928.

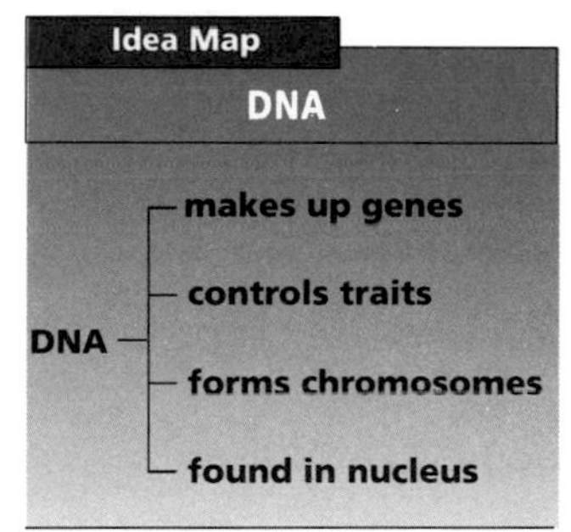


**Study Tip:** Use this idea map as a study guide to the DNA molecule. The features of DNA are shown.

## TEACH

### Concept Development

▶ Each twist of the DNA molecule comprises 10 base pairs.

▶ Have students study Figure 28-3 carefully. Until now, students have learned that the nucleus is the "control center" of the cell. Now they should be able to see that the nucleus controls the cell through its genes and the DNA code making up the chromosomes.

### Independent Practice

**Study Guide**/*Teacher Resource Package*, p. 163. Use the Study Guide worksheet shown at the bottom of this page for independent practice.

### Connection: Math

A scientist has calculated the size of one virus chromosome to be $1.6 \times 10^5$ nitrogen-base pairs long. It takes about 1000 base pairs to make up one gene. How many genes are present on the virus DNA? (Answer is 160: $1.6 \times 10^5 = 160\ 000$; $160\ 000 \div 1000 = 160$)

Note: you may have to review scientific notation to enable students to convert $1.6 \times 10^5$ to standard notation.

### Idea Map

Have students use the idea map as a study guide to the major concepts of this section.

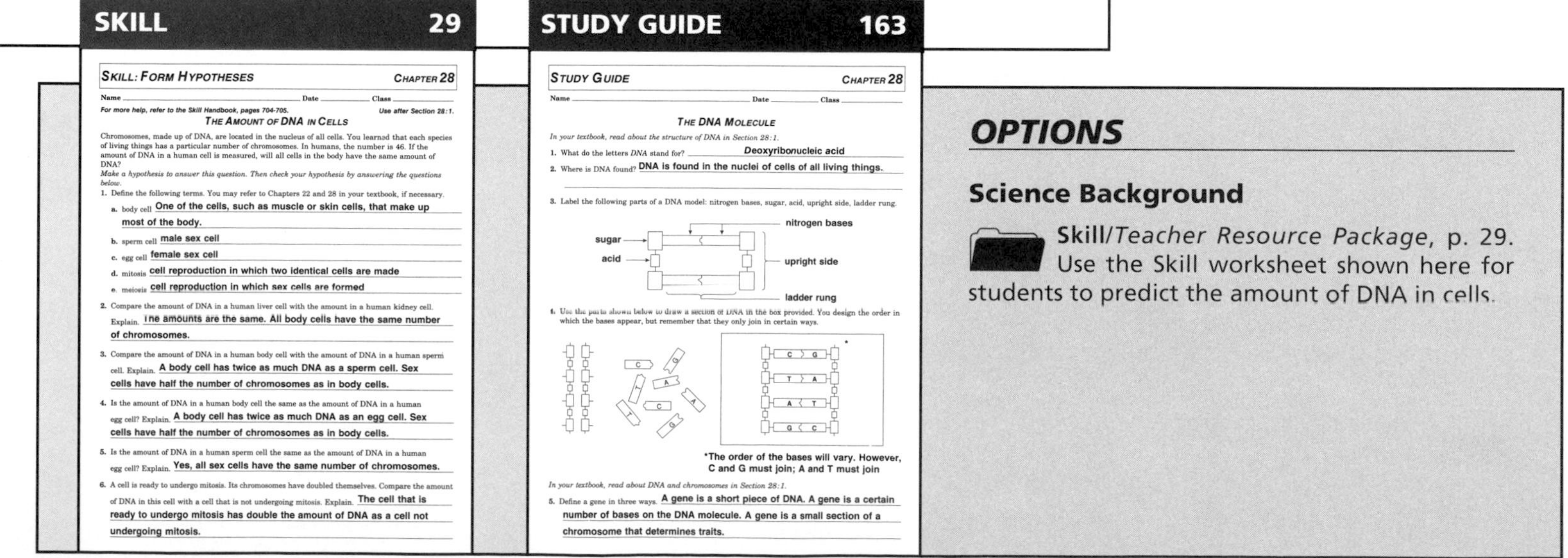

**SKILL 29**

**SKILL: FORM HYPOTHESES** CHAPTER 28

Name ______ Date ______ Class ______

*For more help, refer to the Skill Handbook, pages 704-705.* *Use after Section 28:1.*

**THE AMOUNT OF DNA IN CELLS**

Chromosomes, made up of DNA, are located in the nucleus of all cells. You learned that each species of living things has a particular number of chromosomes. In humans, the number is 46. If the amount of DNA in a human cell is measured, will all cells in the body have the same amount of DNA?

*Make a hypothesis to answer this question. Then check your hypothesis by answering the questions below.*

1. Define the following terms. You may refer to Chapters 22 and 28 in your textbook, if necessary.
   a. body cell **One of the cells, such as muscle or skin cells, that make up most of the body.**
   b. sperm cell **male sex cell**
   c. egg cell **female sex cell**
   d. mitosis **cell reproduction in which two identical cells are made**
   e. meiosis **cell reproduction in which sex cells are formed**
2. Compare the amount of DNA in a human liver cell with the amount in a human kidney cell. Explain. **The amounts are the same. All body cells have the same number of chromosomes.**
3. Compare the amount of DNA in a human body cell with the amount of DNA in a human sperm cell. Explain. **A body cell has twice as much DNA as a sperm cell. Sex cells have half the number of chromosomes as in body cells.**
4. Is the amount of DNA in a human body cell the same as the amount of DNA in a human egg cell? Explain. **A body cell has twice as much DNA as an egg cell. Sex cells have half the number of chromosomes as in body cells.**
5. Is the amount of DNA in a human sperm cell the same as the amount of DNA in a human egg cell? Explain. **Yes, all sex cells have the same number of chromosomes.**
6. A cell is ready to undergo mitosis. Its chromosomes have doubled themselves. Compare the amount of DNA in this cell with a cell that is not undergoing mitosis. Explain. **The cell that is ready to undergo mitosis has double the amount of DNA as a cell not undergoing mitosis.**

**STUDY GUIDE 163**

**STUDY GUIDE** CHAPTER 28

Name ______ Date ______ Class ______

**THE DNA MOLECULE**

*In your textbook, read about the structure of DNA in Section 28:1.*

1. What do the letters *DNA* stand for? **Deoxyribonucleic acid**
2. Where is DNA found? **DNA is found in the nuclei of cells of all living things.**
3. Label the following parts of a DNA model: nitrogen bases, sugar, acid, upright side, ladder rung.



4. Use the parts shown below to draw a section of DNA in the box provided. You design the order in which the bases appear, but remember that they only join in certain ways.



***The order of the bases will vary. However, C and G must join; A and T must join**

*In your textbook, read about DNA and chromosomes in Section 28:1.*

5. Define a gene in three ways. **A gene is a short piece of DNA. A gene is a certain number of bases on the DNA molecule. A gene is a small section of a chromosome that determines traits.**

## OPTIONS

### Science Background

**Skill**/*Teacher Resource Package*, p. 29. Use the Skill worksheet shown here for students to predict the amount of DNA in cells.

## TEACH

### Concept Development

▶ Make sure students realize that as long as the harmful bacteria are killed, they cannot cause a disease.

▶ Explain that the practice of injecting dead bacteria into humans is quite common. It stimulates the body into producing antibodies against a foreign antigen.

▶ An understanding of experiment D is critical. Results are difficult to explain until the student reads on. The last part of the experiment shows that it is indeed DNA, and not other cellular components, that controls traits.

▶ Not all cells can transform or pick up pieces of other cells. This is a peculiarity of bacteria.

▶ Explain that biologists are able to separate different cell parts and then combine the different cell parts experimentally. This requires the use of the ultracentrifuge.

### Independent Practice

**Study Guide**/*Teacher Resource Package*, p. 164. Use the Study Guide worksheet shown at the bottom of this page for independent practice.

### *Student Journal*

Set up the following story line for students. You are a reporter for a science magazine and have just learned about the 1928 experiments dealing with bacteria and mice. Write an article with a catchy headline that reports on the results and significance of these experiments.

Figure 28-4 shows what happened in these experiments. Look at the first experiment, **a.** These results make sense. You would expect that living harmless bacteria would not kill a mouse. The results in experiments **b** and **c** also make sense. Living pneumonia bacteria would be expected to kill the mouse. Dead pneumonia bacteria would be expected to have no effect on the mouse. Do the results in experiment **d** make sense? They didn't at first. Remember from Chapter 1 that a scientist may have to form a new hypothesis if the experiment shows that the old hypothesis is not supported. The scientist who did this experiment had to form a new hypothesis to explain the results of experiment **d.** He hypothesized that the living harmless bacteria picked up a cell part from the dead harmful bacteria. He said that this cell part caused the harmless bacteria to change into harmful bacteria.

What cell part did the harmless bacteria pick up? This question was answered years later by another group of scientists. They injected a mixture of living harmless bacteria and the DNA from pneumonia bacteria into a mouse. The mouse died. Then they injected living harmless bacteria mixed with other chemicals from the pneumonia bacteria. When the harmless bacteria picked up DNA from the harmful bacteria, they were changed into harmful bacteria. When the harmless bacteria picked up any other

**Figure 28-4** How scientists proved that DNA controls traits

| | Experiment a | Experiment b | Experiment c | Experiment d |
|---|---|---|---|---|
| Scientist injects mouse with | Living, harmless bacteria | Living pneumonia bacteria | Dead pneumonia bacteria | Mixture of living, harmless bacteria and dead pneumonia bacteria |
| Results | Mouse lives | Mouse dies | Mouse lives | Mouse dies |
| Meaning | Harmless bacteria cannot kill a mouse | Harmful bacteria can kill a mouse | Dead, harmful bacteria cannot kill a mouse | ? This combination should not kill a mouse |



## *OPTIONS*

### Science Background

In 1928, Fred Griffith conducted experiments in bacterial transformation. During transformation, the harmless form changed into the harmful form. In 1944, Avery, McCarty, and MacLeod showed that the chemical that changed the bacteria was DNA.

**Transparency Master**/*Teacher Resource Package*, p. 141. Use the Transparency Master shown here to show how DNA controls traits.

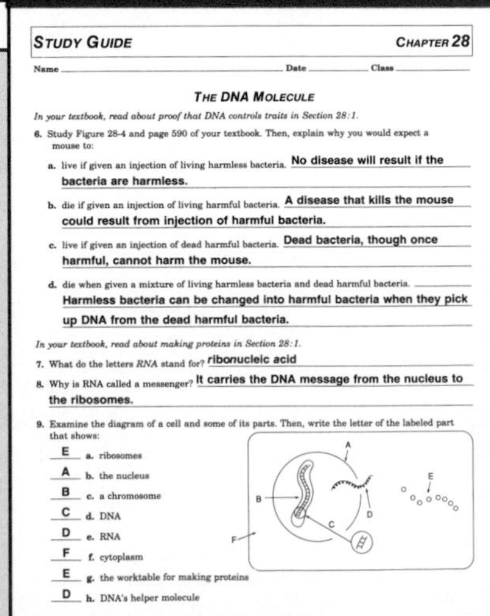

**STUDY GUIDE 164**

STUDY GUIDE — CHAPTER 28

Name ____________ Date ________ Class ________

*THE DNA MOLECULE*

*In your textbook, read about proof that DNA controls traits in Section 28:1.*

6. Study Figure 28-4 and page 590 of your textbook. Then, explain why you would expect a mouse to:
   a. live if given an injection of living harmless bacteria. **No disease will result if the bacteria are harmless.**
   b. die if given an injection of living harmful bacteria. **A disease that kills the mouse could result from injection of harmful bacteria.**
   c. live if given an injection of dead harmful bacteria. **Dead bacteria, though once harmful, cannot harm the mouse.**
   d. die when given a mixture of living harmless bacteria and dead harmful bacteria. **Harmless bacteria can be changed into harmful bacteria when they pick up DNA from the dead harmful bacteria.**

*In your textbook, read about making proteins in Section 28:1.*

7. What do the letters *RNA* stand for? **ribonucleic acid**
8. Why is RNA called a messenger? **It carries the DNA message from the nucleus to the ribosomes.**
9. Examine the diagram of a cell and some of its parts. Then, write the letter of the labeled part that shows:
   **E** a. ribosomes
   **A** b. the nucleus
   **B** c. a chromosome
   **C** d. DNA
   **D** e. RNA
   **F** f. cytoplasm
   **E** g. the worktable for making proteins
   **D** h. DNA's helper molecule



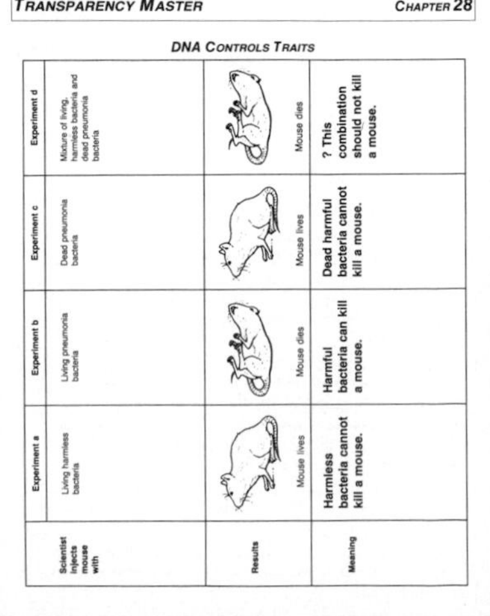

**TRANSPARENCY 141**

TRANSPARENCY MASTER — CHAPTER 28

*DNA CONTROLS TRAITS*

| | Experiment a | Experiment b | Experiment c | Experiment d |
|---|---|---|---|---|
| Scientist injects mouse with | Living harmless bacteria | Living pneumonia bacteria | Dead pneumonia bacteria | Mixture of living, harmless bacteria and dead pneumonia bacteria |
| Results | Mouse lives | Mouse dies | Mouse lives | Mouse dies |
| Meaning | **Harmless bacteria cannot kill a mouse.** | **Harmful bacteria can kill a mouse.** | **Dead harmful bacteria cannot kill a mouse.** | **? This combination should not kill a mouse.** |

| a. Person with normal blood cells | G A G T G A G G C T T C |
|---|---|
| | I I I I I I I I I I I I |
| | C T C A C T C C G A A G |

| b. Person with sickle cells | G A G T G A G G C T A C |
|---|---|
| | I I I I I I I I I I I I |
| | C T C A C T C C G A T G |

**Figure 28-5** A change in just one pair of DNA bases can change a living thing's traits.

chemical from the harmful bacteria, they did not change. The scientists concluded that only DNA caused the bacteria to change. DNA is the chemical that controls traits.

## How DNA Works

How does DNA direct a plant to make chlorophyll? How does DNA direct the formation of sickle-shaped or round red blood cells? How does DNA direct all traits in all living things? Let's compare DNA and how it works to a computer. A computer reads electrical messages stored within its memory bank. These messages are stored in code form. DNA also has a code. The nitrogen bases in the DNA molecule spell out a message that is stored in code form.

**What determines the coded message in DNA?**

The order of the nitrogen bases, A, T, C, and G, is the coded message. Figure 28-5a shows the order of bases somewhere on a section of DNA. Let's suppose this order directs the formation of round red blood cells. Now, let's suppose we look at the same place on the DNA of another person. We see the message in Figure 28-5b. This order of bases directs the formation of sickle cells.

In this example, the orders of the bases in DNA are slightly different. This difference causes the traits to be very different. One person has normal round red blood cells. The other has sickle cells.

New orders of nitrogen bases can make new messages. Each message gives the cell instructions for a different trait. Think about our alphabet for a moment. The letters of the alphabet are always the same. We put the letters into different orders to form different words. The letters w, o, and l form the word *owl*. They also can form the word *low*. Each order codes for a different meaning. How many different letters are in the DNA alphabet?

### Mini Lab

**How Are Bases Paired?**

**Formulate a model:** Mark each of 20 students with A, T, C, or G. Form a row of 10 students to show one side of DNA. Call up a student to pair with your letter. What pairs are possible? *For more help, refer to the **Skill Handbook**, pages 706-711.*



# TEACH

### Concept Development

▶ Explain that reading the dots and dashes of Morse code allows one to form words and sentences. The DNA code is similar. The order of nitrogen bases, rather than the sequence of dots and dashes, forms the code of our cells.

▶ Point out that *code of life* refers to the order of nitrogen bases in DNA.

### Guided Practice

Have students write down their answers to the margin question: the order of the nitrogen bases.

### Mini Lab

The only combinations between bases will be A + T, and C + G.

### ASSESSMENT

**Content:** Give students the following sequence of bases and have them determine the complementary sequence. TAGCGGACTGTA

### *Portfolio*

Have students prepare a chart with columns labeled Normal Blood Cells and Sickle Cells, and rows labeled DNA Bases Present and Appearance of Red Blood Cells. Ask them to complete the chart using Figures 28-5 and 27-7 as guides.

### GLENCOE TECHNOLOGY

**Videodisc**

**Biology: The Dynamics of Life**
Animation: *DNA Transcription*
Disc 1, Side 1, Ch. 32

### ENRICHMENT 30

ENRICHMENT CHAPTER 28

Name ______ Date ______ Class ______

*Use after Section 28:2.*

THE DNA CODE

At some time in your life, you may have enjoyed sending messages in code. DNA works like a code. The nitrogen bases, A, G, T, and C, make up the code letters. Each group of three nitrogen bases represents one code word. Each code word tells the cell to make a particular link in a protein chain. There are 20 different kinds of links that make up proteins. For example, the nitrogen bases in the order ATG tell the cell to make one link. The bases TAG or TGA produce two other links.

The table below shows all the different combinations of nitrogen bases in 3-letter code words. The numbers in the table represent the 20 different kinds of links. Notice that many of the links have more than one code word. The code words marked "Stop" tell the cell to stop adding links.

| | | | |
|---|---|---|---|
| AAA Link 1 | AGA Link 6 | ATA Link 10 | ACA Link 17 |
| AAG Link 1 | AGG Link 6 | ATG Link 10 | ACG Link 17 |
| AAT Link 2 | AGT Link 6 | ATT Stop | ACT Stop |
| AAC Link 2 | AGC Link 6 | ATC Stop | ACC Link 18 |
| GAA Link 2 | GGA Link 7 | GTA Link 11 | GCA Link 19 |
| GAG Link 2 | GGG Link 7 | GTG Link 11 | GCG Link 19 |
| GAT Link 2 | GGT Link 7 | GTT Link 12 | GCT Link 19 |
| GAC Link 2 | GGC Link 7 | GTC Link 12 | GCC Link 19 |
| TAA Link 3 | TGA Link 8 | TTA Link 13 | TCA Link 6 |
| TAG Link 3 | TGG Link 8 | TTG Link 13 | TCG Link 6 |
| TAT Link 3 | TGT Link 8 | TTT Link 14 | TCT Link 19 |
| TAC Link 4 | TGC Link 8 | TTC Link 14 | TCC Link 19 |
| CAA Link 5 | CGA Link 9 | CTA Link 15 | CCA Link 20 |
| CAG Link 5 | CGG Link 9 | CTG Link 15 | CCG Link 20 |
| CAT Link 5 | CGT Link 9 | CTT Link 16 | CCT Link 20 |
| CAC Link 5 | CGC Link 9 | CTC Link 16 | CCC Link 20 |

1. The sequence of nitrogen bases below represents a section of DNA. Use your table to figure out the sequence of links that would be made from this DNA chain. Write the number of the correct link in the blank below each code word.

AAT GCA TCA CGT TTT GGC TAG CAA GTG CAT
2 19 6 9 14 7 3 5 11 5

2. A mutation can occur when one nitrogen base is changed, lost, or added to the chain. The DNA chains below are the same as the chain in question 1, but one mutation has been made to each. Use your table to figure out the new sequence of links. Write the number of the link in the blank below each code word. Notice how the message changes.

nitrogen base T changed to A here
AAT GCA TCA CGT TTA GGC TAG CAA GTG CAT
2 19 6 9 13 7 3 5 11 5

nitrogen base G lost here
AAT CAT CAC GTT TTG GCT AGC AAG TGC AT
2 5 5 12 13 19 6 1 8

## OPTIONS

### ACTIVITY/Challenge

Have students list examples in which the changing of a three-letter sequence can alter the meaning of a word. Suggestions might be pin, nip; pat, tap, apt; or eat, ate, tea.

**Enrichment/*Teacher Resource Package*,** p. 30. Use the Enrichment worksheet shown here to help students understand the DNA code.

## TEACH

### Concept Development

▶ Use the example of hair color to explain that the kind of protein you make shows up as your traits. Explain that black hair color is due to a slightly different protein than is blond hair color.

▶ Figure 28-7 is critical to understanding the role of RNA in the process of protein formation and translation of the code located in DNA.

**Cooperative Learning:** Divide the class into five or six groups. Have each group compare and contrast DNA and RNA using Figure 28-7 as a guide plus any other diagrams or explanations within the chapter.

### Guided Practice

Have students write down their answers to the margin question: It carries the DNA message from the DNA in the nucleus to the ribosomes in the cytoplasm.

**Understand Science Words:** The term *mRNA* is not used. If you feel it is important, explain that the letter *m* represents the word *messenger,* thus the term *messenger RNA.*

**GLENCOE TECHNOLOGY**

**Videodisc**

**The Secret of Life**
*Protein Synthesis*

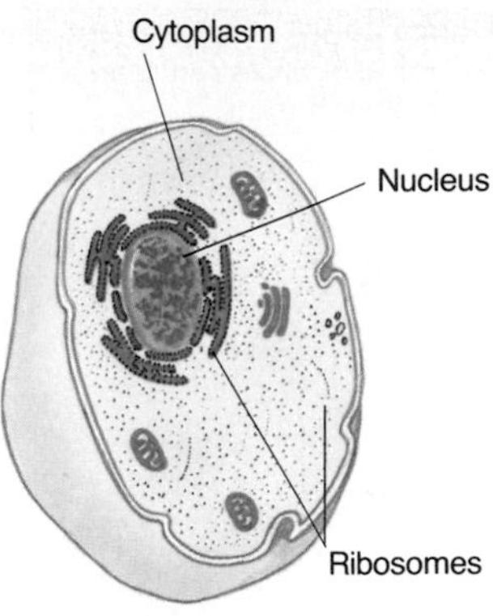


**Figure 28-6** Ribosomes have a part in making proteins. Where are the ribosomes located? in the cytoplasm

## Making Proteins

DNA directs the making of proteins in cells. How do cells make proteins? Remember from Chapter 2 that ribosomes are the cell parts where proteins are made. Look at Figure 28-6. Notice that the ribosomes are in the cytoplasm, not in the nucleus.

**What does RNA do in the making of proteins?**

How can DNA in the nucleus control what goes on at the ribosomes? It has a helper molecule called RNA. These letters are the initials for *ribo*n*ucleic* (ri boh noo KLAY ik) *acid*. **RNA** is a chemical that acts as a messenger for DNA. RNA carries the coded DNA message from the nucleus to the ribosomes. The ribosome acts as a worktable for making proteins. When RNA arrives at the ribosomes, it carries a message that must be decoded before it can direct the formation of a protein. Think of the coded DNA message as being written in a certain language. This is the language of the nitrogen bases A, T, C, and G. A protein is written in another language. Somehow, the DNA language must be translated into the protein language. The **genetic code** is the code that translates the DNA language into the protein language. Once the message has been translated, the protein can be made. Figure 28-7 shows how proteins are made. Traits show up in a cell because of the kinds of proteins being made at the ribosomes. Thus, the genes on the chromosomes have sent their messages to the ribosomes to make certain traits appear.

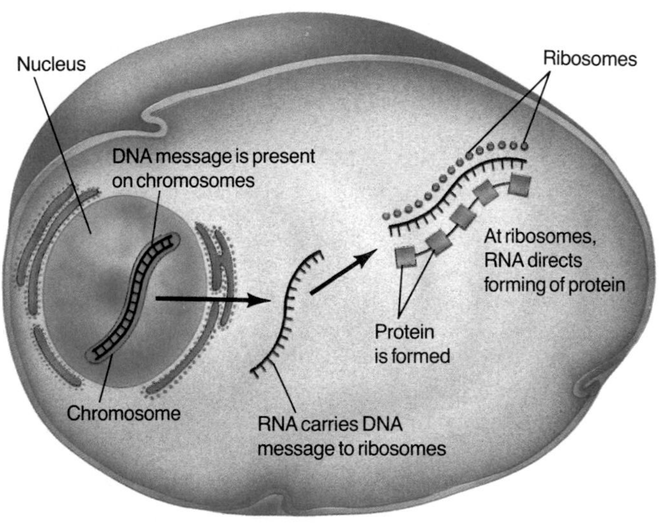


**Figure 28-7** Steps needed in the cell to change the DNA message into a protein. Where are proteins made? at the ribosomes



## OPTIONS

### Science Background

The normal flow of information within a cell is: DNA → mRNA → tRNA → Protein. Certain viruses (the AIDS virus in particular) reverse the flow of information so that it is: viral RNA → DNA → mRNA → tRNA → Protein.

## TECHNOLOGY

### DNA Fingerprinting

In violent crimes such as rape or murder, physical evidence is often left at the scene. Blood, hair, or sperm cells may be the only evidence that the police have. A person can be arrested and found guilty long after a crime is committed. How do police know that this person is the rapist or murderer? They are able to identify the sperm or blood as his.

Every person has a special set of "fingerprints" from the moment he or she is born. These fingerprints are not on the fingertips. They are in each cell in the body. These fingerprints are the order of the bases in a person's DNA.

Today, it is possible to picture a person's DNA by using special laboratory tests. Any cell in the body can be used. In the case of a rapist, police are able to match the DNA of the sperm cells collected at the scene of the crime with the DNA of the arrested man's blood cells. Thus, police have proof that the sperm came from the man arrested. It is impossible for the sperm to have come from any other person because the order of the bases in DNA is different for each person. DNA fingerprinting will be used more and more in the future to solve crimes.

Mobile crime lab

How does the order of nitrogen bases in DNA fit into this story? The order of bases is the coded message that controls what kinds of proteins are made. The kinds of proteins you have determine what traits you have. Thus, all your traits and the traits of all living things result from the order of nitrogen bases in DNA.

## How DNA Copies Itself

You have seen that DNA has many jobs. It carries the genetic messages for all living things. It controls the making of proteins. Remember that in cell reproduction, chromosomes make copies of themselves just before mitosis and meiosis. You know that chromosomes are made mostly of DNA. Now you know that DNA is what is copied. How does DNA copy itself? What is the story behind cell reproduction?



## *OPTIONS*

### Science Background

Nitrogen bases are joined by hydrogen bonds. Hydrogen bonds are weak bonds in that they are easily broken. This property facilitates the unzipping of the DNA molecule for replication and protein synthesis.

**How DNA Copies Itself**/*Transparency Package,* number 28b. Use color transparency number 28b as you teach the steps of DNA replication.

**How Does DNA Make Protein?**/*Lab Manual,* pp. 235-238. Use this lab as an extension to how DNA controls the making of protein.

## TECHNOLOGY

### DNA Fingerprinting

#### Background

Scientists form a profile of chemicals by chopping DNA into short segments that form specific patterns for each person. These segments are then subjected to electrophoresis and can be "sorted out" and categorized according to their molecular weights. The odds of two people (not including identical twins) having the same DNA fingerprints are calculated at 30 billion to 1.

#### Discussion

Typically, sperm cells in a rape case are analyzed for DNA fingerprinting. Sperm cells contain only half the DNA of body cells, but they can still be used to identify a criminal. All the DNA of the collected sperm taken together matches the DNA of the body cells.

#### References

Fackelmann, Kathy A. "Beyond the Genome: The Ethics of DNA Testing." *Science News*, November 5, 1994, pp. 298-299.

Friend, Tim. "Power Tool." *National Wildlife*, October-November 1995, pp. 16-24.

Hawaleshka, Danylo. "A High-Tech Tool for Police." *Maclean's*, March 24, 1997, pp. 52-53.

Nichols, Mark. "DNA on Trial." *Maclean's*, February 6, 1995, pp. 56-60.

Smiley, Brenda. "Fingerprinting the Dead." *Archaeology*, November-December 1996, pp. 66-67.

## TEACH

### Concept Development

▶ Explain that each new molecule of DNA contains exactly one-half of the original DNA.

***Portfolio***

Give students the following portion of a DNA molecule. Have them sequence the steps that take place during copying.

A – T
T – A
C – G
C – G
G – C
A – T

## TEACH

### Independent Practice

**Study Guide**/*Teacher Resource Package*, p. 165. Use the study guide worksheet shown at the bottom of this page for independent practice.

### Check for Understanding

Have students respond to the first three questions in Check Your Understanding.

### Reteach

**Reteaching**/*Teacher Resource Package*, p. 83. Use this worksheet to give students additional practice with DNA structure and how DNA copies itself.

**Extension:** Assign Critical Thinking, Biology and Reading, or some of the **OPTIONS** available with this lesson.

## 3 APPLY

### Brainstorming

Give students this scenario: You are a scientist in the late 1930s. Explain how you will prove that DNA controls our traits.

## 4 CLOSE

### Convergent Question

Tell students that they are made up of half their father's and half their mother's DNA. Ask why brothers and sisters don't look exactly like them.

### Answers to Check Your Understanding

1. It makes up the genes of all living things.
2. The order of bases controls the making of proteins in cells. The kind of protein made shows up as a trait.
3. RNA acts as a messenger and carries the DNA message from the nucleus to the ribosomes. Ribosomes act as worktables for making proteins.
4. the order of the nitrogen bases
5. T-T-C-G-A-G-G-A-C-G

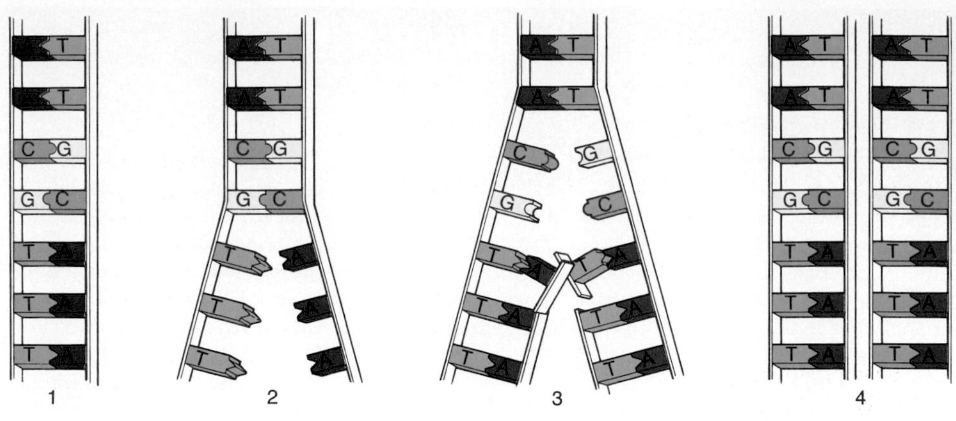


**Figure 28-8** Steps used by DNA as it copies itself

Think about the ladder model of DNA again. Follow the steps in Figure 28-8 as you read how DNA copies itself.

1. DNA is ready to make a copy of itself.
2. The molecule begins to open up along its middle.
3. Loose nitrogen bases with the sugar and acid attached are present in the cell nucleus. These bases are not part of the DNA yet. They join the bases that are on the opened rungs. Look carefully and you will see that A joins only with T. C joins only with G. The loose nitrogen bases continue to join the bases on the DNA molecule.
4. Finally, two DNA molecules have formed.

Look closely at the two newly formed DNA molecules in step 4 of Figure 28-8. They are exactly alike. The order of nitrogen bases is also exactly the same as in the original molecule in step 1 of the figure. The genes on the two chromosomes are exactly the same. When a cell reproduces, the new cells that form have the same genetic message.

### Check Your Understanding

1. Why is DNA important to all living things?
2. How does the order of nitrogen bases in DNA control traits?
3. How do RNA and ribosomes help DNA make proteins?
4. **Critical Thinking:** What features of the DNA structure are important if the genetic message is to be copied exactly?
5. **Biology and Reading:** Suppose one side of a piece of DNA has the bases A-A-G-C-T-C-C-T-G-C. What bases would the other side of the DNA molecule have?

### STUDY GUIDE 165

STUDY GUIDE CHAPTER 28

Name ____ Date ____ Class ____

**THE DNA MOLECULE**

*In your textbook, read about how DNA copies itself in Section 28:1.*

10. The model of DNA on the left is ready to be copied. The copies on the right labeled 1-3 show the resulting molecules as drawn by several students.

DNA to be copied: AT, TA, CG, CG, GC, TA, GC

1: AT, TA, CG, CG, GC, TA, GC and AT, TA, CG, AG, GC, TA, GC

2: AT, TA, CG, CG, GC, TA, GC and AT, TA, CG, CG, GC, TA, GC

3: AA, TT, CC, GC, GG, TT, GG

a. Explain why number 1 is incorrect. **The order of nitrogen bases must be exactly the same as the original. One copy is correct and one is not.**

b. Explain why number 2 is correct. **Both models are identical to each other and to the original.**

c. Explain why number 3 is incorrect. **The base pairs in this model are incorrectly joined. Also, two copies must be made, not just one.**

11. How does the genetic message compare in the original cell and the two new cells that form after the cell reproduces? **It should be the same.**

**HOW THE GENETIC MESSAGE CHANGES**

*In your textbook, read about mutations in Section 28:2.*

1. Complete the diagram on the left. Then, circle the areas in the diagram on the right that show a mutation.

DNA correctly copied: AT, TA, CG, CG, GC, TA, GC, AT → AT, TA, CG, CG, GC, TA, GC, AT and AT, TA, CG, CG, GC, TA, GC, AT

DNA incorrectly copied: AT, TA, CG, CG, GC, TA, GC, AT → AT, TA, CG, CG, GC, TA, GC, AT and AT, TA, CG, CG, GC, (CA), GC, AT

2. List three causes of mutations. **An error may take place when DNA is being copied. The cell may be exposed to certain chemicals. The cell may be exposed to radiation.**

### RETEACHING 83

RETEACHING CHAPTER 28

Name ____ Date ____ Class ____

Use with Section 28:1.

**DNA STRUCTURE AND HOW DNA COPIES ITSELF**

*Complete the diagrams by filling in the correct terms and letters.*

Ladder upright
Sides of DNA molecule
Ladder rung
sugar T A Sugar
**nitrogen bases** form rungs
acid Acid
sugar G C sugar
**G** joins only with **C**
acid acid
sugar A T sugar
**A** joins only with **T**

1 2 3 4

## 28:2 How the Genetic Message Changes

In Section 28:1 you read that newly formed cells have the same genetic message as the cell from which they came. Sometimes they do not. What happens when the message changes? Can scientists change the message of living things?

**Objectives**

4. **Describe** how mutations occur.
5. **Describe** how cloning and breeding produce offspring with desired traits.
6. **Explain** how recombinant DNA and gene therapy can help humans.

**Key Science Words**

mutation
radiation
identical twins
fraternal twins
breeding
recombinant DNA
gene therapy

### Mutations

Sometimes errors happen when chromosomes are copied. The bases A, T, C, and G may join incorrectly. Joining incorrectly results in a change called a mutation (myew TAY shun). A **mutation** is any change in copying the DNA message.

What happens if you hit the wrong key on a computer keyboard? Doesn't the computer get the wrong message? Cells are like the computer. A wrong base in the DNA gives the cell the wrong message. The result is that the wrong type of protein is made. This change may cause a different trait to appear.

Hemophilia is a serious blood disease that can start from a mutation. Look at the pedigree in Figure 28-9. You

**Figure 28-9** A family pedigree showing where a mutation for hemophilia may have occurred

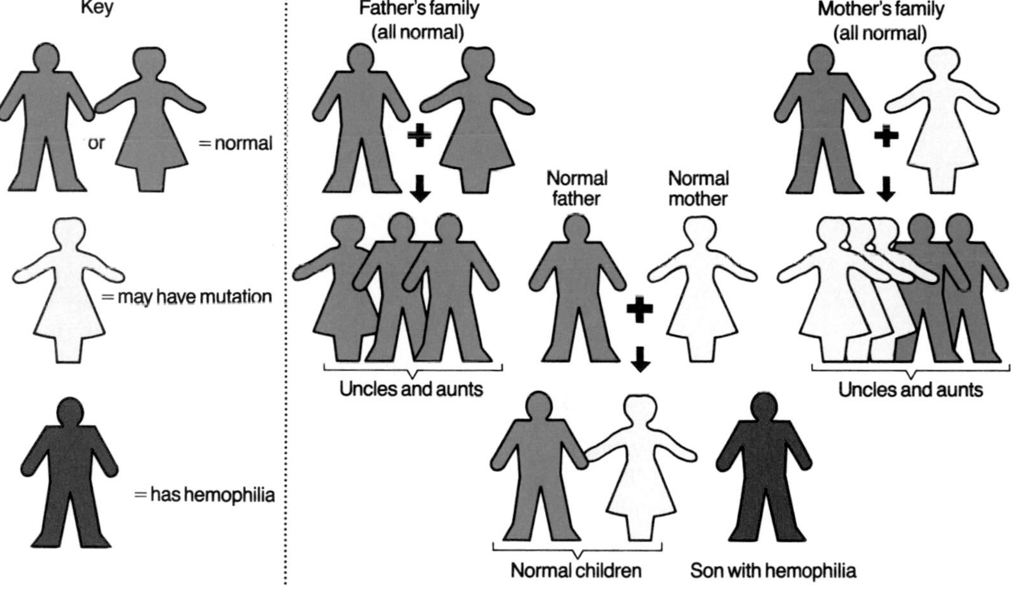




## 28:2 How the Genetic Message Changes

### PREPARATION

#### Materials Needed

Make copies of the Focus, Critical Thinking/Problem Solving, Study Guide, Application, Reteaching, and Lab worksheets in the *Teacher Resource Package*.

▶ Gather *Coleus* plants, plastic cups, soil, and aluminum foil for Lab 28-2.

▶ Have materials like those in Lab 28-1 on hand for the Mini Lab.

#### Key Science Words

| | |
|---|---|
| mutation | fraternal twins |
| radiation | gene therapy |
| identical twins | breeding |
| | recombinant DNA |

#### Process Skills

In the Mini Lab, students will formulate a model. In Lab 28-2, students will observe, infer, and interpret data. In the Skill Check, students will understand science words.

### 1 FOCUS

▶ The objectives are listed on the student page. Remind students to preview these objectives as a guide to this numbered section.

#### Bridging

The problem of skin cancer and radiation was touched upon in Chapter 22. Refer to that chapter and discuss how the development of skin cancer and changes in DNA caused by exposure to the sun might be related.

**Focus**/*Teacher Resource Package*, p. 55. Use the Focus transparency shown at the bottom of this page as an introduction to cloning.

**FOCUS 55**

FOCUS CHAPTER 28

Name Date Class

TWINS

These twins formed from two different fertilized eggs.

These twins formed from the splitting of one fertilized egg.

1. Which twins are exactly alike? Explain.
2. How are the other twins related to each other?

### OPTIONS

#### Science Background

A point mutation is a mutation in which only one base pair is altered. It is also possible for a base pair to be added or deleted. Such a mutation causes a shift in the base sequence and is called a frameshift mutation. Somatic mutations take place in body cells during development. These mutations are not heritable and usually are not noticeable.

# 2 TEACH

## MOTIVATION/Brainstorming

Ask students to name mutation-causing agents. (List may include radiation from sunlight, X rays, atomic bomb explosions, certain drugs, industrial pollutants, insecticides, and even food additives.)

## Concept Development

▶ Tell students that not all mutations are harmful. Some are helpful. The vast majority are probably neither helpful nor harmful and go unnoticed by life-forms.

▶ Alert students to the role that mutations play in evolution. Mutations that provide an organism with some slight advantage will increase that living thing's chances of survival.

## Mini Lab

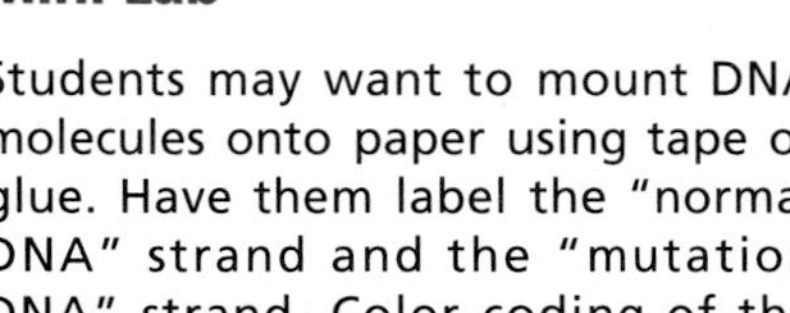

Students may want to mount DNA molecules onto paper using tape or glue. Have them label the "normal DNA" strand and the "mutation DNA" strand. Color coding of the mutation site is suggested.

## ✓ ASSESSMENT

**Content:** Ask students to examine Figures A and B. Ask them to identify the figure that contains the mutation, identify the location of the mutation, and offer a possible explanation for what might have caused it to occur.

| Figure A | Figure B |
|---|---|
| A – T | C – G |
| T – C | G – C |
| C – G | T – A |
| C – G | A – T |
| G – C | G – C |

## Guided Practice

Have students write down their answers to the margin question: a change in the genetic message.

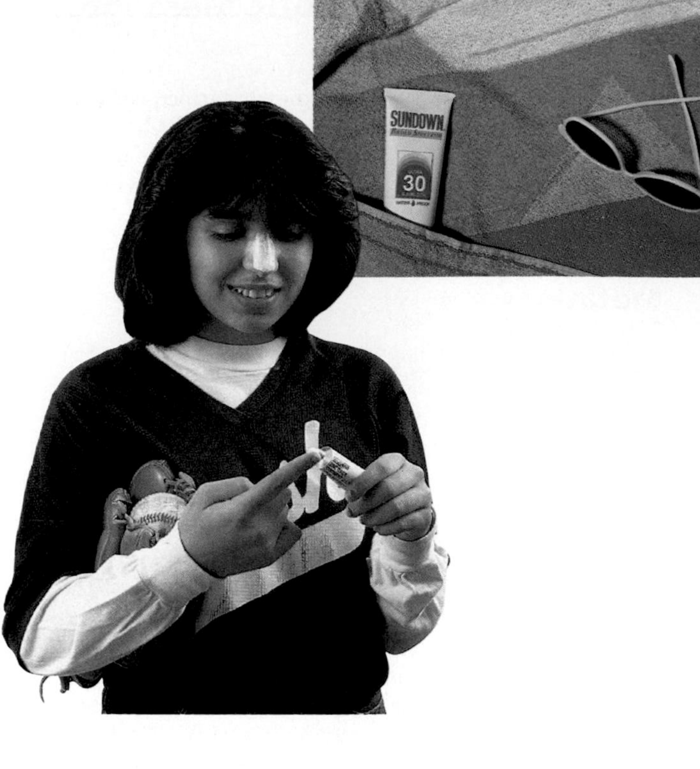



**Figure 28-10** Sunscreens help protect the skin from the sun's radiation.

### Mini Lab

**How Does a Mutation Occur?**

**Formulate a model:** Use parts like those in Lab 28-1 to build DNA as follows: AACGTA. Build a second model but show a mutation. Label the mutation. What can cause it? *For more help, refer to the **Skill Handbook,** pages 706-711.*

can see when the disease first appeared in this family. We expect traits to be passed from parents to children. In this family, the grandparents, uncles and aunts, and parents do not show the trait. It suddenly appears in the son. We can guess that a mutation took place when the sex cells of one of the parents were being made. It may have happened in the mother or grandmother, but it first appears in their child. A mutation causes a change in a child's trait only when it takes place in the parent's sex cells.

What causes mutations?

What causes mutations to occur? Many mutations are simply the results of copying mistakes. An error may take place in the pairing of bases when DNA is copied during cell reproduction. Other mutations may be caused by something from outside the cell. Certain chemicals and some forms of radiation (rayd ee AY shun) can cause mutations. **Radiation** is energy that is given off by atoms. The sun releases a large amount of radiation. X rays, ultraviolet light, and visible light are examples of kinds of radiation from the sun. Very powerful radiation, such as X rays and ultraviolet light, can cause mutations.



## OPTIONS

**Critical Thinking/Problem Solving/** *Teacher Resource Package,* p. 28. Use the Critical Thinking/Problem Solving worksheet shown here to help students understand changes in the genetic code.

**How Can a Mutation in DNA Affect an Organism?/***Lab Manual,* pp. 239-242. Use this lab as an extension to changes in the genetic message.

### CRITICAL THINKING 28

CRITICAL THINKING/PROBLEM SOLVING — CHAPTER 28

Name ______ Date ______ Class ______

*Use after Section 28:2.*

**HOW CAN YOU FIND CHANGES IN THE GENETIC CODE?**

The students in biology looked at the model of DNA. They saw the two upright sides of sugar and acid wound around each other into a spiral. They saw the bases attached to the sides. The names of the most common bases in DNA are adenine (whose symbol is A), thymine (whose symbol is T), guanine (G), and cytosine (C). They saw how the bases pair to make a smooth spiral with no bulges.

The students learned that DNA copies itself when each base pairs with its opposite type. A always pairs with T, and G always pairs with C. In that way, two molecules of DNA are formed that are usually identical. A mistake in copying can cause a mutation.

The students did an exercise to test whether they could spot possible mutations. The teacher explained that a strand of DNA is one side of a DNA molecule. It has one upright side and unpaired bases. The teacher put these sequences of bases on the blackboard:

Strand A. A T T C A T G C A T
Strand B. T A A G T A C G T A
Strand C. A T A C A T G G A T
Strand D. A T T C A T G C A T
Strand E. T A A G T A C G T A
Strand F. T A G T A C G T A C
Strand G. T A A A G T A C G T

*Then the teacher asked a series of questions. Answer the questions below that the teacher asked.*

**Analyzing the Problem**

1. When strand A copies itself to make another DNA strand, will the order of the bases be like strand B, C, or D? **strand B**
2. When strand A was part of a molecule containing two strands, what was the sequence of bases in its opposite strand? Write the symbols here: **T A A G T A C G T A like strand B.**
3. How many copying errors, or mutations, are present in strand C, assuming it was copied from strand B? Write the number and the symbols for each error here: **two—A T A instead of A T T and G G A instead of G C A**

**Solving the Problem**

1. Sometimes errors occur by the loss (or deletion) or the addition of a base. Suppose that Strands E, F, and G are copied from strand D. Which strand shows a deletion error? Explain. **Strand F; the third base, which should be A to pair with the T in strand D, is missing.**
2. Which strand, E, F, or G, shows an addition mutation? **strand G**

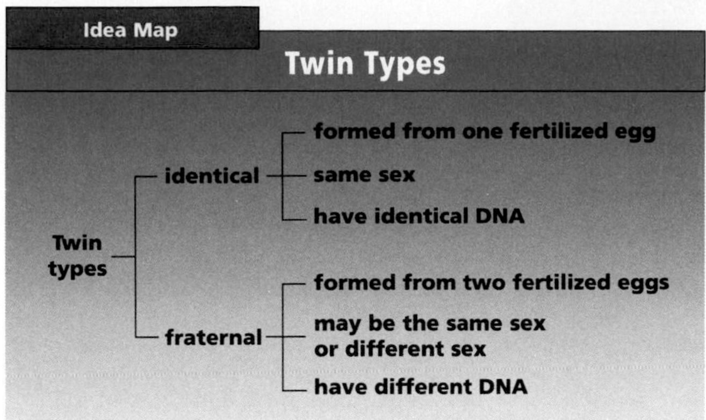


**Study Tip:** Use this idea map as a study guide to twin types. The features of identical and fraternal twins are shown.

## Cloning

Do you have an identical twin? If you do, you have a clone. That's because **identical twins** are two children that form from the splitting of one fertilized egg. They have exactly the same genes. You know that two clones have the same genes. They are exact copies of each other. Since identical twins have the same genes, they must have the same DNA. See if you can tell which diagram in Figure 28-11 will result in identical twins. One of the diagrams shows how fraternal twins are formed. **Fraternal twins** are twins that form from two different fertilized eggs. They are not clones. Fraternal twins are no more alike than two other children in the same family would be. They don't even have to be the same sex.

**Figure 28-11** Differences between identical and fraternal twins. Which diagram shows identical twins? b

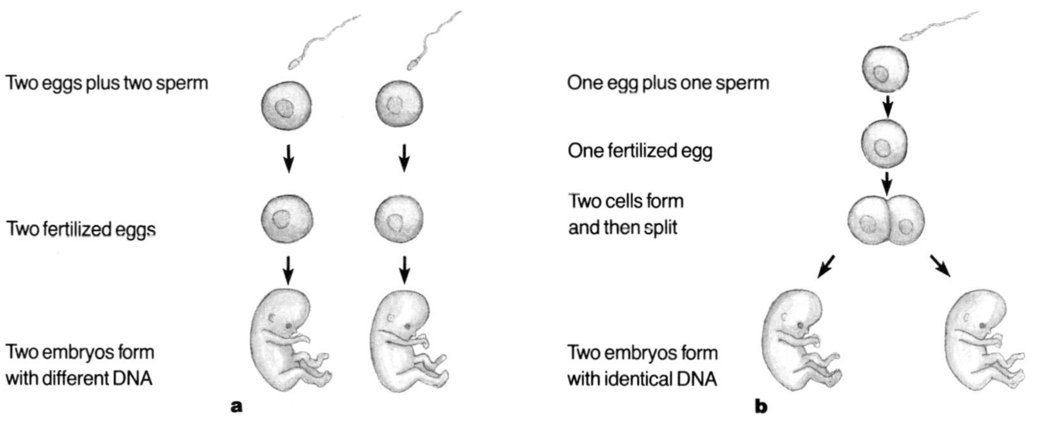




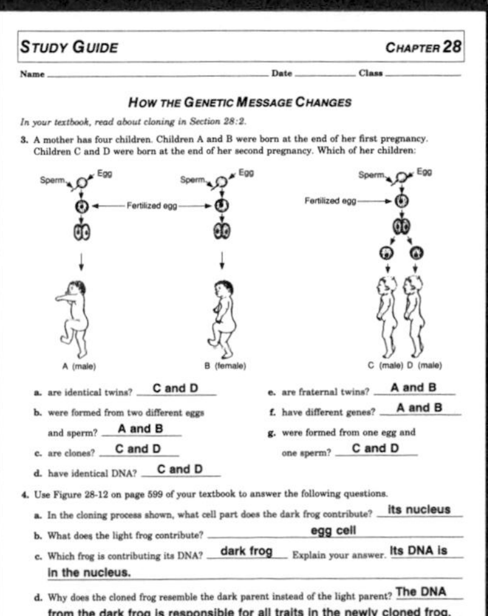

**STUDY GUIDE 166**

STUDY GUIDE — CHAPTER 28

Name ______ Date ______ Class ______

*HOW THE GENETIC MESSAGE CHANGES*

*In your textbook, read about cloning in Section 28:2.*

3. A mother has four children. Children A and B were born at the end of her first pregnancy. Children C and D were born at the end of her second pregnancy. Which of her children:



a. are identical twins? C and D
b. were formed from two different eggs and sperm? A and B
c. are clones? C and D
d. have identical DNA? C and D
e. are fraternal twins? A and B
f. have different genes? A and B
g. were formed from one egg and one sperm? C and D

4. Use Figure 28-12 on page 599 of your textbook to answer the following questions.
a. In the cloning process shown, what cell part does the dark frog contribute? its nucleus
b. What does the light frog contribute? egg cell
c. Which frog is contributing its DNA? dark frog Explain your answer. Its DNA is in the nucleus.
d. Why does the cloned frog resemble the dark parent instead of the light parent? The DNA from the dark frog is responsible for all traits in the newly cloned frog.

## TEACH

### Concept Development

▶ Identical twins will always be the same sex. Fraternal twins may or may not be the same sex.

▶ Students may want to know that twins occur about once in every 85 pregnancies. Two-thirds of these are fraternal. Triplets occur once in every 8000 pregnancies.

### Idea Map

Have students use the idea map as a study guide to the major concepts of this section.

### Independent Practice

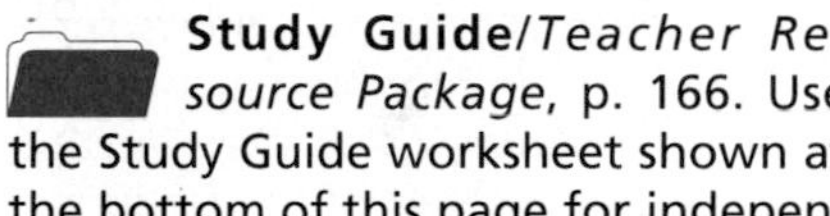

**Study Guide/***Teacher Resource Package*, p. 166. Use the Study Guide worksheet shown at the bottom of this page for independent practice.

### *Student Journal*

Set up the following story line for students. You have just written a science fiction short story about a cloned human. The nucleus of a cell from a tall, blonde, blue-eyed person was inserted into a nucleus-free cell from a short, black-haired, green-eyed person. Write the last page of the short story and describe what the clone looked like and explain why.

### GLENCOE TECHNOLOGY

**Videotape**

**The Secret of Life**
*In the Land of Milk and Money: Biotechnology*

**Videodisc**

**Biology: The Dynamics of Life**
Movie: *Bioengineering*
Disc 1, Side 1, Ch. 37

## Lab 28-2 Cloning

### Overview

In this lab, students are introduced to the concept of cloning. *Coleus* will be used as the experimental organism.

**Objectives:** Upon completion of this lab, students will be able to (1) **demonstrate** the technique of plant cloning, (2) **explain** why the process of cloning yields so many offspring.

**Time Allotment:** first day – 20 minutes; days 2-8 – 10 minutes each

### Preparation

▶ **Alternate Materials:** Plants such as *Geranium* or *Zebrina* may be used in place of *Coleus.* Polystyrene cups may be used instead of plastic cups.

▶ One *Coleus* plant could yield 5 to 8 cuttings.

**Lab 28-2 worksheet**/*Teacher Resource Package*, pp. 111-112.

### Teaching the Lab

▶ The term *asexual reproduction* or *asexual propagation* could be used to describe the process being used here. The correct term is *cloning.*

▶ Steps 11-12 are optional.

### ✓ ASSESSMENT

**Skill:** Provide students with a leaf having a very distinct vein pattern, margin, coloration, or shape. Advise them that this leaf can be cloned into a small plant. Ask them to diagram the appearance of the resulting plant after one month.

# Cloning

# Lab 28–2

## Problem: Can a *Coleus* plant be cloned?

### Skills

observe, infer, interpret data

### Materials

| | |
|---|---|
| clear plastic cup | soil |
| aluminum foil | water |
| scissors | label |
| *Coleus* plant | metric ruler |

### Procedure

1. Copy the data table.
2. Label a clear plastic cup with your name. Fill it with water.
3. Cover the top of the cup with foil.
4. Use scissors to punch a small hole in the center of the foil. **CAUTION:** *Always be careful when using scissors.*
5. Use scissors to cut off a small branch from a *Coleus* plant. The figure shows where to make the cut.
6. Insert the cut end of the branch through the hole in the foil. The bottom of the branch must dip into the water.
7. In your table, write today's date. Draw what the end of the cut branch looks like.
8. Record the number of students who cut a branch from the same plant.
9. **Observe:** Check the branch each day for the next few days. Look for roots that appear at the cut end.
10. When you see roots beginning to appear on your branch, record the date in your table and draw the cut end. Continue to check your branch and draw its appearance.
11. When roots are about two centimeters long, remove the plant from the cup. Pour out the water and fill the cup with soil.
12. Place your plant into the soil, being careful not to break its new roots. Moisten the soil with water.

### Data and Observations

1. How many branches were cut from your original plant?
2. How many days did it take for roots to appear on your branch?

**NUMBER OF STUDENTS USING SAME PLANT** ____

| | | |
|---|---|---|
| Date 4/16 | Date 4/17 | Date 4/18 |
| Date 4/19 | Date 4/20 | Date 4/21 |

### Analyze and Apply

1. Explain the advantage of cloning plants.
2. **Infer:** How would the DNA in plants grown from the same original plant compare? Why?
3. **Apply:** Was the cloning method you used sexual or asexual reproduction? Explain.

### Extension

Garden stores sell chemicals that are supposed to speed up root growth from the cut ends of plants. **Design an experiment** to test this.

598 DNA—Life's Code 28:2

## ANSWERS

### Data and Observations

1. Answers will vary.
2. Answers will vary – possibly 3 to 6 days.

### Analyze and Apply

1. All plants will be identical, and one can clone many plants from one original plant.
2. The DNA is exactly the same because the process is actually asexual reproduction.
3. asexual reproduction; only one parent was involved

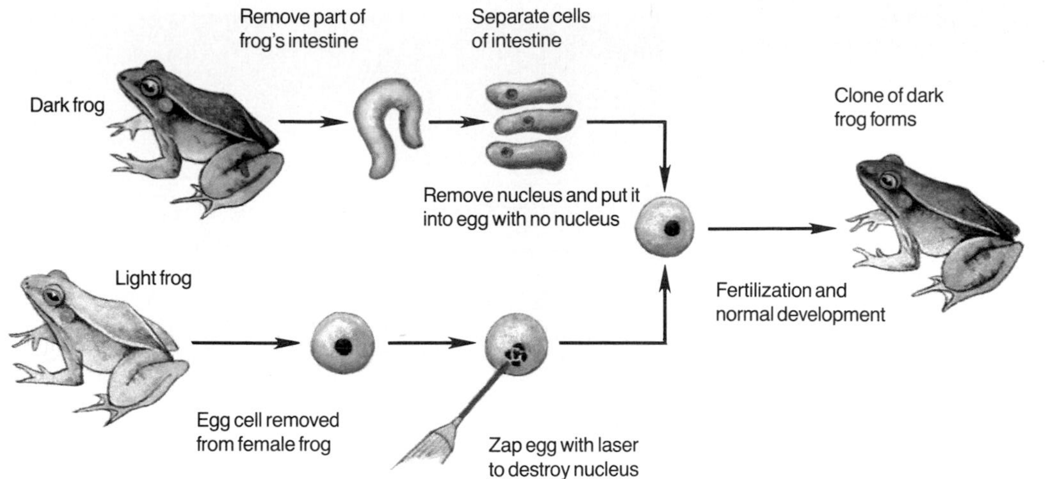


**Figure 28-12** Steps needed to clone a frog

Can humans clone animals? The answer is yes. Figure 28-12 shows how frogs are cloned. It is not science fiction. It has been done. Even a sheep has been cloned. Can humans be cloned in the laboratory? The technology makes it seem possible. More importantly, should it be done?

## Plant and Animal Breeding

Cloning is a way of producing living things with identical desirable traits. How do we get a living thing with the desired trait in the first place?

In Chapters 26 and 27 you worked many genetic problems. They showed what to expect with certain matings. Knowing what to expect tells us which living things to breed for certain traits. **Breeding** is the bringing together of two living things to produce offspring.

Selective breeding can bring out the desired traits of living things. If breeders bring together parents with different desired traits, the result can be offspring with the best traits of both parents.

An example of this can be seen in cotton plants. A certain type of cotton plant grew in Africa, but a plant disease was destroying it. Another type of cotton plant did not get the disease, but this plant could not grow in Africa. It grew only in Central America. The two plants were bred with each other. The result was a new type of cotton plant that would grow in Africa and did not get the disease.

### Bio Tip

**Consumer:** Humans have been breeding dogs for hundreds of years. The evidence is seen in the many different varieties that exist today.

**How has breeding produced better cotton plants?**



## TEACH

### Concept Development

▶ A study of Figure 28-12 will help students understand the process of cloning in animals. Any body cell from within the dark frog could be used as a cell source. The egg cell from the light frog with its nucleus destroyed does not have to be fertilized because the nucleus being added to it from the dark frog already has the full chromosome number. All body cells have the full chromosome number.

▶ Discuss how breeeding has produced more robust varieties of plants and animals.

### MOTIVATION/Brainstorming

Ask students why the new clone has the dark frog trait rather than the light frog trait. (DNA was contributed by the dark frog cell because its nucleus was used. The DNA code for dark skin protein is the code being used. The only thing being contributed by the light frog is the cytoplasm within its egg cell.)

**Analogy:** A signal within the genetic code turns gene action on and off just as a capital letter starts a sentence and a period stops a sentence.

### Guided Practice

Have students write down their answers to the margin question: It has produced a cotton plant that is resistant to disease and that will grow in Africa.

## APPLICATION 28

APPLICATION: CONSUMER — CHAPTER 28

Name ________ Date ________ Class ________

*Use after Section 28:2.*

PLANT AND ANIMAL BREEDING

Since the beginning of agriculture, farmers have tried to improve plants and animals for food. They have mated animals with the traits they wanted. They have saved seeds from the best plants and planted the seeds the following year. Selecting hardy, productive plants and animals to produce the next generation is called selective breeding. Selective breeding is based on the techniques of mass selection, inbreeding, and hybridization.

Mass selection is choosing the best plants and animals from a large number for further breeding. Inbreeding is the mating of related individuals to establish pure lines. Breeders use inbreeding to keep desirable traits in a plant or animal species. In hybridization, two different but related varieties or species of plants or animals are mated. The results of the mating are called hybrids.

*Examples of selective breeding are described below. In the space provided, identify each example as either mass selection, inbreeding, or hybridization.*

1. Luther Burbank grew large numbers of fruits, flowers, vegetables and grains. He selected the best plants to breed for the next generation. He grew thousands of plants trying to produce one improved species.
   mass selection
2. A dog had five puppies. Two of the puppies, a male and a female, were a golden color. When the puppies were mature, they were mated to each other in an attempt to produce puppies with the same golden color.
   inbreeding
3. A male championship race horse was mated to its female offspring to produce a horse with the championship traits of the male horse.
   inbreeding
4. An apple liked by consumers for its taste, crispness, and long storage life was pollinated with the pollen of an apple that resisted disease. The resulting apple has the taste, crispness, and long storage life shoppers want. It also resists diseases.
   hybridization
5. A large pink daisy is discovered in a bed of all white daisies. Seeds from the pink daisy are collected and planted the next year with hopes of producing more pink daisies.
   mass selection
6. The bull mastiff is the result of breeding an aggressive bull dog with a large, fast, gentle mastiff. The bull mastiff has proved to be a good guard dog.
   hybridization

## OPTIONS

**Application**/*Teacher Resource Package*, p. 28. Use the Application worksheet shown here to give students practice with plant and animal breeding.

## GLENCOE TECHNOLOGY

**Videodisc**

**Biology: The Dynamics of Life**
Animation: *Gene Cloning*
Disc 1, Side 1, Ch. 38

**Videotape**

**The Secret of Life**
*Tinkering with Our Genes: Genetic Medicine*

# TEACH

## Concept Development

▶ Explain that other applications of recombinant DNA research include: drug synthesis (insulin, interferon, vaccines) and development of crops that are resistant to heat, cold, disease, drought, and pestilence and have higher nutritional value.

▶ Gene therapy may be helpful in treating such diseases as severe combined immunodeficiency (SCID) as well as certain cancers. It also shows promise with Alzheimers and Parkinson's diseases as well as rheumatoid arthritis. This therapy is still in its infancy, and much research is needed before it becomes a common procedure in the war against human genetic disorders.

## Check for Understanding

Ask students to list several ways that humans have changed the gene code or altered traits of a living thing. Lists should include mutations, cloning, breeding, and gene splicing. Discuss whether the changes were helpful or harmful.

### Reteach

Have students prepare a brief written report on one of their answers to Check Your Understanding.

## Independent Practice

**Study Guide**/*Teacher Resource Package*, p. 167. Use the Study Guide worksheet shown at the bottom of this page for independent practice.

**Figure 28-13** Scientists have selectively bred Brahman bulls for small size.

Brahman cattle are usually raised for meat, Figure 28-13. Recently, scientists have been able to breed smaller Brahmans. Each cow produces less meat than a larger size cow, but it also does not eat as much. This is a real help to those farmers who don't have much land for cattle to graze. Ten mini cows can graze on the same amount of grass as one large cow and can produce almost three times as much beef.

Many crops and livestock are the result of careful selective breeding. The fruits and vegetables that you buy at the grocery store are the products of selective breeding. Breeders are producing cattle that have less fat, and they are trying to breed chickens that lay low-cholesterol eggs.

**Figure 28-14** Steps needed to splice a wire

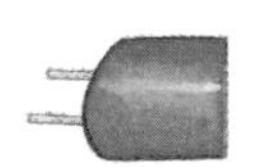


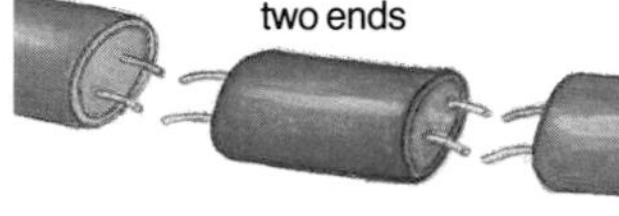


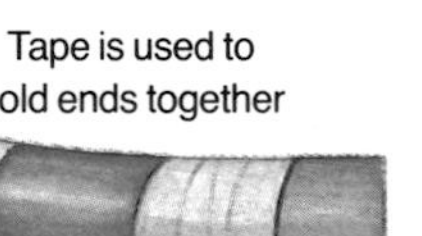


## Splicing Genes and Gene Therapy

Have you ever spliced a wire or broken film? If you have, then you know that the word *splice* means to insert, or join together. Figure 28-14 shows wires that have been joined together.

Today, scientists can splice genes. To do this, a bacterium receives a section of DNA from another organism, such as a human. The new DNA combination that is formed in the bacterium is recombinant (ree KAHM buh nunt) DNA. **Recombinant DNA** is the DNA that is formed when DNA from one organism is put into the DNA of another organism.

Gene splicing produces bacteria that can make certain chemicals. For example, scientists splice human genes that control the making of insulin into bacteria. The bacteria then make insulin in large amounts. Diabetics benefit



## OPTIONS

### ACTIVITY/Challenge

Have some students prepare a debate to be presented to the rest of the class. The topic is the pros and cons of recombinant DNA technology (gene splicing).

### STUDY GUIDE 167

STUDY GUIDE — CHAPTER 28

Name ______ Date ______ Class ______

**HOW THE GENETIC MESSAGE CHANGES**

*In your textbook, read about breeding of plants and animals and splicing genes between organisms in Section 28:2.*

5. This diagram shows two bulls. Below each is a description of that bull's traits.

Bull A: Sleek, clean, solid-colored fur; Long tail, long legs; Milk production of his offspring is low.

Bull B: Rough, spotted fur; Short tail, short legs; Milk production of his offspring is high.

a. Which bull would you choose to breed with a cow to produce a herd of cows that would supply a lot of milk? **B** Why? **The milk production of bull B's offspring is high.**

b. Which bull would you choose for breeding to produce "beautiful" offspring? **A** Why? **Bull A has a sleek, clean, solid-colored fur.**

c. Which bull would you choose for breeding to produce offspring that would not jump over fences? **B** Why? **Bull B has shorter legs.**

d. Explain the value of plant and animal breeding. **The value of breeding is producing offspring with desired traits. It's been used to breed plants that won't get certain diseases and cattle that have less fat or greater milk production.**

6. What is recombinant DNA? **Recombinant DNA is the DNA that is formed when DNA from one organism is put into the DNA of another organism.**

7. Why does gene splicing work? **It works because the genetic code for all living things is the same.**

8. List four ways that gene splicing is of value to humans now or may be of value in the future. **Gene splicing can help make human insulin, growth hormone, and plants that are not harmed by chemical sprays. Someday, it may help to cure certain genetic diseases.**

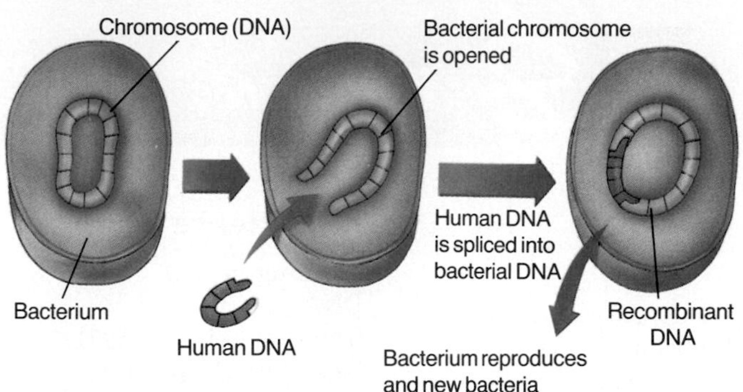


**Figure 28-15** In gene splicing, a bacterium receives DNA from another organism.

because they can't make their own insulin.

Splicing of human genes into bacteria also has helped make human growth hormone. This hormone is given to children when their pituitary gland does not make enough of it. The real future of gene splicing may be in the ability to cure certain genetic diseases.

Gene therapy is the newest hopeful weapon against genetic disorders. **Gene therapy** is the adding of a healthy gene into the body of a person suffering from a disorder caused by a defective or mutated gene. This healthy gene then takes the place of the defective gene on a chromosome. How can healthy genes be delivered to chromosomes? Viruses can carry the healthy DNA or gene into cells. Once there, this healthy gene will begin to function as if it had been present in the defective cell from the start. Disorders that might be cured through gene therapy include cystic fibrosis, hemophilia, and certain muscle disorders.

## Check Your Understanding

6. How does a mutation occur?
7. How could you produce an apple tree that was resistant to disease and produced large fruit?
8. What is being spliced to form recombinant DNA?
9. **Critical Thinking:** Could you breed an animal with exactly the combination of genes you wanted? Explain.
10. **Biology and Writing:** Do you think research on splicing genes should continue? Write a paragraph with at least three reasons that explain your answer.

### Skill Check

**Understand science words: recombinant DNA.** The word part *re* means again. In your dictionary, find three words with the word part *re* in them. *For more help, refer to the* ***Skill Handbook,*** *pages 706-711.*

## TEACH

### Check for Understanding

Have students respond to the first three questions in Check Your Understanding.

### Independent Practice

**Study Guide**/*Teacher Resource Package*, p. 168. Use the Study Guide worksheet shown at the bottom of this page for independent practice.

### Reteach

**Reteaching**/*Teacher Resource Package*, p. 84. Use this worksheet to give students additional practice with how DNA changes.

## 3 APPLY

### ACTIVITY/Challenge

Have students use references to determine how humans breed dogs. Have them report on certain dog breeds.

## 4 CLOSE

### Audiovisual

Show the video *Heredity and Mutation*, Films for the Humanities.

### Answers to Check Your Understanding

6. When DNA copies itself, an error may take place in the pairing of bases.
7. You could breed a tree that produces large fruit with one that is resistant to disease.
8. genes
9. No; many genes are hidden in recessive form, so you would have to breed only for traits you can see.
10. Answers will vary. Example: Splicing genes should not continue because it is dangerous.

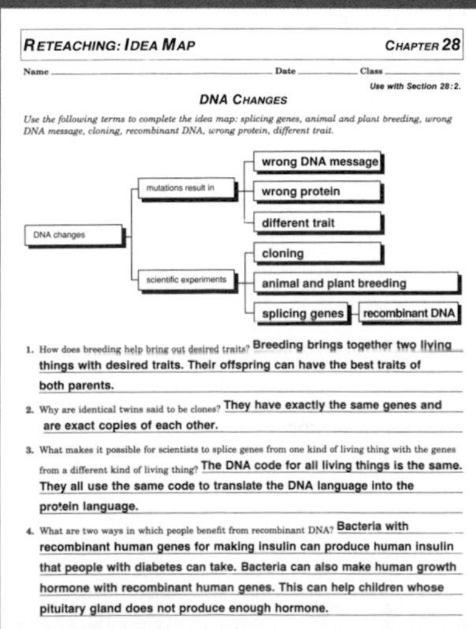

**RETEACHING 84**

RETEACHING: IDEA MAP — CHAPTER 28

Name ______ Date ______ Class ______

*Use with Section 28:2.*

**DNA CHANGES**

*Use the following terms to complete the idea map: splicing genes, animal and plant breeding, wrong DNA message, cloning, recombinant DNA, wrong protein, different trait.*

DNA changes
- mutations result in: wrong DNA message; wrong protein; different trait
- scientific experiments: cloning; animal and plant breeding; splicing genes — recombinant DNA

1. How does breeding help bring out desired traits? **Breeding brings together two living things with desired traits. Their offspring can have the best traits of both parents.**
2. Why are identical twins said to be clones? **They have exactly the same genes and are exact copies of each other.**
3. What makes it possible for scientists to splice genes from one kind of living thing with the genes from a different kind of living thing? **The DNA code for all living things is the same. They all use the same code to translate the DNA language into the protein language.**
4. What are two ways in which people benefit from recombinant DNA? **Bacteria with recombinant human genes for making insulin can produce human insulin that people with diabetes can take. Bacteria can also make human growth hormone with recombinant human genes. This can help children whose pituitary gland does not produce enough hormone.**

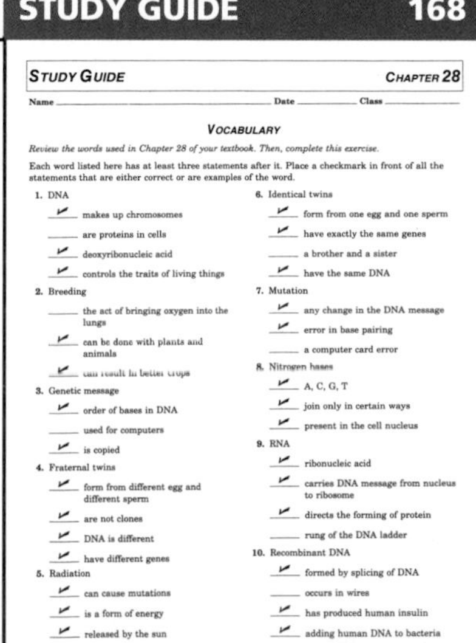

**STUDY GUIDE 168**

STUDY GUIDE — CHAPTER 28

Name ______ Date ______ Class ______

**VOCABULARY**

*Review the words used in Chapter 28 of your textbook. Then, complete this exercise.*

Each word listed here has at least three statements after it. Place a checkmark in front of all the statements that are either correct or are examples of the word.

1. DNA
   - ✓ makes up chromosomes
   - ___ are proteins in cells
   - ✓ deoxyribonucleic acid
   - ✓ controls the traits of living things
2. Breeding
   - ___ the act of bringing oxygen into the lungs
   - ✓ can be done with plants and animals
   - ✓ can result in better crops
3. Genetic message
   - ✓ order of bases in DNA
   - ___ used for computers
   - ✓ is copied
4. Fraternal twins
   - ✓ form from different egg and different sperm
   - ✓ are not clones
   - ✓ DNA is different
   - ✓ have different genes
5. Radiation
   - ✓ can cause mutations
   - ✓ is a form of energy
   - ✓ released by the sun
6. Identical twins
   - ✓ form from one egg and one sperm
   - ✓ have exactly the same genes
   - ___ a brother and a sister
   - ✓ have the same DNA
7. Mutation
   - ✓ any change in the DNA message
   - ✓ error in base pairing
   - ___ a computer card error
8. Nitrogen bases
   - ✓ A, C, G, T
   - ✓ join only in certain ways
   - ✓ present in the cell nucleus
9. RNA
   - ✓ ribonucleic acid
   - ✓ carries DNA message from nucleus to ribosome
   - ✓ directs the forming of protein
   - ___ rung of the DNA ladder
10. Recombinant DNA
    - ✓ formed by splicing of DNA
    - ___ occurs in wires
    - ✓ has produced human insulin
    - ✓ adding human DNA to bacteria

## CHAPTER 28 REVIEW

### Summary

Summary statements can be used by students to review the major concepts of the chapter.

### Key Science Words

All boldfaced terms from the chapter are listed.

### Testing Yourself

#### Using Words

1. DNA
2. breeding
3. nitrogen bases
4. mutation
5. recombinant DNA
6. identical twins
7. fraternal twins
8. RNA
9. genetic code
10. radiation

#### Finding Main Ideas

11. p. 599; Selective breeding can bring out the desired traits of living things.
12. p. 596; the incorrect pairing of bases, certain chemicals, and some forms of radiation
13. p. 601; Healthy genes are inserted into the chromosomes of a person suffering from a genetic disorder.
14. p. 592; A helper molecule called RNA carries the DNA code to the ribosomes.
15. p. 594; First, the molecule opens along the nitrogen bases; second, new nitrogen bases in the nucleus join the bases on the opened rungs.
16. p. 591; The DNA message is determined by the order of the nitrogen bases A, C, G, and T.
17. p. 597; Identical twins have the same genes, the same DNA, and were formed from one egg and one sperm. Fraternal twins have different genes, different DNA, and were formed from two eggs and two sperm.
18. pp. 586-587; It resembles a ladder that has been twisted.
19. p. 589; A gene is a short piece of DNA, a certain number of bases on the DNA molecule, and a small section of chromosome that determines traits.
20. p. 601; Gene splicing has produced bacteria that make human insulin and human growth hormone.

#### Using Main Ideas

21. Bases A and T can join, and bases C and G can join. They fit like pieces of a puzzle.
22. A family pedigree can show that hemophilia shows up without having ever appeared in any family member before.

# Chapter 28 Review

## Summary

### 28:1 The DNA Molecule

1. DNA is a molecule that makes up the chromosomes and genes of living things. The rungs of the ladder-shaped molecule are nitrogen bases.
2. Through experiments with bacteria, scientists showed that DNA controls traits.
3. The order of nitrogen bases A, T, C, and G forms a DNA message. RNA carries the DNA message to the ribosomes where specific proteins are formed. DNA is copied exactly to form new chromosomes just before mitosis and meiosis.

### 28:2 How the Genetic Message Changes

4. The DNA message may change if a mutation takes place. Mutations may be caused by some chemicals and radiation or by mistakes in copying.
5. Identical twins are clones. Clones have the same DNA and traits. Fraternal twins have different DNA. Selective breeding has produced new and better types of plants and animals.
6. Genes from one living thing can be spliced into the DNA of a different living thing. The result is recombinant DNA. Gene therapy adds healthy genes directly to an affected person to help in treating certain genetic disorders.

## Key Science Words

breeding (p. 599)
DNA (p. 586)
fraternal twins (p. 597)
gene therapy (p. 601)
genetic code (p. 592)
identical twins (p. 597)
mutation (p. 595)
nitrogen base (p. 587)
radiation (p. 596)
recombinant DNA (p. 600)
RNA (p. 592)

## Testing Yourself

### Using Words

*Choose the word from the list of Key Science Words that best fits the definition.*

1. molecule that makes up genes
2. causing two living things to mate and produce desired offspring
3. form the rungs of the DNA molecule
4. a change in the DNA message
5. formed when DNA is spliced into the DNA of another organism
6. twins formed from the splitting of one fertilized egg
7. twins formed from two different fertilized eggs
8. molecule that acts as a cell messenger for DNA
9. translates the DNA language into the protein language
10. energy that is given off by atoms

# Review

## Testing Yourself *continued*

**Finding Main Ideas**

*List the page number where each main idea below is found. Then, explain each main idea.*

11. what is gained from selective breeding
12. what causes mutations to occur
13. how gene therapy works
14. how DNA in the nucleus can control what goes on at the ribosomes
15. the steps involved when DNA copies itself
16. what determines the DNA message
17. how identical and fraternal twins differ
18. what a model of DNA looks like
19. three ways to describe a gene
20. examples of how gene splicing has helped humans

**Using Main Ideas**

*Answer the questions by referring to the page number after each question.*

21. Which DNA bases can join with one another? Why? (p. 587)
22. How can one show that hemophilia in a family may result from a mutation? (p. 595)
23. What are the steps in gene splicing? (p. 600)
24. What are the steps in making protein, starting with DNA in the nucleus? (p. 592)
25. How has plant breeding offered a solution to the problem of diseased cotton in Africa? (p. 599)
26. What is copied when new chromosomes form just before mitosis and meiosis? (p. 593)
27. Where would you find DNA in a pine tree? Why does a pine tree have DNA? (pp. 587, 589)

## Skill Review 

*For more help, refer to the* ***Skill Handbook,*** *pages 704-719.*

1. **Observe:** Look at Figure 28-4. Why do dead harmful bacteria injected into a rat not cause any problem to the animal?
2. **Interpret diagrams:** What happens to the original DNA molecule when it copies itself? Look at Figure 28-8 for help.
3. **Interpret data:** What part of a bacterium carries the genetic message? What is the evidence for this, as shown by the scientists who did the experiments described on page 590?
4. **Infer:** Why is it important that mutations be prevented from occurring in testes and ovaries?

## Finding Out More

**Critical Thinking**

1. Why would a mutation that appears in a person's body cell not appear in that person's offspring?
2. Should humans be cloned? Give reasons for your answer.

**Applications**

1. Write a report discussing the effects of the sun's radiation on the skin.
2. You wish to breed sheep in order to sell them as meat. What traits would you look for in the parents? Why?



### Using Main Ideas

23. DNA within a bacterium is split open; a section of another organism's DNA is placed inside the bacterium; the bacterium closes back together, incorporating the new DNA as part of its DNA.
24. DNA in the nucleus contains a code of nitrogen bases; RNA carries the DNA code from the nucleus to ribosomes; the ribosomes make protein according to the code on the RNA.
25. A cotton plant that could grow only in Central America and did not get a disease was bred to a cotton plant that grew in Africa and did get the disease. The result was a plant that would grow in Africa and would not get the disease.
26. DNA is copied.
27. DNA is in the nucleus of each cell in a pine tree. All cells in living things have DNA.

### Skill Review

1. Harmful bacteria that are dead will not kill the animal.
2. It opens along the middle, loose nitrogen bases join the bases on the DNA molecule, and finally two DNA molecules have formed.
3. Nitrogen bases in the DNA molecule carry the genetic message. Harmless bacteria change when they pick up DNA from harmful bacteria, but the offspring's traits are unaffected.
4. A mutation in the sex cells causes a change in the offspring's traits.

### Finding Out More

**Critical Thinking**

1. Mutations cause a change only when they occur in the sex cells.
2. Answers will vary.

**Applications**

1. Reports should include reference to cancer and other skin problems.
2. Answers will vary, but should include large size, little fat, and lots of muscle since the sheep would be a source of food.

 **Chapter Review/***Teacher Resource Package, pp. 148-149.*

 **Chapter Test/***Teacher Resource Package*, pp. 150-152.

**Chapter Test/***Computer Test Bank*

# Evolution

## PLANNING GUIDE

| CONTENT | TEXT FEATURES | TEACHER RESOURCE PACKAGE | OTHER COMPONENTS |
|---|---|---|---|
| (2 1/2 days)<br>29:1 Changes In Living Things<br>Adaptations<br>An Example of Survival<br>Natural Selection<br>Mutations<br>Species Formation<br>Primate and Human Evolution | Skill Check: *Define Words in Context,* p. 609<br>Skill Check: *Understand Science Words,* p. 613<br>Lab 29-1: *Adaptations,* p. 608<br>Check Your Understanding, p. 614 | Enrichment: *Human Evolution,* p. 31<br>Application: *Adaptations of Zoo Animals,* p. 29<br>Reteaching: *How Natural Selection Works,* p. 85<br>Critical Thinking/Problem Solving: *How Could Hollow Hip Bones Help Supersaurus?* p. 29<br>Focus: *Natural Selection,* p. 57<br>Study Guide: *Changes in Living Things,* pp. 169-171<br>Lab 29-1: *Adaptations,* pp. 113-114 | **Laboratory Manual:**<br>*How Do Some Living Things Vary?* p. 243<br>Color Transparency 29: *How Species Are Formed*<br>**STVS:** *Lizard Invasion,* Ecology (Disc 6, Side 1) |
| (3 1/2 days)<br>29:2 Explanations for Evolution<br>Darwin's Work<br>Fossil Evidence<br>Other Evidence | Skill Check: *Interpret Diagrams,* p. 619<br>Mini Lab: *How Do Peas Vary?* p. 616<br>Mini Lab: *How Do Wood and Fossil Wood Compare?* p. 617<br>Lab 29-2: *Fossil Prints,* p. 618<br>Idea Map, p. 621<br>Check Your Understanding, p. 621<br>Applying Technology: *A Fossil Hunt,* p. 622 | Reteaching: *Evolution,* p. 86<br>Skill: *Outline,* p. 30<br>Study Guide: *Explanations for Evolution,* pp. 172-173<br>Study Guide: *Vocabulary,* p. 174<br>Lab 29-2: *Fossil Prints,* pp. 115-116<br>Transparency Master: *The Geologic Time Scale,* p. 147 | **Laboratory Manual:**<br>*How Do Fossils Show Change?* p. 247<br>**STVS:** *Resistance to Pesticides in Cockroaches,* Ecology (Disc 6, Side 2) |
| Chapter Review | Summary<br>Key Science Words<br>Testing Yourself<br>Finding Main Ideas<br>Using Main Ideas<br>Skill Review | **ASSESSMENT RESOURCES**<br>Chapter Review, pp. 153-154<br>Chapter Test, pp. 155-157<br>Performance Assessment in the Biology Classroom<br>Alternate Assessment in the Science Classroom<br>Computer Test Bank | |

## GLENCOE TECHNOLOGY

**Infinite Voyage,** *The Search for Ancient Americans*
**Science and Technology Videodisc Series,** *Lizard Invasion,* Ecology (Disc 6, Side 1)
*Resistance to Pesticides in Cockroaches,* Ecology (Disc 6, Side 2)

**The Secret of Life,** *It's in the Genes: Evolution*

## MATERIALS NEEDED

| LAB 29-1, p. 608 | LAB 29-2, p. 618 | APPLYING TECHNOLOGY, p. 622 | MARGIN FEATURES |
|---|---|---|---|
| white paper<br>black paper<br>30 white dots<br>30 black dots<br>damp paper towel | plastic dish<br>100-mL graduated cylinder<br>warm water<br>metric ruler<br>leaf<br>plastic powder | beads, 3 groups<br>container | Skill Check, p. 613<br>dictionary<br>Skill Check, p. 619<br>pencil<br>paper<br>Mini Lab, p. 616<br>metric ruler<br>pencil<br>paper<br>10 pea pods<br>Mini Lab, p. 617<br>graduated cylinder<br>water<br>piece of wood<br>piece of petrified wood |

## OBJECTIVES

For more information about National Science Standards, see page 5T.

| SECTION | OBJECTIVE | CORRELATION of QUESTIONS to OBJECTIVES: CHECK YOUR UNDERSTANDING | CHAPTER REVIEW | TRP CHAPTER REVIEW | TRP CHAPTER TEST |
|---|---|---|---|---|---|
| 29:1 National Science Stds: UCP.2, UCP.3, UCP.4, UCP.5, C.3, C.4, C.6 | 1. **Give examples** of how adaptations help organisms survive. | 1 | 1, 4, 5, 15, 16, 17 | 3, 7, 9 | 1, 2, 3, 21, 37, 39, 41, 43 |
| | 2. **Explain** how changes in life-forms occur. | 2, 4 | 10, 18, 23 | 11 | 5, 10, 11, 22, 24, 26 |
| | 3. **Describe** the classification and evolution of primates and humans. | 3 | 9, 13, 22 | 1, 24 | 27, 29 |
| 29:2 National Science Stds: UCP.2, UCP.3, UCP.4, UCP.5, A.1, A.2, C.3, C.4, C.5, C.6, D.3, G.1, G.2, G.3 | 4. **Communicate** Darwin's main ideas. | 6, 8 | 2, 12, 20 | 2, 10, 14, 15, 16, 17, 18, 21, 22, 25 | 4, 6, 7, 8, 9, 13, 14, 15, 16, 17, 18, 19, 20, 23, 25, 36, 38, 40, 42 |
| | 5. **Describe** evidence that supports evolution. | 7, 9, 10 | 6, 7, 8, 11, 14, 19, 21 | 4, 5, 6, 8, 12, 13, 19, 20, 23 | 10, 12, 28, 30, 34, 35, 38, 39, 40, 41, 42 |

# Evolution

## CHAPTER OVERVIEW

### Key Concepts

In this chapter, students will be introduced to natural selection through a series of events occurring between a predator and its prey. Mutations and species formation are covered. Primate and human evolution are discussed. The major statements by Darwin and the evidence supporting evolution are examined, including fossil, embryological, biochemical, and vestigial organ evidence.

### Key Science Words

competition
evolution
extinct
fertile
fossil
natural selection
new-world monkey
old-world monkey
primate
sedimentary rock
species
variation
vestigial structure

### Skill Development

In Lab 29-1, students will **form a hypothesis, experiment, interpret data, infer,** and **calculate** while studying color as an adaptation. In Lab 29-2, they will **formulate a model, observe,** and **infer** while making a fossil print. In the Skill Check on page 609, students will **define words in context** relating to natural selection. In the Skill Check on page 613, they will **understand** the **science word** *primate.* In the Skill Check on page 619, they will **interpret diagrams** of rock layers. In the Mini Lab on page 616, students will **make a bar graph** relating to variation in plants. In the Mini Lab on page 617, they will **infer** how petrified wood is formed.

### Bridging

In Chapter 28, students learned that all living organisms are composed of cells and that these cells contain the chemical DNA. This chapter will explain why life-forms are so similar in organization and in chemical composition.



**CHAPTER PREVIEW**

### Chapter Content

Review this outline for Chapter 29 before you read the chapter.



### Skills in this Chapter

The skills that you will use in this chapter are listed below.

- In **Lab 29-1,** you will form hypotheses, experiment, interpret data, infer, and calculate. In **Lab 29-2,** you will formulate a model, observe, and infer.
- In the **Skill Checks,** you will understand science words, define words in context, and interpret diagrams.
- In the **Mini Labs,** you will infer and make a bar graph.

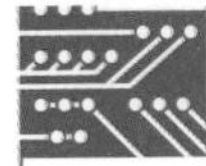

## TECH PREP

For Tech Prep activities in this chapter of the Teacher Wraparound Edition, see especially the Motivation and Portfolio on page 607, the Guided Practice on page 610, the Activity on page 612, the Connection on page 613, and the Options Enrichment on page 616.

See also the Glencoe Homepage at **www.glencoe.com**

Chapter 29

# Evolution

There are millions of different kinds of living things on Earth. Each and every kind of living thing is well suited to where it lives. For example, there are many different kinds of cacti. The smaller photograph below shows a close-up view of a cactus plant. Cactus plants have spines instead of leaves. The spines are an adaptation that helps to reduce water loss. Cactus plants are well suited for growing in areas where there is little water available. How do scientists explain the great variety of living things on the face of Earth? How do they explain the fact that living things are well suited to where they live? This chapter will answer these questions.

**Try This!**

**What Does Opposable Mean?** Tape your thumbs to the insides of your palms. Try to write, button a shirt, pick up a piece of paper, and turn the pages of this book. Most primates have an opposable thumb. What does opposable mean? How helpful is this trait?

**interNET CONNECTION**

For more information about the material in this chapter, follow the link for the chapter on the Glencoe Homepage at **http://www.glencoe.com**



**ASSESSMENT PLANNER**

**Portfolio**

Strategies on the following pages represent student products that can be placed into a best-work portfolio: pp. 607, 613, 616.

**PERFORMANCE ASSESSMENT**

Skill Check, pp. 609, 613
Mini Lab, pp. 616, 617
Lab, pp. 608, 618

**CONTENT ASSESSMENT**

Check for Understanding, pp. 612, 614, 621
Chapter Review, pp. 624-625

**GROUP ASSESSMENT**

Opportunities for group assessment occur with Cooperative Learning Strategies.

**Student Journal**

Strategies on the following pages represent opportunities for writing in a Student Journal: p. 620.

**GETTING STARTED**

## Using the Photos

Discuss other ways that the cactus is adapted to a desert environment. The cells of the plant hold enormous volumes of water. The spines themselves help to reduce water loss through transpiration. Cacti are thick and fleshy, which helps them withstand the extremely dry conditions. Desert plants are widely spaced. This helps to reduce competition for water and nutrients. Have students try to think of adaptations of other desert organisms. Lizards have scaly skin for protection against drying out. Many desert animals are active at night when it is cooler. Many desert animals have larger extremities for getting rid of excess body heat.

## MOTIVATION/Try This!

**What does opposable mean?** Students will observe the difficulty of being able to perform simple tasks when unable to use the thumb. An opposable thumb allows for contact between the thumb and other fingers. Without this trait, primates would not be able to develop such skills as tool making and use, tree climbing, etc.

## Chapter Preview

Have students study the chapter outline before they begin to read the chapter. They should note the overall organization of the chapter, that the first topic describes the general changes in living things that occur over time. This is followed by a section that deals with the evidence that supports evolution.

## Misconception

The following statement is often heard from students, "Humans evolved from apes." Advise students that evolutionary evidence supports the theory that humans and apes evolved from a common early ancestor.

## 29:1 Changes in Living Things

### PREPARATION

#### Materials Needed

Make copies of the Focus, Application, Critical Thinking/Problem Solving, Enrichment, Reteaching, Study Guide, and Lab worksheets in the *Teacher Resource Package*.

▶ Gather black and white paper and prepare paper dots for Lab 29-1.

#### Key Science Words

natural selection
species
fertile
old-world monkey
primate
new-world monkey

#### Process Skills

In Lab 29-1, students will form a hypothesis, experiment, interpret data, infer, and calculate. In the Skill Checks, they will define words in context and understand science words.

### 1 FOCUS

▶ The objectives are listed on the student page. Remind students to preview these objectives as a guide to this numbered section.

#### MOTIVATION/Display

Gather pictures from nature magazines and display them for the class. In particular, use any animal that blends well with its surroundings or mimics objects in its environment. Ask students how each animal benefits from the coloration.

**Focus**/*Teacher Resource Package*, p. 57. Use the Focus transparency shown at the bottom of this page as an introduction to natural selection.



**Objectives**

1. **Give examples** of how adaptations help organisms survive.
2. **Explain** how changes in life-forms occur.
3. **Describe** the classification and evolution of primates and humans.

**Key Science Words**

natural selection
species
fertile
primate
new-world monkey
old-world monkey

## 29:1 Changes in Living Things

Living things are well suited to where they live. What exactly does this statement mean?

### Adaptations

All the living things in the world have certain kinds of adaptations. An adaptation is a trait that makes a living thing able to survive in its surroundings. You first read about adaptations in Chapter 2.

Look at the foot of the bird shown in Figure 29-1a. The bird spends a great deal of its time perched in the long stalks of reeds. Note the bird's long toes. The long toes curl around small shoots and help the bird remain perched in the reeds. The bird shown in Figure 29-1b spends a lot of time in water. How does the webbing between the toes help this bird swim?

### An Example of Survival

How do traits help organisms survive in their environments? What is the outcome when an organism with a certain trait is able to survive? The following example will answer these questions.

**Figure 29-1** Many birds have feet adapted to perching on tree branches (a). Geese have feet adapted to swimming (b).

a

b

**OPTIONS**

**Enrichment:** Students can identify with the word *camouflage*. Ask if the term fits the idea of adaptation.

FOCUS 57

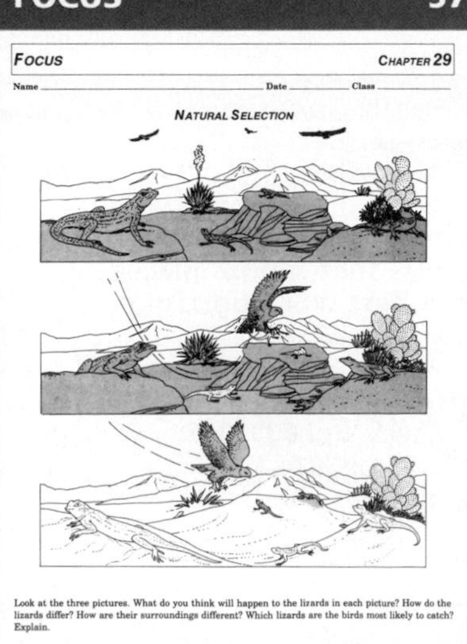

*Focus* — *Chapter* 29

Name ____ Date ____ Class ____

*Natural Selection*

Look at the three pictures. What do you think will happen to the lizards in each picture? How do the lizards differ? How are their surroundings different? Which lizards are the birds most likely to catch? Explain.

A group of mice lives in an area that has dark soil. Owls that eat mice also live in this area. Because dark mice blend well with the dark soil, owls cannot see them easily. Thus, the dark mice are better protected because they blend with the soil color, Figure 29-2. Their color is an adaptation, a trait that helps them survive.

**Figure 29-2** Owls can easily see light-colored mice on dark soil.

Dark mice do not always have offspring that are also dark in color. Every now and then they have light-colored offspring. The light-colored offspring, in turn, have other light-colored mice. Light-colored mice are easy to spot against the dark soil. As a result of their color, the light-colored mice are usually the first to be eaten by the owls. Light-colored mice living on dark soil are poorly adapted to their surroundings. Few light mice survive. As a result, few light mice reproduce. The number of light-colored mice tends to remain low. The dark mice, however, survive and reproduce. The dark mice will continue to outnumber the light mice.

**How do adaptations help living things survive?**

Suppose chemical changes take place in the soil and cause it to change color. The soil becomes lighter. Now owls can spot the dark mice on the ground more easily than they can spot the light mice. As a result, owls eat more dark mice than light mice.

Dark mice are no longer adapted to these surroundings. However, light mice are adapted to the light-colored soil. Dark color has become an unfavorable trait. The balance between the two mouse types begins to change. More light mice survive and reproduce. Any dark mice that are born are more likely to be eaten. Few of them get a chance to survive and reproduce.



## 2 TEACH

### MOTIVATION/Brainstorming

Ask students for a list of animals that blend with their environments. Some examples would be snowshoe hare, tiger, walking stick, and chameleon. All of these animals have protective coloration.

### Guided Practice

Have students write down their answers to the margin question: they make a living thing better suited to its environment.

### Independent Practice

**Study Guide**/*Teacher Resource Package*, p. 169. Use the Study Guide worksheet shown at the bottom of this page for independent practice.

### Portfolio

Give students the following information. You are a scientist who studies rabbit populations within a forest community. You observe a population of rabbits that feed in the forest and in turn are fed upon by wolves. It is summertime and you notice that the population of rabbits is mainly white (albino) with a few dark brown rabbits. You revisit the same site ten years later and observe dark brown rabbits but no white ones. Prepare a report of your findings and explain what has occurred within the past ten years to bring about this change in rabbit coloration.

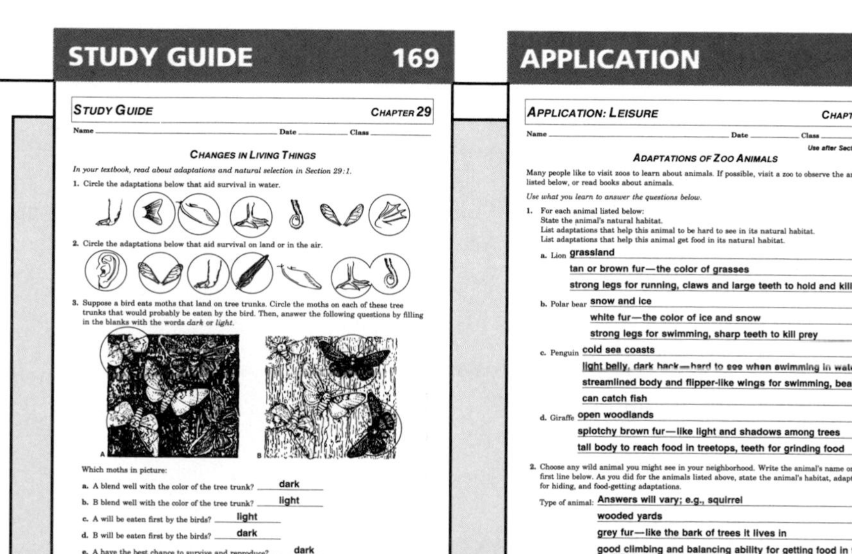

**STUDY GUIDE 169**

STUDY GUIDE — CHAPTER 29

Name ______ Date ______ Class ______

**CHANGES IN LIVING THINGS**

*In your textbook, read about adaptations and natural selection in Section 29:1.*

1. Circle the adaptations below that aid survival in water.
2. Circle the adaptations below that aid survival on land or in the air.
3. Suppose a bird eats moths that land on tree trunks. Circle the moths on each of these tree trunks that would probably be eaten by the bird. Then, answer the following questions by filling in the blanks with the words *dark* or *light*.

A B

Which moths in picture:

a. A blend well with the color of the tree trunk? dark
b. B blend well with the color of the tree trunk? light
c. A will be eaten first by the birds? light
d. B will be eaten first by the birds? dark
e. A have the best chance to survive and reproduce? dark
f. B have the best chance to survive and reproduce? light

**APPLICATION 29**

APPLICATION: LEISURE — CHAPTER 29

Name ______ Date ______ Class ______

*Use after Section 29:1.*

**ADAPTATIONS OF ZOO ANIMALS**

Many people like to visit zoos to learn about animals. If possible, visit a zoo to observe the animals listed below, or read books about animals.

*Use what you learn to answer the questions below.*

1. For each animal listed below:
   State the animal's natural habitat.
   List adaptations that help this animal to be hard to see in its natural habitat.
   List adaptations that help this animal get food in its natural habitat.
   a. Lion grassland
      tan or brown fur—the color of grasses
      strong legs for running, claws and large teeth to hold and kill prey
   b. Polar bear snow and ice
      white fur—the color of ice and snow
      strong legs for swimming, sharp teeth to kill prey
   c. Penguin cold sea coasts
      light belly, dark back—hard to see when swimming in water
      streamlined body and flipper-like wings for swimming, beak that can catch fish
   d. Giraffe open woodlands
      splotchy brown fur—like light and shadows among trees
      tall body to reach food in treetops, teeth for grinding food
2. Choose any wild animal you might see in your neighborhood. Write the animal's name on the first line below. As you did for the animals listed above, state the animal's habitat, adaptations for hiding, and food-getting adaptations.
   Type of animal: Answers will vary; e.g., squirrel
   wooded yards
   grey fur—like the bark of trees it lives in
   good climbing and balancing ability for getting food in trees, gnawing teeth

### OPTIONS

**Application**/*Teacher Resource Package*, p. 29. Use the Application worksheet shown here to give students further practice with adaptations.

# Lab 29-1 Adaptations

### Overview

Through the use of modeling, students will learn that an organism can be protected from its prey by matching its background.

**Objectives:** Upon completion of this lab, students will be able to: (1) **demonstrate** that white dots on a white background are difficult to see, (2) **demonstrate** that black dots on a black background are difficult to see, (3) **conclude** that an organism matching its background can be an adaptation that helps in survival.

**Time Allotment:** 1 class period

### Preparation

▶ Use a paper punch to make the dots prior to class. Store the dots in envelopes.

**Lab 29-1 worksheet**/*Teacher Resource Package*, pp. 113-114.

### Teaching the Lab

▶ Students should spread the dots out evenly but randomly on the paper backgrounds.

**Cooperative Learning:** Divide the class into groups of three students. For more information, see pp. 22T-23T in the Teacher Guide.

**Skill:** Provide students with the following information. Lab 29-1 was performed again outdoors on grass by a group of students. This time, however, the dots used were larger in size and some were green and others were red. The data collected is shown in the following table.

| | Dot X | Dot Y |
|---|---|---|
| Trial | | |
| 1 | 20 | 4 |
| 2 | 18 | 2 |
| 3 | 18 | 1 |
| 4 | 22 | 3 |

Which color was dot X? Dot Y? Support your answer with the data.

# Adaptations

Lab 29–1

## Problem: How is color an adaptation?

### Skills

form hypotheses, experiment, interpret data, infer, calculate

### Materials

white paper
black paper
30 white dots
30 black dots
damp paper towel

### Procedure

1. Copy the data table. It should include space for 10 trials.
2. Spread 30 black and 30 white dots out on a page of white paper.
3. Write down a **hypothesis** as to whether you will be able to pick up more white dots or more black dots in 10 trials.
4. Moisten all your fingers with a damp paper towel. Quickly touch one dot with each finger. The dots should stick to your fingers. Do this as quickly as you can.
5. Count the number of black dots and white dots you picked up.
6. Record your totals as Trial 1.
7. Return all the dots to the white paper.
8. Repeat steps 4 through 7 nine more times.
9. Switch to the black background paper. Spread the dots on the black paper.
10. Write down a new **hypothesis.** Repeat steps 4 through 7 for ten trials.
11. Total each column in your table.

### Data and Observations

1. Which color dots, black or white, were picked up more often on the white background?
2. Which color dots were picked up more often on the black background?

### Analyze and Apply

1. **Check your hypotheses:** Are your hypotheses supported by your data? Why or why not?
2. **Infer:** What was the adaptation in this activity?
3. **Apply:** In nature, how does color help as an adaptation?

### Extension

**Calculate:** Put the data for the class on the board. Calculate the averages for each category.

| | WHITE BACKGROUND | | BLACK BACKGROUND | |
|---|---|---|---|---|
| TRIAL | BLACK DOTS | WHITE DOTS | BLACK DOTS | WHITE DOTS |
| 1 | 7 | 3 | 5 | 5 |
| 2 | 6 | 4 | 3 | 7 |
| 3 | 8 | 2 | 4 | 6 |
| | | | | |
| 9 | 5 | 5 | 3 | 7 |
| 10 | 6 | 4 | 2 | 8 |
| Total | 70 | 30 | 42 | 58 |



## ANSWERS

### Data and Observations

1. black
2. white

### Analyze and Apply

1. Students who hypothesized that they would be able to pick up more dots of the opposite color will say their hypotheses were supported.
2. The adaptation was body color (black or white).
3. Color may help an organism blend in with its surroundings and avoid being captured and eaten.

## Natural Selection

We can now ask an important question using the mouse and owl story. What determined which mouse was better adapted to its surroundings? The owls determined, by eating certain mice, which color was an adaptation for survival. Only mice that are not eaten survive to reproduce. On the dark soil, more dark mice survived because owls did not see them. On the light soil, more light mice escaped the owls. **Natural selection** is the process in which something in a living thing's surroundings determines if it will or will not survive to have offspring. In natural selection, something in nature does the selecting. In our example, the owls did the selecting. When the soil became lighter, the group of mice changed from mostly dark to mostly light. The change was the result of natural selection.

Living things that are suited to their surroundings survive. They will be the ones most likely to reproduce. Their traits will be passed on to their offspring. Living things that are not suited to their surroundings won't survive. They won't reproduce, they won't have offspring, and their traits won't be passed on.

Why wouldn't a bright red frog survive in a muddy pond? Can you explain why more mud-colored frogs are likely to survive in muddy ponds than red frogs? How can web-toed frogs survive in water better than frogs with toe pads?

### Skill Check

**Define words in context:** Read this statement: "The dinosaurs became extinct due to natural selection." What does natural selection mean as used in this sentence? *For more help, refer to the **Skill Handbook**, pages 706-711.*

## Mutations

Adaptations are traits that help living things survive in their environments. Recall from Chapter 26 that traits are controlled by genes. Thus, adaptations are controlled by genes.

What is a source for new traits that help living things survive? Many new traits come from mutations. Remember, a mutation is a change in the DNA code. Mutations may supply living things with sources of new traits. Thus, they may supply new adaptations.

Are all mutations helpful for survival? No, some are harmful. For example, a mutation causes a change in the gene that controls fur color. Inheriting this gene caused the deer in Figure 29-3 to have white fur. The deer no longer blends in with its surroundings. It can be seen more easily by its enemies and may be eaten.

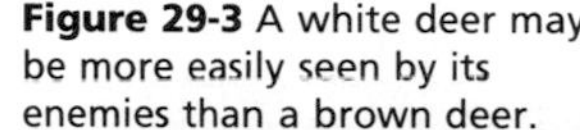

**Figure 29-3** A white deer may be more easily seen by its enemies than a brown deer.



## TEACH

### Videocassette

Show the videocassette *Natural Selection*, Films for the Humanities and Sciences.

### Bridging

Propose these questions to your class: How do you explain a dark-colored mouse producing offspring that are light-colored? Can the dark-colored mouse control the type of coloration in its offspring? These two questions will begin to set the stage for ideas regarding natural selection. An answer to the first question is that genes control mouse coloration. Thus, a certain gene combination in the offspring can result in light color. *Genetic variation* is a term you may want to introduce. The second question will help students realize that living things cannot control the genetic traits that their offspring will inherit. The sorting out of traits is a random event occurring at meiosis.

### ACTIVITY/Brainstorming

Ask students: If few light-colored mice survive, will the genes for this trait be as plentiful in the population as the genes for dark-colored fur?

### Bridging

Ask students to review Section 28:2, which deals with mutations from a genetic perspective. Point out that in this section, mutations are being examined as potential sources of new adaptations.

### CRITICAL THINKING 29

CRITICAL THINKING/PROBLEM SOLVING — CHAPTER 29

Name ______ Date ______ Class ______

Use after Section 29:1.

**HOW COULD HOLLOW HIP BONES HELP SUPERSAURUS?**

*Supersaurus* was one of the largest plant eating dinosaurs that ever lived. It was about 12 meters (40 feet) tall, weighed almost 30 tons and was 39 meters (130 feet) long from head to tail. It lived 135 million years ago. Scientists from Brigham Young University have uncovered the pelvis and vertebrae of the giant dinosaur. Its pelvis is almost 2 meters (six feet) long and about 1⅓ meters (four feet) wide. It was the largest *Supersaurus* bone ever found. And the dinosaur was not even full grown! While preparing and assembling the bones, scientists found that the dinosaur's hip bone is hollow. Some smaller dinosaurs also had hollow bones. How might hollow bones be helpful to dinosaurs?

**Analyzing the Problem**

1. Some scientists are not convinced that the hip bone was naturally hollow. How could the inside of the hip bone have become hollow before the bone was uncovered? **Perhaps the center decayed or was eaten by decomposers.**
2. When architects design support beams for skyscrapers, they use hollow beams because they are light and more flexible. The load they carry is distributed on or nearer the outer edges. How could this help you understand the *Supersaurus* bones? **Hollow bones might support the animal's weight better than solid ones of the same size.**

**Solving the Problem**

1. How do you think hollow bones might have helped *Supersaurus* and other dinosaurs? **Hollow bones would weigh less than solid bones. This would reduce body weight and the dinosaur would need less food.**

### OPTIONS

**How Do Some Living Things Vary?**/*Lab Manual*, pp. 243-246. Use this lab as an extension to variation in living things.

**Critical Thinking/Problem Solving/** *Teacher Resource Package*, p. 29. Use the Critical Thinking/Problem Solving worksheet shown here to extend the concept of adaptations.

### GLENCOE TECHNOLOGY

**Videotape**

**The Secret of Life**

*It's in the Genes: Evolution*

## TEACH

### Concept Development

▶ Mutations are sometimes said to be the "raw material" of adaptations. Ask students if they can explain this idea.

### ACTIVITY/Brainstorming

What kinds of adaptations are seen in animals such as penguins, seals, or whales? Students may describe the long, slim body as an adaptation, enabling these animals to swim easily through the water. They may also describe the adaptation of front legs modified into finlike parts to help with swimming.

### MOTIVATION/Display

Prepare a large diagram of the following fictitious animal. Ask students to describe the adaptations shown that help the animal with (a) food capturing (b) swimming (c) protection from predators. You may wish to point out that the animal feeds on fish and that its underside is gray while its top side is a deep blue except for the green sprigs on its head and tail.

### Guided Practice

Ask students to design an animal that is adapted to blending in with plant branches. They may either: (a) list the desirable traits that would aid in the animal's survival, or (b) make a drawing of the animal and plant. Supply the following hints if necessary: size and color of the animal, location of plant branch, leaf size, and leaf shape.

### Independent Practice

Study Guide/*Teacher Resource Package*, p. 170. Use the Study Guide worksheet shown at the bottom of this page for independent practice.

**Bio Tip**

**Consumer:** Wearing a lead apron when you have dental X rays helps prevent mutations of your reproductive cells.

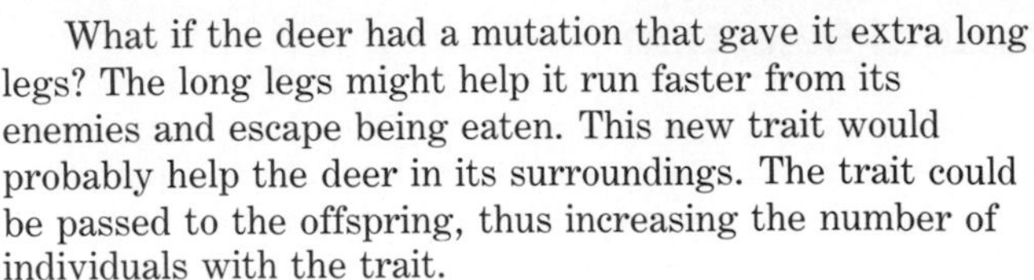

What if the deer had a mutation that gave it extra long legs? The long legs might help it run faster from its enemies and escape being eaten. This new trait would probably help the deer in its surroundings. The trait could be passed to the offspring, thus increasing the number of individuals with the trait.

Mutations are natural events. Mutations appear in every living thing. These changes in genes may be helpful, harmful, or have no effect at all.

## Species Formation

Chapter 3 described a species as the smallest group of living things in classification. There is another way to describe a species. A **species** is a group of living things that can breed with others of the same species and form fertile offspring. **Fertile** means being able to reproduce by forming egg or sperm cells. For example, the plants shown in Figure 29-4c are very similar in appearance. Yet, they can't produce fertile offspring if they breed with one another. Another example is shown in Figures 29-4a and 4b. These macaws look very much alike. They belong to different species, however. If they breed with each other, they won't form fertile offspring.

**Figure 29-4** The macaws (a,b) are different species, as are the plants (c).



**STUDY GUIDE 170**

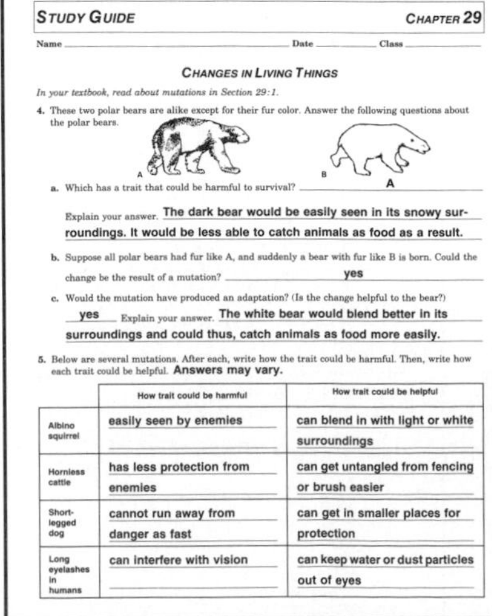

STUDY GUIDE CHAPTER 29

Name ____ Date ____ Class ____

CHANGES IN LIVING THINGS

*In your textbook, read about mutations in Section 29:1.*

4. These two polar bears are alike except for their fur color. Answer the following questions about the polar bears.

A B

a. Which has a trait that could be harmful to survival? **A**

Explain your answer. **The dark bear would be easily seen in its snowy surroundings. It would be less able to catch animals as food as a result.**

b. Suppose all polar bears had fur like A, and suddenly a bear with fur like B is born. Could the change be the result of a mutation? **yes**

c. Would the mutation have produced an adaptation? (Is the change helpful to the bear?) **yes** Explain your answer. **The white bear would blend better in its surroundings and could thus, catch animals as food more easily.**

5. Below are several mutations. After each, write how the trait could be harmful. Then, write how each trait could be helpful. **Answers may vary.**

| | How trait could be harmful | How trait could be helpful |
|---|---|---|
| Albino squirrel | easily seen by enemies | can blend in with light or white surroundings |
| Hornless cattle | has less protection from enemies | can get untangled from fencing or brush easier |
| Short-legged dog | cannot run away from danger as fast | can get in smaller places for protection |
| Long eyelashes in humans | can interfere with vision | can keep water or dust particles out of eyes |

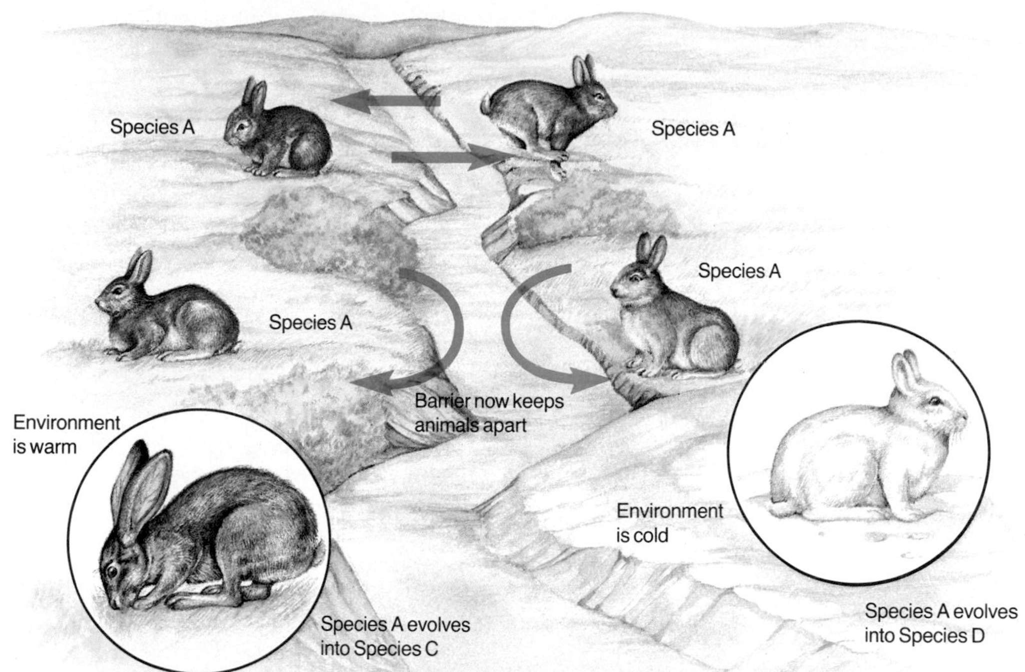


**Figure 29-5** A barrier can lead to the formation of new species.

Now we have a problem. Only members of the same species can breed and form offspring. Yet, new species are constantly appearing on Earth. How can new species appear? Follow the events shown in Figures 29-5 to see how this is possible. Let's start with a group of animals living on either side of a shallow stream. The animals at the top of Figure 29-5 are all of the same species. They can wade across the stream easily, but they are unable to swim. Due to a flood, the stream becomes a very wide river. It remains that way. The animals become separated into two groups. Being unable to swim, they are unable to cross the river. They continue to live apart for thousands of years. Note that the living conditions on each side of the river are different. During the time of separation, natural selection has taken place in each group. In each group, individuals lacking traits favorable for the new environment have died. Individuals with the favorable traits have survived and reproduced. The two groups gradually become different because their environments are different. In time, each group may become a different species.

**How does the formation of a barrier lead to new species?**



## TEACH

### Filmstrip

Show the filmstrip *Evolution III: Speciation,* Carolina Biological Supply Co.

### Guided Practice

Have students write down their answers to the margin question: a barrier separates organisms. If conditions on either side of the barrier are different, new species may evolve.

### GLENCOE TECHNOLOGY

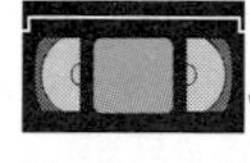

**Videotape**

**The Secret of Life**
*Tinkering with Our Genes: Genetic Medicine*

**Videodisc**

**Biology: The Dynamics of Life**
Animation: *Geographic Isolation*
Disc 1, Side 2, Ch. 6

Movie: *The Galapagos*
Disc 1, Side 2, Ch. 4

## OPTIONS

**Challenge:** Locate information on the Abert and Kaibab squirrel species. They live in different regions of the Grand Canyon. Report on: (a) how they are alike (b) how they differ (c) why they might have originally been the same species (d) how the canyon separated them into different groups (e) why they evolved into different species once separated by a canyon barrier.

### Science Background

The forming of new species via a physical barrier is known as allopatric speciation. Barriers can be formed by volcanic eruptions, glaciers, or new mountain ranges. The term *allopatric* means other (allo) homeland (patric).

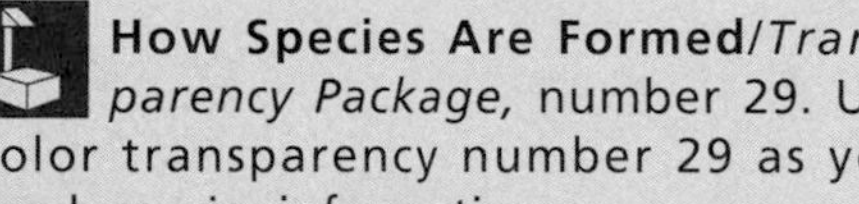

**How Species Are Formed**/*Transparency Package,* number 29. Use color transparency number 29 as you teach species information.

## TEACH

### Check for Understanding

Provide students with the following scenario: Lions will chase their prey at very high speeds. Their prey can escape only if they are faster than the lions. Which of the prey will survive, those with stronger muscles for running or those with weaker muscles? Lions will also give up the chase after a few minutes of pursuing their prey. Which prey will survive, those with short endurance or those with long endurance? Which traits will be passed to future offspring of the prey?

### Reteach

Ask certain students to explain their answers to the questions asked in Check for Understanding. Discuss students' answers and make any corrections or comments.

### ACTIVITY/Field Trip

Visit the produce section of a large grocery store. Note the number of "new" plants available for human use that you did not see several years ago. Explain: (a) why these plants may not have been available a few years ago and what has allowed them to appear now, and (b) what the role of humans has been in making these new plants available.

### Independent Practice

**Study Guide**/*Teacher Resource Package*, p. 171. Use the Study Guide worksheet shown at the bottom of this page for independent practice.

**Figure 29-6** The finches on the Galapagos Islands are adapted to eating different kinds of food.

In the example just given, three events led to the development of two new species. First, a barrier formed that separated members of a species. The barrier could have been a river, ocean, new mountain, glacier, or a lava flow. Second, the animals found themselves living in different environments. Third, the groups began to show different traits as a result of natural selection. The two groups in time became two different species. As a result, they would not be able to breed and form fertile offspring if brought back together.

The finches that live on the Galapagos Islands are a well-known example of the forming of new species. On the Galapagos are several species of finch. Each species has a different beak shape. Some finches have thick beaks and are adapted to eating seeds. Some have small beaks and are adapted to eating insets. How did the finches come to live on the islands? The ancestor of the Galapagos finches probably flew to the islands from the mainland of South America. New species began to evolve when the finches spread out over the islands. The different groups of finches didn't come into contact with one another for a long time. Over time, the different groups became adapted to their new environments. They also became less like one another. A single finch ancestor had evolved into many different species.



## OPTIONS

**Challenge:** The Galapagos Islands have a number of different finch species. Have students research the mechanism that led to the formation of these different bird species.

## STUDY GUIDE 171

STUDY GUIDE — CHAPTER 29

Name ____ Date ____ Class ____

### CHANGES IN LIVING THINGS

*In your textbook, read about species formation in Section 29:1.*

6. a. How can the word *species* be defined as it relates to the classification of living things? **A species is the smallest group of living things.**

   b. How can the word *species* be defined as it relates to breeding? **A species is a group of living things that can breed with others of the same species and form fertile offspring.**

7. Use Figure 29-5 on page 611 of your textbook to help you do this exercise. The following statements involve events that can result in the formation of a new species. Number them from 1-6 in the proper order. Number one is done for you.

   **3** a. animals are separated into two groups and must now live apart

   **4** b. environments on either side of the river change with time

   1 c. animals of the same species living on either side of a stream can move from one side to the other

   **2** d. the stream changes into a river

   **5** e. animals, due to natural selection, undergo change

   **6** f. after thousands of years, two different species of animals form

8. In the example in Figure 29-5, what three events led to the formation of new species? **A barrier formed that separated members of the species. Members of the species found themselves living in different environments. The groups began to show different traits as a result of natural selection.**

9. How do the finches of the Galapagos Islands differ? **Each has a differently shaped beak.**

10. Describe how a single finch ancestor probably evolved into many different species. **The ancestor probably flew to the islands from the mainland of South America. New species began to evolve when the finches spread out over the islands and became adapted to their new environments. Without contact with one another, the groups became less alike.**

## Primate and Human Evolution

Fossils are very important when it comes to tracing the evolution of humans. They provide us with evidence of past life. Before looking at human evolution, a brief review of human classification will help. Some of the categories for humans are as follows: Phylum Chordate, Class Mammal, Order Primate.

The primate order is the one in which monkeys, apes, and humans are classified. **Primates** are mammals with eyes that face forward, a well-developed cerebrum, and thumbs that can be used for grasping. About 45 million years ago, primates evolved into two main groups, new-world monkeys and a second group. **New-world monkeys** have a tail that can grasp like a hand and nostrils that open upward. Howler and spider monkeys are examples of new-world monkeys. The second group that evolved is the ancestor of the old-world monkeys. It is also the ancestor of the group that evolved into apes and humanlike life-forms. **Old-world monkeys** can't grasp with their tails, if they have one, and their nostrils open downward. Baboons are old-world monkeys. Apes don't have tails and include gorillas and chimpanzees. Figure 29-7 shows the old-world monkeys, new-world monkeys, apes, and how they are related.

### Skill Check

**Understand science words: primate.** The word *primate* comes from the word *prime* meaning first. In your dictionary, find three words with the word *prime* or word part *prima* in them. *For more help, refer to the **Skill Handbook**, pages 706-711.*

**Figure 29-7** Primate groups include new-world monkeys, old-world monkeys, and apes.

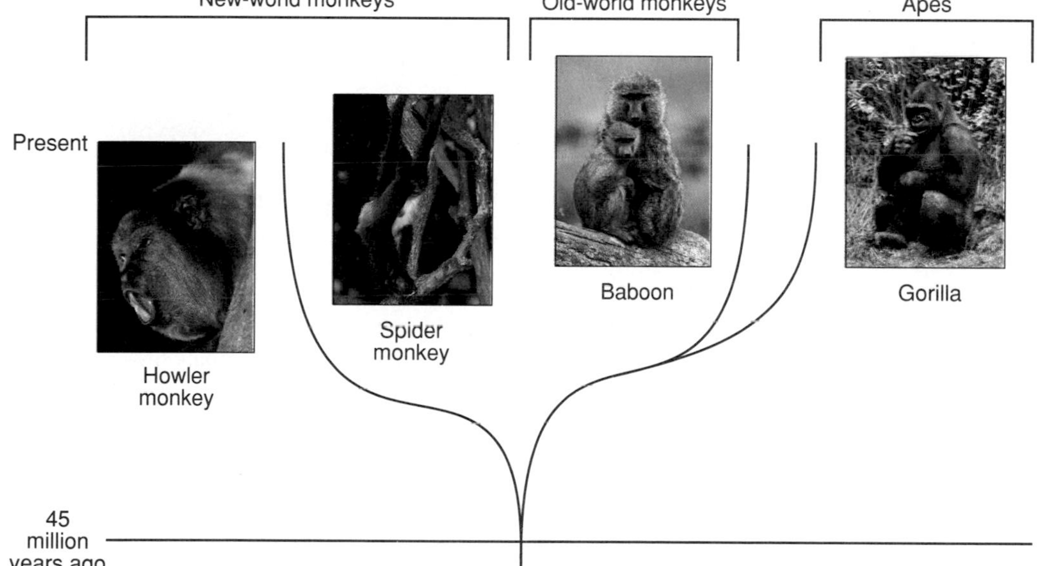




## TEACH

### Concept Development

▶ Lesser known primates include lemurs, tarsiers, marmosets, and galagos (prosimians). More common are monkeys, apes, and humans (anthropoids).

### Videocassette

Show the videocassette *Retracing Man's Steps,* Films for the Humanities and Sciences.

### Bridging

Review the characteristics of each classification category with your students. Phylum chordate–having a chord along the dorsal side sometime during development, dorsal hollow nerve cord, gill slits. Class mammal–body usually covered with hair, constant body temperature, diaphragm, females with mammary glands that secrete milk to nourish young. Order primates–five digits on hand and foot, opposable thumb, usually one young at birth, eyes face forward and are surrounded and protected by bone.

### Connection: English

Ask students to read the play *Inherit the Wind* or rent the video version of the film. Have them report their impressions to the class.

### *Portfolio*

Have students prepare an idea map that compares Old and New World monkeys. They should begin the map with the word *primates.* Traits of primates must be included in the map along with traits and examples for both monkey groups.

### ENRICHMENT 31

ENRICHMENT CHAPTER 29

Name ______ Date ______ Class ______

Use after Section 29:1.

HUMAN EVOLUTION

Fossils indicate that over millions of years, the human species has evolved to its present form. Many humanlike fossils have been discovered.

*Study each of the skulls below. Use reference books to find out the appearance of each, where they were found, what kind of tools they used, and how long ago they lived. Then write this information in the blanks below each skull.*

Australopithecus: receding forehead; Africa; bone and pebble tools; 3.5 million years ago

Zinjanthropus: broad, flat thumb; Africa; simple stone tools; 3 million years ago

Java: heavy bones; massive forehead; Java; used simple stone tools; 1.6 million years ago

Peking: heavy brow ridges; receding chin and protruding jaw; China; tools and fire; 1.5 million years ago

Neandarthal: receding forehead; prominent brow ridges; large brain; Europe; many types of tools; 130 000 years ago

Cro-Magnon: high forehead; strong chin; like modern humans; France; many types of tools; 50 000 years ago

## *OPTIONS*

### ACTIVITY/Challenge

Ask students to use references to report on the appearance of early humans. Ask them to describe their traits, social structures, areas on Earth where they lived, advancements such as development of tools or use of fire. Diagrams would also be helpful.

**Enrichment/***Teacher Resource Package,* p. 31. Use the Enrichment worksheet shown here to extend knowledge of human evolution.

## GLENCOE TECHNOLOGY

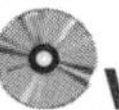

**Videodisc**

**Biology: The Dynamics of Life**
Movie: *Primate Characteristics*
Disc 1, Side 2, Ch. 7

## TEACH

### Guided Practice

Have students write down their answers to the margin question: humans have thumbs for grasping, eyes that face forward, and well-developed cerebrums.

### Check for Understanding

Have students respond to the first three questions in Check Your Understanding.

### Reteach

*Reteaching/Teacher Resource Package*, p. 85. Use this worksheet if students are having trouble understanding natural selection.

**Extension:** Assign Critical Thinking, Biology and Math, or some of the **OPTIONS** available with this lesson.

## 3 APPLY

### Software

*Evolution,* Queue, Inc.

## 4 CLOSE

### Videocassette

Show the videocassette *The Human Influence,* Films for the Humanities and Sciences.

### Answers to Check Your Understanding

1. Mice are protected from their enemies when they blend in with their surroundings.
2. If a barrier splits a population into two groups and the two groups have different environments, natural selection may eventually lead to the formation of two new species.
3. Both have eyes that face forward, a well developed cerebrum, and thumbs that can be used for grasping.
4. No. They can breed with other dog breeds to produce offspring.
5. The probability is about 1, or about 100 percent.

**Figure 29-8** Early humanlike forms include *Australopithecus* (ah stray loh PITH uh cus), *Homo habilis,* and Neanderthals. *Homo habilis* is called "handy man" because they used stone tools.

**How are humans primates?**

Humanlike ancestors first appeared around three million years ago. These ancestors walked upright, just as modern humans do. However, they were much shorter than modern humans. Several different humanlike groups evolved, but they all became extinct. Figure 29-8 shows these humanlike groups. Two groups of *Homo sapiens* have lived on Earth. *Homo sapiens* is the only human life-form alive today. One of the *Homo sapiens* groups was known as Neanderthal (nee AN dur thawl) man. Neanderthal man was shorter than modern humans and had thicker bones. Neanderthal man became extinct, and the second *Homo sapiens* group evolved into modern humans.

### Check Your Understanding

1. How does color of a mouse affect its survival?
2. How can two species develop from one species?
3. How are new-world monkeys and old-world monkeys alike?
4. **Critical Thinking:** People who breed dogs sometimes end up with breeds that are different from any others. Are these dog breeds new species?
5. **Biology and Math:** In humans, the rate of mutation is about 1 in 100 000. You have about 100 000 genes. What is the chance that you have a gene that has a mutation and is, therefore, different from any of the genes your mother or father have?



## OPTIONS

**Challenge:** Ask students to explain, on the basis of mutations and natural selection, how the brain size of humans has increased over time.

## RETEACHING 85

**RETEACHING: IDEA MAP** — **CHAPTER 29**

Name ______ Date ______ Class ______

*Use with Section 29:1.*

**HOW NATURAL SELECTION WORKS**

*Use the following terms to complete the idea map below: poorly, well, easy, hard, feed, do not feed, survive, die, blend, do not blend.*

*Start here and follow the arrows.*

Dark lizards live in an area with dark soil.

Light offspring are born.

Light offspring **do not blend** with surroundings.

Hawks **feed** on light offspring. Light offspring are **easy** to see.

Light lizards are **poorly** suited to surroundings.

Light lizards **die** and cannot reproduce.

Dark offspring are born.

Dark offspring **blend** with the surroundings.

Hawks **do not feed** on dark offspring. Dark offspring are **hard** to see.

Dark lizards are **well** suited to the surroundings.

Dark lizards **survive** and reproduce.

1. What is natural selection? **Natural selection is the process in which something in a living thing's environment determines if it will or will not survive.**
2. Which living things survive? **The living things that are best suited to their surroundings will survive.**
3. What effect do mutations have on survival? **Mutations may be helpful, harmful, or have no effect at all on survival.**

## 29:2 Explanations for Evolution

How can we explain that life-forms have changed with time? What is the evidence that life-forms had a common beginning?

**Objectives**

4. **Communicate** Darwin's main ideas.
5. **Describe** evidence that supports evolution.

**Key Science Words**

variation
competition
evolution
fossil
extinct
sedimentary rock
vestigial structure

### Darwin's Work

Much of what we have stated so far about adaptations and natural selection is not new. It was said over 100 years ago by Charles Darwin. When he was young, Darwin made a voyage around the world on a ship. During the trip, he observed many kinds of plants and animals and gathered examples of them. On some islands he collected living things not found anywhere else on Earth. Darwin saw that these living things were similar to life present in other parts of the world. For 20 years after his trip, Darwin studied the material he had collected. Finally, in 1859, he wrote a book explaining evolution (ev uh LEW shun) and his theory of natural selection.

Darwin made a number of important points in his book. We shall summarize several of his more important ideas.

1. **Living things overproduce.** More offspring are produced than survive. A single cottonwood tree forms thousands of seeds, Figure 29-9a. Frog eggs, shown in Figure 29-9b, are produced in the hundreds.

a

b

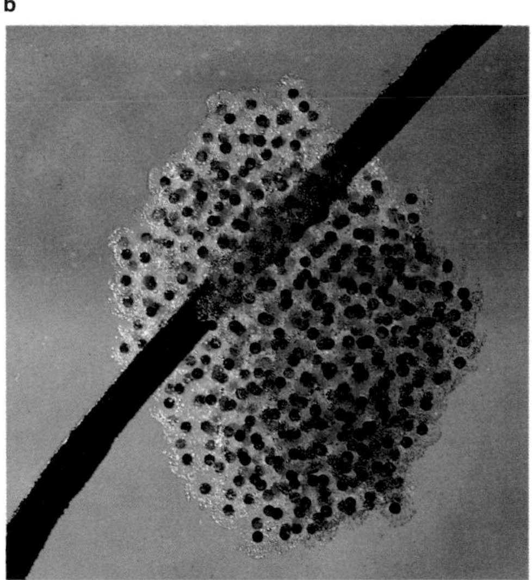

**Figure 29-9** Cottonwood trees (a) and frogs (b) are examples of living things that produce more offspring than can survive.



## 29:2 Explanations for Evolution

### PREPARATION

#### Materials Needed

Make copies of the Transparency Master, Skill, Reteaching, Study Guide, and Lab worksheets in the *Teacher Resource Package*.

▶ Gather pea pods, rulers, wood blocks, petrified wood samples, balances, and graduated cylinders for the Mini Labs.

▶ Purchase the powder needed to form fossil imprints and gather leaves needed for Lab 29-2.

#### Key Science Words

variation
competition
evolution
fossil
extinct
sedimentary rock
vestigial structure

#### Process Skills

In Lab 29-2, students will formulate a model, observe, and infer. In the Skill Check, they will interpret diagrams. In the Mini Labs, they will make a bar graph and infer.

### 1 FOCUS

▶ The objectives are listed on the student page. Remind students to preview these objectives as a guide to this numbered section.

#### MOTIVATION/Brainstorming

Ask students whether living things change because they want to or because they need to. Neither idea is correct. Living things change as a result of natural selection.

### OPTIONS

**Challenge:** Ask students how bacteria becoming resistant to certain antibiotics is an example of Darwin's principles. More bacteria are produced than can survive. There is variation among bacteria. Some are resistant to antibiotics. Some aren't. Bacteria compete for resources. Selection is always taking place. In this case, resistant bacteria survive. Nonresistant bacteria die. Over time, only resistant bacteria will be present in the population.

# 2 TEACH

## Convergent Question

Ask students to provide examples of variation in humans, dogs, or trees.

**Cooperative Learning:** Divide the class into 6 to 8 groups. Present the following: The giraffe today has a long neck that makes it well adapted for reaching its food. Have students explain how a change from short- to long-necked giraffes may have taken place over time using the four major themes of Darwin's work as a guide.

## Guided Practice

Have students write down their answers to the margin question: there aren't enough resources to go around, and living things must compete for them.

## Mini Lab

Help students with their bar graphs if necessary.

**Performance:** Provide students with ten needles from the same species of a conifer and a metric ruler. Ask them to demonstrate, using the materials provided, that variation exists among living things.

## Independent Practice

**Study Guide**/*Teacher Resource Package*, p. 172. Use the Study Guide worksheet shown at the bottom of this page for independent practice.

## Student Journal

Tell students to restate Darwin's four points described on pages 615 and 616. However, ask them to use the tiger moth shown in Figure 29-10 as an example for each point.

**Figure 29-10** Tiger moths show variation in their wing patterns.

**2. There is variation (ver ee AY shun) among the offspring.** A **variation** is a trait that makes an individual different from others of its species. Each living thing does not appear exactly like all the others. Some of the differences between individuals are inherited. Figure 29-10 shows variations in color and pattern of tiger moth wings.

**What does it mean that living things struggle to survive?**

**3. There is a struggle to survive.** There are more living things than there are resources to go around. This results in competition. **Competition** is the struggle among living things to get their needs for life. Young pine trees compete for light, water, and soil nutrients. Rabbits compete with other rabbits for food, shelter, and mates.

**4. Natural selection is always taking place.** Individuals that have less desirable traits are less fit. They reproduce fewer offspring. Individuals that have desirable traits are more fit. They reproduce more offspring. The organisms alive today are the ones that are better suited to their surroundings. The traits, or variations, that make them more fit are the ones they inherited and will, in turn, pass to their offspring.

Darwin realized that species of organisms are always changing. He knew that the changes in species do not occur quickly. Darwin's studies led him to form the theory of natural selection to explain evolution. **Evolution** is a change in the hereditary features of a group of organisms over time. When a species changes through time, it is said to have evolved.

### Mini Lab

**How Do Peas Vary?**

**Make a bar graph:** Find the average diameter of the peas in 10 pea pods. Make a bar graph of your data. How do your data relate to Darwin's statement about variation? *For more help, refer to the* ***Skill Handbook****, pages 715-717.*



## OPTIONS

**Enrichment:** Ask students to provide examples of how term *evolution* is used in a nonbiological sense. That is, how is the term used to describe other events of change? Students may suggest evolving hair styles, clothing trends, auto design, etc.

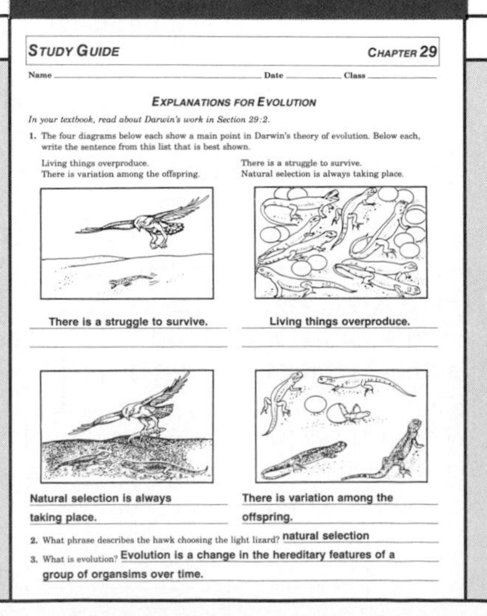

STUDY GUIDE 172

STUDY GUIDE CHAPTER 29

Name ______ Date ______ Class ______

EXPLANATIONS FOR EVOLUTION

*In your textbook, read about Darwin's work in Section 29:2.*

1. The four diagrams below each show a main point in Darwin's theory of evolution. Below each, write the sentence from this list that is best shown.

Living things overproduce.
There is variation among the offspring.
There is a struggle to survive.
Natural selection is always taking place.

There is a struggle to survive.

Living things overproduce.

Natural selection is always taking place.

There is variation among the offspring.

2. What phrase describes the hawk choosing the light lizard? natural selection

3. What is evolution? Evolution is a change in the hereditary features of a group of organsims over time.

## Fossil Evidence

What evidence supports evolution? Some evidence of evolution comes from fossils. **Fossils** are the remains of once-living things from ages past.

A fossil may be a print of a leaf. It may be a footprint of an animal. It could even be a skeleton. A fossil could be an animal trapped and frozen in ice. Or a fossil could be an insect trapped in hardened plant sap.

When living things from the past are compared to living things today, we can see that change has occurred. For example, Figure 29-11 shows the fossil remains of an extinct (ihk STINGT) animal. An **extinct** life-form is one that no longer exists.

Fossils are found in Earth's crust. They are present in sedimentary rocks. **Sedimentary rocks** form from layers of mud, sand, and other fine particles. The mud, sand, and fine particles are called sediments. These sediments form at the bottom of seas. Many animals and plants die and settle to the bottoms of oceans, lakes, or ponds with the sediments. These sediments change into rock over millions of years. Fossils form within these sediment layers. Fossils give us a record of what types of living things were on Earth in the past. Scientists can tell how old fossils are by dating them. Being able to date fossils gives scientists an idea of the history of life on Earth.

### Mini Lab

**How Do Wood and Fossil Wood Compare?**

**Infer:** Compare the density of a piece of wood with the density of a piece of fossil wood. Which has higher density? Why? *For more help, refer to the* ***Skill Handbook,*** *pages 706-711.*

**Figure 29-11** This wooly mammoth is extinct. What present-day animal does it resemble?



## TEACH

### Demonstration

To illustrate the formation of sedimentary rocks, do the following: Place several spoonfuls of loose soil into an empty mayonnaise jar. Add several spoonfuls of clay soil and several spoonfuls of small stones. Add enough water to fill the jar two-thirds full. Cover the jar and shake several times. Allow the soil, clay, and stones to settle. To further illustrate how a fossil is formed, place a clam or oyster shell into the same jar. Remix the contents and allow it to settle once again.

### Mini Lab

Purchase fossils from a biological supply house or from a local rock shop. Ask the industrial shop teacher to supply you with small cubes of wood that fit into a graduated cylinder or beaker. Procedure: (a) Design a suitable data table. (b) Mass the wood and fossil using a balance. (c) Determine fossil and wood volume by measuring displacement. (d) Review the formula for determining density (density = mass divided by volume).

### ASSESSMENT

**Performance:** Provide students with a balance, graduated cylinder, water, and a small piece of bone. Have them determine the density of the bone. Based on their calculations, ask them to predict the density for this same piece of bone if it were fossilized.

### ACTIVITY/Project

Ask students to prepare a time line that illustrates the appearance of different life-forms on Earth. Students may divide their charts into the different eras or periods. The information can be found in most college textbooks.

### TRANSPARENCY 147

TRANSPARENCY MASTER — CHAPTER 29

GEOLOGIC TIME SCALE

| Era | Period | Epoch | Age (years ago) | Representative life forms |
|---|---|---|---|---|
| Cenozoic | Quaternary | Recent | 100 000 | Humans; modern forms of plants and animals |
| | | Pleistocene | 1 000 000 | Extinction of many mammals; primitive humans; grasslands |
| | Tertiary | Pliocene | 10 000 000 | Early humans; other mammals; herbs |
| | | Miocene | 30 000 000 | Mammals; grasses |
| | | Oligocene | 40 000 000 | Primates and other mammals; forests common |
| | | Eocene | 60 000 000 | Primitive horse; other mammals; flowering plants |
| | | Paleocene | 75 000 000 | Mammals predominant; more modern flowering plants |
| Mesozoic | Cretaceous | | 135 000 000 | Extinction of giant reptiles; birds and insects; flowering plants |
| | Jurassic | | 165 000 000 | Dinosaurs dominant; primitive birds and mammals; earliest flowering plants |
| | Triassic | | 205 000 000 | Dinosaurs and other reptiles; early mammals; primitive seed plants |
| Paleozoic | Permian | | 230 000 000 | Rise of insects; early reptiles |
| | Carboniferous | | 280 000 000 | Insects and amphibians; mosses and ferns |
| | Devonian | | 325 000 000 | Age of fishes; early amphibians; early bryophytes; ferns |
| | Silurian | | 360 000 000 | Club mosses; insects and other invertebrates |
| | Ordovician | | 425 000 000 | Primitive mollusks and fish; algae |
| | Cambrian | | 500 000 000 | Protists; sponges, jellyfish; spore-producing plants |
| Precambrian | | | 4 500 000 000 | Monerans; simple protists; fungi; simple invertebrates |

## OPTIONS

### Science Background

The oldest known fossils are deposits of cyanobacteria in sedimentary rock. Their age is estimated to be between 3 and 3.5 billion years.

**How Do Fossils Show Change?**/*Lab Manual*, pp. 247-250. Use this lab as an extension to fossil evidence.

**Transparency Master**/*Teacher Resource Package*, p. 147. Use the Transparency Master shown here to explain the geologic time scale.

## Lab 29-2 Fossil Prints

### Overview

Students will simulate the formation of a fossil in sedimentary rock. A rapid-setting dental powder is used.

**Objectives:** Upon completion of this lab, students will be able to: (1) **define** what a fossil is, (2) **model** the steps involved in forming a fossil, (3) **identify** the steps involved in forming a real fossil.

**Time Allotment:** 30 minutes

### Preparation

▶ Plastic dishes may be cottage cheese containers.

▶ Order the plastic powder from: L.D. Caulk Co., Division of Dentsply, Milford, DE 19963. Request Type 2 Jeltrate or substitute. Another source is: Coe Labs, 3737 W. 127 St., Chicago, IL 60658. Request Type 2 alginate or substitute.

▶ The powder may also be ordered or purchased from your local dentist.

▶ Note: The exact amount of water to powder may vary with each brand used. Adjust accordingly.

**Lab 29-2 worksheet**/*Teacher Resource Package*, pp. 115-116.

### Teaching the Lab

▶ Students can remove the solidified powder by simply flexing the sides of the container. The powder should pop out.

▶ Review the markings on the graduated cylinder with students as part of the prelab.

**Cooperative Learning:** Divide the class into groups of two students. For more information, refer to the Teacher Guide, pp. 22T-23T.

### ASSESSMENT

**Performance:** Provide students with a piece of clay and a mollusk shell. Ask them to demonstrate the steps that occur in nature as a fossil imprint of the shell forms.

# Fossil Prints — Lab 29–2

## Problem: How is a fossil print made in rock?

### Skills

formulate a model, observe, infer

### Materials

plastic dish
100 mL graduated cylinder
leaf
warm water
plastic powder
metric ruler

### Procedure

1. Copy the data table.
2. List as many traits as you can for the actual leaf. You may want to measure certain leaf parts with a ruler.
3. Measure 60 mL of plastic powder with a graduated cylinder. Pour the powder into a small dish as in the figure.
4. Slowly add 36 mL of warm water to the powder. Use your hands to mix the water and powder until it is claylike. The claylike substance represents soft rock.
5. Flatten the soft rock so that it covers the dish bottom. Smooth out the top surface.
6. Place a leaf onto the soft rock's surface.
7. Gently press the leaf into the soft rock.
8. Wait five minutes. Then remove the leaf.
9. Allow the soft rock to harden for a few more minutes.
10. Examine the leaf "fossil" print. Compare it to the original leaf.
11. **Observe:** List as many traits as you can for the leaf "fossil" in your data table.
12. Clean and return your materials to their proper place.

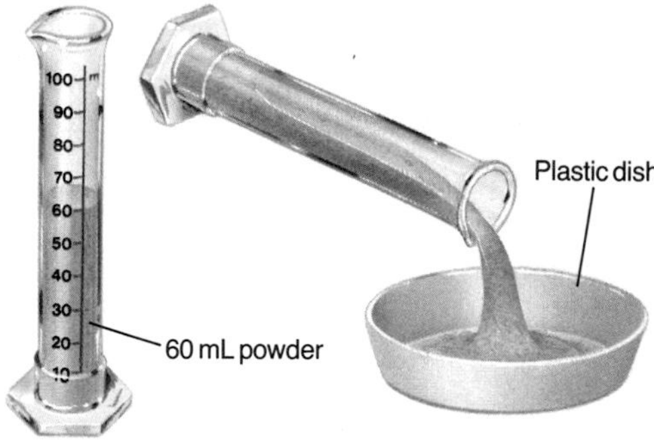




### Data and Observations

1. How are the actual leaf and leaf "fossil" alike? How do they differ?
2. Match the following steps with the numbered steps in the procedure.
   (a) A soft soil such as clay is present.
   (b) A leaf falls onto the clay surface.
   (c) The leaf gets pressed down into the clay.
   (d) The clay hardens. The leaf breaks down and disappears.
   (e) A print of the leaf remains on the hard rock.

| TRAITS | |
|---|---|
| Actual leaf | Leaf "fossil" |
| | |

### Analyze and Apply

1. How do fossils give information about past ages?
2. **Infer:** List several examples of what may become a fossil.
3. **Apply:** How long would steps 8 and 9 take if this were happening in nature?

### Extension

**Observe:** Examine pieces of marble for fossils. How did the fossils get into the marble?

## ANSWERS

### Data and Observations

1. margin of the leaf, vein pattern; different color, thickness, and more detail in the live organism
2. (a) 4 (b) 6 (c) 7 (d) 8, 9 (e) 10

### Analyze and Apply

1. They show the imprint of a living thing from that age.
2. leaf, skeleton of an animal, entire plant, insect
3. millions of years

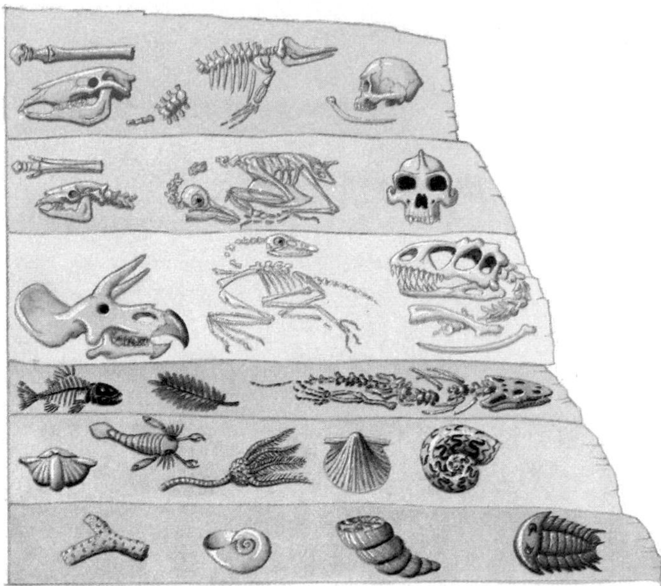

Figure 29-12 is a side view of sedimentary rocks in Earth's crust. Where is the oldest layer of rock located? Because it was the first layer to form, it is on the bottom. Younger layers settle on top of the oldest layer. So, as you move up the layers, the rocks get younger. This is similar to stacking newspapers. Suppose you always put the most recent newspaper on top of the stack. As long as the stack is left alone, the oldest paper will always be on the bottom. The newest paper will be on top. Fossils found in the lower layers of rock are older than those found in the upper layers.

Compare the fossils in the bottom layers to those near the top. How many fossil forms found toward the bottom and middle layers are still alive today? Not too many. How many fossil forms found in the top layers are still alive today? Certainly more than in the middle and bottom layers.

**Figure 29-12** Fossils in the lower layers of sedimentary rock are older than those found near the top layers.

## Other Evidence

In Chapter 3, you studied that comparing the origins of body structures and comparing body chemistries are ways to determine relationships among different species. Each of these comparisons is evidence of evolution.

### Skill Check

**Interpret diagrams:** Study Figure 29-12. In which rock layers are the more simple fossils found? *For more help, refer to the* ***Skill Handbook,*** *pages 706-711.*



## TEACH

### Videocassette

Show the videocassette *The Record of the Rocks,* Films for the Humanities and Sciences.

### Bridging

Ask students who have had a course in earth science the following question: Why does one not find fossil evidence in igneous or metamorphic rocks?

### Filmstrip

Show the filmstrip *The Fossil History of Man,* Carolina Biological Supply Co.

### Independent Practice

**Study Guide**/*Teacher Resource Package*, p. 173. Use the Study Guide worksheet shown at the bottom of this page for independent practice.

## GLENCOE TECHNOLOGY

**Videotape**

**The Secret of Life**

*Gone Before You Know It: The Biodiversity Crisis*

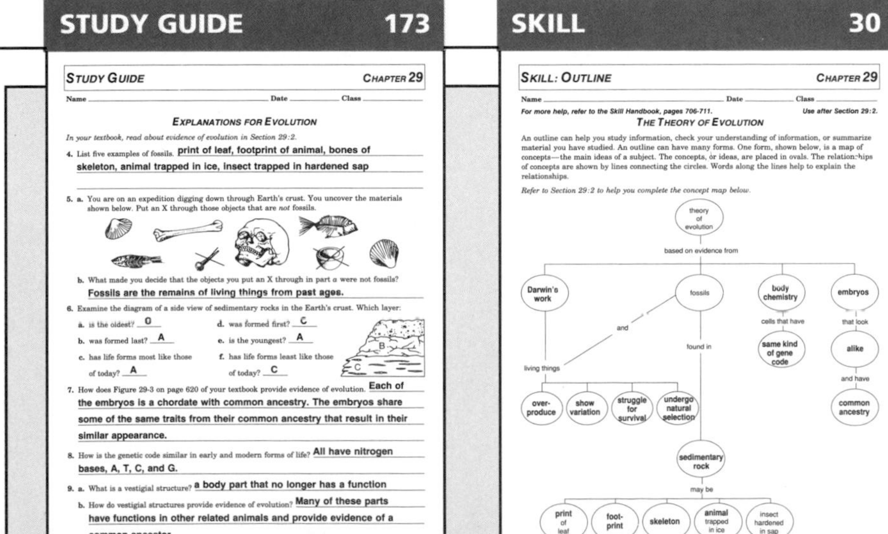

**STUDY GUIDE 173**

STUDY GUIDE CHAPTER 29

Name ______ Date ______ Class ______

EXPLANATIONS FOR EVOLUTION

*In your textbook, read about evidence of evolution in Section 29:2.*

4. List five examples of fossils. print of leaf, footprint of animal, bones of skeleton, animal trapped in ice, insect trapped in hardened sap

5. a. You are on an expedition digging down through Earth's crust. You uncover the materials shown below. Put an X through those objects that are *not* fossils.

b. What made you decide that the objects you put an X through in part *a* were not fossils? Fossils are the remains of living things from past ages.

6. Examine the diagram of a side view of sedimentary rocks in the Earth's crust. Which layer:

a. is the oldest? C
b. was formed last? A
c. has life forms most like those of today? A
d. was formed first? C
e. is the youngest? A
f. has life forms least like those of today? C

7. How does Figure 29-3 on page 620 of your textbook provide evidence of evolution. Each of the embryos is a chordate with common ancestry. The embryos share some of the same traits from their common ancestry that result in their similar appearance.

8. How is the genetic code similar in early and modern forms of life? All have nitrogen bases, A, T, C, and G.

9. a. What is a vestigial structure? a body part that no longer has a function

b. How do vestigial structures provide evidence of evolution? Many of these parts have functions in other related animals and provide evidence of a common ancestor.

**SKILL 30**

SKILL: OUTLINE CHAPTER 29

Name ______ Date ______ Class ______

*For more help, refer to the Skill Handbook, pages 706-711.* *Use after Section 29:2.*

THE THEORY OF EVOLUTION

An outline can help you study information, check your understanding of information, or summarize material you have studied. An outline can have many forms. One form, shown below, is a map of concepts—the main ideas of a subject. The concepts, or ideas, are placed in ovals. The relationships of concepts are shown by lines connecting the circles. Words along the lines help to explain the relationships.

*Refer to Section 29:2 to help you complete the concept map below.*



## OPTIONS

**Skill**/*Teacher Resource Package*, p. 30. Use the Skill worksheet shown here to reinforce the concept of evolution.

# TEACH

## Concept Development

▶ Structures with similar origins and functions are called *homologous structures.* Structures with similar functions but different origins are called analogous structures.

▶ Plant-eating animals typically have a rather long caecum (cecum) along with the appendix that aids in the breakdown and digestion of cellulose.

## ACTIVITY/Demonstration

Ask if any students can wiggle their ears. Ear muscles are considered to be vestigial structures.

## Bridging

Ask how it is possible for human DNA placed within a bacterial cell to work. All living things contain DNA. This is very strong evidence of a common origin.

## Guided Practice

Have students write down their answers to the margin question: a vestigal structure is evidence of a common ancestor.

## Independent Practice

Study Guide/*Teacher Resource Package*, p. 174. Use the Study Guide worksheet shown at the bottom of this page for independent practice.

### *Student Journal*

Provide students with the following information. You are participating in a debate regarding evolution. Your team supports the theory that evolution has occurred and is still occurring. Prepare a summary of the evidence that you could use to defend your position.

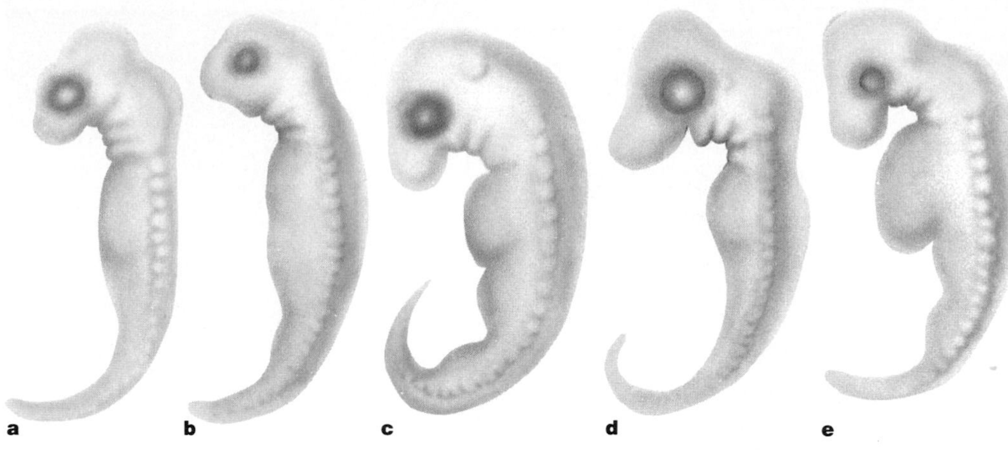


**Figure 29-13** Embryos of a fish (a), frog (b), turtle (c), bird (d), and rabbit (e) look similar.

Another comparison also provides evidence of evolution. Figure 29-13 shows five different animal embryos. Fish, frog, turtle, bird, and rabbit embryos are shown. Which embryo is which? You really can't tell. They all look very much alike at this stage in their development. As they get older, you will be able to tell which embryo is which. How is the similar appearance of embryos evidence of evolution? All five animals are chordates. They have a common ancestry. The embryos share some of the same traits from their common ancestry. These traits result in all the embryos looking similar.

What other evidence is there of evolution? If later life-forms evolved from earlier ones, wouldn't the later forms have something in common with the earlier forms? Read the following examples for the answer to this question. Early life-forms are made of cells. So are later life-forms. Early life-forms have DNA as part of their chromosomes. So do later life-forms. The gene code in early life-forms is made of nitrogen bases, A, T, C, and G. Later life-forms have the same kind of gene code.

Have you ever wondered what that little pink lump is in the corner of your eye? What does it do? What does your appendix do? Both of these body parts are called vestigial (vuh STIJ ee ul) structures. A **vestigial structure** is a body part that no longer has a function. How is a vestigial body part evidence of evolution? Most of these body parts do have jobs in other animals. For example, in many mammals the appendix helps digest food. Rabbits are examples of animals with an appendix that still works to digest food.

**How are vestigial structures evidence of evolution?**



## *OPTIONS*

### Science Background

The small pink lump in the corner of the eye is called the nictitating membrane. Other vestigial structures in humans are: third molars, muscles that move the ears, body hair, coccyx vertebrae (tailbone), appendix.

**STUDY GUIDE 174**

STUDY GUIDE CHAPTER 29

Name ________ Date ________ Class ________

VOCABULARY

*Review the new words in Chapter 29 of your textbook. Then, write a sentence defining each word.*

1. species — a group of living things that can breed with others of the same species and form fertile offspring
2. old-world monkeys — monkeys that can't grasp with their tails and that have nostrils that open downward
3. natural selection — process in which something in the surroundings determines if a living thing survives to have offspring
4. vestigial structure — a body part that no longer has a function
5. sedimentary rocks — rocks that form from layers of mud, sand, and other fine particles
6. primate — mammal with eyes that face forward, a well-developed cerebrum, and grasping thumbs
7. evolution — change in the hereditary features of a group of organisms over time
8. new-world monkeys — monkeys that have a grasping tail and nostrils that open upward
9. extinct — a life-form that no longer exists
10. competition — a struggle among living things to get their needs for life
11. fossils — remains of once-living things from ages past
12. fertile — able to reproduce by forming egg or sperm cells
13. variation — trait that makes an individual different from others in its species

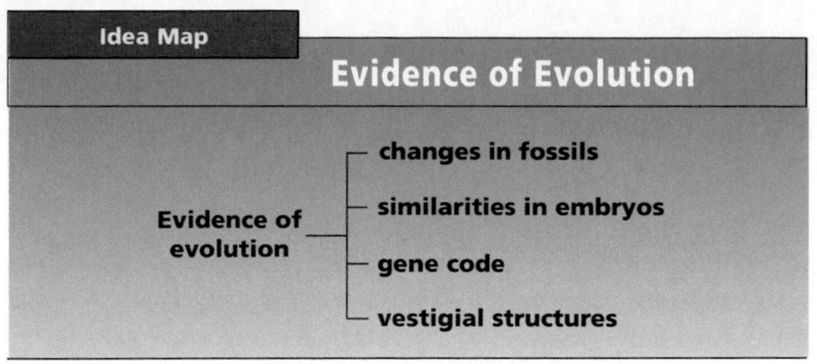


**Study Tip:** Use the idea map as a study guide to the kinds of evidence that show that species evolve over time.

This part of a rabbit intestine helps to break down plant material, the rabbit's chief source of food. The pink lump in your eye is all that is left of a third eyelid. In other animals the third eyelid is usually very thin and covers the entire eye. Frogs and turtles are examples of animals with third eyelids. The third eyelid of these animals protects the eye while the animal is under water. Birds, fish, and reptiles also have third eyelids that protect the eye.

We may not have any use for our appendixes or what remains of our third eyelid. These structures, however, are still useful to related animals. Thus, the presence of vestigial structures is evidence of a common ancestor for us and related animals. We still have the genes for appendixes and third eyelids, even though we don't use these structures. Animals that are related to us also have the genes for these traits.

## Check Your Understanding

6. List and describe the four major points of Darwin's theory of natural selection.
7. Give several examples of fossils and tell where they are formed.
8. Describe how the embryos shown in Figure 29-13 are alike.
9. **Critical Thinking:** An athlete breaks her leg in a high jump. Years later, she has a son who walks with a slight limp. Is this an example of evolution? Why or why not?
10. **Biology and Reading:** What kinds of evidence do scientists have for evolution?

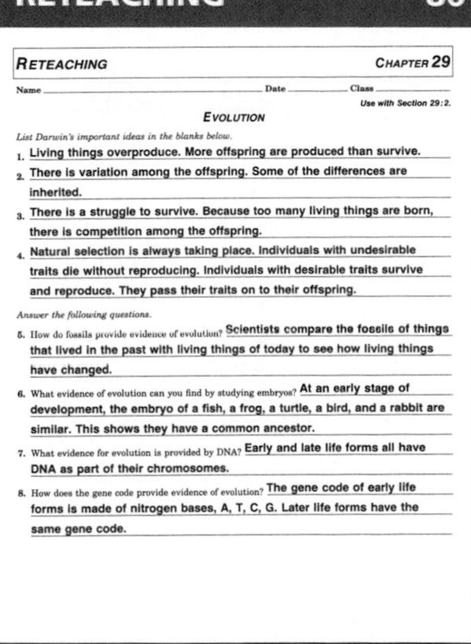
**RETEACHING 86**

*Reteaching* — *Chapter* 29

Name ______ Date ______ Class ______

*Use with Section 29:2.*

*Evolution*

*List Darwin's important ideas in the blanks below.*

1. Living things overproduce. More offspring are produced than survive.
2. There is variation among the offspring. Some of the differences are inherited.
3. There is a struggle to survive. Because too many living things are born, there is competition among the offspring.
4. Natural selection is always taking place. Individuals with undesirable traits die without reproducing. Individuals with desirable traits survive and reproduce. They pass their traits on to their offspring.

*Answer the following questions.*

5. How do fossils provide evidence of evolution? Scientists compare the fossils of things that lived in the past with living things of today to see how living things have changed.
6. What evidence of evolution can you find by studying embryos? At an early stage of development, the embryo of a fish, a frog, a turtle, a bird, and a rabbit are similar. This shows they have a common ancestor.
7. What evidence for evolution is provided by DNA? Early and late life forms all have DNA as part of their chromosomes.
8. How does the gene code provide evidence of evolution? The gene code of early life forms is made of nitrogen bases, A, T, C, G. Later life forms have the same gene code.



# TEACH

## Idea Map

Have students use the idea map as a study guide to the major concepts of this section.

## Check for Understanding

Have students respond to the first three questions in Check Your Understanding.

### Reteach

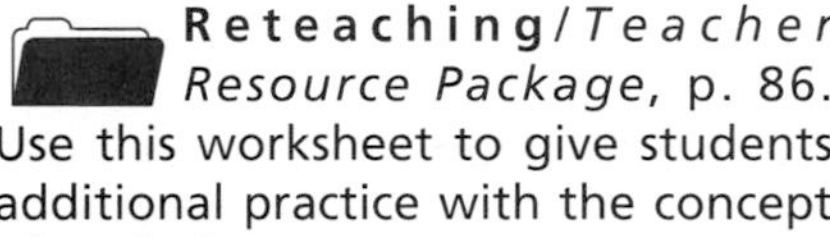
**Reteaching**/*Teacher Resource Package*, p. 86. Use this worksheet to give students additional practice with the concept of evolution.

**Extension:** Assign Critical Thinking, Biology and Reading, or some of the **OPTIONS** available with this lesson.

# 3 APPLY

## ACTIVITY/Brainstorming

Why would you expect the following traits *not* to be passed to offspring: muscle building, tatoos, and plastic surgery to reshape a person's nose?

# 4 CLOSE

## Convergent Question

Have students suggest how the dinosaurs may have become extinct.

## Answers to Check Your Understanding

6. (1) Living things tend to overproduce. (2) Over production leads to competition. (3) Variation occurs among living things. (4) Living things with favorable variations will survive and reproduce more offspring.
7. a leaf print formed in sediments, an animal trapped in ice, a skeleton trapped in sediments
8. The embryos have tails, eyes, slits at the base of the neck.
9. No. There must be a change in the genes of the sex cell.
10. fossils, embryos, chemical evidence, and vestigial structures

### Purpose

Students will use the known age of model index fossils to predict the age of unknown rock samples containing index fossils.

### Process Skills

Formulate models, interpret data, infer, observe, classify, recognize and use spatial relationships

### Time Required

Thirty minutes to explain and conduct the activity

### Possible Problems

Remind students that there may be more than one index fossil in their unknowns. Have students return all beads to the correct container before opening the next unknown. Some students may have difficulty interpreting the table. Prepare a transparency of the table, and project it onto your classroom screen with an overhead projector. Give students practice in using the table by providing them with a few examples. You might even use some of the actual "unknowns" as examples.

### Teaching Strategies

▶ Unknowns can be placed in self-sealing plastic bags or small empty medicine vials with snap-top caps.

▶ Unknowns are to be prepared as follows:

Label bag or vial as #1: add about 100 round beads, 2 of type A, and 3 of type C.

Label bag or vial as #2: add about 100 round beads, 2 of type B, 2 of type C, and 3 of type E.

Label bag or vial as #3: add about 100 round beads, 2 of type D, and 2 of type E.

▶ Beads may be purchased in most craft stores or in specialized bead shops.

# TECH PREP APPLYING TECHNOLOGY

## A Fossil Hunt

Have you ever found a fossil? There is a certain thrill about finding evidence of life long gone. The presence of fossils in rock layers can be more useful than just showing us what life forms looked like millions of years ago. Fossils may also be used by scientists to judge the age of the rock layers in which they are found.

### Identifying the Problem

You are a geologist who works for an oil company. After locating some fossils in sedimentary rock layers in Wyoming, you want to determine the age of these rock layers. Why would you want to know the age of a layer of rocks? As a geologist who is looking for rocks that may contain oil deposits, age is an important clue.

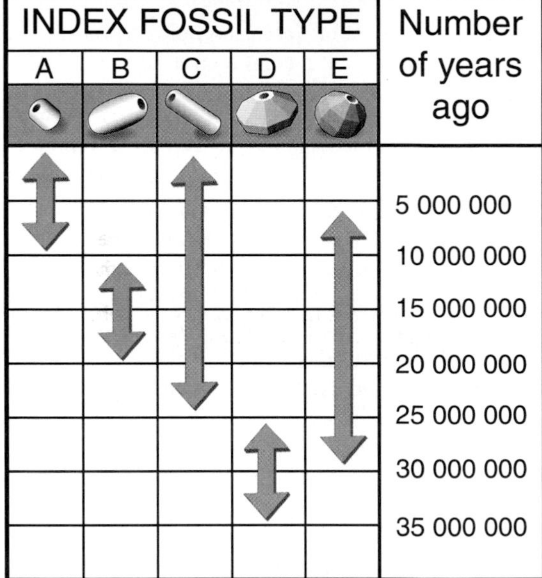

| INDEX FOSSIL TYPE | | | | | Number of years ago |
|---|---|---|---|---|---|
| A | B | C | D | E | |
| | | | | | 5 000 000 |
| | | | | | 10 000 000 |
| | | | | | 15 000 000 |
| | | | | | 20 000 000 |
| | | | | | 25 000 000 |
| | | | | | 30 000 000 |
| | | | | | 35 000 000 |

### Collecting Information

One method for determining age of a rock layer is through the presence of index fossils. Index fossils are certain fossil types that are known to have lived during certain time periods on Earth. Thus, the age of these index fossils has been established. If a certain index fossil is found in a rock layer, the age of that layer can be determined by the presence of the index fossil buried in it.

The table on this page shows the shapes of five beads. Each bead represents a different index fossil. The table gives the ages of the index fossils. If you find one, two, or even three index fossils in a rock layer, you should be able to determine the age of the rock from where the index fossils were found.

#### Technology Connection

Our dependence on oil as fuel to generate electricity may be decreasing. Solar cells have improved so much in the amount of energy they can trap that they can now compete with fossil fuels such as oil. Two large energy companies, Enron and Amoco, have joined together to build a solar plant in Nevada. This plant will generate enough energy to supply a city of 100 000 people.



#### Technology Connection

The facility being built will generate 100 megawatts of electricity. It will use a thin-film, silicon-based photovoltaic cell that transforms into electricity about eight percent of the sunlight that reaches it. One major benefit of solar power is that it is nonpolluting. As world demand for fossil fuels continues to increase, the supply continues to decrease. Solar energy technology may be the solution to this problem.

## Carrying Out an Experiment

You will be given three groups of "fossils" collected from three layers of rock. All the beads represent fossils, but only the special shapes shown in the table represent index fossils.

1. Make a data table in which you will record the number of your fossil group, the shape of the index fossils found, and the age of the rock layer from which the fossils were taken.
2. Examine the first group of fossils labeled #1. Look for the presence of index fossil shapes according to the table.
3. Record which index fossils are present in the group.
4. Using the rock ages on the right side of the table, record the age of the unknown rock sample from which the group of fossils was taken. For example, if you found index fossil shapes A, C, and E in the group, then the rock sample would be between 5 million and 10 million years old. Why? Because this is the only time period in which all three index fossils are found together.
5. Return all fossils to their container before starting to examine the next group of fossils.
6. Repeat steps 2-5 with two more groups of fossils marked #2 and #3.

### Career Connection

- **Petroleum and Natural Gas Exploration Worker** Works for an oil company or an oil drilling company; may work on land or on an ocean platform
- **Field Assistant** Prepares, classifies, and sorts rock and fossil specimens; cleans, washes, and prepares samples for further study
- **Paleontologist** Studies fossil plants and animals in order to trace their evolutionary trends; categorizes fossils as to their age and location; looks for petroleum bearing rock formations

## Assessing Your Results

Record the ages for rock samples #1, #2, and #3 in your data table. Explain why the age for rock sample #1 could *not* be between 5 million and 10 million years old. Explain how it might be possible for a rock sample to not contain an index fossil even though it is known that the fossil should be present.

▶ If beads are not easily available, birdseed can be substituted. Use seeds that are similar in color but slightly different in shape. If a change in index fossil type is made, advise students of the new shapes that correspond to index fossils A-E.

▶ To save time and materials, you can have students work in groups of two or three.

▶ Store the bags or vials for the following year.

▶ If available, show students examples of actual fossils. Ask students to bring in and explain examples that they may have collected.

### Conclusions

Students will be able to understand how index fossils are used to determine the age of rock layers. They will also use modeling to illustrate a concept.

### Answers to Questions

Ages for unknowns: #1 is 5 million years or younger, #2 is 10 to 20 million years old, #3 is 25 to 30 million years old. If sample #1 were between 5 and 10 million years old, index fossil E would have to be present. Student answers may vary but could include the idea that the sample size was too small and the index fossil was not observed, or the index fossil was incorrectly identified.

### ASSESSMENT

**Skill** Have students draw the index fossils that would be expected to be observed in rock layers that are 30 to 35 million years of age and 20 to 35 million years of age.

### Career Connection

**Career Path**
**Petroleum and Natural Gas Exploration Worker** High-school diploma and on-the-job training
**Field Assistant** High-school diploma and some on-the-job training
**Paleontologist** College degree in biology or geology; advanced college study

**Career Issue**
Have students explore other means available to paleontologists for dating the ages of rock layers. (radioactive dating)

## CHAPTER 29 REVIEW

### Summary

Summary statements can be used by students to review the major concepts of the chapter.

### Key Science Words

All boldfaced terms from the chapter are listed.

### Testing Yourself

#### Using Words

1. evolution
2. competition
3. extinct
4. natural selection
5. variation
6. sedimentary rock
7. fossil
8. vestigial structure
9. primate
10. fertile

#### Finding Main Ideas

11. p. 619; The oldest layer is at the bottom because it was the first layer to form. Newer layers settle on top of the oldest layer.
12. p. 615; More offspring are produced than can survive. They aren't able to survive because resources are limited.
13. p. 613; Humans are Homo sapiens. They belong to the Phylum Chordate, the Class Mammal, and the Order Primate. Primates are mammals with eyes that face forward, a well-developed cerebrum, and thumbs that can be used for grasping.
14. p. 620; Since chordate embryos share similar traits, they must share a common ancestry.
15. p. 607; If the color of the background soil changes, the survival of the two different colored mice changes.
16. p. 609; Because of natural selection, some individuals are better adapted to their environment than others. These individuals will survive and produce more offspring than less well adapted individuals.

# Chapter 29

# Review

## Summary

### 29:1 Changes in Living Things

1. All living things have adaptations that help them survive.
2. Changes in life-forms are the result of natural selection, mutations, and changes in the environment.
3. Primates evolved about 45 million years ago into two main groups. One of these groups was the ancestor of apes and humans. Humans evolved about three million years ago.

### 29:2 Explanations for Evolution

4. Darwin described his ideas about evolution over 100 years ago. His four main points were: living things overproduce, variation occurs in living things, there is a struggle to survive, and those that survive to reproduce are the ones that have traits best suited for their surroundings.
5. Evidence of evolution comes from fossil remains, similarity of embryos, similarity between chemical makeup of early and present-day life-forms, and from the presence of vestigial structures.

## Key Science Words

competition (p. 616)
evolution (p. 616)
extinct (p. 617)
fertile (p. 610)
fossil (p. 617)
natural selection (p. 609)
new-world monkey (p. 613)
old-world monkey (p. 613)
primate (p. 613)
sedimentary rock (p. 617)
species (p. 610)
variation (p. 616)
vestigial structure (p. 620)

## Testing Yourself

### Using Words

*Choose the word from the list of Key Science Words that best fits the definition.*

1. a change in hereditary features of a group over time
2. struggle among living things for their needs
3. species that is no longer living
4. when something in the environment determines whether or not a living thing survives
5. a trait that makes a living thing different from others in its species
6. a type of rock formed from layers of mud and sand
7. remains of any once-living thing
8. the appendix or the third eyelid of humans
9. mammal that has eyes that face forward and grasping thumbs
10. being able to reproduce by forming egg or sperm cells



#### Using Main Ideas

17. Dark-colored mice are well adapted to living in areas with dark soil because they can hide more easily from their enemies.
18. First, a barrier forms that separates members of a species. The environment on either side of the barrier changes. Individuals that are well adapted to the new environments will survive and reproduce. Over time new species may evolve.
19. (a) Fossils show that extinct life-forms are related to present-day life-forms.
(b) Earlier and later life-forms have the same DNA code and, therefore, have something in common.
20. Some individuals have inherited adaptations that allow them to survive and reproduce better than individuals without those adaptations.
21. Animals and plants die and settle at the bottom of oceans, lakes, or ponds with sediments of mud, sand, and other fine particles. They are covered with more sediments, and over millions of years, the sediments change into rock. The animal or plant parts covered by the sediment form permanent imprints in the rock.

# Review

## Testing Yourself *continued*

**Finding Main Ideas**

*List the page number where each main idea below is found. Then, explain each main idea.*

11. why the oldest layer of sedimentary rock is found on the bottom
12. what Darwin meant when he said that living things overproduce
13. human classification and traits of the order to which they belong
14. the importance of chordate embryos looking alike
15. what may cause the balance to change in favor of the survival of light-colored or dark-colored mice
16. how natural selection works

**Using Main Ideas**

*Answer the questions by referring to the page number after each question.*

17. Why are dark-colored mice well adapted or not well adapted to living in areas with dark soil? (p. 607)
18. What are three events that lead to the formation of new species? (p. 612)
19. How are the following used as evidence of evolution?
    (a) comparison of fossil life-forms to present day life-forms (p. 617)
    (b) the DNA codes of all life-forms being made up of the same nitrogen bases (p. 620)
20. What did Darwin mean when he indicated that some individuals are more fit than others? (p. 616)
21. How are fossils formed? (p. 617)
22. How do monkeys and apes differ? (p. 613)
23. What are the three effects of mutations? (p. 609)

## Skill Review 

*For more help, refer to the **Skill Handbook,** pages 704-719.*

1. **Infer:** Which flower would have the best chance of being pollinated by a moth, one that opens during the day or one that opens during the night? Why?
2. **Define words in context:** During the evolution of the horse, the number of toes decreased from four to three to one. What does the word evolution mean as used in this sentence?
3. **Interpret data:** During an experiment, a student reported the following distances that the frogs she was raising were able to jump: 1.2 m, 0.5 m, 0.9 m, 1.0 m, 0.6 m, 1.1 m. Are her data evidence of variation?
4. **Interpret diagrams:** Study Figure 29-13. How are the embryos alike?

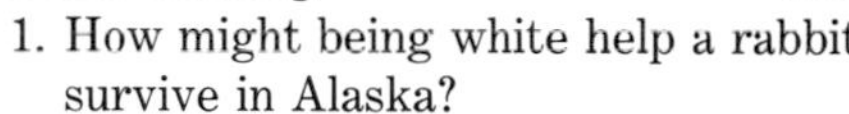


## Finding Out More

**Critical Thinking**

1. How might being white help a rabbit survive in Alaska?
2. Assume the trait in question 1 is a genetic trait. Will it be passed to future generations? Explain.

**Applications**

1. You wish to breed a chicken that produces many large eggs. Outline a breeding program that could accomplish this goal.
2. Use references to determine what the brain size, general appearance, and method of walking (two or four legs) were for early humanlike species.

625

**Chapter Review**/*Teacher Resource Package*, pp. 153-154.

**Chapter Test**/*Teacher Resource Package*, pp. 155-157.

**Chapter Test**/*Computer Test Bank*

**Unit Test**/*Teacher Resource Package*, pp. 158-159.

### Using Main Ideas

22. Monkeys have tails, and apes don't have tails.
23. Mutations change the DNA code, are a source for new traits, and may supply new adaptations.

### Skill Review

1. night; Moths fly at night.
2. At one time, horses had more toes than presently. However, some horses were born that had fewer toes. These horses were able to survive better and produce more young. Over time, horses with fewer toes became more numerous and eventually were the only kind of horse.
3. yes
4. tails, eyes, slits at base of neck, the beginnings of the spine

### Finding Out More

**Critical Thinking**

1. It would blend in with the snowy background and could easily hide from enemies.
2. Yes. Rabbits with this trait are more likely to survive and produce offspring. Their offspring will, in turn, have the trait because it is genetically controlled.

**Applications**

1. Select the chickens that produce the most eggs and the largest eggs and only breed them. Do this each generation.
2. Brain size was generally smaller. Early humans were generally shorter. However, all humanlike species (hominids) walked upright.

# Unit 7 Traits of Living Things

CONNECTIONS

## Biology in Your World

The traits of an organism are determined by its genetic makeup. Scientists have learned to manipulate genetic material to produce desirable traits in plants and animals. Studying fossils helps us to understand the natural evolution of certain traits in all forms of life.

### Literature Connection

Jack Horner was the first person to discover a dinosaur nest. Some nests contained only eggs, but others held the remains of hatched shells and babies. A few nests held larger babies than the other nests. This suggests that the babies stayed in the nest until they reached a certain size.

Horner concluded that the babies were cared for by a parent. Before, scientists thought dinosaurs laid their eggs and abandoned their young. Horner also concluded that the babies must have grown quickly to the size at which they could leave the nest. This fast growth made him wonder if the dinosaurs could have been warm-blooded.

### History Connection

Charles Willson Peale (1741-1827) of Maryland painted many portraits of Revolutionary War heroes. He also founded the first major American museum. The Peale Museum opened in Philadelphia's Independence Hall and contained his paintings, technological gadgets, and stuffed animals. Its most famous exhibit was the complete mastodon skeleton. Peale recorded the excavation in his painting *Exhuming the Mastodon.*

C O N N E C T I O N S

# Biology in Your World

### Variety: The Spice of Life

In this unit, you learned that variations in plants and animals are due to changes in the genetic material. Studying fossils reveals the variety of living things from the past. Today, scientists can use the variations in living things to breed crops and animals with certain traits.

LITERATURE

## Looking for Baby Dinosaurs

John Horner's interest in baby dinosaurs began with a can full of tiny dinosaur bones. He got the bones from the owners of a rock shop in Montana in 1978.

Horner spent the next six years looking for more bones at a site in Montana.

He found the bones of baby dinosaurs in what appeared to be nests. The nests were close together and the babies were all the same age. Horner believes that these dinosaurs lived in groups and took care of their young.

Horner wrote *Digging Dinosaurs* about his findings. He also discusses why dinosaurs succeeded for 140 million years.

HISTORY

## Putting the Pieces Together

In 1801, farmers found some large bones in swamps near New York City. The bones were those of a mastodon. A mastodon is an extinct relative of the elephant. Mastodons lived in North America until about 10 000 years ago.

Charles Willson Peale was an artist and museum owner from Philadelphia who studied natural history. Hoping to find a complete mastodon skeleton, Peale organized an excavation at one of the sites. Most of the bones of an entire mastodon were found. In order to complete the skeleton, missing bones were carved from wood. The skeleton was displayed in Peale's museum.

626

**GLENCOE TECHNOLOGY**

**Videodisc**

Biology: The Dynamics of Life
Movie: *Discovering Dinosaurs*
Disc 1, Side 2, Ch. 2

LEISURE

## Breeding Guppies

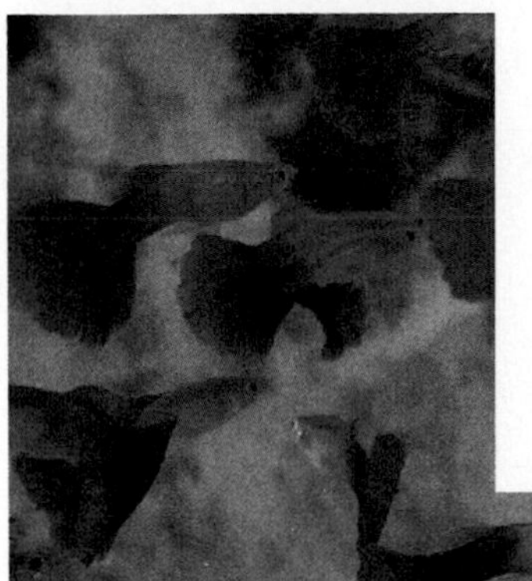

Female guppies are gray, ordinary-looking tropical fish. However, male guppies can be brightly colored. If you would like to see genetics in action, consider breeding guppies. You will need a tank and large jars to use as maternity wards. Check at a library or pet store for books on aquariums and breeding guppies. After each group is born, choose the largest and liveliest fish as the parents of the next generation. You can expect interesting results within a few generations. Keep records—you may succeed in producing a new strain of guppies!

CONSUMER

## Delicious, Nutritious Bananas

Bananas are a very nutritious type of fruit. They are a good source of carbohydrates, phosphorus, and potassium. They are also a natural source of vitamins A and C. Bananas have very little fat.

Bananas grow in the tropics, where the plants produce fruit almost continuously. Thus, fresh bananas are almost always available.

The banana we eat is the Cavendish variety. It is being threatened by a fungus called black sigatoka (sihg uh TOH kuh). The fungus attacks the leaves of banana plants and can spread to the fruits. If not stopped, the fungus can kill the plants. At one time, chemical sprays were able to kill the fungus. But now, some of the fungi have become resistant to the chemicals.

Plant breeders are trying to develop a new hybrid banana that is resistant to black sigatoka. Some wild varieties of bananas are resistant to the fungus. The breeders are crossing these wild, nonedible varieties with the edible varieties.

Maybe in a few years you will see a new banana on your grocery store shelf.



## CULTURAL DIVERSITY

### Digestion of Lactose

The enzyme lactase is needed for the digestion of lactose, the sugar in milk. Lactase is usually present in human infants but absent in most adults. In some people, however, this enzyme is missing altogether due to a genetic mutation. Without it, drinking milk or eating some milk products can result in indigestion, cramps, and diarrhea because the lactose is broken down by bacterial fermentation to produce gas. Members of some African groups have no lactase and therefore use very little milk in their diet. Understandably, they do not herd dairy animals. Other ethnic groups have high percentages of members who cannot digest lactose. Have students research these groups.

## CONNECTIONS

### Leisure Connection

Guppies are prolific breeders, reproducing 30 to 50 live young every four to six weeks. The young guppies are only 3 millimeters long and are often eaten by the larger fish. Therefore, the baby guppies should be separated from the adult fish until they are too big to be eaten.

In the wild, guppies feed on worms, shellfish, and insect larvae. They are sometimes released in warm waters to control mosquitoes. If students breed too many fish, guppies can be released in ponds or lakes that are around 20°C. Sometimes, pet stores will buy surplus fish.

### Consumer Connection

Consumers in the United States eat about 11 billion bananas yearly. Most of these varieties have thick yellow skins like the Cavendish. However, there are other varieties available. Red Jamaicas are small red bananas that have thin skins. Plantains are hard and starchy, and they are usually cooked and eaten as a vegetable.

When creating a new hybrid, some bananas have to be pollinated by hand. Therefore, producing a Cavendish hybrid resistant to black sigatoka will be difficult. In the meantime, consumers may see more Red Jamaicas and Plantains at the grocery store.

### References


Fisher, Daniel. "Tusk Tales." *Discover*, February 1997, pp. 22-23.

Horner, John R. *Dinosaur Lives*. New York: Harper Collins, 1997.

McDonald, Kim A. "The Iconoclastic Fossil Hunter." *The Chronicle of Higher Education*, November 16, 1994, pp. A8-A14.

Monastersky, Richard. "Sex and Violence in the Ice Age World." *Science News*, November 2, 1996, p. 287.

Whitern, Wilfred A. *Guppies*. Neptune, NJ: T.F.H. Publications, Inc., 1980.

# Unit 8 Relationships in the Environment

## UNIT OVERVIEW

Unit 8 focuses on the various levels of ecological relationships. Chapter 30 deals with populations as the basic unit of ecology, and communities as the places where interrelationships occur between populations. Chapter 31 discusses the ecosystem and biome levels of ecological relationships. Chapter 32 focuses on some ecological problems today, such as pollution, acid rain, and endangered species.

## Advance Planning

Audiovisual and computer software suggestions are located within each specific chapter under the heading TEACH.

**To prepare for:**

Lab 30-1: Have graph paper and rulers ready for each team of students.

Lab 30-2: Order owl pellets and get cardboard and glue for students to mount their skeletons.

Lab 31-1: Order 2 *Elodea* sprigs for each team. Prepare the bromthymol blue solution (blue and green) several days in advance.

Lab 31-2: Order brine shrimp eggs. Make the salt solutions a day in advance so all of the salt dissolves.

Mini Lab: Collect dead leaves, jars, and plastic wrap.

Lab 32-1: Collect rainwater and household chemicals several days prior to the lab.

Lab 32-2: Prepare the bromthymol blue solution. Prepare the live and dead yeast mixtures on the day of the lab.

Mini Lab: Collect jars with lids.

# Unit 8

**CONTENTS**

628

# Relationships in the Environment

**What would happen if...**

there were no mosquitoes? You may have memories of hot, itchy summers when you thought the world would be better off without mosquitoes. But, would it? Fish depend on mosquito larvae for food. Many birds, in turn, depend on the fish. Without the mosquito, many fish would starve to death. Many birds would then starve. The balance of nature would be upset by the lack of mosquitoes.

What is the effect of spraying chemicals to kill mosquitoes? In many communities, trucks that spray to kill mosquitoes are a regular sight. The spray that kills mosquitoes kills honeybees, too. What would be the effect of killing honeybees? What is the cost to the environment of getting rid of mosquitoes? Is the cost too high?



UNIT OVERVIEW

## Photo Teaching Tip

Have students read the passage and examine the photographs on these unit opening pages. Ask students to make a list of all the animals and plants they consider to be a nuisance. Put this list on the chalkboard and ask students to brainstorm for a list of reasons why each of these so-called pests is valuable to the balance of nature. Begin the discussion by reviewing the connection between mosquitoes and fish as described in the passage. You may wish to lead the discussion to the importance of protecting species from extinction.

## Theme Development

Living things interact with their environment. Students have studied the different kingdoms of living things and found out how their body parts work. Now, students will go beyond an individual living thing and study how similar and different populations interface with each other and the environment. They should understand that each species of living thing is an important part of the community because other living things may depend on its members for food or even for their habitat. They will learn how living things, humans included, can cause changes in the environment that make it unhealthy for other living things.

**CULTURAL DIVERSITY**

See the Cultural Diversity feature located on the Connections page at the end of this unit.

For Tech Prep activities, see the Applying Technology feature in this unit, the Finding Out More applications questions in the Chapter Reviews, and the Tech Prep Applications booklet in the Teacher Resource Package.

# Populations and Communities

## PLANNING GUIDE

| CONTENT | TEXT FEATURES | TEACHER RESOURCE PACKAGE | OTHER COMPONENTS |
|---|---|---|---|
| (2 days)<br>30:1 Populations<br>Population Size and Arrangement<br>Population Changes<br>Limits on Population Size | Skill Check: *Understand Science Words,* p. 632<br>Skill Check: *Make a Line Graph,* p. 634<br>Lab 30-1: *Competition,* p. 637<br>Science and Society: *Problems With an Increasing Human Population,* p. 635<br>Check Your Understanding, p. 636 | Reteaching: *Limiting Factors,* p. 87<br>Study Guide: *Populations,* pp. 175-176<br>Lab 30-1: *Competition,* pp. 117-118<br>Focus: *Interfering With Fish Populations,* p. 59 | **STVS:** *Fish Survey,* Ecology (Disc 6, Side 1) |
| (2 days)<br>30:2 Communities<br>Parts of a Community<br>Producers<br>Consumers<br>Decomposers | Mini Lab: *What's in a Community,* p. 640<br>Check Your Understanding, p. 640<br>Idea Map, p. 639 | Reteaching: *Populations and Communities,* p. 88<br>Skill: *Infer,* p. 31<br>Study Guide: *Communities,* p. 177 | **STVS:** *Evaluating Artificial Reefs,* Ecology (Disc 6, Side 1) |
| (1 1/2 days)<br>30:3 Energy in a Community<br>Food Chains<br>Energy Flow in a Community | Check Your Understanding, p. 643 | Enrichment: *A Salt Marsh Food Web,* p. 32<br>Transparency Master: *Energy Pyramid,* p. 151<br>Reteaching: *Energy in a Community,* p. 89<br>Study Guide: *Energy in a Community,* p. 178 | **Laboratory Manual:**<br>*What Are Some Parts of a Food Chain and a Food Web?* p. 251<br>Color Transparency 30: *Food Web*<br>**STVS:** *Zooplankton,* Ecology (Disc 6, Side 1) |
| (1 1/2 days)<br>30:4 Relationships in a Community<br>Mutualism<br>Commensalism<br>Parasitism<br>Predation | Skill Check: *Observe,* p. 646<br>Lab 30-2: *Predation,* p. 649<br>Check Your Understanding, p. 648 | Application: *Predators and Prey,* p. 30<br>Reteaching: *Relationships in a Community,* p. 90<br>Critical Thinking/Problem Solving: *What Do Predators Do in a Community?* p. 30<br>Study Guide: *Relationships in a Community,* p. 179<br>Study Guide: *Vocabulary,* p. 180<br>Lab 30-2: *Predation,* pp. 119-120 | **Laboratory Manual:**<br>*How Do Predator and Prey Populations Change?* p. 255<br>**STVS:** *Shorebird Preserves,* Ecology (Disc 6, Side 1) |
| Chapter Review | Summary<br>Key Science Words<br>Testing Yourself<br>Finding Main Ideas<br>Using Main Ideas<br>Skill Review | **ASSESSMENT RESOURCES**<br>Chapter Review, pp. 160-161<br>Chapter Test, pp. 162-164<br>Performance Assessment in the Biology Classroom<br>Alternate Assessment in the Science Classroom<br>Computer Test Bank | |

### GLENCOE TECHNOLOGY

**Infinite Voyage,** *Living with Disaster*

**The Secret of Life,** *Gone Before You Know It: The Biodiversity Crisis*

## MATERIALS NEEDED

| LAB 30-1, p. 637 | LAB 30-2, p. 649 | MARGIN FEATURES |
|---|---|---|
| metric ruler<br>graph paper | light microscope<br>microscope slide<br>coverslip<br>bowl<br>forceps<br>cardboard<br>glue<br>owl pellet<br>water | Skill Check, p. 632<br>dictionary<br>Skill Check, p. 634<br>pencil<br>graph paper<br>Mini Lab, p. 640<br>pencil<br>paper |

## OBJECTIVES

For more information about National Science Standards, see page 5T.

| SECTION | OBJECTIVE | CORRELATION of QUESTIONS to OBJECTIVES | | | |
|---|---|---|---|---|---|
| | | CHECK YOUR UNDERSTANDING | CHAPTER REVIEW | TRP CHAPTER REVIEW | TRP CHAPTER TEST |
| 30:1 National Science Stds: UCP.1, UCP.3, UCP.4, A.2, B.5, B.6, C.4, C.5, C.6, F.2, F.3, F.4 | 1. **Relate** the importance and methods of counting populations. | 1 | 10, 11, 17, 24 | 12, 14, 25 | 1, 50 |
| | 2. **Discuss** why populations change size. | 2, 5 | 6, 8, 25 | 7, 9, 15, 16 | 4, 10, 23, 49, 52 |
| | 3. **Explain** how limiting factors affect a population. | 3, 4 | 3, 18, 22 | 8, 24, 26 | 5, 11, 13, 20 |
| 30:2 National Science Stds: UCP.1, B.5, B.6, C.4, C.5, C.6, F.3, F.4 | 4. **Describe** the different parts of a community. | 6, 9 | 4, 9, 19 | 3, 19, 21, 28 | 9, 16 |
| | 5. **Explain** the importance of producers, consumers, and decomposers. | 7, 8, 10 | 12, 15, 23 | 18, 20 | 8, 19, 33, 34, 35, 36, 37, 38, 39, 40, 41, 42, 43, 44, 45, 46, 47, 48 |
| 30:3 National Science Stds: UCP.1, UCP.2, UCP.3, UCP.4, B.5, B.6, C.4, C.5, C.6, D.1, F.3 | 6. **Trace** the path of energy and materials through a community. | 11 | 1, 5 | 11, 13, 17 | 7, 12, 24, 51 |
| | 7. **Explain** how food chains are connected. | 13, 14 | 13, 16 | 10, 22 | 3, 15 |
| 30:4 National Science Stds: UCP.1, UCP.2, C.4, C.6 | 8. **Give examples** of mutualism. | 16, 20 | 14 | 23 | 18, 26, 29, 31 |
| | 9. **Describe** commensalism. | 18, 19 | 26 | 1, 4, 6 | 22, 27, 32 |
| | 10. **Compare** parasitism with predation. | 17, 20 | 2, 7, 21 | 2, 5, 27 | 2, 6, 14, 17, 21, 25, 28, 30 |

# Populations and Communities

## CHAPTER OVERVIEW

### Key Concepts

In this chapter, students will focus on the basic unit of ecology—the population. Next, the three basic parts of the community—plants, animals, and decomposers—are discussed as to how each fits into food chains and food webs. Mutualism, parasitism, and commensalism are examined.

### Key Science Words

| | |
|---|---|
| commensalism | parasitism |
| community | population |
| emigration | predation |
| energy pyramid | predator |
| food chain | prey |
| food web | primary consumer |
| habitat | secondary consumer |
| immigration | |
| limiting factor | |
| niche | |

### Skill Development

In Lab 30-1, students will **form hypotheses, interpret data, infer,** and **design an experiment** relating to species competition. In Lab 30-2, they will **observe, infer,** and **design an experiment** regarding predation. On page 632, students will **understand** the **science word** *population.* In the Skill Check on page 634, students will **make a line graph** of population size. In the Skill Check on page 646, they will **observe** a lamprey to determine how it gets food. In the Mini Lab on page 640, students will **make and use a table** of producers, consumers, and decomposers.

### Bridging

In Unit 2, students learned how all organisms were classified into kingdoms by structure and function of body parts. In this chapter, students will learn how the organisms in the different kingdoms are grouped into populations and how they interact in a community.



**CHAPTER PREVIEW**

### Chapter Content

Review this outline for Chapter 30 before you read the chapter.

**30:1 Populations**
- Population Size and Arrangement
- Population Changes
- Limits of Population Size

**30:2 Communities**
- Parts of a Community
- Producers
- Consumers
- Decomposers

**30:3 Energy in a Community**
- Food Chains
- Energy Flow in the Community

**30:4 Relationships in a Community**
- Mutualism
- Commensalism
- Parasitism
- Predation

### Skills in this Chapter

The skills that you will use in this chapter are listed below.
- In **Lab 30-1,** you will form hypotheses, interpret data, infer, and design an experiment. In **Lab 30-2,** you will observe, infer, and design an experiment.
- In the **Skill Checks,** you will understand science words, make a line graph, and observe.
- In the **Mini Labs,** you will make and use tables.

## TECH PREP

For Tech Prep activities in this chapter of the Teacher Wraparound Edition, see especially the Cooperative Learning and Options activities and the Portfolio on page 634, the Options Activity on page 639, the Student Journal on page 647, and the Brainstorming activity on page 648.

See also the Glencoe Homepage at **www.glencoe.com**

Chapter 30

# Populations and Communities

Some animals, like the caribou (KAR uh boo) on the left, live together in large groups called herds. There may be 50 to several hundred caribou in a herd. Each year the caribou migrate to a better feeding area. During migration, smaller herds combine to form a larger herd that may number in the thousands. What benefit might there be to living in a herd? What disadvantage is there to living in a herd?

Animals like the wolves shown below live together in small groups called packs. There may be 6 to 10 wolves in one pack. The pack may roam over an area of 25 to 50 square kilometers. Sometimes the wolves follow the caribou on their migration. How do wolves benefit from living in small packs? What might happen if the pack increased to 50 or 60 wolves?

**Try This!**

**How important is each organism in a community?** Write down the name of a plant. Now write the name of an organism that eats the plant. Continue to name an organism that eats the last organism you named. Your list shows relationships among living things.

*inter*NET CONNECTION

For more information about the material in this chapter, follow the link for the chapter on the Glencoe Homepage at **http://www.glencoe.com**



## ASSESSMENT PLANNER

***Portfolio***

Strategies on the following pages represent student products that can be placed into a best-work portfolio: pp. 634, 638, 642.

**PERFORMANCE ASSESSMENT**

Skill Check, pp. 632, 634, 646
Mini Lab, p. 640
Lab, pp. 637, 649

**CONTENT ASSESSMENT**

Check for Understanding, pp. 636, 640, 642, 643, 647, 648
Chapter Review, pp. 650-651

**GROUP ASSESSMENT**

Opportunities for group assessment occur with Cooperative Learning Strategies.

***Student Journal***

Strategies on the following pages represent opportunities for writing in a Student Journal: pp. 636, 647.

GETTING STARTED

### Using the Photos

Most grazing animals travel in large herds. The benefit of a large herd is that the young are protected and cared for by the entire herd. Point out that some herds of grazing animals may number in the thousands. Explain that predators travel in small packs and kill the sick or injured animals in the herd. Point out that if there were as many wolves in a pack as there were caribou in a herd, the wolves would have to kill a large number of caribou to survive. This would eventually destroy the food supply of the wolves.

### MOTIVATION/Try This!

**How important is each organism in a community?** Students will **recognize and use spatial relationships** among living things. They should realize that most relationships form not a linear chain, but a web. Point out to students that some of the organisms may feed at several different levels.

### Chapter Preview

Have students study the chapter outline before they begin to read the chapter. They should note the overall organization of the chapter, that each kind of living thing is a part of a population of a given area. Next, they will study how one population interacts with other populations to form a community. They will then learn how members of each population fit into the food chains and food web of a community and how energy flows through a food web in a community. Last, they will study relationships in a community.

### Misconception

Students may not understand that all living things in a community are important in the food web. They may regard mice and snakes as undesirable or useless animals. Point out that mice supply many predators such as the snake with food, and snakes are food for larger predators.

## 30:1 Populations

### PREPARATION

#### Materials Needed

Make copies of the Focus, Study Guide, Reteaching, and Lab worksheets in the *Teacher Resource Package.*

▶ Collect and arrange the pictures for the Focus activity.

▶ Have graph paper and rulers ready for Lab 30-1.

#### Key Science Words

population, emigration, immigration, limiting factor

#### Process Skills

In Lab 30-1, students will form hypotheses, interpret data, infer, and design an experiment. In the Skill Checks, they will understand science words and make a line graph.

### 1 FOCUS

▶ The objectives are listed on the student page. Remind students to preview these objectives as a guide to this numbered section.

#### MOTIVATION/Bulletin Board

Have pictures of plant and animal populations that show clumped, random, and uniform populations. A good example of a uniform population is trees that were planted in rows by humans. Have students discuss the advantages and disadvantages of the three types of populations in terms of competition for space, water, and sun.

#### MOTIVATION/Videocassette

Show the videocassette *Populations, Communities and Biomes,* Insight Media.

Focus/*Teacher Resource Package,* p. 59. Use the Focus transparency shown at the bottom of the page as an introduction to populations.



## 30:1 Populations

**Objectives**

1. **Relate** the importance and methods of counting populations.
2. **Discuss** why populations change in size.
3. **Explain** how limiting factors affect a population.

**Key Science Words**

population
emigration
immigration
limiting factor

On your way to school each day, you see different kinds of living things. If you stopped to take a closer look, you might see many individuals of the same kind of organism. There could be dozens or hundreds of one kind of living thing in one place. These living things make up a population. A **population** is a group of living things of the same species that live in an area.

### Population Size and Arrangement

Suppose you counted all of the daisies in the field in Figure 30-1. They would make up the daisy population of that field. You can't see the populations of spiders, earthworms, or field mice in the photo, but they are there as well.

A change in the size of one population often causes a change in the size of another population. For this reason, scientists usually want to know of any increases or decreases in populations they study. Finding the size of a population is not always easy. It may be hard to count the number of mice or earthworms in a population because these animals move. Some animals may be counted twice. Some may not be counted at all.

Counting the number of daisies in the small field would be easier because they do not move and the size of the field is small. Sometimes the area where the members of a population are found is very large. Finding out the size of the spruce tree population of Yellowstone National Park would be much harder due to the park's large size.

**Skill Check**

**Understand science words: population.** The word part *popula* means a group of people. In your dictionary, find three words with the word part *popula* in them. *For more help, refer to the **Skill Handbook,** pages 706-711.*

**Figure 30-1** Some populations are easier to see and count than others.



### OPTIONS

#### Science Background

Animal population size can be influenced by the amount of food and space available. Natural populations cannot increase forever; population size is also controlled by environmental and biological factors. Animals immigrating to one area are emigrating from another area. A chief factor influencing the size of a plant population is the amount of daily sunlight and the space available in which roots can grow and spread.

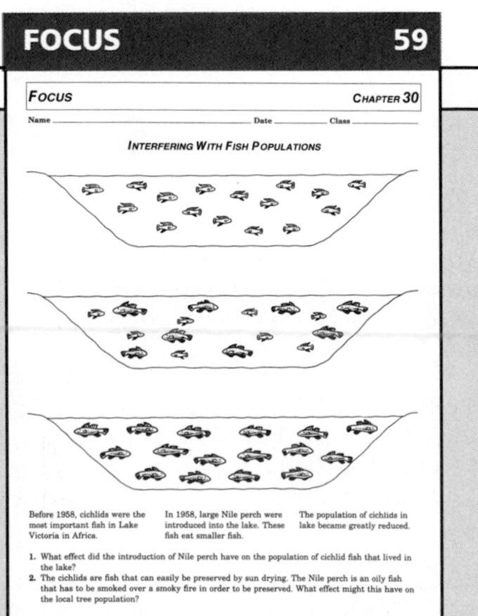

FOCUS 59

FOCUS CHAPTER 30

Name ______ Date ______ Class ______

INTERFERING WITH FISH POPULATIONS

Before 1958, cichlids were the most important fish in Lake Victoria in Africa.

In 1958, large Nile perch were introduced into the lake. These fish eat smaller fish.

The population of cichlids in lake became greatly reduced.

1. What effect did the introduction of Nile perch have on the population of cichlid fish that lived in the lake?
2. The cichlids are fish that can easily be preserved by sun drying. The Nile perch is an oily fish that has to be smoked over a smoky fire in order to be preserved. What effect might this have on the local tree population?

a

b

c

Populations are counted in different ways. Animals such as mountain goats can be marked with ear tags, Figure 30-2. Birds are marked by leg bands. Sometimes radio transmitters are put on animals. The animals are then followed to their nesting places and their offspring are counted. Trees may be marked with paint or ribbons. How is the human population in the United States counted?

Populations are not spread out evenly. Figure 30-3 shows that the population of the United States is clumped in cities. Is there a high or low population where you live?

A clumped population can be a useful plan. Members of the population may help one another find food or shelter. Many species of animals live in groups for protection. What kinds of animals live in groups called herds, flocks, and packs? Plants may be helped by living in clumps. Trees may be protected from strong winds if they grow close together. deer, birds, wolves

**Figure 30-2** Ear tags (a), leg bands (b), and radio transmitters (c) are ways of marking animals for population counts.

**How can an animal population be counted?**

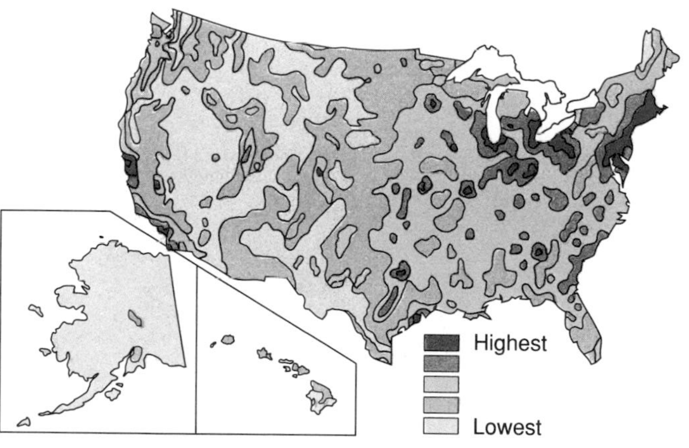


**Figure 30-3** The human population in the United States is clumped. Which area has most people? the eastern half of the country, especially the east coast



## 2 TEACH

### MOTIVATION/Audiovisual

Show about ten 35-mm color slides that depict populations of plants and animals. Some slides may show several different populations. A slide of a wooded area might show an oak tree population or a maple tree population. Explain that two or more interacting populations form a community.

### Concept Development

▶ Explain that populations too large to count are estimated using sampling techniques in which the organisms in a unit area are counted and then this number is multiplied by the total area in which the organisms live. Explain that some populations of plants are uniform in their arrangement because humans plant them. Crop fields and sections of forests may show this uniform arrangement.

### Guided Practice

Have students write down their answers to the margin question: by marking the animals with tags, bands, or transmitters.

### Independent Practice

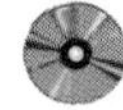
**Study Guide**/*Teacher Resource Package,* p. 175. Use the Study Guide worksheet shown at the bottom of this page for independent practice.

**STUDY GUIDE 175**

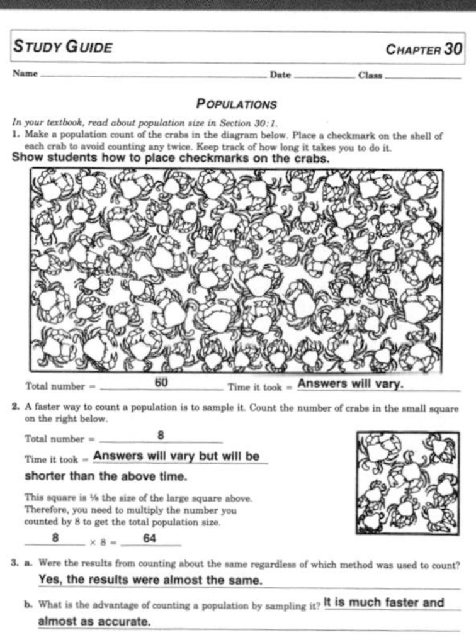

STUDY GUIDE — CHAPTER 30

Name ____ Date ____ Class ____

POPULATIONS

*In your textbook, read about population size in Section 30:1.*

1. Make a population count of the crabs in the diagram below. Place a checkmark on the shell of each crab to avoid counting any twice. Keep track of how long it takes you to do it.
**Show students how to place checkmarks on the crabs.**

Total number = **60** Time it took = **Answers will vary.**

2. A faster way to count a population is to sample it. Count the number of crabs in the small square on the right below.

Total number = **8**

Time it took = **Answers will vary but will be shorter than the above time.**

This square is ⅛ the size of the large square above. Therefore, you need to multiply the number you counted by 8 to get the total population size.

**8** × 8 = **64**

3. a. Were the results from counting about the same regardless of which method was used to count?
**Yes, the results were almost the same.**

b. What is the advantage of counting a population by sampling it? **It is much faster and almost as accurate.**

## OPTIONS

### ACTIVITY/Challenge

Have students determine how many trees are in the school yard or, if there is a fence row, how many trees are along the fence row. Point out that finding the size of a plant population is easier than finding that of an animal population because plants do not move around.

**GLENCOE TECHNOLOGY**

**Videodisc**

**Biology: The Dynamics of Life**
Animation: *Carrying Capacity*
Disc 1, Side 1, Ch. 13

# TEACH

## Skill Check

Data for making the line graph can be found in the *World Almanac and Book of Facts* published by the New York Times. Ask students to find out what factors are causing the human population to change in their state.

**Cooperative Learning:** Divide students into several teams. Have the teams find out when the greatest number of immigrants came to the United States and why people immigrated. Have them make graphs and charts showing their data.

## Guided Practice

Have students write down their answers to the margin question: it increases the population.

## ACTIVITY/Filmstrip

Show the filmstrip *Population Ecology,* Carolina Biological Supply Co.

## Independent Practice

**Study Guide**/*Teacher Resource Package,* p. 176. Use the Study Guide worksheet shown below for independent practice.

## Portfolio

The *World Almanac* lists census data for each state at ten-year intervals. Have each student compare census data for his or her state and another state for the past 100 years. Tell them to plot two lines on a graph, using a different color pencil for each state. They should show the number of people on the vertical axis and the years on the horizontal axis. Have them compare their graph with Figure 30-4 and then write a paragraph explaining the following: which of the two states shows the greater population increase; which line on the graph is similar to the line on Figure 30-4; what effect immigration and emigration had on the population size; and why the population in one state increased faster than the other.

### Skill Check

**Make a line graph:** Find out what the population size was for your state in 1950, 1960, 1970, 1980, and 1990. Plot the data on a line graph. Make a point showing where you think the population for 2000 will be. *For more help, refer to the* ***Skill Handbook,*** *pages 715-717.*

## Population Changes

Let's examine how the human population has changed in the United States in the last forty years. You can see from the figures in Table 30-1 that the population is increasing.

**TABLE 30–1. UNITED STATES POPULATION**

| Year | Number of People |
|---|---|
| 1960 | 180 671 000 |
| 1970 | 205 052 000 |
| 1980 | 227 757 000 |
| 1990 | 249 600 000 |
| 1997 (est) | 268 000 000 |

The graph in Figure 30-4 shows the growth in world population since the year 1500. Several things can cause a population to change in size. If the number of births goes up or the number of deaths goes down, a population will increase. The population will decrease if there are more deaths than births. Animal populations can increase or decrease in other ways. Animals move from place to place. The movement of animals out of an area is called **emigration** (em uh GRAY shun). Emigration causes a decrease in population numbers.

Animals that move out of one population usually enter another population. The movement of animals into a population is called **immigration** (ihm uh GRAY shun). Immigration causes an increase in population numbers.

**How does immigration affect a population?**

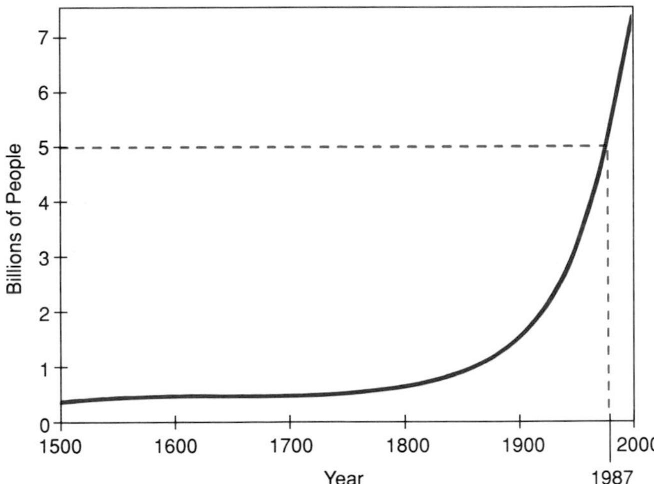


**Figure 30-4** This graph shows how the human population in the world has grown since 1500. What is the population expected to be in the year 2000?



## OPTIONS

**Enrichment:** Explain to students that the change in size of a plant population is dependent on the number of seeds that sprout and grow into new plants and how many plants are destroyed by insects and environmental factors.

### ACTIVITY/Project

Ask students to investigate why the human life span has increased. Booklets are available from life insurance companies that explain why humans are living longer. Survivor charts that show which age groups have higher death rates, and why, are also available.

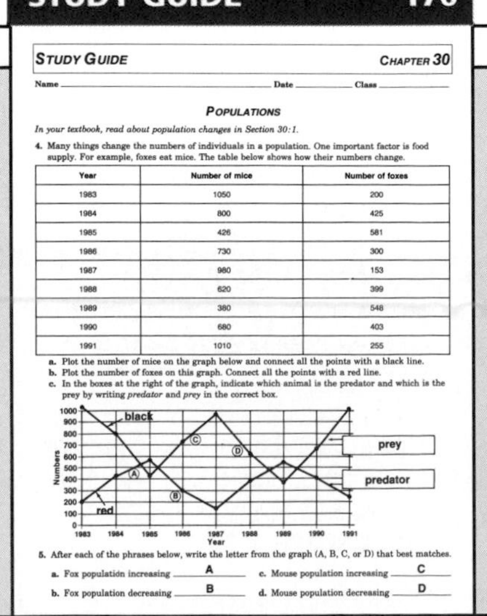

STUDY GUIDE 176

STUDY GUIDE — CHAPTER 30

Name ______ Date ______ Class ______

POPULATIONS

*In your textbook, read about population changes in Section 30:1.*

4. Many things change the numbers of individuals in a population. One important factor is food supply. For example, foxes eat mice. The table below shows how their numbers change.

| Year | Number of mice | Number of foxes |
|---|---|---|
| 1983 | 1050 | 200 |
| 1984 | 800 | 425 |
| 1985 | 426 | 581 |
| 1986 | 730 | 300 |
| 1987 | 980 | 153 |
| 1988 | 620 | 399 |
| 1989 | 380 | 548 |
| 1990 | 680 | 403 |
| 1991 | 1010 | 255 |

a. Plot the number of mice on the graph below and connect all the points with a black line.
b. Plot the number of foxes on this graph. Connect all the points with a red line.
c. In the boxes at the right of the graph, indicate which animal is the predator and which is the prey by writing *predator* and *prey* in the correct box.



5. After each of the phrases below, write the letter from the graph (A, B, C, or D) that best matches.

a. Fox population increasing ___A___
b. Fox population decreasing ___B___
c. Mouse population increasing ___C___
d. Mouse population decreasing ___D___

# Science and Society

## Problems with an Increasing Human Population

In 1975 the human population on Earth was four billion, or double its size in 1930. By 2020 the human population will probably double its size again, to almost 8.1 billion. Many countries today face food shortages each year. In the countries that lack food, one-half of the humans who die each year are children under the age of five. Some of these deaths are caused by diseases, but many of the deaths are caused by starvation.

Much of the land in the United States that was available for growing food crops is no longer available. Many farms have been sold, and the land is used for building homes and new industrial sites. Due to improved farming methods and better fertilizers, however, farmers can now grow more food on less land.

### What Do You Think?

**1.** Thirty years ago, most countries produced their own food. Now, the United States and six other countries are the only ones in the world that produce enough food to feed all of their people. These countries have extra food that they store or sell to other countries.

Some people think that a country with extra food should give it to countries that do not have enough. If they give it away, who will pay the farmer?

**2.** The populations of different countries are growing at different rates. In 1996, the world population grew at a rate of 1.4 percent. In some developing countries in Asia and Africa, however, the growth rate was over 3 percent. These countries are the least able to feed their increasing populations. Should the government control family size? If so, who should decide which families should have children and how many they should have?

In some countries, people have trouble raising enough food to eat.

**3.** Certain sprays that kill weeds and insects are banned in the United States because they are harmful. These sprays are sold to farmers in countries that have trouble raising enough food. If the sprays are not used, food production will decrease because some of the crops will be destroyed by insects or weeds. Should all countries pass laws banning the use of such sprays? Should the companies that make the chemicals be prevented from selling them?

**Conclusion:** Is it better to control population growth or to increase the amount of food produced?

635

### Suggested Readings


Byington, Scott. "Simulating Population Growth." *The American Biology Teacher*, June 1997, p. 353.

Dasgupta, Partha S. "Population, Poverty and the Local Environment." *Scientific American*, February 1995, pp. 40-45.

Elliot, Michael and Christopher Dickey. "Body Politics." *Newsweek*, September 12, 1994, pp. 22-27.

Foster, Rebecca. "The Cairo Conference: The Stakes, The Players." *Ms. Magazine*, September-October, 1994, p. 12.

Kiessling, Kerstin L. and Hans Landberg. *Population, Economic Development, and the Environment*. Oxford, England: Oxford University Press, 1994.

Prosterman, Roy L., Tim Hanstad, Li Ping. "Can China Feed Itself?" *Scientific American*, November 1996, pp. 90-96.



Science and Society


## Problems with an Increasing Human Population

### Background

By the year 2020 there will be eight billion people living on Earth. It has been estimated that food production on farms and ranches will have to increase to two or three times what it is today in order to feed everyone. In some non-industrialized nations, humans spend most of their waking hours gathering food for their families. Even though food production has increased in the past forty years in most of the world, the rate of food production has not increased fast enough to keep up with the human population increase.

### Teaching Suggestions

Have students find out: (a) which countries produce enough food to feed their people, (b) what natural disasters have occurred to keep food production down, and (c) what agricultural practices enable some countries to produce enough food. Students can work in cooperative groups for this assignment.

### What Do You Think?

1. Countries that give away surplus food often end up paying for the food through taxes. Surplus grains often end up not being used at all.
2. Some countries, in fact, do have government-controlled limits on family size. In other countries, infant and child mortality is high, so couples have more children to compensate.
3. Discuss the possibilities of using biological controls instead of pesticides.

**Conclusion:** Allow students to discuss different means of controlling population, from government-controlled limits to voluntary cooperation. Encourage students to discuss increasing crop yields through better farming techniques, fertilizer, and irrigation.

## TEACH

### Connection: Math

Students can get the wind speed and temperature from the local paper and calculate the wind chill for a week. Wind chill charts are found in most encyclopedias.

### Guided Practice

Have students write down their answers to the margin question: it slows down.

### *Student Journal*

After you discuss the limits of population size, take the students on a walk around the school or a field or forest if it is near the school. Have them write in their journals the names of the limiting factors and how they affect specific organisms in the area. Have them explain how the limiting factors would have to be changed to make the conditions better for these organisms.

### Check for Understanding

Have students respond to the first three questions in Check Your Understanding.

**Reteach**

**Reteaching**/*Teacher Resource Package,* p. 87. Use this worksheet to give students additional practice with limiting factors.

**Extension:** Assign Critical Thinking, Biology and Math, or some of the **OPTIONS** available with this lesson.

## 3 APPLY

**ACTIVITY/Software**

*Population Concepts,* Queue, Inc.

## 4 CLOSE

**ACTIVITY/Software**

*Ecology,* Queue, Inc.

**Figure 30-5** Competition affects the size of a population. What is the limiting factor in this photo? food

**What happens to plant growth when space is limited?**

## Limits of Population Size

Why don't populations increase forever? What prevents the world from being overrun with all kinds of living things? Any condition that keeps the size of a population from increasing is called a **limiting factor.** Almost anything that affects the lives of organisms can be a limiting factor. Lack of light, space, water, or food are all limiting factors.

A population uses more sunlight, space, water, and food as it increases in size. Members of the same and similar species have the same needs. Remember from Chapter 29 that the struggle among organisms to get their needs for life is called competition. An increase in the population size causes more competition for the same materials. Some organisms may not get enough of these materials. The lack of needed materials causes population growth to slow down by decreasing the number of births and increasing the number of deaths.

Plant roots growing close to each other compete for nutrients, water, and space. The stems and leaves of different plants compete for light. A decrease in the amount of space and light causes the growth of plants to slow down. Why do gardeners pull out some of their lettuce seedlings?

Look again at the human population graph in Figure 30-4. The line slants upward showing a rapid increase in population. What does this tell you about the limiting factors in a human population? Could you make a prediction about the population size in the year 2000? Do you think human populations are affected by limiting factors? not yet

### Check Your Understanding

1. Explain which is easier to count, plant populations or animal populations.
2. There are very few deaths in a deer population, yet the population is decreasing. Explain what might be happening to cause the decrease.
3. How can limiting factors cause a population change?
4. **Critical Thinking:** What might become limiting factors in a country whose population becomes too large?
5. **Biology and Math:** If there are 10 wolves in one pack roaming over an area of 20 square kilometers, what is the average density of wolves per square kilometer?



### Answers to Check Your Understanding

1. plants, because they do not move about
2. Some of the deer could be emigrating to a new area, there could be fewer births, or limiting factors could be decreasing the population.
3. Living things compete for resources. When these resources are scarce, the population size may decrease. When these resources are abundant, population size may increase.
4. food, space, water, mates, shelter, jobs
5. 1/2 wolf per $km^2$

**RETEACHING 87**

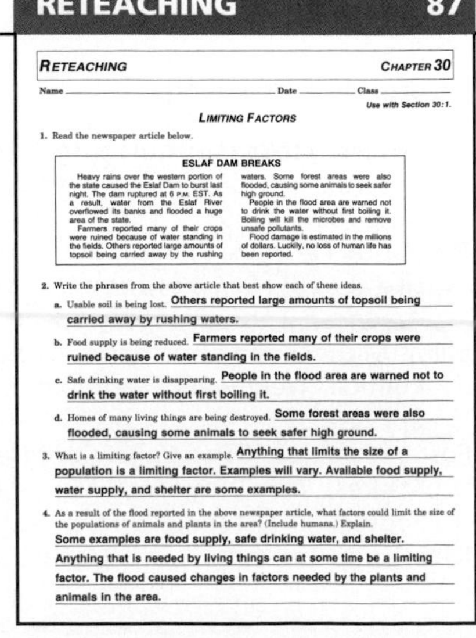

RETEACHING CHAPTER 30

Name ______ Date ______ Class ______

Use with Section 30:1.

LIMITING FACTORS

1. Read the newspaper article below.

ESLAF DAM BREAKS

Heavy rains over the western portion of the state caused the Eslaf Dam to burst last night. The dam ruptured at 6 P.M. EST. As a result, water from the Eslaf River overflowed its banks and flooded a huge area of the state.

Farmers reported many of their crops were ruined because of water standing in the fields. Others reported large amounts of topsoil being carried away by the rushing waters. Some forest areas were also flooded, causing some animals to seek safer high ground.

People in the flood area are warned not to drink the water without first boiling it. Boiling will kill the microbes and remove unsafe pollutants.

Flood damage is estimated in the millions of dollars. Luckily, no loss of human life has been reported.

2. Write the phrases from the above article that best show each of these ideas.
   a. Usable soil is being lost. **Others reported large amounts of topsoil being carried away by rushing waters.**
   b. Food supply is being reduced. **Farmers reported many of their crops were ruined because of water standing in the fields.**
   c. Safe drinking water is disappearing. **People in the flood area are warned not to drink the water without first boiling it.**
   d. Homes of many living things are being destroyed. **Some forest areas were also flooded, causing some animals to seek safer high ground.**
3. What is a limiting factor? Give an example. **Anything that limits the size of a population is a limiting factor. Examples will vary. Available food supply, water supply, and shelter are some examples.**
4. As a result of the flood reported in the above newspaper article, what factors could limit the size of the populations of animals and plants in the area? (Include humans.) Explain. **Some examples are food supply, safe drinking water, and shelter. Anything that is needed by living things can at some time be a limiting factor. The flood caused changes in factors needed by the plants and animals in the area.**

# Lab 30–1 Competition

## Problem: What happens when two animal species compete?

### Skills

form hypotheses, interpret data, infer, design an experiment

### Materials

graph paper
ruler

### Procedure

In 1973, there were no houses in a large, old field. By 1988, there were 250 houses. A biologist found a population of dusky field mice living in the field. The biologist sampled this population from 1977 to 1992. Write a **hypothesis** in your notebook about what you think happened to the mouse population between 1977 and 1992.

1. Prepare a sheet of graph paper as follows. Mark the horizontal axis with the years 1977 to 1994. Label the horizontal axis "Years Sampled."
2. Mark the vertical axis from 0 to 90. Label the vertical axis "Size of Mouse Population per 1000 Square Meters."
3. Study the information in the table.
4. Plot the data for the field mouse population on the graph.
5. Connect the data points with a ruler to make a line graph. Continue the line to the right so you can estimate the size of the mouse population in 1994.

| YEAR | NUMBER OF MICE PER 1000 SQUARE METERS | YEAR | NUMBER OF MICE PER 1000 SQUARE METERS |
|---|---|---|---|
| 1977 | 89 | 1985 | 47 |
| 1978 | 85 | 1986 | 45 |
| 1979 | 82 | 1987 | 36 |
| 1980 | 79 | 1988 | 29 |
| 1981 | 78 | 1989 | 26 |
| 1982 | 51 | 1990 | 17 |
| 1983 | 48 | 1991 | 14 |
| 1984 | 42 | 1992 | 7 |

### Data and Observations

1. Did the number of mice increase or decrease between 1977 and 1988?
2. Between which two years was the greatest change in population seen?

### Analyze and Apply

1. **Infer:** What limiting factors are affecting the population size of the dusky field mouse?
2. With what other living thing do you think these mice are competing?
3. What might be happening to the birthrate of mice?
4. Suppose in 1984, 17 mice emigrated. What do you think would happen to the mouse population?
5. Predict what will happen to the number of mice in 1994.
6. List two factors that can increase the mouse population.
7. What effect would immigration have on the mouse population?
8. **Check your hypothesis:** Was your hypothesis supported by the data? Why or why not?
9. **Apply:** What happens when two animal species compete?

### Extension

**Design an experiment** that would show how a limiting factor, such as size of habitat, affects a plant population.



## ANSWERS

### Data and Observations

1. decrease
2. between 1981 and 1982

### Analyze and Apply

1. space and food
2. humans
3. decreasing
4. The population would sharply decrease. There would be less competition for space, food, and shelter.
5. The mice will probably disappear from the field.
6. immigration and increased birthrate
7. It would increase the size of the mouse population and cause competition for food, space, and shelter.
8. Students who hypothesized that the mouse population would decrease and probably disappear from the field will say that their hypotheses were supported.
9. The species that is more successful in obtaining its needs is able to increase in population size.

## Lab 30-1 Competition

### Overview

This simulation gives the student a chance to see how the size of one population (mouse) can decrease because of an increase in the size of another population (human). The concepts of limiting factor, competition, and population change are reinforced.

**Objectives:** Students will be able to: (1) **demonstrate** how a population changes in size over time, (2) **interpret** a graph, and (3) **explain** the cause and effect of competition.

**Time Allotment:** 45 minutes

### Preparation

▶ Have graph paper and a ruler for each student.

Lab 30-1 worksheet/*Teacher Resource Package,* pp. 117-118.

### Teaching the Lab

▶ Be sure students use sharp pencils when plotting their data. Many students will not know how to set up the graph. Review graphing techniques during the prelab.

▶ **Troubleshooting:** Give students an idea of how big an area 1000 square meters is: about 1/5 the size of a football field.

**Cooperative Learning:** Divide the class into groups of two students. For more information, see pp. 22T-23T in the Teacher Guide.

### ASSESSMENT

**Skill:** Have students determine the effect of immigration on the population of mice from 1977-1992. Tell students that in the even years, eight mice immigrated into the area, and in the odd years, four mice immigrated into the area. Direct students to use this information to calculate the size of the population due to immigration and then graph the population size on the original graph using a different color pen. Have students explain the effect of immigration on the population of mice.

# 30:2 Communities

## PREPARATION

### Materials Needed

Make copies of the Skill, Reteaching, and Study Guide worksheets in the *Teacher Resource Package.*

### Key Science Words

community
habitat
niche
primary consumer
secondary consumer

### Process Skills

In the Mini Lab, students will make and use tables.

## 1 FOCUS

▶ The objectives are listed on the student page. Remind students to preview these objectives as a guide to this numbered section.

### Divergent Question

Ask students to describe some interactions they have observed between two living things. Ask them to tell which of the two living things depended on the other.

### Guided Practice

Have students write down their answers to the margin question: the place where an organism lives.

### Portfolio

Have students make a chart with three headings: organism, habitat, niche. Have them list the names, the habitat, and the niche for each of 10-15 organisms that live in their area. They should include a paragraph to explain what can happen to any organism that shares the same habitat with another organism, or tries to share the same niche with another organism.

# 30:2 Communities

### Objectives

4. **Describe** different parts of a community.
5. **Explain** the importance of producers, consumers, and decomposers.

### Key Science Words

community
habitat
niche
primary consumer
secondary consumer

The different populations in an area depend on each other. How can this be true? Look again at Figure 30-1. The daisies in the field can provide shelter for the spiders. The mice can eat the seeds of the daisy. What do you think the earthworm depends on? How do the mice and the earthworms help the daisies?

## Parts of a Community

The populations of daisies, mice, and earthworms in Figure 30-1 make up a community. A **community** is all of the living things in an area that depend upon each other. Some communities, like the field of daisies, may be easy to see. Other communities may be harder to see. Is there a pond or lake near your home? What living things make up the pond community? Communities can be identified by the kinds of living things found there. What types of communities can you identify?

What is a habitat?

In a community, every living thing has a place to live that best suits it. You know that birds live in trees, fish live in ponds, and people live in houses. The place where a plant or animal lives is its **habitat** (HAB uh tat). A squirrel may use several different trees in the forest as its habitat, while a skunk may use one hollow log and the surrounding area as its habitat. Your skin is the habitat for the millions of bacteria that live there. Figure 30-6 shows a lichen attached to a rock. What do you think the lichen's habitat is?

**Figure 30-6** Some lichens can grow on rocks.

Every living thing in a community also has a function, or job. The job of the organism in the community is its **niche** (NITSH). The job of most green plants is to produce food from sunlight, water, and carbon dioxide by photosynthesis. Animals have jobs, too. Earthworms help break up the soil so that plant roots can get water and nutrients. Bees pollinate flowers when they collect nectar to make honey.

You live in the community, too. Your habitat is your home, your school, and other places where you spend time. Your job is to study and learn. Your niche is that of a student. Think about other niches in your community. What do you think is the niche of an auto mechanic? How do other people depend on the mail carrier or the bus driver? How does your teacher depend upon you? How do you depend on your teacher?



## OPTIONS

### Science Background

The differences between environmental factors in two bordering communities tend to keep organisms from spreading to a new community. Communities are named after the dominant living thing or land feature, for example, a grassland or pond.

**Skill**/*Teacher Resource Package,* p. 31. Use the Skill worksheet shown here to reinforce the concept of community dynamics.

### SKILL 31

SKILL: INFER — CHAPTER 30

Name ______ Date ______ Class ______

*For more help, refer to the Skill Handbook, pages 706-711.* *Use after Section 30:2.*

#### A MEADOW COMMUNITY

A three-year study of the South Fork Meadow showed that 36 kinds of animals lived in the community. Of these, 6 kinds of animals made up the largest populations. These animals are shown in the table below. Grasses and weeds made up the plant populations of the community.

**The Most Abundant Animals at South Fork Meadow**

| Animal | Mouse | Rabbit | Hawk | Deer | Snake | Cricket |
|---|---|---|---|---|---|---|
| Food source in meadow | seeds | plant parts | mice, birds, snakes | plant purts | mice, birds, rabbits | plant parts |

In the summer of the second year of the study, the populations of the crickets, mice, deer, and rabbits increased in size. In contrast, the hawk and snake populations stayed nearly the same sizes as the first year's populations. By the middle of the second summer, the meadow had developed many bare spots where the soil could be seen. By the middle of the third year, the cricket, rabbit, and mouse populations had decreased to about the population sizes of the first year. But, the hawk and snake populations had nearly doubled. The amount of bare soil had decreased.

1. Which organisms are the primary consumers in the meadow? **mouse, rabbit, deer, and cricket**
2. Which organisms are the secondary consumers in the meadow? **hawk and snake**
3. Which organisms are the producers in the meadow? Explain. **grasses and weeds, because they make their own food**
4. In which of the three years do you think there was a better balance in the meadow community? Why? **The first year because there were enough producers to feed the primary consumers**
5. Which populations do you think caused the balance in the community to change for the worse? Why do you think this change came about? **The primary consumers caused the balance to change. Their populations increased more quickly than their food, the plants, could reproduce.**
6. Why did the hawk and snake populations increase during the third year? **Because their food supply was so plentiful, they could grow and reproduce quickly.**

a

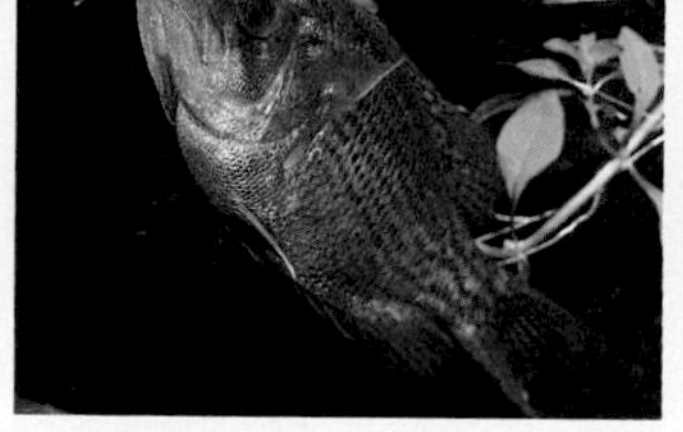

b

c

**Figure 30-7** In a pond community, water plants (a) may be the producers, fish (b) may be the primary consumers, and turtles (c) may be the secondary consumers.

## Producers

All of the organisms in a community are related to each other through their jobs. These jobs may be divided into three types: producers, consumers, and decomposers.

Producers are the organisms that make food in a community. In almost all communities, green plants, algae, or blue-green bacteria are the producers. Producers use some of the food for themselves. The food they make will also help to feed all the animals in the community. Producers also make oxygen during photosynthesis. Other organisms need the oxygen to live.

**What do producers do for a community?**

## Consumers

The animals in a community must have food. Animals can't make their own food, so they are consumers. Consumers are organisms that eat other organisms.

Many animals, such as mice and deer, get food by eating plants. Animals that eat only plants are **primary consumers.**

Some animals eat other animals. Owls eat mice, and turtles eat fish. These animals are secondary consumers. **Secondary consumers** are animals that eat other animals.

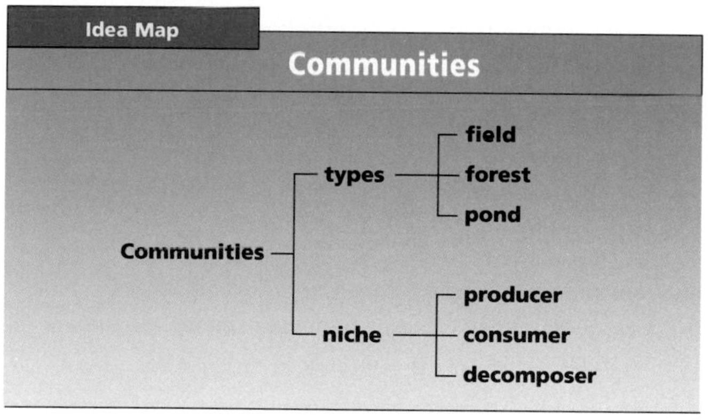


**Study Tip:** Use this idea map as a study guide to communities. Can you think of other types of communities?



# 2 TEACH

### MOTIVATION/Videocassette

Show the videocassette *Matter, Energy, and Ecosystems,* Insight Media.

### Concept Development

▶ Point out that some protists are producers in aquatic communities.

### Guided Practice

Have students write down their answers to the margin question: They make food from the sun for the community.

### Idea Map

Have students use the idea map as a study guide to the major concepts of this section.

**Cooperative Learning:** Have student teams make charts and posters of animals that are primary consumers and animals that are secondary consumers. Provide *National Geographic, Audubon,* and *National Wildlife* magazines for their use. Have each team report on their findings. Have them point out the features that each group of animals has, i.e., hooves on some primary consumers and claws on many secondary consumers.

**Misconception:** Students think that mushrooms get their food from the soil in the same way trees get their minerals from the soil. Point out that mushrooms are using dead matter in the soil as food. Explain that they absorb organic matter by secreting enzymes to break down the dead matter into nutrients.

### ACTIVITY/Videocassette

Show the videocassette *To Know a Pond,* Insight Media.

### Independent Practice

**Study Guide**/*Teacher Resource Package,* p. 177. Use the Study Guide worksheet shown at the bottom of this page for independent practice.

**STUDY GUIDE 177**

STUDY GUIDE — CHAPTER 30

Name ______ Date ______ Class ______

COMMUNITIES

*In your textbook, read about the parts of a community in Section 30:2.*

1. This picture of a community shows many different kinds of living things. Using these colors, shade the following parts.

Green—producers
Red—primary consumers
Yellow—secondary consumers
Blue—decomposers

green green green yellow red red green red yellow blue

2. What is the habitat of the deer? the field
3. What is the niche of the deer? primary consumer
4. List two decomposers in this community. bacteria, fungi (mushrooms and molds)
5. What producers are present in this community? grass, trees, plants
6. List two secondary consumers and tell what they eat. bird—worm; large member of cat family—cattle or deer

## OPTIONS

### ACTIVITY/Challenge

Ask students to design a space station community and to list all the organisms that would be needed to sustain life.

## TEACH

### Mini Lab

Make a chart so each student has the same data. Use the headings Organisms Seen, Producers, Consumers, Decomposers.

### ✓ ASSESSMENT

**Content:** After you have discussed Section 30:3, have students use the organisms from their Table of Organisms to list several food chains that could occur in the community. They may have to use reference books to find out what some of the organisms on their list eat.

## 3 APPLY

### ACTIVITY/Software

*Ecosystems; Coexist: Population Dynamics; Compete: Plant Competition*, Queue, Inc.

## 4 CLOSE

### ACTIVITY/Field Trip

Take a field trip to observe organisms in their environment. Have students list the producers, consumers, and decomposers they find.

### Answers to Check Your Understanding

6. The earthworm's habitat is the soil. Its niche is to break up soil so plants can grow.
7. Producers make their own food from light, while consumers depend on others for food.
8. They recycle nutrients and break down dead matter.
9. producers and decomposers
10. When humans eat only plants, they are primary consumers. When humans eat other animals, they are secondary consumers.

**Figure 30-8** A forest community has many different producers, consumers, and decomposers.

### Mini Lab

**What's in a Community?**

**Make and use tables:** Make a table listing 10 organisms you see in your area and whether each is a producer, consumer, or decomposer. Which appears in the largest number? *For more help, refer to the **Skill Handbook**, pages 715-717.*

## Decomposers

In the pond and many other communities, decomposers form another group. Remember from Chapter 4 that decomposers are living things that get their food from breaking down dead matter. Bacteria and fungi, such as molds and mushrooms, are decomposers. Decomposers break down dead matter into simpler chemicals. Decomposers are very important to a community. They recycle nutrients so other organisms can use them. Without decomposers, Earth would soon be covered by dead material.

The forest community shown in Figure 30-8 has many different populations of producers, consumers, and decomposers. There are populations of ferns, oak trees, squirrels, and mushrooms. The upper left-hand corner of the figure shows a section of soil from the forest community. This soil section has been enlarged about 2000 times to show the small organisms found there. Bacteria, fungi, and protozoan populations fill the soil. Many of these organisms are decomposers.

### Check Your Understanding

6. What is the habitat and niche of an earthworm?
7. How are producers different from consumers?
8. Why are decomposers important to a community?
9. **Critical Thinking:** Which of the following groups are necessary to life on Earth: producers, primary consumers, secondary consumers, decomposers?
10. **Biology and Reading:** Are humans primary consumers or secondary consumers? Explain your answer.



## OPTIONS

### ACTIVITY/Project

Have students bring in articles from newspapers and magazines that report on producers, consumers, and/or decomposers. Have them report on the articles and explain how useful these organisms are.

### RETEACHING 88

**RETEACHING: IDEA MAP** — CHAPTER 30

Name ________ Date ________ Class ________

Use with Section 30:2.

**POPULATIONS AND COMMUNITIES**

*Use the following terms to help you complete the idea map: beneath logs, consumers, decomposers, soil, producers, trees.*

Forest community
- niches
  - producers
  - consumers
  - decomposers
- habitats
  - trees
  - soil
  - beneath logs

1. What would you call a group of living things of the same species in a community? **a population**
2. What two things does an organism obtain from its habitat? **shelter and food**
3. What are the habitats of some of the living things that live in a forest? **trees, soil, and beneath logs**
4. What are three niches in every community? **producers, consumers, and decomposers**
5. How do consumers depend on producers? **Consumers eat the food and use the oxygen made by producers.**
6. What is the niche of a decomposer? **A decomposer breaks down dead matter into simpler chemicals.**

# 30:3 Energy in a Community

Recall that energy is the ability to do work. You learned in Chapter 2 that organisms get energy from food. This food energy can be traced back to the sun. How?

Producers, or green organisms, use light energy to make food. This light energy comes from the sun. To make food, green organisms change energy from the sun into chemical energy (sugar). This chemical energy is, in turn, used by consumers. A community's energy source is the sun.

**Objectives**

6. **Trace** the path of energy and materials through a community.
7. **Explain** how food chains are connected.

**Key Science Words**

food chain
food web
energy pyramid

## Food Chains

When an animal eats a plant and is then eaten by another animal, you have a food chain. A **food chain** is a pathway of energy and materials through a community. Grass→grasshopper→bird is an example of a simple food chain. Each of the living things in it is a link in the food chain. Each depends on the other living things in the food chain. Figure 30-9 shows several important steps in a food chain.

**What is an example of a simple food chain?**

**Figure 30-9** A food chain is a pathway of food through a community.



## 30:3 Energy in a Community

## PREPARATION

### Materials Needed

Make copies of the Enrichment, Transparency Master, Reteaching, and Study Guide worksheets in the *Teacher Resource Package*.

### Key Science Words

food chain
food web
energy pyramid

## 1 FOCUS

▶ The objectives are listed on the student page. Remind students to preview these objectives as a guide to this numbered section.

### Convergent Question

Ask students where they think they get their energy for their body activities. Students should conclude that their energy comes from the food they eat. They should recognize that their energy ultimately comes from the sun.

### Guided Practice

Have students write down their answers to the margin question: grass–grasshopper–toad.

**GLENCOE TECHNOLOGY**

**Videodisc**

**Biology: The Dynamics of Life**
Movie: *How Organisms Interact*
Disc 1, Side 1, Ch. 4

**ENRICHMENT 32**

ENRICHMENT — CHAPTER 30

Name ______ Date ______ Class ______

*Use after Section 30:3.*

**A SALT MARSH FOOD WEB**

A salt marsh is a community that includes plants and animals that live in salt water and those that live on the land nearby. The picture shows a food web in a salt marsh.

*Use the picture to help you answer the questions below.*

1. Name the producers in the salt marsh community. **land plants and water plants**
2. Name four primary consumers in the salt marsh community. **snails, smelt, insects, and crayfish**
3. Name four secondary consumers in the salt marsh community. **vole, heron, sandpiper, rail**
4. Name two animals in the salt marsh community that eat both plants and animals. **rail and vole**
5. Name the two animals at the top of the salt marsh food chains. **marsh hawk and short-eared owl**
6. Why are the decomposers shown as they are in this picture? **The decomposers get their energy from all the organisms in the salt marsh.**

Marsh hawk, Short-eared owl, Vole, Insects, Heron, Sand piper, Rail, Snail, Land plants, Decomposers, Snail, Water plants, Smelt

## OPTIONS

### Science Background

Organisms store some energy for their body processes, but most is lost as heat. Short food chains are more efficient in transferring energy than longer food chains.

**Food Web**/*Transparency Package*, number 30. Use color transparency number 30 as you teach about food webs.

**Enrichment**/*Teacher Resource Package*, p. 32. Use the Enrichment worksheet shown here as an extension to food webs.

# 2 TEACH

### Check for Understanding

Have students make a single food chain of local organisms and list producers, primary and secondary consumers, and decomposers.

### Reteach

List several food chains on an overhead transparency. Connect the food chains into food webs.

### Guided Practice

Have students write down their answers to the margin question: interconnected food chains in a community.

### Independent Practice

**Study Guide**/*Teacher Resource Package,* p. 178. Use the Study Guide worksheet shown at the bottom of this page for independent practice.

### *Portfolio*

Have students make a table with the following headings: producers, primary consumers, secondary consumers, decomposers. Then, have them write the names of the organisms in Figure 30-10 under the correct heading. Ask them to write a paragraph explaining what they think will happen in the community if both mice and rats disappear from the community.

**Figure 30-10** This food web shows the relationships among living things in a meadow community.

> **Bio Tip**
>
> **Consumer:** On a farm, it takes 16 kilograms of plants to produce one kilogram of beef, but only five kilograms of plants to produce one kilogram of chicken. Think about this the next time you help with the shopping.

1. Food chains start with producers such as plants or other green organisms. They are the only organisms able to make their own food.
2. The next links in the food chain are the primary consumers (plant-eating animals).
3. Above the primary consumers are the secondary consumers (meat-eating animals). These animals eat primary consumers.
4. All living things die. They are food for the decomposers.

The arrows in the diagram show the direction in which the energy moves. In our example, plant leaves are eaten by grasshoppers, so an arrow points from the plant to the grasshopper. The next arrow shows that the grasshopper is eaten by a rat. Decomposers, in turn, feed on all organisms after they have died. Decomposers include mushrooms, bacteria, and protists.

Several different primary consumers can eat the same producer. Several secondary consumers can eat the same primary consumer. The same producer or consumer may be part of several different food chains. Several food chains may be connected. Figure 30-10 shows how. Food chains connected in a community are called a **food web.** A food web shows how energy is moved through a community.

**What is a food web?**



## *OPTIONS*

### ACTIVITY/Challenge

Set up a jar of pond water with protozoans, algae, water fleas, and other micro-invertebrates. Have the students make drawings of the organisms they see and use the drawings to make several food chains.

**Transparency Master**/*Teacher Resource Package,* p. 151. Use the Transparency Master shown here to show relationships in an energy pyramid.

**STUDY GUIDE 178**

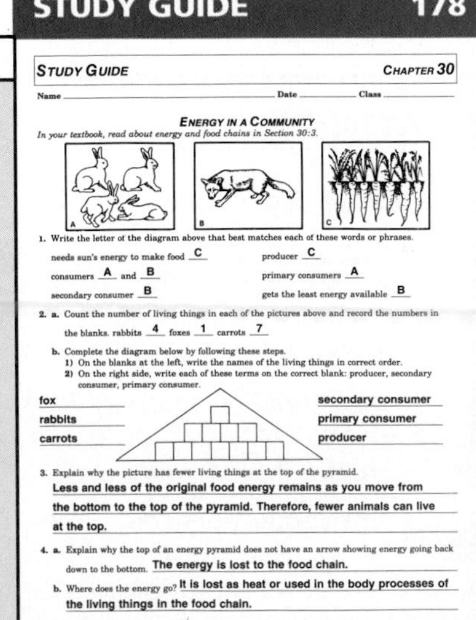
STUDY GUIDE — CHAPTER 30

Name ______ Date ______ Class ______

ENERGY IN A COMMUNITY

*In your textbook, read about energy and food chains in Section 30:3.*

A B C

1. Write the letter of the diagram above that best matches each of these words or phrases.

needs sun's energy to make food C — producer C

consumers A and B — primary consumers A

secondary consumer B — gets the least energy available B

2. a. Count the number of living things in each of the pictures above and record the numbers in the blanks. rabbits 4 foxes 1 carrots 7

b. Complete the diagram below by following these steps.
1) On the blanks at the left, write the names of the living things in correct order.
2) On the right side, write each of these terms on the correct blank: producer, secondary consumer, primary consumer.

fox — secondary consumer
rabbits — primary consumer
carrots — producer

3. Explain why the picture has fewer living things at the top of the pyramid. Less and less of the original food energy remains as you move from the bottom to the top of the pyramid. Therefore, fewer animals can live at the top.

4. a. Explain why the top of an energy pyramid does not have an arrow showing energy going back down to the bottom. The energy is lost to the food chain.

b. Where does the energy go? It is lost as heat or used in the body processes of the living things in the food chain.

**TRANSPARENCY 151**

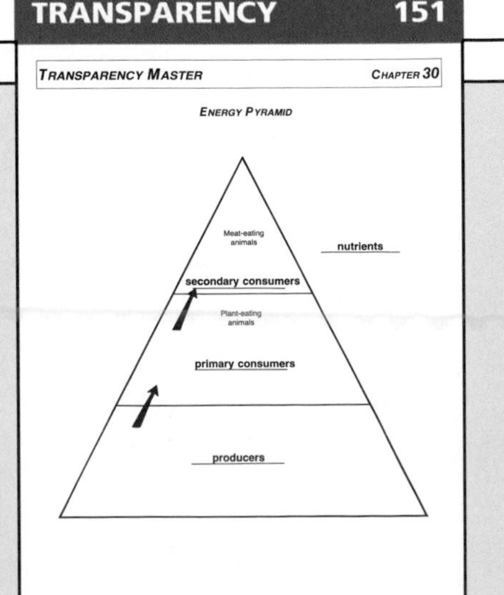

## Energy Flow in the Community

A producer gets energy from the sun and uses it to produce food and oxygen. Much of the food made through photosynthesis is used by the producer for growth and other cell processes. A large amount of the energy is lost as heat. Only a small amount of the food energy produced is available to the primary consumer that eats the producer. Therefore, primary consumers must eat large numbers of plants to get the energy they need for their cell processes and growth. When a secondary consumer eats a primary consumer, again only some of the food energy is available. The secondary consumer must eat several animals to get enough energy. You can see that each step of the food chain has fewer organisms than the step before it.

At each step of the food chain, from producers to primary consumers to secondary consumers, the amount of available energy becomes less and less. This loss of energy through a food chain can be shown in the shape of a pyramid. An **energy pyramid** is a diagram that shows energy loss in the food chain. Figure 30-11 shows an energy pyramid. The pyramid is wider at the bottom than at the top. There is more available energy at the bottom of the pyramid than at the top. Notice in Figure 30-11 that the number of primary consumers is smaller than the number of producers. Notice also that the number of secondary consumers is smaller than the number of primary consumers. There are fewer organisms at the top of the pyramid because less energy is available to them. Figure 30-11 also shows that all living things lose heat to their surroundings.

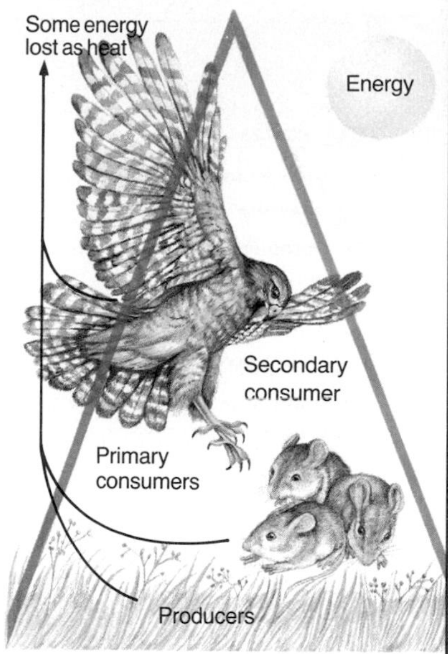


**Figure 30-11** An energy pyramid shows energy loss in a food chain.

### Check Your Understanding

11. Why must primary consumers eat large numbers of plants?
12. Which would be affected more by the loss of a species—a food chain or a food web? Why?
13. Why are plants necessary to the path of energy through a community?
14. **Critical Thinking:** What would happen to the food chains in an area if all plants were to die?
15. **Biology and Reading:** What happens to most of the energy at each level of a food chain?



## TEACH

### Check for Understanding

Have students respond to the first three questions in Check Your Understanding.

### Reteach

**Reteaching**/*Teacher Resource Package*, p. 89. Use this worksheet to give students additional practice with energy flow in a community.

**Extension:** Assign Critical Thinking, Biology and Reading, or some of the **OPTIONS** available with this lesson.

## 3 APPLY

### ACTIVITY/Hands-On

Have pictures of plants, animals, and fungi from magazines mounted on construction paper. Give each team of four students seven to ten pictures. Have them develop a food chain using the pictures. Have them present their food chain to the class. When the teams have presented their food chains, have several teams use their food chains to make a food web to present to the class.

## 4 CLOSE

### ACTIVITY/Software

*Food Chains,* Queue, Inc.

### Answers to Check Your Understanding

11. because only a small amount of the energy in the producers is available to the primary consumer
12. a food chain; The species that is lost represents a significant portion of the food chain, and its consumers will not be likely to have other food sources.
13. Plants are the only organisms that can convert sunlight energy into food energy.
14. The consumers and decomposers would die because they would have no food.
15. At each level of the food chain, most of the energy is lost as heat.

**RETEACHING 89**

RETEACHING: IDEA MAP — CHAPTER 30

Name ______ Date ______ Class ______

Use with Section 30:3.

ENERGY IN A COMMUNITY

*Use the following words or phrases to complete the idea map: secondary consumers, decomposers, producers, cell processes, primary consumers, growth. You may use a word or phrase more than once.*

Energy → transfer of food in a food chain → secondary consumers → decomposers; primary consumers → decomposers; producers → decomposers
Energy → used up as food → cell processes; growth

1. What is the beginning source of energy in a community? the light from the sun
2. Which group of organisms are able to use this energy in the community? producers
3. What does this group of organisms do with the energy they receive? They make food and oxygen.
4. Which groups of organisms in a community cannot make food? consumers and decomposers
5. What is the source of energy for these organisms in the community? living or dead animals and plants
6. What do all organisms need energy for? growth and cell processes
7. How does light energy from the sun eventually reach the decomposers? Plants use the light from the sun to make food, and this is passed on to the consumers. When the producers and consumers die, the decomposers use these organisms for energy.

### OPTIONS

**What Are Some Parts of a Food Chain and a Food Web?**/*Lab Manual,* pp. 251-254. Use this lab as an extension to food webs.

## 30:4 Relationships in a Community

## PREPARATION

### Materials Needed

Make copies of the Application, Critical Thinking/Problem Solving, Reteaching, Study Guide, and Lab worksheets in the *Teacher Resource Package.*

▶ Get owl pellets, cardboard, and glue ready for Lab 30-2. The cardboard may be cut from large cartons.

### Key Science Words

commensalism
parasitism
predation
predator
prey

### Process Skills

In Lab 30-2, students will observe, infer, and design an experiment.

## 1 FOCUS

▶ The objectives are listed on the student page. Remind students to preview these objectives as a guide to this numbered section.

### Divergent Question

Ask students to explain some ways in which two organisms can be helpful or harmful to one another. Have them list the names of specific organisms. You could start with the example of a hawk and a mouse. List answers on the chalkboard.

### Guided Practice

Have students write down their answers to the margin question: a relationship in which two organisms live together and depend on each other for survival.

## 30:4 Relationships in a Community

**Objectives**

8. **Give examples** of mutualism.
9. **Describe** commensalism.
10. **Compare** parasitism with predation.

**Key Science Words**

commensalism
parasitism
predation
predator
prey

There are many types of relationships within a community. Each relationship shows a different way that one living thing affects another. The relationship of one kind of organism to another may be helpful to both. In other relationships, one organism benefits while the second is harmed. There are still other cases where one organism is helped and the other is neither harmed nor helped. Communities could not exist without these different relationships.

### Mutualism

Suppose there is a good rock concert in town and you would like to go. Right now, you do not have enough money for a ticket. Your best friend offers to buy you a ticket for this concert if you will give her a ride to the concert. It sounds like a good deal. Both you and your friend benefit from the arrangement.

What is mutualism?

This is an example of mutualism. Remember from Chapter 5 that mutualism is a relationship in which two organisms live in a community and depend on each other. Both organisms benefit from the relationship. Mutualism occurs among members of all five kingdoms. You first read about this kind of relationship in Chapter 5 when you read about lichens.

An example of a relationship between an animal and a protist can be seen in a dead log that contains termites, Figure 30-12. Termites are insects that eat wood. They are

**Figure 30-12** The relationship among protists (a) living in the intestines of termites (b) is an example of mutualism.



## OPTIONS

### Science Background

Sometimes, two populations have very little effect on each other. Some of the relationships between two populations can be helpful to both populations while some are harmful to one of the populations. In some cases, only one population benefits while the other is unaffected. These relationships between populations can cause one of the populations in the community to increase or decrease in size over a period of time.

**Figure 30-13** The relationship between orchid plants and trees is an example of commensalism.

not able to digest wood. They could not use it for energy if it weren't for a protist that lives inside the termite's intestine. The protist can digest the wood for the termite so that the termite can use it for energy. The protist has a home inside the termite and uses some of the wood for its own energy.

Humans can show mutualism with other organisms. An example is the bacterium that lives in the human intestine. The bacterium makes vitamin $B_{12}$ for the human, who provides a home and food for the bacterium.

**Bio Tip**

**Health:** The bacterium that makes vitamin $B_{12}$ in the human intestine also makes vitamin K, which is needed for blood clotting. Newborn babies have to build up the population of these bacteria before they can get enough vitamin K in their bodies to clot their blood.

## Commensalism

Let's return to the example of the rock concert tickets. Suppose your friend has a free ticket, but can't go to the concert. She gives the ticket to you. You benefit from the free ticket. Whether or not you use the ticket doesn't matter to your friend, does it? She got the ticket free. She won't lose anything whether you use the ticket or not.

What is commensalism?

Commensalism (kuh MEN suh lihz um) is seen in this example. **Commensalism** is a relationship in which two organisms live in a community, and one benefits while the other gets no benefit and is not harmed.

Commensalism occurs between many living things. For example, orchid plants and trees have this kind of relationship. Orchids receive more sunlight for photosynthesis if they grow high on tree branches rather than close to the ground, Figure 30-13. The orchid plants are helped by the tree. The tree gets nothing in return from the orchids.



## 2 TEACH

### MOTIVATION/Filmstrip

Show the filmstrip *Symbiosis,* Carolina Biological Supply Co.

### Concept Development

▶ Explain to the students that *mutual* means having something in common.

▶ Make sure students understand that one organism benefits and the other is not harmed in a commensal relationship.

### Audiovisual

Make an overhead transparency listing the organisms that live as mutualists.

**Analogy:** The analogy of the rock concert will be discussed with the topics of mutualism, commensalism, and parasitism. In each section, the analogy of the ticket and two friends should help to reinforce these three community relationships.

### Guided Practice

Have students write down their answers to the margin question: a relationship in which one organism benefits and the other is unaffected.

### Independent Practice

**Study Guide**/*Teacher Resource Package,* p. 179. Use the Study Guide worksheet shown at the bottom of this page for independent practice.

**STUDY GUIDE 179**

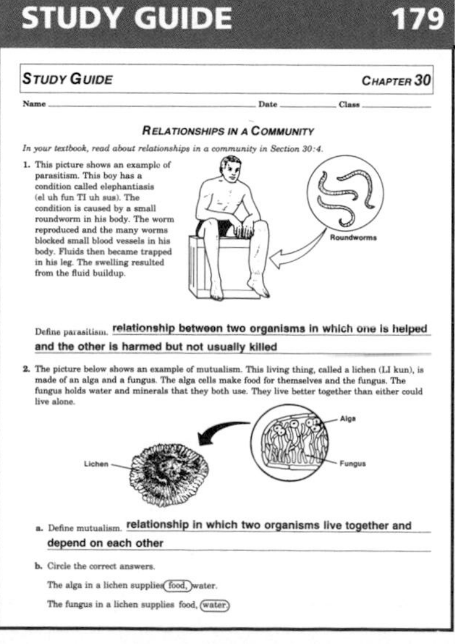

STUDY GUIDE CHAPTER 30

Name ________ Date ________ Class ________

RELATIONSHIPS IN A COMMUNITY

*In your textbook, read about relationships in a community in Section 30:4.*

1. This picture shows an example of parasitism. This boy has a condition called elephantiasis (el uh fun TI uh sus). The condition is caused by a small roundworm in his body. The worm reproduced and the many worms blocked small blood vessels in his body. Fluids then became trapped in his leg. The swelling resulted from the fluid buildup.



Define parasitism. relationship between two organisms in which one is helped and the other is harmed but not usually killed

2. The picture below shows an example of mutualism. This living thing, called a lichen (LI kun), is made of an alga and a fungus. The alga cells make food for themselves and the fungus. The fungus holds water and minerals that they both use. They live better together than either could live alone.



a. Define mutualism. relationship in which two organisms live together and depend on each other

b. Circle the correct answers.

The alga in a lichen supplies (food,) water.

The fungus in a lichen supplies food, (water.)

## TEACH

### Concept Development

▶ Explain to students that the organism a parasite attaches to is called a host.

▶ Point out that parasites can be monerans, protists, fungi, plants, or animals.

▶ Ask students why it is important that parasites not kill their hosts. Have them predict what would happen to the parasites if all of their hosts were to die.

### Audiovisual

Make an overhead transparency listing the organisms that are parasites.

### Skill Check

The lamprey has a mouth that is adapted to piercing flesh and sucking the body fluids of its host.

### Independent Practice

Study Guide/*Teacher Resource Package,* p. 180. Use the Study Guide worksheet shown at the bottom of the page for independent practice.

## GLENCOE TECHNOLOGY

**Videodisc**

**Biology: The Dynamics of Life**
Movie: *Symbiosis*
Disc 1, Side 1, Ch. 5

**The Secret of Life**
*Predator-Prey*

*Mutualism*

### Skill Check

**Observe:** A lamprey is a parasitic fish. Look at the photo of a lamprey on this page and see if you can tell how the lamprey gets its food. Does it have any body parts that are adapted to its way of life? *For more help, refer to the* ***Skill Handbook,*** *pages 704-705.*

## Parasitism

Let's return to the rock concert tickets one more time. Suppose your friend offers to sell you some concert tickets. Later, you find out that she has charged you much more money than the tickets were worth. Your friend is making a profit at your expense.

This is an example of parasitism (PAR uh suh tihz um). **Parasitism** is a relationship between two organisms in which one is helped and the other is harmed. Remember that a parasite lives at the expense of another living thing. The living thing on which the parasite lives is called the host. The host is rarely killed in this kind of relationship. What would happen to the parasite if it killed its host?

Many organisms are parasites. Fleas are parasites with which you may be familiar. Fleas use the blood of dogs or other animals for food. The flea benefits and the host animal is harmed in the relationship.

There are other common examples of parasitism. When you get sick, chances are that bacteria or viruses are using your body as a host. At some time you may have had strep throat. This disease is caused by bacteria living in your throat and using your body as a host. The bacteria grow and reproduce. They give off chemicals that irritate your throat tissues and interfere with the function of other body systems.

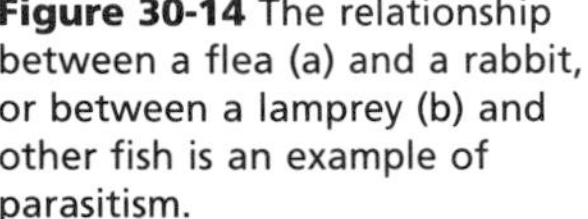

**Figure 30-14** The relationship between a flea (a) and a rabbit, or between a lamprey (b) and other fish is an example of parasitism.

a

b

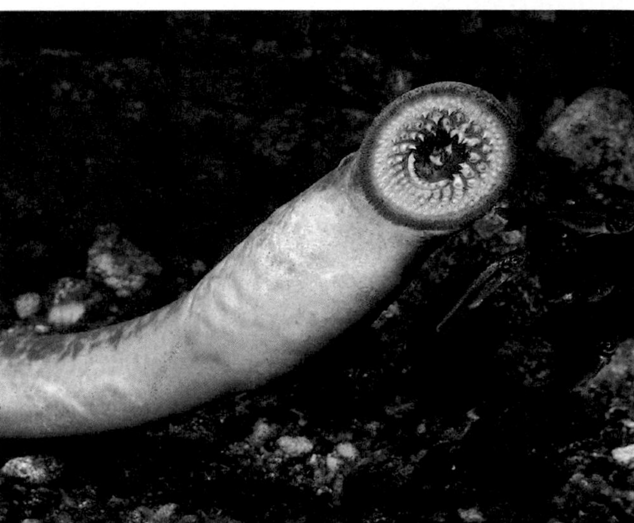

## OPTIONS

**Challenge:** Have students research and present a report to the class on parasites that are harmful to humans.

## STUDY GUIDE 180

STUDY GUIDE CHAPTER 30

Name ______ Date ______ Class ______

VOCABULARY

*Review the new words used in Chapter 30 of your textbook. Then complete this puzzle. Use the words listed below to fill in the blanks in the sentences that follow.*

| | | |
|---|---|---|
| commensalism | population | immigration |
| energy pyramid | habitat | emigration |
| niche | community | limiting factor |
| primary consumers | food chain | secondary consumers |
| prey | food web | predator |
| predation | parasitism | |

A group of organisms of the same species make up a **population**.

The size of this group of organisms can change because new organisms enter and leave. This is called **immigration** and **emigration**.

A condition that keeps the size of a population from increasing is a **limiting factor**.

Living things of different kinds live together in a **community**. The place where a plant or animal lives is its **habitat**. The job of the organism is its **niche**. Animals that eat only plants are **primary consumers**. Animals that eat other animals are **secondary consumers**.

Energy and materials are passed from one living thing to another in a community through a **food chain**. A **food web** shows how food chains are connected in a community. An **energy pyramid** is a diagram that shows energy loss in the food chain.

Living things depend on each other in several ways. In **commensalism**, one organism is helped and the other is not affected. In **parasitism**, one organism is helped and the other is harmed but not usually killed. In **predation**, one animal hunts and kills another animal. The **prey** is the animal that is eaten. The **predator** is the animal that hunts and kills the other animal.

Figure 30-15 A predator hunts and kills other animals for food.

## Predation

There is another relationship in a community in which one organism benefits and the other is harmed. This is **predation,** or the predator-prey relationship. An animal that hunts, kills, and eats another animal is a **predator.** The **prey** is the animal that the predator kills and eats. Unlike parasitism, in predation the predator actually kills the prey.

What is a predator?

Predators are secondary consumers. They can limit the sizes of some populations. Rabbits living near a field of clover become prey for owls that live in the same area. Look at the graph of the rabbit population in Figure 30-16. The graph shows sudden increases and decreases in population size. What are some things that might cause the population to decrease?

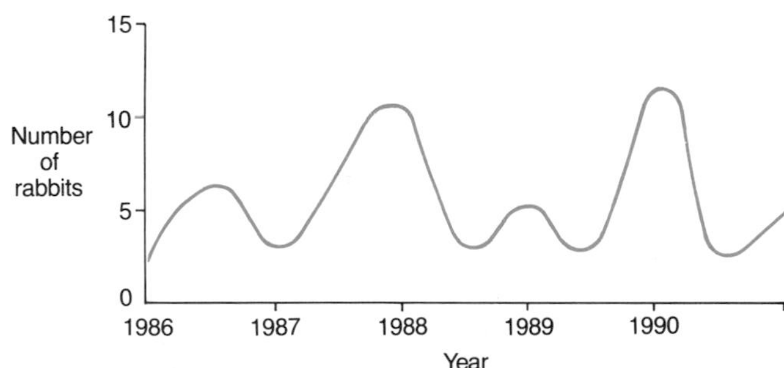


Figure 30-16 Predators can be a limiting factor for a prey population.



## TEACH

### Student Journal

Wolves are being reintroduced into some of the western states, and farmers, ranchers, and hunters state that these predators will destroy livestock and reduce the numbers of game animals. Direct students to write a report in which they defend both sides—why a predator should be reintroduced into an area and why it should not.

### Guided Practice

Have students write a short paragraph to explain whether a predator is a producer, a primary consumer, or a secondary consumer. Have them give a reason for their answer.

### Check for Understanding

Have students use Table 30-2 on page 648 to explain how the four relationships in a community are alike and how they are different.

### Reteach

Divide the students into teams and have them use the library to find several examples of each type of community relationship. Have student teams report their findings to the class.

### Guided Practice

Have students write down their answers to the margin question: An animal that kills and eats another animal.

## CRITICAL THINKING 30

CRITICAL THINKING/PROBLEM SOLVING — CHAPTER 30

Name ____ Date ____ Class ____

*Use after Section 30:4.*

**WHAT DO PREDATORS DO IN A COMMUNITY?**

The game warden in a county noticed many foxes in the woods. She knew that foxes were predators and she thought the foxes were eating too many rabbits. The game warden decided to catch some of the foxes and release them somewhere else. During the following summer, the game warden moved four foxes. She trapped and moved the same number of foxes each year for three more years. During that time, she noticed that the populations of owls and mice increased in size. She was surprised to find there were no more rabbits than before. What is happening to the balance in this community?

**Analyzing the Problem**

1. Which animals are predators? foxes and owls
2. Which animals are prey? rabbits and mice
3. Which of the animals compete with each other? Explain. The foxes and owls may compete with each other for mice and rabbits as food.

**Solving the Problem**

1. How could removal of foxes be related to the increase of mice? Trapping and removing the foxes removed a predator of mice. More mice could reproduce and grow to be adults, who also reproduce.
2. Why do you suppose the rabbit population did not increase, even though foxes were removed? Owls are another predator and must be eating the rabbits.
3. Do the results suggest that a predator can have more than one prey? Give an example. Yes, owls seem to eat both mice and rabbits.
4. Do the results suggest that a prey can be food for more than one predator? Yes, rabbits seem to be food for both owls and foxes.
5. Why do some people think predators are harmful and should be removed from an area? Some people want to keep the prey for human uses.
6. Why do some people think predators have a helpful role in a community? Predators eat prey and keep a balance of animals in the area. If not eaten, the population of prey animals might grow larger and eat plants that humans need.

## APPLICATION 30

APPLICATION: ENVIRONMENT — CHAPTER 30

Name ____ Date ____ Class ____

*Use after Section 30:4.*

**PREDATORS AND PREY**

Many prey animals have behaviors that help them escape from their predators. One way that prey animals protect themselves is to stay together in a group.

1. Give three examples of animals that live, feed, or travel together in groups. Answers might include flocks of geese, schools of fish, or herds of zebras.

Do the experiment described below. You will need a partner and 8-10 table tennis balls. You and your partner should stand about 2 meters apart in an open area, facing each other. One of you will act as the predator. The other person will toss 8-10 balls to the predator. The balls will represent the prey. The balls should be tossed gently, but also quickly, all at once, and without warning. The balls would represent a group of prey animals running past a predator. The predator will try to catch as many balls (prey) as possible. Record on the table below how many balls the predator catches. Repeat the experiment five times.

Now do another experiment. This time, toss only one ball to the predator. The ball would represent what it is like for a predator to try to catch a single prey animal. Repeat this experiment five times. On the table, record whether one or zero balls was caught each time.

| | |
|---|---|
| Number of balls (prey) caught when many balls are tossed: | Answers will vary but students should catch |
| Number of balls (prey) caught when one ball is tossed: | more balls when only one is thrown. |

*Refer to your table to answer the following questions.*

2. Which do you think is harder, catching a ball when it is tossed as part of a group or when it is tossed separately? Explain. It is harder to catch a ball when it is tossed in a group, because it is harder to focus your attention on any single ball.
3. How do you think a prey animal is protected from predators by being part of a group? It might be hard for the predator to catch any single animal due to the movement of all the animals together; animals in the middle of the group would be less likely to be caught than those on the outside: one animal might warn the group if predators are sighted.
4. How would you change the experiment to find out what happens if one member of the prey group is sick or old? All but one of the balls could be thrown at once, and then the last ball could be thrown separately.

## OPTIONS

**Application**/*Teacher Resource Package,* p. 30. Use the Application worksheet shown here as an extension to predator-prey relationships.

**Critical Thinking/Problem Solving**/*Teacher Resource Package,* p. 30. Use the Critical Thinking/Problem Solving worksheet shown here to give additional practice with problems involving predator-prey interactions.

# TEACH

## Idea Map

Have students use the idea map as a study guide to the major concepts of this section.

## Check for Understanding

Have students respond to the first three questions in Check Your Understanding.

## Reteach

**Reteaching**/*Teacher Resource Package,* p. 90. Use this worksheet to give students additional practice with relationships in a community.

**Extension:** Assign Critical Thinking, Biology and Reading, or some of the **OPTIONS** available with this lesson.

# 3 APPLY

## Brainstorming

Have student teams discuss the following question and report to the class on their findings. A shopping center is going to be built in an area that is now a forest. What concern should the developer and citizens of the community have for the plants and animals in the forest?

# 4 CLOSE

## Guest Speaker

Have a veterinarian come to your class to speak about parasites in pets and farm animals.

## Answers to Check Your Understanding

16. The bacteria that make Vitamin $B_{12}$ in the human intestine have a mutualistic relationship with humans.
17. The orchids get more light.
18. parasitism
19. If too many orchid plants cover the tree, the tree might not be able to make food. The tree would be weakened and harmed.
20. (a) mutualism, (b) parasitism, (c) commensalism

**Study Tip:** Use this idea map as study guide to community relationships. Can you give an example of each kind of relationship?

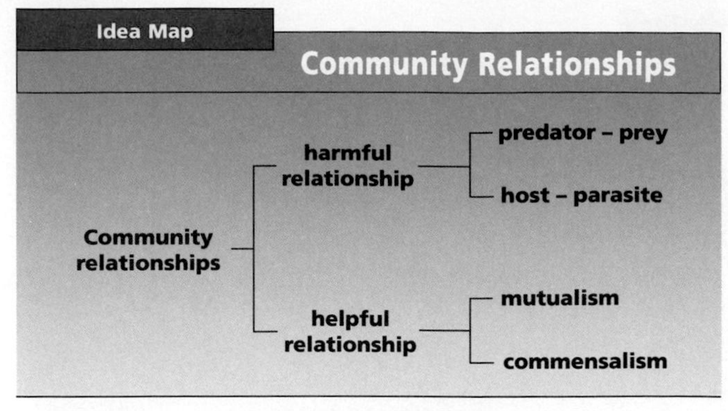


| TABLE 30–2. RELATIONSHIPS | | |
|---|---|---|
| **Type of Relationship** | **Organism 1** | **Organism 2** |
| Mutualism | helped | helped |
| Commensalism | helped | unaffected |
| Parasitism | helped | harmed |
| Predation | helped | harmed (killed) |

Table 30-2 is a summary of the relationships that may occur among populations in a community. Now you can see why the statement "no living thing lives alone" is true.

## Check Your Understanding

16. Give an example of mutualism involving humans.
17. How do orchid plants benefit by living high in trees?
18. What word describes the relationship between fleas and their hosts?
19. **Critical Thinking:** Explain how the relationship between orchid plants and a tree might change from commensalism to parasitism.
20. **Biology and Reading:** For each of the pairs of organisms below, write the word that best describes their community relationship.
    a. Ants protect a tree, which provides food for ants.
    b. A liver fluke lives inside a sheep, which develops liver disease.
    c. An insect eats the hairs that fall out of a deer's skin, but the deer is not affected.



## OPTIONS

### ACTIVITY/Project

Have a group of students research the predator/prey relationship between the lynx and the snowshoe hare. Ask them to prepare a poster illustrating how this relationship results in population cycles.

**How Do Predator and Prey Populations Change?**/*Lab Manual,* pp. 255-258. Use this lab as an extension to population dynamics involving predator-prey interactions.

RETEACHING 90

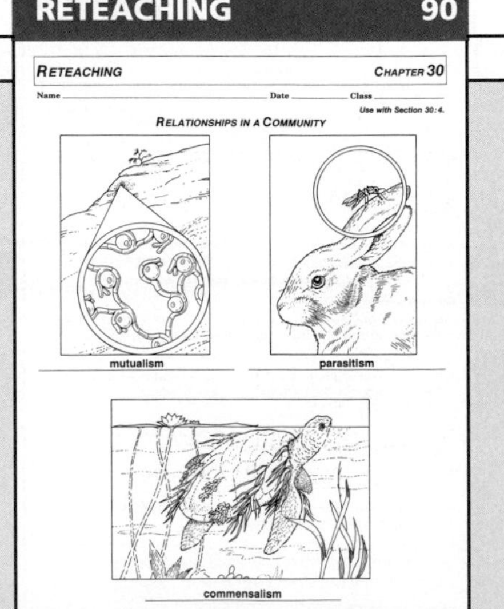

# Lab 30–2 Predation

## Problem: What do owls eat?

### Skills

observe, infer, design an experiment

### Materials

water
bowl
forceps
glue
glass slide
coverslip
light microscope
owl pellet
cardboard

### Procedure

Owl pellets are made of food, bones, and fur an owl has eaten but has not been able to digest. An owl forms pellets in its stomach and then coughs them up. Examining an owl pellet will give you some ideas about what an owl eats.

1. Copy the data table.
2. Place an owl pellet in the bowl. Add some water.
3. Use the forceps to remove the outside covering of the pellet.
4. Place a small piece of this covering on a glass slide and add a coverslip.
5. **Observe:** Examine the covering of the pellet on low power of your microscope. Make a drawing of what you see.
6. Continue to break the pellet open. Save all the parts present in the pellet.
7. Place the parts on a sheet of paper. Try to figure out how many living things the owl ate. Examine the bones and record your observations.
8. Try to assemble the bony parts into a skeleton. You can do this by gluing them to cardboard.

### Data and Observations

1. What was the outside covering of the pellet? Where do you suppose it came from?
2. What types of things did you see in the pellet?

### Analyze and Apply

1. An owl coughs up one pellet a day. How many animals did your owl eat each day?
2. Judging from the size of the remains in the owl pellet, name the kinds of living things that the owl ate.
3. Do owls eat plants or animals?
4. Is an owl a producer or a consumer? Explain your answer.
5. Is an owl a primary consumer or a secondary consumer?
6. **Apply:** Suppose an owl population increases from three to eight. What can happen to the population of organisms eaten by the owls? Why?

### Extension

**Design an experiment** to show what happens to a predator population when the size of the prey population decreases.

Answers will vary.

| NUMBERS OBSERVED | | | | |
|---|---|---|---|---|
| Leg Bone | Rib | Mammal Skull | Bird Skull | Mammal Jawbone |
| 3 | 8 | 0 | 2 | 4 |



## ANSWERS

### Data and Observations

1. feathers or fur from the animals eaten by the owl
2. feathers, fur, bones

### Analyze and Apply

1. Answers will vary. Sometimes 5 or 6 skulls are easily seen.
2. Answers will vary. Owls eat small mammals, such as mice, voles, rabbits, and squirrels. They also eat birds.
3. animals
4. consumer; Owls eat other animals and do not make their own food.
5. secondary consumer
6. They will begin to decrease because more owls are present to prey on the animals.

## Lab 30-2 Predation

### Overview

Students will find out what owls eat by examining an owl pellet.

**Objectives:** Students will be able to (1) **determine** what an owl eats, (2) **classify** an owl as an herbivore or carnivore, and (3) **assemble** the bones from an owl pellet to make a skeleton.

**Time Allotment:** 45 minutes

### Preparation

▶ Obtain zoology reference books that show the skeletons of birds and small mammals.

▶ Cut 10 cm x 10 cm squares for students to use in assembling the skeleton from the owl pellet.

▶ Owl pellets can be found under trees, ordered from a biological supply house, or obtained from a local zoo.

**Lab 30-2 worksheet**/*Teacher Resource Package,* pp. 119-120.

### Teaching the Lab

**Cooperative Learning:** Divide the class into groups of two students. For more information, see pp. 22T-23T in the Teacher Guide.

**Performance:** Direct students to use the leftover bones that have not been glued to cardboard to create a mythical prey that the owl eats. Have them arrange the bones on the cardboard before gluing and give the animal they made a name based on the properties and arrangement pattern of the bones.

## CHAPTER 30 REVIEW

### Summary

Summary statements can be used by students to review the major concepts of the chapter.

### Key Science Words

All boldfaced terms from the chapter are listed.

### Testing Yourself

#### Using Words

1. food chain
2. parasitism
3. limiting factor
4. habitat
5. energy pyramid
6. immigration
7. prey
8. emigration
9. niche
10. population

#### Finding Main Ideas

11. p. 633; Animal herds and flocks are clumped populations.
12. p. 640; Decomposers recycle nutrients for reuse by a community.
13. p. 641; A food chain shows how energy is passed through a community.
14. p. 644; The relationship is called mutualism.
15. p. 639; Plants are the producers, organisms that make food in a community.
16. p. 642; The second link in a food chain is a plant eater.
17. p. 633; Animals are sometimes marked with ear staples, birds with bands on their legs, and trees with paint or ribbons. Animals with radio transmitters can be followed to their homes, where their young can be counted.
18. p. 636; Limiting factors are the amount of food, space, and water.
19. p. 639; Animals that eat only plants are primary consumers. Animals that eat other animals are secondary consumers.
20. p. 643; An energy pyramid shows energy loss in a food chain.

# Chapter 30 Review

## Summary

### 30:1 Populations

1. There are many methods for counting the living things in a population.
2. A change in population size can be caused by a change in birth or death rates.
3. Limiting factors keep populations from increasing forever.

### 30:2 Communities

4. Each kind of organism in a community has its own habitat and niche.
5. Green organisms supply food in a community. Animals are consumers. Bacteria and fungi are decomposers.

### 30:3 Energy in a Community

6. The energy in a community comes originally from the sun. It is passed through the community as food.
7. Food chains are connected to form food webs.

### 30:4 Relationships in a Community

8. In mutualism, two living things depend on each other.
9. Commensalism is a relationship in which one organism benefits and the other is unaffected.
10. In parasitism, the host is harmed but not killed. A predator kills its prey.

## Key Science Words

commensalism (p. 645)
community (p. 638)
emigration (p. 634)
energy pyramid (p. 643)
food chain (p. 641)
food web (p. 642)
habitat (p. 638)
immigration (p. 634)
limiting factor (p. 636)
niche (p. 638)
parasitism (p. 646)
population (p. 632)
predation (p. 647)
predator (p. 647)
prey (p. 647)
primary consumer (p. 639)
secondary consumer (p. 639)

## Testing Yourself

**Using Words**

*Choose the word from the list of Key Science Words that best fits the definition.*

1. simple pathway of energy and materials through a community
2. relationship in which one member is helped and the other is harmed
3. keeps a population from increasing
4. place where an organism lives
5. a diagram that shows energy loss
6. animals moving into a population
7. eaten by a predator



### Using Main Ideas

21. It could increase.
22. The sun is used as energy by producers to make food.
23. Bacteria and fungi are decomposers.
24. Radio transmitters allow animals to be followed to their homes, where the offspring can be counted.
25. Birth increases a population by adding new offspring; immigration increases a population by adding new adults.
26. commensalism

# Review

## Testing Yourself *continued*

8. animals moving out of a population
9. an organism's job in a community
10. a group of the same kind of living things in an area

**Finding Main Ideas**

*List the page number where each main idea below is found. Then, explain each main idea.*

11. examples of clumped populations
12. the job of decomposers in a community
13. what a food chain shows
14. a relationship in which two organisms depend on each other
15. the importance of plants to a community
16. what the second link of a food chain is
17. what methods are used to count populations
18. what the limiting factors for animal populations are
19. the difference between primary consumers and secondary consumers
20. what an energy pyramid shows

**Using Main Ideas**

*Answer the questions by referring to the page number after each question.*

21. If a predator population in an area decreases, what could happen to the size of the prey population? (p. 647)
22. What role does the sun play in a community? (p. 643)
23. What are two examples of decomposers? (p. 640)
24. Why are radio transmitters sometimes used to count a population? (p. 633)
25. What factors can increase a population? (p. 634)
26. What relationship is shown by orchids growing on a tree? (p. 645)

## Skill Review ✓

*For more help, refer to the* ***Skill Handbook,*** *pages 704-719.*

1. **Make a line graph:** Find out from the school office how many students have been in your school each year during the past ten years. Make a line graph to show these data. Is the school population increasing or decreasing?
2. **Infer:** Infer what happens to two different organisms in a community if they have the same needs for food.
3. **Form hypotheses:** Form a hypothesis to explain how you could tell if owls eat small plant-eating animals.
4. **Make and use tables:** Make a table listing 10 organisms as producers, primary consumers, and secondary consumers. List how each organism gets food.

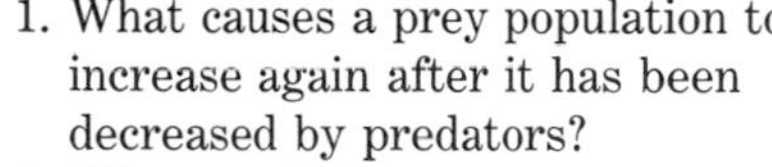


## Finding Out More

**Critical Thinking**

1. What causes a prey population to increase again after it has been decreased by predators?
2. What would happen in a food web if one kind of organism in the web were all killed by a disease?

**Applications**

1. Make a list of 6 to 10 different organisms that live near your school or home. Construct a food web using the organisms on your list.
2. Find out why fungi and bacteria are useful organisms in a farm field.

651

**Chapter Review**/*Teacher Resource Package,* pp. 160-161.

**Chapter Test**/*Teacher Resource Package,* pp. 162-164.

**Chapter Review**/*Computer Test Bank*

## Skill Review

1. Students can also plot the number of students in each grade for ten years to see which grades show an increase or decrease.
2. Students should be able to explain how organisms compete for the same food and that one or both of the populations may show a decrease in number.
3. By examining owl pellets and identifying the skulls of plant eaters and animal eaters, students could form a hypothesis as to the importance of plant eaters to owls.
4. Do this as an in-class assignment and give each student a chart. You may have some magazine and newspaper articles that will help them make their list of organisms.

## Finding Out More

### Critical Thinking

1. When a prey population decreases, the predators do not have enough to eat. Food becomes a limiting factor and the predator population decreases. A decrease in the predator population eventually allows the prey population to increase.
2. The web would be disrupted. Secondary consumers would have to find new prey species.

### Applications

1. Students should show a web that depicts animals feeding at more than one level.
2. Students should realize that these organisms are decomposers and are necessary for recycling materials.

# Ecosystems and Biomes

## PLANNING GUIDE

| CONTENT | TEXT FEATURES | TEACHER RESOURCE PACKAGE | OTHER COMPONENTS |
|---|---|---|---|
| (2 days)<br>31:1 Ecosystems<br>Parts of an Ecosystem<br>Soil and Nitrogen Cycle<br>Water<br>Oxygen-Carbon Dioxide Cycle | Skill Check: *Understand Science Words,* p. 655<br>Mini Lab: *What Are the Parts of the Water Cycle?* p. 657<br>Lab 31-1: *Carbon Dioxide,* p. 659<br>Check Your Understanding, p. 658<br>Career Close-up: *Wildlife Technician,* p. 656 | Application: *Space Invaders,* p. 31<br>Enrichment: *How Fast Can Change Take Place in an Ecosystem,* p. 33<br>Reteaching: *The Water Cycle,* p. 91<br>Critical Thinking/Problem Solving: *How Can a Nonliving Factor Affect Animals in an Ecosystem?* p. 31<br>Study Guide: *Parts of an Ecosystem,* p. 181<br>Lab 31-1: *Carbon Dioxide,* pp. 121-122 | **Laboratory Manual:**<br>*How Much Water Will Soil Hold?* p. 259<br>**Laboratory Manual:**<br>*How Can a Nonliving Part of an Ecosystem Harm Living Things?* p. 263<br>Color Transparency 31a: *The Nitrogen Cycle*<br>Color Transparency 31b: *The Water Cycle*<br>**STVS:** *Unusual Estuary,* Ecology (Disc 6, Side 1) |
| (1 1/2 days)<br>31:2 Succession<br>Succession in a Land Community<br>Succession in a Water Community | Mini Lab: *How Do Leaves Decay?* p. 661<br>Check Your Understanding, p. 662 | Reteaching: *Succession,* p. 92<br>Focus: *Changes in a Community,* p. 61<br>Transparency Master: *Succession—Bare Land to Forest,* p. 155<br>Study Guide: *Succession,* pp. 182-184 | **Laboratory Manual:**<br>*How Can a Nonliving Part of an Ecosystem Help Living Things?* p. 265<br>**STVS:** *Weed-Eating Fish,* Ecology (Disc 6, Side 1) |
| (1 1/2 days)<br>31:3 How Living Things Are Distributed<br>Climate<br>Land Biomes<br>Water Ecosystems | Skill Check: *Sequence,* p. 664<br>Lab 31-2: *Brine Shrimp,* p. 667<br>Idea Map, p. 666<br>Check Your Understanding, p. 666<br>Applying Technology: *Taking a Sidewalk's Temperature,* p. 668 | Reteaching: *Climates and Biomes,* p. 93<br>Skill: *Interpret Diagrams,* p. 32<br>Study Guide: *How Living Things Are Distributed* p. 185<br>Study Guide: *Vocabulary,* p. 186<br>Lab 31-2: *Brine Shrimp,* pp. 123-124 | **STVS:** *Desert in the Antarctic,* Ecology (Disc 6, Side 1) |
| Chapter Review | Summary<br>Key Science Words<br>Testing Yourself<br>Finding Main Ideas<br>Using Main Ideas<br>Skill Review | **ASSESSMENT RESOURCES**<br>Chapter Review, pp. 165-166<br>Chapter Test, pp. 167-169<br>Performance Assessment in the Biology Classroom<br>Alternate Assessment in the Science Classroom<br>Computer Test Bank | |

## GLENCOE TECHNOLOGY

**Infinite Voyage,** *Life in the Balance*
**Science and Technology Videodisc Series,** *Unusual Estuary,* Ecology (Disc 6, Side 1)
*Weed-Eating Fish,* Ecology (Disc 6, Side 1)
*Desert in the Antarctic,* Ecology (Disc 6, Side 1)
**The Secret of Life,** *Gone Before You Know It: The Biodiversity Crisis*

## MATERIALS NEEDED

| LAB 31-1, p. 659 | LAB 31-2, p. 667 | APPLYING TECHNOLOGY, p. 668 | MARGIN FEATURES |
|---|---|---|---|
| 4 test tubes<br>wax pencil<br>4 stoppers<br>lamp<br>blue test liquid<br>green test liquid<br>2 *Elodea* sprigs<br>2 test-tube racks | 3 small plastic cups<br>marking pen<br>hand lens<br>flat toothpick<br>distilled water<br>brine shrimp eggs<br>strong salt solution<br>weak salt solution | thermometer<br>watch or clock<br>graph paper | Skill Check, p. 655<br>dictionary<br>Skill Check, p. 664<br>pencil<br>paper<br>Mini Lab, p. 657<br>pencil<br>paper<br>Mini Lab, p. 661<br>jar<br>plastic wrap<br>soil<br>water<br>dead leaves |

## OBJECTIVES

For more information about National Science Standards, see page 5T.

| SECTION | OBJECTIVE | CORRELATION of QUESTIONS to OBJECTIVES | | | |
|---|---|---|---|---|---|
| | | CHECK YOUR UNDERSTANDING | CHAPTER REVIEW | TRP CHAPTER REVIEW | TRP CHAPTER TEST |
| 31:1 National Science Stds: UCP.1, UCP.3, UCP.4, B.6, C.4, C.5, D.1, D.2, F.3, F.4 | 1. **Describe** the parts of an ecosystem. | 1 | 2, 3, 8, 11, 14 | 2, 26, 28 | 2, 25, 26, 30 |
| | 2. **Describe** how the water cycle affects an ecosystem. | 3, 4 | 1, 10, 19, 21 | 4, 30 | 1, 22, 47, 48, 49, 50 |
| | 3. **Explain** the nitrogen cycle and the oxygen-carbon dioxide cycle. | 2, 5 | 7, 13, 19, 20, 22 | 1, 5, 7, 8, 31 | 3, 4, 6, 9, 10, 13, 16, 17, 20, 21, 23, 24, 27, 31, 42, 43, 44, 45, 46 |
| 31:2 National Science Stds: UCP.1, UCP.2, UCP.3, UCP.4, B.5, B.6, C.4, C.5, D.1, D.2, F.3 | 4. **Describe** succession in a land community. | 7, 9 | 4, 9, 16 | 3, 6, 23, 24, 28, 29 | 5, 11, 12, 14, 29, 32, 33, 34, 35, 36 |
| | 5. **Describe** succession in a water community. | 8, 10 | 4, 9, 18, 24 | 12, 25 | 5, 8, 37, 38, 39, 40, 41 |
| 31:3 National Science Stds: UCP.1, UCP.3, UCP.4, C.4, F.3, F.4 | 6. **Explain** how climate helps determine what living things live in an area. | 11, 14 | 5, 6 | 11, 12, 13, 14, 15, 16, 33, 34 | 18, 19, 28 |
| | 7. **Describe** the major land biomes and water ecosys-tems. | 12, 13, 15 | 12, 15, 23, 25 | 9, 17, 18, 19, 20, 21, 22, 32 | 7, 15 |

# Ecosystems and Biomes

## CHAPTER OVERVIEW

### Key Concepts

This chapter deals with the interaction between the community and the nonliving parts of the environment to form an ecosystem. The major nonliving factors that affect the community are discussed. Changes over a long time (succession) in both aquatic and terrestrial communities are covered. Biomes are discussed as large land areas encompassing numerous ecosystems.

### Key Science Words

| | |
|---|---|
| biome | nitrogen cycle |
| climate | precipitation |
| climax community | succession |
| ecology | water cycle |
| ecosystem | |

### Skill Development

In Lab 31-1 on page 659, students will **observe, form hypotheses, interpret data, infer,** and **design an experiment** to show that plants give off carbon dioxide. In Lab 31-2 on page 667, they will **observe, form hypotheses, interpret data,** and **design an experiment** to see how different concentrations of salt water affect brine shrimp. In the Skill Check on page 655, students will **understand** the **science word** *ecology.* In the Skill Check on page 664, they will **sequence** the climates of the major land biomes. In the Mini Lab on page 657, students will **interpret a diagram** of the water cycle. In the Mini Lab on page 661, they will **experiment** to find out how decomposers break down leaves.

### Bridging

Chapter 30 develops the concept of living factors making up populations and communities and how these living factors interact with each other in the community. Chapter 31 discusses how nonliving factors interact with the living factors in an ecosystem and biome.

**CHAPTER PREVIEW**

### Chapter Content

Review this outline for Chapter 31 before you read the chapter.



### Skills in this Chapter

The skills that you will use in this chapter are listed below.

- In **Lab 31-1,** you will observe, form hypotheses, interpret data, infer, and design an experiment. In **Lab 31-2,** you will observe, form hypotheses, interpret data, and design an experiment.
- In the **Skill Checks,** you will understand science words and sequence.
- In the **Mini Labs,** you will interpret diagrams and experiment.



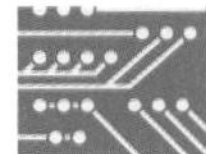

## TECH PREP

For Tech Prep activities in this chapter of the Teacher Wraparound Edition, see especially the Student Journals on pages 655 and 661, the Activity on page 662, the Connection on page 665, and the Guest Speaker activity on page 666.

See also the Glencoe Homepage at **www.glencoe.com**

Chapter 31

# Ecosystems and Biomes

Look at the photo on the left. What part of the environment is causing the trees to bend toward one side? Why are the limbs growing more on one side than on the other? The photo on this page shows a field. Do you see any animals in the field? Where could they be? What happened to the green plants in the field?

In the last chapter, you learned that populations of living things interact with each other in a community. Within a community, living things also interact with their environment. In this and the next chapter, you will see how living things interact with their environments.

**Try This!**

**How do you interact with your environment?** Make a list of all the living and nonliving things in your environment. All of the living things interact with each other and with each nonliving thing. How does each thing affect you? How do you interact with each thing?

**interNET CONNECTION**

For more information about the material in this chapter, follow the link for the chapter on the Glencoe Homepage at **http://www.glencoe.com**



## ASSESSMENT PLANNER

**Portfolio**

Strategies on the following pages represent student products that can be placed into a best-work portfolio: pp. 655, 665.

**PERFORMANCE ASSESSMENT**

Skill Check, pp. 655, 664
Mini Lab, pp. 657, 661
Lab, pp. 659, 667

**CONTENT ASSESSMENT**

Check for Understanding, pp. 657, 658, 662, 666
Chapter Review, pp. 670-671

**GROUP ASSESSMENT**

Opportunities for group assessment occur with Cooperative Learning Strategies.

**Student Journal**

Strategies on the following pages represent opportunities for writing in a Student Journal: pp. 655, 661, 663.

## GETTING STARTED

### Using the Photos

After students tell you that wind and temperature have affected the organisms shown in the photo, point out that these factors make up the nonliving environment that surrounds an organism. Explain to students that the nonliving factors described in their text are the major factors that interact with living things in an ecosystem.

### MOTIVATION/Try This!

**How do you interact with your environment?** Show 5-6 color slides or frames from a filmstrip that show environments in which organisms live (wet, dry, hot, cold, windy). Ask students to identify what effect the environment can have on the organisms not seen in the pictures.

### Chapter Preview

In this chapter, students will learn how living factors interact with nonliving factors to form ecosystems and biomes. They will find out how specific factors such as minerals, water, and gases are recycled through organisms and the environment. They will learn how succession occurs in both land and water communities and compare the types of ecosystems and biomes in the world.

### Misconception

Students may think that all minerals are recycled rapidly through the living and nonliving world. Point out that some minerals end up in rock strata and sediment layers for centuries.

## 31:1 Ecosystems

### PREPARATION

#### Materials Needed

Make copies of the Critical Thinking/Problem Solving, Application, Enrichment, Study Guide, Reteaching, and Lab worksheets in the *Teacher Resource Package*.

▶ Gather rocks for the Focus activity.

▶ Prepare the bromthymol blue and green solutions and purchase *Elodea* sprigs for Lab 31-1.

#### Key Science Words

ecosystem
ecology
nitrogen cycle
water cycle

#### Process Skills

In Lab 31-1, students will observe, form hypotheses, interpret data, infer, and design an experiment. In the Skill Check, students will understand science words. In the Mini Lab, students will interpret diagrams.

### 1 FOCUS

▶ The objectives are listed on the student page. Remind students to preview these objectives as a guide to this numbered section.

#### ACTIVITY/Hands-On

Rub two rocks, either hard or soft, over a white piece of paper. Let students use a hand lens to examine the small particles that come from the rocks. Have students describe the particles as to size, color, and shape. Tell them that these particles are minerals and are important parts of soil.

#### Guided Practice

Have students write down their answers to the margin question: living and nonliving parts.

## 31:1 Ecosystems

**Objectives**

1. **Describe** the parts of an ecosystem.
2. **Describe** how the water cycle affects an ecosystem.
3. **Explain** the nitrogen cycle and the oxygen-carbon dioxide cycle.

**Key Science Words**

ecosystem
ecology
nitrogen cycle
water cycle

A community interacting with the environment is an **ecosystem.** An ecosystem can be as small as a roadside ditch or as large as one of the Great Lakes.

### Parts of an Ecosystem

**What two parts make up an ecosystem?**

How do communities differ from ecosystems? A community includes the living things in an area. When you study a community, you study only how the living things affect each other. An ecosystem includes the nonliving parts as well as the living parts in an area. When you study an ecosystem, you study how the nonliving and living parts affect each other.

The study of how living things interact with each other and with their environment is called **ecology** (ih KAHL uh jee). Ecologists study ecosystems to find out how the different parts interact. They observe living things in nature and in the laboratory. They collect data by making measurements and by carrying out carefully controlled experiments. Sometimes it is difficult or impossible to set up the same conditions in a lab that exist in nature. Then, ecologists may set up a model of an ecosystem on a computer, Figure 31-1a. However, they must realize when they make conclusions based on lab experiments or models that the data did not come from nature.

Look at the seashore ecosystem in Figure 31-1b. The living parts of this ecosystem are the plants, animals, and algae you see, as well as the bacteria, protozoans, and fungi that are too small to be seen. All these organisms are the producers, consumers, and decomposers you read about in Chapter 30. Suppose you could remove the living parts

**Figure 31-1** Ecologists sometimes try to reproduce natural conditions by setting up a computer model (a). A seashore (b) is an ecosystem.

a

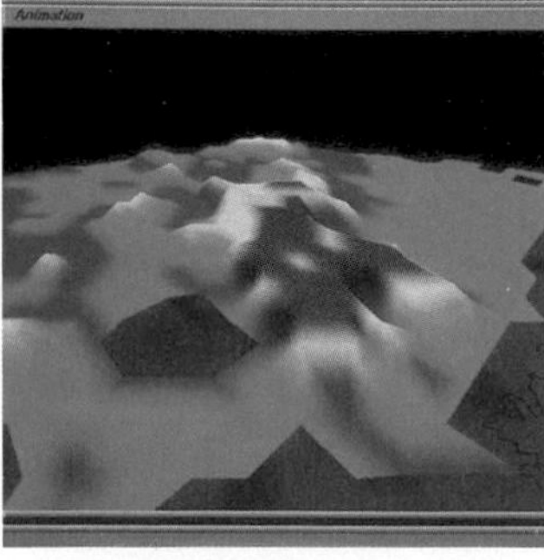

b

### OPTIONS

#### Science Background

Nutrients are passed through living and nonliving parts of an ecosystem as minerals. Since these chemicals pass through living things (bio) and soil and rocks (geo), the cycle is called a biogeochemical cycle.

**Critical Thinking/Problem Solving/** *Teacher Resource Package*, p. 31. Use the Critical Thinking/Problem Solving worksheet shown here as an extension to parts of an ecosystem.

**CRITICAL THINKING 31**

CRITICAL THINKING/PROBLEM SOLVING — CHAPTER 31

Name ______ Date ______ Class ______

*Use after Section 31:1.*

***HOW CAN A NONLIVING FACTOR AFFECT ANIMALS IN AN ECOSYSTEM?***

As students walked past a meadow near the school, they heard crickets chirping. The crickets were noisier in the afternoon than in the morning and were chirping more quickly. What could be the cause?

In their biology class, the students learned that crickets are cold-blooded animals. They become more active as the temperature rises. Warm-blooded animals, such as humans, have the same body temperature all day long. Their cycles of activity are not so sensitive to temperature.

One student read that the temperature of the air can be estimated by measuring the activity of a cold-blooded animal and using a mathematical formula. The formula for crickets is: number of chirps per second divided by 2, plus 8, equals air temperature in degrees Celsius.

The students decided to test the formula by doing an experiment with cricket calls. Each student listened to a cricket and noted the air temperature with a thermometer at a certain time during the day. Students put their results into a table:

| Calculated air temperature | | | Measured air temperature °C |
|---|---|---|---|
| Chirps per second | Divided by 2 | Plus 8 = result | |
| 45 | 22.5 | 30.5 | 30 |
| 30 | 15 | 23 | 25 |
| 43 | 21.5 | 29.5 | 31 |
| 38 | 19 | 27 | 26 |

**Analyzing the Problem**

1. What is a cold-blooded animal? an animal whose body temperature and activity is affected by the temperature of its environment
2. What nonliving part of the ecosystem affected the chirping activity of the crickets? air temperature

**Solving the Problem**

1. Do the data show that air temperature can be estimated by cricket chirps? Yes, the numbers are not exactly the same, but they are close.
2. Suppose you have a tape of crickets chirping at different times of day. How could you tell if the air temperature is increasing or decreasing? Count the chirps at each time. The temperature is increasing if the rate of chirping increases.

**Figure 31-2** The nitrogen cycle. How is nitrogen gas returned to the air?

from Figure 31-1b. What would you see? You would see the nonliving parts. The nonliving parts of an ecosystem include the soil, air, water, light, and temperature. Nonliving parts help to determine where organisms can live.

## Soil and the Nitrogen Cycle

The soil is an important nonliving part of an ecosystem. Soil comes from rocks that have broken down. It has many jobs. Soil serves as a place for plants to anchor their roots. Soil serves as a home for many living things. Ants and earthworms moving through the soil make holes for air and water. Moles, shrews, and other animals that burrow help mix and loosen the soil as they tunnel in all directions.

Soil holds other nonliving parts such as water, nutrients, and gases. Some of the most important soil nutrients needed by plants are called nitrates (NI trayts). Nitrates are chemicals that contain nitrogen. The way that nitrates get into soil is an example of the way living and nonliving parts of the ecosystem work together. Use Figure 31-2 to trace how nitrogen moves through an ecosystem.

1. Plants and animals need nitrogen to make protein. Air is mostly nitrogen, but most living things can't make protein from the nitrogen in the air. Bacteria living on the roots of some plants change the nitrogen in air to a form that plants can use to make protein.
2. When living things die, they are decomposed by bacteria and fungi. Some of the nitrogen in their bodies is changed into nitrates.

### Skill Check

**Understand science words: ecology:** The word part *eco* means environment. In your dictionary, find three words with the word part *eco* in them. *For more help, refer to the **Skill Handbook**, pages 706-711.*



**STUDY GUIDE 181**

*STUDY GUIDE* — *CHAPTER 31*

Name ______ Date ______ Class ______

*ECOSYSTEMS*

*In your textbook, read about cycles in an ecosystem in Section 31:1.*

1. The incomplete diagram below shows the water cycle. Complete the diagram by following these steps.
   a. Use blue to shade the arrow that shows evaporation of water into air from ground or streams.
   b. Use green to shade the arrow that shows evaporation of water into air from plants.
   c. Leave unshaded the arrow that shows water falling from the clouds.

blue green unshaded

2. Complete the diagram below of the oxygen and carbon dioxide cycle by following these steps. Write the symbols for the gases in all the spaces marked A. Use $O_2$ as the symbol for oxygen and $CO_2$ as the symbol for carbon dioxide. Then, identify the process that takes place in all the spaces marked B. Use P as the symbol for photosynthesis and R as the symbol for respiration.

Producers use $CO_2$ (A) during P (B) and give off $CO_2$ during R (B). Producers give off $O_2$ during P (B) and use it during R (B). Consumers give off $CO_2$ (A) during R (B). Consumers use $O_2$ (A) during R (B).

## OPTIONS

**How Can a Nonliving Part of an Ecosystem Harm Living Things?/***Lab Manual*, pp. 263-264. Use this lab as an extension to living and nonliving parts of an ecosystem.

**The Nitrogen Cycle/***Transparency Package*, number 31a. Use color transparency number 31a as you teach the nitrogen cycle.

# 2 TEACH

### MOTIVATION/Demonstration

Demonstrate how water evaporates by boiling water in a beaker as you are discussing the water cycle. Hold a beaker containing a little ice over the beaker of boiling water. Some water should condense on the cold beaker.

### Concept Development

▶ Explain that nitrogen is but one of the many chemical elements that are cycled in an ecosystem.

▶ Point out that when living things die, they decompose and become nutrients that are recycled. Even bacteria and fungi die and are recycled.

### Independent Practice

**Study Guide/***Teacher Resource Package*, p. 181. Use the Study Guide worksheet shown at the bottom of this page for independent practice.

### Audiovisual

Make a list on the chalkboard or an overhead transparency of the organisms that live in the soil. Point out some of their special features, such as burrowing parts.

### *Portfolio*

Have students find out what other important soil nutrients similar to nitrogen are cycled through an ecosystem. Tell them to make a chart similar to Figure 31-2 to show what living and nonliving parts these nutrients pass through in the cycle.

### *Student Journal*

Have students research and write in their journals about the Biosphere II Project. The report should focus on the following: where and why the site was selected; what was hoped to be found about a closed ecosystem; major problems that developed as the project progressed; findings that were unexpected; future research to be conducted.

## Career Close-Up

### Wildlife Technician

**Background**

Throughout the year, wildlife technicians are doing population counts, checking on the health of game animals, building shelters for wildlife, doing age and growth studies, and talking to hunting groups. A wildlife technician must also know about the plants in the area, as most game animals are plant eaters. If specific food plants in the area are depleted or overbrowsed, the wildlife technician sets up a program for reseeding or replanting these plants.

**Related Careers**

fisheries biology technician, animal caretaker in laboratories and zoos

**Job Requirements**

A high school diploma with on-the-job training and some course work at a local community college are required. A general interest in plants and animals is also helpful.

**For More Information**

Check the library for government publications that list opportunities in the field.

## TEACH

### Concept Development

▶ Point out that water not only serves as a habitat for some living things, such as fish, but also makes up over 90% of the body weight of living things.

▶ Point out how important evaporation and condensation are in the water cycle.

**GLENCOE TECHNOLOGY**

**Videodisc**

**The Secret of Life**

*Water Cycle*

## Career Close-Up

### Wildlife Technician

Wildlife technician observing animal populations

**A** wildlife technician is a person who usually assists a wildlife biologist. Wildlife technicians count the numbers of animals in populations such as deer, turkeys, squirrels, and rabbits.

A wildlife technician may be expected to speak to people about wildlife management. During the hunting season, he or she may examine the animals killed by hunters to check for diseases and determine the ages of the animals. These data are used to predict how many surviving animals of each type are in the area. This information will tell if the number of hunting permits issued must be changed or if an area must be restocked.

A wildlife technician usually works for a state wildlife agency. The individual must have an interest in all kinds of living things and some knowledge of populations, communities, and ecosystems. Some wildlife technicians get an Associate of Arts degree to prepare for their jobs.

**Bio Tip**

**Consumer:** Manure rebuilds soil better than chemical fertilizers do. Manure broken down by decomposers provides nitrogen for plants, becomes part of the soil, and helps retain water in the soil. Dry manure can be bought in garden shops.

3. The nitrates are left in the soil. They can be used by plants as a source of nitrogen for making protein. Some plants are then eaten by animals. These animals can use nitrogen compounds found in plants to make proteins. When the plants or animals die, the nitrogen in their bodies is again returned to the soil.

4. Some nitrates in the soil are changed back into nitrogen gas by bacteria. The nitrogen gas is released into the air. The cycle repeats itself over and over again. The reusing of nitrogen in an ecosystem is called the **nitrogen cycle.**

The nonliving parts of the nitrogen cycle are nitrogen and soil. The living parts are the plants, animals, bacteria, and fungi. In every ecosystem, the nonliving and living parts interact with each other. Many other chemical materials are cycled through the ecosystem just as nitrogen is cycled.

656 Ecosystems and Biomes 31:1

## OPTIONS

**Application/***Teacher Resource Package*, p. 31. Use the Application worksheet shown here to give additional practice with balance in ecosystems.

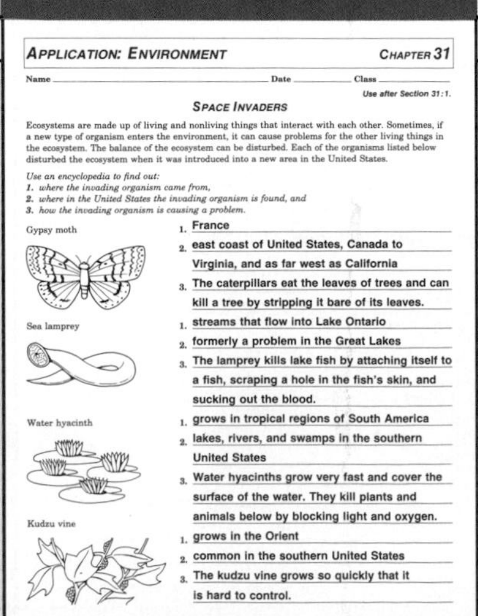

**APPLICATION 31**

*APPLICATION: ENVIRONMENT* — *CHAPTER 31*

Name ______ Date ______ Class ______

*Use after Section 31:1.*

**SPACE INVADERS**

Ecosystems are made up of living and nonliving things that interact with each other. Sometimes, if a new type of organism enters the environment, it can cause problems for the other living things in the ecosystem. The balance of the ecosystem can be disturbed. Each of the organisms listed below disturbed the ecosystem when it was introduced into a new area in the United States.

*Use an encyclopedia to find out:*
*1. where the invading organism came from,*
*2. where in the United States the invading organism is found, and*
*3. how the invading organism is causing a problem.*

Gypsy moth
1. France
2. east coast of United States, Canada to Virginia, and as far west as California
3. The caterpillars eat the leaves of trees and can kill a tree by stripping it bare of its leaves.

Sea lamprey
1. streams that flow into Lake Ontario
2. formerly a problem in the Great Lakes
3. The lamprey kills lake fish by attaching itself to a fish, scraping a hole in the fish's skin, and sucking out the blood.

Water hyacinth
1. grows in tropical regions of South America
2. lakes, rivers, and swamps in the southern United States
3. Water hyacinths grow very fast and cover the surface of the water. They kill plants and animals below by blocking light and oxygen.

Kudzu vine
1. grows in the Orient
2. common in the southern United States
3. The kudzu vine grows so quickly that it is hard to control.

## Water

Water is a nonliving part of an ecosystem that is needed by all living things. The cells of all living things are mostly water. Without water, the cells would die. Plants can't make food without water. What would happen to an ecosystem if plants couldn't make food?

Water, like nitrogen, is cycled through the ecosystem. The path that water takes through an ecosystem is called the **water cycle.** Use Figure 31-3 to help you understand the steps of the water cycle.

1. Water in the air falls to Earth as rain or snow.
2. Some water runs off the land into rivers, ponds, lakes, or oceans.
3. Water from rain may soak into the soil.
4. Some of this water is taken up by plants through their roots. Animals may drink some of the water on the ground. The rest flows into underground lakes and rivers.
5. Excess water passes out of plants through their leaves. Animals lose water through body openings or as wastes are removed. The water evaporates (ee VAP uh raytz) into the air. Evaporate means to change from a liquid to a gas. Water is present in the air as a gas.
6. Water from lakes, rivers, and oceans also evaporates into the air.

### Mini Lab

**What Are the Parts of the Water Cycle?**

**Interpret diagrams:** Look at Figure 31-3. What parts of an ecosystem are important in the water cycle? What role does each part play? Are all living things part of the water cycle? *For more help, refer to the* ***Skill Handbook,*** *pages 706-711.*

**In what form is water found in the air?**

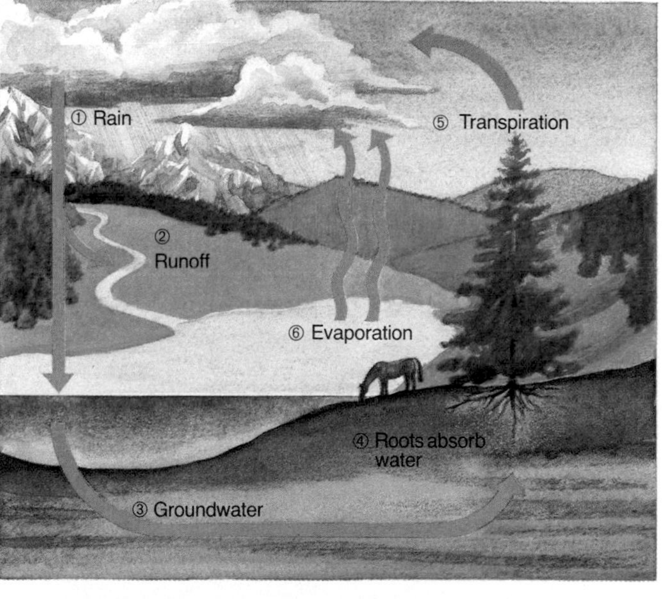


**Figure 31-3** The water cycle. How is water returned to the air? it evaporates



## TEACH

### Concept Development

▶ Reinforce the importance of water in photosynthesis. Explain that energy becomes available to all living things by way of food chains and that plants are the first step of a food chain.

▶ Make sure that students understand that carbon dioxide dissolved in water in aquatic ecosystems is an important raw material for photosynthesis.

### Mini Lab

As students are making their diagrams, make a list of the living and nonliving parts of the community on the chalkboard and have students pick out the most important factors for their water cycle diagram.

### ✓ ASSESSMENT

**Oral:** Have a student do a class report explaining the path water takes in the water cycle. The student should use a poster or projectoral and name each part of the ecosystem water passes through as well as the state of matter of the water at each step.

### Guided Practice

Have students write down their answers to the margin question: gas.

### Check for Understanding

Have students write a paragraph explaining how the organisms in a community are affected by nonliving factors. Have them list examples they have seen in the area where they live.

### Reteach

Make a list of the nonliving factors that you discussed with the students. Ask them to explain how each one or several interact with the members of a community. Make a list of the student responses on the chalkboard. At the end of the discussion, have students decide which of the nonliving factors may be the most important to organisms.

### ENRICHMENT 33

*ENRICHMENT* — *CHAPTER 31*

Name ______ Date ______ Class ______

*Use with Section 31:1.*

***HOW FAST CAN CHANGE TAKE PLACE IN AN ECOSYSTEM?***

Suppose an animal enters an ecosystem where it has never lived before. A small population of gray grass mice entered the tall grasslands area of a state park in the fall of 1984. From 1984 to 1990 a study took place to find out the new places the gray mice moved into each year. The graph below shows how many acres of tall grasslands the mice lived in during each year of the study. The mice continued to live in the places where they had lived in the years before, but as the population grew larger, they spread out into new places.

*Answer the questions that follow based on the information above and in the graph.*



1. On how many acres of grassland did the mouse population live in 1984? 1 in 1985? 5 in 1988? 40 in 1990? 80
2. The gray mouse eats grass seeds. It also kills the roots of grass plants. What will happen to the grasslands if the gray mice stay there for a number of years? The grassland will be destroyed.

   What might happen to the mouse population after a number of years? It will decrease.
3. What can happen to the soil after the mice have destroyed the roots of the grass? Seeds of different plants can begin to grow in the area.
4. What is the name of the change that is occurring in this ecosystem? succession

### OPTIONS

**Enrichment**/*Teacher Resource Package,* p. 33. Use the Enrichment worksheet shown here as an extension to structure of ecosystems.

**The Water Cycle**/*Transparency Package,* number 31b. Use color transparency number 31b as you teach the water cycle.

## TEACH

### Guided Practice

Have students write down their answers to the margin question: for respiration and photosynthesis.

### Check for Understanding

Have students respond to the first three questions in Check Your Understanding.

### Reteach

**Reteaching**/*Teacher Resource Package*, p. 91. Use this worksheet to give students additional practice with the water cycle.

**Extension:** Assign Critical Thinking, Biology and Reading, or some of the **OPTIONS** available with this lesson.

## 3 APPLY

### ACTIVITY/Field Trip

Use the school yard, fence row, or wooded area for a field trip. Have students record on a chart the nonliving factors that affect the organisms they see. Discuss the findings with the class.

## 4 CLOSE

### Convergent Question

Have students answer the question: In what way would the nonliving factor(s) have to be changed so a ________ could live in this area? Fill in the blank with names like polar bears, tropical fish, orchids, etc.

### Answers to Check Your Understanding

1. living and nonliving factors
2. Refer to Figure 31-3.
3. to make proteins
4. Water would not be recycled through the trees, so the amount of transpiration would be less; less oxygen would be produced; less carbon dioxide would be used.
5. It means a recurring series of events. The elements can be used over and over again.

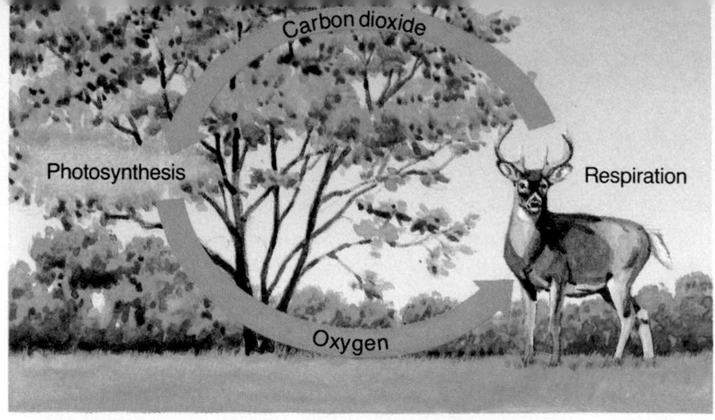


**Figure 31-4** The oxygen-carbon dioxide cycle. What is given off during photosynthesis?

## Oxygen-Carbon Dioxide Cycle

Why do living things need oxygen?

Air is needed in an ecosystem. Air is made up of several gases that are important to living things. You have learned about nitrogen. Two other gases in the air are oxygen and carbon dioxide. Why are these gases important? Without oxygen and carbon dioxide, respiration and photosynthesis can't take place.

Trace the path of oxygen and carbon dioxide through the ecosystem in Figure 31-4. Oxygen given off by producers during photosynthesis is needed by all living things for cellular respiration. Carbon dioxide given off during respiration is used by producers for making food. Oxygen and carbon dioxide can be cycled.

Producers and consumers in a water ecosystem, such as a lake, also have an oxygen-carbon dioxide cycle. Water has air mixed into it. Oxygen in the water is used by fish, plants, and other water life for respiration. Carbon dioxide in the water is used by water plants, blue-green bacteria, and algae for photosynthesis.

### Check Your Understanding

1. What two parts make up an ecosystem?
2. Draw a diagram of the water cycle.
3. Why do plants and animals need nitrogen?
4. **Critical Thinking:** If a large forest was destroyed, how would cycling of materials be affected?
5. **Biology and Reading:** What is meant by the word *cycle* when discussing the nitrogen, water, and oxygen–carbon dioxide cycles?



### OPTIONS

**Enrichment:** Construct a bulletin board that shows some of the nutrient cycles. Show pictures of plants, animals, rocks, streams, and the atmosphere as components of various cycles. Set up a separate section for each cycle.

**How Much Water Will Soil Hold?**/*Lab Manual*, pp. 259-262. Use this lab as an extension to cycles.

**RETEACHING 91**

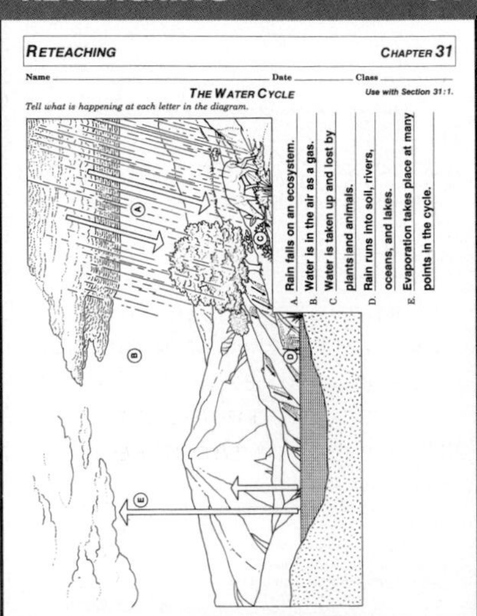
RETEACHING CHAPTER 31

Name Date Class

THE WATER CYCLE

Use with Section 31:1.

*Tell what is happening at each letter in the diagram.*

A. Rain falls on an ecosystem.
B. Water is in the air as a gas.
C. Water is taken up and lost by plants and animals.
D. Rain runs into soil, rivers, oceans, and lakes.
E. Evaporation takes place at many points in the cycle.

# Lab 31–1 Carbon Dioxide

## Problem: Do plants use carbon dioxide or give off carbon dioxide?

### Skills

observe, form hypotheses, interpret data, infer, design an experiment

### Materials

4 test tubes
marking pencil
4 stoppers
lamp
blue test liquid
green test liquid
2 *Elodea* sprigs
2 test-tube racks

### Procedure

1. Read the problem. In your notebook, write a **hypothesis** that answers the question.
2. Copy the data table.
3. Number the tubes 1 to 4 with the marking pencil.
4. Fill tubes 1 and 2 nearly full with green test liquid. Put an *Elodea* sprig in tube 2. Stopper both tubes. Place them in a rack.
5. Record the color of the liquid in each tube.
6. Place both tubes in front of the lamp and turn the lamp on.
7. Keep both tubes in the light overnight.
8. Fill tubes 3 and 4 nearly full with blue test liquid. Put an *Elodea* sprig in tube 4. Stopper both tubes. Place them in a rack.
9. Record the color of the liquid in each tube.
10. Place both tubes in the dark overnight.
11. **Observe:** Record the color of the liquids in all tubes the next day.
12. Dispose of the materials as directed by your teacher.

### Data and Observations

1. In which tubes, 1 or 2, did the green test liquid change color?
2. In which tubes, 3 or 4, did the blue test liquid change color?

| TUBE | PLANT ADDED? | WHERE PLACED (DARK OR LIGHT) | COLOR OF TUBE FIRST DAY | COLOR OF TUBE SECOND DAY |
|---|---|---|---|---|
| 1 | | light | green | green |
| 2 | √ | light | green | blue |
| 3 | | dark | blue | blue |
| 4 | √ | dark | blue | green |

### Analyze and Apply

1. What was the purpose of tube 1?
2. What was the purpose of tube 3?
3. If the green liquid turns blue, carbon dioxide was used in the tube. Which tube, 1 or 2, showed evidence that carbon dioxide was used overnight?
4. If the blue liquid turns green, carbon dioxide was given off in the tube. Which tube, 3 or 4, showed evidence that carbon dioxide was given off overnight? Explain.
5. Do plants use or give off carbon dioxide?
6. **Check your hypothesis:** Was your hypothesis supported by your data? Why or why not?
7. **Apply:** What process occurs in green plants that gives off carbon dioxide? What process in plants uses carbon dioxide?

### Extension

**Design an experiment** to test whether other water plants, such as hornwort, use and give off carbon dioxide.



## ANSWERS

### Data and Observations

1. tube 2–color changed from green to blue
2. tube 4–color changed from blue to green

### Analyze and Apply

1. as a control; to see if the green test liquid would turn blue without the plant
2. as a control; to see if the blue test liquid would turn green without the plant
3. tube 2 because the green test liquid turned blue
4. tube 4 because the blue test liquid turned green
5. both; Carbon dioxide is used during photosynthesis and given off during respiration.
6. Students who hypothesized that plants both use and give off carbon dioxide will say that their hypotheses were supported.
7. respiration; photosynthesis

## Lab 31-1 Carbon Dioxide

### Overview

Students learn that carbon dioxide is used by plants in photosynthesis.

**Objectives:** Upon completing this lab, students will be able to: (1) **demonstrate** the test for $CO_2$, (2) **infer** when a plant uses or gives off $CO_2$, and (3) **separate and control variables** in an experiment.

**Time Allotment:** 45 minutes on day 1 and 10-15 minutes on day 2

### Preparation

▶ For the blue liquid, add 0.1 g bromthymol blue powder to 2 L distilled water.

▶ For the green liquid, crush a seltzer tablet and add small amounts of it to the blue liquid until the liquid turns green.

▶ *Elodea* can be maintained in a gallon jar nearly full of water, with 5 cm of sand in which to anchor the plants. Put the jar in a lighted area.

**Lab 31-1 worksheet**/*Teacher Resource Package*, pp. 121-122.

### Teaching the Lab

▶ Explain that $CO_2$ is constantly produced by plants. However, during photosynthesis, the amount of $CO_2$ used outweighs the amount produced, resulting in a net loss to the environment.

**Cooperative Learning:** Divide the class into groups of two students. For more information see pp. 22T-23T in the Teacher Guide.

### ASSESSMENT

**Content:** Have students make a hypothesis stating what would happen to the color of the liquid in each of the test tubes if the *Elodea* plant were replaced with a pond snail and all other conditions were kept the same. Have students use their knowledge of the oxygen-carbon dioxide cycle in their hypothesis.

## 31:2 Succession

## PREPARATION

### Materials Needed

Make copies of the Focus, Transparency, Study Guide, and Reteaching worksheets in the *Teacher Resource Package*.

▶ Gather two plastic shoe boxes, sand, and soil for the Focus Demonstration.

▶ Have glass jars, soil, dead leaves, and plastic wrap available for the Mini Lab.

### Key Science Words

succession climax community

### Process Skills

In the Mini Lab, students will experiment.

## 1 FOCUS

▶ The objectives are listed on the student page. Remind students to preview these objectives as a guide to this numbered section.

### MOTIVATION/Demonstration

Fill a plastic shoe box with one inch of sand and add tap water until nearly full. Fill a second plastic shoe box with several inches of soil. Ask the students to imagine that the shoe box with water is a large pond and the shoe box with soil is a recently plowed field. Ask them how long it would take for organisms to appear in each. Ask them to explain what kinds of organisms could appear over a period of time.

Focus/*Teacher Resource Package*, p. 61. Use the Focus transparency shown at the bottom of this page to introduce succession.

### Guided Practice

Have students write down their answers to the margin question: changes that take place in a community as it ages.

## 31:2 Succession

**Objectives**

4. **Describe** succession in a land community.
5. **Describe** succession in a water community.

**Key Science Words**

succession
climax community

What is succession?

The living and nonliving parts of an ecosystem may change over a period of time. For example, as trees grow taller in a forest ecosystem, they change the amounts of light, temperature, and water in the forest. Forests that have large trees are cooler, darker, and wetter than forests made up of small trees. Decaying leaves make the soil more acid. Some forest organisms cannot live in these new conditions. They are replaced by organisms that can live in the new conditions. When we refer to the changes in an ecosystem, we often refer just to the changes in the living part, the community. The changes that take place in a community as it gets older are called **succession.** All communities go through succession. You might not notice the changes unless you observe the community for many years, because they happen slowly.

### Succession in a Land Community

To understand the stages of succession in one type of land community, we will start with a plowed field. In Figure 31-5a, there is nothing but bare soil.

After a few weeks or months, weed seeds are carried to the field by wind and animals. Weeds begin to grow over the soil. Worms and grasshoppers may be among the first animals to arrive. Beetles and ants arrive soon after. Figure 31-5b shows the changes that have occurred in the community.

As plants and animals die and decompose, their remains add nutrients to the soil. The soil becomes better suited to a greater variety of larger plants. Bushes and small trees begin to grow. Animal populations also change. Rabbits,

a

b

**Figure 31-5** The first stage of succession on land is bare soil (a). The next stage is the growth of producers (b).



**FOCUS 61**

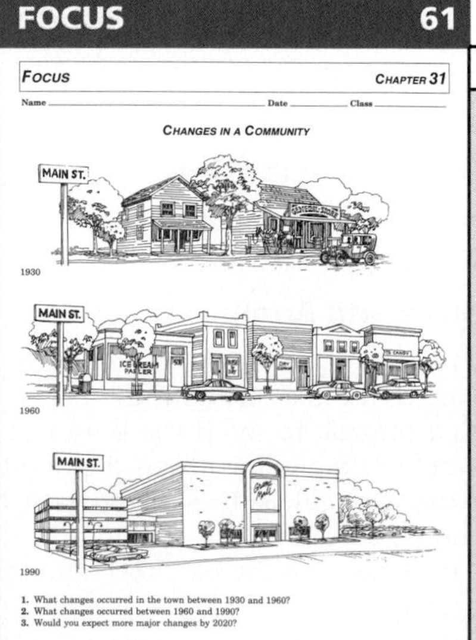

*Focus* *Chapter 31*

Name ______ Date ______ Class ______

*Changes in a Community*



1. What changes occurred in the town between 1930 and 1960?
2. What changes occurred between 1960 and 1990?
3. Would you expect more major changes by 2020?

## OPTIONS

### Science Background

Climax communities can be changed by unusual conditions such as earthquakes, volcanoes, and erosion, or human activities such as lumbering, clearing land, or pollution.

**How Can a Nonliving Part of an Ecosystem Help Living Things?**/*Lab Manual*, pp. 265-266. Use this lab as an extension to relationships between living and nonliving parts of the environment.

a

b

**Figure 31-6** A wide variety of plants provide food for many different animals (a). A climax community is usually a forest (b).

mice, foxes, skunks, and hawks may be found in the community shown in Figure 31-6a.

As more plants and animals die in the community, more nutrients are added to the soil. The soil itself becomes looser and better suited to the growth of larger plants. Now large trees are able to grow. If the community is not disturbed, a forest begins to develop. There are more kinds of animals in the forest than there are in a vacant field.

The last stage in the succession of this community is a forest like the one in Figure 31-6b. We call the forest the climax community. A **climax community** is the last or final stage of succession in a community. It is the final stage because it is stable and can replace itself with little change from then on. It may take 150 years or more for an area to become a climax community.

### Mini Lab

**How Do Leaves Decay?**

**Experiment:** Fill a jar half full with dead leaves and soil. Moisten with water and cover with plastic wrap. Punch holes in the wrap. Record the odor weekly. What causes the change? *For more help, refer to the **Skill Handbook**, pages 704-705.*

## Succession in a Water Community

Could the place you are sitting right now have been a pond or lake a few hundred years ago? It may seem impossible when you think of the living things that make up a water community. Once, however, there might have been a water community where you are sitting.

Let's look at the stages of succession for a pond that changes into a land community. The bottom and sides of a new pond are bare. The only green organisms in the pond are algae floating on the water surface. Snails and a few small fish can be found. After several years, the pond becomes more shallow because dead plants and algae have piled up on the bottom. Larger fish, tadpoles, crayfish, and frogs are now present.

**Why does a pond become more shallow over time?**



# 2 TEACH

### Mini Lab

The odor will have a "moldy earth" smell. This odor is caused by fungi and bacteria decomposing the leaves.

### ASSESSMENT

**Skill:** Have students make and test a hypothesis as to whether the soil in the container is needed to bring about decay in leaves. Students can set up another jar with leaves but without soil and compare the results of the two jars after several weeks.

### Guided Practice

Have students write down their answers to the margin question: dead plants and algae pile up on the bottom, filling the pond.

### *Student Journal*

Make a copy of the transparency on page 155 of the *Transparency Masters* book for students and have them research how many years it takes to complete each stage of succession from bare soil to the mature forest. Tell them to find out what organisms from the five kingdoms are in each stage and are important in bringing about succession. Have them report their findings in their journals.

**TRANSPARENCY 155**

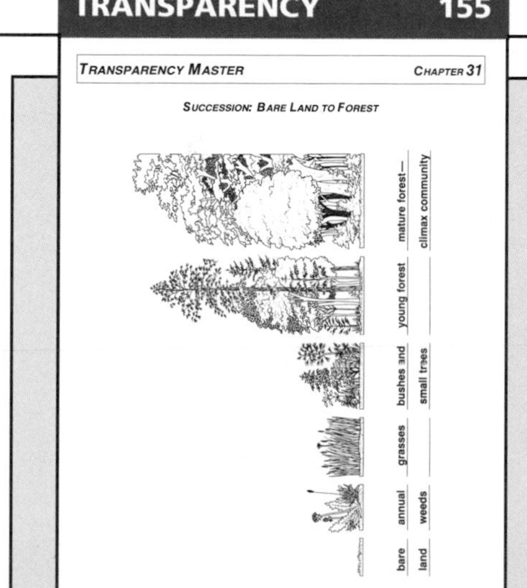

*TRANSPARENCY MASTER* *CHAPTER 31*

*SUCCESSION: BARE LAND TO FOREST*



**STUDY GUIDE 182**

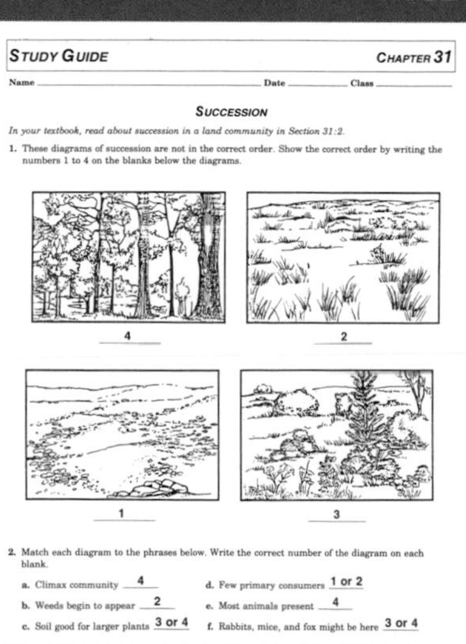

*STUDY GUIDE* *CHAPTER 31*

Name ______ Date ______ Class ______

*SUCCESSION*

*In your textbook, read about succession in a land community in Section 31:2.*

1. These diagrams of succession are not in the correct order. Show the correct order by writing the numbers 1 to 4 on the blanks below the diagrams.

4 2 1 3

2. Match each diagram to the phrases below. Write the correct number of the diagram on each blank.

a. Climax community 4
b. Weeds begin to appear 2
c. Soil good for larger plants 3 or 4
d. Few primary consumers 1 or 2
e. Most animals present 4
f. Rabbits, mice, and fox might be here 3 or 4

## *OPTIONS*

**Enrichment:** Living things change the environment and bring about succession. For example, as small plants fill a meadow they enrich the soil, thus making it more hospitable for shrubs. The shrubs add more nutrients to the soil as they die and decompose, allowing growth of trees. As trees in a forest grow larger, they provide shade and the smaller plants in their shadow may fail to thrive.

**Transparency Master**/*Teacher Resource Package*, p. 155. Use the Transparency Master shown here as you teach land succession.

# TEACH

### Check for Understanding

Have students respond to the first three questions in Check Your Understanding.

### Reteach

**Reteaching**/*Teacher Resource Package*, p. 92. Use this worksheet to give students additional practice with succession.

**Extension:** Assign Critical Thinking, Biology and Reading, or some of the **OPTIONS** available with this lesson.

## 3 APPLY

### ACTIVITY/Project

Have student teams build a model to show ecological succession in a plowed field. Have them fill a plastic box with two inches of soil, sprinkle with a cup of mixed seeds (grass, bird seed, dried beans, and cover with 1/4 inch of soil. Add water until moist. Students should notice that grass seeds sprout and grow rapidly only to be replaced by the beans and larger plants.

## 4 CLOSE

### Divergent Question

Ask students to describe areas around their home that are undergoing succession. How do they know that succession is occurring?

### Answers to Check Your Understanding

6. the series of changes a community goes through as it gets older
7. primary consumers
8. frogs and turtles because they can live out of water
9. Since animals contribute nutrients to the soil when they die, plant succession may proceed more slowly.
10. No, the climax community for any body of water is a land community. That is because the body of water is filled in by soil.

**Figure 31-7** Succession in a pond leads to dry land.

Over the years, more dead plants pile up and soil washes into the pond. The pond becomes more shallow and plants begin to grow around the edge of the pond, Figure 31-7c. Eventually, there will not be enough water for the fish to survive. Animals that can live out of water part of the time, such as turtles and frogs, will increase in number.

A hundred years later, the ground looks nothing like a pond. It may look like what you see in Figure 31-7d. Grass, shrubs, and small trees grow where the old pond was. Much of the animal life is completely different from when there was a young pond. Some day the area the pond occupied may contain a hardwood forest. A forest is not the only type of climax community that can develop. Depending upon soil, water, air, and sunlight, the climax community for an area might be a prairie or grassland, a desert, or a tropical rain forest. What kind of climax community can you identify where you live?

## Check Your Understanding

6. What is succession?
7. If weeds are the first things to appear in a land community, what will the first animals probably be?
8. What animals are found in an older pond that is very shallow? Why?
9. **Critical Thinking:** How would succession be affected if animals did not return to an area after a fire?
10. **Biology and Reading:** Is the climax community for a lake a watery environment? Why or why not?



## *OPTIONS*

**Enrichment:** Have students set up a closed ecosystem in a gallon jar using aquatic organisms. They can observe this for several weeks. Use the article "Sealed-Jar Ecosystem" by John Zuke in *The Science Teacher,* March, 1986: Vol. 53, pp. 46-47 as a reference.

**RETEACHING 92**

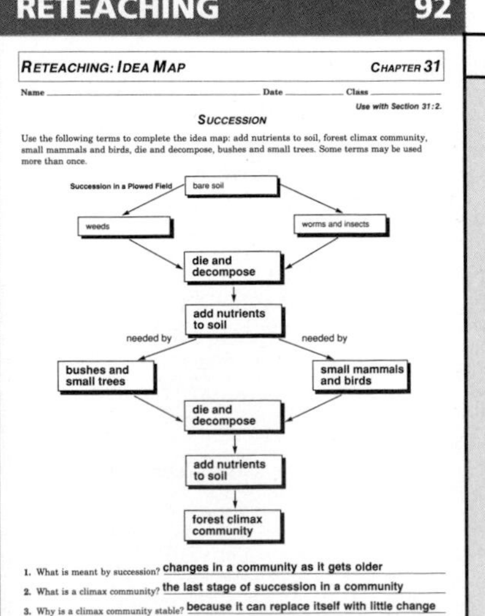

RETEACHING: IDEA MAP — CHAPTER 31

Name ______ Date ______ Class ______

*Use with Section 31:2.*

SUCCESSION

Use the following terms to complete the idea map: add nutrients to soil, forest climax community, small mammals and birds, die and decompose, bushes and small trees. Some terms may be used more than once.



1. What is meant by succession? changes in a community as it gets older
2. What is a climax community? the last stage of succession in a community
3. Why is a climax community stable? because it can replace itself with little change

# 31:3 How Living Things Are Distributed

The nonliving parts of an ecosystem usually control what organisms can live in a certain area. Temperature, sunlight, and water are very important factors in determining which organisms live in a place.

**Objectives**

6. **Explain** how climate helps determine what living things live in an area.
7. **Describe** the major land biomes and water ecosystems.

**Key Science Words**

precipitation
climate
biome

## Climate

Light and temperature are two nonliving parts of an ecosystem that are not cycled. Light from the sun is used by plants to make food. Producers transfer energy to other living things through food chains.

Temperature and light are often related. The soil in a field is warmed by the sun. The soil in a forest is cool because the leaves of trees prevent most of the sun's light from warming the ground.

The temperature of an ecosystem helps determine what organisms live there. Polar bears can live in very cold ecosystems, while lions, elephants, and palm trees live only in warmer ecosystems. Temperatures that are too low for some living things are just right for others.

The water cycle is also related to temperature and light. The amount of sunlight can affect the rate of evaporation. Water evaporates faster at warm temperatures. The temperature of the air affects the type of precipitation (prih sihp uh TAY shun) falling to Earth. **Precipitation** is water in the air that falls to Earth as rain or snow. Figure 31-8 shows temperatures and precipitation for the land biomes. You will learn about biomes on the next page.

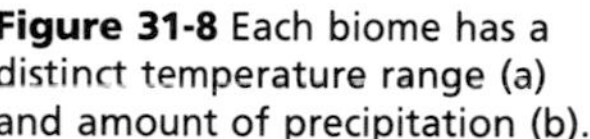

**Figure 31-8** Each biome has a distinct temperature range (a) and amount of precipitation (b).

Temperature Range °C
40
30
20
10
0
-10
-20
-30
Desert Taiga Tundra Temp. Forest Grass-land Rain Forest
a

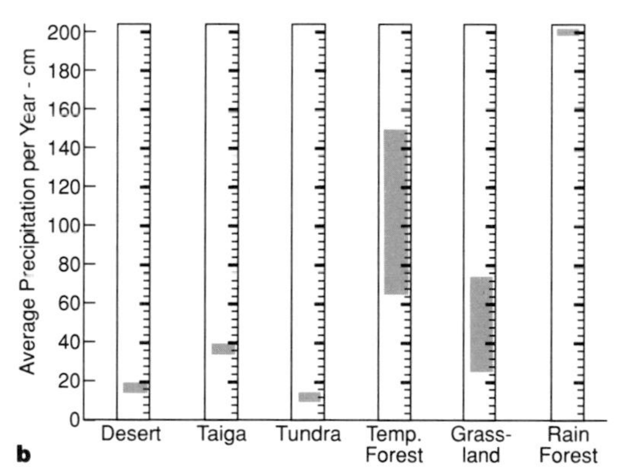




**SKILL 32**

SKILL: INTERPRET DIAGRAMS — CHAPTER 31

Name ______ Date ______ Class ______

*For more help, refer to the Skill Handbook, pages 706-711.* *Use after Section 31:3.*

BIOMES

The world has many kinds of biomes. You would have to travel hundreds of kilometers across North America to see just a few of them. A faster way of seeing what these biomes look like would be to climb a high mountain. As you travel up the mountain, you would notice that the average temperature and precipitation change. As a result, the different areas on the mountain resemble different biomes.

*The diagram below shows the areas found on a particular mountainside at different elevations. Study the diagram, and then answer the questions.*

Snow and Ice
4750 meters
Tundra
4500 meters
Taiga
4000 meters
Temperate Forests

1. At what elevations would you pass through an area that resembles a taiga? **4000 meters to 4500 meters**
2. If you climbed to an elevation of 3000 meters, what biome-like area would you see? **temperate forests**
3. At what elevations would you see mostly small shrubs and lichens? **4500 meters to 4750 meters**
4. Why do you think there is no desert-like area on the mountain? **The mountain gets too much precipitation.**
5. Above what elevation would you expect not to see any plants? **4750 meters**
6. What could you conclude about the areas on the mountain if the west side of the mountain received much more precipitation than those on the east side? **The areas on the west side of the mountain might have different climate, plants, and animals from the areas on the east side.**

## OPTIONS

### Science Background

Many ecosystems in the same area having the same climate and organisms make up a biome. There are six basic biomes in the world.

Skill/*Teacher Resource Package*, p. 32. Use the Skill worksheet shown here to teach climate variations among biomes.

## 31:3 Biomes

## PREPARATION

### Materials Needed

Make copies of the Skill, Study Guide, Reteaching, and Lab worksheets in the *Teacher Resource Package*.

▶ Have on hand a large world map for the Focus activity.

▶ Order brine shrimp eggs for Lab 31-2. Make the salt solutions a day in advance so all of the salt dissolves.

### Key Science Words

precipitation
climate
biome

### Process Skills

In Lab 31-2, students will observe, form hypotheses, interpret data, and design an experiment. In the Skill Check, students will sequence.

## 1 FOCUS

▶ The objectives are listed on the student page. Remind students to preview these objectives as a guide to this numbered section.

### Divergent Question

Have students identify the following areas of the world on a map: polar, temperate, and equatorial. Ask them to identify the types of climate and living things they would expect to find in each area.

**Student Journal**

The tundra biome is sometimes thought of as a northern desert. Tell students to look at the data presented in Figure 31-8 and have them write in their journals why they believe this could be a true statement.

# 2 TEACH

### MOTIVATION/Filmstrip

Show the filmstrip *Biomes Collection,* Carolina Biological Supply, Co.

### Guided Practice

Have students write down their answers to the margin question: the amounts of light and precipitation and the average temperature of an area over a period of years.

### ACTIVITY/Software

*Self-Help: Ecology,* Queue.

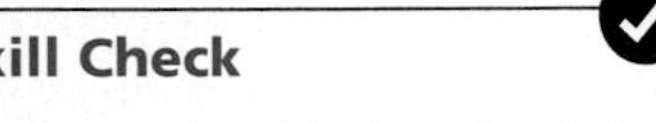

### Skill Check

Students should give the following answers:

a. tundra, desert, taiga, grassland, temperate forest, tropical rain forest (in this order)

b. tundra, taiga, temperate forest, grassland, desert, tropical rain forest (in this order)

wettest/warmest–tropical rain forest

coolest/driest–tundra

### Independent Practice

**Study Guide**/*Teacher Resource Package*, p. 185. Use the Study Guide worksheet shown at the bottom of this page for independent practice.

## GLENCOE TECHNOLOGY

**Videodisc**

**Biology: The Dynamics of Life**

Movie: *Tundra*

Disc 1, Side 1, Ch. 7

Movie: *Taiga*

Disc 1, Side 1, Ch. 8

What is climate?

All of these factors—light, temperature, and precipitation—taken over many years is called the **climate** of an area. Some climates are wet while others are dry. Some are cold while others are warm or hot. The climate of an area helps determine what kinds of plants and animals can live there. You can see how living and nonliving factors act together to make an ecosystem.

### Skill Check

**Sequence:** List the six major land biomes in order from (a) least to most precipitation and (b) from the lowest to highest temperature. Which biome is the wettest and warmest? Which biome is coolest and driest? *For more help, refer to the* ***Skill Handbook****, pages 706-711.*

## Land Biomes

There are large areas on Earth that have similar climates and climax communities. A land area with a distinct climate and with specific types of plants and animals is called a **biome** (BI ohm). Each biome has its own distinct producers, consumers, and decomposers. A biome is made up of all the ecosystems on Earth that have similar climates and organisms. For example, there are many deserts on Earth. They all have very dry climates and similar organisms. All the deserts together make up the desert biome. Figure 31-10 shows the major biomes and some of the living things found in each one. Biomes include tropical rain forests, grasslands, deserts, temperate forests, taiga (TI guh), and tundra (TUN druh).

A biome usually covers parts of several continents. Each continent has several biomes. Look at the map in Figure 31-9. How many different biomes are there in the continent of North America?

**Figure 31-9** The major land biomes



## GLENCOE TECHNOLOGY

**Videodisc**

**Biology: The Dynamics of Life**

Movie: *Desert*

Disc 1, Side 1, Ch. 9

## STUDY GUIDE 185

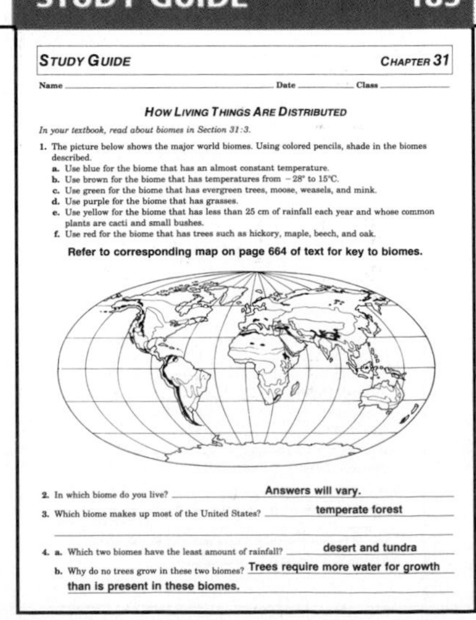

STUDY GUIDE CHAPTER 31

Name ________ Date ________ Class ________

*HOW LIVING THINGS ARE DISTRIBUTED*

*In your textbook, read about biomes in Section 31:3.*

1. The picture below shows the major world biomes. Using colored pencils, shade in the biomes described.
   a. Use blue for the biome that has an almost constant temperature.
   b. Use brown for the biome that has temperatures from −28° to 15°C.
   c. Use green for the biome that has evergreen trees, moose, weasels, and mink.
   d. Use purple for the biome that has grasses.
   e. Use yellow for the biome that has less than 25 cm of rainfall each year and whose common plants are cacti and small bushes.
   f. Use red for the biome that has trees such as hickory, maple, beech, and oak.

**Refer to corresponding map on page 664 of text for key to biomes.**

2. In which biome do you live? **Answers will vary.**
3. Which biome makes up most of the United States? **temperate forest**
4. a. Which two biomes have the least amount of rainfall? **desert and tundra**
   b. Why do no trees grow in these two biomes? **Trees require more water for growth than is present in these biomes.**

**Figure 31-10** Biomes found on land

| COMMON PLANTS | BIOME | COMMON ANIMALS |
|---|---|---|
| Vines<br>Palm trees<br>Orchids<br>Ferns |  Tropical rain forest | Tree frogs<br>Birds<br>Insects<br>Monkeys |
| Grasses |  Grassland | Antelope<br>Gophers<br>Rabbits<br>Prairie dogs |
| Cacti<br>Small bushes |  Desert | Lizards<br>Snakes<br>Scorpions<br>Mice |
| Trees: Hickory<br>Maple<br>Beech<br>Oak |  Temperate forest | Deer<br>Black bears<br>Squirrels<br>Insects |
| Evergreen trees:<br>Spruces<br>Conifers<br>Firs |  Taiga | Moose<br>Weasels<br>Mink |
| Lichens<br>Mosses<br>Grasses<br>Small bushes |  Tundra | Caribou<br>Musk oxen<br>Polar bears |



## TEACH

### Concept Development

▶ Point out that several biomes may be found in one country.

▶ Explain that at one time one-half of the land on Earth was grassland biome and today that land is used by humans to grow crops, raise animals, or for housing.

### Audiovisual

Show the videocassette *Populations, Communities, and Biomes,* Insight Media.

### Connection: Geography

Have students find pictures of other countries showing different biomes. The student can glue these pictures to construction paper and list the name of the country, the continent on which the country is found, and the biome in the picture. Note that some countries may have more than one biome.

### Independent Practice

**Study Guide**/*Teacher Resource Package*, p. 186. Use the Study Guide worksheet shown at the bottom of this page for independent practice.

### Portfolio

Tell students that each biome has its own distinct producers, consumers, and decomposers adapted to living in that biome. Using the organisms listed in Figure 31-10, have them find out what specific adaptation organisms have that allows them to survive in that specific region. Have students relate the adaptations to the climate and land features. Have them present their findings in report form in their journals.

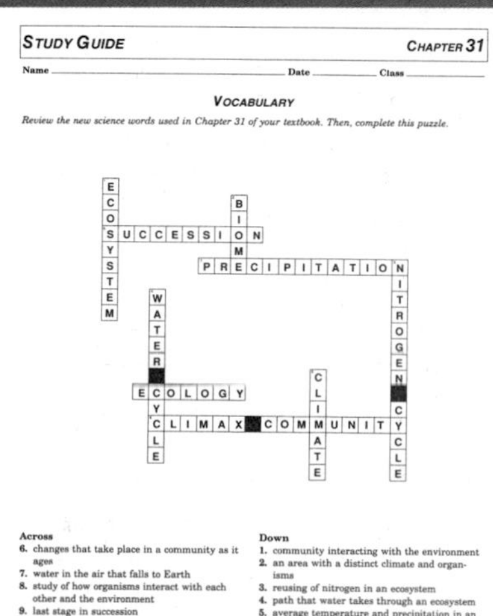

**STUDY GUIDE 186**

STUDY GUIDE CHAPTER 31

Name ______ Date ______ Class ______

*VOCABULARY*

*Review the new science words used in Chapter 31 of your textbook. Then, complete this puzzle.*



**Across**
6. changes that take place in a community as it ages
7. water in the air that falls to Earth
8. study of how organisms interact with each other and the environment
9. last stage in succession

**Down**
1. community interacting with the environment
2. an area with a distinct climate and organisms
3. reusing of nitrogen in an ecosystem
4. path that water takes through an ecosystem
5. average temperature and precipitation in an area

**GLENCOE TECHNOLOGY**

**Videodisc**

**Biology: The Dynamics of Life**
Movie: *Tropical Rain Forest*
Disc 1, Side 1, Ch. 12

**GLENCOE TECHNOLOGY**

**Videodisc**

**Biology: The Dynamics of Life**
Movie: *Temperate Forest*
Disc 1, Side 1, Ch. 11

## Idea Map

Have students use the idea map as a study guide to the major concepts of this section.

## Guided Practice

Have students write down their answers to the margin question: the ocean.

## Check for Understanding

Have students respond to the first three questions in Check Your Understanding.

### Reteach

**Reteaching**/*Teacher Resource Package*, p. 93. Use this worksheet to give students additional practice with the climates of different biomes.

**Extension:** Assign Critical Thinking, Biology and Writing, or some of the **OPTIONS** available with this lesson.

## 3 APPLY

### Guest Speaker

Invite a wildlife biologist to speak to your class about working with communities and ecosystems.

## 4 CLOSE

Ask students to use the words population, community, ecosystem, and biome to explain how each word is related to the others.

### Answers to Check Your Understanding

11. There is too little rainfall, and the temperatures are too cold.
12. tundra; caribou, musk oxen, polar bears
13. The saltwater ecosystem contains a large amount of mineral salts; a freshwater ecosystem does not.
14. The current biomes would change into different biomes as the climate changed. The regions most affected would be ice, tundra, and taiga.
15. Answers will vary.



**Study Tip:** Use this idea map as a study guide to biomes and ecosystems. Give an example of an animal that lives in each one.

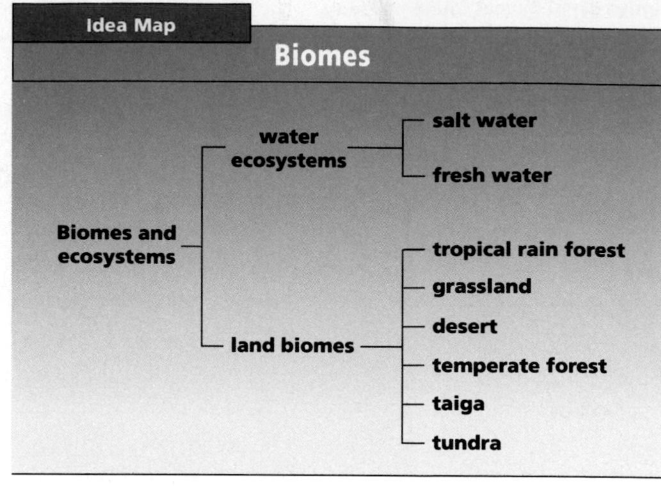


## Water Ecosystems

Water ecosystems are divided into fresh water and salt water. Streams, rivers, and lakes make up the freshwater ecosystem. Streams and rivers may start as water draining from a pond or lake and eventually flow into an ocean. The oceans make up the saltwater ecosystem because they contain large amounts of salts. Most freshwater organisms would die if placed in an ocean.

**What is Earth's largest ecosystem?**

The ocean ecosystem is Earth's largest ecosystem. It can be divided into smaller ecosystems. These include tidal pools, salt marshes, seashores, coral reefs, and the open ocean.

### Check Your Understanding

11. Why can't palm trees live in the tundra?
12. What biome has the lowest average precipitation? Name three organisms that live there.
13. What major difference between freshwater and saltwater ecosystems determines what organisms will live there?
14. **Critical Thinking:** What might be an effect on today's biomes if all climates became warmer?
15. **Biology and Writing:** Write a paragraph of at least five sentences that describes the biome you live in.



## OPTIONS

**Challenge:** Have students locate the biome in which they live and make a list of the smaller ecosystems within this biome. It may include swamps, rivers, streams, fields, fence rows, etc. Make sure they understand the difference between an ecosystem and a biome. Have students identify the biome(s) that border(s) the biome in which they live.

### RETEACHING 93

RETEACHING — CHAPTER 31

Name ______ Date ______ Class ______

*Use with Section 31:3.*

**CLIMATES AND BIOMES**

Rainfall and temperature are the important factors that make up the climate of a biome. Whether or not a plant or an animal survives in a biome depends on the climate of the biome.

**Annual Rainfall and Temperature in Biomes**

| Biome | Annual rainfall (in cm) | Normal temperature range (in °C) | |
|---|---|---|---|
| | | Highest | Lowest |
| Temperate forest | 65-150 | 38 | −24 |
| Tundra | 10-15 | 15 | −28 |
| Tropical rain forest | 200 | almost constant 25 | |
| Taiga | 35-40 | 22 | −24 |
| Desert | Under 25 | 38 | 10 |
| Grassland | 25-75 | 25 | 0 |

*Use the information in the table and in your textbook to answer these questions about biomes.*

1. Which biome has the greatest amount of rainfall in a year? **tropical rain forest**
   The least amount of rainfall in a year? **tundra**
2. Which biome has the highest normal temperature? **desert and temperate forest**
   The lowest normal temperature? **tundra**
3. Is the tropical rain forest or the desert the warmer biome throughout the year? Explain. **The tropical rainforest is warmer throughout the year because there is very little difference between its highest and lowest temperatures.**
4. List the biomes in order from greatest to least amount of rainfall. **tropical rain forest, temperate forest, grassland, taiga, desert, tundra**
5. Which biome has the widest range of temperature from highest to lowest? **temperate forest**
6. Which biome is the warmest and the wettest? **tropical rain forest**
7. Why do you think the tundra is sometimes called the arctic desert? **The tundra biome has a small amount of rainfall, as does the desert biome.**

# Lab 31–2 Brine Shrimp

## Problem: How does salt affect the growth of brine shrimp?

### Skills

observe, form hypotheses, interpret data, design an experiment

### Materials

flat toothpick
brine shrimp eggs
weak salt solution
strong salt solution
distilled water
3 small plastic cups
marking pen
hand lens

### Procedure

1. Copy the data table.
2. Label three cups 1, 2, and 3 with a marking pen.
3. Fill the cups nearly full with the following:
   cup 1—distilled water
   cup 2—weak salt solution
   cup 3—strong salt solution
4. Dip a toothpick into the brine shrimp eggs. Brine shrimp are tiny animals that live in salt water. Place the eggs that cling to the toothpick into cup 1.
5. Repeat step 4 with cups 2 and 3. The eggs will float on the water.
6. Place the cups on a shelf in the classroom.

| CUP | AMOUNT OF SALT | DAY WHEN MOST SHRIMP HATCHED | CUP WITH MOST BRINE SHRIMP |
|---|---|---|---|
| 1 | none | did not hatch | |
| 2 | weak solution | day 3 | |
| 3 | strong solution | day 2 | √ |

7. Write a **hypothesis** in your notebook to explain which cup will have the most brine shrimp at the end of 3 days.
8. **Observe:** Observe the cups for the next three days. Use the hand lens to look for small orange-colored animals swimming in a jerking motion. These objects are the brine shrimp that hatched from the eggs. Note the cup in which you see the first shrimp and the number of shrimp that hatch in each cup on each day. Record your data in the data table.

### Data and Observations

1. How many days did it take for you to see the first brine shrimp?
2. In which of the three cups did they hatch first?
3. In which of these three cups did the most brine shrimp hatch? In which did they grow best?
4. In which of the three cups did the brine shrimp not hatch at all?

### Analyze and Apply

1. What would you call the ecosystems in which you were trying to grow brine shrimp?
2. In which type of ecosystem do you think brine shrimp hatch and grow best?
3. **Check your hypothesis:** Was your hypothesis supported by your data? Why or why not?
4. **Apply:** How does salt affect the growth of brine shrimp?

### Extension

**Design an experiment** to find out if there is a maximum strength of salt water at which brine shrimp will grow.



## ANSWERS

### Data and Observations

1. Answers may vary. Students should see the first shrimp on day 1 or 2.
2. cup 3
3. cup 3; cup 3
4. cup 1

### Analyze and Apply

1. saltwater
2. the ocean or a saltwater ecosystem
3. Students who hypothesized that brine shrimp hatch best in strong salt solution will say that their hypotheses were supported.
4. The greater the salt concentration, the greater the hatching and growth of brine shrimp.

| amount of salt | cup number |
|---|---|
| 1 teaspoon | 1 |
| 2 teaspoons | 2 |
| 4 teaspoons | 3 |
| 8 teaspoons | 4 |
| 16 teaspoons | 5 |

## Lab 31-2 Brine Shrimp

### Overview

Students will determine how salt affects the growth of brine shrimp eggs by placing the eggs in distilled water and two different concentrations of salt water.

**Objectives:** Students will be able to (1) **infer** which solution grows brine shrimp best, (2) **form a hypothesis** to explain which solution will have the most brine shrimp hatch, and (3) **determine** the type of ecosystem in which brine shrimp grow.

**Time Allotment:** Day 0: 45 minutes; days 1, 2, 3: 15 minutes.

### Preparation

▶ Weak salt solution: add a teaspoon of noniodized salt to a quart of water and stir until dissolved.

▶ Strong salt solution: add four teaspoons of noniodized salt to a quart of water and stir until dissolved.

Lab 31-2 worksheet/*Teacher Resource Package*, pp. 123-124.

### Teaching the Lab

▶ Students will notice that the stronger salt solution will hatch the most eggs. This salt solution is nearly the same as ocean water.

**Cooperative Learning:** Divide the class into groups of two or three students. For more information, see p. 22T-23T in the Teacher Guide.

### ASSESSMENT

**Performance:** Direct students to use the extension activity and make salt solutions according to the table at the left. Repeat procedure steps 4-6 and observe the cups for several days. Have students compare the results for the different salt solutions and write a report stating their observations and conclusions.

## APPLYING TECHNOLOGY

### Purpose

Students will learn the meaning of the terms *biotic* and *abiotic.* They will gather and plot temperature changes occurring during the day for a sidewalk and a grassy area ecosystem.

### Process Skills

Observe, communicate, measure, interpret data, form hypotheses, experiment, use numbers

### Time Required

Twenty minutes to record data on first day. Thirty minutes on following day to record data from other classes and to complete graphs

### Possible Problems

Remind students that thermometers are fragile. They should not shake the thermometers down. Thermometers can easily roll off desktops or lab tables.

### Teaching Strategies

Gather materials needed in advance of the trip outdoors. You will need thermometers (make sure they are all either Celsius or Fahrenheit) and chalk.

Visit the site of study in advance of the first class so that you can mark an area for study in chalk on the school sidewalk. All classes should visit the same site. Choose an area that receives maximum sun during the day.

Conduct this activity on a bright sunny day during the spring or fall.

Assign students into groups of four. Student roles will be recorder, timer, thermometer placer, and thermometer reader.

If you have only one or two classes conducting this activity, ask responsible students to be excused from study hall so that they can gather information at different times during the day. Ideally, four readings should be made during the day.

# TECH PREP APPLYING TECHNOLOGY

## Taking a Sidewalk's Temperature

Communities can be found right "under your nose." A parking lot contains a community of living things. So does a concrete sidewalk and the grassy area alongside it. Your backyard, a vacant lot, or a school football field also houses a community. What, then, makes a community different from an ecosystem? Once you begin to study both the living and the nonliving things that are present in a community, you are looking at the bigger picture—an ecosystem.

### Identifying the Problem

What information can you gain by examining the sidewalk outside your school and any grassy areas next to the sidewalk? You could determine temperature changes that occur during the day. You could also conduct an inventory of the animal and plant types that are present. You might want to see if the humidity of the sidewalk ecosystem changes during the day. In this activity, you will explore the temperature changes that occur during the day and survey the plant and animal life in a nearby grassy area. Form a hypothesis about how temperature changes may differ between the sidewalk and the grassy areas during the day.

**Technology Connection**

An important ecosystem worldwide is the land farm. However, farmers are now able to move beyond raising crops only on land. Through advances in technology, one of the newest farming practices is aquaculture. Aquaculture is the farming of certain plants and animals in large tanks of water (see photo). Salmon, shrimp, catfish, and tilapia (a freshwater fish) are animals being raised through this new technology.

### Collecting Information

Using references, look up and define the meaning of the terms *abiotic factors* and *biotic factors.* Give examples of each. Describe some of the biotic and abiotic factors that you might study in a sidewalk or grassy area ecosystem. Describe the biotic and abiotic factors that you actually will be studying in this activity.

668

### Technology Connection

Eighty six million metric tons of fish were caught worldwide in 1993. Twenty two percent of that total was supplied through aquaculture. Within 15 years, that percentage is expected to increase to 40 percent of the total fish consumed in the world. Of course "fish farmers" face ecological problems similar to those farmers growing traditional crops on land. Aquaculturalists must deal with disease, pollution, and space.

## Carrying Out an Experiment

1. Prepare a table in which you will record the temperature of the sidewalk and the grassy area nearby. You should record the data your group collects and the average data from all the groups in your class. Other classes will collect data at different times throughout the day. You will complete your data table the following day by adding the data collected by these classes.
2. Place a thermometer onto the chalk-marked area of the school sidewalk. Wait five minutes, and record the temperature and time.
3. Place the thermometer in the grassy area near the sidewalk. Wait five minutes and record the temperature.
4. Record and describe any animal life present on the sidewalk and in the grassy area.
5. Record and describe any plant life present on the sidewalk (or in any cracks) and in the grassy area.
6. Calculate class averages for the sidewalk temperature and grassy area temperature. Record these numbers for the time that your class meets.
7. At the next class session, complete your data table by adding the information from all other classes.
8. Prepare a graph that plots temperature changes for the sidewalk and grassy area against time in hours.

### Career Connection

- **Fisher** Fishes the oceans or lakes; may also dig for clams and oysters near shore; helps set and empty lobster traps
- **Ecologist** Regulates, controls, and prevents air, land, and water pollution; may work for the federal government or local agencies; solves environmental problems
- **Park Ranger** Makes surveys of plant and animal life; works for the federal government in national parks, forests, and wildlife refuges; studies relationships among living things

## Assessing Your Results

Explain why the sidewalk and the grassy area are ecosystems. How did the temperature of the sidewalk change during the day? How did the temperature of the grassy area change during the day? Which area showed a greater temperature change? Offer an explanation for why this may have been observed. How might this change affect the types of life forms that live there? Was your hypothesis supported by your data?

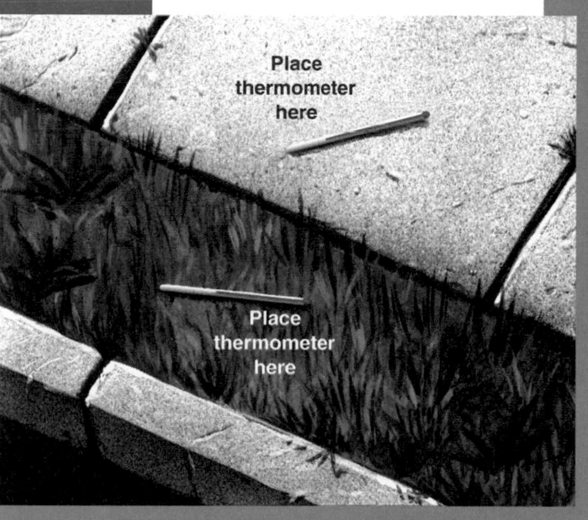


▶ Review the Celsius scale with students. Review averaging and graphing skills with students.

### Conclusions

Students will see a trend of gradual temperature rise during the day for both the sidewalk and the grassy areas. However, the temperature rise for the sidewalk area will be much higher. Plant life on the sidewalk (in cracks) and grassy areas may consist of weeds or grass. Animal life may include ants, beetles, spiders, earthworms, or isopods (sow/pill bugs).

### Answers to Questions

Sidewalks and grassy areas are ecosystems because they include both life forms and physical characteristics. The temperature of the sidewalk and grassy areas rose during the day, but the rise was greater on the sidewalk. Concrete tends to absorb more solar heat than does grass. Soil of the grassy area also contains moisture, which tends to maintain a more constant temperature. Life-forms living in the grassy area are not subjected to wide fluctuations in temperature during the day.

### ASSESSMENT

**Performance** Have students add to their graphs what they might expect if data had been collected for air temperatures above the sidewalk area. (Temperatures will rise during the day but will be below those for the sidewalk and higher than those for the grassy area.)

### Career Connection

**Career Path**
**Fisher** A high-school diploma is recommended.
**Ecologist** High-school diploma and a college degree from a four-year college; degree may be in environmental sciences or ecology.
**Park Ranger** High-school diploma; college degree may be needed for some jobs; on-the-job training

**Career Issue**
Have students discuss why an ecologist would be interested in the greenhouse effect. What impact might rising temperatures have on living things? Have them research why global temperatures are rising.

## CHAPTER 31 REVIEW

### Summary

Summary statements can be used by students to review major concepts of the chapter.

### Key Science Words

All boldfaced terms from the chapter are listed.

### Testing Yourself

#### Using Words

1. water cycle
2. biome
3. ecosystem
4. climax community
5. precipitation
6. climate
7. nitrogen cycle
8. ecology
9. succession

#### Finding Main Ideas

10. p. 657; Water falls to Earth as rain. Some evaporates and some soaks into the ground. Water from the ground, oceans, lakes, plants, and animals, evaporates into the air.
11. p. 654; Ecosystems can be found on land or water.
12. p. 660; Trees in a forest absorb the heat of the sun and shade the soil.
13. p. 658; producers and consumers
14. p. 655; Soil is nonliving.
15. p. 664; tundra, taiga, temperate forest, grassland, and desert
16. p. 660; bare soil
17. p. 658; Respiration and photosynthesis cannot take place without these gases.
18. p. 662; The pond becomes shallower.

#### Using Main Ideas

19. Oxygen, water, nitrogen, and carbon dioxide are recycled through an ecosystem.
20. Oxygen is respired by all organisms and carbon dioxide is used by producers for photosynthesis.
21. It controls what kinds of plants and animals can live in an area.
22. Their remains add nutrients to the soil.
23. an ocean
24. Dead plants pile up and soil washes into the pond.
25. Each biome is different in its climate, plants, and animals.

# Chapter 31 Review

## Summary

### 31:1 Ecosystems

1. Ecosystems are made up of living and nonliving parts.
2. Water is a nonliving part that is cycled through the ecosystem and is necessary for all living things.
3. The nitrogen cycle returns nitrates to the soil so they can be used for plant growth. The oxygen-carbon dioxide cycle provides carbon dioxide for plant photosynthesis and oxygen for plant and animal respiration.

### 31:2 Succession

4. The changes that take place in the living parts of a community as it gets older are called succession. Each stage of succession is represented by different plants and animals.
5. Succession in a water community results in formation of dry land.

### 31:3 How Living Things Are Distributed

6. Light, temperature, and precipitation make up an area's climate. Climate determines what living things live in an area.
7. Biomes are large land areas with distinct climate and living things. Water ecosystems include fresh water and salt water.

## Key Science Words

biome (p. 664)
climate (p. 664)
climax community (p. 661)
ecology (p. 654)
ecosystem (p. 654)
nitrogen cycle (p. 656)
precipitation (p. 663)
succession (p. 660)
water cycle (p. 657)

## Testing Yourself

#### Using Words

*Choose the word from the list of Key Science Words that best fits the definition.*

1. path of water through an ecosystem
2. has a distinct climate and certain plants and animals
3. community interacting with the environment
4. the last stage of succession
5. one factor that makes up the climate of an area
6. the average of temperature and precipitation over many years
7. the reusing of nitrogen in an ecosystem
8. the study of how living things interact with each other and their environment
9. changes in a community over time

# Review

## Testing Yourself *continued*

**Finding Main Ideas**

*List the page number where each main idea below is found. Then, explain each main idea.*

10. how water is recycled in an ecosystem
11. where ecosystems can be found
12. why the soil in a vacant field has a different temperature than the soil in a forest
13. what uses oxygen in an ecosystem
14. whether soil is a living or a nonliving part of an ecosystem
15. what the different biomes are in North America
16. what the first stage of succession on land is
17. why oxygen and carbon dioxide are important gases in an ecosystem
18. what happens to the depth of a pond as the community changes

**Using Main Ideas**

*Answer the questions by referring to the page number after each question.*

19. Which of the following are recycled through an ecosystem: oxygen, water, light, nitrogen, carbon dioxide? (pp. 655, 657, 658, 663)
20. How is the oxygen-carbon dioxide cycle important to an ecosystem? (p. 658)
21. How does climate affect plants and animals? (p. 663)
22. How can the soil be affected by the death of plants and animals? (p. 655)
23. Which has more salts, an ocean or river? (p. 666)
24. Why does a pond become more shallow as it ages? (p. 662)
25. What kinds of things separate one biome from another? (p. 664)

## Skill Review ✓

*For more help, refer to the **Skill Handbook,** pages 704-719.*

1. **Sequence:** List 5 ecosystems in the area where you live. Sequence these ecosystems from smallest to largest.
2. **Infer:** How can boiling water become part of the water cycle?
3. **Infer:** Explain why leaves that fell on the forest floor in autumn disappear during the next summer.
4. **Design an experiment:** Design an experiment to show how you could tell if you are giving off carbon dioxide when you breathe out.

## Finding Out More

TECH PREP

### Critical Thinking

1. Suppose you set up a terrarium with plants, soil, decomposers, and water. You sealed it so it was air-tight. Could the plants live in this sealed environment? What is the one factor you would have to supply?
2. Use the map of biomes on page 664 to find out which biome you live in. What would happen if your biome got half the amount of precipitation it now gets?

### Applications

1. Visit a park or area near your school. Record evidence for how nonliving things affect the living things.
2. Find out about Biosphere II. Describe some of the problems scientists had to solve while planning and building this project.



## Skill Review

1. Answers might include ponds, forests, meadows, and so on. The size of the ecosystem will vary with the area in which the students live.
2. Boiling water condenses and forms water droplets when it cools. These water droplets become heavier than air and form precipitation.
3. Leaves are decomposed by fungi and bacteria over a period of time.
4. Blow into a jar containing blue liquid. If the blue liquid changes to green or yellow, you are breathing out $CO_2$.

## Finding Out More

### Critical Thinking

1. The sealed environment could recycle its materials such as water, nitrogen, oxygen, and carbon dioxide, as long as it was supplied with light as an energy source.
2. The types of plants and animals in the biome would change. Plants that need a lot of moisture would not be able to survive with half the amount of moisture.

### Applications

1. Temperature in autumn causes leaves to change and fall. Plants become dormant. Animals become less active or hibernate in winter. Some plants need more light or moisture (or less) than others. For example, ferns would get scorched in bright sun.
2. Some of the problems have been keeping a balance in the water, oxygen, carbon dioxide, and nitrogen so that cycles would be stable. Producing enough food for the scientists also was a problem, so that food would not need to be brought in. Scientists also needed to find a way to regulate the temperature.

**Chapter Review**/*Teacher Resource Package*, pp. 165-166.

**Chapter Test**/*Teacher Resource Package*, pp. 167-169.

**Chapter Test**/*Computer Test Bank*

# Solving Ecological Problems

## PLANNING GUIDE

| CONTENT | TEXT FEATURES | TEACHER RESOURCE PACKAGE | OTHER COMPONENTS |
|---|---|---|---|
| (1 1/2 days)<br>32:1 Resources and Human Activities<br>Wildlife and Plants<br>Soil<br>Water<br>Fossil Fuels | Skill Check: *Calculate,* p. 677<br>Mini Lab: *How Could You Protect a Species?* p. 675<br>Mini Lab: *What Does Sediment Look Like?* p. 676<br>Check Your Understanding, p. 677 | Application: *How Humans Affect Animal and Plant Resources,* p. 32<br>Reteaching: *Resources and Human Activities,* p. 94<br>Critical Thinking/Problem Solving: *Can Pollution Change Succession?* p. 32<br>Skill: *Use Numbers,* pp. 33-34<br>Study Guide: *Resources and Human Activities,* p. 187 | **STVS:** *Saving the Spotted Owl,* Ecology (Disc 6, Side 1) |
| (2 1/2 days)<br>32:2 Problems From Pollution<br>Air Pollution<br>Water Pollution<br>Acid Rain<br>Land Pollution | Skill Check: *Experiment,* p. 679<br>Skill Check: *Understand Science Words,* p. 681<br>Technology: *The Greenhouse Effect—A Computer Model,* p. 680<br>Lab 32-1: *pH,* p. 682<br>Idea Map, p. 685<br>Check Your Understanding, p. 685 | Enrichment: *Acid Rain and Aquatic Organisms,* pp. 34-35<br>Reteaching: *Air Pollution,* p. 95<br>Focus: *Lichens Monitor Air Pollution,* p. 63<br>Transparency Master: *Acid Rain,* p. 161<br>Study Guide: *Problems From Pollution,* pp. 188-190<br>Lab 32-1: *pH,* pp. 125-126 | **Laboratory Manual:**<br>*How Do Chemical Pollutants Affect Living Things?* p. 271<br>**Laboratory Manual:**<br>*How Does Thermal Pollution Affect Living Things?* p. 267<br>Color Transparency 32: *How Pesticides Are Concentrated in a Food Chain*<br>**STVS:** *Arctic Haze,* Ecology (Disc 6, Side 2) |
| (2 days)<br>32:3 Working Toward Solutions<br>Conserving Our Resources<br>Keeping Our Environment Clean | Check Your Understanding, p. 689<br>Lab 32-2: *Pollution,* p. 686 | Reteaching: *Solutions to Environmental Problems,* p. 96<br>Study Guide: *Working Toward Solutions,* p. 191<br>Study Guide: *Vocabulary,* p. 192<br>Lab 32-2: *Pollution,* pp. 127-128 | **STVS:** *Treating Acid Lakes,* Ecology (Disc 6, Side 2) |
| Chapter Review | Summary<br>Key Science Words<br>Testing Yourself<br>Finding Main Ideas<br>Using Main Ideas<br>Skill Review | **ASSESSMENT RESOURCES**<br>Chapter Review, pp. 170-171<br>Chapter Test, pp. 172-174<br>Performance Assessment in the Biology Classroom<br>Alternate Assessment in the Science Classroom<br>Computer Test Bank | |

## GLENCOE TECHNOLOGY

**Infinite Voyage,** *Crisis in the Atmosphere*
**Science and Technology Videodisc Series,** *Saving the Spotted Owl,* Ecology (Disc 6, Side 1)
*Arctic Haze,* Ecology (Disc 6, Side 2)
*Treating Acid Lakes,* Ecology (Disc 6, Side 2)

**The Secret of Life,** *Gone Before You Know It: The Biodiversity Crisis*

## MATERIALS NEEDED

| LAB 32-1, p. 682 | LAB 32-2, p. 686 | MARGIN FEATURES |
|---|---|---|
| pH paper<br>pH chart<br>forceps<br>marking pencil<br>7 glass slides<br>7 droppers<br>ammonia<br>baking soda<br>cola<br>vinegar<br>2 samples of rainwater<br>distilled water | 4 test tubes<br>marking pencil<br>test-tube rack<br>5 droppers<br>blue test liquid<br>dead yeast mixture<br>live yeast mixture<br>detergent<br>hydrogen peroxide<br>distilled water | Skill Check, p. 677<br>pencil<br>paper<br>Skill Check, p. 679<br>plastic cup<br>distilled water<br>Skill Check, p. 681<br>dictionary<br>Mini Lab, p. 675<br>pencil<br>paper<br>Mini Lab, p. 676<br>jar with lid<br>soil<br>water |

## OBJECTIVES

For more information about National Science Standards, see page 5T.

| SECTION | OBJECTIVE | CORRELATION of QUESTIONS to OBJECTIVES: CHECK YOUR UNDERSTANDING | CHAPTER REVIEW | TRP CHAPTER REVIEW | TRP CHAPTER TEST |
|---|---|---|---|---|---|
| 32:1 National Science Stds: UCP.1, UCP.3, UCP.4, B.5, B.6, C.4, C.5, D.1, D.3, F.2, F.3, F.4, F.5, F.6, G.1 | 1. **Explain** how wildlife and plants are affected by humans. | 1, 4 | 5, 20 | 11 | 3, 29, 35, 49 |
| | 2. **Describe** how water and soil can be lost to the environment. | 2 | 4, 11, 17 | 6, 22, 23, 24 | 8, 20, 24, 42, 43 |
| | 3. **Explain** why fossil fuels are being used up. | 3 | 14, 21, 22 | 5, 29 | 5 |
| 32:2 National Science Stds: UCP.3, UCP.4, F.1, F.3, F.4, F.5, F.6 | 4. **Discuss** the causes of air pollution and acid rain. | 6 | 1, 8, 10, 15, 16, 25 | 1, 3, 7, 9, 12, 13, 14, 16, 18, 19 | 1, 4, 6, 7, 10, 12, 13, 14, 15, 18, 19, 21, 28, 30, 32, 34, 36, 38, 46, 47, 51, 52 |
| | 5. **Discuss** problems that come from nonbiodegradable chemicals that pollute water. | 7, 8, 9 | 2, 3, 6, 7, 23, 26, 27 | 2, 8, 17, 19, 21, 28 | 9, 11, 16, 17, 22, 23, 25, 26, 27, 33, 39, 40, 41, 50 |
| 32:3 National Science Stds: UCP.3, UCP.4, A.2, B.5, B.6, C.5, E.1, E.2, F.1, F.2, F.3, F.4, F.5, F.6, G.1, G.2 | 6. **Explain** methods of conserving resources. | 11, 13 | 13, 18, 19 | 4 | 31, 44 |
| | 7. **Discuss** ways to keep the environment clean. | 12, 14, 15 | 9, 24 | 10, 15 | 2, 37, 45, 48 |

# Solving Ecological Problems

## CHAPTER OVERVIEW

### Key Concepts

In this chapter, students will learn about the major ecological problems in the world today. They will learn how plants, animals, soil, water, and air can be damaged. They will find out about the damage caused by air and water pollution. The chapter closes with ways humans can conserve natural resources and keep the environment clean.

### Key Science Words

| | |
|---|---|
| acid | ozone |
| acid rain | pesticide |
| base | pollution |
| biodegradable | radon |
| endangered | recycling |
| erosion | sediment |
| fossil fuel | smog |
| greenhouse effect | threatened |
| natural resource | toxic |

### Skill Development

In Lab 32-1, students will **observe, experiment, interpret data,** and **classify** to compare the pH of rainwater with that of household chemicals. In Lab 32-2, students will **observe, form hypotheses, interpret data,** and **design an experiment** to determine the effects of pollutants on yeast. In the Skill Check on page 677, students will **calculate** the world population. In the Skill Check on page 679, students will **experiment** with rainwater. On page 681, students will **understand** the **science word** *pesticide*. In the Mini Lab on page 675, students will **outline** protection for an endangered species. In the Mini Lab on page 676, students will **classify** sediment found in soil.

### Bridging

Students learned about the different levels of ecological relationships in populations, communities, ecosystems, and biomes in Chapters 30 and 31. Chapter 32 explores environmental problems and solutions for them.

**CHAPTER PREVIEW**

### Chapter Content

Review this outline for Chapter 32 before you read the chapter.

**32:1 Resources and Human Activities**
- Wildlife and Plants
- Soil
- Water
- Fossil Fuels

**32:2 Problems from Pollution**
- Air Pollution
- Water Pollution
- Acid Rain
- Land Pollution

**32:3 Working Toward Solutions**
- Conserving Our Resources
- Keeping Our Environment Clean

### Skills in this Chapter

The skills that you will use in this chapter are listed below.

- In **Lab 32-1,** you will observe, experiment, interpret data, and classify. In **Lab 32-2,** you will observe, form hypotheses, interpret data, and design an experiment.
- In the **Skill Checks,** you will calculate, experiment, and understand science words.
- In the **Mini Labs,** you will outline and classify.

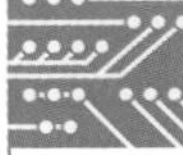

## TECH PREP

For Tech Prep activities in this chapter of the Teacher Wraparound Edition, see especially the Activity on page 677, the Options Activity and Student Journal on page 679, the Portfolio on page 681, the Cooperative Learning activities on pages 687 and 689, and the Connection on page 688.

See also the Glencoe Homepage at **www.glencoe.com**

The following Glencoe resources provide additional opportunities for integrating science and technology.

**Manufacturing Technology: Today and Tomorrow**
Section II: Manufacturing Tools, Materials, and Processes
Activity 1: Making a Product from Recycled Materials

Section IV: The Future in Manufacturing Technology
Activity 1: Designing a Product for the Future

Chapter 32

# Solving Ecological Problems

The increasing human population has brought about vast changes in the environment. Humans have changed their environment by building factories, parks, shopping malls, and by farming. Other living things change the environment, too, but not as fast as humans do. The beavers on the opposite page cut down trees and dam up streams. This activity creates ponds where the beavers can build their dens. The river in the photo on this page has been littered with trash thrown away by humans. Both of these activities have changed the environment.

In this chapter, you will read about the changes humans have made to the environment. Think about some of the ways you may have changed your environment.

**Try This!**

**What changes can occur in natural resources?** Prepare a tray with soil and a tray with grass growing in soil. Tip the trays as you pour equal amounts of water over each. Collect the water as it runs off the soil. Compare the amount of soil that has washed out of each tray. What effect did the grass have?

**interNET CONNECTION**

For more information about the material in this chapter, follow the link for the chapter on the Glencoe Homepage at **http://www.glencoe.com**

673

**ASSESSMENT PLANNER**

***Portfolio***

Strategies on the following pages represent student products that can be placed into a best-work portfolio: pp. 681, 684.

**PERFORMANCE ASSESSMENT**

Skill Check, pp. 677, 679, 681
Mini Lab, pp. 675, 676
Lab, pp. 682, 686

**CONTENT ASSESSMENT**

Check for Understanding, pp. 677, 679, 684, 685, 689
Chapter Review, pp. 690-691

**GROUP ASSESSMENT**

Opportunities for group assessment occur with Cooperative Learning Strategies.

***Student Journal***

Strategies on the following pages represent opportunities for writing in a Student Journal: pp. 679, 683.

**GETTING STARTED**

## Using the Photos

In looking at the photos, students should think of other ways humans have changed the environment. They should think of helpful as well as harmful changes to the environment. They should also think of ways that they could improve their environment.

## MOTIVATION/Try This!

**What changes can occur in natural resources?** Upon completing the chapter-opening activity, students will have **experimented.** They will discover that grass protects soil from washing away.

## Chapter Preview

Have students study the chapter outline before they begin to read the chapter. They should note that the chapter begins with a discussion of endangered species and possible solutions to this problem. The next broad topic is damage to our environment by various types of pollution. The last section focuses on possible ways to solve pollution problems.

## Misconception

Students usually assume that damage to natural resources happens somewhere other than where they live. Point out a few local problems involving damage to natural resources in the community where the students live.

## 32:1 Resources and Human Activities

### PREPARATION

#### Materials Needed

Make copies of the Skill, Application, Critical Thinking/Problem Solving, Study Guide, and Reteaching worksheets in the *Teacher Resource Package*.

▶ Obtain soil and jars for the Mini Lab.

#### Key Science Words

natural resource
endangered
threatened
erosion
sediment
pollution
fossil fuel

#### Process Skills

In the Skill Check, students will calculate. In the Mini Labs, students will outline and classify.

### 1 FOCUS

▶ The objectives are listed on the student page. Remind students to preview these objectives as a guide to this numbered section.

#### MOTIVATION/Bulletin Board

Display pictures of natural resources. *National Wildlife* and *International Wildlife* have excellent photos. Discuss problems that occur when natural resources are not protected.

#### Guided Practice

Have students write down their answers to the margin question: one that can be replaced during a lifetime.



## 32:1 Resources and Human Activities

**Objectives**

1. **Explain** how wildlife and plants are affected by humans.
2. **Describe** how water and soil can be lost to the environment.
3. **Explain** why fossil fuels are being used up.

**Key Science Words**

natural resource
endangered
threatened
erosion
sediment
pollution
fossil fuel

Air, water, soil, plants, and animals are some of our natural resources. A **natural resource** is any part of the environment used by humans. If these resources are used unwisely, the lives of many things can be harmed. For instance, cutting down a forest will destroy many of the animal populations that live there. See Figure 32-1.

### Wildlife and Plants

**What is a renewable resource?**

Wildlife and plants are important resources for humans. They provide food, fibers, and building materials. Nearly half the medicines being used today come from living things. Plants make oxygen, which all living things use during respiration. Soil that is covered with plants absorbs and holds water from precipitation. Thus, the soil is not worn away. Animals may be considered a renewable resource. A renewable resource is a resource that can be replaced within a person's lifetime. Animals can reproduce, keeping the population stable. Plants are a renewable resource, too. A field can regrow in about five years. A forest may take 25 to 100 years or more to regrow.

As the human population gets bigger, it needs more space to grow food and to build homes and factories. Where do we get the space? We use the land around us. As we take over this land, we use more natural resources, make more trash, and take away habitats of animals and plants.

**Figure 32-1** Cutting down a forest destroys many of the plant and animal populations that live there.



### OPTIONS

#### Science Background

There are over 995 animals and 529 plants on the threatened and endangered lists.

a

b

c

**Figure 32-2** The green pitcher plant (a) and whooping crane (b) are endangered species. Elephants (c) are killed illegally so their tusks can be carved into jewelry.

What happens to animal populations that are forced from their habitats? They must find new places to live. If they can't find enough space or food, they die. Unlike animals, plants can't move to a new area when their habitats are used for land development. These plants usually die. When a species of living thing no longer exists anywhere on Earth, it is extinct. Over the past three and one-half centuries, nearly 200 animal species have become extinct in the United States alone!

In the world today, over 1200 animal and plant species are endangered. **Endangered** means that a species is in danger of becoming extinct. Two endangered species are shown in Figure 32-2. Some living things are threatened. **Threatened** means that a species of a living thing is close to being endangered. There are almost 250 threatened species of plants and animals in the world.

How do animals and plants become threatened or endangered? Illegal hunting is one way. For example, African elephants are killed for their ivory tusks. Reptiles such as alligators and crocodiles are hunted illegally so their skins can be made into shoes, belts, and purses.

The rain forests of South America, Africa, and Asia are being cut down at an extremely high rate to make room for farmland. Every year, 200 000 square kilometers of rain forest are destroyed. This is an area equal to 76 football fields being destroyed every minute! Decreasing the forests means there are fewer plants releasing oxygen into the air and more carbon dioxide building up in the air. The ecological balance is upset. Why is this a problem for humans and other living things?

Scientists estimate that over 100 species of plants in these forests are becoming extinct each day. Extinction has always happened as a part of evolution, but by cutting down the rain forests, we speed up the process.

### Mini Lab

**How Could You Protect a Species?**

**Outline:** Pick a plant or animal and assume that it has become endangered. Make an outline showing what you would do to protect this species. What needs does this species have? *For more help, refer to the* ***Skill Handbook****, pages 706-711.*



## 2 TEACH

### MOTIVATION/Display

Write the word *humans* on the chalkboard and then write the words *wildlife, plants, soil, water,* and *fossil fuels* around it. Discuss how humans depend on each of these things, and how they are all interrelated.

### Concept Development

▶ Explain that legal hunting of some animals helps control animal populations that no longer have natural predators.

▶ Discuss how some human needs are causing loss of habitat for animals.

### Mini Lab

This open-ended activity will give students a chance to report on a favorite living thing. Have students compare their ideas in small groups.

### ✓ ASSESSMENT

**Oral:** Have a student do a class report explaining how two species, the American alligator and the bald eagle, are no longer considered endangered species. Tell the student to find out what methods were used to increase the numbers of individuals and how long it has taken for these two species to make a comeback.

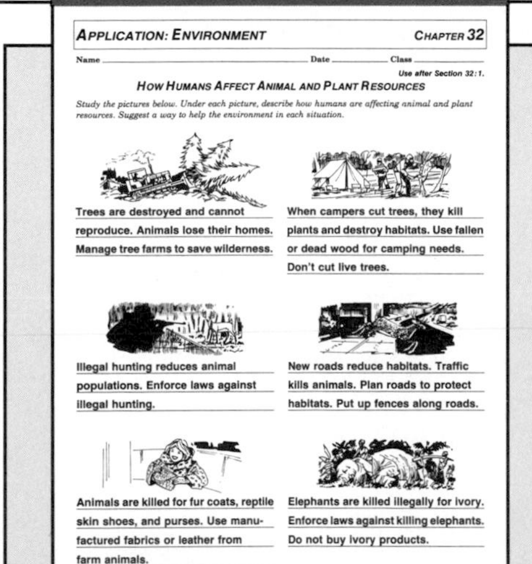

**APPLICATION 32**

APPLICATION: ENVIRONMENT — CHAPTER 32

Name ______ Date ______ Class ______

Use after Section 32:1.

HOW HUMANS AFFECT ANIMAL AND PLANT RESOURCES

*Study the pictures below. Under each picture, describe how humans are affecting animal and plant resources. Suggest a way to help the environment in each situation.*

Trees are destroyed and cannot reproduce. Animals lose their homes. Manage tree farms to save wilderness.

When campers cut trees, they kill plants and destroy habitats. Use fallen or dead wood for camping needs. Don't cut live trees.

Illegal hunting reduces animal populations. Enforce laws against illegal hunting.

New roads reduce habitats. Traffic kills animals. Plan roads to protect habitats. Put up fences along roads.

Animals are killed for fur coats, reptile skin shoes, and purses. Use manufactured fabrics or leather from farm animals.

Elephants are killed illegally for ivory. Enforce laws against killing elephants. Do not buy ivory products.

**SKILL 33**

SKILL: USE NUMBERS, CALCULATE — CHAPTER 32

Name ______ Date ______ Class ______

*For more help, refer to the Skill Handbook, pages 718-719.* — *Use with Section 32:1.*

DISAPPEARANCE OF TROPICAL RAIN FORESTS

Tropical rain forests of the world are being cut down at the rate of 116 500 square kilometers each year. The forests are cut down to make way for farming, or the cut trees are used for fuel or lumber. To understand the rate at which the forests are being destroyed, think about the size of the state of Pennsylvania. The state has a land area of about 117 000 square kilometers. If Pennsylvania were covered with forests, and the forests were destroyed at the rate of 117 000 square kilometers a year, the forests would be gone at the end of one year.

Suppose a tropical rain forest covered all the land of the United States. Then suppose the trees in this forest were being cut down at the same rate as those in the world's rain forests. How long would it take for the forest to disappear completely from the United States? How long would it take for the forest to disappear in particular states?

*Calculate the number of years and days it would take to remove the forest from each of the states listed in Table 2 on the next page. Use Formula A or Formula B below. You can use Table 1 on the next page to change decimal parts of a year into days. For example, 0.2 year equals 73 days. Write the results of your calculations in the last column of Table 2. Two have been done for you.*

Use Formula A for states larger than 100 000 square kilometers

$$\frac{\text{land area in square kilometers}}{\text{square kilometers removed yearly}} = \text{number of years to remove forest}$$

Example: Alabama

$$\frac{\text{132 000 square kilometers}}{\text{117 000 square kilometers removed yearly}} = \text{1.1 years (1 year and 36 days)}$$

Use Formula B for states smaller than 100 000 square kilometers

$$\frac{\text{land area in square kilometers}}{\text{square kilometers removed yearly}} \times \text{365 days} = \text{number of days to remove forest}$$

Example: Delaware

$$\frac{\text{5200 square kilometers}}{\text{117 000 square kilometers removed yearly}} \times \text{365 days} = \text{16 days}$$

*When you have completed Table 2, answer the following questions.*

1. How many square kilometers of tropical forests in the world are being destroyed each day? (Divide the number of square kilometers destroyed each year by 365 days.) **320 square kilometers each day**

2. What problems can you think of that may be caused by destruction of the world's tropical rain forests? **Answers will vary. Habitats of organisms will be destroyed and organisms might become extinct. Bare land might cause serious erosion. Loss of forests might cause climate changes.**

## OPTIONS

**Skill**/*Teacher Resource Package*, p. 33. Use the Skill worksheet shown here for students to use numbers to understand the disappearance of tropical forests.

**Application**/*Teacher Resource Package*, p. 32. Use the Application worksheet shown here to teach students how humans affect animal and plant resources.

## TEACH

### Mini Lab

Water that is not clear after 3 days may have fine clay and silt particles suspended in it. Have students compare the layers of sediment with a hand lens.

### ASSESSMENT

**Skill:** Have students compare and classify the sediments in their area. Tell each student to bring in a pint-sized container and a cup of soil in a plastic bag. Have students add equal amounts of soil and tap water to the jar and shake the jar to break up the soil. After the soil has settled, have students use a ruler to measure the thickness of each layer, record the color of the sediment, and classify the sediments by color and thickness of the layer.

### Guided Practice

Have students write down their answers to the margin question: Wind separates small particles and blows them away.

### Independent Practice

**Study Guide**/*Teacher Resource Package*, p. 187. Use the Study Guide worksheet shown at the bottom of the page for independent practice.

### Mini Lab

**What Does Sediment Look Like?**

**Classify:** Add soil to a jar of water. Cover and shake. Let the soil settle for three days. How many layers of sediment do you see? Describe the layers. Is the water clear? *For more help, refer to the* ***Skill Handbook****, pages 715-717.*

## Soil

Soil is an important natural resource. The roots of plants hold soil in place. When plants are removed and the soil is not protected, erosion can take place. **Erosion** is the wearing away of soil by wind and water. When soil erodes, it is carried away faster than it can be replaced.

Soil that washes into streams and rivers can cause problems. As the current in a stream slows, the floating soil particles settle to the bottom. Material that settles on the bottom of the stream is called **sediment.** Sediment covers the habitats of water plants and animals. Fish eggs on the river bottom die because they don't get enough oxygen.

When bare soil dries out, it can erode quickly. Why? The wind easily separates the small particles of soil and blows them away. Erosion carries away good topsoil. The subsoil layer under the topsoil is exposed. Plants don't grow well in subsoil because it is packed down, contains little oxygen, and lacks the nutrients found in topsoil.

How does dry soil erode?

## Water

Water is the most abundant resource on Earth. It is a renewable resource because it is cycled in the environment and can be reused. Water covers over 70 percent of Earth's surface. Over 70 percent of your body is water. You could not go without water for more than a few days at a time.

Water in the air is an important factor in climate and weather. The more water there is in the air, the wetter the climate will be in an area. Water is used to irrigate crops, for industry, and for cooling in electric power plants.

**Figure 32-3** Soil washes into streams and becomes sediment (a). Wind can erode soil by blowing it away (b). Water is used to irrigate crops (c).

a

b

c

### OPTIONS

**Critical Thinking/Problem Solving/** *Teacher Resource Package*, p. 32. Use the Critical Thinking/Problem Solving worksheet shown here to teach how pollution can change succession.

**CRITICAL THINKING 32**

CRITICAL THINKING/PROBLEM SOLVING — CHAPTER 32

Name ________ Date ________ Class ________

*Use after Section 32:1.*

**CAN POLLUTION CHANGE SUCCESSION?**

The students in a biology class went on a field trip in a rural area. They studied the ecology of two ponds about 50 meters apart. Both received good water supplies from underground springs. Both received about the same amount of runoff from rain.

The appearance of the ponds was quite different. Pond 1 was covered with green and brown pond scum. It had an unpleasant odor. When students took a water sample and looked at it with a microscope, they saw many kinds of single-celled and hairlike algae. This pond was near a cow pasture and near a field of growing crops. The field had been heavily fertilized. Its wire fence was rusty and broken through.

Pond 2 stood farther away from the pasture, on a hill near a pine-hardwood forest. The students could see the bottom of the pond through its clear brownish water. Under the microscope, the students saw a few algae and a few swimming protozoa.

The teacher said, "Each pond is proceeding through its natural succession. However, environmental conditions affect the rate of changes. Pollution can be caused by too much of a good thing, or by harmful substances." What conditions speed up changes in the pond environment?

**Analyzing the Problem**

1. What is succession? the natural process by which a pond ages, or changes over time
2. What is pollution? any factor that causes undesirable changes in the environment
3. What materials flow into Pond 1? wastes from cows and from fertilizers
4. What materials flow into Pond 2? rainwater, pine needles, and leaves

**Solving the Problem**

1. What effect did the runoff have on pond 1? It caused rapid heavy growth of algae.
2. What effect did the runoff have on pond 2? It supported growth of a mixed community of plants and animals.
3. Which pond is more polluted? pond 1
4. Which pond shows a faster rate of succession? pond 1
5. How could succession be slowed in the polluted pond? Decrease the flow of cow wastes and excess fertilizer into the pond.

**STUDY GUIDE 187**

STUDY GUIDE — CHAPTER 32

Name ________ Date ________ Class ________

**RESOURCES AND HUMAN ACTIVITIES**

*In your textbook, read about natural resources in Section 32:1.*

1. Each year during the five years from 1986 through 1990, millions of square kilometers of rain forests were cut down and cleared in Africa and Asia to make room for farmland. Use the information in the table below to help you answer the questions that follow.

| | Number of square kilometers of forest removed | |
|---|---|---|
| Year | Africa | Asia |
| 1990 | 24 000 | 16 000 |
| 1989 | 24 000 | 16 000 |
| 1988 | 24 000 | 16 000 |
| 1987 | 24 000 | 16 000 |
| 1986 | 24 000 | 16 000 |

2. a. How many square kilometers of forests were destroyed in Africa during the last five years? 120 000
   b. How many square kilometers of forests were destroyed in Asia during the last five years? 80 000
3. What happens to the soil when all of the plants are removed from the forests? The soil is not protected and erosion occurs. The soil is worn away by wind and water.
4. What happens to this soil when it is washed into a stream? It settles to the bottom and covers the habitats of plants and animals.
5. How does clearing the forests affect the level of gases in the air? There will be a decrease of oxygen and an increase of carbon dioxide.
6. How does the removal of the forest environments affect the extinction of organisms? It speeds up the process of extinction.

A major problem with lakes and ponds is rapid aging. Remember that lakes and ponds normally go through succession and become dry land. Producers living in the ponds and lakes need nutrients to live. Sometimes more nutrients are added to the water than are needed by these producers. The extra nutrients are in fertilizers washed from nearby farms, and in untreated sewage. These nutrients cause the pond or lake to go through succession faster than normal. Review the stages of aging in a lake as shown in Figure 31-7 on page 662. Aging is normally a slow, natural process, but it is speeded up because of pollution. **Pollution** is anything that makes the surroundings unhealthy or unclean.

## Skill Check

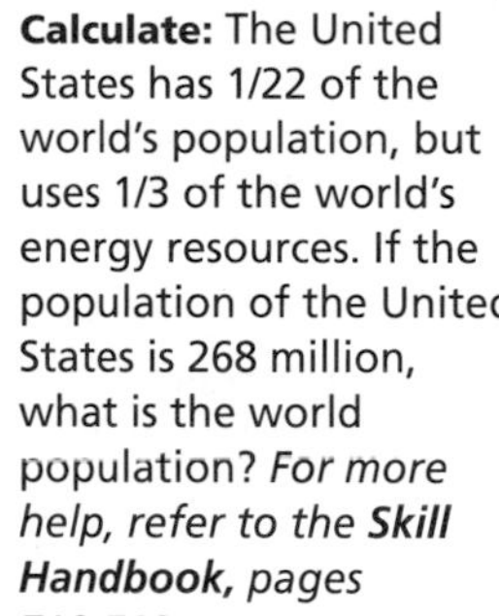

**Calculate:** The United States has 1/22 of the world's population, but uses 1/3 of the world's energy resources. If the population of the United States is 268 million, what is the world population? *For more help, refer to the* ***Skill Handbook****, pages 718-719.*

## Fossil Fuels

Coal, oil, and natural gas are all fossil fuels that humans use to run their cars, heat their homes, and produce electricity. A **fossil fuel** is the remains of organisms that lived millions of years ago. These remains were compressed over long periods of time at tremendous pressure. Because they take so long to form, fossil fuels are not renewable.

**Why are fossil fuels not renewable?**

Americans use fossil fuels for about 85 percent of our energy needs. As the human population continues to grow, more energy is needed to run larger cities and more cars, homes, and factories. The supply of fossil fuels is being used up at an alarming rate. When coal, oil, and natural gas supplies are gone, there will be no more. New energy sources must be found. Governments must help to save our fossil fuel supplies by passing laws limiting their use.

## Check Your Understanding

1. What is the difference among extinct, endangered, and threatened species?
2. How is aging of ponds and lakes affected by pollution?
3. Why must new sources of energy be found?
4. **Critical Thinking:** Why should people who live in cities be concerned about rain forests being destroyed?
5. **Biology and Math:** It is estimated that 100 species of plants become extinct in rain forests each day. How many plant species become extinct in rain forests in a year?



## RETEACHING 94

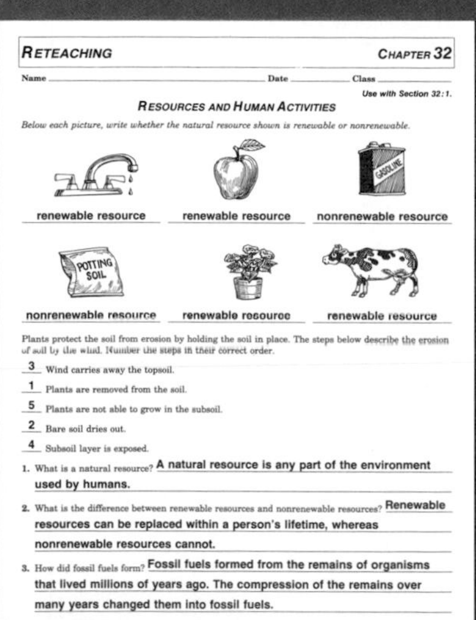

RETEACHING CHAPTER 32

Name ______ Date ______ Class ______

*Use with Section 32:1.*

**RESOURCES AND HUMAN ACTIVITIES**

*Below each picture, write whether the natural resource shown is renewable or nonrenewable.*



renewable resource | renewable resource | nonrenewable resource

nonrenewable resource | renewable resource | renewable resource

Plants protect the soil from erosion by holding the soil in place. The steps below describe the erosion of soil by the wind. Number the steps in their correct order.

3 Wind carries away the topsoil.
1 Plants are removed from the soil.
5 Plants are not able to grow in the subsoil.
2 Bare soil dries out.
4 Subsoil layer is exposed.

1. What is a natural resource? A natural resource is any part of the environment used by humans.
2. What is the difference between renewable resources and nonrenewable resources? Renewable resources can be replaced within a person's lifetime, whereas nonrenewable resources cannot.
3. How did fossil fuels form? Fossil fuels formed from the remains of organisms that lived millions of years ago. The compression of the remains over many years changed them into fossil fuels.

## GLENCOE TECHNOLOGY

**Videodisc**

**Biology: The Dynamics of Life**
Animation: *The Atmosphere*
Disc 1, Side 1, Ch. 14

# TEACH

### Skill Check

5000 million

### Guided Practice

Have students write down their answers to the margin question: They take so long to form.

### Check for Understanding

Have students respond to the first three questions in Check Your Understanding.

### Reteach

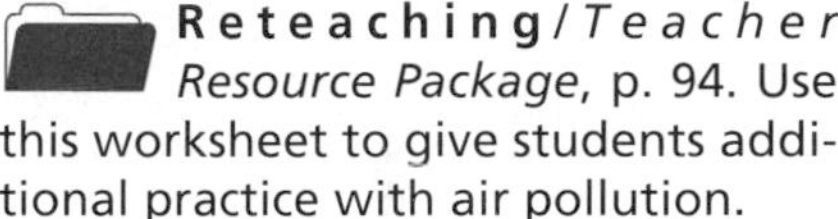

**Reteaching**/*Teacher Resource Package*, p. 94. Use this worksheet to give students additional practice with air pollution.

**Extension:** Assign Critical Thinking, Biology and Math, or some of the **OPTIONS** available with this lesson.

# 3 APPLY

### ACTIVITY/Project

Have each student report on an endangered species and explain how to protect it.

# 4 CLOSE

### Summary

Discuss how living things depend on the quality of their environment.

### Answers to Check Your Understanding

1. Extinct means a species no longer exists; endangered means a species is in danger of becoming extinct; threatened means a species is close to being endangered.
2. It increases the rate at which ponds and lakes age.
3. Fossil fuels are not renewable and are being used up.
4. There will be less oxygen and more carbon dioxide.
5. 100 x 365 = 36 500

## 32:2 Problems from Pollution

## PREPARATION

### Materials Needed

Make copies of the Focus, Study Guide, Transparency Master, Enrichment, Reteaching,and Lab worksheets from the *Teacher Resource Package*.

▶ Obtain balloons and iron filings for the Focus motivation.

▶ Gather ammonia, baking soda, cola, vinegar, and rainwater for Lab 32-1.

### Key Science Words

| | |
|---|---|
| toxic | pesticide |
| smog | biodegradable |
| greenhouse effect | acid |
| ozone | base |
| radon | acid rain |

### Process Skills

In Lab 32-1, students will observe, experiment, interpret data, and classify. In the Skill Checks, students will experiment and understand science words.

## 1 FOCUS

▶ The objectives are listed on the student page. Remind students to preview these objectives as a guide to this numbered section.

### MOTIVATION/Demonstration

Fill a balloon with air to show how a normal alveoli sac looks as air is taken in. Fill a second balloon with air but first add 5-10 mL of water and some iron filings. The fluid shows how emphysema prevents oxygen from entering the blood cells. The iron filings represent particulates taken in during breathing.

**Focus**/*Teacher Resource Package*, p. 63. Use the Focus transparency shown at the bottom of this page as an introduction to studying air pollution.



**Objectives**

4. **Discuss** the causes of air pollution and acid rain.
5. **Discuss** problems that come from nonbiodegradable chemicals that pollute water.

**Key Science Words**

toxic
smog
greenhouse effect
ozone
radon
pesticide
biodegradable
acid
base
acid rain

## 32:2 Problems from Pollution

What is the effect of pollution on our bodies and on our environment? A number of health problems are now known to be caused by pollutants in our environment. Plants and animals are dying from pollutants. Humans are affected by pollution, too.

### Air Pollution

Most air pollution is caused by burning coal, oil, gasoline, or natural gas. In North America, cars, lawn mowers, and other gasoline engines burn a total of 14 million gallons of fuel each hour. Gasoline produces several harmful gases when burned in an engine. These pollutants are given off in toxic amounts. **Toxic** means poisonous.

Many large cities, such as New York and Los Angeles, suffer from smog emergencies. **Smog** is a combination of smoke and fog. It is made thicker by chemical fumes. Most of the smog that blankets our cities is produced when the harmful gases from industry and auto exhausts react with the energy in sunlight. Trace how smog develops, using Figure 32-4b. This reaction produces new chemicals that are irritating to the eyes, nose, lungs and throat. A smog emergency occurs when the air is very still for a long time, allowing dangerous amounts of these gases to build up.

Factories and power plants that burn coal also make a lot of air pollution. Coal contains a small amount of sulfur.

**Figure 32-4** Air pollution is caused by the burning of fossil fuels in engines (a). Smog is formed when harmful gases from burning fossil fuels react with sunlight (b).

a

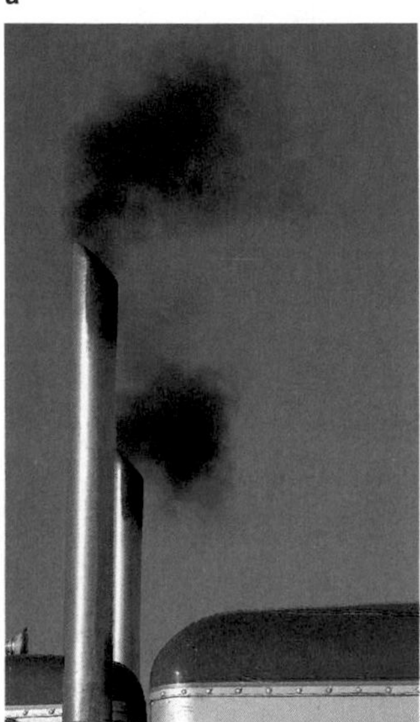

b

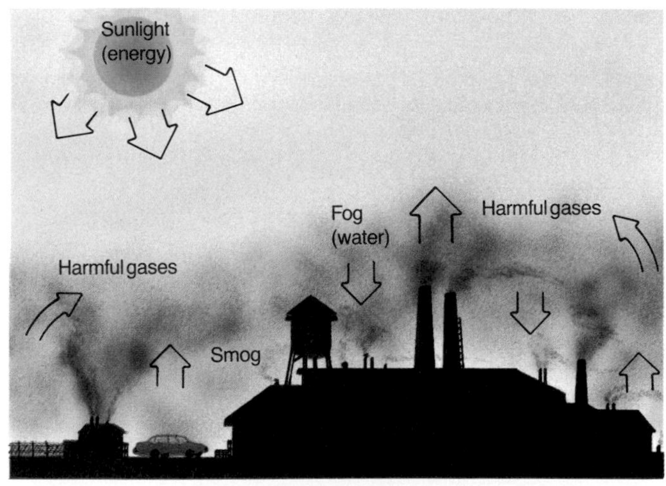




## *OPTIONS*

### Science Background

Mercury from industrial waste enters food chains through freshwater and saltwater ecosystems. When humans eat contaminated fish, the mercury travels to the brain and becomes toxic.

**FOCUS 63**

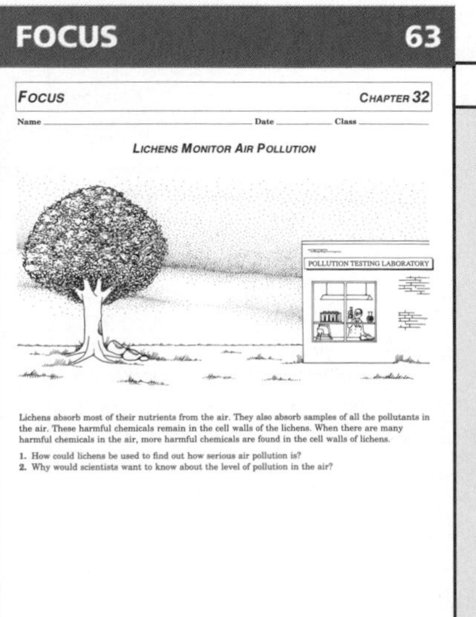
*Focus* — *Chapter 32*

Name ______ Date ______ Class ______

**Lichens Monitor Air Pollution**

POLLUTION TESTING LABORATORY

Lichens absorb most of their nutrients from the air. They also absorb samples of all the pollutants in the air. These harmful chemicals remain in the cell walls of the lichens. When there are many harmful chemicals in the air, more harmful chemicals are found in the cell walls of lichens.

1. How could lichens be used to find out how serious air pollution is?
2. Why would scientists want to know about the level of pollution in the air?

a

b

When coal burns, the sulfur combines with oxygen in the air to form sulfur dioxide. Sulfur dioxide can kill humans. Power plants that burn coal also give off tiny particles of soot or dust. These particles block sunlight and get into our lungs when we breathe.

When fossil fuels are burned by factories and power plants, large amounts of carbon dioxide are made and released into the air. We don't usually think of carbon dioxide as being a pollutant, but in large amounts it can be. This carbon dioxide forms a layer around Earth and traps heat so it can't escape into the atmosphere. The trapped heat may cause temperatures on Earth to rise slowly, a process know as the **greenhouse effect.** If temperatures on Earth rise even a few degrees, polar ice may melt and the habitats of plants and animals may change. Deserts can form in areas that were once forests.

You may have heard or read that Earth's ozone layer is getting thinner. **Ozone** is a molecule made of three oxygen atoms. It forms a layer high above Earth's surface that keeps harmful radiation from reaching Earth.

Scientists have learned that ozone is being destroyed by certain chemicals found in spray cans, refrigerators, and air conditioners. The same chemicals are used to make plastic foam food containers. The chemicals destroy ozone, and harmful radiation can reach Earth's surface.

Some air pollution problems can be found inside your home. You may have heard of **radon**, a gas that gives off radiation. Radon is found naturally in the ground. In large amounts, radon may cause cancer. Radon can build up to dangerous levels in some buildings. Fans can be added to these buildings to blow the polluted air outside.

**Figure 32-5** Factories and power plants that burn coal cause air pollution (a). Chemicals from spray cans and plastic foam containers (b) destroy Earth's ozone layer.

### Skill Check

**Experiment:** Wash and rinse a plastic cup in distilled water. Use the cup to collect rain water, then evaporate the water. Look for a film of tiny particles on the side of the cup. How do particles get into the cup? *For more help, refer to the* ***Skill Handbook,*** *pages 704-705.*

How is ozone destroyed?



## OPTIONS

### ACTIVITY/Challenge

Have students explain how the pollution control devices on an automobile work.

**How Does Thermal Pollution Affect Living Things?** *Lab Manual,* pp. 267-270. Use this lab as an extension to studying pollution problems.

## 2 TEACH

### MOTIVATION/Software

*Pollute,* Opportunities for Learning.

### Concept Development

▶ Explain that some pollutants enter the water cycle from the atmosphere. When it rains, these pollutants enter water systems and can enter plants through the soil.

▶ Remind students they learned about carbon monoxide in Chapter 13.

▶ Point out that over 70 percent of the sulfur in our atmosphere comes from burning coal. Explain that some sulfur enters the atmosphere during decomposition.

### Skill Check

Many particulates in the air dissolve in precipitation. Dissolved solids reappear when the precipitation evaporates.

### Guided Practice

Have students write down their answers to the margin question: by certain chemicals.

### *Student Journal*

Have students write in their journals about the use of ethanol in automobiles as a fuel additive. Tell them to focus on: whether it may reduce the amount of smog in suburban areas, the cost per gallon of producing ethanol compared to gasoline, how ethanol can add to the amount of carbon dioxide in the atmosphere, and why race car drivers use this fuel in their cars.

### *GLENCOE* TECHNOLOGY

**Videodisc**

**Biology: The Dynamics of Life**
Animation: *The Greenhouse Effect*
Disc 1, Side 1, Ch. 15

## TECHNOLOGY

### The Greenhouse Effect—A Computer Model

#### Background

Climatologists have predicted a warming trend that some say is caused by the greenhouse effect. They claim that gases like carbon dioxide, methane, and ozone accumulate in the atmosphere as a result of human activity and agricultural practices and block the exit of heat energy. They claim that temperatures have been rising slowly over the past several decades. Scientists use data to make inferences, predictions, and hypotheses. By looking at increases in temperature around the world they have noticed that warmer temperatures have drifted toward the poles.

#### Discussion

Point out that some fertile areas could become deserts if the climatologists are correct. Explain that scientists have predicted that the melting of glaciers and polar ice would raise the sea level of the world by six feet. Show students a topographical map of towns near the seashore to help them understand how much land could be underwater. Point out that one benefit of increased carbon dioxide could be more photosynthetic activity, thus, more food for humans. This is possible only if enough water is available.

#### References


Doyle, Roger. "Carbon Dioxide Emissions." *Scientific American*, May 1996, p. 24.

Goldberg, Jeff. "Water World?" *Science World*, November 3, 1995, pp. 12-16.

Schneider, David. "The Rising Seas." *Scientific American*, March 1997, pp. 112-117.

Smil, Vaclav. *Cycles of Life: Civilization and the Biosphere*. San Francisco: W. H. Freeman and Company, 1997.


## TEACH

### Guided Practice

Have students write down their answers to the margin question: They cause damage to soil and wildlife.

## TECHNOLOGY

### The Greenhouse Effect — A Computer Model

Many scientists are certain that the hotter than usual weather of 1988 was due to the greenhouse effect. They say that the gases from factories, home fuels, and car exhaust are causing the temperature on Earth to rise. These gases trap the heat energy from the sun. Any increase in the amounts of these gases allows more heat to be trapped in the atmosphere.

People who study weather and climate patterns have developed a computer model to explain the hot, dry summer of 1988. Worldwide climate data from several years were placed into the computer, which showed what climate patterns had looked like in the past.

The scientists added more atmospheric gases to the computer model. The gases were added in the amount in which they had been increasing. The computer showed the expected changes for the next 30 years.

What changes were predicted based on the increase in gases? These changes included warmer winters around the world, increased melting of glaciers and polar ice, a six-foot rise in sea level, some desert biomes becoming wetter, and rain forests becoming drier.

Not all scientists agree on the results of the computer model. Some say that the model has not included the effects of plants, clouds, soil moisture, and mountains on global weather. They also claim that it is too difficult to predict weather for more than a few weeks at a time. They agree that Earth is warming, but not as fast or as much as the model shows. Which group of scientists is correct? More work using the models to predict weather patterns may give us the answers.

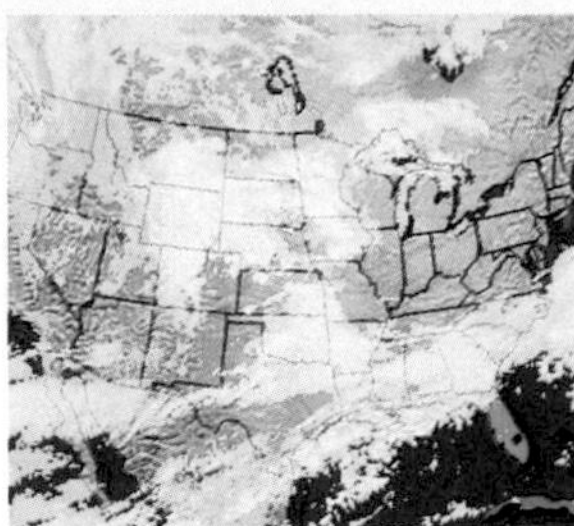

Scientists can use computers to model weather patterns.

### Water Pollution

Why are pesticides a problem to the environment?

Water from streams and rivers supplies this country with about half of its drinking water. However, much of it contains toxic wastes. Our water also contains untreated sewage, and fertilizers and pesticides washed from farmlands. **Pesticides** are chemicals used to kill unwanted pests such as rodents or insects. When pesticides are washed off farmlands into rivers and streams, they pollute the water. Pesticides can cause damage to soil and wildlife. Because pesticides are so dangerous, many of them are no longer allowed to be used in this country.



## *OPTIONS*

**How Pesticides Are Concentrated in a Food Chain**/*Transparency Package*, number 32. Use color transparency number 32 as you teach pollution from chemicals.

The best known of all pesticides is DDT. DDT is a chemical spray used to kill insects. It has been shown to cause cancer. DDT also affects the shells of bird eggs. The shells of birds that have eaten DDT in their food are thin and break easily. These eggs don't hatch, so no offspring are produced. Because of DDT, some bird species have become endangered. DDT is so harmful that the Environmental Protection Agency (EPA) ruled in 1973 that DDT could no longer be used in the United States. However, it is still widely used in other countries.

A serious problem today is that other pesticides similar to DDT are still being used. Many of these chemicals are not biodegradable (bi oh duh GRAYD uh bul.) **Biodegradable** means that something can be broken down by microbes into harmless chemicals and used by other living things. Because many pesticides are not biodegradable, they remain in an ecosystem. DDT, other pesticides, and weed killers also move through food chains. This means that humans and other meat-eating animals can have very high amounts of these chemicals in their bodies.

We use chemicals to rid our crops of pests, to wash our clothes, and to manufacture the many products we use. Many dangerous wastes are left over when chemicals are made. The waste products from chemical manufacturing are often toxic. These wastes may cause pollution.

A big problem with toxic wastes is how to get rid of them. One solution has been to bury them at dump sites. We now know that buried chemical wastes can escape into the air, the surrounding soil, and nearby water. When these wastes escape into the water, the water is polluted. Therefore, burying wastes is not a good solution.

## Skill Check

**Understand science words: pesticide.** The word part *pest* means something that annoys. In your dictionary, find three words with the word part *pest* in them. *For more help, refer to the **Skill Handbook**, pages 706-711.*

a

b

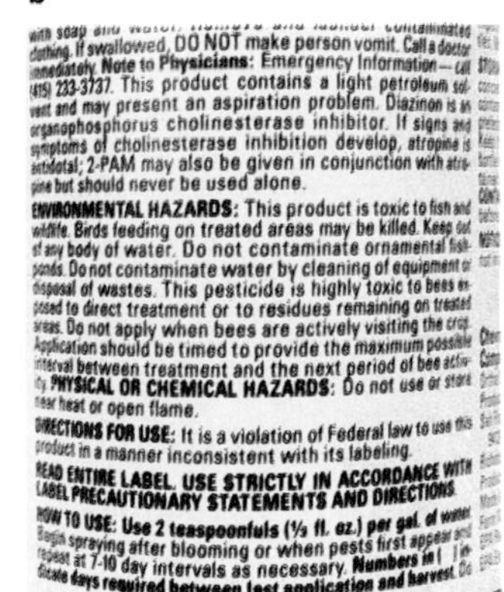


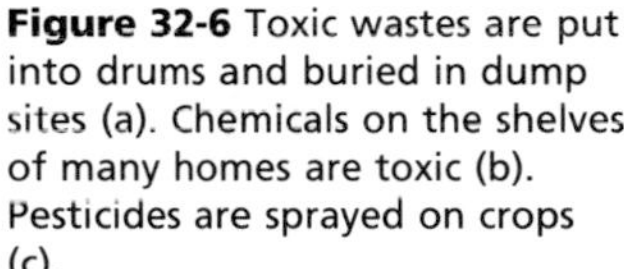

**Figure 32-6** Toxic wastes are put into drums and buried in dump sites (a). Chemicals on the shelves of many homes are toxic (b). Pesticides are sprayed on crops (c).

c

# TEACH

### Concept Development

▶ Emphasize that pollutants on soil surfaces can percolate down and contaminate underground water. This water may be the water supply for many humans.

▶ Explain that many pesticides are applied as sprays. Those not broken down by microorganisms end up in food chains. Here, they block certain physiological processes in living things.

### Independent Practice

**Study Guide**/*Teacher Resource Package*, p. 188. Use the Study Guide worksheet shown at the bottom of this page for independent practice.

### *Portfolio*

Direct students to read the precautionary statements on the labels of at least four pesticides used around the home. Have them make a chart listing what the pesticides are used for and how they are hazards to humans, domestic animals, and the environment. Have them determine whether or not the pesticide is biodegradable.

## STUDY GUIDE 188

STUDY GUIDE — CHAPTER 32

Name ______ Date ______ Class ______

**PROBLEMS FROM POLLUTION**

*In your textbook, read about air and water pollution in Section 32:2.*

1. Fill in the boxes of the pollution table below with the correct words or phrases: burning coal, oil, gasoline, and natural gas; PCBs; harmful gases; heavy metals; pesticides and weed killers; tiny particles; sulfur dioxide. Several are already done for you.

Pollution
- Water
  - toxic wastes, untreated sewage, fertilizer runoff
  - PCBs, heavy metals, pesticides and weed killers
- Air
  - burning coal, oil, gasoline, and natural gas
  - harmful gases, sulfur dioxide, tiny particles

2. How is smog produced? Smog is produced when harmful gases react with the energy in sunlight to produce new, irritating chemicals.

3. What is the greenhouse effect? The carbon dioxide formed by burning fossil fuels forms a layer around Earth and traps heat so it can't escape into the atmosphere. The trapped heat may eventually cause Earth's temperature to rise.

4. Why are chemicals such as PCBs and DDT so harmful to the environment? Because they are not biodegradable and cannot be excreted by animals, they accumulate in the environment and in the bodies of animals.

## OPTIONS

### Science Background

When DDT was developed over 40 years ago, it was called a "miracle" chemical because it killed disease-carrying and crop-destroying insects. Only a few scientists suspected that DDT could also be harmful to other living things. Not only did DDT enter the food chains, it also entered the water cycle. This particular pesticide sometimes becomes concentrated in food eaten by humans.

# Lab 32-1 pH

## Overview

Students will determine whether several household solutions are acidic, basic, or neutral. Students find the pH of the solutions by using the pH (number) scale and their data. They will compare the pH of solutions with that of samples of rain water.

**Objectives:** Students will be able to (1) **determine** the pH of several household chemicals and rainwater, (2) **demonstrate** the use of pH paper, (3) **classify** the solutions as acids or bases by using pH paper.

**Time Allotment:** 45 minutes

## Preparation

▶ Have the known solution in dropper bottles, beakers, or flasks for easy dispensing. Prepare several sets. Place each kind of solution at a different spot in the room so the various solutions do not get mixed together.

▶ Cut the pH paper into 3-cm strips prior to class.

**Lab 32-1 worksheet**/*Teacher Resource Package*, pp. 125-126.

## Teaching the Lab

▶ Have student teams put their data on a class chart on the chalkboard or a transparency so that they can compare their data with that of other teams.

▶ Freshly distilled water, free of carbon dioxide, has a pH of 7. If $CO_2$ is present, the water will have a slightly acidic pH.

▶ Acid rain will have a pH between 1 and 5.5

**Performance:** Give teams of students a plastic cup that was rinsed three times with distilled water and covered with plastic wrap. Tell them to place the cups on the schoolyard or at home as it begins to rain and collect enough rainwater to do the pH test. Have students compare their results. They could compare the pH of rain each time it rains in a one-month period.

# pH

## Lab 32–1

**Problem: How does the pH of rainwater compare with that of some household products?**

### Skills

observe, experiment, interpret data, classify

### Materials

pH paper
pH chart
forceps
marking pencil
7 glass slides
7 droppers
ammonia
baking soda
cola
vinegar
2 samples of rainwater
distilled water

### Procedure

1. Copy the data table.
2. Label 5 glass slides 1 to 5.
3. Put a drop of liquid on each slide, as listed below. **CAUTION:** *If solutions are spilled on skin, rinse with water at once and notify your teacher.*
   Slide 1—baking soda
   Slide 2—distilled water
   Slide 3—vinegar
   Slide 4—cola
   Slide 5—ammonia
4. Pick up a piece of pH paper with the forceps.
5. Touch the pH paper to the liquid on slide 1 and remove the paper.
6. **Observe:** Compare the color of the wet end of the paper with the pH color chart.
7. Record the pH of the liquid on slide 1.
8. Discard the pH paper and rinse the forceps in tap water.
9. Test the liquids on slides 2 through 5 by repeating steps 4 through 8.
10. Put a drop of the first rainwater sample on a clean slide. Put a drop of the second rainwater sample on a second slide.
11. Test the pH of each sample of rainwater. Follow steps 4 through 8.
12. Record the pH of each sample.

| SLIDE | CHECK CORRECT ANSWER | | | |
|---|---|---|---|---|
| | pH | ACID | NEUTRAL | BASE |
| 1. Baking soda | 8 | | | √ |
| 2. Distilled water | 7 | | √ | |
| 3. Vinegar | 2.5–3.5 | √ | | |
| 4. Cola | 6 | √ | | |
| 5. Ammonia | 11 | | | √ |
| 6. Rainwater | Answers | will vary. | | |
| 7. Rainwater | Answers | will vary. | | |

13. **Classify:** Finish filling in the data table by checking the correct column.

### Data and Observations

1. Which of the materials tested was the strongest acid? How do you know?
2. Which of the materials tested was the strongest base? How do you know?

### Analyze and Apply

1. Were your rainwater samples acidic or basic?
2. Why did you handle the pH paper with forceps?
3. What household product had a pH closest to your rainwater samples?
4. **Apply:** Why is acid rain a problem?

### Extension

**Experiment:** Sample several bodies of water in your community and determine if the pH of the water is above or below 5.5.



## ANSWERS

### Data and Observations

1. Vinegar; the lower the pH, the stronger the acid.
2. Ammonia; it had the highest pH.

### Analyze and Apply

1. acidic
2. so substances on the hands did not affect the pH paper
3. Answers will vary. Most students will answer either cola or vinegar.
4. It kills trees and other plants, destroys buildings, and makes lakes too acid to support life.

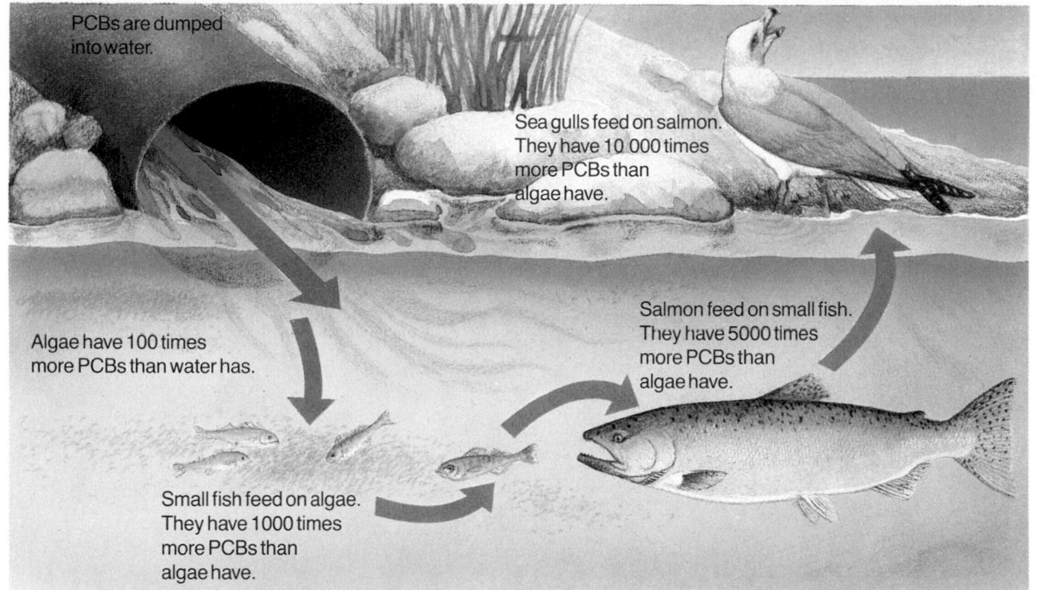


**Figure 32-7** PCBs increase in the bodies of animals at every step of a food chain. Why does this happen?

One class of dangerous chemicals present in water is PCBs. PCBs are toxic wastes produced when paints and inks are made. PCBs cause many serious health problems in people. Many lakes polluted with PCBs have signs posted around them warning people not to eat fish taken from the lake. Fish have large amounts of PCBs in their bodies because of their place in the food chain. Figure 32-7 shows what happens to PCBs in a food chain. Animals can't excrete PCBs from their bodies. At every step up in a food chain, the amount of PCBs in each animal increases.

Other pollutants found in water are heavy metals such as lead and mercury. These metals come from factories that dump their wastes into rivers or lakes. These chemicals, too, can build up in fish and the animals that eat fish. Lead is also a problem in paint found in old houses. Sometimes small children eat this paint and get sick.

## Acid Rain

A very serious problem related to pollution is acid rain. A substance can be an acid, a base, or neutral. Scientists have devised a scale called the pH scale to measure how strong an acid or a base is. The pH scale is numbered from 0 to 14. See Figure 32-8.

**Figure 32-8** pH scale

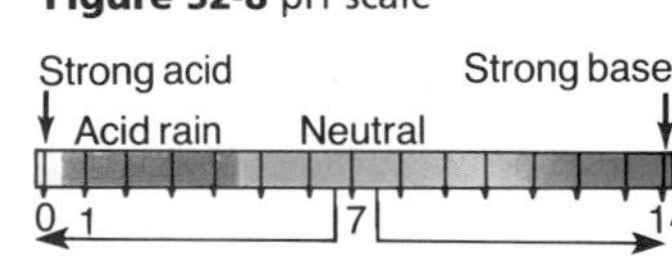




# TEACH

## Concept Development

▶ Explain that some aquatic animals develop tumors and ulcers when they are exposed to chemical pollutants in the water. When scientists find animals in this condition, they know that harmful chemicals are present.

▶ Explain to students that lead in the environment comes mostly from automobile exhaust, and mercury comes from industrial waste. These heavy metals enter food chains and become more concentrated as they move up the chain.

## Guided Practice

Have students diagram an aquatic food chain using algae, water fleas, small fish, and humans to show how PCBs pass through food chains and cause problems for humans.

### Student Journal

Tell students that the author of *Alice in Wonderland* had a reason for having a character called the Mad Hatter in his story. It was known that persons who made hats often became "mad" after a period of time. Find out the connection between mercury and why workers in hat factories suffered from nervous conditions.

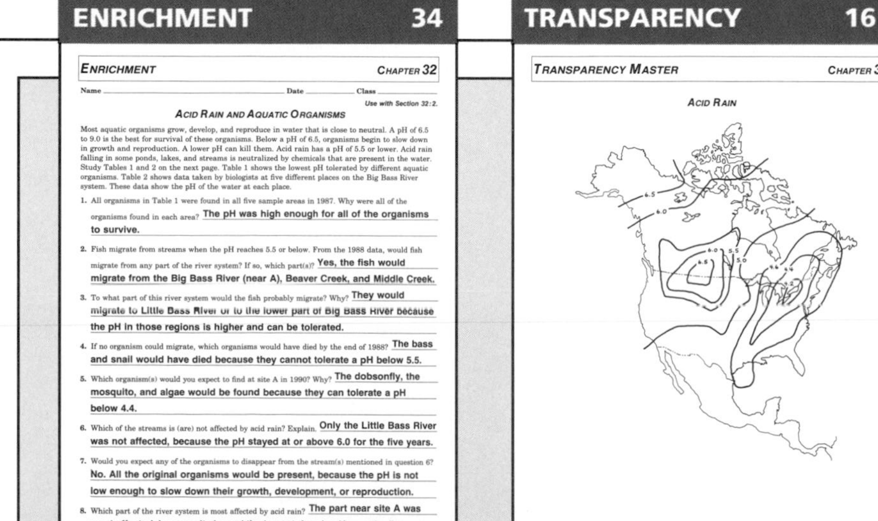

**ENRICHMENT 34**

ENRICHMENT — CHAPTER 32

Name ______ Date ______ Class ______

*Use with Section 32:2.*

ACID RAIN AND AQUATIC ORGANISMS

Most aquatic organisms grow, develop, and reproduce in water that is close to neutral. A pH of 6.5 to 9.0 is the best for survival of these organisms. Below a pH of 6.5, organisms begin to slow down in growth and reproduction. A lower pH can kill them. Acid rain has a pH of 5.5 or lower. Acid rain falling in some ponds, lakes, and streams is neutralized by chemicals that are present in the water. Study Tables 1 and 2 on the next page. Table 1 shows the lowest pH tolerated by different aquatic organisms. Table 2 shows data taken by biologists at five different places on the Big Bass River system. These data show the pH of the water at each place.

1. All organisms in Table 1 were found in all five sample areas in 1987. Why were all of the organisms found in each area? **The pH was high enough for all of the organisms to survive.**
2. Fish migrate from streams when the pH reaches 5.5 or below. From the 1988 data, would fish migrate from any part of the river system? If so, which part(s)? **Yes, the fish would migrate from the Big Bass River (near A), Beaver Creek, and Middle Creek.**
3. To what part of this river system would the fish probably migrate? Why? **They would migrate to Little Bass River or to the lower part of Big Bass River because the pH in those regions is higher and can be tolerated.**
4. If no organism could migrate, which organisms would have died by the end of 1988? **The bass and snail would have died because they cannot tolerate a pH below 5.5.**
5. Which organism(s) would you expect to find at site A in 1990? Why? **The dobsonfly, the mosquito, and algae would be found because they can tolerate a pH below 4.4.**
6. Which of the streams is (are) not affected by acid rain? Explain. **Only the Little Bass River was not affected, because the pH stayed at or above 6.0 for the five years.**
7. Would you expect any of the organisms to disappear from the stream(s) mentioned in question 6? **No. All the original organisms would be present, because the pH is not low enough to slow down their growth, development, or reproduction.**
8. Which part of the river system is most affected by acid rain? **The part near site A was most affected, because it showed the largest drop in pH over the five years.**

**TRANSPARENCY 161**

TRANSPARENCY MASTER — CHAPTER 32

ACID RAIN

## OPTIONS

**Transparency Master**/*Teacher Resource Package,* p. 161. Use the Transparency Master shown here to teach acid rain.

**Enrichment**/*Teacher Resource Package,* p. 34. Use the Enrichment worksheet shown here to teach how acid rain affects aquatic organisms.

# TEACH

## Concept Development

▶ Explain that acids in solution taste sour and release hydrogen gas when reacted with certain metals.

▶ Point out that normal rainfall is not pure water. The water in rain combines with carbon dioxide to form carbonic acid. Normal rain has a pH of 5.5 or above. Tell students that soda water is a weak solution of carbonic acid.

## Independent Practice

**Study Guide**/*Teacher Resource Package*, p. 189. Use the Study Guide worksheet shown at the bottom of this page for independent practice.

## Check for Understanding

Have students make a chart relating pollutants to the problems they cause. Tell them to use the following pollutants for their chart: carbon monoxide, sulfur dioxide, smog, particulates, heavy metals, and pesticides. Help students locate the sections of the text that explain these problems.

### Reteach

Give a short matching quiz using the pollutants and problems from the chart in Check for Understanding.

### *Portfolio*

Tell students that each species of aquatic organism lives within a specific pH range. Some have a wider or narrower range of tolerance to changes in the pH of water. Have students research the pH tolerance ranges of at least ten aquatic plants and animals and explain what has happened to aquatic ecosystems in which acid rain has altered the pH of the water. Have them present their findings in report form in their portfolios.

**Figure 32-9** Acid rain damages buildings (a) as well as forests (b). What gas causes acid rain? sulfur dioxide

a

**Acids,** such as vinegar and lemon juice, are liquids that have pH values lower than 7. The stronger an acid is, the lower its pH. Pure water is neutral, neither acid nor base, and has a pH of 7. **Bases,** such as ammonia and lye, are liquids that have pH values greater than 7. The stronger a base is, the higher its pH. The pH of rain is normally above 5.5, or almost neutral. **Acid rain** is rain that has a pH between 1 and 5.5.

b

How is it possible for rain (or fog or snow) to be acid? Acids form in the air when gases such as sulfur dioxide react with water. Remember that these gases are formed by burning fossil fuels. As acid rain falls onto Earth, it damages forests, crops, soil, and buildings. See Figure 32-9. Acid rain has turned lakes and ponds into bodies of water with a pH value below 4.5. It has been estimated that 15 percent of all Minnesota lakes are now too acid for most living things. Over 200 lakes in New York State are dead due to acid rain. Nothing can live in them.

Acid rain seems to be worse for the east coast of the United States and Canada. Much of the countries' industry is in the east. Winds also carry acid rain from the western and central parts of these countries toward the east.

### Bio Tip

**Ecology:** Each person in the United States produces more than four pounds of garbage each day. Eighty percent of this garbage can be recycled by using modern technology.

## Land Pollution

We live in a plastic and chemical age. Think about all the items we use each day that are made of plastic. You probably have heard that plastic is not biodegradable . Plastics can't be broken down.

684 Solving Ecological Problems 32:2

## *OPTIONS*

**How Do Chemical Pollutants Affect Living Things?**/*Lab Manual,* pp. 271-274. Use this lab as an extension to studying pollution from chemicals.

## STUDY GUIDE 189

STUDY GUIDE CHAPTER 32

Name ______ Date ______ Class ______

PROBLEMS FROM POLLUTION

*In your textbook, read about acid rain in Section 32:2.*

1. Study the map below.



2. Using an atlas and the map above, list the states that have a high sensitivity to acid rain. Place a checkmark by those states that have only small areas of high sensitivity to acid rain.

| West | | East | |
|---|---|---|---|
| Washington | Tennessee | Vermont | Minnesota |
| Oregon | Kentucky | North Carolina | Wisconsin |
| California | West Virginia | South Carolina | ✓ Michigan |
| ✓ Nevada | Pennsylvania | Florida | ✓ Texas |
| Idaho | Maryland | Louisiana | ✓ Oklahoma |
| Montana | New Jersey | Georgia | ✓ Missouri |
| Wyoming | Delaware | Alabama | |
| Utah | New York | Arkansas | |
| Colorado | Massachusetts | Mississippi | |
| New Mexico | Connecticut | Rhode Island | |
| Arizona | District of Columbia | Maine | |
| | New Hampshire | | |

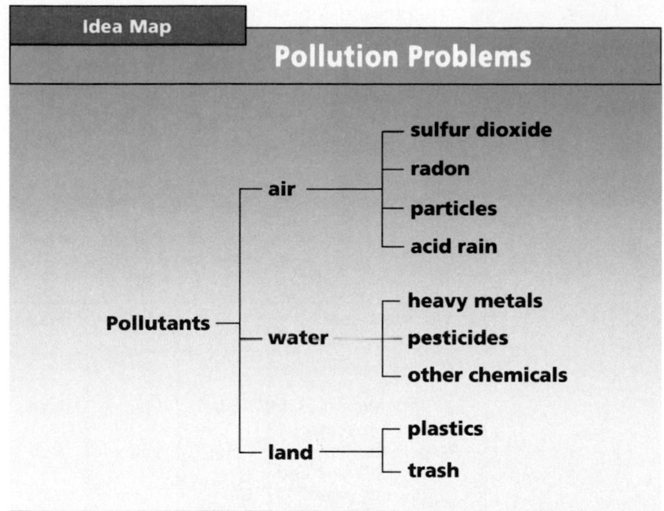


**Study Tip:** Use this idea map as a study guide to pollutants. Can you think of any other pollutants?

Suppose you set a bag filled with bits of food, paper scraps, and plastic jugs and bottles on the curb. The trash collectors forget to come to your house. In a few days, the bits of food begin to rot. The plastic containers, however, remain unchanged. After several weeks, the food and paper have rotted away. The plastic is still there. In fact, if you were to move away and then return in a few years, the plastic would still be unchanged. Think of all the plastic items people throw away each day. None of them are biodegradable. Dumps and landfills are becoming filled with these items. We are running out of places to put them.

## Check Your Understanding

6. What is a major source of air pollution?
7. In what parts of the food chain do pesticides build up?
8. What problems arise from the use of plastics?
9. **Critical Thinking:** Bald eagles almost became extinct because DDT prevented their eggs from hatching. The eagles got the DDT from the fish they ate. How did the fish get the DDT?
10. **Biology and Writing:** In this section, you learned about damage to humans from air, water, and land pollution. Tell about another kind of pollution that damages humans.

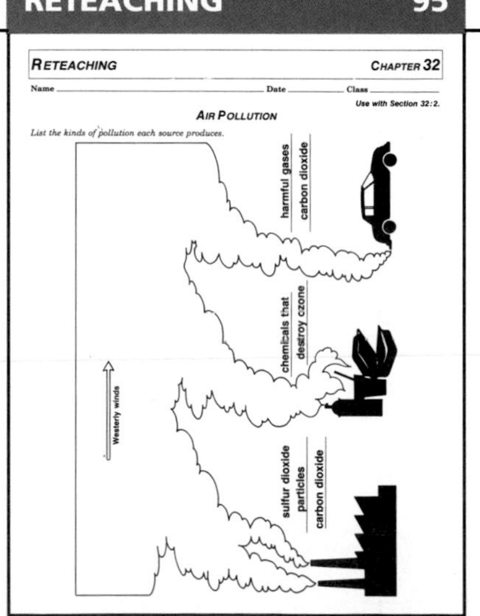

**RETEACHING 95**

RETEACHING CHAPTER 32

Name ______ Date ______ Class ______

*Use with Section 32:2.*

AIR POLLUTION

*List the kinds of pollution each source produces.*



**STUDY GUIDE 190**

STUDY GUIDE CHAPTER 32

Name ______ Date ______ Class ______

PROBLEMS FROM POLLUTION

3. What are two reasons that the eastern part of the United States has a greater acid rain problem than other parts?
   a. Winds carry acid rain from the western and central parts of the United States toward the east.
   b. Most of the industry in the United States is located in the east.
4. What does the pH scale measure? The pH scale measures how strong an acid or base a substance is.
5. What are some examples of acids? vinegar, lemon juice
6. What are some examples of bases? ammonia, lye
7. What pH do acids have? Acids have pH values lower than 7.
8. What pH do bases have? Bases have pH values higher than 7.
9. What is the pH of pure water? The pH of pure water is 7.
10. What is the pH of normal rain? The pH of normal rain is above 5.5.
11. What is the pH of acid rain? The pH of acid rain is between 1 and 5.5.
12. How is acid rain formed? Harmful gases such as sulfur dioxide react with sunlight and water to form acid rain.
13. List three results of acid rain in the United States.
   a. It has been estimated that 15% of all Minnesota lakes are now too acid for life.
   b. Over 200 lakes in the state of New York are "dead".
   c. Some lakes and ponds have a pH value below 4.5.
14. What things other than lakes does acid rain damage? Acid rain damages forests, crops, soil, and buildings.
15. Why must countries work together to solve the acid rain problem? The gases that cause acid rain are carried by the wind from one area to another.

## TEACH

### Idea Map

Have students use the idea map as a study guide to the major concepts of this section.

### Independent Practice

**Study Guide**/*Teacher Resource Package*, p. 190. Use the Study Guide worksheet shown at the bottom of this page for independent practice.

### Check for Understanding

Have students respond to the first three questions in Check your Understanding.

### Reteach

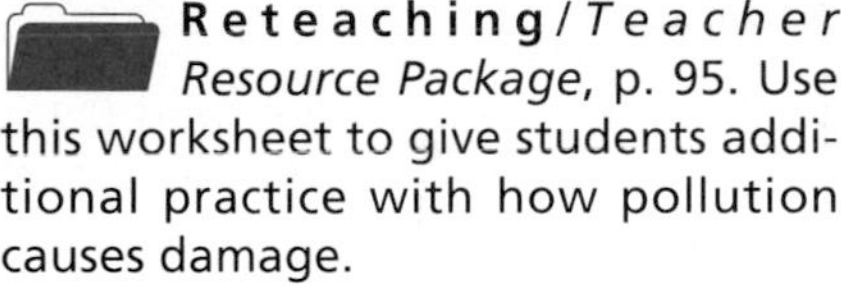

**Reteaching**/*Teacher Resource Package*, p. 95. Use this worksheet to give students additional practice with how pollution causes damage.

**Extension:** Assign Critical Thinking, Biology and Writing, or some of the **OPTIONS** available with this lesson.

### Divergent Question

Ask students which pollutants can be found in the rivers and streams where they live.

### Audiovisual

Show the video *Acid Rain: No Simple Solution*, Insight Media.

### Answers to Check Your Understanding

6. burning of fossil fuels
7. secondary consumers
8. They don't break down, so they fill up landfills.
9. They ate smaller fish or protists that were contaminated with DDT that washed off fields.
10. Answers may include noise pollution, pollution of food sources from pesticides, and so on.

## Lab 32-2 Pollution

### Overview

Students will find out how certain pollutants affect yeast cells.

**Objectives:** Students will be able to (1) **describe** how some pollutants affect yeast cells, (2) **test** for the presence of carbon dioxide as a by-product of respiration, (3) **separate and control** variables in an experiment.

**Time Allotment:** 45 minutes on the first day, and 10-15 minutes on the next day

### Preparation

▶ For bromthymol blue, add 0.1 g bromthymol blue powder to 2 L of distilled water.

▶ For live yeast mixture, add 1/4 cake of yeast and 1 tablespoon of table sugar to 500 mL of distilled water. Stir the mixture until the yeast cake breaks apart.

▶ For dead yeast mixture, boil 100 mL of live yeast mixture for at least 5 minutes on a hot plate.

**Lab 32-2 worksheet**/*Teacher Resource Package,* pp. 127-128.

### ASSESSMENT

**Skill:** Have students make and test a hypothesis as to whether different amounts of a pollutant added to yeast cells will still kill the yeast. Direct students to add different amounts of a pollutant to the yeast cells while keeping the number of drops of yeast cells constant. Have students use water as a control against which to compare the effects of the pollutants.

| amount of yeast | amount of water/pollutant |
|---|---|
| 10 drops | 1 drop |
| 10 drops | 2 drops |
| 10 drops | 4 drops |
| 10 drops | 8 drops |

# Pollution

## Lab 32–2

### Problem: What is the effect of pollutants on yeast?

#### Skills

observe, form hypotheses, interpret data, design an experiment

#### Materials

4 test tubes
marking pencil
test-tube rack
5 droppers
blue test liquid
dead yeast mixture
live yeast mixture
detergent
hydrogen peroxide
water

#### Procedure

Yeast cells are living organisms. In this lab, they will be treated with different pollutants to see if they are killed.

1. Copy the data table.
2. Number four test tubes 1 to 4.
3. Add the following to each tube: Tube 1—10 drops of dead yeast mixture and 10 drops of water; Tube 2—10 drops of live yeast mixture and 10 drops of water; Tube 3—10 drops of detergent and 10 drops of live yeast mixture; Tube 4—10 drops of hydrogen peroxide and 10 drops of live yeast mixture.
4. Fill all tubes almost full with blue test liquid.
5. **Observe:** Record the color in each tube. This is day 1.
6. Write a **hypothesis** to state whether the yeast will be killed by detergent or hydrogen peroxide.
7. Leave the tubes in the rack overnight.
8. Record the color in each tube on day 2.
9. Dispose of your materials as directed by your teacher.

| TUBE | CONTENTS | COLOR IN TUBE DAY 1 | COLOR IN TUBE DAY 2 | YEAST ALIVE OR DEAD |
|---|---|---|---|---|
| 1 | Dead yeast and water | blue | blue | dead |
| 2 | Live yeast and water | blue | green | alive |
| 3 | Detergent and live yeast | blue | blue | dead |
| 4 | Hydrogen peroxide and live yeast | blue | blue | dead |

#### Data and Observations

1. What color was each tube on day 1?
2. What color was each tube on day 2?

#### Analyze and Apply

If the tube contents change from blue to green overnight, the yeast cells are alive. If the tube contents stay blue, the yeast cells are dead. Complete the last column of the data table, then answer the questions.

1. Which tubes contained live organisms the first day? Which tubes contained live organisms the second day?
2. **Interpret data:** The blue liquid changes to green if carbon dioxide gas is present. How is this gas related to living things?
3. What was the purpose of tubes 1 and 2?
4. **Check your hypothesis:** Did your data support your hypothesis? Why or why not?
5. **Apply:** From this lab, what would you say might be the effect of hydrogen peroxide and detergent on some other living things?

#### Extension

**Design an experiment** to test the effect of an acid and a base on yeast.



## ANSWERS

### Data and Observations

1. All tubes were blue on day 1.
2. Tube 2 was green. Tubes 1, 3, and 4 were blue.

### Analyze and Apply

1. tubes 2, 3, 4; tube 2
2. All living things give off carbon dioxide during respiration.
3. as controls to see if dead yeast would change the color of the test liquid and to see what changes live yeast would make
4. Answers will vary. Students who said that detergent and hydrogen peroxide would kill yeast will report that their hypotheses were supported.
5. Hydrogen peroxide and detergent would probably harm or kill other living things.

## 32:3 Working Toward Solutions

Our environment is in trouble. It is being damaged by pollutants in the air, soil, and water. We are using up our resources and making more and more wastes as our population increases. Many resources can't be replaced.

Look at the area where you live. You probably can find examples of water, land, and air pollution. What can be done to solve pollution problems before they begin to limit population growth? How can we use resources wisely?

**Objectives**

6. **Explain** methods of conserving resources.
7. **Discuss** ways to keep the environment clean.

**Key Science Words**

recycling

### Conserving Our Resources

There are several solutions to the problem of endangered plants and animals. One depends on the United States government. Another is up to you as an active, caring individual.

Today, the government has set aside over 500 areas called National Wildlife Refuges. A refuge is an area that protects or shelters living things from being harmed by humans.

In 1973, the United States Congress passed the Endangered Species Act. This law states that anyone found guilty of killing, capturing, or removing any endangered species from its environment can be fined up to $20 000 and jailed for one year. The law also protects the habitat of any endangered species. In order for government refuges and laws to work, everyone must cooperate. This means that everyone must obey the law and make every effort not to reduce our wildlife populations.

**What is the Endangered Species Act?**

Erosion of soil can be slowed by planting crops across the slope of the land, rather than up and down, as shown in Figure 32-10. Strips of grass can be planted between fields to hold the soil. Rows of trees can be planted as windbreaks so the wind doesn't blow away dry soil.

Water can be reused if wastes and harmful chemicals are first removed. Conserving water would also guard the supply of this precious resource. Each time an average-size lawn is watered, up to 2000 gallons of water is used. That's enough water to fill your bathtub 40 times! Now, imagine thousands of people watering their lawns daily. You can see that a lot of water is used on lawns.

Do you allow the water to run as you brush your teeth? If you do, you are sending about five gallons of water down the drain each time you brush. Turning off the tap is a simple solution to this problem.

**Figure 32-10** Erosion of soil can be slowed by planting crops across a slope.

### OPTIONS

**Science Background**

No new water is being made on Earth today. All of the water used by living things is recycled through the water cycle.

## 32:3 Working Toward Solutions

### PREPARATION

**Materials Needed**

Make copies of the Study Guide, Reteaching, and Lab worksheets in the *Teacher Resource Package*.

▶ Prepare bromthymol blue solution for Lab 32-2. Prepare live and dead yeast mixtures on the day of the Lab.

▶ Have students bring in magazine or newspaper articles on solutions to pollution problems for the Focus activity.

**Key Science Words**

recycling

**Process Skills**

In Lab 32-2, students will observe, form hypotheses, interpret data, and design an experiment.

### 1 FOCUS

▶ The objectives are listed on the student page. Remind students to preview these objectives as a guide to this numbered section.

**Cooperative Learning:** Have teams of students bring in newspaper or magazine articles that deal with solutions to pollution problems. Have each team choose one article, tell what pollutant is being discussed, and explain the proposed solution to the problem. When all teams have reported, close with a discussion about the cost of solving these problems.

**Guided Practice**

Have students write down their answers to the margin question: a law that makes it a crime to kill endangered species.

# 2 TEACH

## MOTIVATION/Brainstorming

Ask students to name things that can be recycled.

## Concept Development

- ▶ Point out that more and more pesticides are being replaced by animals, plants, protists, monerans, or fungi that serve as parasites or predators on unwanted organisms.
- ▶ Explain that it costs money to recycle materials, but if materials are not recycled, raw materials may not be available in the future.
- ▶ Explain that each citizen in a community is responsible for keeping the environment clean.

## Guided Practice

Have students write down their answers to the margin question: to remove sulfur dioxide and particles from smoke.

## Connection: Social Studies

Have students find out what local and state laws govern pollution. Have them also find out what state and local agencies are responsible for enforcement and how rigidly the laws are being enforced.

## Independent Practice

**Study Guide**/*Teacher Resource Package*, p. 191. Use the Study Guide worksheet shown at the bottom of this page for independent practice.

**GLENCOE TECHNOLOGY**

**Videodisc**

Biology: The Dynamics of Life
Movie: *Recycling*
Disc 1, Side 1, Ch. 16

## Keeping Our Environment Clean

Why do some power plants use scrubbers?

How can we clean up our environment and keep it clean? Many steps have been taken to improve our air and water. New power plants that use coal must have scrubbers in their smokestacks. The scrubbers help to remove sulfur dioxide and particles from the smoke before it is released into the air. More improvements in air quality could be made if cleaner sources of energy were used. Solar and nuclear energy can be used to make electricity, instead of using coal or oil.

How has water quality improved? Several large bodies of water in the United States have been cleaned up in the last 20 years. Parts of Lake Erie were once so polluted with chemicals that people could not eat Lake Erie fish. Today people can swim and fish in Lake Erie again.

Wise use of technology also can help solve the pollution problem. Scientists have found that insect pests can be destroyed by their natural enemies. Aphids are tiny insects that suck the juices from plants. The ladybird beetle is a natural predator of aphids. Ladybird beetles can be released in an area where there are aphids. The aphids will be destroyed in a natural way without the use of pesticides.

Bacteria and viruses also can destroy some kinds of insects without hurting other living things. For example, bacteria can be used to control gypsy moths. The gypsy moth caterpillar has destroyed many oak forests. A bacterium found in the soil is able to kill this caterpillar. Biologists grow this type of bacterium in large tanks. The bacteria are then sprayed on tree leaves. The gypsy moth caterpillars die several hours after eating leaves with bacteria on them. Biologists are discovering ways to solve problems in our environment without creating new ones.

**Figure 32-11** Soil bacteria can interrupt the life cycle of the gypsy moth (a). A ladybird beetle eats aphids (b).

a

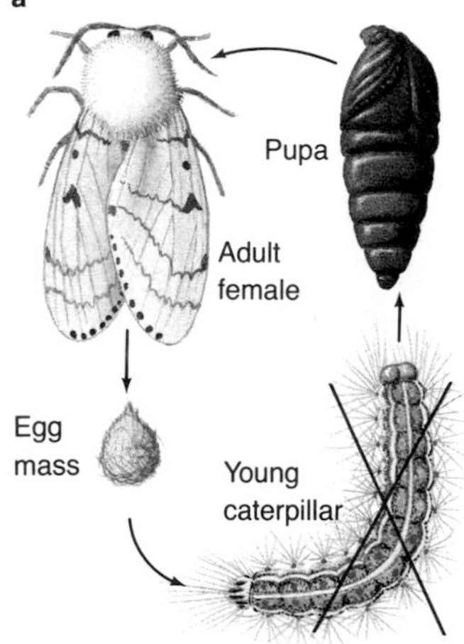


b

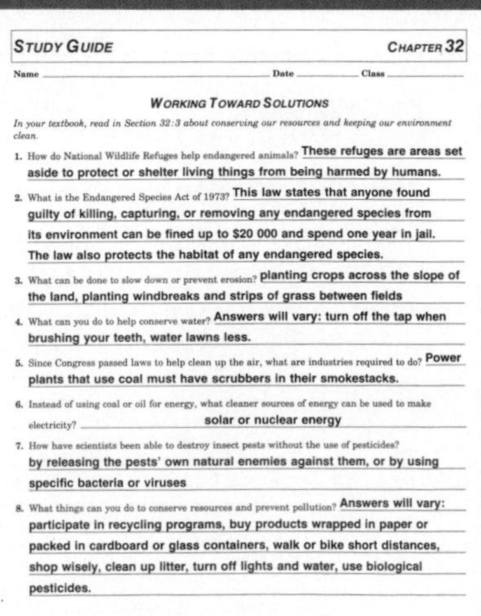

**STUDY GUIDE 191**

STUDY GUIDE — CHAPTER 32

Name ______ Date ______ Class ______

**WORKING TOWARD SOLUTIONS**

*In your textbook, read in Section 32:3 about conserving our resources and keeping our environment clean.*

1. How do National Wildlife Refuges help endangered animals? **These refuges are areas set aside to protect or shelter living things from being harmed by humans.**
2. What is the Endangered Species Act of 1973? **This law states that anyone found guilty of killing, capturing, or removing any endangered species from its environment can be fined up to $20 000 and spend one year in jail. The law also protects the habitat of any endangered species.**
3. What can be done to slow down or prevent erosion? **planting crops across the slope of the land, planting windbreaks and strips of grass between fields**
4. What can you do to help conserve water? **Answers will vary: turn off the tap when brushing your teeth, water lawns less.**
5. Since Congress passed laws to help clean up the air, what are industries required to do? **Power plants that use coal must have scrubbers in their smokestacks.**
6. Instead of using coal or oil for energy, what cleaner sources of energy can be used to make electricity? **solar or nuclear energy**
7. How have scientists been able to destroy insect pests without the use of pesticides? **by releasing the pests' own natural enemies against them, or by using specific bacteria or viruses**
8. What things can you do to conserve resources and prevent pollution? **Answers will vary: participate in recycling programs, buy products wrapped in paper or packed in cardboard or glass containers, walk or bike short distances, shop wisely, clean up litter, turn off lights and water, use biological pesticides.**

a

b

**Figure 32-12** Recycling (a) can reduce the amount of trash that goes into landfills. Over-packaged products (b) cause landfills to fill up. Why? The packaging is not biodegradable.

There are many things that people can do to help prevent pollution. **Recycling** is the reusing of resources. There are recycling centers for glass, plastic, and aluminum. Used paper and cardboard can be made into new paper. Over 500 000 trees could be saved every week if everyone recycled just the Sunday newspapers.

Being a wise consumer can also help. You can buy products wrapped in paper or packed in cardboard or glass. You can avoid buying individually wrapped items such as cheese slices. Every item that is packed in something other than plastic will help to reduce landfills.

How can you help reduce pollution? You can walk or bike short distances instead of driving, shop wisely and carefully for items that won't pollute the environment, recycle, clean up litter, turn off lights and water, and use natural pesticides. The answers to pollution problems are neither easy nor inexpensive. To have a good quality of life, everyone must work to keep the environment clean.

**Bio Tip**

**Ecology:** Using one metric ton of recycled paper for printing is equal to cutting down 17 trees to make into paper pulp.

## Check Your Understanding

11. List one way to conserve each of the following: animals, water, trees.
12. How have laws helped in cleaning up our environment?
13. How can gypsy moths be controlled naturally?
14. **Critical Thinking:** How can producing more fuel-efficient cars help reduce air pollution?
15. **Biology and Reading:** What are at least two steps that have been taken to keep the air clean?

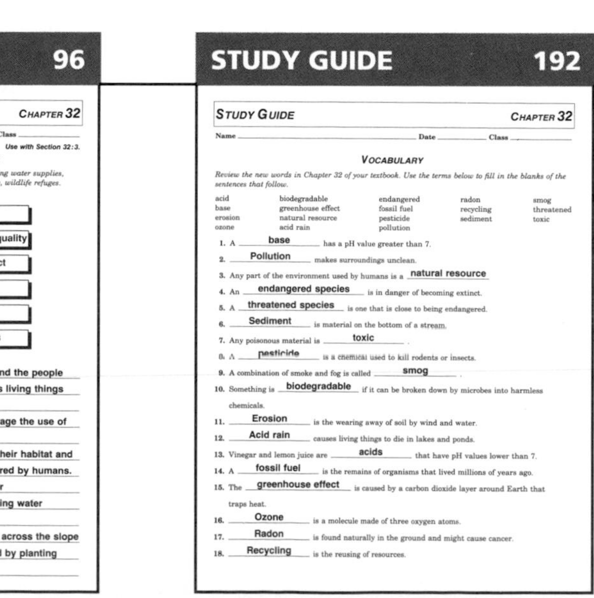

**RETEACHING 96**

RETEACHING: IDEA MAP — CHAPTER 32

Name ______ Date ______ Class ______

Use with Section 32:3.

SOLUTIONS TO ENVIRONMENTAL PROBLEMS

*Use the following terms and phrases to complete the idea map: recycling, cleaning water supplies, Endangered Species Act, good farming practices, laws for improving air quality, wildlife refuges.*

Solutions by the government: wildlife refuges; laws for improving air quality; Endangered Species Act

Solutions by the people: recycling; good farming practices; cleaning water supplies

1. What two groups can help protect the environment? the government and the people
2. How does a wildlife refuge help living things? It protects and shelters living things from being harmed by humans.
3. How do laws for improving air quality help fight pollution? They encourage the use of scrubbers and of cleaner energy sources.
4. How does the Endangered Species Act protect living things? It protects their habitat and limits the kinds of living things that can be killed or captured by humans.
5. What kinds of items can be recycled? water, glass, aluminum, paper
6. How can water be conserved? by not watering lawns, by not running water excessively
7. How can farmers and gardeners slow down erosion? by planting crops across the slope of the land, by planting strips of grass between fields, and by planting windbreaks

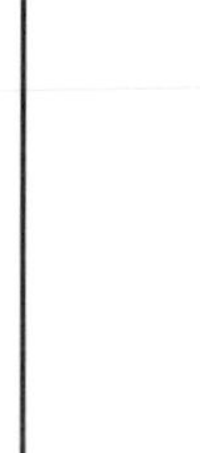

**STUDY GUIDE 192**

STUDY GUIDE — CHAPTER 32

Name ______ Date ______ Class ______

VOCABULARY

*Review the new words in Chapter 32 of your textbook. Use the terms below to fill in the blanks of the sentences that follow.*

| | | | | |
|---|---|---|---|---|
| acid | biodegradable | endangered | radon | smog |
| base | greenhouse effect | fossil fuel | recycling | threatened |
| erosion | natural resource | pesticide | sediment | toxic |
| ozone | acid rain | pollution | | |

1. A base has a pH value greater than 7.
2. Pollution makes surroundings unclean.
3. Any part of the environment used by humans is a natural resource
4. An endangered species is in danger of becoming extinct.
5. A threatened species is one that is close to being endangered.
6. Sediment is material on the bottom of a stream.
7. Any poisonous material is toxic.
8. A pesticide is a chemical used to kill rodents or insects.
9. A combination of smoke and fog is called smog.
10. Something is biodegradable if it can be broken down by microbes into harmless chemicals.
11. Erosion is the wearing away of soil by wind and water.
12. Acid rain causes living things to die in lakes and ponds.
13. Vinegar and lemon juice are acids that have pH values lower than 7.
14. A fossil fuel is the remains of organisms that lived millions of years ago.
15. The greenhouse effect is caused by a carbon dioxide layer around Earth that traps heat.
16. Ozone is a molecule made of three oxygen atoms.
17. Radon is found naturally in the ground and might cause cancer.
18. Recycling is the reusing of resources.

## TEACH

### Independent Practice

**Study Guide**/*Teacher Resource Package*, p. 192. Use the Study Guide worksheet shown at the bottom of this page for independent practice.

### Check for Understanding

Have students respond to the first three questions in Check Your Understanding.

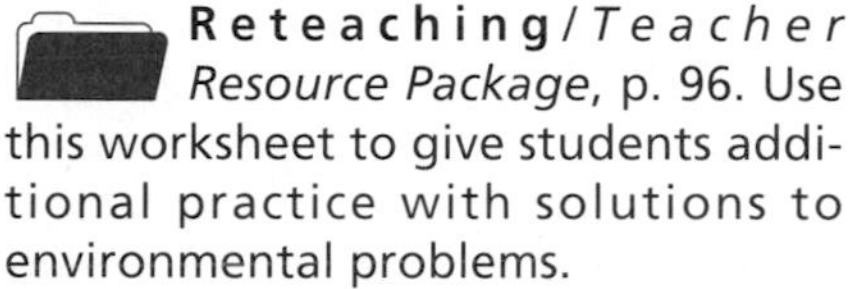

### Reteach

**Reteaching**/*Teacher Resource Package*, p. 96. Use this worksheet to give students additional practice with solutions to environmental problems.

**Extension:** Assign Critical Thinking, Biology and Reading, or some of the **OPTIONS** available with this lesson.

## 3 APPLY

**Cooperative Learning:** Set up student teams to research how garbage, sewage, and litter are handled in your area.

## 4 CLOSE

### Summary

Have students explain how they can help improve the environment.

### Answers to Check Your Understanding

11. animals–don't wear reptile shoes, bags, or belts; water–turn off water when brushing teeth; trees–recycle paper
12. New power plants must have scrubbers.
13. by spraying trees with bacteria that kill gypsy moths
14. More efficient cars use less gasoline, so less poisonous gas and carbon dioxide are produced.
15. New power plants must have scrubbers; alternate energy sources, such as solar energy, are being explored.

## CHAPTER 32 REVIEW

### Summary

Summary statements can be used by students to review the major concepts of the chapter.

### Key Science Words

All boldfaced terms from the chapter are listed.

### Testing Yourself

#### Using Words

1. acid
2. toxic
3. sediment
4. erosion
5. threatened
6. pesticide
7. radon
8. smog
9. biodegradable
10. greenhouse effect

#### Finding Main Ideas

11. p. 677; The pond goes through succession faster than normal.
12. p. 681; They can escape into the air, soil, and nearby water.
13. p. 687; It protects endangered species and their habitats.
14. p. 677; The remains of organisms were compressed under tremendous pressure for long periods of time.
15. p. 679; Certain chemicals found in spray cans, refrigerators, and air conditioners destroy ozone.
16. p. 684; Acid rain forms when certain gases react with water.
17. p. 676; It dries out, and wind easily separates small soil particles and blows them around.
18. p. 674; Plants and animals can be replaced within a person's lifetime.
19. p. 687; Water can be conserved by not watering lawns and by turning off water while brushing teeth.

#### Using Main Ideas

20. People have destroyed the habitats of other living things, hunted illegally, and used animal parts for jewelry, shoes, and purses.



# Chapter 32 Review

## Summary

### 32:1 Resources and Human Activities

1. Wildlife and plants are natural, renewable resources. They become endangered when they die faster than they can renew themselves.
2. Soil erosion occurs when soil is carried away by wind and water faster than it can be replaced. Water is a valuable resource. Pollution can cause ponds and lakes to go through a rapid aging process.
3. Fossil fuels are a nonrenewable source of energy. They are rapidly being used up.

### 32:2 Problems from Pollution

4. Air pollution is caused by burning fossil fuels in large amounts.
5. Heavy metals and pesticides can pollute water and build up in fish and other animals. Plastics and other materials that are not biodegradable fill up landfills and pollute Earth's surface.

### 32:3 Working Toward Solutions

6. Resources can be conserved by effective laws and by each person using resources wisely and carefully.
7. Keeping the environment clean depends on industry, government enforcement of laws, the development of clean energy sources, and the actions of all people.

## Key Science Words

acid (p. 684)
acid rain (p. 684)
base (p. 684)
biodegradable (p. 681)
endangered (p. 675)
erosion (p. 676)
fossil fuel (p. 677)
greenhouse effect (p. 679)
natural resource (p. 674)
ozone (p. 679)
pesticide (p. 680)
pollution (p. 677)
radon (p. 679)
recycling (p. 689)
sediment (p. 676)
smog (p. 678)
threatened (p. 675)
toxic (p. 678)

## Testing Yourself

### Using Words

*Choose the word from the list of Key Science Words that best fits the definition.*

1. has a pH value less than 7
2. poisonous or harmful
3. particles deposited in a stream
4. wearing away of soil
5. close to being endangered
6. a chemical that kills unwanted organisms
7. a gas that gives off radiation
8. a combination of smoke and fog
9. ability of a chemical to be broken down by microbes
10. process that causes heat to be trapped near Earth's surface

# Review

## Testing Yourself *continued*

**Finding Main Ideas**

*List the page number where each main idea below is found. Then, explain each main idea.*

11. what happens to a pond when extra nutrients are washed into it
12. what can happen to toxic waste buried in dump sites
13. what the purpose of the Endangered Species Act is
14. how fossil fuels are formed
15. why the ozone layer is being destroyed
16. how acid rain is formed
17. why bare soil is easily moved about by the wind
18. why plants and animals are renewable resources
19. how water can be conserved

**Using Main Ideas**

*Answer the questions by referring to the page number after each question.*

20. What are two ways that people make living things endangered or threatened? (p. 675)
21. What are two energy sources that are cleaner than fossil fuels and are renewable? (p. 688)
22. What is radon? (p. 679)
23. What effects does DDT have on living things? (p. 681)
24. How can bacteria be used to control plant pests naturally? (p. 688)
25. Where does sulfur dioxide in the air come from? (p. 679)
26. How do pollutants get into our water supply? (p. 680)
27. Why do you suppose water at a water treatment plant is checked for mercury and lead? (p. 683)

## Skill Review ✓

*For more help, refer to the* ***Skill Handbook,*** *pages 704-719.*

1. **Outline:** Make an outline of the different kinds of pollution discussed in this chapter.
2. **Calculate:** Assume there are 4800 species of animals in your state. Pollution is causing 30 species to die each week. Calculate how many years it will be before all animals are gone.
3. **Classify:** Classify the following items as to whether each is biodegradable: banana peel, plastic bottle, newspaper, DDT, plastic foam container, sandwich.
4. **Design an experiment:** Design an experiment to show whether newspapers are biodegradable.

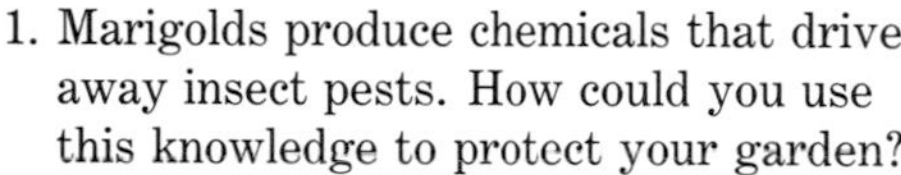


## Finding Out More

**Critical Thinking**

1. Marigolds produce chemicals that drive away insect pests. How could you use this knowledge to protect your garden?
2. Some people think that zoos and refuges are a waste of time and money. Give reasons for and against this statement.

**Applications**

1. Tape cardboard coated with a thin layer of vaseline to various locations in your community. After a few days, remove the cards and examine the particles trapped in the vaseline. Try to find out where they came from.
2. Write a report describing why Love Canal in New York State caused problems in the environment.



## Finding Out More

### Critical Thinking

1. You can plant marigolds around the edge.
2. An argument for would be the protection of species; an argument against would be the large amount of land used.

### Applications

1. Students may choose the front and back doors of a house or apartment, a wall of the school, or perhaps a room in their home.
2. Toxic materials were dumped into the canal. After a period of time, it was noticed that certain cancers appeared in the residents at a higher rate than average. Subsequently, the residents were evacuated and their homes boarded up.

**Chapter Review**/*Teacher Resource Package*, pp. 170-171.

**Chapter Test**/*Teacher Resource Package*, pp. 172-174.

**Chapter Test**/*Computer Test Bank*

**Unit Test**/*Teacher Resource Package*, pp. 175-176.

### Using Main Ideas

21. Solar and nuclear energy can be used.
22. Radon is a natural gas that gives off radiation.
23. It can cause cancer and prevent offspring from being produced.
24. It can be used to destroy some insects without harming other living things.
25. Sulfur from burning coal combines with oxygen in the air to form sulfur dioxide.
26. Pesticides and fertilizers can wash off farmland into streams and rivers.
27. Mercury and lead are pollutants that can build up in animals.

### Skill Review

1. Students will find the second section of the chapter helpful in making their outlines.
2. It will take a little over three years.
3. Banana peel, newspaper, and sandwich are biodegradable. Plastic bottle, DDT, and plastic foam are not.
4. Students should suggest using water, sunlight, and soil as biodegraders.

# Unit 8 Relationships in the Environment

**CONNECTIONS**

## Biology in Your World

Human behavior is interconnected with and makes an impact on all other organisms. Students must make decisions about their own behavior and decide what is most practical for themselves and the rest of the world.

### Literature Connection

Rachel Carson (1907-1964) was a marine biologist and author who worked for the U.S. Fish and Wildlife Service. Her controversial book, *Silent Spring,* sparked debate over the impact of pesticides on the environment.

Carson emphasized that all life is interrelated and that the extermination of one organism has repercussions on other organisms. For example, bald eagles came close to extinction because of loss of habitat and DDT. They ate fish containing DDT residues, which caused the birds to lay eggs with fragile shells. Since the restriction of DDT, successful breeding programs have increased the bald eagle population.

### History Connection

In May 1990, the Environmental Protection Agency declared Love Canal safe. The government spent twelve years and $250 million to contain 22 000 tons of toxic waste and seal the site. The homes closest to the dump were destroyed.

In August 1990, the remaining 236 homes went up for sale. Resettlement officials attracted many prospective buyers with low prices and studies showing the area to be habitable. Environmental groups and some former residents disagree with the government about the safety of the area.



**CONNECTIONS**

# Biology in Your World

### Disrupting a Delicate Balance

As you have seen in this unit, relationships are important in a community. If the survival of a single species is threatened, it may result in a chain reaction. We live in a rather fragile environment. Even well-meaning interference can be disastrous.

**LITERATURE**

## Environmental Awakening

In 1962, biologist Rachel Carson wrote about the dangers of spraying with pesticides in her book, *Silent Spring.* She pointed out that robins died after eating earthworms that had picked up DDT residues. DDT had been sprayed on the leaves of elm trees. It was used to kill the beetles carrying the fungus that caused Dutch Elm disease. Carson also warned that pesticides could contaminate human food supplies.

Because of Carson's book, the use of certain pesticides was restricted in the United States.

**HISTORY**

## A Ticking Time Bomb

In the 1920s, Love Canal was a partly-dug canal in the city of Niagara Falls, New York. It became a disposal site for household trash. Nearby chemical plants also dumped wastes in it. When the site was covered in 1958, about 20 000 metric tons of waste had been buried. A school was built on one edge of the canal. Homes were built on the other side. After very heavy rains in the mid 1970s, chemicals seeped out of the canal. These chemicals have been known to cause cancer, birth defects, miscarriages, and liver damage. Dangerous levels of these chemicals were found in the basements of many homes. In 1978-79, families were told to move from their homes. In 1990, the government allowed people to move into the homes.

LEISURE

## Making Your Own Paper

Recycle old newspapers yourself. First, tear the paper into small pieces. Put the pieces into a blender with five cups of water. Blend until you get a watery pulp. Put a piece of window screen into a dishpan with about three centimeters of water. Pour one cup of the paper pulp over the screen and spread it evenly with your fingers. Lift the screen and drain. Put the screen and pulp between several sheets of newspaper, and press with a board to squeeze out the extra water. Remove the pulp from the screen and leave it on dry newspaper overnight. When the pulp is dry, carefully peel off your handmade paper.

CONSUMER

## Stop the Throw-away Lifestyle

Landfills are overflowing. Some resources are being used up. What can you do to help? Instead of creating more waste, you can make some smart choices as a consumer.

Try not to buy products that are packaged in single-serving sizes or in unnecessary plastic. Whenever you can, buy things in large containers or in bulk. If possible, take your own containers and bags with you when you shop.

Set up a system that encourages recycling. Separate aluminum cans, glass containers, and newspapers and recycle them if you have a recycling center near you.

When you buy a new sweater to replace one that does not fit anymore, do not put the old one in the trash. Give it to someone who can use it or add it to the rag bag.

Substitute paper for plastic whenever you can. The plastic rings connecting a six-pack of canned soft drinks take a long time to decay in a landfill. Also, birds can get caught in the plastic rings and be strangled. Some types of plastic give off toxic fumes when they are burned.

Do Earth a favor. Think of the long-term effects before you buy something.

CULTURAL DIVERSITY

### George Washington Carver

African American agriculturalist George Washington Carver (1865-1943) is best known for his work in establishing crops such as cotton, peanuts, and sweet potatoes in the southern United States. Carver's research was directed toward the poor farmer and showed how reliance on a single cash crop ultimately depleted the soil of nutrients, leaving it open to destructive erosion. Initiate a discussion about the agricultural methods used by farmers in different societies and how such methods are related to protecting soil against erosion. Have students research farming methods and irrigation practices and make models that illustrate the methods used by people of different cultures.

CONNECTIONS

### Leisure Connection

Students may be interested in products made from or packaged in recycled paper. If the back of cardboard packaging is gray, it was made from recycled paper. For more information on recycled paper products, write to Seventh Generation at 10 Farrell St., South Burlington, VT 05403.

### Consumer Connection

In order to cut down on waste, students may need to make decisions about what not to buy. Landfills are brimming with bags of grass clippings and leaves. Organic matter in plastic bags doesn't break down quickly. Grass clippings can be left on the lawn or composted with leaves and kitchen scraps. These alternatives don't require purchasing plastic bags.

### References


Carson, Rachel. *Silent Spring.* Boston: Houghton Mifflin Co., 1962.

The Earth Works Group. *50 Simple Things Kids Can Do to Save the Earth.* Kansas City, MO: Andrews and McMeel, 1990.

Grove, Noel. "Recycling." *National Geographic*, July 1994, pp. 92-115.

Hoffman, Andrew J. "An Uneasy Rebirth of Love Canal, *Environment*, March 1995, pp. 4-17.

Horrigan, Alice and Jim Motavalli. "Talking Trash." *E.*, March-April, 1997, pp. 28-36.

Ingalls, Zoe. "Soak It, Boil It, Mash It: A Chicago Center Focuses on Paper Making." *The Chronicle of Higher Education*, December 8, 1995, pp. B6-B8.

Peart, Karen N. "Three Deadly Legacies." *Scholastic Update*, April 15, 1994, pp. 6-10.

Schwartz, Linda. "Save the Earth: Cleaning Up the Environment Starts With You." *Boy's Life*, May 1997, pp. 8-12.

Tierney, John. "Recycling Is Garbage." *The New York Times Magazine*, June 30, 1996, p. 24.

# Measuring in SI

Careful measurement is an important part of science. But even if you measure very carefully, your figures will have no meaning if your measuring unit means something different each time you use it. To be practical, a measurement unit must always mean the same thing to everybody. A *standard* unit is a definite amount used by everyone when measuring.

Most people in the world and all scientists use the International System (SI) of units. This is a modern form of the metric system.

In SI, all of the units are related to each other by the same set of prefixes. It is easy to change to smaller or larger units simply by multiplying or dividing by ten.

The main units used for measuring are the meter–for distance, gram–for mass, and liter–for volume.

Below is a table of the SI prefixes, their meanings, and rules for changing from one unit to another.

**I. Rules for expressing units of measurement**

A. Changing meters
   1. Meters can be changed into smaller units such as decimeters (dm), centimeters (cm), or millimeters (mm).
   2. Meters can be changed into larger units such as dekameters (dam), hectometers (hm), or kilometers (km).

B. Changing grams
   1. Grams can be changed into smaller units such as decigrams (dg), centigrams (cg), or milligrams (mg).
   2. Grams can be changed into larger units such as dekagrams (dag), hectograms (hg), or kilograms (kg).

C. Changing liters
   1. Liters can be changed into smaller units such as deciliters (dL), centiliters (cL), or milliliters (mL).
   2. Liters can be changed into larger units such as dekaliters (daL), hectoliters (hL), or kiloliters (kL).

**II. Rules for changing from one unit to another**

A. When changing from a *smaller unit to a larger unit* you must *divide.* (This type of change is shown by the arrow that points upward in the table.)

**SI UNITS**

| Prefixes | Symbol | Meaning | | | |
|---|---|---|---|---|---|
| **kilo** | k | 1000 | thousand | Larger unit to Smaller unit ↓ | ↑ |
| **hecto** | h | 100 | hundred | | |
| **deka** | da | 10 | ten | | |
| **gram, meter, liter** | g,m,L | | main unit | | |
| **deci** | d | 0.1 | tenth | | |
| **centi** | c | 0.01 | hundredth | | |
| **milli** | m | 0.001 | thousandth | | Smaller unit to Larger unit |

1. If you move up the table from any prefix to
   one above it, then divide by 10.
   two above it, then divide by 100.
   three above it, then divide by 1000.
   four above it, then divide by 10 000.
   five above it, then divide by 100 000.
   six above it, then divide by 1 000 000.
2. Examples
   Change 13.2 grams (g) to hectograms (hg). (*Hecto* is two places above gram in the table. Divide by 100.)
   13.2 g ÷ 100 = 0.132 hg

   Change 2.6 decimeters (dm) to kilometers (km).
   (*Kilo* is four places above *deci* in the table. Divide by 10 000.)
   2.6 dm ÷ 10 000 = 0.00026 km

   Change 14 milliliters (mL) to liters (L). (Liter is three places above *milli* in the table. Divide by 1000.)
   14 mL ÷ 1000 = 0.014 L

B. When changing from a *larger unit to a smaller unit* you must *multiply.* (This type of change is shown by the arrow that points downward in the table.)

1. If you move down the table from any prefix to
   one below it, then multiply by 10.
   two below it, then multiply by 100.
   three below it, then multiply by 1000.
   four below it, then multiply by 10 000.
   five below it, then multiply by 100 000.
   six below it, then multiply by 1 000 000.

2. Examples
   Change 25 hectograms (hg) to dekagrams (dag). (*Deka* is one place below *hecto* in the table. Multiply by 10.)
   25 hg × 10 = 250 dag

   Change 126 meters (m) to millimeters (mm). (*Milli* is three places below meter in the table. Multiply by 1000.)
   126 m × 1000 = 126 000 mm

   Change 0.08 kiloliters (kL) to dekaliters (daL).
   (*Deka* is two places below *kilo* in the table. Multiply by 100.)
   0.08 kL × 100 = 8.0 daL

# Classification of Living Things

Scientists recognize five kingdoms of living things: the Moneran Kingdom, Protist Kingdom, Fungus Kingdom, Plant Kingdom, and Animal Kingdom. These five kingdoms are divided into smaller groups. Scientists do not always agree about the grouping of living things within a kingdom, or that there should be five kingdoms. This appendix lists the classification groups that are generally accepted by most scientists. Use this appendix as you study Chapters 3-8.

**MONERAN KINGDOM**
One-celled; no nucleus; cell wall and cell membrane present.

**TRUE BACTERIA**
Includes several phyla; may be round, rod-shaped, or spiral-shaped; cannot make food. Example: bacteria that cause sore throats.
**Blue-green Bacteria Phylum**
Contain colored pigments, usually blue-green; can make food. Example: *Anabaena.*

**PROTIST KINGDOM**
Nucleus present; one-celled or multicellular.

**ANIMAL-LIKE PROTISTS**
One-celled; most move about; classification based on type of movement; cannot make food.

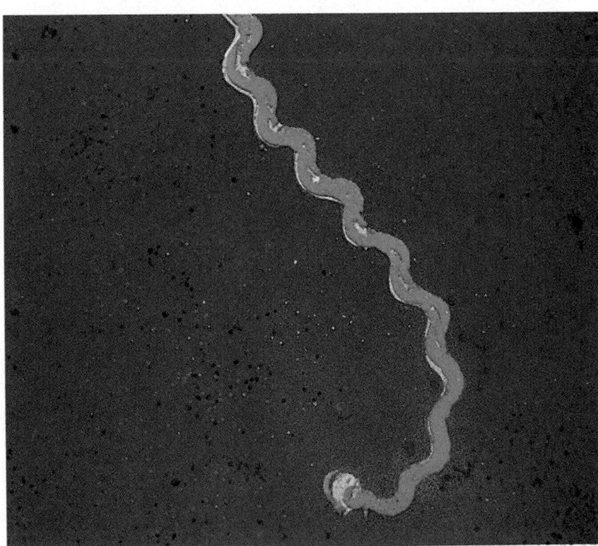

**Amoeba Phylum**
Move with false feet. Example: *Amoeba proteus.*
**Ciliate Phylum**
Move with cilia. Example: *Paramecium.*
**Flagellate Phylum**
Move with flagella. Example: *Trypanosoma.*
**Sporozoan Phylum**
Do not move. Example: *Plasmodium.*

**PLANTLIKE PROTISTS**
Some can move about; one-celled or multicellular; can make food.
**Euglena Phylum**
Move with flagella; one-celled; no cell wall. Example: *Euglena gracilus.*
**Diatom Phylum**
One-celled; golden-brown color; has two-part glass covering. Example: *Navicula.*
**Dinoflagellate Phylum**
One-celled; red or brown color. Example: *Gongaulax.*

**Green Algae Phylum**
One-celled or multicellular. Example: *Spirogyra* (spi ruh GI ruh).

**Red and Brown Algae Phyla**
Multicellular. Example: kelp.

**FUNGUSLIKE PROTISTS**
Have a life cycle that includes a moving, slimy mass, an amoebalike stage, and a spore-forming stage; cannot make food. Examples: two phyla that include the slime molds.

**FUNGUS KINGDOM**
Body consisting of hyphae; reproduce by spores.

**Sporangium Fungus Phylum**
Cannot make food; spores formed in sporangia. Example: bread mold.

**Club Fungus Phylum**
Cannot make food; spores formed in club-shaped structures. Examples: mushrooms, smuts.

**Sac Fungus Phylum**
Cannot make food; spores formed in saclike structures. Examples: yeast, *Penicillium.*

**Lichens**
Can make food; combination of fungus and organism with chlorophyll. Examples: British soldier lichen, reindeer moss.

**PLANT KINGDOM**
Green; chlorophyll in cells; can make food; cell walls present.

**NONVASCULAR PLANTS**
No tubes for carrying materials throughout plant; reproduce by means of spores; no roots, stems, leaves; includes several phyla (botanists use the term *division* instead of *phylum*). Examples: mosses and liverworts.

**VASCULAR PLANTS**
Tubes for carrying materials throughout plant; roots, stems, and leaves present; includes several phyla.

**Fern Phylum**
Reproduce with spores. Example: Boston fern.

**Conifer Phylum**
Reproduce with seeds; seeds in cones. Examples: pines, firs.

**Flowering Plant Phylum**
Reproduce with seeds; seeds in flowers. Examples: maple tree, daisy, wheat.

**ANIMAL KINGDOM**
Cannot make food; can move about; multicellular.

**INVERTEBRATES**
Animals without backbones; includes several phyla.

**Sponge Phylum**
Two cell layers; no symmetry; body with pores and canals; no tissues or organs; reproduce sexually and asexually. Examples: glass sponges, bath sponges.

**Stinging-cell Animal Phylum**
Two cell layers; radial symmetry; one opening (mouth) leading into a hollow body; mouth surrounded by tentacles; stinging cells; first phylum to show tissues. Examples: coral, sea anemone, *Hydra.*

**Flatworm Phylum**
Three cell layers and body organs; flattened body; bilateral symmetry; parasitic or free living; one opening (mouth) if present; first phylum to show a nervous system. Examples: *Planaria,* tapeworms.

**Roundworm Phylum**
Round body; bilateral symmetry; parasitic or free living; mouth and anus; sexual reproduction only. Examples: hookworm, *Ascaris.*

**Segmented Worm Phylum**
Segmented body; bilateral symmetry; free living; mouth and anus; heart and circulatory system. Examples: earthworm, clamworm, leech.

**Soft-bodied Animal Phylum**
Soft body covered by a fleshy mantle; movement by a muscular foot; bilateral symmetry; free living; mouth and anus; reproduce sexually.

**Snail Class**
Large muscular foot; one shell or no shell on outside of body. Examples: snails, slugs.

**Clam Class**
Large muscular foot; two shells. Examples: clams, oysters, scallops.

**Squid Class**
Muscular foot divided into tentacles and arms; shell (if present) inside body. Examples: squid, octopus.

**Jointed-leg Animal Phylum**
Body in sections; bilateral symmetry; free living; mouth and anus; first phylum to show a skeletal system (exoskeleton).
**Crayfish Class**
Body in two sections; five pairs of legs; two pairs of antennae. Examples: crabs, crayfish, shrimp.
**Spider Class**
Body in two sections; four pairs of legs; no antennae. Examples: spiders, scorpions.
**Centipede Class**
Body in two sections; one pair of poisonous claws; body segmented with one pair of legs per body segment. Example: centipedes.
**Millipede Class**
Body in two sections; body segmented with two pairs of legs per body segment. Example: millipedes.
**Insect Class**
Body in three sections; three pairs of legs; one pair of antennae; wings usually present. Examples: bees, ants, moths, beetles.
**Spiny-skin Animal Phylum**
Body with five part design; radial symmetry; covered with spines; tube feet. Examples: sea urchin, starfish.

**VERTEBRATES**
Animals with backbones; includes one phylum.
**Chordate Phylum**
Tough, flexible rod along the back sometime during life; skeleton inside body (endoskeleton); sexual reproduction; bilateral symmetry.
**Jawless Fish Class**
No scales or jaw; skeleton of cartilage; fins not paired; no gill covering; cold blooded; parasitic or free living. Example: lamprey.
**Cartilage Fish Class**
Toothlike scales; jaw; skeleton of cartilage; paired fins; no gill covering; cold blooded. Examples: sharks, rays.
**Bony Fish Class**
First to show bony skeleton; bony scales; paired fins; covered gills; most have a swim bladder; cold blooded. Examples: trout, salmon, goldfish.
**Amphibian Class**
Moist, scaleless skin; young live in water, adults on land; four legs; cold blooded. Examples: frog, toad, salamander.
**Reptile Class**
Dry, scaly skin; first to show ability to lay eggs outside of water; egg with shell and protective membranes; four legs (except snakes); cold blooded. Examples: turtle, snake, alligator.
**Bird Class**
Feathers; wings; scales and claws on legs; beak present; no teeth; warm blooded. Examples: swan, robin, sparrow, owl.
**Mammal Class**
Hair; mammary glands for nursing young; young develop within body of mother; warm blooded. Examples: human, dog, lion, whale.

# Rootwords, Prefixes, and Suffixes

Many words can be broken down into root words, prefixes, and suffixes. A prefix is a word part that, when added to the beginning of a word, changes the meaning of the word. A suffix is a word part that, when added to the end of a word, changes the meaning of the word. Knowing the meanings of certain root words, prefixes, and suffixes can help you understand biology words, and therefore, biology better. Use the table on these two pages to help you understand words used in biology.

P = prefix
R = root
S = suffix

| Word Part | Word | Meaning | Word Example | Definition |
|---|---|---|---|---|
| R | amphi | double | amphibian | double life |
| P | anti | against | antibiotic | against life |
| P | arthro | join/joint | arthritis | disease of joints |
| R | atrium | entrance | left atrium | entrance to heart |
| R | bio | life | biodegradable | change by life forms |
| | | | biology | study of life |
| R | carb | carbon | carbohydrate | having carbon and water |
| | | | carbon dioxide | having carbon and oxygen |
| P | cardio | heart | cardiac muscle | muscle found in heart |
| R | chloro | green | chlorophyll | green pigment in leaf |
| | | | chloroplast | green body in plant cell |
| R | chromo | colored | chromosome | colored body in cell nucleus |
| P | cyto | cell | cytoplasm | liquid within cell |
| R | dermis | covering | dermis | skin |
| | | | epidermis | outer skin covering |
| P | dia | through | diaphragm | through the middle of body |
| P | endo | inside | endoskeleton | skeleton inside body |
| | | | endospore | spore formed inside bacteria |
| P | epi | upon | epidermis | outer skin covering |
| | | | epiglottis | upon the glottis, or throat |
| P | ex | out of | exhaling | push air out of lungs |
| | | | extinct | life forms no longer on Earth |
| P | exo | outer | exoskeleton | skeleton on outside of body |
| R | fertile | to bear | fertilization | process that occurs before bearing young |
| | | | fertilizer | helps soil bear better plants |
| R | flagellum | whip | flagellum | cell part that looks like a whip |
| P | gen | create | genetics | how life traits are created |
| R | genos | race | genus | classification group |

| Word Part | Word | Meaning | Word Example | Definition |
|---|---|---|---|---|
| P | hemo | blood | hemophilia | disease in which the blood does not clot properly |
| | | | hemoglobin | protein in red blood cells |
| R | herb | grass | herb | small grass-like plant |
| | | | herbivore | plant-eating animal |
| R | hydr | water | hydroponics | growing plants in water |
| P | hypo | below | hypothesis | to put under or suppose |
| S | logy | study of | biology | study of life |
| | | | ecology | study of where life forms live |
| P | micro | small | microscope | instrument used to look at small objects |
| P | mito | thread | mitochondrion | thread-like parts in cytoplasm |
| | | | mitosis | changes in cell chromosomes (threads) during reproduction |
| R | morph | shape | metamorphosis | changes in body shape |
| P | multi | many | multicellular | having many cells |
| R | nephros | kidney | nephron | filter unit of kidney |
| R | neuro | nerve | neuron | nerve cell |
| P | non | not | noncommunicable | disease that is not catching |
| R | ov | egg | ovary | organ that makes egg cells |
| | | | oviduct | organ that carries egg from ovary |
| R | para | beside or near | parasite | living beside or inside |
| R | pherein | to carry | pheromone | carries insect messages |
| P | photo | light | photosynthesis | use of light by plants to make food |
| | | | phototropism | plants turning toward light |
| R | phyl | tribe | phylum | group name used to classify |
| R | phyll | leaf | chlorophyll | cell part that gives leaves their green color |
| R | pneuma | breath | pneumonia | disease of lungs that makes breathing difficult |
| P | post | after | posterior | hind part of animal |
| P | pre | before | prenatal | before birth |
| | | | prescription | writing before |
| P | re | back | regenerate | to grow back again |
| P | semi | half | semilunar | half moon-shaped valve |
| R | scop | to look | microscope | instrument you look through |
| S | soma | body | chromosome | colored body in cells |
| R | spor | seed | spore | cell that acts like a seed in reproduction |
| | | | sporozoan | protozoan that forms spores during reproduction |
| R | stoma | mouth | stomate | pore in leaves like a mouth |
| | | | stomach | sac connected to mouth |
| S | trop | turning | phototropism | turning toward light |
| | | | thigmotropism | turning toward something that touches |
| P | zoo | animal | zoology | study of animals |

# Safety in the Laboratory

The biology laboratory is a safe place to work if you are aware of important safety rules and if you are careful. You must be responsible for your own safety and for the safety of others. The safety rules given here will protect you and others from harm in the lab. While carrying out procedures in any of the **Labs,** notice the safety symbols and caution statements. The safety symbols are explained in the chart on the next page.

1. Always obtain your teacher's permission to begin a lab.
2. Study the procedure. If you have questions, ask your teacher. Be sure you understand all safety symbols shown.
3. Use the safety equipment provided for you. Goggles and a safety apron should be worn when any lab calls for using chemicals.
4. When you are heating a test tube, always slant it so the mouth points away from you and others.
5. Never eat or drink in the lab. Never inhale chemicals. Do not taste any substance or draw any material into your mouth.
6. If you spill any chemical, wash it off immediately with water. Report the spill immediately to your teacher.
7. Know the location and proper use of the fire extinguisher, safety shower, fire blanket, first aid kit, and fire alarm.
8. Keep all materials away from open flames. Tie back long hair.
9. If a fire should break out in the classroom, or if your clothing should catch fire, smother it with the fire blanket or a coat, or get under a safety shower. **NEVER RUN.**
10. Report any accident or injury, no matter how small, to your teacher.

Follow these procedures as you clean up your work area.

1. Turn off the water and gas. Disconnect electrical devices.
2. Return materials to their places.
3. Dispose of chemicals and other materials as directed by your teacher. Place broken glass and solid substances in the proper containers. Never discard materials in the sink.
4. Clean your work area.
5. Wash your hands thoroughly after working in the laboratory.

**FIRST AID IN THE LABORATORY**

| Injury | Safe response |
|---|---|
| **Burns** | Apply cold water. Call your teacher immediately. |
| **Cuts and bruises** | Stop any bleeding by applying direct pressure. Cover cuts with a clean dressing. Apply cold compresses to bruises. Call your teacher immediately. |
| **Fainting** | Leave the person lying down. Loosen any tight clothing and keep crowds away. Call your teacher immediately. |
| **Foreign matter in eye** | Flush with plenty of water. Use eyewash bottle or fountain. |
| **Poisoning** | Note the suspected poisoning agent and call your teacher immediately. |
| **Any spills on skin** | Flush with large amounts of water or use safety shower. Call your teacher immediately. |

## SAFETY SYMBOLS

**DISPOSAL ALERT**
This symbol appears when care must be taken to dispose of materials properly.

**BIOLOGICAL HAZARD**
This symbol appears when there is danger involving bacteria, fungi, or protists.

**OPEN FLAME ALERT**
This symbol appears when use of an open flame could cause a fire or an explosion.

**THERMAL SAFETY**
This symbol appears as a reminder to use caution when handling hot objects.

**SHARP OBJECT SAFETY**
This symbol appears when a danger of cuts or punctures caused by the use of sharp objects exists.

**FUME SAFETY**
This symbol appears when chemicals or chemical reactions could cause dangerous fumes.

**ELECTRICAL SAFETY**
This symbol appears when care should be taken when using electrical equipment.

**SKIN PROTECTION SAFETY**
This symbol appears when use of caustic chemicals might irritate the skin or when contact with microorganisms might transmit infection.

**ANIMAL SAFETY**
This symbol appears whenever live animals are studied and the safety of the animals and the students must be ensured.

**RADIOACTIVE SAFETY**
This symbol appears when radioactive materials are used.

**CLOTHING PROTECTION SAFETY**
This symbol appears when substances used could stain or burn clothing.

**FIRE SAFETY**
This symbol appears when care should be taken around open flames.

**EXPLOSION SAFETY**
This symbol appears when the misuse of chemicals could cause an explosion.

**EYE SAFETY**
This symbol appears when a danger to the eyes exists. Safety goggles should be worn when this symbol appears.

**POISON SAFETY**
This symbol appears when poisonous substances are used.

**CHEMICAL SAFETY**
This symbol appears when chemicals used can cause burns or are poisonous if absorbed through the skin.

# 1 Practicing Scientific Method

Scientists use orderly methods to learn new information and solve problems. The methods used in this process of understanding the world include observing, forming a hypothesis, separating and controlling variables, interpreting data, and designing an experiment. Practice scientific methods by following the examples explained below.

## Skill: Observing

When you use your senses, you are observing. What you observe can give you a lot of information about events or things. This information helps you make sense of the world around you.

**Learning the Skill**
There are two kinds of observations, qualitative and quantitative.

(a) Qualitative observations describe something without using numbers. You might use words such as sweet or sour, good or poor.
(b) In quantitative observations, numbers are used to describe something. Measurements are often made. Your height, age, and mass are quantitative observations.

**Example**

1. Qualitative observations for an animal may be any of the following: large or small, tall or short, smooth or furry, brown or black, long ears or short ears, naked or hairy tail.
2. Quantitative observations for the same animal may be: mass–459 g; height–27 cm; ear length–14 mm; age–283 days.

## Skill: Forming a Hypothesis

A hypothesis is a possible explanation.

**Learning the Skill**

1. First, you need to recognize the problem or event that needs explaining. You can then begin to form a hypothesis to explain it.
2. You make a hypothesis in the form of a statement that contains the words *if* and *then.*

**Example**
You are raising African violets and want to know if salt water will kill them. A possible hypothesis is: *If* I give salt water to my African violets, *then* they will continue to grow as usual. Another hypothesis could be: *If* I give salt water to my African violets, *then* they will die.

## Skill: Separating and Controlling Variables

To be sure you know which variable caused a particular outcome in your experiment, you must test only one variable at a time.

1. Determine what variables can affect the outcome of an experiment.
2. Make sure that only one variable is allowed to change during the experiment.
3. Set up an experimental group and a control group. The experimental group is the one in which you expect to see a change. The control group is the one in which you don't expect to see a change.

**Example**

1. You wish to know whether or not caffeine increases heart rate. First, you must find out what other things (variables) might increase heart rate. Is activity important? Is amount of caffeine consumed or time of day important?

2. Keep all variables but one constant. For example, you could:
   (a) have all your subjects sit during the entire experiment.
   (b) give all your subjects the same amount of caffeine.
   (c) collect your data the same time each day.

   The variable that remains constant is the giving of caffeine to all the subjects. Changing any other variable can cause misleading results. For example, you allow your subjects to walk around as they wish. You won't be able to tell if any changes in heart rate are due to the caffeine or the different activities.
3. Compare your results with those from a control group. In the control group, all variables are identical to those in the experimental group with one exception. The members of the control group drink plain water.

## Skill: Interpreting Data

The word *interpret* means to explain the meaning of something. When you interpret data, you explain what the data mean.

**Learning the Skill**
Check the data:
   (a) Compare the control group and the experimental group. Compare them for qualitative and quantitative differences.
   (b) Decide if the variable being tested had any effect. If there is no difference between the control group and the experimental group, then the variable being tested probably had no effect.

**Example**
1. You wish to find out if fertilizer affects plant growth. Your hypothesis is: If I add fertilizer to the soil, then my plants will grow larger.
2. The table shows the data you collect.
3. You compare group A (control group) to groups B and C (experimental groups). In groups B and C, the plants are taller.
4. You decide that adding fertilizer to the soil did have an effect on plant growth. It caused the plants in groups B and C to grow taller. Also, the amount of fertilizer had an effect because the plants in group C grew taller than the plants in group B.

**EFFECT OF FERTILIZER ON PLANTS**

| GROUP | TREATMENT | HEIGHT | AFTER 3 WEEKS |
|---|---|---|---|
| A | no fertilizer added to soil | 16.5 cm | 17 cm |
| B | 3 g fertilizer added to soil | 16.5 cm | 31 cm |
| C | 6 g fertilizer added to soil | 16.5 cm | 48 cm |

## Skill: Designing an Experiment

A successful experiment follows the steps of the scientific method.

**Learning the Skill**
1. Make observations. You observe that your neighbor fertilizes her chrysanthemums and that they are taller than yours.
2. Recognize the problem. Does fertilizer cause plants to grow larger?
3. Research the problem to see what kinds of things could cause plants to grow larger.
4. Form a hypothesis: If I fertilize my plants, then they will grow larger.
5. Run the experiment.
   (a) Gather your materials. For this experiment, you would need fertilizer, water, plants, soil, a ruler, and a light source.
   (b) Set up the procedure. The only difference between the control group and the experimental group should be whether or not fertilizer is received.
   (c) Collect the data in a table.
6. Interpret the data. Decide if the fertilizer had an effect on plant growth.
7. Reach a conclusion. Is your original hypothesis supported by your data?
8. If your original hypothesis is not supported, form another hypothesis and test it.

# 2 Reading Science

As you use this textbook, you will find it useful to practice the following reading skills. Processes in nature generally occur in particular sequences. Learning the skill of sequencing will help you remember these processes. The skills of outlining and summarizing main ideas, understanding science words, and interpreting diagrams will help you study the major concepts in this text. As you learn about biology, you will find the skills of inferring and making models will help you apply your knowledge.

## Skill: Sequencing

Sequencing is the arranging of facts or ideas into a series. The order in which the facts or ideas are arranged is important because one fact must lead to the next in the series.

**Learning the Skill**

1. List all items to be sequenced.
2. Decide which item is the first in the series.
3. Continue to list all items in their proper order.

**Examples**

Place the following steps of sharpening a pencil in their proper sequence. Remember that the steps must be in the proper order for the activity to be carried out successfully. Each item must lead to the next.

(a) Turn the handle of the sharpener several times.
(b) Locate the pencil sharpener.
(c) Insert the pencil into the opening of the sharpener.
(d) Check the pencil point to see if it is properly sharpened.
(e) Remove the pencil from the sharpener.

In this example, the correct sequence is (b), (c), (a), (e), (d).

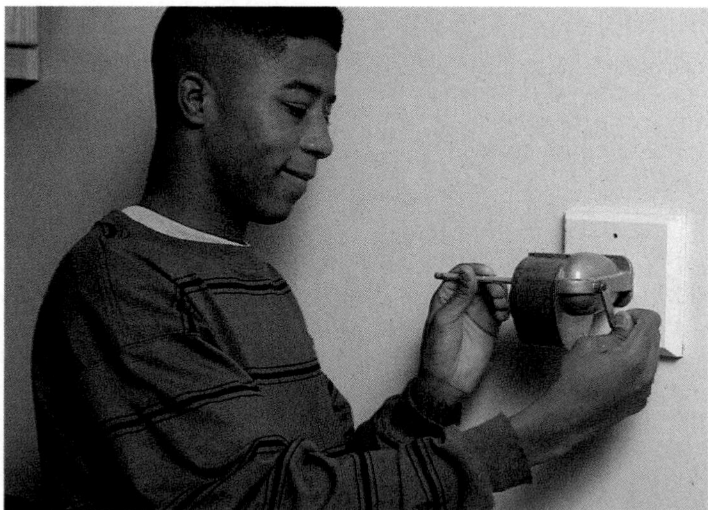

## Skill: Outlining and Summarizing Main Ideas

Being able to outline the material in your textbook will help you learn and understand it and to see how ideas are connected.

**Learning the Skill**

1. Look over the material to be outlined and summarized. Locate the major headings or ideas. These are the numbered sections in each chapter. Use a Roman numeral for each numbered section and write the title of the section next to the Roman numeral.
2. Write a capital letter for each subsection. Subsections are indicated by the headings that follow each numbered section in a chapter.
3. Write the main idea of the subsection next to the capital letter. Leave several lines between all capital letters for supporting ideas.
4. List several supporting ideas below each main idea. These ideas should be in the form of numbered statements.
5. To summarize the main ideas, write the outline in the form of complete sentences.

**Example**

1. Outline numbered Section 22:1.

I. Mitosis (title of numbered section)
   A. Body Growth and Repair (title of subsection)
      1. cell reproduction is called mitosis (supporting idea)
      2. body growth occurs because of mitosis (supporting idea)
      3. body repair occurs because of mitosis (supporting idea)
   B. An Introduction to Mitosis (title of subsection)
      1. one cell forms two during mitosis (supporting idea)
      2. mitosis can be compared to making a photocopy (supporting idea)

2. Summarize the information in the outline. Mitosis is the process of cell reproduction that occurs as our bodies grow and repair themselves. During this process, one cell reproduces to form two. Mitosis can be compared to the making of a photocopy.

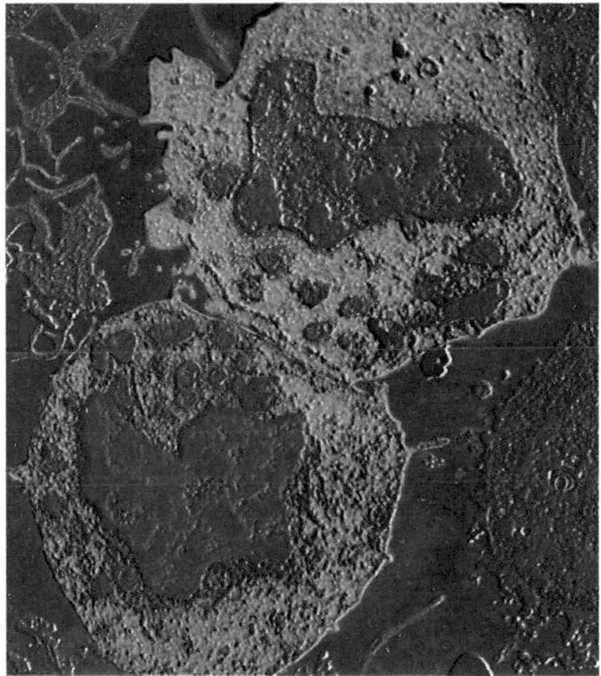



707

## Skill: Understanding Science Words

There are two ways you can understand science words better. One way is to define the word in context. The way the word is used gives you a clue as to its meaning. A second way to understand science words is to look at the parts that make up the word. Each word part can give you a clue as to the meaning of the whole word.

**Learning the Skill: Defining Words in Context**

1. First, read to see if the word is defined directly in the sentence.
2. If the word is not defined directly, read several sentences beyond the one in which the word first appears. These sentences may provide information about the definition of the word.
3. If possible, define the word based on your own past knowledge. You may have learned the word in an earlier chapter, or you may be familiar with it because you hear it day to day.
4. Figure out the meaning of the word by how it is used in the sentence and by the sentences around it.

**Examples**

Find the definitions of the underlined words.

1. *Biology* is the study of life. The word *is* gives you a clue that the word is defined directly in the sentence.
2. A cat is a *mammal.* A mammal is an animal that has body hair, provides its young with milk from mammary glands, and is warm-blooded. The second sentence contains the definition of mammal.
3. All living things can *reproduce. Reproduce* is a familiar word. It is defined in Chapter 2. It is also one you probably hear from day to day.
4. Green plants carry out *photosynthesis.* Thus, they are able to make their own food. These sentences tell you that photosynthesis is the ability of a plant to make food.

**Learning the Skill: Understanding Word Parts**

1. Look at the word to see how many word parts you think it has. The word may have one or more word parts.
2. You may recognize parts of the word from previous lessons. Or, you may recognize parts of the word from other familiar words. Try to define each word part if you can. Then define the whole word.
3. Look for root words and prefixes or suffixes. A root word is the main part of the word. A prefix is a word part added to the front of a root word to change its meaning. A suffix is a word part added to the end of a root word to change its meaning.

**Examples**

1. What does the word *microorganism* mean? The word *microorganism* has two word parts, *micro* and *organism.* You remember the word *microscope* and that *micro* means small. You also remember that the word *organism* means a living thing. Therefore the word *microorganism* means a small living thing.
2. Examples of root words are:
   emia—blood
   vertebrate—animal with a backbone
   bio—life
   zoo—animal
3. The prefixes *an* and *in* mean without. What do the words *anemia* and *invertebrate* mean?
   anemia—having too little blood
   invertebrate—animal without a backbone
4. The suffix *logy* means the study of. What is biology? Zoology?
   biology—the study of life
   zoology—the study of animals

## Skill: Interpreting Diagrams

When you look at a diagram, note first the orientation and symmetry of the organism. Knowing the orientation tells you where the front end is, where the tail end is, where the front side is, and where the back side is. Knowing the symmetry tells you how many sides the organism has. If it is an inside view of an organism, note whether it shows the whole organism or only part of one. Usually, inside views, also called sections, show only parts of organisms. If the diagram is of a process, note the order in which steps of the process occur.

**Learning the Skill: Orientation**

1. First, note whether the diagram, as in Figure 1, shows an internal or external view.
2. Locate the anterior, or head end, posterior, or tail end, dorsal, or back side, ventral, or belly side.

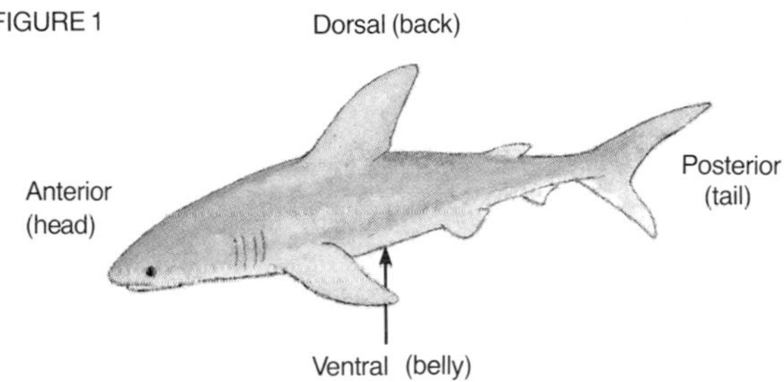


**Learning the Skill: Symmetry**

There are two types of symmetry, radial and bilateral. Bilateral means having two sides alike. Radial means spreading out from the center.

1. Draw an imaginary line through the center of the animal from its anterior end to its posterior end.
2. If the animal forms two mirror images, it probably has bilateral symmetry. You must do step 3 to find out for sure.
3. Draw a second line at right angles to the first. If the animal forms four equal parts, it has radial symmetry. If it doesn't, it has bilateral symmetry.

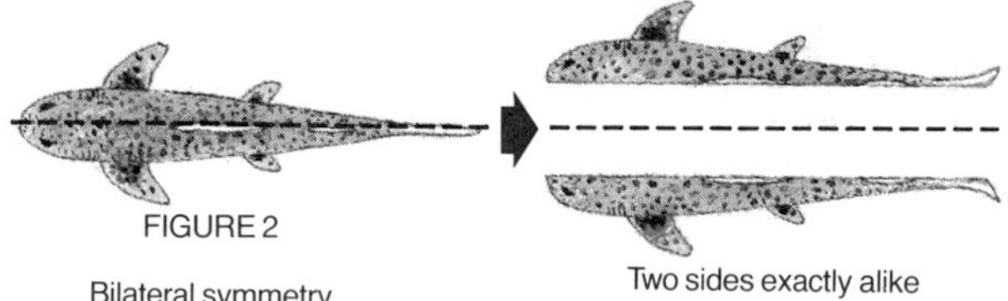


**Example**

1. Study the figure of the shark. A shark has bilateral symmetry. Note that the two sides are mirror images of one another.
2. Study the figure of the bicycle wheel and the sea anemone. A sea anemone has radial symmetry. The parts of the animal spread out from the center, much like the spokes of a wheel.

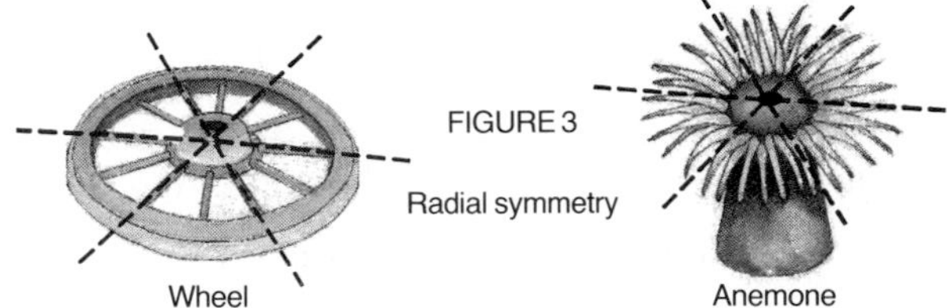


**Learning the Skill: Sections**

There are two kinds of sections, long sections and cross sections.

1. Cross sections are slices made at right angles to the axis. A slice through your waist would give a cross section of you.
2. Long sections are slices that run along the axis. A slice from the top of your head to the floor would give a long section of you.

**Example**

Study the figure of the squash. This figure shows a cross section of the squash. A cross section across the top part shows only the fleshy part of the squash. A cross section across the lower part of the squash shows an interior chamber filled with seeds.

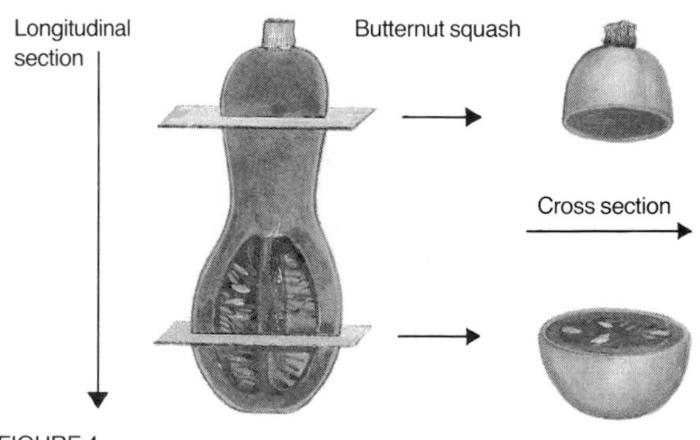

## Skill: Inferring

To infer means to make an assumption based on facts, observations, or experimental data. You make an inference each time you say "I'll bet the school cafeteria is serving pizza for lunch because I can smell it."

### Learning the Skill

1. Define the question. Sometimes you infer something without thinking about the question behind the inference. In the pizza example, what is the question? The question might be "what are we having for lunch today?" Sometimes, the question is asked directly, as in the example that follows.
2. Study the available information. In the pizza example, the available information is the odor of pizza in the air.
3. Look for connecting ideas. Make comparisons. Again in the pizza example, what else could you be smelling? Could it be sub sandwiches? The odor is similar, but your cafeteria has never served subs before.
4. Answer the question. Based on your responses to steps 2 and 3, you infer that pizza is on the menu.
5. Support your answer using the available information.

### Example

1. The following information is about human population growth. One question that might come up regarding this information is "which country might have a food shortage by 2020?"
2. Study the table. What information does it give? It gives information about expected population growth in Japan and Pakistan.
3. Compare the expected increase in population for Japan with the expected increase in population for Pakistan.
4. Based on the data, you could infer that Pakistan will have a food storage by 2020.
5. Supporting evidence is that the population of Pakistan is increasing at a much faster rate than the population of Japan. The larger a population becomes, the more food it requires. What additional information could change your inference? If Japan had to import most of its food, and Pakistan was the largest producer of grain in the world, you might infer that Japan would have a food shortage, or that neither country would have a food shortage.

**POPULATION GROWTH**

| COUNTRY | 1996 | 2020 (ESTIMATED) |
|---|---|---|
| JAPAN | 125 million | 124 million |
| PAKISTAN | 129 million | 199 million |

## Skill: Formulating Models

In a model, ideas in biology are represented by familiar objects. Models help to simplify processes or structures that are often difficult to understand.

**Learning the Skill**

1. Recognize the process or structure that is to be modeled.
2. Research the process or structure before you start your model. Find out how a structure such as an artery is put together. Find out what takes place during a process, such as breathing in and out.
3. Think of a simple way of showing this same process or structure using materials that are readily available.
4. Construct your model. Note if it operates correctly and shows the idea you wish it to show.

**Example**

1. Formulate a model that shows how the sizes of arteries, veins, and capillaries differ.
2. Read what you can about arteries, veins, and capillaries. Arteries are usually round and are the thickest of the three blood vessel types. Veins are flat and thinner than arteries. Capillaries are very small in size. All three vessels are hollow tubes.
3. Materials that you could use include straws to represent capillaries, mostaccioli noodles to represent veins, and manicotti noodles to represent arteries.
4. To construct the model, you might want to cook the noodles first. Cooking will make them softer, and make their texture more like that of arteries and veins. Cut the noodles and straws into short sections and glue them to thick paper so that the openings are facing outward. Label each structure appropriately.

# 3 Using a Microscope

There are many opportunities in this biology text for you to make your own observations and experimentations. In many labs, you will be able to practice your skills of using a compound microscope, making a wet mount, calculating magnification, and making scale drawings.

## Skill: Using a Compound Microscope

The compound microscope allows you to magnify objects 100, 400, and 1000 X (times) their natural size. The word *compound* refers to the fact that a compound microscope has several lenses for magnifying objects.

### Steps Needed to Acquire Skill

1. Review the parts of the compound microscope shown in Figure 1.
2. Place the glass slide and object to be viewed onto the stage. Hold the slide in place with the stage clips.
3. Move the slide so that the object to be viewed is directly over the stage opening.
4. Turn the low-power objective into place until you hear a click. Look to the side and turn the coarse adjustment until the low-power objective is almost touching the slide.
5. Turn on the microscope (if electric) or adjust the mirror toward a light source. Never use direct sunlight.
6. Look through the eyepiece and adjust the diaphragm until you see a bright light. The mirror may also have to be readjusted.
7. Use the coarse adjustment to raise the body tube until you can see the object.
8. Use the fine wheel adjustment to bring the object into sharp focus.
9. To see the object under high power, rotate the objectives until you hear another click. Use only the fine adjustment to sharpen the focus.

### Troubleshooting

**Problem:** Object can't be seen clearly or found under low or high power.

**Solution:** Make sure that the eyepiece, objectives, slide, and coverslip are clean. Check to be sure that the water is not covering the coverslip and if it is, make a new wet mount. If the objective or eyepiece lens is dirty, clean it with lens paper. Check the coarse adjustment. Check the diaphragm opening. If there is too much light, transparent objects, such as amoebas, will be washed out.

**Problem:** Object can't be found under high power.

**Solution:** Check to be sure the object is centered over the stage opening. You may need to return to low power first.

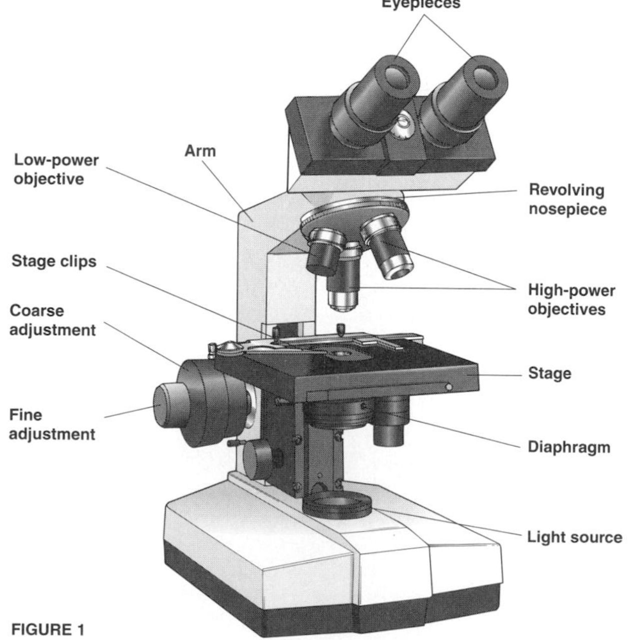


FIGURE 1

## Skill: Making a Wet Mount

All the slides you prepare for observing under the microscope are called wet mounts. They are called wet mounts because the object to be viewed is prepared or mounted in water.

**Learning the Skill**

1. Add a drop or two of water to the center of a clean microscope slide.
2. Place the object to be viewed in the drop of water as shown in Figure 1.
3. Pick up a coverslip by its edges. Do not touch the surface of the coverslip. Stand the coverslip on its edge next to the drop of water.
4. Slowly lower the coverslip over the drop of water and the object to be viewed as shown in Figure 2.
5. Make sure that the object is totally covered with water. If it is not, remove the coverslip, add more water, and replace the coverslip.

**Troubleshooting**

1. Figures 3 and 4 show correctly prepared wet mounts.
2. The directions that follow tell you what to do if problems occur when you are making wet mounts.
   (a) Figure 5 shows what happens when there is not enough water on the wet mount or the coverslip is lowered too quickly. The water doesn't cover the entire object, or air bubbles may form.
   (b) Figure 6 shows what happens when too much water is used or the coverslip is dirty. If too much water is on the slide, place the tip of a paper towel at the edge of the coverslip to absorb some of the excess water. If there is dirt on the coverslip, replace it. If these steps don't work, clean the slide, dry the slide and coverslip, and make a new wet mount.
   (c) Figure 7 shows what happens when the object being viewed is too thick. Prepare a thinner object for viewing and make a new wet mount.

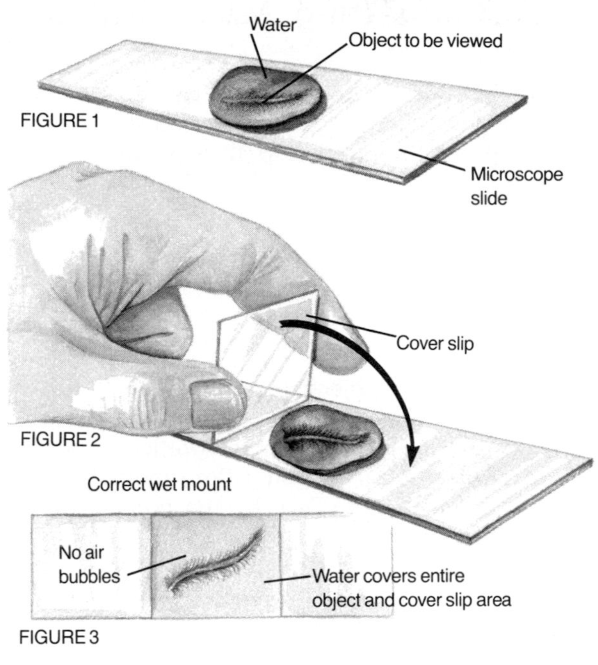

FIGURE 1

FIGURE 2

FIGURE 3

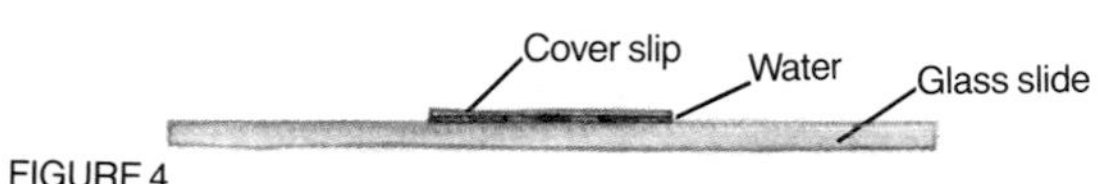

FIGURE 4

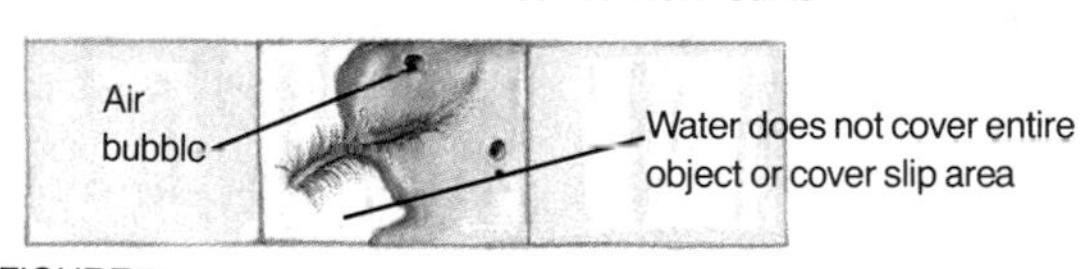

FIGURE 5

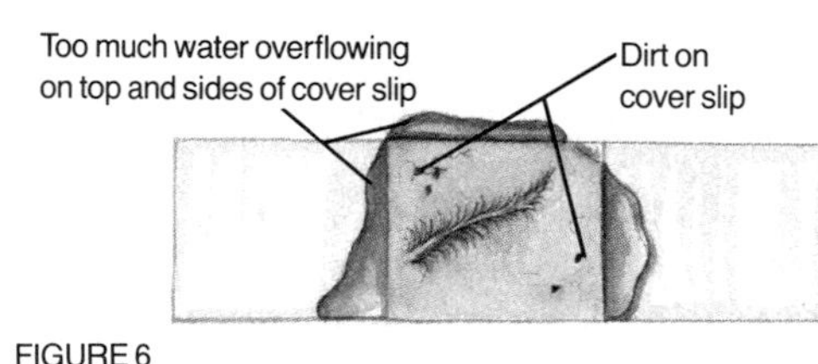

FIGURE 6

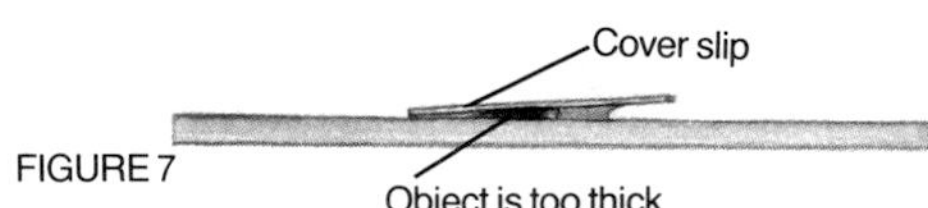

FIGURE 7

## Skill: Calculating Magnification

Objects viewed under the microscope appear larger than normal because they are magnified. Total magnification describes how much larger an object appears when viewed through the microscope.

**Learning the Skill**

1. Look for a number marked with an X on the following:
   (a) eyepiece
   (b) low-power objective
   (c) high-power objective
   The X stands for how many times the lens of the microscope part magnifies an object.
2. To calculate *total* magnification, multiply the number on the eyepiece by the number on the objective.

**Example**

If the eyepiece magnification is 7×, the low-power objective magnification is 10×, and the high-power objective magnification is 40×:
   (a) then total magnification under low power is 7× for the eyepiece × 10× for the low-power objective = 70× (7 × 10 = 70).
   (b) then total magnification under high power is 7× for the eyepiece × 40× for the high-power objective = 280× (7 × 40 = 280).

## Skill: Making Scale Drawings

When you draw objects seen through the microscope, the size that you make your drawing is important. Your drawing should be in proportion to the size the object appears to be when viewed through the microscope. This is called drawing to scale. Drawing to scale allows you to compare the sizes of different objects. It also allows you to form an idea of the actual size of the object being viewed, Figure 1.

**Learning the Skill**

1. Draw a circle on your paper. The circle may be any size.
2. Imagine the circle divided into four equal sections, as in Figure 2.

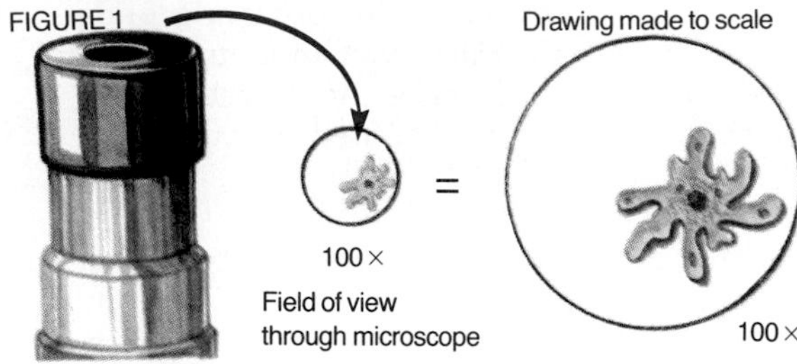


3. Locate an object under low or high power of the microscope. Imagine the field of view also divided into four equal sections.
4. Note how much of the field of view is taken up by the object. Also note what part of the field of view the object is in.
5. Draw the object in the circle. Position the object in about the same part of the circle as it appears in the field of view. Also, draw the object so that it takes up about the same amount of space within the circle as it actually takes up in the field of view, Figure 2.

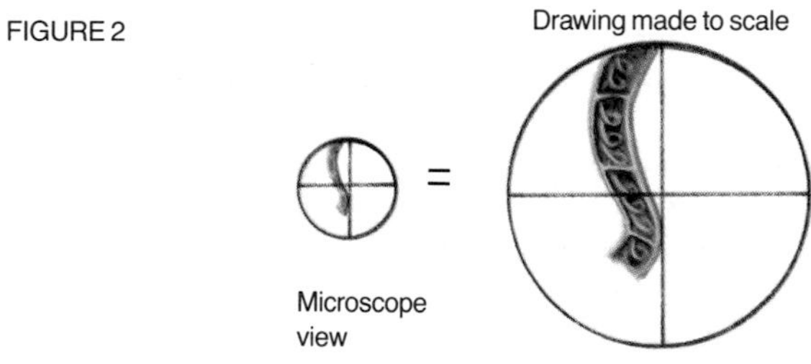


**Troubleshooting**

Figure 3 shows objects that haven't been drawn to scale correctly. One object has been drawn too small. One object has been drawn too large. The object in Figure 2 is drawn to the correct size and in the correct position.

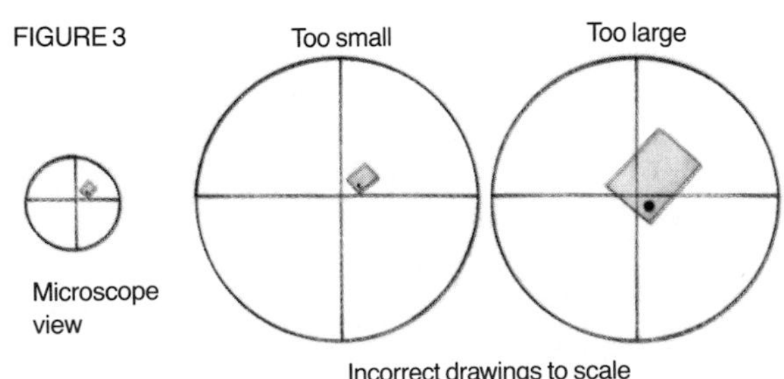

# 4 Organizing Information

As a scientist, you will be expected to collect data and draw conclusions. The skills of classifying, making and using tables, making line graphs, and making bar graphs will help you organize and interpret the information that you gather.

## Skill: Classifying

Everyone, including scientists, groups or classifies to show similarities and differences among things and to put things in order.

**Learning the Skill**

1. Study the things to be classified.
2. Look at their traits and note traits that are similar and those that are different.
3. Pick out one major trait that can be used to separate the objects into at least two (or even three) separate groups.
4. Separate the things to be classified into these two groups. Then, taking each group at a time, look for other traits that can be used to separate each of the two groups into subgroups.
5. Continue to separate each subgroup into smaller and smaller groups until each group contains only one object.

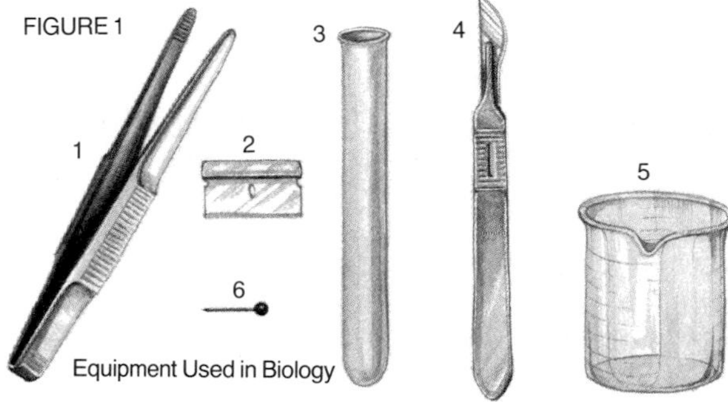


FIGURE 1 Equipment Used in Biology

**Example**

1. Classify equipment used in biology labs.
2. First, separate the equipment into two groups, possibly glass and metal.
3. Separate the glass items into subgroups. Separate metal items into subgroups. The diagram shows some possible subgroups for classifying the equipment.

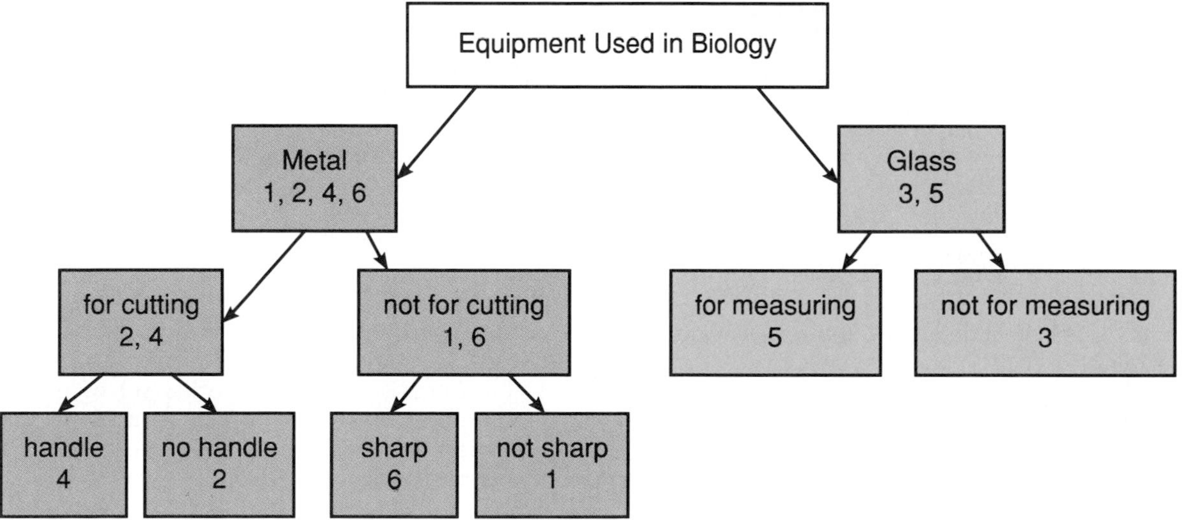

## Skill: Making and Using Tables

Tables organize information in a way that makes it easy to read and interpret. Tables may be used to show qualitative information or quantitative information. (See page 704 of the Skill Handbook.) Tables have two uses. One, they are used to record data. Two, they provide information.

**Learning the Skill**
Tables have the following characteristics:
1. a title that tells what kind of information appears in the table
2. columns that run up and down
3. rows that run left to right
4. a heading for each column that tells what type of information is in that column

**Example**
An experiment was run to see if yeast cells give off different amounts of carbon dioxide gas when provided with different kinds of food. Starch, a food source, was put into a test tube containing yeast. Sugar, a second food source, was put into another test tube containing yeast. A third test tube that contained yeast, the control, had no food added. The test tubes were labeled A (no food), B (sugar added), and C (starch added). For the next 20 minutes, the experimenter counted the number of bubbles given off in each test tube. The number of bubbles was used as a measurement of the amount of carbon dioxide gas being given off by the yeast. The results were that the yeast in tube A gave off 3 bubbles in 20 minutes. The yeast in tube B gave off 104 bubbles. The yeast in tube C gave off 58 bubbles.

The results of this experiment could be summarized as follows:

Tube A, no food added, 3 bubbles given off in 20 minutes.

Tube B, sugar added, 104 bubbles given off in 20 minutes.

Tube C, starch added, 58 bubbles given off in 20 minutes.

Note that in this experiment, quantitative observations were made.

Another way to summarize the results is in a table. Study the table. Note how the information gathered in the yeast experiment is organized.

**BUBBLES GIVEN OFF BY YEAST USING DIFFERENT FOOD TYPES**

| TUBE | FOOD ADDED | BUBBLES GIVEN OFF IN 20 MINUTES |
|---|---|---|
| A | none | 3 |
| B | sugar | 104 |
| C | starch | 58 |

## Skill: Making Line Graphs

A line graph is used to show how quantities of things change. It shows these changes in picture form. Very often, the data that are shown in a line graph are collected in a table first. Then, the data in the table are graphed.

**Learning the Skill**
1. Look at Figure 1 as you follow these directions. On a piece of graph paper draw two lines, one horizontal (A) and one vertical (B). The lines must meet in the lower left corner.
2. On the horizontal line, make regularly spaced marks (C). Label the marks. For example, if data were gathered every hour, then the first mark on the horizontal line would be labeled 1 hour, the second mark 2 hours, and so on.
3. On the vertical line, make regularly spaced marks. Label the marks. These labels must also match the data to be graphed. If the numbers to be graphed are the numbers of yeast cells, then the first mark could be labeled 1 cell, the second mark 2 cells, and so on. The way you label your graph depends on the data. If the quantities you want to graph are large, you will want to start with larger numbers on your graph.
4. Plot the data. Look at the data in the following table. One set of numbers represents the information that goes on the horizontal line (time in hours). The other set of numbers represents the information that goes on the vertical line (number of cells). First, plot the number of hours (zero). Place your pencil on the horizontal line at zero hours. Next plot

| REPRODUCTION IN YEAST | |
|---|---|
| NUMBER OF YEAST CELLS | TIME IN HOURS |
| 5 | 0 |
| 10 | 1 |
| 18 | 2 |
| 45 | 3 |
| 30 | 4 |
| 8 | 5 |

the number of cells. Move your pencil up to the mark labeled 5. Make a dot (D). Continue to plot the data until all the rows of numbers have been plotted.

5. If you have trouble determining where to place your dot, draw dotted lines from the horizontal and vertical lines. Where the dotted lines intersect is where you place your dot (E).
6. Connect each dot with a smooth line (F).
7. If you are graphing two sets of data, be sure to make the two lines different colors so you can tell them apart. Or, you can use a solid line for one and a dashed line for the other. Include a key that tells which color or line represents which set of data (G).

**Example**

Let's see how the data in the table would look in a line graph.

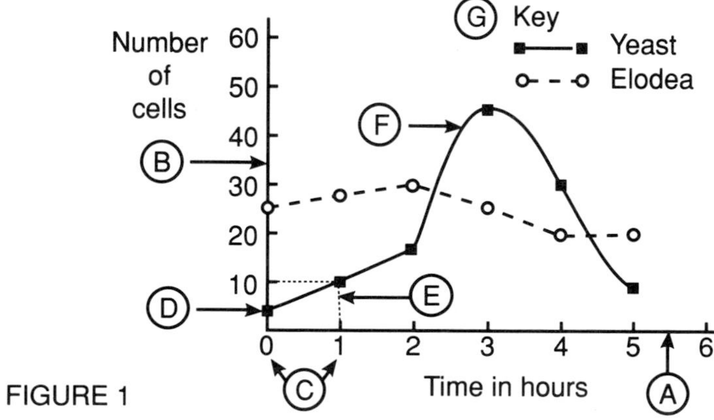


FIGURE 1

## Skill: Making Bar Graphs

A bar graph is used to compare quantities of different kinds of things. It shows these comparisons in picture form.

**Learning the Skill**

1. Repeat steps 1-3 under Making Line Graphs. Label the horizontal and vertical lines with the appropriate labels.
2. Plot the data from the table below.

| BREATH HOLDING ABILITY | |
|---|---|
| AMOUNT OF TIME ABLE TO HOLD BREATH (MINUTES) | ANIMAL |
| 20 | Gray Seal |
| 10 | Walrus |
| 50 | Blue Whale |
| 15 | Beaver |

Place your pencil on the horizontal line where it is marked gray seal. Plot the number of minutes a gray seal can hold its breath by moving your pencil up to the appropriate mark on the vertical line. Place a small mark there. Then, draw a vertical bar up to the mark. Continue to plot the data until the values for all animals have been plotted.

**Example**

1. Let's see how the data in the table would look in a graph.

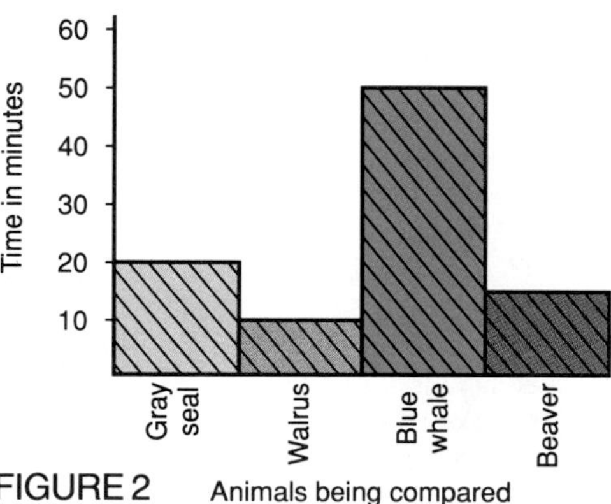


FIGURE 2

# 5 Using Numbers

Two important methods of science are the comparing of information and the repeating of experiments. All scientists use the skills of measuring volume, length, and mass in SI, and calculating in order to confirm hypotheses, results, and conclusions.

## Skill: Measuring in SI—Volume

Volume measurements are made in milliliters (mL) or liters (L). 1000 milliliters = 1 liter. The equipment used for measuring volume is a graduated cylinder.

**Learning the Skill**

1. Locate the units on the graduated cylinder in Figure 1.
2. To read the volume of liquid in a graduated cylinder, do the following:
   (a) Place the cylinder on a flat surface.
   (b) Make sure you read the volume of liquid at eye level.
   (c) The surface of the liquid will form a curve. Find the lowest point of the curve and read the volume.

**Troubleshooting**

1. Figure 1 shows a correct volume reading.
2. The reading in Figure 2 is incorrect because:
   (a) The cylinder is not on a flat surface.
   (b) The top of the liquid is not at eye level.
   (c) The lowest point of the curve is not being used to read the volume.

## Skill: Measuring in SI—Length

Measurements of length, distance, and height are made in millimeters (mm), centimeters (cm), and meters (m). The equipment used for measuring length is a metric ruler.

**Learning the Skill**

1. First decide which units you want to measure in. The units you choose will depend on the size of the object or distance being measured.
2. Note the divisions on the ruler. Figure 3 shows these divisions. One meter is divided into 100 centimeters. Each centimeter is divided into 10 millimeters.
3. Place one end of the object you are measuring next to the zero mark on the ruler. Note where the other end stops and read the number of units shown there.

**Example**

1. Measure the length of a leaf.
2. Place one end of the leaf at the zero mark as shown in Figure 4.
3. The leaf in the figure is 35 mm.

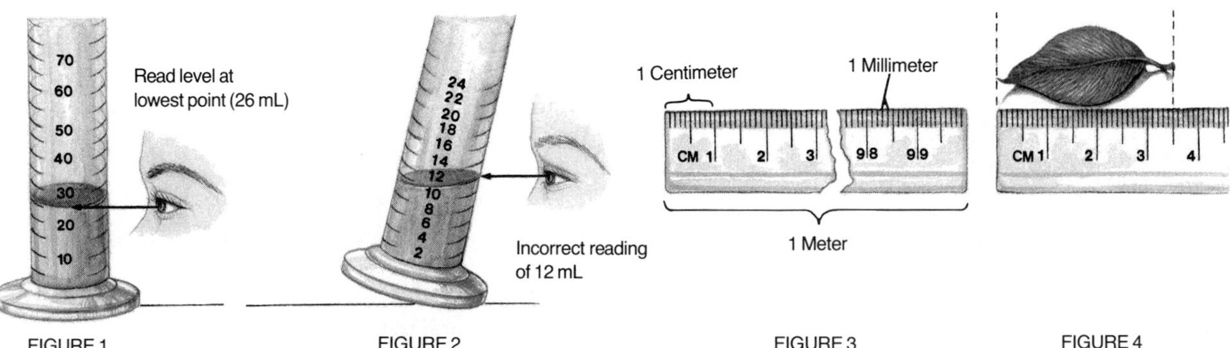


FIGURE 1 FIGURE 2 FIGURE 3 FIGURE 4

## Skill: Measuring in SI—Mass

Mass measurements are made in grams (g), milligrams (mg), or kilograms (kg). The equipment used is a balance.

**Learning the Skill**

1. Place the object to be massed onto the pan. Note: liquids must be in a container; powders or solids must be placed on paper. The mass of the container or paper must later be subtracted from the total mass.
2. Slide all the riders to the left and release the locking mechanism.
3. Figure 1 shows that the back beam measures mass in 10 gram units. The middle beam measures mass in 1 gram units. The front beam measures mass in 0.1 and 0.01 gram units.
4. Starting with the rider on the back beam, move the rider to the right until the pointer drops below the zero point. Then move it back one notch. Repeat this process with the rider on the middle beam. Move the front rider until the pointer lines up with the zero point (center line).
5. The mass of the object will be the total of all the numbers indicated by the riders of the three beams.

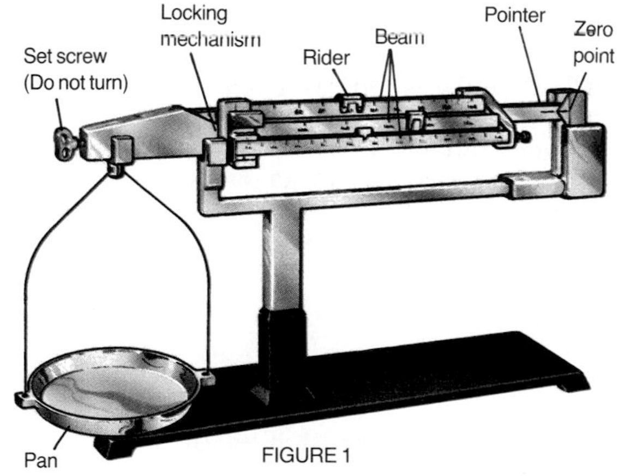

FIGURE 1

## Skill: Calculating

Calculations may involve averaging numbers, comparing sets of numbers, or calculating area or volume.

**Learning the Skill: Averaging**

1. Add together the numbers to be averaged.
2. Divide the total by the number of figures used. The resulting number is the average.

**Learning the Skill: Comparing Numbers**

1. To find out how many times larger something is, *divide the smaller number into the larger.*
2. To find out how many times smaller something is than something else, *divide the larger number into the smaller.*

**Learning the Skill: Calculating Area**

1. Multiply the length times the width.
2. The result is the area in *square* units, for example, square meters or square centimeters.

**Example**

1. Calculate the area of the rectangle shown below.

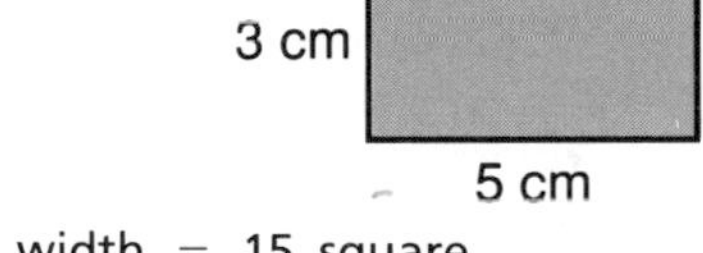


2. Length = 3 cm
   Width = 5 cm
   Area = length × width = 15 square centimeters
   Square units may be written $cm^2$, $m^2$, or $mm^2$.

**Learning the Skill: Calculating Volume**

1. Multiply length times width times height.
2. The result is the volume in *cubic* units, such as cubic meters or cubic centimeters.

**Example**

1. Calculate the volume of the cube shown below.

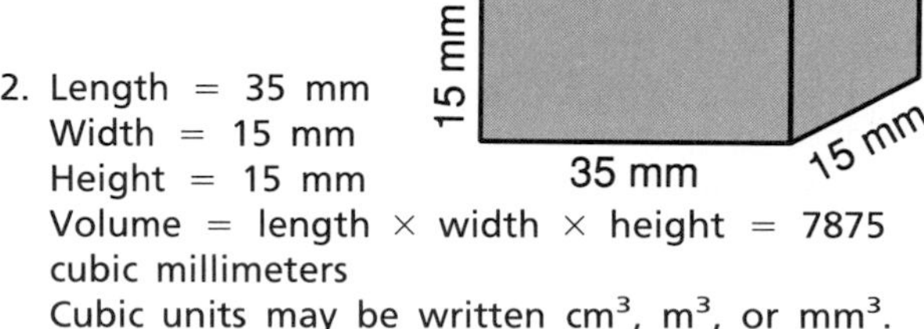


2. Length = 35 mm
   Width = 15 mm
   Height = 15 mm
   Volume = length × width × height = 7875 cubic millimeters
   Cubic units may be written $cm^3$, $m^3$, or $mm^3$.

# GLOSSARY

## PRONUNCIATION KEY

a . . . b**a**ck (bak)
ay . . . d**ay** (day)
ah . . . f**a**ther (fahth ur)
ow . . . fl**ow**er (flow ur)
ar . . . **car** (car)
e . . . l**e**ss (les)
ee . . . l**ea**f (leef)
ih . . . tr**i**p (trihp)
i(i+con+e) . . . **i**dea, l**i**fe (i dee uh, life)

oh . . . g**o** (goh)
aw . . . s**o**ft (sawft)
or . . . **or**bit (or but)
oy . . . c**oi**n (coyn)
oo . . . f**oo**t (foot)
ew . . . f**oo**d (fewd)
yoo . . . p**ure** (pyoor)
yew . . . f**ew** (fyew)
uh . . . comm**a** (cahm uh)
u(+con) . . . flow**er** (flow ur)

sh . . . **sh**elf (shelf)
ch . . . na**t**ure (nay chur)
g . . . **g**ift (gihft)
j . . . **g**em (jem)
ing . . . s**ing** (sing)
zh . . . vi**si**on (vihzh un)
k . . . ca**k**e (kayk)
s . . . **s**eed, **c**ent (seed, sent)
z . . . **z**one, rai**s**e (zohn, rayz)

## A

**Acids:** liquids that have pH values lower than 7 (p. 684)

**Acid rain:** rain that has a pH between 1 and 5.5 (p. 684)

**Acquired Immune Deficiency Snydrome,** or **AIDS:** a disease of the immune system (p. 259)

**Adaptation:** a trait that makes a living thing better able to survive (p. 29)

**Algae** (AL jee): plantlike protists (p. 97)

**Alveoli** (al VEE uh li): the tiny air sacs of the lungs (p. 268)

**Amniocentesis** (am nee oh sen TEE sus): a way of looking at the chromosomes of a fetus (p. 567)

**Amniotic** (am nee AHT ik) **sac:** a tissue filled with liquid that protects the embryo (p. 525)

**Amphibian** (am FIHB ee un): an animal that lives part of its life in water and another part of its life on land (p. 169)

**Anemia:** a condition in which there are too few red blood cells in the blood (p. 248)

**Animals:** organisms that have many cells, can't make their own food, and can move (p. 62)

**Annual** (AN yul): a plant that completes its life cycle within one year (p. 447)

**Annual ring:** each ring of xylem in a woody stem (p. 423)

**Antacid** (ant AS ud): a drug that changes acid into water and a salt (p. 382)

**Antennae** (an TEN ee): appendages of the head that are used for sensing smell and touch (p. 159)

**Antibiotics** (an ti bi AHT iks): chemical substances that kill or slow the growth of bacteria (p. 84)

**Antibodies** (ANT ih bohd eez): chemicals that help destroy bacteria or viruses (p. 257)

**Antigens** (ANT ih junz): foreign substances, usually proteins, that invade the body and cause diseases (p. 257)

**Antihistamine** (ANT i HIHS tuh meen): a drug that reduces swelling of tissues by stopping the leaking of blood plasma from capillaries (p. 380)

**Anus** (AY nus): an opening through which undigested food leaves the body (p. 145)

**Aorta** (ay ORT uh): the largest artery in the body (p. 230)

**Appendage** (uh PEN dihj): a structure that grows out of an animal's body (p. 158)

**Appendix** (uh PEN dihks): a small, fingerlike part of the digestive system found where the small and large intestines meet (p. 212)

**Artery** (ART uh ree): a blood vessel that carries blood away from the heart (p. 226)

**Arthritis** (ar THRIT us): a disease of bone joints (p. 299)

**Asexual** (ay SEK shul) **reproduction:** the reproducing of a living thing from only one parent (p. 80)

**Atria** (AY tree uh): the small top chambers of the heart (p. 225)

**Auditory** (AHD uh tor ee) **nerve:** a nerve that carries messages of sound to the brain (p. 341)

**Autosomes** (AH toh sohmz): chromosomes that do not determine the sex of a person (p. 568)

**Axon** (AK sahn): the part of a neuron that sends messages to surrounding neurons or body organs (p. 313)

## B

**Bacteria:** very small, one-celled monerans (p. 79)

**Balanced diet:** a diet with the right amount of each nutrient (p. 192)

**Ball-and-socket joint:** a joint that allows you to twist and turn the bones in a circle where they meet (p. 291)

**Bases:** liquids that have pH values greater than 7 (p. 684)

**Behavior:** the way an animal acts (p. 352)

**Bicuspid** (bi KUS pud) **valve:** the valve between the left atrium and the left ventricle (p. 227)

**Biennial** (bi EN ee ul): a plant that produces seeds at the end of the second year of growth and then dies (p. 447)

**Bile:** a green liquid that breaks large fat droplets into small fat droplets (p. 210)

**Biodegradable** (bi oh duh GRAYD uh bul): something that can be broken down by microbes into harmless chemicals and used by other living things (p. 681)

**Biology:** the study of living and once-living things (p. 5)

**Biome** (BI ohm): a land area with a distinct climate and with specific types of plants and animals (p. 664)

**Biotechnology** (bi oh tek NAHL uh jee): the use of living things to solve practical problems (p. 85)

**Blade:** the thin, flat part of a leaf (p. 398)

**Blood pressure:** the force created when blood pushes against the walls of vessels (p. 232)

**Blue-green bacteria:** small, one-celled monerans that contain chlorophyll and can make their own food (p. 87)

**Body cells:** cells that make up most of the body, such as the skin, blood, bones, and stomach (p. 464)

**Bone marrow:** the soft center part of the bone (p. 248)

**Bony fish:** fish that have skeletons made mostly of bone (p. 168)

**Brain:** the organ that sends and receives messages to and from all body parts (p. 314)

**Breeding:** the bringing together of two living things to produce offspring (p. 599)

**Bronchi** (BRAUN ki): two short tubes that carry air from the trachea to the left and right lung (p. 267)

**Budding:** reproduction in which a small part of the parent grows into a new organism (p. 106)

**Bulb:** a short, underground stem surrounded by fleshy leaves that contain stored food (p. 486)

## C

**Caffeine:** a stimulant found in coffee and tea, cocoa, chocolate, and some soft drinks (p. 385)

**Calorie:** a measure of the energy in food (p. 193)

**Cambium:** a thin layer of cells that divide to form new phloem on the outside and new xylem on the inside (p. 421)

**Cancer:** a disease in which body cells reproduce at an abnormally fast rate (p. 478)

**Capillary:** the smallest kind of blood vessel (p. 234)

**Capsule:** a sticky outer layer produced by bacteria (p. 80)

**Carbohydrates** (kar boh HI drayts): nutrients that supply you with energy (p. 185)

**Carbon monoxide** (CO): an odorless, colorless gas sometimes found in the air (p. 272)

**Cardiac** (KAR dee ac) **muscle:** the muscle that makes up the heart (p. 294)

**Cartilage** (KART ul ihj): a tough, flexible tissue that supports and shapes the bodies of some fish and some animal parts (p. 167)

**Cartilage fish:** jawed fish in which the entire skeleton is made of cartilage (p. 167)

**Cell:** the basic unit of all living things (p. 27)

**Cell membrane:** the cell part that gives the cell shape and holds the cytoplasm (p. 32)

**Cell wall:** the thick, outer covering outside the cell membrane (p. 35)

**Cellular respiration:** the process by which food is broken down and energy is released (p. 27)

**Celsius** (SEL see us): a scale with which scientists measure temperature (p. 14)

**Centrioles** (SEN tree ohlz): cell parts that help with cell reproduction (p. 35)

**Cerebellum** (ser uh BEL uhm): the brain part that helps make your movements smooth and graceful, rather than robotlike (p. 317)

**Cerebrum** (suh REE brum): the brain part that controls thought, reason, and the senses (p. 316)

**Chemical change:** turns food into a form that cells can use (p. 206)

**Chlorophyll** (KLOHR uh fihl): a chemical that gives plants their green color and traps light energy (p. 114)

**Chloroplasts** (KLOR uh plasts): cell parts that contain the green pigment, chlorophyll (p. 35)

**Cholesterol** (kuh LES tuh ral): a fatlike chemical found in certain foods (p. 237)

**Chordate** (KOR dayt): an animal that, at some time in its life, has a tough, flexible rod along its back (p. 165)

**Chromosomes** (KROH muh sohmz): cell parts with information that determines what traits a living thing will have (p. 32)

**Cilia:** short, hairlike parts on the surface of a cell (p. 96)

**Circulatory system:** the body system made up of your blood, blood vessels, and heart (p. 222)

**Class:** the largest group within a phylum (p. 54)

**Classify:** to group things together based on similarities (p. 48)

**Cleavage** (KLEE vihj): the series of changes that take place to turn one fertilized egg into a hollow ball of many cells (p. 525)

**Climate:** the average light, temperature, and precipitation in an area taken over many years (p. 664)

**Climax community:** the last or final stage of succession in a community (p. 661)

**Clone** (KLOHN): a living thing that comes from one parent and is identical to the parent (p. 496)

**Club fungi:** fungi with club-shaped parts that produce spores (p. 105)

**Cocaine:** a controlled drug used for its stimulant effects (p. 384)

**Cochlea** (KAHK lee uh): a liquid-filled, coiled chamber in the ear that contains nerve cells (p. 341)

**Cold-blooded:** having a body temperature that changes with the temperature of the surroundings (p. 166)

**Colony:** a group of similar cells growing next to each other that do not depend on each other

**Color blindness:** a problem in which red and green are not seen as they should be (p. 576)

**Commensalism** (kuh MEN suh lihz um): a relationship in which two organisms live together, and one benefits while the other gets no benefit and is not harmed (p. 645)

**Communicable** (kuh MYEW nih kuh bul) **diseases:** those that can be passed from one organism to another (p. 83)

**Community:** all of the living things in an area that depend upon each other (p. 638)

**Competition:** the struggle among living things to get their needs for life (p. 616)

**Complete metamorphosis:** a series of changes in an insect in which the young do not look like the adult (p. 536)

**Compound eyes:** eyes with many lenses (p. 159)

**Cones:** nerve cells that can detect color (p. 336)

**Conifer** (KAHN uh fur): a plant that produces seeds in cones (p. 124)

**Consumers:** living things that eat, or consume, other living things (p. 27)

**Control:** a standard for comparing results (p. 18)

**Cork:** the outer layer of a woody stem (p. 421)

**Cornea** (KOR nee uh): the clear outer covering at the front of the eye (p. 335)

**Coronary** (KOR uh ner ee) **vessels:** the blood vessels that carry blood to and from the heart itself (p. 237)

**Cortex:** a food storage tissue in plants (p. 420)

**Courting behaviors:** behaviors used by males and females to attract one another for mating (p. 359)

**Cross pollination:** when pollen from the stamen of one flower is carried to the pistil of a flower on a different plant (p. 490)

**Cutting:** a small section of plant stem that has been removed from a parent plant and planted to form a new plant (p. 487)

**Cyst** (SIHST): a young worm with a protective covering (p. 143)

**Cytoplasm** (SITE uh plaz um): the clear, jellylike material between the cell membrane and the nucleus that makes up most of the cell (p. 32)

# D

**Data:** the recorded facts or measurements from an experiment (p. 18)

**Day-neutral plants:** plants in which flowering doesn't depend on day length (p. 441)

**Decomposers:** living things that get their food from breaking down dead matter into simpler chemicals (p. 81)

**Dendrites:** parts of the neuron that receive messages from nearby neurons (p. 313)

**Dependence:** needing a certain drug in order to carry out normal daily activities (p. 383)

**Depressant** (dih PRES unt): a drug that slows down messages in the nervous system (p. 377)

**Dermis:** a thick layer of cells that form in the inner part of the skin (p. 343)

**Development:** all the changes that occur as a living thing grows (p. 26)

**Diabetes mellitus** (di uh BEET us · MEL uht us): a disease that results when the pancreas doesn't make enough insulin (p. 324)

**Diaphragm:** a sheetlike muscle that separates the inside of your chest from the intestines and other organs of your abdomen (p. 270)

**Diffusion** (dif YEW zhun): the movement of a substance from where there is a large amount of it to where there is a small amount of it (p. 38)

**Digestion:** the changing of food into a usable form (p. 205)

**Digestive system:** a group of organs that take in food and change it into a form the body can use (p. 204)

**DNA:** a molecule that makes up genes and determines the traits of all living things (p. 586)

**Dominant** (DAHM uh nunt) **genes:** those that keep other genes from showing their traits (p. 549)

**Dosage:** how much and how often to take a drug (p. 372)

**Drug:** a chemical that changes the way a living thing functions when it is taken into the body (p. 370)

**Drug abuse:** the incorrect or improper use of a drug (p. 383)

**Dyslexia** (dis LEHK see uh): a genetic disorder in which the person sees and writes some letters or words backward (p. 579)

## E

**Eardrum:** the membrane that vibrates at the end of the ear canal (p. 340)

**Ecology** (ih KAHL uh jee): the study of how living things interact with each other and with their environment (p. 654)

**Ecosystem:** a community interacting with the environment (p. 654)

**Egg:** a female reproductive cell (p. 118)

**Embryo** (EM bree oh): an organism in its earliest stages of growth (p. 124)

**Emigration** (em uh GRAY shun): the movement of animals out of an area (p. 634)

**Emphysema** (em fuh SEE muh): a lung disease that results in the breakdown of alveoli (p. 273)

**Endangered:** a species of living thing that is in danger of becoming extinct (p. 675)

**Endocrine** (EN duh krin) **system:** the body system made up of small glands that make chemicals that carry messages through the body (p. 320)

**Endodermis:** a ring of waxy cells that surrounds the xylem in roots (p. 429)

**Endoskeleton** (EN doh skel uht uhn): a skeleton on the inside of the body (p. 165)

**Endospore** (EN duh spor): a thick-walled structure that forms inside a bacterial cell (p. 81)

**Energy pyramid:** a diagram that shows energy loss in the food chain (p. 643)

**Enzymes** (EN zimes): chemicals that speed up the rate of chemical change (p. 206)

**Epidermis** (ep uh DUR mus): the outer layer of cells of a plant (p. 401)

**Epiglottis** (ep uh GLAHT uhs): a small flap that closes over the windpipe when you swallow (p. 267)

**Erosion:** the wearing away of soil by wind and water (p. 676)

**Esophagus** (ih SAHF uh gus): a tube that connects the mouth to the stomach (p. 209)

**Estrogen** (ES truh jun): a female hormone (p. 516)

**Estrous** (ES trus) **cycle:** a cycle in which a female will mate only at certain times (p. 510)

**Ethyl** (ETH ul) **alcohol:** a drug formed from sugars by yeast that is found in all alcoholic drinks (p. 386)

**Evolution** (ev uh LEW shun): a change in the hereditary features of a group of organisms over time (p. 616)

**Excretory** (EK skruh tor ee) **system:** the body system made up of those organs that rid the body of liquid wastes (p. 274)

**Exoskeleton** (EK soh skel uht uhn): a skeleton on the outside of the body (p. 159)

**Experiment:** testing a hypothesis using a series of steps with controlled conditions (p. 16)

**External fertilization:** the joining of egg and sperm outside the body (p. 508)

**Extinct** (ihk STINGT): a life-form that no longer exists (p. 617)

## F

**Family:** the largest group within an order (p. 54)

**Farsighted:** being able to see clearly far away but not close up (p. 345)

**Fats:** nutrients that are stored by your body and used later as a source of energy (p. 185)

**Fern:** a vascular plant that reproduces with spores (p. 121)

**Fertile:** being able to reproduce by forming egg or sperm cells (p. 610)

**Fertilization:** the joining of the egg and sperm (p. 118)

**Fertilizer:** a substance made of minerals that improves soil for plant growth (p. 449)

**Fetus** (FEET us): an embryo that has all of its body systems (p. 529)

**Fibrous roots:** many-branched roots that grow in clusters (p. 428)

**Fission** (FIHSH un): the process of one organism dividing into two organisms (p. 80)

**Fixed joints:** joints that don't move (p. 291)

**Flagellum** (fluh JEL um): a whiplike thread used for movement by bacteria (p. 80)

**Flatworms:** the simplest worms, they have flattened bodies (p. 142)

**Flower:** the reproductive part of a flowering plant (p. 128)

**Flowering plant:** a vascular plant that produces seeds inside a flower (p. 128)

**Food chain:** a pathway of energy and materials through a community (p. 641)

**Food web:** food chains connected in a community (p. 642)

**Fossil fuel:** the remains of organisms that lived millions of years ago (p. 677)

**Fossils:** the remains of once-living things from ages past (p. 617)

**Fraternal twins:** twins that form from two different fertilized eggs (p. 597)

**Fruit:** an enlarged ovary that contains seeds (p. 492)

**Fungi:** organisms that have cell walls and absorb food from their surroundings (p. 61)

## G

**Gallbladder:** a small, baglike part located under the liver that stores bile (p. 210)

**Gene** (JEEN): a small section of chromosome that determines a specific trait of an organism (p. 547)

**Genetic code:** the code that translates the DNA language into the protein language (p. 592)

**Genetic counseling:** the use of genetics to predict and explain traits in children (p. 580)

**Genetics** (juh NET ihks): the study of how traits are passed from parents to offspring (p. 546)

**Genus** (JEE nus): the largest group within a family (p. 55)

**Germination** (jur muh NAY shun): the first growth of a young plant from a seed (p. 494)

**Gill:** a structure used by fish and some other animals to breathe in water (p. 166)

**Grafting:** the process of joining a stem of one plant to the stem of another plant (p. 487)

**Gravitropism** (grav uh TROH pihz um): the response of a plant to gravity (p. 444)

**Greenhouse effect:** the trapping of heat near Earth's surface by a layer of carbon dioxide (p. 679)

**Guard cells:** green cells that change the size of the stomata in a leaf (p. 402)

## H

**Habitat** (HAB uh tat): the place where a plant or animal lives (p. 638)

**Heart attack:** the death of a section of heart muscle (p. 237)

**Hemoglobin** (HEE muh gloh bun): a protein in red blood cells that joins with oxygen and gives the red cells their color (p. 248)

**Hemophilia** (hee muh FIHL ee uh): a disease in which a person's blood won't clot (p. 252)

**Herbaceous** (hur BAY shus) **stems:** soft, green stems (p. 420)

**Heterozygous** (HET uh roh ZI gus): an individual with a dominant and a recessive gene for a trait (p. 549)

**Hibernation:** the state of being inactive during cold weather (p. 170)

**Hinge joints:** joints that allow bones to move only back and forth (p. 291)

**Hookworm:** a roundworm that is a parasite of humans (p. 145)

**Hormones:** chemicals made in one part of an organism that affect other parts of the organism (p. 320)

**Host:** an organism that provides food for a parasite (p. 73)

**Hydrochloric** (hi druh KLOR ik) **acid:** a chemical often called stomach acid (p. 210)

**Hypertension** (HI pur ten chun): occurs when blood pressure is extremely high (p. 236)

**Hyphae** (HI fee): threadlike structures that make up the bodies of most fungi (p. 102)

**Hypothesis** (hi PAHTH uh sus): a statement that can be tested (p. 16)

## I

**Identical twins:** two children that form from the splitting of one fertilized egg (p. 597)

**Immigration** (ihm uh GRAY shun): the movement of animals into a population (p. 634)

**Immune system:** the body system made up of proteins, cells, and tissues that identify and defend the body against foreign chemicals and organisms (p. 256)

**Immunity:** the ability of a person who once had a disease to be protected from getting the same disease again (p. 258)

**Incomplete dominance:** a case of inheritance in which neither gene is totally dominant over the other (p. 573)

**Incomplete metamorphosis:** a series of changes in an insect from nymph to adult (p. 536)

**Inhalant:** a drug breathed in through the lungs in order to cause a behavior change (p. 379)

**Innate behavior:** a way of responding that does not require learning (p. 353)

**Instinct:** a complex pattern of behavior that an animal is born with (p. 353)

**Insulin** (IHN suh lun): a hormone that lets your body cells take in glucose, a sugar, from your blood (p. 323)

**Interferon** (ihnt ur FIHR ahn): a chemical substance that interferes with the way viruses reproduce (p. 77)

**Internal fertilization:** the joining of egg and sperm inside the female's body (p. 509)

**International System of Units:** a measuring system based on units of 10 (p. 11)

**Invertebrates** (in VERT uh brayts): animals without backbones (p. 134)

**Involuntary muscles:** muscles you can't control (p. 294)

**Iris:** a muscle that controls the amount of light entering the eye (p. 334)

## J

**Jawless fish:** fish that have no jaws and are not covered with scales (p. 167)

**Jointed-leg animal:** an invertebrate with an outside skeleton, bilateral symmetry, and jointed appendages (p. 158)

## K

**Kilogram** (kg): an SI unit of mass (p. 14)

**Kingdom:** the largest group of living things (p. 53)

**Koch's postulates** (KAHKS · PAHS chuh lutz): steps for proving that a disease is caused by a certain microscopic organism (p. 82)

## L

**Labor:** contractions of the uterus (p. 530)

**Large intestine:** a tubelike organ at the end of the digestive tract (p. 212)

**Larva** (LAR vuh): a young animal that looks completely different from the adult (p. 536)

**Lateral buds:** buds along the sides of a stem that give rise to new branches, leaves, or flowers (p. 423)

**Learned behaviors:** behaviors that must be taught (p. 356)

**Lens:** a clear part of the eye that changes shape as you view things at different distances (p. 335)

**Lens muscle:** a muscle that pulls on the lens and changes its shape (p. 335)

**Leukemia** (lew KEE mee uh): a blood cancer in which the number of white blood cells increases at an abnormally fast rate (p. 250)

**Lichen** (LI kun): a fungus and an organism with chlorophyll that live together (p. 107)

**Ligaments** (LIGH uh munts): tough fibers that hold one bone to another (p. 289)

**Light microscope:** light passes through the object being looked at and then through two or more lenses (p. 7)

**Limiting factor:** any condition that keeps the size of a population from increasing (p. 636)

**Liver:** the largest organ in the body, it makes a chemical called bile (p. 210)

**Long-day plants:** plants that flower when the day length rises above 12 to 14 hours (p. 441)

## M

**Mammal** (MAM ul): an animal that has hair and feeds milk to its young (p. 173)

**Mammary** (MAM uh ree) **glands:** body parts that produce milk (p. 173)

**Mantle:** a thin, fleshy tissue that covers a soft-bodied animal (p. 149)

**Medulla** (muh DUL uh): the brain part that controls heartbeat, breathing, and blood pressure (p. 317)

**Meiosis** (mi OH sus): a kind of cell reproduction that forms eggs and sperm (p. 471)

**Menstrual** (MEN struhl) **cycle:** the monthly changes that take place in the female reproductive organs (p. 515)

**Menstruation** (men STRAY shun): a loss of cells from the uterine lining, blood, and egg (p. 516)

**Metamorphosis** (met uh MOR fuh sus): the changes in appearance and lifestyle that occur between the young and the adult stages (p. 535)

**Meter:** an SI unit of length (p. 11)

**Midrib:** the main vein of a leaf (p. 399)

**Migration:** a kind of behavior in which animals move from place to place in response to the season of the year (p. 363)

**Minerals:** nutrients needed to help form different cell parts (p. 190)

**Mitochondria** (mite uh KAHN dree uh): cell parts that produce energy from food that has been digested (p. 34)

**Mitosis** (mi TOH sus): cell reproduction in which two identical cells are made from one cell (p. 464)

**Molting:** shedding an exoskeleton (p. 159)

**Monerans:** one-celled organisms that don't have a nucleus (p. 61)

**Moss:** a small, nonvascular plant that has both stems and leaves but no roots (p. 117)

**Mucus** (MYEW kus): a thick, sticky material that protects the stomach and intestinal linings from enzymes and stomach acid (p. 216)

**Multicellular:** an organism having many different cells that do certain jobs for the organism (p. 97)

**Muscular dystrophy** (MUS kyuh lur · DIHS truh fee): a disease that causes the slow wasting away of skeletal muscle tissue (p. 300)

**Muscular system:** all the muscles in your body (p. 292)

**Mutation** (myew TAY shun): any change in copying the DNA message (p. 595)

**Mutualism:** a living arrangement in which both organisms benefit (p. 107)



725

## N

**Natural resource:** any part of the environment used by humans (p. 674)

**Natural selection:** the process in which something in a living thing's surroundings determines if it will or will not survive to have offspring (p. 609)

**Navel:** where the umbilical cord was attached to the body (p. 530)

**Nearsighted:** being able to see clearly close up but not far away (p. 344)

**Nephron** (NEF rahn): a tiny filter unit of the kidney (p. 276)

**Nerve:** many neurons bunched together (p. 312)

**Nervous system:** the body system made up of cells and organs that let an animal detect changes and respond to them (p. 310)

**Neurons** (NOO rahnz): nerve cells (p. 312)

**New-world monkeys:** those that have a tail that can grasp like a hand and nostrils that open upward (p. 613)

**Niche** (NITSH): the job of the organism in the community (p. 638)

**Nicotine:** a stimulant found in tobacco (p. 385)

**Nitrogen bases:** the chemicals that form the rungs of a DNA molecule (p. 587)

**Nitrogen cycle:** the reusing of nitrogen in an ecosystem (p. 656)

**Nonvascular** (nahn VAS kyuh lur) **plants:** plants that don't have tubelike cells in their stems and leaves (p. 116)

**Nuclear membrane:** a structure that surrounds the nucleus and separates it from the rest of the cell (p. 32)

**Nucleolus** (new KLEE uh lus): the cell part that helps make ribosomes (p. 32)

**Nucleus** (NEW klee us): the cell part that controls most of the cell's activities (p. 32)

**Nutrients** (NEW tree unts): the chemicals in food that cells need (p. 184)

**Nutrition** (new TRISH un): the study of nutrients and how your body uses them (p. 184)

**Nymph** (NIHMF): a young insect that looks similar to the adult (p. 536)

## O

**Old-world monkeys:** those that can't grasp with their tail, if they have one, and whose nostrils open downward (p. 613)

**Olfactory** (ohl FAK tree) **nerve:** a nerve that carries messages from the nose to the brain (p. 339)

**Optic** (AHP tihk) **nerve:** a nerve that carries messages from the retina to the brain (p. 336)

**Order:** the largest group within a class (p. 54)

**Organ:** a group of tissues that work together to do a job (p. 41)

**Organ system:** a group of organs that work together to do a certain job (p. 41)

**Organism:** a living thing (p. 41)

**Osmosis** (ahs MOH sus): the movement of water across the cell membrane (p. 39)

**Ovaries** (OHV uh reez): the female sex organs that produce eggs (p. 474)

**Over-the-counter drug:** one that you can buy legally without a prescription (p. 371)

**Overdose:** the result of too much of a drug in the body (p. 374)

**Oviducts** (OH vuh dukts): tubelike organs that connect the ovaries to the uterus (p. 514)

**Ovules:** tiny, round parts of a seed plant that contain the egg cells (p. 489)

**Ozone:** a molecule made of three oxygen atoms that forms a layer high above Earth's surface (p. 679)

## P

**Palisade layer:** the layer of long, green cells below the upper epidermis of a leaf (p. 402)

**Pancreas** (PAN kree us): an organ located below the stomach that makes three different enzymes (p. 210)

**Parasite** (PAR uh site): an organism that lives in or on another living thing and gets food from it (p. 73)

**Parasitism** (PAR uh suh tihz um): a relationship between two organisms in which one is helped and the other is harmed (p. 646)

**Parental care:** a behavior in which adults give food, protection, and warmth to eggs or young (p. 365)

**Pasteurization** (pas chuh ruh ZAY shun): the process of heating milk to kill harmful bacteria (p. 85)

**Pedigree:** a diagram that can show how a certain trait is passed along in a family (p. 581)

**Penis** (PEE nus): places sperm in vagina during mating (p. 513)

**Perennial** (puh REN ee ul): a plant that doesn't die at the end of one or two years of growth (p. 448)

**Pesticides:** chemicals used to kill unwanted pests (p. 680)

**Petals:** often brightly colored and scented parts of a flower that attract insects to the flower (p. 488)

**Pheromones** (FER uh mohnz): chemicals that affect the behavior of members of the same species (p. 360)

**Phloem** (FLOH em): cells that carry food that is made in the leaves to all parts of the plant (p. 121)

**Photosynthesis** (foht oh SIHN thuh sus): the process in which plants use water, carbon dioxide, and energy from the sun to make food (p. 114)

**Phototropism** (foh toh TROH pihz um): the growth of a plant in response to light (p. 443)

**Phylum** (FI lum): the largest group within a kingdom (p. 54)

**Physical change:** occurs when large food pieces are broken down into smaller pieces (p. 205)

**Pistil** (PIH stul): the female reproductive organ of a flower (p. 489)

**Pituitary** (puh TEW uh ter ee) **gland:** an endocrine gland that forms many different hormones (p. 321)

**Placenta** (pluh SENT uh): an organ that connects the embryo to the mother's uterus (p. 526)

**Planarian** (pluh NAIR ee un): a common freshwater flatworm that is not a parasite (p. 144)

**Plants:** organisms that are made up of many cells, have chlorophyll, and can make their own food (p. 62)

**Plasma** (PLAZ muh): the nonliving, yellow liquid part of blood (p. 247)

**Platelets** (PLAYT lutz): cell parts that aid in forming blood clots (p. 251)

**Pneumonia** (noo MOH nyuh): a lung disease caused by bacteria, a virus, or both (p. 273)

**Polar body:** a small cell formed during meiosis in a female (p. 475)

**Pollen:** the tiny yellow grains of seed plants in which sperm develop (p. 125)

**Pollination** (pahl ih NAY shun): the transfer of pollen from the male part of a seed plant to the female part (p. 490)

**Pollution:** anything that makes the surroundings unhealthy or unclean (p. 677)

**Population:** a group of living things of the same species that live in an area (p. 632)

**Pore:** a small opening in a sponge through which water enters (p. 137)

**Precipitation** (prih sihp uh TAY shun): water in the air that falls to Earth as rain or snow (p. 663)

**Predation:** the predator-prey relationship (p. 647)

**Predator:** an animal that hunts, kills, and eats another animal (p. 647)

**Prescription drug:** one that a doctor must tell you to take (p. 371)

**Prey:** the animal that the predator kills and eats (p. 647)

**Primary consumers:** animals that eat only plants (p. 639)

**Primates:** mammals with eyes that face forward, a well-developed cerebrum, and thumbs that can be used for grasping (p. 613)

**Producers:** living things that make, or produce, their own food (p. 27)

**Proteins** (PROH teenz): nutrients that are used to build and repair body parts (p. 185)

**Protists:** mostly single-celled organisms that have a nucleus and other cell parts (p. 61)

**Protozoans** (proht uh ZOH uhnz): one-celled animal-like organisms with a nucleus (p. 94)

**Psychedelic** (si kuh DEL ihk) **drug:** one that alters the way the mind works and changes the signals we receive from our sense organs (p. 378)

**Puberty** (PEW bur tee): the stage in life when a person begins to develop sex cells (p. 474)

**Pulmonary** (PUL muh ner ee) **artery:** an artery that carries blood away from the heart to the lungs (p. 229)

**Pulmonary veins:** veins that carry blood from the lungs to the left side of the heart (p. 229)

**Punnett** (PUH nuht) **square:** a way to show which genes can combine when egg and sperm join (p. 553)

**Pupa** (PYEW puh): a quiet, non-feeding stage that occurs between larva and adult (p. 537)

**Pupil:** an opening in the center of the iris through which light enters the eye (p. 334)

**Pure dominant:** an organism with two dominant genes for a trait (p. 549)

**Pure recessive:** an organism with two recessive genes for a trait (p. 549)

# R

**Radiation:** energy that is given off by atoms (p. 596)

**Radon:** a gas found in the ground that gives off radiation (p. 679)

**Recessive** (rih SES ihv) **genes:** those that do not show their traits when dominant genes are present (p. 549)

**Recombinant DNA:** the DNA that is formed when DNA from one organism is put into the DNA of another organism (p. 600)

**Recommended daily allowance:** the amount of each vitamin and mineral a person needs each day to stay in good health (p. 189)

**Recycling:** the reusing of resources (p. 689)

**Red blood cells:** cells in the blood that carry oxygen to the body tissues (p. 247)

**Reflexes:** quick, protective reactions that occur within the nervous system (p. 318)

**Regeneration** (rih jen uh RAY shun): reproduction in which the parent separates into two or more pieces and each piece forms a new organism (p. 505)

**Reproduce:** to form offspring similar to the parents (p. 26)

**Reproductive system:** the body system used to produce offspring (p. 513)

**Reptile:** an animal that has dry, scaly skin and can live on land (p. 170)

**Respiratory** (RES pruh tor ee) **system:** the body system made up of body parts that help with the exchange of gases (p. 264)

**Retina** (RET nuh): a structure at the back of the eye made of light-detecting nerve cells (p. 335)

**Ribosomes** (RI buh sohmz): cell parts where proteins are made (p. 34)

**RNA:** a chemical that acts as a messenger for DNA (p. 592)

**Rods:** nerve cells that detect motion and help us to tell if an object is light or dark (p. 336)

**Root hairs:** threadlike cells of the epidermis that absorb water and minerals for a plant (p. 429)

**Roundworms:** worms that have long bodies with pointed ends (p. 145)

**Runner:** a stem that grows along the ground or in the air and forms a new plant at its tip (p. 486)

## S

**Sac fungi:** fungi that produce spores in saclike structures (p. 106)

**Saliva:** a liquid that is formed in the mouth and contains an enzyme (p. 208)

**Salivary** (SAL uh ver ee) **glands:** three pairs of small glands located under the tongue and behind the jaw (p. 208)

**Saprophytes** (SAP ruh fites): organisms that use dead materials for food (p. 81)

**Scientific method:** a series of steps used to solve problems (p. 15)

**Scientific name:** the genus and species names together (p. 59)

**Sclera** (SKLER uh): the tough, white outer covering of the eye (p. 334)

**Scrotum** (SKROH tum): the sac that holds the testes (p. 513)

**Secondary consumers:** animals that eat other animals (p. 639)

**Sediment:** material that settles on the bottom of a body of water (p. 676)

**Sedimentary rocks:** those that form from layers of mud, sand, and other fine particles (p. 617)

**Seed:** a part of a plant that contains a new, young plant and stored food (p. 124)

**Segmented worms:** worms with bodies divided into sections called segments (p. 146)

**Self pollination:** when pollen moves from the stamen of one flower to the pistil of the same flower or of another flower on the same plant (p. 490)

**Semicircular canals:** inner ear parts that help us keep our balance (p. 341)

**Semilunar** (sem ih LEW nur) **valves:** valves located between the ventricles and their arteries (p. 227)

**Sense organs:** parts of the nervous system that tell an animal what is going on around it (p. 332)

**Sepals:** often green, leaflike parts of a flower that protect the young flower while it is still a bud (p. 488)

**Sex cells:** reproductive cells produced in sex organs (p. 471)

**Sex chromosomes:** chromosomes that determine sex (p. 568)

**Sexual reproduction:** the forming of a new organism by the union of two reproductive cells (p. 118)

**Sexually transmitted diseases:** those transmitted through sexual contact (p. 518)

**Short-day plants:** plants that flower when the day length falls below 12 to 14 hours (p. 441)

**Sickle-cell anemia:** a genetic disorder in which all the red blood cells are shaped like sickles (p. 574)

**Side effect:** a change other than the expected change caused by a drug (p. 372)

**Sister chromatids** (KROH muh tidz): two strands of a doubled chromosome (p. 466)

**Skeletal muscles:** muscles that move the bones of the skeleton (p. 293)

**Skeletal system:** the framework of bones in your body (p. 286)

**Slime molds:** funguslike protists that are consumers (p. 99)

**Small intestine:** a long, hollow, tubelike organ where most of the chemical digestion of food takes place (p. 210)

**Smog:** a combination of smoke and fog (p. 678)

**Smooth muscle:** involuntary muscle that makes up the intestines, arteries, and many other body organs (p. 295)

**Social insects:** insects that live in groups, with each individual doing a certain job (p. 362)

**Soft-bodied animals:** animals with a soft body that is usually protected by a hard shell (p. 149)

**Solid bone:** the very compact or hard part of a bone (p. 289)

**Species:** the smallest group of living things (p. 55); a group of living things that can breed with others of the same species and form fertile offspring (p. 610)

**Sperm:** a male reproductive cell (p. 118)

**Spinal cord:** the body part that carries messages from the brain to body nerves or from body nerves to the brain (p. 315)

**Spiny-skin animal:** an invertebrate with a five-part body design, radial symmetry, and spines (p. 163)

**Sponges:** simple invertebrates that have pores (p. 137)

**Spongy bone:** the part of a bone that has many empty spaces, much like those in a sponge (p. 289)

**Spongy layer:** the layer of round, green cells directly below the palisade layer of a leaf (p. 402)

**Sporangia:** structures, found on the tips of hyphae, that make spores (p. 104)

**Sporangium** (spuh RAN jee uhm) **fungi:** fungi that produce spores in sporangia (p. 104)

**Spores:** special cells that develop into new organisms (p. 96)

**Sporozoans** (spor uh ZOH uhnz): protozoans that reproduce by forming spores (p. 96)

**Sprains:** injuries that occur to your ligaments at a joint (p. 300)

**Stamen:** the male reproductive organ of a flower (p. 488)

**Stereomicroscope** (STER ee oh MI kruh skohp): used for viewing large objects and things through which light cannot pass (p. 8)

**Stimulant** (STIHM yuh lunt): a drug that speeds up body activities that are controlled by the nervous system (p. 376)

**Stimulus:** something that causes a reaction in an organism (p. 353)

**Stinging-cell animals:** animals with stinging cells and hollow, sock-shaped bodies that lack organs (p. 139)

**Stoma** (STOH muh): a small pore or opening in the lower epidermis of a leaf (p. 402)

**Stomach:** a baglike, muscular organ that mixes and chemically changes protein (p. 209)

**Succession:** the changes that take place in a community as it gets older (p. 660)

**Symmetry** (SIH muh tree): the balanced arrangement of body parts around a center point or along a center line (p. 134)

**Synapse** (SIN aps): a small space between the axon of one neuron and the dendrite of a nearby neuron (p. 313)

## T

**Taproot:** a large, single root with smaller side roots (p. 428)

**Tapeworm:** a kind of flatworm that has a flattened, ribbonlike body divided into sections (p. 142)

**Taste buds:** nerve cells in the tongue that detect chemical molecules (p. 339)

**Technology:** the use of scientific discoveries to solve everyday problems (p. 20)

**Tendon** (TEN duhn): a tough, fibrous tissue that connects muscle to bone (p. 295)

**Tentacles** (TENT ih kulz): armlike parts of stinging-cell animals (p. 139)

**Terminal bud:** the bud at the tip of a stem (p. 423)

**Testes** (TES teez): the male sex organs that produce sperm (p. 474)

**Theory:** a hypothesis that has been tested again and again by many scientists, with similar results each time (p. 19)

**Thigmotropism:** a plant growth response to contact (p. 444)

**Threatened:** a species of living thing that is close to being endangered (p. 675)

**Thyroid** (THI royd) **gland:** an important endocrine gland that is found near the lower part of your neck and produces thyroxine (p. 322)

**Thyroxine** (thi RAHK sun): the hormone that controls how fast your cells release energy from food (p. 322)

**Tissue:** a group of similar cells that work together to carry out a special job (p. 40)

**Toxic:** poisonous (p. 678)

**Trachea** (TRAY kee uh): a tube about 15 centimeters long that carries air to two shorter tubes called the bronchi (p. 267)

**Trait:** a feature that a thing has (p. 48)

**Transpiration** (trans puh RAY shun): the process of water passing out through the stomata of leaves (p. 403)

**Tricuspid** (tri KUS pud) **valve:** the valve between the right atrium and the right ventricle (p. 227)

**Tropism** (TROH pihz um): a movement of a plant caused by a change in growth as a response to a stimulus (p. 443)

**Tube feet:** parts of a starfish that are like suction cups and help the starfish move, attach to rocks, and get food (p. 163)

**Tuber:** an underground root or stem swollen with stored food (p. 485)

## U

**Umbilical** (uhm BIL ih kuhl) **cord:** made of blood vessels that connect the embryo to the placenta (p. 526)

**Urea** (yoo REE uh): a waste that results from the breakdown of body protein (p. 274)

**Ureter** (YOOR ut ur): a tube that carries wastes from a kidney to the uninary bladder (p. 276)

**Urethra** (yoo REE thruh): a tube that carries liquid wastes from the urinary bladder to outside the body (p. 276)

**Urinary bladder:** a sac that stores liquid wastes removed from the kidneys (p. 276)

**Urine:** the waste liquid that reaches the ureter (p. 277)

**Uterus** (YEWT uh rus): a muscular organ in which the fertilized egg develops (p. 514)

## V

**Vaccines:** substances made from weakened or dead viruses that protect you against certain diseases (p. 78)

**Vacuole** (VAK yuh wol): a liquid-filled space that stores food, water, and minerals (p. 35)

**Vagina** (vuh JI nuh): a muscular tube that leads from outside the female's body to the uterus (p. 515)

**Valves:** flaps in the heart that keep blood flowing in one direction (p. 227)

**Variable:** something that causes the changes observed in an experiment (p. 18)

**Variation** (ver ee AY shun): a trait that makes an individual different from others of its species (p. 616)

**Vas deferens** (VAS · DEF uh runs): carries sperm from testes to urethra (p. 513)

**Vascular** (VAS kyuh lur) **plants:** plants that have tubelike cells in their roots, stems, and leaves to carry food and water (p. 116)

**Vein:** a blood vessel that carries blood back to the heart (p. 226)

**Vena cava** (VEE nuh · KAY vuh): the largest vein in the body (p. 229)

**Ventricles** (VEN trih kulz): the large bottom chambers of the heart (p. 225)

**Vertebrates** (VERT uh brayts): animals with backbones (p. 134)

**Vestigial** (vuh STIJ ee ul) **structure:** a body part that no longer has a function (p. 620)

**Villi** (VIHL i): the fingerlike parts on the lining of the small intestine (p. 214)

**Virus:** a chromosome-like part surrounded by a protein coat (p. 72)

**Vitamins:** chemical compounds needed in very small amounts for growth and tissue repair of the body (p. 188)

**Vitreous** (VI tree us) **humor:** a jelly-like material inside the eye (p. 335)

**Volume:** the amount of space a substance occupies (p. 13)

**Voluntary muscles:** muscles you can control (p. 294)

## W

**Warm-blooded:** having a body temperature that is controlled so that it stays about the same no matter what the temperature of the surroundings (p. 171)

**Water cycle:** the path that water takes through an ecosystem (p. 657)

**White blood cells:** the cells in the blood that destroy harmful microbes, remove dead cells, and make proteins that help prevent disease (p. 249)

**Wilting:** when a plant loses water faster than it can be replaced (p. 403)

**Woody stem:** a nongreen stem that grows to be thick and hard (p. 421)

## X

**X chromosomes:** sex chromosomes of the female (p. 568)

**Xylem** (ZI lum): cells that carry water and dissolved minerals from the roots to the leaves (p. 121)

## Y

**Y chromosome:** a sex chromosome found only in males (p. 568)

# INDEX

## A



## B

## C

# D

# E



# F

## G



## H

## N



## O

# P

# Q



# R



# S

# T

## U



## V



## W



## X



## Y

# PHOTO CREDITS

**Cover photo:** Claude Steelman/Tom Stack & Associates
**iv,** Walter Cunningham; **v**(t) Larry Lefever/Grant Heilman, (inset) Rod Plank/Tom Stack & Associates, (c) Alvin Staffan, (b) Steve Lissau; **vi**(tl) Johnny Johnson/Animals, Animals, (tc) Walt Anderson/Tom Stack & Associates, (b) Aaron Haupt/Glencoe; **vii**(t) David Dennis, (c) a, Larry Lipsky/Tom Stack & Associates, b, Hans Pfletschin/Peter Arnold, Inc., c, National Audubon Society/Photo Researchers, d, Fred Bavendam/Peter Arnold, Inc., e, Spencer Swanger/Tom Stack & Associates, f, Aaron Haupt/Glencoe, g, Tim Courlas, h, H. Gritscher/Peter Arnold, Inc., (b) Lynn Stone; **ix**(tr) a, Donald Dietz/Stock Boston, b, Tim Davis/Photo Researchers, (tr), Hank Morgan/Science Source/Photo Researchers, (bl) a, John Radcliffe Infirmary/SPL/Photo Researchers, b, Lennart Nilsson, (br) First Image; **x**(t) Cobalt Productions, (b) Alan Carey; **xi**(t) R. Feldman/Dan McCoy Rainbow, (b) Harris Biological Supply Company; **xii**(b) W. Gregory Brown/Animals, Animals, (t) Lynn Stone; **xiii**(b) Gerard Photography, (t) Dr Tony Brian/SPL/Photo Researchers; **xiv**(t) David Brownell, (b) Pictures Unlimited; **xv**(b) Grant Heilman/Grant Heilman, (t) Elaine Shay; **xvi,** James Westwater; **xvii**(t) Pictures Unlimited, (b) Alan Carey; **xviii**(t) William Weber, (b) John Giannicchi/SPL/Photo Researchers; **xix**(t) Elaine Shay, (b) David Frazier; **xx**(t) Frank Balthis, (b) Ted Rice; **xxi,** David Dennis; **xxii**(t) Comstock, (b) Bob McKeever/Tom Stack & Associates; **xxiii**(t) Larry Lefever/Grant Heilman, (b) G.C. Lockwood; **xxiv,** Pictures Unlimited; **1,** M. Woodbridge Williams/National Park Service; **2-3,** JPL/NASA, (inset)NASA; **4,** Bob Daemmrich; **5,** Tom Courlas; **6**(l) USDA, (r) Doug Martin; **8**(l) E. Herzog/Photo Researchers, (c) Michael Abbey/Photo Researchers, (r) Martin Rodgers/Taurus Photos; **11**(t) David Dennis, (b) Tom Courlas; **13,** NASA; **20,** Hank Morgan/Photo Researchers; **21**(t) F. Stuart Westmorland/Tom Stack & Associates, (b) Lionel Deleigne/Stock Boston; **24,** Johnny Johnson/Animals, Animals; **25,** Walt Anderson/Tom Stack & Associates; **27**(l) David M. Dennis/Tom Stack & Associates, (r) Dwight Kuhn; **29**(t) John Kaprieliam/Photo Researchers, (b) Larry Hamill; **31,** Cabisco/Visuals Unlimited; **35**(l) K.R. Porter/Photo Researchers, (r) Biophoto Associates/Science Source/Photo Researchers; **37**(tl) John Shaw/Tom Stack & Associates, (bl) Oxford Scientific Films/Animals, Animals, (tr) Fritz Prenzel/Animals, Animals, (br) Juan M. Renyito/Animals, Animals; **42** Studiohio; **43** Grant Heilman/Grant Heilman Photography; **46,** Brian Parker/Tom Stack & Associates; **47,** Peter Parks/Oxford Scientific Films/Animals, Animals; **48,** Bud Fowle, **50,** First Image; **54**(l) Jeffrey W. Lang/Photo Researchers, (tr) John Serrao/Photo Researchers, (br) Soames Summerhay/Photo Researchers; **56,** First Image; **57**(t) Comstock, (c) Lynn Stone, (b) Dave R. Fleethan/Tom Stack & Associates; **59**(l) Anup Shah/Animals, (c) Ralph A. Beinhold/Animals, Animals, (r) Rod Planckt/Tom Stack & Associates, (tr) Jerry L. Ferrana/Photo Researchers; **60**(l) Alan Carey, (c) Michael Collier, (r) David Zosel; **61,** Phil Farnes/Photo Researchers; **66,** Steve Schapiro/Sygma; **67,** Doug Martin; **68-69,** Roger & Donna Aitkenhead/Animals, Animals, (inset) Runk-Schoenberger/Grant Heilman; **70,** Dr. Gopal Murti/Science Photo Library/Custom Medical Stock Photo; **71,** CNR/Science Photo Library/Photo Researchers; **72**(l) Omikron/Science Source/Photo Researchers, (c) CNRI/SPL/Photo Researchers, (r) R. Feldman-Dan McCoy/Rainbow; **73,** Omnikon/Photo Researchers; **74**(t) Prof. Luc Montagnier/CNRI/Science Photo Library/Photo Researchers, (b) James E. Lloyd/Animals, Animals; **78**(l) Pictures Unlimited, (r) Martin Rodgers/Stock Boston; **79**(tl) Dr. Tony Brain & David Parker/Science Photo Library/Photo Researchers, (bl) CNRI/Science Photo Library/Photo Researchers, (r) CNRI/Science Photo Library/Photo Researchers; **81,** John Kohout/Root Resources; **83**(l) NIH/Science Photo Library/Photo Researchers, (r) CNRI/Science Photo Library/Photo Researchers; **84**(l) Doug Martin, (c),(r) First Image; **85,** First Image; **86**(t) Ken W. Davis/Tom Stack & Associates, (b) First Image; **88,** Brian Parker/Tom Stack & Associates; **89**(l) Biophoto Associates-Science Source/Photo Researchers, (r) Arthur M. Siegelman; **92,** Rod Planck/Tom Stack & Associates; **93,** Ward's Natural Science Establishment; **94**(l) Eric Grave/Science Source/Photo Researchers, (c) Oxford Scientific Films/Animals, Animals, (r) John Shaw/Tom Stack & Associates; **96,** Eric Grave/Photo Researchers; **98**(t) Jan Hinsch/SPL/Photo Researchers, (b) Ward's Natural Science Establishment; **99,** Joyce Photographics/Photo Researchers; **102**(r) David M. Dennis/Tom Stack & Associates, (l) Jack Swenson/Tom Stack & Associates; **105,** Cary Wolinsky/Stock Boston; **106**(l) A. McClenaghan/Photo Researchers, (r) Biophoto Associates/Photo Researchers; **107,** Breck Kent; **112,** Ray Bishop/Stock Boston; **113,** Al Staffan; **114,** Courtesy: Ray F. Everett; **117**(l) Rod Planck/Tom Stack & Associates, (r) Jack Wilburn/Earth Series; **118,** Aaron Haupt/Glencoe; **122,** Breck P. Kent/Earth Scenes; **124**(l) Rich Brommer, (c) John Gerlach/Tom Stack & Associates, (r) Doug Sokell/Tom Stack & Associates; **126**(t) Eric Carle/Stock Boston, (b) Bob Daemmrich; **127**(t) Donald Dietz/Stock Boston, (b) Tim Davis/Photo Researchers; **128**(t) Kodansha/Science Software System, (bl) Jack Wilburn/Earth Scenes, (br) Zig Leszczynski/Earth Scenes; **129**(t) Patti Murray/Earth Scenes, (b) First Image; **132,** Nancy Sefton/Photo Researchers; **133,** Brian Parker/Tom Stack & Associates; **137,** Cindy Garoutte/Tom Stack & Associates; **139**(l) William Curtsinge/Photo Researchers, (c) Dave Fleetham/Tom Stack & Associates, (r) Lynn Funkhouser/Peter Arnold, Inc.; **142,** Michael Klein/Peter Arnold, Inc.; **146**(l) Alvin Staffan, (r) Breck Kent; **149**(l) George Holton/Photo Researchers, (r) Garry G. Gibson/Photo Researchers; **150**(t) F.E. Unverhav/Animals, Animals, (b) Ed Robinson/Tom Stack & Associates; **152**(l) Larry Lipsky/Tom Stack & Associates, (lc) Hans Pfletschinger/Peter Arnold, Inc., (rc) National Audubon Society/Photo Researchers, (r) Fred Bavendam/Peter Arnold, Inc.; **153**(rc) Tim Courlas, (r) H. Gritscher/Peter Arnold, Inc., (l) Spencer Swanger/Tom Stack & Associates, (lc) Aaron Haupt/Glencoe; **156,** Mark Newman/Tom Stack & Associates; **157,** Colin Milkens/Oxford Scientific Films/Animals, Animals; **158,** Jack Swenson/Tom Stack & Associates; **159**(t) George Harrison/Grant Heilman, (b) Donald Specker/Animals, Animals; **160,** Tom McHugh/Photo Researchers; **163**(t) Ed Robinson/Tom Stack & Associates, (c) Larry Tackett/Tom Stack & Associates, (r) Runk-Schoenberger/Grant Heilman, (b) Ed Robinson/Tom Stack & Associates; **164,** Brian Parker/Tom Stack & Associates; **165,** Animals, Animals; **168**(t) Fred McConnaughey/Photo Researchers, (b) Al Staffan; **169,** Breck Kent/Animals, Animals; **170**(t) Alvin Staffan, (b) David Dennis; **173,** Wendy Neefus/Animals, Animals; **174**(l) Dave Watts/Tom Stack & Associates, (r) Dave Watts/Tom Stack & Associates; **175**(t) Tom McHugh/Photo Researchers, (b) P.F. Gero/Sygma; **178**(t) Peter Veit/DRK Photo, (b) Jim Zuckerman/Stock Concepts; **179**(t) Runk-Schoenberger/Grant Heilman, (b) Bud Fowle; **180-181,** Doug Martin, (inset) Bob Daemmrich; **182, 183,** ML Uttermohlen; **185,** Tim Courlas; **191,** Studiohio; **196,** Light Source; **197,** Bob Daemmrich; **198,** Mark Burnett; **199,** Glencoe file photo; **202,** Calvin Larsen/Photo Researchers; **203,** Davis Scharf/Peter Arnold, Inc.; **211,** Doug Martin; **217,** ML Uttermohlen; **220,** Steve Elmore/Tom Stack & Associates; **221,** Lennart Nilsson; **225,** David Gifford/SPL/Photo Researchers; **237,** Randy Schieber; **238,** Dan McCoy/Rainbow; **239**(l) Sygma, (r) Dan McCoy/Rainbow; **242,** CNRI/SPL/Photo Researchers; **243,** Aaron Haupt/Glencoe; **244,** Doug Martin; **245,** Geri Murphy; **246,** Carolina Biological Supply Co.; **247**(l) Doug Martin, (r) E.R. Degginger/Bruce Coleman; **249**(l) Runk-Schoenberger/Grant Heilman, (r) Lennart Nilsson; **250**(t) John Radcliffe Infirmary/SPL/Photo Researchers, (r) Lennart Nilsson; **251**(rt) Hickson-Bender, (rc) S.I.U./Bruce Coleman, (l) Lennart Nilsson, (rb) L. Moskowitz/Medichrome; **252,** Doug Martin; **259,** Lennart Nilsson; **262,** Dave B. Fleetham/Tom Stack & Associates; **263,** Doug Martin; **272,** Young/Hoffhines; **273,** Susan McCartney/Photo Researchers;



**742**

**280,** Owen Franklin/Stock Boston; **281**(t) Doug Martin, (b) Hank Morgan/Science Source/Photo Researchers; **284,** Roy Bishop/Stock Boston; **285,288,** First Image; **292,293** Bob Daemmrich; **294,** M. Huberland/Science Source/Photo Researchers; **299,** Comstock, (inset) Alexander Tsiaras/Stock Boston; **300,** Julie Simon/Discover Magazine; **301,** First Image; **308,** Aaron Haupt/Glencoe; **309,** SPL/Science Source/Photo Researchers; **304**(t) Doug Martin, (b) Scala/Art Resource; **305,** Doug Martin; **306-307,** David Parker/SPL/Photo Researchers, (insert) Sheila Terry/SPL/Photo Researchers; **326,** MRI-Cross section of body; **327,** Blair Seitz/Photo Researchers; **330,** Holt Studios, Ltd/Animals, Animals; **331,** Thomas Eisner; **333,** W. Perry Conway/Tom Stack & Associates; **334,** Tim Courias; **336,** Ralph Eagle, Jr. MD/Science Source/Photo Researchers; **339,** Omnikron/Science Source/Photo Researchers; **342,** Doug Martin; **350,** ML Uttermohlen; **351,** Anthony Mercieca/Photo Researchers; **356,** David Frazier; **359**(l) Nik Kleinberg/Stock Boston, (r) J.H. Robinson/Photo Researchers; **361**(l) Cary Wolinsky/Stock Boston, (r) Stephen Krasemann/Peter Arnold, Inc.; **362,** Gary Milburn/Tom Stack & Associates; **363**(l) Stephen Krasemann/Peter Arnold, Inc., (r) K. Brink/Vireo; **364,** John David Brandt/Animals, Animals; **365,** Mac Albin; **368,** Pictures Unlimited; **369,** Aaron Haupt/Glencoe; **370,** Doug Martin; **371**(l) Tim Courlas, (r) Aaron Haupt/Glencoe; **378,** Tom HcHugh/Photo Researchers; **380,** Doug Martin; **381,** ML Uttermohlen; **382,** Pictures Unlimited; **383,** Scott Camazine/Photo Researchers; **384**(l) Christopher Brown/Stock Boston, (r) Hugh Patrick Brown/Sygma; **385**(l) Norm Thomas/Photo Researchers, (r) Pictures Unlimited; **386,** ML Uttermohlen; **388,** Larry Kolvoord/The Image Works; **389**(t) Bob Daemmrich/Stock Boston, (b) Bob Daemmrich/The Image Works; **392**(t) Historical Picture Service, (b) Tate Gallery/Art Resource; **393,** Doug Martin; **394-395,** Michael P. Gadomsli/Photo Researchers, (inset) Sam Bryan/Photo Researchers; **396,** Randy Schieber; **397,** Aaron Haupt/Glencoe; **399**(l) Jack Fields/Photo Researchers, (r) G.I. Bernard/Earth Scenes; **403**(t) Jerry Bauer, (b) Studiohio; **404,** Steve Lissau; **408,** Pictures Unlimited; **411**(l),(c) G.R. Roberts, (r) John Gerlack/Tom Stack & Associates; **412,** Aaron Haupt/Glencoe; **413**(t) G.I. Bernard/Earth Scenes,` (b) Aaron Haupt/Glencoe; **414**(l) Pictures Unlimited, (r) John Shaw/Tom Stack & Associates; **418,** H. Reinhard/Okopia/Photo Researchers; **419,** Doug Martin; **421,** Jeff Lepore/ Photo Researchers; **422,** Doug Martin; **426,** Kjell B. Sandved/Photo Researchers; **427,** First Image; **428**(l) John Kaprielian/Photo Researchers, (r) Michael P. Gadomski/Photo Researchers; **431**(l) Bob Daemmrich, (r) Alan & Sandy Carey; **433**(l) G.R. Roberts, (r) Dwight Kuhn; **435**(l) Breck P. Kent/Earth Scenes, (r) Elaine Braithwaite/Peter Arnold, Inc.; **438,** Lefever-Grushow/Grant Heilman; **439,** Norm Thomas/Photo Researchers; **440,** Pat Lynch/Photo Researchers; **443,** G.R. Roberts; **444,** John Colwell/Grant Heilman; **446**(l) Charles C. Johnson, (r) Nuridsany et Perennou/Photo Researchers; **447**(l) Richard Kolar/Earth Scenes, (r) Richard Schiell/Earth Scenes; **449**(l) David M. Dennis, (c) Greg Vaughn/Tom Stack & Associates, (r) First Image; **450**(l) First Image, (c) Rod Planck/Tom Stack & Associates, (r) Kenneth Murray/Photo Researchers; **451**(l) Alan & Sandy Carey, (r) Bob Daemmrich; **452**(l) Michael S. Thompson/Comstock, (c) Kathy Merrifield/Photo Researchers, (r) G.R. Roberts; **453**(l) Patti Murray/Animals, Animals, (r) John Gerlach/Tom Stack & Associates; **454**(t) Mark Thayer; **454**(b) Morton/White; **455** Doug Martin; **458**(t) Zig Leszczynski/Earth Scenes, (b) Illustrated London News; **459**(t) USDA, (b) David Frazier; **460-461,** Grant Heilman/Grant Heilman, (inset) Jeff Heger/Houston Convention & Visitor Bureau; **462,** IFA/Peter Arnold, Inc., **463,** D.R. Specker/Earth Scenes; **465,** ML Uttermohlen; **467,** Grant Heilman/Grant Heilman; **470,** Photo Researchers; **474,** David Scharf/Peter Arnold, Inc., **478,** Oscar Auerbach, M.D.; **479**(l) Aaron Haupt/Glencoe, (r) Brian Yablonsky/duomo; **482,** Breck Kent/Animals, Animals; **483,** Frank Cezus; **486,** Pictures Unlimited; **487,** Biological Photo Service; **489**(l) Dr. Jeremy Burgess/SPL/Photo Researchers, (r) Runk-Schoenberger/Grant Heilman; **493**(t) Agricultural Research Service/USDA, (b) First Image; **496,** LLT Rhodes/Earth Scenes; **498,** Stephen J. Krasemann/Photo Researchers; **499,** Glencoe file photo; **502,** Peter Menzel/Stock Boston; **503,** John Gerlach/Animals, Animals; **505,** Walker England/Photo Researchers; **506,** Carolina Biological Supply Company; **508,** Mark Stouffer/Animals, Animals; **509,** David M. Dennis; **510,** John Shaw/Tom Stack & Associates; **517**(t) Francis Leroy/Biocosmos/Science Photo Library/Photo Researchers, (b) Hank Morgan/Science Source/Photo Researchers; **522,** Lennart Nilsson; **523,** Studiohio; **527,** Howard Sochurek/Medical Images, Inc.; **528**(l) Lennart Nilsson, (r) Hans Pfletschinger/Peter Arnold, Inc.; **529,** Lennart Nilsson; **532,** Hans Pfletschinger/Peter Arnold, Inc.; **537**(tl), (bl) John Shaw/Tom Stack & Associates, (tr) Michael Fogden/Animals, Animals, (br) Don and Pat Valenti/Tom Stack & Associates; **540**(t) Latent Image, (b) David Frazier; **541**(t) Zig Lesczynski/Animals, Animals; **542-543,** Omnikon/Science Source/Photo Researchers, (inset) Catherine Ursillo/Photo Researchers; **544,545;** Studiohio; **548,** Aaron Haupt/Glencoe; **551**(t) Mavournea Hay/Daemmrich Associates, (b) Aaron Haupt/Glencoe; **555**(t) Doug Martin, (b) First Image; **559,** Bettmann Archives; **564,565,** Tim Courlas; **567,** American Cancer Society; **568**(t) File Photo, (b) Biophoto Associates/Photo Researchers; **569,** Doug Martin; **572**(l) Studio Productions, (r) First Image; **573,** Tim Courlas; **574**(l) Biological Photo Service, (r) Murayama/Biological Photo Service; **576,** First Image; **579,** Richard Hutching/Science Source/Photo Researchers; **580,** Doug Martin; **584,** Runk-Schoenberger/Grant Heilman; **585,593,** First Image; **596**(t) Aaron Haupt/Glencoe, (b) First Image; **600,** John Colwell/Grant Heilman; **604,** D. Cavagnaro/Peter Arnold, Inc.; **605,** First Image; **606**(l) Richard R. Hansen/Photo Researchers, (r) Cary Wolinsky/Stock Boston; **609,** Richard P. Smith/Tom Stack & Associates; **610**(l) Tom McHugh/Photo Researchers, (r) Kenneth W. Fink/Photo Researchers; **613**(l) Tom McHugh/Photo Researchers, (lc) Tom McHugh/Photo Researchers, (rc) Stephen J. Krasemann/Photo Researchers, (r) Tom McHugh/Photo Researchers; **615**(l) Aaron Haupt/Glencoe, (r) John Shaw/Tom Stack & Associates; **616**(l) Ray Coleman/Photo Researchers, (r) Jeff Lepore/Photo Researchers; **619,** National Museum of Natural History; **619,** Bohdan Hrynewych/Stock Boston; **623,** Francois Gohier/Photo Researchers; **626**(t) E.R. Degginger/Earth Scenes, (b) National Museum of Natural History; **627**(t) Studiohio, (b) Bud Fowle; **628-629,** Larry Lefever/Grant Heilman, (inset) Rod Plank/Tom Stack & Associates; **630,** Stephen J. Krasemann/Photo Researchers; **631,** Thomas Kitchin/Tom Stack & Associates; **632,** Tim Cullinan; **633**(l) Joe McDonald/Animals, Animals, (c) Larry Brock/Tom Stack & Associates, (r) Richard P. Smith/Tom Stack & Associates; **635**(t) Bruno J. Zehnder/Peter Arnold, Inc., (r) Eugene Gilliam; **636,** John Shaw/Tom Stack & Associates; **638,** Gerald Corsi/Tom Stack & Associates; **639**(l) C.C. Lockwood/Earth Scenes, (c) E.R. Degginger/Animals, Animals, (r) Zig Leszcynski/Animals, Animals; **640,** David Thompson/Oxford Scientific Films/Earth Scenes; **644**(l) Eric V. Grave/Photo Researchers, (r) Dwight R. Kuhn, **645,** Jacques Jangoux/Peter Arnold, Inc.; **646**(l) Animals, Animals/G.I. Bernard, (r) Zig Leszczynski/Animals, Animals; **647,** Charles G. Summers, Jr/Tom Stack & Associates; **652,** Kevin Schafer/Tom Stack & Associates; **653,** Sydney Thomson/Earth Scenes; **654**(l) Matt Meadows; (r) David C. Fritts/Animals, Animals; **656,** Bob Daemmrich; **660**(l) Doug Martin, (r) Pictures Unlimited; **661**(l) Michael P. Gadomski/Photo Researchers, (r) Michael P. Gadomski/Photo Researchers; **665**(a) Brian Parker/Tom Stack & Associates, (b) Richard R. Thom/Tom Stack & Associates, (c) Michael & Barbara Reed/Earth Scenes, (d) Michael P. Gadomski/Photo Researchers, (e) Charlie Ott/Photo Researchers, (f) Tom Stack/Tom Stack & Associates; **669,** Carl Purcell/Photo Researchers; **672,** Pat & Tom Leeson/Photo Researchers; **673,** David Dennis;



**743**